INDEX BIOLOGORUM

INVESTIGATORES · LABORATORIA
PERIODICA

EDIDIT

G. CHR. HIRSCH
UTRECHT · HOLLAND

EDITIO PRIMA

BERLIN · JULIUS SPRINGER · 1928

ISBN-13: 978-3-642-98565-2   e-ISBN-13: 978-3-642-99380-0
DOI: 10.1007/978-3-642-99380-0

COPYRIGHT 1928 BY JULIUS SPRINGER IN BERLIN
Softcover reprint of the hardcover 1st edition 1928

Praefatio

Biologos intellego omnes investigatores, qui vitae naturam ab omni parte indagant. Quotannis biologi magis magisque in diversas partes discedunt. Singulae enim biologiae disciplinae, etiam quae olim inter se artissime erant connexae, in dies magis secernuntur alia ab alia. Malo inde orto variis remediis obviam ire fortasse possemus; e quibus unum, sit externum tantummodo, in medium proferre propositum habeo. Indicem enim in lucem edere aggressus sum, in quo biologorum omnium non modo nomina enarrantur, sed etiam quae natura sit studiorum investigationumque, quibus illi operam dant.

Ut ea, quae nisi magno damno et incommodo separari nequeunt, denuo inter se conjugantur et recte cohaereant, ad laborem huius operis componendi me accinxi.

Non modo enumerationem specto illorum virorum doctorum nominum et domiciliorum, sed etiam studiorum, quibus incumbunt, et deinde conspectum laboratoriorum eo ordine, quem natura indicat earum rerum, quae in eis navantur, dispositorum; idque hoc consilio, ut biologi omnium gentium inter se quam maxime conspirent. Magnopere gaudeo aedes Julii Springer officium suscepisse divulgandi meum librum.

Lectorem fortasse delectat cognoscere nos operi huic perficiendo unius anni et dimidii laborem impendisse. Labores nostri quattuordecim milia horarum absumpserunt. Triginta fere milia epistularum a nobis ex urbe Utrecht in omnes orbis terrarum partes missae sunt, ut quam plurima indicia colligerentur.

Ubicumque gentium adiutores, quorum nomina infra memorantur et quibus hoc loco gratias agere mihi liceat quam maximas, praesto mihi esse voluerunt. Multi etiam alii viri docti, rescribentes ad epistulas meas, officiosos in me se praestiterunt. Neque tamen ego omnia, quae viri illi docti mecum communicaverunt, in usum meum recipere potuisse doleo. Quae de vita biologorum et de libris ab eis editis dici possunt, omisi, ne libri pretium fieret maius. Systema biologiae studiorumque, quibus biologi singuli operam dant, certo ordine expositorum — quod systema in hac ipsa editione in lucem edere mihi in animo erat — eandem ob causam prelo tradere nondum mihi licebat. Spero fore ut in editione posteriore non desideretur. Interea sufficiat systema laboratoriorum pro systemate biologiae.

Index, qui exstat, continet indicia de quattuordecim fere milibus investigatorum, de plus quam sex milibus laboratoriorum, de trecentis quinquagimta septem periodicis. Quorum periodicorum dispositionem aedes, quae hunc librum divulgant, curaverunt.

Indicia maximam partem e propriis investigatorum laboratoriumque rectorum documentis oriuntur. Nonnulla tamen mihi praebebant adiutores. Alia inveni in annalibus, libris continentibus virorum doctorum laboratoriumque conspectus talibusque commentariis aliis. Haec indicia in conspectibus investigatorum sigillo °, quod ad singula indicia additum est, notata sunt.

Praeterea usus sum his sigillis:
  ⊕ = Permutatio (Tausch, Exchange, Echange, Cambio).
  ⊙ = Promotio doctoris.

Usui, quem specto, index tunc modo inservire poterit, cum aequis temporis spatiis interiectis denuo in lucem prodibit. Ut hoc fieri possit, benevole biologi me adiuvent, cum de supplementis emendationibusque, quae desiderantur, certiorem me faciunt. Collegis librum meum, qui communi rei nostrae servire vult, commendare mihi liceat!

Utrecht (Nederland),
Zoölogisch Laboratorium,
Janskerkhof 3.

G. Chr. Hirsch

Adjutores

Telesforo de Aranzadi, Barcelona (España), Anthropologisches Institut.
Ari-Ibn-Sahav, Jerusalem (Palästina), Hebräische Universität.
Thomas Barbour, Cambridge, Mass. (U.S.A.), Museum of comparative Zoology.
W. Benham, Dunedin (New Zealand), University of Otago, University's Museum.
Ludwig Bretschneider, Utrecht (Nederland), Zoologisches Laboratorium.
E. D. Brown, American Society for Pharmakology and Experimental Therapeutics, Minneapolis, Min. (U.S.A.), University.
P. E. Brown, American Society of Agronomy, Ames, Ia. (U.S.A.), Iowa State College.
N. J. ten Cate Hoedemaker, Utrecht (Nederland), Zoologisches Laboratorium.
C. P. Close, American Society for Horticulture Science, College Park Md. (U.S.A.).
Arthur F. Coca, American Association of Immunologists, New York City (U.S.A.).
L. J. Cole, Woods Hole, Mass. (U.S.A.), Division of Biology and Agriculture.
C. W. Collins, American Association of Economic Entomologists, Melrose Highlands, Mass. (U.S.A.).
R. Cook, American Genetic Association, Washington, D.C. (U.S.A.).
W. W. Cort, Baltimore, Md. (U.S.A.), John Hopkins University, Department of Medical Zoology.
Georges F. Cosmetatos, Athen (Griechenland), Histologisches Institut.
P. J. S. Cramer, Buitenzorg (Nederlandsch-Oost-Indië), General agricultural Experiment Station.
Marie Daiber, Zürich (Schweiz), Zoologisches Institut.
H. Dampf, Mexico, D.F. (Mexico).
J. J. Davis, Lafayette, Ind. (U.S.A.), Agricultural Experiment Station.
D. Deineka, Leningrad (U.d.S.S.R.), Anatomisch-Histologisches Laboratorium der Universität.
Valentin Dogiel, Leningrad (U.d.S.S.R.), Zootomisches Institut der Universität.
F. Dreyer, Bloemfontein (South Africa), Grey University College.
Aemilius Bernhardus Drooglever-Fortuyn, Peking (China), Union Medical College.
B. M. Duggar, National Research Council, Washington, D.C. (U.S.A.).
Scott V. Eaton, American Society of Plant Physiologists, Chicago, Ill. (U.S.A.), Department of Botany (University).
Helene K. Emme, Leningrad (U.d.S.S.R.), Institut für angewandte Botanik und neue Kultur.
L. Freund, Prag II (C.S.R.), Zoologisches Institut.
E. Furrer, Zürich (Schweiz).
F. Garcia del Cid, Barcelona (España), Universidad.
Wright A. Gardener, Auburn, Ala. (U.S.A.), Polytechnical Institut, Department of Botany.
G. E. Gates, Rangoon (British-India), Judson College.
Gesellschaft für kulturelle Verbindung der Sowjetunion mit dem Auslande, Moskau (U.d.S.S.R.).
N. Gist Gee, Peking (China), China Medical Board.
Ernst Hartert, Tring, Herts (England), Zoological Museum.
R. J. Haskell, American Phytopathological Society, Washington, D.C. (U.S.A.), Bureau of Plant Industry.
Wolter Hellén, Helsinki (Finnland), Zoologisches Museum.
L. Herrera, Mexico, D.F. (Mexico), Estudios Biológicos.
Hans Hirsch, Halle a. d. S. (Deutschland).
K. Hofeneder, Innsbruck (Deutsch-Österreich).

N. v. Hofsten, Upsala (Sverige), Zoologisches Institut.
Johan Huus, Bergen (Norge), Bergens Museum.
Raffaele Issel, Genova (Italia), Istituto Zoologico.
Arthur de Jaczewski, Leningrad (U.d.S.S.R.), Institut Jaczewski de Mycologie et de Pathologie végétale.
C. Janicki, Warszawa (Polska), Zoologisches Institut.
David Starr Jordan, Stanford University, Cal. (U.S.A.).
F. C. Koch, Federation of American Societies for Experimental Biology, Chicago, Ill. (U.S.A.), Department of Physiological Chemistry.
A. Kol, Leningrad (U.d.S.S.R.), U.S.S.R.-Institut für angewandte Botanik.
Walter Kotte, Freiburg i. Br. (Deutschland), Weinbau-Institut.
A. Kryshtofovich, Leningrad (U.d.S.S.R.), Geologisches Kommite.
Olga Kuttner, Berlin (Deutschland), Hirschwaldsche Buchhandlung.
Robert Larson, Lund (Sverige).
A. H. Leim, St. Andrews, N.B. (Canada), Atlantic Biological Station.
I. J. Lewis, Botanical Society of America, University, Va. (U.S.A.).
Theodor Lippmaa, Tartu — Dorpat (Eesti).
Alex. Lwoff, Moskau (U.d.S.S.R.).
J. G. van der Maas, Utrecht (Nederland).
Wilhelm Marinelli, Wien (Deutsch-Österreich), II. Zoologisches Institut.
Walter D. Meek, American Physiological Society, Madison, Wis. (U.S.A.).
E. Minnich, American Society of Zoologists, Minneapolis, Minn. (U.S.A.), University Department of Animal Biology.
Th. Moroff, Sofia (Bulgarien), Zoologisches Institut.
F. E. Northup, Rangoon (British-India), Judson College.
T. S. Palmer, American Ornithologists Union, Washington, D.C. (U.S.A.).
G. H. de Paula Souza, São Paulo (Brasil), Hygienisches Institut.
J. Pillet, Lyon (France), Faculté médicine.
N. Podkopaew, Leningrad (U.d.S.S.R.), Physiologisches Institut der Akademie der Wissenschaften.
Albert Policard, Lyon (France), Institut Histologique.
Osvaldo Polimanti, Perugia (Italia), Istituto Fisiologico.
Valère Puskarin, Cluj — Klausenburg (România), Speläologisches Institut.
E. Racovitza, Cluj — Klausenburg (România), Speläologisches Institut.
C. Regel, Kaunas (Litauen), Botanisches Laboratorium der Universität.
R. H. Roberts, Berkeley, Cal. (U.S.A.).
Rudolf Schön, Pillau (Deutschland), Preußische Fischereistation.
Baron M. de Selys Longchamps, Bruxelles — Brussel (Belgique), Université Libre.
James M. Sherman, Society of American Bakteriologists, Ithaka, N.Y. (U.S.A.), Cornel University.
O. Shinoda, Kyoto (Japan), Zoologisches Laboratorium.
A. Franklin Shull, Ann Arbor, Mich. (U.S.A.), University of Michigan.
Hakon R. G. Sparck, København (Danmark), Zoologisches Museum.
Karlis Starzs, Riga (Latvija), Institut für Pflanzenschutz.
Erwin Stresemann, Deutsche Ornithologische Gesellschaft, Berlin (Deutschland).
P. Suzuki, Tokyo (Japan), Pathologisches Institut.
Z. v. Szilady, Budapest 80 (Ungarn), Zoologische Abteilung des Ungarischen National-Museums.
F. C. Unger, Amersfoort (Nederland).
United States Department of Agriculture, Washington, D.C. (U.S.A.).
Lewis H. Weed, American Association of Anatomists, Baltimore, Md. (U.S.A.), John Hopkins Medical School.
A. O. Weese, Ecological Society of America, Norman, Okla. (U.S.A.), University of Oklahoma.
S. C. White, American Dairy Association, Storrs, Conn. (U.S.A.).
O. Wilhelm, Concepción (Chile), Escuela di Medicina, Universidad.
A. Zamels, Riga (Latvija), Botanisches Institut.
B. Zarnik, Zagreb — Agram (S.H.S.), Morphologisch-Biologisches Institut.
M. M. Zawadowski, Moskau (U.d.S.S.R.), Zoo-Park.

Conspectus

	Pag.
I. Investigatores	1
II. Laboratoria	336
Labor. biologiae generalis	336
Institutiones biologicae generales	336
Labor. evolutionis et geneticorum	341
Laboratoria hydrobiologiae	343
1. Laboratoria marina	343
2. Laboratoria limnologiae	347
Musea biologica generalia	350
Institutiones historiae biologiae	355
Laboratoria zoologica	356
Labor. zoologica generalia et Laboratoria morphologiae zoologicae (Anatomiae comparatae)	356
Musea zoologica	370
Horti zoologici (Vivaria)	373
Labor. specialia embryologiae	374
Laboratoria histologiae	375
Laboratoria parasitologica	379
Laboratoria parasitologiae generalis (Zoologiae applicatae)	379
Labor. entomologica specialia	381
Laboratoria malariae	389
Laboratoria helminthologiae	389
Institutiones ornithologiae	389
Laboratoria spelaeologiae	390
Laboratoria botanica	390
Laboratoria botaniae generalis	390
Musea botanica et herbaria	405
Horti botanici	408
Laboratoria et horti botaniae pharmacognosiae	412
Labor. physiologiae plantarum	412
Laboratoria phytopathologiae et mycologiae	415
Laboratoria oecologiae (phytosociologiae)	423
Laboratoria anatomica	423
Labor. anatomiae generalis et experimentalis, fac. med.	423
Laboratoria anatomiae cerebri et nervorum	431
Labor. et Musea anthropologiae	432
Labor. anatomiae animalium domesticorum (fac. veterin.)	433
Laboratoria chemo-physicalia biologica	435
Laboratoria biochemiae generalis	435
Laboratoria biophysica	442
Labor. nutrimentorum specialia	443
Laboratoria fermentorum	443
Laboratoria physiologica	444
Laboratoria physiologiae generalis (fac. medicinae)	444
Labor. physiologiae comparatae (fac. scientiarum natur., fac. veterin., fac. agricult.)	452
Labor. physiologica specialia	454
1. Digestio et nutritio	454
2. Physiologia exercitationis corporis	455
3. Physiologia humana, altitudine montium impressa	455
4. Psychologia experimentalis	456
5. Neurologia physiologica	458
6. Endocrinologia (secretio interna)	458
Laboratoria anatomiae pathologicae	459
Anatomia pathologica humana experimentalis (fac. med.)	459
Anatomia pathologica animalium (fac. veterin.)	467
Laboratoria physiologiae pathologicae	469
Laboratoria physiologiae pathologicae experimentalis generalis	469
Laboratoria pharmacologiae experimentalis	473
Laboratoria serologiae et immunologiae	477
Labor. bacteriologiae hominis	479
Laboratoria bacteriologiae animalium	486
Laboratoria hygienae hominis	487
Laboratoria hygienae animalium	492
Cancer	493
Laboratoria palaeontologica	493
Musea palaeontologica	496
Laboratoria culturae plantarum et animalium	497
Institutiones generales culturae plantarum et animalium	497
Laboratoria culturae plantarum generalia	499
Laboratoria et institutiones agriculturae specialis	507
Laboratoria silviculturae	509
Institutiones horticulturae	513
Institutiones praticulturae	517
Institutiones culturae plantarum specialium	518
1. Viticultura	518
2. Pomologia	519
3. Cultura mori	519
4. Institutiones seminologiae	519
5. Cultura sacchari	521
6. Cultura heveae	522
7. Cultura nicotianae	522
8. Cultura theae	523
9. Cultura solani tuberosi	523
10. Cultura coffeae	523
11. Cultura citri	523
Laboratoria culturae animalium generalia	523
Laboratoria culturae animalium specialia	527
1. Cultura piscium	527
2. Sericicultura	530
3. Apides	531
4. Animalia pelicea	531
Labor. productionis lactariae	531
Labor. microbiologiae agris (Pedologiae). (Bacteriae et protozoa)	533
Laboratoria terrae uliginosae	536
Labor. chemiae et physicae agris	536
III. Periodica	54

I. INVESTIGATORES

Aagaard, Otto C. (1883), Dr. med., Priv.Doc. Copenhagen (Danmark), Rigshospitalet. *Anat. of the lymphatic-system.* ⓟ Univ. of Copenhagen 1924.

Aaltonen, Viktor Toivo (1889), Prof., Bodenkunde, Forstwiss. Versuchsanstalt. Helsinki (Finland), Rauhankatu 4. *Forstliche Bodenkunde.* ⓟ Helsinki 1919.

Aamodt, Olaf S. (1892), Associate Pathologist, Office of Cereal Crops and Diseases, U.S. Dept. of Agriculture cooperating with the Univ. of Minnesota (U.S.A.), Univ. Farm, St. Paul. *Breeding of cereals for disease resistance; resistance in spring wheat and cold resistance in winter wheat.* ⓟ *Wheat varieties of all species, but especially vulgare.*

Aase, Hannah C., Ass. Prof. of Botan., State Coll. of Washington. Pullman, Wash. (U.S.A.). o

Aaser, C., Dr., Abt.-Dir., Vet.-Inst. Oslo (Norge). *Bact.* o

Abad, Jose, Prof., Univ. of St. Tomas. Manila, Philippine Islands (U.S.A.). *Phytol. and Zool. Pharm.* o

Abad, Mariano, Prof., Dr., Labor. Farm. Fac. Med. Valladolid (España). *Pharmacol.* o

Abadjleff, B., Dr., Prosektor Bact. u. Serol. Inst. d. Univ. Sofia (Bulgarien). o

Abajo-Trujillo, José, Preparador del Mus. Nacional de Ciencias Nat. Madrid (España), Corredera Alta 2. *Entomol.* o

Abba, Francesco (1862), Prof., Dir. Ufficio d'Ig. Torino (Italia), Via Schina 8. ⓟ Torino 1904.

Abbado, Michele, Lib. Doc. R.-I. med. vet. Milano (Italia). *Botan.* o

Abbott, Alexander Crever (1860), Prof. Hygiene and Bact.; Dir., School of Hygiene, Univ. of Pennsylvania, Philadelphia, Pa. (U.S.A.). ⓟ Univ. of Maryland 1884.

Abbott, Charles Harlan (1889), Prof. of Zool., Univ. of Redlands, Cal. (U.S.A.), 28 South Univ. St. *Seasonal Succession of Birds in relation to Life-Zones. Migration of the Butterfly Pyrameis cardui. Ecol. Reactions of Land Isopods to Light.* ⓟ Brown Univ. 1918.

Abbott, Clinton Gilbert (1881), Dir., Natural History Museum. Balboa Park, San Diego, California (U.S.A.). *Ornithology; life histories of birds.*

Abbott, Howard Clinton (1895), Prof. of Botany, Univ. of South Dakota. Vermillion (U.S.A.). *Plant Pathol., Plant Physiol.*

Abbott, Walter S. (1879), Entomologist, U.S. Dept. of Agriculture. Vienna, Virginia (U.S.A.). *Testing insecticides in connection with the enforcement of the Federal Insecticide.*

Abdallan, S., Prof. Staatsuniv. Erivan, Armenien (U.d.S.S.R.). *Palaeont.* o

Abderhalden, Emil (1877), Geh.R. Prof. Dr. med. et phil. h. c., Dir. des physiol. Inst. der Univ. Halle a. Saale (Deutschland), Paulusstr. 10. *Stoffwechselphysiol. Physiol. Chemie. Ferment-, Inkret-, Vitaminforschung. Eiweißchemie.* ⓟ Basel 1901.

Abe, Katsuma, Prof. of Keiô-Gijuku Univ. Yotsuya, Tôkyô (Japan), Pharmacol. Inst., Coll. of Med. *Pharm.* o

Abe, Takuji (1896), Ass. Kyoto (Japan), Labor. of Phytopath., Department of Agric., Imper. Univ. *Rot of the rice-seedlings caused by Achlya prolifera Nees. Root and foot rot diseases of some plants. Mycol. flora of the Shikoku Island of Japan.* ⓟ Kyoto 1927.

Abe, Yoshio, Prof. of Hiroshima High Seminary (Kôtô Shihan Gakkô). Hiroshima (Japan), Biol. Labor. *Mammalian Zool.* o

Abegg, Fred Anton (1896), Graduate ass. at the Univ. of Wisconsin, Genetics Dept., Agric. Chem. Bldg. Madison, Wis. (U.S.A.) or 1611 Chadbourne Ave. *Nature of Waxy Gene in Maize. Genetochem. phases.*

Abel, Franz (1878), Leipzig-W 31 (Deutschland), Könneritzstr. 104. *Hesperiden der ganzen Erde.* ⓟ Hesperiden (Lep.).

Abel, John J. (1857), Prof. of Pharmacol., J. Hopkins Univ. Johns Hopkins Medical School, Wolfe and Monument Sts., Baltimore, Md. (U.S.A.). *The composition of animal tissues and fluids, function and isolation of special chemical principles of the animal organism; ethyl sulphid, carbamic acid, pigment of negro's skin, isolation of the blood-pressure raising constituent of the suprarenal capsules in form of a benzoyl derivative; pharmacol. action of alcohol, phthaleins and derivatives; the poisons of the Amanita Phalloides; action of antimonials in experimental trypanosomiasis; action of convulsant dyes; researches on lymph hearts of frogs, vividiffusion and plasmaphaeresis, crystalline poisons of Bufo Agua, albumoses in the tissues; active principle of the pituitary gland, histamine and histamine like substances in the animal tissues, crystalline insulin.* ⓟ Straßburg 1888.

Abel, Othenio (1875), o. Prof. der Palaeobiol., Dir. Palaeobiol. Inst. Univ. Wien XIII/2 (Österreich), Jenullgasse 2. *Palaeobiol., Palaeozool., vergl. Anat. und Stammesgeschichte der Wirbeltiere, fossile Säugetiere, Fährten und Lebensspuren aller Art bei rezenten und fossilen Formen, fossile Wirbeltiere.* ⓟ Wien 1899.

Abel, Rudolf (1868), Dr. med., o. Univ.-Prof., Mitglied d. Reichsgesundheitsrates. Jena (Deutschland), Reichardtstiege 2. *Hygiene, Bact.* o

Abele, Karlis (1896), Priv.Doc. Univ. Riga (Latvija), Botan. Inst. *Pflanzencytol. und -anat. Embryosack. Zellkern.* ⓟ Marburg 1923.

Abelin, J. (1883), Dr. phil. et med., Priv.Doc. an der Univ. Bern (Schweiz), Waldheimstr. 10. *Stoffwechsel, innere Sekretion, Schilddrüse.* ⓟ Bern 1910.

Abeloos, Marcel (1901), Ass. Fac. d. Sc. Paris (France), Labor. d'Evolution, 105, Bd. Raspail. *Biol. expérim.: régénération, pigments.*

Abelous, M., Prof. Fac. Méd. Toulouse (France). *Physiol.* o

Abolin, Robert I., Doc. Univ. Taschkent (U.d. S.S.R.), Ul. Stalina 23. *Geobotan., Boden.* o

Aboliņš, Leo (1895), Priv.Doc. Vergl. Anat. und Experim.-Zool. Inst. der Univ. Riga (Lettland), Albertstr. 10; 1927: Napoli (Italia), Stazione Zool. *Farbwechsel der Fische. Regeneration der Amphibien- und Fischextremitäten. Atmungsphysiol. Adaptationen.* ⓟ Riga 1924.

Abonyi, Alex., Dr., Priv.Doc. der Univ. Budapest (Ungarn). *Histol.* o

Abovian, Sargis, Prof., Univ. Erivan, Armenien (U.d.S.S.R.). *Milchwirtschaftslehre.* o

Ábrahám, András, Dr., Mitarbeiter am Zool. und komp.-anat. Inst. der Univ. Budapest (Ungarn). o

Abramoff, Iwan (1884), Pflanzenschutzstation Fern-Ost-Gebiet. Chabarowsk (U.d.S.S.R.), Primorsk Gouvernement, K. Marksstr. 29. *Trockenbestäubung des Weizens und der chinesischen Kulturpflanzen gegen den Brandpilz. Aussaat und Düngung im Verhältnis zur Brandbefallung. Immunität der Weizenarten gegen den Brandpilz. Krankheiten der Sojabohne.* ⓟ Botan. u. mykol. Obj. d. fernen Ostens.

Abramov, Serge (1875), Dir., Labor. d'analyses med. Rive gauche. Paris VI (France), 48, rue Monsieur le Prince. *Microorganismes pathogènes. Inoculations préventives*. ⓟ Moscou 1899.

Abrams, Le Roy (1874), Prof. of Botan. and Curator of the Dudley Herbarium Stanford Univ., California (U.S.A.). *Taxonomic Botan. (Spermatophyta). Flora of Los Angeles and Vicinity, of the Pacific States.* ⓟ Columbia, 1910. ⓟ Plants, especially of the western United States.

Abrikossoff, Alexey (1875), Prof. Pathol. Anat. I. Staatsuniv. Moskau 69 (U.d.S.S.R.), Troubnikowsky pereulok 26, Log. 5. *Pathol.-anat. Morphol.* ⓟ Moskau 1904.

Absolon, Karl (1877), Prof. nat. der Karls-Univ. in Prag, Kustos der zool. und palaeont. Abteilung am mährischen Landesmus. in Brno (Brünn) C.S.R. *Höhlenfauna, Apterygota orbis terrarum. Arthropoda. Limnobiol. Homo fossilis. Palaeolithicum. Pleistozäne Tierwelt. Karstphänomen (spec. Höhlen). Hydrographie. Tiergeographie.* ⓟ Prag 1904. ⓟ Rezente Höhlentiere: Mollusca, Coleoptera, Diptera, Orthoptera, Myriopoda, Crustacea, Arachnoidea, Vermes. Kollembola.

Achard, Charles (1860), Prof. à la Fac. de Méd. Paris 16e (France), 37, Rue Galilée. *Fonction rénale. Anatomie pathologique.* ⓟ Paris, 1887.

Achard, Geneviève (1898), Moniteur à l'Inst. de Physique Biol. de la Fac. des Sc., Préparateur à l'Inst. d'Hydrol. de la Fac. de Méd. de Strasbourg (France), 31, Avenue des Vosges. *Physique biol. suspensions biol., Action de la chaleur sur le développement de l'œuf d'Oursin. Hydrol.: physico-chim. de sources.*

Achelis, Johann Daniel (1898), Ass. am physiol. Inst. der Univ. Leipzig (Deutschland), Liebigstr. 16. *Sinnes- und Nervenphysiol.* ⓟ Leipzig, 1922.

Ackerknecht, E. Gerhard (1883), Prof. Dr. Zürich 7 (Schweiz), Forchstr. 149. *Vet. Anat. Kreislaufapparat.* o

Ackerman, Lloyd (1892), Instr. of Biol., Western Reserve Univ. Cleveland, Ohio (U. S. A.). *Insect physiol., Aphids and Thysanoptera.* ⓟ Univ. of Michigan 1925.

Ackermann, Dankwart (1878), Prof. Dr. med., Vorstand des Physiol.-Chem. Inst. der Univ. Würzburg (Deutschland), Röntgenring 9. *Tierische Basen, Intermediärer Stoffwechsel, Abbau durch Mikroorganismen.* ⓟ Rostock 1902.

Ackert, James Edward (1879), Prof. of Zool., Kansas State Agricultural Coll., Manhattan, Kansas (U.S.A.). *Helminthology.* ⓟ Univ. of Illinois 1918. ⓟ Hookworms (Necator suillus).

Ackley, Alma B., Instr. in Botan., Detroit City Coll. Detroit, Mich. (U.S.A.), 665 W. Warren St. *Algae.* o

Acqua, Camillo (1863), Dir. Staz. di Gelsicolt. e Bachicolt. Ascoli Piceno (Italia). ⓟ Roma 1920.

Acquarone, Paul (1900), Ass. in Botan., Johns Hopkins Univ. Baltimore, Maryland (U.S.A.). *Morphology of plants; anatomy of the Musaceae; ecology.*

Adair, Henry Steven (1899), Junior Entomol., Bureau of Entomol., U. S. Department of Agric. Brownwood, Texas (U.S.A.). *Pecan Insect.*

Adamberg, Leida (1899), Ass. Tartu (Estland), Physiol. Inst. *Innere Secretion.* ⓟ Tartu.

Adametz, L., Prof. Hochsch. f. Bodenkultur. Wien XIX (Österreich), Eichendorffgasse 7. *Tierzucht.* o

Adamoff, Vladimir Vladimirovitch (1876), Botan. Station Expérim. Paludéenne, Dir. sc. du Jardin botan. Letce-Wielkie (près Vitebsk). Minsk, Russie Blanche (U.d.S.S.R.), 36, A. rue Communale. *Systématique Floristique Phanérogames, Flore de la Russie Blanche, Russie Centrale.*

Adams, A. Elizabeth (1892), Associate Prof. of Zool. Mount Holyoke Coll. South Hadley, Mass. (U.S.A.). *Embryol., Experimental Embryol. Development of the mouth in the amphibian embryo.* ⓟ Yale Univ. 1923.

Adams, Charles C. (1873), Prof. of Forest. Zool. Syracuse, N.Y. (U.S.A.), the New York State Coll. of Forestry. *Entomol., Ecol.* ⓟ phil. Chicago 1908, Sc. Wesleyan 1920. o

Adams, George Edward (1874), Prof. of agronomy. Kingston, R.I. (U.S.A.). *Agronomy horticulture.* o

Adams, James Fowler (1888), Associate Plant Pathologist, Agric. Experiment Station, Univ. of Delaware. Plant Pathologist for State Board. of Agric. Newark, Delaware (U.S.A.), Box 425. *Cytology of Uredinales, Diseases of Orchard and Ornamentals.* ⓟ Columbia Univ. 1918.

Adams, John (1872), Botanist, Experimental Farms Branch. Division of Botany, Central Experimental Farm, Ottawa (Canada). *Morphology and Ecology of Seed-plants. Plant Life, Flowering Plants.* ⓟ Cambridge 1900. ⓟ Seeds from Botan. Garden.

Adams, Leverett Allen (1877), Ass. Prof. of Zool. Univ. of Illinois, Urbana, Ill. (U.S.A.). *Vertebrate comp. Anat. musculature, skull, fossil footprints.* ⓟ Columbia Univ. 1916. ⓟ Skeletal material.

Adams, Walter B., Prof. American Univ. Beirut (Syrien). *Pharmacol. Therapeutics.* o

Adamson, R. S., Prof. Univ. Cape Town (South Africa). *Botan.* o

Adelheim, Roman (1881), Prof., Dir. Pathol. Inst. Univ. Riga (Latvija), Waldemarstr. 34. *Pathol. Anat. der Infektionskrankheiten.* ⓟ Dorpat 1907.

Adelsohn, Joseph (1896), Ass. d. microbiol. Inst. Univ., Leiter der Vaccinenabteilung d. Sanität-Bact. Inst. Woronesch (U.d.S.S.R.), Malaja Bogojawlenskaja N 7 qu. 4. *Med. Bact., Vaccination.*

Addington, Lawrence H. (1901), Research Ass. in Dairy Production. East Lansing, Mich. (U.S.A.), Box 1002 or Dairy Dept. *Cottonseed meal injury to dairy cattle.*

Addis, Thomas (1881), Prof. of Med., Stanford Univ. Medical School, San Francisco (U.S.A.). *Physiol. and Pathology of the Kidney.* ⓟ Edinburgh Univ. 1905.

Addison, William H. F. (1880), Prof. of Normal Histol. and Embryol. Univ. of Pennsylvania. Philadelphia, Pa. (U.S.A.). *Experim. histol. Compar. neuro-anat. Brain and Spinal Cord. Development of the Purkinje cells. Rhinencephalon of Delphinus delphis. The cell-changes in the hypophysis after castration. Histol. study of the spleen of the rabbit, under heightened phagocytic activity, cerebellar tracts in teleosts.* ⓟ Univ. of Toronto 1917.

Addems, Ruth Margery (1896), Instr. in Botan., Univ. of Wisconsin, Madison, Wis. (U.S.A.). *Plant Physiol., structure of protoplasm, plant nutrition.* ⓟ Univ. of Wis., 1926.

Adelmann, Howard Bernhardt (1898), Ass. Prof. of Histol. and Embryol., Cornell Univ. Ithaca, Stimson Hall, N.Y. (U.S.A.). *Embryol., development of the vertebrate head, development of the eye, eye muscles, etc., notochord, prechordal plate.* ⓟ Cornell Univ. 1924.

Adler, Abraham (1891), Ass. an der Med. Univ.-Klinik Leipzig (Deutschland), Liebigstr. 22. *Physiol. und Pathol. der Leber. Stoffwechsel der Gallenbestandteile: Bilirubin, Cholesterin, Urobilin, Gallensäuren. Harnblasen-Innervation.* ⓟ Frankfurt a. M. 1918.

Adler, Bruno Friedrich Wilhelm Karl Adolf (1874), Prof., o. Mitglied des wiss. Forschungsinst. von

Timirjaseff, Priv.Doc. Med. Fac. I. Univ. Moskau (U.d.S.S.R.), Pjatnitzkaja, 48. *Anthrop., Geographie (Anthropogeographie), Völkerkunde.* ⑨ Phil. Leipzig 1900, mag. sc. Moskau 1910.

Adloff, Paul (1870), Dr. phil. med. dent. h. c. o. Univ.Prof., Dir. der Zahnärztl. Univ.Klinik. Königsberg (Deutschland), Steindamm 3. *Entwicklungsgeschichte und vergl. Anat. des Gehirnes des Menschen und der Anthropomorphen.* ⑨ Greifswald. o

Adolph, Edward Frederick (1895), Ass. Prof. of Physiol., School of Med. and Dentistry, Univ. of Rochester, New York (U.S.A.). *Physiol. of living membranes, physiol. of the kidneys, metabolism of water.* ⑨ Harvard 1920.

Adova, Anna Nicolaevna (1890), Chef de labor. de Chim. biol. fac. méd. II. Univ. Moscou (U.d. S.S.R.), Krivoarbatski pereoulok 19—4. *Ferments protéoclastiques; dérivés de la gouanidine.*

Adrian, Edgar Douglas (1889), Lect. in Physiol., Univ. Cambridge (England), Trinity College. *Nervous system.* ⑨ Cambridge, 1912.

Adrianee, Guy Webb (1895), Associate Prof. Horticulture, Agric. and Mech. Coll. Texas (U.S.A.), Coll. Station. *Fruit Growing. Sterility in Pecans, Phenol., Morphol. and Physiol. Developing fruit of the Pecan.* ⑨ Univ. of California 1917.

Adrianow, Paul I., Landwirtschaftl. Akad. nam. Timirjasew. Moskau (U.d.S.S.R.), Petrowsko-Rasumowskoje, Solomenaja Storoshka 8, W. 1. *Agronomie, Nahrungsmittel.*

Aduceo, V., Prof. Univ. Pisa (Italia). *Fisiol. umana.* o

Afanassieff, Wjatscheslaw (1859), Prof. Univ. Woronesh (U.d.S.S.R.), Grusovajastr. 6. *Pathol. Anat. Kefirbakterien. Microorganismen aus der Gruppe der sog. Septicaemia haemorrh. Organe der inneren Sekretion.* ⑨ Mil.-Med. Akad. St. Petersburg 1882.

Africa, Candido, Ass. Prof. Univ. of the Philip. Manila, Philippine Islands (U.S.A.). *Parasit.* o

Afzelius, Karl (1887), Doc. Stockholm (Sverige), Botan. Inst., Stockholms Högskola. *Embryol. und Zytol. der Compositae und Orchidaceae. Flora von Madagaskar.* ⑨ Stockholm 1924.

Agafonowa, M. W., Ass. Labor. d. Embryol. Univ. Leningrad (U.d.S.S.R.). o

Agar, Wilfred Eade (1882), Prof. of Zool., Univ. of Melbourne (Australia). *Cytol., with special reference to the Metazoan Nucleus. Genetics: Inheritance in Asexual Reproduction, Production of inheritable variations.* ⑨ Cambridge 1903.

D'-Agata, Giuseppe, Lib. Doc., Univ. Catania, Sicilia (Italia). *Pathol.* o

Agduhr, Erik (1886), Prof., Dir. anat.-histol. Inst. Tierärztl. Hochsch. Experimentalfältet (Sverige), Veterinärhögskolan. *Postfötale Entwickelung unter verschiedenen Funktions- und Nutritionsverhältnissen.* ⑨ Stockholm 1924. ⑦ Histol. Praep.

Agelskiy, Tichon (1900), Ass. physiol. Labor. d. Vet.-Inst. Kazan (U.d.S.S.R.). *Mechanik der Gelenke.* ⑨ Kazan 1925.

Aggazzotti, Alberto (1877), Prof. st., Dir. Ist. di Fisiol. Sperimentale Univ. Modena (Italia), Piazzale Garibaldi 1. ⑨ Torino 1923.

Agharkar, Shankar Purushottam (1884), Ghose Prof. of Botan. Univ. Calcutta (India), 35, Ballygange Circular Road. *Plant Geography.* ⑨ Berlin 1919. ⑦ Phanerogamic plants.

Ågren, Hugo Andreas (1877), Liz. phil., Fischereiintendant. Gävle (Sverige). *Apterygota Südschwedens und Lapplands. Fischereiwirtschaft.*

Agte, Wladimir N. Jalta (U.d.S.S.R.), Staatl. Nikitskij Botan. Garten. *Forstwirtschaft.* o

de Aguilar-Amat, Juan B. (1882), Dir. du Mus. de Biol. Barcelona (Espagne), Apartado 593. *Malacol.: Mollusques terrestres paléactiques.* ⑨ Barcelona 1911. ⑦ Mollusques terrestres et fluviatiles.

Aguilar Blanch, Romualdo. Valencia (España), Pasaje de Monistrol 4. *Mammalia, Aves.* o

Aguilar y Carmena, Fernando, Dir. de la Estación de Biología vegetal. Illescas, Toledo (España). *Biol. végétale.* o

Agulló y Garsot, Juan. Cambrils, Tarragona (España). *Entomol. agricole.* o

Aguinaga, Enrique González, Doc. Univ. Lima (Peru). *Anat., Physiol. d. Tiere.* o

Ahl, Christoph Gustav Ernst (1898), Verwalter d. herpetologischen Abteilung des Zool. Mus. Berlin N 4 (Deutschland), Invalidenstr. 43. *Systematik, Ökologie, Morphologie und Phylogenie der Reptilien, Amphibien und Fische.* ⑨ Berlin 1921. ⑦ Reptilien und Amphibien.

Ahlgren, Johan Gunnar (1898), Priv.Doc. der Physiol. Univ. Lund (Schweden), Physiologicum. *Regulation des Stoffwechsels durch die Hormone. Pharm. des Stoffwechsels. Gewebsoxydation sowie ihre Beeinflussung durch Insulin, Adrenalin, Thyroxin und Hypophysepraep.* ⑨ Lund 1926.

Aichel, Otto (1871), Dr. phil. et med., Prof., Dir. des Anthrop. Inst. der Univ. Kiel (Deutschland), Caprivistr. 9. *Physische Anthrop., vergl. Anat., Vererbungslehre.* ⑨ phil. Erlangen, med. Würzburg.

Aida, Tatsuo Prof. of Zool. and Anat. in Polytechnique. Kyôto (Japan), Biol. Labor., Kyôto Kôto Kôgei Gakkô. *Genetics of Gold-fishes.*

Aikman, John M. (1893), Associate Prof. of Biol., NebraskaWesleyanUniv.,Lincoln,Nebraska(U.S.A.). *Plant Ecology: Structure of the Deciduous Forest in Nebraska.*

Airaghi, Carlo, Prof. Univ. Fac. sc. Milano (Italia). *Palaeont.* o

Airila, Y., Dr. med., Prof. a. d. Univ., Dir. med.-chem. Labor. u. d. Pharmacol. Anstalt. Helsinki (Finnland), Fjälldalsgatan 5 A. *Pharm.* o

Airoldi, Marco (1900), Ass. à l'Inst. de Géol. Genova (Italie), Mus. Geol., Regia Univ. ⑨ Genova.

Aitken, William Alexander (1895), Associate Prof. of Anat. Ames, Ia. (U.S.A.), Vet. Division, Iowa State Coll. *Oestrous Cycle and Reproductive Phenomena of the Mare: Abdominal Viscera of the Ox.* ⑨ Iowa State Coll. 1917.

Ajello, Luigi (1898), Aiuto cattedra di anatomia patol. R. Università. Palermo (Italia), Via de Spuches 8. *Anatomia-patol. Batteriol. e sierol.* ⑨ med. Palermo.

Ajon, Guido (1891), Dr., Chim. della R. Stazione Sperim. di Agrumicolt. e Frutticoltura. Acireale (Italia). *Biochim. des limons.* ⑨ Palermo 1914. ⑦ Essences de limons.

Akamatsu, Shigeru, Dr. Med., Prof. Chiba (Japan), Labor. of Med. Chem., Med. Univ. *Biochem.*

Akatzuka, Kôzô, Ass. Prof. of Kyôto Imperial Univ. Kyôto (Japan), Zool. Labor., Coll. of Sc. *Phycol. and Planktonkunde.*

Åkerman, Ernst Åke (1887), Abteilungsdir., Weizen- und Haferabteilung des Saatzuchtvereins. Svalöf (Sverige). *Vererbungslehre und Pflanzenphysiol.* ⑨ Lund 1916.

Akimowa, Olga Dmitriewna (1883), Ass. Botan. Inst. Univ. Minsk (U.d.S.S.R.), Karl-Marx-Str. 56. *Algen.*

Aksentjew, Boris N., Doc. Inst. f. Volkswirtschaft u. Landwirtschaftl. Inst. Odessa (U.d.S.S.R.). Ul. Schepkina 7, W. 18. *Pflanzenphysiol. u. Biol.*

Akula, Tukaram Gangaram (1887). Bombay (Br. India), Parel, Haffkine Inst. *Stegomyia, Malaria. Entomol.*

Alabern, Enrique, Dr. Palma de Mallorca (España), Borne-Pelaires 104. *Cytol. gen. Histol.* o

Alaejos y Sanz, Luis. Santander (España), 19, Castelar. *Biol. marine.* o

Albertoni, Pietro (1849), Prof. Physiol. Bologna (Italia), Via della Libertà 5. *Action des bromures sur l'excitabilité de l'écorie cérébrale. Centres cérébraux*

1*

du mouvement. Centres d'arrêt dans le crapaud. Valeur alimentaire du vin, des albumines etc. Action des sucres sur l'organisme. Les sucres et l'action protectrice du foie. Valeur physiol. de l'ammoniaque.

Al'blekij, B. D., Doc. Univ. Nizny Novgorod (U.d.S.S.R.), Sovetskaja Ploščad 8. *Pharmacol.* o

Albrecht, Josef (1888), Dr. phil. Wien 5 (Österreich), Schönbrunner Str. 70. *Revision der pliocänen Säuger des Wiener Beckens. Paläont. Ergebn. der serbischen Forschungsreise.* ① Wien 1924.

Albrecht, William Albert (1888), Associate Prof. of Soils, Ass. Agric. Expt. Sta. Soils Dept. Coll. of Agric. Univ.of Missouri, Columbia, Mo. (U.S.A.). *Soil Bact.: Nitrogen Fixation, Inoculation, Nitrification* ① Univ. of Illinois 1919.

Alden, Charles H. (1894), Ass. Entomologist, U. S. D. A. Fort Valley, Georgia (U.S.A.). *Peach insects, life history and control.*

Aldenburg Bentinck, Godard Adrian, Graf von (1887). Overveen (Holland), Bloemendaalscheweg 196. *Entomol.(Lepidoptera).*) ① Darmstadt 1915.

Aldous, Alfred Evan (1885), Prof. of Pasture Improvement. Manhattan, Kansas (U.S.A.), Kansas State College. *Various tame forage species and mixtures of the different species are also tested to determine the ones best suited for seeding pasture lands in the various portions of Kansas. Selections of native forage grasses are being made to increase the viability of the seed. The aim being to get a strain that will have high enough viability to use in reseeding native pastures.*

Alder, Homer E., Ass. Prof. of Botan., Nebraska Wesleyan Univ. Univ. Place, Neb. (U.S.A.), 724, E. 22 d., St. *Taxonomy.* o

Aldrich, John Merton (1866), Associate Curator, Division of Insects, United States National Mus. Washington D.C. (U.S.A.). *Classification of Diptera of North America. Sarcophaga and allies in North America.* ① Leland Stanford Junior Univ. 1906.

Alechin, Wasilij W. (1882), Prof. I. Univ. et d'Acad. de Mines. Moscou (U.d.S.S.R.), Cabinet de Botan., Dolgorukowskaja 29, W. 98. *Phytosociol., Geographie botan., Systématique des phanérogames, Méthodique des explorations géobotaniques, steppes russes.* ① Moscou 1918.

Alekseev, Evgenij, Prof. Landwirtschaftl. Inst. Kiev (U.d.S.S.R.), Brest-Litovskoe-Chaussee 39. *Allg. Forstkunde.* o

Alekseieff, W., Prof., Dr., Inst. f. Pharm. u. Therapie d. Univ. Sofia (Bulgarien). o

Aleksejew, Aleksej M., Doc. Univ. u. Inst. f. Land- u. Forstwirtschaft. Kasan (U.d.S.S.R.), Ul. Galaktinowa 5, W. 1. *Botan. Zellphysiol. Microbiol.* o

Aleksjejev, A. I., Doc. Univ. Taschkent (U.d. S.S.R.). *Biochem.* o

Alemanno, Augusto (1904), Ass. Osservatorio di Fitopat. per le Puglie. Taranto (Italia), Piazza Ebalia 1. *Insetti.* ① Portici 1926.

Aleschin, Boris (1901), Aspirant Forschungs-Inst. Zool. I. Univ. Moskau 34 (U.d.S.S.R.), Ostoschenka 38. *Experim. und physiol. Zytol. und Histol. Aktuelle Reaktion der Zelle und Gewebe. Chondriom. Normale und artifizielle Metamorphose.* ① Moskau 1925.

Aleskowskij, Michail W., Ass., Saratower Staatl. Inst. f. Landwirtschaft u. Melioration. Saratow (U.d.S.S.R.), Nikolskaja 29, W. 8. *Botan. Anat. Cytol.* o

Alessandri, A., Prof. Univ. Roma (Italia). *Batteriol.* o

Alessandrini, G., Prof. Univ. Roma (Italia). *Parasit.* o

Alessandro, Giovanni, Lib. Doc. Univ. Messina (Italia). *Physiol.* o

Alexander, Charles Paul (1889), Ass. Prof. of Entomol., Massachusetts Agric. Coll. Amherst, Massachusetts (U.S.A.), Fernald Hall. *Systematic Entomology. Tipuloidea (Diptera) of the World.* ① Cornell 1918.' ② 3500 species of Tipuloidea from all regions of the World. Exchange material in this group for species not in the collection or will determine material from remote regions for the privilege of retaining duplicates. Write before sending material.

Alexander, Edward Gordon (1901), Ass. Prof of Biol., Central College. Fayette, Missouri (U.S.A.). *Ecology, Physiol. of Protista, especially Euglena. Ecology of Birds and Mammals.*

Alexander, Gustav (1873), o. Univ.Prof. Wien VIII (Österreich), Skodagasse 75. *Anat. d. Gangl. vestibul. Entwicklungsgesch. d. pars inferior labyr. Hörtheorie.* ① Wien. o

Alexandrov, Wasily G. (1887), Prof., Dir. Physiol. Labor. d. Botan. Gartens. Tiflis, Transkaukasien (U.d.S.S.R.). *Anat. und Physiol. der Pflanzen. Ekol., Wasserhaushalt, quantitative und qualitative Anat.* ① Leningrad 1912.

Alexandrowicz, Jerzy Stanislaw (1886), prof. d'histologie et Dir. de l'inst. d'histologie à l'Univ. Wilno (Pologne). *Système nerveux sympathique chez les invertébrés. Innervation des organes de la vision.* ① phil. Zürich 1910, med. Jena 1913.

Alexejev, Alexius K., Prof. Univ. Odessa (U.d S.S.R.), Geol. Kabinet. *Palaeont. des Tertiärs, speziell Vertebrata.* o

Alezais, M., Prof. Ecole Méd. Marseille (France). *Anat.* o

Alfken, Johann Diedrich (1862), Entomologe am Mus. für Natur-, Völker- und Handelskunde in Bremen. (Deutschland), Delmestr. 18. *Apidae. Die Insekten des Memmert.*

Alföldy, Zoltán, Mitarbeiter am Pharm. Inst. der Univ. Debreczen (Ungarn). o

Aliaga, José M., Prof. Univ. Nac. Córdoba (Argentina). *Anat. descr.* o

Alicante, Marcos (1895), Soil biologist. Bureau of Science, Manila, P. I. (Philippines). ① Univ. of Illinois 1923.

Alissowa, Eugenie N., Dir. d. Botan. Kabinetts d. Süd-Ussurijsker Abtlg. d. Geographischen Gesellschaft. Nikolsko-Ussurijssk (U.d.S.S.R.), Krasnosnamenskaja. *Botan., Anat., Morphol., Phytosoziol.*

Aljakritskij, Vjačeslav V., Prof. Univ. Smolensk (U.d.S.S.R.). *Pathol. Anat.* o

Aljawdina, Ksenija P., Doc. Polytechn. Inst. Iwanowo-Wosnessensk (U.d.S.S.R.), Sewerskaja ul. 22. *Microbiol. Phytopath.* o

Alksnis, Jēkabs, Prof. Dr. Univ. Rīga (Latvija), Brīvibas iela 2, dz. 12. *Pathol. Anat. d. Menschen.* o

Allan, Harry Howard (1882), Dr., Feilding Agric. High School. Feilding, New Zealand (Australie). 4 Camden St. *Wild hybrids in New Zealand. Response to environment in plants. Artificial crossing of wild plants. Analysis of ,,Linnean Species" in New Zealand Plants. New Zealand Trees and Shrubs. Taxonomic status of the Genus Hebe. Vegetation of Mount Peel.* ① Auckland, N.Z., 1924. ② Seeds of grasses. Mosses and Lichens.

Allan Thompson, J., Dr. M. A. Dominion Mus. Wellington (New Zealand). *Palaeont.* o

Allee, Warder Clyde (1885), Associate Prof. of Zool. Univ. of Chicago, Ill. (U.S.A.). *Behavior and distribution of animals, physiol. analysis of animal behavior.* ① Chicago 1912.

Allen, Agnes Louise (1898), Associate, Inst. for biol. Research of Johns Hopkins Univ. Baltimore, Md. (U.S.A.), 605 Park Avenue.

Allen, Bennet M. (1877), Prof. of Zool. Los Angeles, Cal. (U.S.A.), Univ. of Southern California. *Development of germ glands, origin of germ cells, influence of glands of internal secretion upon growth.* ① Chicago 1903. o

Allen, Charles Elmer (1872), Prof. of Botany, Univ. of Wisconsin. Dept. of Botany, Madison, Wis. (U.S.A.). *Cytology and genetics, analysis of inheritance in Sphaerocarpos.* ① Univ. of Wis. 1904.

Allen, Charles L. (1889), Ass. Prof. of Animal Husbandry. Ithaca, N.Y. (U.S.A.). *Dairy cattle, milk.*

Allen, Edgar Johnson (1866), Dir. Marine Biol. Labor. Plymouth, Devon. (England), Citadel Hill. *Marine Biol. Crustacea, Polychaeta.* ⊕ London 1885.

Allen, E. W., Chief, Office of Experim. Stations, U.S. Department of Agric., Washington, D.C. (U.S.A.). o

Allen, Ezra (1870), Prof. of Biology, Ursinus Coll., Research Associate at the Carnegie Inst., Dept. of Genetics, Cold Spring Harbor, N.Y. Collegeville, Pa. (U.S.A.). *Cytology.* ⊕ Univ. of Pennsylvania 1914.

Allen, Frank W. (1887), Ass. Pomol. Berkeley, Cal. (U.S.A.), Agric. Experim. Station. Univ. of California. ⊕ Iowa State Coll. 1913.

Allen, Harry W. (1892), Entomologist. Japanese Beetle Laboratory Riverton, N.J. (U.S.A.). *Parasites of insects, ecology and taxonomy of Tachinidae (Diptera).* ⊕ Ohio State Univ. Columbus, Ohio 1926. ⊕ Tachinidae.

Allen, Paul William (1883), Prof. and Head of Dept. of Bact. at Univ. of Tennessee. Knoxville, Tenn. (U.S.A.). *Microbiology. Industrial Fermentations.* ⊕ Univ. of Illinois.

Allen, William Fitch (1875), Prof. of Anat., Head of the Dept. Univ. of Oregon Medical School. Portland, Oregon (U.S.A.). *Neurol. and Embryol. Exper. and Morphol. Neurol., Blood and Lymphatic systems.* ⊕ Univ. of Minnesota 1915.

Allen, William Ray (1885), Associate Prof. of Zool. Univ. of Kentucky, Lexington, Ky. (U.S.A.). *Limnology, Ichthyology, Feeding habits of Pelecypoda, Neotropical fishes.* ⊕ Indiana Univ., Bloomington.

Allen, Winfred Emory (1873), Assistant Prof. of Biology, Scripps Inst. of Univ. La Jolla, Calif. (U.S.A.). *Marine plankton diatoms and dinoflagellates. Quantitative studies. Ecological research. Accumulation and interpretation of numerical records of seasonal, geographic, topographic, and bathymetric distribution of diatoms and dinoflagellates of the marine plankton.* ⊕ Microplankton.

Allis, Edward Phelps junr. (1851). Menton (France), Palais de Carnolès. *Anat., development of the vertebrate head.* ⊕ Boston, Mass., 1871.

Allison, Franklin E. (1892), Research chemist and bacteriologist, U. S. Department of Agric., Washington, D. C. (U.S.A.). *Relations, particularly chemical, existing between Bacillus radicicola and its host plants. The mechanism of nitrogen fixation by leguminous plants.* ⊕ Rutgers College 1917.

Allison, Robert V., Tropical Plant Research Foundation. Central Baragua, Baragua, Province de Camagney (Cuba). *Physiol.* o

Allyn, Harriett May (1883), Ass. Prof. Poughkeepsie, N.Y. (U.S.A.), Vassar Coll. ⊕ Univ. of Chicago 1912.

Almquist, Ernst Bernhard (1852), Prof. emerit. Stockholm (Sverige), Östermalmsgatan 80. *Bact. Hygiene. Linné. Artbildung.* ⊕ 1882.

Alonskiy, Sergius (1896), Doc. Kazan (U.d. S.S.R.), Veter.-Inst., Arskoje Pole 96. *Physikalische Chemie d. Lebenserscheinungen.* ⊕ Kazan 1920.

Alonzo, Giovanni, Lib. Doc. Catania, Sicilia (Italia), Univ. *Patol. gener. Istol. patol.* o

Aloy, J., Prof. Univ. Toulouse (France). *Chim. biol.* o

Alpatov, Wladimir (1898), Custos Zool. Mus.Univ., Priv.Doc. der Variationsstatistik. Moskau (U.d. S.S.R.) 66, Gozochowski per, Haus 16, W. 2. *Variabilität der Tiere, geographische Variabilität der Honigbiene, Systematik der arktischen Crustaceen und der palaearktischen Ameisen.* ⊕ Arktische Crustaceen u. russische Bienenrassen.

Alpern, D. E., Priv.Doc. Staatl. Psychoneurol. Inst. Charkow (U.d.S.S.R.). *Pathophysiol.* o

van Alphen, Adriaan Jacob Sem. (1877), Bactériologiste à l'Inst. de Serothérapie de l'Etat. Rotterdam (Holland), Schoonderloostraat 82a. *Sera divers.* ⊕ Bern 1910.

Alsberg, Carl Lucas (1877), Dir. Food Research Inst., Stanford Univ., California (U.S.A.). *Agric.* *Chem., Food Chem., Chem. of milking and baking, starch. Colloid Chem. in its application to biol.* ⊕ Columbia Univ., New York City 1900.

Alsterberg, Gustaf (1892), Doc. Univ. Lund (Sverige). *Süßwasserbiol. und Hydrographie, Wasseranalyse (Sauerstoffanalyse), Biochem. (Geschlechtsreaktionen) und Histol. der Nervengewebe.* ⊕ Lund 1922.

van Alstyne, Lewis Morrell (1892), Ass. in Research Geneva, New York (U.S.A.). *Horticulture. Fruit variety tests.*

Alvarado, Salustio (1897), Dr. rer. nat., Prof. der Biol. und Physiol. im Inst. Nacional de II Enseñanza de Tarragona (Spanien). *Chondriosomen der Pflanzen und Entstehung der Plastiden. Jugendformen bei Phanerogamen und Farnen. Mesogloea der Medusen.* ⊕ Madrid 1922.

Alvarez Lopez, Enrique (1897), Prof. à l'Inst. de Cadiz (España), Inst. de 2. Enseñanza. *Zoogeographie.*

Alverdes, Friedrich (1889), a. o. Prof. der Zool. am Zool. Inst. Halle a. S. (Deutschland), Domplatz 4. *Sinnesphysiol. der Wirbellosen, Tierpsychol., Perlen, Copepoden, Vererbungslehre, Infusorien, Zentralnervensystem der Arthropoden. Rassen- und Artbildung. Verhalten der niederen Organismen. Tiersoziologie.* ⊕ Marburg i. H. 1912.

Alves Ribeiro, Benjamin (1899), Assistant at Institute of Hygiene of São Paolo (Brazil), Rua Brigadeiro Tobias 45. *Physiology and Psychology.* ⊕ São Paulo, 1921.

Alway, Frederick James (1874), Prof. soil chem., chief div. of soils, Univ. of Minn. St. Paul, Minn. (U.S.A.), 1386 Grantham Avenue. *Soil chem. Soil physics.* o

Amadori, Leonida (1900), Dr., Ass. in prima nell' Orto Botan. della R. Univ. di Pisa (Italia). *Fisiol. vegetale. Micol.* ⊕ Bologna.

Amaldi, Paola (1897), Dr. Firenze (Italia), Via S. Salvi 12. *Anat. vegetale. Helwingia.* ⊕ Firenze 1921.

Amalitzky, Anna P., Geol. Mus. der Akademie der Wissenschaften. Leningrad, W. O. (U.d.S.S.R.), Tutschkowa Nab. 2. *Palaeont., spez. Wirbeltiere und Wirbellose.* o

Amann, Jean-Jules (1859), Lausanne (Suisse), 45, Avenue de Rumine. *Biochim., Bryologie. Flore des Mousses et Bryogéographie de la Suisse.* ⊕ Lausanne (Univ.) 1900. ⊕ Mousses.

Amantea, Giuseppe (1885), Prof. Fisiol., Dir. Ist. di Fisiol. Univ. Messina (Italia). *Funzione dei centri nervosi. Funzione dell' apparato genitale.* ⊕ Roma 1925.

do Amaral, Afranio (1894), B. Sc. and Lit., M.D., D.P.H., Dir., Antivenin Inst. of America. Glenolden, Penna. (U.S.A.). *Biol. and Immunology applied to poisonous animals, Snakes. Bancroftose. Filariose de Bancroft.* ⊕ M.D. Bahia, (Brazil) 1916, D.P.H. Boston (U.S.A.) 1924. ⊕ Snakes, both Neotropical and Nearctic.

Amari, Shin-ichi, Sc. of Imp. Sericult. Experim. Station. Suginami-chô near Tôkyô (Japan), Pathol. Labor. *Parasit. of silk-worms.*

Amato, Alessandro, Prof. incar. Univ. Bari (Italia). *Bact. Parasitol.* o

D'Amato, Luigi, Prof. Univ. Messina (Italia). *Pathol.* o

Amatucci Mallardo, Cesare (1903), Ass. Ist. di Patol. gen. Univ. Napoli (Italia), S. Andrea delle Dame 21. *Patol. sperimentale.* ⊕ Napoli 1926.

Ambard, M. L., Prof. Fac. Méd. Strasbourg (France), Labor. de Pharmacodynamie. *Physiol.* o

Ambrož, Emil (1881), Prof. Dr., Jihlava (C.S.R.), Stefanikovo nám 19. *Phanérogames, protection des plantes.* o

Amemiya, Ikusaku, Ass. Prof. of Tôkyô Imperial Univ. Komaba near Tôkyô (Japan), Inst. of Hydrobiol., Fac. of Agric. *Aquatic animals.*

Ames, Adeline (1879), Prof. of Botany in Sweet Briar Coll., Va. (U.S.A.). ⊕ Cornell Univ. 1913.

Ames, Oakes (1874), A. M., Ass. Prof. of Botan. and Curator of Botan. Mus of Harvard Univ. North Easton, Mass. (U.S.A.). *Orchid flora of Florida.* ○

van Ameyden, Ubbe Peter (1891), Dr. zool., Landwirtschaftlicher Berater der Fiong Ham Zuckerfabriken. Semarang, Java (Ned.-Indië). *Zuckerrohrkultur.* ① Utrecht 1917.

Amiredžibi, Simon, Prof. Univ. Tiflis (U.d. S.S.R.). *Bact. u. Hygiene.* ○

Ammal, Edavaleth Kakkat Janahi, Prof. of Botan., The Women's Christian Coll., Cathedral P.O. Madras (India). *Genetics.* ○

Ammon, Paul Ludwig Karl (1895), Doc. der Zool. Tula (U.d.S.S.R.), Morosowstr. 18. *Ornithol. im Gouvernement Tula.*

Ammons, Nelle (1889), Instr. in Botany at West Virginia Univ. Morgantown, Va. (U.S.A.), 146, Hunt St.

Amoore, K. L., Harpenden, Herts (England), Inst. of Plant Nutr. and Soils Problems. *Fermentation.* ○

Amos, Felipe R., Ass. Prof. Forest School. Manila, Philippine Islands (U.S.A.). *Forest Engin.* ○

Amoss, Harold L. (1886), Assoc. Prof. of Med., Johns Hopkins Univ., Assoc. Physician, Johns Hopkins Hospital. Baltimore, Md. (U.S.A.). *Experimental Pathology, Bact. Hemolytic Streptococci.* ① M.D. Harvard 1911, Dr.P.H. Harvard 1912, Sc.D. (Hon.) Washington Univ. 1922.

Amschler, Johann Wolfgang (1893), Dr. rer. techn. Priv.Doc. München (Deutschland), Technische Hochsch. *Tierzucht und Züchtungsbiol.* ① 1922. ⓟ Haus- und Wildtiere Rußlands, Sibiriens, Turkestans und der Kaukasusländer.

Amsler, Cäsar (1881), o. Prof. Pharmacol. Univ. Riga (Latvija), Pharm. Inst., Kronwald-Boulevard 9. *Experim. Pharm. und Pathol.* ① Basel 1910.

Ancel, M., Prof. Fac. Méd. Strasbourg (France). *Embryol.* ○

d'Ancona, Umberto (1896), Aiuto e Libero Doc. Ist. di Anat. e Fisiol. comparate della R. Univ. di Roma (Italia), Via Depretis 91. *Biol. des poissons, réproduction, détermination du sexe. Métabolisme.* ① Roma 1920.

Anders, Gustav(1866). Charlottenburg (Deutschland), Königin-Elisabeth-Str. 50. *Pflanzenbiol., Vererbungslehre, Rassenhygiene.*

Anders, H. E. (1886), Dr. med., Priv.Doc. und I. Ass. des Pathologischen Inst. der Univ. Freiburg i. Br. (Deutschland), Albertstr. 19. *Allgemeine und spezielle Pathologie. Teratologie.* ① Berlin 1911. ⓟ Tierische und menschliche Mißbildungen.

Andersen, A. C., Landbohøjskolens Forsøgsvirksomhed. København V (Danmark), Rolighedvej. ○

Andersen, C. W., Dr., Landw. Hochsch., Seruminst. København V (Danmark), Bülowsvej. ○

Andersen, Dorothy Hansine (1901), Ass. in Anatomy. Story Memorial Hospital, Rochester, NY. (U.S.A.). *Lymphatics, lymphatic drainage of the female reproductive tract.* ① Johns Hopkins Medical School 1926.

Andersen, Karl Theodor (1898), a.o. Hochschulprof. Weihenstephan b.München (Deutschland), Zool. Inst. der landwirtschaftlichen Hochsch. *Entwicklungsgeschichte der Mollusken u. Wirbeltiere. Morphologie der Mollusken. Physiol. des Herzens der Reptilien.* ① Erlangen 1922.

Anderson, Arthur (1890), Associate Prof. of physiol. Chem. Pennsylvania State Coll. (U.S.A.). *Agric. biol. Chem.* ① Univ. of Minnesota 1923.

Anderson, Carl George (1897), Ass. in Plant Pathologie, Univ. of Minnesota. Plant Pathology Department Univ. Farm St. Paul (U.S.A.). *Sweet corn diseases, seed treatments, fertilizers, mostly in connection with sweet corn for canning at the canning plants.*

Anderson, Edgar (1897), Ass. Prof. of Botan., Washington Univ. Geneticist, Missouri Botan. Garden. Saint Louis (U.S.A.). *Geographical variation.* ① Harvard Univ. 1922.

Anderson, Emma N., Ass. Prof. of Botan., Univ. of Nebraska. Lincoln, Neb. (U.S.A.), 1507 R. St. ○

Anderson, E. O. (1897), Ass. Prof. Dairy Manufacturing, Ass. Dairy Mannufacturing in Experiment Station. Storrs, Conn. (U.S.A.). *Milk and its products. Some of the factors (physical and chemical) which influence the stability of casein.*

Anderson, Flora C., Ass. Prof. of Botan., Indiana Univ. Bloomington, Ind. (U.S.A.), 327 S. Henderson St. ○

Anderson, Harry Warren (1885), Assoc. Prof. of Pomological Pathol., Dep. of Hortic., Univ. of Illinois, Urbana, Illinois (U.S.A.). *Diseases of horticultural crops, fruit crops.* ① Univ. of Illinois 1917.

Anderson, Jacob Peter (1874), Florist. Juneau, Alaska (U.S.A.), Box 1074. *Herbarium material, Plant diseases, Plant breeding.*

Anderson, John Arlington (1893), Instr. Department of Bact., Rutgers Univ., New Brunswick, N.J. (U.S.A.). *Fermentation enzymes.* ① Univ. of Wisconsin 1926. ⓟ Any cultures of bacteria.

Anderson, John Edward, Ass. prof. in Botan. Cedar Rapids, Iowa (U.S.A.), Coe College. *Genetics, mycol.* ○

Anderson, John F. (1873), Dir., Research and Biol. Labor. of E. R. Squibb & Sons. New Brunswick, N.J. (U.S.A.). *Bacteriology, Immunology. Anaphylaxis, Etiology of Measles, Etiology of Typhoid Fever.* ① Univ. of Virginia 1895.

Anderson, Rudolph Martin (1876), Dr., Zoologist, Chief of Division of Biol., Victoria Memorial Mus., Department of Mines (the National Mus. of Canada), Ottawa, Ontario (Canada). *Mammalogy, systematic of mammals of Canada, taxonomy, distribution, and ecology; Ornithology (Nearctic); Distribution, ecology and life histories of arctic and subarctic animals; Wild Life Conservation.* ① 1906. Univ. of Iowa. ⓟ Plants.

Anderson, Walter Sewell (1867), Prof. of Genetics, Univ. of Kentucky. Lexington, Ky. (U. S. A.). *Coat Colors in Horses. Vitality and Morphology of Spermatozoa. Sterility of Males. Domestic Animals.*

Anderson, William Arthur, jr. (1900), Microscopist, Department of Feed Control, Kentucky Agric. Experiment Station. Lexington, Ky. (U.S.A.). *Systematic botany of the higher plants.* ① Univ. of Kentucky 1923. ⓟ Herbarium specimens.

Andersson, J. G., Dr., Prof. Högskolan. Stockholm (Sverige). *Palaeont.* ○

Andervont, Howard Bancroft (1898). School of Hygiene, Baltimore, Maryland (U.S.A.), 615 North Wolfe Street. *Filterable Viruses, Protozoology, Helminthology, Bact., Immunology.*

Ando, Akira. Nagoya (Japan), Kasuga-chô 19, Naka-ku. *Parasit.: Cestoda.*

Andrásovszky, Josef, Dr., Ampelol. Inst. des Ung. Min. f. Landw. Budapest II (Ungarn), Debröigasse 17. *Ampelographie.* ○

André, Emile-Henri (1870). Prof. ord. Fac. Sc. Univ. Genève (Suisse), 10 Délices. *Biol. lacustre. Parasit. Infusoires, Acanthocéphales, Hirudinées, Invertébrés de la Suisse.* ① Genève 1893.

André, Jean-Marie-Gustave (1876), Prof. membre de l'Inst. de France et de l'Académie d'agriculture. Paris 6e (France), 120, Boulevard Raspail. *Chimie agricole. Chimie du sol. Chimie végétale.* ① 1897.

Andrea, M., Prof. Fac. d. Med. Bahia (Brasil). *Pathol. Anat.* ○

Andrea, Teodosio, Prof. Univ. Nac. d. Litoral. Corrientes (Argentina). *Anat. topogr.* ○

Baron **Andreánszky**, Gábor (1895), Ass. Inst. systematische Botan. und Pflanzengeographie Univ. Budapest (Ungarn). *Flora der mediterranen Gebirge.* ① Budapest 1922.

Andrée, Julius (1889), Dr. phil., Priv.Doc. für Geol. u. Paläont. an der Univ. Münster i.W. (Deutschland), Kampstr. 2. *Paläont. u. Diluvial-Prähistorie.* ① Münster i. W. 1917.

Andrée, Karl Erich (1880), Dr. phil., o.ö. Prof. für Geol. und Paläont. an der Albertus-Univ. zu Königsberg i. Pr. (Deutschland), Brahmsstr. 19 I r. *Rezente und fossile Sedimente des Meeres. Fossile Isopoden. Gebirgsbildung. Bernstein-Inklusen, Pflanzen und Tiere des Alttertiärs.* ⓓ Göttingen 1904. ⓟ Fossilien, Bernstein-Inklusen.

Andreev, N. N., Priv.Doc. Staatl. Inst. f. ärztl. Fortbildung. Leningrad (U.d.S.S.R.), Kiročnaja 41. *Bact.* o

Andreev, V. A., Priv.Doc. Staatl. Inst. f. ärztl. Fortbildung. Leningrad (U.d.S.S.R.), Kiročnaja 41. *Hygiene.* o

Andrejev, Nicolas Hyp. (1881), Doc. Phytopath. Landwirtschaftl. Inst., Phytopath. Kabinett. St. Novočerkassk (U.d.S.S.R.), Chomutowskij 33. *Phytopath.* ⓓ Don. Polyt. Inst. 1916. ⓟ Mycol. Obj.

Andres, Angelo (1851), Prof. Univ. Parma (Italia), Palazzo dell' Ateneo. *Actinia.* ⓓ Pavia.

Andreu y Rubio, José, Prof. de Hist. Nat. en el Sem. de Orihuela, Alicante (España). *Diptères d'Espagne.* o

Andrewes, Sir Frederik William (1859), Prof., Univ. London (England), South Kensington S. W. 7. St. Barthol. Hospital. *Pathol. Bacteriol.* ⓓ Oxford 1887.

Andrews, Ethan Allen (1859), Prof. of Zool. Johns Hopkins Univ. Baltimore, Md. (U.S.A.). *Non-vertebrates. Annelides, Insects, Amphioxus, Crayfish, Protozoa.* ⓓ J. H. Univ. 1887. ⓟ Folliculina.

Andrews, Frank M., Prof. of Botan., Indiana Univ., Desires exchanges in Physiol. Bloomington, Ind. (U.S.A.), 901 East 10 St. *Physiol.* o

Andrews, James C. (1892), Ass. Prof. of Physiol. Chem., School of Med., Univ. of Pennsylvania, Philadelphia, Pa. (U.S.A.). *Physical chem. of amino acids and proteins.* ⓓ Columbia Univ. 1918.

Andrews, Justin Meredith (1902), Dr., Instr. in Protozool., Johns Hopkins Univ. Baltimore, Maryland (U.S.A.), 615 Wolfe Street. *Parasitic protozoa of man and domestic animals; intestinal flagellates, amoebae, and coccidia; immunology of protozoan diseases, physiol. changes in the host during infection of a biochem. nature.* ⓓ Baltimore 1926. ⓟ Parasitic protozoa.

Andrijevškyj, P., Prof. Ukrainische Univ. Praha II (C.S.R.), Stěpánská ul. č. 49. *Biol.* o

Andriska, Viktor, Dr., Mitarb. am Hygien. Inst. Univ. Budapest (Ungarn). o

Andronescu, Demetrius I., Prof. Gov. Agric. Experim. Sta. Bucarest (Romania), Soseau Kiselef 22. *Physiol., cytol., genetics.* o

Andrus, E. Cowles (1896), Associate in Med., Johns Hopkins Univ., and Resident Physician, Johns Hopkins Hospital. Baltimore, Md. (U.S.A.). *Physiol. of cardiac rhythm.* ⓓ Johns Hopkins Univ. Medical School 1921.

Anel, Marcel (1900), Ass. Fac. d. Sc. Paris (France), Labor. d'Evolution, 105, Boul. Raspail. *Biol. experim., Histol.* o

Anelli, Giuseppe (1890), Ass. della Sezione Istol. dell' Ist. Anat. della R. Univ. di Milano (Italia), Via Antonio Rosmini I A

Angel, Clément Fernand (1881), Préparateur d'herpétol. et d'Ichthyol. au Mus. National d'Hist. Nat. de Paris V (France), 57, rue Cuvier *Reptiles. Larves de Poissons.* ⓓ Paris 1909.

de Angelis, Giovanni (1879), Aiuto Ist. di Patol. Generale Univ. Bologna (Italia). *Batt. Micol. Istol. patol. Fisiol. patol.* ⓓ Messina.

Angeloff, Stephan (1878), o. Prof. für Bakter., Serologie und Infektionskrankheiten an der Vet.-Med. Fac. der Univ., Dir. der Vet.-bact. Inst. in Sofia (Bulgarien), Alabinskastr. 3. *Schafpocken, Rotz, Rinder- und Schweinepest.* ⓓ Gießen 1908.

v. Angerer, Karl (1883), Priv.Doc., ao. Prof., Ass. Hygien. Inst. München (Deutschland), Pettendorfer Str. 34. *Hygiene und Bact.: Anwendungen der Physik u. physikalischen Chemie.* ⓓ München 1908.

Anghelowa, Maria, Dr., Prosektor Inst. f. Histol. u. Embryol. des Menschen. Sofia (Bulgarien). o

Angier, Roswell Parker (1874), Prof. of psychol. Univ. of Chicago. New Haven, Conn. (U.S.A.), 140 Edgehill Rd. *Psychol.* o

d'Angremond, Arend (1883), Dr. phil., Dir. Proefstation voor Vorstenlandsche Tabak. Klaten, Java (Ned.-Indië). *Cytologie und Genetik der Gattung Musa, und der Phasmiden.* ⓓ Zürich 1914.

Anguita, Aug., Prof. Univ. of St. Tomas. Manila, Philippine Islands (U.S.A.). *Anat.* o

Angyal, Dezsö. Budapest I (Ungarn), Horthystraße 15 a. *Pomol.* o

Anlčhov, N. N., Prof. Staatl. Inst. f. experim Med. Leningrad (U.d.S.S.R.). *Pathol.* o

Anigstein, Ludwik (1891), Dr. med. et phil., Chef de la Séction de Parasit. de l'Inst. d'Hygiène de l'Etat, Lect. en Parasit. à l'Ecole d'Hygiène de l'Etat Polonais. Varsovie (Pologne), Kujawska 2. *Microbiol., Trypanosomes, spirochètes, encephalite, "Rickettsia".* ⓓ phil. nat. Heidelberg 1913, med. Dorpat 1915. ⓟ Préparations de bact. et protozool.

Anile, Antonino (1869), Prof. di Anat. umana nelle R. Accademia Belle Arti Roma (Italia), Via Ripetta 218. *Anat. humaine, système nerveux. Biol. générale.*

Ankel, Wulf Emmo (1897), Dr. phil. nat., Ass. am Zool. Inst. der Univ. Gießen (Deutschland). *Cytologie der Geschlechtszellen, Spermatozoendimorphismus der Prosobranchia, Biol. der Prosobranchia.* ⓓ Frankfurt a. M. 1923. ⓟ Lebende Prosobranchier, vor allem exotische.

Anker, Jean Thore Højer Jensen (1892), Unterbibliothekar Univ.-Bibliothek København S (Danmark), Amagerbrogade 168, IV. *Tierische und menschliche Biol. und Vererbungslehre. Vererbung der Haarfarbe beim Dachshunde.*

Annenhova-Chlopina, N. P., Leiter einer wiss. Abt. Zool. Mus. Leningrad (U.d.S.S.R.), Univ. nab 1. *Wirbellose.* o

Anrep, Gleb (1890), Lect. in Physiol. Cambridge (England). ⓓ Petrograd 1915.

Anselmi, Raffaello (1893), Libero doc. di Anat. e Fisiol. Compar., Aiuto Anat. Compar., Prof. incaricato di Anat. e Fisiol. Umana per la Fac. di Sc. Nat. Univ. di Genova (Italia). *Microfaune marine. Anat. et physiol. de la respiration des amphibies. Hémogregarines des vertébrés.*

Anserov, N. I., Prof., Dir. Anat. Inst. Univ. Perm (U.d.S.S.R.), Ul. Karl Marksa 26. o

Anthony, Raoul Louis Ferdinand (1874), Prof. au Mus. national d'Histoire naturelle, Prof. à l'Ecole d'Anthrop., Dir. à l'Ecole des Hautes Etudes. Paris 5e, (France) 55, Rue de Buffon. *Anat. comparée surtout des Mammifères, anat. du cerveau des Cétacés, des Primates. Morphogénie. Méthodologie. Facteurs éolutifs.*

Anthony, Roy David (1884), Prof. of Experimental Pomology. Pennsylvania State Coll., State Coll., Pa. (U.S.A.). *Improvement of fruit stocks; fruit fertility; fruit storage.* ⓓ Cornell 1920.

Anthonius, Otto, Priv.Doc. Wien (Österreich), Schönbrunn. *Zool.* o

Antipa, Gregori (1867), Dir. du Mus. d'Hist. nat. Bucarest (Romania), Piatza Victoria. *Zool., Ichthyol. Hydrobiol., Limnol. et Oceanographie. Fischbiol.* ⓓ Jena 1891.

Antoine, G., Prof. Ecole de Méd. vét. Bruxelles (Belgique). *Pharmacol.* o

Antokolčskaja, Maria Pavlovna (1887), Ass. Mykol Inst. Leningrad (U.d.S.S.R.), Zagarodnyi prosp. 33 log. 34. *Mykol. und Phytopath. Pilze.*

Antonini, Attilio, Prof. R. I. Med. vet. Milano (Italia). *Anat.* o

Antonow, Alexis (1900), Ass. Histol. Labor. Univ. Saratow (U.d.S.S.R.). *Gl. pinealis des Menschen und der Tiere. Fett- und Lipoidablagerungen in inkretorischen Organen.*

Antón y Ferrándiz, Manuel, Dir. del Mus. de Antrop., Catedrático jubilado de la Fac. de Ciencias. Madrid (España), Olózaga, 5 y 7. *Antrop.* o

Anufriev, W. N., Ass. Dr., Kubansches Med. Inst., Anat. Labor. Krasnodar (U.d.S.S.R.), Str. Sedin 4. o

Anufriew, Gennadij (1889), Doc. Leningrad (U.d.S.S.R.), Forstinst. *Phytosoziol. der Wälder, Wiesen, Moore. Torfmoor-Stratigraphie. Pollenanalytische Untersuchungen. Paleobotan. Systematik: Sphagnum.* ⊕ Sphagnum.

Anutschin, A. W., Prof., Dir. Zool. Labor. Veter. Inst. Moskau (U.d.S.S.R.).

Aoki, Bun-ichirô, Prof. Zool. of Aichi Med. Univ. Nagoya (Japan), Biol. Labor. *Mammalian Zool.*

Aoki, Kao-ru (1877), Prof. of Tôhoku Imper. Univ. Igaku-Lakushi. Sendai (Japan), Bact. Inst. *Immunisatorische Untersuchung von Bakterien, Variationsforschung, Differenzierung, Vaccinherstellung, Sera. Filtrierbares Virus.* ⊕ Tokio 1905. ⊕ Bakterienstämme, besonders Paratyphus-Gruppe.

Aparicio, F., Prof. Univ. Paraná (Argentina). *Paleont.* o

Aparicio, Julio, Prof. Univ. Nac. Bogotá (Colombia). *Physiol.* o

Apollinaire, Marie (1867), Frère de la Congrégation des Frères des Ecoles Chrétiennes, Ex-Prof. de Zool. à la Fac. de Méd., Ex-Prof. d'Entomol. agricole à la Fac. d'Agronomie. Bogotá (Colombia), Apartado No. 473. *Invertébrés de la Faune Colombienne.* ⊕ Bogotá 1912.

Aportolakis, Georges (1890), Prosecteur à l'Inst. Anat. de l'Univ. Athènes (Grèce), Rue Andrianou 21. *Muscle pyramidale et masseter.* ⊕ Athènes 1911.

Apostoleanu, Eugen (1889), Prof., Ecole Sup. d'Agric. de Bucarest (Romania). *Anat. et Physiol. animale. Système lymphatique.* ⊕ Méd. vét. Bucarest 1916, Paris 1924. ⊕ Préparations microscopiques.

Appel, Gustav Otto (1897), Dr. agr., Ass. am Bot. Inst. Braunschweig (Deutschland), Fasanenstr. 48. *Pflanzenzüchtung.* ⊕ Berlin 1924.

Appel, Otto (1867), Geh.R.-R., Dir. d. Biol. Reichsanstalt für Land- u. Forstwirtschaft Berlin-Dahlem, Hon.-Prof. a. d. Landw. Hochsch. Berlin. Berlin-Dahlem (Deutschland), Königin-Luise-Str. 17—19. *Pflanzenkrankheiten und Pflanzenschutz. Kartoffelkrankheiten, Krankheiten der Zuckerrübe. Landwirtschaftliche Unkräuter.* ⊕ Würzburg 1897.

Appleman, Charles Orville (1878), Prof. Plant Physiol., Dean of Graduate School, Univ. of Maryland; Plant Physiologist, Maryland Agr. Expt. Station. Coll. Park, Maryland (U.S.A.). *Plant Respiration, Physiol. and Biochem. aspects of plant food storage. Pectic substances and Pectic enzymes.* ⊕ Univ. of Chicago 1910.

Appleton, A. B., Lecturer Downing Coll. Cambridge (England). *Anat.* o

Appleton, Joseph Luke Teasdale (1888), Prof. Bacteriopathology. Philadelphia, Pa. (U.S.A.), Dental School, Univ. of Pa., 4001 Spruce St. *Bact. of the mouth, and tooth development.*

Apstein, Carl H. (1862), Prof. Dr., Wiss. Beamter d. Preuß. Akad. d. Wissenschaften. Berlin N 4, (Deutschland), Invalidenstr. 43. Zool. Inst. *Plankton d. Meeres und Süßwassers; Biol. d. Fische, Salpen, Alciopiden, Tomopteriden. Literatur. Nomenclatur.* ⊕ Kiel 1898.

Aptekar, Esfir M., Doc., Jekaterinoslawskij Inst. f. Volksbildung. Dnepropetrowsk (U.d.S.S.R.), Upornaja 11, W. 5. *Botan.* o

Aral, Kwanji, Ass. Prof of Tôhoku Imper. Univ., Pathol. Inst., Coll. of Med. Sendai (Japan). *Pathol.* o

Arakawa, Sachiyo (1901), Division of Agric. and Microbiol. Kurashiki Okayama-ken (Japan), Ohara Inst. of Agric. Res. *Soil Bact.*

Araki, Torasaburô, Prof., Med. Dr., President of Kyôto Imper. Univ. (Japan). *Med. and Biol. Chem.*

Aranda y Millán, Francisco, Prof. Catedrático de Biol. general en la Univ. Zaragoza (España), Coso Nr. 110. o

de Aranzadi, Telesforo (1860), Prof. de Anthrop. Univ. Barcelona (España), Cortes 635. *Anthrop., Ethnographie et Préhistoire des Basques.* ⊕ Pharm. Madrid 1882, sc. nat. 1888.

Arbuthnot, Kenneth D. (1902), Instr. of Entomol. M. S. C. East Lansing, Mich. (U. S. A.). ⊕ Col. Agric. Coll., Ft. Collins 1926.

Arcangeli, Alceste, Lib. doc. R. I. med. vet. Milano (Italia). *Anat. compar.* o

Archangelsky, Andreas D., Dr. Prof. Geol., Univ. u. Bergakademie, Obergeol. am Comité Géol. de Russie. Moskau (U.d.S.S.R.). *Mesozoische Fauna.* o

Archangelsky, Wasiliy Michalovich (1883), Prof., Physiol. Biochem. Dnepropetrovsk, Ukraina (U.d.S.S.R.), Sevastopolskaja 18. *Relationship between the endocrine organs and the activity of cortex cerebri.* ⊕ Moskau 1918.

Archer, Sir Geoffrey. London S.W. 1 (England), 40 Pont Street. *Ornithol.* o

Archey, Gilbert (1890), Secretary and Dir., Auckland Inst. and Mus. Auckland (New Zealand). *Chilopoda and Birds.* ⊕ Christchurch, N.Z., 1913.

Archimowitsch, Alexander S., Leiter, Versuchsstation der Selektion-Station. Belaja Zerkow (U.d.S.S.R.). *Selektion d. Zuckerrübe.* o

Arcidiacono, N., Prof., Univ. Nac. Córdoba (Argentina). *Hygiene.* o

Ardanaz, Félix, General de Estado Mayor. Santander (España), Segismundo Moret, 3. *Entomol.* o

Ardati, Najib M., Prof., American Univ. Beirut (Syrien). *Bact.* o

Arditi, Horacio, Prof., Univ. Nac. La Plata (Argentina). *Zool.* o

Ardois, Juan. Madrid (España), Alberto Aguilera Nr. 60. *Coleoptera.* o

Arefiev, S. P., Doc., Univ. Vostočny j. Pedagog Inst. Kazan, Tatar. Rep. (U.d.S.S.R.), Karl-Marx-Straße 47. *Botan.* o

Arens, L. E., Wiss. „P. F. Lesshaft" Inst. Leningrad (U.d.S.S.R.), Pr. Maklina 32. *Zoopsychol.* o

Arens, Pedro Martin José Trinidad (1884), Dr. phil., Dir. des Wissenschaftlichen Dienstes der Rubber Cultuur Maatschappy Amsterdam. Galang, Sumatra, (Ned. Indië). *Hevea brasiliensis.* ⊕ Bonn a. Rhein 1907. ⊕ Moose.

Arévalo Carretero, Celso. Prof. Inst. du Cardinal Cisueros, Chef Sect. Hydrobiol. au Musée nation. Sc. nat. Madrid (España), Narval 25. *Plancton d'eau douce. Pêche fluviale. Cladocera. Rotifera. Hydrachnida.* ⊕ Madrid 1908.

Argand, E., Prof. Dr. Neuchâtel (Suisse), Chemins de Pavés 9. *Palaeont.* o

Argaud, Prof. Dr., Labor. d'Histol., Fac. de méd. Toulouse (France). o

Arima, Junzô, Lect. of Keiô-Gynku Univ. Tôkyô (Japan), Pharm. Inst., Coll. of Med. *Pharmacol.* o

Ariola, Vincenzo, Lib. doc., R. univ. degli studi. Genova (Italia). *Parasit.* o

Aristorhova, Olga, Dr., Dir. d. Epidemiol. Inst. Novi Sad (S.H.S.), Vojvodina. *Bact.* o

Aristov, Michael (1890), Ass. Leningrad (U.d.S.S.R.), Gercen-st-t. 44. *Biol. Garden injurious insects, Apples trees.*

Aristowsky, Wjatscheslav (1882), Prof., Bact. Inst. Kasan (U.d.S.S.R.), B. Krasnaja 59. *Spirochaeten und Spirochaetosen.* ⊕ 1912.

Arisz, Willem Hendrik (1888), Prof. Dr. Univ. Groningen (Holland), Nieuwe Kijk in't Jat-straat 84. *Botan. Physiol.* ⊕ Utrecht 1914.

Arkwright, Joseph A. (1864), Bact., Lister Inst. London S.W. 1 (England), Chelsea Gardens. *Bact.* ⊕ Cambridge 1895.

Arldt, Theodor Karl Hermann (1878), Prof. Dr. phil. Radeberg (Deutschland), Badstr. 13. *Paläogeographie.* ⊕ Leipzig 1901.

Arloing, M., Prof./Fac. Méd. Lyon (France). *Bact.* o

Armbruster, Ludwig (1886), Prof., Dir. Berlin-Dahlem (Deutschland), Lentzeallee 86. *Bienenkunde.* ⊕ Freiburg i. Br. 1913. ⊕ Apistica.

Armitage, Frederick Louis (1874), Bact. Auckland (New Zealand), General Hospital. *Med. Bact. and Immunol.*

Armstrong, George Miller (1893), Chief, Division of Boll Weevil Control, Clemson College, S.C. Pee Dee Exp. Station, Florence, S. C. (U.S.A.) *Cotton Physiol., relations of vegetativeness to fruit production.* ① Washington Univ. 1921.

Arnandow, Nikola (1887), a.o. Prof. Sofia (Bulgarien), Botan. Inst. der Univ. *Moose, Wasserpilze.* ① Sofia 1909.

Arnaud, Gabriel (1882), Dir. adjoint Station Centr. de Phytopath. Paris 14 (France), 11bis, rue d'Alesia. *Maladies des Plantes horticoles et champignons asterinoides.* ① Paris 1922.

Arnaudi, Carlo (1899), Dr. sc. agr., Aiuto nella Sezione di ricerche di Batteriol. Agraria e Industriale dell' Ist. Sieroterapico Milanese. Milano (Italia). *Batt. Agraria; Microbiol. e patol. vegetale.* ① Milano 1923. ② Hyphomyceta.

Arndt, Charles H. (1894), Dir. Coffee Experiment Station. Port au Prince (Haiti), Service Technique. *Plant Physiol. and Pathol. of the coffee plant.* ① Univ. of Pennsylvania 1921.

Arndt, Hans-Joachim (1900), Dr. med. et med. vet., Priv.-Doz., I. Ass. am Pathol. Inst. d. Univ. Marburg a.d.L. (Deutschland). *Allgemeine Pathol. und pathol. Anat. Vergl. Pathol. Zellstoffwechsel, besonders Fett- und Kohlehydratstoffwechsel (experim.-pathol.). Endokrinol. (morphol.).* ① med. Berlin 1924, med. vet. Berlin 1924.

Arndt, Walther (1891), Dr. med. et phil., Kustos am Zool. Mus. der Univ. Berlin N 4 (Deutschland), Invalidenstr. 43. *Porifera, Turbellaria, Spelaeobiol.* ① med. Breslau 1919, phil. 1920.

Arnesen, Emily, Dr., Konserv., Zool. Mus. Oslo (Norge), Töien. *Zool.* ○

Arnett, Clare Newton (1884), Head of animal husbandry, Mont. State Coll. and Experim. Station. Bozeman, Mtn. (U.S.A.). *Animal husbandry.* ○

Arnhart, Ludwig (1858), Dr. Wien XVIII (Österreich), Anton-Frank-Gasse 8. *Apis.* ① Wien 1922. ② Mikro- und makroskopische Präparate über Apis.

Arnold, George, Dir., Rhodesia Mus. Bulawayo (S. Rhodesia), Box 240. *Entomol., especially Hymenopterol. Formicidae and Sphegidae of South Africa.* ① Liverpool 1916. ② Hymenoptera Fossoria and Formicidae.

Arnold, G. T., Prof., Inst. f. Volksbildung. Charkow (U.d.S.S.R.). *Zool.* ○

Arnoldov, Michajl, Prof.,Vet.Inst. (Arskoe Pole 96). Kazan, Tatar. Rep. (U.d.S.S.R.). *Epizootol.* ○

Arnol'dov, Vladimir Andr., Prof., N. G. Cernyševskij Univ. Ploščad imeni Lenina. Saratov (U.d.S.S.R.). *Experim. Hygiene u. Epizootol.* ○

Arny, Albert Cedric (1877). St. Paul, Minn. (U.S.A.), Univ. Farm. *Agronomy. Farm Crops.*

Aron, Max (1892), Chargé de Cours à la Fac. de Méd. de Strasbourg (France), Inst. d'Histol. *Histophysiol. sexuelle. Histophysiol. des glandes endocrines, spécialement des glandes endocrines embryonnaires (pancréas thyroide). Hématopoïèse chez l'embryon. Développement des os. Embryol. causale: localisations germinales, facteurs de la croissance.* ① Strasbourg 1919.

Aronson, Joseph D. (1877). Associate in Pathology at the Henry Phipps Institute, Philadelphia, Pa. (U.S.A.), 7th and Lombard Sts. *Tuberculosis.* ① Medico-Chirurgical College, Philadelphia, 1908.

Arrhenius, Olof W. (1895), Dr. Fil. Stockholm (Schweden), Gamla Haga. *Ernährungs- und Agrikulturphysiol., Ökol., Bodenkunde.* ① Stockholm 1920.

D'Arrigo, G., Prof. Univ. Napoli (Italia). *Anat.* ○

Arrigoni Degli Oddi, Comte Ettore (1868), Prof. de Zool. Padua (Italia), via Umberto 10. *Ornithol.* ① Padua 1888. ② Oiseaux Palearticques.

D'Arsonval, M.A., Prof., Coll. de France, Nogent-sur-Marne, Seine, (France), 49bis, avenue de la Belle-Gabrielle. *Physiol.* ○

Artega, Francisco, Prof., Univ. Libre. Bogotá (Columbia). *Biol. Anthrop.* ○

v. **Arthaber,** Gustav (1864), Dr. phil., Prof. für Palaeont. am Palaeont. Inst. der Univ. Wien IX, (Österreich), Ferstelgasse 3. *Evertebrata: Cephalopoden, Reptilien.* ① Wien 1892.

Arthur, John M. (1893), Biochemist, Boyce Thompson Inst. for Plant Research, Inc., Yonkers, N. Y. (U.S.A.). *Effect of Light, Carbon Dioxide Supply and other Environmental Factors on the Growth and Reproduction of Plants.* ① Univ. of Chicago 1926.

Arthur, Joseph Charles (1850), Prof. emeritus of Botan., Purdue Univ., Lafayette, Ind. (U.S.A.). *Uredinales. Plant Dissection, Living Plants and their Properties.* ① Iowa State College 1872.

Arthus, Maurice (1862), Prof. de physiol., fac. de méd. de l'Univ. de Lausanne (Suisse), Inst. de Physiol., Champ de l'Air. *Hématol. Digestion. Venins. Anaphylaxie. Protiotonies. Anesthésie. Physiol. microbienne.* ① Paris Sc. physiques 1893, nat. 1890, med. 1896.

Artom, Camillo (1893), Libero doc. di fisiol., di chim. biol., di fisiol. nella R.Univ. di Palermo (Italia), Ist. di Fisiol., Corso Tuköry. *Corrélations fonctionnels entre parathyroide et sécrétion intestinale. Endocrinol., Métabolisme intermédiaire.*

Artom, Cesare (1879), Prof., Zool., Biol. gen., Dir. Ist. Zool., Univ. di Pavia (Italia). *Cellule sessuali. Genetica. Citol. sperimentale. Sesso.*

Artobolevskij, N., Prof., Univ. Drahomanow. Kiev (U.d.S.S.R.). *Biol.* ○

Artschwager, Ernst (1889), Associate Pathologist. Washington, D.C. (U.S.A.). *Pflanzenanat.; Anat. der Zuckerrübe, des Zuckerrohrs, der Kartoffel.* ① Cornell Univ. 1918.

Artzuml, Wahan, Prof. Univ. Erivan, Armenien (U.d.S.S.R.). *Anat.* ○

Aruch, E., Prof. Univ. Perugia (Italia). *Pathol.* ○

Aruncbovskij, V. M., Prof., I. Univ. Moskau (U.d.S.S.R.). *Botan.* ○

Arvay, Sándor, Ass. an Physiol. und allg. pathol. Inst. der Univ. Debreczen (Ungarn). ○

Arwidsson, Ivar (1873), Doc., Konserv., Zool. Mus., Univ. Uppsala (Sverige). *Polychäten: Maldaniden (Systematik u. Faunistik), Sylliden (Anat.)* ① Uppsala 1907.

v. **Arx, Max** (1857), Dr. med. Olten (Schweiz), Bahnhofstr. 43. *Evolution d. organ. Substanz, Konstructionspläne z. Körperbau u. Menschwerdung nach der Ballontheorie u. dem Princip der statischen Gleichgewichtslage.* ① Zürich 1882.

Asada, Jun-ichi, Ass. of Tōkyō Imper. Univ., Zool. Inst., Fac. of Sc. Tōkyō (Japan). *Helminthol.*

Asahina, Yasuhiko, Dr. pharm., Prof., Pharmacetu. Chem. of Imper. Univ. Tōkyō (Japan), Labor. of Pharm. Chem., Coll. of Med. *Analysis of Chinese Med. Plants.*

Asai, Takeshirō, Prof. of Aichi Med. Univ. Nagoya (Japan), Anat. Inst. *Anat*.

Asami, Yoshichi (1894), Ass. Prof. Horticulture, Agric. Coll., Imp. Univ. Tōkyō (Japan), Horticultural Inst. *Pomol.: Crab-Apples and Nectarines of Japan.* ① Tōkyō Imp. Univ. 1923. ② Fruit-trees, vegetables and flowers.

Asbukin, Agafonik Paol, Prof. Univ. Tomsk (U.d.S.S.R.), Prospekt Timirjazeva 3. *Anat.* ○

Aschmarin, Peter Alexandrowitsch (1888), Ass. biochem. Abt., Inst. f. experim. Med. Leningrad (U.d.S.S. R.), Lopuchinskaja 12. *Säure-Basen im Organismus.*

Aschoff, Karl Albert Ludwig (1866), o.ö.Prof. der pathol. Anat. Freiburg i. Br. (Deutschland), Eschenhof. *Thrombose. Verfettung. Histol. d. sog. entzündlichen Reaktionen, retikuloendotheliales System, vitale Färbung, Ikterus. Nierensekretion, Steinbildungen, Reizleitungssystem des Herzens, Kropf, Ovulation, Ulcus ventriculi, Lungenphthise, Wurmfortsatzentzündung, Gallenblasenentzündung, rheumatische Myokarditis. Pyelonephritis. Pathol. und Grund-*

lagen der Herzschwäche. Atherosklerose und andere Sklerosen. Cholelithiasis. Thrombose. Skorbut. Veränderungen der motorischen Ganglienzellen bei Wundstarrkrampf. ④ Bonn 1889.

Ascoli, Alberto, Prof. R. I. med. vet. Milano (Italia). *Farmacol., Toxicol.* ○

Ascoli, Maurizio (1876), Prof. di ruolo di patol. med. nella R. Univ. di Catania (Italia), Viale Margherita 18. *Immunité. Amibiase. Cancer.*

Asdell, Sydney Arthur (1897), Research Ass. Cambridge (England), Animal Nutrition Inst. *Reproduction in Mammals.* ④ Cambridge 1925.

Ashby, Edwin (1861), „Wittunga" Blackwood (South Australia). *Polyplacophora (Chiton). Palaeont. in Polyplacophora. Ornithol., especially Australian.* ⑬ Australian exotic Polyplacophora (Chitons). Seeds of Australian native shrubs.

Ashby, Sydney Francis (1874), Mycol., Imp. Bureau of Mycol. Kew, Surrey (England), 17 The Green. *General Pathol.* ④ Edinburgh 1900.

Asher, Leon (1865), Prof., Dir. Physiol. Inst. (Hellerionum) Univ. Bern (Schweiz), Optingenstr. Nr. 18. *Experim. Physiol., Biochem. Stoffaustausch zwischen Blut u. Gewebe. Anteil einfachster Stoffe an den Lebenserscheinungen. Unregelmäßigkeit des Herzschlags. Physiol. d. Schilddrüse, Physiol. d. Nebenschilddrüse.* ④ Leipzig 1890.

Asheshov, Igor N. (1891), Dir. State Bact. Labor. Dubrovnik (S.H.S.). *Bacteriophagy (d'Herelle's phenomenon). Ultrafiltration. Ultraviruses.* ④ Saratov, Russia 1916. ⑬ Living cultures and photographs of the Bacteriophage.

Ashford, Bailey K. (1873), Prof. Tropical Med. and Mycol., Columbia Univ. N.Y. San Juan, (Porto Rico), Box 887. *Mycology, experimental and classical. Genus Monilia.* ④ Georgetown Univ. Washington, D.C. 1896. ⑬ Malaria. Filaria. Monilia psilosis. Necator americanus and Pathol. slides of organs of experim. animals.

Ashman, Richard (1890), Ass. Prof. of Physiol., Tulane Univ., School of Med. New Orleans, La. (U.S.A.), 376, Millaudon St. *Electrocardiography; heart-block. Internormal phase in cardiac conduction.* ④ Tulane Univ. New Orleans (La.) 1925.

Ashworth, James H., Prof. Edinburgh (Scotland). *Zool.* ○

Asimoff, Grigory Josifowitsch (1894), Labor. experim. Biol. Swerdlow Univ. Moskau (U.d.S.S.R.), 3 Meschanskaja, 8, W. 24. *Antithyreoidin. Bedingte Reflexe und ihr Zusammenhang mit der Schild- und Sexualdrüse. Morphogenetische Prozesse und ihr Zusammenhang mit der Schilddrüse.*

Askanazy, Max (1865), Prof. Dr. med., Dir. des Pathol. Inst. der Univ. Genf. Genève (Suisse), 16, rue de Candolle. *Pathol. des Blutes und der Blutbildungsorgane. Geschwulstlehre, besonders experim. Tierische Parasiten in der menschlichen Krankheitslehre. Regeneration und Transplantation von Muskulatur. Pathol. Histologie.* ④ Königsberg i. Pr. 1890. ⑬ Präparate.

Aslander, J. Alfred O. (1892). Ithaca, N.Y., (U.S.A.), Botan. Department, Cornell Univ. *Effect of poisons on plant growth with aim to find a method for eradicating weeds on arable land.*

Asô, Keijirô, Prof. of Imper. Univ. Tôkyô (Japan), Minami-Teramachi 18, Yotsuya. *Agric. Chem.*

Asplund, Erik (1888). Stockholm 50 (Sverige). *Flora of South America.* ④ Uppsala 1920.

Asselin, Elie, Chargé de cours. Univ. Montreal (Canada), Rue St. Denis 1265. *Physiol. génér.* ○

Assèn, Zlataroff (1885), Prof. extrao. chim. méd. Univ., Dir. Inst. de chim. méd. fac. méd. l'Univ. Sofia (Bulgarie), rue „Tchépino" 3. *Biochim.: chim. des ferments, bromatol., chim. alimentaire.* ④ Grenoble 1908.

Assis Gonçalves, Alfredo de, Prof. Fac. d. Med. Curityba, Paraná (Brasil). *Microbiol.* ○

Astanin, Peter Petrowitsch (1885), Ass. biochem. Abt., Inst. für experim. Med., Prof. Vet.-Inst. Leningrad (U.d.S.S.R.), Lopuchinskaja 12. *Grundgesetze des biochem. Umsatzes.*

Astauroff, Boris (1907), I. Mosk. Univ., Inst. f. Experim. Biol. Moskau (U.d.S.S.R.), Woronzowo-Pole 6. *Genetic der Drosophilaarten.*

Aszódy, Zoltán (1891), Adjunkt physiol. u. pathol.-chem. Inst. Univ. Budapest VIII (Ungarn), Eszterházy-ut 9. *Stoffwechseluntersuchungen.* ④ Budapest 1914.

Atanasoff, Dimitar (1894), Lect. Sofia (Bulgaria), Inst. of Phytopath., The Univ. *Virus diseases of plants.* ④ Ph.D. Univ. of Wisconsin 1920. ⑬ Herbarium material and cultures of plant disease causing organisms.

Atkins, William Ringrose Gelston (1884), Fellow of Inst. of Chem., Fellow of the Royal Society, Head of Department of General Physiol. Plymouth (England), Marine Biol. Labor. *Plant physiol.; osmotic pressure, hydrogen ion concentration, plant distribution, minor constituents of sea water in relation to plant growth.* ④ Dublin 1906.

Atwell, Charles Beach (1855), Prof. of Botany, Coll. of Liberal Arts, Northwestern Univ., Evanston Ill., (U.S.A.), 26, Fisk Hall. *General comparative morphol. of Seed plants. History of Botan.* ④ Syracuse Univ. 1879.

Atwell, Wayne Jason (1889), Prof. of Anat., Univ. of Buffalo, Head of the Dept. of Anat. Buffalo, N.Y. (U.S.A.), 24 High Street. *Compar. morphol., morphogenesis and physiol. of the hypophysis cerebri; description of a young human embryo.* ④ Univ. Michigan 1917.

Atwood, Winfred McKenzie (1883), Prof. of Plant Physiol. Corvallis, Ore. (U.S.A.), Oregon Agric. Coll. *Germination of Avena. Effect of Formaldehyde of Wheat.* ④ Univ. Chicago 1913.

Atzeni Tedesco, Plinio (1897), Dr., Ass. della R. Clinica Med. Univ. di Cagliari (Italia), Via Canelles 11. *Hèmatol.* ④ Cagliari 1921.

Atzler, Edgar (1887), Prof. d. Physiol., Dir. des Kaiser-Wilhelm-Inst. für Arbeitsphysiol., Berlin N 4 (Deutschland), Invalidenstr. 103a. *Arbeitsphysiol.* ④ Heidelberg 1913.

Aub, Joseph Charles (1890), Ass. Prof. of Med., Harvard Med. School. Boston, Mass. (U.S.A.), Mass. General Hospital. *Physiol. of the internal secretions and of inorganic salt metabolism.* ④ Harvard Med. School 1914.

Aubel, Eugène (1884), Prof. de chim. physiol. Bordeaux (France), Fac. des sc. *Fermentations, Oxydo-réductions.* ④ Paris 1925.

Aubert, M., Prof. suppléant d'Anat., Fac. Méd. Marseille (France), 70, Chemin de St. Julien. *Anat.* ○

Aubry, Hector, Prof. agr. Univ. Montreal (Canada), Rue St. Denis 1265. *Bact. theor. et prat.* ○

Auchter, Eugene Curtis (1889), Head, Department of Horticulture Univ. of Maryland and Experim. Station. College Park, Md. (U.S.A.), ④ Cornell Univ. Ithaca (N.Y.) 1923.

Audas, James Wales (1872), Senior Ass. National Herbarium of Victoria (Australia). *Victorian Grasses. Australian Fodder Shrub, Salt-bush. Indigenous Fibre Plants. Clovers Naturalized in Victoria. Flora of Victoria. Weeds Poison Plants.*

Audige, M., Maître de Conférences., Zool. Fac. Sc. Caen (France), Labor. maritime, Luc-sur-Mer. *Zool.* ○

Auer, John (1875), Prof. Pharmacol., Dir. dept., Univ. School of Med. St. Louis, Mo. (U.S.A.), 1402 S. Grand Avenue. *Digestion. Respiration Heart, Physiol. Action of Various Drugs.* ○

Auer, Väinö, Dr. phil., Doc. Univ. Helsinki (Finnland), Fredriksgatan 66 B. *Botan. Mooruntersuchungen.* ○

Auerbach, Max (1879), Dir. der Bad. Landessammlungen für Naturkunde in Karlsruhe i. B. und der Anstalt für Bodensee-Forschung in Staad b. Konstanz, a.o. Prof. a. d. Techn. Hochschule. Karlsruhe i. B. (Deutschland), Landessammlung für Naturkunde, Friedrichsplatz. *Cnidosporidien. Hydrobiol. (Seenforschung); Fischereibiologie.* ℗ Basel 1902. ℗ *Tiere und Pflanzen des Bodensees.*

Augenblick, Maurice Lewis (1892). Wilkes-Barre, Pa. (U.S.A.), 47, So. Washington St. *Botan. and Bact.* ℗ Philadelphia, Pa. (U.S.A.) 1915.

Augier, Marius (1886), Chef du Labor. d'anat. Fac. de Méd. Paris (France). *Embryol., craniol.* ℗ 1913. ℗ *Moulages d'anat. humaine.*

Augustin, Béla (1877), Dr., Leiter der Kgl. Ung. Drogenversuchsstation, Prof. der Apothekersch., Doc. der Volkswirtschaftl. Fac. der Univ. Budapest II (Ungarn), Debröi ut. 15. *Pharmakognosie, pharmazeutische Botanik, Kultur und Analyse der Arzneipflanzen.* ℗ Budapest 1899, Bern 1904. ℗ *Ungarns Arzneipflanzen lebend, getrocknet, Samen.*

Augustine, D. L., Assoc. Prof., Compar. Pathol., Harvard Med. School. Boston, Mass. (U.S.A.). *Helminthol.* o

Aujeszky, Aladár (1869), o.ö. Prof. der Bact. an der Veterinär-Hochsch. in Budapest VII, (Ungarn) Hungária-Körút, 244. ℗ Med. Budapest 1892.

Aulló y Costilla, Manuel, Jefe del Servicio de Estudio y Extinción de plagas forestales. Madrid (España), Ferraz 40. *Entomol.* o

Aulmann, Georg (1884), Dr. phil., Dir. des Löbbecke- und Naturwiss. Mus. und des Zool. Gartens der Stadt Düsseldorf (Deutschland). *Schädlinge kolonialer Kulturpflanzen und Systematik der Psylliden.* ℗ Erlangen 1907.

Aunap, Eduard (1888), Ass. am Histol. Inst. der Univ. Tartu (Estland). *Entwicklung des Knorpels. Histol. der Nasenschleimhaut*

Aurivillius, Christopher (1853), Prof. Stockholm 50 (Sverige). *Entomol.* ℗ Upsala 1880.

Aurivillius, Sven Magnus (1892), Dir., Kristinebergs Zool. Station. Fiskebäckskil (Sverige). *Gorgonarien. Faunistik.* ℗ 1923.

Austen, E. E., Deputy Keeper Dept. of Entomol. British Mus. of Nat. Hist. London, S. W. (England), Cromwell Road. *Entomol.* o

Austin, James Harold (1883), Prof. of Research Med., Univ. of Pennsylvania. Philadelphia, Pa. (U.S.A.). *Physical chem. of blood; acidosis; renal and splenic function. Hydrogen Ion Concentration of the Blood.* ℗ Univ. of Pennsylvania 1908.

Austin, Lloyd (1898), Dir. of Eddy Tree Breeding Station. Placerville, Cal. (U.S.A.), 60, Bedford Ave. *Forest and fruit tree breeding and improvement. Development of rapid-growing strains of forest trees. Forest nursery.* ℗ Univ. of California 1920. ℗ *Pinus ponderosa, lambertiana, sabiniana, jeffreyi, monticola. Pseudotsuga taxifolia. Abies concolor, magnifica. Sequoia sempervirens, gigantea.* Libocedrus decurrens.

Austin, Mary L. (1896), Lect. Zool., Barnard Coll. New York City (U.S.A.), Dept. of Zool. *Protozool.* ℗ Wellesley 1922.

Austin, Richard Sisson (1885), Prof. of Pathol., Univ. of Cincinnati, Ohio (U.S.A.), Cincinnati General Hospital. *Descriptive and experim. pathol.* ℗ Harvard Med. School 1911. ℗ *Pathol.* material (preserved tissues and microscopic sections).

Austmann, Kristján Jónsson (1890), Ass. Prof., Dept. Physiol., Fac. of Med., Univ. of Manitoba. Winnipeg, Man. (U.S.A.), Med. Coll. *Intracranial relationships. Suprarenal glands. Gastrophilus infestation. Human skin.* ℗ Univ. of Manitoba 1921.

Avari, Cooverji Rustomji (1872). Parel, Bombay No. 12 (Br. India), Old Government House. *Bacteriophage.*

Ava Sarria, Pedro, Prof. Univ. Valencia (España). *Anat. descript., Embryol.* o

Averein, Victor Grigorjevich (1885), Prof. appl. Zool., Biol. of mammals a. birds, Kharu Inst. of Agric. a. Forestry; Chief of Division of Protection of Plants Peopl. Kommissar. of Agric. in Ukraine. Charkow (U.d.S.S.R.). *Theor. a. appl. Mammal. a. Ornithol. Entomol.*

Averianoff,Paul (1887),Prof. Physiol. pathol.Univ. Smolensk. Moscou 19 (U.d.S.S.R.), Lenivka, Lebiajy per. 1/35. *Pathol. de l'appareil Cardio-Vasculaire.*

Avery, Priscilla (1899), Research ass. in Genetics, Coll. of Agric., Teaching Fellow in Zool., Univ. of California. Oakland, Cal. (U.S.A.), 29 Westall Avenue. *Cytol., taxonomy, genetics of Crepis and of Nicotiana.* ℗ Berkeley, Cal., 1926.

Avetta, Carlo (1861), Prof. Botan., Dir. Ist. ed Orto Botan. Univ. Parma (Italia), via Farini 174. *Farmacognosia.* ℗ Roma.

Avoni, Aldo, Lib. doc. Bologna (Italia). *Pathol.* o

Avtonomova, E. G., Doc. Gosudarstv. Donskoj Univ. Rostov am Don (U.d.S.S.R.). *Med. Chem.* o

Awdeewa, Marie (1888), Mitarbeiter K. E. P. S. Akademie der Wissenschaften, Moskau (U.d.S.S.R.), Woronzowo Pole 6. *Physiol. des Stoffwechsels (Kohlenhydratstoffwechsel).*

Awerin, W. G., Ass. Zool. Labor. Pädagogisches Inst. (I.N.O.). Charkow (U.d.S.S.R.). o

Awerinzew, Sergius (1875), Leiter d. Ichthyol. Labor., Inst. für Fischereiwirtschaft. Moskau (U.d.S.S.R.), Piatnitzkaja Str. 33. *Protozoa: Cnidosporidia u. Rhizopoda. Pisces: Biol. u. Rassenfragen d.Fische v. Barents- u. Weißen Meere.* ℗ Charkov 1914.

Awerkieff, Dmitrij (1889), Ältester Ass. N.-Nowgorod (U. d. S. S. R.), Univ., Kabinett der Botan. *Geobotan. Steppenfrage, Systematik* ℗ Herbarmaterial.

Awtokratow, Demetrius (1868), Prof. Tierärztl. Hochsch.NowotcherkasskVeterin.Inst. (U.d.S.S.R.). *Vergl. Anat. d. Haustiere, Gefäßlehre.* ℗ med. vet. Kazan 1900.

Axentieff, Boris (1894), Ass., Botan. Labor., Univ. Odessa, Ukraine (U.d.S.S.R.). *Lichtwirkung auf die Samenkeimung. Bacillariaceae.*

Ayers, S. Henry (1884), Research Dir. New York, N.Y. (U.S.A.), 22 East 75th St. *Bact. fermentations in food products.* ℗ Mass. Inst. of Technol. 1905.

Azbukin, A. P., Prof., Dir., Anat. Labor., Univ. Tomsk (U.d.S.S.R.). o

Azéma, Maurice (1905), Ass. à la Fac. des Sc. Labor. de Zool., Paris V (France), 1, rue Victor Cousin. *Histo-physiol. de l'excrétion chez les Tuniciers.* ℗ Paris 1925.

Azevedo Macedo, José de, Prof. Fac. d. Med. Curityba, Paraná (Brasil). *Biochem.* o

de Azevedo Marques, Luiz Augusto (1880), Ass. au Service d'Entomol. Agric. a Inst. Biol. de Defesa Agric. Rio de Janeiro (Brasil.). *Recherches d'entomol. agric.*

Azpeitia y Moros, Florentino (1859). Madrid (España), Calle Príncipe de Vergara 23. *Malacol., Diatomol.* ℗ Mollusques, Diatomées.

Azzi, A., Prof. Univ. Torino (Italia). *Batteriol. e Immunol.* o

Baas-Becking, Lourens Gerhard Marinus (1895), Prof. of Biol., Stanford Univ., California (U.S.A.). *Physiol. of the cell, radiation, autotrophic bacteria, application of mathematics on Biol., Structure of protoplasm. Growth.* ℗ Phil. Stanford 1921, Sc. Utrecht 1925.

Baashuus-Jessen, Johannes (1887), Registrar, The Live-Stock Registration Office of Norway. Oslo (Norway), Parkveien 8 III. *Horsebreeding, particularly colours, natural history, points, selection.*

Babcock, Ernest Brown (1877), Prof. of Genetics in the Experim. Station. Berkeley, Cal. (U.S.A.). Univ. of California. *Genetics and cytol. in relation to taxonomy, phylogeny and evolution of the genus Crepis. Genetics in relation to Agric.* ℗ Univ. of California 1905. ℗ Seeds of Crepis.

Babcock, Orville Gorman (1885), Ass. Entomol. Division of Insects Affecting Health of Animals. Sonora, Tex. (U.S.A.), Box 407. *Lice on animals, blow flies, sheep scab.* ⑬ Fort Collins, Col., 1910, ⑫ Goat lice, Blow flies.

Babić, Krunoslav, Dr., Priv.Doc. d. Morphol. u. Systematik der Wirbellosen (Phil. Fac.), Kustos am Kroat. Nationalmus., Zool. Abt. Zagreb (S.H.S.), Haulikova ul. 4. *Systematik d. Spongien u. Bryozoen.* o

Babiy, Paul Peter (1894), Curator of invertebrate Zool., Dept. of Entomol. N.Y. State Coll. of Agric. Ithaca, N.Y. (U.S.A.), Cornell Univ. *Hymenoptera; Apidae (Bombus und Psithyrus). Psammocharidae (gen. Ceropales Latr.).* ⑬ Wien 1924. ⑫ Hummeln, Ceropales Latr. (Psammocharidae).

Babkin, Boris (1877), Prof. of Physiol., Dalhousie Univ., Halifax, N.S. (Canada), Dept. of Physiol., Med. Sc. Building. *Secretion and movements of the digestive tract.* ⑬ M.D. Mil.-Med. Academy 1904, D.Sc. Univ. of London 1925.

Babor, Josef (1872), Dr., Prof. der allgemeinen Biol. an der med. Fac. der Komenský-Univ. Bratislava Schiffbeck, (C.S.R.), Biol. Inst. *Systematische und morphol. Malakozool., Palaeont. und Palaeanthropol., vergl. Neurol., Morphol. der Wirbeltiere.* ⑬ Prag 1897. ⑫ Recente und fossile Mollusca.

Bach, Fritz Werner (1887), ao. Prof. f. Hygien. Bonn a. Rh. (Deutschland), Rheinbacherstr. 15. *Parasitische Protozoen des Menschen.* ⑬ Leipzig1912.

Bachem, Carl (1880), nichtbeamteter ao. Prof. Dr. med., Bonn a. Rh. (Deutschland), Heerstr. 11. *Pharmakol., Toxikol.* ⑬ Bonn 1904.

Bachmann, Ewald Theodor (1850), Prof. i. R. Radebeul b. Dresden (Deutschland), Moltkestr. 24. *Flechten, bes. Kalkflechten.* ⑬ Leipzig 1881.

Bachmann, Freda M. (1878), Prof. Bact. and Hygiene. Stout Inst. Menomonie, Wis. (U.S.A.). *Bact. of foods. Hemolytic streptococci. Discomycetes. Fertilization in Collema. Migration of the germ cells. Vitamines. Toxicity of Juice extracted from succulent Onion Scales. Toxin-producing anaerobes in Wisconsin Soils.* ⑬ Univ. Wis. 1912. ⑫ Cultures of bacteria, algae and fungi.

Bachmann, Fritz H. V. (1887), Dr. phil., Priv.Doc. für Botan. an der Univ. Leipzig, I. Ass. am botan. Inst. Leipzig (Deutschland), Linnéstr. 1. *Stoffwechselphysiol., Wasserhaushalt der Pflanzen.* ⑬ Leipzig 1912.

Bachmann, Jean George (1877), Prof. of Physiol. Georgia (U.S.A.), Emory Univ., School of Med. *The heart and circulation. Pharmacodynamy.* ⑬ Jefferson Med. Coll. 1907.

Bachofen-Echt, Adolf (1864), Dr. phil. Wien XIX (Österreich), Sprinsiedelgasse 28. *Säugetiere Diluv und Tertiär. Insekten Bernstein.* ⑬ Wien 1925.

Bachtine, Benjamin Sergeevitsch (1888), Chef des travaux, Inst. Jaczewski, et de Phytopath. à l'Inst. Agronomique à Detskoie (Tsarskoie) Selo, Prof. de Phytopath. Inst. de Zool. appliquée et Phytopath. Leningrad (U.d.S.S.R.), rue Torgovaia 28, logem. 8. *Maladies des plantes agricoles et des arbres fruitiers, Maladies du lin. Peronosporées.*

Back, Ernest Adna (1880), Senior Entomologist. Amherst, Mass. (U.S.A.), Bureau of Entomol., Department of Agric. *Insects affecting all kinds of stored commodities and household insects.* ⑬ Amherst, Mass. 1907.

Backman, E. L., Prof. Dr. Uppsala (Sverige). *Pharmacol.* o

Backman, Gaston (1883), Prof., Prosektor der Anat., Kgl. Karolin. Inst. Stockholm (Schweden), Hantverkaregatan 10. *Anthrop., Lungenanat., Variationsstatistik, Körperlänge.* ⑬ Med. Upsala 1913, med. h. c. Riga 1925.

Bądzyński, St., Dr. med. et phil., Prof. Warszawa (Pologne), Univ., Fac. de méd., r. Krakowskie-Przedmieście 26/28. *Biochem.* o

Bäckström, Kåre Mattias (1895), Ass. Univ. Stockholm (Sverige), Zootomiska Inst., Drottninggatan. No. 116. *Forebrain morphol. in Selachians.* ⑬ Stockholms Högskola 1924.

baron de Baehr, Venceslas-Brunon (1873), Prof. de Cytol., Dir. de l'Inst. de Cytol. de l'Univ. de Varsovie (Pologne). *Cytol., Chromosomes, Hérédité, Problème du sexe.*

Bär, J., Dr., Kustos, Inst. f. system. Botan., Botan. Garten. Zürich (Schweiz). o

Baer, Jean Georges (1902), Dr. Sc., Ass. au labor. d'Anat. normale à l'Univ. de Genève (Suisse), 14, Rue de Hollande. *Parasit., Platodes: Cestodes.* ⑬ Neuchâtel 1924.

Baer, William Gustav (1867), Ass., Zool. Inst., Forstl. Hochsch. Tharandt, Sachsen (Deutschland), Cottastr. 161. *Forstlich bedeutungsvolle Insekten (Tachinen) und Wirbeltiere.*

Baerg, William J. (1885), Prof. of Entomol. Fayetteville, Ark. (U.S.A.), Univ. of Arkansas. *Economic Entomol., Poisonous arthropods.* ⑬ Cornell Univ. 1922.

Bagg, Halsey J. (1889), Associate in Anat., Cornell Univ. Med. Coll. and Biol. of the Memorial Hospital. New York City (U.S.A.), 2 West, 106 St. *Experim. Cancer. Biol. effect of radium and X-ray irradiation on animal tissues. Hereditary structural defects in animals associated with irradiation of the germ plasm.* ⑬ Columbia Univ. 1918.

Baginski, Stefan Etienne (1892), Ass. d'Histol. Vilno (Pologne), Inst. d'Histol. de l'Univ. *Microphysiol. des glandes endocrines.* ⑬ Wilno 1924.

Baglioni, S., Prof. Univ. Roma (Italia). *Fisiol.* o

Bahía y Urrutia, Luis, Abogado, ex Senador del Reino. Madrid (España), Almagro 29. *Agric.* o

Bahl, Karm Narayan (1891), Prof., Univ. of Lucknow (India). *Morphol. of the Oligochaeta (Pheretima).* ⑬ Sc. Lahore 1913, Phil. Oxford 1921. ⑫ Indian Animals. Microscopical preparations illustrating the Anat. and Development of Pheretima.

Bahr, Louis (1876), Labor.-Dir. København F (Danmark), Virginievej 9. *Bact. Bienenkrankheiten. Bekämpfung schädlicher Nagetiere. Paratyphus der Honigbiene.* ⑬ Hannover 1922. ⑫ Bakterienstämme: Paratyphus-Gärtner-Gruppe.

Bail, Oskar, Dr., Prof. d. Hygiene (Bact.). Praha II (C.S.R.), Preslova, Hygien. Inst. d. Deutschen Univ. o

Bailey, Clyde H. (1887), Prof. of agric. Biochem. St. Paul, Minn. (U.S.A.), Univ. Farm. *Biochem. and chemical technology of the production, merchandizing, grading, and milling of cereals; fermentation and baking of bread dough.* ⑬ Univ. of Maryland 1921. ⑫ Wheat flour spring wheat varieties.

Bailey, Irving W. (1884), Associate Prof., Bussey Inst. for Research in Applied Biol., Harvard Univ. Boston 30, Mass. (U.S.A.). *Compar., Physiol. and Pathol. Anat. of Angiosperms. Cytol. and Histogenesis of Cambium and its derivative tissues.* ⑬ Harvard 1909.

Bailey, Percival (1892), Instr. in Surgery, Harward Medical School, Associate in Surgery, Peter Bent Brigham Hospital, Boston, Mass. (U.S.A.), 721 Huntington Ave. *Path. anat., more especially of brain tumors. Tumors of the Glioma Group.* ⑬ Univ. of Chicago 1918.

Bailey, Vernon (1864), Biol., Biol. Survey, U.S. Department of Agric. Washington, D.C. (U.S.A.), 1834, Kalorama Road. *Mammals, Birds, Reptiles, Plants: taxonomy, distribution and habits.*

Baird, Alfred Briggs (1891), Entomologist, Entomol. Branch Department Agric. Ottawa, Chatam, Ontario (Canada), Entomol. Labor., 261, Victoria Avenue. *Insect parasitism, European Corn Borer.*

Baird, Edgar Alan, Associate Prof. in Botan. Univ. of North Dakota Grand Forks (U.S.A.), *Cytol. of the Fungi, Taxonomy of Flowering Plants.* ⑬ Univ. of Wisconsin 1920.

Baltsell, George Alfred (1885), Associate Prof. of Biol., Yale Univ. New Haven, Conn. (U.S.A.), Osborn Zool. Labor. Yale Station. *Life History of the Infusoria. Tissue culture. Fibrin, Microdissection. Connective tissue in the embryo and in pathol. conditions.* ⓟ Yale Univ. 1914.

Bajarunas, Michael W., Oberkustos Geolog. und Miner. Mus., Akademie der Wissenschaften. Leningrad (U.d.S.S.R.), Wassily Ostrow, Tutschkowa Nab. *Palaeont. des Tertiärs.* o

Bajkov, Alexandr. (1894), Investigator-Biologist, Biol. Board of Canada (Canada). *Hydrobiol. and Ichthyol.* ⓟ Brno 1923. ⓟ *Fishes and Freshwater Plancton.*

Bakaloff, Peter, ao. Prof. Palaeont., Geol. Inst. d. Univ. Sofia (Bulgarien). *Palaeont.* o

Baker, Albert Wesley (1891), Prof. of Systematic Entomol. and Zool., Ontario Agric. Coll. Guelph (Canada). *Mallophaga.* ⓟ *Mallophaga.*

Baker, Clarence Everett (1896), Ass. i. Pomol. (Research). Lafayette, Ind. (U.S.A.), Purdue Univ. *Cold storage of fruits; orchard soil management; pruning of apples; small fruit culture.*

Baker, E. C. Stuart. London G. E. 19 (England), 6 Harold Road, Upper Norwood. *Ornithol.* o

Baker, Horace Greeley (1873), Associate Prof. of Biol., Southwestern Coll. Winfield, Kan. (U.S.A.). *Euthenics in Eugenic. Human Evolution.*

Baker, H. R. (1896), Ass. Prof. of Biol. Newark, Del. (U.S.A.), Univ. of Delaware. *Systematic Botan.* ⓟ *Lygodium palmatum, Potentilla fruticosa* L., *Viola striata.*

Baker, John Randal (1900), Demonstrator in the Department of Zool. and Compar. Anat., Oxford (England), Univ. Mus. *Sexual abnormalities in Vertebrates; Seasonal changes in the reproductive system of animals; Fauna of the New Hebrides. Sex in Man and Animals.*

Baker, Mary Francis (1876). Florida (U.S.A.), 225 Kentucky Avenue, Winter Park. *Practical botan. Flora of the state of Florida.*

Baker, Merle P. (1899), Ass. Prof. in Dairying, Iowa State Coll. Ames, Ia. (U.S.A.), Dairy Department. *Dairy Bact.*

Bakhuizen v. d. Brink, Reinier Cornelis (1881). Ass. Herbarium Buitenzorg, Java (Nederl.-O.-Indiē), Wigmanweg 15. *Pflanzensystematik.*

Bakke, Arthur Laurence (1886), Prof. of Plant Physiol., Iowa State Coll. Ames, Ia. (U.S.A.). ⓟ Univ. of Chicago 1917.

Bakker, Abel (1898), Ginneken bei Breda (Holland). *Nierensekretion; Allgemeine Zellphysiol.: Protoplasmabewegung, Phagozytose, Chemische Zusammensetzung; Atmung.* ⓟ Groningen 1926.

Bakker, Dirk Leonhard (1876), Dr., Prof. in der Zootechnik an der Landwirtschaftlichen Hochsch. Wageningen (Holland), Nassauweg 1. *Anat., Zootechnik.* ⓟ Bern 1909.

Bakó, Gábor, Mitarb. an der Entomol. Station. Budapest (Ungarn), Kitaibel-Gasse 1. *Entomol.* o

Balachonov, Pierre I., Chef de la Section Phytopath. Station de Défense des Plantes. Samara (U.d.S.S.R.), *Pathol. végétale.* o

Balakirev, P. W., Prof. Dr., Anat. Labor. Inst. f. Med. Wiss. Leningrad (U.d.S.S.R.). o

Balan, Mirtscho, Dr., Prosektor Anat. Inst. d. Univ., Med. Fac. Sofia (Bulgarien). o

Balasova, Olga Nik., Doz. II. Univ. Moskau (U.d.S.S.R.). *Biol. Chem.* o

Balbuena, Felix Fernandez (1877). Gijon (España). *Imprégnation Argentique de la rétine et d'autres centres nerveux.* ⓟ Madrid 1900.

Baldacci, Antonio (1867), Prof. Accademico Benedettino della R. Accademia delle Sc. di Bologna (Italia), Fuori Porta Zamboni. *Botan.: Flora de Montenegro et d'Albania.* ⓟ Bologna 1891.

Baldasseroni, Vincenzo (1884), Prof. incar. Zool. agraria e forestale, Ist. agrario e forestale, Mus.

zool. Univ. Firenze (Italia), Via Romana 19. *Lombrichi, Tifloscolecidi, Chetognati, Biol. marina.*

Baldi, D., Prof. Univ. Pisa (Italia), *Farm. e Tossicol.* o

Baldoni, Alessandro, Prof. Univ. Bari (Italia). *Pharm.* o

Balducci, Enrico (1866), Libero doc. in Zool. e Anat. compar. del Vertebrati, Univ. di Firenze (Italia), Via Mazzetta 8. *Osteol., Morphol. du sternum des oiseaux.*

Balduf, Walter Valentine (1889), Dr., Ass. Prof. of Entomol. Urbana, Ill. (U.S.A.), Natural History Building, Univ. of Illinois. *Entomol.; Chalcidoidea, Eurytomidae.* ⓟ Ohio State Univ. 1922. ⓟ *Chalcidoidea, esp. Eurytomidae.*

Baldus, Karl (1898), Dr. phil. nat. Ass. am Zool. Inst. Heidelberg (Deutschland), Klausenpfad 20. *Reizphysiol. der Insekten.* ⓟ Heidelberg 1922.

Baldwin, Henry Ives (1896), Research Forester, Brown Company, Berlin, N.H. (U.S.A.). Research Department. *Growth and yield of pulpwood forests; reproduction and management of spruce; vegetative periods, periodicity of seed years, effect of source of seed, extraction of seed; thinnings and reproduction cuttings.*

Baldwin, S. Prentiss (1868), Dir., Baldwin Bird Research Labor. Cleveland, Ohio (U.S.A.), 817, Williamson Building. *Live wild birds, living under natural conditions, by means of ringing, and by use of labor. methods of study of the individual.*

Balej, Stephan, Dr., Dir. Inst. f. experim. Psychol. Lwów (Polen), 10, Ringpl. o

Balfour, Arthur James, Earl of, Pres. Nat. Inst. of Industrial Psychol. London W.C. 1 (England), 329 High Holborn. o

Balguerias y Quesada, Eduardo, Prof. aux. Univ., Conserv. Herbarios del Jardin Botán. Madrid (España), Silva, 44, 3. *Botan.* o

Balinsky, Boris (1905), sc. Research on Zool. Kiev (U.d.S.S.R.), ul. Korolenko 37, Biol. Inst. *Experim. Morphol.*

Balkashina, Elisabeth (1899), Sc. worker of the Committee of the Academy of Sc. for investigation. Moscow (U.d.S.S.R.), Voronzoro Pole 6. *Genetic of Drosophila, concerning the genotype of Natural Population.*

Ball, Prof. d'Anat. Pathol., Ecole Vét. Lyon (France). *Anat. Pathol.* o

Ball, Antoine J. (1897), Collaborateur au Musée d'Histoire Naturelle à Bruxelles (Belgique), 160, rue Belliard. *Copéopnthes; Dytiscidae de Belgique.* ⓟ Bruxelles 1924.

Ball, Carleton Roy (1873), Senior Agronomist in Charge, Office of Cereal Crops and Diseases, Bureau of Plant Industry. Washington, D.C. (U.S.A.), U.S. Department of Agric. *Agronomy, botan. and physiol. of cereals; Taxonomy and ecology of North American willows (Salix).* ⓟ Iowa State Coll. 1920.

Ball, Elmer Darwin (1870), Research Entomologist Florida State Plant Board, Sanford, Fla.(U.S.A.). *Grasshoppers, Codling Moth. Insects that transmit Plant Diseases, the Beet leafhopper and the Curly leaf Disease, the Potato leafhopper and Hopperburn of Potatoes. Biol. of the Celery leaf-tyer. Inheritance of Egg laying in Poultry; Classification of the Cercopidae Fulgoridae Membracidae and Cicadellidae. Life histories Food Plants and Economic relations of the leafhoppers.* ⓟ Ohio State Univ. 1907. ⓟ *North American Cercopidae Membracidae Fulgoridae and Cicadellidae.*

Ball, Francis J. (1859), Collaborateur sc. au Mus. Royal d'Histoire nat. Bruxelles (Belgique), 160, rue Belliard. *Les Androconia chez les Rhopalocères; différenciation spécifique et générique. Les Rhopalocères et les Bourdons de Belgique et du Monde.*

Ball, Gordon H. (1899), Instr. in Zool., Univ. of California, Southern Branch. Los Angeles, Cal. (U.S.A.), 855 N. Vermont Avenue. *Life-histories of intestinal Protozoa. Physiol. studies on Paramecium.*

Variation in fresh-water mussels. Ecol. Experiments on the action of various endocrine substances, of liver, and of glycogen on the division rate of Paramecium. The behavior of a conjugating race of Paramecium caudatum. Ⓟ Slides of Parasitic Protozoa.
Ball, Nikolaus (1872), Dir. Pathol.-Anat. Inst., Tierärztl. Hochsch. Leningrad (U.d.S.S.R.), Černigovskaij Str. 5. *Pathol. Anat. der Haustiere. Callus.* Ⓓ Dorpat 1897. Ⓟ Pathol.-anat. Praep. der Tierpathol.
Ball, Oscar Melville (1868), Prof. of Biol. Agric. Coll. Texas. College Station, Texas (U.S.A.). *Plant physiol. and bact.* o
Ballenegger, Robert, Dr., Priv.Doc. der Bodenkunde an der Landwirtschaftl. Univ. Budapest (Ungarn). *Bodenkunde.* o
Ballerstedt, Max (1857) Bückeburg (Deutschland), Hannoversche Str. 12. *Fauna des norddeutschen Wealdensandsteins, insbesondere Schildkröten und Krokodile. Fährten von Dinosauriern.*
Ballou, Henry A., Prof. Imp. Coll. of Tropical Agric. Trinidad (Br.-W.-India). *Zool. and Entomol.* o
Ballowitz, Emil (1859), Prof., Geh.Med.R., Univ. Münster (Deutschland), Hittorfstr. 10. *Anat., Histol., Physiol. der Zelle.* o
Bally, Walter (1882) Botanicus an der Versuchsstation Malang, Java (Nederl. Indien) Proefstation. *Kultur und Krankheiten des Kaffeebaums.* Ⓓ Zürich 1907.
Baló, József, Dr., Mitarb. am Pathol.-anat. Inst. Nr. 1, Univ. Budapest (Ungarn). o
Baló, Lajos, Mitarb. am Anat. Inst. Univ. Pécs, (Ungarn). o
v. Balogh, Ernö (1890), o. Prof. Dir. Pathol.-anat. Inst. Univ. Szeged (Ungarn), Kossuth Str. 40. *Pathol., Anat. u. Histol.* Ⓓ Budapest 1913.
Balss, Heinrich (1886) Prof. Dr. Konserv. der zool. Sammlung. München (Deutschland), Alte Akademie, Neuhauserstr. 51. *Systematik der Decapoden u. Stomatopoden. Palaeont. der Decapoden. Geschichte der Zool.* Ⓓ München 1908. Ⓟ Decapoden u. Stomatopoden.
Balteanu, Jean (1887), Conferancier Fac. med. Univ. Jassy (Roumanie). *Bactériol., Immunol., variations microbiennes.*
Baltzer, Fritz (1884) o. Prof. f. Zool., vergl. Anat. u. allg. Biol. und Dir. des zool. Inst. an der Univ. Bern (Schweiz), Bollwerk 10. *Vermes, Bastardierung, Geschlechtsbestimmung, Physiol. u. Psychol. der Spinnen.* Ⓓ Würzburg 1908.
Bambacioni, Valeria (1895), Ass. Lib. doc. Ist. Botan. Univ. Roma (Italia), Via Milano 75. *Citol., Fisiol. vegetale, Embriol. vegetale.*
Bambarén, Carlos, Prof. Univ. Lima (Peru). *Biol.* o
Bamford, Ronald (1901), Instr. Botan. and Cytol. Burlington, Vt. (U.S.A.), Botany Departm., Univ. of Vermont. *Chromosome behavior in Violet hybrids of F_1 generation during meiosis.* Ⓓ M.S. Univ. of Vermont 1926.
Bancroft, Helen (1887), Lect. in Agric. Botan. Oxford (England), The School of Rural Economy. *Botany of Farm Crops; Anat. of Plants.* Ⓓ London 1915.
Banerjee, Bhosanath (1895), Research Ass. Department of Bio-Chem., Indian Inst. of Sc. Hebbal, Bangalore (India). *Power Alcohol. Fermentation Industries. Biochem. Tropical vegetation. Mucors.* Ⓓ Calcutta Univ. 1918.
Bang, O., Prof. Dr., Landwirtschaftl. Hochsch. København V (Danmark), Bülowsvej. *Pathol.* o
Bangham, Ralph V. (1895), Prof. Biol. Coll. of Wooster Head of Dept. Wooster, Ohio (U.S.A.). *Parasitol. Parasites of fresh water species of fish.* Ⓓ Ohio State Univ. 1923. Ⓟ Parasites of fish.
Banker, Howard James (1866), Research work in Eugenics under the Carnegie Inst. of Washington. Long Island, N.Y. (U.S.A.), Eugenics Record Office, Cold Spring Harbor. *Human heredity. North American Hydnaceae. Genealogy of the Banker or Bancker families of North America. Underwood Families of North America.* Ⓓ Columbia 1906. Ⓟ Mycol.
Banks, Nathan (1868), Cur. insects mus., Comp. zool., Harvard Univ. Holliston, Mass. (U.S.A.). *Entomol.* o
Bannerman, David Armitage (1886), Special Ass., Dept. of Zool., British Mus., Nat. History. London W 14 (England), 60 Addison Road Kensington. *Ornithol. Canary Islands.*
Bannier, J. P., Dr. Advis. Oei Tjung Ham-Unternehmung. Pasaroeang, Java (Nederl.-O.-Indië). *Kultur trop. Gewächse.* o
Bannwarth, Emil (1862), Ass. Berlin N 4 (Deutschland), Invalidenstr., Zool. Mus. *Kalkkorallen. Milz der Katze. Crania helvet. antiqua.* Ⓓ 1887.
Bánsághy, Grg., Dr., Mitarb. am Pathol.-anat. Inst. der Univ. Debreczen (Ungarn). o
Bant, A., Dr. Prof. Lwów (Pologne), Académie Vét., r. Kochanowskiego 40. *Anat.* o
Banta, Arthur M. (1877), Resident Investigator, Station for Experim. Evolution, Carnegie Inst. of Washington. Cold Spring Harbor, L.I., N.Y. (U.S.A.). *Control of sex in Cladocera. Inheritance in parthenogenesis and in sexual reproduction in Cladocera. The Fauna of Mayfield's Cave. Selection in Cladocera on the basis of a physiol. character.* Ⓓ Harvard 1907. Ⓟ Cladocera. Alcoholic specimens.
Banting, Frederick Grant (1891), Prof. Med. Research Univ. of Toronto (Canada). *Internal secretion of Pancreas. The Suprarenales.*
Baptista, J. B. F., Prof. Univ. Rio de Janeiro (Brasil). *Anat.* o
Barackman, Rufus A. (1899), Phosphate Fellow, Div. of Agric. Biochem., Dept. of Agric., Univ. of Minnesota. St. Paul, Minn. (U.S.A.). *Phosphates in bread.*
Baraldi, A., Prof. Univ. Rosario (Argentina). *Patol.* o
Baranoff, Nikolaj (1887), Entomol. an dem Inst. für Tropenkrankheiten. Skoplje (S.H.S.). *Culicidae, Simuliidae, Anthomyidae und Tachinidae. Dipteren-Fauna Mazedoniens.*
Baranov, Paul (1892), Prof., Dir. Botan. Inst., Univ., Doc. Taschkent (U.d.S.S.R.). *Zytol. und Embryol. der Pflanzen.*
Baranov, Wladimir (1889), Prof. Omsk (U.d. S.S.R.). Siberisches Inst. für Land- und Forstwirtschaft. *Geographie und Oekol. der Pflanzen. Oberflächenspannung.* Ⓟ Halophyten und Xerophyten.
Baranskij, Dimitrij J. (1889), Ass. des Odessaer landwirtschaftlichen Inst., Versuchsstation. Odessa, Ukraine (U.d.S.S.R.). *Züchtung, Genetik, Ökol. der Gerste.*
Baráss, Gyula, Mitarb. am Anat.-histol. Inst. der Franz-Josef-Univ. Szeged (Ungarn). o
Baráth, Jenö (1895), Ass. III. Med. Univ.-Klinik Budapest (Ungarn), Ludoviceum-ut 2. *Pathol. u. Pharm. des vegetativen Nervensystems des Menschen. Funktionelle Untersuchung der Blutdruckstörungen.* Ⓓ med. Budapest 1918.
Barbacci, Ottone (1860), Dir., Ist. Anat. patol. Univ. Siena (Italia), Via Camollia 36. *Anat. e Istol. patol. Tumori.*
Barbanti-Silva, Raul, Lib. doc., Univ. Modena (Italia). *Anat. Istol.* o
Barber, Bertram A. (1885), Prof. of Biol., Hillsdale Coll. Mich. (U.S.A.). *Ecol. of Mallophaga.* Ⓟ Freshwater Sponges.
Barber, Harry G. (1871), Roselle, N.T. (U.S.A.), 143 E., 3 Ave. *Taxonomy: Hemiptera-Heteroptera.* Ⓓ Harvard Univ. 1898. Ⓟ Hemiptera-Heteroptera.
Barbey, Auguste (1872), Expert forestier, Lauréat de l'Inst. Moncherand sur Orbe, Vaud (Suisse). *Entomol. forestière, recherches biol. sur l'évolution des insectes renageant le bois et les forêts. Les Scolytides de l'Europe centrale.* Ⓓ Dr. h. c. de l'Ecole poly-

technique fédérale de Zürich 1919. ⑦ Préparations d'insectes.
Barbier, Anne-Marie (1902), Prép., Station Centrale de Phytopath. Paris 14 (France), 11 bis, rue d'Alésia.
Barbosa Sueiro, Manuel Bernardo (1894), First Ass. of Anat., Sec. Ass. of Anthrop. Univ. Lisbon. Monte-Estoril (Portugal), Avenida Castelhana, Vila Beira, 3. *Human anat.: osteol. and angeiol. Anat. anthrop.*
Barbour, Henry Gray (1886), Prof. of Physiol. and Pharmacol. Louisville, Ky. (U.S.A.), 101, W. Chestnut St., Department of Physiol. and Pharmacol., Univ. of Louisville. *Regulation of body temperature, blood specific gravity, operative and anesthetic shock. Endocrinol. and Metabolism.*
Barbour, Thomas (1884), Lect. in Zool. in Harvard Univ. and Curator of Reptiles and Amphibians in Mus. of Compar. Zool., Univ. Harv., President Boston Society of Nat. History. Boston, Mass. (U.S.A.), 278 Clarendon St. *Taxonymy. Exploration in Amer. Tropics.* ⑦ Harvard Univ. 1910.
Barclay-Smith, E., Prof., Univ. London (England), South Kensington, S. W. 7. *Anat.* o
Barcroft, Joseph (1872), Prof. of Physiol. Univ. Cambridge (England), 13 Grange Road. *Physiol. of respiration, respiratory function of blood.* ③ Cambridge 1896.
Bard, Louis (1857), Prof. de Clinique méd. à l'Univ. Lyon (France), 5, Avenue Jules Ferry. *Spécificité cellulaire: classification des tumeurs, pathogénie du cancer, électivité cellulaire des virus. Pathogénie des dilatations idiopathiques des organes tubulés et cavitaires. Physiol. du labyrinthe et Sens de la Gyration. Rythme cardiaque. Audition, vision, sens thermique. Physiol. du cervelet, du sens musculaire, de la glande pinéale.* ①Lyon 1879.
Bardach, Jacques (1857), Prof. de microbiol., Inst. d'Instr. publique, Inst. de Hautes Etudes. Odessa (U.d.S.S.R.), Rue Tolstoi 6. *Microbiol. des Limans (lacs salés). Purification des eaux d'Egout. Immunité. Diphtérie.* ① Moskau 1894. ⑦ Cultures de microbes du sol, des limans et des champs aux égout.
Bardan, Emma (1899), Ass. de Biol. Madrid (España), Alcalá 31. *Ichthyol.* ① Madrid 1922.
Bárdarson, Gudmundur G. (1880), Adjunkt at the „Mentaskóli", Coll., Reykjavik (Iceland). *Malacol. and palaeont.; distrubition of the marine Mollusca at the coasts of Iceland; marine shellbearing fossil layers in Iceland deposited in Holocene, Pleistocene and Pliocene.* ⑦ Marine Mollusca from Iceland.
Bardeen, Charles R. (1871), Prof. of Anat., Univ. of Wisconsin. Madison, Wis. (U.S.A.). ① Johns Hopkins Univ. 1897.
Bardelli, Plinio Carlo (1887), Prof. Patol., Vet., Dir. del Labor. Militare per la preparazione del Siero Antitetanico. Bologna 16 (Italia), Viale Audinot 49. *Semiol. Sierol.: sieri antitossici. Batteriol: anaerobi, ifomiceti patogeni per gli animali utili.* ① Pisa 1919. ⑦ Ceppi di Bacillo tetanico. Colture di Criptococco del Rivolta.
Bardier, Emile Désiré (1870), Prof. à la Fac. de Méd. de Toulouse (France), 10, rue d'Etienne. *Physiol. Pathol.: La syncope adrénalino-chloroformique.*
Bare, Clarence O. (1889), Ass. Entomol., U.S., Dept. of Agric. Tampa, Flo. (U.S.A.), Box 4583. *Parasit. of celery leaf tyer; Notonectidae.* ① Univ. of Kansas 1924. ⑦ Notonectids of genera, Notonecta, Buenoa and Plea.
Barfurth, Dietrich (1849), Univ.-Prof. Dr. phil. et med. et Dr. med. hon. c., Geh.Med.R., Dir. des anat. Inst. in Rostock a. D. Seit 1. Okt. 1921 Prof. emeritus. Rostock i. M. (Deutschland), Schackstr. 1. *Anat. und Embryol., speziell: experim. Untersuchungen über Regeneration und Superregeneration, Vererbung der Hyperdactylie, Entwicklungsmechanik.* ① Bonn phil. 1874. med. 1882; Halle med. h. c. 1921.

Barge, I. A. J., Prof. der Anat., Dir. Anat. Labor. Leiden (Holland). o
Baribeau, Charles Henri Bernard (1898), Inspecteur. Ste. Anne de la Pocatière, Quebec (Canada), Labor. de Pathol. végétale. *Pommes de terre. Maladies.*
Barikine, O., Ass. de l'Inst. Microbiol. de l'Etat. Moskau (U.d.S.S.R.), Pogodinskaja 10. o
Barikine, Wladimir (1879), Prof., Dir. Inst. microbiol. sc. Moscou (U.d.S.S.R.), r. Pogodinskaja 10. *Microbiol. méd. et immunol. Paratyph. Vaccination antityphique. Typhus exanthématique. Immunité comme la fonction de l'état des biocolloides de l'organisme.* ⑦ Petersbourg 1905.
Baringer, John William (1895), Associate Pathol., U.S. Dept. Agric. Columbus, Ohio (U.S.A.), Botan. Dept. O.S.U. *Barberry Eradication.* ① Ohio State Univ. 1924.
Barkan, Georg (1889), Dr. med., planmäßiger Ass. am Pharmakol. Inst. der Univ. Frankfurt a. M. (Deutschland), Myliusstr. 26 II. *Physiol. und Pharmakol. des Eisenstoffwechsels. Bluteisen.* ① München 1914.
Barker, B. T. P., Prof., Univ. Bristol (England). *Agric., Biol.* o
Barker, Henry D. (1893), Botan. and Plant Pathol. Port-au-Prince (Haiti), Service Technique de l'Agric. *Diseases of pineapple, cotton, and other tropical crop plants.* ① Univ. of Minnesota 1922. ⑦ Some Haitian plants and fungi could be offered in exchange for other tropical plants and fungi.
Barker, Lewellys Franklin (1867), Prof. emer., John Hopkins Hospital. Baltimore, Md. (U.S.A.), 208 Stratford Rd. Guilford. *Anat. pathol.* o
Barnard, W. G., Curator of Mus. of Pathol. Anat. London, W.C. (England), Univ. Coll. Hosp. Med. School, University Str. o
Barnes, Bertie Frank (1888), Lect. London E. C. 4 (England), Birkbeck Coll., Univ. of London, Breams Buildings. *Mycol., Structure and development of the fungi.* ① London 1922.
Barnett, Robert John (1874), Prof. of Horticulture, Experim. Station Pomologist. Manhattan, Kan. (U.S.A.), Kansas State Agric. Coll. *Pomology.*
Barnette, Robert Marlin (1900), Ass. Chem., Soils, Agric. Exp. Sta. Univ of Florida. Gainesville Flo. (U.S.A.). *Soil Chem. and Bacteriol., Plant Nutrition.* ① Rutgers Univ. 1923. ⑦ Plants, soils.
Barnhart, John H., M. D., Botan. Garden New York (U.S.A.), Bronx Park. o
Barnsby, P., Prof. Univ. Poitiers (France). *Anat.* o
Baron, Johannes (1877), o. Prof. philol. Fak. Staatl. Akad. Braunsberg, Ostpr. (Deutschland). ① Münster 1902, med. Breslau 1912. ⑦ Mikrosk. Präparate Hist. des Eileiters. Säugetiere.
Baroni, Eugenio (1865), Prof., Botan. Univ. di Ferrara, lib. doc. Bologna (Italia). *Botan.: Anat., Sistematica, Lichenol.* ① Pisa 1890.
Barqmann, Helene Elizabeth (1896), Research student, Natural History Mus. London N.W. 3 (England), 43 Bekize Park. *Morphol. of the Central Nervous system of the Gastropoda pulmonata.*
Barr, Rose Aileen (1902), Ass. keeper in Zool. The Manchester Mus. Manchester (England), Victoria Univ. *Anat. of the Mollusca.* ① Manchester 1925.
de las Barras de Aragón, Francisco, Prof., Catedrático de Antrop. de la Univ. Central. Jefe de la Sección de Etnografía del Mus. Antrop. Madrid (España), Covarrubias, 21. *Anthrop.* o
Barre, Henry W., Dir., S.C. Experim. Station. Clemson College, South Carolina (U.S.A.). *Pathol. Physiol.* o
La Barre, Jean (1896), Ass. aux cours de Pharmacographie et Pharmacodynamie de l'Univ. de Bruxelles (Belgique), 72, rue Mercelis. *Coagulation du sang. Anaphylaxie. Sécrétions internes. Pharm.* ① Bruxelles 1922.

Barreiro y Martinez, P. Augustin Jesus. (1869), Presidente de la Sociedad Española de Anthrop. Etnografia y Prehistoria. Madrid 9 (España), General Portier 6. *Coelenterata: Corallinae. Conchol.*
Barrenscheen, Hermann Karl (1887), Priv.Doc., Ass. Inst. f. med. Chem. Univ. Wien XIII,1 (Deutsch-Österreich), Spohrstr. 51 II. *Physiol. Chem. Kohlehydratstoffwechsel.* ① Wien 1910.
Barrett, James Theophilus (1876), Prof. of Plant Pathol. Univ. of California Graduate School of tropical Agric.; Associate Dir. of the Citrus Experim. Station and Plant Pathol. Riverside, Cal. (U.S.A.). *Diseases of deciduous fruits, Subtropical fruits, Walnuts and truck crops.* ① Cornell Univ. 1910.
Barrington, Frederick James Fitzmaurice (1884), Ass. St. Peters Hospital. Ass. in Surg. Univ. Coll. Hospital. London W.C. 1 (England), Med. School. *Experim. physiol. of nutrition.* ① London 1907.
Barrois, Charles (1851), President Académie des Sc. de Paris, Prof. honoraire Univ. Lille (France), 41, rue Pascal. *Faunes paléozoïques.* ① Paris 1876.
de Barros, D., Prof. Univ. Rio de Janeiro (Brasilien). *Histol.* ○
de Barros, F., Prof. Univ. Porto Alegre (Brasil). *Physiol.* ○
Barroso, Manuel Jerónimo. Salamanca (España), Inst. *Bryozoa.* ○
Barrows, William Morton (1883), Prof. of Zool. and Entomol. Columbus, O. (U.S.A.), Ohio State Univ. *Human Heredity, animal Reactions, Spiders of Ohio.* ① Harvard Univ. 1920.
Barrus, Mortier Franklin (1879), Prof. of Plant Pathol. Ithaca, N.Y. (U.S.A.), Cornell Univ. *Diseases of the potato and Control measures.* ① Cornell Univ. 1911.
Barry, David T., Prof. Cork (Ireland). *Physiol.* ○
Barsakoff, Bojan (1885), Ass. Sofia (Bulgarien), Botan. Inst. der Univ. *Hymenomycetes.*
Barsali, E., Prof. Univ. Pisa (Italia). *Botan.* ○
Barsanov'era, Vera Aleks, Doc. I. Univ. Moskau (U.d.S.S.R.). *Palaeont.* ○
Barss, Howard Phillips (1885), Ass. Prof. of Botan. and Plant Pathol. Head of Department of Botan. and Plant Pathol. Corvallis, Ore. (U.S.A.), Oregon State Agric. Coll. *Diseases of cereals and fruit and control materials (fungicides).* ① Plant diseases occurring in the Northwestern Pacific Coast section of the U.S.A.
Barta, Ede, Dr., Mitarb. am Anat. Inst. Nr. 2. Univ. Budapest (Ungarn). ○
Bartelmez, George W. (1885), Ass. Prof. Anat. Chicago, Ill. (U.S.A.), Hull Labor. of Anat. The Univ. *Anat. of young human embryos. The Histol. of the endometrium.* ① Chicago 1910.
Bartels, Max, Pasir Datar, Halte Tjisaat Preanger, Java (Nederl.-O.-Indië). *Ornithol.* ○
Bartenef, Alexander (1882), Prof. Rostov am Don (U.d.S.S.R.), Nord Caucasisch. Univ., Zool. Kabinett. *Odonata. Darvinismus. Geschichte der Deszendenzlehre.* ① Odonata.
Barth, A. J., Prof. Norges Landbrukshöiskole. Aas b. Oslo (Norge). ○
Barthel, Johan Gottfrid Christian (1873), Prof. und Vorsteher der Bact. Abteilung der Zentralanstalt für landwirtschaftl. Versuchswesen, Experimentalfältet Stockholm (Schweden). *Bodenbact. und Molkereibact.* ① Stockholm 1895. ② Kulturen von Mikroorganismen.
Barthélémy, Charles Henri (1885), Chef de Travaux, Inst. de Zool. et de Biol. Générale. Strasbourg (France), Rue de l'Université. *Analyse de la fécondation. Energétique animale. Parasitol.* ① Strasbourg 1919.
Barthélet, Jean Jules (1903), Préparateur Station Pathol. végétale. Paris (France), 11bis, Rue d'Alésia. *Phytopath.* ① Montpellier 1922.
Bartholomew, Elbert Thomas (1878), Ass. Plant Pathol., Univ. of California. Riverside, Cal. (U.S.A.), Citrus Experim. Station. *Physiol. Diseases of Citrus.* ① Univ. of Wisconsin 1914.

Bartlett, F. C., Prof. Univ. Cambridge (England). *Experim. Psychol.* ○
Bartlett, Harley H. (1886), Prof. of Botan., and Dir. of the Botan. Garden. Ann Arbor, Mich. (U.S.A.), Univ. of Michigan. ○
Bartolomé del Cerro, Abelardo, Prof. Catedrático Univ. Valladolid (España). *Biol.* ○
Bartrina, Jesús, Prof. Univ. Valencia (España). *Anat. descript. Embryol.* ○
Bartrum, J. A., Prof. Univ. Coll. Auckland (New Zealand). *Palaeont.* ○
Bartsch, Otto (1881), Dr. phil., Ass. am Zool. Inst. der Landwirtschaftl. Hochsch. Berlin. Berlin-Niederschönhausen (Deutschland), Kaiser-Wilhelm-Straße 66. *Geflügelzucht.* ① Berlin 1922.
Bartsch, Paul (1871), Dir. Histol. and Physiol. labs. Harvard Univ., med. Dept. Washington, D.C. (U.S.A.), 1456 Belmont Str. N.W. *Med. Zool., pearl mussel.* ○
Bartual, Juan (1883), Prof. Histol., Fac. Med. Valencia (España), Embajador Viela 1. ① Madrid 1889.
Bartucz, Lajos, Dr., Anthrop. der Ethnograph. Abt. des Nationalmus. Budapest (Ungarn). ○
Barulina, Helene (1896), Senior Ass. Inst. of Applied Botany and New Cultures. Leningrad (U.d. S.S.R.), Herzen St. 44. *Systematic of cultivated plants: Ervum Lens L., Vicia sativa L., Vicia Ervilia Schr.*
Barykin, Vlad. Alek., Prof. I. Univ. Moskau (U.d.S.S.R.). *Bact.* ○
Baschinskaja, Maria M., Vorsteherin d. Botan. Abtlg. d. Staatl. Wolynsko-Wissensch. Untersuchungsmus. Shitomir (U.d.S.S.R.), Lermontowstr. 13. *Geobotan. Phytogeographie. Phytosoziol. Systematik.* ○
Bascom, Kellogg F. (1891), Associate Prof. of Anat. Richmond, Va. (U.S.A.), Med. Coll. of Virginia. *Interstitial Cells of the Gonads of Cattle. Cryptorchid testes. Pattern and Total Tubule Length in the Testicles.* ① Univ. of Chicago 1922.
Basile, C., Prof. Univ. Roma (Italia). *Parasit.* ○
Basilevsky, Nina (1902), Ass. of Herbarium Principal Botan. garden, Inst. of Applied Botan. Leningrad (U.d.S.S.R.), Proviantskaja 3, fl. 22. *Flora of Turkestan sanddeserts. Biol. of psammophytes and Classification of genus Astragalus.*
Basinger, Almon T. (1886), Ass. in Entomol. Riverside, Cal. (U.S.A.), Citrus Exper. Stat. *Investig. of life histories of beneficial insects.* ① Cacinellidae.
Baskina, Wera P. (1896). Perm (U.d.S.S.R.), Zaimka-Univ. *Zool. Statistische Zooókol. Hydrobiol. der Salzgewässer.*
Baskina-Muchina, Olga P. (1899). Perm (U.d.S.S.R.), Zaimka-Univ. *Vergl. Histol. der Schwimmblase.*
Basler, Adolf (1878), tit. a. o. Prof., Leiter des Rassenbiol. Inst., Priv.Doc. der Physiol. in Tübingen (Deutschland), Katharinenstr. 20. *Physiol. des Stehens und Gehens. Physiol. der Körperhaare und der äußeren Haut. Verschmelzung von Gesichtsempfindungen. Blutbewegung in den Haargefäßen. Die Modifizierbarkeit des menschlichen Schädels. Vererbung der Zygodaktylie beim Menschen.* ① Tübingen 1902.
Bašmaka, Vlad. Il'ie, Doc. I. Univ. Moskau (U.d.S.S.R.). *Physiol.* ○
Bass, Erwin (1895), Dr. med., Ass.-Arzt an der Med. Univ.-Klinik Greifswald (Deutschland). *Muskelaktionsströme. Sauerstoffverbrauch des Zentralnervensystems. Nervöse Atmungsregulation. Asthma bronchiale. Funktionsprüfung der Atmung. Residualluftbestimmung und Atemvolummessung.* ① Tübingen 1921.
Bassalik, Casimir, Dr. Prof. Warszawa (Pologne), Univ. Inst. de physiol. végét., r. Krakowskie-Przedmieście 26/28. *Botan.* ○
Bassarskaja, Lydia D., Ass. d. Chem.-pharm. Inst. Odessa (U.d.S.S.R.), Staro Porto-Frankowskaja 41, W. 7. *Botan. Systematik. Arzneipflanzen.* ○

Bassler, Ray S. (1878), Curator of Palaeont., U.S. National Mus. Washington, D.C. (U.S.A.). *Fossil microorganisms, Paleozoic stratigraphy, paleont.* ⓟ George Washington Univ. 1905.

Bastert, Christine Elizabeth (1890). Amsterdam (Holland), Meerhuizenplein 34. *Atmungsphysiol. der Amphibien.*

Bastos Monteiro, Hernani(1891), Prof. ord. d'Anat. topographique, Fac. Méd. Porto (Portugal), R. do Pombal, 47. *Anat. humaine et comparée.* ⓟ Porto 1915.

Bataillon, Jean Eugène (1869), Prof. à l'Univ. de Montpellier (Faculté des Sc.), Dir. de la Station marine de Cette. Montpellier (France), 1, Rue Richer de Belleval. *Fécondation et Parthénogénèse, Etudes experim. et analytiques. Métamorphoses des Batraciens et des Insectes. Ontogénèse des Poissons et des Batraciens. Teratogénèse. Cinèse (Ascaris). Parthénogénèse traumatique des Batraciens-Polyspermie.* ⓟ Lyon. ⓟ Matériel marin de la côte Cettoise.

Batalina-Silberminz, Marie A., Privatg. Moskau (U.d.S.S.R.), Leontjewsky Per., 29. *Palaeont., spez. Ostracoden.* o

Bate, Dorothea M. A. London S.W. (England), British Mus., Natural History. *Pleistocene Mammalia and Birds.* o

Bates, Carlos Glazier (1885), Silviculturist and Dir., Rocky Mt. Forest Exp. Sta. Colorado Springs. Col. (U.S.A.), *Growth and reproduction of forest trees; influence of forests upon streamflow; forest ecol. and tree physiol.*

Bates, G. L. Bitye, Ebolowa (Französisch Kamerun). *Ornithol.* o

Bateson, William, Dr. Sc., Dir. John Innes Horticultural Inst. Surrey (England), Merton Park. o

Bather, Francis Arthur (1863), Keeper of the Department of Geol. and Palaeont. in the British Mus., President of the Geological Society of London. London SW 7 (England), British Museum (Natural History). *Organic Evolution: palaeont. evidences. Echinoderma: fossil Pelmatozoa. Annelida fossilia.* ⓟ Oxford 1900.

Batchinsky, Alexandra (1878), Ass. Pflanzenbiol., Med. Inst. Leningrad (U.d.S.S.R.), Gatchinskaja Str. 14. *Biol. der Hefepilze.*

Batchinsky, Peter (1875), Ass. d. Hygiene, Med. Inst. Leningrad (U.d.S.S.R.), Gatchinskaja Str. 14. *Biol. Wasseranalyse.*

Battaerd, P., Prof. Univ. of Stellenbosch (South Africa), Dept. of Bausiol. ⓟ Leiden.

Battaglia, Filippo (1896), Aiuto nell' Istituto di Anat. Patol. R. Univ. Milano (Italia), Ospedale Maggiore. *Reins. Eosinophilie.*

Battaglia, M., Prof. Univ. Napoli (Italia). *Anat.* o

Battelli, F., Prof. Dr. Genève (Suisse), Boulevard des Tranchées 14. *Physiol. Ferments: Oxydase.* o

Battez, M. G., agrégé Physiol., Fac. Méd. Algier (Afrique). *Physiol.* o

Bauch, Robert (1897), Priv.-Doc. Dr. Rostock (Deutschland), Botan. Inst. der Univ. *Sexualitätsproblem bei Pilzen. Ustilagineen und Basidiomyceten.* ⓟ Würzburg 1921. ⓟ Ustilagineen, Hymenomyceten, Mucorineen.

Baudet, Edmond, Arthur, René, Floribert (1886), Dr. med. vet., Konserv. an Inst. für parasitäre und Infektionskrankheiten (Fac. der Veterinär-Med.) Utrecht (Holland), Willem Barentzstr. 57. *Biol. der parasitären Nematoden.* ⓟ Bern 1911.

Baudrimont, Edouard Marcel Albert (1883), Dr. med., Préparateur d'Histol. à la Fac. de Méd. de Bordeaux (France). *Histol. Poumons (Mammifères, Oiseaux, Reptiles, Amphibiens).* ⓟ Bordeaux 1910.

Baudys, Eduard, Dr. ing., Doc. für Phytopath. der Mähr. Landwirtschaftl. Versuchsanstalt. Brno (C.S.R.), Trávniky 5/6. *Phytopath., Zoocecidol., Mykol.* o

Bauer, A. A., Prof. II. Univ. Moskau (U.d.S.S.R.). *Pflanzenzucht.* o

Bauer, Alfred (1883), Weinbaudir. Neustadt a. d. Haardt (Deutschland), Hindenburgstr. 9. *Reblausrassen, Reblausbekämpfung, Rebveredlung.*

Bauer, Karl Heinrich (1890), Prof. Dr., a.o. Prof. Göttingen (Deutschland), Chir. Univ.-Klinik. *Allgemeine Vererbungspathol. des Menschen (Bewegungsapparat, Blutkrankheiten, Mißbildungen), Menschliche Konstitutionslehre (Systemerkrankungen, bes. d. Stützgewebe), Rassenhygiene (biol. Grundlagen).* ⓟ Würzburg 1916. ⓟ Meerschweinchen, Ratten, weiße Mäuse.

Bauer, Victor (1881), Priv.Doc. und Ass. am Zool. Inst. der Univ. Bonn, lehrbeauftragt für Fischerei und Fischzucht an der landwirtschaftlichen Hochschule Bonn-Poppelsdorf. Bonn (Deutschland), Zool. Inst. (Poppelsdorfer Schloß). *Vergl. Physiol. der Sinne, bes. des Lichtsinns.* ⓟ Freiburg i. B. 1903.

Bauman, Louis (1880), Ass. Prof. of Med. New York City (U.S.A.), Presbyterian Hospital, Columbia Univ. *Biol. Chem.*

Baumann, Alexander A., Dir. Selektionsabtlg. d. Dnepro-Petrowsker Provinzversuchsstation. Sinelnikowo (U.d.S.S.R.). *Agronomie. Selektion.* o

Baumann, Emil J. (1891), Chem. New York, N.Y. (U.S.A.), Gun Hill Road near Jerome Avenue. *Physiol. Chem.* ⓟ 1915.

Baumann, Eugen (1868), Dr. phil. Zürich 6 (Schweiz), Sonneggstr. 35. *Hydrobiol., spez. Wasserpflanzen und Sumpfpflanzen.* ⓟ Zürich 1910. ⓟ Wasser und Sumpfpflanzen, spez. Potamogetonaceae, Najadaceae, Myriophyllum, Cyperaceae.

Baumann, Franz (1885), ao. Prof. Zool. Univ., Dir. der Zool. Sammlungen d. Naturhist. Mus. der Stadt Bern (Schweiz), Mühlemattstr. 2. *Parasit., Hydrobiol. und Faunistik der einheimischen Tierwelt.* ⓟ Bern 1909.

Baumgärtel, Traugott (1891), Priv.Doc. für Bact. und Konserv. an der Techn. Hochsch. in München (Deutschland). *Mikrobiol.* ⓟ Gießen 1921. ⓟ Bact.-, Pilz- und Algenkulturen.

von Baumgarten, Paul Clemens (1848), Dr. med., o.ö. Prof. der allgemeinen Path. und der path. Anat. an der Univ. Tübingen und Dir. des path. Inst. (im Ruhestand). Dresden-N. 8 (Deutschland), Radeberger Str. 9 I. *Path. Histologie, Bact. und experim. Path. Spezialgebiet: Erforschung der Wechselwirkung zwischen den Mikroparasiten und den von ihnen ergriffenen Geweben des Wirtsorganismus.* ⓟ Leipzig 1873.

Baunacke, Karl Walther (1883), Zoologe, Dr. phil. Vorstand der Abt. Pflanzenschutz der Staatl. Landwirtschaftl. Versuchsanstalt Dresden-A. 16 (Deutschland), Stübleallee 2. *Landwirtschaftl. Bodenschädlinge: Rübennematoden, Drahtwurm und Wurzelschmarotzer. Schädlinge des Obst-, Wein- und Gemüsebaues und deren Bekämpfung. Tierische Sinnesphysiol. Angewandte Entomol. Vogelschutz. Kartoffelkrebsbekämpfung. Rübenfliege.* ⓟ Greifswald 1912.

Baur, Erwin (1875), Dr. phil. et med., Dr. agr. h. c., o. Prof. an der Landwirtschaftl. Hochsch. zu Berlin, Dir. des Inst. für Vererbungsforschung. Berlin-Dahlem (Deutschland), Albrecht-Thaer-Weg 4. *Vererbungslehre, Mutationsproblem bei Phanerogamen. Wissenschaftliche Grundlagen der Pflanzenzüchtung.* ⓟ med. Kiel 1899, phil. Freiburg 1903, agr. h. c. Wien 1924.

Baur, Max (1893), Priv.Doc. der Pharmacol., z. Z. stellv. Dir. des Pharmacol. Inst. Kiel (Deutschland), Seeblick 2. *Dünndarmperistaltik, Abhängigkeit von physikalischen und pharm. Eingriffen. Die physiol. Grundlagen der Osmotherapie.* ⓟ Köln 1920.

Baurmann, Hermann (1897), Dr. med. Freiburg i. Baden (Deutschland), Zool. Inst. *Entwicklungsmechanik, spez. Determinationsproblem, Milzfunktion.* ⓟ Freiburg 1924.

Bavendamm, Werner Hermann Theodor (1898), Dr.phil.,Hochsch.-Ass.Tharandt i.Sa. (Deutschland), Forstliche Hochsch., Botan. Inst. *Mykol. (holzzerstörende Pilze), Morphol. und Systematik der*

Pflanzen. Vererbungslehre. Farblose und rote Schwefelbact. des Süß- und Salzwassers. ① Berlin 1923.
Baxter, Dow Vawter (1898), Ass. Prof. of Forestry. Ann Arbor, Mich. (U.S.A.), Univ. of Michigan. *Silvics, Forest Path.* ① Ann Arbor (Mich.) 1924.
Baxter, Evelyn Vida (1879), Hon. Mem. B.O.W., The Grove Upper Largo Fife (England), *Scottish Ornithol., Migration, Distribution.*
Bayer, August (1882), Dr. phil., Prof. de Botan. systématique, Ecole Supérieure d'Agric. Brno (C.S.R.), Cerná Pole. *Dendrol.; Anat. du bois; Pyrenomycetes. Morphol. d. Rhizome v. Pteris. Cruciferen. Weibl. Blüten d. Cupressineen.* ① Prague 1905. ℗ Semences, fruits, herbiers; maladies de plantes ligneuses; champignons parasit.; bois; altérations et défauts des bois.
Bayer, Edvin, Ph.D., Dir. de la Section de Botan. du Mus. National, Praha II (C.S.R.), Václavské nám. 74. o
Bayer, Emil (1875), o. Prof. der allgemeinen und landwirtschaftl. Zool. an der Hochsch. für Bodenkultur. Brünn (C.S.R.). *Vergl. Morphol. der Würmer, Biol. der Insekten, Cecidol. (Gallenkunde), Limnol., spez.: tierisches Plankton der Süßwässer.*
Bayer, Gustav (1879), o. Prof., Dir. d. Inst. f. allg. u. experim. Pathol. d. Univ. Innsbruck (Deutsch-Österreich), Müllerstr. 44. *Innere Sekretion, Nebennieren.* ① Wien 1904.
Bayer, Margarete (1896), Dr. phil., wiss. Hilfsarbeiterin am Zool. Inst. der Landwirtschaftl. Hochsch. Berlin. Berlin-Südende (Deutschland), Potsdamer Str. 6. *Biol. der Silphiden.* ① Berlin 1924.
Baylis, Adelaide B. (1883), Ass. in Bact. New York City (U.S.A.), 303 East 20th Street, New York Post Graduate Medical School and Hospital. *Bact., Serol.*
Baylis Ash, E., Lect. Univ. Birmingham (England). *Pathol.* o
Baylis, Harry Arnold (1889), Ass., Dept. of Zool., British Mus. London S.W. 7 (England), Cromwell Road. *Nemathelminthes, Platyhelminthes, Nemertinea.* ① Oxford 1912.
Bayliss, Leonard Ernest (1900), Sharpey Scholar in Physiol., Univ. Coll. London N.W. 3 (England), St. Cuthberts West Heath Road. *Physical Chem. as applied to Physiol.* ④ Cambridge 1922.
Bayne, P. M., Prof. West China Union Univ. Chengtu, Szechuan (China). *Zool.* o
Bayne-Jones, Stanhope (1888), Prof. of Bact. Rochester, N.Y. (U.S.A.), School of Med. and Dentistry Univ. of Rochester, Crittenden Boulevard. *Med. Bact.* ① Johns Hopkins 1914.
Bazarewski, Stefan (1871), Prof. à l'Univ. de Wilno et Dir. de l'Inst. de chem. agricole et de microbiol. agricole. Wilno (Pologne), Podgorna 3—17. *Chimie et microbiol. agricole, Biol. de sol.* ① Riga 1912.
Bazaroff, Nicholas I. (1902), Inst. of General Biol. II. Univ., Collab. Labor. of Experim. Biol. Zoopark. Moscow (U.d.S.S.R.), Pjatnitzkaja 10, Ap. 7. *Mechanics of development. Influence of internal factors on the morphogenesis.*
Baxett, Henry Cuthbert (1885), Prof. of Physiol. Philadelphia, Pa. (U.S.A.), Med. School, Univ. of Pennsylvania. *Physiol. of Temperature Control and of Blood Circulation. Crossed Circulation. Decerebrate Cats. Differential Pressures Baths.* ① Oxford 1919.
Beach, Walter Spurgeon (1890), Ass. Prof. of Plant Pathol., Pennsylvania State Coll. Bustleton, Pa. (U. S. A.), Field laboratory. *Fungous diseases of vegetables, lettuce, rhubarb, lima beans and mushrooms.* ① Univ. Ill. 1918.
Bean, Barton A. (1860), Ichthyol. Washington, D.C. (U.S.A.), U.S. National Mus. o
Bean, Robert Bennett (1874), Prof. of Anat., Univ. of Virginia, Charlottesville, Va. (U.S.A.). *Physical anthrop. Growth in children. Human types. Old Americans of Virginia. The Racial Anat. of the Philippine Islanders.*

Bean, W. J. (1863), Curator. Kew (England), Royal Botan. Gardens. *Botan. and horticultural collections of living plants. Orchids. Trees and Shrubs Hardy in the British Isles.*
Bear, Firman E. (1884), Prof. of Soils, Ohio State Univ. Columbus, Ohio (U.S.A.). *Nitrogen Economy Studies. Soil Management.* ① Univ. of Wisconsin 1917.
Beattie, James Martin (1868), Prof. of Bact. Univ. of Liverpool (England). ① Edinburgh 1894.
Beattie, R. Kent (1875), Pathol. in Charge, Foreign Plant Quarantines, Federal Horticultural Board United States Department of Agric. Washington, D.C. (U.S.A.), 2518—17th St. N. W. *Taxonomy and distribution of the plants in the Pacific Northwest States of the U.S.A. Plant Disease Geography.*
de Beauchamp, M. P., Prof. Fac. Sc. Strasbourg (France). *Zool.* o
de Beaufort, Lieven Ferdinand (1879), Dir. of the Zool. Mus., Amsterdam (Holland), Plantage Middenlaan. *Fishes (anat. and classification), recent and extinct. Birds and mammals (classification). Zoogeography. Fishes of the Indo-Australian Archipelago.* ① Amsterdam 1908.
Beaumont, John Herbert (1894), Ass. Prof. of Horticulture. St. Paul, Minn. (U.S.A.), Univ. Farm. *Genetics and Plant Breeding. Cytol.: Sterility and chromosome behavior in fruits.* ℗ Hardy fruit plants.
Beauverd, Gustave (1867), Dr., Conserv. de l'Herbier Boissier, Inst. botan., Univ. Genève (Suisse), 69, rue Liotard. *Botan. Syst.: Melanipyrum, Species Hepaticorum; Géobotan. (Rhône).*
Beauverie, Jean J. (1874), Prof. à la Fac. des Sc. de l'Univ. de Lyon (France), Labor. de Botan. *Cytol. (organismes inférieurs, notamment Bact.) Phytopath. Polymorphisme des champignons.*
de Beaux, Oscar (1879), Dr., Prof. Univ., Conserv. Mus. Civico di Storia Naturale „Giacomo Doria". Genova 2 (Italia), Via Brigata Liguria 9. *Vertebrati: Uccelli e Mammiferi delle Colonie Italiane, Africa, Italia. Morfol. esterna dei Mammiferi. Morfol. craniale normale e patol. dei Mammiferi. Morfol. esterna dei Neonati dei Mammiferi tanto selvatici che domestici. Sviluppo postembrionale della forma esterna e del cranio nei Mammiferi.* ① Genova 1915. ℗ Micromammiferi.
Beccari, Ludovico, Prof. straord. univ. Ferrara (Italia). *Fisiol. sperim.* o
Beccari, Nello (1883), Prof. st. Anat. e Fisiol. compar., Prof. incar. Istol. e Fisiol. gen. Univ. Firenze (Italia), Via Romana 19. *Anat. del sistema nervoso dell' uomo e dei vertebrati (nervi cranici, mesencefalo, rombencefalo, circonvoluzioni cerebrali, morfol. dell' encefalo di vertebrati fossili). Prima origine delle cellule genitali e differenziamento del sesso negli anfibi. Sviluppo delle ghiandole cutanee nei mammiferi. Corpi Ghiandolari cutanei di alcuni ruminanti. Muscoli del Torace.*
Becherer, Alfred (1897), Dr. phil. Basel (Schweiz), Pfirtergasse 28. *Floristik und Pflanzengeographie der Nordwestschweiz.* ① Basel 1924.
Bechhold, Heinrich (1866), Univ.-Prof., Dir. des Inst. für Kolloidforschung. Frankfurt a. M. (Deutschland), Theodor-Sternhaus. *Kolloide.* ① Berlin 1889.
Bechterev, V. M., Prof., Dir. Inst. z. Studium d. Gehirnes u. d. Psychol. Tätigkeit. Leningrad (U.d.S.S.R.), Petrovskaja ul. 3. o
Beck, Adolf (1863), med. doctor, prof. of physiol. Lwow (Poland), 4 Asnyka. *Physiol. of the central nervous system.* ① Univ. of Cracov 1890.
Beck-Mannagetta, Günther, Dr., Prof. der Botan. Praha II (C.S.R.), Vinična 3a. o
Beck, Max (1862), Dr. med., Prof., Geh.Med.R., Dir. des staatlichen Medizinaluntersuchungsamts. Stade, Prov. Hannover (Deutschland). *Bact.* ① Tübingen 1886. ℗ Plankton.
Beck, Olga (1894), Dr. phil. Wien XIX (Österreich), Hartäckerstr. 26. *Pflanzenpathol.* ① Wien 1918.

Beck, William A. (1883), Prof. of Botan. Dayton, Ohio (U.S.A.), Univ. *Plant Physiol.* ⓓ Univ. of Fribourg 1926.

Bečka, Jan (1889), Ph. et M.U. Dr., Prof. der med. Chem. an der Tierärztl. Hochsch. in Brünn. Brno (C.S.R.), Pražská 69. *Interferometrische und refraktometrische Maßanalyse. Eiweißkörperkoagulation in Tropfen. Toxikol. Mikrochem. Herstellung wirksamer Substanzen aus Organen mit innerer Sekretion.* ⓓ Praha, Ph.Dr. 1912, MU.Dr. 1918.

Becker, Elery R. (1896), Ass. Prof. of Zool. Ames, Ia. (U.S.A.), Iowa State Coll. *Protozoa.* ⓓ Johns Hopkins 1923.

Becker, Jan Gerarders (1887), Lect. Univ. Witwatersrand, Johannesburg, Transvaal (S. Africa), P.O. Box 1038. *Pathol. Bacteriol.* ⓓ Glasgow 1912.

Becker, Joachim Hermann (1869), Leiter der Hauptstelle für Pflanzenschutz. Eutin (Deutschland), Markt 9. *Pflanzenernährung.* ⓓ Leipzig 1895.

Becker, Karl Ernst (1893), Dr. phil., Vorsteher der Botan. Abteilung der Anhaltischen Versuchsstation und Leiter der Hauptstelle für Pflanzenschutz in Anhalt. Bernburg (Deutschland), Junkergasse 3. *Pflanzenkrankheiten.* ⓓ Halle 1920.

Becker, Raymond B. (1892), Associate Prof. in Dairy Husbandry. Stillwater, Okla. (U.S.A.), A. and M. Coll. *Nutrition of dairy cattle.* ⓓ Husbandry 1925.

Becker, Th. Liegnitz, Schlesien (Deutschland), Weißenburger Str. 3. *Entomol., Dipteren.* o

Beckwith, Charles Stewart (1891), Cranberry Specialist, New Jersey Agric. Experim. Station. Pemberton, N.J. (U.S.A.). *Cranberry and blueberry culture.* ⓓ Rutgers Coll. 1924.

Beckwith, Cora J. (1875), Ph. D., Assoc. Prof. of Zool. Poughkeepsie, N.Y. (U.S.A.), Vassar Coll. *Histol. Cytol. Embryol.* ⓓ Columbia Univ. 1914.

Bédart, Prof. Univ. Lille (France). *Pharm.* o

Bedeljan, Joachim L., Prof. Univ. Eriwan (U.d.S.S.R.), Universität 352. *Botan. Pflanzen-Physiol.* o

Bedford, Mary Duchess of Woburn Abbey, Beds. (England). *Ornithol.* o

Bedot, M., Dr., Dir. du Mus. d'Histoire nat. Genève, Kt. Genf (Suisse). o

Bedujagin, F. I., Doc. landwirtschaftl. Inst. Krasnodar, Kaukasus (U.d.S.S.R.), Novaja 107. *Techn. d. Ernährung.* o

Beebe, William (1877), Dir., Department of Tropical Research, New York Zool. Society. New York City (U.S.A.), New York Zool. Park. *Evolution of Fish and Birds, Oceanography.* ⓓ Columbia Univ. 1898.

de Beer, Gavin Rylands (1899), Fellow of Merton Coll., Jenkinson Memorial Lect. in Compar. and Experim. Embryol. Univ. of Oxford (England), 4 Holywell Street. *Compar. Anat. and Embryol. of Vertebrata. Experim. Embryol. Growth. Genetics. Compar. Anat., Histol. and Development of the Pituitary Body.* ⓓ Oxford 1924.

Beers, Charles Dale (1901). Baltimore, Md. (U.S.A.), Zool. Labor., Johns Hopkins Univ. *Life histories and function of conjugation and encystment in fresh-water protozoa.* ⓓ Johns Hopkins Univ. 1925.

Begak, D. A. (1905), Sc. collaborator by Central Sc.-Invest. Peat Inst. Moscow (U.d.S.S.R.), Nikolskystreet 10. *Microbiol. of soil and peat.*

Begemann, Herman, Abteilungsvorsteher in der Proefstation voor Bergcultures Malang, O. Java (Nederl.-O.-Indië), Willemstraat 4. *Vergl. Physiol. des Blutes. Bekämpfung von Kaffeeschädlingen durch Schlupfwespen und durch das Studium der Lebensperiodizität der Schädlinge.* ⓓ Utrecht 1925.

Beghtel, Floyd E., Prof. of Botan., Indiana Central Coll. Indianapolis, Ind. (U.S.A.), Univ. Heights. *Morphol.* o

Begtrup, Erik (1888), Priv.Doc. d. Ernährungslehre, Univ. København (Danmark), Hygien. Inst., Ny Vestergade 11. *Diätetik, Stoffwechsel.*

Béguet, Maurice Eugène (1887), Chef de labor. à l'Inst. Pasteur d'Algérie. Alger (Algérie). *Biol. des bactéries.* ⓓ Alger 1913.

Begutschew, Peter P., Spezialist d. Wolga Kolonisation u. Meliorations-Expedition. Saratow (U.d.S.S.R.), Sowetskaja 24/26. *Agronomie, angew. Botan. Pflanzen-Oekol.* o

Béguinot, Augusto (1875), Prof. d. Botan. Univ. di Modena (Italia), R. Ist. ed Orto Botan. *Botan. ecol.,fitogeografica, sistematica (delle piante superiori), genetica. Flore regionali.* ⓓ Roma 1898.

Behm, Alarik (1871), Dir. des zool. Gartens „Skansen". Stockholm 14 (Sverige). *Zucht u. Bastardzucht (Canidae, Corvidae, Tetraoninae, Strigidae).*

Behn, H. (1876), Reg.R.,Mitglied der Biol. Reichsanstalt für Land- u. Forstwirtschaft. Leiter des Lab. f. prakt. Bodenbact. Berlin-Dahlem (Deutschland). ⓓ Rostock 1900.

Behning, Arvid L. (1890), Dir. der Biol. Wolga-Station, Prof. der Zool., Landwirtsch. Inst. Saratow (U.d.S.S.R.), Biol. Station. *Plankton, Crustaceen, Biol. des Sterlets.* ⓓ Leipzig 1912.

von Behr-Pinnow, Carl F. L.(1864), Privatgelehrter, Vorsitzender des Deutschen Bundes für Volksaufartung und Erbkunde. Berlin W 15 (Deutschland), Sächsische Str. 6. *Menschliche Erbbiol. und deren Anwendung, Eugenik. Geburtenrückgang und Säuglingssterblichkeit. Soziale Kultur und Volkswohlfahrt.* ⓓ jur. Göttingen 1886, med. h. c. Berlin 1909.

Behre, Ellinor H. (1886), Ass. Prof. of Zool. Baton Rouge, La. (U.S.A.), Louisiana State Univ. *Physiol. Zool.; color changes in cold blooded animals, in relation to environ mental factors.* ⓓ Univ. of Chicago 1918.

Behre, Jeanette Allen (1891), Instr. in Chem., Cornell Univ. Med. Coll. New York City (U.S.A.) and Research Chem., Union Central Life Insurance Co., Cincinnati, Ohio (U.S.A.), 3614 Vine St. *Biol. Chem., blood and urine.*

Behrens, Charles August (1885), Prof. bact. and sanitary Sc. Purdue Univ. West-Lafayette, Ind. (U.S.A.), 217 Lutz Ave. *Bact.* o

Behrens, Johannes (1864), Prof. Dr. phil., Geh.-Ob.R., Dir. d. Biol. Reichsanstalt für Land- und Forstw. (l. R.). Hildesheim (Deutschland), Küchenthalstr. 15. ⓓ Kiel 1886.

Behrsin, Ivan And., Doc. Univ. Perm (U.d.S.S.R.), Ul. Karl Marksa 26. *Zootechn.* o

Beilline, Isaac G., Chef de la Section Phytopath., Station Expér. Voronesh (U.d.S.S.R.). *Pathol. végétale, Ustilaginées.* o

Beitzke, Hermann (1875), o. Prof. der pathol. Anat. an der Univ. Graz (Deutsch-Österreich), Waldgasse 28. *Pathol. der Tuberkulose.* ⓓ Kiel 1899.

Beke, Ladislaus, Mitarb. an der Stat. für Pflanzenschutz d. Ung. Min. für Landw. Budapest II (Ungarn), Debröigasse 17. *Kartoffelkrankheiten.* o

Beker, Ernest-Jul. G., Doc. I. Univ. Moskau (U.d.S.S.R.). *Zool.* o

Bekkendorf, C. M., Prof. Astrachan (U.d.S.S.R.), Staatl. Med. Inst. *Normale Histol.* o

Beklemischew, Wladimir (1890), Prof. Zool. Univ., Mitglied des Biol. Forschungsinst., Dir. der Biol. Station des B. F. Inst. (am Kama-Flusse). Perm (U.d.S.S.R.). *Morphol. und Tiergeographie der Turbellaria. Oekol. und Variabilität der Culicidae. Alluviale Zoocoenosen. Biol. Systematik.*

Belajew, Nicolaus (1899), I. Univ., Mitarbeiter d. Inst. f. Experim. Biol. Moskau (U.d.S.S.R.), Woronzowo Pole, 6. *Raupen-Färbung. Genetik der Drosophila-Arten.*

Belák, Alexander (1886), o.ö. Univ.-Prof. Debreczen (Ungarn), Hygien. Inst. *Immunitätsforschung. Kolloidchem. Blutbildung.* ⓓ Budapest 1910. ⓓ Frösche, Kaninchen.

Bělař, Karl I. (1895), Dr. phil., Ass. am K.-W.-I. f. Biol. (Abteilung Hartmann), Priv.Doc.f. Zool. an der F.-W.-Univ. Berlin. Berlin-Dahlem (Deutschland), Kaiser-Wilhelm-Inst. für Biol. *Morphol. und Physiol. der Karyokinese, Protoplasmastruktur, Fortpflanzungsphysiol. (Abhängigkeit der Auslösung von Formwechselprozessen von Außen- und Innenbedingungen).* ⓟ Wien 1919. ⓟ Kulturen von: Amoeba proteus, Myxomyceten, Erdamoeben, Actinosphaerium, Euglypha und andere Süßwasserprotozoen, ferner: Erdnematoden. Präparate von: Protozoen, Nematoden und sonstigen cytologisch interessanten Objekten.

Belavić, Marija (1896), Leiterin der chem. Abteilung des Epidemiol. Inst. Osijek (S.H.S.). *Med. Chem.* ⓟ Graz 1920.

Bělehrádek, Jan (1896), Doc. de Biol. générale, Faculté de Méd., Dir. provisoire de l'Inst. de Biol. Générale, Univ. Masaryk, Brno (C.S.R.), Udolní 73. *Physico-chimie du protoplasma et de la cellule. Action de température sur le protoplasma et sur ses fonctions. L'état colloidal, perméabilité.* ⓟ Univ. Charles, Prague 1921.

Belfanti, Serafino (1860), Prof., Dir. d. Serotherapeutischen Inst. Mailand, Univ. Milano 24 (Italia), Via Darwin 20. *Immunol., Biochem. und Bact.* ⓟ Turin 1886. ⓟ Kulturen, Stämme, Sera, Vakzine.

Belfiore, Ignazio, Lib. Doc. Catania, Sicilia (Italia), Univ. *Biochem.* o

Beling, Demetrius (1882), Prof. der Zool., Inst. für Volksaufklärung, Dir. der Biol. Dnjepr-Station. Kiew, Ukraine (U.d.S.S.R.), Gerschuni Str. 44/2. *Ichthyol, Hydrobiol.* ⓟ Kiew 1916.

Belizin, Vladimir (1905), Ass. Entomol. Plant Protection Station. Stawropol, Caucasus (U.d.S.S.R.), Okrzu.: *Hymenoptera (Cynipidae). Economic Entomol.*

Beljajewa, Elisabeth N., Geol. Mus. d. Akad. d. Wiss. Leningrad (U.d.S.S.R.). *Palaeont. der Wirbeltiere.* o

Beljakow, Eugen, Ass., Univ. Saratow (U.d.S.S.R.), Botan. Kabinett. *Botan. Phytogeographie.* o

Belkin, Grigorij (1896), Prosektor. Leningrad (U.d.S.S.R.), Černigovskaij Str. *Pathol. Anat. der Haustiere.* ⓟ Moskau 1921.

Belkin, Rafael I. (1894), Inst. General Biol. II. Univ., Collab.Labor. of Experim. Biol., Zoopark. Moscow (U.d.S.S.R.), Smolensky boulevard 53, Ap. 100. *Endocrinol.* ⓟ 1924.

Bell, Ernest L. (1876). Flushing, N.Y. (U.S.A.), 438 Amity Street. *Hesperiidae.* ⓟ Lepidoptera (local) for specimens of Hesperidae.

Bell, Frederick Heward (1902), Research Scientist, International Fisheries Commission. Seattle, Wash., (U.S.A.) Univ. of Washington. *Races of Hippoglossus.*

Bell, Hugh Philip (1889), Associate Prof. of Botan. Halifax, Nova Scotia (Canada), Botan. Labor. Dalhousie Univ. *Marine Algae.* ⓟ Toronto 1912.

Bell, Muriel Emma (1898), Univ. of Otago. Med. School, Lecturer. Dunedin (N.-Zealand). *Histol. Physiol.* ⓟ New Zealand 1922.

Bellido, Jesús M. (1880), Prof. Univ., Fac. Méd., Sous-dir. Inst. de Physiol. Barcelona (España), Emancipación, 32. *Electrophysiol., Chim. physique biol.* ⓟ Barcelona 1902.

Bellido y Golferichs, Jesús María, Prof. Catedrático Fac. Med., Dir. Lab. de Fisiol. Barcelona (España). *Physiol.* o

Belling, Dimitrij, Prof. Landwirtschaftl. Inst. Kiew (U.d.S.S.R.). *Ichthyol.* o

Bellocq, Philippe (1888), chargé de cours à la Fac. de Méd. Strasbourg (France), 24, rue Schweighäuser. *Anatomie déscriptive. L'oreille interne osseuse chez l'homme. L'os temporal chez l'homme adulte.* ⓟ Toulouse 1919.

Bellón Uriarte, Luis (1897), Ass., Inst. Español de Oceanografía. Málaga (España), Paseo de la Farola 47. *Systématique d'Algues Marines.* ⓟ Univ. de Madrid 1921.

Bellucci, Luigi (1879), Aiuto nel labor. di Fisiol. della R. Univ. di Siena; Doc. di Oto-rino-laringol. nella R. Univ. di Roma (Italia), Via di Città 13 bis. *Centres corticales et subcorticales. Micoses.*

Belon, Pedro, Prof., Univ. Nac. Buenos-Aires (Argentina). *Anat. descr.* o

Belonovski, Georg (1875), Prof. Bact. und Epidemiol., Dir. Bact. Abteilung d. Inst. für ärztliche Fortbildung. Leningrad (U.d.S.S.R.), 41 Kiroschaja. *Immunität.* ⓟ Leningrad 1902.

Belonsov, Victor (1898), Ass. Entomol. Plant Protection Station. Stawropol, Caucasus (U.d.S.S.R.), Okrzu. *Orthoptera, Economic Entomol.*

Beltrán, Enrique (1903), Prof. es Sc. Naturelles, Chef de la Station Biol. du Golfe. Veracruz, Ver. (República Mexicana), Boîte postale No. 250. *Biol. maritime, Protozoaires.* ⓟ México 1923.

Beltrán Bigorra, François (1886), Prof. de Biol., Univ., Dir. du Jardin Botan. Valencia (España), Pizarro 12-2. *Phytopat.* ⓟ Madrid 1910. ⓟ Phanérogames, Mousses et Uredinées.

van Bemmelen, Johan Frans (1859), o. Prof. der Zool. an der Reichsuniv. Groningen (Holland), Verlengde Heereweg 87. *Farbenmuster der Tiere, spez. der Insekten. Erblichkeitsstudien beim Menschen mittels der Methoden der Bio-Genealogie. Brachiopoden. Die Entwicklung der Farben und Adern auf den Schmetterlingsflügeln. Der Schädelbau der Monotremen. Indische Haustiere.* ⓟ Leiden 1882.

Benazzi, Mario (1902), Aiuto. Torino (Italia), Palazzo Carignano. *Ghiandola tiroide.* ⓟ Bologna 1925.

Benbrook, Edward Antony (1892), Prof. and Head of Department of Veterinary Pathol., Iowa State Coll. of Agric. and Mechanic Arts. Ames, Ia. (U.S.A.). *Vet. Pathol. Vet. Parasitol. Milk Hygiene. Meat Hygiene.* ⓟ Parasites of domesticated animals. Pathol. tissue sections from domesticated animals.

Bencini, Bruno (1896), Aiuto alla cattedre di Zool. e Anat. compar. della R. Univ. di Cagliari, (Italia). *Vitamine. Physiol. comparée.*

Bencze, Józef, Mitarb. am Physiol. Inst. d. Univ. Pécs (Ungarn). o

Benda, Carl (1857), G.S.R., Prof. Univ., Dir. pathol. anat. Abt. Moabit. Berlin NW 40 (Deutschland), Kronprinzenufer 30. *Normale u. pathol. Anat.* o

Bender, Harold Bohn (1902), Head of Biol. Dept., Coll. of Yale in China. Hunan (China), Changsha. *Plant Pathol.: Fungi imperfecti. Parasit.: Mosquitoes.* ⓟ Invertebrate parasit.

Bendl, Walther Ernst (1881), Dr. phil., Prof. Salzburg (Österreich), Bundesrealschule. *Hymenoptera. Landplanarien. Süßwassertrikladen.* ⓟ Graz 1907.

Benedetti, Edoardo (1894), Dr., Tecnico presso l'Ist. di Anat. compar. della R. Univ. di Bologna (Italia). *Anat. compar. du système nerveux centrale des poissons et des amphibies. Influence du champ électromagnétique de haute fréquence sur les germes végétaux.* ⓟ Bologna 1921.

Benedict, Don M., Ass. Prof., New York State Coll. of Forestry at Syracuse Univ. Syracuse, N.Y. (U.S.A.). o

Benedict, Francis Gano (1870), Dir., Nutrition Labor. of the Carnegie Institution of Washington. Boston, Mass. (U.S.A.), 29 Vila Street. *Metabolism of humans and animals. Energy metabolism.* ⓟ Univ. of Maine 1924.

Benedict, Harris Miller (1873), Prof. of Botan. Cincinnati, Ohio (U.S.A.) Univ. *Plant Physiol., Senility in Plants. Mechanism of Water movement in cells.* ⓟ Cornell Univ. 1914.

Benedict, Stanley R. (1884), Prof. of Chem., Cornell Univ. Med. Coll. New York City (U.S.A.), 477 First Avenue. *Biochem. Analytical method. Creatin and creatinine metabolism. Carbohydrate metabolism. Ammoniac formation. Uric acid metabolism.* ⓟ Yale Univ. 1908.

Benedict, Ralph Curtiss (1883), President Investigator, Brooklyn Botan. Garden, Chairman, Biol. Dept. New York City (U.S.A.), High School. *Ferns: Genetics.* Ⓓ Columbia Univ. 1911.

Beneke, Rudolf (1861), Dr. med., Geh.Med.R. o. ö. Prof. der pathol. Anat. an der Univ., Dir. des pathol. Inst. in Halle a./S. (Deutschland), Pathol. Inst. *Beziehungen der pathol. Anat. zur Entwicklungsmechanik. Mißbildungslehre. Geschwulstlehre. Thrombose. Geschichte der Med. Knochen, Gelenke, Gehirn.* Ⓓ Straßburg i. E. 1885.

Beneke, Wilhelm (1868), Dr., Univ.Prof., Dir. d. Botan. Gartens u. Instit. Münster (Deutschland). *Pflanzenphysiol. Bakterien.* o

Benevolenskaja, Sophia V. (1892), Senior Ass. Pathol. Physiol. Tomsk, Siberia (U.d.S.S.R.), Univ.

Benewolenskij, Iwan I., Dir. d. Archangelsker Moor-Versuchsfeldes. Archangelsk (U.d.S.S.R.), Gubsemuprawlenie. *Moor-Kultur.* o

Bengtson, Ida A. (1881), Bact. Rolla, Mo. (U.S.A.), U. S. Public Health Service. *Trachoma.* Ⓓ Univ. of Chicago 1919.

Bengtsson, Knut Nikolaus Ferdinand (1887), Ass. Experimentalfältet (Sverige). *Bodenbact., biochem. Umsetzungen im Ackerboden.*

Bengtsson, Simon Fredrik (1860), Dir. u. Doc. Entomol. Mus. Lund (Sverige). *Ephemerida. Plecoptera. Hummeln (Bombus u. Psithyrus). Braconiden (Hym.). Insectenlarven (Coleopteren).* Ⓓ Univ. Lund 1897. Ⓟ Paläarkt. Hummeln.

Benick, Ludwig (1874), Konserv. am Naturhistorischen Mus. in Lübeck (Deutschland). *Coleopteren, Systematik und Biol.* Ⓟ Coleopteren, Steninae.

Benham, William Blaxland (1860), Prof. of Biol. Univ. of Otago. Dunedin (New Zealand). *Annelida, (Plathyhelmia, Nemertina.)* Ⓓ London 1887. Ⓟ Marine animals.

Benitez, Clemente, Prof. Univ. Nac. d. Litoral. Corrientes (Argentina). *Histol. Embryol.* o

Benito, Fernández Riofrio (1896), Prof. aux. Univ. Barcelona (España). *Mycol.* Ⓓ Madrid 1921.

Benlloch, J., Prof., Dr., Lab. de Farmacol., Fac. Med. Cadiz (España). *Farmacol.* o

Bennet-Clark, Thomas A., Ass. to Prof. of Botan. School of Botan., Trinity Coll. Dublin (Ireland). *Plant Physiol.* Ⓓ Cambridge 1923.

Bennett, Carlyle Wilson (1895), Research Ass. in Plant Pathol., Michigan State Coll. East Lansing, Mich. (U.S.A.). *Fruit Diseases; virus diseases of Rubus.* Ⓓ Univ. of Wisconsin 1926.

Bennett, Edward W. (1899), Ass. Curator, Canterbury Mus. Christchurch (New Zealand). *Freshwater Entomostraca. Geographical Distribution: botan., zool., geol.* Ⓟ Fresh-water Entomostraca, Mollusca (New Zealand).

Bennett, James Percy (1886), Associate Prof. of Pomol. and Associate Pomol. in the Experim. Station. Berkeley, Cal. (U.S.A.), 331 Hilgard Hall, Univ. of California. *Plant Physiol.* Ⓓ Univ. of Wisconsin 1918.

Benninghoff, Alfred (1890), a.o. Prof., Dr. med., I. Prosektor am Anat. Inst. in Kiel (Deutschland). *Reaktionsweisen der Gewebe. Funktionelle Strukturen der Gewebe.* Ⓓ Heidelberg 1919.

Bennitt, Rudolf (1898), Instr. in Zool., Tufts Coll., Mass. (U.S.A.), Barnum Mus. *Invertebrate zool.; general physiol. Physiol. of vision in arthropods.* Ⓓ Harvard Univ. 1923. Ⓟ Hydroids and worms.

Benois, Charles-Louis (1885), Ass. Inst. Jaczewski de Mycol. et Pathol. végétale. Leningrad (U.d. S.S.R.), Perspective Anglaise 29. *Systématique Mycol.* o

Benoit, Charles Pierre Arthur (1901), Chef du Labor. de Chim. clinique de la Fac. de Méd. à l'Hôpital Général de Montpellier (France). *Chim. biol. Comportement de l'Acide picrique dans l'organisme.* Ⓓ Montpellier 1925.

Benoit, Jacques (1896), Ass. à l'Inst. d'Histol. de la Fac. de Méd. de Strasbourg (France), 40, rue Oberlin. *Histophysiol. sur les glandes génitales et sur les caractères sexuels secondaires (Oiseaux). Changement de sexe et l'hermaphrodisme chez les Oiseaux.* Ⓓ Strasbourg 1925.

Bensley, Benjamin Arthur (1875), Prof. of Zool. Univ., Head Dept. of Biol. Univ., Dir. Royal Ontario Mus. of Zool. Toronto (Canada), 37 Admiral Road. *Mammalian morphol., anat. of the rabbit.* Ⓓ Columbia Univ. 1902.

Benson, Gilbert Thereon (1896), Ass. Curator and Librarian, Dudley Herbarium, Stanford Univ. California (U.S.A.), Box 1362. *Trees and shrubs of the Pacific Coast.* Ⓓ Stanford Univ. 1925.

van Benthem Jutting, Woutera Sophie Suzanna (1899), Curator at the Zool. Mus. of Amsterdam (Holland), Plantage Middenlaan. *Invertebrates, especially Mollusca.*

Bentivoglio, Tito (1868), Dr., Bibliotecario della R. Accademia delle Sc. di Modena, Prof. ordinario di Sc. Modena (Italia), Corso Trento Trieste 46. *Libellulidae.*

Benussi, Vittorio (1878), Prof. o. st. Psicol. Sperimentale, Dir. del Labor. di Psicol. Univ. Padova (Italia), Corte Capitaniato 5. *La percezione della forma, del tempo, del movimento e dello spazio. I sintomi respiratori delle situazioni di coscienza. La suggestione e l' ipnosi. La Psicoanalisi.* Ⓓ Graz 1902.

Béraha, Hertzel (1899), Labor. de Biochim. de la Fac. de Méd. de Montpellier (France), Inst. de Biol., rue Martels. *Chim. physique en méd. Fonctions rénales.* Ⓓ Univ. de Montpellier 1926.

Berberich, Joseph (1897), Ass. der Univ.-Ohrenklinik Frankfurt a. M. (Deutschland), Eschenbachstraße 14. *Pathol. Anat. des Ohres. Stoffwechsel, Cholesterin, Harnsäure.* Ⓓ Frankfurt a. M. 1919.

Berbiela Jordana, Marcelino Baldomero (1861), Prof. Anat. descr. Zaragoza (España), San Miguel 6. Ⓓ Zaragoza 1880.

Berblinger, Walther (1882), Dr. med., o. Prof. für allgemeine Pathol. und pathol. Anat., Dir. des Pathol. Inst. der Univ. Jena (Deutschland). *Innere Sekretion, speziell Hypophyse. Zirbel. Keimdrüsen. Sexualität und Nebennieren. Pathol. Anat. des Herzens: Herzmuskelglykogen. Pathol. des Nervensystems, spez. Nervenregeneration. Leukaemie. Traumatische Krankheiten.* Ⓓ Straßburg 1908. Ⓟ Mikroskop. Präparate u. Photographien d. inneren Sekretion.

Berckhemer, Fritz (1890), Dr. rer. nat., Hauptkonserv.; Leiter der geolog. Abteilung der Württ. Naturaliensammlung in Stuttgart (Deutschland), Archivstr. 3. *Ammoniten des Schwäbischen oberen weißen Juras. Diluviale Säugetiere von Württemberg.* Ⓓ Tübingen 1913. Ⓟ Württemb. Fossilien und Abgüsse nach den hauptsächlichen Originalen der geol. Abt. des Mus.

Berczeller, László (1890), Dir. des ernährungsphysiol. Labor. (Josephinum). Wien IX (Österreich), Bleichergasse 6. *Ernährungsphysiol., Eiweißchem., Kataphorese, Antigengehalt menschl. Sera und Blutkuchen, Cholesterin, Chem. der Zellmembranen, biol. Wertung der Nahrungsmittel, Membran der roten Blutkörperchen, Regulation des Blutkreislaufes.* Ⓓ Budapest 1914.

Berdnikov, A. I., Prof. Ass. Ac. Russe. Paris (France), Boulevard Raspail 96. *Bact.* o

Beresin, V. I., Prof., Staatl. Med. Inst. Astrachan (U.d.S.S.R.). *Pharm., Physiol.* o

Berg, Kaj (1899), Ass. Freshwater-biol. Labor. Univ. Copenhagen. Hilleröd (Danmark), Banevej 17. *Cladocera.*

Berg, Leo (1876), Prof. Univ., Dir. Bureau für angewandte Ichthyol. Leningrad (U.d.S.S.R.), Morskaya 42. *Ichthyol., Zoogeographie, spec. Russland.* Ⓓ Moskau 1909.

Berg, Viktor R., Doc. Landw. Inst. d. Bergrepublik. Wladikawkas (U.d.S.S.R.). *Agronomie. Selektion.* o

Bergami, Gino (1903), Ass. nell' Ist. di Fisiol. della R. Univ. di Napoli (Italia), S. Andrea delle Dame 21. *Biochim. Physiol.* ⓟ Napoli 1926.

Bergauer, Vladimír (1898), Dr., Ass. des Inst. für allg. Biol. und experim. Morphol. der med. Fak. in Praha II (C.S.R.), Kateřinská 32. *Physik. Chem. der Lebensprozesse, speziell des Alterns, Vererbung.* ⓟ Prag 1923.

Bergaut, K. F., Dr. Ass. Histol. Labor. Med. Inst. Astrachan (U.d.S.S.R.). o

Bergeim, Olaf (1888), Ass. Prof. of Physiol. Chem., Univ. of Illinois Coll. of Med., Chicago, Ill. (U.S.A.), 1817 W Polk St. *Chem. of Digestion. Nutrition.* ⓟ Jefferson Med. Coll. 1914.

Berger, Edward William (1869), Entomol. State Plant Board of Florida, Gainesville Fla. (U.S.A.) *Injourious insects, principally citrus insects, entomogenous Fungi; Icera purchasi.* ⓟ Baltimore Md. 1899.

Berger, Josef Frz. (1860), Präsident des Österr. Entomologen-Vereins. Wien (Österreich), Lerchenfelderstr. 67/31. *Makro-Lepidopteren.* ⓟ Makro-Lepidopteren.

Berger, Käthe Bertha Charlotte (1897), Dr. phil., zur Zeit Leiterin des Vivariums München (Deutschland), Zool. Inst., Neuhauser Str. 51. *Sinnesphysiol. niederer Wirbeltiere (Reptilien).* ⓟ München 1923.

Bergerhoff, Rudolf Otto Karl, Dr. phil., Ass. als Chem. Berlin-Steglitz (Deutschland), Schloßstr. 53. *Hormone.* ⓟ Berlin 1923.

Bergersen, Birger (1891), Doc. Oslo (Norge), Tandlaegeinstituttet, Gjetemyrsveien. *Koriumuntersuchungen.*

Bergey, David H. (1860), Prof. of Hygiene and Bact., Univ. of Pennsylvania. Philadelphia, Pa. (U.S.A.), 206 South 53 Street. ⓟ Univ. of Pennsylvania 1884.

Bergglas, Bernhard (1899), Univ. Ass. Wien III (Österreich), Ob. Weißgerberstr. 11/10. *Hyperdactylie u. Os intermetatarsale. Adventitia d. Arterien. Gefäßnerven.* ⓟ Wien 1925.

Berglund, Hilding (1887), Prof. and Chief, Department of Med., Univ. of Minnesota. Minneapolis, Minn. (U.S.A.), Univ. Hospital. *Normal and pathol. problems of human metabolism.* ⓟ Stockholm Med. 1916, Sc. 1920.

Bergman, Herbert Floyd (1883), Prof. of Botan., Univ. of Hawaii. Honolulu (Hawaii). *Plant Physiol.: respiration and aeration. Biochem.: plant proteins.* ⓟ Univ. of Minnesota 1918. ⓟ Dried plants for herbarium specimens, or seeds of Hawaiian plants.

Bergold, Marianne (1899), Hilfsass. am Chem. Labor. der Bayr. Landesanstalt für Pflanzenbau und Pflanzenschutz München (Deutschland). *Agrikulturchem. Pharmazeut.-chem. Pflanzenuntersuch.*

Bergs, Janis (1863), Prof. landwirtsch. Fac. Univ. Riga (Latvija), Dzirnavu iela 9, dz. 6. *Pflanzenbau, Pflanzenzüchtung.* ⓟ Riga 1889. ⓟ Winterroggen, Rispenhafer, Fahnenhafer, Kartoffeln, Schließmohn, Wiesenklee.

Bergstrand, Hilding (1886), Prof. General Pathol. Caroline Inst. Stockholm (Sverige). *Pathol. anat., general pathol. and bact.* ⓟ Stockholm 1916.

Bergtold, William Harry (1865), Denver, Colo. (U.S.A.), 1159 Race Street. *Ornithol., Incubation periods of birds.* ⓟ Univ. of Buffalo 1886.

Beritašvili, Ivan, Prof. Univ. Tiflis (U.d.S.S.R.). *Physiol. d. Tiere.* o

Beritoff, Johann Ivan (1884), Prof. Physiol. und Reflexol. Pädagogische Fac. Univ. Tiflis, Georgien (U.d.S.S.R.). *Physiol. des Muskels und Nervensystems: Erregung u. Hemmung; Kontraktionsfähigkeit und Ermüdung; Bau u. Innervation des Muskels; Aktionsströme der Muskeln u. Nerven; Reflextätigkeit des Zentralnervensystems: Abwehrreflexe und tonische Reflexe; Reflexumkehr; individuelle Reflextätigkeit des Zentralnervensystems: motorische individuelle Reflexe u. ihre Entstehung bei verschiedenen Tierarten.* ⓟ Leningrad 1915.

Berkelbach v. d. Sprenkel, Hendrik (1890), Dr., Konserv. des Histol. Labor. der Univ. Utrecht (Holland), Beetsstraat 22. *Mikroskopische Anat. des zentralen und peripheren Nervensystems.* ⓟ Utrecht.

Berkeley, Garven Hugh (1894), Sr. Plant Pathol.-in-charge-of-Dominion Research Labor. St. Catharines, Ont. (Canada), 204 St. Paul St. *Fruit Diseases, vegetable Diseases.* ⓟ Toronto 1923. ⓟ Plant Disease material.

Berkner, Friedrich Wilhelm (1874), Dr., o.ö. Prof. für Pflanzenbau und Pflanzenzüchtung. Breslau 16. (Deutschland) *Getreidezüchtung (Weizen, Hafer).* ⓟ Halle a. S. 1907.

Freiherr von Berlepsch, Hans (1857), Besitzer der Staatlich anerkannten Versuchs- u. Musterstation für Vogelschutz, Burg Seebach, Kr. Langensalza (Deutschland). *Ornithol., angewandte Ornithol., biol. Schädlingsbekämpfung, wissenschaftl. Vogelschutz.* ⓟ Halle a. d. S. 1923.

Berlese, Antonio, Prof. Firenze (Italia), Via Romana 19. *Entomol.* o

Berlin, John Herved (1890), Fil.Dr., Konserv. Zool. Mus. Univ. Lund (Sverige). *Gregarinen und Lumbriciden.* ⓟ Lund 1923.

Bermudez, Andrés, Prof., Univ. Nac. Bogotá (Columbia). *Pharm.* o

Bernales, Juan Noto, Prof. Univ. Lima (Peru). *Pharmacol.* o

Bernard, Charles Jean (1876), Dr., Dir. Theeproefstation. Tjiapoes, Buitenzorg (Java). *Embryol. et cytol., anat., physiol., algues d'eau douce du domaine Malais. Sujets d'agronomie tropicale en particulier le thé et les engrais verts; Maladies et ennemis du théiér.* ⓟ Genève 1902.

Bernardini, Domenico, Prof. R. I. med. vet. Milano (Italia). *Pathol.* o

Bernatsky, Jenö (1873), Landw. Dir. i. P., Doc. an der Univ. Pest Hidegk út. (Ungarn). *Phytopathol., Weinbau, eßbare Pilze.* ⓟ Budapest 1899.

Berndt, Wilhelm (1878), Dr. phil., Prof., Abteilungsvorsteher am Zool. Inst. der Univ. zu Berlin N 4 (Deutschland), Invalidenstr. 43. *Cirripedien, besonders bohrende und parasitische Formen. Alcippe lampas. Cryptophialus. Aerothoracica. Vererbungslehre, speziell Vererbung von Haustier- und Degenerationsmerkmalen bei Knochenfischen (Zierkarauschen, Goldfische, Schleierschwanzrassen usw.).* ⓟ Berlin 1902.

Berner, O., Dr., Prosektor. Oslo (Norge). *Anat., Anthrop.* o

Bernhardt, Georg (1881), Dr. med. Neubabelsberg bei Berlin (Deutschland), Augustastr. 29. *Bact.* ⓟ 1904.

Bernhardt, Harold Frederick (1898), Instr. in Biol., Dir. of Botanical Gardens and Greenhouses, Temple Univ. Philadelphia, Pa. (U.S.A.), Broad St. and Montgomery Ave. *Cytol.: Chromosome Numbers. Mosses: Structure and Identification.* ⓟ Herbarium: Plants collected in Pennsylvania. Mosses-fruiting.

Bernhoft-Osa, Anders (1897). Voss (Norge). *Ornithol. Norwegens.* ⓟ Vogelbälge Norwegens.

Bernied, Prof. Ecole prép. de méd. et de pharm. Reims (France). *Bact.* o

Bernier, A., Prof. Univ. Montreal (Canada), Rue St. Denis 1265. *Pathol. génér.* o

Berrie, A. R., Ass. Bact. Charing Cross Hospital, Inst. of Pathol. London W.C. 2 (England). *Bact. and Haematol.* ⓟ Edinburgh.

Berrill, Norman John (1903), Ass. London W.C. 1 (England), Zool. Dept. Univ. Coll. *Experim. Embryol. and Regeneration.* ⓟ Bristol 1924.

Berry, Edward Wilber (1875), Prof. of Palaeont. Baltimore, Md. (U.S.A.), Johns Hopkins Univ. *Classification and evolution in plants. Palaeobotan. Evolutionary History.* o

Berry, Richard James Arthur, Prof. Univ. Melbourne (Australia). *Anat.* o

Bersa, Egon (1897), Dr. phil., ao. Ass. am pflanzenphysiol. Inst. der Univ. Graz (Österreich), Schubertstr. 53. *Röntgenphysiol. und Elektrophysiol. der Pflanzen; Schwefelbakterien.* ⓟ Graz 1920.

Bertelli, D., Prof. Univ. Padua (Italia). *Anat.* o

Bertels, Alexis O., Conserv. à la Station de Défense des Plantes. Leningrad (U.d.S.S.R.), Rue Tschaikovski 4. *Pathol. végétale.* o

Berthelot, Albert Charles Marie (1881), Chef de Labor. à l'Inst. Pasteur, Licencié ès sc. Paris XV (France), 25, Rue Dutot. *Chim. biol., appliquée à la Microbiol. Flore intestinale. Toxines, ptomatnes, vaccins.* ⓟ Paris 1913.

Bertholt, Lloyd M. (1899), Prof. of Biol., Western Maryland Coll. Westminster, Md. (U.S.A.). *Physiol. and Behavior of the Honeybee.*

Bertoni, Arnoldo de Winkelried (1879), Entomol. et chef du Bureau de Phytopath. de la „Directión d'Agric. et Def. Agricole" du Paraguay; Propriétaire du Mus. Zool. a Puerto Bertoni (Paraguay). Assomption, Azara 538, ou Puerto Bertoni (Paraguay). *Faune du Paraguay.* ⓟ Yaguarasapá, Aut Paraná, Paraguay, 1893. ⓟ Hymenoptères Apoïdées et Vespoidées.

Bertram, Johannes Eduard Ferdinand (1894), Med. wiss. Leiter der Firma C. H. Boehringer Sohn, Hamburg 33 (Deutschland), Wasmannstr. 32. *Zuckerstoffwechsel.* ⓟ Hamburg 1923.

Bertrand, Gabriel (1867), Prof. de Chim. Biol. à la Fac. des sc. (Sorbonne) et à l'Inst. Pasteur. Paris XV (France), 28, Rue Dutot. *Eléments oligosynergiques, diastases, vitamines, hormones. Sucres et tissus végétaux, glucosides. Microbes et fermentations.*

Bertrand, Ivan, Chef de Labor. Fac. méd. Paris (France), Hôpital de la Salpêtrière. *Anat. Pathol.* o

Bertrand-Fenaux, Paul Charles Edouard (1879), Prof. Paléobotan. Univ. Lille (France), 159, Rue Brule Maison. *Flores houillères: Flores primaires, dévonienne, carbonifère et permienne.* ⓟ Lille 1909. ⓟ Empreintes de Fougères houillères, Types de Charbons.

Bervoets d'Oostkerke, Raymond Emile (1890), Dr. ès Sc. Bruxelles (Belgique), 2, Boulevard St. Michel. *Physiol. entomol., vol des insectes.* ⓟ Bruxelles 1913.

Besançon, M. F., Prof. Fac. Méd. Paris XVII (France), 76, Rue de Monceau. *Bact.* o

de Besche, Dr., Doc. Univ. Oslo (Norge). *Bact.* o

Bessey, Ernst Athearn (1877), Prof. of Botan. East Lansing, Mich. (U.S.A.), Department of Botan. Michigan State Coll. *Plant Pathol., Phylogeny of Plants, Mycol.* ⓟ Halle a. S. 1904.

Bessmertny, Sophie (1889), Stellv. Dir. K. A. Timiriasev-Biomus., Mitarb. des Labor. f. exper. Biol. Moskau (U.d.S.S.R.), Sadowaja-Samotetschnaja 7, Wohn. 16. *Genetik, Physiol. der Schilddrüse. Genetische Studien bei Hühnern u. Haustauben. Bestimmung d. Abstandes zwischen den quantitativen Genen.*

Bessubetz, Stephan (1891), Oberass. Tierärztl. Hochsch., Leiter der serol. Abteilung des Veterinärbact. Inst. Leningrad (U.d.S.S.R.), Meschdunarodny Prospekt 83a. *Bact. und Infektionskrankheiten. Kerne bei den Bakterien. Paratyphus der Kälber. Bacillum anthracis in der Milch.* ⓟ Sera und Vakzine.

Best, M. G. London S.W. 10 (England), 123 Cheyne Walk, Chelsea. *Ornithol.* o

Bestužec A. P., Prof. Staatsuniv., Med. Fac. Minsk, Weißr. (U.d.S.S.R.). *Biochem., Pharm.* o

Bethe, Albrecht (1872), Dr. phil. et med., o. Prof. an der Univ., Dir. des Inst. für animalische Physiol. Frankfurt a. M. (Deutschland), Kettenhofweg 126. *Vergl. u. spezielle Physiol. des Nerven- u. Muskelsystems. Histol. des Nervensystems.* ⓟ Phil. München 1895, Med. Straßburg 1898.

Bethke, Roland Martin (1896), Ass. in Animal Industry. Wooster, Ohio (U.S.A.), Ohio Experim. Station. *Nutrition, particularly vitamin and mineral.*

Betrem, Johan George (1899), Ass. an der landwirtschaftlichen Hochsch., Labor. für Entomol. Wageningen (Holland), Bowlespark 14. *Ökonomische Entomol., Hymenoptera, Acarina.* ⓟ Hymenoptera.

Betten, Cornelius (1877), Dir. of Resident Instruction, N. Y. State Coll. of Agric. at Cornell Univ. Ithaca, N.Y. (U.S.A.), Roberts Hall. *Trichoptera, taxonomy.* ⓟ Lake Forest Phil. 1906, Sc. 1923.

Bettencomt, Nicolau (1872), Prof. libre de Bact. à la Fac. de Méd. de Lisbonne (Portugal), Inst. Camara Pestana. *Sérol.; immunité.*

Bettencourt, Annibal (1868), Prof. Bact. et Parasit. Fac. Méd. Univ., Dir. de l'Inst. de Bact. Camara Pestana. Lisboa (Portugal). *Bact., Helminthol. (Schistosomes), Protozool.* ⓟ Sérums antidiphtérique et antitétanique.

Betti, Mario (1875), Dir. dell' Ist. di Chim. generale Univ. Bologna (Italia). *Ossidazioni spontanee, Enantiomorfismo della materia vivente. Biochim. vegetale.* ⓟ Pisa 1897.

Bettini, Plinio (1902), Dr. Cremona (Italia), Via Platini 4. *Osmose des membranes vivantes. Cancer.* ⓟ Parma 1924.

Bettzieche, Fritz Karl Rudolf (1901), Dr. phil., Ass. am Physiol.-chem. Inst. der Univ. Leipzig-R., (Deutschland), Wallwitzstr. 9. *Einwirkung von Grignardreagenz auf Aminosäuren u. Polipeptide. Aminoalkohole. Oxyaminosäuren. Bestimmung freier Carboxylgruppen in Polipeptiden.* ⓟ Leipzig 1924.

Beumée, J. G. B. (1888), Chef des Herbars. Buitenzorg, Java, (Nederl. Indië), Laan v. d. Wijck 1. *Systematik und Verbreitung der höheren Pflanzen des Malaiischen Archipels.* ⓟ Wageningen 1922.

Beutler, Ruth (1897), Dr. phil., Laborantin d. zool. Inst. in München (Deutschland), Neuhauserstr. 51, Alte Akademie. *Vergl. Physiol. d. Ernährung.* ⓟ Rostock 1923.

Beutenmüller, William (1864), Consulting Entomol. Tenafly, N. J. (U.S.A.), 85 Elm St. *Insects Inj. to Forest and Shade trees; Gall Insects (Cynipidae); Lepid., Coleopt.*

Beutner, Reinhard (1885), Ass. Prof. of Pharmacol., School of Med., Univ. of Louisville, Ky. (U.S.A.), 101 West Chestnut Street. *Anwendung der physikalischen Chem. auf biol. Probleme; elektromotor. Kräfte aus synthetischen organischen Substanzen ohne Mitwirkung eines Metalls, künstliche Nachahmung verschiedener anderer biol. Erscheinungen.* ⓟ Ing. Karlsruhe 1908, phil. und med. Berlin 1926.

Beverslais, Jozef Reinville (1890), Ass. an der Landwirtschaftl. Hochsch. in Wageningen (Holland), Landbouwhoogeschool. *Anat. und mikroskopische Bestimmung des Holzes.* ⓟ Wageningen 1925.

Bews, John W. (1884), Prof. of Botan. in the Univ. of Durham. Newcastle-on-Tyne (England), Armstrong Coll. *Ecol., Plant Physiol., Geographical Distribution. Vegetation of S. Africa.* ⓟ Edinburgh 1912.

Bey, Innes, Dr. Cairo (Egypten), 6 Square Halim Pasha. *Ornithol.* o

Beyer, Adolf Bernhard Barnabas (1902), Dr. phil., Hilfsass. Freiburg i. Br. (Deutschland), Botan. Inst. *Reizphysiol. der Pflanzen, Tropismen.* ⓟ Berlin 1925.

Beyer, H. Otley, Prof. Univ. of the Philipp. Manila, Philippine Islands (U.S.A.). *Anthrop.* o

Beyer, J. J., Ass. Labor. für Zwiebelkrankheiten. Lisse (Holland). o

Beylot, Marc (1866), Dr. med., Chef des travaux d'anat. générale et d'Histol. à la Fac. de Méd. de Bordeaux (France). ⓟ 1910.

Beyschlag, Franz (1856), Dr. phil., Ing., Prof. Berlin-Wilmersdorf (Deutschland), Binger Str. 57. *Geol., Palaeont.* o

Bezděk, Martin, Prof. Praha II (C.S.R.), Jungmannova 18. *Amélioration des Plantes cultivées.* o

Bezdíček, Jiří (1905) Ass. Histol.-embryol. Inst. Univ. Brno (C.S.R.) Údolní 73. *Bindegewebe.*

Bézi, István, Dr., Mitarb. am Pathol.-anat. Inst. Univ. Budapest (Ungarn). o

von Beznák, Aladár (1901), Dr. med., I. Ass. an dem Physiol. Inst. der Univ. Debreczen (Ungarn); z. Z.: London W. C. 1 (England), Inst. of Pathol.

Chem., Univ. Coll. Hospital Med. School, Gowerstr. *Oxydative und reduktive Processe. Photooxydation der Proteine; Oxydation der Glutaminsäure im Tiere.* ⓟ 1924.
Bezobrazova, N. F., Doc. Univ. Taschkent (U.d. S.S.R.). *Palaeont.* ○
Bezssonoff, Nikolaï (1885), Dir. Technique labor. de Chim. biol. de Colombes, Conseil en physiol. végétale et bact. agricole des labor. G. Truffaut à Versailles. Colombes, Seine (France), 15, rue Besson. *Préparation de vitamines C, A et B. Mesurations des vitamines. Bact. agric.: du sol, fixation microbiol. de l'azote. Physiol. végétale: besoins des plantes en aliments minéraux.* ⓟ St. Petersburg 1914.
Bhalerao, Govind D. (1896), Prof. of Zool. Nagpur, C.P. (India), Hislop Coll. *Helminthol., Trematoda.* ⓟ Allahabad 1922. ⓟ Helminths from Burma and India.
Bharucha, Kaikhashru Hoshangji (1900), Ass. Chem. Haffkine Inst. Parel. Bombay 7 (India), Khalukdina Terrace Govalia Tank Road. *Analytical Chem.* ⓟ B.A.Sc. (Honours) Bombay 1922.
Bhattacharya, Dakshina Ranjan (1888), Head of the Department of Zool., Dean Fac. of Sc. Univ. Allahabad (India). *Compar. Anat. of Vertebrates, Animal Cytol.* ⓟ Ph.D. Dublin 1924, D.Sc. Paris 1925. ⓟ Local fauna.
Bialaszewicz, Kazimier (1882), Dir. du Labor. de Physiol. à l'Inst. Nencki, Prof. de Physiol. des Animaux à l'Univ. de Varsovie (Pologne), Rue Sniadecki 8. *Métabolisme chimique et énergétique chez les animaux inférieurs. Physiol. du développement embryonnaire.* ⓟ Cracovie 1909.
Bialynicki-Birula, Alexis (wiss. Name: A. Birula) (1864), Chef-Zool., stellv. Dir. des Zool. Mus. Akad. der Wiss., Prof. Univ. Leningrad (U.d.S.S.R.). *Morphol., Systematik und geograph. Verbreitung der Arachniden, Insekten (Hymenoptera), Mammalia, Palaeont. posttertiäre Mammal.*
Bianchedi, Gherardo, Lib. Doc. Ist. super. di Med. vet. Bologna (Italia). *Anat. patol. vet.* ○
Bickerton, William. Watford (England), 21 Oxhey Road. *Ornithol.* ○
Bickij, V. A., Prof. Landwirtschaftl. Inst. Leningrad (U.d.S.S.R.). *Zoohygiene u. Zoopathol.* ○
Bickijs, Jānis, Doc. Univ. Caur Cēsim (Latvija). *Phytopat., Entom.* ○
Bidwell, Edward. London N. 4 (England), 12 Woodberry Grove, Finsbury Park. *Ornithol.* ○
Bieberdorf, Gustav A. (1898), Ass. Entomol. and Ass. Prof. Entomol. Stillwater, Okla. (U.S.A.), Entomol. Dept., Oklahoma A. and M. Coll.
Biedermann, Wilhelm (1852), o. Prof. der Physiol. an der Univ. Jena (Deutschland), Botzstr. 4. *Muskel- u. Nerven-Physiol., peristaltische Bewegungen, vergl. Physiol. der Verdauung, Chemie der Zelle, Fermente, Physiol. der Skelettsubstanzen, Tierfarben. Vergl. Physiol. der irritablen Substanzen. Integument. Elektrophysiol.* ⓟ Prag 1878.
Biedl, Arthur (1869), o. Univ.Prof. deutsche med. Fac., Dir. Inst. für allgemeine u. experim. Pathol. Praha II (C.S.R.), Salmovská 3. *Pathol. Physiol.: Endokrinol.* ⓟ Wien 1892.
Biehler, Wilhelm (1894), Pharm. Labor. der Knoll AG. Ludwigshafen a. Rh. (Deutschland), Schillerstr. 78 II. *Experim. Pharmacol.* ⓟ Med. München 1920, phil. Münster i. W. 1924.
Bieling, Richard (1888), Priv.Doc. für Hygiene an der Univ. Frankfurt a. M. und Leiter der sero-bact. Abteilung der I. G. Farbenindustrie Aktiengesellschaft, Höchst a. M. Soden a. T. (Deutschland), Königsteinerstr. 64. *Immunitätslehre, Chemotherapie, bakterielle Infektionskrankheiten, Hygiene.* ⓟ Freiburg 1912. ⓟ Heilsera und Impfstoffe. Chemo-therapeutische Mittel, z. B. Salvarsan.
Bien, Zoltán (1891), Dr., Konserv. a. d. Reichsuniv. in Leiden (Holland), Rijnsburger Weg 163. *Serologie, Pharmacol.* ⓟ Budapest 1916.

Bierbaum, Kurt (1881), a.o. Prof. der Tierärztl. Hochsch., Vorsteher der Serumabteilung des Hygien. Inst. Berlin NW 6 (Deutschland), Luisenstr. 56. *Bact. und Serologie.* ⓟ Gießen 1906.
Bierens de Haan, J. A. (1883), Dr., Priv-Doc. an der Univ. Amsterdam (Holland), 39 Prins Hendriklaan. *Tierpsychol. und Sinnesphysiol.* ⓟ Utrecht 1913.
Bierry, Prof., Dir. Labor. de Physiol., Fac. Sc. Marseille (France).
Bierry, Henri Georges (1877), Prof. physiol., Fac. de Sc., Marseille (France). *Ferments solubles. Métabolisme des hydrats de carbone.*
Bierry, M. H., Maître de Conf. Ecole des Hautes Etudes. Paris 16 (France), 2, Avenue de la Grande Armée. *Physiol.* ○
Biese, Walter (1895), Dr. phil. Berlin N 4 (Deutschland), Invalidenstr. 43. *Palaeobiol., Phylogenese: Crinoidea, Pisces.* ⓟ Berlin 1925.
Biester, Harry Edward (1892), Ass. Prof. Vet. Investigation. Ames, Ia. (U.S.A.), Iowa State Coll. *Pathol. (animal), Investigation diseases of swine and other species.* ⓟ Gross or micro preparations of animal diseases.
Biffen, Sir R. H., Prof. Univ. Cambridge (England). *Agric. Botan.* ○
Biffi-Gentili, Ugo (1870), Dir. degli Ist. Ospitalieri di Piacenza; Doc. di Igiene Sperimentale nella R. Univ. di Bologna. Piacenza (Italia), Ospedale Civile. *Hematol. Bact.* ⓟ Bologna 1895. ⓟ Préparations microscopiques de Verruga Peruviana.
Bigalke, Rudolph (1896) Dir. of the National Zool. Gardens. Pretoria, Transvaal (South Africa), P. O. Box 754. *Morphol. of animals.* ⓟ Berlin 1926.
Bigelow, Henry Bryant (1879), Research Curator. Cambridge, Mass. (U.S.A.), Mus. Compar. Zool. *Pelagic Coelenterates. Oceanic Biol. Biol. and Physical oceanography especially of Western North Atlantic.* ⓟ Harvard Univ. 1906.
Bigelow, Maurice A. (1872), Ph.D., Prof. of Biol. and Dir. of the School of Practical Arts. New York (U.S.A.), Teachers Coll., Columbia Univ. *Early development of Lepas.* ⓟ Harvard 1901. ○
Bigelow, Robert Payne (1863), Prof. of Zool. and Parasit., Massachusetts Inst. of Technol. Cambridge, Mass. (U.S.A.). *Compar. Anat., Histol., Embryol., General biol., Parasit., Ichthyol. Stomatopoda of the Pacific Ocean, Ecol. of larval fishes of the Atlantic coast of Massachusetts.* ⓟ Johns Hopkins 1892.
Bigger, John Henry (1898), Ass. Entomologist, Illinois Natural History Survey, Urbana, Ill. Jacksonville, Ill. (U.S.A.), 1114 S. Main Street. *Insects affecting fruit, grain and forage crops. Fruit Tree Leaf Roller in Illinois.*
Bigger, Joseph W., Prof. Univ. Trin. Coll. Dublin (Ireland). *Bact.* ○
Bignami, A., Prof. Univ. Roma (Italia). *Patol. gener.* ○
Bijhouwer, Jan Tijs Pieter (1898), Ass. Landb. Hoogesch. Abt. Syst. u. Pflanzengeogr. Wageningen (Holland). *Dendrologie, Untersuchung der Niederländischen Pflanzengemeinschaften.* ⓟ Wageningen 1926.
Bik, W. I., Ass., Dr. Anat. Inst. Univ. Kasan (U.d.S.S.R.). ○
Bikk, Franz, Dr., Prof. der allg. und speziellen Tierzucht. Praha XII (C.S.R.), Inst. f. allg. und spec. Tierzucht d. čech. techn. Hochsch. ○
Billard, Armand Eugène (1871), Prof. de Zool. à la Fac. des Sc. de l'Univ. de Poitiers (France). *Hydroides.* ⓟ Hydroides.
Billet, H., Prof. Univ. Lille (France). *Anat.* ○
Billings, Frederick H. (1869), Prof. of Botany and Bact. Univ. of Redlands Cal. (U.S.A.) 948 East Colton Ave. ⓟ München 1900.
Bilousev, M. F., Prof. Inst. f. Volksbildung. Charkow (U.d.S.S.R.). *Physiol.* ○

Bimes, D., Prof. Ecole Nat. Vétér. Toulouse (France). *Anat. pathol.* ○

Binard, Alice (1896). Bruxelles (Belgique), Rue Potagère 101, St. Josse. *Morphol. du Système nerveux des Invertébrés, spécialement des Polychètes.*

Binet, Léon René (1891), Prof. agrégé de physiol. à la Fac. de Méd. de Paris VII (France), 5, Avenue Bosquet. *Physiol. du poumon. Physiol. normale et pathol. du nourrisson.* ⓟ Collection de plantes murales pour l'enseignement de la Physiol.

Bingham, Marjorie J. (1895), Detroit, Mich. (U.S.A.), 9617 N. Martindale Ave. *Senility in plants. Relation of Age to Size in certain Root Cells and in Vein-islets of the Leaves of Salix nigra Marsh* ⓟ Herbarium specimens of the flora of the middle western United States.

Binz, August (1870), Kustos der Herbarien Univ. Basel (Schweiz), Gundeldinger Str.175. *Floristik der Schweiz.* ⓓ Zürich 1892.

Bioletti, Frederic Theodore (1865), Prof. of Viticulture, Coll. of Agric., Univ. of California, and Viticulturist, Agric. Experim. Station California. Berkeley, Cal. (U.S.A.), 336 Hilgard Hall, U.C. *Yeasts; Microbiol. of alcohol and alcoholic products; Grapes on the Pacific Coast; Olives.* ⓟ Gigas mutants of Vitis vinifera.

Bioret, Georges Félix Marie (1880), Prof. de Botan. Univ. Catholique. Angers (France), Labor. de Botan. *Lichens. Algues d'eau douce (plancton).* ⓓ Angers.

Blourge, Philibert Melchior Joseph Ehislam (1864), Prof. ordinaire à l'Univ. catholique. Louvain (Belgique), Rue du Canal, Inst. Carnoy. *Fungi imperfecti, Phytopath., Fermentations.* ⓓ Louvain 1891.

Bird, Henry (1869), Entomologist. Rye, N.Y (U.S.A.), 600 Milton Road. *Biol. of Noctuid Moths particularly the genus Papaipema; Artificial Acidulation of Soils.*

Birge, Edward Asahel (1851), in charge, Natural History Division, Wisconsin Geological and Natural History Survey. Madison, Wis. (U.S.A.), 2011 Van Hise Ave. *Limnol.; temperature of lakes; penetration of sun's radiation; productivity in organic matter.* ⓟ Ph. Harvard 1878, Hon. Sc. Univ. of Pittsburgh 1897, LL. Williams Coll. 1903, Univ. of Wisconsin 1915, Univ. of Missouri 1919, Ph. Renselaer Polytechnic Inst. 1924.

Birkhaug, Konrad Elias (1892), Ass. Prof. of Bact. Rochester, N.Y. (U.S.A.), The Univ. of Rochester School of Med. and Dentistry. *Streptococcal Infections: their bact. and immunol. Erysipelas and Scarlet Fever.* ⓓ Baltimore, Md. 1924. ⓟ Streptococcal cultures.

Birnie, Sjuwke Maria (1894), Hauptass. für die Allgemeine Botan. an der Univ. Utrecht (Holland), Maliebaan 131. *Reizphysiol. der Pflanzen.*

Biró, István, Mitarb. am Pharmacol. Inst. Univ. Budapest (Ungarn). ○

Biró, Lajos, Dr., Zool. Abt. d. Ung. National Mus. Budapest (Ungarn). *Hymenopterol.* ○

Bisbee, Ruth C. (1889), Lect. in Zool. Liverpool (England), Univ. *Genetics, Embryol.* ⓓ Liverpool 1914.

Bisby, Guy Richard (1889), Prof. of Plant Pathol., Ph.D. Manitoba Agric. Coll. (U.S.A.). *Mycol. Cereal Diseases. Potato Diseases.* ⓓ Univ. of Minnesota 1918. ⓟ Fungi.

Bischoff, Wilhelm Carl Maria (1890), Dr. phil.,₁ Ass. am forstzool. Inst. d. Univ. Freiburg i. Br. (Deutschland). *Biol. u. Morphologie d. Dipterenlarven, intrazellulare Symbioseforschung.* ⓓ Greifswald 1920. ⓟ Blepharoceridenmetamorphosen.

Bishop, Sherman C. (1887), Zoologist, New York State Mus., Albany, N.Y. (U.S.A.). *Reptiles and amphibians; lifehistories of North American newts and salamanders; taxonomy of arachnids; spiders and phalangids; life histories of spiders.* ⓓ Cornell Univ. 1925. ⓟ North American salamanders and spiders; phalangids.

Bishopp, Fred Corry (1884), Entomologist in Charge, Insects Affecting Health of Animals. Dallas, Tex. (U.S.A.), P. O. Box 208. *Insects which attack live stock, poultry, game birds and animals, and man. Ixodidae and Siphonaptera. Chemotropic and other responses of insects.* ⓟ North American Ixodoidae.

Bitancourt, Agesilau Antonio (1899), Prof. de Botan. à l'Ecole Agricole „Luiz de Queirez" à Piracicaba, Etat de São Paulo (Brasil), Rua de Commercio, 309. *Botan. et Phytopath.* ⓓ Paris 1919.

Bitter, Georg (1873), Dr., Univ.Prof. Göttingen (Deutschland), Unt. Karspüle 2a. *Botan.* ○

Bittner, Heinrich (1897), ○. Prof. Anat., Histol. und Embryol. der Haustiere Univ. Sofia (Bulgarien), Ulitza Regentska 11. *Hausgeflügel; Anat., Histol. und Embryol. der Zehe der Haustiere; Sektionstechnik der Haustiere.* ⓓ Berlin 1921.

Bjarnhjeðinsson, Saemundur, Prof. Háskóli Islands. Reykjavík (Island). ○

Bjelichow, Demetrius Wassili (1898). Kasan (U.d.S.S.R.), Zootomisches Kabinett Univ. *Vergl. Morphol. der Wirbellosen: Enteropneusta. Hydrobiol.: Biocoenotik und Plankton des Wolga-Flusses.*

Bjelajeva-Sankova, Anna Ivanovna (1892), Laborant Botan. Garten. Leningrad (U.d.S.S.R.), Kirocznaja 27, Wohn. 5. *Bryol.*

Blaauw, A. H., Prof. der Botan., Dr., Landw. Hochsch. Wageningen (Holland). *Pflanzenphysiol., spez. Phototropismus; Einfluß der Temperatur auf Periodizität der Kulturpflanzen.* ○

Blacher, Leonidas J. (1900). Ass. Inst. of General Biol. II. Univ., I. Ass. Labor. of Experim. Biol. of the Zoopark. Moscou (U.d.S.S.R.). *Mechanics of development. The influence of internal factors (glands of internal secretion) upon the morphogenesis. Genetics of Lebistes reticulatus.* ⓓ 1925.

Black, Davidson (1884), Prof. of Anat. and Head of the Dept. of Anat. Peking Union Med. Coll. Peking (China). *Physical Anthrop. and Compar. Neurol.* ⓓ Univ. of Toronto 1923.

Black, Otis Fisher (1867), Biochemist: Bur. Plant Industry, U. S. Dept. Agr. Washington, D.C. (U.S.A.). *Biochem. and pharm. investigation of poisonous and med. plants.*

Blackburn, Kathleen Bever (1892), Lect. in Botan. Newcastle-on-Tyne (England), Armstrong Coll. (Univ. of Durham). *Cytol., chromosome number in Rosa, Populus etc. Inheritance of Sex in members of the Family Caryophyllaceae with especial reference to the occurrence of Sex chromosomes.* ⓓ London 1924.

Blackman, Maulsby Willett (1876), Prof. of Forest Entomol., N. Y. State Coll. of Forestry, Syracuse Univ. Syracuse, N.Y (U.S.A.). *Forest Entomol., boring insects, Cerambycidae, Buprestidae, Apidae. Taxonomy and Biol. of Bark-beetles.* ⓓ Harvard Univ. 1905. ⓟ Opidae and Cerambycidae.

Blackman, Vernon Herbert (1872), Prof. of Plant Physiol. and Pathol. London S.W. 7 (England), Imperial Coll. of Sc. and Technol. ⓓ Cambridge1895.

Blagoweschenski, Andrey Vassilievich (1889), Prof. of Plant Physiol. Univ. Taschkent (U.d. S.S.R.), Balikschinskaja 8. *Specific and synthetic actions of plant ferments (proteases). Osmotic pressures of desert and mountainous plants.*

Blagoweschtensky, Nicolai (1893), Ass. Bact. Inst. der Univ. Kazan (U.d.S.S.R.), B. Krasnaja 59. *Bact. Immunitätsforschung.* ⓓ Rostov a. D. 1919.

Blaha, Josef (1900), Ass. Mähr. Landw. Versuchsanstalt. Brno (C.S.R.), Schwarze Felder. *Physiol. d. Rebe, Mykol. d. Mostes.* ⓓ Brünn 1925.

Blain, Walter L. (1892), Associate Plant Pathologist and Associate Prof. of Plant Pathol. Alabama Agric. Experim. Station. Auburn, Ala. (U.S.A.), Botan. Department, Alabama Polytechnic Inst. *Disease Resistance in Sweet Potatoes. Life History and Control of Pecan Scab. Diseases of Southern Crops. Truck Crops.* ⓓ Univ. of Illinois 1926.

Blair, Duncan Maccallum (1896), Lect. Anat. Univ. Glasgow (England). *Human anat. Histol.* ⓓ Univ. of Glasgow.
Blair, John Edward (1899), Instr. in Bact. Stanford Univ., Calif. (U.S.A.), Box 1711. *Bact.: The Bacteriophage Phenomenon.* ⓓ Brown Univ. 1923.
Blair, Joseph Cullen (1871), Prof. chief Hortic. Agric. Exper. Station. Urbana, Ill. (U.S.A.). o
Blair, Mary Constance (1877), Labor. Ass. Evanston, Ill. (U.S.A.), 1011 Grove St. *Embryol. of Reboulia hemisphaerica.* ⓓ Northwestern Univ. 1922.
Blaisdell, Frank Ellsworth (1862), Prof. of Surgery, Stanford Med. School, Dir. of the Labor. of Experim. Surgery. San Francisco, Cal. (U.S.A.), 1520 Lake Street. *Coleoptera of the Tenebrionid Tribe Eleodiini.* ⓓ San Francisco, Cal., 1889.
Blake, Charles Henry (1901), Ass. Cambridge, Mass. (U.S.A.), Dept. Biol. and Public Health, Mass. Inst. of Technol. *Crustacea.*
Blake, Sidney F. (1892), Associate Botanist, U. S. Dept. Agric. Washington, D. C. (U.S.A.). *Systematic Botan.: Compositae.* ⓓ Harvard Univ. 1917. ⓟ Compositae.
Blakeslee, Albert Francis (1874), Ass. Dir., Department of Genetics, Carnegie Inst. of Washington. Cold Spring Harbor, N.Y. (U.S.A.). *Sexuality in Mucorales, Genetics of Datura.* ⓓ Harvard Univ. 1904.
Blanc, Georges (1884), Dir. de l'Inst. Pasteur Hellénique. Athènes (Grèce). *Pathol. expérim., parasit. (Helminthes). Etude des virus filtrants.* ⓟ Prép. de Leishmanias. Mollusques d'eau douce, parasit. intestinaux de l'homme et des animaux. Cultures microbiennes et virus filtrants.
Blanc, Henri (1859), Prof. Zool., d'Anat. comparée, physiol. gén. Lausanne (Suisse), 36, Avenue des Alpes, Mus. zool. *Poissons, Phalangidea, Isopoda, Amphipoda, Ceratium, Rhizopodes, Taenia, Bothriocephalus, maturation et fécondation de l'œuf, Plankton du lac Léman, Asellus, caprellidae, anomalies de l'appareil génital hermaphrodite, Dixippus.*
Blanchard, Frank Nelson (1888), Ass. Prof. of Zool. Ann Arbor, Mich. (U.S.A.), Univ. of Michigan. *Herpetol. Life histories and systematic relations of North American snakes, life histories of salamanders.* ⓓ Univ. of Michigan 1919. ⓟ North American snakes.
Blanchard, Frieda Cobb, Ass. Dir. Botan. Gardens, Univ. of Michigan. Ann Arbor, Mich. (U.S.A.). *Genetics, systematic botan.* o
Blanchard, Ralph Arthur (1900), Ass. Entomol. Monroe, Mich. (U.S.A.). *European Corn Borer.* ⓓ Washington Univ. 1924.
Blanchetiere, Alexandre (1875), Prof. agrégé à la Fac. de méd., Paris XIV (France), 2, rue Marié-Davy. *Biochim.: chim. des matières albuminoides, chim. pathol.* ⓓ Paris 1909.
Blanck, Edwin (1877), o. Prof. und Dir. des Agrikulturchem. und Bodenkundl. Inst. d. Univ. Göttingen (Deutschland), Wilhelm-Weber-Str. 40. *Bodenlehre u. Pflanzenernährungslehre. Die chem. Verwitterung in der ägyptischen Wüste.* ⓓ Heidelberg 1900.
Blanco, Ramón (1891), Chef Section Agron. Lérida (España). *Génétique: blés et bétail bovin. Citol.: chromosomes dans le bétail.* ⓓ Madrid 1917.
Blaringhem, Louis F. T. (1878), Prof. de Botan. à la Fac. des Sc. et de l'Ecole Normale Supérieure. Paris V (France), 45, rue d'Ulm. *Biol. florale et génétique des plantes cultivées. Mutations et hybrides des Céréales, Lins etc. plantes horticoles alimentaires et ornementales.* ⓓ Paris 1907. ⓟ Conifères, Chênes, Arbres à feuilles caduques, Rosiers. Plantes mutantes, et de lignées controlées de Céréales, Légumineuses, plantes horticoles (graines).
Blatný, Ctibor Eugen Maria Karel (1897), Dr. Ing., Phytopath. Inst. der Staatlichen Versuchsanstalten für Pflanzenproduktion. Prag II (C.S.R.), Václavské nám 44. *Viruskrankheiten der Kulturpflanzen, Virus-*

krankheiten übertragende Insekten, Krankheiten und Schädlinge des Hopfens; durch Protozoen verursachte Krankheiten der Pflanzen, die landwirtschaftlich wichtigen Milben, Scydmaeniden und Pselaphidae der Welt (Coleoptera). ⓓ Prag 1924. ⓟ Praep. der kranken Pflanzen, auch getrocknete kranke Pflanzen.
Blaydes, Glenn W. (1900), Ass. Prof. of Botan., Ohio Wesleyan Univ., Delaware, Ohio (U.S.A.), 95 Oak Hill Avenue. *Transpiration as shown by the cobalt chloride test.*
Bledowski, Richard (1886), Dr. phil. Prof. de Zool. à l'Univ. Libre de Pologne à Varsovie (Pologne), Labor. de Zool., rue Polna 30. *Biol. systématique et embryol. des Hyménoptères parasites.* ⓓ Berne 1909. ⓟ Hyménoptères parasites.
Bledsoe, Rosewell Page (1888), Agronomist, Georgia Exp. Sta. (U.S.A.). *Crops, soils, plant breeding, fertilizers, crop insects.*
Blegvad, Harald (1886), Dr. phil., Ass. Danske Biol. Station. København S. (Danmark), Fischersgade 19. *Marine Ecol.: Bottom Fauna, Food of Animals. Biol. of Amphipoda, Cumacea, Polychaeta.* ⓓ Copenhagen 1921.
Bleier, Hubert (1896), Dr. phil., Ass. an der Lehrkanzel für Phytopath. der Hochsch. für Bodenkultur. Wien XVIII (Österreich). *Pflanzen-Cytol., Pflanzenzüchtung.* ⓓ Kiel 1925.
Bleuler, Paul Eugen (1857), Dr. med., o. Prof. der Psychiatrie, Univ. Zürich (Schweiz), Burghölzli. *Psychoide als Princip der organischen Entwicklung. Entwicklungslehre im Zusammenhang mit der Mneme.* ⓓ Bern 1884.
Bleyer, Benno (1885), Prof., Dr. phil. Weihenstephan bei München (Deutschland). *Biochem. der Kohlehydrate (Gärung, Ernährung) und der Milch.* ⓓ München 1910.
Bleyer, Leo (1893), I. Ass. hygien. Inst. Basel (Schweiz), Petersplatz 10. *Einfluß von Metallsalzen auf die Tetanolysinhämolyse u. auf die Blutkatalase. Serodiagnostik der Schwangerschaft. Wirkung von Alkylresorzincarbonsäuren auf Bakterien und bakterielle Toxine.* ⓓ Innsbruck 1919.
de Blieck, Levinus (1878), Prof. à l'Univ. Fac. vét. Utrecht, (Holland), Biltstraat 168. *Maladies parasitaires et contagieuses des animaux domestiques. immunité spéciale.* ⓓ Bern 1904. ⓟ Cultures de Bactéries.
Bliss, Eleanor Albert (1899), Jacques Loeb Fellow in Med. Johns Hopkins Univ. Baltimore, Md. (U.S.A.), 1026 North Calvert Street. *Bact.; haemolytic streptococci; Erysipelas.* ⓓ Johns Hopkins Univ. 1925.
Blizzard, Alpheus Wesley, Ass. Prof. of Biol. New York City (U.S.A.), New York Univ., Univ. Heights. *Cytol., basidiomycetes.* o
Bljaher, Leonid Jak., Doc. II. Univ. Moskau (U.d.S.S.R.). *Allg. Biol.* o
Bloch, Bruno (1878), Dir. Dermatol. Univ.Klinik. Zürich (Schweiz), Plattenstr. 49. *Pigmentforschung. Beziehungen der Hautkrankheiten zu Stoffwechsel. Endocrinol. und Immunbiol. Biol. der Haut. Experim. Krebsforschung.* ⓓ Basel 1903.
Bloch, Robert (1898), Dr. phil. Charlottenburg 2 (Deutschland), Knesebeckstr. 83. *Pflanzenphysiol.* ⓓ Berlin 1925.
Blochmann, Friedrich (1858), Prof. der Zool. im Ruhestand. Tübingen (Deutschland), Wilhelmstr.11. *Weiterherausgabe von Bütschlis Vergl. Anat. Mikroskopische Tierwelt des Süßwassers. Brachiopoden. Cestoden u. Trematoden.* ⓓ Heidelberg 1881.
Blochwitz, Adalbert (1875). Berlin N 20 (Deutschland), Buttmannstr. 7 I. *Schimmelpilze (Systematik, Physiol., Vererbung).*
Blodgett, Forest Milo (1885), Ass. Prof. in Plant Pathol. in Cornell Univ. Ithaca, N.Y. (U.S.A.), Coll. of Agric. *Virus diseases of Plants: Potatoes.* ⓓ Cornell Univ. 1914.
Blohm, Ernst (1868), Dr. phil., stellvertr. Vorsteher der Agrikulturchem. Versuchsstation der

Landwirtschaftskammer für die Prov. Schlesw.-Holstein, Kiel (Deutschland), Kronshagener Weg 3. *Agrikulturchemie.* ⓓ Kiel 1895.

Blom, Jakob Hansen (1898), Dr. phil., Ass. at the Plantphysiol. Labor. of the Royal Vet.- and Agric. Coll. København (Danmark), Roligbedsvej 23. *Biochem. of soils.* ⓓ Kiel 1922.

Blomberg, Eduardo, Prof. Univ. Nac. La Plata (Argentina). *Anat. Pathol. Hygiene.* o

Blomquist, Hugo L., Prof. of Botan., Duke Univ. Durham N.C. (U.S.A.). o

Blood, H. Loran (1900), Instr. in Botan. and Plant Pathol. Logan, Utah (U.S.A.), Utah Agric. Coll. *Diseases of Utah crops and their controls.* ⓓ Utah Agric. Coll. 1925.

Bloom, William (1899), Douglas Smith Foundation Fellow in Anat., Univ. of Chicago, Ill. (U.S.A.), Dept. of Anat. *Experim. histol. and tissue culture, immune phenomena, developmental potencies of connective tissue, blood cells and the blood forming organs.* ⓓ Johns Hopkins Univ. 1923.

Bloor, Walter Ray (1877), Prof. of Biochem., School of Med. and Dentistry Univ. of Rochester, N.Y. (U.S.A.). *Metabolism of Fat.*

Blotevogel, Wilhelm (1894), Priv.Doc. an der Univ. Hamburg, II. Prosektor am Anat. Inst. Hamburg 20 (Deutschland), Erikastr. 1. *Vitale Färbungen. Sexualhormone und Uterusnervensystem.* ⓓ Hamburg 1920.

Bluhm, Agnes (1862), Dr. med., Volontärin am Kaiser-Wilh.-Inst. für Biol., Berlin-Dahlem. Berlin-Lichterfelde (Deutschland), Unter den Eichen 54. *Experim. Vererbungsforschung: Säugetiere (Maus), menschliche Vererbungslehre, Rassen- und Sozialhygiene, Keimgiftstudien.* ⓓ Zürich 1889.

Blum, Gebhard (1888), Priv.Doc., Ass. am Botan. Inst., Freiburg (Schweiz). *Physikalische Pflanzenphysiol; osmotische Untersuchungen, ökol. Pflanzengeographie.* ⓓ Freiburg (Schweiz) 1914.

Blumberg, John (1861), Doc., Dir. chir.-anat. Inst. Tartu (Estland). *Topographische Anat.* ⓓ Dorpat 1889.

Blume, Wilhelm (1893), Ass. am Pharmakol. Inst. der Univ. Bonn a. Rh. (Deutschland), Wilhelmstr. 33. *Experim. Pharmakol.: Zentralnervensystem der Säugetiere.* ⓓ Berlin 1921.

Blumenthal, Ferdinand (1877), Geh.San.R., a. o. Prof., Dir. Inst. f. Krebsforschung. Berlin W 10 (Deutschland), Bendlerstr. 20. o

Blunck, Hans (1885), Dr. phil., Reg.R. bei der Biol. Reichsanstalt (Leiter der Zweigstelle Kiel), Priv.Doc. an der Univ. Kiel (Deutschland). *Phytopath., angewandte Entomol.* ⓓ Marburg a. d. L. 1912.

Bluntschli, Hans (1877), Dr. med., Dir. der Dr. Senckenbergischen Anat. Inst. der Univ. Frankfurt a. M. (Deutschland), Gärtnerweg 54. *Funktionelle Anat. des Schädels, vergl. Anat. namentlich der Primaten.* ⓓ Heidelberg 1903.

Boas, Friedrich (1886), Dr. phil., Prof. der Botan. und der Phytopath. Hochsch. Weihenstephan-Freising, Bayern (Deutschland). *Physiol. der Zelle, chem. Physiol.; Mikrobiol. Physiol. der Salze (Hylergographie). Lupine. Getreidekrankheiten.* ⓓ München 1911.

Boas, Johan Erik Vesti (1855), Prof. Zool. K. Vet.- u. Landwirtschaftl. Hochsch. København V (Danmark), Gl. Kongevej 84. *Morphol. der Wirbeltiere, spez. Blutgefäßsystem, Crustaceen, Phylogenie.* ⓓ København 1881.

Bobilioff, W., Dr. Buitenzorg, Java (Nederl.-O.-Indië). *Physiol. von Hevea.* o

Bobin, W. W., Dir. Anat. Labor., Kubansches Med. Inst. Krasnodar (U.d.S.S.R.), Str. Sedin 4. o

Bobko, Eugen (1890), Prof. Omsk (U.d.S.S.R.), Inst. für Land- und Forstwirtschaft. *Agric.-Chem. Bodendüngung und Züchtung der technisch wichtigen landwirtschaftl. Pflanzen.* ⓓ Samen von Lein, Hanf, Sonnenblumen, Rüben, Tabak und anderen Öl- und Textilpflanzen.

Bobrinskoy, Count Alexis (1893). London W. 1 (England), Sr. James' Club. *Ornithol. Birds of the river Dniepr basin.* ⓓ Oxford 1921.

Bobrinskoy, Nicolas (1890), Ass. Zool. Mus. Univ. Moscow (U.d.S.S.R.), Trubnikovsky per., 26, 2. *Systematic of Birds, Reptiles (Snakes), Mammals (Bats). Faunistic, specially of Palearctic Asia.*

Bobrow, Eugen (1902), Ass. Leningrad (U.d.S.S.R.), Botan. Hauptgarten. *Botan. Geographie.*

Bocchia, I., Doc. Univ. Parma (Italia). *Igiene.* o

Bocci, Balduino (1852), Prof. Fisiol. Univ. Siena (Italia), Labor. di Fisiol. *Fisiol. Psicol. sperimentale.*

Bock, Friedrich (1899) Dr. phil. planm. Ass. Tübingen (Deutschland), Zool. Inst. *Koloniebildende Protozoen. Fische.* ⓓ Marburg 1922.

Bock, Joseph Carl (1884), Prof. physiol. chem., Dir. of Dept., Marquette Univ. School of Med. Milwaukee, Wis. (U.S.A.), 638, 4 Street. *Microanalysis in bloodchem. Colorimetry.* ⓓ Wien, Techn. Hochsch.

Bock, Karl Alfred Sixten (1884), Doc. Zool. Univ., gegenwärtig stellv. Prof. Uppsala (Sverige), Ö. Ågatan 13. *Turbellarien (Polycladen, Tricladen, Alloeocoelen, Rhabdocoelen), Nemertinen, Nematomorphen und Cyclostomen.* ⓓ Uppsala 1914.

Bodansky, Meyer (1896), Associate Prof. of Biol. Chem. Galveston, Tex. (U.S.A.), Univ. of Texas, Med. Coll. *Experim. Anemia.* ⓓ Cornell Univ. 1923.

Bode, Hans Robert (1899) Dr. phil. Ass. Bonn a. Rh. (Deutschland), Botan. Inst. d. Univ. *Pflanzenphysiol.: Wasserhaushalt der Pflanze.* ⓓ Jena 1922.

Bodnár, János (1889), o. Univ.Prof. der med. Chem. Univ. Debrecen, Dir. Pflanzenbiochem. Inst. Budapest II (Ungarn), Debrői ut 15; Debrecen, Simonyi ut 16. *Biochem., analytische, med. und gerichtliche Chem.* ⓓ Kolozsvár 1911.

Bodó, Richárd (1895), Priv.Doc., Ass. Prof. physiol. Inst. Univ. Pécs (Ungarn), Rákóczi-Str. 80. *Physikal. Chem. Physiol. der Harnabsonderung.* ⓓ Budapest 1922.

van Boeckel, Louis J. J. M. (1886), Insp. principal d'hygiène du Gouvernement, Dir. du labor. Central de l'administration de l'hygiène de l'Etat. Bruxelles (Belgique), Avenue Clays, 25. *Bact. et Hygiène.* ⓓ Louvain 1909, Gand 1913.

Boecker, Eduard (1886), Abteilungsleiter und Prof. am Preußischen Inst. für Infektionskrankheiten Robert Koch. Berlin N 39 (Deutschland), Föhrer Str. 2. *Diagnostische Bact. Zool., Anat., Pathol., Chemotherapie.* ⓓ Göttingen 1910.

Boedecker, Friedrich (1883), Dr. phil., Leiter der wiss. Labor. der J. D. Riedel A.G. Berlin-Britz (Deutschland), Riedelstr. 1—32. *Synthese und Pharmakol. therapeutisch wirksamer Substanzen, Schlafmittel und Analgetica, Lecithin, Cholesterin, Gallensäuren.* ⓓ Freiburg I. Br. 1907.

Boedijn, Karel Bernard (1893), Botanist at the experim. station of the A.V.R.O.S. Medan, Deli (Sumatra, Niederl. Indien). *Phytopath. and general Mycol. Genetics and Cytol.: oilpalm.* ⓓ Amsterdam 1925. ⓟ Fungi.

Böhm, Leopold Karl (1886), Prof. der Allgemeinen Zool. und Parasitenkunde an der Tierärztl. Hochsch. Wien III (Österreich). *Tierische Parasiten: Kokzidien, Cestoden, Nematoden, Milben.* ⓓ Wien Phil. 1910, med. vet. 1916. ⓟ Tierische Parasiten aus Säugetieren und Vögeln.

Boehm, Rudolf (1844), Dr. med., Dr. phil. h. c., o. Prof., Geh.Med.R. Leipzig (Deutschland), Seeburgstr. 100 II. *Pharmacol., Biochemie.* ⓓ Würzburg 1867.

Böhmer, J., Prof. Dr., Landwirtschaftl. Hochsch. Aas (Norge). *Forstwiss.* o

Böhmig, Ludwig (1858), Univ.Prof. Graz-Groisbach, Steiermark (Österreich). *Zool.* o

Böhne, Arnold (1860), wiss. Hilfsarbeiter St. Mus. Bremen (Deutschland), Sedanstr. 78. *Conchylien.*

Boeke, J., Prof. Dr., Dir. des Histol. und Embryol. Labor. der Univ. Utrecht (Holland). *Nervensystem, Sinnesorgane, Bindegewebe.* o

Böker, Hans (1886), Prof. Dr. med., I. Prosektor Anat. Inst. Freiburg i. Br. (Deutschland), Jacobistr. 58. *Vergl. biol. Anat. der Wirbeltiere, Phylogenese der Lebensäußerungen, Ornithol., Anat. des Vogelflugs.* ⓓ Freiburg i. Br. 1913.

Bölsche, Wilhelm (1861). Schreiberhau i. R. (Deutschland), Villa Carmen. *Volkstümliche Naturkunde, Naturschutz.*

Börner, Aloys (1868), Dr., Prof. für angewandte Chem. a. d. Univ. und Vorsteher der Landwirtschaftl. Versuchsstation Münster i. W. (Deutschland), Südstr. 74. *Agrikulturchemie, Lebensmittelchemie.* ⓓ Münster i. W. 1891.

de Boer, Siebe (1876), Priv.Doc. der Physiol. der Univ. Amsterdam (Holland), Lomanstraat 51. *Innervation des Muskeltonus, Herzphysiol., segmentale Innervation der Muskeln. Electrophysiol., Pharm. des Herzens und des C. N. S., die glatten Muskeln (Milz). Physiol. der Niere. Physiol. und Pharm. des Flimmerns.* ⓓ Amsterdam 1914.

Börgesen, Frederik Christian Emil (1866). København (Danmark), Botan. Mus. *Algol., Plantgeography.* ⓓ København 1904. ⓟ Algae.

Börner, Carl (1880), Oberreg.R., Dr. phil., Mitglied der Biol. Reichsanstalt für Land- und Forstwirtschaft, Leiter der Zweigstelle Naumburg a. d. S. (Deutschland), Luisenstr. 26. *Angewandte Entomol., Aphiden (Systematik, Biol., Bekämpfung, Karyol.). Genetik der Reblausimmunität in der Gattung Vitis. Generelle Morphol., Metamorphose und Systematik der Arthropoden, vornehmlich Insekten; generelle Tokontol.* ⓓ Marburg a. d. L. 1903. ⓟ Apterygoten, Aphiden.

Börnstein, Paul Heinrich (1886), Dr., Med.R., Kreisarzt Berlin-Spandau. Berlin W 35 (Deutschland), Blumeshof 14 II. *Seuchenbekämpfung, Hygiene, Bact.* ⓓ 1914.

Boeshore, Irwin (1883), Ass. Prof. of Botan. Philadelphia, Pa. (U. S.A.), MacFarlane Hall, Univ. of Pennsylvania. *Pteridophyta, Gymnospermia. Paleobotan.* ⓓ Univ. of Pennsylvania 1920.

Böttcher, Georg (1900), Wissenschaftl. Mitarb. der Preußischen Landesanstalt für Wasser-, Boden- und Lufthygiene. Berlin-Dahlem (Deutschland). *Wasser und Abwasser. Hygiene.* ⓓ Marburg a. d. L. 1925.

Boettger, Caesar Rudolf (1888). Frankfurt a. M. (Deutschland), Humboldtstr. 42. *Mollusca. Zoogeographie: Mittelmeergebiet, Westafrika, Kanarische Inseln.* ⓓ Bonn 1911.

Bofill, José Maria, Dr. med. Barcelona (España), Aragón, 281. *Entomol.* o

Bogačev, Vladimir (1881), Prof. de Géol. et Paléont. à l'Ecole Polytechnique à l'Univ. Baku, Adzerbaydjan, Caucase (U.d.S.S.R.). *Paléont.: Mollusques, Poissons, Mammifères. Géol. du Caucase. Zoogéographie.* ⓓ Dorpat 1910.

Bogdán, József, Dr., Mitarb. am Anat.-histol. Inst. der Franz-Josef-Univ. Szeged (Ungarn). o

Bogdan, Wassilij S., Prof. d. Kubansker Landwirtschaftl. Inst. Krasnodar (U.d.S.S.R.), Nowaja-Str. 107. *Agronomie. Selektion. Botan. Systematik. Landwirtschaftl. Pflanzen.* o

Bogdanow, Ellius A. (1872), Prof. der allgemeinen Tierzucht Landw. Timiriazew'schen Akademie, Dir. Zootechn. Versuchsstation. Moskau (U.d.S.S.R.). *Ernährungsphysiol. des Milch-, Mast-, Arbeits- und Jungviehs. Züchtungslehre: Veränderlichkeit, Vererbung, Zuchtwahl. Genetik: künstliche Mutationen d. Calliphora vomitoria.* ⓓ Leningrad 1910 Mag. Zool. und Physiol.

Bogdanow-Katkow, N. N., Dir. Inst. d. angew. Zool. Univ. Leningrad (U.d.S.S.R.). *Entomol.* o

Bogert, Lotta Jean (1888), Dr., Research Chemist, Henry Ford Hospital, Detroit, Mich. (U.S.A.). *Metabolism of calcium and magnesium.* ⓓ Yale Univ. 1916.

Bogojawlensky, Konstantin (1899), Ass. Kabinet Histol. und Embryol. Moskau (U.d.S.S.R.), Gerzenstr. 6. *Allgemeine Zytol.: Blut der Wirbellosen.*

Bogojawlensky, N. W., Prof., Dir. Histol.-Embryol. Kab. d. I. Univ. Moskau (U.d.S.S.R.). o

Bogojawlensky, Sergej Georgiewitsch (1888), Ass. Kabinet d. Zool. und Entomol., Vorsteher d. Versuchs-Bienenstandes. Woronesch-Gouv. (U.d.S. S.R.), Landwirtschaftl. Inst. *Biol. d. schädlichen Insekten, chem. Bekämpfungsmittel, Biol. d. Bienen.*

Bogolepov, A., Prof. à l'Univ. Tomsk, Sibérie (U.d.S.S.R.). *Dermatophytes.* o

Bogoljubow, Nikolaus (1872), Prof. Univ. Woronesch (U.d.S.S.R.). *Palaeont.: mesozoische Reptilien, Mammalia der Quartärfauna.* ⓓ Moskau 1912.

Bogoljabsky, Sergius (1885), Doc. Moskau (U.d.S.S.R.), I. Univ., Inst. für vergl. Anat. *Vergl. Anat. d. Wirbeltiere. Naturgeschichte der Haustiere: Canis, Ovis, Bos.* ⓟ Embryonen.

Bogomolez, Alexandre (1881), Prof. de Physiol. Pathol. Univ. II. Moscou (U.d.S.S.R.), Bolchaja Jakimanca 40, 37. *Corrélation neuro-iono-endocrinienne dans l'organisme. Problème et prophylaxie du fatigue neuro-musculaire. Sécrétions internes. Specifité des centres nerveux du métabolisme.*

Bogoslovska, Marie A., Ass. Inst. Agronomique. Voronesh (U.d.S.S.R.). *Maladies des pommes de terre.* o

Bogoslowsky, Alexandre (1900), Forschungsinst. für Zool. Physik.-Mathem. Fac. der I. Univ. Moskau, II. Ass. an der Biol. Station zu Bolschewo (U.d. S.S.R.). *Embryol., Histol. und Biol. der Rotatorien.*

Bogucki, Mieczysław (1884), Ass. Labor. de Physiol. Inst. Nencki, Prof. d'Embryol. Univ. Warszawa (Pologne), 8, rue Śniadecki. *Physiol. du développement.* ⓓ Cracovie 1916.

Bohl, Boris (1897), Prosektor. Kasan (U.d. S.S.R.), Vet.-Inst., Arskoe Pole 96. *Rotz der Lungen.*

Bohl, Carl (1871), Prof. Kasan (U.d.S.S.R.), Vet.-Inst., Arskoe Pole 96. *Pathol. Anat. und Histol. Bindegewebe.* ⓓ Kasan 1895.

. Bohlin, Anders Birger (1898), Fil. lic. Upsala (Sverige), Zool. Inst. *Cavicornier der chinesischen Hipparionfauna (Antilopen). Giraffidae.* ⓓ Uppsala 1926.

Bohn, Georges (1868), Dir. de labor. à l'Ecole des Hautes-Etudes et à la Fac. Sc. Paris 5 (France), 2, rue des Arènes. *Biol. expér. et chim. Reproduction, croissance.*

Boino-Rodzevič, M. A. (1890), Wiss. P. T. Lesshaft Inst. Leningrad (U.d.S.S.R.), Persp. Maklin 32. *Anatomie comp.*

Bois, Désiré Georges Jean Marie, (1856) Prof. de culture au Mus. national d'histoire naturelle. Paris V (France), 57, Rue Cuvier. *Plantes alimentaires.*

du Bois, Eugene Floyd (1882), Med. Dir. Russell Sage Inst. of Pathol.; Associate Prof. of Med., Cornell Univ. Med. Coll.; Dir. Second Med. Division, Bellevue Hospital. New York, N.Y. (U.S.A.), 1215 Park Avenue. *Normal and pathol. metabolism, especially basal metabolism.* ⓓ Columbia 1906.

du Bois-Reymond, René (1863), Dr. med., a.o. Univ.Prof. Berlin-Nicolassee (Deutschland), Paul-Krause-Str. 3. *Physiol. der Bewegung.* o

Bojarkin, Al. Nik., Doc. I. Univ. Moskau (U.d. S.S.R.). *Botan.* o

Bokor, Elemér (1887). Budapest VII (Ungarn), Damjanich-u. 36. I/1. *Käfer Ungarns. Höhlen- und Subterranfauna.* ⓓ Budapest 1925. ⓟ Mitteleuropäische Käfer.

Bokor, Rudolf (1898), dipl. Forsting., Ass. a. d. k. ung. Hochsch. für Berg- und Forsting. Sopron (Ungarn), Botan. Inst. *Biol. u. Bact. des Waldbodens.* ⓓ Budapest 1924.

Boicato, Virgilio (1901), Ass. Ist. di Fisiol. Univ. Parma (Italia). *Ricambio intermedio degli idrati di carbonio.* ⓟ Parma.

Boichowitinowa, Marie A., Doc. für Geol. und Palaeont. Moskau (U.d.S.S.R.), Bergakademie. *Spongia.* o

Boldingh, Isaak, Dr. Buitenzorg, Java (Nederl.-O.-Indië). *System der Pflanzen.* o

Boldyreff, Vasily Nikolaevich (1873), Dir. of Pavlov Physiol. Inst., Battle Creek, Mich. (U.S.A.), Battle Creek Sanitarium. *Physiol. of digestion, blood, and cerebral function. Internal secretion and metabolism (pancreas). Regulation of Heat (Thyreoid gland). Fundamental Laws of Cerebral Function.* ⓓ Petersburg 1898. ⓟ All digestive juices of dog: gastric juice, pancreatic juice, intestinal juice.

Boldyrev, Vassilij Fedorovič (1883), Prof. angew. Zool., Landwirtsch. Akademie, Labor. d. angew. Zool. Moskau (U.d.S.S.R.), Petrovsko-Razumovskoe. *Schädliche Insekten. Biol. der Insekten: Orthoptera, Spermatophorbefruchtung. Neuropt., Isoptera, Coleoptera, Diptera. Fauna hiemalis. Fauna rossica (g. Chionea; Orthoptera).*

Bolin, Pehr Karl Vilhelm (1865), Oberass. Centralanstalt für landwirtsch. Versuchswesen. Experimentalfältet (Sverige). *Pflanzenzucht.*

Bolin, Zera Exley (1888), Dr. of Pharmacy, Dr. of Med., Instr. in pathol. Univ. of California, Med. school. San Francisco, Cal. (U.S.A.), 40 Edge Hill Way. *Bact., Pathol., Biol. chem.*

Bolivar, Ignacio (1850), Dir. Mus. Nacional de Ciencias Nat., Palacco del Hipódromo. Madrid (Espagne). *Orthoptères.* ⓓ Madrid 1901. ⓟ Orthoptères, coléoptères, hyménoptères.

Bolivar y Pieltain, Cándido (1897), Sección de Entomol. del Mus. Nacional de Ciencias Natur., Prof. en la Univ. de Madrid 6 (España), Dique de Leon, 27. *Entomol. géneralé. Orthoptera, sp. Acrididae, Carabidae palearcticae, Chalcidoidea Eupelmidae.* ⓓ Madrid.

Bolk, Lodewijk (1866), Prof. der Anat. Univ. Amsterdam (Holland), Mauritskade 61. *Menschliche Anat. Odontol. Abstammungslehre. Muskelsegmentation.* ⓓ H. c. Leiden.

Bolkay, Stephen Joseph (1887), Dr. phil., Keeper of the Zool. Dept. of the Bosnian-Hercegovinian State Mus. Sarajevo 2 (S.H.S.). *Theoretical and practical Zool.; Osteol.; Palaeont.; Phylogeny; Zoogeography.* ⓓ Budapest 1909. ⓟ Living Amphibians and Reptiles from Bosnia, Hercegovina and Dalmatia and Skins of small Mammals (with skulls) from Bosnia and Hercegovina.

Bolle, P. C., Dr., Proefstation. Cheribon (Nederl.-O.-Indië). *Phytopath., Zucker.* o

Bolley, Henry Luke (1865), Botan. and plant physiol. of Gov. Expt. Sta. for N.D., Dean div. of Biol. Fargo, N.D. (U.S.A.). *Plant pathol.* o

Bolognesi, Giuseppe, Prof. Univ. Siena (Italia). *Patol.* o

Bolot, A., Prof. Univ. Besançon (France). *Physiol.* o

Bolotov, Vladimir (1893), Biochem. Groupe Académique Russe. Paris V (France), 11, Rue Guy-de-la-Brosse.

Bolton, Lloyd L. (1899), Research Sc., International Fisheries Commission, Univ. of Washington (U.S.A.). *Spawning and early larval history of halibut.*

Boltunov, A. P., Prof. Staatl. Pädagog. „Herzen"-Inst. Leningrad (U.d.S.S.R.). *Psychol.* o

Bommer, Charles (1866), Prof. de botan. à l'Univ., Dir. du Labor. de Morphol. et de botan. systématique, Conserv. au Jardin botan. de l'Etat. Bruxelles (Belgique), 47, rue Hobbema. *Morphol. végétale, paléobotan., botan. systématique, dendrol., biol. forestière, Gymnospermes.* ⓓ Bruxelles 1889.

Bonar, Lee (1891), Ass. Prof. of Botan., Univ. of California. Berkeley, Cal. (U.S.A.), Dept. of Botan. *Mycol., Ascomycetes.* ⓓ Univ. Michigan 1922.

Bonasewicz, Felix (1898), Ass.Warszawa (Poland), Polna 70. *Identification of wood.* ⓓ Warsaw 1923. ⓟ Wood growed in Poland.

Bonaventura, Enzo (1891), Dott., Prof., Dir. Labor. di Psicol. sperimentale e Prof. di Psicol. Univ. Firenze (Italia). *Percezione dello spazio e del tempo.*

Bonazzi, Augusto (1893), Dir. Chaparra Agric. Experim. Station. San Manuel, Ote (Cuba). *Bact., mycol., pathol.*

Bondarzew, A. S. (1877), Vorst. d. Abtlg. f. Pflanzenpathol. Botan. Garten. Leningrad (U.d.S.S.R.), Pessotschnaja, 2a. lg. 5. *Polyporaceae, Thelephoraceae und Hydnaceae.* ⓓ Polytechnikum Riga 1908. ⓟ Pilzexsiccate.

Bondarzewa-Monteverde, V. N. (1889), Ass. d. Abtlg. f. Pflanzenpathol. Botan. Garten. Leningrad (U.d.S.S.R.), Pessotschnaja, 2a. lg. 5. *Phytopath. und Mycol. (Discomycetes).*

von Bonde, Cecil (1895), Dir. of the Fisheries Survey. Dii. Aquarium St. James. *Ichthyology, Gen. Marine Biol.* ⓓ Capetown 1923. ⓟ S. Afric. Marine Animals.

Bonde, Reiner, Ass. Plant Pathol. Orono, Maine (U.S.A.), Experim. Station. *Virus Diseases of the Potato. Alternaria Solani as a Potato Tuber Rot. Strains of the Fungi.*

Bondorf, C., Prof. Dr., Landw. Hochsch. København V (Danmark), Bülowsvej. *Bact. des Bodens.* o

Bondorff, Karl Andreas Christian (1891), Prof. of agronomy at The Royal Agric. Coll., Copenhagen (Danmark). *Quantitative relation between nutrition and production by plants (law of diminishing return). The chem. composition of cultivated plants in relation to nutrition.* ⓓ Copenhagen 1913.

Bondroit, Jean (1882), Dessinateur. Ixelles (Belgique), 4, rue Maes. *Fourmis. Biol. générale.*

Bonet, Federico Marco (1906), Lic. Sc. Nat. Madrid (España), Calle de Atocha 78, I. *Entomol.: Neuroptères et Collemboles d'Espagne.* ⓓ Madrid 1926. ⓟ Collembola, Thysanura, Neuroptera, Coleoptera, Orthoptera, Lepidoptera.

Bongert, Jakob (1865), Dr. med. vet., o. Prof. der Nahrungsmittelhygiene und Dir. des Inst. für Nahrungsmittelkunde an der Tierärztl. Hochsch. zu Berlin W 50, (Deutschland), Prager Str. 11. *Tierische Parasiten. Hygiene der von Tieren stammenden Nahrungsmittel (Fleisch und Milch). Bact., Fleischuntersuchung. Drüsen der Pferde.* ⓓ Berlin 1923.

Bonne, Curt (1897), Saatzuchtleiter. Schlanstedt, Kr. Oschersleben, Prov. Sachsen (Deutschland). *Physiol., Morphol., Pflanzenpathol.*

Bonnema, Jan (1864), Prof. of geol. Groningen (Holland), Herm. Coll. str. 13. *Fossile Ostracoden.* ⓓ Groningen 1909. ⓟ Fossile Ostracoden.

Bonnet, M. A., Chef de travaux de Zool., Fac. Sc. Lyon (France). *Zool.* o

Bonnet, Pierre Numa Louis (1897), Ass. Zool. Fac. Sc. Toulouse (France). *Biol. et Anat. des Araignées.* ⓓ Toulouse 1922.

Bonnevie, Kristine E. H. (1872), Prof. der Zool. an der Univ. Oslo (Norwegen). *Embryol. Vererbungsforschung.* ⓓ Oslo 1906.

Bonnier, Gert (1890), Doc. Univ. Stockholm (Sverige), Högskola, Drottninggatan 116. *Genetics, Experim. studies on the determination of Sex.* ⓓ Stockholm 1924.

Bonrath, Wilhelm (1890), Dr. phil., wiss. Mitarbeiter d. Landw. Abt. der J. G. A.G. Frankfurt a. M., Werk Leverkusen (Deutschland), Biol. Inst. *Schädlingsbekämpfungsmittel: Beizfrage, Weinbaumittel.* ⓓ Gießen 1913.

Bonvicini, Mario (1899), Primo Ass. Ist. Allevamento vegetale per la Cerealicoltura. Bologna (Italia), Via Toscana 121. *Selezione e coltivazione dei cereali.* ⓓ Bologna.

Bookman, Samuel (1869), Physiol. Chem., Mt. Sinai Hospital. New York City (U.S.A.), 48 E. 82 St. *Biochem.* ⓓ 1895.

Boothby, Walter Meredith (1880), Associate Prof. of Med., Head of Section of Clinical Metabolism, Mayo Clinic, Rochester, Minn. (U.S.A.). *Anaesthetica Haemoglobin. Regulation of the circulation. Thyroid. Thyroxin. Parathyroid gland. Insulin. Nitrogenous metabolism.* ⓟ Harvard 1906.

Borchert, Alfred (1886), Dr. med. vet., Reg.R. bei der Biol. Reichsanstalt für Land- und Forstwirtschaft in Berlin-Dahlem, ao. Prof. an der Tierärztl. Hochsch. in Berlin. Berlin-Steglitz (Deutschland), Vionvillestr. 20 III. *Bienenseuchen.* ⓟ Berlin 1914.

Borchmann, Fritz Heinrich Christian (1870), Koleopterol. am Zool. Staats-Inst. und Zool. Mus. Hamburg 34 (Deutschland), Bauerberg 71 I. *Insecta: Coleoptera Lagriidae, Alleculidae, Meloidae, Nilionidae, Othniidae.*

Borelć, Berislav (1891), Dir. staatl. epidemiol. Inst. Zagreb (S.H.S.), Mlinarska cesta 14. *Bact. und Epidemiol.* ⓟ Med. Bern 1919, med. vet. 1924.

Bordas, F., Prof., Dir. Labor. Univ. Paris (France). *Hyg. génér. et expér.* o

Bordas, Léonard (1864), Dr., Prof. de Zool. Fac. Sc., Dir. Station entomol. de Bretagne. Rennes (France), 36, rue de Châteaudun. *Arthropodes: Insectes (Anat., Histol., Embryol., Physiol). Hyménoptères, Coléoptères, Lépidoptères, Orthoptères, Hémiptères. Larves aquatiques (Anat. et Biol.) des Coléoptères, des Nèvroptères et des Hémiptères.*

Bordás Celma, Manuel (1880). Barcelona (España), Villanueva y Geltrú. *Ovogénèse et Spermatogénèse chez la Sagitta. Stigmenstruktur u. Funktion bei Blatta, Spermiogen., Oogenese bei Dendrocoelum.* ⓟ Univ. Central de Madrid 1913.

Bordet, Jules (1870), Dir. de l'Inst. Pasteur de Bruxelles, Prof. à la Fac. de Med. de l'Univ. de Bruxelles (Belgique) *Bact. méd., immunité contre les maladies infectieuses, sérol. et physiol. du sang.* ⓟ Bruxelles 1892.

Bordsilowskij, Eugen I., Doc. d. Inst. f. Volksbildung u. des Staatl. Pharm. Technikums. Kiew (U.d.S.S.R.), Dmitriewskaja 52, W. 4. *Botan. Systematik.* o

Boresch, Karl (1886), Dr., Prof. an der landwirtschaftl. Abteilung der Prager deutschen techn. Hochsch. in Tetschen-Liebwerd. Tetschen a. E. Nr. 800 (C.S.R.). *Agrikulturchem., Pflanzenernährung, Pflanzenphysiol.* ⓟ Prag 1909.

Borg, Folke (1892), Lect. Wexiö (Sverige), Esplanade 16. *Bryozoa.* ⓟ Univ. of Upsala 1926.

Borg, John (1873), Prof. Nat. Hist., Superint. of Agric. Attard (Malta). *Systematic and Economic Botan. (Phanerogams). Cultivation and Diseases of the Orange-tribe. Banano in Malta. Fruit trees in the Maltese Islands. The Pig. American Vines. Flora of the Maltese Islands.* ⓟ Malta 1918.

Borg, Paul (1888), Plant Pathol., Dept. of Agric. Malta. *Plant pests and diseases.* ⓟ Malta 1909.

Borgatti, Giuseppe (1899), Ass. chargé à l'Inst. de physiol. Bologne (Italie). *Chém. biol.: Cholestérine, Acétone.* ⓟ Bologne 1923.

Borgert, Adolf Hermann Constant (1868), Prof., Dr., stellvertr. Dir. des Zool. und vergl. anat. Inst. der Univ. Bonn (Deutschland), Poppelsdorfer Schloß. *Protozool., Planktol., Tunicata, Meeresforschung, Faunistik. Tripylee Radiolarien.* ⓟ Kiel 1891.

Borges, Ildefonso (1865), Prof. Zool. et Pathol. exotique, Ecole de Méd. Vét., Chef de service de l'Inst. de Bact. Lisboa (Portugal), Inst. Camara Pestana. *Bact. Helminthol.* ⓟ 1912.

Borggardt, Alexandre I., Dir. Sect. Phytopath., Station Expérim. Agronomique. Ekatherinoslav (U.d.S.S.R.). *Pathol. végétale, Ustilaginées.* o

Borgh, Tage (1895). Uppsala (Sverige), Zool. Inst. *Fauna der fließenden Gewässer in den schwedischen Hochgebirgen (Verteilungsprobleme).* ⓟ Uppsala.

Borgherini, Alessandro (1857), Prof. pareggiato di Patol., Univ. Padova (Italia). *Sistema Nervoso.*

Borghi, Bruno (1900), Dr. med., Ist. di Pathol., Generale Univ. Milano (Italia), Via G. Strambio 31. *Appareil réticulo-endothélial. Tumeurs.* ⓟ Pavia 1924.

Boring, Alice M. (1883), Ph.D., Prof. of Zool. Peking (China), Yenching Univ., Wellesley Coll. *Cytol. Histol.* ⓟ Bryn M. Coll. 1910.

Borisevitsch, Georges F., Conserv. à Stat. Expérim. Agronomique. Vinnitsa, Gouv. Kiev (U.d.S.S.R.). *Maladies de la betterave, Ascomycétes.* o

Borisowna-Lepeskinskaja, Olga (1871), Prof. staatl. Timirjaseff-Inst. f. wiss. Forsch. Moskau (U.d.S.S.R.), Piatnizkaja 48, Histol. Labor. *Histol. Erythrozyten membran. Knochengewebe, Trypanosomen.*

Borissiak, Alexis (1872), Prof., Ecole des Mines. Léningrad (U.d.S.S.R.), W. O., 21 Linie, 2. *Mammifères fossiles; Rhinocéridées et les Mastodonts du miocène de Tourkestan.*

Borissov, P. G., Prof. Polytechn. Inst. Ivanovo, Voznessensk (U.d.S.S.R.). *Zootechn.* o

Borja, H., Prof. Univ. Quito (Ecuador). *Physiol.* o

Borkowski, Roman (1882), Dr., Adjunkt an der Techn. Hochsch. in Lemberg (Inst. für Pflanzenbau und Pflanzenzucht). Dublany bei Lemberg (Polen). *Pflanzenzucht.* ⓟ Neuchâtel (Schweiz) 1913.

Borland, Andrew Allen (1878), Prof. of Dairy Husbandry. State College, Pa. (U.S.A.), Dairy Department. *Dairy production, bact.*

Borman, Frank William (1877), Dir. Egyptian Government Zool. Service. Giza (Egypt), Zool. Gardens. *Ornithol.* ⓟ Oological specimens.

Bormans, A., Doc. Univ. Torino (Italia). *Batteriol.* o

Born, Axel (1887), Dr., o. ö. Prof. Charlottenburg (Deutschland), Techn. Hochsch. *Geol. u. Palaeont.* o

Born, Paul (1859), Dr. phil. h. c. Herzogenbuchsee, Ct. Bern (Schweiz). *Carabus, Cychrus und Calosoma der ganzen Welt.* ⓟ Bern 1920. ⓟ Carabus, Cychrus, Calosoma.

Bornebusch, Carl Heinrich (1886), Laborant. Springforbi (Danmark). *Biol., chem. und mechan. Untersuchungen von Waldböden. Bodenflora im Walde. Azotobacter.*

Bornstein, Arthur (1881), Dr., Univ. Prof. Hamburg (Deutschland), Hartwicusstr. 10. *Pharm.* o

Boros, Adam (1900), Phil. Dr., Ass. des landwirtschaftl. Versuchswesens. Budapest II (Ungarn), Aldás- u. 4. *Pflanzengeographie der ungarischen Tiefebene. Kryptogamen. Entwickelungsstudien, Moor und Steppenprobleme, Phytopalaeontol. des Diluviums. Drogenkunde.* ⓟ Budapest 1922. ⓟ Ungarische Herbarpflanzen, auch Moose, evtl. Petrefakten.

Borodin, D. N., President Russian Agric. Agency in America. New York, N.Y. (U.S.A.), Room 309, Liberty Street 136. o

Borodin, Iwan Parfeniewicz (1847), President d. Russ. Botan. Ges., Dir. d. Botan. Mus. d. Akademie. Leningrad (U.d.S.S.R.), Academie d. Sc., Mus. Botan., Wass. Ostrow, quai de l'Université 5. *Anat., Physiol., Floristik.* ⓟ St. Pétersbourg 1876, h. c. Odessa 1886.

Borodkina, Zin. Vikt., Doc. II. Univ. Moskau (U.d.S.S.R.). *Biochem.* o

Boros, Ludwig (1896), Univ. Ass. Szeged (Ungarn), Egyetem Allattani Iutézet. *Chirocephalus und Branchypus. Darmkanal.* ⓟ Chirocephalus und Branchypus.

Borovskij, V. M., Doc. II. Univ. Moskau (U.d.S.S.R.). *Psychol.* o

Borowikow, Georg Andreewitsch (1881), Prof., Dir. Botan. Labor., Landwirtschaftl. Inst., Dir. Landwirtschaftl. Versuchsstation bei Wosnesensk. Odessa (U.d.S.S.R.). *Anat., Physiol. und Biol. d. Pflanzen. Pflanzenwachstum.* ⓟ Odessa 1915.

Borradaile, Lancelot Alexander (1872), Lect. Zool. Univ. of Cambridge, Fellow and Tutor of Selwyn Coll. Cambridge (England). *Crustacea.* ⓟ Cambridge 1922.

Borrel, M., Prof. Fac. Méd. Strasbourg. Paris 15 (France), 207, rue de Vaugirard. *Bact.* o

Borrer, C. D. London S. W. 3 (England), 6 Durham Place, Chelsea. *Ornithol.* o

Borries, Gustav Valdemar Theodor (1887), Dr. med., Priv.Doc. København (Danmark), Holsteinsgade 49. *Physiol. and Pathol. of the stato-cinetic Labyrinth, of the Eye-movements, of Fixation.* ⊕ Copenhagen 1920.

Borrmann, Robert (1870), Prof., Dr. med., Dir. des Pathol. Inst. der Krankenanstalt in Bremen (Deutschland), Lüneburger Str. 9. *Pathol. Anat., Geschwülste, Krebs.* ⊕ Göttingen 1895.

Borsieri, C., Prof. Univ. Roma (Italia). *Zool.* o

Borst, Max (1869), o.ö. Prof. der allg. Pathol. u. der pathol. Anat. a. d. Univ. München (Deutschland), Prinzregentenstr. 11. *Geschwulstforschung, Entzündung, Transplantation.* ⊕ Würzburg 1892.

Boruchin, Moses (1893), Prosektor Anat. Inst. Minsk (U.d.S.S.R.), Moskauer Str. 38. *Individuelle Eigenschaften des menschlichen Gehirns.*

Borza, Alexander (1887), Prof. systemat. Botan., Univ., Dir. des Inst. für syst. Botan., Botan. Mus., Botan. Garten. Cluj (Romañía). *Systematik der Phanerogamen (Cerastium, Dianthus, Centaurea, Melampyrum, Artemisia, Asyneuma). Pflanzenökol.: Steppenvegetation.* ⊕ Budapest 1913.

Borzęcki, Konstantin Marian (1883). Lwów (Polen), Szeptyckichgasse 16. *Desmidiaceen.* ⊕ Phil. Lemberg 1918, med. 1926.

Bos, Hemmo (1857), Dr. d. Biol., Vorstand Kulturkontrolle (Echtheitsprüfung) an der Staatl. Versuchsstation für Samenkontrolle. Wageningen (Holland), Hinkeloordscheweg 11. *Varietätsmale der Kulturpflanzen, Gartenbaubotan., Phaenol.* ⊕ Groningen 1885.

Bosaeus, Wilhelm (1891), Doc. pathol. Anat. Univ. Uppsala (Sverige). *Experim. Pathogenesis und Ephebogenesis bei Wirbeltieren. Genese der Ovarialembryome bei Amphibien.*

Bosc, Edouard (1863), Dr., Chef des Travaux d'Anat. Pathol., Fac. de Méd. Montpellier (France). *Tumeurs.* ⊕ Montpellier 1899.

Boschhardt, Dr. med., chem. Abt. f. Schiffs- u. Tropenkrankheiten. Hamburg 4 (Deutschland). o

Boschma, Hilbrand (1893), Dr. zool. et bot. Hauptass., Priv.Doc. Oegstgeest (Holland), de Kempenaerstraat 66. *Rhizocephalen (systematisch). Ernährung und Entwicklungsgesch. von Riffkorallen.* ⊕ Amsterdam 1920.

Bosco, J., Prof. Univ. Rosario (Argentina). *Anat.* o

Bose, M. N., Dr., Prof., Med. Coll. Belgachia, Bengalen (India). *Anat.* o

Bose, S. R., Prof., Med. Coll. Belgachia, Bengalen (India). *Botan.* o

Boshart, Karl (1887), Reg.R. an der Bayer. Landesanstalt für Pflanzenbau und Pflanzenschutz. München (Deutschland), Kaulbachstr. 63a. *Gemüsebau, Ernährung der Pflanzen. Arzneipflanzenkultur.* ⊕ München 1911.

Boshi, Sen (1888). Calcutta (India), Vivekananda Labor., 8 Bosepara Lane. *Electric changes in living tissues.* ⊕ Calcutta 1911. ⊕ *Seeds of Mimosa pudica, and Desmodium gyrans.*

Bosman, Louis Pierre (1895), Lect. Physiol. Chem. Univ. Capetown (South-Africa). *Biochem. Hydrol. properties of certain amino acids. Nature of co-enzyme of lipase. Calcium metabolism and ovarian hormone.* ⊕ Edinburgh 1922.

Boss, Andrew (1867), Prof. of agronomy Dir.expt. Station. St. Paul, Minn. (U.S.A.), 1443 Raymond Av. *Animal husbandry.* o

Bossé, Georges G. (1887), Inst. de recherches sc. de Timiriaseff, Chef de division. Moscou (U.d.S.S.R.), Piatnizkaia 48. *Microbiol. agric. Procès osmotiques. Origine du caoutchouc.* ⊕ Papillons et Herbier de Colombie, Mexique, Guatemala, Cuba, Panama, Vénézuela.

Boswell, Victor Rickman (1900) Associate in Olericulture. College Park, Md. (U.S.A.). *Fertilizer and cultural practices with Pisum sativum, growth and yield of peas. Temperature influence upon quality of peas (chemical studies). Environmental factors to Brassica oleracea. Physiol., chem. and morphol. conditions.* ⊕ Univ. of Maryland 1926.

Botey y Mateu, Thimotée (1892) Licencié ès sc. nat. et Pharm. Prof. de Botan. agricole à l'Ecole supérieure d'Agric. de Barcelona (España) calle Claris, 113, entlo., 2a. *Anat. et physiol. végétale, classification des plantes cultivées.* ⊕ Barcelona 1916, Madrid 1918.

Botez, Joan G. (1892), Maître de conf. Paléont. et d'Anthrop. Univ. Jassy (Roumanie), 12 strada Sperantei. *La faune des Mammifères quaternaires de la Bessarabie. Mécanisme adaptatif dans les variations du tissu osseux et musculaire. Tortues eocènes du bassin parisien.*

Botez, Mihail A., Prof. Univ. Cluj (România). *Hygiene.* o

Botezat, Eugen (1871), Univ.Prof. Cernăuți (România), Zool. Inst., Univ. *Vergl. Histol., Sinnesorgane, periphere Nervenendigungen, Morphol. der Geweihe von Cerviden, Säugetiere, Vögel, Tardigrada.* ⊕ Cernăuți 1898.

Both, Jolán (1898), Ass. pharm. Inst. Szeged (Ungarn), Kálvária 5. *Blutzuckerbestimmungen.* ⊕ Szeged 1925.

Bottazzi, Filippo (1867), Prof. Fisiol. Napoli (Italia), S. Andrea delle Dame 21. *Fisiol. normale. Biochem.* ⊕ Geneva 1903.

Bottini, Antonio (1850), Aiuto e Libero Doc. di Botan. nella R. Univ. di Pisa (Italia), Lung' Arno Regio No. 17. *Musci, Sphagnum.* ⊕ Pisa 1888.

Bottu, Prof. Ecole prép. de méd. et de pharm. Reims (France). *Toxicol.* o

Bouček, Bohuslav (1886), Prof. agrégé. Brno (C.S.R.), Uvoz 33. *Pharm. du système nerveux. Respiration. Digitale et les cardiotoniques. Standardisation biol. de méd. héroiques. Strychnine.* ⊕ Prague 1911.

Bouckaert, Jean Jacques Philippe (1901), Dr. med. Ostende (Belgique), Rue Christine 112. *Physiol. et Pharm. des Hypnotiques et des Narcotiques.* ⊕ Gand 1925.

Boudrina, Catherine P., Conserv. à l'Inst. Jaczewski. Leningrad (U.d.S.S.R.), Perspective Anglaise 2°. *Systématique générale, Myxomycètes.* o

Boudrina-Voloschinova, Anna P., Chef des trav. Sta. de défense des plantes. Leningrad (U.d.S.S.R.), rue Tschaikovski 7. *Pathol. végétale. Expertise phytopath. des graines.* o

Boughton, Frederick Seymour (1859). Rochester, N.Y. (U.S.A.). *Plant systematic.*

Boughton, Ivan Bertrand (1893), Chief, Vet. Sc., Service Technique de l'Agric., Haiti. Port au Prince (Haiti). *Vet. Sc., bact.* ⊕ Ohio State Univ. 1916.

Bouhey, R., Dir. Adj. de Labor., Ecole pratique des hautes Etudes. Paris (France). *Morphol. expérim.* o

Bouin, Maurice, Prof. Univ. Nancy (France). *Zool. appliquée.* o

Bouin, Pol André (1870), Prof. d'Histol. à la fac. de méd. de Strasbourg (France), 28, Rue Erckmann-Chatrian. *Histol. générale et expérim. Cytol. Glandes sexuelles, endocrinol. Déterminisme des caractères sexuels secondaires. Histophysiol. des organes de la reproduction chez la femelle.*

Boulavkina, A. A., Conserv., Section des Plantes Vivantes, Jardin Botan. principal. Leningrad (U.d.S.S.R.), Aplekarski Ostrov, Pessotchnaia ¹/₅. *Systématisation: Patrinia, Flore de la Région Lacustre.* o

Boule, Pierre-Marcellin (1861), Prof. au Mus. national d'Histoire nat., Dir. de l'Inst. de Paléontol. humaine. Paris 5 (France), 3 place Valhubert. *Paléontol.; Paléontol. humaine.*

Boulenger, Charles, L. (1885), Prof. of Zool. in the Univ. London, N.W. (Egl.),Bedford Coll.*Paraits., particularly Nematoda; Myzostomida.*⊕Birmingham1913.

Boulenger, George Albert (1858), Attaché au Jardin Botan. de l'Etat Bruxelles (Belgique). *Botan. (Rhodol.).* ⓓ LL: St. Andrews, Phil. Gießen, Sc. Louvain.

Bouloumoy, P., Prof., St. Josephs-Univ. Beirut (Syrien). *Pathol.* ○

Boulton, Wolfrid Rudyerd, jr. (1901), Associate, American Mus. Natural History. Long Island, N.Y. (U.S.A.), Cold Spring Harbor. *Ornithol.: Anat., Physiol., Distribution.* ⓟ Bird skeletons and birds in spirit, particularly Fringillidae.

Bounhiol, Jean-Paul (1870), Prof. Fac. Sc. Univ. Bordeaux (France), 11, rue Gaston Lespiault. *Zool. et Physiol. animale. Biol. et Pêches maritimes.*

Bourcart, Jacques Paul (1891), Dr. sc. nat., Chef des travaux de Geographie Physique à la Fac. des Sc. de Paris, Chef de la mission géol. permanente à l'Inst. Sc. Chérifien (Maroc.). Paris XIII (France), 224, rue de Tolbiac. *Géol., géographie et botan. de la Péninsule des Balkans, spécialement de l'Albanie, géol. et paléont. du Néogène et du Quaternaire marocain.* ⓓ Paris 1922.

Bourne, Benjamin Arthur, Prof. of Plant Pathol., Coll. of Agric. and Mechan. Arts, Univ. of Porto Rico. Mayaguez (Porto Rico). ○

Bourov, Serge S., Dir. Stat. de défense des plantes. Moscou (U.d.S.S.R.), Rue Sadovaia-Trioumphalnaia 10. *Pathol. végétale.* ○

Boutan, M. L., Prof. Zool. Fac. Sc. Alger, (Afrique). *Zool.* ○

Bouvied, Prof. Ecole prép. de méd. et de pharm. Reims (France). *Anat.* ○

Bouvier, M., Prof. au Mus. Paris 5 (France), 55, rue de Buffon. *Zool.* ○

Bouwens, Henriette (1881). Nymegen (Holland), De Zandkuil, Berg en Dalsche weg 371. *Phytopath.: Erysipheen.*

Bovie, William T. (1882), Ph.D., Ass. Prof. of Biophysics, Instr. in Bact., M. Douglas Flattery Research Fellow. Boston 17, Mass. (U.S.A.), Harvard Med. School. *Photobiol.* ⓓ Harvard 1914.

Bovien, Prosper (1894), Unterbibliothekar d. kgl. Tierärztl. u. Landwirtsch. Hochsch. København (Danmark), Classensgade 48. *Ascidiae from the Auckland and Campbell islands. Caryophyllaeidae from Java. Pentastomen. Trematoden.* ⓓ København 1919.

Bovšík, G. A. (1894), Wiss. P. F. Leshaft Inst. Leningrad (U.d.S.S.R.) Persp. Maklin 32. *Mikrobiol. Ferments, champignons.* ⓓ 1915.

Bowen, Robert Hall (1892), Ass. Prof. of Zool., Columbia Univ. New York City (U.S.A.), Dept. of Zool. *Cellular Biol.: Structurs of protoplasm, spermatogenesis, Golgi apparatus.* ⓓ Columbia 1920.

Bowkiewicz, Jan (1896), Dr. phil. Wilno (Pologne), ul. Zakretowa 15, Zakład Biol. Ogólnej U.S.B. *Hydrobiol. Cladocera. Euphyllopoda.* ⓓ Wilno 1924. ⓟ Cladocera (Crustacea).

Bowman, Howard Hiestand Minich (1886), Dr. of Phil., Prof. of Biol. Toledo, Ohio (U.S.A.), Univ. of the City. *Botan.; Plant Physiol. and Anat.; Taxonomy; Ecol. Botan. Ecol. of the Dry Tortugas and of the Red Mangrove.* ⓓ Univ. of Pennsylvania 1917.

Box, Harold Edmund (1898), Entomol., Central Aguirre Sugar Company. Central Aguirre (Porto Rico). *Sugar Cane Insect, biol. control of Porto Rican sugar-cane pests.* ⓟ Neotropical Insects (coleoptera and hymenoptera)

Boyce, John Shaw (1889), Pathol., U. S. Department of Agric. Portland, Ore. (U.S.A.), 630 New Post Office Building. *Pathol. of trees and shrubs, decays and defects in wood.* ⓓ Stanford 1917. ⓟ Specimens of fungi parasitic on forest trees.

Boycott, A. C., Prof., Univ. London S.W. 7 (England), South Kensington. *Pathol.* ○

Boyd, Mark F. (1889) Dir. Station for Field Studies of Malaria International Health Board, Rockefeller Foundation. Edenton, N.C. (U.S.A.) and New York City (U.S.A.), 61 Broadway. *Field investigations in epidemiol. of malaria including ecol. of anophelines, Epidemiol. and Parasitol. Mycetoma.* ⓓ Iowa 1911.

Boyd, Theodore Elliott (1893), Associate Prof. of Physiol. Loyola Univ. School of Med. Chicago, Ill. (U.S.A.), 706 S. Lincoln Street. *Gastro-Intestinal Physiol. Physiol. of the Parathyroids.* ⓓ Univ. of Chicago 1923.

Boyden, Alan Arthur (1897), Ass. Prof. of Zool. New Brunswick, N.J. (U.S.A.), Rutgers Univ. *Serol. Precipitin Reaction in the Study of Animal Relationships.* ⓓ Univ. of Wisconsin 1925.

Boyden, Edward Allen (1886), Associate Prof. of Anat., Coll. of Med., Univ. of Illinois. Chicago, Ill. (U.S.A.), 1817 West Polk St. *Experim. study of embryonic nephric system. Physiol. and anat. of gall bladder.* ⓓ Harvard Univ. 1916.

Boyer, A. J., Research Lab., Bureau of Dairy Industry, U.S. Dept. of Agric., Washington, D.C. (U.S.A.). *Abnormal fermentation.* ○

Boynton, Lyman Crowell (1893), Ass. Prof. Biochem., Univ. of Washington. Seattle, Wash. (U.S.A.), Dept. of Chem. *Quantitative studies of the storage and distribution of Vitamin A in the body, and of content of the fat-soluble vitamins in the body, oils of various species of Pacific Coast salmon. Digestive enzymes of woodboring molluscs.* ⓓ Columbia Univ. 1924.

Boysen-Jensen, Peter (1883), Dr. phil., Lektor der Pflanzenphysiol. an der Univ. Kopenhagen (Dänemark), Gothersgade 140. *Zymasegärung; Reizleitung bei Avena; Stoffproduktion bei Pflanzen.* ⓓ Kopenhagen 1910.

Bozler, Emil (1901), Dr. phil., Priv.Doc. an der Univ. München (Deutschland), Zool. Inst., Neuhauser Str. 51. *Physiol. der Sinnesorgane der Arthropoden, Reizphysiol. der Protozoa, Physiol. und Morphol. des Nervensystems der niedersten Metazoen (Medusen).* ⓓ München 1923.

Bozza, Giorgio (1901), Aiuto dell' Ist. Anat. di Firenze (Italia), Fac. di Med., Via degli Alfani 33. *Développement et histol. de la glande pinéale.*

Braafladt, Louis H., Prof., Chantung Christian Univ. Tsinan (China). *Pathol.* ○

Bracher, Rose (1894), Ass. Lect. in Botan. Univ. of Bristol (England). *Euglena. Rhytisma.* ⓓ Sc. Wisconsin 1920, med. Bristol 1918.

Brachet, Albert (1869), Prof. d'Anat. et d'Embryol. à l'Univ. de Bruxelles (Belgique), 32, rue Léonard de Vinci. *Embryol. des Vertébrés; Amphibiens et Reptiles. Embryol. experim.: analyse de la Fécondation, segmentation, localisations germinales de l'œuf.* ⓓ Liège 1894.

Bradford, Frederick Charles (1887), Associate Prof. of Horticulture, Michigan State Coll., East Lansing, Mich. (U.S.A.). *Pomol., nutrition of trees, root-stocks and pomol. history.* ⓓ Maine 1911.

Bradley, George Hirst (1893), Ass. Entomol. U. S. Bureau of Entomol. Mound, La. (U.S.A.). *Biol. and Control of Mosquitoes.* ⓟ Mounted mosquitoes. Southern U. S. species.

Bradley, Harold Cornelius (1878), Research Dir. Woods Hole, Marine Biol. Labor. Madison, Wis. (U.S.A.), Coll. Hills. *Biochem.* ○

Bradley, James Chester (1884), Dr., Prof. of Entomol. and Curator of Invertebrate Zool. on Cornell Univ. Ithaca, N.Y. (U.S.A.), Roberts Hall. *Taxonomy of Hymenoptera: Siricidae, Evaniidae, Mutillidae, Masaridae, Aculeata.* ⓓ Cornell Univ. Ithaca 1910. ⓟ Hymenoptera of the world.

Bradley, O. Charnoch, M.D., Prof., R. Veter. Coll. Edinburgh (Scotland). *Anat., Histol.* ○

Braendegaard, Jens (1887). København (Danmark), Østersøgade 30. *Araneae.* ⓓ 1924.

Bragg, Arthur N. (1897), Ass. Prof. of Zool., Marquette Univ. Milwaukee, Wis. (U.S.A.). *General Zool.* ⓓ Bates Coll. 1924.

Bragin, Ev. Alex., Doc. II. Univ. Moskau (U.d. S.S.R.). *Experim. Hygiene.* o

Bragina, Anna P., Entomol. de l'Inst. Central de l'Hygiène à Belgrade (S.H.S.), Centralni Higijenski Zavod. *Entomol. appliquée: Biol. des Insectes, nuisibles à l'agric., et de leurs parasit. Entomol. méd.: Biol. et systématique des Moustiques des environs de Belgrade et des Phlébotomes. Morphol. des larves de Culex, Theobaldia et Aedes.* Ⓟ Larves des Moustiques. Moustiques secs, insectes nuisibles à l'Agric., carnassiers, ennemis des Moustiques, Diptères.

Bragone, Giuseppina (1899), Ass. Ist. Botan. R. Univ. di Roma (Italia), Via Panisperma 193. *Physiol. végétale. Nutrition.* Ⓓ Roma 1920.

Brahm, Carl (1872), *Abteilungsvorsteher Tierphysiol. Inst. Landwirtschaftl. Hochsch. Berlin N 4 (Deutschland), Invalidenstr. 42. Physiol. Chem. Tierische Ernährung und Stoffwechsel.* Ⓓ 1901.

Brailowsky, Sergej (1899), Doc. Univ. N.-Nowgorod (U.d.S.S.R.), Gogolstr., Tupik, 6, 7. *Physiol. Chem., Hematol.* Ⓓ Moskau 1913.

Brain, C. K., Prof. Univ. of Stellenbosch (South Africa), Dept. of Entomol. Ⓓ M. A. Ohio, Sc. Birmingham. o

Brakenhoff, Hermann (1873), Wissenschaftl. Hilfsarb. St. Mus. Bremen (Deutschland), Reederstraße 11. *Floristik. Hölzer.*

Bramwell, John Crighton (1889), Manchester R.-Infirmary; Honorary Ass. Lect. in Experim. Physiol. Univ. Manchester (England). *Cardiac Pathol. Pulse of Arteries.* Ⓓ Manchester 1923.

Branca, Albert Jean (1868), Prof. à la Fac. de Med. Paris VI (France), 5, rue Palatine. *Histol. et Embryol. des organes sexuels, des annexes embryonnaires.* Ⓓ Paris 1926. Ⓟ Embryons d'Axolote, de Petromyzon. Uterus gravides de Chiroptères.

v. Brand, Frh. Theodor (1899), Dr. phil., Ass. am Physiol. Inst. der Univ. Erlangen i. Bayern (Deutschland). *Vergl. Stoffwechselphysiol.* Ⓓ München 1922.

van der Branden, Frédéric, Dir. Adj. Lab. Bact. Léopoldsville (Congo belge). o

Brandes, Elmer W. (1891), Senior Pathol. in Charge, Sugar-Plant Investigations, U. S. Department of Agric. Washington, D.C. (U.S.A.), 3404 Fulton St., N.W. Ⓓ Michigan Univ. 1920.

Brandes, Gustav (1862), Honorarprof. d. Zool. an d. Techn. Hochsch. mit Lehrauftrag f. Allgem. Zool. u. vergl. Anat., Dir. des Zool. Gartens. Dresden-A. 1 (Deutschland), Tiergartenstr. 1. *Vogelatmung, Hirudineen.* Ⓓ Leipzig 1888.

Brandt, K. (1854), Univ.Prof. emer. Kiel (Deutschland), Düppelstr. 3. *Zool., spez. des Meeres (Plankton, Mollusken).* o

Brandt, Max (1890), I. Ass. Pathol. Inst. Univ. Riga (Latvija), Taubenstr. 30. *Pathol. der Lungen primäre Tumoren). Teerkrebs.* Ⓓ Riga 1926.

Brandt, Walter (1889), Prof. der Anat. Köln a.Rh. (Deutschland), Anat. Inst. *Konstitutionsanat. Entwicklungsmechanik. Extremitätentransplantationen.* Ⓓ Würzburg 1920. Ⓟ *Amphibien.*

Brandwijk, Marie Gerdina (1899), Dr. phil., Junior Lect. in Pharm., Univ. of the Witwatersrand. Johannesburg (South Africa), Pharm. Labor. *Botan. and Chem. investigation of South African indigenous med. and poisonous plants.* Ⓓ Utrecht 1923.

Branham, Sara Elizabeth, Instr. in Bact., The Univ. of Chicago, Ill. (U.S.A.), Dept. Hygiene and Bact. *Immunol., Physiol. of Bact. Toxins of the paratyphoid group of bacteria.* Ⓓ Univ. of Chicago 1923.

Branner, John Casper, Ph.D., LL.D., Sc.D., President Emer. of Leland Stanford Junior Univ., California (U.S.A.). o

Brannon, James M., Research Prof. in Bact., Dept. of Dairy Husbandry, Univ. of Illinois. Urbana, Ill. (U.S.A.). *Physiol., cytol., bact.* o

Brannon, Melvin Amos (1865), Chancellor of the Univ. of Montana. Helena, Mtn. (U.S.A.), State Capitol. *Physiol. Botan.* Ⓓ Univ. of Chicago 1912.

Brash, James Couper (1886), Prof. of Anat., Univ. Birmingham (England), Arden Lodge, Dorridge. *Human Embryol. Physical Anthrop. The Growth of the Jaws and the Skull.* Ⓓ Edinburgh 1910.

Brassler, Karl (1897). Berlin O 34 (Deutschland), Kopernikusstr. 14 b. Kressler. *Melolonthidae. Angewandte Entomol. Pelzschädlingsbekämpfung. Dasselfliegenplage.* Ⓟ *Melolonthiden. Schädlinge und entspr. biol. Objekte.*

Bratley, Cyril Oliver (1903). Ithaca, N.Y.(U.S.A.), Cornell Univ. *Small fruit diseases.*

Braucher, Ralph Waldo (1871), Entomol. Chicago, Ill. (U.S.A.), 10 North Clark Street. *Fruit and shade tree insects.*

Braun, Alfred (1894), Dr. med., Wissenschaftl. Hilfsarbeiter im Reichsgesundheitsamt. Charlottenburg (Deutschland), Clausewitzstr. 3. *Heilmittel, Nahrungsmittel, Genußmittel.* Ⓓ Leipzig 1922.

Braun, Emma L., Ass. Prof. of Botan., Univ. of Cincinnati, O. (U.S.A.), 2702. May St. *Ecol.* o

Braun, Hans (1896), Ass., Biol. Reichsanstalt Berlin-Dahlem, Labor. f. angewandte Vererbungslehre. Berlin-Steglitz (Deutschland), Ganghoferstr. 2. *Morphol., Physiol. u. Züchtung der Kartoffel.* Ⓓ Berlin 1925.

Braun, Harry, Ass. Pathol., Bureau of Plant Industry, U. S. Dept. of Agric. Washington, D.C. (U.S.A.). *Physiol., Mycol., Pathol.* o

Braun, Hugo (1881), ao. Prof., Dr. med., Vorsteher der bact.-hygien. Abteilung des Hygien. Inst. der Univ. Frankfurt a. M. (Deutschland), Paul-Ehrlich-Str. 40. *Bact., Parasitol., Immunitätslehre, Hygiene. Serumfestigkeit der Bact. Immunität gegen Trypanosomen. Tierische Trypanosomenkrankheiten in Deutsch-Ostafrika. Ekto- und Endoplasma der Bact. Verwendungsstoffwechsel pathogener Bact. Prüfungsmethoden der Wundantiseptica im Tierexperim. Theorie der Wassermannschen Reaktion. Colitisbazillen.* Ⓓ Prag 1907. Ⓟ Bacterienkulturen.

Braun, John William, Ass. Prof. of Plant Pathol., Univ. of Wisconsin, Horticult. Bldg. Madison, Wis. (U.S.A.). o

Braun, Karl (1870), Reg.R. Prof. Dr., Leiter der Zweigstelle der Biol. Reichsanstalt f. Land- und Forstwirtschaft. Stade, Prov. Hannover (Deutschland), Harsefelder Str. 57a. *Pflanzenkrankheiten und ihre Bekämpfung. Tropische Kulturpflanzen speziell aus dem früheren Deutsch-Ostafrika. Geschichte der Kultur- und Medizinalpflanzen.* Ⓓ Basel 1900.

Braun, Maximilian Gustav Christian Carl (1850), Prof. emer. der Zool. und vergl. Anat. Königsberg, Pr. (Deutschland), Cäcilienallee 7. *Helminthol. Trematodes. Cestodes.* Ⓓ Würzburg med. 1874, phil. 1877.

Braun, Wilhelm (1898), Ass. Weihenstephan bei München (Deutschland). *Biochem. der Kohlehydrate.* Ⓓ Techn. Hochsch. München.

Braun-Blanquet, Josias (1884), Priv.Doc. für Botan. an der Eidgenöss. Techn. Hochsch. in Zürich (Schweiz). Montpellier, Herault (Frankreich), 44, Rue Pont de Lattes. *Pflanzensoziol. Systematik der Phanerogamen. Florengeschichte. Vegetationsverhältnisse der Schneestufe. Flores dans le Massif Central de France.* Ⓟ Phanerogamen der Alpen und der Mittelmeerländer.

Brauner, Alexander (1857), Prof. Zool. Odessa (U.d.S.S.R.), Landwirtschaftl. Inst. *Palaeont. der quartären Wirbeltiere. Vergl. Anat. der Wirbeltiere. Fauna Südrußl.* Ⓓ Odessa 1920 h. c.

Brauner, Leo (1898), Dr. phil., Priv.Doc. für Botan., 1. Ass. am Botan. Inst. der Univ. Jena (Deutschland). *Reizphysiol. der Pflanzen, Elektrophysiol. der Pflanzen.* Ⓓ Jena 1922.

Bravo, Helia (1903), Prof. de Botan., Ass. de Histol. México, D.F. (Mexico), Dr. Carmona y Valle 1. *Protozoares Mexicaines: Flagellata.* Ⓓ Mexico 1922. Ⓟ Préparations microscopiques de Protozoaires.

Bray, William L. Prof. of Botan. and Head of the Department of Botan., Syracuse, N.Y. (U.S.A.), Univ. *Geographical distribution of plants. Tissues of plants. The development of the vegetation of New York State and Texas.* ⓐ Chicago 1898.

Brecher, Leonore (1886), Dr. phil. Chem. Abt. des Pathol. Inst. Univ. Berlin. Berlin-Charlottenburg (Deutschland), Fredericiastr. 13 III, bei Dr. H. H. Weber. *Experim. Biol.: Regeneration, Farbanpassung bei Schmetterlingspuppen (physico-chem., experim.-biol.).* ⓐ Wien 1916.

Bredemann, Gustav (1880), Dr. phil., Prof. und Dir. des Inst. für Pflanzenzüchtung der Preußischen Landwirtschaftl. Versuchs- und Forschungs-Anstalten. Landsberg a. d. Warthe (Deutschland), Theaterstr. 25. *Vererbungsstudien für praktische Züchtung. Vervollkommnung der bei den Züchtungsarbeiten anzuwendenden Methoden.* ⓐ Marburg a. L. 1908. ⓟ Sämereien fast aller landw. Kulturpflanzen in ihren versch. Sorten u. Varietäten.

Breder, Charles M. (1897), Research Associate, New York Aquarium. Research Associate in Ichthyol. (Honorary) American Mus. of Natural History. New York City (U.S.A.), Battery Park. *Problems in the adaptations of fishes to their environment. Life history studies. Amphibia. Food of Cyprinids. Life history of Sciaenidae. Locomotion of Fishes. Marine fishes of N.Y. and Southern New England. Fishes of the Rio Chucunaque.*

Breed, Robert Stanley (1877), Chief in Research (Bact.) New York State Agric. Experim. Station. Geneva, N.Y. (U.S.A.) *Taxonomy and Dairy Bact.* ⓐ Harvard 1902. ⓟ Bacteria of the Bacillus prodigiosus type.

v. Breemen, P. J. (1875). Chief Sugar experim. Station Pasoeroean (Nederl.-O.-Indië). *Entomol.* ⓐ Amsterdam 1905.

Bregger, John T. (1896), Chief of Research Dept., Stark Bro's Nurseries Louisiana, Mo. (U.S.A.). *Fruit varieties; propagation; Scion-rooting. Diseases of fruits. Propagation of Citrus fruits (California) and compar. value of stocks (citrus).*

Brehm, Vincenz (1879). Eger (C.S.R.). *Systematik der Süßwasser-Entomostraken. Tiergeographie des Süßwassers. Copepoda und Branchiura.* ⓐ Innsbruck 1902.

v. Brehmer, W. (1888), Laboratoriumsvorsteher der Biol. Reichsanstalt für Land- und Forstwirtschaft, Pflanzenanat. Labor. Berlin-Dahlem (Deutschland), Altensteinstr. 30. *Pflanzenanat. u. Mikrochem.* ⓐ Berlin 1915.

Breier, S., Dr., Dir. d. Inst. für Serol., Staatl. Bact. Inst. Sofia (Bulgarien), Buleward Makedonia. o

Breindl, Václav (Wenzel) (1890), Dr., Priv.Doc. der Zool. an der Karls-Univ. in Prag II (C.S.R.), Ú. Karlova 3. *Protozool. und Parasitol. Parasitische Protozoen, vor allem Trypanosomen, Malariaplasmodien und Entamoeben. Plathelminthes und Nemathelmintes.* ⓐ Prag 1912. ⓟ Tierische Parasiten.

Breinl, Friedrich (1888), Priv.Doc., Leiter serol. Abt. Hygien. Inst. deutsch. Univ. Praha II (C.S.R.), Albertov Presslová 7. *Mutation der Bakterien. Invisible Virusarten.* ⓐ Prag 1913.

Breisacher, Leo (1873), Consulting Gastro-Enterologist, Harper Hospital. Detroit, Mich. (U.S.A.), 723 D. Whitney Bldg. *Gastro-Enterol., Ernährungskrankheiten. Schilddrüse; Eiweißbedarf d. gesunden Menschen; Physiol. des Schlafes; Nervus laryngeus.* ⓐ 1892.

Breitenbecher, Joseph Kumler (1884), Ph.D., Lect. in Zool., Dept. of Zool., Montreal, Quebec (Canada), McGill Univ. *Insects and their variation and heredity, behavior, embryol., cytol., evolution, reaction to environment. Bruchus quadrimaculatus sex-limited inheritance, polymorphism, allelomorphism, somatic mutations, elytral mosaics, unilateral inheritance, bilateral asymmetry, piebal traits, fecundity, semisterility, abnormal sex ratios, mutations, responses to environment, desert insects. Leptinotarsa decemlineata.* ⓐ Univ. of Chicago 1913. ⓟ Living cultures of Bruchus quadrimaculatus Fabr., of Bruchus chiensis and Zabrotes pectoralis, Bruchidae (Mylabridae).

Breitfuss, Leonid (1864). Berlin-Schöneberg (Deutschland), Stubenrauchstr. 1. *Marine Biol. und Fischerei, Spongiol. (Calcarea), Ozeanographie.*

Bremekamp, Cornelis Elisa Bertus (1888), Ph.D., Prof. of Botan., Transvaal Univ. Coll. Pretoria (South Africa). *Movement of plants.* ⓐ Utrecht 1912.

Bremer, Friedrich Wilhelm (1894), Dr. med., Ass.-Arzt an der I. Med. Klinik München (Deutschland), Ziemssenstr. *Neurol., Vererbungslehre a. d. Gebiet der Neurol.* ⓐ Göttingen 1921.

Bremer, Gustav (1891), Dr. in agric. sc., Botan. at the Sugar experim. station Pasoeroean, Java (Nederl.-Indië). *Cytol. of sugar-cane.* ⓐ Wageningen 1921.

Bremer, Hans (1891), Dr. phil., wissenschaftl. Angestellter der Biol. Reichsanstalt für Land- und Forstwirtschaft. Kiel (Deutschland), Niemannsweg 13. *Angewandte Entomol.: Rübenschädlinge.* ⓐ Breslau 1922.

Bremer, John Lewis (1874), Associate Prof. of Histol. Boston, Mass. (U.S.A.), Harvard Med. School. *Development of blood vessels, of nerves. Mechanics of development.* Harvard 1901.

Brenčēns, Voldemars, Dr., Doc. Univ. Rīga (Latvija), Brīvības ielā 144, dz. 3. *Anat. d. Tiere.* o

Brenchley, Winifred Elsie (1883), Dr., Head of Botan. Department Rothamsted Labor., Harpenden, Herts. (England), Rothamsted Experim. Station. *Plant nutrition. Effect of special elements on growth. Plant poisons.* ⓐ London 1911.

Brenner, Magnus Widar (1887), Doc. an der Univ. Agrogeol. am staatl. Inst. für Bodenforschung. Helsingfors (Finnland), Bergmansgatan 15. *Pflanzenphysiol., besonders der Zelle. Pflanzenökol.: Verhältnis zwischen Vegetation und Boden.* ⓐ Helsingfors 1915.

Brentzel, Wanda Weniger (1895), Department of Plant Pathol. State Coll. Station. Fargo, N.D. (U.S.A.). *Cytol. of rusts. Cereal Diseases.* ⓐ Univ. of Chicago 1918.

Brentzel, William Edward (1889), Plant Pathol. Fargo, N.D. (U.S.A.), North Dakota Experim. Station, State Coll. Station. *Flax diseases, Cereal smuts.* ⓐ State Coll. Station 1924.

Breslavec, Lidya P., Doc. II. Univ. Moskau (U.d. S.S.R.). *Botan.* o

Bressau, Ernst (1877), Dr. med. et phil., o.ö. Univ.Prof. der Zool., Dir. des zool. Inst. der Univ. Köln. Köln-Bayental (Deutschland), Hölderlinstr. 7. *Vergl. Entwicklungsgeschichte der Turbellarien u. der Mammalia (Milchdrüsenapparat). Biol. der Stechmücken und anderer medizinischer Schädlinge. Bedeutung der physikochem. Umweltfaktoren für das Leben der niederen Organismen (speziell der Protozoen), Wasserstoffionenkonzentration.* ⓐ Straßburg, Elsaß 1902.

Bressou, Clément Jean Pierre (1887), Prof. Ecole Nation. Vét. d'Alfort, Seine (France). *Anat.; Histol. et Embryol. compar.* ⓐ Toulouse 1917. ⓟ Prép. anat. et histol. de Mammifères et d'oiseaux.

Breteau, M. P., Prof., Ecole du Val-de Grâce. Paris 5 (France), 277bis, rue St. Jacques. *Pharm.* o

Brethes, Juan, Prof. Univ. Nac. La Plata (Argentina). *Zool.* o

Bretnall, George Herbert (1871), Head of the Department of Biol., Baker Univ. Baldwin City, Kan. (U.S.A.), Box 227. *Zool., Compar. Anat., Physiol., Geol.* ⓐ Cornell Coll. 1897.

Bretscher, Konrad (1858), Dr. Zürich 6, (Schweiz), Weinbergstr. 146. *Vogelzug, Oligochäten der Schweiz.* ⓐ Zürich 1899.

Bretschneider, Ludwig Hermann (1899), Ass. a. d. Abt. f. experim. Histol. d. zool. Labor. Utrecht (Holland), Janskerkhof 3. *Sexualität bei Protisten. Cytol. des Flimmerepithels. Cytol. der Ciliaten. Plasma. Plasmosomen.* ⓐ Cytol. Praep.

Breuil, Henri Edouard Prosper (1877), Prof. d'Ethnographie, Inst. de Paléont. humaine. Paris (France), 1, Rue René Pernhard. *Civilisations paléolithiques. Art paléolithique et néolithique.*

von Breuning, Stephan (1894), Dr. phil. Wien IV (Österreich), Prinz-Eugen-Str. 18. *Carabini (Fam. Carabidae, Insecten, Coleoptera).* ⓟ Wien 1920. ⓟ *Carabini der Welt (Fam. Carabidae, Col.).*

Brewbaker, Harvey E. (1895), Instr. and Ass. St. Paul, Minn. (U.S.A.), Univ. Farm. *Corn and rye breeding and genetic.* Univ. of Minnesota 1926.

Brezina, Ernst, ao. Univ. Prof. Wien I (Deutsch-Österreich), Börseplatz 3. *Hygiene.* o

Brian, Alessandro G. G. (1873), Aiuto volontarie dell'Istit. Zool. R. Univ. di Genova (Italia), 5 Corso Firenze 5. *Copépodes parasit. et libres. Isopodes.* ⓟ Genova 1897.

Brick, Marie J., Ass., Univ. zu Taschkent, Turkestan (U.d.S.S.R.), Geol. Kabinett der Mittelasiat. Univ. *Juraflora.* o

Bridges, Calvin Blackman (1889), Ph.D., Research Ass., Carnegie Inst. Leonia, N.J. (U.S.A.), 475 Grand Ave. *Sex-limited inheritance of Drosophila, mutation, Mendelian inheritance, lineage, determination of sex.* ⓟ Columbia Univ. 1916. o

Brien, Paul (1894), Dr. sc. nat., Ass. de Zool. à la Fac. des Sc., Chargé du cours de Biol. à l'école des sc. politiques et sociales de l'Univ. de Bruxelles (Belgique), Labor. de Zool., rue Ad. Buyl, Univ. libre. *Tuniciers; Embryol. Bourgeonnement, Morphol.* ⓟ Bruxelles 1922.

Brierley, Philip (1899), Junior Pathol. Washington, D.C. (U.S.A.), U.S. Bureau Plant Industry. *Diseases of Irish Potato, of Herbaceous Ornamental Plants.*

Brierley, Wilfrid Gordon (1885), Associate Prof. of Horticulture. St. Paul, Minn. (U.S.A.), Univ. Farm. *Pomol.: economic and physiol. phases.* ⓟ Washington State Univ. 1913.

Brierley, William Broadhurst (1889), Head of the Department of Mycol., Harpenden (England), Rothamsted Experim. Station. *Mycol. and Plant Pathol., Genetics. Fungi: Actinomycetes of the Soil, genetics of Fungi and Bacteria.* ⓟ Manchester 1920.

Briest, Franz (1864), Ehrensenator der Univ. Greifswald. Boltenhagen, Post Grimmen, V.-Pommern (Deutschland). *Tierische Schädlinge der Landwirtschaft.* ⓟ 1923.

Briffa, J., Ass. Prof. Univ. of Malta. La Valletta (Malta). *Anat.* o

Briggs, Fred N. (1896), Ass. Pathol. and Associate in Agronomy in the Experim. Station, Univ. of California. Berkeley, Cal. (U.S.A.), 124 Hilgard Hall. *The inheritance of resistance in wheat.* ⓟ Univ. of California 1925.

Briggs, George Edward (1893), Univ. Lect. in Plant Physiol. Cambridge (England), Botan. School. *Photosynthesis, Plant respiration, Enzyme Action.* ⓟ Cambridge 1915.

Brigham, Carl Campbell (1890), Associate Prof. of Psychol. Princeton, N.J. (U.S.A.), Univ. *Mental Tests.* ⓟ Princeton Univ. 1916.

Brigl, Percy (1885), Prof. f. Agric.-Chem. a. d. landw. Hochsch., Dr. phil., Vorstand d. landw. Versuchsstation. Hohenheim b. Stuttgart (Deutschland). *Tierernährung, Chem. der Kohlenhydrate u. des Eiweiß.* ⓟ Berlin 1909.

Brilliant, Warwara Alexandrowna (1888), Ass. Sektion der Pflanzenphysiol., Botan. Garten. Leningrad (U.d.S.S.R.), Snamenskaja 19. *Physiol. und Anat. der Pflanzen. Photosynthese und Wassergehalt der Pflanze. Permeabilität der Zelle. Farbstoffe der Pflanzen.*

Brimble, Lionel John Farnham (1904), Lect. Botan. Univ. Glasgow (Scotland). *Plant Physiol.: Permeability and Photosynthesis.* ⓟ London 1925.

Brimley, Clement Samuel (1863), Ass., Division of Entomol., North Carolina Department of Agric. (in charge of the Insect Collection). Raleigh, N.C. (U.S.A.). *Fauna of North Carolina, Amphibians, Reptiles, Birds and Insects: Odonata, Diptera and Hymenoptera (Wasps).*

Brindley, Harold Hulme (1865), Lect. app. by the spec. Board of Indian Civ. Service Stud. Cambridge (England), St. John's Coll. *Regeneration and variation in Insects.* ⓟ Cambr. 1887.

Brink, Royal Alexander (1897), Ass. Prof. of Genetics, Univ. of Wisconsin. Madison, Wis. (U.S.A.). *Nature of the gene; role of nucleus and cytoplasm in heredity; pollen tube growth; plant breeding.* ⓟ Harvard Univ. 1923.

Brinkmann, August (1878), Dr. phil., Prof. der Zool. (Inhaber der Sundtches zool. Lehrkanzel des Mus. zu Bergen), Dir. des zool. Mus. und der biol. Station des Mus. Bergen (Norwegen), Bergens Mus. *Kompar. Anat. der Haut der Säugetiere, Osteol. und Rassengeschichte der Haustiere. Systematik, Anat. und Biol. der Nemertinen und Turbellarien. Parasit. der Vögel.* ⓟ Kopenhagen 1911.

Briot, Augustin (1873), Prof. de Physiol., Inst. Catholique de Paris 6 (France), 74, rue de Vaugirard. *Ferments digestifs, (ferments coagulants), antidiastases.* ⓟ Paris 1908.

Briquet, John Isaac (1870), Dir. du Conservatoire et du Jardin botan. de la Ville de Genève (Suisse), Conservatoire botan., 192, Route de Lausanne. *Systématique des Phanérogames: Labiées; géobotan. et floristique des Alpes occidentales, des Alpes maritimes et de la Corse; règles internationales de nomenclature systématique Galeopsis, Cytides; Buplèvres; Centaurées.* ⓟ Genève 1891.

Briscoe, Chas. F. (1868), Prof. of Bact. and Chief in Bact. in the Agric. Experim. Station. A and M College, Miss. (U.S.A.). ⓟ Univ. of Illinois 1912.

Bristol, Charles L. (1859), Prof. of Biol., Emeritus. New York City (U.S.A.), New York Univ., Univ. Heights. ⓟ Chicago Univ. 1894.

de Brito Fontes, António (1893), Ass. Fac. Med. Inst. d'Anat. Lisboa (Portugal), Rua do Passadiço, 128-3° Dto. *Achondroplasie.* ⓟ Lisbonne 1922.

Britton, Wilton Everett (1868), Entomologist Conn. Agric. Experim. Station; State Entomol.; Superintendent State Geol. and Natural History Survey. New Haven, Conn. (U.S.A.). *Injurious insects of Conn.; Hemiptera of Conn.* ⓟ Yale Univ. 1903. *Insects.*

Britzian, Khoren, Prof. Staatsuniv. Erivan, Armenien (U.d.S.S.R.). *Tierzuchtlehre.* o

Brizi, Ugo, Prof. Univ. Milano (Italia). *Botan.* o

Broadbent, Bessie M. (1895), Junior Entomol. U.S.Department of Agric. Washington, D.C. (U.S.A.), Apartment 36, 1448 Park Road N.W. *Greenhouse Insects, Tropical and Subtropical Fruit Insects.*

Broadfoot, R. C., Dir. Agric. Coll. and Research Inst. Coimbatore (Br. India). o

Broch, Hjalmar (1882), Prof. aggrégé, Doc. für Zool. an der Univ. und Dir. der marinebiol. Univ. Station (Dröbek). Oslo (Norwegen), Sophus Liesgate 10 III. *Systematik, Biogeographie und Biol. von Coelenteraten (bes. Hydrozoen und Oktokorallen) und Cirripedien. Plankton (Peridineen). Fischereiuntersuchungen (Hering). Allgemeine marine Faunistik und Biogeographie.* ⓟ Oslo 1910.

Brockmann-Jerosch, Heinrich (1879), Prof., Dr. phil., Prof. an der Univ. Zürich (Schweiz). *Botan., Pflanzengeographie, Geographie. Flora des Puschlavs. Glaciale Pflanzenreste. Vegetation der Schweiz.* ⓟ Zürich 1905.

Brocq-Rousseu, M., Dir. du Labor. des Rech. Vét. Paris 14e (France), 21, Rue Montbrun. *Pathol.* o

Brode, Howard Stidham (1866), Prof. of Biol. and Curator of Whitman Coll. Mus., Whitman Coll., Walla Walla, Wash. (U.S.A.), 433 E. Alder Street. *Annelids.* ⓟ Univ. of Chicago 1896.

Broderick, Ralph Alexander (1888), Lect. in Bact. to Dental Students, Univ. of Birmingham (England), 141 Great Charles Street. *Bact. of the Mouth.* ⊕ Birmingham 1924.

Brodersen, Johannes Hermann (1878), Prof. extraordinarius am Anat. Inst. der Univ. Hamburg (Deutschland), Uhlenhorster Weg 31. *Experim. Histol. von Blut, Knorpel, Epithel.* ⊕ Rostock 1904.

Brodsky, Jacob (1882), Ass. Allgem. Pathol. Univ. Rostow am Don (U.d.S.S.R.), Dmitrijewskaja Str. 126. *Resorptionsfähigkeit der Pleurahöhle.*

Brodsky, Lvovitsch Abráham (1883), Prof. d. Fac. Sc. phys. et nat. et Fac. Sc. agr. et Fac. med. Univ. Taschkent, Usbekistan (U.d.S.S.R.). *Protozoaires; Rhizopodes et Infusories ciliées, Insects de Torrent; Protozoaire et insects du sol.* ⊕ Sc. Genève 1908, Moscou 1909. ℗ Invertébrés de l'Asie centrale (Insectes, Araignés, Myriapodes).

Brody, Samuel (1890), Ass. Prof. Columbia, Mo. (U.S.A.). *Growth, senescence, energy. Metabolism, Milk Secretion.* ⊕ Univ. California 1917.

v. d. Broek, A. J. P., o. Prof., Dir. d. Anat. Labor. Utrecht (Holland), Janskerkhof 3. *Genitalorgane.* ⊕ Med. h. c. Amsterdam 1923. o

TenBroeck, Carl (1885), Prof. of Bact. and Head of the Dept. of Pathol. Peking (China), Union Med. Coll. *Bact. Tetanus, paratyphoid bacilli, dysentery, and diseases of animals.* ⊕ Harvard Univ. 1913. ℗ Cultures of pathogenic organisms.

Broemser, Philipp (1886), Dr. med., o.ö. Prof. für Physiol., Dir. der physiol. Anstalt der Univ. Basel (Schweiz), Vesalgasse 1. *Allgemeine Nervenphysiol., Methodik der Kreislaufforschung, Dynamik des Blutkreislaufs, Physiol. Akustik.* ⊕ Marburg a. d. L. 1911.

Brofeldt, Pekka (1884), Fischereirat. Helsinki (Finland), Gengatan 1 A. *Fischereibiol.*

Brohmer, Paul (1885), Dr. phil., Doc. an der Pädagogischen Akademie Kiel (Deutschland), Karlstr. 38 III. *Säugetiere. Fauna von Deutschland.* ⊕ Jena 1909.

Brolli, Ferdinand (1874), Dr., o. Prof. a. d. Univ. u. Dir. der bayer. Staatssammlung für Palaeont. u. histor. Geol. München (Deutschland), Neuhauser Str. 51. *Fossile Amphibien (Stegocephalen) u. Reptilien.* ⊕ München 1898.

Broman, Ivar (1868), Prof., Dir. Anat. Inst. Lund (Sverige), Rosenvillan. *Embryol. der Wirbeltiere, speziell die Menschen.* ⊕ Lund 1899. ℗ Wirbeltierembryonen, besonders jüngere Stadien.

Bromley, Stanley W. (1899), Entomol. New York City (U.S.A.), 511, Fifth Avenue. *Asilidae: Diptera. Odonata.* ⊕ Amherst, Mass., 1924. ℗ Asilidae.

Brøndsted, Holger Valdemar (1893), Magister. Birkerød (Danmark). *Zool. Porifera.* ⊕ Copenhagen.

Bronfenbrenner, Jacques J. (1883), Associate Member of the Rockefeller Inst. New York (U.S.A.). *terum diagnosis of Echinococcosis. Syphilis and Suberculosis. Lysis of Tubercle bac. in vivo. Ex-Terim. Syphilis in rabbits. Serum ferments. Abderpalden reaction (Mechanism). Nature of Anaphylahoxin. Methods of study of bact. fermentations. Botulism. Nature of the Bacteriophage phenomenon. Respiration of filtrable viruses.*

Bronsow, Alexander Ja., Doc. MoskauerVet.-Inst., Dir. Staatl. Inst. f. Wiesenbau. Dorf Sucharewo, Moskauer Gouv. (U.d.S.S.R.), Station Katuar, Sawel. sh. d. *Pflanzen-Geographie. Geobotan.* o

Brooks, Albert Nelson (1897), Ass. Plant Pathol. in charge of Strawberry Disease Investigations. Plant City, Florida (U.S.A.). *Plant pathol. strawberry diseases in Florida.* ⊕ Univ. of Wisconsin 1296.

Brooks, Allan (1869). Okanagan Landing, Brit. Col. (Canada). *Ornithol.* ℗ Bird and Mammal Skins.

Brooks, Charles (1872), Senior Plant Pathol. U.S. Dept. Agric. Washington. D.C. (U.S.A.). *Fruit Diseases. Storage. Transportation Diseases. Non-parasitic Diseases of Apples.* ⊕ Univ. Missouri 1908.

Brooks, Frederick Tom (1882), Univ. Lect. in Botan., Cambridge (England), Botan. School. *Plant Pathol.* ⊕ Cambridge 1905.

Brooks, Matilda Moldenhauer, Associate Biol. Washington, D.C. (U.S.A.), Hygien. Labor. *Cellular physiol.: respiration, permeability of living cells, oxidation-reduction phenomena.* ⊕ Harvard Univ.1920.

Brooks, Sumner Cushing (1888), Prof. of Physiol. and Biochem., Rutgers Univ. New Brunswick, N.J. (U.S.A.), New Jersey Hall. *Growth, permeability, electrical conductance of cells.* ⊕ Harvard Univ. 1916.

Brosch, Nikolai (1892), Versuchs- u. Pflanzenzuchtstation des Lettländischen Landwirtschaftl. Zentralvereins. Priekulu selekcijas stacija bei Cēsis (Latvija). *Gerste, Hafer.* ⊕ Moskau 1917.

Brotherus, Viktor Ferdinand (1849), Prof. im Ruhestand. Helsingfors (Finland). *Systematik der Laubmoose.* ⊕ Helsingfors 1873. ℗ Laubmoose.

Brotzu, Giuseppe (1895), Aiuto dell' Ist. di Igiene della R. Univ. di Bologna (Italia). *Hygiene. Malaria. Bact.* ⊕ Cagliari 1919.

Broun, Goronwy Owen (1895), Associate Prof. of Med., St. Louis Univ. School of Med. 1402 S. Grand Ave. St. Louis, Mo. (U.S.A.), St. Mary's Infirmary, 1536 Papin St. *Pathol. and Physiol. of the Blood, Liver.* ⊕ St. Louis, Mo. 1918.

Brouwer, Gerrit (1900), Dr. rer. nat. Haag (Holland), van Weede van Dijkveldstraat 33. *Psychol. der höheren Tiere.* ⊕ Utrecht 1926.

Browicz, Thaddée, Prof., Dr. med., Dir. Inst. d'Anat. pathol. Univ. Kraków (Pologne). o

Browiński, J., Dr. phil., Prof. Poznań (Pologne), Univ., r. św. Pawła 10. *Biochim.* o

Brown, B. A., Doc. in Botan., Bernice Panahi Bishop Mus. Honolulu (Hawaii). *Plant Anat., Ecol.* o

Brown, Gordon G. (1887), Horticulturist, Hood River Experim. Station. Hood River, Ore. (U.S.A.). *Influence of shade crops and fertilizers on growth and production, apples and pears. Small fruits.*

Brown, Harold Duke (1892), Teacher on Staff Univ., Biol. Dept. Chengtu, Szechwan (China), West China Union Univ. *Genetics, Plant Pathol.* ⊕ Cornell.

Brown, Harry Bales (1876), Prof. of Cotton Investigations, La. State Univ. Baton Rouge, La. (U.S.A.), 910 America St. *Plant Breeding, Cotton.* ⊕ Cornell Univ. 1910.

Brown, Harry Philip (1887), Prof. of Wood Technol. Syracuse, N. Y. (U.S.A.), New York State Coll. of Forestry. *Structure and Identification of Wood, Physical Properties of Wood. Dendrol. Physiol. of Tree Growth.* ⊕ Cornell Univ., Ithaca, N.Y. (U.S.A.), 1914. ℗ Authentic Wood samples. Herbarium Material, trees of N.E. United States.

Brown, Henry Coddington, Bact. Wellcome Bureau of Sc. Research. London W.C. 1 (England), 25/28, Endsleigh Gardens. ⊕ Cambridge 1902.

Brown, James Greenlief (1880), Prof. of Plant Pathol., Univ. of Arizona; Agric. Experim. Station. Tucson, Ariz. (U.S.A.), Univ. Station. *Plant Pathol.: resistance of cotton to certain diseases.* ⊕ Univ. of Chicago 1925.

Brown, Mabel M., Ass. Prof. of Botan., Univ. of New Hampshire. Durham, N.H. (U.S.A.). *Morphol., cytol.* o

Brown, Marguerite Wessels (1895), Staff of Med. Fac. Chengtu, Szechwan (China), West China Union Univ. *Bact.* ⊕ Toronto.

Brown, Nathan Clifford (1856). Portland, Maine (U.S.A.), 218 Middle Street. *Habits and distribution of North American birds.* ⊕ Baudoin (honorary) 1885.

Brown, Nellie Adalesa (1877), Associate Plant Pathol., Department of Agric. Washington, D.C. (U.S.A.), Bureau of Plant Industry. *Plant pathol.: bact. diseases of plants.* ⊕ Univ. of Michigan 1901.

Brown, Orville Harry (1875), Supt. public health of Arizona. Phoenix Ariz. (U.S.A.), 2026 N. 3 St. *Physiol. Pharm.* o

Brown, Percy Edgar (1885), Prof. of Soils, Iowa State Coll.; and Chief in Soil Chem. and Bact., and Associate in Charge, Soil Survey, Iowa Agric. Experim. Station. Ames, Ia. (U.S.A.), Soils Dept. *Soil Bact.: Nitrification, Ammonification, Nonsymbiotic and Symbiotic Nitrogen fixation, Sulfofication, Molds, Actinomycetes, Rhiz. Leg. Studies, Bact. Changes in Field Soils. Soil Chem.: Phosphorus Availability, Potassium Changes in Soils, Acidity. Classification Soils; Soil Fertility.* ⊕ Rutgers Coll., New Brunswick, N.J. 1912.
Brown, Robert Ankeny (1899), Collector of Marine specimens. Des Moines, Ia. (U.S.A.), Brown Marine Mus. *The study and collection of Marine life from sea coasts of Labrador, Nova Scotia, Maine, Florida, West Indies, Brazil and Argentina, Alaska, British Columbia, Washington, California, Gulf of Mexico and West coast of Mexico, Hawaiian Islands.* ⊕ Iowa 1923. ⊕ Mounted fish.
Brown, T. Graham, Prof. Univ. Coll. of S. Wales and M. Cardiff, Wales (England). *Physiol.* o
Brown, William (1888), Ass. Prof. (Plant Path.), Department of Botan. London (England), South Kensington, Imperial Coll. of Sc. and Technol. *Physiol. of Parasitism, Enzymes of Fungi, Technique of Culture of Fungi, Dialysis.* ⊕ London 1915.
Brown, William H., Dir., Bureau of Sc. Manila (Philip. Islands). *Physiol., Ecol., Morphol., Economic Botan.* o
Brown, Williamson J. (1901), Instr. in Entomol. Oklahoma Agric. and Mechanical Coll. Stillwater, Okla. (U.S.A.). *Taxonomy of Scarabaeidae.* ⊕ Univ. of Kansas 1923. ⊕ Coleoptera.
Browne, William Ward (1884), Ass. Prof. Biol. The Coll. of the City of NewYork (U.S.A.), 139th St. *Bact., Pathogenic, Diagnostic, Serol., Water, Oysters, Eggs, Fish.* ⊕ Brown Univ. Prov., R. I. (U.S.A.) 1912.
Brownell, Stanley J. (1893), Ass. Prof. Dairy Extension. State Coll., Penna. (U.S.A.). *Inbreeding sterility, fertility, and its relation to dairy production.*
Browning, C. H., Prof. Univ. Glasgow (Scotland). *Bact.* o
Browning, Harold William (1893), Prof. of Botan., Rhode Island State Coll., Kingston (U. S.A.). *Plant Physiol.: physiol. of the fungi.* ⊕ Univ. of Wisconsin 1920.
Brožek, Arthur (1882), Lect. in Genetics, Eugenics and Biometrics, Plantphysiol. Inst. Univ. Praha II (C.S.R.), 433 Benátecká Street, Charles Univ. *Eugenics, Biol. Statistics, Genetics in Mimulusplants.* ⊕ Prague 1907.
Brshesitzkij, Michail W., Prof. am Azerbaidshansker Polytechnikum u. an d. Univ. Vorsteher d. Samenkontrollstation. Baku (U.d.S.S.R.), Kooperatiwnaja 5. *Angew. Botan. Reis.* o
Bruce, John Ronald. Port Erin, Isle of Man (England), Biol. Station. *Physiol. and Biochem. of Marine organisms.* ⊕ Univ. of Liverpool.
Brüchatowa, Anna (1898), Laborant Inst. für experim. Biol. Moskau (U.d.S.S.R.), Woronzowo Pole 6. *Einfluß der physikalisch-chem. Faktoren des Außenmediums auf die phototaktischen Bewegungen der Süßwassertiere.*
Brücke, Ernst Th. (1880), o.ö. Prof., Dr. med. Innsbruck (Österreich), Schöpfstr. 41. *Nerv. Muskelphysiol.* ⊕ Wien 1904.
Brückner, Gerhard (1902), Dr. phil., Wiss. Hilfsarbeiter im Inst. für Müllerei an der Preußischen Versuchs- und Forschungsanstalt für Getreideverarbeitung und Futterveredelung zu Berlin. Berlin-Lichterfelde (Deutschland), Auguststr. 35. *Botan. Systematik. Getreideverarbeitung.* ⊕ Berlin 1926.
Brüel, Ludwig (1871), a. o. Prof. f. Zool. a. d. Univ. Halle-Wittenberg u. Kustos d. zool. u. entomol. Sammlungen d. zool. Inst. d. Univ. Halle a. d. S. (Deutschland), Domplatz 4. *Morphol. u. System. d. Heteropoden; experim. Entwicklungsgesch.*

d. Pulmonaten; System. (Rassenfragen) b. Lepidopteren (Rhopaloceren). ⊕ Leipzig 1897.
Brühl, P., Prof. Univ. Calcutta (Engl. India). *Botan.* o
Brues, Charles T. (1879), Associate Prof. of Economic Entomol. Boston, Mass. (U.S.A.), Bussey Inst. Harvard Univ., Forest Hills. *Entomol.* ⊕ Univ. of Texas 1901.
Brugués y Escuder, Casimiro, Dr. en Farm. y en Cien., Prof. aux. de la Fac. de Farm. Barcelona (España), Bruch, 44, 2. *Histol. végétale.* o
Brumfield, Mary E. (1902), Ass. Seed Analyst, New Jersey State Seed Labor.; Ass. in Botan. Department, New Jersey Coll. for Women. New Brunswick, N. J. (U.S.A.), 108 George Street. *Analysis of seed samples for purity.* ⊕ Virginia Polytechnic Inst. 1925.
Brumpt, Emile (1877), Prof. de Parasit. à la Fac. de Méd. Paris (France), 15, Rue de l'Ecole de Médecine. *Parasit. humaine et comparée.* ⊕ Paris.
Brun, L., Prof. Ecole d. Service d. Santé de la Marine. Bordeaux (France). *Anat. pathol. Bact.* o
de Brun, P., Prof. St. Joseph-Univ. Beirut (Syrien). *Histol.* o
Brun, Rudolf (1885), Dr. med., Priv.Doc. für Neurobiol. und Neurol. Zürich 1 (Schweiz), Theaterstraße 14. *Instinktbiol., vergl. Psychol. der Ameisen und der Insekten überhaupt, speziell der sozialen Insekten, der sozialen Hymenopteren. Vergl. Anat. des Insektengehirns. Die Raumorientierung der Ameisen.* ⊕ Zürich 1911. ⊕ Ameisen.
Bruner, Lawrence (1856), Prof. of Entomol., Univ. of Nebraska. Lincoln, Neb. (U.S.A.). *Entomol.*
Bruner, Stephan Cole (1891), Chief, Dept. of Phytopath. and Entomol., Agric. Experim. Station. Santiago de las Vegas (Cuba). *Phytopath.; economic entomol.; systematic entomol: Hemiptera.* ⊕ West Raleigh, N.C., 1912. ⊕ Hemiptera.
Brunett, Earl Louis (1898), Prof. Avian pathol., New York State Vet. Coll. at Cornell Univ. Ithaca, N.Y. (U.S.A.). *Diseases of poultry.*
Bruni, Angelo Cesare (1884), Prof. stabile, Dir. Ist. Anat. ed Istol. Ist. Sup. di Med. vet. Milano (Italia), Viale Monte Nero, 70. *Morfol. dell' apparecchio di conduzione, eccitamento del cuore dei mammiferi. Corpi genitali considerati come ghiandole endocrine. Sistema reticoloendoteliale.*
Brunies, Stefan (1877), Oberaufseher des Schweizerischen Nationalparkes. Basel (Schweiz), Oberalpstr. 11. *Pflanzengeographie, Flora Graubündens, Naturschutz.* ⊕ Breslau 1900.
v. Brunn, Karl Oskar Max (1852), Prof. Dr. phil., Kustos an Zool. Staatsinst. u. Zool. Mus., im Ruhestand. Hamburg (Deutschland), Sechslingspforte 6. *Orthopteren-Systematik.* ⊕ Leipzig 1884.
Bruno, Francesco (1897), V.-Dir. Giardino Coloniale di Palermo (Italia). ⊕ Palermo 1921.
Bruno, Giovanni (1893), Prof. Anat. Umana Univ. Camerino (Italia), Ist. Anat. *Anat. ed Embriol. umana e comparata: Miocardio. Istol. patol.* ⊕ Palermo1920.
Bruno, P., Prof. Univ. Napoli (Italia). *Bact.* o
Bruns, Hayo (1872), Prof. Dr. med., Dir. des Inst. für Hygiene u. Bact., Gelsenkirchen (Deutschland). *Trinkwasserhygiene, Abwasserhygiene, Infektionskrankheiten, Bakt., Ankylostomiasis, Bergwerkshygiene.* ⊕ Straßburg 1895.
Bruns, Oskar (1878), Prof., Dir. Med. Univ.-Poliklinik. Königsberg, Pr. (Deutschland), Paradeplatz 19 u. Drummstr. 31a. *Physiol. und Pathol. der Atmung und der Herztätigkeit.* ⊕ Tübingen 1903.
Brunson, Arthur Maxwell (1891), Associate Agronomist, U. S. Department of Agric. Manhattan, Kan. (U.S.A.), Agronomy Department, Kansas State Agric. Coll. *Corn genetics, selection within the selfed lines.* ⊕ Cornell Univ. 1923. ⊕ Seed corn.
Brunst, Victor (1903), Aspirant sc., research on Zool. Kiev (U.d.S.S.R.), Ul. Korolenko 48a-17. *Regeneration in Amphibia. The influence of the nervous system on regeneration.*

Brunton, Charles Eason (1892), Demonstrator in Physiol. Univ. Manchester (England), Med. School. *Experim. Physiol. and Pathol.*

Bruntz, Louis (1877), Prof. de Pharm. et de Zool. pharm. à la Fac. de Pharmacie de l'Univ. de Nancy (France). *Histophysiol.: organes excréteurs, phagocytaires et globuligènes des Arthropodes. Plantes officinales et Plantes à drogues médicamenteuses.* ① Nancy 1911.

Bruschettini, Alessandro (1868), Lib. Doc., Dir. e proprietario del Labor. di Terapia Sperimentale. Genova (Italia), Piazza Savonarola 7. *Terapia sperimentale: Preparazione di Sieri e Vaccini.*

Brussin, Alexander M., Doc. II. Univ. Moskau (U.d.S.S.R.). *Bact.* o

Bruun, Anton Frederik (1901), Ass. at „The Danish Commission for the Study of the Sea". Hellerup (Danmark), Strandvej 34 B. *Biol. and System of Eggs and young stages of Teleostean Fishes, esp. Heterosomata; Physiol. and Anat. of poisonous fishes; physiol. and morphol. influence of fresh water upon salt water organisms.* ① Copenhagen 1926.

de Bruyn, Helena L. G. (1885), Botanist at the Inst. voor Phytopath. Wageningen (Holland), Hinkeloordsche Weg 5.

Bruynoghe, Richard (1881), Prof. d'Hygiène et de Bact.. à l'Univ. de Louvain (Belgique), 96, Rue Vital Decorter 96. *L'immunité: La constitution de l'alexine et de l'opsonine. Agglutinogène et agglutinines — précipitogène, précipitines. Pluralité et complexicité des bactériophages. L'action du Radium en biol. Trypanosomes.* ① Louvain 1907.

Bryan, Edwin H. Jr. (1898), Entomol., Bernice P. Bishop Mus., Instr. in Entomol., Univ. Honolulu (Hawaii). *Mounting, Labelling and Identifying collections of insects made in connection with an entomol. survey of the Pacific Islands.* ③ Univ. of Hawaii 1924.

Bryan, George Smith (1879), Assoc. Prof. Botan., Univ. of Wisconsin. Madison, Wis. (U.S.A.), 803 State St. *Morphol. of plants.* ① Univ. of Chicago 1914.

Bryan, Mary Katherine (1877), Ass. Pathol. Washington, D.C. (U.S.A.), U. S. Department of Agric. *Bact. diseases of plants.* ① Stanford Univ. 1908.

Bryan, William Alanson (1875), Dir. Los Angeles Mus. Los Angeles Cal. (U.S.A.), Exposition Park. *Polynesian Natural History and Ethnol.* ③ Iowa State Coll. 1896.

Bryant, Harold Child (1886), Economic Ornithol. Univ. of California Mus. of Vertebrate Zool.: Dir. Bureau of Education and Research, California Fish and Game Commission. Berkeley, Cal. (U.S.A.). *Economic ornithology and mammal.* ① Univ. of Calif. 1913.

Bryce, Thomas H. (1862), Prof. Anat., Univ. Glasgow (England). *Human anat. Embryol. Early Development and Imbedding of the Human ovum. Physical Anthrop.* ① Univ. of Edinburgh 1886.

Bryn, Halfdan (1864), Dr. Trondhjem (Norge). *Rassenanthropol.*

Bryson, Harry R. (1892), Instr. in Entomol., Kansas State Agric. Coll. Manhattan, Kan. (U.S.A.). *Insects attacking the roots of staple crops, sorghums.* ⑦ Elateridae.

Bryiz, C. V., Prof. Dr., Landw. Hochsch. København (Danmark) Bülowsvej. *Forstwissenschaft.* o

Brzeziński, Joseph, Dr., Prof. Kraków (Pologne), Prądnik Czerwony. *Botan.* o

Bubák, Franz, Dr., Prof. der Pflanzenpathol. Praha-Vršovice 28 (C.S.R.), Phytopath. Inst. d. Cech. techn. Hochsch. o

Bublitschenko, Nikolaus L., Bergingenieur, Comité Géol. de Russie. Leningrad (U.d.S.S.R.), W.O., Sredny Prospekt, 72-B. *Palaeont.; speziell Brachiopoda.* o

Bucciardi, Giulio (1899), Aiuto Univ. Modena (Italia), Piazza S. Eufemia 4. *Fisiol. sperimentale.* ⑦ Modena 1925.

Buch, Hans Robert Viktor (1884), Doc. an der Univ., Amanuensis an dem Botan. Mus. Helsingfors (Finnland). *Systematik der Lebermoose.* ⑩ Helsingfors 1912. ⑫ Moose.

Buchalova, V. I., Doc. Univ. Voronez (U.d.S.S.R.). Universitätskaja 5. *Zool.* o

Buchanan, George (1882), Superintendent, Routine Division, South African Inst. for Med. Research. Johannesburg (S. Africa), Box 1038. *Spirochaetal Diseases. Streptococcal Nephritis. Bacillary Dysenterie in the Sub-tropics. Immunisation with Gold Chloride treated Vaccines.* ① Edinburgh 1918.

Buchanan, James William (1888), Ass. Prof. of Biol., Yale Univ. New Haven, Conn. (U.S.A.). *Physiol. of Regeneration: Planaria. Action of cyanide and anesthetics on oxidative metabolism.* ① Univ. of Chicago 1921.

Buchanan, L. L. (1893), Ass. Biol. Washington, D.C. (U.S.A.), U.S. Dept. of Agric. *Taxonomy of Curculionidae (Coleoptera).* ① Univ. of Iowa 1916. ⑫ Curculionidae.

Buchanan, Robert Earle (1883), Prof. of Bact. and Bacteriol. Agric. Experim. Station, Ames, Ia, Dean of Graduate Coll., Iowa State Coll. of Agric. and Mechanic Arts (U.S.A.). *Nomenclature and classification of bacteria. Physiol. of Bacteria.* ① Univ. of Chicago 1908 ⑫ Pure cultures of certain species of bacteria, particularly members of the colontyphoid series of bacteria.

Buch Andersen, Erik (1892). København (Danmark), Techn. Hochsch. *Chemo-dynamie bei Cancer.* ① København 1917.

Buchheim, Alexander (1888), Ass. f. Phytopath., Landwirtschaftl. Akademie. Moskau (U.d.S.S.R.), Petrowskoje-Rasumowskoje. *Biol. der parasit. Pilze: Uredineen, Erysiphaceen.* ① Bern 1915.

Buchholz, John Theodore (1888), Prof. of Botan., Univ. of Texas. Austin, Tex. (U.S.A.). *Embryol. of Seed Plants: Coniferales. Pollen-tube growth.* ① Chicago 1917. ⑫ Dissected pine embryo preparations.

Buchmann, Wilhelm Walter (1900), Dr. phil. Berlin-Dahlem (Deutschland), Biol. Reichsanstalt. *Systematik und Morphol. der Heteropoden. Histo-Physiol.: Oxydasen des tierischen Gewebes. Resorption und Sekretion.* ① Halle 1923.

Buchner, Paul Ernst Christof (1886), o. Prof. an der Univ., Dir. des zool. Inst. Breslau (Deutschland), Sternstr. *Geschlechtszellen, Symbiose zwischen Tieren und Pflanzen, Salpenentwicklung, Bryozoen, Brackwasserprobleme.* ① München 1909.

Buchstab, Lasar (1868), Prof. Forschungskatheder d. Morphol. u. Physiol. Odessa(U.d.S.S.R.), Troizkajastr. 32. ① 1891, 1925.

Buchwald, Johannes (1869), Institutsdir. und Prof. an der Preußischen Versuchs- und Forschungsanstalt für Getreideverarbeitung, Dir. des Inst. fü Müllerei, Berlin N 65 (Deutschland), Seestr. 11. ① Berlin 1893.

Buchwald, Niels Fabritius (1898), Ass., pflanzenpathol. Abt. København V (Danmark), Rolighedsvej 23. *Mykol. und Pflanzenpathol.* ⑫ Pilze, speziell Hymenomyceten.

Buckell, Edward Ronald (1889), Entomol. Vernon, B. C. (Canada). Court House, *Life histories and devise control measures for injurious insects Orthoptera.* ⑫ Canadian Orthoptera.

Buckman, Harry O. (1883), Prof. of Soil Technol., N.Y. State Coll. of Agricult., Cornell Univ., Ithaca, N.Y. (U.S.A.). ① Cornell Univ. 1912.

Buckmann, Adolf (1900), Dr. rer. nat., Ass. der deutsch. wissensch. Kommission f. Meeresforschung. Helgoland (Deutschland), Biol. Anstalt. *Fischereibiol. des Meeres. Biol. u. Systematik der Appendicularien.* ① Hamburg 1923.

Buckner, Garrett Davis (1885), Research Prof. of Chem., Kentucky Agric. Experim. Station. Lexington, Ky. (U.S.A.). *Biol. chem.* ① Princeton Univ. 1912. ⑫ White Leghorn chickens.

Budaházy, Imre, Dr., Stat. für Pflanzenschutz d. Ung. Min. f. Landw. Budapest II (Ungarn), Debröigasse 17. *Kartoffelkrankheiten.* ○

Buday, Kálmán, Prof. Dr., Dir. des Pathol. anat. Inst. No. I, Univ. Budapest (Ungarn). *Pathol.* ○

Budde, Hermann (1890), Dr. phil. Dahl bei Hagen, Westfalen (Deutschland). *Süßwasseralgen.* ⊕ Kiel 1923. ⊕ Algen.

Frh. v. Buddenbrock, Wolfgang (1884), o. Prof. der Zool. an der Univ. Kiel (Deutschland), Karolinenweg 9. *Vergl. Physiol.: statische Sinnesorgane, Lichtsinn, Atmung.* ⊕ Heidelberg 1910.

Buder, Johannes (1884), Dr. phil., o. Prof. an der Univ., Dir. des Botan. Inst. u. Gartens. Greifswald (Deutschland), Grimmer Str. 88. *Pfropfbastarde. Phototropismus und Geotropismus. Taxien. Purpurbakterien.* ⊕ Berlin 1908.

Budington, Robert A. (1872), Prof. of Zool., Oberlin Coll., Oberlin, Ohio (U.S.A.). *Endocrine Glands.* ⊕ Williams Coll. 1896.

Bücher, Hermann (1882), Berlin-Steglitz (Deutschland), Beymestr. 10. *Heuschreckenplage und ihre Bekämpfung.* ⊕ Leipzig; oec. publ. h. c. München.

Buell, Mary Van Rensselaer (1893), Associate, Dept. Physiol. Chem. Baltimore, Md. (U.S.A.), Johns Hopkins Medical School. *Metabolism, blood chem. and nutrition.* ⊕ Madison, Wis. 1919.

De Buen, Fernando (1896), Chef du Dept. de Biol. de la Dir. Général des Pêches. Madrid (España), Alcalá 31. *Ichthyol.* ⊕ Madrid 1917.

De Buen, Odón (1863), Prof., Biol. générale, Univ., Président Section Océanographie (Union international Geodésique et Géophysique). Madrid (España), Alcalá 31. *Océanographie biol. et appliqué à la pêche.* ⊕ Madrid 1885.

Buen y del Cos, Odón de, ex Senador, Prof., Dir. del Inst. Español de Oceanografiá, Catedrático de Biol. general de la Univ. Central. Madrid (España), Lagasca, 116. *Biol. marina.* ○

Buen y Lozano, Sadí de, Jefe de Sección del Inst. Alfonso XIII, Auxiliar de la Fac. de Med., Secretario de la Comisión Central de Trabajos Antipalúdicos. Madrid (España), Serrano 112. *Parasit. humaine.* ○

Bünger, Heinrich (1880), Dr. phil., Prof., Dir. d. Inst. f. Milcherz. a. d. Pr. Versuchsanst. f. Milchwirtsch. Kiel (Deutschland), Forstweg 16. *Tierzucht. Milcherzeugung. Tierernährung.* ○

von Büren, Günther (1889), Dr. phil., Beauftragter Doc. für Botan. und Mikrotechnik an der Hochsch. Bern (Schweiz), Marktgasse 46/II. *Pflanzliche Cytol. und Embryol.* ⊕ Bern 1915.

Bürger, Josef Bernhard (1886), Abteilungsleiter, Prof. Preuß. Landesanstalt für Wasser-, Boden- und Lufthygiene. Berlin-Dahlem (Deutschland), U. d. Eichen 77. *Wasser-Reinigung durch Filterung und dergl. Hygiene der Abwasser-Beseitigung.* ⊕ Bonn a. Rh. 1911.

Bürgers, Th. Josef, o. Prof. Med. Akad., Dir. d. Hygien. Inst. Düsseldorf (Deutschland). *Hygiene u. Bact.* ○

Bürgi, Emil (1872), o.ö. Prof. für med. Chem. und Pharm. Bern (Schweiz), Freiburgstr. 30. *Arzneikombinationen, Chlorophyll-, Schwefelwirkungen.* ⊕ Bern 1896.

Bürker, Karl (1872), o. Prof. der Physiol., Dir. des Physiol. Inst. Univ. Gießen (Deutschland), Friedrichstr. 17. *Resorption in der Leber, Blut, speziell körperliche Bestandteile, Haemoglobin, Kolorimetrie, Blutplättchen und Blutgerinnung, Wirkungen des Höhenklimas auf das Blut, Verteilung des Haemoglobins auf die Oberfläche von Erythrocyten, Thermodynamik des Muskels, Blut- u. Muskelgerinnung, Elektrotonus, Theorie der Narkose.* ⊕ Tübingen: scient. nat. 1897; med. 1900.

Büttokofer, Johann (1850), Dir. des Zool. Garten Rotterdam emer. Bern (Schweiz), Hallwylstr. 32. *Ornithol.* ⊕ Phil. h. c. Bern 1895.

Buglia, Cj., Prof. Univ. Pisa (Italia). *Fisiol. speriment.* ○

Buice, W. A., Prof. Univ. Waco, Tex. (U.S.A.). *Biol.* ○

Bujard, E., Prof. Dr. Genève (Suisse), Place Claparède 4. *Histol. Embryol.* ○

Bujas, Ramiro, Dr., Priv.Doc. f. experim. Psychol., Adjunkt am Physiol. Inst. Zagreb (S.H.S.), Salata. *Experim. Psychol.* ○

Bujor, Paul (1862), Prof. de Morphol. animale Fac. Sc. Univ., Dir. du Labor. de Morphol. Jassy (Roumănie), Strada Carol 41. *Anat. comparée. Faune des Lacs Salés de Roumanie.* ⊕ Genève.

Bujorean, Georg (1893), Wissenschaftl. Präparator Inst. für systematische Botan. Univ. Cluj (Romania), Grădina botanică, Str. Regală 26. *Pflanzensukzessionen. Oekol. der Epiphyten. Pflanzen- und Tierphänol. Pflanzenteratol.* ⊕ 1925.

Bukowski, G., Dr., Prof. Bochnia (Pologne). *Paléont.* ○

Bulatav, Emilio, Assoc. Prof. Univ. of the Philipp. Manila, Philippine Islands (U.S.A.). *Physiol.* ○

Bulger, Jacob W. (1887), Ass. Sc. (Insect Physiol.). Washington, D. C. (U.S.A.), Bureau of Entomol. *Physiol. of the honey-bee.* ⊕ Ohio State Univ. 1924.

Bull, Carroll Gideon (1884), Prof. of immunol., School of Hygiene and Pub. Health, John Hopkins Univ. Baltimore (U.S.A.), Windsor Hills, Montgomery Road. *Agglutination of bacteria in Vivo. Toxin, anti-toxin.* ○

Bull, Herbert O. (1903), Analyt. chemist. Plymouth (England), Marine Biol. Labor. *Fisheries. Biol. of Fishes.*

Bull, Lionel Batley (1889), Dir., South Australian Government Labor. of Pathol. and Bact. Adelaide (South Australia), Adelaide Hospital. *Pathol. and Bact. Compar. Pathol.* ⊕ Melbourne 1919. ⊕ Pathol. specimens; Bact. cultures.

Buller, Arthur H. K., Prof. of Botan., Univ. of Manitoba. Winnipeg, Manitoba (Canada). *Fungi, physiol. problems.* ○

Bulliard, Henri (1885), Préparateur d'Histol. à la Fac. de Méd., chef de labor. à la clinique gynécol. de l'hôpital Broca. Paris (France). *Cancer, caractères sexuels, sexualité.* ⊕ Paris 1910. ⊕ Préparations microscopiques d'Histol. ou d'Anat. pathol. de l'appareil génital.

Bulloch, William, Prof. Univ. London (England), South Kensington, S. W. 7. *Bact.* ○

Bumpus, Hermon O. (1862), consulting Dir. Buffalo Mus. of Sc., Buffalo, N.Y. (U.S.A.). *Embryol., Oecol., Variation.* ⊕ Clark Univ. 1890.

Bunak, Victor-Valerian (1891), Prof. der Anthropol. Univ., Dir. anthropol. Sektion d. Kab. z. Studium der Person. Moskau (U.d.S.S.R.), Mochovaja, 11. *Anthropometrie, Anthropogenetik, Rassenkunde, Morphol. des Schädels, des Gehirns. Anthrop. Typus der Tscheremissen und Moskvinen.* ⊕ Moskauer Univ. 1918.

Bungenberg de Jong, Hendrik Gerard (1893), Dr., o. Prof., Med. Chemie, Univ. Leiden (Holland). ⊕ Utrecht 1921.

Bunker, John Wymond Miller (1886), Associate Prof. of Physiol. and Biochem. at the Massachusetts Inst. of Technol. Cambridge, Mass. (U.S.A.). *Vitamines, enzymes, protein derivatives. Proteins from blood.* ⊕ Brown Univ. 1912.

Bunting, Charles Henry (1875), Prof. of Pathol., Univ. of Wisconsin, Medical School. Madison, Wis. (U.S.A.), Sc. Hall. *Haematol.* ⊕ Johns Hopkins 1901.

Bunyard, Percy F. (1870). Croydon, Surrey (England), 57 Kidderminster Rd. *Compar. Ool. Nidol. Economic Ornithol. Down and Nestfeathers of Anatidæ.* ⊕ Ool. Specimens.

Buresch, Iwan (1885), Dir. der wissenschaftl. Anstalten des Königs von Bulgarien (Zool. Garten, Naturhistorisches Mus., Entomol. Station, wissenschaftl. Bibliothek). Sofia (Bulgarien), königl.

Palais. *Lepidopteren-Fauna Bulgariens und Höhlen-Fauna Bulgariens.* ⓟ München 1911. ⓟ Höhlenfauna Bulgariens.

Burfield, Stanley Thomas (1889), Senior Lect. Univ. Liverpool (England), Zool. Dept. *Chaetognatha; Compar. Anat. of the Vertebrata.* ⓟ Liverpool 1920.

Van der Burg, Boke (1879), Prof. Lact. and Dairy sc. Wageningen (Holland).

von Burg, Gustav (1871). Olten (Schweiz). *Ornithol. und Mammal. von Mitteleuropa: Vogelzug. Säugetiere des Alpengebiets, alpine Subspecies.* ⓟ Bälge und Stopfpraep. Vögel und Kleinsäuger des paläarktischen Faunengebietes, Schädel, Skelette gleicher Provenienz.

Burgeff, Hans (1883), Dr. phil., Prof. der Botan. u. Pharmacognosie. Würzburg (Deutschland), Botan. Inst. *Pilze, Vererbung bei Haplositen. Mycorhiza. Systematik der Lepidopterengattung Zygaena. Wurzelpilze der Orchideen.* ⓟ Jena 1909. ⓟ Schmetterlinge der Gattung Zygaena Fab.

Burger, Owen Francis, Plant Pathol., Agric. Experim. Stat. Univ. of Florida. Gainesville, Flo. (U.S.A.), 325 N. Colson St. o

Burgess, Albert Franklin (1873), Entomol. Melrose Highlands, Mass. (U.S.A.). *Preventing spread of the gipsy moth.* o

Burgess, Edward Sandford (1855), Prof. (retired) of Hunten Coll., New York City. Yonkers, N.Y. (U.S.A.), 61 Ridge Road. *Botan., American Asters, Anthropol., Paleontol., Human Descent.* ⓟ Sc. Hamilton Coll. 1904, Ph. Columbia Univ., 1899. ⓟ Asters (herbarium specimens of Asters of the northeastern U.S.).

Burgwitz, Georg (1889), Ass. Botan. Garten, Inst. für Experim. Agronomie. Leningrad (U.d.S.S.R.). *Bakteriosen der Pflanzen u. Bodenbact.*

Burian, Richard (1871), o. Univ.Prof., Dir. physiol. Inst. Med. Fac. Beograd (S.H.S.), 92 Zrinskoga ulica. *Chem. u. physiol. Chem. der Purine u. d. Harnsäure, Exkretion, Nierenfunktion, Nerven- u. Muskelphysiol.* ⓟ Wien 1894.

Burkard, Otto (1876), Univ.Prof. Graz (Österreich), Freiheitsplatz 2. *Hygiene.* o

Burke, Victor (1882), Prof. and Head — Dept. Bact. — State Coll. of Washington. Pullman, Wash. (U.S.A.). *Bact. — variations and adaptation of the bacteria — antiseptic dyes.* ⓟ Stanford Univ. 1912.

Burkewitsch, Wladimir St., Landwirtschaftl. Akademie Timirjasew, Station f. Pflanzenernährung. Moskau (U.d.S.S.R.), Ermolajewskij per. 6, W. 7. *Pflanzen-Physiol., Physiol. d. Stoffwechsels.*

Burkey, Lloyd Allen (1896), Bact. Department. Ames, Ia. (U.S.A.). *Dairy or Soil Bact. Variations of the Legume Bacteria.* ⓟ Culture of Legume Bacteria (Alfalfa Organism).

Burkholder, Walter H. (1891), Ass. Prof. of Plant Pathol., Cornell Univ. Ithaca, N.Y. (U.S.A.). *Diseases of vegetables (Phaseolus vulgaris L.).* ⓟ Cornell Univ. 1917.

Burky, Earl L. (1898), Ass. in Ophthalmol., Johns Hopkins Med. School. Baltimore, Md. (U.S.A.), Wilmer Inst. of Ophthalmol. *Chem. and Serol. studies of the protein of crystalline lens of the eye. Bact. of the eye.*

Burlingame, Leonas L. (1876), Prof. Biol., Cytol., Genetics. Stanford Univ., Cal. (U.S.A.), Box 1455. *Genetics and Cytol. of Clarkia species.* ⓟ Chicago 1908. ⓟ Cytol. slides.

Burlingham, Gertrude S. Ass. Biology, Eastern district High School, Brooklyn, N.Y. (U.S.A.) 556 Lafayette Ave. *Russula, Lactaria. Dairy Science.* o

Burlison, William Leonidas (1882), Prof. Crop Production, Head Department of Agronomy Univ. of Illinois. Urbana, Ill. (U.S.A.), 212 Old Agricultural Building. *Crop Production. Plant Physiol.* ⓟ Univ. of Illinois 1915.

Burn, J. H. (1892), Dir. Pharmacol. Labor., Pharm. Soc. of Gt. Britain. London W.C. 1 (England), 17 Bloomsbury Square. *Biol. Assay.*

Burnett, Lyman C. (1881), Chief in Cereal Breeding Iowa Agric. Experim. Station. Ames, Ia. (U.S.A.). *Production of new and more economical sorts of small grains and maize.* ⓟ Seeds of oats, Winter Wheat, Maize.

Burnett, Theodore C. (1861), Ass. Prof. of Physiol. Univ. of California (U.S.A.), Faculty Club. *Physiol.*

Burnham, Stewart Henry (1870), Ass. Curator of Botan. Ithaca, N.Y. (U.S.A.), Department of Botan., Cornell Univ.

Burns, David, Prof., Univ. Durham Coll. of Med. Newcastle upon Tyne (England). *Physiol.* o

Burns, George Plumer, Prof. of Botan., Univ. of Vermont. Burlington, Vermt. (U.S.A.). *Silvicultural requirements of forest trees, plant succession.* o

Buromskij, Iwan D., Doc. d. Landwirtschaftl. Akademie Timirjasew u. des Inst. f. Volkswirtschaft Plechanow. Moskau (U.d.S.S.R.), Petrowsko-Rasumowskoje, Chem. Abtlg. *Pflanzen-Physiol., Landwirtschaftl. Bact.* o

Burova, L. F., Doc. Mittelasiat. Univ. Taschkent (U.d.S.S.R.). *Protist.* o

Burr, William Wesley (1880), Agronomist and Ass. Dir. of Experim. Station, Univ. of Nebraska. Lincoln, Neb. (U.S.A.). *Soil investigations. Loss of water in the vapor stage, effect of climatic and cultural conditions on nitrification, general problem of the restoration of the organic content of the soil.*

Burr, Prof., Dir. chem. Inst. Preuß. Vers.-Anst. f. Milchwirtschaft. Kiel (Deutschland).

Burri, Robert, Prof. Dr., Dir. Schweiz. milchwirtschaftl. u. bact. Anstalt. Liebefeld bei Bern (Schweiz). *Bact., Bienenbrut.* o

Burrill, Alfred C. Burrill (1881), Curator. Jefferson City, Mo. (U.S.A.), Missouri State Resources Mus. *Formicidae, Aphidae.* ⓟ Yale 1903, Harvard 1905. ⓟ Formicida, Aphididae.

Burrows, Montrose Thomas (1884)), Associate prof. of Experim. Surgery, Washington Univ. School of Med. and Dir. of the Research Labor. of the Barnard Free Skin and Cancer Hospital, St. Louis, Mo. (U.S.A.). *Experim. Cancer. Pathol.* ⓟ John Hopkins Medical School 1909.

Burstein, Moses I., Prof. d. Weißrussischen Staatl. Landw. Akademie, Vorsteher d. Versuchsstation f. Gartenbau. Gorki (U.d.S.S.R.). *Gartenbau, Selektion.* o

Burt, Edward A., Mycol., Missouri Botan. Garden, and Prof. in the Hemy Shaw School of Botan. of Washington Univ., Retired. St. Louis, Mo. (U.S.A.), 4542 Tower Grove Place. *Mycol.* o

Burton, Georgia Winifred (1894), Head of Home Economics Research, Exper. Sta. Auburn, Ala. (U.S.A.), Alabama Polytechnic Inst. *Nutrition.* ⓟ Columbia Univ. 1925.

Burton, Maurice (1898), Ass., Natural History Mus. (British Mus.). London S.W. 7 (England), Cromwell Rd. *Sponges: Systematics, morphol., Embryol.* ⓟ London Univ. 1921.

Busacca, Archimede (1893), Dr. med., Libero Doc. di Clinica Oculistica. Firenze (Italia), Via Bonifacio Lupi 2. *Anat. et pathol. de l'œil.* ⓟ Palermo.

Buscalioni, Luigi (1863), Dir. Orto Botan. e del giardino Coloniale, Prof. Botan. Univ. Palermo (Italia). *Botan. generale.* ⓟ Torino.

Busch, Elisabeth Alexandrowna (1886), Mitarb. Botan. Mus. d. Akademie d. Wiss. Leningrad (U.d.S.S.R.), Karpowka 19, Qu. 48. *Flora des Kaukasus, Sibiriens und des Fernen Ostens. Systematik der Primulaceen, Ericaceen, Empetraceen, Staphyleaceen, Celastraceen, Butaceen, Anacardiaceen, Aquifoliaceen und Aceraceen.*

Busch, Nicolai Adolfowitsch (1869), Prof. Univ. Leningrad (U.d.S.S.R), Karpowka 19, Qu. 48. *Flora des Kaukasus, Sibiriens, des Fernen Ostens. Pflanzengeographie des Kaukasus, der Krim und des Europäischen Rußlands. Systematik der Rhoeadales und Ranales; insbesondere d. Cruciferen.* ⓟ Dorpat 1911.

Busch, Werner Ernst Wilhelm (1889), Dr. med. et phil. Magdeburg (Deutschland), Augustastr. 37. *Meeres- und Binnengewässerhydrobiol. Marine Ciliaten und Flagellaten. Marine Copepoden, Eihüllenbildung, geographische und Tiefenverteilung, Einfluß der wechselnden Umweltfaktoren auf die Körpergestalt. Peridineen, Diatomeen.* Ⓓ Med. Kiel 1917, phil. Kiel 1920.

Buschkowitsch, W. O. (1878), Ass. d. Anat. Inst. Mitarb. Psychiatrisch. Klinik Leningrad. Odessa (U.d.S.S.R.), Baranova 9. *Anthrop. der Blutgruppen.* Ⓓ Kiew 1903.

Buschmakin, N. D., Prof., Dir. d. Anat. Inst. Univ. Irkutsk (U.d.S.S.R.). o

Buschmann, Arnold (1873), Prof. landwirtsch. Fac. Univ. Riga (Latvia), Freiheitstr. 84—6. *Ernährung der landwirtschaftl. Nutztiere. Specifische Wirkungen einzelner Nährstoffe u. ganzer Futtermittel. Nährstoffgehalt.*

Busck, August (1870), Associate Entomol., U. S. Bureau of Entomol. Washington, D.C. (U.S.A.), U. S. National Mus. *Micro-Lepidoptera.*

Bushnell, John (1893), Ass. in Horticulture, Ohio Agric. Experim. Station. Wooster, Ohio (U.S.A). *Physiol. of the potato plant.* Ⓓ Univ. of Chicago 1925.

Bushnell, Leland D. (1880), Prof. and Head of the Department of Bact. Kansas State Agric. Coll. Manhattan, Kan. (U.S.A.). Ⓓ Harvard Univ. 1921.

Businco, Armando (1886), Prof. Anat. Patol. Perugia (Italia), Univ. *Anat. patol. umana e vet.*

De Busk, Ezra Franklin (1884), Extension Citrus Pathol.-Entomol., Univ. of Florida. Gainesville, Flo. (U.S.A.). *Practical control of Diseases and Insect pests of Citrus Fruits Trees.*

Bușnița, Th., Chef de Travaux Inst. de Zool. Cluj (România), Str. Miko 5. *Zool. théoritique.* o

Busquet, M. H., Agr. Fac. Méd. Paris 10 (France), 2, Rue Condorcet. *Physiol.* o

Busse, Walter (1865), Dr. phil., Geh.Ob.Reg.R., Doc. an der Landwirtschaftl. Hochsch. Berlin. Berlin-Wilmersdorf (Deutschland), Hildegardstr. 3. *Nutzpflanzen, besonders der warmen Länder.* Ⓓ Freiburg i. Br. 1892.

Busson, Bruno (1880), Dir. der bundesstaatlichen Schutzimpfungsanstalt gegen Wut und der bundesstaatlichen Serumkontrollstelle am staatl. Serotherapeutischen Inst. Wien (Österreich). *Bact.,Serol., Experim. Pathol.* Ⓓ Czernowitz 1901.

Busuttill, R., Prof. Univ. of Malta. La Valletta (Malta). *Anat. and Histol.* o

Butjagin, Paul Vass., Prof. Tomsker Staats-Univ. Tomsk (U.d.S.S.R.), Prospekt Timirjazeva 3. *Bact.* o

Butkevič, V. C., Prof. II. Univ. Moskau (U.d. S.S.R.). *Physiol. d. Pflanzen.* o

Butkov, P. I., Staatl. Med. Inst. Astrachan (U.d.S.S.R.). *Physiol. Chem.* o

Butler, Amos William (1860). Indianapolis, Ind. (U.S.A.), 52 Downey Ave. *Ornithol. Mammal. Herpetol.* Ⓓ Hanover Coll. 1915, Indiana Univ. 1922.

Butler, Arthur Lennox (1873). Horsham (England), St. Leonard's Park. *Ornithol.: Trochilidae.* Ⓟ Trochilidae.

Butler, Charles St. John (1875), Captain, Med. Corps, U.S. Navy. Ingénieur chargé du Service National d'Hygiène Publique d'Haïti. Washington, D.C. (U.S.A.), U. S. Navy Department, or Port-au-Prince (Haïti). *Bact. and Tropical Med.*

Butler, Edwin John (1874), Dir., Imperial Bureau of Mycol. Kew, Surrey (England), 17 The Green. *Plant pathol.; Mycol.* Ⓓ Dublin 1898.

Butler, Elmer G. (1900), Instr. in Compar. Anat., Univ. Princeton, N.J. (U.S.A.), 24 Dickinson St. *Exbryol. of the vascular system in vertebrates.* Ⓓ Princeton Univ. 1926.

Butler, James Bayley (1884), Prof. Zool. Univ. Coll. Dublin (Ireland), 81 Ranelach Road. *Marine Zool.* Ⓓ M.A. Dublin 1907, M.B. 1909. Ⓟ Marine Zoological specimens.

Butler, Lowell F. (1894), Ass. Pathol., United States Department of Agric. Chicago, Ill. (U.S.A.), 1425 So. Racine Ave. *Diseases of fruits.*

Butler, Ormond R., Prof. of Bot. New Hampshire Coll. of Agric. and Bot., New Hampshire Agric. Exper. Stat. Durham, N.H. (U.S.A.). o

Butning, Pavel Ernestovič (1887), Ass. Zool. Inst. Tierärztl. Hochsch., Abteilung f. allgem. Biol. des Inst. f. med. Wissensch. Leningrad (U.d.S.S.R.), Sovietski, 4. *Experim. Morphol. und Physiol. der Cestodenentwicklung.*

Butt, Richard V. (1906), Research Sc., International Fisheries Commission. Seattle, Wash. (U.S.A.), Univ. of Washington. *Growth of the Halibut (Hippoglossus).*

von Buttel-Reepen, Hugo (1860), Prof., Dr. phil., Leiter des Naturhistor. Mus. und der Denkmalspflege Oldenburg i. O., Oldenburg (Deutschland), Bismarckstr. 32. *Biol. und Physiol. der staatenbild. Insekten, Tierpsychol., descendenztheoretische und prähistorische Fragen.*

Butterfield, C. T. (1890), Bac., U.S. Public. Health Service. Cincinnati, Ohio (U.S.A.), 3 and Kilgour Sts. *Bact. Water Purification.*

Butters, Frederic King (1878), Associate Prof. of Botan., Univ. of Minnesota. Minneapolis, Minn. (U.S.A.). *Taxonomy of Ferns. Plant geography.* Ⓓ Harvard 1917.

Buxbaum, Franz (1900), Dr. phil. Neubistritz (C.S.R.). *Systematische Pflanzenanat., Floristik und systematische Botan., physiol. Anat. Melanthioideae. Flora von Tunesien.* Ⓓ Graz 1922. Ⓟ Herbarpflanzen, alpine Flora.

Buxton, J. B., Prof., Dir. of Inst. of Animal Pathol., Univ. Cambridge (England). o

Buxton, L. H. D., Lect. Univ. Oxford (England). *Physic. Anthrop.* o

Buxton, Patrick Alfred (1892), Dir., Department of Entomol., School of Hygiene and Tropical Med. London W. C. 1 (England), 23 Endsleigh Gardens. *Med. Entomol. Insect Ecol. Conditions of life in deserts.*

Buxtorf, August (1877), Prof. Geol., Palaeont. Univ. Basel (Schweiz), Bernoullianum. *Jura der Alpen.* Ⓓ Basel 1900.

Buy, J., Prof. Univ. Clermont-Ferrand, Puy-de-Dôme (France). *Anat.* o

Buys, John L. (1897), Prof. of Biol. and Head of Biol. Department at St. Lawrence Univ. Canton, N.Y. (U.S.A.). *Cicadellidae of the order Homoptera. Systematic.* Ⓓ Cornell Univ., Ithaca, N.Y., 1922.

Buytendyk, Jacobus Johannes (1887), Prof. der Physiol. Reichsuniv. Dir. des Physiol. Inst. Groningen (Holland), Emmasingel 14. *Stoffund Gaswechselphysiol. Tierpsychol. Eigenschaften der Kohlensäure.* Ⓓ Utrecht 1918.

Byck, Wadim (1894), Prosektorvertreter, Katheder der normalen Anat. des Menschen. Kazan (U.d. S.S.R.), Anat. Inst. der Staatsuniv. *Morphol. des Gehirns, Rasseneigentümlichkeiten.*

Byerly, Theodore C. (1902), Instr. in Zool., Univ. of Michigan. Ann Arbor, Mich. (U.S.A.), 1028 E. Vaughn. *Experim. embryol.; relation between respiration of the cell and growth.* Ⓓ Univ. of Iowa 1926.

Byers, Charles Francis (1902). Johnstown, Pa. (U.S.A.), 139 Sec. Ave. Westmont. *Taxonomy of the Odonata (Insects).* Ⓟ North American Odonata.

Bykoff, C. M. (1886), L'aide de dir. labor. de Physiol. de l'Inst. Experim. Méd. Leningrad (U.d. S.S.R.), Lopuchinskij 12. *Physiol.* Ⓓ Kasan 1912.

Bykowski, Jaxa Ludwik, confer **Jaxa.**

van der Byl, Paul Andries (1888), Prof. of Plant Pathol. and Mycol. (incl. Bact.), Univ. Stellenbosch (Union of South Africa), Bosmanstraat. *Systematic Mycol.: Polyporaceae, Auriculariaceae, Tremellaceae, Dacryomycetaceae, Thelephoraceae, Gasteromycetes, Fungi.* Ⓓ Univ. of the Cape of Good Hope 1915. Ⓟ Herbarium of South African Fungi.

Byrne, G. T., Prof. Univ. Hong-Kong (China). *Biochem.* o

Bytel, Johanna Hubertha, Ass. am anat. und embryol. Inst. Groningen (Holland), Herman Colleniusstraat 21. *Experim. Embryol., insbes. der Amphibia.*

C... confer etiam Tsch...

Caballero, Arturo, Prof., Catedrático Univ., Jefe de la Sección de Herbarios del Jardín Botán. Madrid (España), Alvarez de Castro 16, 3. *Botan., Micol.* o

Cabannes, Prof. Univ. Montpellier (France). *Hist. nat. méd.* o

Cabrera, Angel (1879), Dr. Phil., Curator in Chief Department of Paleontol., Mus. de La Plata, Prof. of Paleontol. in the Inst. del Mus., Univ. of La Plata (Argentine), Calle 3, No. 1034. *Vertebrata; Mammals, fossil and living.* ① Madrid 1900.

Cabrera y Díaz, Agustín, Prof., Dr. en Ciencias. Laguna de Tenerife (Canarias). *Entomol.* o

Cagnetto, Giovanni (1874), Prof. stabile nella R. Univ., Dir. dell' Ist. di Anat. Patol. Padova (Italia). *Patol. delle ossa e del Polmone. Anat. Patol. dell' Apparato Sessuale Maschile.* ① Padova 1924.

Cain, Stanley Adair (1902), Instr. in Botan., Butler Univ., Curator of the Herbarium. Indianapolis, Ind. (U.S.A.). *Ecol.* ① Butler Univ. 1924. ⑦ Herbarium mounts of Indiana plants.

Di Cairano, Vitale (1896), Ass. Osserv. Fitopat. per le Puglie. Taranto (Italia), Piazza Ebalia 1. *Insetti.* ① Portici 1921.

Cairo da Silva, Nilo, Prof. Fac. d Med. Curityba, Paraná (Brasil). *Pathol.* o

Caius, John Fernand (1877), Pharm. Labor., Haffkine Inst. Bombay (India), St. Xavier's Coll. *Med. Chem. Ptomaines, urine concentration, saliva of nonvenomous snakes, Indian fish-oil, anthelmintics.*

Cajander, Aimo Kaarlo (1879), Prof. (Waldbau) an der Univ. Helsinki, Gen. Dir. der Forstverwaltung Finnlands. Helsinki (Finnland), Museokatu 18. *Pflanzengeographie. Pflanzengeographische Grundlagen des Waldbaus.* ① Helsinki 1915.

Calabresi, Enrica (1891), Libera doc. in Zool., Aiuto presso l'Ist. di Zool. Univ. di Firenze (Italia), Via Romana 19. *Hexacorallae, systématique et anat. Brenthidae (Coleoptera). Amphibia, Reptilia.* ① Firenze 1914.

Calabresi, Massimo (1903), Ass. dell Ist. Anat. Firenze (Italia), Via degli Alfani 33. ① Firenze 1926.

Calderwood, William L. (1865), Inspector of Salmon Fisheries of Scotland, Member of Council Royal Soc. of Edinburgh (Scotland), Fishery Board for Scotland. *Fishes: Salmonidae.*

Caldis, Panos Demetrius (1896), Pathol., Haytian Pineapple Co. (California Packing Corporation). Cape Haitien (Haiti). *Pineapple Diseases. Etiol. and transmission of Endosepsis in the fruit of Ficus carica L.* ① Univ. of California 1926.

Caldwell, George Thornhill (1892) Prof. of Zool., Univ. of Arizona. Tucson, Arizona (U.S.A.), Biology Department. *Physiol., respiration in relation to water metabolism.* ① Univ. of Chicago 1923.

Caldwell, Mary Letitia. Instr. in Chem., Columbia Univ. New York City (U.S.A.), Havemeyer Hall. *Chem. nature of Enzymes (Especially Amylases). Separation of Amino Acids.* ① Columbia Univ., New York 1921.

Caldwell, Otis William (1869), Prof. of Education and Dir. of the Lincoln Experim. School, Teachers' Coll., Columbia Univ. New York, N.Y. (U.S.A.), 646 Park Ave. *Plant Morphol. Practical botan.* ① Chicago 1898. o

Cale, William Earle (1904), Graduate Ass. in Bact. Knoxville, Tenn. (U.S.A.), Department of Bact. Univ. of Tennessee. *Public health, Food and Dairy Bact.* ① Univ. of Tennessee 1926. ⑦ Bact. Cultures. Yeast Cultures.

Calendoli, E., Prof. Univ. Napoli (Italia). *Batteriol.* o

Califano, Luigi (1901), Ass. Ist. Pathol. Generale, R. Univ. Napoli (Italia), S. Andrea Dame 21. *Pathol. expérim.* ① Napoli.

Calkins, Sary Nathan (1869), Prof. of Protozool. Columbia Univ. New York, N.Y. (U.S.A.). *Protozool. esp. Ciliata.* ① Columbia 1896. o

Call, Leland Everett (1881), Dean of Agric. and Dir. of Agric. Experim. Station, Kansas State Agric. Coll. Manhattan, Kan. (U.S.A.). *Agronomy.* ① Ohio State Univ. 1912.

Callier, Alfons (1866). Bunzlau, Schlesien (Deutschland), Adler-Apotheke. *Systematische Botan.: Alnus.* ⑦ Phanerogamen und Gefäßkryptogamen. Flora silesiaca exsiccata.

Calman, William Thomas (1871), Deputy-Keeper of Zool. London, S. W. 7 (England), Cromwell Road, British Mus. (Natural History). *Systematic Zool., Crustacea.* ① Dundee 1900.

Calmette, Albert 'Léon Charles (1863), Prof. honoraire, Sous-Dir. de l'Inst. Pasteur de Paris XV (France), 25, rue Dutot. *Bact. et Physiol. Ankylostomiase. Animaux venineux.* ① Paris 1886.

Călugăreanŭ, D., o. Prof. Univ. Cluj (România). *Allg. Physiol.* o

Calvert, Philip Powell (1871), Prof. of Zool., Zool. Labor., Univ. of Pennsylvania. Philadelphia, Pa. (U.S.A.). *Odonata (Taxonomy, Morphol., Development, Ecol., Distribution), Seasonal Distribution of Organisms in the Tropics, History of Entomol.* ① Univ. of Pennsylvania 1895.

Calvet, Louis (1868), Prof. de Zool. générale et appliquée à la Fac. des Sc. de l'Univ. de Clermont-Ferrand (France), 3, Cours Sablon. *Bryozoaires marins.* ① Montpellier 1912.

Calvino, Mameli Eva (1886), Prof. de Botan., Univ. de Cagliari, Sardaigne (Italie), Jardin Botanique. *Anat. et Physiol de la canne à sucre, du pollen. Biol. de la fleur. Lichenol.* ① Pavia 1907.

Camero, Gabriel, Univ. Nac. Bogotá (Columbia). *Anat. Pathol.* o

Cameron, Alexander T. (1882), Prof. of Biochem., Fac. of Med., Univ. of Manitoba. (Honorary) Associate in Biochem., Winnipeg General Hospital. Winnipeg (Canada), Department of Biochem. Med. Coll. *Chem. of the internal secretions.* ① Edinburgh 1925.

Cameron, Gordon Roy (1899), Ass. Dir. Hall Research Inst. Melbourne, Victoria (Australia), Walter and Eliza Hall Inst., Melbourne Hospital. *Pancreatic regeneration. Compensatory Hypertrophy. General morbid anat. and histol.*

Cameron, J., Prof. Dalhousie Univ. Halifax, Nova Scotia (Canada). *Anat.* o

Cameron, Sidney H. (1897), Ass. in Subtropical Horticulture. Berkeley, Cal. (U.S.A.), 339 Hilgard Hall. *Carbohydrate Cycle in Citrus.* ① Univ. of California 1922.

Caminopetros, Jean (1893). Athènes (Grèce), Inst. Pasteur Hellénique. *Mollusques. Leishmania.* ① Athènes. ⑦ Mollusques. Préparations de Leishmania tropica et Leishmania Donovani.

Camis, Mario (1878), Prof. di Fisiol. Parma Italia), Roma Univ. *Physiol. du système nerveux. Sang, hémoglobine.* ① Roma 1902.

Cammerloher, Hermann (1885), Priv. Doc. an der Univ., Ass. am Botan. Inst. der Univ. Wien III (Österreich), Rennweg 14. *Anat., Biol. der Pflanzen.* ① Wien 1910.

Camp, Charles Lewis (1893), Ass. Prof. of Zool. and Curator of Reptiles and Amphibians in the Mus. of Paleont. Berkeley, Cal. (U.S.A.), Dept. Geol., Univ. California. *Fossil vertebrates of the Triassic of Southwestern North America. Morphol. and Paleontol. of Reptiles and Amphibians.* ① Columbia 1923.

Del Campana, Domenico, Prof. incar. R. univ. Firenze (Italia). *Paleont.* o

Campbell, Douglas Houghton (1859), Prof. emer. Stanford University, Cal. (U.S.A.). *Morphol. and*

Plant-evolution. Plant-distribution. Bryophytes, Pteridophytes, lawer Phanerogams. ⓓ Michigan 1886.
Campbell, J.W., Lect. Sea View Road, New Brighton. Christchurch (N. Zealand). *Zool.* o
Campbell, James Argyll (1884), Member of Research State National Inst. for Med. Research. London (England). *Tissue O_2- and CO_2-Tensions. Metabolism. Action of light radiations. Effects of altered O_2 pressures in air.* ⓓ Edinburgh, Scotland, Med. 1912, Sc. 1914.
Campbell, Roy Elliott (1890), Associate Entomol., U.S. Bureau of Entomol. Alhambra, Cal. (U.S.A.), 200 So. 3rd. St., P. O. Box 297. *Biol. of insects affecting truck crops.* ⓓ Univ. of California 1913.
Campbell, Roy Jones (1896), Prof. of Biol. Sabattus, Maine, (U.S.A.). *Bact. and immunity* ⓟ Harvard School of Public Health 1921. ⓟ *Plants and Animals of the New England coast especially Maine.*
Campbell, William (1889), Prof. Bacteriol. Dir. Labor. Serol. Immuniol. and med. Protozool. New Medical. Buildgs. Univ Cape Town (South Africa). New Medical School Buildgs. *Bact., Anaphylaxis.*
del Campo, Estanislao, Prof., Dir. Lab. de Fisiol. Fac. Med. Sevilla (España). o
Campos Fillol, Juan (1875), Prof., Catedratico Higiene, Bact., Fac. Med. Valencia (España),Lauria No. 2. *Oftalmol.* ⓓ Med. Valencia 1899, farm. 1919. ⓟ Prepar. farm.
Campos Fillol, Rafael, Dr. med., Prof. aux. Fac. Valencia (España), Pí y Margall 1. *Histol.* o
Camus, M. J., Prof. agr., Fac. Méd. Paris (France). *Physiol.* o
Canavan, William Paul (1897), Instr. of Zool. Univ. of Pennsylvania. Philadelphia, Pa. (U.S.A.), Zool. Labor. *Parasitol.: Trematoda, Cestoda, Nematoda.* ⓓ Leigh Univ. 1924. ⓟ *Round Worms ex Horse, Cat, Dog, Fish tapeworms, Trematoda ex Frogs, Necturus, etc. Alcoholics. Parasites of Zool. Garden Animals.*
Čančík, Josef, Dr., Priv.Doc. d. Hygiene. Praha II (C.S.R.), Bojiště 3. o
Candura, Giuseppe Salvatore (1899), Ass. Labor. Zool. ed Entomol. agraria Ist. Sup. Agr. Portici, Napoli (Italia). *Entomol. dei prodotti agrari: biol. microlepidotteri.* ⓓ Portici 1923.
Canestrini, Alessandro (1879), Preside del R. Ist. Tecnico di Rovereto (Italia). *Biol. generale.* ⓓ Innsbruck 1903.
Canizarez, Miguel, Ass. Prof. Univ. of the Philipp. Manila, Philippine Islands (U.S.A.). *Anat.* o
Cannan, Robert Keith (1894), Senior Ass., Department of Biochem., Univ. Coll. London, W.C. 1 (England), Gower Street. *Chem. of Metabolic Processes: Mechanism of Biol. Oxidation, Reduction.* ⓓ London 1914.
Cannon, Walter Bradford (1871), Prof. of Physiol Harvard Univ. Med. School. Boston 17, Mass. (U.S.A.). *Mechanical Aspects of Digestion. Traumatic Shock. Physiol. Aspects of Emotional Excitements. Functions of the Sympathetic Nervous System. Nervous Control of Internal Secretion.* ⓓ Harvard Med. School 1900.
Cannon, William A., Staff Member, Dept. of Botan. Research, Carnegie Inst. of Washington. Stanford Univ., Cal. (U.S.A.), Box 809. o
Cantacuzène, Jean (1863), Prof. Fac. Méd. (Méd. expérim.), Dir. de Inst. de Serusi st Vaccinusis, Dir. du Labor. de Méd. expérim. de la Fac. de Méd. Bucarest (România), 2 Piatza Alex. Lahovary. *Epidémiol., Pathol. expérim. et Immunité chez les Invertébrés.* ⓓ Paris 1894.
Capitan, M., Chargé de Cours Coll. de France. Paris 5 (France), 5, rue de Ursulines. *Anat.* o
Capocaccia, Mario (1900), Aiuto nell' Ist. di Pathol. generale Univ. Genova 18 (Italia), Viale Benedetto XV. *Histopathol. Physiopathol.* ⓓ Univ. di Genova 1923.

di Caporiacco, Lodovico (1900), Dr., Ass. Firenze (Italia), Via Romana 19. *Arachnidae. Pisces.* ⓓ Firenze 1920.
Cappe de Baillon, Pierre (1874). Lille, Nord (France), 73, rue des Stations. *Embryol., Anat., Biol., Tératol. des Insectes.* ⓓ Nancy 1920.
Cappelletti, Carlo (1900), Aiuto, Inst. Botan. de Padova (Italia), Via Orto Botanico 15. *Anat., Cytol., Mycol., Hépatiques. Immunol. dans les plantes.* ⓓ Torino 1921.
Capra, Felice (1896), Conserv. Mus. Civico di Sloria Nat. Genova 102 (Italia), Via Brigala Liguria 9. *Sistematica e zoogeografia dei Coleotteri: Coccinellidae.* ⓓ 1923.
Car, Lazar (1860), o. Prof. vergl. Anat. Univ. Zagreb (S.H.S.), Gajgasse 47. *Systematik der Crustazeen: Copepoden. Bewegungen bei den Tieren.* ⓓ Jena 1881.
Caradonna, Giambattista (1867), Prof. o. stabile di Anat., normale Vet., Dir. Ist. Sup. di Med. Vet. Perugia (Italia), Villino proprio. *Anat. Veterinaria.*
Carano, Enrico (1877), Prof. di Botan., Dir. Ist. Botanico. Univ. di Roma 3 (Italia), Via Milano 75. *Anat. végétale, embryol., cytol.*
Carbonaro, Giuseppe (1891), Ass., Ist. di Farm., Univ. di Messina (Italia), Via Risorgimento, 199. *Amoebiasis.*
Carboncini, Guglielmo (1882), Ass. Ist. di Cerealicoltura. Bologna (Italia), Via Milazzo 4. *Sélection des céréales.* ⓓ Bologna 1925.
Carbone, Domenico (1880), Libero doc. di batt. agraria e libero doc. d'igiene, Univ. di Milano 121 (Italia), Via Mancini 15. *Decomposizione dei vegetali in genere. Reazioni nelle piante. Ferment. lattica.* ⓓ Pisa 1904. ⓟ *Culture di microrganismi.*
Carbonieri, Pier Luigi (1902), Aiuto nell' Ist. d'Igiene, Univ. Modena (Italia). *Saccaromiceti del Latte acido.*
Cardot, Henry (1886), Prof. agrégé de l'Univ., docteur ès sc., Chef de lLabor. (physiol.) à la Fac. de Méd. de Paris XIII (France), 164, rue Jeanne d'Arc prol. *Lois polaires de l'excitation et électrotonus. Réflexe linguo-maxillaire et excitabilité des centres nerveux. Physiol. générale de la cellule: variations et hérédité des caractères acquis. Cœur et Sinus contractiles des Mollusques, flagelle des Hélices. Faune malacologique du Nord-Est de la France.* ⓓ Paris 1912. ⓟ *Mollusques terrestres et fluviatiles de France.*
Carey, Cornelia Lee (1892), Instr. in Botan., Barnard Coll. New York, N.Y. (U.S.A.), Columbia Univ. *Physiol. Botan., Bact.* ⓓ Columbia Univ. 1923.
Carey, Eben James (1889), Dir. of the Department and Prof. of Anat.; Dean of Students; Acting Dean of the Fac., Marquette Univ., School of Med. Milwaukee, Wis. (U.S.A.), 638/4 Street. *Anat., Embryol.* ⓓ Chicago 1925.
Cargill, Miriam Ayer (1901), Instr. in Botan. Kingston, R.I. (U.S.A.), R.I. State Coll. ⓓ Wisconsin 1925.
Carl, Johann (1877), Dr. phil., 1er Ass. au Mus. d'histoire nat. (Entomol.), Priv.Doc. à l'Univ., Fac. des Sc. et Fac. des Sc. économiques et sociales, Genève (Suisse), Mus. d'histoire nat. *Diplopoda, Orthoptera (Systématique), morphol. et répartition géographique). Hyménoptera helvetica. Crustacea: Copepoda; Isopoda, Amphipoda et Decapoda helvetica. Problèmes de zoogéographie.* ⓓ Berne 1898.
Carleton, Alice, Demonstrator f. Human Anat. Univ. Oxford (England), 45 Baubury Rd.
Carleton, Harry Montgomerie (1896), Univ. Lect. in Histol. Oxford (England), Physiol. Labor. *Histol. in relation to Physiol. and experim. Med.: Tissue Culture, Pneumococcosis, capillary contractility.* ⓓ 1919.
Carlgren, O. H., Prof., Dr. Lund (Sverige). *Zool.* o

Carlier, Edmond William Wace (1861), Prof. of Physiol. Dorridge near Birmingham (England). Morningside. *Experim. Histol. Physiol. Chem. Experim. Physiol.* ⊕ 1891.

De Carlo, John (1888), Demonstrator of Topographic and applied Anat. Philadelphia, Pa. (U.S.A.), 1629 S. Broad St. ⊕ Philadelphia, Pa., 1911.

Carlson, Anton Julius (1875), Prof. of Physiol., Univ. of Chicago (U.S.A.). *Physiol. of the Endocrine glands, Alimentary tract.* ⊕ Stanford Univ. 1903.

Carlson, Margery C., Fellow in Microchem., Ph.D. Yonkers, N.Y. (U.S.A.), Boyce Thompson Inst. for Plant Research. *Plant propagation, Microchem. and histol.* ⊕ Univ. of Wisconsin 1925.

Carmichael, Leonard (1898), Ass. Prof. of Psychol., Univ. Princeton, N.J. (U.S.A.), Eno Hall. *Physiol. and Experim. Psychol.* ⊕ Harvard Univ. 1924.

Carminati, Valentino (1899), Ass. volontario, Ist. di Patol. generale, Univ. di Milano (Italia), Via G. Strambio, 31. *Ormone sessuale. Avitaminosi.* ⊕ Pavia 1926.

Carneiro, Abdon Petit, Prof. Fac. d. Med. Univ. Curityba, Paraná (Brasil). *Histol.* o

Carnevale, Pasquale (1875), Prof. mat. sc., RR. Scuole Med. Capracotta Campobasso (Italia). *Enzime.* ⊕ Torino. ℗ Radiolarie.

De Caro, Luigi (1901), Aiuto di fisiol., Univ. Napoli (Italia), S. Andrea delle Dame 21. *Fisiol. e Chim. Biol.*

Caroli, Angelo (1887), Ass., Ist. di Zool. e Anat. comparata della R. Univ. Siena (Italia), Piazza Giordano Bruno 3. *Anat. degli anfibi anuri.* ⊕ Napoli 1913.

Caron, Omer (1893), Botan. Quebec (Canada), Ministère de l'Agric. *Botan. systématique (Phanérogames). Pathol. végétale (Mycol., Espèces nuisibles). Physiol. végétale (surtout dans ses rapports avec la pathol).*

Carothers, E. Eleanor (1882), Lect. in Zool. Philadelphia, Pa. (U.S.A.), Zool. Bldg. Univ. of Pennsylvania. *Acrididae (Orthoptera), Cytol. and Genetic.* ⊕ Univ. of Pennsylvania 1916.

Carpelan, Jarl Otto Casimir (1897). Uleåborg, A/B Uleå (Finnland). *Pflanzenkrankheiten. Lappländische Vögel (Leben, Eier, Nester usw.), besonders Petsamo-Gebiet am Eismeer. Insekten: Longicornes (Käfer) und Bombycidae (Schmetterlinge), finnische Arten.* ℗ Finnländische Insekten, Vogeleier und Bälge. Eier von paläarktischen Sumpfvögeln, paläarktische Longicornes und Bombycides.

Carpenter, Charles Milton (1895), Ass. Prof. Labor. of Labor. Diagnosis. Cornell Univ. Ithaca, N.Y. (U.S.A.). *Compar. Pathol. and Bact.* ⊕ Med. vet. Cornell Univ. Ithaca, N.Y., 1917, phil. 1921.

Carpenter, George Herbert (1865), Keeper of the Mus., Victoria Univ. Manchester (England). *Arachnida (Araneida; Pycnogonida). Insecta (Apterygota; Larvae of Diptera and Coleoptera). Geographical Distribution. Insect Transformation.* ⊕ M.Sc. Belfast 1914, D. Sc. 1918.

Carpenter, Thorne Martin (1878), Physiol. Chem., Nutrition Labor. Boston, Mass. (U.S.A.), 29 Vila St. *Gas analysis, respiratory exchange in relation to intermediary metabolism.* ⊕ Harvard Univ. 1915.

Carpentier, A., Prof. Univ. Lille (France). *Botan.* o

Carpentier, Fritz M. J. (1890), Conservat. des collections de Zool. à l'Univ. de Liége (Belgique), 10, rue Vivegnis. *Morphol. et anat. des Insectes.* ⊕ Liége 1919.

Carr, Lancelot Ashbourn (1882). Lichfield (England), White House, Market St. *Invertebrata fauna. Lepidoptera. Araneae. Ichneumonidae. Braconidae. Chalastogastra. Hemiptera and Aculeata.*

Carr, Ralph Howard (1878), Prof. of Agric. Chem., Purdue Univ. Lafayette, Ind. (U.S.A.). *Plant and Soil Chem., Nutrition.* ⊕ Univ. of Wisconsin 1913.

Carrão, Manoel Lustoza, Prof. Fac. d. Med. Univ. Curityba, Paraná (Brasil). *Physiol.* o

Carrasco Formiguera, Rossend (1892). Barcelona (España), Copérnico, Bonanova. *Carbohidrat metabolism. Insuline.* ⊕ Barcelona 1914.

Carrel, Alexis (1873), Member, Rockefeller Inst. for Med. Research. New York, N.Y. (U.S.A.), 66th Street and Avenue A. *Experim. surgery. Cicatrization, regeneration, growth, old age, tumor formation.* ⊕ Lyons, France 1900.

Carren, C., Prof. Univ. Lille (France). *Pharm.* o

Carriker, Melbourne Armstrong (1879). Santa Marta, (Colombia, S.A.), Apartado 51. *Central and South American Ornithol. Insect family ,,Mallophaga".* ℗ Birds, Mammals, Reptiles, Amphibians and Bird's Eggs, Mallophaga.

Carrisso, Luiz Wittnich (1886), Prof. de Botan., Dir. Inst. Botan. Univ. Coimbra (Portugal). *Geographie botan. du Portugal: regions côtières (sables et marecages marins).* ⊕ Coimbra 1918. ℗ Graines et échantillons d'herbier. Prépar. microscop.

Carroll, Mitchel (1885), Prof. of Biol.,Franklin and Marshall Coll. Lancaster, Pa. (U.S.A.). *Abdominal tracheae and respiration in larva of Mecistogaster modestus. Mosquito. Spermatogenesis in Camnula pellucida. Numerical variations in the chromosome complex within the individual.* ⊕ Univ. of Pennsylvania 1919.

Carroll, Robert Patrick (1903), Instr. in Biol. at Washington and Lee Univ. Lexington, Va., until June 1927 (U.S.A.), Winfall, Va., permanently. *Hepaticae.* ⊕ Univ. of Virginia 1925. ℗ Hepaticae of Virginia.

Carruthers, A. D., Barmer Hall, King's Lynn Norfolk (England). *Ornithol.* o

Carsner, Eubanks (1891), Associate Pathol., U. S. Dept. of Agric. Citrus Experim. Station Riverside, Cal. (U.S.A.). *Disease of Sugar beets called curly-top.* ⊕ Wisconsin 1917.

Carter, Nellie, Lect. in Botan, East London Coll. London E. 1 (England), Mile End Rd. *Algol. British Desmidiaceae.* ⊕ Birmingham 1919.

Carter, Price Wallator (1898), Ass. Lect. in Botan. Univ. Coll. of Wales, Aberystwyth (England). *Algol.* ⊕ Aberystwyth 1924.

Carter, Thomas, Wensleydale, Mulgrave Road, Sutton, Surrey (England). *Ornithol.* o

Carter, Walter (1897), Assoc. Entomol., U. S. Bureau of Entomol. Irvin Halls, Id. (U.S.A.). *Ecol. of sugar beet leafhopper.*

Carter, William Spencer (1869), Dir., div. of med. educ., Rockefeller Foundation. New York, N.Y. (U.S.A.), 61 Broadway. *Comp. Physiol.* o

Cartledge, Joseph Lincoln (1895), Instr., Botan., Univ. of Pittsburgh, Pa. (U.S.A.). *Sex in Fungi; Genetics, pollen sterility.* ⊕ Univ. of Pennsylvania 1921.

Cartwright, Oscar L., Ass. Entomol., Experim. Station. Clemson College, South Carolina (U.S.A.). *Injurious Insects.* ℗ Coleoptera, Cicindelidae.

Cartwright, William Bell (1894), Associate Entomol., U. S. Bureau of Entomol. Sacramento, Cal. (U.S.A.), 600, 26th Street. *Cereal and Forage Insects.* ⊕ Nashville, Tenn. 1917.

Carver, William Angus (1895), Ass. Cotton Specialist. Gainsville, Flo. (U.S.A.), Agric. Experim. Station. *Cotton variety, Cotton breeding, genetic study.* ⊕ Iowa State Coll. 1925.

Cary, Lewis R., Ass. Prof. of Biol., Univ. Princeton, N.J. (U.S.A.). o

Casan, Ignacio, Prof. Colège des P. P. Escolapios de Castellón de la Plana (España). *Hydrobiol., Conchyliol.*

Casares-Gil, Antonio (1871), Prof. agrégé Mus. Nat. Sc. Nat. Madrid (España). *Flore bryol. de la Péninsule Ibérique.* ℗ Mousses et Hépatiques.

Case, Ermine Cowles (1871), Prof. Historical Geol. and Paleont., Dir. of the Geol. Mus. Dept. Geol. Univ. of Michigan. Ann Arbor, Mich. (U.S.A.) *Vertebrate Paleont. and Paleogeography: Permian and Triassic. Stratigraphy of the Paleozoic of Michi-*

gan (U.S.A.). ④ Univ. of Chicago 1896. ⑨ Invertebrate fossils from the Ordovician, Silurian and Devonian of Michigan.
Case, Isabella Macgregor (1900), Ass. Botan. Univ. Glasgow (Scotland). ④ Glasgow 1925.
Cash, James Robert (1893), Associate Prof. of Pathol. Peking Union Med. Coll. Peking (China). *Pathol. Anat. Experim. Hypertension. Histogenesis of the Tubercle. Parasit. Diseases. Kalar Azar.* ④ Johns Hopkins Univ. 1919.
Casper, Max (1866), Dr. med., o.ö. Prof., Dir. des Vet.-Inst. und der Tierklinik Breslau 16 (Deutschland), Uferzeile 11. *Veterinärmed. Geschwülste.* ④ Freiburg i. Br. 1897.
Cassal, Godofredo, Prof. Univ. Nac. Buenos Aires (Argentina). *Anat. Physiol. anim.* ○
Cassanello, R., Prof. Univ. Pisa (Italia). *Patol.* ○
Cassano, Cosimo Bernardo (1904), Dr.Ass. Osservatorio Fitopat. Taranto (Italia), Piazza Ebalia 1. *Insetti.* ⑨ Perugia 1926.
Casselberry, Russell David (1892), Ass. Prof. Zool. Penn. State Coll., Pa. (U.S.A.). *Entomol.* ⑨ Penn. State Coll. 1922. ⑨ Insects.
Castagnari, Giovanni (1897), Aiuto Ist. di Fisiol. Univ. di Bologna (Italia), Via Galliera 24. *Chim. biol. Acetone. Funziona epatica. Urotropina.*
Castaldi, Luigi (1890), Dir. Ist. anat. umana di Cagliari, Sardegna (Italia). *Fegato, Mesencefalo, Glandole endocrine, Costituzioni.* ⑨ Firenze 1926.
Castejón y Martínez Arizala, Raphael (1893), Prof. Córdoba (España), Escuela de Vet. *Bact. Parasit. animal.* ④ Univ. de Sevilla 1913.
Castellani, Aldo (1876), Prof. of Tropical Med., Tulane Univ., New Orleans, La. (U.S.A.) 1551 Canal St. N.O. *Tropical Med., Tropical Dermatol., Bact. and Mycol.* ④ Florence, Italy 1899.
Castelli, Agostino (1879), Aiuto Ist. d'Igiene Univ., Prof. incaricato di Batteriol. Cagliari, Sardegna (Italia). *Batteriol. e parissotol.* ⑨ Preparati di sangue malarico.
Castetter, Edward F. (1892), Ass. Prof. of Botan. Ames, Ia. (U.S.A.), Department of Botan. *Morphol., Histol., Cytol. and Genetics of the Cucurbitaceae. Species crosses in the genus cucurbita, Cytol. of the hybrids.* ④ Iowa State Coll. 1924.
Castiglioni, Giovanni, Prof. incar. Univ. Milano (Italia). *Pathol.* ○
Castille, A., Prof. Univ. Cath. Louvain (Belgique). *Pharm.* ○
Castillo, Odena, Prof. Univ. Nac. d. Litoral. Corrientes (Argentina). *Physiol.* ○
Castle, William Ernest (1867), Prof. of Genetics, Bussey Inst., Harvard Univ. Boston, Mass. (U.S.A.), Forest Hills. *Genetics of mammals: domestic rodents, rots, mice, guinea-pigs, rabbits.* ④ Harvard Univ. 1896.
Castrén, Harry (1891), Dr. med., Doc. der pathol. Anat., stellvertretender Prof. der pathol. Anat. an der Univ. Helsingfors (Finnland), Pathol. Inst. *Histol. pathol. Zellen. Geschwulstforschung. Pathol. des Zentralnervensystems.* ④ Helsingfors 1923.
Catalano, Giuseppe (1888), Priv.Doc. Botan. Univ Palermo (Italia), Botan. Garten. *Xerophylie, Anat. der Graminaceen.* ④ 1911.
Cataneı, A. (1895), Chef de labor. à l'Inst. Pasteur d'Algérie, Alger (Algérie), 59, rue de Lyon. *Protozoaires. Paludisme.* ④ Alger 1922. ⑨ Champignons pathogènes.
ten Cate Hoedemaker, Nicolaus Jurjan (1897), Ass. an der Abteilg. f. experim. Histol. d. Zool. Labor. Utrecht (Holland), *Vergl. Histophysiol.* ⑨ Histol. Praep.
Caterini, Francesco (1895), Prof. incaricato di Paleont., Ass. alla cattedra di Geol., Libero doc. Univ. Pisa (Italia), Ist. di Geol. *Mammiferi pliocenici e quaternari. Lamellibranchi postpliocenici.* ④ Pisa 1917.
Cathcart, Edward Provan (1877), Prof. of Physiol. Chem. Univ. Glasgow (England), Inst. of Physiol.

Protein metabolism. Muscular work. ④ Glasgow Univ. 1900.
Catsaras, Johannes (1883), Prof. Physiol. pathol. Univ., Dir. du Labor. de Pathol. expérim. Athènes (Grèce), Rue Mavromichali 5a. *Pathol. Experim. Epith. Geschwülste, Lympho-granulomatose Leukämien, Myzetome.* ④ Athen 1905. ⑨ Malaria Myzetome.
Cattaneo, D., Prof. Univ. Roma (Italia). *Istol. oftalm.* ○
Cattaneo, Giacomo, Prof. R. univ. degli studi. Genova (Italia). *Anat. comp.* ○
Cattaneo, Pasquale, Lib. doc. R. univ. degli studi. Genova (Italia). *Patol.* ○
Cattell, Y. McKeen (1860), President of the Psychol. Corporation. Garrison-on-Hudson, N.Y. (U.S.A.). *Psychol. measurements, individual differences, application of psychol.* ④ Ph., LL., Sc. Leipzig 1886. ○
Caudell, Andrew Nelson (1872), Custodian of Orthoptera, U. S. National Mus. Washington, D.C. (U.S.A.). *Systematics of Orthopterous insects.* ④ Oklahoma Agric. Coll. 1897. ⑨ Orthoptera.
Caullery, Maurice (1868), Prof. de Zool. à la Fac. de Sc. de l'Univ. de Paris, Dir. de la Station Zool. de Wimereux. Paris (France), 105 Bould. Raspail *Zool. marine: Protozoaires, Orthonectides, Cructacés' Tuniciers, Annélides. Sexualité. Parasitisme et Symbiose. Histoire des sc. biol. en France.* ④ Paris 1895. ⑨ Animaux de Wimereux.
Causey, David (1899), Instr. in Biol. Department of Biol. Univ. Princeton, N.J. (U.S.A.). *Parasit., animal, particularly protozoan parasit. Mitochondria.* ④ Univ. of California 1925.
Cavallini, Francesca (1900), Ass., Ist. Zool., Univ. di Pavia (Italia). *Biol. dei Protozoi.*
Cavara, Fridiano (1857), Prof. o. Univ., Dir. de l'Inst. et Jardin botan. Napoli (Italia), via Foria 223. *Botan. et Biol. expérimentale. Paléobotan. Mycol. Phytopath.* ⑨ Exsiccata des Phanérogames italiennes et de l'Afrique du Nord (Libie).
Cavazza, Filippo, Lib. doc. Univ. Bologna (Italia). *Fisiol.* ○
Cavazzani, Emilio, Lib. doc. Univ. Bologna (Italia). *Farm.* ○
Cave, Alexander J. E. (1898), Demonstrator in Anat. Univ. of Leeds (England), Anat. Department Med. School. *Developmental errors of the axial skeleton. Compar. anat. of cervical region.* ④ Manchester Univ. 1922.
Cavrilescu, N., Chef de Travaux de l'Inst. de Physiol. générale, Univ. Cluj (România). ○
Cayley, Dorothy M. (1874), John Innes Horticultural Inst. London S.W. 19 (England), Mostyn Road, Merton Park, Wimbledon. *Plant Pathol. Mycol. Genetics of Fungi.*
Cehernov, Alexandre (1877), Prof. I. et II. Univ. Moscou, Mohovaja (U.d.S.S.R.), Inst. géol. de la I. Univ. de l'Etat. *Ammonites.* ⑨ Ammonites.
Ceballos, Gonzalo (1895). Cadiz (España), Calderón de la Barca 8. *Himenopteros. Ichneumonidae.* ④ Madrid 1917.
Čečulin, Sergej Ionovič, Doc. I. Univ. Moskau (U.d.S.S.R.). *Pathol.* ○
Cedercreutz, Carl Wilhelm (1893). Helsingfors (Finland), Unionsgatan 4. *Algol. (Zygnemales), floristische Pflanzengeographie.*
Cekovschi, Constantin (1881), Prof., Chef der praktischen Arbeiten, Zool. Inst., Univ. Cernăuti (România). *Säugetier- und Vogelfauna, Histol. der Sinnesorgane. Periphere Nervenendigungen.*
Celestino da Costa, Augusto (1884), Prof. o. Fac. Méd. Univ. Lisbonne (Portugal), Rua nova de Santo Antonio 33. *Histophysiol. des glandes à sécrétion interne: appareil surrénal et hypophyse. Embryol. des Mammifères.*
Centanni, Eugenio (1873), Prof. Patol. Generale Univ. Modena (Italia), Via Ganaceto 62. *Biochem. Immunol. Tumeurs.*

Centi, Mancia Gio Battista, Lib. doc. Ist. sup. di Med. vet. Bologna (Italia). *Zootechnica.* ○

Čerbačer, Dmitrij Mich., Prof. II. Univ. Moskau (U.d.S.S.R.). *Pharmakognosie.* ○

Ceredi, Antonio (1899), Ass. Ist. di Igiene, Univ. di Bologna (Italia), Regia Univ. *Igiene. Batteriol.* (U.d.S.S.R.), *Bact.* ○

Cerikover, Revekka Linov, Doc. II. Univ. Moskau (U.d.S.S.R.), *Bact.* ○

Cernik, Leo Franz. Olmütz (C.S.R.), Pohlgasse 2. *Phytopath. Niedere Pilze als Krankheitserreger.* ⓓ Med. Wien 1915, Brünn 1922.

Černišev, A. S., Doc. II. Univ. Moskau (U.d. S.S.R.). *Physiol. d. Nervensystems.* ○

Černjavsky, Paule (1892), Ass. Botan. Inst. Univ. Beograd (S.H.S.), Takovska 41. *Pflanzengeographie von Südserbien (Macedonien).*

Černozubov, N., Dr., Dir. d. Bact. Stat. Novi Pazar (S.H.S.). ○

Cerny, Adolf (1885), Leiter der Hydrobiol. Donaustation. Wien III (Deutsch-Österreich), Petrusgasse 11. *Hydrobiol. der Binnengewässer. Plankton.* ⓟ Planktonproben.

Cerruti, Carlo Francesco (1899), Ass. Effettivo, Ist. di Igiene della R. Univ. di Torino (Italia), Via Bidone 37. *Batteriol. e Sierol., febbre melitense, aborto epizootico.*

Cerruti, Tomás (1871), Prof. de Histol. Normal y Embriol. Rosario de Santa Fé (Argentina), Calle 3 de Febrero No. 923. *Histol. (Organografía). Cytol. Formation des cellules. Embryol.* ⓓ Buenos Aires 1897.

Cerva, Friedrich Attila (1856), Oberinspektor des Zool.-botan. Gartens Budapest (Ungarn). *Ornithol. faunistica et oecol.; Ool. Avicultura. Entomol. biol.*

Cervera Moltó, Augusto, Dr. en Med., Prof. ayudante de Histol. de la Fac. de Med. Valencia (España), Colón. 2. *Histol.* ○

Cesa-Bianchi, Dominico, Prof. incar. Univ. Milano (Italia). *Pathol.* ○

Cesaris Demel, Antonio (1866), Dir. Ist. di Anat.-Patol. Univ. Pisa (Italia), Via Solferino 37. *Fisio. patol. del cuore isolato. Colorazione vitale (Leucociti, eritrociti, piastrine emoconi). Origine delle piastrine dai megacariociti. Circolazione della milza, splenomi. Teoria anafilattica dell' ulcera rotonda dello stomaco. Apparato Uropoietico.* ⓓ Torino 1890. ⓟ Preparati stol. dimostranti l' origine delle piastrine dai megacariociti.

Cesaris Demel, Venceslao (1897), Ass. di Anat. Patol. Univ. di Pisa (Italia), 37, Via Solferino. *Anat. ed istol. patol. generale e speciale. Patol. del fascio di His. Istochim. dei tumori maligni.* ⓓ Pisa 1921.

Četverikov, Sergej Serg., Doc. I. Univ. Moskau (U.d.S.S.R.). *Zool., Embryol.* ○

Ceyp, Karl (1900), Ass. botan. Inst. Univ. Praha II (C.S.R.), Na Slupi 433. *Vergl. Morphol. der Pflanzen, der Blüten. Mykol.: Hydnaceae, Mycena. Pathol. der Pflanzen, Pflanzenteratol.* ⓓ Prag 1923. ⓟ Exsiccate von Pilzen, Pflanzenpathol. Exsiccate.

Chabakov, Alexander Wassilievicz (1904), Palaeont. Section d. Russ. Geol. Kommitee. Leningrad W.O. (U.d.S.S.R.), Ssredny pr. 726. *Fossile Fische Rußlands (Palaeozoicum: Carbon, Perm; Mesozoicum: Trias, Jura). Selachier u. Ganoiden. Palaeont. d. Permformation d. Wjatka-Gebietes.* ⓟ Fossile Fische aus d. Palaeozoicum u. Mesozoicum Rußlands (Palaeonisciden u. Plaxysomiden).

Chabanaud, Paul (1876), Préparateur à l'Ecole des Hautes-Etudes, Correspondant du Mus. National d'Histoire natur. Paris 5 (France), 57, rue Cuvier. *Poissons: Téléostéens marins des colonies françaises (morphol., anat., systématique).* ⓟ Poissons de mer du littoral français. (Tunisie, Algérie, Maroc, Côte occidentale d'Afrique tropicale, Madagascar, Indo-Chine).

Chabanier, M. H., Chef de Labor., Hôpital Necker. Paris 15 (France), 151, Rue de Sèvres. *Pharm.* ○

Chabrol, Maurice (1893), Chef des Travaux de physiol., Fac. de Méd. Alger (Algérie), 80, rue Michelet. *Physiol. des capsules surrénales. Physiol. de la circulation: régulation de la pression artérielle.*

Chachloff, W. A., Doc. d. Technol. Inst. Tomsk (U.d.S.S.R.),Geol. Kabinett.*Paläophytol. Sibiriens.* ○

Chachlov, V. A., Prof. Tomsker Staats-Univ. Tomsk (U.d.S.S.R.), Prospekt Timirjazeva 3: *Experim. Zool. u. Embryol.* ○

Chadani, Ryô, Ass. of Med. Univ. Kanazawa (Japan), Pathol. Inst. *Pathol.* ○

Chadwick, Herbert Clifton, Naturalist, Univ. of Liverpool. Port Erin, Isle of Man (England), Biol. Sea Station. *Systematic Zool., Echinodermata.* ⓓ Manchester.

Chahovitch, Xénophon (1897), Doc. de Pathol. expérim. à la Fac. de Méd. Beograd (S.H.S.), Choumadiska 18. *Immunité chez les invertébrés. Coagulabilité du sang. Métabolisme de base. Avitaminoses. Cancer expérim.* ⓓ Lyon 1922.

Chaine, Joseph. Prof. d'Anat. comparée à la Fac. des Sc. de Bordeaux, Dir. du Mus. d'histoire nat. de Bordeaux (France), 247, Cours de l'Argonne. *Anat. compar. des Vertébrés: la myol., le squelette, les otolithes des poissons, etc.* ⓟ Peaux de lionnes, dogues de Bordeaux avec crane; oiseaux montés; fossiles du bassin de l'Aquitaine.

Chait, Sarah (1897), Aspirant wissenschaftl. Forschungsinst. in Odessa, Ukraine (U.d.S.S.R.), I.N.O., Kominternstr. 2. *Bact.: Sulfatreduktion. Microbiol. des aktivierten Schlammes.*

Chalatow, Simon (1884), Prof., Dir. Abteilung für Pathol. Physiol., Med. Inst. Leningrad (U.d.S.S.R.). *Pathol. des Stoffwechsels. Anisotrope Verfettung. Gallesekretion. Diathesenlehre. Harnausscheidung.* ⓓ St. Petersburg 1911.

Chamberlain, Charles Joseph (1863), Prof. of Botan., The Univ. of Chicago, Ill. (U.S.A.), Dept. of Botan. *Morphol. and Cytol., especially of Cycads.* ⓓ Univ. of Chicago 1897.

Chamberlin, Joseph Conrad (1898), Instr. in Biol. and Zool. at San Jose State Teachers Coll. San Jose, Cal. (U.S.A.). *Systematic Arachnol. and Coccidol. (Entomol.). Compar. morphol. of Arthropods. Chelonethida (Arachnida) and Coccidæ (Hemiptera).* ⓓ Chelonethida (Pseudoscorpiones).

Chamberlin, Ralph V. (1879), Prof. of Zool., Univ. of Utah. Salt Lake City, Utah (U.S.A.). *Arachnida; Myriopoda; Annelida.* ⓓ Cornell Univ. 1905. ⓟ Myriopoda and Arachnida.

Chamberlin, Thomas R. (1889), Ass. Entomol., U. S. Department of Agric. Forest Grove, Ore. (U.S.A.). *Cereal and forage insects, insect parasitism.* ⓓ Univ. of Utah 1920.

Chamberlin, W. S. (1890), Forest Entomol. Oregon State Coll. Corvallis, Ore. (U.S.A.). *Forest Insects. Coleoptera: Buprestidae, Cerambycidae, Scolytidae, Cleridae.* ⓓ Oregon 1921. ⓟ Buprestidae and Cerambycidae of North America.

Chambers, Ernest Leslie (1896), Ass. State Entomol. Madison, Wis. (U.S.A.), State Department of Agric.

Chambers, Robert (1881), Prof. Microscopic Anat. New York City, N.Y. (U.S.A.), Cornell Univ. Med. Coll. *Cytol., Experim. cell physiol. Experim. embryol., Physical structure of protoplasm microdissection and microinjection. Apparatus invented for microdissection and injection of living Cells. Action of salts on the plasma membrane and on the internal protoplasm, Hydrogen ionconcentration of the cytoplasm and nucleus.* ⓓ Univ. of Munich 1908.

Chambliss, Charles Edward (1871), Agronomist in charge of rice investigations U.S. Dept. of Agric. Washington, D.C. (U.S.A.), 1833 Kilbourne Pl. ○

Chamot, Emile M. (1868), Prof of Chem. Microscopy. Ithaca, N.Y. (U.S.A.), Cornell Univ., Baker Labor. of Chem. *Micro-Chem.* ⓓ Cornell Univ.1897.

Champlain, Alfred B. (1882), Entomol. Harrisburg, Pa. (U.S.A.), Department of Agric., Bureau of Plant Industry. *General Entomol. Coleoptera. Larvae of the Family Cleridae.* ⓟ Coleoptera.

Champlin, Manley (1885), Sr. Prof. of Field Husbandry, Univ. of Saskatchewan. Saskatoon, Saskatchewan (Canada). *Crop improvement.*

Champy, Christian (1885), Prof. agrégé à la Fac. de Méd. Paris V (France), 55, rue Geoffroy Saint Hilaire. *Cytol. générale. Cultures de tissus. Biol. générale: sexualité, action des hormons.* ⓟ Paris 1913.

Chanamirow, Alexander (1901). Ass. Anat. Inst. Rostow a. Don (U.d.S.S.R.). *Variet. d. A. subclavia dextra.*

Chandler, Asa C. (1891), Prof. of Biol., Rice Inst., Officer in charge, Hookworm Research Labor., Calcutta School of Tropical Med. Calcutta (Br.-India), Rice Inst., Houston, Texas (U.S.A.). School of Tropical Med. *Parasit., Med. Entomol.* ⓘ Univ. of California 1914. ⓟ Parasites.

Chandler, Simon B. (1887), Ass. Prof. of Gross Anat. Loyola Univ. School of Med. Chicago, Ill. (U.S.A.), 706 So. Lincoln St. *Histol. of the Parathyroid gland, Accessory Glands, Relation to Sexual Maturity.*

Chandler, Stewart Curtis (1889), Field Entomol. for Southern Illinois, Illinois State Natural History Survey. Urbana, Ill. (U.S.A.). *Fruit insects, peach insects.* ⓘ Univ. of Wisconsin, Madison, 1915.

Chandler, William Henry (1878), Prof. of Pomol. Berkeley, Cal. (U.S.A.), Division of Pomol., Univ. *Killing of plant tissue by low temperature. The effect of blooming and of setting and maturing fruit on growth and other behavior of trees; the effect of the California climate on the hardiness of fruit trees.* ⓘ Univ. of Missouri 1914.

Chaney, Ralph W. (1890), Research Associate, Carnegie Inst. of Washington. Berkeley, Cal. (U.S.A.), Bacon Hall. *Tertiary palaeobotan. of western North America.* ⓘ Univ. of Chicago 1919. ⓟ Tertiary plant fossils.

Chang, Ching-Yueh (1896), Prof. of Plant Morphol. Southeastern Univ. Nanking (China), Department of Botan. *Morphol. and anat. of Vascular Plants. Anat. of Fossil Plants.* ⓘ Chicago 1926. ⓟ Preserved plant material, microscopic slides of plants.

Chang, F. L., Yale Univ. in China. Changsha (China). *Botan.* o

Chang, H. S. (1896), Prof. of Entomol. Univ., Entomol. of the Bureau of Entomol. of Kiangsi Province. Nanking (China), Coll. of Agric. *Insect taxonomy, Locust Control.* ⓘ Univ. of California 1921.

Channon, Harold John (1897), Biochem., Dept. of Experim. Pathol., Univ. of Leeds (England). *Fats.* ⓘ London 1919.

Chapchal, Elias, o. Prof., Dir. Anat. Inst. Univ., Med. Fac. Sofia (Bulgarien). o

Chapman, Abel (1851). Houxty-Wark, S.O., Northumberland (England). *General Zool. Ornithol.* ⓘ M.A. hon. c. Durham 1921.

Chapman, F., Keeper Nat. Mus. of Nat. Hist. Melbourne (Australia). *Palaeont.* o

Chapman, Frank M. (1864), Curator of Birds. New York City (U.S.A.), Am. Mus. nat. Hist. *Geographical distribution and speciation of Birds.* ⓟ Birds.

Chapman, Henry G., Prof. Univ. Sydney (Australia). *Physiol.* o

Chapman, Henry Leonard Russell (1890), Superintendent, Beal Botan. Garden. East Lansing, Mich. (U.S.A.), Michigan State Coll. *Plant breeding, plant immigrants.* ⓟ Plants and seeds.

Chapman, Paul Jones (1900), Instr. in Entomol. at Cornell Univ. Ithaca, N.Y. (U.S.A.), Roberts Hall. *Economic entomol., Corrodentia.* ⓟ Corrodentia.

Chapman, Robert Edward (1804), Lect. in Botan., King's Coll., Univ. London S. E. 5 (England), The Platanes, Champion Hill. *Photosynthesis of water plants. Carbohydrate enzymes of monocotyledon leaves.* ⓘ 1921.

Chapman, Royal Norton (1889), Prof. and Head of the Division of Entomol. and Economic Zool. St. Paul, Minn. (U.S.A.), Univ. Farm. *Animal Ecol.; wood-boring beetles; wings of insects; insects in stored food products; pelvis of burrowing mammals.* ⓘ Univ. of Minnesota 1914.

Chappelear, George Warren (1889), Head, Department of Biol., State Teachers Coll. Harrisonburg, Va. (U.S.A.). *Physiol. Botan. Plants of the Shenandeah Valley of Virginia.* ⓘ Virginia Polytechnic Inst. 1913. ⓟ Plant specimens.

Chappius, Pierre Alfred (1891), Dr. phil., Adjoint à la direction de l'Inst. de Speol. de l'Univ. Cluj (Roumanie), Căsuţa postală 158. *Copepodes d'eau douce: Harpacticides et Cyclopides. Syncarides.* ⓘ Basel 1919. ⓟ Harpacticides d'eaux douces. Syncarides.

Charbonnier, A., Prof. Univ. Caen (France). *Anat.* o

Charitonow, Demetrius (1896), Ass. Zool. Univ., wissensch. Mitarbeiter d. Inst. für biol. Forschung Univ. Perm (U.d.S.S.R.), Zaimka, Zool. Labor. d. Univ. *Geographie, Oekol., Systematik u. Biol. d. Araneae. Entomol.: Coleoptera, schädl. Insekten. Hydrobiol.: Copepoda.* ⓘ Perm 1926. ⓟ Araneina.

Charlemagne, Nikolaus (1887), Zool. a. Zool. Mus. Ukrainische Akademie der Wissenschaften. Kiew (U.d.S.S.R.), Tschudnowskystr. 2. *Ornithol.*

Charles, Vera K., Mycol., Bureau of Plant Industry, U.S. Dept. of Agric. Washington, D.C. (U.S.A.). *Mycol., Pathol.* o

Charusin, Oleg A. (1899). Moskau (U.d.S.S.R.), Zool. Mus. d. I. Moskauer Univ. *Systematische Ornithol.* ⓘ Moskau 1926. ⓟ Vogel-Bälge aus Zentralrußland und N.-Kaukasus.

Chase, Agnes (1869), Associate agrostol., U. S. Department of Agric. Washington, D.C. (U.S.A.). *Taxonomy and morphol. of Gramineae.* ⓟ Grasses (Gramineae) American.

Chase, Ethel W. B., Dean of Women and Ass. Prof. of Botan., Coll. of the City of Detroit, Mich. (U.S.A.), 4041 Cass. Ave. *Ecol., Plant Histol., and Physiol.* o

Chase, R. W. Bewdley, Worcestershire (England), Herne's Nest. *Ornithol.* o

Chase, Samuel Wood (1892), Associate Prof. of Histol. and Embryol., Western Reserve Univ. Cleveland, Ohio (U.S.A.), Labor. of Histol. and Embryol., Western Reserve Univ. School of Med., 2109 Adelbert Road. *Histol. of Urogenital System. Histol. and Histogenesis of Teeth.* ⓘ Harvard Univ. 1921.

Chasen, Frederick Nutter (1896), Curator, Raffles Mus., Singapore (India). *Zool. of Malaysia.*

Chatton, Edouard Pierre Léon (1883), Prof. de Biol. générale à l'Univ. de Strasbourg (France), Inst. de Biol. générale. Fac. des Sc. *Protistol. Flagellés libres et parasit. (Dinoflagellés, Volvocinées, Monadines, Tetramitidés, Trypanosomides etc.). Rhizopodes (Amibes libres et parasit. de l'homme ou des animaux. Protéomyxés. Radiolaires. Haplosporidies. Schizosporidies). Sporozoaires (Grégarines, Coccidies, Plasmodidés, Sarcosporidies). Cnidiés (Microsporidies, Paramyxidies). Ciliés (Nicollellidae. Thigmotriches. Foettingeriidae). Protophytes (Oscillospiracées, Bactériacées). Champignons (Chytridinées, Saccharomycétées, Laboulbéniacées) Biol. générale des Protistes.* ⓘ Paris 1922. ⓟ Préparations microscopiques et cultures de Protistes.

Chauchard, Albert Emmanuel (1874), Dr. ès sc., Dr. en méd., Dir. du Labor. de Physiol. à l'Ecole pratique des Hautes Etudes (Sorbonne). Paris 5e (France), 6, place de la Sorbonne. (Labor.: Paris 5e, 1, rue Victor Cousin.) *Excitabilité. Physiol. du système nerveux.* ⓘ Paris.

Chauchard, Bertha Elise Marie (1884), Dr. ès sc., Dr. en méd., Dir.-adjoint du Labor. de Physiol. à l'Ecole pratique des Hautes Etudes (Sorbonne). Paris 5e (France), 6, place de la Sorbonne. (Labor.: 1, rue Victor Cousin). *Excitabilité. Physiol. du*

Système nerveux. *Analyse de l'excitabilité d'un nerf sécrétoire: la corde du tympan.* ① Paris 1925.

Chaussin, Jules Paul (1871), Dr. ès Sc. Physiques, préparateur de Physiol. générale à la Sorbonne. Paris V (France), 1, rue Victor Cousin. *Physiol. urinaire chez l'homme et chez les herbivores. Physiol. végétale: milieu intérieur dans les plantes.*

Chaves, Pedro Roberto (1887), Ass. et Prof. Libre Fac. med. Lisbôa (Portugal), Rua da Penha de França (à Graça) 230. *Cytol. des glandes: pancréas et foie. Cytogenèse des cellules.*

Cheatum, Elmer Philip (1901), Ass. Prof. of Biol. in the Southern Methodist Univ., Dallas, Tex. (U.S.A.). *Ecol.: Aquatic and terrestrial Snails, ecol. aspects of their life history.* ② Terrestrial and aquatic Snails.

Checchia-Rispoli, Giuseppe (1877), Prof. di Geol. e Paleontol. Univ. di Cagliari. Roma 11 (Italia), Via dei Pianellari. *Foraminiferi, Echinidi viventi e fossili, Crostacei, Molluschi.* ① Roma 1926.

Chelmer-Fainsilber, Rose (1894), Ass. der Anat. Rostow a. Don (U.d.S.S.R.), Anat. Inst. *Anomalie d. Nieren. Verdoppelung d. A. maxillaris interna. Rechtsseitige Lage d. Aortenbogen. Polydactylie.*

Chen, Chao Chi (1898), Ass. in Biochem., Peking (China), Department of Biochem., Peking Union Med. Coll. ① Peking National Univ. 1920.

Chen, E. T., Prof. Univ. Amoy (China). *Psychol.* o

Chen, K. K. (1898), Graduate, Johns Hopkins Med. School, Baltimore, Md. (U.S.A.). *Pharmacol., Biochem.* ① Ph. Wisconsin 1923, med. Johns Hopkins Med. School 1927.

Ch'en, N. S., Doc. Univ. Nanking (China). *Zool.* o

Chen, Shisan C. (1894), Prof. of Zool., National Southeastern Univ. Nanking (China). *Genetics and Experim. Embryol. of Carassius auratus.* ① Nanking 1918. ② Preserved specimens of Chinese goldfishes with many varieties.

Chen, W. Y., Lect. Southeastern Univ. Nanking (China). *Zool.* o

Cher, Sofia Dimitr., Doc. Gosudarstv. Donskoj Univ. Rostov a. Don (U.d.S.S.R.). *Pathol. Anat.* o

Chèrlè Lignière, Massimo (1877), Lib. doc. Anat. Topografica Univ., Già incaricato dell' Insegnamento e della Dir. dell' Ist. anat., Aiuto. Parma (Italia), Ist. di Anat. umana normale della Regia. *Anat. umana normale e topografica.* ① Parma 1903.

Chermezon, Henri (1885), Maître de conférences de Botan. à la Fac. des Sc. Strasbourg, Bas-Rhin (France), Inst. Botan., 7, rue de l'Univ. *Systématique et anat. des Cypéracées.* ① Paris 1911.

Cherrie, George Kruck (1865). Newfane, Vt. (U.S.A.). *Ornithol.* o

Cherry, Thomas (1861), Cancer Research Fellow. Melbourne, Victoria (Australia), Walter and Eliza Hall Inst., Melbourne Hospital. *Relation of Cancer to Tuberculosis.* ① Melbourne 1893.

Chester, Wayland Morgan (1870), Prof. of Biol. Colgate Univ. Hamilton, N.Y. (U.S.A.). *Coelenterata.* o

Chesters, Charles G. C. (1904), Ass. Lect. in Botan., Univ. Birmingham (England), The Botan. Department. *Plant Physiol., factors conditioning growth of Coprophilous Fungi. Ecol.* ① Glasgow Univ. 1926.

Cheung Shui Ngaan, Ass. in Pathol., Anat., Embryol., Hackett Med. School. Canton (China). o

Chevalier, A., Dir. Labor. Univ. Paris (France). *Agron. colon.* o

Chevey, Pierre (1900), Préparateur au Mus. national d'Histoire natur. Paris X (France), 112, rue du Faubourg St. Denis. *Ichthyol. (Biol. des eaux douces, Océanographie). Embryol. des Poissons.* ① Paris 1925.

Cheymol, Jean Henri (1896), Préparateur à la chaire de Biol. générale Coll. de France. Paris (France), Place Marcelin Berthelot.

Cheyney, Edward G. (1878), Prof. of Forestry. St. Paul, Minn. (U.S.A.). Univ. Farm. *Silviculture.*

① Cornell Univ. 1900. ② Native tree seedlings of Minnesota.

Chiappella, A. Riccardo (1868), Libero Doc. in Igiene. Pistoia (Italia), Ufficio d'Igiene. *Microbiol. delle acque.* ① Pisa 1893.

Chiarugi, Alberto (1901), Ass. Ist. Botan. Univ. Firenze (Italia) 14 Via Lamarmora 4. *Citol. ed Embriol. vegetale. Morfol. e anat. delle Fanerogame.* ① Firenze 1924.

Chiarugi, Giulio (1859), Prof. o., Dir. Ist. Anat. Firenze (Italia), Via Alfani, 33. *Anat. dell' uomo, Istol., Embriol.*

Chiatellino, Antonio (1899), Fellow Rockefeller Found. Torino (Italia), Via Thesauro 2. *Fisiol. della ghiandola tiroide.* ① Torino 1924.

Chibnall, Albert Charles (1894). London (England), Univ. Coll. *Plant Biochem.* ① Cambridge 1915.

Chichkoff, Georg (1865), Prof. ordinaire. Sofia (Bulgarie), Ulitza Cepino, 9. *Faune d'eau douce: Copepoda, Turbellaria, Pisces.* ① Genève 1892. ② Poissons d'eau douce: Triclades et Copépodes.

Chien, Sung Shu (1885), Ass. Prof. in Botan. Peking (China), Tsing Hua Coll. *Plant ecol. physiol.*

Child, Charles Manning (1869), Prof. of Zool. in the Univ. of Chicago, Ill. (U.S.A.), Zool. Labor. *Physiol. Integration and Individuation in the simpler organisms: the Nature of Polarity and Symmetry. Physiol. Isolation von Teilen des Organismus. Senescence and Rejuvenescence. Individuality in Organisms. Origin and Development of the Nervous System. Physiol. Foundations of Behavior.* ① Univ. of Leipzig 1894.

Childs, Leroy (1888), Supt. Hood River, Ore. (U.S.A.), Branch Experim. Station. *Entomol. and Plant pathol., control of insect pests and plant diseases of apples and pears.* ① Stanford Univ., California, 1914.

Chilow, Konstantin (1893), Ass. Klinik für Hals, Nasen u. Ohrenkrankheiten, Milit. Med. Akademie. Leningrad (U.d.S.S.R.), Klinitscheskaja 5. *Bau des Labyrinthes. Otolithenfunktion bei Amphibien u. Fischen.*

Chilton, Charles (1860), Rector, Prof. of Biol. Canterbury Coll. Christchurch (New Zealand). *Crustacea: Amphipoda, Isopoda.* ① New Zealand D. Sc. 1893; Hon.L.L.D. Aberdeen 1912. ② Amphipoda, Isopoda.

Chimenkow, Victor G., Obergeol. Comité Géol. de Russie. Moskau (U.d.S.S.R.), Jamskaja, 1. *Palaeont.* o

Chin, H. F., Shantung Christian Univ. Tsinan (China). *Zool.* o

Chino, Mitsusige, Ass. Zool. Labor. Kyôto (Japan), Imperial Univ. *Genetics.* o

Ch'in Yao-T'ing, Prof. Shantung Christian Univ. Tsinan (China). *Biol.* o

Chio, Mario (1880), Dir. Inst. Pharm. Univ. de Sassari (Italia). *Oxide de carbone, cyanure, coagulation du sang.*

Chiodi, Valentino (1898), Ass. di Anat. Ist. superiore di Med. vet. Milano (Italia), Città degli Studi. *Placenta dei ruminanti. Istol. del sistema reticolo-endoteliale.* ① Milano 1925.

Chipp, Thomas Ford (1886), Ass. Dir. Royal Botan. Gardens. London (England), Kew. *Botan. in relation to economics, ecol. and forestry.* ① London 1922.

Chitre, Ganpat Damodar (1883), Haffkine Inst., Parel. Bombay (India). *Indian Plague Research.* ① Bombay Univ. 1904.

Chitrow, Wladimir N., Leiter d. Muratowschen Botan. Gartens u. d. Schapilowschen Landwirtsch. Prov. - Versuchsstation. Muratowo, Orlowsches Gouv. (U.d.S.S.R.). *Angew. Botan.* o

Chittenden, Alfred Knight (1879), Prof. of Forestry, Michigan State Coll. East Lansing, Mich. (U.S.A.). *Forestry. Silviculture. Forest Conditions in Northern New Hampshire. The Red Gum.* ① Yale Univ. 1900.

Chittenden, Frank Hurlbät (1858), Entomol.,Truck Crop Insect Investigations, Bureau of Entomol., U. S. Department of Agric. Washington, D.C. (U.S.A.), 1323 Vermont Ave. *Systematic work on Coleoptera, chiefly Rhynchophora and Chrysomelidae.* ⑨ Univ. of Pittsburgh 1904.

Chittenden, Frederick James (1873), Dir., Royal Horticultural Society's Gardens, Wisley. Ripley, Surrey (England). *Garden plants.* ⑫ Garden plants and seeds.

Chittenden, Russell Henry (1856), Dir. emer. of the Sheffield Sc. School. New Haven, Conn. (U.S.A.), Yale Univ., 83 Trumbull St. *Digestive proteolysis. Physiol. economy in nutrition of man.* ⑨ Yale Univ. o

Chivers, Arthur H., Prof. of Botan., Dartmouth Coll. Hanover, N.H. (U.S.A.). *Fungi.* o

Chlebaroff, Georg (1886), Prof. der speziellen Tierzucht. Agronomische Fac. Univ., Dir. Inst. für spezielle Tierzucht, Leiter der Zentral-Versuchsstation für Geflügelzucht. Sofia (Bulgarien). *Rassengeschichtl. u. kraniol. Forschungen bei Rindern, Schweinen u. Schafen. Biol. Beobachtungen über das Wachstum der Kälber und Ferkel der bulgarischen Landrassen. Wolle-Untersuchungen.* ⑫ Rinder- und Schweineschädel; Schafwolle verschiedener bulgarischer Schafrassen; Photographien verschiedener Haustiere Bulgariens.

Chlebnikowa, Marie (1892), Ass. Zool. Kabinett der Univ. Tomsk (U.d.S.S.R.), Preobrashensk Str. Nr. 25, Kb. 1. *Entomol. Biol. der schädlichen Insekten; Systematik der Hemiptera-Heteroptera.*

Chlopin, G. N., Prof. Med. Inst. Leningrad (U.d.S.S.R.), Petr. st., ul L'va Tolstogo 6—8. *Hygiene.* o

Chlopin, Nikolaus Grigoriewitsch (1897), Priv.-Doc. Inst. Histol. u. Embryol. Med. Militär-Akad. Leningrad (U.d.S.S.R.), Prospekt von Karl Marx, 7 A, W. 2. *Experim. Zell- und Gewebeforschung, Explantation. Vitalfärbung. Bindegewebe, Blut und Epithel.* ⑨ Phil. Petersburg 1922; med. 1923.

Chmelař, František (1891), Doc. der Landwirtschaftl. Hochsch., Vorstand der Sektion für Samenprüfung der Landwirtschaftl. Landesversuchsanstalt in Brünn. Brno (C.S.R.), Květná 19. *Methoden der Samenprüfung und Methoden der Sortenprüfung, Samenbau.* ⑨ 1917. ⑫ Pflanzen und speziell Samen.

Chmelewsky, Ludmila V., Doc. Univ. Rostow a. Don (U.d.S.S.R.), Geol. Kabinett. *Palaeont.* o

Choate, Helen Ashhurst (1882), Associate Prof. of Botan. Northampton, Mass. (U.S.A.), Smith Coll. *Plant Physiol. History of Botan.* ⑨ Univ. of Chicago 1920.

Chochlow, Benedict A., Doc. d. Staatl. Univ. Tomsk (U.d.S.S.R.), Nikitskaja ul. 26. *Palaeobotan.*

Chodat, Fernand François Louis (1900), Priv.Doc. Univ., Ass. Inst. Botan. Univ. Genève (Suisse), 5, Chemin du Square. *Ecol., Génétique; Microbiol. générale et Enzymol.* ⑨ Genève 1921.

Chodat, Robert Hippolyte (1865), Prof. Botan. et Microbiol. Genève (Suisse), Inst. *Cytol., Anat. des lianes, Anat. des fossiles. Algol. Enzymol. Mycètes de fermentation. Physiol. de la nutrition et de la Sexualité. Monographia Polygalacearum.* ⑨ Genève 1887. ⑫ Algues, Champignons.

Cholewa, Josef (1885), Leiter d. staatl. Krankenhauses. Brežice (S.H.S.). *Experim. Cancer.* ⑨ Graz 1917.

von Cholnoky, Béla Eugen (1899). Szeged (Ungarn), Dugonics-Platz 2. *Algen, Diatomeen, Flagellaten, niedere Chlorophyceen und Conjugaten, besonders in ihren öcol. und zytol. Beziehungen (Karyol.).* ⑨ Szeged (Ungarn) 1924.

Cholnoky, Lázló (1899), Ass. Pécs (Ungarn), Chem. Inst., Univ. *Pflanzenfarbstoffe.* ⑨ Univ. Budapest.

Cholodnij, Nikolaj G. (1882). Prof. d. Kiewschen Inst. f. Volksbildung. Kiew (U.d.S.S.R.). *Pflanzen-Physiol. Microbiol, Eisen- und Schwefelbakterien.* ⑨ Kiew 1907. ⑫ Mikr. Präp. von Eisen- und Schwefelbakterien.

Chopra, Bashambar Nath. (1898), Ass., Superintendent Zool. Survey of India. Calcutta (India), Indian Mus. *Entomol. and Carcinol. Survey of Indian Fauna.* ⑨ Lahore, India, 1923.

Choronshitzky, Borys (1871), Dr. med., Leiter der privaten Oto-laryngol. Klinik „Opieka" in Warschau, Sekretär der polnischen Oto-laryngol. Gesellschaft.Warszawa (Polen), Boduena 5. *Entwicklung der Milz, Leber u. des Pfortadersystems; Anat. der Tonsillen. Otol. und Laryngol.* ⑨ Moskau 1898.

Chou, Tsan-Quo (1887), Associate in Pharm., P.U.M.C. Peking (China). *Chem. study of Chinese drugs.* ⑨ Genève 1914.

Choudjakov, Nikolaj Nik., Prof. Timirjasev Akad. Moskau (U.d.S.S.R.), Petrovsko'e-Razumovsko'e. *Pflanzenphysiol. u. Mikrobiol.* o

Chow, H. F., Keeper, Educ. Ass. Mus. Changsha, Hunan (China). *Zool.* o

Chranowa, Anna (1899), Ass. am zool. Inst. Sofia (Bulgarien), Univ., Oboriste 13. *Microbiol., Variationslehre.* ⑨ Greifswald 1926.

Chranewytsch, Wasyl (1887), Prof. Sil'sho Inst. Kamjanec-Podilsky, Ukraine (U.d.S.S.R.), Schewczenkostr. 23. *Theoretische und angewandte Zool.: Entomol. u. besonders Lepidopterol. Säugetiere. Vögel.* ⑫ Schmetterlinge und Puppen aus der Ukraine.

Chranilov, Nicolas (1900), Naturwissenschaftl. Inst. Peterhof (U.d.S.S.R.). *Vergl. Anat. der Fische: Skelettsystem, Kopf, Darmkanal, Schwimmblase (Anat., Physiol. und Entwicklungsgeschichte).*

Chrebtow, Aristokles (1876), Prof. Perm (U.d.S.S.R.), Univ. *Angewandte Botan. (Unkrautflora, Nutzpflanzen).* ⑨ Riga 1908. ⑫ Samen, Herbarien.

Christ, John Henry (1896), In charge Sandpoint Substation, Univ. of Idaho. Sandpoint, Id. (U.S.A.). *Crop adaptation.* ⑨ Iowa State Coll. 1920.

Christeller, Erwin (1889), Dr. med., Dir. der pathol.-anat. Abteilung des Rudolf-Virchow-Krankenhauses Berlin N 65. Berlin-Charlottenburg (Deutschland), Kaiserdamm 84. *Methode der histotopographischen Organgefrierschnitte. Histol.-mikrochem. und mikrophotographische Technik. Pathol. Anat. der Knochenerkrankungen, der Urogenitalorgane, des Typhus abdominalis.* ⑨ Berlin 1914. ⑫ Histotopographische Organgefrierschnitte. Diapositive zur pathol. Anat. und Histol. Pathol.-histol. Praep.

Christensen, Carl (1872), Magister sc., Curator of the Botan. Mus. of the Univ. Copenhagen (Denmark). *Taxonomy of ferns; History of Botan.* ⑨ Copenhagen 1900.

Christensen, Jonas J. (1892), Ass. Plant Pathol., Univ. of Minnesota, Agent, Cereal Invest. U.S. Dept. of Agric. St. Paul, Minn. (U.S.A.), Univ. Farm. *Root rot of cereals; Imperfecti of cereals; corn smut.* ⑨ Univ. of Minnesota 1925.

Christensen, N. P. C., Prof. Dr., Landw. Hochsch. København V (Danmark), Bülowsvej. *Bact.* o

Christian, Henry Asbury (1876), Carney Hosp. and Peter Bent Brigham Hospital. Brooklyn, Mass (U.S.A.), 68 Monmouth St. *Pathol.* o

Christiansen, M., Prof. Dr., Landw. Hochsch. København V (Danmark), Bülowsvej. *Allg. Bact.*

Christiansen, Werner Friedrich (1900), Dr. phil., Wiss. Hilfsarbeiter am Bact. Inst. der Preuß. Versuchs- und Forschungsanstalt für Milchwirtschaft. Kiel (Deutschland), Winterbeker Weg 20. *Bact. Systematik, Pflanzengeographie und Pflanzensoziol. der Phanerogamen und Gefäßkryptogamen, besonders von Schleswig-Holstein.* ⑨ Kiel 1924. ⑫ Phanerogamen und Gefäßkryptogamen.

Christiansen, Willi (1885). Kiel-Gaarden (Deutschland), Brommystr. 36. *Floristik und Pflanzengeographie von Schleswig-Holstein.*

Christiansen-Weniger, Friedrich Johann Georg (1897), Priv.Doc. und Ass. Schwoitsch bei Breslau

(Deutschland). *Vererbungsstudien an Getreide, Variabilität der Ährenform, Entwicklungsrhythmus bei Weizen, Immunitätszüchtung.* ⓟ Breslau 1922. ⓔ Sämereien landw. Nutzpflanzen und Unkräuter.
Christie, Arthur W. (1892), Ass. Prof. of Fruit Products, Univ. of California. Berkeley, Cal. (U.S.A.), 336 Hilgard Hall. *Dehydration.* ⓟ Univ. of California 1915.
Christie, Jesse Roy (1889), Associate Nematol., U. S. Department of Agric. Bureau of Plant Industry. Stationed at Eastern Field Station, East Falls Church, Va. (U.S.A.). *Free-living nematodes, plant infecting nematodes, and nematodes parasit. in insects and other invertebrates.* ⓟ Univ. of Illinois 1918.
Christie, Werner Hosewinkel (1877), Prof., Dir. der Saatzuchtwirtschaft des „Falleskjöpet" Oslo. Hjellum (Norwegen). *Pflanzenzüchtung: Rotklee, Hafer, Gerste. Vererbungslehre: Haustauben, Hühner.* ⓟ Oslo 1915.
Christmann, Georg (1874), Ministerialrat, Dir. der Bayerischen Landesanstalt für Pflanzenbau u. Pflanzenschutz. München (Deutschland), Liebigstraße 20.
Christophersen, Erling (1898), Ph.D., Amanuensis. Oslo (Norge), Botan. Mus. *Plant ecol.* ⓟ Yale 1924.
Chrobák, Adolf (1892), Adj. Mähr. Landw. Versuchsanstalt. Brno (C.S.R.), Schwarze Felder. *Wein, Obst. Mykol.* ⓟ Brünn 1920.
Chrowley, Mark Thomas (1904), Prof. Botany, Assoc. Prof. of Histol., Embryol., Physiol. Fordham Univ. New York City (U.S.A.). *Mammalian Anat. Sexuality in Plants.*
Chrustalev, Alexander Aleks., Doc. I. Univ. Moskau (U.d.S.S.R.). *Hygiene.* ○
Chrustschoff, Grigory (1897), Doc. Moskau (U.d.S.S.R.), I. Staatsuniv. *Histol.: das Bindegewebe der niederen Vertebrata, Gewebezüchtung.*
Chrysler, Mintin A. (1871), Ass. Prof. of Botan., Rutgers Univ. New Brunswick, N.J. (U.S.A.). *Anat. of vascular plants, especially monocotyledons and gymnosperms. Plant life of Maryland.* ⓟ Chicago 1909.
Chu, Hai-Ju (1898), Department of Biol. Yale in China. Changsha, Hunan (China). *Parasit. Helminthes.* ⓟ Soochow 1923. ⓔ *Dipylidium canium with scolex, Clonorchis sinensis, Ancylostoma duodenale.*
Chu, Yuanting T. (1896), Prof. of Biol. Shanghai (China), St. John's Univ. *Plecoptera. Chinese Fishes.* ⓟ Cornell Univ. 1926. ⓔ Chinese Fishes.
Chuang, Chai-Hsuan, Doc. Tsing Hua Univ. Peking (China). *Psychol.* ○
Chubb, Laurence J., M.Sc., Ass. in the Dept. of Geol., Univ. Coll., Univ. of London W.C. 1 (England), Gower St., *Palaeont.*
Chudjakow, Nikolai N., Prof., Landwirtschaftl. Akademie Timirjasew. Moskau (U.d.S.S.R.), Petrowsko-Rasumowskoje. *Pflanzen-Physiol. Bact.* ○
Chun, Woon-Young (1895), Prof. of Dendrol., National Southeastern Univ. Nanking (China), Coll. of Agric. *Taxonomy of Woody Plants.* ⓟ Harvard Univ. 1919. ⓔ Chinese Plants; Chinese Tree, seeds.
Chung, C. C., Prof. Univ., Dir. Botan. Mus. Amoy (China). *Botan.* ○
Chung, Hsin Hsuan (1892), Prof. of Botan., Univ. Amoy (China). *Flora of Southeastern China. Chinese water fungi. Trees and Shrubs.* ⓟ Harvard 1918. ⓔ Botan. specimens.
Chupp, Charles David (1886), Ass. Prof. of Plant Pathol. at Cornell Univ. Ithaca, N.Y. (U.S.A.). *Diseases of garden-vegetables. Classification of the Cercosporas.* ⓟ Cornell Univ. Ithaca, N.Y. 1916. ⓔ Fungous species, Cercosporas.
Church, A. H., Lect. Univ. Oxford (England). *Botan.* ○

Church, Margaret Brooks (1889), Ass. mycol. Washington D.C. (U.S.A.), Bureau of Chem., U.S. Department of Agric. *Saprophytic molds in foods and drugs. Aspergilli and Penicillia.* ⓟ Brown Univ., Providence, R.I. 1918.
Churchman, John Woolman (1877), Prof. of Experim. Therapeutics, Cornell Univ. Med. School, New York City, N.Y. (U.S.A.), 28 St. and 1 Ave. *The Selective Bacteriostatic Effects of Aniline Dyes.* ⓟ Johns Hopkins 1902.
Chythiv, M., Lect. Inst. Narodnoi Osviti. Kamjanec-Podilsky, Ukraine (U.d.S.S.R.), Zatonsky Str. 31. *Zool.* ○
Ciaccio, Carmelo, Prof. Univ. Messina (Italia). *Pathol. gen. Lipoidea. Cellule.* ○
Ciechanowski, Stanisław (1869), Dr., o. Prof., Dir. des Inst. für pathol. Anat. der Univ. Kraków (Polen), Grzegorzeckagasse 16. *Krebs. Appendicitis. Prostatahypertrophie u. verwandte Zustände.* ⓟ Kraków 1894.
Ciferri, Raffaele (1897), Dr. é. Sc. Agric., Dir. de la Station National Agronomique et Coll. d'Agric. de Moca (République Dominicaine). *Mycol.: Ustilaginales, Torulopsidaceae. Maladies du Cacaoyer, Cafetier et Canne à sucre. Biol.: Mycorrhizes, plantes tropicales. Agric.: Acajou.* ⓟ Bologne (Italie) 1920.
Cillero y Angulo, Marcelino (1884), Prof. Catedrático Inst. Burgos (España). *Embriol. Genetica.* ⓟ Madrid 1906.
Cimmermann, S. E., Prof. Mittelasiat. Univ. Taskent (U.d.S.S.R.). *Anat.* ○
Cipollone, T., Prof. Univ. Roma (Italia). *Anat. patol.* ○
Cipriani, Carlo (1895), Ass. vol. de la Clinique Méd. de Turin. Torino (Italia), Via Accademia Albertina 1. *Biochim., nephritiques et hépatiques.* ⓟ Turin 1920.
Cipriani, Lidio (1894), Libero Doc. in Antrop. presso la R. Univ. Aiuto nel Mus. Nazionale d'Antrop. e Etnol. Firenze 3 (Italia), Via del Proconsolo, 12. *Antrop. razziale. Evoluzione biol.*
Ciro, Bortolotti (1878), Insegnante di Sc. Naturali nei R. Licei. Cividale del Friuli (Italia), R. Liceo-Ginnasio. *Zool.: Vermi, Oligocheti terricoli. Paleontol.: Mammiferi.* ⓟ Bologna 1901. ⓔ Animali e piante del Friuli.
Čistović, F. J., o. Prof. St. Inst. f. ärztl. Fortbildung. Leningrad (U.d.S.S.R.), Kiročnaja 41. *Pathol. Anat.* ○
Citterio, Vittorio (1895), Ass. all' Ist. di Anat. e Fisiol. Comparate, Univ. Pavia (Italia), Palazzo Botta. *Anat. dei Vertebrati.* ⓟ Pavia.
Ciuca, Michel (1883), Prof. d'Hygiène Fac. Med. Jassy (România). *Bact. et Hygiène, Immunol., physiol. microb.* ⓟ 1908.
Ciurea, Ioan (1878), Prof. de Parasitol. à la Fac. de Méd. Vét. Bucarest (Roumanie), Splaiul Dr. C. Davila No. 8. *Vers parasites des poissons. Insectes: Simulies.*
Civalleri, A., Doc. Univ. Torino (Italia). *Anat.* ○
Claassen, Peter Walter (1886), Prof. of Biol., Cornell Univ. Ithaca, N.Y. (U.S.A.). *Plecoptera; Stream Pollution Problems; Insect Ecol.* ⓟ Cornell Univ. 1918. ⓔ Plecoptera.
Clach, A. J., Prof. Univ. London S. W. 7 (England), South Kensington. *Pharmacol.* ○
Claessen, Gunnlaugur, Prof. extrao. Univ. Islands. Reykjavik (Island). *Physiol.* ○
Clark, Alfred Joseph (1885), Prof. of Pharm. Edinburgh (England), Univ. *Action of drugs and ions on isolated organs.* ⓟ Cambridge 1910.
Clark, Austin Hobart (1880), Curator, Division of Echinoderms, U.S. Nation. Mus. Washington, D.C. (U.S.A.). *Aves, Insecta, Echinodermata.* ⓟ Harvard Univ. 1903.
Clark, Charles Frederick (1873), Associate Horticulturist United States Department of Agric.

Washington, D.C. (U.S.A.), Bureau of Plant Industry. *Potato breeding: development of new varieties resistant to the virus diseases.* ⓓ Cornell Univ. 1909.

Clark, Eliot Round (1881), Dr. (M.D.), Prof. of Anat., Dir. of the Dept. of Anat., Univ. of Pennsylvania. Philadelphia, Pa. (U.S.A.), Med. School. *Experim. Anat. and Embryol. Living Cells and Tissues. Reactions and Growth of lymphatics, bloodvessels and bone. Development and reactions of wandering cells.* ⓓ Johns Hopkins Med. School 1907.

Clark, Eugene Sumner (1897), Bact. and Chem. Springfield, Ill. (U.S.A.), State Dept. of Health. *Water, sewage.* ⓟ Bacteria from water supplies.

Clark, George Albert (1894), Lect. in Physiol., Univ. Sheffield (England). *Experim. Physiol.* ⓓ Durham Univ. 1917.

Clark, Guy Wendell (1887), Ass. Prof. Biochem. and Pharm. Berkeley, Cal. (U.S.A.), Dept. of Biochem. and Pharm., Univ. of California. *Mineral metabolism.* ⓓ Univ. of California. 1918.

Clark, Hubert Lyman (1870), Ph.D., Curator of Echinoderms, Mus. of Compar. Zool. Cambridge, Mass. (U.S.A.). *Echinoderms. Apodous Holothurians. Ophiurans. Hawaiian and other Pacific Echini. Echinoderms of Iones Strait and South Africa.* ⓓ Johns Hopkins Univ. 1897. ⓟ Echinoderms.

Clark, I. Harold (1899), Ass. Pomol., New Jersey agric. Experim. Station and Ass. Prof. of Pomol., Rutgers Univ. New Brunswick, N. J. (U.S.A.). *Pomol.*

Clark, Lawrence T. (1881), Junior Dir., Med. Research and Biol. Labor., Parke, Davis & Company. Detroit, Mich. (U. S. A.). *Bact. and Biol. products, serums and vaccines, production of med. plants.*

Clark, Lois. Ass. Prof. of Botan. Moscow, Id. (U.S.A.), Univ. of Idaho. *Taxonomy of Hepaticae.* ⓓ Univ. of Minnesota 1919.

Clark, Orton Loring (1887). Ass. Prof. of Botan., Massachusetts Agric. Coll. Amherst, Mass. (U.S.A.). *Plant Physiol.*

Clark, Paul Franklin (1882), Prof. of Bact. Univ. of Wisc., Med. School. Madison, Wis. (U.S.A.), Van Hise St. *Bact.* o

Clark, W. E. le G., Reader Univ. London S.W. 7 (England), South Kensington. *Anat.* o

Clark, W. Mansfield (1884), Prof. of Chem., Hygien. Labor., U.S. Public Health Service. Washington, D.C. (U.S.A.), 25 and E Streets. *Oxidation-reduction potentials and acid-base equilibria. Biochem.* ⓓ Johns Hopkins 1910.

Clarke, James Alexander, Jr. (1891), Chief of Department of Applied Immunol. of Jefferson Hospital Philadelphia, Pa. (U.S.A.), 334 South 21 Street. *Immunol. (Atopy).* ⓓ Jefferson Med. Coll. 1916. ⓟ Extracts (Protein) for use in skin testing and therepeutics of asthma etc.

Clarke, Stephenson. Cuckfield, Sussex (England), Borde Hill, *Ornithol.* o

Clarke, William Eagle. Edinburgh (Scotland), 8 Grosvenor Street. *Ornithol.* o

Classen, Theodor (1882), Prof., z. Z. als Fischereiexpert in Walfischbay (S.A. Union, Südafrika). *Ichtyol., Hydrobiol., Hydrographie.* ⓓ Moskau 1907.

Clausen, Jens (1891), Ass. Royal Vet. and Agric. Coll. København (Danmark), Genetic Labor., Rolighedsvej 23. *Genetics, Cytol. and Evolution. Genus Viola. Taxonomy.* ⓓ Copenhagen 1926. ⓟ Seeds from Viola species.

Clausen, Peter Heinrich (1886), Dr. phil. Kiel (Deutschland), Kleiststr. 27. *Cytol.* ⓓ Kiel 1926.

Clausen, R. E., Prof., Dir. Botan. Inst. Högskola Stockholms (Sverige). o

Clausen, Roy E., (1891) Ph.D., Associate Prof. of Gene tics, Univ. of California. Berkeley, Cal. (U.S.A.), 315 Hilgard Hall. *Genetics of Nicotiana. Chromosomes, hybridisation.* ⓓ Univ of Calif. 1914. ⓟ Seeds of Nicotiana.

Claußen, Peter (1877), o. Prof. der Botan., Dir. des botan. u. pharmacognostischen Inst. u. d. botan. Gartens. Marburg a.d. Lahn (Deutschland), Deutschhausstr. 28 I. *Entwicklungsgeschichte der Pflanzen.* ⓓ Berlin 1901.

Clay, William Marion (1906), Ass. in Biol. Transylvania Univ. Lexington, Ky. (U.S.A.), 440 Columbia Avenue. *Genetica. Eugenics. Phytopath. Heredity of immunity and susceptibility of plants to diseases.*

Clayton, Edward Eastman (1895), Associate in Research, New York Agr. Experim. Station. Riverhead, N.Y. (U.S.A.). *Diseases of vegetable crops.* ⓓ Univ. of Wisconsin 1920.

Clayton, Walter Frederick (1844), Prof. of Pathol. and Histol., and Pathol. Richmond, Va. (U.S.A.), 21 East Main St. *Pathol. pharm. Cellular Pathol. Physical Origin of Nerve Function.* ⓓ Med. Birmingham Med. Coll., phil. Wittenberg 1921. ⓟ Histol. and Pathol. microscopical slides of Human Diseases. Animals and Plants of the United States of America.

Van Cleave, Harley Jones (1886), Associate Prof. of Zool., Univ. of Illinois. Urbana, Ill. (U.S.A.), 318 Natural History. *Morphol. and taxonomy of the Acanthocephala.* ⓓ Univ. of Illinois 1913. ⓟ Acanthocephala.

Cleland, John Burton (1878), Marks Prof. of Pathol., Univ. of Adelaide (S. Australia). *Pathol. Anat. and Histol. Epidemiol. Diseases of natives. Basidiomycetous Fungi.* ⓓ Sydney 1902.

Cleland, Ralph E. (1892), Associate Prof. of Biol., Goucher Coll. Baltimore, Md. (U.S.A.). *Cytol. and genetics of Oenothera.* ⓓ Univ. of Pennsylvania 1919.

Clemens, Wilbert Amie (1887), Dir. Pacific Biol. Sta. Nanaimo, B. C. (Canada). *General Marine and Fisheries Investigations.* ⓓ Cornell, 1915.

Clément, Hugues Emile (1881), Chargé de Cours, Dir. par interim labor. physiol. Fac. des sc. Lyon (France), 37, Quai Gailleton. *Physiol. générale et comparée, Action de la force centrifuge sur larves, vision colorée, l'oxyhémoglobine. Népenthes.* ⓓ Lyon 1921.

Clemente, Leopoldo S. (1896), Ass. Prof. Zool. Manila (Philip. Islands), Dept. of Zool., Univ. of the Philip. *Genetics and allied subjects. Entomol., classification. Mutation and inheritance in Drosophila, mice, rats and guinea pigs.* ⓓ Urbana, Ill., 1923. ⓟ Mutant forms of Philippine Drosophila, Philippine mice and insects.

Clements, Frederic Edward (1874), Associate in Ecol., Carnegie Inst. of Washington. Santa Barbara, Cal. (U.S.A.), Mission Canyon (October—May); Manitou, Colo.(U.S.A.), Alpine Labor.(June—Sept.). *Ecol.: Mycol.* ⓓ Univ. of Nebraska 1898.

Clench, William James (1897), Curator, Mollusks, Mus. of Compar. Zool., Harvard Univ. Cambridge, Mass. (U.S.A.). *Freshwater mollusks: Physa, Limnaea, Physopsis, Isidora and Bulimus.* ⓟ Mollusks of all families.

Le Clerg, Erwin Louis (1901), Ass. Plant Pathol. Colorado Agric. Experim. Station, Instr. in Botan. Fort Collins, Col. (U.S.A.), Botan. Department. *Bacterial Root-rot of Alfalfa; Leaf temperature of Lettuce Tip-burn; Crown Rust Spore Germination; Surface tension in relation to spore germination.* ⓓ Iowa State Coll. 1925.

Clermont, Dominique (1880), Agrégé d'Anat. à la Fac. de Méd. Toulouse (France), 19, Rue Ninau. *Lymphatiques des voies biliaires. Le chordome sacrococcygien.*

Cleveland, Clarence R. (1889), Ass. Entomol., Purdue Univ., Agric. Experim. Station. West Lafayette, Ind. (U.S.A.). *Economic Entomol.* ⓓ Univ. of Wisconsin 1912.

Cleveland, L. R. (1892), Ass. Prof. of Protozool. Boston, Mass. (U.S.A.), Harvard Med. School. *Protozool. Relation of protozoa to their hosts. Symbiosis in insects. Cellulose digestion.* ⓟ Protozoa.

Clifford, Eddy O. Associate Entomol., Experim. Station. Clemson Coll., South Carolina (U.S.A.).

Clinton, George Perkins (1867), Botan., Conn. Agr. Exp. Sta.; Research Associate, Yale Univ. New Haven, Conn. (U.S.A.). *Plant Pathol.; Mycol. Ustilaginales.* ⓓ Harvard Univ. 1902. ⓟ Ustilaginales.

Cloetta, Max (1868), o. Prof. Zürich (Schweiz), Plattenstr. 58. *Digitalisforschung.* ⓓ Zürich 1891.

Close, Charles Philip (1868), Extension Horticulturist, U. S. Dept. of Agric., Washington, D.C. (U.S.A.). *Fruits, nuts, vegetables and ornamental plants.*

Clouston, David (1901), Junior Lect. in Agric. Botan., in Queen's Univ. in Charge of Seed Testing Department, Ministry of Agric. for Northern Ireland. Belfast (N. Ireland). *Physiol. of Germination. Lethal temperatures. Pasture problems.* ⓓ Edinburgh MA. 1923.

Clum, Harold H., Prof. of Botan. and Plant Pathol., Coll. of Agric., Univ. of Porto Rico. Mayaguez (Porto Rico). *Physiol.. Pathol.* o

Clutterbuck, P. W. (1897), Demonstrator Chem. Physiol. Manchester (England), Department of Physiol., Univ. *Biochem.* ⓓ Leeds 1922.

Cluver, Eustace Henry (1894), Ass. Health Officer. Pretoria (South Africa), Union Buildings. *Physiol. Hygiene.* ⓓ London 1925.

Cobau, Roberto, Lib. Doc. R. I. Med. vet. Milano (Italia). *Botan.* o

Cobb, John Nathan (1868), Prof. of Fisheries and Dean of the Coll. of Fisheries. Seattle, Wash. (U.S.A.), Univ. *Fish and Fisheries, biochem.* ⓟ Fishes.

Cobb, Stanley (1887), Prof. of Neuropathol. Boston 17, Mass. (U.S.A.), Harvard Med. School. *Neuropathol. and Neurophysiol.: cerebral circulation.* ⓓ Harvard 1914. ⓟ Neuropathol. microscopical specimens.

Cobbett, Louis (1862), Lect. in Pathol. Univ. Cambridge (England), Pathol. Department, Med. School. *Bact.* ⓓ Cambridge 1884.

Coca, Arthur F. (1875), Prof. of Immunol., Cornell Univ. (Med. Coll.). New York City, N.Y. (U.S.A.), 8 West 16 Street. *Immunol., Hypersensitiveness. Anaphylaxis. Hay Fever and Asthma. Idiosyncrasies.* ⓓ Univ. of Pennsylvania 1900.

Cochran, Doris Mable (1898), Aid, Division of Reptiles and Amphibians, U.S. National Mus. Washington, D.C. (U.S.A.), 2027 First Street, N.W. *Herpetol. Systematic classification.* ⓟ Reptiles and amphibians of the United States.

Cochrau, Samuel, Prof. Shantung Christian Univ. Tsinan (China). *Bact.* o

Cockayne, Leonhard, Ph.D. Ngaio, Wellington (N. Zealand). *Botan. syst.* o

Cockerell, Theodore Dru Alison (1866), Prof. of Zool., Univ. of Colorado. Boulder, Col. (U.S.A.), *Entomol., Apoidea (bees) and fossil insects. Land mollusca, Plants, living and fossil.* ⓓ Honorary D.Sc. Colorado Coll. 1913.

Cody, L. R. (1877), Horticultural Commissioner of Santa Clara County. San Jose, Cal. (U.S.A.), Hall of Justice. *Economic entomol. and pathol.*

Cody, Madison D., Prof. of Botan. and Bact., Univ. of Florida. Gainesville, Flo. (U.S.A.). *Ecol., physiol.* o

Coe, Dana Griswold (1893), Agric. for American Cyanamid Company. Homeaddress: Marysville, Ohio (U.S.A.), 560 North Maine Street; Businessaddress: Oklahoma City, Okla. (U.S.A.), Hotel Skirvin, or American Cyanamid Company, New York City, N.Y. (U.S.A.). *Cotton (and other crops) various soil types and climatic conditions, rates and mixtures of fertilizers.* ⓓ Iowa State Coll., Ames, Ia. 1924.

Coe, Wesley Roswell (1869), Prof. of Biol. and Curator of the Zool. Collections, Yale Univ. New Haven, Conn. (U.S.A.). *Morphol. of Nemerteans. Marine Invertebrate Zool. Cytol. and Genetics. Embryol. of Invertebrates.* ⓓ Yale Univ. 1895.

Coffman, Franklin Arthur (1892), Associate Agronomist in Oat Investigations, U. S. Dept. of Agric. Washington, D.C. (U.S.A.). *Agronomic, Genetic, Oat Plant.* ⓓ Kansas State Agric. Coll. 1922.

Cognetti de Martiis, Luigi (1878), Prof. tit. Genova (Italia), R. Univ., Via Balbi 5. *Sistematica e anat. degli Oligocheti; istol. Platelminti, Molluschi, Ditteri. Vertebrati: Cellule sessuali, retina, pelle, org. pariet. Gregarine Monocistidee, Lymphosporidium.* ⓓ Bari. 1925.

Coghill, George Ellett (1872), Prof. of compar. Anat., The Wistar Inst. of Anat. and Biol. Philadelphia, Pa. (U.S.A.), 36 and Woodland Avenue. *The growth of the nervous system in relation to the development of behavior.* ⓓ Brown Univ. 1902. ⓟ Amblystoma and other amphibians.

Cohen, Barnett (1891), Chem. Washington, D.C. (U.S.A.), Hygien. Labor., U.S. Public Health Service. *Physical chem. of bacterial and tissue reductions. The antiscorbutic vitamine.* ⓓ Yale Univ. 1921.

Cohen, Seymour J. (1896), Associate in Pharm., Univ. of Illinois, 1817 West Polk St., Chicago, Ill. (U.S.A.). ⓓ Rush Med. Coll. 1919.

Cohen-Kysper, Adolf (1863), Dr. med., Arzt. Hamburg (Deutschland), Esplanade 39. *Theoretische Biol. Mechan. Analyse der Veränderungen vitaler Systeme.* ⓓ München 1887.

Cohen Stuart, C. P. (1889), Dr., Botan. der Versuchsstation für Tee. Buitenzorg, Java (Ned. O. Indië), Proefstation voor Thee. *Züchtung, Cytol., experim. Morphol. der Teepflanze; Systematik der Theaceae.* ⓓ Utrecht 1916.

Cohey, Charles (1881), Agrégé à l'Univ., attaché à l'Inst. Pasteur. Bruxelles (Belgique). *Bact. Maladies infectieuses. Pédiatrie.* ⓓ Bruxelles 1903.

Cohn, Alfred Einstein (1879), Member Rockefeller Inst. for Med. Research. New York (U.S.A.), 66 Street and Avenue ,,A". *Physiol. and pharm. of the heart and circulation.* ⓓ Columbia Univ. 1904.

Cohn, Ludwig (1873), Dr. phil., Abteilungsvorsteher am Städtischen Mus. Bremen (Deutschland). *Vergl. Anat. der Wirbeltiere, Reptilien.* ⓓ Königsberg 1896.

Cohn, Philip, Ass. in Botan., Columbia Univ. New York City (U.S.A.), 239 C. Mosholu Parkway. *Histol., Physiol.* o

Coit, John Eliot (1880), Head of Coit Agric. Service. Pasadena, Cal. (U.S.A.), 1880 Linda Vesta Avenue. *Horticult.* o

Coker, Robert E. (1876), Prof. of Zool., Univ. of N.C. Chapel Hill, N.C. (U.S.A.). *Hydrobiol.* ⓓ Johns Hopkins Univ.

Coker, William Chambers (1872), Kenan Prof. of Botan. Chapel Hill, N.C. (U.S.A.), Univ. of North Carolina. *Fungi.* ⓓ Johns Hopkins Univ. 1901. ⓟ Fungi.

Colas-Belcour, Jacques Charles Georges Joseph (1894), Lic., attaché à l'Inst. Pasteur (Labor. de Mr. Mesnil). Paris (France), 96, Rue Talguière. *Biol. des Insectes. Protozoaires. Transmission et cultures des Flagellés sanguicoles.* ⓓ Paris 1923.

Cole, Arch Evan (1895), Ass. Prof. in Zool. Northwestern Univ. Evanston, Ill. (U.S.A.), Univ. Hall. *Invertebrate Physiol. Respiration of Animals that can live in low Oxygen tensions.* ⓓ Univ. of Wisconsin 1920.

Cole, Elbert C. (1891), Ass. Prof. of Biol., William Coll. Williamstown, Mass. (U.S.A.), 7 Southworth Street. *Neurol. particularly of the digestive tract. Methylene blue technique.* ⓓ Harvard Univ. 1924.

Cole, Frank J., Prof. Univ. Reading (England). *Zool.* o

Cole, L. J. (1877), Prof. of Genetics, Univ. of Wisconsin. Washington, D.C. (U.S.A.). *Animal behavior and genetics.* ⓓ Harvard 1906. o

Cole, S. W., Lect. Univ. Cambridge (England). *Biochem.* o

Cole, William Earle (1904), Graduate Ass. in Bact. Knoxville, Tenn. (U.S.A.), Dept. of Bact., Univ. of Tenn. *Food and Dairy Bact.* ⓓ Univ. of Tenn. 1926. ⓟ Bacterial Cultures, Yeasts.

Cole, William Harder (1892), Prof. of Biol., Clark Univ. Worcester, Mass. (U.S.A.). *Experim. Zool.: skin transplantation; tropisms.* ⓓ Harvard Univ., Cambridge, Mass. 1921.

Coleman, George Edward (1867), Research Associate, Hooper Foundation for Med. Research, Univ. of California, San Francisco, Cal. Santa Barbara, Cal. (U.S.A.), Box 119, Hot Springs Road. *Serum prophylaxis and therapy.* ⓓ Univ. of California 1891.

Coleman, Leslie Charles (1878), Prof. of Plant Pathol., Univ. of Toronto (Canada), Botan. Department. *Diseases of Plants.* ⓓ Univ. Göttingen 1907.

Colin, H., Prof. Inst. Cath. Paris (France), Rue de Vaugirard 74. *Physiol. végét.* o

Colizza, Corrado (1888), Ispettore per le malattie delle piante. Roma (Italia), Ministero Economia Nazionale. *Biol. degli insetti dannosi all' agric.* ⓓ Portici 1918.

Colla, Silvia (1902), Ass. Ist. di Fisiol. Univ. Torino (Italia), Via Montebello W. 4. *Fisiol. vegetale. Micol.* ⓓ Torino 1924. ⓟ Fungi parasit.

Collander, Paul Runar (1894), Dr. phil., Ass. am Botan. Inst. der Univ. Helsingfors, Doc. für Pflanzenphysiol. an der Univ. Helsingfors (Finnland), Auroragatan 13. *Zellphysiol., Protoplasmapermeabilität.* ⓓ Helsingfors 1920.

Colle, G., Prof. Univ. Padua (Italia). *Anat.* o

Collett, Mary E. (1888), Associate Prof. of Physiol. School of Nursing, Western Reserve Univ. Cleveland, Ohio (U.S.A.). *General physiol.: toxicity of acids and salts to Infusoria; intracellular enzyme experim. Human physiol.: metabolism and minute volume of the heart.* ⓓ Univ. of Pennsylvania 1919.

Collevati, Umberto (1894), Ass. Volontario. Pavia (Italia), Ist. di Patol. Generale, R. Univ. *Trapianti di occhi.* ⓓ Bologna 1921.

Colley, Reginald H., Pathol. in Charge, Office of Forest Pathol., Bureau of Plant Industry, U. S. Dept. of Agric. Madison, Wis. (U.S.A.), 1712 Madison St. *Mycol., Pathol.* o

Collin, Anton (1863), Kustos und Prof. am Zool. Mus. der Univ. Berlin N 4 (Deutschland), Invalidenstr. 43. *Vermes.* ⓓ Berlin 1888.

Collin, Remy (1880), Prof. d'histol. à la Fac. de Méd. Nancy (France), 31, rue Lionnois. *Histol. de l'hypophyse. Histol. générale du Système nerveux.* ⓓ Nancy 1920.

Collinge, Walter Edward. Keeper of the Yorkshire Mus., York (England). *Economic Biol., Ornithol.; Terrestrial Isopoda. Injurious Insects.* ⓓ St. Andrews 1916.

Collingwood, B. J., Prof. Univ. London (England), South Kensington, S.W. 7. *Physiol.* o

Collins, C. Walter (1882), Entomol., Gipsy Moth and Brown-tail Moth Investigations, Bureau of Entomol., U. S. Dept. of Agric. Melrose Highlands, Mass. (U.S.A₁), Gipsy Moth Labor. *Parasit. and predators of the Gipsy Moth, Brown-tail Moth and Satin Moth.* ⓓ Univ. of Delaware 1913.

Collins, Ernest J., Botan. The John Innes Horticultural Inst. London S.W. 19 (England), Merton Park. *Plant Genetics and Physiol.* ⓓ London.

Collins, Guy U. (1872), Sr. Botan. in charge, Biophysical Labor., Bureau Plant Industry, Lauham, Md. (U.S.A.). *Genetics of maize.*

Collins, Henry B. Jr. (1899), Ass. Curator, Division of Ethnol., U.S. National Mus. Washington, D.C. (U.S.A.). *Ethnol. Physical Anthrop., North American Indians; Craniol., Apes and Man; Anthropometry.* ⓓ George Washington Univ. 1924.

Collins, James F., Pathol., U.S. Dept. of Agric., Curator of the Herbarium and Demonstrator in Botan., Brown Univ. Providence (Rhode Island), 13 Brown St. *Fungous disease of woody plants, trees, mosses.* o

Collins, Julius Lloyd (1889), Ass. Prof. of Genetics. Berkeley, Cal. (U.S.A.), Univ. of Calif. *Genetics of Crepis species; Drosophila melanogaster. Origin of Chromosome variations in Crepis.* ⓓ Univ. of California 1923.

Collip, James Bertram (1892), Prof. of Biochem. Edmonton, Alberta (Canada), Univ. of Alberta. *Biochem. and Physiol. of the organs of Internal Secretion.* ⓓ Toronto 1912.

Collison, Reginald C. (1884), Chief in Research in Agronomy, New York Agric. Experim. Station. Geneva, N.Y. (U.S.A.). *Plant and Soil.*

Colosi, Giuseppe (1892), Prof. tit. Zool. e anat. compar. Univ., Dir. Ist. di zool. ed anat. compar. Univ. di Siena (Italia). *Sistematica e anat. dei Crostacei, Molluschi gasteropodi. Distribuzione degli animali. Fisiol. generale della respirazione.*

Colton, Harold Sellers (1881), Prof. of Zool., Univ. of Pennsylvania, Zool. Labor. Philadelphia (U.S.A.). *Ecol. of Mollusca.* ⓓ Univ. of Pennsylvania 1908.

Columbino, Carlo, Prof. incar. Univ. Milano (Italia). *Physiol.* o

Coman, Francis Dana (1895), Instr. in Anat., Johns Hopkins Univ. Baltimore, Md. (U.S.A.), Johns Hopkins Med. School. *Nervous System: Sympathetic innervation of cross-striated musculature, and of blood vessels. Plurisegmental innervation of skeletal musculature. Cerebellum.* ⓓ Johns Hopkins 1924.

Comandon, Jean (1877), Dr. en méd., Président de la section de cinématographie technique à l'Office National des Recherches sc. et des Inventions. Sèvres près Paris (France), 7, rue Avice. *Cytol., hématol., cinématographie biol. Histophysiol.* ⓓ Paris.

Combemale, Pierre (1893), Prof. agrégé Fac. Méd., Labor. de Physiol. Lille (France), 93, rue d'Esquermes. *Psychol. endocrinienne. Physiol. du système nerveux central.* ⓓ 1919.

Combes, Raoul Pierre Emil (1883), Chargé de cours à la Fac. des Sc. (Sorbonne), Prof. à l'Ecole nationale d'Horticulture et à l'Inst. d'Agronomie Coloniale. Paris X (France), 15 bis, Rue Alexandre Parodi. *Biochim.: Pigments anthocyaniques. Biochim. du jaunissement automnal des arbres. La protéolyse chez les arbres. Composés pectiques. Excrétions par les racines.*

Comel, Marcello (1902), Fellow Rockefeller Foundation. Torino (Italia), Via Rosa Govone 4. *Fisiol. del ricambio dei tessuti. Fisiol. dell' eccitabilità dei tessuti.* ⓓ Torino 1926.

Comes, Salvatore (1880), Lib. Doc. Zool. Anat. Compar. Incaricato di Anat. e Fisiol. compar. Univ. Palermo (Italia), Ist. Zool. *Tropismi negli Insetti. Fisiol. gen. del sistema nervoso. Condrioma delle cellule cartil. Morphol. e Fisiol. dell' oocite dei pesci e dei Mammiferi, Flagellati. Termiti. Infusori. Malaria. Rigenerazione.* ⓓ Catania.

Commeret, Armand Louis Léon (1903), Préparateur de Physiol. à l'Ecole de Med. Marseille (France), 22, rue Docteur Escat.

Compere, Harold (1896), Ass. in Entomol. Riverside, Cal. (U.S.A.), Citrus Experim. Station. *Chalcidoid Parasit. of Coccidae. Taxonomy and Chalcidoid parasit. of Coccidae.* ⓟ

Compton, Leroy Everett (1897), Ass. Pathol. Lafayette, Ind. (U.S.A.), Purdue Agric. Expt. Station. *Cereal Leaf Rust.* ⓓ Purdue Univ. 1926.

Compton, Robert Harold (1886), Prof. of Botan. in the Univ. of Cape Town, Dir. of the National Botan. Gardens of South Africa. Kirstenbosch near Cape Town and Whitehill near Matjesfontein (South Africa). *Systematics distribution and ecol. of the South African flora (Angiospermae, Gymnospermae and Pteridophyta); morphol. and anat. of Filicineae and the seedlings of Angiospermae.* ⓓ Cambridge 1911.

Comstock, John H. (1849), B.S., Prof. emer. of Entomol., Cornell Univ. Ithaca, N.Y. (U.S.A.), 123 Roberts Pl. *Insects-life, Cotton Insects, the wings of insects, Butterflies.* o

Comte, Henri Marie Joseph (1899), Fac. Méd., Inst. d'Anat. Cours Gambetta (France). *Anat.* ⓛ Lyon 1926.

Conard, Henry Shoemaker (1874), Prof. of Botan., Grinnell, Ia. (U.S.A.), Grinnell Coll. *Flora of Iowa. Waterlilies: Genus Nymphaea, Structure and development of the Hay-scented Fern, Dennstaedtia punctilobula.* ⓛ Univ. of Pennsylvania 1901. ⓟ Herbarium material of the Flora of Iowa. Desiderata: Specimens of species occurring in Iowa from „loci classici".

Concepcion, Isabelo (1886), Prof. of Biochem. Manila (Philippines), Coll. of Med., Univ. of the Philippine. *Nutrition and metabolism.* ⓛ Coll. of Med., Manila 1912.

Conde, Otto (1905), Volontär des Pflanzenschutzinst. Riga (Latvija), Kirchenstr. 4a, Wohn. 10. *Tenthrediniden.* ⓟ Tenthrediniden, Cynipiden und Formiciden.

Conde Andreu, José (1895), Prof. agr. d'Anat. descriptive. Zaragoza (España), Paseo Maria Agustin 5. *Anat. du Rachis. Viskosität des Blutes.* ⓛ Zaragoza 1921.

Condelli, Sebastiano (1873), Chef Chim. principal, Dir. du Labor. chim. des Douanes et I. I. de Livourne, Prof. de Chim. générale à l'Univ. de Pise. Livourne (Italia). *Biochim.* ⓛ Rome 1900.

Condorelli, Luigi (1899), Prof. agrégé de Chim. Physiol. et de Pathol. Méd., R. Univ. Napoli (Italia), II. Clinique Méd. *Pathol. et Physiopathol. des échanges nutritifs: hydrates de C (sucre combiné du sang), lipoides, et sels minéraux. Microchém.*

Condorelli, Mario, Prof. incar. Catania, Sicilia (Italia). *Parasit.* o

Cone, Sydney M. (1869), Associate Prof. of Pathol., Univ. of Maryland and Coll. of Physicians and Surgeons. Pikesville, Md. (U.S.A.). *Bone pathol. Anat. and pathol. of the peripheral nerves.* ⓛ Univ. of Pennsylvania 1893.

Conger, Allen Clifton (1887), Assoc. Prof. Zcol. Ohio Wesleyan Univ. Delaware, Ohio (U.S.A.). *Ornithol., Ecol.*

Conklin, Edwin Grant (1863), Prof. of Biol. Princeton, N.J. (U.S.A.), Princeton Univ. *Embryol. of Gastropoda, Brachiopoda, Tunicata; Promorphol. of Ovum; Cell-Lineage of Embryonic Organs; Localization of Morphogenetic Substances in Eggs; Cell-Size, Nuclear-Size, Body-Size; Mosaic Development, Equipotentiality in Mollusca, Tunicata, Amphioxus; Nuclear and Cell-Division; Mitosis and Amitosis; Organic Adaptation.* ⓛ Johns Hopkins 1891, Sc.D. h. c. Ohio Wesleyan Univ., Univ. of Pennsylvania, L.L.D. h. c. Western Reserve Univ.

Conn, Harold Joel (1926), Chief in Research (Soil Bact.) in New York, Agric. Experim. Station, Geneva, N. Y. (U.S.A.). *Bact. of the soil.* ⓛ Cornell Univ. 1911. ⓟ Cultures of various soil bacteria, as for example: Bacillus cereus, Bacillus megatherium, and other spore-forming bacteria as well as certain non-spore-formers not yet named.

Connell, J. S. M., Curator Pathol. Mus. Birmingham (England). o

Conner, Samuel Dicken (1872), Research Chem. Purdue Univ. Agric. Experim. Station. La Fayette Ind. (U.S.A.). *Soil chem. Soil Acidity. Fertilizer availibility. Soil deficiencies.* ⓛ Purdue Univ. 1907.

Connolly, C. J., Prof., Univ. St. Francis Xavier. Antigonish, N.Sc. (Canada). *Biol.* o

Connors, Charles Henry (1884), Associate in Plant Breeding; Head, Department of Floriculture. New Brunswick, N.J. (U.S.A.), New Jersey Agric. Experim. Station. *Breeding of peaches and carnations.*

Conrad, Carl M. (1897), Ass. Plant Physiol. College Park, Maryland (U.S.A.), Univ. *Pectic Constituents of Plants.* ⓛ Univ. of Maryland 1925.

Conrad, John P. (1893), Research Ass. in Agronomy, Univ. of California. Davis, Cal. (U.S.A.). *Agronomy.*

Conradi, Heinrich, Univ.Doc. a. d. Techn. Hochschule. Dresden-Blasewitz (Deutschland), Waldparkstr. 5. *Hygiene. Bact.* o

Conrey, Guy Woolard (1887), Associate in Agronomy, in charge Ohio Soil Survey. Wooster, Ohio. (U.S.A.). Ohio Agric. Experim. Station. *Soil origin and development, and agric. relationships.* ⓛ Ohio State Univ. 1921.

Consoli, L., Prof. Univ. Palermo (Italia). *Zool.* o

Constantineanu, Joan, (1860), Prof. Physiol. végét. Univ. Jassy (Romañia), Labor. *Physiol. végét. Botan. sistématique: Chytridinées, Urédinées, Ustilaginées.* ⓛ Halle a. S. 1906. ⓟ Plantes: Urédinées, Ustilaginées.

Constantinesco, Georges K. (1888), Prof. de zootechnie Fac. méd. vét. Bucarest (Romãnia), Str. Logofatul Taut 7. *Biometrie, Génétique, Vie sexuelle des animaux domestiques. Stérilité.* ⓛ Bucarest 1913.

Convers, Daniel, Prof. Escuela Nac. d. Vet. Bogotá (Columbia), Carrera ya Calle 32. *Histol.* o

Cook, Melville T. (1869), Botan. and Plant Pathol., Insular Experim. Station. Rio Piedras (Porto Rico). *Diseases of Tropical Plants.* ⓛ Ohio State Univ. 1904.

Cook, Orator Fuller (1867), Bionomist, Botan., Bureau of Plant Industry, U. S. Dept. of Agric. Washington, D.C. (U.S.A.). *Tropical agric., botan., zool., evolution.* o

Cook, Walter Robert Ivimey (1901), Demonstrator in Botan. London (England), King's Coll. *Plasmodiophorales, Chytridiales and primitive fungi.* ⓛ Univ. of London 1925. ⓟ Ligniera, Olpidium.

Cook, William C. (1895), Associate Prof. of Entomol., Univ. of Minnesota. St. Paul, Minn. (U.S.A.), Univ. Farm. *Ecol. of arid lands, especially the Noctuidae (Lep.). Insecticide Chem. Attractants and Repellants.* ⓛ Cornell Univ. 1917. ⓟ Noctuidae of western U.S.

Cool, Catharina (1874), Ass. Herbarium Leiden (Holland), Witterozenstr. 34. *Kryptogamen: Mycol.* ⓟ Kryptogamen, besonders höhere Pilze.

Cooley, Jacquelin Smith (1883), Plant Pathol., Department of Agric. Washington, D.C. (U.S.A.). *Storage and transportation diseases of pome and stone fruits. Heterosporium variabile. Cedar Rust of apples. Apple leaves, Gymnosporium. Transpiration of potted potatoes.* ⓛ Washington Univ. St. Louis, Missouri.

Cooley, Robert Allen (1873), Prof. of Entomol., Experim. Station Entomol., State Entomol. Univ. of Montana. Bozeman, Mtn. (U.S.A.), State Coll. *Insects that affect health of man and animals.* ⓟ Different groups of insects. Noctuidae, Orthoptera; Heteroptera.

Coolhaas, Caspar (1895), landwirtschaftl. Ingenieur. Wageningen (Holland), Veluviaweg 8. *Bodenbact.* ⓛ Wageningen 1921.

Coons, George Herbert (1885), Prof. of Botan., Michigan State Coll. East Lansing, Mich. (U.S.A.). *Phytopath.* Univ. of Michigan 1915.

Cooper, Clive Forster (1880), Superintendent, Univ. Mus. of Zool. Cambridge (England). *Vertebrate Palaeont.* ⓛ Cambridge.

Cooper, Delmer C. (1896), Instr. in Biol., Purdue Univ. W. Lafayette, Ind. (U.S.A.), 21 N. Salisbury. *Plant Pathol.*

Cooper, Hugh (1896), Pathol. Imperial Inst. of Vet. Research, Muktsar, Ritani, U.P (Brit. India).

Cooper, John Ralph (1884), Prof. of Horticulture (in charge of Horticulture, Forestry and Landscape Gardening). Fayetteville, Ark. (U.S.A.), Univ. of Arkansas. *Physiol. phases of Horticulture.*

Cooper, William Skinner (1884), Associate Prof. of Botan., Univ. of Minnesota. Minneapolis, Minn. (U.S.A.). *Plant ecol.: plant succession. The Broad-sclerophyll Vegetation of California.* ⓛ Univ. of Chicago.

Coordt, Werner (1900), Ass. Mykol. Inst. Forstl. Hochsch. Hann.-Münden (Deutschland), Philo-

sophenweg 16. *Biochem.: Stoffwechselprodukte von Fadenpilzen.* ⓓ Köln 1924.
Copeland, Edwin Bingham (1873), Retired Prof. Chico, Cal. (U.S.A.). *Tropical agric., and Pteridol.* ⓓ Halle 1896. ⓟ Pteridophyta.
Copeland, William F. (1872), Prof. of Agric. Ohio Univ., Athens, O. (U.S.A.). *Forestry.* ⓓ Ohio Univ. 1907. o
Copenhaver, Wilfred Monroe (1898), Instr. in Anat., Rochester, N.Y. (U.S.A.), School. of Med., Univ. *Experim. Embryol.,* early development of the heart. ⓓ Yale Univ., New Haven, Conn. 1925.
Copson, Godfrey Vernon (1884), Prof. of Bact. at Oregon Agric. Coll. Corvallis, Ore. (U.S.A.). *Bact.*
Corbett, George Hamblin (1887), Government Entomol., Straits Settlements and Federated Malay States. Kuala Lumpur (F.M.S.). *Malayan Insects. Aleurodidae: Heliothrips, Rhynchophorus, Scotinophara, Batrachedra.* ⓓ Edinburgh 1909. ⓟ Aleurodids.
Corbett, Lee Cleveland (1867), Senior Horticulturist in Charge, Office of Horticultural Investigations. Washington, D.C. (U.S.A.), Bureau of Plant Industry, U. S. Dept. of Agric. ⓓ Cornell Univ. 1896.
Cordero, Narciso, Ass. Prof. Univ. of the Philip. Manila, Philippine Islands (U.S.A.). *Physiol.* o
Cori, Carl F. (1896), Biol. Chem. Buffalo, N.Y. (U.S.A.), State Inst. for the Study of Malignant Disease. *Biol. chem.: Carbohydrate metabolism.* ⓓ Prague, C.S.R., 1920.
Cori, Carl J. (1865), o. Univ.Prof., Dir. zool. Inst. deutsche Univ. Praha II (C.S.R.), Viničná 8. *Morphol., Entwicklungsgeschichte, Biol. der Entoprocta, Bryozoa,Phoronidea, Biol. der marinen Fauna, Limnol.* ⓓ Phil. Leipzig 1890, med. Deutsche Univ. Prag 1891. ⓟ Entoprocta, Phoronidea.
Cori, Gerty T. (1896), Ass. Buffalo, N.Y. (U.S.A.), State Inst. for the Study of Malignant Disease. *Carbohydrate metabolism.* ⓓ Prague, C.S.R. 1920.
Corkins, Clifford Leon (1896), Head, Dept. of Entomol. Univ. of Wyo. Laramie, Wyo. (U.S.A.). ⓟ Orthoptera.
Corner, George Washington (1889), Prof. of Anat., Univ. of Rochester, School of Med. and Dentistry. Rochester, N.Y. (U.S.A.), Crittenden Boul. *Histol. and Physiol. of reproductive Organs. Early Mammalian Embryol. History of Anat.* ⓓ Johns Hopkins Univ., Baltimore 1913.
Cornet, Jules, Prof. géol., Ecole des Mines. Mons (Belgique), 12, boul. Elisabeth. *Paléontol.* o
Cornil, Lucien (1888), Prof. agrégé à la Fac. de Méd. de Nancy, Chef des travaux d'Anat. Pathol. Nancy (France), 2, rue Girardet. *Anat. et physiol. pathol.: Système nerveux. Tumeurs cérébrales.* ⓓ Paris 1923. ⓟ Préparations histol. pathol. du Système Nerveux]. Tumeurs du Système Nerveux.
Cornwall, Ira E. (1875), F. R. G. S. Quarantine Station William Head, B.C. (Canada), *Recent and fossil Cirripedia; systematic, ecol., and morphol.* ⓟ Recent Cirripedia. Fossil Cirripedia.
Cornwall, Leon Hastings (1887), Associate in Neurol., Coll. of Physicians, Columbia Univ., Attending Neurol., New York City Hospital and Vanderbilt Clinic. New York, N.Y. (U.S.A.), 55 East 76 Street. *Neuropathol.: Origin of myelin. Multiple Sclerosis. Serol. and Chem. of the spinal fluid.* ⓓ Syracuse 1910.
Corporaal, Johannes Bastiaan (1880), Curator (Keeper) of Entomol. of the R. Zool. Soc. ,,Natura Artis Magistra". Amsterdam (Holland). *Coleoptera: Cleridae orb. terr.* ⓟ Coleoptera: Cleridae.
de Corral, José María (1889), Prof. aux. Univ., Ass. Labor. d. Physiol. d. Junta para Ampliación de Estud. Madrid (España), Sagasta 28. *Wasserstoffionenkonzentrationen. Gasstoffwechsel. Chem. Blutanalyse.* ⓓ Valladolid 1912.

Correia, Ant. A. M., Prof. Univ. Porto (Portugal). *Biol.* o
Correns, Carl Erich (1864), I. Dir. des Kaiser-Wilhelm-Inst. für Biol., Hon.Prof. an der Univ. Berlin-Dahlem (Deutschland), Bolzmannstr. *Die ungeschlechtliche Vermehrung der Laubmoose. Bestimmung und Vererbung des Geschlechtes. Vererbungswissenschaft.* ⓓ München 1889.
Cort, William Walter (1887), Prof. of Helminthol., School of Hygiene and Public Health, Johns Hopkins Univ. Baltimore, Md. (U.S.A.), 615 N. Wolfe Street. *Helminthol.; Epidemiol. of Hookworm disease, Biol. of Hookworms, Trematodes, Ascariasis.* ⓓ Univ. of Illinois 1914. ⓟ Human Helminths.
Cortelezzi, Emilio Daniel (1886), Prof. de Anat. et Fisiol. Comparada en la Fac. de Agron. La Plata (Rep. Argentina), Calle 60 No. 324. *Anat. et physiol. comp. Parasit. (Cecidiol.), Eriofidi. Apparato digerente dei Vertebrati, Istol., Citol., Apparato interno del Golgi, Lacunoma.* ⓓ Med. Pavia 1902, Torino 1925.
Cortés Latorre, Cayetano (1896), Prof. Pharm. Univ. Barcelona (España). *Algues d'eau douce.* ⓓ Santiago 1922.
Corti, Alfredo (1880), Prof. e Dir. Ist. Anat., Fisiol. compar. e Istol. Univ. Torino (Italia), Palazzo Carignano. *Parassit. (Cecidiol.), Eriofidi. Apparato digerente dei Vertebrati, Istol., Citol., Apparato interno del Golgi, Lacunoma.* ⓓ Med. Pavia 1902, Torino 1925.
Corti, Arnold (1873), Dr. phil., Dir. der Chem. Fabrik Flora, Dübendorf (Schweiz). *Palaearktische Agrotinae, Biol., Anat., Genetik.* ⓓ München 1897.
Corti, Emilio (1865), Aiuto nell' Ist. Zool. della Univ. di Pavia (Italia). ⓓ Pavia 1893.
Cory, Ernest Neale (1886), State Entomol., Prof. of Entomol. College Park, Md. (U.S.A.), Univ. *Economic entomol.* ⓓ American Univ. 1926.
Cosma, Joan (1895), Chief of works. Cluj (România), Physiol. Labor. *The acid. base equilibrium of the blood. Blood-gases.* ⓓ Cluj 1923.
Cosmetatos, F. Georges (1876), Prof. Histol. et Embryol. Athènes (Grèce), Rue Scoufa 30. *Embryol. de l'appareil de la vision. Teratogenie spéciale.* ⓓ Paris 1898.
Cosmovici, Nicolas L. (1889), Prof. de Physiol. générale et compar. à l'Univ. Fac. des Sc. Jassy (Roumanie), Rue Sărărie 76. *Physiol. et chim. physiol.: coagulation du sang et du lait. Protéiques du lait. Poisons de l'extrait acqueux des tentacules et des nématocistes d'Adamsia palliata. Venin de Cobra. Physiol. du cœur des Crustacées. Antithrombine. Zool.: Protozoaires, Rotifères de Roumanie. Paléont.: Proidotea Haugi (Isopod). Lamellibranches et Gastéropodes de l'Oligocène moyen de Roumanie.* ⓓ Paris ,,Sorbonne" 1915.
le Cosquino de Bussy, Louis Philibert (1879), Dir. der Abt. Handelsmus. des Kolonial-Inst., Amsterdam, Prof. a. d. Univ. Utrecht. Baarn (Holland), Westeinde 7. *Economische Entomol. Ornithol. Entwicklung von Megalobatrachus.* ⓓ Amsterdam 1904.
Cossio, R., Doc. Univ. Tucumán (Argentina). *Bact.* o
Costa, Antonio (1902), Ass. effettivo. Firenze (Italia), Via dei Malcontenti 4. *Anat. e istol. patol.* ⓓ Firenze 1926.
Costa, Sauveur (1871), Prof. de Bact. à l'Ecole de Méd. de l'Univ. d'Aix. Marseille (France), 89, Boulevard Périer. ⓓ Marseille 1921.
da Costa Lima, Angelo (1887), Prof. d'Entomol. Agric. à Escola Superior de Agric. e Med. Vet., Chef du Service de Vigilance Sanitaire Vegetale à Inst. Bilogico de Defesa Agric. Rio de Janeiro (Brasil), Casa Luiz de Rezende, Rua de Ouvider 116. *Entomol. agric.* ⓓ Rio de Janeiro 1909.
Costantin, J. (1857), Prof. Mus. d'Histoire nat. Paris 5 (France), 200, Rue St. Jacques. *Mycorhizes*

des plantes montagnardes; Pleurobes du Eryngium. ⓓ 1901.

Costantino, Giorgio (1903), Ass. presso il R. Labor. di Entomol. Agraria i. Portici, Napoli (Italia). *Studio morfo-biol.: Ceratitis. Entomol. economica. La coltivazione dell' Anona Cherimolia.* ⓓ 1925. ⓟ Ceratitis capitata.

Costerus, Jan Constantijn (1849). Hilversum (Holland), 21 Ministerpark. *Pflanzen-Teratol.* ⓓ Utrecht 1875.

Cotner, Frank B., Prof. of Botan. and Bact., Montana State Coll. Bozeman, Mtn. (U.S.A.). *Mycol.* o

Cotoni, Louis Alexandre Joseph (1884), Chef de labor. à l'Inst. Pasteur. Paris 2 (France), rue Léon-Vandoyer. *Bact. Préparation du sérum antipneumococcique.* ⓓ Paris.

Cotronei, Giulio, Prof. Univ. Siena (Italia). *Zool., Fisiol., Anat.* o

Cotte, Henri Jules (1873), Prof. à la Fac. des Sc. et à l'École de Méd. de Marseille (France), 213, Rue d'Endoume. *Spongiaires, Cécidies. Sexualité et Microcalorimétrie.* ⓓ Marseille.

Cotter, Ralph (1896), Ass. Pathol., Bureau of Plant Industry, Office of Cereal Crops and Diseases, U.S. Dept. of Agric. St. Paul, Minn. (U.S.A.), Univ. Farm. *Parasitism and specialization of Puccinia graminis.*

Cotterina, G., Doc. Univ. Padua (Italia). *Batteriol.* o

Cotton, A. D. (1879), Keeper of Herbarium, Royal Botan. Gardens. London (England), Kew. *General Systematic Botan. Algol. Lichenol. Mycol. Phytopath.*

Cotton, C. A., M. Sc. Prof. Victoria Coll. Wellington (New Zealand). *Palaeont.* o

Cotton, Richard Thomas (1893), Entomol., U.S. Bureau of Entomol. Washington, D.C. (U.S.A.). *Insects attacking stored products. Effect of high and low temperatures on insects, new gases for fumigation work. Classification of the Curculionid larvae.* ⓓ George Washington Univ. 1924. ⓟ Weevils of the subfamily Calandrinae.

Cottreau, Jean (1877), Ass. au Mus. National d'Histoire Nat. Paris I (France), 252, rue de Rivoli. *Paléont.*

Cougdon, Edgar Davidson (1879), Prof. of Anat. Chulalongkara Med. School Bangkok (Siam). *Development of the Arterial System. Mechanism of Development.* ⓓ Syracuse Univ. 1901.

Coulaeff, Stephan (1896), Wissenschaftl. Mitarbeiter u. Präparator an der Biol. Station zu Bolschewo. Moskau (U.d.S.S.R.), I. Univ., Zool. Mus., Gerzenstr. *Biol. und Histol. der Hirudineen.*

de Coulon, André Sämes (1890), Chef rech. expér. du Centre anticancereux Romand. Lausanne (Suisse), Avenue de Beaumont. *Physico-chim. du cancer.* ⓓ Neuchâtel 1916.

Coulongeat, Cl., Prof. Univ. Poitiers (France). *Hist. Nat.* o

Coulter, John Merle (1851), Boyce Thompson Inst. for Plant Research. Yonkers, N.Y. (U.S.A.). *Plant Morphol.; Angiosperms and Gymnosperms. Rocky Mountain Botany; Plant structures; Plantbreeding; Evolution of sex in plants; Spermatophytae.* ⓓ Hanover Coll., Indiana 1870.

Coulter, Stanley (1853), Dean of School of Sc. and Dir. of the Biol. Labor. Lafayette, Ind. (U.S.A.). Purdue Univ. *Forest trees of Indiana.* ⓓ Hanover Coll. 1889. o

Coupin, Fernande-Henriette, Préparateur au Mus. d'Histoire natur. Paris V (France), 55, rue de Buffon. *Anat. comparée des Mammifères.*

Courmont, M. Paul, Prof. Fac. Méd. Lyon (France). *Bact.* o

Courrier, Robert (1895), Dr., Ass. à l'Inst. d'Histol. de la Fac. de Méd. de Strasbourg (France). *Histophysiol. de l'appareil génital femelle. Déterminisme des Caractères sexuels secondaires chez les Poissons et chez les Mammifères à spermatogenèse périodique. L'oeuf d'Oursin.* ⓓ Strasbourg 1924.

De Coursey, Russell M. (1900), Research Fellow in Entomol. Urbana, Ill. (U.S.A.), Department of Entomol., Univ. of Illinois. *Pollenia rudis Fab.* ⓓ Univ. of Illinois 1925.

Courtade, M. D., Chef de Labor. Fac. Méd. Paris 15 (France), 9, Place Beaugrenelle. *Physiol.* o

Courtland, S. Mudge (1888), Associate Prof. Bacteriol., Univ. of California. Davis, Cal. (U.S.A.), 109/1 Street. *Dairy Bacteriol.* ⓓ Brown Univ. 1915.

Courtois, Père S. J., Dir. des Sikawei Mus. Shanghai (China). *Ornithol.* o

Cousin, E., Prof. Univ. Quito (Ecuador). *Bact.* o

Coutiere, M. H., Prof. Fac. Pharm. Paris 6 (France), 4, Avenue de l'Observatoire. *Zool.* o

Couvreur, M., Prof. Fac. Sc. à Lyon (France). *Physiol.* o

Coventry, Frances A. (1897), Research Instr., Univ. of Chicago, Ill. (U.S.A.), Dept. of Hygiene and Bacteriol. *Parasitol. immunol. aspects.* ⓓ Johns Hopkins Univ. 1924.

Coville, Frederick Vernon (1867), Botan., U.S. Dept. of Agric., Curator, U.S. National Herbarium. Washington, D.C. (U.S.A.), Bureau of Plant Industry. *Taxonomy; Ribes, Juncus, Ericaceae. Breeding and culture of acid-soil plants.* ⓓ Cornell 1887.

Coville, Perkins (1898), Instr. in Forestry. Department of Forestry, Iowa State Coll. Ames, Ia. (U.S.A.). *Forest products, logging, lumbering, wood preservation, kiln drying, lumber markets.*

Coward, Katharine H. (1885), Biochem. in charge Vitamin-testing Department, Pharm. Labor. London W.C. 1 (England), 17 Bloomsbury Square. *Vitamin A and the Lipochromes.* ⓓ M. Sc. Manchester 1908, London 1923.

Cowdry, Edmund V. (1888), Associate Member Rockefeller Inst. for Med. Research. New York, N.Y. (U.S.A.), 66 Street and Avenue A. *Mitochondria, Golgi apparatus, structure of nerve cell, Rickettsiae, Herpes, Vaccinia, Encephalitis, Botalinus poisoning, Heartwater.* ⓓ Chicago 1913.

Cowgill, George Raymond (1893), Ass. Prof. of Physiol. Chem., Yale Univ. New Haven, Conn. (U.S.A.), 333 Cedar Street. *Physiol. of Vitamins; Gastro-intestinal Physiol.* ⓓ Yale Univ. 1921.

Cowles, Henry Chandler. Prof. of Botan., Univ. of Chicago, Ill. (U.S.A.). *Ecol.* o

Cowles, Rheinart Parker (1872), Associate Prof. of Zool. Baltimore, Md. (U.S.A.), Johns Hopkins Univ. *Limnol., Oceanography, Ecol.* ⓓ Johns Hopkins Univ. 1904.

Cox, H. A., Lect. Univ. Cambridge (England). *Forestry.* o

Cox, Leslie Reginald (1897), Ass., Dept. of Biol., British Mus., Nat. Hist. London S.W. 7 (England), Cromwell Road. *Fossil Mollusca.* ⓓ Cambridge 1925.

Cox, Sir Percy Zachariah. London S. W. 1 (England), Whitehall Court 3. *Ornithol.* o

Craig, Charles Franklin (1872), Dir. of Department of Preventive Med. and Clinical Pathol., Army Med. School, Washington, D.C. (U.S.A.), Army Med. Center. *Protozool. and Serol.* ⓓ Yale Univ., New Haven, Conn. 1894.

Craigie, Edward Horne (1894), Ass. Prof. of Comparative Anat. and Neurol. Toronto 5 (Canada), Dept. of Biol., Univ. Compar. neurol. *Vascularity of brain. Olfactory sense in migration of fish, respiratory reflexes on chem. stimulation, rate of growth, weightlength relation, fishes.* ⓓ Toronto 1920.

Cram, Eloise B. (1896), Associate Zool., Bureau of Animal Industry, U.S. Department of Agric., Washington, D.C. (U.S.A.). *Systematic, morphol., pathol. and life history studies of helminths parasit. in domestic animals.* ⓓ Washington 1925.

Cramer, P. J. S., Dr., Dir. Gen. agric. experim. Station. Buitenzorg, Java (Nederl.-O.-Indië). *Selection, Kaffee, Hevea.* o

Crampton, Guy Chester (1881), Prof. of Insect Morphol. Amherst, Mass. (U.S.A.), Massachusetts Agric. Coll. *Insect Morphol., Phylogeny, behavior.* Ⓓ Berlin 1908. Ⓟ Insects of all orders.

Crampton, Henry E. (1875), Prof. of Zool., Columbia Univ. New York, N.Y. (U.S.A.), Barnard Coll. *Evolution, embryol. Variation, heredity.* Ⓓ Columbia 1899.

Crandall, Lee Saunders (1887), Curatur of Birds, New York (U.S.A.), Zool. Park. *Birds in captivity; gen. ornithol.*

Crandoll, Linton Brown (1873), Prof. of Agric. and Specialist in Bee Culture. Storrs, Conn. (U.S.A.). Ⓓ Alfred Univ., N.Y. 1904.

Crane, Harley Lucius (1891), Associate Prof. of Horticulture, West Virginia Univ. and Associate Horticulturist, W.Va., Agr. Exp. Sta. Morgantown, W.Va. (U.S.A.), Coll. of Agric. *Pomol.*

Cranwell, Daniel J., Prof. Univ. Nac. Buenos Aires (Argentina). *Pathol.* o

Crawford, D. L., Prof. Univ. Honolulu (Hawaian Islands). *Entomol.* o

Crawford, James Chamberlain (1880), Ass. Entomol., North Carolina State Department of Agric. Raleigh, N.C. (U.S.A.). *Economic Entomol., Hymenoptera, especially Aculeata.* Ⓓ Washington. Ⓟ Aculeate Hymenoptera.

Creaser, Charles William (1897), Prof. of Zool. Detroit, Mich. (U.S.A.), Coll. of the City, 4841 Cass Ave. *The Taxonomy of Lampreys. Life-history of fishes, species of the Great Lakes.* Ⓓ Univ., Ann Arbor, Mich. 1923. Ⓟ Fresh water Fishes Lampreys.

Creed, Richard Stephen (1898), Lect. Oxford (England), New Coll. *Physiol. of Nervous System and Special Senses.*

Cremer, Max (1865), Dr., Prof. Physiol. Tierärztl. Hochsch. Berlin NW6 (Deutschland), Physiol. Inst., Luisenstr. 56. Ⓓ Würzburg 1887.

Cremonese, Guido (1875), Lib. Doc. d'Igiene Univ., Dir. Ist Biol., Malariol. Roma 34 (Italia), Via Dalmazia 36. *Malaria.* Ⓓ 1901.

Crespi Jaume, Luis (1889), Dr., Prof. d'Agric. Madrid (España), Mus. Nac. de Ciencias Nat., Hipodromo. *Lichens. Physiol. vegetale agricole.*

Cresson, Ezra Townsend, Jr. (1876), Ass. Curator Academy of Nat. Sc. of Philadelphia, Pa. Swarthmore, Pa. (U.S.A.). *Entomol., Diptera, Acalyptrate: Ephydridae and Micropezidae.*

Crèvecoeur, Adolphe (1895). Bruxelles (Belgique), 253, Chaussée d'Ixelles. *Hyménoptères mellifères et vespiformes de la faune de Belgique.*

Cribbs, James Elias (1891), Prof. of Biol., Drury Coll. Springfield, Mo. (U.S.A.). *Botany.* Ⓓ Univ. of Chicago 1918.

Criddle, Norman (1875), Entomol. Canadian Department of Agric., in charge for Manitoba. (Canada), Treesbank. *Economic Entomol.: field crop insects.*

Crile, W. (1864), Prof. of Surgery, Western Reserve Univ., Med. School. Cleveland, Ohio. (U.S.A.), Clinic, Euclid Ave. at E. 93 rd St. *Respiratory System, Bloodpressure, Transfusion, nature of the emotions, Kinetik, Thyroid gland, bipolar interpretation of living processes.* Ⓓ Wooster Univ. Cleveland 1887.

de Crinis, Maximinian (1889), ao. Univ.Prof. für Neurol. u. Psychiatrie. Graz (Deutsch-Österreich), Parkstr. 17. *Stoffwechseluntersuchungen bei Psychosen, im speziellen Epilepsie. Humoralpathol. bei Gehirnkrankheiten. Histochem.* Ⓓ Karl-Franzens-Univ. in Graz 1912.

Crist, John William (1887), Associate Prof. of Horticulture, Michigan State Coll. East Lansing, Mich. (U.S.A.). Ⓓ Univ. of Nebraska, 1923.

Cristiani, H., Prof. Dr. Genève (Suisse), Corraterie No. 15. *Hyg., Bact.* o

Cristol, Paul (1899), Chef des travaux de Chim. Biol. à la Fac. de Méd. Montpellier, Hérault (France), 4, rue de la Barralerie. *Chim. Biol. Chim. du Sang du point de vue physiopath.* Ⓓ Montpellier 1922.

Crocker, William (1876), Dir., Boyce Thompson Inst. for Plant Research. Yonkers, N.Y. (U.S.A.), 1086 North Broadway. *Delayed germination; Effects of noxious gases on plants; Gravity as a form stimulus; injury in greenhouses from illuminating gas.* Ⓓ Univ. Chicago 1906.

Crofts, Doris R. (1893), Lect. in Biol., King's Coll. for Women, Univ. London W. 8 (England), Campden Hill Road. *Rectal gland of Elasmobranch Fishes. Haliotis.* Ⓓ London 1924. Ⓟ Axolotls.

Crosby, Cyrus Richard (1879), Prof. of Entomol. at Cornell Univ. Ithaca, N.Y. (U.S.A.), Roberts Hall. *Economic entomol., arachnol. (spiders and opilinnes).* Ⓟ Spiders and Opiliones.

Crossland, Cyril (1878), Prof. Zool. Labor. Cambridge (England). *Marine Ecol. and Coral Reefs. Cultivation of Mother of Pearl Shell. Polychaete Worms. Exped. Zanzibar, Cape Verde, Panama, Pacific Islands.* Ⓓ Cambridge 1900. London 1900.

Crossman, Samuel Sutton (1886), Entomol., First Ass., Preventing Spread of Moths, Bureau of Entomol., U. S. Dept. of Agric. Melrose Highlands, Mass. (U.S.A.), 964 Maine St. *Biol. Control of Insects.*

Crow, William Bernard (1895), Lect. in Botan., Univ. Coll. of South Wales and Monmouthshire, Univ. of Wales. Cardiff (England), Newport Road. *Principles of Phylogeny; Homol.; Morphol. of Colonial Algae.* Ⓓ Phil. London 1923, Sc. Wales 1926.

Crowden, Guy Pascoe (1894), Ass. Inst. of Physiol. Univ. Coll. London W.C. 1 (England) Gower Street. *Industriae fatigue. Physiol.*

Crowell, Milton Frederick (1895), Field Ass., New Hampshire Department of Agric. Durham, N.H. (U.S.A.). *Life history and habits of Pyrausta nubilalis Hubn. Protozoa and Porifera in New Hampshire.* Ⓓ Univ. of New Hampshire 1925.

Crozier, William John (1892), Associate Prof. of General Physiol., Harvard Univ. Cambridge, Mass. (U.S.A.). *Analytical studies of behavior and sensory physiol. of invertebrates: echinoderms, molluscs; excitation of sense organs; tropisms; temperature characteristics of vital processes.* Ⓓ Harvard Univ. 1915.

Cruess, William S. (1886), Associate Prof. of Fruit Products and Chem. in experim. Sta. Berkeley, Cal. (U.S.A.), 1466 Scenic Ave. *Bact. and physics applied to Fruit Products Industries.* Ⓓ Univ. of Cal., Berkeley, 1911.

Crüger, Otto (1888), Dr. phil., wiss. Mitarbeiter der I. G. Farbenindustrie A.G., Abt. Schädlingsbekämpfung. Landsberg a. d.Warthe (Deutschland), Heinersdorfer Str. 56 III. *Praktischer Pflanzenschutz und Schädlingsbekämpfung, Pflanzenphysiol. (bes. Mikrochem. der Pflanze).* Ⓓ Marburg a. d. L. 1920.

Cruikshank, Douglas, Prof. American Univ. Beirut (Syria). *Pathol. Anat.* o

Cruickshank, George A. (1903), Instr. in Bact., Ass. in Animal Breeding and Pathol. Kingston, R.I. (U.S.A.), Rhode Island State Coll. *Diseases of Poultry: especially Bacillary white Diarrhea, Paralysis, Blackhead of Turkeys, Inheritance of Body Weight in Poultry.* Ⓟ Turkeys. Chickens.

Csiki, Ernö, Dir. d. Zool. Abt. d. Ung. National-Mus. Budapest (Ungarn). *Coleopterol.* o

Csörgey, Titus, Dir. d. Ung. Ornithol. Inst. Budapest II (Ungarn), Debröigasse 15. *Ökonom. Ornithol.* o

Csonka, Frank A. (1889), Protein Investigation Labor., U. S. Dept. Agric., Bureau of Chem. Washington, D.C. (U.S.A.). *Biochem. Nutrition.* Ⓓ Cornell 1924.

Csontos, Josef (1889), Dr., Priv.Doc., Ass. des Bact. Inst. der Veterinär-Hochsch. in Budapest

VII (Ungarn), Hungaria krt. 244. *Bact., Pathol. des Geflügels.* ⑨ Budapest 1912.

Cuatrecasas Arumí, José (1903), Pharm., Prof. aux. de Botan., Fac. Pharm. Univ. Barcelona (España). *Sistematique phanerogamique iberique et Myxomycètes.* ⑨ Barcelona. ⑫ Plantes espagnoles d'herbier.

Cuatrecasas Arumí, Juan (1899), Prof. Fac. Med. Barcelona (España). *Pathol. interne humaine et Endocrinol.* ⑨ Barcelone.

Cuboni, Ettore (1898), Ass. Labor. della Direz., Ist. Sieroterapico Milanese. Milano 24 (Italia), Via Darwin N. 20. *Microbiol. ed Immunol.* ⑨ Bologna 1922.

Cučmarev, Z. I., Priv.Doc., Staatl. Psychoneurol. Inst. Charkow (U.d.S.S.R.). *Psychophysiol.* o

Cuenot, M., Prof. Fac. Sc. Nancy (France). *Zool.* o

Cullen, Glenn Ernest (1890), Prof. of Biochem., Vanderbilt Med. School. Nashville, Tenn. (U.S.A.). *Biochem. Hydrogen Ion Concentration of blood. Blood Electrolytes. Blood gases.* ⑨ Columbia 1917.

Cullinan, Frank P. (1895), Research Associate in Pomol. Lafayette, Ind. (U.S.A.), Department of Horticulture, Purdue Agric. Experim. Station. *Behavior of native fruits; soil management including, fertilizers, causes of differences in growth.* ⑨ Cornell Univ. 1917. ⑫ Scions of fruit trees now growing at this station.

Cullis, W. C., Prof., Univ. London, S.W. 7 (England), S. Kensington. *Physiol.* o

Cullmann, Carl Phil. Herm. (1893), Studienrat. Idar a. d. Nahe, Rheinland (Deutschland). *Bewegungsformen niederer Organismen. Anat. u. morphol. Besonderheiten im Bau der Blüten.*

Cummings, Marshall Baxter (1875), Coll. Prof. of Horticult., Univ. of Vt. Burlington, Vt. (U.S.A.), 236 Loomis St. o

Cummins, Harold (1893), Associate Prof. of Anat., School of Med. Tulane Univ. New Orleans, La. (U.S.A.), Dept. of Anat. *Dermatoglyphics: configurations of epidermal ridges of the palmar and plantar surfaces, human and compar. Vestibular ear: physiol. anat.* ⑨ Tulane Univ. 1925.

Cummins, H. A., Prof., Univ. Coll. Cork (Ireland). *Botany.* o

Cuñat, Sch. P., Prof. de hist. nat. Valencia (España). *Flora de Alcira.* o

Da Cunha, L., Prof. Univ. Rio de Janeiro (Brasil). *Anat.* o

Cunliffe, Norman (1889), Advisory Officer, Ministry of Agric. and Fisheries, Christopher Welch Lecturer in Economic Zool., Univ. Oxford (England), School of Rural Economy. *Resistance of plants to insect attack.*

Cunningham, Bert (1883), Prof. Biol., Head of Biol. Dept. Duke Univ. Durham, N.C. (U.S.A.). *Embryol., Protozoa.* ⑨ Univ. of Wisconsin 1920.

Cunningham, Gordon Herriot (1892), Government Mycol. Dept. of Agric. Wellington (New Zealand), Private Bag. *Diseases of plants, Fungous, bacterial and non-parasit. Systematics of New Zealand Fungi.* ⑨ Wellington, N.Z. 1927. ⑫ Fungi.

Cunningham, Robert Sydney (1891), Prof. of Anat. Nashville, Tenn. (U.S.A.), Vanderbilt Univ. School of Med. *Histo-pathol. of blood and tuberculosis.* ⑨ Johns Hopkins Med. School 1915.

Cunnington, William Alfred (1877), Lect. of Biol., St. Bartholomew's Hospital Med. Coll. London E.C.1 (England). *Crustacea. Limnol.* ⑨ Phil. Jena 1902, Sc. London 1921.

Cunze, Reinhard (1898), Dr. phil. nat. Braunschweig (Deutschland), Fasanenstr. 52. *Moore des Harzes.* ⑨ Freiburg i. B. 1924.

Curran, C. Howard (1894), Entomol. Ottawa (Canada), Entomol. Branch, Dept. of Agric. *Diptera; Stored Product Insects.* ⑫ Diptera.

Currence, Iroy M. (1897), Graduate Ass. Cornell Univ. Ithaca, N.Y. (U.S.A.), Veg. Garding Dept. *Inheritance, Phaseolus vulgaris.*

Curtis, Carlton C., Prof. of Botan. Columbia Univ. New York (U.S.A.). *Nature and development of plants.* ⑨ Syracuse (U.S.A.) 1893. o

Curtis, Francis Roy (1902), Research Ass. Physiol. Middlesex Hospital Med. School. London (England). *Experim. Physiol.* ⑨ Leeds Univ. 1925.

Curtis, George M. (1890), Associate Prof., Univ. of Chicago Hospital Chicago, Ill. (U.S.A.). *Physiol., Pathol. and Anat. of the Thyroid. Endocrine Glands, Metabolism, Kidney Function.* ⑨ Phil. Univ. of Michigan 1914, Med. Rush Med. Coll. 1920. ⑫ Microscopic preparations of the human thyroid gland. Histol. sections of human normal and pathol. glands.

Curtis, M. F., Prof. d'Anat. pathol., Fac. Méd. Lille (France). *Pathol.* o

Curtis, Maynie R. (1880), Associate in Cancer Research, Inst. of Cancer Research Columbia Univ. New York City (U.S.A.), 1145 Amsterdam Ave. *Cancer.* ⑨ Univ. of Michigan 1913.

Curtis, Otis F., Prof. of Plant Physiol., New York State Coll. of Agric., Cornell Univ. Ithaca, N.Y. (U.S.A.). *Plant Physiol.* o

Curtis, Winterton Conway (1875), Prof. of Zool., Univ. of Missouri, Columbia, Mo. (U.S.A.), Lefèvre Hall. *Embryol. of Unionidae, Anat. of Turbellaria, Effect of X-rays upon Regeneration.* ⑨ Johns Hopkins 1901.

Curtois, Father, Keeper, Mus. Siccawei, Shanghai (China). *Zool.* o

Curzi, Mario (1898), Ass. ordinario Labor. Crittogamico di Pavia (Italia), R. Orto Botanico. *Patol. vegetale e Micol.* ⑨ Pisa 1922.

Cusi Ventades, Ernesto (1892), Prof. aux., Univ. Central, Conserv. del Mus. Nac. de Ciencias Nat. Madrid (España), Noviciado 10. *Herpetol.* ⑨ Madrid.

Cutler, Donald Ward (1890), Head of the General Microbiol. Department, Rothamsted Experim. Station. Harpenden (England). *Investigations on the Soil population, relationships between the bacteria and protozoa. Evolution: Heredity and Variation. Cytol.: Marine Biol.: parasit. protozoa and soil protozoa.* ⑨ Cambridge 1913.

Cutler, John Sumner (1899), Ass. in Agronomy. Wooster, Ohio (U.S.A.), Ohio Agric. Experim. Station. ⑨ Ohio State Univ. 1923.

Cutore, Gaetano (1869), Prof. Anat. Univ. Catania (Italia), via Gesuiti 45. *Ghiandola endocrina dell' ampolla deferenziale degli Equidi. Cellule interstiziali del testicolo senescente.* ⑨ Catania 1926.

Cutright, Clifford R. (1893), Ass. Entomol. Wooster, Ohio (U.S.A.), Ohio Agric. Experim. Station. *Fruit insects. Subterranean aphids.* ⑫ Slides of Subterranean aphids.

Cutter, Robert K. (1898), Member of staff, Cutter Labor. Berkeley, Cal. (U.S.A.), 819 Ensenada Ave. *Allergy.* ⑨ Yale Univ. 1923.

Cutting, Ernest Melville (1882), Senior Ass. in Botan. London W. S. 1 (England), Univ. Coll., Gower Str. *Cytol. Fungi.* ⑨ Cambridge 1906.

Cuzner, Harold, Prof. Forest School. Manila, Philippine Islands (U.S.A.). *Dendrol.* o

Czaja, Alphons Theodor (1894), Dr. phil., Priv.-Doc. für Botan. an der Univ. Berlin. Berlin-Dahlem (Deutschland), Pflanzenphysiol. Inst. *Physikalische Chem. der Zelle, Membranen, Entwicklungsmechanik und Genetik der leptosporangiaten Farne.* ⑨ Würzburg 1920.

Czarnecki, Edward Stanisław (1892), Ass. Pathol. Générale. Wilno (Pologne), Zakretowa 15. *Endocrinol.* ⑨ Wilno 1925.

Czarnocki, Wilhelm-Czesław (1886), Pros. Anat. Patol. Fac. Med. Warszawa (Pologne), Chalubinski 5. *Atrophie jaune du foie.* ⑨ Cracovie 1913.

Czekanowski, Jan (1882), Prof. d'Anthrop. et d'Ethnol. Lwów (Pologne), Rue Długosza 8.

Anthropol., Biol. générale, Biométrie. ⊕ Zürich 1906. ℗ Objets craniol.

Czellitzer, Paul Arthur (1871), Dr., Augenarzt. Berlin W 9 (Deutschland), Potsdamer Str. 5. *Familienforschung und Vererbung. Augenärztliche Erbkunde, Astigmatismus- und Myopieerbgang.* Jüdische Familienforschung. ⊕ Breslau 1895.

Czerneckl, Wicenty, Dr. med., Priv.Doc. Fac. Med. Lwów (Pologne), Univ., Inst. de pathol. gén. et expérim. o

Czerniakowska, Katherine (1892), Ass. des Herbars, Botan. Hauptgarten. Leningrad (U.d.S.S.R.), *Systematik, Floristik Persiens und Turkestans, Acantholimon (Plumbaginaceae).* ℗ Herbarmaterial Persiens.

Czerski, S., Dr. Prof. Lwów (Polska), R. Kochanowskiego 40. *Histol. Embryol.* o

Czógler, Kálmán, (1884), Prof. Dr., Szeged (Ungarn), Horvath Mihaly Gasse 2. *Mollusken.* ℗ Mollusken, bes. Najaden.

Czubalski, François (1884), Prof. de l'Univ. de Varsovie (Pologne), Inst. de Physiol. Krakowskie, Przedmiescie 26/28. *Digestion, fonctions sécrétoires, corrélations fonctionnelles.* ⊕ Lwow 1910.

Czurda, Viktor (1897), Dr. rer. nat., Ass. am pflanzenphysiol. Inst. der deutschen Univ. Prag II (C.S.R.), Vinicná 3a. *Entwicklungs-, Ernährungs- und Fortpflanzungsphysiol. der Algen, spez. der Konjugaten. Physiol der Stärkebildung.* ⊕ Deutsche Univ. in Prag 1922.

Dabelow, Adolf (1899) Dr. med. et phil. nat., Prosector, Ontleedkundig Labor. Univ. Amsterdam (Holland), Mauritskade 61. *Vergl. Anat. und Ontogenie des Skeletts.* ⊕ Freiburg i. Br. 1924 und 1926.

Dabrowski, E., Dr. med., Prof. Univ., Dir. biochem. Labor. Poznań (Polska), r. Ś. Pawla 10. *Biochem.* o

Dąbrowski, Wacław, Dr. ingr. Warszawa (Pologne), École Sup. Agric. *Botan.* o

Däniker, Albert Ulrich (1894), Dr. phil. Küsnacht, Zürich (Schweiz), Im Dillilee. *Theoretische Botan. Systematik u.Pflanzengeographie,Oekol.* ⊕ Zürich 1923.

Dafert, Franz (1863), Prof. Wien II/1 (Deutsch-Österreich), Marinellig. 4. *Agrikulturchem.* ⊕ Gießen 1884, Wien 1904.

Dafert, Otto (1888), Ass. pharmakognostischen Inst. Univ. Wien II/1 (Österreich), Marinelligasse 4. *Agrikulturchem, Terephtalydiharnstoff, Gehalt d. Digitalisarten a. wirksamen Stoffen, Studien über Saponine, Düngg. mit Chlorcalcium, Chem. Untersuchg. am Höhlendünger v. Mixnitz.* ⊕ Wien, phil. 1913, jur. 1922.

Dahl, Friedrich (1856), Prof. Dr., Kustos a. D. des Zool. Univ.-Mus. Berlin. Falkenhagen West, Post Seegefeld, (Deutschland). *Systematisch-biol. und tiergeographische Forschungsmethode. Spinnen. Isopoden.* ⊕ Kiel.

Dahl, Knut (1871), Prof. Agric. Coll., Aas near Oslo, Dir. of experim. Freshwater stat. Oslo (Norge). Zool. Mus., Trondhjemsveien 23. *Freshwater biol., Fishery.* ⊕ Oslo 1911.

Dahl, Ore, Kons. Botan. Mus. Oslo (Norge), Töien. o

Dahlgren, Karl Vilhelm Ossian (1888), Priv.Doc. Univ. Uppsala (Sverige), Botan. Inst. *Embryol, der Angiospermen. Botan. Vererbungsforschung.* ⊕ Uppsala 1916.

Dahlgren, Ulric (1870), Prof. of Biol. Princeton Univ. Princeton, N.J., Dir. and trustee of the Mount Desert Island Biol. Labor. Salisbury Cove Maine. Princeton, N.J. (U.S.A.). *Animal Histol., Production of light by animals, Electric fishes, invertebrate Zool.* ⊕ Princeton Univ. 1894.

Dahm, Paul (1895), Dr., Ass. am Botan. Inst. der Univ. Bonn (Deutschland), Bremener Str. 54. *Salz- und Ionenaufnahme der Pflanzen.* ⊕ Bonn 1924.

Dahmer, Georg (1879), Dr. phil., Korrespondent der Preußischen Geologischen Landesanstalt in Berlin. Höchst a. Main (Deutschland), Zeilsheimer Weg 21. *Devon. Palaeont.* ⊕ Marburg a. d. L. 1901.

Dalber, Marie (1868), Tit.Prof. für Zool. an der Univ. Zürich, Prosektor am Zool.- vergl. anat. Labor. beider Hochsch. in Zürich 7 (Schweiz), Krähbühlstr. 6. *Vögel. Trilobita. Merostomata. Arachnoidea. Protracheata. Myriapoda. Pantopoda. Tardigrada.* ⊕ Zürich 1904.

Daikubara, Gintarô, Prof. of Kyûshû Imperial Univ., Chief of Govern. Agric., Experim. Stat. of Suigen, Korea. Fukuoka (Japan), Chem. Inst., Coll. of Agric. *Agric. Chem.* o

Daille, A., Prof. Ecole Nat. vét. Toulouse (France). *Pathol. gén., Microbiol.* o

Dainelli, Giotto (1878), Prof. Geol., Geografia Fisica, Univ., Dir. Ist. Geol. e Paleont. Firenze 14 (Italia), Via Lamarmora 4. *Paleont.: Invertebrati del Giura, Creta, Eocene, Miocene, Quaternario. Geol., Geografia Fisica e Morfol. Terrestre.*

Dakin, William John (1883), Prof. of Zool., Dir. of the Zool. Department, Univ. of Liverpool (England). *Zool., compar. Physiol. Animal sense organs and physiol. of aquatic organisms. Pecten, Buccinum, Pearls, Sex.* ⊕ Liverpool 1906.

Dalbis, Louis Janvier, Prof. de Biol. et d'Histol. à l'Univ. de Montréal (Canada). *Biol. de Ostraea virginica. Asclepias syriaca. Spirochètes du Canada. Développement de l'Osyris alba L. L'immigration des espèces florales dans l'Amérique du Nord.* ⊕ Paris 1921. ℗ Plantes et préparations microscopiques diverses, faune et flore canadiennes.

Dale, Ernest E. (1888), Instr. in Botan., Univ. of Michigan. Ann Arbor, Mich. (U.S.A.), Department of Botan. *Genetics of peppers (Capsicum annuum).* ⊕ Kansas State Agric. Coll. 1920.

Dale, H. H., Dir., Dept. of Biochem. and Pharm., Nation. Inst. for Med. Research. London W.C. 2 (England), 15 York Buildings Adelphi o

Daleg, Albert (1893), Chef des travaux d'Anat. et Embryol. Univ. Bruxelles. Uccle (Belgique), 349 av. Brugmann. *Embryol., Physiol. du développement.* ⊕ Bruxelles 1919.

Dall, William Healey (1845), Honorary Curator of Mollusca, U.S. Nat. Mus. Prof. Palaeont., Wagner Free Inst. of Sc., Philadelphia. Washington, D.C. (U.S.A.),[1] U.S. National Mus. *Sc. exploration of Alaska; geographical distribution, description and classification of mollusca, especially marine forms, and the bivalves. Tertiary Fauna of Florida.* ⊕ Sc. Univ. of Pennsylvania 1904, L.L. Washington Univ. 1915.

Dalla Fior, Giuseppe (1884), Prof. presso il R. Ist. tecnico of Trento (Italia), Via della Collina 8. *Floristica.* ⊕ Wien 1907. ℗ Piante del Trentino.

Dallimore, William (1871), Keeper of Mus. of Economic Botan. London (England), Royal Botan. Gardens, Kew. *Economic Botan., Horticulture, Forestry.*

Damas, D., Prof. Univ. Inst. Ed. v. Beneden d'Anat. comp. et Embryol. Liège (Belgique). *Zool. générale.* o

Damiani, Giacomo (1871), R. Liceo „Morgagni" Forlì (Italia). *Ornithol.; Ichthyol.; Biol. marina; Cetol. del Mediterraneo; Fauna dell' Isola d' Elba.* ⊕ Genova.

Dammerman, Karel Willem (1885), Dir. Zool. Mus. and Labor. Buitenzorg, Java (Ned. Indië). *Zoogeography of the Indo-Australian Archipelago. Mammals.* ⊕ Utrecht 1910.

Dammert, Augusto, Prof. Univ. Lima (Peru). *Anat.* o

Damon, Samuel Reed (1893), Associate in Bact., Johns Hopkins School of Hygien. and Public Health. Baltimore, Md. (U.S.A.), 615 N. Wolfe St. *The production of vitaminly bacteria; relation of certain members of the paratyphoid group to food poisoning; distribution of Clostridium botulinum in the soil; Thermophilice bacteria.* ⊕ Brown Univ. 1921. ℗ Bact. slide preparations and cultures.

Damperow, Demetrius I., Comité Géol. de Russie. Leningrad (U.d.S.S.R.), W.O., Sredny Prospekt, 72-B. *Palaeobotan.* ○

Dampf, Alfons (1884), Dr. phil., Reg.R., Prof. und Regierungsentomol. Mexiko, D.F. (Mexiko), Avenida Insurgentes 171. *Morphol. und Anat. der Lepidopterenlarven. Generationsorgane der Lepidopteren; Systematik der Aphaniptera; Wanderheuschrecke; zirkumpolare Biogeographie; Biocoenose der Hochmoore. Landwirtschaftliche und med. Entomol.; Biocoenotik.* ① Königsberg i. Pr. 1909. ② Aphaniptera.

Dana, Bliss F. (1891), Ass. Plant Pathol., Ass. Prof. Plant Pathol., State Coll. of Washington. Bullmann, Wash. (U.S.A.). *Plant pathol. Virus diseases of field crops.*

Dana, Samuel Trask (1883), Dir., Northeastern Forest Experim. Station. Amherst, Mass. (U.S.A.). *Protection of forests from fire.* ① Yale Univ. 1907.

Danchakoff, Vera, Ass. Prof. of Anat., Columbia Univ. New York (U.S.A.), 437 West, 59 St. ○

Danforth, Charles Haskell (1883), Prof. of Anat. Stanford Univ., Cal. (U.S.A.). *Human heredity, Anthrop. (physical), Anat. hair, hypertrichosis.* ① Washington Univ. 1912.

Danforth, Stuart T. (1900), Prof. of Zool., Coll. of Agric. of the Univ. of Porto Rico. Mayagüez (Porto Rico). *Ornithol. of the Western Hemisphere, particularly Neotropical. Antillean Herpetol. Entomology.* ① Cornell 1925. ② West Indian Birds, Reptiles and Insects.

Dangeard, P. A. (1862), Prof. à la Sarbonne, Membre de l'Inst. (Académie des sc. de Paris). Paris (France), 1 rue Victor Cousin. *Sexualité des Champignons, Sexualité générale. Structure et développement des organismes inférieurs; Protozoaires, Protophytes. La symbiose chez les Légumineuses. La structure cellulaire.*

Daniel, John Franklin (1873), Prof. of Zool. Univ. of California. Berkeley, Cal. (U.S.A.). *Elasmobranch Fishes. Morphogenesis.* ① Johns Hopkins Univ. 1909.

Daniel, Lucien-Louis (1856), Prof. de Botan. appliquée à l'Univ. de Rennes (France), 6, rue de la Palestine. *Horticulture scientifique, la greffe, sa théorie et ses applications. Les bractées de l'involucre des Composées. Les greffes herbacées et ligneuses. Les hybrides de greffe.* ① Rennes 1903.

Daniel, O., Doc. Univ. Tartu (Estland), Aia tän 46. *Botan. Forstw.* ○

Danielewicz, Vladislas, Dir. Jardin de Saxe. Warszawa (Pologne). *Botan.* ○

Danielopolu, Daniel (1884), Prof. Fac. Méd. Univ. Bucarest (România), Strada L. Catargi, 5 bis. *Système nerveux végétatif.* ① Bucarest 1910.

Danilewsky, Wasilij (1852), Ukrainische Akad. der Wissensch. Kieff. Charkow, Ukraina (U.d.S.S.R.), Mironossitz-Straße, 68. *Arbeitsphysiol. u. Elektrophysiol. Parasit. comparée du sang.*

Danilov, Athanasius (1879), Ober-Konserv. des Botan. Hauptgartens. Leningrad (U.d.S.S.R.), Pessotschnaja 2. *Biol. und Physiol. niederer Pflanzen. Algen, Pilze, Flechten und Moose.* ① Leningrad 1911.

Danilov, N. P., Prof. Med. Inst. Moskau (U.d.S.S.R.), Ul. Spartakov 2. *Hygiene.* ○

Danilov, Nicolaus Vasiljevitsch (1900), Ass. Physiol. Labor. Univ. Rostow a. Don (U.d.S.S.R.), Suvorowskaja 41. *Nervenregeneration, Physiol. des Herzmuskels.* ① Rostow a. Don.

Danilow, Simeon (1898), Ass. der Anat. Rostow a. Don (U.d.S.S.R.), Anat. Inst. *Mißbildung des Uterus. Morphol. der Brustbeines. Varietäten d. A. brachialis. Topographie d. Brustorgane.*

Danilowsky, Johann W., Comité Géol. de Russie. Leningrad (U.d.S.S.R.), W.O., Sredny Prospekt, 72-B. *Palaeont.; speziell Quartärmollusken und Devonfauna.* ○

Dannenberg, Arthur (1865), Prof. Techn. Hochsch. Aachen (Deutschland), Krefelder Str. 4. *Geol. d. Steinkohlenlager.* ○

Dannevig, Alf (1886), Cand. real. Vorsteher der Brutanstalt für Seefische und Hummern bei Flødevigen. Arendal (Norwegen). *Fischkultur (marine), Hummerkultur, Biol. der Fische und Hummern.*

Danser, Benedictus Hubertus (1891), Doctor Botan. et Zool., Ass. am Herbarium des Botan. Gartens. Buitenzorg, Java (Nederl.-Indië). *Beschreibung, Systematik und Floristik der Polygonaceen (Rumex und Polygonum), besonders für Europa und Niederländisch-Ostindien. Systematik, Beschreibung und Floristik von Nepenthes. Artbastarde (Pflanzen und Tiere) und deren Eigenschaften (besonders Fertilität und Sterilität).* ① Amsterdam 1921.

Dantan, J., Maître de Conf. Alger (Afrique), Univ. *Zool. et Parasit.* ○

von Darányi, Gyula (1888), Doc. Univ., Oberbact. Budapest IV (Ungarn), Városház-ut 10. *Bact.: bes. pyogene Arten, Sporenbildende Bazillen, Immunitätsreaktionen, kolloidale Eigenschaften des Blutserums.* ① Budapest 1912. ② Bakterienkulturen.

Darbaker, Leasure Kline (1879), Prof. Histol. Pharm. and Bact., Univ. Pittsburgh, Dept. Pharm., Pittsburgh Coll. of Pharm. Pittsburgh, Pa. (U.S.A.), 7025 Hamilton Ave., Homewood Station. *Plant Path., Bact., Histol. Pharm.* ① Univ. Pittsburgh 1910.

Darbishire, Otto Vernon (1870), Prof. of Botan., Head of Botan. Labor. and Univ. Gardens. Bristol (England), The Univ. *Structure and development of Lichens. Marine Algae.* ① Kiel 1895.

Darini, Eugène S. (1894), Priv.Doc., Ass. Histol. Labor., Mitarbeiter Biol. Forschungsinst. Perm (U.d.S.S.R.), Zaïmka, Univ. *Übergangsepithel, Morphol. des Blutes und Bindegewebes bei Crustaceen, Beziehungen zwischen Bindegewebe und Muskelfaser, Experim. Untersuchung der Blutgefäße bei Crustaceen.*

Darling, Chester A. (1880), Prof. of Biol., Allegheny Coll. Meadville, Pa. (U.S.A.). *Bact. Spoilage in Fresh Fruits and Vegetables.* ① Columbia Univ., New York City 1909.

Darlington, Henry Townsend (1875), Associate Prof. of Botan., Dir. of the Beal Botan. Garden. East Lansing, Mich. (U.S.A.), 224 Bailey Street. *Ecol. and Taxonomy of higher plants.* ① Univ. of Chicago 1923. ② Plants and seeds from North America.

Darrow, George McMillan (1889), Associate Pomol., U.S. Dept. of Agric. Washington, D.C. (U.S.A.). *Physiol. of small fruits.* ① Johns Hopkins 1926. ② Fragaria, Rubus, Ribes, Prunus.

Darrow, William Hinds (1890), Extension Prof. of Pomol. Storrs, Conn. (U.S.A.), Conn. Agll. Coll. *Fruit Growing.* ① Univ. of Maine, Orono, Me. 1913.

Darst, Williard Holden (1884), Prof. of Agronomy, North Carolina State Coll. Raleigh, N.C. (U.S.A.). *Field Crops.* ① Cornell Univ. 1918.

Dart, Raymond Arthur (1893), Prof. of Anat. Johannesburg (South Africa), Univ. of the Witwatersrand. *Neurol., Anthrop.* ① Sydney 1926.

Darwin Wen, Prof. Dr., Dir. Labor. des sc. sericicult. Univ. Nantung (China). *Seidenraupenzucht.* ○

Darzine, Egone (1894), Ass. Univ. Riga (Latvija). *Sérol.* ① Paris.

Das, Balkishan, Recogn. teacher, Univ. Delhi (Br.-India). *Biol.* ○

Dasso, Héctor, Prof. Univ. Nac. La Plata (Argentina). *Microbiol.* ○

Daubigny, Théodule, Prof. tit. Ecole Méd. Vét. Montreal (Canada). *Anat. génér.* ○

Daukes, Sidney Herbert, Dir. Wellcome Mus. of Med. Sc. London, W.C. 1 (England), 25/28 Endsleigh Gardens. *Pathol. and Parasit.* ① Cambridge 1905.

Daut, Karl (1863). Bern (Schweiz). *Ornithol. der Schweiz.*

Dauth, Albert, Prof. tit. Ecole Méd. Vét. Montreal (Canada). *Pathol. génér., histol., bact.* o
Dautrebande, Lucien (1894), Ass. Fondation Reine Elisabeth, Bruxelles (Belgique), Hôpital Brugmann. *Physiol. et Physiopath. de la Respiration. Equilibre acido-base.*
Dauvart, Anna (1892), Ass. am Vergl. anat. und experim. zool. Inst. der Latvija-Univ. Riga (Lettland), Kr. Barona ielā 52, dz. 5. *Geschlechtliche Saisonvariation am Froschskelett. Heterotopie des Froschhodens. Selektionsstudien an Mäusen.*
Davenport, Charles Benedict (1866), Dir. Dept. of Genetics, Carnegie Inst. of Washington. Long Island, N.Y. (U.S.A.), Cold Spring Harbor. *Genetics, especially of human (and mammalia) growth and development. Experim. Morphol. Inheritance in Poultry.* ⓓ Harvard Univ. 1893.
Davenport, Eugene (1856). Woodland, Mich. (U.S.A.). *Agric., Breeding.* ⓓ Michigan Agric. Coll. 1878.
Davey, A. J., Lect. in Botan. Bangor (Gt. Britain), Memorial Buildings, Univ. Coll. of N.Wales. *Seedling Anat. Cytol. of Rhodophyceae.* ⓓ London 1912.
Davida, Eugen (1884), Univ.Prof., Dir. anat. Inst. Univ. Szeged (Ungarn), Kossuth S. sugárút 40. *Absolutes u. relatives Volumen der menschlichen Knochen, Obliteration der Schädelnähte.* ⓓ Budapest 1907.
Davidoff, Božimir (1870), Prof., Jardin botan. Sofia (Bulgarie). *Flore, écol., géographie botan.* ⓓ Sofia 1893.
Davidovitch, Stanislav (1891), Senior Ass. Genetics Stat. of Acclimatation. Detskoe Selo (U.d. S.S.R.), 2 Kolpinskaja St. *Genetics of different characters in Nicotiana.*
Davidson, David (1900), Ass., Rockefeller Inst. for Med. Research, New York City (U.S.A.), 66 St. and Ave. A. *Biochem. Oxydations, Syntheses.* ⓓ Yale Univ. 1924.
Davidson, Hamish Reid (1893), Univ. Demonstrator in Agric., Ass., Animal Nutrition Inst. Cambridge (England), School of Agric. *Swine Husbandry.* ⓓ Cambridge 1920.
Davidson, J. (1885), Entomol. Harpenden, Herts. (England), Rothamsted Exp. Station. *Aphides; Ixodoidea; Mosquitoes; Flies and Lice.* ⓓ Liverpool 1915.
Davidson, Jehiel, Associate Chem., Bureau of Chem., U.S. Department of Agric. Washington, D.C. (U.S.A.). *Plant nutrition, chemical composition of cultivated plants.* ⓓ Cornell Univ., Ithaca, N.Y. 1914.
Davidson, John (1878), Ass. Prof. of Botan., and Botan. in charge of the Botan. Gardens and Herbarium of the Province of B.C. Vancouver (British Columbia), Dept. of Botany, Univ. of B.C. *Economic Botan., Dendrol., Botan. Survey of British Columbia, Taxonomy.* ⓟ Herbarium specimens of B.C. Flora, Garden seeds of Native plants.
Davidson, Viola M., during summer at Atlantic Biol. Sta. St. Andrew's, N. B. Can.; during winter at Biol. labor., Univ. of Toronto (Canada), 338 Albany Avenue. *Plankton: Ostracoda, Diatom.* ⓓ Toronto Univ. 1925.
Davidson, William M. (1887), Associate Entomol., I and F Board, U.S. Dept. Agric. Vienna, Va. (U.S.A.). *Insecticides.* ⓓ Stanford Univ., Calif. 1910. ⓟ Syrphidae of North America.
Davies, Arthur Morley (1869), Ass. Prof. of Palaeont. London S.W. 7 (England), Dept. of Geol., Imperial Coll. of Sc. and Technol. *Foraminifera, Mollusca.* ⓓ London 1907. ⓟ Fossils.
Davies, William (1899), Research Agronomist in charge of Herbage Plants. Aberystwyth, Wales (England), Agric. Buildings, U. C. W. *Seeds mixture problems.* ⓓ Univ. of Wales. 1925.
Davies, William Maldwyn (1902), Research Entomol. Harpenden, Herts. (England), Rothamsted Experim. Station. *Collembola, Control of Weeds by insect pests in New Zealand and Australia.* ⓓ Univ. Coll. of North Wales 1924.
Davis, Bradley Moore (1871), Prof. of Botan., Univ. of Michigan. Ann Arbor, Mich. (U.S.A.). *Genetics and cytol. of Oenothera.* ⓓ Harvard Univ. 1895.
Davis, Charles Deforest (1878), Ass. Prof. of Farm Crops. Manhattan, Kan. (U.S.A.), Kansas State Agric. Coll. *Grain and forage crops.* ⓓ Kansas St. Agric. Coll. 1926.
Davis, Donald Walton (1882), Prof. of Biol., Coll. of William and Mary. Williamsburg, Va. (U.S.A.). *Genetics of Genus Impatiens.* ⓓ Harvard Univ. 1913.
Davis, Edgar W. (1895), Junior Entomol. Richfield, Utah (U.S.A.), Box 342. *Eutettix tenella.* ⓟ Hemiptera.
Davis, Everett F. (1901), Fellow-in-Biochem. Boyce Thompson Inst. for Plant Research. Yonkers, N.Y. (U.S.A.). *Plant Physiol., Plant Path., Plant Chem.* ⓓ Rutgers Univ. 1923.
Davis, Hallowell (1896), Tutor and Instr. in Biochem. Sc. Harvard Univ. Boston, Mass. (U.S.A.), Labor. of Physiol. Harvard Med. School. *Physiol. of the Neuro-Muscular system; phenomenon of conduction.* ⓓ Harvard Med. School 1922.
Davis, Herbert Perry (1889), Prof. of Dairy Husbandry-Univ. of Nebraska, Chairman of Department. Lincoln, Neb. (U.S.A.). *Dairy cattle genetics, and nutrition.* ⓓ Pennsylvania State Coll. 1914. ⓟ Dairy cattle: Holstein-Friesian, Jersey, Ayrshire, Guernsey. Rats, Albinos and mixed.
Davis, John Henry (1901), Ass. Prof. of Biol. Davidson, N.C. (U.S.A.). *Plant Ecol. of the Western North Carolina Mountains.* ⓓ Davidson Coll. 1924.
Davis, John Jefferson (1852), Curator of Herbarium, Univ. of Wisconsin. Madison, Wis. (U.S.A.), Biol. Bldg. *Parasitic fungi of Wisconsin.* ⓓ Hahneman Med. Coll., Chicago 1875. ⓟ Parasitic fungi.
Davis, John June (1885), Head Dept. Entomol. Purdue Univ. and Purdue Agric. Exper. Sta. Lafayette, Ind. (U.S.A.). *Insdcts of economic importance.* ⓓ Univ. of Illinois 1907.
Davis, Leslie John, Ass. Bact., Wellcome Bureau of Sc. London W.C. 1 (England), 25/28 Endsleigh Gardens. *Bact.* ⓓ Edinburgh 1924.
Davis, Nelson F. (1872), Prof. of Biol. Lewisburg, Pa. (U.S.A.), Bucknell Univ. *Apple-tree tent caterpillar, chestnut culture.* ⓓ Lewisburg, Bucknell Univ., 1903. o
Davis, Robert Lesley (1891), Associate Agronomist, U.S. Agric. Experim. Station. Mayaguez (Porto Rico). *Genetics. Sugar cane and field corn (maize). Self-fertilized lines.* ⓓ Illinois 1916.
Davis, Ward B. (1893), Associate Chem. Los Angeles, Cal. (U.S.A.), 148 So. Mission Road. *Plant physiol. and biochem.* ⓓ Univ. of Chicago 1924.
Davis, William Harold (1876), Ass. Prof. of Botan. at the Massachusetts Agric. Coll. Amherst, Mass. (U.S.A.). *Mycol. and Plant Pathol.* ⓓ Cornell Univ. 1912, Wisconsin Univ. 1922.
Davis, Wilmer E. (1867), Associate Prof. of Botan, Kansas State Agric. Coll. Manhattan, Kan. (U.S.A.). *Seed germination.* ⓓ Univ. of Illinois 1903.
Davy, Joseph Burtt (1870), Hon.M.A. Ph.D. Lecturer in Tropical Forest Botan., Imperial Forestry Inst. Univ. of Oxford (England), Parks Rd. *Systematic Botan. (Phanerogamae); Ecol.; Forest Botan.* ⓟ Herbarium specimens.
Davydov, Paul N., Dir. Stat. Défense des Plantes. Omsk, Siberie (U.d.S.S.R.). *Pathol. végétale, Ustilaginées, Fungicides.* o
Dawidow, Waldemar (1899), Aspirant Forschungsinst. für Zool. bei Phys. Mathemat. Fac. I. Univ., Ass. Biol. Station. Bolschewo (U.d.S.S.R.). *Allgemeine Zytol.: Vergl. Morphol. der Zellkernstrukturen bei Wirbellosen.*

Dawson, Alden Benjamin (1892), Associate Prof. of Biol., Coll. of Arts and Pure Sc., New York Univ. New York, N.Y. (U.S.A.), Dept. of Biol., Univ. Heights. *Vertebrate Zool. Experim. and histol. study of glands. Development of the Vertebrate Skeleton.* ⓟ Harvard Univ. 1918.

Dawson, Arthur James (1888), Instr. in Zool. Cambridge, Mass. (U.S.A.), Zool. Labor., Harvard Univ. *Protozool. Physiol., Inheritance, Life Cycles.* ⓟ Yale 1918.

Dawson, Percy Millard (1873), Prof. of physiol., Univ. of Wisc. Madison, Wis. (U.S.A.). *Physiol.* o

Dawson, William Leon (1873), Dir. International Mus. Compar. Oology. Santa Barbara, Cal. (U.S.A.), „Los Colibris" Mission Caxugon. *Ornithol.* o

Dawydoff, Constantin (1880), attaché au Labor. Arago. Banyuls s. Mer, Pyr. Or. (France). *Réduction, restitution des animaux; embryol. des Invertébrés.*

Dawydowskie, Hyppolyte (1887), Prosektor Inst. Path. Anat. I. Univ. Moskau (U.d.S.S.R.), Troubezkoi pereulok. 10, Log. 1. *Entzündungslehre und Infektionskrankheiten. Fleckfieber.* ⓟ Moskau 1911.

Day, Albert M. (1897), in Charge Rodent Control. Wyoming District, Bureau of Biol. Survey, U.S. Department of Agric. Laramie, Wyo. (U.S.A.), Univ. of Wyo. *Rodents injurious to crops.* ⓟ Laramie Wyo. 1922. ⓟ Prairie dogs, ground squirrels, pocket gophers, mice.

Day, Alfred E., Prof. American Univ. Beirut (Syria). *Botan.* o

Day, Cameron Donald (1889), Prof., and Dir. Department of Biol. in Trinity Univ., Waxahachie, Tex. (U.S.A.). *On the Descent of the Testes and Cryptorchism.* ⓟ Kansas Univ. 1924.

Dayton, William Adams (1885), Associate Plant Ecol., in charge Forage Investigations, Forest Service, U.S. Department of Agric. Washington, D.C. (U.S.A.). *Ecol. conditions of plants.* ⓟ Williams Coll. 1908. ⓟ Range plant material.

Deam, Chas. C. (1865), State Forester of Indiana. Bluffton, Ind. (U.S.A.). *Grasses and Sedges of Indiana, Flora of Indiana.* ⓟ Honary M.A. at Wabash Coll. ⓟ Herbarium specimens of the Indiana flora.

Dean, Bashford (1867), Prof. of Vertebrate Zool. Columbia Univ. New York (U.S.A.), Riverdale-on-Hudson. *Palaeichthyol. and embryol. of fishes.* *(Myxinoid, Chimaeroid, Ganoid). Bibliography of fishes.* ⓟ Columbia 1890. o

Dean, Henry Roy (1879), Prof. of Pathol., Univ. of Cambridge (England), Department of Pathol. Univ. *Pathol., Serol.* ⓟ Oxford 1902.

Dean, Horace S. (1899), Plant Pathol., Tela Railroad Company, Tela (Honduras), Research Department. *The „Panama Disease" of bananas.*

Dean, J. Atlee (1888), Dir. Analytical and Biol. Labor., Pres. Dean Labor., Inc. Dir. Hospital Labs. Philadelphia, Pa. (U.S.A.), 614 S. 48 Street. *Analytical Chem., Toxicol., Bacteriol., Serol., Pharm.* ⓟ 1908. ⓟ Colloidal Gold, Solutions in Ampuls, Reagents and Test Solutions.

Deane, Ruth Ven (1851). Chicago, Ill. (U.S.A.), 223 N. State St. *Ornithol.* o

Deane, Walter (1848). Cambridge, Mass. (U.S.A.), 29 Brewster St. *Botan.* o

Dearborn, George van Ness (1869), Med. Officer Expert, Psychol. and Psychiatrist, U.S. Veterans' Bureau. New York City (U.S.A.), U.S. Veterans' Hospital, Kingsbridge. *Med. psychol., intellectual regression in the psychotic and psychoneurotic. Physiol., biol.* ⓟ Columbia M.D. 1893, Ph.D. 1899.

Dearstyne, Roy Styring (1891), Associate in Poultry Disease Investigations. Raleigh, N.C. (U.S.A.), *Poultry Diseases from bacteriol., serol. and pathol. standpoint.* ⓟ State Coll. of North Carolina 1922.

Debaisieux, Paul (1886), Prof. de Zool. et de Biol. générale, Univ. de Louvain (Belgique), 95 Rue de Namur, Inst. de Zool. de l'Univ. *Cytol. des Sporozoaires (Cnidosporidies).* ⓟ Univ. de Louvain Méd. 1910, Sc. Zool. 1913.

Debbarman, P. M., Curator, Herbarium. Royal Botan. Garden, Sibpore near Calcutta (India). *Taxonomy.* o

Debeyre, Albert Pierre (1877), Prof. agrégé, chef des Travaux d'Histol. et Embryol. Fac. Méd. Lille (France), 43, Rue Henri Kolb. *Histol. et Embryol., Pancréas, Foie; Morphol. du lobule hépatique. Circulation porte; Etude de jeunes œufs ou embryons humains; Premières cellules adipeuses.* ⓟ Lille 1904.

Debierre, M. Ch., Prof. Fac. Méd. Lille (France). *Anat.* o

Debré, Robert (1882), Prof. agrégé à la Fac. de Méd. de Paris (France), 5 Rue de l'Univ. *Maladies infectieuses. Immunol.*

Dechaume, Jean (1896), Préparateur de recherches au labor. d'anat. pathol. de la Fac. de méd. de Lyon (France), 13, quai de la Guillotière. *Système nerveux.* ⓟ Lyon 1926.

Deckenbach, Constantin (1866), Prof., Dir. phytopath. Abteil. der Südkrimer Station für Pflanzenschutz. Jalta, Krim (U.d.S.S.R), Postfach 39. *Pilze und Algen des Schwarzen Meeres, Pilzkrankheiten der landwirtschaftl. Kulturpflanzen der Krim: Erysiphaceen, Krankheiten der Reben, Cucurbitaceen und des Tabaks. Oidium.* ⓟ Kasan 1903. ⓟ Meeresalgen des Schwarzen Meeres. Pflanzengallen. Herbarexemplare von Pflanzenkrankheiten der landwirtschaftl. Kulturpflanzen.

Deckert, Richard Frederic (1879). Miami, Flo. (U.S.A.), 553 N.W., 64. St. *Life histories and habits of Florida Amphibians and Reptiles.* ⓟ Various living Amphibians and Reptiles.

Decksbach, Nikolai (1891), Hydrobiol. Biol. Station Kossino; Doc. landwirtschaftl. Akademie Timirjasev zu Petrowsko-Rasumowskoje. Moskau (U.d.S.S.R.), Twerskoi Boulevard 9, 38. *Hydrobiol., speziell Limnol.: biol. Seetypen; Quantitative Bodenuntersuchungen, Rotatorien, Cladoceren, Apusiden, Studien an Wasserinsecten (Larven und Imago), Quellstudien.*

Decoud, J. T., Prof. Univ. Nation. Asunción (Paraguay). *Anat.* o

Decrock, E., Prof. Univ. Fac. Sc. Marseille (France). *Botan. agric.* o

Dederer, Pauline Hamilton (1878), Prof. of Zool., Connecticut Coll., New London, Conn. (U.S.A.). *Spermatogenesis and Oogenesis in Lepidoptera.* ⓟ Columbia Univ. 1915.

Deegener, Paul (1875), Prof. Dr., beamteter a.o. Prof. an der Univ. Berlin. Berlin-Charlottenburg (Deutschland), Schillerstr. 114. *Anat., Histol., Metamorphose der Insecten. Sociol. der Insecten. Allgem. Tiersociol. Ethol. der Insecten.* ⓟ Berlin 1900.

Defrise, Aldo (1901), Aiuto dell' Ist. Anat. della R. Univ. di Milano; Incaricato dell' insegnamento dell' Anat. umana alla Fac. di Sc. Naturali. Milano (Italia), Viale Romagna 23. *Apparato Uro-genitale maschile e femminile. Sistema nervoso vegetativo. Ricerche microscopiche in cellule ed organi viventi.* ⓟ Pavia 1924.

von Degen, Arpad (1866), Dr. med., K. ung. Hofrat, Oberdir., Doc. a. d. Univ. Budapest VI (Ungarn), Vilma királynő ut. 26. *Systematik, Pflanzengeographie, Pflanzenphysiol. und Biol.* ⓟ Budapest 1889. ⓟ Getrocknete Pflanzen.

Degener, Otto (1899), Instr. in Botan., Univ. Honolulu, T.H. Hawaii, 2220 Van Couver Highway. *Vascular Plants of Hawaii.* ⓟ Univ. Hawaii 1923. ⓟ Hawaiian Vascular Plants.

Degerből, Magnus Anton (1895), Museumsamanuensis. København (Danmark), Univ.-zool. Mus. *Mammal. Glauconia, praehistor. Hunde.* ⓟ 1922.

Degner, Eduard (1886), Dr. phil., Kustos am Zool. Staatsinst. u. Zool. Mus. Hamburg 1 (Deutschland), Steintorwall. *Cephalopoden. Farbenwechsel der Krebse. Gastropoden. Anat. u. Systemat.* ⓟ Leipzig 1912. ⓟ Mollusken.

Degtiarev, P. T., Doc. Med. Inst. Odessa (U.d. S.S.R.), Valichowskij per. 2. *Physiol. Chem.* o

Dehée, René (1898), Ass. Géol. et Minéralogie Univ. Lille (France), 159, Rue Brule, Maison. *Faunes des Terrains Paléozoïques.*

Dehnel, G., Univ., Inst. d'Anat. Comp. Warszawa (Pologne), r. Krak.-Przedmieście 26/28. *Histol., Embryol.* o

Dehong, Walter (1894), Ass. in Horticulture. St. Paul, Minn. (U.S.A.), Univ. of Minnesota, Univ. Farm. *Pentosans, glucosids, hemicelluloses, Respiration of the apple.* ⓓ McGill, Montreal.

Dehorne, Prof. Univ. Lille (France). *Histol. comp. et biol. marit.* o

Dehorne, M. L., Prép. Zool., Fac. Sc. Paris 5 (France), Rue de la Sorbonne. *Zool.* o

Delhun, B. I. (1881), Prosector d. Kathed. Histol. und Embryol., Histol. Labor. Kiew, Ukraina (U.d. S.S.R.), Leninstr. 37. *Chondriosomen u. Bindegewebe.* ⓓ Kiew 1908.

Delnega, Ass. Priv.Doz. Inst. d. vergl. Anat., I. Univ. Moskau (U.d.S.S.R.), Gerzenstr. 6. *Zool. Histol.* o

Delneka, Dimitry (1876), Prof. Univ. Leningrad (U.d.S.S.R.), Anat.-Histol. Labor. *Histogenese des Knochen- und Knorpelgewebes, Chondriosomen, Golgi-Apparat u. andere Plasmabestandteile.*

Dekaprelewitsch, Leonard L., Doc., Staatl. Polytechnikum, Vorsteher d. Selektions-Abt. d. Botan. Gartens. Tiflis (U.d.S.S.R.), Ul. Kamo 73. *Selektion u. Genetik d. landw. Pflanzen.* o

Delabarre, Edmund Burke, Prof. of Psychol., Brown Univ. Providence, R.I. (U.S.A.), 9 Arlington Ave. *Bewegungsempfindungen.* ⓓ Freiburg i. Br. 1891. o

Delachaux, Théodore (1879), Ass. au Labor. de Zool., Univ. Neuchâtel (Suisse). *Faune d'eau douce et cavernicole, Copépodes (Harpacticides), Cladocères, Ostracodes.*

Delacour, Jean (1890), Président de la Section d'Ornithol. de la Société Nationale d'Acclimatation de France, et de la Ligue Française pour la Protection des Oiseaux, Chef de la mission d'exploration zool. (Oiseaux et Mammifères) en Indochine. Château de Clères, Seine Inférieure (France) et Huê, Annam (Indochine). *Ornithol. systématique; recherche, déterminations et description des Oiseaux, principalement de l'Indochine. Protection des Animaux.* ⓓ Lille 1914. ⓟ Des jeunes Mammifères et Oiseaux vivants.

Delattre, A., Prof. Lille (France). *Anat. pathol.* o

Delaunay, Henri Marie Eugène (1881), Prof. agrégé de Physiol. à la Fac. de Méd. de Bordeaux (France), 25, Rue Rolland. *Métabolisme des substances protéiques dans l'organisme animal. Biochim. des échanges azotés (Physiol. comparée).* ⓓ Bordeaux 1910.

Delcourt-Bernard, Emile (1893), Méd.-hygiéniste de la Fac. de Bruxelles (Belgique), 78, Rue de Trèves. *Physiol. générale. Physio-pathol. de la Respiration. Septicémie expérim. Metabolisme basal. L'Antisepsie.* ⓓ Liège 1921.

Delezenne, M. C., Prof. Inst. Pasteur. Paris 15 (France), 6, Rue Mizon. *Physiol.* o

Dell, E. M., Reader Univ. London S.W. 7 (England), S. Kensington. *Botan.* o

Dell' Oro, B., Prof. Univ. Rosario (Argentina). *Anat.* o

Delmas, Robert (1900), Ass. à la Fac. des Sc. de Toulouse (France). *Tenthrédes. Myriapodes.* ⓓ Toulouse 1920. ⓟ Tenthrédes.

Delmé-Radcliffe. Headcorn, Kent (England), Shenley House. *Ornithol.* o

Delphy, Jean (1887), Chef de travaux à l'Ecole pratique des Hautes-Etudes, Ass. à la Fac. des Sc. de l'Univ. de Paris. Clamart, Seine (France), 74, Avenue Schneider. *Zool. (spécialement: Oligochètes), Anat., Physiol., Embryol., Protist. et Hydrobiol., Histoire de la Biol.* ⓓ Paris 1921.

Delsman, Hendricus Christoffel (1886), Dr., Dir Labor. f. Meeresuntersuchungen Batavia (Ned. Indië). *Fischentwicklung. Marine Hydrobiol.* ⓓ Amsterdam 1912.

Delwiche, Edmond J. (1874), Prof. of Agronomy. Green Bay, Wis. (U.S.A.), 302 Sheridan Bldg. *Production of high quality and disease resistant peas for canning purposes; hardy winter wheat; rust resisting spring wheats; early varieties of yellow maize.* ⓓ Madison, Wis., 1909. ⓟ Several new varieties of peas, wheat, maize and soybeans.

Demant, Pierre, Dr. med., Univ., Inst. de pathol. générale et expérim. Warszawa (Pologne), Rue Krakowskie Przedmieście 26/28. o

Demaree, Delzie (1889), Ass. Prof. of Botan., Univ. of Arkansas. Fayetteville, Ark. (U.S.A.). *Physiographie, Ecol.* ⓓ Chicago 1921.

Dembowska, Stanislawa (1892), Agrégée de biol. à l'Inst. Nencki à Varsovie. (Pologne), Śniadeckich 8. *Régénération des protozoaires.* ⓓ Moscou 1916.

Dembowski, Jan (1889), Dr. phil., Prof. der Biol. an der Freien Polnischen Univ., Priv.Doc. f. Zool. an der Staatsuniv. Warschau (Polen), Śniadeckich 8. *Experim. Untersuchungen über das Verhalten der Tiere, insbesondere der Infusorien: Nahrungsaufnahme, Nahrungswahl, Bewegung, Tropismen, Bauinstinkte der Insektenlarven, der Landkrabben, Plastizität der Instinkte.* ⓓ Warschau 1921.

Demel, Cesaris, Dott., Ass. Anat. Patol. Univ. Pisa (Italia). o

Demel, Kazimierz (1889), Adjoint au Labor. de Pêche marine à Hel (Pologne) *Ethol. animale, associations animales nat., Biol. de la Baltique. La Faune des cavernes d'Ojców (Pologne).*

Dementjeva, Tatjana (1904), Wiss. Inst. für Meereskunde. Moskau (U.d.S.S.R.). *Variabilität der niederen Crustaceen, arctische Amphipoden.*

Demerec, Milislav (1895), Dr., Investigator, Department of Genetics, Carnegie Inst. of Washington. Cold Spring Harbor, N.Y. (U.S.A.). *Inheritance of chlorophyll in Maize, Highly mutable genes in Delphinium and Drosophila.* ⓓ Cornell Univ., Ithaca, N.Y., 1923.

Demers, L. J., Prof. agr. Ecole Med. Vet. Montreal (Canada). *Parasit. et anat. pathol.* o

Demeter, Hans (1888), Dr. med. vet., Conserv am tieranat. Inst. der Univ. München (Deutschland), Veterinärstr. 6. *Descriptive und topographische Anat., Histol. und Embryol. der Haustiere.* ⓓ München 1913. ⓟ Osteol. und splanchnol. Praepar. der Haustiere, histol. und embryol. Praepar.

Demeter, Karl Joseph (1892), Leiter der Bact. Abteilung der Südd. Forschungsanstalt für Milchwirtschaft an der Hochsch. in Weihenstephan (Deutschland). *Landwirtschaftl. Microbiol., speziell Molkereibact.* ⓓ München 1921.

Demetriades, Theodor D. Athen (Griechenland), Karageorgistr. 14; derzeit Wien VIII, Alserstr. 41. *Otorhinolaryngol. Labyrinth und vegetatives Nervensystem. Schädelresonanz. Labyrinthprüfung mit Minimalreizen. Hirntumoren und Gehörorgan. Optischer Nystagmus. cochleare Reflexe. Iontophorese der Nase.* ⓓ Athen 1916.

Demianowsky, Sergey Iakowlewitsch (1883), Moskau 69 (U.d.S.S.R.), Trubnikowsky 26, W. 19. *Gärung und der intermediäre Kohlenstoffwechsel im tierisch. Organism. Wechsel des natürl. Pigmentsim. lebend. Organismus.* ⓓ Moskau 1921.

Demidenko, Tit Trofimowitsch (1894), Ass. für spezielle Pflanzenbaulehre. Moskau (U.d.S.S.R.), Denejny 3/14. *Leinmüdigkeit. Einfluß des osmotischen Druckes auf den Pflanzenstand. Biol. des Mais.*

Demidov, Doc. Uralsche Univ. Sverdlovsk (U.d.S.S.R.), Ul. Dekabristov 1. *Forstkunde, Technol. d. Holzes.* o

Demjanov, Nik. Jak., Prof. Timirjasev Akad. Moskau (U.d.S.S.R.). *Biochem.* o

Dem'janovskij, Sergej Jak., Prof. Univ. Smolensk (U.d.S.S.R.). *Biol. Chem.* o

Demoll, Reinhard (1882), o. Univ.Prof. Dr., Vorstand des Zool. Inst. der Tierärztl. Fac. der Univ. München; Vorst. der Biol. Versuchsanst. München; Vorst. des Hofer-Inst. Wielenbach; Dir. des Inst. f. Seenforschung u. Seenbewirtschaft. in Langenargen am Bodensee. München (Deutschland), Veterinärstr. 6. *Biol. der Fische, Inzucht, Flug der Vögel, Atmung.* ⓟ Freiburg 1907.

Demoussy, François Emile (1866), Prof. à l'Inst. Agronomique. Paris (France), 16, rue Claude Bernard. *Physiol. Végétale.* ⓟ Paris 1899.

Dempwolff, Carl Friedrich Wilhelm (1867), Dr. phil., Abteilungsleiter Landwirtschaftl. Versuchsstation. Hildesheim (Deutschland), Göbenstr. 16 II. *Nahrungsmittelchem. und Wasserhygiene.* ⓟ Rostock 1903.

Dengler, Alfred (1874), Prof. an der Forstl. Hochsch. Eberswalde i. Preußen (Deutschland). *Biol. des Waldes und der deutschen Waldbäume.* ⓟ München 1903. ⓟ Photographien über Wald und Forstwirtschaft.

Denigès, Georges Noël Fort (1859), Prof. de chim. biol. et méd. à l'Univ. de Bordeaux. (France), 53, rue d'Alzon. *Biochim. analytique et microchim.* ⓟ Bordeaux 1898.

Denis, Jean Marcel Robert (1893), Agrégé de l'Univ., Ass. au Labor. Arago de Banyuls s. Mer, Pyr. Or. (France). *Aptérygotes. Systématique et Anat.*

Denis, Marcel (1897), Ass. à la Fac. de Sc. de Clermont Ferrand Puy de Dôme (France). *Botan. Systématique (Euphorbiacées africano-melgaches).Phytogéographie. Systématique et biol. des Algues d'eau douce.*

Denis, Willey (1879), Prof. of Bio-Chem. New Orleans, La. (U.S.A.), Tulane Univ. *Animal metabolism.* ⓟ Chicago 1907.

Denissow, Sachar N., Ass. d. Weißruss. Landwirtschaftl. Akademie. Gorki (U.d.S.S.R.). *Botan. Phytopath. Systematik.*

Denniston, Rollin (1874), Ass. Prof. Botan. Madison, Wis. (U.S.A.), Biol. Bldg., Univ. of Wisconsin. *Plant anat. Taxonomy of flowering plants. Lichens.* ⓟ Univ. of Wisconsin 1904. ⓟ Lichens.

Denny, Frank Earl (1883), Plant Physiol. at Boyce Thompson Inst., Inc. Yonkers, N.Y. (U.S.A.), 1086 N. Broadway. *Effect of stimulative chem. on plants. Methods of shortening the rest period of buds. Chem. treatments for hastening the coloration of plants. Ripening processes in plants.* ⓟ Chicago 1916.

Denny-Brown, Derek Ernest (1901), Beit Memorial Research Fellow. Oxford (England), Labor. of Physiol. *Physiol. of the Nervous System.* ⓟ New Zealand 1923.

Densch, Alfred Gustav (1874), Prof. Dr., Dir. des Inst. für Bodenkunde und Pflanzenernährung an den Preußischen landwirtschaftl. Versuchs- und Forschungsanstalten. Landsberg a. d. Warthe (Deutschland). *Ausnutzung der Nährstoffe durch die Pflanze, Ermittlung des Nährstoffbedarfs der Pflanzen durch chem. und biol. Methoden. Bodenreaktionen und deren Bestimmung auf chem. und bact. Wege.* ⓟ Würzburg 1901.

Densmore, Hiram D. (1862), Prof. in Botany. Beloit, Wis. (U.S.A.), 718 Clury St. *Cytol.* o

Dentici, Salvatore, Lib. doc. Univ. Messina (Italia). *Pathol.* o

Deperet, M. C., Prof. Fac. Sc. de Lyon (France). *Paléont.* o

Derenne, Franz (1873), Dir. de la Revue Mensuelle de l'Union des Entomol. belges: „Lambillionea". Bruxelles, Ixelles (Belgique), Avenue Louis Lepoutre 69. *Macrolépidoptères de la faune paléarctique et principalement de la faune belge.* ⓟ Macrolépidoptères de la faune paléarctique.

Derick, Clifford Lambie (1894), Ass., Rockefeller Inst. for Med. Research, and Ass. Resident Physician, Hospital of the Rockefeller Inst. for Med. Research. New York (U.S.A.), 66 Street and Avenue A. *Allergic nature of rheumatic fever.* ⓟ McGill Univ. 1918.

Derjavin, (1878), Prof. Dir. Pacific Ocean fishery Research Sta. Vladivostok (U.d.S.S.R.). *Biol. Geogr. of fishes. Fisheries fishculture. Amphipoda, Mysidacca, Cumacea.* ⓟ Crustaca.

Derjugin, Konst. Mich. (1878), Dir. Hydrobiol. Abt. d. St. Hydrol. Inst., Prof. d. Univ., Dir. d. Naturwiss. Inst. zu Peterhof. Leningrad (U.d.S.S.R.), Was. Ostr., 16 L., H. 29, W. 8. *Hydrobiol., Ichthyol., Ornithol., Vergl. Anat. d. Wirbeltiere, Zoogeographie.* ⓟ Petersburg 1915.

Dern, August (1858), Bayerischer Landesinspektor für Weinbau a. D., Ob.Reg.R. Würzburg (Deutschland), Sanderring 19 I. *Rebschädlingsbekämpfung.*

Dernby, K. G. E., Dr., Doc. Univ. Stockholm (Sverige). *Biochem.* o

Derrien, Eugène Louis François (1879), Prof. de Chim. biol. et méd. à la Fac. de Méd. de l'Univ. de Montpellier. (France), Inst. de Biol., rue Montels. *Biol. des porphyrines nat. et des pigments y apparentés.* ⓟ Montpellier 1906.

von Derschau, Max (1866), Dr. phil. Auerbach, Hessen (Deutschland). *Cytol. der Pflanzen.* ⓟ Leipzig 1893.

Desbouis, G., Prof. Univ. Caen (France). *Physiol.* o

Deschiens, Robert Edouard André (1896), Attaché à l'Inst. Pasteur. Paris (France), rue Dutot. *Parasit., Protozool., Protozoaires intestinaux, leur morphol. et cultures.* ⓟ 1921.

Deschin, A. A., Prof., Dir. Anat. Inst., II. Univ. Moskau (U.d.S.S.R.), Dewitschje Pole, Trubezkoi pereulok. *Anat.* o

Desderi, Paul (1886), Priv.Doc. Hygiène et Bact. Turin (Italia), Place St. Martin 3. *Bact.: sero- et vaccinothérapie, parasit., histol. pathol.* ⓟ 1910. ⓟ Cultures de microorganismes (bact. typh., parat., melitensis, bact. de l'Ozéna). Stock-vaccins antiozena.

Desceő, Dezső (1893), a.o. Prof. der Physiol. Budapest (Ungarn), Physiol. Inst. der kgl. ung. tierärztl. Hochsch., Rottenbiller-ut 23. *Atmung, Resorption.* ⓟ Budapest 1915.

Desgrez, Alexandre (1863), Prof. à la Fac. de Méd. de Paris, Membre de l'Inst. et de l'Académie de Méd. Paris VI (France), 21, rue de l'Ecole de Méd. *Chim. biol. et méd.; Hydrol.* ⓟ Paris 1912.

Desoil, Paul (1870), Prof. de Zool. Méd. à la Fac. de Méd. de Lille,Nord (France), 13,Place Philippe Le Bon. *Parasit.* ⓟ Lille 1895. ⓟ Préparations de parasit.

Despott, Giuseppe (1879), Superintendent of fisheries, Curator Natural History section, National Mus. Valletta (Malta). *Pisces, Reptiles, Birds of Malta. Mollusca. Palaeont.* ⓟ Sheds of Marine or Terrestrial Mollusca. Palaeont. specimens, especially Pleistocene.

Dessy, Giorgio (1900), Interno nei Labor. della Dir. dello Ist. Sieroterapico Milanese. Milano 24 (Italia), Via Darwin 20. *Patol. Generale.*

Detjen, Louis Reinhold (1884), Associate Prof. of Horticulture and Assoc. Horticulturist. Newark, Del. (U.S.A.). *Physiol. Dropping of Fruits while young. Cabbage Characters and Heredity. Influence of Nitrogen, Potash and Phosphoric Acid on Apple Production.* ⓟ North Carolina Coll. of Agric. 1911.

Detlefsen, John A. (1883), Wistar Inst. of Anat. and Biol. Philadelphia, Pa. (U.S.A.). *Heredity, variations, species crosses, animal breeding.* ⓟ Harvard 1912. o

Detmer, Wilhelm (1850), Dr. phil., Prof. d. Botan. a. d. Univ. Jena (Deutschland), Sonnenberg-Str. 1a. *Pflanzenphysiol. Bodenkunde.* ⓟ Leipzig 1871.

Detmers, Fredericka (1867), Ass. Botan. Wooster, Ohio (U.S.A.), Dept. of Botan.OhioAgr. Exp.Station.

Weeds and poisonous plants of Ohio. ⑨ Ohio State Univ. 1912.
Detwiler, Samuel R. (1890), Associate Prof. of Zool. Harvard Univ. Cambridge, Mass. (U.S.A.), Department of Zool. Compar. *Embryol. and Neurol.* ⑨ Yale Univ. 1918.
Deuber, Carl G. (1898), Instr. of Plant Physiol., Yale Univ. New Haven, Conn. (U.S.A.), Osborn Botan. Labor. *Physiol. of the chloroplast pigments, quantitative determinations of pigment contents of normal and abnormal plant material.* ⑨ Univ. of Missouri 1925.
Deuel, Harry James Jr. (1897), Instr. in Physiol. New York City (U.S.A.), Cornell Univ. Med. Coll. 477/1 Avenue. *Digestibility of foodstuffs, intermediary metabolism of carbohydrates, physiol. of vitamins.* ⑨ Yale Univ. 1923.
Dexler, Hermann (1866), o. Prof. der Tierseuchenlehre, Deutsch. Univ., Dir. des Tierärztl. Inst. Praha II (C.S.R.), Legerova 48. *Vergl. Anat. d. Zentralnervensystems, Tierpsychol.* ⑨ Tierärztl. Hochsch. Wien 1887.
Dezani, Serafino (1884), Prof. incaricato di Farm. e Libero Doc. di Chim. Fisiol., Univ. di Torino (Italia), Corso Raffaello 30. *Piante med. delle Colonie italiane. Genesi dell' Acido solfocianico, dell' Acido tiosolforico, della Colesterina negli animali. Utilizzazione dell' Acido cianidrico per parte delle piante.* ⑨ Torino 1914.
Dhéré, Charles Joseph (1876), Prof. de Physiol. et de Chim. physiol. Fac. sc. Univ. Fribourg (Suisse). *Chim. physiol.: Pigments, spectrochim., physiochim. des colloides, action des rayons ultraviolets.* ⑨ Méd. Paris 1898, sc. 1909.
Diakonova, Hélène A., Conserv. Station de Défense des Plantes. Leningrad (U.d.S.S.R.), rue Tschaikovski 7. *Maladies des arbres fruitiers, Pathol. végétale.* o
Diakonow, Peter Petrowitsch (1882), Chef Biol. Abt. des Sanitäts-Inst., Prof. für Mikrobiol. an der physico-mathematischen Fac. der I. Univ. Moskau (U.d.S.S.R.), Zoubovsky Boulevard 29, Wohn. 13. *Dynamische Anthropometrie. Konstitutionslehre. Hygiene der Arbeit.* ⑨ Moskau 1907.
Diamare, Vincenzo, Prof. Univ. Napoli (Italia). *Istol., fisiol. gen.* o
Diamare, Vincenzo, Prof. Univ. Pisa (Italia). *Zool., anat. e fisiol.* o
Dias Amado, Luiz E. (1901), Ass. Histol. et Embryol., Fac. Méd. Lisbôa (Portugal), Rua D. Estefania 75, 2d. *Cytol. animale: Spermatogénèse.* ⑨ Lisbôa 1925.
Diaz del Villar, Juan Manuel, Prof., Dr. med., Catedrático en la Escuela de Vet. Madrid (España), Atocha, 114 duplicado. *Epizoarios y Entomozoarios.* o
Dibble, Charles Bradford (1901), Instr. Entomol. Entomol. Dept. Mich. St. Coll. East Lansing, Mich. (U.S.A.). *Economic Pests. Forest Insects.* ⑨ Mich. St. Coll. 1925.
Dible, J. Henry (1889), Prof. of Pathol., Univ. London W.C. 1 (England), The Royal Free Hospital. *Pathol. Bact.* ⑨ Glasgow 1912.
Dice, James A. (1885), Chairman Department of Dairy Husbandry, North Dakota Agric. Coll. Fargo, N.D. (U.S.A.), State Coll. Station. *Dairy cattle, feeding, breeding and management of the heard.*
Dice, Lee Raymond (1887), Curator of Mammals, Mus. of Zool., Univ. of Michigan. Ann Arbor, Mich. (U.S.A.). *Biogeography, evolution, mammals.* ⑨ Univ. of California 1915. ⑦ *Hares or pikas.*
Dicenty, Desiderius, Dir. d. Ampetol. Inst. d. Ung. Min. f. Landw. Budapest II (Ungarn), Debröigasse No. 17.
Dickens, Albert (1867), Prof. Horticult., Kan. State Agric. Coll. Manhattan, Kan. (U.S.A.). o
Dickey, DonaldRyder (1887) Research Associate in Vertebrate Zool. California Inst. Pasadena Cal. (U.S.A.), 514 Lester Avenue. *Birds and Mammals of North and Central America, distributional, systematic, and life history, fauna of the southwestern United States and of Salvador.* ⑨ M.A., Hon., Occidental Coll. Los Angeles (Cal.) 1925. ⑦ *Birds and mammals.*
Dickinson, J., B. Sc., Chem., Animal Nutrition, Research Div. Monistry of Agric. Northern (Ireland). o
Dickinson, Sydney (1898), Research worker. Rothamsted Experim. Sta.Harpenden, Herts. (England). *The physiol. and genetics of the Smut Fungi.*
Dickson, Bertram Thomas, Prof. of Botan. and Head of Dept., Macdonald Coll. P. 6. Quebec (Canada). o
Dickson, Ernest Charles (1881), Prof. of Public Health and Preventive Med., Stanford Univ. School of Med. San Francisco, Cal. (U.S.A.), Sacramento and Webster Streets. *Food infection, Food intoxication. Botulism.* ⑨ Univ. of Toronto 1917.
Dickson, Frank, Ass. Prof. of Botan., Univ. of British Columbia. Vancouver, B.C. (Canada). o
Dickson, James G. (1891), Prof. Plant Pathol., Univ. of Wisconsin and Agent, U.S. Dept. Agr. Office of Cereal Investigations. Madison, Wis. (U.S.A.), Dept. of Plant Pathol. *Cereal and Field Crops Pathol., pathol. and physiol. of parasitism, nature of resistance to disease.* ⑨ Univ. of Wisconsin 1920. ⑦ *Disease specimens of cereal crops.*
Didier, Pierre Augustin (1873), Dir. du Jardin zool. de la Ville de Lyon (France), Parc de la Tête d'or. ⑨ Lyon 1897.
Diéguez, Salustiano (1891), Philosophical Seminary of the Vincentians Priests. Leon (España), Paules-Villafranca del Bierzo. *Chromosomes, Ontogenetics in the flowers in order to know the variety and species of the plants.* ⑨ Barcelona.
Diehl, William W. (1891), Ass. Mycol., Bureau Plant Industry, U.S. Dept. Agric. Washington, D.C. (U.S.A.). *Mycol.: systematic, taxonomic, morphol., geographical.* ⑨ Iowa State Coll. 1915. ⑦ Specimens of fungi, especially pyrenomycetes.
Diels, Ludwig (1874), Prof. an der Univ. Berlin, Dir. des Botan. Gartens und Mus. zu Berlin-Dahlem (Deutschland), Königin-Luise-Str. 6—8. *Botan. und Pflanzengeographie. Systematik der Phanerogamen. Oekol. der Pflanzen. Pflanzengeographie. Rhythmik der Pflanzen. Flora von Australien und China.* ⑨ Berlin 1896.
Dienaide, F. R., Prof., Dr., Head Dept of Med., Union Med. Coll. Peking (China). o
Dieterich, Victor (1879), Dr., Oberforstrat, früher Vorstand der Forst-Versuchsanstalt Tübingen. Stuttgart (Deutschland), Relenbergstr. 51 p. *Forstliche Ertragskunde, ertragskundlich-waldbauliche Untersuchungen, Oekonomik des Waldbaus, Forstbenutzung.* ⑨ Tübingen 1911.
Dietrich, Albert (1873), Prof. Dir. pathol. Inst. Köln-Lindenthal ⊰(Deutschland), Weyerthal 121. *Pathol. anat.* o
Dietrich, Wilhelm Otto (1881), Ass. Geol.-palaeont. Inst. u. Mus. Univ. Berlin N 4 (Deutschland), Invalidenstr. 43. *Stratigraphische Palaeont. Mammalia fossilia.* ⑨ Tübingen 1903.
Dietsch, Werner (1892), Chem. am Labor. für physiol. Chem. und Ernährungsforschung an Dr. Lahmanns Sanatorium, Dresden-Weißer Hirsch. Dresden-N. (Deutschland), Theresienstr. *Vitamine.* ⑨ Techn. Hochsch. Dresden 1923.
Diettert, Reuben Arthur (1901), Ass. in Botan. East Lansing, Mich. (U.S.A.), Botan. Dept., Michigan State Coll. *Bacterial spot of Lima Beans caused by Bacterium viridifaciens.* ⑨ De Pauw Univ. 1925.
Diez Tortosa, Juan Luis, Prof., Decano y Catedrático en la Fac. de Farm. Granada (España), Reyes Católicos, 47. *Botán.* o
Dijkstra, Onno Hendrik (1899), Prosector-conserv. bei der Pathol. Anat. an der Reichs-Univ. in Groningen (Holland), Pathol.-Anat. Labor. *Anat. Pathol. Anat.*

Dik, Knut, Prof., Dr., Norges Landwirtschaftl. Hochsch. Aas (Norge). *Pflanzenkultur, Ackerbauversuche.* ○

van Diller, Louis Rhijnvis (1888), Besoekisch Proefstation. Djember, Java (Ned.-O.-Indië). *Hevea Brasiliensis, Fermentation des Tabaks.*

van Dillewijn, Cornelius (1899), Ass. am Botan. Labor. der Univ. Utrecht (Holland), Vondelkade 26. *Phototropismus.*

Dilling, Walter James (1886), Associate Prof. of Pharm. and General Therapeutics Liverpool Univ. Aigburth, Liverpool (England), ,,Kareol", Mines Avenue. *Pharm.*

Dimitrakoff, M., Dr., Ass. Inst. f. Pharm. u. Therapie Univ. Sofia (Bulgarien). ○

Dimitrijević-Speth, Vojin (1889), Chef des Bact. Inst. Vel. Bečkerek (S.H.S.), Paschitsch-Str. 43. *Hydromechanik der Bakterienzelle: Verhalten der Mikroorganismen in Gallerten, Schleimen, in flüssigkeitsgefüllten porösen Körpern. Anzüchtung der Leptothrix. Techn. der Isolierung von Anaerob. aus Gemischen mit Aerobiern.* ⑨ Tübingen 1920.

Dimitroff, Théodor (1884), Doc. ordinaire de Sylviculture à la Fac. agronomique et forestière de l'Univ. Sofia (Bulgarie). *Pinus. Champignons. Plant ligneuses exotiques.* ⑨ Sofia, Bulgarie, 1924.

Dimitrowa, Ariada (1897), Ass. am Biol. Inst. der Univ. Sofia (Bulgarien), Dunaw 3. *Myrmeleoniden in Bulgarien, Thrazien und Mazedonien. Stimulierung der Regenerationsprozesse bei Hydra viridis.* ⑨ Rostock i.M. 1926.

Dimmock, George (1852). Springfield, Mass. (U.S.A.), 531 Berkshire Ave. *Anat. and Physiol. of Insects; Early Stages of Coleoptera.* ⑨ Leipzig 1881.

Dimo, Nikolaj A., Prof. Univ. Taschkent (U.d. S.S.R.), Inst. f. Bodenkunde u. Geobotan. Obuchowstr. *Geobotan. Bodenkunde.* ○

Dimock, William Wallace (1880), Head dept. vet. Sc., Ky. Agric. Experim. Stat. Lexington, Ky. (U.S.A.). *Pathol. Bact.* ○

Dinan, Thomas John (1890), Ass. to the Prof. of Zool. Dublin (Ireland), Univ. Coll., Merrion Str. *Marine and general Zool.* ⑨ National Univ. of Ireland 1924.

Dingelstedt, Theodor (1890), Rektor u. Prof. Leningrad (U.d.S.S.R.), Forstinst. *Pflanzengeographie, Bodenkunde, Phytosoziol., Forstwesen.*

Dingemanse, Mej. E. (1886), 1. Chem. Ass. Pharm. therapeut. Labor. Univ. Amsterdam (Holland). *Biochem.* ⑨ Zürich 1920.

Dingler, Hermann (1846), o. Prof. der Botan. in Pension, Dr. med. u. phil. Aschaffenburg (Deutschland), Grünewaldstr. 15. *Biol. Periodizität der Laubblätter der Holzgewächse und Systematik des Genus Rosa, Mechanik der pflanzlichen Flugorgane.* ⑨ Med. München 1870, phil. Leipzig 1883. ⑫ Herbarexemplare von Rosen.

Dingler, Max (1883), Dr. phil., außeretatmäßiger ao. Prof. für Forstzool. und Schädlingsbekämpfung. Gießen, Hessen (Deutschland), Wilhelmstr. 9. *Hausinsekten, Cocciden.* ⑨ Würzburg 1909.

Dinis, Ed., Prof., Fac. d. Med. Bahia (Brasil). *Anat.* ○

Dionisi, A., Prof. Univ. Roma (Italia). *Anat. Patol.* ○

Dirsch, Vitalius, Zool. Zool. Mus. Ukrain. Akademie der Wissenschaften. Kiew (U.d.S.S.R.), Tschodnowskystr. 2. *Orthoptera.*

Dsche, Zacharias (1895). Wien VIII (Deutsch-Österreich), Laudongasse 42. *Kohlehydratchem., Stoffwechsel beim Carcinom.* ⑨ Wien 1921.

Disselhorst, Rudolf (1854), Geh.R., Dr. med. et sc. nat., o.ö.Prof. Halle a. d. S. (Deutschland), Wettinerstr. 37. *Vergl. Anat. und Physiol. Emigration farbloser Zellen aus dem Blute; der Harnleiter der Wirbeltiere; die accessorischen Geschlechtsdrüsen der Wirbeltiere mit bes. Berücksichtigung d. Menschen; Anhangsdrüsen und Ausführungsgänge der Geschlechtsorgane bei Monotremen und Marsupialen; Anhangsdrüsen und Ausführungsgänge d. m. Geschlechtsorg. b. d. Wirbeltieren; vergl. Anat. u. Physiol. d. Haussäuger.* ⑨ Med. Halle a. d. S. 1887, sc. nat. Tübingen 1897.

Ditlevsen, Hjalmar, Thomas (1864), Museumsamanuensis des Zool. Mus. der Univ. København (Danmark). *Polychaeta. Freilebende Nematoden.* ⑨ København 1903.

Ditmars, Raymond Lee (1876), Cur. Dept. of Mammals. New York, N.Y. (U.S.A.), Zool. Park. ○

Dittler, Rudolf (1881), o.ö. Prof. der Physiol., Dir. des Physiol. Inst. zu Marburg a. d. L. (Deutschland). *Nerv-Muskelphysiol., physiol. Optik, Physiol. der inneren Sekretion.* ⑨ Leipzig 1907.

Dittrich, Gustav (1875), Prof. Dr., Studienrat. Breslau 16 (Deutschland), Uferzeile 14. *Mykol.: Höhere Pilze, Pilzvergiftungen.* ⑨ Breslau 1898.

Dixey, Frederick Augustus (1855), Lect. of Zool., Wadham Coll. Oxford (England). *Lepidoptera.* ⑨ Oxford 1879.

Dixit, D. L., Prof. Univ. of Bombay. Poona (Br.-India). *Botan., Zool.* ○

Dixon, Andrew Francis (1868), Univ. Prof. of Anat. Dublin (Scotland), School of Anat., Trinity Coll. *Human Anat. and Embryol.* ⑨ 1889.

Dixon, Boris (1873), Adjoint de l'École Générale d'agric. à Varsovie, Chef de la Station Expérim. de Pisciculture à Ruda Maleniecka (Pologne). *Ichthyol. appliquée.* ⑨ 1897.

Dixon, Henry Horatio, Prof. of Botan. Univ., of Plant Biol. in Trinity Coll., Dir. of Trinity Coll. Botan. Garden. Dublin (Scotland), School of Botan. *Plant Physiol., Cytol. Transpiration and the ascent of Sap in Plants.* ⑨ Dublin 1897. ⑫ Pteridophytes and Seed-Plants.

Dixon, W. E., Prof. Univ. Downing Coll. Cambridge (England). *Pharm.* ○

Djakonoff, Fedor F. (1899), II. Ass. Biol. Wolga-Station, Spezialist für Fischerei, Gouv. Saratow (U.d.S.S.R.), Biol. Station. *Bewuchs, Fischerei.* ⑨ Biol. Fischmaterial.

Djakonov, A.M., Konserv. Zool. Mus. Leningrad (U.d.S.S.R.), Univ. nab. 1. *Wirbellose Tiere.* ○

Dmitriev, Sergej Fed., Prof. I. Univ. Moskau (U.d.S.S.R.). *Bact.* ○

Dmitriew, Andrej M., Prof., Timirjasew Landwirtschaftl. Akademie, Inst. f. Wiesenbau. Lobnja, Moskauer Gouv. (U.d.S.S.R.). *Botan. Geobotan.* ○

Dmochowski, Antoine (1896), Dr. phil., Chef des travaux au Labor. de Chim. Biol. Univ. Varsovie (Pologne). *Verdauungshämatin. Oxyproteinsäuren. Buscainoreaktion. Purinchem. und Purinstoffwechsel.* ⑨ Varsovie 1924.

Doak, Kenneth Davis (1903), Ass. in Botan. Lafayette, Ind. (U.S.A.), Purdue Agric. Experim. Station. *Cereal Leaf Rust.* ⑨ Lafayette, Ind. 1926.

Doan, Charles Austin (1896), Associate, Rockefeller Inst., New York City (U.S.A.), 66 St. and Ave. A. *Pathol. and Bact.* ⑨ Johns Hopkins 1923.

Doan, Francis Janney (1896), Ass. Prof. of Dairy Manufactures. State Coll., Pennsylvania (U.S.A.), Dairy Department. *Dairy Bacteriol.* ⑨ Pennsylvania State Coll. 1922.

Doane, Rennie W. (1871), Prof. of Entomol. Stanford Univ., Cal. (U.S.A.). *Economic Entomol.: Insects and Disease.* ⑨ Stanford Univ. 1896.

Dobbins, R. A. (1895), Ass. Prof. of Botan., Ohio Northern Univ. Ada, O. (U.S.A.). *General Botan. and Mycol. Clavariaceae, Polyporiaceae.* ⑨ Ohio State Univ. 1922. ⑫ Powdery Mildews and Polypores, Ohio.

Dobell, Clifford (1886), Protist., Med. Research Council. London N.W. 3 (England), National Inst. for Med. Research, Hampstead. *Protozool. and Bact. Intestinal Protozoa of Man.* ⑨ Cambridge 1906.

Dobreff, Minko (1898), Pleven (Bulgarien). *Normale und pathol. Physiol.: Verdauung, die Verdauungssäfte, die Absonderung der Verdauungssäfte, Vergl. Physiol. der Verdauung, Secretine, Harnabsonderung, Avitaminose.* ⑨ Berlin 1925.

Dik — Dollo

Dobrinin, N. F., Doc. II. Univ. Moskau (U.d.S.S.R.). *Psychol.* o

Dobroljubowa, Tatiana (1889), Ass. géol. II. Univ. Moskau (U.d.S.S.R.). *Faune du carbonifère de l'Oural du Nord.*

Dobroscky, Irene Dorothy (1899), Ass.Plant Pathol. and Entomol. Yonkers, N.Y. (U.S.A.), Boyce Thompson Inst. for Plant Research. *Plant diseases, insect carriers of Plant Diseases., Pathol. Entomol.* ℗ Cornell Univ. 1923.

Dobrov, Sergej Aleks, Konserv. Geol. Inst. Moskau (U.d.S.S.R.). *Palaeont.* o

Dobrovolsky, M. E., o. Prof. Inst. f. ärztl. Fortbildung. Leningrad (U.d.S.S.R.), Piročnoja 41. *Hygiene.* o

Dobrowitsky, Peisach Josifowitsch (1897), Nervenarzt, Aspirant Labor. experim. Biol., Swerdlow Univ. Moskau (U.d.S.S.R.), Pretschistensky Boulevard 29, Wohn. 49. *Nervös-muskuläre Ermüdung. Histol. und Physiol. der endokrinen Drüsen.*

Dobrowolski, Jean, Prof. Poznań (Pologne), Univ. *Botan.* o

Dobrozrakowa, Taissa Leonidovna (1895), Chef de travaux, Station Pathol. végétale de l'Inst. Agronomique de Leningrad (U.d.S.S.R.), Detskoe Selo. *Mycol. et pathol. végétale: bouillies fungicides, maladies dites „mosaïque", „enroulement".* ℗ Champignons.

Doby, Géza Karl (1877), Dr., o.ö. Univ.Prof. Budapest IV (Ungarn), Szerb-ut 23. *Pflanzenbiochem., physiol. und pathol. Enzymol., Biochem. der Pflanzenernährung, Bodenchem. (Agrochem.).* ℗ Budapest 1902.

Dobzhansky, Theodosy G. (1900), Oberass. Leningrad (U.d.S.S.R.), W. O. Univ., Labor. f. Genetik. *Vererbungslehre: pleiotrope Wirkung der Gene, geographische und individuelle Variabilität.* ℗ Coccinellidae.

Docters van Leeuwen, Willem Marius (1880), Dir. des Botan. Gartens Buitenzorg, Java (Ned.-Indië). *Die Gallen (Ätiol., Anat., Physiol., Morphol.). Biol. der Flora auf den Gipfeln der Vulkane in Nied.-Ost-Indien.* ℗ Amsterdam 1907.

Dodd, David Rollin (1889), Agronomy Specialist. Morgantown, W.Va. (U.S.A.), West Virginia Univ., Coll. of Agric. *Influence of soybeans on crops and on nitrate content of soil.* ℗ West Virginia Univ. 1914.

Dodds, Clifford Ten Eyck (1896), Entomol., Citrus Fruit Association. Santa Paula, Cal. (U.S.A.), 837 Ojai Road. *Economic entomol.; Citrus pests.* ℗ Univ. of California 1922.

Dodds, E. C., Prof., Univ. London S.W. 7 (England), S. Kensington. *Biochem.* o

Dodds-Parker, A. P., Lect. Univ. Oxford (England). *Appl. Anat.* o

Dodge, Bernard (1872), Pathol., Bureau Plant Industry. Washington, D.C. (U.S.A.). *Cytol. and Pathol. of the Fungi.* ℗ Columbia Univ. 1912.

Dodge, Carroll William (1895), Ass. Prof. of Botan., Harvard Univ., Curator Farlow Reference Library of Cryptogamic Botan., Acting Curator, Farlow Herbarium. Cambridge 38, Mass. (U.S.A.), 20 Divinity Ave. *Mycol., higher basidiomycetes and fungi which cause human disease. Lichens. Physiol. of fungi; fermentations. Bibliography, biography and history of biol. Systematic treatment of algae.* ℗ Washington Univ. 1918. ℗ Cultures of fungi which cause human disease. Dried material, algae, fungi, lichens, bryophytes.

Dodge, Charles Wright (1863), M.S., Prof. Univ. of Rochester, N.Y. (U.S.A.). *General Zool.* o

Dodson, William Rufus (1867), Dean Coll. of Agric. and Dir. Experim. Stations, La. State Univ. Baton Rouge, La. (U.S.A.). *Agric. Analysis of feedstuffs and fertilizers.* o

Döderlein, Ludwig (1855), Geh.R.R., Univ.Prof., Dir. Zool. Staats-Mus. München (Deutschland), Herzogstr. 64. *Zool. Zoogeographie, Echinodermata.* o

Dömmel, Martha (1902), Anat. Pathol. Inst. Univ. Pécs (Ungarn), Inczédy-ut 1. *Histol., Bact., Serol.* ℗ Pécs 1926.

Dömsödy, Péter (1903), Univ.Ass. Debreczen (Ungarn), Szent Anna-ut 60. *Biochem.*

Doerr, Robert (1871), o. Univ.Prof. Hygiene, Dir. hygien. Inst. Univ. Basel (Schweiz), Riehentorstr. Nr. 33. *Ätiol. der Infektionskrankheiten: bazilläre Dysenterie, filtrierbare Virusarten, insbes. Phlebotomiusfieber, Hühnerpest, Herpes febrilis, Encephalitis epidemica. Immunität: Anaphylaxie, Allergie, Idiosynkrasien, Immunochemie der Eiweißkörper, Chem. und physiol. Dynamik der Bakterientoxine.* ℗ Wien 1897. ℗ Bakterienkulturen, virushaltige Organe.

Dörries, Wilhelm (1886), Dr. phil., Studienrat, wiss. Mitarbeiter der Preuß. Landesanstalt für Wasserhygiene in Berlin-Dahlem. Berlin-Zehlendorf (Deutschland), Gertraudstr. 10. *Chem. Physiol.; Ernährungsphysiol. der Wasserpilze; abnormales Dickenwachstum.* ℗ Göttingen 1910.

Dogiel, Valentin (1882), Prof. der Zool. Univ., Vizedir. des Naturwiss. Inst. zu Peterhof. Leningrad (U.d.S.S.R.), Zool. Labor. *Parasit. Protozoa, besonders Infusorien (Morphol., Systematik, Vermehrungsprozesse, Biol. der genannten Gruppen). Ökol. Untersuchungen an Land-Evertebraten. Quantitative Analyse der Landfauna. Fauna Zentralafrikas.* ℗ Leningrad 1913.

Dognon, André (1900), Ass. à l'Inst. de Physique biol. Strasbourg (France), 21, rue Goethe. *Electrol.: excitation électrique des nerfs et muscles. Chronaxie. Rayons X. Actions biol.* ℗ Strasbourg 1925.

Dohnal, Theodor, Dr., Prof. der Anat. und Physiol. der Haustiere. Brno (C.S.R.), Černa pole 102, Landwirtschaftl. Hochsch. o

Dohrn, Max (1874), Abteilungsleiter im wiss. Labor. der Chem. Fabrik auf Aktien vorm. E. Schering. Berlin-Charlottenburg (Deutschland), Schloßstr. 67. *Purinstoffwechsel (Atophan). Sexualhormone.* ℗ Heidelberg 1898.

Dohrn, Reinhard (1880), Prof., Dir. della Stazione Zool. di Napoli (Italia). ℗ Marburg 1904.

Doisy, Edward Adelbert (1893), Prof. of Biol. Chem. at St. Louis Univ. School of Med. St. Louis, Mo. (U.S.A.), 1402 South Grand Boulevard. *Biol. Chem. Insulin, ovarian hormone.* ℗ Harvard Univ. 1920.

Dojarenko, A. G., Prof. I. Univ. Moskau (U.d.S.S.R.). *Agric. Chem.* o

Dokturowsky, Wladimir, Prof., Dir. d. Geobotan. Kabinetts d.Wissensch.-experim. Torfinst. Moskau 2 (U.d.S.S.R.), Arbat-Str. 51, quart. 42. *Geobotan. Untersuchungen im Torfmoorgebiet. Pflanzengeogr. und Ökol., Torfanalyse, Moorstratigraphie und Pollenanalyse im Torf.*

Dokukin, Michail W., Dir. d. Moor-Versuchsstation. Minsk (U.d.S.S.R.), Kommunalnaja 36a, W. 1. *Moorforschung.* o

Dold, Hermann (1882), a.o. Prof. für Hygiene an der Univ. Marburg u. Dir. des Inst. „E. v. Behring". Marburg a. d. L. (Deutschland). *Hygiene u. Bact. Serol.* ℗ Tübingen 1905.

Dole, Eleazer Johnson (1888), Ass. Prof. Botan., Univ. of Vermont, Ass. Botan., Vermont Agric. Experim. Station. Burlington, Vt. (U.S.A.), 433 South Prospect Street. *Plant Morphol. Light and Water Relations of Plants.* ℗ Univ. of Vermont 1923.

Dolk, H. E. (1904), Ass. Botan. Labor. Utrecht. Bilthoven (Holland), v. Dycklaan 23. *Phototropie.* o

Dolifus, Robert Ph. (1887), Secrétaire général de l'Office de Faunistique et de Parasit. marocaines de l'Inst. sc. chérifien. Paris V (France), Mus. National, 57 rue Cuvier, et Rabat (Maroc), Inst. sc. Chérifien. *Faunistique et Parasit.* ℗ Paris.

Dollo, Louis, Prof. Fac. sc., Dir. Mus. de paléont. Bruxelles (Belgique), 31, rue Vautier. *Paléont. Biol. gén.* o

5*

Domac, Julije, Dr.,o. Prof. emer. Pharmakognosie. Zagreb (S.H.S.), Trg I/8. *Anat. d. Pflanzen, Pharmakognosie.* o

Domagk, Gerhard (1895), Priv.Doc., Dr. med. Münster i. W. (Deutschland), Pathol. Inst. der Univ. *Retikuloendothel, Amyloid, Anaphylaxie, Eiweißstoffwechsel. Gewebsschädigungen durch Röntgenstrahlen.* ① Kiel 1921.

v. Domaniewski, Janusz (1891). Warschau (Polen), Polnisches Naturhistorisches Staatsmus., Krakowskie Przedmieście 26/28. *Ornithol., Mammalol., Zoogeographie.*

Domantowitsch, Michail K., Landw. Akademie Timirjasew. Moskau (U.d.S.S.R.), Petrowsko-Rasumowskoje Iwanowskaja 7. *Pflanzenphysiol.* o

Dombrovska, Jeanne (1890), Prosektor Anat. pathol. Fac. Med. Warszawa (Pologne), Chalubinski 5. *Ulcus gastrique, atherosclerose, anaemie pernicieuse.* ① Pétersbourg 1914.

Dombrovsky, Boris S. (1887), Prof. Vladivostok, Siberia (U.d.S.S.R.), Univ. d'Etat à l'Extrême-Orient. *Paléont. Cétacés fossiles. Mollusques fossiles.* ② Fossiles invertébrés: mollusques, brachiopodes, plantes fossiles des âges Permien, triasique, jurassique, tertiaire.

Domin, Karl, Dr., Prof. der Botan., Dir. des Pharm.-botan. Inst. der Karls-Univ. Praha II (C.S.R.), Benátecká 433. *Systématique, Géobotan., Morphol., Tératol. | Plantes méd. et cultivées, Protection des plantes.* o

Domizio, Giovanni di, Lib. doc. Univ. Bologna (Italia). *Pathol. trop. vet.* o

Domogalla, Bernhard (1894), Biol. Chem. for the City of Madison, Wis. (U.S.A.), Univ. Club. *Biochem. Lake Investigations. Biochem. Sewage Treatment. Food Problems.* ① Univ. of Wisconsin 1925.

Domontowitsch,Michael Konstantinowitsch(1884), Station für Pflanzenernährung und Düngung, Landwirtsch. Akadem. Moskau 8 (U.d.S.S.R.). *Physiol. der Pflanzenernährung. Wasserstoffionenkonzentration.*

Donadson, R., Prof. St. George Hospital Med. School. London, S.W. (England), Hyde Park Corner. *Pathol.* o

Donaldson, Henry Herbert (1857), Prof. of Neurol., The Wistar Inst. Philadelphia, Pa. (U.S.A.). *The postnatal growth of the mammalian nervous system.* ① Johns Hopkins Univ. 1885.

Donaldson, John C. (1888), Associate Prof. of Anat. Univ. of Pittsburgh, Pa. (U.S.A.). *Endocrine glands. Adrenal.* ① Johns Hopkins 1914.

Donat, Artur (1893), Dr. phil., wiss. Ass. am Botan. Inst. der Univ. Freiburg i. Br. (Deutschland), Roßkopfstr. 2 III. *Limnol., Pflanzengeographie (Süßwasseralgen).* ① Berlin 1925.

Donatien, André Louis (1889), Chef de labor., Inst. Pasteur d'Algérie. Alger (Algérie). *Maladies microbiennes de l'algérie. Piroplasmes.* ① 1926. ② Préparations d'hématozoaires des animaux domestiques de l'Afrique du Nord.

Doncieux, M., Préparateur Fac. Sc. Lyon (France) *Paléont.* o

Donegan, J. T., Prof. Univ. Coll. Galway (Irland). *Physiol.* o

Donges, Rudolph (1880), Dr. med., Generaloberarzt. Hannover (Deutschland), Adolfstr. 9. *Bact.* ① Würzburg 1903.

Doniselli, Casimiro, Prof. incar. Univ. Milano (Italia). *Psychol. experim.* o

Dons, Carl (1882), Custos zool. Abt. d. Mus. (Kgl. Norske Videnskabers Selskab.) Trondhjem (Norge). *Decapoden, Hydroiden u. sessile Protozoen.*

Donshov, Vladimir, Prof. Univ. Irkutsk (U.d. S.S.R.). *Pathol. Anat.* o

Dontas, Spiridon (1878), o. Prof. der experim. Pharm. der Univ. zu Athen. Athènes (Griechenland), Rue du Phalère 47. *Pharm. des Muskel- und Nervensystems.* ① Athen 1898.

Dooley, Marion Sylvester (1879), Prof. of Materia Med. and Pharmacol., Coll. of Med., Syracuse Univ., Syracuse, N.Y. (U.S.A.), 309 So. McBride St. *Digitalis, Problems in Anaesthesia, Respiratory Stimulants on a Carbon Dioxide basis, Correlation of emptying of the Stomach with Gastric Peristalsis.* ① Syracuse Univ. 1914.

Doolittle, Sears P. (1890), Plant Pathol., Office of Vegetable and Forage Diseases, U. S. Dept. of Agric. Dept. of Plant Pathol., Univ. of Wisconsin. Madison, Wis. (U.S.A.). *Mosaic diseases of truck crops, cucumber, tomatoes and lettuce.* ① Univ. of Wisconsin 1918.

Dop, P., Prof. Univ. Toulouse (France). *Botan.* o

Doppelmair, Georg G. (1880), Prof. Forst-Inst., Doc. Univ. Leningrad (U.d.S.S.R.). *Biol. der Säugetiere und Vögel, angewandte und Jagdzool., Jagdwirtschaft.* ① Petersburg 1912.

Doran, William Leonard, Ass. Prof. of botan., Mass. Agric. Coll. Amherst, Mass. (U.S.A.). o

Dorello, Primo (1872), Prof. di Anat. Normale presso la R. Univ. di Perugia (Italia), Ist. Anat., S. Francesco. *Embriol. del Sistema Nervoso Centrale. Mecanico articol. (Articol. temporo-mandibolare). Sviluppo dell' articol. temporo-mandibolare. Biol. sessuale del Gen. Helix.*

de Dorlodot, Henry, Prof. de Paléont. à l'Univ. Louvain (Belgique), 42, rue de Bériot. o

Dorn, Karl Alfred Ferdinand (1884), Leipzig-Schleußig (Deutschland), Könneritzstr. 5 I. *Coleoptera palaearctica: Elateridae.* ② Coleoptera.

Dorner, Herman Bernard (1878), Prof. and Chief in Floriculture. Urbano,Ill. (U.S.A.), Univ. of Illinois, 100 Floriculture Building. ① Purdue Univ. 1901.

Dorning, Heinrich (1880), ordentl. Beobachter des Königl. ung. Ornithol. Inst. in Budapest X (Ungarn), Simor-ut 13. *Ornithol. (Oekol. der Vögel u. Aviphaen.).*

Dorogin, Georg (1878), Mykol. Labor. des Inst. für experim. Agronomie. Leningrad (U.d.S.S.R.). *Phytopath.: Krankheiten der Gemüsepflanzen u. der Samen.*

Dorogostajsky, Vitaly, Prof. Naučno-Inst. Irkutsk (U.d.S.S.R.). *Zool.* o

Doroschenko, Assia B. (1889), Ass. Plant Physiol., Inst. of Applied Botan. and New Cultures. Leningrad (U.d.S.S.R.), 44 Herzenstr. *Photoperiodism and Physiol. of Pollen.*

Dorsey, Henry (1882), Prof. of Agronomy. Storrs, Conn. (U.S.A.). *Farm crops, soil technol.* ① Iowa State Coll. 1926.

Dorsey, Maxwell Jay (1880), Chief in Pomol. Univ. of Illinois. Urbana, Ill. (U.S.A.). *Horticulture, cytol., genetics.* ① Cornell Univ. 1913.

van Dorsselaer, René (1895), Préparateur. Wesembeek, Br. (Belgique), Avenue Oscar de Burbure. *Etude systématique et morphol. spécialisée des Haliplides. Dytiscides et Gyrinides belges et mondiaux.* ② Dytiscides.

Dosdall, Louise (1893), Mycol., Minnesota Agric. Experim. Sta. St. Paul, Minn. (U.S.A.), Univ. Farm. *Mycol., Plant Pathol.* ① Univ. of Minnesota 1922.

Dosrenko, Aleksej G., Prof., Landwirtschaftl. Akademie namens Timirjasew. Moskau (U.d.S.S.R.), Petrowsko-Rasumowskoje. *Agronomie.* o

Dostál, Rudolf (1885), Ph.D., Prof. de Botan. à l'École Vét. Brno. (C.S.R.), Pražská 69. *Morphol. expérim. des plantes, sérol. botan. Plantes pathogènes des animaux.*. ① Prague 1908. ② Plantes méd. et vénéneuses.

Doten, Samuel Bradford (1875), Prof. of entomol. Univ. of Nevada, Dir. Nevada Experim. Sta. Reno, Nev. (U.S.A.). o

Doty, Hiram S., Prof. of biol., Simpson Coll. Indianola, Ia. (U.S.A.), 911 N. Buxton St. *Physiol., Ecol.* o

Doubrow, Serge (1893), Ass. labor. d'histol. Lyon (France), 18, Quai Claude Bernard. *Anat. pathol. de la Tuberculose et de la Syphilis. Cancer.* ① Lyon 1924.

Doubt, Sarah L., Head of the Dept. of Botan., Washburn Coll. Topeka, Kan. (U.S.A.), *Physiol., Genetics.* ○

Douglas, Claude Gordon (1882), Fellow of St. John's Coll., Demonstrator in Physiol. Univ. Oxford (England). *Human Physiol., physiol. of respiration, respiratory exchange, circulation, blood.* ⓓ Oxford Univ. 1913.

Douglas, J. S. C., Prof. Univ. Fac. Med. Sheffield (England). *Pathol.* ○

Douglas, S. B., Dir. Dept. Bact. and experim. Pathol. London W.C. 2 (England), 15 York Buildings Adelphi. ○

Douglass, James Robert, Ass. Entomol., U.S. Bureau of Entomol., U.S. Department of Agric. Estancia, N. Mex. (U.S.A.), P.O. Box 353. *Bean Beetle Investigations.* ⓟ *Chrysomelidae and Coccinellidae of Southwest.*

Douin, Robert Charles Victor (1892), Maître de Conférences à la Fac. des Sc. de Lyon (France), Labor. de Botan., 16, Quai Claude Bernard. *Les Muscinées; les Hépatiques (embryol., anat., cytol., systématique et géographie botan., paléobotanique).* ⓓ Paris 1925.

Douris, R., Prof. Univ. Nancy (France). *Chim. biol.* ○

Dove, Walter E. (1894), Research-Bureau of Entomol., U.S. Department of Agric. Washington, D.C. Dallas, Tex. (U.S.A.), Box 208. *Parasit. Insects affecting health. Nematodes causing skin affections of man.*

Dove, William Franklin (1897), Associate Biol. Orono, Me. (U.S.A.), Univ. of Maine, Experim. Station. *Differentiation of tissues by means of Transplantation. Relation of Endocrine glands to development. Effect of Age of Parents on the heritable traits. Interrelations between growth in size of organs and development of body form.* ⓓ Wisconsin 1923.

Dowden, Philip Berry (1901), Ass. Entomol. U.S. Dept. of Agric., Bureau of Entomol. Melrose Highlands, Mass. (U.S.A.), 17 E. Highland Ave. *Entomol. Parasit. Hymenoptera.* ⓓ Amherst, Mass., 1923.

Dowding, Eleanor Silver (1901), Lect. Edmonton, Alberta (Canada), Univ. *Ecol. Morphol.* ⓓ Univ. of Alberta 1923. ⓟ *Alpine plants from Rocky Mts. Insectivorous plants. Algae.*

Dowell, Philip (1864), Teacher of Biol., Curtis High School. Port Richmond, N.Y. (U.S.A.), 86 Bond St. *Systematic Botan.: Ferns, violets, local floras.* ⓓ Augustana 1900. ⓟ *Herbarium specimens. Histol. slides.*

Dowgiallo, N. D. (1898), Ass. Anat. Inst. Odessa (U.d.S.S.R.), Olgiewskaja 4. *Vegetatives Nervensystem.*

Down, E. E. (1892), Ass. Prof. East Lansing, Mich. (U.S.A.), Michigan State Coll. *Plant Breeding.* ⓓ Michigan State Coll. 1915. ⓟ *Wheat, Oats, Barley, Alfalfa, Sugar Beets, Field Beans, Sweet Clover.*

Downes, William (1874), Ass. Entomol. Victoria (British Columbia), Dominion Entomol. Labor. *Economic insect problems, affecting small fruits, systematic studies of Hemiptera.* ⓟ *Hemiptera of Western Canada.*

Downey, Hal, Prof. of Histol., Univ. of Minnesota. Minneapolis, Minn. (U.S.A.), Department of Animal Biol. *Morphol. Hematol., Blood-forming organs, Cells of the connective tissue.* ⓓ Univ. of Minnesota 1909. ⓟ *Hematol. preparations.*

Downie, Allan Watt, Ass. Lect. Manchester (England). Pathol. Department, Victoria Univ. *Med. Bact. Tissue culture and Experim. Cancer.*

Downs, Paul Andrew (1891), Associate Prof. in the Department of Dairy Husbandry and Experim. Station. In charge of Dairy Bact. and Manufacture. Lincoln, Neb. (U.S.A.), Univ. of Nebraska. *Dairy Bact.* ⓓ Cornell Univ. 1923.

Dox, Arthur Wayland (1882), Research Chem., Parke, Davis & Co. Detroit, Mich. (U.S.A.). *Biochem., Pharm.* ⓓ Yale Univ. 1909.

Doyer, Catharina Magdalena (1898), Dr. Ass. Centralbureau für Schimmelkulturen. Baarn (Holland), Javalaan 6. *Pilzkulturen.* ⓓ Utrecht 1925.

Doyer, Lucie Christina (1883), Mykol. a. d. Reichsversuchsstation für Samenkontrolle Wageningen (Holland), Hoogstraat 71. *Samenkrankheiten.* ⓓ Utrecht 1914.

Doyle, Joseph (1891), Prof. of Botan. in Univ. Coll. Dublin (Ireland). *Enzymes of Conifers in relation to starch formation: Conifer physiol. in general. Optimum conditions for germination and growth of wheat and cereals. Farm problems in general.* ⓓ Dublin 1915.

Doyon, M. M., Prof. Fac. Méd. Lyon (France). *Physiol.* ○

Dozier, Herbert Lawrence (1895), Entomol. Newark, Del. (U.S.A.), Del. Agr. Expt. Station. *Codling Moth. Grope Leafhopper (Erythroneura vitia).* ⓓ Ohio State Univ. 1922. ⓟ *Whiteflies (Aleyrodidae), and chalcid parasit. of Coccidae and Aleyrodidae.*

Dragendorff, Otto (1877), Dr. med., Univ.Prof. Greifswald (Deutschland), Wolgaster Str. 30. *Anat.* ○

Drago, Umberto, Lib. doc., Univ. Catania, Sicilia (Italia). *Zool., Fisiol., Anat. compar.* ○

Drăgoiu, Jean (1878), Prof. d'Histol. et d'Embryol. à la Fac. de Med. Cluj (Roumanie), Rue Pasteur 6. *Cellules sexuelles.* ⓓ Bucarest 1903.

Dragstedt, Lester R. (1893), Ass. Prof. Surgery Univ. Chicago, Ill. (U.S.A.). *Physiol. of the Parathyroid Glands, of Stomach and Duodenum.*

Drahorad, Friedrich (1891), Adjunkt, diplom. Ing., Dr. der Bodenkultur. Bundesanstalt für Pflanzenbau u. Samenprüfung. Wien II (Österreich), Prater, Lagerhausstr. 174. *Biol. der landw. Kulturpflanzen, Samenkunde, Pflanzenbaulehre, Vererbungslehre, speziell Pflanzenzüchtung und Getreidebau.* ⓓ Wien 1923.

Drahovzal, Frant (1898), Ass. Inst. für experim. Pathol. Univ. Brno (C.S.R.), Úvoz 73. *Chem. Analyse der Zelle.* ⓓ Böhm. techn Hochsch. Brünn 1925.

Drain, Brooks Daniel (1891), Ass.Prof. of Pomology. Amherst, Mass. (U.S.A.). ⓓ Univ. of Chicago 1925.

Drastich, Ludvik (1887), Med. Univ. Dr., Ass. im Inst. für allgemeine Biol. d. med. Fac. d. Masaryk-Univ. Brno (C.S.R.), Úðolní 73. *Aphidol. Wirkung der Sauerstoffkonzentration auf Stoffwechsel und Morphogenese.* ⓓ Brno 1925.

Drechsel, Joseph (1897), Ass. am anat. Inst. der Univ. Sofia (Bulgarien), Marin Drinoff 18. *Architektur (makroskopische) der Herzventrikel bei Wirbeltieren. Hautempfindungen bei Reizung mit galvanischem Strome.* ⓓ Würzburg 1923.

Drechsler, Charles (1892), Associate Pathol. in the Bureau of Plant Industry, United States Department of Agric., Washington, D.C. (U.S.A.). *Classification and morphol. of fungi parasitic on fruits, vegetable and forage crops; taxonomic work on the genera Pythium, Phytophthora, Helminthosporium, and terrestrial Saprolegniaceae.* ⓓ HarvardUniv. 1917.

Drennan, A. M., Prof. Univ. of Otago. Dunedin, New Zealand (Australia). *Pathol.* ○

Drennan, Matthew Robertson (1885), Prof. of Anat., in the Univ. of Capetown (South Africa), Anat. Dept. *Human Anat., Embryol., Physical Anthrop.* ⓓ Edinburgh 1910.

Drenowski, Alexander K. (1879), Custos am Schulmus. beim Unterrichtsministerium. Sofia (Bulgarien), Bulevard Chr. Botjew 50. *Lepidopterenfauna in den Hochbergen Bulgariens, vertikale Verbreitung und Verteilung der Lepidopteren auf denselben, biol. Erscheinungen einiger Arten der Rhopalocera.* ⓟ *Lepidopteren.*

Drensky, Konstantin, Dr., Dir. d. Inst. f. d. Bekämpfung der Malaria. Burgas (Bulgarien). ○

Drensky, Pentscho (1887), Ass. an der Königl. Entomol. Station in Sofia (Bulgarien), Königl.

Naturhistorisch Mus. *Aranaea, Pisces der bulgarischen Fauna.* ③ Sofia 1911.
Dresbach, Melvin (1874), Prof. of Physiol. Albany, N.Y. (U.S.A.), Albany Med. Coll. *Toxicity of normal human urine, elliptical human erythrocytes, blood pressure of the sheep, pancreatic bladder in cat, phenolphthalein action on animals, interpolated extrasystoles, general physiol. of inhibition, digitalis, pharm. action of k-strophanthidin.* ③ Ohio Med. Univ. 1903.
Dresel, Ernst Gerhard (1885), Dr. med et phil., o.ö. Prof. der Hygiene, Dir. des Hygien. Inst. in Greifswald (Deutschland), Martin-Luther-Str. 6. *Hygiene, Bact., Serol.* ③ Heidelberg med. 1910, phil. 1913.
Drevermann, Fritz (1875), Prof. der Geol. und Palaeont. an der Univ. Frankfurt a. M. Schönberg i. Taunus (Deutschland), Parkstr. 9. *Palaeont. bes. Wirbeltiere. Museumskunde.* ③ Marburg 1900.
Drew, Gilman A. (1868), Ass. Dir. Marine Biol. Labor. Woods Hole, Mass. (U.S.A.). ③ Johns Hopkins 1898. ○
Drew, Kathleen, Ass. Lect. in Botan. Manchester (England), Univ. *Algae.*
Dreyer, G., Prof. Univ. Fac. Med. Oxford (England). *Pathol.* ○
Dreyer, George Peter (1866), Head dept. of physiol. Coll. of Med. Univ. of Ill. La Grange, Ill. (U.S.A.), W. Cossit Ave. *Secretory nerves of the suprarenal glands.* ○
Dreyer, T. F., Prof. Univ. Bloemfontein, Orange Free State (South Africa). *Zool.* ○
Dreżepolski, Roman (1884), Dr. phil., Prof. an dem Marien-Gymnasium in Poznań 3 (Pologne), rue Podolska 21. *Flagellata spec. Eugleninae.* ③ Univ. Léopol 1908.
Driesch, Hans A. E. (1867), Dr. phil., Dr. iur. h. c. (Aberdeen), Dr. med. h. c. (Hamburg), Dr. Sc. h. c. (Nanking, China), Ordinarius der Philosophie in Leipzig (Deutschland), Zöllnerstr. 1. *Experim. zool. Morphol. Allgemeine Philosophie mit besonderer Berücksichtigung der Philosophie des Lebendigen.* ③ Jena 1889.
Driggers, Byrley Floyd (1901), Associate Cranberry Specialist, New Jersey Agricultural Experim. Station. New Brunswick, N.J. (U.S.A.). *Economic Entomol.; insects of Cranberry and Blueberry (Vaccinium spp).* ③ Rutgers Coll. 1924. ⑰ *Insects of Vaccinium spp.*
Drinkard, Alfred Washington Jr. (1883), Dir., Virginia Agric. Experim. Station. Blackeburg, Virginia (U.S.A.). *Pomol., physiol. problems on the nutrition of apple and peach trees.* ③ Cornell Univ. 1913.
Drobow, Wassilij P., Prof. Univ. Taschkent (U.d.S.S.R.), Per. Nogina 6. *Botan. Systematik. Phytosoziol.* ○
Droogleever Fortuyn, Aemilius Bernardus (18)86, Ass. Prof. of anat., Peking Union Med. Coll. Peking (China). *Histol. of cerebral cortex of mammals. Embryol. of the Chinese race. Histol. and microscopic anat. of the nervous system of Invertebrates. Phaenogenetics.* ③ Amsterdam 1911.
Droogleever Fortuyn-van Leyden, Cornelia Elisabeth (1885), Histol. Labor. Peking (China), Union Med. Coll. *Conditions of mitosis.*
Drosdow, Nikolaj A., Dir. Wissenschaftl. Schwarzmeer-Forschungs-Inst. Station Kubanskaja (U.d.S.S.R.), Sew. Kawkas. sh. d. „Chutorok". *Pflanzenkunde. Selektion d. Mais.* ○
Drost, Rudolf (1892), Dr. phil., Leiter der Vogelwarte der Staatl. Biol. Anstalt Helgoland (Deutschland). *Vogelzugsforschung.* ③ Göttingen 1923. ⑰ Vögel.
Drude, Karl Georg Oscar (1852), Geh.R., Prof. der Botan. i. R. (emer.) der Techn. Hochsch. in Dresden. Dresden-Bühlau (Deutschland), Thorner Str. 6. *Pflanzengeographie und insbesondere Oekol., Phänol.,*

Entwicklungsgeschichte der Florenreiche, Flora und Pflanzengeogr. von Mitteleuropa. ③ Göttingen 1874.
Druet, Jules (1892), Prof. Dampremy, Charleroi (Belgique), 50, Rue de Moscou. *Lépidoptères (macros et micros) du monde entier, spécialement noctuidae. Etude des premiers états, variations, mélanisme, hermaphroditisme, parasites des chenilles, Botan. (plantes de Belgique) au point de vue tératologique.* ③ Couvin 1912. ⑰ Macrolépidoptères d'Europ. Lépidoptères du Congo, Mongolie, Tonkin, Maroc, Amérique du Sud.
Drummond, Jack Cecil (1891), Prof. Biochem. Univ. Coll. London (England). *Biochem. of Animal Nutrition.* ③ London 1912.
Drummond, James Montagu Frank (1881), Regius Prof. of Botan., Univ. Glasgow (Scotland), Botan. Department. *Plant Physiol. (Photosynthesis, Hydathodes). Genetics (Crop-breeding, Brassica). Taxonomy of Angiosperms.* ③ Cambridge 1925. ⑰ Angiosperm Flowers, Genetical Material.
Drushinin, Dimitry (1894), Ass. Moskau (U.d.S.S.R.), Petrowko-Rasumowski, 17. *Agronomische Chem.*
Dry, Francis William (1891), Ackroyd Memorial Research Fellow. Headingley, Leeds (England), 9 Canterbury Drive. *Inheritance of colour in the Wensleydale Breed of sheep, attempt to establish a flock of pure-breeding white animals. Mammalian hair, especially that of the Mouse (Mus musculus), with special reference to problems in Genetics.* ③ Leeds 1925. ⑰ Wool of British Breeds of sheep.
Dryerre, H., Prof. R. Vet. Coll. Edinburgh (Scotland). *Physiol* ○
Drzewicki, Stefan (1898), Ass. des Zool. Inst. an der Univ. Jan Kazimierz in Lwów (Polen), ul. Mikołaja 4. *Innere Secretion bei Amphibien und Reptilien. Kreuzungsversuche an Reptilien.*
Dšanelidze, Alexander, Doc. Univ. Tiflis (U.d.S.S.R.). *Palaeont.* ○
Dšavachišvili, Georg, Doc. Univ. Tiflis (U.d.S.S.R.). *Anthrop., Ethnographie u. Zool.* ○
Dsewanowskij, Sergej A., Krimer Forschungs-Inst. Simferopol (U.d.S.S.R.), Turgenewskaja 4. *Botan. Geobotan.* ○
Dubin, Harry Ennis (1891), Dir. of the Biochem. Department of the H. A. Metz Labor. New York, N.Y. (U.S.A.), 122 Hudson St. *Research on Odd-Carbon Fats, Suprarenin and Vitamines.* ③ Univ. of Pennsylvania 1916. ⑰ Suprarenin. Active Principle of Suprarenal Gland Oscodal. A Concentrate Containing the Active Principles of Cod Liver Oil, Antirachitic and Antiophthalmic Vitamines.
Dubjanskij, V. A., Doc. I. Univ. Moskau (U.d.S.S.R.). *Botan.* ○
Dubois, Emile-Adolphe (1867), Dir. de l'école communale, Lépidopterol. Menin (Belgique), 32, rue de l'Emancipation. *Lépidoptères: Rhopalocères du Globe.* ⑰ Rhopalocères de l'Indochine française, de l'Amérique du Sud (Guyane française et Colombie).
Dubois, Eugène, Prof., Dr., Dir. Geol. Labor. Univ. Amsterdam. Haarlem (Holland). *Geol., Pitecanthropus, Larynx.* ○
Dubois, Georges (1890), Chargé de cours Fac. Sc. Univ., Labor. de Géol. Lille (France), 159, Rue Brule-Maison. *Faunes quaternaires.* ③ Lille 1924.
Dubos, René Jules (1901), Research Ass. in Soil Microbiol., Ass. in Sanitary Sc. New-Brunswick, N.J. (U.S.A.), N.J. Agric. Experim. Station. *Cellulose decomposition by bacteria, their occurence, abundance, and actual importance in natural processes. Bact., Biochem.* ③ Rutgers Univ. 1927. ⑰ Pure Cultures of cellulose decomposing bacteria.
Duboscq, Octave (1868), Dir. du Labor. Arago, Prof. à la Sorbonne. Banyuls sur mer, Pyr. Orient (France). *Protist* ○
Dubreuil, Georges (1879), Prof. d'Anat. générale et d'histol. à la Fac. de Méd. de Bordeaux, Dir. de l'Inst. Pasteur municipal de Bordeaux (France).

Histol: *Poumon. Tissus conjonctifs. Os et ossification.* Ⓟ Clichés microphotographiques, se rapportant à l'histologie des Mammifères.

Ducceschi, Virgilio, Prof. uff. Fisiol. Fac. Med. Univ. Padova (Italia), Viale Loredau, 6.

Ducháček, François (1875), Prof. Ecole polytechnique tchèque, Dir. Inst. des fermentations. Brno (C.S.R.). *Microbiol.: les fermentations et les enzymes.* Ⓓ Prague 1903.

Duchan, František, Ass. Inst. expérim. pour la production végétale. Praha-Žižkov (C.S.R.), Táboritská 31. *Physiol. de la nourriture des plantes cultivées.* o

Duchoň, François (1897), Ass. de l'inst. biochim. d'Etat. Praha I (C.S.R.), Valentinská 51. *Nutrition des plantes, les engrais, la chim. agricole, la production végétale. La nutrition de la beterave à sucre.* Ⓓ Prague 1920.

Duchworth, W. L. H., Prof. Univ. Jesus Coll. Cambridge (England). *Human Anat.* o

Dudgeon, Leonard Stanley (1876), Prof. of Pathol. Univ. of London, Dir. of Pathol. St. Thomas Hospital London. London W. (England), 6. Stanhope Street. *Pathol. and Bacteriol.* Ⓓ London 1897.

Dudgeon, Winfield (1886), Prof. of Botan., Ewing Christian Coll. Allahabad (India). *Morphol., diclinous angiosperms, Ecol. of India. Epiphyta, Algae.* Ⓓ Chicago 1917. Ⓟ Algae, Epiphyta.

Dudich, Endre, Curator Zool. Abt. d. Ung. National Mus. Budapest (Ungarn). *Crustacea, niedere Tiere.* o

Dudley jr., John E. (1886). Associate Entomol., Bureau of Entomol., U. S. Dep. of Agric. Madison, Wis. (U.S.A.), 1532 Univ. Ave. *Truck Crop Insect, pea aphis on canning peas, striped cucumber beetle, onion maggot, potato leafhopper.* Ⓓ Univ. of Wisconsin, Madison 1926.

Dudley Lamson, Paul (1884), Prof. of Pharm., Vanderbilt Univ. Med. School. Nashville, Tenn. (U.S.A.). *Pharm. Polycythaemia, Blood Volume methods. The Liver as Regulator of erythrocyte content of the Blood, and Blood Fluids. Carbon Tetrachloride. Anthelmintics. Heart Rhythme.* Ⓓ Harvard Univ. 1909.

Düggeli, M. (1878), Prof., Dr. a. d. Eidg. Techn. Hochsch. Zürich 7 (Schweiz), Hofstr. 75. *Pflanzengeographie u. wirtschaftl. Monographie d. Sihltales bei Einsiedeln.* o

Dürck, Hermann (1869), Geh.Med.R., o. Honorarprof. a. d. Univ. München, Vorstand des Pathol. Inst. des städt. Krankenhauses München rechts der Isar. München 27 (Deutschland), Geibelstr. 1 II. *Pathol. Anat. der Tropenkrankheiten und des Nervensystems.* Ⓓ München 1892.

Dürcken, Bernhard (1881), Dr. phil., o. Univ.Prof., Vorsteher der Abtlg. für Entwicklungsmechanik und Vererbung der Anatomie, Univ. Breslau XVI (Deutschland), Maxstr. 6. *Entwicklungsmechanik u. Vererbung.* Ⓓ Göttingen 1907.

Duerden, Herbert (1891), Lect. Department of Botan., Birkbeck Coll., Univ. London, E.C. 4 (England), Fetter Lane, Breams Buildings. *Anat. of plants. Vascular Cryptogams. Lycopodiales.* Ⓓ London Univ. 1925.

Duerden, J. E., Prof. Rhodes Univ. Coll. Grahamstown, Cape Col. (S. Africa). *Zool.* o

Dürigen, Bruno (1853), Prof. Landw. Schule. Berlin SO 16 (Deutschland), Schmidstr. 8a. *Zool. Geflügelkunde.* o

Duerst, Joh. Ulrich, Prof. Dr., Vet. med. Fac. Bern (Schweiz), Neubrückstr. 10. *Tierzucht, Hygiene.* o

Duesberg, Jules (1881), Prof. d'Anat. à l'Univ. de Liège (Belgique), 22, quai Mativa. *Cytol. L'œuf et ses localisations germinales.* Ⓓ Liège 1905.

Dufraisse, Charles Robert (1885), Sous-Dir. Labor. de Chim. Organique. Paris 5 (France), Coll. de France, Rue des Ecoles. *Catalyse d'oxydation par l'oxygène libre: actions antioxygènes et prooxygènes.* Ⓓ Paris 1923.

Dufrenoy, Jean (1894), Chef de Travaux, Chargé de la dir. Stat. Pathol. Végétale. Brive Corrèze (France), Coll. *Cellules migratrices. Sphériacés parasit. Géographie Botan. Actinomycés. Tumeurs cancreuses. Cellule géante chez les végétaux. Endotrophic Mycorhiza. L'excrétion des colorants vitaux.* Ⓓ Paris. Ⓟ Cultures de Champignons phytopath.

Dufresne, G. A. H., Prof. tit. Univ. Montreal (Canada), Rue St. Denis 1265. *Physiol.* o

Duggar, Benjamin Minge (1872), Research prof. plant physiol., Washington Univ., In Charge of graduate labor., Shaw School of Botan., Plant Physiol. to the Missouri Botan. Garden. St. Louis, Mo. (U.S.A.). *Plant nutrition, bact. filtration, virus diseases of plants.* Ⓓ Cornell Univ. 1898.

Duke, Dr., Dir. Inst. of bact. Research. Entebbe, Uganda (Br. East Africa). o

Dulanto, Abraham Rodriguez, Prof. Univ. Lima (Peru). *Botan.* o

Dulzetto, Filippo (1894), Ass. Ist. Zool. ed Anat. compar. Univ. Catania (Italia), Via Androne 35. *Ghiandole a secrezione interna; Istofisiol.* Ⓓ Catania.

Dunavan, David, Ass. Prof. of Entomol. and Zool. Clemson Coll., South Carolina (U.S.A.). Ⓓ Agric. Coll. of Oregon, U.S.A. Ⓟ Coleoptera, Carabidae.

Duncan, Carl Dudley (1895), Instr. in Botan. and General Biol. Stanford University Cal. (U.S.A.) Box 4, or San Jose, Cal.(U.S.A.), State Teachers Coll. *Systematic: Diploptera (Vespidae, Eumenidae, Masaridae). Bembicidae. Other Hymenoptera Aculeata. Insect Ecol.* Ⓟ Hymenoptera Aculeata. Vespidae and Bembicidae.

Duncker, Hans (1881), Studienrat und Verwalter des naturwiss. Kabinetts des Realgymnasiums Bremen (Deutschland), Wernigeroder Str. 22. *Vererbungslehre: Kanarienvögel, Wellensittiche, Hühner u. Enten; Letalfaktoren, Physiol. der Erbfaktoren; Farbenproblem der Vogelfeder. Rassenhygiene: Wirkung der Domestikation auf Phaenotyp u. Genotyp.* Ⓓ Göttingen 1905. Ⓟ Vögel.

Duncker, P. Georg E. (1870), Dr. phil., Kustos am Zool. Mus. Hamburg I (Deutschland), Steintorwall. *Biostatistik (Biometrie); Ichthyol. (Systematik, Faunistik, Variation).* Ⓓ Kiel 1895. Ⓟ Norddeutsche Salz- und Süßwasserfische.

Dunegan, John Clymer (1898), Ass. Pathol., Office of Fruit Disease Investigations, U. S. Dept. Agric. Fort Valley, Ga. (U.S.A.). *Fungus and bact. disease of drupaceous fruits. Diseases caused by members of the genus Sclerotinia.* Ⓓ Pennsylvania State Coll. 1921.

Dungan, George H. (1881), Ass. Prof. of Crop Production, Agronomy Dept., Univ. of Illinois. Urbana, Ill. (U.S.A.). *Physiol. of the corn plant.* Ⓓ Univ. of Wisconsin 1925.

Baron von Dungern, Emil (1867), Prof. Dr. med. Ludwigshafen am Bodensee (Deutschland). *Bact., Serol.* Ⓓ Freiburg i. Br. 1892.

Dunham, Henry G. (1895), Dir., Bact. Labor. Detroit, Mich. (U.S.A.), Digestive Ferments Co., 920 Henry St. *Bact. Reagents and Culture Media in the dehydrated form.* Ⓓ Mass. Agric. Coll. 1917. Ⓟ Bact. Reagents and dehydrated Culture Med.

Dunkerly, John S. (1881), Prof. Manchester (England), Zool. Dept., Univ. *Parasit., Protozool.* Ⓟ Parasit. Protozoa.

Dunkin, George William (1886), Vet. Superintendent. London, N.W. 7 (England), Med. Research Council, Farm Labor., Mill Hill. *Immunisation against dog Distemper.* Ⓓ Liverpool 1910.

Dunlap, Knight (1875), Prof. of Experim. Psychol. Johns Hopkins Univ. Baltimore, Md. (U.S.A.), 500 W 33 St. *Compar. Psychol.* Ⓓ Harvard 1902. o

Dunlop, Henry Adams (1898), Ass. Dir. Seattle, Wash. (U.S.A.), Internat. Fisheries Comm. Univ. of

Washington. *Age, growth and migration of Hippoglossus.* ⊕ Univ. of Br. Columbia, Vancouver 1912.

Dunn, Emmett Reid (1894). Northampton, Mass. (U.S.A.). *Herpetol., Amphibia. Salamanders.* ⊕ Harvard Univ. 1921. ℗ *Salamanders.*

Dunn, Halbert Louis (1896), Associate Prof. of Biometry and Vital Statistics. Baltimore, Md. (U.S.A.), School of Hygiene, Johns Hopkins. *Variation, Statistical methods of analysis in the field of medicine and Biol.* ⊕ Univ. of Minnesota 1923.

Dunn, John Shaw (1883), Prof. of Pathol. Manchester (England), Dept. of Pathol., Victoria Univ. *Morbid Anat. and Histol., Experim. Pathol.* ⊕ Glasgow 1912.

Dunn, Leslie C. (1893), Geneticist, Storrs Agric. Experim. Station. Storrs, Conn. (U.S.A.). *Inheritance in Poultry, of Color and Pattern in Mice.* ⊕ Harvard Univ. 1920. ℗ Morphol. Varieties of Fowls. Material on Embryo Abnormalities in Fowls.

Dunn, Marin S., Ass. Prof. of Botan., Philadelphia Coll. of Pharm. and Sc. Philadelphia, Pa. (U.S.A.), 145 N. 10th St. *Physiol., Morphol.* o

Dunn, Max S. (1895), Ass. Prof. of Biochem. Philadelphia, Pa. (U.S.A.). *Chem. of the Proteins.* ⊕ Univ. of Illinois, Urbana 1921.

Dunn, Stuart (1900), Instr. in Botan., Univ. of New Hampshire and Ass. Botan., New Hampshire Agric. Experim. Station. Durham, N.H. (U.S.A.). *PlantPhysiol.: Resistance of the apple to winter injury. The relation of time of harvest to yield and carbohydrate content of the potato.* ⊕ The Univ. of Minnesota, Minneapolis 1923.

Dunn, T. Avery (1900), Strietman Fellow in Biochem., Univ. of Minnesota. St. Paul, Minn. (U.S.A.), Univ. Farm. *Cereal Chem.* ⊕ Columbia 1927.

Dunton, Leila Elisabeth (1883), Ass. Prof. Food Economics and Nutrition. Manhattan, Kan. (U.S.A.), Kansas State Agric. Coll. *Relation of variety and growth conditions to the food value of wheat. Vitamins of foods and food composition in relation to nutrition.* ⊕ Wisconsin, Univ. 1922. ℗ Albino rats.

Duparque, André (1892), Ass. Géol. Fac. Sc. Univ. Lille (France), 159, Rue Brûle-Maison. *Paléont. de la houille.*

Dupray, Martin (1890), Dir., the Dupray Labor. Hutchinson, Kan. (U.S.A.), Rms. 306—309 Hoke Bldg. *Bact., Physiol. Chem. and Pathol.* ⊕ Univ. of Wisconsin 1914.

Dupuis, Paul (1869), Conserv. au Mus. Royal d'Hist. Natur. Bruxelles. Ixelles (Belgique), 33, Rue l'Abbaye. *Mollusques vivants.*

Durañona, Lucio, Prof. Univ. Nac. Buenos Aires (Argentina). *Botan.* o

Durham, George Benjamin (1896), Instr. in Botan. and Genetics. Storrs, Conn. (U.S.A.), Conn. Agric. Coll. *Genetics and morphol. of crop plants.* ⊕ Conn. Agric. Coll. 1919. ℗ Slides on tomato blossoms, buds and on squash tissue.

Durieux, C., Inst. Pasteur. Dakar, Sénégal (Afrique). *Bact.* o

Durig, Arnold (1872), Hofrat, Univ. Prof. Wien XVIII (Deutsch-Österreich), Lazaristengasse 10. *Physiol., Arbeits- u. Stoffwechselphysiol. Appetit.* ⊕ Innsbruck 1898.

Durmaškin, V. M., Doc. Univ. Nizny Novgorod (U.d.S.S.R.), Sovetskaja Ploščad 8. *Anat.* o

Durrell, Laurence Wood (1888), Botan. and Chief of Botan. Section State Experim. Station. Fort Collins, Col. (U.S.A.). *Plant Pathol. and Physiol. of Fungi.* ⊕ Iowa State Agric. Coll.

Durst, Charles Elmer (1884). Chicago, Ill. (U.S.A.), 53 West Jackson Blvd. *Selection.* ⊕ 1924.

Duruz, Willis P. (1896), Ass. Pomol. Experim. Station Univ. of Calif. Davis, Cal. (U.S.A.). *Diseases and insect control.*

Dusham, Edward Henry (1887), Prof. of Entomol. Pennsylvania State Coll., Pa. (U.S.A.), Coll. Heights. *Economic Entomol.* ⊕ Cornell Univ. 1924.

Dusmet Alonso, Jose Ma. (1869), Naturalista agr. Mus. Madrid I (España), Rue de Claudio Coello 19. *Hyménoptères de l'Espagne et paléarctiques. Vespides et Apides.* ⊕ Univ. de Madrid 1893. ℗ *Aculeata.*

Dusser de Barenne, Joannes Gregorius (1885), Dr. Priv.Doc. der Physiol. an der Reichsuniv. Utrecht (Holland), Wilhelminapark 28a. *Animal. Physiol., Zentralnervensystem, Muskel- und Nervenphysiol., Kreislauf, Atmung, Stoffwechsel.* ⊕ Amsterdam 1919.

Dustin, Albert-Pierre-Jean (1884), Prof. d'Anat. pathol. à la Fac. de Méd. de Bruxelles, Membre Academie de Méd., Chef des Services d'Anat. pathol. des Hôpitaux de Bruxelles (Belgique), 62, rue Berckmans. *Anat. pathol. et pathol. expérim. Cancérol. — Déterminisme de la Caryocinèse — Les poisons caryoclasiques — Biol. et pathol. du thymus — Mode d'action des radiations sur la cellule vivante.* ⊕ Méd. 1907, Sc. anat. 1909.

Dutcher, R. Adams (1886), Head of Department of Agric. and Biol. Chem. State College, Pennsylvania. (U.S.A.). *Vitamins and Deficiency Diseases.* ⊕ Univ. of Missouri 1912.

Duthie, A. V., Lect. Stellenbosch (South Africa), Dept. of Botan., Univ. ⊕ Cape.

Dutoit, Daniel (1894), Dr. ès sc., Ass. de botan. à la fac. des sc., Lausanne. Corsier sur Vevey (Suisse), Ct. de Vaud. *Géographie botan., systématique.* ⊕ Lausanne 1923.

Dutton, Loraine Orr. (1898), Dir. of labor., Methodist hospital, Memphis, Tenn. (U.S.A.). *Bact. The role of the bacteriophage in streptococcus infections.* ⊕ Univ. of Tennessee 1920. ℗ Strains of Bacteriophage.

Duval, Charles Warren (1876), Dir. Labor., Charity and Presbyn hosps. New Orleans, La (U.S.A.), Richmond Pl. *Pathol., bact.* o

Duvauchelle, Prof. Ecole Colon. d'Agric. Tunis (Afrique). *Anat. et physiol. anim.* o

van Duzee, Edward Payson (1861), Curator, Department of Entomol., California Acad. of Sc. San Francisco, Cal. (U.S.A.). *Hemiptera.*

Dwight, Jonathan (1858), Research Ass. New York (U.S.A.), American Mus. of Nat. Hist. *Ornithol.* ⊕ Columbia Univ. 1893.

Dyckerhoff, Fritz (1884), Dr. phil., wiss. Angestellter an der Zweigstelle der Biol. Reichsanstalt Aschersleben (Deutschland). *Tierische Schädlinge der Zier- und Gemüsepflanzen.* ⊕ Halle 1920. ℗ Tierische Schädlinge der Zier- u. Gemüsepflanzen.

Dyk, Anton, Prof. des Forstschutzes und Jagdwesens. Brno (C.S.R.), Černa pole 102, Landwirtschaftl. Hochsch. o

van Dyk, William T., Prof. American Univ. Zool. Mus. Beirut (Syrien). *Zool.* o

van Dyke, Edwin Cooper (1869), Associate Prof. of Entomol., Univ. of California. Berkeley, Cal. (U.S.A.), Agric. Hall. *Systematic Entomol. (Coleoptera), Forest Entomol., Geographical Distribution of Insects, Insect Biol.* ⊕ Cooper Med. Coll. 1895.

van Dyke, Harry B. (1895), Associate Prof. of Pharm., Univ. of Chicago (U.S.A.), Department of Pharm. *Pharmacol. and Physiol.* ⊕ Ph. Univ. of Chicago 1921, M. Rush Med. Coll. 1922.

Dykstra, Theodore Peter (1896), Junior Plant pathol. Botan. and Plantpathol., Dept. Corvallis, Ore. (U.S.A.). *Potato diseases.* ⊕ Univ. of Wisconsin 1925.

Dymond, John Richardson (1887), Ass. Prof. of Systematic Zool. Toronto (Canada), Dept. of Biol., Univ. of Toronto. *Ichthyol.* ⊕ Univ. of Toronto 1912. ℗ Fishes.

Dyrdowska, Marie (1888), Ass. Univ., Inst. Anat. Biol. Poznań (Polen), ul. Wjazdowa 3. *Molluskenfauna der Provinz Posen. Oribatiden von Polen.* ⊕ Poznań 1926. ℗ Mollusken u. Oribatiden von Polen.

Dziedzic, Josef (1892), Leiter der Saatzuchtanstalt. Przeworsk (Polen). *Getreide-, Kartoffel-*

und Gräserzüchtung. Ⓟ *Getreide- und Wiesengräserpflanzen.*

Dziubałtowski, Seweryn (1883), Prof. de botan. de l'Ecole supérieur d'Agric. à Varsovie (Pologne), rue Miodowa, 23. *Phytosociol. et botan. forestière.* Ⓟ Neuchâtel.

Dzunkovski, E., Dr., Dir. d. Labor. f. Tropenkrankheiten. Beograd (S.H.S.). o

Eames, Arthur Johnson (1881), Prof. of Botan. Ithaca, N.Y. (U.S.A.), New York State Coll. of Agric., Cornell Univ. *Morphol., Anat. of Vascular Plants (Angiospermae).* Ⓟ Harvard Univ. 1912.

Earl, Rollo Othwell (1892), Assistant Prof. of Biol. Kingston, Ont. (Canada), Queen's Univ. *Cytol. of chromosomes and their relationship to heredity; chromosomes of interspecific hybrids.* Ⓟ Chicago 1926.

Earle, Franklin Sumner (1856), Sugar Cane Technol., Tropical Plant Research Foundation. Herradura (Cuba). *Tropical Agric., Sugar Cane Technol. (including Pathol.).* Ⓟ Alabama Polytechnic Inst. 1900.

Earle, H. G., Prof. Univ. Hong-kong (China). *Physiol.* o

East, Edward Murray (1879), Prof. of Genetics, Harvard Univ. Boston 30, Mass. (U.S.A.), Bussey Inst. Ⓟ Univ. of Illinois 1907.

Eastham, John Wm. (1879), Provincial Plant Pathol. for British Columbia. Court House, Vancouver, B.C. (Canada). *Potato diseases. Diseases of cultivated plants in general.*

Eastham, Leonard (1893), Lect. in Entomol. and Zool. Birmingham (England), Univ., Edmund Street. *Embryol. of Lepidoptera. Development of Imaginal Discs in the Head of Lepidoptera.*

Eastwood, Alice, Curator Botan. Dept., Calif., Academy of Sc. San Francisco, Cal. (U.S.A.), Golden Gate Park. *California native and exotic phanerogams.* o

Eaton, Elon Howard (1866), Prof., State ornithol. Geneva, N.Y. (U.S.A.). o

Eaton, Frank Morris (1893), Ass. Physiol., Bureau of Plant Industry, U.S. Department of Agric. Sacaton, Ariz. (U.S.A.). *Physiol. of the Cotton Plant. Water Requirements and Transpiration. Alkali and Drought Resistant Relationships.* Ⓟ Univ. of Minnesota 1926.

Eaton, George F. (1872), Univ.-Prof., Secretary of Connecticut Acad. of Arts and Sc. New Haven, Conn. (U.S.A.), 70 Sachem St. *Compar. osteol., ralaeont.* Ⓟ Harvard 1900. o

Eaton, Scott Verne (1885), Ass. Prof. of Plant Physiol. Univ. of Chicago (U.S.A.), Department of Botan. *Relation of the essential elements to plant nutrition, especially sulphur.* Ⓟ Univ. of Chicago 1920.

Ebbecke, Ulrich (1883), o. ö. Prof. d. Physiol., Dir. d. Physiol. Inst. Bonn (Deutschland), Nußallee 11. *Gefäß-, Sinnes-, Nerven- und Elektrophysiol.* Ⓟ Kiel 1907.

Eber, August (1865), Dr. phil. et med. vet., o. Prof. und Dir. des Tierseucheninst. (früheren Veterinärinst.) und Inst. für animalische Nahrungsmittelkunde an der Univ. Leipzig (Deutschland), Linnéstr. 11. *Beziehungen zwischen Tier- und Menschentuberkulose; Bekämpfung der Tiertuberkulose; Erkennung und Bekämpfung der Tierseuchen; Fleischvergiftungen des Menschen, verursacht durch Bakterien der Coli-Typhus-Gruppe; seuchenhafte Geflügelkrankheiten; vergl. Morphol. des Unpaarzeher- und Paarzeherfußes.* Ⓟ Phil. Leipzig 1895, med. vet. Leipzig 1923.

Eberhardt, M., Prof. Univ. Besançon, Doubs (France). *Botan. Agric.* o

Eberle, Georg (1899), Dr. phil., Ass. Lübeck (Deutschland), Victoriastr. 19. *Prakt. Pflanzenschutz. Bodenazidität. Naturdenkmalpflege, insbes. Trapa natans und Ciconia ciconia. Naturkundliche Photographie.* Ⓟ Bonn a. Rhein 1922. Ⓟ Naturkundliche Aufnahmen.

Eberson, Frederick (1892), Ass. Prof. of Med., Epidemiol., Univ. Hospital, San Francisco, Cal. (U.S.A.), Univ. of California, School of Med. *Immunol.* Ⓟ Ph.D. Columbia Univ. 1918, M.D. Univ. of Minnesota 1924.

Eberstaller, Oskar (1851), Priv.Doc., ao. Prof. Univ. Graz (Österreich), Waltendorf. *Anat. des zentr. Nervensystems.* Ⓟ Graz 1876.

Ebert, Boris (1882), Prof. Mikrobiol. Inst. f. Med. Wiss. Leningrad (U.d.S.S.R.). Tschernyscheff per. 7, Wohnung 23. *Mikrobiol. Colityphus bei Vögeln u. Nagern. Bacterielle Bekämpfung. Immunität.* Ⓟ Petersburg 1922.

Eberth, Karl (1835), Univ.Prof. G.M.R. Berlin-Halensee (Deutschland), Seesener Str. 20. *Pathol. Anat. Bact. Histol. Nematoden.* o

Ebner, Richard (1885), Dr. phil., Mittelschul-Prof. Wien IX (Deutsch-Österreich), Beethovengasse 3. *Orthoptera (Systematik, Biol., geographische Verbreitung).* Ⓟ Wien 1922. Ⓟ Orthopteren.

Echerson, Sophia H., Boyce Thompson Inst. for Plant Research. Yonkers, N.Y. (U.S.A.). *Plant Microchem., physiol.* o

Echeverria, Pedro M., Escuela Nac. d. Vet. Bogotá (Columbia), Carrera 7a, Calle 32. *Anat.* o

Ecker, Enrique Eduardo (1887), Ass. Prof. of Immunol., Western Reserve Univ., Cleveland, Ohio (U.S.A.), 2109 Adalbert Road. *Bact., Immunol. and Experim. Pathol.* Ⓟ Univ. of Chicago 1917.

Eckles, Clarence Henry (1875), Prof. of Dairy Husbandry, Univ. of Minnesota. St. Paul, Minn. (U.S.A.), Univ. Farm. *The growth and nutrition of cattle* Ⓟ Iowa State Coll. 1897.

Eckstein, Erich (1888), Ass. am Pathol. Inst. des Städt. Krankenhauses Ludwigshafen a.Rh. (Deutschland). *Untersuchungen über Hämolyse.* Ⓟ Heidelberg 1921.

Eckstein, Fritz (1890), Dr. phil. nat., Abteilungsvorstand. Rathenow a. d. H. (Deutschland). *Angewandte Entomol., Bact., Biochem.* Ⓟ Straßburg 1913.

Eckstein, Henry Charles (1890), Ass. Prof. in Physiol. Chem. Ann Arbor, Mich. (U.S.A.), Univ. Michigan. *Chem. of the fats and proteins.*

Eckstein, Karl (1859), Geh.Reg.R., Prof. der Zool. am 1. Zool. Inst. der forstlichen Hochsch. in Eberswalde (Deutschland), Neue Schweizer Str. 24. *Angewandte Zool.: Forstzool., Biol., Schädlingsbekämpfung. Fischereiwiss. u. Fischereiwirtschaft.* Ⓟ Gießen 1884.

Eddy, C. O. (1894), Associate Entomol., Experim. Station, Clemson College, South Carolina (U.S.A.). *Econcmic Entomol.: Toxicity of Insecticides, Mexican Bean Beetle, Cotton Flea Hopper.* Ⓟ Ohio State Univ. 1920.

Eddy, Nathan B. (1890), Ass. Prof. of Physiol. and Pharm., Univ. of Alberta, Edmonton, Alta. (Canada). Department of Physiol. *Internal secretions and Pharm.* Ⓟ Cornell Univ. 1911.

Eddy, Samuel (1897), Ass. in Zool., Univ. of Illinois. Urbana, Ill. (U.S.A.), 301 Nat. Hist. Bldg. *Ecol., Protozool. and Planktol.* Ⓟ Univ. of Illinois 1925.

Edelstein, Sophie J., Doc. des Twerschen Pädagog. Inst. Twer (U.d.S.S.R.), Ul. Uritzkogo. *Botan.* o

Edelstein, Witalij I., Prof., Landwirtschaftl. Akademie Timirjasew. Moskau (U.d.S.S.R.), Petrowsko-Rasumowskoje. *Gartenbau.* o

Edgell, Beatrice (1871), Prof. of Psychol., Bedford Coll., Univ. of London N.W. 1 (England), Regents Park. *Psychol. Problems of memory.* Ⓟ London 1894, Wales 1897, Würzburg 1900.

Edgerton, Claude W., Prof. of Botan., Louisiana State Univ. Baton Rouge, La. (U.S.A.). o

Edie, E. S., Prof. Univ. Cape Town (South Africa). *Bio-Chem.* o

Edinger, Tilly (1897), Dr. phil. nat., Volontärass. Frankfurt a. M. (Deutschland), Geol.-Palaeont. Inst. der Univ., Robert-Mayer-Str. 6. *Palaeoneurol. Fossile Reptilien.* ⑨ Frankfurt a. M. 1921.

Edkins, J. S., Reader Univ. London SW 7. (England), S. Kensington. *Physiol.* o

Edlbacher, Siegfried (1886), Univ.Prof. Heidelberg (Deutschland), Mittelstr. 18. *Physiol. Chem.* o

Edson, Howard A. (1875), Senior Pathol. in Charge, Office of Vegetable and Forage Diseases, Bureau of Plant Industry, U.S. Dept. of Agric. Washington, D. C. (U.S.A.), 14th and B. Sts., S. W. *Diseases of vegetable and forage crops.* ⑨ Univ. of Wisconsin 1913.

Edwards, Charles L. (1863), Dir. of Nature Study, Los Angeles, Cal. (U.S.A.), 611 Braun Bldg., 1240 South Main St. ⑨ Leipzig 1890. o

Edwards, Francis Rees (1897), Head of Dept. of Chim. Industry, Georgia Experim. Station, Ga. (U.S.A.). *Infectious Abortion in Live Stock. Climatic Effects.* ⑨ Ohio State Univ. 1921.

Edwards, James Thomas (1889), Dir., Imperial Inst. of Vet. Research. Muktesar (Brit.-India), Ritani, U. P. *Trypanosoma.* ⑨ London 1926.

Edwards, Wilfred Norman (1890), Ass., Geol. Department, British Mus. (Natural History). London S.W. 7 (England), Cromwell Road. *Palaeobotan.* ⑨ Cambridge 1911.

van Eecke, Rudolf (1886), Konserv. am Ryks Mus. van Natuurlyke Historie te Leiden (Holland), Maredijk 161. *Systematik der Lepidoptera, Indo-Australische Heterocera, Thysanoptera Neerlandica. Genitalorgane der Lepidopteren, Limacodidae.*

Efflatoun, Hassan C. Bey. Dir. Entomol. Res. Plant Protection Section, Ministry of Agric. Cairo (Egypt.).*Entomol., spec. Egyptian Diptera.* ⑫ Diptera palaearct. region.

Efimoff, Wasilii W. (1890), Priv.Doc. Univ. Leningrad, 1. Ass. Inst. für Physik und Biophysik zu Moskau (U.d.S.S.R.), Krapotkinsstr., Haus 32, Wohnung 5. *Physiko-chem. Biol. Ionenlehre. Arbeitsphysiol.*

Efimov, Alexander Efim., Prof. Omsky Med. Inst. Omsk (U.d.S.S.R.), Ul. Lenina 9. *Histol.* o

von Eggeling, Jakob Friedrich Georg Burghard Heinrich (1869), o.ö. Prof. der Anat., Dr. med., Dir. der Anat. Anstalt der Univ. Breslau XVI (Deutschland), Maxstr. 6. *Anat. des Menschen. Vergl. Anat. der Wirbeltiere (Muskeln, Skelett, Verdauungsorgane, Haut).* ⑨ Heidelberg 1895.

Eggers, Friedrich (1888), Dr. phil., Priv.Doc. der Zool. an der Univ. Kiel (Deutschland). *Morphol. und Physiol. der Sinnesorgane der Insekten; tympanale und chordotonale Sinnesapparate.* ⑨ Gießen 1919.

Eggers, Hans (1873), Forstrat. Stolberg, Harz (Deutschland). *Biol. und Systematik der Borkenkäfer (Ipidae).* ⑫ Ipidae.

Eggleston, Willard Webster (1863), Ass. Botan., Bureau of Plant Industry, U.S. Department of Agric. Washington, D.C. (U.S.A.), 612 Randolph St. N.W. *Plants that poison man, horses, cattle, sheep, goats etc. Lupinus, Astragalus, Oxytropis, Delphinium, Asclepiadaceae, Compositae, Ericaceae, Umbelliferae, Liliaceae, Solanaceae. Plants of the Apple family. Crataegus.*

Eggleton, Frank E. (1893), Instr. in Zool., Univ. of Michigan. Ann Arbor, Mich. (U.S.A.), Department of Zool. *Animal Ecol., Hydrobiol. (Limnol.), Bottom Fauna of Fresh-water Lakes.* ⑨ Michigan 1923.

Eghis, Samuel (1873), Dir. Akklimatisations-Stat. Detskoje Sselo, Prof. Landwirtsch. Inst. Leningrad. Detskoje Sselo (U.d.S.S.R.), Kolpinskaja 2. *Vererbung: Nicotiana. Interspezifische Kreuzung in Genus Nicotiana. Züchtung des Buchweizens und Kartoffel. Nachwirkungs- und Mutationserscheinungen bei der künstlichen Erziehung der Kulturpflanzen.* ⑨ Halle a. d. S. 1901. ⑫ Samen der verschiedenen Rassen der N. Tabacum und N. rustica.

Eglit, Edgar (1890), Dir. Versuchs- u. Pflanzenzuchtstat., Landwirtsch. Zentralver. Priekulu selekcijas stacija über Cēsis (Lettland). *Kartoffel, Roggen, Flachs.* ⑨ Moskau 1918.

Eglīts, Makšis (1892), Priv.Doc. Univ., Dir. Inst. für Pflanzenschutz. Riga (Latvia), Baznīcas ielā No. 4a dz. 10. *Kartoffelkrankheiten. Pflanzenschutzmittel.* ⑨ Moskow 1918.

Egorowa, Alexandra (1899), Ass. Mikrobiol. Landwirtsch. Inst.; Ass. Haupt Botan. Garten. Leningrad (U.d.S.S.R.). *Mikrobiol. des Wassers.*

Eguren y Bengoa, Enrique de, Prof., Vicerrector y Catedrático Univ. Oviedo (España). *Anthrop.* o

Ehik, Julius (1891), Dr., Custos an dem Ungarischen National-Mus., Leiter der mammal. Sammlung. Budapest 80 (Ungarn). *Systematik der Nagetiere. Diluviale Säugetiere.* ⑨ Budapest 1914. ⑫ Kleine Säugetiere.

Ehlers, George Marion (1891), Ass. Prof. of Geol., Department of Geol., Univ. of Michigan. Ann Arbor, Mich. (U.S.A.). *Invertebrate Paleont. and Stratigraphy: Palaeont. and Stratigraphy of Paleozoic rocks of Michigan.* ⑨ Michigan 1913. ⑫ Invertebrate fossils from the Paleozoic rocks of Michigan.

Ehlers, Heinrich Wilhelm Ewald (1875), Dr. med., Prosector, Dir. der pathol.-anat. und bact. Abteilung des städtischen Krankenhauses Berlin-Neukölln (Deutschland), Innstr. 29. *Pathol. Anat.: Geschwulstlehre.* ⑨ Bonn 1904. ⑫ Mikroskopische Praep. der pathol. Histologie.

Ehlers, John H., Ass. Prof. of botan. and Curator of the Phanerogamic Herbarium, Univ. of Michigan. Ann Arbor, Mich. (U.S.A.). *Physiol., Systematic botan.* o

Ehrenbaum, Ernst M. E. (1861), Prof. Dr., Leiter der Fischereibiol. Abteilung des Zool. Staatsinst., Mitglied der Deutschen Wiss. Kommission für Meeresforschung. Hamburg (Deutschland), Kirchenallee 47. *Fischereibiol., Eier und Larven von Fischen, Lebensgeschichte der Fische, Makrele, Aal, Hering, Elasmobranchier; Fische der Arktis, von Westafrika, Fischereiverhältnisse der Vereinigten Staaten von Nordamerika, Türkische Seefischerei.* ⑨ Kiel 1884.

Ehrenberg, Kurt (1896), Dr. phil., Priv.Doc., Ass. am paläobiol. Inst. der Univ. Wien I (Deutsch-Österreich). *Paläobiol.: Pelmatozoa; Pleistocäne Säugetiere (besonders Ursus spelaeus); allgemeine Paläobiol. der Evertebrata (besonders Anpassungen an Sessilität); Erhaltungsbedingungen und Erhaltungszustände fossiler Tierreste; Gebißentwicklung und Zahnwechsel bei Säugetieren.* ⑨ Wien 1921.

Ehrenberg, Paul Richard Rudolf (1875), o. ö. Prof. an der Schlesischen Friedrich-Wilhelm-Univ., Dir. des Agrikulturchem. und Bact. Inst. der Univ. Breslau 6 (Deutschland), Nikolaistadtgraben 9 I. *Landwirtschaftl. Tier- und Pflanzenernährung, Bodenforschung, Bodenbact. Kolloidchem. des Erdbodens.* ⑨ Jena 1898.

Ehrenberg, Rudolf (1884), a.o. Prof. der Physiol. Göttingen (Deutschland), Am weißen Stein 16. *Theoretische Biol. vom Standpunkt der Irreversibilität des elementaren Lebensvorganges. Chem. der Alterungsveränderungen, Chem. der tryptischen Fermente, Radiometrische Mikroanalyse und Kolloidanalyse, Mikroanalyse der Entwicklungsvorgänge.* ⑨ Göttingen 1909.

Ehrhorn, Edward Macfarlane (1862), Consulting Entomol. Honolulu (Hawaii), Box 2456. *Termite control. Coccidae.* ⑫ Coccidae.

Ehrlich, Curt (1888), Dr. med. vet., Abteilungsvorsteher an Tierseucheninst. der Landwirtschaftskammer für die Provinz Hannover. Hannover (Deutschland), Vahrenwalder Str. 58. *Tuberkulose, Jungtierkrankheiten, bact. Fleischbeschau.* ⑨ Berlin 1912.

Ehrlich, Felix (1877), Dr. phil., o. Prof., Dir. des Inst. für Biochem. u. landwirtschaftl.

Technol. d. Univ. Breslau 16 (Deutschland), Fürstenstr. 102. *Biochem., Pflanzenphysiol., Organische Chem. (Kohlenhydrate, Eiweißstoffe), Gärung.* Ⓓ Berlin 1900. Ⓟ Lebende Mikroorganismen (Hefen, Schimmelpilze, Bact.). Organisch-chem. Praep.

Ehrmann, Hermann Felix Paul (1868), Studienrat. Leipzig-Gohlis (Deutschland), Eisenacher Str. 15 III. *Morphol., Physiol., Biol., Systematik und Verbreitung der Land- und Süßwassermollusken. Zoogeographie.* Ⓟ Land- und Süßwassermollusken.

Eichfeld, Johann G., Leiter d. Landwirtsch. u. Kolonisations-Versuchsabt. d. Murman-Eisenbahn. Chibiny, Murmansches Gouv. (U.d.S.S.R.), Versuchsfeld. *Selektion.* ○

Eichler, Oskar (1898), Dr. med., Ass. Düsseldorf (Deutschland), Pharmacol. Inst., Moorenstr. 5. *Pharm. des Kreislaufs.* Ⓓ Königsberg 1924.

Eichler, Paul (1888), Dr., Leiter des Biol. Inst. des Annen-Realgymn., Doc. an der Volkshochsch. Dresden-A., (Deutschland), Hohe Str. 38. *Experim. Tierphysiol.* Ⓓ Tübingen 1910.

Eichler, Witold (1874), Dr. med. Pabjanice (Pologne). *Coleopterol. palearctique.* Ⓓ Dorpat 1899. Ⓟ Coleoptères palearctiques.

Eide, Dir. Statens Skogforsöksstation Landbrukshöiskolen. Aas (Norge). ○

Eidmann, Hermann A. (1897), Dr. phil., Prof. für angewandte Zool. und vergl. Anat. der Forstlichen Versuchsanstalt München. Tung Chi (China), Univ. Shanghai. *Entomol., speciell Biol. und Psychol. der Ameisen.* Ⓓ München 1921.

Eigenmann, Carl H. (1863), Prof. of Zool., and Dean of the Graduate School. Bloomington, Ind. (U.S.A.), Indiana Univ. *Vertebrates of North America, american Characidae, fresh-water fishes of British Guiana, fishes of western south America and the Doradidae.* Ⓓ 1889. ○

Eiger, Marius (1873), o. Prof. Wilno (Polen), Physiol. Inst., Zakretowastr. 15. *Herznervation. Elektrokardiographie. Innere Sekretion. Interferometrie. Blutkolloide. Vegetatives System. Retina. Cortisches Organ.* Ⓓ Moskau 1909.

Eikenberry, William L., Head of Sc. Dept., State Normal School. East Stroudsburg, Pa. (U.S.A.). *Plant Physiol, Ecol.* ○

Eiman, John (1886), Ass. Prof. in Pathol., Univ. of Pennsylvania; Dir. Dept. of Pathol., Presbyterian Hosp. Philadelphia, Pa. (U.S.A.). *General Pathol.* Ⓓ Univ. of Pennsylvania 1918.

Eisentraut, Bruno Martin (1902), Ass. am Mus. für Naturkunde. Berlin N 4 (Deutschland), Invalidenstr. 43, Zool. Mus. *Vererbungscytol., Biol.: Orthopteren u. a. (Anpassungsbiol.), Systematik der Orthopteren, Hemipteren, Ascidien.* Ⓓ Halle a. d. S. 1925.

Eisler, M., Univ.Prof. Wien IX (Österreich), Zimmermannsgasse 4. *Serol. Serotherapie.* ○

Eisler, Paul (1862), o. Prof. Univ., Prosektor. Halle (Deutschland), Magdeburger Str. 26. ○

Eismond, J., Dr. Prof. Warszawa (Pologne), Inst. Nencki, r. Śniadeckich 8. *Histol. Embryol.* ○

Eitingen, Gregor N., Prof. Dr., Moskauer Forst-Inst. Moskau (U.d.S.S.R.), Rushejnyj per. 4, W. 7. *Waldbau.* ○

Ejsmont, Leopold (1902), Ass., Inst. für Zool. und Parasit. der tierärztl. Hochsch. Warschau (Polen), Grochowska 77. *Helminthol. parasitär, Trematodes.* ○

Ekblom, Tore (1896), Ass.Stat. Bakt.Labor. Stockholm (Sverige),Vasagatan 15-17. *Entomol.: Hemiptera, Diptera, specially Culicidae.* Ⓓ Upsala 1926.

Ekman, Gunnar Henrik Julius (1883), Doc. Univ. Helsinki (Sverige), Zool. Inst. *Anat., Physiol. und Embryol. der Vertebraten; Entwicklungsmechanik, speziell Kiemen-, Augen- und Herzentwicklung der Anuren.* Ⓓ Helsingfors 1914.

Ekman, Sven (1876), Dr. phil., Doc. a. d. Univ. Uppsala (Schweden), Univ. *Euphyllopoda, Cladocera, Copepoda, Ostracoda des süßen Wassers; Amphipoda, Isopoda u. Schizopoda; Cordylophora in Schweden; Holothurioidea aus Australien, Antarktis und Sub-antarktis; Lemmus; Geographie u. Ökol. der Land- u. Süßwasservertebraten u. der Süßwasserentomostraken Skandinaviens, der Alpen u. der Arktis; marine Relikte in nordeuropäischen Binnenseen, in der Ostsee u. im Kaspisee; Reliktenfrage; Artbildung durch Milieuinduktion; Tiefenfauna u. Tiefenablagerungen d. Binnenseen; Methodik der Tiefseeforschung; Vogelzug.* Ⓓ Uppsala 1904.

Ekman, T. T., Dr., Fiskeriintendent. Södertälje (Sverige). ○

Elagin, P. N., Prof. Pädagog. „Herzen"-Inst. Leningrad (U.d.S.S.R.), Moika 48-50-52. *Biol.* ○

Elbert, Wolfgang (1896), Diplomlandwirt, Abteilungsvorsteher der Abteilung für Pflanzenbau und Pflanzenschutz an der Landw. Versuchsanstalt. Harleshausen b. Kassel (Deutschland). *Pflanzenbau und Pflanzenschutz.* Ⓓ Bonn-Poppelsdorf 1924.

Elberth, B. I., Doc. Univ. Med. Fac. Minsk, Weißr. (U.d.S.S.R.). *Mikrobiol. Bact.* ○

Elbrecht, Berend (1883), Senior Lect. Botan. Pretoria (S. Africa), Transvaal-Univ. Coll. *Transvaal Algae.*

Elcock, Harry A. (1901), Ass. Botan. East Lansing, Mich. (U.S.A.), Botan. Dept., Mich. State Coll. *Tuberculosis. Bacterium beticolum.* Ⓓ Washington 1926.

Elek, László (1896), Hilfsass. Debrecen (Ungarn), Med. Klinik. *Intermediärer Stoffwechsel.* Ⓓ Budapest 1921.

Elenev, Paul (1877), Sachverständiger an der Abteilung Mykol. und Phytopath. der Staatsanstalt für experim. Agronomie. Leningrad (U.d.S.S.R.) Englischer Prospekt, 29, Labor. Jaczewski. *Mycol. Biol. und Floristic. Phytopath. der Ackerpflanzen, spez. Auswinterung des Getreides.* Ⓓ Moskau 1899.

Elenewskij, Richard A., Exped.-Leiter. Station Lobnja, Sawel. sh. d., Moskauer Gouv. (U.d.S.S.R.), Staatl. Inst. f. Wiesenbau. *Geobotan. Wiesenbau.* ○

Elenkin, Alexandr Alexandrovicz (1873), Dir. Kryptogamen-Inst. des Botan. Gartens, Doc. Univ. Leningrad (U.d.S.S.R.). *Kryptogamen.* Ⓟ Cryptogamen.

Eleonsky, Alexander (1886), Prof. Ichthyol. Labor. Landw. Akad. Moskau (U.d.S.S.R.), Petrowsko-Rasumowskoje. *Allg. u. spez. Fischzucht. Hydrobiol.*

Elfstrand, M., Prof. Dr. Uppsala (Sverige). *Pharm.* ○

Elfving, Fredrik Emil Wolmar (1854), Prof. Univ. emer. Helsinki (Finland), Köpmansgatan 10. *Entwickelungsgeschichte der Flechten.* Ⓓ Helsingfors 1880.

Eliasberg, Paul (1885), Labor. f. Pflanzenchem. u. Pflanzenphysiol. der Akad. d. Wissensch. Leningrad (U.d.S.S.R.), Wassiliewsky Ostrow, I. Linie, 12. *Pflanzenchem.*

Elkeles, Gerhard (1889), Dr. med., Leiter des Untersuchungsamts für ansteckende Krankheiten, Berlin-Westend. Charlottenburg 9 (Deutschland), Neuer Fürstenbrunner Weg 13/15. *Klinische Bact. und Serol. der bacteriellen Erkrankungen.* Ⓓ Berlin 1914. Ⓟ Bakterien-Stämme und Sera.

Elkin, D., Doc., Pädagog. Hochsch. Odessa (U.d.S.S.R.), Ul. Kominterna 2. *Psychol.* ○

Elkind, Amélie (1889), Ass. au Labor. d'Histol. et d'Embryol. à l'Ecole de Méd. à Lausanne (Suisse), Av. des Mousquines 2. *Biol. et Anat. microscopique des Phasmides; Carausins et ses glandes sexuelles. Anat. pathol.*

Elkner, Alexandre (1895), Ass. Univ., Inst. Histol. et Embryol. Warszawa (Pologne), Rue Chałubinskiego 5. *Tissu conjonctif.* Ⓓ Varsovie 1926.

Ellenberger, Wilhelm (1848), Dr. med. et phil. et med. vet., Prof. emer., Geh.R. Dresden-A. 24 (Deutschland), Schweizerstr. 11. *Physiol., Histol., Anat. der Tiere.* Ⓓ Göttingen.

Ellinger, Philipp (1887), Dr. med. et phil., a.o. Prof. der Pharmacol. an d. Univ. Heidelberg (Deutschland), Mozartstr. 7. *Nierenphysiol., Methämoglobin, Zellmutung, Pharm. der Röntgen-Strahlen, physi-*

kalisch-chem. Probleme der Pharm. ⊕ Phil. Greifswald 1911, med. Heidelberg 1913.

Ellinger, Tage Ulrich Holten (1892), Dir., Dept. of Live Stock Economics, International Live Stock Exposition, Chicago Lect., Univ. of Chicago, Ill. (U.S.A.), 120 Exchange Building, Union Stock Yards. *Animal Genetics, Animal Breeding. Physiol. of Reproduction. Agric. Economics.* ⊕ Harvard 1923.

Ellinghaus, Josef (1894), Ass. am Physiol. Inst. der Univ. Berlin N 4. Berlin-Wittenau (Deutschland), Lindenweg 49. *Chem. Physiol. (Stoffwechsel).* ⊕ Berlin 1921.

Elliot, Charlotte (1883), Associate Pathol., U.S. Department of Agric. Washington, D. C. (U.S.A.), Room 304, West Wing. *Bact. diseases of plants.* ⊕ Univ. of Wisconsin 1918.

Elliott, Jessie Sproat, Lect. in Botan., Univ. of Birmingham (England), Botan. Department. *Mycol.* ⊕ Sc. Birmingham.

Ellis, David (1874), Prof. of Bact., Lect. in Botan. Milngaire, Glasgow (Scotland), Birchtor, Strathblane Road. *Bact. Botan. Wild flowers of Scotland and Wales.* ⊕ Phil. Marburg 1902, sc. London 1905.

Ellis, Max Mapes (1887), Assoc. Prof. Univ. of Missouri, Dept. State Entomol. of Ind. Columbia, Mo. (U.S.A.), *Biol., physiol.* o

Ellisonas, Doc. Vergl.-anat. Inst. Kaunas (Litauen), Wilnastr. 2. o

Elmer, Otto Herman (1891), Ass. Chief in Plant Pathol. Ames, Ia. (U.S.A.). *Virus diseases, Cucurbitaceae diseases.* ⊕ Ames 1924.

Elmhirst, K., Superint., Maine Biol. Station. Millport, Scotl. (England). o

Elmore, Clarence Jerome (1870), Prof. of Biol., William Jewell Coll. Liberty, Mo. (U.S.A.), 429 Wilson Street. *Algae, Diatoms of Nebraska. Spring Flora of Liberty, Missouri and Vicinity.* ⊕ Univ. of Nebraska 1915.

Elmore, John Clifford (1896), Junior Entomol., U.S. Dept. of Agric., Bureau of Entomol. Alhambra, Cal. (U.S.A.), P. O. Box 297. *Truck Crop Insect.*

Elpatievsky, Vladimir (1877), Prof. at the Univ., Chef of the Helminthol. Section at the Inst. of Microbiol. of Azerbayan, Baku (U.d.S.S.R.). *Helminthol. (Nematodes parasit.); Meduses and Plankton of Caspian Sea. Entw. d. Sexualzellen.* ⊕ Warsaw 1915.

Elrod, Morton John (1863), Dir. Biol. Stat. Univ. of Montana. Missoula, Mtn. (U.S.A.). o

Elsbach, Eduard Maximiliaan (1899). Groningen (Holland), Brugstr. 7 A. *Pathol. Anat.*

Elßmann, Emil (1889), Dr. phil., Leiter der wissenschaftl. Abteilung an der Höheren Staatslehranstalt für Gartenbau, Weihenstephan. Freising b. München (Deutschland), Meichelbeckstr. 6. *Blütenentwicklung der Obstgehölze. Keimfähigkeit und Fertilität des Pollens bei den Stein- u. Kernobstsorten.* ⊕ München 1921.

v. d. Elst, P., Dr., Chief of Botan. Labor. Gen. Agric. Experim. Sta. Buitenzorg, Java (Nederl.-O.-Indië). *Physiol. of rice plant. Mentek-disease. Nutrition.* o

Elting, Erwin C. (1901), Instr., Dept. of Dairy Husbandry, Univ. of Missouri. Columbia, Miss. (U.S.A.). *Dairy Cattle Production.* ⊕ Univ. of Missouri 1925.

Elton, Charles (1900), Demonstrator Zool., Department of Zool. and Compar. Anat., Univ. Mus. Oxford (England). *Ecol. evolution: Periodic fluctuations in numbers of wild mice, with special reference to epidemics, parasit. and other ecol. Animal communities.* ⊕ Oxford 1922.

Elvehjem, Conrad Arnold (1901), Instr. in Agric. Chem. Agric. Chem. Bldg. Madison, Wis. (U.S.A.). *Animal nutrition, iron assimilation.* ⊕ Univ. of Wisconsin 1924.

Elwyn, Adolph, Ass. Prof. of Anat. New York, N.Y. (U.S.A.), 437 West 59 St. *Histol. of prioreceptive organs.* ⊕ Columbia Univ. 1917.

Elze, Curt (1885), Prof. d. Anat., Dir. d. Anat. Inst. Rostock (Deutschland), Georgstr. 49. *Nervensystem.* ⊕ Freiburg und Wien.

Embden, Gustav Georg (1874), Dr. med., Prof. o. Prof. der Physiol., Dir. des Univ.Inst. für vegetative Physiol. Frankfurt a. M. (Deutschland), Souchaystr. 3. *Physiol. und Pathol. des intermediären Stoffwechsels. Intermediärer Stoffwechsel beim Diabetes mellitus. Herkunft der Acetonkörper. Physiol. der Leber. Chemismus der Muskeltätigkeit.* ⊕ Straßburg 1899.

Emberger, Marie Louis (1897), Botan. de l'Inst. Sc. chérifien et Prof. à l'Inst. des hautes études marocaines. Rabat (Maroc), Avenue Moulay-Youssef. *Mitochondries. Géographie botan. méditerranéenne et spécialement Nordafricaine.* ⊕ Lyon 1921.

Embody, George Charles (1876), Prof. of Agric., Cornell Univ. Ithaca, N.Y. (U.S.A.), Roberts Hall. *Ecol. of Fishes; Fish Culture; Pond Fertilization. Biol. of Lakes and Streams; Fisheries. Habits of Crustacea; Amphipoda.* ⊕ Cornell Univ. 1910.

van Emden, Fritz (1898), Dr. phil. Halle a. d. S. (Deutschland), Thomasiusstr. 8. *Larvensystematik der Insekten, Speicherschädlinge, Biol. d. Carabiden, Systematik der Sandaliden.* ⊕ Leipzig 1921.

Emdin, Paul (1883), Prof., Dir. d. Nervenklinik Univ. Rostow a. Don (U.d.S.S.R.), Puschkinskaja Nr. 159. *Neurol. u. Pathol. der Muskeln. Chirurgie des Nervensystems. Encephalographie.* ⊕ Kazan 1909.

Emeljanoff, Nina (1902), Zool. Mus. Univ. u. Timiriaseff-Inst. Moskau (U.d.S.S.R.). *Entwicklungsmechanik des Geschlechts; Einwirkung d. äußeren Faktoren auf den Organismus; Systematik der Conopidae (Diptera).* ⊕ Conopidae (Diptera).

Emerson, Alfred Edwards (1896), Associate Prof. of Zool., Univ. of Pittsburgh, Pa. (U. S.A.), Dept. of Zool. *Termites, Ecol. Relationships, Ontogenetic and Phylogenetic origin of Polymorphism, Taxonomy.* ⊕ Ithaca, N.Y. 1925. ⊕ Termites and Termitophiles from any locality in the world.

Emerson, Fred Wilbert (1886), Prof. of Botan., Penn Coll., Oskaloosa, Ia. (U.S.A.). *Distribution of plants as related to root systems; environment and plant form.* ⊕ Chicago 1921. ⊕ Various botan. materials in preservative and microscopical preparations. Materials characteristic of inland North America.

Emerson, Herbert William (1880), Ass. Prof. of Bact. and Dir. of Pasteur Inst. Univ. of Michigan Ann Arbor, Mich. (U.S.A.). *Bact. and Hygiene.* ⊕ Univ. of Michigan 1915.

Emerson, Rollins Adams (1873), Dean of the Graduate School and Prof. of Plant Breeding, Cornell Univ. Ithaca, N.Y. (U.S.A.). *Genetics of Zea Mays.* ⊕ Harvard Univ. 1913.

Emerson, Sterling H. (1900), Instr. in Botan., Univ. of Michigan. Ann Arbor, Mich. (U.S.A.), Dept. of Botan. *Genetics and Cytol. of Oenothera.* ⊕ M.A. Michigan 1924.

Emig, William H., Ass. Prof. of botan., Univ. of Pittsburgh, Pa. (U.S.A.). *Bryophyta, histol.* o

Eminett, P. P., Prof. Staatl. Med. Inst. Astrachan (U.d.S.S.R.). *Bact.* o

Emme, Helene (1885), Ass. am Feder. Inst. für angew. Botan. u. neue Kultur, Herzenstr. 44. Leningrad (U.d.S.S.R.), 4, Linie 31, W. 6. *Phytocytol. Spez. Kulturpflanzen.*

Emmel, Victor E. (1878), Prof. of Anat., Coll. of Med., Univ. of Illinois. Chicago, Ill. (U.S.A.), 1817 St. Polk St. *Regeneration of tissues. Origin, cytol. and function of the cellular elements of the blood. Factors influencing growth of tissues and organs.* ⊕ Brown Univ. 1907.

Emmerich, Emil (1882), Prosektor, Dr. med., Dir. des Pathol. Inst. der städt. Krankenanstalt. Kiel (Deutschland). *Pathol. Anat. der Infektionskrank-*

heiten, Reticulo-endotheliales System und Infektionserreger, Experim. Schrumpfnieren durch Röntgenstrahlen, Röntgenstrahlen und Komplement, Bact. u. Serol. ⓓ München 1907. ⓟ Histol. und bact. Praep.
Emoto, Osamu, Ass. Prof. of Tôkyô Imperial Univ. Komaba near Tôkyô (Japan), Vet. Inst., Fac. of Agric. *Vet. med.* o
Emoto, Yoshikadzu (1891), Tokugawa Biol. Inst., Ass., Botan. Inst., Kaiserl. Univ. Tôkyô. Hiratsuka-Machi, Ebara-Gun, Tôkyô-Fu (Japan). *Bact., Mykol. und Pflanzenphysiol. Bact.: Schwefeloxydierende Bakterien. Mykol.: Myxomyceten, Schimmelpilze.* ⓓ Tôkyô 1917.
Enander, Sven Johan (1847). Lillherrdal (Sverige). *Salix, Hieracium.* ⓓ Lund 1918.
Enderlein, Günther (1872), Prof. und Kustos am Zool. Mus. der Univ. Berlin; Verwalter der Centralstelle für blutsaugende Insekten. Berlin SW 11 (Deutschland), Hafenplatz 3 II. *Vergl. Anat. und Morphol. der Insekten; Klassifikation der Insekten; blutsaugende Insekten; Dipteren; vergl. Morphol. der Bacterien-Cyclogenie. Bau, geschlechtl. und ungeschlechtl. Fortpflanzung und Entwicklung der Bacterien.* ⓓ Leipzig 1898. ⓟ Copeognathen der ganzen Erde.
Enders, Howard Edwin (1877), Head Department of Biol., Prof. of Zool., Purdue Univ. West Lafayette, Ind. (U.S.A.), 249 Littleton Street. *External and internal parasit. of the domesticated and wild vertebrated animals; Mallophaga, Anoplura, Acarida, Nemathelminthes and Cestodes.* ⓓ Johns Hopkins Univ. 1906.
Endo, Shigeru (1906), Ass., Dept. of Agric., Labor. of Phytopath., Imperial Univ. Kyoto (Japan). *Riceplant diseases, Sclerotial diseases.* ⓓ Tottori 1926.
Endo, Yasutarô (1887), Prof. of Imper. Sericult. Coll. Ueda, Naganoken (Japan). *Pathol. of mulberry-trees and mycol.* ⓓ Tôkyô 1915.
Engel, E., Doc. Univ. Palermo (Italia). *Anat.* o
Engel, Hendrik (1898), Curator at Zool. Mus. Amsterdam (Holland), Aquariumgebouw, Pl. Middenlaan. *Opisthobranchiate Molluscs.* ⓓ Amsterdam 1925.
Engel, Martha S., Ass. prof. of botan., Parsons Coll. Fairfield, Ia. (U.S.A.). o
Engelhardt, Victor M. (1884), Leiter Pflanzenschutzstation. Charbarowsk (U.d.S.S.R.), K.-Marx-Straße 29. *Biol., Ökol. d. schädlichen Insekten. Schädlingsbekämpfung.* ⓓ Leipzig 1910. ⓟ Orthoptera Liberiae Orientalis, Schädliche Insekten.
Engelhardt, Wladimir (1894), Ass. Biochem. Inst. Moskau 64 (U.d.S.S.R.), Woronzowo Pole 8. *Biochem. der Immunität; Enzymol. des intermediären Stoffwechsels.* ⓓ Moskau 1919.
Enger, Rudolf (1897), Dr. med., Ass. am physiol.-chem. Inst. d. Univ. Leipzig (Deutschland), Liebigstr. 16. *Eiweißchem. (Arginin- u. Cadaverinderivate).* ⓓ Leipzig 1923.
Engfeldt, Nils Olof (1881), Labor. der Chem. Tierärztl. Hochsch. Stockholm (Sverige). *Biochem.: Azetonkörper.* ⓓ Stockholm 1921.
Engle, Earl Theron (1896), Instr. Anat., Stanford Univ., California (U.S.A.). *Ovary of albino mouse; experim. and quantitative follicular atresia, growth of follicle, Corpus luteum atreticum, the senile ovary.* ⓓ Stanford 1925.
Engler, Heinrich Gustav Adolf (1844), o. Prof. emer. Univ. Berlin, ehem. Dir. des botan. Gartens und Mus. Berlin-Dahlem (Deutschland), Königin-Luisen-Str. 6—8. *Systematik und Pflanzengeographie, besonders von Afrika. Entwicklungsgeschichte d. Hochgebirgsfloren.* ⓓ Breslau 1866.
Englis, Duane Taylor (1891), Ass. Prof. of Chem., Univ. of Illinois. Urbana, Ill. (U.S.A.), 264 Chem. Labor. *Quantitative organic analysis. Carbohydrate metabolism of Plants.* ⓓ Univ. of Illinois 1916.
English, Lester Lamar (1899), Research Fellow, Crop Protection Inst. Urbana, Ill. (U.S.A.), State Entomol. Bldg. *Economic Entomol., Insecticides.* ⓓ Iowa State Coll. 1924.
Enken, Boris K., Vorsitzender d. Bureau neuer Kulturen u. Introduktionen. Charkow (U.d.S.S.R.), Tscherkasskaja 25. *Selektion landw. Pflanzen.* o
Enlows, Ella M. A. (1889), Associate Bact. Washington, D.C. (U.S.A.), Hygien. Labor. *Bact., serol.; antibacteria serums. Respiratory and intestinal diseases due to bacterial infection.* ⓓ George Wash. Univ. 1923. ⓟ Cultures of bacteria.
Enomoto, Suzuo (1887), Ass. of Phytopath., Botan. Inst., Hokkaido Imperial Univ. Sapporo (Japan). ⓓ Hokkaido Imperial Univ. 1922.
Enomoto, Yoshiki. Wakayamaken (Japan), Kôyasan 374, Ito-gun. *Ecol. of Birds.*
Enriques, P., Prof. d. Zool., Univ. Padova (Italia). *Leber der Mollusken. Zellehre. Vergl. Anat.* o
Ensign, M. R. (1890), Ass. Horticulturist. Gainesville, Fla. (U.S.A.), Exp. Sta. Univ. of Florida. *Variety tests, water relations, nutritive balance, temperature and growth correlations and hybridization.* ⓓ Cornell Univ. 1917.
Enslin, Eduard (1879), San.R., Dr. med. Fürth i. B. (Deutschland), Schließfach 27. *Biol. und Systematik der Hymenoptera. Tenthredinoidea, Blatt- und Holzwespen Mitteleuropas.* ⓓ Erlangen 1902. ⓟ Hymenopteren-Nester.
Entz, Béla (1877), o. Prof. Univ. Pécs (Ungarn), Dischka-Györzö-Gasse 5. *Pathol. Histol. Trypanosomen. Spirochaeten. Malaria. Balantidium.* ⓓ Budapest 1900. ⓟ Mikroskopische Praep. pathol. Objekte.
Entz, Géza (1875), Conserv. Univ. Utrecht (Holland), Zool. Labor., Janskerkhof 3. *Kernbau und Teilung der Ciliaten. Fibrillensystem bei Ciliaten. Kernbau und Teilung bei Peridineen. Phagocytose bei Protozoen.* ⓓ Budapest 1902. ⓟ Mikr. Praep. von Protisten.
Epple, W. F., Dairy Dept., Purdue Univ., Agric. Experim. Sta. Lafayette, Ind. (U.S.A.). *Action of specific organisms on milk proteins.*
Epstein, Albert K. (1890), Consulting chem. technologist of the firm of Epstein & Harris, Consulting Chem. Chicago, Ill. (U.S.A.), 5 South Wabash Avenue. *Biochem. analytical methods; chem. of proteins, carbohydrates and fats; vitamines, enzymes; industrial bact.* ⓓ Univ. of Chicago 1912.
Epstein, Emil (1875), Priv.Doc. allgemeine u. experim. Pathol., Univ. Wien VI (Deutsch-Österreich), Dreihufeisengasse 3. *Pathol. Biochem., Kolloidchem. des menschlichen Blutserums und der menschlichen Organgewebe.* ⓓ Wien 1900. ⓟ Histol., bact. Praep. und Organextrakte.
Epstein, G. W., Ass., Dr., Histol.-Embryol. Kab. d. I. Univ. Moskau (U.d.S.S.R.). *Histol. Embryol.* o
Eransquin, Rodolfo, Prof. Univ. Nac. Buenos Aires (Argentina), Calle Viamonte 430. *Anat. pathol. Bact.* o
D'Erasmo, G., Prof. Univ. Napoli (Italia). *Paleont.* o
Erazo, A., Prof. Univ. Tegucigalpa, Honduras (Zentralamerika), Plaza de la Merced. *Biol.* o
Ercegović, Anton (1895). Split (S.H.S.), sjemenište. *Lithophytische Algen, besonders Cyanophyceen.* ⓓ Zagreb 1924.
Erdman, Lewis Wilson (1895), Ass. Chief in Soil Bact. and Ass. Prof. of Soils, Iowa Agric. Experim. Station and Iowa State Coll. Ames, Ia. (U.S.A.), Soils Department. *Soil Bact.* ⓓ Iowa State Coll. 1922. ⓟ Cultures of Rhizobium (Legume bacteria).
Erdmann, Anna Marie Rhoda (1870), a.o. Prof. an der Med. Fac. der Univ. Berlin und Abteilungsleiter der Abteilung für experim. Zellforschung des Univ.Inst. für Krebsforschung, Berlin NW 6, Luisenstr. 9, Vdh. II. Berlin-Wilmersdorf (Deutschland), Nassauische Str. 17 II. *Experim. Zellforschung, experim. Protozool. und Gewebezüchtung. Explantation.* ⓓ München 1908.

Erdtman, Otto Gunnar Elias (1897), Department of Botan., Univ. Stockholm Va. (Sverige). *History of the vegetation of Europe. Pollenanalyse von Torfmooren und marinen Sedimenten.* ⓓ Stockholm 1922.

Eremeeve, Antoinette M., Conserv. Stat. Phytopath. du Jardin Botan. Leningrad (U.d.S.S.R.). *Uredinées.* ○

Erhard, Hubert (1883), Dr., a.o. Prof. an der Univ. Gießen (Deutschland), Zool. Inst. *Zellforschung, Tierphysiol., Einheimische Tierwelt. Protoplasmabewegung. Flimmer- u. Geißelbewegung.* ⓓ München 1909.

Eri, Megumu, Prof. in Zool. of Girls' High Seminary (Joshi Kôtô Shihan Gakkô). Nara (Japan). Biol. Labor.

Erickson, Eugene Thaorin, Ass., Plant Pathol. and Botan., Univ. of Minnesota. Minneapolis, Minn. (U.S.A.), 988/15 th ave. S.E. *Plant Physiol. Germination of Seeds.*

Eriksson, Erik Gösta Astle (1874), Leiter Algotsholms Versuchsstation. Kneippbaden (Sverige). *Physiol. Klee, Hafer, Gemüse.* ⓓ Alnarp 1897. ⓟ Samen von Klee, Gartenerbsen und Bohnen, Wrucken, Herbstrüben, Möhren, Pastinaken, Felderbsen und Futterwicken.

Erlacher, Philipp Johann (1886), a.o. Univ.Prof. für orthopädische Chirurgie a. d. Univ. Graz (Österreich), Opernring 4. *Pathol., Physiol., Histol. der peripheren Nerven.* ⓓ Graz 1910.

Erlanger, Joseph (1874), Prof. of Physiol., Washington Univ. School of Med. (U.S.A.). *Physiol. of the heart and Circulation. Active potential in nerve.* ⓓ Johns Hopkins Univ. 1899.

Erlanson, Carl Oscar (1901), Instr. Botan. Univ. of Michigan. Ann Arbor, Mich: (U.S.A.). *Genetical Ecol. Botan. Plant Distribution.* ⓓ Michigan 1925.

Erlanson, Eileen Whitehead (1899), Grad. Ass., Univ. Phanerogamic Herbarium, Univ. of Michigan. Ann Arbor, Mich. (U.S.A.). *Experim. taxonomic botan.; Genus Rosa.* ⓓ Michigan 1924. ⓟ Rosa.

Ermoljewa, Sinaida (1899), Ass. Biochem. Inst. Moskau 64 (U.d.S.S.R.), Woronzowo Pole 8. *Bact., Biol. der Vibrionen und Immunität.* ⓓ Rostow 1921. ⓟ 450 Stämme Vibrionen, besonders leuchtender.

Ernst, Paul (1859), o. Prof., pathol. Anat., allg. Pathol., Dir. pathol. Inst. Heidelberg (Deutschland), Albert-Ueberle-Str. 20. *Pathol. Anat. des Nervensystems. Degeneration, Metamorphose, Nekrose. Kolloide Struktur der Sekrete.* ⓓ Zürich 1884.

Eröss, Gedeon (1898), Ass. pathol. anat. Univ.- Inst. Pécs (Ungarn). *Immunität der Tuberkulose.* ⓓ Budapest 1923.

Esaki, Shirô, Ass. Prof. of Keiô-Gijuku Univ., Anat. Inst., Coll. of Med. Yotsuya, Tôkyô (Japan). *Anat.* ○

Esben-Petersen, Peter (1869). Silkeborg (Danmark). *Neuroptera Planipennia.*

de la Escalera, Fernando (1895), Mus. de Ciencias Nat. Madrid (España), Claudio Coello 115. *Lepidoptera.* ⓟ Lepidoptères d'Espagne.

Escherich, Karl Leopold (1871), o. Prof. für angewandte Zool. a. d. Univ. München (Deutschland), Prinzenstr. 26. *Angewandte Entomol. (Beziehungen zwischen der modernen Pflanzenkultur u. Schädlingsvermehrung); Biol. u. Bekämpfung der Forstschädlinge; Blattschneiderameisen (Attax) in Brasilien.* ⓓ Med. Würzburg 1893, phil. Leipzig 1896. ⓟ Fraßstücke forstlicher Insekten.

Escobio Franco, Jesús. Santander (España), Gaboya, 6, 4. *Antrop.* ○

Escribano, Cayetano, Conserv. del Mus. Nacional de Ciencias Nat., Prof. Auxiliar de la Fac. de Ciencias. Madrid (España), Colmenares 6. *Histol.* ○

Escudier-Donnadieu, Henri (1903), Préparateur d'histol. Bordeaux (France), Ecole de Méd. Navale. *Embryol. Tissu osseux.* ⓟ Préparations histol. Homme et Animaux.

Esdaile, Philippa Chicheley (1888), Reader Biol. Univ., Head of the Dept. of Biol., King's Coll. for Women. London W. 8 (England), Campden Hill Road. *Scales of Salmon. Embryonic skull and laryngeal cartilages of Perameles. Economic Biol. Animal Vegetable Products.* ⓓ Manchester 1917.

Esdorn, Ilse (1897), Dr. phil., Ass. am Botan. Inst. der Techn. Hochsch. zu Braunschweig (Deutschland), Humboldtstr. 1. *Pflanzenpathol., botan. Röntgenol.* ⓓ Kiel 1924.

Esguerra, Carlos, Prof. Univ. Nac. Bogotá (Colombia). *Pathol.* ○

Eshowa, Elena (1883), Ass. Moskau (U.d.S.S.R.), Werchne-Syromjatnitscheskajastr. 4, 8. *Tierstoffwechsel.*

Esmarch, Ferdinand (1886), Dr. phil., wissenschaftl. Mitarbeiter der Staatl. Landwirtschaftl. Versuchsanstalt Dresden-A. 16 (Deutschland), Stübelallee 2. *Pflanzenpathol., Pilzkrankheiten (Morphol., Physiol. und Biol. parasit. Pilze).* ⓓ Kiel 1914.

Espinosa Ventura, Manuel, Conserv. del Mus. de Anat. de la Fac. de Med. Valencia (España). *Anat.* ○

Esplugues Armengol, Julio, Lic. Ciencias Nat., Prof. auxiliar del Inst., Jardinero 2. del Botán. Valencia (España), Hospital, 12. *Botán.* ○

Espregneira Mendes, João, Ass. libre Fac. méd. Porto (Portugal). *Anat. humaine.* ⓓ Porto 1926.

Esser, Peter Hans Heinrich (1859), Dr. phil., Prof. der Botan., Dir. des botan. Gartens und des botan. Inst. der Univ. Köln a. Rh. (Deutschland), Vorgebirgstr. 37. *Physiol.; Phytopath. Blütenbildung am alten Holze. Durch Pilze verursachte Pflanzenkrankheiten.* ⓓ Bonn 1887.

Essey, Hiram Eli (1893), Fellow in Zool. Univ. of Illinois. Urbana, Ill. (U.S.A.), 503 W. California St. *Parasit., Helminthol. (Trematoda and Cestoda).*

Essig, Edward Oliver (1884), Associate Prof. of Entomol., Univ. of California. Berkeley, Cal. (U.S.A.), 200 Agric. Hall. *Economic entomol., Aphididae, Coccidae.* ⓓ Pomona Coll., Claremont, Cal., 1912. ⓟ Aphididae and Coccidae.

Esten, William Merrill (1862), Prof. of Bact., Connecticut Agric. Coll. Storrs, Conn. (U.S.A.). *Dairy Bact. Fermentation. Soil Bact. Bact. function in crop production.* ⓓ Middletown, Conn. 1894. ⓟ About 250 kinds of living bacteria. Many new varieties from Milk and Soil.

Esterly, Calvin O. (1879), Prof. of Zool., Occidental Coll., Zool., Scripps Inst. of Oceanography. Los Angeles, Cal. (U.S.A.). *Distribution and occurrence of marine plankton, pelagic Copepoda. Food of copepoda. Reactions of plankton animals to external conditions, especially with reference to diurnal migration.* ⓓ Harward Univ. 1907.

Estienne, Victor Martial François Joseph (1895), Chargé de cours à l'Univ. cathol. Dép. (Inst.) agronomique de Louvain (Belgique), Inst. Carnoy. *Biochim. végétale. Microchim.* ⓓ Louvain 1922.

Estrada, Adelaide Augusta Fernandes (1900), Préparateur du Labor. d'Analyses Cliniques et Ass. libre de l'Inst. d'Histol. de la Fac. de Méd. de Porto (Portugal). *Histol., Hématol.*

Esty, James Russell (1893), Bact., National Canners Association, Western Branch, San Francisco, Cal. (U.S.A.), 322 Battery Street. *Food poisoning investigations.* ⓓ R. I. State Coll. 1914.

von Euler-Chelpin, Hans Karl August Simon (1873), o. Prof. der allgemeinen und organischen Chem. a. d. Univ. Stockholm (Schweden). *Enzymchem. Vitaminchem. Chem. der Hefe und der alkoholischen Gärung.* ⓓ Berlin 1894.

Evans, A. H., Dr. Cambridge (England), 9 Harvey Road. *Ornithol.* ○

Evans, Alex. W. (1868), Prof. of Botan., Yale Univ. New Haven, Conn. (U.S.A.), 180 Livingston St. *Botan.* ⓓ Yale 1899. ○

Evans, Alice Catherine (1881), Associate Bact., U.S. Public Health Service. Washington, D.C. (U.S.A.). *Hygien. Labor. Brucella melitensis.*

ⓓ Univ. of Wisconsin 1910. ⓟ The various serol. groups of Brucella melitensis.

Evans, Arthur Thompson (1888), Dr., Prof. of Botan. and Plant Pathol., Head of Botan. Department, South Dakota State Coll. Brookings, S.D. (U.S.A.). *Morphol.* ⓓ Univ. of Chicago 1918.

Evans, Charles Lovatt (1884), Jodrell Prof. of Physiol. London (England), Univ. Coll. *Experim. physiol.* ⓓ London 1912.

Evans, Herbert McLean (1882), Prof. of Anat., Head of the Department of Anat. Berkeley, Cal. (U.S.A.), Univ. of California. *Histol., embryol., physiol. of reproduction, endocrinol.* ⓓ Johns Hopkins 1908.

Evans, Mary Jardine (1896), Junior Ass. Bact., Philadelphia General Hospital. Rutledge, Pa. (U.S.A.), Box 229. *Med. Bact.* ⓓ Univ. of Pennsylvania 1918.

Evans, Philip S. jr., Prof. Shantung Christian Univ. Tsinan (China). *Physiol.* o

Evans, T. J., Prof. Guy's Hospital Med. School. London S.E. 1 (England), St. Thomas St., Borough. *Biol.* o

Evenden, James Cawston (1889), Associate Entomol., United States Department of Agric., Bureau of Entomol., Forest Entomol. Idaho (U.S.A.), Cœur d'Alene. *Ravages of bark beetles and defoliating insects of the pine forests.* ⓓ Oregon 1914.

Evenius, Christa (1894), Versuchs- und Lehranstalt f. Bienenzucht. Stettin (Deutschland), Werderstr.32. *Biol., Anat. und Embryol. d. Hymenopteren.* ⓓ Münster 1920.

Evenius, Joachim (1896), Dr. phil., Leiter der Versuchs- und Lehranst. für Bienenzucht d. Landw.-Kammer, Stettin (Deutschland), Werderstr. 32. *Physiol., Anat. und Entwicklungsgeschichte des Verdauungssystems der Insekten; Physiol., Anat. und Entwicklungsgeschichte der Honigbiene (Apis mellifica L.).* ⓓ Münster 1923. ⓟ Praep. zur Biol. der Gattung Apis.

Evermann, Barton Warren (1853), Dir. Mus. California Academy of Sc. and Dir. Steinhart Aquarium, Lect. on conservation of marine life, Stanford Univ. San Francisco, Cal. (U.S.A.), Golden Gate Park. *Ichthyol., Geographic distribution of fishes.* ⓓ Ph. Indiana Univ. 1891, LL. Univ. of Utah, 1922.

Everett, Mark (1899), Prof. and Head of Dept. of Biochem. and Pharmacol., Univ. of Oklahoma Med. School. Norman, Okla. (U.S.A.). *Determination of Sugars and Phosphates in Biol. Material. Nature of Blood and Urine Sugar in Normal. Nature of Organic Acid-Soluble Phosphores of Blood. Determination of Potassium in Small Amounts. Specific Dynamic Action of Protein.* ⓓ Harvard 1924.

Everts, Jhr. Edouard, Jacques, Guilleaume (1849), Dr. phil. Haag (Holland), Emmastraat 28. *Coleopteren Europas und der Niederlande.* ⓓ Erlangen 1873.

Evreïnov, Vladimir (1887). Praha (C.S.R.), Inst. de physiol. végétale, Univ. *Pomol.*

Ewart, Alfred James, Prof. Botan. and Plant Physiol., Univ. Chairman Forest Exam. Board. Melbourne (Australia). *Plant Physiol., systematic Botan. of Victoria and of Central Australia.* ⓓ Phil. Leipzig 1896, sc. London 1897, Melbourne 1907, Oxford 1910, Adelaide 1926.

Ewart, James Cossar, Prof. Univ. Fac. of Med. Edinburgh (Scotland). *Zool.* o

Ewing, James (1888), Dr., Ass. Lect. in Botan. Leeds (Scotland), Univ. *Plant Ecol., Plant Physiol. amphoteric properties of proteins and plant tissues.* ⓓ Aberdeen 1916.

Ewing, Ky Pepper (1898), Junior Entomol., U.S. Dept. of Agric., Bureau of Entomol. Tallulah, La. (U.S.A.), Delta Labor. *New cotton insect pest, the cotton flea hopper (Psallus seriatus).*

Exell, Arthur Wallis (1901), Ass., Dept. of Botan., British Mus., Natural History. London S.W. 7 (England), Cromwell Road. *Systematic Botan. of Extra-European Polypetalae.* ⓓ Cambridge Univ. 1923.

Ext, Werner (1893), Dr. phil., Leiter des Biol. Labor. Wolfen der I. G. Farbenindustrie Aktiengesellschaft, Wolfen, Kreis Bitterfeld. Dessau, Anhalt (Deutschland), Stiftsstr. 19a. *Schädlingsbekämpfung, Saatgutbeizung, Insektenphysiol. und -toxikol.* ⓓ Berlin 1921.

Eyre, J. W. H., Prof. Univ. London S.W. 7 (England), S. Kensington. *Bact.* o

Eyster, William H., Prof. of Botan., Univ. of Maine. Orono, Me. (U.S.A.). *Genetics of maize.* o

Ezaki, Teizô, Ass. Prof. of Kyûshû Imperial Univ. Fukuoka (Japan), Entomol. Labor., Coll. of Agr. *Entomol.*

Ezekiel, Walter Naphtali (1901), National Research Fellow in Botan. St. Paul, Minn. (U.S.A.), Univ. of Minnesota, Univ. Farm. *Taxonomy, Physiol. and pathol. relations of Fruit-rotting Sclerotinias. Physiol. of resistance to Uredinales; physiol. of Uredinales.* ⓓ Maryland 1924.

Ezikov, J. J., Doz. I. Univ. Moskau (U.d.S.S.R.). *Zool. Histol. Embryol.* o

Ezra, Alfred. Cobham, Surrey (England), Foxwarren Park. *Ornithol.* o

Faas, Alexander (1872), Chefgeol. Leningrad (U.d.S.S.R.), Comité géol. de Russie, Wassl. Ostrow, Sredny Prosp. 72 b. *Geol. u. Palaeont.: Echinodermata.*

von Faber, Friedrich Carl (1880), Dir. der Botan. Labor. (Treub-Labor.) des Botan. Gartens zu Buitenzorg, Java (Ned. O.-Indië). *Pflanzenphysiol. und -œkol. der Tropen.* ⓓ Heidelberg 1903.

Fabian, Frederick William (1888), Associate Prof. and Research Associate in Bact. and Hygien. East Lansing, Mich. (U.S.A.), Michigan State Coll. *Fermentation.* ⓓ Michigan State Coll. 1924. ⓟ Bacteria.

Fabiani, R., Prof. Univ. Palermo (Italia). *Paleont.* o

Fabre, Philippe Joseph (1892), Préparateur de Physique biol. à la Fac. de Méd. de Paris (France), Labor. de Physique biol. *Mécanique sphygmomanométrique. Electrobiol.*

Fabris, Aldo, Prof. o. Univ. Genova (Italia). *Anat. patol.* o

Fabris, Angiolo (1895), Aiuto nell' Ist. di Anat. Patol., Univ. di Padova (Italia), Via S. Massimo 18. *Anat. patol. e patol. sperimentale delle emopatie. Cancro sperimentale. Antagonismo batterico.* ⓓ 1920.

Fabritius Buchwald, Niels (1898), Ass. København (Danmark), Ved Klosteret 13. *Mycol. und Pflanzenpathol., forstschädliche Pilze, Systematik der Polyporaceen.* ⓟ Hymenomyceten: Polyporeen, Thelephoreen und Hydnaceen.

La Face, Lidia (1891), Ass. Ist. Anat. Compar. Univ., Lib. doc. Entomol. appl. Roma (Italia), Via Depretis 91. *Morfol. e biol. delle Coccinigle. Formiche. Anofelini.* ⓓ Roma 1915.

Fackenthall, Philip F., Prof. of Pharmacognosy, Med. Coll of Virginia, School of Pharm. Richmond, Va. (U.S.A.). *Drug plants.* o

Fackler, Harry Lee (1894), Associate Entomol., Agric. Experim. Station. Knoxville, Tenn. (U.S.A.). *Entomol.* ⓓ Univ. of Kansas 1918.

Faelli, Ferruccio (1862), Prof. di zootecnia ed Igien zootecnica. Ist. Sup. Med. Vet. Torino (Italia), Via Nizza 52. *Eugenetica. Latte. Alimentazione degli animali.*

Faes, Henry (1878), Dir. de la Station fédérale d'essais viticoles et arboricoles de Lausanne. Montagibert, Lausanne (Suisse). *Entomol. appliquée (Viticulture et arboriculture fruitière). Maladies des plantes cultivées.*

Fahmy, Tewfik (1889), Chief Mycol. Res. Div. Ministry of Agric. Giza (Egypt), Plant Protection Section Cotton Res. Board. *Res. in fungies bacterial diseases of plants.* ⓓ London 1920.

Fåhraeus, R. S., Dr., Doc., Karolinska Inst. Stockholm (Sverige). *Pathol.* o

Fahrenholz, Curt (1890), II. Prosektor am Anat Inst. Leipzig (Deutschland), Liebigstr. 13. *Entwickelungsgeschichte u. vergl. Anat. der Wirbeltiere.* Ⓓ Phil. Jena 1915, med. 1922.

Fahrenholz, Heinrich (1882), Senator. Hildesheim (Deutschland), Humboldtstr. 14. *Ectoparasit. der Säugetiere und Vögel, insbesondere Anopluren.* Ⓟ Anopluren.

Fahringer, Josef (1876), Dr. phil., Prof. Wien XVIII (Deutsch-Österreich), Jörgerstr. 4. *Angewandte Insektenkunde (speziell: Hymenoptera). Systematische Insektenkunde (speziell: Braconidae).* Ⓓ Univ. Wien 1904. Ⓟ Hymenoptera (Braconidae et Ichneumonidae).

Fairchild, David Grandison (1869), Senior Agric. Explorer in Charge, Office of Foreign Seed and Plant Introduction, U. S. Bureau of Plant Industry. Washington, D.C. (U.S.A.), 1331 Conn. Ave. *Botan.* Ⓓ Oberlin Coll. 1915.

Fairhall, Lawrence Turner (1888), Instr., Harvard Univ., School of Public Health. Cambridge, Mass. (U.S.A.), 82 Ellery Street. *Nutrition. The metabolism of inorganic salts.* Ⓓ Harvard 1918.

Falck, August (1848), Univ. Prof., Geh.M.R. Kiel (Deutschland), Feldstr. 58. *Toxikol.* o

Falck, Richard (1873), o. Prof. der technischen Mycol. an der Forstl. Hochsch. in Hann.-Münden, Leiter des mykol. Inst. der Hochsch. Hann.-Münden (Deutschland). *Hausschwamm, holzzerstörende Pilze, Holzschutz, Baumkrankheiten, Humifikation, Sporenverbreitung bei den Fadenpilzen. Biol. Systematik der Pilze, Brandkrankheiten des Getreides, Mutterkorn, Pflanzenschutz und Pflanzenschutzstoffe, Desinfection, Hautdesinfection und Hautreinigung. Lenzitesfäule des Coniferenholzes, Merulinsfäule des Bauholzes.* Ⓓ Breslau 1901. Ⓟ Hausschwamm und holzzerstörende Pilze in Praep.

Falcone, Cesare, Prof. Univ. Napoli (Italia). *Anat.* o

Falk, Isidore Sydney (1899), Associate Prof. of Hygien. and Bact. Chicago, Ill. (U.S.A.), Univ. *Physiol. of the bacteria ; physical chem. of bact. metabolism and of immunol. reactions.* Ⓓ Yale Univ. 1923. Ⓟ Bacteries.

Falkenstein, R. B., Lect., Christian Coll. Canton (China). *Zool.* o

Falkowska, Hélène (1898), Ingénieur des Sc. Horticoles, Ass. á l'Inst. de Génétique de l'École Supérieure d'Agric. à Varsovie. (Pologne), Miodowa 23, Szkota Gt. Gosp. Wiejs. *Mendelisme: Recherches sur les hybrides de la fève et de la féverole.* Ⓓ 1925.

Fall, Henry Clinton (1862), Entomol. Tyngsboro, Mass. (U.S.A.). *Coleoptera, Acmaeodera, Apion.* o

Falqui, Giuseppe, Lib. doc. Univ. Cagliari, Sardinia (Italia). *Botan.* o

Da Fano, Corrado Donato (1879), Reader in Histol. King's Coll., Univ. London W.C. 2 (England), Strand. *Golgi's internal apparatus, Neuroglia. Impregnation Methods. Experim. Encephalitis and Inflammatory Processes of the Central Nervous System in general.* Ⓓ Pavia.

Fano, G., Prof. Univ. Roma (Italia). *Istol., Fisiol.* o

Fant, G. W. (1899), Extension Plant Pathol. Raleigh, N.C. (U.S.A.), State Coll. Station. *Plant diseases. Tobacco and fruit disease.* Ⓟ Local plant disease specimens.

Fantham, H. B., Prof. Univ. Johannesburg, Transvaal (S.Africa). *Zool., Comp. Anat.* o

Farenholtz, Hermann (1884), Dr. phil., Vorsteher der botan. Abteilung am Städtischen Mus. für Natur-, Völker- und Handelskunde in Bremen; Leiter der Bremischen Stelle für Pflanzenschutz. Bremen (Deutschland). *Pflanzenkrankheiten, botan. Warenkunde.* Ⓓ Kiel 1913. Ⓟ Herbarpflanzen.

Farkas, Béla (1884), ao. Prof. der Syst. Zool. Szeged (Ungarn), Szukováthy tér 1. *Protist.* Ⓢ *Spongiol., Histol. und Embryol.* Ⓓ Kolozsvár 1906. Ⓟ Tiere für das Mus.; histol. u. embryol. Praep.

Farkas, Géza (1872), o. Prof., Dir. physiol. Inst. Univ. Budapest VIII (Ungarn), Eszterházy-ut 9. *Physikal.-chem. Biol., Innere Sekretion, Stoffwechsel, Arbeitsphysiol., Sinnesphysiol.* Ⓓ Budapest 1897.

Farmer, Eric (1888), Investigator to the Industrial Fatigue, Research Board. London W.C. 2 (England), 15 York Buildings. *Industrial fatigue. Motion. Experim. psychol.* Ⓓ Cambridge 1914.

Farmer, J. B. (1865), Prof. Imp. Coll. of Sc. and Techn. London S.W. 7 (England), S. Kensington. *Botan.* Ⓓ Oxford 1902.

Farquhar, H., Keeper Dominion Mus. Wellington (New Zealand). *Zool.* o

Farr, Clifford H. (1888), Associate Prof. of Botan., Washington Univ. St. Louis, Mo. (U.S.A.). *Cell Physiol., Cytol., Cytokinesis; Cell enlargement; Effect of chemicals on root hair elongation.* Ⓓ Columbia Univ. 1916.

Farran, George Philip (1876), Inspector. Dublin (Ireland), 3 Kildare Place, Department of Fisheries. *Marine Zool.*

Farrar, Milton D. (1901), Instr. in Zool. and Entomol. Brookings, S.D. (U.S.A.), S.D. Agric. Coll. *Beekeeping.* Ⓟ Solitary bees native to South Dakota. Bremus sp. of S.D.

Farrell, James Irving (1900), Instr. in Physiol. and Pharm., Northwestern Univ. Chicago, Ill. (U.S.A.), 303 East Chicago Avenue. *Gastric and Pancreatic Secretion.* Ⓓ Chicago 1925.

Farsky, Octavianus (1893). Brno (C.S.R.), Cerná pole, U Kopličky č. 29. *Forstschutz, Waldkrankheiten und Schädlinge, Vogelschutz.* Ⓓ Dipl.-Ing. agr. Praha 1920, Dipl.-Ing. Forest Brno 1923.

Farwell, Oliver Atkins (1867), Consulting Botan., Curator of Herbarium and Drug Inspector for Parke, Davis & Co. Detroit, Mich. (U.S.A.), P.O.B. 488. *Systematic Botan.; Plant Histol.; Pharmacognosy.*

Fasiani, G. M., Prof. Univ. Padua (Italia). *Patol.* o

Faßbender, Paul (1892), Ass. an der Württembergischen Landesanstalt für Pflanzenschutz der Landw. Hochsch. Hohenheim-Stuttgart. Stuttgart (Deutschland), Seidenstr. 63. *Pflanzenphysiol.: Keimungs- und Reizphysiol.* Ⓓ Tübingen 1924.

Fassett, Norman Carter (1900), Instr. botan., Biol. Bldg., Univ. of Wisconsin. Madison, Wis. (U.S.A.). *Classification and distribution of flowering plants.* Ⓓ Harvard Univ. 1925.

Fasten, Nathan (1887), Prof. of Zool. and head of department of Zool. and Physiol. Oregon State Agric. Coll., Corvallis, Ore. (U.S.A.). *Cytol. of Decapoda. Game Fishery Problems. Copepod Parasit. of Trout. Animal Behavior.* Ⓓ Madison, Wis., 1914.

Faucheron, Louis Marie (1873), Ass. à la Fac. des Sc. de Lyon, Dir. du Service des Cultures de la Ville de Lyon, Dir. du Jardin et Collections botaniques de la Ville de Lyon au Parc de la Tête d'Or. Lyon 3 (France), 23, Rue Alfred de Musset. *Botan. générale et Horticulture.* Ⓓ Univ. de Lyon 1894. Ⓟ Des plantes et des graines.

Faucon, Prof. Univ. Montpellier (France). *Botan.* o

Faulkner, Gwen H. (1902), Demonstrator Zool. Univ. London N.W. 1 (England), Bedford Coll., Regent's Park. *Serpulid anat.* Ⓓ 1922.

Faull, Joseph Horace (1870), Prof. of Botan. Toronto (Canada), Univ. *Mycol.*

Faure, Charles Louis Alexandre (1886), Dr. en Méd., Licencié ès Sc. nat., Chargé du Cours de Zool. Méd. à la Fac. de Méd. de Toulouse (France), 3, place du Capitole. *Histol. des Arthropodes et notamment des Ixodidés.*

Faure, J. C., Prof. Transvaal Univ. Coll. Pretoria (South Africa). *Entomol.* o

Fauré-Fremiet, Emmanuel (1883), Dr. ès Sc., Sous-Dir. du Labor. d'Embryogénie comparée du Coll. de France, Paris V (France), 46, rue des Ecoles. *Cytol. comparée et systématique des Infusoires ciliés. Cytol., Histo-Chim., Chim. des constituants de la Cellule en général et de l'œuf en particulier. Physiol. de la Fécondation et des premiers stades du développement (transformations chim. et énergétiques). Etude analytique de la croissance des tissus et des organes. Cultures de Tissus ,,in vitro"; ses conditions physico-chim.; conditions expérim. de la formation des complexes cellulaires et des tissus.* ⓟ Paris 1924.

Faust, E. C., Prof. Dr., Dept. of Pathol., Union Med. Coll. Peking (China). o

Faust, Mildred Elizabeth (1899), Instr. in Botan., Syracuse Univ., Syracuse, N.Y. (U.S.A.), Box 62, Univ. Station. *Ecol. of Plants.* ⓟ Univ. of Chicago 1923. ⓔ *Biota of north central New York.*

Favaro, Giuseppe (1877), Prof. o., Dir. dell' Ist. Anat. Univ. Modena (Italia), Via Foro Boario 5. *Vestibolo orale; Sviluppo dei muscoli; Volta del diencefalo; Vasi, seni e cuori caudali e canale caudale; Ciclostomi; Pleura e cavità pleurali retrocardiache (borsa e seno); Miocardio polmonare; Cuore dei vertebrati; Nervo terminale e regione etmoidale mediana; Ginocchio; Ipofisi caudale.*

Favilli, Giovanni (1901), Ass. volontario, Rockefeller Foundation. Firenze, B. 8 (Italia), Viale G. B. Morgagni 18. *Fisiol. patol.: Itterizie (ricerche sperimentali); Istol.: Tessuti di granulazione; Batt.: Undulant fieber (Febbre maltese).* ⓟ Firenze 1924.

Favilli, Narciso (1885), Aiuto Anat. normale Ist. Superiore Med. Vet., Lib. doc., Prof. incar. Anat. e Fisiol. degli animali rurali Ist. Sup. Agr. Pisa (Italia), Via Cuppori 22. *Anat., istol. dell' apparato digerente nei diversi animali domestici.*

Favre, Maurice Jules (1876), Prof. agrégé à la Fac. de Méd., Lyon, Rhône (France). *Histol. de la peau et des muqueuses malpighiennes. Chondriome de la peau. Anat. pathol. du système lymphatique. Maladies ganglionnaires.* ⓟ Lyon 1904. ⓔ *Préparations relatives au Chondriome de la peau, des muqueuses malpighiennes et à l'anat. pathol. du système ganglionnaire.*

Fawcett, E., Prof. Univ. Bristol (England). *Anat.* o

Fawcett, Howard S. (1877), Prof. of Plant Pathol., Univ. of California, Graduate School of Tropical Agric. Riverside, Cal. (U.S.A.), Citrus Experim. Station. *Citrus diseases, temperature relations.* ⓟ Johns Hopkins Univ.

Faworow, Alexej M., Agronom Landw. Prov. Versuchsstation. Odessa (U.d.S.S.R.), Puschkinstraße 44. *Agronomie. Selektion u. Genetik landw. Pflanzen.* o

Fay, Arthur C. (1896), Ass. Prof. of Bact. and Dairy Bact. for the Kansas Experim. Station. Manhattan, Kan. (U.S.A.), Bact. Department K.S.A.C. *Bact.* ⓟ Univ. of Wisconsin 1921.

Fazzari, Ignazio (1899), Ass. alla Cattedra di Anat. umana normale, Univ. di Palermo (Italia), Ist. di Anat. umana normale, Bastioni di Porta Carini. *Morfol. e sviluppo dei vasi del cervelletto; Culture di tessuti in vitro; Sviluppo del diencefalo in ovis.* ⓟ Palermo 1922.

Fedak, Ludwik (1892), Ass. zool., Acad.Vet. Lwów (Polen), r. Lyczakowska 5. *Parasit., Ichtyol.* ⓟ Lwów 1924.

Fedde, Friedrich Karl Georg (1873), Prof. Dr.phil. Dahlem b. Berlin (Deutschland), Fabeckstr. 49. *Pflanzengeographie und -biol. Monographie der Papaveraceae (einschl. Fumarioideae).* ⓟ Breslau 1896. ⓔ *Lichtbilder zur Pflanzengeographie und -biol. Bisher 1500 Diapositive.*

Fedders, G., Dr., Ass. Univ. Rīga (Latvija), J. Jelgavas šoseja 31. *Libero doc.* o

Fedele, Marco, Libero doc. di Zool. nella R. Univ., Capo del Reparto di Zool. della Stazione Zool. di Napoli (Italia). *Tunicata: biol., sistematica, sistema nervoso, fisiol. della nutrizione, distribuzione, ecc. Movimento ciliare nei metazoi. Molluschi opistobranchi: Nutrizione. Innervazione cardiaca. Planctol.* ⓟ Napoli 1924. ⓔ *Animali marini del Golfo di Napoli, Thaliacea (Salpidae, Doliolidae, e Pirosomidae) degli altri mari.*

Federley, Harry (1879), Prof. der Genetik an der Univ. zu Helsingfors (Finnland), Västra Chaussén 31. *Speziesbastarde, Chromosomenuntersuchungen an Mischlingen.* ⓟ Helsingfors 1900.

Fedjakowa, Paula D., wiss. Mitarbeiterin Moor-Versuchsstation Nowgorod. (U.d.S.S.R.). *Botan.* o

Fedjušin, A. V., Doc. Univ., Ass. Zool. Kab. Minsk, Weißrußland (U.d.S.S.R.). *Zool.* o

Fedorov, Stefan Mitrofanovitch (1888), Entomol., Dir. de la Station de la défense des Plantes sur les côtes meridionales de la Crimée, Jalta (U.d.S.S.R.). *Biol. des insectes nuisibles de la vigne et de tabac.* ⓔ *Insectes nuisibles de la vigne et matériaux sur les Orthoptères de la Crimée.*

Fedorovsky, Alexandre S. (1885), Prof. Inst. Narodnoj Osvity. Charkow, Ukraine (U.d.S.S.R.), Vul. Vilnoi Akademii 35. *Paléont. des vertébrés. Cetacea. Proboscidea. Poissons tertiaires. Paléont. humaine.* ⓟ Charkow 1917.

Fedorow, Alexander I. (1895), Doc. d. Mittelasiat. Staatl. Univ. Taschkent (U.d.S.S.R.), Ul. I, Maja 77. *Forstwirtschaft. Genetic.* ⓟ Holzrassen.

Fedorow, Boris Gabriel (1904), Ass. Kasan (U.d.S.S.R.), Zootomisches Kabinett Univ. *Vergl. Morphol. der Wirbellosen: Polychaeten u. Onychophoren. Biocoenotik d. Meere.*

Fedotov, Dmitry (1888), Prof., Chief Zool. Zool. Labor., Academy of Sc., Geol.-Collaborator of the Zool. Committee. Leningrad (U.d.S.S.R.). *Compar. morphol. of Echinodermata, recent and fossil. Paras. Annelides (Aryestomidae) and Arthropoda. Systematik of Carbonif. Lamellibranchiata.* ⓟ St. Petersburg 1916.

Fedtschenko, Boris (1872), Prof. Univ., Dir. Inst. de phytogeographie, Botan. en chef au Jardin Botan. Principal. Leningrad (U.d.S.S.R.). *Géographie botan.; Systématique des plantes: Liliaceae, Ericaceae. Flore de l'Asie Centrale et de l'Amérique de Sud.* ⓟ Jurjev 1905. ⓔ *Plantes de la Russie, Asie Centrale.*

van der Feen, Pieter Jacobus (1892). Domburg, Zeeland (Holland). *Vergl. Physiol. u. Anat. d. Nervensystems u. d. Muskeln bei Invertebraten und Vertebraten.*

Fehér, Dániel (1890), Dr. phil., Diplom-Forstingenieur, o.ö. Hochschulprof., Vorstand des Botan. Inst. und Gartens der k. ung. Hochsch. für Berg- und Forstingenieure. Sopron (Ungarn). *Biochem. und biophysikalische Erforschung der Physiol. der Waldbäume (CO_2-Assimilation, N-Ernährung, Licht usw.). Biol. und Bact. des Waldbodens im Zusammenhange mit dem Wachstum der Waldbestände. Keimungsphysiol. Probleme (Hitzeresistenz der Samen). Einwirkung von Na_2CO_3 und Nitriten auf das Pflanzenwachstum. Wurzelbakterien der Leguminosenhölzer und ihre serol. Diagnose. Fruchtabfall der Laubhölzer.* ⓟ Wien 1920. ⓔ *Samen von Waldbäumen aus dem Botan. Garten. Pflänzlinge aus den Versuchs-Baumschulen des Botan. Gartens.*

Fehlmann, Werner (1887), Dr. phil., Prof. an der Eidg. Techn. Hochsch. Schaffhausen (Schweiz). *Biol. der Süßwasserorganismen, Ichthyol. Biol. Abwasserreinigung, biol. Wasseranalyse, Vogelschutz als biol. Schädlingsbekämpfung.* ⓟ Basel 1911.

Fehr, A., Prof. Versuchs- und Forschungsanstalt f. Milchwirtschaft. Weihenstephan b. Freising (Deutschland). o

Feirer, William Anthony (1900), Instr. in Bact., Johns Hopkins Univ. Baltimore, Md. (U.S.A.). *Bact.; desinfection.* ⓟ Johns Hopkins Univ. 1925.

von Fejérváry, Baron, Geza Dr., Konserv., Zool. Abt. d. Ung. National Mus. Budapest (Ungarn). *Herpetol., rec. u. fossil.*

von Fejérváry Lángh, Aranka, Dr. phil., Custos am Nation. Mus., Abt. Zool. Budapest I (Ungarn), Döbrenteygasse 12. *Geograph. Verbreitung u. Skelett der Reptilien.*

Feldbausch, Karl (1887), Dr. phil. Mannheim, Baden (Deutschland), Parkring 35. *Psychol. und Ökol. der Tiere. Systematische Anat. der Pflanzengattung Viola.* ① Erlangen 1913. ⑫ Höhere Pflanzen, Lichenes.

Feldberg, Wilhelm (1900), Volontärass. am Physiol. Inst. Berlin (Deutschland), Hessische Str. 3/4. *Autonomes Nervensystem.* ① Berlin 1925.

Feldman, Horace Wenger (1899), Research Fellow of the National Research Council in Zool. Boston, Mass. (U.S.A.), Bussey Institution Forest Hills. *Genetics of Mammals, Physiol. of Reproduction, Fertilization.* ① Harvard Univ. 1925. ⑫ Rats and mice of known hereditary composition.

Felice, Givelli (1901), Ass. Ist. Botan. Univ. Pavia (Italia). *Fisiol. e Patol. Vegetale.* ① Torino 1922.

Felín, Enriqueta Ortega, Prof. Univ. Barcelona (España). *Biol.* ○

Felix, Johannes (1859), Vorstand des Geol.-Palaeontol. Mus. der Univ. Leipzig (Deutschland), Gellertstr. 3. *Fossile Anthozoa.* ① Leipzig 1882. ⑫ Fossile Anthozoa.

Felix, Kurt Alfred Arthur Oskar (1888), a.o. Prof., Leiter des Labor. der II. Med. Klinik München 19 (Deutschland), Ruffinistr. 14. *Physiol. Chem. (Konstitution und intermediärer Stoffwechsel der Proteine).* ① München 1913.

Felix, Walther (1860), Dr. med., o. Prof., Dir. des Anat. Inst. Zürich 7 (Schweiz), Köllikerstr. 7. *Harnapparat d. Wirbeltiere, Keimdrüsen, Urogenitalsystem des Menschen, Anat. d. Brustkorbes, der Lungen u. Brustfells, Topograph. Anat. d. Mittelfellraumes u. seiner Organe, Anat. d. Atmungsorgane.* ○

Fellers, Carl Raymond (1893), Research Prof. of Horticultural Manufactures, Mass. Agric. Coll. Amherst, Mass. (U.S.A.). *Agric. Microbiol. and Chem.* ① Rutgers Univ. 1917.

Fellows, Hurley (1891), Associate Pathol., U. S. D. A. Manhattan, Kan. (U.S.A.), Botan. Dept., K. S. A. C. *Foot-rots of wheat.* ① Univ. of Wisconsin 1923.

Felt, E. Porter (1868), State Entomol. of New York. Albany, N.Y. (U.S.A.), State Mus. *Economic Entomol., gall midges (Itonididae), distribution of insects by wind.* ① Cornell Univ. 1894. ⑫ Gall insects and insect galls.

Felton, Lloyd Derr (1885), Ass. Prof. of Preventive Med. and Hygien., Harvard Med. School, Boston Mass. Cambridge, Mass. (U.S.A.), 7 Trowbridge Place. *Antibody of pneumococcus.* ⑪ M. Johns Hopkins Med. School 1916, Sc. Wooster Coll. 1925.

Fenaroli, Luigi (1899), Libero doc. di Botan. Sistematica e Fitogeografia. Milano, XVI, Via S. Vincenzo 38; Tavernola s. Lago d'Iseo, Prov. Bergamo (Italia). *Botan. sistematica e fitogeografia; Flora alpina; Macromiceti; Fotografia botan.* ⑪ Milano 1921. ⑫ Flora alpina, Felci, Carex, Euphrasia, Hieracium.

Feng, Chih-tung (1886), Ass. Pharm., Union Med. Coll., Dept. of Pharm. Peking (China). *Pharm. Chem.* ⑫ Ephedrine, ethylesters.

Feng, H. K., Lect. Nankai Univ. Tientsin (China). *Botan.* ○

Fenn, Wallace Osgood (1893), Prof. of Physiol., Rochester Univ. Med. School. Rochester, N.Y. (U.S.A.), *Gas exchange of nerve, heat production of muscle.* ① Harvard Univ. 1919.

Fennel, Eric A. (1887) Dir., Labor. (Clinical pathol.) The Clinic, Honolulu, T.H. (U.S.A.). *Clinical pathol. (Pathol., Bact., Serol., Chem.).* ① Univ. of Cincinnati 1908.

Fenton, Frederick Azel (1892), Entomol., U. S. Dept. of Agric. Florence, S.C. (U.S.A.), Pee Dee Experim. Stat. *Cotton Insects: boll weevil Anthonomus grandis.* ① Ohio 1918. ⑫ Hymenoptera, Dryinidae.

Ferdinandsen, Carl (1879), Prof. der Pflanzenpathol. København V (Danmark), Rolighedsvej 23. *Mykol., Pflanzenpathol., Unkräuter.* ① Kopenhagen 1920.

Fergus, Ernest Newton (1892), Ass. Prof. of Farm Crops, Ass. Agronomist (in Agric. Experim. Station). Lexington, Ky. (U.S.A.), Univ. of Kentucky. *Breeding and physiol.* ① Ohio State Univ. 1916.

Ferguson, Margaret Clay (1863), Prof. of Botan. and Dir. of Botan. Dept. Wellesley, Mass. (U.S.A.), 46 Dover Road, Wellesley Coll. *Plant physiol., cytol., genetics, compar. morphol.* ① Cornell 1901. ○

Feringer, T. B., Doc. Univ. Nizny Novgorod (U.d.S.S.R.), Sovetskaja Ploščad 8. *Anat.* ○

Ferle, Friedrich Rudolf (1877), Doc. f. Züchtungslehre u. Pflanzenbau am Herderinst. Riga (Latvija), Antonienstr. 1. *Angewandte Botan. Pflanzensiedlung u. -wanderung. Aufnahme nicht obligatorischer mineral. Nahrung d. d. Pflanzen. Rostepidemie. Hanf. Lein.* ① Riga 1904.

Fermi, Claudio (1862), o. Prof., Dir. Inst. de Igiene Univ. Sassari Roma (Italia),Via National 237.

Fermor, K. A., Ass. Lab. d. Genetik Univ. Leningrad (U.d.S.S.R.). ○

Fernald, Evelyn I., Ass. Prof. of Botan. Rockford Coll. New London, Conn. (U.S.A.), Box 233, Connecticut Coll. *Physiol.* ○

Fernald, Henry T. (1866), Prof. of Entomol., Massachusetts Agric. Coll. Amherst, Mass. (U.S.A.). *Applied entomol.* ① Johns Hopkins 1890. ○

Fernald, Merritt L., Fisher Prof. of Nat. history, Harvard Univ. Cambridge, Mass. (U.S.A.), Gray Herbarium. *Classification and distribution of vascular plants, especially of northeastern America.* ○

Fernandes, Daniel Salomon (1886), Gouv.Entomol. Paramaribo, Suriname (Südamerika), Princessestraat 19. *Krankheiten des Cacao.* ① Utrecht 1923.

Fernández, Ambrosio O. S. A. (1882), Prof. Salamanca (España), Col. de Calatrava. *Lépidoptères.* ① Valladolid 1905. ⑫ Papillons de l'Espagne.

Fernandez, Miguel, Prof. Univ. Nac. La Plata (Argentina). *Anat. comp.* ○

Fernandez, Samuel, Prof. Univ. Nac. d. Litoral. Corrientes (Argentina). *Botan.* ○

Fernández Galiano, Emilio (1885), Prof. Histol., Organographie et Physiol. compar. Univ. Barcelona (España), Enrique Granados, 108, 2, 1a. *Mouvements cellulaires. Chimiotactisme des cellules libres, Protozoaires. Tissus conjonctif et musculaire.* ① Madrid 1909.

Fernandez Gonzalez, Manuel, Prof. Univ. St. Tomas. Manila, Philippine Islands (U.S.A.). *Botan.* ○

Fernández Riofrio, Benito (1896), Prof. aux. Univ. Barcelona (España). *Micol.* ① Madrid 1920.

Fernbach, Auguste (1860), à l'Inst. Pasteur, Prof. honor. à la Fac. des Sc. de l'Univ. de Paris XV (France), 26, Rue Dutot. *Fermentations.* ① Paris 1889.

Fernow, Karl H. (1893), Ass. Extension Prof. Ithaca, N.Y. (U.S.A.), Department of Plant Pathol., N.Y. State Coll. of Agric. *Virus Diseases of Plants, especially those affecting Potatoes.* ① Cornell 1925.

Ferraresi, O., Prov. Univ. Roma (Italia). *Anat. patol.* ○

Ferrari, Giulio Cesare, Prof. incar. Univ. Bologna (Italia). *Psicol.* ○

Ferraris, Teodoro, Lib. doc. Univ. Genova (Italia). *Patol. veget.* ○

Ferraro, Pasquale, Lib. doc. Univ. Messina (Italia). *Anat. pathol.* ○

Ferrata, A., Prof. Univ. Napoli (Italia). *Istol.* ○

Ferraz, D., Prof. Univ. Porto Alegre (Brasil). *Pathol.* ○

Ferreira, Eduardo Leite Leal, Prof. Fac. d. Med. Curityba, Paraná (Brasil). *Anat. descript.* ○
Ferreira, Paulo, Prof. Univ. Porto (Portugal). *Biol.* ○
Ferreira de Mira, Mathias (1875), Prof. libre et Premier ass. Fac. Méd. de Lisbonne, chargé du Cours de Chim. Physiol., Dir. Inst. Bento da Rocha Cabral pour la recherche sc. Lisboa (Portugal), Calcada Bento da Rocha Cabral 28. *Physiol. normale. Capsules surrénales.*
Ferrer Galdiano, Manuel, Cons. Hidrobiol. Mus. Nacional de Sc. Nat. Madrid (España), Paseo de Recoletos 37. *Crustáceos.* ○
Ferrière, Charles (1888), Dr. és Sc., Conserv. pour l'entomol. au Mus. d'Histoire Naturelle à Berne (Suisse), Steinerstr. 4. *Biol. et Systématique des Hyménoptères parasites.* ⓟ Genève 1914.
Ferris, Roxana S. (1895), Ass. Curator, Dudley Herbarium, Stanford Univ., Cal. (U.S.A.). *Taxonomic Botan.* ⓟ Stanford Univ. 1916.
Ferris, Harry B. (1865), E. K. Hunt Prof. of Anat., Med. Dept., Yale Univ. New Haven, Conn. (U.S.A.), 395 St. Ronan St. *Anthropol.* ⓟ Jale Univ. 1891.
Ferroux, René (1892), Chef des Travaux de physique appliquée à la biol. Paris V (France), Inst. du Radium, 26, rue d'Ulm. *Applications biol. et thérapeutiques des radiations de Röntgen et des corps radioactifs.* ⓟ Paris 1919.
Ferry, Ronald M. (1891), Instr. and Tutor in Premed. Sc., Harvard Coll., Instr. in Physiol. Chem. Boston, Mass. (U.S.A.), Harvard Med. School. *Biol. Chem., Physical Chem. of Hemoglobin, Physical Chem. of Immunity.* ⓟ Columbia 1916.
Festa, Enrico (1868), Vice-Dir. onorario del Regio Mus. Zool. Torino (Italia), Palazzo Carignano. *Mammiferi, uccelli, Pesci di acqua dolce italiani.*
Fetscher, Rainer (1895), Dr. med., Priv.Doc. für Hygiene. Dresden-N. 23 (Deutschland), Weinbergstraße 96. *Menschliche Erbbiol., Kriminalbiol. Soziale Hygiene und Eugenik. Bact.* ⓟ Tübingen 1921.
Feuerborn, Heinrich Jacob (1883), Dr. phil., Priv.Doc., z. Z. stellv. Dir. des Zool. Inst. Münster i. Westf. (Deutschland), Johannisstr. *Vergl. Morphol. u. Entwicklungsgeschichte der Insekten. Diptera. Psychodidae.* ⓟ Münster i. W. 1908.
Feulgen, Robert (1884), Dr. med., Prof. der physiol. Chemie a. d. Univ. Gießen (Deutschland), Ludwigstr. 46. *Biochem. des Zellkerns und spezieller Lipoide einschl. ihrer Mikrochem. Chem. und Physiol. der Nukleinstoffe.* ⓟ Kiel 1912.
Fewkes, J. Walter (1850), Chief, Bureau American Ethnol. Forest Glen, Md. (U.S.A.). *Ethnol., archaeol., Medusae, Echinoderms.* ⓟ Harvard 1877. ○
Fey, K. J. (1897), Entomol. in Charge. Kashing, Chekiang (China). *Applied entomol.: Paddy Borer (Chilo simplex, Schoenobius incertellus), Aphid.* ⓟ Living pupae of Philosamia, Actias, Anthrera. Eggs of Bombyx mori, Mantid.
Fiala, Anton (1895), Dir. Abteilung für Wein und Obstbau bei der Landw. Versuchsstat. Kosice (C.S.R.), Komenského 22. *Chemie des Wein- u. Obstbaus.* ⓟ Prag 1918.
Fiallos, J. M., Prof. Univ. Tegucigalpa, Honduras (Zentralamerika), Plaza de la Merced. *Bact.*
Fichtenholz, Sophie S. (1885), Ass. Section der Pflanzenphysiol. Botan. Garten. Leningrad (U.d.S.S.R.), Kamennoostrovskij 73/75. *Bedeutung des Blattnervensystems.*
Fick, Ludwig (1868), Dr. med., Honorarprof. in der med. Fac. der Univ., Abt.-Dir. der pathol.-anat. Abt. des städt. Krankenhauses im Friedrichshain, Berlin NO 18 (Deutschland). *Pathol. Anat. und pathol. Histol.* ⓟ Königsberg i. Pr. 1892.
Fick, Rudolf Armin (1866), o.ö. Prof. der Anat., Vorstand der Anat. Anstalt. Berlin NW 23 (Deutschland), Siegmundshof 21 II. *Gelenk-Muskelmechanik,*

Anthropoide, theoretische Vererbungslehre. ⓟ Würzburg 1888.
Fickendey, Ernst (1878), GRR. Prof. Dr., Berater der Rubber CultuurMaatschappij „Amsterdam". Poeloe Radja, Post Tandjong Balei, Sumatra, O.K. (Ned.-Indië). *Kultur der Ölpalme.* ⓟ Leipzig 1904.
Fiener, E. N., Doc. Univ. Voronez (U.d.S.S.R.). *Botan.* ○
Fidrovskij, V., Prof. Univ. Nikolaev (U.d.S.S.R.), Nikol'ska ul. 12/2. *Psychol.* ○
Fiebiger, Josef (1870), Dr., o.ö. Prof. für Histol., Embryol. u. Fischkunde an der Wiener tierärztl. Hochsch., Priv.Doc. f. Parasit. an d. med. Fac. der Univ. Wien XII (Deutsch-Österreich), Ruckergasse 12. *Bau der Delphinlunge, Verknöcherung. Fischhistol. u. -anat., Fischkrankheiten; Protozoen (Coccidien), Räudemilben.* ⓟ Wien 1894. ⓟ *Fische, Parasiten, Histol. Praep.*
Field, Cyrus W. (1878), Dir. Med. Labor. New York City, N.Y. (U.S.A.), 126 E. 64. St. *Clinical pathol., bact., seriol. and biochem.* ⓟ Yale Univ. 1900.
Fiessinger, Noël (1881), Prof. agrégé Fac. de Méd., Paris (France), 16 Boulevard Raspail. *Hématol. chim. et morphol. Ferments des leucocytes. Physiol. pathol. du foie. Exploration fonctionelle. Cirrhoses.* ⓟ Paris 1908.
Figdor, Wilhelm (1866), a.o. Prof. an d. Univ. Wien, Dr. phil., Vorstand der pflanzenphysiol. Abt. der Biol. Versuchsanstalt der Akademie der Wissenschaften in Wien IV (Deutsch-Österreich), Wahllehengasse 9. *Anat. u. Physiol. der Pflanzen (Anisophyllie, Regenerationserscheinungen). Angewandte Botan. (Gallen).* ⓟ Wien 1891. ⓟ *Pflanzen (lebende).*
Figueira, Luís (1894), Dr., Ass. de la Fac. de Méd. de Lisbonne et de l'Inst. Bact. Camara Pestana, service anti-rabique. Lisbonne (Portugal). *Bact., Parasit., diagnostic et traitement de la rage, analyses bact. et sérol., bacille dysentérique Shiga, les Bacilles diphtérimorphes et pseudo-diphtériques; foyer de bilharziose dans l'Algarbe-Portugal; Cobaye dans le diagnostic bact. de la rage; lèpre; colibacillémie.* ⓟ Lisbonne 1920.
Filarszky, Nándor (1858), Dr., Dir. der Botan. Abt. des Ungarischen National-Mus. Budapest V (Ungarn), Akademia u. 2 sz. II. *Pflanzenmorphol. (Organographia, Anat.), Systematik, Algol., Charal.* ⓟ Budapest 1884. ⓟ *Charaphyten aus Ungarn für fremdländische Charaphyten.*
Filatow, Dimitry (1876), Ass. Inst. Experim. Biol. Moskau (U.d.S.S.R.), Woronzowo Pole 6. *Transplantierung und Explantierung der Ohrblase u. der Augenkeimes bei Amphibien.* ⓟ Moskauer Univ.1900.
Fildes, Paul (1882), Ass. Bact., London Hospital. London E. 1 (England). *Med. Bact. Tetanus.* ⓟ Cambridge Univ. 1909.
Filewicz, Wladyslaw (1876), Dr., Privatgelehrter, Präsident der polnischen Obstgartenbesitzer-Gesellschaft. Sinoteka, Eisenbahnstat. Mrozy, Post Katuszyn (Polska). *Regenerationsprozesse bei Pflanzen, Tieren u. Menschen. Chirurg.-therapeutische Experim. an alten Obstbäumen u. edlen schwachen Sorten. Einpfropfen der Testikel junger Tiere i. d. Muskulatur der alten. Verjüngungsexperim. bei Tieren.* ⓟ Wien 1901.
Filho, P., Prof. Univ. Porto Alegre (Brasil). *Microbiol.* ○
Filipčenko, Ju. A., Prof. Univ. Leningrad (U.d.S.S.R.). *Genetik.* ○
Filipjev, Ivan Nik. (1889), Doc. d. angewandten Entomol. Univ., Leiter d. Zool. im Forstinst., Abt. f. Entomol. des Staatsinst. für experim. Agronomie. Leningrad (U.d.S.S.R.), W.O. 7, Lin. 13, Kv. 23. *Angewandte Entomol. (spez. Lepidoptera u. Acridiodea). Oekol. und Zoogeographie (Lepidopteren). Nematoden (spez. freilebende und pflanzenparasit.).* ⓟ Univ. Leningrad 1923. ⓟ *Freilebende Nematoden. Lepidopteren.*
Filippi, D., Prof. Univ. Parma (Italia). *Botan.* ○
Filter, Paul (1880), Vorsteher der botan. Abt. der Landwirtschaftl. Kontrollstation der Landwirt-

schafts-Kammer, Berlin NW 40 (Deutschland), Kronprinzenufer 4. *Samenkontrolle, Futtermittelmikroskopie.* Ⓓ Berlin 1914.

Fine, Morris S. (1886), Dir. Research Dept. Postum Cereal Co. Battle Creek, Mich. (U.S.A.). *Nutrition, Digestion, Blood Chemistry.*

Fingerling, Gustav (1876), Prof. Dr. phil. et agr. h. c., Dir. der Staatl. Sächs. landw. Versuchsanstalt Möckern. Leipzig-Möckern (Deutschland), Gustav-Kühn-Str. 8. *Tierphysiol.; Ernährungsfragen.* Ⓓ Göttingen 1904.

de Finis, J. O., Prof. Univ. Nation. Asunción (Paraguay). *Physiol.*

Fink, Bruce (1861), Prof. of Botan. in Miami Univ. Oxford, Ohio (U.S.A.). *Nature and taxonomy of lichens.* Ⓓ Univ. of Minnesota 1899. Ⓟ *Fungi.*

Fink, David E. (1881), Associate entomol., U. S. Department of Agric. Philadelphia, Pa. (U.S.A.), Department of Zool., Univ. of Pennsylvania. *Insect Physiol., Economic entomol effect of Arsenicals.* Ⓓ Univ. Pennsylvania 1925.

Fink, Gail J. (1887), Chem. Dir., National Lime Association. Washington, D.C. (U.S.A.), 918 G. Street, N.W. *Biol. problems in connection with soil treatment. Relation of lime to dietetic problems, calcium metabolism, assimilation, etc.* Ⓓ Cornell Univ. 1914.

Fink, Nikola (1894), Adjunkt am zool.-zootomischen Inst. der Univ. Zagreb (S.H.S.), Račkoga 11. *Entomol.* Ⓓ Zagreb 1917.

Finkelstein, Eugen (1897), Katheder d. Zool. Charkow, Ukraine (U.d.S.S.R.), Basseinastr. 33. Ⓟ *Süßwasserpolypen.*

Finlay, Harold John (1901), National Research Scholar in Palaeont. Dunedin (New Zealand), 14 Pine Hill Terrace. *Tertiary and recent marine mollusca (excluding Cephalopoda) of New Zealand and Australia.* Ⓓ Otago 1922. Ⓟ *New Zealand and Australian tertiary and recent mollusca.* Gasteropoda.

Finley, William Lovell (1876), Dir. Wild Life Conservation. Jennings Lodge, Oregon. Washington, D.C. (U.S.A.), Nature Magazine. *Wild birds and Mammals.* Ⓟ Berkeley, Cal. 1903.

Finn, Donovan Bartley (1900), Acting Dir., Pacific Experim. Station for Fisheries of the Biol. Board of Canada. Prince Rupert, B.C. (Canada). *The biochem. aspect of fish curing and processing.* Ⓓ Winnipeg 1924.

Finn, Wladimir W. (1878), Prof. d. Kiewschen Landw. Inst. Kiew (U.d.S.S.R.), Ul. Tolstogo 17. *Botan., Morphol., Embryol., Cytol., Systematik.* Ⓓ Kiew 1901.

Finnell, Henry Howard (1894), Dir., Panhandle Agric. Experim. Station, and Associate Agronomist, Okla. Agric. Exp. Station. Goodwell, Okla. (U.S.A.). *Fungous diseases of the sorghums. Mycol. and bact.; seedbed organisms.* Ⓓ Oklahoma Agric. and Mechanical Coll. 1917. Ⓟ *Normal and diseased forms of semi-arid crop plants, Wheat, Barley, Sorghums.*

Fiolle, Jean, Prof. Univ. Marseille (France). *Pathol.* o

Fiori, Adriano (1865), Prof. Botanica sistematica e Geografia botan. Firenze 9 R. (Italia), R. Ist. Superiore agrario e forestale. *Sistematica delle Fanerogame. Botanica forestale, agraria.* Ⓓ Firenze 1913. Ⓟ *Flora Italica Exsiccata.* Xylotomotheca Italica.

Fiori, Paolo, Prof. Univ. Modena (Italia), *Pathol.* o

Fiorito, Giuseppe, Lib. doc. Univ. Catania, Sicilia (Italia). *Batteriol.* o

Fippin, Elmez Otterbein (1879), Dir. agric. Exp. Stations Republic of Haiti, Port au Prince (Haiti). o

Firbas, Franz (1902), Dr., wissenschaftl. Hilfskraft am botan. Inst. der deutschen Univ. in Prag II (C.S.R.), Viničná 3a. *Oekol. und historische Pflanzengeographie, Phytosociol., Phytopalaeont.* (*Moorforschung*). Ⓓ Prag 1924.

Firket, Ch., Prof. emer. Univ. Liège (Belgique). *Anat. pathol.* o

Fischel, Alfred (1868), o.ö. Prof. der Embryol. und Vorsteher des embryol. Inst. der Univ. in Wien 9/3 (Österreich). *Deskriptive und experim. Embryol. (Determination, Differenzierung. Potenzen der Zellen, Formen u. Ursachen der normalen und abnormen Entwicklung, Entwicklungsursachen des Auges, des Pigmentes).* Ⓓ Prag 1894.

Fischer, Albert (1891), z. Zt. wissensch. Gast d. Kaiser-Wilhelm-Ges., Priv.Doc. allgem. Pathol. København (Danmark). Berlin-Dahlem (Deutschland), Thielallee 69/73. *Krebsforschung. Gewebekultur.* Ⓓ Kopenhagen. Ⓟ *Tumoren von Hühnern,* Ratten, Mäusen.

Fischer, A. F., Prof. Forest School. Manila, Philippine Islands (U.S.A.). *Tropic. Forestry.* o

Fischer, Bernhard (1877), o. Prof. der allgemeinen Pathol. u. pathol. Anat., Dir. des Pathol. Inst. der Univ. Frankfurt a. M. (Deutschland), Gartenstr. 229. *Allgemeine Geschwulstlehre, allgemeine Pathol.* Ⓓ Bonn a. Rh. 1900.

Fischer, Eduard (1861), Prof. der Botan. an der Univ. und Dir. des Botan. Gartens in Bern (Schweiz), Kirchenfeldstr. 14. *Biol. der parasitischen Pilze. Morphol. der Fruchtkörper der Suberineen und Gastromyceten.* Ⓓ Straßburg 1883.

Fischer, Emil (1868), Dr. med., Prakt. Arzt u. Biol. Zürich 6 (Schweiz), Bolleystr. 19. *Temperatur-Experim. und Hybridation mit Lepidopteren; Aufstellung und Ausbildung besonderer Methoden: Paarung und Weiterzucht von Tagfaltern in der Gefangenschaft; Züchtung von F_1- und F_2-Generationen von Artbastarden des Genus Celerio Oken.* Ⓓ Zürich 1897. Ⓟ *Temperatur-Aberrationen und Hybriden von Lepidopteren.*

Fischer, Eugen (1874), Dr. med., o. Prof. der Anat. und Anthrop., Dir. des anat. Inst. Freiburg i. B. (Deutschland), Mozartstr. 20. *Menschliche Vererbung, Anthrop. Die Rehobother Bastards.* Ⓓ Freiburg 1898. Ⓟ *Anthrop. Photographien.*

Fischer, Gustav (1889), Dr. phil., Reg.- und Landesökonomierat im Preußischen Ministerium für Landwirtschaft, Domänen und Forsten. Berlin W 9 (Deutschland), Leipziger Platz 10. *Züchtung landwirtschaftl. Nutzpflanzen, Rebenzüchtung.* Ⓓ Halle (Saale) 1913.

Fischer, Hans (1881), Geh.R. Prof. Dr. phil. et med., Dir. des Organisch-chem. Inst. der Techn. Hochsch. in München (Deutschland), Arcisstr. 21. *Blut- und Gallenfarbstoff.* Ⓓ Phil. Marburg 1904, med. München 1908.

Fischer, Hermann (1884), Studienprof., Priv.Doc. a. d. Techn. Hochsch. München (Deutschland), Herzogstr. 58 III. *Bodenbiol., Biol. der Stickstoffbact., Edaphon, Aquakulturchemie, Teichdüngungslehre, Mittelalterliche Botan.* Ⓓ Würzburg 1908.

Fischer, Hugo (1865), Dr. phil. Berlin W 15 (Deutschland), Meierottostr. 1 III. *Pollenmorphol., Kolloid- u. Färbungstheorie, Bodenbact., Theorie der Blütenbildung, Kohlensäureernährung der Pflanzen, Fortpflanzung bei Farnbastarden, Theorie der Abstammungslehre.* Ⓓ Halle a. d. S. 1890.

Fischer, K. A. Wilhelm (1884), Dr. phil., Vorsteher der Hauptstelle für Pflanzenschutz der Landwirtschaftskammer f. d. Prov. Hannover. Göttingen (Deutschland), Nikolausberger Weg 7. *Pflanzenschutz, angew. Botan., Landwirtschaft.* Ⓓ Göttingen 1901.

Fischer, Martin H. (1879), Prof. of Physiol., Coll. of Med., Univ. of Cincinnati, O. (U.S.A.), General Hospital. *Physiol. and pathol. of water absorption and secretion, emulsions theoretically and practically considered; theory of the lyophilic colloids.* Ⓓ Rush Med. Coll., Chicago 1901.

Fischer, Max Heinrich (1892), Dr. med., Priv.Doc. und Ass. am physiol. Inst. der deutschen Univ. Prag VI (C.S.R.), Albertov 5. *Physiol. des Labyrinthes* (*Vestibularapparates*), *d. Gleichgewichtes (Körperstellung und Körperhaltung); Physiol. Optik.* Ⓓ Prag 1919.

Fischer, Walther (1882), o. Prof. der Pathol. und Dir. des Pathol. Inst. der Univ. Rostock (Deutschland), St. Georgstr. 34. *Allg. u. spez. Pathol., Tropenpathol., Parasit.* ⓘ Tübingen 1907.

Fischer, Wilhelm Karl Reinhold (1858), Prof. Dr. phil. Bergedorf bei Hamburg (Deutschland), Augustastr. 3. *Echinoiden, Sipunculiden, Priapuliden.* ⓘ Kiel 1884.

Fisher, Anna Bathsheba (1899), Instr. of Bact. Evanston Hospital, Evanston, Ill. Ass. Dept. of Hygiene and Bact., Univ. of Chicago. Chicago, Ill. (U.S.A.). *Pathogenic Bact. and Protozoa: Malaria.* ⓘ Univ. of Chicago 1924.

Fisher, Charles King (1892), Junior Entomol., U.S. Bureau of Entomol. Alhambra, Cal. (U.S.A.), P. O. Box 297. *Stored Product Insect Investigations.*

Fisher, Clyde (1878), Curator of Visual Instr. in American Mus. of Natural History. New York City (U.S.A.), 77. St. and Central Park West. ⓘ Johns Hopkins Univ. 1913.

Fisher, Durward Frederick, Pathol., Bur. of Plant Ind., U.S. Dept. of Agric. Wenatchee, Wash. (U.S.A.). o

Fisher, Martin Luther (1871), Prof. of Agronomy. Lafayette, Ind. (U.S.A.). *Field Crops; Forage Crops.* ⓘ Wisconsin 1911.

Fisher, Richard Thornton (1876), Dir. of the Harvard Forest. Petersham, Mass. (U.S.A.). *Forestry, including allied biol. sc.* ⓘ Yale Univ., School of Forestry 1902.

Fisher, Walter Kenrick (1878), Dir. and Prof. of Zool., Hopkins Marine Station of Stanford Univ. Pacific Grove, Cal. (U.S.A.). *Echinodermata,* ⓘ Stanford Univ. 1906.

Fisher, Warren S. (1878), Specialist on Forest Coleoptera, U.S. Department of Agric. Washington, D.C. (U.S.A.), U.S. National Mus. *Cerambycidae, Buprestidae (Coleoptera).* ⓟ Buprestidae and Cerambycidae.

Fisk, Emma Luella (1892), Ass. Prof. Univ. of Wisconsin. Madison (U.S.A.), Dept. of Botan. *Morphol. and cytol.: Zea Mays.* ⓘ Wisconsin 1925.

Fiske, Jessie G. (1895), State Seed Analyst o New Jersey and Ass. Prof. of Botan. Rutgers Univ. New Brunswick, N.J. (U.S.A.), Agric. Experim. Station. *Analyses of seeds, seed problems.* ⓘ 1920. ⓟ Seed varieties.

Fister, Wilbur Earl (1892), Dir. of Clinical Labor. Cincinnati, O. (U.S.A.), 580 Doctors Bldg. *Bact. — Pathol. — Blood Chem. — Serol.* ⓘ Cincinnati 1926.

Fitch, Clifford Penny (1884), Chief, Division of Vet. Med., and Prof. of Animal Pathol. and Bact. Univ. of Minnesota. St. Paul, Minn. (U.S.A.), Univ. Farm. *Diseases of animals: poultry, abortion of cattle and horses.* ⓘ Cornell Univ. 1911.

Fitschen, Jost (1869), Altona a. d. Elbe (Deutschland), Lenbachstr. 11. *Koniferen, Rubus.* ⓟ Rubus.

Fitting, Johannes (1877), Prof., Dir. d. Botan. Inst. a. d. Univ. Bonn a. Rh. (Deutschland), Poppelsdorfer Schloß. *Pflanzenphysiol. und -ökol.* ⓘ Straßburg i. Els. 1900.

Fitzgerald, D. P., Prof. Univ. Coll. Cork (Ireland). *Anat.* o

Fitz Gerald, John Gerald (1882), Prof. of Hygiene and Preventive Med., Dir., School of Hygiene and Connaught Labor. Toronto (Canada), Univ. of Toronto. *Bact., Immunol., Preventive Med.* ⓘ M.B. Toronto 1920, L.L. Queen's Univ., Kingston 1925.

Fitz Gerald, Mabel Purefoy, Lect. School of Med. R. Coll. Edinburgh (Scotland). *Bact.* o

Fitzpatrick, Harry Morton (1886), Prof. of Plant Pathol., Cornell Univ. Ithaca, N.Y. (U.S.A.), 220 Bryant Avenue. *Fungi, especially pyrenomycetes.* ⓘ Cornell Univ. 1913.

Fitz Simons, F. W. (1881), Dir., Mus. Port Elizabeth (South Africa). *Snakes, Mammals, Birds of South Africa.* o

Fitz Simons, V. (1901), Ass. Zool. State Mus. Pretoria, Transvaal (South Africa). ·*Lower vertebrates. Fishes.* ⓘ Rhodes Univ. Grahamstown, Cape Colony, 1923.

Fiwejskaja, Ass. Zool. Labor. Univ. Irkutsk (U.d.S.S.R.). *Gewebekultur, Histol.* o

Fjervoll, Dir. Statens forsöksstation for Troms og Finmark. Holt, pr. Troms (Norge). o

Flachs, Karl (1888), Dr., Phytopath., Reg.R. an der B. Landesanstalt für Pflanzenbau u. Pflanzenschutz. München (Deutschland), Klugstr. 30. *Tierische und pilzliche Parasiten landwirtschaftl. u. gärtnerischer Kulturgewächse.* ⓘ München 1916.

Flahault, M. C., Prof. Botan. Fac. Sc. Montpellier (France). *Botan.* o

Flaksberger, Constantin Andreevitsch (1880), Prof., Chief of Wheat Section, Inst. of Applied Botan. and New Cultures. Leningrad (U.d.S.S.R.), 44 Herzen St. *Classification and geography of the wheats of the world.* ⓘ Dorpat. ⓟ Wheat samples from all parts of the world.

Flanders, Stanley E. (1894), Economic Entomol., Saticoy Walnut Growers Association. Saticoy, Cal. (U.S.A.). *Codling moth infestation. The lima bean pod-borer (Etiella schisticolor Zeller).* ⓘ Berkeley, Cal. 1923. ⓟ Life-history (Riker) mounts of local insects.

Flaschenträger, Bonifaz (1894), Ass. am Physiol.-chem. Inst. der Univ. Leipzig C 1, (Deutschland), Liebigstr. 16. *Intermediärer Stoffwechsel. Chem. und Pharm. des Krotonöls. Organische Mikroanalyse.* ⓘ Ing. München 1921, med. Leipzig 1922.

Fleisch, Alfred (1892), Dr. med., Priv.Doc. für Physiol. Zürich (Schweiz), Rämistr. 69. *Physiol. des Blutkreislaufes, der Atmung, des Höhenklimas. Labyrinth. Blutlipoide, Oxydationstheorie. Physiol. der Gehbewegung.* ⓘ Zürich 1918. ⓟ Pneumotachograph.

Fleischer, Bruno (1874), Dr. med., o. Univ.-Prof. der Augenheilkunde. Erlangen (Deutschland), Univ.-Augenklinik. *Vererbung.* ⓘ Tübingen 1898.

Fleischer, R. P. Max (1861), Prof., Dr. h. c. Den Haag (Holland), Columbusstraat 201. *Laubmoose (Musci frondosi) Italiens, Javas, Neu-Guineas, Moosflora von Java, Entwicklung der Zwergmännchen bei den Laubmoosen.* ⓘ Utrecht 1923. ⓟ Musci frondosi (Laubmoose).

Fleischmann, Albert (1862), o.ö. Prof. der Zool. u. vergl. Anat. an der Univ. Erlangen (Deutschland), Puchtaplatz 9. *Vergl. Embryol. der Amnioten. Vergl. Morphol. der Mollusken.* ⓘ Erlangen 1885. ⓟ Amniotenembryonen, Molluskenschalen.

Fleischmann, Walter (1896). Wien XIX (Österreich), Silbergasse 2. *Physiol. des Blutes, insbesondere der Leukocyten.* ⓘ Wien 1922.

Fleisher, Moyer S. (1884), Prof. of Bact. and Hygiene, St. Louis Univ., School of Med., St. Louis, Mo. (U.S.A.). *Physiol. Phenomena in Anaphylactic Shock. Fungi (Monilia) in relation to diseases in man. Staining reactions of Bact. Filterable toxins.* ⓘ Univ. of Pennsylvania 1907.

Fleming, A., Prof. St. Marys Hosp. Med. School. London W. 2 (England), Paddington. *Bact.* o

Fleming, James Henry (1872), Honorary Curator of ornithol., Canadian National Mus. Ottawa. Toronto 4 (Canada), 267 Rusholme Road. *Ornithol., private collection of 32000 birds, island faunas.* ⓟ Collections of birds.

Flerow, Alexander F. (1872), Prof. d. Donschen Polytechn. Inst. Nowotscherkask (U.d.S.S.R.), Pr. Bakunina 33. *Botan., Microbiol. Geobotan. Moorkultur.* ⓘ Dorpat 1901.

Flerow, Boris (1896), Doc. d. Botan. Moskau (U.d.S.S.R.), Rue Gerzeno 6, Botan. Inst. *Cytol. d. Pilze (Ustilagineae), Cytol. d. Algen (Diatomaceae). System. u. Oekol. d. Meeresalgen.*

Flesch, Max (1852), Prof., Dr. med., Arzt. Hochwaldhausen, Oberhessen (Deutschland). *Anthrop., spez. Kriminalanthrop.* ⓘ Würzburg 1872.

Fletcher, Frank C. (1901), Ass. in Systematic Entomol. Ithaca, N.Y. (U.S.A.), Dept. of Entomol.,

Cornell Univ. *Coleoptera, Pselaphidae.* ⓛ Cornell Univ. 1925. ⓟ Coleoptera for Pselaphidae of the world.

Fletcher, Stevenson Whitcomb (1875), Prof. Horticult. State College, Pennsylvania (U.S.A.). o

Fletcher, William (1872), Dir. Inst. for Med. Research. Kuala Lumpur, Federated Malay States (India). *Bact.* ⓛ Cambridge Univ. 1896. ⓟ Cultures of Micro-organisme.

Fleury, Paul (1885), Chef du Labor. de Chim. Biol. à la Fac. de Pharm. de Paris (France), 4, Avenue de l'Observatoire. *Chem. analytique appliquée à la biol. Recherches sur les diastases.* ⓛ Paris 1912.

Flexner, Simon (1863), Dir., The Rockefeller Inst. for Med. Research. New York, N.Y. (U.S.A.), 66. Street and Avenue A. *Bact. and pathol. Pathol. of toxalbumin intoxication, biochem. constitution of snake venoms, experim. pancreatitis and fat necrosis; epidemic cerebrospinal meningitis and its serum treatment; poliomyelitis, its cause and mode of transmission, serum treatment; experim. epidemiol.; epidemic encephalitis.* ⓛ D.Ss. Harvard Univ. 1906, D.Sc. Yale Univ. 1910, D.Sc. Princeton Univ. 1913, LL.D. Univ. of Maryland 1907, LL.D. Brown Univ. 1915, LL.D. Johns Hopkins Univ. 1915, LL.D. Washington Univ. 1915, LL.D. Cambridge Univ., England, 1920, Dr. Univ. of Strasbourg, France, 1923, M.D. Univ. of Louisville 1889.

Florell, Victor H. (1885), Associate Agronomist, Office of Cereal Crops and Diseases, United States Department of Agric. Davis, Cal. (U.S.A.), Univ. Farm. *Plant Breeding and Genetics. Crops-Wheat, Barley, Oats.* ⓛ Univ. of California 1923. ⓟ Varieties of Cereals (Wheat, Barley, Oats).

Florey, Horvard (1898), Research worker. London (England), London Hospital, Whitechapel. *Blood Capillaries and lymphatics.* ⓛ Oxford 1923.

Florin, Carl Rudolf (1894), M.A., Ass. Curator, Palaeobotan. Dept., Naturhistoriska Riksmus. Stockholm 50 (Sverige). *Palaeobotan. of the Coniferales.* ⓛ Stockholm 1920.

Florin Elsa Henrietta (1896), Research Ass., Horticultural Department, R. Academy of Agric., Experimentalfältet. Stocksund (Sverige), Stocksundstorp. *Sterility Problem in Fruit Production. Horticulture.* ⓛ Stockholm 1922.

Florschütz, Frans (1887), Sekretär d. Coll. d. Curatoren der Landbouwhoogeschool Wageningen. Velp Prov. Gelderland (Holland). *Palaeobotan (tertiär en quartär). Lichenes.* ⓛ Leiden 1916.

Flower, Stanley Smyth (1871), Major, Herpetol., Zool. Record, London. Tring, Herts. (England). Spencersgreen End. *Practical Zool. and normal physiol. of Reptilia and Amphibia.* ⓛ London 1893.

Flu, Paul Christian (1884), o. Prof. tropische Hygiene und Parasit., Univ. Leiden (Holland), Witte Singel 40. *Bact. Parasit. Bacteriophagen. Filtrierbares Virus. Sarcoma.* ⓛ Utrecht 1906.

Fluke, Charles L. (1891), Prof. in Economic Entomol. Madison, Wis. (U.S.A.), Univ. of Wisconsin. *Syrphidae (Diptera). Life history of truck crop and fruit insects. Codling moth (Carpocapsa pomonella), pea moth (Laspeyresia nigricana S.), pea aphid (Illinoia pisi Kalt.).* ⓛ Univ. of Wisconsin 1918. ⓟ Determined Syrphidae (Diptera).

Flury, Ferdinand (1877), Dr. phil. et med., o. Prof. der Pharmacol. Würzburg (Deutschland), Pharmacol. Inst. *Toxikol. Tierische Gifte.* ⓛ Chem. Erlangen 1902, med. Würzburg 1910.

Flynn, John Edward (1897), Instr. Ithaca, N.Y. (U.S.A.), Department of Plant Pathol., Cornell Univ. *Plant Pathol.*

Flynn, T. T., Prof. Univ. Hobart, Tasmania (Australia). *Biol.* o

Foà, Anna (1876), Prof. di Bachicoltura Ist. Superiore Agrario. Portici, Napoli (Italia). *Baco da seta. Genetica.* ⓛ Roma 1920.

Foà, Carlo, Prof. Univ. Milano (Italia). *Chim. biol.* o

Fodor, Andor (1884), Prof., Dir. Inst. für Biochem. u. Kolloidchem., Hebräische Univ. Jerusalem (Palästina). *Chem. der Eiweißkörper, Aminosäuren und Polypeptide, Fermente, Kolloidchemie der Eiweiße u. a. kolloider Systeme, insbes. der Biokolloide.* ⓛ Zürich 1909.

Förster, Johann Karl (1903), Dr. Leipzig-R. (Deutschland), Kohlgartenstr. 11I. *Pflanzliche Entwicklungsphysiol. (Ursachen der Formbildung.)* ⓛ Leipzig 1926.

Foerster, R. Earle (1899), Research Biol., Biol. Board of Canada, in charge of Pacific Salmon Research Station. Vedder Crossing P.O., B.C. (Canada). *Life history of the Pacific Salmon (Oncorhynchus), especially the Sockeye salmon (O. nerka) in fresh water: food migration to and from salt water; rate of growth. Efficiency of artificial propagation as compared with natural propagation. Optimum conditions (Oxygen, temperatures and pH) for development and hatch. Changed environment versus inherited characters as regulating habits of salmon when removed as fry or eggs to new areas.*

Föyn, Björn (1898), Cand. mag., Ass. am Zool. Labor. der Univ. Oslo (Norge). *Geschlecht und Geschlechtszellen bei Hydroiden.*

Folbort, G. P., Prof. Staatl. Univ. Leningrad (U.d.S.S.R.). *Physiol.* o

Foley, Henry (1871), Inst. Pasteur d'Algérie. Alger, Algérie (Afrique). *Parasit.: régions sahariennes. Mycol.: champignons supérieurs.* ⓛ Lyon 1895. ⓟ Parasit. animaux, Plantes du Sahara algérien.

Følger, Af. F., Prof. Patol.-Anat. Labor. København V (Danmark), Bülowsvej. o

Folin, Otto (1867), Prof. Biol. Chem., Harvard Med. School, Boston (U.S.A.). *Blood and urine analysis; Metabolism of carbohydrates and of nitrogenous products.* ⓛ Univ. of Chicago 1898.

Folsom, Donald (1891), Plant Pathol., Maine Agric. Experim. Station. Orono, Me. (U.S.A.). *Virus diseases; potato and apple diseases.* ⓛ Univ. of Minnesota 1917. ⓟ Potato tubers perpetuating virus diseases.

Folsom, Justus Watson (1871), Entomol., U.S. Bur. of Entom. Tallulah, La. (U.S.A.). *Cotton Insect, Collembola, Thysanura, Entomol.* ⓟ Collembola and Thysanura.

Fomin, Alexander (1867), Prof. Univ., Dir. Botan. Garten u. Labor. Kiew (U.d.S.S.R.), Kominternstr. 1. *Pteridophyta, Geobotan. und Acclimatisation. Pteridophyta Sibiriens und des weiten Ostens.* ⓛ Dorpat 1913. ⓟ Pflanzen des Kaukasus und der Ukraine.

Fomin, W. E., Prosektor Histol. Inst. d. I. Univ., Prof. Med. Inst. Moskau (U.d.S.S.R.), ul. Spartaka 2. o

Fomitschew, Woldemar D., Com. Géol. d. Russie. Leningrad (U.d.S.S.R.), W.O., Sredny Pr., 72-B. *Palaeont.; Rugosa.* o

Fomitschow, Alexander (1876), Ass. Forstinst. Leningrad (U.d.S.S.R.), Forstinst. Q. N. *Forstkulturwesen.*

Fontès, Georges Jean Paul (1893), Chargé de cours de chimie biol. Fac. de méd. Strasbourg (France), Inst. de Chimie biol. I, Place de l'Hopital. *Microchimie et métabolisme du fer, du glycose et de l'ammoniaque.* ⓛ Montpellier 1919.

Fontes, Joaquim (1892), Prof. libre à la Fac. de Méd. de Lisbonne et Ass. de Physiol. Lisboa (Portugal), Inst. de Fisiol., Fac. de Med. *Physiol. du tissu musculaire.*

Font Quer, Pio (1888), Conserv. Section Botan. du Mus. d'Hist. Nat. Barcelona (España), Apartado 593. *Phanérogames ibero-marroccaines. Sideritis. Fitogeographie ibérique.* ⓛ Barcelona 1910. ⓟ Phanérogames Ibéro-maroccaines.

Foot, Nathan Chandler (1881), Associate Prof. of Pathol., Coll. of Med. of the Univ. of Cincinnati, O. (U.S.A.), Cincinnati General Hospital. *Histogenesis of the mononuclear phagocytes and of the reticulum*

fiber, various problems in general pathol., histogenesis of tumors, thymic tumors. ⒹColumbia 1907.
Forbes, Alexander (1882), Associate Prof. of Physiol. Boston 17, Mass. (U.S.A.), Department of Physiol., Harvard Med. School. *Physiol. of the Nervous System.* ⒹHarvard Univ. 1904.
Forbes, Ernest Browning (1876), Dir., Inst. of Animal Nutrition, Pa. State Coll. State College, Pennsylvania (U.S.A.). *Animal nutrition; energy, mineral and protein metabolism.* ⒹUniv. of Illinois 1908.
Forbes, Reginald D. (1891), Dir., Southern Forest Experim. Station, U.S. Forest Service. New Orleans, La. (U.S.A.), 326 Customhouse. *Forest (silvical) research.* ⒹYale 1913.
Forbes, Stephen Alfred (1844), Chief, Illinois State Natural History Survey. Urbana, Ill. (U.S.A.), Univ. of Ill. *Biol. survey of Illinois.*
Forbes, William T. M. (1885), Instr. in Entomol., Cornell Univ. Ithaca, N.Y. (U.S.A.), Dept. of Entomol. *Lepidoptera of Northeastern U.S.A., larvae, Ithomiinae. Wings of Coleoptera.* ⒹClark Univ. 1910.
Forbush, Edward Houle (1858), State ornithol. Boston (U.S.A.), State House. ○
Ford, Ebenezer (1890), Plymouth Labor. of the Marine Biol. Association of the U. K. Plymouth (England), Citadel Hill. *Fishes and Fisheries. Statistical studies in the biol. of the Herring. Bottom fauna.*
Ford, Edmund Brisco (1901), Research Worker Dept. of Zool. and Compar. Anat., Univ. Mus. Oxford (England), Wadham Coll. *Genetics: Rates of development. Epidemics in wild animals: Parasit. Protozoa. Variation in the Lepidoptera (sexual dimorphism).* ⒹOxford 1924.
Ford, William W., Prof. of Bact. Baltimore, Md. (U.S.A.), School of Hygiene and Public Health, Johns Hopkins Univ. ○
Fordham, Mahalak Glen Clark (1898), Lect. Univ. Liverpool (England), Zool. Dept. *Invertebrata (Aphrodite), Parasit.* ⒹWest Australia 1919.
Forel, August Heinrich (1848), Prof. emer. Univ. Zürich. Yvorne, Kant. Waadt (Schweiz). *Psychiatrie, Neurol., Hypnotismus, Psychoanalyse, Hirnanat., Psychol., Soziol., Formicidae.* ⒹZürich 1872.
De Forest, David McClellan (1903), Instr. in Biol. Schenectady, N.Y. (U.S.A.), Union Coll. *Biol. effects of X-rays.* ⒹUnion Coll., Schenectady, N.Y. 1925.
De Forest, Howard, Chairman of the Botan. Division, Dept. of Biol., Univ. of S. Calif. Los Angeles, Cal. (U.S.A.). ○
Forlini, Euclide (1899), Ass. Anat. Patol. Univ. di Parma (Italia), Ospedale Maggiore. *Forsette retroperitoneale.* ⒹParma 1924.
Formanek, Emanuel, Dr., Prof. d. med. Chem. Praha II (C.S.R.), Kateřinska 32, Med. chem. Inst. d. Karls-Univ. ○
Formosov, Alexandr Nikolaevich (1899), Aspirant Inst. of sc. Investigation Zool. Section, I. Univ. Moscow (U.d.S.S.R.), Krapotkinsky per., d. 21, k. 2. *Zool.; mammals of Palaearctic Region, biol., geographical distribution.* ⒹMoscow 1925. ⓅSkins and skulls of mammals from Europ. Russia, North Caucasus and Siberia.
Fornet, Walter (1877), Dir. des Inst. für Mikrobiol. Saarbrücken (Deutschland), Wittenbergstr. Nr. 20a. *Organotherapie: eßbares Insulin. Bact.: Kultur entfetteten Tuberkelbazillen.* ⒹBerlin 1903. ⓅOrganotherapeutische Praep. (Insulin) und bact. Praep. (Tebephagin).
Forni, Giuseppe Ghevardo (1885), Prof. incar. di patol. Bologna (Italia), Univ. *Istol. Anat. Patol. patol. chirurgica sperimentale.* ⒹBologna 1924.
Forsaith, Carl Cheswell (1888), Ass. Prof., Wood Technol., New York State Coll. of Forestry at Syracuse Univ. Syracuse, N.Y. (U.S.A.). *Morphol. of wood; In Relation to Physical and Mechanical Properties, and to Evolution. The Physical Properties of Timber.* ⒹHarvard Univ. 1917.
Forsius, Runar (1884), Med. licentiatus, Krankenhausarzt, Fredriksberg (Finnland), Epidemiekrankenhaus. *Tenthredinoidea der Welt, Hymenopteren Finlands (Epidemiekrankheiten).* ⒹTenthredinoiden (sowie auch andere Hymenopteren der ganzen Welt).
Forssman, John (1868), o. Prof. Pathol. Bact. und allgemeine Hygiene, Univ. Lund (Sverige), Frimagatan 16. *Heterogenetische Antigene und Antikörper. Wassermannreaktion. Neurotropismus.* ⒹLund 1898.
Forster, George F. (1895), Prof. of Biol., Olivet Coll. Olivet, Mich. (U.S.A.), Department of Biol. *Immunol. Precipitins. Hypersensitiveness to animal products.* ⒹUniv. of Wisconsin 1922.
Forster, M., Prof. Fac. Méd. Strasbourg (France). *Anat.* ○
Fořt, Jindřich, M.S., Chef Station d'Etat de Pomol. et de Culture de la Vigne. Horni Černošice (C.S.R.). *Pomol., culture de la vigne.* ○
Foss, H., Dir. Statens forsøksstation for fjeldbygolerne. Voldbu (Norge). ○
Foster, Harry E. (1881), Dir. Cutter Labor. Berkeley, Cal. (U.S.A.). *Specific antisera, vaccines, toxins and viruses for human use.* ⒹSan Francisco 1908. ⓅAntitoxins; Antisera; Vaccine virus; Rabies vaccine, Tuberculins; Bact. vaccines; Diphtheria Toxin; Antitoxin mixture.
Foster, Nellis B. (1875), Associate Prof. of Med., Cornell Univ., Med. Coll. New York City (U.S.A.), New York Hospital. *Pathol. chem. and metabolism, Endocrine System.* ⒹJohns Hopkins 1902.
Fourneau, Ernest (1872), Chef de service Inst. Pasteur. Paris VII (France), 28, Rue Barbet de Jouy.
Fournes, Alfred (1897), Volontär ornithol. Abt. Naturhist. Mus. Wien XIII (Österreich), Sechshauser Str. 89. ⒹWien 1922. ⓅVogeleier.
Foweracker, C. E., Lect. M.A. Cant. Coll. Christchurch (New Zealand). *Botan.* ○
Fowweather, Frank Scott (1892), Lect. in Chem. Pathol., Univ. of Leeds (England), Med. School. ⒹLiverpool 1925.
Fox, Francis William (1894), Biochem. Johannesburg (S. Africa), S.A. Inst. for Med. Research, Box 1038. *Sterol Metabolism.* ⒹLondon 1918.
Fox, Henry (1875), Associate Entomol., U.S. Bureau of Entomol. Riverton, N.J. (U.S.A.), Japanese Beetle Labor. *Taxonomy, Distribution and Ecology of Eastern North American Orthoptera. Life History and Ecol. of Japanese Beetle. Influence of Climatic Factors on Spread and Distribution of Insects.* ⒹUniv. of Pennsylvania 1905.
Fox, Rolland David (1899), Dir., Bact. Research Labor. Municipal Univ., Dir. Public Health Labor., Akron, Ohio (U.S.A.). *Pathogenic Bact., Cultivation and Immunol., Pleomorphism.* ⒹMunicipal Univ. of Akron 1923. ⓅCultures and microscopic preparations of microorganisms.
Foxworthy, Fred W., Forest Research Officer, F. H. S., Dept. of Forestry. Kuala Lumpur (Federated Malay States). *Systematic botan., dendrol., silviculture.* ○
Fracassi, Humberto, Prof. Univ. Nac. Córdoba (Argentina). *Anat. topogr.* ○
Fracker, Stanley Black (1889), State Entomol. of Wisconsin. Madison, Wis. (U.S.A.), Room 14, State Capitol Annex. *Taxonomy of Heteroptera, and of immature Lepidoptera; bee disease; insect and plant disease control.* ⒹPh.D. Univ. of Illinois 1914. ⓅNearctic and Neotropical Heteroptera.
Frada, Attilio, Lib. doc. Univ. Messina (Italia). *Pathol.* ○
Fraipont, Ch., Prof. Univ. Dir. Labor. Liège (Belgique). *Paléont.* ○
La France, S., Prof. Univ. Palermo (Italia). *Patol.* ○
Franceschini, Piero (1900), aiuto onorario di Anat. um. Firenze (Italia), Via Alfani 33. ⒹFirenze 1926.

Franchini, Giuseppe, Lib. doc. Univ. Bologna (Italia). *Patol.* ○

Francis, Edward (1872), Surgeon, United States Public Health Service. Washington, D.C. (U.S.A.), Hygien. Labor., 25. and E Streets, N.W. *Bact.* ⓓ Univ. of Cincinnati 1897.

Franck, Wilhelm Jacques (1886), Diplomingenieur, Chem. in den technischen Wissenschaften, Dir. der Reichsversuchsstation für Samenkontrolle. Wageningen (Holland), Rijksstraatweg 71 A. *Samenkontrolle und Untersuchung.* ⓓ Delft.

Franco, Enrico Emilio (1881), Prof., Dir. Ist. di Anat. Patol. e Labor. di indagini cliniche dell' ospedale Civile di Venezia (Italia). *Anat. Patol. Tumori. Tessuto adiposo. Kalaazar. Sarcosporidiae. Leishmania. Hémoblastes. Mastzellen. Storia della med.* ⓓ 1906. ⓟ Preparati isto-patol.

Francotte, H., Prof. Univ. Liège (Belgique). *Pathol. génér.* ○

Frandsen, Julius H. (1877), Head, Dept. of Animal and Dairy Husbandry, M. A. C. Amherst, Mass. (U.S.A.). ⓓ Iowa State Coll., Ames, Ia. 1904.

Frandsen, Peter (1876), Prof. of Biol., Head of dept. of biol. Reno, Nev. (U.S.A.), Univ. of Nevada. ⓓ Univ. of Nevada 1895; Harvard Univ. 1898.

Frank, Antoine (1892), Prof. suppléant d'anat. humaine. Bratislava (C.S.R.), Sasinkova 2. *Système de conduction du cœur, ostéol. humaine, système sympathique.* ⓓ Prague 1921. ⓟ Lepidoptera, Coleoptera.

Frank, Anton (1883), Dr., Leiter des pathol. Inst. am Augusta-Hospital, Priv.Doc. Köln (Deutschland), Rhöndorfer Str. 74. *Pathol. Anatomie.* ⓓ Rostock 1908.

Frank, Arthur (1885), Plant Pathol. Puyallup, Wash. (U.S.A.), Experim. Station. *Phytopath. Mycol., Bact., Entomol.* ⓓ Oregon Agric. Coll., Corvallis, Ore., 1915.

Frank, Otto (1865), Geh.HofR., Univ.Prof. München (Deutschland), Haydnstr. 5. *Physiol.* ○

Frank, Robert T. (1875). New York City (U.S.A.), 10 East 85. Street. *Physiol., pharm., and biochem. of the female sex hormone as obtained from the ovarian follicle, corpus luteum and the placenta.* ⓓ Columbia 1900.

Frank-Kamenetzky, Albert (1875), Prof. Univ. Irkutsk, Sibirien (U.d.S.S.R.), Nabereschnaja 34. *Hydrochem. Untersuchungen. Baikalsee.* ⓓ Basel (Schweiz) 1899.

Franke, Adolf (1860), Arnstadt, Thüringen (Deutschland), Stadtilmer Str. 1. *Fossile Foraminiferen und Ostracoden.* ⓟ Foraminiferen und Ostracoden.

Franke, Marjan, Prof. Dr. med. Lwów (Pologne), Univ., Inst. de pathol. gén. et expérim. ○

von Frankenberg, Gerhard (1892), Dr. phil., Mus.-Inspektor, Leiter des Naturhistorischen Mus. in Braunschweig (Deutschland), Rankestr. 5. *Atmung der Insekten.* ⓓ Leipzig 1915.

Frankenberger, Zdenko (1892), Prof. der Histol. und Embryol. an der med. Fac. der Komenský-Univ. in Bratislava (C.S.R.), Prayova 28. *Allgemeine Zytol., Geschlechtsdrüsen.* ⓓ Prag 1917.

Franklin, Henry James (1883), Ph.D., Research Prof., in charge of Cranberrystation of the Massachusetts Agric. Experim. Station. East Wareham, Mass. (U.S.A.). *Cranberry problems, connected with frost prediction, insect and disease control. Weed control, variety studies.* ⓓ Mass. Agric. Coll. 1908. ⓟ Bumblebees.

Franklin, Kenneth James (1897), Fellow and Lect. in Physiol., Oriel Coll., Oxford, Lect., Brasenose Coll., Oxford, Demonstrator of Pharmacol., Univ. of Oxford (England). *Anat., Physiol. and Pharm. of Veins.* ⓓ Oxford 1924.

Franquet, Robert Fernand (1897), Préparateur Mus. nat. d'Hist. nat. Chaire de Culture. Paris VI (France), 59, Rue de Rennes. *Biochim. Chimisme de greffe végétale. Genèse des hydrates de carbone.* ⓓ Paris 1921.

Franz, Shepherd Ivory (1874), Prof. of Psychol., Univ. of California, Southern Branch, Chief of Division of Psychol. and Education, Childrens Hospital, Hollywood, Cal. Los Angeles, Cal. (U.S.A.), 855 North Vermont Avenue. *Functions of the Brain.* ⓓ Columbia 1899.

Franz, Victor (1883), a.o. Prof. der Zool., Kustos des Zool. Inst. und Phyletischen Mus. der Univ. Jena (Deutschland), Westbahnhofstr. 9. *Vergl. Anat. der Lichtsinnesorgane. Amphioxus. Phylogenie.* ⓓ Breslau 1905.

Fraps, R. M., Ass. Tucson, Ariz. (U.S.A.), Desert Labor. *General Physiol. Influence of thyroid on metabolism.*

Fraser, Allan Cameron (1890), Ass. Prof. of Plant Breeding, Cornell Univ. Ithaca, N.Y. (U.S.A.). *Genetics of maize and Rosa.* ⓓ Cornell Univ. 1918.

Fraser, C., Prof. Univ. Vancouver (Brit. Columbia). *Zool.* ○

Fraser, Elizabeth Alice (1880), Senior Lect., Department of Zool., Univ. Coll. London W.C. 1 (England). *Vertebrate Embryol.: Development of Eye Muscles and Urogenital System.* ⓓ Univ. Coll. London 1916.

Fraser, R., Prof. Univ. Sackville (Canada). *Biol.* ○

Fraser-Harris, David (1867), Prof. emer. of Physiol., Halifax. London S.W. 1 (England), The Author's Club, 2. Whitehall Court. *Experim. study of neuro-muscular rhythms. Bio-chem. researches into milk and the reducing ferment of Tissues. Contribution to the Theory of living matter.* ⓓ Med. Univ. of Glasgow 1897; Sc. Birmingham 1911.

Frassi, A., Doc. Univ. Parma (Italia). *Igiene.* ○

Fray, George Franklin (1898), Ass. Horticulturist, Univ. of Delaware. Newark, Del. (U.S.A.). *Nitrogen.*

Frazer, J. E. S., Prof. St. Marys Hosp. Med. School and Univ. London S.W. 7. (England), S. Kens. *Anat.* ○

Frechkop, Serge (1894), Dr. en Sc.; travailleur libre au labor. de Zool. de l'Univ. de Bruxelles (Belgique), 3, Avenue de l'Armée. *Zool. théorique morphol. des Vertébrés.*

Fred, Edwin Broum (1887), Prof. in Agric. Bact. Madison, Wis. (U.S.A.), Univ. of Wisconsin. *Soil Bact. and Fermentation.* ⓓ Goettingen 1911.

Baron **Fredericks,** George Nikolaevitch (1889), Géol. du Comité Géol. de Russ. Leningrad (U.d. S.S.R.), Vasili Ostrov, Sredny Pr. 72 B. *Brachiopoda du paléozoique supérieur (Spiriferidae et Productidae), Bryozoa (Fenestelcidae), Cephalopoda du carbonifère supérieur et permien.* ⓓ Charkov 1917. ⓟ Brachiopoda, Carbonifère supérieur et permien.

Fredericq, Henri (1887), Prof. de Physiol. Fac. Méd. Univ., Dir. Inst. de Physiol. Liége (Belgique), 20, rue de Pitteurs. *Physiol. du cœur: Contraction cardiaque. Electrocardiographie. Nerfs du Cœur. Pouls alternant. Bradycardies. Excitabilité du cœur, des nerfs et des muscles mesurée par les méthodes chronaximéthriques. Pharmacodynamie: caféine. Transmission humorale des actions nerveuses.* ⓓ Liége 1912.

Fredericq, Léon (1851), Prof. émérite de l'Univ. de Liège (Belgique), Boulevard Frère-Orban 3 bis. *Physiol. Physiol. comparée.* ⓓ Gand méd. 1873, Sc. 1874.

Freeborn, Stanley Barron (1891), Associate Prof. of Entomol. Davis, Cal. (U.S.A.), Univ. Farm. *Vet. Parasit., Mosquitoes and Malaria, Poultry Parasit.* ⓓ Mass. Agric. Coll. 1924.

Freedman, Louis (1894), Research chem. and biochem., H. A. Metz Labor., Inc. Brooklyn, N.Y. (U.S.A.), 644 Pacific Street. *Chem., biochem. and pharm. investigations in synthetic pharmaceutical preparations. Chemotherapy: Synthetic epinephrine and derivatives, salvarsan, neo-salvarsan, and other organo arsenic compounds; hydroxyphenylalanines and*

tyrosine derivatives; synthetic organic sulfur compounds. ⓐ Columbia Univ., New York, N.Y., 1922.
Freeman, Edward Monroe (1875), Head, Dept. of Plant Pathol. and Botan., Dean, Coll. of Agric., Forestry, and Home Economics, Univ. of Minnesota. St. Paul, Minn. (U.S.A.), Univ. Farm. *Minnesota Plant Diseases.* ⓐ Univ. of Minnesota 1905.
Freeman, George Fouche (1876), Dir. du Service Technique de l'Agric. et de l'Enseignement Professionel. Port-au-Prince (Haiti). *Botan., alfalfa, tepary beans, papago, sweet corn, cotton, plant genetics.* ⓐ Harvard 1917.
Freeman, J., Prof. St. Marys Hosp. Med. School. London W. 2 (England), Paddington. *Bact.* o
Freer, Ruskin Skidmore, Prof. of Biol. Lynchburg, Va. (U.S.A.), Lynchburg Coll. *Ecol.* o
Frei, Walter (1882), o. Prof. für Pathol. und Physiol. an der vet. med. Fac. der Univ. Zürich, Dir. des vet.-pathol. Inst. Zollikon b. Zürich (Schweiz), Höhestr. 68, *Pathol. und Physiol. der weiblichen Geschlechtsorgane. Sterilität der Haustiere.* ⓐ Zürich 1906. ⓟ Tierpathogene, Bakterienkulturen, pathol. anat. Praep. von Tieren.
Freiberger, Genady, Prof. Tierärztl. Hochsch. Leningrad (U.d.S.S.R.), Cernigovskaja 5. *Pathol. Anat.* o
Freidenfelt, Magnus Fredrik Teodor (1872), Fiskeriintendent. Karlstad (Sverige). *Biometrie, Coregonen-Systematik.* ⓐ Lund 1908.
Freire, Carlos (1892), Chef de Travaux, Prof. compl. Inst. de Bact., Ecole Vet. Montevideo (Uruguay). *Bact.* ⓐ Montevideo 1916.
Freitel, Paul, Ass. au Mus. Paris (France), 61, rue de Buffon. *Paléobot.: flores tertiaires. Paléozool.*
La Frenais, Hugh Maurice (1884), Anti-Rabic Dept., Haffkine Inst. Parel, Bombay (India). *Action of synthetic drugs in rabies.* ⓐ Madras 1907.
Frenckell, Georg (1899), Ass. med. Inst., Leiter der klinischen Stat. d. Abt. für pathol. Physiol. Leningrad (U.d.S.S.R.), Straße des 3. Juli, 77, W. 7. *Pathol. Physiol. des Blutes, der blutbildenden Organe, der Kreislauforgane.* ⓐ Leningrad 1925.
Freis, G. P. (1879), Dr., Physician and Prosector of hospital for mental and neural diseases. Poortugaal near Rotterdam (Holland). *Heredity of Man (Headform, Eyecolor).* ⓐ Amsterdam 1924.
Freudenberg, Wilhelm (1881), Prof. Dr. phil nat. Heidelberg-Schlierbach (Deutschland). *Palaeont. des Menschen.* ⓐ Wien 1905. ⓟ Fossile Säugetiere, besonders aus dem Fundniveau des Homo Heidelbergensis. Kiefer des Neanderthal-Menschen.
Freudl, Eligius, Prof. f. Pflanzenzüchtung, Wiesen- u. Hopfenbau. Děčin-Libverd (C.S.R.), Landwirtschaftl. Abt. d. Deutsch. Techn. Hochsch. Prag. o
Freund, Hans Otto Friedrich (1885), Halle a. d. S. (Deutschland), Blumenstr. 19 pt. *Fortpflanzungsphysiol. d. Algen. Zellphysiol.* ⓐ Halle 1907.
Freund, Hermann, Univ.Prof., Dir.des pharmacol. Inst. Münster (Deutschland), Westring 12. o
Freund, Jules (1890), Philadelphia, Pa. (U.S.A.), The Henry Phipps Inst. of the Univ. of Pennsylvania, 4931 Locurt Str. *Immunol.* ⓐ Budapest 1913.
Freund, Ludwig (1878), ao. Prof. der Zool. Deutsche Univ. Praha II (C.S.R.), Legerova 48. *Cetacea; vergl. Anat. d. Säugetiere u. Vögel; Anoplura. Pathol. piscium.* ⓐ Prag 1904.
Frey, Charles N. (1886), In charge of Fleischmann Labor. New York (U.S.A.), 158 St. and Mott Ave. *Nutrition of Fungi.* ⓐ Univ. of Wisconsin 1919.
Frey, Eduard (1888), Dr. phil., Bern (Schweiz), Hubelmattstr. 42a. *Flechten, Systematik, Floristik, Morphol. Pflanzensoziol.* ⓐ Bern 1920. ⓟ Flechten.
Frey, Ernst (1878), Dr. med., o.ö. Prof. für Pharm., Dir. Pharmacol. Inst. Univ. Rostock i. M. (Deutschland), Kossfelder Str. 21. *Pharm. der Nierentätigkeit, des Muskels, des Herzens.* ⓐ Breslau 1901.
Frey, Hedwig, Prosektor und Prof. am Anat. Inst. der Univ. Zürich (Schweiz), Mommsenstr. 17.

Vergl. Anat. des Skelett- und Muskelsystems: Muskeln der unteren Extremität, Knochen des Rumpfskeletts, Biol. der Scapula. ⓐ Zürich 1912.
Frey, Lucie, Dr. med., Warszawa (Polska), Novogrodska 59. *Pathol., Anat. pathol.*
von Frey, Max (1852), Prof. Physiol., Dir. physiol. Inst. Univ. Würzburg (Deutschland), Röntgenring 9. *Sinnesphysiol.* ⓐ Wien 1877.
Frey, Richard Karl Hjalmar (1886), Kustos entomol. Mus. Univ. Helsinski (Finland), Zool. Mus. *Diptera brachycera, Morphol. der Insekten.* ⓐ Helsingfors 1924. ⓟ Diptera.
Freystadtl, Béla (1883), Dr., Oberarzt Elisabeth-Sanatorium. Budapest (Ungarn), Dorotheengasse 3. *Neurol. des Kehlkopfs und Rachens. Sensibilität der Zunge.* ⓐ Budapest 1906.
Friauf, Alexander P., Doc., Landwirtschaftl. Akademie nam. Timirjasew. Moskau (U.d.S.S.R.), Petrowsko-Rasumowskoje, Iwanowskaja 14. *Gartenbau. Obstbau.* o
Fricke, Max (1882), Kustos der Julius Richter-Sammlung (Geol.-palaeont. Abtlg.) des König-Albert-Mus. Zwickau i. Sa. (Deutschland), Siedlung Weißenborn, Damaschke-Weg 4. *Palaeozool. und Palaeobotan. Vergl. Anat. Die silurischen Ablagerungen am Südrande des Zwickauer Kohlenbeckens, Graptolithenfauna.* ⓐ Tier- und Pflanzenpetrefakten, insbes. aus dem Silur, Karbon und der Dyas.
Frickhinger, Hans Walter (1889), Dr., Leiter der Beratungsstelle für Schädlingsbekämpfung der I. G. Farbenindustrie-Aktiengesellschaft. München (Deutschland), Habsburger Platz 2. *Schädlingsbekämpfung. Haus- und Magazininsekten.* ⓐ Freiburg i. Br. 1914.
Fridericia, Louis Sigurd (1881), o. Prof. der Hygiene an der Univ. Copenhagen B (Dänemark), Ny Vestergade 11. *Ernährung, Vitaminuntersuchungen.* ⓐ Copenhagen 1906.
Fried, Robert (1892), Vorstand der Stalna bakterioloska stanica in Tuzla (S.H.S.). *Hygiene, Serol., Bact., Epidemiol.* ⓐ Wien 1917.
von Friedberg, Wilhelm (1873), Dr., o. Prof. der Palaeont. Poznań (Polen), Palaeont. Inst., Słowackigasse 4/6. *Miozäne Mollusken.* ⓐ Lwów 1899. ⓟ Miozäne Mollusken.
Friedberger, Ernst (1875), Prof. Dr. med., Dir. des Forschungs-Inst. für Hygiene und Immunitätslehre. Berlin-Dahlem (Deutschland), Thielallee 63. *Hygiene, Bact. und Immunitätslehre.* ⓐ Würzburg 1899.
Friede, Xenia (1894), Priv.Doc. II. Univ. Moskau (U.d.S.S.R.), Malaja Nikitskaja 10, Wohn. 12. *Immunität. Anaphylaxie. Heterogenetische Antigene und Antikörper.*
Friedel, Arthur (1880), Dr. med., Ass. der Anat. Anstalt der Univ. Berlin NW 6 (Deutschland), Luisenstr. 56. *Anat. des Menschen, Bewegungsapparat.* ⓐ Berlin 1912.
Friedel, Jean (1874), Maître de Conférences, adjoint à la Fac. des Sc. de Nancy (France), 42, avenue de France. *Physiol. et Anat. végétales. Morphol. et d'anat. comparées. Symétries.* ⓐ Nancy 1924.
Friedenthal, Hans (1870), Dr. med., a.o. Prof. für Physiol. an der Univ. Berlin NW 7, (Deutschland), Dorotheenstr. 13. *Menschheitskunde. Physiognomik. Verwandtschaftslehre.* ⓐ Bonn 1893. ⓟ Menschliche Embryonen durchsichtig gemacht und Affenfoeten. Totenmasken und Hände. Anthrop. Rassenbilder und Aufnahmen.
Friederichs, Karl Paul Theodor (1878), nichtplanmäßiger a. o. Prof. mit Lehrauftrag für angewandte Zool., Reg.- und Ökonomierat a. D. Rostock (Deutschland), Zool. Inst. der Univ. *Angewandte Entomol. (der Tropen und Europas). Embidae. Simuliidae. Lamellicornia.* ⓐ Rostock 1905. ⓟ Lebende Embiiden (Spinnfüßler).
Friedland, Michael (1888), Prof., Dir. orthopäd. Klinik Staatl. Inst. zur Ärztefortbildung. Kasan

(U.d.S.S.R.), Bolschaja Krasnaja 45. *Pathol. Anat. u. Physiol., Bact., Statik u. Dynamik d. menschlichen Körpers: Biomechanik des Pneumothorax.* Ⓓ Kasan 1911.

Friedmann, Herbert (1900), Instr. in Zool., Brown Univ. Providence, R.I. (U.S.A.). *Ornithol.; Vertebrate Zool., Compar. Anat. Parasitism in Birds; the Origin and Evolution of the Instincts Associated with Reproduction in Birds.* Ⓓ Cornell Univ. 1923.

Friedmann, Jode M. (1896), Aspirant des Biol. Forschungs-Inst. Univ. Perm (U.d.S.S.R.), Zaĭmka. *Statistische Zoöökol.: Turbellaria des Baikalsee (Tricladen).*

Friedrichs, Gustav (1890), Ass. Anstalt f. Pflanzenschutz und Samenuntersuchung d. Landwirtschaftskammer Provinz Westfalen. Münster i. Westf. (Deutschland), Südstr. 76. *Chondriosomen u. Chromatophoren. Methodik cytol. Forschung. Pilzliche Schädlinge des Getreides, bes. Brandpilze u. Schneeschimmel. Prüfung von Pflanzenschutzmitteln, vor allem Beizmitteln. Saatgutveredelung. Vergl. Keimprüfungen.* Ⓓ Münster i. W. 1922.

Frierson, Jr., Lorraine Screven (1903), Herpetol. Robson, La. (U.S.A.), Caddo Parish. *Amphibians and Reptiles from all parts of the World.* Ⓟ Amphibians and Reptiles living and preserved.

Fries, Rob. E. (1876), Prof., Dir. d. Bergianischen Gartens. Stockholm 50 (Sverige). *Systematik und Morphol. der Phanerogamen, Pflanzengeographie, Blütenbiol.* Ⓓ Upsala 1905.

Fries, Sigmund Heinrich Stephan (1850), Dr. med., Geh.San.R. Göttingen (Deutschland), Baurat-Gerber-Str. 7. *Biol. der einheimischen Säugetiere, Vögel, Amphibien; Höhlenfauna.* Ⓓ Tübingen 1875.

Fries, T. C. E., Doc. Univ. Uppsala (Sverige). *Botan.* ○

Friese, Heinrich Friedrich August (1860), Prof. und Leiter der Abteilung VI (für biol. Forschung) am Landes-Gesundheitsamt. Schwerin, Mecklenburg (Deutschland), Kirchenstr. 1. *Hymenoptera.* Ⓓ Gießen i. Hessen 1907. Ⓟ Hymenoptera, auch deren Nester und Strepsiptera.

Friesé, Vadime (1881), Sous-Dir. Inst. Microbiol. Moscou (U.d.S.S.R.), Troubnikowsky 26, N. 28. *Immunol. et Bact.* Ⓓ Kazan 1907.

Friesner, Ray Clarence (1894), Prof. of Botan., Butler Univ. Indianapolis, Ind. (U.S.A.). *Cytol.* Ⓓ Univ. of Michigan.

Frimmel, Franz (1888), Leiter des Fürst Liechtenstein Pflanzenzüchtungs-Inst. Eisgrub, Mähren (C.S.R.). *Theoretische und praktische Pflanzenzüchtung.* Ⓓ Wien Univ. 1912.

v. Frisch, Karl (1886), o. Univ.Prof., Dir. des Zool. Inst. der Univ. München (Deutschland), Giselastr. 5. *Farbwechsel, Sinnesphysiol. Farbensinn, Formensinn und Geruchsinn der Biene.* Ⓓ Wien 1910.

De Frise, Aldo, Prof. Univ. Milano (Italia). *Anat. umana.* ○

Frison, Theodore Henry (1895), Systematic Entomol. and Curator. Urbana, Ill. (U.S.A.), Illinois State Natural History Survey. *Insects of Illinois. Classification of Evaniidae and Bremidae (Bombidae), biol. studies of Bremidae (Bombidae).* Ⓓ Univ. of Illinois 1923. Ⓟ Bremidae (Bombidae).

Fritsch, Felix Eugen (1879), Prof. of Botan. Univ., Head of Botan. Dept., East London Coll. London E. 1 (England), Mile End Rd. *Freshwater Algae, taxonomy and biol. Plant Ecol.* Ⓓ Phil. München 1901, sc. London 1905.

Fritsch, Gustav (1838), Geh.M.R., Hon.Prof. Berlin-Lichterfelde (Deutschland), Berliner Str. 30. *Anthrop.: Eingeborene Südafrikas. Anat.: elektrische Fische, Area centralis, Fischgehirn.* ○

Fritsch, Karl (1864), Dr. phil., o.ö. Prof. der systematischen Botan. und Dir. des Botan. Gartens der Univ. Graz (Österreich), Alberstr. 19. *Systematik der Gesneriaceen. Blütenbiol. Flora von Österreich.* Ⓓ Wien 1886.

Fritz, Emanuel (1886), Associate Prof. of Forestry, Univ. of California. Berkeley, Cal. (U.S.A.), 305 Hilgard Hall. *Microscopic structure of economic woods.* Ⓓ Yale Univ. Connecticut 1914. Ⓟ American woods.

Fritze, Adolf (1860), Dr. phil., Prof., Mus.Dir. i. R. Hannover (Deutschland), Vellchenstr. 3 B III. *Angewandte Zool. Orthopterol.* Ⓓ Freiburg i. Br. 1888. Ⓟ Lebende Reptilien, Amphibien, Mollusken und Insekten, besonders Wasserinsekten.

Froblsher, Jr., Martin (1896), Instr. in Bact., The Johns Hopkins Univ. School of Med. Baltimore, Md. (U.S.A.), 1833 E. Monument St. *Bact.* Ⓓ Johns Hopkins Univ. 1925.

Froboese, Curt, Siegfried Waldemar (1891), Dr. med., Priv.Doc. für allg. Pathol. und pathol. Anat. Heidelberg (Deutschland), Gaisbergstr. 101. *Fettstoffwechsel (Intoxikation).* Ⓓ Berlin 1916.

Frodl, Bedřich, Dr.-Ing., Ass. l'Inst. des Recherches agricoles et Forestière. Brno (C.S.R.), Květná ul. *Biochem.* ○

Fröhlich, Friedrich Wilhelm (1879), a.o. Prof. für Physiol. Bonn a. Rh. (Deutschland), Physiol. Inst. der Univ. *Sinnesphysiol. Allgemeine Physiol. des Nervensystems.* Ⓓ Wien 1906.

Frölich, Gustav (1879), a. ö. Prof., Dir. Inst. f. Tierzucht u. Molkereiwesen. Halle a. d. S. (Deutschland), Sophienstr. 15. ○

Frolowa, Sophia (1884), Ass. I. u. II. Univ., Inst. f. Experim. Biol. Moskau (U.d.S.S.R.), Woronzowo Pole 6. *Cytol.: Chromosomengarnituren, normale und poliploide, der Insekten.*

Fromherz, Konrad (1883), Pharmacol. wissenschaftl. Mitarbeiter der Firma Hoffmann-Laroche, Basel. München (Deutschland), Mozartstr. 14 II. *Pharm., Narkotica, Anaesthetica und Hormone.* Ⓓ Phil. Freiburg 1906, med. Freiburg 1908.

Fromm, Emil (1865), o.ö. Prof., Dr. phil., Vorstand des Univ.-Inst. für med. Chem. zu Wien IX (Deutsch-Österreich), Währinger Str. 25. *Schutz des Tierkörpers bei Vergiftungen.* Ⓓ Erlangen 1888.

Fromme, Fred Denton (1886), Prof. of Botan. and Plant Pathol., Virginia Polytechnic Inst., Plant Pathol., Virginia Agric. Experim. Station. Blacksburg, Va. (U.S.A.). *Phytopath. diseases of tobacco, of fruits and of the cereals.* Ⓓ Columbia Univ. 1914.

Fromme, Karl Walther (1879), Dr. med., Prof., Stadtarzt, Leiter des Gesundheitsamtes der Stadt Witten a. d. Ruhr (Deutschland). *Hygiene und Bact.* Ⓓ Berlin 1904.

Fromols-Rakowski, R. J. Danzig, Langgasse 57. *Ornithol.* ○

Fron, Georges (1870), Prof. à l'Inst. National Agronomique. Paris 5 W (France), 16, Rue Claude Bernard. *Cryptogamie. Pathol. végétale. Bois industriels indigénes et coloniaux. Maladies des plantes cultivées tempérées et tropicales. Plantes nuisibles à l'agric.* Ⓓ Paris.

Fronda, Francisco M. (1896), Ass. Prof. of Poultry Husbandry, Univ. of the Philippines, Los Baños Coll. Laguna (Philippine Islands). *Genetics, nutrition and pathol.* Ⓓ Cornell Univ. 1922. Ⓟ Poultry, Cantonese.

Fronius, Richard (1884), Ass. Tierzuchtlabor. Academia de Agr. Cluj (România). *Abstammung, Blutlinien, Milchbact., Chem. der Milch.* Ⓓ Hohenheim,Württ., 1907. Ⓟ Photographien von Haustieren.

Frost, Charles Albert (1872). Framingham, Mass. (U.S.A.), 67 Henry St. *North American Coleoptera. Buprestidae, Cerambycidae, Cicindelidae.* Ⓓ Univ. of Maine 1895. Ⓟ Coleoptera of North America.

Frost, Florence M. (1885), Dir. of the Labor. of the Polyclinic. Memphis, Tenn. (U.S.A.), 20 South Dunlap Street. *Bact. Serol.* Ⓓ Univ. of Wisconsin 1912. Ⓟ Malaria slides.

Frost, Stuart Ward (1891), Associate Prof. Entomol. The Pennsylvania State Coll. Pa. Arendtsville, Pa. (U.S.A.). *Systematic Entomol., genus Phytomyza.*

Apple leaf rollers and Oriental. Fruit moth. Leaf-mining Diptera. Ⓓ Cornell Univ. 1924.
Frost, William Dodge (1867), Prof. of Agric. Bact., Univ. of Wisconsin. Madison, Wis. (U.S.A.). *Food bact. Milk.* Ⓓ Univ. Wis. 1903.
Fruwirth, Carl (1862), HofR., o.ö. Prof., Dr., Dr. h. c. Wien IV (Österreich), Techn. Hochsch., und Waldhof, Amstetten (N.-Österreich). *Landwirtschaftl. Pflanzenzüchtung. Technik des landw. Pflanzenbaues.* Ⓓ h. c. Hohenheim und h. c. Wien.
Fry, Eva Jennie (1893). London, N.W. 3 (England), Westfield Coll., Hampstead. *Effect of Lichens on different substrata on which they live. Relations between the algal and fungal constituents.* Ⓓ Aberystwyth 1919.
Frye, Theodore C. (1869), Prof. of Botan., Univ. of Washington; Dir., Puget Sound Biol. Station. Seattle, Wash. (U.S.A.). *Morphol. of Algae, Taxonomy of Mosses.* Ⓓ Chicago 1902.
Fuchs, Alfred (1872). Augsburg (Deutschland), Maxstr. C 8/III. *Systematik d. Entwicklungsgesch. u. Biol. mitteleuropäischer Orchideen.*
Fuchs, Franz (1874), Ob.Reg.R. Bayreuth (Deutschland), Friedrichstr. 36. *Biol. sämtlicher Insekten, spez. der schädlichen Forstinsekten.* Ⓟ Insekten und Biol. von Insekten.
Fuchs, Richard Friedrich (1870), a.o. Prof. der Physiol. an der Univ. Breslau 16 (Deutschland), Wilhelmsruh. *Blutkreislauf, funktionelle Anpassung, Einfluß d. Höhenklimas, allgemeine Nerven- und Muskelphysiol., Farbenwechsel der Tiere.* Ⓓ Prag 1897.
Fuchs, W., Doc., Biochem. Abt. der Deutschen Techn. Hochsch. Brno (C.S.R.). ○
Fuchsig, Heinrich (1887), Prof. der Naturgesch. an der Bundeserziehungsanstalt Wien XIII; Mitarbeiter der biol. Station in Lunz (N.-Ö.). Wien XIII (Deutsch-Österreich), Dampierrestr. 1. *Physiol. Pflanzenanat.; Beeinflussung des anat. Baues der Pflanzen (speziell der Moose) durch die verschiedenen Standortsfaktoren. Biol. u. Ökol. (der Moose).* Ⓓ Graz 1910. Ⓟ Moos-Herbarstücke.
Fucini, Alberto, Prof. o., Dir. Geol. Ist. Univ. Catania, Sicilia (Italia). ○
von Fudakowski, Josef (1893), Dr. phil., Kustos der zool. Abteilung des Physiographischen Mus. der Polnischen Akademie der Wissenschaften in Kraków (Polen), Sławkowska 17. *Systemat. und Biol. der Odonata und Hymenoptera (Vespidae, Mutillidae, Chryseididae). Vergl. Anat. (Darmkanal und Lymphsystem der Fische).* Ⓓ Kraków 1924. Ⓟ Odonata von Polen (spec. der Tatra-Gebirge),
Fühner, Hermann (1871), Prof., Dr. phil et med., Dir. des Pharmacol. Inst. der Univ. Bonn (Deutschland), Wilhelmstr. 33. *Pharm.; Synergismus und Antagonismus, Nachweis und Bestimmung von Giften auf pharm. Wege. Narcotica: Alkohole.* Ⓓ Phil. Genf 1896, med. Straßburg 1902.
de la Fuente y Morales, José Maria (1855). Pozuelo de Calatrava, province de Ciudad Real (España). *Coléoptères d'Europe.* Ⓟ Coléoptères.
Fülleborn, Friedrich (1866), Dr. med., Prof., GMR., Abteilungsvorst. am Inst. für Schiffs- und Tropenkrankheiten. Hamburg 4 (Deutschland). *Tropenmed., Biol. der Helminthen. Physische Anthrop. des Nord-Nyassa-Gebiets.* Ⓓ Berlin 1894. Ⓟ Helminthol. Material.
Fürst, Carl Magnus (1854), Prof. emer. Lund (Sverige). *Anthrop., Anat.* Ⓓ Lund 1887.
Fürst, Friedrich (1894), Assessor an der Landesanstalt für Pflanzenbau und Pflanzenschutz. München (Deutschland), Theresienstr. 68/0. *Landwirtschaftlicher Pflanzenbau, Pflanzenernährung und Düngung, Bodenbearbeitung, Unkrautbekämpfung.*
Fürst, Pius (1868), Bürgerschuldir. i. P. Wien XV (Österreich), Markgraf-Rüdiger-Str. 23. *Musci.*
Fürth, Otto, a.o. Univ.Prof. mit Titel eines o. Prof., Vorstand der Abteilung für physiol. Chem. im Wiener Physiol. Univ.-Inst. Wien XIX (Deutsch-Österreich), Hasenauer Str. 32. *Eiweißchem.*

Muskelchem. und Muskelphysiol. Intermediärer Stoffwechsel. Vergl. chem. Physiol. der Wirbellosen. Melaninbildung. Ⓓ Wien 1894.
Fuhrmann, Franz (1877), Univ.Prof. Waltendorf b. Graz (Österreich). *Mykol., Bact., Nahrungsmittelchem.* ○
Fuhrmann, Otto (1871), Dr., Prof. der Zool. Neuchâtel (Schweiz), Univ. *Hydrobiol., Parasit. (Platodes).* Ⓓ Basel 1894. Ⓟ Plankton, Cestoden.
Fuji, Teikichi, Lect. Imper. Univ., Inst. of Hygiene Fac. Med. Kyôtô (Japan).
Fujita, Sukeyo (1878), Dir. Experim. Stat for Pearl Culture. Boeton, Celebes (Ned.O.Indië). *Biol. of pearl-bearing molluscs, their cultivation and enforced pearl production.* Ⓓ Tokio 1909.
Fujita, Toshihiko (1877), Prof. Physiol. Tôhoku Reichsuniv. Sendai (Japan), Physiol. Inst. *Allg. Muskel- und Nervenphysiol, insbes. Sinnesphysiol., Physiol. der Gesichts- und Hautsinne u. der Stimmlaute.* Ⓓ Tôkyô 1914.
Fuliński, Benedykt (1881), Prof. d. Zool. u. vergl. Anat. der Haustiere. Lwów (Polen), Zool. Inst. a. d. Politechnik. *Embryol. d. Würmer u. Arthropoden. Turbellaria (Tricladida u. Rhabdocoelida).* Ⓓ Lwow 1908.
Fuller, Albert M., Ass. Curator of Botan., Public Mus. Milwaukee, Wis. (U.S.A.). ○
Fuller, George Damon (1869), Associate Prof. of Plant Ecol. Univ. of Chicago, Ill. (U.S.A.), Department of Botan. *Plant Ecol. The measurement of ecol. factors. Sand dune vegetation. Transitions between grasslands and forests.* Ⓓ Univ. of Chicago 1913.
Fulmek, Leopold (1883), R.R. Bundesanstalt f. Pflanzenschutz. Wien II (Deutsch-Österreich), Trunnerstr. 1. *Angewandte Entomol., spez. Mallophaga.* Ⓓ Wien 1907. Ⓟ Mallophaga.
Fulmer, H. L., Prof. Agric. Coll. Guelph, Ontario (Canada). *Biochem.* ○
Fulton, Bentley Ball (1889), Ass. Entomol. Iowa State Coll. Ames, Ia (U.S.A.). *EconomicEntomol., Ecol. of Orthoptera.* Ⓓ Iowa State 1926. Ⓟ Orthoptera.
Fulton, John Farquhar (1899). Brookline, Mass. (U.S.A.), 126 Longwood Avenue. *The intimate nat. of muscular contraction. Muscle tonus. The kneejerk. The reflex coordination of movement. Electrical variations of living tissues. Physiol. of nerve and muscle. The origin and nature of animal and plant pigments.* Ⓓ Ph.D. Oxford 1925, M.D. Harvard Med. School 1927.
Funk, Casimir (1884), Head of Biochem. Dept. State School of Hygiene. Warszawa (Polska), 24 Chocimska. *Biochem. and Nutrition. Chem. and physiol. of vitamines and vitasterols, chem. and pharm. of insulin, pituitrin etc., chem. of ferments, etiol. of cancer.* Ⓓ Ph.D. Berne 1904, D.Sc. London 1913.
Funk, Georg (1886), ao. Univ.Prof. der Botan. Gießen (Deutschland), Bleichstr. 4. ○
Funkhouser, William Delbert (1881), Head Department of Zool., Dean of the Graduate School, Univ. of Kentucky. Lexington, Ky. (U.S.A.). *Membracidae (Homoptera). Entomol.* Ⓓ Cornell 1916. Ⓟ Membracidae.
Funkquist, Hermann Per Anton (1870), Prof. der Zootechnik an der Landwirtschaftl. Hochsch. zu Alnarp, Åkarp (Schweden). *Erblichkeit der Haarfarbe des Rindviehs, Histogenese der Geschlechtsdrüsen des Rinderbryos. Pinealdrüse.* Ⓓ Lund 1920.
Furlani, Johannes (1881), Dr., Prof. Wien VII (Deutsch-Österreich), Kandlgasse 39. *Klimatol., Standortslehre, Bodenkunde, Pflanzengeographie.* Ⓓ Wien 1904.
Furrer, Ernst (1888), Dr. phil. Zürich 2 (Schweiz), Seestr. 301. *Pflanzengeographie.* Ⓓ Zürich 1914.
Furrow, Clarence Lee (1896), Prof. of Biol., Knox Coll., Dept. of Biol. Galesburg, Ill. (U.S.A.). *Cytol., Physiol. and general biol.* Ⓓ Oklahoma 1922.

Fursikov, Prof. II. Univ. Moskau (U.d.S.S.R.). *Physiol. d. Zentralnervensystems.* ○

Furssenko, Alexander (1903), Inst. of Nat. Hist. Leningrad (U.d.S.S.R.), Old-Peterhof. *Evertebrate zool. Protist. Development and morphol. of the Infusoria, Peritriche colonies. System. morphol. of the Peritricha.*

Fuschini, Carlo (1880), Prof. di Entomol. Agraria, Sericicoltura e Apicoltura nel R. Ist. Superiore Agrario di Perugia (Italia). *Insetti dannosi alle piante coltivate; Sericicoltura. Apicoltura.*

Fuset Tubiá, J. (1871), Prof. de Biol. générale et de Zoografie des Vertébrés. Fac. des Sc. à l'Univ. Barcelona (España). *Aves, Pisces.* ⓟ Poissons des côtes de Catalogne et des îles Baléares.

Gaál, Béla (1896), Ass. Pharmakognost. Inst. Budapest VIII (Ungarn), Üllői-ut 26. ⓓ Budapest 1926.

von Gaál, Stefan (1877), Dr., Priv.Doc. an der Univ. Szeged (Kolozsvár). Budapest VII (Ungarn), Szentkirályi-ut. 7. *Palaeont.; obermiozäne Brackwasser- und Landschnecken.* ⓓ Budapest 1905. ⓟ Obermiozäne Brackwasser- und Landschnecken.

Gaarder, Torbjørn (1885), Dir. Biochem. Labor., Bergens Mus. Bergen (Norge). *Biochem.* ⓓ Oslo 1916.

Gabaricollo, Prof. Univ. Montpellier (France). *Hist. nat. méd.* ○

Gabel, Charles Ernst (1877), Dir., West Virginia State Hygien. Labor. Charleston, W.Va. (U.S.A.), 1902 Washington St. *Med. bact.* ⓓ Wien 1903.

Gabinsky, A. M. (1898), Aspirant des Katheders f. Morphol. u. Physiol. Odessa (U.d.S.S.R.), Pasteurstraße 36. *Craniol. der Juden.* ⓓ Odessa 1926.

Gabrão, A., Prof. Univ. Porto Alegre (Brasil). *Pharmacol.* ○

Gabričevskij, Evgenij Georgievič, Doc. I. Univ. Moskau (U.d.S.S.R.). *Zool. experim. Anat.* ○

Gabriel, C., Prof. Univ. et méd. Marseille (France). *Botan.* ○

Gabrielson, Ira Noel (1889), Ass. Biol., Biol. Survey, U.S. Department of Agric. Portland, Ore. (U.S.A.), 515 Post Office Building. *Noxious rodents. Distribution and economic relations of Birds.*

Gabunia, K. E., Sibir. Abt. des Com. Géol. Tomsk (U.d.S.S.R.). *Spezialität: Korallen des Carbon.* ○

Gadow, Hans Friedrich (1855), Strickland Curator, Reader in Morphol. of Vertebrata, Univ. of Cambridge. Cleramendi, Great Shelford (England), Mus. of Zool. *Morphol., systematic and geographical distribution of Vertebrates. Muskulatur des Beckens u. d. Extremitäten der Ratitae. Osteol. of the Mammalia, Birds. Amphibia and Reptiles. Wanderings of Animals.* ⓓ Phil. Jena 1878, M. A. h. c. Cantab 1884.

Gärtner, August (1848), Dr. med. et Dr.phil. h. c., GRR., ○. Prof. emer. der Hygiene Univ. Jena (Deutschland). *Bact. enteritidis, Gärtner-Bazillen. Trinkwasser in hygien. Beziehung.* ⓓ Berlin 1872.

Gaertner, Gustav (1855), tit. ○. Prof. emer., f. allg. u. experim. Pathol. Univ. Wien XIX (Deutsch-Österreich), Dittesgasse 48. *Ernährungslehre.* ⓓ Wien 1879.

De Gaetani, Luigi, Prof. incar. Univ. Messina (Italia). *Anat. hum. et topogr.* ○

Gaetano, Martino (1900), Aiuto Ist. Fisiol. Univ. Messina (Italia). ⓓ Roma 1925.

Gaetano, Sampietro (1875), Prof. incar. di Igiene delle. abstazioni nella R. Scuola Sup. di Architettura. Dirigente il Labor. Micrografico del Govern. Roma (Italia), Viale Regina Margherita 214. *Batteriol.*

Gäumann, Ernst Albert (1893), Dr. phil., Doc. an der Eidg. Techn. Hochsch. Zürich 6 (Schweiz), Seminarstr. 21. *Mykol., Pflanzenpathol.* ⓓ Bern 1917. ⓟ Kryptogamen, Pflanzenkrankheiten.

Gage, George Edward (1883), Head of Department and Prof. of Animal Pathol. Amherst, Mass. (U.S.A.). *Histol., Physiol., Bact., Serol.* Yale Univ. 1909.

Gage, Simon Henry (1851), Prof. emer. Histol. and Embryol., Cornell Univ. Ithaca, N.Y. (U.S.A.), Stimson Hall. *Histol., Embryol., Physiol. of Vertebrates.*

Gager, C. Stuart (1872), Dr., Dir. Brooklyn Botan. Garden, New York City. Brooklyn, N.Y. (U.S.A.), 1000 Washington Avenue. *Plant Physiol., the effect of radium rays on plants.*

Gahan, Arthur Burton (1880). Entomol. U. S. Bur. of Entomol. U. S. Nat. Mus. Washington D. C. (U.S.A.). *Hymenoptera, Chalcidoidae.*

Gahan, C. J., Keeper of Dept. of Entomol., British Mus. London S.W. (England), Cromwellroad. *Nat. History.* ○

Galdukow, Nikolaj Michalowitsch (1874), Prof. Botan., Dir. Botan. Inst. Staatl. Weißruss. Univ. Minsk (U.d.S.S.R.), Sowjetskaja Str. 33. *Algen, Biol. Analyse d. Wassers, komplementäre chromatische Adaptation, Kolloide d. Pflanzenzellen, Protoplasma, Ultramikroskopie, Samenkunde, Spinnfasern, Unkräuter, offizinelle Pflanzen, Phylogenie der Pflanzen, Konvergenzen, Komplikationen.* ⓓ Kiew 1904.

Gaiger, Sydney Herbert (1884), Prof. of Vet. Pathol., Royal Coll. of vet. Surgeous. Univ. of Liverpool (England). *Infections. Diseases of Animals.* ⓓ Liverpool 1905.

Gail, Floyd Whitney (1884), Dr., Prof. of Botan. Head of Department. Moscow, Ida. (U.S.A.), Dept. of Botan. Univ. of Idaho. *Plant physiol.* ⓓ Washington 1920.

Gaiļīts, Laimons (1885), Lect. der angewandten Entomol. und Leiter des Entomol. Kabinetts an der Univ. Rīgā (Latvija), Kronvalda bulvarī 1. *Entomol. und angewandte Zool. Lepidopteren. Spinnen.*

Gain, Edmond (1868), Prof. de Botan. Fac. Sc. Univ., Dir. de l'Inst. agric. Colon., Dir. du Jardin Botan. de la Ville et du Jardin alpin Univ. Nancy (France). *Physiol. végétale et Botan. agric. Graines. Ecol.*

Gaines, Edward F. (1886), Associate Prof. of Farm Crops; Cerealist. Pullman, Wash. (U.S.A.). *Plant breeding: resistance to covered smut in oats and stinking smut in wheat.* ⓓ Harvard Univ. 1921. ⓟ About 500 varieties of wheat, oats and barley.

Gaines, Walter Lee (1881), Prof. of Milk Production, Coll. of Agric. and Chief, Agric. Experim. Station. Urbana, Ill. (U.S.A.), Univ. of Illinois. *Milk production: physiol. aspects of same.* ⓓ Univ. of Chicago 1915.

Gainey, Percy Leigh (1887), Prof. Bact., Kansas State Agric. Coll.; Soil Bact., Kansas Agric. Experim. Station. Manhattan, Kan. (U.S.A.). *Soil bact., Nitrogenfixation in Soil.* ⓓ Washington Univ. 1910.

Gaiser, Lulu Odell, Lect., Mc.Master Univ. Toronto, Ontario (Canada). *Cytol., Pathol.* ○

Gajl, Kazimierz (1896), Dr. phil., Ass. de l'Univ. de Varsovie (Pologne), Kr. Prz. 26, Inst. de Zool. *Associations et morphol. des Crustacés des eaux douces (Phyllopoda et Copepoda excl. Harpacticidae et Cop. parasit.). Hydrobiol.* ⓓ Varsovie 1924.

Galati Mosella, Rosario (1891), Prof. ord. Treviso (Italia). *Organi di senso dei Molluschi e degli Onellidi, Protozoi.*

Galbo, C., Prof. Univ. Napoli (Italia). *Anat. patol.* ○

Galdiano, Manuel Ferrer, Dr. Madrid (España), Paseo Recoletos 37. *Hydrobiol.* ○

Galenieks, Paul (1891), Priv.Doc., Univ. Riga (Lettland), Botan. Inst. der Univ. *Palaeobotan. und Moorbotan., insbesondere Diluvialflora.* ⓟ Moose, speziell Sphagnales und Bryales.

Galiano, Emilio Fernándes, Univ.Prof. Barcelona (España). *Histol., Protist.* ○

Le Gall, Jean Joseph (1894), Prof. Agr. Univ., Dir. Labor. de l'Office Sc. des Pêches Maritimes.

Boulogne-sur-Mer (France), 17, Boulevard de Chatillon. *Biol. des Poissons, espèces marines. Plancton. Océanographie. Hydrol.*

Gallagher, Bernard A. (1880), Bact., Board of Agric. and Forestry, Honolulu, T.H. (Hawaii), Box 3319. *Diseases of Animals; esp. dom. birds.* ⑨ Cornell Univ. 1901.

Gallardo, Angel, Prof. Univ. Nac. Buenos Aires (Argentina). *Zool.* o

Gallástegui, Cruz Angel (1891), Dir. de la Misión biol. de Galicia. Santiago (España). *Vererbungsprobleme: Mais, Kastanie.* ⑨ Hohenheim 1914. ⑫ Mais, Kastanie.

Galli, Eugenio, Prof. Univ. Nac. La Plata (Argentina). *Anat. topograph.* o

Galli-Valerio, Bruno (1867), Dir. de l'Inst. d'Hygiène et Parasit. de l'Univ. Lausanne (Suisse), Solitude 19. *Hygiène, Parasit., Epidemiol.* ⑨ Med. et med. vet. 1892. ⑫ Parasit. végétaux et animaux.

Galliard, Henri (1891), Préparateur de parasit. Fac. de Méd. Paris (France), 15, rue de l'Ecole de Med. *Parasit. humaine et compar.* ⑨ Paris 1921.

Gallöe, Olaf. Kongens Lyngby (Danmark). *Lichens.* ⑨ Copenhagen.

Galloway, Beverly Thomas (1863), Pathol., United States Department of Agric. Washington, D.C. (U.S.A.). *Theoretical and practical botan.: plant introduction and acclimatisation.* ⑨ Univ. of Missouri, Columbia, Mo., 1884.

Gmaleja, N. F., Priv.Doc. Staatl. Inst. f. ärztl. Fortbildung. Leningrad (U.d.S.S.R.), Kiročnaja 41. *Bact.* o

Gambetta, Laura (1901), Ass. all'Ist. di Zool. della R. Univ. di Torino 108 (Italia), Palazzo Carignano. *Gasteropodi polmonati. Endosimbiosi di Coleotteri silofagi.* ⑨ Torino 1924.

Gamble, F. W., Univ.Prof. Birmingham (England). *Zool.* o

Gamble, James L. (1883), Associate Prof. of Pediatrics, Harvard Med. School. Brookline, Mass. (U.S.A.), 33 Edge Hill Rd. *Physiol. of nutrition, the metabolism of the inorganic elements in terms of acidbase and osmotic pressure equilibriums.* ⑨ Leland Stanford Jr. Univ. 1906.

Gambrell, Foster Lee (1900), Ass. in Research of Entomol. Geneva, N.Y. (U.S.A.), N.Y. State Agr. Expt. Station. *Orchard and vegetable insects. Genus Rhagoletis, particularily, the cherry maggot, the walnut husk maggot, and the apple maggot.* ⑨ Ohio State Univ. 1925. ⑫ Jassidae.

Gamô, Toshioki, Prof. of Ueda Imperial Coll. of Sericulture. Ueda, Nagano-ken (Japan), *Anat. Inst. Anat. of Insects.* o

Gamorak, Nestor T., Prof. Landw. Inst. Kamenetz-Podolsk (U.d.S.S.R.), Ul. Michailitschenko 5. *Botan. Physiol. Pflanzenanat.* o

Gams, Helmut (1893), Dr. phil., Leiter der Biogeol. Station Mooslachen. Wasserburg a. Bodensee, Bayern (Deutschland). *Biocönotik, Systematik, Ökol. und Geographie der Gefäßpflanzen, Bryophyten und Algen, Stratigraphie und Palaeont. des Quartärs, Hydrobiol.* ⑨ Zürich 1918. ⑫ Moose und Gefäßpflanzen der Alpen, Pflanzen und Tiere des Bodensees.

Gandolfi-Hornyold, Alfonso (1879), Priv.Doc. Univ. de Fribourg (Suisse). San Sebastian (España), Museo Naval. *Biol. de l'Anguille.* ⑨ Fribourg, Suisse, 1907.

Gandrup, Johannes (1882), Dir. Besoekish Proefstation. Djember; Java (Ned.-O.-Indië). *Tabaccodiseases, rubber-diseases, coffee-diseases, growth of microorganisms on rubbersheets.* ⑨ Copenhagen 1919.

Gano, Laura (1874), Investigator. Thonotosassa, Flo. (U.S.A.). *Ecol. Botan.* ⑨ Univ. of Chicago 1912.

Ganeschin, Sergius (1879), Botan. Botan. Mus. Akademie d. Wissensch., Doc. Univ. Leningrad (U.d.S.S.R.), Univ.-Quai 5. *Systematik und Geographie d. Pflanzen. Systematik d. Gattung Euphrasia; Ruderalpflanzen.*

Ganfini, C., Prof. Univ. degli Studi, Lib. doc. Univ. Sassari (Italia). *Anat. umana norm.* o

Ganja, Boris (1883), Moorversuchsstation. Minsk (U.d.S.S.R.). *Chem. der Moorkultur.*

Ganong, William Francis (1864), Prof. of Botan. in Smith Coll. Northampton, Mass. (U.S.A.). *Botan. morphol., physiol. and ecol.* ⑨ München 1894.

Gans, Abraham (1885), Chef des pathol.-anat. Labor. und Arzt in der Neurol. und Psychiatrie'an dem Provinciaal Ziekenhuis in Santpoort (Holland)., Priv.Doc. der Neurol. in Leiden (Holland). *Die lokalisatorischen und histol. Befunde des Gehirns.* ⑨ München 1915. ⑫ Hirnpraep.

Gans, Oscar (1888), a.o. Prof., Oberarzt d. Hautklinik. Heidelberg, Baden (Deutschland), Univ.-Hautklinik. *Histopathol. und pathol. Physiol. d. Haut.* ⑨ Freiburg 1912. ⑫ Histol. Schnitte bzw. Material v. Hautkrankheiten.

Ganter, Georg (1885), Univ.Prof., Dir. der med. Poliklinik für innere Med. Rostock (Deutschland). *Physiol., pharmacol. u. pathol. physiol. Untersuchungen: Herz, Gefäße, Darm und glatte Muskulatur überhaupt.* ⑨ Freiburg i. Br. 1912.

Gantscheff, Jelü, o. Prof., Dr., Inst. f. allg. Tierzucht d. Landwirtschaftl. Fac. Sofia (Bulgarien). o

Gaponov, Euphemius A., Ass. Geol. Kabinett Univ. Odessa (U.d.S.S.R.). *Palaeobotan.: fossile Diatomeen.* o

Garbar, Jurij (1895). Selekcijas Stacija c. Stendi (Latvija). *Sommer- u. Winterweizen.* ⑫ Samen.

Garbat, Abraham Leon (1885), Associate Visiting Physician. Lenox Hill Hospital. New York City (U.S.A.). *Bact. Immunity.* ⑨ Cornell Med. School. 1906.

Garber, Ralph John (1890), Head, Dept. of Agronomy. West Virginia Univ. Morgantown, W.Va. (U.S.A.). *Corn, wheat, soybeans, oats, buckwheat, and tobacco.* ⑨ Univ. of Minn. 1922. ⑫ Seed of our resistant strains.

Garbini, Adriano (1857), Prof. Verona (Italia), Leoncino 38. *Istiol. e Idrobiol. Zoonomastica demol.* ⑨ Padova 1882. ⑫ Animali acquatici.

Garbowski, Ludwik (1872), Dr., Prof., Chef de la Section des Maladies des Plantes à l'Inst. agronomique de l'Etat. Bydgoszcz (Pologne). *Phytopath. des plantes cultivées.* ⑨ München 1907.

Garcez do Nascimento, Euripedes, Prof. Fac. d. Med. Curityba, Paraná (Brasil). *Pharm.* o

Garcia, A., Prof. Univ. of the Philipp. Manila, Philippine Islands (U.S.A.). *Anat.* o

Garcia, F., Assoc. Prof. Univ. of the Philipp. Manila, Philippine Islands (U.S.A.). *Pharmacol.* o

Garcia, H., Prof. Univ. Central. Caracas (Venezuela). *Hydrol.* o

Garcia Castelló, Cayetano (1878). Gandia, Valencia (España), Plaza del Cabo Pastor, 12. *Flore du distrit de Gandia.* ⑨ Madrid 1903. ⑫ Espèces classifiées du distrit de Gandia.

Garcia del Cid, Francisco, Dr. en Ciencias Nat., Prof. auxiliar en la Fac. de Ciencias. Barcelona (España), Cortes 513. *Biol.* o

Garcia Fresca y Tolosana, Antonio, Dr. en Ciencias Nat., Catedrático del Inst. Pamplona (España). *Entomol.* o

Garcia Varela, Antonio, Prof. Cat. de Organografia y Fisiol. vegetal, Vicedir. del Jardín Botán. y Jefe de la Sección de cultivos. Madrid (España), Espalter, 11. *Hemíptera et Botan.* o

Garcia Viñals y Busto, José, Dr. en Med., Prof. auxiliar en la Fac. Madrid (España), Jordán 2. *Pathol.* o

Gard, M. Dr. ès sc., Dir. Stat. de pathol. végétale. Bordeaux (France), 20 cours Pasteur. *Maladies et dépérissement des Noyers. Maladies de la vigne.*

Gardiner, John Stanley (1872), Prof. Univ. of Cambridge (England), Zool. Labor. Bredon House, Selwyn Gardens. *Zool. and compar. Anat. Tropical faunae and coral reefs.* ⑨ Cambridge 1895.

Gardner, J. A., Reader Univ., Lect. St. Georges Hospital, Med. School. London S.W. 7 (England), S. Kensington. ' *Physiol. Chem.* o

Gardner, Leon L. (1894). Washington, D. C. (U.S.A.), War Department. *Anat. of Birds.* ⓟ Univ. of Pennsylvania 1920.

Gardner, Max William (1890), Associate Botan., Purdue Univ. Agric. Experim. Station. La Fayette, Ind. (U.S.A.). *Diseases of tomato, potato, apple, soybean, cowpea, cucumber. Virus diseases, bact. diseases, pathol. anat., dissemination.* ⓟ Univ. of Wisconsin 1918.

Gardner, Theodore R. (1899), Ass. Entomol. U.S. Dept. of Agric., Bureau of Entomol. Yokohama (Japan), Am. Consulate. *Parasit. of the Japanese Beetle.* ⓟ Conn. Agric. Coll., Storrs, 1922.

Gardner, Wright A., Head of Dept. of Botan., Alabama Polytechnic Inst. Auburn, Ala. (U.S.A.). *Morphol., Physiol., Soil toxins.*

Garin, Charles-Pierre (1883), Prof. agr. Fac. méd. Lyon (France), 47, cours Morand. *Parasit. Maladies des pays chauds.* ⓟ Lyon 1913.

Garino-Canina, Ettore (1883), Ass. R. Stazione Enol. Sperim. Asti (Italia). *Chim. del vino e fisiol. della fermentazione alcoolica.* ⓟ 1912.

De Garis, Charles Francis (1886), Instr. in Anat., Johns Hopkins Univ. Baltimore, Md. (U.S.A.), Dept. of Anat. *Arm arteries in White and Negro Stocks. Aortic stems in mammals. Localization within somatic motor nucleus of Facial nerve (cat). Physiol. and Morphogenesis of Paramecium.* ⓟ Washington Univ. 1912.

Garlock-Sella, C. Evelyn (1901). Bact. Florence, Wis. (U.S.A.). *Bact. Blood Chem. Pathol. Sectioning.*

Garman, Harrison (1858), Prof., Head of Department, State Entomol. of Kentucky. Lexington, Ky. (U.S.A.). *Ecol., Entomol.* ⓟ Insects of Kentucky.

Garman, Philip (1891), Ass. Entomol. New Haven, Conn. (U.S.A.), Conn. Agr. Exper. Stat. *Research in Entomol.* ⓟ Univ. Illinois 1916.

Garner, Chester Alexander (1897), Associate in Olericulture. Univ. of Illinois, Urba (U.S.A.). *Culture of vegetables on peat or muchsoils. Varieties of vegetables, especially of sweet potatoes (Ipomoea Batatas), sweet corn (Zea mais, saccharata) and their regional adapt ability.* ⓟ *Rare seeds and plants. Seeds of improved varieties of vegetables especially adapted to greenhouseculture and to much soils.*

Garner, Wightman W. (1875), Senior Physiol. in Charge of Tobacco and Plant Nutrition, U.S. Bureau of Plant Industry. Washington, D.C. (U.S.A.), U.S. Department of Agric. *Plant physiol., plant nutrition, tobacco plant. Effect of length of day on plant growth.* ⓟ Johns Hopkins Univ. 1900.

von Garnier, Hubertus (1901). Goslau b. Pitschen, Kr. Kreuzburg. O.-S. (Deutschland). *Angewandte Entomol.*

Garnier, Marcel (1870), Prof., agrégé Physiol, à la Fac. de méd. de Paris, méd. de l'hôpital Lariboisière. Paris 8 (France), 1, rue d'Argenson. *Glandes endocrines: thyroïde et hypophyse. Foie. Reinalbuminurie.* ⓟ Paris 1899.

Garratt, George A. (1898), Ass. Prof. of Forest Products, Yale Univ. New Haven, Conn. (U.S.A.), 205 Prospect St. *Forest Products.*

Garrett, Albert Osbun (1870), Head Department of Biol., East High School. Salt Lake City, Utah (U.S.A.), 791 Ninth Avenue. *Life histories of smuts and rusts; algae, fungi of Utah: mosses of halophytic regions: spermatophyt of the west.* ⓟ Univ. of Kansas, Lawrence, Kan., 1894.ⓟ *Fungi and Spermatophytes.*

Garrey, Walter Eugene (1873), Prof. of Physiol., Med. School, Vanderbilt Univ. Nashville, Tenn. (U.S.A.). *Normal and Pathol. Physiol. of the Heart and Circulations. Neurodynamics. Annual Behavior (the Tropisms).* ⓟ Ph.D. Univ. of Chicago 1900, M.D. Rush Med. Coll. 1909.

Garry, Robert Campbell (1900), Lect. Glasgow (Scotland), Inst. of Physiol., Univ. *Experim. Physiol.*

Garstang, Sylvia Lucy (1898), Ass. London (England), Dept. of Zool. Univ. Coll., Cower Str. *Ascidian Embryol. Genetics.*

Gartkiewicz, Stanisław (1892), Ass. am Physiol. Inst., Univ. Warschau (Polen). *Sinnesphysiol. Respiration des Anodontes à l'état d'activité et repos. Coeur d'Anodonte.*

Garver, Samuel (1883), Associate Agronomist, U. S. Dept. of Agric. Redfield, S.D. (U.S.A.). *Testing of forage crops under field conditions.*

Garvey, Mary Ellen Monica (1896), Instr. of microbiol. and physiol. Amherst, Mass. (U.S.A.), 29 South Prospect Street. *Dairy and food bact.*

Garwood, Edmund Johnston, Yales Goldsmid Prof. of Geol. London (England), Univ. Coll. ⓟ Cambridge.

Gascón y Marin, Joaquín (1870), Prof. d'Anat. déscriptive, Zaragoza (España), Paseo de Sagasta, 39. ⓟ Madrid 1892.

Gąsiorowski, Napol., Priv.Doc. Univ. Lwów (Polska). *Bact.* o

Gaskell, John Ioster (1878), Dir. of the Bonnett Labor. of Pathol. Addenbrook's Hospital Cambridge and Honorary Ass. Physician. Cambridge (England), Uplands Great Shelford. *The Pathol. of Pneumonia. Pathol. of diseases of the Kidney. Previous researches on the origin of the vascular sympathetic nervous systems. Malaria, blackwater fever, cystinaria, the action of X-rays on growing tissues. Cerebrospinal Fever.* ⓟ Cambridge 1908.

Gasow, Heinrich Theodor Johannes (1899), Dr. phil., Vorsteher der Zool. Abt. der Anstalt für Pflanzenschutz u. Samenuntersuchung der Landwirtschaftskammer für die Provinz Westfalen. Münster i. W. (Deutschland), Südstr. 76. *Landwirtschaftliche und forstliche Entomol. (Tipulidenbekämpfung; Bekämpfung des Roggendälchens; Biol. und Bekämpfung von T. viridana, Evetria buoliana, Bupalus piniarius.) Tierische Nützlinge und Vogelschutz.* ⓟ Münster i. W. 1925.

De Gasparis, A., Doc. Univ. Napoli (Italia). *Botan.* o

Gasser, Herbert S. (1888), Prof. of Pharm. Washington Univ. Saint Louis, Mo. (U.S.A.), Kingshighway and Euclid Ave. *Electrophysiol. of Nerve. Physiol. and Pharm. of Muscle.* ⓟ Johns Hopkins Univ. 1915.

Gassner, Gustav (1881), o. Prof. der Botan. an der Techn. Hochsch. Dir. des Botan. Inst. und Gartens Braunschweig (Deutschland), Humboldtstr. 1. *Pflanzenphysiol. und -pathol.* ⓟ Berlin 1905.

Gast, Paul Rupert (1897), Instr. Harvard Univ., Cambridge, Mass. (U.S.A.); *Forest Physiol. Biophysics: Insolation and its relation to silviculture.* ⓟ N.Y. State Coll. of Forestry 1922.

Gaston, Marcel (1894), Prép. Marseille (France), 16, Bd. du Jardin Zool. *Physiol. du Travail.* ⓟ Montpellier 1925.

Gatenby, J. B., Prof., Univ. Trin. Coll. Dublin (Ireland). *Zool.* o

Gater, Bossley Alan Rex (1896), Malaria Research Officer, Federated Malay States. Kuala Lumpur (F.M.S.), Malaria Bureau, Inst. for Med. Research. *Mosquitoes in relation to malaria; insects of medical importance; Malayan economic insects; derris as an insecticide; Artona catoxantha Hamps. and its parasit. and hyperparasit.; introduction of Ptychomyia remota Ald. into Fiji for control of Levuana iridescens, Beth. Baker.* ⓟ *Culicidae of Malaya; Diptera and insects of medical importance; Malayan insects; aquatic organisms.*

Gates, Frank C. (1887), Associate Prof. Botan., Kansas State Agric. Coll. Manhattan, Kan. (U.S.A.). *Plant Ecolog.* ⓟ Univ. of Michigan 1912. ⓟ *Higher plants from Michigan and Kansas, grasses, poisonous plants and weeds.*

Gates, Frederick Lamont (1886), Associate Member, The Rockefeller Inst. for Med. Research. New York City (U.S.A.). *Biophysics: Ultra violet light.* ⓓ Johns Hopkins Univ. 1913.

Gates, Gordon E. (1897), Head, Dept. of Biol., Judson Coll. Rangoon, Burma (Engl. India). *Classification, Morphol. and Physiol. of Oligochaeta.* ⓓ Harvard Univ. 1920. ⓟ *Earthworms of Burma.*

Gates, R. Ruggles (1882), Prof. of Botan., Univ. of London, W.C. 2 (England), King's Coll., Strand. *Cytol. and genetics of Oenothera. Cytol. of Lactuca, Lattoraea. Heredity in man; interracial crosses and inheritance of abnormalities.* ⓓ Chicago 1908. ⓟ Cytol. preparations.

Gates, William H. (1883), Prof.: Zool. and Entomol., Louisiana State Univ. Louisiana (U.S.A.), Univ. Station, Baton Rouge. *Growth. Japanese waltzing mouse.* ⓓ Harvard 1926.

Gatti, A., Prof. Univ. Rosario (Argentina). *Microbiol.* o

Gatunev, Sergius A., Kustos der Geol. Abt.; Geol. und Mineral. Mus. der Akad. der Wiss. Leningrad (U.d.S.S.R.), W.O., Tutschkova Nab. 2. *Tertiäre Palaeont.; speciell Mollusken des Neogens.* o

Gaudineau, Marguerite Anne-Marie (1901), Ing. agr.; Prép. Stat. centr. Phytopath. Paris 14 (France), 11 bis Rue d'Alésia. *Pathol. végétale. Piétin des céréales.*

Gaudron, Julio, Doc. Univ. Lima (Peru). *Botan., Pathol., Zool.* o

Gauthier, Henri Charles (1896), Licencié ès-sc. nat., Ass. à la Fac. des sc. d'Alger (Algérie), Univ. d'Alger. *Faune des eaux continentales de l'Afrique du Nord: Hydrobiol. Systématique: Crustacea: Ostracoda, Cladocera, Phyllopoda.* ⓓ Alger 1920. ⓟ Crustacés des eaux douces.

Gauthier-Lièvre, Lucienne Emilienne (1897), Ass. au Labor. de Botan., Fac. des Sc. Alger (Algérie). *Flore phanérogamique et algol. des eaux douces de l'Afrique du Nord.* ⓓ Univ. d'Alger 1919.

Gautrelet, Jean (1878), Dir. du labor. de Biol. expérim. à l'Ecole des Hautes-Etudes à la Fac. de Méd. de Paris (France). *Sécrétions internes (Surrénale), Réflexe oculo-cardiaque, Sympathique, Choline.*

Gav'alov, Ivan (1900), Entomol. Stawropolian Plant Protection Stat. Stawropol am Caucasus (U.d.S.S.R.), Okrzu. *Coccidae, Economik, Entomol.* ⓟ Coccidae.

van Gaver, Ferdinand (1874), Chef des Travaux prat. Zool. Marseille (France), La Paule, Sainte Marte. *Diptères.* ⓟ Diptères; Mollusques.

Gawrilenko, Anatole (1885), Ass., Priv.Doc. Univ., Labor. de Zool. Leningrad (U.d.S.S.R.), Vass. Ostr. *Morphol. et développement du cerveau intermèd., de l'hypophyse, de l'extremité ant., des cavités prémandibulaires. Poissons. Lézards.*

Gawriloff, Lydie G. (1889), Laborant Section der Pflanzenphysiol. Botan. Garten; Ass. der Botan. im Med. Inst. Leningrad (U.d.S.S.R.), Ordinarnaja, 11/85. *Wasserhaushalt der Pflanze.* ⓓ Leningrad 1923.

Gay, Frederick Parker (1874), Prof. Columbia Univ. New York, N.Y. (U.S.A.), 437 W 59 St. *Pathol., Bact.* o

Gayda, Tullio (1882), Prof. Fisiol., Dir. Ist. di Fisiol. Univ. Pavia (Italia). *Organi sopravviventi, Elettrofisiol., Produzione di calore nell' organismo, Ricambio dei singoli organi.*

Gáyer, Gyula (1883), Kg. Gerichtsrat, Priv.-Doc. an der Szegeder Univ. Szombathely (Ungarn). *Pflanzengeographie von Ungarn, besonders Westungarn. Systematik der Phanerogamen: Aconitum, Rubus, Viola.* ⓓ Kolozsvár 1907.

v. Gaza, Wilhelm (1883), Prof., Priv.Doc. für Chirurgie, Oberarzt der Chir. Univ.-Klinik Göttingen (Deutschland). *Stoffwechsel im Wundgewebe und im Transplantat, Entzündung (spez. Schmerz), Vitalfärbung am Menschen.* ⓓ Greifswald.

Gazzara, Pasquale, Lib. doc. Univ. Messina (Italia). *Anat.* o

Gebbing, Johannes (1874), Dr., Dir. d. zool. Gartens. Leipzig (Deutschland). *Seidenraupenzucht.* o

Gebhardt, Anton (1887), Dr., Ausschußmitglied der ,,Ungar. Entomol. Gesellsch.". Dombovár, Kom. Tolna (Ungarn). *Biol. und Systematik der palearkt. und exot. Buprestiden (Col.).* ⓓ Kolozovár 1910, 1911. ⓟ Buprestiden der Erde. Über 3000 Arten Coleoptera, besonders Ungarns.

Geblen, Hans (1874), Coleopterol. des Zool. Staatsinst. u. Mus. Hamburg 34 (Deutschland), Herlogestraße 10. *Insecta, Coleoptera, Tenebrionidae der Welt.*

Gedoelst, L., Prof. école vét. Cureghem (Belgique), rue Meyerbeer 15. *Parasit.* o

Geduly von Felsötömös, Oliver (1889), Budapest IX (Ungarn), Bakátsgasse 3. *Herpetol. und Amphibiol. (oekol., systematisch und osteol.).* ⓟ Ungarische Reptilien und Amphibien, lebende Tiere wie in Alkohol konserviert. Tropidonotus natrix, tessellatus, Anguis fragilis, Lacerta viridis, agilis, taurica, muralis, Ablepharus pannonicus, Bufo vulgaris, viridis, Rana arvalis, agilis, esculenta, ridibunda, Bombinator igneus, Pelobates fuscus, Hyla arborea, Triton cristatus und vulgaris.

Gee, A. Haldane (1901), Dr. of phil., Fellow of the National Research Council of United States. New Haven, Conn. (U.S.A.), Department of Bact., Sheffield Hall, Yale Univ. *Putrefaction of Haddock Muscle.* ⓓ Univ. of Toronto 1922.

Gee, Nathaniel Gist (1876), Ass. Resident, Dir. of the China Med. Board of the Rockefeller Foundation. Peking (China). *Invertebrate fauna spec. Fresh Water Sponges of China.* ⓓ Wofford Coll. Spartanburg, M.A. 1898, Honorary LL. 1926. ⓟ Chinese Fresh Water Sponges.

Geelmuyden, Hans Christian (1861), Erster Ass. an dem physiol. Inst. der Univ. in Oslo (Norwegen), Bygdöallée 8. *Kohlehydratstoffwechsel. Stoffwechsel bei Diabetes mellitus.* ⓓ Oslo 1895.

Gehring, Alfred (1892), Dr. phil., Leiter d. Landw. Versuchsstation Braunschweig (Deutschland), Kaiser-Wilhelm-Str. 60. *Bodenbact., Kalkfragen des Bodens.* ⓓ Göttingen 1914.

Geier, Ferdinand K., Prof., Inst. f. Volksbildung. Nikolaew (U.d.S.S.R.), I. N. O. *Botan.* o

Geiger, Ernest (1896), Priv.Doc., Ass. am Inst. f. experim. Pathol. d. Univ. Pécs (Ungarn). *Physiol. und Pathol. des Stoffwechsels. Pharm. des Herzens.* ⓓ Graz 1920.

Geiger, Wilhelm (1894), Dr., Technischer Leiter des Inst. der Bekämpfung der Virusschweinepest, Eystrup a. d. Weser (Deutschland). *Virusschweinepest.* ⓓ Hannover 1920.

Geiling, Eugene M. K. (1891), Assoc. Prof. of Pharmacol., Johns Hopkins Univ. Baltimore, Md. (U.S.A.), Johns Hopkins Med. School, Wolfe and Monument Sts. *Nutrition, blood regeneration. Protein hydrolysis, histamine. Glands of internal secretion, insulin, pituitary and adrenaline.* ⓓ Illinois 1915.

Geinitz, Bruno (1889), Priv.Doc. Dr. und Ass. am Zool. Inst., Freiburg i. Br. (Deutschland). *Entwicklungsmechanik, Bienenkunde, angewandte Entomologie.* ⓓ Würzburg 1914.

Geise, Fred W. (1893), Prof. of Olericulture. College Park, Md. (U.S.A.). *Vegetable crop production with particular reference to canning crops.* ⓟ Truck crops.

Geiser, Samuel Wood (1890), Prof. and Head of the Department of Biol., Southern Methodist Univ. Dallas, Tex. (U.S.A.). *Physiol. problems of sex, differential mortality of the two sexes, and factors affecting same; animal orientation to stimuli of a directive sort.* ⓓ Johns Hopkins 1922. ⓟ Land isopods.

Geissler, Karl Paul (1902), Ass. an der zool. Abt. der staatl. Versuchsanstalt für Wein- und Obstbau. Neustadt a. d. Haardt, Rheinpfalz (Deutschland), Karolinenstr. 8. *Anthonomus pomorum u. cinctus. Contarinia pyrivora. Hoplocampa fulvicornis.*
Geist, F., Prof. Univ. Nikolaev (U.d.S.S.R.), Nikol'ska ul. 12/2. *Botan.* ○
Geitler, Lothar (1899), Dr., Ass. am Botan. Inst. der Univ. Wien III (Deutsch-Österreich), Rennweg Nr. 14. *Protist., Cyanophycea.* ⓠ Wien 1922.
v. Gelei, József (1885), Dr., Prof. der Zool. u. Allg. u. Vergl. Anat. Szeged (Ungarn), Tisza Lajos-Körút 6. *Chromosomenlehre, allg. Zytol. Histol. von Cnidarien und Turbellarien. Mikroskopische Anat. der Ciliaten.* ⓠ Kolozsvár 1907. ⓟ Spinnen, Schnecken, Turbellaria, Hydrae, Ciliata d. Umg. v. Szeged. Mikroskopische Praep. von Turbellarien, Hydren und Ciliaten, Knochenschliffe von verschiedenen Vertebraten.
Gellhorn, Ernst (1893), ao. Prof. d. Physiol., Dr. phil. et med. Halle a. d. S (Deutschland),Tiergartenstraße 10. *Physiol. der Sinnesorgane; physico-chem. Physiol. (spez. Osmose, Quellung, Permeabilität, Ionenwirkungen). Physiol. d. glatten Muskulatur. Adrenalin.* ⓠ Phil. Münster 1919, med. Heidelberg 1919.
Gelmer, Oskar F., Dir. Landwirtschaftl. Provinz-Versuchsstation, Selektions-Abtlg. Charkow (U.d.S.S.R.), Postfach 266. *Selektion landw. Pflanzen.* ○
Gemeinhardt, Ernst Berthold Konrad (1883), Polizei-Pharmazierat, Dr. phil. nat., freiw. wissenschaftl. Mitarbeiter an der Preuß. Landesanstalt für Wasser-, Boden- u. Lufthygiene. Berlin-Dahlem (Deutschland). *Systematik, Zytol. und Oekol. der Diatomeen.* ⓠ Jena 1925. ⓟ Diatomeen.
Gemelli, Agostino (1878), Rettore Univ. cattol. del sacro Cuore, Prof. o. di psicol. sperim. Membro del Consiglio sup. della Pubblica Instruzione per il Regno d'Italia. Milano (Italia), Via S. Agnese 4. *Pensiero e dei sentimenti.* ⓠ 1926.
Gemmill, James Fairlie, Prof., Univ. Dundee and Med. School. St. Andrews (Scotland). *Nat. Hist.* ○
Gemünd, Wilhelm (1878), Dr. med., Prof. für Hygiene und Bact. a. d. Techn. Hochsch. Aachen (Deutschland). *Soziale Hygiene, soziale Biol. und Psychobiol.* ⓠ München 1895.
Genaux, Charles (1904), Ass. in Forestry, Washington State Coll. Pullman, Wash. (U.S.A.). *Ecol., theoretical and practical, and Plant Pathol., practical as applied in Silviculture.*
Généreux, Damase, Prof. tit. Ecole Méd. Vét. Montreal (Canada). *Pathol. int., ext. du cheval.* ○
Gengler, Josef (1863). Erlangen (Deutschland), Nürnberger Str. 16 I. *Ornithol.: Faunistik, Systematik, Biol.* ⓠ Erlangen 1894.
Genleys, Paul Victor Laurent (1896), Prép. Insectarium de Mentore, Inst. des Recherches Agr. Tarn (France), Saint-Urcisse. *Insectes Entomophages, larves d'Hyménoptères Braconidae.*
Gennadiew, Alexis (1892), Prosector Anat., Inst. Univ. Kasan (U.d.S.S.R.). *Arterien der Unterextremität.*
Gennaro, Teodoro (1886), Vicedir. R. Sta. bacol. sperimentale, Lib. doc. Zool., Anat. e Fisiol. comp. Univ. Padova (Italia). *Anat., Istol. ed Embriol. degli Insetti: Cocciniglie, Istol. e patol. del Bombyx mori, Flagellosi degli insetti, Rotatoria, Plancton d'acqua dolce, Tardigrada.* ⓠ Padova 1910. ⓟ Cocciniglie. Emitteri eterotteri.
Gennerich, Johannes Wilhelm Theodor Otto (1894), Dr. phil., Oberfischmeister für die Provinz Niederschlesien. Breslau I, Oberpräsidium, Neumarkt 1—8. Breslau 13 (Deutschland), Kaiser-Wilhelm-Str. 115. *Fischereibiol., Abwasserbiol.* ⓠ Berlin 1921.
Gentner, Georg (1877), RR., Dir. Abteilung für Samenkontrolle Landesanstalt für Pflanzenbau u. Pflanzenschutz, Doc. Techn. Hochsch. München 22 (Deutschland), Liebigstr. 25. *Samenkontrolle, Krankheiten des Saatgutes.* ⓠ München 1905.

Gentner, Louis G. (1892), Research Ass. in Entomol., Experim. Station, Michigan State Coll. East Lansing, Mich. (U.S.A.), 225 Baily Street. *Habits and control of insect pests injurious to fruits: aphides, codling moth, pear psylla, the mint flea-beetle. Systematic of North American Halticinae (Coleoptera).* ⓠ Univ. of Wis., Madison 1918. ⓟ North American Halticinae or flea-beetles.
Genung, Elizabeth F. (1883), Associate Prof. of Bact. Smith Coll. Northampton, Mass. (U.S.A.), 2 West St. *Bact.* ⓠ Cornell Univ. 1914.
George, D. C. (1887), Plant Pathol., Arizona Commission of Agric. and Horticulture. Phoenix, Ariz. (U.S.A.), Box 1857. *Plant Disease.* ⓠ State Coll. of Washington 1916.
St. George, R. A. (1894), Ass. Entomol., Division of Forest Insect Investigations, Bureau of Entomol., Washington, D. C. (U.S.A.). *Insects affecting living forest trees, biol., morphol. and taxonomy of immature stages of Coleoptera; Tenebrionidae.* ⓠ George Washington Univ., Washington 1923.
George, Wesley Critz (1888), Prof. of Histol. and Embryol., Univ. of North Carolina. Chapel Hill, N.C. (U.S.A.), Box 410. *Comparative histol. and development of the vascular system.* ⓠ Univ. of North Carolina, Chapel Hill 1918.
Georgévitch, Pierre (1874), o. Prof. der Botan. an der Univ. zu Belgrad (S.H.S.) Studenička 54. *Cytol., Pflanzenpathol. und Bact.* ⓠ Bonn a. Rh. 1906. ⓟ Einheimische Pflanzen.
Georgieff, Thoma (1883), Ass. Tzar Assen N 38 (Bulgarien). *Floristique: Flore de la Bulgarie.* ⓠ Sofia 1911.
Georgiewskaja-Petrunkina, Anna Michailowna (1892), Biochem. Abt. Inst. für experim. Med. Leningrad (U.d.S.S.R.), Lepuchinskaja 12. *Lipoide; Bedingungen, die die Verbindung der Eiweißstoffe mit Alkaloiden, tierischen Basen, Säuren begünstigen.*
Georgopoulos, M., Prof., agrégé à la Fac. de Méd. Athènes (Grèce), Rue Charilaon Trikonpi 1. *Pathol.* ○
Gepp, Antony (1862), Ass. Keeper, Department of Botan., British Mus. (Natural History). London, S.W. 7 (England), Cromwell Road. *Systematic Botan.: Cryptogams, especially Pteridophyta, Bryophyta, Algae.* ⓠ Cambridge Univ. 1889.
Geppert, Julius (1856), Prof. Univ., G.M.R. Gießen (Deutschland), Liebigstr. 34. *Pharmacol.* ○
Geraldino, Brites (1882), Prof. de la Fac. de Méd. Univ. Coïmbre (Portugal). *Histol. normale et pathol. Système nerveux des Aranidae et des Pulmonatae.*
Gérard, Ernest, Prof. Univ. Lille (France). *Pharmacol.* ○
Gerard, M. G., Prof. Fac. Méd. Lille (France), 48, Rue Nocolas Leblanc. *Anat.* ○
Gerard, Pol (1886), Prof. d'Histol. à l'Univ. de Bruxelles (Fac. de Méd.). Uccle (Belgique), 67, rue Joseph Stallaert. *Spermato- et ovogénèse. Glandes germinatives. Placentation.* ⓠ Bruxelles 1909.
Gerassimow, Dmitry (1895), Botan. Wissensch.-Experim. Torfinst. Moskau 1 (U.d.S.S.R.), Kudrinskaja-Ssadowaja, 7, qu. 11. *Stratigraphie der Torfmoore, Ecol. der Moose, Systematik der Sphagna.* ⓟ Sphagna.
Gerber, C., Prof. Univ. Toulouse (France). *Botan.* ○
Gerber, Herman (1891), Bact., Research Labor., Health Dept. New York City (U.S.A.), Ft. E. 16. *St. Anthrax, plague, tetanus, botulinus, rabies.*
Gercenberg, Elena Jakovlevna, Doc. I. Univ. Moskau (U.d.S.S.R.). *Pathol. Anat.* ○
Gerhardt, Ulrich (1875), Dr. med. et phil., o.ö. Prof. an der Univ. Halle a. d. Saale (Deutschland), Wilhelmstr. 27/28. *Morphol. und Biol. der Araneen, Sexualbiol., Mammalia.* ⓠ Med. Berlin 1899, phil. Breslau 1903. ⓟ Spinnen (leb.).
Gericke, William F., Ass. Prof. of Soil Chem., Univ. of Calif. Berkeley, Cal. (U.S.A.). *Plant Nutrition.* ○

Gerlach, Franz (1891), Dr., Dir. der staatl. Tierimpfstoffgewinnungs-Anstalt und der Station für Tierseuchendiagnostik. Mödling bei Wien (Deutsch-Österreich), Friedrichstr. *Bact., Serol. und pathol. Anat.* ③ Wien 1915. ⑫ Photographien und Diapositive von Infektionskrankheiten (namentlich solche von Tieren) und an dem Gesamtgebiete der Bact. und Serol., mikroskopische und pathol.-anat. Praep.

Gerlach, Max (1861), Prof. Dr. Geh.Reg.R., Dir. der Preuß. Versuchs- u. Forschungsanstalt für Getreideverarbeitung u. Futterveredelung. Berlin N (Deutschland), Seestr. 11. *Pflanzenernährung, Anbau, Konservierung u. Verbesserung der Futterpflanzen, Fütterungsversuche.* ③ Halle a. d. Saale 1887.

Gerlach, Werner (1891), Dr. med., Prosektor am Krankenhaus Barmbeck-Hamburg (Deutschland), Hofweg 53. *Entzündungslehre. Beziehungen zwischen Infektion und Immunität.* ③ Tübingen 1917.

Germain, Louis Alfred Pierre, Ass. au Labor. de Biol. Marine. Paris (France), 55, rue de Buffon. *Zoogéographie. Mollusques terrestres et fluviatiles. Chétognathes.* ③ Paris.

Gerónimo Barroso, Manuel (1887), Prof. Salamanca (España). *Bryozoaires.* ③ Madrid.

Geronimus, Efim Solom., Doc. II. Univ. Moskau (U.d.S.S.R.). *Bact.* ○

Gerould, John H. (1868), Prof. of Zool. Hanover, N.H. (U.S.A.), Dartmouth Coll. *Genetics and general physiol. of insects, especially Lepidoptera. Physiol. of coloration in Colias and other butterflies. Periodic reversal of circulation in Lepidoptera. Physiol of sex in Lepidoptera. Anat. and histol. of Holothurians and Sipunculids, the development of Phascolosoma, distribution and taxonomy of Sipunculids, hybridization of species in Colias, mimicry.* ③ Harvard Univ. 1895.

Gerretsen, Frederik Charles (1889), Dir. Microbiol. Abt. d. Rijkslandb. Proefstation Groningen, Priv.-Doc. Microbiol. Rijks-Univ. Groningen (Holland), Zuidzijde Eemskanal 1. *Microbiol. Prozesse in der Agric.* ③ Delft 1921.

Gerry, Eloise (1885), Microscopist in Forest Products, Lect. Univ. of Wisconsin U.S.A. Forest Service. Madison, Wis. (U.S.A.), Forest Products Labor. *Physiol. Anat. and Microchem. of Forest Products.* ③ Univ. of Wisconsin 1921. ⑫ Wood specimens.

Gersbach, Alfons (1892), Dr. med., Kreismedizinalrat. Erkelenz, Rheinland (Deutschland). *Kropfentstehung und -bekämpfung. Wasser- und Bäderbact. Bakteriophage. Anthropometrie.* ③ Frankfurt a. Main 1920.

Gerschunij, Boris (1867), Prosector Anat. Staatsinst. med. Wissensch. Leningrad (U.d.S.S.R.), Tschaikowskistr. 60. *Normale Anat. des Menschen.* ③ Kiew 1897.

Gershenfeld, Louis (1895), Dir. of the Gershenfeld Labor., Prof. of Bact. and Hygiene at the Philadelphia Coll. of Pharm. and Sc. Philadelphia, Pa. (U.S.A.), 1831 Chestnut St. *Bact., Bio-Chem.* ③ Univ. of Pennsylvania 1915.

Gershoy, Alexander (1896), Instr. in Botan.,Univ. of Vermont, Ass. Botan., Agric. Expt. Sta., Vermont. Burlington, Vt. (U.S.A.), 82 Hungerford St. *Genetics and Cytol. of Viola. Phytogeographie.*

Gerth, Heinrich (1884), Prof., Conservator Rijks Geolog. Mineralog. Mus. Leiden (Holland), Wasstr. No. 17. *Palaeont. der Invertebrata besonders Coelenterata.* ③ Bonn 1908.

Gertz, Otto Daniel (1878), Priv.Doc. der Botan. Univ., Botan. Inst., Bibliothekar. Lund (Sverige). *Pflanzenanat., Pflanzenphysiol., Biochem., Cecidol., Geschichte der Botan., Naturschutz.* ③ Lund 1906.

Gerver, A. V., Pädagog. „Herzen"-Inst. Leningrad (U.d.S.S.R.), Moika 48-50-52. *Anat., Physiol.* ○

Gery, Louis (1883), Prof. d'Anat. Pathol. à la Fac. de Méd. d'Asuncion (Paraguy). *Anat. Pathol.* ③ Strasbourg 1919.

Gesell, Robert (1886), Prof. of Physiol., Univ. of Michigan. Ann Arbor, Mich. (U.S.A.). *Electrical phenomena of living tissues; Salivary secretion; Cardiodynamics; Hemorrhage and shock; Volume-flow of blood; Regulation of circulation; Regulation of respiration.* ③ Washington Univ. Med. School 1914.

Gessner, Albert (1888), Dr. phil., Regierungsbotan. am bad. Weinbauinst., Freiburg i. Baden (Deutschland), Goethestr. 9. *Pflanzenschutz, Schädlingsbekämpfung im Weinbau.* ③ Freiburg i. Baden 1920.

Gessner, Otto (1895), Dr. med., Ass. Marburg, Lahn (Deutschland), Wolffstr. 5. *Pharm. (Amphibiengifte, Hormonwirkungen).* ③ Marburg 1920.

Gettler, Alexander O. (1883), Associate Prof. of Chem. at Univ. and Bellevue Hospital Med. Coll., Pathol. Chem. to Bellevue and Allied Hospitals, Toxicologist to New York City (U.S.A.), 338 East 26. St. *Toxicol. and Physiol. Chem.* ③ Columbia 1912.

Getty, Robert E. (1891), Associate Agronomist, U.S. Dept. of Agric. Hays, Kan. (U.S.A.). *Forage Crops for dry lands; sorghum.* ⑫ Seed of leading sorghum varieties.

Gex, Marie-Madeleine (1902), Ass. à l'Inst. de Physique Biol. de la Fac. de Méd. de Strasbourg (France). *Physico-chim. des solutions. Spectres d'absorption ultraviolette.* ③ Strasbourg 1923.

Geyr v. Schweppenburg, Hans Freiherr (1884), a.o. Prof. Forstlich. Hochsch. Hann. Münden (Deutschland). *Ornithol.: Vogelzug. Systematik.* ③ Hann. Münden 1923.

Gheorghiu, Ioan (1889), Chéf des trav., Labor. de bact., Fac. Med. Jassy (România). ③ Jassy 1917.

Gherardini, Pietro, Prof., Ist. sup. d. Med. vet. Bologna (Italia). *Anat. patol.* ○

Ghigi, Alessandro, Prof. o. Univ. Bologna (Italia). *Zool.* ○

Ghirardi, Giordano Emilio (1898), Ass., Ist. di Fisiol., Univ. di Milano (Italia), Città degli studi. *Fisiol. dei nervi e dei muscoli.* ③ Milano 1925.

Ghon, Anton, Prof. d. pathol. Anat. Praha II (C.S.R.), u. nemocnice 4, Pathol.-anat. Inst. der Deutschen Univ. ○

Giacomini, Ercole (1864), Prof. o. Anat. comparata Univ. Bologna (Italia), Ist. di Anat. comparata. *Glandole salivari degli Uccelli; ovidotto dei Sauropsidi; annessi embrionali dei Vertebrati, ovario dei Selaci; espansioni nervose nelle estremità delle fibre muscolari striate dei Pesci e degli Anfibii; fusi neuromuscolari dei Sauropsidi; nervo terminale dei Teleostei; capsule surrenali degli Anfibii, dei Petromizonti, dei Teleostei, dei Ganoidi. Azione della gl. tiroidea e dello jodio sulla metamorfosi degli Anfibii e sulla rigenerazione; azione della tiroide sullo sviluppo e sul piumaggio degli Uccelli.* ③ Siena 1890. ⑫ Serie di embrioni e di larve di Acipenser.

Giacosa, Piero (1853), Prof. o. Farmacol. sperim. R. Univ. Torino (Italia), Via Pallamaglio 31. *Farmacol. e chimica fisiol. Sostanze aromatiche, ossido di Carbonio.* ③ Torino, LLD. Cambridge h. c.

Giaja, J. (1884), Prof. de physiol. a la fac. de philosophie de l'Univ. de Belgrade (S.H.S.). *Métabolisme énergétique comparé. Ferments. Chim. physiol.* ③ Paris, Sorbonne, 1909.

Gianferrari, Luisa, Prof. Univ. Milano (Italia). *Biol. gen.* ○

Giang, Fredericka, Ass. Yenching Univ., Zool. Labor. Peking (China). *Zool.* ○

Giannelli, Luigi, Prof. o. Univ. Ferrara (Italia). *Anat. umana norm.* ○

Gianturco, V., Prof. Univ. Napoli (Italia). *Anat.* ○

Giardina, A., Prof. Univ. Roma (Italia). *Anat. compar.* ○

Gibbs, Charles Shelby (1888), Research in Animal and Silkworm Diseases. Nanking, Kiangsu (China). *Control of Rinderpest in Cattle; Pebrine and Flacherie in Silkworms. Infectious Abortion in Cattle.* ③ Yale

Univ. 1921. ℗ Cultures of Bacteria Isolated in the Orient.

Gibbs, Owen Stanley (1898), Prof. of Pharm., Dalhousie Univ. Halifax, Nova Scotia (Canada). *Kidney function (with especial reference to Uric Acid) excretion in the fowl. Quinine distribution in blood.* ⓓ Edinburgh 1921.

Gibbs, Ronald Darnley (1904), Demonstrator in Botan., McGill Univ. Montreal (Canada). *Plant Physiol.; Light.* ⓓ McGill Univ., Montreal 1926.

Gibson, Archie L. (1892), Junior Entomol. Coeur D'Alene, Id. (U.S.A.), Box 386. *Epidemics of bark beetles.* ⓓ Syracuse 1920.

Giddings, N. J., Prof. of Plant Pathol., West Virginia Univ. Morgantown, W.Va. (U.S.A.). o

Gidley, James Williams (1866), Ass. Curator of Fossil Mammals. Washington, D.C. (U.S.A.), U. S. National Mus. *Vertebrate paleont.; mammalian branch.* ⓓ Princeton Univ. 1898. ℗ Models of the evolution of the horse, of the Eocene amblipod, Uintatherium.

Gidley, William Francis (1887), Dean of the Coll. of Pharm., Univ. of Texas, Galveston, Tex. (U.S.A.). *Chem. of Plant constituents.* ⓓ Ann Arbor, Mich., Univ. of Michigan 1908.

Gidon, F., Prof. Univ. Caen (France). *Histol.* o

Giemsa, Gustav Berthold Carl (1867), Prof., Dir. chem. Abt. Inst. für Schiffs- und Tropenkrankheiten. Hamburg 4 (Deutschland), Bernhardstr. 74. *Chemotherapie, Chem., Färbemethoden für Blut- und Gewebeparasit.* ⓓ Hamburg 1925.

von Gierke, Edgar (1877), Prosektor am Städtischen Krankenhaus, a.o. Prof. für Bact. an der Techn. Hochsch. Karlsruhe i. B. (Deutschland), Maxaustraße 11. *Allgemeine Pathol.; Histol. Morphol. des Zellstoffwechsels: Glykogenablagerung, Verfettung, intercelluläre Fermente, Innersekretorische Drüsen.* ⓓ Heidelberg 1902.

Giersberg, Hermann (1890), Dr. phil., Priv.Doc. am Zool. Inst. Breslau (Deutschland), Zool. Inst., Sternstr. 21. *Plasmabau, Entwicklungsmechanik, Sinnes- und Hirnphysiol.*

Gies, William John (1872). New York, N.Y. (U.S.A.), 609 W, 115 St. *Biol. chem., pathol.* o

Giese, Fritz (1890), Priv.Doc. a. d. Technischen Hochsch. Stuttgart. Stuttgart-Cannstatt (Deutschland), Paulinenstr. 17. *Biol. der Berufsarbeit und des Sports, Arbeitswissenschaft, Körperkultur.* ⓓ Leipzig 1914. ℗ Schautafeln, Photos und Diapositive für Biol. der industriellen Arbeit und des Sports.

Giesen, John (1891), Prof. of Biol. and Chairman of the Department. Worcester, Mass. (U.S.A.), Holy Cross Coll. *Parasit. Zool.: Evolution.* ⓓ Univ. of Dallas, Tex. 1917.

Giesenhagen, Karl (1860), Geh. Reg.R., Dr. phil., o. Univ.-Prof., Vorstand des tierärztlichen botan. Inst. der Univ. und Vorstand des botan. Inst. der Technischen Hochsch. München (Deutschland), Schackstr. 2II. *Morphol. und Entwicklungsgeschichte der Kryptogamen. Angewandte Botan.* ⓓ Marburg 1889.

Gleysztor, Marjan (1901), Inst. de Zool. de l'Univ. Varsovie (Pologne), Krakowskie Przedmieście 26. *Systématique et étol. des Turbellaria-Rhabdocoelida et Lepidoptera (,,Macrolépidoptères").*

van Giffen, Albert Egges (1884), Dir., Biol.-archaeol. Inst., Conserv. der Zool., Reichsuniv. Groningen (Holland), Pertstraat 6. *Tiere u. Pflanzen prae- u. frühhistorischer Zeit. Wüstenfauna.*

Giglio-Tos, Ermanno, Prof. o. Univ. Cagliari, Sardinia (Italia). *Zool., Anat. e Fisiol. compar.* o

Gigon, Alfred (1888), a.o. Prof. Univ. Basel (Schweiz), Feierabendstr. 15. *Physiol. und pathol. Stoffwechsel, Kohlenhydratstoffwechsel, Ernährung. Constitutionslehre (Behaarung, Riesenwuchs u. Zwergwuchs).* ⓓ Basel 1906.

Gilbert, Alfred Holley, Ass. Prof. Botan., Univ. of Vermont, Plant Pathol., Vermont Dept. of Agric. Burlington, Vt. (U.S.A.), 191 Loomis Street. o

Gilbert, Basil E. (1892). Kingston, R.I. (U.S.A.), R.I. Agric. Experim. Station. *Agric. and Bio-Chem.* ⓓ Univ. of Chicago 1925.

Gilbert, Edward M. (1875), Prof. of Botan. and Plant Pathol., Univ. of Wisconsin. Madison, Wis. (U.S.A.). *Mycol., cytol. of fungi, pathol.* ⓓ Wisconsin 1914.

Gilbert, Ruth (1883), Bact. in Charge of Diagnostic Labor. Business: Albany, N.Y. (U.S.A.), New Scotland Ave. Home: Albany, N.Y., 116 North Allen St. *Bact.* ⓓ Albany Med. Coll. 1923.

Gilbert, William W., Plant Pathol., U.S. Dept. of Agric. Washington, D.C. (U.S.A.), Bureau of Plant Industry. *Diseases of cotton and truck crops, cucurbit diseases.* o

Gilchrist, Grace Gertrude (1892), Lect. in Botan. Univ. of Bristol (England). *Bark Canker Disease of Apple Trees caused by Myxosporium corticolum. Footrot caused by Axochyta.* ⓓ Wisconsin 1924.

Gilchrist, J. D. F., Prof. Univ. Cape Town (South Africa). *Zool. comp. Anat.* o

Gil Collado, Juan, Conserv. de Entomol. del Mus. Nac. de Ciencias Nat. Madrid (España), Travesía de Fúcar, 19. o

Gildemeister, Eugen (1878), Prof. Dr. med., Ober-reg.R. und Mitglied des Reichsgesundheitsamts. Berlin-Lichterfelde W (Deutschland), Viktoriastr. 7. *Bact. und Serol. (Bakterienvariabilität).* ⓓ Breslau 1902. ℗ Bakterienkulturen.

Gildemeister, Martin (1876), Dr. med., Prof. der Physiol. Leipzig (Deutschland), Liebigstr. 16. *Animalische Physiol.* ⓓ Berlin 1898.

Gilding, Henry Percy (1895), Joint Senior Demonstrator Physiol., Physiol. Labor., St. Bartholomew's Hospital, Med. Coll. London E.C. 1 (England). *Physiol., Histol., Biochem.*

Gil Fagoaga, Lucio (1896), Prof. Fac. Phil. Univ. Madrid (España), Rios Rosas, 8. *Physiol. psychol.* ⓓ Madrid 1916.

Gilg, Ernst Friedrich (1867), Prof. Botan., Pharmakognosie Univ.; Kustos, Botan. Mus. Berlin-Dahlem (Deutschland). *Systematische Botan. Pharmakognosie. Syllabus der Pflanzenfamilien.* ⓓ Berlin 1891.

Gillis, Paul, Prof. Fac. Méd. Montpellier (France), 5, Rue de l'Observance. *Anat. Embryol. Region inguino-abdom. et canal abdom. Anat. des centres nerveux sympath.* o

Giljarowskij, Iwan P., Prof. Landw. Inst. Samara (U.d.S.S.R.), Rabotschaja ul. 3. *Botan. Microbiol. Mykol.* o

Gilkey, Helen Margaret (1886), Ass. Prof. of Botan. and Curator of Herbarium, O. A. C. Corvallis, Ore. (U.S.A.), 136 N. 30. St. *Hypogaeous Ascomycetes; Weeds. Tuberales of North America.* ⓓ Univ. of California 1915. ℗ Pressed plants from Oregon.

Gillespie, Thomas Haining (1876), Dir. Edinburgh, W. (Scotland), Zool. Park.

Gillet, Joseph Jean Edouard (1865), Prof. honoraire de l'Athénée Royal de Bruxelles (Belgique), 106, rue St. Bernard. *Entomol. Coléoptères: Lamellicornes coprophages.* ⓓ Liège 1888. ℗ Coléoptères, spec. Lamellicornes non coprophages.

Gillette, Clarance P. (1859), Head, Dept. of Zool. and Entomol., Dir. of Agric. Experim. Station, Colorado Agric. Coll., State Entomol. of Colorado. Ft. Collins, Colo. (U.S.A.). *Aphididae.* ⓓ State Univ., Urbana, Ill. 1917. ℗ Aphididae.

Gillis, Merl Conrad (1893), Associate in Olericulture. Urbana, Ill. (U.S.A.), Univ. of Illinois. *Canning crops: sweet corn and tomatoes.* ⓓ Cornell Univ. 1924.

Gilman, Herbert L. (1895), Ass. Prof. of Research in the Diseases of Breeding Cattle. Ithaca, N.Y. (U.S.A.), New York State Veterinary Coll. *Anat. pathol. and bact. of the genital organs of cattle.* ⓓ Vet. med. Cornell Univ. 1917, phil. 1922. ℗ Bact. cultures.

Gilman, Joseph C. (1890), Associate Prof. of Botan. Ames, Ia. (U.S.A.), Iowa State Coll. *Plant pathol., mycol.* ⓓ Washington Univ. 1915.

Gilmore, Charles Whitney (1874), Curator, Vertebrate Paleont., U.S. National Mus. Washington, D.C. (U.S.A.). *Vertebrate paleont., fossil Reptilia, Dinosauria.* ⒹUniv. of Wyoming, Laramie 1901. ⓅModel restoration: Stegosaurus, Triceratops, Diplodocus, Trachodon, Brachyceratops, Dimetrodon, Ceratosaurus.

Gilmore, John Washington (1872), Prof. of Agronomy, Coll. of Agric., Univ. of California. Berkeley, Cal. (U.S.A.). *Field Crop Production and Soil Management.* Ⓓ Cornell Univ. 1905.

Gilson, Gustave (1859), Prof. Univ. Louvain (Belgique). 95 rue de Namur. *Zool. génér. et marine Cytol.* Ⓓ Lonvain 1884. Ⓟ Crustacea.

Giltay, Louis Pierre Oscar (1903), Dr. en Sc. naturelles, Aide-naturaliste au Mus. Royal d'Histoire naturelle de Belgique. Bruxelles (Belgique), 31, rue Vautier. *Arachnides (Morphologie et Phylogénie). Poissons (Morphol. — Ethol.).* Ⓓ Bruxelles 1925 Ⓟ Animaux: Arachnides, Poissons. Plantes: Lichens.

Giltaz, L., Nat. Mus. R. d'Hist. nat. Bruxelles (Belgique), 31, rue Vautier. *Vertébrés.* ○

Giltner, Ward (1882), Prof. Bact. and Hygien., Bact., Dean of Vet. Med. East Lansing, Mich. (U.S.A.), Michigan State Coll. *Infectious diseases of animals, infectious abortion, poultry diseases and parasit. diseases; dairy bact., influence of diseases of the cow on the milk, ice cream bact.; soil bact., nitrogen fixing bact., decomposition of peat; flax retting; vinegar fermentation; sanitary bact., rural sewage disposal.* Ⓟ Nitrogen fixing bact., vinegar yeasts and bact.; Bact. abortus, Salmonella pullora.

Giménez de Aguilar y Cano, Juan, Prof., Catedrático. Casa Blanca, Cuenca (España). *Lepidópteros.* ○

Ginsburg, Joseph M. (1894), Biochem. in Entomol. New Brunswick, N.J. (U.S.A.), Agric. Experim. Station. *Insecticides and Insect Physiol.*

Ginzberger, August (1873), PrivDoc., Vizedir. emer. des Botan. Gartens und Inst. der Univ. Wien XIII/5 (Deutsch-Österreich), Lorenz-Weiss-Gasse 9. *Pflanzengeographie der ostadriatischen Küstengebiete; Vegetation der Stadt (Straßen, Mauern, Dächer). Latyrus. Moore. Centaurea.* Ⓓ Wien 1896.

Di Giovanni, Ant. (1901), Ass. Agr. Staz. Chim.-Agr. Torino (Italia), Via Ormea 47.

Girard, Georges (1888), Dir. de l'Inst. Pasteur, Prof. d'hygiene à l'école de méd. Tananarive (Madagascar). *Peste, Rage, Hygiene soc.*

Girard, Henri Valentin René (1894), Prép. Bordeaux (France), 21, Rue Adrien Bayssellance. *Zool., Parasit., Microbiol.* Ⓓ Bordeaux 1920.

Girard, Pierre (1880), Dir. du labor. de Chimie Physique biol. de la Sorbonne. Paris (France), 87 Boulevard Saint Michel. *Chimie physique biol. Facteurs électriques de l'Osmose. La perméabilité sélective des parois vivantes et des parois inertes polarisées aux différents ions. Conséquences chimiques que comporte cette perméabilité sélective; modifications colloidales du plasma sanguin par l'injection intraveineuse de certains colorants fluorescents. Suppression ou l'atténuation de certains phénomènes de ,,choc''.*

Girmounsky, Alexandre (1887), Prof. Académie des Mines, Géol. du Comité géol. russe. Moscou (U.d.S.S.R.), Kaluschskaja 14, N. 70. *Paléont.: Cephalopoda et Lamellibranchiata.* Ⓓ Moscou 1913.

Girndt, Otto (1895),Univ.Ass., Pharmacol. Inst. der Univ. Frankfurt a. Main (Deutschland), Weigertstr. 3. *Physiol. und Pharm. des Zentralnervensystems.* Ⓓ Tübingen 1921.

Girometta, Umberto, Prof. Dr., Dir. d. Naturw. Mus. Split (S.H.S.). *Höhlenfauna.* ○

Girzitska, Zoe (1895), Aspirant of the Botan. Garden. Kiew (U.d.S.S.R.), Nesterovskaja str. No. 4/1. *Mycol.* Ⓟ Fungi.

Gislén, Torsten E. (1893), Doc. Uppsala (Sverige), Zool. Inst. *Zool. Echinoderms: Taxonomy,*

Morphol., Biol., Palaeont. Insectivorous plants. Ⓓ 1924.

Gistl, Rudolf (1891) Dr., Univ. Ass. München (Deutschland), Gabelsbergerstr. 51. *Kryptogamen.* Ⓓ München 1914.

Lo Giudice, Pietro (1881), Dir. Stabilimente Ittiogenico. Brescia (Italia). *Biol. dei Protozoi e dei Pesci. Piscicultura.* Ⓓ Roma 1924.

Giudiceandrea, Vincenzo, Prof. Univ. Roma (Italia). *Patol.* ○

Gizelt, Adolf, Prof. Tierärztl. Akad. Lwów (Polen), ul. Kochanowskiego. *Pharm.* ○

Gjaja, Ivan, Dr., o. Prof. d. vergl. Physiol., Dir. d. Physiol. Seminars Novi Univ. Beograd (S.H.S.). *Vergl. Physiol., insbes. Fermente.* ○

Gjorgjević, Zivojin, Dr., Prof. Univ. d. Zool. u. vergl. Anat., Dir. d. Zool. Inst. Univ. Beograd (S.H.S.). *Cytol., Sporozoa, Entomol.* ○

Gjurašin, Stjepan (1867), ao. Univ.Prof. d. Botan. Zagreb (S.H.S.), Pantovčak 80. *Systematische Botan., Mykol., Cytol.* Ⓓ Zagreb 1895.

Gjurić, Peter Mihajlov (1891), o. Univ.Prof., Dir. physiol. Inst. Univ., vet.-med. Fac. Zagreb (S.H.S.), Savska cesta 14a. *Physiol. d. Herzens, Kaltblütler.* Ⓓ Wien 1910.

Gladwin, Fred E. (1877), Associate in Research (Horticulture). Fredonia, N.Y. (U.S.A.), Vinegard Labor. of N.Y., Agric. Experim. Station. *Grape. Soil studies, the influence of pruning and training, Cold resistance, testing of new varieties, insect and disease control.*

Glaser, Otto (1880), Stone Prof. of Biol. Winter: Amherst, Mass. (U.S.A.), Amherst Coll. Summer: Woods Hole, Mass. (U.S.A.), Marine Biol. Labor. *Physiol. of development, growth of oysters. Development of Fasciolaria; Rats: Energy of development. Mechanism of folding in Vertebrate nervous System. Chem. of Eggs, secretion and changes in egg at instant of fertilization. Temperature and vital processes. Movement of paramaecium.*

Glasewald, Konrad (1889), Dr. phil. Berlin-Charlottenburg 9 (Deutschland), Lindenallee 12. *Ornithol. u. Naturdenkmalpflege.* Ⓓ Halle a.d. S. 1923.

Glasgow, Hugh (1885), Ph.D., Associate in Research (Entomol.). New York Agric. Experim. Station. Geneva, N.Y. (U.S.A.). *Applied Entomol.* Ⓓ Univ. of Illinois 1913.

Gleason, Henry Allan (1882), Curator, New York Botan. Garden. New York City (U.S.A.), Bronx Park. *Phytogeography; taxonomy of flowering plants, especially of northern South America.* Ⓓ Columbia Univ. 1906.

Glegg, William Edwin (1878). London E. 1 (England), The House, Albion Brewery, Whitechapel Road. *Ornithol.*

Gleisberg, Walther (1891), Dr. phil., Leiter des Pflanzenphysiol. und des Inst. für Baumschulkrankheiten Ketzin a.d. H. (Deutschland). *Unterlagenzüchtung der Obstgehölze und der Rebe. Stecklingsbewurzelung. Stimulationsfragen. Obst-, Gemüse- und Gehölzkrankheiten.* Ⓓ Breslau 1921.

Gley, Eugène (1857), Prof. au Coll. de France (chaire de Biol. générale). Paris VI (France), rue Monsieur le Prince 16. *Sécrétions internes.*

Glinka-Tschernorutzky, Helene (1887), Ass. Med. Inst. Leningrad (U.d.S.S.R.), Malaja Posadskaja 15, W. 112. *Biol. Chem. u. Mikrobiol.: Fermentforschung.* Ⓓ Moskau 1909.

Glišić, Ljubiša (1888), Dr., Doc. für Botan. an der Univ. Belgrad (S.H.S.), Botanischer Garten, Takovska 41. *Embryol., Cytol. der Angiospermen.* Ⓓ Belgrad 1924.

Globus, Joseph H. (1885), Associate Prof. of Neuro-Pathol. and Neuro-Anat., New York Univ. Med. School. New York, N.Y. (U.S.A.), 1125 Park Ave. *Organic neuro-pathol., in close relation to neurol.* Ⓓ Cornell Univ. 1917.

Glod-Werschuk, Valentine (1894), Ass. Pathol.-Anat. Inst. Weißrussisch. Univ. Minsk I (U.d.S.S.R.), Gospitalny per. N. 8, Kb. 1.

Gloerum, Dir. Statens forsöksstation på Möystad. Hjellum, pr. Hamar (Norge). o

Gloyer, Walter O. (1886), Associate in Research. Plant Pathol. Geneva, N.Y. (U.S.A.), N.Y. Agric. Experim. Station. *Diseases of fruits, asters, beans, and soil sterilization by means of fungicides.* ① Wisconsin 1910.

Glucksmann, Sigismond (1870), Dr. med., o. Prof. der Hygiene u. Bact. an der Univ., Dir. des Inst. für Hygiene u. Bact. in Fribourg (Schweiz). *Bact. (diagnostisch).* ① Zürich 1896.

Glück, Christian Maximilian Hugo (1868), Prof. für systematische Botan. und Pharmakognosie. Heidelberg (Deutschland), Lutherstr. 63. *Morphol., Biol., Systematik. Wasser- u. Sumpfgewächse.* ① München 1895. ② Wasser- u. Sumpfgewächse aus allen Gegenden; lebendes Material u. Alkohol- u. Herbarmaterial. Europäische Herbarpflanzen resp. lebendes Material (Hydrophyten).

Glynn, E. E., Prof., Univ. Fac. Med. Liverpool (England). *Pathol.* o

Glynne, Mary D. (1895), Ass. Mycol., Department of Mycol., Univ. of Wales. Harpenden, Herts. (England), Rothamsted Experim. Station. *Disease of Potatoes.* ① Bangor 1917.

Gminder, Adolf (1887), Dr., Reg.R., stellvertretender Vorstand des württemb. Tierärztlichen Landesuntersuchungsamts in Stuttgart (Deutschland), Azenbergstr. 14a. *Bact. Serol. Herstellung von Impfstoffen.* ① Stuttgart 1910.

Gnedovskij, A. S., Doc. Mittelasiat. Univ. Taschkent (U.d.S.S.R.). *Pharmacol.* o

Gôda, Tokusuke (1901), Zool. Inst., Sc. fac., Imp. Univ. of Tokyo (Japan). *Histogenesis of amphibia. Osteol. of mammalia.* ① Tokyo 1925.

Goddard, E. J., Prof. Univ. Brisbane, Queensland (Australia). *Biol.* o

Godlewski, Emile (senior), Dr., Prof. Putawy (Pologne), Inst. d'Agric. de l'Etat. *Botan.* o

Godlewski, Emil (1875), o. Prof. der Embryol. und allg. Biol. Univ. Kraków (Polen), Biol.-embryol. Inst., St. Jana 20. *Experim. Embryol.* ① 1899. ② Axolotl.

Godneff, Tichon (1893), Prof. für Pflanzenphysiol. und Agrochem. Iwanowo-Wosnessensk (U.d.S.S.R.). *Chem. der Pflanzenpigmente.* ① Moskau 1917.

Godwin, Clarence Hurdman, Plant Disease Investigator, Dominion Government Plant Pathol. Labor. Fredericton, N.B. (Canada). o

Godwin, Harry (1901), Univ. Demonstrator Botan. Cambridge (England), Botan. School. *Plant physiol. Ecol.* ① Cambridge 1926.

Goebel, Carl von (1855), Univ.Prof., Dir. Botan. Inst. u. Garten. München (Deutschland), Menzinger Str. 15. *System d. Pflanzen. Morphol. Organographie. Experim.Morphol. Entfaltungsbewegungen.* o

Goebel, François (1896), Ass. à l'Inst. de Chim. physiol. de l'Univ. à Varsovie (Pologne), rue Polna N 52—12. *La cholestérine, l'acidose.* ① Varsovie 1922.

Gödel, Alfred, Priv.Doc. Graz (Österreich), Univ. *Physiol. Anat.* o

Goedewaagen, Matthys Arnoldus Jan (1892), Botan., Rijkslandbouwproefstation. Groningen (Holland), Eemskanaal Z.Z. No. 1.

Görnitz, Karl (1895), Dr. phil., Leiter der Versuchsstelle für Pflanzenschutz der Chemischen Fabrik auf Actien (vorm. E. Schering). Berlin-Zehlendorf-Schoenow (Deutschland), Goerzallee 3. *Angewandte Entomol., Phytopath. Ornithol.: spez. Federfarben.* ① Halle a. d. S. 1921.

Görög, Dénes (1901), Ass. pathol.-anat. Inst. Univ. Pécs (Ungarn). ① Pécs 1924.

Goertler, Kurt (1898), Priv.Doc. für Anat., Univ. München, Ass. an der histol.-embryol. Abteilung der Anat. Anstalt. München-Laim (Deutschland),

Agnes-Bernauer-Str. 112. *Formbildungsvorgänge während der Primitiventwicklung von Amphibieneiern spez. Gestaltung der Medullaranlage und des Herzens unter besonderer Berücksichtigung entwicklungs-physiol. Fragen.* ① Hamburg 1922.

Goethart, J. W. C., Dr., Dir., Reichsherbarium Leiden (Holland), Nonnensteg. *Pflanzensystematik.* o

Göthlin, Gustaf Fredrik (1874), Prof. Physiol., Dir. des Physiol. Inst. Univ. Upsala (Sverige). *Elektrophysiol. des isolierten Muskels; Farbenempfindungen; Entoptischer Nystagmus. Die doppelbrechenden Eigenschaften des Nervengewebes. Energieschwelle für die Empfindung Rot in ihrer Abhängigkeit von der Wellenlänge der Lichtstrahlung.* ① Upsala 1907.

Goetsch, Wilhelm (1887), ao. Prof. für Zool. und vergl. Anat. mit Lehrauftrag für Entwicklungsgeschichte. Univ. München (Deutschland), Zool. Inst., Neuhauser Str. 51. *Entwicklungsmechanik, Algensymbiose.* ① Straßburg (Els.) 1914.

Goffart, Hans (1900), Dr. phil., wissenschaftlicher Hilfsarbeiter. Berlin-Dahlem (Deutschland), Biol. Reichsanstalt für Land- und Forstwirtschaft. *Nematoden.* ① Münster i. Westf. 1923.

Gogelis, Prof., Dir. Vet.-Bact. Inst. Kaunas (Litauen), Keistucio Gatve 18. o

Goggio, E., Doc. Univ. Napoli (Italia). *Anat.* o

Gogol-Janowskij, Georg J., Doc., Landwirtschaftl. Akademie Timirjasew. Moskau (U.d.S.S.R.), Trubnikowskij per. 19, W. 26. *Weinbau.* o

van Goidsenhoven, Ch., Prof., Ecole de Med. vet. Bruxelles (Belgique). *Bact. Maladies contagieuses.* o

Gokhale, Ramchandra Krishnaji (1895), Anti-Rabic Department. Parel, Bombay (India), Haffkine Inst. *Bact.* ① Coll. Bombay 1918.

Gokhale, Shankar Kashinath (1898), Ass. Chem., Biochem. Unit. Haffkine Inst. Post Vile Parle near Bombay (India), B. B. & C. I. Rly. *Physiol. Chem.* ① Bombay 1919.

Gola, Giuseppe (1877), Prof. Botan., Dir. Inst. Botan. Univ. Padova (Italia), Via Orto Botan. 13. *Physiol. végétale, Biochim., Ecol., Hépatiques.* ① Turin 1920.

Goldberg, I. M., Priv.Doc. Baku (U.d.S.S.R.). *Allg. Pathol.* o

Goldenberg, Eugène E. (1896), Priv.Doc. physiol., Chef du séminaire physico-chim. Inst. de Méd. Odessa (U.d.S.S.R.), Rue Léon Tolstoï 10, lg. 1. *Chim. physique du système nerveux, physiol. générale des tissues excitables. Action des ions sur la tissue nerveuse; phénomènes de ,,l'escalier''.*

Goldfederová, Anna, Ass. ,Lab. de Fysiol. l'Univ. Masaryk. Brno (C.S.R.), Údolní 73. o

Goldforb, A. J. (1881), Assoc. Prof. of Biol. Coll. of the City of New York, N.Y. (U.S.A.). *Regeneration, growth, physiol. experim. embryol.* ① Columbia 1909.

Goldman, Jan, Dr. med. Inst. de pathol. générale et expérim. Warszawa (Polska), rue Krakowskie Przedmieście 26/28. o

Goldmann, Franz (1895), Dr. med., wissenschaftliches Mitglied des Hauptgesundheitsamtes. Berlin C (Deutschland), Fischerstr. 39/42. *Soziale Pathol. und Hygiene. Gesundheitsfürsorge.* ① Berlin 1920.

Goldner, Jacques Wilhelm (1893), Chef des Travaux Labor. d'Histol. Fac. méd. Jassy (România), strada Mărzescu 17. *Cytol. et Histophysiol. des glandes endocrines. Correlations interglandulaires. Action de l'Adrénaline sur le thymus, Histophysiol. de la thyroïde et du thymus au cours des Fractures.* ① Cluj 1922.

Goldschmid, Edgar (1881), Univ.Prof., Prosektor d. Senkenberg. Pathol. Inst. Frankfurt a.M. (Deutschland), Mainzerlandstr. 2. *Pathol.* o

Goldschmidt, Richard (1878), Prof. Dr., II. Dir. des Kaiser-Wilhelm-Instituts für Biol. Berlin-

Dahlem (Deutschland). *Genetik, Geschlechtsbestimmung. Vererbungstheorie.* ① Heidelberg 1902.
Goldsmith, Glenn Warren (1886), Investigator with Carnegie Inst. of Washington. Colorado Springs, Col. (U.S.A.), 123 E. Washington St. *Water and gas relations of plants. Physiol. and physics of flower movement. Ecol. relations of the larger soil organisms. The Phytometer Method in Ecol.* ① Univ. of Nebraska 1924. ② *Larger soil organisms, insects native to Colorado.*
Goldsmith, William Marion (1888), Prof. Biol., Southwestern Coll. Winfield, Kans. (U.S.A.), 304 Seward Street. *Cytol., Genetics, Eugenics.* ① Indiana State Univ. 1920.
Golenkin, Michail J., Prof. Dr. I. Univ. Moskau (U.d.S.S.R.), I. Grashdanskaja, Botan. Garten. *Botan., Morphol., Pflanzengeographie. Angew. Botan.* o
Golikova, Sofija Mitrofanovna, Doc. II. Univ. Moskau (U.d.S.S.R.). *Bact.* o
Golińska, Hédvige (1893), Ass. à l'Ecole Supérieure d'Agric. (chaire de culture potagère). Varsovie (Pologne), Skierniewice Palais, L'inst. de culture potagère de l'Ecole Super. d'Agric. *Sélection et croissance des pommes de terre.* ① Pétersbourg 1916.
Golla, Frederick (1877), Dir. Central Pathol. Labor. London S.E. 5 (England), Maudsley Hospital, Denmark Hill. *Physiol. Neurol.* ① Oxford 1900, London 1904.
Gollwitzer-Meier, Klothilde (1894), Priv.Doc. Greifswald (Deutschland), Med. Klinik. *Reaktionsregulation im Körper. Mineralstoffwechsel.* ② München 1919.
Golodnoff, Michel (1886), Inst. Pasteur. Paris XVI (France), 22, rue Copernic. *Immunité locale, Anaphylaxie.* ① Petrograd 1913.
Golovanov, V. A., Doc. Univ. Nizny-Novgorod (U.d.S.S.R.), Sovetskaja Ploščad 8. *Bact.* o
Golubev, Boris Alexandrovitsch (1893), Ass. Labor. für Düngerlehre. Moskau 8 (U.d.S.S.R.), Landwirtschaftl. Akademie. *Boden-Biol. und Chem.: Bodenacidität, Kalkung, Rohphosphatdüngung.* o
Gombocz, Endre, Dr., Priv.Doc. Budapest I (Ungarn), Attilagasse 14. *Floristik: Populus.* o
Gómen, A. Enrique, Prof. Univ. Nac. Bogotá (Columbia). *Botan.* o
Gomez, Angel K. (1891), Associate Prof. Pathol. and Bact. Coll. of Vet. Sc., Los Baños Coll. Laguna, Philippine Islands (U.S.A.). ① Manila 1914. ② *Pathol. slides, Bact. collections.*
Gomez, Liborio (1887), Prof. Pathol. and Bact. Coll. of Med. Univ. of the Philipp. Manila, Philippine Islands (U.S.A.). *Pathol. anatomy, bact., tropic. med.* ① Rush Med. Coll. 1908. ② *Pathol. specimen.*
Gomez-Menor y Ortega, Juan (1899), Lic. Sc. Nat. Prép. microgr. Almeria (España), Estacion de Patol. Vegetal, Calle de Murcia 2. *Hemiptères, Lygeidae et Coccides: biol. et parasit.* ② *Encirtides, Afelinines* (Hyménoptères, Calcidides), Coccides.
Gomolao, Nicolas I., Conserv. Station de Défense des Plantes. Kiev (U.d.S.S.R.), Foundoukleevskaia No. 46. *Maladies de la betterave.* o
Gonçalves, Paulo, Ass. d'Histol. Fac. méd. Porto (Portugal). *Histol.* ① Porto 1924.
Gonzalez, Bienvenido M. (1893), Prof., Head Dept. of Animal Husbandry, Coll. of Agric., Univ. Philippines. Los Baños College (Philippine Islands). *Zootechny, genetics as applied to improvement of farm animals.* ① Univ. of Johns Hopkins 1922. ② *Farm animals.*
v. Gonzenbach, Willi (1880), Prof. für Bact. und Hygiene an der eidgenössisch-technischen Hochsch. Zürich (Schweiz), Zürichbergstr. 4. *Desinfektionslehre. Wunddesinfektion.* ① Zürich 1910.
Gooch, Marjorie (1896), Technician. Princeton, N.J. (U.S.A.), Rockefeller Inst. *Animal genetics and physiol., especially Drosophila, Poultry, Cattle, Rats.* ① Univ. of Maine 1919.

Good, Henry George (1897), Ass. Prof. Ent. and Zool., Ass. Entomol. to Experim. Station. Auburn, Ala. (U.S.A.), Ala. Poly. Inst. *Coleoptera, systematic; Bird migration. Wing-venation of the Coleoptera.* ① Cornell Univ. 1923
Good, Ronald D'Oyley (1896), Ass., Dept. of Botan. London S.W. 7 (England), British Mus. (Nat. Hist.), Cromwell Rd. *Taxonomy of flowering plants. Geographical distribution.* ① 1921.
Goodale, Hubert Dana (1879), Biol., Mount Hope Farm. Williamstown, Mass. (U.S.A.). *Inheritance and physiol. of reproduction in domestic fowl; influence of ultra violet light on growth and reproduction in domestic fowl. Inbreeding in domestic fowl.* ① Columbia Univ. 1913. ② *Pedigreed poultry.*
Goodall, Alexander, Lect. School of Med. R. Coll. Edinburgh (Scotland). *Physiol.* o
Goodhart, G. W., Lect. Univ. Coll. Hosp. Med. School. London W.C. (England), Univ. Str. *Pathol.* o
Gooding, J. Hunter (1891), Agric. E. I. du Pont de Nemours & Co., Inc. Wilmington, Del. (U.S.A.). *Development of organic mercury and other type of seed disinfectants, fungicides etc.* ② *Plants. Seed disinfectants.*
Goodpasture, Ernest William (1886), Prof. of Pathol., Vanderbilt Univ., Med. School. Nashville, Tenn. (U.S.A.). *Pathol. of diseases due to filterable viruses. Inclusion bodies.* ① Johns Hopkins 1912.
Goodrich, Edwin Stephen, Linean Prof. Univ. Oxford (England). *Zool.comp.Anat.Coelcm.Nephridia.* o
Goodrich, Hubert Baker (1887), Prof. Wesleyan Inst., Instr. of embryol., Marine Biol. Labor. Woodshole. Middletown, Conn. (U.S.A.). *Zool.* o
Goodspeed, Thomas H. (1887), Associate Prof. of Botan. and Curator of the Botan. Garden, Univ. of California. Berkeley, Cal.(U.S.A.), BotanyBuildg. *Cytol Genetics of hybrid plants.* ① 1912.
Goormaghtigh, Norbert Oscar Jean (1890), Prof. d'Anat. pathol. Univ. de Gand (Belgique), 53, Bd. Ch. du Kerchove. *Histol. des tumeurs. Histopathol. des glandes endocrines (surrénale, ovaire, thyroide).* ① Gand 1913 et 1922.
van der Goot, P., Dept. v. Landbouw. Buitenzorg, Java (Nederl.-O.-Indië). *Aphididae, Coccidae, Reisbohrer.* o
Gorbunov, Prof. II. Univ. Moskau (U.d.S.S.R.). *Zool.* o
Gordilho, Ad., Prof. Fac. Med. Bahia (Brasil) *Histol.* o
Gordjagin, Andrej Ja., Prof., Univ., Inst. f. Land- u. Forstwirtschaft. Kasan (U.d.S.S.R.), Ul. K. Marksa 43. *Geobotan.* o
Gordon, John (1895), Dr., Lect. in Bact. Leeds (Scotland), Med. School. ① Leeds 1918.
Gordon, Mervyn Henry (1872), Consulting Bact. to St. Bartholomenis Hospital London E.C. 1 (England). *Bact. and Immunol. Streptococci. Vaccinia.* ① Oxford 1903.
Gordon, W. T., Prof. Univ. London S.W. 7 (England), S. Kensington. *Geol.* o
Gore, Shamroo Narayan (1876), Ass. Dir. Haffkine Inst., Parel, Bombay (India). *Faecal, water and soil bacteria. Faecal fats. Hydrocyanic acid gas fumigation.* ① Bombay 1900.
Gorham, Frederic P. (1871), Prof. of Bact. and Head of the Biol. Department of Brown Univ Providence, R.I. (U.S.A.). *Bact., experim. Zool. diphtheria bacillus, bact. of shellfish, diseases of fishes.* ① Brown Univ. 1894.
Gorin, Alexander P., Agronom, Landw. Versuchsstation, Univ. Perm (U.d.S.S.R.), Ul. Karla Marksa 26, W. 1. *Selektion.* o
Gorini, Constantino, Prof. R. Scuola sup. di agric. Milano (Italia). *Batteriol. agraria.* o
Goris, Paul (1895), Ing. Paris (France). *Pharmacol.* o
Gorjaczkowski, Vladimir, Dr. Prof. Warszawa (Pologne), Ecole Sup. Agric., r. Miodowa 23. *Botan.* o

Gorjanović-Kramberger, Karl (1856), o.ö. Univ.-Prof. emer., Dir. emer. des geol.-palaeont. Nationalmus. Zagreb (S.H.S.), Lisinski Gasse 2. *Palaeoichthyol., Palaeomalacol. und Palaeoanthrop.* ① Tübingen 1879.

v. Gorka, Alexander (1878), Dr., Prof. d. Biol. a. d. med. Fac. d. Univ. u. Dir. d. Biol. Inst. d. Univ. Pécs (Ungarn), Rákóczistr. 80. *Vergl. Physiol. d. Verdauungserscheinungen wirbelloser Tiere (spec. d. Mollusken u. Insecten).* ① Budapest 1901.

Gorman, Martin W., Custodian, Forestry Building. Portland, Ore. (U.S.A.). *Systematic botan.* o

Gorman, Michael, J. (1890), Lect. on Agric. Botan. Dublin (Scotland), Coll. of Sc. *Botan. studies of Agric. Grassland.* ① Dublin 1911.

Gorodissky, Henriette (1900), Ukrainisch. Biochem. Inst., Ass. d. Med. Fac. 'Charkow (U.d.S.S.R.), Skripnitzkaja Str. 7. *Biochem. d. Zentralnervensystems.*

Gorodkov, Boris Nikolaevič (1890), Botan. Mus. d. Akad. d. Wissensch. Leningrad (U.d.S.S.R.), Universitätskaja naber., 5. *Pflanzengeographie der westsibirischen Tiefebene und des nördlichen Ural. Systematik der Carexarten.*

Gorovic, Vladova, Prof. Med. Inst. Ekaterinoslav, Ukraine (U.d.S.S.R.). *Hygiene.* c

Gorpintschenko, Catherine (1895), Ass. Station Séricicole. Taschkent (U.d.S.S.R.), rue Kafanow 1. *Sériciculture.*

Gorschkova, S. G., Ass. l'Herbier, Jardin Botan. princ. Leningrad (U.d.S.S.R.), Aptekarski Ostrov, Pessolchnaia ¹/₂. *Systématisation, Tamaricaceae.* o

Goršenin, K. P., Prof. Sibir. Inst. f. Land- u. Forstw. Omsk (U.d.S.S.R.), Usadba za Staroj Zagorodnoj Koščej. *Bodenkunde.* o

Gorshkova, Lydia (1901), Ass. Physiol. Labor. Univ. Rostov a. Don (U.d.S.S.R.), Souvorovskaia No. 41. *The influence of the pituitary body on growth.* ① Rostov a. Don 1925.

Gorsky, Jean (1893), Ass. de Paléont., Inst. des Mines, Collabor. Comité Geol. Leningrad (U.d. S.S.R.), Vassily Ostrov, 21 Linie 2, 31. *Faunes carbonifères, permo-carbonifères et dévoniennes: Rugosa, Brachiopoda, Gastropoda.*

Górski, Marjan (1886), Mitglied-Korrespondent der Technischen Akademie, o. Prof. landwirtsch. Hochsch. Warszawa (Polen), Miodowa 23. *Bodenazidität u. Wachstum der Pflanzen. Der Einfluß der Pflanzen auf die Bodenazidität.* ① Leipzig 1911.

Gortani, Michele (1883), Prof. o. Geol. e Paleont. Univ. Bologna (Italia), R. Mus. Geol. ,, G. Capellini", Via Zamboni 63. *Faune paleozoiche.* ① Bologna 1904. ② Graptoliti della Sardegna. Modelli in gesso di Vertebrati fossili italiani.

Gortner, Ross Aiken (1885), Chief of the Division of Agric. Biochem., Univ. of Minnesota, and Prof. of Agric. Biochem. St. Paul, Minn. (U.S.A.). *The colloidal constituents of wheat flour and their influence on bread manufacture. Chem. and physicochem. studies on proteins. The role of colloids in biol. phenomena, to drought resistance and winter hardiness of plants. Physico-chem. properties of plant tissue fluids in relation to geographical distribution.* ① Columbia Univ. 1909.

Goss, Robert W. (1891), Associate Prof. and Associate Plant Pathol. Univ. of Nebraska and Agric. Experim. Station. Lincoln, Neb. (U.S.A.), Dept. Plant Pathol. Coll. of Agric. *Potato Diseases.* ① Univ. of Wisconsin 1922.

Gothan, Walther (1879), Prof., Dr., Kustos a. d. Preuß. Geol. Landesanstalt, Doc. a. d. Techn. Hochsch. Berlin N 4 (Deutschland), Invalidenstr. 44. *Paläobotan.* ① Jena 1904. ② Fossile Pflanzen.

Gotô, Motonosuke, Prof. of Kyûshû Imper. Univ. Fukuoka (Japan), Chem. Inst., Coll. of Med. *Med. Chem.*

Gotô, Seitarô (1867), Prof. Zool., Sc. Fac., Imperial Univ. Tokyo (Japan). *Trematoda, Asteroidea,*

Hydrozoa. ① Tokyo Imp. Univ. 1890. ② *Trematoda, Cestoda.*

Gotschev, Peter, Ass. Geol. Inst. d. Univ. Sofia (Bulgarien). *Palaeont.* o

Gottberg, F. G., Kalastusneuvos, Mataloushallitus Kalatolousosasto. Helsinki (Finnland), Mariegatan No. 23. *Fischereibiol.* o

Gottlieb, Erik (1894), Dr. København (Danmark), Rigshospitalet. *Fermente und Ammoniak im Blut und Harn.* ② Copenhagen.

Gould, Harley Nathan (1887), Prof. of Biol., Newcomb Coll. of Tulane Univ. New Orleans, La. (U.S.A.). *Origin and differentiation of Germ Cells.* ① Princeton Univ. 1916.

Gould, Harris Perley (1871) Pomol., U. S. Dept. of Agric. Washington, D.C. (U.S.A.). *Peach growing.* ① Cornell Univ. 1897.

Goulden, Cyril H. (1897), Cereal Specialist. Winnipeg, Manitoba (Canada), Rust Research Labor. Manitoba Agric. Coll. *Breeding of cereals, resistance.* ① Univ. of Minnesota 1925.

Gourley, Joseph Harvey (1883), Chief Department of Horticulture, Ohio Agric. Experim. Station. Wooster, Ohio (U.S.A.). *Fertilizer and cultural requirements of fruit trees. Pruning, ringing.* ① Ohio State Univ. 1915.

Gourvitsch, Victor (1899), Collabor. sc. fac. sc, phys. nat. Univ. Taschkent, Usbekistan (U.d.S.S.R.). *Protozoaires.*

Goutner, Lily (1903), Stagiaire, Station de Pathol. végétale de l'Inst. Agronomique. Leningrad (U.d. S.S.R.), Detskoe Selo. *Mycol.: Deuteromycètes, leur ontogenie, cultures pures.* ② Cultures de champignons.

Govaerts, Paul (1889), Adjoint de clinique méd. Bruxelles (Belgique), 24, Rue Marie-Thérèse. *Physiol. Pathol. Sérol.* ① Bruxelles 1914.

Govorov, Alexander, Dr., Prosektor Pathol.-anat. Inst. Zagreb (S.H.S.), Vočarska cesta. *Pathol. Anat.* o

Gowanloch, J. N., Ass. Prof. Dalhousie Univ Halifax, Nova Scotia (Canada). *Zool.* o

Gowland, W. P., Prof. Univ. of Otago. Dunedin, New Zealand (Australia). *Anat.* o

Goworow, Leonid I., Doc. Landwirtschaftl. Akademie Timirjasew, Leiter d. Steppen-Versuchsstat. d. Staatl. Inst. f. Agron. Talowaja, Ju-W. sh. dor. Moskau (U.d.S.S.R.), Petrowsko-Rasumowskoje, Selektionsstation. *Selektion u. Biol. landwirtschaftl. Pflanzen.* o

Goy, Sam (1879), Dir. der Landw. Versuchsstation, Prof. an der Univ. Königsberg i. Pr. (Deutschland), Lange Reihe 3. *Pflanzenernährung und -Düngung, Bodenkunde, Nahrungsmittelchem.* ① Marburg 1908.

Gózony, Lajos (1886), Priv.Doc. Univ. Budapest IX (Ungarn), Rákosgasse 9. *Mikrobiol., Protozool.* ① Budapest 1909.

de Graaff, Willem Cornelis (1877), Prof. en Matière méd., Microbiol., Chem. pathol. Utrecht (Holland) Labor. pharm. *Propriétés biochim. des bact., spécialement la fermentation. Les plantes utiles, les plantes médicinales et à essence, les plantes alimentaires, fourragières, vénéneuses.*

Grabau, Amadeus William (1870), Chief Paleont. Geol. Survey of China, Prof. Paleont. Nat. Univ., Research Associate Central Asiatic Expedition. Peking (China), 5 Tou Ya Tsai Hutung, West City. *Invertebrate Paleont. of China.* ① Cambridge 1900.

Graber, Laurence F. (1887), Prof. of Agronomy, Univ. of Wisconsin, Madison, Wis. (U.S.A.). *Root reserves of herbaceous plants; hay and pasture crops.* ① Univ. of Wisconsin, Madison, Wis. 1910.

Grabfield, G. Philip (1892), Instr. in Pharm., Harvard Med. School, Associate in Med., Peter Bent Brigham Hospital. Boston, Mass. (U.S.A.), 23 Bay State Road. *Pharm., the effect of drugs on the Nitrogen metabolism; pulmonary disease; glands of internal secretion and the nitrogen metabolism.* ① Harvard Med. School 1915.

Grabner, Emil, Dir. d. Versuchs-Anst. für Pflanzenkultur d. Landwirtschafts-Akad. Magyar-Óvár (Ungarn). ○

Grabowski, T., Dr. Prof. Kraków (Pologne), Univ. Zool. ○

Gračanin, Mihovil (1901), Dr., Ass. Poljoprivredna Ogledna i Kontrolna Stanica Osijek I (S.H.S.). *Biochem. Katalasewirkung bei autotrophen Pflanzen. Influence de la lumière sur l'absorption de l'acide phosphorique et du potassium par les plantes.* ⓓ Prag 1925.

Gradmann, Hans (1892), Priv.Doc. der Botan. Erlangen (Deutschland), Essenbacher Str. 6. *Reizphysiol. der Pflanzen. Wasserhaushalt der Pflanzen. Bodenkunde.* ⓓ Tübingen 1920.

Gradojević, Michailo (1887), Doc. de l'Univ., l'Entomol. forestière et la Protection des forêts. Belgrade (S.H.S.), Studenička ulica 55. *Lepidoptères de la Serbie (Macro et Micro). Insectes nuisibles aux forêts.* ⓓ Prague (C.S.R.) 1924.

Graebner, Heinrich Robert Paul (1900), Dr. phil., wiss. Hilfsarbeiter am Westfälischen Provinzial-Mus. für Naturkunde. Münster i. W. (Deutschland), Zool. Garten. *Naturdenkmalpflege (botan.). Bearbeitung der Museumssammlungen. Systematische und pflanzengeographische Untersuchungen in Westfalen.* ⓓ Berlin 1924.

Gräff, Siegfried Wilhelm (1887), ao. Prof. Heidelberg (Deutschland), Pathol. Inst. der Univ. *Allgemeine Pathol. und pathol. Anat. und Histol.; physikalisch-chem. und physiol. Zellforschung, histol. Fermentforschung.* ⓓ Freiburg 1911.

Gräper, Ludwig Ernst (1882), Prof. ordinarius, Dr. med., Prosektor am Anat. Inst. Jena (Deutschland), Kasernenstr. 5. *Topographische Anat. (kindliche Brusthöhle). Entwicklungsmechanik (Extremitäten).* ⓓ Leipzig 1909.

Graf, Franz (1865), Prof., O.R.Chem., Dir. staatl. Bayer. Abwasserstation. München (Deutschland), Veterinärstr. 6. *Hydrochem. und Hydrobiol.; Biol. des Abwassers und der Vorflutgewässer.* ⓓ Erlangen 1896.

Graf, Hans (1898), Dr. med. vet., wiss. Hilfsarbeiter am Pharm. Inst. der Tierärztl. Hochsch. Berlin NW 6 (Deutschland), Philippstr. 13. *Adsorptionsprophylaxis und -therapie, Pharm. der Genitalorgane speziell des Rindes; Veterinär-Pharm.; Lokalanästhesie.* ⓓ Zürich 1923. ⓟ Schweizerischer Pflanzen des Mittellandes. Arzneipflanzen.

Graf, Jacob (1891), Dr. phil. nat. Rüsselsheim a. M., Hessen (Deutschland), Haßlocher Str. 48. *Botanik (Entwicklungsgeschichte, Vererbungslehre).* ⓓ Frankfurt a. M. 1921.

Grafe, Erich (1881), o. Prof. f. innere Med., Dir. der Med. Klin. Würzburg (Deutschland), Luitpoldkrankenhaus, Bau 17. *Physiol. und Pathol. des Stoffwechsels, bes. des respiratorischen.* ⓓ Bonn 1904.

Grafe, Viktor (1878), ao. Prof. für Biochem. a. d. Univ. in Wien VIII (Österreich), Hamerlingplatz 9. *Biochem. der Pflanzen.* ⓓ Wien 1901.

Graff, Paul W., Ass. Prof. of Botan., Univ. of Montana, Biol. Sta. Missoula, Mtn. (U.S.A.). *Cryptogamic botan.* ○

Graham, Edw. H. (1902), Ass. in Herbarium, Carnegie Mus. Pittsburgh, Pa. (U.S.A.). *Classification of trees of Pennsylvania by appearance of bark; Flora of British Guiana.* ⓓ Univ. of Pittsburgh 1927.

Graham, John Young (1869), Prof. of Biol., Univ. of Alabama. Tuscaloosa, Ala. (U.S.A.). ○

Graham, Samuel A. (1891), Ass.Prof. of Entomol., Univ. of Minnesota. St Paul, Minn. (U.S.A.), Univ. Farm. *Ecol. of forest insects.* ⓓ Univ. of Minnesota 1921.

Gram, Hans Christian Joachim (1853), Prof. emer. Pharmacol. Univ. København (Danmark), Aaboulevard 40. *Bact., normale u. pathol. Anat. des Blutes, Pharm., angewandte Zool. (tierische Parasit).* · ⓓ Copenhagen 1883.

Gram, Ernst (1891), Dir. Lyngby (Dänemark), Statens plantepatologiske Forsøg. *Phytopath. Seed desinfection, filtrable vira.* ⓓ 1915.

Gramenicky, M. J., Prof., Inst. f. Med. Wiss. Leningrad (U.d.S.S.R.), 2. Sovetsky Str. 4. *Pharm.* ○

Gramenitska-Tovstoless, Tatiana A., Préparateur Inst. Jaczewski. Leningrad (U.d.S.S.R.), Perspective Anglaise 29. *Maladies du lin.* ○

Gran, H. H., Prof. Dr., Botan. Labor. Univ. Oslo (Norge). ○

Granata, Leopoldo (1885), Prof. Zool., anat., fisiol. compar. Univ. ,Dir. della Staz. biol. Univ. Cagliari, S. Bartolomeo (Italia). *Protist. (protozoi parasiti). Citol. Istofisiol.* ⓓ 1925.

Le Grand, Prof., Ecole prép. de méd. et de pharm. Rouen (France). *Anat. descriptive.* ○

Grandi, Guido (1887), Prof. ufficiale di Entomol., Ist. Superiore Agrario di Bologna (Italia). *Biol., morfol. e sistematica generale degli Insetti. Imenotteri melliferi e predatori. Insetti dei Fichi di tutto il mondo.* ⓓ Bologna 1926.

Grandis, Valentino, Prof. o. univ. degli studi. Genova (Italia). *Fisiol.* ○

Grandoci, Remo, Prof. Univ. Padua e Prof. incar. Univ. Camerino (Italia). *Anat. Fisiol. compar.* ○

Granel, François (1888), Chef de Labor. d'Histol. à la Fac. de Méd. de Montpellier (France), 1, rue Saint-Firmin. *Histol., Embryol. La cellule pulmonaire, l'ossification, la pseudobranchie des Poissons.*

Granit, Ragnar Arthur (1900), Ass. Physiol. Inst., Univ. Helsinki (Finland). *Experim. psychol. Physiol. of the sensorgans.* ⓓ Helsingfors 1927.

Granovsky, Alexander Anastacevitch (1887), Ass.Prof. of Economic Entomol., Univ. of Wisconsin. Madison, Wis. (U.S.A.), 1532 Univ. Avenue. *Grasshoppers. Insects in relation to transmission and dissemination of plant diseases. Symbiotic relationship between the insects and micro-organisms. Cherry Aphid. Taxonomic studies of Aphididae.* ⓓ Univ. of Wisconsin 1926. ⓟ Aphididae.

Grant, Adele Lewis (1881), Senior Lect. Botan. Wellington (South Africa), Huguenot Univ. Coll. *Taxonomy. Monographing South African genera in Scrophulariaceae.* ⓓ Washington Univ. 1923. ⓟ Plants, angiosperms.

Grant, J. B., Prof. Dr., Head of Dept. of Hygiene, Union Med. Coll. Peking (China). ○

Grapmans, Rudolfs, Dr., Doc. Univ. Riga (Latvija), Karlīnes ielā 15, dz. 21 u. Pērnavas ielā 19. *Pathol. Anat. d. Tiere.* ○

Gravatt, Annie Rathbun (1894), Junior Pathol., Office of Forest Pathol. Washington, D.C. (U.S.A.), U.S. Department of Agric. *White pine blister rust and damping off of conifers.* ⓓ Brown Univ. 1918.

Gravatt, George F. (1891), Associate Pathol., Office of Forest Pathol. Washington, D. C. (U.S.A.), U.S. Department of Agric. *Disease resistance, hybridization and testing of introduced species of chestnut. Maple wilt (Verticillium). Nectria canker of trees.* ⓟ Mycol. specimens.

Grave, Benjamin H. (1878), Prof. of Zool. Crawfordsville, Ind. (U.S.A.), Wabash Coll. *General Embryol., breeding seasons, lunar periodicity, rate of growth, age at sexual maturity, duration of life of marine invertebrates.* ⓓ Johns Hopkins Univ. 1910. ○

Grave, Caswell (1870), Prof. of Zool., Washington Univ. St. Louis, Mo. (U.S.A.). ⓓ Johns Hopkins 1899. ○

Graver, Frederick O. (1868), Prof. of Botan. Oberlin Coll. Oberlin Ohio (U.S.A.), 270 Elm St. ⓓ Harvard 1896. ○

Le Graverend, Eugène (1889), Dir. jardins botan. Rouen (France), 114, rue d'Elbeuf. ⓟ Graines et bulbes.

Graves, Arthur Harmount (1879), Curator of Public Instr. Brooklyn, N.Y. (U.S.A.), Botan. Garden. *Diseases of Forest Trees.* ⓓ Yale 1907.

Graves, Stuart (1879), Prof. of pathol. and bact., Univ., School of Med. Louisville, Ky. (U.S.A.), 2500 Longest Ave. ○

Gravier, M. Ch., Prof. au Mus. Paris 5 (France), 55, Rue de Buffo. *Zool.* ○

Gravis, Jean-Joseph-Auguste (1857), Prof. de Botan. à l'Univ. de Liège (Belgique), 22, rue Fusch. *Anat. végétale. Urtica dioica, Tradescantia virginica, Chlorophytum, Amarantacées, Commélinées.* ℗ Bruxelles 1880. ℗ Des échantillons de plantes cultivés au Jardin botan. de Liège.

Gray, James (1891), Lect. experim. Zool. Univ. Cambridge (England), Kings Coll. *Experim. embryol. and cytol.* ℗ Cambridge 1912.

Gray, P. H. H. (1891), Ass. Bact., Experim. Stat. Harpenden, Herts. (England), Wellcott Close, Welwyn Garden City. *Soil Bact. Researches on bact. that decompose aromatic compounds, cellulose etc.* ℗ Oxford Univ. 1920.

Gray, William L. (1898), Inspector, State Plant Board. Natchez, Miss. (U.S.A.). *Insect Pest and Plant Diseases injurious to Farm Crops, Orchards, Ornamentals, plant Quarentines.* ℗ Iridomyrmex Humilis, Anthonomus Grandis, Dialeurodes Citri, Tetranychus-telarius, Heliothis Obsoleta, Pulvinaria vitis, Chrysobothris Femorata, Phyllophaga, Chrysomphalus Obscurus, Coccus Hesperidum L., Aulacaspis pentagona, Chrysomphalus tenebriocosus Eomst., and Termites.

Graz, Otto, Dir. milchwirtschaftl. Forschungsstation. Magyaróvár (Ungarn). ○

Graziadei, George (1894), Ass. et Bact. chez l'Inst. d'Hygiène de l'Univ. de Turin (Italia), Rue Montevecchio 4. *Hygiène et bact. Protozoaire intestinaux; la biol. bacterienne; méthodes de coloration en bact.; la culture du sperme dans les gonorrhées croniques; la Radioemanation en bact.;propagande hygienique etc.* ℗ Turin, Italie, 1920.

Graziani, Alberto (1880), Prof., Libero doc., Dir. Padova (Italia), Via Umberto I 8. *Igiene.* ℗ Padova 1904.

Greaves, Joseph James (1880), Prof. Bact. and Physiol. Utah Agric. Coll., Chem. and Bact. Experim. Station. Logan, Utah (U.S.A.). *Factors influencing nitrogenfixation by bact., influence of inorganic salts on bact.* ℗ Univ. of California 1911.

Green, H. H., Prof. Transvaal Univ. Coll. Pretoria South Africa). *Biochem.* ○

Green, Robert Gladding (1895), Associate Prof. of Bact. Minneapolis, Minn. (U.S.A.), 225 Millard Hall. *Stability of Disperse Systems. Diseases of Carnivora. Filterable Viruses.* ℗ Univ. of Minnesota 1921.

Greenbank, George R., Research Lab., Bureau of Dairy Industry, U.S. Dept. of Agric. Washington, D.C. (U.S.A.). *Fat in dry milks.* ○

Greene, Charles Wilson (1866), Prof. physiol. and pharmacol. Univ. of Missouri. Columbia, Mo. (U.S.A.), 814 Virginia Av. *Heart and circulation, oxygen. want, nitrous oxide anesthetica, gen. physiol.* ℗ Johns Hopkins 1898.

Greenman, Jesse More (1867), Curator of Herbarium, Missouri Botan. Garden, and Prof. of Botan., Washington Univ. St. Louis, Mo. (U.S.A.). *Systematic Botan. (Taxonomy) and Plant Geography. The North American Species of the Genus Senecio.* ℗ Harvard Univ. 1899, Berlin 1901. ℗ Vascular plants.

Greenman, Milton J. (1866), Anat., Dir. of the Wistar Inst. Philadelphia, Pa. (U.S.A.), 3618 Woodland Ave. ℗ Pittsburgh 1912.

Greenwald, Isidor (1887), Biochem., Harriman Research Labor., The Roosevelt Hospital. New York, N.Y. (U.S.A.), 428 West 59 St. *Physiol. of parathyroid glands. Chem. of blood.* ℗ Columbia Univ. 1911.

Greenway, Daniel, Prof. Univ. Nac. La Plata (Argentina). *Parasit.* ○

Greer, Frank E. (1896), Principal Bact., Chicago Dept. of Health. In charge of routine and research work on water analysis. Chicago, Ill. (U.S.A.), 5458 Kimbark Ave. *Anaerobes in sewage. Spiriall from oil waters.* ℗ Kalamazoo Coll. 1921.

Greeves-Carpenter, Cyril Frederick (1897), Entomol., The F. A. Bartlett Tree Expert Co. Philadelphia, Pa. (U.S.A.), 316 S. Juniper St. *Economic Entomol. in relation to shade trees.* ℗ Bablake Coll., Coventry, England, 1914.

Greger, Justin (1886), Dr. phil., Ass. und Supplent der Lehrkanzel für Botan., Warenkunde und technische Mikroskopie an der Deutschen Techn. Hochsch. in Prag (C.S.R.), I, Husgasse 5 (Husova 5). *Floristik und Biol. der Grünalgen. Anat. der landwirtschaftlichen Unkrautsämereien.* ℗ Prag 1914.

Gregory, Charles T. (1887), Associate in Botan. Lafayette, Ind. (U.S.A.), Agric. Experim. Station. *Plant pathol.* ℗ Cornell Univ. 1910.

Gregory, Louise H. (1880), Assoc. Prof. Zool. New York City (U.S.A.), Barnard Coll. *Physiol., Protozool.* ℗ Med. und Ph. Columbia Univ. 1909.

Gregory, William K. (1876), Prof. of Vertebrate Palaeont., Columbia Univ., Curator, Dept. of Comp. Anat., American Mus of Nat. History. New York (U.S.A.), 77 St. and Central Park W. *Origin and Evolutions of the Human Dentition.* ℗ Columbia 1910.

Greguss, Paul (1890), Dr., Hochsch.Ass. der Botan. Budapest (Ungarn), I., Győri-út 13. *Biol. der Sexualorgane, der Samen, Sporen und Pollen. Entwicklungsgeschichtliche Bedeutung der Sexualorgane.* ℗ Budapest 1919.

Greisenegger, Ignaz Karl (1874), Dr., Prof. Mödling bei Wien (Österreich), Elisabethstr. 13. *Pflanzenbau und Agrikulturchem. Verhalten von Superphosphat im Boden.* ℗ Wien 1909.

Grejdich, Evgraf Oskarović, Doc. I. Univ. Moskau (U.d.S.S.R.). *Anat.* ○

Grekov, A. D., Prof. Univ. Taskent (U.d.S.S.R.). *Bact.* ○

Gremjatsky, Michel (1888), Ass. Inst. anthrop. Univ., Membre de l'Inst. des Sc. Biol. Timirjaseff. Moscou (U.d.S.S.R.), B. Karetnyj 15, W. 5. *Anthrop. somatique (Osteol.). Physiol. generale.*

Greschik, Jenö, Dr. Zool. Abt. d. Ung. National Mus. Budapest (Ungarn). *Anat. d. Vögel.* ○

Grevenstuk, Antonie Tzn (1892), Ass. Pharmacol.-therapeut. Labor. Univ. Amsterdam. Baambrugge (Holland) und Amsterdam (Holland), Polderweg 20. *Biochem. u. Pharm. Insulin.* ℗ Seltenere Pflanzen (natürliche Spezies, keine Gärtnereiprodukte), Samen, Zwiebeln.

Greving, Bernhard, Prof., Dr. pharm. Reval (Estland), Schmiedestr. 41. *Enzyme. Milchkunde. Butterkunde. Pathol. Anat. d. Harnorgane.*

Greving, Richard (1887), Priv.Doc. für innere Med., ao. Prof. Erlangen (Deutschland), Med. Klinik. *Vegetatives Nervensystem. Anat., Physiol. u. Pathol. des Zwischenhirns.* ℗ Würzburg 1919. ℗ Histol. Praep. vom Zwischenhirn, Mittelhirn, Medulla oblongata, Rückenmark.

Greze, B. S., Prof. Jaroslavsky Pedagog. Inst. Jaroslavl (U.d.S.S.R.), 122 Respublikanskaja.*Zool.* ○

Gridelli, Edoardo (1895), Conserv. Mus. Civico di Storia Nat. Genova 2 (Italia), Piazza di Francia 9. *Sistematica e Zoogeografia dei coleotteri: Staphylinidae (Philonthus, Quedius).* ℗ Padova 1919.

Grieg, J. A., Kons. Bergens Mus Bergen (Norge). *Zool.* ○

Grier, Norman MacDowell (1890), Prof. of Biol., Des Moines, Ia. (U.S.A.), Univ. *Ecol. and Variation of fresh water mussels (Naiades). Systematic Botan. of Spermatophytes. Paleobotan. of Pittsburgh Cool.* ℗ Univ. of Pittsburgh 1919.

Griesbach, Walter Edwin (1888), Dr. med., Priv.Doc. Univ. Hamburg (Deutschland), Rothenbaumchaussee 30. *Pathol. des Stoffwechsels, speziell*

intermediärer Zuckerstoffwechsel. Blutmengenbestimmung. Ⓓ Freiburg i. B. 1913.
Grießmann, Karl (1890), Dr. phil. nat., Abteilungsvorsteher d. Agric.-Chem. Kontrollstation. Halle a. d. S. (Deutschland), Ulestr. 17 II. *Angewandte Botanik, Samenkunde.* Ⓓ Heidelberg 1913.
Griffin, Lawrence Edmonds (1874), Prof. of Biol. Portland, Ore. (U.S.A.), Reed Coll. *Hydrozoa. Nautilus, Euplotes, Aclesia, Philippine and South American Reptiles.* Ⓓ Johns Hopkins Univ. 1900. Ⓟ Pacific Coast invertebrates generally.
Griffith,Wendell H. (1895), Ass. Prof. of Biochem. St. Louis Univ. School of Med. St. Louis, Mo. (U.S.A.), 1402 South Grand Blvd. *Biochem. Intermediary metabolism of amino acids.* Ⓓ Univ. of Illinois 1923.
Griffiths, David (1867), Horticulturist, U.S. Dept. of Agric. Washington, D.C.Home address: Takoma Park, D.C. (U.S.A.), 6961 Maple St.; *Culture, development and improvement of bulbous plants, and tuberous rooted and related ornamental stocks; Opuntia.* Ⓓ Columbia Univ., New York City. 1900.
Griffiths, H. E., Prof., St. Bartholem. Hosp. Coll. London E.C. 1 (England), West Smithfield. *Anat.* o
Grigaut, Adrien (1884), Chef de labor. Fac. méd. Paris (France), 21, Rue du Vieux Colombier. *Biochem.* Ⓓ 1911.
Griggs, Leland, Ph.D., Prof. of Biol. Hanover, N.H. (U.S A.), Dartmouth Coll. o
Griggs, Robert Fiske (1881), Prof. of Botan. (and head of the Dept. of Botan.), George Washington Univ. Washington, D. C. (U.S.A.), 39; Bradley Lane Chevy Chase Md. *The Flowering Plants.* Ⓓ Harvard 1911.
Grigorieff, Leonidas M. (1900), Prep. Labor. of Experim. Biol. Zoopark. Moscow (U.d.S.S.R.), Dolgy per. 14, Ap. 5. *Pathol. Anat.* Ⓓ Moscow.
Grigorovitsch, Alexandrine, Conserv. Station Phytopath., Inst. Agron. Moscou (U.d.S.S.R.). *Pharmacol.* o
Grigorovič, Nik. Aleksandrovič, Doc. II. Univ. Moskau (U.d.S.S.R.). *Pharmacol.* o
Grigorovitch-Beresovsky, Nicolaj Alexandrovitch (1876), Prof. géol. et paléont. Univ. Rostov sur Don (U.d.S.S.R.), Nicolskaja 57. *Mollusques marines du postpliocène et d'eau douce du pliocène (faune des terrains ,,levantins") et de l'horizon de tschokrak (miocène).* Ⓓ Odessa. Ⓟ Mollusques tertiaires du Caucase et de la Nouvelle Russie.
Grijns, Gerrit (1865), Med. Dr. u. Arzt, Prof. der Physiol. der Tiere a. d. Landwirtschaftl. Hochsch. Wageningen (Holland), Marktstr. 33. *Ernährung der Haustiere. Nahrung und Fruchtbarkeit. Farbensinn. Rote Blutzellen. Polyneuritis gallinarum. Biol. Prozesse im Boden.* Ⓓ Utrecht 1891.
Grimbert, Léon Louis (1860), Prof. de Chim. biol. à la Fac. de Pharm. Paris VI (France), 4, avenue de l'Observatoire. *Chim. biol. appliquée à la pathol.*
Grimes, Michael (1888), Dairy bact., Dairy Research Department, Univ. Coll. Cork (Irish Free State). *Dairy bact. Problems affecting creameries in the Irish Free State.* Ⓓ Iowa State Coll. 1923.
Grimm, Jay John, Prof. of Biol., Canoll Coll. Waukesha, Wis. (U.S.A.). *Plant Physiol.* o
Grimmer, Walter (1878), Dr. phil., Prof. für Milchwirtschaft Univ. Königsberg i. Pr. (Deutschland), Tragheimer Kirchenstr. 83. *Biochem. der Milch und der Molkereiprodukte.* Ⓓ Göttingen 1904.
Grimpe, Johann Georg (1889), Dr. phil.,Priv.Doc. Univ. Leipzig, Kustos der Sammlungen des Zool. Inst. Leipzig C 1 (Deutschland), Talstr. 33. *Mollusca (spez. Cephalopoda), Mammalia.* Ⓓ Leipzig 1913.
Grimshaw, Percy Hall (1869), Ass. Keeper, Natural History Department, The Royal Scottish Mus. Edinburgh (Scotland). *Entomol., British Diptera.*
Grindee, Boris Konstantinovič, Doc. I. Univ. Moskau (U.d.S.S.R.). *Anat.* o
Grinnell, Joseph (1877), Dir., California Mus. of Vertebrate Zool. and Prof. of Zool. Univ. of California. Berkeley, Cal. (U.S.A.). *Distribution of vertebrate animals in California.* Ⓓ Leland Stanford Junior Univ. 1913.
Grintescu, Johann, Prof. Acad. d. Agric. Cluj (România). *Pflanzenanat., Pflanzenphysiol.* o
Griswold, Grace Hall (1872), Research Instr. in Economic Entomol., Cornell Univ. Ithaca, N.Y. (U.S.A.), Care Dept. of Entomol. *Insects that attack ornamental plants. Hymenopterous parasit. of greenhouse aphids and aleyrodids.* Ⓓ Cornell Univ. 1925.
Groat, William Avery (1876), Prof. Clinical Pathol., Coll. of Med. Univ. Syracuse, N.Y. (U.S.A.), 608 East Genesee St. *Biol. Chem., particularly Metabolism.* Ⓓ Syracuse Univ. 1900. Ⓟ Stained Smears Rare Pathol. Bloods (Human).
Grobbelaar, Coert Smit (1886), Lect. Zool., Univ. Stellenbosch, Cape Province (South Africa), Riebeek Street. *Myol. of the Amphibia. Entomol.: Diptera (Tabanidae), Genetics.* Ⓓ 1923. Ⓟ Amphibia, Insecta, Mollusca.
Grobben, Karl (1854), o. Prof. emerit. der Zool. der Univ. Wien XVIII (Deutsch-Österreich), Sternwartestr. 49. *Morphol. Crustacea, Mollusca, Echinoderma.* Ⓓ Wien 1877.
Grochmalicki, Jan (1883), Univ.Prof. Poznań (Polska), Univ. *Faunistik, Crustaceen, Mollusken; Embryol. der Wirbeltiere.* Ⓓ Lwów 1908.
Grodziński, Zygmunt (1896), Ass. vergl. Anat. Univ. Kraków (Polen), św. Anny 6. *Anat. und Entwicklungsgeschichte des Lymph- und Blutgefäßsystems.* Ⓓ Kraków 1923.
Groebbels, Franz (1888), Prof., Priv.Doc. der Physiol. Hamburg 20 (Deutschland), Physiol. Inst. der Univ., Krhs. Eppendorf. *Physiol. der Wirbeltiere, Histophysiol., Vogelflug.* Ⓓ Heidelberg 1913.
v. **Gröer,** Franz (1887), o.ö. Prof. für Kinderheilkunde und Dir. der Kinderklinik an der Univ. zu Lwów (Polen), 5, Senatorska. *Serol. und Immunitätswiss., Physiol. und physik. Chem.* Ⓓ Breslau 1911, St. Petersburg 1912, Wien 1914.
Grönberg, G., Doc. Univ. Stockholm (Sverige). *Zool.* o
Grönblad, Rolf Leo (1895), Odontol. Licentiat Zahnarzt. Elisenvaara (Finland). *Desmidiaceae.* Ⓓ Helsingfors 1922.
Groeneveld, Gerrit (1897), Prosector Anat. Labor. Groningen (Holland), Oostersingel 69. *Anat. und Embryol.*
Grönroos, Hjalmar, Dr. phil. et med., Prof. Univ., Dir. anat. Inst. Helsinki (Finnland), Konstantinsgatan 12. *Anat., Embryol.* o
Gröntved, Johannes (1882). Museumsamanuensis, M. sc. København K (Danmark), Botanisches Mus. *Systematische Botan., Floristik (baltische Strand- pflanzen).* Ⓓ Kopenhagen 1917. Ⓟ Pflanzen: dänische und aus dem Ostbalticum.
Grönwall, Karl Anders Axel (1869), Prof. Geol. Univ. Lund (Sverige), Geol.-Mineralog. Inst. *Paleont. der in Schonen und im baltischen Gebiet vorkommenden Formationen.* Ⓓ Lund 1897.
Groff, George Weidman (1884), Prof. of Horticulture, Lingnan Agric. Coll., Univ. Canton (China). Berkeley, Cal. (U.S.A.), Box 117, Univ. of California. *Systematic and distributional botan., reciprocal plant contributions between China and other countries.* Ⓓ Pennsylvania State Coll. 1918. Ⓟ Herbarium specimens from South China. Seeds and Plants.
Gróh, Julius (1886), Dr., o.ö. Prof. der Chem. an der Kgl. Ungarischen Tierärztl. Hochsch. Budapest VII (Ungarn), Rottenbiller-ut. 23. *Kolloidchem., Spektrochem., Reaktionskinetik, mit Rücksicht auf die biochem. Forschung.* Ⓓ Budapest 1909.
Groll, Hermann (1888), a.o. Univ.Prof., Prosektor am Pathol. Inst. der Univ. München (Deutschland), Schubertstr. 8 I. *Allgemeine Pathol. und spezielle pathol. Anat. Entzündung, Gewebsatmung, Lymphatismus.* Ⓓ München 1911.
Gromowa-Lwowa, Ljubow A., Priv.Doc. u. Ass. II. Univ. Moskau (U.d.S.S.R.), Kadaschewskaja

naber., M. Tolmatschewskij per. 5, W. 5. *Botan., Pflanzen-Anat.* o
Grønlien, Nils (1874). Voss (Norge). *Microlepidoptera, mines.* ③ Oslo. ⑨ *Micros, mining lep. and mines.*
Gropengiesser, Curt (1882), Dr. Bern (Schweiz), Brunnadernstr. 38 A. *Bekämpfung von Pflanzenkrankheiten mit chem. Mitteln. Herstellung von pharm. Praep.* ③ Köln a. Rh. 1925.
Grošelj, Pavel, Dr., Doc. allg. Biol., Dir. Biol. Inst. Med. Fac. Univ. Ljubljana (S.H.S.), Zaloška cesla. *Coelent.* o
Gross, Alfred Otto (1883), Prof. of Biol., Bowdoin Coll. Brunswick, Me. (U.S.A.). *Ornithol.* ③ Harvard, Cambridge, Massachusetts.
Gross, Eberhard (1888), Dr., Priv.Doc. Heidelberg (Deutschland), Handschuhsheimer Str. 45 b. *Physiol.* o
Gross, Emanuel (1868), Dr. der Bodenkultur, Ingenieur, o. ö. Hochsch. Prof. der deutschen techn. Hochsch. in Prag, landwirtschaftl. Abteilung. Tetschen-Liebwerd, Böhmen (C.S.R.), Landw. Hochsch.-Abtlg. *Pflanzenproduktionslehre, Obst- und Gemüsebau, landwirtschaftl. Versuchswesen, mechanische Bearbeitung des Bodens als wesentlicher Produktionsfaktor.* ③ Wien 1911. ⑨ *Landwirtschaftl. Kulturpflanzen Mitteleuropas aller Art in Samen. Einschlägige graphische Darstellungen über Erträge derselben.*
Gross, Erwin George (1892), Ass. Prof. of Pharm. and Toxicol. New Haven, Conn. (U.S.A.), 333 Cedar Street, Sterling Hall of Med. *Chem. Pharm.* ③ Univ. of Wisconsin 1917.
Gross, Louis (1895), Acting Dir. of Labor. New York City (U.S.A.), Mount Sinai Hospital. *Pathol.* ③ McGill Univ. 1916.
Gross, Walter (1878), o. Prof., Dir. pathol. Inst. Univ. Münster i. Westfalen (Deutschland), Westring 17. *Vitale Färbung bes. ihre Verwendung zur Feststellung pathol. Zellveränderungen. Pathol. der Nieren. Pathol. Histol. des Nervensystems.* ③ Heidelberg 1903.
Grosser, Otto (1873), o. Prof. der Anat. an der deutschen Univ. Prag II (C.S.R.), Salmovska 5. *Entwicklungsgeschichte (menschliche und vergl.).* ③ Wien 1900.
Grossheim, Alexander A., Botan. Botan. Garten Tiflis (U.d.S.S.R.). *Pflanzen-Geographie. Systematik.* o
Grosso, Giacomo (1880), Lib. Doc. Batt. Univ. Lib. Doc. di Patol. Gen. Vet. Scuola Sup. di Med.- Vet. Torino (Italia). *Batt. Sierol. Istopatol.* ③ Med. vet. Torino 1902, med. Genova 1916. ⑨ *Microphotographien, histol. Praep. (Vet.- und Humanmed.).*
Groß, Julius (1869), Bibliotecario della Stazione Zool. di Napoli (Italia). *Istol. degli Insetti, Citol., Mendelismo, Spirochaeta, Hydra.* ③ Jena 1901.
Grote, Hermann (1882), Ornithol. Berlin-Charlottenburg (Deutschland), Trendelenburgstr. 16. *Ornithol. Afrika, Kirkisensteppe, Südrussland, Finnland.*
Grote, Louis Radcliffe (1886), Prof. Dr. med., Chefarzt von Dr. Lahmanns Sanatorium. Dresden-Weißer Hirsch (Deutschland). *Klinische Konstitutionspathol. Vererbungslehre. Stoffwechsel.* ③ Berlin 1912.
Grotjahn, Alfred (1869), o. Prof. der sozialen Hygiene. Berlin W 35 (Deutschland), Derfflingerstr. 24. *Anwendung der Vererbungsbiol. und Vererbungspathol. auf den Menschen.* ③ 1894.
Grout, Abel J. (1867), First Ass. in Biol., Curtis High School. New York City (U.S.A.), New Brighton, Vine St. *Biol. of Mosses.* ③ Columbia 1897. ⑨ *American Mosses.*
Grove, William Bywater (1848), Lect. in Botan. Municipal technical School Birmingham (England), 46 Duchess Road. *Mycol.; Bacteria and yeast-fungi;*
Rust-fungi (uredinales); British Flowering Plants. ③ Cambridge 1889. ⑨ *Dried specimens of fungi.*
Grover, Frederick Orville, Prof. of Botan., Oberlin Coll. Oberlin, O. (U.S.A.). *Genetics, Taxonomy.* o
Groves, James F. (1879), Prof. of Botan. Ripon Coll. also U.S. Plant Disease Survey. Ripon, Wis. (U.S.A.). *Plant pathol., Plant physiol. in relation to temperature Effects.* ③ Univ. of Chicago 1915.
Groysbeck, Johann Ludwig (1884), Sekretär der Landwirtschaftskammer für Niederösterreich. Loosdorf an der Westbahn (Österreich). *Biochem., Biol. landwirtschaftl. Nutzpflanzen und Nutzanwendung auf die Gesundheitslehre. Arzneipflanzenkultur.* ⑨ *Arzneipflanzenstecklinge, Samen, Drogen.*
Gruber, August (1853), Dr. phil., Prof. der Zool. a. D. Lindau, Bodensee (Deutschland), Lindenhof. *Protozoa.* ③ Leipzig 1877.
Gruber, Charles M. (1887), Associate Prof. on Pharm. St. Louis, Mo. (U.S.A.), Washington Univ., School of Med. *Endocrinol., Pharm., Physiol.* ③ Ph. Harvard Univ. 1914, M. Washington Univ., School of Med., St. Louis, Mo.
Gruber, Georg Benno (1884), o.ö. Prof. für pathol. Anat. an der Univ. Innsbruck (Deutsch-Österreich). *Pathol. Anat. und Histol., Mißbildungslehre. Morphol. Untersuchungen zur Klärung klinischer Krankheitserscheinungen. Ursachenforschung für Krankheitsbilder.* ③ München 1909. ⑨ *Pathol. anat. Praep.*
von Gruber, Max (1853), Präsident Bayr. Akademie der Wissenschaften, o. Prof. emer. Hygiene und Bact. Univ. München (Deutschland), Prinzenstraße 10. *Physiol. d. Stoffwechsels, Ernährung, Bact. Epidemiol. Cholera. Infektion. Immunität. Entd. d. spez. Agglutination. Leucin. Vererbung. Eugenetik. Alkohol.* ③ Wien 1876.
Grüß, Johannes (1860), Dr. phil., Prof., Wissenschaftl. Mitarbeiter am Inst. für Gärungsgewerbe. Friedrichshagen (Deutschland), Bruno-Wille-Str. 56. *Gärungsphysiol., Palaeomykol. Spezielle Arbeitsgebiete: Biol. der Nectarorganismen; Cytase- und Hydrogenaseuntersuchungen; Erforschung fossiler Pilze; Physiol. Umwandlungen an Kohlenhydrate.* ③ Berlin 1885. ⑨ *Nematophora, fossil auf Sandsteinplatte aus Spitzbergen. Cyclostoma kilthorkense von der Bäreninsel. Dünnschliffe: Nematophora, devonische Pilze, Kieselkohlen. Alpenpflanzen. Fossilien.*
Gruhl, Kurt (1888), Dr. phil. Grünberg, Schlesien (Deutschland), Klietestr. 11. *Paarungsvorspiele der Insekten. Faunistische Erforschung der engeren Heimat (Wirbeltiere, Insekten).* ③ Breslau 1911. ⑨ *Dipteren.*
Gruvel, Jean Abel (1870), Prof. au Mus. national. Paris Ve (France), 57, Rue Cuvier. *Cirripèdes et la faune marine et d'eaux douces des Colonies françaises.* ③ Paris 1894.
Grynfeltt, Edouard (1871), Prof. d'Anat. pathol. Fac. méd. Montpellier (France), 8 Place St. Côme. *Histol. normale et pathol.* ③ Montpellier 1922.
Grysez, Victor-Maurice (1875), Chef de Labor. Inst. Pasteur. Lille (France), 87, Rue Fréderic Moltez. *Bact.* ③ Lyon 1897.
Grzybowski, Alexander Genrichowitsch (1885), Smolensk (U.d.S.S.R.), Meerowskoe Chaussee N. 1, Qu. 9. *Ornithol. Vogelzug.* ③ Med. Moskau 1914.
Gualdi, Antonio (1901), Ass. Ist. Patol. gen. Univ. Napoli (Italia), 21 S. Andrea delle Dame. ③ Napoli 1926.
Guba, Emil Frederick (1897), Ass. Research Prof. of Botan., Market Garden Field Station, Massachusetts Agric. Coll. Waltham, Mass. (U.S.A.), 240 Beaver St. *Plant pathol.: investigations of the diseases of economic plants.* ③ Univ. of Illinois 1923. ⑨ *Genus Pestalozzia.*
Gubányi, Emil (1898), Pflanzenbiochem. Inst. Budapest (Ungarn), Debröi-ut 15. *Agric.- u. Biochem.*
Guberlet, John E. (1887), Prof., Associate Prof. of Zool. Seattle, Wash. (U.S.A.), Univ. of Washington. *Animal parasites of fish and birds. Parasit*

of poultry and sheep, life history studies on chicken cestodes. ⒹIllinois 1914.

Gudger, Eugene Willis (1866), Bibliographer and Associate in Ichthyol., American Mus. of Natural History. New York City (U.S.A.). *Breeding habits and embryol. of fishes; unusual methods of fishing practiced by primitive people; history and bibliography of ichthyol.* ⒹHopkins 1905.

Guéguen, Edouard (1885), Prof. de botan. à l'Ecole de Méd. de Nantes (France), Loire-Inférieure. *Chim. biol., Algues.* ⒹParis. ⓅAlgues.

Günther, Carl (1854), Univ.Prof. G.Med.R. Berlin-Lichterfelde (Deutschland), Hindenburgdamm 122. *Hygiene. Bact.* ⒹBerlin 1879.

Günther, Gustav (1868), Dr. med., Tierarzt, Mag. Pharm., o.ö. Prof. der Arzneimittellehre an der Wiener Tierärztl. Hochsch. Wien III (Deutsch-Österreich), Linke Bahngasse 11. *Pharm. der Gefäße und der Haut.* ⒹWien 1889, 1898, 1901.

Guenther, Konrad (1874), a.o. Prof. Freiburg i. Br. (Deutschland), Reichsgrafenstr. 18. *Ornithol., Naturschutz.* o

Gürber, August (1864), Dr. med., Dr. phil., Dr. med. dent. h. c., ö.o. Prof. der Med., Dir. des pharm. Inst., Marburg a. d. Lahn (Deutschland), Marbacher Weg 5. *Chem. und Physiol. des Blutes (Blutsalze, Serumalbuminkristalle, Hämoglobinkristalle, Stoffaustausch zwischen Plasma und Blutkörperchen, Sauerstoffmangel und Blutkörperchen, endozelluläre Hämoglobinkristalle, Vakuolenbildung in Blutkörperchen); Verdauung (Pepsin und Säure, Pepsin und Trypsin, Änderung des Drehungsvermögens der Eiweiße); Stoffwechsel (Alkohol, Nahrung, Aderlaß, Schilddrüse, Phosphor); Pharm. (chem. Konstitution und Wirkung, Quecksilbervergiftung, Uzara, Suprarenin, Lokalanästhetica, Diuretica, Campher, Narcotica, Antithyreodin, Amphibiengifte).* ⒹZürich 1890, Würzburg 1893, h. c. Marburg 1924.

Gürich, Georg (1859), Univ.Prof. Hamburg (Deutschland), Lübecker Str. 22. *Vertebr. fossil., palaeoz. Crust.* ⒹBreslau.

Guérin, Paul Emile Alexis (1868), Prof. Agrégé Fac. Pharm., Prof. à l'Inst. nat. agr. Paris 6 (France), 4, Avenue de l'Observatoire. *Anat. végétale. Cytol.*

Guerreiro, Luís (1891), Chef des Travaux d'Anat. Fac. Méd., Chargé du Cours d'Anat. topogr. Lisboa (Portugal), R. Cidade Cardiff n. 21—1. *Variations dans l'Anat. physiol. macroscopique humaine.* ⒹLisbonne 1920.

Guerrini, Guido (1878), Prof. Stab. Patol. gen., Dir. Ist. Patol. compar. Milano (Italia), Corso Buenos Aires 48. *Patol. sperimentale, Fisiol. patol., Fisica biol., Chim. biol., Immunol. Anafilassi. Ghiandole endocrine. Fisiopatol. dei muscoli degenerati. Azione patogena dei parassiti animali. Vitamine. Avitaminosi. Patogenesi delle emorragie.* ⒹMilano 1916.

Guevara, Romulo, Ass. Prof. Univ. of the Philipp. Manila, Philippine Islands (U.S.A.). *Pharm.* o

Guggiani, P. B., Prof. Univ. Asunción (Paraguay). *Biol.* o

Guggisberg, Hans (1880), Prof., Dir. der Univ. Frauenklinik. Bern (Schweiz). *Innere Secretion der Placenta. Stoffwechsel in der Schwangerschaft. Wachstumsprobleme.* ⒹBern 1904.

Gugnoni, Cesare (1876), Prof. straord. di Zootecnia, Labor. di Zootecnia. Perugia (Italia), R. Ist. Sup. Agr.

Gul, Harry L. (1890), Ass. Entomol. Wooster, O. (U.S.A.), Ohio Agric. Experim. Station. *Vegetable crop pest investigation.* ⒹKansas State Agric. Coll. 1926.

Guiart, Jules (1870), Prof. Parasit. et Hist. nat. méd., Chargé du cours d'Histoire Méd. Fac. Méd. Lyon, Rhône (France), 58, Boulevard de la Croix-Rousse. *Helminthol. Glande thyroïde dans la série des Vertébrés et en particulier chez les Sélaciens.*

Gastéropodes Opisthobranches. Céphalaspides. ⒹMéd. 1896, sc. nat. 1901.

Guiart, M., Prof. Fac. Méd. Lyon (France). *Zool.* o

Guibert, Hermann Louis Joseph (1898), Chef de labor. des cliniques (Service d'Anat. Pathol.) à la Fac. de Méd. de Montpellier, Hérault (France), 16, rue des Carmes. *Tissu conjonctif, Réactions dans les tumeurs épitheliales. Tumeurs du Tissu conjonctif.* Ⓓ1924. ⓅCobayes.

Guieysse-Pellissier, Ma., Fac. Méd. Paris 5 (France), 26, Rue Vavin. *Histol.* o

Guignard, Jean Louis Léon (1852), Prof. Fac. Pharm. Paris 6 (France), 4, Avenue de l'Observatoire. *Cytol.*

Guilarovski, Ivan P., Prof. Botan. Inst. Agron. Samara (U.d.S.S.R.). *Mycol.* o

Guillaumin, André (1885), Ass. Lab. Mus. Nat. d'Hist. nat. Paris 5 (France), 61, Rue du Buffon. *Flore de la Nouvelle-Calédonie.* ⒹParis 1920.

Guillemard, Francis Henry Hill (1852). Cambridge (England), Old Mill House. *Ornithol.: Birds of the Netherlands India and New Guinea, also of the Philippine Islands.* ⒹCambridge 1881.

Guillemard, M., Prof. Fac. Méd. Alger (Afrique). *Biochem.* o

Guillén García, José Maria de., Dir. la Realde Estación de Patol. veg. Barcelona (España), Avenida Príncipe de Asturias 4. o

Guilliermond, Alexandre (1876), Chargé du Cours de Botanique P.C.N. à la Sorbonne. Paris (France), 12, Rue Cuvier. *Cytol. végétale. Développement et classification des Levures.* ⒹParis 1923.

Guillouet, Prof. Ecole prép. méd. et pharm. Rouen (France). *Anat., Physiol.* o

Guimarães, Ant. Luís M., Prof. Univ. Porto (Portugal). *Biol.* o

Guimarães, F. P., Prof. Univ. Rio de Janeiro (Brasil). *Pathol.* o

Guise, Cedric H. (1890), Ass. Prof. of Forest Management, Cornell Univ. Ithaca, N.Y. (U.S.A.). *Wood preservation and wood seasoning.* ⒹCornell Univ. 1915.

Guitman, Lia S., Chef des trav. Stat. Défense des Plantes. Toula (U.d.S.S.R.). *Pathol. végétale, Ustilaginées.* o

Guizzetti, Pietro (1862), Prof. Anat. Patol. Univ. Parma (Italia), Via Bixio 82. *Ipofisi cerebrale dell' uomo. Ferro nel sistema nervoso centrale.*

Gulick, Addison, A.M., Ph.D., Assoc. Prof. of Physiol. Columbia, Mo. (U.S.A.), Univ. of Missouri. o

Gundersen, Alfred, Curator of plants, Brooklyn Botan. Garden. Brooklyn, N.Y. (U.S.A.), Washington Avenue 1000. *Classification and evolution of Dicotyledons.* o

Gunjco, Gregory (1887), Manager of the Section Technical and officinal Plants of the Botan. Garden. Nikita—Yalta, Crimée (U.d.S.S.R.). *Dipsacus fullonum, Luffa, Plants which give essential oils and alcaloids.* ⓅPlants, Seeds.

Gunn, J. A., Prof. Univ. Fac. Med. Oxford (England). *Pharm.* o

Gunnarsson, Johan Gottfrid (1866), Hvellinge (Sverige). *Gattung Betula.* ⓅBetula.

Gunthorp, Horace (1881), Prof. of Zool., Mills Coll. Oakland, Cal. (U.S.A.). *Myriapoda.* ⒹKansas 1923.

Gurevič, Mich. Osipovič, Prof. II. Univ. Moskau (U.d.S.S.R.). *Neurol.* o

Gurin, Gavril Ivanovič, Prof. Timirjasev Akad. Moskau (U.d.S.S.R.). *Anat. d. Haustiere.* o

Gurney, G. H. Norwich (England), Keswick Hall. *Ornithol.* o

Gurwitsch, Alexander (1874), Prof. Histol., Dir. histol. Inst. I. Univ. Moskau (U.d.S.S.R.), Mochovaja. *Ursachen und Faktoren der Zellteilung (speziell mitogenetische Strahlung). Analytische Embryol. (embryonale Felder). Histophysiol. d. Muskels.* ⒹMünchen 1898, Petersburg 1906.

Gurwitsch, Lydia, Ass. Dr. Histol. Inst. I. Univ. Moskau (U.d.S.S.R.), Mochowaja. *Entwicklungsmechan. Embryonale Felderung. Hormone.* o

Guselʼnikova, E. P., Doc. Univ. Taskent (U.d. S.S.R.). *Mikrobiol.* o

Gušić, Branimir (1901), Ass. der Oto-rhino-laryngologischen Klinik Zagreb (S.H.S.), Piverska ulica N. 3/II. *Lepidoptereol. Faunistische Verbreitung der Rhopaloceren des Balkans und der jugoslavischen Hochgebirge.* ① Zagreb 1926. ⑫ Rhopaloceren.

Gussewa, Kapitolina A., Doc. Moskauer Forst-Inst. Moskau (U.d.S.S.R.), B. Spasskaja, Dokutschajew per. 11, W. 9. *Botan., Mykol.* o

Gustafson, Felix Gustav (1889), Ass. Prof. of Plant Physiol., Univ. of Michigan. Ann Arbor, Mich. (U.S.A.). *Hydrogen ion concentration. Growth and Respiration.* ① Wisconsin 1921.

Gustavson, Reuben Gilbert (1892), Associate Prof. of Chem., Univ. Denver, Col. (U.S.A.). *Bio-chem. of the female sex hormone, isolation and purification of the female sex hormone, structure of the female sex hormone. Concentration of the antirachitic vitamine.* ① Univ. of Chicago 1925.

Gușuleac, Michail (1887), Prof. Univ., Dir. botan. Inst. u. Garten. Cernăuți (România), Grădina Botan. *Systematische Botan.: Flora der Bukowina. Anchuseae. Bothriospermum- und Thyrocarpusfrüchte.* ① Cernăuți 1917. ⑫ Herbarexsiccata Rumäniens.

Guterman, Carl Edward Frederick (1903), Ass. in Department of Plant Pathol., Cornell Univ. Ithaca, N.Y. (U.S.A.). *Plant Pathol.* ① Mass. Agric. Coll. Amherst 1925.

von Gutfeld, Fritz (1888), Dr. med., Dir. der bact. Abteilung des Städtischen Krankenhauses Am Urban. Berlin S (Deutschland). *Serol. der Syphilis, der Tuberkulose. Serodiagnostik der Gravidität. Vakzinetherapie. Theoretische Immunitätsfragen.* ① Berlin 1913.

Gutherz, Siegfried (1881), Dr. med., Priv.Doc. an der Univ. Berlin NW 6 (Deutschland), Anat.-biol. Inst., Luisenstr. 56. *Chromosomen, Spermien-Dimorphismus, Reizwirkung zellulärer Abbaustoffe, Sexualhormone, Ursprung tierischer Keimzellen, Muskelhistol.* ① Berlin 1906.

Guthrie, Charles Claude (1880), Prof. of Physiol. and Pharm. Pittsburgh, Pa. (U.S.A.), School of Med., Univ. of Pittsburgh. *Hemolysis, cerebral anemia, vascular suture and organ transplantation, heredity in fowls with transplanted ovaries, fundamental properties of skeletal and heart muscle, pharm. of magnesium salts, digitalis.* ① M. Univ. of Missouri 1901, ph. Univ. of Chicago 1907.

Guthrie, John Daulton (1903), Ass. Biochem. Yonkers, N.Y. (U.S.A.), Boyce Thompson Inst. *Effect of various enviromental conditions on the growth and chem. composition of plants. Chloroplast pigments.* ① Ohio State Univ. 1925.

Guthrie, Joseph Edward (1871), Prof. of Zool. Ames, Ia. (U.S.A.), Iowa State Coll. *Herpetol., Snakes of Iowa. Collembola of Minnesota.* ① Univ. of Minnesota 1901.

Guthrie, Mary J. (1895.), Ass. Prof. Zool., Univ. of Missouri. Columbia, Mo. (U.S.A.), Lefevre Hall. *Cytol., Cytoplasmic Inclusions.* ① Bryn Maur Coll. 1922.

Gutierrez, Avelino, Prof. Univ. Nac. Buenos Aires (Argentina). *Anat. topogr.* o

Gutmann, M. J. (1894), Dr. med., prakt. Arzt. München (Deutschland), Karlsplatz 8. *Tuberkulose, Zusammenhang mit Konstitution und Rassenfrage, Verbreitung der Tuberkulose. Vererbungspathol. Biol. und Pathol., Psychol. der Juden.* ① München 1919.

Gutner, Rosa Abramowna (1898), Labor. physiol. Chem. Univ. Leningrad (U.d.S.S.R.), Wassily Ostrow Linie 16, Haus 29. *Biogeochem., Zoochem., Phytochem. (Algen).*

Gutowski, Bolesław (1888), Ass. Inst. de Physiol. Univ. Warszawa (Pologne). *Digestion.* ① Varsovie 1922. ⑫ Des Chiens opérés avec la methode de Pawlov.

Gutschy, Ludwig. (1874), Prof. für Mikrobiol. an der landwirtschaftl. und technischen Fak. der Univ. in Zagreb (S.H.S.), Wilsonplatz 2. *Mikrobiol. des Bodens. Chem. und Mikrobiol. der Gärung.* ① Graz 1900.

Gutsell, James Squier (1887), Associate Biol., U.S. Bureau of Fisheries. Beaufort, N.C. (U.S.A.), U.S. Fisheries Labor. *Life-histories and ecological studies of marine bivalves.*

von Guttenberg, Hermann (1881), o. Prof. a. d. Univ., Dir. des Botan. Inst. und Gartens. Rostock (Deutschland), Botan. Inst. *Anat. und Physiol. der Pflanzen, Bewegungserscheinungen. Physiol.Anat. der Pilzgallen.* ① Graz 1904.

Guyénot, Emile Louis Charles (1885), Prof. de Zool. et Anat. comparée, Dir. de la Station de Zool. expérim., Univ. Genève (Suisse). Dir. adjoint de la Station de Zool. maritime de Wimereux, Pas de Calais (France). Genève (Suisse), Univ., Inst. de Zool. *Régénération des Amphibiens; déterminisme du sexe; caractères sexuels secondaires; génétique et hérédité en général; greffe des yeux; toutes questions relatives aux chromosomes. Protozoaires parasites (Coccidies, Microsporidies, Myxosporidies, Grégarines); Cestodes et Trématodes.* ① Genève 1918.

Guyer, Michael F. (1874), Prof. of Zool., Univ. of Wisconsin. Madison, Wis. (U.S.A.). *Experim. evolution; Genetics, Cytol. Experim. embryol.* ① PH.D. Univ. of Missouri 1900, LL.D. 1924.

Guyon, Prép. Coll. de France. Paris 5 (France), 9, Place Marcelin-Berthelot. *Histol.* o

Guyot, Henry (1891), Dr. ès sc. Basel (Schweiz), 35 Wettsteinallée. *Géographie botan., biol. des plantes alpines.* ① Genève. ⑫ Plantes supérieures de l'Europe moyenne et orientale.

Guyton, Fay E. (1893), Ass. Prof. Zool.-Entomol. Auburn, Ala. (U.S.A.). ① Ohio State Univ. 1920.

Guyton, Thomas Lee (1884), Chief Entomol., Pennsylvania Department of Agric. Harrisburg, Pa. (U.S.A.), Bureau of Plant Industry. *Insect. Insecticide treats.* ① Ohio State Univ. 1923.

Gwatkin, Ronald (1890), Dr., Lect. in bact., milk hygiene and poultry diseases at the Ontario Vet. Coll. Guelph, Ontario (Canada). *Animal diseases. Diseases of poultry.* ① Univ. of Toronto 1919.

Gwynne-Vaughan, Helen Charlotte Isabella (1879), Prof. of Botan. in the Univ., Head of the Department of Botan. London, E.C. 4 (England), Birkbeck Coll. *Cytol. and mycol., especially the cytol. of the Fungi. Ascomycetes, Ustilaginales, Uredinales.* ① Univ. of London 1907.

Gyelnik, Vilmos (1906). Budapest I (Ungarn), Görsz-ut 53/a. *Floristik u. Systematik der Flechten. Monographia peltigeraceae.* ⑫ Flechten. o

Gyldenstolpe, Count Nils Carl Gustaf Fersen (1886), Ass., Vertebrate Dept. Stockholm 50 (Sverige), Nat. History Mus. *Ornithol., Mammal., Herpethol.* ① Lund 1924.

von Győrffy, Eugen (1882), Adjunkt der Königl. ungarischen Entomol. Station. Budapest II (Ungarn), Kitaibel Pál Gasse 1. *Coleopterol., Apionidae der Erde.* ⑫ Apionidae der Erde.

Győrffy, István (1880), Dr., o.ö. Prof. der allgem. Botan. Szeged (Ungarn), Tisza Ring 61 I. *Oekol. der Bryophyta, die Flora der Hohen Tatra.* ① Kolozsvár (Siebenbürgen) 1904. ⑫ Moose.

Haagnér, Alwin Karl (1880), Dir.Nat.Zool.Gardens Pretoria, Transvaal (South Africa), Box 754. *Ornithol., Mammal, Protection of Wild Life.* ① Pittsburgh Univ., 1922.

Haaland, Knut Magnus (1876), Prosektor. Bergen (Norge), F. G. Gades Pathol. Inst. *Pathol. Anat. und Bact., experim. Pathol., Cancer.* ① Oslo 1909. ⑫ Pathol. Praep., Mäusetumoren.

Haaland, Margit Oddgjerd (1882), Ass. Bergen (Norge), F. G. Gade's pathol. Inst. *Typhus abdominalis, Cancer. Blutpar. d. Norw. Schneehuhnes.* ⓓ Oslo 1913.

de Haan, Klaas (1900), Ingenieur, Ass. am Labor. für die Physiol. der Tiere. Wageningen (Holland), Dijkgraaf 2. *Vitamine. Blutkörperchen. Farben.*

Haas, Emile (1881), Préparateur au labor. de Physique Méd. de la Fac. de Paris (France), 100 bis. Avenue Kléber. *Optique physiol.* ⓓ Paris 1907.

Haas, Fritz (1886), Dr. phil. nat., Kustos für Zool. am Senckenberg-Mus. Frankfurt a. M. (Deutschland), Victoriaallee 7. *Lamellibranchier, Najaden; geographische Verbreitung der Mollusken.* ⓓ Heidelberg 1910. ⓟ Land- und Süßwassermollusken.

Haas, Jos., Ing., Ass. l'Inst. des Recherches agric. et Forestiére. Brno (C.S.R.), Květná ul. *Biochem.* o

Haas, Paul (1877), Reader in Plant Chem. in the Univ. London W.C. 1 (England), Univ. Coll. *Chem. and Physiol. of Plant Products.* ⓓ London 1899, Freiburg 1901.

Haase-Bessell, Gertraud (1876). Dresden-N. (Deutschland), Hospitalstr. 3 II. *Erbphysiol. Untersuchungen in Verbindung mit zytol. Studien der Gattung Digitalis.* ⓟ Zytol. Digitalispraep.

Hassis, Ferdinand W. (1889), Ass., Department of Plant Physiol., The Johns Hopkins Univ., on leave from Branch of Research, U.S. Forest Service. Baltimore, Md. U.S. Forest Service, Washington, D.C. (U.S.A.). *Plant Physiol.* ⓓ Yale 1913.

Haber, Ernest S. (1896), Ass. Chief, Vegetable Crops Section. Ames, Ia. (U.S.A.), Iowa State Coll. *Experim. and physiol. work with vegetable crops.* ⓓ Iowa State Coll. 1922. ⓟ Squash, sweet corn, popcorn. Tomatoes.

Haber, Julia Moesel (1888), Personal Research Worker, Botan. Dept. State College, Pennsylvania (U.S.A.), 227 W. Beaver Ave. or Botan. Dept. Pennsylvania State Coll. *Morphol. and Anat. of the Euphorbiaceae.* ⓓ Cornell Univ. 1924. ⓟ Euphorbiaceae. Geraniales.

Haber, Vernon Raymond (1887), Associate Prof. of Zool. and Entomol., State College, Pennsylvania (U.S.A.). *Economic Entomol., Vertebrate Zool.* ⓓ Cornell Univ. 1924. ⓟ Cockroaches common to eastern continental America.

Haberlandt, Gottlieb (1854), Dr., Geh.Reg.R., o. Prof. der Botan. an der Univ. Berlin. Berlin-Wilmersdorf (Deutschland), Berliner Str. 66. *Anat. und Physiol. der Pflanzen; Zellenlehre.* ⓓ Tübingen 1876.

Haberlandt, Ludwig (1885), Dr. med., ao. Prof. der Physiol. an der Univ. Innsbruck. Innsbruck-Mühlau Nr. 135, Tirol (Deutsch-Österreich). *Allgemeine Herzphysiol. Innere Sekretion der weiblichen Keimdrüse.* ⓓ Graz 1909.

Hach, P. D. Iw. W. (1889), Dir. Variola-Vakzine-Abteilung Bact. Inst., Priv.Doc. der Mikrobiol. Med. Staatsinst. Kiew (U.d.S.S.R.), Stenka Rasinstrasse 4. *Fleckfieber. Variola-Vakzine. Scharlach. Physiol. der Mikroben. Gewebsimmunität.* ⓓ Kiew 1913. ⓟ Mikrobenkulturen.

Hachisuka, Masa Uji (1903). London W. 1 (England), 37 Portman Square. *Systematic Ornithol.*

Hacker, Henry Pollard (1885), Hon. Research Ass., Zool. Dept. London W.C. 1 (England), Univ. Coll. *Malaria.* ⓓ London 1912.

Hadding, Assar Robert (1886), Doc. Univ. Lund (Sverige). *Paleozoische und Mesozoische Evertebraten, besonders ordovicische Graptoliten und Trilobiten. Senon Brachiopoden.* ⓓ Lund 1913.

Haddow, John Reid (1896), Vet. Research Off. Muktesar P. O. Ritani, U.P. (Brit. India), Imperial Inst. of Vet. Research. *Production of Sera.* ⓓ Royal (Dick) Vet. Coll., Edinburgh, 1923.

Hadley, Faith Palmerlee (1898), Research Ass. in Dental Bact. Ann Arbor, Mich. (U.S.A.), Coll. of Dental Surgery, Univ. of Michigan. *The bact. of dental caries. Root-canal and periapical infections* ⓓ Univ. of Michigan 1920.

Hadley, Frederick B. (1880), Prof. of Vet. Sc., Univ. of Wisconsin. Madison, Wis. (U.S.A.). *Animal Diseases.* ⓓ Ohio State Univ. 1907. ⓟ Parasites of the domesticated animals.

Hadley, Philip Bardwell (1881), Ass.Prof. of Bact., Univ. of Michigan, Coll. of Med. Ann Arbor, Mich. (U.S.A.). *Pathogenic bact. and protozool. Morphol. development and behaviorism in Homarus. Genetics in birds and rabbits; Avian diseases, bact. and protozoan; Bacteriophage; Microbic dissociation; Bibliography. Instability of Bact. species, active dissociation and transmissible autolysis.* ⓓ Brown Univ. 1908.

Hadži, Jovan (1884), o. Prof. Zool., Dir. des Zool. Inst. Univ. Ljubljana (S.H.S.), Gorupova ulica 16. *Morphol., Entwicklungsgeschichte und Systematik der Cnidarier, Spongien. Meeresbiol. Tierische Plankton der Adria.* ⓓ Wien 1907. ⓟ Höhlentiere (besonders Proteus).

Haeckel, Eduard Heinrich Werner (1886), Studienrat, Lehrer am Staatl. Gymnasium, Osterode, Ostpr. (Deutschland). *Anat. der Schnecken.* ⓓ Jena 1911. ⓟ Ostpreußische Schnecken und Muschelschalen.

Haecker, Roman. Th., Geol. Mus. der Akad. der Wiss. Leningrad (U.d.S.S.R.). *Silurische Faunen des europ. Rußlands.* o

Haecker, Valentin (1864), Dr. rer. nat., Dr. med. h. c., o. Prof. der Zool. an der Univ. Halle a. d. S. (Deutschland), Mozartstr. 20. *Zellenlehre, Radiolarien, Vererbungsforschung, Phaenogenetik, Ornithol. Gesang der Vögel, seine anat. u. physiol. Grundlagen.* ⓓ Rer. nat. Tübingen 1889, med. h. c. Halle a. d. S. 1925.

Hägg, R., Ass., Riksmus. Stockholm (Sverige). *Palaeont.* o

Hägglund, Erik Karl Mauritz (1887), Dr. phil. o. Prof. a. d. Akad. Åbo (Finnland). *Holzchem., Gärungschem.* ⓓ Stockholm 1914.

Häggqvist, Gösta Per Engelbert (1891), Prof., Dir. d. histol. Abt. d. Karolinischen Inst. Stockholm (Sverige). *Muskelhistol. Aktiver Bewegungsapparat. Hauthistol.* ⓓ Lund 1919.

Haehn, Hugo (1880), Dr. phil., Abteilungsvorsteher am Inst. f. Gärungsgewerbe der Landwirtschaftl. Hochsch. Berlin N 65 (Deutschland), Seestr. 13. *Biochem. der Hefe, Enzyme, Enzymmodelle.* ⓓ Heidelberg 1904.

Hähne, Hans Karl August (1900), Wissenschaftl. Hilfsarbeiter bei der Biol. Reichsanstalt für Land- und Forstwirtschaft, Zweigstelle Kiel (Deutschland), Niemannsweg 11. *Pflanzenkrankheiten.* ⓓ Jena 1925.

Hämmerle, Juan Andres (1876), Dr. phil. Cuxhaven (Deutschland), Süderwisch 51. *Die Pflanzen des Amtes Ritzebüttel. Phanerogamen und Gefäßkryptogamen, Moose.* ⓓ Göttingen 1898.

Hämmerling, J., Dr., Ass. Kais.-Wilh.-Inst. f. Biol. Berlin-Dahlem (Deutschland). *Protozool.* o

Haempel, Oskar (1882), o. Prof. der Hydrobiol. und Fischereiwirtschaftslehre an der Hochsch. für Bodenkultur in Wien XVIII/1 (Deutsch-Österreich), Hochschulstr. 17. *Angewandte Hydrobiol. Fischzucht in Seen, Teichen und Flüssen. Moderne Fischereibetriebslehre. Fischereibiol. der Alpenseen.* ⓓ Wien 1906.

Haendel, Ludwig (1869), Prof., Dr. med., Geh. Reg.R., Dir. der Bact. Abteilung des Reichsgesundheitsamtes. Berlin-Dahlem (Deutschland), Bötticherstr. 14. *Hygiene und Bact.* ⓓ Freiburg 1896.

Haenel, Karl (1874), Forstmeister und Reg.R. in Bayern. Bamberg (Deutschland), Markusplatz 6. *Biol. Schädlingsbekämpfung, besonders durch praktischen Vogelschutz; Bedeutung der Pardiae und Raptatores für die Forst- und Landwirtschaft; Bedeu-*

tung der Vogelwelt im allgemeinen für den Natur- und Heimatschutz.

Hänsel, Karl Egon Siegfried (1891), Dr. phil. Seesen a. Harz (Deutschland), Bergstr. 1. *Theoretische Zool.; Deszendenztheorie: Entstehung der Arten.* ⑨ Berlin 1913.

Haenseler, Conrad Martin (1888), Ass., Ass.Prof. Plant Pathol. New Brunswick, N.J. (U.S.A.), Agric. Experim. Sta. *Plant diseases. Root rots of peas. Verticillium wilts.* ⑨ Rutgers Coll. 1921.

Häyrén, Ernst Fredrik (1878), Dr. phil., Doc. der Botan. an der Univ. Helsingfors; Amanuensis am Geographischen Inst. derselben Univ. Helsingfors (Finnland), Univ. (oder: W. Chaussén 33 A). *Finnlands Pflanzengeographie. Finnlands Meeresalgen (in der Ostsee und im nördl. Eismeer).* ⑨ Helsingfors. ⑫ Algen.

Hafferl, Anton (1886), Priv.Doc. Ass. Wien IX (Deutsch-Österreich), Porzellangasse 36. *Vergl. Anat.* o

Haffner, Felix (1886), Dr. med., o. Prof. und Dir. des Pharmacol. Inst. der Univ. Königsberg i. Pr. (Deutschland), Kopernikusstr. 3/4. *Pharm., Kolloidchem. der Zelle.* ⑨ München 1912.

von Haffner, Konstantin (1895), Priv.Doc. der Zool. Marburg a. d. L. (Deutschland), Zool. Inst. *Pulmonata, Linguatulida, Symbiose.* ⑨ 1920.

Hafner, Johann (1867). Ljubljana (S.H.S.), Mišiča c. 23. *Schmetterlinge von Krain.* ⑫ Großund Kleinschmetterlinge.

Hafner, Maté (1865). Ljubljana (S.H.S.), Prisojna ulica 6. *Schmetterlinge und Käfer Krains, der Julischen Alpen, Grottenkäfer.* ⑫ Schmetterlinge der Julischen Alpen und Grottenkäfer Krains.

Hagan, Harold Raymond (1886), Associate Prof. Zool. and Head of Department. Salt Lake City, Utah (U.S.A.), Univ. of Utah. *Insect embryol.* ⑨ S.M. Harvard 1917.

Hagem, Oscar (1885), Prof. allg. Botan. Bergens Mus., Dir. botan. Labor., Dir. d. forstlichenVersuchsanstalt für West-Norwegen. Bergen (Norge). *Pflanzenphysiol., Vererbungsforschung, Mykol. und Bact. des Bodens, Pflanzenkrankheiten und Forstbotan.* ⑨ Oslo 1917.

Hagen, Werner Hugo Adolf (1884). Lübeck (Deutschland), Paulstr. 22a I. *Ornithologie.* ⑫ Vögel.

Hagerup, Olaf (1889), Mag. sc. København (Danmark), Botan. Labor. of the Univ. *Empetraceae, Bicornes, Moose, Morphol., Cytol., Periodicity.* ⑨ Copenhagen 1923.

Hagiwara, Tokio (1896), Tokyo (Japan), Kamimeguro 586. *Genetic of the Japanese morning glory (Pharbitis nil), the Balsam (Impatiens balsamina).*

Hagmann, Gottfried (1874), Dr. phil., Chef der Meteorol. Station von Taperinha-Santarem. Taperinha-Santarem-Pará (Brazil). *Fauna amazonica.* ⑨ Basel 1898. ⑫ Fauna amazonica.

Hagmeier, Ludwig Arthur (1886), Kustos und Prof. für Zool. an der Staatlichen Biol. Anstalt. Helgoland (Deutschland). *Oekol. der marinen Bodenfauna. Aquariumstechnik. Biol. der Auster.* ⑨ Heidelberg 1911.

Hague, Florence Sander (1889), Ass. Prof. of Biol., Sweet Briar Coll. Sweet Briar, Va. (U.S.A.). *Oligochaeta.* ⑨ Univ. of Illinois 1921.

von Hahn, Friedrich-Vincenz (1897), Leiter der kolloidbiol. Station am Eppendorfer Krankenhaus in Hamburg. Volksdorf b. Hamburg (Deutschland), Holthusenstr. 6. *Chem. und Biol. der oberflächenaktiven Substanzen. Kolloidchem. Untersuchungen von med. und biol. Substanzen, Körperbestandteilen usw. Vitaminuntersuchungen von kolloidbiol. Standpunkt aus. Untersuchungen über die Anwendung der Kolloidwissenschaft auf die Med. im allgemeinen und ihre Spezialgebiete, hauptsächlich Dermatol.* ⑨ Leipzig 1921. ⑫ Alle kolloidchem. in bezug. Praep.

Hahn, Jaroslav (1897), Priv.Doc. der Zool. an der Karls-Univ. in Prag II (C.S.R.), U Karlova 3. *Anat. und Physiol. der Insekten. Gregarinen der Oligochaeten.* ⑨ Prag 1921.

Hahn, Martin John (1865), o. Prof. der Hygiene an der Univ., Dir. des Hygien. Inst. Berlin NW 7 (Deutschland), Dorotheenstr. 28a. *Hygiene u. Bact.* ⑨ München 1889. ⑫ Bacterienkulturen.

Hahne, August (1873). Stettin-Nt. (Deutschland), Dunkerstr. 19. *Pflanzengeographie Mitteleuropas, Palaeont. des Devons.* ⑫ Insekten (europäische und exotische) und andere biol. Gegenstände.

Haibe, Dir. Inst. prov. bact. Namur (Belgique). o

Hailer, Ekkehard Eugen Reinhold (1877), Dr. rer. nat., Reg.R., Mitglied des Reichsgesundheitsamtes. Berlin-Dahlem (Deutschland), Unter den Eichen 82. *Biol. der Mikroorganismen, Verhalten einzelliger Organismen gegenüber Giften, Desinfektion u. Chemotherapie; gewerbehygien. Fragen.* ⑨ Tübingen 1901.

Haines, Frederick Merlin (1898), Lect., East London Coll., Univ. London N. 6 (England), 26 Talbot Road, Highgate. *Plant Physiol.* ⑨ London 1925.

van Haitsma, John P. (1884), Prof. of Organic Sc. of Calvin Coll. Grand Rapids, Mich. (U.S.A.). *Trematodes of Birds. Crassiphiala bulboglossa.* ⑨ Univ. of Michigan 1912.

Hajdu, István (1897), Dr., I. Ass. bei dem Hygien. Inst. der Stefan Tisza Univ. Debrecen (Ungarn), *Immunitätsforschung.* ⑨ Budapest 1922.

Hajek, August, Univ.Prof. Wien (Deutsch-Österreich), Hochsch. f. Bodenkultur. *Pflanzengeographie.* o

Hajek, Markus, Univ.Prof. Wien IX (Deutsch-Österreich), Beethovengasse 6. *Physiol. Sinnesorgane.* o

Hajós, Károly (1891), Ass. III. med. Univ.-Klinik. Budapest IV (Ungarn), Múz. kövüt 39. *Bact., Serol. Immunitätsforschung. Pathophysiol. der allergischen Erkrankungen und des vegetativen Nervensystems.* ⑨ Budapest 1913.

Håkansson, Artur (1896), Priv.Doc. Univ., Botan. Inst. Lund (Sverige). *Embryol. und Zytol. der Pflanzen.* ⑨ Lund (Schweden) 1923.

Hakki, Ismaíl (1877), Prof. Zool. parasit., Fac. méd. et Ecole vét. sup. Constantinople (Turquie). *Parasit. compar.* ⑨ Constantinople.

Hakobian, Pajlak, Prof. Staatsuniv. Erivan, Armenien (U.d.S.S.R.). *Histol.* o

Hale, James Rashleigh (1874). Kent (England), Boxley Vicarage Maidstone. *Ornithol. Ool. Lepidoptera. Botan.* ⑨ Oxford 1895. ⑫ Eggs.

Hale, Joseph Daniels (1898), Chief of Div. of Timber Physics. Forest Products Labor. of Canada. McGill Univ. Montreal (Canada). *Anat. characters of timber.*

Halid-Bey, Mehmed (1887), Prof. organ. Chem., Nahrungsmittelkunde. Konstantinopel (Türkei), Tierärztl. Hochsch., Skutaric-Sélimlée. *Leberegelseuche.*

Halipré, A., Prof Ecole prép. méd. et pharm. Rouen (France). *Histol.* o

Hall, Ada Roberta (1890), Prof. of Biol., Rome, Ga. (U.S.A.), Shorter Coll. *Resistance of Sea Urchin Eggs to Sulphurous Acid. Regeneration in the Annelid Nerve Cord. The Effect of Oxygen and Carbon Dioxide on the development of the Toad.* ⑨ Univ. of Illinois 1921.

van Hall, Constant Johan Jacob (1875). Zutphen (Holland), Brugstraat 13. *Agronomie und Phytopathol. der tropischen Kulturpflanzen. Krankheiten von Kaffee und Kakao, Krankheiten der Kulturpflanzen von Niederländisch-Ostindien; Krankheiten und Beschädigungen der Kakaopflanze in den verschiedenen Ländern.* ⑨ Amsterdam 1902.

Hall, David G. jr. (1902), Ass. Entomol., Curator of Diptera, Dept. of Entomol., Univ. of Arkansas. Fayetteville, Ark. (U.S.A.). *Diptera, Muscidae, Sarcophagidae, Taxonomy, Biol.* ⑨ North American Diptera.

Hall, E. Raymond (1902), Research Ass. in Mus. of Vertebrate Zool. Berkeley, Cal. (U.S.A.), Univ.

of California. *Mustelidae; anat., palaeont., distribution and systematics.* Ⓓ Univ. of California 1925.
Hall, Frank Gregory (1896), Ass. Prof. of Zool. Duke Univ. Durham, N.C. (U.S.A.). *Physiol. of Cold-blooded vertebrates (fish): Respiration, blood, swimbladder.* Ⓓ Univ. of Wisconsin 1923.
Hall, Harvey Monroe (1874), Investigator with the Carnegie Institution of Washington. Berkeley, Cal. (U.S.A.), Univ. of California. *Experim. taxonomy. Special studies in Haplopappus, the Madinae, and other groups of Spermaphytes.* Ⓓ Univ. of California 1901.
Hall, Maurice Crowther (1881), Senior Zool. and Chief of the Zool. Division, Bureau of Animal Industry, U.S. Department of Agric. Washington, D.C. (U.S.A.). *Vet. parasit., as regards taxonomy, life histories, pathol., treatment and prophylaxis; anthelmintics; chlorinated hydrocarbons, such as carbon tetrachloride and tetrachlorethylene, as anthelmintics.* Ⓓ Ph. D., Georg Wash. Univ. 1915, D.V.M., Georg Wash. Univ. 1916, D.Sc. (honorary), Coll. 1925. Ⓟ Specimens of parasites.
Hall, Richard P. (1900), Ass.Prof. of Microbiol., Department of Biol., Univ. Coll., New York City (U.S.A.), Univ. Heights. *Protozool., cytol. of Mastigophora.* Ⓓ Univ. of California 1924. Ⓟ Slides of Mastigophora (Euglenoids especially).
Hall, Robert, Tasmanian Mus. Hobart. Tasmania (Australia). *Ornithol.* ○
Hall, R. William (1872), Prof. of Biol. Univ. Bethlehem, Pa. (U.S.A.). Ⓓ Harvard 1901. ○
Hall, Walker (1868), Prof. of Pathol. Bristol (England), Univ. *Pathol. and Bact. Purin bodies.* Ⓓ Manchester 1899.
Hall, Winfield Scott (1861), Prof. emeritus on Med. Fac. Northwestern Univ., Chicago, Ill. Berwyn III, Ill. (U.S.A.). *General Biol. Physiol. Anthrop., Sociol.* Ⓓ Northwestern Univ. 1888, Leipzig med. 1894, phil. 1895.
Halla, František, Dir. Ecole d'Agric. Litovel na Moravĕ (C.S.R.). *Phanérogames, botan. appliquée.* ○
Halle, T. G. (1884), Prof., Keeper Palaeobotan. dept. Naturhistoriska Riksmus. Stockholm 50 (Sverige). *Fossil plants from the Palaeozoic of China. Plant-remains of the oldest land-flora.* Ⓓ Upsala, Sweden, 1911.
Haller, Oskar-Sigward (1891), Prof., Dir. Pestlabor. Urda, Uralsk. Gouv. (U.d.S.S.R.). *Bact. Pest.* Ⓟ Nagetiere des Südosten Rußlands.
Graf Haller v. Hallerstein, Victor, (1887), ao. Prof. Univ., Anat. Inst. Berlin NW 6 (Deutschland), Luisenstr. 56. *Anat. des Menschen und vergl. Anat. der Wirbeltiere: Gehirn, Hypophyse, Mund, Nase.* Ⓓ Berlin 1914.
Hallier-Schleiden, Hans (1868), Dr. phil. Oegstgeest bei Leiden (Holland), Dorpstraat 30. *System der Windengewächse; Morphol., Anat., vergl. Phytochem., Pflanzenverbreitung, Entwicklungslehre der Blütenpflanzen; Aufstellung eines natürl. Stammbaumes; Flora von Niederl.-Ostindien; Nomenklaturfrage.* Ⓓ Jena 1892.
Hallion, M., Dir. adj. Ecole des Hautes Etudes. Paris 8 (France), 54, rue du Faubourg St. Honoré. *Physiol.* ○
Hallock, Harold C. (1891), Ass. Entomol. Riverton, N.J. (U.S.A.), Japanese Beetle Labor. *Biol. study of insect parasites. Taxonomy of Diptera (Sarcophagidae).* Ⓓ Cornell Univ. 1922. Ⓟ Diptera (esp. Sarcophagidae).
Hallqvist, C. A., Doc. Univ. Lund (Sverige). *Genetik.* ○
d'Halluin, M., Prof. Univ. Lille (France). *Physique biol.* ○
Halpert, Béla (1896), Instr. in Anat. Baltimore, Md. (U.S.A.), Johns Hopkins Med. School. *Gall-Bladder and Bile-Ducts. Duodenum. Peritoneal abnormalities and topography of the abdominal organs. Vascular (venous) anomalies.* Ⓓ German Univ. of Prague, 1921.

Halversen, William V. (1893), Associate Prof. in Bact., Oregon State Agric. Coll. and Associate Bact., Oregon Agric. Experim. Station. Corvallis, Ore. (U.S.A.). *Agric. Bact., Seasonal conditions and soil treatment on Bacteria and Molds in the Soil.* Ⓓ Iowa State Coll., Ames, Iowa, 1924. Ⓟ Rhizobium leguminosarum.
Hamann, Carl A. (1868), Prof. of Applied Anat., Med. School, Western Reserve Univ. Cleveland, Ohio (U.S.A.), 416 Osborn Bldg. Ⓓ Univ. of Pennsylvania 1890. ○
Hamazaki, Yukio (1895), Prof. Dr., Doc. pathol. Inst. d. Univ. Okayama (Japan). *Milchflecke, Omentum, Lunge.* Ⓓ Okayama 1921.
Hamburger, Franz Anton (1874), Univ.Prof. f. Kinderheilkunde. Graz (Deutsch-Österreich), Mozartgasse 12. *Ausscheidung artfremden antitoxischen Serums, Antitoxin u. Eiweiß, biol. Inkubationszeit, Tuberkuloseimmunität, Schwankungen der Krankheitsdisposition. Arteigenheit und Assimilation.* Ⓓ Graz 1898.
Hamburger, Viktor (1900), Dr., Ass. am Kaiser-Wilhelm-Inst. für Biol., Abt. O. Mangold, Berlin-Dahlem (Deutschland). *Entwicklungsmechanik der Amphibienextremität. Farbensinn der Tiere.* Ⓓ Freiburg i. B. 1924.
Hamdi, Ahmed (1881), Prof. vergl. u. topographische Anat. u. med. Physik. Konstantinopel (Türkei), Tierärztl. Hochsch., Skutaric-Sélimiée. *Nerven- u. Gefäßsystem.*
Hamdi, H. (1874), Prof. pathol. Anat. Med. Fak. Haidar-Pascha. Univ.Constantinopel (Türkei), Kadi-Keuy. Moda Djaddesi 15. *Pathol. Anat.* Ⓓ Constantinopel, Leipzig. Ⓟ Pathol.-histol. Präp.
Hamilton, Clyde C. (1890), Associate Entomol., N.J. Agric. Experim. Station and Associate Prof. of Entomol., Rutgers Coll. New Brunswick, N.J. (U.S.A.). *Economic investigations of fruit insects and insects affecting floricultural plants. Relation of lead arsenate upon the foliage of apple trees to leaf growth. Systematic: larval stages of tigerbeetles and Carabids.* Ⓓ Ph.D. Cornell Univ.
Hamilton, Harold, Dominion Mus. Wellington (New Zealand). *Zool.* ○
Hamilton, William Ferguson (1893), Associate Prof. of Physiol., Univ. Louisville, Ky. (U.S.A.), 101 W. Chestnut Street. *Color vision; cardiac output; water exchange.* Ⓓ Univ. of California 1920. Ⓟ Apparatus for determining specific gravity of small quantities of body fluids.
Hamisky, Gregorio (1891), Prof. de la Fac. de Med. del Litoral. Rosario. (Argentina). *Odontol. Histol. des dents.* Ⓓ Argentina.
Hamlin, John Calhoun (1896), Associate Entomol. (in Field Charge, Dried-Fruit Insect), Bureau of Entomol., U.S. Dept. of Agric. Fresno, Cal. (U.S.A.), Dried-Fruit Labor., 712 Elizabeth Street. *Biol. and economic studies of the insect pests of dried fruits. Feeding habits of insects.* Ⓓ Ohio State Univ. 1918.
Hamm, Friedrich (1891), Dr. phil. nat., Dir.-Ass. für Geol., Mineral. und Palaeont. am Provinzial-Mus. zu Hannover (Deutschland), Krausenstr. 51. *Museol. Darstellung von Geol., Mineral. und Palaeont.* Ⓓ Heidelberg 1922.
Hammar, J. Aug. (1861), Prof. emer. Univ. Upsala (Sverige). *Konstitutionsanat., Inkretorgane, Thymus.* Ⓓ Upsala 1892.
Hammarlund, Carl Theodor Waldemar (1884), Doc. Phytopath. u. Mykol. Univ. Lund. Veredelungsleiter der Weibullsholms Saatzuchtanstalt. Landskrona (Sverige). *Genetik der Erbsen, Plantago usw. Phytopath. der Kulturpflanzen. Veredelung von Gemüsen.* Ⓓ Lund 1924.
Hammarsten, E., Dr., Karolinska Inst. Stockholm (Sverige). *Pharmacol.* ○
Hammarsten, Olof (1841), Prof. emer. Upsala (Sverige). *Milch und Labgerinnung, Blut und Blutgerinnung, Verdauung, Gallensäuren und Galle,*

Proteinstoffe. ① Med. 1869, phil. h. c. Upsala 1893, med. h. c. Christiania 1911.
Hammer, B. W. (1886), Prof. of Dairy Bact., Iowa State Coll., Chief, in Dairy Bact., Iowa Agric. Experim. Station, Iowa State Coll. Ames, Ia. (U.S.A.), Dairy Building. *Dairy Bact., milk fermentations.* ① Univ. of Chicago 1920. ⓟ Cultures of various milk bact.
Hammer, Ernst (1893), Anat.-Pathol. du Wilhelmina Gasthuis, Amsterdam (Holland), 1. Const. Huijgensstraat 101. *Anat. pathol.: Thrombose. Ulcère gastrique et duodénal. Statistique de l'Anat. pathol.* ① Amsterdam 1918. ⓟ Prép. d'Anat. pathol.
Hammermann, Adele (1888), Ass. Mus. d. Botan. Gartens, Med. Inst. Leningrad (U.d.S.S.R.), Pessotschnaja 2. *Pharmakognosie.*
Hammerschmidt, Johann (1876), ao. Prof. Hygiene Univ., Ass. Inst. für Hygiene. Graz (Deutsch-Österreich), Universitätsplatz 4. *Med. Bact. Septicämie, Reticuloendothel.* ① Med. Wien 1900, phil. Graz 1911.
Hammett, Frederick Simonds (1885), Ass. Prof. of Biochem. Philadelphia, Pa. (U.S.A.), The Wistar Inst. of Anat. *Growth and the Relation of the Internal Secretion.* ① Harvard Univ. 1915. ⓟ Albino rats.
Hammond, John (1889), Physiol., Animal Nutrition Inst. Univ. of Cambridge (England), School of Agric. *Biol. and Physiol. of Reproduction and Growth: Fertility and Sterility, Milk and Meat production. Reproduction in the Rabbit.* ① Cambridge 1914.
Hamorak, Nestor (1892), Prof. Botan. u. Phytopath. Landwirtschaftl. Inst. Kamenetz-Podolsk, Ukraine (U.d.S.S.R.), Mychajtytschenkostr. 5. *Anat. und Physiol. der Pflanzen. Mikrochem. und Physiol. des Spaltöffnungsapparates. Wasserhaushalt der Pflanze.* ① Wien 1915.
Hamsík, Antonín (1878), Dr., Prof. der med. Chem. an der Univ. Brno (C.S.R.), Údolní 73. *Lipase, Blutfarbstoffderivate (Oxyhämin, Hämin, Porphyrin).* ① Prag 1902.
Hamshaw Thomas, Hugh (1885), Lect. Botan. Univ. Cambridge (England), Botan. School. *Palaeobotan., especially Mesozoic. Ecol. Recent Pteridophyta.* ① Cambridge 1907. ⓟ Fossil Plants from the Jurassic rocks of Yorkshire.
Hanák, Antonín (1889), Prof. der Physiol. Bratislava (C.S.R.), Comenius Univ. *Anaphylaxie.* ① 1916.
Hance, Robert Theodore (1892), Associate Rockefeller Inst. for Med. Research. New York, N.Y. (U.S.A.), 66. Street and Avenue A. *Cytol., Chromosomes, X-ray effects.* ① Univ. of Pennsylvania 1917.
Handa, M. R. (1900), Ass. Lect. Biol., Univ. Coll. Rangoon, Burma (India), Biol. Department. *Zygnemales. Green Algae, Charales (Chara).* ⓟ Zygnema; Spirogyra; Mougeotia; Utricularia; Anthoceros; Riccia; Various bracket-form parasit. and saprophytic fungi; preparations of Mougeotia and Zygnema showing one-celled and two-celled rhizoidal outgrowths; all stages of conjugation in Spirogyra.
Handel-Mazzetti, Heinrich (1882), Dr. phil., Kustos an der botan. Abteilung des Naturhistorischen Mus. in Wien VIII (Deutsch-Österreich), Zeltgasse 1. *Systematische Botan. und Pflanzengeographie (Taraxacum, Leontopodium, Onobrychis, Haplophyllum. — Vorderasien, Ostasien).* ① Wien 1907.
Handlirsch, Anton (1865), Hofrat, Dr. phil., Mag. pharm., Dir. emer. am naturhist. Staatsmus. in Wien, Priv.Doc. an der Univ. Wien, wirkl. Mitglied der Akad. der Wissenschaften in Wien, Präsident der zool. botan. Gesellschaft. Wien IV (Deutsch-Österreich), Rubensgasse 5. *Allgem. Entomol. Phylogenie und Palaeont. der Arthropoden. Hemiptera, Hymenoptera (Systematik).* ① Graz 1923.
Handovsky, Hans (1888), ao. Prof., Med. Fac. Göttingen (Deutschland), Herzberger Landstr. 38. *Pharmacol. Kolloidchem. der Eiweißkörper.* ① Wien 1912.

Handschin, Eduard (1894), Dr. phil., Priv.Doc. Basel (Schweiz), Zool. Anstalt, Univ. *Allgemeine und angewandte Entomol. Oekol. und Zoogeographie. Apterygota, Coleoptera. Palaeont. der Collembola Subterranformen derselben (Myrmecophile, termitophile, micro- und macrocavernicole Formen).* ① Basel 1918.
Hanicke, Eugen (1869), Dir. Physikalisch-Physiol. Abt. Inst. für experim. Med. Leningrad (U.d.S.S.R.), Ulitza Krasnych Zor. 65. *Vererbung der bedingten Reflexe.* ① St. Petersburg 1892.
Haniel, Curt B. (1886), Dr. phil., Besitzer des Biol. Inst. in Schlederlohe, Post Wolfratshausen, Oberbayern (Deutschland). *Experim. Untersuchungen über die Parthenogenese der Schmetterlinge.* ① München 1920.
Hankinson, Thomas Leroy (1876), Prof. of Zool., Michigan State Normal Coll. Ypsilanti, Mich. (U.S.A.), 96 Oakwood Ave. ① Cornell 1900. ⓟ Fishes, fresh water forms of Eastern U.S.
Hankó, Béla (1886), Dr. phil., Prof. und Priv.Doc., Dir. der Biol. Balatonsee Station. Révfülöp a.-Balaton (Ungarn). *Experim. Morphol., Hydrobiol., Ichthyol.* ① Budapest 1910.
Hanneson, G., Prof., Dr., Reykjavík, Island (Danmark). *Fischereizool.* o
Hannesson, Pálmi (1898), Staatsschule in Akureyri (Island). *Biol. der Süßwasserfische Islands. Ornithol. Ökol. der Vögel.* ① Kopenhagen 1926. ⓟ Isländische Süßwassertiere, Vögel, Eier.
Hannevart, Germaine (1887), Dr. Sc. nat. Bruxelles (Belgique), 109, rue G. Grahy. *Physiol. animale: permeabilité.* Bruxelles 1922.
Hannig, Emil (1872), o. Prof. der Botan. Münster i. Westf. (Deutschland), Langenstr. 17. *Anat., Physiol., Pflanzengeographie.* ① Straßburg 1898.
Hanriot, M., Dr. agr. à la Fac. Méd. Paris 6 (France). *Biochem.* o
Hansen, Albert A. (1891), Associate Botan., Purdue Univ., Agric. Experim. Station. Lafayette, Ind. (U.S.A.). *Weeds, Poisonous plants, seeds.* ① Pennsylvania State Coll. 1915.
Hansen, C. H., Prof., Farmalogisk Labor., Vet. Landw. Hochsch. København V (Danmark), Bülowsvej. *Pharm.* o
Hansen, Hans Jakob (1855), Dr. phil. Gjentofte near København (Danmark), Søgaardsvej 24. *Arthropoda: systematic, comparative morphol., geographical distribution, anat., postembryonic development.* ① Copenhagen 1883.
Hansen, Karen Marie (1881). København (Danmark). *Blood sugar in man.* ① Med. København.
Hansen, Klaus Gustav (1895), Dr. med., Erster Amanuensis an dem Pharmacol. Inst. der Univ. Oslo (Norwegen). *Alkohol. Narkotica. Milchsäure. Vitale Oxydationsprozesse bei erhöhtem Sauerstoffpartialdruck. Phosphorvergiftung.* ① Oslo 1925.
Hansen, P., Dr. med., Pasteur-Inst. d. Univ. Tartu (Estland), Venet. 28. o
Hansen, Victor Georg (1889). København (Danmark), I. E. Ohlsensgade 10 II. *Danish Coleoptera.* ⓟ Danish Coleoptera.
Hanser, Robert (1884), Prof., Pathol. Anatom. Ludwigshafen a. Rhein (Deutschland), Städt. Krankenhaus. *Pathol. Anat., Bact., Serol.* ① Heidelberg 1908.
Hanson, Frank Blair (1886), Prof. of Zool., Washington Univ. St. Louis, Mo. (U.S.A.), Dept. of Zool. *Genetic. Effect of alcohol on germ cells. Modification of the germ plasm.* ① The American Univ. 1918.
Hanström, Bertil (1891), Doc. Zool. Univ. Lund. Landskrona (Sverige), St. Norregadan 120. *Vergl. Anat. und Histol. des Nervensystems und der Sinnesorgane von Turbellarien, Mollusken, Polychaeten, Xiphosuren, Arachnoiden, Pantopoden, Myriapoden, Crustaceen und Insekten. Vergl. Physiol. des Geruchs und Geschmacks.* ① Stockholm 1920.

Hanzawa, M., Prof. of Hokkaido Imper. Univ. Botan. Inst., Fac. of Agric. Sapporo (Japan). *Applied Mycol. and Agric. Microorganism.*

Hanzelka, František (1899), Ass. Mähr. Versuchsanstalt. Brno (C.S.R.), Schwarze Felder. *Wein. Obst.* ⓓ Brünn 1925.

Hanzlik,Paul J.(1885),Prof.of Pharmacol., Stanford Univ. School of Med. San Francisco, Cal. (U.S.A.), Sacramento and Webster Sts. *Pharm., biochem., therapeutics.* ⓓ Western Reserve 1912.

Happe, Heinrich (1890), Dr. med., wissenschaftliches Mitglied im Hauptgesundheitsamte der Stadt Berlin. Berlin C 2 (Deutschland). *Bact. Serol. Serol. der Lues.* ⓓ Halle a. d. S. 1925.

Happold, Frank Charles (1902), Demonstr. Bact., Med. School, Univ. Leeds (England), 37 Hilton Rd., Harchills. ⓓ Manchester 1923.

Hara, Hiroshi, Prof. of Tôkyô Imper. Univ. Horticult. Inst., Fac. of Agric. Komaba near Tôkyô (Japan). *Horticulture.*

Hara, Jûta, Prof. of Tôkyô Imper. Univ., Inst. of Hydrobiol., Fac. of Agric. Komaba near Tôkyô (Japan). *Economic oceanography.*

Hara, Kanesuke (1885), Expert in the Agric. Society of Perfecture. Shizuoka, Jônai (Japan). *Plant Pathol., Mycol. (Pyrenomycetes, Shaeropsidales, Gymnosporangium).*

Hara, Tôru, Lect. of Imper. Univ., Matsuo's Clinique of Inner Med., Coll. of Med. Kyôto (Japan). *Experim. Med. Pathol. Chem.*

Harbitz, Francis, Prof. Dr., Pathol.-Anat. Inst., Reichshospital. Oslo (Norge). *Pathol. Anat.* o

Harden, Arthur, Prof., Univ. London S.W. 7 (England), S. Kensington. *Biochem.* o

Hardenberg, Christian Bernhardus (1874), Chief, Division of Entomol., Dept. of Agric. Lourenço Marques (Port. East Africa), Box 250. *Economic Entomol.* ⓓ Wisconsin 1906. ⓟ Coleoptera, especially Coccinellidae in all stages.

Hardenberg, Johann Dietrich Franz (1902), Ass. Embryol. Inst. des Hubrechtfonds. Utrecht (Holland), Lange Jansstr. 6. *Embryol. der Insecten. Nervenmuskelphysiol.*

Harder, Nikolaus Richard (1888), Dr. phil., o. Prof. der Botan., Vorstand des botan. Inst. und Gartens der Technischen Hochschule in Stuttgart (Deutschland), Seestr. 70. *Gaswechselphysiol.; Mykol.; Rolle von Kern und Protoplasma bei der Vererbung; Reizphysiol.* ⓓ Kiel 1910.

Harding, Harry Alexis (1871), Chief, Dairy Research Division, Mathews Industries, Inc. Detroit, Mich. (U.S.A.), Box 834. *Biol. and Hygien. problems of Milk.* ⓓ Cornell Univ. 1910.

Hardy, George Austin (1888), Ass. Biol. Victoria, British Columbia (Canada), Provincial Mus. *Coleoptera: Cerambycidae.* ⓟ Cerambycidae.

Hardy, Sir.W. B., Gonville Coll., Chairman Food Invest. Board. Cambridge (England). *Pathol.* o

Hareyama, Seigo (1893), Lect. General Zool. of Imper. Univ., Zool. Labor., Coll. of Sc. Kyôtô Japan). *Spermatogenesis of Insects.* ⓓ Kyoto Imp. Univ. 1924.

Hargitt, Charles Weslay (1852), Research Prof. of Zool., Univ. Syracuse, N.Y. (U.S.A.). *General Biol. Morphol. and development of Coelenterata, Behavior of Annelids, Monographic descriptions of Medusae, Actinozoa, Hydrozoa.* ⓓ Phil. Ohio Univ. 1890, sc. (Hon.) Syracuse Univ. 1922.

Hargitt, George Thomas (1881), Prof. of Zool., Syracuse Univ., Syracuse, N.Y. (U.S.A.), Lyman Hall. *Embryol. and cytol., origin of germ cells and sex glands. Coelenterates, amphibia and mammals, origin and history of their germ cells. Germ plasm theory. Feeding of protozoa, effect of the food upon rate of reproduction.* ⓓ Harvard Univ. 1909.

Hárl, Paul (1869), o. Prof. der physiol. und pathol. Chem., Univ. Budapest VIII (Ungarn), Eszterházygasse 9. *Biochem., Tierische Kalorimetrie; Tierische Farbstoffe.* ⓓ Univ. Wien 1894. ⓟ Physiol.-chem. Praep.

Harkness, William John Knox (1896), Lect. in Limnobiol., Univ. of Toronto, Toronto (Canada). *Interrelationships among aquatic forms of life, fresh water fisheries.* ⓓ Univ. of Toronto 1922.

Harlan, Harry V. (1882), Senior Agronomist in Charge of Barley Investigations, U.S.D.A. Washington, D.C. (U.S.A.), 1306 B Street, SW. *Varieties of barley. Inheritance. Kernel development.* ⓓ Univ. of Minnesota 1914.

Harland, Sydney Cross, Prof. Imp. Coll. of Tropical Agric. Trinidad (Br.-W.-India). *Botan., Genetics.* o

Harman, Mary Theresa (1877), Prof. of Zool., Kansas State Agric. Coll., Manhattan, Kan. (U.S.A.). Zool. Department. *Regeneration: Planaria, Thinodrilus. Cytol.: Paratettix, Apotettix, Cavia cobaya, Taenia tenaeformis. Embryol.: teratol., normal Embryol.* ⓓ Indiana Univ., Bloomington, Ind. 1912.

Harmer, Sir Sidney Frederic (1862), Dir. emer. Natural History Depts., British Mus. Melbourn, Royston, Herts. (England), Old Manor House. *Zool.: Polyzoa, Pterobranchia, Cetacea. Protection of Animals.* ⓓ Cambridge 1897.

Harms, Jürgen Wilhelm (1884), o. Prof. der Zool. und vergl. Anat., Vorstand des Zool. Inst. Tübingen (Deutschland). *Experim. Geschlechtsdifferenzierung und Vererbung des Geschlechts. Lebensablauf, Altern und Tod. Sekretion. Seidenraupenzucht.* ⓓ Marburg a. d. L. 1907.

Harms, M., Dr., Konserv. Zool. Mus. Univ. Tartu (Estland), Aia t. 46. o

Harms, Wilhelm Max Bruno (1890), Stadtarzt. Berlin W 62 (Deutschland), Landgrafenstr. 7. *Aphaniptera.* ⓓ Berlin Dr. phil. 1912, Dr. med. 1920. ⓟ Aphaniptera. Dytisciden, Gyriniden, Hydrophiliden.

Harned, Horace H. (1886), Assoc. Prof. Bact., Mississippi Agric. and Mechanical Coll. A. and M. College, Miss. (U.S.A.). *Bact. of green manures.* ⓓ Univ. Wisconsin 1921.

Harned, Joseph E., Lect. Botan., High School. Oakland, Md. (U.S.A.). *Systematic and agric. botan.* o

Harned, Robey Wentworth (1884), Prof. of Zool. and Entomol. of the Mississippi A. and M. Coll., Entomol. of the Mississippi Experim. Station and of the State Plant Board of Mississippi. A. and M. College, Miss. (U.S.A.). *Insects of Mississippi; insects affecting pecans; Argentine ant control and eradication and Cicadidae.* ⓓ Ohio State Univ. 1906.

Harnisch, Otto E. R. (1901), Dr. phil., Ass. am Zool. Inst. der Univ. Köln a. Rh. (Deutschland), Stapelhaus. *Hydrobiol. (Fließende Gewässer, Moore), Atmungsphysiol. wirbelloser Tiere.* ⓓ Breslau 1924.

Harper, Robert Almer (1862), Prof. Botan., Columbia Univ. New York City, N.Y. (U.S.A.). *Cytol.* ⓓ Bonn 1896.

Harrassowitz, Hermann (1885), o. Prof. der Geol. und Palaeont., Dir. des Geol. und Palaeont. Inst. der Univ. Gießen (Deutschland), Ludwigstr. 30. *Fossile Schildkröten.* ⓓ Freiburg i. Br. 1909.

van Harreveld (1879), Dir. of the Bureau of Sugar. Den Haag (Holland), Houtweg 12. *Biol. and culture of Sugar Cane.* ⓓ Groningen 1907. ⓟ Sugar cane varieties.

Harrington, James Bishop (1894), Ass. Prof. of Field Husbandry, in charge of cereal breeding and investigations. Saskatoon, Saskatchewan (Canada), Univ. of Saskatchewan. *Cereal Crops. Plant Breeding; production of improved varieties; cereal variety testing. Breeding for a suitable variety of wheat resistant to black stem rust and the root and foot rots; breeding for a higher yielding more winter-hardy variety of fall rye; amount of natural crossing in wheat, oats and barley under controlled conditions.* ⓓ St. Paul, Minn. 1924.

Harris, Coleman J. (1889), Prof. of Biol., Pennsylvania State Forest School. Mont Alto, Pa. (U.S.A.). *Amphibia, Reptilia, Bryophyta.* ⓓ Bucknell 1912.
Harris, Daniel Thomas (1883), Ass. Prof. Physiol. London W.C. 1 (England), Univ. Coll. *Biol. Action of Light. Histol. Physiol.* ⓓ London 1926.
Harris, Halbert M. (1900), Ass. Prof. in Entomol. Ames, Ia. (U.S.A.), Dept. Zool. and Entomol., Iowa State Coll. *Systematic Entomol.* — *Hemiptera. Nahididae. Anthocoridae. Henicocephalidae. Saldidae.* ⓓ Iowa State Coll. 1925. ⓟ *Hemiptera.*
Harris, Henry Albert (1886), Senior Demonstrator and Curator of Anat. Mus., Univ. Coll., Member of Board of Examiners, Royal Coll. of Surgeons. London W.C.1 (England), Inst. of Anat., Univ. Coll., Gower St. *The Growth of Bone in Health and Disease. The morphol. aspects of Teratol. The special pathol. of the Embryo. The function of the spleen.*
Harris, James A., Prof. of Botan. and Head, Dept. of Botan., Univ. of Minnesota. Minneapolis, Minn. (U.S.A.). *Physiol., ecol., biometry, genetics.* o
Harrison, Francis C. (1871), Prof. of Bact., Principal Macdonald College, Montreal, P. Q. (Canada), Mc Gill Univ. *Bact., Dairy.* ⓓ Toronto 1893.
Harrison, Herbert Spencer (1872), Curator Horniman Mus., Forest Hill. London S.E. (England). *Anthrop.* ⓓ London 1901.
Harrison, L., Prof. Univ. Sydney (Australia). *Zool.* o
Harrison, Roland Wendell (1897), Ass. Prof. of Biol., Southern Methodist Univ. Dallas, Tex. (U.S.A.). *Bact., Immunol.* ⓓ Univ. of Chicago 1925.
Harrison, Ross Granville (1870), Prof. of Compar. Anat. New Haven, Conn. (U.S.A.), Yale Univ. *Embryol. of Vertebrates; Experim. Embryol., Transplantation, Tissue Culture, Regeneration.* ⓓ Ph. D. Johns Hopkins 1894, M. D. Bonn 1899.
Harrison, Thomas J. (1885). Prof. of Field Husbandry, Univ. of Manitoba. Winnipeg, Man. (Canada). *Environmental factors influencing quality in Cereal grains.* ⓓ Univ. of Manitoba 1911. ⓟ *Cereal varieties,* and *forage crop varieties.*
Harrower, G., Prof. King Edward VII Coll. of Med. Singapore (India). *Anat.* o
Harshberger, John W. (1869), Prof. of Botan., Univ. of Pennsylvania. Philadelphia, Pa. (U.S.A.). *Ecol., Fungi.* ⓓ Univ. of Pennsylvania 1893.
Hart, Nelson Collins (1888), Prof. of Botan., Univ. of Western Ontario. London (Canada). *Anat. of Plants. Ecol. Genetics.* ⓓ Univ. of Toronto 1916. ⓟ *Spermatophyta.*
Hart, P. C. (1901), Ass. Pharmacol.-therapeut. Labor. Univ. Amsterdam (Holland), Palestinastr. 12. *Biochem. Vergl. Anat. Embryol.* ⓓ Amsterdam 1925.
Harter, Leonard Lee (1875), Plant Pathol., U. S. Dept. of Agric. Washington, D.C. (U.S.A.). *Diseases Phaseolus and Ipomoea batatas.* ⓓ George Washington Univ. 1917.
Hartert, Ernst Johann Otto (1859), Dr. phil., Dir. des Zool. Mus. in Tring (England). *Ornithol., Systematik, Biol., Geographische Verbreitung.* ⓓ Dr. honoris causa Marburg a. d. L.
Hartings, Marquis of Tavistock, William Russell (1888). Havant, Hampshire (England), Warblington House. *Ruminant mammals, parrots, waterfowl.* ⓟ *Parrakeets.*
Hartlaub, Clemens (1858), Prof., Dr., Custos a. D. Mölln, Lbg. (Deutschland), Villa Annemarie. *Coelenteraten (Hydroiden und Medusen). Echinodermen (Crinoiden). Mammalia (Sirenen).* ⓓ Freiburg 1884.
Hartley, Carl (1887), Pathol. Washington, D.C. (U.S.A.), U.S. Bureau of Plant Industry. *Diseases of forest trees.* ⓓ California 1920.
Hartley, Edwin Adolphus (1893), Ass. Prof. of forest entomol. Syracuse, N.Y. (U.S.A.), N.Y. State Coll. of Forestry. *Hymenoptera — Ichneumonidae. Insect Ecol., factors effecting insect distribution; Parasites of insects, their biol. and use in the control of insect outbreaks.* ⓓ Ohio State Univ. 1921. ⓟ *Ichneumonidae, Tryphoninae. Hymenoptera.*
Hartley, P., Nation. Inst. for Med. Research. London W.C. 2 (England), 15 York Buildings. *Bact. and experim. Pathol.* o
Hartman, Carl Gottfried (1879), Research Associate. Baltimore, Md. (U.S.A.), Carnegie Inst. of Washington, Department of Embryol., Wolfe and Madison Streets. *Physiol. of Reproduction (Mammals) with special reference to Menstruation in Primates. Embryol. (Mammals).* ⓓ Univ. of Texas 1915.
Hartman, Ernest (1896), Dr., Instr. in Zool. Urbana, Ill. (U.S.A.), Univ. of Illinois. *Parasit.; Malaria.* ⓓ Johns Hopkins 1926. ⓟ *Slides of Bird Malaria.*
Hartman, Frank Alexander (1884), Prof. and Head of Department of Physiol., Univ. Buffalo, N.Y. (U.S.A.), 24 High Street. *Physiol. of the Adrenal Gland.* ⓓ Univ. of Kansas 1905.
Hartman, Henry (1889), Associate Prof. of Pomol., Oregon Agric. Coll. Corvallis, Ore. (U.S.A.). ⓓ Iowa State Coll. 1921.
Hartmann, Adele Auguste Karoline (1881), ao. Prof. der Univ., Ass. an der Anat. Anstalt, Abt. für Histol. und Embryol. München (Deutschland), Mozartstr. 17. *Blutbildung und blutbildende Organe (Milz, Thymus, Knochenmark). Gefäßsystem. Biol. Strahlenwirkung (Röntgen- u. Kathodenstrahlen).* ⓓ München 1911.
Hartmann, Max (1876), Hon. Prof. a. d. Univ. Berlin. Mitglied des Kaiser-Wilhelm-Inst. für Biol., Berlin-Dahlem (Deutschland). *Cytol. und Entwicklung der Protisten, Physiol. der Fortpflanzung bei Protisten und niederen Wirbellosen, Physiol. der Befruchtung und Sexualität bei Protisten und Algen, Problem des Todes, Vererbung bei Haploiden.* ⓓ München 1901.
Hartmann, Otto, Univ.Prof. Graz (Deutsch-Österreich). *Zool.* o
Hartmann-Weinberg, Alexandra P., Dr., Kustos d. Geol. Mus. der Akademie der Wissenschaften. Leningrad (U.d.S.S.R.). *Röntgenographie in der Palaeont., Anat. der Wirbeltiere.* o
Hartree, William (1870). Cambridge (England), Newton Road. *Heat production in muscle.* ⓓ Cambridge 1895.
Hartridge, Hamilton (1886), Univ. Lect., Organs of Special Sense, King's Coll. Cambridge (England). *Haemoglobin. Viscin.* ⓓ Cambridge 1908.
Hartsema, Anna Martha (1896), Dr., Hauptass. Labor. f. Techn. Botan. Delft (Holland). *Experim. Morphologie.* ⓓ Utrecht 1924.
Hartt, Constance Endicott (1900), Instr. in Biol. Canton, N.Y. (U.S.A.), St. Lawrence Univ. *Plant physiol. and ecol.* ⓓ Univ. of Chicago 1924.
Hartwell, Burt Laws (1865), Dir. and Agronomist, Agric. Exp. Station, Rhode Island, State Coll., and Prof. Agr. Chem. in the Coll. Kingston, R.I. (U.S.A.) *The relative fertilizer and nutrient needs of agric. crops, including lime and organic matter. The effect of crops on those succeeding them. The toxic effect of aluminum and the growth-inhibiting effect of a field deficiency of manganese.* ⓓ Univ. of Pa. 1903.
Hartwell, Rhoda Alice (1901), Instr. of Zool., Univ. of Vermont. Burlington, Vt. (U.S.A.), 144 South Willard St. *Olfactory Sense of Insects. Muscle Development in Amphibia.* ⓓ Yale Univ. 1923.
Hartzell, Albert (1891), Entomol. Yonkers, N.Y. (U.S.A.), Boyce Thompson Inst. *Insect transmission of plant diseases.* ⓓ Ohio State Univ. 1923. ⓟ *Insect carriers of plant diseases.*
Hartzell, Frederick Zeller (1879), Associate in Research (Entomol.), New York Agric. Experim. Station, Geneva, N.Y. (U.S.A.). *Grape, Pear and Apple Insects. Biometrical Studies of Entomol. Physiol. of Insects and their injuries to plants. Ecol. phases of entomol.* ⓓ Cornell Univ. 1909.

Harukawa, Chûkichi (1887), Entomol. to The Ohara Inst. for Agric. Research. Kurashiki, Okayama-ken (Japan). *Agric. entomol., especially ecol., life-history and methods of control of injurious insects.* ⒹUniv. of Tokyo 1913.

Harvey, Basil C. H. (1875), Prof. of Anat., Dean of Med. Students, Chicago Ill. (U.S.A.), Univ. of Chicago. *Gastric Glands and Glands of Internal Secretion.* Ⓓ Toronto 1898.

Harvey, Edward Maris (1888), Prof. of Horticultural Research, Oregon Agric. Coll. Corvallis, Ore. (U.S.A.). *Plant nutrition, chem. activity of fruit trees and growth correlations.* Ⓓ Univ. of Chicago 1914.

Harvey, E. Newton (1887), Prof. of Physiol., Princeton Univ. Princeton, N.J. (U.S.A.), Guyot Hall. *Bioluminescence, Cell oxydations, Cell permeability.* Ⓓ Columbia Univ. 1911.

Harvey, Ethel Browne (1886). Princeton, N.J. (U.S.A.), *Cytol.* Ⓓ Columbia 1913.

Harvey, Joseph le Roy, Lect., Hackett Med. School. Canton (China). *Pathol., Bact.* o

Harvey, Rodney Beecher (1890), Associate Prof. of Plant Physiol. and Botan., Univ. Minnesota, and Head, Section of Plant Physiol., Minnesota Expt. Station. St. Paul, Minn. (U.S.A.), Univ. Farm. *Hardiness in plants and winter resistance of varieties; freezing points of tissues, effects of low temperature. Respiratory enzymes. Ripening of fruits and vegetables. Storage problems.* Ⓓ Chicago 1918.

Harwey, Hildebrande Wolfe (1887), Hydrographer. Plymouth (England), Biol. Labor., Citadel Hill. *Variable physical and chem. conditions in the sea, animal and vegetable life.* Ⓓ Cambridge 1909.

Haša, Rudolf, Dr., Prof. d. Forstwissensch. Brno (C.S.R.), Černa pole 102, Landwirtsch. Hochsch. o

Hasan, Syed Hadi, Reader, Univ. Aligarh, Unit. Prov. of Agra and Oudh (India), *Botan., Zool.* o

Hase, Albrecht (1882), Dr. phil., Prof. der Zool., Mitglied der Biol. Reichsanstalt für Land- u. Forstwirtschaft. Berlin-Dahlem (Deutschland). *Zool., Entomol.: insbesondere Parasit. (blutsaugende Insekten); angewandte Entomol., physiol. Grundfragen derselben; Temperaturproblem; Geruchsproblem; Ernährungsproblem.* Ⓓ Jena 1907.

Hase, Olga Theodorovna (1898), Ass. Landwirtschaftl. Inst. Leningrad (U.d.S.S.R.). *Geobotan. und Bryol., insbesondere der Torfmoose (Sphagnum). Geobotan. Untersuchungen des Gouvernements Wologda.*

Hasebe, Kotondo, Prof. of Tôhoku Imper. Univ. Anat. Inst., Coll. of Med. Sapporo (Japan). *Anat.*

Haselhoff, Emil (1862), Prof., Dir. landw. Versuchsanstalt. Harlshausen bei Cassel (Deutschland). *Agrikulturchem., Pflanzenphysiol., Bodenkunde.* Ⓓ Marburg (Lahn) 1888.

Haseman, Leonard (1884), Prof. of Entomol., Univ. of Missouri. Columbia, Mo. (U.S.A.). Ⓓ Cornell Univ. 1910.

Hasenbäumer, Julius (1874), Abteilungs-Vorsteher an der Abt. für Bodenforschung und Pflanzenernährung der Landw. Versuchsstation der Landw.-Kammer für Westfalen. Münster i. Westf. (Deutschland). *Agriculturchem. Düngebedürfnis der Böden und Nährstoff-Aufnahme durch die Pflanzen.* Ⓓ Rostock 1898.

Hashda, Kunihiko, Prof. of Imper. Univ., Physiol. Inst., Fac. of Med. Tôkyô (Japan).

Hashimoto, Kinzi, Ass., Bact. Inst., Univ. Sendai (Japan). *Agglutinatorische Analyse von Typhusbazillen.* Ⓓ Sendai.

Haskell, Royal Joyslin (1890), Associate Pathol. in charge of Plant Disease Survey, Secretary-Treasurer of American Phytopath. Society and Business Manager of Phytopath. Washington, D.C. (U.S.A.), Bureau of Plant Industry, U. S. Department of Agric. *Plant disease geography; Seed treatment for oat smut. Potato diseases.* Ⓓ Cornell Univ. 1917.

Haskell, Sidney B. (1881), Dir., Massachusetts Agric. Coll. Experim. Station, Acting Head, Division of Agric. Amherst, Mass. (U.S.A.). *Plant nutrition and fertilizer use.*

Hassall, Albert (1862), Zool. Division, Bureau of Animal Industry, U. S. Dept. of Agric. Washington, D.C. (U.S.A.). *Parasit. worms and hosts.* Ⓓ London 1886. Ⓟ Specimens of parasit.

Hasselbalch, Karl Albert (1874). Snekkersten (Danmark). *Wasserstoffionenkonzentration des Bodens. Reaktion des Blutes u. der Gewebe.* Ⓓ Kopenhagen 1899.

Hasselwander, Albert (1877), o.Univ.Prof.Erlangen (Deutschland), Bismarckstr. 24. *Anat., Röntgenol.* o

Hastings, Edwin George (1872), Prof. of Agric. Bact., Univ. of Wisconsin. Madison, Wis. (U.S.A.), Coll. of Agric. *Dairy Bact.* Ⓓ Univ. of Wisconsin 1899.

Haswell, Wilson G., Prof., Univ. Birmingham (England). *Pathol.* o

Hata, Sahachirô, Prof. of Keiô-Gijuku Univ., Member of Kitasato Inst. for Infect. Diseases. Bact. Inst., Coll. of Med. Yotsuya, Tôkyô (Japan). *Bact.*

Hata, Senshô, Prof. of Kyôto Imper. Coll. of Sericulture, Chem.'Inst. Hanazono near Kyôto (Japan). *Physiol. Chem.*

Hatai, Shinkishi, Prof. in Compar. Physiol. of Tôhoku Imper. Univ., Zool. Labor., Coll. of Sc. Sendai (Japan). *Compar. Physiol. of Nervous System. Physiol. of Respiration.*

Hatch, Melville Harrison (1898), Instr. Department of Animal Biol., Univ. of Minnesota. Minneapolis, Minn. (U.S.A.). *Coleopterol.* Ⓓ Univ. of Michigan 1925. Ⓟ Coleoptera.

Hatcher, Robert A. (1868), Prof. of Pharmacol., Cornell Univ. Med. Coll. New York City (U.S.A.), 414 East 26th Street. Ⓓ Tulane Univ., New Orleans 1898.

Hatori, Shinkichi, Ass., Labor. of Horticulture, Coll. of Agric., Imper. Univ. Kyôto (Japan). o

Hatschek, Berthold, Univ.Prof. emer. Wien VIII (Deutsch-Österreich), Rangegasse 8. *Zool., Entwicklungsgesch. der Anneliden.* o

Hatt, Pierre (1897). Labor. Arago, Banyuls s. M. (France). *Histol. des Lamellibranches.*

Hatt, Robert Torrens (1902), Instr. in Biol., New York Univ. and Field Naturalist, Roosevelt Wild Life Experim. Station. New York City (U.S.A.), New York Univ., Univ. Heights. *Myol., Ecol. and Life-histories of Vertebrates.* Ⓓ Columbia Univ. 1925.

Hatta, Saburô (1864), Prof. Hokkaidô Imper. Univ., Zool. Inst. Sapporo (Japan). *Vertebrate embryol., Zoogeography.* Ⓓ Univ. Tôkyô 1891.

Hattori, Hirotaro (1876), Lect. Tôkyô Imper. Univ., Dir. Tokugawa Inst. for Biol. Research. Kanda, Tôkyô (Japan), 23, Suzukicho. *Microbiol.* Ⓓ Tôkyô 1899.

Hattori, Osamu (1902), Ass. of Zool. Inst. Sc., Coll. Imper. Univ. Tôkyô (Japan). *Histol. of the cardiac nerves of Invertebrates.* Ⓓ Tôkyô Imper. Univ. 1926.

Hattori, S., Prof. Dr. Tôkyô (Japan), Botan. Inst., Fac. of Sc., Imperial Univ. Koishikawa-Ku. *Biochem.*

Hauchecorne, Friedrich (1894), Dr., Dir. d. Zool. Gartens. Halle a. d. S. (Deutschland). *Wirbeltiere, bes. Säugetiere u. Vögel Deutschlands.* Ⓓ Berlin 1924.

Hauck, Emil (1879), Dr. iur., Diplom. Tierarzt, Staats-Veterinäroberinspektor. Wien III (Deutsch-Österreich), Landstraße, Hauptstr. 109. *Kynol. Hippol.* Ⓓ Wien Jur. 1906, Tierarzt 1910.

Hauduroy, Paul (1897), Chef de Lab. Fac. Méd., Agrégé des Fac. Méd. Paris (France), 21, Rue de l'Ecole de Méd. *Bact. Bacteriophage de d'Hérelle.*

v. Hauer, Fritz (1889), ao. Prof., Doc. Wien III (Deutsch-Österreich), linke Bahngasse 11. *Physikal. Eigenschaften des Blutes u. d. Harnes. Leitung von Elektrizität und Wärme in tierischen Geweben.* Ⓓ Wien 1909.

Hauge, Sigfred Melanchton (1895), Research Associate in Biochem., Purdue Univ. Agric. Exper.

Stat. Lafayette, Ind. (U.S.A.). *Animal Nutrition and Biochem.* ⓕ Univ. of Minnesota 1926.

Hauman, Lucien, Prof. Univ. Nac. Buenos Aires (Argentina). *Botan. Pathol. Microbiol.* ○

Haun, Friedrich (1889), Dr. phil., Abteilungs-Vorsteher an der Landwirtschaftl. Versuchsanstalt in Harleshausen b. Cassel. Cassel (Deutschland), Holländische Str. 43¹/₄. *Nahrungsmittel- und Agrikultur-Chem.* ⓕ Leipzig 1915.

Haupt, Arthur Wing (1894), Ass. Prof. of Botan., Univ. of California, Southern Branch. Los Angeles, Cal. (U.S.A.). *Morphol. and cytol. of Hepaticae; cytol. of Cyanophyceae.* ⓕ Univ. of Chicago 1919. ⓟ Microscopical slides and preserved material of liverworts.

Haupt, Hermann (1873), Halle a. d. S. (Deutschland), Burgstr. 19. *Systematik und Biol. der palaearktischen Psammocharidae und Homoptera.* ⓟ Deutsche Homoptera.

Haupt, Hugo (1874), Prof. Dr. phil., Bautzen (Deutschland), Mättigstr. 35, Anstalt für Wasseruntersuchungen. *Wasseruntersuchungen, Flußverunreinigungsfragen, Verunreinigung von Oberflächen sowie von industriellen Brauchwässern.* ⓕ Dresden 1901.

Haurowitz, Felix (1896), Priv.Doc., Dr. med., Dr. rer. nat., I. Ass. am med.-chem. Inst. der deutschen Univ. Prag II (C.S.R.), Havlíčkovo nám. 15. *Chem. des Blutfarbstoffes. Chem. der Fermente. Physikal.-chem. Biol.* ⓕ Prag med. 1922, nat. 1923.

Hauser, Gustav Chr. F. (1856), Dr. phil. et med., Geh. Hofrat, o. Prof. der Pathol. Anat. an der Univ. Erlangen (Deutschland), Rathsberger Str. 21. *Allgemeine Pathol. u. Pathol. Anat. Entomol.* ⓕ Erlangen phil. 1878, med. 1881. ⓟ Coleopteren.

Hausman, Leon Augustus (1888), Ass. Prof. of Zool., Rutgers Coll. and the New Jersey Coll. for Women. New Brunswick, N.J. (U.S.A.). *Structure of human and lower mammalian hairs.* ⓕ Cornell Univ. Ithaca 1919.

Hausrath, Hans (1866), o. Prof. d. Forstwissenschaft Univ. Freiburg i. Br. (Deutschland), Bad. forstl. Versuchsanstalt. *Pflanzengeographie, Forstgeschichte, Waldbau.* ⓕ Freiburg i. Br.

Havet, J., Prof. Univ. Cath. Louvain (Belgique). *Histol., Embryol.*

Havinga, Berend (1892), Dr., Biol. to the Coast-fisheries. Amsterdam (Holland), Rijksadministratiegebouw. *Biol. of Mollusca, Crustacea and Pisces important for the Coast-fisheries.* ⓕ Groningen 1919.

Hawk, Grover C. (1885), Prof. of Zool., Penn. Coll. Oskaloosa, Ia. (U.S.A.). ⓕ Univ. of Chicago 1919.

Hawk, Philip Bovier (1874), Dir. Food Research Labs. of New York. Forest Hills, N.Y. (U.S.A.), 710 Burns St. *Physiol. chem. and toxicol.* ○

Hawkins, Herbert Leader (1887), Prof. of Geol., Palaeont. Univ. Reading (England), 38 The Mount. *Main researches on Fossil Echinoidea.* ⓕ Manchester 1920.

Hawkins, Kennith (1890), Specialist in bee culture, GBLewis Co Watertown, Wis. (U.S.A.). *Production of honey.* ⓕ Washington, D.C., 1919.

Hawkins, Lon A. (1880), Physiol., U. S. Dept. of Agric. Washington, D.C. (U.S.A.), Bureau of Plant Industry. *Physiol. fruit and vegetable storage and transportation.* ⓕ Johns Hopkins Univ. 1913.

Hawley, Ira Myron (1884), Entomol., U. S. D. A. Bureau of Entomol., Cereal and Forage Insects, Alfalfa Weevil (Phytonomus posticus) Investigations. Salt Lake City, Utah (U.S.A.), 413 Fourth Ave. *Economic Entomol. Ecol. Insect Parasit.* ⓕ Cornell Univ. 1916.

Hawley, Ralph Chipman (1880), Prof. Forestry Yale Univ. New Haven, Conn. (U.S.A.), Yale School of Forestry, *Silviculture.* ⓕ Yale 1904.

Hawthorn, Leslie Rushton (1902), Ass. in Research (Horticulture). Geneva, N.Y. (U.S.A.). *The Vegetables of New York.* ⓕ Cornell Univ. 1924.

Hay, Oliver Berry (1846), Assoc. Carnegie Inst. Washington, D.C. (U.S.A.), 1211 Harvard St. *Vertebrate palaeont.* ○

Hay, William Perry (1872), Head dept. Biol. and chem. Washington High Schools. Kensington, Md. (U.S.A.). ○

Hayasaka, Ichirô, Ass. Prof. of Tôhoku Imper. Univ. Sendai (Japan), Geol. Inst., Coll. of Sc. *Palaeont.* ○

Hayashi, Haruo, Prof. of Tôkyô Imperial Univ. Tôkyô (Japan), Pharmacol.Inst., Fac.of Med. *Pharm.*

Hayashi, Ikuhiko, Prof. of Nagasaki Med. Univ. Nagasaki (Japan), Pathol. Inst. *Pathol.*

Hayashi, Inosuke, Prof. of Aichi Med. Univ. Nagoya (Japan), Pharmacol. Inst. *Pharmacol.*

Hayashi, Naosuke, Prof. of Aichi Med. Univ. Nagoya (Japan), Pathol. Inst. *Pathol.*

Hayashi, Takashi, Lect. of Tôhoku Imper. Univ. Sendai (Japan), Chem. Inst., Coll. of Med. *Biochem.*

Hayashi, Teijirô (1901), Nôgakushi, Ass. Hokkaidô Imperial Univ. Sapporo (Japan), Zool. Inst. *Physiol. of animal cells.* ⓕ Hokkaidô Imperial Univ. Sapporo, Japan, 1926.

Hayashi, Terutoshi, Ass. of Tôhoku Imp. Univ. Sendai (Japan), Bact. Inst. Univ. *Agglutinatorische Analyse von Paratyphusgruppe-Bazillen.*

Hayata, Bunzô (1874), Prof. of Systematic Botan. and Dir. of the Botan. Garden, Fac. Sc., Imper. Univ. of Tôkyô. Tôkyôfuka (Japan), No. 2570, Nishisugamomachi. *Systematic Botan.; Nat. Classification of Plants, according to Dynamic System; Selection and Evolution; Plants of Formosa; Vegetation of Mt. Fuji.* ⓕ Tôkyô 1903.

Hayden, Ada, Ass. Prof. of Botan. Ames, Ia. (U.S.A.), Iowa State Coll., Dept. of Botan. *Ecol., Systematic Botan., Morphol., Prairie plants, Distribution of weeds.* ⓕ Iowa State Coll., Ames, Iowa, 1918.

Hayden, Margaret Alger (1884), Ass. Prof. of Zool., Department of Zool. and Physiol. Wellesley, Mass. (U.S.A.), Wellesley Coll. *Cytol. Karyosphere formation in Scarabaeidae. Sex ratios in Diaspidinae.* ⓕ Columbia Univ. 1924.

Hayek, August (1871), OberMed.R. a.o. Prof. Univ. Wien V (Deutsch-Österreich), Margarethenstr. 82. *Systemat. Botan., Pflanzengeographie, Orientflora.* ⓕ Wien med. 1895, phil. 1905. ⓟ Herbarpflanzen aus Europa, Nordafrika und Westasien.

Hayes, Fred Montreville (1885), Associate Prof. of Vet. Sc. Davis, Cal. (U.S.A.), Univ. of Calif., Univ. Farm. *Comparat. Pathol. and Bact. infectious abortion of cattle and dogs.* ⓕ Kansas State Agric. Coll. 1908.

Hayes, Herbert K. (1884), Prof. of Plant Breeding, Univ. of Minnesota. St. Paul, Minn. (U.S.A.), Univ. Farm. *Genetics and Plant Breeding. Disease resistance in small grains and maize. Genetics and breeding of maize. Breeding small grains.* ⓕ Bussey Inst., Boston, Mass., 1921.

Hayes, William P. (1887), Ass. Prof. Entomol. Urbana, Ill. (U.S.A.), Univ. Illinois. *Insect Morphol. and Immature Insects.* ⓕ Cornell Univ. 1923. ⓠ Rhynchophora, Scarabaeidae.

Haymaker, Herbert Henley, Assoc. Prof., Kansas State Agric. Coll. Manhattan, Kan. (U.S.A.), 315 N 16 Street. ○

Haynes, Frederic (1896), Demonstrator in Histol. Univ. Oxford (England), Physiol. Labor, *Pneumokonioses.* ⓕ Oxford 1922.

Hays, Frank A. (1888), Research Prof. in Poultry Investigations, Mass. Agr. Coll. Amherst, Mass. (U.S.A.). *Genetic studies with poultry.* ⓕ Iowa 1917. ⓟ Poultry and eggs.

Hazaniuk, Michael (1892), Ass. physiol. Chem. Kiew (U.d.S.S.R.), Bolshay-Shitomirska N. 18. *Vitamines, biol. value of the various components of nutrition substances.*

Hazelhoff, Engel Hendrik (1900), Entomol. Proefstation Javasuikercultuur. Pasoeroean, Java (Nederl.-O.-Indië). *Atmungsregulierung bei Insekten und*

Spinnen *(Regulierung der Diffusion; Funktion der Stigmenverschlußapparate). Angewandte Entomol. (Zuckerschädlinge und deren Bekämpfung).* ⓓ Utrecht 1926.

Hazen, Tracy Elliot, Ass. Prof. of botan., Barnard Coll., Columbia Univ. New York City (U.S.A.). *Morphol. of Chlorophyceae.* o

Hazlitt, Victoria (1887), Lect. Psychol. London N.W. 1 (England), Bedford Coll., Regent's Park. *Animal Psychol., Mental Tests.* ⓓ London Univ. 1910.

Hazslinszky, Bertalan (1902), Ass. Allg. Bot. Inst. Univ. Budapest IX (Ungarn), Ráday-ut 53, II, 2. *Pflanzenanat.*

Headlee, T. J., Prof. of Entomol. Rutgers Coll. Entomol. New Jersey Agric. Experim. Station, State Entomol. New Jersey Department of Agric. New Brunswick, N.J. (U.S.A.). ⓓ Cornell 1906. o

Heald, Frederick De Forest (1872), Prof. Plant Pathol. and Plant Pathol. of the Agric. Experim. Station, State Coll. of Washington. Pullman, Wash. (U.S.A.), Coll. Station. *Wheat smut (Tilletia tritici). Apple rots and associated fungi. Silver leaf disease of fruit trees. Winter injury of fruits.* ⓓ Leipzig 1897. ⓟ *Mycol. and pathol. specimens.*

Heath, Eugene Schofield, Acting Associate Prof. in Botan., Agnes Scott Coll. Scottdale, Ga. (U.S.A.). *Paleobotan., phyletics, flora of Stone Mountain.* o

Heath, Harold (1868), Prof. of Zool., Hopkins Marine Station of Stanford Univ. Pacific Grove, Cal. (U.S.A.), 181 Ocean View. *Marine Invertebrates, their development.* ⓓ Ohio Wesleyan Univ. 1893.

Heath, Harry Colson (1877), Prof. of Biol. in Westminster Coll. Fulton, Mo. (U.S.A.), 804 Jefferson St. *Plant Physiol. The Use of Potash Shale as a source of potassium for growing plants.* ⓓ Univ. of Chicago 1922.

Heaton, Trevor Braby (1886), Dr. Lee's Reader Anat., Student of Christ Church. Oxford (England), 37 St. Giles. *Pharmacol., Tissue culture.* ⓓ Oxford 1911.

Heberer, Gerhard (1901), Wissenschaftl. Hilfsarbeiter. Halle a. d. S. (Deutschland), Röpziger Str. 4. *Vererbungscytol., Entomostraken, Primaten (recente und fossile).* ⓓ Halle a. d. S. 1924.

Hecht, Gerhard (1900), Dr. med., Pharmacol. I.-G. Farbenindustrie A.-G., Werk Elberfeld (Deutschland). *Pharm., Toxikol.* ⓓ Göttingen 1925.

Hecht, Otto (1900), Dr. phil., wissenschaftl. Angestellter in der chemischen Industrie. Aussig a. d. Elbe (C.S.R.), Biol. Labor. des Vereins für chem. und metallurgische Produktion. *Angewandte Entomol. Ausarbeitung und biol. Prüfung von Bekämpfungsmitteln. Blausäuredurchgasungen.* ⓓ München 1923.

Hecht, Selig (1892), Associate Prof. of Biophysics. New York City (U.S.A.), Columbia Univ. *Physiol. and Photochem. of the visual process in lower animals and man.* ⓓ Harvard Univ. 1917.

Heck, Ludwig (1860), Dr. phil., Prof., Geh.R., Wissenschaftl. Dir. des Zool. Gartens. Berlin W 62 (Deutschland), Budapester Str. 9. *Naturgeschichte der Säugetiere.* ⓓ Leipzig 1884. Med. vet. h. c. Berlin 1926.

Heck, Lutz (1892), Dr. phil., Dir. Ass. am Zool. Garten. Berlin W 62 (Deutschland), Budapester Str. 9. *Säugetierkunde.* ⓓ Berlin 1921.

Hecke, Ludwig, Univ.Prof. Wien XIX (Österreich), Hochsch. f. Bodenkultur. *Phytopath.* o

Hecker, Hilmar (1891), Dr. phil., B. Landesanstalt für Pflanzenbau u. -schutz. München (Deutschland), Steinsdorfstr. 5 I. *Beizfrage.* ⓓ Erlangen 1890.

Heckscher, Hans (1893). København (Danmark), Kommunehospitalet. *Bact.; Pathol. und physiol. Chem.* ⓓ Copenhagen 1918.

Hector, J. M., Prof. Transvaal Univ. Coll. Pretoria (South Africa). *Agric. Botan.* o

Hedgcock, George Grant (1863), Pathol., Investigations in Forest Pathol., U.S. Dept. of Agric. Washington, D.C. (U.S.A.), Bureau Plant Industry. *Life history of fungous and plant parasit. of forest trees and shrubs, the relation of the physiol. water in the soil to that in plants; Colletotrichum on Agaves; Sclerotinia on Caul. flower; Crown gall of the apple tree and grape etc.; Diseases of trees in our national forests; Leaf cone, and stem rusts of conifers; Wood staining fungi.* ⓓ Washington Univ. 1906.

Hedges, Florence (1878), Ass. Pathol., Bur. Plant Industry. Washington, D.C. (U.S.A.). ⓓ Univ. of Michigan 1901.

Hedicke, Hans (1891), Dr. phil., Hilfsarbeiter der Preußischen Akademie der Wissenschaften. Berlin N 4 (Deutschland), Invalidenstr. 43. *Cecidol., Systematik der Apiden und Cynipiden, Biocoenotik.* ⓓ Berlin 1920. ⓟ *Apiden und Cynipiden.*

Hedin, Sven Gustaf (1859), Prof. emer. med., physiol. Chem. Upsala (Schweden). *Spaltungsprodukte des Eiweißes. Osmotische Verhältnisse der Blutkörperchen. Proteolytische Enzyme verschiedener animalen Organe und des Harns.* ⓓ Lund phil. 1886, med. 1893.

Hedlund, Johan Teodor (1861), Prof. landw. Inst. Alnarp, Akarp (Sverige). *Nahrungsaufnahme der Wurzel; Stofftransport innerhalb der Pflanze; Bedingungen des Wachstums bei Gerste und Hafer. Mutationen bei Malva parviflora.* ⓓ Upsala 1892.

Hédon, Emmanuel (1863), Prof. de Physiol. à la Fac. de méd. Montpellier (France), Inst. de Biol., Rue Montels. *Diabète pancréatique expérim. Insuline. Métabolisme des Hydrates de Carbone.* ⓓ Montpellier 1894.

Hédon, Louis (1895), Prof. agr. Physiol. Fac. Méd. Montpellier (France), Labor. de Physiol. Inst. de Biol., Rue Montels. *Diabète expérim. Métabolisme des hydrates de carbone.* ⓓ Montpellier 1923.

Hedrick, Ulysses Prentiss (1870), Vice-Dir., Chief in Research (Horticulture). Geneva, N.Y. (U.S.A.). *Horticulture and Genetics.* ⓓ Hobart Coll., Geneva, N.Y., 1913.

Hée, Alexandre Auguste Alfred (1894), Ass. à l'Inst. Botan., Fac. des Sc. Strasbourg (France), 7, Rue de l'Univ. *Physiol. végétale, Respiration des plantes.*

Heeres, Pieter Anton (1901), Arzt; Ass. am Pathol. Labor. der Reichs-Univ. Groningen (Holland), Turfsingel 1. *Pathol. Physiol.: Haematol.*

Hefnawy, Mahmoud Tewfik (1894), Lect. of Botan., School of Agric. Giza (Egypt), Horticultural Section. *Plant breeding.* ⓓ Nat. Sc. Tripos 1922.

Hegh, Emile Marie Joseph (1877), Ingénieur agric., Chef de bureau au Ministère des colonies de Belgique. Bruxelles (Belgique). *Zooblol. économique. Les moustiques. Les Termites.* ⓓ Louvain 1898.

Hegi, Gustav (1876), a.o. Univ.Prof. München (Deutschland), Tengstr. 18/0. *Flora von Mittel-Europa, Blütenbiol., Kulturpflanzen.* ⓓ Zürich 1900.

Hegner, Robert William (1880), Prof. of Protozool. Head of the Department of Med. Zool., School of Hygiene and Public Health, The Johns Hopkins Univ. Baltimore, Md. (U.S.A.), 615 N. Wolfe Street. *Parasit. protozoa in man: intestinal flagellates, bird malaria, host-parasit. relations among human protozoa.* ⓓ Univ. of Wisconsin 1908. ⓟ *Parasit. protozoa.*

Helberg, K. A. (1880), Pathol. der klinischen Abteilungen des Finsen-Inst. København (Danmark), Frederiksborggade 44. *Pathol. (und normale) Anat. des lymphoiden Gewebes, der Tuberkulose, der Haut, der Schleimhäute, des Tumorgewebes (sp. Kerngröße), Haut des elastischen Gewebes.*

von der Heide, Carl (1872), Prof. Dr. phil., Vorstand der weinchem. Versuchsstation. Geisenheim a. Rhein (Deutschland). *Weinchem.* ⓓ München 1896.

Heidenhain, Martin (1864), o.ö. Prof. der Anat., Dir. des anat. Inst. Tübingen (Deutschland). *Synthetische Morphol.* ⓓ Freiburg.

Heider, Karl (1856), Univ.Prof. emer., Geh. Reg.R. Berlin NW 15 (Deutschland), Schaperstr. 15. *Anneliden. Vergl. Embryol.: Hydrophilus. Salpae.* ○

Helderich, Friedrich (1878), o. Prof., Dir. des anat. Inst. der Univ. Münster, Westfalen (Deutschland), Hittorfstr. 46. *Histol., Anat. des Kindesalters.* ⊕ Göttingen 1903.

Heidemanns, Curt (1894), Ass. zool. Inst. Bonn (Deutschland). *Chem. u. Anat. des Tierkörpers.* ⊕ Bonn 1922.

Heidrich, Herbert (1894), Ass. Deutsche wissenschaftl. Kommission für Meeresforschung. Hamburg 5 (Deutschland), Kirchenallee 47 II. *Fischereibiol. Fragen der Ostsee inkl. Beltsee. Plattfische und Gadiden.* ⊕ Kiel 1922.

Heidsieck, Erich (1888), Dr. med., Prosektor am Anat. Inst. Breslau 16 (Deutschland), Maxstr. 6. *Wirbelsäule. Knochenstruktur.* ⊕ Kiel 1917.

Heikertinger, Franz (1876), Sekretär der zool.botan. Gesellschaft in Wien. XII/2, Hetzendorf (Deutsch-Österreich), Thunhofg. 8. *Systematik und Bionomie der Halticinen (Coleopt., Chrysom.) des paläarktischen (und nearktischen) Gebietes. Mimikry und sonstige Tiertrachthypothesen (kritisch). Nomenklatur.*

Heikinheimo, Olli (1882), Prof., Dir. Forstwissenschaftl. Versuchsanstalt. Helsinki (Finland), Rauhankatu 4. *Waldbau.* ⊕ Helsinki 1916.

Heil, Hans Albrecht (1899), Dr., Priv.Doc. für Botan., Ass. am botan. und am pharmakognostischen Inst. der Techn. Hochsch. Darmstadt (Deutschland), Roquetteweg 3 p. *Phanerogame Parasit.: Sporangienentwicklung der Dictyotaceen, Chamaegigas intrepidus und verwandte Arten.* ⊕ 1922.

Heilborn, Otto (1892), Doc. der Zytol. an Stockholms Högskola. Stockholm (Schweden), Botan. Inst. *Genetische Experimentalzytol., spez. Chromosomenzahlen und -dimensionen der Gattung Carex; Ananas, Carica, Draba, Viola. Zytol. und Artbildung, Systematik. Pflanzengeographie von Ecuador (Polsterpflanzen).* ⊕ Stockholm 1924.

Heilbronn, Afred (1885), Dr. phil., a.o. Univ.Prof. Münster i. W. (Deutschland), Steinfurter Str. 39. *Reizphysiol. der Pflanzen, Genetik der Farne.* ⊕ München 1909.

Heilbrunn, Lewis Victor (1892), Ass. Prof. of Zool. Univ. of Michigan, Ann Arbor, Mich. (U.S.A.), Zool. Dept. *General physiol., colloid chem. of protoplasm, artificial parthenogenesis, cell division, anesthesia.* ⊕ Univ. of Chicago 1914.

Heilmann, Paul (1888), Dir. des Pathol. Inst. am staatl. Krankenstift. Zwickau, Sachsen (Deutschland). *Pathol. Anat. u. allgemeine Pathol.* ⊕ Leipzig 1915.

Heim, Ludwig (1857), Dr., Geh.Med.R., o.ö. Prof. der Hygien. und Bact. Erlangen (Deutschland), Wasserturmstr. 2¹/₂. *Hygien. und Bact.* ⊕ 1880.

Heimerl, Anton (1857), R.R., Priv.Doc. Univ. Wien XIII/2 (Österreich), Hadikgasse 34. *Systematische Botan., Nyctaginaceen u. Phytolaccaceen, Achillea und Artemisia.* ⊕ Wien 1888.

Heimlich, Louis Frederick (1890), Prof. of Botan., Head Dept. of Biol., Univ. Valparaiso, Ind. (U.S.A.), 302 E. Jefferson st. *Plant Cytol., Morphol. and Taxonomy.* ⊕ Univ. of Wisconsin 1926.

Hein, Illo (1893), Prof of Biol. Pennsylvania State Forestry School, Mont Alto, Pennsylvania. Columbia University, N.Y. (U.S.A.), Department of Botan. *Cytol. of Ascomycetes, Erysiphaceae. Mycol. Physiol. species. Variagation.* ⊕ Columbia Univ. 1927.

Heincke, Friedrich (1852), G.R.R. Prof. Dr., Dir. emerit. der staatl. Biol. Anstalt auf Helgoland (Deutschland). *Biol. der Nutzfische. Molluskenfauna der Nordsee.* ⊕ Leipzig 1873.

Heine-Geldern, Robert (1885), Priv.Doc. für Völkerkunde. Univ. Wien I (Österreich), Kolowratring 7. *Völkerkunde Indiens und Südostasiens.* ⊕ Wien 1914.

Heineman, Paul G. (1864), Dir. of Research and Labor., Cook Labor., Inc. Chicago, Ill. (U.S.A.), 536 Lake Shore Drive. *Bacterial Vaccines, Serums, Pharmaceuticals, Anesthetics.* ⊕ Chicago 1907. ⊕ *Bacterial Vaccines.*

Heinemann, Robert (1868), Mitarbeiter am Naturhistorischen Mus. Braunschweig (Deutschland), Göttinger Str. 21. *Systematik und Biol. der deutschen Käfer.*

Heinicke, Arthur John (1892), Prof. Pomol. Ithaca, N.Y. (U.S.A.), Cornell Univ. Dept. Pomol. *Nutrition of fruit trees, pruning, soil management, fruit setting.* ⊕ Cornell 1916.

Heinrich, G., Dr. Serum Lab. d. Univ. Tartu (Estland), Vene t. 28. ○

Heinricher, Emil Johann Lambert (1856), o. Prof. der Botan. Univ. Innsbruck, Dir. d. Botan. Inst. und Botan. Gartens. Innsbruck-Hötting (Österreich), Sternwartstraße 13. *Biol. der parasit. Samenpflanzen (Entwicklungsgeschichte, Aufzucht, Anat. und Physiol.) Vererbung im allgem., besonders von Atavismen. Anat. Physiol.* ⊕ Graz 1879.

Heinroth, Oskar (1871), Dr. med., Leiter des Berliner Aquariums. Berlin W 62 (Deutschland) *Aquarien- u. Terrarienkunde mit zugehöriger Tierpflege. Vogelkunde, insbesondere Lebensäußerungen, geistiges Verhalten, Mauser.* ⊕ Kiel 1895.

Heinsius, Hein Willems (1868). Amsterdam (Holland), Hoofstraat 144. *Botan.* ⊕ Amsterdam 1890.

Heintz, Anatol (1898), Konserv. Palaeont. Mus. Oslo (Norge), Trondhjemsveien 23. *Devonische Fischfossilien (spez. Arthrodirae).* ⊕ Oslo.

Heintz, Louis (1886). Vitoria (España), Colegio de Santa María. *Spéléol., Préhistoire.* ⊕ Madrid 1907.

Heinze, Hermann Berthold (1872), Abteilungsvorsteher Versuchsstat. für Pflanzenkrankheiten u. Pflanzenschutz der Landw. Kammer Prov. Sachsen (Bact. Abteilung). Halle a. d. Saale (Deutschland), Karlstr. 10. *Bodenmykol. und -bact.; normale und pathol. Physiol. und angewandte Botan.: Bodenpilze und Bodenbakterien, stickstoffsammelnde Azetobacterorganismen, Knöllchenbildner der Leguminosen, Veränderlichkeit von Mikroorganismen und Vererbung von neu erworbenen Eigenschaften. Akklimatisierung: Soja hispida, Mais, Serradella, Lupinen.* ⊕ Tübingen 1897. ⊕ *Samen von Soja hispida, Körnermais und Grünfuttermais sowie von Lupinen.*

Heiss, Robert (1884), o. Prof. Univ., Dir. Anat. Inst. Königsberg i. Pr. (Deutschland), Wartenburgstr. 2. ○

Heitz, Emil (1892), Priv.Doc. an der Univ. Hamburg (Deutschland), Inst. für allgemeine Botan. *Morphol. u. Physiol. der Pflanzen-Chloroplasten. Geschlechtschromosomen bei Pflanzen. Chromosomenstudien in Verbindung mit genetisch-systematischen Fragen. Intrazelluläre Symbiose bei Insekten. Zoosporen: Reizphysiol.* ⊕ Heidelberg 1921.

Hekma, Ebel (1868), Arzt, Dir. der physiol. Abteilung der Landwirtschaftl. Versuchsstation Hoorn (Holland). *Physiol. der Milch und Milchprodukte. Tierphysiol. Ernährungs- und Nahrungsprobleme.* ⊕ Groningen 1896.

Hektoen, Ludwig (1863), Head Dept. of Pathol. Univ. Chicago (U.S.A.), 637 S. Wood St. *Bact., Immunol.* ○

Heldt, Thomas Hammann (1883), Physician in Charge Division of Neuropsychiatry. Detroit, Mich. (U.S.A.), Henry Ford Hospital. *Neurol., Neuropsychiatry.* ⊕ John Hopkins Med. Sch., Baltimore, Md., 1916.

Hele, Thomas Shirley (1881), Lect. in Biochem., Univ. of Cambridge (England), Emmanuel Coll. *Intermediary metabolism in animals, Protective synthesis against poisonous substances.* ⊕ Cambridge 1911.

Helff, Otto Maximilian (1897), Associate in Zool., Dept. of Zool., Univ. of Iowa. Iowa City, Ia.

(U.S.A.). *Endocrinol., General Physiol. Amphibian metamorphosis and metabolism.* ① Yale Univ. 1925.
Hellén, Wolter Edvard (1890), Mag. phil., Amanuensis b. d. Univ. zu Helsingfors (Finnland), Zool. Mus. d. Univ. *Coleoptera und Hymenoptera von Nordeuropa. Crustacea, Diptera, Hemiptera, Neuroptera.* ⓟ Paläarktische Hymenoptera parasit.
von Hellens, O. J., Dr. med., Prof. Univ., Dir. Hygien. Inst. Helsinki (Finnland), Parkgatan 11. *Hygiene, Bact.* o
Heller, Józef (1896), Ass. Med.-chem. Inst. Lwów (Polen), Piekarska 52. *Metamorphose der Insekten, physiol. u. chem.* ① Lwów 1922.
Hellevaara, E. Kalastusneuvos, Ph.M., Mataloushallitus Kalatolousosasto. Helsinki (Finland). *Fischzool.* o
Hellman, Torsten (1878), Prof. der Anat. a. d. Univ. Lund (Schweden). *Lymphatisches Gewebe.* ① Upsala 1914.
Hellwig, Christian Alexander (1889), Dr. med., Dir. of labor., membre of staff St. Francis Hospital. Wichita, Kans. (U.S.A.), 1017 North St. Francis Ave. *Thyroid gland.* ① Bonn 1916.
Helmer-Fainsilberg, R. B., Dr., Ass. Anat. Inst. Univ. Rostow am Don (U.d.S.S.R.), Str. F. Engels 141. o
Helms, J., Prof., Dr., Landwirtschaftl. Hochsch. København (Danmark), Bülowsvej. *Botan.* o
Helms, Otto (1866), Chefarzt der Lungenheilstätte Nakkebøllefjord. Pejrup (Danmark). *Die Vögel Grönlands, Biol. der dänischen Vögel, Röntgenaufnahmen von Vögeln.*
Hemenway, Ansel F. (1879), Assoc. Prof. Biol., Univ. of Arizona. Tucson, Ariz. (U.S.A.), 545 N. Vine Ave. *Plant Morphol.; Phloem of Dicotyledons, Late frost injury to woody plants. Desert Plants.* ① Univ. of Chicago 1912.
Hemingway, Albert (1902), Senior Ass., Univ. Coll. London W.C. 1 (England), Gower St. *Physiol.*
Hemmi, Takewo (1890), Prof. of Phytopath. and Mycol., Dept. of Agric. Kyôto (Japan), Imperial Univ. *Pathol. of the rice plant. Wood-rotting fungi. Anthracnoses of plants. Septorioses of plants. Effect of environmental factors to plant diseases. Mycol. flora of the southern part of Honshū in Japan.* ① Hokkaido Imper. Univ., Sapporo 1915.
Hemmingsen, A., Mag.Sc. Insulinlabor. Hellerup (Danmark), Onsgoarding. *Biochem. Synthese d. Insulins.* o
Hempelmann, Friedrich Albert (1878), Dr. phil., a.o. Prof. d. Zool. u. vergl. Anat., Oberass. Leipzig (Deutschland), Talstr. 33, Zool. Inst. d. Univ. *Oligochaeten, Polychaeten, Tierpsychol.* ① Leipzig 1906.
Hempt, Adolphe (1874), Dir. Staatl. Pasteur-Inst. Novi-Sad (S.H.S.). *Bact., speziell Rabiol. Immunität.* ① Graz 1898. ⓟ Rabische Vira. Antirabische Vakzinen.
Henckel, Alexander Hermannowitsch (1872), Prof. Botan. (Morphol., Systematik u. Phytopath.) Univ. Perm (U.d.S.S.R.), Botan. Inst. *Phytoplankton des Karameeres. Oekol. der niederen Pflanzen u. ihre Entwicklungsgeschichte (Plasmodiophora). Biol. der Meeresalgen. Phytoplankton des Kaspischen Meeres.* ① Mag. bot. Leningrad 1902, bot. Odessa 1911. ⓟ Mikroskopische Planktonpraep. (Diatomeen des Karameeres), Pilz- u. Bakterienkulturen.
Hendel, Friedrich (1874), Dr. phil. Wien II/1 (Deutsch - Österreich), Darwingasse 30. *Dipterol. Muscidae acalypteratae u. Blattminenkunde. Paläarkt. Sciomyziden. Bohrfliegen Südamerikas. Die Arten der Platystomidae. Die paläarkt. Agromyzidae.* ① Wien. ⓟ Acalyptrate Musciden.
Henderson, Harry Oram (1889), Associate Prof. Dairy Husbandry in Experim. Station. Morgantown, W.Va. (U.S.A.), Dairy Dept. *Nutrition and biochem.* ① Penns. State Coll. 1916.

Henderson, Jean Tasker (1902), Lect. in Zool. Montreal, P.Q. (Canada), McGill Univ. ① McGill Univ., Montreal 1922.
Henderson, Junius (1865), Prof. of Nat. History and Curator of Mus., Univ. of Colorado. Boulder, Col. (U.S.A.), 1305 Euclid Avenue. *Recent mollusks and fossil invertebrates.* ① Univ. of Colorado 1908. ⓟ Recent mollusks and fossil invertebrates.
Henderson, Lawrence J. (1878), Prof. of Biol. Chem., Harvard Univ. Cambridge, Mass. (U.S.A.), 4 Willard St. *Physical Chem.* ① Harvard 1902.
Henderson, Lena Bondurant (1880), Ass. Prof. of Botan., Rockford Coll. Rockford, Ill. (U.S.A.). *Taxonomy, Ecol.* ① Univ. of Tennessee 1908. ⓟ Flowering plants, Green Algae.
Henderson, Velyien Ewart (1877), Prof. of Pharmacy-Pharmacol., Univ. Toronto (Canada), Med. Building. *Pharm.* ① Univ. of Toronto 1903.
Henderson, Yandell (1873), Prof. of applied Physiol., Yale Univ. New Haven, Conn. (U.S.A.). *The physiol. of the circulation and respiration, biochem. of the blood gases, respiratory metabolism, toxicol. hygien. and therapy of the noxious gases of industry, and the pharm. of anaesthesia.* ① Yale 1898.
Hendrickson, Arthur H. (1890), Associate Pomol., Univ. of California. Davis, Cal. (U.S.A.), Univ. Farm. *Plant Physiol.; Water relations of Plants; Irrigation requirements of Orchard Trees.* ① Leland Stanford Univ. 1926.
Hendrickson, George (1890), Instr., Dept. of Zool. and Entomol. Ames, Ia. (U.S.A.), Iowa State Coll. *Ecol. of prairie insects.* ① Iowa State Coll. of Agric. 1926. ⓟ Prairie insects.
Hendrix, Byron M. (1886), Prof. of Biol. Chem., School of Med. Univ. of Texas. Galveston, Tex. (U.S.A.). *The function of the kidney in relation to the acid-base balance of the animal body.* ① Yale 1915.
Hendry, George W. (1885), Associate Agronomist, Univ. of California. Berkeley, Cal. (U.S.A.), 309 Hilgard Hall. *Agronomic Botan.* ① Cornell 1913.
Hengl, Franz (1894), Oberkommissär der Bundesanstalt für Pflanzenschutz. Wien II (Deutsch-Österreich), Trunnerstr. 1. *Krankheiten und Schädlinge des Weinbaues, Pflanzenschutz-Chem.* ① Wien 1917.
Hening, James Courtenay (1891), Ass. in dairy research. New York (U.S.A.), State Agr. Exp. Station. ① Univ. of Minnesota 1923.
Henke, Karl (1895), Dr. phil., Ass. am zool. Inst. der Univ. Göttingen (Deutschland). *Vererbung bei Schmetterlingen. Lichtsinn bei Wirbellosen.* ① Göttingen 1923.
Henkel, Alexander G., Prof. Univ. Perm II (U.d.S.S.R.), Salmka. *Botan., Systematik.* o
Henkel, Theodor (1855), Prof. Dr., Geh.Reg.R., o. Prof. für Agric.Chem. u. Milchwirtschaft, Vorstand der Bayerischen Haupt-Versuchs-Anstalt für Landwirtschaft an der Techn. Hochsch. München (Deutschland), Luisenstr. 36. *Pflanzenernährung, Tierernährung, Milchwirtschaft. Landw. Versuchs- u. Untersuchungswesen.* ① Erlangen. ⓟ Futter- u. Düngemittel.
Henking, Hermann (1858), Prof., Geh.Reg.R., Generalsekretär des Deutschen Seefischerei-Vereins, Mitglied der Deutschen Wissenschaftl. Kommission für Meeresforschung. Berlin W 9 (Deutschland), Potsdamer Str. 22a III. *Milben, Befruchtung bei den Insekten. Seefischerei, Statistik der Seefischerei (Nordsee, Ostsee). Schollen, Schellfisch, Hering, Lachs, Miesmuschel, Austern.* ① Göttingen 1882.
Henn, Arthur Wilbur (1890), Curator of Ichthyol., Carnegie Mus. Pittsburgh, Pa. (U.S.A.). *Fishes of North and South America.* ① Indiana Univ. 1915.
Henneberg, Bruno (1867), Dr. med., o.ö. Prof., Dir. des Anat. Inst. der Univ. Gießen (Deutschland), Friedrichstr. 6. *Vergl. Embryol. Glatte Muskulatur. Ohrmuschel. Autotomie. Sinushaare. Ontogenie der Genitalorgane.* ① Berlin 1894.

Henneberg, Wilhelm (1871), Dir. des Bact. Inst. der Preußischen Versuchs- und Forschungsanstalt für Milchwirtschaft, Hon.Prof. der Univ. Kiel (Deutschland). *Gärungsbact., Hefen, Milchsäure- und Essigsäurebact., sämtliche Pilze der Milch und Milcherzeugnisse.* ⑨ Rostock 1896. ⑨ Die Pilze (Hefen, Schimmelpilze, Bact.) der Milch u. der Milcherzeugnisse.

Henneguy, Louis Felix (1870), Prof. au Coll. de France. Paris V. (France), 9, rue Thénard. *Cytol., Embryol. des Vertébrés et des Insectes.* ⑨ Paris 1900.

Hennequin, Louise (1901), Préparateur Physiol. Fac. Méd. Nancy (France), Inst. de Physiol., 30, Rue Lionnois. *Physiol.*

Hennicke, Carl Richard (1865), Prof. Dr. med., Augen- und Ohrenarzt, Schriftleiter der Ornithol. Monatsschrift. Gera-Untermhaus (Deutschland), Ernseerweg. *Ornithol.* ⑨ Leipzig 1891.

Hennig, Edwin (1882), o. Prof. für Geol., Palaeont. a. d. Univ. Tübingen (Deutschland), Schloßbergstr. Nr. 15. *Fossile Fische und Dinosaurier, Mesozoische Säuger (sowie auch Muscheln und Arthropoden), Allgemeine Palaeont. (Abstammungslehre).* ⑨ Berlin 1906.

de Hennin de Boussu Walcourt, Emmanuel Joseph Eugène Marie (1876). Ganshoren (Belgique), 3, rue Beeckmans. *Lépidoptères belges y compris les micros.* ⑫ Microlépidoptères belges.

de Hennin de Boussu Walcourt, (Dom) Guy (1879). Abbaye de Maredsous, par Maredret (pr. Namur) (Belgique). *Entomol.: Lépidoptérol.: Biol. des Lépidoptères de Belgique.* ⑫ Lépidoptères de Belgique; insectes; Plantes de la région calcaire (Bassin entre Sambre-et-Meuse).

Henrijean, François (1860), Prof. à l'Univ. de Liège (Belgique), Membre de l'Academie de Méd. de Belgique, Id. de Paris, Docteur Honoris Causa de Lyon, Toulouse. Liège (Belgique), Rue Fabry 11. *Physiol. et Pharmacodynamie.* ⑨ Liège 1885.

Henriksen, Kai Ludvig (1888), Mag. sc., Amanuensis b. d. Zool. Mus. København (Danmark). *Vergl. Morphol. der Insekten. Käferlarven. Biol. der Wasserhymenopteren. Fossile Insekten Dänemarks (tertiäre und Moorinsekten).* ⑨ København 1912.

Henriques, Júlio Augusto (1838), Prof. emer. Univ., Dir. de l'Herbier Inst. Botan. Coimbra (Portugal). *Systématique de Phanerogames. Agric. colonial.*

Henry, Fred G., Prof. McGill Univ. Montreal (Canada). *Pathol.* ○

Henschen, Folke (1881), Prof. pathol. Anat., Dir. pathol. Abteilung d. Karolinischen Inst. Stockholm (Sverige). *Allgemeine und pathol. Anat., experim. und vergl. Pathol. Hirntumoren. Pseudotuberkulose, Nierenkrankheiten, Blutkrankheiten. Retikuloendothel.* ⑨ Stockholm 1910. ⑫ Pathol. Praep.

Henseler, Heinz (1885), o. Prof. u. Dir. des Inst. und Labor. für Tierzucht u. Züchtungsbiol. Techn. Hochsch. München (Deutschland), Arcisstr. *Wissenschaftl. Tierzucht, Vererbungsbiol.* ⑨ Halle (Saale) 1910.

Henshaw, Samuel, Curator of Mus. of Compar. Zool. ⑨ Cambridge, Mass. (U.S.A.), 28 Fayerweather St. ⑨ Harvard 1903. ○

Hentschel, Christopher Carl (1899), Demonstrator in Biol. and Compar. Anat. at St. Barthol. Hospital Med. Coll., Ass. in Zool. to the Univ. of London W. 8 (England), 70 Wynnstay Gardens. *Parasit. Ciliates and Gregarines, the relation of parasit. to host; Cytol.* ⑨ Univ. of London 1926.

Hentschel, Ernst Ludwig (1876), Prof. Dr., Priv.-Doc., Leiter der Hydrobiol. Abt. des Zool. Staatsinst. und Zool. Mus. Hamburg 23 (Deutschland), Jordanstr. 5 IV. *Hydrobiol., Spongien.* ⑨ München. ⑫ Spongien.

Hepburn, David, Prof. Univ. Coll of S. Wales and M. Cardiff, Wales (England). *Human Anat.* ○

Heppner, Myer J. (1899), Research Ass. in Pomol. Davis, Cal. (U.S.A.), Univ. Farm. *Rootstock. Black- and red* ⑨ Univ. of California, Berkeley, Cal. 1922.

Heptner, W. Georgievich (1901), Moskau (U.d.S. S.R.) Zool. Mus. Univ. *Tiergeographie (Paläarktik); Syst. der Wirbeltiere, sp. Vögel u. Säugetiere Rußlands. Syst. der Gerbillinae (Mammalia).*

Heraščenko, M., Prof. Silsko-Hospod-Inst. Kamjanec, Podilsky, Ukraine (U.d.S.S.R.), Levčenko-Str. 23-25. *Physiol.* ○

Herberg, Martin Karl Richard (1893), Dr., Abteilungsleiter (Biol.), Staatl. Hauptstelle für den naturwiss. Unterricht zu Berlin (Deutschland). *Anat. der Schildläuse (Coccidae).* ⑨ Berlin 1918.

Herbert, Paul Anthony (1899), Associate Forester, Forest Taxation Inquiry, U.S. Forest Service. New Haven, Conn. (U.S.A.), Marsh Hall, Yale Univ. *Breeding of forest trees by selection, hybridizing, influence of several chem. elements on seedling and tree development. The rest period of forest tree seed, influence of climatic factors on the sap flow in maple (Acer) trees.* ⑨ Cornell Univ. 1922.

Herbst, Curt (1866), Prof. der Zool. und Dir. des Zool. Inst. Heidelberg, Baden (Deutschland). *Einfluß anorganischer Stoffe auf die Entwicklung der Tiere. Formative u. Richtungsreize in der tierischen Ontogenese. Regeneration. Einfluß des Nervensystems auf dieselbe. Vererbungslehre vom entwicklungsmechanischen Standpunkte aus, besonders Quantitätsgesetz der Wirkung der Kernsubstanzen. Färbung von Salamandra maculosa.* ⑨ Jena 1889.

Herbst, Robert (1895), Dr. med., Ass. am Kaiser-Wilhelm-Inst. f. Arbeitsphysiol. und am physiol. Labor. der Deutschen Hochsch. für Leibesübungen. Berlin N 4 (Deutschland), Invalidenstr. 103 a. *Arbeits- und Sportphysiol.* ⑨ Würzburg 1921.

Hercelles, Oswaldo, Prof. Univ. Lima (Peru). *Anat. Pathol.* ○

Herčík, Ferdinand (1905), Ass. Brno (C.S.R.), 63 Kounicova. *Plant growth in relation to surface energy. Biophysics and biochem.*

Hercus, C. E., Prof. Univ. of Otago. Dunedin, New Zealand (Australia). *Bact.* ○

Herfs, Adolf (1895), Dr., Leiter des Zool. Labor. der I. G. Farbenindustrie Leverkusen (Deutschland). *Oekol. der Milben. Morphol. des Gastropoden-Integuments.* ⑨ Bonn 1919.

Heribert-Nilsson, Nils (1883), Prof., Saatzuchtleiter der Abteilung für Roggen, Hafer und Kartoffel an der Saatzuchtanstalt Weibullsholm bei Landskrona (Schweden). *Oenothera, Salix; Vererbungs- und Artbildungsstudien. Roggen: Züchtungsmethodische Probleme.* ⑨ Lund 1915.

Héricourt, Jules (1850), Chef de Labor. honoraire de la Fac. de Méd. de Paris (France), 12, rue de Douai. *Physiol.; Méd. experim.: Sérothérapie, Zomothérapie.* ⑨ Laur. Ac. Med. Paris 1874.

Hering, Martin (1893), Dr. phil., Kustos am Zool. Mus. der Univ. Berlin N 4 (Deutschland), Invalidenstraße 43. *Blattminen und blattminierende Insekten.* ⑨ 1921. ⑫ Blattminen und blattminierende Insekten.

Heringa, Gerard Carel (1890), o. Prof. der Histol. Univ., Dir. des Histol. Labor. Amsterdam (Holland). *Bindegewebe: Zellen und Zwischensubstanz. Ultrastructur fibrillärer Gewebselemente.* ⑨ Leiden 1916.

Herlitzka, Amedeo (1872), Prof. o. fisiol., Dir. Ist. Fisiol. Univ. Torino (Italia), Corso Raffaello 30. *Fisiol. dell' Aviazione. Elettrofisiol. Fisiol. dei muscoli. Fisiol. dell' uomo in varie condizioni climatiche e di lavoro.* ⑨ Roma 1897.

Herman, Martin, Prof., Dir. Inst. prov. bact. Mons (Belgique), 57, Boul. Sainctelette. ○

Hermann, Henri Xavier (1892), Prof. agrégé de Physiol. à la Fac. de Méd. d'Alger (Afrique), 10, Rue Thuillier. *Physiol. du Pancréas (exocrine et endocrine). Circulation (innervation vasomotrice des differents organes splanchniques). Respiration (L'autoregulation respiratoire). Métabolisme de base.*

Herms, William Brodbeck (1876), Prof. of Parasit. and Entomol., Agric. Experim. Station. Head of

Division of Entomol. and Parasit. Berkeley, Cal. (U.S.A.), Agric. Hall, Univ. California. *Med. Entomol., Parasit., Ecol.* ⌾ Ohio State Univ. 1906.

Hernández, Enrique Gamarra, Prof. Univ. Lima (Peru). *Botan.* o

Hernandez, J. M., Prof. Univ. Tegucigalpa, Honduras (Zentralamerika), Plaza de la Merced. *Anat.* o

Hernández, Parmenio, Prof. Univ. Nac. Bogotá (Columbia). *Anat. Topog. y Med. Operat.* o

Hernandez-Pacheco, Eduardo (1872), Prof. Geol. Univ., Dir. Comisión de Investigaciones paleont. y prehistóricas. Madrid (España), Calle Eloy Gonzalo 19. *Paleont. de vertebrados neogénos.* ⌾ Madrid 1898.

Hernando, Teófilo, Prof. Dr. Lab. de Farmacol. Fac. Med. Madrid (España). *Pharm.* o

Herold, Werner (1886), Swinemünde(Deutschland), Bedastr.4. *Isopoden, Mammalia. AngewandteEntomol.* ⌾ Greifswald 1912. ⌾ Isopoden.

Herpin, Alexandre (1880), Prof. à l'Ecole française de Stomatol., Secrét. gén.Association Stomatol. Internat. Paris 8 (France), 79, Boulevard Haussmann. *Stomatol.; Anat. des dents; Orthodontie; Anthrop.* ⌾ Paris 1907.

Herrenschmidt, André (1874), Chef de Labor. Fac. Méd., Conserv. du Mus. Dupuytren. Paris (France), 23, Rue Franklin. *Histol. pathol. du Cancer.* ⌾ Paris 1904.

Herrera, Alfonso L. (1868), Farmaceutico, Dir. des Etudes Biol., Prof. du Coll. Militaire. Mexico, D.F. (Mexico), 2a Cipres 64. *Plasmogénie, l'origine du protoplasma.* ⌾ Mexico 1889. ⌾ Plantes, minéraux et animaux du Mexique.

Herrera, Fortunato L. (1873), Dir. du Mus. de Histoire Nat. et Prof. de Botan. d' Univ. du Cuzco (Perou), Casilla N. 14. *Botan. systématique et etnol.* ⌾ Perou 1911. ⌾ Plantes.

Herrera, Moisés (1885), Oficial 1º Tècnio de la Dir. de Arqueol. Mexico, D.F. (Mexico), 17. Av. Presidentes 176. Col Portales. *Acrididae.* ⌾ Mexico 1918. ⌾ Acridideos del Valle de Mexico.

Herrick, Charles Judson (1868), Prof. of Neurol. Chicago, Ill. (U.S.A.), Department of Anat., Univ. of Chicago. *Comparative anat. of the vertebrate nervous system.* ⌾ Ph. Columbia Univ. 1900, Sc. Univ. Cincinnati 1926 (Hon.).

Herrick, Chester A. (1893), Ass. Prof. Madison, Wis. (U.S.A.), Zool. Labor., Univ. of Wisconsin. *Helminthol.; Hookworm.* ⌾ John Hopkins Univ. 1925.

Herrick, Francis Hobart (1858), Prof. of Biol., Western Reserve Univ. Cleveland, Ohio (U.S.A.). *Animal Behavior; Life and Instincts of Birds; Origin of Instincts. Crustacea. The American Lobster.* ⌾ Johns Hopkins Univ. 1888.

Herrick, Glenn W. (1870), Prof. of Economic Entomol. and Entomol. of Cornell Univ. Experim. Station. Ithaca, N.Y. (U.S.A.), 219 Kelvin Place. *Injurious insects. Thysanoptera.* ⌾ Cornell Univ. 1896.

Herriott, M. A., Lect. Cant. Coll. Christchurch (New Zealand). *Zool.* o

Herris, G. H. (1893), Ass. Prof. of Zool., Stanford Univ. California (U.S.A.). *Coccidae, Chermidae, Mallophaga, Anoplura, Diptera, Pupipara.* ⌾ Stanford Univ. 1918.

Herrmann, Edmund (1875), Priv.Doc. Gynäkol. Wien IV (Österreich), Schleifmühlgasse 2. *Normales, hypoplastisches und pathol. Ovar; Ovarial- und Plazentarlipoide; Biochem. der Blutlipoide in- und außerhalb der Schwangerschaft.* ⌾ Wien 1899.

Herrmann, Eugen (1863), Doc. für Forstwissenschaft Univ., GORR. u. Forstrat. Breslau VIII (Deutschland), Forckenbeckstr. 8 II. *Morphol., Physiol. (besonders Rassenbildung) und Pathol. der Waldbäume.*

Herrmann, George R. (1894), Ass. Prof. of Med., Tulane Univ. of La. New Orleans, La. (U.S.A.), 1551 Canal St. *Cardiac Physiol. and Pathol. Electrocardiography.* ⌾ Med. Univ. of Michigan 1918; ph. 1922. ⌾ Electrocardiograms.

Hersh, Amos Henry (1891), Ass. Prof. of Biol., Western Reserve Univ. Cleveland, Ohio (U.S.A.), 10940 Euclid Ave. *Dominance in Bar series of Drosophila. Temperature effects upon development in Drosophila.* ⌾ Univ. of Illinois 1922.

Herter, Konrad (1891), Dr. phil., Priv.Doc. f. Zool. an der Univ. Berlin (Lehrauftr. f. vergl. Sinnesphysiol. u. Tierpsychol.). Berlin-Steglitz (Deutschland), Wrangelstr. 5. *Sinnesphysiol. der Wirbellosen.* ⌾ Berlin 1921.

Hertling, Helmuth (1891), Etatsmäßiger Ass. der Biol. Anstalt auf Helgoland (Deutschland), Leuchtturmstr. *Oligochaeten (Morphol.), Fische (Oekol., Ernährung), Nordseemollusken (Oekol., Physiol., Systematik).* ⌾ Göttingen 1919.

Hertwig, Günther (1888), ao. Prof., I. Prosektor, Anat. Inst. Rostock i. Mecklenburg (Deutschland). *Keimzellforschung, embryonale Transplantation.* ⌾ Berlin 1912.

Hertwig, Paula (1889), Dr. phil., Priv.Doc. Univ. Berlin. Berlin-Grunewald (Deutschland), Wangenheimstr. 28. *Entwicklungsmechanik: Experim. Parthenogenese bei Fröschen. Vererbungsversuche (mendelistischer Art) mit Ratten, Mäusen, Hühnern, Antirrhinum.* ⌾ Berlin 1916. ⌾ Ratten und Mäuse in verschiedenen Färbungen und Zeichnungen.

Hertwig, Richard (1850), Prof. emeritus der Zool. u. vergl. Anat. Univ. München (Deutschland), Zool. Inst., Neuhauser Str. 51 (Alte Akademie). *Protozoen, Befruchtungslehre, Geschlechtsbestimmung. Histol. der Radiolarien. Das Nervensystem und die Sinnesorgane der Medusen.* ⌾ Bonn 1872.

Hérubel, Marcel Adolphe (1879), Ass. Labor. de Zool. à la Sorbonne, Paris 5 (France), 1, rue Victor Cousin. *Zool.: Géphyriens. Physiol.: vision.* ⌾ 1908. ⌾ Sipunculides et Echiurides (Thalassema).

van Herwerden, Maria Anna (1874), tit. Lect. an der Univ. Utrecht (Holland), Parkstraat 47. *Allgemeine Cytol. und Erblichkeitslehre. Protoplasmastudien auf physiol. und physiol.-chem. Gebiete bei Invertebrata.* ⌾ 1905.

Herzenberg, Helene (1886), Ass. Pathol. anat. Inst. I. Univ. Moskau (U.d.S.S.R.), Antipiewsky pereulok. N. 10, Log. 20. *Myeloide Metaplasie. Amyloiderzeugung und seine vitale Färbung. Eisenstoffwechsel.* ⌾ Moskau 1921.

Herzfeld, Stephanie (1868), Dr. Wien (Deutsch-Österreich), Botan. Inst., III. Rennweg 14. *Morphol. und Entwicklungsgeschichte, Gymnospermen.* ⌾ Wien.

Herzog, Ernst (1898), Dr. med., Ass.-Arzt am Pathol. Inst. der Univ. Heidelberg (Deutschland). *Histopathol. des peripheren Nervensystems u. Sympathikus.* ⌾ Heidelberg 1921. ⌾ Histol. Praep.

Herzog, Georg (1884), Dr. med., o.Prof. der allgem. Pathol. und der pathol. Anat., Dir. des Pathol. Inst. der Univ. Gießen (Deutschland), Klinikstr. 32. *Bindegewebe, „Entzündung", Geschwülste, Beziehungen der experim. Zool. zur Pathol.* ⌾ Leipzig 1909.

Hescheler, Karl (1868), Prof. Dr., Dir. des zool.-vergl. anat. Inst. beider Hochschulen und Dir. des zool. Mus. der Univ. Zürich (Schweiz). *Vergl. Anat. der Anneliden. Fauna des Diluviums und der Pfahlbauten der Schweiz.* ⌾ Zürich 1895.

Hesler, Lexemuel Ray (1888), Prof. of Botan. Knoxville, Tenn. (U.S.A.), Univ. of Tennessee. *Fruit Diseases: apple and peach diseases; Fungicides, Fungous Flora of Tennessee; Sexual Stages of certain Ascomycetes.* ⌾ Cornell Univ. 1914. ⌾ Fungi.

Hesnard, A., Prof. Ecole d. Service d. Santé de la Marine. Bordeaux (France). *Physiol., Hygiene.* o

Hess, Friedrich Albert (1876), Präsident der Schweizer. Gesellschaft für Vogelkunde. Bern (Schweiz), Spitalgasse 26. *Theoretische und angewandte Ornithol.* ⌾ Palaeärktische Vögel.

Hess, Walter Norton (1890), Prof. of Zool. and Head of the Biol. Department of DePauw Univ.

Greencastle, Ind. (U.S.A.). *Physiol., Animal Behavior.* ⊕ Cornell Univ. 1919.
Hess, W. R., Prof. Dr. Zürich 6 (Schweiz), Susenbergstr. 198. *Physiol. der Blutgefäße.* o
Hesse, Edmond (1872), Prof. de Zool. à l'Univ. de Dijon, Cote d'Or (France), Fac. des Sc. *Sporozoaires, particulièrement Microsporidies. Diptères.* ⊕ Paris 1909.
Hesse, Erich (1895), Dr., Priv.Doc. für Pharmacol. und Toxikol. an der Univ. Breslau (Deutschland), Maxstr. 12. *Pharm. und Toxikol.* ⊕ 1921.
Hesse, Richard (1868), o.ö. Prof. der Zool. u. Dir. des Zool. Inst. der Univ. Berlin N 4 (Deutschland), Invalidenstr. 43. *Sehorgane, Ökol. Abstammungslehre. Tiergeographie.* ⊕ Halle 1892.
Hesselink, Engbertus (1879), Dir. der holl. forstlichen Versuchsanstalt. Amersfoort (Holland), Koninginnelaan 19. *Waldbauliche Probleme, wie Wurzelentwicklung, Einfluß der Herkunft des Kieferssamens.*
Hesselman, Henrik (1874), Prof., Dir. d. forstlichen Versuchsanstalt u. der naturwissenschaftl. Abteilung, Experimentalfältet (Sverige). *Forstliche Ökol. und speziell forstliche Bodenkunde.* ⊕ Uppsala 1904.
Hesser, A. C. H., Prof. Dr., Karolinska Inst. Stockholm (Sverige). *Anat.* o
Heßler, Hugo (1879), Hessische Lehranstalt für Obstbau und Landwirtschaft. Friedberg i. Hessen (Deutschland), Dieffenbachstr. 5. *Obst- u. Gartenbauschädlinge. Pflanzenschutzmittel. Phänol.* ⊕ Gießen 1912.
Hetényi, Géza (1894), Ass. III. med. Klinik. Budapest VI (Ungarn), Andrássy-ut 27. *Kohlehydratstoffwechsel: Insulin; Blutzuckerregulation; Respiratorischer Gaswechsel. Kalkstoffwechsel: Regulation des Blutkalkes.* ⊕ Budapest 1919.
Hetherington, Duncan Charteris (1895), Instr. in Anat. Nashville, Tenn. (U.S.A.), Vanderbilt Univ. School of Med. *Anat., tissue culture (sympathetics).* ⊕ Johns Hopkins Med. School, Baltimore 1926.
Hetsch, Heinrich (1873), Prof. am Staatsinst. für experim. Therapie. Frankfurt a. M. (Deutschland), Paul-Ehrlich-Str. 44. *Prüfung der Schutz- und Heilsera sowie der Tuberkuline.* ⊕ Berlin 1895.
Hett, Johannes (1894), Oberass. am Anat. Inst. Priv.Doc. Univ. Halle a. d. S. (Deutschland), Gr. Steinstr. 52. *Histol., Embryologie (Corpus luteum, Nebenniere, Pankreas, Ovar).* ⊕ Leipzig 1920. ⊕ Histol. Praep.
Hett, Mary L. (1880), Lect. in Zool., Bedford Coll., Univ. London N.W. (England). *Linguatulidae. Mammalian Tonsils.* ⊕ London 1913.
Hettwer, Joseph P. (1895), Ass. Prof. of Physiol. Milwaukee, Wis. (U.S.A.), Marquette Univ. Med. School. *Absorption of undigested protein and Anaphylaxis.* ⊕ Harvard Univ. 1921.
Heubner, Wolfgang Otto Leonhard (1877), Prof. der Pharmacol. u. Toxikol. Göttingen (Deutschland), Haussenstr. 26. *Blutgifte (besonders Methämoglobinbildner); Calciumwirkungen; Schwermetallwirkungen, bes. Eisen, Gold, Quecksilber.* ⊕ Straßburg i. Elsaß 1903.
van Heurn, Willem Cornelis Jhr. (1887), Zool. am Inst. für Pflanzenkrankheiten (Departement für Landwirtschaft usw.) zu Buitenzorg, Java (Ned.-O.-Indië), Bataviascheweg 56a. *Bekämpfung von Schädlingen der Culturpflanzen, von tierischen Feinden des Haushaltes.* ⊕ Bälge der javanischen Kleinsäugetiere und Vögel.
Hewitt, James Arthur (1889), Lect. in Physiol. London (England), Univ. of London, Kings Coll. *Biochem. of Carbohydrates.* ⊕ St. Andrews 1919, London 1922.
Heybowicz-Kulesza, Stella Teresa (1897), Ass. Anat. Fac. Méd. Warszawa (Polska), Chalubinski 5, Inst. d'Anat. Patol. *Tumeurs.* ⊕ Varsovie 1924.

van der Heyde, Henri Christiaan (1898), Conserv. in charge of teaching of physiol. and physiol. chem. Groenekan near Utrecht (Holland). *Comparative physiol., physiol. of the domestic animals. Electrogastrograms.* ⊕ Amsterdam 1922, Paris 1923, Utrecht 1926.
Heyder, Ernst Richard (1884), Öderan, Sachsen (Deutschland), Badgasse 8. *Ornithogeographie, Aviöcol.; Avifauna Sachsens.*
Heymanowitsch, Alexander (1882), Prof., Dir. Psychoneurol. Inst., Leiter d. Forschungskatheder f. Psychoneurol. d. Charkow. Med. Inst. u. Leiter d. Section f. experim. Psychoneurol. u. Morphol. d. Nervensyst. Charkow, Ukraine (U.d.S.S.R.), Karl-Liebknecht-Str. 4. *Neuropathol.; Histopathol. d. Nervensystems.*
Heylemans, Franz (1894), Prof. Forest. Bruxelles (Belgique), Avenue des 7 Bonniers 271. *Entomol. Coléoptères; Coléoptères Paléarctiques. Carabidae. Cerambicidae. Lucanidae.* ⊕ Bruxelles 1914. ⊕ Coléoptères du Globe. Carabidae.
Heymann, Bruno (1871), beamteter ao. Prof. in der Med. Fac., Abteilungsvorsteher am Hygien. Inst. der Univ. Berlin NW 7 (Deutschland), Dorotheenstr. 28a. *Hygien., klimatische Fragen, Avitaminose, Tuberkulose, Trachom, Vaccination, Tumoren, Krankheiten übertragende Insekten, Tollwut.* ⊕ 1897.
Heymann, Paul (1892), Dr. med., Ass.Arzt. Wiesbaden (Deutschland), Ruhbergstr. 15. *Experim. Pharm. Nieren-Physiol.* ⊕ Frankfurt a. M. 1920.
Heymans, Corneille-Jean-François (1892), Dr. med. Prof. de pharmacodynamie à l'Univ. Gand (Belgique), 57, quai des moines. *Physiol. et pharmacol.: la régulation cardio-vasculaire et respiratoire, la technique de la tête ,,isolée'' du chien.* ⊕ Gand 1920.
Heymans, Jean-François (1859), Dr. med., Prof. de thérapeutique générale et éléments de pharmacol. à l'Univ. Gand (Belgique), 49, Boulev. de Kerchove. *Recherches pharm. concernant la régulation cardio-vasculaire et respiratoire. Tuberculose.* ⊕ Louvain-Gand 1891.
Heymons, Richard (1867), Dir. des Zool. Inst. der Landwirtschaftl. Hochsch. Berlin N (Deutschland), Invalidenstr. 42. *Biol., Anat., Entwicklung der Insekten. Pentastomida (Linguatulida).* ⊕ Berlin 1891. ⊕ Pentastomida (Linguatulida).
Heys, Florence M. (1903), Ass. in Zool. Saint Louis, Mo. (U.S.A.), Washington Univ., Department of Zool. *Experim. evolution; Genetics.* ⊕ Washington Univ. 1926.
Hibbard, Rufus Percival (1875), Plant Physiol., Experim. Station, Michigan State Coll. (U.S.A.). *Phytochem., Nutrition.* Univ. of Michigan 1906.
Hibino, Shin-ichi (1888), Prof. Plant Physiol. of Tôhoku Imp. Univ. Sendai (Japan), Biol. Inst. *Plant-physiol., Physiol. of Microorganisms.* ⊕ Tôkyô Imperial Univ. 1912.
Hicken, Cristobal M. (1875), Prof. der Botan., Univ. Buenos Aires (Argentina), Casilla de Correo 1606. *Botan. Systematik.* ⊕ Buenos Aires 1906. ⊕ Herbarmaterial.
Hicks, Cedric Stanton (1892), Prof. Human Physiol. and Pharmacol. Univ. Sheridan Res.Fellow of Adelaide. Adelaide (South Australia), Univ. *Ultraviolet absorption spectrum study of labile systems. And especially thyroid junction from chem. standpoint.* ⊕ New Zealand 1914. ⊕ Pedigree hooded (black and white) rat, from the mottram and hartwell stock of 10 years breeding.
Hicks, Charles Henry (1899), Instr. in Biol., Univ. of Colorado. Boulder, Col. (U.S.A.), Biol. Building. *Habits and parasit. of Hymenoptera.* ⊕ Bees and wasps.
Hicks, J. A. B., Prof., Westminster Hosp. Med. School. London S.W. 1 (England), 12 Caxton Str. *Pathol.* o
Hickson, Sydney John (1859), Prof. emer. Univ. of Manchester. Cambridge (England), 26 Barton

Road. *Corals and Alcyonaria. Fauna of the Deep Sea.* ⓓ M.A. Cambridge 1884, D.Sc. London 1883.

Hidén, Henrik Ilmari Augustus (1898), ao. Ass. der geographischen Institution der Univ. zu Helsinki (Finnland), Kasarmink. 14 B. *Pflanzengeographie und Phanerogamensystematik. Die Phanerogamen Finnlands, Systematik der Gattung Salix. Beschaffenheit und Verbreitung der Hainenvegetation des südöstlichen Finnlands.* ⓓ Helsinki 1926. ⓟ Pflanzen Finnlands.

Hieronymi, Erich (1884), o.ö. Prof., philosophische Fak. d. Univ., Dir. d. Tierärztl. Inst. d. Univ. Königsberg i. Pr. (Deutschland), Hagenstr. 9. *Pathol. Anat. der Haut der Haustiere.* ⓓ Berlin 1911.

Higashi, Mitsuharu (1895), Lect. Zool. Inst., Sc. Coll., Kyôto Imperial Univ. Kyôto(Japan). *Embryol. and Ecol. of Onchididae (Mollusca). Animal Ecol.* ⓓ Tôkyô, Japan, 1920. ⓟ Onchididae, Animals of Sand areas.

Higgins, Bascombe Britt (1887), Botan., Georgia Agric. Experim. Station. Georgia (U.S.A.). *Mycol., pathol. anat., normal and pathol. physiol.* ⓓ Cornell 1913.

Higgins, George Marsh (1890), Ass. Division of Experim. Surgery and Pathol., Mayo Foundation. Rochester, Minn. (U.S.A.). *Experim. Biol.; including embryol., physiol. anat., with especial reference to the liver. Growth and differentiation of tissue.* ⓓ Illinois 1919.

Hikida, Toyoharu, Lect. School of Fishery, Hokkaido Imp. Univ. Sapporo (Japan).

Hilario Atanacio, Roxas (1896), Dr. phil. Ass. Prof. Zool. Univ. of the Philippines. Manila (Philippine Islands), Dept. of Zool. *Worker of Sex Problems.* ⓓ Chicago 1927. ⓟ All sorts of tropical marine and land animals for exchange or purchase.

Hildebrand, Samuel Frederick (1883), Dir. U.S. Fisheries Biol. Station, Beaufort, N.C. (U.S.A.). *Taxonomic studies of fishes. Life history studies of fishes. Fishes in relation to mosquito control. Diamond-back terrapin culture.* ⓓ Univ. of Georgia, Med. Dept., Augusta, Ga.1925. ⓟ Aquatic biol. specimens.

Hildebrandt, Fritz (1887), Dr. med., o. Prof. der Pharmacol. an der Med. Akademie. Düsseldorf (Deutschland), Moorenstr. 5. *Stoffwechsel. Thyroxin. Pentamethylentetrazol (Cardiazol).* ⓓ Heidelberg 1913.

Hildén, Kaarlo Thorsten Oskar (1893), Prof., Dr. phil., Doc. für Anthrop. an der Univ. Helsingfors (Finnland), Västra Chaussén 33. *Anthrop. und Anthropogenetik (menschliche Vererbungslehre).* ⓓ Helsingfors 1920.

Hildreth, Aubrey C. (1893), Instr. and Ass. in Horticulture, Univ. of Minnesota. St. Paul, Minn. (U.S.A.), Univ. Farm. *Cold resistance in fruit trees.* ⓓ Univ. of Minnesota 1926.

Hiley, Wilfrid Edward (1886), Lect. Forest Economics, Imper. Forestry Inst. Oxford (England). *Fungal Diseases of the Common Larch.* ⓓ Oxford 1911.

Hilgendorff, Gustav (1875), Dr. phil., Reg.R. Berlin-Dahlem (Deutschland), Biol. Reichsanstalt für Land- u. Forstwirtschaft. *Pflanzenschutzmittel.* ⓓ Greifswald 1901.

Hilgermann, Robert (1874), Prof., Dr. med., Dir. des Preuß. Hygien. Inst. Landsberg a. d. Warthe (Deutschland), Zechower Str. 48. *Bact. und Hygien., im besonderen Infektionskrankheiten.* ⓓ Breslau 1900.

Hilgers, Wilhelm Edmund (1882), ao. Prof., Priv.-Doc., Hygien. Univ.Inst. Königsberg i. Pr. (Deutschland), Steindamm 9 B. *Bact.; Bacterien der Milch und der Zahnerkrankungen, Ruhrbacillenrassen.* ⓓ Straßburg, Els. 1910. ⓟ Bakterienkulturen; besonders Milchbakterien und Bakterien der Mundhöhle.

Hill, Albert Frederick (1889), Instr. in Botan. and Ass. Curator Botan. Collections, Yale Univ. New Haven, Conn. (U.S.A.), Osborn Botan. Labor. *Ecol., Classification and Distribution of Flowering Plants.* ⓓ Yale 1921. ⓟ Herbarium specimens.

Hill, Archibald Vivian (1886), Foulerton Research Prof. of the Royal Society, working at the Physiol. Labor., Univ. Coll. London N 6 (England), 16 Bishopswood Rd., Highgate. *The intimate mechanism of muscle from the standpoint of energetics. Muscular exercise in man. Energetics of the nervous impulse. Physico-chem. studies of blood, muscle and other tissues.*

Hill, Arthur W. (1875), Dir., Royal Botan. Gardens. Kew, Surrey (England). *Systematic and Economic Botan.* ⓓ Cambridge Sc.D. 1897.

Hill, Charles Chase (1890), Ass. Entomol., Cereal and Forage Insect Investigations, Bureau of Entomol., U.S. Department of Agric. Carlisle, Pa. (U.S.A.), U.S. Entomol. Labor., 337 Franklin Street. *Parasitic insects.* ⓓ Amherst, Mass. 1914.

Hill, G. F., Keeper Nation. Mus. Nat. Hist. Melbourne (Australia) *Entomol.* ○

Hill, George Richard, Jr. (1884), Dir., Dept. of Agric. Research, American Smelting and Refining Company, Salt Lake City, Utah (U.S.A.), 700 McCornick Building. *Plant Physiol. Effects of smelter gases and fumes on plants, crops and animals.* ⓓ Cornell 1912.

Hill, H. W., Prof. Univ. Vancouver (Brit. Columbia). *Bact.* ○

Hill, J. P., Prof., Univ. London S.W. 7 (England), S. Kensington. *Embryol.* ○

Hill, J. Ben (1879), Prof. of Botan., the Pennsylvania State College, Pa. (U.S.A.). *Botan. and Genetics. Epiphytic species of Lycopodium. Inheritance of Cotyledon Form and Size in Digitalis.* ⓓ Univ. of Missouri 1908.

Hill, Leonard Erskine (1866), Dir. Dept. of Applied Physiol. National Inst. of Med. Research, Mount Vernon Hampstead. London N.W. 3. Loughton, Essex (England), Osborne House.

Hill, Reuben L. (1888), Prof. of Chem., Utah Agric. Coll., Head Dept. Chem., in charge of Human Nutrition Investigations, U.A.C. Experim. Station. Logan, Utah (U.S.A.), 645 North 8 East. *Physiol. and Chem. of Milk Secretion. Digestibility and Nutritive Value of Various Milks. Other general Phases of Nutrition. Physiol. and Organic Chem.* ⓓ Cornell Univ. 1915.

Hill, Thomas George (1876), Reader in Plant Physiol. in the Univ. of London (England), Univ. Coll. *Plant physiol.* ⓓ London 1898.

Hill, William Charles Osman (1901), Ass. Lect., Demonstrator of Anat., Univ. Birmingham (England), 25 Galton Road, Warley. *Development of the Pancreas and the Suprarenal. Histol.* ⓓ Univ. of Birmingham 1924.

Hiller, A. G., Doc. Bergakademie. Moskau (U.d.S.S.R.). *Palaeont.* ○

Hiller, Alma (1892), Ass., Rockefeller Inst. for Med. Research. New York City (U.S.A.), 66. St. and Ave. A. *Physiol. Chem., blood, urine, Nephritis.* ⓓ Columbia Univ. 1926.

Hiller, Stanisław (1891), Dr. med., Ass. à l'Univ. de Cracovie. Kraków (Pologne), Lenartowicza 11. *Embryogénie des Vertébrés, Régénération.* ⓓ Cracovie 1921.

Hiller, Waldemar (1897), Dr. phil. Stettin (Deutschland), Friedrichstr. 3. *Holzanat., pollenanalyt. Mooruntersuchung.* ⓓ Greifswald 1921.

Hillrod, Curtis M. (1887), Prof., Biol. and Health, Simmons Coll. Boston, Mass. (U.S.A.). *Public Hygiene, Bact.* ⓓ Dartmouth 1909.

Hilmi, Mehmed (1882), Prof. Anat. u. Teratol. des Pferdes. Konstantinopel (Türkei), Tierärztl. Hochschule, Skutarie-Sélimiée. *Muskulatur.*

Hiltner, Erhard (1893), Landwirtschaftsassessor der Bayr. Landesanstalt für Pflanzenbau und Pflanzenschutz. München (Deutschland), Osterwaldstraße 9 F. *Ernährungsphysiol. und Stoffwechsel-*

pathol. der Pflanzen. Stimulationsforschung. Bodenbiol., Bodenreaktion. Phänol. ⒟ München 1922.
Hilton, William Atwood (1879), Prof. Zool. Pomona Coll., Claremont, Dir., Laguna Marine Labor., Laguna Beach, Cal. Claremont, Cal. (U.S.A.), 1263 Dartmouth Ave. *Neurol. and Histol., Marine Zool. Nervous System and sense organs of Invertebrates. Marine distribution.* ⒟ Cornell 1902. ⒫ *Marine Invertebrates.*
Hilzheimer, Max (1877), Dr. phil., Dir. der naturwiss. Abteilung des Märkischen Mus. Berlin S 14 (Deutschland). *Recente und fossile Säugetiere, besonders Haustiere, Faunistik und Tiergeographie.* ⒟ München 1903.
Himmel, Walter J., Prof. of Botan., Macalester Coll. St. Paul, Minn. (U.S.A.), 1611 James St. *Physiol.* o
Himmelbaur, Wolfgang (1886), Dr. phil., Priv.-Doc., Referent über Arzneipflanzenkultur. Wien II (Deutsch-Österreich), Schuettelstr. 71. *Arzneipflanzenkultur. Mycol. Pflanzenkrankheiten. Embryol.* ⒟ 1909. ⒫ *Arzneipflanzensamen und Setzlinge. Drogen.*
Himmer, Anton (1886), Dr., Ass. an der Landesanstalt für Bienenzucht. Erlangen (Deutschland), Gabelsbergerstr. 5 I. *Bienenkunde und Bienenzucht.* ⒟ München 1922.
Himwich, Harold Edwin (1894), Ass. Prof. in Physiol. New Haven, Conn. (U.S.A.), Yale Univ. Med. School. *Metabolism, especially in regard to carbohydrate metabolism of muscle.* ⒟ Cornell Univ. Med. Coll. 1919.
Hindersson, Elis Richard (1877), Dir. veterinärbact. Inst. des Finnischen Staates. Helsinki (Finland). *Piroplasma.* ⒟ Leipzig 1912. ⒫ *Bakterienkulturen, Sera, Impfstoffe, Tuberkuline.*
Hindhede, Mikkel (1862), Dir. Staatl. Labor. für Ernährungsuntersuchungen. København V (Danmark), Frederiksberg Alle 28. *Ernährungsuntersuchungen mit Menschen: Eiweißminimum, Fettminimum, Versuche mit Kartoffeln, verschiedenen Brotsorten, Gerstengrütze, Hafergrütze usw.* ⒟ Copenhagen 1888.
Hindle, Edward (1886), Royal Society's Kala Agar Commission China, Milner Research Fellow, School of Hygiene and Tropical Med. London N.W. 1 (England), Endsleigh Gardens. *Protozool.: insects in the transmission of disease. Life cycle of spirochaetes.* ⒟ Cambridge 1912.
Hindorf, Richard (1863), Dr. phil., Dir. kolonialer Gesellschaften, die Pflanzenbetriebe in Kamerun und Ostafrika haben. Geschäftsadresse: Berlin W 35 (Deutschland), Flottwellstr. 3. *Agric. der Tropen.* ⒟ Halle 1886.
Hinds, Warren Elmer (1876), Entomol., Louisiana Experim. Station and Extension Service. Baton Rouge, La, (U.S.A.), Louisiana State Univ. *Mexican cotton boll weevil; insects infesting stored grain; the grass worm (Laphygma frugiperda), the grass thrips (Anaphothrips striatus), Thysanoptera of North America; the sugar cane moth borer (Diatraea saccharalis), insect control by means of dust distributed by airplanes; carbon disulphid as an insecticide; the cotton leaf worm (Alabama argillacea); the Mexican bean beetle (Epilachna corrupta).* ⒟ Massachusetts Agric. Coll. 1902. ⒫ *Insect species of economic importance.*
Hine, James S. (1866), Associate Prof. of Zool. and Entomol., Ohio State Univ., Curator of Natural History, Ohio State Mus. Columbus, O. (U.S.A.). *Diptera, families Tabanidae and Asilidae. Mammals and Birds.* ⒟ Ohio State Univ. 1893. ⒫ *Tabanidae.*
Hinman, Jack Jones, Junior (1888), Ass. Prof. of Sanitation, Univ. of Iowa, and Chief, Water Labor. Division, Iowa State Hygienic Labor. Iowa City, Ia. (U.S.A.), 121 Melrose Avenue. *Water Supply and Water Purification, Swimming Pools, Field Water Supply, Treatment of Sewage and Industrial Wastes.* (*Bact., Chem.*) ⒟ Butler Coll. 1915.

Hinrichs, Marie (1892), Research. Chicago, Ill. (U.S.A.), Dept. of Physiol., Univ. *Physiol. of Light Action; Modification of embryol. development.* ⒟ The Univ. of Chicago 1923.
Hinshaw, Horton Corwin (1902), Ass. Zool., Department of Zool., Univ. of California. Berkeley, Cal. (U.S.A.). *Protozool.* ⒟ Univ. of California 1927.
Hinshow, W. R. (1896), Instr. in Bact. and Poultry Disease Investigator, Kansas State Agric. Experim. Station (U.S.A.). *Poultry diseases bacillary white diarrhea.* ⒟ Kansas State Agric. Coll. 1926.
Hinton, William Augustus (1883), Dir. Wassermann Labor., Massachusetts Dept. of Public Health; Instr. in Bact., Harvard Med. School. Boston, Mass. (U.S.A.), 240 Longwood Avenue. *Bact.: Serol.* ⒟ Harvard Med. School, Cambridge, Mass., 1912. ⒫ *Standardized antigen and amboceptor.*
Hintzsche, Erich G. W. (1900), Ass. am Anat. Inst. Halle a. d. S. (Deutschland), Große Steinstr. 52. *Histol. der Stützsubstanzen, besonders der Knochenbildung.* ⒟ Halle a. d. S. 1925.
Hinze, Gustav (1879), Dr. phil., Mus.Dir. Zerbst, Anhalt (Deutschland), Friedrichsholzallee 42. *Schwefelbakterien.* ⒟ 1901.
Hion, Viktor (1902), Dr., Ass. Physiol. Inst. Tartu (Estland). *Innere Sekretion, Ermüdung, Carcinom.* ⒟ Tartu, Estland.
Hirasaka, Kyôsuke (1887), Prof. Zool., Imper. Univ. of Formosa. Tôkyô (Japan), 629 Yoyogi. *Morphol. and ecol.researches on Marine Invertebrates, especially Mollusca. Dorsal Eyes of Onchidium.* ⒟ Tôkyô 1911.
Hirase, Shintarô, Zool. Labor., Coll. of Sc., Imper. Univ. Tôkyô (Japan). *Systematic Chonchiol.*
Hirata, Kenji (1888), Agric. Dept., Hokkaido Imper. Univ. Sapporo (Japan). *Mechanism of sex determination, hemp plant.* ⒟ Hokkaido Imper. Univ. 1923.
Hiratsuka, Eikichi, Chief of Imperial Sericult. Experim. Sta. Nakano near Tôkyô (Japan). *Physiol. (of Nutrition) of Silk-worms. Biochem. of Mulberry.*
Hiratsuka, Naoharu (1878). Sapporo No. 7 (Japan), Higashi-2 chome, Kita 6-jo. *Melampsoraceae (Fungi), Vegetable fibers (especially flax).* ⒟ Sapporo 1896.
Hiratsuka, Naohide (1903). Botan. Inst., Hokkaido Imper. Univ. Sapporo (Japan). *Phytopath. and Mycol.* ⒟ Hokkaido Imper. Univ. 1926.
Hirmer, Max (1893), Dr., Priv.Doc., München-Nymphenburg (Deutschland), Maria-Ward-Str. 14. *Systematische Morphol., Pflanzengeographie und Palaeobotan.; Entwicklungsphysiol. Probleme der Blattstellung.* ⒟ München 1917.
Hirsch, Gottwalt Christian (1888), Dr. rer. nat., Lect. für prop. Zool. und Leiter der Abteil. für experim. Histol. am Zool. Labor. Utrecht (Holland), Janskerkhof 3. *Zellpermeabilität, Phagocytose, Rhythmus der Secretion, Fermente, Mitochondrien, Golgiapparat, Zellanalyse.* ⒟ Tübingen 1914. ⒫ *Mikroskop. Praep. auf den oben genannten Arbeitsgebieten.*
Hirsch, Slavko (1893), Dir. Epidemiol. Inst. u. Primararzt Infektionsabt. des Landeskrankenhauses. Osijek (S.H.S.). *Bact., Serol. u. Klinik der Infektionskrankheiten.* ⒟ Innsbruck 1917.
Hirschfeld, Hans (1873), ao. Prof. Univ., Abteilungsvorsteher am Inst. für Krebsforschung an der Charité. Berlin NW 40 (Deutschland), Alt-Moabit 110. *Haematol., Krebsforschung.* ⒟ Berlin 1897.
Hirschfeld, Magnus (1868), Dr. med., San.R., Leiter des Inst. f. Sexualwiss. Berlin NW 40 (Deutschland), In den Zelten 10. *Sexualbiol., Sexualwiss.* ⒟ Berlin 1892.
Hirschler, Jan (1883), Dr. phil., o.ö. Prof. der Zool. und vergl. Anat. und Dir. d. Zool. Inst. an der Jan Kazimierz Univ. in Lwów (Polen), Sanct Mikolaj-Gasse 4. *Embryol. der Insekten, Cytol. des Zellenplasmas: Golgischer Apparat und*

Mitochondrien, Experim. Analyse der Amphibien-metamorphose, *Theoretische Vererbungslehre, Regenerationsvorgänge bei Tieren.* ① Lwów 1905.

Hirszfeld, Ludwig (1884), Dir. am Hygiene-Inst., Abteilung für Bact. und experim. Med., Priv.Doc. an der Univ. u. Prof. an der Freien Hochsch. Warschau (Polen), Hygiene-Inst., Kujawska 2. *Konstitutionsserol., Blutgruppenforschung. Serodiagnostik der Lues.* ① Berlin 1907. ② Sera.

Hirt, August (1898), Priv.Doc., II. Prosektor. Heidelberg (Deutschland). *Vergl. Anat. des Sympathicus. Innervation der Niere. Faserverlauf im Sympathicus.* ① Heidelberg 1922.

Hirt, Ray R. (1893), Instr. in Forest Botan. New York (U.S.A.), N.Y. State Coll. of Forestry. *Fungi.* ① Syracuse Univ. 1924. ② Fungi.

Hirtz, Friedrich (1878), Prof. der kgl. Forstakademie, Kustos der zool. Abt. des Nationalmus. in Zagreb (S.H.S.), Mesnička Gasse 39. *Ornithol., Herpetol. und Jagdzool., Faunistik Jugoslaviens.* ① Zagreb 1907.

Hisata, Katsujirô, Biol. in Provincial Agric. Experim. Station of Kyôto. Shimogamo, Kyôto (Japan). *Phytopath.*

Hisaw, Frederick Lee (1891), Ass. Prof. of Zool. Madison, Wis. (U.S.A.), Biol. Bldg. Univ. of Wisconsin. *Embryol. and Physiol.* ① Univ. of Wisconsin 1924.

Hitchcock, Albert Spear (1865), Senior Botan. in Charge of Systematic Agrostol., U.S. Department of Agric. Washington, D.C. (U.S.A.), Home: 1867 Park Road; Office: Smithsonian Institution. *Classification and identification of grasses (Gramineae).* ① Iowa State Coll., Ames, Ia. 1920.

Hitchcock, Charles Hanchett (1896), Ass., Rockefeller Inst. for Med. Research, and Ass. Resident Physician, Hospital of the Rockefeller Inst. for Med. Research. New York, N.Y. (U.S.A.), 66 th St. and Avenue A. *Rheumatic fever, problem of the allergic nature of this disease.* ① Johns Hopkins Univ. 1921.

Hitchcock, David Ingersoll (1893), Associate Prof. Physiol. and Biochem., Bryn Mawr Coll., Bryn Mawr, Pa. (U.S.A.). *Colloidal behavior of proteins.* ① Columbia Univ. 1922.

Hite, Bertha Courtright (1880), Ass. Botan. U.S. Department of Agric. Corvallis, Ore. (U.S.A.), Branch Seed Labor. *Seed testing: after-ripening and germination of seeds and selection of 1000 headed kale plants from seedlings.* ② Grass, crop and weed seeds from the Pacific Coast, Northwestern U.S.

Hitti, Yusuf, Prof. American Univ. Beirut (Syrien) *Anat.* o

Hixon, Ralph M. (1895), Assoc. Prof. Chem. Ames, Ia. (U.S.A.), Chem. Department. *Biochem.: Carbohydrates and other poly-hydroxides.* ① Wisconsin 1922.

Hjort, Axel M. (1889), Prof. of Pharmacol., Dartmouth Med. School. Hanover, N.H. (U.S.A.). *Chem. and Physiol. of the Parathyroid Glands. Comparative Local Anesthetics and Hypnotics.* ① Ph. Yale Univ. 1918, med. 1921.

Ho, William T. H. (1893), Prof. in Plant Pathol., National Southeastern Univ. Nanking (China). *Plant Pathol. and Mycol.* ② Specimens of diseased plants and larger fungi.

Hoadley, Leigh (1895), Ass. Prof. of Zool. Providence, R.I. (U.S.A.), Zool. Department, Brown Univ. *Embryol., experim. and descriptive. The localization of potential areas in the germ. Factors active in developmental period. Fertilization.* ① Univ. of Chicago 1923.

Hoagland, Dennis Robert (1884), Associate Prof., Plant Nutrition, Chem. Agric. Experim. Station. Berkeley, Cal. (U.S.A.), Univ. of California. *The absorption of mineral elements by plants, soil solution.* ① Stanford Univ. 1907.

Hoar, Carl S., Ass. Prof. of biol., Williams Coll. Williamstown, Mass. (U.S.A.). *Plant Morphol., evolutionary, bact.* o

Hoare, Cecil Arthur, Wellcome Bureau of Sc. Research. London W.C.1 (England), 25/28 Endsleigh Gardens. *Protozool.* ① Petrograd 1916.

Hobbs, Joseph (1904), Ass. in Filterable Viruses, Johns Hopkins Univ. Baltimore, Md. (U.S.A.), 615 N. Wolfe Str. *Myxomatosis of Rabbits, Herpes and Infectious Tumors of Fowls.* ③ M.I.T. 1925. ② Cell-inclusion preparations.

Hobmaier, M., Prof. Univ., Kab. f. Vet. Wiss. Tartu (Estland), Vene t. 38. *Pathol.* o

Hobstetter, Karl (1875), Dr. med. vet., G.R.R. Univ.-Prof. Jena (Deutschland). *Tierheilkunde, Anat. der Haustiere, Bact., pathol. Anat.* ① Gießen 1901.

Hochapfel, Hans Heinz (1897), wissenschaftlicher Hilfsarbeiter. Berlin-Dahlem (Deutschland), Biol. Reichsanstalt. *Obstfäule.* ① Bonn 1925.

. **Hoche,** Léon Claude Adolphe (1869), Prof. d'Anat. pathol. Fac. de Méd. de l'Univ. de Nancy (France), 16, rue Emile Gallé. *Anat. pathol. des Tumeurs. Pathogénie.* ① 1910. ② Préparations d'histol. pathol.

Hochstetter, Ferdinand (1861), o. ö. Prof. der Anat. und Vorstand der II. anat. Lehrkanzel an der Univ. Wien XIX (Deutsch-Österreich), Pokornygasse 23. *Vergl. Anat. und Entwicklungsgeschichte des Blutgefäßsystems der Wirbeltiere. Entwicklungsgeschichte des Gehirns.* ① Wien 1885.

Hockey, John Frederick (1895), Plant Pathol.-in-charge, Labor. of Plant Pathol. Kentville, Nova Scotia (Canada). *Fruit and vegetable diseases, diseases of apples.* ① McGill Univ., Montreal, Que., 1921. ② Specimens and cultures of local apple diseases.

Hoder, Friedrich (1900), Ass. Hygien. Inst. deutsche Univ. Praha (C.S.R.), Preslová 7. *Bacteriophagen und Bacterienmutation.* ① Prag 1926.

Hodgson, Benjamin Earl (1888), Ass. Entomol.: Sc. investigations at the European Corn Borer Labor. Arlington, Mass. (U.S.A.), 10 Court Street. *Biol. and economic: European Corn Borer (Pyrausta nubilalis Hubn.) in maize and other host plants. The extent and intensity of infestation and the spread of the European Corn Borer in New England.*

Hodgson, Robert Willard (1893), Associate Prof. of Horticulture and Associate Citriculturist in the Agric. Experim. Station, Univ. of California. Berkeley, Cal. (U.S.A.), 339 Hilgard Hall. *Subtropical Horticulture. Plant physiol. as applied to horticultural plants.* ① Univ. of California 1916.

Höber, Rudolf (1873), Dr. med., o. ö. Prof. der Physiol., Dir. des physiol. Inst. der Univ. Kiel (Deutschland). *Physikalische Chem. der Zellen und der Gewebe.* ① Erlangen 1896.

Höeg, Ove Arbo (1898), Konserv., Kgl. Norske Videnskabers Selskab (Trondhjems Mus.), Botan. Abt. Trondhjem (Norge), Videnskabsselskabet. *Paläobotan.: Fossile arktische Floren; fossile Algen. Lichenol.: Systematik und Geographie der Gatt. Pertusaria.* ① Oslo 1923. ② Flechten (Pertusaria).

Höfker, Hinrich (1859), Prof. Dortmund (Deutschland), Limburger Str. 31. *Dendrol.: Ligustrum Cedrus. Kohlensäuredüngung.* ① Jena 1892.

Höfker, Jan (1898), Dr. phil. Haag (Holland), 3e Braamstraat 35. *Protozool., Foraminifera.* ① Leiden. ② Foraminiferenschalen.

Höfler, Karl (1893), Dr. phil., Priv.Doc. der Anat. und Physiol. der Pflanzen an der Univ. Wien XIII/2 (Deutsch-Österreich), Onno Kloppgasse 15. *Zellphysiol., Protoplasmakunde, Permeabilität, osmotische Zustandsgrößen der Pflanzenzelle.* ① Wien 1919.

Höflich, Dr., Prof., Inst. f. Anat., Physiol. u. Pathol. der Haustiere. Weihenstephan b. Freising (Deutschland).

Hoek, Françoise Juriana (1902), Ass. am Zool. Inst. d. Univ. (Abt. Morphol.) Utrecht (Holland), Prins Hendriklaan 11. *Protozoenernährung und -bau.*

Hoeltzer, Rudolf (1890), Ass. Bact. Inst. Univ. Kazan (U.d.S.S.R.), B. Krasnaja 59. *Bact. Kultivierung der Spirochaeta Obermeieri u. pallida.* ⊕ St. Petersburg 1914.

Hönningstad, Dir. Statens forsöksstation Vestenfjelds. Forus, Jederen (Norge). o

Hoepke, Hermann Leonard Theodor August (1889), Priv.Doc., Abteilungsvorsteher Anat. Inst. Heidelberg (Deutschland). *Anat. der Haut. Biddersches Organ bei Bufo.* ⊕ Greifswald 1918.

Hoeppli, Reinhard (1893), Doc. Pathol. (beurlaubter Ass. des Tropeninst. Hamburg, Priv.Doc. Univ.). Univ. of Amoy (China). *Tropenpathol. und Helminthol.* ⊕ Med. Kiel 1919, rer. nat. Hamburg 1924.

Höstermann, Gustav (1872), Dr. phil., Doc. der Botan. und Abteilungsvorsteher, Leiter der pflanzenphysiol. Versuchsstation der Lehr- und Forschungsanstalt für Gartenbau Berlin-Dahlem. Berlin-Steglitz (Deutschland), Schloßstr. 32. *Anat., Organographie, Pflanzenphysiol.* (*Obstbaumphysiol.*). *Krankheiten der gärtnerischen Kulturgewächse, gärtnerischer Pflanzenschutz und Pflanzenzüchtung.* ⊕ Königsberg i. Pr. 1902.

Hofeneder, Karl (1878), Dr. phil. Innsbruck, Tirol (Österreich), Innrain 47. *Strepsipteren. Biol. tierischer Parasit., Insekten.* ⊕ Innsbruck 1912. ⊕ Strepsipteren.

Hoffer, George N., Pathol. U. S. Dept. of Agric. and Assoc. Botan., Purdue Univ. Agric. Experim. Stat. W. Lafayette, Ind. (U.S.A.), 434 Littleton Street. o

Hoffman, Prof. Dr. med. Poznań (Polska), Univ., Inst. de pathol. gén. et expérim. o

Hoffman, Ira Curtis (1888), Ass. in Olericulture. Lafayette, Ind. (U.S.A.), Purdue Univ., Agric. Experim. Stat. *Breeding Vegetable Crops.* ⊕ Purdue Univ. 1916. ⊕ Inbred lines of sweet corn.

Hoffman, William Albert (1894), Ass. Prof. of Parasit. San Juan (Porto Rico), School of Tropical Med. *Anopheles larvae, Transmitters of filaria, Life history and distribution of Schistosoma mansoni in Porto Rico. Taxonomy and biol. of Ceratopogoninae* (*Chironomidae*). ⊕ Johns Hopkins 1924.

Hoffmann, Alfred, Dr. Wien II (Deutsch-Österreich), Trumerstr. 7. *Pflanzenbau. Samenprüfung.* o

Hoffmann, Bernhard August (1860), Prof. Dresden-A. (Deutschland), Uhlandstr. 16 III. *Vogelstimmen, Faunistik u. Ökol.* ⊕ Leipzig 1883.

Hoffmann, Curt (1898), Dr. phil., Ass. am botan. Inst. der Univ. Kiel (Deutschland). *Permeabilität bei Algen.* ⊕ Leipzig 1924. ⊕ Makro-Lepidopteren.

Hoffmann, Erich (1868), o. Prof. an der Univ. und Dir. der Hautklinik in Bonn a. Rh. (Deutschland), Meckenheimer Allee 18. *Syphilis, Syphilisspirochäte, Leptospira dentium im Munde des Menschen. Spirochäten* (*Sp. balanitidis*). ⊕ Berlin 1892.

Hoffmann, Hans (1896), Dr. phil., Priv.Doc. für Zool. und vergl. Anat., I. Ass. am Zool. Inst. Jena, Thüringen (Deutschland), Neugasse 25. *Gastropoda. Zoogeographie.* ⊕ Leipzig 1919.

Hoffmann, Heinrich Fritz August Wilhelm (1875), Pathologist Covadonga Hospital, Habana (Cuba). *Tropical Diseases* (*Spirochaetal diseases; yellow fever*). ⊕ Berlin 1899.

Hoffmann, Hermann (1891), Priv.Doc. für Psychiatrie und Neurol., Oberarzt der Univ.-Klinik für Gemüts- und Nervenkrankheiten. Tübingen (Deutschland). *Psychiatrie, Neurol. Charakterol. und psychiatrischen Konstitutions- und Vererbungslehre.* ⊕ Tübingen 1916.

Hoffmann, Paul Albin (1884), Prof. der Physiol. Freiburg i. Br. (Deutschland), Hebelstr. 33, Physiol. Inst. *Physiol. des Nervensystems. Elektrobiol.* ⊕ Leipzig 1909.

Hoffmann, Richard (1872), a.o. Univ.-Prof. Göttingen (Deutschland), Risdorfer Weg 8. *Zool.* o

Hoffmann, William Edwin, Lect. Hackett Med. Coll. Canton (China). *Zool.* o

Hoffstadt, Rachel E., Ass. Prof. of bact., Univ. of Washington. Seattle, Wash. (U.S.A.). *Plant pathol., bact.* o

Hofmann, Julius V., Prof. of Forestry, Pennsylvania State Forest School. Mont Alto, Pa. (U.S.A.). *Ecol., forestry.* o

von Hofsten, Nils (1881), Prof. der vergl. Anat. an der Univ. Uppsala (Schweden). *Turbellarien: Anat., Systematik, Histol., Oekol. der Rhabdocoelen und Alloeocoelen, Rotatorien* (*Faunistik und Systematik*). *Süßwasserfauna* (*Bodenfauna der Seen*). *Marine arktische Tiergeographie* (*Echinodermen, decapode Crustaceen, Fische*). *Geschichte der Biol.* ⊕ Upsala 1907.

Hogan, Albert G. (1884), Prof. of Animal Nutrition, and Chairman, Department of Agric. Chem., Univ. of Missouri. Columbia, Mo. (U.S.A.), 105 Schweitzer Hall. *Nutritional requirements for growth, reproduction, and lactation of mammals; chick. Vitamines.* ⊕ Yale Univ. 1914.

Hogarth, Alfred Moore (1876), Chairman and Hon. Dir., Coll. of Pestol. London W.C. 1 (England), 52 Bedford Square. *Entomol. Appl. Biol.*

Hogben, Lancelot Thomas (1895), Ass. Prof. of Zool., McGill Univ. Montreal (Canada). *Cytol., Compar. physiol. and biochem. of the pituitary gland. Color response. Haemocyanin.* ⊕ London 1921.

Hogg, Ira Dwight (1892), Ass. Prof. of Anat., Med. School, Univ. of Pittsburgh, Pa. (U.S.A.), Department of Anat. *Growth of the motor nuclei of the brain-stem of the albino rat.* ⊕ 1926.

Hogue, Mary Jane (1883), Instr. of Histol. Philadelphia, Pa. (U.S.A.), School of Med., Univ. of Pennsylvania. *Parasit. protozoa. Tissue culture.* ⊕ Würzburg 1909.

Hohen, E. J., Doc. Med. Inst. Odessa (U.d.S.S.R.), Valichovsky per. 2. *Bact.* o

Hohn, Joseph (1877), Dr. med., Leiter des bact.-serol. Labor. der Stadt Essen, Krankenanstalten. Essen (Deutschland), Gemarkenstr. 47. *Bact. und Serol.* ⊕ 1901.

Hojer, Johan Axel (1890), Doc. in experim. Pathol. Lund (Schweden). *Vitamine. Nahrungshygiene. Kinderhygiene. Maul- und Klauenseuche.* ⊕ 1924.

Hoke, Gladys (1895), Ass. Entomol., State Plant Board of Miss. Como, Miss. (U.S.A.). *Coccidae, Homoptera.* ⊕ Coccidae, Homoptera.

Holbert, James Ransom (1890), Agronomist in the Office of Cereal Investigations, Bureau of Plant Industry. Bloomington, Ill. (U.S.A.). *Corn* (*Zea mays indentata*) *root, stalk and ear-rot diseases, seed selection, seed treatment and breeding. The development of disease-resistant inbred strains and the recombination of these into crosses, double crosses, and synthetic varieties that combine disease resistance with other agronomic qualities.* ⊕ Univ. of Illinois 1926. ⊕ Inbred strains of corn.

Holch, Arthur E. (1891), Head of Biol. Department, Nebr. State Teachers Coll. Peru, Neb. (U.S.A.). *Plant Ecol.; root studies.* ⊕ Univ. of Colorado 1923.

Holdaway, Charles William (1880), Dairy Husbandman, Virginia Agric. Experim. Station, Prof. of Dairy Husbandry, State Coll. of Agric. Blacksburg, Va. (U.S.A.). *Dairy Cattle Nutrition.* ⊕ Blacksburg 1913.

Holdefleiss, Paul (1865), o. Prof., Dir. Abt. Pflanzenbau. Halle a. d. S. (Deutschland), Hohe Weg 31. *Tierzucht, Pflanzenzüchtung.*

Holl, Frederick J. (1898), Ass. in Zool. Durham, N.C. (U.S.A.), Duke Univ. *Ecol. and Parasit.* ⊕ Univ. of Wisconsin 1926.

Holland, John Henry (1869), Ass. Mus., Royal Botan. Gardens, Kew. London (England). *The Useful Plants of Nigeria: Plants suitable for Cultivation in West Africa and other Tropical Dependencies of the British Empire.*

Holland, William Jacob (1848), Pres. Carnegie Hero Fund Commission. Pittsburgh, Pa. (U.S.A), Carnegie Inst. *Zool., Paleont.* o

Hollande, A.-Charles (1881), Prof. de Zool. et Microbiol. à la fac. de Pharmacie de Montpellier, Hérault (France). *Histochim. et cytol. des Insectes; flagellés et sporozoaires; technique histol. et microbiol*

Hollendonner, Ferenc (1882), Priv.Doc. Univ. u. Polytechnikum. Budapest I (Ungarn), Gellért t. 4. *Holz. Praehist. Holzkohle. Lignithistol.* Ⓓ Budapest 1907.

Holler, Gottfried (1886), Ass. II. med. Univ.-Klinik. Wien VIII (Österreich), Langeg. 67. *Morphol. Hämatol. Mineralstoffwechsel.* Ⓓ Wien 1912. Ⓟ Haematol. Praep., Gewebschnitte.

Hollick, Arthur (1857), Paleobotan., New York Botan. Garden, Geol., U.S. Geol. Survey. New York, N.Y. (U.S.A.). *Cretaceous, tertiary, and quaternary fossil plants.* Ⓓ Columbian Univ., Washington, 1897.

Hollingshead, E. Lillian (1900), Research Ass., Genetics. Berkeley, Cal. (U.S.A.), Univ. of California, Genetics Department. *Cytol. and genetic investigations of plants, the genus Crepis, its species and species hybrids.* Ⓓ Univ. of Saskatchewan 1926.

Holloway, John, D.Sc. Lect. Otago Univ. Dunedin (New Zealand). *Botan.* o

Holloway, Thomas E. (1886), Associate Entomol., in charge of sugar cane and rice insect invest., Bureau of Entomol., U.S. Dept. Agr. New Orleans, La. (U.S.A.), 8203 Oak St. *Sugar cane moth borer, Diatraea saccharalis Fab.* Ⓓ Agric. and Mech. Col. of Texas 1908.

Hollstein, Wilhelm (1898), Dr. phil., Ass. am geol.-mineralog. Inst. d. Techn. Hochsch. Danzig. Danzig-Langfuhr (Freistaat Danzig). *Palaeont. der Wirbeltiere.* Ⓓ Münster i. W. 1922.

Holm, George Elmer (1891), Biochem., U.S. Dept. of Agric., Washington, D.C. (U.S.A.), Research Labor., Bureau of Dairy Industry. *Deterioration of fats and oils, butterfat. Autoxidation of fats. Heat coagulation of proteins.* Ⓓ Univ. of Minnesota 1919.

Holm, Otto Ejler (1887), København (Danmark), Østergade 16. *Physiol. Optik, Dunkeladaptation. Experim. Xerophthalmie und Hemeral. Pathogenese u. Aetiol. der Myopie. Vererbung der Myopie und des Nystagmus.* Ⓓ Kopenhagen 1922.

Holman, Richard M. (1886), Associate Prof. of Botan. Berkeley, Cal. (U.S.A.), Department of Botan., Univ. of California. *Plant Physiol. and General Botan. Tropisms. Pollen longevity and germination, conditions for reproduction in green algae.* Ⓓ Univ. of California 1915.

Holman, William Ludlow (1879), Associate Prof. of Bact. Toronto, Ont. (Canada), Univ. of Toronto. *Medical Bact.* Ⓓ Montreal (McGill) 1907.

Holmberg, Otto Rudolf (1874), Konserv. Botan. Mus. Univ., Tauschleiter des Lunds Botan. Förening, Dir. staatl. unterstützter Samenkontrollstation. Lund (Sverige). *Botan. Systematik und Nomenklatur. Skandinavische Phanerogamen. Gramineae, Cyperaceae. Samenkontrolle.*

Holmboe, Jens (1880), o. Prof. der Botan. an der Univ. Oslo (Norwegen), Botan. Garten. *Pflanzengeographie.* Ⓓ Oslo. Ⓟ Phanerogamen, Krypto--gamen.

Holmdahl, David Edvard (1887), Priv.Doc. Anat. Lund (Sverige). *Embryol.: Spinalnervensystem, Coelom, kaudale Partie.* Ⓓ Lund 1918.

Holmes, Francis O. (1899), Dr. Sc., Protozool. of the Boyce Thompson Inst. for Plant Research, Yonkers, N.Y. (U.S.A.), 1086 N. Broadway. *Latex inhabiting protozoa. Insect protozoa. Virus diseases of plants.* Ⓟ Latex inhabiting protozoa.

Holmes, Harlan B. (1898), Junior Aquatic Biol. U.S. Bur. of Fisheries. Stanford University, California (U.S.A.). *Life-history of the Pacific salmons.*

Holmes, Samuel J. (1868), Prof. of Zool., Univ. of California. Berkeley, Cal. (U.S.A.). *Compar.*
Psychol., Genetics, Eugenics. Ⓓ Univ. of California. 1897.

Holmgren, Ivar (1889), Priv.Doc. der Botan. Stockholm (Sverige), Botan. Inst. Univ. *Zytol., Embryol.* Ⓓ Stockholm 1919.

Holmgren, Nils Frithiof (1877), Prof. Zool., Dir. Zootom. Inst. Hochsch. zu Stockholm (Sverige), Drottninggatan 116. *Vergl. Anat., Histol. und Embryol. der Wirbeltiere: Nervensystem, Skelettsystem (Kranium der niederen Vertebraten), Vergl. Anat. der Wirbellosen, speziell Nervensystem der Articulaten, Termiten, systematisch. und vergl. anat.* Ⓓ Stockholm 1907.

Holmquist, Albert Martinius (1891), Prof. of zool., St. Olaf Coll. Northfield, Minn. (U.S.A.). *Ecol. of animals, winter ecol. and hibernation.* Ⓓ Univ. of Chicago 1925.

Holroyd, Roland (1896), Senior Prof. of Biol., La Salle Coll., Instr. in Botan., Univ. of Pennsylvania. Philadelphia, Pa. (U.S.A.), Macfarlane Hall. *Morphol. and Economic Botan.* Ⓓ Univ. of Pennsylvania 1918.

Holste, Arnold (1869), o. Prof. Med. Fac. Univ., Dir. Pharmacol. Inst. Beograd (S.H.S.), Šumadiska No. 18. *Experim. Pharm.: überlebende Organe.* Ⓓ Göttingen 1889.

Holt, Caroline M. (1878), Ass. Prof. of Biol. Simmons Coll. Boston, Mass. Brookline, Mass. (U.S.A.), 101 St. Paul Street. *Anat. and Embryol.* Ⓓ Univ. of Pennsylvania 1915.

Holtedahl, Olaf (1885), Prof. Historical Geol. Oslo (Norge), Geol. Mus. *Invertebrate paleont. of Paleozoic formations of Norway and Arctic regions (Spitsbergen, Bear Island, Novaya Zemlya). Quaternary geol. of Norway.* Ⓓ Oslo 1913.

Holth, H., Prof. Dr. Vet.-Inst. Oslo (Norge). *Bact.* o

Holtz, Friedrich (1898), Ass. Göttingen (Deutschland), Chem. Labor. *Tierische Extractstoffe. Chem. Physiol. der Blutbestandteile. Quantitative biol. Mikromethoden.* Ⓓ Würzburg phil. 1923, med. 1924.

Holtz, Henry F. (1880), Associate in Soils. Pullman, Wash. (U.S.A.). *Soils, physical, chemical and biol.* Ⓓ Washington State Coll. 1913.

Holtzinger, Johannes Hermann Alexander Georg (1884). Tenever, Post Hemelingen, Bremen (Deutschland). *Biol. der Reptilien. Systemat. Herpetol. Anat. Pathol. und Therapie der inneren Sekretion.* Ⓟ Reptilien.

Holzmann, Willi (1878), Nervenarzt. Hamburg (Deutschland), An der Alster 63. *Serol., Vererbungslehre.* Ⓓ München 1903.

Holzmayer, Herbert-Konrad (1898), Ass. zool. Kabinett Univ. Kasan (U.d.S.S.R.). *Ichthyol. (spez. Wachstum d. Fische).*

Homma, Yasu (1892), Ass. of Botan., Coll. of Agric., Hokkaidô Imper. Univ. Sapporo (Japan), Botan. Inst. *Erysiphaceae.* Ⓓ 1916, 1923.

Honcamp, Franz (1875), Prof., Dir. d. Landw. Versuchsanstalt. Rostock i. Meckl. (Deutschland), Graf-Lipper-Str. 4. *Agric.-Chem.* o

Honda, Masaji (1897), Ass. Tôkyô Imper. Univ. Tôkyô-shi-gai (Japan), 2364, Miyanaka Nishisugamo. *Systematic Botan., especially the classification of the Japanese Gramineae.* Ⓓ Tôkyô 1921.

Honda, Seiroku, Prof. of Tôkyô Imper. Univ. Komaba near. Tôkyô (Japan), Inst. of Forestry, Fac. of Agric. *Forestry.*

Honey, Edwin Earl (1891), Ass. Prof. of Extension, Dept. of Plant Pathol., N.Y. State Coll. of Agric., Cornell Univ. Ithaca, N.Y. (U.S.A.), Bailey Hall. *Plant pathol. or mycol. Discomycetes.* Ⓓ Univ. of Illinois 1922. Ⓟ Mycol. or plant pathol. specimens.

Honigmann, Hans Leo (1889), Leiter der Abteilung für Pflanzenschutz u. Schädlingsbekämpfung Saccharinfabrik, Magdeburg-Südost. Magdeburg (Deutschland), Bismarckstr. 36. *Angewandte Zool. u. Botan., Land- u. Süßwassermollusken. Plankton.*

Honing, Jan Antonie (1880), Prof., Landw. Hochschule. Wageningen (Holland), Rijksstraatweg 79. *Genetics: Canna, Nicotiana.* ⊕ Amsterdam 1909.

Honl, Iwan, Dr., Prof. d. Bact. u. Serol., Pathol. anat. Inst. d. Karls-Univ. Praha II (C.S.R.), Preslova. o

Hons, Vilém (1890), Ass. physiol. Inst. Med. Fac. Praha (C.S.R.), Kateřinská 32. *Metabolism. Vitamines.* ⊕ Prague 1915.

Hoogerwerf, Simon (1896), M.D., Priv.Doc. für Physiol. Leiden (Holland), Physiol. Inst. *Elektrophysiol.; Phonetik.* ⊕ Leiden 1924.

de Hoogh, Jerphaas (1898), Adjunct-Phytopath. Wageningen (Holland), Bowlespark 2. *Phytopath., Economic entomol.* ⊕ Wageningen 1925.

Hoogland, Hendrik Jacobus Marinus (1884), Conserv. at the pathol. inst. of the vet. fac. of the univ. of Utrecht (Holland). *Animal pathol., tumours, diseases of the liver.* ⊕ Utrecht 1907.

v. Hook, James M., Prof. of Botan., Indiana Univ. Bloomington, Ind. (U.S.A.). *Mycol., plant pathol.* o

Hooker, Davenport (1887), Prof. of Anat., School of Med., Univ. of Pittsburgh, Pa. (U.S.A.). *Experim. embryol. and regeneration, nervous system. Anat., human and compar.* ⊕ Yale Univ. 1912.

Hooker, Donald Rossell (1876), M.S., M.D., Lect. in Physiol. Hygien., Johns Hopkins Univ. Baltimore, Md. (U.S.A.), 19 West Chase St. *Effect of carbon monoxide and other noxious Gases on growth and metabolism.* ⊕ Johns Hopkins 1905.

Hooker, Henry D. (1892), Associate Prof. of Horticulture, Univ. of Missouri. Columbia, Mo. (U.S.A.). *Carbohydrate-Nitrogen relations, fruitfulness, hardiness to cold, nitrogen fertilizers.* ⊕ Yale 1915.

Hooker, Sanford Burton (1888), Associate Prof. of Immunol., Boston Univ. School of Med., Member of Evans Memorial Inst. for Clinical Research and Preventive Med. Boston 18, Mass. (U.S.A.), 80 E. Concord St. *Immunol. Antigenic specificity. Hypersensitiveness.* ⊕ Boston Univ. 1913.

Hopkins, Andrew Delmar (1857), Senior Entomol. in Charge of Bioclimatic Research. Parkersburg, W.Va. (U.S.A.), 1708 Washington Ave. *The Bioclimatic Law of latitude, longitude and altitude relative to the phenomena of terrestrial life and climate.* ⊕ Hon. Ph. West Virginia Univ. 1894.

Hopkins, Elizabeth Franus (1896), Ass, in Research, Botan. Geneva, N.Y. (U.S.A.), N.Y. Agric. Experim. Station. *Seed Testing. Analyses of Purity, Germination testing of seeds.* ⊕ Mass. Agric. Coll. 1926.

Hopkins, Edwin Fraser (1891), Ass. Prof. of Botan. Ithaca, N.Y. (U.S.A.), Labor. of Plant Physiol., Cornell Univ. *Plant Physiol.* ⊕ Cornell Univ. 1915.

Hopkins, Sir F. G., Prof. Univ. Trin. Coll. Cambridge (England). *Bio-chem.* o

Hopkinson, Emilius. Bathurst, Gambia (West Africa). *Ornithol.* o

Hopwood, Arthur Jindell (1897), Ass. Department of Geol., British Mus. (Nat. Hist.). London S.W. 7 (England). *Fossil Mammalia.* ⊕ Manchester 1924.

Hora, Frant. (1902), Dr., Ass. Inst. allgemeine und experim. Pathol. Univ. Brno (C.S.R.), Französische Str. 12. *Cytolyse.* ⊕ Brünn 1925.

Hora, Sunder Lal (1896), Off. Superintendent, Zool. Survey of India. Calcutta (India), Indian Mus. *Taxonomy of Indian Fishes. Animal life in torrential streams. Fauna of India.* ⊕ Lahore, India, 1922. ⊕ *Animals.*

Hori, Hiroshi, Entomol. Labor., Coll. of Agric., Kûshû Imper. Univ. Fukuoka (Japan). *Entomol., Lepidoptera.*

Hormuzaki, Baron Constantin (1863), Honorary member Roumanian Academy. Cernăuți (România), Shada Zotta 8. *Lepidoptera, Coleoptera, Genitalarmature of lepidoptera.* ⊕ Cernăuți 1888.

Horn, Walther (1871), Dr., Dir. des Deutschen Entomol. Inst. der Kaiser-Wilhelm-Gesellschaft; ständiger Sekretär der Wanderversammlungen Deutscher Entomol. Berlin-Dahlem (Deutschland), Gosslerstraße 20. *Cicindeliden der Welt (Insecta: Coleoptera). Geschichte der Entomol. Entomol. Bibliographie.* ⊕ Berlin 1892. ⊕ *Cicindeliden der Welt* (Ins. Col.). Biol. Objekte über Insecten. Neuroptera, Orthoptera und Coccidae.

Horne, William Titus, Assoc. Prof. of Plant Pathol., Univ. of Calif. Berkeley, Cal. (U.S.A.), 209 Agr. Hall. o

Horowitz-Wlassowa, Aimée (1879), Prof. d'Hygiène Inst. Méd., Chef du Service d'Hygiène à l'Inst. Bact. Ekaterinoslaw (U.d.S.S.R.). *Bact., Epidemiol., Immunité. Microbiol. du sol. Epuration biol. des eaux d'égoût.* ⊕ Paris 1902, Charkow 1902, Petersburg 1906.

Horring, K. (1875). København V (Danmark), Rahbeks Allé 32. *Ornithol.* ⊕ Københavns Univ.

Horsfall, James G. (1905), Ass. Plant Pathol., Cornell Univ. Ithaca, N.Y. (U.S.A.). ⊕ Univ. of Arkansas 1925.

van den Horst, Cornelius Jan (1889), Dr., Vice-dir. Central Inst. for Brain Research. Amsterdam (Holland), Mauritskade 61. *Compar. Neurol. Enteropneusta. Madreporaria.* ⊕ Amsterdam 1916.

Horst, Rutgerus (1849), Conserv. emer. Leiden (Holland), Rijks Mus. van Natuurlijke Historie. *Annelida.* ⊕ Utrecht 1876.

Hortling, Ivar Johannes (1876), Dr. phil., Generalsekretär der Ornithol. Vereinigung in Finnland. Helsingfors (Finnland), Brändö. *Ornithol.: Vogelbiol., Vogelstimmen, Rassen, Zug, lokale ökol. Untersuchungen.* ⊕ Helsingfors 1900.

Horton, John Raymond, Associate Entomol., wichita Labor., cereal and forage insect investigations, U.S. Bureau of Entomol. Wichita, Kan. (U.S.A.), 126 S. Minneapolis Avenue. *Insect pests of cereal and forage crops. Formicidae, Phyllophaga (Fam. Scarabaeidae). Citrus and other subtropical fruit insect pests.* ⊕ Utah Agric. Coll. 1909. ⊕ *Formicidae; Beetles: Scarabaeidae, Tenebrionidae: Phyllophaga, Eleodes, Blapstinus, etc.*

Horvat, Ivo (1897). Dr. phil., Kustos des Botan. Inst. der Univ., Hon.Doc. Techn. Fac. Zagreb (S.H.S.), Mažuranićev trg 29 II. *Systematik der Filicineen; Geobotan.: Bryophyten Jugoslawiens, Kroatiens, Sloveniens, Bosniens.* ⊕ Zagreb 1920.

von Horváth, Géza, Dr., Dir. emer. Zool. Abt. d. Ung. National Mus. Budapest (Ungarn). *Hemiptera.* o

Horvatić, Stjepan (1899), Ass. des botan. Inst. Univ. Zagreb (S.H.S.), Mažuranićev trg 29 II. *Geobotan. der quarnerischen und norddalmatinischen Inseln.* ⊕ Zagreb.

Horwood, Murray Philip (1892), Ass. Prof., Department of Biol. and Public Health, Mass. Inst. of Technol. Cambridge, Mass. (U.S.A.). *Bact., Immunol.* ⊕ Mass. Inst. of Technol. 1921.

Hoschek, Anton, Dr. Sevnica, Slovenien (S.H.S.). *Buprestidae.* o

Hoshino, T., Prof. Med. Univ. Kanazawa (Japan), Hygien. Inst. *Bact. and Hygiene.*

Hoskins, Roy Graham (1880), Prof. of Physiol., O. State Univ. Columbus, Ohio (U.S.A.), Hamilton Hall, O. State Univ. *The internal secretions: adrenals and testes.* ⊕ Johns Hopkins Univ. 1920.

Hosmer, Ralph Sheldon (1874), Prof. of Forestry, Head, Dept. of Forestry, Cornell Univ. Ithaca, N.Y. (U.S.A.). *Forest. Hist. forest policy, forest protection.*

Hosoya, Yûji (1897), Ass. Prof. Tôhoku Imper. Univ. Sendai (Japan), Physiol. Inst. *Physiol. der Netzhaut.* ⊕ Sendai 1923.

Hosseus, Carlos C., Prof. Univ. Nac. Córdoba (Argentina). *Botan.* o

Hotson, John W. (1870), Ass. Prof. of Botan. Univ. of Washington (U.S.A.). *Mycol. and Plant Pathol. Uredinales.* ⊕ Ph. Harvard 1913.

Hotta, Katsuo, Lect. of Aichi Med. Univ. Nagoya (Japan), Sero-chem. Inst. *Sero-chem.*

Hottes, Alfred Carl (1891), Prof. of Floriculture. Columbus, Ohio (U.S.A.). *Floriculture.* ⓓ Cornell Univ. 1913.

Hottes, Charles F., Prof. of Plant Physiol., Univ. of Illinois. Urbana, Ill. (U.S.A.). o

Hou, Hsiang-Chuan (1899), Ass. in Physiol. Peking (China), Union Med. Coll. *Physiol.: Gastric secretion, Splenic functions.* ⓓ Peking Union Med. Coll. 1924.

Houard, Clodomir (1873), Prof. à la Fac. des Sc. Dir. de l'Inst. Botan. et du Jardin Botan. Strasbourg (France), 7, rue de l'Univ. *Les Zoocécidies des Plantes du Globe (Morphol., Anat., Physiol., Répartition géographique, Histoire, Bibliographie).*

Houard, Jeanne Maria, Conserv. des Collections Inst. botan. Univ. Strasbourg (France), 11, Allée de la Robertsau. *Collections vivantes et des herbiers. Biol. et de pathol. végétales.* ⓓ Strasbourg 1919.

Houben, Heinrich Hubert Maria Josef (1875), Dr. phil., ORR., o. Mitglied der Bioi. Reichsanstalt Berlin-Dahlem, a.o. Prof. der Chem. an der Univ. Berlin. Berlin-Zehlendorf-Mitte (Deutschland), Hauptstr. 29. *Organische Chem., Chem. der Pflanzenstoffe, Pflanzenschutz- und -heilstoffe, lichtchem. Auf- und Abbau im Pflanzenreich vorkommender Kohlenstoffverbindungen.* ⓓ 1898.

Houette, Charles (1899), Préparateur, Inst. d'Anat. pathol., Fac. de Méd. Strasbourg (France), 5, rue Edouard Teutsch. ⓓ Strasbourg 1925.

Hough, Walter (1859), Head Curator of Anthrop., U.S. National Museum, Smithsonian Instit. Washington, D.C. (U.S.A.). *Ethnol.; archeol.; ethnography.* ⓓ West Virginia Univ., Morgantown 1894.

Hough, Walter S. (1893), Associate Entomol., Virginia Agric. Experim. Station. Winchester, Va. (U.S.A.). *Biol. and control of apple insects: codling moth and leaf-roller larvae.* ⓓ Ohio State Univ. 1925.

Houghton, E. Mark (1867), Dir. Medical Research and Biol. Labor., Parke, Davis & Co. Detroit, Mich. (U.S.A.), 680 Longfellow Ave. *Pharm. and Bact.* ⓓ Univ. of Michigan 1894.

Hou-Jensen, Hans Marius (1893), Prosector anat. København (Danmark), Nyvej 10. *Vasa renalia.* ⓓ Copenhagen 1921.

House, Homer Doliver (1878), N.Y. State Botan. Albany, N.Y (U.S.A.), N.Y. State Mus. *Flora of New York State; sc. economic and pathol.* ⓓ Columbia 1908.

House, Margaret Chandler (1900), Purnell Research Expert. Ames, Ia. (U.S.A.), Home Economics Division, Ia. State Coll. *Factors influencing the vitamin content of certain vegetables. The effect of storage and cultural conditions. The effect of a high protein diet on reproduction and fertility of rats.* ⓓ Ia. State Coll. 1925.

Houser, Gilbert L. (1866), Prof. of Animal Biol., Dir. of the Labor. of Animal Biol., State Univ. of Iowa. Iowa City, Ia. (U.S.A.), 430 Iowa Ave. ⓓ John Hopkins 1901. o

Houser, John Samuel (1881), Entomol. Ohio Agr. Exp. Station. Wooster, Ohio (U.S.A.). *Economic Entomol.* ⓓ Cornell 1911.

Housiaux, Arsène, Chef du Labor. d'Inspection med. scolaire. Mohlenbeek (Belgique), 94, rue Le Lorrain. *Entomol. et parasit. med.* o

Houssay, Bernardo Alberto (1887), Prof. Physiol., Dir. Physiol. Inst. d. Med. Fac. Univ. Buenos Aires (Argentina), Viamonte 2790. *Innere Sekretion. Gifte der Schlangen, Spinnen u. Skorpione.* ⓓ Buenos Aires 1911.

Hovasse, Raymond (1895), Prof. de Zool. à la Fac. des Sc. de Bayérid, Stamboul (Turquie). *Cytol. (Chromosomes), Protistol. (Zooxanthelles, Ellobiopsidés). Faune de la Turquie.* o

Hovelaque, M., Prof. agr. Fac. Méd. Paris 6 (France), 13, Avenue de l'Observatoire. *Anat.* o

Hover, John Milton, Prof. of Agric., Michigan State Normal Coll. Ypsilanti, Mich. (U.S.A.), 924 Sheridan Ave. *Genetics.* o

Howard, Charles W. (1882), Prof. of Sericulture, Dir. Government Bureau of Sericulture. Canton, (China), Lingnan Univ. (Permanent address Wheaton, Ill. [U.S.A.]). *Sericulture, med. entomol. and parasit.* ⓓ Univ. of Minnesota 1913.

Howard, Henry Eliot. Clarelands near Stourport, Worcestershire (England). *Ornithol.* o

Howard, Leland O. (1857), Chief, Bureau of Entomol., U.S. Department of Agric. Washington, D.C. (U.S.A.). *Economic entomol.; Med. entomol.; Parasit. Hymenoptera.* ⓓ Cornell Univ. Ithaca, N.Y. (U.S.A.) 1877.

Howard, Walter L. (1872), Prof. of Pomol., Dir. Branch of the Coll. of Agric., Univ. of California. Davis, Cal. (U.S.A.), Univ. Farm. *Plant Physiol.: The rest period of plants; rootstocks for deciduous fruit trees.* ⓓ Univ. Halle-Wittenberg 1906.

Howarth, Willis Openshaw (1890), Lect. in Botan., The Univ. Manchester. Lancashire (England), 18 Denmark Road, Southport. *Ecol. Systematic of Gramineae, espec. genus Festuca. Mycol.* ⓓ Manchester 1918.

Howden, R., Prof. Univ. Durham Coll. of Med. Newcastle upon Tyne (England). *Anat.* o

Howe, Clifton D., Dean, Fac. of Forestry, Univ. Toronto, Ontario (Canada). *Forest ecol.* o

Howe, George Henry (1888), Associate in Research, Division of Horticulture, N.Y. State Agric. Experim. Station, Geneva, N.Y. (U.S.A.). *Various phases of variety tests with apples, pears, plums, peaches, cherries, and small fruits, including pruning, propagation and fertilizer experim. plant breeding experim. with several fruits.* ⓓ Univ. of Vermont 1910.

Howe, Marshall Avery (1867), Ass. Dir., The New York Botan. Garden. New York, N.Y. (U.S.A.), Bronx Park. *Taxonomy and Morphol. of Marine Algae.* ⓓ Hon. Sc. D. Univ. of Vermont 1919.

Howe, Paul Edward (1885), Biol. Chem., In charge of Nutrition Investigations. Animal Husbandry Division, U.S. Department of Agric. Washington, D.C. (U.S.A.). *Biol. Chem.: nutrition, fasting and proteins.* ⓓ Univ. of Illinois 1910.

Howell, Alfred Brazier (1886), Collaborator, Biol. Survey, U.S. Dept. Agric. Washington, D.C. (U.S.A.), U.S. National Mus. *Myol. and osteol. Mammalian anat., with special reference to adaptation; geographic variation of mammals. Birds of the California Islands. Anat. of the wood rat.*

Howell, William Henry (1860), Ass. Dir. and Prof. of Physiol. in the School of Hygiene and Public Health, Johns Hopkins Univ. Baltimore, Md. (U.S.A.). *Physiol. of blood, the problem of Coagulation; the effect of temperature, humidity and light on the animal body, its reactions and its resistance to infections.* ⓓ Ph. Johns Hopkins Univ. 1884, Hon. med. Michigan Univ. 1890, Hon. Sc. Yale Univ. 1911, Hon. L. L. Trinity 1901, Michigan Univ. 1912.

Howes, Frank Norman (1901), Ass., Royal Botan. Gardens, Mus. Kew, Surrey (England), Priory Rd., The Priory. *Economic Botan.* ⓓ Pretoria, South Africa 1922.

Howes, Paul Griswold (1892), Curator of Nat. History. Greenwich, Conn. (U.S.A.), Bruce Memorial Mus. of Nat. History, History and Art. Bruce Park. *Causes of animal behavior and reactions.* ⓔ West Indian faunae, Color-preparations of fishes.

Howitt, J. E., Prof. of Botan., Ontario Agric. Coll. Guelph, Ontario (Canada). o

Howland, Ruth B. (1887), Ass. Prof. of Biol. New York City, N.Y. (U.S.A.), Washington Square Coll., Univ. *Experim. Embryol. on Pronephros and Mesonephros of Amblystoma. Contractile Vacuoles of Paramecium, Amoeba verrucosa, Microinjection.* ⓓ Yale Univ. 1920.

Howlett, Freeman (1900), Ass. in Horticulture. Wooster, Ohio (U.S.A.), Ohio Agric. Experiment Sta. *Physiol. investigations with apples, fruit setting and pollination.* ⓓ Cornell Univ. 1925.

Hoyer, Heinrich (1864), o. Prof. der vergl. Anat. an der philos. Fak. der Jagellonischen Univ. zu Kraków (Polska), Annagasse 6. *Entwicklung und vergl. Anat. der Blut- und Lymphgefäße.* ⓟ Straßburg 1892.

Baron Hoyningen, genannt **Huene,** Friedrich Richard (1875), a.o. Prof. Dr., Konserv. der Geol. u. Palaeont. Univ.-Sammlung in Tübingen (Deutschland), Geol. Inst. der Univ. *Fossile Reptilien u. Amphibien des Palaeozoicum u. Mesozoicum. Kontinentale Ablagerungen Karbon bis Kreide. Geol. von Südafrika u. Südamerika.* ⓟ Tübingen 1898.

de Hoyos Sainz, Luis (1869), Prof., Dr. C. Nat. Dir. Labor. d'Authropol. physiol. Madrid (España), Montalban 20. *Croissance et développement de l'enfant Anthropol. physiol. de l'Espagne et de l'Amérique du Sud.*

Hoyt, William Dana (1880), Prof. of Biol. at Washington and Lee Univ. Lexington, Va. (U.S.A.). *Periodicity in reproduction; physiol. aspects and effects of fertilization in ferns; effects of colloidal metals; algae.* ⓟ Johns Hopkins Univ. 1909.

Hôzawa, S., Ass. Prof. of Hokkaidô Imper. Univ. Sapporo (Japan), Physiol. Inst., Fac. of Med. *Physiol.*

Hozawa, Sanji, Prof. in Zool. of Tôhoku Imper. Univ., Rigakushi, Rigaku-hakushi. Sendai (Japan), Zool. Labor., Coll. of Sc. *Kalkschwämme.* ⓟ Tôkyô.

Hrabowski, Johanna Annemarie (1898), Ass. Hauptstelle für Forstlichen Pflanzenschutz. Tharandt bei Dresden (Deutschland), Kirchgasse 147. *Angewandte Zool.* ⓟ Leipzig 1926.

Hrdlička, Aleš, Curator, Division of Physical Anthrop., U.S. National Mus. Washington, D.C. (U.S.A.). *Man's origin, evolution, variation American aborgines.* ⓟ Med. New York 1802, sc. hon. Prag 1921.

Hryniewiecki, Boleslas, Dr., Prof., Dir. du jardin bot. de l'Univ. de Varsovie. Warszawa (Polska), R. Al. Ujazdowskie 6/8. *Botan.* ○

Hsu, N., Lect. Prov. Med. Coll. Soochow (China). *Zool.* ○

Hu, Hsen-Hsu (1894), Head of Dept. of Botan., Prof. of Syst. Botan., National Southeastern Univ., Head of Botan. Division, Biol. Labor., Sc. Society. Nanking (China), Wonder Lane. *Dendrol. and general systematic Botan.* ⓟ Harvard 1925. ⓔ Herbarium of Southeastern China.

Huang, Tsefang F. (1899), Chief, Department of General Adm. and Public Health, National Epidemic Prevention Bureau. Peking (China), Temple of Heaven. *Biol. products. Typhoid Bacilli. Vaccines and Serums and their Production in China.* ⓟ Univ. of Chicago 1924. ⓔ Sera. Vaccines. Diagnostic reagents.

Huapaya, K. Luis, Prof. Univ. Lima (Peru). *Anat., Physiol.* ○

Huard, Chanoine Victor-Alphonse (1853), Dir. du Naturaliste canadien, Conservateur du Mus. d'Histoire naturelle de l'Instruction publique. Québec (Canada). 2, rue Richelieu. *Entomol., Lépidoptères.* ⓟ Laval Univ. Quebec.

Hubbard, C. Andresen (1898), Prof. of Biol. Forest Grove, Ore. (U.S.A.), 406/3. Avenue. *Evolutionary Biol. Mammol., Aquarist, Aviarist.* ⓔ Tropical and Temporata fishes, aviary birds. Specimens of local Flora and Fauna.

Hubbel, Theodore Huntington (1897), Ass. Prof. of Biol., Univ. of Florida. Gainesville, Flo. (U.S.A.), Department of Biol., *Entomol.: Taxonomy, distribution, ecol. of Orthoptera, Rhaphidophorinae, genus Ceuthophilus.* ⓔ North and Central American Dermaptera and Orthoptera.

Hubbs, Carl L. (1894), Curator of Fishes and Instr. in Zool., Mus. of Zool., Univ. of Michigan. Ann Arbor, Mich. (U.S.A.). *Systematic ichthyol.; fisheries biol.; evolution.* ⓟ Stanford 1916. ⓔ Fishes.

Huber, Ernst (1892), Dr. phil., Associate Prof. of Anat. Baltimore, Md. (U.S.A.), Dept. of Anat., Johns Hopkins Med. School. *Vergl. Anat. des Muskelsystems, Gesichtsmuskulatur: (niedere Vertebr. bis Säuger inkl. Mensch). Rassenanthrop. des Muskel- und Skelettsystemes: (Neger, Weiße, Chinesen). Embryol. des menschl. Muskelsystemes. Experim. Untersuchungen über den N. facialis: Neurol.* ⓟ Zürich 1917.

Huber (seit 1911: **Huber-Pestalozzi**), Gottfried Eduard (1877), Dr. med. et phil., Arzt. Zürich (Schweiz), Englischviertelstr. 61. *Algol. (Systematik, Physiol., Süßwasser-Planktol., Pflanzengeographie, mit Ausnahme der Kieselalgen und der Meeresalgen).* ⓟ Phil. Zürich 1904, med. Zürich 1911.

Huber, Gotthelf Carl (1865), Prof. of Anat., Dir. of the Anat. Labor. Univ. of Michigan. Ann. Arbor, Mich. (U.S.A.), 1330 Hill St. *Histol., Physiol. Embryol.* ⓟ Univ. of Michigan 1887. ○

Huber, Harvey Evert, Dean of Coll. of Liberal Arts and Prof. of Biol., Ohio Northern Univ. Ada, Ohio (U.S.A.). ○

Huber, Josef Anton (1899), Dr. phil, Wissenschaftlicher Ass. am Inst. für Pflanzenzüchtung und Pflanzenbau der bayer. Hochsch. für Landwirtschaft und Brauerei in Weihenstephan, Post Freising (Deutschland). *Landwirtschaftliche Kulturpflanzen: Vererbungswiss. und systematisch-Morphol. Gattung Sedum, L. (Crassulaceae)* ⓟ München 1923. ⓔ Lebende Pflanzen der Gattung Sedum (Fam. Crassulaceae).

Huber, Laurence Lester (1893), Assoc. Entomol., Ohio Agric. Experim. Stat. Wooster, O. (U.S.A.). *Economic entomol., European cornborer.* ⓟ Ohio Univ. 1922.

Hubert, Ernest Everett (1887), Prof. of Forest Products. School of Forestry. Moscow, Id. (U.S.A.), Univ. of Idaho. *Forest pathol.: Diseases of trees, studies on wood rots, relative durability of woods, fungous defects in wood products.* ⓟ Univ. of Wisconsin 1923. ⓔ Wood rots, wood inhabiting fungi.

Hublard, Emile, Dir. Mus. d'Histoire natur. Mons (Belgique), Rue de Houdain.

Hucker, George James (1893), Associate in Research, New York Agric. Experim. Station, and Ass. Prof., Cornell Univ. Geneva, N.Y. (U.S.A.). *Bact.: Systematic and physiol.* ⓟ Yale Univ. 1924.

Huckett, Hugh Cecil (1890), Associate in Research, Riverhead, N.Y. (U.S.A.), Long Island Vegetable Research Farm. *Economic Entomol.* ⓟ Toronto, Canada, 1919. ⓔ Muscidae, Anthomylidae (Diptera).

Huddleson, I. Forrest (1893), Research Associate in Bact. East Lansing, Mich. (U.S.A.). *Investigation in infectious abortion in cattle.* ⓟ Cornell Univ. 1925. ⓔ Cultures of bact. abortus.

Hudson, G. V., Hill View. Karori. Wellington (New Zealand). *Entomol.* ○

Hudson, George Henry (1855), Retired Head of Sc. Dept. State Normal School Plattsburgh, N.Y. (U.S.A.), 39 Broad Street. *Paleont., Echinoderms. Eublastoidea, Astrocystites (Steganoblastus).* ⓔ Ordovician fossils of Champlain Valley.

Hudson, Jessie B. (1870), Dir. of branch labor. for the Iowa State Board of Health. Sheffield, Ia. (U.S.A.). *Bact. of the ear. Staining of tubercle bacilli, typhoid carriers.* ⓟ Iowa City 1908.

Hudson, Robert George Spencer (1895), Ass. Lect., Dept. of Geol. Leeds (England), Univ. *Invertebrate Palaeozool., of lower Carboniferous.* ⓟ London 1920.

Hue, F., Prof. École prep. méd. et pharm. Rouen (France). *Pathol.* ○

Hübenett, E. R., (1882), Prof. Chef Labor. Botan. de l'Inst. scientifique Lesshaft. Ass. de l'Inst. de l'instruction physique Lesshaft. Leningrad (U.d.S.S.R.), Torgovaja 25 log. 6. *Les pigments des Algues; Formation de la Chlorophylle.* ⓟ Leningrad 1910.

Huebschmann, Paul Karl Kurt (1878), o. Prof. für Allgemeine Pathol. und pathol. Anat. und Dir. des Pathol. Inst. der Med. Akademie in Düsseldorf (Deutschland), Kronprinzenstr. 49. *Pathol. Anat. der Tuberkulose.* ⓟ Würzburg 1903.

Hueck, Max Werner (1882), Dr. med., o. Prof. für pathol. Anat. u. Dir. des pathol. Inst. Leipzig (Deutschland), Rob.-Schumann-Str. 12b. *Pathol. Histol. d. Mesenchyms (Gefäße, Knochengelenke, Milz), Cholesterin-Stoffwechsel.* ⑨ Rostock 1905.

Hüeber, Theodor (1848), Dr., Generaloberarzt a. D. Ulm a. d. Donau (Deutschland), Heimstr. 7. *Entomol., Halbflügler (Hemiptera heteroptera).* ⑨ München 1874 bzw. Würzburg 1883.

Huelsen, Walter A. (1892), Ass. Chief in Olericulture. Urbana, Ill. (U.S.A.), Dept. of Horticulture, Univ. of Ill. *Crops for canning; sweet corn and tomatoes, plant nutrition; effects of strain of fertilizer treatment on sulfide discoloration.* ⑨ Univ. of Illinois 1926. ⑩ Seeds of sweet corn (Zea Mays var. saccharata) and tomatoes (Lycopersicum esculentum).

Hueppe, Ferdinand (1852), Prof. emer. der Hygiene. Dresden-Loschwitz (Deutschland), Pillnitzer Landstraße 15. *Bact., Biochem., Konstitutionsforschung und Vererbung, Kausalgesetz in Biol. und Pathol.* ⑨ Med. Berlin 1876, iur. (LLD.) h. c. Aberdeen 1906.

Huergo, José M., Prof. Univ. Nac. Buenos Aires (Argentina). *Zool. agric.* o

Huertas, José Vicente, Prof. Univ. Nac. Bogotá (Colombia). *Pathol.* o

Hürthle, Karl (1860), Dr. med., Geh. Med. R. Tübingen (Deutschland), Eugenstr. 2. *Physiol. der Blutbewegung. Struktur der Muskeln. Struktur der Schilddrüse.* ⑨ Tübingen 1884.

Hueter, Carl (1862), Prof., Prosektor am städt. Krankenhaus Altona. Altona-Ottensen (Deutschland), Hohenzollernring 27. *Pathol. Anat.* ⑨ Göttingen 1887.

Hüttig, Carl Albert (1901). Kiel (Deutschland), Knooper Weg 125. *Bodenbact. Pflanzengeographie Schleswig-Holsteins.* ⑨ Kiel 1927.

Huff, Clay G. (1900), Ass. in Med. Entomol., School of Hygiene and Public Health, The Johns Hopkins Univ. Baltimore, Md. (U.S.A.), 615 N. Wolfe St. *Insect Transmission of Disease (Bird Malaria). Insect Immunity.* ⑨ Southwestern Coll. 1924. ⑩ Blood Smears of Various Parasit. of Bird Malaria.

Huff, N. L., Ass. Prof. of botan., Univ. of Minnesota. Minneapolis, Minn. (U.S.A.), 1219/7th St., S. C. *Morphol.* o

Huggins, John Robinson (1883), Ass. Zool. Univ. of Pennsylvania. Philadelphia, Pa. (U.S.A.). *Insect morphol.* ⑨ Univ. of Pennsylvania 1921.

Hughes, Elmer Howard (1881), Ass. Prof. of Animal Husbandry, Univ. of California. Davis, Cal. (U.S.A.), Univ. Farm. *Swine.* ⑨ Univ. of Missouri 1916. ⑩ Swine, Breeding stock.

Hughes, Harold De Mott (1882), Chief in Farm Crops. IowaAgric.Expt. Station Ames, Ia. (U.S.A.). *Production and breeding of Farm Crops. Corn breedings.* ⑨ Univ. of Mo. 1908.

Hughes-Schrader, Sally (1895), Instructor Biol., Bryn Mawr, Pa. (U.S.A.), Coll. *Cytol. Sex determination. Entomol.* ⑨ Columbia Univ. 1924.

Hugouneng, Louis (1860), Prof. de Chim. biol. et méd. à la Fac. de Méd. de l'Univ. de Lyon (France). *Chim. Biol. générale, Chim. méd. Poisons.*

Huguenin, B. (1876), o. Prof. Bern (Schweiz), 6 Engehaldenstr. *Vergl. Pathol.* ⑨ Bern 1905.

Huguet-del-Villar, Emile (1871), Secrétaire général et technique de la Commission d'Edaphol. et Géobotan. (Section Espagnole de la Association Internationale de la Sc. du Sol). Madrid (España), Lista 62,3. der. *Région méditerranéenne.* ⑩ Plantes.

Huidobro y Hernández, José, Conserv. Mus. Nac. de Ciencias Nat. Madrid (España), Ruiz, 12. *Entomol.* o

Huji, Nobuzo, Ass. Prof. of Tôkyô Imperial Univ. Tôkyô (Japan), Chem. Inst., Fac. of Med. *Med. Chem.*

Hujii, Kenjirô, D.Sc., Prof. of Botan., Botan. Inst. Fac. of Sc., Tôkyô Imperial Univ. Tôkyô (Japan), Koishikawa-Ku, Botan. Garden. *General Cytol. and Genetics.*⁴

Hujinami, Akira, Prof. of Kyôto Imper. Univ. Kyôto (Japan), Pathol. Inst., Coll. of Med. *Pathol. Pathol. Anat. and Parasit.*

Hujita, Atsushi (1899), Prof. Zool. High School. Toyama (Japan), Biol. Labor. *Histol.* ⑨ Tôkyô 1923.

Hujita, Masao, Ass. des pharmacol. Inst. Okayama (Japan), Med. Univ. *Experim. Pharm.* ⑨ Med. Akademie Okayama 1920.

Hujita, Tsunenobu (1868), Prof. of Fish-Culture. Sapporo (Japan), Hokkaidô Imperial Univ. *Parasit. Invertebrate Embryol.* ⑨ Tôkyô Imperial Univ. 1893.

Hukaki, Sadayoshi, Ass. der Univ. Fukuoka (Japan), Botan. Inst., Kaiserl. Kyushu Univ. *Ernährung der Pflanzen.* ⑨ Momioka 1921.

Hukamachi, R., Lect. of Keiô-Gijuku Univ. Tôkyô (Japan), Pathol. Inst., Coll. of Med *Pathol.*

Hukkinen, Yrjö Armas (1886), Mag. phil., I. Ass. an der Landwirtschaftl. Versuchsanstalt, Abt. für Schädlinge. Tikkurila, (Finnland). *Schädlinge der Kulturpflanzen; Thysanoptera. Biol. und Verbreitung der Dasychira selenitica Esp., Johannisbeerengallmilbe (Eriophyes ribis Nal.).* ⑨ Helsinki 1919. ⑩ Thysanopteren, Tenthrediniden.

Hukuda, Kunizô (1896), Ass. Prof. Physiol., Tôkyô Imperial Univ. (Japan), Physiol. Labor. *Physiol. problems from a physical, chem. and physicochem. stand point of view. Permeability of frog-skin.* ⑨ Tôkyô Imperial Univ. 1922.

Hukuda, Tokushi, Prof. of Med. Univ., Chiba near Tôkyô (Japan), Pharmacol. Inst. *Pharm.*

Hukuda, Yasona (1895), Prof. Biol., Premedical Course, Manchuria Med. Coll. Mukden, Manchuria (China), 15 Hagimachi, Japanese city. *Cytol. and Genetics, Study of behavior of chromosomes of Plants. Biochem. (Coloring matters of Fungus).*

Hukuhara, Yoshimoto, Prof. of Ôsaka Provincial Med. Univ. Ôsaka (Japan), Hygien. Inst. *Hygiene and Bact.*

Hukui, Tamao (1891), Prof. biol., Shizuoka Higher School, Shizuoka (Japan). *Helminthol. Classification, anat., histol., embryol. of amphistomes and other trematodes.* ⑨ Tôkyô 1920. ⑩ Amphistome parasit. of Japan, and some other parasit.

Hulin, Wilbur Schofield (1899), Instructor Psychol., Univ. Princeton, N.J. (U.S.A.), Eno Hall. *Experim. Psychol.* ⑨ Princeton 1926.

Hull, Thomas Gordon (1889), Chief, Division of Labor. Illinois State Department of Public Health. Springfield, Ill. (U.S.A.), Capitol Building. *Bact. and Serol., as Wassermann tests, Widal tests, cultures for typhoid, diphtheria, examination of specimens for gonococci, tubercle bacilli.* ⑨ Yale 1916.

Hulst, Jean Pierre Louis (1875), Lect., Inst. tropische geneeskunde Rotterdam-Leiden. Leiden (Holland), Morschsingel 6. *Pathol. Anat.* ⑨ Leiden 1898.

Hultkrantz, Johan Wilhelm (1862), o. Prof. Anat. Univ. Dir. d. Anat. Inst. Uppsala (Sverige). *Osteol. Arthrol., Phys. Anthropol. u. Rassenhyg.* ⑨ Stockholm.

Humbert, Eugene P., M.S., Ph.D., Prof. of Genetics, Agric. and Mechanical Coll. of Texas (U.S.A.). College Statione, Tex. o

Humbert, Henri (1887), Chef des travaux, Labor. de Botan., Fac. des Sc. d'Alger (Algérie). *Systématique: Phanérogames, Composées de Madagascar. Phytogéographie: Afrique du Nord et Madagascar. Anat.: Adaptation aux facteurs écol.* ⑨ Alger 1922. ⑩ Plantes de Madagascar.

Hume, H. Harold (1875), State Horticulturist. Mary, Florida (U.S.A.), Glen St. o

Hume, J. B., Prof. St. Barthol. Hosp. Med. School. London E.C. 1 (England), West Smithfield. *Anat.* o

Hummel, Karl (1889), Prof. Dr., a.o. Prof. an der Univ. Gießen (Deutschland), Geol. Inst. *Palaeontol., fossile Trionychia.* ⓓ Freiburg i. Br. 1913.

Humphrey, Clarence John (1882), Mycol., Bureau of Sc., In Charge, Division of Mycol. and Plant Pathol. Manila (P. I.). *Forest mycol. and pathol.* ⓓ Univ. of Wisconsin 1920. ⓟ Forest fungi, basidiomycetes.

Humphrey, Harry B., Senior pathol., in Charge of Cereal Disease Invest., Bureau of Plant Industry, U.S. Dept. of Agric. Washington, D.C. (U.S.A.). o

Humphrey, Rufus Richard (1892), Associate Prof. of Anat., School of Med., Univ. Buffalo, N.Y. (U.S.A.), 24 High St. *Interstitial cells of the testis in Urodeles. Primordial germ cells of Amphibia. Multiple testes of Urodeles.* ⓓ Cornell Univ. 1923.

Hunaoka, Seigo (1890), Prof. Anat. Kaiserl. Univ. Kyôto (Japan). *Panaschierte Blätter. Peripheres Nervensystem.* ⓓ Kyôto, Japan, 1914.

Hunger, F. W. T. (1874). Amsterdam (Holland), van Eeghenstraat 52. *Tropische Kulturpflanzen: Kokos- und Ölpalme. Geschichte der Botan.* ⓓ Jena 1898.

Hungerford, Charles William (1885), Prof. of Plant Pathol., Univ. of Idaho. Plant Pathol. Agric. Experim. Stat. Moscow,ⓘId. (U.S.A.). *Diseases of grains and potatoes, Plant Pathol.* ⓓ Iowa Univ. 1910.

Hungerford, Herbert Barker (1885), Head of Department of Entomol., Curator of Entomol. Mus., State Entomol. Univ. of Kansas. Lawrence, Kan. (U.S.A.), 202 Mus. Bldg. *Biol. and taxonomy of aquatic and semiaquatic Hemiptera.* ⓓ Cornell Univ. 1918. ⓟ American Homopterous and Heteropterous insects.

Hunt, E. L., Prof. St. Georges Hosp. Med. School. London S.W. (England), Hyde Park Corner. *Bact.* o

Hunt, Harrison Randall (1889), Head of Department of Zool. and Geol., Prof. East Lansing, Mich. (U.S.A.), 501 Sunset Lane. *Zool.; Genetics; Eugenics.* ⓓ Harvard 1916.

Hunt, Nicholas Rex (1885), Pathol., Federal Horticultural Board, U.S. Dept. of Agric. Washington, D.C. (U.S.A.). *Forest tree diseases and potato disease.* ⓓ Univ. of California 1909.

Hunt, Reid (1870), Prof. of Pharmacol., Harvard Univ. Boston, Mass. (U.S.A.), Harvard Med. School. *Pharm. and Toxicol. Thyroid, alcohols, nitriles, choline compounds; physiol., pharm. of circulation.* ⓓ Johns Hopkins Univ. 1896.

Hunt, Willis Roberts (1893), Sc. Ass. in Botany, Conn. Agr. Expt. Sta. New Haven, Conn. (U.S.A.). *Uredinales.* ⓓ Yale Univ. 1925. ⓟ Uredinales of New England.

Huntemüller, Otto Werner Gustav (1878), a.o. Prof. der Hygiene und Bact., Leiter der Med. Abteilung des Inst. für Körperkultur an der Univ. Gießen (Deutschland). *Die biol. Grundlagen der Körperbildung.* ⓓ München 1905.

Hunter, Albert Clayton (1893), Bact., U.S. Bureau of Chem. Washington, D.C. (U.S.A), U.S. Dept. of Agric. *Bact. of food spoilage, food preservation and food poisoning. Bact. of fish, shellfish and other sea foods.* ⓓ Brown Univ. 1918.

Hunter, Charles A. (1893), Ass. Dir. State Health Labor., Prof. of Bact. Univ. of South Dakota. Vermillion, S.D. (U.S.A.). *Bact. and Hygiene.* ⓓ Univ. of Wisconsin 1916.

Hunter, George (1894), Prof., Ass. Prof. of Pathol. Chem. Toronto, Ontario (Canada), Univ. *Iminazoles in muscle, urine and blood, the non-protein constituents of blood. Diazo-Reaction.*

Hunter, George William (1902), Fellow in Zool., Univ. of Illinois. Champaign, Ill. (U.S.A.), 1002 South 2. St. *Parasitol. Cestodaria, Cestoda, Trematoda.* ⓓ Univ. of Illinois 1924.

Hunter, Samuel John (1866), Curator entomol. Collections Univ. of Kansas. Lawrence, Kan.

(U.S.A.), 1145 W. Campus Rd. *Comp. Zool. Entomol.* o

Hunter, Walter S., Stanley Hall Prof. of Genetic Psychol., Clark Univ. Worcester, Mass. (U.S.A.). *Animal behavior; space perception; compar. psychol.* ⓓ Chicago 1912. o

Huntington, Ellsworth (1876), Research Associate in Geography in Yale Univ. New Haven, Conn. (U.S.A.). *Relations between climate, together with other geographical factors, and the distribution of crops, domestic animals, racial characteristics, human health, and various human activities.* ⓓ Yale Univ. 1909.

Hurd-Karrer, Annie May (1893), Associate Physiol., U.S. Dept. of Agric. Washington, D.C. (U.S.A.), Bureau of Plant Industry. *Physiol. of cereal plants. Chem. and physical properties of the sap of various cereals grown under different conditions, disease resistance.* ⓓ Univ. of California 1918.

Hursh, Charles Raymond (1895), Associate Forest Ecol., Forest Service, U.S. Dept. of Agric. Asheville, N.C. (U.S.A.), Appalachian Forest Experim. Station. ⓓ Minnesota 1923. ⓟ Botan. and Plant Disease Specimens.

Hurst, Clarence T. (1894), Ass. Prof. of Zool. Mills College, California (U.S.A.). *Morphol. Trematode larvae. Effect of Copper Sulphate on Ducks. Pathol. effects of Trematode larvae on Gasteropod Molluscs.* ⓓ Univ. of California 1926. ⓟ Prepared slides of Trichonympha campanula (Hypermastigote flagellate).

Hurst, Richard Rankin (1895), Ass. Plant Pathol. Charlottetown, P.E.I. (Canada), Experim. Farm. *General plant pathol. Potato, cereal and fruit diseases. General systematic.* ⓓ Toronto 1922.

Hus, Pieter (1894), Agric. Engineer, Phytopath. of the Phytopath. Service. Bennekom, Gemeente Ede (Holland). *Phytopath. and Economic Entomol.* ⓓ Wageningen 1919.

Huse, Gennosuke, Prof. of Tôhoku Imper. Univ. Sendai (Japan), Anat. Inst., Coll. of Med. *Anat.*

Huse, Shin-ichirô, Ass. of Kanasawa Med. Univ. Kanasawa (Japan), Pathol. Inst. *Pathol.*

Husfeld, Bernhard (1900), Ass. am Inst. für Vererbungsforschung der Landwirtschaftl. Hochsch. Berlin-Friedenau (Deutschland), Lauterstr. 16. *Pflanzenzüchtung.*

Huskins, C. Leonard (1894), Exhibition Sc. Research Scholar. London (England), Department of Botan., King's Coll. Strand. *Genetics and Cytol., combined genetical and cytol. study of the origin of fatuoids in cultivated oats, and speltoids in cultivated wheats.* ⓓ Alberta 1925. ⓟ Seeds of many species of Avena.

Husmann, George C. F. (1861), Pomol. in charge of Viticultural Investigations, U.S. Dept. of Agric. Washington, D.C. (U.S.A.). *Viticultural matters U.S.A.* ⓓ Missouri State Univ. 1885.

Huss, Harald Axel (1875), Vorsteher der biol. Labor. des städt. Gesundheitsamtes u. des Wasserwerkes, Lehrer an der pharmaz. Hochsch. (Bact.). Stockholm (Sverige), Vasagatan 13. *Mikroskopische Untersuchung der Lebensmittel. Bact. Untersuchung des Wassers. Pharm. Sterilisationstechnik.* ⓓ 1905.

Hussey, Russell Claudius (1888), Ass. Prof. of Geol., Univ. of Michigan. Ann Arbor, Mich. (U.S.A.), *Invertebrate Paleont., Ordovician period of Michigan.* ⓓ Michigan 1925.

Hustedt, Friedrich Carl (1886). Bremen 4 (Deutschland), Ingelheimer Str. 7. *Diatomeen.* ⓓ Sc. nat. h. c. Halle 1927.

Husz, Béla (1892), Adjunkt am kgl. Ung. Inst. für Pflanzenphysiol. und Pathol. Budapest I (Ungarn), Greguss ut 5. I. 2. *Krankheiten der landwirtschaftl. Kulturpflanzen. Mykologie.* ⓓ Budapest 1919.

Hutcherm, Thomas B. (1882), Agronomist Va. Poly. Inst. and Va. Exp. Station. Blacksburg, Va. (U.S.A.). *Crop, soils investigations.* ⓓ Cornell 1913.

Hutchens, H. J., Prof. Univ. Durham, Coll. of Med. Newcastle upon Tyne (England). *Compar. Pathol. and Bact.* ○

Hutchins, Lee M. (1888), Associate Pathol. Washington, D.C. (U.S.A.), Bureau of Plant Industry. *Physiol. diseases of plants, fruit trees. Oxygen-supplying power of the soil. Plant physiol.* ⓓ Johns Hopkins Univ. Baltimore, Md., 1924.

Hutchinson, Andrew Henderson (1888), Prof. of Botan. and Head of the Department. Vancouver (Canada), Univ. of British Columbia. *Plant Morphol. and Cytol., Forest Ecol. Effect of Radiant Energy on Protoplasm (unicellular).* ⓓ Chicago 1915.

Hutchison, Claude Burton (1885), Dir. for Europe, Division of Agric. Education, International Education Board. Paris (France), 19, Rue Louis-le-Grand. *Plant Production and Plant Genetics.* ⓓ Harvard Univ. 1917.

Hutson, Ray (1896), Ass. Entomol. New Brunswick, N.J. (U.S.A.), Ent. Bldg. N.J. Agric. Experim. Sta. *Bee diseases and Bee ecol.* ⓓ West Virginia Univ. 1922.

Hutton, Joseph Gladden (1873), Associate Agronomist, Associate Prof. of Agronomy, in Charge of Soil Investigations and South Dakota Soil Survey. Brookings, S.D. (U.S.A.), South Dakota State Coll. *The Classification of Soils and the Relation of Soil Fertility to Crop Production.* ⓓ Univ. of Illinois 1910. ⓟ Samples of soil.

Huus, Johan (1892), Amanuensis. Bergens Mus. Bergen (Norge). *Ascidien, embryol. und systematisch. Marine Nematoden, systematisch.*

Huxley, Julian Sorell (1887), Prof. of Zool., Univ. London WC 2 (England), King's Coll., Strand. *Amphibian metamorphosis. Bird behaviour and Ecol., with special reference to Courtship behaviour. Neomendelism, especially relation between Mendelian factors and development. Relative growth-rates of organs. Experim. Embryol., with special reference to the problems of physiol. gradients and differentiation. Dedifferentiation and resorption.* ⓓ Oxford 1909.

Huzella, Theodor (1886), o. Prof., Dir. Anat.-biol. Inst. Debreczen (Ungarn). *Drüsen und Kapillarmechanismus, experim. Zellforschung, Mikrochirurgie, Soziol.* ⓓ Budapest 1911.

Hyde, Roscoe Raymond (1884), Associate Prof., Immunol., Johns Hopkins Univ. Baltimore, Md. (U.S.A.). ⓓ Columbia Univ. 1915. ⓟ Guinea pigs.

Hykeš, Oldrich Vilém (1895), Ph.Dr., Priv.Doc. de Biol. générale a l'École vétérinaire. Brno (C.S.R.), Pražská 69. *Biol. et pathol. des abeilles, des poissons. Hydrobiol., Endokrinol., Respiration des animaux, Influence des actions physiques sur êtres vivants.* ⓓ Karlova univ. Praha 1919.

Hyland, Jay (1900), Instr. in Biol. Orono, Me. (U.S.A.), Biol. Department, Univ. of Maine. *Prevention of germination of weed seeds in Forestry nursery seed beds by application of chem. to the soil.* ⓓ Michigan State Coll., East Lansing, Mich. 1925.

Hyman, Libbie Henrietta (1888), Research Ass., Department of Zool., Univ. of Chicago (U.S.A.). *Physiol. zool. Respiration Comparative vertebrate anat.* ⓓ Univ. of Chicago 1915. ⓟ Turbellaria Tricladida.

Hyman, Orren Williams (1890), Prof. of Histol., Department of Med., Univ. of Tennessee. Memphis, Tenn. (U.S.A.), 879 Madison Ave. *Decapod larval histories. Cytol. of Prosobranch germ cells and fertilization.* ⓓ Princeton 1921.

Hynes, Harold John (1900), Ass. Biol. Sydney (Australia), Biol. Branch, Dept. of Agric. *Plant diseases, Foot-Rot (Helminthosporium sativum) of Cereals. Take-All (Ophiobolus graminis) of Cereals. Reaction of hybrid wheats to Puccinia graministritici. Stalk Rots of Corn; Pea Root Rot.* ⓓ Minnesota Univ. 1925.

Hyslop, James Augustus (1884), Entomol. in Charge, Insect Pest Survey, U.S. Dept. Agric. Washington, D.C. (U.S.A.), Bureau of Entomol. *Elateridae. Statistical data on insect abundance from month to month and year to year; in relation to the climatol. and other factors that might influence the variation in prevalence of insects.* ⓓ Wash. State Coll. 1912.

Ibsen, Heman L. (1886), Prof. of Genetics, Animal Husbandry Department Kansas State Agric. Coll. Manhattan, Kan. (U.S.A.). *Heredity and physiol. of reproduction in guinea-pigs, rabbits and rats.* ⓓ Univ. of Wisconsin 1916. ⓟ Guinea-pigs.

Ichikawa, Sanroku, Prof. of Kyôto Imper. Univ. Kyôto (Japan), Inst. of Forestry, Coll. of Agric. *Forestry.*

Ichimura, Tsutsumi, Prof. of the 4 High School. Kanazawa (Japan), Horomachi 15. *Med. plants of Japan.* ⓓ Tôkyô imperial Univ. 1895.

Ide, Manille (1866), Prof. à l'univ. de Louvain (Belgique), Bd. de Diest 30. *Pharmacodynamie, Anatomie, organes isolés.* ⓓ 1890.

Ideta, Arata (1870), Principal of the Ogôri Agric. School, Yamaguchi Prefecture (Japan). *Phytopath. in general.* ⓓ Sapporo Imperial Coll. of Agric. 1893. ⓟ Specimens of Plant diseases in Japan.

Iehara, Takeo, Lect. Imper. Univ. Kyôto (Japan), Inst. of Hygiene, Fac. Med.

Iglesia, Mario, Prof. Univ. Nac. d. Litoral. Corrientes (Argentina). *Zool, Entomol.* ○

Iglesias é Iglesias, Luis (1895), Prof. Fac. Sc. Santiago de Compostela (España), Rua del Villar 37-39, 2º Galicia. *Coleoptères de la région gallega.* ⓓ Madrid 1917. ⓟ Insectes.

Ignatov, Nik. Konst., Prof. II. Univ. Moskau (U.d.S.S.R.). *Experim. Hyg.* ○

Igoschina, K. N., Botan. Inst. Perm (U.d.S.S.R.), Záimko. *Oekol. der Pflanzen.* ○

v. Ihering, Hermann (1840), o. Hon.Prof. Palaeont. Univ. Göttingen. Bädingen, Oberhessen (Deutschland). *Zool. und Palaeont., Zoogeographie, Faunistik von Brasilien. Morphol. und System der Mollusken.* ⓓ Göttingen med. 1872, phil. 1876. ⓟ Conchylien von Brasilien, zumal marine, recent. Cretaceo-tertiäre Zähne von Selachiern.

Ihle, Johan Egbert Willem (1879), Prof. der Zool. a. d. Univ. Amsterdam. Utrecht (Holland), A. Numankade 9. *Nematoden, Brachyuren, Appendicularien, Thaliaceen.* ⓓ Amsterdam 1906.

Ihsan, Sami (1885), Dir. section de Vaccin à l'Inst. bact. Stamboul (Turquie), Rue Matbaa. ⓓ Constantinople 1916.

Iizuka, Akira, Prof. of Zool. Tôkyô (Japan), Zool. Labor., Gakushû-in. *Annelida.* ⓓ Tôkyô.

Ikari, Jirô, Ass. in Seto Marine Biol. Labor. of Kyôto Imperial Univ. Wakayama-Ken (Japan), Seto-Kanayama, Nishi-Muro-Gun. *Planktonkunde: Diatomeen.*

Ikeda, Sakujirô, Prof. emer. Niigata-Ken (Japan), 74 Zaô-chô, Nagaoka. *Amphibia, Cephalopoda.*

Ikeno, Seiitirô (1855), Prof. Botan. kaiserl. Univ. Tôkyô. Botan. Inst. landw. Fac. Komaba, Tôkyô (Japan). *Vererbungsstudien der japanischen wilden und Kulturpflanzen.* ⓓ 1879.

Ikoma, Yoshihiro (1891), Prof. Biol. Commercial School, and the Adjustment Committee of Plants Prep. Agric. Coll. Tottori (Japan). *Algae, Fungi, Lichen, Mosses, Fern and Higher Plants.* ⓓ Tottori 1914. ⓟ Algae, Fungi, Lichen, Mosses, Fern, Higher Plants.

Iljin, Catherine (1888), Adjoint de Chim. biol. II. Univ. Moscou (U.d.S.S.R.), Rue Malaja Piragowskaja N. 1. *Fermentol.*

Iljin, Boris Sergejewitsch (1889), Ass. der Schwarzmeer-Azowschen Expedition. Moskau 17 (U.d.S.S.R.), Pjatnitzkaja Str. 33, W. 15. *Ichthyol., Gobiidae, Systematik und Biol.* ⓓ St. Petersburg 1913. ⓟ Gobiidae des Schwarzen, Asowschen Meeres und des Kaspisees.

Iljin, Modeste (1889), Conserv. Leningrad (U.d.S.S.R.), Jardin Botan. Principal. *Composées Cynareae; Flore de la Sibérie.* ⓓ Leningrad 1916.

Iljin, Nicholas A. (1903), Ass. Labor. of Experim. Biol. Zoopark, Ass. Inst. General Biol. II. Univ. Moscow (U.d.S.S.R.). *Morphogenetics of animal pigmentation. External factors of morphogenesis. Inheritance and inheritable realizations. Influence of the temperature on the morphogenesis. Causes of the season and age dimorphism. Inheritance of the spotting in mammals.* ⑨ Don-Univ. 1923.

Iljin-Kakneff, Boris (1886), Ass. Moskau (U.d. S.S.R.), Nikolojamskajastr., Lysschikowgasse 5, 1. *Kohlenhydratstoffwechsel.* ⑨ Moskau.

Iljina, Valentina N. (1904), Labor. Experim. Biol. Zoopark. Moscow (U.d.S.S.R.). *External factors of the morphogenesis.*

Iljinsky, Ass. Entomol. Station. Tula (U.d.S.S.R.). *Zool.*

Iljinskij, Alefej (1888), Prof., Ass. Leningrad (U.d.S.S.R.), Anth. Ostrov Glavnyj Botanitschesky Sad N. 1. *Geobotan. Flora von Mittel-Rußland.* Geoyr. *Verbreitung der Holzpflanzen.* ⑭ Herbarium from Midell Russia.

Iljinskij, Nikolaj W., Doc. Milchwirtschaftl. Inst. Wologda (U.d.S.S.R.), Sowetskij per. 15, W. 2. *Geobotan. Wiesenbau.* o

Illick, John Theron (1888), Prof. of Zool., Head Dept. of Biol. Univ. Nanking (China). *Zool. and Genetics.* ⑨ Syracuse Univ. 1913. ⑭ *Plants.*

Illingworth, James Franklin (1870) Entomol., Association of Hawaiian Pineapple Canners, also, Research Associate in Entomol., Bishop Mus. Honolulu, T.H. (Hawaii). *Distribution of pests in the Pacific.* ⑨ Cornell Univ. 1912.

Illuvieff, Victorin (1889). Nikita Yalta, Crimée U(.d.S.S.R.), Gov. Bot. Garden. *Soil chem.* ⑨ Petersburg 1912.

Ilowaisky, David J., Prof. Bergakad., Sekretär der Moskauer Abt. des Comité Géol. de Russie. Moskau (U.d.S.S.R.), Geol. Inst. *Palaeont.; speziell Ammoniten des Jura und der Kreide.* o

Iltis, Hugo (1882), Dr., Prof., Priv.Doc. an der deutschen techn. Hochsch. in Brünn (C.S.R.). *Biologiegeschichte. Biol. der Pflanzen. Pflanzengeographie.* ⑨ Prag, Deutsche Univ. 1905. ⑭ *Pflanzengeographische Photographien von Mitteleuropa.*

Ilvento, Arcangelo (1897), Doc. Igiene Univ. Roma (Italia), Via Dalmazia 24. *Igiene.*

Ilvessalo, Lauri (1887), Forstmeister, Lehrer des Waldbaus a. d. Univ. Helsinki (Finnland), Creutzkatu 7. *Biol. der einheimischen Holzarten, Waldbau.* ⑨ Helsinki 1923.

Ilvessalo, Yrid (1892), Prof. der Waldabschätzung an der forstlichen Versuchsanstalt Finnlands. Helsinki (Finnland), Creutzkatu 7. *Zuwachsverhältnisse der Wälder Finnlands. Vegetationsstatistische Untersuchungen der Waldpflanzenvereine. Boden u. Waldbestand.* ⑨ Helsinki 1921

Imahashi, Tetsuzô (1883). Pharmacol. Labor.,Med. Univ. Okayama (Japan). *Experim. Pharmacol.* o

Imai, Sanshi (1900), Ass. of Phytopath., Coll. of Agric., Hokkaido Imperial Univ. Sapporo (Japan), Botan. Inst. *Mycol. and Phytopath.* ⑨ 1924.

Imas, Rosa (1900), Ass. Inst. de Chim. biol. Cluj (România), Strada Pasteur. *Elimination rénale.*

Imbert, René-Armand-Marie (1901), Aide prép. de physique méd. Univ. Montpellier (France), 9, Rue des Carmélites. *Electrocardiol. Potentiels de contact (indice de nutrition) entre les tissus de l'homme et des animaux ou végétaux vivants et les milieux extérieurs liquides, solides; ou gazeux.*

Imes, Marion (1873), Vet., Zool. Division, Bureau of Animal Industry, U.S. Dept. of Agric. Kansas City, Kan. (U.S.A.), 23 Federal Bldg. *External animal parasit. of livestock.* ⑨ New Hampshire 1900.

Imlah, Helen Woodbridge (1899), Zool. Department, Harvard Univ. Cambridge, Mass. (U.S.A.). *Ascidians.* ⑨ Washington Univ. 1922.

Immendorff, Heinrich (1860), Univ.Prof., Dir. Agric.-chem. Labor. Jena (Deutschland), Am Stelger 9. o

Immer, Forrest R. (1899), St. Paul, Minn. (U.S.A.), Univ. of Minnesota, Univ. Farm. *Plant Genetics, particulary Inheritance in Maize.* ⑨ Univ. of Minnesota 1927.

Imms, A. D. (1880), Chief Entomol., Rothamsted Experim. Station. Harpenden, Herts. (England). *General entomol., morphol. and biol.* ⑨ 1907.

Imrie, Cyril Gray (1890), Lect. on Physiol. Sheffield (England), Univ. *Chem. and Applied Physiol.* ⑨ London 1911.

Ingebrigtsen, Olaf (1890), Lect. Bergen (Norwegen), Katedralschule. *Kraniometrie d. Mammalia (Canis, Cervus).*

Ingram, Alexander (1868), Med. Entomol., South African Inst. for Med. Research. Johannesburg (S.-Africa), Box 1038. *Bionomics of Insects carrying Human diseases.* ⑨ Edinburgh 1892.

Ingram, Collingwood. Cranbrook, Kent (England), The Grange, Benenden. *Ornithol.* o

Ingvar, Sven (1889), Doc. Neurol. Univ. Lund (Sverige). *The comp. anat. of the nervous system.* ⑨ Lund 1919.

Inkowa, Wera (1904), Aspirant Inst. der normalen Anat. Univ. Kasan (U.d.S.S.R.), Popowa Gora 16/3. *Vegetatives Nervensystem.*

Inkster, Robert G. (1897), Demonstrator in Anat., Univ. Leeds (England), Anat. Dept., Med. School. *Talus of the Australian native. Embryol.* ⑨ Edinburgh 1922.

Inman, Ondess Lamar (1890), Prof. of Biol., Antioch Coll. Yellow Springs, Ohio (U.S.A.). *Luminescent Bacteria and biochem. studies on Chlorella sp. Respiration. Iron-depositing Bacteria.* ⑨ Harvard Univ. 1921.

Inomata, Shūjirō (1893), Prof. of Zool. Tottori (Japan), Imperial Agric. Coll. *Economic Zool.: Life cycles and control methods of thrips of crops and trees. Japanese palolo-worm injurious to rice plant.* ⑨ Tôkyô Imperial Univ. 1918.

Inoue, Katsuji, Prof. of Tôhoku Imper. Univ. Sendai (Japan), Chem. Inst., Coll. of Med. *Biochem.*

Inoue, Michio, Prof. of Tôkyô Imperial Univ. Tôkyô (Japan), Anat. Inst., Fac. of Med. *Anat.*

Inouye, Ryûgo, Prof. of Uyeda Imper. Sericult. Coll. Uyeda, Nagano-Ken (Japan), Chem. Labor. *Chem. of Silks.*

Introzzi, Paolo (1898), Ass. Clin. med. Univ. Pavia (Italia). *Istopatol. Ematol.* ⑨ Pavia 1923.

Inukai, Tetsuo (1897), Ass. Prof. Hokkaidô Imperial Univ. Sapporo (Japan), Zool. Inst. *Experim. zool., embryol., variation.* ⑨ Hokkaido Imperial Univ., Sapporo 1922.

Iredale, Tom. Sydney (Australia), Australian Mus. *Ornithol.* o

Ireland, Joseph C., Prof. of Biol. Durant, Okla. (U.S.A.). *Rots of Corn, pathol., physiol.* o

Irger, Julius (1897), Prosektor Anat. Inst. Univ. Minsk (U.d.S.S.R.). *Blutgefäße d. parenchymatösen Organe d. Menschen.*

Irish, John H. (1885), Junior Chem. in the Experim. Station of the Coll. of Agric., Univ. of California. Berkeley, Cal. (U.S.A.), 996 Cragmont Ave.; Univ. of California, 336 Hilgard Hall. *Fruit products in their relation to chem. and bact.*

Irmscher, Edgar (1887), Prof., Dr., Kustos des Herbariums des Inst. für allgem. Botan. Hamburg 36 (Deutschland), Jungiusstr. *Systematik und Pflanzengeographie. Begoniaceae. Musci.* ⑨ Leipzig 1911.

Irwin, Marian (1889), Associate in the Division of General Physiol., Rockefeller Inst. for Med. Research. New York City, N.Y. (U.S.A.), Avenue A and East 66 th Street. *Mechanism of permeability of cells to dyes.* ⑨ Radcliffe Coll. 1919.

Isaachsen, Haakon (1867), Prof. der Tierernährung a. d. Landwirtschaftl. Hochsch. Norwegens

Ås (Norge), Landbrukshöiskolen. *Fütterung d. Milchviehes. Mineralbalance d. Milchkühe. Aufzucht d. Kälber, Ersatz d. Vollmilch mit Centrifugenm. u. Lebertran u. a. Fettarten. Wertbestimmung inländ. Kraftfuttermitt. Wert d. Weidengrases. Mineral- u. Vitaminfragen in d. Schweinezucht. Wert d. norwegischen animal. Futtermitt. in der Schweinefütterung. Proteinbedarf d. Schweines. GewisseAufzuchtfragen i. d. Schweinezucht. Euterphysiol. d. Kuh. Ensilagefragen. Ersatz für Milch b. Ferkel u. Jungschwein, norwegische animal. Futtermitt. Verschiedene Lammungszeiten u. ihre ökonomischen Beziehungen. Wachstumsuntersuchungen b. Rindvieh u. Pferd. Bedeutung d. Tranbeigabe b. Kalb u. Jungschwein.* ⓓ Phil. h. c. Syracuse, N.Y. (U.S.A.) 1923.

Isaacs, Raphael (1891), Instr. in Med., Harvard Med. School, Ass. Physician, Collis P. Huntington Memorial Hospital of the Cancer Commission of Harvard Univ. Boston, Mass. (U.S.A.), 695 Huntington Ave. *Clinical and experim.: Hematol. and Cancer. Development of Blood Cells. Effects of Roentgen Rays and Radium on Blood Cells. Fate of Blood Cells. Salivary Leucopedisis. Statistical studies of Leukemia, Lymphoblastoma, Cancer.* ⓓ Cincinnati 1918.

Isabolinsky, M. (1880), Prof. Bact., Univ., Bact. Inst. Smolensk (U.d.S.S.R.). *Immunität. Serotherapie und Serol.* ⓓ 1907.

Isai, Chiao (1898), Prof. of Physiol., Fu Tan Univ. Shanghai (China), School of Biol. Sc. *Physiol. of brain and glands of internal secretion.* ⓓ Univ. of Chicago 1924.

Ischreyt, Gottfried (1868). Libau (Lettland), Scheunenstr. 22. *Limnol. des Ostbaltikums, Cladoceren.* ⓟ Plankton.

Isely, Dwiht (1887), Associate Prof. of Entomol., Coll. of Agric., Univ. of Arkansas; and Associate Entomol., Arkansas State Agric. Experim. Station. Fayetteville, Ark. (U.S.A.), 3 North Duncan Street. *Economic Entomol.* ⓓ Univ. of Kansas 1913.

Isely, Frederick B. (1873), Dean and Prof. of biol. Culver-Stockton Coll. Worth, Tex. (U.S.A.), 2732 W.E. Ft. o

Isenschmid, Robert (1882), Priv.Doc. Univ. Bern (Schweiz), Hirschengraben 6. *Physiol. menschl. Schilddrüse in Kindesalter. Lokalisation d. Wärmeregulationszentrums im Zwischenhirn.* ⓓ Bern. o

Ishibashi, Matsuzô, Prof. of Chiba Med. Univ. Chiba near Tôkyô (Japan), Pathol. Inst. *Pathol.*

Ishidoya, Tsutomu (1891), Lect. of Botan. Keijô Imper. Univ. Chôsen, Korea (Japan). *Practical Botan.* ⓓ Sapporo 1906.

Ishihara, Makoto, Prof. of Kyûshû Imper. Univ. Fukuoka (Japan), Physiol. Inst., Coll. of Med. *Physiol.*

Ishii, Shigemi. Suginami-chô near Tôkyô (Japan), Kôenji 556. *Diplopoda, Protozoa.*

Ishii, Shirô (1886), Biol. of Government Fishery Experim. St. Rakuma, Karafuto (Japan). *Marine biol. Ommastrephes sloanii pacificus.* ⓟ Teleostei specimen, especially Salmonidae.

Ishii, Tei (1894), Entomol. Nagasaki-Customs (Japan), Imperial Plant Quarantine Service. *Classification and Biol. of the Chalcidoidea (Hymenoptera Insecta), parasit. of Coccids.* ⓓ Tôkyô Imp. Univ. 1918.

Ishikawa, Chiyomatsu, Prof. emer. Imper. Univ. Yotsuya, Tôkyô (Japan), Ôbanchô 19.

Ishikawa, Kintarô, Teacher of Yabe Agric. School. Hamachô near Kumamoto (Japan). *Physiol. and Anat. of Silk-worms.*

Ishikawa, Masashi, Lect. of Tôkyô Imper. Univ. Komaba near Tôkyô (Japan), Fishery Inst., Fac. of Agric. *Cephalopoda.*

Ishikawa, Mitsuharu, Prof. in Botan. of I. High School. Hongô, Tôkyô (Japan), Biol. Labor. *Cytol. of Plant.*

Ishimori, Naoto (1890), Prof. Imper. Coll. of Agric. and Forestry, Entomol. Labor. Morioka (Japan). *Anat. and Physiol. of lepidopterous larvae.* ⓓ Tôkyô Imp. Univ. 1925. ⓟ Silkworm.

Ishiwata, Shigetane, Prof. emer. of Imper. Univ., Prof. of Agric. Univ. Yodobashi near Tôkyô (Japan), Kashiwagi 350. *Anat. and Physiol. of Insects, especially of Silk-worms.*

Ishizaka, Shinkichi, Prof. of Med. Univ. Kanazawa (Japan), Pharmacol. Inst. *Pharmacol.*

Ishizaka, Tomotarô, Prof. of Kyûshû Imper.Univ. Fukuoka (Japan), Pharmacol. Inst., Coll. of Med. *Pharmacol.* o

Israel, Anton (1895), Doc. Inst. Physiol. Taskent (U.d.S.S.R.), Univ. *Anthrop. Turkestans. Biochem. d. Blutes.* ⓓ 1925.

Israels de Jong, S.J. (1878), Prof., agrégé d'Anat. Pathol. à la Fac. de Méd., Méd. des Hôpitaux. Paris (France), 75, rue de Conscelles. *Anat. pathol. de l'appareil respiratoire. Hématol., eosinophilie.*

Isralisky, Wladimir (1887), Ass. bact.-agronom. Stat. Moskau (U.d.S.S.R.), Konjuschkowskaja Str. 31. *Landwirtschaftl. Bact. Bakterielle Krankheiten der Pflanzen, Bodenbact.* ⓟ Phytopath. Bakterien, holzzerstörende Pilze, Bakterien für Mäuseu. Rattenvertilgung, Knöllchenbakterien für Bohnenpflanzen.

Issatschenko, Boris (1871), Dir. Hauptbotan. Garten, Prof. Univ., Prof. des Landwirtschaftl. Inst., Dir. d. Microbiol. Labor. d. Staatl. Hydrol. Inst. Leningrad (U.d.S.S.R.). *Microbiol. des Wassers und Samenkunde (Physiol. d. Keimung).* ⓓ Univ. Leningrad 1895.

von Issekutz, Béla (1886), Dr. o. Prof. Pharmacol., Dir. des pharmacol. Inst. Prof. der Pharmakognosie und Dir. des pharmakognostischen Inst. Szeged (Ungarn), Kálvária-tér 5. *Experim. Pharm.; Insulinforschung.* ⓓ Kolozsvár 1908. ⓟ Rana esculenta.

Issel, Raffaele (1878), Prof. o. Zool. Univ. Genova (Italia), Ist. Zool., Via Balbi 5. *Ecol. marina, Plancton, Sviluppe larvale dei molluschi cefalopodi ed eteropodi. Fauna termale, morfol. e sviluppo larvale dei crostacei decapodi.* ⓓ Genova 1900.

Isshiki, Shûchi, Prof. of High Agric. School. Taihoku, Formosa (Japan), Entomol. Labor. *Systematics of Micro-Lepidoptera. Histol. of Insects.*

Istvánffi, Gyula, Dr., Prof. d. Botan. a. d. Techn. Hochsch. Budapest (Ungarn). *Mykol.* o

Itagaki, Masamitsu, Prof. of Kyûshû Imper. Univ. Fukuoka (Japan), Physiol. Inst., Coll. of Med. *Physiol.*

van Iterson, Gerrit (1878), Prof.Technische Hochsch. in Delft, Dir. des Inst. technische Botan. Delft (Holland), Poortlandlaan 35. *Mikroskopie und Mikrochem. der Faserstoffe. Pflanzliche Rohstofflehre. Mathematische und mikroskopisch-anat. Studien über Blattstellungen.* ⓓ Delft 1907.

Itano, Arao, Sc. in Ôhara, Inst. for Agric. Research. Kurashiki near Okayama (Japan). *Soil Bact.*

Itô, Hirowo (1888), Dr., Prof. Zool. Univ., Tôkyô Imperial Sericultural Coll. Nishigahara, Tôkyô (Japan). *Anat., Histol., Cytol. of Insects, Lepidoptera and Orthoptera.* ⓓ Tôkyô 1913. ⓟ Insects, Orthoptera and Lepidoptera.

Itô, Seiji, Prof. of Imper. Seric. Coll. Hanazono near Kyôto (Japan), Entomol. Labor., *Entomol.*

Itô, Seiya (1883), Prof. Phytopath. and Mycol., Coll. of Agric., Hokkaidô Imperial Univ. Sapporo (Japan), Botan. Inst. *Phytopath. and mycol.* ⓓ 1908.

Itô, Tokutarô, Lect. of Tôhoku Imper. Univ. Sendai (Japan), Botan. Inst., Coll. of Sc. *Botan.*

Itriage, L. G., Prof. Univ. Centr. Caracas (Venezuela). *Hygiene.* o

Ivakin, A. A., Prof. Med. Inst., Dir. Anat. Labor. Kiev (U.d.S.S.R.), Str. Levaschowskaja N. 39. o

Ivanauskas, Prof., Dir. Zool. Labor. Kaunas (Litauen), Wilnastr. 2. o

Ivancich, Antonio (1880), Prof. Liceo Sc. G. Oberolan di Trieste (Italia). *Briol.; Muschi; forme cavernicole.* ⊕ Wien 1907. ⊕ Muschi.
Ivanić, Momčilo (1886). Beograd (S.H.S.), Studeničkа 9 I. *Protozoen, Embryol. der Arachniden, Cytol.* ⊕ München 1913.
Ivanić, Stevan (1884), Chef section bact. et epidemiol., Dir. de l'Inst. Beograd (S.H.S.), Inst. Central d'Hygiene. *Bact. et serol. Parasit., Epidemiol.* ⊕ 1910.
Ivanizky, Michael Th. (1894), Priv.Doc. Anat. Med. Fak. I. Staatsuniv. Moskau (U.d.S.S.R.), Mohowaja, 11; 105. *Pelvis renalis, Glandula thyreoidea.* ⊕ 1925.
Ivanov, Jurij, Prof. Tierärztl. Hochsch. Leningrad (U.d.S.S.R.), Cernigovskaja 5. *Zool. u. vergl. Anat.* o
Ivanov, Vladimir, Doc. Univ. Irkutsk (U.d.S.S. R.). *Physiol. Chem.* o
Ivanov-Jadin, I. I., Prof. landwirtschaftl. Inst. Krasnodar, Kaukasus (U.d.S.S.R.), Novaja 107. *Anat. u. Hygiene d. Haustiere.* o
Ivanov-Smolenskij, A. G., Prof. Pädagog. „Herzen" Inst. Leningrad (U.d.S.S.R.). *Anat. u. Physiol.* o
Ivanow, Sergius (1880), Prof. der Pflanzenphysiol. II. Univ., Prof. der Chem. und Technol. der Fette, Öle und Wachs in der Chem.-Techn. Hochsch., Inst. Mendeleew. Moskau 34 (U.d.S.S.R.), Krapotkinstr. 15, 1. *Einfluß des Klimas auf die chem. Tätigkeit der Pflanzen. Tierische Fette, Pflanzenöle, Lipase, Eiweißreservestoffe.* ⊕ Moskau 1913.
Iversen, Poul (1889), Priv.Doc. København, L. (Danmark), Bispebjerg Hospital. *Normale und pathol. Physiol. Pharm.* ⊕ København 1918.
Ives, Sumner A., Dean of the School of Sc. and Prof. of botan., Howard Coll. Birmingham, Ala. U.S.A.). *Germination of Cassina seeds.* o
Ivy, Andrew Conway (1893), Prof. of Physiol. and Pharmacol., Northwestern Univ., Med. School. Chicago, Ill. (U. S.A.), 303 East Chicago Avenue. *Gastro-Intestinal Tract.* ⊕ Univ. of Chicago 1918.
Iwakawa, Tomotarô, Prof. emer. of Tôkyô Girls' High Seminary. Nakano near Tôkyô (Japan), Arai 543. *Chonchyol.*
Iwakin, A. A. (1893), Doc. Univ., Dir. Anat. Inst. Kiew (U.d.S.S.R.), Theatralna, N. 48 A, 7. *Nervensystem der Blase.*
Iwanoff, Boris (1880), Vorsteher der Pflanzenschutzabteilung. Sofia (Bulgarien), Solunplatz 3. *Pilzkrankheiten — Uredineae.* ⊕ Bern 1906.
Iwanoff, Boris (1898), Ass. Kasan (U.d.S.S.R.), Vet. Inst., Asskoe Pole 96. *Rotz der Lymphdrüsen.*
Iwanoff, Georg (1893), Ass. Anat. Inst., Militär-Med. Akademie. Leningrad (U.d.S.S.R.), Kronwerksky prosp. H. N. 29, W. 19. *Chromaffines System; Interrenalsystem; Arterien d. menschlich. Körpers.*
Iwanoff, Leonid Alexandrowitsch (1871), Prof. Physiol. und Anat. d. Pflanzen Leningrader Forstinst. Leningrad (U.d.S.S.R.). *Physiol. und Oekol. d. Pflanzen; Methodik d. Lichtmessung im Dienste der Pflanzenphysiol. und Pflanzenkultur. Bestimmung d. Lichtgenusses der Holzgewächse. Photosynthese und Transpiration.* ⊕ Petersburg 1906.
Iwanoff, Michael F. (1871), Prof. Tierzucht Landwirtschaftl. Akademie. Moskau (U.d.S.S.R.), Petrowskoe-Rasumowskoe. *Schaf-, Schweine-, Geflügelzucht.* ⊕ Zürich.
Iwanoff, Nicolai Nicolaewitsch (1884), Prof., Dir. botan. u. microbiol. Labor. Leningrad (U.d.S.S.R.), Technol. Inst. *Physiol. und Chem. der Pflanzen (Pilze, Bakterien).*
Ivanov, P. P., Prof., Dir. Embryol. Labor. Univ. Leningrad (U.d.S.S.R.). o
Iwanowsky, Wladimir Nikolajevitsch (1867), Prof. Psychol. Univ. Minsk (U.d.S.S.R.), Schirokaja, 28. ⊕ Kazan 1910.

Iwanow, Eugen W., Bergingenieur, Com. Géol. de Russie. Leningrad (U.d.S.S.R.), W.O., Sredny Pr. 78-B. *Palaeont.; Mesozoische Mollusken.* o
Iwanow, Johann (1895), Ass. der Anat. Rostow a. Don (U.d.S.S.R.), Anat. Inst. *Verdoppelung der unteren Hohlvene. Blutgefäße des Diverticulum Meckeli. Blutgefäße d. Dünn- und Dickdarmes.* ⊕ Moskau 1916.
Iwanowska, Anne (1891), Wissenschaftl. Forschungs-Inst. Odessa (U.d.S.S.R.), Barjatinsky 1-9. *Hemotropische und traumatotropische Reizleitungsvorgänge.*
Iwanowsky, Alexius A., Prof. Univ. Charkow (U.d.S.S.R.). *Anthrop.* o
Iwanowsky, Georg (1901), Labor. Taschkent (U.d.S.S.R.), Puschkinskaja N. 53. *Physiol. der Blutzirkulation.*
Iwanzoff, Nikolaus A. (1863), Prof. II. Univ., Mitglied d. wissenschaftl. Inst. d. Zool. d. I. Univ., Mitglied d. wissenschaftl. Inst. K. A. Timiriaseff. Moskau (U.d.S.S.R.), Sretensky Bulwar, 6-37. *Physiol.; Reflexol.; Evolutionslehre.* ⊕ Moskau 1897.
Iwanzov, S. L., Prof. II. Univ. Moskau (U.d. S.S.R.). *Botan.* o
Iwaoka, Suehiko, Chief sc. of Chem. Labor., Gunze Silk Factory. Ayahe near Kyôto (Japan). *Physiol. of silk-worms and Mulberry.*
Iwirbliss, Nicolas (1898), Chef Section de Technol. des cocons et de la soie Station de Sériciculture d'Asie Centrale. Taschkent (U.d.S.S.R.), Rue Kafanow, 1. *Sériciculture.* ⊕ Cocons de diverses races du ver à soie. Echantillons de la soie grège de diverses races.
Izabolinskij, Michajl Petr., Prof. Univ. Smolensk (U.d.S.S.R.). *Bact.* o
Izar, Guido (1883), Dir. R. Clinica Med. Generale dell' Univ. Messina (Italia), Via dei Mille 224. *Serol. dei tumori maligni. Chemoterapia tubercolosi, tumori maligni, ecc.*
Izawa, T., Lect. of Keiô-Gijuku Univ. Tôkyô (Japan), Bact. Inst., Coll. of Med. *Bact.*
Izquierdo, J., Prof. Univ. Centr. Caracas (Venezuela). *Anat. norm.* o
Izquierdo, José Joaquín (1893), Dr., Prof. of Physiol. in the National School of Med. and the Military Med. School. Mexico City (Mexico), San Luis Potosi 173. *Physiol. of the high altitudes; Labyrinth.* ⊕ Colegio del Estado de Puebla 1917.

Jablokow, Nicolaus (1897), Ass. protozool. Abt. Metschnikow-Inst. Moskau (U.d.S.S.R.), Possowka 44. *Hämatol. Mückenbiol. Malaria.*
Jablonowski, Josef (1863), Generaldir. im kgl. ung. Landwirtschaftlichen Versuchswesen, Dir. der Kgl. ung. staatlichen Entomol. Station. Budapest (Ungarn), II. Kitaibel Pál u. 1. *Angewandte landwirtschaftliche Entomol.* ⊕ Budapest 1887. ⊕ Landwirtschaftlich-entomol. Gegenstände: schädliche Tiere, ihre Verwandlungsformen und Fraßstücke.
Jabouille, Pierre (1875). Bac-ninh, Tonkin (Indo-China), *Ornithol. de l'Indochine.*
Jaccard, Paul (1868), Prof. der Botan. an der Eidg. Techn. Hochsch. Zürich (Schweiz). *Pflanzenphysiol. Dickenwachstum. Geotropismus. Gesetze der Pflanzenverteilung. Embryol. Elektrokultur. Holzanat.* ⊕ Zürich 1893.
Jačevskij, A. A., Prof. Forstinst. Leningrad (U.d.S.S.R.), Lesnoj Inst. per. 5. *Phytopath.* o
Jachimowicz, Jan (1900), Dr. med. Warszawa (Polska), Chałubinskiego 5, Zakład Anat. *Beschreibende, vergl. und pathol. Anat., Zool.* ⊕ Warschau 1925.
Jachontowa, Olga Vlad., Doc. II. Univ., Ass. Anat. Inst. Moskau (U.d.S.S.R.), Dewitschje Pole, Trubezkoi pereulok. *Norm. Anat.* o
Jack, John G., Ass. Prof. of Dendrol., Harvard Univ. East Walpole, Mass. (U.S.A.). *Forest botan., dendrol.* o
Jackson, Cicero Floyd (1882), Prof. of Zool., Univ. of New Hampshire. Durham, N.H. (U.S.A.).

Ecology and Systematic: Vertebrates. Ⓓ Ohio State Univ. 1907. Ⓟ Various groups of Vertebrates.

Jackson, Clarence Martin (1875), Prof. and Head of the Department of Anat., Univ. of Minnesota. Minneapolis, Minn. (U.S.A.), Inst. of Anat. *Effects of malnutrition upon growth·and structure.* Ⓓ Univ. of Missouri 1900, LL.D. 1923.

Jackson, Daniel Dana (1870), Executive Officer and Prof., Department of Chem. Engineering, Columbia Univ., New York City (U.S.A.), Havemeyer Hall. *Bact. and microscopy of water and sewage.* Ⓓ Univ. of Pittsburg 1924.

Jackson, Sir Frederick, ao. H. S. Holt Esq. London W.C. 1 (England), 6 Gray's Inn. *Ornithol.* o

Jackson, Harold Gordon (1888), Head of Zool. Dept., Birkbeck Coll., Univ. London E.C. 4 (England), Breams Buildings. *Crustacea: Morphol. and Systematics of Isopoda.* Ⓓ Liverpool 1926.

Jackson, Hartley Harrad Thompson (1881), Biol., in charge div. Biol. Investigations, U.S. Biol. Survey. Washington, D.C. (U.S.A.). *Mammals.* Ⓓ Washington Univ. 1914.

Jackson, Herbert Spencer (1883), Prof. of Botan., Chief in Botan., Purdue Univ., Agric. Experim. Station. West La Fayette, Ind. (U.S.A.). *Plant Pathol., Mycol.. Uredinales and Ustilaginales.* Ⓓ Cornell Univ. 1905. Ⓟ Ustilaginales and Uredinales.

Jackson, Holmes C. (1875), Prof. of Physiol., Univ. of New York City and Bellevue Hospital Med. Coll. East Orange, N.Y. (U.S.A.), 47 Ashland Ave. *Physiol. chem.* Ⓓ Yale 1899. o

Jackson, John Wilfrid (1880), Senior Ass. Keeper, Mus., the Univ. Manchester (England), „Standene" Goulden Road, Withington. *Palaeont. Recent Mollusca and Brachiopoda.* Ⓓ Manchester 1921. Ⓟ Mollusca (especially Cypraea) and Brachiopoda.

Jackson, Robert Tracy (1861). Peterborough, N.H. (U.S.A.). *Palaeont. and Zool.: Echini.*

Jackson, Vincent W. (1876), Prof. of Botan. and Biol.Manitoba Agric. Coll. Winnipeg, Man. (Canada). *Ecol. of the Prairies (plant and animal).* Ⓓ Univ. of Minnesota 1923.

Jacob, Charles (1878), Prof. de Géol. Toulouse (France), 29, Rue André Délieux. *Ammonites du Crétacé inférieur et moyen. Brachiopodes.*

Jacobi, Arnold Friedrich Victor (1870), Mus.-Dir., Prof. Technische Hochsch. Dresden (Deutschland), Zwinger. *Tiergeographie; Systematik und Vergl. Anat. der Säugetiere; Ornithol.; Systematik der Homoptera (Insecta, Rhynchota).* Ⓓ Leipzig 1895.

Jacoblitz, Ernst (1868), Prof., Abteilungsvorsteher Preuß. Hygien. Inst. Beuthen, Oberschlesien (Deutschland), Gymnasialstr. 6. *Hygiene. Bact Serol.* Ⓓ Berlin 1893.

Jacobj, Carl (1857), Geh.Med.R., o. Prof. und Vorstand des pharmacol.Inst. zu Tübingen (Deutschland). *Pharmacol. und Physiol.* Ⓓ Straßburg 1887.

Jacobs, Merkel Henry (1884), Prof. of General Physiol., Univ. of Pennsylvania, Dir. of the Marine Biol. Labor., Woods Hole, Mass. Philadelphia, Pa. (U.S.A.), Department of Physiol. *General Physiol. especially Cell Permeability.* Ⓓ Univ. of Pennsylvania 1908.

Jacobs, Stanley Edward (1905), Demonstrator. London S.W. 17 (England), 21 Rectory Lane, Tooting Common. *Bact. in its relation to Agric. and Industry.* Ⓓ London Univ. 1926.

Jacobs, Walter A. (1883), Member, Rockefeller Inst. for Med. Research. New York City, N.Y. (U.S.A.), *Structural studies on cardiac glucosides and saponins. Organic arsenic compounds. Alkaloids. Chemotherapy.* Ⓓ Berlin 1907.

Jacobs, Werner Friedrich Christian Gustav (1901), Dr., Ass. am zool. Inst. München (Deutschland), Neuhauser Str. 51, Alte Akademie. *Cytol., Golgiapparat, Sekretion.* Ⓓ Rostock 1924.

Jacobsen, William Cornelius (1894), Chief, Bureau of Plant quarantine and Pest control, State Dept. of Agric., Sacramento, Cal. (U.S.A.), 1341—43. St. *Plant diseases, animal pests.* Ⓓ Berkeley, Cal. 1916.

Jacobshagen, Karl Konrad Eduard (1886), Dr. med., Prof. ord. der Anat. Marburg a. d. Lahn (Deutschland), Anat. *Vergl. Anat. (allgemeine, spezielle).* Ⓓ Jena 1912.

Jacobsohn-Lask, Ludwig (1863), Priv.Doc. und Univ.-Prof. Berlin-Lichterfelde (Deutschland), Mittelstr. 11. *Gestaltsprobleme des Gehirns und vergl. Anat. des Zentralnervensystems.* Ⓓ Berlin 1889.

Jacobson, Edward Richard (1870). Fort de Kock, Sumatra (Ned.-Indië). *Biol. aller Arthropoden, Erforschung ihrer Lebensweise (Gewohnheiten, Nahrung, Nestbau, Fortpflanzung, Verwandlung, Aufenthaltsort) und ihrer geographischen Verbreitung (horizontal und vertikal).* Ⓟ Sumatranische Insekten und andere Arthropoden.

Jacobson, Georgij (1871), Prof. Landwirtschaftl. Inst., Chef-Zool. des Zool. Mus. d. Akademie d. Wissenschaften. Leningrad (U.d.S.S.R.). *Systematik und Zoogeographie der Insekten. Käfer Rußlands. Orthopteren u. Pseudoneuropteren.* Ⓟ Insecta, Coleoptera, spec. Chrysomelidae palaearcticiae.

Jacobsthal, Erwin Jakob Wolfgang (1879), Priv.-Doc., Leitender Oberarzt des Bact.-serol. Inst. am Allg. Krankenh. St. Georg, Hamburg 5 (Deutschland). *Bact., Serol. Leichenbact., Bacteriemmutation, Tuberkulose; Kolloidchem. in der Serol.* Ⓓ Straßburg i. E. 1901. Ⓟ Bacterienkulturen.

Jacoby, Martin (1872), Prof., Dr. med., Dir. der biochem. Abteilung des Krankenhauses Moabit, Berlin W 35 (Deutschland), Derfflingerstr. 19. *Biochem., spez. Fermentforschung. Pharmacol. und Toxikol., spez. der Antigene.* Ⓓ Berlin 1895.

Jacot, Arthur Paul, Prof. Shantung Christian Univ. Tsinan (China). *Biol.* o

Jacquinet, R., Prof. Ecole prép. de méd. et de pharmacie. Reims (France). *Clinique méd.* o

Jaczewski, A. J., Dir., Inst. of Phytopath., Leningrad (U.d.S.S.R). NewYork City (U.S.A.), Care of W. P. Anderson, 512, Fifth Ave.

de Jaczewski, Arthur (1863), Dir. de l'Inst. Jaczewsko de Mycol. et de Pathol. Végétale. Leningrad (U.d.S.S.R.), Anglijskij Prospekt 29. *Mycol. et Pathol. Végétale: Peronosporées, Myxomycètes, champignons parasit. des Espèces forestières. Exoascées.* Ⓟ Prep. mycol. et phytopath.

Jaczewski, Tadeusz (1899), Keeper of the Entomol. Section of the Polish Mus. of Natur. History at Warszawa (Polska), Krakowskie Przedmieście 26/28. *Hemiptera-Heteroptera, aquatic and semiaquatic groups, systematics, ecol., geographical distribution and phylogeny; fossil forms.* Ⓓ Poznań (Poland) 1925. Ⓟ Heteroptera.

Jääskeläinen, Viljo (1886), Mag. phil., Biol. am Bureau für Fischereiuntersuchungen. Helsingfors (Finnland), Lantbruksstyrelsen. *Ökol. der Fische und Fischparasiten des Ladogasees.* Ⓓ Helsingfors 1919.

Jaeger, Maria (1899). Münster i. W. (Deutschland), Krummer Timpen 20. *Anat. und Physiol. der Verholzung.* Ⓓ Münster 1926.

Jägersköld, L. A. (1867), Prof., Intendent der Zool. am Mus. Göteborg 11 (Sverige), Naturhistoriska mus. *Nordische Vögel. Vogelzug. Nematoden. Trematoden.* Ⓓ Upsala 1893. Ⓟ Helminthen. Nordische Vögel.

Jäggli, Mario (1888), Dr., Dir. der höher. Handelsschule. Bellinzona, Tessin (Schweiz). *Laub- und Lebermoose (Pflanzengeographie). Floristica del monte Camoghè presso Bellinzona. Il Delta della Maggia e la sua vegetazione.* Ⓓ Zürich 1908. Ⓟ Moose des Kantons Tessins.

Jaekel, Otto (1863), Geh.Reg.R., Dr., o. Univ.-Prof. für Geol. und Palaeont. in Greifswald (Deutschland). *Morphogenie der Wirbeltiere und Echinodermen. Entwicklungstheorie. Eiszeiten und Diluvium*

Norddeutschlands. Morphogenie der ältesten Wirbeltiere. Ⓓ München 1886.

Järnefelt, Heikki Arvid (1891), Doc. der angewandten Limnol. an der Univ. Helsinki (Finnland), Ritarikatu 9 B. *Produktionsbiol.: Untersuchung der Bodentierproduktion, des Seebodens, des Gashaushalts und der Temperatur des Wassers. Wasserdüngungslehre. Lyncodaphniden, Chydoriden, Fische und ihre Nahrung.* Ⓓ Helsinki 1922.

Järvi, Toivo Henrik (1877), Prof. Dr. Helsingfors (Finnland). *Ökol. der Fische, Araneae.* Ⓓ Helsingfors 1914.

Jaffa, Myer Edward (1857), Consulting Nutrition expert. Calif. State Bd. Health. Berkeley, Cal. (U.S.A.). *Nutrition. Agric. chem.* ○

Jaffe, Henry Lewis (1896), Dir. Labor. Hospital Joint Diseases. New York City, N.Y. (U.S.A.), 1919 Madison Ave. *Pathol. physiol. of the endocrine glands, bones and joints.* Ⓓ New York 1920.

Jaffe, Richard Hermann, Associate Prof. of Pathol., Coll. of med., Univ. of Illinois, Chicago, Ill. (U.S.A.), Research and Educational Hospital, Polk Street. *Pathol. Anat. and Experim. Pathol. Histopathol. of glands with internal secretion, Tuberculosis, Nephritis and malignant Neoplasmas.* Ⓓ Wien 1913.

Jaffé, Rudolf (1885), Prof. Dr. med., Dir. des Pathol.-anat. Inst. am Krankenhaus Moabit-Berlin. Berlin-Zehlendorf (Deutschland), Gertraudstr. 18. *Allgemeine Pathol. und spezielle pathol. Anat., speziell Endocrines System, Lipoidstoffwechsel.* Ⓓ Freiburg i. B. 1909. Ⓟ *Pathol.-anat. Praep.*

Jahn, Eduard (1871), Prof. an der Forstl. Hochschule in Hann.-Münden (Deutschland), Botan. Inst. *Mykol. Forstbotan.* Ⓓ Berlin 1904. Ⓟ *Myxomyceten.*

Jahn, Friedrich (1888), Dr., praktischer Arzt. Schmalkalden, Thüringen (Deutschland). *Pharm.: Org. Sauerstoffverbindungen, Magnesiumnarkose und spezifische Immunität und Verwandtes. Biochem.: Eisenstoffwechsel (Bestimmungsmethode).* Ⓓ 1914.

Jáki, Gyula (1898), I. Ass. des pathol.-anat. Inst. d. Stephan-Tisza-Univ. Debreczen (Ungarn). *Pathol. Anat. und Histol.* Ⓓ Budapest 1922.

Jakimov, V. L., Priv.Doc. Inst. f. ärztl. Fortbildung. Leningrad (U.d.S.S.R.), Kiročnaja 41. *Bact.* ○

Jakob, Alfons (1884), Prof., Vorstand der hirnanat. Abteilung der Staatskrankenanstalt und psychiatrischen Univ.-Klinik Hamburg-Friedrichsberg. Hamburg 22 (Deutschland), Friedrichsberg, Anat.-pathol. Abteilung. *Normale, pathol Anat. und Histol. des Nervensystems.* Ⓓ Straßburg 1909. Ⓟ *Praep. des Nervensystems.*

Jakob, Christfried (1866), Prof. der Biol. an der Univ. Buenos Aires und La Plata. Buenos Aires (Argentina), Charcas 1240. *Neurobiol. (Hirnrinde und Hirnganglien vergl. histol. und histopathol.).* Ⓓ Erlangen 1890.

Jakob, Heinrich (1874), o. Prof. an der Reichsuniv., vet. Fac. Utrecht (Holland), J. W. Frisostraat 30. *Pharmacol. und Toxikol. der Haustiere.* Ⓓ Bern 1902.

Jakobházy, Zsigmond (1867), Univ.Prof., Dir. Pharmakognost. Inst. Budapest VIII (Ungarn), Üllői-ut 26. *Pharm.* Ⓓ 1891.

Jakooljević, Stevan (1891), Dr., Ass. für Botan. Univ.Beograd (S.H.S.), Takovska 41, Botan. Garten. *Pflanzenanat., speziell: Zellmembran.* Ⓓ Belgrad, S.H.S., 1925.

Jakovlev, Sergius A., Prof. Geol., Palaeont. Forstinst. Leningrad (U.d.S.S.R.), Lesnoj, Geol. Kabinett. *Posttertiäre Flora und Fauna.* ○

Jakowatz, Anton, Dr., Prof. d. Botan. u. Pflanzenschutz. Děčin-Liběverd (C.S.R.), Landwirtsch. Abt. d. Deutsch. techn. Hochsch. Prag. ○

Jakowlew, Nikolaus N. (1870), Chefgeol. des russischen geol. Comités, Prof. d. Palaeont. Berginst. Leningrad (U.d.S.S.R.), Wassili Ostrow. *Oberpalaeozoische Fauna d. Evertebrata, insbesondere Korallen (Rugosa), Brachiopoden, Pelmatozoen. Fossile Amphibien und Reptilien. Palaeobiol.*

Jakowlew, Paul I., Stellvertr. Dir. d. Salgirsker Versuchsstation f. Obstbau. Simferopol (U.d.S.S.R.). *Gartenbau.* ○

Jakubski, Antoni Władysław (1885), Prof. ord. d'anat. Comp. et biol., Univ. Poznań (Poiska), Wjazdowa 3. *Histol. du système nerveux des invertébrés, système. et biol. des Rotifères de Pologne, biol. de la mer baltique, zoogéographie de Pologne, zoogéographie générale, rapport de l'hérédité et évolutisme.* Ⓓ Lwów 1908.

Jakuschkin, Iwan W., Prof. Landw. Inst., Dir. d. Versuchsstation. Woronesh (U.d.S.S.R.). *Pflanzenkunde. Kultur d. Zuckerrübe.* ○

Jamieson, John Kay, Prof. of Anat., Univ. of Leeds (England), Anat. Dept., Med. School. *Lymphatic system in Man.* Ⓓ Edinburgh 1894.

Jamieson, Walter A. (1890), Dir., Biol. Labor., Eli Lilly and Company. Indianapolis (U.S.A.). *Immunol. Antitoxins, immune sera, bact. vaccines viruses.* Ⓓ Trinity Coll. 1911.

Jammes, M. L., Prof. de Zool. appl., Fac. Sc., Toulouse (France). *Zool.* ○

Janata, Alexander A., Prof. d. Landwirtsch. Inst. Charkow (U.d.S.S.R.), Lermontowskaja 35, W. 10. *Botan. Geobotan. Samenzucht.*

Janata, O., Prof., Dir. Inst. f. angewandte Botan. Charkow (U.d.S.S.R.). *Botan.* ○

Janchen-Michel, Erwin (1882), Dr. phil., Prof. für systematische Botan. (mit Einschluß der angewandten Botan.) an der Univ. Wien, Vizedir. des Botan. Gartens der Univ.. Amt: Wien III (Deutsch-Österreich), Rennweg 14; Wohnung: Wien III, Ungargasse 71. *Systematik der Blütenpflanzen; Verbreitungsmittel der Früchte und Samen; Nutzpflanzen.* Ⓓ Wien 1907.

Jancke, Oldwig (1901), Ass. an der Zweigstelle Naumburg der Biol. Reichsanstalt. Naumburg a. d. S. (Deutschland), Burckhardtbrücke 2. *Schädlingsbekämpfung im Obstbau.* Ⓓ Greifswald 1923.

Janda, Viktor (1880), Ph.D., Doc. der experim. Zool. an der böhm. Univ. Prag-Letná (C.S.R.), Cechova 16, Zool. Inst., Univ. Karlova 3. *Experim. Zool. (Regeneration).* Ⓓ Prag.

Janek, Alexander (1891), Doc. Univ. Riga (Lettland), Chem. Labor. *Viskosität lyophiler kolloider Lösungen und Prozesse in gelatinierenden Systemen.*

Janensch, Werner (1878), Prof., Kustos geol.-palaeont. Inst. u. Mus. Univ. Berlin N 4 (Deutschland), Invalidenstr. 43. *Fossile Reptilien.* Ⓓ Straßburg i. Elsaß 1901.

Jang, Sham Sher (1888), Ass., Plague Enquiry, Jemadar Indian Med. Dept. Bombay (India), Haffkine Inst. Ⓓ Agra 1907.

Janicki, Constantin (1876), Dr. phil., o. Prof. der Zool. und Dir. des Zool. Inst., Univ. Warschau (Polen), Krak.-Przedm. 26. *Cestoden der Säugetiere und Fische. Embryonale und postembryonale Entwicklung der Cestoden. Flagellaten des Insektendarmes (Hypermastigina).* Ⓓ Basel 1905.

Janisch, Ernst (1892), Dr. phil., wissenschaftlicher Hilfsarbeiter an der Biol. Reichsanstalt, Berlin-Dahlem (Deutschland), *Vergl. Biol.; allgemeine Bekämpfungslehre, angewandte Entomol.* Ⓓ Marburg a. d. L. 1919.

Janisch, Rudolf (1897), Dr. phil., Entomol. Magdeburg (Deutschland), Halberstädter Str. 6. *Pflanzenschutzmittel, Biol. u. Systematik der Blattläuse.* Ⓓ Jena 1923.

Janischewskij, Dmitrij E., Prof. Univ., Dir. Inst. f. Landwirtschaft u. Melioration. Saratow (U.d. S.S.R.), Kropiwnaja 18, W. 3. *Botan. Systematik. Morphol.* ○

Janke, Alexander (1887), Priv.Doc., ao. Prof. Wien VI (Deutsch-Österreich), Getreidemarkt 9 (Inst.). *Mikrobiol., Biochem. Bact., Mykol.* Ⓓ Wien 1915.

Jankowski, Edmond (1849), Prof. à l'Ecole Centrale Agric. et à l'Univ. Libre de Pologne. Warszawa (Polska), r. Warecka 14. *Pomol.*

Janney, Nelson W. (1881). Hermosa Beach, Cal. (U.S.A.), 402 Strand. *Endocrinol., Metabolism.* ⓓ Med. Univ. of Pennsylvania 1906, München 1912, Dr. chem. Marburg 1918.

Janošik, Johann, Dr., Prof. d. Anat. Praha II (C.S.R.), Kateřinska 32, Anat. Inst. d. Karls-Univ. o.

Janowski, Bronislas (1875), Prof. extr. de Botan. et d'Agric. Lwów (Polska), Kochanowskiego 63, L'Académie de la Méd. Vét. *Culture des plantes fourragères, des prairies. Culture des graminées.* ⓓ Dublany 1897. ⓟ *Herbiers des plantes fourragères.*

Janse, Jacobus Marinus (1860), Dr., Prof. der Botan., Dir. Botan. Inst. und Botan. Garten, Univ. Leiden (Holland), Witte Singel 30. *Wasserbewegung der Pflanzen. Permeabilität des Protoplasmas. Polarität bei niederen (Caulerpa) und bei höheren Pflanzen. Bewegungen der Pflanzen. Mycorrhiza. Tropische Pflanzenkrankheiten.* ⓓ Amsterdam 1885.

Janssen, Sigurd (1891), Priv.Doc. für Pharmacol. u. Ass. am Pharmacol. Inst. der Univ. Freiburg i. Br. (Deutschland), Katharinenstr. 27. *Pharmacol.* ⓓ Heidelberg 1921.

Janssonius, Hindrik Haijo (1874), Regierungsbotan. für die Mikrographie der indischen Hölzer, Labor. der Abteilung Handelsmus. des Koloniaal Inst. Amsterdam (Holland), Nicolaas Maesstraat 141 I. *Mikrographie der indischen Hölzer. Anat. der Simplicia.* ⓓ Groningen 1918.

Januškevič, Paul, Prof., Tierärztl. Hochsch. Leningrad (U.d.S.S.R.), Černigovskaja 5. *Anat.* o

Janusz, Victor (1894), Ass. Inst. Anat. pathol. Lwów (Pologne), Rue Piekarska 52. *Système nerveuse.* ⓓ Varsovie 1921.

Japha, Arnold (1877), Dr. med., Dr. phil., ao.Prof. an der Univ. Halle a.d. S. (Deutschland), Schwuchtstr. 17. *Helminthol., Vererbungslehre beim Menschen.* ⓓ Med. Königsberg 1901, phil. Königsberg 1907.

Japp, Gilbert (1886), Prof. Olmütz (Č.S.R.), Sokolenstr. 44. *Hydrobiol.: Desmidiaceen, Rotatorien, Entomostraken, höhere Pilze, Amphibien.* ⓓ Wien 1909. ⓟ *Mährische Amphibien und Reptilien, Planktonfänge, höhere Pilze, Desmidiaceen Mährens und Nord-Böhmens.*

Jappelli, Antonio (1886), Ass. Ist. di Fisiol. Univ., Libero Doc. di Farmacol. Napoli (Italia), 1, Via dei Mille. *Fisiol., Farm.*

Jappelli, Gaetano (1859), Lib. Doc. Fisiol. Univ. Socio ord. Accad. med.-chir. Napoli (Italia), 1, Via del Mille. *Fisiol. dell' alimentazione.* ⓓ Napoli 1883.

Jaquet, Alfred (1865), Prof. Dr., Leiter d. Sanatoriums La Charmille. Riehen, Basel (Schweiz). *Pharmacol. Kreislaufstörungen, respiratorischer Gaswechsel, physiol. Wirkung d. Höhenklimas.* o

Jardim de Vilhena, Henrique, Prof. Univ. Lisbôa (Portugal). *Anat. descript.* o

Jarisch, Adolf, Univ.Prof. Innsbruck (Deutsch-Österreich). *Pharmacol.* o

Jármai, Karl (1887), o. Prof. Pathol. Anat. Budapest (Ungarn), Tierärztl. Hochsch., Rottenbillergasse 25. *Allgemeine Pathol., Pathol. Anat. und Histol. der Tiere.* ⓓ Budapest 1909.

Jarnagin, Milton Preston (1881), Prof. of animal husbandry, Univ. of Ga. Athens, Ga. (U.S.A.), 630 Ulilledge Circle. o

Jarocki, Jerzy (1898), Dr. phil., Ass. à l'Inst. de Zool. de l'Univ. de Warszawa (Polska), Zaklad Zool. Univ. Krakowskie-Przedmieście 26. *Protozool. (Flagellata, Rhizopoda, Mycetozoa); Crustacea (Amphipoda).* ⓓ Warszawa 1926. ⓟ *Mycetozoa de Pologne contre les mêmes des pays étrangers.*

Jaroslaff, Siméon (1888), Prof. de physiol. Inst. de l'économie rurale, Ass. Labor. physiol. Inst. de méd. Kiew (U.d.S.S.R.). *Anesthésie, Fonction de l'apophyse vermiforme de l'intestin aveugle.* ⓓ Kiew 1922.

Jarosz, Jan (1877), Dr., Prof. der Palaeont. an der Bergakademie Kraków (Polska), Loretańska 18. *Palaeont. und Stratigraphie.* ⓓ Kraków 1913.

Jarussova, Natalie (1893), Ass. Moskau 2 (U.d. S.S.R.), Korowy Gasse 10, N. 39. *Tierstoffwechsel, Avitaminose-C.*

Jarvis, Norman Donald (1899), Instr. in Fisheries. Seattle, Wash. (U.S.A.), National Conners Assoc. *Jodine Content of the Pacific Coost Salmon.* ⓓ Univ. of Washington 1925.

Jaschnoff, Leonidas (1860), Prof. der Forstwissenschaft. Kazan (U.d.S.S.R.), Inst. der Land- u. Forstwirtschaft. *Dendrol. u. Waldbau d. Wolga-Gebietes.* ⓟ *Forstpflanzen.*

Jasnitzkij, Wladislaw N., Doc. a. d. Univ. Irkutsk (U.d.S.S.R.), Nabereshn. Angary 24/1. *Botan.* o

Jassilov, G. M., Prof., Dir. Anat. Inst. d. Univ. Woronesch (U.d.S.S.R.), Prosp. d. Revolution 2. o

Jasswoin, Gregor (1884), wissenschaftl. Mitarbeiter Staatsinst. f. Röntgenol. u. Radiol. Leningrad (U.d.S.S.R.), Biol. Labor., Röntgenstr. 6. *Biol. Wirkung des Radiums und der Röntgenstrahlen.* ⓓ Jurjew 1909.

Jastrow, Joseph (1863), Prof. of Psychol., Univ. of Wisconsin. Madison, Wis. (U.S.A.). ⓓ John Hopkins 1886. o

Jasudasen, Francis (1887), Research Department. Haffkine Inst. Parel, Bombay (India). *Bact.* ⓓ Madras Med. School 1914.

Jatschewa, Zdrawka (1896), Ass. Bact. Inst. Med. Fac. Sofia (Bulgarien), Zar Schischman 11. *Allgemeine u. spezielle Mikrobiol.* ⓓ Bern 1920.

Jatsynina, Claudine N., Chef des travaux Station de Défense des Plantes. Moscou (U.d.S.S.R.), Sadovaia-Trioumphalnaia 10. *Maladies des pommes de terre.* o

Javier, Ramon Q. (1894), Instr. Bact., Pathol. Los Baños College, Laguna (Philipp. Islands), College of Vet. Science. ⓓ Agric. Coll., Manhattan, Kan. 1924. ⓟ *Mus. collections; Pathol. slides.*

Javillier, Jean Maurice (1875), Maître de Conférences de Chimie biol. à la Fac. des Sc. de Paris; Dir. de Labor. à l'Inst. des Recherches agronomiques (Station centrale de Recherches sur l'Alimentation). Paris 15 (France), 16, Rue de l'Estrapade. *Les catalyseurs biochimiques (Diastases, Vitamines); les éléments catalytiques (Zinc, manganèse); métabolisme du phosphore.* ⓓ Paris 1919.

Jávorka, Sándor (1883), Dr., Abt.-Dir. a. d. Botan. Abt. des Ungarischen National-Mus. Budapest V (Ungarn), Akademia-ut 2. *Floristik: speciell Ungarische und Balkanische Flora.* ⓟ *Ungarische und balkanische Pflanzen.*

Jawłowski, Hieronim (1887), Ass. der Univ. Wilno (Polen), Zakretowa 15. *Myriopoda.* ⓓ Wilno 1926.

Jaworski, Erich (1890), ao. Prof. d. Univ. Bonn (Deutschland), Geol.-Palaeontol. Inst., Nußallee 2. *Fossile Lamellibranchiaten. Mesozoikum von Südamerika.* ⓓ Bonn 1912.

Jaworski, Alexander L., Leiter d. Botan. Abt. d. Mus. d. Jenissej-Gebiets. Krasnojarsk (U.d. S.S.R.), Kirowa 69. *Mykol.* o

Jaxa Bykowski, Ludwik (1881), o. Prof. de zool. et biol. générale à l'Acad. Vet. Lwów (Pologne). r. Bonifratrow 14. *Biometrie, Hydrobiol.* ⓓ Lwów 1904. ⓟ *Animaux parasit.*

Jaynes, Harold A. (1900), Ass. Entomol., U. S. Dept. of Agric. Shanghai (China), Box 1559. *Parasit. Japanese Beetle Control.*

Jazenko, Alexandra (1888), Ass. Inst. für experim. Biol. Moskau (U.d.S.S.R.), Woronzowo Pole 6. *Hydrophiol.: physiko-chem. Untersuchungen der Mantelhöhlenflüssigkeit der Mollusken. Physiol. der Lamellibranchiata. Einfluß der Faktoren des Außenmediums (Temperatur, pH, Licht, Alkohol) auf Rotatorien und Cladoceren.* ⓓ Frauenhochschule Moskau 1910.

Jazentkovskij, Evgenij Vladimirevič, Prof., Weißr. Inst. f. Landw. Minsk, Weißr. (U.d.S.S.R.). *Entomol.* o

Jazuta, Konstantin (1876), Prof. Anat. Univ. Rostow a. Don (U.d.S.S.R.), Anat. Inst., Fr. Engelsstraße 141. *Arteria mening. media bei d. Menschen u. Säugetieren. Os intermetatarseum. Abhängigkeit der Nierenlage vom Dickdarmgekröse. Hufeisenniere beim menschl. Embryo. Topographie der Carotiden. Krümmung des Jochbogens. Lageanomalie der Art. thyr. inf. Entwicklung der Flossenmuskulatur bei den Selachiern. Peritonealtaschen.* ⑰ Milit.-Med. Akad. zu Petersburg 1901.

Jeannel, René Gabriel (1879), Dir. Vivarium, Mus. Nat. d'Histoire nat. Paris, Prof. de Biol. générale à l'Univ. de Cluj (Roumanie). Paris V (France), 57, rue Cuvier. *Faune cavernicole, spécialement Coléoptères. Elevage d'Invertébrés vivants de tous ordres. Bathysciinae. Trechinae.* ⑰ Méd. Paris 1907, sc. 1911. ⑭ Coléoptères Cavernicoles.

Jedlička, Václav (1893), MUDr., Priv.Doc. der böhm. med. Fac. Prag, Ass. des pathol.-anat. Inst. Praha VI (C.S.R.), Albertov 2089. *Pathol. Anat., Endokrinol.* ⑰ Praha 1918. ⑭ Praep. aus der pathol. Histol. des endokrinen Apparates.

Jedliński, Ladislaus (1886), o.ö. Prof. der Hochsch. für Bodenkultur. Leiter der Lehrkanzel für Forsteinrichtung und Holzmeßkunde. Warszawa (Polen), ul. Hoża 74. *Forstbetriebseinrichtung mit besonderer Berücksichtigung ihrer naturwiss. Grundlagen und Zuwachslehre.*

Jedlitschka, Franz (1901), Ass. Pflanzenphysiol. Inst. deutsch. Univ. Praha XII (C.S.R.), Manesova Nr. 92. *Biochem.* ⑰ Prag 1926.

Jeffrey, Edward Ch., Prof. of Plant Morphol., Harvard Univ. Cambridge, Mass. (U.S.A.), 47 Lake View Avenue. *Anat. and morphol. of hybrids. Palaeontol., coal, anat. of extinct plants.* o

Jeffs, Royal E., Ass. Prof. of Botan., Univ. of Oklahoma. Norman, Okla. (U.S.A.), 427 Coll. Ave. *Physiol. mycol.* o

Jegin, Georg (1882), Entomol. Schweiz. Versuchsanstalt f. Obst-, Wein- u. Gartenbau. Wädenswil (Schweiz). *Parasit. Angewandte Entomol. Protist. Pathol. Anat.* ⑰ Basel 1905.

Jegozov, Michel (1892), Ass. Inst. de Biol. experim. Moscou (U.d.S.S.R.), Grande Jakimanka 47. *Liquide testiculair de Kravkov. Transplantations des glandes endocrines.* ⑰ Moscou 1914.

Jehle, Robert A. (1882), Ass. Plant Pathol., Univ. of Maryland. College Park, Md. (U.S.A.). *Plant Pathol.* ⑰ Cornell Univ. 1914.

Jeliaskowa, Anastassia (1897), Ass. am Zool. Inst. der Univ. Sofia (Bulgarien), Oboriste 13. *Entwicklungsgeschichte, Zellstimulation. Glaskörper der niederen Wirbeltiere. Beschleunigung der Teilungsrate von Paramaecium.* ⑰ Sofia 1921.

Jelin, Wladimir (1883), Prof. Odessa, Ukraine (U.d.S.S.R.), Med. Inst., Bact. Labor. *Biol. der säurefesten Mikroben.* ⑰ Odessa 1911.

Jelinek, Jan (1879), Dr., Prof. de l'amélioration des plantes. Praha-Bubeneč (C.S.R.), Belcredihov 260. *Agrobotan., amélioration des plantes. Qualité meunière et boulangère du froment.* ⑭ Semens des pl. cult.

Jellinek, Ant., Doc.Ing., Forstwirtschaftl. Abt. d. Deutschen Technischen Hochsch. Brno (C.S.R.). o

Jendrassik, Loránd (1896), Ass., Leiter der experim. Pathol. Labor. der med. Klinik, Univ. Pécs (Ungarn), Belklinika. *Allgemeine Pharmacol.; Nerven u. Muskelphysiol.; Physikalisch-chem. Physiol. und Pathol.* ⑰ Budapest 1921.

v. Jeney, Andreas (1891), Dr., Univ.Doc., Adjunkt am Inst. für allgemeine Pathol. und Therapie der Univ. Szeged (Ungarn), Riga u. 24/c. *Bact., Serol., Endokrinol., Biol. (Bakterienvariation, Bakterienantagonismen, lokale Immunität, Bakteriophage-Frage, Seuchenbekämpfung; Blutregeneration; eiweißfreie Leber- und Milzextrakte; Grundumsatz bei Anämie; hormonale Regelung der Magenbewegungen, Rassenbiol., Haemagglutination).* ⑰ Koloscvár (Transylvania) 1914.

Jenkins, Merle Truman (1895), Associate Agronomist, Bureau of Plant Industry, U.S. Department of Agric. Ames, Ia. (U.S.A.), Department of Farm Crops, Iowa State Coll. *Corn (maize) breeding and corn genetics.* ⑰ Iowa State Coll. 1925.

Jenkins, Oliver Peebles (1850), Coll. Prof. Stanford Univ., California (U.S.A.). *Physiol. nervous system Invertebrates.* o

Jenks, Albert Ernest (1869), Prof. Anthrop. Minneapolis, Minn. (U.S.A.), 1930 Emerson Av. S. o

Jennings, Herbert S. (1868), Henry Walters Prof. of Zool. and Dir. of the Zool. Labor., The Johns Hopkins Univ. Baltimore, Md. (U.S.A.). *Genetics of Protozoa and of Rotifera. Heredity in Vegetative Reproduction. The effect of Environment and Manner of Life on Later Generations.* ⑰ Ph. Harvard 1896, L.L. Clark 1909, Sc. Michigan 1918.

Jennings, Otto E., Curator of Botan. (in charge of Botan. and Paleobotan.), Carnegie Mus., Prof. of Botan., Univ. of Pittsburgh, Pa. (U.S.A.). *Systematic and ecol. botan.* o

Jennison, Harry Milliken (1885), Prof. of Botan., Univ. of Tennessee. Knoxville, Tenn. (U.S.A.). *Plant pathol., physiol.* ⑰ Massachusetts Agric. Coll. 1908. ⑭ Plants of Eastern Tennessee.

Jensen, Adolf Severin (1866), Prof. Zool. Univ., Dir. der Wirbeltierabt. des Zool. Mus. København (Danmark). *Arktische u. boreale Fische. Mollusken. Wissenschaftliche Fischereiuntersuchungen bei Grönland. Dänische Nagetiere.* ⑰ Lund 1918.

Jensen, C. E. O. (1859). København N. (Danmark), Nörrebrogade 22. *Moose.*

Jensen, C. O., Prof. Dr., Landw. Hochsch., Seruminst. Bülowsvej, København V (Danmark). o

Jensen, Hjalmar (1865), Lect. a. d. landw. Hochsch. København (Danmark), Rolighedsvej 23. *Pflanzenphysiol. Tabak.* ⑰ 1889.

Jensen, J. K. (1876), U.S.A." government Employee in Indian Educational Work. Santa Fé, N.M. (U.S.A.). *Nidification of birds, Experim. with Nestingboxes.* ⑰ 1894. ⑭ Birds Eggs in Clutches.

Jensen, Lloyd B. (1896), Instr., Hygiene and Bact. Univ. Chicago, Ill. (U.S.A.). *Bact.* ⑰ Univ. of Chicago 1924. ⑭ Bacteria.

Jensen, Paul (1868), o. Prof. der Physiol. u. Dir. des Physiol. Inst. der Univ. Göttingen (Deutschland), Wilhelm-Weber-Str. 39. *Allgemeine Physiol., Zellphysiol., Physiol. der Muskeln und der Protoplasmabewegung. Psychophysik.* ⑰ Jena 1892.

Jensen, Vilhelm Peter Hedlof (1870), Lect. der Bact. und Parasit. an der Univ. København (Danmark), Juliane Mariesvej 22. *Bact., Parasit. Pathogene Hefen.* ⑰ 1903. ⑭ Bact. und parasit. Praep.

Jensen-Haarup, Anders Christian (1863). Silkeborg (Danmark). *Hemiptera, Heteroptera and Homoptera.*

Jepson, Willis Linn (1867), Prof. of Botan., Univ. of California. Berkeley, Cal. (U.S.A.), 9 Mosswood Road. *Variation in Californian plants in relation to phylogeny, geographic distribution.* ⑰ California 1898.

Jermolajew, Wladimir (1905), Conserv. Zool. Mus. Univ. Tomsk, Sibirien (U.d.S.S.R.). *Spinnen.*

Jesenko, Fran (1875), o. Prof., Univ., Dir. pflanzenphysiol. Inst. und botan. Garten. Ljubljana (S.H.S.) univerza. *Cytol. der Gattungsbastarde, Saftsteigen der Pflanzen Plasmakolloide. Turgeszenzdauer abgeschnittener Pflanzensprosse.* ⑰ Wien 1902.

Jespersen, Poul (1891), Ass. Plankton Labor. of the Kommissionen for Danmarks Fiskeri- og Havundersögelser. København (Danmark), Hellerup, 34 Strandvej. *Marine Zooplankton: Geographical distribution and quantity, variation of Crustacea (Copepods, Mysids, Euphausids). Biol. of fishes: Clupea, Hippoglossus, Anguillidae and Sternopty-*

chidae. *Ornithol.: Geographical distribution and migration of birds.* ⓓ Copenhagen 1917. ⓟ Copepods, Mysids, Euphausids and other plankton organisms.

Jessen, Knud (1884), Dr. phil. København (Danmark), Gammelmønt 14, Geol. Survey. *Phytopaleont. of the interglacial and postglacial periods. History of the Anthophytes of Denmark. Phytogeography of Denmark. Arctic Flowering Plants. Ranunculaceae. Rosaceae. Immigration of trees and shrubs, history of the vegetation.* ⓓ Copenhagen 1920.

De Jesus, Zacarias (1896), Dr., Instr. in Vet. Hygiene, Coll. of Vet. Sc., Univ. of the Philippines. Los Baños Coll., Laguna (Philippine Islands). *Vet. Bact. and Meat Hygiene.* ⓓ Univ. of the Philippines 1924. ⓟ *Animal parasit. and cultures of bacteria.*

Jeswlet, Jacob (1879), Prof. für Botan. Systematik, Pflanzengeographie und Dendrol. a. d. Landwirtschaftl. Hochsch., Dir. des Arboretums. Wageningen (Holland), Dijkstraat 2. ⓓ Zürich 1913. ⓟ *Pflanzen, Samen, Diapositive bezüglich Pflanzengeographie und Systematik.*

Jezhikov, J. J. (1893), Priv.Doc., I. Univ., Zool. Mus. Moskau (U.d.S.S.R.). *Nachembryonale Entwicklung der Insekten, vollkommene Verwandlung. Polymorphismus und Variabilität der Ameisen.*

Jilford, Paul E. (1900), Ass. Plant Pathol. Wooster, Ohio (U.S.A.), Experim. Station. *Potato and Floral crop diseases.* ⓓ Ohio State 1926.

Jillnger, George Albert (1897), Ass. Entomol. Wooster, O. (U.S.A.), Ohio Agric. Experim. Station. *Greenhouse insects.* ⓓ Kansas State agric. Coll. 1925.

Jimbô, Tadao (1897), Member Tokugawa Inst. for Biol. Research. Hiratsukamachi, Ebaragun, Tôkyô (Japan). *Plant physiol. Nodulebacteria.*

Jirago, P. I., Doc. II. Univ. Moskau (U.d.S.S.R.). *Histol.* o

Joachimoglu, Georg (1887), ao. Prof., stellv. Dir. des Pharmacol. Inst. d. Univ. Berlin. Charlottenburg 2 (Deutschland), Hardenbergstr. 38. *Pharm.: Arsen, Benzol, Kampfer, Digitalis, Selen und Tellur, Chlorderivate des Methans, Äthans und Äthylens, Sublimat, Zinn- und Germaniumwasserstoff, Insulin.* ⓓ Berlin 1911.

Joakimoff, Dimitri (1864), Priv.Doc., Ass. Sofia (Bulgarien), Ul. Oboriste 13, Zool. Inst. der Univ. *Entomol.* ⓓ Genf 1895. ⓟ *Hemiptera aus Bulgarien.*

Joannidès, Georges (1893), Inst. Pasteur Hellénique. Athènes, Ambelokipi (Grèce), Rue Astériou No. 17. *Réactions d'immunité, cellulaires et humorales.* ⓓ 1914.

Joannović, Gjorgje, Dr., o. Prof. pathol. Anat. Beograd (S.H.S.), Višegradska ulica. *Pathol. Histol.* o

Job, Thesle Theodore (1885), Prof. of Anat. Chicago, Ill. (U.S.A.), 706 So. Lincoln St. *Anat. of Lymphatic syst. Anat. of Common Bile Duct. Bony changes in jaw s.* ⓓ State Univ. of Iowa 1917.

Jobling, Boris, General Sc. Worker, Wellcome Bureau of Sc. Research. London W.C. 1 (England), 25/28 Endsleigh Gardens. *Entomol.*

Jobling, James Wesley (1876), Prof. pathol. Columbia Univ. New York, N.Y. (U.S.A.), 228 W., 71 St. *Bact.* o

Jochems, Sarah Cornelis Johannes (1891), Dr. phil., Erste Botan., Deli Proefstation (Tabak). Medan, Sumatra (Nederl.-O.-Indië), Manggalaan 30. *Phytopath. u. Selection des Tabaks. Bodensterilisation.* ⓓ Amsterdam 1919. ⓟ *Tabakparasit. Pilze u. Bakterien. Sumatranische Pflanzen.*

Jochims, Johannes (1899), Chem. Ass. am Pathol. Inst. der Univ. Freiburg i. Br. (Deutschland), Stadtstr. 41 I. *Physikalisch-chem. Biol.* ⓓ München 1924.

Jodidi, S. L., Biochem., Bureau of Plant Industry, U. S. Dept. of Agric. Washington, D.C. (U.S.A.). *Plant Physiol., biochem.* o

Jodlbauer, Albert (1871), o. Univ.Prof., Dir. pharmacol. Inst. der tierärztl. Fac. der Univ.

München (Deutschland). *Pharmacol.* ⓓ München 1896.

Jörgensen, Carl Adolf (1899), Botan. Lyngby (Danmark). *Plant-pathol.: Mycol. and Bact. Genetics: Resistance, Sports and Chimaeras. Cytol.: Chromosomes, Heteroploidy.* ⓓ Copenhagen 1923.

Jörgensen, Eugen (1862), Lect. Fjösanger b. Bergen (Norge) oder Bergens Mus., Bergen. *Marines Plankton: Dinoflagellaten und Tintinniden; Hepaticae (Lebermoose).* ⓟ *Hepaticae.*

Jötten, Karl Wilhelm (1886), o. Univ.Prof. Hygiene, Dir. hygien. Inst. der Westfälischen Wilhelms-Univ. Münster i.W. (Deutschland), Westring 10. *Tuberkuloseforschung. Gonococcen und Meningococcen.* ⓓ Berlin 1912.

Joffe, Jacob S. (1887), Chem. and Bact., Soil Research. New Brunswick, N.J. (U.S.A.), N.J. Agric. Experim. Station. *Soil colloids; availability of nutrient cations from soil complex capable of base exchange.* ⓓ Rutgers Univ. 1922.

Johan, Béla (1889), Priv.Doc., Dir. Staatl. Hygien. Inst. Budapest I (Ungarn), Kelenhegyi-ut 33. *Pathohistol., Bact., Hygiene.* ⓓ Budapest 1912. ⓟ *Pathol.-histol. Praep.*

Johannsen, Oskar Augustus (1870), Prof. of Entomol. in the Coll. of Agric. of Cornell Univ. Ithaca, N.Y. (U.S.A.). *Insect Morphol. and Embryol. Biol. and Systematics of the Diptera.* ⓓ Cornell Univ. 1902.

Johannsen, Wilhelm Ludwig (1857), Dr., Prof. ordinarius der Pflanzenphysiol. an der Univ. Kopenhagen (Danmark), Gotersgade 140. *Genetik und Pflanzenphysiol. Geschichte der Biol.* ⓓ Kopenhagen und Lund, phil. Freiburg i. Br., bot. et zool. Groningen.

Johansen, Anders Cornelius (1867), Dir., Danish Biol. Station. Hellerup-København (Danmark) Strandvej 34, B. *Biol. and Oecol. of fishes and marine Invertebrates. Influence of Current upon the distribution.* ⓓ Copenhagen.

Johansen, Hans Christian (1897), Ass. am Kabinett f. vergl. Anat. und Zool. d. Staatsuniv. in Tomsk (U.d.S.S.R.). Reval (Estland), Dänisches Konsulat. *Ornithol. u. Zoogeographie des Süd-Ussurilandes in Ostsibirien. Physiographie und Biogeographie des Baikalsees.* ⓓ München 1924. ⓟ *Vogelbälge und Schmetterlinge.*

Johansen, Hermann Eduardowitsch (1866), Prof. vergl. Anat. u. Zool. der Wirbeltiere, Dir. Zool. Mus. Univ. Tomsk, Westsibirien (U.d.S.S.R.). *Ornithol., Ichthyol., Vertebrata Sibiriens. Lepidoptera. Phänol. Parasit. Copepoden (Lernaeopoda).* ⓟ *Vogelbälge, -eler.*

Johansson, Johan Erik (1862), Prof. der Physiol., Dir. Physiol. Labor., Karolinska Inst. Stockholm (Sverige), Kungsklippan 17 A. *Physiol. des Stoffwechsels, Muskelphysiol.* ⓓ Uppsala 1890, Laws Edinburgh 1923.

Johansson, Karl Ivar (1891), Instr. in Animal Husbandry Agric. Coll. Ultuna, Uppsala (Sverige). *Genetics in its relation to Animal Breeding; The shape of the lactation curve as a mendelian character.*

John, Friedrich (1894), Diplomlandwirt. Militsch, Schlesien (Deutschland). *Tierische Schädlinge an Kulturpflanzen.* ⓟ *Pflanzliche und tierische Schädlinge an Kulturpflanzen.*

St. John, Harold (1892), Associate Prof. of Botan., and Curator of Herbarium. Pullman, Wash. (U.S.A.), Botan. Dept., State Coll. of Washington. *Taxonomy of the vascular plants of northern North America. Plant ecol.: relation between the chem. nature of the soil and plant distribution.* ⓓ Harvard 1917. ⓟ *Herbarium specimens of vascular plants of northwest America.*

Johnsen, Sigurd (1884), Konserv. Bergens Mus. zool. avdeling (Vertebraten). Bergen (Norge). *Marine fishes: syst., eccl., distribution. Vertebrate fauna of Norway.*

Johnson, Aaron Guy (1880), Plant Pathol., Bureau of Plant Industry, U.S. Dept. Agric. Washington, D.C. (U.S.A.), Office of Cereal Crops and Diseases. *Cereal diseases, caused by Bacteria, Ascomycetes, and Fungi Imperfecti.* ① Univ. of Wisconsin 1914. ⓟ Helminthosporium diseases of barley.

Johnson, Charles Eugene (1880), Prof. of Forest Zool. and Acting Dir. of the Roosevelt Wild Life Forest Experim. Station, N.Y. State Coll. of Forestry, Syracuse, N.Y. (U.S.A.). *Vertebrate morphol. and ecol., mammals, birds, and reptiles.* ① Univ. of Minnesota 1912.

Johnson, Charles Willison (1863), Curator Boston Soc. Natural History. Brookline, Mass. (U.S.A.). *Mollusca, diptera.* o

Johnson, Duncan Starr (1867), Prof. of Botan. and Dir. of Botan. Garden, Johns Hopkins Univ. Baltimore, Md. (U.S.A.). *Morphol. of Archegoniates and Spermatophytes. Ecol. of maratime plants and desert plants.* ① Johns Hopkins Univ. 1897. ⓟ Jamaican Filicales, Hepaticae, and Algae.

Johnson, Edna Louise, Ass. Prof. of Biol., Univ. of Colorado. Boulder, Col. (U.S.A.). *Plant Physiol., Influence of X-rays upon plants.* ① Chicago 1926.

Johnson, Edward C., Dean and Dir., Agric. Coll. and Experim. Station, Washington State Coll. Pullman, Wash. (U.S.A.). *Agric. economics.* o

Johnson, F., Prof. Dep. of Educ. Coll. of Sc. Dublin (Ireland). *Botan.* o

Johnson, George Edwin (1889), Associate Prof. of Zool. and Experim. Station Mammal., Kansas State Agric. Coll. Manhattan, Kan. (U.S.A.), Zool. Department. *Giant Nerve Fibers of Crustaceans. Hibernation of Ground Squirrel. Habits and Methods of Control or eradication of Injurious Mammals.* ① Harvard Univ. 1923.

Johnson, James (1886), Prof. of Horticulture, Univ. of Wisconsin and Agent Bureau of Plant Industry, U.S. Department of Agric. Madison, Wis. (U.S.A.). *Plant Pathol. Diseases of Tobacco. Virus diseases of plants.* ① Univ. of Wisconsin 1918.

Johnson, John Christopher (1891), Prof. of Biol., Western State Coll. of Colorado, Dir. of the Rocky Mountain Biol. Station. Gunnison, Col. (U.S.A.). *Trematodes.* ① Univ. of California 1919. ⓟ High Mountain flora and fauna.

Johnson, John Thomas (1904), Ass. in Botan., Univ. of Illinois. Urbana, Ill. (U.S.A.), 308 Natural History. *Plant Pathol.*

Johnson, Maynard S. (1900), Ass. Prof. of Economic Zool. and Animal Biol. Univ. of Minnesota, Minneapolis, Minn. (U.S.A.). *Vertebrate compar. anat., economic zool.* ① Univ. of Illinois.

Johnson, Samuel Arthur (1866), Associate Prof. of Zool. Ft. Collins, Col. (U.S.A.), Colorado Agric. Coll. *Economic Entomol.* ① Rutgers Coll. 1895.

Johnson, Thomas (1863), Prof. of Botan., Univ. Coll. Dublin (Ireland), Glenmore, Dewell Park, Rathgar. *Palaeobotan. (especially Devonian and Tertiary Plants of Ireland); Algol.* ① London 1895. ⓟ Fossil Plants (Devonian and Tertiary).

Johnson, Thomas (1870), Dir., Virginia Truck Experim. Station, Prof. of Vegetable Gardening, Virginia Polytechnic Inst. Norfolk, Va. (U.S.A.). *Nutrition and production of vegetable plants.* ① West Virginia Univ. 1900.

Johnston, Charles George (1899), National Research Fellow of Med., Ass. Instr. in Physiol.Chem., School of Med., Univ. of Pennsylvania. Philadelphia, Pa. (U.S.A.). *Tumors, Ovaries and Vitamins. Studies on Blood.* ① Washington Univ. 1926.

Johnston, Charles Otis (1893), Ass. Pathol., Bureau of Plant Industry, U.S. Dept. in cooperation with the Kansas Agric. Experim. Station. Manhattan, Kan. (U.S.A.), Kansas State Agric. Coll. *Leaf rust of wheat, varietal resistance, inheritance of resistance, distribution of physiol. forms, overwintering, effect of rust on yield, and infection phenomena.* ① Kansas State Agric. Coll. 1924.

Johnston, Earl S., Associate Prof. of plant Physiol. and Associate Plant Physiol., Univ. of Maryland, and Maryland Agric. Experim. Station. College Park, Md. (U.S.A.). *Biophysics.* o

Johnston, John B. (1868), Prof. of Compar. Neurol. Minneapolis, Minn. (U.S.A.), Univ. of Minnesota. *Compar. anat. and physiol. of the nervous system of vertebrates; evolution of the forebrain.* ① Univ. of Michigan 1899.

Johnston, Thomas Baillie (1883), Prof. Anat., Guy's Hospital Med. School, Univ. London S.E.1 (England). *Human and compar. Anat. Embryol.* ① Edinburgh.

Johnstone, George R. (1888), Ass. Prof. of Botan. and Chairman of the Botan. Department. Univ. of Southern California (U.S.A.). *Plant Physiol.* ① Univ. of Chicago 1924. ⓟ Selaginella bigelovii. Ephedra.

Johnstone, James, Prof. of Oceanography, Univ. of Liverpool (England). *Parasit. and Pathol. of Fishes.* ① Univ. London.

Jolles, Adolf (1863), Dr. phil., Prof., Hon.Doc. an der Hochsch. für Welthandel in Wien, Leiter und Inhaber eines behördlich autorisierten chem.-mikroskopischen Labor. für med.-chem. und hygien. Untersuchungen. Wien IX (Deutsch-Österreich), Türkenstr. 9. *Chem. der menschlichen Se- und Excrete. Chem. der Nahrungs- und Genußmittel. Biochem. Arbeiten auf dem Gebiete der Fette, Kohlenhydrate und Eiweißkörper.* ① Breslau 1887.

Jollos, Victor (1887), Prof. Dr., o. Prof. der Zool. und Dir. des Zool. Inst. an der Egyptian Univ. Cairo (Ägypten), Department of Zool. (Vom 1.6. bis 15. 9.): Berlin-Dahlem (Deutschland), Kaiser-Wilhelm-Inst. f. Biol.). *Experim. Vererbungslehre u. Geschlechtsbestimmung, Protistenkunde.* ① München 1910.

Jolly, Justin Marie Jules (1870), Prof. Coll. de France. Paris XVI (France), 16, rue Copernic. *Histophysiol. des Tissus et organes formateurs du sang.* ① Paris 1925.

Jolly, William Adam, Prof. of Physiol., Univ. Cape Town (South Africa). *Reflex Action. Electrocardiogram. Lymphatic system.* ① Edinburgh 1906.

Jones, Charles R. (1879), Ass. State Entomol. Associate Prof., Dept. of Zool. and Entomol., Fort Collins, Col. (U.S.A.), Colorado Experim. Station. *Syrphidae. Formicidae. Aphididae.* ① Colorado Agric. Coll. 1926. ⓟ Syrphidae.

Jones, Dan Herbert (1875), B.S.A., Prof. of Bact. Guelph, Ontario (Canada), Ontario Agric. Coll. ① Univ. of Toronto 1908. ⓟ Bact. cultures.

Jones, David Breese (1879), Chem. in Charge, Protein Investigation Labor., Bureau of Chem. U.S. Dept. of Agric. Washington, D.C. (U.S.A.), 216—13th St., S.W. *Proteins, their isolation from various natural sources, their properties, amino acid composition and nutritive value: Methods of protein analyses and amino acid determination: Determination of the nutritive value of different proteins by feeding experim. with small animals (rats). Estimation of vitamins in foodstuffs.* ① Yale Univ. 1910.

Jones, David T., Chairman of Fishery Board for Scotland. Edinburgh (Scotland). o

Jones, Dettmar W. (1890), Associate Entomol., U.S. Dept. Agric. Arlington, Mass. (U.S.A.), 10 Court St. *Parasit. of European Corn Borer.* ① Massachusetts Agric. Coll. 1914.

Jones, Donald Forsha (1890), Geneticist, Connecticut Agric. Experim. Station. New Haven, Conn. (U. S.A.), P.O. Drawer 1106. *Heredity, its application to plant improvement.* ① Harvard Univ. 1918.

Jones, Edward N., Prof. of Botan., Baylor Univ. Waco, Tex. (U.S.A.). *Morphol. and biol. of aquatic seed plants.* o

Jones, Elizabeth (1898), Research Ass. Boston, Mass. (U.S.A.), 695 Huntington Ave. *Cancer. Inheritance of susceptibility factors in mice; experim. production of cancer in animals. Genetics of mammals; experim. modification of the germ plasm.* ⒹUniv. of Maine 1924.

Jones, Evan Thomas (1892), Univ. Coll. of Wales. Plant Breeding Stat. Aberystwyth (England). *Genetic: Oats.* ⒹUniv. Coll. of Wales 1921.

Jones, Fred Reuel (1884), Pathol., Bureau of Plant Industry, Washington, D.C. Madison, Wis. (U.S.A.), Univ. of Wisconsin, Horticulture Building. *Diseases of Leguminous Forage Plants, especially Alfalfa.* ⒹUniv. of Wisconsin 1917.

Jones, George Tallmon (1897), Instr. Botan. Oberlin Coll. Oberlin, Ohio (U.S.A.), 322 West Coll. Street. *Ecol., Mycol.*

Jones, Henry A. (1889), Associate Prof. Truck Crops, Univ. of California. Davis, Cal. (U.S.A.). *Sex Studies on asparagus; curly top resistance in Sugar beets.*

Jones, J. B., Prof. McGill Univ. Montreal (Canada). *Hygien.* o

Jones, James Paul (1898), Ass. Prof. of Zool., Department of Zool., Univ. of Tennessee. Knoxville, Tenn. (U.S.A.). *Herpetol., Lacertilia.*

Jones, J. Share, Prof., Univ. Liverpool (England). *Vet. Anat.* o

Jones, Jenkin William (1888), Associate Agronomist, Superintendent U.S. Rice Field Stat. Biggs, Cal. (U.S.A.). *Genetics of rice.* ⒹCalifornia 1924.

Jones, John McKinley (1886), Chief, Division Range Animal Husbandry, Texas Agric. Experim. Station. College Station, Texas (U.S.A.). *Nutrition and breeding of sheep, cattle and Angora goats. Inheritance of quality or fineness in Wool and Mohair. Shrinkage variation in wool and mohair fleeces.* ⒹUniv. of Missouri 1912.

Jones, John Paul (1897), Research Prof. of Agronomy, Massachusetts Agric. Coll. Amherst, Mass. (U.S.A.). *Tobacco and onions. Soil fertility, plant breeding.* ⒹCornell Univ. 1926.

Jones, Leon K. (1895), Associate in Research, New York Agric. Experim. Station. Geneva, N.Y. (U.S.A.). *Plant Pathol., diseases of peas, beans and tomatoes.* ⒹUniv. of Wisconsin 1922.

Jones, Lewis R., Prof. of Plant Pathol., Univ. of Wisconsin. Madison, Wis. (U.S.A.).

Jones, Lynds (1865), Prof. of Animal Ecol. in Oberlin Coll. Oberlin, Ohio (U.S.A.). *Animal Ecol., Ornithol.* ⒹUniv. of Chicago 1905.

Jones, Marcus E.Claremont,Cal.(U.S.A.) *Syst. bot.* o

Jones, Merlin Perry (1895), Instr. in Extension Entomol. Columbus, Ohio (U.S.A.), Botan. and Zool. Bldg. O.S.U. *Insecticides. Life history, systematic, habits and control of various insects.* ⒹOhio State Univ. 1924.

Jones, O. F. Prof. Vict. Univ., Manchester (England). *Geol. Palaeontol.* o

Jones, Samuel Griffith (1884), Lect. in Botan , Univ. Coll. of Wales. Aberystwyth (England). *Diseases of Plants.* ⒹBangor 1908.

Jones, W. Neilson, Prof. Bedford Coll. for Women. London N.W. 1 (England), Inner Circle Regents Park. *Botan.* o

Jones, Walter (1865), De Lamar Prof. physiol. chem., Johns Hopkins Univ.Baltimore,Md. (U.S.A.), Hopkins Apartments. o

Jonesco-Mihaïesti, Constantin (1883), Prof. Pathol. gén. Fac. Méd. Jassy, Dir. adjoint Inst. de Seruzi si Vaccinusi ,,Dr. J. Cautacuzino". Bucarest (România), 4 Splaiul C. Davila. *Pathol. expérim., Immunol., Sérol., Hématol.* ⒹBucarest 1907.

Jonescu, Constantin N. (1878), Maître de conférences, Fac. Sc., Labor. de Morphol. animale. Univ. de Jassy (Roumanie). *Faune des Insectes Collemboles de Roumanie, Terrestres et cavernicoles.* ⒹJena 1910.

Jonescu, Stan F., Doc. Univ., Maître de Conf. Fac. Sc. Bucarest (România), Labor. de Physiol. végétale, str. Cotroceni 38. *Anthocyanes des plantes. Tanins des fleurs de diverses plantes.* ⒹBucarest 1912.

de Jongh, S. E. (1898), Ass. Pharmaco-therapeut. Labor. Univ. Amsterdam (Holland), Röntgenstraat 15huis. *Pharmacol.* ⒹUtrecht 1921.

Jonnesco, Démètre (1898), Ass. Inst. V. Babes. Bucarest (România), Splaiul Davila 4. *Bact.* ⒹBucarest 1923.

Jons, J., Prof. Univ. Rosario (Argentina), *Hygiene.* o

v. **Joós,** Ilona (1893), Ass. Budapest (Ungarn). Hauptst. Hyg. u. Bact. Inst. *Bact. u. Serol.* ⒹBudapest 1918.

Jooss, Carlo H. (1883), Palaeont., Geol. Stuttgart (Deutschland), Seestr. 64. *Land- und Süßwasser-Mollusken aus Vergangenheit und Gegenwart.* ⒶFossile u. recente Land- u. Süßwasser-Mollusken. ⒹPrinceton Univ. 1907.

Jordan, David Starr (1851), Chancellor Emeritus of Stanford Univ., President from 1891 to 1916. Stanford Univ., California (U.S.A.). *Ichthyol.* ⒹL.L. Cornell Univ. 1888, L.L. Johns Hopkins Univ.

Jordan, Edwin Oakes (1866), Prof. of Bact. and Chairman of the Department of Bact. and Hygiene, The Univ. of Chicago (U.S.A.).

Jordan, Harvey Ernest (1878), Prof. of Histol. and Embryol., Med. School, Univ. of Virginia. Charlottesville, Va. (U.S.A.), 34 Univ. Place. *Striped muscle structure; blood; development of blood; Yolk. Mitochondria.* ⒹPrinceton Univ. 1907.

Jordan, Hermann Jacques (1877), o. Prof. der vergl. Physiol. an der Univ. Leiter der Abt. f. Vergl. Physiol. des Zool. Labor. Utrecht (Holland), Frans Halsstraat 19. *Vergl. Physiol. des Zentralnervensystems u. des Muskelsystems wirbelloser Tiere (Tonus glatter Muskeln). Atmung, Blut, Ernährung, Psychol., Naturphilosophie.* ⒹBonn 1901.

Jordan, Karl (1861), Curator of Entomol. in Zool. Mus. Tring, Herts. (England). *Entomol.: Lepidoptera, Siphonaptera, Anthribidae.* ⒹGöttingen 1885.

Jordanoff, Daki (1893), Ass. f. Botan. Sofia (Bulgarien), Botan. Inst. der Univ. *Systematik der Pflanzen und Pflanzengeographie. Pflanzenphysiol.*

von Jordans, Adolf (1892). Bonn a. Rh. (Deutschland), Marienstr. 13. *Ornithol. Vogelfauna Mallorcas. Sturnus vulgaris.* ⒹBonn 1914.

Jores, Leonhard (1866), o. Prof. allgem. Pathol. u. pathol. Anat., Dir. pathol. Inst. Univ. Kiel (Deutschland), Düppelstr. 25. *Arterien- und Nierenpathol.* Ⓓ1889.

Jörgensen, Pedro (1870). Villarrica (Paraguay). *Biol. of Lepidoptera (systématique).* ⒹCopenhagen 1892. ⒶInsekten und Pflanzen aus Paraguay und Argentina.

Jørstad, Ivar (1887), Government Mycol. Oslo (Norge), Botan. Mus. *General plant pathol., systematic mycol., especially rust fungi.* ⒹUniv. of Wisconsin 1920. ⒶParasit. fungi, especially rusts.

Jorstad, Louis Helmar (1896), Ass. in Pathol. and Research; Ass. in Surgery, Washington Univ. School of Med.; Ass. Bact., Washington Univ. School of Dentistry. St. Louis, Mo. (U.S.A.), Barnard Free Skin and Cancer Hospital. *Experim. Cancer and biol. research. Pathol., Bact.* ⒹWashington Univ. School of Med. 1924.

Joseph, Don Rosco (1881), Vice dean and Prof. physiol. and Dir. dept. U. School of Med. St. Louis, Mo. (U.S.A.), 5929 Julian Av. o

Joseph, Heinrich (1875), ao. Prof. der Zool. (Titel ord. Prof.), Univ. Wien IX (Österreich), Mariannengasse 32. *Vergl. Histol., Protozoenkunde, Zellenlehre, Hydrozoa.* ⒹPrag 1898.

Josephson, K. O., Doc. Univ. Stockholm (Sverige). *Biochem.* o

Josephy, Hermann (1887), Priv.Doc. an der Univ. Abt.-Arzt a. d. Staatskrankenanstalt Friedrichsberg. Hamburg 22 (Deutschland). *Pathol. Anat. des Gehirns.* ⒹRostock 1911.

Josifov, Gordei Maksimovitz, Prof., Dir. Anat. Inst. Univ. St. Voronesch (U.d.S.S.R.), Prospect Revolutii N. 2.

Josifov, J. M., Doc. Pädag. Inst. d. Bergrepublik. Vladikavkaz (U.d.S.S.R.), Ul. Marksa 20. *Anat. u. Physiol. d. Menschen.* o

Josifović, Mladen (1897), Maître de conferences de Pathol. Végétale à la Fac. d'Agric. et Silviculture de l'Univ. Beograd (S.H.S.), 54, Studenička. *Maladies des plantes cultivées et des arbres des forêts. Dégénerescence de la pomme de terre et les parasit. du chêne (Quercus sp.).* ⑨ Toulouse 1923. ⑰ Polyporaceae de Quercus pedunculata.

de Josselin de Jong, Rodolph (1868), o. Prof., Prof. der allgemeinen Pathol. und pathol. Anat., Dir. d. Pathol. Inst. Univ. Utrecht (Holland), Pasteurstr. 2. *Allgemeine Pathol. und Pathol. Anat.* ⑨ Leiden 1895. ⑰ Pathol.-anat. Praep.

Jost, Ludwig (1865), o. Prof. a. d. Univ. Heidelberg (Deutschland), Botan. Inst. *Pflanzenphysiol.* ⑨ Straßburg 1887.

Joszt, Dr. phil., Prof. Lwów (Polska), Dublany. *Biochim.* o

Joubin, Louis (1861), Prof. à l'Inst. Océanographique. Paris (France), 36, rue Geoffroy Saint Hilaire. *Zool. marine. Invertébrés: Mollusques, Echinodermes, Coelenterata.* ⑨ Paris-Sorbonne 1903. ⑰ Mollusques. Echinodermes. Coelentérés.

Joukow, Boris (1892), Doc. I. Univ. Moscou (U.d.S.S.R.), Boulevard de Zoubow 15, log. 22. *Paléoethnol. et Anthrop. préhistorique.* ⑨ Moscou 1921.

Jourdain, Francis C. R. (1865). Ditchingham, Norfolk (England), Waveney Lodge. *Ornithol. and Ool. Birds of the Palaearctic Region.* ⑨ Oxford 1890. ⑰ Eggs of Palaearctic Birds: Skins of birds.

Jourdan, E., Prof. Univ. et méd. Marseille (France). *Histol.* o

Jourdan, M., Prof. Fac. Sc. Marseille (France). *Zool.* o

Joyeux, Charles (1881), Dr. en Méd., Dr. ès Sc., Prof. agrégé à la Fac. de Méd. Paris, 6 (France), 15, rue de l'Ecole de Médecine. *Parasit.*

Jucci, Carlo (1897), Prof., Lib. Doc. Univ. Napoli (Italia), Ist. di Fisiol., S. Andrea delle Dame. *Genetica e Fisiol. dei Insetti: Termiti. Fisiol. e genetica dei bachi da seta (Bombyx mori).* ⑨ Sc. nat. Roma 1920, med. Napoli 1925.

Juday, Chancey (1871), Wisconsin Geol. and Natural History Survey; Lect. in Zool., Univ. of Wisconsin. Madison, Wis. (U.S.A.), Biol. Building. *Limnol., Zool.* ⑨ Indiana Univ. 1896. ⑰ Plankton material.

Juden, G. G., Prof., Dir. Anat. Inst. Univ. Smolensk (U.d.S.S.R.). o

Judin, A. A., Prof. Med. Inst. Moskau (U.d.S.S.R.) ul. Spartaka 2. *Physiol.* o

Juel, Hans Oscar (1863), Prof. der Botan. an der Univ. Upsala (Schweden), Botan. Garten. *Morphol. und Zytol. der Phanerogamen, Zytol. der Pilze. Typische u. parthenogenetische Fortpflanzung der Gattung Antennaria. Tetradenteilung. Entwicklungsgeschichte von Hippuris und Saxifraga. Blütenanat., Systematik der Rosaceen. Kernteilungen in den Basidien, Phylogenie der Basidiomyceten. Rheotropismus der Wurzeln.* ⑨ Upsala 1890.

Jullien, Antoine (1891), Ass. Physiol. Fac. Sc. Lyon, Rhône (France), St. Cyr au Mont d'Or. *Histol. physiol.* ⑨ Lyon 1923.

Jullien, M. A., Prép. Fac. Sc. Univ. Lyon (France). *Zool.* o

Jumelle, Henri, Prof. à la Fac. des Sc., Dir. du Mus. Colonial et du Jardin Botan. Marseille (France), 105, rue Edmond Rostand. *Flore tropicale et produits coloniaux.*

Jung, Frederic Theodore (1898), Ass. Prof. of physiol., Department of Physiol. and Pharmacol., Northwestern Univ. Med. School. Chicago, Ill. (U.S.A.), 303 East Chicago Avenue. *Glandulae parathyreoideae; nature of acidosis.* ⑨ Chicago 1925.

Jungeblut, Claus W. (1897), Ass. Prof. Bact. Stanford Univ. Palo Alto, Cal. (U.S.A.). *Bact. and Immunol.* ⑨ Bern 1921.

Jungmann, Wilhelm (1887), Dr. phil. nat., Botan. Erfurt (Deutschland), Dammweg 11. *Pflanzenphysiol., Samenzucht.* ⑨ Frankfurt a. Main 1920.

Junitski, Alexandre A., Prof. Phytopath. Univ. Kazan (U.d.S.S.R.). *Maladies de sarbres forestiers.* o

Junker, Hermann (1894), Dr. phil. nat. et cand. med., Ass. an der kolloidbiol. Station am Eppendorfer Krankenhaus. Hamburg 20 (Deutschland). *Biol. u. physiol. Wirkung extremer Verdünnungen. Vitaminforschung. Angewandte Kolloidchem.* ⑨ Freiburg i. Breisgau 1922.

Junkersdorf, Peter (1878), Dr. phil. et med., ao. Prof. d. Physiol. an der Univ. Bonn (Deutschland), Mozartstr. 34. *Physiol. Chem., Stoffwechselphysiol.* ⑨ Bonn phil. 1907, med. 1914.

Jurak, Ludwig, Dr., o. Prof. pathol. Anat. d. Haustiere, Dir. pathol. Inst. Zagreb (S.H.S.), Gundulićeva ulica 22. *Pathol. Histol.* o

Jurgeliunas, Prof. Univ., Dir. Hygien.-Bact. Inst. Kaunas (Litauen), Gedimino Gatve 29. o

Jurica, Edmund Joseph (1900), Prof. of Zool., St. Procopius Coll. Lisle, Ill. (U.S.A.). *Motility of the denervated mammalian Esophagus.* ⑨ Chicago 1926.

Jurica, Hilary Stanislaus (1892), Prof. of Botan. and Head of the Department of Biol., St. Procopius Coll. Lisle, Ill. (U.S.A.). *Morphol., Ecol. Dipsacus, Umbelliferae.* ⑨ Chicago 1922.

Jurisch, August Chr., Dr. med., Prosektor Anat. Mus. København (Danmark), Bredgade 62. o

Jurišić, Petar, Dr., Adjunkt Inst. f. allgem. u. experim. Pathol. u. Pharmacol. Zagreb (S.H.S.), Voćarska c. 97. *Pflanzenphysiol., physik. Chem. d. Zelle.* o

Jurriaanse, J. H. (1866). Rotterdam (Holland), Schiekade 75. *Systematik der exotischen Lepidoptera, Rhopalocera und Heterocera.* ⑰ Lepidoptera.

Jurukoff, Bogoja (1903), Ass. Bact. Inst. Med. Fac. Sofia (Bulgarien), Kresna 14. *Mikrobiol.* ⑨ Sofia 1926.

Just, Ernest E., A.B., Ph.D., Prof. of Zool. Howard Univ. Washington, D.C. (U.S.A.), 412 T Street, N.W. o

Just, Günther (1892), Dr. phil., Priv.Doc. für Zool. an der Univ. (Lehrauftrag für allgemeine Biol. und Vererbungslehre). Greifswald (Deutschland), Zool. Inst. *Vererbungslehre: Methoden, Drosophila, Crossing-over, Mensch (spez. Farbenblindheit, Methoden), Mutationen. Lichtsinn niederer Tiere, Tropismen. Biol. Begriffsbildung.* ⑨ Berlin 1919. ⑰ Drosophila-Mutanten.

Just, Jaroslav (1883), Prof. à l'Ecole polytechnique supérieure, fac. d'agric., et chef de l'Inst. expérim. agric. d'Etat pour la biotechnol. animale. Praha II (C.S.R.), Mezibranská 4. *L'alimentation des animaux domestiques, l'influence des differentes vitamines sur la production de lait des vaches laitières.* ⑨ Praha Sc. techn. 1906, phil. 1907.

Justov, N. A., Prof., Dir. Histol. Labor. Vet. Inst. Leningrad (U.d.S.S.R.). o

Jutras, Lorenzo Joseph (1882), Anat.-pathol., Inst. du Radium, Prof. d'Histol. et d'Embryol. à la Fac. dentaire de l'Univ. de Montréal (Canada), 2636 rue Ontario Est. *Cancer. Culture de tissu. Histopathol. avant, pendant et après le traitement du cancer par le Radium et les Rayons-X.* ⑨ Univ. Laval de Montréal 1907. ⑰ Préparations histol. et histopathol.

Juvová, Marie (1898), Ass. Ecole polytechnique Inst. des Fermentations. Brno (C.S.R.). *Microbiol. techn.*

Juzbašjan, S. M., Polytechn. „Lenin"-Inst., Techn. Hochsch. Tiflis (U.d.S.S.R.), Ul. Čavčavadze 3. *Zool. u. Anat. d. Tiere.* o

Juzepczuk, S. V., Cons. l'Herbier, Jardin Botan. princ. Leningrad (U.d.S.S.R.), Aplebarski Ostrov, Pessotchnaia $^1/_2$. *Rosacea, Alchimilla, Rubus.* o

Kabelik, Jan (1891), Prof. agrégé de l'Univ. „Masaryk" à Brno et prosecteur de l'Hôpital provinci 1 à Olomouc (C.S.R.), 24 Presslova. *Bact., sérol., colloide-chim., bactériophagie, la sérol. de la Tuberculose et Syphilis, la néphélométrie du sérum.* ⑨ L'univ. tchèque „Charles IV" à Prague 1918.

Kabonov, Nikolai (1902), Biol. Oka-Station. Murom, Gouvern. Wladimir (U.d.S.S.R.), Perwomajskaja, 4. *Algol.*

Kabrhel, Gustav, Dr., Prof. d. Hygiene u. Bact. Univ. Praha II (C.S.R.), Na bojišti 3, Hygien. Inst. o*

Kaburaki, Tokio, Ass. Prof. in Zool. of Coll. of Agric. Tôkyô Imper. Univ. Komaba near Tôkyô (Japan), Zool. Labor. *Turbellaria.* ⑨ Tôkyô.

Kac, Nik. Jakov, Doc. I. Univ. Moskau (U.d.S.S.R.). *Botan.* o

Kachiani, Nikolaus, Prof. Univ. Tiflis (U.d.S.R.). *Topograph., anat. u. operat. Chir.* o

Kadić, Ottokar (1876), Chefgeol., Doc. für Karstgeol. und Palaeont. der Wirbeltiere an der Univ. Budapest VII. (Ungarn), Stefánia-ut 14. *Palaeont. der Säugetiere und des Menschen (Palaeanthrop.).* ⑨ München 1900.

Kadocsa, Gyula (1880), Stationsleiter an der königl. ungar. Entomol. Station. Budapest II. (Ungarn), Kitaibel Pál-Gasse 1. *Angewandte Entomol. (Schädlinge und Nützlinge der Landwirtschaft, des Garten- und Weinbaues, Schädlinge des Haushaltes). Spezialstudium: Macro- und Microlepidopteren.* ⓟ Ungarische Schmetterlinge mit Schmetterlingen des palaearktischen Faunengebietes.

Kadykov, Josif Iljič, Doc. Inst. f. Landwirtschaft u. Melioration. Saratov (U.d.S.S.R.), Teatral'naja Ploščad. *Anat. u. Histol. d. Tiere.* o

Kändler, Rudolf (1899), wiss. Ass. an der Preuß. Biol. Anstalt Helgoland (Deutschland). *Biol. des Wattenmeeres; speziell Untersuchungen über die Biol. von Ostrea edulis und künstliche Austernzucht.*

Käsebier, Ants (1896), Tartu, (Estland), Raadi möis. *Phytophthora infestans, Schneeschimmel und Keimlingsbrand der Runkelrüben. Pflanzenpathol. Statistik.* ⓟ Pflanzenpathol. Praep.

Kästner, Alfred (1901). Leipzig C 1 (Deutschland), Talstr. 33, Zool. Inst. *Araneae, Pseudoscorpiones, Opiliones, anat. und physiol.*

Kahn, Eugen (1887), Priv.Doc. für Psychiatrie. München (Deutschland), Psychiatrische und Nervenklinik. *Psychiatrie, Erbbiol.* ⑨ München 1911.

Kahn, Reuben L. (1887), Immunol. and Ass. Dir. of Labor., Mich. Dept. of Health. Lansing, Mich. (U.S.A.). *Immunol. and Bact.* ⑨ New York 1916.

Kahn, Richard H., Dr., Prof. d. Physiol. Praha II (C.S.R.), Albertov 5, Physiol. Inst. d. Deutsch. Univ. o

Kaho, Hugo (1885), o. Prof. Pflanzenphysiol., Dir. Botan. Inst. Univ. Tartu (Estonia), Ülikool. Botan. Inst. *Verhalten der Pflanzenzelle gegen Elektrolyte: antagonistische Salzwirkungen; Permeabilität der Pflanzenzelle für Elektrolyte.* ⑨ Univ. Tartu 1923.

Kaiser, Johannes E. (1859), Prof. Dr., Leipzig-Li. (Deutschland), Kanzlerstr. 11. *Helminthol. Acanthocephalen.* ⑨ 1889.

Kaiserling, Carl (1869), Prof. d. Allgem. Pathol. u. pathol. Anat., Dir. Pathol. Inst. Univ. Königsberg i. Pr. (Deutschland), Kopernikusstr. 3/4. *Physik der Zellen u. Gewebe. Haftkrankheit. Krankheiten der Mundhöhle, Naevi.* ⑨ Berlin 1893.

Kajava, Yrjö Henrik (1884), Prof. der Anat. an der Univ. Helsinki (Finnland), Rauhank. 11. *Vergl. Myol. (Antmuskulatur des Säugetiere), Anthrop. (Somatol. der Finnen und Lappen).* ⑨ Helsinki 1914.

Kajiyama, Eiji (1886), Expert of Agric. and Forest. Expert of Imper. Fisheries Inst. (Suisan koshuje). Hiroshima (Japan), Ohcho Fishcult. Sta. *Seafish cult. and Seafish hatching.* ⑨ Tokio 1911.

Kakiuchi, Saburô, Prof. Biochem. of Coll. of Med., Imper. Univ. Tôkyô (Japan). *Biochem.* ⑨ Tôkyô.

Kalaitan, H. St., Doc. landwirtsch. Inst. Krasnodar, Kaukasus (U.d.S.S.R.), Novaja 107. *Bienenzucht.* o

Kalajda, F. K., Dir., Nikitsky Bot. Garten. Jalta (U.d.S.S.R.). *Akklimatisation u. Pflanzenkultur.* o

Kalantarian, Papa, Prof. Staatsuniv. Erivan, Armenien (U.d.S.S.R.). *Landwirtschaftslehre u. Bact.* o

Kalaschnikow, Leonid Nik., Ass. Inst. f. Landwirtschaft u. Melioration. Saratow (U.d.S.S.R.), Prigorskaja 31. *Geobotan.* o

Kalaschnikow, P. P., Ass. Zool. Inst. Univ. Leningrad (U.d.S.S.R.). o

Kalberg, W. A., Prof. Dr., Dir., Anat. Labor. Inst. f. Med. Wiss. Leningrad (U.d.S.S.R.). o

Kalitsky, Kasimir P., Obergeol. Com. Géol. d. Rus. Leningrad (U.d.S.S.R.), W. O., Sredny Pr. N. 72-B. *Entstehung des Naphta.* o

Kalkus, Julius Wilbur (1886), Superintendent of Western Washington Experim. Station of the State Coll. of Wash. and Vet. Pathol. of the Station. Washington (U.S.A.). *Animal Diseases.* ⑨ Kansas City 1909.

Kallenbach, Franz Joseph (1893), Lehrer. Darmstadt (Deutschland), Frankfurter Str. 57. *Mykol., speziell Boletaceae.* ⓟ Pilzexsikkate, insbesondere Boletaeceae, Polyporacea und Hydacenae.

Kallius, Erich (1867), Prof. d. Univ., Dir. des Anat. Inst. Heidelberg (Deutschland). ⑨ Berlin 1892.

Kalm, Prof. Dr., Milchwirtschaftl. Inst. Kiel (Deutschland). o

Kalmykov, Michail Proch., Prof. Omsky Med. Inst. Omsk (U.d.S.S.R.), Ul. Lenina 9. *Physiol.* o

Kalninš, Alfred (1894), Ass. Technical Mycol. Riga (Latvija), Univ., Fac. of Agric. *Soil Microbiol.: decomposition of cellulose by bacteria.* ⑨ Riga 1923.

Kalninš, Eduards, Dr., Doc. Univ. Riga (Latvija), Elizabetes ielā 103, dz. 24. *Neurol.* o

Kalocsay-Kalusza, Boguchwal Joseph (1884), Dr. phil. Poznań (Polska), Gimnazjum im. Bergera ul. Strzelecka 4. *Embryol., physiol. de la croissance et la mécanique du développement, l'influence des facteurs extérieurs et du milieu; le mécanisme de la division de la cellule.* ⑨ Cracovie 1923.

Kalshoven, Louis George Edmund (1892), Zool., Inst. for the Research of Plantdiseases. Buitenzorg, Java (Nederl.-O.-Indië). *Forest insect pests: insect enemies of teak (Tectona); teak termite (Calotermes tectonae); the mahogony twigborer (Hypsipyla robusta).* ⓟ Forest insect specimens of Java. Indo-malayan Coleoptera and Termitidae.

Kaltenbach, Paul (1878), Studienrat und Doc. an den akademischen Kursen. Düsseldorf (Deutschland), Hoffeldstr. 3 III. *Naturdenkmäler Deutschlands, insbesondere des Niederrheins.* ⓟ Photographische Aufnahmen von Naturdenkmälern, Baumaufnahmen (besonders Eichen).

Kalvaitytė, Paulina (1886), Prosektor allgem. Pathol. u. pathol. Anat. Univ. Kaunas (Litauen). *Wirkung der Roentgenstrahlen auf das Lymphogranulom.*

Kalwaryjski, Bernard Eugeniusz (1890), I. Ass. an dem Histol. Embryol. Inst. der Univ. Lwów (Polska), Piekarska 52. *Plexus chorioideus, Liquor cerebrospinalis. Einfluß chem. Agentien auf die Geschlechtselemente in frühen Entwicklungsstadien der Anuren. Sinneszellen und Mundhöhlendrüsen. Morphogenese.* ⑨ Lwów 1922. ⓟ Histol. Praep.

Kamada, Takeo (1901), Ass. of Zool. Inst., Sc. Fac., Imperial Univ. Tôkyô (Japan). *Physiol. of the animal, vegetative functions in the unicellular organism.* ⑨ Imp. Univ. Tôkyô 1925.

Kamat, Shamrao (1900). Bombay (India), Haffkine Inst., Parel. *Chemotherapy of Plague.* ⑨ Berlin 1925.

Kamel, Senji (1893), Ass. Prof. of Dendropathol. Coll. of Agric., Hokkaidô Imperial Univ. Sapporo (Japan), Botan. Inst. *Rust-fungi parasit. on Ferns, Polyporaceae.* ⓓ 1917.

Kamenický, Charles (1894), Chef Station pomol. Prâhonice u. Prahy, Inspecteur des Inst. des Recherches Agronomiques. Praha XII (C.S.R.), Korunní 60. *Biol. des arbres fruitiers, des fruits. Ontogenie, variabilité. Croissement. Pomol.* ⓓ Prague 1923. ⓟ *Des variétés fruitières (greffes).*

Kamensky, Constantin W. (1884), Inst. d'Essais des Sémences du Jardin Botan. Princ. Leningrad (U.d.S.S.R.). *Morphol. des sémences et methodol. d'essais des sémences.*

Kamerling, Zeno (1872), Landbouwkundige. Leiden (Holland), Croneskinkade 2. *Physiol. und Biol. der tropischen Pflanzenwelt; Verdunstung und Wasseraufnahme.* ⓓ Jena 1897.

Kaminsky, G., Prof. Univ. Rosario (Argentina). *Anat.* o

van Kampen, P. N. (1878), o. Prof., Univ. Leiden (Holland), Dir.Zool.Labor. *Vergl.Anat. und Ontogenie des Schädels. Systematik der Amphibien des Indischen Archipels. Zoogeographie.* ⓓ Amsterdam 1904.

Kampmeier, Otto Frederic (1888), Associate Prof. of Anat. Coll. of Med., Univ. of Illinois, Chicago (U.S.A.). *Embryol. of Vascular System, Kidneys.* ⓓ Phil. Princeton Univ. 1912, med. München 1924.

Kan, J. Lw., Doc. I. Univ. Moskau (U.d.S.S.R.). *Zool. Embryol.* o

Kanai, Toshio, Ass. Utsunomiya (Japan), Coll. of Agric. *Botan.*

Kanajew, Iwan (1893), Naturwiss. Inst. Peterhof bei Leningrad (U.d.S.S.R.). *Experim. Zool. u. Genetik. Histol. Vorgänge bei der Regeneration von Pelmatohydra.* ⓓ Petersburg 1918. ⓟ Hydren aus der Umgegend von Peterhof.

Kanaseki, J., Ass. Prof. of Imp. Univ. Kyôto (Japan), Anat. Inst., Coll. of Med. *Anat.*

Kančaveli, Zacharias, Doc. Univ. Tiflis (U.d. S.S.R.). *Botan.* o

Kanda, Sakyô (1874). Tôkyô (Japan), 468 Kami-Ochiai. *Physico-Chem. Studies on Bioluminescence, Cypridina hilgendorfii (Japanese Ostracoda).* ⓓ Univ. of Minnesota 1915. ⓟ Cypridina Hilgendorfii; Sergestes similis Hansen; Cavernularia habereri Moroff; Watasenea Scintillans Berry; Chaetopterus and other luminous animals in Japan.

Kangro, Karl, Dr., Doc. Univ. Rīga (Latvija), Dzīrnavu ielā 59, dz. 2. *Anat. d. Tiere.* o

Kankaanpää, Wäinö (1879), Doc. der Haustier-hygiene an der Univ. Helsingfors (Finnland), Med.-Oberverwaltung. *Tuberkelbazillen. Pathol. Anat. d. Knochenmarkes u. d. Lymphdrüsen.* ⓓ Phil. Leipzig 1911; Helsingfors med. vet. 1916.

Kanouse, Bessie Bernice (1889), Curator of Cryptogamic Collections, Herbarium of the Univ. of Michigan. Ann Arbor, Mich. (U.S.A.), 1236 Prospect Street. *Fungi: Phycomycetes, Ascomycetes.* ⓓ Univ. of Michigan 1926. ⓟ Phycomycetes and Ascomycetes.

Kant, Anton (1875), Niederl.-Ind. Ärzteschule. Soerabaja, Java (Nederl.-O.-Indië), Sumatrastraat 23. *Fungi, Phytopath., Erblichkeitslehre der Pflanzen.* ⓓ Amsterdam 1906.

Kantardjieff, Assen (1898), Ass. an dem Inst. für Allgem. Tierzucht bei der Landw. Fac. der Univ. Sofia (Bulgarien), Bulevard Skobelew 11a. *Molkereibact.*

Kapthammer, Joseph (1888), Dr. phil., Dr. med., Priv.Doc., Ass. am Physiol.-chem. Inst. der Univ. Leipzig (Deutschland), Liebigstr. 16. *Eiweiß-Konstitution, Eiweiß-Bausteine, intermediärer Stoffwechsel.* ⓟ Phil. Erlangen 1916, med. Berlin 1921.

Kaplanskij, Sam. J., Doc. I. Univ. Moskau (U.d.S.S.R.). *Pathol.* o

Kappen, Hubert (1878), Prof. Dir. d. chem. Inst. d. landw. Hochsch. Bonn (Deutschland), Hohenzollernstr. 6. *Agrikulturchem.* o

Kapper, Woldemar (1885), Ass. des Forstinst. Leningrad (U.d.S.S.R.), Forstinst. q. 16. *Samenkunde, Forstkulturwesen. Keimprüfungen.* ⓟ Waldsamen.

Kappert, Hans (1890), Dr. phil., Saatzuchtleiter der Gebr. Dippe AG. Quedlinburg a. Harz (Deutschland), Neuer Weg 17. *Koppelungsphaenomene bei Pisum.* ⓓ 1914.

Karakulin, Boris P. (1888), Botan. Garten. Leningrad (U.d.S.S.R.), Pessotschnaja 2. *Fungi imperfecti: Melanconiales.* ⓓ Leningrad 1912.

Karaman, Stanko (1889), Leiter d. zool. Mus. Skoplje (S.H.S.), Prosvetni Dom. *Vertebrata, spez. Ichthyol., Herpetol.* ⓓ Zagreb 1921.

Karamnickij, Nic. Alex., Doc. I. Univ. Moskau (U.d.S.S.R.). *Botan.* o

Karásek, François (1902), Ass. à la Fac. de Méd. de Prague. Praha II (C.S.R.), Kateřinská 32. *Physiol. humaine. Sécrétions internes. Physiol. du cœur et du muscle.* ⓓ Prague 1926.

Karawajew, Wladimir (1864), Dir. Zool. Mus. Ukrain. Akad. d. Wiss. Kiew (U.d.S.S.R.), Tschudnowskystr. 2. *Systematik und Biol. der Ameisen des palaearktischen und indo-australischen Faunengebietes.* ⓓ Kiew 1890.

Karch, Kristian, Prof. Dr., Volkswirtschaftl. Univ. Budapest (Ungarn). *Pflanzenzucht.* o

Kardassewitsch, Boleslaus Josephus (1895), Ass. Histol. und Emybrol. Med. Fac. Odessa, Ukraina (U.d.S.S.R.), Wnieszniaja 66/1. *Speicheldrüsen, Pankreas.*

Karell, Antal (1899), Ass. Debrecen (Ungarn), Simonyi-ut 16. *Biochem.* ⓓ Polytechnikum Budapest 1924.

Karelskaja, Anna F., Forschungs-Inst. d. Landw. Woronesh (U.d.S.S.R.). *Pflanzenphysiol. Microbiol.* o

Karling, John Sidney (1898). New York City (U.S.A.), Columbia Univ. *Mycol. Nuclear and Cell Division. Influence of Light and Temperature on Growth and Reproduction. Characeae.* ⓓ Univ. of Texas, Austin, Tex., 1919. ⓟ Characeae.

Karlsen, Astrid (1889), Amanuensis. Bergens Mus. Bergen (Norge). *Soil bact., denitrification.* ⓓ Oslo 1922.

Karny, Heinrich Hugo (1886), Ass. am Zool. Mus. Buitenzorg, Java (Nederl.-O.-Indië). *Thysanopteren, Psociden und Orthopteren (spez. Tettigoniiden) des malayischen und austro-malayischen Faunengebietes; Gryllacriden der ganzen Welt.* ⓓ Phil. Wien 1909; med. 1915. ⓟ Insekten.

Karo, Ferdinand (1845), Maître en pharmacie. Bibliothécaire de Mus. de la Société pharmaceut. de Pologne. Warszawa (Polska), r. Hoża 22. *Botan. systématique.* ⓓ Varsovie 1867.

Karper, Robert Earl (1888), Ass. Dir., Texas Agric. Experim. Station, Agronomist in Charge Small Grain Investigations. College Station, Tex. (U.S.A.). *Inheritance of characters in grain sorghums.* ⓓ Kansas State Agric. Coll. 1914. ⓟ Improved varieties, strains and hybrid material of grain sorghums.

Karpetschenko, George Dmitriewitch (1899), Dir. Genetic Labor. at the Inst. of Applied Botan. and New Cultures. Leningrad (U.d.S.S.R.), Detskoe Selo. *Interspecific crosses in Gramineae and Cruciferae and experim. studies on polyploidy.*

Karpinsky, Alexander P., Präsident der Akad. d. Wissenschaft. U.d.S.S.R., Prof., Bergingenieur, Ehrendir. des Com. Géol. de Russie. Leningrad (U.d. S.S.R.). *Palaeont. des Palaeozoicum. Edestidae, Trocholiten etc.* o

Karpoff-Benois, Elisabethe (1892), Ass. Inst. de Jaczewski de Mycol. et Pathol. végétale. Leningrad (U.d.S.S.R.), Perspective Anglaise 29. *Phytopath. forestière.*

Karpov, V., Prof. Med. Inst. Ekaterinoslav, Ukraine (U.d.S.S.R.). *Norm. Histol. u. Embryol.* o

Karpow, Wladimir (1888), Ass. Moskau 8 (U.d. S.S.R.), Timirjasewskaja Akademia, Korpus Ribowedenja. *Landschaftliche Bedeutung unserer Vogelwelt. Beziehungen zwischen Ectoparasit. und ihren Wirten. Biol. der Schädlinge (Nacktschnecken). Winter- (Schnee-) Fauna. Naturschutz.* Ⓟ Moskauer Carabiden. Ectoparasit.

Karr, Walter G. (1892), Ass. Prof. Biochem., Graduate School of Med., Univ. of Pennsylvania, Chief Chem., Philadelphia General Hospital, Philadelphia, Pa. (U.S.A.), 34 th and Pine Sts. *Biochem.* Ⓓ Yale Univ. 1920.

Karsinkin, Georg (1900), Hydrobiol. Station am See „Glubokoje". Moskau (U.d.S.S.R.), Chlebnij per. N. 28; N. 2. *Hydrobiol. bzw. Biocoenol.*

Karsner, Howard T. (1879), Prof. of Pathol., School of Med., Western Reserve Univ. Cleveland, Ohio (U.S.A.), 2109AdelbertRoad. *Pathol. of kidney, of circulatory apparatus. Anaphylaxis.* Ⓓ Univ. of Pennsylvania 1903.

Karsten, Fritz (1884), Dr., Dir. des Tierseucheninst. der Landwirtschafts-Kammer Hannover (Deutschland), Vahrenwalderstr. 58. *Bekämpfung der Tuberkulose des Rindviehes.* Ⓓ Gießen 1909.

Karsten, George H. H. (1863), Dr. phil., Univ.-Prof., Dir. des Botan. Inst. der Univ. Halle-Wittenberg. Halle a. d. S. (Deutschland), Am Kirchtor 1. *Embryol. der Gnetumarten. Mangroveentwicklung und Oekol. Conjugatenentwicklung und andere Algen. Diatomeen : Entwicklung, Verbreitung und Oekol.* Ⓓ 1885.

Kartaševskij, Je. A., Prof. Inst. f. Med. Wiss. Leningrad (U.d.S.S.R.), 2 Sovetsky Str. 4. *Allgem. Pathol.* ○

Karusin, Peter Iv. (1864), Prof. d. syst. Anat., Dir. Anat. Inst. I. Univ. Moskau (U.d.S.S.R.), Granowskystr. 4; 15 oder Inst. d. normalen Anat. *Anat. d. Zentralnervensystems, plastische Anat., Geschichte d. Anat.*

Káš, Václav (1899), Ass. de l'Inst. biochim., Section de l'Inst. pour la production végétale. Praha II (C.S.R.), Na Struze 5. *Bact. agric., Microbiol. du sol.* Ⓓ Prague 1924.

Kasai, Mikio, Sc. of Ôhara Inst. for Agric. Research. Kurashiki near Okayama (Japan). *Phytopath.*

Kasanskij, Alexander S., Prof. Ural. Polytechn. Inst. Swerdlowsk (U.d.S.S.R.), Ul. Trotzkogo 45, W. 1. *Pflanzenphysiol.* ○

Kasanskij, A. N., Prof. Polytechn. Inst. Ivanovo, Voznessensk (U.d.S.S.R.). *Entomol.* c

Kasanzeff, Wladimir (1872), Prof. Leningrad (U.d.S.S.R.), Pesotschnaja 4, Wohn. 26. *Zool., Protozoa.* Ⓓ Zürich 1901.

Kaschenko, Nikolaj F., Akad. Ukrain. Akademie d. Wissensch. Kiew (U.d.S.S.R.). *Akklimatisation.* ○

Kasesák, Ödön (1901), Ass. Allg. Botan. Inst. Univ. Budapest VIII (Ungarn), Múzeum-Körút 4/a. *Pflanzenanat.*

Kaserer, Hermann (1877), o.ö. Prof. für Pflanzenbau an der Hochsch. für Bodenkultur. Wien XVIII (Deutsch-Österreich), Gerdhofer Str. 73. *Düngerlehre, Bodenbact., Ernährungsphysiol. der Bakterien, Ökol. der Kulturpflanzen, Fruchtfolgen, Weinbau (Schädlingsbekämpfung und Betriebslehre).* Ⓓ Wien 1901.

Kashkarov, Daniel N. (1878), Prof. Vertebrate Zool. and Compar. Anat. Univ. Taschkent (U.d. S.S.R.), Novaja 25. *Systematics and ecol. of Vertebrates in Middle Asia. Systematic of Rodents and zonal distribution of vertebrates (Life Zones, habitats, Associations). Histol. of fish bone tissue, and of „blasiges Stützgewebe".* Ⓓ Moskau 1917. Ⓟ Rodents, Reptiles, Birds of Turkestan.

Kaßmann, Franziska (1896), Dr. phil. Münster i. Westf. (Deutschland), Kampstr. 12. *Botan.: Chondriosomen und Chloroplasten.* Ⓓ Münster i. Westf. 1925.

Katagi, Ryûzô, Mitarbeiter im Labor. Med. Univ. Okayama (Japan), Pharmacol. Inst. *Experim. Pharmacol.* Ⓓ Med. Akad. Okayama 1910.

Katagiri, Hideo, Ass. Prof. Imper. Univ. Kyôto (Japan), Inst. of Agric. Chem., Fac. Agric.

Katase, Awashi, Prof. of Prov. Med. Univ. Ôsaka (Japan), Pathol. Inst. *Pathol. and experim. Pathol.*

Katić, Danilo, Dr., Kustos der botan. Abteilung d. Serbischen Landesmus. Beograd (S.H.S.), Miloša Velikog ulica 31. *Flora Serbiens.* ○

Katô, Gen-ichi, Prof. of Keiô-Gijuku Univ. Yotsuya, Tôkyô (Japan), Physiol. Inst., Coll. of Med. *Physiol.*

Katô, Takeo, Prof. of Imper. Univ. Tôkyô (Japan), Geol. Inst., Fac. of Sc. *Geol. Palaeont.*

Katschioni-Walther, Lydia (1889), Pflanzenphysiol. Versuchsstat. Detskoje Sseló. Leningrad (U.d.S.S.R.), Bolsch. Possadskaja 9, Wohng. 9. *Keimung in Abhängigkeit von p_H. Bedeutung des Chlors f. Buchweizen. Einfluß verschiedener Stickstoffquellen auf die Entwickelung und chemische Zusammensetzung der Pflanzen. Adsorption von Farbstoffen an Alumin. hydroxyd.*

Katsu, Yoshitaka (1897), Ass. Prof. Anat. Kyôto (Japan), Inst. of Anat., Coll. of Med., Furitsu Ikadaigaku. *Biol. Physicochem.*

Katsurashima, T., Ass. Prof. of Tôhoku Imper. Univ. Sendai (Japan), Pathol. Inst., Coll. of Med. *Pathol. and pathol. Anat.* Ⓓ Med. Kyôto 1919.

Katterfeldt, Nicolas O., Prép. Station Phytopath. du Jardin Botan. Leningrad (U.d.S.S.R.). *Mycol., Pathol. végétale.* ○

Kaucki, Tytus (1874). Lwów (Polska), r. Zdrovic 10 I st. *Macrolepidoptera Poloniae.*

Kauffman, C. H. (1869), Prof. of Botan. and Dir. of Univ. Herbarium, Univ. of Michigan. Ann Arbor, Mich. (U.S.A.), 1236 Prospect St. *Cryptogamic Botan. Taxonomy, physiol. and pathol. of Fungi.* Ⓓ Michigan 1907. Ⓟ Phanerogams, Fungi, Mosses.

Kaufman, Laura (1889), Dr. phil., Ass. au Département de Morphol. expérim. des animaux à l'Inst. National Polonais d'Économie Rurale à Pulawy (Pologne). *Métamorphose de l'axolotle. Avitaminose. Phénogénétique de la pigmentation chez les Lapins. Croissance. Culture des tissus.* Ⓓ Cracovie 1916.

Kaufmann, Otto (1896), Dr. rer. nat., wissenschaftl. Ass. an der Biol. Reichsanstalt für L. u. F. Rosenthal b. Breslau (Deutschland), Fliegende Station der Biol. Reichsanstalt (Zuckerfabrik). *Lebensgeschichte und Bekämpfung der Rübenfliege. Weißährigkeit der Wiesengräser und ihre Bekämpfung.* Ⓓ Heidelberg 1923.

Kaunitz, Paul (1884), Dir. Epidemiol. Inst. Sarajevo (S.H.S.), Epidemiol. Zavod. *Bact. Serol. Pathol. Histol.* Ⓓ Wien 1909. Ⓟ Bakterienstämme und histol. Praep.

Kaupp, Benjamin F. (1874), Avian Pathol. and Dir. of the Labor., North Carolina State Coll. of Agric. Raleigh, N.C. (U.S.A.), State Coll. Station. *Poultry.* Ⓟ Colorado Agric. Coll. 1909.

Kausky, Eugen, Dr., ao. Prof., Dir. Physiol. Inst. d. Med.Fac.Univ. Ljubljana (S.H.S.),Zaloška cesta. ○

Kavina, Karel (1890), Dr. der Botan. an der Polytechn. Hochsch., landwirtsch. Fac. Prag-Vinohrady 58 (C. S. R.), Villa Grébovka. *Mykol., Biol. u. Systematik der Hymenomyceten. Torfmoose. Lebermoose. Verzweigung der Bryophyten.* Ⓓ Prag 1912. Ⓟ Pilze und Bryophyta.

Kawaguchi, Eisaku, Ass. Prof. of Kyûshû Imper. Univ. Fukuoka (Japan), Seric. Labor., Coll. of Agric. *Insect Histol.*

Kawaguchi, Magojirô, Teacher biol. Fukuoka-Ken (Japan), Meizen-Kô, Kurume. *Ornithol.*

Kawai, Sakyô, Ass. Prof. of Aichi Med. Coll. Nagoya (Japan), Physiol. Inst. *Physiol.*

Kawakami, Zen, Prof. of Keiô-Gijuku Univ. Yotsuya, Tôkyô (Japan), Pathol. Inst., Coll. of Med. *Pathol.*

10*

Kawakita, Motozô, Prof. of Med. Univ. Niigata (Japan), Chem. Inst.

Kawamura, Rinya, Prof. of Med. Univ. Niigata (Japan), Pathol. Inst. *Pathol.*

Kawamura, Sei-Ichi. Takinogawa near Tôkyô (Japan), Kami-nakazato 11. *Mycol.*

Kawamura, Tamiji (1883), Prof. in Compar. Physiol. and Ecol. of Imperial Univ. Kyôto (Japan), Dept. of Sc. *Systematics of Syphonophora and Spongilla. Ecol., Physiol. of Nervous System, and Animal Psychol. Fresh-water Biol.* ① Tôkyô 1907.

Kawano, Usaburô, Prof. of 8 High School. Nagoya (Japan), Zool. Labor. *Compar. Anat.*

Kawase, Sôjirô (1885), Prof. Fac. Agric. Imperial Univ. Tôkyô (Japan). *Biochem. of carbohydrates.* ① Tôkyô Imperial Univ. 1910.

Kawashima, Katsujirô, Prof. Nagasaki-Ken (Japan), Shinjô, Nishi-ômura, Higashi-Sonoki-gûn. *Nutrition Physiol. of Silk-Worms.*

Kay, Herbert Davenport (1893), Dr., Beit memorial Research Fellow, Biochem. to Med. Univ., London Hospital. London E. 1 (England). *Biochem.* ① Cambridge 1925.

Kay-Movat, T. R., Prof. King Edward VII Coll. of Med. Singapore (India). *Physiol.* o

Kayser, Heinrich (1876), Generalarzt. Weimar (Deutschland), Deinhardtsgasse 17. *Immunitätsforschung. Biol. der Bakterien. Wasserbact. Epidemiol. und klinische Bact. des Paratyphus u. Typhus. Systematik d. Paratyphusbakterien.* ① Straßburg i. Els. 1900.

Kazanskij, V. I., Prof. Staatl. Med. Lunačarskij-Inst. Astrachan (U.d.S.S.R.). *Zool.* o

Kazancev, A. I., Prosektor, Dr. Anat. Inst. Univ. Irkutsk (U.d.S.S.R.). o

Kazina, Olga N., Prép. Inst. Jaczewski. Leningrad (U.d.S.S.R.), Perspective Anglaise 29. *Maladies du lin.* o

Kaznowski, Lucjau (1890), Leiter Abteilung für Pflanzenzüchtung Staatl. Inst. f. Landwirtschaft, Doc. f. Pflanzenzüchtung an d. Hochsch. Warszawa. Puławy (Polska), Inst. *Futterpflanzen.* ① Kiew 1915. ℗ Samengräser, -klee, -erbsen.

Kchelmer, Rosa Bor., Doc. Gosudarstv. Donskoj Univ. Rostov an Don (U.d.S.S.R.). *Anat.* o

Kearney, Thomas H. (1874), Physiol., Bureau of Plant Industry, U.S. Dept. of Agric. Washington, D.C. (U.S.A.). *Crop plants adapted to alkali soils and dry climate.* ① Arizona 1920. o

Kedrowsky, Boris (1898), Ass. Med. Zool. Univ., I. Univ., Zool. Mus. Moskau (U.d.S.S.R.), Herzenstr. 6. *Experim. Zytol. Struktur und Physiol. der Nahrungsaufnahme bei Opalina.*

Kedrówsky, Wasily Iwánowitsch (1865), Prof. Pathol. Anat. II. Univ. Moskau (U.d.S.S.R.), Pirogoffstr. 57. *Pathol. Anat., Mikrobiol. der Lepra. Tuberkulose.* ① Moskau 1896. ℗ Leprakulturen.

Keeble, Sir Frederick, Prof. Univ. Fac. Nat. Sc. Oxford (England). *Botan.* o

Keelshoven, L. G. E., Entomol. Inst. f. Pflanzenkrankheiten Buitenzorg, Java (Nederl.-O.-Indië). *Insekten Chinas.* o

Keene, M. F. L., Prof. Univ. London S.W. 7 (England), S. Kensington. *Anat.* o

Keeser, Eduard (1892), Dr. med., Ass. am pharmakol. Inst. Berlin NW 7 (Deutschland), Dorotheenstr. 28. *Adsorptionsvorgänge, Wirkungsbedingungen von Arzneimitteln im Organismus, spez. von optischen Isomeren u. Alkohol.* ① Berlin 1919. o

Kegel, Werner (1882), Dr. phil., Bremen (Deutschland), Braunschweiger Str. 5. *Pflanzenphysiol.* ① Göttingen 1904.

Kehl, Hermann (1886), Prof., Dr. med., Dir. des städtischen Krankenhauses. Siegen i. Westfalen (Deutschland). *Pathol. Anat.* ① Leipzig 1911.

Kehrmann, Friedrich (1864), Dr., Prof. an der Univ. Lausanne (Schweiz), Villa Electa. *Biol. der „Lycaena" (Bläulinge, Lepidoptera).* ① Basel 1887. ℗ Schmetterlinge der Schweiz.

Keibel, Franz (1861), Prof., Dir. Anat. biol. Inst. Univ. Berlin NW 6 (Deutschland), Philippstr. 12. *Entwicklungsgeschichte der Wirbeltiere.*

Keifer, Hartford H. (1902), Ass. Curator. San Francisco, Cal. (U.S.A.), Calif. Academy of Sc., Golden Gate Park. *Microlepidoptera.* ① Berkeley 1924.

Keilin, D., Lect. Univ. Magd. Coll. Cambridge (England). *Parasit.* o

Keiser, Nicolaus (1892), Ass. Zool. der wirbellosen Tiere und Hydrobiol. Univ. Taschkent (U.d.S.S.R.), Usbekistan. *Cladocera und Copepoda, Hydrobiol., Plankton.* ① Leningrad 1916.

Keissler, Karl (1872), Dr., Hofrat, Dir. der botan. Abt. des Naturhistorischen Mus. Wien I/1 (Deutsch-Österreich), Burgring 7. *Fungi (Flechtenparasit. und lichenoide Pilze). Phytoplankton.* ① Wien 1895. ℗ Fungi u. andere Kryptogamen.

Keith, Arthur (1866), Hunterian Prof. Conserv. of Mus., Royal Coll. of Surgeons of England. London W. C. 2 (England). *Anat. and Evolution of Higher Primates.* ① Aberdeen 1888.

Keith, Norman Macdonnell (1885), Associate Prof. of Med., Mayo Foundation, Univ. of Minnesota. Rochester, Minn. (U.S.A.), Mayo Clinic. *Pathol. Physiol.* ① Johns Hopkins 1911.

Keitt, George Wannamaker (1889), Prof. of Plant Pathol., Univ. of Wisconsin. Madison, Wis. (U.S.A.), Department of Plant Pathol. *Plant Pathol.*

Kekčeev, Krikor Chačaturovič, Doc. I. Univ. Moskau (U.d.S.S.R.). *Physiol.* o

Keler, Stefan (1897), Entomol. de la Section des Maladies des Plantes d'Inst. sc. d'agric. Bydgoszcz (Pologne), Zacisze 8. *Entomol. forestière.*

Kelham, Henry Robert. Instow near Barnstaple, N. Devon (England). *Ornithol.* o

Kellaway, Charles Halliley (1889), Dir. Walter and Eliza Hall Inst. Melbourne, Victoria (Australia), Hospital. *Compensatory Hypertrophy. Streptococcal Infection.* ① Melbourne Univ. 1911.

Kelle, Arthur (1882), Prof. Hochsch. für Berg- und Forstingenieure, Leiter der Lehrkanzel für Forstschutz. Sopron (Ungarn). *Forstentomol.* ① Budapest 1908.

Keller, Boris (1874), Prof. Botan. Landwirtschaftl. Hochsch. Woronesh (U.d.S.S.R.). *Oekol. Anat. und Physiol. (besonders der Xerophyten und Halophyten). Vergl. Physiol. Flora und Geobotan. d. russischen Steppen, Halbwüsten und Wüsten.* ℗ Pflanzen d. russischen Steppen, Halbwüsten und Wüsten.

Keller, Conrad (1848), o.ö. Prof. der Zool. an der Techn. Hochsch. (Bundeshochsch.). Zürich (Schweiz), Schaffhauser Str. 83. *Forstzool., Tiergeographie, Haustierzool., Spongiol.* ① Jena 1874.

Keller, Emilie (1898), Botan., Landwirtschaftl. Hochsch. Woronesh (U.d.S.S.R.). *Oekol. Anat. und Physiol. d. Pflanzen: Xerophyten und Halophyten. Geobotan.; russische Steppen, Halbwüsten, Wüsten.*

Keller, Karl, Univ. Prof. Wien III (Österreich), Neulinggasse 32. *Tierzucht.* o

Keller, Robert (1854), Konserv. der naturwiss. u. geographischen Sammlungen der Stadt Winterthur (Schweiz), Trollstr. 32. *Rosa und Rubus; Hypericum.* ① Jena 1877.

Keller, Rudolf (1855). Praha (C.S.R.), Panská 12. *Mikroskopischer Elektrizitätsnachweis, Elektrobiochem. Elektrohistol. Untersuchungen an Pflanzen und Tieren.*

Keller-Ponomarewa, Wanda R., Doc. Univ., Vet.-Inst. Kasan II (U.d.S.S.R.), Soldatskaja 47, W. 2. *Pflanzen-Physiol.* o

Kellerman, Karl Frederic (1879), Associate Chief, Bureau of Plant Industry, U. S. Dept. of Agric. Washington, D.C. (U.S.A.), Room 207, West Wing. *Bact. Plant Physiol.*

Kelley, Arthur Pierson (1897), Ass. Prof. of Botan. New Brunswick, N.J. (U.S.A.), Department of Botan., Rutgers Univ. *Plant distribution in relation*

to edaphic factors. Ⓓ Univ. of Pennsylvania 1923. Ⓟ Herbarium material, plants of New Jersey.
Kelley, John G. (1898), Labor. Ass. in Plant Pathol., at the Tobacco Station of the Florida Agric. Experim. Station. Quincy, Fla. (U.S.A.). *Tobacco diseases.* Ⓓ Univ. of Florida 1923. Ⓟ Cultures of the various diseases of tobacco.
Kellogg, Arthur Remington (1892), Associate Biol., Bureau Biol. Survey, U. S. Dept. Agric. Washington, D.C. (U.S.A.). *Pelagic Mammals (Sirenia, Cetacea and Pinnipedia), living and fossil. Economic and systematic Herpetol.* Ⓓ Univ. of Kansas 1915.
Kellogg, Chester Elijak (1888), Associate Prof. of Psychol., Gill Univ. Montreal (Canada). Arts Building. *Compar. Psychol.* Ⓓ Harvard Univ. 1914.
Kellogg, Claude R. (1886), Prof. of Zool., Fukien Christian Univ. Foochow (China). *Sericulture. Economic Entomol.* Ⓓ Univ. of Wisconsin 1918. Ⓟ Entomol. specimens.
Kellogg, James L. (1866), Prof. of Biol., Williams Coll. Williamstown, Mass. (U.S.A.). *Anat., habits of Lamellibranchiata, Mollusca.* Ⓓ John Hopkins 1892. o
Kellogg, Vernon (1867), Permanent Secretary, National Research Council. Washington, D.C. (U.S.A.). *American Insects. Evolution. Heredity.* Ⓓ Univ. of Kansas 1889.
Kelly, Harry M. (1867), Prof. of Biol. Mt. Vernon, Ia. (U.S.A.), Cornell Coll. *Trematodes. Mollusca.* Ⓓ A.M. Harvard 1893, L.L.D. John B. Stetson U. De Land 1911. o
Kelly, James P. (1885), Prof. of Botan., Pennsylvania State College, State College, Pa. (U.S.A.) *Phlox in cultivation. Genetics of Phlox drummondii, Solanum tuberosum.* Ⓓ Princeton Univ. 1920. Ⓟ Seeds of Phlox Drummondii. Vascular plants of central Pennsylvania.
Kelsheimer, Eugene Gillespie (1901), Ass. Entomol. Wooster, O. (U.S.A.), Ohio Agric. Experim. Station. *Economic Entomol., European Corn Borer Research.* Ⓟ Ames, Iowa 1925.
Kémal-Djénab, Halil (1876), Prof. de Physiol. Univ. Constantinople (Turquie), Rue Mutávelli 2, Sur la grand' Rue de Moda-Cadikeny. *Sécrétions internes, cardiol. physiol.* Ⓓ Constantinople 1909.
Kemp, Tage (1896), M.D., Ass. of the Inst. of general Pathol. of Univ. København P. (Danmark), Juliane Mariesvej 22. *Sexual characters in embryos. The appearance of bacteriophages in chicken embryos.* Ⓓ Copenhagen 1921.
Kempton, Forrest E., Associate Pathol., Office of Cereal Investigations, Bur. of Plant Ind., U.S. Dept. of Agric. Washington, D.C. (U.S.A.). *Plant Pathol.* o
Kendall, Edward Calvin (1886), Prof. of Biochem. Univ. of Minn. Rochester, Minn. (U.S.A.), 627, 8 Av. S.W. *Biochem., active constituent of the thyroid.* o
Kendall, William Converse (1861), Ichthyol., U.S. Bur. of Fisheries. Freeport, Me. (U.S.A.). o
Kendeigh, S. Charles (1904), Research ass., Baldwin Bird Research Labor. Oberlin, Ohio (U.S.A.), 136 Woodland Ave. *Life Histories of Birds, the house wren. Physiological Processes in the Live Bird. Plant and Animal Ecol. Plant Succession.*
Kendrick, James B. (1893), Ass. Botan., Purdue Agr. Exp. Station. Lafayette, Ind. (U.S.A.). *Plant Pathol. Vegetable and Truck Crop Diseases and Diseases of Cereals.* Ⓓ Iowa State Coll., Ames, 1925.
Kendrick,PearlL.(1890), Ass.Dir.Labor., in charge: Western Mich. Division Labor., Michigan Dept. of Health. Grand Rapids, Mich. (U.S.A.), Fuller Ave., N.E. *Bact. and Serol.*
Kenk, Roman (1898), Dr. phil., Univ.Doc. Ljubljana (S.H.S.), Zool. Inst. Univ. *Systematik, Morphol. und Regeneration der Tricladida (Turbellaria), Spelaeol. (zool.).* Ⓓ Graz 1921.
Kennedy, Clarence Hamilton (1879), Ass. Prof. of Entomol. Columbus, O. (U.S.A.), Zoo. and Ent. Dept., Ohio State Univ. *Morphol., Ecol. and Evolution of Insects. Odonata. Ephemera. Sensitivity of insects. Phylogeny and distribution.* Ⓓ Cornell Univ. 1919. Ⓟ Odonata.
Kennedy, Cornelia, Ass. Prof. Agric. biochem., St. Paul, Minn. (U.S.A.), Univ. Farm. *Nutrition of the animal. vitamins and growth.* Ⓓ Johns Hopkins 1919.
Kennedy, Patrick Beveridge (1874), Prof. of Agronomy and Agrostol. in the Experim. Station. Berkéley, Cal. (U.S.A.), 118 Hilgard Hall, Univ. of California. *Trifolium. Grasses, Wild Flowers of California.* Ⓓ Cornell Univ. 1899. Ⓟ Grasses and legumes, particularly Trifolium.
von Kennel, J., Prof. emer. Univ. Tartu (Estland), Aia t. 46. *Zool.* o
Kenoyer, Leslie Alva (1883), Prof. of Biol., Western State Normal School. Kalamazoo, Mich. (U.S.A.). *Plant Geography of Asia. Ecol. of Lake and Swamp Plants.* Ⓓ Univ. of Chicago 1916.
Képinow, Léon I. (1881), Inst. Pasteur, Paris, Labor. de Physiol., Section russe, Fac. des Sc. à la Sorbonne, Prof. agrégé à l'Académie de Méd. de Petersbourg. Paris (France), Boulevard Raspail 96. *Pathol. expérim.: Physiol. Synergismus von Hypophysisextract und Adrenalin.* Ⓓ Moscou 1912.
Kepner, William Allison (1875), Prof. of Biol., Univ. of Virginia. Charlottesville, Va. (U.S.A.), 29 University Place. *Animal histol., behavior of lower achordata.* Ⓓ Univ. of Virginia 1908.
Kerb, Johannes Wolfgang (1884), Dr. phil. et med., Ass.-Arzt am Städtischen Krankenhaus zu Danzig (Freistaat Danzig). *Physiol. Chem. Chem. und Biol. des intermediären Kohlenhydratstoffwechsels im gesunden und diabetischen Organismus.* Ⓓ Phil. Freiburg i. Br. 1907, med. Berlin 1919.
Kerbert, Coenraad (1849), Dir. der Königlichen Zool. Gesellschaft „Natura Artis Magistra" Amsterdam (Holland). *Tiergärtnerei.* Ⓓ Leipzig 1876, Upsala (med. h. c.) 1907. † Sept. 1927.
Kerbosch, M. G. J., Dr., Dir. Gouv. Chinaonderneming. Bandoeng, Java (Ned.-O.-Indië), Tjihjiroean. *Selection v. Chinabaum.* o
Kerdovsky, V. J., Prof. Med. Inst. Moskau (U.d.S.S.R.), Ul. Spartaka 2. *Pathol. Anat.* o
Kerling, Louise C. P. (1900), Ass. Botan. Labor. Univ. Leiden. Haag (Holland), 2. Schuytstraat 150. *Pathol. Anat. und Physiol. der Pflanzen.*
Kern, Frank Dunn (1883), Prof. of Botan., Dean of the Graduate School, The Pennsylvania State College, State College, Pa. (U.S.A.), 116 W. Fairmount Avenue. *Mycol., Uredinol.* Ⓓ Ph. Columbia 1911, Sc. Univ. of Porto Rico 1926. Ⓟ Uredinales.
Kern, Hermann, Dir. d. Stat. für Pflanzenschutz d. Min. f. Landw. Budapest II (Ungarn), Debröigasse 17. o
von Kerpely, Koloman (1864), Prof., Volkswirtschaftl. Univ. Budapest VIII (Ungarn), Eszterhazygasse 3. *Landwirtschaftl. Pflanzenbau. Anat. u. Physiol. des Weizenkorns. Tabakbau.* Ⓓ Keszthely 1884. Ⓟ Landwirtschaftl. Kulturpflanzen, Samen und Fruchtstände der Getreidearten, Mais, Sojabohnen, Baumwolle, Reis usw.
Kerr, Abram Tucker (1873), Prof. of Anat. and Secretary of the Ithaca Division of Cornell Univ. Med. Coll. Ithaca, N.Y. (U.S.A.), 116 Kelvin Place. *Anat.* Ⓓ Buffalo 1897.
Kerr, John Graham (1869), I.R.S., Regius Prof. of Zool., Univ. of Glasgow (England). *Morphol. of lower Vertebrates: Crossopterygii and Dipnoi. Morphol. of Siphonopoda (Cephalopoda); Nautilus and Spirula.* Ⓓ Cambridge 1896.
Kerridge, Phyllis Margaret Tookey (1901), Ass. Department of Physiol. and Biochem., Univ. Coll. London W.C. 1 (England), Gower Street. *Physical Chem. applied to Physiol.* Ⓓ London 1921.
Kerschner, Theodor (1885), Dir. naturwissenschaftl. Abteilung des oberösterreich. Landes-Mus. Linz a. d. Donau (Österreich), Prunerstr. 18. *Recente*

und diluviale Wirbeltiere. Ontogenese der Insekten. Copulationsapparat von Tenebrio. Ⓓ Graz 1913.
Kersten, Hans Ewald (1880), Med.R., Prof. Dr. med., Kreisarzt des Kreisarztbezirkes Gelnhausen-Schlüchtern. Gelnhausen bei Frankfurt a. Main (Deutschland). *Bact., soziale und Tropenhygien.* Ⓓ Kiel 1905.
Kessel, John F. (1894), Associate in Parasit., Union Med. Coll. Peking (China). *Med. Zool. Intestinal Protozoa of Man, Monkeys, Rats and Pigs.* Ⓓ Univ. of California 1923.
Kessler, Ad., Dr. med., Dir. Pharmacol. Inst. d. Tung Chi Univ.Shanghai (China), 22 A Burkill Road.
Kestner, Otto Heinrich (1873), o. Prof. der Physiol. Univ. Hamburg 20 (Deutschland), Krankenhaus Eppendorf. *Physiol. der Ernährung, Stoffwechsel, Klimatol.* Ⓓ Heidelberg 1896.
Kettle, E. H., Prof. Welsh Nat. School of Med. Cardiff, Wales (England). *Pathol. Bact.* o
Keuchenius, Pieter Emile (1886), Phytopath. Kisaran, Sumatra O.-K. (Ned.-O.-Indië). *Phytopath. d. Gummibaumes.* Ⓓ Univ. Utrecht 1910.
Kezer, Alvin (1877), Prof. of agronomy and farm mgr. Ft. Collins, Col. (U.S.A.). o
Khakhlof, Vitaly Andrew (1890), Prof. Experim. Zool. Tomsk (U.d.S.S.R.), State Univ. *Genetics, Experim. Morphol. Theoretic bee-keeping.* Ⓟ Skins of Siberian birds.
Khazanoff, Amram (1890), Dept. of Experim. Agric. and Dune Afforestation, P. J. C. A., Advisor, Division of Hortic., Agric. Experim. Stat., P. Z. E. Haifa (Palestine), Palestine Jewish Colonization Association. *Systematic experim. in Viticulture, Horticulture and Dune fixation and afforestation.* Ⓓ California 1916. Ⓟ Seeds, cuttings of native trees and vines.
Khodjaew, Baschirulla (1887), Ass. Station Séricicole. Taschkent (U.d.S.S.R.), Rue de Kafanow 1. *Sériciculture.*
Khomenko, Johann P., Prof., Geol. am Comité Géol. de Russie (Abt. des Fernen Ostens). Leningrad (U.d.S.S.R.), W.O., Sredny Pr., N. 72B. *Palaeont.; speziell Vertebrata, Mammalia, tertiäre Molluskenfauna des Sachalins.* o
Kiär, Johan (1869), Prof. der Palaeont. Univ. in Oslo (Norge), Geol. Mus. *Älteste Fische, Trilobiten, Korallen.* Ⓓ München 1896. Ⓟ Paläozoische Fossilien aus Norwegen. Silurische u. devonische Fische. Fossilien aus den verschiedenen Formationen Spitzbergens.
Kiær, Sven (1894), Chef de clinique à la division chirurgicales (C.) de l'Univ. de Copenhague. København (Danmark), Rigshospitalet. *Cultures of tissues. Action de la lumière ultra-violette sur les cultures-tissulaires in vitro.* Ⓓ Copenhague 1925.
Kiang, Peter Ch'ing, Prof. Shantung Christian Univ. Tsinan (China). *Bio-Chem.* o
Kibaltschitsch, W. P. (1891), Ass. med. Inst., Anat. Labor. Kiew (U.d.S.S.R.), Gogolenskaja Str. N. 49, W. 2. *Nervensystem der Blase bei Cavia.*
Kibbe, Alice L., Prof. of Biol., Carthage Coll. Carthage, Ill. (U.S.A.). *Taxonomy.* o
Kidd, Franklin (1890), Research Officer Department of Sc. and Industrial Research of Great Britain and Principal Ass. at the Low Temperature Research Station. Cambridge (England), St. Johns Coll. *Plant physiol.; plant growth, cellpermeability. Respiration of plants, physiol. diseases of fruit. Temperature and metabolic balance in plants. Effect of oxygen and Carbondioxides upon metabolism.* Ⓓ London 1915.
Kienholz, Raymond (1894), Instr., Dept. of Botan., Univ. of Ill. Urbana, Ill. (U.S.A.). *Beach vegetation and marine phanerogams of the tropics. Plant physiol. in relation to silviculture.* Ⓓ Univ. of Illinois 1922.
Kiesel, Alexander (1882), Prof. I. Univ., Wirkl. Mitglied des Botan. Inst. d. I. Univ., des Timiriaseff Inst., d. Forschungsinst. Polytechn. Mus., d. Chem.-

Pharmaz. Inst. Moskau (U.d.S.S.R.), Piatnitzkaja N. 48. *Biochem. u. Chem. Physiol. d. Pflanzen. Chem. Bestandteile d. Protoplasmas; pflanzl. Eiweißstoffe u. deren Umbau; pflanzl. Fortpflanzungszellen; Polysaccharide in Pflanzen.* Ⓓ Moskau 1904.
Kießling, Ludwig (1875), GMR., Dr., o. Prof. Techn. Hochsch. München (Deutschland). *Ackerbau, Landwirtschaftl. Pflanzenbau, Landwirtschaftl. Kulturpflanzenzüchtung einschl. Vererbungslehre.*
Kihara, Hitoshi, Ass., Prof. in Genetics of Coll. of Agric., Imper. Univ. Kyôto (Japan), Labor. of Agric. Biol. *Cytol. and genetics of plant.* Ⓓ Kyôto.
Kihara, Takusaburô, Ass., Prof. of Imper. Univ. Kyôto (Japan), Anat. Inst., Coll. of Med. *Anat.*
Kikuchi, Akio, Prof. Imper. Univ. Kyôto (Japan), Inst. of Agronomy, Fac. Agric. *Horticulture.*
Killian, Charles (1887), Maître de conférences de Botan. Alger (Algérie), Fac. des Sc. *Développement des Algues et Champignons. Pathol. végétale.* Ⓓ Fribourg 1911. Ⓟ Prépar. de Cryptogames.
Killip, Ellsworth P. (1890), Aid, Division of Plants, U.S.National Mus. Washington, D.C. (U.S.A.). *Taxonomy of phanerogams, Passifloraceae and Urticaceae of South America.*
Kimla, Rudolf, Dr., Prof. d. pathol. Anat. Praha II (C.S.R.), Preslova, Pathol. anat. Inst. d. Karls-Univ. o
Kimura, Arika (1900). Botan. Inst. der kaiserlichen Univ. zu Tôkyô (Japan). *Systematik der japanischen Salices.* Ⓓ Kais. Univ. Tôkyô 1925. Ⓟ Weidenexemplare.
Kimura, Onari (1885), Prof., Dir. Pathol. Inst. Sendai (Japan), Kita 6, Bancho 230. *Histol. und Histopathol. des peripherischen Nervensystems (bei Menschen und Säugetieren). Experim. Histopathol des Zentralnervensystems bei Säugetieren.* Ⓓ Tôkyô 1910.
Kimura, Ren, Ass. Prof. of Imper. Univ. Kyôto (Japan), Microbiol. Inst., Coll. of Med. *Microbiol. and Serochem.* Ⓓ Kyôto.
Kimura, Tokuzô (1875), Prof. of Nat. Sc. Univ. of Commerce, Lect. of Biol. of the Keiô Univ., Tôkyô Tôkyô-fu (Japan), 1649 Kami-igusa, Jogi-machi. *Organic Evolution.* Ⓓ Agric. Coll., Hokkaidô, Japan, 1901; Stanford Univ., U.S.A., 1906.
Kinashi, Entarô. Wakayama-Ken (Japan), Matsue-mura 998, Kaisô-gun. *Systematic botan.*
Kincaid, Irevor (1872), Executive Chairman, Department of Zool., Univ. of Washington. Seattle, Wash. (U.S.A.). *Entomol., Ichthyol.* Ⓓ Univ. of Washington 1901.
Kindred, James Ernest (1893), Associate Prof., Histol. and Embryol., Med. School, Univ. of Virginia. Charlottesville, Va. (U.S.A.). *Invertebrate hematol.* Ⓓ Univ. of Illinois 1918.
Kinel, Jan (1886), Secrétaire du Mus. Lwów (Polska), Rutowskiego 18. *Entomol., Coléoptères aquatiques (systématique et faunistique).* Ⓓ 1923. Ⓟ Coléoptères aquatiques.
King, Chalmers Jackson (1893), Associate Agronomist and Superintendent, U.S. Field Station. Sacaton, Ariz. (U.S.A.). *Acclimatisation of Cotton, rubber and sub-tropical plants. Breeding corn for alkali resistance.* Ⓓ Clemson Agric. Coll. 1913.
King, Charlotte M., Ass. Chief, Botan. Section, Iowa Agric. Experim. Station. Ames, Ia. (U.S.A.), Station A. *Ecol., Pathol.* o
King, H., Dep. Biochem. Nation. Inst. for Med. Research. London W.C. 2 (England), 15 York Buildings. *Biochem. and Pharm.* o
King, Helen Dran (1869), Ass. Prof. of Embryol. at the Wistar Inst. of Anat. and Biol. Philadelphia, Pa. (U.S.A.). *Sex determination. Inbreeding; domestication; hybridisation; Physiol.: growth and reproduction.* Ⓓ Bryn Mawr Coll. 1899.
King, J. Lyonel (1888), Entomol., U. S. Dept. Agric. Riverton, N.J. (U.S.A.). *Insect Parasit.* Ⓓ Illinois 1916.

King, Jessie L. (1881), Fellow in Anat., Univ. Rochester med. School. Rochester, N.Y. (U.S.A.), Strory memorial Hospital. *Mammalian reproduction.* ⓟ Cornell Univ. 1911.

King, Kenneth M. (1896), Entomol., Entomol. Branch, Dominion of Canada, Department of Agric. Saskatoon, Sask. (Canada), Dominion Entomol. Labor. *Cutworms (Noctuidae), Wireworms (Elateridae), Animal ecol.* ⓟ Univ. of Saskatchewan 1926.

King, Sohtsu Gee (1886), Curator Department of Conchol. Founder and Custodian of the Peking Labor. of Natural History. Peking (China), 11, Kaka Hutung Tungsze Pailou. *Marine Conchol.* ⓟ Molluscan shells.

King, Walter E. (1877), Ass. Dir., Med. Research and Biol. Depts., Parke, Davis & Co., Detroit, Mich. (U.S.A.), 2951 Atwater Street. *Bact.* ⓟ Detroit Coll. of Med. and Surgery, Detroit, Mich., 1914. ⓟ Cultures of Bact., slides, antiserum preparations, vaccines.

King, Willard V. (1888), Entomol., Bureau of Entomol.,.U. S. Department of Agric. Mound, La. (U.S.A.). *Mosquitoes, malaria, insects affecting the health of man; parasit. of sucking insects.* ⓟ Tulane Univ., New Orleans, La., 1915.

Kingsbury, A. N., Inst. Med. Research. Kuala-Lumpur, Fed. Malay States (India). *Pathol.* o

Kingsbury, Francis B. (1886), Chem. in Charge, Biochem. Labor. Metropolitan Life Insurance Co. New York, N.Y. (U.S.A.), 1 Madison Ave. *Blood and Urine Proteins, Carbohydrates. Excretion of Hippuric Acid as a test for renal function.* ⓟ Harvard Univ. 1914.

Kingsley, J. Sterling (1853), Prof. Emer. of Zool-Univ. of Illinois. Berkeley, Cal. (U.S.A.), 2500 Cedar St. *Compar. Zool. Vertebratae. Morphol.* ⓟ Princeton 1885. o

Kinnear, Norman Boyd, British Mus., Natural History. London S.W. 7 (England), Cromwell Road. *Ornithol.* o

Kinoshita, Hironô, Demonstrator of Tôkyô Imper. Univ. Koishikawa-Ku, Tôkyô (Japan), Botan. Inst., Fac. of Sc. *Plant Physiol.*

Kinoshita, Kumao. Takamatsu (Japan), Gobanchô 56. *Corals.*

Kinoshita, R., Ass. Prof. of Hokkaidô Imper. Univ. Sapporo (Japan), Pathol. Inst., Fac. of Med. *Pathol.*

Kinoshita, Shûta, Entomol. of Imper. Agric. Experim. Station. Nishigahara, Tôkyô (Japan), Entomol. Labor. *Entomol. (Thysanura).*

Kinoshita, Tôsaku, Prof. of Ôsaka Provincial Med. Univ. Ôsaka (Japan), Physiol. Inst. *Physiol.*

Kinsey, Alfred C. (1894), B. S., Sc. D., Assoc. Prof. Zool., Indiana Univ. Bloomington, Ind. (U.S.A.). *Taxonomy, phylogeny, life histories, distribution of gall wasps (Cynipidae).* ⓟ Harvard Univ. 1920. ⓟ Cynipidae and their galls.

Kinsley, Albert Thomas (1877), Kinsley Vet. Biol. Labor. Kansas City, Mo. (U.S.A.), 400 New Centre Bldg. ⓟ Kansas City Vet. Coll. 1904.

Kinzel, Wilhelm Joh. Al. Herm. (1863), Prof., RR. Bayerische Landesanstalt für Pflanzenbau u. Pflanzenschutz. München (Deutschland), Liebigstr. 25. *Biol. der Samenkeimung. Mikroskopische Kontrolle von Futter- und Nahrungsmitteln.* ⓟ Erlangen 1891.

Kionka, Heinrich (1868), o. Prof. Pharmacol., Dir. pharmacol. Inst. Univ. Jena (Deutschland), Beethovenstr. 32. *Narkose, Alkohol, Colloidchem. Toxikol.* ⓟ Breslau 1893.

Kiparissowa, Alexandra (1888), Ass. Anat. Inst. Univ. Kasan (U.d.S.S.R.). *Variationen des Gefäßsystems d. Eingeweide u. d. Halses.*

Kipen, Alexandre (1870), Prof. viticulture Inst. agronomique. Odessa (U.d.S.S.R.), Rue Elisabetinskaja (Ul. Schepkina) 9, log. 1. *Viticulture.*

Hybridation et selection de la vigne. Gréffage. Taille. ⓟ Montpellier 1890.

Kirby, Harold, Jr. (1900), Instr. in biol., Yale Univ., Osborn Zool. Labor. New Haven, Conn. (U.S.A.). *Protozool.; Protozoa in termites.* ⓟ Univ. of California 1925. ⓟ Slides of protozoa, especially those of ceratomia; termites.

Kirby, Robert Stearns (1892), Ass. Prof. Extension Plant Pathol. Pennsylvania State College, State College, Pa. (U.S.A.)., Botan. Department. ⓟ Ithaca, N.Y., 1923.

Kirchensteiņš, Auguste (1872), Prof. Mikrobiol. Riga (Latvija), Univ., Mikrobiol. Inst., 1 Kronvalda bulvd. *Entwicklung, Symbiose, Vitaminfrage bei d. Bakterien. Infektionskrankheiten.* ⓟ Riga 1922.

Kirchner, Reinhold (1873), Landwirtschaftsrat an der Staatl. Lehr- u. Versuchsanstalt für Weinu. Obstbau in Neustadt an der Haardt (Deutschland), Haardter Str. 6. *Theoretische u. angewandte Botan., Mykol., Bact. des Weinstockes u. der Obstgewächse. Gärungsphysiol.* ⓟ Breslau 1904.

Kirillowa, Natalie Vlatschislawowna (1882), Extern-Mitarb. Labor. f. experim. Biol. Swerdlov-Univ., Prof. und Vize-Präsident d. biol. Sektion d. Lehrstuhls f. Naturwissenschaft an derselben Univ. Moskau (U.d.S.S.R.), Malaja-Nikitskaja, Vspolny per. 7, Wohn. 1. *Morphogenetische Bedeutung der Schild- und Keimdrüse. Evolutionslehre.*

Kiritchenko, A. N., Konserv. Zool. Mus. Akad. d. Wissenschaften. Leningrad (U.d.S.S.R.). o

Kirk, H. B., Prof., M.A. Victoria Coll. Wellington (New Zealand). *Zool.* o

Kirk, Lawrence E. (1886), Prof. of Field Husbandry. Univ. of Saskatchewan. Saskatoon, Sask. (Canada). *Potato Varieties. Sweet Clover. Red Clover.* ⓟ Univ. of Saskatchewan 1922.

Kirkham, William B. (1882), Prof. of Biol., International Y.M.C.A. Coll. Springfield, Mass. (U.S.A.), 100 Mill Street. *Early mammalian embryol.* ⓟ Yale Univ. 1907.

Kirkwood, Joseph Edward (1872), Prof. of Botan. and Head of Department, State Univ. Missoula, Mont. (U.S.A.). *Ecol. of the Rocky Mountain flora.* ⓟ Columbia Univ. 1903.

Kirschstein, Wilhelm (1863). Berlin-Pankow (Deutschland), Neue Schönholzer Str. 13. *Mykol., Askomyzeten.* ⓟ Askomyzeten.

Kisch, Bruno (1890), o.ö. Prof. der Physiol. an der Univ. Köln a. Rh.-Lindenthal (Deutschland), Lindenburg. *Normale und pathol. Physiol. Kreislauf, Nervensystem. Physiko-chem. Methoden.* ⓟ Deutsche Univ. Prag 1913.

Kischkin, Michail Nikolaevitsch (1892), Kustos des K. A. Timiriasev-Biomus. Moskau (U.d.S.S.R.), Malaja Bronnaja 20a, W. 10. ⓟ Felle und Spiritus-praep. aus dem Nördlichen Ozean, Japanischen Meere, See Chanka, Mittelasien (Turkestan), Krim und Gouvernement Moskau.

Kisel, Alexander R., Prof. Dr., I. Univ., Vorsteher d. Abtlg. f. Pflanzen-Physiol. der Biol. Station. Moskau (U.d.S.S.R.), Pjatnizkaja 48. *Pflanzen-Physiol. u. Biochem.* o

Kiselev, A. O., Doc. Univ. Nizny-Novgorod (U.d.S.S.R.), Sovetskaja Ploščad 8. *Histol.* o

Kiselev, N. J., Prof. Vostočny y Pedagog Inst. Kazan, Tatar. Rep. (U.d.S.S.R.), Karl-Marx-Str. 47. *Zool.* o

Kiselev, N. N., Doc. I. Univ. Moskau (U.d.S.S.R.). *Botan.* o

Kishida, Kyûkichi, Biol. of Bureau of Agric., Dept. of Agric. and Forestry. Takinogawa near Tôkyô (Japan), Nakazato 132. *Spiders, Mammalia of Japan.*

Kishinoue, Kamakichi, Prof. Imper. Univ. Tôkyô (Japan), Hydrobiol. Inst., Fac. of Agric. *Ichthiol., Fishery.*

Kishitani, Teijirô (1896), Member Tokugawa Inst. for Biol. Research. Tôkyô (Japan), Hiratsukamura-

Koyama, Ebaragun. *Plant physiol.; Photobact., Myxobact.* Tôkyô 1925.

Kisker, Georg (1896), Dr. phil., Oberfischmeister für die Provinz Sachsen. Magdeburg (Deutschland), Oberpräsidium. *Fischereiliche Verhältnisse der Provinz Sachsen, Flußfischerei.* ⑨ Marburg 1920.

Kiss von Zilah, Endre (1873). Cehul-Silvanei (România), Jud. Salaj. *Hymenoptera: Ichneumonidae.* ⑨ Budapest 1901. ⑨ Palaearctische Hymenopteren.

Kissel, Z. M., Dr. Ass., Anat. Lab. Med. Inst. Leningrad (U.d.S.S.R.), Str. Lew Tolstoy. ○

Kisselew, N. N. (1886), Priv.Doc. I. Univ. Moskau (U.d.S.S.R.), Botan. Inst. *Pflanzenphysiol.*

Kisselewa, Elisabetha (1891), Ass. zool. Kabinett Univ. Tomsk (U.d.S.S.R.), Krassnoarmeyskaja N. 19, n. 2, oder Univ. *Entomol.: Coleoptera (Fam. Cerambycidae und Ipidae), Diptera (Fam. Culicidae), Ameisen.* ⑨ Tomsk 1917.

Kisser, Josef (1899), Ass. am pflanzenphysiol. Inst. der Univ. Wien XIII/4 (Österreich), Baumgartenstr. 93. *Botan. Mikrotechnik. Chem. Physiol. der Pflanzen. Anat. und Histol. der Pflanzen. Sterile Kultur höherer Pflanzen.* ⑨ Wien 1922.

Kisskalt, Karl (1875), o.ö. Prof. der Hygiene. München (Deutschland), Hygien. Inst. *Epidemiol., Disposition, Wasser, Abwasser, Gewerbehygiene.* ⑨ Würzburg 1898.

Kister, Julius (1870), Prof., Abteilungsvorsteher und stellvertret. Dir. am hygien. Staatsinst.; ao. Prof. der Hygiene an der Univ. Hamburg (Deutschland). *Infektionskrankheiten.* ⑨ Kiel 1894.

Kistiakowsky, Alexander (1904), Zool. Mus. der Ukrainischen Akademie der Wissensch. Kiew (U.d.S.S.R.), Tschudnowskystr. 2. *Ornithol. und Mallophaga von U.d.S.S.R.* ⑨ Mallophaga von U.d.S.S.R.

Kitajima, Taichi, Prof. of Kelô Gijuku Univ., Member of Kitasato Inst. of Infect. Diseases. Tôkyô (Japan), Bact. Inst., Coll. of Med., Yotsuya. *Bact. and Hygiene.*

Kitamura, Naomi, Ass. Prof. in Human Physiol. of Imperial Univ. Kyôto (Japan), Physiol. Labor., Coll. of Med. *Stoffwechselphysiol.*

Kitasato, Yenjirô, Dr. Ass. Tôkyô (Japan), Botan. Inst., Fac. of Sc., Imperial Univ., Koishikawa-Ku. *Biochem.*

Kittredge, Joseph, Jr. (1890), Silviculturist, Lake States Forest Experim. Station, U. S. Forest Service. St. Paul, Minn. (U.S.A.), Univ. Farm. *Forest research in the fields of silviculture and management.* ⑨ Harvard 1913.

Kivirikko, Karl Emil (1870). Helsinki (Finnland), Nervandergatan 11. *Cladocera; Ornithol.* ⑨ Helsingfors 1900.

Kiyohara, Kin, Demonstrator. Tôkyô (Japan), Botan. Inst., Fac. of Sc., Imperial Univ. Koishikawa-Ku.

Kiyono, Kenji, Prof. of Imp. Univ. Kyôto (Japan), Pathol. Inst., Coll. of Med. *Pathol., Pathol. Anat. and Microbiol.*

Kizel, Alex. Rob., Prof. I. Univ. Moskau (U.d. S.S.R.), *Botan.* ○

van der Klaauw, Cornelis Jakob (1893), Conserv. Zool. Labor. Univ. und Priv.Doc. Leiden (Holland). *Bulla auditiva rezenter und fossiler Mammalia, Ossicula auditoria rezenter und fossiler Amnioten und Amphibia: Entwickl. und erwachsener Zustand. Geschichte der Zool.* ⑨ Leiden 1922.

Klages, Karl Henry (1898), Ass. Prof. of Agronomy Oklahoma Agric. and Mechanical Coll. Stillwater, Okla. (U.S.A.). *Forage Crops, Crop Ecol. Adaptation of forage crops to Oklahoma conditions.* ⑨ Univ. of Illinois 1925. ⑨ Heads and Grain of the sorghums.

Klarenbeek, Arie (1888), Konserv. und Priv.-Doc. Tierärztl. Fac., Univ. Utrecht (Holland), C. Houtmanstr. 18. *Krankheiten kleiner Haustiere, Spirochaetenkrankheiten, klinisch, pharmacol., chirurgisch und röntgenol.* ⑨ Bern 1915.

Klarin, Eugen (1895), Vorst. d. Diagnost. Abt. d. Veterinärbakteriol. Staatsinst. Stockholm-Experimentalfältet (Sverige). *Bakt. u. Pathol. Anat.* ⑨ Stockholm 1921.

Kláštersky, Ivan (1901), Conserv.-Adjoint. Praha II (C.S.R.), Václavské nám. 1700. *Systématique des Phanérogames: Pedicularis, Achillea, Campanula. Sociol. végétale: Carpathes.* ⑨ Praha 1926.

Klatt, Berthold (1885), Prof. Dr. phil., Priv.Doc. der Zool., Custos der Säugetierabteilung am Zool. Mus. Hamburg I (Deutschland), Steintorwall. *Probleme der Domestikation. Fütterungsversuche an Tritonen. Vererbungsversuche am Schwammspinner.* ⑨ Berlin 1908.

Klautke, Doc. Univ. Woosung (China). *Botan. Zool.* ○

Klebahn, Henrich (1859), Hon.Prof. an der Univ. Hamburg (Deutschland), Inst. für allgem. Botan., Curschmannstr. 27. *Biol. der Pilze (Uredineen, Phycomyceten, Ascomyceten, Fungi imperfecti) und Pflanzenkrankheiten.* ⑨ Jena 1884.

v. Klebelsberg, Raimund (1886), o. Univ.Prof. Geol. u. Palaeont., Dir. Geol.-palaeont. Inst. Univ. Innsbruck (Österreich), Universitätstr. 4. *Palaeont. (Evertebrata, Faunistik): Fauna der produktiven Steinkohlenformation. Ammoniten. Evertebraten des alpinen Carbon.* ⑨ Wien 1910.

Klechetow, Anatoliy Nikol. (1894), Ass. Agronomist. Moscow (U.d.S.S.R.), Flax Exper. Station, Agric. Academia. *Plant Industry. Flax Diseases, culture, sick. Soil.*

Klein, Boris (1872), Abteilungsvorsteher Bact. Inst. Kiew (U.d.S.S.R.), Proreskaja 14. *Gärungserscheinungen und Immunität: Gärungsagglutination, bact. Nachweis von Zuckerarten und Kohlehydraten. Immunität. Serol. der Krebskrankheit. Malaria.* ⑨ Kiew 1900.

Klein, Edmund Jos. (1866), Prof. Luxemburg Äußerer Ring 20. *Pflanzengeographie von Luxemburg.* ⑨ Hymenophyllum.

Klein, Gustav (1892), ao. Prof. für Anat. u. Physiol. d. Pflanze, I. Ass. am pflanzenphys. Inst. d. Univ. Wien XIX/3 (Deutsch-Österreich), Kahlenbergerdorf. *Intermediärer Stoffwechsel der Pflanze, Histochem.* ⑨ Univ. Wien 1918.

Klein, Karl (1889), Dr. med., MR., Dir. des Med. Unters.-Amtes Düsseldorf (Deutschland), Gneisenaustr. 28. *Bakt. u. Hygiene, Bekämpfung der humanen Infektionskrankheiten.* ⑨ Bonn 1917.

Klein, Ludwig, o. Prof. Techn. Hochsch. G.Hof R. Karlsruhe (Deutschland), Kaiserstr. 2. *Botan.* ○

Klein, Marc (1905), Préparateur-stagiaire à l'Inst. d'histol. de la Fac. de Méd. de Strasbourg (France). *Histol., Histophysiol. des organes génitaux.*

Klein, Otto (1891), Priv.Doc., Ass. d. II. deutsch. med. Univ.-Klinik. Praha II (C.S.R.), Allgem. Krankenhaus. *Wasserhaushalt, Kohlehydratstoffwechsel, Insulinwirkung.* ⑨ Prag 1915.

Klein, Stanislaus (1863), Doc. an der Univ. Warschau. Primarzt der inneren Abteilung des israelitischen Krankenhauses in Warschau (Polen), Nowogrodzka 46. *Haematol. des Menschen, normale und pathol.* ⑨ Warschau 1889. ⑨ Blutpraep. des Menschen, normale und pathol.

Kleine, Friedrich Karl (1869), GMR., Prof., Dr. med., Hon.Prof. Univ. Berlin, Abteilungsdir. am Preuß. Inst. für Infektionskrankheiten „Robert Koch" in Berlin. Mitglied der Schlafkrankheitskommission des Völkerbundes. Berlin N 39 (Deutschland), Pohrerstr. 2. *Bact., Tropenkrankheiten, Schlafkrankheit, Rückfallfieber, Pferdeseuche, Küstenfieber, Tuberkulose.* ⑨ Halle a. d. S.

Kleine, Richard (1874), Abteilungsvorsteher an der Landwirtschaftskammer der Provinz Pommern, Vorsteher der Samenprüfungsstelle und der Hauptstelle für Pflanzenschutz. Stettin (Deutschland), Werderstr. 31. *Ernährungsphysiol. europäischer Chrysomeliden, Fraßbildstudien. Biol. der Bren-*

thidae. Angewandte Entomol. ⑨ Material über Fraßbildstudien, exotische Ipidae, Brenthidae gesucht.

Kleiner, Israel S. (1885), Prof. and Head of the Department of Physiol. Chem., Homoeopathic Med. Coll. and Flower Hospital. New York, N.Y. (U.S.A.), 64 St. and Av. A. *Blood sugar; diabetes mellitus; nutrition.* ⑨ Yale Univ. 1909.

Kleinhoonte, Anthonia (1887), Conserv. Labor. f. Techn. Botan. Delft (Holland), Poortlandlaan 35. *Nyktinastische Bewegungen. Flora von Surinam.*

Kleitman, Nathaniel (1895), Ass. Prof. of Physiol., Univ. of Chicago (U.S.A.), *Physiol. of the Nervous System (Sleep, conditioned reflexes).* ⑨ Univ. of Chicago 1923.

Klencke, Heinrich (1889), Dr. phil. Essen-Ruhr (Deutschland), Bismarckstr. 21 II. *Mikrochem. der Pflanzen, insbesondere aromatische Substanzen; Variabilität niederer Organismen, Grünalgen.* ⑨ Göttingen 1912.

Kler, Prof. Uralsche Univ. Sverdlovsk (U.d.S.S.R.), Ul. Dekabristov I. *Zool.* o

Kleneberger, Emmy (1892), Bact. am Städtisch. Hygien. Univ.-Inst. Frankfurt a. M. (Deutschland), Paul-Ehrlich-Str. 40. *Bakterienmodifikationen; Bakterielle Zersetzung von Kohlehydraten.* ⑨ Frankfurt a. M. 1917.

Kligler, Israel J. (1889), Prof. Jerusalem (Palestine), Hebrew Univ. *Malaria. Protozoan Immunity. Oral Bacteria. Soil Pollution. Intestinal Infections.* ⑨ Columbia Univ. 1915.

Klika, Jaromir (1888), Prof., agrégé de l'Ecole polytechnique, Fac. de l'agric. Praha (C.S.R.), Košíře 333. *Biol. des Discomycètes, des Ascomycètes. Phytogéographie.* ⑨ Praha 1919. ⑨ *Fungi* (Ascomycètes, Uredinées, Ustilaginées).

Klimentova, A. A., Doc. Gosudarstv. Donskoj Univ. Rostov am Don (U.d.S.S.R.). *Pathol.* o

Klimmer, Martin (1873), Dr. phil. et med. vet., Ob.Med.R., o. Univ.-Prof. und Dir. des Vet.-Hygien. Inst. der Univ. Leipzig und Dir. des Vet.-Hygien. Inst. der Univ. Leipzig und Genußmittel in Speyer a. Rh. (Deutschland), Denkmalsallee 110. *Vet.-Hygien. und Seuchenforschung (Tuberkulose, Jungtierkrankheiten, Abortus, Euterkrankheiten usw.).* ⑨ Phil. Bern; med. vet. Dresden-Leipzig.

Klimov, Prof. Tierzucht-Inst. Moskau (U.d.S.S.R.), Smolenski bulv. 57. *Anat. Haustiere.* o

Klimov, K. M., Dr. Prosektor, Anat. Kabinett, Med. Inst. Astrachan (U.d.S.S.R.). *o*

Kling, C. A., Prof. Dr., Karolinska Inst. Stockholm (Sverige). *Serol.* o

Kling, Max (1874), Prof. Dr., Vorstand der Landwirtschaftl. Abteilung der Landwirtschaftl. Kreisversuchsstation und Öffentl. Untersuchungsanstalt für Nahrungs- und Genußmittel in Speyer a. Rh. (Deutschland). *Landwirtschaftl. Versuchs- und Untersuchungswesen (Dünger-, Fütterungs-, Bodenkunde).* ⑨ Breslau 1898.

Klingner, Heinrich (1881), bayerischer Landwirtschaftsrat für Weinbau. Neustadt a. d. Haardt (Deutschland), Moltkestr. 14. *Weinbau.*

Klingstedt, Torsten Holger (1900), Ass. Helsinki (Finnland), Zool. Inst. Univ. *Gametogenese der Trichopteren.* ⑨ Helsinki.

Klisiecki, Andreas (1895), Dr. med., Ass. des Physiol. Inst. der Univ. Lwów (Polen), Piekarska 52. *Verlauf der Strömungsschwindigkeit des Blutes vom Herzen bis zu d. Kapillaren. Die Harnstoffrage im Organismus.* ⑨ Lwów 1921.

Kljuchin, Stepan Mich., Doc. I. Univ. Moskau (U.d.S.S.R.). *Bact.* o

v. Klobusitzky, Dionys (1900), Ass. physiol. Inst. Univ. Pécs (Ungarn), Rákoczi-Str. 80. *Physikalische Chem. u. Physiol.* ⑨ Pécs 1926.

Klöti, Eugen (1891), Leiter Abteilung für Pflanzenschutz- u. Schädlingsbekämpfungsmittel der Chem. Fabrik Flora, Dübendorf Zch. Wallisellen, Zch. (Schweiz). *Angewandte Entomol.* ⑨ Zürich 1920.

Kloos, Abraham Willem Jr. (1880), Doc. Middelbare Techn. School Dordrecht (Holland), Krispijnscheweg 105 zw. *Niederl. Adventiv-Flora, Violen, Euphrasien, Thymi.* ⑨ Delft 1903. ⑨ Niederl. Herbar-Material.

Kloss, Cecil Boden (1877), Dir. Mus., Straits Settlements and Federated Malay States Singapore (India), Raffles Mus. *Zool., Anthrop., Ethnol. of Malaysia.* ⑨ Malaysian zool. specimens.

Klotz, L. Joseph (1895), Research Associate, Plant Pathol. Riverside, Cal. (U.S.A.), Citrus Experim. Station. *Physiol., Biochem. of plant diseases. Fungeus pathogens.* ⑨ Missouri 1923.

Klotz, Oskar (1878), M.D.C.M., Prof. of Pathol. and Bact., Univ. of Toronto (Canada). *Pathol. Arteries, Heart and Kidneys; Yellow Fever.* ⑨ McGill Univ. 1906.

Klukhine, Etienne (1886), Prof. agrégé Fac. méd. Univ., Ass. chargé Inst. Microbiol. Moscou (U.d.S.S.R.), Ulansky 24, ap. 14. *Immunol. et Bact.* ⑨ Moscou 1914.

Klugh, A. Brooker (1882), Ass. Prof. of Biol. Kingston (Canada), Queen's Univ. *Animals Ecol., Limnol., Light as an ecol. Factor.* ⑨ Cornell 1925.

Klut, Hartwig (1875), Prof., Dr. phil., wissensch. Mitglied der Preuß. Landesanstalt für Wasser-, Boden- und Lufthygiene zu Berlin-Dahlem. Berlin-Lichterfelde-W. (Deutschland), Tulpenstr. 6 II. *Hygiene des Trink- und Brauchwassers.* ⑨ Berlin 1901.

Kluyver, Albert Jan (1888), Prof. der Mikrobiol. an der Techn. Hochsch. Delft (Holland), Nieuwe Laan 3. *Biochem. der Mikroorganismen (Atmungs- und Gärungsprozesse).* ⑨ Techn. Hochsch. Delft 1914. ⑨ Bakterien- und Hefenkulturen.

Knauer, Paul (1874), Dir. des Bact. und Serum-Inst. der Landwirtschaftskammer für die Provinz Ostpreußen. Königsberg, Pr. (Deutschland), Beethovenstr. 24/26. *Bact. Tuberkuloseforschung.* ⑨ Gießen 1903.

Kneucker, Joh. Andreas (1862), Kustos der badischen Landessammlung für Naturkunde in Karlsruhe, Baden (Deutschland), Werderplatz 48. *Systematische Botan., Floristik u. Pflanzengeographie, das Florengebiet von Mitteleuropa u. des petraïschen Arabien (Sinai). Glumaceae: das Genus Carex.* ⑨ Glumaceae exsiccatae: Carices exsiccatae, Cyperaceae (excl. Carices) et Juncaceae exsiccatae, Gramineae exsiccatae. Exsiccaten aus dem Sinaigebiet. Lehrsammlungen der angewandten Botan.

Knlep, Hans (1881), o. Prof. d. Botan., Dir. d. Pflanzenphysiol. Inst. der Univ. Berlin. Berlin-Dahlem (Deutschland), Haderslebener Str. 9. *Reizphysiol., Sexualität der Pilze, Geschlechtsbestimmung.* ⑨ Jena 1904.

Kniepkamp, Wilhelm (1894), Dr. med., Ass. d. orthopäd. Univ.-Klinik Berlin. Berlin-Charlottenburg (Deutschland), Bredtschneiderstr. 13. *Vergl. Gelenk- u. Muskelmechanik.* ⑨ Berlin 1922.

Knipowitsch, N. M., Prof., Dir. Zool. Labor. Med. Inst. Leningrad (U.d.S.S.R.), Ul. L'va Tolstoyo 6-8. o

Knight, Harry Hazeeton (1889), Associate Prof. of Entomol., Iowa State Coll. Ames, Ia. (U.S.A.), Dept. Zool. and Entomol. *Entomol. Hemiptera, biol. and systematic studies on family Miridae.* ⑨ Cornell Univ., Ithaca, N.Y., 1920. ⑨ Hemiptera, family Miridae.

Knight, Hugh (1877), Research Entomol. California Spray-Chem. Co. Glendora, Cal. (U.S.A.), Box 111. *Fumigation with HCN gas. Oil sprays. Insect pests infesting citrus trees.*

Knight, Paul (1903), Instr. in Entomol. College Park, Md. (U.S.A.), Univ. of Maryland, Department of Entomol. *Structure and life history of Achroia grisella. Morphol. of the head of the order Isoptera.* ⑨ Univ. of Maryland 1927.

Knight, Robert Cedric (1891), Research Plant Physiol., Ministry of Agric. East Malling, Kent

(England), Research Station. *Physiol. of fruit production.* ⓓ London 1917.

Knoll, Fritz (1883), Dr. phil., o.ö. Univ.Prof., Vorstand des Botan. Inst. und Dir. des Botan. Gartens der Deutschen Univ. in Prag II (C.S.R.), Viničná 3a. *Oekol. der Pflanzen, besonders der Blüte, Wechselbeziehungen zwischen Pflanzen und Tieren. Experim. Methoden. Systematik, Morphol. und Oekol. der Araceen, verschiedener Pilzgruppen. Physiol. und systematische Anat. der Pflanzen, Paläobotan. Morphol. der Blütenpflanzen, besonders der Blüte.* ⓓ Graz (Österreich) 1906.

Knoll, Willy (1876), Dr. med., Chefarzt der bündner Heilstätte Arosa (Schweiz). *Embryol. Hämatol. Atmung und Energieverbrauch. Hochgebirgsflora (spez. Primelbastarde).* ⓓ 1904. ⓟ Praep. von embryol. Blutmaterial.

Knoop, Franz (1875), o. Prof. f. physiol. Chem. Freiburg i. Br. (Deutschland), Burgunderstr. 22. *Intermediärer Stoffwechsel.* ⓓ Freiburg i. Br. 1900.

Knopfli, Walter (1889), Dr. phil., Ass. am Zool. Inst. beider Hochsch. (Eidgen. techn. Hochsch. und Zürch. Univ.). Zürich 4 (Schweiz), Stauffacherstraße 9. *Ornithol. und Tiergeographie.* ⓓ Zürich 1916.

Knorr, Maximilian (1895), Priv.Doc. für Hygiene und Bact. München (Deutschland), Matthias-Pschorr-Ring 1/II. *Hämophile Bakterien, Bact. und Vitamine, Typhus Coli Gruppe, Anaerobiose, experim. Epidemiol., Bact.* ⓓ Würzburg 1921.

Knorring - Neustruewa, O. E., Ass. l'Herbier, Jardin Botan. princ. Leningrad (U.d.S.S.R.), Aplekarski Ostrov, Pessotchnaia ¹/₂. *Systém. et géographie des plantes du Turkestan.* o

Knott, James Edward (1897), Associate Prof., Head of Division of Vegetable Gardening. Pennsylvania State Coll., State College, Pa. (U.S.A.). *Vegetable crops. Catalase activity in relation to growth changes. Green manure fertilization of asparagus.* ⓓ Cornell Univ. 1926.

Knower, Henry McE. (1868), Prof. of Anat., Med. School. Cincinnati, Ohio. (U.S.A.), Univ. of Cincinnati. *Termites Lymphatic and vascular Systems of frog embryos. Muscles of human heart.* ⓓ John Hopkins 1896. o

Knowlton, Frank Hall (1860), Prof. U.S. geol. Survey. Washington, D.C. (U.S.A.), U.S. Nat. Mus. *Botan. palaeont., geol.* o

Knowlton, George F. (1901), Ass. Entomol. Logan, Utah (U.S.A.), Utah Agric. Experim. Station. *Beet Leafhopper, Curly-top. Economic Entomol. Biol. and Taxonomy of the Aphididae. Ecol.* ⓓ Utah Agric. Coll. 1925. ⓟ Aphididae.

Knowlton, Harry Edward (1890), Associate Prof. of Horticulture, Associate Horticulturist, West Virginia Agric. Experim. Station. Morgantown, W.Va. (U.S.A.). *Investigator in fruit pollination, orchard fertilization.* ⓓ Cornell Univ. 1920.

Knox, Leila Charlton (1883), Resident Pathol. New York (U.S.A.), St. Luke's Hospital. *Ossification.* ⓓ Cornell 1918.

Knudson, Arthur (1889), Prof. biol. chem. Union Univ. Medical Dept. Albany, N.Y. (U.S.A.), Albany Med. Coll. *Metabolism of Lipins. Cholesterol. Nutrition.* ⓓ Columbia Univ. 1915.

Knudson, Lewis, Prof. of Plant Physiol., Coll. of Agric., Cornell Univ. Ithaca, N.Y. (U.S.A.). o

Knull, Josef Nissley (1891), Entomol., Pennsylvania Bureau Plant Industry. Hanisbury, Pa. (U.S.A.), 1120N.17thSt.*Entomol.* ⓓ Ohio State Univ. 1924. ⓟ North American Cleridae, Buprestidae, Cerambycidae and Scolytidae.

Knuth, Paul (1866), Prof. Dr. phil., Dir. des Inst. für Tierhygiene der Preuß. landwirtschaftl. Versuchs- und Forschungsanstalten. Landsberg a. d. Warthe (Deutschland), Theaterstr. 26. *Tropenkrankheiten der Haustiere. Tierseuchen: Protozoenkrankheiten, Leukocytozoon.* ⓓ Leipzig. ⓟ Ausstrichpraep. Leukocytozoon anseris.

Knutsen, Martin H. (1887), Associate Prof. of Bact. State College, Penn. (U.S.A.), Pennsylvania State Coll. *Dairy Bact.* ⓓ Univ. of Wisconsin 1916.

Kobayashi, Hajime, Lect. Inper. Univ. Kyôto (Japan), Pathol. Inst., Fac. Med.

Kobayashi, Haruhej, Lect. of Imper. Univ. Kyôto (Japan), Anat. Inst., Coll. of Med. *Anat.*

Kobayashi, Harujirô (1884), Chief of the Research Department, Chosen Government General Hospital, and Prof. of Parasit., Keijo Imperial Univ., Med. Fac. Keijo (Seoul), Chosen, Korea (Japan). *Parasit. and med. Zool., Helminthol. (Flukes and roundworms). Med. Entomol. (Flies, Mosquitos).* ⓓ Tôkyô Imperial Univ., Coll. of Sc. 1909. ⓟ Helminthes, Flies and Mosquitoes.

Kobayashi, Rokuzô, Prof. of Keiô-Gijuku Univ. Tôkyô (Japan), Bact. Inst., Coll. of Med., Yotsuya. *Bact.*

Kobayashi, Shumpei, Lect. Imper. Univ. Kyôto (Japan), Anat. Inst., Fac. Med.

Kobel, Fritz (1896), Pflanzenphysiol. an der Schweiz. Versuchsanstalt für Obst-, Wein- und Gartenbau. Wädenswil, Kt. Zürich (Schweiz). *Befruchtung der Obstarten (Cytol., Pollenkeimung, Parthenokarpie, Apogamie, Selbst- und Fremdbefruchtung, Interlsterilität). Rebenzüchtung und Cytol. der Reben. Kohlensäuredüngung; künstliche Belichtung.* ⓓ Bern 1920.

Kobendza, Roman (1886), Jardin Botan. de l'Univ. Warszawa (Polska), Rue Aleje Mjazdowskie N. 6/8. *Phytosociol. et biol. des plantes.* ⓓ Varsovie 1926.

Kobranow, Nikolaus P. (1883), Prof. Forstinst. Leningrad (U.d.S.S.R.), Forstinst. Q. N. 47. *Forstliche Samenkunde und Forstkulturwesen. Sumpfkiefer. Selection der Eiche.*

Kobuski, Clarence E. (1900), Rufus J. Lackland Research Fellow. St. Louis, Mo. (U. S. A.), Missouri Botan. Garden. *Morphol. and Taxonomy of Higher Plants.* ⓓ Washington 1925.

Kočergin, S. M., Prof. Sibir. Inst. f. Land- u. Forstwirtschaft. Omsk (U.d.S.S.R.), Usadba za Staroj Zagorodnoj. Roščej. *Vererbung.* o

Koch, Anton (1901), Ass. Greifswald (Deutschland), Zool. Inst. *Eiwachstum, Dotterkernproblem. Tierisches Leuchten.* ⓓ München 1924.

Koch, Eberhard (1892), Priv.-Doc. für normale und pathol. Physiol. Köln-Rhein (Deutschland), Lindenthal. *Normale und pathol. Physiol. des Kreislaufes, des Nervensystems, Elektrophysiol.* ⓓ Bonn 1818.

Koch, Ferdo, Prof. emer. Geol. u. Palaeont. Zagreb (S.H.S.), Demetrova ulica 1. o

Koch, Fred Conrad (1876), Prof. of Physiol. Chem., Chairman of the Department. Univ. of Chicago, Ill. (U.S.A.). *Biochem. of Internal Secretion, Blood analysis, Phospholipins, Secretin, Gastrin, Thyroid.* ⓓ Univ. of Chicago 1912.

Koch, Louis (1890), Chief Plant Breeding Station for Animal Crops. Buitenzorg, Java (Ned.-O.-Indië). *Applied Botan., Breeding Oryza Sativa, Zea mays, Manihot utillissima, Arachis hypogaea, Ipomoea batatas, Soya Max, several grasses, Hibiscus sabdoriffa altissima, Sorghum vulgare, Mimosa invisa.* ⓓ Wageningen 1912. ⓟ Planting material from: Oryza sativa, Zea mays, Manihot utillissima, Arachis hypogaea, Ipomoea batatas, Soya max, Panicum maximum, Pennisetum purpureum, Hibiscus sabdoriffa altissima, Sorghum vulgare, Mimosa invisa.

Koch, Walo (1896). Zürich (Schweiz), Kantonsapotheke. *Pflanzensoziol., Floristik.* ⓓ Zürich 1925.

Koch, Walter (1880), ao. Prof. der allgemeinen Pathol. und Pathol. Anat. Univ. Berlin, Dir. Pathol. Inst. Krankenhaus Berlin-Westend. Geschäftsführer der Sozialhygien. Akademie Berlin-Charlottenburg. Berlin-Wilmersdorf (Deutschland), Landauer Str. 4. *Pathol. Anat. mit besonderer Berücksichtigung der klinisch-topographischen Pathol. Herzanat. Tuberkulose. Konstitution.*

Drüsen mit innerer Sekretion. Spezifische Muskelsysteme des Herzens. ⒹFreiburg 1906.
Koch, Walter (1902), Ass. München (Deutschland), Jagdstr. 9 I. *Vergl. Osteol. der Säugetiere.* ⒹMünchen 1925.
Koch, Wilhelm (1887), Dr. phil., Reg.R., Landes-Fischerei-Sachverständiger im Badisch. Ministerium des Innern. Karlsruhe (Deutschland), Ettlinger Str. 15. *Fischereibiol.* ⒹMünchen 1911.
Kochmann, Martin (1878), o.ö. Prof. u. Dir. des pharmacol. Inst. der Univ. Halle a. d. S. (Deutschland), Magdeburger Str. 22a. *Pharm. der Narkose, Lokalanästhesie, Ionen.* ⒹJena 1902.
Kochs, Julius (1871), Dr. phil., Stud.R., Doc. Berlin-Dahlem (Deutschland), Lehr- u. Forschungsanstalt für Gartenbau. *Chem.-physiol. Untersuchungen an lagerndem Kernobst. Nährstoffbestand der Beerensträucher während der Vegetationsperiode.* Ⓓ 1899.
de Kock, G. v. d. W., Prof. Transvaal Univ. Coll. Pretoria (South Africa). *Pathol.* o
Kočnar, Karel, Dr. Ing. l'Inst. des Recherches agric. et Forestière. Brno (C.S.R.), Květná ul. *Mendelismus.* o
Kodama, Sakuji (1893), Ass. Prof. Tôhoku Reichs-Univ. Sendai (Japan), Physiol. Inst. *Physik. Chem. d. Wachstums der Zelle u. d. Gewebe in vitro.* ⒹSendai 1919.
Kodama, Sakuzaemon, Ass. Prof. Tôhoku Imper. Univ. Sendai (Japan), Anat. Inst., Fac. Med.
Köck, Gustav (1879), Dr., Hofrat an der Bundesanstalt für Pflanzenschutz, Priv.Doc. an der Hochschule für Bodenkultur in Wien (Österreich), III/4, Aspangstr.· 17. *Phytopath.* ⒹWien 1903.
Koefoed, Einar (1875), Mag. sc., Ass. und Bibliothekar an der wissenschaftl. Abteilung des Bureaus des Fischereidir. Bergen (Norwegen). *Fischeier und Jungfische. Systematik der Fische, insbesondere der Tiefseefische.* ⒹKopenhagen 1901.
Koegel, Anton (1889), Dr., Priv.Doc. der techn. Hochsch. zu München (Deutschland), Eggernstr. 7 II. *Allgemeine Pathol.: Parasit., tierärztl. Entomol.* Ⓓ Gießen 1914.
Koehler, Alfred Edward (1896), Physician, Henry Ford Hospital. Detroit, Mich. (U.S.A.). *Metabolism, Acid-base equilibrium.* Ⓓ Ph.D. Univ. of Wisconsin 1921; M.D. Harvard Med. School 1923.
Köhler, Erich (1889), Dr. phil., wiss. Hilfsarbeiter an der Biol. Reichsanstalt für Land- u. Forstwirtschaft. Berlin-Dahlem (Deutschland), Königin-Luise-Str. 19. *Pflanzenpathol. Pathol. und Physiol. der Kartoffel.* ⒹMünchen 1919.
Koehler, Otto (1889), o. Prof., Vorstand des zool. Inst. und Mus. Königsberg, Pr. (Deutschland). *Vergl. Sinnesphysiol. Vererbungslehre.* ⒹMünchen 1911.
Koehler, René (1860), Prof. à la Fac. des Sc. de l'Univ. Lyon (France), 29, Rue Guillond. *Echinodermes vivants.*
Koenig, Alexander Ferdinand (1858), Dr. med. h. c., o. Hon.Prof. an der Univ. Bonn a. Rhein (Deutschland), Coblenzer Str. 164. *Ornithol.* ⒹMarburg 1884.
Körbler, Juraj Georg (1900), Ass. an der chirurgischen Klinik der Univ. Zagreb (S.H.S.), Ulica ,,C" 17. *Experim. Krebsforschung. Rattensarkom. Explantation und Gewebezüchtung. Geschichte der Med. bei den Südslaven.* ⒹFreiburg i. Br. 1923.
Koernicke, Max (1874), o. Prof. der Botan. a. d. Landwirtschaftl. Hochsch. Bonn-Poppelsdorf, Hon.-Prof. a. d. Univ. Bonn, Dir. des botan. Inst. der Landwirtschaftl. Hochsch. Bonn-Poppelsdorf. Bonn (Deutschland), Bonner Talweg 45, *Pflanzliche Cytol. Wirkung verschiedener Strahlengattungen auf den pflanzl. Organismus.* ⒹBonn 1897.
von Kőrösy, Kornél (1879), Priv.Doc., Prof. Univ. Budapest VI (Ungarn), Lendvaygasse 25.
Physikalisch-chem. Biol., spezielle Physiol. der Verdauung und Resorption, Mikrobiol., Vererbungslehre. ⒹBudapest 1903.
Kövessi, François (1875), Prof. à l'École d'Ingenieur des Mines et des Forêts. Sopron (Ungarn). *Anat., Physiol. et Pathol. des plantes forestières et agric. Biophysique. Energetique des plantes.* Ⓓ Paris 1901.
Kofler, Ludwig (1891), Prof. für Pharmakognosie, Vorstand des pharmacognostischen Inst. der Univ. Innsbruck (Österreich), Anatomiestr. 1. *Saponine.* ⒹPhil. Wien 1914, med. 1920.
Kofoid, Charles Atwood (1865), Prof. of Zool., Chairman of Department, Consulting Parasit. of California State Board of Health. Berkeley, Cal. (U.S.A.), Univ. of California. *Cellular Biol., Protozool., Parasit.* ⒹPh.D. Harvard 1894, Sc.D. Oberlin College 1915, Sc.D. Univ. of Wales 1920.
Kogan, E. N., Prof. Astrachan (U.d.S.S.R.), Staatl. Med. Lunačarsky Inst. *Allg. Pathol., pathol. Anat.* o
Koganei, Yoshikiyo (1858), Prof. emer. Anat. Kaiserl. Univ. Tôkyô (Japan). *Physische Anthrop.* Ⓓ 1882.
Kogoj, Franjo (1894), Univ.Doc., Supplent an der Dermatol. Klinik. Zagreb (S.H.S.), Salata, Dermatologische Klinik. *Infektion mit pathogenen Pilzen. Pathol. Histol. der Haut: Atrophien, Dyskeratosen.* ⒹKarls-Univ. Praha 1920.
Kohan, Johann (1877), Ass. Inst. experim. Biol. Moskau (U.d.S.S.R.), Kursowaj Querstr. 10. *Innere Secretion d. Hoden u. Ovarien. Transplantation der Geschlechtsdrüsen bei Tieren und Menschen.*
Kohl, Edwin Jacob, Ass. Prof. of Botan., Purdue Univ. West Lafayette, Ind. (U.S.A.), 218 Fowler Ave. o
Kohlbrugge, J. H. F., Prof. Univ. Utrecht (Holland). *Ethnol. Anthropol. Geschichte der Biol.*
Kohlrausch, Arnt (1884), o. Prof. Physiol. Univ. Greifswald (Deutschland), Rubenowstr. 3. *Nerven- und Muskelphysiol., Sinnes-Physiol., Elektrophysiol.* ⒹMarburg 1911.
Kohn, Alfred, Dr., Prof. d. Histol. Praha II (C.S.R.), Salmovska 5, Histol. Inst. d. Deutschen Univ. o
Kohn, Lawrence A. (1894), Instr. in med. Rochester, N.Y. (U.S.A.), Univ. *Bact.* ⒹJohns Hopkins Univ. 1923.
Kohn-Speyer, Alice Charlotte (1902), Ass. Cancer Researcher. London S.W. 1 (England), Lister Inst. *Tissue Culture in relation to immunity ito tumours.*
Kolde, H., Prof. of Hokkaidô Imperial Univ. Sapporo (Japan), Inst. of Forestry, Fac. of Agric. *Forest Management and Statistics.*
Koike, Keiji, Prof. of Chiba Med. Univ. Chiba near Tôkyô (Japan), Anat. Inst. *Anat.*
Koizumi, Gen-ichi (1883), Prof. of Botan., Imper. Univ. Kyôto (Japan), Botan. Inst., Department of Sc. *Systematic Botan. Flora and Phytogeography of Japan. Salicol., Rosaceae, Morus.* ⒹTôkyô 1923.
Koizumi, Tan, Prof. of Keiô-Gijuku Univ. Tôkyô (Japan), Pathol. Labor., Coll. of Med. *Parasit. Protist.* o
Kojima, Hitoshi (1895), Ass. Prof. Kyûshû Imper. Univ. Fukuoka (Japan), Botan. Inst. *Physiol. and cytol. of the relation between growth and cell-division. Cytol. in Solanaceae. Heredity in Papaver somniferum L. and Celosia cristata L.* ⒹKyûshû Imperial Univ. 1918.
Kôketsu, Riichiro (1886), Prof. der Pflanzenphysiol. an der Univ. Fukuoka (Japan), Botan. Inst., Kaiserliche Kyûshû Univ. *Serodiagnostische Untersuchungen der Pflanzen. Elektrophysiol. der Pflanzen (Elektro-Reizphysiol.). Wasserhaushalt der Pflanzen (Transpiration, Welken usw.).* Ⓓ Tôkyô Univ. 1912.

Kokina, Susanna (1894), Laborant des Labor. experim. Ökol., Botan. Garten. Leningrad (U.d. S.S.R.). *Wasserhaushalt der Pflanzen.*

Kokine, Abraham J. (1890), Prép. à la Section de Physiol. Végétale du Jardin Botan. Leningrad (U.d.S.S.R.), Pessotchnaja 24 B. *Physiol. végétale: les produits de la photosynthèse.* Ⓓ Leningrad 1924.

Kokita, Harno, Lect. Imper. Univ. Kyôto (Japan), Pathol. Inst., Fac. Med.

Kokubo, Seiji (1889), Ass. Prof. Tôhoku Imper. Univ. Sendai. Asamushi near Aomori (Japan), Marine Biol. Sta. *Hydrobiol., Physiol.* Ⓓ Sapporo 1907.

Kol, Alexander Karlowitch (1877), Chief of Bur. of Introduction of the U.d.S.S.R. Inst. of Applied Botan. and New Cultures. Leningrad (U.d.S.S.R.), 44 Herzen street. *Introduction of new useful plants and their varieties.* Ⓓ Leipzig, Agric. Inst., Univ. 1900. Ⓟ Seeds of plants of Russian agric. crops.

Kolbe, Hermann Julius (1855), Prof., Kustos a. D. am Zool. Mus. der Univ. Berlin. Berlin-Lichterfelde-West (Deutschland), Steinäckerstr. 12. *Biol. der Coleopteren, Neuropteren und Psociden. Metamorphose und Cecidiol. der Coleopteren. Palaeontol. und Tiergeographie der Insekten.*

Kolbe, Robert William (1882), Dr. phil., wiss. Mitarbeiter an der Preuß. Landesanstalt für Wasser-, Boden- und Lufthygiene. Berlin-Lichterfelde (Deutschland), Ringstr. 68. *Algol.; Diatomeen (Systematik, Oekol., Cytol., Entwicklungsgeschichte), Chlorophyceen, Florideen.* Ⓓ Berlin 1926. Ⓟ Materialien und Präp. von Diatomeen.

Koleneeva, Val. Fed., Doc. II. Univ. Moskau (U.d.S.S.R.). *Morphol. Syst. d. Pflanzen.* o

Kolesnikov, N. W., Ass. Dr. Kubansches Med. Inst., Anat. Labor. Krasnodar (U.d.S.S.R.), Str. Sedin N. 4. o

Kolkunow, Wladimir W., Prof., Landwirt. Inst. Kiew (U.d.S.S.R.), Polytechnikum 2, W. 18. *Pflanzenbau. Geobotan. u. Selektion.* o

Kolkwitz, Richard (1873), Prof. Dr. phil. et med. h. c., ao. Prof. an der Univ. Berlin für Botan., Abt.-Leiter an der Preuß. Landesanstalt f. Wasser-, Boden- u. Lufthygiene in Berlin-Dahlem. Berlin-Steglitz (Deutschland), Rothenburgstr. 30. *Pflanzenphysiol. Hydrobiol.* Ⓓ Berlin 1895.

Kollán, Sándor, Dr., extraord. Dir. d. Inst. f. Zool. u. Parasit. d. Ung. Tierärztl. Hochsch. Budapest (Ungarn), Rottenbillergasse. *Zool., Parasit.* o

Kolle, Wilhelm (1868), Prof. Dr., Dir. Staats-Inst. für Experim. Therapie u. des chemotherap. Forschungs-Inst. ,,Georg-Speyer-Haus". Frankfurt a. Main (Deutschland), Paul-Ehrlich-Str. 44. *Experim. Forschungen über Immunität und Chemotherapie von Infektionskrankheiten und Wertbestimmung serol. und chemotherapeut. Praeparate.* Ⓓ Würzburg 1892.

Koller, Gottfried (1902), Dr. phil., Hilfsass. bei der Preuß. Kommission für Meeresforschung. Kiel (Deutschland), Zool. Inst. *Vergl. Physiol.* (*Farbwechsel, Farbensinn, Ernährungsphysiol.*). Ⓓ Kiel 1926.

Kolli, V. A., Doc. II. Univ. Moskau (U.d.S.S.R.). *Anat.* o

Kollmann, Max (1880), Prof. Fac. Sc., Dir. du Labor. Marion. Marseille (France). *Régénération, Pancréas.*

Kollog, C. R., Lect. Fukien Christian Univ. Foochow (China). *Zool.* o

Kolmer, John Albert (1886), Prof. path. and bact., Head dept. Grad. School of Med., Univ. of Pa. Cynwyd, Pa. (U.S.A.). *Pathol. bact. Immunol.* o

Kolmer, Walther Arthur (1879), ao. Prof., Abteilungsvorstand Physiol. Inst. Univ. Wien XIX (Österreich), Vegagasse 15. *Vergl. Histol. der Sinnesorgane: Innere Sekretion, Zytol. Geschmacksorgan, Geruchsorgan, Gehörorgan.* Ⓓ Wien 1903. Ⓟ Histol. Diapositive: Sinnesorgane, Protozoen, Zytol.

Kołodziejczyk, January (1889), Chargé de cours à l'Ecole Superieure d'Agric. Prof. de botan. systématique l'Univ. Libre de Pologne. Warszawa, (Polska), Krucza 40. *Morphologie des plantes, geographie.* Ⓟ Cracovie 1919.

Kołodziejski, Zygmunt (1893), Ass. Inst. zool. Univ. Kraków (Polska), św. Anny 6. *Changement de matière chez les Invertébrés inférieurs; Transplantation et régénération chez les Batraciens; Metabolism of the budding Hydra.* Ⓓ Kraków 1924. Ⓟ Quelques races de ,,Drosophila melanogaster". Jeunes axolotls (blancs et noirs).

Kolokolnikov, A. I., Doc. II. Univ. Moskau (U.d.S.S.R.). *Method. d. Biol.* o

Kolosov, Doc. Uralsche Univ. Sverdlovsk (U.d.S.S.R.), Ul. Dekabristov 1. *Entomol.* o

Kolossow, Alexander (1862), Prof. Histol. Univ. Rostow am Don (U.d.S.S.R.), Krasnoarmejskaja 67. *Beziehungen der Zellen zueinander und über die Saftkanälchen in Deck- und Drüsenepithelien und im glatten Muskelgewebe. Struktur der Blutkapillarenwand. Bau und Entwickelung des lockeren Bindegewebes.* Ⓓ Moskau 1892. Ⓟ Histol. Praep.

Kolossow, Nicolai (1897), Histol. Inst. der Staats-Univ. Kazan (U.d.S.S.R.), Podluschnaja 48. *Ursprung der Fettsubstanzen in Nebennieren-Rinde. Einfluß des Pilocarpin auf Rindenelemente der Nebennieren. Morphol. Bedeutung der Langerhansschen Inseln.*

von Kolosváry, Gabriel (1901), Dr., Univ.Ass. Szeged (Ungarn), Tisza Lajos körút No. 6, Zool. Inst. *Araneae verae; morphol. u. biol. Spinnenfauna der ungarischen Tiefebene und des Balatonsees, Tierpsychol.* Ⓓ 1925. Ⓟ Spinnen, kleine Säuger.

Kolpakowa, F. A., Ass. Protist. Labor. Leningrad (U.d.S.S.R.), Med. Inst. o

Koltzoff, Nikolai Konstantinovicz (1872), Dir. Inst. Experim. Biol. der Commiss. f. öff. Gesundheitsschutz, Dir. Zentrale genetische Station des Volkskommissariats für Landwirtschaft, Dir. der genetischen Abteilung d. K.E.P.S. der Akademie der Wissenschaft, o. Prof. der I. u. II. Univ. Moskau (U.d.S.S.R.), Voronzovo Pole 6. *Genetische Analyse der erblichen chem. Eigenschaften des Blutes. Gestalt der Zelle (Pigmentzelle).* Ⓓ Moskau 1894.

Komai, Taku (1886), Prof. Zool., Imper. Univ. Kyôto (Japan). *Genetic of Insects, morphol. and taxanomy of Crustacea and Coelenterata. Aberrant Ctenophores, Coeloplana and Gastrodes.* Ⓓ Tôkyô Imperial Univ. 1927.

Komárek, Julius (1892), Prof. Karls Univ., Dir. des Inst. für allgemeine und systematische Zool., Vorstand der staatl. Versuchsanstalt für Forstschutz. Doc. für Landwirtsch. Entomol. an d. technischen Hochsch. Praha II (C.S.R.), u Karlova 3. *Systematik der Süßwassertricladen, ,,Chlamydozoen" filtrierbare Krankheitserreger, Hydrobiol., Forstentomol., Landwirtschaftl. Entomol. Blepharocerиden von Kaukasus und Armenien. Wipfelkrankheit der Nonne und der Erreger derselben. Hydracarina. Ephemeridenlarven.* Ⓓ Prag 1916.

Komarnitzkij, Nikolaj A. (1888), Doc. I. Univ. u. Landwirtschaftl. Akademie Timirjasew. Moskau (U.d.S.S.R.), Tichwinskij per. 9, W. 13. *Morphol. u. Entwicklung d. Pflanzen. Flechten. Pilze.* o

Komarov, Vladimir (1869), Prof. Univ., Vice-Dir. des Botan. Hauptgartens u. Chef der Abteilung der lebenden Pflanzen des Botan. Hauptgartens. Leningrad (U.d.S.S.R.), Pessotschnaia 1/2. *Flora Mandschuriens und Chinas.* Ⓓ Moskau 1911.

Komarow, Semjon, Dr., Prosektor Univ. Rīga (Latvija), Pulkv. Brieža ielā 7, dz. 6. *Physiol. u. physiol. Chem.* o

Komatsu, Shigeru, Prof. of Imperial Univ. Kyôto (Japan), Biochem. Labor., Coll. of Sc. *Organic Chem.; Phyto-Chem.* Ⓓ Kyôto.

Kominami, Kiyoshi, Ass. Prof. of Imper. Univ. Komaba near Tôkyô (Japan), Botan. Labor., Coll. of Agric. *Dendrol.*

Komine, Shigeyuki. Sendai (Japan), Zool. Inst., Coll. of Sc., Tôhoku Imperial Univ. *Reizphysiol.*
Komm, Ernst (1899), Priv.Doc. biol. Chem. Techn. Hochsch. Dresden, Vorstand des Labor. für physiol. Chem. und Ernährungsforschung an Dr. Lahmanns Sanatorium. Dresden-Weißer Hirsch (Deutschland), Rißweg 61. *Eiweißchemie, Vitamine.* ⓓ Gießen 1923.
Komocki, Witold (1875), Dr. med. Warszawa (Polska), 5 rue Skorupka. *Anat. et Histol. pathol., Bact. Les problèmes hématol.* ⓓ Kiew 1898.
Kompancev, N. N., Doc. Univ. Taskent (U.d. S.S.R.). *Pharmaz. Chem., Pharmacol.* o
Kompanejetz, Salomon (1872), Prof. Oto-Rhino-Laryngol. Dnepropetrowsk (U.d.S.S.R.), Prospect Karl Marx 17. *Beziehungen zwischen Tongehör und Sprachgehör; Tonleitung im Schädel; Lage der Sprachlaute in der Tonscala; Physiol. des Otolithenapparates beim Menschen; Methodik der Untersuchung der Funktionen des Otolithenapparates.* ⓓ Kiew 1897.
Kompanejez, Anna (1892), Ass. Microbiol. Inst. d. Volksgesundheitskommissariats. Moskau (U.d. S.S.R.), Pogodinskaja 10. *Microbiol.*
Komura, Taiji (1889), Animal husbandry man. Kung-chu-ling, South Manchuria (China), Agr. Exp. Stat., South Manchuria Railway Co. *Heredity of live-stock, sheep and pig. Parasit.: nematoda. History, distribution and classification of live-stock in china. Sheep-farming in South Manchuria. Pig.* ⓓ Coll. of Agric., Tôhoku Imperial Univ. 1913.
Kon, Y., Prof. of Hokkaidô Imperial Univ. Sapporo (Japan), Pathol. Inst., Fac. of Med. *Pathol.*
Konaschko, P. I., Ass. Dr. Med. Inst. Anat. Lab. Kiev (U.d.S.S.R.), Str. Levaschowskaja N. 39. o
Kondô, Heizaburô, Prof. of Imperial Univ. Tôkyô (Japan), Inst. of Pharmaceut. Chem., Fac. of Med. *Pharmaceut. Chem.*
Kondô, Kinsuke, Prof. of Imper. Univ. Kyôto (Japan), Chem. Labor., Coll. of Agric. *Nutrition Chem., Chem. of Protein.*
Kondô, Mantarō (1883), Dir. des Ōhara-Inst. für Landwirtschaftl. Forschungen. Okayama-Ken (Japan), Kurashiki. *Landwirtschaftl. Samenkunde.* ⓓ Kais. Univ. Tokiô.
Kondô, Shôji, Ass. Prof. of Tôhoku Imper. Univ. Sendai (Japan), Bact. Inst., Coll. of Med. *Bact.*
Kondretjew, N. S. (1887), Prof. d. Anat. Odessa (U.d.S.S.R.), Olgiewskaja 4. *Nervensystem.* ⓓ Charkow 1912.
Koningsberger, J. C., Dr., früher: Dir. Botan. Garten Buitenzorg, jetzt: Hollands Kolonialminister. Utrecht (Holland), Alex. Numankade 6. o
Koningsberger, Victor Jacob (1895), Dr. der Versuchsstation für die Zuckerrohrkultur (Kulturabteilung). Pasoeroean, Java (Nederl.-Indië), Heerenstraat. *Pflanzenphysiol. Angewandte Botan.* ⓓ Utrecht (Holland) 1922.
Konishi, Kametarô, Ass. Prof. Imper. Univ. Kyôto (Japan), Inst. of Agric. Chem., Fac. Agric.
Konjev, Dimitrije, Dr., o. Prof. der landwirtsch. Bact., Dir. d. Inst. f. landw. Bact. u. Tierseuchenlehre. Beograd (S.H.S.), Birčaninova ulica 36. o
Kôno, Hiromichi (1905). Sapporo (Japan), N. 4, W. 7. *Entomol.: Japanische Attelabiden, Coleopteren aus Korea und N.-Sachalin, Argynnis.*
Kôno, Sôichi (1879). Keijô (Japan), Goverment Normal School. *Insects, shell-fishes and fossils in Korea.* ⓓ Hiroshima 1914. ⓟ Specimens of animals (Insects and shell fishes) and fossils in Korea.
Konokotina, Anastasia (1888), Ass. d. Biol. d. Pflanzen im Med. Inst. Leningrad (U.d.S.S.R.), Gattschinskaja Str. 14. *Biol. der Hefepilze.*
Kononow, Wladimir Nikolaewitsch (1885), Chef d. Hydrobiol. Abteilung des Sanitäts-Inst. Moskau (U.d.S.S.R.), Blaguscha, 6, W. 1. *Sanitäre Hydrobiol.* ⓓ Moskau 1910.

Konopacka, Bronisława (1886), Volontaire à l'Inst. d'Histol. et d'Embryol. de l'Univ. de Varsovie. Warszawa (Polska),rue Chałubinskiego 5. *Mécanique et histochim. du développement.* ⓓ Cracovie 1908.
Konopacka, Wanda (1886), Ass. de l'Inst. de Phytopath. de l'Ecole supérieure d'Agric. Skierniewice (Pologne). *Mycol., Phytopath.*
Konopacki, Mieczysław (1880), Prof. Dr., Dir. de l'Inst. d'Histol. et Embryol. Fac. de Med. Univ. Warszawa (Polska), r. Chałubinskiego N. 5. *La cinétique du développement et la microchim. du développement.* ⓓ Lwów 1909.
Konowalow, Iwan N., Prof. Landw. Inst. Woronesh (U.d.S.S.R.). *Agronomie. Feldbau u. Gemüsegartenbau.* o
Konrich, Friedrich (1878), ao. Prof. Med. Fac. d. Univ. Berlin, ORR., Mitglied des Deutschen Reichsgesundheitsamtes. Berlin-Charlottenburg 5 (Deutschland), Kuno-Fischer-Str. 18. *Hygiene, Militär- und Gewerbehygiene.* ⓓ Berlin 1905. ⓟ Mikrobiol. Praep.
Konšel, Josef, Ing., Prof. d. Forstproduktion. Brno (C.S.R.), Černapole 102, Landwirtschaftl. Hochsch. o
Konstanty, Ewald (1899), Biol. Reichsanstalt für Land- u. Forstwirtschaft, Pflanzenanat. Labor. Berlin-Dahlem (Deutschland), Königin-Luise-Str.19. *Pflanzenanat.* ⓓ Berlin 1925.
Konsuloff, Stefan (1885), Dr. phil., ao. Prof. Sofia (Bulgarien),,,Zibra"-Str., beim Boris-Garten. *Zellularphysiol., Parasit. (Protozoen), Biol. der Mücken.* ⓓ Sofia 1921.
Kooiman, Havik Nicolaas (1893), Ass. Libr., Agric. Coll. Wageningen (Holland), Rijksstraatweg 52. *Bibliography of Genetic Literature.* ⓓ Utrecht 1920.
Kooper, Willem Johannes Cornelis (1888), Landbauberater Internationale Crediet- en Handelsvereeniging ,,Rotterdam", Semarang, Java (Nederl.-O.-Indië). *Pflanzensoziol.; Zuckerkultur.* ⓓ Utrecht 1927.
Koornneef, Jan (1868), Amsterdam (Holland), 1, Const. Huygensstr. 67. *Floristik der niederländischen Phanerogamen; Faunistik und Biol. der niederländischen Hymenoptera: Tenthredinoidea und Aculeata.* ⓟ Niederländische Insekten gegen allgemein palaearktische Hymenoptera.
Koosnetzow, Anatolius (1897), Ass. section of Pharmacol., Inst. of Experim. Med. Leningrad (U.d. S.S.R.), 10 te Sowjetskaja 14, Log. 36. *Endocrinol.* o
Kopatschewskaja, Marija A., Leiter d. Landwirtschaftl. Meteorol. Sektion d. Wissenschaftl. Landwirtschaftl. Komitee. Kiew (U.d.S.S.R.), Sretenskaja ul. 7, W. 1. *Geobotan.* o
Kopeć, Stefan (1888), Dr. phil., Priv.Doc. à l'Univ.deVarsovie, Chef duDepartement de Morphol. expérim. des animaux à l'Inst. National Polonais d'Économie Rurale à Pulawy (Pologne). *Castration, transplantation, régénération, métamorphose chez les insectes, batraciens et poissons. Longuévité, influence du jeûne. Hérédité du poids chez les lapins. Variabilité et hérédité de couleur des œufs de poules. Problème de la valeur morphogénétique de nouveau-nés. La xénie chez les oiseaux.* ⓓ Cracovie 1913.
Kopecky, Otakar (1876), o. Prof. Dr., Ing. chem. Brno (C.S.R.), Akademická ul. 42. *Landwirtsch. Technol.: Zuckerfabrikation, Raffination, Malzfabrik., Bierbrauerei, Spiritusindustrie, Hefe-, Essig-, Stärkefabrikation.* ⓓ Prag 1907.
Kopeloff, Nicholas (1890), Ph.D., Associate in Bact., Psychiatric Inst. Ward's Island, N.Y. (U.S.A.). *Bact. with special reference to psychiatric problems, studies in focal infections, malaria, gastrointestinal tract, lactobacillus acidophilus. Fermentation.* ⓓ Rutgers Univ. 1916. ⓟ Lactobacillus acidophilus.
Kopke, Ayres (1866), Prof. à l'Ecole de Méd. Tropicale et Chef du Service de la Tuberculose à l'Inst. Bact. Camara Pestana. Lisbôa (Portugal),

18, Rue Antonio Enes. *Protozoaires pathogènes. Maladies tropicales.* ⓟ Préparations des protozoaires.

Kopp, P., Prof. Univ. Tartu (Estland), Vene t. 38. *Agr.* o

Koppányi, Theodore (1901), Instr. in Physiol., Univ. of Chicago, Ill. (U.S.A.), Department of Physiol. *Physiol. morphol. of the central nervous system, regeneration and transplantation, glands with internal secretions, eye and pupillary reaction, sex.* ⓓ Wien 1923.

Koppe, Fritz (1896), Dr. phil. Kiel (Deutschland), Herzog-Friedrich-Str. 56 II. *Bryol., Algol.* ⓓ Kiel 1922. ⓟ Moose.

Kopsch, Friedrich (1868), Dr. med., beamteter ao. Prof. der Anat., I. Prosektor am Anat. Inst. der Univ. Berlin. Berlin-Wilmersdorf (Deutschland), Kaiserplatz 2. *Experim. Entwicklungsgeschichte der Wirbeltiere. Binnengerüst (= Apparato reticolare interno von Golgi).* ⓓ Berlin 1892.

Koræn, G. M., Dr., Karolinska Inst. Stockholm (Sverige). *Bact.* o

Korczewski, Michal (1889), Prof. in Szkoła Główna Gosp.Wiejskiego (Ecole supérieure d'Agric.). Warszawa (Polska), Miodowa 23. *Plant Physiol. esp. Growth process, physical. and physiochem., Physiol. organisation of cell and protoplasm (colloidal structure, permeability, biochem.).* ⓓ Cracow 1916.

Korde, Nina (1895), Ass., Polytechn. Inst., Landwirtsch. Fac., Zool. Kabinet. Iwanowo-Wosnessensk (U.d.S.S.R.). *Hydrobiol. Rotatoria, Cladocera, Turbellaria.*

Kordes, Herbert (1898), Dr. phil., Ass. u. wiss. Mitarbeiter an der Pflanzenphysiol. Versuchsstation der Lehr- u. Forschungsanstalt f. Gartenbau Berlin-Dahlem. Berlin-Lichterfelde-W. (Deutschland), Lortzingstr. 37 III. *Landwirtsch. Microbiol. und Phytopath. gärtnerischer Kulturpflanzen.* ⓓ Würzburg 1922. ⓟ Phytopath. Praep. gärtnerischer Kulturpflanzen.

Korentchevsky, Vladimir (1880), Prof., Dr. med., Research worker for the Med. Research Council of Great Britain and Ireland. London SW. 1 (England), Lister Inst. of Preventive Med. *Endocrine glands and metabolism, male sexual glands. Experim. tumours. Vitamins and Rickets.* ⓓ Moscow 1907.

Korff, Gustav (1872), Dr. phil., Prof., Vorstand der Pflanzenschutzabteilung an der Bayer. Landesanstalt für Pflanzenbau u. Pflanzenschutz, Doc. für Pflanzenkrankheiten u. Pflanzenschutz an der Landwirtschaftl. Abteilung der Techn. Hochsch. München (Deutschland), Liebigstr. 25. *Pflanzenkrankheiten und Pflanzenschutz.* ⓓ Erlangen 1898.

Korff-Petersen, Arthur (1882), Prof., Dr. med., Dir. des Hygien. Inst. der Univ. Kiel (Deutschland), Hohenbergstr. 20. *Chem. des Tuberkelbacillus.* ⓓ Leipzig 1909.

Kōriba, Kwan (1882), Prof. Botan., Kaiserl. Univ. Kioto (Japan), Botan. Inst. *Physiol., Ökol. und Morphol. der Pflanzen.* ⓓ Tokio 1912.

Kořínek, Jean (1889), Maître des Conférences à l'Univ. Charles. Praha II (C.S.R.), Benátská 433. *Biol. des microbes, autolyse, maladies microbiennes des plantes, microbes de terre.* ⓓ Prague 1916.

Kormos, Tivadar (1881). Budapest VII (Ungarn), Gizella ut 47. *Paläomammal., Eiszeitforschung.* ⓓ Budapest 1906. ⓟ Fossilien.

Kornfeld, Werner (1892), Hilfsarzt an der Wiener Kinderklinik, ehem. Ass. am Wiener Embryol. Inst. Wien XVIII (Deutsch-Österreich), Weimarer Str. 7. *Proportionsverschiebungen in der embryonalen und postembryonalen Entwicklung des Menschen. Zellteilungsrhythmus. Menschliche Mißbildungslehre, spez. Urogenitalsystem. Hypothyreosen. Vergl. Anat. wirbelloser Tiere, Polychaeten; vergl. Histogenese der Anuren. Entwicklungsmechanik (Transplantationen).* ⓓ Phil. Univ. Wien 1914; med. 1924.

Kornhauser, Sidney Isaac (1887), Prof. of Anat. Louisville, Ky. (U.S.A.), Univ., Med. Dept. *Cytol. Sex Determination.* ⓓ Harvard Univ. 1912. ⓟ Preparations of human tissues (normal) or cytol. slides of Anisolabis maritima.

Kornilow, K. N., Prof. II. Univ. Moskau (U.d. S.S.R.). *Psychol.* o

Korns, J. H., Prof. Dr., Dept. of Med., Union Med. Coll. Peking (China). *Pharmacol.* o

Kornylowitsch, N. P., Prof., Dir. Anat. Labor. Med. Inst. Leningrad (U.d.S.S.R.), Str. Lew Tolstoy. o

Korolew, Serg. Alexandrowitsch (1874), Prof. technisch. Lehranstalt, Vorstand der Versuchsstat. Vologda (U.d.S.S.R.), Molotschno-Chosjaistwennyi Inst. (Milchwirtschaftliches Inst.). *Mikrobiol. v. Milch, Käse u. Butter, Milchsäurebakterien.*

Korotaewa, A. G., Doc., Ass. Anat. Inst. II. Univ. Moskau (U.d.S.S.R.), Dewitschje Pole, Trubezkoi pereulok. o

Korotneff, Nicolas (1865), Prof. agrégé. Moscou 34 (U.d.S.S.R.), Barykovsky per. 9. *Oecol. des coléoptères. Régime alimentaire des oiseaux.* ⓓ Moscou 1890. ⓟ Coléoptères.

Korovin, Eugène (1891), Priv.Doc. Univ. Taschkent (U.d.S.S.R.), Michaylovskaja 14. *Botan., géographique et systématique. Angiospermes de l'Asie Centrale. Umbelliferae et Salsolaceae.* ⓓ Moscou 1921. ⓟ Herbiers de la Flore de l'Asie Centrale.

Korovin, I., Prof. Med. Univ. Ekaterinoslav, Ukraine (U.d.S.S.R.). *Pathol. Anat.* o

Korovin, Michel Kalinikovitch (1883), Prof. géol., paléont. Inst. Technol. de Sibérie, géol. en chef du Comité Géol. de Sibérie. Tomsk, Sibérie (U.d. S.S.R.). *Géol. des gisements houillers, paléont. des invertébrés du Paléozoïque, spécialement des coraux dévoniens et des Brachiopoda.* ⓓ Tomsk 1914. ⓟ Coraux et Brachiopodes du Silurien de la Podkamennaya Toungouska, du Dévonien de Libédiansk, du Carbonifèrien de Tomsk, flore jurassique d'Irkoutsk, restes fossiles des Mammalia de Sibérie: Elephas, Rhinoceros, Bison.

Korsakowa, Marie (1881), Labor. f. Pflanzenchem. u. Pflanzenphysiol. der Akademie d. Wiss. Lenin. grad (U.d.S.S.R.), Wassiliewsky Ostrow, Bolschoi Prospekt 24. *Biochem. des Bodens, Pflanzenphysiol., Fermentforschung.* ⓓ St. Petersburg 1903.

Korschelt, Eugen (1858), Prof. der Zool. und vergl. Anat., Dir. des Zool. Inst. an der Univ. Marburg a. d. L. (Deutschland). *Cytol., Morphol., Entwicklungsgeschichte, Biol. der Tiere.* ⓓ 1882.

Korschikoff, Alexander Arkadjewitsch (1889). Prof. der Botan. (Morphol. u. System.) u. Dir. d. Botan. Labor., Doc. d. Technol. Inst. Charkow (U.d.S.S.R.), Seminarskaja 5. *Mikrobiol. Algol. (Morphol. u. System.), insbesondere Flagellatae, Volvocales, Protococcales.* ⓓ Univ. Charkow.

Korsmo, Emil (1863), Prof. Oslo (Norge), Det kgl. Landbruksdepartement, Victoriaterasse 3. *Biol. der Unkräuter.*

Korstian, Clarence F. (1889), Associate Silviculturist, Appalachian Forest Experim. Station. Asheville, N.C. (U.S.A.). *Silviculture.*

Korteweg, Remmert (1884), Arzt, II. Ass. pathol. Anat. Wilhelminagasthuis. Amsterdam (Holland), Prins Hendriklaan 16. *Pathol. Anat. (Tuberkulose).*

Korvenkontio, Valio Armas (1889), Amanuensis Zool. Mus. Univ. Helsinki (Finnland), Katajanokankatu 4 F. *Mammalia, Histol. der Zähne, Jagdkunde, Pelztierzucht, Bisamratte.* ⓓ Helsinki 1914. ⓟ Schädel und Bälge, Rodentia, Insectivora und Carnivora, Spiritusexempl. oder lebendige Tiere.

Kos, Franc (1885), Kustos. Ljubljana (S.H.S.), Narodni muzej. *Zool.* ⓓ 1920.

Kōsaka, Yūshō, Prof. of Med. Univ. Okayama (Japan), Anat. Inst. *Anat.*

Košanin, Nedelyko (1874), o. Prof. für Botan. Dir. des Botan. Gartens u. Inst. der Univ. Beograd (S.H.S.), Takovska 41. *Flora und Pflanzengeo-*

graphie von Mazedonien und der Balkanländer. Ⓓ Leipzig 1905. Ⓟ Balkanpflanzen, speziell aus Mazedonien.

Koschkina, K. A., Dr., Ass. Anat. Labor. Univ. Tomsk (U.d.S.S.R.). o

Koser, Stewart A. (1894), Ass. Prof. of Bact. Univ. of Illinois. Urbana, Ill. (U.S.A.). *Colon Group of bacteria, food poisoning, botulism, and the bacteriophage.* Ⓓ Yale Univ. 1918. Ⓟ Colon Group of bacteria.

Kosin, Nikolaj I., Doc., Inst. f. Volkserziehung Plechanow. Moskau (U.d.S.S.R.), Walowaja 29. *Botan., Phytopath.* o

Kosinski, Karol Charles (1887), I. Ass. Labor. d'Anat. topographique, chargé de cours d'anat. topogr. Wilno (Polska), Rue Zakretowa, II a. *Innervation cutanée. Veines superficielles des membres inférieures. Variations artérielles.* Ⓓ Cracovie 1914.

Koŝir, Alija, Dr., Doc. Histol., Anat. Inst. d. Med. Fac. Univ. Ljubljana (S.H.S.), Zaloška cesta. *Histol.* o

Kosmin, Natalie P. (1905), Inst. of General Biol., II. Univ., Labor. of Experim. Biol. of the Zoopark. Moscow (U.d.S.S.R.). *Physiol. of development. Metabolism of the eggs of Ascaris.* Ⓓ Moscow 1925.

Kosminsky, Peter (1888), Doc. I. Univ., Wissenschaftl. Mitarbeiter von K.E.P.S. Akademie der Wiss. Moskau (U.d.S.S.R.), Starokonjuschenny 18. *Intersexualität und Gynandromorphismus bei den Schmetterlingen. Temperaturexperim. mit Insekten. Vererbung der Färbung und Zeichnung der Raupen.* o

Koso-Poljanskij, Boris M., Prof. Univ. Woronesh (U.d.S.S.R.), B. Dewizkaja 21. *Botan.* o

Kossel, Albrecht (1853), Geh.R., Univ.Prof. im Ruhestand, Leiter des Inst. für Eiweißforschung der Univ. Heidelberg (Deutschland), Landfriedstr. 7. *Chem. der Eiweißkörper.* Ⓓ Rostock 1878.

Kossinskaja, Ekaterina Konstantinovna (1900), Préparateur du Jardin Botan. Leningrad (U.d. S.S.R.), Perspective Aptekarsky $^1/_3$, ap. 93/b. *Algol.* o

Kossowitsch, Nadeschda Zwowna (1890), Lehrkanzel für Physiol. und Anat. d. Pflanzen. Leningrad (U.d.S.S.R.). *Photosynthese der Holzarten.*

Kostanecki, K., Dr., Prof. Kraków (Polska), Univ. Inst. d'Anat., r. Kopernika 13. o

Kostecki, Edouard (1886), Dr.Sc., Dir. sc. de la Société pour l'amélioration des plantes „Ouditsch". Warszawa (Polska), 25, rue Hoża. *L'amélioration des graines de la betterave à sucre des céreales, des graminées, du trèfle rouge et de la pomme de terre.* Ⓓ Fribourg (Suisse) 1910.

Kostinowitsch, L. S. (1891), Ass. Katheder für normale Anat. Odessa (U.d.S.S.R.), Sadikom-Str. N. 2, 18. *Craniol.* Ⓓ Odessa 1916.

Kostitch, Alexandre George (1893), Prof. d'Histol. et d'Embryol. Fac. Méd., Dir. Inst. d'Histol. Beograd (S.H.S.), 29/III Deligradska. *Glandes génitales: différents facteurs blastophthoriques: action de l'alcool, de l'iode. de l'arsenic et du brome. Hématol.: chondriome des leucocytes, phagocytose.* Ⓓ Strasbourg. Ⓟ Préparations et microphotographies.

Kostjamin, N. N., Prof. Chem.-Pharm. Inst. Odessa (U.d.S.S.R.), Ul. Krasnoj Gvardii 17. *Hygiene.* o

Kostoff, Dontcho (1897), Ass. an der Univ., Agronomische Fac. Sofia (Bulgaria), Wranja 58. *Genetik u. Cytol. mit Berücksichtigung der Pflanzenzüchtung.* Ⓓ Halle a. d. S. 1924.

Kostylew, Nicolaus (1887), Ass. d. Zool. Inst. der Militär-medizin. Akademie. Leningrad (U.d.S.S.R.). *Parasit., Acanthocephalen.* Ⓓ Mil.-Med. Akademie 1912. Ⓟ Helminthen, Acanthocephalen.

Kostytschew, Sergius (1877), Prof. d. Univ., Leiter d. Bureaus f. landwirtschaftl. Mikrobiol. d. Inst. f. experim. Landwirtschaft. Leningrad (U.d.S.S.R.), Akad. d. Wissensch., Labor. f. Biochem. d. Pflanzen. *Pflanzenchem., Pflanzenphysiol. Mikrobiol. des Bodens. Gärungen. Anaerobe Atmung der Pflanzen.* Ⓓ St. Petersburg 1900.

Kotake, Yashiro, Prof. of Provincial Med. Univ. Ōsaka (Japan), Chem. Inst. *Biochem.: protein and amino-acids.*

Kotikova, E. A., Wiss. Lesshaft-Inst. Leningrad (U.d.S.S.R.), Pr. Maklina 32. *Anat.* o

Kotila, John Ernest, Research Ass. in Plant Pathol., Michigan Agric. Coll. East Lansing, Mich. (U.S.A.). o

Kotow, Michail I., Wissenschaftl. Landwirtschaftl. Komitee d. Landwirtschaftl. Akademie. Charkow (U.d.S.S.R.), Karasinskaja 17, W. 1. *Geobotan.* o

Kotowski, Felix (1895), Ph.D., Prof., Coll. of Agric. (Szkoła Główna Gospodarstwa Wiejskiego). Warszawa-Skierniewice (Polska). *Physiol. and morphol. of economic plants; breeding and genetic work on truck crops (problems on germination, growth, seed stimulation, correlations-biometrically).* Ⓓ Cracov, Poland, 1919. Ⓟ Collections of seeds of economic plants, cultivated in Poland.

Kotsaftis, A., Chef du Labor. de Physiol. à l'Univ. Athènes (Grèce), Rue Scoufa 5. o

Kotte, Walter (1893), Dr. phil., Regierungsbotaniker. Freiburg i. Br. (Deutschland), Badisches Weinbau-Inst. *Pflanzenphysiol., Phytopath., Pflanzenschutz.* Ⓓ Berlin 1920. Ⓟ Phytopath. Praep. und Photographien.

Kotthoff, Peter (1883), Abteilungsvorsteher an der Anstalt für Pflanzenschutz und Samenkontrolle. Münster i. W. (Deutschland), Südstr. 76. *Phytopath.* Ⓓ Münster i. W. 1913.

Kouda, Jacob M., Chef des travaux, Inst. Agronomique. Kiev (U.d.S.S.R.). *Mycol.* o

Kovačević, Željko (1893), Prof., Chef der Entomol. Phytopath. Abt. Poljoprivredna Ogledna i Kontrolna Stanica Osijek I (S.H.S.). *Myriapoden. Glomeriden Kroatiens: Biol. Bekämpfung der Schädlinge. Schwammspinner u. Ringelspinner und ihre Parasit.* Ⓓ Zagreb 1922.

Kovács, András (1902), Ass. Pécs (Ungarn), Baranya Comitat Irgalmasok-ut 18. *Pathol. des Herzens und des Gefäßsystems.* Ⓓ Pécs 1926.

Kovács, Karl, Dr. med., Ass. am Anat.-biol. Inst. d. Univ. Debreczen (Ungarn). o

Kowalski, Władysław (1891), Ass. chir. Univ.-Klinik. Poznań (Polska), ul. Długa 1. *Innere Sekretion: Schilddrüse, Nebenschilddrüse, Thymus.* Breslau 1920.

Kowalsky, Wictor (1899), Ass. d. Chem.-Pharmaz. Inst., Aspirant d. wissenschaftl. Forschungs-Inst. f. Biol. Odessa (U.d.S.S.R.), Wneschnaja 72. *Giftigkeit d. Salzlösungen für Organismen.*

Kowesschnikowa, A. K., Abt.-Leiter, Lesshaft-Inst., Anat. Labor. Leningrad (U.d.S.S.R.), Torgowoja 12. o

Koževnikov, Grigorij Alek., Prof. I. Univ. Moskau (U.d.S.S.R.), *Zool., Embryol.* o

Kozhautschikow, Wassily Dmitrievitsch (1866), Dir. d. Martjanovschen Staatsmus. Minussinsk, Sibirien (U.d.S.S.R.), Martjanov-Str. 6. *Aphadiini (Coleoptera, Lamellicornia) und Lepidoptera. Faunistik.* Ⓓ St. Petersburg 1892. Ⓟ Coleoptera und Lepidoptera.

Kozikowski, A., Prof. Lwów (Polska), r. Nabielaka 22. *Zool.* o

Kozin, N. J., Doc. L.-V.-Plechanov-Inst. f. Volkswirtschaft. Moskau (U.d.S.S.R.), Pereulok 28. *Phytopath.* o

Kozłowska, Aniela (1898), Ass. de l'Univ. Jagiellonienne Kraków (Polska), L'Inst. botan., Lubicz 46. *Géographie botan., Paleobotan., Systematique.* Ⓓ Cracovie 1921. Ⓟ Herbier.

Kozlowski, Anton (1889), Ph.D., Chem. in State Dept. of Health in Albany, N.Y. (U.S.A.). *Plant cytol., plant- and animalbiochem. Bact., Plantphysiol.* Ⓓ Lwów 1919.

Kozłowski, Roman (1889), Prof. Paléont. et Géol. Univ., Chargé des cours de Paléont. Univ. Warszawa

(Polska), Zakład Geologiczny. *Brachiopodes fossiles. Fossiles dévoniens de Parana.* ① Paris (Sorbonne) 1910. ② Brachiopodes carbonifères de Bolivie.

Koźmiński, Zygmunt (1902), Ass. Stat. Hydrobiol. de Wigry Suwałki (Polska). *Ecol. et systématique des Orthoptères, Entomostracés, Amphibiens et Reptiles.*

Kozo-Polianski, Boris M., Prof. Botan. à l'Inst. Agronomique. Voronèje (U.d.S.S.R.), Universitätskaja 5. *Mycol.* o

Krabbe, Knud Haraldsen (1885), Priv.Doc. København (Danmark), Osterbrogade 21. *Anat. of the pineal gland and subcommissural organs, endocrine glands and neurol.* ① København 1915.

Kracheninnikov, Hippolyte (1884), Conserv. Leningrad (U.d.S.S.R.), Jardin Botan. Principal. *Geobotan., Systématique: Composées surtout Artemisia, végétation des Steppes et déserts.* ① Moscou 1913.

Kraemer, Hermann (1872), o. Prof. an der Univ. Gießen, Dir. des Inst. für Tierzucht. Oberer Hardthof bei Gießen (Deutschland). *Tierzucht, Beurteilung der Tiere, Vererbungsfragen, Abstammungslehre, Rassengeschichte der Haustiere.* ① Zürich 1899.

Kräusel, Richard (1890), Dr. phil., Priv.Doc. an der Univ. Frankfurt a. M. (Deutschland), Hohenzollernplatz 24. *Palaeobotan., Anat. der Pflanzen. Kohlenbildung.* ① Breslau 1913.

Krainskij, Sergej W., Prof., Kubansches Landw. Inst. Krasnodar (U.d.S.S.R.). *Angew. Botan. Gartenbau. Experim. Morphol. u. Biol.* o

Krajina,Vladimir, Ass.Inst.de Botan. pharmaceut. Univ. Praha VI (C.S.R.), Benátská 433. *Phanérogames, géobotan.* o

Kram, Kyril M., Ass., Landw. Bezirks-Versuchsstation. Tulun (U.d.S.S.R.), Irkutsker Gouv. *Agronomie. Selektion landw. Pflanzen.* o

Kramer, Otto (1891), Dr. rer. nat., Ökonomierat, Vorstand der württ. Versuchsanstalt für Wein- und Obstbau in Weinsberg (Deutschland), Urbanstr. *Schädlingsbekämpfung im Weinbau und Obstbau, Gärungsphysiol. und -bact., Weinchem.* ① Greifswald 1919.

Kramp, Paul Lassenius (1887), Mag.sc., Ass. curator at the Zool. Mus. of the Univ. København (Danmark), Krystalgade. *Coelenterata, Medusae and Hydroids, compar. morphol. and classification, geographical distribution, seasonal occurrence, life history, and relation to physical conditions. Biol. of marine wood-boring animals.* ① Copenhagen 1911. ② Coelenterata, espec. from North European seas. Medusae from all parts of the world.

Krantz, Fred A. (1890), Ass. Prof. in Horticulture, Univ. of Minnesota, Ass. Horticulturist, Minnesota Agric. Experim. Station. St. Paul, Minn. (U.S.A.), Univ. Farm. *Potato Breeding, Vegetable Breeding.* ① Univ. of Minnesota 1924. ② Potato seed or tubers of seedlings selected in self-fertilized lines.

Krapivin, Vera, (1900), Ass. Station. of Acclimatation of Agric. Inst. Leningrad (U.d.S.S.R.), Detskoe Selo, 2 Kolpinskayast. *Genetical and selectionary study of the Tobacco plant.*

Krascheninnikoff, Fedor N. (1869), Prof. I. et II. Univ. Moscou (U.d.S.S.R.), Labor. botan., Gerzena, Univ. log. 66. *Physiol. végétale, Photosynthèse, Chim. végétale, Echange gazeux.* ① Moscou 1901.

Krascheninnikow, Sergius (1895), Priv.Doc. Vet.-Zootechn. Inst. Kiew, Ukraine (U.d.S.S.R.), Fundukliwskastraße 62-9. *Protistol. Cytol. der Ophryoscoleciden. Systematik der Gattung Trachelomonas.* ① Kyjiw 1923.

Krasinsky, Nik. Petr, Doc. I. Univ. Moskau (U.d.S.S.R.). *Botan.* o

Krasnikov, Vasilij Vas., Prof. I. Univ. Moskau (U.d.S.S.R.). *Agric. Chem.* o

Krasnogorskij, N. J., Prof. Staatl. Pädagog. „Herzen" Inst. Naturwiss. Abt. Leningrad (U.d. S.S.R.), Moika 48. *Anat. u. Physiol. d. Kindes.* o

Krasnosselsky-Maximow, Tatiana A. (1884), Prof. botan. School of pharmacy, Ass. Prof.of Plant Physiol.

Inst. of Applied Botan. and New Cultures. Leningrad (U.d.S.S.R.), 44 Herzenstr. *Waterrelations of plants, physiol. of germination.* ① St. Petersburg 1904.

Krassawin, Mstislaw I., Dir. Abt. f. Gartenbau d. Versuchsstation f. Gartenbau u. Landwirtschaft. Sotschi (U.d.S.S.R.), Komsomolskaja ul. 12. o

Krassinskij, Nikolaj P., Doc. I. Univ. Moskau (U.d.S.S.R.), B. Archangelskij per. 12, W. 7. *Botan., Pflanzen-Anat.*

Krassovsky, Irene W. (1896), Ass. Plant Physiol. Inst. of Applied Botan. and New Cultures. Leningrad (U.d.S.S.R.), Herzen Street 44. *Plant Physiol.: Root systems.*

Krassowski, P. N., Ass. am Botan. Inst. u. d. Biol. Station der Univ. Záimko, Perm II (U.d.S.S.R.). *Sumpfforschung.* o

Krassuskaja, A. A. (1854), Prof., Dir. Anat. Labor. Leningrad (U.d.S.S.R.), Wissenschaftl. Inst. Lesshaft, Pr. Maklina 32. *Anat. d. Menschen.*

Krastin, Lucija, Dr., Ass. Univ. Riga (Latvija), Kalpaka bulv. 10, dz. 26 u. *Anat.* o

Kratinoff, Alexander (1900), Labor. experim. Biol. an der Swerdlov Univ. Moskau 55 (U.d.S.S.R.), Mjusskaja Pl. 3, Biolabor. *Motorische Tätigkeit des Verdauungstraktus. Schilddrüse.*

Kraus, Erik Johannes (1887), Dr., Priv.Doc. für pathol. Anat. an der Deutschen Univ. und I. Ass. am Pathol. Inst. Prag II (C.S.R.), Trojická 4. *Pathol. Anat. und Histol. des endocrinen Systems. Milz, haematopoetisches System, Pigmente, Bact.* (*Hühnertyphus, Faecalis-Gruppe, Paratyphus β, Kapselbacterien*). ① 1911. ② Histopathol. Praep.

Kraus, Ezra Jacob (1885), Prof. of Applied Botan., Univ. of Wisconsin. Madison, Wis. (U.S.A.), Biol. Building. *Interrelations of vegetative and reproductive functions of flowering plants; morphol. and anat. of agric. plants.* ① Univ. of Chicago 1917. ② Microscopical preparations of various cultivated plants.

Kraus, Friedrich (1858), Geh.Med.R., o.ö. Prof. für innere Med. an der Univ. Berlin NW (Deutschland), Brückenallee 7. *Persönlichkeitslehre.* ① Prag 1882.

Kraus, Rudolf (1868), Univ.Prof. an der med. Fac., Dir. des staatl. serotherapeutischen Inst. Wien IX (Deutsch-Österreich), Zimmermanngasse 3. *Microbiol., Sero- u. Immunobiol.* ① Prag 1893.

Krause, Kurt (1883), Dr. phil., Kustos und Prof. am Botan. Mus. der Univ. Berlin. Berlin-Dahlem (Deutschland), Königin-Luisen-Str. 6—8. *Systematische Botan.: Araceae, Liliaceae, Loranthaceae, Rutaceae, Sapotaceae, Rubiaceae. Floristik: orientalische Flora.* ① Berlin 1905.

Krause, Rudolf (1865), ao. Prof. der Anat. und Prosektor am anat.-biol. Inst. der Univ. Berlin-Halensee (Deutschland), Nestorstr. 1. *Bau und Entwicklung der Sinnesorgane und der Drüsen.* ① Berlin 1891.

Krauspe, Carl August (1895), Ass. am Pathol. Inst. der Univ. Leipzig (Deutschland), Liebigstr. 26. *Pathol. Anat. der Infektionskrankheiten. Pathogenese der Infektionskrankheiten. Allgemeine Immunitätslehre. Pathol. Anatomie der Kinderkrankheiten.* ① Königsberg I. Pr. 1921.

Krauss, Frederick George (1870), Prof. of Agronomy and Genetics, Univ. of Hawaii, and Consulting Geneticist, Experim. Station of the Association of Hawaiian Canners, Honolulu (Hawai), 2447 Parker Street. *Breeding of Pineapples (Ananas sativus) and Leguminous Field Crops. Genetical analysis.* ① Univ. of Hawaii 1921. ② Seeds of Crop Plants.

Krausse, Anton Hermann (1878), Zool. an der Forstlichen Hochsch. in Eberswalde (Deutschland), Hohenzollernstr. 1. *Systematik und Biol.; Forstzool.; angewandte Entomol.; Säugetierkunde und Myrmekol.* ① Phil. Jena, sc. nat. Cagliari. ② Kleinsäuger und Ameisen.

Kraybill, Henry Reist (1891), Prof of. Agric. Chem. and State Chem., Purdue Univ. Lafayette, Ind. (U.S.A.). *Chem. of growth and reproduction in plants. Role of phosphorus in plant metabolism. Biochem. of virus diseases of plants. Chem. of the apple tree. Protein synthesis in plants.* ③ The Univ. of Chicago 1917.

Krečatovič, Lev. Melch., Prof. I. u. II. Univ. Moskau (U.d.S.S.R.). *Botan.* ○

Krecker, Frederick H. (1881), Prof. of Zool., Ass. Dir., Franz Theodore Stone Labor. Columbus, O. (U.S.A.), Department of Zool., Ohio State Univ. *Regeneration and animal Ecol. Annelida. Aquatic animals.* ④ Princeton 1909.

Krečman, R. J., Doc. Univ. Nizny-Novgorod (U.d.S.S.R.), Sovetskaja Ploščad 8. *Pharm.* ○

Krediet, Gerrit (1886), Prof. der Vet.-anat., Univ. Utrecht (Holland), Fred. Hendrikstraat 82. *Hermaphroditismus bei Haustieren. Biol. des Geschlechtsapparates. Teratol. der Haustiere.* ③ Bern 1910.

Krehan, Max (1889). Děčin-Libverd (Tetschen a. d. Elbe) (C.S.R.), Agrikulturchem. Labor. der landwirtsch. Hochsch. *Biol. u. Physiol. der Essigbakterien, Bedeutung der Salze für Wachstum und enzymatische Prozesse in der Essigbakterienzelle. Photokatalyse. Stoffaufnahme in die lebende Pflanzenzelle.* ① Prag.

Krehl, Ludolf (1861), Prof., Dir. der med. Univ.-Klinik. Heidelberg (Deutschland), Berstr. 106. *Pathol. Physiol.* ③ 1887.

Kreidl, Alois, Univ.Prof., Dir. des Inst. für Vergl. Physiol. Wien VIII (Deutsch-Österreich), Schlösselgasse 13. *Physiol.: Reiznachwirkung. Irreziprozität. Stimm- u. Musikapparate bei Tieren. Sinnesorgane.* ○

Kreier, Georgij K., Dir. Versuchsstat. f. Arzneipfl. Mogilew (U.d.S.S.R.). *Botan., Arzneipflanzen.* ○

Kreitmair, Hanns (1894), Dr. med., Leiter der pharmacol.Abteilung derChemischenFabrikE.Merck. Darmstadt (Deutschland), Liebigstr. 24. *Pharm. (Wirkung von Giften, Wertbestimmung von Arzneimitteln).* ① Erlangen.

Kremer, Johann Paul (1883), II. Prosektor am Anat. Inst. zu Bonn a. Rh. (Deutschland). *Die Metamorphose der Insekten und der Wirbeltiere in ihrer Beziehung zur modernen Zellforschung.* ③ Berlin phil. 1914, med. 1919.

Kremky, Jerzy (1897), Ass. de la section entomol. du Mus. d'Histoire Natur. Warszawa (Polska), rue Krakowskie Przedmieście 26/28. *Macrolépidoptères. Systématique et morphol. La faune de la Pologne, les Danaïdides de l'Amérique méridionale.*

Krempf, Armand (1879), Dr. ès sc., Dir. de l'Inst. océanograph. de l'Indochine. Cauda par Nhatrang, Annam (Indochine françoise), Labor. *Coelentérés: Anthozoaires et Cténophores, Applications industrielles: Forines de Poisson et produits d'autolyse riches en acides amidés.* ① Paris. ② Forines de poissons. *Autolysots de poissons préparés. Animaux marins.*

Krenner, Endre (1900), Ass. Pflanzenphysiol. u. Phytopath. Inst. Budapest II (Ungarn), Debröi-ut N. 15. *Krankheiten der Champignons, Getreiderost, Getreidebrand, Kartoffelkrankheiten, Rübenkrankheiten. Pflanzensystematik: Bacillarien, Süßwasseralgen; Oscillarien, Bewegung.* ④ Budapest 1924.

Krepuska, Gyula (1893). Budapest X (Ungarn), Hédervary-ut 21. *Protistol. Fauna hungara.* ○

Krestovnikov, A. N., Vize-Dir. Wiss. ,,P.-O.-Lesshaft-Inst.''. Leningrad (U.d.S.S.R.), Pr. Maklina 32. *Physiol.* ○

Kretschetowitsch, Lew M., Prof. Dr. I. u. II. Univ. u. Höheren Zootechn. Inst. Moskau (U.d.S.S.R.), Ul. Gerzena 6, Univ., W. 58. *Pflanzen-Morphol. Systematik. Pflanzen-Geographie. Phytopath. Flora i. Devon u. d. Unterkreide.* ○

Kreuter, Eugenia (1886), Ass., State Inst. of Experim. Agronomy, Bureau of Applied Entomol. Leningrad (U.d.S.S.R.), 44 Herzen Street. *Part of temperature in biol. of Diptera.* ④ 1921. ② *Diptera; Chloropidae.*

Kribs, David Alson (1896), Instr. of Forest Products. St. Paul, Minn. (U.S.A.), Division of Forestry, Univ. Farm. *Compar. anat. of wood, natural classification of plants.*

Kričevskij, Il'ja L'vovič, Prof. II. Univ. u. Med. Inst. Moskau (U.d.S.S.R.), ul. Spartaka 2. *Bact.* ○

Krieg, Hans (1891), Dr., Entomol. der W. Güttler A.-G. Hamburg 21 (Deutschland), Hofweg 50 II. *Angewandte Entomol. Arsenmittel und ihre Wirkung auf fressende Insekten unter Berücksichtigung biol., physiol. und meteorol. Verhältnisse. Flugzeugbekämpfung. Systematik: Termiten.* ① Bonn 1922. ② *Wirtschaftlich wichtige fressende Schädlinge.*

Krieger, Louis Charles Christopher (1873), Botan. mycol. in The Howard A. Kelly Mycol. Library, Baltimore, Md. (U.S.A.), 2114 N. Calvert St. *Mycol., Taxonomy of the Basidiomycetes; Higher Fungi of the world.* ② North American Basidiomycetes.

von Kries, Johannes (1853), Geh.R., Prof. emeritus. Freiburg i. Br. (Deutschland), Goethestr. 42. *Physiol.: Muskeln, Kreislauf und Sinnesorgane. Psychol.* ① Leipzig 1875.

Krige, A. V., Lect. Univ. of Stellenbosch (South Africa), Dept. of Geol. ① Stellenbosch. ○

Krigl, Kálmán (1902), Path.-anat. Inst. Univ. Pécs (Ungarn), Perczel ucca 24. *Pathol. Histol.* ③ Pécs 1927.

Krijgsman, Berend Jan (1901), Ass. am Inst. f. Parasit. und Infektionskrankheiten, Leiter des Labor. f. Tropenkrankheiten a. d. Vet. Fac. d. Univ. Utrecht (Holland), Biltstraat 172. *Parasit. Protozoen. Histol. d. Secretion und Resorption.* ② Praep. von parasit. Protozoen.

Krimberg, Eugenija, Dr., Ass. Univ. Rīga (Latvija), Brīvības ielā 15, dz. 12. *Physiol. u. physiol. Chem.* ○

Krinizkij, Chalva Jos., Prof. Gosudarstv. Donskoj Univ. Rostov a. Don (U.d.S.S.R.). *Pathol. Anat.* ○

Krischtafowitsch, Nikolaus J., Prof. Landwirtschaftl. Inst. Charkow (U.d.S.S.R.). *Stratigraphie des Quartärs.* ○

Kristensen, Martin Kristian (1888), Abteilungsvorsteher Statens Seruninst. København (Danmark). *Bact. Classification of the haemoglobinophilic Bacteria.* ① Kopenhagen 1922.

Kristofferson, Karl Birger (1887). Landskrona (Sverige). *Genetic: Aquilegia vulgaris. Brassica oleracea, Daucus carota, Malva sp. Phaseolus vulgaris, Viola tricolor.* ① Lund 1926.

Křiženecký, Jaroslav (1896), Doc. der allgemeinen Biol. an der Landwirtsch. Hochsch., Vorstand der Sektion für Züchtungsbiol. im Zoocentr. Landes-Forschungs-Inst. Brno (C.S.R.), Černá pole. *Innere Sekretion, Sexualbiol., Vitamine, Entwicklungsmechanik, landwirtschaftliche Tierzucht, Vererbung erworbener Eigenschaften, Wirkung der äußeren Verhältnisse auf den Tierorganismus, Physiol. und Vererbung der Produktionseigenschaften der landwirtschaftlichen Tiere.* ① Prag 1919.

Krjukot, Alexander (1878), Prof. Univ. Taschkent (U.d.S.S.R.), Djukowskaja 36. *Morphol. des Blutes, normale und pathol.Morphol. der blutbildenden Organe.* ① Moskau 1909.

Kröber, Otto Th. (1882), Dipterol. am Zool. Staats-Inst. und Zool. Mus. Hamburg 24 (Deutschland), Graumannsweg 63. *Diptera; Conopidae, Therevidae, Omphralidae, Tabanidae der Welt.*

Krölling, Otto (1891), Priv.Doc., Dr. med.vet., diplom. Tierarzt (o. Hochsch. Ass.). Wien III (Deutsch-Österreich), Linke Bahngasse 11, Tierärztl. Hochsch. *Embryol.: Plazenta. Histol.: Entwicklung der Hautdrüsen bei Säugetieren.* ① Wien 1920. ② Embryonen, histol. Praep.

Kroemer, Karl (1871), Prof. Dr., Dir. der Pflanzenphysiol. Versuchsstation und der Wissenschaftlichen Abteilung der Rebenversuchsstation an der Lehr- und Forschungsanstalt für Wein-, Obst- und Gartenbau in Geisenheim a. Rh. (Deutschland).

Physiol. der Rebe und der gärtnerischen Nutzpflanzen, Mykol. der Weinbereitung. Ⓓ Marburg a. d. L. 1903. Ⓟ Hefen und Weinorganismen.

Kröning, Friedrich Johann Karl (1897), Dr. phil., Ass. Göttingen (Deutschland), Bahnhofstr. 28, Zool. Inst. *Keimzellreifung. Vererbung Säuger.* Ⓓ Göttingen 1922.

Kroetz, Christian Albert (1894), Priv.Doc. innere Med. Greifswald (Deutschland), Med. Univ.-Klinik. *Mineralstoffwechsel; Säurebasengleichgewicht. Biochem. und Biophysik der Strahlenwirkungen. Hämodynamik.* Ⓓ 1920.

Krogh, Marie (1874), Dr. med., Priv.Doc. København (Danmark), Ny Vestergade 11. *Stoffwechseluntersuchungen (respiratorische). Wertbestimmungen (biol.) von Arzneimitteln, Digitalis, Strophanthin u. a.* Ⓓ Copenhagen 1907.

Krogh, Schack August Steenberg (1874), Prof. of Zool. physiol., (Sc. Fac.). Univ. København B (Danmark), 11 Ny Vestergade. *Physiol. of respiration and respiratory metabolism in vertebrates and invertebrates. Physiol. of the blood capillaries and capillary circulation.* Ⓓ Phil. Copenhagen 1903, LL. D. h.c. Edinburgh 1921.

Krogius, August (1871), Prof. Psychol. Staatsuniv. Saratow (U.d.S.S.R.).

Krohn, Válnó (1891), Ph.D. Hamina (Finnland). *Botan. Nahrungsphysiol.* ⒹHelsinki 1923. ⓅPflanzen.

Krokos, Woldemar J., Prof. Geol., Miner. Inst. für Landwirtschaft. Odessa (U.d.S.S.R.). *Palaeont. des Tertiärs und fossile Böden Südrußlands.* o

Kromayer, Ernst (1862), Prof. Dr. med., Arzt. Berlin W 15 (Deutschland), Meinekestr. 27. *Normale und pathol. Anat. der Haut.* Ⓓ Bonn 1885.

Kronacher, Carl (1871), o. Prof. und Dir. des Inst. für Tierzucht und Vererbungsforschung der Tierärztlichen Hochsch. Hannover (Deutschland), Hohenzollernstr. 31 II. *Haar- und Wolleforschung. Haustiergenetik (Schweine, Ziegen, Pferde), Konstitutionsforschung, Inzuchtforschung, Körperbauehre, Pferdezucht, Schweinezucht, Ziegenzucht.* Ⓓ Med. vet. Bern 1903, Med. vet. h. c. Leipzig, Dr. der Landwirtschaft e. h. Hohenheim, Dr. der Bodenkultur e. h. Wien. Ⓜ Mikroskopische und sonstige Praep. von Haar und Wolle. Tierschädel.

Kronberger, Max (1886), R.R., Bayr. Landesanst. Pflanzenbau u. Pflanzenschutz. München (Deutschland), Mauerkirchenstr. 2/III. *Landw. Bact.*

Kronfeld, Peter (1899), Ass. 1. Univ.-Augenklinik. Wien IX (Deutsch-Österreich), Alserstr. 4. *Physiol. des Auges, insbesondere des Flüssigkeitswechsels im Auge.* Ⓓ Wien 1923. Ⓟ Histol. Praep. tierischer und menschlicher Augen.

Krontowski, Alexej (1885), Prof., Vorstand der Abt. f. experim. Med. des Bact. Inst. und der Abt. für Biol. und experim. Med. des Röntgeninst. Kiew, Ukraine (U.d.S.S.R.), Bact. Inst. oder Kiew, Röntgeninst., Tolstoi-Str. 7. *Pathol. Physiol., Bact.; Gewebskulturen, Explantation, physikalisch-chem. u. mikrochem. Untersuchungen des Stoffwechsels in Gewebskulturen aus normalen und Krebsgeweben; deren Beeinflussung durch Röntgenstrahlen; innere Sekretion. Analyse der Immunitätserscheinungen mittels Explantation. Experim. und vergl. Pathol. der bösartigen Neubildungen. Erblichkeits- und Konstitutionslehre.* Ⓓ Kiew 1911.

Kroon, Henri Margarethus (1868), Prof., Reichsuniv. Utrecht (Holland), J. W. Frisostraat 1. *Zootechnik und Tierhygiene. Beurteilung und Rassen der Haustiere. Züchtungsbiol. Tierzucht (Eigenschaftsanalyse bei den Haustieren). Futtermittelkunde und Fütterung der Haustiere. Stallhygiene. Pflege der Haustiere und Milchgewinnung.* Ⓓ Bern 1915.

Krosby, Peter (1893), Doc. in Pflanzenkultur. Aas (Norge), Landw. Hochsch. *Samenbeizung. Brassica rapa und Br. napus.*

Krotkina, Marie A. (1888), Labor. of Plant Physiol. at the Inst. of Applied Botan. and New Cultures. Leningrad (U.d.S.S.R.), 44 Herzenstr. *Waterrelations of plants, physiol. of crop plants.*

Kroulik, Alois, Dr. ing., Priv.Doc. f. allg. u. landwirtsch. Bact. Praha XII (C.S.R.), Chodska 2. o

Krügel, Curt (1876), Dr., Dir. Landwirtsch. Versuchsstat. Hamburg-Horn. Hamburg 26 (Deutschland). Ⓓ Breslau 1897.

von Krüger, Friedrich (1862), Russ. Wirkl. Staatsrat, Exzellenz, ao. Prof. der physiol. Chem. an der Univ. Rostock i. M. (Deutschland), Patriotischer Weg 24. *Physiol. Chem.: Blut und Blutfarbstoff. Verdauungsfermente.* Ⓓ Dorpat 1886.

Krüger, Heinrich Wilhelm (1857), Prof. Dr., Dir. der Versuchsstation Bernburg i. Anhalt (Deutschland). *Biol. und Kultur der Zuckerrübe. Mikrobiol.* Ⓓ Halle a. d. S. 1883. Ⓟ Samen.

Krüger, Leopold (1901), Ass. am Inst. für Tierzucht u. Züchtungsbiol. der Techn. Hochsch. München (Deutschland). *Tierzucht und Züchtungsbiol., Haarfarben.* Ⓓ München, 1926.

Krüger, Paul (1886), ao. Prof. an der Univ. Berlin N 4 (Deutschland), Invalidenstr. 48, Zool. Inst. *Cirripedien, vergl. Physiol. der Tiere: Stoffwechsel.* Ⓓ Halle a. d. S. 1910.

Kruizinga, Pieter (1885), Conserv. Mineral. Geol. Mus. Inst. f. Bergbaukunde der Techn. Hochsch. Delft. Rijswijk (Holland), Julianastr. 21. *Palaeont. und Geol.* Ⓓ Groningen 1918.

Krukowski, Marjan (1879), Dipl. Ingenieur der Landwirtsch., Leiter der Subabteilung für Pflanzenzucht des Staatlichen Wissenschaftlichen Inst. für Landwirtschaft zu Bydgoszcz (Polen), Krakowska 2a. *Genetik (Saatzucht).* Ⓓ Riga 1913. Ⓟ Pflanzen.

Krull, Christian (1895), Priv.Doc. und wissenschaftl. Ass. der Mitscherlich-Gesellschaft am Pflanzenbauinst. der Albertus-Univ. in Königsberg, Pr. (Deutschland). *Pflanzenphysiol., Reaktion der Pflanzen gegen Säure und Alkalität ihres Wachstumsmediums.* Ⓓ Königsberg, Pr., 1923.

Krull, Rudolf (1861). Breslau X (Deutschland), Rosenthaler Str. 45. *Holzzerstörende Pilze und Pilzschäden am verbauten Holz (Hausschwamm, Trockenfäule etc.).* Ⓓ 1890. Ⓟ Schausammlungen der „Holzzerstörung durch Haus- und Baumpilze".

Krumbach, Thilo (1874), Prof. Berlin NW 7 (Deutschland), Georgenstr. 34—36. *Chaetopoden, Ctenophoren. Biol. Meereskunde. Allg. Biol.* Ⓓ Breslau 1896.

Krumbeck, Lothar (1878), Dr. phil. Univ. Prof. Erlangen (Deutschland), Min.-Geol. Inst. *Palaeogeographie von Nordbayern. Stratigraphie d. nordbayrischen Lias.* Ⓓ München.

Krumbhaar, Edward Bell (1882), Associate Prof. of Pathol., Univ. of Pennsylvania, Grad. School of Med., Dir. of Labor., Philadelphia Gen. Hospital. Philadelphia, Pa. (U.S.A.). *Pathol. and Experim. Med.* Ⓓ Univ. of Pennsylvania M.D. 1908, Ph.D. 1916.

Krummacher, Otto (1864), Dr. med., o. Prof. der Physiol. und Abteilungsvorsteher am physiol. Inst. Münster i. W. (Deutschland), Warendorfer Str. 76. *Stoffwechselphysiol. Krystallographie der Hämoglobinarten.* Ⓓ Bonn 1890.

Krumwiede, Charles (1879), Prof. of Bact. and Hygiene Univ. and Bellevue Med. Coll. New York, N.Y. (U.S.A.), Foot E, 16 St. o

Krunoslav, Babić (1875), Kustos am Kroatischen Zool. Landes-Mus. und Hon.Doc. für Zool. der Univ. Zagreb (S.H.S.), Demeter-Gasse 1. *Morphol. der Spongien der kroatischen Gewässer und des Adriatischen Meeres. Biol. und bionomische Verhältnisse im Adriatischen Meere.* Ⓓ Zagreb 1899.

Krupko, Stefan (1890), Ass. de l'Univ. Warszawa (Polska), Jardin Botan., Al. Ujazdowskie 6/8. *Mitochondries et Plastides. Embryol. des Phanérogames.* Ⓓ Kieff 1918.

Kruse, Walther (1864), o. Prof. der Hygiene, Dir. Hygien. Inst. Leipzig (Deutschland), Liebigstr. 24. *Bact. Ernährung. Hygiene.* ⓓ Berlin 1888.

Kruyt, Hugo Rudolph (1882), o. Prof., Univ. Utrecht (Holland), Wilhelminapark 37. *Kolloidchem.* ⓓ Utrecht 1908.

Krylov, Dimitry (1879), o. Prof. Univ. Sofia (Bulgarien), Pathol. Anat Inst. Alexandrova Bolnica. ⓓ St. Petersburg 1903.

Krylov, Porphyry Krylov (1850), Prof. of Botan., Dir. of the Herbarium of Univ. Tomsk, West Siberia (U.d.S.S.R.), Timirjasewskij prosp. 3. *Phytogeography, Floristic, Systematic of the floral plants.* ⓓ Univ. of Kazan 1906. ⓟ Dry Siberian plants.

Kryński, L., Dr., Prof. Warszawa (Polska), Inst. d'Anat. Topogr., r. Chałubińskiego 5. o

Kryshtofovich, Africanus (1885), Obergeol. Comité Géol., Prof. Univ. für Phytopaleontol. und Kaustobioliten, Doc. Bergakademie f. Paleobotan., Konserv. d. Botan. Gartens, Herbarium. Leningrad (U.d.S.S.R.), Srednij Prospekt, 72. *General Paleobotan. Quaternary flora of Sibiria and European Russia; Tertiary flora of U.d.S.S.R., especially Ukraina and the Far East; Cretaceous flora of Sakhalin and Amurland, and the Jurassic flora of the whole U.d.S.S.R. Paleoclimatol. Correlation of the American and East Asiatic floral phases.* ⓓ Odessa 1908. ⓟ Cretaceous plantimpressions.

Krzemieniewska, Helène (1878). Lwów (Polska), r. Długosza 5. *Microbiol.*

Krzemieniewski, Seweryn (1871), Dr. phil., Prof., Fac. des Sc. à l'Univ. Lwów (Polska), r. Mikotaja 4. *Microbiol. (Myxobacteria).* ⓓ Cracovie 1901.

Krzysik, Stanislaw Marjan (1891). Warszawa (Polska), Inst. zool. Univ., rue Krakowskie Przedmieście 26/28. *Hydrobiol., Limnol.: Tricladida, Bryozoa, Spongillidae, Amphipoda d'eau douce.* ⓓ Varsovie 1927.

Krzyszlalowicz, François (1868), Dr. med., Prof. de dermatol. et vénérol. à l'Univ. Warszawa (Polska), rue Koszykowa 82a. *Histol. normale et pathol. de la peau bact. des maladies cutanées — spirochaete pallida — mycoses de la peau (teignes).* ⓓ Cracovie (Pologne) 1892. ⓟ Cultures des champignons de teignes (trichophyton, microsporon, favus).

Krzyszkowsky, Konstantin Nicolaewitsch (1877), Prof. Dir. d. Tierphysiol. Labor. im Agronom. Inst. Leningrad (U.d.S.S.R.), Detskoe Ssèlo. *Bedingte Reflexe u. bedingte Reaktionen, Tierpsychol., Physiol. d. Vögel: Verdauung, Fortpflanzung, Verdauungsphysiol.* ⓓ Leningrad (Mil. Acad.) 1906.

Krzywanek, Fr. Wilhelm (1896), Dr. med. vet., Priv.Doc. für Tierphysiol., Ass. am vet.-physiol. Inst. der Univ. Leipzig (Deutschland), Tiroler Str. 6. *Physiol. der Atmung, des Stoffwechsels und der Verdauung, Milz.* ⓓ Dresden 1921.

Kubart, Bruno (1882), Dr. ao. Prof. an der Univ. Graz, Vorstand des phytopalaeont. Labor. der Univ. Graz (Deutsch-Österreich), Holteigasse 6. *Phytopalaeont.* ⓓ Wien 1906. ⓟ Dünnschliffe.

Kubichek, W. F. (1892), Instr. in Zool., Curator of B. H. Bailey Mus. Cedar Rapids, Ia. (U.S.A.), Coe Coll. *Economic Ornithol.* ⓓ Iowa 1917. ⓟ Birds, skins and mounted.

Kubo, Inokichi (1874), Prof. Kaiserl. Kyūshū-Univ. Fukuoka (Japan), Nr. 105 Daimyō-Str. *Physiol. und Pathol. des Bogengangapparates. Lepidopterol., besonders Rhopalocera.* ⓓ 1908.

Kubo, Seitoku, Ass. Prof. of Keiō-Gijuku Univ. Yotsuya, Tōkyō (Japan), Physiol. Inst., Coll. of Med.

Kučera, Petr (1893), Ing.agr., Inst. des Recherches agr. Košice (C.S.R.) *Pédol. Agrochim. Analyse méchanique.*

Kuchler, Ludwig Franz (1892), Assessor, stellv. Abteilungsleiter, Futtermittelkontrollabteilung der bayr. Landesanstalt für Pflanzenbau u. Pflanzenschutz, München (Deutschland). *Biol. Futtermitteluntersuchungen, tierische Ernährungsphysiol.* ⓟ Futtermittelprobe, mikroskopische Futtermittelpraep.

Kuczynski, Max Hans (1890), Dr. phil. et med., Abteilungsvorsteher am Pathol. Inst. der Univ. Berlin, beamt. ao. Prof. für Pathol. Berlin W 15 (Deutschland), Kurfürstendamm 213. *Analytische und experim. Pathogenese. Pathogenese der Infektionskrankheiten, Rickettsiosen. Funktionelle Analyse pathol. Gewebsreaktionen. Geographische Pathol.* ⓓ Phil. Berlin 1913, med. Berlin 1919.

Kudelka, Ladislaus (1879). Poznań (Polska), Sołacz Sleska 16. *Vergl. Anat. der Pflanzen, Arzneipflanzen.* ⓓ Jagiell. Univ. in Krakau 1910.

Kudo, Roksabro (1886), D.Sc., Ass. Prof. of Zool. of the Univ. of Illinois, Urbana, Ill. (U.S.A.), Department of Zool. *Parasit., Parasit. Protozool., Sporozoa, Insect Protozoa, Fishparasites, Cytol.* ⓓ Imperial Univ. of Tokio, Japan, 1910. ⓟ Protozoa, especially parasit. forms and Sporozoa.

Kudô, Tokuan, Prof. of Med. Univ. Niigata (Japan), Anat. Inst. *Anat.*

Kudô, Yûshô, Prof. of Hokkaidô Imper. Univ. Sapporo (Japan), Botan. Labor., Coll. of Agric. *Usefull Trees of Japan.*

Kudrjawzew, Nicolai (1898), Prosektor d. Physiol. Inst. Univ., Dir. des Arbeitsphysiol. Labor. des Psychoneurol. Inst. Charkow, Ukraine (U.d.S.S.R.), Puschkinskaja 45, W. 3. *Ermüdung des Nervensystems.* ⓓ Charkow.

Kudrjawzewa, Anna (1893), Ass. Med. Inst., Mitarbeiter d. Biochem. Inst. Charkow (U.d.S.S.R.), Lermontowskaja Str. 19. *Stoffwechsel u. Ernährungstörungen.*

Kühn, Alfred (1885), o. Prof. der Zool. und Vergl. Anat., Dir. des Zool. Inst. der Univ. Göttingen (Deutschland), Prinz-Albrecht-Str. 24. *Farbensinn der Tiere, Taxien, Vererbung bei Säugetieren (Nagetieren, Katzen) und Insekten (Hymenopteren, Lepidopteren), Temperaturmodifikation, Protozoenkernteilung, Hydroidenentwicklung.* ⓓ Freiburg i. Br. 1908. ⓟ Temperaturaberrationen u. geographische Rassen von Schmetterlingen.

Kühn, Othmar (1892), Dr. phil., Prof. an der Bundes-Erziehungsanstalt Wien, wissensch. Mitarbeiter des Naturhistor. Mus. und der Geol. Bundesanstalt. Wien I (Deutsch-Österreich), Burgring 7. *Palaeont. der Coelenteraten.* ⓓ Wien 1919. ⓟ Korallen, rezent und fossil.

Kühnholtz-Lordat, Georges (1888), Prof. Botan. Ecole nation. d'Agric., Dir. Stat. de physiol. et pathol. végétales. Montpellier (France). *Biol. des plantes littorales, leur géographie, sociol. et physiol. Biol. des plantes nuisibles à l'Agric. Pathol. méditerranéenne. L'association à Corynephorus canescens et Helichrysum Stoechas (Corynephoretum atlanticum). L'association à Lavandula latifolia et Thymus vulgaris en Provence.* ⓟ Végétation des Dunes. Plantes nuisibles aux cultures.

Külz, Fritz (1887), Prof., Dr. med., Dir. des pharmacol. Inst. der Univ. Kiel (Deutschland), Hospitalstr. 20. *Pharm.* ⓓ Berlin 1914.

Kümmerle, Eugen Béla (1876), Dr., Abteilungsleiter an der botan. Abteilung des Ungarischen National-Mus. Budapest V (Ungarn), Akadémia-ut 2. *Pteridol. Lycopodiaceae, Selaginellaceae, Psilotaceae.* ⓓ Budapest 1903. ⓟ Pteridophyten aller Weltteile.

Künkel, Karl (1861), Prof. Heidelberg (Deutschland), Mittelstr. 44. *Landpulmonaten: Züchtung, Vermehrung, Fremd- und Selbstbefruchtung, Embryonalentwicklung, Lebensdauer, Vererbung.* ⓓ Dr. h. c. Freiburg in Baden 1921.

Künne, Clemens (1901), Ass. der deutschen wissenschaftl. Kommission für Meeresforschung an der Biol. Anstalt Helgoland (Deutschland). *Marines Plankton. Rädertiere.* ⓓ Göttingen 1925.

Küpfer, Max (1888), Dr., Prof. Eidg. Techn. Hochsch. Zürich 8 (Schweiz), Klausstr. 20. *Anat. und Physiol. der Haussäugetiere, spez. Genitalsystem.*

Kürsteiner, J., Dr., Adj. Schweiz. Milchwirtschaftl. u. Bact. Inst. Liebefeld bei Bern (Schweiz). *Bact.* o

Küster, Emil (1877), Prof. Dr. med. und med. vet., z. Z. Volontär am Inst. für experim. Therapie, Frankfurt a. M. Oberursel a. T. (Deutschland). *Hygiene; Bact., Serol., Fermentforschung.* ⓛ Med. Würzburg 1901, med. vet. Gießen 1909. ⓟ Testpraep. zum Nachweis der Abderhalden-Abbaufermente.

Küster, Ernst (1874), o. Prof. d. Botan. Univ., Dir. des botan. Gartens. Gießen (Deutschland), Brandpl. 4. *Pathol. Pflanzenanat., Gallen der Pflanzen, Panaschierung, Physiol. der Pflanzenzelle. Alter u. Tod. Zonenbildung u. Pflanzenrhythmus.* ⓛ München 1896.

Küster, William (1863), Prof. Techn. Hochsch., Dir. Labor. für organ. u. pharmaz. Chem. Stuttgart (Deutschland), Kernerplatz 1. *Blut- und Gallenfarbstoff nebst Derivaten (Porphyrine). Synthese der durch oxydative Eingriffe aus ihnen erhaltenen Spaltprodukte.* ⓛ Leipzig 1889.

Kufferath, Hubert (1882), Dir. du Labor. Intercommunal de Chim. et de Bact. de l'Agglomération Bruxelloise à Bruxelles (Belgique), 20, rue Joseph II. *Bact. du lait et des denrées alimentaires. Levures. Microscopie des denrées alimentaires. Algol. systématique, géographique. Cultures pures des algues.* ⓛ Gembloux 1903, Univ. de Bruxelles 1912.

Kuhl, Willi (1892), Dr. phil., Ass. am Zool. Inst. der Univ. Frankfurt a. M. (Deutschland), Robert-Mayer-Str. 6. *Histol.; Cytol.; vergl. Anat.; Variabilitätsforschung. Insecta: Cirkulationssystem (Coleoptera). Chaetognatha: Kopf histol.; spec. Retrocerebralorgan. Orthoptera: Forficula auricularia: Variabilität der Abdominalanhänge. Amphibia: Kalksäckchenproblem bei Rana.* ⓛ Marburg 1920.

Kuhn, K. K., Cons. Section des Plantes vivantes, Jardin Botan. principal. Leningrad (U.d.S.S.R.), Aplekarski Ostrov, Pessotchnaia ¹/₂. *Culture des plantes.* o

Kuhn, Otto (1896), Dr. rer. nat., Ass. Göttingen (Deutschland), Zool. Inst. *Vererbung bei Vögeln. Innere Sekretion.* ⓛ Tübingen 1922.

Kuhn, Philalethes (1870), o. ö. Prof. für Hygiene an der Univ. Gießen a. d. L. (Deutschland), Hygien. Inst. *Biol. der Bakterien. Ätiol. der multiplen Sklerose. Abwasserfragen. Rassenhygiene.* ⓛ Berlin 1894.

Kuiper, Koenraad (1888), Dir. Zool. Gardens. Rotterdam (Holland), Kruisstraat 21. *Mammol., Genetics.* ⓛ Amsterdam. ⓟ Vertebrates and plants.

Kujala, Viljo Vilho (1891), Ass. Forstwissenschaftl. Versuchsanstalt Finnlands. Helsinki (Finnland), Rauhankatu 4. *Aut- und Synökol. der Waldpflanzen; Waldvegetationstypen.*

Kukuk, Paul (1877), Dr. phil., Bergassessor, Leiter der geol. Abteilung der Westfälischen Berggewerkschaftskasse in Bochum, Priv.Doc. für angewandte Geol. an der Univ. Münster. Bochum (Deutschland), Bergstr. 135. *Geol. der Kohle. Die Calamariaceen des Rheinisch-westfälischen Steinkohlenbeckens.* ⓛ Bonn 1920. ⓟ Pflanzliche und tierische Reste aus dem Steinkohlengebirge Westfalens.

Kulagin, N. M., Prof. I. Univ., Dir. Zool. Labor. Landw. Akad. Moskau (U.d.S.S.R.), Petrowsko-Rasumowskoje. *Embryol.* o

Kulczycki, Włodzimierz (1862), o. Prof. der Vergl. Anat der Haustiere, Vet. Med. Akad. Lwów (Polska), ul. Kochanowskiego 67. *Insecta, Protozoa als Parasit. Anat.* ⓛ Lwów 1887.

Kulczyński, Stanisław (1895), Prof. Univ. Lwów (Polska), Kochanowskiego 8. *Systematik, Geographie der Pflanzen, Pflanzensoziol., Paleobotan. (Diluvium).* ⓛ Krakau 1919.

Kulesza, Witold (1891), Priv.Doc. Univ. Poznań (Polska), Sołacz Dwór. *Bryol., Forstbotan.* ⓛ Poznań 1924.

Kulik, Nestor A., Oberkustos des Geol. Mus. der Akad. d. Wissensch. Leningrad (U.d.S.S.R.). *Palaeont. Fauna d. nördlichen Urals.* o

Kulikow, Wenceslas (1889), Chef du labor. des biocolloïdes à l'Inst. de Microbiol., Chargé de Cours de Chimie Biol. à l'Ecole Supérieure Technique. Moscou (U.d.S.S.R.), Centre, Maschkow pereulok 1, log. 31. *Chim. des colloïdes biol. Toxines, ferments. Chim. biol. technique.*

Kulke, Joachim (1892). Goldberg i. Schles. (Deutschland), Trotzendorfplatz 5. *Floristik der Phanerogamen Schlesiens; Pflanzengeographie Schlesiens.* ⓟ Phanerogamen-Exsikkate Schlesiens.

Kull, Harry A. (1886), Prof. der Histol. und Embryol., Dir. Histol. Inst. der Univ. Tartu (Estland), Wallgraben 14. *Zytol., innere Sekretion. Mitochondrien, Apparato reticolare Golgi, Zellteilung. Pancreas, Langerhanssche Inseln, chromaffines Gewebe, Darmepithel. Mitochondrienfärbung, Kupferkarmin, Blutfärbung zum Ersatz der Injectionen. Panethsche Zellen.* ⓛ 1924.

Kulmatycki, Wladimir Julian (1895), Leiter der Abteilung für Binnenfischerei a. d. Staatl. Landwirtschaftl. Versuchsinst. in Bydgoszcz (Polen), Zacisze 7. *Fischkunde, Binnenfischerei.*

Kulp, Walter L. (1890), Ass. Prof., General Bact. New Haven, Conn. (U.S.A.), 12 Sheffield Hall, Yale Univ. ⓛ Yale Univ. 1923.

Kultiassor, Michael (1891), Doc. Univ. Inst. f. Bodenkunde u. Geobotan. Taschkent (U.d.S.S.R.), Botan. Garten, Postfach 122. *Pflanzengeographie. Flora u. Vegetation Mittelasiens (West-Turkestan). Phytosoziol.*

Kumada, Asao (1898), Ass. Imp. Piscicultural Experim. Station. Muroyoshida near Toyohashi, Aichi-Ken (Japan). *Biol. of fresh-water animals. Parasitol.* ⓛ Tōkyō 1922. ⓟ Parasits of aquatic animals.

Kumada, Tōsiro (1884), Dir. Hayatomo Fisheries Labor. Simonoseki (Japan). *Practical Zool. Practical Botan. Algae.* ⓛ Tōkyō 1914. ⓟ Japanese Trawl-fishes.

Kumagai, Kyōsuke, Prof. of Aichi Med. Univ. Nagoya (Japan), Physiol. Inst. *Physiol.*

Kumagawa, Hachirō, Ass. Prof. of Keiō-Gijuku Univ. Hongō, Tōkyō (Japan), Mukōgaoka-Yayoichō No. 3. *Med. Chem.*

Kumaris, Jean (1879), Prof. d'anthrop. Univ., Dir. du Mus. d'Anthrop. Athènes (Grèce), Académie. *Physische Anthrop.* ⓛ Athènes 1902.

Kume, Matazō, Prof. of Tōkyō Girl's High Seminary. Ochanomizu, Tōkyō (Japan), Biol. Labor.

Kuminowa, Tarija W., Doc. Staatl. Univ. Tomsk (U.d.S.S.R.), Ul. Belinskogo 38, W. 1. *Botan. Geobotan.* o

Kumm, Paul Maria Joseph Karl (1866), Provinzial-Mus.-Dir. a. W. Danzig (Freie Stadt), Thornscher Weg 131. *Botan. und Zool. der Provinz Westpreußen; Naturdenkmalpflege.* ⓛ Breslau 1889.

Kummer, Georg (1885). Schaffhausen (Schweiz), Korallenstr. 11. *Flora Kanton Schaffhausen.*

Kunitomo, Kana, Prof. of Med. Univ. Nagasaki (Japan), Anat. Inst. *Anat.*

Kunstler, M. J., Prof. Fac. Sc. Bordeaux (France). *Zool.* o

Kunstman, C. I., Wiss. „P.-F.-Lesshaft"-Inst. Leningrad (U.d.S.S.R.), Pr. Maklina 32. *Physiol.'* o

Kuntz, Albert (1879), Prof. of Anat., St. Louis Univ. School of Med. St. Louis, Mo. (U.S.A.), 1402 South Grand Blvd. *Development, anat. and physiol. of the sympathetic nervous system. Anat. of the central nervous system.* ⓛ Ph.D. State Univ. of Iowa 1910, M.D. St.Louis Univ. School of Med. 1918.

Kuntz, János (1891). Budapest IX (Ungarn), Ranolder u. 3. I. 8. *Phytochem. und Pharmacognosie. Drogenversuchsstation.* ⓛ Budapest 1918.

Kuntz, William Abraham (1896), Ass. Plant Pathol. Lake Alfred, Flo. (U.S.A.). *Diseases of*

insects; diseases of citrus. Aphids. Ⓟ Pennsylvania State Coll. 1923.
Kuntze, Roman (1902), Ass. des Inst. für Forstschutz der Polytechnik in Lwów (Polska), ul. Nabielaka 22. *Systematik und Faunistik: Coleoptera Poloniae spec. Carabus, Halticini, Ipidae. Zoogeographie südöstlichen Polens (Podolisches Plateau). Genetik und Biometrik, ihre Beziehungen zur Systematik. Forstzool.: Nager, Borkenkäfer.* Ⓟ Lwów 1925. Ⓟ Carabus und Halticinen Polens, außerdem charakteristische Käfersorten Podoliens und der Karpathen.
Kuo, Y. P., Lect. Normal Univ. Peking (China). *Zool.* o
Kupalov, Peter Stepanovich (1888), Ass. Physiol. Dept., Inst. of Experim.Med.Leningrad(U.d.S.S.R.), Barmaleeva 9, 10. *Physiol. of the Brain.*
Kupelwieser, E., Priv.Doc. Physiol. Inst. d. Univ. Wien IX (Österreich), Schwarzspanierstr. 17. o
Kupelwieser, Hans, Priv.Doc. Univ. München, Inh. d. Biol. Station Lunz am See (Österreich). *Zool.* o
Kupfer, Karl Reinhold, Prof. Herder-Inst. Riga (Latvija), Stabu ielā 23. *Pflanzengeographie u. system. Botan.* o
Kurbatov, Nikolai Ivanovich (1885), Prof. Univ., Dir. of Experim. Sta. for studying of Fertilisers. Taschkent (U.d.S.S.R.). *Soil and Fertilizer.*
Kurdiani, Salomon, Prof. Univ. Tiflis (U.d. S.S.R.). *Forstwiss.* o
Kurdiumov, N. A. (1898), Ass. St. Woronesch (U.d.S.S.R.), Kolzovskaja 22. *Anat.*
Kurisaki, Masumi. Fukuoka (Japan), Zool. Inst., Coll. of Agric., Kyûshû Imper. Univ. *Entomol.*
Kurkiewicz, Thadée (1885), Prof. Histol. et Embryol. Univ. Poznań (Polska), rue Fredry 10. *Histogenèse du muscle cardiaque. Cellules de Paneth. Développement des muscles striés.* Ⓟ 1911.
Kuroda, Genji, Prof. South Manchulian Med. Univ. Mukden, Manchuria (China), Physiol. Inst., S.M.M.V. *Psychol., Sinnesphysiol.* o
Kuroda, Nagamichi (1889). Akasaka, Tôkyô (Japan), No. 1, Fukuyoshi Chô. *Ornithol. and Mammal. Geese, Swans, Ducks of the world. Phaesants of Japan. Migration of Birds.* Ⓟ Tôkyô Imperial Univ. 1915. Ⓟ Bird-skins. Anatidae.
Kuroda, Tokubei, Ass. of Imper. Univ. Kyôto (Japan), Geol. Inst., Coll. of Sc. *Conchyliol.*
Kurokawa, K., Lect. of Keiô-Gijuku Univ. Tôkyô (Japan), Pathol. Inst., Coll. of Med. *Pathol.*
Kurosawa, R., Ass. Prof. of Hokkaidô Imper. Univ. Sapporo (Japan), Inst. of Vet. sc., Fac. of Agric.
Kurosawa, Rokurô, Ass. Prof. of Imper. Univ. Kyôto (Japan), Pharmacol. Inst., Coll. of Med. *Pharmacol.*
Kursanov, Leo (1877), Prof., Dir. d. Kabinett der Morphol. und Systematik der Pflanzen. Moskau (U.d.S.S.R.), Gr. Nikitskaya Str. 6. *Morphol. u. Cytol. d. Algen u. Pilze. Phytopath.* Ⓟ Moskau 1901.
Kurz, Albert (1886), Dr. rer. nat. Bern (Schweiz), Sandrainstr. 56. *Hydrobiol.; Süßwasseralgen (Systematik, Oekol., Geobotan.); Desmidiaceen und Diatomaceen.* Ⓟ Süßwasseralgen.
Kurzius, Bernard, Dir. d. Königl. Zool. Gartens. Sofia (Bulgarien), Bulevard Christo Botew 47. o
Kusama, Shigeru, Prof. of Keiô-Gijuku Univ., Member of Kitasato Inst. of Infect. Diseases. Yotsuya, Tôkyô (Japan), Pathol. Inst., Coll. of Med. *Pathol.*
Kusano, Shunsuke, Prof. of Tôkyô Imper. Univ. Komaba near Tôkyô (Japan), Botan. Labor., Coll. *Botan.*
Kuščer, Ludwig (1891), Zool. Mus. in Zagreb. Kranj (S.H.S.). *Mollusken, Systematik u. Faunistik. Struktur des Gastropodenschale.* Ⓟ Wien 1919. Ⓟ Mollusken.
Kuschke, J. Ed., Doc. Landwirtschaftl. Inst. Krasnodar, Kaukasus (U.d.S.S.R.), Novaja 107. *Phytopath.* o

Kuseneva-Prochorova, O. I., Cons. l'Herbier, Jardin Botan. principal. Leningrad (U.d.S.S.R.), Aplekarski Ostrov, Pessotchnaia $^1/_8$. *Géobotan.: végétation de l'Extrême Orient et de la Région arctique.* o
Kusin, Boris (1903), Zool. Mus. d. Univ., Ass. Timiriaseff Inst. Moskau (U.d.S.S.R.). *Abstammungslehre, experim. Ökol. Systematik u. Faunistik d. Wasserkäfer; geographische Variabilität v. Mylabris (Coleoptera); experim. œkol. Studien an Calliphora (Dipt.); Geschlechtssystem d. Dytiscidae u. Hydrophilidae (Col.).* Ⓟ Insekten (Coleoptera, spez. Aquatica u. Meloïdae, bes. Mylabris).
Kusmin, Sergius P. (1896), Ass. Labor. of plant physiol. Inst. of Applied Botan. and New Cultures at Leningrad. Mardakiani near Baku, Caucasus (U.d.S.S.R.). *Waterrelations of plants, root system investigations.*
Kusnetzowsky, Nikolaus (1891), Priv.Doc. Militär-Med. Akademie. Leningrad (U.d.S.S.R.), Prosp. v. Karl Marx 7a, W. 1. *Pathol. Anat.*
Kusnezov, Nicholas Jak. (1873), Chief Zool. Academy of Sc., Doc. of Compar. Physiol. Univ. Leningrad (U.d.S.S.R.), Zool. Mus. *Morphol. of Lepidoptera, Pigmentation of Insects. Gynandromorphism. Geographical distribution of Lepidoptera. Classification of Lepidoptera. Fauna of Russia. Physiol. of Invertebrata and of pigmentation in Insects.*
Kutin, Adolf, Prof. Ecole d'Agric. Tábor (C.S.R.). *Culture des plantes médicinales.* o
Kutter, Heinrich (1896). Zürich 6 (Schweiz), Weiherstr. 33. *Myrmekol.* Ⓟ Ameisen.
Kuttner, Olga, Dr. phil., Wissenschaftl. Mitarbeiterin der Hirschwaldschen Buchhandlung. Berlin (Deutschland), Schlüterstr. 56. *Hydrobiol., Cladoceren.* Ⓟ Freiburg i. Br. 1909.
Kuwabara, Kagemasa, Ass. Prof. Imper. Univ. Kyôto (Japan), Pharmacol. Inst., Fac. of Med.
Kuwada, Yoshinari, Prof. Anat. and Cytol. of Plants, Botan. Labor., Coll. of Sc., Imperial Univ. Kyôto (Japan). *Plant Cytol.* Ⓟ Tôkyô.
Kuwana, Inokichi, Chief Biol. of Plant Quarantee St. Yokohama (Japan), Custom. *Insecta: Coccidae.* Ⓟ Tôkyô.
Kuwayama, Satoru, Entomol. of Governm. Agric. Exp. St. Kotoni-mura near Sapporo (Japan), Entomol. Labor. *Lepidopterol.* o
Kuyper, Jan (1884), Dr., Dir. der Versuchsstation für Tabak (Deliproefstation). Medan (Nederl.-O.-Indië). *Physiol. und Pathol. der Tabakpflanze.* Ⓟ Utrecht 1909.
Kuznecov, Sergej Ivan, Doc. I. Univ. Moskau (U.d.S.S.R.). *Botan.*
Kuznetzov, Nicolas (1864), Chef section géobotan. du Jardin botan. principal, Prof. de botan. à l'Univ. Leningrad (U.d.S.S.R.), Jardin botan. *Géographie botan. de l'Europe orientale. Flore du Caucase.* Ⓟ Odessa 1911.
Kuznetzov, Nikolaj Nikolaevitsch (1898), Doc. Univ., Agronomische Fac. Taschkent, Uzbekistan (U.d.S.S.R.). *Tiergeographie. Systematik d. Ameisen. Blattwespen und Aculeata.*
Kwaan Seung Woh, Dean Hackett Med. Coll. Canton (China). *Anat. comp., Histol., Embryol.* o
Kwaschnin-Ssamarin, Mikolajus (1887), Priv.Doc. für Palaeont., angewandte Zool. und angewandte Botan. Kaunas (Litauen) Keistucio N 46—2. *Einfluß des Bodeneises auf die Pflanzenwelt Sibiriens. Einfluß des Gipses auf die Pflanzenformation. Abstammung des litauischen Pferdes.* Ⓟ Zähne, Füße und Schädel der rezenten und fossilen Equiden.
Kyas, Otto, Doc., Inst. des Recherches agricoles et Forestière. Brno (C.S.R.), Květná ul. *Biochem.* o
Kyes, Preston (1875), Doc., Prof. Univ. Chicago, Ill. (U.S.A.), 5717 Kimbark Av. *Experim. Pathol.* o
Kyle, Curtis H. (1878), Agronomist in the U.S. Dep. of Agric. Washington D.C. (U.S.A.). *Maize breeding.*

Kyle, Edwin Jackson (1869), Dean, Prof. School of Agric. College Station. Texas (U.S.A.). *Horticulture.* ○

Kyle, Harry Macdonald (1872). Peterley Corner, Gt. Missenden (England). *Fishes: anat. and biol.* ① St. Andrews 1895.

Kylin, Eskil (1889), Dir. d. inneren Abt. d. allgem. Krankenhauses. Jönköping (Sverige). *Mineralstoffwechsel: speziell Ca- und K-Stoffwechsel. Innere Sekretion: Adrenalinwirkung, Hormonwirkung, innere Sekretion d. Sexualdrüsen.* ① Lund 1920.

Kylin, Harald (1879), Prof. Univ. Lund (Sverige), Botan. Labor. *Entwicklungsgeschichte der Meeresalgen.* ① Upsala 1907.

Kyriasides, Kyriakos (1887), Dir. Labor. de Biol. et Bact. de l'Hôpital Evangelismos. Athènes (Grèce), Rue Marsseille 3a. *Biol. med. Vaccins et serums diagnostiques.* ① Athènes 1907. ℗ Vaccin thérapeuthique et prophylactique.

Kytmanow, Nikolaj A., Leiter d. Naturwiss. u. Kulturhistor. Mus. Jenissejsk (U.d.S.S.R.). *Agronomie. Botan.* ○

Laake, Ernest William (1887), Associate Entomol., Bureau of Entomol., U.S. Dept. of Agric. Dallas, Tex. (U.S.A.), P.O. Box 208. *Insects affecting the health of domestic animals. Med. entomol.* ① Texas A. and M. Coll. 1913. ℗ Imature stages of various Muscids and Oestrids.

Laanes, Theophil (1898), Magister pharmaciae; Ass. am Pharmakognostischen Inst. der Univ. Tartu (Estland), Ritterstr. 2. *Gehalt der Arzneipflanzen an ätherischem Öl.*

Labbé, Alphonse (1869), Prof. de Zool. à l'Ecole de méd. de Nantes, Loire Inférieure (France), Dir. du Labor. de Biol. Marine du Croisic. Nantes (France), Ecole de méd. *Faune des marais salants. Adaptation; origine des espèces. Copépodes; Nudibranches; Flagellés.* ℗ Paris 1909.

Lacassagne, Antoine Marcelin Bernard (1884), Sous-Dir. du Labor. Pasteur à l'Inst. du Radium de l'Univ. de Paris V (France), 26, rue d'Ulm. *Radiophysiol., Histopathol. des cancers.*

Lachi, Pilade, Prof. o. Univ. Genova (Italia). *Anat. umana norm.* ○

Lachowski, Petr Jakob (1889), Prof., Dir. Histol. u. Embryol. Labor. d. Kuban. Med. Inst. Krasnodar (U.d.S.S.R.), Pospolitakinskaja Str. 40. *Organismustheorie, Kolloidchemie und speciell Ultramikroskopie des Protoplasmus (Paramaec.). Physiol. Theorie der Kariokinese. Intravitale Untersuchung. Vakuolenverdauung.* ① Ekaterinoslav 1919, Krasnodar 1923.

Lackey, Charles Franklin (1899), Junior Plant Pathol., Sugar-Plant Investigations, U.S. Dept. of Agric. Citrus Experiment Station Riverside, Cal. (U.S.A.). *Curly-top of sugar beets and breeding strains resistant to the disease.* ℗ Rusts and mildews from California, flowering plants, lizards and butterflies.

Lackschewitz, Paul (1865), Dr. med. Libau (Lettland), Ulichstr. 48. *Neuropteren und Trichopteren der palaearkt. Region. Diptera, Nematocera polyneura der palaearkt. Region. Botan.: Salicaceae Eurasiens.* ① Dorpat 1893. ℗ Insecta: Neuroptera, Trichoptera u. Nematocera polyneura des ostbaltischen Gebietes. Phanerogamae (spez. Salicaceae) des Ostbaltikums.

Lacoste, André Pierre (1885), Agrégé d'Histol. à la Fac. de Méd. Bordeaux (France), Labor. d'Histol. *Adaptations functionnelles de l'appareil respiratoire des mammifères aquatiques. Développement du crâne. Les greffes de la cornée (Autoplastie. Hétéroplastie).* ① Med. Bordeaux 1913, sc. Paris 1923. ℗ Pièces sur les adaptations de l'appareil respiratoire de mammifères aquatiques (Cétacès, Siréniens, Hippopotame, Phoque etc.).

Lämmermayr, Ludwig (1877), Prof. Graz (Deutsch-Österreich), Lichtenfelsgasse Realgymnasium. *Oekol. der Pflanzen, speziell der grünen Höhlenvegetation. Entwicklung der Buchenassociation seit dem Tertiär. Serpentinpflanzen.* ① Wien 1900.

van Laer, Marc Henri (1893), Prof. à l'Inst. Supérieur des Fermentations (chim. Biol.) de Gand. Bruxelles (Belgique), 83, rue Berckmans. *Fermentation alcool. Diastases: Réaction du milieu.* ① Bruxelles 1919.

Lafon, J., Prof. Ecole Nation. Vét. Toulouse (France). *Physiol.* ○

Lagarde, Joannès Joseph (1866), Maître de Conférences à la Univ. de Strasbourg (France), 12, rue de l'Observatoire. *Discomycètes.* ① Montpellier.

Lagassè, Felix Scott (1898), Ass. Horticulturist. Newark, Delaware (U.S.A.), Horticultural Dept. *Nutritional studies with the apple and peach.* ① Maryland, Coll. Park, Maryland U.S.A. 1923.

Lagerberg, Torsten (1882), Prof. der Botan. an der Forstl. Hochsch. Experimentalfältet (Schweden). *Blaufäule des Nadelholzes. Pilzschädlinge und Pilzschäden der Waldbäume und des Holzes.* ① Uppsala 1909. ℗ Forstschädliche Pilze.

Laguesse, Gustave François Antoine (di Edouard) (1861), Prof. d'Histol. à la Fac. de Méd. de Lille (France), rue d'Artois 50. *Histol., histophysiol., histogénie, spécialement rate, glandes salivaires, pancréas, poumon, tissu conjonctif.* ① Lille 1896.

Lahille, Fernando, Prof. Univ. Nac. Buenos Aires (Argentina). *Zool.* ○

Laibach, Friedrich (1885), Dr. phil., Privatdoc. für Botan. Frankfurt a. M. (Deutschland), Vogelweidstr. 14. *Entwicklungsgeschichte der Pflanzen (Pilze) und Vererbungslehre (Heterostylie, Artbastarde).* ① Bonn 1907.

Laidlaw, P. P., Med. Research council. London, W.C. 2 (England), 15 York Buildings, Adelphi. *Bact. and Experim. Pathol.* ○

Laing, Gordon Dacomb (1899), Ass. Pathol., South African Inst. for Med. Research. Johannesburg (S. Africa), Box 1038. *Pathol. and Bact.* ① St. Andrews, Scotland, 1923.

Laing, Robert Malcolm (1865), M.A., Lect. Univ. Christchurch (New Zealand), 37 Macmillan Av. Cashmere Hills. *Marine Algol.* ① New Zealand 1884. ℗ Seaweeds (Algae).

Laitakari, Erkki (1892), Forstmeister; ao. Doc. der forstl. Betriebslehre Univ. Helsinki (Finnland), Kirkkokatu 4. *Morphol. der Wurzelsysteme der Waldbäume.* ① Helsinki 1921.

Laja, F., Ass. bacteriol. Versuchsstat. Univ. Tartu (Eesti) Venetan 34. ○

Lakari, Oiva Johannes (1883), Prof., Chef d. Abt. für Forstbetriebseinrichtung Staatsforstdirektion. Helsinki (Finnland), Metsähallitus. *Waldbau.* ① Helsinki 1915.

Lakin, C. E., Prof. Middlesex Hospit. Med. School. London W. (England), Mortimer Str., Oxford Str. *Pathol.* ○

Lakon, Georg (1882), Dr., a.o. Prof. für Botan. an der Technischen Hochsch. Stuttgart, Doc. für landw. Samenkunde und Abteilungsvorsteher an der Landw. Hochsch. Hohenheim. (Vorstand der Württ. Landesanstalt für Samenprüfung.) Hohenheim bei Stuttgart (Deutschland). *Periodizität; Entwicklungsphysiol.; Keimungsphysiol. Samenkunde und Samenprüfung; Enzyme (speziell Katalase); Panaschüre; Pilzliche Parasit. der Insekten (Entmophthoreen usw.); Pflanzenkrankheiten; Teratol.* ① Athen 1904.

Lakowitz, Konrad (1859), Dr., Prof. Danzig (Freie Stadt), Brabank 3. *Meeres- und Süßwasseralgen; Pilze.* ① Breslau 1881. ℗ Ostsee-Meeresalgen.

Lallemand, Victor Joseph Thomas (1880), Dr. en méd. Uccle (Belgique), 8, rue du Pacifique. *Insectes, homoptères: Cereopides et Eufulgorides. Systématique et biol.* ① Bruxelles 1905. ℗ Homoptères; Cereopides, fulgorides.

Lalou, Socrate D. (1875), Prof., Dir. Labor. Pharmacol. Fac. Méd. Bucarest (România), 5, Rue

Victor Emmanuel III. *Secrétion pancréatique.* ⓂMéd. Paris, 1898, sc. Paris, Sorbonne, 1912.
Lam, Herman Johannes (1892), Ph.D.; Sc. Ass. at the Herbarium and Mus. for Systematical Botan. at Buitenzorg, Java (Nederl.-O.-Indië). *Systematical Botan. flora of Malesia: Verbenaceae, Sapotaceae and Burseraceae, plant-geography.* Ⓓ Utrecht 1919.
Lamarque, Jean-François Paul (1894), Prof. agrégé de physique o-méd. Fac. Méd. Univ. Montpellier (France), 4 Passage Lonjon. *Actions des radiations (rayons X et rayonnement du radium) sur les tissus vivants. Action du courant électrique sur les êtres vivants.*
Lamauri, Julius N., Dir. d. Botan. Gartens. Tiflis (U.d.S.S.R.). o
Lamb, Alvin R. (1890), Dir. of Research, Moorman experiment Station, Moorman Mfg. Co., Quincy, Ill. (U.S.A.). *Animal nutrition and fermentation.* Ⓓ Univ. of Wisconsin 1913.
Lamb, Francis William (1874), Reader in Human Physiol. and Ass. Dir., Physiol. Labor. Univ. of Manchester (England), Med. School. *Physiol.* Ⓓ Dublin Univ. 1903.
Lambert, Mayer Simon (1870), Prof. Physiol. Fac. Méd. Nancy (France), 30, Rue Lionnois, Inst. de Physiol. Fd. Ol. *Physiol. nerveuse: fatigue des nerfs. Gangl. sympath. Metabolisme. Nutrition. Respiration. Hyperglycémie. Sécrétion interne de l'ovaire, Insuline et pancréas. Circulation. Sécrétion urinaire. Action des venins. Vision. Transport humoral de l'excitation nerveuse.* Ⓓ Nancy 1895.
Lambert, Robert Archibald (1883), Dir. and Prof. of Pathol., School of Tropical Med. of the Univ. of Porto Rico. San Juan (Porto Rico), School of Tropical Med. *Pathol.* Ⓓ Tulane Univ. Med. School, New Orleans, 1907.
Lambrecht, Kálmán (1889), Priv.Doc. Univ. zu Pécs, Ungarn. Budapest VIII (Ungarn), Pratergasse 59/D III. 32. *Palaeont. der Vertebraten (Palaeoornithol.)..* Ⓓ Budapest 1913. Ⓟ *Vertebraten, Vogelskelette und Fossilien, besonders Aves.*
Lameere, Auguste Alfred Lucien Gaston (1864), Prof. de Zool. à l'Univ. de Bruxelles (Belgique), rue Defacgoz, 74. *Phylog.; Prionides; Insectes fossiles.* Ⓓ Bruxelles 1887.
Lamhey, Ernest M. K., Prof. of Plant Physiol. and Plant Physiol. Wilmington, Del. (U.S.A.), 806 West St. o
Lammering, Dietrich (1893), Dr. phil. nat., Chem. und Biol. Leiter des wissenschaftlichen Labor. einer Fabrik für Pflanzenschutzmittel. Braunschweig (Deutschland), Rebenstr. 25. *Angewandte Botan. und angewandte Zool. (Pflanzenschutz und Schädlingsbekämpfung).* Ⓓ Frankfurt a. M. 1922.
Lampe, Lois, Instr. in Botan., Ohio State Univ. Columbus, Ohio (U.S.A.). *Plant Microchemistry and Plant Physiol.* Ⓓ Ohio State Univ. 1922.
Lamprecht, Herbert Anton Karl (1889), Versuchsleiter staatl. Versuchsanstalt für Gemüsebau Alnarp (Sverige), Åkarp. *Physiol. (Stoffwechsel) u. Biochem. der Pflanzen. Biol. von Dacne bipustulata. Wasserabgabe grüner Erbsen. Anthocyan und Zuckergehalt in Beta vulg. Chem. Zusammensetzung und biol. Eigenschaften einiger Gemüsearten.* Ⓓ Graz 1917.
Lams, Honoré Julien Charles (1883), Dr. en méd., Prof. d'Histol. générale et d'Embryol. à l'Univ. de l'Etat à Gand (Belgique), 292, Chaussée de Courtrai. *Ovogenèse et premiers stades de l'embryogenèse. Histogenèse et structure de la dent.* Ⓓ Gand méd. 1906, spéc. 1913.
Lamy, Louis Edouard (1866), Ass. au Mus. nat. d'histoire nat. Paris V (France), 55, rue de Buffon. *Mollusques marins.* Ⓓ Paris 1911.
Lancaster, T. L., Lect. Univ. Coll. Auckland (New Zealand). *Botan.* o
Lancefield, Rebecca Craighill (1895), Ass. Rockefeller Inst. for Med. Research, Department of Rheumatic Fever. New York (U.S.A.), 66 Street and Avenue A. *Work on the antigens contained in the streptococci. Drosophila, non-disjunction.* Ⓓ Columbia Univ. 1925.
Landacre, Francis Leroy (1867), Prof. of Anat. Ohio State Univ. Columbus, O. (U.S.A.), 2026 Inka Av. *Zool., embryol., entomol., histol.* o
Landau, Eber (1878), o. Prof., Dir. Lehrstuhl für Histol. u. Embryol. Kaunas (Litauen), Kauko g-vė 5. *Hirnanat.* Ⓓ Dorpat 1907.
Landauer, Walter (1896), Ass. Geneticist. Storrs, Conn. (U.S.A.), Storrs Agric. Experiment Station. *Chondrodystrophy in chicken embryos; Experiments concerning the mortality of chicken embryos; Causation of the hair slope in mammals.* Ⓓ Heidelberg 1921.
Landry, Prof. Ecole prép. de méd. et de pharmacie Reims (France). *Anat.* o
Landsteiner, Karl (1868), Member of the Rockefeller Inst. for Med. Research, New York City (U.S.A.), 66 Street and Avenue A. *Serol., Chem. of Antigens, Blood Groups, Bact.* Ⓓ Wien 1891.
Lane, Henry Higgins (1878), Univ.Prof. Lawrence, Kan. (U.S.A.). *Zool., embryol., the correlation between structure and function in the development of the special senses in mammals.* o
Lane, Merton C. (1893), Ass. Entomol., U.S. Bureau of Entomol. Toppenish, Washington (U.S.A.), Box 448, *Elateridae (Coleoptera), economical and systematic.* Ⓟ Elateridae.
Lang, Franz Josef (1894), Dr. med. a.o. Prof. für pathol. Anat., o. Ass. am pathol. anat. Inst. Innsbruck (Deutsch-Österreich), Müllerstr. 44, Pathol. anat. Inst. der Univ. *Knochen, Lungen, Blut, Gewebskulturen.* Ⓓ Innsbruck 1918.
Lang, Maxime Jomé David (1902), Chef du Labor. de Chim. clinique de la Fac. de Méd. à l'Hôpital Suburbain de Montpellier (France). *Chim. biol.*
Láng, Sándor (1902), Ass. Physiol. Inst. Budapest VIII (Ungarn), Eszterházy-ut 9. *Innere Secretion.* o
Lang, W. H., Prof. Vict. Univ. Manchester (England). *Botan. Cryptogams.* o
Lang, Wilhelm (1876), Dr., Vorstand der Württ. Landesanstalt für Pflanzenschutz. Hohenheim bei Stuttgart (Deutschland). *Pflanzenkrankheiten und Pflanzenschutz; Krankheiten der Getreide, ihre Bekämpfung durch Beizen; Bodenschädlinge; Krankheiten am Hopfen.* Ⓓ Tübingen 1904.
Lang, William Dickson (1878), Ass. Keeper, Department of Geol., British Mus. (Nat. Hist.), London S.W. 7 (England), Cromwell Road. *Fossil Invertebrate Animals; especially Corals and Polyzoa Bryozoa. Recent Mosquitoes.* Ⓓ Cambridge 1919.
Langdow, La Dema Mary, Ass. Prof. of Biol., Goucher Coll. Baltimore, Md. (U.S.A.), Allston Apts., N. Charles St. *Morphol., anat., ecol.* o
Lange, Axel (1871), Curator of the Botan. Garden. København (Danmark). *History of botan. Garden, Alpine plants.*
Lange, Bernhard (1895), Dr. med., Ass. am Anat. Inst. der Univ. Breslau 9 (Deutschland), Sternstraße 54. *Rassen- und Artunterschiede (Schädel, Haut).* Ⓓ Breslau 1923.
de Lange, Daniel (1878), Dr., Dir. des Embryol. Inst. der Hubrecht-Stiftung in Utrecht (Holland), Janskerkhof 2. *Normale, vergl. Embryol. der Vertebraten; die ersten Entwicklungsstadien (Keimblattbildung, Organanlage), Bildung der Eihüllen. Plazentation. Entwicklung der Amphibien und Säugetiere, der holoblastischen Anamnia und Selachier. Kopfproblem.* Ⓓ Amsterdam 1906.
Lange, Friedrich (1897), Ass. d. Arzneimittelprüfungsamts. Tartu (Estland), Pharmakol. Inst. *Ovarialhormon.* Ⓓ Dorpat 1923.
Lange, Hermann (1893), Priv.Doc. Univ. Frankfurt a. M. (Deutschland) *Physiol.* o
Lange, Ludwig (1873), Dr. med., Prof., ORR., Mitglied im Reichsgesundheitsamt. Berlin-Dahlem

(Deutschland), Boettlcherstraße 10. *Bact.; Tuberkuloseforschung; Immunitätsforschung.* ⓟ München 1896.
Lange, Octavian K., Prof. Geol. Univ. Taschkent, Turkestan (U.d.S.S.R.). Geol. Kabinett. *Kreidefauna.* o
Lange, Siegfried (1891), Dr. phil., Ass. am Botan. Inst. der Univ. Greifswald, Pommern (Deutschland), Bleichstr. 13/14I. *Botan. Reizphysiol.* ⓟ Greifswald 1921. ⓟ Algenpraep. des Greifswalder Boddens.
Langecker, Hedwig (1894), Priv.Doc. an der deutschen Univ., Ass. am pharmacol.-pharmakognostischen Inst. Praha II (C.S.R.), Albertov 7. *Experim. Pharm. Eiweißchem., Insulin, vegetatives Nervensystem.* ⓟ Med. et rer. nat. Prag.
Langelaan, Jacob Willem (1872), Prof. emer., Amsterdam Univ. Baaurn (Holland), 15 Amsterd. Str. Weg. *Muskel und Nerv. Morphol. des Zentralnervensystems.* ⓟ Amsterdam 1900.
Langer, Arturo (1900), Ass. d'Anat. umana normale. Firenze (Italia), Via Alfani 33. ⓟ Firenze 1926.
Langer, Hans (1887), Oberarzt des Kais.-Auguste-Viktoria-Haus. Berlin-Charlottenburg (Deutschland), Mommsenstr. 12. *Bact.-serol. Fragen der Kinderheilkunde. Immunisierung gegen Tuberkulose.* ⓟ Berlin 1911.
Langeron, Maurice Charles Pierre (1874), Chef de lab. Fac. méd., Chef des travaux de parasit. Inst. méd. colon. Paris (France), 15, Rue de l'Ecole de méd. *Parasit. animale, Malariol., Mycol.* ⓟ Paris 1902.
Langfeld, Herbert Sidney (1874), Prof. of Psychol., Dir. Psychol. Labor., Univ. Princeton, N.J. (U.S.A.), Eno Hall. *Experim. Psychol.* ⓟ Univ. of Berlin 1909.
Langfeldt, Einar, Dr., Prof. Physiol. Inst. Univ. Oslo (Norge). *Biochem.* o
Langford, George S. (1901), Deputy State Entomol. Fort Collins, Col. (U.S.A.). *Applied and experim. entomol.* ⓟ Maryland 1924.
Langhans, Viktor, Dr., Prof. d. landwirtsch. Zool., Fischzucht u. Teichwirtschaft Doksy (Hirschberg i. Böhmen) (C.S.R.), Forschungsstat. f. Hydrobiol. u. Fischzucht. o
Langhoffer, August (1861), Prof., Univ. Zagreb (S.H.S.), Demetergasse 1. *Entomol., Zool., Spelaeol.* ⓟ Jena. o
Langier, Henri (1888), Chef de travaux au labor. de Physiol. de la Sorbonne. Paris VI (France), 1, rue Hautefeuille. *Physiol. du système nerveux.* ⓟ Paris 1923.
Langworthy, Orthello Richardson (1897), Associate in Anat., Johns Hopkins Med. School. Baltimore, Md. (U.S.A.). *Correlated morphol. and physiol. studies of the central nervous system, its development in young animals.* ⓟ Johns Hopkins Med. School 1922.
Lanjouw, Joseph (1902), Conserv. Botan. Mus. and Herb., Utrecht (Holland), Lange Nieuwstraat 106. *Systematical Botan.: Euphorbiaceae from Surinam.*
Lantz, Harvey Lee (1888), Ass. Chief, Pomol. Section, Iowa Agric. Experim. Station. Ames, Ia. (U.S.A.). *Fruit breeding. Apples, Pears, Plums.* ⓟ Iowa State Coll. 1918.
Lanuza, Vicente Rodriguez, Prof. Univ. St. Tomas. Manila (Philippine Islands). *Parasit., Bact.* o
von Lanz, Titus (1897), Priv.Doc., Ass. an der Anat. Anstalt München (Deutschland), Pettenkofer Straße 11. *Zelle, Vitalfärbung, Nebenhoden.* ⓟ München 1922.
Lanza, Domenico (1868), Libero doc. di Botan. Univ. Palermo; Cons. dell' Erbario del R. Ist. Botan. R. Orto Botanico Palermo (Italia). *Sistematica. Flora sicula. Genetica.* ⓟ Piante di Sicilia.

Lanzoni, Francesco (1881), Aiuto di Botan. Univ. Parma (Italia). *Farmacobotan. Storia della Botan.*
Lapage, Geoffrey (1888), Lect. in Zool. Zool. Department, The Victoria Univ. of Manchester (England). *Protozool.* ⓟ Manchester 1912.
Lapicque, Louis (1866), Prof. de Physiol. générale à la Fac. des Sc. de l'Univ. de Paris (Sorbonne); Prof. honoraire au Mus. national d'Histoire nat. Paris 4 (France), 21. Boulevard Henri IV. *Electrophysiol., excitabilité électrique. Echanges cellulaires; propriétés physicochim. de la matière vivante. Le fer dans l'organisme, la ration alimentaire, le poids du cerveau, la race nègre, spécialement les Negritos et les Parias.* ⓟ Méd. et Sc. Paris 1919, hon. ès sc., Oxford.
Lapicque, Marcelle (1873), Dir. adjoint Labor. Physiol. Ecole des Hautes Etudes (Sorbonne). Paris (France), 21, Boulevard Henri IV. *Excitation électrique des nerfs et des muscles. Conduction dans le cœur. Poisons. Alimentation.* ⓟ Paris 1905.
Lapidari, Mario (1902), Ass. vol. Pavia (Italia), Ist. Patol. gen. Univ. *Tumori sperim.* ⓟ Pavia 1925.
Lapiner, Michail (1898), Prof. d. Physiol. a. d. Swerdlov Univ., Labor. f. experim. Biol. Moskau (U.d.S.S.R.), 55, Miusskaja Pl. 3. *Biochem. der Schilddrüse und des Hodens.*
Laporte, F., Prof. Univ. Toulouse (France). *Pathol.* o
Lappi-Seppälä, Martti (1900), Ass. Forstwiss. Versuchsanstalt. Helsinki (Finnland), Lutherstr. 4. *Zuwachslehre der gemischten Bestände.* o Helsinki 1924.
Lapschina, Eustolie (1889), Naturwissenschaftl. Inst. in Peterhof. Leningrad (U.d.S.S.R.), Estlandskaja Str. 7, Qu. 14. *Überwinterung und vegetative Vermehrung der Pflanzen. Flora des Europäischen Rußlands.*
Laptev, M. K., Doc. Mittelasiat. Univ. Taskent (U.d.S.S.R.). *Zool. d. Wirbeltiere.* o
Laquer, Fritz (1888), Dr. med., Priv.Doc. an der Univ. Frankfurt. Beurlaubt. Leiter der N. V. Organon. Oss. (Holland). Nymwegen (Holland), Van Spaenstr. 16. *Intermediärer Stoffwechsel, Physiol. Chem. der Kohlenhydrate, Chem. der Hormone. Klimatol.* ⓟ Heidelberg 1912.
Laqueur, Ernst (1880), o. Prof., Dir. des Pharmacotherapeutischen Labor. der Univ. Amsterdam (Holland), Polderweg 20. *Fermente, Hormone, Adsorption, Lunge, Geschlechtshormone, chemotherapeutische Praeparate aus der Chininreihe und Akridinreihe, Insulin.* ⓟ Breslau 1905.
Large, Thomas (1871), U.S. Bureau of Plant Industry, Office of White Pine Blister Rust Control. Spokane, Wash. (U.S.A.), East 1528 18th Avenue. *Animal ecol. of the „Inland Empire", Ecol. of the Genus Ribes in relation to White Pine and Blister Rust.*
Larionow, Leonid F. (1902), Ass. Pathol. Physiol. Tomsk, Siberia (U.d.S.S.R.), Univ. *Cancer, exper. Tar-Cancer, cultivation of tumors in vitro.*
Larmer, Finley G. (1902), Ass. in Botan. Dept. of Botan., Mich. State Coll. East Lansing, Mich. (U.S.A.). *Cercospora beticola.*
Larrimer, Walter H. (1889), Senior Entomol., in Charge Cereal and Forage Insect Investigations U. S. Dept. of Agric. Washington, D.C. (U.S.A.). ⓟ Ohio Staate 1925.
Larrousse, Fernand Lucien (1888), Préparateur, Fac. de Méd. (Labor. de Parasit.) Paris (France), 3, place St. Michel. *Parasit., Tiques.* ⓟ Paris 1923. ⓟ Arthropodes parasit. Carabidés.
Larsell, Olof (1886), Prof. of Anat. Portland, Ore. (U.S.A.), Univ. of Oregon Medical School. *Compar. Neurol. Histol.* ⓟ Northwestern Univ. 1918.
Larsen, Carl Sophus (1874). Rislebrek pr. Faaborg (Danmark). *Dänische Lepidoptera, spec. Microlepidoptera. Südamerikanische Papiloniden:*

Aeneas u. *Lysander, Dismorphia, Catasticta, Helicarius, Catagramma, Prepona, Agrias, Eryciniden, Lycaeniden, Hesperiden, Castnia und Pericopis.*
Larsen, Carl Syrach (1898). Forest Botan. Garden. Charlottenlund (Danmark). *Forest Botan.* Ⓓ Vet. and Agric. Coll. København 1923.
Larsen, Julius Ansgar (1877), Ass. Prof. of Forestry. Ames, Ia. (U.S.A.). *Silviculture; Seed, reforestation, growth, distribution. Ecol. of forest trees. Forest types and natural reproduction of trees.* Ⓓ Yale Forest School 1910.
Larson, Andrew Olof (1887), Associate Entomol. Alhambra, Cal. (U.S.A.), Box 297. *Investigations of life habits and control of the bean weevils (Bruchids).*
Larson, Carl W. (1881), Chief, Bureau of Dairy Industry, U.S. Dept. of Agric. Washington, D.C. (U.S.A.), 1209 Delafield Place, N.W. *Bact. and chem. of milk; breeding and nutrition of dairy cattle; dairy plant management.* Ⓓ Columbia Univ. 1916.
Larson, Winford Porter (1880), Prof. of Bact. and Immunol., Univ. of Minnesota. Minneapolis, Minn. (U.S.A.). Ⓓ Univ. of Illinois 1904.
Lasareff, Peter (1878), Dir. Inst. Biophysik. Moskau VI (U.d.S.S.R.), Miusskaja 3. *Photochem. Biophysik: Ionentheorie der Reizung.* Ⓓ Moskau 1907.
Lašas, Vladas (1892), Prof., Dir. Labor. für Physiol. u. physiol. Chem. Kaunas (Litauen). *Anaphylaxie.* Ⓓ Dorpat 1922.
Laschewskaja, Wladislawa I., Doc. Univ. Woronesh (U.d.S.S.R.), Str. d. 9. Januar 21, W. 1. *Botan. Anat. d. Pflanzen.* o
Laske, Carl (1883), Leiter der Hauptstelle für Pflanzenschutz bei der Landwirtschaftskammer Schlesien. Breslau (Deutschland), Gustav-Freytag-Str. 29 I. *Abbau- und sog. Viruskrankheiten der Kartoffel.* Ⓓ Breslau 1915. Ⓟ *Tierische Parasit. und Krankheitsbilder von Kulturpflanzen.*
Laskowski, Joseph (1900), Ass. d'Anat. Patol. Fac. Méd. Warszawa (Polska), Chalubiński 5. *Tumeurs expérim. Ferments du foie.* Ⓓ Varsovie 1924.
Lasowsky, Julius (1903), Aspirant Inst. für Pathol. Anat. Moskau (U.d.S.S.R.), I. Univ., Inst. für Pathol. Anat. *Drüsen mit innerer Sekretion.*
Lassablière, M. P., Chef de Labor., Fac. Méd. Paris (France). *Pathol.* o
Lasseur, Philippe Antoine (1882), Prof. Microbiol. Fac. Pharmacie. Nancy (France). *Bactéries chromogènes, Pigments bactériens, Morphol. de la cellule bactérienne. Sérol.: Fixation de l'alexine. Bacillus Chloraphis. Bacillus Le Monnieri.* Ⓓ Nancy 1923.
Lastočkin, Dimitri (1890), Prof. Zool., Dir. Zool. Kabinett, Polytechn. Inst. Iwanowo-Wossnesensk (U.d.S.S.R.). *Anat. u. Physiol. der Oligochaeten u. Holothuroidea; Regeneration; Systematik u. Faunistik der Oligochaeta limicola u. Eucopepoda. Hydrol. der stehenden Gewässer. Tierassoziationen in Seen.*
Latarjet, André (1877), Prof. d'Anat. à la Fac. de méd. de Lyon (France), 1, Cours de Verdun. *Anat. humaine: le Sympathique (anat., descriptive et experim.), les lymphatiques.* Ⓓ Lyon 1921.
Lataste, Fernand (1847), Prof. honoraire d'Univ. Cadillac-sur-Garonne, Gironde (France). *Herpétol., Mammal., Zooéthique. Tératol., Psychisme animal.* Ⓓ Paris 1876.
Lathouwers, Victor (1880), Chef des Travaux à la Station de Recherches pour l'Amélioration des Plantes de l'Etat, Chargé des Cours de Génétique et d'Amélioration des Plantes cultivées, à l'Inst. Agronomique de l'Etat, à Gembloux (Belgique). *Génétique: Mendelisme, Cytol., Nombre chromosomiques, croisements interspécifiques. Biol. florale (autogamie, allogamie). Méthodes d'Amélioration. Variétés.* Ⓓ Louvain 1903. Ⓟ Semences de variétés sélectionnées.
Lathrop, Frank H. (1891), Entomol., U. S. Bureau of Entomol. Washington, D.C. (U.S.A.). *Deciduous fruit insects. Cicadellidae (Jassidae).* Ⓓ Ohio State Univ. 1923. Ⓟ *North American species of Cicadellidae (Jassidae).*
Latimer, Homer Barker (1882), Prof. of Anat., Med. School, Univ. of Kansas. Lawrence, Kan. (U.S.A.). *Anat., growth and quantitative anat.* Ⓓ Univ. of Minnesota 1921.
Latta, John Stephens, Associate Prof. of Anat., Univ. of Nebraska Med. Coll. Omaha, Neb. (U.S.A.), 4311 Wakeley Street. *Histogenesis of the blood and the blood-forming Organs. Correlation of structure and function in the Lymphatic System.* Ⓓ Cornell 1920.
Latter, Margaret Yoan (1901), Demonstrator in Botan. London (England), King's Coll. Strand. *Cytol. of Lathyrus. Microdissection. Nuclei.* Ⓓ London 1925.
Laubenheimer, Kurt (1877), Prof. Dr. med., Mitglied des Staatsinst. für experim. Therapie. Frankfurt a. M. (Deutschland), Paul-Ehrlich-Str. 44. *Serol., Immunitätswissenschaft.* Ⓓ Gießen 1903.
Laubert, Richard (1870), Dr. phil., Reg.R. und Vorsteher der wissenschaftl. Sammlungen der Biol. Reichsanstalt für Land- und Forstwirtschaft in Berlin-Dahlem. Berlin-Zehlendorf (Deutschland), Elfriedenstr. 5. *Botan., Phytopathol., parasit. Pilze, Krankheiten der Kulturgewächse des Gartenbaues.* Ⓓ Erlangen 1896. Ⓟ *Pflanzenkrankheiten und parasit. Pilze.*
Laubmann, Alfred Louis (1886), Konserv. Abteilungsvorstand Zool. Staatssammlung. München (Deutschland), Neuhauser Str. 51. *Ornithol.: Nomenklatur, Faunistik und Systematik. Alcedinidae.* Ⓓ München 1911. Ⓟ *Vogelbälge.*
Lauche, Arnold (1890), Dr. med., Priv.Doc. an der Univ. Bonn (Deutschland), Pathol. Inst. der Univ. *Pathol. der Lungen und der weiblichen Genitalien (endometrioide Wucherungen), Biol. und Systematik der Rädertiere (Rotatoria).* Ⓓ Bonn 1914.
Laude, Hilmer Henry (1887), Associate Prof. of Agronomy. Manhattan, Kan. (U.S.A.). *Crops and soils.* Ⓓ Texas Agric. and Mechanical Coll. 1918.
Laufberger, Vilém (1890), Prof. med. Fac. Univ., Dir. Inst. für allgemeine u. experim. Pathol. Brno (C.S.R.), Úvoz 33. *Mechanismus des Insulinwirkung. Ultrastruktur der Zelle.* Ⓓ Prag 1917.
Laufer, Berthold, Curator of Anthrop., Field Mus. of Natural History. Chicago, Ill. (U.S.A.). o
Laufer, M., Dir. Adj. de Labor. Paris (France). *Physiol.* o
Lauffer, Jorge, Agregado al Mus. Nac. de Ciencias Nat. Madrid (España), Juan de Mena 5. *Coleópteros y Lepidópteros.*
Laughlin, Harry Hamilton (1880), Ass. Dir., Eugenics Record Office, Carnegie Inst. of Washington. Long Island, N.Y. (U.S.A.), Cold Spring Harbor. *Human heredity. The biol. aspects of immigration. Eugenical aspects of sexual sterilization. Genetics of the thoroughbred horse.* Ⓓ Princeton Univ. 1917.
Laughton, Nelles Boyd (1896), Assoc. Prof. in Physiol., Univ. of Western Ontario, Med. School. London, Ont. (Canada). *Nervous System. Internal Secretion. Blood pressure.* Ⓓ Univ. of Toronto 1921.
Launoy, Léon Louis (1876), Prof. agrégé, Fac. Pharmacie. Paris, Saint-Germain en Laye (France), 17, rue de Lorraine, *Ferments, venins. Cellule hépatique. Spirilloses (Syphilis du lapin et Spirillose des poules). Thyroïdes. Sécrétions gastrique et pancréatique. Sérums sanguins, anaphylaxie. Pharmacodynamie.* Ⓓ Paris 1903.
Laurens, Henry (1885), Prof. of Physiol. New Orleans, La. (U.S.A.), Dept. of Physiol., Tulane Univ. *Reactions of animals to light. Physiol. and anat. of the heart. Melanophores. Spectral sensitivity and visibility. Radiation on metabolism and growth, blood.* Ⓓ Harvard 1911.
Laurie, Robert Douglas (1874), Prof. Dept. of Zool. Univ. of Wales. Aberystwyth, Wales (Eng-

land), Univ. Coll. *Crustacea. Animal Ecol.* ⑨ Oxford 1905.
Lautenschläger, Carl Ludwig (1888), Prof., Dr. med., Dr. ing., Leiter der pharmaz.-wissenschaftl. Abt. und Labor. der I.-G. Farbenindustrie Aktiengesellschaft Höchst a. M., Doc. a. d. Univ. Frankfurt a.M. (Deutschland), Schumannstr. 7. *Pharmacol., Biochem., angewandte Botan., pharmazeutische Chem.* ⑨ Ing. Karlsruhe i. B. 1913, med. Freiburg i. Br. 1919.
Lautner, Franz Karl (1892), Dr. phil. Erlangen, Bayern (Deutschland), Harfenstr. 7. *Biol. der Apis mellifica. Biol. und Systematik der heimischen Mollusken und Flora.* ⑨ Erlangen 1917. ⑫ Mollusken mit Angabe der Bodenbeschaffenheit der Fundorte; mikroskopische Praep. der Honigbiene, ihrer Krankheiten und Feinde. Tiere der Nordsee.
Lavauden, Louis (1881), Ingénieur agronome, Inspecteur des Eaux et Forêts. Tunis (Afrique), 12, Rue de Cronstadt. *Ornithol. Mammal., spécialement de l'Afrique du nord, et des Alpes.* ⑨ Paris 1905.
Lavin, Rodrigo Leonardo (1867), Prof. Physiol. Fac. méd., Doyen de la Fac. Pres. de l'Acad. royale de méd. Cadix (España). *Syst. nerveux. Applications sanit. de la physiol.* ⑨ 1889.
Lavoreria, Daniel E., Prof. Univ. Lima (Peru). *Histol.* o
Lavrov, Nicolas N., Doc. Univ., Doc. Technol. Inst. Tomsk, Sibérie (U.d.S.S.R.), Timirjazevskij Prosp. 9. *Flore mycol. de Sibérie. Mycol. systématique.* o
Lavrov, Sergej Dm., Prof. Med. Inst. Omsk (U.d. S.S.R.), Ul. Lenina 9 u. Inst. f. Land- u. Forstw., Usadba za Staroj Zagorodnoj Roščej. *Zool., allg. Biol.* o
Lavy, William Henry (1874), Barkor Prof. of Cryptogamic Botan. Manchester (England), Univ. ⑨ Glasgow 1895.
Lawrence, William Evans (1888), Associate Prof. of Plant Ecol., Oregon State Agric. Coll. Corvallis, Ore. (U.S.A.). ⑫ Herbarium specimens of native flora and other native material.
Lawrenko, Eugen M., Dir. Herbarium d. Botan. Gartens. Charkow (U.d.S.S.R.), Klotschkowskaja No. 52. *Geobotan. Floristik.* o
Lawrentiev, A. P. (1898), Prosektor. Odessa, Ukraine (U.d.S.S.R.), Olgiewskaja 4, Anat. Inst. *Nervensystem.* o
Lawrentiev, B. N., Prosektor, Dr., Histol. Labor. Univ. Kazan (U.d.S.S.R.). o
Lawson, A., Prof. Univ. Sydney (Australia). *Botan.* o
Lawrow, Boris A. (1884), Prof. Tierphysiol., Zootechn. Inst. Moskau (U.d.S.S.R.), Bolschaja Ssadowaja Str. 25, W. 50. *Stoffwechsel; Gaswechsel bei der B-Avitaminose. Stickstoffwechsel bei Vögeln.*
Lawrow, Dawid (1867), Prof. Odessa, Ukraina (U.d.S.S.R.), Olgiewskaja 4. *Experim. Pharmacol. Intoxikationen. Lipoide.* ⑨ Sc. nat. Moskau 1890, med. 1893.
Lawrow, M., Geol. Mus. der Akademie der Wiss. Leningrad (U.d.S.S.R.). *Quartäre Mollusken des Nordens.* o
Lawson, Caesar, Provincial Entomol. and Prof. of Economic Entomol. Guelph, Ontario (Canada), Ontario Agric. Coll.
Laxa, Otakar (1874), Prof. Lactol. Praha-Vršovice (Č.S.R.), Havlíčkovy sady 28. *Laiterie, bact. d'agric.* ⑨ Prague 1901.
Layman, Prof., II. Helminthol. Inst. Moskau (U.d.S.S.R.), Pimenowsky per. 5. o
Lazarenko, Andreas (1901), Botan. Kabinett Ukrainische Akademie der Wissensch. Kiew (U.d. S.S.R.). *Bryophyta.* o
Lazier, Edgar L. (1899), Associate in Zool., Univ. of California. Berkeley, Cal. (U.S.A.), Department of Zool. *Anat. and physiol. of the Teredinidae.* ⑨ Univ. of California 1923.

Lazzaro, C., Prof. Univ. Palermo (Italia). *Farmacol.* o
Leach, John Albert, Education Department. Melbourne (Australia). *Ornithol.* o
Leach, Julian Gilbert (1894), Ass. Prof. in Plant Pathol., Univ. of Minnesota. St. Paul, Minn. (U.S.A.), Univ. Farm. *Diseases of potatoes and vegetables. Insect Dissemination of plant pathogens.* ⑨ Minnesota 1922.
Leach, William (1891), M.Sc., Lect. in Botan., The Botan. Dept., The Univ. Birmingham (England). *Ecol. Experim. Morphol. and Physiol. of Bryophyta. General Physiol. Plant Anat.* ⑨ Victoria Univ. Manchester 1922.
Leake, Chauncey D. (1896), Associate Prof. of Pharmacol., Univ. of Wisconsin. Madison, Wis. (U.S.A.). *Action of anesthetics on blood p_H; Regulation of red blood cell production; Pharmacol. of the nitrites. History of Med. and Sc.* ⑨ Univ. of Wisconsin 1923.
Learn, C. D. (1875), Ass. Prof. Botan. Colorado Agric. Coll. Fort Collins, Col. (U.S.A.). *Pleurotus.* ⑨ Cornell Univ. 1912.
Leathes, John Beresford (1864), Prof. of Physiol. Univ. of Sheffield (England). *Physiol. Chem.* ⑨ Oxford 1891.
Lebbe, D. B. (1879), Prof. Dr. Abbaye de Maredsous (Belgique). *Zool.* o
Lebedev, Alexander (1874), Prof. der Zool. und Entomol. Kiew, Ukraine (U.d.S.S.R.), Zool. Labor. des Landwirtschaftl. Inst. *Anat. und Systematik d. Coleopteren und Hymenopteren.* ⑨ Kazan 1899. ⑫ Coleoptera (Mylabris, Lethrus, Cleoninae), Apidae.
Lebedev, Vladimir Nik., Prof. II. Univ. Moskau (U.d.S.S.R.). *Zool.* o
Lebedeva, Lydia (1871), Ass. Leningrad (U.d. S.S.R.), Jardin Botan. Princ. *Mycol., Agaricineae, Basidiomycetae.* ⑫ Herbarium mycol.
Lebedew, Alexander (1881), Prof. I. Univ. Moskau (U.d.S.S.R.), Belinskistr. 1, W. 14. *Biochem. und Agrikulturchem. Alkoholgärung auf Grund extracellulärer Fermente.* ⑨ Moskau 1914.
Lebedew, Alexander F., Prof. Univ. Rostow a.Don (U.d.S.S.R.). *Pflanzen-Physiol. Hydrol.*
Lebedew, Nicolas J. (1863), Prof. der Geol. Berg-Inst. Dnepropetrowsk, Ukraine (U.d.S.S.R.). *Palaeont. der paläozoischen (hauptsächlich carbonischen) Invertebrata, historische Geol. des Carbons. Obersilurische Fauna des Timan. Korallen im Devon.* ⑫ Fossile Tiere und Pflanzen des Carbons.
Lebedincev, Elisabeth V. (1884), Ass. Labor. experim. Ökol. d. Botan. Gartens. Leningrad (U.d. S.S.R.). *Wasserhaushalt der Pflanzen. Experim. Anat. der Pflanzen.*
Lebedinsky, Naum Gregor (1888), o. Prof., Dir. vergl.-anat. und experim.-zool. Inst. Univ. Riga (Latvija), Albertstr. 10. *Selektionsstudien an Mäusen, geschlechtliche Zuchtwahl, Homoeosis- und Isopotenzerscheinungen. Palaeornithol. Frankreichs.* ⑨ Zürich 1913.
Lebedkin, Sergej J. (1886), Prof., Dir. Anat. Inst. Minsk (U.d.S.S.R.), Univ. *Labyrinth der Säuger.* ⑨ Moskau 1914.
Lebel, Michel Ludwig (1893), Ass. im Pathol. Inst. zu Sofia (Bulgarien), Bulevard „Chisto Botew" No. 183. *Lipoide in der Nebennierenrinde.* ⑨ Moskau 1917.
Leblanc, Ely (1871), Prof. d'Anat. à la Fac. de Méd. Univ. d'Alger (Algérie). *Anat. Anthrop. Anat.* ⑨ Paris 1925. ⑫ Animaux de l'Afrique du Nord (reptiles, petits mammifères) contre documents anthrop. (squelettes, moulages).
Leboucq, Georges (1880), Prof. d'Anat. à l'Univ. de Gand (Belgique), 11, Boulevard Léopold. *Œil et centres nerveux.* ⑨ Sc. nat. Gand 1901, méd. 1904.
Lebour, Marie Victoire, Naturalist at the Plymouth Marine Labor. Plymouth (England), Citadel Hill. *Marine Zool., Plankton. Dinoflagellates. Helminthol.* ⑨ Durham 1917.

Lebrun, Hector (1866), Prof. d'anat. comparée, Paléont. Gand (Belgique), 43, Boulevard Léopold. *Anat. comparée, Cytol.* ⓓ Med. Louvain 1890, sc. 1893.

Lebzelter, Viktor (1889), Custos der anthrop. Sammlung am Naturhist. Staatsmus. Wien I (Deutsch-Österreich), Burgring 7. *Rassenanthrop. Konstitutionsanthrop.* ⓓ 1914.

Lécaillon, Albert (1863), Prof. Zool. Fac. Sc. Univ. Toulouse (France). *Génétique chez les Oiseaux (Canards, Poules, Pigeons).* ⓓ Paris 1911.

Leche, Wilhelm (1850), Prof. Stockholm (Schweden). *Morphol. der Säugetiere.*

Lechler, Hermann (1898), Ass. am Inst. f. Seenforschung. Langenargen a. B. (Deutschland). ⓓ Tübingen 1924.

Lecloux, Jules Michel Joseph (1896), Ass. d'Anatomopathol. à l'Univ. de Liège (Belgique), 1 rue des Bonnes Villes. *Cancérol. expérim. Action des composés d'acides gras sur l'évolution du cancer au goudron chez la souris. Physiol. de la cellule cancereuse.* ⓓ Liège 1925.

Lecomte, H., Prof. Univ. Paris (France). *Phanérogames.* o

Lecomte du Noüy, Pierre (1883), Associate Member of the Rockefeller Inst. New York, N.Y. (U.S.A.), 66th Street and Avenue A. *Physico-chem. Biol. Molecular Physics in Relation to Biol. and Med. Molec. Physics of Colloids. Surface Equilibria of Colloids.* ⓓ Paris 1916.

Lederer, Max (1885), Pathol., Jewish Hospital. Brooklyn, N.Y. (U.S.A.). *Hematol. Pathol. Anat. and Histol.* ⓓ Columbia Univ. 1906. ⓟ Pathol. Specimens of Tissues.

Ledingham, J. C. G. (1875), Chief Bact. Lister Inst., Prof. Bact. Univ. London S.W. 1 (England), Chelsea Gardens. *Bact., Immunity, Pathol.* ⓓ Aberdeen 1902.

Ledoux, E., Prof. Ecole de Méd. Besançon (France). *Pathol.*

Ledoux, Huberte Henriette (1901), Ass. de Géol. à l'Univ. libre de Bruxelles (Belgique), 139, Rue Masui. *Paleophytol.: Flore de la Période dévonienne* ⓓ Bruxelles 1925.

Ledoux, Paul Vincent Désiré (1898), Ass. du Labor. de Morphol. et de Botan. Systematique, Inst. botan. Léo Errera, Univ. Bruxelles (Belgique), 139, rue Masui. *Phytomorphol. speciale: Cormophyta (Rubiaceae; Euphorbiaceae). Phytogeographie floristique et oecol. de Belgique: Flagellatae, Algae (Ulotrichales, Oedogoniales), Cormophyta.* ⓓ Bruxelles 1923. ⓟ Cormophyta: matériaux vivants, fixés ou secs.

Leduc, M. S., Prof., Ecole Méd. Pharmacie. Nantes (France). *Physiol.* o

Lee, Frederic Schiller (1859), Dalton Prof. of Physiol. Columbia Univ. New York (U.S.A.), 437 W., 59 St. *Reproduction. Fatigue and occupation.* ⓓ Ph. John Hopkins 1885, L.L. St. Lawrenc 1918. o

Lee, Henry Atherton (1894), Pathol., Experim. of the Hawaiian Sugar Planters' Association. Honolulu (Hawaii), Experim. Station. *Diseases of Sugar cane. The normal roots of sugar cane.* ⓓ Univ. of California 1916.

Lee, Milton O. (1901), Instr., Department of Physiol., Ohio State Univ. Columbus, O. (U.S.A.). *Endocrinol.* ⓓ Ohio State Univ. 1925.

Lee, S. C., Lect.National Normal Univ. Peking (China) *Botan.* o

Lee, Thomas G. (1860), Prof. of Compar. Anat., Inst. of Anat. Univ. of Minnesota. Minneapolis, Minn. (U.S.A.). *Anat. Histol.* ⓓ Univ. of Pennsylvania 1886. o

Leeder, Friedrich (1862), Hofrat, Oberforstmeister i. R. Gmunden (Ober-Österreich), Georgstr. 16. *Floristiker der österreichischen Alpenländer Nieder- und Oberösterreich, Steiermark, Salzburg, Kärnten.*

Leeder, Karl (1864), Hofrat, Prof., Ingenieur, Hon.Doc. ao. Prof. für Wildkunde und Jagdbetrieb an der Hochsch. für Bodenkultur.Wien VII (Deutsch-Österreich), Zeismannsbrunngasse 4. *Allgemeine Biol. der Jagdtiere.*

Leefmans, S. (1884), Dir. Zoöl. Labor., Inst. f. Pflanzenkrankheiten. Buitenzorg, Java (Nederl.-O.-Indië), Tjikeumeukweg 86. *Angewandte Entomol.* ⓟ Lamellicornia N.-O.-Indiens.

Leegaard, Caroline (1885), Ass. to the Botan. Labor. of the Univ. Oslo (Norge), Keysers Gate 9. *Plankton (Ciliaten). Plant Anat.*

Leendertz, Karel (1892), Botan., Samenkontrollstation. Wageningen (Holland), Rijksstraatweg 41. *Samenkunde.* ⓓ 1918. ⓟ Gräsersamen.

Lefever, Rufus H. (1895), Independent Nat. York, Pa. (U.S.A.), 1001 East King St. *Ornithol. of Shantung, China General collecting.* ⓓ Lebanon Valley Coll. 1917. ⓟ Birds of China . Shells of Shingtau.

Lefevre, Edwin (1859), Ass. Bact., Dept. of Agric. Washington, D.C. (U.S.A.), 1420 Newton Str. *Physiol.* o

Lefevre, M. J., Prof. Lycée Pasteur. Neuilly-sur-Seine (France). *Physiol.* o

Leffingwell, Dana Jackson (1901), Curator of the Charles R. Conner Mus. State Coll. of Washington. Pullman, Wash. (U.S.A.). *Ornithol.* ⓟ Cornell Univ. 1926. ⓟ Birdskins and Mammal skins.

Legagneux, Henri (1872), Chef labor. inst. océanographique. Mus. du Havre (France). *Biol., bact.* ⓓ Paris 1898.

Legendre, René André (1880), Dir. du Labor. de Physiol. comparée à l'Ecole des Hautes Etudes, Sous-Dir. du Labor. maritime du Coll. de France, à Concarneau (Finistère). Paris 14e (France), 27, rue d'Alésia. *Physiol. comparée, biol. marine.* ⓓ Paris 1910.

Léger, Louis Urbain (1866), Prof. de Zool. à la Fac. des Sc. de Grenoble (France), Rue Hébert. *Protist. et Parasit. Hydrobiol., Pisciculture.* ⓟ Faune des eaux de montagne.

Legrand, A., Prof. adj. Univ. Fac. cath. Lille (France). *Physiol.* o

Lehbert, Rudolph (1858). Reval (Estland), Ratsapotheke, W. 3. *Flora des ostbaltischen Gebiets. Calamagrostis. Betula. Kieselhaargebilde phanerogamer Pflanzen.*

Lehenbauer, Philip A. (1885), Prof. of Botan. and Horticulture. Univ. of Nevada. Reno, Nev. (U.S.A.). *Water in Relation to Plant activity. Growth of Maize in Relation to Temperature.* ⓓ Univ. of Illinois 1914.

Lehman, Samuel G. (1887), Prof. Plant Pathol., North Carolina State Coll. of Agric. and Eng., Plant Pathol. North Carolina Agric. Experim. Sta. Raleigh, N.C. (U.S.A.), State Coll. *Plant Pathol. Soybean diseases.* ⓓ Washington Univ. 1922.

Lehmann, Conrad (1898), Dr. phil., Vorsteher der Biol. Abt. für Fischerei an der Landwirtschaftl. Versuchsstation in Münster i. Westf. (Deutschland), Südstr. 72. *Fischereibiol., Abwasserbiol.* ⓓ Berlin 1921.

Lehmann, Ernst (1880), o. Prof. Botan. Univ., Dir. des botan. Gartens. Tübingen (Deutschland), Wilhelmstr. 5. *Vererbungslehre. Keimungsphysiol. Veronica. Oenotheraforschung.* ⓓ Straßburg

Lehmann, Gunther (1897), Dr. med., 1. Ass. am Kaiser-Wilhelm-Inst. für Arbeitsphysiol. Berlin N 4 (Deutschland), Invalidenstr. 103a. *Energetik des menschlichen Körpers. Ermüdungsforschung. Physikalische Chem. des Blutes.* ⓓ Berlin 1922.

Lehmann, Hans (1889), Dr. phil. nat. Luxemburg (Luxemburg), Boulevard Henderstr. 13. *Angewandte Zool., Schädlinge der Obstbäume und des Rebstockes.* ⓓ Freiburg i. Br. 1915.

Lehmann, Karl (1858), Univ.Prof., G. HofR., Dir. Hygien. Inst. Würzburg (Deutschland), Schellingstraße 20. *Hygien. Toxikol. Bact.* o

Lehmann, Otto (1865), Prof. Dr., Dir. des Altonaer Mus. Altona a. d. Elbe (Deutschland), Museumstr.

Vererbungslehre, Landeskunde von Schl.-Holstein. ① Jena 1887.

Lehmann-Facius, Hermann (1899), Dr. med. Mannheim (Deutschland), Pathol. Inst. der städt. Krankenanstalten. *Serol. (Serodiagnostik von Karzinom, Tuberkulose, Gravidität).* ⓟ Heidelberg 1923.

Lehmberg, Karl (1897), Dr. phil. Gandersheim a. Harz (Deutschland). *Entwicklung der Wasserleitungsbahnen b. Pflanzen.* ⓟ Göttingen 1923.

Lehner, Josef (1882), Dr. med., Priv.Doc. für Histol. und Ass. am Histol. Inst. der med. Fac. der Univ. Wien IX (Deutsch-Österreich), Schwarzspanierstr. 17. *Histol. und Histogenese von Zahn, Magen, Hoden und Nebenhoden, Bindegewebe und Blut.* ⓟ Wien 1907.

Lehnert, Berthold Edwin (1884), Vorst. Serol. Abt. d. Veterin. Bakteriol. Staatsinst. Stockholm (Sverige), Experimentalfältet. *Impfstoffe, Serodiagnose.* ⓟ Hannover 1921.

Lei, Hei Kit (1902), Lignan Univ. Canton (China). *Anat. and pathol. of silk worm.* ⓟ Lingnan Univ. Canton 1926.

Leibensohn, Alexis (1868), Ass.d.Pharmacol. Inst. Odessa (U.d.S.S.R.), Olgijewskaja 4. *Intoxikationen.* ⓟ Odessa 1924.

Leiby, Rowland Willis (1892), Chief in Entomol., North Carolina Department of Agric. Raleigh, N.C. (U.S.A.). *General Entomol. Polyembryony of Insects.* ⓟ Cornell 1921.

Leick, Erich (1882), Dr. phil., ao. Prof. d. Botan. u. Pharmakognosie a. d. Univ. Greifswald i. Pomm. (Deutschland), Arndtstr. 31. *Wärmehaushalt der Pflanzen; Stomatärbewegung; Transpiration.* ⓟ Greifswald 1910.

Leick-Schultz, Marie (1880), Dr. phil. Greifswald i. Pomm. (Deutschland), Arndtstr. 31. *Süß- und Brackwasseralgen.* ⓟ Greifswald 1914.

Leidenfrost, Gyula (1885), Dr., Hochsch.Prof., Priv.Doc. Budapest IX (Ungarn), Boráros tér 1. *Ichthyol., Thalassozool. (Echinodermen).* ⓟ Budapest 1916.

Leigh-Sharpe, William Harold (1881), Lect. in Zool. Chelsea Polytechnic, St. Mary's Hospital Med. School London S.W. 10 (England), 17 Clyde St. *Parasit. Parasitic Copepoda.* ⓟ London 1920.

Leighty, Clyde Evert (1882), Agronomist in Charge of Eastern U. S. Wheat Investigations. Bureau Plant Industry. Washington, D.C. (U.S.A.). *Agronomic, genetic, physiol. and breeding investigations on wheat, rye and buckwheat.* ⓟ Cornell Univ. 1912. Ⓟ *Triticum, Aegilops, Secale, Fagopyrum.*

Leim, Alexander Henry (1897), Ass. Dir., Atlantic Biol. Station. St. Andrews, N.B. (Canada). *Fisheries. Life History of the Shad (Alosa sapidissima). Effect of various physical conditions on the development of eggs and larvae of various fishes: cod, smelt, Atlantic salmon, whitefish, lake trout. Marine and fresh water biol.* ⓟ Toronto 1924.

Leininger, Hermann (1885), Dr., Prof., Konserv. an den Landessammlungen für Naturkunde, Zool. Abt. Karlsruhe i. B. (Deutschland), Kaiserallee 115. *Entomol., Hymenoptera aculeata. Botan.* ⓟ Heidelberg 1911.

Leinzinger, Mária (1900), Ass. pharmakogn. Inst. Szeged (Ungarn), Kálvária tér 5. *Biol. Titrieren des Mutterkorns. Stoffwechsel im überlebenden Gewebe.* ⓟ Szeged 1924.

Leiper, R. T., Prof. School of Hygiene and Tropic. Med. London N.W. (England), Endsleigh Gardens. *Helminthol.* o

Leira, Lamberto, Assoc. Prof. Univ. of the Philipp. Manila (Philippine Islands). *Parasit.* o

Leisering, Bruno (1878), Dr., Prof., Studienrat. Berlin NO 43 (Deutschland), Am Friedrichshain 15. *Pflanzenanat. und -morphol.* ⓟ Berlin 1899.

Leitão da Cunha, Raul (1881), Prof. cathedr. d'Anat. Pathol. Fac. Méd. Univ. Rio de Janeiro (Brasil), 52 Rua das Palmeiras. *Anat. pathol., immunité, cytol. du liquide céphalo-rachidien.* ⓟ Rio de Janeiro 1903.

Leitch, James Muil (1901), Demonstrator in Bact., Royal Technical Coll. Glasgow (England), 1, Ormonde Mount, Muirend. *Booked Foodstuffs.*

van der Lek, Hendrik Adrianus Abraham (1881), Dr., Botan. at the labor. for horticultural research of the Agric. Univ. Wageningen (Holland), Zoomweg 10. *Applied botan.* ⓟ 1925.

Lelep, Doc. Univ. Fac. méd. Poznań (Polska), r. ś. Pawła 10. *Biochim.* o

Lelievre, Dr., Prép. Labor. d'Histol., Fac. Méd. Paris (France), 6, rue Leclerc. *Histol.* o

Lemann, Isaac Ivan (1877), Prof., Clinical Med., Tulane Univ. New Orleans, La. (U.S.A.). *Metabolism.* ⓟ Tulane Univ. 1900.

Lemcke, Alfred (1864), Dir. des Samenuntersuchungsamtes und der Hauptstelle für Pflanzenschutz an der Landwirtschaftskammer. Königsberg i. Pr. (Deutschland), Beethovenstr. 24/26. *Pflanzenkrankheiten. Carex. Moorkunde.* ⓟ Königsberg i. Pr. 1892.

Lemesle, Robert August Joseph (1894), Preparateur Inst. des Recherches Agron. Paris (France). *Détermination des champignons. Systématique e Structure des Phanérogames.* ⓟ Méd. Laur. Fac. méd., sc. n. Paris.

Lemière, G., Prof. Univ. Fac. cath. Lille (France). *Hyg. Microbiol.* o

Lemierre, André Alfred (1875), Prof. Bact. méd. Fac. Méd. Paris (France), 217, Faubourg Saint-Honoré. *Bact., méd.* ⓟ 1926.

Lemmermann, Otto (1869), o. Prof. für Agrikulturchem. u. Bact. an der Landwirtschaftl. Hochschule Berlin und Oberleiter der Landwirtschaftl. Versuchsstation der Landwirtschaftskammer für die Provinz Brandenburg. Berlin-Dahlem (Deutschland), Albrecht-Thaer-Weg 1. *Agrikulturchem. (Pflanzenernährung), Agrikulturbact., Bodenkunde.* ⓟ Jena.

Lemoigne, Maurice (1883), Chef labor. Inst. Pasteur. Lille (France), Bd. Louis XIV. *Fermentations.*

Lendl, Adolf (1862), Prof. emer., Dir. des Zool. Gartens emer. Budapest II (Ungarn), Jégverem-ut 6. *Biol., Ornithol., Entomol., Araneol.* ⓟ Budapest 1883.

Lendner, Alfred (1873), Prof. Pharmacognosie Univ. Genève (Suisse), 6, Rue Emile Yung. *Champignons Mucorinées. Botan. appliquée: Pathol. végétale. Pharmacognosie.* ⓟ Genève 1896.

Lengerich, Hanns (1893), Dr. phil., Leiter des Forschungsinst. für die Fischindustrie e. V. Altona a. d. Elbe (Deutschland), Flottbecker Chaussee 92. *Fischereibiol. und Fischverarbeitungstechnik. Hydroidensystematik und -entwicklungsgeschichte.* ⓟ Berlin 1920.

v. Lengerken, Hanns (1889), Dr. phil., ao. Prof. für Zool. an der Landwirtschaftl. Hochsch. zu Berlin. Berlin-Schöneberg (Deutschland), Hauptstr. 130. *Insektenbiol., Lebenserscheinungen der Käfer, Brutpflegeinstinkte der Käfer, Teratol. Erscheinungen der Käfer, Metatelie der Insekten, Biol. der Silphini, Gynandromorphismus bei Lucanus. Extraintestinale Verdauung.* ⓟ 1914.

Lengyel, Géza (1884), landwirtschaftl. Versuchsstationsleiter. Budapest II (Ungarn), Kisrókus-ut 15. *Floristik, Systematik der Blütenpflanzen. Angewandte Botan.* ⓟ Budapest 1907. Ⓟ *Pflanzen.*

Lenhossék, Mihály, Prof. Dr., Dir. I. Anat. Inst. Univ. Budapest (Ungarn). *Anat. Histol. Cytol.* o

Lenjkow, Peter W., Botan. d. Landwirtschaftl. Akademie Timirjasew. Moskau (U.d.S.S.R.), Petrowsko-Rasumowskoje. *Angew. Botan.* o

Lenoir, Maurice (1888), Ass. de Botan. Fac. Sc. Nancy, Meurthe et Moselle (France), 31,Rue de Paris. *Cytol. Evolution du tissu vasculaire. Chromatines.* ⓟ Nancy 1919.

Lentz, Otto (1873), Geh. Obermedizinalrat, Prof., Dr. med., Ministerialrat im Preußischen Mini-

sterium für Volkswohlfahrt. Berlin-Wilmersdorf (Deutschland), Rüdesheimer Platz 6. *Bact., Seuchenbekämpfung*. ⓓ 1895.
Lenz, Friedrich (1889), Dr., wissenschaftl. Ass. an der Hydrobiol. Anstalt der Kaiser-Wilhelm-Gesellschaft zu Plön, Holstein (Deutschland), Prinzenstraße 12. *Hydrobiol. (Limnol.); Systematik, Morphol. und Ökol. der Dipterenlarven und -puppen, Chironomiden.* ⓓ Kiel 1919.
Lenz, Fritz (1887), Dr., Prof. der Rassenhygiene (Eugenik) an der Univ. München. Herrsching bei München (Deutschland). *Rassenhygien. (Eugenik). Menschliche Erblichkeitslehre. Allgemeine Genetik.* ⓓ Freiburg i. B. 1912. ⓟ Speziesbastarde von Schmetterlingen.
Leo, Hans (1854), o. Univ.Prof. d. Pharmacol. in Bonn (Deutschland), Coblenzer Str. 93 I. *Physiol. u. pathol. Chem. Pharm.* ⓓ Phil. Bonn 1878; med. 1882.
Leővey, Ferenc (1903), Ass. Physiol. Inst. Univ. Budapest VIII (Ungarn), Eszterházy-ut 9. *Physiol. der Niere.*
Léon, N., Prof. Univ. Jassy (România). *Parasit.* o
De Leon, Walfrido, Assoc. Prof. Univ. of the Philippine, Manila (Philippine Islands). *Pathol. and Bact.* o
Leonard, Emery C. (1892), Aid, Division of Plants, U. S. National Mus. Washington, D.C. (U.S.A.). *Taxonomic studies of West Indian phanerogams, Acanthaceae of tropical America.* ⓓ Ohio State Univ. 1916.
Leonard, Lewis Thompson (1885), Associate Physiol. in the United States Department of Agric. Washington, D.C. (U.S.A.). *Soil bact., the legumes and their nodule organisms.* ⓓ Washington, D.C. 1921. ⓟ Commercial legume bacteria culture.
Leonhardt, Otto (1864). Nossen i. Sa. (Deutschland), Meißner Str. 7. *Phanerogamen, Cryptogamen.* ⓟ Pflanzen (Cryptog. Phanerog.).
Leonian, Leon H., Ass. prof. of Plant Pathol., West Virginia Univ., Ass. Plant Pathol., West Virginia Agric. Experim. Station. Morgantown, W.Va. (U.S.A.). *Physiol. of fungi.* o
Leonov, N. D., Doc. Univ. Taskent (U.d.S.S.R.). *Physiol. d. Pflanzen.* o
Leontowitsch, Alexander Wassiljewitsch (1869), Prof. Physiol. der Tiere. Moskau (U.d.S.S.R.), 8 Petrovsko-Razumovskoje, Landwirtsch. Inst., Timirjasewsche Landw. Akad., Haus 9. *Vitale Methylenblau-Färbung der Nerven.* ⓓ Kiew 1900.
Lepechin, A. L., Doc. Univ. Med. Fac. Anat. Inst. Taschkent (U.d.S.S.R.). o
Lepeschkin, Wladimir W. (1876), Prof. Dr. Naturwiss. Abt. d. Russ. Volks-Univ. Praha II (C.S.R.), Benatska 433. *Physiol. d. Zelle. Permeabilität d. Protoplasmas. Kolloidchem. d. leb. Materie. Stärke, Eiweißkörper, Hämolyse.* ⓓ Zürich 1901.
Lepešinskaja, Olga Borisovna, Doc. I. Univ. Moskau (U.d.S.S.R.). *Histol.* o
de Lépiney, Jacques (1896), Entomol. agricole de l'Inst. Sc. Chérifien Rabat (Maroc). Youssef (N. Africa), Avenue Moulay. *Biol. des insectes utiles ou nuisibles, hyménoptères parasit.* ⓟ Insectes nuisibles du Maroc; hyménoptères parasit.
Leplat, Georges (1890), Agrégé spécial à l'Univ. de Liège (Belgique), 19, Rue des Anges. *Histol., embryol., teratol. et anat. comparée de l'organe visuel. Physiol. normale et pathol. de l'œil.* ⓓ Liège 1925.
Lepnewa, G. S., Ass. Hydrobiol. Labor. Univ. Leningrad (U.d.S.S.R.). o
Lepri, G., Prof. Univ. Roma (Italia). *Zool.* o
Leptschenko, Jacob Ch., Botan., Botan. Sektion d. Wissenschaftl. Kom. d. Ukrain. „Narkomsema". Kiew (U.d.S.S.R.), Korolenko-Str. 21, W. 3. *Botan. Floristik.* o
Lerche, Martin (1892), Dr., stellvertr. Dir. des Bact. Inst. der Landwirtschaftskammer Schlesien. Breslau 16 (Deutschland), Kaiserstr. 55. *Bact. und pathol. Anat. (Veterinärmed.). Abortus, Geflügel-* *krankheiten.* ⓓ Hannover 1920. ⓟ Bakterien.-Kulturen.
Leriche, Maurice Henri Charles (1875), Prof. de Géol. à l'Univ. de Bruxelles. Uccle (Belgique), 123, avenue Montjoie. *Poissons fossiles.* ⓓ Bruxelles 1910.
Lermontova, Katharina (1889), Mitarbeiter des Comité Géol. de Russie. Leningrad (U.d.S.S.R.), V. O. Sredny Prosp. 72 b. *Palaeont. Kambrische Trilobiten des asiatischen Rußlands (Sibirien, Turkestan). Larven der Trilobiten. Kambr. Fauna Turkestans.* ⓟ Kambrische Trilobiten.
Leroux, Paul Henri Roger (1892), Chef des Travaux Pratiques d'Anat. Pathol. Fac. Méd. Paris (France), 21, Rue de l'Ecole de Médicine. *Anat. Pathol. Cancer expérim.* ⓓ Paris 1921. ⓟ Films d'enseignement de l'Anat. Pathol.
Lesage, P., Prof. Univ. Rennes (France). *Botan.* o
Lesbre, M., Dir. Ecole Nat. Vét. Lyon (France). *Anat.* o
Leschke, Erich (1888), ao. Prof. f. Innere Med., Abteilungsarzt d. II. med. Klinik d. Charité. Berlin W 15 (Deutschland), Kurfürstendamm 66. *Vegetatives System und Stoffwechsel, Histochem., Chemotherapie. Wechselseitige Beziehungen der Drüsen mit innerer Sekretion.* ⓓ Bonn 1911.
Lesley, James Wyvill (1888), Ass. in Genetics. Riverside, Cal. (U.S.A.), Citrus Experim. Station of Univ. of California. *Tomato breeding, disease resistance. Cyto-genetics of tomato.* ⓓ Univ. of Cambridge 1919. ⓟ Tomato seeds.
Lesley, Margaret Mann (1891). Riverside, Cal. (U.S.A.), 129 Nogales St. *Cytol. in relation to Genetics.* ⓓ Univ. of California 1921.
Lesnikowa, Aldona Th. (1889), Geol. Comité Rußlands, Ass. Univ. Leningrad (U.d.S.S.R.), Wass. Ost. 12 Linie U. 33, Qu. 18. *Trilobiten und Brachiopoden des Baltischen Silurs. Quartäre Pflanzen.* ⓟ Versteinerung des Silurs der Umgegend von Leningrad.
Lesser, Ernst Josef (1879), Dr. med., Vorstand des Labor. der städtischen Krankenanstalten in Mannheim (Deutschland). *Kohlehydratstoffwechsel. Anoxybiose. Diabetes. Fermente. Innere Secretion des Pancreas.* ⓓ München 1903.
Lestage, J. A. (1879), Ass. à la Station de Biol. lacustre d'Overmeire (Belgique). Uccle (Belgique), 10, Avenue de la Floride. *Larves aquatiques (Ephemeroptera, Plecoptera, Trichoptera). Hydrobiol. Adultes des mêmes Ordres. Planipennia, Megaloptera, Mecoptera.*
Lester, Vera Esther (1895), Ass. Bact. at the State Serum Inst. København (Danmark), v. Amagerport 8. *Bact.*
Lestoquard, Félix (1897), Chef de labor. Inst. Pasteur d'Algérie. Alger (Algérie). *Maladies microbiennes et parasitaires des animaux. Piroplasmes.* ⓓ Toulouse 1922. ⓟ Préparations de protozoaires agents des maladies des animaux.
de Leszczyński, Roman Jean (1891), Ass. Univ. Poznań (Polska), Fredry 10. *Pharm. expérim.* ⓓ Poznań 1924.
Lettan, Georg (1878), Dr., Augenarzt. Lörrach. Baden (Deutschland), Markus-Pflüger-Str. 11. *Lichenol. Europas (Flechten).* ⓓ Heidelberg 1904
Letterer, Erich (1895), Priv.Doc. Würzburg (Deutschland), Pathol. Inst. *Amyloid.* ⓓ Würzburg 1921.
Letulle, Maurice E. J. L. (1853), Prof. honoraire d'anat. pathol. à la Fac. de méd. Paris XVI (France), 24, rue Boissière. *Anat. pathol., Poumon.* ⓟ Préparations d'histol. pathol.
Leuenberger, Fritz (1860), Dr. h. c. Bern (Schweiz) Marzilistr. *Bienenkunde: Anat., Biol., Bienenkrankheiten und deren Bekämpfung.* ⓓ Bern 1926. ⓟ Bienen mit Acarapis woodi.
Leukel, Robert Whilmer (1888), Associate Pathol., Office of Cereal Crops and Diseases, Bureau of Plant Industry, U.S. Dept. of Agric. Washington, D.C.

(U.S.A.). *Diseases of cereals and other grasses.* ⓓ Univ. of Wisconsin 1921. ⓟ *Diseases on cereals, as nematode galls, smuts, rusts.*

Leupold, Ernst (1884), Prof. allg. Pathol. u. Pathol. Anat., Dir. Pathol. Inst. Univ. Greifswald (Deutschland). *Beziehungen zwischen Nebennieren und männlichen Keimdrüsen. Bedeutung des Cholesterin-Phosphatidstoffwechsels für die Geschlechtsbestimmung.* ⓓ München 1910.

van Leuven Osterhout, Winthrop John (1871), Member Rockefeller Inst. for Med. Research. New York, N.Y. (U.S.A.), 66th St. and Ave. A. *Permeability of cells: bioelectrical phenomena.* ⓓ Univ. of California 1899.

Leuzinger, Hans (1897), Dr. phil., Chef de la Station cantonale d'entomol. appliquée. Châteauneuf. Sion, Valais (Schweiz). *Angewandte Entomol. Wein- und Obstbaumschädlinge Blutlaus, Reblaus, Traubenwickler. Simaethis pariana. Entwicklungsgeschichte v. Carausius morosus.* ⓓ Zürich 1925.

Leuzzi, F., Prof. Univ. Napoli (Italia). Anat. o

Levaditi, Constantin (1879), Prof. Inst. Pasteur. Paris XV (France), 54, rue des Volontaires. *Neurovaccin, rage, encephalite, herpes. Rage. Syphilis.* ⓓ Paris 1902.

Levander, K. M., Dr. phil., o. Prof. Univ. Helsinki N. (Finnland), Järnvägsgatan 13. *Zool., Plankton. Fischbiol.* o

Levene, Phoebus Aaron Theodor (1869), Member Rockefeller Inst. for Med. Research. New York City (U.S.A.), 66th Street and Avenue A. *Structural chem. of biol. important substances (proteins, carbohydrates, nitrogenous sugars, nucleic acids, lipoids) and stereochem.* ⓓ St. Petersburg 1891.

Levi, A., Prof. Univ. Padova (Italia). *Istol.* o

Levi, Giuseppe (1872), Prof. d'Anat. humaine, Dir. Inst. d'Anat. Univ. Torino (Italia), Corso Massime D'Areglio 52. *Histol., Embryol. gén. Cultivation des tissus.*

Levi, Jackson Horlacher (1896), Assoc. Prof., Animal Husbandry, Kentucky Experim. Station. Lexington, Ky. (U.S.A.). *Sheep investigations.* ⓓ Kansas 1919.

Levíček, Jan, Ass. Anat. Lab. Univ. Masaryk, Brno (C.S.R.), Údolní 73. *Histologie.* o

Levin, E. I., Doc., Univ. Karolinska Inst. Stockholm (Sverige). *Bact.* o

Levine, Harold (1900), Food Research Labor. Inc. New York, N.Y. (U.S.A.), 39 W 38th St. *Ketosis in the rat. Vitamins.* ⓓ Yale Univ. 1926.

Levine, Michael (1886), Cancer Research Labor. New York City (U.S.A.), Montefiore Hospital. *Plant cancer, crown gall; cytol.* ⓓ Columbia Univ. 1913.

Levine, Moses ben Naphtali (1886), Associate Pathol., U.S. Dept. of Agric. St. Paul, Minn. (U.S.A.), Biol. Club, Univ. of Minnesota. *Phytopath. and genetics, ecol. and physiol.; Cereal rusts, smuts and mildew. Physiol. specialization. Morphol. idiosyncrasies. Genetic nature.* ⓓ Univ. of Minnesota 1924.

Levine, Philip (1900), Ass. at the Rockefeller Inst. New York. Brooklyn, N.Y. (U.S.A.), 609 Pennsylvania Avenue. *Bact. and Serol.* ⓓ Cornell Univ. Med. Coll. 1923.

Levine, Victor E., Prof. of Biol. Chem. and Nutrition, Creighton Univ. School of Med. Omaha, Neb. (U.S.A.). *Chem. of metabolism and nutrition.* o

Levinson, Samuel A. (1895), Associate Department of Pathol., Univ. of Illinois, Coll. of Med. Chicago, Ill. (U.S.A.). *Experim. Pathol., Immunol.*

Levinthal, Walter (1886), Dr. med., Ass. am preuß. Inst. für Infektionskrankheiten („Robert Koch"). Berlin N 39 (Deutschland), Föhrer Str. 2. *Bact. und Epidemiol. der Influenza. Filtrierbare Virusarten. Umwandlung von Bakterien mit Einzelkulturen.* ⓓ München 1912.

Levitt, Michael M. (1903), Doc. angewandte Zool. Landwirtschaftl. Inst., Aspirant d. Katheders d.

Zool. Kiew (U.d.S.S.R.), Leninstr. 82-5. *Entomol.: Anat. d. inneren Geschlechtsapp. der Blattkäfer.*

Levy, Fritz (1887), Dr. phil. et med., Privatlabor. Berlin W 57 (Deutschland), Winterfeldstr. 35 II. *Zellteilungsphysiol., Pathol. der Zelle, Parthenogenese.* ⓓ Phil. Berlin 1915; med. 1916.

Lévy, Robert (1886), Maître de Conférences Fac. Sc. Univ., Dir. Labor. Ecole pratique des Hautes-Etudes. Paris (France), 45, Rue d'Ulm. *Toxines d'origine animale.* ⓓ Paris 1921.

Lew, T. T., Prof. Yenching Univ. Peking (China). *Psychol.* o

Lewaschoff, Michael M. (1899), Ass. Biol. Wolga-Station. Saratow (U.d.S.S.R.). *Parasit. und freilebende Nematoden.* ⓟ *Nematoden.*

Lewicki, Stefan (1890), Dr. d'agric., Priv.Doc. de génétique de l'Univ. de Cracovie, chef de section de l'amélioration des plantes de l'Inst. National de l'Economie Rurale. Pulawy (Polska). *Génétique appliquée, l'amélioration des blés.* ⓓ Cracovie 1925. ⓟ *Varietés et races du froment, de l'orge, de l'avoine et du millet.*

Lewin, Louis (1850), Hon.Prof. Techn. Hochsch. Berlin NW 40 (Deutschland), Hindersinstr. 2. *Pharmakol. Toxikal.* o

Lewiński, Jan (1876), Dr. phil., Prof. o. de Géol. et de Paléont. à l'Univ. de Varsovie (Pologne). Labor. Géol. Univ. *Ammonites du Jurassique supérieur (Cardiocératidés, Appellíidées, Virgatites).* ⓓ Lwów. ⓟ *Ammonites du Jurassique supérieur de la Pologne.*

Lewis, Francis John, Prof. of Botan. Univ. of Alberta. Edmonton (Canada). *Plant Ecol.and Physiol. Pleistocene Plant Deposits.* ⓓ Liverpool 1911.

Lewis, Frederic Thomas (1875), Associate Prof. of Embryol., Harvard Med. School. Boston, Mass. (U.S.A.). *Vertebrate embryol.; Morphol. of the cell.* ⓓ Harvard 1901.

Lewis, Harrison Flint (1893), Chief Federal Migratory Bird Officer, Ontario and Quebec. Ottawa (Canada), Canadian National Parks. *Ornithol.* ⓓ Toronto, Ontario, 1926.

Lewis, Howard Bishop (1887), Prof. Physiol. Chem. Med. School, Univ. of Michigan. Ann Arbor, Mich. (U.S.A.). *Nutrition, Metabolism of Sulfur, Intermediary Metabolism of the Amino Acids, Protein Metabolism.* ⓓ Yale Univ. 1913.

Lewis, H. C. (1902), Ass. Entomol., California Dept. of Agric. Sacramento, Cal. (U.S.A.). ⓓ Ohio State Univ. 1925.

Lewis, Isaac M., Prof. of Botan., Univ. of Texas. Austin, Tex. (U.S.A.). *Pathol., Cytol.* o

Lewis, Ivey Foreman, Prof. of Biol., Univ. of Virginia. Charlottesville, Va. (U.S.A.). *Algae.* o

Lewis, Julian Herman (1891), Ass. Prof. of Pathol., Member of the Otho S. A. Sprague Memorial Inst. for Med. Research. Chicago, Ill. (U.S.A.), Univ. *Immunol.* ⓓ Phil. Chicago 1915, med. 1917.

Lewis, Margaret R. (1881), Associate, Carnegie Inst. of Washington. Baltimore, Md. (U.S.A.), Johns Hopkins Med. School. *Cytol., tissue culture.*

Lewis, Paul A. (1879), Member Rockefeller Inst. f. Med. Research. Princeton, N.J. (U.S.A.), Dept. for Animal Pathol. *Comp. Pathol. Bact.* o

Lewis, Robert Donald (1897), Ass. Extension Prof. of Plant Breeding. Ithaca, N.Y. (U.S.A.), Coll. of Agric. *Plant Breeding: Genetics of Cereals.* ⓓ Cornell 1926. ⓟ *Dwarfs in oats.*

Lewis, Warren Harmon (1870), Prof. Physiol. Anat., Johns Hopkins Univ., Research associate, Carnegie Inst. of Washington. Baltimore, Md. (U.S.A.), Johns Hopkins Med. School. *Cytol., Tissue Culture, Experim. Embryol.* ⓓ Johns Hopkins Univ. 1900.

Lewitsky, Gregor A. (1878), Vorstand der Cytol. Abteilung des Inst. für angewandte Botan. Leningrad (U.d.S.S.R.), Herzenstr. 44. *Genetische und systematische Cytol.; Chondriosomen. Karyol. der*

Gattung Festuca. ⓓ Kiew 1904. ⓟ Gattung Festuca (Herbarexemplare u. Samen).
Lewy, F. H. (1885), a.o. Prof. a. d. Univ. Berlin W 10 (Deutschland), Matthäikirchstr. 8. *Nervensystem des Menschen und der Tiere. Norm. u. pathol. Anat. und Physiol. Infektiöse Erkrankungen des Nervensystems.* ⓓ Berlin 1910.
Ley, Auguste, o. Prof. Univ. libre. ,,Bruxelles (Belgique). *Psychol.* ○
L'hoëst, Michel (1869), Dir. Jardin Zool. Anvers (Belgique). 20, Place de la Gare. ⓓ Bruxelles 1891.
Lhoták, Kamil, Prof. Univ. Charles, chef de l'Inst. de Pharmacognosie et Pharmacol. Praha II (C.S.R.), Legerova 38. ○
Li, Ju Chi (1896), Prof. of Biol. Shanghai (China), School of Biol. Sc., Fuh Tan Univ. *Genetic, experim. embryol.* ⓓ Columbia 1926.
Liachowetzky, A. M., Ass., Dr. Histol. Embryol. Kab. II. Univ. Moskau (U.d.S.S.R.), Pogodinskaja ul. 6. ○
Librowitch, Leonid S., Com. Géol. de Russie. Leningrad (U.d.S.S.R.), W. O., Sredny Pr. 72-B. *Palaeont.; Spongia.* ○
Liburnau, Ludwig, Ritter von (1856), ao. Prof. Hochsch. f. Bodenkultur, Hofrat, Dir. zool. Abteilung Naturhistorisches Mus. Wien VII (Deutsch-Österreich), Burggasse 7. *Zool.: niedere Tiere. Plathelminten, Bryozoen, Vögel (Systematik und Vogelzug), Säugetiere (Systematik), ausgestorbene Lemuren. Polypomedusen. Ornis Südarabiens. Pipridae. Steinböcke Innerasiens. Affen, Halbaffen und Huftiere von Zentral-Afrika.* ⓓ Wien 1879.
Lichačev, A. A., Prof. Med. Inst. Leningrad (U.d.S.S.R.), Ul. L'va Tolstogo 6-8. *Pharmacol.* ○
Licharew, Boris K. (1887), Geol. des Russischen Geol. Comités. Leningrad (U.d.S.S.R.), 15 Linia, 48. *Paleont. Permische und obercarbonische Faunen d. europäischen Rußlands. Besonders: Brachiopoden; Fusulinen.*
Lichatschev, L. J., Dr., Ass. Histol. Labor. Tierärztl. Inst. Saratow (U.d.S.S.R.). ○
Lichtenheld, Georg (1877), Geh.Vet.R. Wiesbaden (Deutschland), Panoramaweg 9. *Tropische Tierseuchen.* ⓓ Leipzig 1903.
Lid, Johannes (1886), Kunserv. Oslo (Norge), Botanisk Mus. *Pflanzengeographie, spez. Svalbard (Spitzbergen) und arctischen Gebieten. Sphagnol. Pflanzenoekol. und Pflanzensociol.* ⓟ Sphagnum.
Liddell, Edward George Tandy (1895), Fellow of Trinity Coll. and Univ., Demonstrator in Physiol. Oxford (England). *Physiol. of the nervous system.*
Lieb, Charles C. (1880), Prof. of Pharmacol., Coll. of Physicians and Surgeons, Columbia Univ. New York City (U.S.A.), 437 West 59 Street. *Pharmacol.*
Lieb, Hans, Univ.Prof. Graz (Österreich), Kirchengasse 13. *Biochem. Pharm.* ○
Lieben, Fritz (1890), Priv.Doc. f. physiol. Chem. u. Ass. an d. Abt. f. physiol. Chem., Physiol. Univ.- Inst. Wien I (Deutsch-Österreich), Mölkerbastei 5. *Probleme des Eiweiß- und Kohlehydratstoffwechsels: Einwirkung von Salpetersäure, Halogenen auf Eiweiß.* ⓓ Wien 1917.
Lieberfarb, Alexander Sigismundowitsch (1885), Labor. f. experim. Biol. Swerdlov Univ. Moskau 55 (U.d.S.S.R.), Miusskaja pl. 3. *Physiol. d. Schilddrüse u. d. Ovariums.* ⓓ Odessa 1911.
Lieberkind, Ingvald (1897), Ass. Univ. Zool. Studiensammlung. København (Danmark), Nórregade 10. *Anat. and Embryol. and Biol., Echinodermata.* ⓓ København 1925.
Liebermann, Albert Ernst (1899), Dr. phil. München 19 (Deutschland), Prinzenstr. 20. *Angewandte Entomol. und Ornithol.* ⓓ München 1924.
Liebermeister, Gustav (1879), Dr., leitender Arzt der Inneren Abteilung des Städt. Krankenhauses und der Städt. Tuberkulose-Fürsorgestelle. Düren, Rheinpreußen (Deutschland), Städt. Krankenhaus. *Pathol. Physiol. innerer Krankheiten, Infektions-krankheiten. Atmungsmechanik. Tuberkulose.* ⓓ Tübingen 1901.
Liebetanz, Bernard (1892), Ass. Inst. für Botan. und Phytopath. Univ. Poznań (Polska), ul. Grunwaldzka 5. *Algol.: halophile Mikroflora. Färbungsmethoden pflanzlicher Gewebe.* ⓓ Poznań 1924. ⓟ *Anat. und systematische, spez. doppelgefärbte Praep. von Holzgewächsen.*
Liebtag, Charlotte E. (1905), Grad. Ass. Botan. Univ. of Illinois. Urbana, Ill. (U.S.A.), 912 W. Illinois St. *Plant Ecol.*
Liebus, Adalbert (1876), Dr. phil., a.o. Prof. Prag II (C.S.R.), Viničná 3. *Palaeont.: Foraminiferen.* ⓓ Prag, deutsche Univ. 1900. ⓟ Foraminiferen mit Bestimmungen.
Lielmanis, Janis (1895). c. Stendi (Latvija), Selekcijas stacijā. *Hafer und Erbsenzüchtung.* ⓟ Samenproben.
Liepin, Tenis (1895). Leningrad (U.d.S.S.R.), Academy of Sc., Bureau of Genetics and Eugenics. *Variability of Rotatoria and of the chrysomelid beetle Phaedon Cochleariae. Vererbung von Augen- und Haarfarben bei Menschen. Vererbung von Kurzsichtigkeit.*
Liese, Johannes (1891), Dr. phil., Priv.Doc. Eberswalde (Deutschland), Kaiser-Friedrich-Str. 25. *Forstliche Botan.; Mykol., Holzaufbau, Holzerstörung, Imprägnierung.* ⓓ Berlin 1920. ⓟ Holzzerstörende Pilze.
Lieske, Rudolf (1886), Prof. Dr. Berlin-Dahlem (Deutschland), Biol. Reichsanstalt. *Pflanzenphysiol. allgemeine Bact.* ⓓ Leipzig 1909. ⓟ Bakterienkulturen, Strahlenpilze.
Light, S. F. (1886), Ass. Prof. of Zool., Univ. of California. Berkeley, Cal. (U.S.A.), Department of Zool. *Zool. Research on the termites (Isoptera) and the Protozoa of the termites.* ⓓ Univ. of California 1926. ⓟ Protozoa from various termites. Oriental and American termites (Isoptera).
Lignac, Georg Otto Emile (1891), Priv.Doc. und Konserv., Prosektor am Boerhaave-Labor. zu Leiden (Holland), Rynsburgerweg 54. *Pigmente (Hautpigment, Blutpigment — besonders Hämatoidin —, Malariapigment, Pigment bei ,,Melanosis" Coli, Porphyrine, Pigment der Zirbeldrüse, Stoffwechselstörungen (Cystinstoffwechsel bei Kindern), Appendixcarcinoid, Maçensyphilis, Thrombo-arteriitis multiplex luetica, Tuberkulose.* ⓓ Leiden 1922.
Lignau, Nicolai (1873), Prof. Zool. am Landwirtschaftl. Inst., Leiter der Sektion für Zool. d. wissenschaftl.-biol. Forschungs-Inst. Odessa (U.d.S.S.R.), Mering-Str. 52, W. 22. *Embryol. (Polydesmus, Juliden, Diplopoda). Systematisch-faunistische Untersuchungen an Myriapoden. Hydrobiol.: Untersuch. üb. d. Bewuchs v. Seeorganismen.* ⓓ Petersburg Univ. 1912. ⓟ Evertebrata d. Landfauna Südrußlands, Fauna d. Schwarzen Meeres (namentlich d. Odessaer Hafens).
Liljestrand, Göran (1886), Doc. für Physiol. und Pharmacol. Stockholm (Schweden), Karolinska Inst. *Lungengaswechsel, Blutkreislauf. bes. beim Menschen.* ⓓ Stockholm 1917.
Lillie, Frank Rattray (1870), Prof. of Embryol. and Chairman of the Department of Zool. Univ. of Chicago, Ill. (U.S.A.). *Biol. of Sex in Fowl. Cattle. Fertilization. Development of the Chick.* ⓓ Chicago 1894.
Lillie, Ralph Stayner (1875), Prof. of General Physiol. Univ. of Chicago, Ill. (U.S.A.), Physiol. Labor. *Stimulation Processes, Nervous and other Transmissions in Living Protoplasm, Narcosis, Salt Action, Physiol. of Development, Physiol. Effects of Radiation.* ⓓ Univ. of Chicago 1901.
Lim, Robert Kho-Seng (1897), Prof. of Physiol. Department of Physiol., Peking Union Med. Coll. Peking (China). *Physiol. of digestive secretions.* ⓓ Edinburgh Univ. Phil. 1920, Sc. 1924.
de Lima, A. P., Prof. Sc. Univ. Porto (Portugal). *Biol.* ○

de Lima, J. A. Pires, Prof. Univ. Porto (Portugal). *Anat.* ○

de Lima Salazar, Abel, Prof. Univ. Porto (Portugal). *Histol.* ○

Limpricht, Wolfgang Dr., Priv.Doc. a. d. Univ. Breslau 10 (Deutschland), Waisenhausstr. 12. *Pflanzengeographie (Ost-Europa, Z.- u. Ost-Asien), Systematik d. Phanerogamen (Gattg. Pedicularis, Taccaceae) u. Bryophyten (Schlesiens Laub- u. Lebermoose).* ⊕ Breslau 1902.

Limson, Marciano, Ass. Prof. Univ. of the Philipp. Manila (Philippine Islands). *Anat.* ○

Lin, Kuo-Hao (1898), Instr. Biochem. Union Med. Coll. Peking (China), Department of Biochem. *Biochem.* ⊕ Harvard 1924.

Lincoln, Frederick Charles (1892), Associate Biol., Bureau of Biol. Survey, United States Department of Agric. Washington, D.C. (U.S.A.), 114 Maple Ave., Takoma Park. *Migrations and life histories of birds.* ⊕ Denver, Colorado, 1910.

Lind, Sigfrid (1899), I. Ass. des Hygien. Inst. der Univ. Tartu (Eesti), Lepiku tän. N 14 kr 3. *Nahrungsmittelchem. u. -bact.*

Lindberg, Harald, Dr. phil., Custos Botan. Mus. d. Univ. Helsinki (Finnland), Berggatan 20 E. *Botan., Pflanzensystematik.*

Lindberg, Håkan (1898), Amanuens. am Zool. Mus. der Univ. zu Helsingfors (Finnland), Berggatan 20. *Hemipterol., Insektenökol.* ⊕ Helsingfors 1926. ⊕ Coleoptera palaearctica, Hemiptera orbis terrarum.

Lindblad, Gerhard (1889). Kristianstad (Sverige). *Blutzucker und Wasserstoffionenkonz. des Blutes. Temperaturregelung bei Fischen.* ⊕ Lund.

Lindeman, Woldemar (1867), Prof. der Univ. Warschau (Polen), Ludna 11. *Allgemeine Pathol. und Toxikol.* ⊕ Moskau 1896.

Lindemuth, Karl (1892), Dr. der Landwirtschaft, Landwirtschaftslehrer. Stolp, Pommern (Deutschland), Bahnhofstr. 15. *Biol. der Arten der Gattung Vicia. Biol. landwirtschaftl. schädlicher Insekten.* ⊕ Berlin 1923. ⊕ Blütenbesucher und Schmarotzer der Arten der Gattung Vicia. Landwirtschaftl. schädliche Insekten und deren Schädigungen.

Linden, Bernard A. (1892), Dr. Bact. Microbiol. Labor., U. S. Bureau of Chem., U. S. Dept. of Agric. Washington, D.C. (U.S.A.). *Food bact.; food poisoning.*

von Linden, Gräfin Maria (1869), Vorsteher des Parasit. Labor. der Univ. Bonn a. Rhein (Deutschland), Quantiusstr. 13. *Parasit. Nematodenerkrankung. Untersuchung über die Chemotherap. Wirkungen des Kupfersalzes bei parasit. Erkrankungen und Infektionskrankheiten.* ⊕ Tübingen 1895. ⊕ Praep. über Lungenstrongylose. Tuberculose.

Linders, Frans Josua (1882), Doc. Univ., Vize-Dir. und Statistiker des schwedischen Staatsinst. für Rassenbiol. Uppsala (Sverige). *Biometrie. Bevölkerungsstatistik.* ⊕ Uppsala 1926.

Lindfors, K. M. Thore (1889), Laborator. Experimentalfältet Stockholm (Sverige). *Pflanzenkrankheiten. Morphol., Systematik, Biol. und Zytol. der Pilze, bes. der Rostpilze.* ⊕ Uppsala 1924. ⊕ Parasitische Pilze.

Lindhard, E., Prof. Dr., Landw. Hochsch. København V (Danmark), Bülowsvej. *Forstwissenschaft.* ○

Lindhard, F. P. F. (1870), Dr. med., Prof. a. d. Univ. København K (Danmark), Studiestraede 6. *Respirations- u. Kreislaufsphysiol., Muskelphysiol., Physiol. d. Muskelarbeit.* ⊕ Kopenhagen 1914.

Lindholm, Wilhelm Alexander Adolf (1874), Leiter des Malakozool. Abteilung Zool. Mus. der Akademie der Wissensch., wissensch. Mitarb. Geol. Comité. Leningrad (U.d.S.S.R.). *Land- und Süßwassermollusken (rezent und pleistozän) der palaearktischen Region in systematischer, zoogeographischer und ökol. Beziehung.*

Lindinger, Leonhard (1879), Kustos am Inst. für angewandte Botan. zu Hamburg, Vorstand der Amtlichen Pflanzenbeschau, Hamburg-Freihafen.

Neu-Rahlstedt b. Hamburg (Deutschland), Schillerstraße 13. *Schildläuse (Coccidae). Monokotylen-Anat. u. Morphol. Kanarenflora.* ⊕ Erlangen 1902.

Lindman, Carl Axel Magnus (1856), Prof., Dir. Botan. Abteilg. Naturhistor. Reichsmus. Stockholm (Schweden), Tegnérgatan 34. *Morphol., Biol.; Pflanzengeogr., botan. Reisen; deskriptive Botan.: südamerikan. Flora und schwedische Phanerogamflora.* ⊕ Uppsala 1886.

Lindner, Erwin (1888), Dr. phil., Hauptkonserv. an der Württ. Naturaliensammlung. Stuttgart (Deutschland), Archivstr. 3. *Entomol., speziell Diptera; Ostracoda.* ⊕ München 1913.

Lindner, Paul (1861), Prof., Vorsteher der biol. Abt. am Inst. für Gärungsgewerbe der Landwirtschaftl. Hochsch. Berlin N 65 (Deutschland), Seestraße 13. *Gärungsmikroben. Gär- und Assimilationsversuche. Mikrophotographische Aufnahmen. Allgemeine Naturgeschichte der Gärung. Naturgärungen. Mikroflora der Pulquegärung (Mexiko).* ⊕ Berlin 1888. ⊕ Reinkulturen von Gärungsmikroben.

Lindstrom, Ernest W. (1891), Prof. and Head of Genetics, Iowa State Coll. Ames, Ia. (U.S.A.), Department of Genetics. *Genetics of maize (Zea mays), Tomato (lycopersicum).* ⊕ Cornell Univ., Ithaca, N.Y. 1917. ⊕ Tomato seed, Maize seed.

Lineburg, Bruce (1888), Prof. of Biol., Lake Forest Coll., Lake Forest, Ill. (U.S.A.). *Bees, Animal behavior, Bee diseases. Growth and feeding of larvae.* ⊕ Johns Hopkins Univ. 1924. ⊕ Various species of bees with nests etc. from different countries.

Ling, Schmorl M. (1893), Ass. chem. labor., Dept. of Med. Peking (China), Union Med. Coll. *Metabolism. Phlorhizin Diabetes. Basal metabolism.* ⊕ Med. St. John's Univ. Shanghai 1922, sc. Univ. of Pennsylvania 1925.

v. Lingelsheim, Alexander (1874), Doc. für Botan. an der Techn. Hochsch., Priv.Doc. für Pharmakognosie an der Univ., Ass. am Botan. Garten und Mus. der Univ. Breslau XVI (Deutschland), Inststr. 11. *Angewandte Botan., Pharmakognosie, Phytochem.* ⊕ Rostock 1907.

v. Lingelsheim, Walter (1866), Dir. Staatl. Hygien. Inst. Beuthen, Oberschles. (Deutschland). *Bact., insbesondere pathogene Kokken.* ⊕ Berlin 1891.

Linin, Marija, Ass. Univ. Rīga (Latvija), Antonijas ielā 14-b, dz. 3 u. Kronvalda bulv. 1. *Moorforschung u. Bryol.* ○

Link, George K. K., Ass. Prof. of Plant Pathol. Chicago, Ill. (U.S.A.), Univ. ○

Linkola, Kaarlo (1888), Prof. Botan., Dir. Botan. Garten u. Botan. Inst. Univ. Helsinki (Finnland). *Pflanzengeographie, Pflanzenökol. Einfluß der Kultur auf die Flora. Überwinterung der Unkräuter.* ⊕ Helsinki 1919.

Linnaniemi, Walter Mikael (1876), o. Prof. der Zool. Univ. Turku (Finnland). *Collembola (fennicaborealia), Pflanzenminen, Tierökol., Phytökol. der Insekten.* ⊕ Helsinki 1908. ⊕ Collembola fennica, Rhynchophora.

Linsbauer, Karl (1872), o. Prof. d. Anat. und Physiol. der Pflanzen a. d. Univ. Graz (Österreich), Liebiggasse 7. *Entwicklungsphysiol., Strahlenphysiol. u. Reizphysiol. der Pflanzen.* ⊕ Wien 1898.

Linsbauer, Ludwig (1869), Dir. der Bundeslehr- und Versuchsanstalt für Wein-, Obst- und Gartenbau in Klosterneuburg bei Wien (Deutsch-Österreich). *Physiol. der Kulturpflanzen. Phänol. Pflanzenpathol. Physiol.* ⊕ Wien 1893.

Linton, Edwin (1855), Prof. Med. Dept. Univ. of Georgia. Augusta, Ga. (U.S.A.), 1104 Milledge Road. *Parasit.* ⊕ Yale 1890. ○

Liouville, Jacques (1879), Dir. Inst. sc. Cherifien et des Archives sc. du Protectorat français, Chef du Bureau d'Afrique de la Commission de l'Atlantique de la Section Océanographique du Conseil International de Recherches. Rabat (Maroc) et Paris VII (France), 35, rue de l'univ. *Océanographie physique: littoral des fonds marins. Océanographie biol. mammi-*

fères marins: phoques et Cétacés. ⓓ Paris 1906. ⓔ Oiseaux, insectes, poissons, tuniciers, vermidiens, mollusques, vers, échinodermes, coelentérés, spongiaires du Maroc littoral et territorial.

Lipin, Alexander (1877), Dir. Hydrobiol. Abteilung des Inst. für Fischereiwirtschaft. Moskau (U.d.S.S.R.), Pjatnitzkaja Str. 33. *Plankton; Ernährung der Fische.* ⓓ Kasan.

Lipin, Nina (1883), Ass. der hydrobiol. Abteilung des Inst. für Fischereiwirtschaft. Moskau (U.d. S.S.R.), Piatnitzkaja 33. *Bodentiere, Chironomidenlarven.* ⓓ Tübingen 1909.

Lipkin, Boris Ja., Doc. Weißruss. landw. Akad. Gorki (U.d.S.S.R.). *Forstwirtschaft. Biol.* o

Lipman, Charles Bernard (1883), Prof. of Plant Physiol. and Dean of the Graduate Division Univ. of California. Berkeley, Cal. (U.S.A.), 113 California Hall. *Nitrogen fixation; the identity of the essential chem. elements for plants; the relation of salts in the root medium to plant growth; injection of trees as a solution for problems in nutrition and in the physiol. and parasitic diseases of plants; bact. flora of sea water and on the chem. composition of the latter.* ⓓ California 1910.

Lipman, Jacob Goodale (1874), Dean of Coll. of Agric., Rutgers Univ., Dir. New Jersey Agric. Experim. Station. New Brunswick, N.J. (U.S.A.). *Soil chem., microbiol., fertility.* ⓓ Rutgers 1903, D.Sc. (hon.) 1923. ⓔ *Cultures of soil microorganisms.*

Lipolla, F., Prof. Univ. Palermo (Italia). *Anthrop.* o

Lipop, Jerzy (1888), Dr., Kustodian of the Physiogr. Muz. of the Polish Acad of Sc. Kraków (Polska), Sławkowska 17. *Paleobotan.* ⓓ Cracow 1926.

Lipowskij, Wladimir I., Akademiker. Kiew, Ukraine (U.d.S.S.R.), Akademie d. Wissenschaften. *Botan.* o

Lippich, Friedrich, Dr., Prof. d. physiol. Chem. Praha II (C.S.R.), Salmova 3. *Medizinchem. Inst. der Deutschen Univ.* o

Lippincott, Leon Stanley (1888), Dir. of Labor., Vicksburg Sanitarium and Crawford Street Hospital. Vicksburg, Miss. (U.S.A.). *Bact., Serol., Pathol., Haematol., Biol. Chem., Metabolism.*

Lippincott, William Adams (1882), Prof. of Poultry Husbandry. Berkeley, Cal. (U.S.A.), Room 7. Agric. Hall. *Genetics of poultry.* ⓓ Univ. of Wisconsin 1920.

Lippmaa, Theodor (1892), Dr. Bot. Tartu (Estland), Lossi tän. 15—8. *Biochem. Pflanzenpigmente (Rhodoxanthin). Anat. und Synökol. der Pflanzen. Pigmenttypen bei Pteridophyta und Anthophyta. Systematik.* ⓓ Univ. Tartu 1926. ⓔ *Antophyta vom Altai, Resedaceen und Droseraceen.*

Lipschitz, Werner Ludwig (1892), o. Prof. u. Dir. d. Pharmacol. Inst. d. Univ. Frankfurt a. M.-Süd (Deutschland), Weigertstr. 3. *Pharmacol. der Zellatmung u. des Camphers. Toxikol. d. arom. Nitro- u. Aminokörper. Eiweißstoffwechsel u. Wärmehaushalt. Ausscheidung von Substanzen aus dem Blut. Arzneipotenzierung.* ⓓ Phil. Berlin 1915, med. Leipzig 1916.

Lipschütz, Alexander (1883), Prof. Physiol., Dir. Physiol. Inst. Univ. Concepcion (Chile), Caupolican 17. *Innere Sekretion der Geschlechtsdrüsen. Partialkastration und kompensatorische Reaktionen der Geschlechtsdrüsen. Lokalisation der Hormonproduktion in Testikel und Ovarium. Ovarialtransplantation. Experim. Hermaphroditismus. Antagonismus der Geschlechtsdrüsen. Isolierung der Ovarialhormone.* ⓓ Göttingen 1907.

Lipska, Irène (1882), Chef de la sous-section des fermentations à l'Inst. National Sc. d'Agric. de Bydgoszcz (Polska), Zacisze 8. *Microbiol. appliquée: fermentation alcoolique (vinification), fermentation lactique (lactobacilles de lactose et de maltose); Citromyces.* ⓓ Genève 1910. ⓔ *Cultures de levures et de moisissures (citromyces).*

Lipták, Pál (1887), Priv.Doc., Adjunkt Pharmakognost. Inst. Budapest VIII (Ungarn), Üllői-ut 26. ⓓ Budapest 1910.

Liro, I. J., Dr. phil., ao. Prof. Univ., Dir. Pflanzenpathol. Inst. Helsinki (Finnland), Gengatan 5 M. *Pflanzenbiol. u. -pathol., Mykol.* o

Lisbonne, Marcel Paul (1883), Prof. de Microbiol., Fac. de Méd. Montpellier (France), 14, Avenue du Stand. *Actions bactériophagiques. Immunité. Actions diastasiques. Microbes du tube digestif.* ⓔ *Préparations de Microbes. Cultures microbiennes.*

Lisenkov, N. K., Prof. Med. Inst., Anat. Labor. Odessa (U.d.S.S.R.). o

Lisowski, Joseph Stephan (1872), Prof. d'histoire naturelle au lycée d'Etat. Puławy (Polska), Lycée de garçons. *Cycle annuaire de la succession des phénomènes de la nature.* ⓓ Kieff 1896.

Lisse, Martin W. (1891), Associate Prof. of biophysical chem., Dept. of Agric. and Biol. Chem. State College, Pennsylvania (U.S.A.), 222 S. Atherton St. *Bact. Cataphoresis.*

Lissitsin, Konstantin J., Prof. Geol. am Don schen Polytechn. Inst. Nowotscherkassk (U.d. S.S.R.), Geol. Kabinet. *Carbonische Fauna.* o

Lissner, Helmuth (1895), Dr., Ass. der Deutschen wiss. Kommission für Meeresforschung. Hamburg 5 (Deutschland), Kirchenallee 47 I. *Fischereibiol., Hering, Makroplankton.* ⓓ Leipzig 1922.

Lissón, Carlos I., Prof. Univ. Lima (Peru). *Paléontol.* o

List, George M. (1885), Chief Deputy State Entomol. of Colorado, Associate Entomol., Colorado Agric. Coll. Experim. Station. Fort Collins, Col. (U.S.A.). *Economic entomol. Codling moth and other fruit insects. Cimicidae.* ⓓ Colorado Agric. Coll. 1924.

Lister, Sir Frederick Spencer (1876), Dir. South African Inst. for Med. Research. Johannesburg, Transvaal (South Africa). Box 1038. *Bact., Lobar Pneumonia, Influenza, Cerebrospinal Meningitis, Filterable Viruses.* ⓓ London 1905.

Litschauer, Viktor (1879), Prof. für Naturgeschichte, Chem. u. Warenkunde an der Handelsakademie. Innsbruck, Tirol (Deutsch-Österreich), Mandelsberger Str. 9 I. *Systematik der Pilze, Thelephoraceen, Hydnaceen und Polyporaceen.* ⓔ *Exsiccate von Pilzen.*

Litschkow, Boris L. (1888), Prof. de géol. et paléont. Univ. Comité géol. de Russie. Kiew (U.d. S.S.R.), boulevard de Schewtschenko 36-9. *Paléont. der Kreide.*

Little, Clarence Cook (1888), President Univ. of Michigan. Ann Arbor, Mich. (U.S.A.). *Genetics, cancer, modification of the germplasm in mammals.* ⓓ Haward 1914. ⓔ *Mus. faroensis, M. muralis.*

Litvak, Josif (1900), Ass. gener. pathol. Inst. of Chem. and Pharmacy. Odessa (U.d.S.S.R.), Alexandrovskaja ploschad 3/4. *Avitaminose.*

Litvinov, Démetrius Ivanovicz (1854), Botan. en Chef de Mus. botan. de l'Académie des Sc. Leningrad (U.d.S.S.R.), Quai de l'Univ. 5. *Gramineae de la flore de la Russie asiatique.* ⓓ Moscau 1879.

Lityński, Alfred (1880), Dr. phil., Dir. de la Station hydrobiol. de Wigry. Suwałki (Polska). *Hydrographie des lacs, Phylloppoda, Copepoda, Salmonidae.* ⓓ Cracovie 1912.

Liu, Chung Lo (1901), Ass. Prof. in Biol. Tsing Hua Coll. Peking (China). *Aculeate Hymenoptera. Vespidae and Bombidae.* ⓓ Cornell Univ. 1926.

Liu, Ju-ch'iang (1895), Ass., Department of Pharmacol. P. U. M. C. Peking (China). *Systematic Botan.* ⓓ Univ. of Wisconsin 1924. ⓔ *Chinese plants.*

Livanow, Nikolaus (1876), Prof. Zool., Dir. Zootomisches Kabinett d. Univ. Kasan (U.d.S.S.R.). *Vergl. anat. und Histol. d. Wirbellosen, Annelides (Polychaeta u. Hirudinea). Biocoenotik d. Meere.* ⓓ St. Petersburg 1904.

Livingston, Burton Edward (1875), Prof. of Plant Physiol. and Dir. of the Labor. of Plant Physiol. of the Johns Hopkins Univ. Baltimore, Md. (U.S.A.),

Homewood. *Relations of water, mineral-salt, oxygen, temperature, sunlight; physiol. ecol., atmometry. Diffusion and Osmotic Pressure in Plants. Relation of Desert Plants to Soil Moisture and to Evaporation. Distribution of Vegetation in the United States.* ① Univ. of Chicago 1901.

Livini, Ferdinando (1868), Prof. o. Anat. Umana Normale, Dir. Ist. Anat. Univ. Milano (Italia), Viale Regina Margherita 85. *Embriol. generale, Istogenesi.* ① Firenze.

Livon, M. J., Chef des trav. prat. Histol., Ecole Méd. et de Pharmacie, Marseille (France). *Histol.* o

Ljachoveckij, Anatolij Mich., Doc. II. Univ. Moskau (U.d.S.S.R.). *Histol. Embryol.* o

Ljajman, Ed. Konst., Doc. I. Univ. Moskau (U.d.S.S.R.). *Zool. Histol.* o

Ljubinski, Vsevolod, Dr., Dir. d. Epidemiol. Inst. Sarajevo (S.H.S.). *Bact.* o

Llambias, Joaquin, Prof. Univ. Nac. Buenos Aires (Argentina). *Anat. Physiol. Pathol.* o

Llenas, M., Prof. Esc. super. de agric. Barcelona (España). *Botan.* o

Lleras, A. Federico, Prof. Univ. Nac. Bogotá (Columbia). *Bact. Parasit.* o

Lloyd, Blodwen (1901), Ass. Lect. in Botan., Royal Technical Coll. Glasgow C. 1 (Scotland). *Macsyfferwyd,* Aberdare, Belmont Terrace. *Marine Plankton.* ① Aberystwyth, Wales, 1922.

Lloyd, Francis Ernest (1868), Prof. of Botan. Montreal (Canada), McGill Univ. *Embryol.; Cellular physiol.; Stomata and Transpiration; Abscission; Rubber; Tannin; Fluorescence; Pollen.* ① Princeton 1895.

Lloyd, John William (1876), Chief in olericulture Agric. Experim. Station. Urbana, Ill. (U.S.A.). o

Lloyd, Joseph Henry (1892), Lect. in Zool. and Compar. Anat. in the Univ. Coll. of South Wales (Univ. of Wales). Cardiff (England), Newport Road. ① Birmingham 1914.

Loader, Freda (1893), Lect. in Botan. Southhampton (England), Univ. Coll. ① London 1916.

Lobik, Alexis (1888), Dir. Stat. Défense des Plantes du gouvernement du Térek. Essentouky, Caucase du Nord (U.d.S.S.R.). *Flore mycol. Sclerotinia Libertiana Fuck sur l'hélianthe; sa biol., son expansion géographique. Nielle des céréales; influence du degré de la souillure des grains par les spores sur le degré de la contamination des blés dans les champs. Effets nuisibles de la rouille sur le froment. Rapport entre l'action du mildiou sur la vigne et la température de l'air et les phénomènes aqueux. Dégénérescence des pommes de terre, son expansion et sa portée économique. Russthau dans les conditions des cultures pures.* ① Petersburg 1914. ℗ Flore mycol. du gouvernement du Térek.

Lobo, B., Prof. Univ. Rio de Janeiro (Brasil). *Microbiol.* o

Locatelli, Piera (1900), Ass. Labor. de Pathol. Générale Fac. Méd. Univ. Pavia (Italia), Palazzo Botta. *Système nerveux, régénération, métabolisme des graisses.*

Lochhead, Allan Grant (1890), Dominion Agric. Bact. Ottawa (Canada), Central Experim. Farm. *Soil and Dairy Bact.* ① McGill 1919, Leipzig 1914.

Locke, Arthur (1897), Seymour Coman Fellow of the Univ. of Chicago, Ill. (U.S.A.), Favill Labor., St. Luke's Hospital. *Isolation of Immune Substances (Antitoxins, Hemolysins etc.). Mechanism of Immunity.* ① Chicago 1922.

Lockemann, Georg (1871), Geh.Reg.R., Prof., Dr. phil., Dir. der chem. Abteilung am Inst. Robert Koch, Berlin, ao. Prof. a. d. Univ. Berlin-Grunewald (Deutschland), Königsweg 126. *Arsennachweis. Adsorptionsvorgänge. Desinfectionsprüfung. Wachstumsverhältnisse von Tuberkelbacillen auf eiweißfreien Nährlösungen.* ① Heidelberg 1896.

Locklin, Harrison D. (1891), Horticulturist, Western Washington Experim. Station of the State Coll. of Washington. Puyallup, Wash. (U.S.A.). *Small Fruits or Berries. Pruning, Training, Soil Management. Bulbs: Variety testing, Planting methods, Fertilizers.* ① Pullman, Wash. 1915.

Lockwood, A. T. Stewart (1892), Associate Entomol. in charge Billings Labor. of the U.S. Dept. of Agric. Billings, Mtn. (U.S.A.), Box 1094. *Ecol. and control of Orthopterous insects. Biometry.* ① Univ. of Minnesota 1926. ℗ Orthoptera of the northern plains of the U.S.

de Lodijensky, Vera (1887), Labor. zool. académie des Sc. Leningrad (U.d.S.S.R.), 19, rue Rose Luxemburg. *Zool. expérim.: régénération et transplantation.*

Loeb, Leo (1869), Prof. of Pathol. St. Louis, Mo. (U.S.A.), Washington Univ. Med. School. *Growth phenomena, Sexual cycle (Placentomata), Aetiol. of Tumors. Cell pathol. (Amoebocyte tissue). Analysis of Individuality differential.* ① Zürich 1896/97.

Löhner, Leopold Robert Josef (1884), Dr. med. et phil., ao. Prof. der Physiol. an der Univ. Graz (Österreich), III, Halbärthgasse 6. *Physiol. der Gallenwege und Galle, Geruchsphysiol. des Hundes (Polizeihunde-Arbeit und -Dressur), Inzucht und Bastardierung (Umstimmungsphänomene, parakinetische Faktoren.* ① Med. Graz 1908, phil. 1910.

Löhnis, Felix (1874), Dr., o. Prof. d. Univ., Dir. des Inst. für landwirtschaftl. Bact. und Bodenkunde. Leipzig (Deutschland), Johannisallee 21. *Landwirtschaftliche Bact.* ① Leipzig 1901. ℗ Landwirtschaftlich wichtige Bakterien.

Löhnis, Maria Petronella (1888). Scheveningen (Holland). *Plant pathol. and bact.* ① Utrecht 1922.

Loehwing, Walter Ferdinand (1896), Ass. Prof. of Plant Physiol. Iowa City, Ia (U.S.A.), State Univ. *Soil Fertility, influence of nutrients on internal metabolism, effect of nutrients on carbohydrate-nitrogen ratio.* ① Chicago 1925.

Lönnberg, Einar (1865), Prof., Dir. Abteilung für Vertebraten im Naturhistorischen Reichsmus. Stockholm 50 (Sverige). *Systematik und Verbreitung der Vertebraten.* ① Upsala 1891.

Lönnroth, Erik Johannes (1883), Doc. Univ. Helsinki (Finnland), Elisabethstr. 21. *Dendrometrie, Waldbiol.* ① Helsinki 1919.

Löns, Max (1886), Dir. des Hygien.-bact. Inst. der Stadt Dortmund (Deutschland). *Bact. der Typhus- und Paratyphus-Erkrankungen.* ① Göttingen 1913.

Loescheke, Hermann (1882), Dr. med., Dir. des Pathol., Bact. und Serol. Inst. der Städtischen Krankenanstalten, Doc. an der Handels-Hochsch. Mannheim (Deutschland). *Lungenpathol. (Tuberkulose, Emphysem, Bronchiektasen), Innere Sekretion. Knochenpathol. Mißbildungen. Serol. (Carcinom- u. Tuberkulosediagnostik).* ① Bonn 1907.

Loesener, Theodor (1865), Prov., Dr. phil., Kustos am Botan. Mus. in Berlin-Dahlem, pensioniert. Berlin-Steglitz (Deutschland), Humboldtstr. 28. *Systematische Botan., Zingiberaceae, Marantaceae, Aquifoliaceae, Celastraceae, Hippocrateaceae.* ① Berlin 1890.

Löte, Józef, Prof. Dr., Dir. d. Allg.Pathol.Inst. o Univ. Szeged (Ungarn). *Allg. Pathol., Pathophysiol.*

Loevenhart, Arthur Solomon (1878), Prof. of Pharmacol.andToxicol.,Univ. ofWisconsin. Madison, Wis. (U.S.A.), Science Hall. *Enzyme action, biol. oxidation, blood formation, antiseptics, anti-syphilitic drugs, local anesthetics, acidosis.* ① Johns Hopkins Univ. 1903.

Lövö, P., Dir. Statens forsöksstation Nordenfjelds. Mohaltan, pr. Trondhjem (Norge). o

Loewe, Siegfried (1884), o. Prof. Univ., Dir. Pharmacol. Inst., ao. Prof. Univ. Göttingen. Tartu (Estland), Wallgraben 12. *Arzneiwirkungsprüfung, Pharmacol. u. Biochem. d. Sexualhormone, Pharm.der Arzneikombinationen, Seitenkettendthylamine.* ① Straßburg i. E. 1908.

Loewenstein, Ernst (1878), Prof. Univ. Wien XVIII (Deutsch-Österreich), Bastiengasse 61. *Immunität. Tetanus, Diphtherie.* ⓓ 1902.

Loewenthal, Hans (1879), Dr. med. et phil. Berlin W 50 (Deutschland), Achenbachstr. 4. *Gewebezüchtung.* ⓓ Phil. Würzburg 1922, med. Berlin 1924.

Löwenthal, Karl (1892), Dr. med., Oberarzt am Pathol. Inst. des städtischen Krankenhauses Moabit in Berlin (Deutschland), NW 21, Turmstr. 21. *Pathol. Anat. und experim. Pathol., experim. Geschwulstforschung, Lipoidstoffwechsel (Arteriosklerose, Nephrose usw.), Konstitutionspathol., Drüsen mit innerer Sekretion.* ⓓ Freiburg i. Br. 1915.

Loewenthal, Waldemar (1874), Priv.Doc. für Hygiene und Bact. an der Univ. Bern (Schweiz), Inst. f. Hygiene u. Bact. *Protozoen; Spirochäten. Diphtherie-, Typhus-, Paratyphus-, Ruhrbazillen; Seuchenbekämpfung. Mutation.* ⓓ Erlangen 1897.

Loewi, Otto (1873), Univ. Prof. Graz (Österreich), Joh.-Fux-Gasse 35. *Pharmacol.* o

Loewinson-Lessing, Franz (1861), Prof. Univ. u. Polytechn. Inst., Dir. d. Geol. Mus. d. Akad. d. Wiss. Leningrad (U.d.S.S.R.), Quart. 12. Polytechn. Inst. *Palaeont.* ⓓ Petersburg 1898.

Loewy, Adolf (1862), Prof., Leiter d. Schweizerischen Forschungsinst. Davos (Schweiz). *Physiol. des Hochgebirges; Tuberculoseforschung.* ⓓ Berlin 1885. ⓟ Murmeltiere.

Löyning, Paul (1895), Konserv. beim zool. Mus. Oslo (Norge), N. Abbedingen, V. Aker. *Nudibranchien.*

Loginoff, Viktor (1870), Prof. Vet. Inst. Kazan, Tatar. Rep. (U.d.S.S.R.), Arskoe Pole 96. *Histol. Embryol.*

Logvinovič-Miller, Natalija G., Doc. II. Univ. Moskau (U.d.S.S.R.). *Histol. Embryol.* o

Lohman, Marion Lee (1903), Instr. in Botan., Miami Univ., Oxford, Ohio (U.S.A.). *Mycorrhiza of the higher plants, forest plants of Iowa. Taxonomy of the genus Lycoperdon.* ⓓ State Univ. of Iowa 1926.

Lohmann, Hans Theodor (1863), o. Prof. für Zool. a. d. Univ., Dir. d. Zool. Staatsinst. und Zool. Mus. in Hamburg (Deutschland), Uhlenhorster Weg 26 II. *Planktonforschung, Appendicularien.* ⓓ Kiel 1889.

Lohwag, Heinrich (1884), Dr. Wien III (Österreich), Rennweg 2. *Morphol. der Hymenomyzeten.* ⓓ Wien 1907.

Loir, Adrien (1862), Dir. Inst. océanographique. Le Havre (France), Mus. d'Histoire Nat. *Biol., Bact.* ⓓ Paris 1892.

Loisel, Gustave Antoine (1864), Dir. Ecole pratique des Hauts-Etudes, Ass. Zool. Fac. Sc. Univ. Paris VI (France), 6, rue de l'Ecole de Méd. *Mammal.* ⓓ Méd. Paris 1893, sc. 1896.

Loman, Jan Cornelis Christiaan (1856). Amsterdam (Holland), Van Baerlestraat 158. *Pycnogoniden. Opilioniden. Tardigraden.* ⓓ Amsterdam 1881.

Lombana, Joaquin, Prof. Univ. Nac. Bogotá (Columbia). *Anat.* o

Lombard, Warren P. (1855), Prof. Emer. of Physiol. Univ. of Michigan. Ann Arbor, Mich. (U.S.A.), 805 Oxford Road. *General Physiol. of Muscle and Nerve.* ⓓ Med. Harvard 1878, sc. Geneva, N.Y., 1909. o

Lombardi, P. Lorenza (1890), Vice-Dir. Staz. di Gelsicolt. e Bachicoltura. Ascoli Piceno (Italia). *Nuove razze del baco da seta mediante. Fissazione di incroci. Selezione-studio delle principali varietà del gelso.*

Lombardo, Giacomo, Lib. Doc. Univ. Catania, Sicilia (Italia). *Anat. patol.* o

Lombroso, Ugo (1877), Prof. stab., Dir. Ist. Fisiol. Univ. Palermo (Italia), Corso Tukory. *Ricambio materiale, metabolismo intermedio dei grassi, funzione interna ed esterna del pancreas, funzione dell'intestino, endocrinol., enzimol.* ⓓ Roma.

Lomer, Gerhard R. (1882), Univ. Librarian. Montreal (Canada), McGill Univ. Library. *Ornithol.* ⓓ Columbia Univ., N.Y., 1910.

Lommatzsch, Modest (1879), Leiter der Abteilung für Demonstration von lebendem Material im K. A. Timirjasev-Bio-Mus. Moskau (U.d.S.S.R.), Malaja Dmitrovka 6. *Biol. (Ökol.) der Tiere.* ⓟ Lebende Reptilien, Amphibien, Fische, Insekten, hydrobiol. Material.

Łomnicki, Jaroslaw Ludomir Marjan (1873), Dir. des Dziedukycki'schen Mus. Lwów (Polska), Rutowskigasse 18. *Formicidae (Hymenopt.) von Europa. Coleoptera des polnischen Gebietes. Foraminifera fossilia miocaenica Poloniae.* ⓟ Coleoptera und Formicidae.

London, Efim (1869), Leiter der Abteilung für allgemeine Pathol. des Inst. für experim. Med. Leningrad (U.d.S.S.R.). *Intermediäre Stoffwechsel. Radium in der Biol. und Med. Physiol. und pathol. Chymol. Experim. Physiol. und Pathol. der Verdauung.* ⓓ Warschau 1895.

Long, Esmond R. (1890), Associate Prof. of Pathol., Univ. of Chicago, Ill. (U.S.A.), Department of Pathol. *General pathol., and chem. and immunity of tuberculosis.* ⓓ Ph.D. Chicago 1919, M.D. Rush Med. Coll. 1926.

Long, Joseph A. (1879), Assoc. Prof. of Embryol. Univ. of California. Berkeley, Cal. (U.S.A.), 854 Regal Road. *Embryol. (Early development of Mammals) and Oestrous Cycle.* ⓓ Harvard Univ. 1904.

Longcope, Warfield T. (1877), Prof. of Med., Johns Hopkins Med. School. Baltimore, Md. (U.S.A.), Johns Hopkins Hospital. *Bact. immunol. anaphylaxis, allergy and Hodgkin's disease.* ⓓ Johns Hopkins Univ. 1901.

Longley, Albert Edward (1893), Ass. Botan. Washington, D.C. (U.S.A.), Bureau of Plant Industry. *Cytol. Polyploidy, Polyspory and Hybridism in the Angiosperma. Rubus and Crataegus. Chromosomes in Maize. Diploid and Polyploid Forms in Raspberries. Triploid Citrus.* ⓓ Harvard Univ. 1923.

Longley, William H., Prof. of Biol. Baltimore Md. (U.S.A.), Goucher Coll. o

Longo, Biagio (1872), Prof. stabile Botan., Dir. Orto Botan. Univ. Pisa (Italia). ⓓ Pisa 1915. o

Longstaff, Tom George. Picket Hill, Ringwood, Hants. (England). *Ornithol.* o

Loomis, Harold Frederick (1896), Ass. Agronomist in U.S. Department of Agric. Sacaton, Ariz. (U.S.A.). *Culture and breeding of cotton. Systematic studies of Millipedes.* ⓟ Millipeds.

Loomis, Leverett Mills (1857). San Francisco, Cal. (U.S.A.), Care, California Academy of Sc., Golden Gate Park. *Bird migration, Tubinares. The Albatrosses, Petrels, and Diving Petrels.*

Loomis, Walter E. (1898), Dr., Ass. Prof. of Horticulture. Fayetteville, Ark. (U.S.A.), Department of Horticulture, Univ. of Arkan. *Physiol. of horticulture plants, it's chem. phases.* ⓓ Cornell Univ. 1924.

Looney, Joseph M. (1896), Dr., Ass. Prof. of Physiol. Chem. Jefferson Med. Coll. and Biochem. Jefferson Hospital. Philadelphia, Pa. (U.S.A.). *Protein Chem., determination of Amino acids, amines in blood. Spinal fluids.* ⓓ Harvard 1920.

Loos, Kurt (1859), Ing., Forstmeister. Liboch a. d. E. (C.S.R.). *Ornithol.: Nutzen und Schaden der Vögel; Brutgeschäft der Spechte und anderer Vögel. Zug der Vögel, Ringversuche. Entomol.: Lärchenminiermotte, Borkenkäfer, die Nonne. Reptilien: Verbreitung und Lebensweise. Säugetiere, Amphibien. Botan.*

Lopez-Figueroa, J., Prof. Univ. Nac. Buenos-Aires (Argentina). *Anat. descr.* o

López Mendigutia, Fernando, Dr. C. Nat., Prof. aux., Fac. de Cienc. Barcelona (España). *Entomol.* o

Lopo de Canath, Fausto (1891), Prof. Fac. Méd. Univ. Coimbra. Lisbôa (Portugal), Praca do Rio de Janeiro 26. *Bact. de la Tuberculose.*

Loppens, Karel (1875), Biol. Coxyde village (Belgique), Labor. de biol. *Biol. marine et d'eau douce. Biol. des dunes littorales belges.*

Lorch, Wilhelm (1867), Prof., Dr. Berlin-Friedenau (Deutschland), Fregestr. 7 III. *Anat. und Physiol. der Laubmoose.* ① München 1894.

Lord, Clive Errol (1889), Dir. Tasmanian Mus. Hobart, Tasmania (Australia). *Vertebratefauna of Tasmania.*

Lord, Earll Leslie (1881), Prof. of Horticulture, Univ. of Florida. Gainesville, Fla. (U.S.A.), 1340 Lake St. *Pomol. (Systematic and Orcharding). Citriculture. Tropical and Subtropical Horticulture, Viticulture.* ① Cornell Univ. 1909. ⑨ Tropical and subtropical fruits, particularly citrous fruits.

Lord, Frederic Pomeray (1876), Prof. of Anat., Dartmouth Med. School. Hanover, N.H. (U.S.A.). *Temporo-mandibular articulation, mechanics of body motions.* ① Hanover, N.H. 1903.

Lorenz, Gustav Friedrich (1901), Dr. phil. et med. Freiburg i. B. (Deutschland), Ludwigstr. 3. *Nervenphysiol.: Strychnintetanus und seine Actionsströme, Ermüdung bei tetanischer Reizung. Sinnesphysiol.: Empfindung kleiner Zeiten, Schule des Rhythmus. Gefäßphysiol.: Venenblutbewegung.* ① Phil. Münster i. W. 1924, med. Freiburg i. Br. 1926.

Loro y Gómez del Pulgar, Manuel, Prof., Catedrático del Inst. Badajoz (España). *Biol.* o

Losa España, Faurino Mariano (1893). Miranda de Ebro, Burgos (España). *Flore de la Province de Burgos.* ① Madrid 1915. ⑨ Plantes d'Espagne.

Losarenko, Andrej S., Ukrain. Akademie d. Wissensch., Botan. Garten. Kiew (U.d.S.S.R.), Ul. Tolstogo 38 b/17. *Botan. Biol.* o

Losch, Hermann (1888), Dr. phil., Biol. bei der Landw. Versuchsstation der J. G. Farbenindustrie A.G. Limburgerhof, Post Mutterstadt II, Rheinpfalz (Deutschland). *Pflanzenkrankheiten und Schädlingsbekämpfung.* ① Göttingen 1913.

Losino-Losinskij, L. K. (1899), Ass. Wiss. ,,P. F. Lesshaft" Inst. Leningrad (U.d.S.S.R.), Ekateringovsky pr. 1072. *Protistol. physiol. Pantopoda.* ① Leningrad 1924.

Loth, Edward (1884), Prof. d'Anat. descriptive à la Fac. de Méd. à Varsovie (Pologne), Rue Chałubinski 5. *Anat. descriptive macroscopique: variations. L'anthrop. des parties molles: muscles, intestins, artères.* ① Phil. Zürich 1907, med. Heidelberg 1912.

Loth, Jadwiga (1884), Ass. honoraire à l'Inst. d'Anat. descriptive de Varsovie (Pologne), Rue Chałubinski 5. *Variations du rachis.*

Lotka, Alfred James (1880), Supervisor of Mathematical Research. New York, N.Y. (U.S.A.), 1 Madison Avenue. *Vital Statistics of population. Physical Biol.* ① Birmingham (England) 1912.

Lottier, Henri (1884), Prof. Zootechnic Ecole Nat. d'agric. Montpellier (France), 23, Avenue de Toulouse. *Alimentation des animaux.* ① Montpellier 1921. ⑨ Race des Causses. Mérinos d'arles.

Loudon, Julian (1881), Chief Physician, St. Michael's Hospital, Associate in Med., Univ. Toronto (Canada), 254 Sherbourne Street. *Cancer Research.* ① Univ. of Toronto 1906.

Lounsbury, James A. (1896), Ass. Prof. of Botan., Marquette Univ. Milwaukee, Wis. (U.S.A.), Dept. of Botan. *Morphol. and taxonomy of the Saprolegniaceae.* ① Univ. of Wisconsin 1923. ⑨ Saprolegniaceae.

Loustau Gómez de Membrillera, José (1889), Prof. Biol., Rector of Univ. Murcia (España). *Genetics, Philosophical Biol.* ① Madrid 1914.

Love, Harry H. (1880), Prof. of Plant Breeding. Ithaca, N.Y. (U.S.A.), Coll. of Agric. *Genetics and breeding of small grains. Biometric and experim. technic.* ① Cornell Univ. 1909.

Low, Alexander (1868), Prof. of Anat. Aberdeen (Scotland), Univ. *Human Embryol. Physical Anthrop.* ① Aberdeen 1894.

Lowdermilk, Walter Clay (1887), Research Prof. Forestry Univ. Nanking (China). *Methods in forest, soil and water conservation applicable in measures of famine prevention in China. Factors controlling absorption.* ① Oxford 1919. ⑨ Seeds of drought resistant herbs, shrubs and trees.

Lowe, Charles William (1885), Lect. in Botan., Univ. of Manitoba. Winnipeg (Canada). *Freshwater Algae. Ecol.* ① Birmingham 1914.

Lowe, Percy Roycroft, British Mus. (Nat. Hist.). London S.W. 7 (England), Cromwell Road. *Ornithol.* o

Lowry, Philip R. (1896), Ass. Entomol., New Hampshire Agric. Experim. Station, Ass. Prof. of Economic Entomol., Univ. of New Hampshire. Durham, N.H. (U.S.A.). *Economic insects. European corn borer (Pyrausta nubilalis) and stalk borer (Papaipema nebris).* ① M.S. Ohio State Univ. 1921.

Lowther, Florence de Loiselle (1884), Ass. Prof. of Zool. New York, N.Y. (U.S.A.), 21 Claremont Ave. *Compar. vertebrate anat. Cytol. of Protozoa.* ① Columbia 1926.

Lozada, Echenique B., Doc. Univ. Tucumán (Argentina). *Fisiol.* o

Lozano Rey, Luis, Mus. Ciencias Nat. Madrid (España). *Ichthyol.* o

Łoziński, Paul (1883), Dr. phil., Doc. der vergl. Anat. und Histol. Univ. in Krakau (Polen), Karmelickagasse 9. *Histol., Zytol. Systematik der aculeaten Hymenopteren.* ① Krakau 1907.

Lu, Paul K. Y. (1899), Instr. of Botan., Dept. of Biol., Yenching Univ. Haitien, Peking (China). *Local Flora Peking.* ① Grinnell Coll., Iowa, 1924. ⑨ Flowering Plant in North China.

Lubarsch, Otto (1860), Geh.Med.R., o. Prof. Univ. Berlin, Dir. pathol. Inst. u. Mus. Univ., Mitgl. d. Reichsgesundheitsamtes. Charlottenburg 2 (Deutschland), Bismarckstr. 111. *Allg. Pathol. u. pathol. Anat.* ① Straßburg i. E. 1883.

Lubimenko, Vladimir Nikolajewisch (1873), Chef Section de Physiol. végétale au Jardin Botan. principal, Prof. de Botan. Académie Milit. de Méd. et à l'Inst. de Méd., Chef de la Section de Botan. à l'Inst. Sc. de Lesshaft. Leningrad (U.d.S.S.R.), Aptekarsky Ostrov. *Anat. et physiol. végétale, floristique et cytol., pigments des plastes, photosynthèse, influence de la lumière sur les échanges des matières dans l'organisme végétal.* ① Leningrad 1910.

Lubimoff, Methodius P. (1893), Aspirant-Parasit., Labor. of Experim. Biol., Zoopark. Moscow (U.d.S.S.R.). *Parasit., Helminthol.*

Lubischew, Alexander (1890), Doc. Univ., Sekretär des Biol. Inst. Perm (U.d.S.S.R.), Zaīmka. *Systematik der Halticinen (Coleoptera, Chrysomelidae). Schädliche Insekten, speziell Halticini und Apion, Biometrik: Klassifikation der Verteilungskurven. Theoretische Biol. Natur der Vererbungsfaktoren.*

Lubosch, Wilhelm (1875), o. ö. Prof. der Anat., Vorstand der Abteilung für topographische und angewandte Anat. Würzburg (Deutschland), Röntgenring 5. *Anat. und vergl. Anat.* ① Berlin 1898.

von Lucanus, Friedrich (1869). Berlin (Deutschland), Lessingstr. 32. *Ornithol.; besonders Biol. und Psychol. der Vögel. Vogelflug.*

Lucas, Frederic Augustus (1852), hon. Dir. Am. Mus. of Nat. Hist. New York, N.Y. (U.S.A.). *Anat. of birds. Fossil vertebrates.* o

Lucas Keene, Mary Frances, Prof. London W.C. 1 (England), 8 Hunter St. *Human embryol.: development of the ductless glands, appearance of function in these glands and in other organs of the human embryo.* ① London Univ. 1911.

Lucien, Maurice (1880), Dr., Prof. d'Anat. à la Fac. de Méd. Nancy, Meurthe et Moselle (France), 16, Rue de Verdun. *Anat. et Endocrinol. Développement des coulisses fibreuses, des gaines synoviales et des aponévroses du poignet et de la main. Surrénales et organes chromaffines.*

Luckhardt, Arno Benedict (1885), Prof. Univ. of Chicago, Ill. (U.S.A.), 5216 Greenwood Ave. *Physiol. of the parathyroid glands, gastric and pancreatic secretion.* o

Lucksch, Franz (1872), ao. Prof. für pathol. Anat. der deutsch. Univ. Prag II, 497 (C.S.R.). *Encephalitis epidemica, Vaccineencephalitis, perniciöse Anaemie, Mißbildungen, Vererbung derselben.* ⓓ Deutsche Univ. in Prag 1895. ⓟ Praep. von Encephalitis.

Ludewig, Karl (1897), wissenschaftlicher Hilfsarbeiter bei der Biol. Reichsanstalt, Zweigstelle Kiel (Deutschland), Niemannsweg 13. *Fritfliege.* ⓓ Berlin 1922.

Ludford, Reginald James (1895), Dr. of Sc. and Dr. of Phil., Lect. in Cytol. in the Department of Anat., Univ. Coll., London (England), 1, Oakfield Road, Southgate. *General and Pathol. Cytol. in Relation to the problems of Med. Sc.* ⓓ London Univ. 1918.

Ludwig, Alfred (1879), Dr. phil. nat. Siegen i. Westf. (Deutschland), Sandstr. 30. *Floristik (Südwest-Westfalen und Westerwald) einschließlich Gallen und parasitische Pilze. Systematik: Gattung Chenopodium.* ⓓ Straßburg i. Els. 1905. ⓟ Phanerogamen, parasitische Pilze, Gallen.

Ludwig, Clinton Albert (1886), Associate Botan. and Plant Pathol., S. Car. Agric. Experim. Station. Clemson College, South Carolina (U.S.A.). *Physiol. and pathol. of cotton plant.* ⓓ Univ. of Michigan 1917.

Ludwig, Oskar (1888), Dr. phil., Ass. am Inst. für landwirtschaftlichen Bact. Göttingen (Deutschland), Planckstr. 1. *Bact., Phytopath.* ⓓ 1923.

Ludwig, Wilhelm (1901), Ass. Zool. Inst. Leipzig (Deutschland), Talstr. 33. *Hemiptera: Kopulationsorgane. Paramaecium: Physiol.* ⓓ Leipzig 1925.

Ludwigs, Karl (1879), Prof. Dr., Dir. der Hauptstelle für Pflanzenschutz der Landwirtschaftskammer für die Provinz Brandenburg und für Berlin in Berlin-Dahlem (Deutschland), Königin-Luise-Str. 19. *Pflanzenschutz und Schädlingsbekämpfung. Biol. der Pilzkrankheiten.* ⓓ München 1911.

Lüdi, Werner (1888), Priv.Doc. Univ. Bern (Schweiz), Brunnmattstr. 70. *Pflanzengeographie.* ⓓ Bern 1917.

Lueg, Werner (1893), Priv.Doc. für innere Med. an der Univ. Berlin, Ass. der 1. Med. Univ.-Klinik. Berlin NW 6 (Deutschland), Charité. *Kreislauf, Elektrophysiol.* ⓓ Berlin 1921.

Lührs, Erich (1887), Inst.-Leiter des Bact. Inst. der Oldenburgischen Landwirtschaftskammer. Oldenburg i. O. (Deutschland), Marslatourstr. *Bact., Serol., Entomol.* ⓓ Berlin.

Luengo, Emilio (1898), Ass. Inst. Nacional de Higiene de Alfonso XIII, Prof. in charge Labor. de Investigaciones Clinicas. Madrid (España), Rodriguez San Pedro 46. *Parasit. (malaria). Epidemiol. diseases.* ⓓ Madrid 1920. ⓟ Slides of blood-diseases, malaria, protozoa and worms.

Lüpfle, Karl (1880), Univ.Prof., Dir. Tierhygien. Inst. München (Deutschland), Lachnerstr. 3. *Hygiene. Path.* o

Lüstner, Gustav (1869), Prof. Dr. phil., Vorsteher der Pflanzenpathol. Versuchstation der Lehr- und Forschungsanstalt Geisenheim a. Rh. (Deutschland). *Pflanzenpathol. Naturdenkmalpflege. Heimatkunde.* ⓓ 1897.

Lüttschwager, Hans (1889), Dr. phil., Abteilungsdir. am Staatlichen Mus. in Danzig (Freistaat Danzig), Grünes Tor, Langer Markt 24. *Heimische Faunistik, Ornithol. und Mammal., vor allem Kleinsäuger.* ⓓ Breslau 1915.

Lützkendorf, Erna (1897), Ass. Anat. Inst. Univ. Kasan (U.d.S.S.R.). *Vegetatives Nervensystem.*

Luginbill, Philip (1884), Associate Entomol., U.S. Bureau of Entomol. Monroe, Mich. (U.S.A.), Corn Borer Station, Drawer 359. *Cereal and forage insects. Southern Corn Rootworm, Fall Army worm, Chicing bug, Corn Earworm, Southern Corn Stalkborer, Bill bugs and European corn borer.* ⓓ George Washington Univ. 1924. ⓟ Phyllophaga (Lachnosterna).

Łukaszewicz, Joseph, Dr. Prof. Kraków (Polska), Acad. Minière, r. Tomasza 9. *Botan.* o

Lukeš, Franz, Dr., Prof. d. pathol. Anat. Bratislava (C.S.R.), Pathol. Inst. d. Komensky Univ. o

Lull, Richard Swann (1867), Dir., Curator of Vertebrate Palaeont., Peabody Mus., Prof. of Palaeont., Yale Univ. New Haven, Conn. (U.S.A.). *Vertebrate Palaeont. and Evolution.* ⓓ Rutgers 1893.

Lumbau, Delio (1898), Aiuto all' Ist. di Igiene di Sassari (Italia). *Rabbia e malaria.* ⓓ Sassari 1926.

Lumière, Auguste (1862), Corresp. nat. Académie de Méd. de Paris. Lyon (France), 49, Rue Villon. *Pharmacodynamie. Relations entre les fonctions chim. et les propriétés pharmacodynamiques. Biocolloïdes.*

Lumsden, Thomas (1874), Cancer researcher. London S.W. 1 (England), Lister Inst. *Immunity in relation to transplantable tumours of vertebrates.* ⓓ Aberdeen 1897.

Luna, Emerico (1882), Prof. o. Anat. umana normale Univ. Palermo (Italia). ⓓ Palermo 1922.

Lund, Elmer Julius (1884), Prof. of Zool., in charge of Physiol. Austin, Tex. (U.S.A.), Univ. of Texas. *Cell physiol.* ⓓ Johns Hopkins 1914.

Lund, J., Dir. Statens forsöksstation for Sörlandet. Treit nr. Kristiansand S. (Norge). o

Lund, Mogens (1898). København (Danmark), Slagelsegade 7. *Lichens: Cladoniacées.*

Lundbeck, Johannes (1901), Dr. phil., Ass. Neukuhren, Samland, Ostpr. (Deutschland), Seefischereistation. *Hydrobiol., Fischereibiol., quantitative Bodenteruntersuchungen.* ⓓ Kiel 1925.

Lundbeck, William (1863), Dir. der Arthropodenabteilung des Zool. Mus. København (Danmark), Krystalgade. *Diptera.*

Lundblad, Carl Olov (1890). Experimentalfältet Stockholm (Schweden). *Hydracarinen. Wasserhemipteren (speziell Corixidae). Phyllopoden. Angewandte Entomol.* ⓓ Uppsala. ⓟ Hydracarinen.

Lundborg, Herman Bernhard (1868), Prof., Dir. Statens Inst. för Rassenbiol. Uppsala (Schweden). *Neurol. und Psychiatrie, Anthrop. und Rassenbiol. Med.-biol. Anthrop. der Finnen und Lappen Schwedens.* ⓓ Lund 1902. ⓟ Photographien von Rassentypen aus Schweden (Schweden, Finnen, Lappen und andere).

Lundegårdh, Henrik Gunnar (1888), o. Prof., Chef der Botan. Abt. und des Botan. Labor. der Zentralanstalt für Landwirtschaftl. Versuchswesen in Stockholm-Experimentalfältet, Dir. ökol. Station der Hallands Väderö. Experimentalfältet (Sverige). *Ökol. Pflanzenphysiol. (experim. Ökol.) und ihre Anwendung auf landwirtschaftliche Probleme. Kohlensäureassimilation, Probleme der Stoffaufnahme.* ⓓ Stockholm 1917.

Lunghetti, Bernardino, e.o. Prof. Univ. Cagliari, Sardinia (Italia). *Anat. e Istol. patol.* o

Lungrew, E. A. (1899), Ass. Plant Pathol. Ft. Collins, Col. (U.S.A.). *Cereal diseases: wheat rust, wheat Smut.* ⓓ Ft. Collins, Col. 1923.

Lunt, Herbert A. (1898), First Ass. in Soil Fertility and Soil Experim. Fields. Urbana, Ill. (U.S.A.), Dept. of Agronomy, Univ. of Illinois. *Soil and plant analyses. Chem. studies of soils subjected to different treatments in the fields. Effect of straw upon nitrification in the soil.* ⓓ Washington 1923.

Luntz, Albert (1900), Kaiser-Wilhelm-Inst. für Biol. Berlin-Dahlem (Deutschland). *Generationswechsel der Rädertiere und Cladoceren.* ⓓ Berlin 1926.

Lupi Noguera, Raul, Prof. ord. Univ. Lisbôa (Portugal). *Chim. farm. e biol.* o

Lupu, Hélène (1883), Chef des travaux au Labor. de Physiol. Univ. de Jassy (Roumanie). *Histophysiol. de l'intestin des poissons. Régénération de l'épithélium intestinal du Cobitis fossilis. Respiration intestinale. Sang de cobitis fossilis.* ⓓ Jassy 1913.

Lurini, Lydia (1896), Ass. di Anat. e Fisiol. comparate nella R. Univ. Firenze (Italia), Via Romana 19. *Riproduzione cellulare.* ⓓ Firenze 1918.

De Lury, Ralph Emerson (1881). Ottawa (Canada), Dominion Observatory. *Ornithol. and mammal.* ⓓ Toronto Univ. 1907.

Lus, Janis (1897), Ass. Univ., wissensch. Mitarbeiter der Akademie d. Wissenschaft. Leningrad (U.d.S.S.R.), Karl-Liebknechts-Prosp. 106, W. 19. *Regeneration und Transplantation: Planaria, Polarität und Heteromorphose. Vererbungsstudien bei Menschen (Eugenik): Vererbungsstudien an Coccinelliden. Variabilität bei Macrosiphum solidaginis.*

Lusena, Marcello (1896), Aiuto, Lib. doc. Batteriol. e Immunol., Labor. d'Igiene. Firenze (Italia), Viale G. B. Morgagni 18. *Virus filtrabili (Epitelioma contagioso). Immunità locale. Agglutinazione aspecifica. Vaccinazione orale. Mutazioni batteriche. Batteriolisi. Anticorpi batteriol. nella infezione (setticemia) eberthiana.* ⓓ 1921.

Lush, Jay Laurence (1896), Animal Husbandman, in charge of breeding investigations, Texas Agric. Experim. Station. College Station, Tex. (U.S.A.). *Genetics as applied to animal husbandry problems.* ⓓ Univ. of Wisconsin 1922.

Lusk, Graham, Prof. of Physiol., Cornell Univ., Med. Coll. New York City (U.S.A.). *Nutrition, Colorimetry.* ⓓ Columbia 1887, München 1891.

Lustig, Alessandro (1857), Prof. o. di Patol. gen., Sperimentale e di Batteriol., Immunol. Firenze (Italia). *Malattie infettive: Colera, peste, tifo exantematico, febbre ondulante (malta). Nucleo-proteidi bacterici. Patol. cellulare. Virus filtrabili.* ⓓ Vienna 1882 e Firenze.

Lute, Anna M., State Seed Analyst. Fort Collins, Col. (U.S.A.). *Seed purity, germination testing.* ⓟ Weed seeds.

Luther, Alexander Ferdinand (1877), ao. Prof. der Zool. an der Univ. Helsingfors und Vorstand der Zool. Station Tvärminne. Helsingfors (Finnland), Djurgårdsvillan 8. *Morphol. der Vertebraten, spez. Pisces (Kopf, Myol.); Entwicklungsmechanik (Amphibien). Turbellaria, Mollusken Finnlands.* ⓓ 1904.

Lutman, Benjamin F. (1879), Prof. of Plant pathol., Univ. of Vermont, Plant pathol., Vermont Agric. Experim. Station. Burlington, Vt. (U.S.A.), 111 North Prospect St. *Diseases of potato. The cytol. effect of disease and old age in plants.* ⓓ Univ. of Wisconsin 1909.

Lutshnik, Victor (1892), Dir. Mus. and Plant Protection Sta. Stauropol en Caucases (U.d.S.S.R.), Boxpost N. 76. *Entomol.: Carabidae orbis terrarum. Economic Entomol.* ⓟ Carabidae (inkl. Cicindelidae).

Lutz, Frank Eugen (1879), Curator of Entomol., American Mus. of Natural History. New York (U.S.A.), 77 St. and Central Park W. *Insects, variation, heredity, assortive mating.* ⓓ Chicago 1902. o

Lutz, Georg (1892), Vorstand der bact. Abteilung des Katharinenhospitals Stuttgart und Priv.Doc. für Hygiene an der technischen Hochsch. Stuttgart, Württbg. (Deutschland), Kriegsbergstr. 60. *Bact. und Gewerbehygiene.* ⓓ Tübingen 1916.

Lutz, H., Chargés de cours Fac. de Pharmacie. Paris (France), 4, Avenue de l'Observatoire. *Cryptog.* o

Luxenburg, Anna (1890), Dr. en phil., Ass. à l'Inst. de Botan. générale de l'Univ. de Varsovie (Pologne), Koszykowa 11 m. 16. *Recherches cytol. sur les grains de pollen des Malvacées. L'évolution du sac pollinique et des grains de pollen chez les Conifères.* ⓓ Varsovie 1922.

van Luyk, Abraham (1874), Ass. Phytopath. Labor. Baarn (Holland), Marisstraat 18. *Fungi imperfecti. Ustilaginales.* ⓟ Ustilaginales.

Lvoff, Sergius (1879), Sub-chef Section de Physiol. végétale au Jardin Botan. princ., Prof. adjoint Univ., Prof. de Botan. Inst. vét. Leningrad (U.d. S.S.R.). *Ferments: Fermentation alcoholiques. Physiol. végétale.* ⓓ Leningrad 1911.

Lwoff, André (1902), Ass. à l'Inst. Pasteur. Paris XV (France), 96, Rue Falguière. *Protozool. Infusoires parasit. Morphol. Cytol. Histophysiol. Cycle. Infusoires libres: physiol. de la nutrition.* ⓓ Paris 1921.

Lwoff, Marguerite (1905), Inst. Pasteur. Paris (France), 96, rue Falquière. *Biol., physiol. de la nutrition, des Copépodes.* ⓓ Paris 1926.

Lyka, Károly (1869), Hochschullehrer. Budapest VI (Ungarn), Nagy János-u. 12. *Systematik der Gattung Thymus.* ⓟ Exsiccata der Gattung Thymus.

Lyman, George R. (1871), Dean of the Coll. of Agric., Dept. of Plant pathol., Univ. of West-Virginia. Morgantown, W.Va. (U.S.A.). *Mycol. plant diseases.* ⓓ Harvard 1906. o

Lyman, John F. (1881), Prof. of Agric. Chem. Columbus, O. (U.S.A.), Ohio State Univ. *Biochem., Animal Nutrition.* ⓓ Yale 1909.

Lynch, Kenneth Merrill (1887), Prof. of Pathol., Med. Coll. of the State of South Carolina. Charleston, S.C. (U.S.A.). *Protozoa of Man. Morbid Anat. of Negro in disease.* ⓓ Univ. of Texas 1910.

Lynge, Bernt (1884), Doc. Oslo (Norge), Botan. Mus. *Flechten. Arktische Phanerogamen.* ⓓ Univ. Oslo 1917. ⓟ Flechten und arktische Phanerogamen.

Lyon, Elias Potter (1867), Prof. of Physiol. and Dean, Med. School, Univ. of Minnesota. Minneapolis, Minn. (U.S.A.). *Marine Physiol.* ⓓ Chicago 1897.

Lyon, Marcus Ward, jr. (1875), Pathol., The Clinic. South Bend, Ind. (U.S.A.). *Pathol., Bact., human Parasit. Classification and geographic distribution of Mammals. Mammals and Flowering Plants of northern Indiana.* ⓓ Washington Univ. M.D. 1902, Ph.D. 1913. ⓟ Flowering plants of Northern Indiana.

Lyon, Thomas Lyttleton (1869), Prof. of Soil Technol. Ithaca, N.Y. (U.S.A.), Cornell Univ. *Chem. and biol. research in soils.* ⓓ Cornell Univ. 1904.

Lyssenkow, Nikolaus (1865), Dir. Forschungskatheder f. Morphol. u. Physiol. Odessa (U.d.S.S.R.), Baranowa 12. *Neurol.: Anat. des Kleinhirns.* ⓓ Moskau 1897.

Lythgoe, Richard James (1896), Beit Memorial Research Fellow, Investigations-Med. Research Council. London W.C. 1 (England), Dept. of Physiol., Univ. Coll., Gower St. *Physiol. of Vision.* ⓟ Honey bees.

Ma, W. C., Prof. Dr., Head Dept. Anat. Union Med. Coll. Peking (China). o

Maarschalk, Hendrik (1876), Agric. engineer. Wageningen (Holland), Nassauweg 4. *Phytopath. and economic entomol.* ⓓ Wageningen 1913.

McAllister, Frederick (1876), Prof. of Botan., Univ. of Texas. Austin, Tex. (U.S.A.). *Cytol. pertaining to pyrenoids and plastids.* ⓓ Univ. of Wisconsin 1910.

Macallum, Archibald Bruce (1885), Prof. of Biochem., Univ. of Western Ontario. London (Canada), Ottaway Ave. ⓓ M.D. Univ. of Toronto 1910, Ph.D. 1919.

MacAloney, Harvey T. (1896), Ass. Entomol., Bureau of Entomol., U. S. D. A. Amherst, Mass. (U.S.A.), Northeastern Forest Experim. Station. *Forest Insects.* ⓓ Syracuse Univ. 1925. ⓟ Forest Insects.

MacAndrews, A. H. (1897), Ass. Prof., Forest Entomol. Syracuse, N.Y. (U.S.A.), Coll. Forestry. *Forest Entomol. Scolytids.* ⓓ Syracuse Univ. 1926. ⓟ Scolytids of N. A.

McArdle, Richard E. (1899), Junior Forester, Pacific Northwest Forest Experim. Station, U.S. Forest Service. Portland, Ore. (U.S.A.), 514 Lewis Bldg. ⓓ Univ. of Michigan 1924.

MacArthur, John Wood (1889), Ass. Prof. of Genetics Univ. of Toronto and Prof. of Genetics, Ontario Agric. Coll. Toronto (Canada), Dept. of

Biol., Univ. *Genetics: Lycopersicum, Poultry. Eugenics. Experim. Biol.: Daphnia magna.* ⓓ Univ. of Chicago 1920.

McAtee, Waldo L. (1883), Ass. in Charge Food Habits Research, U. S. Biol. Survey; also Custodian of Hemiptera, U. S. National Mus. Washington, D.C. (U.S.A.), U. S. Dept. Agric. *Food habits of vertebrates; taxonomy of Hemiptera, Typhlocybinae.* ⓓ Univ. of Indiana 1906. ⓟ *Hemiptera.*

McAvoy, Blanche (1885), Instr. of Biol., Illinois State Normal Univ. Normal, Ill. (U.S.A.). *Cytol. Ecol. Interglacial fossil.* ⓓ Ohio State Univ. 1912.

MacBride, E.W., Prof. Imperial Coll. London S.W.7 (England), S. Kensington *Zool.* o

Macbride, Thomas H., President Emeritus, State Univ. Iowa City, Ia. (U.S.A.). *Mycol., Myxomycetes.* o

McBryde, Charles Neil (1872), Bact., Bureau of Animal Industry, U. S. Dept. of Agric. Ames, Ia. (U.S.A.), Box 175. *Diseases of domestic animals, swine diseases. Immunity, serums and vaccines.* ⓓ Med. Johns Hopkins Univ. 1897, phil. George Washington Univ. 1911.

McCall, Arthur Gillett (1874), Prof. of Geol. and Soils, Univ. of Maryland, in Charge of Soil Investigations, Maryland Experim. Station College Park, Md. (U.S.A.). *Soils and Plant Nutrition.* ⓓ Johns Hopkins 1916.

McCall, Max Adams (1888), Agronomist in Charge of Cereal Agronomy, Office of Cereal Crops and Diseases. Washington, D. Cl (U.S.A.), Bureau of Plant Industry, U. S. Department of Agric. *Cereals.* ⓓ State Coll. of Washington, Pullman, Wash., 1922.

MacCallum, Peter, Prof. Univ. Melbourne (Australia). *Pathol.* o

M'Candlish, Andrew Corrie (1890), Advisery Officer in Milk Production, West of Scotland Agric. Coll. Wigtownshire, Scotland (England), Claunch, Sorbie. *Factors influencing milk production. Influence of breeding, feeding and management on milk production.* ⓓ Iowa State Coll. 1915.

McCann, William S. (1889), Prof. of Med., Univ. of Rochester, School of Med. Rochester, N.Y. (U.S.A.). *Pathol. Physiol.: metabolism.* ⓓ Cornell Univ. 1915.

McClean, Alan Percy Douglas (1902), Mycol., Department of Agric. Durban (Union of South Africa), Natal Herbarium. *Plant pathol.* ⓓ Pretoria 1924.

McClelland, John Robert (1893), Dir. of Research, The Kolynos Company. New Haven, Conn. (U.S.A.), 186 Willard St. *Oral Bact.*

McClelland, Thomas B. (1886), Horticulturist, U.S. Agric. Experim. Station. Mayaguez (Porto Rico). *Horticultural plants in the tropics: coffee, coconuts, vanilla, vegetables, photoperiodism.* ⓓ Kentucky State Univ. 1907.

MacClement, W. T., Prof. of Biol., Queens Univ. Kingston, Ontario (Canada). o

McClendon, Jesse Francis (1880), Prof. of Physiol. Chem., Univ. of Minnesota, Med. School. Minneapolis, Minn. (U.S.A.), 815 Fulton St., S.E. *Ionic equilibria in blood; iodine metabolism.* ⓓ Pennsylvania 1906.

McClung, Clarence Erwin (1870), Dir., Zool. Labor., Univ. of Pennsylvania and Prof. of Zool. Swarthmore, Pa. (U.S.A.), 417 Harvard St. *Maturation phenomena in animals.* ⓓ Univ. of Kansas 1892.

McClure, Charles F. W. (1865), Prof. of Compar. Anat. in Princeton Univ. Princeton, N.J. (U.S.A.). *Morphol. and physiol. of the vascular system, the lymphatic and venous systems. Oedema.* ⓓ Princeton Univ. 1888.

McClure, Floyd Alonzo, Prof., Christian Coll. Canton (China). *Botan.* o

McClymonds, Arthur E. (1891), Superintendent of the Aberdeen Cooperative Substation. Aberdeen, Id. (U.S.A.). *Breeding of Cereals and Forage Crops. Selection of potatoes and breeding up adapted strains of corn.* ⓓ Manhattan, Kan. 1915. ⓟ *Varieties of various irrigated grains.*

Di Macco, Gennaro (1895), Libero doc. di Patol. generale, Aiuto all' Ist. di Patol. generale della Univ. Napoli (Italia), 21, S. Andrea delle Dame. *Fisiopatol.-batteriol. Fisico-chim. del sangue, proteinoterapia, fisiopatol. della pressione arteriosa, immunità, lipasi, azione biol. dell' alcool.*

McCollam, Millard E. (1894), Agronomist, Western Wash. Experim. Sta. Puyallup, Wash. (U.S.A.). *Pastures, Grasses and Forage Crops.* ⓓ Berkeley, Cal., 1917.

McColloch, James Walker (1889), Prof. of Entomol. Manhattan, Kan. (U.S.A.). *Cereal Crop Insects; Soil inhabiting insects.* ⓓ Kansas State agric. Coll. 1923.

McCollum, Elmer V. (1879), Prof. of Chem. Hygiene, Sch. Hygiene and Pub. Health, Johns Hopkins Univ. Baltimore, Md. (U.S.A.). *Physiol. and Biochem.* ⓓ Yale Univ. 1906.

McComas, Henry Clay (1875), Assoc. Prof. of Psychol., Univ. Princeton, N.J. (U.S.A.), 109 Broadmead. *Types of Attention.* ⓓ Harvard Univ. 1910.

McCombs, Lois Ferree (1899), Ass. technol. librarian, Carnegie Library of Pittsburgh, Forbes Street. Pittsburgh, Pa. (U.S.A.), 1205 Farragut Street. *Food bact., bibliographies.* ⓓ Univ. of Wisconsin 1922.

McConkey, Oswald (1891), Lect. in Agronomy and research practical plant-breeding with field crops. Guelph, Ontario (Canada), Ontario Agric. Coll. *Crop Ecol. and practical breeding and improvement. Grasses and clovers.* ⓓ Univ. of Illinois 1922.

McCormack, James Mines (1879), Physician St. Michael's Hospital. Toronto (Canada). *Cancer Research.* ⓓ Trinity Univ. 1901.

McCrackan, Robert Franklin (1882), Associate Prof. of Biochem. Richmond, Va. (U.S.A.), Med. Coll. *Biochem.* ⓓ Columbia Univ. 1910.

McCray, Francis A. (1888), Prof. of Agronomy, Sam Houston State Teachers Coll. Huntsville, Tex. (U.S.A.). *Fiber crops, forage crops, cereals, root crops and plant genetics.* ⓓ Wisconsin 1914.

McCrudden, Francis (1879), Prof. Therap. Med. School. Boston, Mass. (U.S.A.), 520 Commonwealth avenue. *Biol. chem.* ⓓ Harvard Univ. 1908.

McCubbin, Walter Alex., Chief Plant Pathol., Pennsylvania Bureau of Plant Industry. Harrisburg, Pa. (U.S.A.). o

McCue, Charles Andrew (1879), Prof. of horticulture, Del. Agric. Experim. Station. Newark, Del. (U.S.A.). o

McCulloch, Ernest C. (1899), Instr., Coll. of Agric., Dept. Animal Pathol. and Hygien., Univ. of Illinois, Ass. in Animal Pathol. and Hygien., Experim. Station. Urbana, Ill. (U.S.A.), Labor. *Compar. Pathol. and Parasit., Parasit. Zool.* ⓓ Kansas State Agric. Coll. 1924.

McCulloch, Lucia (1873), Ass. Pathol., Labor. of Plant Pathol., Bureau of Plant Industry, United States Department of Agric. Washington, D.C. (U.S.A.). *Bact. Diseases of Plants.*

MacCurdy, Hansford M. (1868), Prof. of Biol. and Geol., Head of Departments. Alma, Mich. (U.S.A.), Coll. *Zool., theoretical and practical, heredity. Selection and Cross Breeding, inheritance of coat-pigments and Coat-patterns in Rats and Guinea-pigs. Degeneration of Ganglion Cells.* ⓓ Harvard Univ. 1906.

McCutcheon, Morton (1888), Ass. Prof. of Pathol. Philadelphia, Pa. (U.S.A.), Univ. of Pennsylvania. *Osmosis in cells. Kinetics of growth. Mechanism of vital staining. Locomotion of leucocytes.* ⓓ Philadelphia 1917.

McDaniel, Eugenia Inez (1884), Associate Prof. of Entomol., Research Ass. in Entomol., Exp. Station, Mich. State Coll. East Lansing, Mich. (U.S.A.). *Economic Entomol.*

MacDaniels, Laurence H., Prof. of Pomol., State Coll. of Agric., Cornell Univ. Ithaca, N.Y. (U.S.A.). o

McDermott, Frank Alexander (1885), Chief Chem., Eastern Alcohol Corporation, Pennsgrove, N.J. Claymont, Del. (U.S.A.), 16 Cathedral Ave. *Fermentation, bact. and chem. of Entomol., Lampyridae (Coleoptera). Biol. and chem. of luminescence.* ⓓ Pittsburgh 1914.

Macdonald, Adam Davidson (1895), Lect. Experim. Physiol. Univ. Manchester (England). *Actions of Glandular Extracts.* ⓓ Edinburgh 1923.

MacDonald, Gilmour Byers (1883), Prof. Forestry, Chief forestry sect. Experim. Sta., Iowa State Coll. Ames, Ia. (U.S.A.), Forestry Department. *Farm forestry.* ⓓ Univ. of Nebraska, Lincoln, 1914. ⓟ Materials from the native trees of Iowa.

Macdonald, J. S., Prof. Univ. Liverpool (England). *Physiol. and Histol.* o

McDonald, Stuart, Prof. Univ. Durham Coll. of Med. Newcastle upon Tyne (England). *Pathol.* o

McDonough, Frank L. (1888), Crop Pest Specialist, Niagara Sprayer Co. Middleport, N.Y. (U.S.A.). ⓓ Conn. Agr. Coll. 1911.

MacDougal, Daniel Trembly, Dir. of the Labor. for Plant Physiol., Carnegie Inst. of Washington. Tucson, Ariz. (U.S.A.), Desert Labor. o

Macdougall, R. Stewart, Prof. R. Veter. Coll. Edinburgh (Scotland). *Biol.* o

McDougall, W. B. (1883), Ass. Prof. of Botan., Univ. of Illinois. Urbana, Ill. (U.S.A.). *Plant ecol., Mycorrhiza and other symbiotic phenomena.* ⓓ Univ. of Michigan 1913.

McDowall, Robert John Stewart (1892), Prof. of Physiol., Univ. London (England), Kings Coll. *Experim. Physiol. with special reference to the Circulation and Respiration.* ⓓ Edinburgh 1921.

McDowell, E. Carleton (1887), Resident Investigator in Department of Genetics of the Carnegie Inst. of Washington. Long Island, N.Y. (U.S.A.), Carnegie Inst., Cold Spring Harbor. *Physiol. of reproduction and genetics in mice.* ⓓ Swarthmore Coll. 1909.

Mace, M. E., Prof. Fac. Méd. Nancy (France). *Bact.* o

McEwen, Robert Stanley (1888), Associate Prof. in Oberlin Coll. Oberlin, O. (U.S.A.), 208 Forest St. *Embryol., Histol., Genetics.* ⓓ Columbia Univ. 1917.

MacFall, Claude Matthews (1880), Department of Physiol. Harvard. Boston, Mass. (U.S.A.), Harvard Med. School. *Endocrines.* ⓓ Univ. of Virginia 1926.

McFarland, Frank Theodore (1886), Prof., Head Dept. of Botan., Univ. Kentucky. Lexington, Ky. (U.S.A.). ⓓ Univ. of Wisconsin 1921.

McFarlane, Andrew (1898), Ass., Pharmacol. Dept., Univ. Edinburgh (Scotland). *Anti-diuretic action of Pituitary. Standardisation of Digitalis.* ⓓ MD. Edinburgh 1923.

MacFarlane, John Muirhead, Emeritus Prof. Botany. Univ. of Pennsylvania. Philadelphia, Pa. (U.S.A.). o

MacGillivray, William David Kerr (1867), Ornithol. Broken Hill (Australia). *Ornithol., anthrop., botan.* ⓓ Melbourne Univ. 1891.

MacGillivray, John Henry (1899), Research Ass. n Horticulture. W. Lafayette, Ind. (U.S.A.), Purdue Agric. Experim. Station. *Chem. Physiol. of Vegetable Plants.* ⓓ Univ. of Wisconsin 1925.

McGinty, Rupert A. (1886), Associate Horticulturist, Col. Experim. Sta., Col. Agric. Coll. Fort Collins, Col. (U.S.A.). *Physiol. and breeding of Vegetable Crops.* ⓓ Washington Univ. 1919.

Macgregor, Agnes, Lect. School of Med. of the R. Coll. Edinburgh (Scotland). *Pathol., Bact.* o

McGregor, James H. (1872), Prof. Zool. Columbia Univ. New York (U.S.A.). *Fossil races of man, reptilian and primates.* ⓓ Columbia 1899. o

MacGregor, Malcolm Evan (1889), Wellcome Entomol. Field Labor. Wisley, Ripley, Surrey (England). *Med. Entomol. Biol. and bionomics of the insect vectors of disease, under natural field conditions.* ⓓ Univ. of Harvard (U.S.A.). ⓟ Culicidae (Mosquitoes).

McGregor, Richard Crittenden (1871), Ornithol., Bureau of Sc. Manila (Philippine Islands). *Philippine ornithol.* ⓓ Stanford 1898.

McGulgan, Hugh Alister (1874), Prof. Univ. of Illinois, Coll. of Med. Evanston, Ill. (U.S.A.), 2375 Sherman Av. *Pharmacol.* o

Mach, Felix (1868), Prof. Dr., Nahrungsmittelchem. Dir. d. Staatl. Landwirtsch. Versuchsanstalt Augustenberg, Post Grötzingen i. Baden (Deutschland). *Agrikulturchem.* ⓓ Berlin 1892.

McHargue, James Spencer (1878), Research Chem. Lexington, Ky. (U.S.A.), Kentucky Agric. Experim. Station, Univ. of Kentucky. *Distribution and function of Copper, Manganese, Zinc, Boron, Nickel and Cobalt in soils, plants and animals.* ⓓ 1921.

McHatton, Thomas Hubbard (1883), Prof. of Horticulture, Univ. of Georgia. Athens, Ga. (U.S.A.), State Coll. of Agric. ⓓ Spring Hill 1907.

Machebœuf, Michel Alexandre (1900), Prép. chim. biol. fac. sc. Paris XV (France), Inst. Pasteur, Rue Dutot. *Chim. biol. et physicochim. biol. React. du benjoin colloid. dans liquide cephal. Insuline. Phosphor sanguine.* ⓓ Paris.

Machens, Andreas (1883), Leiter Bact. Anstalt d. Landw.-Kammer. Braunschweig (Deutschland), Hochstr. 17/18. *Biol. Milchuntersuchungen.* ⓓ Gießen 1908.

Machida, Sakukichi, Prof. Imper. Univ., Labor. of Hydro-biol., Coll. of Agric. Tôkyô (Japan). *Fishery.*

Machida, Tirô, Ass. Prof. Tôkyô Imper. Univ., Zool. Labor., Coll. of Agric. Komaba near Tôkyô (Japan). *Cytol. and genetics of Lepidoptera.*

Machle, F. P., Dir. Bact. Labor. Bombay (Brit.-India). o

Machle, Edward C., Lect. Hackett Med. Coll. Canton (China). *Pharmac.* o

Macht, David I. (1882), Lect. in Pharmacol., Johns Hopkins Univ., and Dir. of Pharmacol. Research Labor., Hynson, Westcott, and Dunning. Baltimore, Md. (U.S.A.). *Pharmacol.* ⓓ Johns Hopkins Univ. 1906, Phar.D. (Honorary) Univ. of Maryland 1914.

McIndoo, Norman Eugene (1881), Entomol. Washington, D.C. (U.S.A.), Bureau of Entomol. Dept. Agric. *Behavior and senses of insects, insect histol., physiol. effects of insecticides, and discovery of new insecticides.* ⓓ Univ. of Pennsylvania 1911.

McInteer, Berthus Boston (1887), Ass. Prof. Botan. Univ. of Kentucky. Lexington, Ky. (U.S.A.). ⓓ Univ. of Kentucky 1926.

McIntosh, J., Prof. Univ. London S.W. 7 (England), S. Kensington. *Pathol.* o

McIntosh, Allen (1893), Ass., Division of Animal Biol., Univ. of Minnesota. Minneapolis, Minn. (U.S.A.). *Trematodes of birds.* ⓟ Trematodes of birds, reptils and mammals.

McJunkin, Frank A. (1882), Associate Prof. of Pathol., Washington Univ. St. Louis, Mo. (U.S.A.), 1245 Lyndover Place. *Blood and tissue Phagocytes.* ⓓ Univ. of Michigan 1906.

Mack, Margaret Elizabeth (1871), Associate Prof. Biol. Reno, Nev. (U.S.A.), Univ. Nevada. *Hygiene.*

MacKay, Alexander Howard (1848). Dartmouth, Nova Scotia (Canada), 61 Queen St. *Diatomaceae of Nova Scotia and Canada.* ⓓ Dalhousia Univ. 1892.

McKay, Marion Bertice (1887), Plant Pathol., Oregon Experim. Sta., Agent Bureau of Plant Industry, U. S. Dept. of Agric. Corvallis, Ore. (U.S.A.), Oregon Agric. Coll. *Potato diseases, diseases of tulip and narcissus, sugar beet curly top on truck crops.* ⓓ Univ. of Wisconsin, Madison, 1915.

MacKeith, Malcolm Henry (1895), Demonstrator of Pharmacol., Univ., Fellow and Tutor, Magdalen Coll. Oxford (England). ⓓ Oxford 1921.

Mackersie, William G. (1894), Lect. in Pharmacol. Winnipeg, Manitoba (Canada), Med. Coll. *Pharm.*

Mackevičaitė-Lašienė, Janina (1897), I. Ass. des Lehrstuhls f. allgemeine Pathol. u. pathol. Anat. Kaunas (Latvija), Aukštaičių g-vė 17 No. *Fingertuberkulose.* ⓓ Berlin 1922.

McKibben, Paul Stilwell (1886), Prof. of Anat., The Univ. of Western Ontario Med. School. London, Ontario (Canada). *Anat. of the Nervous System.* ⓓ Univ. of Chicago 1911.

Mackie, Frederic Percival (1875), Dir., Haffkine Inst. Parel, Bombay (India). *Tropical plagues, Pathol. Bact. Kala Azar. Toxikol.* ⓓ Bristol 1897. ⓟ Snake Venoms.

Mackie, Thomas Jones (1888), Prof. of Bact., Univ. of Edinburgh (Scottland). *Med. Bact., Immunol. and Serol.* ⓓ Univ. of Glasgow 1910. ⓟ Bact. Cultures and Microscopic Preparations of bact. nature.

Mackie, Wm. W., Ass. Prof., Coll. of Agric., Univ. of California. Berkeley, Cal. (U.S.A.), 124 Hilgard Hall. o

McKinney, Harold Hall (1889), Pathol. in Charge of Virus Disease Investigations on Cereals. Washington, D.C. (U.S.A.), Office of Cereal Crops and Diseases, U.S. Dept. of Agric. *Viruses, their properties, their relations of each other and their specific nature. Tobacco mosaic. Quantitative methods of the purification of the virus. The mosaics of the grasses occurring on winter wheat, winter rye and winter barley, resistant varieties, soil transmission of the virus, cell inclusions and the organs of the cytoplasm.* ⓓ 1919. ⓟ Cytol. preparations dealing with the cell inclusions associated with the virus diseases of plants and animals; virus extracts from mosaic affected plants.

Mackinnon, Doris L. (1883), Reader Zool. Univ., Kings Coll., Zool. Dept. London W.C. 2 (England), Strand. *Protozool.* ⓓ Aberdeen 1914.

Macku, Johann, Dr., Prof. f. Botan. Brno (C.S.R.), Tschech. techn. Hochsch. *Botan. appliquée. Champignons.* o

Mackworth-Praed, Cyril Winthrop. London S.W. 7 (England), 51 Onslow Gardens. *African Ornithol.* ⓓ Trinity Coll., Cambridge, 1914.

McLaine, Leonard Septimus (1887), Chief, Division of Foreign Pests suppression Secretary, Destructive Insect and Pest Act Advisory Board. Ottawa (Canada), Department of Agric. *Insect and pests, new and introduced pests.* ⓓ Mass. Agric. Coll. 1912.

Maclaren, Norman H. W. (1880), Lect. Embryol. Univ. Glasgow (Scotland). *Mammalian embryol.* ⓓ Jena 1904.

McLarty, Harold Ross (1891), Pathol. in Charge. Summerland, British Columbia (Canada), Dominion Labor. of Plant Pathol. *Physiol. diseases of fruit trees.* ⓓ McMaster Univ. 1920.

McLaughlin, Frederick A., Ass. Prof. Botan., Mass. Agric. Coll. Amherst, Mass. (U.S.A.). o

MacLean, H., Prof. St. Thomas Hospit. Med. School. London (England). Albert Embankment, Lambeth S.E. *Pathol.* o

MacLean, Mary Winifred (1904), Ass. Psychol., McGill Univ., Montreal. Souris, Prince Edward Island (Canada). ⓓ Montreal 1926.

McLean, Robert Colquhoun (1890), Prof. Botan., Univ. Coll. of South Wales. Cardiff (England), 3 Chargot Road. *Marine ecol. and Palaeobotan.* ⓓ Cambridge 1914.

McLean Thompson, J., Prof. Univ. Liverpool (England). *Botan.* o

MacLeod, Donald John (1894), Pathol. in charge of Plant Pathol. Labor. Fredericton, N.B. (Canada), Experim. Station. *Phytopath.* ⓓ Queens Univ., Kingston, Ont. (Canada) 1922.

McLeod, James Walter (1887), Prof. Leeds (England), 18 Springfield Mount. *Bact. Biochem. of bacteria growth. Bacterial respiration.* ⓓ Glasgow 1908. ⓟ Bacterial cultures.

MacLeod, Robert Brodie (1901), Ass. Psychol., McGill Univ. Montreal (Canada), 462 Union Ave. ⓓ Montreal 1926.

McLuckie, John (1890), Ass. Prof. Botan., Univ. Sydney (Australia). *Mycol., Plant Physiol., Ecol., Parasit. Flora of Mt. Wilson.* ⓓ Sydney 1922.

MacMillan, Howard Gove (1890), Pathol., U.S. Department of Agric. Greeley, Col. (U.S.A.), 1211 Eighth Street. *Potato diseases in western United States. Injurious effects of light.* ⓓ Univ. of Wisconsin 1919.

MacMillan, Warren B. (1894), Instr. Forestry. State College, Pennsylvania (U.S.A.). *Silviculture and utilization.* ⓟ Different native species of trees-seedling form.

McMurphy, James (1871), Associate Prof. of Botan. Stanford University, California (U.S.A.). *Local Fleshy Fungi and Plant Diseases.* ⓓ Stanford 1909.

McMurrich, James P., Dr., Prof. of Anat. Univ. of Toronto (Canada). o

McMurtrey, James Edward Jr. (1893), Ass. Physiol., Tobacco and Plant Nutrition, Bureau of Plant Industry, U.S. Department of Agric. Washington, D.C. (U.S.A.). *Nutrition of the tobacco plant, development, soil nutrition diseases of tobacco.* ⓓ Univ. of Kentucky, Lexington, Ky. 1917.

McNair, James Birtley (1889), Associate in Economic Botan. Chicago, Ill. (U.S.A.), Field Mus. of Natural History. *Rhus Dermatitis. Citrus Products.* ⓓ Univ. of California 1916.

Mac Neal, Ward J. (1881), Prof. of Pathol. and Bact. and Dir. of the Labor. in the New York Post-Graduate Med. School and Hospital. Office: New York City (U.S.A.), 303 E. 20 th St.; Residence: Forest Hills, N.Y. (U.S.A.), 82 Rockrose Place. *Finer structure of the Spleen. Flocculation tests in Syphilis and in Tuberculosis by methods of Vernes. Pathogenic effects of intestinal anaerobes, especially in relation to the blood and blood-forming organs. Methylene Azure B as a tissue stain.* ⓓ Pub. Univ. of Michigan, Ann Arbor, Mich. 1904; med. 1905.

McNeel, Travis E. (1900), Jr. Entomol., U. S. Bureau of Entomol. Mound, La. (U.S.A.). *Bionomics of Anopheles Mosquitoes.* ⓓ Miss. A. and M. Coll. 1923.

Mac Nider, William de B. (1881), Kenan Research Prof. of Pharmacol., Univ. of North Carolina. Chapel Hill, N.C. (U.S.A.). *Kidney-action of diuretics. Urine formation. Disturbance in acid-base equilibrium of blood and tissue juices.* ⓓ Univ. of North Carolina 1903.

McNutt, G. W. (1894), Assoc. Prof. Vet. Anat. and Histol. Ames, Ia. (U.S.A.), Vet. Div. Iowa State Coll. ⓓ Iowa State Coll. 1917.

Macoun, William Terrill (1869), Dominion Horticulturist. Ottawa (Canada), Experim. Farm. ⓟ Horticultural Plants, Fruits.

McPhee, Hugh C. (1896), Animal Husbandman in Genetics. Washington, D.C. (U.S.A.), United States Department of Agric., Bureau of Animal Industry. *Inbreeding and crossbreeding in livestock. The Genetics of Swine.* ⓓ Cambridge, Mass. 1923.

Macrinov, John (1874), Ass. of the Dir. Microbiol. Section Inst. of Experim. Med. Leningrad, Aptekarsky Island (U.d.S.S.R.), Lopouchinskaiastr. 12. *General microbiol. Bact. of the soil. Milk and dairy bact. Retting of spinning plants; fermentation of dough, cabbage. Injurers of wood.*

McSwiney, Bryan Austin (1894), Prof. of Physiol., Univ. of Leeds (England), Med. School. *Movement and function of smooth muscle. Pulse wave and the elasticity of arteries.* ⓓ Trinity Coll. Dublin 1917.

McTaggart, Alexander (1888), Ass. Prof. of Agronomy, Macdonald Coll., McGill Univ., Montreal, Que. Macdonald College, Prov. Quebec (Canada). *Breeding and selection, forage crops. Soils.* ⓓ Cornell Univ. 1921. ⓟ Seeds of timothy, orchard grass, hardy alfalfas, red clover.

McWhorter, Frank P. (1896), Plant Pathol. Valruck Experim. Station. Norfolk, Va. (U.S.A.). *Diseases of vegetables, flowers, Carica papaya.* ① Chicago 1920. ⓟ Cultures of Fusarium lycopersici. Carica papaya diseases.

Macy, Harold (1895), Ass. Prof. of Dairy Bact., Division of Dairy Husbandry, Univ. of Minnesota. St. Paul, Minn. (U.S.A.), Univ. Farm. *Microbiol. of butter, milk and powdered milk.* ① Cornell Univ. 1917. ⓟ Certain cultures of micro-organisms isolated from dairy products.

Macy, Icie Gertrude (1892), In charge of Nutrition Research, Merrill-Palmer School. Detroit, Mich. (U.S.A.), 71 Ferry Avenue, East. *Nutrition Research.* ① Yale Univ. 1920.

v. **Madarász,** Gyula (1858), Dirig. Custos emer. Nat. Mus. Budapest (Ungarn), Mátyás-tér 14. *Ornithol. Ungarns, Deutsch-Ost-Afrikas.* ① Budapest 1880.

Mader, Leopold (1886). Wien XIX/2 (Österreich), Schätzgasse 3. *Coleopterol., palaearkt. Coccinellidae. Parasitismus.* ⓟ Palaearkt. Insekten. Farne.

Madrid, Samuel Bernardo (1872), Prof. der Histol. an der Med. Fac., Buenos Aires (Argentinien), Luipacha 128. *Physikalische Chemie des Blutes.* ① Buenos Aires 1895. ⓟ Préparations microscopiques de tous les organs et tissus des vertébrés supérieurs.

Madrid Moreno, José, Prof., Vicedir. del Mus. Nac. de Ciencias Nat., Jefe Sección de Microbiol., Subjefe del Labor. Municipal, Catedr. de Técnica micrográfica e Histol. vegetal y animal Fac. Cienc. Madrid (España), Serrano 40. *Microbiol.* o

Maeda, Takeshige, Ass. in Botan. Labor., Coll. of Sc., Imper. Univ. Kyôto (Japan). *Cytol.*

Mägi, Jaan (1883), Prof. Tierzuchtlehre Univ. Tartu (Estland), Leiter d. Zootechn. Versuchsstat. in der Versuchswirtschaft Raadi bei Tartu. *Fütterungslehre.* ① Tartu 1925.

Maekawa, Tokujirô (1886), ao. Prof., Hokkaidô Imper. Univ. Sapporo (Japan), Inst. für Gartenbau. *Geschlechtliche und ungeschlechtliche Fortpflanzung der Pflanzen, genetisch u. physiol.* ① Univ. Sapporo 1913.

Märtens, Max (1870), Dr. med., Augenarzt. Braunschweig (Deutschland), Wilhelmi-Torwall 17. *Kehlkopfentwicklung der Amphibien. Anat. und pathol. Anat. des Auges.* ① Göttingen 1895. ⓟ Praep. aus der Anat. und pathol. Anat. des Auges.

Maestrini, D., Prof. Univ. Roma (Italia). *Fisiol.* o

Maevsky, Vladimir (1888), Katheder of Morphol. and Physiol. Odessa (U.d.S.S.R.), Pasteur-Street 24. *Physiol. of salivary secretion and of autonomic nervous system.* ① Odessa 1912.

Maffei, Siro Luigi (1874), Aiuto Ist. Botan. Pavia (Italia). *Micol., Fitopat., Sistematica, Fitopaleont.*

Magalhães, A. de (1873), Chef de service Inst. Bact. Camara Pestana, Prof. d'Hygiène Inst. Commercial et Industrial. Lisboa (Portugal). *Bact.* ① Lissabon 1913.

Magarinos Torres, Antonio Francisco (1896), Ingénieur agronome, Ass. du Service de Vigilence Sanitaire Végetale, de l'Inst. Biol. de Defesa Agric. Rio de Janeiro (Brasil), Rue Raymundo Corrêa 36. *Entomol. agric.*

Magarinos Torres, Carlos (1891), Ass. do Inst. Oswaldo Cruz. Rio de Janeiro (Brasil), Caixa Postal 926. *Anat.-Pathol.* ① Rio de Janeiro 1918. ⓟ Tissu de schizotrypanosomiasis américaine (molestia de Chagas), de granulome habronémique, de dermatite verruqueuse (chromoblastomycose).

Magdeburg, Paul Albert Wilhelm (1900), Dr. phil. nat. Leipzig N 22 (Deutschland), Eisenacher Str. 40. *Algenfloristik. Soziol. der Pflanzen, spez. der niederen Kryptogamen.* ① Freiburg i. Br. 1924. ⓟ Algen.

Maggiore, L., Prof. Univ. Roma (Italia). *Physiol. Optik.* o

Magliano, Arturo (1889), Ass. Zootecnia ed igiene zootecnica, Ist. sup. Med. Vet. Torino (Italia). *Alimentazione.*

Maglio, Carlo (1878), Doc. R. Liceo di Sondrio (Italia). *Idracarini. Rigenerazione dei Teleostei.* ① Pavia 1901. ⓟ Preparazioni d'idracarini.

Magnan, L., Dir. Labor., Cours méd. des hautes Etudes. Paris (France). *Morphol. expérim.* o

Magnus, Rudolf (1873), o. Prof. der Pharmacol., Dir. des pharmacol. Inst. der Reichs-Univ. Utrecht (Holland), Koningslaan 76. *Expérim. Pharm. Pathol. Physiol. Normale Physiol.: Zentralnervensystem, Körperstellung, Labyrinth (Vestibularorgan).* ① Heidelberg 1898. † 25. Juli 1927.

Magnus, Werner (1876), Dr. phil., ao. Prof. an der Univ. Berlin (Deutschland), Am Karlsbad 4 a. *Formbildung der Pflanzen.* ① Bonn 1901.

Mágocsy-Dietz, Sándor (1855), Dr., ö.o. Prof. der botan. Morphol. und Physiol. Budapest VIII (Ungarn), Muzeumring 4. *Pflanzenanat. und Physiol.* ① Budapest 1883. ⓟ Pflanzenoekol., teratol. Fälle.

Magoon, Charles Alden (1883), Associate Bact., Office of Horticulture, Bureau of Plant Industry, United States Department of Agric. Washington, D.C. (U.S.A.). *Bact. and physiol. research in fruit and vegetable utilization.* ① American Univ. 1924.

Magrou, M., Chef Labor., Hospice de la Salpétrière. Paris (France), 47, Boulevard de l'Hôpital. *Botan.* o

Magruder, Roy (1900), Ass. in Horticulture. Wooster, O. (U.S.A.), Ohio Agric. Experim. Station. *Plant breeding, variety testing and soil fertilization.* ① Purdue Univ. 1922.

Maguire, C. W., Lect. Univ., Curator Pathol. Mus. Birmingham (England). *Pathol.* o

v. **Magyary-Kossa,** Gyula (1865), Prof. Pharmacol., Toxikol. Budapest VII (Ungarn), Rottenbiller-u. 23, Tierärztl. Hochsch. *Toxikol.; Geschichte der Med.* ① Budapest 1889.

Maharadzé-Lomoouri, Nathalie G., Conserv. Station Expérim. Viticole. Ouriatoubani, Kahetie, Caucase (U.d.S.S.R.). *Maladies de la vigne.* o

Maheux, Georges (1889), Prof. d'Entomol., Univ. Laval, Québec, Dir. du Labor. de Biol. du Ministère de l'Agric. Quebec (Canada). *Entomol. appliquée: insectes et animaux nuisibles à l'agric., à l'horticulture, aux forêts, aux maisons, hommes et animaux.* ① Univ. Laval, Québec 1920. ⓟ Coleoptères, Orthoptères.

Mahnert, Alphons (1892), Ass. Univ.-Frauenklinik. Graz (Österreich), Goethestr. 48. *Stoffwechsel, Innere Sekretion der Geschlechtsdrüsen.* ① Graz 1916.

Mahoudeau, M. P. G., Prof. Ecole d'Anthrop. Paris (France), 15, rue de l'Ecole de Méd. *Paléont.* o

Maidl, Franz (1887), Dr. phil., Kustos am Naturhistorischen Mus. Wien I (Deutsch-Österreich), Burgring 7, *Entomol., Hymenopterol., med. Entomol.* ① Wien 1911. ⓟ Hymenoptera der Welt. Hymenopternester.

Maige, L., Prof. Univ. Lille (France). *Botan.* o

Maignon, François (1877), Prof. de Physiol. à l'Ecole Nationale Vét. d'Alfort (Seine). Paris XIII (France), 171, rue de Colbiac. *Système nerveux: pneumogastrique et spinal. Nutrition: Influence des saisons sur la glycogénie, sur les combustions respiratoires. Rôle des graisses dans la glycogénie, l'utilisation des protéines alimentaires. Diastases tissulaires. Acidose physiol. Anaphylaxie.*

Mail, George Allen (1894), Ass. Division of Entomol. Univ. of Minnesota. St. Paul, Minn. (U.S.A.), Univ. Farm. *Wireworm. Ecol.*

Maillard, M., Prof. Fac. Méd. Alger (Afrique). *Biochem.* o

Maillefer, Arthur (1880), Prof. extraordinaire. Lausanne (Suisse), Av. Montagibert 22. *Tropismes. Anat. végétale.* ① Lausanne 1906.

Mains, Edwin B. (1890), Associate Prof., Associate Botan., Purdue Univ. Agric. Experim. Sta., Agent,

Cereal Disease Investigations, U.S. Dept. of Agric. La Fayette, Ind. (U.S.A.). *Rusts (Uredinales). The physiol. specialization of the rusts and the breeding of rust resistant plants specially cereals. The nature of relation of the rusts and their hosts (resistance, susceptibility).* ⓓ Univ. of Michigan 1916.

Mainx, Felix (1900), Dr. rer. nat., Ass. am pflanzenphysiol. Inst. der deutschen Univ. Prag II (C.S.R.), Viničná 3a. *Ernährungs- und Reizphysiol. der niederen Algen, speziell der Eugleninen; experim. Physiol. der Kernteilung.* ⓓ Prag 1923.

Mainzer, Fritz (1895), Dr. med., Ass.Arzt. Altona a. d. Elbe (Deutschland), Städtisches Krankenhaus. *Physikalische Chem. des Blutes und ihre Pathol.; Reaktionsregulation, Atmungsregulation.* ⓓ Frankfurt a. M. 1923.

Mair, Rudolf (1889), Ass. der Anat. Anstalt zu Berlin NW 6 (Deutschland), Luisenstr. 56. *Bau und Wachstum des menschlichen Schädels, mit besonderer Berücksichtigung des fossilen Menschen.* ⓓ Innsbruck 1918.

Maire, René Charles Joseph Ernest (1878), Prof. de Botan. à l'Univ. (Algérie), 3, rue de Linné. *Flore de l'Afrique du Nord. Géographie Botan. de l'Afrique du Nord. Mycol. (particulièrement Basidiomycètes).* ⓓ Alger 1911. ⓟ *Plantes et Champignons de l'Afrique du Nord.*

Maisin, J., Prof. Univ. Cath. Louvain (Belgique). *Anat. Pathol.* o

Maitland, Hugh Bethune (1895), Ass. Bact. London S.W. 1 (England), Lister Inst., Chelsea Gardens. *Bact.* ⓓ Toronto 1922.

Maiwald, Kurt F. G. (1899), Dr. phil. Ass. am Agric.-chem. und bact. Inst. der Univ. Breslau 16 (Deutschland), Hansastr. 25. *Agric.-chem. Ernährung der Kulturpflanzen, Biochem., Bodenkunde.* ⓓ Breslau 1923.

Maizit, János (1883), Doc. der chem. Fac. der Lettländischen Univ. Riga (Latvija), Kurmanovstr. 11, Qu. 14. *Pharmazeutische Chem., Pharmakognosie. Filices.* ⓓ Dorpat 1910.

Majima, Rikô, Prof. Tôhoku Imper. Univ., Sc. of Inst. of Physical and Chem. Research. Sendai (Japan), Chem. Labor., Coll. of Sc. *Organic and Bio- (Phyto-) Chem.*

Major, Ralph Hermon (1884), Prof. Med., Univ. Kansas City, Kan. (U.S.A.), Bell Memorial Hospital. *Pathol. Thyroid gland. Pancreas. Renal function. Insulin.* ⓓ Johns Hopkins 1910.

Major, Thomas Grant (1898), Prof. Univ. Ottawa, Ontario (Canada), Central Experim. Farm. *Diseases of tobacco.* ⓓ McGill Univ. 1921.

Maki, Moichirô (1886), Zool. Inst., Coll. of Sc., Imper. Univ. Kyôto (Japan). *Faunistic of Formosa. Japanese Reptiles.* ⓟ Snakes and lizards in Japan esp. in Formosa.

Makino, Tomitarô, Lect., Botan. Inst., Fac. of Sc., Imper. Univ. Koishikawa-Ku, Tôkyô (Japan). *Systematic Botan.*

Makiyama, Jirô, Ass. Prof. of Imper. Univ., Geol. Inst., Coll. of Sc. Kyôto (Japan). *Paleo-zool.*

Makrinov, Ivan A., Chef des trav. Inst. Méd. Expérim. Leningrad (U.d.S.S.R.), rue Lopouchinskaia. *Champignons lignicoles.* o

Maksimov, N. A., Doc. Univ., Abt. f. Biol. Leningrad (U.d.S.S.R.). *Botan.* o

Makušok, Marhel Emel., Doc. I. Univ., Ass. Anat. Labor. Moskau (U.d.S.S.R.). *Zool., Embryol.* o

Malan, D. E., Prof. Transvaal Univ. Coll. Pretoria (South Africa). *Zool.* o

Malandkar, Mangesh Anant (1892), Ass. Biochem., Biochem. Unit., Haffkine Inst. Parel, Bombay (India). *Biochem.* ⓓ Bombay 1915.

Malaquin, M., Prof. Fac. Sc. Lille (France). *Zool.* o

Malarski, Henryk (1887), Dr. phil., Leiter der Abteilung für die Fütterung der Tiere in dem wissenschaftl. agric. Inst. i. Puławy (Polska). *Chlorophyllgruppe. Biochem., Fütterung der Tiere, Analysen der Futterstoffe mit neuen mikrochem. Methoden. Berechnung der Futterrationen nach den neuen wissenschaftlichen Grundlagen speziell für das Geflügel. (Eiweißqualität, Mineralstoffe, Vitamine.)* ⓓ Kraków 1910.

Malcolm, John (1873), Prof. of Physiol. Univ. of Otago, Dunedin (New Zealand). *Nutritive value, composition, vitamin content of food stuffs especially N. Z. fishes.* ⓓ Edinburgh 1899.

Malczewski, Wladimir Pawlowitsch (1886), Ass. Pflanzenanat. und Physiol. des Forstinst. Leningrad (U.d.S.S.R.). *Anat. und Physiol. (Photosynthese) der Holzgewächse.* ⓓ St. Petersburg 1912.

Maleev, Vladimir (1894), Ass. Botan. Yalta (U.d. S.S.R.), Botan. Garden of Nikita. *Sistematic Botan. and Geobotan.*

Maleew, Wladimir P., Doc., Industriel. Technikum Suchum (U.d.S.S.R.), Puschkinstr., Haus Schabranow. *Botan. Pflanzen-Physiol.* o

Malenchini, Fernando, Prof. Univ. Nac. La Plata (Argentina). *Anat. experim., Embryol. Histol.* o

Malengreau, Fernand (1880), Prof. de Biochim. à la Fac. de Méd. de l'Univ. de Louvain (Belgique). *Synthèse et propriétés physiol. des aminoalcool. Vitamines B-Bios.* ⓓ Louvain.

Malenotti, Ettore (1887), Lib. Doc. Entomol. Agraria Ist. Sup. Agr. Milano, Dir. Osserv. Region. di Fitopatol. Verona (Italia), Via G. Mameli 3. *Insetti dannosi alle piante coltivate. Agriotidi; Grillotalpe; Anomala vitis.* ⓓ Firenze 1920.

Maleš, Branimir (1897), Ass. Inst. d. vergl. Physiol. Univ., Chef des Labor. für experim. Psychol. Centr. Hygien. Inst. Beograd (S.H.S.), Vojvode Milenka 7. *Stoffwechsel der Poikilothermen; vergl. Psychol. d. Protozoen.*

Malfatti, Hans, Univ.Prof. Innsbruck (Deutsch-Österreich). *Pharmacol.* o

Malinowski, Edmund (1885), Dr. Sc., Prof. of Genetics, Coll. of Agric. Warszawa (Polska), 23 Miodowastr. *Genetics.* ⓓ Geneva, 1910.

Maljutitzkij, Nikolaj K., Prof. d. Landwirtsch. Inst. Kiew (U.d.S.S.R.), Polytechn. Inst. No. 1, W. 8. *Pflanzen-Physiol. Genetic. Selektion.*

Malkovský, Karel Maria (1898), Conserv. of the Department of Botan., National Mus. Praha II (C.S.R.), Václavské nám. 1700. *Plant Anat. Cytol., Histol. Plant Physiol. Regeneration, Influence of Centrifugal Force. Microtechnical Methods.* ⓓ Praha 1922. ⓟ Microscopical Preparates on Plant Anat.

von Mallász, Josef (1875), Mus.Dir. Deva, Siebenbürgen (România). *Blütenbiol. Coleoptera, Mundwerkzeuge u. Kopulationsorgane der Caraben. Hydrobiol.* ⓟ Siebenbürgische Caraben.

Mallet-Guy, Pierre Albert (1897), Prosecteur Fac. méd. Inst. d'anat. Lyon (France), 7, Rue Servient. ⓓ Lyon 1926.

Mallmann, Walter Le Roy (1895), Ass. Prof. of Bact. East Lansing, Mich. (U.S.A.), Dept. of Bact., Michigan State Coll. *Bact. of water and sewage, Bacteriophage and diseases of chickens, particularly bacillary White Diarrhea.* ⓓ Michigan State Coll. 1922.

Mallory, Frank Burr (1862), Pathol. to the Boston City Hospital. Boston, Mass. (U.S.A.). *Morphol. pathol.* ⓓ Harvard 1890. ⓟ Microscopic sections of interesting lesions.

Malméjac, Jean Léopold Henri (1903), Prépar. de recherches labor. de Physiol. Fac. Méd. Alger (Afrique), 21, Boulevard Baudin. *Mechan. de l'inhibition du cœur par excitation du vague. Surrénale.*

Maloch, Franz (1862), Prof. Plzeň (C.S.R.), Karlova 27. *Flora von Pilsen.*

Malone, R. H., Ass. Dir., Bact. Labor Bombay (Brit.-India). o

Malowitschko, Eugenia Eustachiewna (1882), Oberass. Histol. Univ. Odessa (U.d.S.S.R.), Staroportofrancusskaja 41. *Protoplasmastruktur. Nebennieren u. Schilddrüse.* ⓓ Odessa 1917.

Malsburg, K., Dr. Prof. Lwów (Polska), Dublany. *Histol. Embryol.* o

Malta, Nikolajs (1890), Doc. Riga (Latvija), Kronvalda bulv. 4. *Systematik und Geographie der Bryophyten. Ökol. der Felsenpflanzen, spez. Kryptogamen.* ⒟ Riga 1925. ⒠ Moose.

Malte, Oscar (1880), Chief Botan., National Herbarium of Canada. Ottawa, Ont. (Canada). *The Phanerogamic Flora of Canada, with specialization in grasses.* ⒟ Lund, 1910. ⒠ Plants.

Freiherr von Maltzan, Friedrich Franz (1886), Dr. Neuhof bei Penzlin, Mecklenburg (Deutschland). *Bekämpfung der Rübenfliege und des Maikäfers.* ⒟ Berlin 1919.

Malvor, E., Prof., Fac. med. Dir. Inst. bact. Liège (Belgique), 1, rue des Bonnes-Villes. o

Maly, Karlo, Kustos d. Botan. Abt. des Bosnischherzegowischen Landesmus. Sarajevo (S.H.S.). *Pteridophyta.* o

Malyshev, Sergius J. (1884), Dir. Zoopsychol., Experim. Sta. at Borissovka Kursk gov. (the branch of P. F. Lesshaft Inst. of Sc.), Doc. of Univ. Leningrad (U.d.S.S.R.), Anglijsky 31, Lesshaft Inst. *Insec behavior, esp. habits of bees and wesps. Zoopsychol., esp. Instincts. Honeybees, their nat. hist. and culture.* ⒟ Dorpat 1914. ⒠ Solitary bees and wesps and their nests.

de Man, Johannes Govertus (1850). Yerseke, Provinz Zeeland (Holland). *Die Decapoda Macrura der Siboga-Expedition, Callianassidae, Atyidae, Hippolytidae, Palaemonidae, Cyclometopa. Freilebende Nematoden.* ⒟ Leiden 1873.

Manaresi, Angelo (1881), Prof. d'Arboriculture et d'Horticulture à l'Ist. superiore agrario. Bologna (Italia). *Biol. florale des arbres fruitiers et de la vigne. Maturation des fruits. Pathol. des arbres.*

Manasse, Paul (1866), o.ö. Prof. an der Univ. Würzburg (Deutschland), Frauenlandstr. 12. *Pathol. Anat. des Ohres, der Nase und des Kehlkopfes. Pathol. Anat. der Tuberkulose der oberen Luftwege. Physiol.: die Empfindlichkeit des Trommelfells und der Paukenhöhlenschleimhaut für äußere Reize.* ⒟ Straßburg i. Elsaß 1890.

Mancini, Mario Ajazzi (1887), Libero Doc. di Farmacol. presso la Univ. Firenze 15 (Italia), Via Alfani 33.

Mandel, John Alfred (1865), Prof. New York Coll. Vet. Yonkers, N.Y. (U.S.A.), 496 Warburton Ave. *Physiol. chem.* o

Mandelbaum, Martin (1881), Dr., Oberarzt und Leiter der serol. und bact. Abteilung des Krankenhauses München-Schwabing (Deutschland), Kölner Platz 1 oder Elisabethstr 7 II. *Structur des Plasmas und Serums; Diphtherie, Masern, Scharlach.* ⒟ München 1910.

Mandereau, L., Prof. Ecole Méd. Besançon (France). *Anat.*

Mandl, Karl (1891), Ass. f. Botan. u. Warenkunde, Techn. Hochsch. Wien V (Deutsch-Österreich), Einsiedlergasse 60. *Mineralische Einlagerungen in pflanzlichen Geweben. Cicindelae.*

Mandoul, Antoine Henri (1872), Prof. de Zool. et Parasit., Fac. de Méd. de Bordeaux. Talence (France), „Amata" rue Pierre Curie. *Parasit Coloration tégumentaire.* ⒟ Bordeaux 1920.

Maneval, Willis Edgar (1877), Ass. Prof. of Botan., Univ. of Missouri. Columbia, Mo. (U.S.A.), 305 Hicks Ave. *Plant Pathol. and Mycol. Longevity and conditions for germination of spores of rusts and other fungi; growth of excised root tips under sterile conditions. Plant Disease.* ⒟ Johns Hopkins Univ. 1912. ⒠ Fungi.

Maney, Thomas Joseph (1888), Chief, Pomology Section, Iowa State Coll. Ames, Ia. (U.S.A.), ⒠ Plants and scion wood of various Pomological fruits.

Mangelsdorf, Paul C. (1899), Ass. Geneticist, Conn. Agric. Experim. Station. New Haven, Conn. (U.S.A.). *Heredity in Maize.* ⒟ Harvard Univ. 1925.

Mangin, M., Dir. Mus. Paris (France), 57, rue Cuvier. *Botan.* o

Mangold, Ernst (1879), Dr. med. et phil., o. Prof. der Tierphysiol. an der Landwirtschaftl. Hochsch. Berlin, Dir. des Tierphysiol. Inst. der Landwirtsch. Hochschule. Berlin N 4 (Deutschland), Invalidenstraße 42. *Allg. und vergl. Herzphysiol.; Physiol. der Verdauungsorgane und ihrer Innervation; Starrezustände der Muskulatur, Mechanism. und Chemism.; Physiol. der glatten Muskulatur; Härtemessung der Organe; respirat. Stoffwechsel; Hungerstoffwechsel; Blutgerinnung.* ⒟ Jena med. 1903, phil. 1905.

Mangold, Otto (1891), Dr., Priv.Doc. Univ. Berlin, Abt.-Leiter Kaiser-Wilhelm-Inst. für Biol. Dahlem bei Berlin (Deutschland), Boltzmannstr. 2. *Entwicklungsmechanik: Determinationsproblem. Physiol.: Niedere Sinne niederer Tiere.* ⒟ Freiburg i. Br. 1919.

Manguat, P., Prof. Univ. Angers (France). *Zool. des vertebr.* o

Manguikian, Tigran, Aspirant Wiss. Forschungs-Inst. Odessa (U.d.S.S.R.), Remeslenstr. 19. *Süßwasser-Molluskenfauna aus dem tertiären und quartären Becken der Südukraine.* ⒠ Conchylien aus den Kujalnikablagerungen bei Odessa.

Mankowski, Alexandre (1868), Prof. ord. à la Fac. de Méd. à l'Univ. de Sofia (Bulgarie). *Morphol. de la Sécrétion interne; spécialement du pancréas.* ⒟ 1894.

Mankowsky, Marie (1899), Dr., Ass. Sofia (Bulgarie), Rue Dobronelja N 2. *Histol. et Embryol.* ⒟ Sofia 1924.

Mann, Frank Charles (1887), Dir., Inst. of Experim. Med., Mayo Foundation, Univ. of Minnesota. Rochester, Minn. (U.S.A.). *Physiol., pathol.* ⒟ Indiana Univ. 1913.

Mann, Sydney Andrew (1882), Biochem. Central Pathol. Labor. London S.E. 5 (England), Mandsley Hospital. ⒟ London 1918.

Mann, William (1886), Dir. Nat. Zool. Park., Hon. Custodian N. S. National Mus. Washington, D.C. (U.S.A.). *Formicidae, Myrmecophiles, Zoögeographie.* ⒟ Harvard 1915.

Manninger, Rudolf, ao. Prof., Dr., Dir. d. Inst. für Seuchenlehre, Tierärztl. Hochsch. Budapest (Ungarn), Rottenbillergasse. o

Manns, Thomas Franklin (1876), Plant Pathol. and Soil Bact. Univ. of Delaware. Newark, Del. (U.S.A.). *Relation of soil flora to crop production.* ⒟ Philadelphia, Pennsylvania, 1914. ⒠ Diseases of Sweet Potato.

Mannsfeld, Géza, Prof. Dr., Dir. d. Pathol. u. des Pharm. Inst. Univ. Pécs (Ungarn). *Pathol.* o

Mannsfeld, Wilhelm (1903), Leiter Ichtyol. Arbeiten am Fischerei-Amt, Lettländ. Landwirtsch.-Minist. Riga (Latvija), Antonienstr. 8, W. 8. *Systematik und Geographie der Fische Lettlands (spez. Clupea harengus, Salmo salar, Coregonus albula). Hirudinea.* ⒠ Plankton Lettländischer Gewässer. Hirudinea Lettlands.

Mannu, Andrea (1873), Prof. Anat. Vet. Parma (Italia), Borgo Carissimi 10. *Morfol. e Embriol. del Sistema arterioso.* ⒟ Parma 1923.

Manoliu, Eugénie (1898), Ass. Labor. d'Hygiène Fac. Med. Jassy (Roumanie). *Bact., Variations microbiennes.* ⒟ Jassy 1924.

Manouvrier, M., Dir. du Labor. d'Anthrop. Paris (France), 1, rue Clovis. *Anat.* o

Mansfeld, Karl (1897), Dr. phil., Leiter der Versuchs- und Musterstation für Vogelschutz, Burg Seebach, Kr. Langensalza (Deutschland). *Angewandte Zool., Vogelschutz.* ⒟ Berlin 1922.

Mansfeld, Rudolf (1901), Dr. phil., Ass. am Botan. Mus. Berlin-Dahlem (Deutschland). *Systematik der Phanerogamen, Orchideen, Euphorbiaceen u. a. Dicotylenfamilien.* ⒟ Berlin 1924.

Manskaja, Sophie M., Ass., Krimsches Pädagog. Inst. Simferopol (U.d.S.S.R.), Bitakskaja 2, W. 8. *Botan., Pflanzen-Physiol.* o

Manson-Bahr, Philip Henry (1881), Lect. School of Hygiene and Trop. Med. London (England), 32 Weymouth Street W. 1. *Filaria. Schistosoma.* ⓓ Cambridge 1914.

Mantegazza, Ambrogio (1898), Ass. Ist. Farmacol., Pavia (Italia), Via L. Mascheroni 6. *Distribuzione dell'arsenico negli organi. Eliminazione di bromuri e joduri.* ⓓ Pavia 1926.

Manter, Harold W. (1898), Ass. Prof. of Zool., Univ. of Nebraska. Lincoln, Neb. (U.S.A.), 204 Bessey Hall. *Trematodes. Parasit. of domestic and lower animals.*

Manteufel, Alexandra Ja., Ass., Forst-Inst. Moskau (U.d.S.S.R.), Ostoshenka 8, W. 4. *Botan.* ○

Manteufel, Paul (1879), Prof. Dr. med., ORR., Mitglied im Reichsgesundheitsamt. Berlin-Dahlem (Deutschland), Reichsgesundheitsamt, Bact. Abt. *Serol., Chemotherapie,* ⓓ Halle a. S. 1904.

Mantovani, Mario, Prof. incar. Univ. Bologna (Italia). *Bact.* ○

Manujlov, T. M., Prof. Landwirtschaftl. Inst. Voronez (U.d.S.S.R.). *Anat. u. Physiol. d. Tiere.* ○

Manwaring, Wilfred H. (1871), Prof. Bact. and Exper. Pathol., Stanford Univ. Cal. (U.S.A.), 3 Bact. Bldg. (Pt. 900 Local 19), 364 Kingsley, Palo Alto (PA 1500). *Experim. functional pathol., Immunol.* ⓓ Johns Hopkins Univ. 1904.

Marburg, Otto (1874), a. ö. Prof. der Wiener Univ., Vorstand des Neurol. Inst. der Univ. Wien I (Deutsch-Österreich), Operngasse 4. *Normale, vergl. und pathol. Anat. des Nervensystems.* ⓓ 1899.

Marceau, Francis (1871), Prof. de Zool. à la Fac. des Sc. et à l'Ecole de Méd. de Besançon (France), 10 bis rue de la Convention. *Histol. et Physiol. du système musculaire.* ⓓ Besançon 1923. ⓟ Prép. d'histol. (système musculaire).

March, A. W., Lect. Christian Coll. Hangchow (China). *Zool.* ○

Marchal, Emile, Julius Jos. (1871), Prof. Inst. agron. Etat. Dir. Stat. de Phytopathol. de l'Etat Gembloux (Belgique), 46, Chaussée de Namur. *Physiol. et Pathol. végét.: Champignons paras. des plantes cultivées.*

Marchal, Paul (1862), Membre de l'Inst., Prof. Zool. appliquée à l'Agric. Inst. Nation. Agron., Dir. Stat. Entomol. Paris (France), 16, rue Claude Bernard. *Entomol. agric. Aphidés. Hyménoptères parasit. Entomol. générale. Crustacea.* ⓓ Paris.

Marchand, Felix (1846), GMR. Univ. Prof. emer. Leipzig (Deutschland), Goethestr. 6. *Pathol. Anat.* ○

Marchese, Liborio, Lib. doc. Univ. Catania Sicilia (Italia). *Anat. umana topogr.* ○

Marchesini, R., Prof. Univ. Roma (Italia). *Istol. e tecnica microl.* ○

Marchiafava, E., Prof. Univ. Roma (Italia). *Anat. patol.* ○

Márcis, Árpád, Dr., Priv.Doc. d. Serol., Inst. f. Seuchenlehre, Tierärztl. Hochsch. Budapest (Ungarn), Rottenbillergasse.

Marchlewski, L., Dr. phil., Prof. Univ. Kraków (Polska), Univ., Fac. méd., r. Kopernika 7. *Biochem.* ○

Marchoux, Emile, (1862), Prof. Inst. Pasteur. Paris (France), 96, rue Falguière. *Bact., parasit., épidémiol. et thérapeutique tropicales. Paludisme, Lèpre, Spirochétoses.*

Marcotte, Léon (1882), Prof. de Sc. nat. l'Univ. de Montréal. Sherbrooke (Canada), Séminaire. *Batraciens, Urodèles.* ⓓ Montréal 1926. ⓟ Oiseaux de la Province de Québec.

Marcovitch, André (1874). Sofia 8 (Bulgarie), Victor Grigorevitch No. 5. *Biol. des insectes nuisible et systématique des lépidoptera.* ⓓ Sofia 1894.

Marcovitch, Simon (1890), Entomol. Knoxville, Tenn. (U.S.A.), Univ. of Tennessee. *Insects and insecticides.* ⓓ Minnesota 1916.

Marcu, Orest (1898), Ass. Zool. Inst. Univ. Cernăuți (România). ⓓ 1926.

Marcucci, Ermete (1876), Aiuto e Lib. doc. Anat. e Fisiol. compar. Napoli (Italia), Palazzo Medievale a Mezzocannone. *Fenomeni di rigenerazione nei Rettili e negli Anfibi.*

Marcus, Ernst Gustav Gotthelf (1893), Priv.Doc. a. d. Univ., Ass. a. Zool. Inst. d. Univ. Berlin N 4 (Deutschland), Invalidenstr. 43. *Tardigrada. Tiergeographie, Coleoptera. Bryozoa (Systematik, Verbreitung, Embryol., Physiol.).* ⓓ Berlin 1919.

Marcus, Harry (1880), a.o. Universitätsprof. und Konserv. an der Anat. München (Deutschland). *Muskelhistol. und vergl. Entwicklungsgeschichte: Gymnophiona. Vergl. Anat. der Lungen.* ⓓ München 1904.

Marden, Aaron (1889). So. Harpswell, Maine. (U.S.A.). *Oology, variation in coloration of eggs: relation of the arrival of birds to the time of nesting, the cause of migration, ornithol.*

Maréchal, Paul (1889), Dr. en sc. nat., Prof. à l'Athénée Royal de Liège (Belgique), rue de Campine 48. *Biol. des Insectes, Hyménoptères.*

Marek, Emil, Ing., Ass. l'Inst. des Recherches agric. et Forestière. Brno (C.S.R.), Květná ul. *Biochem.*

Mares, Franz, Dr., Prof. d. Physiol. Praha II (C.S.R.), Katerinska 32, Physiol. Inst. d. Karls-Univ. ○

Maresch, Rudolf (1868), o. ö. Prof., Leiter des Pathol.-anat. Inst. der Univ. Wien IX (Deutsch-Österreich), Währingerstr. 6/8. *Pathol. Anat.* ⓓ Prag 1895.

Margaria, Rodolfo (1901), Labor. Fisiol. Univ. Torino (Italia), Via Carlo Alberto 35. *Fisiol. delle arteria.* ⓓ Torino 1924.

Marginowsky, Eugène (1874), Dir. de l'Inst. tropical. Prof. Univ. II. Moscou (U.d.S.S.R.), B. Tulskaya 77a. *Protoplasma, Leishmania, Malaria, Spirochaeta.* ⓓ Moscou 1899. ⓟ Prép. des Protozoa. Microphotographies des Protozoa, helmints.

Marl, Mario (1870), Prof. Arezzo (Italia), R. Liceo Petrarca. *Istol. degli organi generativi, spermatogeneti dei Decapodei. Teratol.: La derodimia e la deradelfia dei mammiferi.* ⓓ Roma.

Maria, Hermano Apolinar, Prof. Univ. Nac. Bogotá (Colombia). *Zool. gen.* ○

Mariani, E., Prof. R. Scuola d'ingegn. Milano (Italia). *Palaeont.* ○

Mariante, T., Prof. Univ. Porto Alegre (Brasil). *Pathol.* ○

Mariconda, P., Prof. Univ. Sassari (Italia) e Roma (Italia). *Anat. patol.* ○

Marie, A., Prof., Service de microbiol. de l'Inst. Pasteur. Paris (France), Rue Dutot 21-25. ○

Marie-Victorin, Frère (1885), Membre de la Société Royale du Canada, Prof. de Botan. à la Fac. des Sc. de l'Univ. Montréal (Canada), 1265, rue Saint-Denis. *Botan. générale; systématique. Flore phanérogamique du Québec. Filicinées, Lycopodinées.* ⓓ Québec 1901. ⓟ Plantes phanérogames du Québec.

Marine, David (1880), Dir. of Labor., Montefiore Hospital, Ass. Prof. Pathol., Columbia Univ. New York City (U.S.A.), Montefiore Hospital, Gun Hill Road near Jerome Ave. *Physiol. and pathol. of the ductless glands.* ⓓ Johns Hopkins Univ. 1905. ⓟ Collection illustrating most of the morphol. changes in the thyroid gland.

Mark, Edward Laurens (1847), Hersey Prof. emeritus of Anat., Harvard Univ., and Dir. of the Bermuda Biol. Station for Research. Cambridge, Mass. (U.S.A.), Mus. of Compar. Anat., or 109 Irving Street. *Vergl. Anatomie u. Biol.* ⓓ Leipzig 1876.

Markelov, I. I., Prof. Univ. Taschkent (U.d.S.S.R.). *Pharmazeut. Chem., Pharmacol.* ○

Markgraf, Friedrich (1897), Dr. phil., Ass. am Botan. Mus. Berlin-Dahlem (Deutschland), Königin-Luise-Str. 6—8. *Vegetationskunde, Pflanzengeographie, Systematik (besonders Apocynaceae).* ⓓ Berlin 1921.

Markle, Millard S., Prof. of Botan., Earlham Coll. Earlham, Ind. (U.S.A.). *Microtechnique.* ○

Markley, Klare Stephen (1895), Ass. Biochem. Washington, D.C. (U.S.A.), U. S. Dept. of Agric., Bureau of Plant Industry. *Biochem. study of the activity of soil flora under the influence of crop rotation.* ⓓ Washington, D.C., 1925.

Markoff, Jovan P. (1872), o. Prof. Agr. Fac. Univ. Beograd (S.H.S.), Zootechn. Inst., Dobračina 16. *Anat. u. Physiol. d. landw. Nutztiere; Allgemeine Tierzucht; Zoohygiene.* ⓓ Warschau 1908.

Markoff, Wladimir (1883), Prof., Dir. Bact. Inst. Med. Fac. Univ. Sofia (Bulgarien), ,,6. September" No. 15a. *Serol.* ⓓ Berlin 1908. ⓟ Bakterien- u. Hefenkulturen.

Markowa, Marie Fedorowna (1901), Aspirant Station de Pathol. végétale de l'Inst. Agronomique Leningrad (U.d.S.S.R.), Detskoe Selo. *Mycol. et pathol. végétale: biol. des Ustilaginées. Bouillies fongicides (toxicité).* ⓟ Champignons.

Markowski, J., Dr. Prof. Lwów (Polska), Univ., Inst. d'Anat., r. Piekarska 52. o

Markowskij, Jurij M., Dnepropetrowsker Biol. Station d. All-Ukrain. Akad. d. Wissensch. Kiew (U.d. S.S.R.), Ul. Lenina 9, W. 4. *Hydrobiol.* o

Markus, František, Dir. de l'Inst. de Pomol. Bohonice u Brno (C.S.R.). *Pomol.* o

Marlatt, Charles Lester (1863), Assoc. Chief in charge regulatory work Bur. of Entomol. Washington, D.C. (U.S.A.), Dept. of Agric. *Entomol.* o

Marmier, Louis Jean Alexis (1865), Dir. adjoint de l'Inst. Pasteur. Lille (France), 18, Bd. Louis XIV. *Physique biol.* ⓓ Paris Sc. 1895, méd. 1896.

von Marochino, Vincenz (1882), Doc. Univ. Zagreb (S.H.S.), Palmotićeva 28. *Protozoa.* ⓓ Wien 1909.

v. Maros, Tibor (1893), Dr., Ass. Univ. Szeged (Ungarn), Kalvarienplatz 5. *Biochem.*

Marquard, Otto (1890), Dr., Preuß. Oberfischmeister für die ostpommersche Hochsee- und Küstenfischerei. Stolpmünde i. Pommern (Deutschland), Villa Eldorado, Eldoradoweg 3. *Hydrobiol., Biol. der Meeresfische und ihrer Nahrungstier- und Pflanzenwelt. Limnol. Fischereibiol.* ⓓ Münster (Westf.) 1923. ⓟ Meeresfische (Ostsee); Nahrungstiere der Süßwasserfische.

Marr, J. E., Prof. Univ. Cambridge (England). *Geol., Palaeont.* o

Marro, G., Doc. Univ. Torino (Italia). *Antrop.* o

Marsh, Charles Dwight (1855), Physiol. in Charge of Investigations of Stock Poisoning by Plants. Washington, D.C. (U.S.A.), U.S. Dept. of Agric. *Poisoning by plants. Copepoda.* ⓓ Univ. of Chicago 1904.

Marsh, Millard C. (1872), Biol., State Inst. for the Study of Malignant Disease, Buffalo, N.Y. Springville, Erie County, N.Y. (U.S.A.), Biol. Station. *Experim. Cancer research.* ⓓ George Washington Univ. 1905.

Marsh, Ray Stanley (1894), Ass. Prof. of Horticulture. Urbana, Ill. (U.S.A.), Univ. of Illinois. *Nutrition fruit crops.* ⓓ Univ. of Missouri 1921.

Marsh, Robert P., Head, Dept. of Biol., Gettysburg Coll. Gettysburg, Pa. (U.S.A.). *Physiol.* o

Marshall, Charles Edward (1866), Prof. of Microbiol. Dir. of the Graduate School, Massachusetts Agric. Coll. Amherst, Mass. (U.S.A.). *Microbiol.* ⓓ Univ. of Michigan 1902.

Marshall, E. Kennerly Jr. (1889), Prof. of Physiol., The Johns Hopkins Univ., School of Med. Baltimore, Md. (U.S.A.), The Johns Hopkins Med. School. *Urinary secretion, circulation.* ⓓ Johns Hopkins Univ. 1917.

Marshall, Francis Hugh Adam (1878), Reader Agric. Physiol., S. Dir. Inst. of Animal Nutrition. Cambridge (England), School of Agric. *Physiol. of Reproduction.* ⓓ Sc. Edinburgh 1904, Cambridge 1912.

Marshall, John F., Dir. British Mosquito Control Inst. Hayling Island, Hants. (England). *Mosquitoes.* o

Marshall, Max Skidmore (1895), Ph.D., Ass. Dir., in Charge of Investigation, of the Bureau of Labor. Michigan Dept. of Health. Lansing, Mich. (U.S.A.). *Biol. Products, bact.* ⓓ Ann Arbor, Michigan 1925.

Marshall, P., Dr., M.A., Lands and Survey Dept. Wellington (New Zealand). *Palaeont.* o

Marshall, Roy Edgar (1891), Associate Prof. and Research Assoc. in Horticulture. East Lansing, Mich. (U.S.A.) Michigan State Coll. *Fruit Ripening and Cold Storage, Pruning Investigations, Pollination, Systematic Pomol.* ⓓ Univ. of Nebraska 1913.

Marshall, Rush P. (1891), Ass. Pathol., Forest Pathol. New Haven, Conn. (U.S.A.), Osborn Botan. Labor., Yale Univ. *Healing and treatment of tree wounds. Chestnut blight, potato wart.* ⓓ Univ. Wisc. in Plant Pathol. 1924.

Marshall, Ruth (1869), Prof. of Zool. Rockford, Ill. (U.S.A.) Rockford Coll. *Hydracarina.* ⓓ Univ. of Nebraska 1907. ⓟ Hydrachnids.

Marshall, William Stanley (1866), Prof. of entomol. Univ. of Wisconsin. Madison, Wis. (U.S.A.), 139 E., Gilman St. o

Martelli, Giovanni (1877), Ispettore per le malattie delle piante, Dir. R. Osservatorio Fitopat. per le Puglie. Taranto (Italia), Piazza Ebalia 1. *Insetti.* ⓓ Portici 1920.

Martelli, Ugolino (1860), Prof., lib. doc., ass. Botan. Univ. Pisa. Firenze (Italia), Via di Soffiano No. 15. *Sistematica. Biol. Fanerogame. Pandanacee.*

Martelli, V., Prof. Univ., Lib. doc. Univ. Cagliari. Sassari (Italia). *Botan.* o

Martens, Pierre (1895), Prof. à l'Univ. (Fac. des Sc., Dépt. Botan.). Louvain (Belgique), 23, rue Marie Thérèse. *Cytol., chromosomes végétaux, Noyau. Mycol. générale: systématique, développement, reproduction, cytol. des champignons. Relations entre fleurs et insectes, Paléobotan. Ptéridophytes et Gymnospermes. Morphol., anat., développement, cytol., évolution.* ⓓ Louvain 1920.

Martin, A., Prof. Ecole Nat. Vét. Toulouse (France). *Botan. méd.* o

Martin, C. J., Prof. Univ., Dir. Dep. of Pathol. Lister Inst. London S.W. 7 (England), S. Kensington. *Experim. Pathol.*

Martin, Ch., Prof. Ecole prép. de Méd. et de Pharmacie Angers (France). *Anat.* o

Martin, Ella M., Ass. Prof. of Botan. Illinois Wesleyan Univ. Bloomington, Ill. (U.S.A.). *Mycol.* o

Martin, George Hamilton (1887), Ass. Pathol., Bureau of Plant Industry, Department of Agric. Washington, D.C. (U.S.A.). *Plant diseases.* ⓓ Washington State Coll. of Agric. 1916.

Martin, George Willard (1886), Associate Prof. of Botan., Univ. of Iowa. Iowa City, Ia. (U.S.A.). *Mycol., Marine ecol.* ⓓ Chicago 1922. ⓟ Fungi.

Martin, Harry Mathias (1892), Ass. Prof. of. Animal Pathol. and Hygiene, Univ. of Nebraska Lincoln, Neb. (U.S.A.), Dept. of An. Pathol. and Hygiene. *Parasitol. of domestic animals.* ⓓ Univ. of Nebraska, Lincoln, Neb. 1923.

Martin, John H. (1893), Associate Agronomist in Charge of Grain Sòrghum and Broomcorn Investigations. Washington, D.C. (U.S.A.), Bureau of Plant Industry, U. S. Department of Agric. *Breeding, culture, physiol., genetics, taxonomy, and adaptation of grain sorghums and broomcorn varieties.* ⓓ Univ. of Minnesota 1926. ⓟ Seed of grain sorghum and broomcorn varieties.

Martin, John N., Prof. of Morphol. and Cytol., Iowa State Coll. Ames, Ia. (U.S.A.), 507 Welch Ave. o

Martin, Joseph F. (1887), Dr. en Méd., Préparateur d'Anat. Pathol. à la Fac. de Méd. de Lyon (VI. Arrondissement) (France), 3, Rue Pierre Cornellie. *Anat. pathol., Histopathol., Cancérol.* ⓟ Prép. histol. de biopsies de tumeurs humaines ou animales.

Martin, Paul (1861), Geh.Med.R., o. Univ.-Prof. für Tieranat. Gießen (Deutschland), Keplerstr. 5 II. *Magen- u. Darmentwicklung.* ① Zürich 1894.

Martin, William Hope (1890), Prof. of Plant Pathol., Rutgers Univ. New Brunswick, N.J. (U.S.A.), New Jersey Agric. Experim. Station. *Potato and fruit diseases.* ① Rutgers Coll. 1919.

Martin Lecumberri, Esteban, Prof., Dir. y Catedrático. Figueras (España). *Diatomáceas, Microfotografía.* o

Martin-Rosset. Lyon (France), Labor. Botan., Fac. des Sc. *Action du p_H sur la végétation et chim. végétale.*

Martinescou, George (1874), Prof. à la Fac. de Méd. Cluj (Roumanie), Rue Dr. V. Babés 8. *Pharmacodynamie (Digitale et action cardio-vasculaire. Action de la Coféine sur l'excitabilité musculaire.* ① 1897.

Martinez, F. M., Prof. Univ. Central. Caracas (Venezuela). *Zool.* o

Martinez de la Escalera, Manuel (1867), Prof. agr. Labor. Entomol. du Mus. Nat. de Hist. Natur. Madrid (España). *Coléoptères d'Espagne, d'Afrique, Ténébrionides.* ℗ Coléoptères d'Espagne et du Maroc.

Martínez y Fernández Castillo, Antonio, Prof., Catedrático en el Inst. de San Isidro. Madrid (España), Ferraz 84. *Entomol. e Histol.* o

Martinez Gámez, Vicente, Prof. Catedrático. Cartagena (España). *Ornitol. de España.* o

Martini, Erich (1880), Abt.-Vorsteher am Tropeninst., Univ.-Prof. Hamburg 20 (Deutschland), Tarpenbekstr. 96 I. *Med. Entomol. Zellkonstanz.* ℗ Phil. Rostock 1902, med. 1905. ℗ Culicidae.

Martinotti, Giovanni (1857), Prof. o. di Anat. patol. Univ. Bologna 17 (Italia), Via Dante 2. *Tubercolosi. Vaccinazione antitubercolare. Storia della Med.* ① Torino 1880.

Martynov, Avenir Nik., Doc. I. Univ. Moskau (U.d.S.S.R.). *Zool. u. Embryol.* o

Martynov, W. F., Prof., Dir. Histol. Labor. Med. Inst. Leningrad (U.d.S.S.R.), Str. Lew Tolstoy. o

Martynow, Andreas B. (1879), Conserv. Zool. Mus. Acad. Sc. Leningrad (U.d.S.S.R.). *Entomol.: Systemat. of Trichoptera, Neuroptera, morphol. of Insects: wings of insects, thorax, genital structure. Paléoentomol.: Fossil insects (mesozoic and permian). Crustacea: Amphipoda (Gammaridea), Isopoda, Schizopoda, Cumacea. Faune of Russia. Ecol. of Crustacea of the rivers.* ① Warszawa 1911. ℗ Trichoptères de la Russie. Malacostracés des rivières Don et Dnjepr.

Marumo, Nobukatsu, Ass. Prof. Tôkyô Imper. Univ. Komaba near Tôkyô (Japan), Zool. Labor., Coll. of Agric. *Systematic Entomol. (Lepidoptera, Heterocera).*

Marvasi, Antonio (1900). Macerata (Italia), Ist. Tecnico Nazionale Crugoli. *Malacol. d'acqua dolce.* ① Modena 1922.

Marwick, J., M. A. Geol. Survey. Wellington (New Zealand). *Palaeont.* o

Marzocchi, V., Doc. Univ. Torino (Italia). *Parasit.* o

Masai, Yasunaga, Prof. Ôsaka Prov. Med. Univ. Ôsaka (Japan), Labour-Physiol. Inst. *Arbeitsphysiol.*

Masarowitsch, W. A., Doc. Bergakademie. Moskau (U.d.S.S.R.), Geol. Kabinett. *Palaeont.* o

Mascarenhas, Antonio Constancio de Espectação Brás, Ass. d'Anat. Fac. Méd. Porto (Portugal). *Anthrop.* ℗ Porto 1924.

Maschkowzeff, Alexander (1891), Priv.Doc. Univ. Ass. Inst. f. vergl. Anat. Moskau (U.d.S.S.R.). *Embryol., vergl. Anat., Urogenitalsyst. d. Ganoiden.* ① Gießen 1914.

Masefield, John Richard Beech (1850). Rosehill, Cheadle Staffordshire (England). *Wild Bird protection.*

Masing, Ernst (1879), Prof. Dir. Med. Klinik. Tartu (Estland), Gartenstr. 28. *Morphol., Physiol., Chemie u. Pathol. d. Blutes.* ℗ Dorpat 1908. ℗ Dibotriocephalus latus.

Masj, Luigi, Lib. doc. Univ. Genova (Italia). *Zool.* o

Maskell, Francis Gerard (1899), Lect. and Demonstrator in Zool. Dept. of Biol., Victoria Univ. Coll. Wellington (New Zealand). *Cyclostomata. Development, histol. and anat. of Geotria australis.* ① Auckland 1924.

Maskovec, Alek. A., Doc. I. Univ. Moskau (U.d.S.S.R.). *Zool. Embryol.* o

Masobrio, G., Doc. Univ. Torino. (Italia). *Patol.* o

Mason, Arthur Charles (1891), Ass. Entomol., Bureau of Entomol., U.S. Depart. of Agric. Honolulu (Hawaii), Box 340. *Mediterranean Fruit Fly. Biol. of the Papaya Fruit Fly. Camphor Thrips. Rust Mite of Citrus. Citrus Thrips.* ① Univ. of Florida 1915.

Mason, Edmund William (1890), Ass. Mycol., Imper. Bureau of Mycol. Kew, Surrey (England), 17, The Green. *Systematic mycol.*

Mason, Karl Ernest (1900), Instr. in Anat., Vanderbilt Med. School. Nashville, Tenn. (U.S.A.), Department of Anat. *Dietary deficiency, and its effect on the histol. of the reproductive and other endocrine glands, albino rat. Tuberculous infection of the male reproductive system; guinea pig.* ① Yale Univ. 1925. ℗ Prep. histol.

Mason, Preston Walter (1889), Associate Entomol., U. S. Bureau Entomol. Washington, D.C. (U.S.A.). *Aphiidae, toxonomy.* ① George Washington Univ. 1925. ℗ Aphids.

Mason, Silas Cheever (1857), Arboriculturist U.S. Dept. Agric. Indio, Cal. (U.S.A.). o

Massalskij, Alexander P., Ass., Pädagog. Inst. Jaroslawl (U.d.S.S.R.), Oktoberstr. 22, W. 2. *Botan. Geobotan.* o

Massazza, Adolfo (1898), Ass. effettivo Ist. di Fisiol. Genova (Italia), Via Assarotti 19/3. *Neurol.*

Massey, A. B., Associate Prof. Virginia Polytechnic Inst. and Associate Bact. and Plant Pathol., Agric. Experim. Stat. Blacksburg, Va. (U.S.A.). o

Massey, Herbert, Ivy Lea, Burnage. Didsbury, Manchester (England). *Ornithol.* o

Massey, Louis Melville (1889), Prof. of Plant Pathol. and head of Plant Pathol. Dept. Ithaca, N.Y. (U.S.A.), New York State Coll. of Agric., Cornell Univ. *Plant diseases.* ① Cornell Univ. 1916.

Massia, Georges (1882), Chef de Travaux Parasit. Fac. Méd. Lyon, Rhône (France), 10, Rue de la Barre. *Parasit. et Dermatol. Mycol. parasitaire. Moisissures.* ① Lyon 1911. ℗ Prép. de mycol. Cultures. Prép. microscopiques.

Massini, Carlo Luigi, Lib. doc. Univ. Genova (Italia). *Patol.* o

Massino, Boris (1889), Prof. Parasit. Inst. Vét. Kazan (U.d.S.S.R.). *Helminthofauna de l'homme et de toutes les classes des animaux vertébrés: Anat., Biol., Patol.* ① 1924. ℗ Trematodes, Cestodes, Nematodes et Acanthocephales chez Mammalia, Aves, Pisces, Reptilia et Amphibia de région Volga-Kama.

Masson, Claude Laurent Pierre (1880), Prof. d'Anat. pathol. à la Fac. de Méd. de l'Univ. Strasbourg (France), Inst. d'Anat. pathol. *Anat. pathol., Histol. pathol. Les lésions, l'histophysiol. générale. Tumeurs, des lésions de tissus in vivo: Naevi, Carcinoïdes, Tumeurs des glomus.* ℗ Paris 1919. ℗ Préparations microscopiques.

Mast, Samuel Ottmar, Prof. of Zool. Baltimore, Md. (U.S.A.), Johns Hopkins Univ. *Physiol. of Invertebrates and general physiol. Structure and physiol. of Flowering Plants. Light and the behavior of organisms.* ① Harvard 1906.

Mastauskis, Stanislovas (1890), Lect. der Entomol. Landw. Akad. Dotnuva (Litauen), Žemės Ūkio Akad., Entomol. Kabinetas. *Angewandte Entomol.*

Masucci, Peter (1892), Chief Chem., H. K. Mulford Company. Philadelphia, Pa. (U.S.A.), 640 No. Broad Street. *Endocrine products, therapeutic serums.* ① Mass. Inst. of Technol. 1915.

Masui, Kiyoshi, Prof. Tôkyô Imper. Univ. Komaba near Tôkyô (Japan), Vet. Labor., Coll. of Agric. *Genetics of Cattles (Horse). Cytol.*

Masul, Kôki (1894), Lect. in Plant Physiol. Kyôto (Japan), Botan. Inst., Coll. of Sc., Imp. Univ. *Mycorrhiza of Abies firma. Growth of the mycorrhizal root. Mycorrhiza of Quercus, of Alnus.* ① Kyôto Imperial Univ. 1923.

Matcudaira, Yoritaka. Tôkyô (Japan), 74 Koishikawa-Hisakata-machi. *Ornithol.* ○

Mateescu, Stefan (1888), Chef de travaux pratiques Inst. de Géol. Univ. Cluj (România). *Tertiaire et spécialement Pliocène des Sous-Carpathes. Paléogène du bassin de la Transylvanie.* ① Bucarest 1925. ② Fossiles paléogènes du bassin de la Transylvanie.

Materna, Alois, Dr., Priv.Doc. d. pathol. Anat. Opava (C.S.R.), Prosektur d. Landeskrankenhauses. ○

Matheson, D. C., Prof. R. Vet. Coll. Edinburgh (Scotland). *Pathol. Bact.* ○

Matheson, K. J., Research Labor., Bureau of Dairy Industry, U.S. Dept. of Agric. Washington, D. C. (U.S.A.). *Abnormal fermentation in swiss cheese.* ○

Matheson, Robert (1881), Prof. of Entomol., Cornell Univ., Roberts Hall, Ithaca, N.Y. (U.S.A.)- *Med. Entomol. and Parasit.* ① Cornell Univ. 1906. ② *Parasit. Insects, especially those, attacking man and domestic animals.*

Mathews, Albert Prescott (1871), Prof. of Biochem. Cincinnati, Ohio (U.S.A.), Med.Coll., EdenAve. *Cellular respiration. Blood coagulation.* ① Columbia 1898.

Mathews, Clarence Wentworth (1861), Prof. of horticulture. Lexington, Ky. (U.S.A.). ○

Mathews, Gregory Macalister (1876), Foulis Court fair oak Hants (England). *Ornithol., Nomenclature.*

Mathias, Ernst (1886), I. Ass. am Pathol. Inst. der Univ. Breslau (Deutschland), Tiergartenstr. 25/27. *Allgemeine Pathol. und pathol. Anat., Geschwulstforschung und Krankheiten der Nebennieren, Beziehungen zwischen innerer Sekretion und Geschlechtsmerkmalen.* ① Königsberg 1912.

Mathis, Constant Jean (1871), Dr., Dir. Inst. Pasteur de l'Afrique occidentale française. Dakar (Sénégal).

Mathis, Paul (1892). Klein-Schwein b. Gramschütz, Kr. Glogau (Deutschland). *Kreuzungen der Kartoffeln, Nackt- und Spelzhafer, Weizen, Mohrrüben und anderen landwirtschaftlichen Pflanzen; Auftreten von Mutationen, besonders bei Kartoffeln.* ① Berlin 1925.

Matiegka, Heinrich, Dr., Prof. d. Anthrop. u. Demographie. Praha II (C.S.R.) u. Karlova 3. Anthrop. Inst. d. Karls-Univ. ○

Matjuschenko, Viktor (1893), Botan. Torf-Experiment-Inst. Moskau 11 (U.d.S.S.R.), Pokrowsky-Bulvar 8, W. 8. *Geobotan. d. Moore, Torfanalyse, speziell Carexarten.*

Matjuschev, I. W., Dr., Ass. Anat. Inst. Univ. Perm (U.d.S.S.R.). ○

Matouschek, Franz (1871), Univ.-Lect. Wien XVIII (Deutsch-Österreich), Kutschkergasse 40. *Gallen der Moose; Biol. der Moose.* ② Moose.

Matoušek, Alois, Dr.-Ing., Priv.Doc. Pflanzenproduktion. Praha-Smichow (C.S.R.), Švedska 34. *Botan. appl.* ○

Matsuda, Takeshi, Chief of Ishigami Research Labor. of Infectious Deseases. Hamadera near Ôsaka (Japan). *Experim. Pathol. of Tuberculosis.*

Matsui, Yoshiichi (1891), Dir. Imp. Piscicultural Experim. Stat. (Branch of the Imp. Fisheries Inst.). Muroyoshida near Toyohashi, Aichi-Ken (Japan). *Normal and pathol. physiol. of fishes. Genetics of fishes. Practical aquatic zool.* ① Tôkyô 1914. ② *Freshwater animals and plants. Various kinds of Goldfish. Preparations of fish-disease.*

Matsumoto, Hikohichirô, Prof. of Tôhôku Imper. Univ. Sendai (Japan), Geol. Labor., Coll. of Sc. *Ophiuroidea.*

Matsumoto, Takashi, Prof., Imperial Coll. of Agric. and Forestry. Morioka (Japan). *Plant Pathol.*

Matsumura, Shônen, Prof. Hokkaidô Imper. Univ. Sapporo (Japan), Entomol. Labor., Coll. of Agric. *Systematic Entomol.*

Matsumura, Tsutomu, Prof. Chiba Med. Univ. Chiba near Tôkyô (Japan), Hygien. Inst. *Hygiene and Bact.*

Matsuoka, Zenji, Ass. Prof. of Prov. Med. Univ. Ôsaka (Japan), Chem. Inst. *Biochem.*

Matsuura, Isamu (1905), Ass. Kyôto (Japan), Labor. of Phytopath., Dept. of Agric., Imper. Univ. *Diseases of the Rice-seedlings.* Tottori 1925.

Mattei, Giovanni Ettore (1865), Prof. di Botan. Univ., Dir. Ist. ed Orto Botan. Messina (Italia). *Morfol. e Biol. Vegetale. Filogenesi. Flora dell'Eritrea e della Somalia Italiana. Tulipa, Calligonum.*

Mattfeld, Johannes (1895), Dr. phil., Kustos am Botan. Garten. Berlin-Dahlem (Deutschland), Botan. Mus., Königin-Luise-Str. 6—8. *Systematische Botan.: Phanerogamen (Compositae, Caryophyllaceae, Casuarinaceae, Cyperaceae, Abies). Pflanzengeographie (Floristik, Genetische Pflanzengeographie). Floristische Kartierung Mitteleuropas, Flora der Balkanhalbinsel.* ① Berlin 1920.

Matthaei, Rupprecht (1895), Dr. med., Priv.Doc. f. Physiol., Ass. am Physiol. Inst. der Univ. Bonn (Deutschland), Am Botan. Garten 26. *Gestaltlehre. Farbenlehre.* ① Bonn 1919. ② Farbenkreis.

Matthes, Ernst (1889), o. Prof. der Zool. Dir. Zool. Instit. Univ. Greifswald (Deutschland). *Entwicklungsgeschichte des Schädels. Vergl. Morphologie des Schädels. Wassersäugetiere (besonders Sirenia). Physiol des chemischen Sinnes.* ① Breslau 1912.

Matthew, Charles White (1902), Research Chem. Washington (U.S.A.), Henry Ford Hospital. *Chem. changes in the blood during pregnancy, Normal and pathol. Physico-Chem. Studies of oxidation and reduction.* ① Marquette Univ. 1919.

Matthew, William D. (1871), Curator of Vertebrates Paleont. and Curator in Chief of Division I, Am. Mus. of Nat. Hist. N.Y. Hastings-on-Hudson, N.Y. (U.S.A.). ① Columbia 1895. ○

Matthey, Robert (1900). Genève (Suisse), Labor. de Zool., Univ. *Greffe de l'œil. Conditions de la sexualité chez les Reptiles. Trypanoplasma helicis.* ① Lausanne 1923.

Mattill, Henry A. (1883), Prof. of Biochem., Department of Vital Economics. Univ. of Rochester, N.Y. (U.S.A.). *Metabolism. Vitamins: Vitamins and Reproduction in the Rat.* ① Univ. of Illinois 1910.

Mattirolo, Oreste (1856), Prof. o. Botan. Univ. Torino 6 (Italia), R. Ist. botan., al Valentino. *Micol. Anat. Fisiol. Licheni. Storia delle Botan.* ① Med. e sc. nat. Torino 1894. ② Tuberaceae. Hymenogastreae. Podascineae. Sclerodermataceae. Lycopodaceae e affini.

Matulionis, Prof. Landwirtschaftl. Akad. Dotnava (Litauen). *Forstwiss.* ○

Matveev, Boris Step., Doc., Ass. Inst. Vergl. Anat. I. Univ. Moskau (U.d.S.S.R.). *Zool., Embryol.* ○

Matveev, V. N., Gos. Inst. Eksp. Med., Leiter der Epizool. Abt. Staatl. Inst. f. experim. Med. Leningrad (U.d.S.S.R.), Lopuchinskaja ul 12. ○

Matwejew, Valérian P. (1886), Ass. Fac. méd. Univ. Taschkent (U.d.S.S.R.). *Anat. normale.* ① Paris 1912.

Matz, Julius (1886), Plant Pathol. Fortuna (Porto Rico). *Sugar cane diseases and insect pests. Cane varieties. Experiments with fertilizers. Mosaic disease of sugar cane. Root rot and Top rot of sugar cane. Varietal resistance to disease in cane. Plasmodiophora vascularum, a sugar cane parasite.* Ⓓ Massachusetts Agric. Coll., Amherst, Mass., 1913.

Matz, Lew Josefowitsch (1895), Biol. Abt. Sanitäts-Inst., Leiter d. prakt. Arbeiten Phys.-Mathem. Fac. I. Univ. Moskau 34 (U.d.S.S.R.), Kleine Lewschinsky-Gasse 3, Wohn. 15. *Nahrungsmittelbact. Desinfektion des Milzbrandes. Bact. der Luft.* Ⓓ Moskau 1924.

Matzdorff, Carl (1859), Prof. Berlin NW 7 (Deutschland), Dorotheenstr. 12. *Bryozoen.* Ⓓ Kiel 1882.

Mauch, Oscar (1886), Dr. med., Arzt, Zentralpräsident der Schweizerischen Ornithol. Gesellschaft. Zofingen (Schweiz), Kanton Aargau. *Biol. der Haustaube. Rassenkunde.* Ⓓ Zürich 1911. Ⓟ Rassentaube.

Mauer, Fedor M., Mittelasiat. Univ. Taschkent (U.d.S.S.R.), Gogolstr. 32/34. *Selektion. Samenzucht. Flachsbau.*

Maurer, Friedrich (1859), Geh.Hofr., Prof., Dr. med., o. Prof. der Anat. und Dir. der Anat. Anstalt. Jena, Thüringen (Deutschland), Ob. Philosophenweg 20. *Schilddrüse, Thymus und Kiemenspaltenderivate der Wirbeltiere. Rumpfmuskelsystem der Wirbeltiere. Integument der Wirbeltiere (Säugetierhaare). Histol. der Wirbellosen und der Wirbeltiere.* Ⓓ Heidelberg 1883.

Mauriac, Pierre (1882), Prof. méd. expérim. Fac. Méd. Univ. Bordeaux (France), 12, Rue Vauban. *Méd. expérim. Bact.*

Maurizio, Adam (1862), Dr. phil., o. Prof. der Botan. u. Warenkunde der Techn. Hochsch. in Lwów (Lemberg), pensioniert. Bydgoszcz (Polska), Wawrzyniaka 14. *Botan. u. Chem. des Getreides, Geschichte der Nahrung u. Landwirtschaft, früher Biol. der Pilze.* Ⓓ Bern 1894.

Mauwaring, Wilfred Hamilton (1871), Prof. Stanford Univ., California (U.S.A.). *Experim. pathol., bact., immunol.* ○

Mavor, James Watt (1883), Prof. of Biol. Union Coll., Schenectady, N.Y. (U.S.A.). *Effects of X-rays on the mechanism of inheritance in Drosophila; Myxosporidia; Development of the coral Agaricia; Circulation of the water in the Bay of Fundy.* Ⓓ Harvard Univ. 1913.

Mavrogordato, Anthony (1873), Fellow in Industrial Hygiene, South African Inst. for Med. Research. Johannesburg (S. Africa), Box 1038. *Tuberculosis, Miners' Phthisis, Ventilation.* Ⓓ Oxford 1896.

Mawas, M. J., Dir. sc. Fondation Rothschild. Paris 5 (France), 141, Boulevard St. Michel. *Pathol.* ○

Mawrodiadi, P. A., Prof., Dir. Zool. Kab. Landu. Forstwirtschaftl. Inst. Minsk (U.d.S.S.R.). *Histol.* ○

Maximovič, R., Conserv. au Ministère d'Enseignement public. Praha III (C.S.R.), Ministerstvo školstvi. *Protection de la nature, Phanérogames.* ○

Maximow, Alexander (1874), Prof. of Anat., Univ. of Chicago, Ill. (U.S.A.), Department of Anat. *Experim. Histol. of blood and connective tissue. Tissue Culture.* Ⓓ St. Petersburg 1896.

Maximow, Nicolas Alexandrowitsch (1880), Prof., Head Dept. of Plant Physiol. in the Inst. of Applied Botan. and New Cultures, Head Dept. of Botan. at the State Pedagogical Inst. Leningrad (U.d. S.S.R.), 44 Herzenstr. *Physiol. and experim. ecol. of plants, water relations, drought and frost-resistance.* Ⓓ Petersburg 1902.

Maxon, William R. (1877), Associate Curator, Division of Plants, U.S. National Mus., Washington, D.C. (U.S.A.). *Pteridophyta of tropical America.* Ⓓ Syracuse 1898.

Maxwell, John Preston (1871), Prof., Head of the Dept. of Obstetrics and Gynecol., Union Med. Coll. Peking (China). *Parasit., Tropical Med.* Ⓓ Tropical Med. 1910.

Maxwell, Leila Delliber, Bact., clinical. Pathol., Georgia Baptist Hospital, Atlanta, Ga. Brookhauen, Ga. (U.S.A.), 124 Stewart Drive. *Blood Grouping, Culture Media, Sectioning, Staining tissue, Blood Chemistry, Blood Counts, Malaria.* Ⓓ Columbia Univ., New York City, 1918. Ⓟ Sections of tissue, stained slides, Organisms causing various diseases, stained slides of these.

May, Curtis (1897), Ass. in Botan., Ohio Agric. Experim. Station. Wooster, O. (U.S.A.). *Phytopath. Physiol. Resistance and Susceptibility to Disease.* Ⓓ Ohio State Univ. 1923.

May, Henry G. (1886), Chief of Animal Breeding and Pathol. in the Agric. Experim. Station, Prof. of Bact. in Rhode Island State Coll. Kingston, R.I. (U.S.A.). *Inheritance in poultry. Poultry diseases.* Ⓓ Univ. of Illinois 1917.

May, Raoul Michel (1900), Research Fellow, American Field Service. Paris 5 (France), Labor. d'anat. comparée Sorbonne. *Zool. expérim.: transplantation, greffe. Application de la microchim. à la biol.* Ⓓ Harvard Univ. 1924.

Maydell, Ernest (1878), Dr. med., Prof. für Physiol. an der Jagiellon. Univ. Krakau (Polen), Zyblikiewicz-Str. N. 5. *Physiol. der Verdauungsorgane, Verdauungshormone.* Ⓓ Kijewsche Univ. St. Włodzimier.

Mayer, André (1875), Prof. au Coll. de France. Paris 9 (France), 33, Faubourg Poissonnière. *Physiol. générale. Physicochim. biol.: oxydations, respiration, composition cellulaire.* Ⓓ Paris 1900.

Mayer, Edmund (1889), Dr. med., Prosektor am Krankenhaus Lankwitz. Berlin-Wilmersdorf (Deutschland), Motzstr. 46. *Pathol. des lymphohämatopoetischen Systems, Grenzgebiete der Morphol. und Physiol., besonders in methodol. Hinsicht, Bact. an der Leiche.* Ⓓ Berlin 1914.

Mayer, Konrad (1898), Ass. Weihenstephan bei München (Deutschland). *Accessorische Nährstoffe.* Ⓓ Weihenstephan 1925.

Mayer, Martin (1875), Abteilungsvorsteher am Inst. für Schiffs- u. Tropenkrankheiten. Hamburg 4 (Deutschland). *Bact., tierische Parasit., insbes. Trypanosomen, Leishmanien, Plasmodien und andere Blutparasit. (neu entdeckte Anämieerreger, sog. Bartonella muris und Bartonella bacilliformis des Menschen).* Ⓓ Heidelberg 1900.

Mayer-Gmelin, Hugo Karl Hans Adolf (1873), Prof. an der Landbouwhoogeschool. Labor.-Adr.: Afdeeling Landbouwplantenteelt der Landbouwhoogeschool, Heerenstraat, Wageningen (Holland). Priv.-Adr.: Wageningen, Diedenweg 2. *Landwirtschaftlicher Pflanzenbau. Inzucht- und Züchtungsversuche mit Roggen und Rotklee. Züchtung verschiedener landwirtschaftlicher Kulturgewächse, Getreide und Hülsenfrüchte. Ursachen des Zurückganges bei wiederholtem Nachbau von Flachs aus dem Ostseegebiet. Bedeutung der Grannen beim Getreide. Ungleichwertigkeit der Nachkommenschaften einzelner Individuen bei Gräsern inländischen Ursprungs, in Zusammenhang mit der Züchtung von Gräsern.* Ⓟ Rassen landwirtschaftlicher Kulturpflanzen.

Mayer-Hirschlerowa, Zofja (1892), Dr. phil., Mitarbeiterin des Zool. Inst. an der Jan Kazimierz-Univ. Lwów (Lemberg) (Polen), Sanct-Mikolaj-Gasse 4. *Amphibienmetamorphose, Explantation und Transplantation bei Tieren, Golgischer Apparat.* Ⓓ Lwów (Lemberg) 1924.

Mayet, Lucien (1874), Prof. Anthrop. et Paléont. humaine, Fac. Sc. Univ. Lyon (France), 16, quai Claude-Bernard. *Anthrop. générale. Anthrop. anat. Paléont. humaine.* Ⓓ Med. Lyon 1909, sc. Paris.

Mayhew, Roy L. (1890), Ass. Prof. of Zool. Baton Rouge, La. (U.S.A.), Dept. Zool. La. State Univ.

Parasit. Cestode parasit. of Birds. ① Univ. of Illinois 1924.
Maymone, Bartolo (1884), Dir. Ist. Sperim. Zootecnico e Liber. Doc. Univ. Roma (Italia). *Genetica animale. Alimentazione.*
Maynard, Edward Jackson (1892), Associal Animal Husbandman, Colorado Experim. Station. Fort Collins, Col. (U.S.A.). *Nutritional experim.*
Maynard, Leonard A. (1887), Prof. of Animal Nutrition, Cornell Univ. Ithaca, N.Y. (U.S.A.). *Calcium and phosphorus nutrition. Protein nutrition, metabolism.* ① Cornell Univ. 1915.
Mayné, Raymond (1887), Ingénieur Agronome, Prof. Inst. Agronom. Gembloux, Dir. Stat. Entomol. Gembloux. Ixelles-Bruxelles (Belgique), 17, avenue Macau. *Entomol. économique agricole belge et du Congo Belge.* ① Gembloux 1909. ② Insectes et autres animaux nuisibles à l'Agric.
Mayr, Ernst (1904), Ass. am Zool. Mus. Berlin N 4 (Deutschland), Invalidenstr. 43. *Zoogeographie der Landtiere, Ornithol.* ① Berlin 1926.
Mayr, Erwin (1899). Salzburg (Deutsch-Österreich), Faberstr. 13. *Pflanzenbau und Pflanzenzüchtung, Sortengeographie der Getreidearten.* ① Wien 1923.
Mazaki, Takeo, Prof. Med. Univ. Niigata (Japan), Pharm. Inst. *Pharm.* o
Mazarovič, Alek. Nik., Prof. I. Univ. Moskau (U.d.S.S.R.). *Geol., Palaeont.* o
Maze, M. P., Chef de serv. Inst. Pasteur. Paris 15 (France), 26, rue Dutot. *Botan.* o
Maziarski, Stanislas (1873), Dr. en méd., Prof. d'Histol. à la Fac. de Méd. de l'Univ. des Jagellons Cracovie (Pologne), Labor. d'Histol. de l'Univ. (privat.: 4, rue Łobzowska. *Recherches histol. et cytol. sur les glandes chez les Vertebrés et Invertebrés. Relation du cytoplasme et du noyau dans les cellules glandulaires. Histophysiol. des glandes filières des Lépidoptères. Résidu fusorial pendant la caryocinèse. Divers problèmes de la cytol. Tissu musculaire des Insectes.* ① Cracovie 1897.
Mazurkiewicz, Ladislas, Dr., Prof. Warszawa (Polska), Univ. Inst. Pharmacogn., r. Brzozowa 12. *Botan.* o
Mazza, Felice, Lib. doc. Univ. Cagliari, Sardinia (Italia). *Zool. e Anat. compar.* o
Mazzarelli, Giuseppe, Prof. Univ. Messina (Italia). *Anat. Physiol. compar.* o
Mecartney, John Lupton (1901), Ass. in Pomol., New York State Coll. of Agric. at Cornell Univ. Ithaca, N.Y. (U.S.A.), 214 Thurston Avenue. *Factors associated with and influencing blossom-bud formation in the apple.*
Mechanik, Naum (1896), Prosektor Anat. Labor. Inst. für Med. Wissensch. Leningrad (U.d.S.S.R.). *Osteol. d. unteren Extremitäten.* ① 1921.
Medalia, David Bernard (1888), Dr. med. Boston, Mass. (U.S.A.), 484 Commonwealth Ave. *Bact., Anaphylaxis.* ① Tufts Coll. Med. School, Boston 1917.
Medinger, Pierre (1879), Chim. du Labor. pratique de Bact. Luxembourg (Luxembourg), avenue de la gare 8. *Chim. physiol. et biol., denrées alimentaires.* ① Zürich 1905.
Medisch, Mark Nikolaevič, Chef des trav. de Phytopath., Doc. Inst. Agronomique. Minsk (U.d.S.S.R.). *Pathol. végétale, Mikrobiol.* o
Medvedkova, Lidya Ivan, Doc. I. Univ. Moskau (U.d.S.S.R.). *Pharmacol.* o
Medwedewa, Nina (1898), Prof. agrégé. Moskau (U.d.S.S.R.), B. Kalushskaja 22. *Métabolisme des hydrates de carbone. Sécrétions internes. Système nerveux végétatif. Tumeurs.*
Meek, Walter J. (1878), Prof. of Physiol., Univ. of Wisconsin Med. School, Ass. Dean of the Med. School. Madison, Wis. (U.S.A.), Sc. Hall. *Heart and circulation.* ① Univ. of Chicago 1909.
van der Meer, Jikke H. H. (1893), Botan. of the Inst. for Phytopath., Labor. for Mycol. and Potato Research, Wageningen (Holland), Haarweg 3a.

Phytopath. Fungi: Verticillium, Olpidium. ①Amsterdam 1925.
van der Meer Mohr, J.C., Ass. (1892), Deli Proefstation. Medan, Sumatra (Nederl. - O. - Indië). *Economic Entomol.*
Meffert, Boris F., Obergeol. Com. Géol. de Russie. Leningrad (U.d.S.S.R.), W. O., Sredny Pr. 72-B. *Foraminifera.* o
Meggitt, Frederick Joseph (1890), Prof. of Biol., Univ. of Rangoon (India), Univ. Coll. *Cestoda: Cyclophyllidea. Trematoda. Acanthocephala.* ① Univ. of Birmingham Sc. 1913, Phil. 1919. ② Cestoda, Trematoda, Algae.
Meglitzky, Peter (1894), Ass. Abteilung Pathol. Physiol. Med. Inst. Leningrad (U.d.S.S.R.), Simbirskaja Str. 13, W. 10. *Salzstoffwechsel.*
Megrail, Emerson (1890), Asst. Prof. of Hygien. and Bact., W. R. U. Med. School. Cleveland, Ohio (U.S.A.). ① Western Reserve Univ. Med. School 1915.
Méhes, Gyula (1881). Budapest II (Ungarn), Zsigmond-ut 9. *Ostracoden (recent und palaeont.), Eichengallen, Gallwespen.* ① Budapest 1908. ② Ungarische Eichengallen und Gallwespen.
Méhes, Gyula (1897), Ass. Pharmacol. Inst. Szeged (Ungarn), Kálvária tér 5. *Physiol. der Gefäße der überlebenden Warmblüterleber, Angriffspunkt einiger Narcotica, Schlafzentrum, Diuresezentrum.* ① Szeged 1924.
Méhelÿ, Lajos, Prof. Dr., Dir. d. Zool. u. kompar.-anat. Inst. Univ., Dir. d. Anthrop. Inst. Budapest (Ungarn). *Herpet. Mammalia.* o
Mei, C. S., Lect. National Normal Univ. Peking (China). *Zool.* o
Meier, Elena I., Ass. Landwirtschaftl. Akademie Timirjasew I. Univ. Moskau I (U.d.S.S.R.), Grashdanskaja 28. *Botan. Systematik d. Pflanzen.* o
Meier, Fred Campbell (1893), Extension Plant Pathol. Washington, D.C. (U.S.A.), Bureau of Plant Industry, U.S. Dept. of Agric. ① Harvard Univ. 1917.
Meier, Henry J. A. (1881), Prof. of Botan., Syracuse Univ. Syracuse, N.Y. (U.S.A.). *Plant Physiol.; salt relations; H-ion concentration; protoplasm; Regeneration.* ① Columbia Univ. 1920.
Meier, Konstantin I., Doc. Landwirtschaftl. Akademie Timirjasew. Moskau (U.d.S.S.R.), Sowetskaja Plosch. 10/36, W. 3. *Pflanzenphysiol.* o
Meier, Rolf (1897), Dr. med., Ass. a. Pharmacol. Inst. Göttingen (Deutschland). *Wirkung von local wirksamen chem. definierten Substanzen auf den Stoffwechsel von Einzelzellen.* ① Göttingen 1923.
Meigen, Friedrich (1864), Prof., Dr. phil. Dresden-A. 16 (Deutschland), Holbeinstr. 107. *Pflanzengeographie and Moose.* ① Marburg 1886.
Meigs, Edward Browning (1879), Physiol., Bureau of Dairy Industry, United States Department of Agric. Beltsville, Md. (U.S.A.), Dairy Experim. Station. *Nutrition of dairy cattle and physiol. of milk secretion.* ① Univ. of Pennsylvania 1904.
Meiklejohn, Richard. British Consul. Reval (Estland). *Ornithol.* o
Meinecke, Emilio P. M., Bureau of Plant Industry, and Consulting Pathol., Forest Service, District 5. San Francisco, Cal. (U.S.A.), Ferry Bldg. *Mycol., phytopath., forest pathol.* o
Meinertzhagen, Annie Constance. London W. 11 (England), 17 Kensington Park Garden. *Ornithol.*
Meinertzhagen, Richard (1878). London W. 11 (England), 17 Kensington Park Garden. *Ornithol.*
Meisenheimer, Johannes (1873), o. Univ. Prof., Dir. des Zool. Inst. der Univ. Leipzig (Deutschland), Talstr. 33 II. *Sexualprobleme.* ① Marburg 1897.
Meissel, Maxim (1901), Histol. Labor. Med. Inst. Leningrad (U.d.S.S.R.), Ul. Gerzena 29. *Histol. d. Nervensystems.*
Meissner, Prof. Inst. d. Fischzucht. Moskau (U.d.S.S.R.), Piatnizkaja Str. o

Meissner, Max (1872), Dir. des Königsberger Tiergartens. Königsberg i. Pr. (Deutschland), Hufenallee 32. *Mammalia.*

Meixner, Adolf (1883), Vorstand Zool.-botan. Abteilung Steiermärkisches Landesmus. Joanneum. Graz I (Österreich) Raubergasse 10. *Kleinschmetterlinge Steiermarks.* ⓓ Graz 1907.

Meixner, Josef (1889), Dr. phil., Priv.Doc. u. o. Ass. am Zool.-zootomischen Inst. der Univ. Graz III, Steiermark (Österreich), Universitätsplatz 2. *Turbellaria, Coleoptera (Morphol., Physiol. u. Oekol.).* ⓓ Graz 1913. ⓟ Coleoptera (Trechini).

Mekel, Johannes Christoffer (1904), Ass. am Zool. Labor. Groningen (Holland), Petrus Hendrikszstr. 80. *Zeichnung und Farbe von Raupen, Puppen und Schmetterlingen. Vererbung der Blumenfarbe bei der Kornblume (Centaurea).* ⓟ Samen der verschiedenen .Centaurea-Arten vom natürlichen Standorte.

Mekseev, Michaïl V., Doc. I. Univ. Moskau (U.d.S.S.R.). *Pathol. Anat.* o

Melander, Axel Leonard (1878), Head Prof. of Biol., Coll. of the City of New York (U.S.A.). *Diptera.* ⓓ Harvard 1914. ⓟ Diptera.

Melander, Leonard W. (1893), Associate Pathol., Bureau of Plant Industry, U.S. of Agric. St. Paul, Minn. (U.S.A.), Univ. Farm. ⓓ Univ. of Minnesota 1924.

Melanidès, I., Chef de Labor. Inst. Pasteur Héllénique. Athènes (Grèce), Rue Kifissias. o

Melchers, Leo Edward (1887), Prof., Head of the Department Botan. and Plant Pathol. Manhattan, Kan. (U.S.A.), Kansas State Agric. Coll. *Diseases of Cereals, Forage Crops. Vegetables. Disease Resistance, smut of maize.* ⓟ Flora of Kansas, Cultures of fungi.

Melchior, Hans (1894), Dr. phil., Ass. am Botan. Mus. zu Berlin-Dahlem (Deutschland), Königin-Luise-Str. 6/8. *Systematische Botan.: Violaceae, Theaceae, Bignoniaceae und andere Familien; Süßwasseralgen; Phylogenie der Phanerogamen.* ⓓ Berlin 1920.

Melchior, Lauritz (1871), Prosektor, Dir. Pathol. Inst. Kommunehospitalet. København 7 (Danmark), N. Farimagpade. *Pathol. Anat.* ⓓ Copenhagen 1904.

Melczev, Miklós (1891), Ass. Budapest VIII (Ungarn), József-ut 56. *Biol. der Haut.* ⓓ Budapest 1918. ⓟ Normale u. pathohistol. Praep. aus der Haut.

Melhus, Irving E., Prof. of Plant Pathol., Iowa State Coll. Ames, Ia. (U.S.A.). *Plant Pathol., Mycol.* o

Melichowa, Eugenie (1892), Ass. Rostow a. Don (U.d.S.S.R.), Anat. Inst. *M. sternalis. Obliteration d. Nähte bei Menschen u. Affen. Muskeln d. Frosches.*

Melik-Megrabow, Awak (1881), Prof., Dir. Physiol. Inst. Med. Inst. Odessa (U.d.S.S.R.), Olgiewskaja Str. 4. *Anaphylaxie. Motor. u. sekretor. Funktionen des Verdauungsapparates.* ⓓ Odessa 1919.

Melik-Paschaev, Nikol. Samll., Doc. II. Univ.,Ass., Anat. Inst. Moskau (U.d.S.S.R.), Dewitschje Pole, Trubezkoi pereulok. o

Melin, D. E., Doc. Univ. Uppsala (Sverige). *Entomol.* o

Melissinos, C., Prof. d'Anat. pathol. Univ., Dir. Inst. d'Anat. pathol. Athènes (Grèce), Rue Agathopoleos 4. o

Mĕlka, Jaroslav (1904), Ass. Bratislava (C.S.R.), Sasinkova 3. *Physiol.*

Melkich, Aleksander, Prof. Univ. Irkutsk (U.d.S.S.R.). *Pathol.* o

Mell, Rudolf Emil (1878). Berlin-Steglitz (Deutschland), Rueckertstr. 4. *Biol., Zoogeographie, Systematik der Reptilien und Macrolepidopteren Chinas. Sphingiden, Trichopteren, Ephemeropteren, Gryllacriden, Tettigoniden, Phasmoiden.* ⓓ Rostock 1924. ⓟ Macrolepidopteren Chinas.

Mellanby, J., Prof., Univ. London S.W. 7. (England), S. Kensington. *Physiol.* o

Mellon, Ralph Robertson, Dir. of Labor., Highland Hospital, Rochester N.Y. (U.S.A.). *Microbic heredity. Biol. of the bacteria: complex life cycle. Sexual cycle. Mutation changes in bacteria growth stages, change under influence of suitable environment.* ⓓ M.D. Univ. of Michigan, Ann Arbor, Mich., 1909, Dr.Ph. Harvard Med. School, Boston, Mass., 1916.

Melnik, Stepan P., Prof. an der Weißruss. Staatl. Landw. Akademie Gorki (U.d.S.S.R.), Landwirtschaftliche Akademie 25. *Forstwirtschaft.* o

Melnikow-Raswedenkow, Nikolaus Fedotovicz (1866), Dir. Univ. Pathol.-anat. Inst. Charkow (U.d.S.S.R.), K.-Liebknecht-Str. 41. *Milzpathol. u. pathol. Morphol., Splenomegalien: leukämische, tuberkulöse, Gaucher. Echinococcus, Trichinen, Distomen, besonders die Kapselstruktur d. Echinococcusträgers an der Leber und der Niere. Hamartome, Choristome; Cholesteatome.* ⓓ Moskau 1895.

Mendel, Lafayette Benedict (1872), Prof. of Physiol. Chem., Yale Univ. New Haven, Conn. (U.S.A.), 333 Cedar Street. *Nutrition and Growth.* ⓓ Phil. Yale Univ. 1893, Sc. (Hon.) Univ. of Michigan 1913.

Mendenhall, Eugene Warren (1873), Inspector, Div. of Plant Industry, Ohio Department of Agric. Columbus, O. (U.S.A.), 97 Brighton Rd. *Entomol. Phytopath.*

Mendonça, Francisco d'Ascensão (1889), Inspecteur Jardin Botan. Univ. Coimbra (Portugal), Inst. Botân. *Flore Portugaise, plantes vasculaires.*

Mendigutia, Fernando L. (1884), Prof. d'anat. et physiol. animal à la Fac. des Sc. de l'Univ. de Barcelona (España). *Physiol. générale et Hygiene.* ⓓ Madrid 1910.

Mendoza Guazon, Maria Paz., Assoc. Prof. Univ. Philipp. Manila (Philippine Islands). *Pathol. and Bact.* o

Mendy, Juan, Prof. Univ. Nacional. La Plata (Argentina). *Parasit.* o

Menegaux, Henri August, Mus. d'Hist. Nat. Paris (France). *Ornithol.* o

Menetrier, M. P., Prof., Fac. Méd. Paris (France). *Pathol.* o

Menezes, M., Prof. Univ. Porto Alegre (Brasil). *Anat.* o

Menge, Edward John von Komorowski (1882), Dir. of the Department of Zool., Marquette Univ. Milwaukee, Wis. (U.S.A.). *Animal psychol.; spleen development and comparisons.* ⓓ Univ. of Dallas 1916, Sc.D. Hon. De Paul Univ. 1926.

Menghin, Oswald (1888), o. ö. Prof. für Urgeschichte des Menschen an der Univ. in Wien IX (Deutsch-Österreich), Wasagasse 4, Inst. der Univ. *Urgeschichte des Menschen in allen Zweigen.* ⓓ Wien 1910.

Menten, Maud Leonora (1879), Associate Prof. of Pathol., Univ. of Pittsburgh, Pa. (U.S.A.), 5614 Walnut Street. ⓓ M.D. Univ. of Toronto 1911; Ph.D. Univ. of Chicago 1916.

Mentz, Aug., Prof. Dr., Dir. Botan. Labor., Vet. u. Landw. Hochsch. København (Danmark), Bülowsvej. o

Menzbir, Michael Alexandr., Prof. Inst. d. vergl. Anat., I. Univ. Moskau (U.d.S.S.R.), Herzenstr. 6. *Ornithol. Embryol.* o

Menzel, Paul Julius (1864), Dr. med, Sanitätsrat. Dresden-A. (Deutschland), Mathildenstr. 46. *Palaeobotan. (Tertiärpflanzen).* ⓓ Leipzig 1889.

Menzel, Richard (1890), Zool. an der Teeversuchsstation in Buitenzorg, Java (Nederl.-O.-Indië). *Studium und Bekämpfung der Schädlinge des Tees (speziell biol. Bekämpfung; Parasit.: Hymenopteren und Tachinen). Moosfauna, Bromeliaceen- und Nepenthesfauna (Harpacticiden, Cyclopiden, Ostracoden, Nematoden).* ⓓ Basel 1914.

La Mer, Victor Kuhn (1895), Ass. Prof. of Chem., Columbia Univ. New York City (U.S.A.). *Physical Chem. Vitamins; Oxydation-reduction potentials; Neutral salt effects; Calcium deposition.* ① Columbia 1921.

Mercer, Stephen Pascal (1891), Prof. Agric. Botan. Queens Univ., Head of Seed testing and plant disease division, Ministry of Agric. for Northern Ireland. Belfast (Northern Ireland). *Pasture problems. Biol. of grasses. Seed (grass and clover) production.* ① London 1914. ② Weed seeds.

Mercier, M., Prof. Fac. Sc. Caen. Labor. de Zool. Luc-sur-Mer, Calvados (France). *Zool.* o

Mercier, Pierre Jean Jacques Henri (1890), Chef des recherches physiques au Centre Anticancereux. Lausanne (Suisse), 24 Avenue de Florimont. *Radioactivité.* ① Genève 1921.

Mereshkowsky, Sergius (1863), Vorst. d. Labor. für Bekämpfung d. Schädlinge d. Landwirtschaft d. Abt. d. landwirtschaftl. Mikrobiol., Inst. für Experim. Agronomie. Leningrad (U.d.S.S.R.), Ulitza Herzena 42. *Vernichtung der Nagetiere durch Bakterienkulturen.*

Merikallio, Einari Fredrik (1888), Rektor, Mag. Phil. Kerava (Finnland). *Aves (geogr. Verbreitung, Phaenol., Vogelnamen).* ② Vogeleier.

Merkel, Friedrich (1881), Dr. phil., Geschäftsführer der Saatzucht- und Kolonialabteilung sowie des Sonderausschusses für Versuchsringwesen bei der Deutschen Landwirtschafts-Gesellschaft. Berlin SW 11 (Deutschland), Dessauer Str. 14. *Sortenversuchswesen, Saatenanerkennung, Hochzuchtregister für landwirtschaftl. Pflanzenzuchten, Auskünfte über Saatzucht und Saatbau, koloniale Landwirtschaft, Förderung des Versuchsringwesens.* ① Breslau 1906. ② Samenproben, Getreide, Futterpflanzen.

Merkel, Hermann (1873), Univ.Prof., Dir. Gericht. Med. Inst. München (Deutschland), Lachnerstr. 18. *Pathol. Anat.* o

Merkenschlager, Fritz (1892), Priv.Doc. Kiel (Deutschland), Univ. *Agrikulturbotan. (Physiol. und Stoffwechselpathol. der Kulturpflanzen).* ① München 1920.

Merker, Karl Ernst (1888), Dr. phil., Priv.Doc. und Ass. am Zool. Inst. Gießen (Deutschland), Bahnhofstr. 84. *Kalkkörper der Echinodermen und Lichtbiol. der Tiere.* ① Gießen 1914.

Merkle, Fred G. (1892), Associate Prof. of Soil Technol. State College, Pennsylvania (U.S.A.). *Geol. Physics, Chem. and Biol. of Soils.* ① Mass. Agric. Coll. 1914.

Merklen, Pierre Charles Louis (1896), Prép. Travaux Physiol. Fac. Méd., Chargé du cours d'Education Physique et Organisation du Travail et du Sport. Nancy (France), Inst. de Physiol., 30, rue Lionnois. *Physiol.* ② *Le Rhythme du Cœur.*

Merl, Edmund Maria (1889), R.R., Botan. Bayer. Landesanstalt Pflanzenbau u. Pflanzenschutz. München (Deutschland), Destouchesstr. 45 II. *Samenkontrolle. Biol., Morphol. u. Systematik der Gattungen Utricularia u. Genlisea.* ① München 1914.

Merle, Paul, Prof. Ecole prép. d. Méd. et Pharm. Clermont-Ferrand (France). *Histol.* o

Merlet, Anne-Marie (1899), Préparateur d'Histol. Fac. de Méd. de Bordeaux (France). ① Bordeaux 1919. ② Préparations histol. humaines et animales.

Merrell, William D., Prof. of Biol., Botan., Univ. of Rochester, N.Y. (U.S.A.). *Morphol., Ecol.* o

Merriam, C. Hart (1855), Research Associate, Smithsonian Inst., under Harriman Fund. Washington, D.C. (U.S.A.), 1919,16. St. *Botan. Geogr. distrib. of trees shrubs, cactuses, yuccas. Zool.: Ornithol., Mammals, Geomyidae. Ethnographie.* ② Columbia 1879. o

Merrill, Elmer D. (1876), Dean, Coll. of Agric., Univ. of California, and Dir. of the Agric. Experim. Station. Berkeley, Cal. (U.S.A.). *Taxonomy of vascular plants and phytogeography of the Philippine, Malaysian, Polynesian and Chinese floras.* ③ (Honorary) 1926. ② Botan. material from Borneo, Sumatra, Philippines, China for similar material from the entire Indo-Malaysian and Polynesian regions, China and Japan.

Merrill, Geo B. (1886). Assoc. Entomol. State Plant Board of Florida, Gainesville Fla. (U.S.A.). *Econom., Taxonom.: Coccidae, Aleycodidae,* ② Coccidae, Aleycodidae.

Merrill, Melvin Clarence (1884), Editorial Chief of Publications, United States Department of Agric. Washington, D.C. (U.S.A.), 215 Thirteenth Street, SW. *Plant Physiol.* ① Washington Univ. 1915.

Merriman, Mabel L., Associate Prof. of Botan., Hunter Coll. New York City (U.S.A.), Park Ave. and 60 St. *Cytol., algae.* o

Mertens, August (1864), Prof. Dr., Dir. Städt. Mus. f. Heimatkunde. Magdeburg (Deutschland), Domplatz. *Zool., Botan., Geol.* o

Mertens, Hermann (1895), Preußischer Oberfischmeister für die Küstengewässer des Regierungsbezirkes Stettin in Swinemünde (Deutschland), Augustastr. 11. *Fischereibiol.* ② Münster 1920.

Mertens, Robert (1894), Kustos der Wirbeltierabteilung am Senckenberg-Mus., Frankfurt a. M. (Deutschland), Viktoriaallee 7. *Ethol., Ökol., Verbreitung und Systematik der Amphibien, Reptilien und Säugetiere.* ① Leipzig 1915. ② Amphibien und Reptilien.

Merton, Hugo (1879), ao. Prof. der Zool. an der Univ. Heidelberg (Deutschland), Philosophenweg 16. *Spermien und Flimmerzellen (morphol. und physiol.).* ① Heidelberg 1904.

Merton Yarwood, Williams (1883), Prof. of Palaeont., Univ. of British Columbia. Vancouver (Canada). *Palaeont. of the Silurian of eastern Canada. Palaeozoic of Mackenzie Valley. Mesozoic of western plains of Canada.* ① Yale, New Haven, Conn., 1912.

Mescherjakow, D. P., Doc. d. Timirjasew Landw. Akademie. Station Lobnja, Moskauer Gouv. (U.d.S.S.R.) Sawelow. sh. d. Staatl. Inst. f. Wiesenbau. *Geobotan., Moorforschung.* o

Mesnil, Felix Etienne Pierre (1868), Prof. à l'Inst. Pasteur, Membre de l'Inst. de France. Paris 15 (France), 21, rue Ernest Renan. *Biol. des Annélides marines. Parasitol. marine. Trypanosomes pathogènes.*

Messerle, Nikolaus (1897). Zürich (Schweiz), a. Beckenhofstr. 58. *Vegetatives Nervensystem, Elektrokardiographie, Psychogalvanisches.* ① Zürich 1923.

Messerschmidt, Theodor (1886), ao. Prof. techn. Hochsch. Hannover (Deutschland), Baumstr. 3 A. *Hygiene. Bact.* o

Messineo, G., Doc. Univ. Torino (Italia).) *Parasit.* o

Messing, Per Richard (1868), Prof. Stockholm (Sverige). *Botan. (Mycol.)* ① Stockholm 1911.

Měšťan, François (1896), Ass. Ecole polytechn. Inst. des Fermentations. Brno (C.S.R.). *Microbiol. techn.*

Mestayer, Marjorie Katharine (1879), Ass. Dominion Mus. Wellington (New Zealand). *Recent and Fossil Mollusca.* ② Recent and Fossil Mollusca.

Mestrezat, William (1883), Prof. agr. Fac. Méd., Chef de Labor. à l'Inst. Pasteur, Maître de conférence de Physiol. pathol. à l'Ecole des Hautes Etudes (Coll. de France); Dir. du Labor. de Chim. de la Clinique Chir. Paris 15 (France), 25, rue Dutot. *Chim. et physiol. du Liquide Céphalorachidien et des sécrétions. Perméabilité sélective des membranes artificielles et de la Cellule. Equilibres de membrane.* ① Paris 1920.

Metalnikov, Serge (1871), Prof. Inst. Pasteur. Paris (France), 96, Rue Falguière. *Physiol. normale et pathol. des Invertébrés. Immunité, Digestion intracellulaire, phagocytose; Immortalité des unicellulaires, spermatozoines. Immunité des invertébrés. Tuberculose chez les chénilles de Galleria mell.*

Metcalf, F. P., Prof., Fukien Christian Univ. Foochow (China). *Botan.* o

Metcalf, Haven (1875), Senior Pathol. in Charge, Dir. in the U.S. Department of Agric. Washington, D.C. (U.S.A.). *Diseases of forest trees.* ⓓ Univ. of Nebraska 1902.

Metcalf, Maynard Mayo (1868), Prof., Research Associate in Zool. Baltimore, Md. (U.S.A.), Johns Hopkins Univ. *Protozoa: Opalinidae (cytol. and taxonomy); geographical distribution and paleogeography from host-parasite. data. Embryol. of Chiton. The Eyes and Subneural Gland in Salpidaes Morphol. of the Tunicata.* ⓓ Ph.D. Johns Hopkin. Univ. 1893, Sc.D. (honorary) Oberlin Coll. 1914.

Metcalf, Woodbridge (1888), Associate Prof. of Forestry. Berkeley, Cal. (U.S.A.), 305 Hilgard Hall, Univ. of California. *Dendrol., Forest Protection and Silviculture.* ⓓ Univ. of Michigan 1912. ⓟ Tree specimens and seeds.

Metcalf, Zeno Payne (1885), Dir. of Instruction, Head Dept. of Zool. and Entomol., N.C. State Coll., Raleigh, N.C. (U.S.A.). *Economic Entomol., Morphol. and Taxonomy of the Homoptera.* ⓓ Harvard Univ. 1924. ⓟ Homoptera.

Metge, Gustav (1877), Abteilungsvorsteher, Agric.-chem. Kontrollstation der Landwirtschaftskammer f. d. Prov. Sachsen. Halle a. d. S. (Deutschland), Kaiserplatz 19. *Agrikulturchem.; Rauchschäden; Wasser und Abwasser.* ⓓ 1904.

Metjolkin, Anatol (1894). Ass. Inst. Tropenkrankheiten, Moskau (U.d.S.S.R.) Pogodkinskaja 10. *Piroplasmosen.*

Mettam, K.W., Prof. Transvaal Univ. Coll. Pretoria (South Africa). *Anat.* ○

Metz, Chas. W. (1889), Member of Staff., Department of Genetics, Carnegie Inst. of Washington. Cold Spring Harbor, N.Y. (U.S.A.). *Cytol. and genetics of Diptera.* ⓓ Columbia Univ. 1916.

Metzelaar, Jan (1891), Fisheries expert to State of Michigan and Curator of Michigan fishes, Mus. of Zool., Ann Arbor, Mich. (U.S.A.). *Fish culture. Heredity in domestic Pigeons.* ⓓ Amsterdam 1919.

Metzger Ch., Prof. suppl., Ecole préparat. de Méd. et de Pharm. Angers (France). *Anat. et physiol.* ○

Metzger, Frederick William (1901), Junior Entomol., Physiol. Division, Japanese Beetle Labor. Riverton, N.J. (U.S.A.), U.S. Department of Agric. *Japanese beetle.*

Metzner, Herbert (1900), Labor.-Vorstand Forschungsinst. f. d. Fischindustrie. Altona (Deutschland), Flottbecker Chaussee 92. *Physiol. Chem., Biochem. Fischverwertung.* ⓓ Leipzig 1925.

Metzner, Paul Karl (1926), Priv.Doc. und Ass. am Pflanzenbiol. Inst., Berlin-Dahlem (Deutschland), Königin-Luise-Str. 1. *Reizphysiol. der Pflanzen, Geißel- und Flimmerbewegung. Photodynamische Erscheinung. Elektrophysiol. der Pflanzen (bes. Galvanotaxis, Elektrizitätsproduktion).* ⓓ Leipzig 1920.

Metzner, Rudolf (1858), Prof. Dr. med. Riehen b. Basel (Schweiz), Wenkenstr. 30. *Physiol. der Drüsen.* ○

Meurman, Mauri Olavi (1893), Dr. phil., Dir. of Experim. Station. Messukylä (Finnland). *Plantcytol., chromosome behavior and genetical investigations by Avena and Pisum.* ⓓ Helsinki 1919. ⓟ Seeds of Avena, Pisum and several flowering plants.

Mevius, Walter (1893), Dr. phil., Priv.Doc. für Botan. und Ass. am Botan. Inst. der Univ. Münster i. Westf. (Deutschland), Annenstr. 11. *Ernährungsphysiol. der Pflanzen, bes. Nährstoffaufnahme.* ⓓ Münster i. Westf. 1920.

von Meyenburg, Hanns (1887), Prof., Dr. med., Dir. des Pathol. Inst. der Univ. Zürich 7 (Schweiz), Freie Str. 89. *Pathol. Anat. und allgemeine Pathol., speziell Tuberkulose und Krebs.* ⓓ 1914.

Meyer, Adolf (1893), Bibliotheksrat an der Hamburgischen Staats- und Univ.-Bibliothek, Priv.Doc. an der Univ. Hamburg 26 (Deutschland), Sievekingsallee 10 I. *Phil. Grundprobleme und Geschichte der Biol., sowie theoretische Biol.* ⓓ Jena 1916. ⓟ Diapositive zur Geschichte der Biol.

Meyer, Alfred Henry (1888), Ass. in Botan. Madison, Wis. (U.S.A.), 206 N. Archard St. *Toxic effects of Magnesium salts on peas.* ⓓ Univ. of Wisconsin 1911.

Meyer, Arthur William (1873), Prof. of Anat. Stanford Univ., California (U.S.A.). *Pre-natal pathol.; embryol.; the lymphatics; dietary morphol. changes; growth; variations; nutrition effects.* ⓓ Johns Hopkins Univ. 1905.

Meyer, Bernard Sandler (1901), Instr. in Botan., Ohio State Univ. Columbus, O. (U.S.A.), Dept. of Botan. *Plant Physiol., Ecol.* ⓓ Ohio State Univ. 1926.

Meyer, Constantin Ign. (1871), Prof. Moskau 10 (U.d.S.S.R.), 1 Mestschanskaja 28. *Embryol. d. Pflanzen. Algol.*

Meyer, Eduard (1859), Prof. d. Zool., Kubanisch. Landwirtsch., Pädagog. und Med. Inst. Krasnodar, Kuban-Gebiet (U.d.S.S.R.), Basarnaja 11. *Anneliden, Morphol. und Entwickelungsgeschichte. Ameisen, Biol.* ⓓ St. Petersburg 1898.

Meyer, Eduard (1888), Ass. Mannheim-Käfertal (Deutschland), Rüdesheimer Str. 69. *Psychol. der Werte und der Persönlichkeit.* ⓓ 1921.

Meyer, Erich (1874), o. ö. Prof. der inneren Med., Univ. Göttingen, Dir. der med. Klinik. Göttingen (Deutschland). *Pathol. Physiol. Pathol. des Wasserhaushaltes. Blut.* ⓓ Halle 1898.

Meyer, Fritz Jürgen (1891), Dr. phil., Priv.Doc. für Botan. an der Technischen Hochsch. in Braunschweig (Deutschland), Damm 34. *Pflanzenanat., Anat. der Leitbündel.* ⓓ Marburg 1915.

Meyer, Hans (1892), Ass. Abt. Schiemann, Staatliches Inst.für Infektionskrankheiten „Robert Koch", Berlin. Berlin-Grunewald (Deutschland), Franzensbader Str. 13. *Bact. Immunitätsforschung. Pneumokokken. Streptokokken. Anaphylaxie. Reticuloendothel, Immunität und Anaphylaxie.* ⓓ Berlin 1923. ⓟ Pneumokokken, Streptokokken, Meningokokken. Antipneumokokkenserum.

Meyer, Hans Horst (1853), Prof. d. Pharmacol. i. R., Med. Fac., Univ. Wien XIX (Deutsch-Österreich), Weimarer Str. 83. *Experim. Pharmacol.: Narkose, Diurese. Experim. Pathol.: Tetanus, Entzündung. Experim. Physiol.: Wärmeregulation, Sensibilität, Muskeltätigkeit.* ⓓ Königsberg 1877.

Meyer, Heinrich J. (1886), Dr., Pflanzenpathol. in der Abteilung für Schädlingsbekämpfung der Höchster Farbwerke. Höchst a.M.-Sindlingen (Deutschland), Am Wasserwerk. *Insecta noxia, Aphidae.* ⓓ Bonn 1912.

Meyer, Jean (1878), Prof. de Pathol. générale à l'Univ. de Bruxelles (Belgique), 9, rue Thérésienne. *Physico-Pathol. du cœur et des reins. Circulation et endocrinol. en rapport avec la circulation.* ⓓ Bruxelles 1905.

Meyer, Karl F. (1884), Prof. of Bact., Dir. of the George Williams Hooper Foundation for Med. Research and Dir. of Labor. for Research in the Canning Industries. San Francisco, Cal. (U.S.A.), Hooper Foundation, 2. and Parnassus Ave. *Bact. and Immunol.; Compar. Pathol.; Foods; Piroplasmoses. Botulism, tetanus, typhoid and parathyphoid, Brucella group of bacteria; hypersensitiveness; plague.* ⓓ Zürich 1924. ⓟ Bact. preparations, cultures and sera.

Meyer, Kurt (1882), Dir. Bact. Abt. des Rudolf-Virchow-Krankenhauses. Berlin N 65 (Deutschland). *Antikörperbildung gegen Lipoide. Heterogenetisches Antigen. Streptokokken. Enterococcengruppe. Anwendung der Gewebekultur auf Fragen der Infektion u. Immunität.* ⓓ Straßburg 1911.

Meyer, Nikolai (1889), Bureau für Angewandte Entomol., Reichsinst. f. Experim. Agronomie. Leningrad (U.d.S.S.R.), Herzerstr. 44. *Ichneumonidae, Biol. und biol. Bekämpfungsmethode.*

Meyer, Paul (1902), Dr. med., Ass. am Pharmacol. Inst. der Univ. Frankfurt a. M. (Deutschland), Cronberger Str. 25. *Pharmacol., Physiol. der Zellatmung.* Ⓓ Frankfurt a. M. 1925.

Meyer, Reinhold (1892), Dr. phil., Biol. der Chemischen Fabrik E. Merck. Darmstadt (Deutschland), Heinrichstr. 78 pt. *Angewandte Zool. und Botan. Hymenoptera aculeata.* Ⓓ Jena 1919. Ⓟ Hymenoptera aculeata.

Meyer, Robert (1864), Prof. Dr., Vorstand des Pathol. Inst. der Univ.-Frauenklinik. Berlin N 24 (Deutschland), Artilleriestr. 18. *Entwicklungsgeschichte, Histol. und Pathol. der weiblichen Genitalien. Embryonale Gewebsmißbildungen. — Ovulation und Menstruation. Über epitheliale Gebilde im fötalen und kindlichen Uterus.* Ⓓ Straßburg 1889. Ⓟ Histol. und pathol. Praep. und Mikrophotogramme der weiblichen Genitalien.

Meyer, Walter Huber (1896), Ass. Silviculturist, U.S. Forest Service. Portland, Ore. (U.S.A.), Pacific Northwest Forest Experim. Station, 514 Lewis Building. *Forestry.*

Meyer, Wolfgang (1868), Dr. phil., Anstaltsleiter, Landwirtschaftliche Untersuchungsanstalt. Augsburg (Deutschland), Liebigplatz 11. *Düngemittel, Futtermittel, Sämereien.* Ⓓ Leipzig 1895.

de Meyere, Johannes Cornelis Hendrik (1866), Prof. of applied Zool. and heredity at the Univ. of Amsterdam (Holland), Sarphatistraat 76. *Diptera, especially of Holland and the East Indies. Biol. of Agromyrines (Diptera). Echinoidea. Genetics. Hairs and feathers.* Ⓓ Amsterdam 1893.

Meyerhof, Otto (1884), Prof. Dr. med., wissenschaftliches Mitglied des Kaiser-Wilhelm-Inst. für Biol. Berlin-Dahlem (Deutschland), Boltzmannstr. *Allgemeine Physiol. und Biochem., Oxydation Tumoren.* Ⓓ Heidelberg 1909.

Meyerson, Ignace (1888), Dir.-adjoinct du labor. de Psychol. expérim. à la Sorbonne. Paris XIII (France), 23, rue St. Hippolyte. *Psycho-physiol. de la vision. Psychol. de la perception en général, et en particulier chez l'enfant.*

Mezö, Imre (1896), Ass. Allg. Botan. Inst. Budapest VIII (Ungarn), Muz.-krt 4a III. *Pflanzenanat.*

Mhaskar, Krishnarao Shripat (1879), Ass. Dir., Haffkine Inst. Parel, Bombay (India). *Pharmacol. Malaria. Vaccine lymph. Ankylostomiasis.* Ⓓ Bombay Univ. 1906.

Micatovich, Giovanni (1896), Ass. Ist. Botan. Univ. Firenze (Italia), Via Lamarmora 4. *Fisiol. vegetale.* Ⓓ 1926.

Michaelis, Leonor (1875), Resident Lect. at Johns Hopkins Univ. Baltimore, Md. (U.S.A.), Johns Hopkins Hospital. *Physikalische Chem., angewandt auf Biol. und Med. Dynamik der Oberflächen. Wasserstoffionenkonzentration.* Ⓓ Berlin 1897.

Michaelis, Peter (1900), Dr. phil., Ass. am Botan. Inst. der Univ. Jena (Deutschland). *Vererbung: reziproke Bastarde; experim. Erzeugung heteroploider Formen.* Ⓓ München 1923.

Michaelow, Al., Dr., II. Dir. d. Inst. f. Mikrobiol., Staatl. Bact. Inst. Sofia (Bulgarien), Buleward Makedonia. o

Michaelsen, Johann Wilhelm (1860), Prof. Dr., Hauptkustos a. D. Hamburg 1 (Deutschland), Zool. Staatsinst. *Morphol., Systematik, Stammesgeschichte und geographische Verbreitung der Oligochäten und der Ascidien.* Ⓓ 1886.

Michailoff, Boris (1897), Ass. Vet. Inst. Kasan (U.d.S.S.R.), Arskoe Pole 96. *Veränderungen des Nervensystems bei Lyssa.*

Michajlovskij, Jean-Pierre (1877), Prof. Tachkent (U.d.S.S.R.), Postale boîte 68. *Physiol. du sang et du cœur.* Ⓓ Charkov 1899. Ⓟ *Animaux momifiés.*

Michalowsky, Ivan Osip, Prof., Dir. Histol. Labor. d. Univ. Smolensk (U.d.S.S.R.).

Michejew, Alexander A., Prof. Aserbaidshanker Polytechn. Inst. Baku (U.d.S.S.R.), Asiatische Str. 133, W. 33, Ecke Marienstr. *Bodenkunde. Botan. Erdkunde.* o

Michel-Durand, Emile (1885), Dir.-Adjoint labor. de Biol. végétale de Fontainebleau, Seine et Marne (France). *Physiol. végétale.*

Michels, Franz (1891), Dr. phil. nat., Geol. an der Preuß. Geol. Landesanstalt. Berlin N 4 (Deutschland), Invalidenstr. 44. *Tektonik und Stratigraphie des Rheinischen Schiefergebirges. Faunen des Devons.* Ⓓ Frankfurt a. M. 1921.

Michels, Nicholas E. (1891), Ass. Prof. of biol. Saint Louis (U.S.A.), Univ. of Med. *Haematol., compar. histol.* Ⓓ Univ. of Louvain 1922.

Michin, Boris (1893), Doc. Dir. d. Mosk. Stat. f. Seidenraupenzucht. Moskau (U.d.S.S.R.), Svenigorod Chaussee 9. *Biol. von Bombyx mori: Verdauungsapparat, Mech. d. Seidenabsonderung, Raupenfütterung mit Scorzonerablättern, Pebrinebestimmung in der Graine.* Ⓟ Cytol. u. biol. Praep. von Bombyx mori L.

Michin, W. S., Ichthyol. Abt., Inst. d. Experim. Agronomie. Leningrad (U.d.S.S.R.). o

Michotte, A., Prof. Univ. Cath. Louvain (Belgique). *Physiol. experim.* o

Mickel, Clarence E. (1892), Ass. Prof. of Entomol., Univ. of Minnesota. Curator of the insect collections of the Univ. St. Paul, Minn. (U.S.A.), Division of Entomol., Univ. Farm. *Biol. taxonomy of the Mutillidae of North America.* Ⓓ Univ. of Minnesota 1925.

Micoletzky, Heinrich (1883), a.o. Prof. der Univ. Innsbruck (Deutsch-Österreich), Universitätsstr. 4. Zool. Inst. *Freilebende Nematoden, namentlich Ökol. und Systematik. Turbellarien. Hydrobiol.* Ⓓ Graz 1907. Ⓟ Mikroskop. Praep. freilebender Nematoden.

Miculicich, Miroslav (1883), o.Univ.Prof., Dir. Inst. experim. Pathol. (pathol. Physiol.) und Pharmacol. Univ. Zagreb (S.H.S.), Vočarska cesta 97. *Biol., norm., experim. u. pathol., in Anwendung auf die experim. Med. (pathol. Physiol.).* Ⓓ Phil. Jena 1905, med. Graz 1912.

Miczynski, Kazimierz (1899), Adjunkt des Labor. für Pflanzenzüchtung der Techn. Hochschule Lemberg. Dublany bei Lemberg (Polen). *Genetik der Gattung Aegilops und des Weizens. Getreidezüchtung.* Ⓓ Krakau 1925. Ⓟ Arten der Gattung Aegilops.

Middleton, Austin Ralph (1881), Prof. of Zool. and Dir. of the Biol. Station., Univ. of Louisville, Ky. (U.S.A.). *Genetics of the Protozoa. Animal Behavior.* Ⓓ Johns Hopkins Univ. 1915.

Miehe, Hugo (1875), Dr. phil., o. Prof. der Botan. und Dir. des Inst. für Botan. an der Landwirtschaftl. Hochschule zu Berlin N 4 (Deutschland), Invalidenstraße 42. *Mikrobiol., Symbiose, Entwicklungsgeschichte.* Ⓓ Bonn 1899.

Mielck, Wilhelm Walter Otto (1878), Prof., Dir. Staatl. Biol. Anstalt, Mitglied der Deutschen Wissenschaftl. Kommission für Meeresforschung. Helgoland (Deutschland). *Marine Fischerei-Biol. u. Planktonforschung. Fischbrut. Protozoen, spez. Radiolarien. Schizopoden, Copepoden, Chaetognathen. Plankton-Diatomeen u. Peridineen.* Ⓓ Kiel 1907.

Miermod, Gaston (1885), Dr. ès sc., Ass. de Malacol. Genève (Suisse), Mus. d'Histoire nat. *Anat. des organes reproducteurs et digestifs des Mollusques pulmonés.* Ⓓ Genève 1913. Ⓟ Mollusques terrestres paléarctiques.

Mierzejewski, Wł., Prof. Dr. Wilno (Pologne), Inst. Anat. Compar., Univ. Zool. o

Mießner, Hermann (1870), o. Prof. für Hygiene, Seuchenlehre und Veterinärpolizei. Dir. des Hygien. Inst. der Tierärztl. Hochschule zu Hannover (Deutschland), Misburger Damm 16. *Aufzuchtkrankheiten; Paratyphosen; Rotz; Paratuberkulose; Gasödeme; Tollwut; Parasiten.* Ⓟ med. vet. h. c. Greifswald 1924.

Miezis, V., Dr., Chef der Fischerei-Abt. d. Lettländischen Landwirtschaftsministeriums. Riga (Latvija), Colpaka bulv. 6. *Fischzool.* o

Migone, L. E., Prof. Univ. Nation. Asunción (Paraguay). *Bact.* o
Migot, André (1892), Ass. à la Fac. des Sc. de Paris, Labor. Arago à Banyuls sur Mer, Pyr. Or. (France). *Zool.: Hydracarides et Halacarides.* ⓓ Paris 1918. ⓟ Hydracarides et Halacarides.
Migula, Walter, Dr. Prof. Eisenach (Deutschland), R.-Wagner-Str. 3. *Botan.: Pflanzenbiol.* o
Mihollć, Stanko (1891), Head Chem. dept. of State Inst. Hygiene. Zagreb (S.H.S.), 45 Jurjevska. *Biochem.: Analyses of drugs.* ⓓ Zagreb 1918.
Mihovil, Gračanin (1901), Chef der Agrochem. Abteilung der Landwirtschaftl. Versuchsstation Osijek 1 (S.H.S.), Reiznergasse 72. *Biochem. der Pflanzen: Stoffwechsel. Nährstoffresorption durch die Pflanzen. Bodenazidität, Katalasetätigkeit bei den autotrophen Pflanzen.* ⓓ Karls-Univ. Prag 1925.
van der Mijll Dekker, Wilhelmina Maria (1885), Botanikerin, Rijkslandbouwproefstat. voor Veevoederonderzoek. Wageningen (Holland). *Mikroskop. Untersuchungen der Viehfuttermittel.*
Mikami, Shôzo, Lect. Tôhoku Imper. Univ., Physiol. Inst., Coll. of Med. Sendai (Japan). *Physiol.*
Miki, Shigeru (1901), Ass. Botan. Imper. Univ. Botan. Inst. Dept. Sc. Kyôto (Japan). *Plant Ecol.* ⓓ 1925. ⓟ Plants and Preparations.
Miklaschewskaja, H. P., Leiterin d. Landw. Abt. d. Staatl. Mus. d. Jeniss. Gebiets. Krasnojarsk (U.d.S.S.R.), Ploschad Prosweschenija 2. *Flora d. Jenisseischen Gebiets.* o
Miklaszewski, Jean (1874), Dir. Départ. Forêts au Ministère de l'Agric. et des Domaines de l'Etat. Warszawa (Polska), 16, rue Poznańska. *Sylviculture.* ⓓ St. Petersbourg 1898.
Mikutowicz, Johann Mathias (1872), Bibliothekar. Rīga (Latvija), Pastendes ielā 9. *Bryol.: Sphagnum, Bryum, Drepanocladus).* ⓟ Bryotheca baltica.
Mikyška, Rudolf, Prof. Hradec Králové (C.S.R.), Karlova 441. *Phanérogames. Géobotan.* o
Milad, Yousif (1894), Physiol., Divis. of Hortic. Giza (Egypt). *Lime-induced Chlorosis.* ⓓ Univ. of California.
Milani, Alfons (1864), Dr. phil. Eltville, Rhein (Deutschland). *Vergl. Anat. der Wirbeltiere. Biol. der Forstinsekten. Angewandte Entomol.* ⓓ Gießen 1894.
Milanowsky, E. V., Doc. Bergakademie u. I. Univ. Moskau (U.d.S.S.R.). *Palaeont.* o
Milbrath, David Gallus (1880), Plant Pathol., California State Dept. of Agric. Sacramento, Cal. (U.S.A.). *Resistance in herbaceous ornamentals to plant diseases. Downy mildews on Allium. Sclerotinia cinerea. Prunus avium.* ⓓ Wisconsin 1904.
Mildbraed, Johannes (1879), Dr. phil., Kustos und Prof. am Botan. Mus. zu Berlin-Dahlem (Deutschland). *Blütenpflanzen; Acanthaceen. Floristik des tropischen Afrikas.* ⓓ Berlin 1904.
Miles, George Herbert (1880), Ass., Resistence Dir. National Inst. of Industrial Psychol. London W.C. 1 (England), 329 High Holborn. *Industrial Psychol.* ⓓ London 1905.
Miles, Lee Ellis (1890), Extension Plant Pathol. Auburn, Ala. (U.S.A.), Polytechn. Inst. *Plant Pathol. and Mycol.* ⓓ Univ. of Illinois 1920. ⓟ Mycol. and plant disease specimens.
Miles, Walter R. (1885), Prof. Experim. Psychol., Stanford Univ., California (U.S.A.), Box 1408. *Physiol. Psychol.: Influence of Alcohol on Human Subjects; Motor Functions of the Eyes; Effects of Reduced Diet; Motor Coordination; Static Equilibrium of Man.* ⓓ Iowa State Univ. 1913.
Miliutin, Wladimir (1879), Prof. Univ. Nishny-Nowgorod (U.d.S.S.R.), Studenaja 34. *Histol. und Embryol. Untersuchung der Gewebe im polarisierten Lichte.* ⓓ Petersburg. ⓟ Histol. und embryol. Praep.
Miliutina, Helene (1893), Doc. Univ. Nishny-Nowgorod (U.d.S.S.R.), Studenaja 34. *Histol. und Embryol. Gehirnentwickelung.*

Miljan, August (1889), Ass. Pflanzenbiol. Versuchsstation der Univ. Tartu (Estland). *Wiesenbau.* ⓓ Tartu.
Miljan, Erika (1898), Ass. an der Pflanzenbiol. Versuchsstation der Univ. Tartu (Estland).
Miljanić, Niko, Dr., Hon.Prof. für Anat., Dir. Anat. Inst. Beograd (S.H.S.), Miloša Pocerca ulica Kasama. *Topogr. u. chirurg. Anat.* o
Millais, John Guille (1865), Zool. and Botan. Horsham, Sussex (England). *Birds. Wildfowler in Scotland. Mammalia. Big Game.*
Millar, Charles J. (1898), Instr. in Biol. Rolla, Miss. (U.S.A.). *Bact.*
Millar, William Gilbert (1897), Ass. in the Dept. of Pathol., Univ., Ass. Pathol., Royal Infirmary, Edinburgh (Crichton Scholarship in Pathol.). Edinburgh (Scotland). *Erythrocyte. Anaemia.*
Millard, Wilfrid Arthur (1880), Lect. in Agric. Botan. and Adviser in Mycol. Leeds (England), The Univ. *Potato disease, Actinomycosis Leaf Roll. Disease of Rhabarb., Crown Rot.*
Miller, A. E., Demonstrator Univ. Dept. of Zool. Glasgow (Scotland). o
Miller, August Edward (1899), Res. Entomol. State Nat. History Survey Division. Urbana, Ill. (U.S.A.). *Termites of Illinois. Acarina of the United States.* ⓓ Ohio State Univ. 1925. ⓟ Acarina.
Miller, B. V., Prof. Polytechn. Inst. Ivanovo-Voznessensk (U.d.S.S.R.). *Botan.* o
Miller, David (1890), Dominion Entomol., Biol. Labor., Private Buy. Wellington (New Zealand). *Diptera.*
Miller, David F. (1892), Instr., Dept. of Zool. and Entomol., Ohio State Univ., Columbus, O. (U.S.A.). *Animal behavior, reactions of Insects and the effects of heat as a stimulus. Determination of sex in lower organisms.* ⓓ Ohio State Univ. 1923. ⓟ Cimex lectularis (eggs, nymphs, adults) preserved or on microscope slides. Gastrophilus equi (eggs, larvae, adults both sexes).
Miller, Edgar Calvin LeRoy (1867), Prof. of Bact. and Biochem. Richmond, Va. (U.S.A.), 2915 Seminary Avenue. ⓓ Michigan 1894.
Miller, Edwin Cyrus (1878), Prof. Plant Physiol., Plant Physiol. Kansas Exp. Station. Manhattan, Kan. (U.S.A.), Dept. Botan. K.S.A.C. *Water relations of plants. Chem. of plants.* ⓓ Yale Univ. 1910.
Miller, Frederick Robert, Prof. of Physiol., Med. School, Univ. of Western Ontario. London (Canada). *Nervous System, Cerebellum and Reflexes; Viscero-motor Reflexes.* ⓓ München 1911.
Miller, Gerrit S. (1869), Curator, Division of Mammals, U.S. National Mus. Washington, D.C. (U.S.A.). *Taxonomy of Mammals.* ⓓ Cambridge, Mass., 1894.
Miller, Henry Farrand (1889), Dir. State Cooperative Labor. Kenosha, Wis. (U.S.A.), City Hall. ⓓ Madison, Wis., 1916.
Miller, Hugh, Lect. School of Med. R. Coll. Edinburgh (Scotland). *Zool., Botan.* o
Miller, James (1875), Prof. of Pathol., Queen's Univ. (Canada). *Morbid Anat. and Histol. Practical Pathol. and Post Mortem Technique.* ⓓ Edinburgh 1899.
Miller, John Willoughby (1880), Prof., Dr., Prosektor an den Städtischen Krankenanstalten in Barmen. Leiter des Städtischen Untersuchungsamtes. Elberfeld (Deutschland), Jaegerstr. 2. *Normale und pathol. Histol. des weiblichen Genitals, besonders des Ovariums.* ⓓ Berlin 1904.
Miller, Julian C. (1895). Clemson College, South Carolina (U.S.A.), Hort. Dept. *Vegetable Crops.* ⓓ Cornell Univ. 1926.
Miller, Lawrence P. (1901), Ass. Chemist Agric. Experim. Sta. Lafayette, Ind. (U.S.A.). *Plant chem.*
Miller, Loye (1874), Prof. of Biol., Univ. of California, Los Angeles, Cal. (U.S.A.). *Fossil Birds. Geographic distribution of Birds and Mammals.* ⓓ Univ. of California 1912. ⓟ Skins and skeletons

of recent birds from California and Salvador, fossil Birds and Mammals from Pleistocene of Rancho La Brea.

Miller, Maria (1900), Landwirtschaftl. Inst. Detskoje Sseló (U.d.S.S.R.). *Pflanzenphysiol. Stickstoffernährung der Pflanzen.*

Miller, Paul Robert (1899), Extension Specialist, Agronomy. East Lansing, Mich. (U.S.A.), Michigan State Coll. *Disseminating crops.* ⓟ Michigan State Coll., East Lansing, Mich. 1924.

Miller, Pierre Alphonse (1897), Research Ass. in Plant Pathol. Riverside, Cal. (U.S.A.), Citrus Experim. Station. *Erysiphe cichoracearum.*

Miller, Ralph Lester (1902), Ass. Entomol., Florida Agric. Exp. Sta. Lake Alfred, Fla. (U.S.A.), Citrus Experim. Station. *Citrus Aphid and other citrus insects. Morphol. of Pentatomidae.*

Miller, Raymond J. (1889), Associate Prof., Agric. and Biol.Chem. State College, Pennsylvania (U.S.A.). *Nutrition and metabolism.* ⓟ Jeffersen Med.Coll.1918.

Miller, Robert C. (1899), Ass. Prof. of Zool., Univ. of Washington. Seattle (U.S.A.). *Experim. Zool., marine Invertebrates, Biol. of wood-boring Mollusks: Ornithol.: flight of birds.* ⓟ California 1923.

Miller, Shirley Putnam (1878), Instr. in Anat., Dept. of Anat., Univ. of Minnesota. Minneapolis, Minn. (U.S.A.), Inst. of Anat. *Mitochondria in the vertebrate cell, cell activity. Compar. anat. of fishes.* ⓟ Univ. of Minnesota 1922.

Miller, Viktor (1880)., Prof. Botan. Landw. Fak. am Iwanowo-Wosnessensker Polytechn. Inst.Moskau (U.d.S.S.R.), Ostojenka 40, W. 2. *Botan., Algol., Blütenbiol. Phytopathol.*

Miller, Ward (1892), Prof. of Biol., Grove City Coll. Grove City, Pa. (U.S.A.). *Plant morphol.*

Miller, William Byron (1895), Ass. Range Examiner. Fairbanks (Alaska), U.S. Biol. Surrey. *Reindeer grazing.* ⓟ Univ. of California 1925.

Mills, Ralph Garfield (1881), Prof. of Pathol., Univ. of Minnesota, in the Mayo Foundation. Rochester, Minn. (U.S.A.). *Tissue immunity, Parasitology.* ⓟ Northwestern Univ. Med. Sch. 1907.

Mills, Wilfred Douglas (1895), Extension Instr. in Plant Pathol. Ithaca, N.Y. (U.S.A.), Dept. Plant Pathol. Cornell Univ. *Factors conditioning the Velocity of disinfection, seed-borne pathogens. Plant diseases and insects.* ⓟ Mich. State Coll. 1922.

Milojević, Borivoje D., Dr., ao. Prof. Zool. Beograd (S.H.S.), Zool. Inst. *Protoz., Entwicklungsmech.* o

Miloslavskij, Valerian Vladimirovič, Prof. Univ. Kazan, Tatar.Rep. (U.d.S.S.R.). *Experim. Hygiene.* o

Milovanov, Doc. II. Univ. Moskau (U.d.S.S.R.). *Geobotan.* o

Milovidov, Pierre (1896), Inst. physiol. de plantes de l'Univ. Charles. Praha (C.S.R.). *Botan.: anat. et cytol. Panachure végétale. Tubercules radicaux. Symbiosis.* ⓟ Prague, Univ. Charles, 1924.

Milroy, Thomas H., Prof. Fac. of Sc. Univ. Belfast (Ireland). *Physiol.* o

Mimuroto, Yoshimitsu, Prof. of Tôkyô Imper. Seric. Coll. Nôgakushi. Nishigahara, Tôkyô (Japan), Chem. Labor. *Biochem. and Chem. physiol. of Silkworms.*

Mina, Iwan D., Kiewsches Wissensch. Inst. für Selektion. Kiew (U.d.S.S.R.), Turgenewskaja 57, W. 9. *Selektion landw. Pflanzen.* o

Minami, Sadaham, Prof. Kyôto Imper. Sericult. Coll. Hanazono near Kyôto (Japan), Genetical Labor. *Genetics of Silk-worms.*

Minami, T., Prof. Hokkaidô Imper. Univ., Fac. of Agric. Sapporo (Japan). *Food Crops.*

Mineff, Michael, Dr., Doc. Anat. Inst. d. Univ. Sofia (Bulgarien). *Topograph. Anat.* o

Minenkow, Alexander R., Prof. a. d. Staatl. Univ. inNishny-Nowgorod(U.d.S.S.R.),. *Pflanzenphysiol.* o

Miner, John Rice (1892), Associate Member, Inst. for Biol. Research of Johns Hopkins Univ. Baltimore, Md. (U.S.A.), 1901 East Madison Street. *Biometry.* ⓟ Hopkins 1922.

Minervin, Sergei (1888), Ass. Microbiol. Inst. d. Volksgesundheits-Kommissariats. Moskau (U.d. S.S.R.), Pogodinskaja 10. *Mikrobiol.* ⓟ Varsovie 1914.

Minerwin, Walentin W., Bact., Landwirtschaftl. Akademie Timirjasew. Moskau (U.d.S.S.R.), Petrowsko-Rasumowskoje 5, W. 5. *Landwirtschaftl. Bact. Flachs und andere Faserpflanzen.* o

Mingazzini, Giovanni (1859), Prof. di Clinica neuropsichiatrica nelle R. Univ. di Roma (Italia). *Anat. patol. del Cervello. Il Cervello in relazione con i fenomeni psichici. Centri nervosi. Der Balken.* Zentrale Hypoglossusbahn. *Le afasie.* ⓟ Roma 1883.

Minio, Michelangelo (1872), Prof., Dir. incar. Mus. Civico di Storia naturale. Venezia 108 (Italia), S. Cassiano-Ponte Raspi 1557. *Floristica e Fitogeografia. Fitofenol.* ⓟ Padova 1896.

Minkiewicz, Romuald (1878), Dir. du Labor. de Biol. générale à l'Inst. Nencki, Président Inst., Prof. de Biol. générale Univ. Varsovie (Pologne), rue Sniadeckich 8. *Analyse expérim. de l'instinct, de l'habitude et de la mémoire des animaux inférieurs. Synchromatisme, chromotropisme et vision chromatique. Le polybolisme nerveux fondamental: excitabilité polybolique variable et conductibilité qualitative. Morphodynamique générale (protist. et végétaux).* ⓟ St. Pétersbourg 1900.

Minkowski, Mieczyslaw (1884), Dr., Prof., Oberass. am Hirnanat. Inst. der Univ. Zürich (Schweiz), Physikstr. 6. *Neurol., Anat., Physiol., pathol. Anat. und Klinik des zentralen Nervensystems. Zentrale optische Bahnen. Anat. Verbindungen des Großhirns. Studium der Reflexe, Athetose, Aphasie. Neurobiol. Untersuchungen an menschlichen Foeten. Bewegungen, Reflexe und idio-muskuläre Phänomene.* ⓟ Breslau 1907, Kasan (Rußland) 1908.

Minnich, Dwight Elmer (1890), Associate Prof. of Animal Biol., Univ. of Minnesota. Minneapolis, Minn. (U.S.A.), Dept. of Animal Biol. *Sensory Physiol. and Animal Conduct of insects.* ⓟ Harvard 1917.

Minobé, Hiromu (1898), Ass. Zool. Inst. Fac. Agric. Imper. Univ. Tôkyô (Japan), No.1, Shichichome, Aoyama-kita machi. *Development of the thyroid gland, food-habits of Tachydromus tachydromoides.* ⓟ Tôkyô Imper. Univ. 1925.

Minod, Marcel-Maurice (1887), Ass. Inst. de Botan. Univ. Genève (Suisse), Grand Rue 17. *Anat. et biol. botan.; systématique des Scrophulariacées; ampelographie et pathol. de la vigne; vaisseaux laticifères.* ⓟ Genève 1917. ⓟ Stemodia.

Minoshima, T., Ass. Prof. of Hokkaidô Imper. Univ. Sapporo (Japan), Physiol. Inst., Fac. of Med. *Physiol.*

Minot, Ann Stone (1894), Research Associate in Pharmacol. Nashville, Tenn. (U.S.A.), Vanderbilt Univ. *Toxicity of Carbontetrachloride and its use as an anthelmintic.* ⓟ Harvard Univ. 1923.

Minouchi, Osamu (1896), Lect. of Imper. Univ. in courses of Histol. and Cytol. Kyôto (Japan), Zool. Labor., Coll. of Sc. *Spermatogenesis of vertebrates. Mitochondria and other cellular organs.* ⓟ Tokyo 1922.

Minoura, Tadanaru, Prof. Kyôto Med. Univ. (Provincial). Hanazono near Kyôto (Japan), Biol. Labor. *Experim. Embryol.* ⓟ Kyôto.

Mir y Lhambías, Antonio, Prof., Catedrático de Agric. Mahón (España). o

Miram, Emilie (1870), Zool. Mus. der Akademie der Wissenschaften. Leningrad (U.d.S.S.R.). *Entomol.: Systematik der Orthopteren.*

Miram, Konst. Rud., Prof. Gosudarstv. Donskoj Univ. Rostov a. Don (U.d.S.S.R.). *Pathol.* o

Mirande, Marcel (1864), Prof. Fac. Sc. Univ. Grenoble, Dir.-fondateur de l'Inst. Botan. Alpin du Lautaret, Hautes-Alpes (France). *Botan. alpine, Cuscutacées, Cassytacées. Anat., Physiol., Cytol., Chim. végétales. Histoire de la Botan.* ⓟ Graines de plantes alpines.

de Miranda y Rivera, Alvaro (1896), Dir. Inst. Español de Oceanografía. Málaga (España). *Crustacés décapodes et Cirripédes.* Ⓓ Univ. de Madrid 1922. Ⓟ *Crustacea, Poissons.*

Mirbt, Carl Alexander (1902), Diplomlandwirt. Berlin-Steglitz (Deutschland), Grunewaldstr. 34. *Tierzucht; Geschlechtsbestimmung; Tierernährung.*

Mire, Ch., Prof. Ecole prép. de méd. et de pharmacie. Reims (France). *Botan.* o

Miroljubov, Viktor Pavl, Prof. Staats-Univ. Tomsk (U.d.S.S.R.), Prospekt Timirjazeva 3. *Pathol. Anat.* o

Mirto, G., Doc. Univ. Palermo (Italia). *Elettrobiol.* o

Mirza, M. B. (1900). Frankfurt a. M. (Deutschland), Robert-Mayer-Str. 6, Zool. Inst. der Univ. *Nematodes: Dracunculus medinensis.* Ⓓ London1923.

Mischenko, Paul J., Prof., Kubansches Landw. Inst. Krasnodar (U.d.S.S.R.). *Botan., Systematik, Geographie.* o

Mischustin, Eugen (1901), Ass. Bact. Agronomical Stat. Moscow (U.d.S.S.R.), Konjuschkovskaja 31. *Soil bact.*

Misikov, M. A., Prof. Pädag. Inst. d. Bergrepublik. Vladikavkaz (U.d.S.S.R.), Ul. Marksa 20. *Anthrop.* o

Mislawsky, Nikolaj Aleksandrovič, Prof., Dir. Histol. Labor. Univ. Kazan (U.d.S.S.R.), Černyševskaja 18. o

Misuri, Alfredo, Lib. doc. Univ. Messina (Italia). *Zool.* o

Misvoer, Hans, Prof. Dr., Landwirtschaftl. Hochsch. Aas (Norge). *Gartenbau.* o

Mitamura, Tokushirô, Ass. Prof. Imper. Univ. Tôkyô (Japan), Pathol. Inst., Fac. of Med. *Experim. Pathol.*

Mitchell, Harold H. (1886), Prof. of Animal Nutrition, Univ. of Illinois. Urbana, Ill. (U.S.A.), 556 Old Agric. Building. *Protein metabolism. The protein value of animal foods and farm feeds. The basal metabolism of farm animals, maintenance, growth and fattening.* Ⓓ Univ. of Illinois 1915.

Mitchell, Helen S. (1895), Prof. of Nutrition and Dir. of Nutrition Research at Battle Creek Sanitarium and Coll. Battle Creek, Mich. (U.S.A.). *Vitamins and minerals.* Ⓓ Yale 1921.

Mitchell, Peter Chalmers (1864), Dr., Secretary Zool. Society of London (England), Regent's Park. *Vertebrate Anat.; Biol. Philosophy.* Ⓓ Aberdeen 1884, Oxford 1888.

Mitchell, Philip H. (1883), Prof. of Physiol. in Brown Univ. Providence, R.I. (U.S.A.). *Enzymes of purin metabolism, carbohydrate utilization, absorption of potassium by animal cells, nutrition and oxygen requirements of shell fish, studies in shad culture.* Ⓓ Yale 1907.

Mitchell, Theodore B. (1890), Ass. Prof. Zool. and Entomol. Raleigh, N.C. (U.S.A.), Dept. Zool. and Entomol., N.C. State Coll. *Systematic study of the Apoidea, the genus Megachile.* Ⓓ N.C. State Coll. 1924.

Mitscherlich, Eilh. Alfred (1874), Univ.Prof. Dr. phil., Dir. des Pflanzenbau-Inst. und geschäftsführender Dir. des Landwirtschaftl. Inst. der Univ. Königsberg i. Pr. (Deutschland), Tragheimer Kirchenstr. 83. *Pflanzenphysiol. Bodenkunde.* Ⓓ Kiel 1898.

Mitsuhashi, Shinji, Ass. Tôkyô Imper. Univ. Komaba near Tôkyô (Japan), Zool. Labor., Coll. of Agric. *Systematic Entomol.*

Mitrophanova, Julia (1899), Laborant Entomol. Dept. of the Bact. Inst. Perm (U.d.S.S.R.), Solikamskaja 17/34. *Oekol., Systematics, Variability of the Culicidae.* Ⓟ Culicidae.

Mitrović, Tanasije (1892), Doc. für Zootechnic Agr. Fac. Univ. Beograd (S.H.S.), Zootechnisches Inst.; Dobračina 16. *Tierzucht.* Ⓓ Wien 1921.

Mittasseau, Jean (1901), Inst. agricole, prép. Station de pathol. végétale. Paris (France), 54, rue St. Georges. *Céréales.*

Mitter, Julian Herron (1881), Reader in Botan. Botan. Labor. Univ. Allahabad (India). *Plant Pathol. and Physiol.* Ⓓ Lahore, Panjab, 1906.

Miura, Tsunesuke, Lect. Imper. Univ. Kyôto (Japan), Inst. of Microbiol., Fac. Med. o

Miwa, M., Prof. Hokkaidô Imper. Univ. Sapporo (Japan), Pharmacol. Inst., Fac. of Med. *Pharmacol.*

Mix, Arthur J., Prof. of Botan., Univ. of Kansas. Lawrence, Kan. (U.S.A.), *Pathol., Physiol.* o

Miyabe, Kingo (1860), Prof. of Botan., Dir. Botan. Garden, Hokkaidô Imper. Univ. Sapporo (Japan), Kita 6-jo, Nishi 13-chome. *Phytopath., Mycol., Flora of Northern Japan. Laminariaceae. Flora of the Kurile Islands and Sachalin. Laminariaceae of Hokkaido.* Ⓓ Harward 1889.

Miyai, Shigetsugu, Prof. Med. Univ. Niigata (Japan), Inst. of Bact. and Hygiene. *Microbiol.*

Miyaji, Denzaburô, Lect. Kyôto Imper. Univ. Ôtsu, Shiga-ken (Japan), Lake-side Biol. Station. *Physiol.*

Miyajima, Mikinosuke, Member Kitasato Inst. of Infectious Diseases, Lect. of Keiô-Gijuku Univ. Shiba, Tôkyô (Japan). *Lepidoptera (Rhopalocera). Pathol. (Human).* Ⓓ Med. and sc. Tôkyô.

Miyake, Chûichi (1894), Division of Phytopath. Ohara Inst. for Agric. Research. Kurashiki, Okayama (Japan). *Phytopath.: Brown Shot hol disease of cherry leaves caused by Mycosphaerella cerasella. Gibberella aubinetii causal fungus of the wilt-disease of Horse-bean.*

Miyake, Kiichi, Ass. Prof. Tôkyô Imper. Univ. Komaba near Tôkyô (Japan), Botan. Labor., Coll. of Agric. *Genetics.*

Miyake, S., Ass. Prof. Hokkaidô Imper. Univ. Sapporo (Japan), Chem. Inst., Fac. of Agric. *Organic and Biochem.*

Miyake, Tsutome (1880), Chief Plant Pathol., Dept. of Agric., Government Research Inst. Taihoku, Formosa (Japan). *Plant Pathol., chiefly for Sugar Cane and Pineapple.* Ⓓ Sapporo Agric. Coll. 1904.

Miyashita, Yoshinobu, Lect. of Jikeikai Med. Univ. Shiba, Tôkyô (Japan), Biol. Labor. *Protozool. Enteropneusta.*

Miyazaki, H., Prof. Hokkaidô Imper. Univ. Sapporo (Japan), Physiol. Inst., Fac. of Med. *Physiol.*

Miyazaki, Saburô, Ass. Prof. of Keiô-Gijuku Univ. Yotsuya, Tôkyô (Japan), Pharmacol. Inst., Coll. of Med.

Miyoshi, Manabu, Prof. emer. of Tôkyô Imper. Univ. Koishikawa, Tôkyô (Japan), Botan. Inst., Botan. Garden. *Systematic and Ecol. of Plant.*

Mjasnikow, Boris J., Vorsteher der Selektions-Abt. der Tulunsk. Versuchsstation. Tulun, Irkutsker Gouv. (U.d.S.S.R.). *Agronomie, Selektion.* o

Mjassoiedov, S. W., Prof., Dir. Histol.-Embryol. Labor. Univ. Tomsk (U.d.S.S.R.). o

Mjöen, J. A. H., Dr. Vindern, pr. Oslo (Norge). *Rassenbiol.* o

Mó, A., Prof. Univ. Rosario (Argentina). *Psychol.* o

Mochi, Aldobrandino (1874), Prof., Dir. Museo Nazionale d'Antrop. e Etnol. Florence (Italia). *Anthrop., Ethnol., Paléont.* Ⓓ Pise 1911. Ⓟ Objets anthrop. et etnographiques.

Mochizuki, Shûsaburô, Prof. Keiô-Gijuku Univ. Yotsuya, Tôkyô (Japan), Anat. Inst., Coll. of Med. *Anat.*

Mockeridge, Florence Annie (1889), Dr., Head of Dept. of Biol. at Univ. Coll. of Swansea, S.Wales (England). *Physiol. Botan. and Plant Biochem.* Ⓓ London 1917.

Mockus, Sigismundus (1880), Doc., Leiter Zootom. Inst. d. Vet.-Abt. Univ. Kaunas (Litauen). *Haustiere.* o

Moczarski, Z., Dr., Prof. Poznań (Polska), Univ. *Zool.* o

Modestow, Valery Wladimirowitsch (1890), Doc. Landwirtschaftl. Entomol., Entomol. Labor. d.

Landwirtschaftsak. Timirjaseff, Gouvernementsspezialist der Forstentomol. Moskau (U.d.S.S.R.), Petrowsko-Rasumowskoje. *Kreuzfeldheuschrecke (Stethophyma flavicosta Fisch). Biol. der verkürzten Birnenblattwespe (Micronematus abbreviatus Het.). Forstschädlinge und ihre Bekämpfung.*

Modilewski, Jakob (1883), Prof. Handelstechn. Hochsch., Mitglied des botan. wissenschaftl. Erforschungskatheder am Botan. Garten. Kiew, Ukraine (U.d.S.S.R.), Karawajewskaja 17, W. 6. *Botan. Embryol. und Zytol.* ⓓ München.

Modrakowski, Georges (1875), Prof. Univ., Dir. Inst. Pharmacol. experim. Warszawa (Polska), Krakowskie. Przedmieście 26/28. *Cœur et vaisseaux, reins, l'appareil digestif.* ⓓ Leipzig 1898.

Möbius, Martin (1859), o.ö. Prof. der Botan. an der Univ. Frankfurt a. M., Geh.R., seit Ostern 1926 emeritiert. Frankfurt a. M. (Deutschland), Königsteiner Str. 52. *Anat. und Physiol. der Pflanzen, Algen, Pflanzenfarben.* ⓓ Heidelberg 1883.

von Möllendorff, Wilhelm (1887), Prof., Dir. Anat. Inst. Kiel (Deutschland), Prinz-Heinrich-Str. 40. *Allgemeine Histol. und Histophysiol., vitale Färbung, Niere, Blutgefäß-Bindegewebsapparat.* ⓓ Heidelberg 1911.

Möller, Thure, Dr. med., ao. Prof., Adjunkt Univ. Helsinki (Finnland), Fredsgatan 11 B. *Anat.* o

Möllers, Bernhard (1876), ORR., ao. Prof. Univ., Mitgl. des Reichsgesundheitsamts. Berlin NW 87, (Deutschland), Klopstockstr. 18. *Bact.* ⓓ Berlin 1902.

Moen, Olav (1875), Prof. Norges Landbrukshöiskole. Aas b. Oslo (Norge). *Gartenbau, Gemüsepflanzen.* ⓓ Aas 1913.

Moenkhaus, William J. (1871), Prof. Physiol. Inst., Bloomington, Ind. (U.S.A.), Indiana Univ. ⓓ Chicago 1913.

Mörner, Carl Thore (1864), Prof. der med. und physiol. Chem. an der Univ. Upsala (Schweden). *Proteine.* ⓓ Upsala 1892.

v. Moesz, Gustav (1873), Dr., Abteilungsdir. des Ungarischen National Mus. Budapest V. (Ungarn), Akadémia u. 2. *Mykol., Cecidiol., Pflanzenpathol. Pilzflora Ungarns.* ⓓ Kolozsvár 1908.

Baron de Moffarts, Paul (1869). Château de Botassart par Noirefontaine (Belgique). *Lépidoptères paléarctiques. Pectinicornes, Lamellicornes, Longicornes du globe.* ⓓ Liège 1891. ⓓ Lépidoptères paléarctiques. Pectinicornes, Lamellicornes, Longicornes.

Moffett, Lacy Irvine (1878), Ornithol. Kiangyin, Ku. (China). *Birds of China.* ⓓ Bird skins from China.

Mogiljanskij, Nikolaj K., Wissensch. Landwirtsch. Komitee der Ukrain. „Narkomsema". Charkow (U.d.S.S.R.), Ul. Liebknechta 82. *Angew. Botan.* o

Mogil'nickij, Boris Nester, Prof. Univ. Nizny Novgorod (U.d.S.S.R.), Sovetskaja Ploščad 8. *Pathol. Anat.* o

Mohamed-Sharif, Lect. Univ. Aligarh, Unit. Prov. of Agra and Oudh. (India). *Zool.* o

Mohler, John Robbins (1875), Chief of Bureau of Animal Industry. Washington, D.C. (U.S.A.), Dept. of Agric. *Livestock production, animal disease.* ⓓ V.M. Univ. of Pennsylvania 1896, Sc. Iowa State Coll. 1920, Univ. of Pennsylvania 1925. ⓟ Bacterial cultures of causative agents of infectious diseases of livestock.

Mohr, Erna W. (1894), Freiwillige wissenschaftl. Hilfsarbeiterin. Hamburg (Deutschland), Zool. Mus. *Ichthyol., Fischereibiol., Mammal., Wirbeltier-Faunistik.* ⓓ Kleinsäugetiere, Amphibien, Fische von Schleswig-Holstein.

Mohr, Otto Lous (1886), Prof. Anat. Univ. Oslo (Norge), Anat. Inst. *Zytol.; experim. Genetik.* ⓓ Univ. Oslo 1917.

Moissejew, Alexander S. Leningrad (U.d.S.S.R.), Com. Géol. de Russie. *Jura- und Kreidefaunen und Juraflora der Krim.* o

Moissejewa, Maria (1889), Ass. Pflanzenphysiol. Labor. Univ. Kiew, Ukraine (U.d.S.S.R.). INO. *Physiol. der Zellteilung.*

Mokejewa, Ekaterina A., Doc. an der Mittel-Asiatischen Staatl. Univ. Taschkent (U.d.S.S.R.), Str. I, Maja 4. *Botan., Agronomie.* o

Mokin, Nikita Nicitich (1891), Doc. Univ. Taschkent (U.d.S.S.R.). *Soil and Fertilizer.*

Mokrzecki, Sigismund Athanasius (1865), o.Prof. der Entomol. und des Forstschutzes an der Hochsch. für Bodenkultur in Warschau (Polen), Skierniewice-Palac, Entomol.-Inst. *Biol. der Insekten, Schädlinge der Pflanzen, Innere Therapie der Pflanzen.* ⓓ Hon. c. Simferopol (Crim) 1918. ⓟ Chalcidoidea.

Mola, Pasquale (1872), Dir. della Scuola di pesca di Cagliari (Italia), Gabinetto di Sc. *Parassit., Idrobiol., Botan. Acquicoltura e Piscicoltura.* ⓓ Napoli 1902. ⓟ Cestodi, rotiferi e piante sarde.

Molander, Arvid Ragnar (1886). Fiskebäckskil (Sverige). *Fischereibiol.: Clupea, Gadus, Pleuronectes, Alcyonarien, Gorgonarien.* ⓓ Upsala 1916.

Moldenhawer, Constantin (1890), Chargé de Cours à l'Univ. de Cracovie. Poznań (Polska), 53, rue Matejki. *Botan. agricole, génétique des plantes. Bastarde zwischen Weizen und Aegilops. Croisement de Raphanus avec Brassica.* ⓓ Poznanie 1921.

Moldovan, Juliu, o. Prof. Univ. Cluj (România), *Hygiene.* o

Molina, Ricardo, Prof. Univ. of St. Tomas. Manila (Philippine Islands). *Anat., Histol., Embryol.* o

Molina, Wenceslao, Prof. Univ. Lima (Peru). *Anat. Physiol.* o

Molinari, Léon Lucien (1901), Prép. Insectarium Inst. des Recherches agronomiques. Menton, A. M. (France), Avenue Cernuschi. *Entomol.*

Molisch, Hans (1856), o.ö. Prof. und Dir. des Pflanzenphysiol. Inst. an der Wiener Univ. Wien VIII (Deutsch-Österreich), Zeltgasse 2. *Anat. und Physiol. der Pflanze. Mikrochem., Mikrobiol.* ⓓ Wien 1880.

Moljakov, L. I., Prof. Pedagog. Inst. Jaroslavl (U.d.S.S.R.), Respublikanskaja 122. *Pflanzenzucht.* o

Moll, Jan Willem (1851), Prof. emer. der Botan. Groningen (Holland), Oranjesingel 3. *Pflanzenanat.* ⓓ Leiden 1876.

Møller, Knud Ove (1896), Priv.Doc. Kopenhagen (Dänemark), Univ. Farmacol. Inst. *Experim. Pharmacol., spec. Nierenphysiol., Wirkungsweise der Diuretica. Wasser- und Salzwechsel, Theophyllindiurese.* ⓓ Kopenhagen 1926.

Møller Sørensen, A., Prosector Patol.-Anat. Labor. Vet. Landw. Hochsch. København V (Danmark), Bülowsvej. o

Møllgaard, Holger (1885), Prof. of physiol. København Charlottelund (Danmark), 15 Esperanceallé. *Animal nutrition; experim. therapy. Tuberculosis.*

Mollard, M., Prof. Fac. Sc. Paris 5 (France), 16, rue Vauquelin. *Botan.* o

Mollier, Siegfried (1866), o.ö. Univ.Prof., GMR., Dir. der Anat. Anstalt München (Deutschland), Vilshofener Str. 10. *Embryol., Histol. mechanische Fragen aus dem Gebiet des Bewegungsapparates.* ⓓ München 1889. ⓟ Diapositive von Aktaufnahmen des lebenden menschlichen Körpers.

Mollison, Theodor (1874), Dr., o. Prof. für Anthrop. an der Univ. München (Deutschland), Friedrichstraße 19. *Primatenmorphol., Fossile Hominiden, Serol. Verwandtschaftsreaktionen.* ⓓ Freiburg i. B. 1898.

Moltschanowa, Olga (1886), Abteilungsvorsteher. Moskau (U.d.S.S.R.), Pogodinskajastr. 8, 6. *Tierstoffwechsel, Gaswechsel.*

Molz, Emil (1876), Dr. phil., stellv. Dir. der Versuchsstation für Pflanzenkrankheiten. Halle a. d. S. (Deutschland), Bismarckstr. 18. *Pflanzenpathol.: Immunitätszüchtung und Pflanzentherapie.* ⓓ Jena 1907.

Molz, Francis J. (1890), Prof. of Biol. Dayton, O. (U.S.A.). Univ. *Ascent of Sap.* ⓓ Freiburg (Schweiz) 1924.

Momiyama, Tokutarô (Japan), Sasazuka 1146. *Ornithol.*

Lo Monaco, Domenico (1863), Dir. Ist. Chim. Fisiol. Univ., Prof. Fac. med. Roma (Italia), Via Depretis 92.

von Monakow, Constantin (1853), Prof. hon., Dir. d. Hirnanat. Inst. u. d. Nervenpoliklinik d. Univ. Zürich (Schweiz), Dufourstr. 116. *Neuronenlehre: Hirnanat., -physiol., -pathol., vergl., experim. Sehsphäre u. die opt. subkortikalen Zentren, Lokal. im Großhirn, der rote Kern d. Haube, Schizophrenie u. Plexus chorioidei, Biol. d. Instinktwelt.* ⓓ Zürich 1880.

Monara, Albert (1886), Dr. ès sc., Priv.Doc. à l'Univ. de Neuchâtel (Suisse), La Chaux-de-fonds, Nord 31. *Harpacticides marins et d'eau douce. — Faune des eaux douces.* ⓓ Neuchâtel 1919.

Monari, Dino, Lib. Doc. Ist. super. di Med. vet. Bologna (Italia). *Anat. patol. vet.* ○

Mond, Rudolf (1894), Priv.Doc., Ass. am Physiol. Inst. Kiel (Deutschland), Hegewischstr. 5. *Physiol. des Bluts, Allgemeine physikalisch-chem. Physiol.* ⓓ Kiel 1921.

Mondonnedo, Mariano (1887), Ass. Prof. of Animal Husbandry. Los Baños, Laguna (Philippines), Coll. of Agric. *Swine husbandry.* ⓓ Coll. of Vet. Sc. 1926.

Monia, G., Prof. Fac. Med. Bahia (Brasilia). *Allg. Pathol.* ○

Monjuschko, Voldemar (1903), Ass. expéd. du prof. B. A. Fedtschenko, l'Oural. Leningrad (U.d.S.S.R.), Jardin Botan. *Scrophulariaceae. Végétation de l'Ingrie et du gouvernem. Astrakhan.* ○

Monné, Ludwik (1904). Lwów (Polska), ul. Wiśniowieckich 1. *Heterochromosomen bei Hermaphroditen. Genetik von Lymantria dispar.*

Monod, André Théodore (1902), Préparateur au Mus. National d'histoire nat. Paris 5 (France), 57, rue Cuvier. *Carcinol.: Isopodes aquatiques. Gnathiidae. Océanographie biol., ichthyol. appliquée, poissons du Cameroun.* ⓓ Paris 1926.

Monro, Charles Carmichael Arthur (1894), Ass. Zool. Dept. of the British Mus., Nat. Hist. London (England), Cromwell Road. *Taxonomy of the Annelida Polychaeta.* ⓓ Oxford 1920.

Monserrat, Carlos, Ass. Prof. Univ. of the Philippines. Manila (Philippine Islands). *Pathol. and Bact.* ○

Montana, Eliseo, Prof. Univ. Nac. Bogotá (Columbia). *Histol. e Embriol.* ○

Monte Santos, Nicolaus (1882), Prof. Forstl. Hochsch. u. Landwirtsch. Hochsch. Palaion Phaleron (Griechenland). *Morphol. der Pflanzen.* ⓓ München 1912.

Montell, Justus (1869), Forstmeister. Muonio (Finnland). *Gefäßpflanzen, Vogeleier, Vogelbölge, Schmetterlinge.* ⓟ Pflanzen (nur Gefäßpflanzen), Vogeleier und Schmetterlinge.

Montemartini, Luigi (1869), Prof. Botan. Univ. e Dir. Labor. Crittogamico. Pavia (Italia), Orto Botan. *Anat., fisiol. e patol. vegetali. Ficol.* ⓓ Pavia 1920. ⓟ *Funghi.*

Monterosso, Bruno (1887), Libero doc. in Zool., Anat. e Fisiol. comparate; Aiuto Zool. e Anat. comparata univ. di Catania; Prof. incaricato di Istol. e Fisiol. generale. Catania, Sicilia (Italia), Ist. Zool. della R. Univ. *Struttura e funzione delle ghiandole sessuali (Mammiferi). Anat. e istol. dei Cirripedi (Balanus) e dei Copepodi parassiti (Peroderma). Biol. sessuale dei Ragni (Salticus). Ghiandola uropoietica degli Uccelli (Linota). Teratol. (Vertebrati: Sus). Pesca e biol. marina.* ⓓ R. Univ. di Capania 1911.

Monteverde, Nicolaus A., Vorstand Mus. Botan. Hauptgarten. Leningrad (U.d.S.S.R.), Apotheker. insel, Pessotschnaja $^1/_2$. *Physiol. der Pflanzen und Arzneipflanzen. Salpeter. Calcium, Magnesiumoxalat in Pflanzen. Chlorophyll. Lichteinfluß. Aromatische Pflanzen und ätherische Öle.* ⓓ Petersburg 1889.

Montfort, Camille, Prof. Univ. Halle a. d. S. (Deutschland), Gütchenstr. 20b. *Botan.* ○

Montgomery, B. Elwood (1899). Poseyville, Ind. (U.S.A.), R. 3. *Entomol., Odonata, taxonomy, life history, distribution. Hibernating insects.* ⓓ Purdue Univ. 1925. ⓟ *American Odonata.*

Monti, Achille (1863), Prof. o., Dir. Ist. Anat. Patol. Univ. Pavia (Italia). *Malattie infettive e parassitarie. Malaria.* ⓓ Pavia 1887.

Monti, Rina (1872), Prof. o., Dir. Ist. Anat. e fisiol. compar. Univ. Milano (Italia), via Gadio, 2. *Istol. compar. e sperimentale del sistema nervoso degli animali inferiori. Istol. e fisiol. sperimentale degli animali ibernanti. Limnol. degli alti laghi alpini italiani. Biol. lacustre dei grandi laghi marginali delle alpi italiane. Plankton d' acqua dolce. Protozoi. Idracnidi. Cladoceri.* ⓓ Pavia 1892.

Monticelli, F. S., Prof. Univ. Napoli (Italia). *Zool.* ○

Montoussis, C., Dir. du Lab. central de l'Hygiène. Athènes (Grèce), rue Scoufa 73. *Bact.* ○

Moody, Robert Orton (1864), Associate Prof. of Anat., Univ. of California. Berkeley, Cal. (U.S.A), Department of Anat. ⓓ Yale 1894.

Moomaw, Leroy (1889), Associate Agronomist, U.S. Department of Agric. and Supt. Dickinson Substation. Dickinson, N.D. (U.S.A.). *Tillage and Rotation, Plant Physiol.* ⓓ Univ. of Missouri 1915. ⓟ *Seeds of improved and standard varieties of cereal and forage crops.*

Moore, A. R. (1882), Prof. of General Physiol. and Head of the Department of Zool. Eugene, Ore. (U.S.A.), Univ. of Oregon. *Nervous system of invertebrates, luminescence.* ⓓ California 1911. ⓟ *Marine invertebrates (littoral).*

Moore, Andrew Charles, Prof. of Biol., Univ. of South Carolina. Columbia, S.C. (U.S.A.). ○

Moore, A. E., Prof., Univ. Coll. Cork (Ireland). *Pathol.* ○

Moore, Barrington (1883), New York, N.Y. (U.S.A.), 925 Park Avenue. *Forestry and plant ecol.* ⓓ Yale 1908.

Moore, Carl R. (1892), Associate Prof. in Zool., The Univ. of Chicago, Ill. (U.S.A.). *Sex in Vertebrates, Sex modifications, gonad transplantation, Physiol. of the scrotum, Histol. of sex glands. Organs of internal secretion.* ⓓ Univ. of Chicago 1916.

Moore, Caroline S., Associate Prof. of Biol., Univ. Redlands, Cal. (U.S.A.), 32 S. University St. *Genetics, physiol.* ○

Moore, Dwight Munson (1891), Prof. and Head of the Botan. Department, Univ. of Arkansas. Fayetteville, Ark. (U.S.A.). *Plant Physiol.: the Rest Period of Bulbs.* ⓓ Ohio State Univ. 1924.

Moore, Emmeline (1872), Dir. State Biol. Surveys. Albany, N.Y. (U.S.A.), Conservation Dept. *Fish Culture. Fish diseases and fish food. Biol. survey of Lake George. Survey of Genesee river system.* ⓓ Cornell 1914.

Moore, Ernest John (1897), Lect. in Botan., Municipal Technical School Birmingham (England). ⓓ Manchester 1920.

Moore, George A. (1899), Instr. of Zool., Okla. A. and M. Coll. Stillwater, Okla. (U.S.A.), Zool. Dept. *Taxidermy.*

Moore, George Thomas (1871), Dir., Missouri Botan. Garden and Engelmann Prof. of Botan., Washington Univ. St. Louis, Mo. (U.S.A.). *Bacteriol of Soil, Fresh-water Algae.* ⓓ Harvard Univ. 1900.

Moore, J. Percy (1867), Prof. of Zool., Univ. of Pennsylvania. Philadelphia, Pa. (U.S.A.). *Systematics of Annelida. Systematics, anat. and bionomics of Hirudinea. Control of mosquitoes by means of fishes* ⓓ Univ. of Pennsylvania 1896.

Moore, Ransom Asa (1861), Prof. of agronomy Dept. of Agronomy (Wisc.). Madison, Wis. (U.S.A.), 38 Virginia Terrace. *Agronomy.* o

Mooser, Hermann (1891), Dr. med. Pathol. to the American Hospital and to the Lord Cowdry Sanatorium, Mexico-City. D.F. (Mexico). *Immunity in diseases produced by Spirochaetes.* ⊕ Zürich 1917.

Moraczewski, Wacław Damian (1867), Dr. phil. et med., o. Prof. biol. Chem. vet.-med. Akademie (h. t. Rector). Lwów (Polska), Akademja Medycyny Weter, Kochanowskiego 61. *Pathol. u. klinische Chem.* ⊕ Phil. Zürich 1894; med. 1895.

de Moraes Sannento, Antonio Luis, o. Prof. Univ. Coimbra (Portugal). *Pathol.* o

Moraev, Petr (1864), Ass. Leningrad (U.d.S.S.R.), Černigovskaij Str. 5. *Pathol. Anat. der Haustiere.* ⊕ Tierärztl. Hochsch. Kazan 1899.

Morán Bayo, Juan, Prof., Catedrático de Agric. Córdoba (España). o

Mordvilko, Alexander K. (1867), Chef-Zool., Zool. Mus. Akademie der Wissensch. Leningrad (U.d. S.S.R.). *Aphidodea, Vermes.* ⊕ Univ. Kiev 1901.

Mordwnkina, Alexandra Ivanovna (1893), Ass. Hafer-Sektion Feder. Inst. angewandte Botan. u. neue Kulturen. Leningrad (U.d.S.S.R.), Herzenstr. 44. *Botan. und genetische Studien am kultivierten Hafer.*

Moreau, Fernand (1886), Prof. de botan. pure et appliquée à la Fac. des Sc. de Clermont, Dir. de la Station biol. de Besse (Puy-de-Dôme). Clermont (France). *Organisation et développement des Champignons, structure et biol. des Lichens, systématique et écol. des Mousses et Lichens. Phytopath., sélection des Céréales.*

Moreau, Valentine (1886). Clermont, Puy-de-Dôme (France), 28, rue Vermenouze. *Cryptogamie, particulièrement Lichens. Sexualité chez les Urédinées.*

Moreira, Carlos (1869), Dir. de l'Inst. Biol. et chef d' Entomol. Agricole. Rio de Janeiro (Brasil), Rua Prudente de Moraes 63, Ipanema. *Insectes nuisibles, crustacés, Oligochaetae. Tomaspis liturata. Stephanoderes coffeae. Hemiptères capsides du tabac.*

Morel, Jules Pierre (1893), Aide-préparateur d'Histol. Lille (France), Fac. de Méd. *Modifications du tissu conjonctif dans l'oedème. Anastomoses artério-veineuses. Tests endocriniens.* ⊕ Lille 1921.

Morel, M. L. E., Chef de Labor. Fac. Méd. Paris 7 (France), 31, Boulevard Raspail. *Physiol.* o

Morel, Pierre Victor Albert (1875), Prof. de Chim. organique et Toxicol. Fac. Méd. et Pharmacie. Lyon (France), 13, qual Claude Bernard. *Biochem. Pharmacol.* ⊕ Lyon 1910.

Morelli, Elisa (1902), Ass. univ. Ist. di Patol. generale di Firenze (Italia), Viale Morgagni 18. *Tiroide umana.* ⊕ Firenze 1925.

Morelli, Ferdinando, Lib. doc. Univ. Genova (Italia). *Patol.* o

Morenas, Léon (1891), Prép. labor. de Parasit. Lyon (France), 28, quai de la Guillotière. *Parasit.* ⊕ Lyon 1922.

Morency, Henry Lloyd (1896), Extension vet. (Col. Agric. Coll.), Animal Pathol. Fort Collins, Col. (U.S.A.), Box 238. *Animal Diseases.* ⊕ Colorado Agric. Coll. 1925.

Morera, Antonio Benitez (1905), Entomol. Cádix (España), Rosario no. 10, Dpd. 3. *Lépidopt. et coléopt. Batraciens, Reptiles.* ⊕ Reptiles et Batraciens.

Morgan, Agnes Fay (1884), Prof. of Household Sc. Univ. of California. Berkeley, Cal. (U.S.A.). *Chem. of Nutrition.* ⊕ Univ. of Chicago 1914.

Morgan, Ann Haven (1882), Prof. of Zool. South Hadley, Mass. (U.S.A.), Mount Holyoke Coll. *Experim. Zool., Ephemeridae, natural history.* ⊕ Cornell Univ. 1912. ⊕ Ephemeridae.

Morgan, Thomas H. (1866), Prof. of Experim. Zool. Univ. New York, N.Y. (U.S.A.), Columbia Univ., 409 W., 117 st. *Heredity, Regeneration.*

Theory of the Gene. ⊕ Phil. John Hopkins 1890, L.L. 1917. o

Morgenstern, L. I., Doc. Univ. Nizny Novgorod (U.d.S.S.R.), Sovetskaja Ploščad 8. *Pathol. Anat.* o

Morgenthaler, Otto (1886), Dr. phil., Bact. an der Schweizerischen Milchwirtschaftl. und Bact. Anstalt. Liebefeld b. Bern (Schweiz). *Krankheiten der Honigbiene.* ⊕ Bern 1910. ⊕ Bienen-Parasit.

Morgulis, Sergius (1885), Prof. of Biochem., Chairman of the Department of Biochem., Univ. of Nebraska Coll. of Med. Omaha, Neb. (U.S.A.), 42 and Dewey Ave. *Metabolism, blood chem., enzymes.* ⊕ Harvard 1910.

Mori, Kikuo, Ass. Med. Univ. Kanazawa (Japan), Pathol. Inst. *Anat.*

Mori, Nello (1880), Lib. doc. batt. Fac. di Med. Univ., Dir. Ist. neoimmunitario italiano. Palermo (Italia), Via Girgenti, 5. *Citol. dei batteri. Biol. ed abitudini di vita dei topi campagnoli. Mezzi per combattere le loro infestioni (invasioni). Natura dei virus filtrabili. Etiol. e cura del cancro umano.* ⊕ R. Univ. di Pisa 1903. ⊕ Culture microbiche. Isopatina antineoplastica mori.

Mori, Otto (1889), ao. Prof. Tôkyô (Japan), Anat. Inst. der Kaiserl. Univ. *Vergl. Anat. u. Entwicklungsgeschichte. Entwicklung des Schädelskeletts des Dornhaies.* ⊕ Med. Tôkyô 1914, phil. 1919.

Mori, Shigeko, Ass. Prof. Imper. Univ. Kyôto (Japan), Pathol. Inst., Coll. of Med. *Pathol. and Pathol. Anat.*

Morin, Sergei (1863), Prof. Prakt. Zool. Univ. Odessa (U.d.S.S.R.), Cominternstr. 2, INO., Zool. Labor. *Embryol. der Spinnen, Hydrobiol. Teredo, Zoogeographie d. Spinnen u. Arachnoidea.*

Morini, Fausto, o. Prof. Univ. Bologna (Italia). *Botan.*

Morini, Lorenzo (1887), Prof., Priv.Doc. R. Univ. Modena (Ialia). *Batt. ed Istol. patol.* ⊕ Modena 1923.

Morishima, Kurata, Prof. Imper. Univ. Kyôto (Japan), Pharmacol. Inst., Coll. of Med. *Experim. Pharmacol.*

Morishita, Kaoru, Biol. Govern. Central Experim. Stat. Hygiene Section. Taihoku, Formosa (Japan). *Parasit.* (Nematol.).

Morison, J., Ass.-Dir. Bact. Labor. Bombay (Brit.-India). o

Morita, Jun-ichi (1892), Prof. of Zool. at the Kotogakko. Osaka (Japan). (At present: Paris 13, 111, Rue Broca, Labor. clinique gynecol. Hôpital Broca.) *Mechanisme of development with special reference to heredity. Cytol. (plastosomes (mitochondries) and chromosomes) of the sexual cells. Tissue culture. Regeneration and thyroidisation.* ⊕ Tôkyô. ⊕ Prep. of Cytol. of the gonad of Rana esculenta.

Morland, D. M. T. (1892), Ass. Biol. in charge of Bee Research Inst. of Plant Pathol. Harpenden, Herts. (England), Rothamsted Experim. Station. *Biol. of the Bee. Temperature and Metabolism in the Beehive.* ⊕ Cambridge.

Mornac, G., Prof. Ecole prép. de Méd. et Pharm. Clermont-Ferrand, Puy-de-Dôme (France). *Pathol.* o

Morochin, Dmitrij J., Prof. an Inst. für Land- und Forstwirtschaft. Kasan (U.d.S.S.R.), Ul. Komlewa 20, W. 5. *Forstwirtschaft.* o

Moroder y Sala, Emilio, Cons. del Mus. de Hist. Nat. de la Fac. de Ciencias. Valencia (España), Maestro Chapi, 12. *Coléopteros y Hemípteros.* o

Moroff, Theodor (1877), o. Prof. der Vergl. Anat., Entwicklungsgeschichte und Histol. Sofia (Bulgarien), Zool. Inst. der Univ., Ul. Oboriste 13. *Cytol., Entwickelungsgeschichte, Protozoa. Systematik. Entwicklung der Kiemen bei Knochenfischen Octocorallia. Flagellaten. Coccidien. Sarkosporidien.* ⊕ München 1901.

Morosow, Wladimir A., Versuchsstation für Flachsbau. Moskau (U.d.S.S.R.), Petrowsko-Rasumowskoje, Landwirtschaftl. Akademie 72. *Botan., Ackerbau.* ·o

Morosowa-Wodjanizkaja, Nina W., Wissensch. Mitarbeiterin der Noworossijsker Biol. Station. Noworossijsk (U.d.S.S.R.), Slepzowskaja 3. *Botan. Hydrobiol.* ○

Morote y Greus, Francisco, Prof., Dr. en Ciencias, Dir. y Catedrático de Agric. del Inst. Valencia (España), Plaza de San Pablo, 3. *Patol. vegetal.* ○

Morozov, Boris (1893), Phytopath. Stauropolian Plant Protection Stations, Chief of the Division of Phytopath. Stauropol au Caucasus (U.d.S.S.R.). *Phytopath. and Mycol. Erysiphaceae et Ustilagineae.* ⓟ Mycol. objects.

Morpurgo, Benedetto (1861), Prof. di ruolo dell' Univ. di Torino (Italia), Corso Raffaello 30. *Istol. patol.; patol. sperimentale della nutrizione, della rigenerazione, dei tumori; parabiosi.* ⒹⒹ 1884. ⓟ Preparati d'istol. patol.

Morrill, Albro David (1854), Prof. of Biol., Hamilton Coll. Clinton, N.Y. (U.S.A.), Oneida Co. *Biochem. Geol.* Ⓓ Dartmouth 1879. ○

Morrill, Austin Winfield (1880), Consulting Entomol., Dir. of Agric. and Entomol. investigation. Los Angeles, Cal. E. U. A. (U.S.A.). Ⓓ Mass. Agric. Coll. 1903.

Morrill, Charles V. (1884), Associate Prof. of Anat., Cornell Univ. Med. Coll. New York City (U.S.A.), 477/1st Avenue. *Regeneration; Chromosomes; Teratol.; Amphibia.* Ⓓ Columbia Univ. 1910.

Morris, Harry Elwood, Associate Prof. Botan., Bact., Montana Experim. Station. Bozeman, Mont. (U.S.A.). ○

Morris, H. M. (1896), Ass. Entomol. Harpenden (England), Rothamsted Experim. Station. *Insect Fauna of the soil. Insecticides.*

Morris, Oscar Matison (1874), Prof., Head dept. Hortic. State Coll. of Wash. Pullman, Wash. (U.S.A.). *Hortic. Botan.* ○

Morris, Patrick Francis (1896), Ass. National Herbarium, Melbourne, Victoria (Australia). *Plants Weeds. General and system. Botan.* ⓟ Australian plants.

Morrow, Kenneth Sinclair (1897), Ass. Dairyman. Agric. Coll., Clemson College,S.C.(U.S.A.). *Nutrition, mineral metabolism.* Univ. of Minnesota 1925. ○

Morse, Albert P. (1863), Curator Natural History, Peabody Mus. of Salem, Mass., Zool. Mus., Wellesley Coll. Wellesley, Mass. (U.S.A.), 16 Upland Road. *Insects, particularly North Amer. Orthoptera. Birds.*

Morse, Warner Jackson (1872), Dir. Maine Agric. Experim. Stat. Orono, Me. (U.S.A.). *Phytopath., bact.* Ⓓ Wisconsin 1912.

Morse, William Joseph (1884), Agronomist, Bureau Plant Industry, United States Dept. of Agric. Washington, D.C. (U.S.A.). *The Soybean.*

Morse, Withrow (1880), Prof. and Chief of Biochem. Philadelphia, Pa. (U.S.A.), Jefferson Med. Coll. and Hospital. *Pathol. chem. Enzymes of animal fluids and tissues; autolysis; uric acid in the organism.* Ⓓ Columbia Univ.

Morstadt, Hermann (1877), Prof., Dr., RR. Biol. Reichsanstalt. Berlin-Dahlem (Deutschland), Königin-Luise-Str. 17. *Allgemeine Pflanzenpathol.; Termiten.* Ⓓ Heidelberg 1902.

Mortensen, Ernest (1902), Ass. Entomol., Prickly-Pear Investigations. Uvalde, Tex. (U.S.A.), Box 509. *Insects affecting Opuntias. Feeding adaptations.*

Mortensen, Theodor (1868), Head-Curator Dept. of Invertebrates. København (Danmark), Zool. Mus. *Echinoderms, Ctenophores, Mimetics, Marine ecol.* Ⓓ Copenhagen. ⓟ Echinoderms.

de Mortillet, A., Prof. éc. d'anthrop. Paris (France), Rue de l'Ecole de Méd. *Ethnogr. comp.* ○

Morton, George Edwin (1879), Head of dept. of animal husbandry Colorado. Agric. Coll. Fort Collins, Col. (U.S.A.). ○

Moschini, A., Prof. Univ. Pavia (Italia). *Fisiol.* ○

Moschkoff, Anna (1882), Ass. Univ. Leningrad (U.d.S.S.R.), Millionnais 10, log. 6. *Chim. microbiol.* Ⓓ Sorbonne 1913.

Moseley, Edwin Lincoln (1865), Prof. of Biol. Sc. Bowling Green, Ohio (U.S.A.), State Normal Coll. *Changes of level in the Great Lake Region. Flora of Northern Ohio. Bird migration. Habits of N.A. Mammals. Cause of Trembles and Milk-Sickness.* Ⓓ Univ. of Michigan 1885. ⓟ Spermatophyta of Northern Ohio (birds, shells and minerals).

Moser, Elek (1903), Ass. Physiol. Inst. Budapest (Ungarn), Eszterházy-ut 9. *Überlebende Organe.*

Moser, Johannes (1892), Dr. phil., Kustos am Zool. Mus. der Univ. Berlin N 4 (Deutschland), Invalidenstr. 43. *Coelenteraten.* Ⓓ Breslau 1917.

Moshkovski, Shabsai (1895), Ass. Tropical Inst. Moscow (U.d.S.S.R.), Telegrafny 7/14. *General pathol., immunol., haematol.; med. protozool., chemotherapeutics.* Ⓓ Moscow 1919.

Mosonyi, János (1898), I. Ass. Physiol. Inst. Budapest (Ungarn), Eszterházy-ut 9. *Physiol. der Niere.* Ⓓ Budapest 1924.

Mosquera, A. N., Prof. Univ. Quito (Ecuador). *Pathol.* ○

Moss, C. E., Prof. Univ. Johannesburg, Transvaal (S. Africa). *Botan.* ○

Moss, Ezra Henry (1892), Ass. Prof. of Botan., Univ. of Alberta. Edmonton, Alberta (Canada). *Coniferous Rusts. Vegetation of Alberta. Ecol. of Grasslands.* Ⓓ Univ. of Toronto 1925.

Mossesvili, Varlaam, Prof. Univ. Tiflis (U.d.S.S.R.). *Pharmacol.* ○

Most, August (1867), Prof. der Chirurgie an der Univ., leitender Arzt des St. Georg-Krankenhauses. Breslau 16 (Deutschland), Friedrich-Ebert-Str. 12. *Anat., topographische Anat. u. Chirurgie des Lymphgefäßsystems.* Ⓓ Würzburg 1892.

Mosyka, Józef (1900), Ass. Inst. de Botan. Univ. Kraków (Polska), Rue Lubicz 46. *Lichenol., Floristik, Systematik, Geographie, Oekol.* Ⓓ Univ. Krakow 1925. ⓟ Lichenes exsiccati aus Polen.

Moszkowski, Max (1873), Dr., Privatgelehrter. Berlin-Grunewald (Deutschland), Herthastr. 2a. *Zool. und Anthrop.* Ⓓ Breslau 1896.

Moszyński, Ambroise (1894), Dr. phil., Ass. à l'Inst. Zool. de l'Univ. Poznań (Polska), Wjazdowa 5. *Oligochaeta terricola et limicola.* Ⓓ Poznań 1925.

Motăş, Constantin (1891), Chef des travaux de Zool. descriptive à la Fac. des Sc. de Jassy (Roumanie), Labor. de Zool. Univ.; Grenoble (France), Cours Berriat 20. *Hydracariens, systématique, faunistique et zoogéographique.* ⓟ Hydracariens des Alpes du Dauphiné et du sud de la France.

Mothes, A. Kurt (1900), Dr. phil., Ass. am Botan. Inst. der Univ. Halle a. d. S. (Deutschland), Kirchtor 1. *Stoffwechselphysiol. der Pflanzen, Eiweißaufund -abbau, Amide, Pflanzenbasen.* Ⓓ Leipzig 1925.

Motorin, A. A., Prof. II. Univ. Moskau (U.d.S.S.R.). *Tierzucht.* ○

Motte, Jean (1897), Ass. Fac. Sc. Montpellier (France), Inst. Botan. *Cytol. végétale.*

Mottier, David M. (1864), Prof. of Botan. and Head of Department, Indiana Univ. Bloomington, Ind. (U.S.A.). *Morphol. and Cytol.* Ⓓ Bonn 1897.

Mottram, V. H., Prof., Univ. London S.W. 7 (England), S. Kensington. *Physiol.* ○

Mouchet, Rene Libert Joseph (1884), Méd. en Chef Adjoint au Congo Belge. Boma (Congo-Belge). *Epidemiol. Anat. pathol.* Ⓓ 1907.

Mouchtak, Ahmed, Prof. Univ. Stamboul. Konstantinopel (Türkei). *Palaeont.* ○

de Moulin, Frederik Willem Karel (1891), Doc., Niederl.-Indische Tierärzteschule, Java. Buitenzorg (Nederl.-O.-Indië), Batavascheweg 50. *Protoplasmastrukturen.* Ⓓ Utrecht (Holland) 1918.

Moureau, Mare Jean (1890), Prép. du Labor. de Méd. experim. et Bact. Bordeaux (France), 8, Rue Poquelin Molière. *Leucocytes. Analyses biol.* Ⓓ Bordeaux 1919. ⓟ Cultures microbiennes.

Mouri, Osman, Prof. Inst. Bact. Stamboul (Turquie), Rue Matbaa. ○

Mourier des Gayets, M., Prép. Fac. Sc. Lyon (France). *Botan.* ○

Moussu, M., Prof. Ecole Vét. Alfort (Seine), Prof. Inst. Nation. agron. Paris 5 (France), 16, R. Claude Bernard. *Physiol.* ○

Moutafowa, Ripsimia (1890), Adjunct Hydrol. Inst. Leningrad (U.d.S.S.R.), 2 Linie, 23. *Microbiol. des Wassers.*

Mouton, M., Maître de conf. Fac. Sc. Paris 15 (France), 42, Rue Mathurin-Régnier. *Physiol.* ○

Moyano y Moyano, Pedro (1863), Dir., Prof. de Physiol. et Hygiène. Zaragoza (España), Escuela de Vet. *Hygiène, Zootechnique.* ⓟ Madrid 1886.

Moycho, Wacław (1884), Dr. ès sc. nat., chef des travaux pratiques de botan. à l'école Supérieure d'Agric. Varsovie (Pologne), rue Polna 32. *Enzymes protéol. d'origine bactérienne.* ⓟ Paris 1915.

Mozołowski, Włodzimierz (1895), Ass. Med.-chem. Inst. Univ. Lwów (Polska), Tarnowskiego 74. *Antipepsin, Chinhydronelektrode, Blutammoniak.* ⓟ Lwów 1922.

Mršič, Vilim (1896), Adjunkt für Biol. Zagreb (S.H.S.), Morfološko-Biol. Inst. Šalata. *Theoretische und experim. Zool., insbes. Fischereibiol.* ⓟ München 1922.

Much, Hans (1880), Prof. Univ. u. Dir. Inst. für Immunität (pathol. Biol.). Hamburg (Deutschland), Alsterkamp 12. *Pathol. Biol. u. experim. Therapie.* ⓟ Würzburg 1903.

Much, Rudolf, Univ.Prof. Wien I (Österreich), Burgring 7. *Anat. Anthrop.* ○

Muchin, Michael W. (1897), Ass. Physiol. der Tiere. Perm (U.d.S.S.R.), Zaimka, Univ. *Physiol. Wirkung der Hormone auf das Warmblüterherz.*

Muckermann, Hermann (1877), Leiter der Abteilung für Eugenik, Inst. für Anthropol., Erblichkeitsforschung u. Eugenik der Kaiser-Wilhelm-Gesellschaft zur Förderung der Wissenschaften. Berlin-Schlachtensee (Deutschland), Waldemarstraße 81. *Keimzellforschung, Erforschung der biol. Grundlagen und Bedingungen für den Aufbau von Familie und Volk, Eugenik.* ⓟ 1902.

Mudd, Stuart (1893), Ass. Prof. of Experim. Pathol., Associate in Pathol., Villa Nova, Pa. (U.S.A.), St. Thomas Road. *Cell surfaces; immunochem.* ⓟ Harvard 1920.

Mudge, G. P., Prof., East L. Coll. London (England). *Zool.* ○

Mühldorf, Anton (1890). Cernăuți (România), Piața. *Anat. u. Physiol. der Pflanzen, physiol. Anat. der Moose, Trennungsgewebe, Moose Rumäniens.* ⓟ Cernăuți 1913.

Mühlenbachs, Viktors (1898), Dr., Ass. des Hygien. Inst., Univ. Riga (Latvija), Dzirnavu ielā 47 dz. 2. *Carex, adventive Flora Lettlands.* ⓟ Dorpat

Müllegger, Sebastian (1886), Privatgelehrter. Büsum (Deutschland), Zool. Station. *Wissenschaftl. Meeresforschung in bezug auf Tier- u. Pflanzenwelt.* ⓟ Meerestiere, lebend und praep.

Müller, Adolf (1888), Dr. phil. nat., angewandter Entomol. Frankfurt a. Main (Deutschland), Brüder-Grimm-Str. 26 I. *Angewandte und physiol. Entomol. Ferner Systematik, Zool. und Anat. der Weberknechte (Opilioniden).* ⓟ Frankfurt a. Main 1921.

Müller, Arno (1879), Dr. phil., Reg.R. am Reichsgesundheitsamt. Berlin-Friedenau (Deutschland), Retzdorffpromenade 2 II. *Wasser- und Abwasserbiol. (Sauerstoffzehrung).* ⓟ Jena 1904.

Müller, Detlev (1899), Mag. sc. Kopenhagen (Dänemark), Pflanzenphysiol. Labor. d. Univ., Gothersgde 140. *Atmungsenzyme der Pflanzen. Stoffproduktion u. Assimilation d. Pflanzen.* ⓟ 1923.

Müller, Erich (1898), Dr. med., Ass. Berlin N 4 (Deutschland), Invalidenstr. 103a, z. Zt.: London, W. C. 1, 58, Torrington Square. *Herz- und Gefäßphysiol. Gasstoffwechsel. Arbeitsphysiol. (Energetik).* ⓟ Berlin 1924.

Müller, Franz (1871), a.o. Prof. an der Univ. Berlin. Charlottenburg 9-Westend (Deutschland), Kastanien-Allee 39. *Experim. Pharmacol. Wirkung des Klimas auf den Menschen (Blutgase, Stoffwechsel, Atmung, Blutfarbstoffe). Diagnostik.* ⓟ Heidelberg rer. nat. 1892, med. 1898.

Müller, Friedrich, Univ.Prof. Graz (Österreich). *Anat.* ○

Müller, Gustav Wilhelm (1857), Geh.Reg.R., o. Prof. emer. Zool. u. vergl. Anat. Greifswald (Deutschland), Roonstr. 3. *Ostracoden, Insekten-, besonders Fliegenlarven, Gordius, Hydrobiol.* ⓟ Greifswald 1880. ⓟ Fliegenlarven und Gordius.

Müller, Hans (1894), Dr. phil. Braunschweig (Deutschland), Salzdahlumer Str. 111 I. *Biol. der Myriopoden.* ⓟ Göttingen 1922.

Müller, Helmut (1896), Dr. med., Priv.Doc., Ass. am Physiol. Inst. der Univ. Königsberg, Preußen (Deutschland), Copernicusstr. 1-2. *Biogene Amine.* ⓟ Königsberg 1923.

Mueller, Herbert Constantin (1891). Bombay (British India), Post Box 199. *Entwickelungsmechanik, Regeneration an Tieren. Hydroiden. Embryol.* ⓟ Leipzig 1913. ⓟ Meerestiere der Brit.-Indischen Westküste, Wirbeltiere und Wasserbewohner aus der Umgebung von Bombay.

Mueller, Justus Frederick (1902), Research Ass., Univ. of Illinois. Urbana, Ill. (U.S.A.), Department of Zool. *Parasit. Nematodes. Helminthol.* ⓟ Univ. of Illinois 1926.

Müller, Karl (1881), Dr., Dir. des Badischen Weinbauinst. und der Hauptstelle für Pflanzenschutz in Baden. Freiburg i. Br. (Deutschland), Badisches Weinbauinst., Peterhof. *Lebermoose, angewandte Botan., speziell Weinbau.* ⓟ Freiburg i. Br. 1906.

Müller, Karl Otto (1897), Wiss. Angestellter (Labor.-Vorsteher) an der Biol. Reichsanstalt für Land- und Forstwirtschaft, Priv. u. Hon.Doc. a. d. Landwirtschaftl. Hochsch. Berlin-Steglitz (Deutschland), Friedrichsruher Str. 41a. *Kartoffel (Sol. tuberosum); Kiefer (Pinus silvestris) und Pilze. Immunitätszüchtung und -physiol.; außerdem Pathol. der höheren Pflanzen und Entwicklungsgeschichte der Pilze.* ⓟ Berlin 1921. ⓟ Kartoffelsorten, Pilzkulturen.

Müller, László, Ass. Pathol. Inst. d. Ung. Elisabeth Univ. Pécs-Fünfkirchen (Ungarn). *Pathol.: Tumores. Nekrosen.* ○

Müller, Lene (1894), Rhein. Obst- u. Gartenbauschule für Frauen. Godesberg a. Rhein. Bonn, Rhein (Deutschland), Sechenstr. 38. *Phytopath.* ⓟ Wien 1919. ⓟ Pflanzenschädlinge.

Müller, Reiner (1879), Dr. med., o. Prof. für Hygien. und Bact.; Dir. des Hygien.-Inst. der Univ. Köln. Köln-Lindenthal (Deutschland), Weyertal 123. *Geschichte und Geographie der Hygien. und Bact. Bakterienmutationen. Abtrennung des „Regnum bacteriorum" vom Regnum plantarum. Bact. des Typhus abdom., Paratyphus und der Fleischvergiftungen. Mund-Oscillarien. Mikrophotographie.* ⓟ Kiel 1903. ⓟ Kulturen pathogener Bakterien, krankheitsübertragende Arthropoden.

Müller, Rudolf, Univ.Prof. Graz (Österreich). *Pharmacol.* ○

Müller-Thurgau, Hermann (1850), Prof., Dr., Privatgelehrter. Wädenswil b. Zürich (Schweiz). *Eiweißbildung in Pflanzen; Ursachen der Blüten- und Fruchtbildung. Lebensvorgänge und Wirkungsweise der Plasmopara viticola. Gärungsvorgänge und Bact. des Weines.* ⓟ Würzburg 1874.

Muende, J., Demonstrater in Pathol., Charing Cross Hospital Inst. of Pathol. London W.C. 2 (England). ⓟ London 1926.

Münch, Ernst (1876), Dr. phil. oec. publ., o. Prof., Vorstand der Botan. Abteilung der Forstlichen Hochsch. Tharandt i. Sa. (Deutschland). *Physiol. Baumrassen, Saftstrombahnen der Pflanzen, Krankheiten der Bäume. Immunität u. Krankheitsempfänglichkeit d. Holzpflanzen. Hitzeschäden an Wald-*

pflanzen. *Knospenentfaltung der Fichte u. Spätfrostgefahr.* Ⓓ München 1909. Ⓟ Holzpilze.
Muenscher, Walter C. (1891), Ass. Prof. of Botan. Ithaca, N.Y. (U.S.A.), Department of Botan. *Weeds, life history, control methods, distribution. Myxomycetes. Transpiration. Marine algae. Ecol. of pollution algae.* Ⓓ Cornell Univ., Ithaca, N.Y., 1921. Ⓟ Myxomycetes. Algae.
Muesebeck, Carl F. W. (1894), Entomol., U. S. Department of Agric. Melrose Highlands, Mass. (U.S.A.), U. S. Bureau of Entomol. *Biol. and interrelationships of insect parasit., parasit. Hymenoptera, Braconidae.* Ⓓ Cornell Univ. 1915.
Müntzing, Arne (1903), Amanuensis, Svalöf (Schweden). *Genetik: Art- und Linienkreuzungen der Gattungen Galeopsis, Potentilla Stachis, Anthericum, Triticum und Lamium und ihre Cytol.*
Muffel, Paul P. (1881), Ass. Bact. Inst. Univ., Leiter d. Malariastation. Woronesch (U.d.S.S.R.), Wedenskaja str. N. 13. *Angeborene Malaria. Epidemiol. d. Malaria.* Ⓓ Dorpat 1910.
Muir, James, Prof., R. Technical Coll. Glasgow (Scotland). *Natural Philosophy.* ○
Muir, Robert, Prof., Univ. Glasgow (Scotland). *Pathol.* ○
Muir-Wood, Helen Marguerite (1895), London S.W. 7 (England), British Mus., Natural History. *Fossil Brachiopoda.* Ⓓ London 1920.
Mukerji, Sushil Kumar (1896), Univ. Lect. in Botan., Lucknow Univ. Lucknow (India). London W.C. 1 (England), c/o Prof. I.W. Oliver, I. R. S., Univ. Coll. *Plant Ecol. Systematic Morphol. of Angiosperms. Alpine and Himalayan Vegetation. Economic and Med. Plants. The Genus Mercurialis (Euphorbiacea).* Ⓓ Allahabad, India, 1916. Ⓟ Himalayan plants from Kashmir and the unexplored regions beyond, photographs of the vegetation.
Mukherji, D. B., o. Prof. Ravenshaw Coll. Cuttack, Bihar-Orissa (India). *Botan.* ○
Mulford, Walter (1877), Acting dean Coll. of Agric. Univ. of California. Berkeley, Cal. (U.S.A.), 305 Hilgard Hall. *Forestry.* ○
Mullens, William Herbert, M.A. Kewhurst Manor, Little Common, Bexhill-on-Sea (England). *Ornithol.* ○
Muller, Hermann J. (1890), Prof. of Zool. Univ. of Texas. Austin, Tex. (U.S.A.). *Heredity, variation, and evolution; linear linkage and crossing over, the theory of the gene, mutation, complex characters, heredity and environment in the determination of mental traits, the genetic effects of radiation, photomicromanipulation, effects of selection and non-selection, chromosome aberrancies.* Ⓓ Columbia Univ. 1915. Ⓟ Cultures of Drosophila melanogaster, mutant races.
Muller, Joseph Gh. G. (1885), Pharmacien, Entomol. Visé (Belgique), rue Haute. *Insectes: Coléoptères, hémiptères, orthoptères, diptères, hyménoptères et lépidoptères surtout de l'Est de la Belgique.* Ⓓ Liège 1909. Ⓟ Insectes Coléoptères, hémiptères, orthoptères, lépidoptères, diptères etc. de Belgique.
Muller, M. R., Cons. Section d'Acclimatation, Jardin Botan. princ. Leningrad (U.d.S.S.R.), Aptekarski Ostrov, Pessotchnaia ¹/₂. *Dendrol.* ○
Mulon, M., Prof. agr. Fac. Méd. Paris (France), 27, Avenue Bugeaud. *Histol.* ○
Mumford, Herbert Windsor (1871), Dir. Agric. Expt. Station and Extension Service Univ. of Ill. Urbana, Ill. (U.S.A.), *Animal husbandry.* ○
Muncie, J. H. (1890), Agent U.S. Dept. Agr. Ames, Ia. (U.S.A.), Botan. and Plant Pathol. Section. *Plant Pathol. Bact. diseases.* Ⓓ Iowa State Coll. 1925.
Munda, August, Dr. Ljubljana (S.H.S.), Nunska ulica. *Ichthyol.* ○
Mundinger, Frederick George (1891), Associate in entomol. research. Hudson Valley Horticultural Investigations. Vassar College (U.S.A.), Box 51. *Economic insects.* Ⓓ Syracuse 1914.

Munesada, Tetsuji, Sc. Ohara Inst. for Agric. Research. Kurashiki near, Okayama (Japan). *Pharmacol. and Phyto-chem.*
Munger, Thornton T. (1883), Dir., Pacific Northwest Forest Experim. Station of the U. S. Forest Service. Portland, Ore. (U.S.A.), 514 Lewis Bldg. *Silviculture, heredity of trees, forest ecol., growth of trees and forests.* Ⓓ Yale Forest School 1908. Ⓟ Tree seeds.
Munn, Mancel T. (1887), Associate Botan. (Ass. Prof.) New York State Agric. Experim. Station. Geneva, N.Y. (U.S.A.). *Seed testing and analysis. Seed-borne plant diseases: Physiol. of germination; Practical means of identification.* Ⓟ Seed herbarium.
Munn, Philip Winchester. Puerto Alcudia, Majorca, Balearic Isles (España). *Ornithol.* ○
Muñoz Cobo, Luis, Prof., Dr. en Ciencias. Málaga (España). *Malacol. y Mineral.* ○
Muñoz y Medina, José María (1895), Prof. agregé Fac. de Pharmacie Univ., Granada (España). Parasitol. helminthol. et Protozool. Ⓓ Pharm. Madrid. Ⓟ Vers parasites.
Muñoz Rivero, Emilio, Prof. Dr. Lab. de Farmacol. Fac. Med. Sevilla (España). ○
Munro, Hugh Kenneth (1894), Entomol., Division of Entomol., Dept. of Agric., Union of South Africa. Pretoria (South Africa), Box 513. *Systematics and biol. of the family Trypetidae, Diptera. Larvae of the muscoid groups.* Ⓓ Univ. of South Africa 1919.
Munro, J. A. (1896), State Entomol. Fargo, N.D. (U.S.A.), State Coll. Station. *Entomol. and Beekeeping.* Ⓓ Kansas State Agric. Coll. 1925. Ⓟ Insects.
Munson, John P. (1870), Head of Department of Biol., Investigator on Elizabeth Thompson Sc. Foundation. Ellensburg, Wash. (U.S.A.), 706 North Anderson Street. *Minute structure of the chelonian brain; Origin of germ cells; cell division and cell differentiation; Cytol. of Collaterial glands; cancer; centrosome.* Ⓓ Univ. of Chicago 1897.
Munteanu, Anastase, Prof. Acad. d'Agric. Cluj (România). *Pflanzenzüchtung.* ○
Munz, Philip Alexander (1892), Prof. of Botan., Pomona Coll. Claremont, Cal. (U.S.A.), 1165 Indian Hill Blvd. *Taxonomy of plants of western North America.* Ⓓ Cornell 1917. Ⓟ Plants of Cal.
Muralewicz, Wenceslaus (1881), Ass. Katheder Zool. I. Univ. Moskau (U.d.S.S.R.), Wosstannja Platz 1, Qu. 23. *Anat., Systematik, Verbreitung der Myriopoden und Insecten.* Ⓓ Moskau 1910.
Murashima, Taiichi, Ass. Prof. Tôhoku Imper. Univ. Sendai (Japan), Pharmacol. Inst., Coll. of Med. *Pharmacol.*
Murashkinsky, Konstant (1884), Prof. of Phytopath. of Siberian Agric. Academy. Omsk (U.d. S.S.R.). Ⓟ Pilze.
Murata, Miakichi, Prof. Provincal Med. Univ. Osaka (Japan), Pathol. Inst. *Pathol. and pathol. Anat.*
Murawjowa, Helene (1896), Labor. Leningrad (U.d.S.S.R.), Haupt Botan. Garten. *Samen der Kulturgewächse und Unkräuter, Nahrungsprodukte. Möglichkeit der Kultur des Chenopodium Quinoa in dem nordwestlichen Bezirk Rußlands.* Ⓟ Mikroskop. Präp., Gramineae Cruciferae, Convolvulaceae, Papilionaceae.
Murbeck, Svante (1859), Prof. emer. Lund (Schweden). *Systematik der Gattung Verbascum. Flora von Südbosnien und der Herzegovina. Parthenogenet. Embryobildung. Blütenbau der Papaveraceen. Staminale Pseudopetalie und Herkunft der Blütenkrone. Wüstenpflanzen. Flore du Maroc. Celsia.* Ⓓ Lund. Ⓟ Herbarmaterial und Samen von Verbascum. Arten aus der Mittelmeerregion und dem Orient.
Murisier, Paul, Chef des travaux de Zool. à l'Univ. de Lausanne (Suisse), Palais de Rumine. *Pigments. Sexualité chez les Oiseaux. Greffe. Biol. des Poissons. Pisciculture experim.*

Murlin, John R. (1874), Prof. of Physiol. Univ. of Rochester N.Y. (U.S.A.). *Nutrition: metabolism of infancy, of pregnancy and influence of the endocrine organs, pancreas.* ① Univ. of Pennsylvania 1901.
Murneek, Andrew Edward (1888), Ass. Prof. of Horticulture. Univ. of Missouri. Columbia, Mo. (U.S.A.). *Physiol.* ① Univ. of Wisconsin 1924.
Murphy, James B. (1884), Member of the Rockefeller Inst., in charge of the Division of Bio-Physics. New York City, N.Y. (U.S.A.), 66 St. and Ave. A. *Cancer.* ① Johns Hopkins Univ. 1909. ② Chicken Sarcoma.
Murphy, Margaret Agnes (1889), Women's Med. Service. Parel, Bombay (India), Haffkine Inst. *Bact. Anaemia of Pregnancy.* ① Edinburgh 1914.
Murphy, Paul A. (1887), Head of Plant Diseases Division, Department of Agric. Dublin (Ireland). *Diseases of Economic Plants. Diseases of Potatoes. Virus diseases.* ① Trinity Coll., Dublin, 1922.
Murphy, Robert Cushman (1887), Ass. Dir., American Mus. of Natural History. New York City (U.S.A.). *Oceanography. Distribution of life in the sea. Marine Ornithol.* ② Polynesian birds.
Murr, Erich (1898), Ass. Zool. Inst. u. Mus. der Univ. Königsberg, Pr. (Deutschland), Sternwartstraße 1. *Morphol. und Physiol. des Wirbeltierauges. Inkretorische Organe.* ① München 1924. ② Augen von Wirbeltieren, auch embryonale.
Murray, A., Prof. Univ. Otago. Dunedin (New Zealand). *Pathol.* o
Murray, Everitt George Dunne (1890), Univ. Lect. in Pathol., Fellow of Christs Coll. Cambridge (England), Department of Pathol., Univ. *Bact.* ① London 1916.
Murray, Harold R. (1900), Ass. Dept. Genetic. New Haven, Conn. (U.S.A.), Experim. Station. *Breeding of Vegetable Crops.*
Murray, Thomas J. (1891), Prof. of Bact., Rutgers Univ. New Brunswick, N.J. (U.S.A.). *Bacterium coli and aerogenes. Biol. Soil Processes. Fixation of Atmospheric Nitrogen. Angular-Leafspot of Tobacco. Food Accessory Substances and the Nitrite Bacteria* ① Purdue Univ. 1915.
Murto, Johan Anders (1884), Med. u. Chirurg. Dr., Dir. des Staatlichen Serumlabor. u. Pasteur-Inst. in Helsingfors (Finnland). *Serol. u. bact. Diagnostik. Antibakterielle Vaccine-Therapie. Bakterienkonglutination, Theorie der Seroreaktionen bei Lues und Lepra, Oligodynamische Untersuchungen an Bacterien.* ① Helsingfors 1914. ② Verschiedene Bact. Kleine Laboratoriumstierchen.
Muscatello, Giuseppe, Lib. doc. Univ. Catania, Sicilia (Italia). *Botan.* o
Mušeghian, Figran, Prof. Staatsuniv. Erivan, Armenien (U.d.S.S.R.). *Anat. d. Haustiere.* o
Muschold, Paul (1861), Dr., Obergeneralarzt a.D. Berlin-Lichterfelde (Deutschland), Sternstr. 43, I. *Hygiene. Bact. (Mikrobiol).* ① Berlin 1883.
Muskett, Arthur Edmund (1900), Lect. in Agric. Botan., the Queen's Univ., in Charge of Plant Disease Work in the Seed Testing and Plant Disease Division of the Ministry of Agric. for N. Ireland. Belfast (Ireland). *Mycol. and Plant Pathol. Diseases of Potatoes, Flax, Fruit, Cereales. Celery leaf sput. Forest Nursery Diseases.* ① London 1922. ② Cultures of Parasit. fungi.
Mussatowa, Alexandra Ja., Doc. a. Inst. f. Volksbildung. Dnepropetrowsk (U.d.S.S.R.), Torgowaja 10, W. 3. *Botan. Systematik.* o
Mustafa, Mahmud (1893), Lect. Botan. Cairo (Egypte), The Egyptian Univ. ① Cambridge-Tripos 1924.
Musunecl, Abele, Lib. doc. Univ. Catania, Sicilia (Italia). *Patol.* o
Muszyński, Jean Casimir (1884), Prof. de Pharmacognosie (Matière méd.) et de Culture des plantes méd. Univ. Etienne Batory. Vilno (Pologne). *Matière méd. (Anat., Microchem.), Cécidiol., Maladies des plantes méd., Histoire des drogues.* ① Dorpat 1917. ② Plantes méd., sémences des plantes indigènes en Pologne, Cécidies.
Mutel, Ismar Tavares, Prof. Fac. Med. Curityba, Paraná (Brasilia). *Anat., Physiol. pathol.* o
Mutel, M., Prof. agr. d'Anat. Nancy (France), Rue Lionnois. *Anat.* o
Mutermilch, Stéfan (1878), Chef de Labor. Inst. Pasteur. Paris 15 (France), 25, Rue Dutot. *Sérodiagnostic de la Syphilis.* ① Varsovie 1902.
Muth, Franz (1869), Prof. Dr. rer. nat., Dir. der Lehr- und Forschungsanstalt für Wein-, Obst- und Gartenbau. Geisenheim a. Rh. (Deutschland), Straße 15. *Stimulation (Schwefelkohlenstoff), Kohlensäure-Assimilation, Pflanzenpathol.* ① Tübingen 1898.
Muttkowski, Richard Anthony (1887), Prof. of Biol. and Head of Department of Biol. Univ. of Detroit, Mich. (U.S.A.). *Ecol. of aquatic animals, Arthropoda. Insect Physiol. and morphol. Taxonomy of Odonata.* ① Univ. of Wisconsin 1916.
Muuteanu, Anastase (1895), Prof. Cluj (România), Landw. Akademie. *Pflanzenzüchtung.* ① Cluj 1921.
Muzyha, Maxim, Vorst. Bact. Inst. Lwów (Polska). o
Myers, Charles Emory (1882), Prof. of Plant Breeding, Pennsylvania State College, Pa. (U.S.A.), 304 W. Fairmount Ave. ① Cornell Univ. 1922. ② Seed of new varieties of tomatoes and cabbage.
Myers, Charles Samuel (1873), Dir. National Inst. of Industrial Psychol., Member of the Industrial Fatigue Research Board. London, W.C. 1 (England), 329 High Holborn. *Worker's movements; Reduction of monotony and fatigue; Vocational and Industrial Psychol.* ① Cambridge 1895.
Myers, George Sprague (1905), Ass. Curator of Fishes, Stanford Univ., California (U.S.A.). *Ichthyol. and Herpetol., fresh-water fishes of South America, Africa and tropical Asia. Taxonomy of Poeciliidae (Cyprinodontidae) and Characidae.*
Myers, Harold B. (1886), Prof. of Pharmacol., Univ. of Oregon Med. School. Portland, Ore. (U.S.A.). *In toleration of drugs.* ① Western Reserve Univ. 1911.
Myers, J. G., B.Sc. Agric. Dept. Wellington (New Zealand). *Entomol.* o
Myers, Victor Caryl (1883), Prof. and Head of Biochem., The State Univ. of Iowa. Iowa City, Ia. (U.S.A.). *Metabolism of creatine and creatinine, chem. of the blood in health and disease.* ① Yale Univ. 1909.
Myrbäck, Karl David Reinhold (1900), Doc Stockholm (Sverige), Univ. *Biochem. Organische Chem. Chem. d. Enzyme.* ① Stockholm 1926.

Nábělek, Albert (1894). Bratislava (C.S.R.), Hluboká cesta 4. *Anat., histol. of the Basidiomycetes. Neoplasma of the phanerog.* ① Phil. Vienna 1918, med. Bratislava 1927. ② Exsikkata.
Nábělek, François (1884), Dr. de phil., Doc. à l'Univ. Masaryk à Brno (C.S.R.), Král. Pole. *Flore Orientale (Palestine, Syrie, Mésopotamie, Perse, Kurdistan, Arménie)* ① Vienne 1907. ② Specim. d'herbier, photographies de plantes vivantes.
Nabours, Robert Kirkland (1875), Prof. of Zool. and Zool. of the Agric. Experim. Station. Manhattan, Kan. (U.S.A.). *Heredity and Parthenogenesis in Orthoptera, Tettigidae.* ① The Univ. of Chicago 1911.
Nachtsheim, Hans (1890), Dr. phil., ao. Prof. an der Landwirtschaftl. Hochsch. Berlin, Abteilungsvorsteher am Inst. für Vererbungsforschung. Berlin-Dahlem (Deutschland), Schorlemer Allee. *Geschlechtsbestimmung. Haustiergenetik. Drosophila-Genetik.* ① München 1913.
Nachtwey, Robert Christian Reinhard (1893), Dr. phil. nat. Bremen (Deutschland), Realschule Sögestraße. *Entwicklungsgeschichte und Anat. der Rotatorien.* ① Frankfurt a. M. 1924.
Nadson, Georges (1867), Prof. Inst. Méd., Chef du Labor. biol. Inst. de Roentgenol. et de Radiol.

Leningrad (U.d.S.S.R.), Jardin botan. *Physiol. des plantes et Microbiol. Morphol. des bactéries.* ⓐ Varsovie 1903.

Nael, Adolf (1883), ao. Prof. Napoli (Italia), Acquario. *Allg. Morphol. u. Stammesgeschichte; Cephalopoden, leb. u. fossil; vergl. Anat. u. Embryol. der Wirbeltiere; vergl. Embryol. u. Anat. der Wirbellosen, insbes. Mollusken. Theoretische Morphol. Phylogenie.* ⓐ Zürich 1909.

Naegeli, O., Prof. Dr. Zürich 7 (Schweiz), Schmelzbergstr. 40. *Blut.* ○

Närr, Johan, Dr., o. Prof. Pharmacol., Vet. Fac. Sofia (Bulgarien). ○

Naeslund, Carl Albert (1892), Associate Prof. in bact. and hygiene. Uppsala (Sverige). *Actinomyces. Formation of concretions in the saliva. Toxicity of arsenic compounds.* ⓐ Uppsala 1922.

Naeslund, John (1894), Doc. pathol. Anat. Univ. Upsala (Sverige). *Fonction du corps jaune, gestation et développement du foetus.* ⓐ Upsala 1925.

Nagai, Hisomu, Prof. Imperial Univ. Tôkyô (Japan), Physiol. Inst., Fac. of Med. *Physiol. and general biol.*

Nagai, Nagayoshi, Ph.D., Prof.emer. Tokyo Imper. Univ. Shibuya near Tôkyô (Japan), Aoyama-Minamichô 7-Chôme 3. *Pharmacol. and Organic Chem.*

Nagamatsu, Hideichi, Ass. Prof. Aichi Med. Univ. Nagoya (Japan), Anat. Inst. *Anat.*

Nagamori, Shinzaburô, Ass. Komaba near Tôkyô (Japan), Zool. Labor., Coll. of Agric., Imper. Univ. *Sériculture.*

Nagasaki, Sentarô, Prof. Provincial Med. Univ. Ôsaka (Japan), Pharmacol. Inst. *Pharmacol.*

Nagayo, Matao, Prof. Imper. Univ. Tôkyô (Japan), Pathol. Inst., Fac. of Med. *Pathol., Anat.*

Nagel, Walter (1888), Dr. phil. Frankfurt a. M. (Deutschland), Königstr. 38 II. *Schädlingsbekämpfung tierischer und pflanzlicher Parasiten. Desinfektion.* ⓐ Kiel 1916.

Nageotte, M., Prof. Coll. de France. Paris 6 (France), 82, Rue Notre-Dame des Champs. *Histol.* ○

Nagibina, Maria P. Dmitrijew, Kursker Gouv. (U.d.S.S.R.), Staro-Perschinskaja Biol. Station d. Moskauer Ges. d. Naturforscher. *Geobotan.*

Nagorny, Alexander W. (1887), Prof. Univ., Inst. f. Landwirtschaft. Charkow (U.d.S.S.R.), Gospitalnyj per. 5. *Physiol. u. kolloid. Chemie. Alterungserscheinungen. Atmung bei Insekten. Sauerstoffdepot im Organismus. Protoplasmastructur. Wasserbindung.*

Nagorny, Pantaléon, I. Dir. Stat. Phytopath. Jardin Botan. Tiflis, Caucase (U.d.S.S.R.). *Mycol. et Pathol. génér.* ○

Nagy, Ladislaus (1901). Debrecen (Ungarn), Simonyi-ut 36/c. *Tabakchem.*

Naidu, B. P. Balakrishna (1882), Special Research Officer on Plague Inquiry. Parel, Bombay (India). Haffkine Inst. ⓐ Edinburgh 1921. ⓑ *Pathol. and Bact.* specimens on Plague.

Nakadai, Motoji (1894), Ass.-Prof. Kumamoto (Japan), The Inst. of Anat., Med. Coll. *Cytol.* ⓐ Kumamoto Med. Coll. 1916.

Nakagawa, Tomoichi, Prof. Provincial Med. Univ. Ôsaka (Japan), Physiol. Inst. *Physiol.*

Nakahara, Warô. New York City, N.Y. (U.S.A.), Rockfeller Inst., 66 at. A av. *Cytol. of Insects.*

Nakai, Takenoshin, Prof. of Botan. Koishikawa, Tôkyô (Japan), Botan Inst. Tôkyô Inperial Univ. *Cryptogamic Botan. East Asiatic plants.*

Nakamura, Hachitarô (1881), Prof. allg. Pathol. u. pathol. Anat. Kanazawa (Japan), Nagamachi Rokutanchô 2. *Pathol. Anat. u. experim. Pathol.: innere Sekretion.* ⓐ 1905.

Nakamura, Hisao (1904), Ass. Kyôto (Japan), Labor. of Phytopath., Department of Agric., Imper. Univ. *Septorioses of plants.* ⓐ Kyôto 1927.

Nakamura, Masao (1867), Prof. Utsunomiya (Japan), Agric. Coll. *Theoretical and practical botan.*

Nakamura, Y., Prof. Hokkaidô Imper. Univ. Sapporo (Japan), Bact. Inst., Fac. of Med. *Bact.*

Nakane, Shin-ichi, Prof. Kyôto Imper. Coll. of Sericulture Nôgakushi. Hanazono near Kyôto (Japan), Chem. Inst. *Biochem.*

Nakano, Haruhusa (1883), Ass. Prof. Imper. Univ. Tôkyô (Japan), Botan. Inst. *Physiol. der Algen. Pflanzenökol.* ⓐ Tôkyô 1909.

Nakao, Manzô, Chief Sc. of Central Research Labor. South Manchurian Railway Co. Dairen, Manchuria (China). ○

Nakarai, Boku, Lect. Imper. Univ. Kyôto (Japan), Pathol. Inst., Fac. Med.

Nakata, Kakugorô (1886), Prof. Plant Pathol. Agric. Dept. Kyûshû Imper. Univ. Fukuoka (Japan). *Bact. disease of plants, specially tobacco diseases.* ⓐ Imper. Univ. Tôkyô 1911. ⓑ *Pathogenic bacteria of plants. Sclerotium Rolfsii Sacc. and its allied cultures.*

Nakatomi, Sadao (1893), Agronomist, Specialist of Cotton plant. Kinshu (Manchuria), Kwan-tung Agric. Experim. Station. *Breeding of Cotton Plant. Chromosome numbers of mutants in rice plant. Piricularia-disease resistance in rice plant.* ⓐ Tôkyô 1919.

Nakayama, Heijirô, Prof. Kyûshû Imper. Univ. Fukuoka (Japan), Pathol. Inst., Coll. of Med. *Pathol.*

Nakayama, Shônosuke (1886), Entomol., Imper. Plant Quarantine Service. Yokohama Customs (Japan). *Seed and fruit insect. Stored product insect.* ⓐ Stanford Univ. 1914.

Nakazawa, Kiichi. Yamanashi-ken (Japan), Kanoiwamura, Higashi-Yamanashi-gun. *Schizopoda.*

Nakazawa, Ryôji (1878), Mycol., Chief of Section of Fermentation Industrial Department, Government Research Inst. Taiwan, Formosa (Japan), Taipeh. *Microorganisms of fermentation.* ⓐ Tôkyô Imper. Univ. 1905. ⓑ *Microorganisms of fermentation.*

Nakazawa, T., Lect. Keiô-Gijuku Univ. Tôkyô (Japan), Physiol. Inst., Coll. of Med. *Physiol.*

Nalęcz Dybowski, Benedykt Tadeusz (1833), Prof. emer. der Zool. und Vergl. Anat. an der Univ. Lwów (Polska), Zascianekstraße 12, Villa Biały Dworek. *Crustacea. Mollusca. Pisces. Batrachia. Aves. Mammalia, Fauna Ostsibiriens und Kamtschatkas. Fauna des Baikalsees.* ⓐ Berlin 1860, Dorpat 1862.

Nalepa, Alfred (1856), Reg. R., Prof. i. R. Baden b. Wien (Oesterreich), Epsteingasse 5. *Eriophyiden (Anat., System., Oekol.).* ⓐ Wien 1884.

Nalivkin, Dmitry V. (1889), Prof. Stratigraphie Mining Inst. Leningrad (U.d.S.S.R.), W. O., 21 Linie, 2. *Paleont. and Paleozoogeography. Variation statistics in the paleoni. Paleoecol.* ⓐ Leningrad 1915.

van Name, Willard G. (1872), Associate Curator, American Mus. of Nat. History. New York City (U.S.A.), 77. St. and Central Park West. *Tunicata and Crustacea (Isopoda).* ⓐ Yale Univ. 1898.

Namikawa, Isao, Prof. in Horticulture, Coll. of Agric., Imper. Univ. Kyôto (Japan), Labor. of Horticult. *Pomol.* ⓐ Hokaidô Imper. Univ., Sapporo.

Namysłowski, Bolesław (1882), Prof. d. Botan. u. Phytopath. Fac. Agric. u. Sylvikultur. Poznań (Polska), Mazowiecka 15. *Mykol., Algol., Anat. d. Pflanzen. Mineralquellenflora. Geschichte d. Botan. in Polen.* ⓐ Krakau 1907.

Nañagas, Juan Cancio (1889), Associate Prof. of Anat. Coll. of Med., Univ. of the Philippines. Manila (Philippine Islands). *Neuro-Anat., Surgical Anat., Anthropometry. Anencephalic Human Fetuses. Motor Cortex of Macacus rhesus.* ⓐ Univ. of the Philippines 1915. ⓑ *Brains of human fetuses and adults, Central Nervous System of Macacus rhesus, Human Skeletons (Filipinos), Human stillborn fetuses.*

Napravil, Emil, Ing., Chef de l'Inst. de Contrôle des graines. Bratislava (C.S.R.), Kapitulná 2. *Agrobotan.* ○

Nardelli, G., Prof. Univ. Roma (Italia). *Pharmacol. sperim.* ○

Naryschkina, Maria Aleksseyewna (1902), Biogeochem. Labor. des Radium-Inst. Akad. d. Wiss. Leningrad (U.d.S.S.R.), Röntgenstr. 1. *Chem. des lebendigen Substrats. Seetiere.* ⓓ Leningrad 1925.

Nash, Thomas Palmer, Jr. (1890), Prof. of Chem., Univ. of Tennessee, Coll. of Med. Memphis, Tenn. (U.S.A.). *Ammonia production in the animal. Mechanism of phlorhizin action. Insulin.* ⓓ Cornell Univ. 1922.

Nassonov, Dimitry (1895), Ass. Histol. Labor. Univ. Leningrad (U.d.S.S.R.), Anat.-Histol. Labor. *Zellsekretion, Cytol.*

Nassonov, Nikolaus (1855), Dir. Zool. Labor. Biol. Station in Sebastopol. Leningrad (U.d.S.S.R.), Tutschkova Naberejnaja 2a. *Morphol., experim. Zool., Zoogeographie. Regeneration, Transplantation.* ⓓ Moskauer Univ. 1879.

Nassonov, Sophie (1899), Peterhoffsches Naturwissenschaftl. Inst. Leningrad (U.d.S.S.R.), Tutschkowa naberejnaia 2a. *Cytol.*

Nassonowa, S. N. Histol. Labor. Univ. Leningrad (U.d.S.S.R.) *Histol.* ○

Nasu, Shôzaburô, Prof. Tôhoku Imper. Univ. Sendai (Japan), Pathol. Inst., Coll. of Med. *Pathol.*

Natali, Claudio, Ass. Ist. d'Anat. patol. Firenze (Italia), 33 via degli Alfani. *Anat. e istol. patol.*

Natali, W. F., Doc. II. Univ. Moskau (U.d. S.S.R.). *Theorie d. Entwicklung.* ○

Nathišvili, Aleks., Prof. Univ. Tiflis (U.d.S.S.R.). *Anat., Histol.* ○

von Nathusius, Gottlob (1884). Hundisburg, Bez. Magdeburg (Deutschland). *Ornithol. Vogelschutz u. Beringung.*

Natvig, Leif Reinhardt (1894), Konserv. am Zool. Mus., Entomol. Abt. Oslo (Norge). *Parasitische Insekten auf warmblütigen Tieren in Norwegen. Renntierparasit., Renntierzucht und Biol. des zahmen Renntieres.*

Nauck, Ernst Theodor (1896), Dr. med., Priv.Doc. der Anat., II. Prosektor am Anat. Inst. der Univ. Marburg a. d. Lahn (Deutschland). *Skelett der paarigen Gliedmaßen der Wirbeltiere, Gelenke. Anpassungen im Blutgefäßsystem an Haemodynamik, Wangenontogenie, Parotisinnervation).* ⓓ Marburg 1921.

Naumann, Einar (1891), Doc.f. Botan. und Limnol. an der K. Univ. Lund (Schweden). *Limnol.; Tierund Pflanzenwelt des Süßwassers.* ⓓ Lund 1917.

Naumann, Karl Arno (1862), ao. Prof., Dr., Dipl. Ing. Chemie, Studiendir. Höhere Staatslehranst. für Gartenbau Pillnitz (Deutschland). *Botan. Versuchswesen, Pflanzengeographie, Phytopath.* ⓓ Leipzig 1885.

Naumenko, Pavel, Doc. Landwirtschaftl. Inst. Kiev (U.d.S.S.R.), Brest-Litovskoe-Chaussee 39. *Physiol. der Tiere.* ○

Naumov, Nicolas Alexandrowitch (1888), Prof. Inst. Agron., Prof. Univ., Chef des travaux Inst. Jaczevski de Mycol. et Pathol. végétale. Leningrad (U.d.S.S.R.), Angliskij Prosp. 29. *Mycol. et Pathol. végétale.* Mucorinées, Deuteromycètes, Plasmodiophora. Tératol. Parasit. cryptogames et moyens d'évaluation des dégats.* ⓓ St. Pétersbourg 1910. ⓟ Champignons.

Naumova, Nadejda Alexandrowna (1895), Chef de labor. Station experim. de viticulture. Odessa (U.d.S.S.R.). *Mycol. et Pathol. végétale: maladies de la vigne, bouillies fongicides.*

Navarro, Regino J., Ass. Prof. Univ. of the Philip. Manila, Philippine Islands (U.S.A.). *Pathol. and Bact.* ○

Navarro Martin, Francisco de P. (1898), Dir. du Labor. Biol. Marino de Baleares. Palma de Mallorca, Iles Baléares (España), Terreno. *Biol. des poissons comestibles de la Méditerranée, Clupeidae et Engraulidae.* ⓓ Madrid 1919.

Navarro y Gil, Vicente, Prof. Univ. Valencia (España). *Anat. Topogr.* ○

Navás, Longin S. J. (1858), Prof. Zaragoza (España), Apartado 32. *Insectes, Neuroptères.* ⓓ Madrid 1904. ⓟ Neuroptères.

Navaschine, Serge G., Prof., Dir. de l'Inst. Timiriazev. Moscou (U.d.S.S.R.), Piatnitskaia 48. *Cytol.* ○

Naveau, Georges Raymont Léonard (1889), Conserv. du Mus., sc. d'Anvers, Prof. à l'Ecole supérièure d'Agric. in Belvers, Co-Administrateur de l'Ecole d'Horticulture de l'Etat à Vilvondes. Anvers (Belgique), 272, Beeldekensstraat. *Biol. botan. Mycol. Bryol.* ⓟ Mousses exotiques.

Naylor, Ernst E. (1899), Instr. in General Botan., Univ. of Missouri. Columbia, Mo. (U.S.A.), 104 Lefevre Hall. *Cytol. Staining of Plant Tissues with Acid and Basic Dyes.* ⓓ Univ. of Missouri 1924.

Naylor, Nellie May (1885), Ass. Prof., Chem. Dept. Iowa State Coll. Ames, Ia. (U.S.A.), Station A. *Enzymes developed during growth of the mold. P. Roqueforti in cheese. Enzymes in germinating grains, wheat, rye, corn.* ⓓ Columbia Univ.

Nazari, V., Prof. Univ. Roma (Italia). *Agron.* ○

Neal, David Carleton, Plant Pathol., Mississippi Experim. Stat. and State Plant Board. A. and M. College, Miss. (U.S.A.), Box 95. ○

Neal, Herbert Vincent (1869), Prof. in Zool. Tufts College, Massachusetts (U.S.A.). *Embryol. and morphol. of the nervous system.* ⓓ Harvard 1893.

Neal, Marcus Pinson (1887), Prof. of Pathol., Univ. of Missouri, School of Med. and Pathol. to the Univ. of Missouri and Boone Country General Hospital. Columbia, Mo. (U.S.A.). *Tuberculosis, hematol.* ⓓ Univ. Coll. of Med., Richmond, Va. 1912.

Neave, Jerris (1901), Lect. in Zool., Univ. of Manitoba. Winnipeg, Man. (Canada). *Entomol.* ⓓ Manchester Univ. 1923.

Necchi, Ludovico, Prof. incar. Univ. catt. del sacro cuore. Milano (Italia). *Biol.* ○

Necheles, Heinrich (1897), Associate in Physiol., Union Med. Coll. Peking (China). *Allgem. Physiol.* ⓓ Med. Hamburg 1923, phil. 1923.

Nechoroschew, Basil P. (1893), Geol., geol. Comm. Leningrad (U.d.S.S.R.), W. O., Srednij Pr. 72-B. *Paleozoic Bryozoa.* ⓟ Paloozoic Bryozoa.

Nedocučaev, Nicolaus (1872), Prof. d. Landw. u. Polytechn. Hoschsch. Leningrad (U.d.S.S.R.). *Allg. u. spez. Pflanzenbau. Nitratspeicherung.* ⓓ Moskau 1905. ⓟ Samen, Herbarien d. landw. Pflanzen, Tabellen, Modelle.

Née, Prof. Ecole prép. de méd. et de pharmacie, Rouen (France). *Pathol.* ○

Needham, Dorothy Mary (née Moyle) (1896), Beit Memorial Research Fellow. Cambridge (England), Conduit Head Road. *Chem. of muscle. Micro-injection of pH and rH-indicators.* ⓓ Cambridge 1926.

Needham, James George (1868), Head of Dept. of Entomol. and Limnol., Cornell Univ. Ithaca, N.Y. (U.S.A.). *Fresh water biol., aquatic insects.* ⓓ Cornell 1898. ⓟ Aquatic Neuropteroid Insects.

Needham, Joseph (1900), Fellow of Gonville and Caius Coll. Cambridge (England), Conduit Head Road. *Chem. Embryol. Physiol. of the cycloses. Micro-injection of pH and rH indicators. Biochem. of the developing embryo.* ⓓ Cambridge 1925.

Needham, Paul Robert (1902), Instr. in Ecol. and Limnol., Cornell Univ. Ithaca, N.Y. (U.S.A.), Dept. of Entomol. *Insect hydrobiol., ecol.* ⓓ Cornell Univ., Ithaca, N.Y. 1924.

von Neergaard, Kurt (1887), Priv.Doc. f. Inn. Med. an der Univ. Basel, Oberarzt am Univ.-Inst. f. physikal. Therapie in Zürich. Zürich 7 (Schweiz), Merkurstr. 70. *Blutkörperchensenkungsreaktion; periphere Zirkulation; Atmungspathol.* ⓓ Zürich 1916.

Neethling, J. H., Prof. Stellenbosch (South Africa), Dept. of Genetics, Univ. *Genetics.* ⓓ Cornell. o

Negodi, Giorgio Carlo (1900), Aiuto all' Ist. ed Orto Botan. della R. Univ. di Cagliari (Italia). *Polimorfismo morfol. e sessuale. Flora della Sardegne. Istofisiol. vegetale.* ⓓ Padova 1922.

Nègre, M., Prof. St. Joseph Univ. Beirut (Syrien). *Anat.* o

Negrette, L., Prof. Univ. Rosario (Argentina). *Farmacol.* o

Negri, Giovanni (1877), Dir. Inst. Botan. Univ. Firenze (Italia), via Lamarmora 4. *Ecol. Bryol. Geographie botan. de l'Italie et de l'Afrique.* ⓓ Torino 1925.

Negrín, Juan, Prof. Dr., Dir. Labor. Fisiol. de la ·Residencia de Estudiantes. Madrid (España), Pinar 17. *Fisiol.* o

Negrín y López, Juan, Prof., Catedrático de la Fac. de Med. Madrid (España), Serrano 73. *Physiol. humaine.* o

Negrini, Francesco, Prof. o. Ist. super. di Med. vet. Bologna (Italia). *Anat. descript. e topograf.* o

Nehrling, Henry (1853), Collaborator Bur. of Plant Industry U.S. Dept. of Agric. Naples, Fla. (U.S.A.). *Ornithol., botan., hortic., ecol. of N. Am. Birds.* o

Neilson-Jones, W., Prof. Univ. London S.W.7 (England), S. Kensington. *Botan.* o

Neioloffs, Wladimir (1874), Ass. of the Dir. Microbiol. Section Inst. of the Experim. Med. Aptekarsky Island. Leningrad (U.d.S.S.R.), Lopouchinskaja Str. 12. *Biol. Chem.*

Neisser, Max (1869), o.ö. Prof. der Hygiene und Bact. an der Univ., Dir. des städt. Hygien. Univ.-Inst. Frankfurt a. M. (Deutschland), Paul-Ehrlich-Straße 40. *Bact. und Serol.* ⓓ Berlin 1893. ⓔ Reinkulturen von Bact. und Praep. von Infektionskrankheiten.

Neiswander, Claud R. (1893), Ass. Entomol., Ohio Agric. Experim. Station. Wooster, O. (U.S.A.). *Corn borer.* ⓓ Ohio State Univ. 1926.

Neiswestnowa-Shadina, Katherine (1897), Ass. Biol. Oka-Station. Murom, GouvernementWladimir (U.d.S.S.R.), Perwomajskaja 4. *Süßwasser-Zooplankton. Rotatoria.*

Neitschenko, Grigorij N., Botan. Sektion d. Landw. Wissenschaftl. Komitee d. Ukraine. Charkow (U.d.S.S.R.), Dergatschewskij per. 5, W. 1. *Angew. Botan.* o

Nekrasov, Sergej Alek., Prof. I. Univ. Moskau (U.d.S.S.R.). *Agric. chem.* o

Nekrassow, Alexey (1874), Prof., Mitglied d. Timirjasewsch. Inst. der wissensch. Forschung. Moskau 8 (U.d.S.S.R.), Petrowskoje-Rasumowskoje, Iwanowskaja ulitza 16. *Vergl. Morphol. des Laichs der Süßwassermollusken. Biol. der Parasit. des Laichs der Süßwassermollusken. Geschichte des Darwinismus und der Deszendenztheorie. Befruchtung und Reifung des tierischen Eies. Geschlechtliche Zuchtwahl. Reifung und Befruchtung der Pteropoden (Cymbulia). Knospung und geschl. Vermehrung d. Hydromedusen (Eleutheria).* ⓓ Moskau 1900.

Nekrassowa, Wera (1884), Ass. Leningrad (U.d. S.S.R.), Jardin Botan. Principal, Herbarium. *Systématique et géographie botan. Saxifragaceae.* o

Nel, G. C., Prof. Stellenbosch (South Africa), Dept. of Botan., Univ. ⓓ B.A. Cape, Phil. Berlin. o

Nelidov, N. P., Prof., Dir. Anat. Kabinett Med. Inst. Astrachan (U.d.S.S.R.). o

Nelis, C., Prof. Univ. Cath. Louvain (Belgique). *Anat.* o

Neller, Joseph R. (1891), Associate Chem., State Agric. Experim. Station. Pullman, Wash. (U.S.A.). *Plant Biochem., Agric. Chem.* ⓓ Rutgers Univ. 1920.

Neloubov, D. N., Cons. Section de sémences et Station d'essais de Sémences, Jardin Botan. principal. Leningrad (U.d.S.S.R.), Aptekarski Ostrov, Pessotchnaia ¹/₂. *Physiol. des plantes.* o

Nelson, Aven (1859), President emeritus and Prof. of Botan., Univ. of Wyoming. Laramie, Wyo. (U.S.A.). ⓓ Ph. Univ. of Denver, Denver, Col. 1904, Sc. Univ. of Colorado, Boulder, Col. 1926. ⓔ Herbarium specimens of Rocky Mountain Seed Plants.

Nelson, Casper Irving (1886), Prof. of Bact., Senior Bact. Fargo, N.D. (U.S.A.), State College. *Genetics of Bacteria; Plant serol., Flax, intracellular protein fractions (especially the globulins).* ⓓ River Falls, Wis., 1926.

Nelson, Edward William (1855), Chief, Bureau of Biol. Survey, U. S. Dept. Agric. Washington, D.C. (U.S.A.). *Mammals, birds.* ⓓ Hon. Sc. George Washington Univ. 1920.

Nelson, Erwin E. (1891), Ass. Prof. of Pharmacol., Univ. of Michigan Med. School. Ann Arbor, Mich. (U.S.A.), 1258 Ferdon Road. *Pharm., Bioassay of drugs.* ⓓ Ph. Univ. of Missouri 1920, M. Univ. of Michigan 1926.

Nelson, James Allen (1875). Mount Vernon, Ohio (U.S.A.), Rural Route 3. *Insect Development.* ⓓ Univ. of Pennsylvania 1903.

Nelson, James C., Principal, Senior High School. Salem, Ore. (U.S.A.), 104, E. Wilson St. *Grasses, sedges.* o

Nelson, John A. (1890), Prof. of Dairying. Bozeman, Mtn. (U.S.A.). *Dairy Bact.* ⓓ Iowa State Coll. 1922.

Nelson, John M. (1876), Prof. of Organic Chem., Columbia Univ. New York City (U.S.A.). *Electron conception of Valence. Oxidation-reduction in organic chem. Chem. nature of enzymes.* ⓓ Columbia Univ. 1907.

Nelson, Julius Richards (1900), Oystes Research Specialist, New Jersey Agric. Experim. Station. New Brunswick, N.J. (U.S.A.). *Biol. of the Oyster.* ⓓ Rutgers Univ. 1923. ⓔ All stages in life history and ecol. types of the American oyster.

Nelson, Nels C. (1875), Associate Curator of Anthrop., American Mus. of Natural History. New York City (U.S.A.). ⓓ California 1908. o

Nelson, Thurlow Christian (1890), Prof. of Zool., Rutgers Univ.; Biologist. N. J. Experim. Station and Board of Shellfisheries. New Brunswick, N.J. (U.S.A.). *Biol. of the Oyster. Physiol. of lamellibranch molluscs. Parasit.* ⓓ Wisconsin 1917. ⓔ Stages in lifehistory, specimens etc. of the American Oyster for oysters of other species.

Němec, Antonín (1894), Chef de l'Inst. Biochim. des Inst. pour la Production de Plantes et de l'Inst. Biochim. de recherches Forestières. Praha-Vinohrady (C.S.R.), Halvíčkovy sady 58. *Biochim. de Plantes. Enzymol. de graines. Biochim. du sol agric. et forestier, détermination de l'exigences pour engrais des sols agric. par voie microchim. Acidité du sol agric. et forestier. Nutrition des arbres forestiers.* ⓓ Prague 1916.

Němec, Bohumil (1873), o. Prof. d. Anat. und Physiol. der Pflanzen an der Karls-Univ. Praha II (C.S.R.), Benátská 433. *Cytol., Physiol. d. Wachstums, Regeneration, Reizphysiol., Mykol. Befruchtungsvorgänge.* ⓓ Prag 1896.

Nemeczek, Albin (1864), Dir. emer. Veterinäramt der Stadt Wien. Purkersdorf bei Wien (Österreich), Postfach 8. *Protistol.* ⓓ Wien 1910.

Němejc, František (1901), Adjunkt of the geol.-palaeont. sect. of the National Mus. Praha II (C.S.R.), Příčná 7/IIIp. *Fossil flora of the youngest beds: pliocene, pleistocene and holocene and of the carboniferous beds of the Czechoslovak Republic.* ⓓ Prague 1924. ⓔ Fossil plants of Bohemia.

Nemiloff, Anton W. (1879), Prof. Anat., Histol. Landwirtsch. Inst., Prof. der Cytol. und der physiol, Histol. Univ., Vorstand Histol. Labor. Zootechn. Abteilung im Inst. für experim. Agronomie. Leningrad (U.d.S.S.R.), Ulitza Tschaikowskogo Haus 1, log. 5. *Histophysiol. der Geschlechts- und Milchdrüsen.* ⓓ Univ. von Leningrad 1902.

Nemiloff, Nadeschda K. (1880), Arzt, Ass. Labor. Anat. und Histol. der Haussäugetiere Landwirtsch. Inst. Leningrad (U.d.S.S.R.), Ulitza Tschaikowskogo H. 1, Log. 5. *Anat. und Histol. der endokrinen Organe.* ⓓ Petersburg 1911.

Nenjukov, Dim. Vasil., Doc. I. Univ. Moskau (U.d.S.S.R.). *Zool., Embryol.* o

Neoral, Karel (1889), Dir. Wein- u. Obst-Sektion Mähr. Landw. Versuchsanstalt. Brno (C.S.R.), Schwarze Felder. *Physiol. d. Rebe u. d. Obstes, Mycol.* ⓓ Prag 1912.

Neppi, Valeria (1877), Prof. del R. Ist. Magistrale. Trieste (Italia), via Milano 3. *Polipi idroidi, sistematica, riproduzione, rigenerazione nelle meduse.* ⓓ 1920.

Neresheimer, Eugen Robert (1876), Reg.R., Dr., Priv.Doc. für Zool. an der Univ. Wien, Fischereifachreferent im Bundesministerium für Land- und Forstwirtschaft. Wien IX (Deutsch-Österreich), Borschkegasse 7. *Protozoen, Fischparasiten, Fischkrankheiten, Ichthyol., Fischereiwissenschaft, Abwasserkunde.* ⓓ München 1902.

Nešković, Milutin, Dr., ao. Prof. der Physiol., Physiol. Inst. Beograd (S.H.S.), Resavska ulica. *Fermente, experim. Pathol.* o

Nesom, George H. (1874), Ass. Prof. St. Paul, Minn. (U.S.A.), Univ. Farm. *Sandy and peat soils.* ⓓ Univ. of Minnesota 1916.

Nesterchuck, Gregory (1884), Ass. Academy of Agric. Gorky, White-Russia (U.d.S.S.R.), Krassninskaja ul. 14. *Phytopath. Vegetable parasit. of the pine cultures. Symbiosis. Fungi.* ⓓ Gorky 1925.

Nesterov, Nik. Step., Prof. Timirjasev Akad. Petrovskoje Razumovskoje. Moskau (U.d.S.S.R.). *Forstwesen.* o

Nestervodskij, Doc. Landwirtschaftl. Inst. Kiev (U.d.S.S.R.), Brest-Litovskoe-Chaussee 39. *Bienenzucht.* o

Nestler, Anton, Dr., Prof. der allgem. Botan. Praha II (C.S.R.), Preslova, Lebensmittel-Untersuch.-Anstalt. o

Nestorer, Kyb. C. (1898), Ass. in Horticulture. Experim. Station Morgantown, W.Va. (U.S.A.). *Physiol. of Vegetable Crops.* ⓓ Cornell Univ.

Nestšadimenko, M.P. (1869), Prof. Med. Inst. Univ., Leiter d. wissenschaftl. Erforschungskatheder der theoretischen Med., Dir. d. Bact. Staatl. Inst. Kiew (U.d.S.S.R.), Bajkowa gora. *Microbiol. Epidemiol.* ⓓ Kiew 1910.

Netolitzky, Fritz (1875), o. Prof. Univ. Cernăuți (România). *Anat., Physiol. der Pflanzen. Pharmakognosie, Anbau von Heilpflanzen. Spezielle und angewandte Entomol. Dicotyledonenblätter. Vegetabilien in den Faeces.* ⓓ Wien 1899. ⓟ *Heilpflanzen. Bembidiini (Carabidae, Insecta).*

Netoušek, Miloš, Dr., Prof. d. allg. u. experim. Pathol. Bratislawa (C.S.R.). Inst. f. allg. u. experim. Pathol. d. Komensky-Univ. o

Netschaewa, Natalie (1887), Hydrol. Inst., Ass. Univ. Leningrad (U.d.S.S.R.), Wassili Ostrow, 6 Linie, 17. *Microbiol. Denitrification im Newa-Flusse.* ⓓ Leningrad 1915.

Netschajewa-Diakonowa, Aglaja Kirillowna (1879), Inst. der descriptiven Anat. II. Univ. Moskau (U.d.S.S.R.), Zoubowsky Boulevard 29, Wohn. 13. *Nervenendigungen, Gelenkevolution.*

Netter, Hans Carl Bernhard (1899), Dr. med., Ass. am Physiol. Inst. Kiel (Deutschland), Hegewischstr. 5. *Physikalisch-chem. Biol. Elektrokinetische Vorgänge an Oberflächen und Membranen. Eiweißkörper. Permeabilität von Pflanzenzellen; Membraneigenschaften der Nervenfasern.*

Neubauer, Wilhelm, o. Prof. Wien XVIII (Österreich), Scheibengerg. 20. *Forstwirtschaft.* o

Neubaur, Rudolf (1886), Dr. phil., Staatlicher Oberfischmeister für die Provinz Schleswig-Holstein. Kiel (Deutschland), Schloß. *Fischereibiol. der Ostsee und der Binnengewässer. Ornithol. Copepoden.* ⓓ Halle a. d. S. 1912.

Neuberg, Carl (1877), o. Prof. und Dir. des Kaiser-Wilhelm.-Inst. für Biochem. Berlin-Dahlem (Deutschland), Hittorfstr. 18. *Biochem der Pflanzen und der Tiere, Mikrobiol.* ⓓ Berlin 1900. ⓟ *Mikroorganismen.*

Neubert, Kurt (1898), Ass. Anat. Inst. Univ. Tübingen (Deutschland). *Histogenese der Speicheldrüsen spez. des menschlichen Pankreas. Entwicklung der Langerhansschen Inseln im menschlichen Pankreas. Der Übergang der arteriellen in die venöse Blutbahn bei der Milz.* ⓓ Tübingen 1924.

Neubürger, Karl (1890), Dr., Prosektor bei den oberbayrischen Heil- und Pflegeanstalten. Haar bei München (Deutschland), Heilanstalt. *Pathol. Anat. des Nervensystems.* ⓓ Freiburg i. B. 1914.

Neuburg, Marie Th., Kustos des Geol. Mus. der Akademie der Wissenschaften. Leningrad (U.d.S.S.R.), Wassily Ostrow, Tutschkowa Nab. 2. *Perm und Juraflora von Sibirien.* o

Neuchet, Enrique, Prof. Univ. Nac. Buenos Aires (Argentina). *Psychol.* o

Neufeld, Fred (1869), Prof., Dir. Preuß. Inst. f. Infektionskrankh. Berlin-Wilmersdorf (Deutschland), Nassauische Str. 53. *Bact.* o

Neuhaus, Ernst (1902), Ass. an der preuß. Versuchs- u. Forschungsanstalt f. Milchwirtschaft in Kiel (Deutschland), Prüne 48.

Neumann, Max-Paul (1874), Dr., Prof., Inst.Dir. Berlin W 30 (Deutschland), Luitpoldstr. 43. *Technol. des Brotes und Gärung des Mehlteiges.* ⓓ Leipzig 1900. ⓟ *Getreide, Mehl, Brot nebst tierischen und pflanzlichem Zubehör.*

Neumann, Olga Th., Prof. Berginst. Swerdlowsk (U.d.S.S.R.). *Palaeont.* o

Neumann, Rudolf Otto (1868), Prof., Dr. med. et phil., Geh.Med.Rat., o. Prof. der Hygiene an der Univ. und Dir. des Hygien. Staatsinst. Hamburg 36 (Deutschland), Jungiusstr. 1. *Hygiene und Bact.* ⓓ Phil. Erlangen 1894. med. Würzburg 1899.

Neumann, Václav (1884), Doc. Univ. Brno (C.S.R.), Liliová 2a. *Anat. pathol.* ⓓ Prague 1908.

Neumark, Eugen (1883), Dr. med. vet., Abteilungsleiter im Hauptgesundheitsamt der Stadt Berlin. Berlin-Schöneberg (Deutschland), Bozener Str. 22. *Bact., Serol., hygien. Nahrungsmittelkontrolle, Schädlingsbekämpfung.* ⓓ Gießen 1907.

Neumayer, Hans (1887), Ass. am Botan. Inst. der Univ. Wien III (Deutsch-Österreich), Rennweg 14. *Phylogenetische Systematik und Organographie, Floristik, Pflanzengeographie.* ⓓ Wien 1918.

Neumayer, Ludwig (1866), Dr., ao. Univ.Prof. München (Deutschland), Pettenkoferstr. 11. *Entwicklung des Nervensystems, Histol. des Nervensystems, vergl. Anat. des Nervensystems und Kopfskelettes, Entwicklung und vergl. Anat. des Darmkanals. Histol. Technik.* ⓓ München 1891.

Neunzig, Rudolf (1896). Berlin-Hermsdorf (Deutschland), Neue Bismarckstr. 42. *Systematische, biol. und vererbungswissenschaftliche Forschungen auf ornithol. Gebiet, Biol., Erblichkeit und Systematik der Ploceiden.* ⓟ *Gesucht Nestjunge der Spermestinae und Ploceinae konserviert.*

Neuschloß, M., Prof. Univ. Rosario (Argentina). *Physik. Biol.* o

Neuville, Henri Emmanuel (1872), Ass. au Labor. d'Anat. compar. du Mus. national d'Histoire naturelle de Paris, Secrétaire de l'Inst. de Paléont. humaine. Paris 5 (France), 55 rue de Buffon. *Anat. compar. des Vertébrés, des Mammifères supérieurs.* ⓓ Paris 1901.

Neuwirth, František, Prof. Bučovice na Moravě (C.S.R.), Tyršova 34. *Mycol.* o

Neveu-Lemaire, Maurice (1872), Prof. agrégé, Chef des travaux de Parasit. à la Fac. de méd. de l'Univ. Paris VI (France), 15, rue de l'Ecole de Méd. *Parasit.: Protozoaires, Vers, Arthropodes parasit. de l'homme et des animaux domestiques, spécialement Nématodes. Pathol. tropicale.*

Nevinny, Hans (1900), II. Ass. Pathol.-anat. Inst. Univ. Innsbruck (Österreich), Müllerstr. 44. *Geschwülste der Nieren und der weiblichen Genitalorgane, Knorpelerkrankungen, pathol. Anat. der Trichinose.* Ⓓ Innsbruck 1925.

Nevole, Jan (1878), Prof. Brno (C.S.R.), Huterův rybník 5. *Pflanzengeographie, Floristik (Alpen).*

Nevskij, Walerian Pawlowich (1893), Uzbekistans Stat. of Plant Protection, Orchard's Division, Chief. Tashkent (U.d.S.S.R.), Pushkinskaja 37. *Systematics of Aphididae and Eluteridae. Applied Entomol.* Ⓓ Aphididae.

Nevzad, Moustapha, Doc. Agric. Univ. Stambul (Türkei). *Pharmacol.* o

Newcomb, W. D., Prof. St. Mary's Hospit. Med. School. London W. 2 (England), Paddington. *Pathol.* o

Newcombe, Frederick Charles (1858), Emeritus Prof. of Botan., Univ. of Michigan. Honolulu (Hawaii), 1928 Vancouver Highway. *Plant Physiol.: Plant tropisms.* Ⓓ Leipzig 1893.

Newcomer, Erval Jackson (1890), Associate Entomol., U.S. Dept. of Agric. Yakima, Wash. (U.S.A.), Box 243. *Deciduous Fruit Insects.* Ⓓ Stanford Univ. 1911.

Newell, Wilmon (1878), Dean and Dir. Experim. Station and Agric. Extension Div. Univ. of Florida. Gainesville, Fla. (U.S.A.). *Entomol.* o

Newhall, Allan Goodrich (1894), Ass. pathol. Ohio Agric. Experim. Station. Wooster, O. (U.S.A.). *Diseases of vegetable crops, truck and greenhouse.*

Newham, H. B., Prof. School of Hygiene and Trop. Med. London N.W. (England), Endsleigh Gardens. *Pathol.* o

Newman, George Burgess (1890), Associate Prof. of Zool., Penn. State Coll. State College, Pennsylvania (U.S.A.), 142 S. Frazier St. *Histol., Histol. technique, Embryol., Entomol.*

Newman, Horatio Hackett (1875), Prof. of Zool., Univ. of Chicago, Ill. (U.S.A.). *The biol. of twins. Hybridol. Organic evolution. Vertebrate zool. Experim. embryol. (especially marine).* Ⓓ Univ. of Chicago 1905.

Newman, Leslie Frank (1882), Dir. of Natural Sc. in Coll. Cambridge (England), St. Catharines. *Ecol. and biol. of soil.* Ⓓ Cambridge 1908.

Newodowski, Gabriel St. Kiev, Ukraina (U.d.S.S.R.), Lvovskaja 55-10. *Mycoflora ukrainica: Pyrenomycetes et Fungi imperfecti. Mycoflora Leguminosarum et Graminearum. Morbi Betae Sacchariferae (putriditas radicis).* Ⓓ Pilze.

Newsom, Isaac Ernest (1883), Prof. of Vet. Pathol., Colorado Agric. Coll., Vet. Pathol., Colorado Experim. Station. Fort Collins, Col. (U.S.A.). *Sheep diseases, contagious abortion, cattle.*

Newth, H. G., Lect. Univ. Birmingham (England). *Zool.* o

Newton, Margaret, Plant Pathol. Dom. Rust Labor., Agric. Coll. Winnipeg, Man. (Canada). *Biol. forms of wheat stem rust.* Ⓓ Minnesota 1922.

Newton, Robert (1889), Prof. of Field Husbandry and Plant Biochem. Edmonton (Canada), Univ. of Alberta. *Physiol. and biochem. of crop plants, especially wheat; resistance to frost, drought and disease.* Ⓓ Minnesota 1923.

Newton, William, Sc. Ass. Carnegie Inst. Coastal Labor. Carmel, Cal. (U.S.A.). *Metabolism of nitrogen compounds.* o

Neysse, Arthur Nisswald (1867), Prof. of Biol., Boston Univ. Boston, Mass. (U.S.A.), 688 Boylston Street. *Mammalian embryol.* Ⓓ Ph.D. Harvard Univ. 1894, M.D. Basel 1907.

Nezlobinski, Nikola, Dr., Leiter der Malariastation, Stanica za Malariju. Struga na Ohridskom jezeru, Macedonija (S.H.S.). *Entomol.* o

Ni, Tsang-Gi, Instr. in physiol. Peking Union Med. Coll. Peking (China). *Metabolism of stomach.* Ⓓ Chekiang Med. Coll. 1916, Univ. of Mich. 1919 and 1922.

Niant, James Stewart (1900), Plant Pathol. Ithaca, N.Y. (U.S.A.), Department of Plant Pathol., Cornell Univ. *Damping-off studies with conifers. Use of organic mercury compounds as soil disinfectants.* Ⓓ Pennsylvania State Coll. 1924.

Nibe, Tominosuke, Biol. Riku- u. Branch of Imper. Agric. Experim. St. Hanadate, Senpokugun, Akita-ken (Japan). *Ornithol.*

La Nicca, Richard (1867). Bern (Schweiz), Berna Str. 6. *Schweizerische Floristik.* Ⓓ Zürich 1895. Ⓟ Schweizer Phanerogamen.

Niccolini, Pietro Maria (1895), Prof. Dott., Ass. eff. Ist. di materia medica Univ. Firenze 11 (Italia), Via Faenza 58. *Farm. dell' ematina. Azione di alcuni narcotici, ipnotici, ed antispasmodici sull' utero. Viscum album. Renotirina. Sinergismo tra farm. ed ormoni: digitalici e tiroide.* Ⓓ Firenze 1920.

Nice, Leonard Blaine (1882), Prof. of Physiol., Univ. of Oklahoma. Norman, Okla. (U.S.A.). *Effects of Alcohol, Nicotine and Caffeine on White Mice. Respiration. The Blood of Complement Deficient Guinea Pigs. Removal of Semi-circular canals in young Chicks. Oklahoma Birds.* Ⓓ Clark Univ. 1911.

Nicholas, John Spangler (1895), Ass. Prof. of Biol., Yale Univ. New Haven, Conn. (U.S.A.), Osborn Zool. Labor. *Experim. Morphol., Experim. Embryol.* Ⓓ Yale Univ. 1921.

Nicholls, A. G., Prof. Dalhousie Univ. Halifax, Nova Scotia (Canada). *Pathol. and Bact.* o

Nicholls, George Edward (1878), Prof. of Biol. Univ. of Western Australia, Perth (Australia). *System. of Crustacea.* Ⓓ London 1901.

Nichols, George Elwood (1882), Prof. of Botan., Yale Univ. New Haven, Conn. (U.S.A.), 439 Edgewood Ave. *Ecol. (Geobotan.). The vegetation of northeastern America and geobotan. classification; dendrol.; taxonomy of bryophytes of northeastern America.* Ⓓ Yale 1909.

Nichols, Henry T. (1877), Med. Inspector, Panama Canal Department, U.S. Army. Quarry Heights (Panama Canal Zone). *Pathol. Experim. Syphilis, Typhoid carriers; Thyphoid Vaccine.* Ⓓ M.D. Univ. of Pennsylvania 1904.

Nichols, John Treadwell (1883), Ass. Curator of recent Fishes, American Mus. of Natural History. New York City (U.S.A.). *Ichthyol., Ornithol., Herpetol., Mammal. Systematic: The fresh-water fishes of China. Fauna of Long Island, N.Y.; Alaska. Migration and distribution. Carangidae (Fishes); Chelonia (Reptiles); Limicolae (Birds); Sciuridae.*

Nichols, M. Starr (1887), Ass. Prof., Sanitary Chem., Univ. of Wisconsin, Chem. and Bact. to Wis.State Labor. Hygiene. Madison, Wis. (U.S.A.). *Water supply biol. and bact. Chem.* Ⓓ Univ. of Wisconsin 1926.

Nichols, Susan Persival (1873), Prof. of Botan., Oberlin Coll. Oberlin, Ohio (U.S.A.). *Healing of punctured cells in some Algae.* Ⓓ Univ. of Wisconsin 1902.

Nicholson, G. W. de P., Reader Univ. London S.W. 7 (England), South Kensington. *Histol. Pathol. Anat.* o

Nicloux, M., Prof. Fac. Méd. Strasbourg (France). *Biochem.* o

Nicod, Jean-Louis (1895), Prof. d'anat. pathol. à l'Univ. de Lausanne (Suisse), Inst. pathol.

Nicolai, Leon (1862), Prof. de Parasit. à la Fac. de Med. de Jassy (Roumanie). *Plathelminthes: Cestodes. Insectes: Culicidés.* Ⓓ Jena 1887. Ⓟ Dibothriocephalus latus et Diplogonoporus Brauni Leon et autres parasit. de Roumanie.

Nicolaïdès, P., Prof. de Physiol. Univ., Dir. de l'Inst. Physiol. Athènes (Grèce), Rue Phidias 14. o

Nicolas, A., Prof. Univ. Paris (France). *Anat.* o

Nicolas, Léon Marie Joseph Gustave (1879), Prof. de Botan. à la Fac. des Sc., Dir. de l'Inst. Agric. de l'Univ. Toulouse (France). *Physiol. végétale. Tératol. végétale. Amélioration des plantes cultivées.* Ⓓ Toulouse 1921.

Nicolau, Georges (1885), Prof. à l'Acad. d'Agric. Cluj (Roumanie). *Anat. et Physiol. animale. Pehade achromique chez le cheval. Insuffisance alimentaire. Croissance. Insuline et sécrétion du lait. Hémolysines. Hypodermose des bovidés. Effets de l'injection de différentes substances chez les plantes. Formation du lait.* ① Bucarest 1910. ② *Crânes des carnivores. Parasit. des animaux domestiques.*

Nicoll, Luigi (1900), Ass., Ist. di Anat. patol., R. Univ. d. Pavia (Italia). ① Pavia 1924.

Nicolle, M. Charles, Dir. Inst. Pasteur. Tunis (Afrique). *Bact.* ○

Nicolski, A. (1858), Prof. Zool. Charkow (U.d.S.S.R.), Seminarskaja 34. *Zoogeographie, Ichthyol. Herpetol.*

Niedenzu, Franz Josef (1857). GRR. Dr., Prof. emer. Mathem. und Naturwiss. an der Staatl. Akademie. Braunsberg, Ostpreußen (Deutschland), Stadtpark 3. *System. Botan., Anat. und Pflanzengeographie. Malpighiaceae.* ① Breslau 1889.

Niedra, Dāgmara Anna Eleonore (1892), Ass. Landwirtschaftl. Fac. d. Univ. Rīga (Latvija), Bunginieka ielā 19a. *Mikrobiol., Milchwirtschaftliche Nahrungsmittelbact.* ① Rīga 1924.

Niedziałkowski, Wacław (1892), Ass. l'Inst. de l'aménagement des forêts de l'Ecole Sup. d'Agric. Varsovie (Pologne). *Phytosociol. des forêts.* ① Varsovie 1924.

van Niel, Cornelis Bernardus (1897), Konserv. des Labor. für Mikrobiol. der technischen Hochsch. Delft (Holland), Oldenbarneveldstraat 1. *Biochem. der Mikroorganismen.* ① 1923. ② Kulturen von Hefen- und Bakterienarten.

Nielsen, Kristian Brünnich (1872). København (Danmark), Amagerbrogade 51 I. *Palaeont.; Senon, Danian, Brachiopoden, Bryozoen, Foraminifera, Serpulidae, Korallen, Hydrokorallen, Crinoiden, Asteroiden.* ① Copenhagen 1914. ② *Versteinerungen aus dänischem Senon und Danien.*

Nielsen, Niels (1900), Ass., Plantphysiol. Labor. of the Royal Vet. and Agric. Coll. København (Danmark), Rolighedsvej 23. *Stimulus phenomenas in plants (Avena-Coleoptile), Sexuality in Mucors, Mouldflora of Soils.*

Nielsen, Peder Kristian. Silkeborg (Danmark). *Diptera (Tipulidae).*

Niemczycki, St., Dr. phil., Prof. Lwów (Polska), Acad. Vétérinaire, r. Krapewskiego 52. *Biochem.* ○

Niemeyer, Ludwig (1898), Dr. phil., wissenschaftlicher Angestellter. Berncastel-Cues a. d. Mosel (Deutschland). *Rebenkrankheiten.* ① Münster i. Westf. 1923.

Nienburg, O. K. Wilhelm (1882), o. Prof. der Botan. an der Univ. Kiel (Deutschland), Adolfstr. 52. *Ökol. der Meeresalgen. Entwicklungsgeschichte der Pilze und Flechten.* ① 1907.

Nierenstein, Maximilian (1877), Prof. of Biochem. Bristol (England), The Univ. *Plantchem. Tannins.* ① Berne 1901.

Nierstrasz, Hugo Fredrik (1872), o. Prof. der Zool. an der Reichsuniv. zu Utrecht (Holland), W. Barentzstraat 7. *Nemertinen. Solenogastres, Isopoden, Chitonen. Geographische Verbreitung der Tiere. Vergl. Anat.* ① Utrecht 1902. ② *Solenogastren, Chitonen, Isopoden, parasit. Nemertinen.*

Nieschulz, Otto Christian Henry (1899), Zool.-Parasit. des Tierärztlichen Labor. Buitenzorg, Java (Nederl.-O.-Indië), Veeartsenijkundig Labor. *Veterinärmed. Parasit. Surraübertragung.* ① Hamburg 1922. ② Tropische Parasit. von Haustieren.

Niesiołowski, Witold (1866), Kustos Lepidopterol. Abt. im Physiographischen Mus., Poln. Akad. d. Wiss. Kraków (Polska), Sławkowska 17. *Macrolepidoptera, Systematik.* ② Macrolepidoptera.

Niethammer, Anneliese (1901), Univ.Ass. Praha (C.S.R.), Badenho 291. *Chem. Reizwirkungen bei höheren und niederen Pflanzen.* ① Ing. Techn. Hochschule Prag 1923, rer. nat. Deutsche Univ. Prag 1925.

Nieuwejaar, Otto (1895), Priv.Doc. Univ. Rīga (Latvija), Rainis Boulev. 5, W. 4, *Forstwissensch.*

Nieuwenhuijse, Pieter (1883), Prosektor, Pathol. Anat., Univ. Utrecht (Holland), Frederik Hendrikstraat 49. *Pathol. Anat., Muskeln.*

Nieuwland, Johanna Adriana (1888), Apotheker, Botan. an der Reichsversuchsstation für Samenkontrolle in Wageningen (Holland), Nudestraat 15. *Reinheitsbestimmung der Samen.*

Niezabitowski-Lubicz, Edward (1875), Dr., Prof. Allgem. Biol. med. Fac. Univ. Poznań (Polska). *Allgemeine Biol. und Palaeobiol. (Mammalia foss.). Braconidae, Crustacea, Hippolyte.* ① Kraków 1900.

Nightingale, Gordon T. (1898), Biochem. in Horticulture. New Brunswick, N.J. (U.S.A.), N.J. Experim. Station, Hort. Dept. *Nitrogen and carbohydrate metabolism.* ① Univ. of Wisconsin.

Niijima, Y., Prof. Hokkaidō Imper. Univ. Sapporo (Japan), Inst. of Forestry, Fac. of Agric. *Sylviculture and Forest Protection.*

Nikanorov, S. M., N. G. Černijševskij Univ. Saratov (U.d.S.S.R.), Ploščod imeni Lenina. *Bact.* ○

Nikiforovskij, P. M., Prof. Univ. Voronesch (U.d.S.S.R.), Universitätskaja 5. *Physiol.* ○

Nikitin, Peter (1890), Botan. Landwirtschaftl. Hochsch. Woronesch (U.d.S.S.R.). *Palaeophytol. d. posttertiär. Zeit.* ② Posttertiäre Pflanzenreste.

Nikitin, V. A., Doc. Staatl. Univ. Leningrad (U.d.S.S.R.). *Zootechn.* ○

Nikitin, W. N., Dir. Biol. Station. Sewastopol (U.d.S.S.R.). ○

Nikitinskij, Jakob Ja., Prof. Inst. f. Volkswirtschaft Plechanow. Moskau (U.d.S.S.R.), Sazepa 44. *Botan., Mikrobiol.* ○

Niklas, Prof. Dr., Agrikulturchem. Inst. Weihenstephan b. Freising (Deutschland). ○

Niklewski, Bronislas, Dr. Prof. Poznań (Polska), Univ. Sołacz-Palais, r. Mazowiecka 15. *Botan.* ○

Nikolaefi, Alexander, Dr., Ass. Inst. f. Pharmacol. u. Therapie Univ. Sofia (Bulgarien). ○

Nikolaeff, Michael (1893), Ass. Pharmacol. Inst. Militär-Med. Akademie. Leningrad (U.d.S.S.R.), Tschaikowskistr. 25. *Funktion der endokrinen Drüsen unter dem Einfluß pharm. Agenzien.* ① St. Petersburg 1914.

Nikolaev, Vladimir Vas., Prof. I. Univ. Moskau und Univ. Smolensk (U.d.S.S.R.). *Pharmacol.* ○

Nikolaewa, Helene (1899), Ass. Physiol. Inst. Univ. Minsk (U.d.S.S.R.), Universitätskaja 5. *Sekretion d. Darmsaftes.*

Nikoljukin, Nicolas (1896), Doc., Dir. Inst. Vergl. Anat. Univ. Woronesch (U.d.S.S.R.). *Morphol. des Schädels der Anuren u. der Teleostier. Entwickelung des Rostralapparates bei den Anurenlarven.*

Nikoloff, A. P., Kustos Botan. Inst. d. Univ. Sofia (Bulgarien). *Lichenol.* ○

Nikoloff, Eugen (1899), Ass. Physiol. Inst. Med. Fac. Sofia (Bulgarien), Struga 12. *Biochem.* ① Sofia 1925.

Nikol'skaja, Serafima Vasil., Doc. II. Univ. Moskau (U.d.S.S.R.). *Anat. d. Menschen.* ○

Nikolsky, Alexander Mihailovich (1858), Prof. Univ. Charkow (U.d.S.S.R.), Seminarskaja 34. *Zool., Herpetol., Zoogeographie.* ① Petersburg 1889.

Nikschitsch, Johann H., Com. Géol. de Russie. Leningrad (U.d.S.S.R.), W. O., Sredny Pr. 72-B. *Palaeont.; Mesozoische Ammoniten des Kaukasus.* ○

Nilsson, Fredrik Wilhelm (1903), Agronom. Landskrona (Sverige). *Genetical investigations in grasses.*

Nilsson-Cantell, Carl-August (1893), Prof. Wisby (Schweden), Allmänna läroverket, Skövde oder Skepparegatan 16. *Crustacea: Cirripedien, Geschlechtsmerkmale bei Decapoden.* ① Upsala 1921.

Nilsson-Ehle, N. H., Prof. Univ. Lund (Sverige). *Genetik.* ○

Nilsson-Leissner, Gunnar (1895), Ass. Svalöf (Sverige). *Genetics and breeding of forage crops.* ① Lund 1926.

Ninman, Herman J., Ass. Plant Pathol.; State Department of Agric., State Capitol Annex. Madison, Wis. (U.S.A.). o
Nire, Kageo. Komaba near Tôkyô (Japan), Entomol. Labor., Fac. of Agr., Imper. Univ. *Lepidoptera. Rhopalocera.*
Nishi, Seiho (1885), Prof. Anat. Med. Fac., Tokio Imper. Univ. Meziro-Bunkwa-Mura 57 near Tokio (Japan). *Vergl. Myol. und Anthrop.* ① Tokio Imper. Univ. 1908.
Nishimura, Genkichi (1897), Prof. of Biol., Premedical Course Manchuria Med. Coll. Mukden, Manchuria (China). *Cytol. of insecta.*
Nishishita, Masami (1898), Ass. Med. Univ. Okayama (Japan), Pharmacol. Inst. *Experim. Pharmacol.* ① Med. Akademie Okayama 1923.
Nishiwaki, Yasukichi (1880), Prof. Zymomycol. and Fermentol. at the Technol. Coll., Dir. fermentol. dept. in the Coll. Ôsaka (Japan), Technol. Coll. (Koto-Kogyo-Gakko). *Zymomycol. Sc. of fermentation organisms. Fermentol., especially of the Japanese flora of yeasts and moulds, in relation to the fermentation industries.* ① Ôsaka Technol. Coll. 1901. ② Japanese flora of fermentation organisms.
Nisikado, Yosikazu (1892), Plant pathol., The Ohara Inst. for Agric. Research. Kurashiki, Okayama-ken (Japan). *Plant pathol. and mycol. Rice diseases. Helminthosporium. Seed treatment for cereal diseases.* ① Morioka 1913. ② Specimens and pure-cultures of fungi causing plant diseases, Helminthosporium.
Nissen, Rudolf (1895), Priv.Doc. f. Chirurgie, Ass. Chir. Univ.-Klinik. München (Deutschland), Nußbaumstr. 20—22. *Experim. Lungenpathol. Blutkrankheiten. Reticulo-endotheliales Stoffwechselsystem. Pathol. des Mediastinums.* ① Breslau 1920.
Nißle, Alfred (1874), Dr. med., a.o. Prof., Leiter des Untersuchungsamts für ansteckende Krankheiten. Freiburg i. Br. (Deutschland), Johanniterstr. 15. *Normale und pathogene Darmbakterien. Bakterielle Darmtherapie. Praktische Rassenhygiene. Cariesproblem. Krebsproblem.* ① Freiburg i. Br. 1899.
Nitsche, Max (1875), Dr. ing., Prof. der Abteilung für Landwirtschaft der deutschen technischen Hochschule Prag in Tetschen-Liebwerd. Tetschen a. d. Elbe (Č.S.R.), Lausitzer Str. 844. *Züchtungsbiol. (landwirtschaftliche Tierzucht). Kraniol. Forschung auf dem Gebiete prähistorischer und rezenter Haustierformen sowie Wachstumsuntersuchungen.* ① Wien 1908.
Nitzescu, J. Joan (1844), Prof. tit., Dir. Inst. de Physiol. Fac. Med. Cluj (România), Strada Mico 1. *Sécrétion endocrine du pancreas; thyroïde et glycogène; vitamines.* ① Bucuresti 1913.
Nitzschke, Hans (1890). Wilhelmshaven (Deutschland), Hollmannstr. 13. *Pflanzengeographie. Halophyten im Marschgebiet der Jade.* ① Halle a. d. S. 1913.
Noack, Konrad Ludwig (1891), Dr. phil. nat., o. Prof. der Botan. an der Forstlichen Hochsch. Eberswalde (Deutschland), Botan. Inst. *Panaschierte Gewächse.* ① Freiburg i. Br. 1913. ② Panaschierte Gewächse.
Noack, Kurt (1888), Dr. phil., o. Prof., Dir. des Botan. Inst. Erlangen (Deutschland), Botan. Garten. *Biochem. der Pflanzen.* ① Leipzig 1912.
Noack, Martin (1888), Dr. phil., Botan. an der Biol. Reichsanstalt für Land- und Forstwirtschaft. Berlin-Südende (Deutschland), Steglitzer Str. 39. *Phytopath.* ① Zürich 1921. ② Pilzliche Exsiccate.
Noback, Gustave J. (1890), Associate Prof. of Anat., New York Univ. and Bellevue Hospital, Med. Coll. New York City, N.Y. (U.S.A.), 338 East 26 Street. *Developmental Anat. Fetus, Infant, Child, Early Human Embryo. Thymus. Anthropometry.* ① Univ. of Minnesota 1923.

Noble, Gladwyn Kingsley (1894), Curator, Department of Herpetol. New York City (U.S.A.), American Mus. of Nat. History, 77 Street and Central Park West. *Morphol., distribution, and biol. of reptiles and amphibians.* ① Columbia 1922. ② Reptiles and amphibians.
Noble, Robert Jackson (1894), Principal Ass. Biol., Department of Agric. Sidney N.S.W. (Australia). *Diseases of Wheat, of pome and stone fruits, citrus fruits, field crops, sugar cane, potatoes.* ① Minnesota, U.S.A., 1923. ② Herbarium (dried specimens) of fungous and bacterial diseases of plants.
Noble, Willis Bernard (1898), Ass. Entomol., Cereal and Forage Insect Division, Bureau of Entomol., U.S. Dept. of Agric. West Lafayette, Ind. (U.S.A.), Box 95. *Insects attacking cereal and forage plants. Phytophaga destructor Say. and its parasit.* ① Ohio State Univ. 1922.
Nobre, Augusto Pereira, Prof. Univ. Porto (Portugal). *Biol.* o
Nobuyoshi, Towasa (1899). Osaka (Japan) Komatzubaracho Kitaku. *Hymenoptera.*
Nocht, Bernhard (1857), Dir. d. Inst. f. Schiffsu. Tropenkrankheiten. Hamburg (Deutschland), Brahmsallee 107. *Tropenkrankheiten. Tropenhygiene. Malaria.* o
Noé, Adolf C., Associate Prof. of Paleobotan., Univ. of Chicago, Ill. (U.S.A.). o
Noël, Robert (1893), Prof. Agrégé d'Histol. à la Fac. de Méd. de Lyon, Rhône (France), 18, quai Claude Bernard. *Cytol. gé-érale et histophysiol. Cellule hépatique. Plaques motrices.* ① Méd. Lyon 1920, sc. nat. Paris 1922.
Noël-Paton, Diarmid (1859), Regius Prof. of Physiol., Dir. of Inst. of Physiol. Univ. Glasgow (Scotland). *Experim. and chem. Physiol. Physiol. of continuity of live. Human Physiol. Vet. Physiol. Regulations of Metabolisme.* ① Edinburgh 1886.
Nöller, Prof. Tierärztl. Hochsch. Berlin (Deutschland). *Vet.-Med. Protozoa.* o
Noelli, Alberto (1873), Prof. Torino (Italia), R. Liceo Gioberti. *Micol. Flora del Piemonte. Anat. e fisiol. vegetale.* ① Torino 1907. ② Piante del Piemonte.
Nömmik, Anton (1882), ao. Prof. an der Univ. Tartu (Esthonia), Landwirtschaftl. Versuchsstation Raadi. *Bodenchem., Bodenfruchtbarkeit, Düngung.* ① U.S.A. 1926. ② Böden.
Noether, Paul (1880), wissenschaftl. Mitarbeiter am Pharmacol. Inst. der Univ. Freiburg i. Br. (Deutschland), Thurneestr. 64. *Nicotin, Adrenalin, Guanidin, Allium sativum.* ① 1910.
Nogteff, Wassilij (1890), Priv.Doc. Botan. Kabinett. Univ. Nijny-Nowgorod (U.d.S.S.R.). *Geobotan., Theorie der Phytosoziol., Pflanzen-Ökol.* ② Herbarpflanzen.
Noguchi, Hideyo (1876), Member, Rockefeller Inst. for Med. Research. New York, N.Y. (U.S.A.), Ave. A and 66 St. *Bact., Parasit.; Immunol.; Pathol. Spirochetes; Treponema pallidum in the brain.* ① Tôkyô Med. Coll. 1897.
Noinsky, E., Prof. Geol. Univ. Kazan (U.d.S.S.R.) *Palaeont. des Palaeozoicum.* o
Nojima, Tomoo (1902), Ass. Kyôto (Japan), Labor. of Phytopath., Dept. of Agric., Imper. Univ. *Diseases of the leguminous plants caused by Fusarium wood-rotting fungi.* ① Kagoshima 1924.
Noland, Lowell Evan (1896), Ass. Prof. of Zool., Univ. of Wisconsin. Madison, Wis. (U.S.A.), Biol. Building. *Protozool.; ciliates.* ① Univ. of Wisconsin 1924.
Nolf, Pierre Adrien Emile Louis (1873), Prof. pathol. et thérapeut. gén. Univ. Liége, Dir. du labor. de recherche méd. ,,Reine Elisabeth". Bruxelles (Belgique), 51, rue Stevens Delauroy. *Physiol. normale et pathol.* ① Liége 1901.
Noll, Alfred (1870), a.o. Prof. für Physiol. Jena (Deutschland), Wildstr. 17. *Pupille (vergl. neurol.); Granulalehre (Histophysiol.).* ① Marburg 1895.

Noll, Hans (1885), Dr. hon. c., Lehrer der Naturwiss. am Landerziehungsheim Schloß Glarisegg, Steckborn (Schweiz). *Ornithol., biol. Fragen. Sumpfvögel.* ⒹHon. c. Basel. ⓅPhotographische Aufnahmen aus dem Sumpfvogelleben.

Nolle, Jakov Christ., Doc. I. Univ. Moskau (U.d.S.S.R.). *Pharmacol.* o

Nolla, Jose A. B. (1902), Plant Breeder, Attached Plant Pathol. Dept. Rio Piedras (Porto Rico), Insular Experim. Station. *Tobacco diseases. Onion diseases. Mycol.: Imperfect fungi: Macrosporium, Alternaria and Stemphylium. Genetics: Eggplant breeding and tobacco breeding.* ⒹCornell Univ. Ithaca, N.Y. (U.S.A.) 1926.

Nomals, Pēters, Doc. Univ. Rīga (Latvija), Stērlnieku ielā 11, dz. 6 u. Kronvalda bulv. 1. *Moorforschung.* o

Nomura, Ekitarô (1887), Prof. Zool. Sendai (Japan), Biol. Labor., Tôhoku Imper. Univ. *Aquatic Oligochaetes, Embryol., Experim. Zool.* ⒹImper. Univ. Tôkyô 1912.

Nomura, Shichiroku, Lect. of Imper. Univ. Tôhoku (Japan), Zool. Labor., Coll. of Sc. *Experim. Biol.*

Nonell, I., Prof. Esc. super. de Agric. Barcelona (España). *Biol., Pathol.* o

Nonevitsch, Elias (1863), Prof., Dir. Pathol.anat. Inst. d. Vet.-Abt. Kaunas (Litauen). *Fleischbeschau, Rotz, Ichthyol. Parasitol.* ⒹDorpat 1889.

Nonidez, José Fernandez (1892), Associate in Anat. New York City (U.S.A.), Cornell Univ. Med. Coll., 477 First Avenue. *Cytol. and experim. Endocrinol.* ⒹMadrid 1914.

Nonnenbruch, Wilhelm (1887), Prof. und Chefarzt innere Abtlg. Krankenhaus Frankfurt a. d. Oder (Deutschland). *Stoffwechsel.* Ⓓ 1911.

Nordgård, Ole (1862), Dir. Biol. Station, Kustos am Mus. d. Kgl. Norske Videnskabers Selskab. Trondhjem (Norge). *Systematik, Verbreitung der Bryozoen.*

Nordhagen, Rolf (1894), Prof. für systematische Botan. u. Pflanzengeographie. Bergen (Norwegen), Bergens Museum. *Pflanzengeographie, spez. Pflanzensoziol., Floristik, Waldgrenzen; Postglazialgeol., Torfmoore, Kalktuffe, Klimaschwankungen.*

Nordhausen, Max (1876), o. Prof. der Botan. an der Univ. Marburg a. d. L. (Deutschland), Marbacher Weg 20. *Pflanzenphysiol., Anat.* ⒹBerlin 1897.

Nordmann, Otto (1877), Provinzialobstbauinsp., Obst- und Gartenbaulehrer. Bad Kreuznach (Deutschland). *Obst-, Gemüse- und Gartenbau. Pflanzenpathol.*

Nordqvist, Harald Oscar (1884), Priv.Doc. der Fischereibiol. Univ. Lund, Vorstand der Fischerei-Versuchsstation. Aneboda, Ugglehult (Sverige). *Fischereibiol., Teichwirtschaft.* Ⓓ Lund 1921.

Noriega del Aguila, Miguel, Prof. Univ. Lima (Peru). *Bact. Histol.* o

Norman, John Roxbrough (1898), Ass., Department of Zool., British Mus. (Natural History). London S.W. 7 (England). *Fishes; Systematic.*

de Noronha, Cordato (1900), Ecole méd. de Gôa (Indes Portugaises). *Histol. L'Energétique Cellulaire.* Ⓓ Gôa 1924.

Norris, Harry W. (1862), Prof. of Zool. Grinnell Coll., Grinnell, Ia. (U.S.A.). *Compar. Anat. of nervous system.* ⒹA.M., Iowa Coll. 1889; sc. 1924. o

Norris, Roland Victor (1887), Prof. of Biochem., Indian Inst. of Sc. Bangalore (British India), Hebbal P.O. *Agric. Chem. Soil organisms. Chem. of enzyme action. Indian food materials.* ⒹLondon 1914.

Northcroft, Earle Fead (1896), Botan. Biol. Labor. Dept. of Agric. Wellington (New Zealand), Bay View Road Napier. *Blackberry Pest. Ecol. of the Plants of the Lawyers. Head region near Dunedin. Ecol. Classification of the Vegetation of the Chatham Islands. Gleicheniaceae: syst., anat. and embryol.* Ⓓ Otago Univ. 1924.

Northup, Flora Eleanor (1901), Lect., Dept. of Biol., Judson Coll. Rangoon, Burma (India). Ⓓ Washington Univ. 1925.

Norton, John Bitting Smith (1872), Plant Pathol., Md. Experim. Station, Prof. Systematic Botan. and Mycol., Univ. of Maryland. College Park, Md. (U.S.A.). *Mycol., Plant diseases, especially non parasit. Dahlias.* Ⓓ Hon. Univ. of Maryland 1923.

Noskiewicz, J., Dr., Konserv. Lwów (Polska), Mus. Dzieduszyckich. *Zool.* o

Noureddine, Prof.Univ.Stambul (Türkei). *Anat.* o

Novák, František (1892), Priv.Doc. der Karls-Univ. und Ass. des Botan. Inst. Praha II, 433 (C.S.R.), Benátská 2. *Serpentinpflanzen. (Oekol. und geobotan. Untersuchungen). Dianthus, Centaurea (Monographische Studien). Floristik und Systematik der Pflanzen aus Tschechoslowakei und Balkanhalbinsel.* p_H *in Böden. Pflanzenassoziationen.* Ⓓ Praha 1919.

Novák, Josef, Inspecteur au l'Inst. de Phytopath. Brno (C.S.R.), Trávnicky 8. *Phytopath.*

Novak, Peter (1879), Beamter bei der Entomol.-Phytopat. Sektion der Versuchsanstalt. Split (S.H.S.). *Käfer, für die Landwirtschaft schädliche Insekten.*

Novák, Stanislav, Inspecteur aux Inst. expérim. pour la production végétale. Praha-Vinohrady (C.S.R.), Havlíčkovy sady 58. *Phytopath.* o

Novák, Václav (1888), Prof. Hochsch. für Bodenkultur, Dir. bodenkundl. Sektion Landwirtschaftl. Versuchsanstalt. Brno (C.S.R.), Zemřdělska 1. *Phaenol.* Ⓓ Prague 1912.

Novák, Vladimir, Prof. Dr. Univ. Brno (C.S.R.), Veveří 95. *Botan. physique.* o

Novic, Katherina (1898), Berginst., Geol. Kabinétt, Ekaterinoslaw (U.d.S.S.R.). *Mikrofauna der Kohlenkalkablagerungen Rußlands: Protozoa.* ⓅProtozoa d. Kohlenkalkablagerungen.

Novicki, Semen, Dr., Bakteriološka Stanica. Šabac (S.H.S.). *Allgem. Pathol.* o

Novikov, Sergej Aleks., Prof. Timirjasev Akad. Moskau (U.d.S.S.R.), Petrovsko'e-Razumovsko'e. *Allg. Biol.* o

Novis, A., Prof. Fac. d. Med. Bahia (Brasil). *Physiol.* o

Novy, Frederick G. (1864), Prof. of Bact., Dir. Hygien. Labor., Med. School, Univ. of Michigan. Ann Arbor, Mich. (U.S.A.). *Cocaine. Cellular Toxins. Physiol. chem.* Ⓓ Sc. Michigan 1890, M. 1891, LL. Cincinnati 1920. o

Nový, Josef (1899). Ass. Samenkontrollanstalt Košice, (C.S.R.). Ⓓ Praha 1922.

Nowak, Jan (1880), Prof., Dr. ès sc. Kraków (Polska), 53, rue Grodzka. *Paléont. systématique et paléobiol. des céphalopodes supracrétacés. Cephalopoden, Seeigel, Ammoniten.* ⒹLwów (Pologne) 1907. ⓅBrachiopodes du Carbonifère infér. Cephalopodes du Malm.

Nowicki, Witold, Prof. Dr. med., Dir. Inst. d'Anat. pathol. Lwów (Polska), rue piekarska 52. o

Nowiński, Maryan (1897), Dr. phil., Ingenieur d'agronomie. Pryńcza-Matopolska (Polska). *Géobotan., Sociol. des plantes.* Ⓓ Cracovie 1921.

Nowopokrowskij, IwanW. (1880), Prof. a. Donschen Inst. f. Landwirtschaft u. Melioration. Nowotscherkassk (U.d.S.S.R.), Sowetskaja 31. *Botan. Geobotan. Systematik. Palaeobotan.*

Noyons, Adriaan Karel (1878), Prof. de Physiol. experim. à l'Univ. de Louvain (Belgique), Boulevard de Tirlemont extérieur 35. *La combustion organique, Calorimetrie directe et indirecte, Cardiol., Electrophysiol., Glandes à sécrétion interne.* Ⓓ Utrecht 1908.

Nucho, Ni'meh K., Prof. American Univ. Beirut (Syrien). *Histol.* o

Nuckols, Samuel B. (1887), Associate Agronomist, U.S. Dept. of Agric., Sugar Plant Investigations. Salt Lake City, Utah (U.S.A.). *Sugar beets.* Ⓓ Univ. of Missouri 1912.

Nuernbergk, Erich (1900), Dr. sc. nat., Ass. am Pflanzenphysiol. Inst. Univ. München-Nymphenburg (Deutschland), Menzinger Str. 13. *Physikalische Reizphysiol. der Pflanzen, spez. Phototropismus. Pflanzenökol. Tagesschlaf der Pflanzen.* Ⓓ Halle a. d. Saale 1924.

Nusdin, Peter (1896), Ass. Labor. biol. Chem. Vet.-Inst. Kazan (U.d.S.S.R.). *Katalase.* Ⓓ Kazan 1923.

Nuttall, G. H. F., Prof. Magdalenes Coll. Cambridge (England). *Biol.* o

Nutting, Charles C. (1858), Prof. of Zool. Iowa City (U.S.A.), State Univ. *Gorgoniacea. American Hydroids.* Ⓓ Blackburn, Ill., 1882. o

Nyárády, Erasmus Julius (1881), Konserv. Univ.-u. Siebenbürgisches Landes-Herbarium. Cluj (România). *Phanerogamen, Genus Alyssum.* Ⓟ Budapest 1904. Ⓟ Tatra-Pflanzen.

Nybelin, Dr., Konserv. Hydrograph. Stat. Bornö b. Holma (Sverige). *Plankton.* o

Nyberg, Carl (1879), Dr. med. et chir., Doc. an der Univ. Helsingfors Grankulla, (Finnland). *Bact. Variationserscheinungen bei den Bakterien. Ornithol., Ool.* Ⓓ Helsingfors 1914. Ⓟ Vogeleier in ganzen Gelegen.

Nydam, Frans Egbert (1901), Landbau-Ingenieur an der Kolonialen Landbauschule. Deventer (Holland). *Genetische Zusammensetzung einiger Kulturpflanzen.*

Nye, Robert N. (1892), Ass., Thorndike Memorial Labor., Boston City Hospital; Instr., Dept. Bact., Harvard Med. School, Boston, Mass. Chestnut Hill, Mass. (U.S.A.), 32 Lawrence Road. *Bact.* Ⓓ Harvard 1918. Ⓟ Stock cultures of various yeast like fungi.

Oakley, R. A., Agronomist, U. S. Bureau of Plant Industry. Washington. D.C. (U.S.A.). o

Ôba, Shirô, Prof. Aichi Med. Univ. Nagoya (Japan), Inst. of Bact. and Hygiene. *Bact.*

Obenberger, Johann, Dr., Adjkt. d. Entomol. Praha II (C.S.R.), Vaclavské, National-Mus. o

Obenchain, Jeannette Brown (1876), Research Associate in Anat. Univ. of Chicago (U.S.A.), Dept. of Anat. *Compar. Neurol., Marsupialia.* Ⓓ Univ. of Chicago 1924.

Oberholser, Harry Church (1870), President, Acad. of Sc. Washington, D.C. (U.S.A.), U. S. Nat. Mus. *Zool. Ornithol.* o

Oberling, Charles (1895), Chef de Travaux à l'Inst. d'Anat. pathol. Strasbourg (France). *Anat. pathol. Pathol. rénale.* Ⓓ Strasbourg 1920. Ⓟ Prép. histol.

Oberndorfer, Siegfried (1876), Univ.Prof., Dir. Pathol.Inst. München (Deutschland), Prinzregentenstraße 48. *Pathol. Anat.* o

Oberstein, Paul Wilhelm Otto (1884), Dr. phil., Saatzuchtinspektor der Landwirtschaftskammer Schlesien, Breslau X, Matthiasplatz 5. Breslau XVI (Deutschland), Novastr. 13. *Bekämpfung tierischer und pflanzlicher Schädlinge. Saatenanerkennung und Sortenkunde bei Kartoffeln, Getreide, Futtersämereien.* Ⓓ Breslau 1910.

Oboldujev, Prof. Univ. Taschkent (U.d.S.S.R.). *Mikrobiol.* o

Obraszova, Aleks. And., Doc. Inst. f. Landwirtschaft u. Melioration. Saratov (U.d.S.S.R.), Teatral'naja Ploščad. *Landw. Mikrobiol.* o

Obreshkove, Vasil (1890), Head of Department of Biol., Saint Stephan's Coll. Annandale-on-Hudson, N.J. (U.S.A.). *Photic reactions. Accessory testicular lobes.* Ⓓ Harvard Univ. 1918.

Obrutshev, Dmitri Vl. (1900), Geol. Comités. Leningrad (U.d.S.S.R.), W. O., Sredn. pr. 72. *Palaeoichthyol.: Devonische Ichthyofauna von Sibirien, Anat. u. Systematik der Arthrodira u. Antiarcha, Selachia d. russischen Kreide.*

Ocaranza, Fernando (1876), Dir. de la Fac. de Méd., Prof. de Physiol. de la même Fac. et de l'école méd. militaire. México, D.F. (Mexico), Calle de San Juan de Letrán núm. 19. *Physiol., hématol. normales et pathol. Sécrétion interne. Histol. du testicule.* Ⓓ México 1900.

Ochoterena, Isaac (1885), Prof. Univ., Fac. de Philosophie. Tacubaya, D.F. (México), Xicotencatl 3. *Histol. compar. du système nerveux. Mycol.: Dermatophytes. Cactacées mexicaines.* Ⓓ Univ.National du Mexique 1910. Ⓟ Préparations microscopiques du Système Nerveux; préparations Mycol.; Cactacées mexicaines vivantes.

Ochsner, Fritz (1899), Dr. Winterthur (Schweiz), Wülflingerstr. 65. *Ökol. und soziol. Studien an Epiphytengesellschaften (Algen, Moose, Flechten).* Ⓓ Zürich 1926. Ⓟ Moose, Flechten.

O'Connor, James M., Prof. Univ. Coll. Dublin (Ireland). *Physiol., Histol.* o

Oderkirk, Galen C. (1900), Junior Biol., U. S. Department of Entomol., Purdue Univ. Lafayette, Ind. (U.S.A.). *Injurious animal pests in Illinois and Indiana.* Ⓓ North Dakota Agric. Coll. 1925.

Odhner, Nils Johan Teodor (1879), Prof., Abt.-Dir. Naturhistoriska Riksmus. Stockholm 50 (Sverige). *Trematoden, Stomatopoden, Fam. Hanthidae unter den cyclometopen Krabben.* Ⓓ Uppsala 1905. Ⓟ Trematoden.

Odland, Theodore Eugene (1892), Associate Prof. of Agronomy, West Virginia Univ. and Associate Agronomist, West Virgina Agric. Experim. Station. Morgantown, W.Va. (U.S.A.), West Virginia Univ. *Farm crops.* Ⓓ Cornell 1926. Ⓟ Farm Crops.

Odnoralow, Nikolaus (1897). Rostow a. Don (U.d.S.S.R.), Anat. Inst. *Die Assimilation u. Manifestation d. Atlas. Topographie d. Lig. hepato-duodenale. Topographie d. Leber u. d. Magens.*

O'Donoghue, Charles Henry (1885), Prof. of Zool. Univ. of Manitoba. Winnipeg Man. (Canada). *Bryozoa. Opisthobranchiata. Circulatory system of vertebrata.* Ⓓ London 1912.

Ødum, Hilmar (1900), Ass. of „Danmarks geol. Undersøgelse". København (Danmark), Gl. Mønt 14. *Invertebrate Palaeont. and Palaeobiol.* Ⓓ Copenhagen 1926.

Oechslin, Max (1893), Forstmeister, Adjunkt des Kantonsforstamtes Uri. Altdorf-Uri (Schweiz), Zum Birkenhof. *Botan. und Pflanzengeographie.*

Oehlkers, Friedrich (1890), a.o. Prof. für Botan. an der Univ. Tübingen (Deutschland), Botan. Inst. *Botan. Vererbungsforschung (Gattung Oenothera). Reizphysiol. (Phototropismus).* Ⓓ München 1917.

Oehme, Curt (1883), Prof. Dr. med., Oberarzt Med. Univ.-Klinik Bonn a. Rh. (Deutschland), Haendelstr. 9. *Pathol. Physiol., insbesondere von physikalisch chem. Gesichtspunkten. Mineralstoffwechsel, Wasserhaushalt. Stoffwechsel. Knochensystem. Hypophyse (Rachitis, Liquor und Hypophysensekretion). Histaminwirkung (Theorie der Potentialgifte).* Ⓓ Leipzig 1908.

Oertel, Everett (1897), Ass. in Agric., Cornell Univ., Coll. of Agric. Ithaca, N.Y. (U.S.A.), 404 Eddy St. *Agric., and Economic Entomol.* Ⓓ Univ. of Wisconsin 1924.

Oertel, Horst (1873), Prof. of Pathol. in McGill Univ. and Pathol. in Chief to the Royal Victoria Hospital, Montreal, P.Q. (Canada), Dir. of the pathol. Inst. *Pathol. life.* Ⓓ Yale Univ. New Haven, Conn. 1894

Oertel, Otto (1891), o. Prof. der Anat., gleichzeitig I. Prosektor der Anat. Tübingen (Deutschland), Neckarhalde 48. *Topographische Anat.* Ⓓ 1918.

Øhrvall, H. A., Prof. Dr. Uppsala (Sverige). *Physiol.* o

Oekland, Fridthjof (1893), Doc. für Zool. an der Landwirtschaftl. Hochsch. Norwegens. Oslo (Norwegen), Majorstuveien 15. *Verbreitung, Oekol. und Genetik der Landschnecken. Land- und Süßwasserfauna von Nowaja Semlja.* Ⓓ Oslo.

Ökrös, Sándor (1902), Ass. Pathol. Anat. Univ. Debreczen (Ungarn). *Histol. d. Tuberkulose. Histol. d. Verätzungen.*

Oelkers, Prof. d. Forstl. Hochsch. Hann.-Münden (Deutschland). *Forstwirtschaft.* ○

von Oettingen, Wolfgang Felix (1888), Instr. in Pharmacol. Med.School Western Reserve Univ.Cleveland, O. (U.S.A.). *Toxicol., Pharm. of the Intestine.* ① Ph. Göttingen 1913, med. Heidelberg 1916.

Ôga, Ichirô (1883), Prof. of Botan., Educational Coll. Mukden, Manchuria (China). ① Tôkyô Imper. Univ. 1909.

Ogata, Tomosakurô, o. Prof. Imper. Univ. Tôkyô (Japan), Pathol. Inst., Fac. of Med. *Pathol., pathol. Anat., Avitaminose Geschwülste.* ① Tôkyô 1907.

Ogawa, Masanaga, Prof. Kyûshû Imper. Univ. Fukuoka (Japan), Bact. Inst., Coll. of Med. *Bact.*

Ogawa, Mutsunosuke, Prof. Imper. Univ. Kyôto (Japan), Anat. Inst., Coll. of Med. *Anat. des Menschen.*

Ogden, William Butler (1901), Research Ass. in Horticulture. Madison, Wis. (U.S.A.), Horticulture Bldg, Univ. of Wisconsin. *Fertilizer requirements of tobacco. Previous crop effect upon growth of tobacco. Over wintering of tobacco mosaic. Properties of tobacco mosaic virus.*

Ogilvie, Lawrence (1898), Gov. Plant Pathol. Dept. Agric. Bermuda (Brit. America). *Diseases of truck crops. Economic entomol.* ① Cambridge 1923.

Ogloblin, Aleksandr Alekseevich (1897), Ass. of the Entomol. department of the National Mus. Praha II (C.S.R.), U. Karlova 3, Zool.Ustav. *Strepsiptera and Hymenoptera (Systematic, biol. and morphol.). Anat. of lower insects, Collembola; the nervous and excretory systems. Parasit. castration at Homoptera: sex-determination.* ① Prague 1923. ⓟ Microscopical preparations of Strepsipterae, Bethylidae and Serphodea (Hymenoptera parasit.).

Ognell, Johann (1855), o. Prof. emer. Univ. Moscau (U.d.S.S.R.), Scheremetjewski per. *Entwicklungsgeschichte, Histol.*

Ognev, Sergey Ivanovich (1886), Prof. Moskau (U.d.S.S.R.), Zool. Mus., I. Univ. *Systematische Mammaliol., Ornithol.* ① 1910.

Ogrizek, Albert (1891), Dr., ao. Prof. der Tierzucht u. Milchwirtschaft, Adjunkt des Inst. für Tierzucht. Zagreb (S.H.S.), Petrova ul. 2a. *Biometrik u. Variationslehre. Weiden, Wolle, Abstammung, Serologie.* ① Wien 1914.

Oguchi, T., Lect. Keiô-Gijuku Univ. Tôkyô (Japan), Bact. Inst., Coll. of Med. *Bact.*

Oguma, Kan (1885), Prof. in the Hokkaidô Imper. Univ. Sapporo (Japan), Zool. Inst. *Morphol. of chromosomes, Animal Histol. Systematic of Odonata.* ① Hokkaidô Imper. Univ., Sapporo, 1911.

Ogura, K., Prof. Hokkaidô Imper. Univ. Sapporo (Japan), Inst. of Vet. Sc., Fac. of Agr. *Vet. Surgery. Animal Anat.*

Ogura, Yudzuru (1895), Lect. Botan., Fac. Sc., Tôkyô Imper. Univ. Koishikawa, Tôkyô (Japan), Botan. Inst., Botan. Gardens. *Anat. of Plant: Ferns. Palaeobotan.: internal Structure.* ① Taishô 1919.

Oguro, K., Prof. Hokkaidô Imper. Univ. Sapporo (Japan), Biochem. Inst., Fac. of Med. *Bio- and physiol. Chem.*

Ôgushi, Kikutarô, Prof. Provincial Med. Univ. Ôsaka (Japan), Anat. Inst. *Anat.*

Ogyu, Kikuo, Ass. Prof. Imper. Univ. Kyôto (Japan), Pharmacol. Inst., Fac. Med.

O'Hanlon, Sister Mary Ellen (1882), Prof. of Biol., Rosary Coll., River Forest, Ill. (U.S.A.). *Morphol., especially Hepatics.* ① The Univ. of Chicago 1925. ⓟ Microscopic slides of structures of the Marchantiaceae.

Ohe, Ivan (1884), Haupt Botan. Garten. Leningrad (U.d.S.S.R.), Pessotschnaja 2. *Phytopath., Mykol.* ① 1913.

Ohga, Ichiro, Prof. of Botan. Dairen, Manchuria (China).

Ohlsson, Carl Erik (1891), Prof. de chim. à l'inst. d'agric. à Alnarp. Lund (Sverige). *Chim. biol. Amylas du malt.* ① Lund 1917.

Ohmachi, Fumiye (1898), Member Tokugawa Inst. for Biol. Research. Ebaragun, Tôkyô (Japan), 220 Koyama, Hiratsuka-machi. *Cytol. of Grylloidea, Orthoptera; Heredity, Sex and Taxonomy.* ① Tôkyô Imper. Univ. 1921.

O'Honnell, Frank Getchell (1896), Ass. Plant Quarantine Inspector. Washington, D.C. (U.S.A.), Federal Horticultural Board. *Plant Pathol. specializing in seed and bulb sterilization.*

Oikawa, Shu (1893), Prof., Dir. Labor. of hygiene, Med. Coll. Niigata (Japan), 740 Asahimachi-dori, 2-bancho. *Hygien. application of bio-physical chem.* ① Tôkyô Imper. Univ. 1919.

Oinuma, Shôroku, Prof. Med. Univ. Okayama (Japan), Physiol. Inst. *Physiol.*

Oka, Asajirô (1868), Prof. Zool., Kôtô-Shihan-Gakkô (High Ceminary). Usigome, Tôkyô (Japan), 17 Kawada-mati. *Morphol. and Classification of Hirudinea and of Ascidiae. Theoretical Biol., Evolution.* ① Leipzig 1894.

Okada, M., Ass. Prof. Hokkaidô Imper. Univ. Sapporo (Japan), Anat. Inst., Fac. of Med. *Anat.*

Okada, Ryôsuke, Ass. Med. Univ. Kanazawa (Japan), Pathol. Inst. *Pathol.*

Okada, Yaichirô (1892), Lect. Tôkyô Normal Coll. and Fishery Coll. Koishikawa, Tôkyô (Japan), Tôkyô Kôtô Shihan Gakkô. *Zoogeographical distribution of Amphibia and Reptilia in Japan. Japanese Bryozoa, Hexactinellid sponges, and Myzostoma.* ① Fishery coll. 1915. ⓟ Amphibia, Reptilia, Bryozoa specimens.

Okada, Yônosuke (1895), Ass. Prof. Sendai (Japan), Biol. Inst., Tohoku Imper. Univ. *Plant Physiol.* ① Tôkyô Imper. Univ. 1919.

Okagawa, Masayuki, Prof. Provincial Univ. Ôsaka (Japan), Pharmacol. Inst. *Pharmacol.*

Okajima, Ginji (1878), Prof. Entomol. and Zool. Imper. Coll. of Agric. and Forestry. Kagoshima (Japan), 460, Ue-Arata-chô. *Anat. and Oecol. of Collembola and Aphididae (Rhynchota).* ① Tôkyô Imper. Univ. 1901.

Okajima, Keiji (1882), Prof., Dir. Yotsuya, Tôkyô (Japan), Anat. Inst., Keiô Univ. *Vergl. Anat. der Sinnesorgane der Wirbeltiere, besonders der Amphibien. Histol. Technik: Färbemethoden, Rekonstruktionsmethoden. Myol. der Japaner. Anat. des Menschen.* ① Kanazawa 1902.

Okamoto, Kikuo (1892), o. Prof. Anat.,· Med. Akad. Kanazawa (Japan), Anat. Inst. *Physische Anthrop.* ① Kanazawa 1915. ⓟ Anat. anthrop. Praep., besonders Skelette der Japaner und Primaten Japans.

Okamura, Kintarô, Chief, Prof. of Imper. Inst. for Fishery. Fukagawa, Tôkyô (Japan). *Diatom; Plankton in general.*

O'Kane, Walter Collins (1877), Prof. of Economic Entomol., Univ. of New Hampshire; Deputy Commissioner Agric., State of New Hampshire. Durham, N.H. (U.S.A.). *Life history of European corn borer, black flies, apple maggot.* ① Ohio State Univ. 1909.

Okazaki, Tsunetarô, Tôkyô (Japan), Gakushûin. *Entomol.*

Okey, Ruth E. (1893), Associate Prof. Household Sc. (Nutrition) Univ. of California. Berkeley, Cal. (U.S.A.), 209 Home Econ. Building. *Metabolism of Women (Cyclic Variations), Nitrogenous Compounds, Carbohydrate, Lipoids, Minerals. Ca., P. Intermediary Metabolism: Carbohydrates.* ① Univ. of Illinois 1918.

Okkelberg, Peter (1880), Ass. Prof. of Zool. Univ. of Michigan. Ann Arbor, Mich. (U.S.A.), Zool. Dept. *Embryol., Compar. Anat. Germ cell in lower vertebrates.* ① Univ. of Michigan 1918. ⓟ Land and fresh water mollusca.

Oksijuk, Peter F., Doc. d. Landwirtsch. Inst. in Kiew (U.d.S.S.R.), Tarassowskaja 19, W. 8. *Botan. Floristik u. Cytol.*

Oksner, Alfred N., Konserv. d. Kiewschen Botan. Gartens. Kiew (U.d.S.S.R.), Ul. Kominterna Nr. 1. *Botan.* o

Okuneff, Nicolaus (1896), Ass. Zool. Labor., Akademie der Wissenschaften, Doc. der allgem. Pathol. Univ. Kasan. Leningrad (U.d.S.S.R.), Moschaiskaja 15, W. 9. *Parenterale Resorption. Histol. d. Hungerzustandes (Lipoid, Chondriom), Vitalfärbung der Aorta. Leukozyten und Lipoide, Schicksal intravenöser Farbstoffe.* ⓟ Leningrad 1922.

Okuni, Tadashi, Biol. Govern. Centr. Research Inst. Taihoku, Formosa (Japan), Dept. of Agric. *Phytopath.*

Okushima, Kwan-ichirō, Prof. Med. Univ. Okayama (Japan), Pharmacol. Inst. *Pharmacol.*

Olaru, Dimitrie (1881), Pharmacien-chim. de l'Univ. de Bucarest, Chef-chim. Cluj (Roumanie), Labor. d'hygiène, 6, Str. N. Bălcescu. *Physiol. animale et végétale, normale et pathol. Pharm., Biochim. (Spéc. Tuberculose, Rachitisme, Cancer). Mycol., Bact. du sol, Influence du Manganès.* ⓟ Paris 1920.

Oldham, Charles. Berkhamsted, Herts. (England), The Bollin, Shrublands Road. *Ornithol.* o

Olitsky, Peter K. (1886), Associate Member, The Rockefeller Inst. for Med. Research. New York City, N.Y. (U.S.A.), 66 Street and Avenue A. *Cerebrospinal meningitis, bacillary dysentery, typhus, vesicular stomatitis, mosaic disease, filter-passing viruses, mitochondria, anaerobiosis, serol. and immunity.* ⓟ Cornell Univ. New York 1909.

Oliveira, João Duartede, o. Prof. Univ. Coimbra (Portugal). *Histol., Chem., Top. Anat.* o

Oliveira Reis, Jose, (1898) Ass. Fac. Med. de Coimbra (Portugal). *Histol.*

Oliver, Francis Wall (1864), Prof. of Botan. u. Univ. Coll. London W. C. 1 (England), Gower Street. *Plant ecol., especially of maritime vegetation. Palaeobotan.* ⓟ Sc. London 1888, M.A. Cambridge 1888.

Oliver, Jean (1889), Prof. of Pathol. Stanford Univ. San Francisco (U.S.A.), Lane Hospital. *Tuberculosis, lepra, immunol., Lethargica. Phys.-chem. und vital phenom.* ⓟ Stanford Univ. 1915.

Oliver, Wade Wright (1890), Prof. of Bact., Long Island Coll. Hospital. Brooklyn, N.Y. (U.S.A.), Hoagland Labor. 335, Henry St. *Bact. adaptation to diminished oxygen tension; Sprue; Pneumococcus.* ⓟ Cincinnati 1915.

Oliver, Walter Reginald Brook (1883), Dominion Mus. Wellington (New Zealand). *Pteridophytes and Spermophytes. Plant Ecol. Birds.* ⓟ Wellington 1927. ⓟ New Zealand Pteroidphytes and Spermophytes.

Olivier, Prof. agr. Anat. Fac. Méd. Lille (France). *Anat.* o

Olivo, Oliviero Mario (1896), Aiuto, Ist. Anat. Univ. Torino 16 (Italia), Corso Massimo d'Azeglio 52. *Coltura di tessuti in vitro. Differenziazione e sdifferenziazione dei tessuti. Fisiol. del cuore embrionale. Degenerazione e rigenerazione sperimentale di organi sensitivi. Citol.* ⓟ Torino 1925.

Olmer, David, Prof. Univ. Fac. méd. Marseille (France). *Pathol.* o

Olofsson, Ossian (1886), Fiskeriintendent. Luleå (Sverige). *Hydrobiol., Fischereibiol.* ⓟ Uppsala 1918.

Olsen, Carsten (1891), Ass. am Carlsberg Labor. København (Danmark). *Pflanzengeographie, Pflanzenphysiol. und Bodenkunde. Bodenchem. und bodenbiol. Untersuchungen. Pflanzen in ihrer Abhängigkeit vom Boden. Ustica. H-ion concentration of the soil.* ⓟ Kopenhagen 1921.

Olsen, Ørjan (1885), Dr. phil. Oslo (Norwegen). Zur Zeit: Consulat de Norvége, Papeete (Tahiti). *Ornithol. External characters and biol. of Bryde's whale (Balaenoptera Brydei). Pycnogonida.* ⓟ Innsbruck 1921.

Olson, P. J. (1887), Ass. North Dakota Agric. Experim. Inst. Fargo, N.D. (U.S.A.), State Coll. Station. *Corn and sweet Clover Breeding.* ⓟ Univ. of Illinois 1913.

Olstad, Ola (1885), Fischereiass. Oslo (Norge), Zool. Mus., Trondhjemsveien 23. *Süßwasserbiol., Odonaten.* ⓟ Oslo 1925.

Olszewski, Wolf Adolf (1886), Stadtamtsrat, Vorsteher der Labor. der Wasserwerke der Stadt Dresden. Dresden-N. (Deutschland), Wasserwerk Saloppe. *Wasserbact., Flora und Fauna des Wassers.*

Oltmanns, Friedrich (1860), o. Prof. der Botan. Freiburg i. Br. (Deutschland), Jakobistr. 23. *Algen, Pflanzengeographie.* ⓟ Straßburg 1884.

Oman, Andrew E. (1877), Ass. Biol. (U.S. Biol. Survey). Manhattan, Kan. (U.S.A.), Care State Coll. *Eradication of Rodents.*

Omang, Simen Oscar Fredrik (1867). V. Aker pr. Oslo (Norge), Sognsveien 37, *Hieracium.*

Omeliansky, Basil (1867), Dir. Section of the General Microbiol., Inst. of Exper. Med. Leningrad (U.d.S.S.R.), Aptekarsky Island, Lopouchinskaia Street 12. *General Microbiol. Chem. on the microbiol. processes. Fermentations. Microbiol. of the Soil.* ⓟ Cultures of microbes.

Omer Cooper, Joseph (1893), Zool. Labor. Boscombe, Bournemouth (England), 6, Queensland Road. *Syst. of Crustacea Isopoda. Fresh water fauna of Abyssinia. Distribution, Oecol. and Variation.* ⓟ Cambridge 1924.

Ondroušek, Vladimír (1895), Ass. Mähr. Landw. Versuchsanstalt. Brno (C.S.R.), Schwarze Felder. *Wein. Obst.* ⓟ Brünn 1922.

O'Neal, Claude Edgar (1884), Prof. of Botan., Head of Dept. of Botan., Wesleyan Univ. Delaware, Ohio (U.S.A.), 95 Oak Hill Avenue. *Plant Cytol., Plant Morphol., Bact.* ⓟ Indiana Univ. 1922.

de Ong, Elmer Ralph (1882), Ass. Entomol., California Agric. Experim. Station. Berkeley, Cal. (U.S.A.), Univ. of California. *Insecticides.* ⓟ Univ. of California 1915.

Ono, Shun-ichi. Koishikawa, Tōkyō (Japan), Kobinata-Daimachi 2 chome, 35. *Cytol.*

Ono, Kotaro (1877). Tropical Plantation of MBGK. *Physiol. of Cocos nucifera.* ⓟ Tokyo.

Ontscharov, Alexei Alexandrowitsch (1888), Ass. am Histol. Labor. Krasnodar (U.d.S.S.R.), Griwenskaja 30/32. *Histol. der Decidualzellen.* ⓟ Charkow, 1923. ⓟ Mikroskopische Praeparate.

Ontschukowa, Miliza (1896), Laborant, Botan. Hauptgarten. Leningrad (U.d.S.S.R.). *Microbiol. des Wassers.*

van Oordt, Gregorius Johannes (1892), Dr. zool., Conserv. Rijksuniv.Utrecht (Holland), Zool. Labor., Abt. prop. Zool. u. experim. Histol., Biltstraat 172. *Experimentelle Zool., Sexualität, Ornithol.* ⓟ Utrecht 1921.

van Oort, E. D., Prof. extrao. a. d. Univ., Dir. des Zool. Reichsmus. Leiden (Holland), Witte Singel. *Ornithol., Carabidi.* o

Oparin, Alexander I., Doc. I. Univ. Moskau (U.d.S.S.R.), B. Kosichinskij per. 4, W. 14. *Botan., Pflanzen-Physiol.* o

Oparina-Charitonova, Natalie (1898), Ass. Zool. Univ., Ass. d. Biol. Kamo-Station. Perm (U.d.S.S.R.), Zaimka, Zool. Labor. Univ. *Zooplankton: Rotatoria und Cladocera des Süß- und Salzwassers.* ⓟ Perm 1926.

Operman, François (1885), Chef de la Section entomo-phytopat. dans la Station agric. Zagreb (S.H.S.), rue Kačić 9. *Entomol. agric. et forestière; Tenebrionidae (Coleopt.) de la Péninsule balcanique.* ⓟ Tenebrionidae (Coléopt.) balcan.

Opie, Eugene L. (1873), Prof. of Experim. Pathol., Univ. of Pennsylvania; Dir. of Labor., Henry Phipps Inst. Philadelphia, Pa. (U.S.A.), 2203 St. James Place. *Tuberculosis and immunity. Pathol. Bact.* ⓟ M.D. Johns Hopkins Univ. 1897.

Opieńska, J., Dr. phil. Warszawa (Polska), Univ. Fac. méd., r. Krakowskie-Przedmieście 26/28. *Biochim.* o

Opitz, Kurt (1877), Dr., o. Prof. an der Landwirtschaftl. Hochsch. Berlin und Dir. des Inst. für Acker- und Pflanzenbau Dahlem. Berlin-Dahlem

(Deutschland), Albrecht-Thaer-Weg 3. *Sorten- und Saatgutfrage; Abbau der Kartoffel; physikalische Bodenuntersuchungen im Anschluß an verschiedene Bodenbearbeitungsmethoden; Bestimmung des Düngebedürfnisses der Böden; Flachszüchtung.* ⓓ Breslau 1904.

Opletal, Jos. (1863), Prof. Ing. Forest. l'Ecole Supérieure d'Agric. et de Silvic. Brno (C.S.R.), Černa pole, Lesmická ulice. *Forstbenutzung.*

Opoczyński, Kazimierz (1877), Prof. d'anat. pathol. à la fac. de méd. de l'Univ. d'Etienne (Stefan) Batory à Wilno (Pologne), Wielka 17, m. 4. *Anat. pathol. La maladie de Hodgkin, la transplantation des néoplasmes, l'anat. pathol. des glandes endocriniennes.* ⓓ Odessa 1915.

Oppenheim, Doc. Univ. Woosung (China). *Allg. Pathol., Histol.* o

Oppenheim, Jacob David (1899), Ass. of Prof. Dr. O. Warburg, Dept. of Botan. of the Agric. Experim. Stat. and Hebrew Univ. Rehoboth (Palestine). *Physiol. and breeding of: Citrus species and var., Prunus communis var., Vitis spec. and var. Periodicity and water-supply.*

Oppenheim, Paul (1863), Univ.Prof. Berlin-Lichterfelde-W. (Deutschland), Sternstr. 19. *Palaeontol.* o

Oppenheimer, Carl (1874), Prof., Dr. phil. et med. Berlin-Wilmersdorf (Deutschland), Güntzelstr. 49. *Biochem., Fermentlehre, Stoffwechselphysiol.* ⓓ Phil. Berlin 1894, med. 1898.

Oppenheimer, Ernst (1888), Dr., Leiter der wiss. Abteilung u. Pharmacol. Labor. der Krause-Medico-Gesellschaft. München (Deutschland), Äußere Prinzregentenstr. 69. *Pharmacol. und Pharmakotherapie.* ⓓ Freiburg i. Br. 1913.

Oppenheimer, Heinz Reinhard (1899), Dr. phil., Leiter einer Baumschule. Sichron Jacob (Palaestina). *Physiol. und Pathol. der gärtnerischen Kulturpflanzen (subtropischer Obstbäume).* ⓓ Wien 1922. Ⓟ *Lebende Pflanzen (Stauden oder Bäume).*

Oppenheimer, Joseph (1883), Dr., Priv.Doc. der Palaeont. an der deutschen techn. Hochsch. Brünn (CSR.), Schillergasse 10. *Ammoniten des Oberjura.* ⓓ Wien 1906.

Oppermann, Adolf (1861), Prof. der Forstwissenschaft, Dir. des forstlichen Versuchswesens in Dänemark. Springforbi (Danmark). *Biol. der Waldbäume; Wachstumsgesetze der Waldbäume; Erblichkeit und Rassen der Waldbäume.* ⓓ h. c. Hochsch. für Bodenkultur Wien 1922.

Orator, Victor (1894), Dr., Priv.Doc. für Chirurgie, Ass. der chirurg. Univ. Klinik Graz (Deutsch-Österreich), Landeskrankenhaus. *Pathol. und Physiol. der Prostata (Hypertrophie), der Schilddrüse '(Struma), des Magens (Ulcus und Carcinom).* ⓓ Wien 1919.

Orbeli, C. A., Dir. Wiss. „P. F. Lesshaft" Inst. Leningrad (U.d.S.S.R.), Pr. Maklina 32. *Physiol.* o

Orbison, Agnes M. (1897), Ass. Prof. of Biol., Elmira Coll. Elmira, N.Y. (U.S.A.). *Cytol. Cytoplasmic inclusions.*

d'Orchymont, Armand (1881). Mont St. Amand lez Gand (Belgique), Rue de l'Industrie 106. *Systématique de la famille des Hydrophilides (Coléoptères) pour le globe entier. Morphol., ontogénèse (premiers états, métamorphoses), bionomie, histologie, physiol., zoogéographie, paléontol., phylogénie. Nervation alaire de l'ordre des Coléoptères.* Ⓟ Animaux ou plantes.

Ordman, David (1896), Ass. Bact., South African Inst. for Med. Research. Johannesburg (S. Africa), Box 1038. *Bact.* ⓓ Cape Town 1924.

Orla-Jensen, Prof. Dr., Dir. Bioteknisk Labor. Sølvtorvet, København (Danmark). o

Orlandi, Cesare (1888), Aiuto Dir. Ist. di Cerealicoltura. Bologna (Italia), Piazza Galvano 4. *Selezione dei cereali.* ⓓ 1922.

Orlov, Jurij Alexandrowitsch (1893), Ass. Inst. für Histol. u. Embryol. Med. Militärakad. Leningrad (U.d.S.S.R.), Wassiljewskij Ostrow Kubanskij pereulok 2. *Vergl. Histol. des sympathischen Nervensystems: das viscerale Nervensystem der Insekten, Crustaceen.* ⓓ St. Petersburg.

Orlova-Nesterchuck, Anna (1892), Ass. Academy of Agric. Gorky, White-Russia (U.d.S.S.R.), Krassninskaja ul. 14. *Entomol. Parthenogenesis of Hemiteles areator. Biol. of Apanteles glomeratus.* Ⓟ Russian Apides.

Orman, Emile (1882), Prof. de botan., Inst. agronomique. Louvain (Belgique), 24, Rue du Canal. *Cytol.* ⓓ Louvain 1912.

O'Roke, Earl C. (1881), Ass. Prof. Zool., South Dakota State Coll. Brookings, S.D. (U.S.A.). *Animal Ecol. Parasit.* ⓓ Univ. of Kansas 1912.

Orr, Thomas G. (1884), Prof. of Surgery, Univ. of Kansas. Kansas City, Kan. (U.S.A.), Bell Memorial Hospital. *Intestinal Obstruction and toxic conditions arising from intestinal tract. Relationship of Blood Chem. to Surgery.* ⓓ Johns Hopkins Hospital 1910.

Orrù, Efisio (1861), Aiuto Ist. anat. Cagliari (Italia), Via Dettori 4. *Sviluppo pancreas e fegato; sistema nervoso dei pesci; Terminazioni nervose.* ⓓ Cagliari 1889.

Orsós, Ferenc (1879), o. Prof. pathol. anat. Univ. Debreczen (Ungarn), Pathol. anat. Inst. *Pathol. Anat. und Histol., gerichtliche Med. Tuberkulose-Disposition, lymphatische Organe, Knochenmark.* ⓓ Budapest 1903.

Ortenburger, Arthur Irving (1898), Ass. Prof. of Zool., in Charge of the Mus. of Zool. Norman, Okla. (U.S.A.), Univ. of Oklahoma. *Herpetol., reptiles of the U. S. The Genera Masticophis and Coluber: the Whip snakes and Racers.* ⓓ Univ. Mich. 1925. Ⓟ Reptiles and amphibians from Okla.

Ortmann, Arnold Edward (1863), Curator of Invertebrate Zool., Carnegie Mus., and Prof. of Zool., Univ. of Pittsburgh, Pa. (U.S.A.), Carnegie Mus. *Freshwater Invertebrates, Crustacea and Mollusca (Naiades) of North America. Their taxonomy, anat., ecol. and geographical distribution. History of development of distribution.* ⓓ Phil. Jena 1886, Sc. (hon.) Pittsburgh 1911. Ⓟ Invertebrates, preferably Mollusks. Freshwater Mussels (Naiades) of North America.

Orton, Clayton Roberts (1885), Prof. Plant Pathol. State Coll., Research Division, Agric. Dept. The Bayer Company, Inc. Labor. at The Boyce Thompson Inst. for Plant Research,Yonkers, N.Y. (U.S.A.). *Seed borne parasites, disinfectants. Mycol. (Uredinales, Dothideaceae).* ⓓ Columbia Univ. 1924. Ⓟ Species of Phyllochora, Uredinales of Pennsylvania.

Orton, James Herbert (1884), Chief Naturalist, Labor., Marine Biol. Association of the United Kingdom. Plymouth (England). *Marine Bionomics and Biol. Environmental conditions and rate of growth and other reactions. Sex and sex-change. Ciliary mechanisms.* ⓓ London Univ. 1914.

Orton, William Allen (1877), Sc. Dir. and General Manager, Tropical Plant Research Foundation. Washington, D.C. (U.S.A.), 600 Cedar St.,Takoma Park. *Tropical Agric. Plant Pathol.* ⓓ Burlington 1897.

Orzechowski, Casimir (1878), Prof. de neurol. de l'Univ., Dir. de la Clinique neurol. de l'Univ. à Varsovie (Pologne), Pl. Napoleon 6. *Anat. et Histol. pathol. du système nerveux de l'homme. Neurol. clinique.* ⓓ Lwów 1900.

Osborn, Henry Fairfield (1857). New York, N.Y. (U.S.A.), Am. Mus. of Natural History. *Zool. Comp. anat. Paleont. Geol.* o

Osborn, Henry L. (1857), Dean of the Fac. and Prof. of Biol. Hamline Univ. St. Paul, Minn. (U.S.A.), 1599 Hewitt Ave. *Morphol. of Trematodes.* ⓓ John Hopkins 1884. o

Osborn, Herbert (1856), Research Prof., Ohio State Univ., Dir., Ohio Biol. Survey. Columbus, O. (U.S.A.), Ohio State Univ. *Economic Entomol., Meadow Insects. Taxonomy of Homoptera. Homoptera of North America and Neotropics.* ⓓ Iowa State Coll. 1880, Sc. (hon.) 1916. Ⓟ Homoptera of America.

Osborne, Thomas Burr (1859), Research Chem., Connecticut Agric. Experim. Station, Carnegie Inst. of Washington, Biochem., Yale Univ. New Haven, Conn. (U.S.A.), 58 Huntington St. *Protein chem. and nutrition.* ⓓ Ph. Yale 1885, hon. Sc. Yale 1910.

Osburn, Raymond C. (1872), Head, Department of Zool. and Entomol., Ohio State Univ.; Dir., Franz Theodore Stone Labor. Columbus, O.(U.S.A.), Ohio State Univ. In Summer: Put-in-Bay, O. (U.S.A.), Stone Labor. *Aquatic biol., Fishes. Marine Bryozoa.* ⓓ Columbia Univ. 1906.

Oschanin, Leo (1884), Pr.Doc. d. Anthrop. Univ. Taschkent (U.d.S.S.R.), Observatorskaja 16. *Rassenbestand der eingeborenen Bevölkerung Mittelasiens: Uzbeken aus China, Kezmine, Samarkand; Tadschiken aus Kara-Tegin, Samarkand, Buchara; Kirgisen het Isseyk-kul's; mittelasiatische eingeb. Juden. Herkunft der Turkmenen.* ⓟ Photographien anthropol. Typen Mittelasiens.

Oschkaderov, Wassilij (1895), Prosektor, Inst. für normale Anat., Univ. Woronesch (U.d.S.S.R.). *Lage und Befestigung des Dickdarms. Knochen, Nierenlage.* ⓓ Voronesch 1920.

Osgood, Wilfred Holmes (1875), Curator of Zool., Field Mus. of National History. Chicago, Ill. (U.S.A.). *Mammal. Ornithol. Zool.* o

Ôshima, Hiroshi (1885), Prof. zool., Kyûshû Imper. Univ. Fukuoka (Japan), Zool. Labor., Department of Agric. *Taxonomy of Holothurioidea. Embryol. of Echinoderms, especially of Holothurioidea. Morphol. and life-history of the Pantopoda found parasit. in the bivalve, Tapes.* ⓓ Tokio 1909.

Ôshima, Hukuzô, Ass. Prof. Aichi Med. Univ. Nagoya (Japan), Pathol. Inst. *Pathol.* o

Ôshima, K., Prof. Hokkaidô Imper. Univ. Sapporo (Japan), Chem. Inst., Fac. of Agric. *Agric. Chem.* o

Ôshima, Masamitsu. Hiratsuka near Tôkyô (Japan), Togoshi 282. *Herpetol.* o

Oslund, Robert M. (1893), Ass. Prof. of Physiol. Chicago, Ill. (U.S.A.), Univ. of Ill., Coll. of Med. *Endocrinol. Testicular hormone production.* ⓓ Univ. of Chicago 1923.

Osmaston, Bertram Beresford (1868). Oxford (England), 116 Banbury Road. *Ornithol. East Indian Birds and eggs.*

Osmun, Albert Vincent, Prof. of Bot., Mass. Agric. Coll., Head of Dept. of Bot., Agric. Exp. Stat. Amherst, Mass. (U.S.A.). o

Osolin, Viktor (1885), Adjunkt Hydrobiol. Stat. Univ. Riga (Lettland), Kirchenstr. 5. *Hydracarinen.*

Ostenfeld, Carl Emil Hansen (1873), Prof. at the Univ. and Dir. of the Botan. Gardens. København, (Danmark). *Taxonomy and geographical distribution of spermatophytes (especially arctic and northern regions); genetical and experim. taxonomy; phytoplankton; marine spermatophytes.* ⓓ København 1906.

Osterberger, Bullion A. (1899), Ass. Entomol. Baton Rouge, La. (U.S.A.), Agric. Experim. Station, Louisiana State Univ.*Insect enemies of sugar cane.*

Osterhout, Winthrop J. V. (1871), Member Rockefeller Inst. for Med. Research. New York (U.S.A.), 66 St. and Ave. A. *Plantphysiol. Injury, Recovery and Death, in relation to Conductivity and Permeability.* ⓓ California 1899. o

Ostertag, Berthold (1895), Leiter des Neuropath. Labor. III. Heil- u. Pflege-Anstalt. Berlin-Buch (Deutschland). *Pathol. des Zentralnervensystems, insbesondere Histopathol.* ⓓ Tübingen 1920.

v. **Ostertag,** Robert (1864), Ministerialrat, Veterinärreferent im württembergischen Ministerium des Innern, Dir. des württembergischen Tierärztl. Landes-Untersuchungsamts. Stuttgart (Deutschland), Dorotheenstr. 1. *Veterinärhygiene., Bact. der Tierseuchen, Fleischhygiene, Milchhygiene.* ⓓ Med. Freiburg i. Br. 1888, med. vet. h. c. Gießen, Wien,Berlin.

Ostroumowa, Marianne Wladislaw (1883), Ass. Kasan (U.d.S.S.R.), Zool. Kabinett, Univ. *Morphol. u. Regeneration der Coelenteraten.*

Ostrowskaja, Manepha K. (1889), I. Ass. Labor. f. Anat. u. Physiol. d. Pflanz., Landwirtsch. Inst., Pflanzenphysiol. Versuchsstat. Detskoje Sseló. Leningrad (U.d.S.S.R.), Uliza Krasnich Sorj 75, W. 38. *Einfluß der p_H auf die Entwicklung von Gemüsepflanzen. Periodizität der Beleuchtung in ihrem Einfluß auf Wachstum und Ernteertrag einiger Kulturpflanzen. Physiol. Bedeutung der Grannen.* ⓟ Mikroskopische Praep. Pflanzenanat.

Ostrowski, Thaddèe, Prof. Dr. med., Dir. Inst. d'Anat. pathol. Univ. Lwów (Polska), Rue Piekarska 52. o

Ôsugi, Shigeru, Prof. Imper. Univ. Kyôto (Japan), Chem. Labor., Coll. of Agric. *Agric. Chem., Chem. of Fertilizer.*

Osvald, Karl, Ing., Ass. à l'Inst. d'Amélioration des Plantes. Praha-Vršovice (C.S.R.), Nádražin ul. 13. *Botan. appliquée. Hopfenzüchtung.* ⓓ Prag 1925.

Osvald, Karl Hugo (1892), Priv.Doc. der Pflanzenbiol. Univ. Upsala, Dir. des Schwed. Moorkulturvereins. Jönköping (Schweden). *Skandinavische Vegetation. Vegetation und Stratigraphie der Moore. Bodenreaktion und Pflanzenwachstum. Reaktion und Kalkbedürfnis der Ackerböden. Wurzelsysteme der Pflanzen. Düngungsversuche und andere Kulturversuche, hauptsächlich auf Moorböden.* ⓓ Upsala 1923.

Oswald, Adolf (1870), Prof. Univ., Dr. Zürich 7 (Schweiz), Hofstr. 78. *Thyreoglobulin, Physiol. u. Pathol. d. Schilddrüse, endemischer Kretinismus, Myxödem, innere Sekretion.* o

Otani, Sajûrô, Ass. Prof. Imper. Univ. Kyôto (Japan), Hygien. Inst., Fac. Med. o

Otis, Charles Herbert (1886), Ass. Prof. of Biol. (in charge of Botan.), Western Reserve Univ. Cleveland, Ohio (U.S.A.). Acting Ass.Prof. of Botan. Univ. of Wisconsin. Madison, Wis. (U.S.A.). *Field and forest botan.; plant physiol.; bryophytes. Shrubs.* ⓓ Univ. of Michigan 1913. ⓟ Microscopical slides.

Otruba, Josef (1889). Olomouc, Mähren (C.S.R.). *Carex, Potentilla, Alchemilla, Viola.*

Ottenberg, Reuben (1882), Associate Physician. Mount Sinai Hospital New York (U.S.A.). *Human Blood Groups.* ⓓ Columbia Univ. 1905.

Otterstrøm, CarlVilhelm Theodor (1881), Fischereibiol. Frederiksdal b. Lyngby (Danmark). *Süßwasserfische, faunistisch und biol. Fischkrankheiten. Limnol.* ⓓ København 1906.

Ottley, Alice M., Associate Prof. of Botan., Curator Herbarium, Coll. Wellesley, Mass. (U.S.A.), 46 Dover Rod. *Taxonomy.* o

Otto, Hermann (1889), Dr. phil. Berlin (Deutschland), Am Schlesischen Bhf. 2. *Normale Physiol., theoretische u. angewandte Botan., Mykol.* ⓓ Berlin 1914.

Otto, Jan Pieter (1901), Ass. Zool. Labor. Groningen (Holland), Reitemakersrijge 14. *Limnol.*

Otto, Richard Ernst Wilhelm (1872), Abteilungsdir. im Inst. „Robert Koch". Berlin N 39 (Deutschland), Föhrerstr. 2. *Hygiene, Bact. und Immunität. Pest, Cholera und Staphylokokken; Wertbemessung der Heilsera; Anaphylaxie (Analyse des Th. Smithschen Phänomens, Auffindungen der Antianaphylaxie und der passiven Anaphylaxie); Diphtherie- und Typhusimpfung; Händedesinfektion; experim. Fleckfieber; d'Herelle'sches Phänomen.* ⓓ Berlin1895.

Ottoleghi, Donato (1874), Prof. o. Igien. Univ. Bologna (Italia), Ist. d'Igiene. *Batteriol. (eziol. della peste suina, dell' encefalite letargica. Batteriol. e immunol. del carbonchio, riproduzione del tripanosomi. Disinfettanti. Tecnica batteriol.). Igiene. Vaccinazione contro la tifoide e contro la tubercolosi.*

Purificazione delle acque potabili. Malaria. ⓓ Laureato all' Univ. di Torino 1897.

Oudemans, Anthonie Cornelis (1858). Arnhem (Holland), Burgemeester Weertsstraat 65. *Acari, Mallophaga, Pediculi, Suctoria.* ⓓ Utrecht 1885.

Oudemans, J. Th., Dr. Putten (Holland), Schovenhorst. *Lepidoptera, Hymenoptera, Thysanura, Collembola.* o

Oudendal, Adrianus Jacob François (1887), Lect., Pathol. Med. School Batavia, Java. Weltevreden, Java (Nederl.-O.-Indië), Kramat 99. *Pathol. Histol., human intestinal parasit., tumours.* ⓓ 1919. ⓟ Tropical normal and pathol. preparations.

Outkine, Maxime S., Chef des trav. Inst. Agron. Moscou (U.d.S.S.R.). *Maladies des pommes de terre. Pathol. végétale.* o

Ovazza, Vittorio Emanuele (1878), med. Roma (Italia), Corso d'Italia 43. *Malariol.* ⓓ Turin 1913.

Overbeck, Fritz (1898), Dr., Priv.Doc., Ass. am Botan. Inst. der Univ. Frankfurt a. M. (Deutschland), Viktoriaallee 9. *Physiol. Pflanzenanat. (Turgeszenz-Mechanismen) und Moorkunde.* ⓓ Heidelberg 1922.

Overholser, Earle Long (1888), Ass. Prof. of Pomol. and Associate Pomol., Univ. of California. Davis, Cal. (U.S.A.), Univ. Farm., Division of Pomol. *Physiol. studies of fruit plants.* ⓓ Cornell Univ. 1926.

Overholts, Lee O., Prof. of Botan., State College, Pennsylvania (U.S.A.). *Taxonomy of higher fungi, forest pathol.* o

Overton, C. E., Prof. Univ. Lund (Sverige). *Pharmacol.* o

Overton, James Bertram (1869), Prof. Plant Physiol., Chairman of Dept. of Botan., Univ. of Wisconsin. Madison, Wis. (U.S.A.), 512 Wisconsin Ave. *Chromosome reduction and chromosome structure in plants. Physiol. of reproduction, sap flow and root pressure.* ⓓ Univ. of Chicago 1901.

Owen, Forrest Vern, Associate Biol. Orono, Me. (U.S.A.), Maine Agric. Expt. Stat. *Breeding of potatoes and clover. Theoretical genetic.* ⓓ Wisconsin 1926.

Owen, William Ludwell (1884); Research Bact., La. Sugar Experim. Station, La. State Univ. Baton Rouge, La. (U.S.A.). *Soil and Sugar Bact.*

Owens, Charles E., Associate Prof. of Plant pathol. Oregon Agric. Coll. Corvallis, Ore. (U.S.A.). *Plant pathol.* o

Oxley, Arthur D., Prof. of Biol., Lambuth Coll. Jackson, Tenn. (U.S.A.). o

Oxner, Mieczyslaw (1879), Ass. au Mus. Océanographique de Monaco (Principauté). *Régénération chez les vers marins (Nemertinea). Expériences sur la mémoire chez les poissons.* ⓓ Zürich 1905.

Oyama, Junji (1894), Ass. Prof. Kyûshû Imper. Univ. Fukuoka (Japan), Zool. Labor. *Biol. and anat. of Urodela: Diemictylus and Hynobius.* ⓓ Tôkyô 1918. ⓟ Japanese common newt, Diemictylus pyrrhogaster.

Ôyama, Kashiwa. Sendagaya near Tôkyô (Japan), Onden. *Paleont.*

Oyama, Ryôtoku, Prof. Kyûshû Imper. Univ. Fukuoka (Japan), Anat. Inst., Coll. of Med. *Anat.*

van Oye, Paul Herman Gustave (1886), Doc. an der Univ. Gent (Belgique), Tentoonstellingslaan. *Hydrobiol.* ⓓ Sc. et med. Gent. ⓟ Planctonproben aus den Tropen.

van Oyen, C. F., Dr., Prof. der Tierärztl. Fac. d. Univ. Utrecht. Groenekan bei Utrecht (Holland). *Hygiene der Tiere.* o

Ozaki, Katsumi, Ass. Prof. of Tôhoku Imper. Univ. Sendai (Japan), Biol. Inst., Coll. of Sc. *Biol.*

Ozaki, M., Ass. Prof. of Hokkaidô Impér. Univ. Sapporo (Japan), Pharmacol. Inst., Fac. of Med. *Pharmacol.* o

Ozaki, Yoshimasa (1891), Ass. in Zool., Zool. Inst., Sc. Fac., Imper. Univ. Tôkyô (Japan). *Parasit.: Trematoda, Cestoda, Protozoa.* ⓓ Hiroshima Normal Coll. 1913. ⓟ Preparations and Specimens of Trematodes.

Ozaki, Yoshizumi, Prof. Imper. Univ. Kyôtô (Japan), Pharmacol. Inst., Coll. of Med. *Experim. Pharmacol.*

Ozaplewski, Eugen, Dr. med., Univ.Prof. Köln (Deutschland), Vorgebirgsstr. 19. *Hygiene.* o

Ozawa, Yoshiaki (1899), Ass. Prof. Geol. Inst. Imper. Univ. Tôkyô (Japan). *Palaeozoic fossils: late Palaeozoic Foraminifera. Brachiopoda and corals. Tertiary and Quarternary Foraminifera. Fusulinidae.* ⓓ Imper. Univ. Tôkyô 1923. ⓟ Fossil Foraminifera and their thin slides. Late Paleozoic Corals and Brachiopods.

Ozoliņš, Pāvils (1895), Ass. Univ. Riga (Lettland), Zellu ielā 25, dz. 2. *Entwicklung des Gefäßsystems.*

Ozols, Edgars (1899), Subass. a. d. Univ. Lettland, Entomol. a. d. Inst. für Pflanzenschutz. Riga (Latvija), Baznicas iela No. 4a dz. 10. *Entomol., Ichneumonol.* ⓟ Ichneumonidae.

Ozols, Peters (1897), Bact. labor. vét. Riga (Latvija), Kronvalda bulv. 1, Inst. de Microbiol. *Rage; bact. de lait.*

Ozorov, M. N., Conserv. Station de Défense des Plantes. Tambov (U.d.S.S.R.). o

Paál, Árpád (1889), Priv.Doc. Univ., Adjunkt Kgl. Versuchsstation für Pflanzenphysiol. u. Pathol. Budapest 11 (Ungarn), Debröi-ut 17. *Geotropismus, Phototropismus, Reizleitung, Hormone bei Pflanzen. Krankheiten des Tabaks.* ⓓ Budapest 1911.

Pabisch, Heinrich (1877), Prof., Doc. für technische Rohstofflehre und Mikroskopie. Wien VI (Deutsch-Österreich), Grasgasse 5. *Angewandte Botan. und Zool., Pharmakognosie, technische Rohstofflehre und Mikroskopie, Ethnotoxikol.*

Pace, Domenico (1870), Priv.Doc. de Clinique Méd. à l'Univ. Napoli (Italia). *Histol. du tissu spécifique du coeur. Electrocardiol.*

Pacella, Guido, Prof. Univ. Nac. La Plata (Argentina). *Physic. Biol. Physiol.* o

Pachon, M., Prof. Fac. Méd. Bordeaux (France). *Physiol.* o

Pacinotti, Giuseppe, Prof. incarn. Univ. Camerino (Italia). *Patol. gener.* o

Pack, Dean A. (1889), Associate Agronomist, U. S, Department of Agric. Salt Lake City, Utah (U.S.A.). 1876 South Main Street. *Plant Physiol. and Plant Breeding, sugar beets.* ⓓ Chicago 1920.

Pack, Herbert John (1892), Prof. Entomol., Utah Agric. Coll., State Experim. Sta. Entomol. Logan, Utah (U.S.A.). *Economic insects of Utah.* ⓓ Utah Agric. Coll. 1913.

Packard, Clyde Monroe (1889), Associate Entomol., U. S. Department of Agric., Bureau of Entomol. West Lafayette, Ind. (U.S.A.), Box 495. *Cereal and Forage Crop Insects, especially Hessian Fly.* ⓓ Univ. of Cal. 1925.

Paczoski, Joseph (1864), Dr. phil., Dir. du Labor. de Systematique et Sociol. des plantes de l'Univ. à Poznań (Polska). *Sociol. et geographie des plantes.*

Paddock, Wendell (1866), Prof. Horticulture, Ohio State Univ. Columbus, Ohio (U.S.A.), 1077 Westwood Ave. o

Paechtner, Johannes (1881), Dr. vet. Physiol. Inst. d. Tierärztl. Hochsch. Hannover (Deutschland), Podbielskystr. 17. *Physiol. d. Nutztiere. Wirtschaftsphysiol.* o

Paeckelmann, Wolfgang (1882), Vorsitzender der bergischen Landschaftsstelle für Naturdenkmalpflege. Barmen (Deutschland), Bleicherstr. 3. *Naturschutz. Pflanzengeographie und Geschichte der Botan. des bergischen Landes.*

Paczoski, Joseph (1864), Prof., Dir. Labor. de Systématique et Sociol. Végétales Univ. Poznań (Polska). *Sociol. et géographie végétales.* ⓓ Poznań.

Pagano, G., Doc. Univ. Palermo (Italia). *Fisiol.* o

Page, Harold James (1890), Chief Chem., Rothamsted Experim. Station, Harpenden, Herts (England).

The Chem. and Biochem. of soils, manures and crops. ⑨ Univ. Coll., London 1910.

Pagliani, Luigi (1896), Dr., Ass. del Labor. di Botan. del R. Ist. Inferiore Agrario e della R. Univ., Fac. di Sc. Milano (Italia), Via Giuseppe Colombo 64, Città degli Studii. *Botan. generale, coltivazione delle piante med. ed aromatiche.*

Pagliano, Théophile Charles Louis (1893), Prof. Zool. et Entomol. agric. Tunis, Tunisie (Afrique), Ecole coloniale d'Agric. ⑫ Insectes du Protectorat tunisien.

Paillot, André (1885), Dir. Station entomol. du Sud-Est. Saint-Genis-Laval, Rhône (France), 22, Avenue Clémenceau. *Biol. des Insectes nuisibles; traitements de destruction. Maladies microbiennes des Insectes utiles et nuisibles. Cytol. normale et pathol. des Insectes.* ⑨ 1910. ⑫ Insectes nuisibles à l'Agric.

Paine, Sydney Gross (1881), Ass. Prof. of Bact. Imperial Coll. of Sc. and Technol. South Kensington, London (England). *Bact. in its relation to diseases of Plants. Agric. Fermentation Industries.* ⑨ Univ. of London.

Painter, Henry Raymond (1888), Ass. Entomol., U. S. Bureau of Entomol. Westlafayette (U.S.A.). *Life histories and control methods of cereal and forage insects.* ⑨ Okla. A. and M. Coll. 1916. ⑫ Acrididae.

Painter, Reginald Henry (1901), Ass. Prof. of Entomol., Kansas State Agric. College, and Collaborator, U. S. Dept. Agric., Bureau of Entomol. Manhattan, Kan. (U.S.A.), Entomol. Department. *Taxonomy and Biol. of Bombyliidae and Asilidae (Diptera). Insect Physiol. Resistance of Plants to Insect attack, especially Hessian Fly and Chinch Bug on wheat and Corn.* ⑨ Ohio State Univ. 1926. ⑫ Diptera, especially, Bombyliidae and Asilidae.

Painter, Theophilus S. (1889), Prof. of Zool., Univ. of Texas. Austin, Tex. (U.S.A.). *Mammalian cytol.* ⑨ Yale Univ. 1913.

Painvin, Georges Jean (1886), Ingénieur en Chef des Mines, Prof. des Paléont. à l'Ecole Nationale Supérieure des Mines, Conserv. du Mus. de Paléont. de l'Ecole Nationale Supérieure des Mines. Paris 5 (France), 60, Boulevard Saint Michel. *Paléont. Invertébrés: application à la Géol.*

Pak, Chubyung (1902), Ass. Peking (China), Kanmen Hutung 91. *Pharmacol.* ⑨ Seoul 1922, Freiburg i. Br. 1926.

Pal, Jacob (1863), o. Univ. Prov., Dir. I. med. Abt. am Allgemeinen Krankenh. Wien IX (Deutsch-Österreich), Garnisong. 3. *Glatte Muskulatur: Physiol., Pathol., Pharmacol. der Gefäße wie der anderen Hohlorgane.* ⑨ Wien 1886.

Palacios, Isidro (1859), Prof. de l'Univ. de Saint Louis, Mo. (U.S.A.), 8, Rue de Zaragoza 61. *Chimie générale, Zool. et Botan.* ⑨ San Luis Potosí 1884. ⑫ Plantes diseques.

Palamartschuk, Andrej I., Leiter d. Selektions-Abt. d. Nikitskij Botan. Gartens. Jalta (U.d.S.S.R.). *Selektion. Tabakzucht.* o

Paldrock, Alexander (1871), Prof. f. Dermatol., Univ. Dorpat (Estland), Sternstr. 7. *Gonokokken. Lepra.* ⑨ Dorpat 1898.

Palibin, Jean W. (1872), Conserv. en chef du Mus. de Jardin botan. principal, Spécialiste de l'Inst. Fédéral de Botan. appl., Paléobotan. du Comité Geol. Leningrad (U.d.S.S.R.). *Phanérogames de la Russie et du Caucase. Les flores de Manchourie, Corée et de la Chine boréale; Mongolie. Les flores tertiaires paléarctiques et quaternaires.* ⑨ Genève 1908.

Palik, Piroska (1895), Dr. Inst. für systematische Botan. u. Pflanzengeogr. Univ. Budapest (Ungarn). *Hydrodictyon. Saxifraga. Juniperus.* ⑨ Budapest 1920.

Palladin, Alexander (1885), Prof. physiol. Chem. Med. Fac., Dir. Biochem. Inst. Charkow (U.d. S.S.R.), Veterinärplatz. *Biochem. d. Kreatinstoffwechsels, Biochem. der Avitaminosen, Biochem. des Gehirns.* ⑨ Petersburg 1916.

Palladin, Lydia (1896), Ukrain. Biochem. Inst. Charkow (U.d.S.S.R.), Veterinärplatz. *Biochem. d. Milz, Biochem. d. experim. Tetanie.* ⑨ Charkow 1926.

Palladin, Olga (1896), Ass. Nordwestliches milchwirtschaftliches Untersuch.-Labor. Leningrad (U.d. S.S.R.), pr. Wolodarsky 37. *Untersuchungen der Milchsäurebakterien und Schimmelpilze.*

Palm, Bjorn T., Dir., Deli Experim. Station. Medan, S.O.K. (Ned.-O.-Indië). *Embryol., cytol. of Angiosperms.* o

Palma, Ricardo, Prof. Univ. Lima (Peru). *Anat.* o

Palmer, Charles Mervin (1900), Ass. Prof. of Botan., Butler Coll., Indianapolis, Ind. (U.S.A.). *Bact. (Microbiol.). Plant Pathol. Phycol.*

Palmer, E. T., Dir. of the Ontario Horticultural Experim. Station, Vineland Station, Ontario (Canada). *Experim. Horticulture and Plant Breeding.* o

Palmer, Ephraim Laurence (1888), Prof. of Rural Education, Cornell Univ., Dir. of Nature Education. Ithaca, N.Y. (U.S.A.), Cornell Univ. *Seed impurities.* ⑨ Cornell 1917.

Palmer, Leroy S. (1887), Prof. of Agric. Biochem. Univ. of Minnesota. St. Paul, Minn. (U.S.A.), Univ. Farm. *Dairy Chem., Animal Nutrition, Plant and Animal Coloring Matters, particularly carotinoids.* ⑨ Univ. of Missouri 1913.

Palmer, Miriam Augusta (1878), Insect Delineator, Ass. Prof. in Entomol. Ft. Collins, Colo. (U.S.A.), Dept. of Entomol., C.A.C. *Aphididae. Heredity in the Coccinellid. Life History of Ladybeetles. Lachnus.* ⑨ Colorado State Agric. Coll. 1926.

Palmer, Theodore S. (1868), Expert in Game Conservation, Biol. Survey, U. S. Dept. of Agric. Washington, D.C. (U.S.A.), 1939 Biltmore St., N.W. *Ornithol., birds and game protection.* ⑨ Georgetown 1895. o

Palmer, Walter Walker (1882), Bard Prof. of Med., Columbia Univ., and Dir. of the Med. Service, Presbyterian Hospital. New York City (U.S.A.). *Metabolism, Therapy, urinary acidity, colloids, hydrogen ion concentrat., acid excretion.* ⑨ Harvard Med. School, Boston, Mass. 1910.

Palmgren, Alvar (1880), Adjunkt der Botan. an der Univ. Helsingfors (Finland), Andrégatan 19. *Pflanzengeographie. Pflanzensystematik: Hieracia, Taraxaca, Carices fulvellae.* ⑨ Helsingfors 1914.

Palmgren, Axel Jakob (1892), Prosektor at the anat. Inst. of the Vet. Highsch. Stockholm (Sverige), Veterinärhögskolan. *Arthrol. of the domestic animals, innervation of the joints.* ⑨ Stockholm 1921.

Palmgren, Rolf, Mag. phil., Dir. Tiergarten, Staatl. Inspektor für Naturschutz. Helsinki (Finnand), Högholmen. *Naturdenkmalpflege, Ornithol.* o

Palombi, Arturo (1899), Prof. di Sc. naturali nel R. Liceo Sc. di Avellino. Napoli (Italia), Ist. Zool. della R. Univ., Rampe del Salvatore 4. *Policladi. Alloioceli. Trematodi.*

Paloncimo, Lauri, Ass. Inst. f. Haustierlehre. Helsinki (Finnland). *Immunität und Stoffwechsel.* o

Pammel, Louis Herman (1862), Prof. of Botan., Iowa State Coll. and Botan., Iowa Agric. Experim. Station. Ames, Ia. (U.S.A.), Department of Botan. *Economic Botan., Weeds, Honey plants, Conservation of wild flowers and medicinal plants, plant life.* ⑨ Ph. Washington Univ. 1899, Sc. Univ. of Wisconsin 1925. ⑫ Plants.

Pampanini, Renato (1875), Aiuto e Conserv. del Mus. Botan. dell'Univ. Firenze 14 (Italia), Via Lamarmora 4. *Sistematica botan.* ⑨ 1907.

Pan, M. T., Dr., Lect. Dept. of anat., Union Med. Coll. Peking (China). o

Pancotto, Ettore (1894), Ass. Incaricato, nel Labor. di Anat. Patol. Bologna (Italia), R. Univ. ⑨ Firenze 1919.

Pane, Nicola (1855), Prof. ord. stabile d'Univ. Napoli (Italia), S. Andrea delle Dame n. 2. *Tuberculosi dell' uomo e sperimentale, chemioterapia e immuno-*

terapia. Costituzione morfol. del siero di sangue (biol. dei granuli plasmatici).
Panfilow, Ssosont (1889), Ass. Wologda (U.d. S.S.R.), Milchwirtschaftliches Inst. *Microbiol. von Käse, Butter. Milchsäurebakterien.*
Panini, Francesco (1885), Aiuto all' Ist. di Chim. Farmaceut. Doc. di Tecnica e Legislazione Farmaceut. Univ. Modena (Italia). *Fermentazione e fermenti in rapporto alla stabilizzazione delle piante e droghe vegetali.* ⓓ Modena.
Panisset, Lucien (1880), Prof. à l'Ecole nationale vét. d'Alfort, Seine (France). *Maladies infectieuses des animaux.* ⓓ Lyon 1908. ⓟ Cultures et préparations.
Pankow, Michail M., Prof. d. Landw. Inst. d. Bergrepublik. Wladikawkas (U.d.S.S.R.), Alexandrowskij per. 3. *Pflanzen-Physiol.* o
Panning, Albert (1894), Zool. Mus. Hamburg 1 (Deutschland), Steintorwall. *Asseln, Seewalzen, systematisch.* ⓓ Marburg 1920.
Panormov, Aleksej Aleksandrovič, Prof. Univ. Kazan, Tatar Rep. (U.d.S.S.R.). *Biochem.* o
Pantin, Carl Frederick Abel (1899), Physiol. to the Marine Biol. Ass. of the United Kingdom. Plymouth (England), Marine Biol. Labor., The Hoe. *Cell Physiol. Amoeboid movement. The physical and chem. structures of protoplasm. The action of ions on protoplasm. Oxidation in tissues. The physiol. of Haemocyanin.* ⓓ Cambridge 1921.
Paoli, Guido (1881), Capo per le malattie delle piante, Dir. Osservatorio di Fitopatol. Chiavari, Prov. Genova (Italia), Corso Umberto 38. *Entomol. agraria.* ⓓ 1914. ⓟ *Insetti agrari; parti di piante danneggiate da insetti.*
Papadopolos, Alexandar (1898), Méd. Ass. Nich, Serbie (S.H.S.), Inst. Epidemiol. *Traitement antirabique. Virus filtrant, propagation, incubation, virulence. Immunisation préventive et infection par le virus fix.* ⓓ Bordeaux 1924.
Papanicolaou, George N. (1883), Ass. Prof. Anat., Cornell Univ. Med. Coll. New York City (U.S.A.), 1 st Ave. and 28th St. *Sex and reproduction; Oestrus in mammals; morphol. and physiol. of ovaries; ovarial and thyroid hormones; human physiol. of reproduction.* ⓓ Med. Athènes (Grèce) 1904; phil. München 1910.
Pape, Heinrich (1891), Dr. phil., RR. Biol. Reichsanstalt für Land- und Forstwirtschaft. Berlin-Dahlem (Deutschland), Königin-Luise-Str. 19. *Phytopath., angewandte Botan., Mycologie.* ⓓ Berlin 1917.
Papenheim, Caecilius (1890), Dr. Essen-Ruhr (Deutschland), Franziskanerkloster, Franziskanerstr. 69. *Rassenbiol.* ⓓ Münster i. W. 1924.
Papilian, Victor (1888), Prof., Dir. Inst. d'Anat. descriptive et topogr. Cluj (Roumanie), Str. Minerva 9. *Anat., physiol., histol. du système végétativ.* ⓓ Bucarest 1916.
Papin, P., Prof. Ecole prép. de Méd. et de Pharmacie Angers (France). *Histol.* o
Papp, Constantin (1896), Ass. au Labor. de Botan. Jassy (Roumanie), Str. Păcurari 79 bis. *Mousses et hépatiques de la Moldavie.* ⓓ Jassy 1926. ⓟ *Mousses et hépatiques de la Moldavie.*
Papp, Károly, Prof. Dr., Dir. d. Palaeont. Inst. Univ. Budapest (Ungarn). o
Pappenheim, Paul (1878), Dr. phil. Prof. 2. Dir. am Zool. Mus. der Univ. Berlin N 4 (Deutschland), Invalidenstr. 43. *Embryol. der Spinnen; Systematik, vergl. Morphol., Osteol. der Fische.* ⓓ Berlin 1902.
Parabucev, A. V., Doc. Gosudarstv. Donskoj Univ. Rostov a. Don (U.d.S.S.R.). *Pathol. Anat.* o
Paraf, Jean (1888), Chef de Clinique à la Fac. de méd. Paris (France), 35 bis, Rue Jouffroy. *Tuberculose. Maladies infectieuses. Gonococcie. Meningococcie*
Paramonow, Sergius (1894), Zool. Mus. d. Ukrainischen Akad. d. Wissensch. Kiew (U.d.S.S.R.), Tschidnowskystr. 2. *Dipteren von U.d.S.S.R.* ⓓ Kiew 1917.

Paraschtschuk, Simeon (1873), Prof. Landwirtschaftl. Inst., Dir. des Nord-West-Distriktes milchwirtsch. Untersuch.-Labor. Leningrad (U.d.S.S.R.), Prospect Wolodarsky 37. ⓓ Nowaja Alexandrija. ⓟ *Schimmelpilze aus Butter, Milchbakterien.*
Parat, Maurice (1899), Ass. d'Anat. et d'Histol. comparées à la Sorbonne. Paris (France), 6, Rue d'Ulm. *Cytol.: Appareil de Golgi, vacuome. Physiol. cellulaire: Colorations vitales, P_H et R_H cytoplasmiques.* ⓓ Paris 1922. ⓟ Préparations microscopiques.
Paravicini, Eugen (1889), Dr. Basel (Schweiz), Laufenstr. 25. *Kulturpflanzen, speziell tropische.* ⓓ Zürich 1916.
Pardo, Agostín, Prof. Univ. Nac. La Plata (Argentina). *Pathol.* o
Pardo García, Louis (1897), Ajudant du Labor. de Hydrobiol. Espagnole et Conserv. dans l'Inst. Nacionel de 2 Enseñance. Valencia (España), Gran Via del Marqués del Turla 65. *Hydrobiol.: pêche et pisciculture.* ⓓ Barcelona 1919.
Pardubský, Karel, Prof. Dr., Ecole Sup. Méd. Vét. Brno (C.S.R.), Praźská 69. o
Paredes, J. N., Prof. Univ. Quito (Ecuador). *Botan.* o
Paredes, Jose, Prof. Univ. of St. Tomas. Manila (Philippine Islands). *Physiol. u. Pharmacol.* o
Pareil von Wold Kjerschow Agersborg, Helmer (1881), Wheeler Prof. of Biol., James Millikin Univ. Decatur, Ill. (U.S.A.), 1428 West Riverview Ave. *The physiol. and the behavior of the Twenty-rayed starfish. Classification, histol. and cytol. structures of Nudibranchs and Pteropods. The sensory receptors and the reaction of nudibranchs to physico-chem. stimuli. The structure of the sensory receptors in Opisthobranchs. Germ-cell secretions in nudibranchs. Cytopathol.: the effect of parasitism upon the tissues. Variation in prosobranchiate molluscs from the coast of Norway.* ⓓ Univ. of Illinois 1913.
Partentiev, Ivan Aleks., Doc. I. Univ. Moskau (U.d.S.S.R.). *Zool., Embryol.* o
Parfitt, Elliott Hill (1886), Dairy Bact. Lafayette, Ind. (U.S.A.), Purdue Univ. *Bact. studies of butter and cream.* ⓓ Purdue Univ. 1923.
Parhon, Constantin I. (1874), Prof. de Clin. neuropsychiatrique Univ. Jassy, Dir. hôpital pour les maladies nerveuses Socola (Jassy). Bucarest (România), 3, Rue Luterane. *Endocrinol. Ilikibiol. (Biol. des âges). Histobiol.: tissu musculaire, centres nerveux. Biochim.: Teneur en eau des organes et tissus, Electrolytes. Anat. pathol. du système nerveux et des glandes endocrines.* ⓓ München 1900.
Pariser, Käte (1893), Dr. phil., Kaiser-Wilhelm-Inst. für Biol. Berlin W 62 (Deutschland), Kurfürstenstr. 59. *Spermatogenese bei Saturniden. Triploide Intersexualität bei Saturniden.* ⓓ Berlin 1919.
Parish, Samuel Bonsall (1838), Honorary Curator Herbarium, Univ., Lect. on Calif. Botan., Stanford Univ. Berkeley, Cal. (U.S.A.), 1668 Scenic Ave. *Flora of Southern California.* ⓓ New York Univ. 1858.
Park, James, Prof. Univ. Otago. Dunedin (New Zealand). *Palaeont.* o
Park, Jay Boardman (1884), Prof. of Farm Crops and Head of the Department of Farm Crops, The Ohio State Univ. Columbus, O. (U.S.A.). *Genetic studies in barley, oats and soybeans. Breeding experim. in maize, oats, wheat, barley, soybeans and sweet clover (Melilotus).* ⓓ Harvard Univ. 1916. ⓟ *Pure lines, hybrids and varieties of the crops.*
Park, William Hallock (1863), Dir. of the Bureau of Labor. of the Department of Health of the City of New York, Prof. of Bact. and Hygien. med. Department of New York Univ. New York City, N.Y. (U.S.A.), 315 West 76 Street. *Bact.* ⓓ Columbia 1886. ⓟ Pathogenic bact. antitoxins, toxins.
Parker, Chas. S. (1883), Head Dept. Botan., Howard Univ. Washington, D.C. (U.S.A.), 321 — 11 St. N. E. *Plant Pathol., Systematic Botan.* ⓓ Washington 1923. ⓟ Vascular Plants of the

Parker, Frank Wilson (1897), Soil Chem., Alabama Agric. Experim. Station. Auburn, Ala. (U.S.A.). *Soil solution and factors influencing its composition, mineral nutrition of crop plants.* Ⓓ Univ. of Wisconsin 1921.

Parker, George Howard (1864), Prof. of Zool. and Dir. of the Zool. Labor., Harvard Univ. Cambridge, Mass. (U.S.A.), 16 Berkeley St. *Animal Reactions. Sense Organs and Nervous System.* Ⓓ Harvard Univ. 1887.

Parker, Harry Lament (1893), Ass. Entomol., U. S. Department of Agric. Hyères, Var. (France), Domaine du Mont Fenouillet. *Biol., habits, morphol. of parasit. Insects. Morphol. of the larvae of the Chalcidae.* Ⓓ Paris 1925.

Parker, John Bernard (1870), Prof. of Biol., Catholic Univ. of America. Washington, D.C. (U.S.A.), 1217 Lawrence St., N. E. *Entomol. Bembicinae (Fossorial wasps).* Ⓓ Ohio State Univ. 1915. Ⓟ Wasps.

Parker, John Huntington (1891), Prof. of Plant Breeding, Agronomy Dept., Kansas State Agric. Coll. Manhattan, Kan. (U.S.A.). *Crop improvement, wheat, oats, barley, sorghums.* Ⓟ Seeds of crop plant varieties.

Parker, John Robert (1884), Entomol., Montana Experim. Station. Bozeman, Mtn. (U.S.A.). *Effects of temperature and moisture upon grass-hoppers; life-history and systematic studies of aphids; fruit insects; oil sprays.* Ⓓ Univ. of Minnesota 1922. Ⓟ Aphids slides.

Parker, Ralph Langley (1892), Associate Prof. of Entomol. and Agric., Ass. Entomol., Kans. Agric. Experim. Station, State Apiarist, Kansas Entomol. Commission. Manhattan, Kan. (U.S.A.), Department of Entomol., Kans. State Agric. Coll. *Apiculture, horticultural Entomol., Physiol.* Ⓓ Cornell Univ. 1925.

Parker, Ralph R. (1888), Special Expert, U. S. Public Health Service. Hamilton, Mtn. (U.S.A.). *Med. entomol. Rocky Mountain spotted fever and tularemia. Sarcophagidae (Diptera).* Ⓓ Harvard Med. School 1915. Ⓟ Sarcophagidae.

Parkes, Alan Sterling (1900), Beit Memorial Research Fellow. London W.C. 1 (England), Dept. of Physiol. and Biochem., Univ. Coll. *Physiol. of Reproduction.* Ⓓ Manchester 1923.

Parkin, Thomas, Fairseat, High Wickham. Hastings, Sussex (England). *Ornithol.* o

Parman, Daniel Cleveland (1885), Ass. Entomol., Bureau of Entomol., U. S. Dept. Agric. Uvalde, Tex. (U.S.A.). *Insects affecting the Health of Animals. Insecticides: Toxicity of Chemicals to Insects; Chemotropism of Insects. Life History and Ecol. of Insects attacking Animals.*

Parmenter, Charles Leroy (1882), Ass. Prof. Univ. of Pennsylvania (U.S.A.), Zool. Bldg. *Cytol.; Chromosomes of Parthenogenetic Frog material.* Ⓓ Univ. Pennsylvania. Ⓟ Amphibian Cytol. Material.

Parmentier, Paul (1860), Prof., Dir. Jardin botan. Besançon (France), Inst. botan. *Anat. appliquée à la systématique. Botan. agricole. Les Noyers et les Carya en France.* Ⓓ Besançon 1898.

Parnas, Jakob Karol (1884), Dr. phil., Prof. ord. und Dir. des Med.-chem. Inst. der Univ. Lwów (Polska), Piekarska 52. *Chemie des Blutes: Bildung und Schicksal des in Stickstoffverbindungen ausgeschiedenen Ammoniaks, Fragen des intermediären Stoffwechsels der einfachsten Stickstoffverbindungen. Stoffwechsel der Kohlehydrate, Chem. der Muskelprozesse; Stoffwechsel der embryonalen Entwicklung. Chem. der Lipoidstoffe.* Ⓓ München 1907.

Parodi, Lorenzo, Prof. Univ. Nac. La Plata (Argentina). *Botan. agric.* o

Parodi, Umberto (1879), Prof. tit. di Anat. Patol. Univ., Dir. Ist. di Anat. Patol. Catania (Italia), Via Biblioteca 4. *Cancro sperimentale.*

Parow, Karl Edmund (1870), Prof., Dr., Vorsteher des Forschungsinst. für Stärkefabrikation und Kartoffeltrocknung. Berlin N 65 (Deutschland), Seestr. 13. *Stärkegewerbe und Kartoffeltrocknung.* Ⓓ Greifswald 1896.

Parr, Leland Wilbur, (Assoc. Prof. Bact. and Hygiene, Chairman Dept. of Pathol., Bact., Hygiene and Parasit., American Univ. Beirut, Grand Liban (Syria). *Bact. and Hygiene, Immunol. Isohemagglutination. Intestinal spirochetes.* Ⓓ Chicago 1923.

Parreira, Henrique, Dir. de l'Inst. d'Anat. Pathol. Fac. de Méd. Lisbôa (Portugal). o

Parrot, Louis Michel (1883), Chef de Labor. Inst. Pasteur d'Algérie. Alger (Algérie), Inst. Pasteur. *Parasit. et pathol. exotique.* Ⓓ Paris 1908. Ⓟ Phlébotomes.

Parrott, Percival John (1874), Chief of Research, New York State Agric. Experim. Station. Geneva, N.Y. (U.S.A.). *Economic Entomol.* Ⓓ Kansas State Univ. 1897.

Parshley, Howard Madison (1884), Prof. of Zool. in Smith Coll. Northampton, Mass. (U.S.A.), Department of Zool. *Hemiptera-Heteroptera. Sociol. Biol.* Ⓓ Harvard 1917. Ⓟ North American Hemiptera-Heteroptera.

Parsons, Frederick Gymer (1863), Prof. of Anat., Univ. London S. E. I. (England), Med. School, Dr. Thomas's Hospital. *Anat., Anthrop.*

Partridge, Newton Lyman (1890), Ass. Prof. of Horticulture and Research Ass. in Horticulture, Michigan State Coll. Paw Paw, Mich. (U.S.A.), P.O. Box 133. *Viticulture. Physiol. studies of American bunch grapes and the application of the results to viticulture.* Ⓓ Univ. of Illinois 1917.

van de Pas, Ludwig Gerard Hermann Georg (1874), Dr. med. vet., o. Prof. Univ., Tierärztl. Fac. Buenos Aires (Brasilien), Olaguer 2649. *Vergl. descriptive und topographische Anat.* Ⓓ Utrecht 1898; Bern 1912.

Pasanisi, Ettore (1873), Aiuto ord. nell' Ist. di Anat. Chir. nell' Univ. Napoli (Italia), Corso Vittorio Emmanuele 655. *Anat.* Ⓓ Napoli 1899.

Pascal, C., Prof. Univ. Rosario (Argentina). *Botan.* o

Paschen, Enrique Federico Mauricio (1860), Prof., Dr. med., leitender Oberarzt der Staatsimpfanstalt. Hamburg 13 (Deutschland), Alte Rabenstraße 14. *Mikrobiol. Variola, Vaccine, filtrierbare Erreger.* Ⓓ Heidelberg 1885.

Pascher, Adolf A. (1881), Prof. Botan. Deutsche Univ., Leiter Abt. pharmaz. Botan. u. Kryptogamenkunde, Leiter Botan. Abt. hydrobiol. Station Hirschberg i. B. Praha II (C.S.R.), Vinčina 3a. *Protist.; Niedere Pflanzen (bes. Algen), Hydrobiol., Morphol. und Systematik der höheren Pflanzen. Monocotyl. Solanaceae. Vererbungslehre. Pflanzengeographie. Flagellaten.* Ⓓ 1905. Ⓟ Pflanzengeographische und mikrobiol. Diapositive und Bilder.

Paschkis, Karl (1896), Arzt II. Med. Univ.-Klinik. Wien XVIII (Deutsch-Österreich), Klostergasse 16. *Retikulo-endothelialer Stoffwechselapparat und Vitalfärbung.* Ⓓ Wien 1919.

Pascichi, Luigi, Lib. doc. Univ. Bologna (Italia). *Patol. gener.* o

Pascual Dodero, Julian (1893), Ingenieur Agronome, Membre de l'Inst. Géographique, Agrégé au Labor. de Mycol. du Mus. de Ciencias Naturales. Madrid (España), Cervantes 23. *Mycol. Pathol. végétale. Botan. phanérogamique.* Ⓓ Madrid 1916. Ⓟ Plantes. Exsiccata.

Pásková, Ludmila, Ass. Inst. de Botan. générale et systématique. Brno (C.S.R.), Kounicova 63. *Algol.* o

Paspaleff, Georg (1895), Ass. am Zool. Inst. der Univ. Sofia (Bulgarien). *Zellphysiol. Stimulationsversuche an Polygonum. Protoplasmareifung bei Seeigeleiern.* Ⓓ Sofia 1920, Greifswald 1925.

Pasquini, Pasquale (1901), Aiuto Ist. di Zool. Univ. Roma (Italia), Via Domenico Cimarosa 18.

Embriol. dei vertebrati, embriol. sperimentale. ⓘ Roma 1921.
Passamaneck, Emanuel (1902), Ass. in Biochem. Med. Coll. of Virginia, Charlottesville, Va. (U.S.A.). *Biochem.*
Pasternazkaja, Wera F., Prof. d. Chem.-Pharmaz. Inst. Odessa (U.d.S.S.R.), Staroportofrankowskaja 41, W. 7. *Botan., Systematik. Geobotan.* o
Paszkiewicz, Ludwik Antoni (1878), Prof. d'Anat. pathol. Fac. Med. Inst. d'Anat. pathol. Warszawa (Polska), rue Chatubinski 5. *Anat. Pathol. de plexus choroïdes.* ⓘ 1904.
Patch, Edith Marion (1876), Entomol., Maine Agric. Experim. Station; Prof. of Entomol., Fac. of Graduate Studies, Univ. of Maine. Orono, Me. (U.S.A.). *Aphididae, alternation of food plants of migratory species.* ⓘ Cornell Univ. 1911.
Patch, Lawrence H. (1896), Ass. Entomol., U.S. Bureau of Entomol. Sandusky, O. (U.S.A.), P.O. Box 976. *European Corn Borer.* ⓟ Amherst, Mass. 1919.
Pateff, Paul (1889), Ass. der Pflanzenschutzabteilung d. Landwirtsch. Versuchsstation. Sofia (Bulgarien), Solun Pl. 3. *Bact. Pflanzenkrankheiten; Rhizopoda (System u. Biol.).*
Pater, Béla (1860), Leiter der Arzneipflanzen-Versuchsstation. Cluj (România). *Arzneipflanzen, Krankheiten der Pflanzen.* ⓘ Budapest 1883, Klausenburg 1907.
Patkanian, Aschchene R., Préparat. Inst. Jaczewski. Leningrad (U.d.S.S.R.), Angliskij Prospekt 29. *Fungicides.* o
Paton, D. Noël, Prof. Univ. Glasgow (Scotland). *Physiol.* o
Paton, Stewart, Lect. in Biol. Univ. Princeton, N.J. (U.S.A.), Greenlands. o
von Patow, Freiherr Carl (1880), Dr. phil. nat. Rittergut Calberwisch b. Osterburg, Altmark (Deutschland). *Vererbung der Milchleistung beim Rind.* ⓘ Jena 1925.
Patrick, Austin Lathrop (1889), Prof. of Soil Technol. State College, Pennsylvania (U.S.A.). ⓘ Pennsylvania State Coll. 1913.
Patrizi, Luigi Mariano, e.o. Prof., Inst. super. Med. vet. Bologna (Italia). *Fisiol.* o
Patschovsky, Norbert (1892). Ettal bei Oberammergau, Ob.-Bayern (Deutschland). *Mikrochem. der Pflanzen. Physiol. der Entwicklung u. Fortpflanzung (spez. Moose).* ⓘ Jena 1917. ⓟ Lebende Pflanzen der alpinen Region.
Patta, Aldo (1879), Doc. de Pharmacol. experim., Prof. chargé Pharmacothérapie Univ. Milano (Italia), Rue A. Verga 4. *Pharmacol. des composés arsénicaux organ. Pharmacol. de l'or.* ⓘ 1903.
Pattarin, Piero (1899), Ass. Ist. Anat. Univ. Milano (Italia). ⓘ Pavia 1924.
Patten, Bradley Merrill (1887), Associate Prof. of Histol. and Embryol., Med. School, Western Reserve Univ. Cleveland, Ohio (U.S.A.), 2109 Adelbert Rd. *Histol. and Embryol., development of the cardiovascular system.* ⓘ Harvard 1914.
Patten, C. J. (1870), M.D., Sc.D., Prof. of Anat. Sheffield (England), Univ. *Ornithol., Bird-Migration at Irish Light-Stations.* ⓘ Trinity Coll., Dublin Univ. 1896. ⓟ Bird Skins.
Patten, William (1861), Prof. of Biol. (Zool.), Dir. of Course in Evolution. Hanover, N.H. (U.S.A.), Dartmouth Coll. *Evolution of Vertebrates. Anat. of Ostracoderms and Palaeozoic Fishes. Anat. and Embryol. of Arachnids. Limulus, Scorpions, Eurypterids. Insects. Types of mouches and arthropods* ⓘ Leipzig 1885. ⓟ Palaeozoic Fishes, Ostracoderms.
Patterson, Cecil F., Prof. of Horticulture, Univ. of Saskatchewan. Saskatoon, Saskatchewan (Canada). *Physiol.* o
Patterson, Flora Wambaugh (1847), Mycol. Brooklyn, N.J. (U.S.A.), 1215 Ocean Ave. *Mycol. Pathol. Taxonomy of fungi.* o

Patterson, John Thomas (1878), Prof. of Zool. Univ. of Texas. Austin, Tex. (U.S.A.). *Embryol. polyembryony.* ⓘ Univ. of Chicago, Ill. 1908 ⓟ Armadillo embryos.
Patterson, Jorelyn, Biochem., Charmy Cross Hospital Inst. of Pathol. London W.C. 2 (England). *Biochem.*
Patterson, Thomas Leon (1884), Prof. of Physiol., Detroit Coll. of Med. and Surgery. Detroit, Mich. (U.S.A.), 1516 St. Antoine Street. *Influence of age on variations in the hunger contractions of the empty stomach; Cause of the variations; Compar. studies on the physiol. of the empty and filled stomach of mammals, birds, reptiles, amphibians and marine and terrestrial molluscs; Effect of stimulation of sensory nerves on gastric motility; Gastric and pulmonary tonus; Endocrine studies on pituitary gland.* ⓘ Univ. of Chicago 1920.
Patteson, George Walker (1890), Extension Agronomist, State Agric. Coll. Blacksburg, Va. (U.S.A.) *Field crops, lime fertilizer.* ⓟ Wheat, corn, soybeans, cowpeas, cotton, peanuts, red clover seed, barley, rye, oats, and sweet clover.
Patton, Reuben Tom (1883), Senior Lect. Agric. Botan. Melbourne, Victoria (Australia), Univ. *Anat. of Woods, Plant Distribution.* ⓘ Harward 1922. ⓟ Victorian forest plants.
Patzelt, Viktor (1887), Priv.Doc., o. Ass. am Histol. Inst. der Univ. Wien IX (Deutsch-Österreich), Schwarzspanierstr. 17. *Histol. (Kehlkopf, Haut, Darm, normale und abnorme Geschlechtsentwicklung).* ⓘ Graz 1913.
Paukul, Ernst (1872), Prof., Dir. Inst. der allgemeinen Pathol. Univ. Riga (Latvija), Kronwald. Boulev. 10, W. 12. *Pathol. Physiol.* ⓘ Bern 1903.
Paul, Hermann (1876), Dr. phil., Reg.R. I. Kl., Leiter der Botan. Abteilung der Bayer. Landesanstalt für Moorwirtschaft. München (Deutschland), Hedwigstr. 3 I. *Botan. und Geol. (Stratigraphie) der Moose; Wiesen- und Weidenkunde; Systematik und Biol. der Moose, besonders der Sphagna.* ⓘ Berlin 1902. ⓟ Bryophyten.
Paul, John Rodman (1893), Dir. of the Ayer Clinical Labor. Pennsylvania Hosp. (U.S.A.), 8th and Spruce Sts. *Bact. and chem. pathol.* ⓘ Johns Hopkins Med. School 1919.
de Paula Souza, Geraldo Horacio (1889), Chief Health Officer for the State of S. Paulo, Dir. of Inst. of Hygiene, Prof. of Hygiene, Med. School. S. Paulo (Brazil), Caixa Postal 1985. *Biochem., Hygiene.* ⓘ 1920.
Paul-Boncour, Georges (1866), Prof. d'Anthrop. Paris (France). *Anthrop. pathol.* ⓘ Paris 1916.
Paulescu, N., Prof. Univ. Bucuresti (România). *Physiol.* o
Paulin, Alfons (1853), Hon.Prof. system. Botan. u. Dir. des botan. Gartens der Univ. Ljubljana (S.H.S.), Mikloŝičstr. 28 II. *Systematik der Pteridophyten u. Phanerogamen: Flora von Slovenien (Krain u. Südsteiermark); Pflanzengeographie von Slovenien.* ⓘ Graz 1880. ⓟ Pteridophyta und Phanerogamen, namentlich Alpenpflanzen.
Paulli, S., Prof. Dr., Anat. Labor., Vet. Landw. Hochsch. København V (Danmark), Bülowsvej. o
Paulsen, Ove Vilhelm (1874), Prof. København (Danmark), Botan. garden. *Plant geography of Central -Asia, of the West Indies. Phytoplankton: Peridineales.* ⓘ Copenhagen 1911.
Pauson-Herzfelder, Helene (1895), Dr. phil. Bamberg (Deutschland), Ottostr. 7. *Experim. Beeinflussung von Moosen physiol. und morphol. Art.* ⓘ München 1920.
Pavari, Aldo (1888), Dir. Staz. Sperim. di Selvicoltura, Ist. Superiore Agrario e Forestale. Firenze (Italia). *Botan. applicata.* ⓟ Semi e piante forestali.
Pavlot, Jean (1866), Prof. ord. tit., Prof. d'anat. pathol. à l'Univ. Lyon (France), 23, quai Gailleton. *Histol. pathol. Inflammations, tumeurs.*ⓟ Prép. histol.

Pavlov, Nikolai Vassiljevitsch (1893), I. Univ. Moskau (U.d.S.S.R.), Danilovsky val. Haus 22/3, W. 21. *Systematik der Pflanzen, Floristik der Kirgisischen Steppen und Mongolei. Botan. Geographie Asiens.* ⑨ Pflanzen Sibiriens und der kirgisischen Steppe.

Pavlov, Vjačeslav Alexandrovič (1895), Ass. Zootom. Inst. Univ., Ass., Priv.Doc. f. allgem. Biol. Inst. f. Med. Wissensch. Leningrad (U.d.S.S.R.), Univ., Zootom. Inst. *Physik. Chemie des Protoplasmas. Regeneration u. Wachstum (Protozoen u. niedere Tiere).*

Pavlović, Drago, Dr., Centralni Higijenski zavod, IV. sprat. Beograd (S.H.S.). *Bact.* o

Pavlović, Pelar S., provis. Dir., Kustos d. Geol.-paläontol. Abt. d. Serbischen Landesmus. Beograd (S.H.S.), Miloša Velikoy ulica 31. *Malakol.* o

Pavlovskij, Konstantin, Doc. Landwirtschaftl. Inst. Kiev (U.d.S.S.R.), Brest-Litovskoe-Chaussee 39. *Zoohygien.* o

Pawlow, Alexius P., Dr., Prof. Geol. I. Univ. Moskau (U.d.S.S.R.). *Palaeont., speziell mesozoische Pelecypoden, fossiler Mensch.* o

Pawlow, Marie W., Prof. I. Univ. Moskau (U.d.S.S.R.). *Tertiäre Palaeont. der Wirbeltiere* o

Pawlow, W. A., Prof., Dir. Histol. Labor. Univ. Saratow (U.d.S.S.R.). o

Pawłowski, Bogumil (1898), Dr. phil., Adjoint au Jardin Botan. de l'Univ. Jagellonienne Kraków (Polska), 45, rue Lubicz. *Floristique, geographie botan., surtout des Carpathes. Phytosociol.* ⑨ Kraków 1922. ⑨ Plantes d'herbier de la Pologne.

Pawlowskiy, Eugeniy (1904), Prosektor Physiol. Labor. d. Vet.-Inst. Kazan (U.d.S.S.R.). *Physiol. d. Atmung.* ⑨ Kazan 1924.

Pawlowsky, Eugen (1884), Prof. der Zool. u. vergl. Anat. Militärmed. Akad. Leningrad (U.d. S.S.R.), Prospect Karla Marksa 7 A, log. 3 A. *Vergl. Anat. der Arthropoden (Scorpiones, Ixodoidea, Insekten). Parasit. experim. Wirkung der Parasit. auf ihren Wirt. Wirkung der Giftarthropoden auf Menschen und Tiere.* ⑨ St. Petersburg 1913. ⑨ *Parasit. Tiere. Giftitiere. Zecken. Blutsaugende Insekten. Tropische Skorpione.*

Pax, Ferdinand (1885), beamteter a.o. Prof. der Zool. an der Univ. und Kustos des Zool. Mus. Breslau 9 (Deutschland), Sternstr. 21. *Anthozoen, allgemeine Faunistik und Tiergeographie, Naturschutz.* ⑨ Breslau 1907.

Payn, William Arthur (1871). Grantham (England), Garden Cottage, Willoughley Hall. *Birds.*

Payne, Fernandus (1881), Prof. of Zool. Bloomington, Ind. (U.S.A.), Indiana Univ. *Chromosomes. Cytoplasmic inclusions, heredity in Drosophila. Fresh-water hydroids. Young human embryos.* ⑨ Columbia Univ. 1909.

Payne, Nellie M. (1900), National Research Fellow in Zool. Philadelphia, Pa. (U.S.A.), Zool. Labor., 38 Street and Woodland Avenue. *Inheritance of Physical Traits in Human Beings. Cold-hardiness in insects. Insect Physiol.: Nutrition. Low temperature effects on marine eggs.* ⑨ Minnesota 1925.

Payson, Edwin Blake (1893), Prof. of Botan. Laramie, Wyo. (U.S.A.), Univ. of Wyoming. *Taxonomy of Angiosperms. Cruciferae, plants of the Rocky Mountain Region. Phylogenetic taxonomy.* ⑨ Washington Univ. 1921. ⑨ Herbarium specimens of flowering plants from the Rocky Mountain Region.

Payton, Carrick Gordon (1897), Lect. and Senior Demonstrator of Anat. Univ. of Birmingham (England). *Regional Anat. The Growth of Bone.* ⑨ Univ. Edinburgh 1921.

De la Paz, Dniel, Prof. Univ. of the Philipp. Manila (Philippine Islands). *Pharmacol.* o

Pazos y Caballero, José Hipólito (1867), Dr. méd. et chir., Chef de santé. Antonio de los Baños (Isla de Cuba, Provincia de La Habana), Martí 46 S. *Mosquitos de Cuba.* ⑨ Habana 1894.

Peairs, Leonard Marion (1886), Prof. of Entomol.; Entomol., Agr. Experim. Station. Morgantown, W.Va. (U.S.A.). *Fruit-insects. Influence of climatic factors on development of insects.* ⑨ Kansas 1905.

Pearl, Raymond (1879), Dir. of the Inst. for Biol. Research of the Johns Hopkins Univ., Prof. of Biol. Johns Hopkins Med. School, Research Prof. Johns Hopkins Univ. Baltimore, Md. (U.S.A.). *Biol., Biometry, Vital Statistics. Genetics. Diseases of Poultry.* ⑨ Ph.D. Michigan 1902; Sc.D. Dartmouth 1919; L.L.D. Maine 1919.

Pearsall, W. H., Reader Univ. Leeds (England). *Botan.* o

Pearse, Arthur Sperry (1877), Graduate Prof. of Zool. Durham, N.C. (U.S.A.), Duke Univ. *Ecol., Parasit.* ⑨ Harvard Univ. 1908.

Pearson, A. B., Dr., Ass. Univ. Christchurch (New Zealand). *Pathol.* o

Pearson, Helga Sharpe (1898), Ass. London (England), Gower St., Dept. of Zool. Univ. Coll. *Zool. Vertebrate Palaeont.* ⑨ London 1919.

Pearson, Thomas Gilbert (1873). New York, N.Y. (U.S.A.), 2257 Loring Pl. *Ornithol.* o

Peau, Etienne (1878), Chef des Travaux Biol. Inst. Océanographique Rédact. Sc. Le Havre (France), Mus., Place du Vieux-Marché. *Biol. et applications industrielles de l'Océanographie.* ⑨ Le Havre 1918. ⑨ Animaux marins, Plancton, Algues.

Pečerskij, Georgij Mak., Doc. I. Univ. Moskau (U.d.S.S.R.). *Histol.* o

Pech, Jacques Louis (1889), Prof. de physique méd. Univ. Montpellier (France), Inst. de Biol., rue Montels. *Fluorescences de tissus des êtres vivants. Variations du champ électrique ambiant sur les végétaux. Potentiels de contact (indice de nutrition) entre les tissus de l'homme et des animaux ou végétaux vivants et les milieux extérieurs liquides solides ou gazeux.*

Péchy, Koloman (1902). Pécs (Ungarn), Anna-ut 4. *Pathol. Histol.*

Peck, Mary Ellen (1904), Research Ass. Brooklyn, N.Y. (U.S.A.), Brooklyn Botan. Garden. *Genetics.*

Pedašenko, D. D., Prof. Polytechn. Inst. Novočerkassk (U.d.S.S.R.). *Zool.* o

Pederson, Carl S. (1897), Ass. Bact. Geneva, N.Y. (U.S.A.), New York Agric. Experim. Station. *The Forms of Lactic Acid Produced by Pure and Mixed Cultures of Bacteria.* ⑨ Univ. of Wisconsin 1924.

Peebles, Florence (1874), Investigator. Woods Hole, Mass. (U.S.A.). *Regeneration, experim. Embryol. Transplantation.* ⑨ Bryn Mawr Coll. 1900.

Peeler, William B., Head of Sc. Dept., Hardin Junior Coll. Mexico, Mo. (U.S.A.), 201 W. Boulevard. *General botan.* o

Peglion, Vittorio, Prof. uff. Ist. super. agrar. Bologna (Italia). *Biol. agr.* o

Peirce, George James (1868), Prof. of Botan. and Plant Physiol. Stanford Univ., California (U.S.A.). *Physiol. of Plants.* ⑨ Leipzig 1894.

Peirson, Henry (1894), State Forest Entomol. Augusta, Me. (U.S.A.), Maine Forest Service. *Life history and control of forest and shade tree Insects.* ⑨ Harvard 1920.

Peiser, Elisabeth (1892), Ass. am Physiol. Inst. der Univ. Berlin (Deutschland), Winterfeldtstr. 23. *Biochem. Chem. der Polysaccharide.* ⑨ Berlin 1919.

Peixoto, A., Prof. Univ. Río de Janeiro (Brasil). *Hygiene.* o

Pekár, Mihály (1871), o. Univ.Prof., Dir. Physiol. Inst. Univ. Pécs (Ungarn), Rákóczi-Str. 80. *Physiol. des Zentralnervensystems.* ⑨ Budapest 1897.

Pekelharing, N. R., Dr., Adviseur Intern. Cred. Handels Vereen. Rotterdam. Semarang, Java (Nederl.-O.-Indië). *Kultur v. Hevea u. anderen trop. Gewächsen.* o

Peklo, Jaroslav, Dr., Prof. d. angewandt. Botan. Praha II (C.S.R.), Karlovo nám. 288, Inst. f. angew. Botan. d. čech. techn. Hochsch. *Physiol.*

végétale, phytopath., bâtardation, génétique, botan. appliquée. o
Pellegrin, Jacques (1873), Ass. au Mus. national d'histoire natur. Paris V (France), 57, rue Cuvier. *Ichtyol., Herpétol., Pisciculture. Poissons vénéneux. Cichlidés. Poissons d'Afrique.* ⑨ Méd. Paris 1879, Sc. 1904.
Pellett, Frank C. (1879). Hamilton, Ill. (U.S.A.). *Beekeeping and Plants which are the source of honey.*
Pelosse, Jean Paul Louis (1885), Agrégé de l'Univ., chargé de cours à la Fac. des Sc. Lyon (France), 18, rue du Béguin. *Zool. théorique et pratique: Entomostracés d'eau douce, plankton d'eau douce. Sériciculture.*
Pelseneer, Paul (1863), Secrétaire perpétuel de l'Académie royale de Belgique. Bruxelles (Belgique), 23, rue Longue Haie. *L'Organisation, l'embryon. et l'éthol. des Mollusques, Crustacés, Trématodes et Echinodermes.* ⑨ D.Sc. Bruxelles 1884; D.Sc. h. c. Leeds 1906.
Peltier, George Leo (1888), Prof. of Plant Pathol., Univ. of Nebr. and Plant Pathol., Nebr. Agric. Experim. Station. Lincoln, Neb. (U.S.A.), Dept. of Plant Pathol., Coll. of Agric. *Diseases of floricultural and ornamental plants, Rhizoctonia, citrus canker and citrus scab, stem rust of wheat, Ozonium rootrot.* ⑨ Univ. of Illinois 1915.
Pelzig, Leonid A., Dir. Landw. Versuchsstation. Tschakino, Tambow. Gouv. (U.d.S.S.R.). *Agronom. Angew. Botan.* o
Pembrey, M. S., Prof. Univ. London S.W. 7 (England), Kensingt. *Physiol.* o
Penard, Thomas Edward (1878). Arlington, Mass. (U.S.A.), 12 Norfolk Road. *South American Ornithol.* ⑨ Boston 1900.
Penecke, Richard (1886), Dr. med., Leiter der Landesprosektur. Cieszyń (Polska), Rynek 11. *Pathol. Anat. und Histol., Bact., Serol., Hämatol., med.-chem. Untersuchungen. Liquor-Diagnostik.* ⑨ Graz 1911.
Penfound, Wm. Theodore (1897), Ass. in Botan. Urbana, Ill. (U.S.A.), Univ. of Illinois. *Plant ecol. distribution, ecol. anat. of plants.*
Penland, C. W., Ass. Prof. of Biol., Colorado Coll. Colorado Springs, Col. (U.S.A.). o
Pennell, Francis W. (1886), Curator of Plants, Academy of Nat. Sc. Philadelphia, Pa. (U.S.A.), 19th and Race Sts. *Plant distribution and taxonomy: Scrophulariaceae.* ⑫ Plants of U.S. and Andes of Peru and Chile.
Penners, Andreas (1890), Dr. phil., Priv.Doc. für Zool. und vergl. Anat. Würzburg (Deutschland), Zool. Inst. *Entwicklungsgeschichte (Oligochaeten). Entwicklungsmechanik (Oligochaeten, Amphibien).* ⑨ Würzburg 1920.
Pennington, Leigh H. (1877), Prof. of Botan. in the New York State Coll. of Forestry at Syracuse Univ. Syracuse, N.Y. (U.S.A.), 219 Clarendon Street. *Forest Pathol.* ⑨ Univ. of Michigan 1909.
Pennisi, Alessandro, Lib. doc. Univ. Messina (Italia). *Pathol.* o
Pennock, Charles J. (1857), Lect. Chester County, Pa. (U.S.A.), Kennett Square. *North American Birds, Geographical range and economic value of Birds.* ⑫ Bird skins.
Penrose, Francis George (1857). Bournemouth (England), 51 Surrey Road. *Ornithol.* ⑨ London 1885.
Pensa, Antonio (1874), Prof. de l'Univ. Parma (Italia), Inst. d'Anat. humaine. *Distribution des nerfs dans les glandes. Angiol. humaine et comparée. Pancréas, thymus (anat. et developpement). Cytol. animale et végétale. Structure interne des protozoaires.* ⑨ 1920.
Pentchew, Angel (1894), Dr. med., Prosektor des Pathol. Inst. der Univ. Sofia (Bulgarien), Neofit Rilsky 12. *Pathol. Anat.* ⑨ Berlin 1921.
Pentegov, B. P. (1887), Prof. of chem. Univ., member of Pacific maritim. stat., Member of Research Inst. Vladivostok (U.d.S.S.R.), Univ. *Dispersoidol. Seawater. Soil.* ⑨ St. Petersburg 1913.
Pentman, Izrail S., Prof. Med. Inst. Omsk (U.d.S.S.R.), Ul. Lenina 9. *Allg. Pathol.* o
Penzig, Otto (1856), Prof. o. di Botan. Univ., Dir. Orto ed Ist. Botan. Genova (Italia), Corso Dogali 1, B. *Teratol. vegetale; Morfol., Mycol., Flora di Liguria, Nomenclatura volgare delle piante italiane. Storia della Botan.* ⑨ Breslau 1877.
Péola, Paolo (1869), Prof. di Sc. nel R. Ist. tecnico V. E. II. Genova (Italia). *Paleofitol. Zool. Botan. Igiene. Euphorbia. Olivo. Fauna valdostava.*
Pepere, Alberto (1875), Prof. stab. Anat. e istol. patol. R. Univ. Milano (Italia), Ospedale Maggiore. *Anat. patol. Tumori primitivi del fegato. Ghiandola paratiroidea. Secrezione interna. Batteriol.*
Pepeu, Francesco (1897), Dirigente la Sezione ,,Sieri umani" Ist. Sieroterapico. Milano (Italia), Via Darwin 20. *Immunol. e Profilassi.* ⑨ Vienna 1910.
Pérard, Charles Henri (1885), Ass. Inst. Pasteur. Paris 15 (France), 106, rue Brancion. *Sporozaires: Coccidies.* ⑨ Alfort 1908.
Percival, E., Lect. Univ. Leeds (England). *Zool.* o
Percival, George Hector (1901), Ass. Edinburgh (Scotland), Pharmacol. Dept. Univ. *Calcium metabolism.* ⑨ Edinburgh 1924.
Percival, John (1863), Prof. Agric. Botany. Univ. Reading (England). *Cultivated Farm Plants.*
Peredelsky, Anatolius A. (1904), Prep. Inst. of General Biol. II. Univ., Labor. of Experim. Biol. of the Zoopark. Moscow (U.d.S.S.R.), Dewitschje Pole, Pogodinskaja 6. *Experim., Cytol. Factors of pigmentation.* ⑨ Moscow.
Pereira, M., Prof. Univ. Porto Alegre (Brasil). *Histol.* o
Pereira de Macedo, José, Prof. Fac. d. Med. Curityba, Paraná (Brasil). *Anat.* o
Pereira da Silva, Estevão (1877), Chef de service. Lisbôa (Portugal), Inst. Camara Pestana. *Rage.* ⑨ Lisbonne 1926.
Perelmann, Leonid Ruvimovič, Doc. II. Univ. Moskau (U.d.S.S.R.). *Allg. Pathol.* o
Perevezentsev, Alexandre S., Chef des trav. Stat. Défense des Plantes. Moscou (U.d.S.S.R.), Sadovaia-Trioumphainaia 10. o
Pérez, Charles (1873), Prof. Zool. Fac. Sc. Univ., Dir. Stat. Biol. de Roscoff (Finistère). Paris 5 (France), Sorbonne, 1, rue Victor Cousin. *Processus histol. de la métamorphose. Cytol. des gamètes. Protozoaires parasit. Hydrozoaires. Crustacés. Décapodes, Epicarides, Rhizocéphales.* ⑨ Paris 1902. ⑫ Animaux et plantes marines de la région de Roscoff.
Perez Serrano, Mariano, Prof. esc. esp. de ing. de montes. Madrid (España), Rey Francisco 4. *Zool. Entomol.* o
Perfiliew, Peter (1897), Ass. Militär-Med. Akademie Leningrad (U.d.S.S.R.), Fontanka 2, log. 280. *Vergl. Anat. der Arthropoden (Insekten).* ⑨ Petrograd 1923. ⑫ Amphibien und Reptilien.
Perfiljew, Boris S., Vorsteher d. Stadt. Inst. für Wiesenbau. Sowchos. ,,Katschalkino", Station Lobnja, Moskauer Gouv. (U.d.S.S.R.), Sawel sh. d. *Wiesenbau.* o
Pergola, Mazzini (1880), Prof., Libero Doc. in Igiene e Batt., Coadiutore Med. presso il Laborat. di Micrografia e Batt. Roma 32 (Italia), Via di Porta Maggiore, 132. *Bacillo difterico. Bacillo di tipo bovino. Vibrioni. Intossicazione alimentare.*
Perichanjanz, Jakob (1879), Prof. Dir. Pharmacol. Labor. Univ. Perm, Saïmka (U.d.S.S.R.). *Innere Sekretion. Peripheres Nervensystem. Stoffwechsel.* ⑨ Charkow 1911.
Périot, Maurice (1891), Prof. suppl. physiol. Ecole d. Med., Chargé du cours de Physiol. du Travail. Marseille (France), 3, Bd. Salvator. *Physiol. humaine.* ⑨ Montpellier 1922.

Perkins, George Henry (1844), State geologist. Burlington, Vt. (U.S.A.). *Geol. Entomol.* o
Perkins, Henry Farnham (1877), Prof. Univ. of Vermont. Burlington, Vt. (U.S.A.). *Zool. Fresh water invertebrates.* o
Perkins, Roger Griswold (1874), Prof. of Hygiene and Preventive Med. Cleveland, O. (U.S.A.), W.R.U. School of Med. *Bact., Epidemiol. and Prevention of communicable Diseases.* ⑨ Johns Hopkins 1898.
Perlzweig, William Alexandre (1891), Associate in Med. and Chem. to Med. Clinic, Johns Hopkins Univ. Med. School and Hospital. Baltimore, Md. (U.S.A.). *Pathol. chem., immuno-chem.; metabolism.* ⑨ Columbia Univ., New York, 1915.
Perna, Giovanni, Lib. doc. Univ. Bologna (Italia). *Anat. umana.* o
Pernetti, Joseph (1899), Aide ord. dans l'Inst. de Pharmacol. et Thérapie dans l'Univ. Napoli (Italia), Rue Enrique Pessina N. 73. *Empoisonnements, actions photodynamiques et pharmacodynamiques.* ⑨ Naples 1923.
Pernice, B., Doc. Univ. Palermo (Italia). *Anat. patol.* o
Pernigotti, Mario (1880), Prof. ord. Liceo di Lodi, Milano (Italia). *Ciprinidi.* ⑨ Roma 1915.
Pernkopf, Eduard (1888), a.o. Univ.Prof., Dr. med., Priv.Doc. u. Ass. am II. Anat. Inst. Wien IV (Österreich), Schönbrunner Str. 1. *Anat. und Embryol. Entwicklung des Magen-Darmtraktus. Inversion der Eingeweide.* ⑨ Wien 1912.
Perović, Drago, Dr., o. Prof. Anat., Dir. Anat. Inst. Zagreb (S.H.S.), Voćarska cesta 97. o
Perotti, Renato (1879), Prof. à l'Inst. supérieur d'Agric. Pisa (Italia). *Pathol. végétale. Bact. agric.* o
Perrin, G., Prof. Ecole prép. de Méd. et de Pharmacie. Clermont-Ferrand (France). *Histol.* o
Perrin, M. L., Chef de Trav. Zool. Fac. Sc. Grenoble (France). *Zool.* o
Perrier, Rémy (1861), Prof. Fac. Sc. Paris (France), 12, rue Perrier. *Anat. comparée. Faune de France: Mollusques Gastéropodes. Holothuries.*
Perroncito, Aldo (1882), Prof. o. Patol. Gen. Pavia (Italia), Corso Vittorio Emanuele 77. *Rigenerazioni. Innesti. Patol. sperimentale del fegato.*
Perrot, E., Prof. Univ. Paris (France). *Hist. natur. d. méd.* o
Perrow, Mosby Garland (1876), Dir. Lynchburg, Va. (U.S.A.). *Biochem., Bact.* ⑨ Washington and Lee 1906.
Perry, W. J., Reader Univ. London S.W. 7 (England), S. Kensingt. *Anthrop.* o
Persidsky, Boris, Botan. Wissenschaftl. Meeresinst. Botan. Garten. Moskau (U.d.S.S.R.), Rue Herzen 6. ⑨ Moskau.
Perušek, Milena, Dr., Ass. Landwirtsch. Versuchsstation. Ljubljana (S.H.S.). *Phytopath.* o
van Pesch, Adrianus Jacobus (1879). Amsterdam (Holland), Marinewerfkade 8. *Antipatharia.* ⑨ Amsterdam 1910.
Pesch, Karl Ludwig (1889), Priv.Doc. f. Hygiene u. Bact., Oberarzt a. Hygien. Inst. d. Univ. Köln (Deutschland), Gereonskloster 20. *Diphtherie, Paratyphus, physikalisch-chem. Fragen der Serol., soziale Hygien.* ⑨ Greifswald 1920.
Peschkowskaja, Ludmilla (1885), Ass. Inst. für Experim. Biol. Moskau 6 (U.d.S.S.R.), Worontzowo Polje. *Protistol. Fibrilläre Strukturen der Ciliaten.*
Peserico, Enoch (1897), Aiuto, Prof. incar. Milano (Italia), Ist. di Fisiol., Via G. Strambio. *Ricambio dei carboidrati. Conduttività elettrica dei tessuti.* ⑨ Padova 1925.
Peskett, G. L., Demonstr. Univ. Oxford (England). *Biochem.* o
Pesola, Vilho Aleksanteri (1892), Ass., Abt. Pflanzenzüchtung Staatl. Landwirtschaftl. Versuchsanstalt. Tikkurila (Finnland). *Angewandte Botan. und Vererbungslehre, Pflanzenzüchtung: Getreidearten und Gräser.* ⑨ Samenproben kultivierter Pflanzen.

Pessin, Louis J. (1897), Associate Forest Ecol. New Orleans, La. (U.S.A.), 7822 St. Charles Ave. ⑨ Johns Hopkins Univ., Baltimore, Md. 1923.
Pesta, Otto (1885), Dr. phil., Kustos am Naturhistor. Mus. (ehem. Hofmus.), Priv.Doc. an der Hochsch. für Bodenkultur. Wien I (Österreich), Burgring 7. *Crustaceenkunde: Meeres- u. Süßwassercopepoden, Decapoden, Süßwasser-Cladoceren. Hydrobiol. des Süßwassers (Limnol.). Planktonkunde.* ⑨ Innsbruck 1907.
Peter, Karl (1870), o Prof. und Dir. des Anat. Inst. Greifswald (Deutschland). *Anat. und Entwicklung der Niere. Zellteilung. Biol. Betrachtung der Entwicklung.* ⑨ Freiburg i. Br. 1894.
Péterfi, Tibor (1883), Prof., Dr., wissensch. Berater der Firma C. Zeiß, Berlin. Berlin-Dahlem (Deutschland), Kaiser-Wilh.-Inst. f. Biol. *Experim. Cytol., Mikrol., Mikrotechnik.* ⑨ Kolozsvár 1906.
Peterhans, Emile (1899), Ass. Lausanne (Suisse), Labor. de Géol., Palais de Rumine. *Paléont.* ⑨ Lausanne 1926.
Peters, Harold S. (1902), Ass. Entomol. of Virginia Truck Experim. Station. Norfolk, Va. (U.S.A.). *Insects of Truck and Garden crops of the Eastern and Southern parts of Virginia. Mallophaga of the United States, Genus Degeeriella, from birds.* ⑫ *Mallophaga of Birds of the Eastern United States, Genus Degeeriella.*
Peters, Peter Martin Nicolaus (1900), Zool. Staatsinst. u. Zool. Mus. in Hamburg. Hamburg-Finkenwärder (Deutschland). *Fischereibiol. Regenerationsstudien an Chaetopoden. Protozoen Plankton.* ⑨ Hamburg 1922.
Peters, R. A., Prof. Univ. Oxford (England). *Biochem.* o
Peters, Theodor (1888), Dr. phil. Braunschweig (Deutschland), Helmstedter Str. 91 II. *Phyllodine Acacien, Lichtkeimung.* ⑨ Kiel 1912.
Peters, Wilhelm (1880), Dr. phil., o. ö. Prof. der Psychol. und Vorstand der Psychol. Anstalt der Univ. Jena, Thüringen (Deutschland), Pfaffenstieg Nr. 3 a. *Grenzfragen der Biol. und Psychol.: Vererbung psychischer Eigenschaften, psychische Entwicklung, psychische Konstitution. Probleme der experim. Psychol.: Gedächtnis, Intelligenz, Gefühlsleben. Entwicklung der Leistungsfähigkeit des Kindes, abnorme Kinder und ihre psychischen Leistungsfähigkeit.* ⑨ Leipzig 1904.
Petersen, Axel (1877), Vet. Ringsted (Danmark). *Danish Diptera: Simuliida.* ⑨ Copenhagen 1901. ⑫ *Danish Simuliida.*
Petersen, Carl Georg Johannes (1860), Retired Dir. of The Danish Biol. Station. København (Danmark). *Fish-Biol. Quantitative valuation of animals on the sea-bottom. Theoretical Biol.* ⑨ Copenhagen Univ. 1888.
Petersen, Erik J. (1894), Abteilungsvorsteher, mag. sc. Statens Planteavls Labor. Lyngby-København (Danmark), Peter Bangsvej 59 St. *Bact. CO_2-production of the soil. Cytol. of Bact. Azotobact. Nodule bact.*
Petersen, Grace Agnes (1905), Ass. in Plant Pathol. at Cornell Univ. Ithaca, N.Y. (U.S.A.), Department of Plant Pathol. *Mycol.; blue-green algae of the region about Ithaca.* ⑨ Cornell Univ. 1926.
Petersen, Hans (1885), Dr. phil. et med., o. ö. Prof. der Anat., Vorstand der Anat. Anstalt der Univ. Würzburg (Deutschland), Köllikerstraße. *Histol., Histol. des Knochens, der Skelettorgane, Knochen- und Gelenkmechanik, physikalische Anat.* ⑨ Jena phil. 1907, med. 1911. ⑫ *Mikroskop. Praep.*
Petersen, Henning Eiler (1877), Ass. Botan. Labor. Univ. Lect. Polytechn. Hochsch. Kopenhagen (Dänemark), Gothersgade 140. *Algol.; Ceratium. Phanerogamen: Polymorphie. Anat. Mycol.: Phycomyceten; Merulius.* ⑨ Kopenhagen 1914.

Petersen, Johannes Boye (1887), Amanuensis am Botan. Garten der Univ. København K. (Danmark), Gothersgade 130. *Süßwasseralgen, Luftalgen, Diatomeen, Cyanophyceen, Flagellaten.*

Petersen, Niels Frederick (1877), Botan. Ass. in the Department of Agronomy, Univ. of Nebraska. Lincoln, Neb. (U.S.A.), Station A. *Plant Morphol. and Physiol. Systematic Botan. of the Flowering Plants.* ⓓ Univ. of Nebraska 1911.

Petersen, Wilhelm (1854), Magister zool. Nömme bei Reval (Estland). *Morphol. der Generationsorgane der Lepidoptera. Zoogeographie: Lepidopteren. Fauna des arktischen Gebietes von Europa und die Eiszeit.* ⓓ Dorpat 1887.

Petersen, William F. (1887), Prof., Department of Pathol., Univ. of Illinois, Coll. of Med. Chicago, Ill. (U.S.A.), 508 S. Honore St. *Functional Pathol.* ⓓ Rush Med. Coll. 1912.

Petersen, Wm. Earl (1892), Ass. Prof. Dairy Husbandry. St. Paul, Minn. (U.S.A.), Dairy Dept., Univ. Farm. *Physiol. of Milk Secretion.* ⓓ Univ. of Minnesota 1916.

Peterson, Daniel (1893). Malmö (Sverige), Rēalskolan. *Inzucht bei Pflanzen.* ⓓ Lund.

Peterson, Paul Donald (1894), Ass. Plant Pathol. Univ. Farm. St. Paul, Minn. (U.S.A.). *Sweet Clover Hay Poisoning; Corn Root Rots. Raspberry Mosaic Control.* ⓓ Univ. of Minnesota 1926.

Peterson, William Harold (1880), Prof. Agric. Chem., Univ. of Wisconsin. Madison, Wis. (U.S.A.). *Fermentation, mineral elements in plants and nutrition.* ⓓ Middletown, Conn., 1907.

Petersons, Jānis, Dr., Ass. Univ. Rīga (Latvija), Pernavas ielā 19. *Anat. d. Tiere.* o

Petkoff, Peter, Doc. f. Anat. u. Physiol. d. Haustiere Inst. f. allg. Tierzucht der Landw. Fac. Sofia (Bulgarien). *Histol., Insektenbiol.* o

Petkoff, Stéphan (1866), Prof. ord. de Botan. à la Fac. des Sc. de l'Univ. Sofia (Bulgarie), Inst. de Botan. *Botan. spéciale, Cryptogamie: Schizophyceae, Zygophyceae, Algues vertes, Phaeophyceae, Rhodophyceae, Charophyta.* ⓓ Gand (Belgique) 1894.

Petlach, Stanislav (1889), Ass. Labor. Pharmacol. Univ. Brno (C.S.R.), Tučkova 13. *Saponines. Effets physiol. de Drosera rotundifolia.* ⓓ Brno 1924.

Petragnani, Giovanni, Prof. Univ. Siena (Italia). *Batteriol.* o

Petrascheck, Wilhelm (1876), Dr., o. Prof. Geol., Palaeont. und Lagerstättenlehre, Montanistische Hochsch. Leoben (Österreich). *Palaeont. Kreidefaunen. Geol. der Kohle.* ⓓ Leipzig 1899.

Petrbok, Jaroslav, Národní Mus. Praha (C.S.R.), Pod Slovany No. 9. *Biol. u. Bionomie der rezenten Najaden, rezenten, holocaenen u. pleistrocaenen Gastropoden.* ⓟ Recente u. fossile Land- u. Süßwassermollusken der ganzen Welt. o

Petrén, Karl (1868), Prof. innere Med. Univ. Lund (Sverige). *Neurol.: sensorische Bahnen des zentralen Nervensystems. Stoffwechselstörungen bei Diabetes. Faktoren des Blutzuckerwertes. Faktoren der Azidose bez. Ketose.* ⓓ Lund 1896.

Petrescu, Constantin (1879), Ass.-Chef des travaux au Labor. de Botan., Ass. provis. au Labor. de Physiol. méd. Jassy (Roumanie), Str. Carol. Univ. *Phanérogames systématiques, associations des phanérogames avec les champignons.* ⓟ Phanérogames en associations avec des champignons.

Petri, Else (1887), Ass. Berlin (Deutschland), Pathol. Inst. Univ. Charité, Schumannstr. *Pathol. Anat. u. Histol. der Vergiftungen.* ⓓ Berlin 1918.

Petri, Lionello (1875), Dir. Staz. di Patol. veg. Roma 30 (Italia), Via S. Susanna 13. *Patol. e Fisiol. vegetale. Malattie dell'olivo.* ⓓ Roma 1925.

Petrie, George Ford (1874), Bact.-in-Charge, Serum Department Lister Inst. Elstree, Herts. (England). *Therapeutic sera, immunol.* ⓓ Aberdeen Univ. 1898.

Petřík, Josef (1894), Priv.Doc., interimistischer Leiter des Physiol. Inst. der med. Fac. in Brno (C.S.R.), Údolní 73. *Serol. der Winterschläfer, H-Ionenkonzentrationsregulierung bei Evertebraten, Röntgenstrahlenwirkungen.* ⓓ Praha 1917.

Petrik, Milivoj, (1894), Sanitary Engineer Inst. and School of Hygiene. Zagreb (S.H.S.), 14 Mlinarska cesta. *Analysis of water.* ⓓ Cambridge 1926.

Petronievics, Branislav (1875), Prof. an der Univ. Beograd (S.H.S.), Novi Univ. *Palaeont. Archaeopteryx, säugetierähnliche Reptilien (Cynodontia), primitive Proboscidia (Moeritherium und Palaeomastodon); Abstammungslehre (Gesetze der organischen Entwicklung).* ⓓ Leipzig 1898.

Petropawlowsky, W. P. (1889), Doc. Physiol. Labor. Univ. Perm (U.d.S.S.R.). *Electrophysiol. Cathodendepression am Nervenstamm. Empfindlichkeit und Bewegung unter der Wirkung der Nervenpolarisation.*

Petrov, Gabriel G. (1881), Prof. Omsk (U.d. S.S.R.), Sibirisches Inst. für Land- und Forstwirtschaft. *Physiol. der Ernährung höherer Pflanzen, Assimilation des Stickstoffwechsels.* ⓓ Univ. Moskau 1917.

Petrović, Joso, Dr., Musealkustos Zem. Muzej. Sarajevo (S.H.S.). *Anthrop.* o

Petrunkevitch, Alexander (1875), Prof. of Zool., Hon. A. M., Ph.D., Hon. D.Sc. New Haven, Conn. (U.S.A.), Osborn Zool. Labor., Yale Univ. *Arachnol; Spiders, their structure, relationship, systematics, geographical distribution, paleont., instincts. Microphotography.* ⓓ Moscow 1898.

Petrunkin, Michail Leontiewitsch (1891), Áss. biochem. Abteilung Inst. für experim. Med. Leningrad (U.d.S.S.R.), Lopuchinskaja 12. *Fermente: Verbindung der Eiweißstoffe mit Alkaloiden, tierischen Basen, Säuren.*

Petruschky, Johannes (1863), Prof., o. Hon.Prof., Dir. Hygien. Inst. der techn. Hochsch. Danzig. Danzig-Langfuhr, Baumbachallee 5. *Hygiene, Bact., Immunität.* ⓓ Königsberg i. Pr. 1888. ⓟ Bakterienkulturen.

Petry, Edward J., South Dakota Biol. and Geol. Survey. Brookings, S.D. (U.S.A.), 625, Twelfth Ave. *Economic botan., plant breeding, plant pathol., physiol., taxonomy.* o

Petry, Loren C., Prof. of Botan., Coll. of Agric., Cornell Univ. Ithaca, N.Y. (U.S.A.). *Paleobotan.* o

Petschacher, Ludwig (1889), Priv.Doc. innere Med. Univ. Innsbruck (Österreich), Templestr. 2. *Serumeiweißkörper, Stoffwechsel.* ⓓ Wien 1913.

Pettinari, Vittorio (1901), Ass. Ist. Patol. chir. Univ. Milano 14 (Italia), Viale Regina Margherita 77. *Tozicité des champignons supérieurs. Histol., physiol. de l'ovaire tel que glande endocrine. Greffe des glandes de l'ovaire et des glandes surrénales. Défenses de l'organisme vis-à-vis à certaines infections mycol.*

Pettinger, Nicholas Albert (1901). Urbana, Ill. (U.S.A.), 110 Old Agric. Bldg., Univ. of Illinois. *Plant Breeding and Genetics.* ⓓ Univ. of Illinois 1924.

Pettit, Auguste (1869), Membre de l'Académie de méd., Prof. à l'Inst. Pasteur, Paris (France), 70 rue Jullien Vanves. Délégué auprès du Gouvernement Canadien (mission temporaire). *Spirochètes, cancer, immunité cellulaire.* ⓓ D.Sc. Paris 1896, D.M. Paris 1899. ⓟ Spirochæta icterohemorragiæ.

Pettit, Rufus H. (1869), Prof. of Entomol., Entomol. of Michigan Agric. Experim. Station. East Lansing, Mich. (U.S.A.). *Control of injurious insects.* ⓓ Cornell Univ. 1895.

Pevalek, Ivo (1893), Prof. extraord. an der Forst- und Landwirtschaftl. Fac. der Univ. und Priv.Doc. der philosophischen Fac. (Systematik u. Geobotan.). Zagreb (S.H.S) Jurišić 14. *Süßwasseralgen (spec. Desmidiaceen). Flora und Vegetation von Jugoslavien.* ⓓ Zagreb 1917. ⓟ Exsiccate aus Jugoslawien.

Pevzner, Moïsej Isaak, Prof. Univ. Smolensk (U.d.S.S.R.). *Pathol.* o

Peyer, B., Dr., Priv.Doc. d. Univ. Zürich 1 (Schweiz). Wohllebgasse 4. *Zool.* o

Peyron, Dr., Prof. Ecole d. Méd. Marseille (France). *Anat. Pathol.* ○

Peyronel, Beniamino (1890), Vice Dir. Stazione di Patol. Vegetale. Roma 30 (Italia), Via S. Susanna 13. *Patol. vegetale, Micol.*

Pezzali, Giulio, Lib. doc. R. univ. degli studi. Genova (Italia). *Patol.* ○

Pfannenstiel, Wilhelm H. J. (1890), Dr. med., Ass. am Hygien. Inst., Bact. Abteilung. Münster i. Westf. (Deutschland), Roxelerstr. 20. *Bact. und Serol. der säurefesten Bazillen. Stabilitätsveränderungen des Serums, Serum-Bact.* ⓓ München 1914.

Pfeffer, Johann Georg (1854), Hon. Prof. der Zool. an der Univ. Hamburg (Deutschland), Zool. Staatsinst. *Zoogeographie; Wirbeltiere, Mollusken; Cephalopoden.* ⓓ Berlin 1877.

Pfeiffer, Hans (1890), Dr. phil. Bremen I (Deutschland), Wilhelmstr. 7. *Systematik der Cyperaceen; kausale Gewebeanat. der Pflanzen (Entwicklungsmechanik der Pflanzengewebe), Entwicklungsphysiol., physikochem. Erforschung der Protoplasten, spec. pH- und rH-Messungen lebender Gewebe.* ⓓ Washington, D.C. 1919. ⓟ Herbarpflanzen: Cyperaceen, hauptsächlich Südamerikas.

Pfeiffer, Hermann (1877), o. ö. Prof. für allg. und experim. Pathol. und Vorstand ders. Lehrkanzel, Univ. Graz (Österreich), Universitätsplatz 4. *Serol., Bact., Pathol. general.* ⓓ Wien 1901.

Pfeiffer, Norma E., Morphol., Boyce Thompson Inst. Yonkers, N.Y. (U.S.A.). *Microchem. of plants; morphol.; effect of light.* ⓓ Univ. of Chicago 1913.

Pfeiffer, Richard (1858), Prof. Dr., GMR., Univ. Breslau (Deutschland), Tiergartenstr. 74. *Hygien. und Bact. Infektion und Immunität, Entdeckung der spezifisch bakterienlösenden Immunstoffe, deren Benutzung zur Serodiagnostik. Aktive Immunisierung bei Cholera, Typhus, Entdeckung der Aetiol. der Influenza und der Endotoxine.* ⓓ Berlin 1880.

Pfeil, Erich (1894), Biol. Reichsanst., Lab. f. prakt. Bodenkunde. Berlin-Steglitz (Deutschland), Bergstr. 11. *Bodenchem. u. -physik, Bodenbact.* ⓓ 1918.

Pfeiler, Willy (1881), ao. Prof. für Tierhygiene, Leiter der Virusforschungsanstalt. Jena (Deutschland). *Züchtung des Maul- und Klauenseuche-Erregers; Versuche zur Heilung der Bornaischen Krankheit; experim. Therapie auf dem Gebiete der Maul- und Klauenseuche. Heilung der Blutvergiftungen mit Jodcer-Verbindungen; septischer Abort, Puerperalfieber, Krebs. Zellulartherapie bei spezifischen Infektionskrankheiten. Ansteckende Euterentzündung in Schleswig-Holstein; Hühnertyphus.* ⓓ Gießen 1904. ⓟ Bakterienstämme. Praep. aus dem Gebiete der experim. Therapie.

Pfister, Hilta Ines Christina (1899), Ass. Lect., Physiol. Department, Univ. Birmingham (England), 116 Moseley Road. *Histol. Experim. Physiol.*

Pfuhl, Wilhelm (1889), ao. Prof. Greifswald (Deutschland), St. Georgsfeld 1. *Anat. Histol. der Leber; Wachstum; Muskelmechanik.* ⓓ Berlin 1914.

Pfurtscheller, Paul (1855), Prof. emer. Wien III (Deutsch-Österreich), Streichergasse 10. *Zool. Wandtafeln: Anat. des Tierkörpers.* ⓓ Wien 1878.

Phelps, Lillian Aline (1902), Instr. in Zool. Ithaca, N.Y. (U.S.A.), Zool. Labor., Cornell Univ. *Microdissection of ameba and of ameban nuclei. Regeneration in ameba.* ⓓ Univ. of Kansas 1924.

Philibert, André Paul Henri (1875), Prof. Agrégé, Fac. de Méd., Paris (France), 4 avenue Hoche. *Bacille tuberculeux. Conditions de l'Infection tuberculeuse. Virus Cytotropes.* ⓓ Paris 1908.

Philip, Cornelius B. (1900), Ass. Entomol. Bozeman, Mtn. (U.S.A.), Montana Agric. Experim. Station. *Cutworm: population, development, ecol. study. Experim. with oil emulsions as insecticidal sprays. Developmental and taxonomical studies on the Tabanidae, or horseflies and deerflies. Hydrogen-ion activity in natural waters.* ⓓ Univ. of Minnesota 1925. ⓟ Horseflies and deerflies, especially from Minnesota and Montana.

Philippson, Maurice, Dr. sc., Prof. Univ., Dir. Lab. de physiol. des animaux. Bruxelles (Belgique), 57, rue d'Arlon. *Physiol. générale.* ○

Philiptschenko, Jurius (1882), Prof. Univ., Chef Bureau für Genetik u. Eugenik Russ. Akad. d. Wiss. Leningrad (U.d.S.S.R.). *Variabilitätslehre, Vererbung der quantitativen Merkmale; Morphol. und Systematik der Apterygoten.* ⓓ 1917.

Phillips, Charles (1890), Prof. of Pathol. Med. Coll. Richmond, Va. (U.S.A.). *General pathol., histol., anat., physiol.* ⓓ Virginia 1916.

Phillips, Everett Franklin (1878), Prof. of Apic., New York State Coll. of Agric., Cornell Univ., Ithaca, N.Y. (U.S.A.), Roberts Hall. *Apiculture and related subjects.* ⓓ Univ. of Pennsylvania 1904.

Phillips, Thomas G. (1887), Prof. of Agric. and Biol. Chem. and Chem. in the Experim. Station, Univ. of New Hampshire. Durham, N.H. (U.S.A.). *Plant Biochem.* ⓓ Univ. of Chicago 1918.

Phillipps, William John (1893), Mus. Sc. Ass. Wellington (New Zealand), Dominion Mus. *Fishes.*

Philp, Guy L. (1890), Ass. Pomol. in Experim. Station. Davis, Cal. (U.S.A.), Univ. Farm. *Pollination studies with deciduous fruits. Deciduous fruit varieties.* ⓓ Ore. Agric. Coll. 1916. ⓟ Fruit variety collections for scion or grafting wood.

Philpott, Alfred (1871). Nelson (New Zealand), Cawthron Inst. *Lepidoptera: Hepialoidea and Micropterygoidea.* ⓟ New Zealand Lepidoptera.

Phipps, Ivan Francis (1902), Geneticist and Plant Breeder at the Waite Research Inst., Univ. of Adelaide, South Australia. Ithaca, N.Y. (U.S.A.), Dept. of Plant Breeding, Cornell Univ. *Inheritance in Zea Mais and of virescent seedlings in Maize.* ⓓ Melbourne 1925. ⓟ Genetical material of Maize.

Phisalix, Marie, Labor. d'Herpétol. du Mus. National. Paris V (France), 57, rue Cuvier. *Venins et les animaux vénineux. Protozoaires pathogènes.* ⓓ Paris 1900.

Phocas, Alexandre (1873), Dr. en Méd., Pharmacien de 1a Classe, Prof. agrégé de Chim. biol. à l'Univ. d'Athènes, Chef de Labor. de Pharmacol. Athène (Grèce), Rue Hippocrate 101. *Chim. biol., Pharmacol.* ⓓ Paris et Athènes.

Phokine, Alexandre D., Mus. d'Histoire Nat. Viatka (U.d.S.S.R.). *Mycol.* ○

Photakis, B., Chef du Lab. d'Anat. pathol. à l'Univ. Athènes, Charocopon, près Athènes (Grèce), Rue Nicolara ○

Pia, Julius (1887), Kustos am Naturhistorischen Mus. und Priv.Doc. für Palaeont. an der Univ. Wien I (Deutsch-Österreich), Burgring 7. *Fossile Thallophyten, fossile Kephalopoden, Gesteinsbildung durch Organismen, Variabilität der fossilen Organismen, biol. Grundlagen der Palaeont., Fauna der Trias und des Lias.* ⓓ Wien 1911. ⓟ Fossile Algen.

Pianese, G., Prof. Univ. Napoli (Italia). *Anat.* ○

Dal Piaz, G., Prof. Univ. Padua (Italia). *Paleont.* ○

Picard, François (1879), Prof. à la Fac. Sc. Paris (France), Labor. d'Evolution des êtes organ. 105, Bd Raspail. *Biol. des Insectes.*

Picbauer, Richard (1886), Mykol. der phytopath. Sektion der Mähr. Landes-Versuchsanstalt. Brno (C.S.R.), Akademická 6. *Mykol., mikroskopische Pilze, ihre geographische Verbreitung.* ⓓ Brno 1925. ⓟ Mikroskopische Pilze.

Picchi, Luigi (1868), Incaricato di istol. patol. e aiuto nell'Ist. di Anat. Pat. Firenze (Italia), 26 Via Pandolfini. *Tumori.*

Piccinini, Guido Maria (1879), Prof. tit. cattedra di Farmacol., Dir. del Labor. annesso Univ. Modena (Italia), Via S. Eufemia. *Farm. sperimentale, Chimica Biol.* ⓓ Bologna 1925.

Picciolie, Lodovico (1867), Prof. Selvicoltura, alpicoltura e tecnol. del legna. Firenze (Italia), R. Ist. superiore agrario e forestale. *Biol. fiorale. ecol. delle piante foraggere.* ⓓ Firenze 1887.

Pick, Ernst P. (1872), Dr. med, o. Prof. Pharmacol. an der Univ., Vorstand des Pharmacol. Inst. der Univ. Wien VIII (Deutsch-Österreich), Albertgasse 34. *Diurese, Leberfunktion für den Kreislauf und die Diurese, Angriffspunkte der Schlafmittel.* ⓓ Prag, deutsche Univ. 1896.

Pick, Ludwig (1868), Hon.Prof. med. Fac. Univ., Abteilungsdir. pathol.-anat. Abteilung Städtisches Krankenhaus am Friedrichshain. Berlin NW 6 (Deutschland), Philippstr. 21. *Pathol. Anat. und Histol., allgemeine Pathol.* ⓓ Leipzig 1892.

Pickel, V. O., Doc. Landwirtschaftl. Inst. Kraznodar, Kaukasus (U.d.S.S.R.), Novaja 107. *Entomol., Bekämpfung der Schädlinge.* o

Pickens, Andrew Lee (1890). Greenville, S.C (U.S.A.), 202 Grove Street. *Vertebrates of Southeastern United States. Trichodina steinii.* ⓓ Univ. of Virginia 1924.

Pickett, Fermen Layton (1881), Prof. and Head of Department of Botan., State Coll. of Washington. Pullman, Wash. (U.S.A.). *Taxonomy and Ecol. of Ferns and Mosses; Effects of Extreme Desiccation.* ⓓ Indiana Univ. 1915.

Pico Estrada, O., Prof. Univ. Rosario (Argentina). *Fisiol.* o

Picqué, Robert Léon (1877), Prof. d'Anat., Fac. de Méd. Bordeaux. Talence, Gironde (France), 47 Cours Merlin. *Anat. chirurgicale, Embryol.*

Pidoplitschko, Nikolaj M. (1904), wiss. Mitarbeiter d. Botan. Mus. u. d. Herbariums d. Ukrain. Akad. d. Wissenschaften. Kiew (U.d.S.S.R.). *Mycol. Floristik.*

Piech, Kazimierz (1893), Adjunkt Labor. Botan. Janczewskianum Univ. Kraków (Polska), Aleja Mickiewicza 21. *Physiol. Pflanzenanat., Zytol. der Pflanzen.* ⓓ Kraków 1922.

Pieper, Ernst (1887), Dr. med., wissensch. Mitglied Hauptgesundheitsamt der Stadt Berlin. Berlin-Tempelhof (Deutschland), Hohenzollernkorso 37 b. *Bact. Diagnostik, Serol. Wa. R., Hygiene.* ⓓ Greifswald 1914.

Pieper, John Jacob (1886), Ass. Prof. of Agronomy. Urbana, Ill. (U.S.A.), Agronomy Dept., Univ. of Illinois. *Farm Crops.* ⓓ Univ. of Illinois 1916.

Pieraguoli, L., Prof. Univ. Pisa (Italia). *Paleont.* o

Pierantoni, Umberto (1876), Prof. o. Anat. e Fisiol. compar. Univ. Napoli (Italia), Galleria Umberto I, 27. *Simbiosi ed organi simbiotici nei metazoi: Animali luminosi e biofotogenesi. Protodrilus.*

Pierce, William Dwight (1881), Consulting Entomol. Banning, Cal. (U.S.A.), *Rhynchophora, Strepsiptera, life history, habits, classification. Cotton Boll Weevil. Economics of Cotton. Relations of climate and environment to the reactions of life. Sanitary Entomol. Reptilia, desert fauna and flora.* ⓓ George Washington Univ. 1917. ⓟ Speciments of desert and mountain reptiles, spiders, scorpions, insects, plants.

Piéron, Henri (1881), Prof. de Physiol. des Sensations au Coll. de France. Le Vésinet, Seine et Oise (France), 52 Route de la Plaine. *Physiol. des sensations et psychophysiol.*

Pierret, Prof. Univ. Lille (France). *Hygiene, Bact.* o

Piersanti, Carlo (1888), Prof. Jesi, Ancona (Italia), R. Ist. Tecnico. *Molluschi terrestri. Letargo delle Helix e resistenza all'inanizione.* ⓓ Bologna 1912. ⓟ Conchiglie terrestri e marine.

Piersol, William Hunter (1873), Prof. of Histol. and Embryol., Univ. of Toronto (Canada). *Vertebrate Histol. and Embryol., especially human. Embryol. of Amphibia.* ⓓ Toronto 1899. ⓟ Eggs and larvae of Amblystoma maculatum, Amb. Jeffersonianum, Plethodon erythronotus for eggs and larvae of other Urodeles.

Pierson, Charles J. (1866), Prof. of Zool. and Head of Department of Zool. and Agric., Univ. of Maryland. College Park, Md. (U.S.A.). *Vertebrate Embryol.* ⓓ Univ. of California 1916.

Pieters, Adrian John (1866), Agronomist, Clover Investigations, U S. Department of Agric. Home: Washington, D.C. (U.S.A.), 7206 Blair Road. Office: Washington, U.S. Dept. of Agric. *The clovers and related legumes.* ⓓ Univ. of Michigan 1915.

La Pietra, Michele (1881). Sansevero (Italia), Via Montebello 19. *Staurogamia.* ⓓ Portici 1921.

Pietsch, Albert (1889), Kommissar für Naturdenkmalpflege. Wensickendorf bei Berlin (Deutschland). *Pflanzenphysiol. und Naturschutz. Reizphysiol. Wirkung allerkleinster Reize (Arndt-Schulzsche Reizregel).*

Pietschmann, Victor (1881), Dr., Reg.R., Kustos I. Klasse am Naturhistorischen Mus. Wien IV (Deutsch-Österreich), Favoritenstr. 38. *Systematik; Ichthyol. Selachii. Japanische Plagiostomen; Fische.* ⓓ Wien 1904.

Pigorini, Luciano (1882), Dir. R. Sta. Bacol. Sperim., Lib. Doc. Chim. Fisiol. Padova (Italia). *Fisiol. e chimica fisiol. degli insetti, normale e patol., sviluppo embrionale.* ⓟ Preparati anat. e anat.-patol. del Bombyx mori.

Pigulewsky, Sergius Wladimirowitch (1899), Ass. Histol. Inst. Taschkent (U.d.S.S.R.). *Parasit. Chromosomen, Crossing over.* ⓟ Helminthenpraep.

Piiper, Johannes (1882), Prof. Zool., Dir. des Zool. Inst. u. Mus. Univ. Tartu (Estland). *Embryol. Entwickelung der Wirbelsäule der Vögel.* ⓓ Leningrad 1913. ⓟ Vogelembryonen.

Pike, Frank Henry, Coll. Prof. New York, N.Y. (U.S.A.), 510 Audubon Ave. *Physiol., experim. Biol.* o

Pilát, Albert (1903), Ass. des Botan. Inst. der Karls-Univ. Prag II (C.S.R.), Na Slupi 433. *Mykol.: Polyporaceae, Corticiaceae, Cyphellaceae, Phytopathol.* ⓟ Pilzexsiccate (Hymenomycetes): Polyporaceae, Hydnaceae, Corticiaceae, Cyphellaceae, Phylacteriaceae.

Pilat, M. W., Prof., Dir. Histol. Labor. Med. Inst. Astrachan (U.d.S.S.R.). o

Pilger, Robert (1876), Dr., ao. Prof. an der Univ. Zweiter Dir. des Botan. Gartens und Mus. in Berlin-Dahlem (Deutschland). *Botan., Systematik (Gramineae, Coniferae, Plantago) und vergl. Morphol.* ⓓ Berlin 1898.

Pilla, R., Prof. Univ. Porto Alegre (Brasil). *Physiol.* o

v. Pillich, Franz (1876), Magister pharm. Simontornya (Ungarn). *Insekten: Lepidopt., Coleopt., Hymenopt., Hemiptera, Diptera. Plantae: Phanerogam.*

Pillsbury, Albert Elliot (1903), Ass. in Entomol. N.H. Agric. Experim. Station. Durham, N.H. (U.S.A.). *Life history of Pissodes strobi.* ⓓ Univ. of New Hampshire 1926.

Pilon, Henri, Prof. tit. Ecole Pharm. Montreal (Canada). *Botan.* o

Pinckney, John Stuart (1901), Junior Entomol.: U.S. Dept. of Agric., Bureau of Entomol. Wichita, Kan. (U.S.A.), U.S. Entomol. Labor., 126 South Minneapolis Ave. *Sereal and Forage Crop Insects.* ⓓ Clemson Agric. Coll. 1921.

Pincussen, Ludwig (1873), Dr. med. et phil., Dir. der Biol.-chem. Abteilung des städtischen Krankenhauses am Urban. Berlin-Wilmersdorf (Deutschland), Uhlandstr. 110. *Biochem. des Stoffwechsels, Fermentmethoden, Biol. der Lichtwirkung.* ⓓ Phil. 1897, med. 1910.

Pindar, L. Otley (1870). Versailles, Ky. (U.S.A.). *Ornithol., geographic distribution and migration.* ⓓ Med. Coll. of Ohio, Cincinnati, Ohio 1891.

Pinewitsch, Lydia (1888), Ass. Labor. f. Anat. u. Physiol. der Pflanzen Landwirtsch. Inst., pflanzenphysiol. Abt. der Station f. Akklimatisation der Pflanzen (Detskoje Sseló), Leningrad (U.d.S.S.R.), Uliza Krasnjich Zorj 75, Wohn. 51. *Chlorose bei Pflanzen. Quantitative Untersuchung üb. d. Farb-*

stoffgehalt d. Pflanzen im Zusammenhang mit Ernährungsbedingungen.
Piney, Alfred (1896), Dir. Inst. of Pathol. Charing Cross Hospital, London W.C. 2. Hendon, London N.W. 4 (England), 53 Foscote Road. *Haematol.* ⓓ Birmingham 1918.
Ping, C., Southeastern Univ. Lect. Nanking (China). *Zool.* ○
Pinhey, Kathleen Frances (1901), Demonstrator, Dept. of Zool., McGill Univ. Montreal (Canada). *Physical chem. of haemocyanin.* ⓓ McGill Univ. Montreal 1922.
Pinkhof, Meijer (1892), Dr., 1. Ass. für Pflanzenphysiol. an der Univ. von Amsterdam (Holland), Plantage Muidergracht 27. *Pflanzenphysiol., Reizphysiol.; Theoretische Biol.; Systematische Zool.; Araneae und Opiliones in Westeuropa.* ⓓ Amsterdam 1925.
Pinkus, Felix (1868), Dr., a.o. Prof. an der Univ. Berlin (Deutschland), Lützowstr. 65. *Anat. der Haut. Syphilidol.* ⓓ Freiburg i. Br. 1894.
Pinoy, Pierre Ernest (1873), Prof. d'Histoire natur. méd. et de parasit. à l'Univ. d'Alger (Afrique). *Cryptogamie (Myxomycètes, Champignons pathogènes). Bact. Les champignons dans le sang Séro.—Diagnostic desmycoses. Myxomycètes.* ⓓ Paris.
Pinto, Cezar (1894), Ass. Inst. Oswaldo Cruz. Rio de Janeiro (Brasil), Caixa Postal 926. *Med. Entomol.; Reduviidae; Culicidae; Phlebotomus species; Hirudinea.* ⓓ Rio de Janeiro 1919. ⓟ *Material of Med. Entomol.*
Pinto de Aguiar, Alb. Pereira, Prof. Univ. Porto (Portugal). *Fisiol., Patol.* ○
Pinto Nunes, Joaquim (1901), Ass. d'Histol. Fac. Med. Porto (Portugal), R. Herois de Chaves 812. *Histol. de l'Ovaire.* ⓓ Porto 1925.
Pintner, Theodor (1857), Dr., o.ö. Prof. an der Univ. in Wien; derzeit provisorischer Leiter des I. Zool. Inst. der Wiener Univ. Wien IX/1 (Österreich), Liechtensteinstr. 61. *Cestoden; parasitische Plathelminthen; Parasiten.* ⓓ Graz 1888.
Piquemal, Paul Lucien, Moniteur de Pathol. générale et expérim. Toulouse (France), 102, Allée St. Agne. ⓓ Toulouse 1925.
Pires de Lima, Joaquim Alberto (1877), Prof. o., Dir. Inst. d'Anat. Fac. Méd. Porto (Portugal), R. Alvares Cabral 348. *Anat., Tératol., Anthrop.* ⓓ Porto 1903.
Pirie, James Hunter Harvey (1878), Bact. and Pathol. (Research), South African Inst. for Med. Research. Johannesburg (S. Africa), Box 1038. *Bact. and Pathol., African Diseases.* ⓓ Edinburgh 1902.
Pirotta, Pietro Rosswaldo (1853), Prof. Fisiol. veget. Univ., Dir. Ist. Orto Botan. Roma (Italia), Via Milano 75. *Fisiol. veget., micol. fitopatol. fisiol.*
Pirquet, Clemens (1874), Prof. der Kinderheilkunde Univ. Wien VIII (Deutsch-Österreich), Alserstraße 21. *Tuberkulose; Allergie; Ernährungskunde.* ⓓ Graz 1900.
Pisarev, V. E., Doc. Staatl. Univ. Inst. f. angew. Botan. Leningrad (U.d.S.S.R.), ul. Herzena 44. *Selektion des Getreides.* ○
Pisek, Arthur (1894), Priv.Doc. Ass. Botan. Inst. Univ. Innsbruck (Deutsch-Österreich), Sternwartestraße. *Karyol. und Fortpflanzung der Kormophyten. Phototropismus.* ⓓ Innsbruck 1920.
Piskernik, Angela (1886), Prof. der Botan. Ljubljana (S.H.S.), Janez Trdinova 8/III. *Pflanzenbiol. und Oekol.* ⓓ 1914.
Pi-Suner, Augusto (1879), Prof. Physiol. Fac. Med. Barcelona (España), Cameros 3-10. *Sensibilité interne, regulation trophique.* ⓓ Barcelona 1899.
Pi-Sunner, Jaume (1903), Adjoint à la Chaire de Thérapeutique. Barcelona (España), Cameros 3. *Metabolisme. Pharmacol.* ⓓ Paris 1925.
Pittaluga, Gustavo (1876), Prof. o. Fac. Méd. Univ. Madrid (España), Blanca de Navarra 4. *Parasit., Protozoaires pathogènes, Hématol. Diptera.* ⓓ Rome 1901. ⓟ *Préparations de sang (Hématol., Parasit. du sang) et des organes hématopoïétiques.*
Pittaluga y Fattorini, Gustavo, Priv.Doc. Univ. Fac. med. Madrid (España). *Parasit. Pathol. tropica.* ○
Pittard, E., Prof. Dr. Genève (Suisse), Chemins des Cottage 36. *Anthrop.* ○
Pittier, Henri François (1857), Dir. du Mus. Commercial. Caracas (Vénézuéla, Amérique du Sud). *Botan. systématique et technique, Ecol.* ⓓ Lausanne 1885. ⓟ *Plantes de Venezuela.*
Pittioni, Emanuel (1879). Wien XVIII Österreich), Simonygasse 2. *Lepidopterol., Faunistik.*
Pittman, Don Warren (1891), Associate Agronomist, Utah Agric. Coll. Logan, Utah (U.S.A.). *Soils in relation to plant growth.* ⓓ Utah Agric. Coll. 1916.
Piveteau, Jacques Henri Jean (1898). Paris (France), 77, rue Notre-Dame des Champs. *Paléontol. des Vertébrés.*
Plagge, Homer Henry (1894), Ass., Pomol. section, Iowa Agric. Experim. Station. Ames, Ia. (U.S.A.), Iowa State Coll. *Eperim. pomol., preming, fertilizing, spraying, fruit breeding.* ⓓ Ames, Iowa 1916.
Plakidas, Antonios, G. (1895), Research Ass., Dept. of Plant Pathol., Univ. of California. Berkeley, Cal. (U.S.A.). *Diseases of the strawberry (Fragaria). ,,Yellows", a virus disease of the strawberry.* ⓓ Univ. of California 1924. ⓟ *Strawberry plants.*
Plančić, Josip (1888), Ass. am Inst. für angew. Zool., Leiter der ornithol. Abteilung. Zagreb (S.H.S.), Kačićeva ul. 9. *Vogelzug durch das Balkangebiet. Nutzen und Schaden der Vögel.*
Plank, Harold Kauffman (1891), Associate Entomol. in Charge of Sugar Cane Borer Work for The Cuba Sugar Club Experim. Station of The Tropical Plant Research Foundation. Central Baraguá, Baraguá (Provincia de Camaguey, Cuba). *Sugar cane moth stalkborer, Diatraea saccharalis Fabr., under Cuban conditions, and its control by parasites and by poisonous dusts.* ⓓ The Pennsylvania State Coll. 1914. ⓟ *Eggs, larvae, pupae, and adults of the Cuban variety of Diatraea saccharalis Fabr.*
Plantefol, Lucien (1891), Préparateur au Coll. de France. Paris (France), Place Marcellin Berthelot. *Physiol. végétale: influence des phénomènes d'hydratation sur la forme et la vie des végétaux. Ecol. des Mousses.* ⓓ Paris 1923.
Planteneux, Edmond (1882), Chef de labor. à l'Inst. Pasteur d'Algérie. Alger (Algérie). *Rage et maladies microbiennes des animaux.* ⓓ Toulouse 1926.
Plasaj, Stjepan, Dr., o. Prof. der Seuchenlehre d. Haustiere, Dir. des Inst. Zagreb (S.H.S.), Ciglana, gradska kuća VI. *Bact.* ○
Plate, Ludwig Hermann (1862), o. Prof. der Zool. und Dir. des Phyletischen Mus. in Jena (Deutschland), Zool. Inst. *Theoretische Zool., Abstammungslehre. Mollusken, spez. Chitonen. Fauna chilensis. Fauna ceylanica. Vergl. Anat. Vererbung bei Hunden.* ⓓ Jena 1885. ⓟ *Konservierte Mollusken von Chile, dem Roten Meer, Westindien, Ceylon.*
Plattner, Friedrich (1896), Priv.Doc., Ass. am Physiol. Inst. Innsbruck (Österreich), Schöpfstr. 41. *Physiol. der Herznerven-Stoffe.* ⓓ Innsbruck 1922.
Plaut, Menko (1885), Dr., Saatzuchtdir. der Aug.-Knoche-Wallnitz G. m. b. H. Hamersleben, Kr. Oschersleben (Deutschland). *Pflanzenanat. (Wurzelperiodizität); Keimpflanzung; Samendesinfection (Beizung); Rüben- und Getreidezucht; Vegetationsversuche.* ⓓ Marburg 1909. ⓟ *Getreidesorten.*
Plečnik, Janez, o.ö. Prof., Dir. des Anat. Inst. Med. Fac. Univ. Ljubljana (S.H.S.), Zaloška cesta. *Pathol. Anat.* ○
Plehn, Albert (1861), a.o. Prof. an der Univ. Berlin; Ärztl. Dir. des Urbankrankenhauses. Berlin W 62 (Deutschland), Burggrafenstr. 4. *Blutlehre; Parasitol. der Malaria; Parasitol. der tropischen Hautkrankheiten.* ⓓ Kiel 1886. ⓟ *Malariapraep.*

Plehn, Marianne (1863), Prof., Dr. phil., Konserv. an der Bayer. Biol. Versuchsanstalt für Fischerei in München (Deutschland), Veterinärstr. 6. *Fischkrankheiten.* Ⓓ Zürich, 1896.

Plenk, Hanns (1887), Priv.Doc. f. Histol., Ass. Histol. Inst. Univ. Wien IX (Deutsch-Österreich), Schwarzspanierstr. 17. *Zellgrößen; Brachiopodenentwicklung; Histol. der Muskelfasern bei Mollusken, Anneliden, Scoleciden; Argyrophiles Bindegewebe (Gitterfasern).* Ⓓ Phil. Wien 1910, med. 1919.

Plevako, D., Prof. Sil'sko-Hospod. Inst. Kamjanec-Podilsky, Ukraine (U.d.S.S.R.), Sevčento-Straße 23—25. *Pflanzenzucht.* o

Plitt, Charles C. (1869), Prof. of Botan., School of Pharmacie, Univ. of Maryland. Baltimore, Md. (U.S.A.), 3933 Lowndes Avenue. *Lichens and Pharmacognosy.* Ⓓ Sc. Academy of Sc., Baltimore 1921. Ⓟ Lichens, herbarium specimens of Med. Plants.

Plješakov, Vladimir (1893), Prosektor f. Histol. am Morphol.-biol. Inst. Univ. Zagreb (S.H.S.), *Gewebekulturen, Placenta.* Ⓓ Charkow 1919.

Ploetz, Alfred (1860), Dr. med. Herrsching bei München (Deutschland). *Ursachen von Mutationen der Erbanlagen, besonders durch Alkohol bei Kaninchen. Rassenhygiene.* Ⓓ Zürich 1890.

Plońskier, Maurycy (1897), Prosektor des jüdischen Krankenhauses in Warszawa (Polska), Prosta 8. w. 41. *Tumorforschung. Experim. Karzinome. Heterotransplantation der menschlichen Karzinome.* Ⓓ Warschau 1925.

Plotnikov, Wassily (1878), Dir. Station d. Pflanzenschutzes, Doc. Univ. Taschkent (U.d.S.S.R.), Puschkin-Str. 37. *Heuschrecken, schädliche Insekten des Ackerbaues.* Ⓓ Petersburg 1901.

Plotz, Harry (1890), Attaché à l'Inst. Pasteur, Chief of Labor., American Hospital. Paris (France). 25, Rue Dutot. *Bact. Etiol. of typhus, fever. Mechanism of infection and immunity.* Ⓓ New York 1913.

Plough, Harold H. (1892), Prof. of Biol., Amherst Coll., Amherst, Mass. (U.S.A.), Dept. of Biol. *Animal Genetics (Drosophila). Experim. Embryol. (Echinodermata).* Ⓓ Columbia Univ. 1917.

Plunkett, Orda Allen, Ass. Prof. of Botan., Illinois State Normal Univ. Normal, Ill. (U.S.A.). o

Poche, Franz (1879), Dr. phil., Privatgelehrter. Wien VI (Deutsch-Österreich), Gumpendorfer Str. 36. *Systematik der Protozoa und Cnidaria; Platodaria; Allgemeine zool. Systematik; Nomenklatur.* Ⓓ Wien 1915.

Pochorecka-Lelep, Dr. phil., Priv.Doc. Poznań (Polska), Univ. Fac. de méd., r. ś. Pawła 10. *Biochim.* o

Podjapolsky, Peter P., Dr., Priv.Doc. Univ. Saratow (U.d.S.S.R.). *Spektroskopische Untersuchungen des Chlorophylls und Hämoglobins im fossilen Zustande.* o

Podkopaev, Nicolas (1892), Priv.Doc., Senior Physiol. Academy of Sc. Leningrad (U.d.S.S.R., Tuchkova nab., 2 A. *Physiol. of the central nervous system.* Ⓓ Leningrad 1914.

Podpěra, Josef (1878), Ph.D. ö.o. Prof. der allgemeinen und systematischen Botan. Naturw. Fak. der Masaryk-Univ. Brno (C.S.R.), Kounicova 63. *Systematische und phylogenetische Botan. Pflanzengeographie. Bryophyta.* Ⓓ Prag 1903.

Pöch, Hella (1893), Dr. phil. Wien IX/3 (Österreich), Freiheitsplatz 10. *Physische Anthrop. Morphol. Rassenanat. der Buschmänner.* Ⓓ Wien 1919.

van Poeteren, Nicolaas (1882), Chief of the Phytopath. Service. Wageningen (Holland). *Phytopath. and economic entomol.* Ⓓ Wageningen 1908. Ⓟ Insects, Fungi. Preparations of plant diseases.

Pogodina-Lebedewa, Barbara J. (1893), Ass. Berginst. Dnepropetrowsk (U.d.S.S.R.). *Lamellibranchiaten des Carbons.*

Pogrebow, Nikolaus Th., Comité Géol. de Russie. Leningrad (U.d.S.S.R.), W. O., Sredny Pr. 72-B. *Kaustobiolithen.* o

Pohl, Erwin Robert (1904), Division of Stratigraphic Paleont., U.S.National Mus., Washington, D.C. (U.S.A.). *Devonian Paleont.*

Pohl, Fritz (1897), Dr. phil. Cottbus (Deutschland), Kaiser-Friedrich-Str. 28 I. *Rassenkunde.* Ⓓ Greifswald 1922. Ⓟ Rassenkundliche Photographien.

Pohle, Ernst Albert (1895), Dr. med., Ass. Prof. of Roentgenol., Univ. of Michigan, Ann Arbor, Mich. (U.S.A.), Dept. of Roentgenol., Univ. Hospital. *Biol. of radiation.* Ⓓ Frankfurt a. M. 1920.

Pohle, Hermann Ernst (1892), Dr. phil., Kustos der Säugetierabteilung des Zool. Mus. der Univ. Berlin N 4 (Deutschland), Invalidenstr. 43. *Systematik und Geographie der Säugetiere.* Ⓓ Berlin 1920.

Pojariskaja, A. P., Doc. Gosudarstv. Donskoj Univ. Rostov a. Don (U.d.S.S.R.). *Anat.* o

Pojarkowa, A. I., Ass. Section des Plantes Vivantes Jardin Botan. principal. Leningrad (U.d. S.S.R.), Aptekarski Ostrov, Pessotchnaia $^1/_2$. *Oecol. des plantes, Ribes de la Russie asiatique.* o

Pokrowsky, Gabriel (1899), Ass. sericiculture Fac. Agr. Univ., Ass. Sta. Sericicole. Taschkent (U.d. S.S.R.), rue de Kafanow 1. *Agronomie. Culture du Murus. Sericiculture.* Ⓟ Morus, Maclura, Scorzonera.

Pokrowski, Georg (1901), Ass. Physikal. Inst. Techn. Hochsch. Moskau (U.d.S.S.R.). *Optik der Kolloide, Photophysik und Photochem. der Pflanzen. Lichtzerstreuung im Auge.* Ⓓ Moskau 1924.

Pokrowsky, S. W., Keeper Zool. Mus. Moskau (U.d.S.S.R.), Gerzenstr. 6. *Zool.* o

Pol, Rudolf (1879), a.o. Prof. für Pathol. an der Univ., Prosektor am Pathol. Inst. Rostock (Deutschland), Friedrich-Franz-Str. 106. *Teratol. der Extremitäten und Knochen. Die Vertebraten-Hypermelie. „Brachydaktylie" — „Klinodaktylie" — Hyperphalangie und ihre Grundlagen.* Ⓓ Heidelberg 1904.

Polák, Bohuslav, Dr., Prof. d. Pharmacol. Bratislava (C.S.R.), Pharmacol. Inst. d. Komensky-Univ. o

Polettini, Bruno (1891), Prof. Patol. Generale, Dir. Ist. di Patol. Generale Univ. Sassari (Italia). *Piastrine. Tubercolosi. Batteriol. Anafilassi.*

Polgár, Sándor (1876), Dr., Prof. Györ (Ungarn), Bisinger-Park 4. *Floristik und Pflanzengeographie. Phanerogamen-Systematik, Morphol. und Biol. Adventivpflanzen. Süßwasserflora.* Ⓓ Budapest 1913. Ⓟ Getrocknete Gefäßpflanzen und Characeen, besonders Adventivpflanzen.

Poliakov, Grigory Ivanovich (1876). Moskau (U.d. S.S.R.), Pokrovka, Lialin Pereulok 27, Kw. 1. *Ornithol.*

Poliakova, A. N., Doc. Univ. Nizny-Novgorod (U.d.S.S.R.), Sovetskaja Ploščad 8. *Pathol.* o

Poliakow, Elias (1905), Mitarb. Katheder d. Zool. der Wirbellosen. Charkow, Ukraine (U.d.S.S.R.), Artjemstr. 46. *Experim. Zool. d. Wirbellosen. Hydrozoa und Protozoa.* Ⓟ Hydroiden.

Polianska, Olga (1892), Ass. section géo-botan. du Jardin Botan. principal. Leningrad (U.d.S.S.R.). *Géographie botan. de la Russie Blanche.*

Polianskij, I. I., Prof. Staatl. Pädagog. „Herzen"-Inst. Leningrad (U.d.S.S.R.), Mofka 48-50-52. *Biol.* o

Policard, Albert (1881), Prof. Fac. Méd., Dir. Inst. d'Histol. Lyon (France), Inst. d'Histol. *Histophysiol. et histol. experim., Cytol., Culture des tissus, Histochim. normale et pathol. Le tube urinaire.* Ⓓ Paris.

Polimanti, Osvaldo (1869), o. Prof., Dir. Physiol. Instit. Univ. Perugia (Italia); Dir. Hydrobiol. Stat. am Trasimeno-See. Monte del Lago (Umbria), R. Statione Idrobiol. *Physiol. Zentralnervensystem der Wirbeltiere u. Wirbellosen. Winterschlaf. Ethol. Limnol. Chem. der Seen, Stoffwechsel.*

Poliński, Władysław (1885), Doc. de Zool. et d'anat. comparée Univ. de Cracovie et à l'Ecole Sup. Agr. et Forest. à Varsovie, Conserv. au Mus. Polonais d'Histoire Nat. Varsovie. Warszawa (Polska), Krakowskie Przedmieście 26-28, Pol. Państwowe Muz. Przyrodnicze. *Anat., systématique et distri-*

bution *géographique des Mollusques terrestres et fluviatiles. Isopoda, Tricladida, Odonata, Amphibia, Reptilia. Zoogéographie.* Ⓓ Cracovie 1911.
Politis, I., Prof. de Botan. à l'Univ. Athènes (Grèce), Rue Caraiscou 57, Le Pirée. ○
Politzer, Georg (1898), Ass. Embryol. Inst. Wien I (Deutsch-Österreich), Laurenserberg 4. *Einfluß der Röntgenstrahlen, des ultravioletten Lichtes und der Vitalfarbstoffe auf die Karyokinese und den Zellteilungsrhythmus. Zellwanderung, Pigmentwanderung. Doppelbildungen, Cyklopie, Atresien des Gehörgangs. Traubesche Membranen. Röntgenstrahlen und Regeneration.* Ⓓ Wien 1922.
Poljak, Stephan (1889), Ass. an der Neurol.-psychiatr. Klinik zu Zagreb, Kroatien (S.H.S.), Marulić. ul. 1. *Histol., Anat., Physiol. des Nervensystems. Rückenmark der Säugetiere, Sympathicuszentren; Oktavussystem (Cochlearis) der Säugetiere u. die mit ihm koordinierten motorischen Apparate des Hirnstammes, Bauprinzipien des Cochlearissystems, Projektion der Cochlea an die primären Cochleariszentren des Bulbus, Innervation der Vestibularisendstellen, Assoziations- und Projektionsfasern des Säugetierhirns.* Ⓓ Odessa 1916, Zagreb 1920.
Poll, Heinrich (1877), Prof. Dr., Dir. des Anat. Inst. der Univ. Hamburg (Deutschland). *Vererbungslehre, Anat.* Ⓓ Berlin 1900.
Pollacci, Gino, Prof. Univ. Siena (Italia). *Botan.* ○
Poller, Konrad (1896), Dr. phil., Ass. am Physiol.-chem. Inst. Univ. Würzburg (Deutschland), Kantstraße 54. *Tierische Alkaloide.* Ⓓ Würzburg 1924.
Pollock, James B., Prof. of Botan., Univ. of Michigan. Ann Arbor, Mich. (U.S.A.). *Pathol., soil biol., bact., calcareous algae.* ○
Polson, Cyril (1901), Ass. Lect. in Chem. Pathol. Manchester (England), Department of Pathol., Victoria Univ. *Morbid Anat. and Histol.: Nephritis and the function of the kidney. The pathol. of the Central Nervous System.*
Połtorzycka, Stanislasse (1893), Ass. d'Anat. Pathol. Fac. Med. Warszawa (Polska), Inst. d'Anat. Pathol., Chalubiński 5. *Fibres grillagées dans les tumeurs.* Ⓓ Petersbourg 1916.
Poluszyński, Gustav (1887), Dr. phil., Adjunkt des Zool. Inst. der Univ. Lwów (Polska), ul. Mikołaja 4. *Embryol. der parasitischen Insekten, Spermato- und Ovogenese; Vererbungsversuche an den Insekten.* Ⓓ Lwów 1922.
Pomeroy, Carl S. (1882), Associate Pomol., U. S. Dept. of Agric. Riverside, Cal. (U.S.A.), Box 586. *Selection, citrus and deciduous fruits.*
Pommer, Gustav, Univ.Prof. Innsbruck (Deutsch-Österreich). *Pathol. Anat.* ○
Pompeckj, Josef Felix (1867), Univ.Prof., G.Berg-Rat, Dir. Geol. Inst. Univ. Berlin-Schmargendorf (Deutschland), Aug.-Viktoria-Str. 64. *Palaeontol.* ○
Poncy, Robert (1875), Prof., Bibliothécaire de la Société Zool. Genève (Suisse), Rhône, 59, *Ornithol. Biol. des Echassiers et Palmipèdes du bassin du Léman.*
Pond, Samuel Ernest (1890), Ass. Prof., Physiol., School of Med. Univ. of Pennsylvania. Drexel Hill, Pa. (U.S.A.), 3433 Brighton Avenue. *Effects of radiation on living cell, effects on processes of calcification.* Ⓓ Univ. 1921.
Ponebšek, Janko (1861). Ljubljana (S.H.S.), Križevnska ulica 2. *Ornithol.* Ⓓ Graz 1891. Ⓟ Vogelbälge, Vogeleierschalen.
Ponomarev, Alexej P., Prof. Inst. f. Land- u. Forstwirtschaft, Doc. Univ. Kasan II (U.d.S.S.R.), Soldatskaja 47, W. 2. *Pflanzen-Physiol. Microbiol.* ○
Pontano, Thomas (1882), Med. en chef Prof., Priv.-Doc. de pathol. et de clinique méd. dans l'univ. de Roma (Italia), Castelfidardo 8. *Maladies d'infection. Paludisme, disenterie amibien, foie, syphilis.*
Ponzo, Antonino (1875), Lib. doc. botan. Univ. Milano (Italia), R. Ist. tecnico. *Biol. vegetale, floristica e ontogenesi nelle piante.*

Pool, Raymond J. (1882), Prof. of Botan., Chairman of the Dept. of Botan., The Univ. of Nebraska, Station A. Lincoln, Neb. (U.S.A.). *Ecol., Systematic, Phytopath. Vegetation of the Nebraska Sand Dunes. Nebraska Trees.* Ⓓ Nebraska 1913.
Pool, William Arthur (1889), Dir. Research Inst., Animal Diseases Research Association of Scotland. Edinburgh (Scotland), Movedun Inst., Gilmerton. *Animal Pathol.* Ⓓ London 1911.
Poole, Robert Franklin (1893), Associate Plant Pathol. and Associate Prof. Raleigh, N.C. (U.S.A.), State Coll. *Plant diseases. Sweet Potatoes.* Ⓓ Rutgers Univ. 1921. Ⓟ *Illustrations of sweet potato diseases.*
Poos, Frederick W. (1891), Entomol., Virginia Truck Experim. Station. Norfolk, Va. (U.S.A.), Box 881. *Insects injurious to truck crops.* Ⓓ Ohio State Univ. 1926.
Pop, Emil (1897), Ass., Inst. f. systematische Botan. Univ. Cluj (România), Str. Regala 26. *Ökol. und stratigrafische Untersuchung der rumänischen Torfmoore.* Ⓓ Cluj 1923.
Pope, Merritt N. (1883), Associate Agronomist, Barley Investigations U.S.D.A. Washington, D.C. (U.S.A.), 1306 B Street, SW. *Breeding and testing of varieties of barley. Morphol. and physiol. of the barley plant. Inheritance.*
Pope, Philip Huntley (1887), Dr. Walla Walla, Wash. (U.S.A.), U.S. Veterans' Hospital. *Herpetol., the habits, life history and systematic status of the Amphibia. Compar. anat. of the Vertebrates.* Ⓓ Pittsburgh, Pa. 1921. Ⓟ *Models for the demonstration of mitotic cell division.*
Popescu, C., Prof. Scuala sup. de agric. Bucarest (România). *Pflanzenanat., Pflanzenphysiol.* ○
Popescu-Voiteşti, Ion (1876), Prof. de Géol. et Paléont. à l'Univ. de Cluj (Roumanie). *Le Tertiaire, Nummulites des régions carpathiques roumaines.* Ⓓ Paris 1910. Ⓟ *Nummulites du bassin de la Transylvanie. Nummulites perforatus.*
Poplavsky, Hélène (1895), Ass. Inst. Histol. Wilna (Polen), Zakretowa 15. *Système nerveux, l'œil.* Ⓓ Wilno 1926.
Poplewski, Roman (1894), Chef trav. anat. Univ. Warszawa (Polska), rue Chatubirisori 5. *Anat. du visage.* Ⓟ *Photographies, concernants l'expression du visage.*
Popoff, Methodi (1881), o. Prof. der Biol. an der Med. Fak. der Univ. Sofia; z. Zt. Kgl. Bulgarischer Gesandter und Bevollmächtigter Minister in Berlin. Sofia (Bulgarien), Biol. Inst. d. Univ. *Zellphysiol., speziell Zellstimulationsfragen.* Ⓓ München 1906.
Popoff, Nicolai Wladimirowitsch (1894), Prof., Ass. der eugenetischen Abt. Inst. experim. Biol. Moskau 11 (U.d.S.S.R.), Woronzowo Pole 6. *Vererbung des Menschen (Ohr, Blut, musikalisches Talent usw.); Spektrale Untersuchungen des Blutes und anderer Pigmente, Toxikol.* Ⓓ Moskau 1917.
Popoff, Nicolas (1881), Prof. d'Histol. et d'Embryol. à la Fac. de Méd. de l'Univ. de Lausanne (Suisse), Ecole de Méd. *Ovogénèse, Spermatogénèse.*
Popoff, Porphirii Ivanowitsch (1892), Prof. Pharmacol. Kasan (U.d.S.S.R.), Vet.-Inst., Pharmacol. Labor. *Wirkung der Herzmittel auf das Froschherz bei Reizung des N. vagus.* Ⓓ Kasan 1923.
Popov, N. A., Prof. Staats-Univ. Tomsk (U.d. S.S.R.), Prospekt Timirjazewa 3. *Physiol.* ○
Popov, Peter (1888), Ass. Tropic. Instit. Moskau (U d.S.S.R.) Pogodinskaja 10. *Schädliche Insekten. Phlebotomus.*
Popović, Jovo (1882), Ingenieur, Leiter der Phytopath. Anstalt, hon. Kustos des Landesmus. Sarajevo (S.H.S.), Zemaljski Muzej. *Bekämpfung landwirtschaftl. Schädlinge.* Ⓓ Wien 1906.
Popovici, Alexandru (1866), Prof. Botan. Univ. Jassy (România), Alea Copou. *Anat. végétale et Mycol.* Ⓓ Bonn a. Rh. 1893. Ⓟ *Champignons, Ascomycetes, Basidiomycetes.*

Popovici, Elena (1897), Ass. au Labor. de Botan. de Jassy (Roumanie), Alea Copou. *Cytol. et anat. végétale.* ① Jassy 1924.

Popovici Baznosanu, A., Prof. Univ. Bucarest (România). *Zool.* ○

Popoviciu, Georg (1895), Priv.Doc., I. Ass. des Pharmacol. Inst., Cluj (Rumänien), Str.V. Babeş, 8. *Pharmacol., Biochem., Rassenfrage. Innere Sekretion. Ionen. Autonomes Nervensystem. Alkaloide. Blutchemismus.* ① Budapest 1919.

Popoviteb, Dragolioub (1877), Dir. Inst. Epidemiol. Nische (S.H.S.). *Bact., serol.* ① Paris 1905.

Popow, Iwan P., Doc. Landwirtschaftl. Akademie Timirjasew. Moskau (U.d.S.S.R.), Petrowsko-Rasumowskoje, Wysselki 30, W. 2. *Gartenbau. Pflanzenzucht.* ○

Popow, Woldemar (1899), Ass. der Anat. Rostow a. Don (U.d.S.S.R.), Anat. Inst. *Kehlkopfsäcke b. Mensch. Oligodaktylie. Proc. trochl. calcanei. Trigonum Grinfeldi.* ① Rostow a. Don 1922.

Popowa, Ariadna N. (1897), Ass. Biol. Wolga-Station. Saratow (U.d.S.S.R.). *Süßwasserinsekten, Odonata.* ⑫ Odonata.

Popowa, Galja M., Doc. der Mittelasiatischen Staatl. Univ. Taschkent (U.d.S.S.R.), Stratonowskij Tupik 6. *Agronomie. Selektion.* ○

Popowa, Helene M., Doc. am Pharmazeut. Technikum. Simferopol (U.d.S.S.R.), Feodossejskaja 2, W. 8. *Pflanzenphysiol., Mikrobiol.* ○

Popowa, Vera L. (1904), labor. of plant physiol. Inst. of Applied Botan. and New Cultures. Leningrad (U.d.S.S.R.), 44 Herzen-Str. *Waterrelations of plants, physiol. of crop plants.*

Popp, Henry William (1892), Ass. Prof. of Botan., State College, Pennsylvania (U.S.A.), Dept. of Botan. *Plant Physiol.; effect of light on plants; growth.* ① Univ. of Chicago 1926.

Popp, Otto Max (1878), Prof., Dr., Vorsteher der Versuchs- und Kontrollstation der Oldenburgischen Landwirtschaftskammer. Oldenburg i. O. (Deutschland), Sedanstr. 27. *Phosphorsäureernährung der Kulturpflanzen. Säurekrankheiten der Haustiere (Lecksucht).* ① Halle a. d. S. 1900.

Poppe, Kurt (1880), o. Prof. der med. Fak. Univ. Rostock (Deutschland), Blücherplatz (Palais). *Bact. Tierhygiene.* ⑫ Phil. Leipzig 1902, med. vet. München 1922. ⑫ Bacterienkulturen und Praep.

Poppelbaum, Hermann (1891), Dr. phil. Langen bei Frankfurt a. M. (Deutschland), Friedrich-Ebert-Straße 12. *Theoretische Vererbungslehre. Anthrop. und Zool. Untersuchungen an gynandromorphen Schmetterlingen.* ① München 1913.

Popta, Canna Maria Louisa, Dr. phil., Konserv. am Naturhistorischen Reichsmus. in Leiden, Abteilung Fische. Leiden (Holland), Vliet 41. *Ichthyol.: systematisch geographisch; anat.: embryol. und physiol. (Die Funktion der Schwimmblase der Fische.)* ① Bern 1898.

Porcher, Charles Casimir Toussaint (1872), Prof. de chim., Dir. de l'Ecole Nationale Vét. Lyon, Rhône (France), 2. Quai Chauveau. *Chim. biol.*

Porchunow, Alexander J., Prof. der Staatl. Univ. Nishnij-Nowgorod (U.d.S.S.R.), Univ. *Geobotan.* ○

Poretzky, Artemir Sergejevicz (1901), Wissenschaftl. Mitarbeiter Botan. Labor. Univ., Botan. Garten (Geobotan. Abt.). Leningrad (U.d.S.S.R.), Was. Ostr. 1, Linie 28. *Pflanzenmorphol. und -systematik, Geobotan.* ① Leningrad 1925.

Poretzky, Wadim S. (1893), Ass. Hydrobiol. Labor. Botan. Garten Univ., Adjunkt-Hydrol. Hydrol. Inst. Leningrad (U.d.S.S.R.), Wass. Ostr. 1, Linie 28, W. 4. *Bazillarien.*

Porodko, Theodor (1877), Prof. der Anat. und Physiol. der Pflanzen Univ. Botan. Labor. Odessa, Ukraine (U.d.S.S.R.), Kominternstr. 2. *Kinetik der Absterbeprozesse der Pflanzen. Geotropische Stimmungsänderungen bei den Keimlingen. Chemotropismus und Traumatotropismus der Pflanzenwurzeln.* ① Warschau 1903.

Porpeta y Llorente, Julian, Prof. Univ. fac. de med. Madrid (España). *Anat. Embryol.* ○

Porsch, Otto (1875), o. Prof. der Botan. Hochsch. für Bodenkultur. Wien VIII (Deutsch-Österreich), Zeltg. 6. *Systematik der Blütenpflanzen, Stammesgeschichte, Biol. Anat. Galeopsisarten. Spaltöffnungsapparat im Lichte der Phylogenie. Phylogenet. Erklärung des Embryosackes. Blütenbiol.* ① Wien 1901. ⑫ Diapositive Photos von Tropenpflanzen.

Porsild, Merten P. (1872), Dir. Danish Arctic Statisko, Greenland (Danmark). *Arctic plants, Eskimos.* ① København 1900. ⑫ Arctic vascul. plants.

Porta, Antonio (1874), Libero doc. in Anat. Compar. e Zool., ed in Clinica Dermosifilopatica, Univ. di Parma. San Remo (Italia), Corso O. Raimondo, 6. *Coleotteri paleartici.* ① Parma. ⑫ Coleotteri paleartici.

Porte, William S. (1891), Ass. Pathol. U.S. Dept. of Agric. Washington, D.C. (U.S.A.), Bureau of Plant Industry. *Breeding for disease resistance, tomatoes resistant to Fusarium wilt, Macrosporium Solani.* ① Univ. New Jersey, New Brunswick, N.J. 1915.

Porter, Annie, South African Inst. for Med. Research, Senior Lect. in Parasit., Univ. of the Witwatersrand. Johannesburg (S. Africa), Box 1038. *Parasit. (Med. Zool.), Protozool. and Helminthol.* ① London 1910.

Porter, Bennet A. (1892), Associate Entomol., Bureau of Entomol., U.S. Dept. of Agric. Vincennes, Ind. (U.S.A.), 2 E. Locust St. *Fruit insects. Oil sprays for the control of the San Jose scale. The codling moth, a deformity of peach fruit caused by the tarnished plant bug (Lygus pratensis) and other insects.* ① Massachusetts Agric. Coll., Amherst, Mass. 1914.

Porter, Carlos, Prof.Univ., Dir. Mus. y Labor. de Zool. appliquée. Santiago (Chile), Casilla 2974. *Zool. appl.* ○

Porter, Charles Lyman (1889), Assoc. Prof. Plant Pathol. and Plant Physiol. Purdue Univ. W. Lafayette, Ind. (U.S.A.), 484 Northwestern Ave. *Ecol. relationships of fungi.* ① Univ. of Illinois, Urbana, Ill. 1923.

Porter, Eugene L. (1881), Prof. of Physiol. Med. School Univ. of Texas. Galveston, Tex. (U.S.A.). *Reflex action.* ① Harvard 1912.

Porter, James Pertice (1873), Prof. Ohio Univ. Athens, O. (U.S.A.). *Psychol. intelligence and imitation in birds.* ○

Porter, R. Howard (1892), Prof. of Plant Pathol. Univ., Coll. of agric. and forestry. Nanking (China). *Diseases of Economic plants caused by members of the genus Cercospora; Kaoliany.* ① Ames, Ia., 1920. ⑫ Fungi in East China.

Porter, William Townsend (1862), Prof. Harvard Univ. Boston, Mass. (U.S.A.), Harvard Med. School. *Physiol.* ○

Porterfield, Willand M. jr., Prof. of Biol., St. John's Univ. Shanghai (China). ○

Portheim, Leopold, Dir. Biol. Versuchsanstalt, Abt. Botan. Wien II (Deutsch-Österreich), Prater. *Experim. Botan.* ○

Portier, Paul Jules (1866), Dr. Prof de Physiol. compar. à la Fac. des Sc. de Paris (Sorbonne), Prof. de Physiol. des Etres marins à l'Inst. océanographique, Paris 5 (France), 195, Rue St. Jacques. *Physiol. compar. Physiol. des Animaux aquatiques.* ① Paris 1923.

Portmann, Adolf (1897), I. Ass. an der Zool. Anstalt der Univ. Basel (Schweiz). *Ernährungsphysiol. während der Embryonalperiode (Blutkreislauf und Dotterresorption). Dedifferenzierung von Zellen und Geweben unter Einfluß schwacher Giftlösungen.* ① Basel 1921.

Pospelow, Wladimir (1872), Dir. Bureau für angewandte Entomol. Inst. für experim. Agronomie. Leningrad (U.d.S.S.R.), 44 Herzenstr. *Metamorphose der Insekten (Lepidoptera). Intrazellulaere Symbionten* *der Insekten.* ⓓ Univ. Moskau 1896. ⓟ Lepidoptera: Noctuidae. Coleoptera: Curculionidae. Orthoptera: Acrididae.

Posselt, Adolf (1867), Univ.Prof. Innsbruck (Deutsch-Österreich), Museumstr. 9 I. *Echinococcus.* ⓓ Innsbruck 1891.

Possipkin, Arkadi (1899), Ass. Anat. Inst. Univ. Kasan (U.d.S.S.R.), 2. Soldatenstr. 7. *Innervation der Lungen.*

Posthumus, Oene (1898). Haren bei Groningen (Holland), Rijksstraatweg A 64a. *Stelar morphol.; Pteridophytes; systematics; Filicales; Palaeobotan.: Palaeozoic and Mesozoic; Palaeozool.: otoliths of fishes; Foraminifera* ⓓ Groningen 1924.

Postma, Nicola (1901), Ass. bei der Abt. für Vergl. Physiol. des Zool. Inst. der Staatsuniv. Utrecht (Holland), Oosterstraat 6. *Zool., vergl. Physiol., Respiration, Herz, Exkretion.*

Postrigan, Sawwa A., Wiss. Mitarbeiter der Botan. Sektion d. Wissenschaftl. Komitee d. Ukraine. Charkow (U.d.S.S.R.), Jumowskaja 8. *Botan., Geobotan.* o

Potapenko, Georgij J., Doc. des Odessaer Inst. für Volksbildung. Odessa (U.d.S.S.R.), Gogolstr. 9, W. 16. *Geobotan.* o

Potenciano, Conrado, Prof. Univ. of St. Tomas. Manila (Philippine Islands). *Biochem.* o

Potonié, Horst Willi (1893),Volontärass. Preuß. Landesanstalt für Fischerei. Berlin-Friedrichshagen (Deutschland), Seestr. 109a. *Fischereibiol., Tierphysiol. Temperatur und Atmungsgeschwindigkeit von Kaltblütern. Herztätigkeit des Flußkrebses. Totenstarre glatter Muskulatur.* ⓓ Berlin 1923.

Potter, Dorothy (1896), Ass. Prof. of Physiol. Philadelphia, Pa. (U.S.A.), Women's Med. Coll. of Pennsylvania. *Biochem. Changes in the Blood in Anaesthesia.* ⓓ Edinburgh 1922.

Potter, Fredric, M. D., Dir. Pathol. Labor. Peoples-Hospital. Akron, Ohio (U.S.A.). o

Potter, George Frederick (1891), Prof. of Horticulture, Univ. of New Hampshire; Horticulturist, New Hampshire Experim. Station. Durham, N.H. (U.S.A.). *Pomol., fruit bud formation, winter injury, economics of apple production.* ⓓ Univ. of Wisconsin 1916.

Potthoff, Heinz (1895). Marburg a. d. Lahn (Deutschland), Biegenstr. 38 I. *Bact., Befruchtungsvorgänge bei Bacterien. Entwicklungsgeschichte der Algen.* ⓓ Münster i. W. 1922.

Pottier, Jacques Georges (1892), Maître de conférences adjoint, chef de travaux à la Fac. des sc. de Besançon, Doubs (France), 22, rue de Dôle, ou Chamars, l'Inst. botan. de la Fac. des sc., Jardin botan., Rue Girod de Chantrans. *Anat. et histol. des Muscinées.* ⓓ Paris 1920.

Potts, F. A., Lect., Trinity Hall's Coll. Cambridge (England). *Zool.* o

Potts, G., Prof. Univ. Bloemfontein, Orange Free States (South Africa). *Botan.* o

Pouchet, G., Dir. Adj. de Labor. Paris (France). *Physiol.* o

Poujol, Paulin Eugène Gustave (1867), Prof. d'anat. Pathol. à la Fac. mixte de Méd. et de Pharmacie de l'Univ. d'Alger (Algérie).

Poulsen, Erik Mellentin (1900). København K. (Danmark), St. Kannikestraede 9. *Ecol. of Freshwater Crustaceans.* ⓓ Copenhagen 1925.

Poulsson, Leif Tütein (1898), Amanuensis. Oslo (Norge), Physiol. Inst., Univ. *Physiol. Endokrine Organe (Hypophyse).* ⓓ Oslo 1923.

Poulsson, Poul Edvard (1858), Prof. Pharmacol. Toxikol. Oslo (Norge). *Pharmacol., Vitamine.* ⓓ Christiania 1892.

Poulton, E. B., Prof. Univ. Oxford (England). *Zool.* o

Poulton, Ethel Maud (1889), Lect. Biol. Education Dept. Univ. Birmingham (England). Beech Holme, 53 Beeches Road, West Bromwich, Staffordshire. *Lichens. Algae. Structure, Life-History and Systematic position.* ⓓ Geneva 1925.

Poutiers, Raymond Julien (1886), Lic. Sc., Dir. Insectarium. Menton (France). *Zool. agricole.* ⓟ Insectes.

Povah, Alfred Hubert William (1889), Associate Prof. of Botan., Northwestern Univ. Evanston, Ill. (U.S.A.), 9 Fisk Hall. *Mycol. Mucorales. Plant Pathol. Diseases of Populus.* ⓓ Univ. of Michigan1916.

Powers, Edwin B. (1880), Prof. and Head of the Department of Zool., Univ. of Tennessee. Knoxville, Tenn. (U.S.A.). *Ecol. and Physiol. of marine and fresh Water Fishes.* ⓓ Univ. of Chicago.

Poyarkoff, Eraste (1886), Prof. physiol. animale Univ., Dir. Sta. Sericicole. Taschkent (U.d.S.S.R.). *Protophysiol. Physiol. de l'immunité. Métamorphose des Insectes. Physiol. du ver à soie.* ⓓ Paris 1910.

Poynter, Charles W. M. (1875), Chairman Anat. Dept., Univ. of Nebraska, Coll. of Med. Omaha, Neb. (U.S.A.); 42 and Dewey Ave. *Development and Growth Disturbances (Teratol.). Cell Nutrition.* ⓓ Univ. of Nebraska 1898 and 1902.

Pożerski, Edouard Alexandre (1875), Dr. és sc., Dr. en méd., Chef de labor. à l'Inst. Pasteur. Paris XVII (France), 16, rue Sauffroy. *Physiol.: Digestion. Ferments digestifs. Immunité. Physiol. des microbes. Hygiène alimentaire.*

Prado y Sáinz, Salvador, Prof., Dr. en Cienc. Nat., Catedrático y Dir. del Inst. Guadalajara (España). *Biol.*

Praeger, William Emilius (1863), Prof. of Biol., Kalamazoo Coll. Kalamazoo, Mich. (U.S.A.). *Botan., Ecol. Botan.* ⓓ (Hon.) Kalamazoo 1925.

Prašek, Emil, Dr., o. Prof. Hygiene, Dir. Hygien. Inst. Zagreb (S.H.S.), Gajeva ulica 40. *Bact.* o

Prashad, Baini (1894), Superintendent, Zool. Survey of India. Calcutta (India), Indian Mus. *Indian Molluscs. Fauna of India.* ⓓ Lahore, India, 1917.

Prát, Silvestr (1895), Ph.D., Priv.Doc., Ass. of the Plant Physiol. Labor., Charles Univ. Praha (C.S.R.), Benátská 433. *Plasmolysis, permeability. Plant Nutrition. Cyanophyceae, Calcareous Algae.* ⓓ Charles Univ., Prague 1919. ⓟ Cyanophyceae cultures and preparations.

Pratje, Andre (1892), Dr. phil. nat. et med., Priv.-Doc. für Anat. und Anthrop., Prosektor an der Anat. Anst. der Univ. Erlangen (Deutschland). *Anat. Röntgenuntersuchungen am Lebenden. Zellchem. Das Leuchten der Organismen.* ⓓ Phil. nat. Freiburg i. B., med. Breslau.

Pratje, Otto (1890), Priv.Doc. für Geol. und Palaeont. Königsberg i. Pr. (Deutschland), Geol. Inst. der Univ., Lange Reihe 4. *Organogene Meeressedimente und Küstensedimente.* ⓓ Freiburg i. Br. 1921. ⓟ Organogene Sedimente.

Pratt, Dudley J., Associate Prof. of botan. and plant physiol., Texas Agric. and Mechanical Coll., Univ. of Texas. College Station, Tex. (U.S.A.). *Cytol.* o

Pratt, Henry Sherring (1859), Prof. of Biol., Haverford Coll. Haverford, Pa. (U.S.A.). *Parasit., classification and life history of Trematodes.* ⓓ Univ. of Leipzig 1892.

Pratt, Joseph Hersey (1872). Boston, Mass. (U.S.A.), 270 Commonwealth Ave. *Pathol. physiol. and Pathol. anat. Histol. of pneumonia. Bloodplatelets. Absorption of fats from intestine in pancreatic disease. Condition of vasomotor center in diphtheria intoxications. Uric acid in gout. The emptying of the stomach when pancreatic secretion is absent from the duodenum. Toxic cirrhosis of the liver.* ⓓ J. Hopkins Univ. 1898.

Praum, Auguste (1870), Dir. Labor. pratique de Bact. du Grand-Duché. Luxembourg, G. D. (Luxembourg), 4, Rue Dicks. *Hygiène et Bact.* ⓓ Luxembourg 1894, h. c. Nancy 1924.

Prausnitz, Carl (1876), o. Prof. der Hygiene Breslau XVI (Deutschland), Maxstr. 4. *Bact., Serol., Hygien.* ⓓ Breslau 1903.

Pravoslaview, Paul A., Dr., Prof. Geol. Militärmed. Akademie. Leningrad (U.d.S.S.R.), Geol. Inst.

Palaeont. der Wirbeltiere; palaeozoische und mesozoische Reptilien. o
Preble, Edward Alexander (1871). Washington, D.C. (U.S.A.), 3027 Newark St. *Geogr. distribution of birds mammals and plants.* o
Predtečevskij, S., Prof. Med. Inst. Ekaterinoslav, Ukraine (U.d.S.S.R.). *Bact.* o
Predtetshenskij, Sergej (1898), Ass. Bureau of applied Entomol. Leningrad (U.d.S.S.R.), Ul. Hertzena 44. *Acrididea, Biol.* ℗ Acrididea.
Pregl, Fritz (1869), o.ö. Prof. der med. Chem. an der Univ., med. Fac. Graz III (Deutsch-Österreich), Universitätsplatz 2. *Mikroanalyse.* ① Graz 1890, med. et phil. h. c. Göttingen.
Preisz, Hugo, Prof. Dr., Dir. Allg. Pathol. Inst. u. Dir. Bact. Inst. Univ. Budapest (Ungarn).)*Pathol. Bact.* o
Prell, Adrienne Renée, geb. Koehler (1892). Tharandt i. Sa. (Deutschland), Sidonienstr. 174c. *Zool., Bienenkunde.* ① Bern 1917.
Prell, Heinrich Bernward (1888), o. Prof. Zool., Dir. Zool. Inst. d.Forstlichen Hochsch., Vorstand der Zool. Abt. der Sächsischen Forstlichen Versuchsanstalt; Leiter der Hauptstelle Tharandt für Forstlichen Pflanzenschutz, Leiter der Forschungsstelle für Pelztierkunde. Tharandt i. Sa. (Deutschland), Forstliche Hochsch. *Allgemeine und angewandte Zool., Entomol., Vererbungslehre.* ① Marburg a.d.L. 1913.
Prenant, Auguste (1861), Prof. d'Histol. Fac. Méd. Paris (France). *Histol., Cytol. et Embryol.* ℗ Préparations histol.
Prenant, Marcel (1893), Chef de travaux à la Station Biol. de Roscoff, Finistère (France). *Zool.* ① Paris 1924.
Prenn, Friedrich (1878). Kufstein, Tirol (Deutsch-Österreich). *Libellenbiol., Vogelzug.*
Preti, Giacomo (1893), R. Ispettore per le malattie delle piante. Ventimiglia (Italia), Via Cavour 26. *Micol. e fitopat.* ℗ Portici.
Preudel, Alexander (1888), Ass. of Biol. Labor. Med. Inst. Odessa (U.d.S.S.R.), Street Pasteur 2. *Hydrobiol., Hirudinea; Entomol.: Diptera, Nematocera.* ℗ Culicidea, Crustacea (fresh-water), Hirudinea.
Prianischnikow, Dimitry Nikolaewitsch (1865), Prof. Landwirtsch. Akademie Timirjasew und Univ. Moskau 8 (U.d.S.S.R.), Petrowsko-Rasumowskoje. *Stickstoffumsatz in der Pflanze, Aufnahme von Aschenbestandteilen, Einfluß der Reaktion des Mediums. Verhältnisse zwischen Pflanze, Boden und Düngemittel.* ① Phil. Moskau 1889, Ehrendoktor Breslau 1923.
Pribram, Bruno Oskar (1887), Dir. der chirurgischen Abt. des St. Hildegard-Krankenhauses, Priv.-Doc. f. Chirurgie an der Univ. Berlin NW 40 (Deutschland), Kronprinzenufer 14. *Allgemeine und spezielle Chirurgie, allgem. Biol. und Physiol.* ① Med. Wien 1913, chem. 1909.
Pribram, Ernst (1879), ao. Prof. der Univ. Wien, Ass. Prof. of Pathol. at the Chicago Univ., Rush Med. Coll. Wien IX (Deutsch-Österreich), Michelbeuergasse 1a; Chicago, Ill. (U.S.A.), 1433, N. Claremont Avenue. *Systemat. Bact., Fermentlehre, Serol., pathol. Physiol., Cellularbiol. und Cellularpathol. Mykol., Pharmacol., Physikalisch-chem. Biol.* ① Prag 1903, Madison, Wisconsin, 1926. ℗ Bakterien, Hefen, Pilze, lebend in Reinkulturen, mikroskop. und pathol.-histol. Praep.
Price, Emma Käte (1900), Dir. of Orange County Labor. Orange, Cal. (U.S.A.), County Hospital. *Bact.*
Price, Emmett William (1896), Associate Parasit. Zool. Division, Bureau of Animal Industry, U.S. Dept. of Agric. Washington, D.C. (U.S.A.). *Parasit. of domestic animals and the diseases produced by them.* ① Washington 1918.
Priestley, J. G., Lect. Univ. Oxford (England) *Physiol.* o
Priestley, Joseph Hubert (1883), Prof. of Botan. Univ. of Leeds (England). *Physiol. of Growth and Development of Plants. Causal Anat.* ① London 1903.

Prieur, F., Prof. Ecole de Méd. Besançon (France). *Histol.* o
Prigorowsky, Michael M , Obergeol. am Com. Géol. de Russie. Leningrad (U.d.S.S.R.), W. O., Sredny Pr. 72-B. o
Primans, Jekabs (1892), Prosektor der norm. Anat. Riga (Lettland), Valdemara ielá 41. *Craniol. der Letten. Anat. und Embryol. der Harnorgane.* ① St. Petersburg 1918.
Prince, Arthur Reginald (1900), Prof. of Biol. and Provincial Entomol. at the Agric. Coll., Department of Nat. Resources, Nova Scotia Agric. Coll. Truro, Nova Scotia (Canada), Box 427. *Plant Pathol., Mycol. (Phycomycetes). Bryol., Mosses, Hepatics and Lichens. Entomol., Coleoptera. Systematic Botan. (Phanerogams, Cryptogams).* ℗ Phanerogams, Cryptogams (Fungi, Mosses, Hepatics, Lichens, Ferns etc.), Insects (especially Coleoptera) of Nova Scotia.
Principi, Paolo (1884), Prof. incaricato di Geol. applicata, Univ. d. Genova (Italia), Ist. Geol. *Dicotyledoni fossili. Flora oligocaenica.*
Pringle, Harold, Prof. Univ. Trinity Coll. Dublin (Ireland). *Physiol.* o
Pringsheim, Ernst Georg (1881), o. Prof. an der Deutschen Univ. Prag II (C.S.R.), Viničná 3a. *Reizphysiol., Mikrobiol., insbesondere der Algen und Bakterien.* ① Leipzig 1905. ℗ Algenreinkulturen.
Printz, Karl Henrik Oppegaard (1888), Prof., Dir. Botan. Inst. Norw. Landw. Hochsch. Aas bei Oslo (Norge). *Algol., Chlorophyceen und Meeresalgen Nordeuropas, Flora von Asien, Pflanzenanat.* ① Oslo 1921. ℗ Algen.
Pritchett, Ida (1891), Associate, Department of Pathol. and Bact. Rockefeller Inst. for Med. Research. New York City (U.S.A.), 66 Street and Avenue A. *Epidemiol. of Mouse Typhoid.* ① Johns Hopkins Univ. 1922.
Probst, Eugen (1893), Hochschulass. Freising b. München (Deutschland), Hochsch. Weihenstephan. *Kleintierzucht. Geflügelzucht. Wolluntersuchung.* ① Hochsch. Weihenstephan 1925.
Proca, G., Prof. Univ. Bucaresti (România) *Allg. Pathol.* o
Procházka, Jan Svatopluk (1891), Ph.Dr., Adjunkt der geol.-paleont. Abt. des Nationalmus., Lect. für Naturdenkmalpflege u. Naturschutz an der Karlsuniv. u. d. Hochsch. f. Land- u. Forstwirtschaft. Praha XII (C.S.R.), Slovenská 11. *Naturschutz, Naturdenkmalpflege. Ausrottung von Tieren und Pflanzen durch den Menschen.* ① Prag 1915.
Prochorow, Alexander N., Prof. des Iwanowo-Wosnessensker Polytechn. Inst. Iwanowo-Wosnessensk (U.d.S.S.R.). *Agronomie. Flachsbau. Nahrungspflanzen.* o
Procter, Joan Beauchamp (1897), Curator of Reptiles to the Zool. Soc., Secretary to the World List of sc. Periodicals. London (England), Zool. Soc., Regent's Park. *Herpetol., systematic, morphol., living Reptiles and Batrachians.* ℗ Reptiles and Batrachians.
Prodan, Julius, Prof. Acad. d. Agric. Cluj (România). *Syst. Botan. Phytopath.* o
Proebsting, Edward Louis (1897), Ass. Prof. of Pomol., Univ. of California. Davis, Cal. (U.S.A.), Univ. Farm. *Nutrition of tree fruits. Histol. of the graft union.* ① Cornell Univ. 1924.
Pröscholdt, Oskar (1878), Dir. Gesundheitsamt der Landwirtschaftskammer für Pommern. Züllchow-Stettin, (Deutschland) Bachstr. 5a. *Pathol. Anat., Bact. der Tiere. Serol. d. Infektionskrankheiten. Parasit. d. Tiere. Rindertuberkulose. Impfstoffe (Piroplasma).* ① Univ. Bern 1907. ℗ Diapositive von pathol.-anat. Veränderungen v. Tierkrankheiten u. von klinischen Krankheitsbildern.
Proida, Platon Artemovitsch (1900), Phytopath. Charkow (U.d.S.S.R.), Versuchsstation der Landwirtschaftl. Abteilung Phytopath. *Weizenbrand und Flugbrand. Mosaikkrankheiten der Zuckerrübe.* ℗ Herbarienmaterial der Ackerkulturen.

Promptov, Alexandre (1898), Aspirant I. Univ. Moscow (U.d.S.S.R.), Herzen's Str. 6, Labor. of the Experim. Zool. *Genetics of birds. Drosophila.*
Propatschikh, George (1899), Ass. Fergana (U.d. S.S.R.), Station de Séricicult. *Séricicult.*
Proschkina-Lawrenko, Anastasija J., Ass. des Charkow. Inst. für Volksbildung. Charkow (U.d.S.S.R.), Klotschkowskaja 50/52, Botan. Garten. *Botan., Algen.* ○
Proskorjakow, Eugeny (1895), Ass. d. Botan., Landwirtschaftl. Hochsch. Woronesh (U.d.S.S.R.). *Oekol. Physiol. Physiol. d. Samenkeimung.*
Protić, Georg (1864), Zemaljski Muzej (Landesmuseum) Sarajevo (S.H.S.). *Pflanz. Anat. und Physiol.: Geotropismus der Axenorgane und Blätter von Euphorbia. Hydrobiol. und Plankton der Seen Bosniens und der Hercegovina. Pilzparasitäre Krankheiten der Kulturpflanzen Bosniens und der Hercegovina.* ⓓ Univ. Wien 1891.
Protopopov, V. P, Prof. Inst. f. Volksbildung. Charkow (U.d.S.S.R.) *Reflexol.* ○
Provinciali, Celso (1897), Aiuto Ist. Farmacol., Tossicol. e Terapia sperim. della R. Univ. di Milano (Italia), Città degli studii, Via G. Strambio. *Cerio, Sparteina, Basi ammoniche quaternarie, Olio di Chaulmoogra, Organoterapia.* ⓓ Parma 1921.
Prüffer, Jan (1890), Acting head department of Zool. Univ. Wilno (Polska), Zakretowa 15. *Systematical and experim. Entomol. Systematic of the Lepidoptera and Odonata; experim. morphol. of the organs of lower senses in butterflies.* ⓓ Kraków 1920. ⓟ Lepidoptera and Odonata.
Prunet, A., Prof. Univ. Toulouse (France). *Botan. appl.* ○
Pruvost, Pierre, Eugène (1890), Prof. de Géol. et Minéral. à l'Univ. de Lille (France), 159, rue Brule-Maison. *Faunes carbonifères: Arthropodes, Insectes, Mollusques. Fossiles du terrain houiller.* ⓓ Lille 1919.
Prym, Paul Eugen (1881), Prof. Dr. med., stellvertr. Dir. des Pathol. Inst. der Univ. und I. Ass. Bonn, Rhein (Deutschland), Meckenheimer Allee 75. *Allgemeine Pathol. und pathol. Anat. Geschwulstlehre. Röntgenstrahlenwirkungen.* ⓓ Bonn 1906.
Prynada, Basil (1897), Comité Géol. de Russie. Leningrad (U.d.S.S.R.), Wasiljewski ostrow, Gredny prosp. 72 b. *Mesozoische Flora. Triassische wirbellose Fauna des Ussurigebietes.*
Przibram, Hans (1874), a.ö. Prof. f. experim. Zool. an der Univ. u. Leiter d. Biol. Versuchsanstalt d. Akad. d. Wissensch. Wien II (Deutsch-Österreich), Prater, Vivarium. *Entwicklungsmechanik, Regeneration und Transplantation, experim. Morphol. (Einfluß äußerer Faktoren, Farbphysiol.). Quantitative und theoretische Biol.* ⓓ Wien 1899.
v. Przyborowski, Josef (1895), Dr. phil., Adjunkt an der Lehrkanzel f. Pflanzenzüchtung an der Univ., 2. wissenschaftl. Leiter der Saatzuchtabt. der Landwirtschaftsgesellschaft. Krakau (Polen), Łobzowska No. 24. *Vererbungslehre: Moment der Spaltung der Anlagen. Methoden der Feldversuche.* ⓓ 1922. ⓟ Samen von verschiedenen Sorten landwirtschaftl. Pflanzen, Varietäten von Epilobium hirsutum.
Przylecki, Stanislas Jean (1891), Dr., Prof. ordinaire. Varsovie (Pologne), 35, rue Marszalkowska. *Excrétion et sécretion. Métabolisme intermédiaire des corps azotés. Enzymes. Physiol. comparée.* ⓓ Cracovie 1920.
Ptschelintzew, Woldemar F., Prof. Univ., Com. Géol. de Russie. Leningrad (U.d.S.S.R.), Geol. Kabinett der Univ. *Jurassische und cretacische Faunen; Fucoiden und Nerineen der Krim.* ○
Ptschelkin, Wassilij M., Ass. des Timirjasew Wissenschaftl. Versuchsinst. Iwanowo-Wosnessensk (U.d.S.S.R.), Moskauer Str. 14. *Botan., Geobotan.* ○
Puccinelli, Enrico (1901), Aiuto nell'Ist. di Anat. Patol. dell'Univ. di Pisa (Italia), Via Solforino 37. *Alterazioni del vasi cerebrali. Pigmenti contenenti ferro. Splenectomia.* ⓓ Pisa 1924.

Puccioni, Nello (1881), Prof. incar. Antrop., Dir. Ist. di Antrop. Pavia (Italia), R. Univ. *Antrop. e etnografia della Somalia. Paleont. umana in Italia.*
Puche, José (1896), Ass. Inst. Physiol. Doc. de la Fac. de Med. de Barcelona (España), Cortes 638, ou Inst. de Physiol. *Regulation glycemie, Sensibilité viscérale.* ⓓ Madrid 1926.
Puchner, Heinrich (1865), Dr. phil., o. Hochsch.-Prof. a. d. Landw. Hochsch. Weihenstephan. Freising, Bayern (Deutschland). *Torfkunde, Kartoffelforschung.* ⓓ Leipzig 1890.
Pučkovskij, S. E., Prof. Univ. Voronez (U.d. S.S.R.), Universitetskaja 5. *Histol., Embryol.* ○
Pütter, August (1879), Prof. der Physiol., Dir. des Physiol. Inst. der Univ. Heidelberg (Deutschland), Akademiestr. 3. *Vergl. Physiol. Physiol. der Ernährung. Physiol. der Reizbeantwortungen.* ⓓ Phil. Breslau 1901, med. Göttingen 1903.
Pugliese, Angelo (1866), Prof. o. Fisiol., Dir. Labor. di Fisiol. Sperim. Ist. Sup. Med. Vet. Milano (Italia). *Secrezioni interne. Vitamine.*
Puglisi, Stefano, Lib. doc. Univ. Messina (Italia). *Pathol.* ○
Pujiula, Jaques G., Dir. del Labor. Biol. de Sarriá. Barcelona (España), Col. de S. Ignacio, Sarriá. *Histol., Embryol.* ○
Pulcher, Claudio (1889), Aiuto Ist. Patol. Gen. Torino (Italia), via del Mille 7. *Fisiopat. e fisicochim. dei mycol., tubercolosi.* ⓓ Vienna 1913.
Pulewka, Paul (1896), 1. Ass. im Pharmacol. Inst. der Univ. Königsberg i. Pr. (Deutschland), Lange Reihe 14. *Giftempfindlichkeit, Pharmacol. des Stoffwechsels.* ⓓ Königsberg i. Pr. 1923.
Pulido Valente, Francisco, Prof. ord. Univ. Lisbôa (Portugal). *Patol.* ○
Pulikovsky, Nodezhda (1895), Ass. Zootom. Inst. Univ. Leningrad (U.d.S.S.R.). *Vergl. Anat. der Insekten. Respiration der Wasserinsekten.*
Pulle, August Adriaan (1878), Dr. phil., o.Prof. der speziellen Botan. und Pflanzengeographie a. d. Univ. Utrecht, Dir. des Botan. Gartens „Cantonspark" in Baarn. (Holland), Javalaan 5. *Floristik und Pflanzengeographie der Niederländischen Kolonien, Flora von Surinam und Niederl.-Neu-Guinea.* ⓓ Utrecht 1906.
Pulling, Howard E., Prof. of Botan., Wellesley Coll. Wellesley, S. 1, Mass. (U.S.A.), 201. Weston Road. *Plant physiol.* ○
Pullinger, B. D., Ass. Univ. Inst. for Med. Res. Johannesburg, Transvaal (S. Africa). *Pathol.* ○
Puntoni, V., Prof. Univ. Roma (Italia). *Batt.* ○
Purdy, Helen Alice, Dep. of Pathol., Boyce Thompson Inst. for Plant Research. Croton-on-Hudson, N.Y. (U.S.A.). ○
Purdy, W. J., Med. Research Council. London W.C. 2 (England), 15 York Buildings Adelphi. *Bact. Experim. Pathol.* ○
Puscariu, Emile (1859), Prof. tit. Histol. normale Fac. Méd., Dir. Inst. antirabique. Jassy (România), Strada Gându 3. *Histol. humaine normale et pathol.* ⓓ Budapest 1886.
Puşcariu, Valeriu (1896), Ass. de l'Inst. de Spéol. Cluj (Roumanie). *Spéol. Amphipodes des eaux douces.* ⓓ Jassy 1923. ⓟ Amphipodes.
Puschnig, Roman (1875). Klagenfurt (Österreich), Römerbad. *Systematica und Faunistik der Orthopteren und Odonaten (ostalpines Gebiet). Insektenbiol.: Schutzfärbung und verwandte Erscheinungen.* ⓓ Graz 1899. ⓟ Odonata, Mimikry- und Mimeseobjekte.
Pustet, August (1891), Ass. an der Landesanstalt für Pflanzenbau und Pflanzenschutz München. Pasing bei München (Deutschland), Riemerschmidstraße 47. *Angewandte Zool., Bekämpfung tierischer Schädlinge, der Bisamratte und der übrigen Nager, schädlicher Vögel und Insekten, besonders Bodenschädlinge.* ⓓ München 1921. ⓟ Praep. con. Bisamratte.

Pustowoj, Wassilij St., Leiter d. Selektions-Abt. Landw. Inst. Krasnodar, Kuban (U.d.S.S.R.). *Selektion d. Weizens u. d. Sonnenblume.* o

Putza, Francesco, Prof. incar. Univ. Cagliari, Sardinia (Italia). *Patol.* o

Puusepp, Ludvig (1875), Prof. der Neuropathol. Univ. Tartu, Dir. der Univ.-Nervenklinik. Tartu, (Estland), Küütrit.. 2. *Physiol. und Pathol. des Nervensystems, Endokrinol., Gehirntumoren, Physiol. der Epiphyse.* [D] Petersburg 1902. [P] Mikroskopische und makroskopische pathol. Praep.

de Puymaly, André (1883), Ass. de Botan. à la Fac. des Sc. de Bordeaux (France), 20, Cours Pasteur. *Algol.* [D] Paris 1925.

Puzanow, Johann (1885), Prof., Dir. Pädagogisch. Inst. Simferopol, Krim (U.d.S.S.R.), Bitakstr. 5. *Vergl. Anat. Ichthyol. Tiergeographie. Landmollusken d. Pontusländer.*

Pyatakov, Michael Leonid (1886), Chief zool. of the Pacific-Ocean Sc. Fishery Research Station. Wladiwostok (U.d.S.S.R.), Str. 1 Maja, Dalriba. *Embryol. of Argulus foliaceus; Systematic of decapod Crustaceans of the Ochotsk- and Japan Seas.* [P] Decapoda Crustaceans of the Japan Sea.

Pyle, Norman J. (1897), Ass. Research Prof. of Avian Pathol. Amherst, Mass. (U.S.A.), Department of Vet. Sc., Animal and Pathol., Massachusetts Agric. Coll. [D] Univ. of Pennsylvania 1918. [P] Bact. strains, animal and avian diseases.

Quagliariello, Gaetano (1883), Prof. di Chim. Biol. Napoli (Italia), S. Andrea delle Dame, 21. *Biochim.* [D] Catania 1925.

Quaintance, Altus Lacy (1870), Ass., Chief of Bureau of Entomol. Washington, D.C. (U.S.A.), Cosmos Club. o

Quanjer, Hendrik Marius (1879), Prof. of the Univ. of Agric. Wageningen (Holland), Lowicksche Allée. *Virus-diseases of plants. Potato-diseases.* [D] Amsterdam 1906. [P] Virus-diseases of potatoes, potato-varieties, plant-diseases.

Quarta, Gaetano (1897), Ass. Oss. di Fitopat. per Puglie. Taranto (Italia), Piazza Ebalia 1. *Insetti.* [D] Bologna 1924.

Quarti, Giacomo (1896), Aiuto-Ist. di Anat. Umana Normale R. Univ. di Pavia (Italia). [D] Pavia 1921.

Quastel, Juda Hirsch (1899), Research worker in Biochem. Labor. Cambridge (England), Trinity Coll. *Chem. of bact. metabolism and growth, changes induced by 'resting' or non-proliferating bacteria; chem. of biol. oxidation phenomena.* [D] Cambridge Univ. 1924.

Quensel, Ulrik (1863), o. Prof. Univ. Uppsala (Sverige). *Pathol. Anat.* [D] Lund 1894.

Quenstedt, Werner (1893), Ass. Berlin N 4 (Deutschland), Invalidenstr. 43. *Devonische Mollusken aus Spitzbergen, Sedimentation und Fossilisation.* [D] München 1922.

Ritter von Querner, Friedrich (1900), Dr., Ass. des I. Zool. Inst. der Univ. Wien I (Deutsch-Österreich). *Cestoden, Cephalopoden.* [D] Wien 1923.

de Quervain, Fritz (1868), o. Prof. der Chirurgie, Vorsteher der Univ.-Klinik. Bern (Schweiz), Kirchenfeldstr. 60. *Pathol. Physiol. der Schilddrüse.* [D] Bern 1891.

Quesada, Fortunato, Prof. Univ. Lima (Peru). *Anat. topogr.* o

Quiel, Günther (1890), Dr. phil., Oberfischmeister für die Prov. Brandenburg und die Stadtgemeinde Berlin. Berlin-Lichterfelde (Deutschland), Sternstr. Nr. 2. *Fischereibiol. angewandte Hydrobiol.* [D] Berlin 1914.

Quilis Perez, Modesto (1904), Lic. Pharmacie. Valencia (España), R. Maldonado 54. *Hymenoptera: Bombus, Osmia.* [P] Hymenoptera.

Quillian, Marvin Clarke (1874), Prof. of Biol., Wesleyan Coll., Macon, Ga. (U.S.A.). [D] Vanderbilt Univ. 1900.

Quinby, John Roy (1902), Superintendent, Substation No. 12, Texas Agric. Experim. Station. Chillicothe, Tex. (U.S.A.). *Sorghums and cotton.* [D] Texas A. and M. Coll. 1924. [P] Field crops: sorghums, cotton, small grains.

Quin Lan, Christine Elizabeth (1894), Ass. Lect. Botan. Cardiff, Wales (England), Univ. Coll. Newport Rd. *Vascular system of the Magnoliaceae, especially of the node. Compar. anat. of the Ranales.* [D] Nation. Univ. Ireland 1918.

Quinn, Jeremiah Thomas (1894), Ass. Prof. of Horticulture. Missouri Univ. Columbia, Mo. (U.S.A.) *Nutrition.* [D] Missouri Univ. 1924.

Quinquaud, Prof. Ecole prép. de méd. et de pharmacie. Reims (France). *Physiol.* o

Quinquaud, Alfred (1884), Ass. au labor. de M. Gley. Paris (France), Coll. de France. *Endocrinol.*

Quintanilha, Aurélio Pereira da Silva (1892), Ass. Inst. Botan. Univ. Coimbra (Portugal). *Cytol. végétale; sécretion et cellules secrétrices; plantes carnivores, cytol. des phénomènes de digestion. Cytol. des Champignons. Synchytrium.* [D] Coimbra.

Quintaret, Gustave François (1870), Ass. à la Fac. des Sc. et Prof. à l'Ecole de Méd. et de Pharmacie de Marseille (France), 69, Cours Lieutaud. *Cécidies des Racines (Rhizocécidies).*

Quisumbing, Eduardo (1895), Fellow, National Research Council, Research Associate, Dept. of Botan., Univ. of California. Berkeley, Cal. (U.S.A.); (Home address: Los Baños, Laguna, Philippines). *Piperaceae of the Philippines. Philippine Flora. Morphol., Ecol.* [D] Chicago 1922. [P] Plants of the Philippines.

Raab, Wilhelm (1895), Ass. an der propaedeutischen Klinik der deutschen Univ. Prag II (C.S.R.), Allgemeines Krankenhaus, Klinik Biedl. *Einfluß der innersekretorischen Organe auf den Stoffwechsel, insbesondere den Fettstoffwechsel. Bedeutung der Hypophyse für Fettstoffwechsel und chem. Wärmeregulation. Hypophysäre und zerebrale Fettsucht.* [D] Wien 1920.

Raabe, H., Dr., Priv.Doc. Warszawa (Polska). 2 Mianowskiego 8. *Zool.* o

Rabanus, Adolf (1890), Dr. phil. nat. Uerdingen a. Niederrhein (Deutschland), Am Röttgen 30. *Konservierung des Holzes.* [D] Freiburg i. Br. 1914.

Rabaud, Etienne (1868), Prof. Biol. experim. Fac. Sc., Dir. labor. Recherches biol. Paris 5 (France), 3, Rue Vauquelin. *Signification des formes en fonction de l'éthol.*

Rabbeno, Angelo (1890), Doc. fisiol., Aiuto di farmacol. Univ., Laureato in Med. e Chir. Pavia (Italia), Palazzo Botta. *Sintesi protettive. Grassi. Saponi. Colesterina. Catalasi. Sali di Magnesio. Canale deferente. Tiroide. Secrezione sebacea. Età, clima, lavoro, narcosi in alta montagna. Meccanica respiratoria, ricambio respiratorio, reazione del sangue nel bagno di mare. Iniezioni endovenose di acqua di mare. Ricambio respiratorio dei tessuti.*

Rabbethge, Oscar (1880), Dr., Leiter der Abt. Pflanzenzucht der Zuckerfabrik Klein-Wanzleben vorm. Rabbethge & Giesecke. Klein-Wanzleben, Bez. Magdeburg (Deutschland). *Zuckerrübenzucht.* [D] Basel 1904.

Rabe, Hans, Univ.Prof. Graz (Österreich). *Histol.* o

Raber, Oran L. (1893), Ass., Prof. of Botan., Univ. Tucson, Arizona (U.S.A.), 901, Speedway. *Plant physiol.* [D] Harvard 1920.

Rabien, Herbert (1900), Dr. rer. techn., Ass. am Botan. Inst. der Techn. Hochsch. Braunschweig (Deutschland), Humboldtstr. 1. *Pflanzenphysiol. und -pathol.* [D] Braunschweig 1926.

Rabinerson, Alexander (1896), Ass. Staatsinst. experim. Agronomie, Abteilung angewandte Ichthyol. Leningrad (U.d.S.S.R.), Morskaja 42. *Vergl. Anat. des Viszeralskeletts und der Wirbelsäule der Fische; Systematik und Biol. (Altersbestimmung) der Heringe der russischen Gewässer; Wechselwirkung zwischen kolloiden Lösungen; Wechselwirkung von Kolloiden und Eiweißstoffen aus Normal- und Immunseris.* [D] Kiew 1921.

Rabl, Hans (1868), o. Prof. Dir. Inst. f. Histol. u. Embryol. Univ. Graz (Österreich), Universitätsplatz 4. *Histol. u. Embryol. d. Haut, Ovarium, Vorniere, Kiementaschen, Schilddrüse, branchiogene Organe, Zunge.* Ⓓ Wien 1893.

Racek, Jan, Ass. l'Inst. des Recherches agric. et Forestière. Brno (C.S.R.), Květná ul. *Biochem.*

Racicot, Homere Noe (1893), Plant Pathol. Ste. Anne de la Pocatiere, Que. (Canada), Labor. of Plant Pathol. *Vegetable and fruit diseases.* Ⓓ Toronto 1921. Ⓟ Plant and fungus disease specimens.

Racovitza, Emile G. (1868), Prof. Univ., Dir. Inst. de Spéol. Cluj (Roumanie), Case postale 158. *Spéol. Isopodes.* Ⓓ Paris 1896.

Radais, M.P., Prof. Fac. Pharmacie. Paris 6 (France), 4, Avenue de l'Observatoire. *Botan. Cryptogam. et Microbiol.* o

Radasch, Henry Erdmann (1874), Prof. of Histol. and Embryol. in the Jefferson Med. Coll., Philadelphia, Pa. (U.S.A.), Daniel Baugh Inst. of Anat. and Biol., 11. and Clinton Sts. *Acid Cells of the Stomach. Composition of Compact Bone. Effect of Ligation on Arterial Vessels.* Ⓓ Philadelphia, Pa., 1901.

Radcliffe, Lewis (1880), Deputy Commissioner, U.S. Bureau of Fisheries. Washington, D.C. (U.S.A.). *Systematic ichthyol., fishes of Philippine Islands, Peru, Eastern Pacific and United States. Economic fisheries.* Ⓓ George Washington 1915.

Radeff, Nenko (1899), Ass. am Kgl. Naturhistorischen Mus. in Sofia (Bulgarien). *Wirbeltiere Bulgariens, Speleol.* Ⓓ Sofia 1923.

Radermacher, Arnold (1886), Dr. rer. nat., Botaniker der Untersuchungsstelle Java, Zuckerindustrie Pasoeroean. Java (Nederl.-O.-Indië). *Systematik d. Angiospermen. Pflanzenembryol.* Ⓓ Utrecht 1925.

Radian, S., Prof. Univ. Bucarest (România). *Botan.* o

Radkević, O. N., Prof. Univ. Taschkent (U.d. S.S.R.). *Botan.* o

Rádl, Emanuel (1873), o. Prof. d. Philosophie, Naturwiss. Fac. d. tschech. Univ. Praha IV (C.S.R.), 279. *Geschichte der Wissenschaft, vergl. Physiol., Phil. Resultate der Naturwiss. und ihre Beziehungen zum zeitgenössischen Kulturleben.* Ⓓ Prag 1898.

Radlkofer, Ludwig (1829), Geh. Hofrat, Univ.-Prof. u. Dir. des Botan. Mus. (Herbariums) München-Nymphenburg, Menzinger Str. 13. München 2, SW 3 (Deutschland), Sonnenstr. 7 I. *Systematische Botan. Befruchtungsprozeß im Pflanzenreiche u. im Tierreiche, Verhältnis d. Parthenogenesis zu d. and. Fortpflanzungsarten. Kristalle proteinartiger Körper pflanzlichen u. tierischen Ursprungs. Sapindaceae: Serjania, Allophylus.* Ⓓ Med. München 1854, phil. Jena 1855.

Radovanovič, Svetolik, Dr., o. Prof. Univ., Dir. Geol. u. palaeont. Inst. Beograd (S.H.S.). *Palaeont.* o

Radsimovskij, V., Prof. Univ. Kiev (U.d.S.S.R.). *Physiol.* o

Radugin, Konstantin Wladimiriwisch (1899), Sibirische Abt. Geol. Comité. Tomsk (U.d.S.S.R.). *Palaeont. der Molluscoidea und Corallen. Tabulata aus dem Silur von Mittelsibirien.* Ⓟ Corallen aus dem Devon.

Radzivilovskij, G. L., Prof. Univ. Taschkent (U.d.S.S.R.). *Physiol. d. Haustiere.* o

Rae, Frederick James (1883), Governm. Botan. of Victoria, Dir. of Botan. Gardens and National Herbarium. Domain South Yarra, Melbourne, Victoria (Australia). Ⓓ Melbourne Univ. Ⓟ Australian plants.

Raebiger, Hans (1871), Prof., Dir. des Bact. Inst. der Landwirtschaftskammer für die Provinz Sachsen. Halle a. d. S. (Deutschland), Freiimfelderstr. 68. *Tierseuchenforschung, Bact., pathol. Anat. und Histol. Schädlingsbekämpfung (Hamster, Bisamratten, Ratten und Mäuse). Angewandte Botan.: Verwertung der Pilze unter besonderer Berücksichtigung der giftigen und giftverdächtigen.* Ⓟ Phil. Leipzig 1904, med. vet. h. c. der Tierärztl. Hochsch. Hannover 1925. Ⓟ Tierpathogene Bakterienkulturen, pathol.-histol. Praep.

Raeder, J. Milford (1892), Ass. Plant Pathol., Idaho Agric. Experim. Station. Moscow, Id. (U.S.A.). *Plant Diseases.* Ⓓ 1920.

Raffaele, Federico (1862), Prof. o. Zool. Univ. Roma 10 (Italia), Via Ferdinando di Savoia 3. *Embriol. dei vertebrati, uova e stadi larvali dei Teleostei, Embriol. sperim.* Ⓓ Napoli 1884.

Raffaella, Franceschi (1894), Pisa (Italia), Via G. Giusti N. 1. *Paleont. Brachiopodi del lias medio dell'Apennino centrale. Natura dei Fuccoidi.* Ⓓ Pisa 1919.

Rafiq, Ahmad, Lect. Univ. Aligarh, Unit. Prov. of Agra and Oudh (India). *Botan.* o

Ragsdale, Arthur Chester (1890), Prof. of Dairy Husbandry (Department Chairman). Columbia, Mo. (U.S.A.), Room 1, Dairy Building. *Dairy cattle, growth, breeding.* Ⓓ Univ. of Wisconsin 1925.

Rahm, Gilbert Franz (1885), Prof. Dr. Salzburg (Deutsch-Österreich). Univ. *Physiol. des Stoffwechsels der Tiere im latenten Lebenszustand, Tardigraden, Nematoden und Rotatorien.* Ⓓ 1921. Ⓟ Praep. von Tardigraden.

Rahn, Otto (1881), Prof. of Bact., Cornell Univ. Ithaca, N.Y. (U.S.A.), Dairy Dept. *Landwirtschaftl. Bact. (Milch, Boden, Nahrungsmittel). Physiol. und Systematik der Bakterien.* Ⓓ Göttingen 1902.

Raikes, H. R., Lect. Univ. Oxford (England). *Anthrop.* o

Raikow, B. E., Prof. Zool. Labor. Pädagog. Inst. Leningrad (U.d.S.S.R.). o

Raikowa, Hilaria (1896), Botan. Garten Univ., Ass. am Katheder f. Botan. Taschkent (U.d.S.S.R.), Mittelasiatische Staatsuniv. *Geobotan. u. Lichenol.* Ⓓ Mittelas. Univ. 1920.

Railliet, M., Prof. Hon. Ecole Vét. St. Germain-sur-Morin, Seine-et-Marne (France), 19, rue de Melun. *Zool.* o

Raiment, P. C., Demonstr. Univ. Oxford (England). *Biochem.* o

Raimondi, Carlo (1854), Prof. ordin. mat. med. e farmacol. Univ. di Siena (Italia). *Farm. speriment. Tossicol.* Ⓓ Pavia.

Rainer, Fr., Prof. Univ. Bucarest (România). *Zool.* o

Raineri, Rita (1896), Chef des travaux. Torino (Italia), Via Ormea 110. *Algues.* Ⓓ Torino 1919. Ⓟ Phanerogames.

Raitsits, Emil (1882), ao. Prof., Dir. Poliklinik Tierärztl. Hochsch. Budapest VII (Ungarn), Rottenbiller Gasse 23-25. Ⓓ Budapest 1904. Ⓟ Ungarische Hunderassen: Komondor, Puli, Pumi und Kuvasz.

Raiziss, George W. (1884), Prof. of Chemotherapy Graduate School of Med., Univ. of Pennsylvania; Dir. of the Dermatol. Research Labor. Philadelphia, Pa. (U.S.A.). Business address: 1720 Lombard St., Home address: 303 Pine Manor Apts. 49 th and Pine Sts. *The chem. of organic compounds of arsenic, mercury, bismuth, etc.; organic chem., experim. chemotherapy; advanced therapeutics.* Ⓓ Freiburg i. Br. 1909. Ⓟ Various organic compounds of arsenic and mercury, rats infected with trypanosoma equiperdum, rabbits infected with the treponema pallidum.

Rajkowa, Ilarija A., Doc. der Mittelasiat. Staatl. Univ., Botan. am Botan. Garten. Taschkent (U.d.S.S.R.), Ul. I. Maja 43. *Botan., Geobotan.* o

Rakestraw, Norris W. (1895), Ass. Prof. of Chem., Brown Univ. Providence, R.I. (U.S.A.), Metcalf Chem. Labor. *Biol. Chem. Chem. analysis of blood. Enzymes. Biol. behavior of colloids.* Ⓓ Stanford Univ. 1921.

Raleigh, George Joseph (1898), Chicago, Ill. (U.S.A.), 5731 Kenwood Ave. *Plant physiol.* Ⓓ Univ. of Nebraska 1923.

Raleigh, Walter P. (1901), Ass. Ames, Ia. (U.S.A.). *Corn Diseases.* ① Kansas State Agric. Coll. 1923.

Rall, James (1887), Prof. Botan. Ackerbauinst. Charkow (U.d.S.S.R.), Pouschkinskaja 80. *Desmediaceae, Hydrobiol.*

Ramaley, Francis (1870), Prof. of Biol., Univ. of Colorado. Boulder, Col. (U.S.A.). *Plant ecol. and plant geography of Colorado.* ② Univ. of Minnesota 1899.

Ramalho, Alfredo (1894), Dir.,,Aquário Vasco da Gama", Station de Biol. maritime. Dàfundo (Portugal). *Faune ichthyol. du Portugal, œufs et larves de Poissons, écol. de la Sardine et du Thynnus thynnus et autres Scombriformes.* ① Lissabon 1924.

Ramaswami, C. V. (1895), Research Ass. Indian Inst. of Sc. Hebbal, Bangalore (India). *South Indian soils.* ① 1917.

Rambousek, François G. (1886), Dr. phil., Vorstand der phytopath. Abteilung des Forschungsinst. für die tschechoslowakische Zuckerindustrie. Praha XVIII (C.S.R.), Střešovice, Vořechovka 112. *Rübenschädlinge u. Rübenkrankheiten. System: Staphylinidae (Col.).* ① Prague 1913. ② Rübenschädlinge u. Krankheiten.

Ramenskij, L. G., Doc. der Staatl. Univ. in Woronesh (U.d.S.S.R.). *Geobotan., Bodenkunde, Wiesenbau.* o

Ramires, Adolfo Augusto Baptista, Prof. o. Inst. super. de Agronomia Trapada da Ajuda. Lisbôa (Portugal). *Mikrobiol. agric.* o

Ramírez, R. R., Prof. Univ. Tegucigalpa (Honduras), Plaza de la Merced. *Physiol.* o

Ramme, Willy (1887), Dr. phil., Kustos am Zool. Mus. der Univ. Berlin N 4 (Deutschland), Invalidenstraße 43. *Systematik und Biol. der Orthoptera.* ① Berlin 1910. ② Orthoptera.

Rammler, Fritz, Dir. der Zool. Stat. Kajana, Suvenniemi (Finnland). o

Rammul, Alexander (1875), Prof. Tartu (Estonia). Hygien. Inst. *Biochem. der säurefesten Bakterien.* ① Tartu.

Ramoino, Gio. Batta, Lib. doc. Univ. Genova (Italia). *Anat. Patol.* o

Ramon, Gaston Léon (1886), Vét., Chef de l'Inst. Pasteur à Garches, S. o. (France). *Sérothérapie.* ① Alfort 1910.

Ramon Vila, Barbera, Prof. Univ. Valencia (España). *Patol. Gen.* o

Ramón y Cajal, Pedro, Prof., Catedrático jubilado de la Fac. de Med. Zaragoza (España), Sitios, 6. *Histol. des cellules nerveuses.* o

Rams Bottom, John (1885), Ass. Keeper, Botan. Dept., British Mus., Nat. Hist. London S.W. 7., (England). *Mycol., British Fungi, Enemies of the Rose.* ① Cambridge 1908.

Ramsden, Charles Theodore (1876), Dr. of Nat. Sc. Guantánamo (Cuba), P.O. Box 146. *Cuban: Macrolepidoptera, diurnal and nocturnal; Coleoptera; Ornithol.; Ool.; Herpetol.; Conchol.; Mammal.* ① Univ. Nacional, Habana 1908.

Ramsden, Walter, Prof. of Biochem., Univ. of Liverpool and Fellow of Pembroke Coll. Oxford (England). *Capillary chem.* ① Oxford 1892.

Ramsey, Glen Blaine (1889), Associate Plant Pathol., United States Department of Agric. Chicago, Ill. (U.S.A.), 1425 So., Racine Ave. *Diseases of vegetables.* ① Univ. of Chicago 1925.

Ramström, Oskar Martin (1861), Prof. der Anat. der Univ. Upsala (Schweden). *Nervenanat.: Bauchwand, Diaphragma, Peritoneum. Palaeont.: Palaeolithicum (Piltdown-Fund, Java-Trinil-Fund, Wadjak-Fund).* ① Lund 1905.

Ramult, Miroslaw (1890), Ass. Zool. Inst. Univ., Mitarbeiter der Physiographischen Kommission der Akademie d. Wissenschaften. Kraków (Polska), St. Anna 6. *Physiol. der niederen Crustaceen (Blutkreisläufe, Wachstum, osmotische Erscheinungen.*

Anpassung); Wachstums- u. Anpassungserscheinungen bei Fischen. Faunistik. Cladocera. ① Cracovie 1920. ② Cladoceren.

Ramzin, Sergije, Dr., Bact. Stanica. Šabac, Serbien (S.H.S.). *Bact., Chemother., Serol.* o

Rancken, Dodo Emil Artur (1872), Adjunkt der Physiol. Univ. Helsinki (Finnland). *Körpertemper. Wasserstoffwechsel. Blutkreislauf.* ① Helsingfors 1909.

Rand, Frederick Vernon (1883), Associate Editor, Biol. Abstracts; Lect. in Bact., Johns Hopkins Univ.; Collaborator in Plant Pathol., U.S. Dept. of Agric. Baltimore, Md. (U.S.A.). *Plant pathol., fungous and bact. diseases, insect transmission of disease, cytol. of plant diseases.* ① Columbia Univ. 1920.

Rands, Robert Delafield (1890), Pathol., in sugar cane investigations, U.S. Dept. of Agric. Washington, D.C. (U.S.A.), Bureau Plant Industry. *Sugar cane diseases and varietal resistance.* ① Univ. Wisconsin 1917.

Ranke, Otto (1899), III. Ass. am Pathol. Inst. Freiburg i. Br. (Deutschland), Albertstr. 19. *Artherosklerose und senile Ektasie der Aorta.* ① Freiburg i. Br. 1923.

Ranker, Emery Romain (1896), Associate Physiol. Washington, D.C. (U.S.A.), Bureau of Plant Industry, U. S. D. A. *Physiol. of disease resistance; Methods for the determination of nitrogen; Plant nutrition; Fixation of atmospheric nitrogen by higher plants; Horticulture.* ① Washington Univ., St. Louis, Mo. 1926.

Rankin, Walter M. (1857), Prof. emer. of Biol. Univ. Princeton, N.J. (U.S.A.). ① Munich 1886.

Rankin, William Howard (1888), Associate in Research (Plant Pathol.), New York State Agric. Experim. Station. Geneva, N.Y. (U.S.A.). *Diseases of shade and forest trees. Diseases of small fruits, especially of raspberries (Rubus).* ① Cornell Univ. 1914.

Rankov, Milivoj, Dr., Dir. Inst. f. Tropenkrankheiten. Skoplje (S.H.S.). *Bact.* o

Ransch, Stephen Walter (1880), Prof. and head Dept. of Neuroanat. and Histol. Washington Univ. Med. School. St. Louis, Mo. (U.S.A.), 5875 Maple Ave. o

Ranson, Gilbert (1899), Préparateur sc. au Mus. National d'Histoire Nat. Paris 5. (France), 99, rue de Buffon. *Histol., Zool. experim.* ① Paris 1923.

Rant, A., Dr., Doc. a. d. Niederl. Ärzte-Schule. Soerabaja, Java (Nederl.-O.-Indië), Embong Malang. *Krankheiten des Chinabaumes. Blumenbiol.* o

Ranzi, Silvio (1902), Ass. Staz. Zool. Napoli (Italia), Aquarium. *Embriol. dei Vertebrati: Ductus endolymphaticus; Processi generali di sviluppo; Placodi ed organo di senso spiracolare. Embriol. dei Cefalopodi: normale e sperimentale. Sistematica dei Policladi.* ① Roma 1924.

Rao, H. Srinivasa (1894), Ass. Superintendent, Zool. Survey of India. Calcutta (India), Indian Mus. *Molluscs and Invertebrate. Fauna of India.* ① Madras, India, 1925.

Rapales von Ruhmwerth, Raymund (1885), Prof. der Landwirtschaftl. Akademie. Budapest (Ungarn), Üllői-ut 119. *Theorien der Biosoziol., Geschichte der Biol. (Neuzeit), Pflanzengeographie des Ungarischen Tieflandes.* ① 1907.

Raper, H. S. (1882), Prof., Dir. Physiol. Labor. Univ. Manchester (England). *Physiol.* ① Leeds 1903.

Räsänen, Veli J. P. B. (1888), Mag. phil., Lect. im Landwirtschaftl. Inst. Kúrkijoki (Finnland). *Botan., Lichenes. Stridulationsapparate bei Ameisen.* ① Helsinki 1914. ② Flechten Finnlands.

Rascanu, V., Prof. Univ. Jassy (Romania). *Physiol.* o

Rasch, Walter Ernst (1891), Dr. phil., Wissenschaftl. Mitarbeiter der ,,Deutschen Gesellschaft für Schädlingsbekämpfung m. b. H.". Frankfurt a. M.-Eschersheim (Deutschland), Lindenring 13. *Desinfektionen mit Blausäure und Zyklon.* ① Berlin 1915.

Rasdorskij, Wladimir F., Prof. des Landwirtsch. Inst. der Bergrepublik, Wissensch. Mitarbeiter des Nord-Kaukasischen Inst. für Landeskunde. Wladikawkas (U.d.S.S.R.), Ul. Butirina 26. *Pflanzen-Anat., Flora von Nord-Kaukasus.* o

Rašek, Jaroslav (1898), Ass. Landwirtschaftl. Versuchsanstalt, Phytopath. Sektion und Inst. für die angew. Ornithol. M.A.P. Brno (C.S.R.), Černá pole 201. *Bekämpfung von Insektenplagen: Liparis monacha. Chemische Mittel zur Bekämpfung der Pflanzenschädlinge. Vogelschutz.* ⓓ Brno 1923. ⓟ *Pathol. Beschädigungen der Forstpflanzen, Gallen.*

Rasmussen, Andrew Theodore (1883), Prof. of Neurol., Department of Anat., Univ. of Minnesota. Minneapolis, Minn. (U.S.A.), 1055/14 th Ave. S. E. *Hypophysis. Interstitial cells of testis and ovary. Fiber tracts in central nervous system.* ⓓ Cornell Univ. 1916.

Rasmussen, Rasmus (1871). Thorshavn (Färö, Island), Föroya Fólka Máskuli. *Phenol.: bloming plants on the Färö Isl.*

Rasumow, Alexander S., Doc. Moskauer Techn. Hochsch. Moskau (U.d.S.S.R.), Daew per. 7, W. 1. *Botan., Microchem. Mycol.* o

Rasumow, Victor I. (1902), Labor. of plant physiol., Inst. of applied Botan. and New Cultures. Leningrad (U.d.S.S.R.), 44 Herzenstr. *Waterrelations of plants, photoperiodism.* ⓓ Leningrad 1926.

Rath, Floyd Cecil (1894), City Chem. and Bact. Madison, Wis. (U.S.A.), 603 Prospect Ave. *Foods and milk examinations.* ⓓ Madison, Wis., 1920.

Rathbun, Mary Jane (1860), Associate in Zool., U.S. National Mus. Washington, D.C. (U.S.A.). *Classification of Decapod Crustaceans, recent and fossil.* ⓓ George Washington Univ. 1917.

Rathery, Prof. Univ. Paris (France). *Pathol. experim. et compar.* o

Raudonikis, P., Prof. Dir. Pharmacol. Inst. Kaunas (Litauen), Gedimino Gatve 29. o

Rauff, Hermann (1853), Dr. phil., Geh.BergR., o. Prof. emer. der Geol. und Palaeont. an der Techn. Hochsch. Berlin. Charlottenbg 2 (Deutschland), Leibnizstr. 91. *Palaeospongiol.; Receptaculitidae; Mikroorganismen; Pseudoorganismen (durch Gesteinsdeformation) und Problematica. Geol. Aufnahmen am Rhein und in der Eifel. Rheinisches Devon, Eifelkalkmulden.* ⓓ Bonn 1878.

Raúl de Miranda (1902), Ass. Fac. Sc. Univ. Coïmbra (Portugal), Praca da Republica 35. *Paléont. Flora triassica.* ⓓ Coïmbra 1926.

Raum, Hans (1883), a.o. Prof. an der Hochsch. für Landwirtschaft und Brauerei. Weihenstephan, Bayern (Deutschland). *Erbanalyse der landw. Kulturpflanzen, besonders Getreide. Züchtung von Hopfen. Pflanzenbestand und seine Veränderungen auf Wiesen und Weiden.* ⓓ München 1906.

Rauther, Theodor Eugen Max (1879), Prof., Dr. phil., Dir. der Württ. Naturaliensammlung, Doc. an der Techn. Hochsch. Stuttgart (Deutschland), Archivstr. 3. *Vergl. Anat. der Fische; Nematoden und Nematomorpha; Genitalorgane der Säugetiere; Theorie des zool. Systems.* ⓓ Jena 1903.

Rautmann, Hugo (1879), Vorsteher der Tuberkulose-Abteilung des Inst. Halle a. S. (Deutschland), Freiimfelderstr. 68. *Rindertuberkulosetilgungsverfahren.* ⓓ Bern 1904.

Raventos, Jacques (1905), Ass. au Inst. de Physiol. Barcelona (España), Carril 103. *Regulation de la glycemie.* o

Ravn, Jesper P. J. (1866), Inspecteur du Mus. de Min. et de Géol. de l'Univ., Prof. de Paléont. à l'Univ. København K. (Danmark), Oestervoldgade 7. *Faune crétacique et tertiaire du Danemark.* ⓓ Copenhagen 1892.

Rawitscher, Felix (1890), Dr., a.o. Prof. für Botan. Freiburg i. Br. (Deutschland), Kronenstr. 18. *Reizphysiol. der Pflanzen. Mykol. Pflanzengeographie.* ⓓ 1912.

Raybaud, Laurent (1876), Maître de conférences de Botan. à la Fac. des sc. Nancy, Meurthe et Moselle (France), Jardin Botan. *Pathol. végétale, Biol. Végétale, Mycol., action des radiations sur les êtres vivants.*

Raymond, Lee Carleton (1889), Ass. Prof. of Agronomy. Quebec (Canada), Macdonald Coll., P. O., P. Q. *Corn and root crops.* ⓓ Univ. of Wisconsin 1924.

Razdorsky, V. T., Prof. Pädag. Inst. d. Bergrepublik. Vladikavkaz (U.d.S.S.R.), Ul. Marksa 20. *Anat. u. Morphol. d. Pflanzen.* o

Razowski, Franz (1865), Leiter chem. Centrallabor. mit biol. Abt. für Reinkulturen, Ass. für Biochem. Univ. Nižnij Novgorod (U.d.S.S.R.), Octjaberskaja 25. *Biochem. Torf. Gärungschem. Holzdestill.* ⓓ Dorpat 1892.

Rea, George Harold (1880), Extension Ass. Prof. in Apic. Ithaca, N.Y. (U.S.A.), Department of Entomol., New York State Coll. of Agric., Cornell Univ.

Reach, Felix (1872), Priv.Doc. Univ. Wien IX (Deutsch-Österreich), Alserbachstr. 5. *Physiol. der Gallenwege.* ⓓ Prag, Deutsche Univ. 1895.

Read, Bernard Emms (1887), Prof., Head Dept. of Pharmacol. Peking (China), Peking Union Med. Coll. *Chaulmoogra and Ephedrine. Flora sinensis.* ⓓ Univ. 1924. ⓟ *Flora of North China.*

Reade, John Moore, Prof. of Botan., Univ. of Georgia, Dir. of the Biol. Labor. Athens, Ga. (U.S.A.).

Reagan, F. T., Ph.D., Hon. Research Ass. Dept. of Anat. and Histol. Univ. Coll. London (England). *Embryol.* o

Rebagliati, Raul, Prof. Univ. Lima (Peru). *Bact.* o

Rebel, Hans (1861), Dr., Prof., HofR., Erster Dir. des Naturhistor. Mus. Wien I (Österreich), Burgring 7. *Allgemeine Entomol. Lepidopteren, insbesondere Saturniidae und palaearktische Mikrolepidoptera.* ⓓ Jur. Wien 1886, phil. 1895.

Rebelo da Silva, Luis Antonio, Prof. o. Inst. super. de Agronomia, Trapada da Ajuda. Lisbôa (Portugal). *Chim. agric.* o

Reche, Carl Otto (1879), o. Univ.Prof. Anthrop. u. Ethnographie, Dir. Anthrop.-Ethn. Inst. d. Univ. Wien IX (Deutsch-Österreich), Van-Swieten-Gasse 1. *Biol. Anthrop., Rassenhygiene, Ethnographie Afrikas und der Südsee.* ⓓ Breslau 1904.

Rechinger, Karl (1867), Reg.R., Kustos an der botan. Abteilung des Naturhistorischen Mus. in Wien. Wien I (Deutsch-Österreich), Friedrichstr. 6. *Systematik der Phanerogamen. Flora der Südseeinseln. Pflanzengeographie.* ⓓ Wien 1893. ⓟ *Europäische Herbarpflanzen (Phanerogamen), besonders: Rumex, Salix, Potamogeton.*

Reck, Hans (1886), Prof., Dr. Berlin N 4 (Deutschland), Invalidenstr. 43. *Diluvialfauna Ost- und Zentralafrikas, Oberjurassische Flugsaurier Deutsch-Ostafrikas.* ⓓ München 1910.

Recknagel, Arthur Bernhard (1883), Prof. of forestry Cornell Univ. Ithaca, N.Y. (U.S.A.), 523 Highland Rd. *Forestry.* o

Record, Samuel James (1881), Prof. of Forest Products, Yale Univ. New Haven, Conn. (U.S.A.), 205 Prospect Street. *Structure, properties of woods.* ⓓ Yale Univ., New Haven, Conn., 1906.

Records, Edward (1886), Research Prof. of Vet. Sc., Univ. of Nevada. Reno, Nev. (U.S.A.). *Contagious and infectious diseases of domesticated animals.* ⓓ Univ. of Pennsylvania, Philadelphia 1909. ⓟ *Cultures of pathogenic bacteria; pathol. specimens.*

Redaelli, Piero (1898), Aiuto e lib. doc. Pavia (Italia), Ist. di Anat. Patol. Univ. *Patol. del testicolo e del cancro. Micol. Micropatol. umana e sperimentale.*

Reddick, Donald (1883), Prof. of Plant Pathol., New York State Coll. of Agric., Cornell Univ. Ithaca, N.Y. (U.S.A.). *Conditions of Parasitism.*

Development of disease resistance. Resistance in beans (Phaseolus vulgaris) for mosaic and for anthracnose (Colletotrichum). Resistance in potato (Solanum) for Phytophthora. ⓟ Ph.D. Cornell Univ. 1909. ⓟ Phytophthora resistant potatoes, Anthracnose resistant beans.

Redeke, Heinrich Carl (1873), Dr., Dir. des Rijksinst. voor Biol. Visscherijonderzoek, Dir. der Zool. Station Den Helder, Priv.Doc. an der Univ. Amsterdam (Holland). *Fische und Fischerei, Hydrobiol.* ⓟ Amsterdam 1898.

Redenz, Ernst (1898), Prosektor d. Anat., Abtlg. f. Mikroskopie. Würzburg (Deutschland), Köllickerstr. 6. *Nebenhoden, Spermienbewegung, Mechanisches Verhalten verschiedener Gewebe.* ⓟ Heidelberg 1922.

Redfield, Alfred C. (1890), Ass. Prof. of Physiol., Harvard Univ. Readville, Mass. (U.S.A.). ⓟ Harvard 1917.

Redikorzev, Vladimir (1873), Chef-Zool. Leningrad (U.d.S.S.R.), Zool. Mus. Akad. d. Wiss. *Systematik der Tunicaten und Pseudoscorpione. Anat., Morphol. und Systematik der Coelenteraten, Arthropoden und Tunicaten.* ⓟ Heidelberg 1901. ⓟ Palaearktische Pseudoscorpione.

Redina, Lydia (1893), Ass. Moskau 7, (U.d.S.S.R.), Rostowsky G. 15, W. 44. *Mineralstoffwechsel.*

Rédjeb, Tewfik, Prof. Univ. Stambul (Türkei). *Histol., Embryol.* o

Reed, Bessie Price, Independent ornithol. Dallas, Tex. (U.S.A.), Reagan St. 2635. ⓟ Kansas 1923.

Reed, Carlos Isaac (1887), Associate Prof. of Physiol., Coll. of Med., Baylor Univ. Dallas, Tex. (U.S.A.), 2635 Reagan St. *Photophysiol. Thyroids and parathyroids; adrenalin; physiol. and pharmacol. of heparin.* ⓟ Chicago 1925.

Reed, Edward L., Prof. of Biol., John Tarleton Coll. Stephenville, Tex. (U.S.A.). o

Reed, Guilford B., Prof. Bact., Queens Univ. Kingston, Ontario (Canada), 210 Albert St. *Bact., Physiol.* o

Reed, Howard S. (1876), Prof. of Plant Physiol., Univ. of California, Citrus Experim. Station. Riverside, Cal. (U.S.A.). *Plant nutrition, bact., the nature of the growth process, biometry.* ⓟ Missouri 1907.

Reed, Hugh Daniel (1875), Prof. Zool. Cornell Univ. Ithaca, N.Y. (U.S.A.), 107 Brandon Pl. *Columella auris of Amphibia. Integument Exoskeleton of Fishes. Metamorphosis and Neoteny, especially the morphol. aspect.* ⓟ Cornell Univ. 1893. ⓟ Larval stages of Anura.

Recher, Max Moore (1892), Ass. Entomol., U. S. Dept. of Agric. U.S. Entomol. Labor. Forest Grove, Ore. (U.S.A.). *Economic Entomol., Cereal and Forage Insects.* ⓟ Leland Stanford Junior Univ. 1918.

van Rees, J., Dr., Prof. emer. Hilversum (Holland), Godelindenweg 20. *Histol.* o

Reese, Albert Moore (1872), Prof. of Zool. in West Virginia Univ. Morgantown, W.Va. (U.S.A.). *Anat., embryol., habits etc. of lower vertebrates, alligator.* ⓟ Johns Hopkins Univ. 1900.

Reeves, George I. (1879), Associate Entomol., U.S. Dept. of Agric. Salt Lake City, Utah (U.S.A.), 473/4 Avenue. *Ecol. and economic relations of Phytonomus posticus Gyl. Alfalfa Weevil.*

Reeves, Harry Gordon (1896), Senior Demonstrator, St. Bartholomeus Hospital, Med. Coll., Univ. of London E.C. 1, 6 Giltspur St. London S.W.1. (England), 8, Claverton St. *Physiol. and Biochem.* ⓟ Birmingham 1925.

Reeves, Joseph Amos (1900), Ass. Entomol., Florida State Plant Board. Sanford, Fla. (U.S.A.), Entomol. Labor. *Life History of the Celery Leaftyer; Chrysomelidae, Genitalia of the Genus Gypona.* ⓟ Ohio State Univ. 1924.

Reeves, Robert G. (1898), Instr. in Botan., Iowa State Coll. Ames, Ia. (U.S.A.), Dept. of Botan. *Plant Cytol.*

Réfik, Ahmed, Prof. ad. Univ. Stambul (Türkei). *Bact., Hygien.* o

Regan, William Michael (1884), Associate Prof. of Animal Husbandry, Univ. of California. Davis, Cal. (U.S.A.). *Dairy cattle, genetic, nutrition, growth and milk secretion.* ⓟ Univ. of Missouri, Columbia 1912.

Regaud, Claude (1870), Prof. Inst. Pasteur, Dir. Labor. de Radiophysiol. Univ. Paris (France). *Applications biol. et thérapeutiques des radiations.* ⓟ Lyon.

Regel, Constantin (1890), o. Prof. Univ., Dir. Botan. Garten. Kaunas (Litauen). *Floristik der Arktis. Pflanzengeographie, Soziol. der Kola Lappmark. Vegetation der Wiesen.* ⓟ Würzburg 1921.

Regen, Johannes (1868), Prof. emer. Wien IX (Deutsch-Österreich), Grundlstr. 2. *Stridulation und Gehörsinn der Lokustiden und Grylliden, speziell von Thamnotrizon. Liogryllus. Winterschlaf. Atmung der Insekten. Spermatophoren von Liogryllus.* ⓟ Wien 1897.

Regnier, Robert Marie Gustave (1894), Dir. Station entomol. du Nord-Ouest et du Mus. de Rouen (France), 16, Rue Dufay. *Biol. des insectes et animaux utiles ou nuisibles à l'Agric.* ⓟ Paris 1913 et 1914.

Reh, Ludwig (1867), Kustos. Hamburg (Deutschland), Zool. Staatsinst. u. zool. Mus. *Angewandte Zool.* ⓟ Jena 1892.

Rehberg, Paul Brandt (1895), Ass. of the zoophysiol. Labor., Univ. of København (Danmark), Dyrefysiologisk Labor. *Circulation: Capillaries. Kidney. Microchem.* ⓟ København 1926.

Rehder, Alfred (1863), Curator of the Herbarium, Arnold Arboretum, Harvard Univ. Jamaica Plain, Mass. (U.S.A.). *Dendrol., taxonomy and phytogeography of ligneous plants. Lonicera, Azaleas.* ⓟ Ligneous plants.

Reibisch, F., Univ.Prof. Charakt. Ordinarius, Leiter des Zool. Mus. Univ. Kiel (Deutschland), Feldstr. 96. o

Reich, Edward (1885), Generalsekr.Tschech. Akad. Landw., Doc. Landw. Hochsch., Vorstand Schulwesenabt. Ministerium d. Landwirtsch. Brno. Praha XII (C.S.R.), Benešovská 1888. *Pflanzenernährung und -kultur. Wiesenbau.* ⓟ Tschech. techn. Hochsch. Prag 1918.

Reichardt, Axel (1891), Wissensch. Mitarbeiter Zool. Mus. Russische Akad. d. Wissensch. Leningrad (U.d.S.S.R.). *Systematik, Biol. der Histeriden (Coleoptera). Anat. des Copulationsapparats. Kohlmotte (Plutella mac.).*

Reichel, John (1886), Dir. Biol. Labor. of the H. K. Mulford Co. Philadelphia, Pa. (U.S.A.). *Tuberculous Infection, Bacillus Vaccine, Cholera virus.* ⓟ Univ. of Pennsylvania 1906.

Reichelt, Max Wilhelm Ludwig (1900), Dr. phil. Leipzig C1 (Deutschland), Hohe Str. 13I. *Pigmentbildung, Schuppen- und Flügelentwicklung der Schmetterlinge.* ⓟ Leipzig 1924.

Reichenbach, Hans, Prof. Univ. Göttingen (Deutschland), Herzberger Landstr. 57. *Hygiene.* o

Reichenow, Anton (1847), Prof., Dr. phil., Geh. RegR. Hamburg (Deutschland), Wrangelstr. 16. *Theor. Zool. (Systematik, Faunistik), Vögel.* ⓟ Rostock 1871.

Reichenow, Johann Eduard (1883), Prof. Dr., Abteilungsvorsteher im Inst. für Schiffs- und Tropenkrankheiten. Hamburg 4 (Deutschland), Bernhardstr. 74. *Protozool., bes. parasitische Protozoen.* ⓟ München 1908.

Reichensperger, August (1878), Dr. phil., Prof. für Zool. u. vergl. Anat., Dir. des Zool. Inst. Freiburg (Schweiz), *Anat. und Biol., Echinodermen, Insekten, Myrmekophilen u. Termitophilen.* ⓟ Bonn 1905. ⓟ Myrmekophilen (sp. Paussidae, Clavigeridae, Histeridae) und Termitophilen (sp. Staphylinidae, Rysopaussidae etc.), ferner Ameisen, Termiten.

Reicher, M., Dr. Prof. Wilno (Polska), Inst. d'Anat., R. Słowackiego. *Anat.* o

Reid, D. G., Lect., Gouville and Caius Coll. Cambridge (England). *Anat.* o
Reid, P. E. (1897), Prof. of Botan. and Head of the Dept. of Biol. in Austin College, Sherman, Tex. (U.S.A.). *Phytopath.*
Reighard, Jacob Ellsworth (1861), Prof. of Zool., Dir. Biol. Stat. Univ. Mich. Natur Sc. Bldg. Ann Arbor, Mich. (U.S.A.). *Fresh-water Biol. Evolution, Development, behavior and habits of fishes. Subaquatic photography. Anat. of the bat.* Ⓓ Michigan 1882. o
Reihling, Karl (1888), Reg.R., Dr. rer. nat., Vorstand der Bodenkundlichen Abteilung der Württ. Forstl. Versuchsanstalt. Stuttgart (Deutschland), Arminstr. 6 III. *Symbiosen und Biocönosen.* Ⓓ Tübingen 1914.
Reiling, Hans (1886), Dr. phil., Saatzuchtleiter. Mittelstendorf b. Soltau, Hannover (Deutschland). *Pflanzenzüchtung, Kartoffelzüchtung.* Ⓓ Jena 1912.
Reimers, Hermann J. O. (1893), Ass. Botan. Mus. Berlin-Dahlem (Deutschland), Königin-Luise-Str. 6-8. *Systematik, Geographie, Morphol. der Laub- u. Lebermoose. Europäische Moore. Pflanzliche Textilstoffe.* Ⓓ Hamburg 1922.
Reinberg, Samuel (1897), Doc. Inst. für Röntgenol. u. Radiol. Leningrad (U.d.S.S.R.),Roentgen-Str. 6. *Physiol. der Bronchien, des Uterus und der Eileiter, Röntgendiagnostik in der Palaeont.* Ⓓ Leningrad 1921.
v. Reinbold, Béla (1875), Prof., Dr., Univ. in Szeged (Ungarn), Népkert sor 3. *Biochem. des Blutfarbstoffes.* Ⓓ Kolozsvár 1900.
Reinders, Eildert (1885), Prof. Landwirtschaftl. Hochsch. in Wageningen (Holland), Nassauweg 5. *Anat. tropischer Nutzpflanzen.* Ⓓ Groningen 1913.
Reinhardt, Adolf (1878), Leiter des Pathol. Inst. beim Krankenhaus St. Georg in Leipzig-Eutritzsch. Leipzig (Deutschland), Asterstr. 17. *Geschwülste. Innere Sekretion. Bakterienlatenz. Leishmaniosen. Serodiagnose der Tuberkulose.* Ⓓ Marburg 1904. Ⓟ *Pathol.-anat. u. mikroskop. Praep.*
Reinhardt, Alexander W., Prof., Inst. für Volksbildung. Dnepropetrowsk (U.d.S.S.R.), Nagornaja 12. *Pflanzenphysiol. und Microbiol.* o
Reinhardt, Max Otto (1854), ao. Prof. Univ. Berlin W 50 (Deutschland), Augsburger Str. 9. *Pflanzenphysiol.* Ⓓ Berlin 1884.
Reinhardt, Richard (1874), Prof. Univ. Leipzig (Deutschland), Denkmals-Allee 110. *Infektionskrankheiten der Haustiere.* o
Reinhold, John Gunther (1900), Ass. Biochem., Philadelphia General Hospital and the Graduate School of Med. of the Univ. of Pennsylvania. Philadelphia, Pa. (U.S.A.), Biochem. Labor., *H-ion determination. Inorganic salt metabolism. Carbohydrate metabolism.* Ⓓ Yale Univ. 1926.
Reinitzer, Friedrich (1857), Prof., Ingenieur, Vorstand des Botan. Inst. der Techn. Hochsch. Graz, Steiermark (Österreich). *Harze (chem. und biochem.). Zellwandchem. Pflanzliche Enzyme.* Ⓟ *Holzzerstörende Pilze.*
Reinke, Johannes (1849), Dr., Prof. emer. der Botan. an der Univ. Kiel. Preetz bei Kiel (Deutschland). *Quellung. Chem. und Physiol. des Protoplasma. Biodynamik. Descendenztheorie. Algen. Lichenen. Morphol. der Vegetationsorgane bei Phanerogamen.* Ⓓ Rostock 1871.
Reinking, Otto August (1890), Plant Pathol. Boston, Mass. (U.S.A.), United Fruit Company. Ⓓ Univ. of Wisconsin 1922.
Reinmuth, Ernst Friedrich (1901), Ass. für Phytopathol. Landwirtschaftl. Versuchsstation in Rostock (Deutschland).*Pflanzliche u. tierische Feinde d.Kulturpflanzen.* Ⓟ *Pilzliche Erkrankungen an Pflanzen.*
Reinsch, Friedrich Kurt (1896), Dr., Ass. an der Lehrkanzel für Hydrobiol. und Fischereiwirtschaftslehre der Hochsch. für Bodenkultur in Wien XVIII (Deutsch-Österreich). *Hydrobiol. und Fischereibiol.* Ⓓ München 1922.

Reinwaldt, Edwin (1890), Univ.-Ass. Tartu (Estland), Zool. Inst., Aia 46. *Systematik und Verbreitung der estländischen Muriden.* Ⓓ Dorpat Ⓟ *Bälge und Schädel estländischer Muriden.*
Reis, Karoline (1886), Dr., Ass. an dem Biol. Inst. der Univ. in Lemberg (Polen), Plac Akademicki 3. *Hauttransplantationen. Histol. der Gasdrüse und die Oval in der Schwimmblase der Knochenfische. Embryonale Entwicklung der Knochenfische.* Ⓓ Lemberg 1909.
Reiser, Otmar (1861), Museumskustos, R.R. emer. Pickern, Post Lembach b. Marburg (S.H.S.). Ⓓ Wien, Hochsch. f. Bodenkultur 1887.
Reisin, Eduard (1890),Central-Paedagogisches Inst. in Mitau. Riga (Latvija), Martinstr. 8. *Distomen der Frösche. Entwicklung der Pneumonoeces variegatus.* Ⓟ *Distomen und andere Parasit. der Frösche, Fische und Vögel.*
Reisinger, Erich (1900), Priv.Doc. Univ. Graz (Österreich), Ass. Zool. Inst. d. Univ. Köln a. Rh. (Deutschland), Zool. Inst., Stapelhaus. *Vergl. Anat. und Entwicklungsgeschichte der Turbellaria.* Ⓓ Graz 1922.
Reis Martins, Miguel A. (1864), Prof. Bact. et de l'Histoire de la Méd. Ecole de Méd. Vét., Chef de service à l'Inst. de Bact. Camara Pestana. Lisbôa (Portugal). *Bact.* Ⓓ Lisbonne 1911.
Reiss, Paul (1901), Chef de labor. au Centre Régional de Lutte contre le Cancer. Strasbourg (France), 16, Allée de la Robertsau. *Physicochim. de la cellule. Le p_H intérieur cellulaire.* Ⓓ Strasbourg 1926.
Reiter, Hans Conrad Julius (1881), Prof. an der Univ. Rostock, Dir. des Mecklenburgischen Landesgesundheitsamtes in Schwerin i. M. (Deutschland). *Aetiol. und Epidemiol. der Weilschen Krankheit (Icterus infectiosis). Immunitätsverhältnisse bei d. experim. Kaninchensyphilis, Züchtungsverfahren d. Spirochaeta pallida und der Spiroch. dentium. Perorale Immunisierung gegen Infektionen und Intoxicationen. Studien über die ,,Stumme Infektion". Das Infektionsproblem. Chemotherapie. Neue spezifische Behandlung der Paralyse mit lebender Spirochaeta pallida.* Ⓓ Leipzig 1906.
Reith, Ferdinand (1900), Ass. an der Abteilung für Entwicklungsmechanik und Vererbung, Anat. Inst. Breslau 16 (Deutschland), Maxstr. 6. *Entwicklungsmechanik und Vererbung bei Säugetieren.* Ⓓ Würzburg 1925.
Rekling, Felix Alf Eigil (1896), Wissenschaftl. Ass. des Labor. Finsens med. Lichtinst. København (Danmark). *Lichtbiol.*
Reko, Victor A. (1880), Prof., Chefe del dept. cientifico de la caro Bayer. Mexico, D. F. (Mexico), Aptdo 208 bis. *Farmacol. et phon. experim. Toxicol.*
Remane, Adolf (1898), Priv.Doc. für Zool. an der Univ. Kiel (Deutschland), Zool. Inst., Hegewischstraße 3. *Morphol. der Gastrotrichen und Archianneliden. Morphol. und Rassen der Anthropoiden. Allgemeine Rassenkunde des Menschen. Biol. der Wirbeltiere.* Ⓓ 1921.
Remlinger, M., Dir. de l'Inst. Pasteur. Tanger (Maroc). *Bact.* o
Remotti, Ettore (1893), Libero doc. in Fisiol., Generale Aiuto nell' Ist. di Anat. compar. R. Univ. Bologna (Italia). *Biol. embrionale dei teleostei: migrazioni, resistenza della capsula, processi biochim. della fecondazione, fisiol. della goccia oleosa, adattamenti difensivi, cromatofori, metabolismo, metabolismo dei celenterati. Vesica natatoria dei pesci. Migrazioni dei pesci. Fisiomorfologia del sesso nei pesci e negli uccelli. Anat. dei pesci. Fisiomorfol. del sacco vitellino degli uccelli.* Ⓓ Messina 1920
Remy, Eduard (1881), Dr. phil., Abteilungsvorstand am Hygien. Inst. der Univ. Freiburg, Chem. Abteilung. Freiburg i. Breisgau (Deutschland), Hebelstr. 12. *Biochem., Grenzgebiet der Chem. und Bact. (Adsorptionserscheinungen), Ernährungslehre, Hygiene der Abwässer, Desinfektion.* Ⓓ Rostock 1909.

Remy, Theodor (1868), Geh. Reg.-R. Dr. phil., o. Prof. an der Landwirtschaftl. Hochsch. Bonn-Poppelsdorf. Bonn (Deutschland), Haydnstr. 3. *Bodenkultur- und Pflanzenbaulehre. Bodenbiol., Chem. des Bodens, Düngungsfragen, Kartoffel- und Zuckerrübenbau.* ① Kiel 1896.

Renard, Konstantin G., Prof. der Weißruss. landwirtschaftl. Akademie. Gorki (U.d.S.S.R.). *Selektion landwirtschaftl. Pflanzen.* o

Renaux, Ernest (1885), Sous-Dir. Inst. Pasteur, Bruxelles (Belgique), 95, rue Belliard. *Bact.* ① Bruxelles 1908.

Rendahl, C.H., Doc.Univ.Stockholm(Sverige).*Zool.* o

Rendle, Alfred Barton (1865), Keeper of Botan., British Mus., President of Linnean Society. London S.W. 7 (England), British Mus., Nat. Hist., Cromwell Rd. *Taxonomy of flowering Plants. Flora of Jamaica. African Plants.* ① London 1898.

Renich, Mary E. (1877), Ass. Prof. Biol. Alma Coll. Alma, Mich. (U.S.A.). *Plant Physiol. Growth Problems.* ① Univ. of Illinois 1920.

Renier, Armand, Dir. service géol. Bruxelles (Belgique), 71, avenue de l'Armée. *Paléont.* o

Renner, Otto (1883), Dr. phil., o.ö. Prof. an der Univ., Dir. des Botan. Inst. und Gartens. Jena (Deutschland). *Vererbung der Gattung Oenothera, Wasserversorgung der Pflanzen.* ① München 1906.

Renngarten, Vladimir (1882), Géol. en chef du Commité géol. de Russie, Ingénieur des Mines. Leningrad (U.d.S.S.R.), 10 Wassil. Ostr., Tutschkowskaja nabershn. *Paléont.: Céphalopodes, Lamellibranches, Brachiopodes, Echinodermes, Spongiaires, Foraminifères des terrains crétacés du Caucase.*

Renouf, Louis Percy Watts (1887), Prof. of Zool. Cork National Univ. Coll. (Ireland), *Evolution. Faunistics especially in connection with fisheries and agric.* ① Cambridge (England) 1912.

Renqvist, Yujo (1894), Priv.Doc., Ass. Physiol. Inst. Helsinki (Finnland). *Bewegungs- und Geschmacksempfindungen.* ① 1921.

Rensch, Bernhard (1900), Dr. phil., Verwalter der Molluskenabteilung und der Schausammlung des Zool. Mus. Berlin N 4 (Deutschland), Invalidenstr. 43. *Phylogenetische, tiergeographische und systematische Untersuchungen an Vögeln und Mollusken, geographische Variation. Färbung der Vogelfedern. Sinnesphysiol. der Vögel. Biol. und Sinnesphysiol. der pflanzenparasitären Nematoden.* ① Halle a. d. S. 1922. ② Mollusken, besonders ostasiatische, indomalayische und polynesische.

Renvall, K. Thorsten G. (1868). Helsinki (Finnland), Elisabethstr. 21 A. *Ornithol.* ① 1911.

Resser, Charles E. (1889), Associate Curator, U.S. National Mus. and Ass. Prof. Geol., George Washington Univ. Washington, D.C. (U.S.A.), Smithsonian Inst. *Cambrian Palaeont. and Stratigraphy. Ozarkian Fossils of Novaya Zemlya.* ① Geo. Washington Univ. 1917.

Ressler, Barton C. V. (1898), Ass. Prof. of Zool. Knoxville, Tenn. (U.S.A.), Univ. of Tenn. *Genetics. Invertebrate Embryol.* ① Iowa State Coll 1923.

Resvoll, Thekla Ragnhild (1871), Amanuensis Botan. Labor. Univ. Oslo. Bestun (Norge). *Pflanzenanat., Biol. arktischer und der Gebirgspflanzen.* ① Oslo 1918.

Reswoj, Peter (1887), Zool. Mus. Russ. Akad. d. Wissensch., Ass. Univ. Leningrad (U.d.S.S.R.). *Systematik und Biol. der Spongien. Hydrobiol., Süßwasserplankton, Systematik u. Biol. d. Rotatoria.* ① Univ. Leningrad 1915.

Retinger, Julius Maria (1885), o. Prof., Head Dept. of Physiol. Chem. Stefan Batory Univ. Wilno (Polska). *Nature of proteol. ferments of the intestinal tract, the organs and the blood. Metabolism of milk formation.* ① Leipzig 1913.

Retterer, M., Prof. agr. Fac. Méd. Paris 6 (France), 15, Rue de l'école de Méd. *Histol.* o

Rettger, Leo Frederick (1874), Prof. of Bact., Yale Univ. New Haven, Conn. (U.S.A.), 70 Ogden Street. *Bacterial nutrition and metabolism; Variation; Intestinal flora, transformation and influence on diet. Bacillary white diarrhea of chicks; infectious abortion of cattle. Carbon dioxide requirements of bacteria.* ① Yale Univ. 1902. ② Bact. cultures.

Reuszer, Herbert W. (1903), Instr. in Agronomy. New Brunswick, N.J. (U.S.A.), Agric. Experim. Station. *Soil Microbiol., probably concerning the microbiol. content of permanent pasture soils.* ① Univ. of Missouri 1925.

Reuter, Enzio, Dr. phil., Prof. Univ., Dir. Zool. Inst. und Mus. Helsinki (Finnland), Fabriksgatan 21 C. *Zool., mikrosk. Anat.* o

Reuterwall, O. G. M., Dr., Karolinska Inst. Stockholm (Sverige). *Pathol. Anat.* o

Reverdin, Jaques Louis (1842), Prof. honoraire. Genève (Suisse), Rive de Pregny. *Etude anat. des papillons, système génitale male et femelle.* ① Paris 1870.

Révy, Desiderius (1900), Ass. Landw. Akademie. Magyaróvár (Ungarn). *Keimungsphysiol. der Samen der Blütenpflanzen. Parasit. Pflanzenkrankheiten.* ① Magyaróvár 1922.

Rewerdatto, Viktor W., Prof. d. Tomsker Staatl. Univ. Tomsk (U.d.S.S.R.), Timirjasew Pr. 3. *Geobotan.* o

Rex, Edgar George. Ass. in Plant Pathol., Pennsylvania State Coll. State College, Pa. (U.S.A.), Botan. Bldg. o

Rex, Hugo (1861), ao. Prof. der Anat. (Tit. o. Prof.), Deutsche Univ. Praha XII (C.S.R.), Mánesova 10. *Vergl. Anat. der Wirbeltiere. Cephalogenese der Vögel.* ① Prag 1883.

Rey, Luis Lozano, Conserv. Mus. Ciencias Natur. Madrid (España). *Ichthyol.* o

Reyes, Alicia E. (1893). Mexico, D.F. (México), Madero 72. *Paléont.: Eléphants du Quaternaire.* ① Univ. National de México 1919. ② Photografies de specimens paléont. mexicanies.

Reyes Calvo, Manuel. Lic. Calza (España). *Botan.* o

Reyne, Adriaan (1890), Utrecht (Holland), Oude Kerkstraat 12. *Schädliche Insekten der tropischen Kulturen, speziell von Kakao und Kaffee. Anat. und Biol. der Thysanopteren.* ① Utrecht 1926.

Reynès, H., Prof. Univ., Ecole méd. Marseille (France). *Pathol.* o

Reymann, Georg Christian (1878), Labor. am Statens Seruminst. København C (Danmark). *Immunobiol. (spec. Tetanus, Anaerobenzüchtung, antitoxische Eiweißuntersuchungen).*

Reynolds, Ernest Shaw (1884), Plant Physiol. of the Missouri Botan. Garden and Assoc. Prof. of Plant Physiol. at Washington Univ. St. Louis, Mo. (U.S.A.). *Nutritive Relationships of Fungi, parasitism, and the resistance of hosts to parasitic fungi.* ① Univ. of Illinois 1909.

Reznikoff, Paul (1896), Associate in Anat., Instr. in Med., Cornell Univ. Med. Coll. New York City, N.Y. (U.S.A.), 28. St. and 1st Ave. *Experim. cell biol.* ① Cornell Univ. Med. Coll. 1920.

Reznitschenko, Michael S. (1900), Ass. Inst. of General Biol. II. Univ., Supernum., Collaborator Labor. of Experim. Biol. Zoopark. Moscow (U.d. S.S.R.). *Physical-chem. Biol. Internal factors of development. Action of hormones and iones.* ① Rostow on the Don 1924.

Rezzesi, Francesco Domenico (1902), Ass., Ist. di Patol. Generale della R. Univ. di Bologna (Italia). *Istol. Patol., Tumori.* ① Bologna 1925.

Rhoade, Arthur Stevens (1893), Ass. Plant Pathol. Cocoa, Flo. (U.S.A.), Florida Agric. Experim Station, Cocoa Labor. *Diseases of woody plants. Woodrotting fungi and timber decay. Fruit diseases, grapes and citrus and other subtropical fruits.* ① Syracuse Univ. 1917.

Rhode, Heinz (1896), Leiter med. Abteilung der I. G. Farbenindustrie A.G., Werk Leverkusen b. Köln a. Rh. Köln-Mülheim (Deutschland), Berliner Str. 109. *Lokalanaesthetika, Schwefel, Giftwirkung in verschiedenem Milieu.* ① Göttingen 1920.

Rhodes, Robert Clinton (1887), Prof. of Biol., Emory Univ. Georgia (U.S.A.). *Protozool., Cytol., Eugenics.* ⓓ Univ. of California 1917. ⓟ Protozoa, live, preserved or prepared slides.

Rhumbler, Ludwig (1864) Dr. phil. Prof. an der Forstlichen Hochsch. Hannöversch-Münden (Deutschland), Veckerhagener Str. 73. *Zellmechanik. Foraminiferen. Forstzool.: Coleoptera (Lamellicornia), Rhynchota. Geweihkunde.* ⓓ Straßburg i. E. 1888.

Riabinin, Anatol Valentinus N. (1874), Géol. en chef du Comité Géol. de Russie; Prof. paléont. Inst. des Mines. Leningrad W.O. (U.d.S.S.R.), Sredny Prospect, 72-b. *Amphibiens paléozoïques. Reptiles mésozoiques. Mammifères tertiaires et posttertiaires.*

Riaboff, John (1897), Ass. Pomol. Section Government Botan. Garden, Nikita. Jalta Crimes (U.d. S.S.R.). *Biol. of Fruit tree Varieties: Blossoming, Pollination, Ripening, Immunity.* ⓟ Cuttings, Seeds, Leafs, Flowers of Fruit trees.

Riabuschinsky, Nadine (1886), Ass. Inst. für Ernährungsphysiol. Kom. für Volksgesundheit. Moskau (U.d.S.S.R.), Maly Charitoniewsky per. 12, W. 1. *Stoffwechsel, Gaswechsel und Microchem. des Blutes.* ⓓ Moskau 1916.

Ribbing, L., Prof., Dr. Akad. Stockholm (Sverige). *Anat.* ○

Ribet, Marcel René (1894), Prof. agrégé d'anat. à la Fac. de Méd., Chef des Travaux pratiques. Univ. d'Alger (Algérie). *Anat.* ⓓ Alger 1923.

Ribeyro, Ramon (1878), Prof. de Parasit. de la Fac. de Med. Lima (Peru), Casilla 1102. *Parasit. bact., Vaccine, Dysenterie, Typhus.* ⓟ Préparations parasit. de l'homme.

Ricca, Ubaldo (1874), Lib. doc. univ. Genova (Italia), Via Corsica 6-10. *Fisiol. vegetale fisica. La propagazione di Stimolo nella Mimosa.* ⓓ Genova 1897.

Riccardo, Salvatore (1892), Aiuto Labor. Batt Ist. Superiore Agrario. Portici (Italia). *Nitrificazione, denitrificazione.* ⓓ Portici 1919.

Rice, Edward Loranus (1871), Prof of Zool., Ohio Wesleyan Univ. Delaware, O. (U.S.A.) 265 N. Sandusky St. *Embryology of Vertebrate Skull.* ⓓ Munich 1895.

Rice, Mabel A., Ass. Prof. in Biol. Wheaton Coll. Norton, Mass. (U.S.A.). ○

Rice, Waldo Silas (1895), Associate Prof. Animal Husbandry. Athens, Ga. (U.S.A.), Georgia State Coll. of Agric. ⓓ Univ. of Nebraska 1921.

Rich, Willis H. (1885), Chief Investigator, Salmon Fisheries, U.S. Bureau of Fisheries. Stanford Univ., California (U.S.A.). *Life Histories and Ecol. of the Pacific Coast Salmons.* ⓓ Stanford Univ. 1924.

Richard, Jules (1863), Dir. Mus. océanographique. Monaco (Principauté). *Biol. marine.* ⓓ Paris.

Richards, Alfred Newton (1876), Prof. of Pharmacol., Univ. of Pennsylvania. Philadelphia, Pa. (U.S.A.). *Physiol. Action of Histamine. Physiol. and Pharm. of the Kidney.* ⓓ Columbia 1901.

Richards, Aute (1885), Prof. of Zool., Head of Zool. Dept., Dir. of Zool. Mus. Univ. of Oklahoma. Norman, Okla. (U.S.A.), 434 Chautauqua Ave. *Cytol. and Cellular Embryol. Origin of Germ Cells. Mechanics of Cell Division.* ⓓ Princeton 1911.

Richards, Bert Lorin (1886), Prof. of Botan. and Plant Pathol. Logan, Utah (U.S.A.), Utah Agric. Coll. *Virus diseases, Rhizoctonia of various crops.* ⓓ Univ. of Wisconsin 1920.

Richards, C. Andrey, Ass. Pathol., Labor. of Forest Pathol., Bureau of Plant Industry. Madison, Wis. (U.S.A.), 417 Sterling Place. *Woodpulp deterioration, toxicity of wood extracts and wood preservatives.* ○

Richards, E. H., Prof. Rothamsted Exper. Stat. Harpenden, Herts. (England). *Ferments.* ○

Richards, Herbert Maule (1871), Prof. of Botan. Barnard Coll., Columbia Univ. New York City, N.Y. (U.S.A.). *Plant Physiol., Chem.* ⓓ Harvard Univ. 1895.

Richards, Oscar W. (1902). Ralph Sanger Scholar at the Labor. of General Physiol., Harvard Univ. Cambridge, Mass. (U.S.A.). *General Physiol.* ⓓ Univ. of Oregon 1925.

Richards, Owain Westmacott (1901). Oxford (England), Dept. of Compar. Anat., Univ. Mus. *Systematics and habits of European bees and wasps. European aquatic hemiptera. British Diptera.* ⓓ Oxford 1924.

Richardsen, August (1873), Prof., Dr. phil., Dir. des Inst. für Tierzucht und Molkereiwesen an der Landwirtschaftlichen Hochsch. Bonn-Poppelsdorf (Deutschland). *Tierzucht.* ⓓ Jena 1902.

Richardson, James Keith (1899), Ass. Plant Pathol. Fredericton, N.B. (Canada), Box 867. *Plant diseases.* ⓓ McGell Univ. Montreal 1922.

Richet, Charles (1850), Prof. hon. Physiol. Fac., méd., Dir. Inst. Marey. Paris VII (France), 15, Rue de l'Univ. *Physiol. et Psychophysiol.* ⓓ Méd. 1877, sc. 1879, h. c. 1887, Prix Nobel 1913.

Richey, Frederick D. (1884), Agronomist in Charge of Corn Investigations, Bureau of Plant Industry, U.S. Dept. of Agric. Washington, D.C. (U.S.A.). *Genetics and agronomics of maize.* ⓓ Univ. of Missouri 1909.

Richmond, Charles Wallace (1868), Assoc. Curator U.S. Nat. mus. Washington, D.C. (U.S.A.), 1929 Park Road N.W. *Ornithol.* ○

Richmond, Edward Avery (1887), U.S. Dept. of Agric., New Jersey State Dept. of Agric. Riverton, N.J. (U.S.A.). *Physiol. Studies on the Japanese Beetle including chemotropism. Phototropism, basic Physiol. and Anat.* ⓓ Cornell 1924. ⓟ Hydrophilidae.

Richter, Aladár, Dr., Univ.Prof. emer. Budapest VIII (Ungarn), Nepozinházgasse 32. *Pflanzenanat.* ○

Richter, Andreas (1871), Prof. Univ., Dir. Abteilung für angewandte Botan. Landwirtschaftl. Versuchsstation. Saratow (U.d.S.S.R.). *Pflanzenphysiol. u. Microbiol.* ⓓ St. Petersburg 1904.

Richter, Curt Paul (1894), Associate in Psychobiol., Johns Hopkins Med. School. Baltimore, Md. (U.S.A.). *Human and animal behavior problems. Electrophysiol. Sleep and stupors.* ⓓ Johns Hopkins 1921.

Richter, Hans (1880), o. Prof. Anat., Embryol., Histol., vergl. Anat. vet.-med. Fac. Univ. Tartu (Estonia), Russischestr. 22. *Iris und Pupille. Luftsack (Divertese tubae auditivae). Gelenkfunktionen spez. federnde od. Schnappgelenke. Vererbung: Inzuchtwirkung, Beziehung der Funktion zur Vererbung. Erklärung der Wirkung der Vitamine. Anat., Physiol. u. Biol. vom Rentier.* ⓓ Zürich 1909.

Richter, Irène (1895), Ass. Labor. Histol. der Haussäugetiere Landwirtschaftl. Inst. Leningrad (U.d.S.S.R.), Gagarinskaja 8, W. 2. *Histol. der Milchdrüse u. d. endokrinen Organe.*

Richter, Karl (1897), wissenschaftl. Hilfsarbeiter am Bact. Inst. der Preuß. Versuchs- u. Forschungsanstalt für Milchwirtschaft. Kiel (Deutschland). *Milchbact.; Pflanzenpathol.; arktische Dipteren.* ⓓ 1920. ⓟ Arktische Dipteren, Hymenopteren und Lepidopteren.

Richter, Oswald, Dr., der Botan. Brno (C.S R), Krmenského nám., Deutsche techn. Hochschule. ○

Richter, Rudolf (1881), Dr., a.o. Prof. der Geol. und Palaeont. an der Univ. Frankfurt a. M. (Deutschland), Senckenberg-Mus., Viktoria-Allee 7. *Evertebraten des Palaeozoikums, Palaeobiol. auf Grund heutiger Flachseebeobachtungen.* ⓓ Marburg 1909. ⓟ Fossilien des Palaeozoikums.

Rickett, H. W. (1896), Ass., Prof. of Botan. Department of Botan., Univ. of Missouri. Columbia, Mo. (U.S.A.). *Cytol. study of fertilization in Angiosperms, reactions of gametes.* ⓓ Wisconsin 1922.

Rickman, Wiktor (1887), Ass. Physiol. Inst. of Academy of Sc. Leningrad (U.d.S.S.R.), 34, Sredny prosp. Wass. Ostr. *Central nervous system.*

Ricôme, H., Prof. Univ. Poitiers (France). *Botan.* ○

Riddle, Oscar (1877), Investigator, Carnegie Inst. of Washington. Cold Spring Harbor, L.I., N.Y. (U.S.A.), Station for Experim. Evolution. *Sexuality and reproduction.* ① Unif. of Chicago 1907.

Ridgway, Robert (1850), Curator Division of Birds, U.S. National Mus. Washington (U.S.A.). *Birds of North and Middle America.*

Riech, Fritz (1896), Ass. Zool. Inst. u. Mus. Königsberg, Pr. (Deutschland), Zool. Inst., Sternwartstr. 1. *Entwicklungsgeschichte, Morphol. u. Physiol. der Parietalorgane. Faunistik, Oekol. und Experimentalphysiol. der Brackwassergebiete.* ② Königsberg 1924.

Riede, Wilhelm (1891), Dr., Priv.Doc. an der Landwirtschaftl. Hochsch. für Botan. und Vererbungslehre; Ass. am Botan. Inst. der Landwirtschaftl. Hochsch. Bonn a. Rhein (Deutschland), Meckenheimer Allee 106 II. *Kohlensäureversuche. Vererbungsversuche mit versch. Kulturpflanzen, Weiden.* ① München 1919.

Riedl, Franz (1887), Reg.R. der Bayr. Landesanstalt für Pflanzenbau u. Pflanzenschutz. München (Deutschland), Hildegardstr. 16/0. *Futterbau.*

Riegler, Paul, Prof. Univ. Bucarest (România). *Pathol. Anat. u. Mikrobiol.* ○

Riehl, Julius, Prof. Dr., Prof. Praha II (C.S.R.), Salmova 3. Experim. pathol. Inst.d. Deutsch.Univ. ○

Riehm, Eduard (1882), Reg.R. und Mitglied der Biol. Reichsanstalt Berlin-Dahlem (Deutschland). *Phytopath.* ① Halle a. d. S. 1925.

Ries, Donald T. (1903), Instr. in Entomol. M.S.C. East Lansing, Mich. (U.S.A.). ① Cornell. Univ. 1925.

Ries, Victor H. (1894), Ass. Prof. of Floriculture. Columbus, Ohio (U.S.A.). *Floriculture.* ① M.S. Cornell Univ. 1916.

Riesser, Otto (1882), Dr. phil. nat. et med., o. Prof. der Pharmacol.; Dir. des Pharmacol. Inst. der Univ. Greifswald (Deutschland). *Physiol., physiol. Chem. u. Pharm. der Muskeln, Physiol. und Pharm. des Stoffwechsels.* ① Phil. nat. Heidelberg 1906, med. Königsberg 1913.

van der Riet, B. de St. J., Prof. of organic chem. and Biochem. Stellenbosch (South Africa), Dept. of Chem., Univ. ② M.A. Cape, Ph. Halle. ○

Du Rietz, Gustav Einar, Doc. Univ. Uppsala (Sverige), Växtbiol. Inst. *Pflanzenbiol. Lichenol. Plantsociol. and -geography. Vascular plants. New Zealand.*

Rigg, George B. (1872), Associate Prof. of Botan. Univ. of Washington. Seattle, Wash. (U.S.A.), Dept. of Botan. *Physiol. and ecol. of sphagnum bogs. Physiol. of evergreenness.* ① Chicago 1914.

Riggs, Lloyd Kendrick (1888), Dir. of Research, New Jersey Coll. of Pharmacy. New Brunswick, N.J. (U.S.A.), 39 North 5 Avenue. *Pharm. of unsaturated Hydrocarbons. Physiol. Action of Anesthetics. Antiseptics.* ① Univ. of Chicago 1918.

Rigler, Rudolf (1898), Ass. Inst. allg. u. exp. Pathol. Univ. Wien IX (Österreich), Kinderspitalgasse 15. *Atmung von Bakterien. Physiol. der tierischen Wärmeregulierung. Pharm. Wirkung biogener Amine.* ① Wien 1922.

Riikoja, Heinrich (1891), ao. Prof. der Zool. der Evertebraten an der Univ. Tartu (Estland), Jakobi t. 62. *Limnol., spez. Bodenfauna, Plankton (quantitative Untersuchung): Rotatorien.* ① Tartu. ② Planktonproben.

Van Rijnberk, Gérard Abraham (1875), Prof. ord. de Physiol. dans la Fac. de Méd. de l'Univ. d'Amsterdam. Blaricum (Holland). *Système nerveux central.* ① Rome 1901.

van Rijssel, Evert Cornelis (1889), ProsektorBact. der städt. Krankenhäuser von Rotterdam (Holland), Beukelsdyk 166. *Pathol. Anat.; Bact.; Serol.*

Rikli, Martin (1867), Prof. Dr., Konserv. und Doc. an der Eidg. Technischen Hochsch. Zürich (Schweiz), Herzogstr. 10. *Vegetation der Mittelmeerländer und der Polargebiete.* ① Basel 1894. ② Phanerogamen der Mittelmeerländer und der Arktis.

Riley, William Albert (1876), Head, Dept. of Animal Biol., Univ. of Minnesota. Minneapolis, Minn. (U.S.A.). *Parasit.; Insect histol.* ① Cornell Univ. 1903.

Rimbaud, L., Prof. Univ. Montpellier (France). *Pathol.* ○

Rimpau, W. (1877), Prof. Dr. med., Dir. der Staatl. Bact. Unters. Anstalt München. Solln bei München (Deutschland), Sohncke-Str. 23. *Seuchenbekämpfung, Hygien., Bact.* ① München 1901.

Rimsky-Korsakow, Michael (1873), Prof., Dir. Entomol. Kabinett Staatl. Univ. Leningrad (U.d.S.S.R.), Forstinst., W. 14. *Allgemeine Zool. Angewandte Entomol. Schädliche Schlupfwespen. Wasserhymenopteren.* ① St. Petersburg 1895.

Rindone, Alfredo (1900), Ass. Ist. Anat. umana normale. Palermo (Italia). *Degenerazione e rigenerazione nervosa, App. tireo-paratireoideo. Ipofisi.* ① Palermo 1923. ② Embrioni.

Ring, Gordon Clark (1901), Fellow at Harvard Med. School. Westfield, Mass. (U.S.A.), 32 Jefferson St. *Thyroid.*

Ringoen, Adolph R. (1887), Ass. Prof. of Animal Biol. Minneapolis, Minn. (U.S.A.), Univ. of Minnesota, Biol. Dept. *Animal biol.* ① Univ. of Minnesota 1919.

Rintoul, Leonora Jeffrey (1878). Lahill, Largo Fife (Scotland). *Scottish Ornithol. Migration. Distribution.*

Riofrío, Benito Fernández, Prof. Univ. Barcelona (España). *Botán.* ○

Rio-Hortega, Pio del, Prof. Dr. Med. Madrid (España). Conde de Aranda, 4, 2., *Histol.: Gliazellen, Nervenzellen.* ○

Rioja Lo Bianco, Enrique (1895), Prof. Madrid (España), Mus. Nacional de Ciencias Nat. *Système, anat. et biol. des Annélides polychètes. Mollusques.* ① Madrid 1917. ② Annelida polychaeta. Mollusca.

Rioja y Martín, José, Prof., Catedrático de Zoografía de animales inferiores y moluscos de la Univ. Central. Madrid (España). *Anat. de animales inferiores.* ○

Rippa, G., Prof. Univ. Napoli (Italia). *Botan.* ○

Rippel, August (1888), Prof. Univ., Dir. Inst. f. landwirtsch. Bact. Göttingen (Deutschland), Bergstraße 2. *Landwirtsch. Bact.* ○

Ris, Friedrich (1867), Dr. med., Dir. der kantonalen Irrenanstalt in Rheinau, Kt. Zürich (Schweiz). *Entomol.: Taxonomie, Faunistik etc. der Odonata orbis terrarum, der Neuroptera und Pseudoneuroptera (sensu lato) der Schweiz.* ② Zürich 1890. ② Odonata.

Risa-Bey, Ismaïl (1884), Prof. Bact., Serol. u. Infektionskrankheiten Tierärztl. Hochsch., Dir. des Bact. Labor. des Städt. Schlachthofes. Konstantinopel (Türkei), Skutarie Sélimiée. *Lungenseuche der Ziege. Geflügelcholera.*

Risch, Carl (1878). Bärwalde, Neumark (Deutschland). *Hydrobiol.: Chemismus des Wassers.*

Rischkow, Vitolij (1896), Prof., Katheder der Botan. und Inst. f. Pädagogik. Charkow, Ukraine (U.d.S.S.R.), 54 Artiom Str. *Anat. und Cytol. der buntblättrigen Pflanzen.* ② Buntblättrige Pflanzen.

Risga, Peter (1883), Ass. of Agric. Technol. Labor. Riga (Latvija), Kronvalda bulv. 1, Univ. *Experim. apiary.* ① Boston Univ. 1920.

Rispal, L., Prof. Univ. Toulouse (France). *Bact.* ○

Risquez, J. R., Prof. Univ. Central. Caracas (Venezuela). *Bact., Parasit., Anat. pathol.* ○

Riße, Karl (1895). Ülzen b. Unna i. Westfalen (Deutschland). *Chromosomenzählung und Auflösung der Tapetenzellen bei den Dipsacaceen.*

Risskaltschrek, Apollinaria Terentjewka (1891), Ass. Labor. physiol. Chem. Univ. Leningrad (U.d.S.S.R.), Wassil. Ostrow, 16 Linie, Haus 29. *Eiweiß-Chem.*

Ritchie, Arthur David (1891), Lect. in Chem., Physiol. Univ. Manchester (England). *Chem. of Muscular Processes.*

Ritchie, James (1882), Keeper, Natur. History Department, Royal Scottish Mus., Edinburgh (Scotland). *Development of faunas; anat. and systematics of marine invertebrates, especially hydroid zoophytes; teratol.* ⓓ Aberdeen 1912.

Ritter, William Emerson (1856), President, Sc. Service; Prof. emer. of Zool., Univ. of California; Dir. emer., Scripps Inst. of Oceanography. La Jolla, Cal. Berkeley, Cal. (U.S.A.). *Anat. and embryol. of tunicates and enteropneusts. Psychobiol. studies on ants, salamanders, woodpeckers, beavers, monkeys and human infants.* ⓓ Harvard Univ. 1893. ⓟ Tunicates of the Pacific Coast of North America.

Rittershofer, August (1870), Gerichtlicher Sachverständiger für Ungezieferbekämpfung und Stalldesinfektion, Berlin W 50 (Deutschland), Augsburger Str. 22. *Ungezieferbekämpfung und Stalldesinfektion.* ⓟ Ungeziefer und Schädlinge.

Ritzema Bos, Jan (1850), Prof. emer. der Landwirtschaftlich. Hochsch., Wageningen (Holland), Bowlespark 3. *Phytophat.: die für Land- und Forstwirtschaft sowie für den Gartenbau schädlichen Tiere.* ⓓ Groningen 1874.

Rivas Mateos, Marcelo, Prof., Catedrático de la Fac. de Farmacia. Madrid (España), Hortaleza 85. *Botán.* ○

Rivas Merizalde, Luis Maria, Prof. Univ. Nac. Bogotá (Colombia). *Anat.* ○

Rivera Campanile, Giulia (1892), Ass. Stazione di Patol. Vegetale. Roma 30 (Italia), Via S. Susanna 13. *Fisiol. e Patol. vegetale.* ⓓ Roma 1921.

Rivera dei Ducchi, Vincens, Prof. incaricato Univ. Bari (Italia). *Botan.* ○

Rivera Gallo, Victoriano (1899), Prof., Catedrático del Inst. Oceanograf. Huesca (España). *Animaux marins Echinodermata.*

Rivero, M. J., Prof. Univ. Central. Caracas (Venezuela). *Histol. Physiol.* ○

Rivers, Thomas Milton (1888), Associate Member Rockefeller Inst. New York City (U.S.A.), 66 Street and Avenue A. *Bact.* ⓓ Johns Hopkins 1915.

Rivier, André (1894), Chef de travaux de botan. à l'école nationale d'agric. de Montpellier (France). *Pathol. végétale et agricole: étude de l'efficacité comparée des différents produits anticryptogamiques. Phytogénétique: l'hérédité des caractères acquis sous l'influence du milieu.* ⓓ Montpellier.

Riviere, Bernard B. (1880). Norwich (England), Poringland. *Ornithol.*

Rizzatti, Ferruccio (1862), Prof. ordinario di sc. nat. nel R. Liceo Cavour di Torino (Italia). *Storia delle scienze.* ⓓ Bologna 1900.

Rjabov, Michail (1890), Ass. Bureau of Appl. Entomol. Leningrad (U.d.S.S.R.), ul. Herzena 44. *Lepidoptera.* ⓟ Lepidoptera.

Rjabow, Iwan N., Ass. d. Nikitskij Botan. Gartens. Jalta, Krim (U.d.S.S.R.). *Obstbau.* ○

Roach, Blanche Muriel Bristol (1888), Algol., Rothamsted Agric. Experim. Station. Harpenden, Herts. (England). *Nature, distribution and physiol. of the algae of cultivated soils.* ⓓ Birmingham 1919.

Roach, W. A. (1895), Ass. Research Chem., Insecticides and Fungicides Dept. Harpenden, Herts. (England), Rothamsted Expt. Stat. *Fungicides, mode of action of sulphur as a soil fungicide. Biochem. of disease resistance in plants.* ⓓ London 1917.

Roaf, Herbert Eldon (1881), Prof. of Physiol. at the London Hospital, Med. Coll., Univ. London E. 1 (England). *Physiol. Biol. Chem.* ⓓ Toronto 1902.

Robbins, Wilfred William (1884), Associate Prof. Botan., Coll. of Agric., Univ. of California, and Botan. in Agric. Experim. Station. Davis, Cal. (U.S.A.). *Experim. botan.* ⓓ Univ. of Chicago 1917.

Robbins, William Jacob (1890), Prof. of Botan. Columbia, Mo. (U.S.A.), Dept. of Botan., Univ. of Missouri. *Physiol. of microorganisms, hydrogen-ion concentration, absorption of water and dissolved materials, tissue culture.* ⓓ Cornell Univ. 1915.

Robbins, William Rei (1899), Ass. Olericulturist. New Brunswick, N.J. (U.S.A.), New Jersey Agric Experim. Station. ⓓ Rutgers Univ. 1926.

Robel, Jan Zygmund (1886), Chef des Trav. prat. Inst. Chim. méd. Univ. Kraków (Polska), 7, Rue Kopernik. *Chim. des colorants pyrrol. synthetiques et naturels (chlorophylles, l'hémoglobines). Toxicol. chim.* ⓓ Kraków 1910.

Roberts, Edith A. (1881), Prof. and Chairman, Department of Botan. Poughkeepsie, N.Y. (U.S.A.), Vassar Coll. *Plant Physiol. and Ecol.* ⓓ Univ. of Chicago 1916. ⓟ Seeds.

Roberts, Elmer (1886), Associate Prof. in Animal Breeding, Coll. of Agric., Ass. Chief in Animal Breeding, Agric. Experim. Station, Univ. of Illinois, Urbana, Ill. (U.S.A.), 508 West Iowa St. *Genetics, inheritance of resistance to disease, hypotrichosis, and domestic animals.* ⓓ Univ. of Illinois 1917.

Roberts, F., Lect. Clares Coll. Cambridge (England). *Physiol.* ○

Roberts, Herbert T., Ass. Prof. of Botan., Univ. of Manitoba. Winnipeg (Canada). *Genetics, plant physiol.* ○

Roberts, John William (1882), Pathol., Fruit Diseases, Bureau Plant Industry, U.S. Department of Agric. Washington, D.C. (U.S.A.). *Diseases of the apple, peach, plum and cherry.* ⓓ Univ. of Nebraska 1904.

Roberts, Ray Harland (1890), Associate Prof. of Horticulture, Univ. of Wisconsin. Madison, Wis. (U.S.A.), Hort. Dept. *Physiol. studies with tree fruits: biennial bearing, fruit set, blossom but formation, stock and scion relations.* ⓓ Univ. of Wisconsin 1924.

Roberts, Raymond (1899), Instr. in Entomol., Univ. of Nebraska. Lincoln, Neb. (U.S.A.). *Tenthredinoidea.* ⓓ Colorado Agric. Coll. 1924.

Roberts, Thomas Sadler (1858), Prof. Ornithol., Dir. Zool. Mus., Univ. of Minnesota. Minneapolis, Minn. (U.S.A.). *Ornithol.* ⓓ Univ. of Pennsylvania 1885.

Roberts, Walter Morrell (1890), Dir. Labor. for Clinical Invest. and Research. Manchester (England), Royal Infirmary. *Gastric function. Fractional test meal method. Jaundice, with special reference to the van den Bergh reaction.* ⓓ Manchester Univ. 1911.

Robertson, A., Prof. School of Hygien. and Tropic. Med. London N.W. (England), Endsleigh Gardens. *Protozool.* ○

Robertson, Albert Duncan (1876), Prof. and Head Zool. Department, Univ. of Western Ontario. London (Canada). *Compar. Anat. Genetics. Oyster Culture.* ⓓ Univ. of Toronto 1910. ⓟ Vertebrates and Mollusca.

Robertson, David W. (1894), Associate Agronomist. Fort Collins, Col. (U.S.A.), Colorado Agric. Coll. *Plant Breeding.* ⓓ Univ. of Minnesota 1920.

Robertson, G. Scott, Head Chem. a. Animal, Prof. Nutrition Research, Dir. Mon. of Agric. Northern (Ireland). ○

Robertson, Muriel (1883), Protozool. to the Lister Inst. Chelsea Gardens. London S.W. 1 (England). *Protozool.: Trypanosomiasis. Anaerobic Bacteria. The spore bearing anaerobes.* ⓓ Glasgow 1923.

Robertson, Thorburn Brailsford (1884), Prof. of Biochem. and General Physiol., Univ. of Adelaide (South Australia). *Physical Chem. of Proteins. Growth and senescence of Animals.* ⓓ Phil. California 1907, sc. Adelaide 1908.

Robertson, William R. B., Ph.D., Assoc. Prof. of Zool. Columbia, Mo. (U.S.A.), Univ. of Missouri. ○

Robinsohn, Isak (1874). Wien IX (Deutsch-Österreich), Glasergasse 27. *Pflanzenbiol., hauptsächlich Blütenbiol.* ⓓ Wien 1903.

Robinson, Arthur, Prof. Univ. Edinburgh (Scotland). *Anat.* ○

Robinson, Basil N., Comité Géol. de Russie. Leningrad (U.d.S.S.R.), W.O., Sredny Pr., 72-B. *Palaeont. des neolit. Kaukasus.* ○

Robinson, Byron L. (1892), Prof. of Microscopic Anat., Univ. of Arkansas, Med. School. Little Rock,

Ark. (U.S.A.). *Macrophages and Monocytes in Adrenalectomized Albino Rats. Sweat Glands of the Dog.* ⓊUniv. of Minnesota 1920.

Robinson, Charles Summers (1885), Research Associate Chem., Michigan Agric. Experim. Station. East Lansing, Mich. (U.S.A.). *Mineral metabolism of dairy animals; composition of plant juices; organic constituents of peat; methods of analysis of plant and animal materials.* ⓊUniv. of Michigan 1917.

Robinson, Elliott S. (1894), Ass. Dir., Antitoxin and Vaccine Labor., Massachusetts, and Instr., Harvard Med. School, Departments of Bact. Jamaica Plain, Mass. (U.S.A.), 375 South Street. ⓊYale Univ. 1922.

Robinson, G. Canby, Dean and Prof. of Med., Vanderbilt Univ., School of Med. Nashville, Tenn. (U.S.A.). *Problems on the human circulation.* ⓊJohns Hopkins 1903.

Robinson, Jesse Mathews (1889), Acting Head Dept. Zool. Entomol., Acting Entomol Alabama Experim. Station. Auburn, Ala. (U.S.A.). *Cotton Insects, Vegetable Insects, Coccidae, Life histories and control.* ⓊOhio State Univ. 1916.

Robinson, Montgomery Valient (1888), Dir., Clinical Labor. of Chicago, Ill. (U S A), 4655 South MichiganAvenue. *Pathol.* ⓊSt. Louis1918. ⓅPathol. specimens.

Robinson, T. Ralph, Physiol., Bur. of Plant Ind., U.S. Dept. of Agric. Washington, D.C. (U.S.A.). o

Robinson, Wilfrid (1884), Prof. of Botan., Univ. coll. of Wales. Aberystwyth, Wales (England). *Plant Physiol., Mycol. and general Morphol.* ⓅManchester 1919.

Robinson, William Lipsett (1885), Ass. Prof. of Pathol., Univ. of Toronto, Ont. (Canada). ⓊUniv. of Toronto 1913.

Robinson, Winifred J. (1867), Dean, Women's Coll. Newark, Del. (U.S.A.). *Plant Morphol. Pteridophyta of the Hawaiian Islands.* ⓊColumbia 1912. o

Robinson, Wirt (1864), Prof. of Chem. West Point, N.Y. (U.S.A.). *Birds, Mammals, Insects* ⓊU.S. Military Academy 1887. ⓅColeoptera.

Robson, Guy Coburn (1888), Ass. Keeper, Dept. of Zool. London S.W. 7 (England), British Mus. (Natural History), Cromwell Road. *Mollusca.* ⓊOxford 1910.

Robyns, Walter (1901), Ass. au Jardin Botan. de l'Etat à Bruxelles, chargé de Cours à l'Univ. de Louvain (Belgique), Rue du Canal 110. *Cytol. (figure achromatique). Systématique générale. Flore Congolaise Géobotan.* ⓊLouvain 1923.

Rocasolano, Antonio Gregorio, Prof. Catedrático Fac. Sc. Univ. Zaragoza (España). *Biochem.* o

Roček, Josef, Dr., Prof. d. Hygiene. Brno (C.S.R.), Hygien. Inst. d. Masaryk-Univ. o

Roch, Felix (1901), Dr. phil. Wissenschaftlicher Mitarbeiter der Preußischen Landesanstalt für Wasser-, Boden- und Lufthygiene zu Berlin-Dahlem. Berlin NW 21 (Deutschland), Wilsnacker Str. 9. *Hydrobiol., besonders wasserhygien. Fragen (Süßwasser, Brackwasser und Seewasser). Hydroiden (Morphol. und Ökol.).* ⓊBerlin 1924.

Rocha Pereira, Alfredo (188 6), Prof. athol. Méd. Fac. Méd., Ass. au Labor. Méd. Porto (Portugal). *Serol.* ⓊPorto 1911.

Rock, Joseph F. Ch., Botan. Explorer, Arnold Arboretum, Harvard Univ., and Collaborator, Bur. of Plant Industry, U.S. Dept. of Agric. Jamaica Plain, Mass. (U.S.A.). *Systematic botan., plant geography.* o

Rockwood, Elbert William (1860), Prof. of Chem. and Toxicol., State Univ. of Iowa. Iowa City, Ia. (U.S.A.). *Enzymes: Uric acid.* ⓊPh.D. Yale Univ. 1904, M.D. State Univ. of Iowa 1894.

Rockwood, Lawrence Peck (1886), Associate Entomol., Bureau of Entomol., U.S. Department of Agric., Division of Cereal and Forage Crop, Insect Investigations. Forest Grove, Ore. (U.S.A.). *Insect pests of cereal and forage crops; especially Acrididae,*

Noctuidae (larvae). ⓊMassachusetts Agric. Coll. 1912. ⓅAcrididae or Noctuidae of the Pacific Northwest.

Rodenhiser, Herman A., Ass. in Plant Pathol., Agric. Experim. Stat. St. Paul, Minn. (U.S.A.), Univ. Farm. o

Rodhain, Jérôme, Dir. Labor. bact. Léopoldsville (Congo belge). o

Rodi, Giuseppe, Prof. incar. Univ. Genova (Italia). *Anat. topograf.* o

Rodriguez Rosillo, Abile (1891), Prof. à l'Inst. de 2.Ensegnement de Cáceres (España), Plaza Mayor 17. *Nutrition des Plantes et Phytoplancton d'eau douce.* ⓊMadrid 1912.

Rodríguez y López Neyra, Carlos (1885), Prof., Catedrático de Farmacia Univ. Granada (España), San José 1.*Parasit.* ⓊMadrid 1905.ⓅVermes parasit.

Rodschanow, Sarkis (1899), Aspirant der Anat. Anat. Inst. Rostow, Don (U.d.S.S.R.). *Musc. humero-scapularis. Innervation d. Fußrückens.*

Roedel, Hugo (1858), Prof. Frankfurt a. d. Oder (Deutschland). *Paleocängeschiebe.* ⓊHalle a. d. S. 1881. ⓅSedimentärgeschiebe.

Roehl, Wilhelm (1881), Vorstand der chemotherapeut. Abt. der I.-G. Farbenindustrie. A.-G. Elberfeld (Deutschland), Königstr. 124. *Chemotherapie der Infektionskrankheiten, besonders der Tropenkrankheiten (Trypanosomiasis, Malaria).* ⓊHeidelberg. ⓅPathogene Parasit. und deren Überträger.

Roemer, Theodor Ernst Martin (1883), o. Prof., Dir. des Inst. für Pflanzenbau und Pflanzenzüchtung der Univ. Halle a. d. S. (Deutschland), Ludwig-Wucherer-Str. 2. *Pflanzenzüchtung.* ⓊJena 1910.

Roepke, Walter Karl Johann (1882), o. Prof. für Pflanzenkrankheiten an der Landw. Hochsch. in Wageningen (Holland), Rijksstraatweg 16. *Angew. Entomol., Biol. und Faunistik der Indomalaiischen Insekten.* ⓊZürich 1907. ⓅInsekten.

Rösch, Gustav Adolf (1902), Dr. phil., Ass. München (Deutschland), Zool. Inst. der Univ. *Biol., Physiol. and Psychol. der staatenbildenden Insekten, speziell Honigbiene.* ⓊBreslau 1925.

Roeser, Jacob Jr. (1894), Ass. Silviculturist, Rocky Mt. Experim. Station, U.S. Forest Service. Colorado Springs, Col. (U. S. A.), P.O. Box 1068. *Source of seed and influence of parentage on growth of forest trees; relation of methods of cutting of forest stands to growth and reproduction; heat resistance of coniferous seedlings.*

Rösler, Otto A. (1889), Dr., ao. Prof. für innere Med. an der Univ. in Graz (Österreich), Riesstr. 1, Med. Univ.-Klinik. *Haematol. Studien, Kernstruktur unter pathol. und künstlich erzeugten Krankheitszuständen.* ⓊGraz 1912.

Rössle, Robert (1876), Dr. med., o. ö. Prof. der Pathol., Vorsteher des Pathol. Anstalt der Univ. Basel (Schweiz), Hebelstr. 24. *Allgemeine und spezielle Anat. pathol. Immunitätsforschung, Entzündungslehre. Pathol. des Wachstums.* ⓊMünchen 1900.

Rössler, Erwin (1876), Prof. Dr., Dir. Inst. angew. Zool. Zagreb (S.H.S.), Kačićgasse 9. *Hydrobiol., speziell in Bezug auf die fischereiliche und teichwirtschaftl. Praxis. Vogelzug. Lacertilia.* ⓊZagreb 1900.

Rössler Hubert (1876), Prof. Dr. phil., Dir. der Hessischen Landwirtschaftlichen Versuchsstation. Darmstadt (Deutschland), Rheinstr. 91. *Pflanzenernährung und Düngung.* ⓊBonn 1902.

Röthig, Paul (1874), Prof. Dr. med., Stadtrat a. D. Berlin-Charlottenburg 2 (Deutschland), Grolmanstr. 4/5. *Vergl. Anat. des Centralnervensystems der Wirbeltiere, Embryol., Mikroskopische Anat.* ⓊBerlin 1898. ⓅPraep. aus dem Gebiete der Embryol. und der vergl. Anat. des Zentralnervensystems der Wirbeltiere

Roewer, Carl-Friedrich (1881), Prof Dr. phil. Bremen (Deutschland), Am Weidedamm 5. *Arachnoiden, besonders Araneen, Opilioniden (ausschließlich Acarinen).* ⓊJena 1906. ⓅArachnoiden (exkl. Acarinen), besonders Opilioniden und Araneen.

Rogalski, T., Dr. Doc. Kraków (Polska), Univ. Inst. d'Anat., R. Kopernika 13. ○
Rogenhofer, Emanuel (1879), Dr. phil., Reg.R. an der Bundesanstalt für Pflanzenbau und Samenprüfung. Wien II (Deutsch-Österreich), Lagerhausstr. 174. *Samenkontrolle.* ℗ Wien 1905.
Roger, Prof., Doyen Fac. Méd. Paris (France), 85, Boulevard St. Germain. *Physiol.* ○
Rogers, Charles Fletcher (1902), Ass. Botan., Colorado Agric. Experim. Station, Instr. in Botan., Colorado Agric. Coll. Fort Collins, Col. (U.S.A.), Department of Botan. *Carbohydrate metabolism, Mineral nutrients, effects of toxic substances on plants, photosynthesis.* ℗ Weed samples from 3600 feet to 14000 feet altitude.
Rogers, Charles G. (1875), Prof. of Compar. Physiol., Coll. Oberlin, O. (U.S.A.), 378 Reamer Pl. *Effect of salts upon heart action. Nerve cells; Relations of temperatures upon heart action and of lower organisms. Heat product by eggs of Echinoderms before, during and after fertilization.* ℗ Ph. Univ. of California 1904, sc. h. c. Syracuse 1904. ○
Rogers, Charles Henry (1888), Curator of the Princeton Mus. of Zool. Princeton, N.J. (U.S.A.), Box 63. *Ornithol.* ⓓ Princeton Univ. 1909.
Rogers, Fred Terry (1889), Prof. of Physiol., Baylor Univ., Dallas, Tex. (U.S.A.). *Physiol. of the stomach, body temperature regulation; Basal nuclei of the brain.* ⓓ Ph.D. Chicago 1916.
Rogers, James Speed (1892), Prof. of Biol., Univ. of Florida. Gainesville, Fla. (U.S.A.). *Animal Ecol. Life-history and distribution of the Tipuloidea (Diptera).* ⓓ Univ. of Michigan 1915. ℗ Adult and immature stages of Tipuloidea.
Rogers, Lore Alford (1875), Dir., Research Labor. Bureau of Dairy Industry, U.S. Department of Agric. Washington, D.C. (U.S.A.). *Bact. of dairy products.* ⓓ Univ. of Maine 1896.
Rogoff, Julius Moses (1884), Associate Prof., Experim. Med. Cleveland, Ohio (U.S.A.), Western Reserve Univ. *Endocrinol.* ⓓ Western Reserve Univ. 1908.
Rogóyski, Casimir, Dr. Prof. Wilno (Polska), R. Objadowa 2, Univ. *Botan.* ○
Rogozina, Marie Stepanovna (1897), Ass. Histol. Labor. Univ. Perm (U.d.S.S.R.). *Darmepithelzellen bei Fischen. Eléments géants, periphere Nerven der Aeschnalarven.*
Rogoziński, F., Dr. Prof. Kraków (Polska), R. Siemiradskiego 12. *Physiol.* ○
Rohde, Emil, Prof. Univ., G.R.R. Breslau 16 (Deutschland), Parkstr. 1. *Zool. Histol. Allgem. Zellenlehre.* ○
Rohdenburg, George Louis (1883), Dir. of Labor. Lenox Hill, Hospital N.Y.C., Associate in Cancer Research Columbia Univ. New York City (U.S.A.), 905 West End Avenue. *Cancer.* ⓓ New York 1905.
Rohleder, Herbert P. T. (1902), Dr. phil., Research Geol., Royal School of Mines (Geol.). London S. W. 7 (England), Prince Consort Road. *Allgemeine und historische Geol., Palaeont. Geol. und Palaeont. von Nord-Irland.* ⓓ München 1926.
Rohlena, Josef (1874). Praha VI (C.S.R.), Botan. Inst. der Böhmisch. Univ. *Floristik, Balkanländer.*
Rohweder, Max Ferdinand Theodor (1892), Saatzuchtleiter der Fa. C. Braune, G. m. b. H., Saatzuchtwirtschaft. Bernburg a. d. S. (Deutschland), Kaiserstr 11a I. *Pflanzenzucht, spez. Zuckerrüben, Weizen und Gerste. Mikroskopische und physiol. Unterscheidung der Sorten. Pflanzengeographische Charakterisierung. Definition der Erbfaktoren.* ℗ Genetisch definierte Formen der Spezies: Beta, Tritikum, Hordeum, gezogen durch Bastardierung, spontane oder künstliche Mutation.
Rohwer, Sievert Allen (1887), Entomol. in Charge of Taxonomic Investigations of U.S. Bureau of Entomol. Washington, D.C. (U.S.A.). *Hymenoptera. Chalastogastra.*

Roig Binimelis, Jeronimo, Lic. en Ciencias, Prof. auxiliar de la Fac. de Ciencias. Barcelona (España), Casonovas 117. *Biol.* ○
Rojanskij, Nikolaj App., Prof. Gosudarstv. Donskoj Univ. Rostov am Don (U.d.S.S.R.). *Physiol.*
Rojdestvenski, Nicolas A., Conserv. à l'Inst. Jaczewski, Chef du Labor. de Phytopath. à la Station Expérim., Korenevo, Gouv. Moscou. Leningrad (U.S.S.R.), Perspective Anglaise, 29. *Maladies des pommes de terre.* ○
Rolants, Edmond, Chef Inst. Pasteur. Lille (France). *Hygiène.*
Rolfs, Fred M. (1875), Head of the Department of Botan. and Plant Pathol., Oklahoma Agric. and Med. Coll. Stillwater, Okla. (U.S.A.). *Potato Failures. Fruit Tree Diseases and Fungicides, Winter Killing of Twigs, Cankers and Sun Scald of Peach Twigs, Disease of Stone Fruits, Angular Leaf Spot of Cotton, Cotton Anthracnose.* ⓓ Cornell Univ. 1913.
Rolfs, Peter Henry (1865), Dir. (President) Escola Superior de Agric. e Vet. do Estado de Minas Geraes. Viçosa, Minas Geraes (Brasil). *Plant Life.* ⓓ Florida Agric. Coll.
Rolleisen, Gunnar (1899), Fiskeristipendiat. Bergen (Norge), Fiskeridirektoratet. *Süßwasseroligochaeten.*
Rollow, Adolf Ch. Prof., Staatl. Polytechn. Inst. Tiflis (U.d.S.S.R.), Anastasjewskaja 14. *Gartenbau.* ○
Roman, Per Abraham (1872), Ass., Naturhist. Riksmus., entomol. avdeln. Stockholm 50 (Sverige). *Ichneumoniden.* ⓓ Upsala 1909. ℗ Nordpaläarkt. Ichneumoniden.
Roman, Frederic, Maitre de conférences de Géol. à l'Univ. Lyon (France), A Quai St. Clair. *Paléont. des Céphalopodes jurassiques et Vertébrés tertiaires.* ⓓ Lyon 1926.
Romaschow, Demetrius (1873), Stellvertr. Dir. Inst. f. Ernährungsphysiol. Moskau (U.d.S.S.R.), Mal. Wlassjewsky ker., H. 3, W. 4. *Tierstoffwechsel.* ⓓ Moskau 1900.
Romeis, Benno (1888), ao. Prof. Univ., Leiter der Abt. für experim. Biol., Anat. Anstalt. München (Deutschland), Ferdinand-Miller-Pl. 3 III r. *Innere Sekretion, speziell Schilddrüse, Keimdrüse, Thymus, Epithelkörper. Mikrotechnik.* ℗München1911.
Romell, Lars-Gunnar Torgny (1891), Prof. agrégé à Stockholms Högskola, Ass. à l'inst. de l'Etat pour experim. forestière (Experimentalfältet). Djursholm (Sverige). *Physiol. végétale: périodicité diurne, échange gazeux; écol. végétale: aération du sol, teneur en CO_2 de l'air comme facteur de production, période de l'accroissement annuel des arbres; phytogéographie; statistique phytosociol.; protection de la nature.* ℗ Stockholm 1922.
Romer, Alfred S. (1894), Associate Prof. of Vertebrate Paleont. and Curator of Vertebrate Paleont., Walker Mus. Univ. of Chicago, Ill. (U.S.A.). *Vertebrate Paleont.: Palaeozoic Amphibia and Reptilia; Compar. Myol.* ⓓ Columbia 1921.
Romien, Marc Louis (1889), Agrégé d'histol. des Fac. de Méd. et Dr. sc. nat., Prof. d'histol. à la Fac. de Méd. Marseille (France), Palais du Pharo. *Histol. et embryol., la cytol., l'hématol. et l'histochimie.* ⓓ Montpellier 1923. ℗ Prép. histol.
Romijn, Gijsbert (1868), Pharmaceut. Inspector für die Volksgezondheit. Bloemendaal (Holland), Jepenlaan 26. *Cladoceren, Pharmacie, Hygiene des Bodens, des Wassers und der Luft, Ostracoden, Oligochaeten, einzellige Pflanzen u. Tiere, Rotatoria, Hydracarina, Insecta larvae, Hydrobiol.* ⓓ Leiden 1893.
Rona, Peter (1871), Dr. med. et phil., beamteter ao. Prof. der med. Chem. an der Univ., Vorsteher der chem. Abteilung des Pathol. Inst. der Univ. Berlin W 62 (Deutschland), Landgrafenstr. 8. *Physiol. Chem., Kolloidchem., Fermentlehre.* ⓓ Wien med. 1896; phil. 1903.
Roncato, A., Prof. Univ. Padua (Italia). *Fisiol.* ○

Rondelli, Maria (1899), Ass. ist. Zool. Univ. Torino (Italia), Palazzo Carignano. *Simbiosi fisiol. ereditaria Ixodoidea.* Ⓓ 1921. Ⓟ Ixodoidea.

Rondoni, Pietro (1882), Prof. o. Patol. gen. Milano (Italia), Viale Romagna 33. *Biol. dei tumori, avitaminosi e malattie da carenza, tubercolosi, reazioni immunitarie, chim. delle melanine, colorazioni vitali.* Ⓓ Firenze 1906.

Ronnicke, Paul (1867). Graz (Deutsch-Österreich), Maigasse 19. *Europ. Macrolepidopteren.* Ⓟ Europ. Macrolepidopteren in jedem Entwicklungsstadium.

Ronniger, Karl (1871). Wien XII/2 (Deutsch-Österreich), Strohberggasse 29. *Systematik: Thymus, Gentiana, Galium, Alectorolophus, Euphrasia.* Ⓟ Herbarpflanzen der europäischen Flora (Gefäßpflanzen).

Ronzani, Enrico, Prof. incar. Univ. Milano (Italia). *Igiene.* o

Roosa, Nikolai (1899), Agronom. Administrator d. Moorversuchsstation ,,Tooma". Tooma Sookatsejaam (Estland), Vägeva kaudu. *Biol. der wichtigsten Unkräuter.*

Root, Francis Metcalf (1889), Associate Prof. of Med. Entomol., School of Hygien. and Public Health, the Johns Hopkins Univ. Baltimore, Md. (U.S.A.), 615 N. Wolfe Street. *Culicidae, Anopheles.* Ⓓ Johns Hopkins 1917.

Roots, Walter (1903), Ass. am Pflanzenbaukabinett. Pflanzenbiol. Versuchsstation der Univ. Tartu (Estland).

Rootsi, Nikolai (1888), Doc. der Pflanzenbaulehre. Kabinett für Pflanzenbaulehre, Univ. Tartu (Estland). *Untersuchungen über Wurzelrückstände, Verhältnisse zwischen Vegetationsfaktoren und anat. Bau der Kulturpflanzen.*

Rophille, Francisco, Prof. Univ. Nac. La Plata (Argentina). *Anat. descr.* o

Roquette-Pinto, Edgard (1884), Prof. d'Anthrop. au Mus. Nacional de Rio de Janeiro. Doc. de Physiol. a la Fac. de Med. Rio de Janeiro (Brasil.), Mus. Nacional. *Anthrop., Ethnol., Physiol., Psychol.*

Rorer, James B., Mycol., Chief Dept. of Agron., Asoc. de Agric. del Ecuador. Guayaquil (Ecuador), Casilla de Correo. *Diseases of tropical plants.* o

Rosa, Daniele (1857), Prof. o. Zool. ed Anat. compar. Univ. Modena (Italia). *Oligocheti, Policheti (Tomopteridi), Evoluzione.* Ⓓ Sanari 1899.

De Rosa, F., Doc. Univ. Napoli (Italia). *Botan.* o

Rosa, J. T. (1895), Ass. Prof. of Truck Crops, Univ. of California. Davis, Cal. (U.S.A.), Univ. Farm. *Physiol., plant breeding, truck crop production.* Ⓓ Univ. of Missouri 1922.

Rosa, Karel, Ass. Inst. Botan., Ecole Supp. Agric. Praha-Smichow (C.S.R.), Hřebenky 1178. *Algol.* o

Rosa, O., Prof. Univ. Porto Alegre (Brasil). *Anat.* o

Rosanow, Anatol (1883), Mitgl. d. Section wissenschaftl. Arbeiter. Oster, Kr. Tschernigow, Ukraine (U.d.S.S.R.). *Systematik der Macrolepidopteren (Bombyces).* Ⓟ Paläarktische Macrolepidopteren.

Roschdestwensky, Konstantin (1890), Prosektor. Rostow, Don (U.d.S.S.R.), Anat. Inst. *Rechtsseitige Aortenbogen. Sirenenbildung. Asymmetrie d. Extremitätenknochen. Proportion d. Skelettes.* Ⓓ Charkow 1915.

Rose, Anton Richard (1877), Chem., Prudential Insurance Company of America. Edgewater, N.J. (U.S.A.), P. O. Box 376. *Biochem. problems in the ,,Longevity Service".* Ⓓ Columbia Univ. 1913.

Rose, Dean H. (1878), Associate Pathol., Office of Fruit Disease Investigations. Washington, D.C. (U.S.A.), Bureau of Plant Industry. *Diseases of fruits.* Ⓓ Chicago, Ill. 1917.

Rose, Mary S. (1874), Prof. of Nutrition, Teachers Coll., Columbia Univ. New York City (U.S.A.). *Nutrition biochem.* Ⓓ Yale Univ. 1909.

Rose, Raymond Charles (1889), Instr. St. Paul, Minn. (U. S. A.), Coll. of Agric. *Plant Pathol.* Ⓓ Univ. of Minn.

Rose, Waldemar (1895), Ass. am Bact. und Serum-Inst. Landsberg a. W. (Deutschland), Heinersdorfer Straße 14. *Pathol., Bact., Serol., Parasit.* Ⓓ Tierärztl. Hochsch. Hannover 1924.

Rose, William Cumming (1887), Prof. of Physiol. Chem., Univ. of Illinois. Urbana, Ill. (U.S.A.), Chem. Department. *Nutrition, intermediary metabolism of creatine, creatinine, uric acid; relation of the amino acids to growth; synthesis of amino acids in the animal organism.* Ⓓ Yale Univ. 1911.

Rosemann, Rudolf (1870), o. Prof. der Physiol., Dir. des Physiol. Inst. Univ. Münster, Westfalen (Deutschland), Grevener Str. 31. *Stoffwechselphysiol., besonders Alkoholstoffwechsel. Physiol. des Magens. Registrierung der menschlichen Stimme.* Ⓓ Med. Greifswald 1893, phil. h. c. Münster i. W. 1925.

Rosen, Harry Robert (1889), Associate Prof. of Plant Pathol., Univ. of Arkansas, Associate Plant Pathol., Arkansas Agric. Experim. Station. Fayetteville, Arkan. (U.S.A.). *Bact. diseases of plants, cotton wilt, mosaic diseases.* Ⓓ Washington Univ. 1922.

Rosén, Nils Walfrid (1882), Dr. phil., Staatl. Fischereintendent. Göteborg (Schweden). *Ichthyol., Lachs, Fischerei.* Ⓓ Lund 1910.

Rosenau, Milton Joseph (1869), Prof. of Preventive Med. and Hygien. Boston, Mass. (U.S.A.), Harvard Med. School. *Hygiene.* Ⓓ Univ. of Pennsylvania 1889.

Rosenberg, Hans (1890), wissenschaftl. Ass. am Physiol. Inst. der Tierärztl. Hochsch. Berlin NW 6 (Deutschland), Luisenstr. 56. *Elektrophysiol., besonders allgemeine Nervenphysiol. (Aktionsstrom, Leitungsgeschwindigkeit, Verstärker, Oszillograph). Kolloidoklasie; Transfusion, Blutmengenbestimmung.* Ⓓ Berlin 1914.

Rosenberg, Leonid S., Prof. Timirjasev Akad. Moskau (U.d.S.S.R.), Petrovskoje-Razumovskoje. *Technol. d. Fischereiprod.* o

Rosenberg, Otto Gustaf (1872), Prof., Dr. phil., Dir. des botan. Inst. der Univ. Stockholm (Schweden). *Genetische Zytol., Bastard-Zytol. und Parthenogenesis der höheren Pflanzen. Physiol.-zytol. Studien über Drosera rotundifolia. Zytol. und morphol. Studien an Drosera longifolia, rotundifolia.* Ⓓ Bonn 1899. Ⓟ Zytol. Praep.

Rosenberg, V. A., Doc. Univ. Nizny Novgorod (U.d.S.S.R.), Sovetskaja Ploščad 8. *Bact.* o

Rosenblatt, Mélanie (1879), Ass. du labor. de Chim. Biol. à l'Inst. Pasteur de Paris (France), 28, Rue Dutot. *Elements oligosynergiques, diastases.* Ⓓ 1913.

Rosenblatt, Michel, Prof. Techn. f. angew. Chem. Odessa (U.d.S.S.R.), Ostrovidovstr. 4. *Biochem.* o

Rosendahl, Carl Otto (1875), Prof. of Botan. Univ. of Minnesota. Minneapolis, Minn. (U.S.A.). *Systematic Botan., Geographical distribution. Minnesota Trees and Shrubs. Spring Flowers, Ferns.* Ⓓ Univ. Berlin 1905. Ⓟ Ferns and Flowering Plants.

Rosenfeld, Arthur Hinton (1886), Consulting Technol. to the American Sugar Cane League. New Orleans, La. (U.S.A.), 1005 New Orleans Bank Bldg. *Insects and diseases; adaptation of varieties of cane tolerant to Mosaic Disease to Louisiana conditions.*

Rosenfeld, Franz (1871), std. wissenschaftl. Mitglied der landwirtschaftl. Kontrollstation der Landwirtschaftskammer. Berlin-Reinickendorf (Deutschland), Residenzstr. 116/5. *Untersuchung von Futter- u. Düngemitteln.* Ⓓ Heidelberg 1900.

Rosenfeld, Lasar (1877), Priv.Doc. Charkow, Ukraine (U.d.S.S.R.), Tschernoglosowskaja 4a. *Fermente. Nucleinstoffwechsel. Chem. der Plasteine.* Ⓓ Charkow 1907.

Rosenkranz, Friedrich (1900), Dr. phil. Wien VII (Österreich), Neubaugürtel 54. *Floristische u. ökol. Pflanzengeographie, spez. von Niederösterreich; Phänol. von Österreich.* Ⓓ Wien 1923.

Rosenow, Edward Carl (1875), Prof. Experim. Bact., Mayo Foundation, Univ. of Minnesota. Rochester, Minn. (U.S.A.). *Poliomyelitis, Encepha-*

litis, Local Infection and Elective Localization. ⒹUniv. of Chicago, Rush Med. School 1902. ⒫ Poliomyelitis antistreptococcus serum.

Rosenow, Leo (1888), Prof. Physiol., Dir. Physiol. Inst. Weißruss. Univ. Minsk (U.d.S.S.R.). *Protease d. Pankreassaftes. Ionentheorie d. Reizung. CO_2-Bildung durch d. Hefe unter verschiedenen Bedingungen. Bedingungsreflexe.* Ⓓ Moskau 1911.

Rosental, Joseph (1884), Ass. Physiol. Labor. Inst. f. exper. Med. Leningrad (U.d.S.S.R.), Lopuchinskaja 12. *Physiol. des Zentral-Nervensystems (Großhirn).*

Rosenthaler, Leopold (1875), ao. Prof. für gerichtliche Chem. und Pharm.-Chem. Bern (Schweiz), Pharmazeut. Inst. der Univ. *Glykoside, Enzyme, Analyse, Biochem. der Pflanzen (Blausäure-Problem).* Ⓓ Straßburg i. E. 1901.

Rosewall, Oscar Waldemar (1889), Prof. of Entomol., Louisiana State Univ. Baton Rouge, La. (U.S.A.). *Family Pentatomidae (Hemiptera).* Ⓓ Louisiana State Univ. 1916. ⒫ Insects of Family Pentatomidae (Hemiptera) and Coleoptera.

Rosher, Arthur B., Bact., Charing Cross Hospital Inst. of Pathol. London W. C. 2 (England). *Bact. and Serol.* Ⓓ 1916.

Roshevitz, Romain (1882), Sous Chef de l'Herbier. Leningrad (U.d.S.S.R.), Jardin Botan. Principal. *Systématique et géographie des Graminées. Flore du Turkestan.* Ⓓ Leningrad 1910.

Rosinski, Bolesław (1885), Doc. Inst. Anthrop. Univ. Lwów (Polska), Długosza 8. *Hérédité chez l'homme. Selection sexuelle.* Ⓓ Lwów 1921.

Rosiński, Joseph, Prof. Warszawa (Polska), R. Poznańska 16. *Botan.* ○

Roskam, Jacques (1890), Titulaire, en qualité de Chargé de cours, de la Chaire de Pathol. méd. et de Thérapeutique spéciale des maladies internes à l'Univ. Chef des travaux de la Clinique méd. Liège (Belgique), rue des Carmes, 10. *Hématol.: physiol. normale et pathol. des humeurs et des éléments figurés du Sang, des plaquettes sanguines.* Ⓓ Liège 1914.

Roskin, Grigory (1892), Priv.Doc. Univ. Moskau (U.d.S.S.R.), Woronzevo Pole 6. *Zytol. der Protist., Muskelzelle, Krebszelle.*

Rosnatovsky, Jakob P. (1890), Prosektor Kath. d. allgem. Pathol. und Chef d. Serum-Abt. des Bact. Inst. Krasnodar, Kooban (U.d.S.S.R.), Raschpilewskaja 104. *Typhus. Maltafieber. Masern. Hogdsons Krankheit. Serol.* Ⓓ Rostow am Don 1923.

Rospigliosi Vigil, C., Dr. Prof. Univ. Lima (Peru). *Hist. Nat.* ○

Ross, Anna (1872), Lect. in Physiol. and Psychol. Guelph, Ontario (Canada), Ontario Agric. Coll. Ⓓ Univ. of Toronto 1902.

Ross, Hermann (1862), Hauptkonserv. u. Abteilungsleiter am Botan. Mus. München-Nymphenburg (Deutschland), Botan. Inst. *Mittelmeerflora, Physiol. Anat. der Pflanzen, Gallenkunde (Cecidol.).* Ⓓ Freiburg i. Br. 1887.

Ross, Howed Ellis (1881), Prof. of Dairy Industry, Cornell Univ. Ithaca, N.Y. (U.S.A.), Dairy Building. *Dairy Labor. Guide. Labor. Guide in Market Milk. The Cow and Handling of Milk.* Ⓓ Cornell Univ. 1909.

Ross, J. C., Prof. Transvaal Univ. Coll. Pretoria (South Africa). *Agric. Chem.* ○

Rosselet, Alfred (1887), Dr., Prof. à la Fac. de Méd., Chef du Service de radiol. à l'Hopital de Lausanne (Suisse), 18, Avenue Secrétan. *Radiol. (Rayons x, Radium, Lumière). Radiobiol.*

Rossell y Vila, M., Prof. Esc. super. de Agric. Barcelona (España). *Zootechn.* ○

Rossi, Giacomo Marco Maria (1872), Prof. Batt. Agraria Ist. Sup. Agrario Portici, Dir. Labor. relativi e delle annesse Stazione di Microbiol. Industriale e Stazione Agric.-Antimalarica. Napoli (Italia). *Batt. del terreno agrario. Terreni sterili. Fermentazione pectica. Paludismo ed anofelismo senza malaria.*

Rossi, Gilberto, Prof. straord. Univ. Firenze (Italia). *Fisiol.* ○

De' Rossi, Gino (1874), Prof. o. di Microbiol. agraria e Tecnica, Ist. superiore Agrario di Perugia (Italia). ⒫ Culture, Prep. microscopici e Diapositive di Microorganismi.

Rossi, Giovanni, Lib. doc. R. univ. degli studi. Genova (Italia). *Anat. patol.* ○

Rossijskij, D. M., Prof. Med. Inst. Moskau (U.d.S.S.R.), ul. Spartaka 2. *Pharmacol.* ○

Rossinsky, Dmitry Mihailovich (1865), Prof. de l'Inst. Textile de Moscou. Moscou-Centre (U.d.S.S.R.), Tschistijé Proudy, Dome 3-A, Qu. 27. *Migration des oiseaux. Insectes utiles: séricigènes et cirigènes; Bacol. expérim.; Sériciculture; Bombyciens sauvages; Araignés.*

Rossolimo, Leonid (1894), Priv.Doc. Univ. Moskau (U.d.S.S.R.), Zool. Mus., Herzenstr. 6. *Biol., Morphol. und Systematik der Infusorien. Hydrol. und Plankton der Binnengewässer.* Ⓓ Moskau 1926.

Rossner, Ferdinand (1900), Dr. phil. Strausberg b. Berlin (Deutschland), Wilhelmstr. 97 I. *Blütenbiol.* Ⓓ Greifswald 1922.

Rost, Eugen (1870), Dr. med., nichtbeamteter ao. Prof. an der Univ. und Geh.Reg.R. im Reichsgesundheitsamt. Berlin NW 87 (Deutschland), Klopstockstr. 18. *Pharmakol., Toxikol. und Ernährungsphysiol.* Ⓓ Heidelberg 1896.

v. Rostafiński, Jan (1882), ao. Prof. an der Landw. Hochsch. in Warszawa, V.-Präses der Polnischen Zootechn. Gesellschaft. Warszawa (Polska), rue Bracka 5/m. 21. *Stoffwechselversuche: Flora des Pansen beim Wiederkäuer, Amide als Eiweißersatz auf der Weide bei Milchkühen. Craniol. und biometrische Arbeiten. Milch- u. Fettproduktion u. Vererbung derselben.* Ⓓ Kraków 1907.

Rostom-Bek, Doc. Univ. Tiflis (U.d.S.S.R.). *Zool.* ○

Rostrup, Ove Georg Frederik (1864), Museumskonserv. København V (Danmark), Paludan-Müllersvej 5. *Pilze: Ascomyceten, Fungi imperfecti.* Ⓓ København 1890. ⒫ Pilze.

Rostrup, Sofie (1857), Abteilungsvorsteher der zool. Abteilung der Statens plantepatol. Forsög. Lyngby. København V (Danmark), Paludan-Müllersvej 5. *Angewandte Zool. (schädliche Tiere von Land- und Gartenbau).* Ⓓ Kopenhagen 1889.

Roszkowski, Waclaw (1886), Prof. extr. de l'Anat. des Animaux Domestiques à l'Inst. Vét. de l'Univ. Varsovie (Pologne), Grochowska 77. *Tube digéstive des Vertébrés. Amphibiens et Reptiles paléarctiques (Faunistique); Mollusques paléarctiques; Lymnacidae (du Monde entier); Anat., Systematique.* Ⓓ Lausanne 1913.

Rotarides, Michael (1893), Dr., Ass. an dem allgem. Zool. Inst. d. Univ. Szeged (Ungarn), Tisza Lajos körut (Ring) 6. *Land- und Süßwassermollusken, hauptsächlich: Landschnecken (Faunistik, histol. Anat., Palaeont.).* Ⓓ 1920. ⒫ Land- und Süßwasser-Mollusken, Sonderabdrucke.

Roth, Emil (1867). Reutlingen (Deutschland), Kaiserstr. 56. *Palaeontol.*

Roth, Franz (1881), staatl. Kommissar für Naturdenkmalpflege im Regierungsbezirk Aachen, Rheinprovinz (Deutschland), Försterstr. 18 II. *Pflanzenzytol.; Histol.; Pflanzengeographie.* Ⓓ Bonn a. Rh. 1907.

Roth, George Byron (1879), Prof. of Physiol. and Pharmacol. George Washington Univ., Med. School. Washington, D.C. (U.S.A.), 1335 H St. N. W. Ⓓ Univ. of Michigan 1909.

Roth, Gyula (1873), Prof. der Hochsch. für Berg- und Forstingenieure, Leiter der forstl. Versuchsanstalt. Sopron (Ungarn). *Waldbau.*

Rothberger, Carl Julius (1871), ao. Univ.Prof., stellv. Leiter Inst. für allgemeine und experim. Pathol. Wien IX (Österreich), Kinderspitalgasse 15. *Normale und pathol. Physiol. des Blutkreislaufes (Elektrokardiogramm). Pharm. d. Gefäße.* ⒹWien 1897.

Rothers, Boris (1890), Phytopath. Sewero-Dwiner Pflanzenschutzstation. Veliki-Ousting, Sewero-Dwinsk. Gouvern. (U.d.S.S.R.). *Mykol. und Phytopath.* ⓓ Petersburg 1912 ⓟ Mykol. Flora des Sewero-Dwina-Gouvern.

Rothfeld, Jakób (1884), Doc. für Neurol. an der Jan Kazimir'schen Univ. Lwów (Polska), Pańskagasse 3. *Physiol. und Pathol. des Vestibularapparates, Koordinationsstörungen. Hals- und Labyrinthreflexe in der menschlichen Pathol.* ⓓ Lwów 1909.

Rothlin, Ernst (1888), Priv.Doc. für Physiol. Basel (Schweiz), Leimenstr. 41. *Physiol. und Pharmacol. des autonomen Nervensystems.* ⓓ 1915.

Rothschild, Lord Lionel Walter (1868), Dr. h. c. (Gießen), Fellow of the Royal Society Trustee of the British Mus. Tring (England), Zool. Mus. *Ornithol., Lepidoptera.*

Roubal, Jan (1880), Prof., Bańská Bystrica (C.S.R.), Haditel' štát. diev. gymn. *Systématique, zoogéographie des Coléoptères paléarctiques. Myrmecophiles (Col.). Coleoptera microcavernicola (Coléopt. dans nids).* ⓟ Coléoptères paléarctiques.

Roubaud, M. E., Chef de Labor. à l'Inst. Pasteur. Paris 15 (France), 96, Rue Falguière. *Zool.* ○

Roud, A., Prof. Univ. Lausanne, Kt. Waadt (Suisse), Le Verger, Pontaise. *Anat. descript.* ○

Rougebief, Henriette, Licencier ès Sc., Ass. à l'Inst. Pasteur d' Algérie. Alger (Algérie). *Microbiol.*

Roughton, Francis John Worsley (1899), Univ. Lect. in Biochem., Trinity Coll. Cambridge (England), 27 Millington Road. *Applications of physical Chem. to Biochem. and Physiol. Especially the Measurement of the Velocity of the very rapid chem. Reactions between (I) Haemoglobin and dissolved cases (II) the red Blood Corpuscle and its Fluid Environment.* ⓓ Cambridge 1926.

Roule, Louis (1861), Prof. Mus. National d'Hist. Nat. Paris 5 (France), 57, Rue Cuvier. *Embryol.; Biol. générale; Poissons migrateurs.* ⓓ Sc. Paris 1881, méd. 1891. ⓟ Poissons, Reptiles.

Roullard, Fred P. (1884), County Horticultural Commissioner. Fresno, Cal. (U.S.A.), Holland Bldg. 320. *Plant disease.* ⓓ Univ. of Idaho 1908.

Rouppert, Kazimierz Stefan (1885), Prof. de Botan. Univ., Fac. Agr. Kraków (Polska), Al. Mickiewicza 21. *Anat. physiol. et pathol. des plantes.* ⓓ Jagellone à Cracovie 1909.

Rous, Francis Peyton (1879), Member Rockefeller Inst. med. Research. New York, N.Y. (U.S.A.), 125 E., 24 St. *Pathol. Bact.* ○

Rouslacroix, Albert (1878), Chef du Labor. Central des Cliniques à l'École de Méd. Chargé de Cours de Microscopie clinique, Méd. Chef des Hopitaux; Chef de Labor. au centre Anti-Cancéreux. Marseille (France), 119, Cours Lientaud. *Anat. Pathol. et Cancer. Accessoirement Bact. clinique.* ⓓ 1903. ⓟ Coupes anat.-pathol. concernant les lésions Cancéreuses.

Roussakov, Leonid (1897), Inst. Jaczewski de Mycol. et Pathol. végétale. Leningrad (U.d.S.S.R.), Englischer Prosp. 29, 12. *Rouilles des céréales.* ⓟ Rouille.

Rousseau, Prof. Ecole Colon. d'Agric. Tunis (Afrique). *Microbiol. Agric.* ○

Roussy, B., Dir. adj. de Labor. de Phys. Paris (France). *Physique Biol.* ○

Roussy, Gustave (1874), Prof. d'Anat. pathol., Dir. de l'Inst. du Cancer de la Fac. de Méd. Paris (France), 31, Avenue Victor Emmanuel III. *Anat. pathol. et Cancer.*

Rouvière, Henri (1875), Chef des Travaux anat. à la Fac. de méd. Paris (France), 55, rue Geoffroy St. Hilaire. *Anat.*

de Rouville, M. E., Chef de trav. Fac. Sc. Montpellier (France). *Zool.* ○

Roux, Jean (1876), Dr. phil., Custos am Naturhist. Mus. Basel (Schweiz), Augustinergasse 2. *Herpetol. (Reptilien und Amphibien), Süßwasser dekapode Crustaceen.* ⓓ Genève 1899.

Roux, Pierre Paul Emile (1873), Dir. de l'Inst. Pasteur. Paris 15 (France), 25, Rue Dutot. *Microbiol. génér. Diphtérie. Charbon. Rage.*

Rovelli, A., Doc. Univ. Tucumán (Argentina). *Farmacia práctica, Pharmacol.* ○

Rovida, Giulio (1895), Ass. Firenze (R. 8) (Italia), Viale Morgogni 18. *Reazioni immunitarie della sifilide. Vaccinazioni enteriche e immunita locale. Ricerche sperimentali sui prodotti tossici gassosi.* ⓓ Firenze 1923.

Rowe, Allan Winter (1879), Chief of Research, Evans Memorial Prof. of Chem., Univ. Boston, Mass. (U.S.A.), 80 East Concord Street. *Human Metabolism; biochem. and biophysical methods for the objective diagnosis of disease; Endocrine diseases. Disorders of calcium metabolism, anaesthesia, renal function.* ⓓ Göttingen 1906.

Roxas, Hilarie A. (1896), Instr. in Zool., Univ. of the Philippines. Manila (Philipp. Islands), Dept. of Zool. *General problems of sex.* ⓓ Chicago 1926.

Roy Fraser (1889), Prof. of Biol. in Mount Allison Univ. Sackville, New Brunswick (Canada). *Streptococcal pathogenesis; factors promoting streptococcal growth in cultures; various improvements in bact. and immunol. technique.* ⓓ Univ. of Kansas.

Rozanov, L. P., Prof. Staatsuniv., Med. Fac. Minsk, Weißr. (U.d.S.S.R.). *Physiol.* ○

Rozanov, Michail P., Doz. I. Univ. Moskau (U.d.S.S.R.), *Zool. Embryol.* ○

Rozanova, Marie (1888), Ober-Ass. Univ. Leningrad (U.d.S.S.R.), Was. Ostr., 16 Linie 29. *Exper. Pflanzensystematik der Ranunculaceen. Rubus. Geobotan.* ⓟ Pflanzen und Samen der reinen Linien (Ranunculus).

Rožansky, Nikolaus (1884), Prof. Physiol., Allgem. Biol. Rostov a. Don (U.d.S.S.R.), Suworowskaja-Str. 41. *Bluteiweißregeneration. Lymphabsonderung durch chronische Lymphfistel. Nervenbeeinflussung der Flimmerbewegung in der Trachea (Hemmungsnerven).* ⓓ Kriegs-Med. Akademie Petrograd 1913.

Roznatovskij, J. P., Doc. Gosudarstv. Donskoj Univ. Rostov am Don (U.d.S.S.R.). *Mikrobiol.* ○

Rozsypal, Jan, Insp. à l'Inst. de Phytopath. Brno (C.S.R.), Cerna pole 201. ○

Różychi, K., Doc. Tierärztl. Akad. Lwów (Polska). *Ernährung d. Haustiere.* ○

Różycki, Str., Dr., Prof. Univ., Dir. Inst. d'Anat. Poznań (Polska), Dolna Wilda. ○

Rubaschkin, W. I., Prof., Dir. Histol. Labor. Med. Inst. Charkow (U.d.S.S.R.), Str. K. Liebknecht 41. ○

Rubay, P. Prof. Ecole méd. vét. Bruxelles (Belgique), 141, Av. Molière. *Physiol. et Chem. physiol. expérim.* ○

Rubeli, Theodor Oskar (1861), Prof. Anat., Histol., and Embryol., Dir. Vet.-anat. Inst. Univ. Bern, (Schweiz), Engehaldenstr. 6. *Milchdrüsen, Zehen der Wiederkäuer, Poly- u. Hyperdaktylie.* ⓓ Bern 1889.

Rubentschik, Lev (1895), Wissenschaftl. Forschungsinst. Odessa, Ukraine (U.d.S.S.R.), I.N.O., Kominternstr. 2. *Harnstoffgärung, Cellulosegärung, Sulfatreduktion, Bact. der Rieselfelder, des Bodens.* ⓓ Odessa 1926.

Rubinsky, Dmitrij (1881), Prep., Katheder der Morphol. und Systematik der Pflanzen, Univ. N.-Nowgorod (U.d.S.S.R.). ⓓ Kasan.

Rubinstein, Dmitri Leonidovich (1893), Prof. der Biol. Med. Inst., Mitgl. Wissenschaftl. Forschungsinst. für Biol. Odessa (U.d.S.S.R.), Preobrajenskaja str. 5. *Physiol. Ionenwirkungen, Ionenantagonismus. Biol. Wirkungen der Röntgenstrahlen.*

Rublewa, Raissa (1897), Forschungskatheder f. Morphol. u. Physiol. Odessa (U.d.S.S.R.), Bebelstraße 20. *Pathol. Anat.*

Rubner, Max (1854), Prof. Univ. Berlin-Großlichterfelde-W (Deutschland), Drakestr. 69. *Physiol. Gesetze des Energieverbrauches. Ernährungsphysiol. d. Hefezelle.* ○

Rudakov, Kyrill (1901), Leader the Section Soil Bact. at Bact.-Agronomical Stat. Moskau (U.d. S.S.R.).
Rudbārds, Jānis, Lektor Univ. Rīga (Latvija), Elizabetes ielā 91/93, Dz. 14, u. Kronvalda bulv. 1. *Angew. Ichthyol.* ○
Rude, Clifford Symes (1894). Extension Entomol., Oklahoma A. and M. Coll. Stillwater, Okla. (U.S.A.). *Parasitic insects.* Ⓟ Ticks and mites.
Rudolfs, Willem (1886), Chief, Sewage research and Prof. of Water and Sewage. New Brunswick, N.J. (U.S.A.), State Univ. of New Jersey. *Biol. of sewage disposal, stream pollution and potable water.* Ⓓ State Univ. of New Jersey 1920.
Rudolph, Bert Alexander (1889), Ass. Plant Pathol., Univ. of California; in charge of the Univ., Deciduous Fruit Station. San Jose, Cal. (U.S.A.). *Diseases of economic fruits and vegetable.* Ⓓ Stanford Univ. 1925.
Rudolph, Karl (1881), Dr. phil., Priv.Doc. tit. ao. Univ.Prof. f. systemat.Botan. Praha II (C.S.R.), Viniční 3a, Botan. Inst. d. Deutschen Univ. *Palaeobotan., besonders Quartär (Moorforschung), Anat. fossiler Pflanzen, historische Pflanzengeographie.* Ⓓ 1905.
Rudzinski, Dionis (1866), Prof., Dir. Dotnuvische Versuchsstation für Pflanzenzüchtung. Dotnuva (Litauen). *Morphol. und Biol. der Pflanzen, Einfluß mineralischer Nahrung und Bodenfeuchtigkeit während verschiedener Wachstumsphasen.* Ⓟ Samen verschiedener Sortenzüchtungen.
Rudzsky, M. D., Prof., Dir. Zool. Labor. Univ. Tomsk (U.d.S.S.R.). ○
La Rue, Carl Downey (1888), Ass. Prof. of Botan., Univ. of Michigan. Ann Arbor, Mich. (U.S.A.), Department of Botan. *Rubber-bearing plants, pure line selection, Genetics of Cryptogamic plants.* Ⓓ Univ. of Michigan 1921.
La Rue, George R. (1882), Associate Prof. of Zool. and Dir. of the Biol. Station, Univ. of Michigan. Ann Arbor, Mich. (U.S.A.), Zool. Department. *Compar. Anat. and life histories of Cestoda and Nematoda.* Ⓓ Doone Coll. 1907.
Rübel, Eduard (1876), Prof. Dr. Doc. an der Eidg. Techn. Hochsch. in Zürich, Präsident der Schweizer. Pflanzengeographischen Kommission usw. Zürich (Schweiz), Zürichbergstr. 30. *Geobotan., speziell Pflanzensoziol.* Ⓓ Zürich 1901.
Rüdin, Ernst (1874), o. Prof. für Psychiatrie an der Univ. Basel u. Leiter der Psychiatrischen Klinik in Basel, Leiter der genealogischen Abt. der Deutschen Forschungs-Anstalt für Psychiatrie, Kaiser-Wilhelm-Inst. in München. Basel-Friedmatt (Schweiz). *Psychiatrische Erblichkeitslehre, Rassenbiol. und Rassenhygiene (Eugenik).* Ⓓ Zürich.
Ruehe, Harrison August (1888), Prof. of Dairy Manufactures and Head of Dairy Dept., Univ. of Illinois. Urbana, Ill. (U.S.A.), 808 Iowa St. *Chem., bact., physical and economic studies of dairy products, dairy cattle breeding, milk secretion.* Ⓓ Cornell Univ. 1921.
Ruehle, Godfrey Leonard Alvin (1884), Prof. of Bact., Univ. of Idaho and Bact., Experim. Sta., Univ. of Idaho. Moscow, Id. (U.S.A.). *Keeping Quality of Butter.* Ⓓ Univ. of Washington 1910.
von Rümker, Kurt Heinrich Theodor (1859), Dr. phil., Geh.Reg.R., Prof. emer. der Landw. Emersleben, Kr. Halberstadt (Deutschland). *Landwirtschaft und landw. Pflanzenzüchtung.* Ⓓ Halle a. d. S. 1888, agric. h.c.Wien 1919. Ⓟ Hochzuchten von Getreide, speziell Roggen und Weizen.
Rüster, Paul (1897), Dr. phil. Breslau I (Deutschland), Schweidnitzer Str. 32. *Phytopath.* Ⓓ Breslau 1921.
Rüter, Elisabeth (1883), Dr. phil. Hamburg 23 (Deutschland), Hagenau 62. *Vorblattbildung bei Monokotylen. Glykosung und Milchsäurebildung im Vogelblute. Einfluß von Alkaloiden und Salzen auf die Vitalfärbung.* Ⓓ München 1917.

Rütimeyer, Leopold (1856), Prof. Dr. med. Univ. Basel (Schweiz), Sociusstr. 25. *Ethnographie.* ○
Ruffin, Winford A. (1902), Specialist Entomol. Alabama Extension service. Auburn, Ala. (U.S.A.). *Bees.* Ⓓ Iowa State Coll. 1924. Ⓟ Insects.
Ruffini, Angelo, o. Prof. Univ. Bologna (Italia). *Istol., Fisiol. gener.* ○
Ruge, Reinhold Friedrich (1862), Marinegeneralstabsarzt a. D., Prof. Dr. med., Klotzsche bei Dresden (Deutschland), Bahnhofstr. 6. *Malariaparasiten, Ruhramöben.* Ⓓ Berlin 1885.
Ruhemann, Ernst (1897), Dr. med., Ass. am Pathol. Inst. der Univ. Leipzig (Deutschland), Promenadenstr. 11 I. *Normale Anat. des peripheren u. sympathischen Nervensystems. Angewandte Anat. des N. phrenicus: anat. Beiträge zur artificiellen Zwerchfellähmung; Pericardinnervation. Teratol. Thoraxmißbildungen; Muskeldefekte; Hermaphroditismus. Pathol. Anat. der Leber (Ikterus).* Ⓓ München 1921.
Ruhland, Wilhelm (1878), o. Prof. an der Univ., Dir. des Botan. Inst. und Gartens. Leipzig (Deutschland), Linnéstr. 1. *Pflanzenphysiol., physikal. Chem. der Zelle, Stoffwechselphysiol.* Ⓓ Berlin 1899.
Ruhmann, Max Hermann (1880), Provincial Entomologist. Vernon (British Columbia). *Economic Entomol.* Ⓟ Entomol. specimens of British Columbia.
Ruickoldt, Ernst (1892), Dr. med., Ass. des Pharmacol. Inst. Rostock i. M. (Deutschland). *Pharmacol.* Ⓓ München 1920.
Ruiz, F., Prof. Univ. Rosario (Argentina). *Anat.* ○
Rumbold, Caroline Thomas, Ass. Pathol. Madison, Wis. (U.S.A.), Old Soils Building. *Blue stain of wood.* Ⓓ München 1907.
Rumjantzew, Alexis W. (1889), Doc. I. Univ., Dir. Hydrobiol. Station am See Glubokoje, Ass. Inst. für experim. Biol. Moskau (U.d.S.S.R.), Kabinett f. Histol., I. Univ., Herzenstr. 6. *Experim. Zytol. Struktur des Protoplasmas. Vitalfärbung. Reaktion der Zelle und Gewebe. Gewebszüchtung. Hydrobiol.* Ⓓ Moskau 1913.
Rumphorst, Hermann (1896), Preuß. Oberfischmeister. Stralsund i. Pomm. (Deutschland). *Fischereibiol.* Ⓓ Münster 1923.
Runge, Stanislaw (1888), Prof. d. landwirtsch. Vet.-med. Univ. Poznań (Polska). *Sterilität der Haustiere, infektiöser Abortus, Einfluß der Endokrinaldrüsen auf den Organismus der Haustiere, Im- und Transplantation der Geschlechtsdrüsen.* Ⓓ Lwów 1915.
Runnström, John A. M. (1888), Prosektor des Zootomischen Inst., stellvertr. Doc. experim. Zool. Hochsch. Stockholm (Sverige), Högskolan. *Kolloidchem. Bau der Zelle. Physiol. der Befruchtung u. Reifung. Elektrolytenwirkung. Permeabilität.* Ⓓ Stockholm 1914.
Runnström, Sven Valdemar (1896), Amanuensis Biol. Station Mus. Bergen. Herdla (Norge). *Entwicklungsgeschichte der Echinodermen und Cirripedien, Planktonstudien.* Ⓓ Lund 1925.
Runow, Ephim Wassiliewitsch (1901), Ass. Bact.-agronomical Station. Moskau 10 (U.d.S.S.R.), Perejaslavka 52, 3. *Boden-Bact.*
Ruoff, Selma (1887), wissensch. Ass. an der Bayerischen Landesanstalt für Moorwirtschaft. München (Deutschland), Amalienstr. 53 III. *Pflanzensoziol. der Moorvegetation. Moorstratigraphie, pollenanalyt. Untersuchungen.*
Ruppert, Fritz (1887), Prof. Univ. Nac. La Plata, Vicente Lopez (Argentina), Calle Monstrio 1420. *Bact. ultravis. Erreger, Microbiol.* Ⓓ Gießen 1910.
Ruppin, Arthur (1876), Doz. Soziol. Hebräische Univ. Jerusalem (Palestine). *Anthrop. der Juden.* Ⓓ Halle a. d. S. 1902.
Rusconi, Mario (1900), Ass. Ist. di Anat. Umana Normale R. Univ. di Pavia (Italia). Ⓓ Pavia 1925.
Rušencov, Dmitrij, Prof. Tierärztl. Hochsch. Leningrad (U.d.S.S.R.), Černigovskaja 5. *Mikrobiol.* ○

Rusjaev, Boris A., Doc. I. Univ. Moskau (U.d. S.S.R.). *Anat.* o

Rusk, H. P. (1884), Prof. of Cattle Husbandry and Head of the Department of Animal Husbandry, Univ. of Illinois. Urbana, Ill. (U.S.A.). ⊕ Univ. of Missouri 1911. o

Russ, Karl, Prof. Univ. Wien III (Deutsch-Österreich), Hauptstr. 146. *Bact. Serol.* o

Russell, Frederick Fuller (1876), Mem. Publ. Health Council State N.Y. New York, N.Y. (U.S.A.), 7 W., 43 St. *Pathol. Bact.* o

Russell, Frederick Stratten (1897), Ass. Naturalist to the Marine Biol. Association of the United Kingdom. Plymouth, Devon. (England), The Labor., Citadel Hill. *Fishery Research; vertical distribution and movements of plankton organisms, pelagic young of fish.* ⊕ Cambridge 1922.

Russell, Lilian (1886), Mus. Ass. and Demonstrator in Zool. Univ. London E.C. 4 (England), Birkbeck Coll., Bream's Buildings. *Mollusca. Morphol. Systematics of Nudibranchs.* ⊕ London 1923.

Russell, Paul (1889), Ass. Botan., Office of Foreign Plant Introduction, Bureau of Plant Industry, U. S. Department of Agric. Washington (U.S.A.). *Prunus.* ⊕ George Washington Univ. 1924.

Russell, Paul Farr (1894), Field Staff, International Health Board, Rockefeller Foundation, Dir., Straits Settlements Rural Sanitation. New York, N.Y. (U.S.A.), 61 Broadway. *Mosquitoes.* ⊕ Cornell 1921.

Russinoff, Victor (1891), Doc. Naturwiss. Abt. Pädagog. Fac. N.-Nowgorod (U.d.S.S.R.), Chalatny per. 10, 2. *Reflexol.* ⊕ Kasan 1914.

Russo, Achille (1866), Prof. titolare di Anat., Fisiol. compar. e Zool. R. Univ. di Catania (Italia). *Anat. ed Embriol. degli Echinodermi. Turbellarii. Sistematica e Ciclo di sviluppo dei Ciliati (Cryptochilum echini Maupas). Istol. dei Mammiferi. Pesca delle Sardine, delle Seppie, nel Golfo di Catania.*

Rustia, Constantio, Ass. Prof. Univ. of the Philippines. Manila (Philippine Islands). *Zool.* o

Ruszkowski, Jan Władysław (1889), Dr. et ingenieur d'agric.; adjunct de l'Univ. Poznań (Polska), Sołacka 3. *Entomol. appliquée, Biol. des Insectes, Hymenoptera, Phytophaga.* ⊕ Poznań.

Ruszkowski, Jerzy Stanisław (1887), Ass. du Labor. de Zool. Warszawa (Polska), 26, Krakowskie Przedmieście Univ., Labor. de Zool. *Evolution et systématique des Plathelminthes (Trematodes et Cestodes).* ⊕ Varsovie 1923.

Rutgers, A. A. L. (1884), Dr., Dir. des Landbau-Ministeriums. Buitenzorg, Java (Nederl.-O.-Indiē). *Reizphysiol. der Pflanzen. Phytopath. Hevea. Selection, Hevea, Ölpalme.* ⊕ Utrecht 1910.

Ruth, Warren A., Associate Prof. of Pomol. Physiol., Univ. of Illinois. Urbana, Ill. (U.S.A.). o

Rutherford, Andrew, Lect. School of Med. R. coll. Edinburgh (Scotland). *Pathol.* o

Ruthven, Alexander G. (1882), Prof. of Zool., Dir. Univ. Mus., Dir. Mus. of Zool. Ann Arbor, Mich. (U.S.A.). *Herpetol.* ⊕ Michigan 1906. ⊕ Reptiles and Amphibians.

Rutkiewicz, Bohdan (1887), Prof. suppl. Philosophie Univ., Prof. de Génétique aux Cours supérieures d'Horticulture. Lublin (Polska). *Individualité, finalité, évolution, vitalism.* ⊕ Grenoble1921.

Rutten, Louis Martin Robert (1884), Prof. de Geol., Mineral., Krystallographie und Palaeont. an der Univ. Utrecht (Holland), C. Evertsenstr. 7. *Fossile Foraminiferen.* ⊕ 1909.

Ruttner, Franz (1882), Dr. phil., Priv.Doc. an der Univ. Wien, Leiter der Biol. Station in Lunz am See (Nieder-Österreich). *Hydrobiol.: Stoffhaushalt der Gewässer; Verteilung des Planktons; Wanderungen; Bioćnotik des Süßwassers. Pflanzenphysiol.: Kohlensäureassimilation der Wasserpflanzen.* ⊕ Deutsche Univ. Prag 1906.

Ruud, Gudrun (1882), Zool. Labor. Univ. Oslo (Norge). *Entwicklungsmechanik.* ⊕ Oslo 1913.

Ružička, Stanislav (1872), Prof. d'Hygiène et Dir. de l'Inst. d'Hygiène de l'Univ. de Bratislava (C.S.R.). *Hygiène basée sur la nouvelle doctrine „eubiotique".* ⊕ Praha 1897.

Růžička, Vladislav (1870), o. Prof. d. med. Fac. Karls-Univ., Dir. Inst. f. allg. Biol. u. experim. Morphol. Praha II (C.S.R.), Kateřinská ul. 32. *Altersforschung und Lebensprozesse vom Standpunkte d. phys. Chem., Vererbung. Struktur u. Plasma. Vererbungslehre. Eugenik.*

Ruzsky, Michael (1864), Prof., Dir. Zool. Kabinett Univ. Tomsk, Sibirien (U.d.S.S.R.), Monastyrskaja Str. 18, 2 (oder Zool. Kabinett d. Univ.). *Fische Sibiriens; Entomol. (Allgemeine Biol. und Zoogeographie). Myrmicidae. Zoogeographie Sibiriens. Vögel d. westlichen Sibiriens.* ⊕ Charkow. ⊕ Ameisen.

Ryan, Andrew Howard, Assoc. physiol. Waterbury, Conn. (U.S.A.). *Industrial physiol. pharm.* o

Rybak, Ottokar, Dr., Prof. d. Pharmacol. u. Pharmakognosie, Tierärztl. Hochsch. Brno (C.S.R.), Pražska 67. *Plantes méd. Pharmakognosie.* o

Rybinsky, Sergeous Basil (1898), Ass. Bact. Inst. (malarian Dept.). Kiew, Ukraine (U.d.S.S.R.), Lenin-street 53 b, 1. *Parasit. and toxicol.*

Rydberg, Per A., Curator, New York Botanical Garden. New York City (U.S.A.). *Flora of the Rocky Mountain region.* o

Rydzewski, B., Dr. Prof. Wilno (Polska), Inst. Géol. Univ., r. Zackretowa 15. *Paléont.* o

Ryffel, J. H., Reader Univ. London S.W. 7 (England), S. Kensingt. *Biochem.* o

Rylkowa, E. W., Doc. Univ., Ass. Inst. Vergl. Anat. I. Univ. Moskau (U.d.S.S.R.), Herzenstr. 6. *Zool. Embryol.* o

Rylov, Wjatscheslav (1889), Priv.Doc. Univ., Hydrobiol. Inst. Leningrad (U.d.S.S.R.), Zool. Mus. d. Russ. Akad. d. Wissensch. *Hydrobiol. (hauptsächlich Zooplankton), Systematik u. Biol. d. Süßwassers. Eucopopoden. Hydroidea. Alcyonaria, bes. der russ. Arktis.* ⊕ Leningr. Univ.

Ryrie, B. J., Prof. Univ. Cape Town (South Africa). *Pathol.* o

Rytz, August Rudolf Walther (1882), Prof. extraord. a. d. Univ., Konserv. am Botan. Inst. Bern (Schweiz), Ländtweg 5. *Pflanzengeographie; Biol. und Geographie der parasitischen Pilze; Synchytrium; Flora der Eiszeit, der postglazialen und prähistorischen Zeit.* ⊕ Bern 1907. ⊕ Synchytrium.

Ryžkow, P. M., Doc. Tomsky Technol. Inst. Tomsk (U.d.S.S.R.), Timirjazewskij Prosp. 9. *Palaeont.* o

Rzóska, Juljan (1900), Dr. phil., Ass. am Zool. Inst. der Univ. Poznań (Polska), Matejki 49. *Copepoda: Cyclopidae u. Centropagidae. Faunistik u. Biol.* ⊕ Poznań 1925.

Š... confer etiam Sch. Sh. Tsch.

Saalas, Uunio (1882), Dr. phil., Prof. der Landwirtschafts- und Forstzool. an der Univ. Helsinki (Finnland), Cygnaeuksenkatu 4. *Schädliche und nützliche Insekten (besonders Coleopteren und ihre frühen Entwicklungsstadien; Ipiden, Elateriden etc.).* ⊕ Helsinki 1919.

Sabalitschka, Theodor (1889), Priv.Doc. für pharmaz. Chem. an der Univ. Berlin. Berlin-Steglitz (Deutschland), Elisenstr. 7. *Chem. des pflanzlichen Stoffwechsels, Pharmazeutische Chem., Chem. u. Wertbeurteilung der Nahrungsmittel, entwicklungshemmende u. abtötende Wirkung chem. Stoffe gegenüber Bact. u. Pilzen, Phytochem., Kultur von Arzneipflanzen.* ⊕ Phil. Berlin 1918, rer. pol. Kiel 1920.

Sabaschnikow, Wladimir W., Dir. d. Jenisseisker landw. Prov.-Versuchsstation. Krasnojarsk (U.d. S.S.R.). *Agronomie. Selektion.* o

Sabbatani, Luigi (1883), Prof. di Farmacol. Univ. Padova 14 (Italia), Viale Loredan 2. *Chim.-Fisica e Metalli colloidali in Farm.*

Sabet, Younis Salem (1898), Lect. in Botan. The Egyptian Univ. Cairo (Egypte).

Sabin, Florence Rena (1871), Member of the Rockefeller Inst. for Med. Research. New York City, N.Y. (U.S.A.), 66 Street and Ave. A. *Blood, Lymphatics. Pathol. Bloodvessel. Tuberculosis.* ① Johns Hopkins Med. School 1900.

Sabinin, Dmitrij A., Prof. a. d. Staatl. Univ. Perm (U.d.S.S.R.). *Pflanzenphysiol.* o

Sablin, B., Doc. Univ. Taschkent (U.d.S.S.R.). *Bact.* o

Sabrazès, Jean (1867), Prof. anat. path. et de Microscopie clinique Fac. Méd. Univ. Bordeaux (France), 50, rue Ferrère. *Cysticerque, Hématol. Glands digestives. Sang. Hernier diaphrague.* ① Bordeaux 1893. ② Préparation cytol. hémat. anat.-pathol.

Sabri, Hussein (1885), Prof. Fac. Sc. Univ. et à l'Ecole Supérieure Vét. Stamboul (Turquie). *Physiol. du cœur.*

Sabussow, Georg (1899), Ass. Histol. Inst. Univ. Kazan (U.d.S.S.R.), Zweite Bergstr. 42. *Veränderungen in der Ohrspeicheldrüse bei Unterbindung ihres Ausführungsganges. Sekretion des Plexus chorioideus.* ① Kazan 1926.

Sabussowa, Zoe Hippolit (1901), Ass. Kasan (U.d.S.S.R.), Zootom. Kabinett, Univ. *Vergl. Morphol. der Wirbellosen. Turbellaria. Biocoenotik.*

Saccardi, Pietro (1889), Prof. st. Chim.-Farmaceut. Univ. Camerino (Italia). *Melanine. Adrenale.*

Sacco, Federico (1864), Prof. di Paleont., Univ. di Torino (Italia), Castello del Valentino. *Paléont.: Molluschi e Vertebrati Mammiferi.* ① Torino 1886.

Van Saceghem, René (1886), Dr. med. Vét., Dir. du Labor. de recherches Vét. du Congo, Inspecteur principal, Chef du Service. Elisabethville (Congo Belge). *Bact. Vaccina. Trypanosomes.*

Sacerdotti, Cesare (1868), Prof. ord. (stabile) di Patol. generale. Pisa (Italia), via Trieste 14. *Istol. normale e patol. Ematol.: le piastrine del sangue. Fisiopatol.: sviluppo eteroplastico dei tessuti. Batteriol. Anafilassi.* ① Pavia 1892.

Sacharov Gavril Petr., Prof. I. Univ. Moskau, (U.d.S.S.R.). *Pathol.* o

Sacharov, Nikolaj Loovič., Doc. Inst. f. Landwirtschaft u. Melioration. Saratov (U.d.S.S.R.), Teatral'naja Ploščad. *Entomol.* o

Sacharov, Taissa M., Doc. Forst-Inst. Moskau (U.d.S.S.R.), Sofijskaja naber. 24, W. 2. *Botan., Pflanzen-Physiol.* o

Sachs, Hans (1877), Dr. med., o. Prof. a. d. Univ., Dir. der wissenschaftl. Abt. des Inst. f. experim. Krebsforschung. Heidelberg (Deutschland), Bergstr. 55. *Bact., Immunitäts- und Serumforschung, Serodiagnostik der Syphilis. Antikörper gegen Lipoide.* ① Leipzig 1900.

Sachße, Hans Friedrich (1890), Oberförster und Ass. Forstl. Hochsch. Tharandt, Sachsen (Deutschland), Sidonienstr. 174 p. *Forstliche Standortslehre (insbes. Düngung im Forstbetrieb und Standortsflora).*

Sachtleben, Hans (1893), Dr. phil., wissenschaftl. Hilfsarbeiter und Leiter des Forstzool. Labor. Berlin-Dahlem (Deutschland), Biol. Reichsanstalt für Land- und Forstwirtschaft. *Forstzool. Ornithol. Mammal.* ① München 1917.

Sachweh, Paul (1885), Dr., Dir. Bact. Inst. der Landwirtschaftskammer für die Provinz Westfalen. Münster i. W. (Deutschland), Kronprinzenstr. 15. *Rindertuberkulose.* ① Gießen 1910.

Sack, Alexander Lwovitsch (1898), Labor. f. experim. Biol. Swerdlow Univ., Timiriasevs Wissenschaftl. Inst. Moskau (U.d.S.S.R.), Granatny per. 7. *Physiol. der bedingten Reflexe, der endocrinen Drüsen, des Erbrechens.*

Sack, Pius (1865), Prof. Dr. phil. Frankfurt a. M. (Deutschland), Klettenbergstr. 9. *Diptera, speciell Syrphidae orbis terrarum.* ① Jena 1891.

Sackett, Walter George (1880), Bact., Colorado Agric. Experim. Station. Fort Collins, Col. (U.S.A.). *Soil Bact., Plant Diseases, Food Poisoning.* ① Univ. of Chicago 1918. ② Cultures of bact. and yeasts.

Sadikov, V. S., confer Ssadikow.

Sadovnikova-Koltzova, Maria, Ass. Univ. Inst. of Experim. Biol. Moscow (U.d.S.S.R.), Woronzowo Pole 6. *Genetics of Animal Behavior (Rats). Leben der Ameisen.* ① Moscow 1910.

Saeger, Albert (1895), Head of the Department of Biol. Junior Coll. of Kansas City, Mo. (U.S.A.). *Anat. and physiol. of the Lemnaceae (Duckweeds).* ① Univ. of Missouri 1924. ② Living or preserved specimens of Lemnaceae.

Saelhof, Clarence Charles (1897), Research Associate, John McCormick Inst. for Infectious Diseases, 629 S. Wood St. Chicago, Ill. (U.S.A.), 4458 W. Madison Street. *Bact. Serol.* ① Med. Univ. of Illinois 1921, phil. Univ. of Chicago 1924.

Saemundsson, Bjarni (1867), Wissenschaftlicher Fischereikonsulent. Reykjavik (Island). *Biol. d. Fische.*

Safford, William E. (1859), Botanist. Washington, D.C. (U.S.A.), U. S. Dept. of Agric. *Tropical plants. Cactacea of Northeastern and Central Mexico. Annona. Lignumnephr. Narcotics and stimulants. Chenopium. Dahlia. Paradise key. Datura.* ① Washington Univ. 1920.

Safro, V. I. (1888), Chief Entomol., California Cyanide Co. Los Angeles, Cal. (U.S.A.), Box 250, Arcade Station. *Destruction of insect and rodent pests by means of cyanide fumigation.* ① Cornell Univ. 1909. ② Insects attacking stored products.

Sagastume, Alfredo, Prof. Univ., Dir. Inst. Pathol. Tegucigalpa (Honduras), Plaza de la Merced. *Pathol. Anat.*

Sager, James L. (1877), Head of Biol. Department. Exeter (England), Univ. Coll. of the South West. *Plant Ecol. and Physiol. Flora of Devon. Soil Acidity. Importance of the Light Factor. Causes of Rhythm in ,,Vital" Phenomena.* ① Cambridge Univ. 1900.

von Sághy, Franz (1893), Univ.Ass. Hygien. Inst. Budapest VII (Hungary), Péterffy Sándor-ut 40. *Bact. und Serol. Milz u. Eisenwirkung. Pilocarpin u. Immunkörperbildung.* ① Budapest 1921.

Saguchi, Sakae, Prof. Med. Univ. Kanazawa (Japan), Anat. Inst. *Anat.*

Sahli, Hermann (1856), Dr. der Med. Univ.-Klinik u. Prof. Med. Univ. Bern (Schweiz), Seftigenstraße 11. *Hämodynamik. Immunität. Antikörperlehre. Blut. Vererbung erworbener Eigenschaften. Allgemeine Pathol. u. Therapie der Tuberkulose. Genetische Beziehungen zu Infektionskrankheiten (z. B. Beziehung zwischen Variole u. Varicelle). Wesen der Neurosen.* ① Bern 1881.

Sahlstedt, A. V., Prof. Veterinärhögskolan, experimentalfältet Stockholm (Sverige). *Physiol.* o

Sahović, Gjorgje, Dr., ao. Prof. pathol. Anat. Beograd (S.H.S.), Višegradska ulica. *Experim. Pathol.* o

Saint-Hilaire, Constantin (1866), Prof.Zool. Univ. Woronesch (U.d.S.S.R.). *Histol., Histo-physiol.; Hydrobiol. Nutrition de la cellule, phagocytose, phénomènes de la sécrétion, morphogénèse des téguments etc.; faune de la mer Blanche, faune d'eau douce de Gouv. Voronech.* ① St. Pétersbourg 1897.

de Saint Rat, Louis (1891), Ass. de Chim. biol. Fac. Sc. Univ. Paris (France), 28, Rue Dutot. *Analyse méd. Sucres.* ① Paris 1921.

Saitô, Katsuhisa, Lect. Prov. Med. Univ. Ôsaka (Japan), Anat. Inst. *Anat.*

Saito, Kendô (1878), Dir. Zentrale Unters.-Anstalt, Südmandschurische Eisenbahngesellschaft. Dairen (Mandschurei). *Gärungsorganismen.* ① Tokio 1900.

Saitô, Tamao (1880), Head Neuro-Biol. Station, Prof. Psychiatry Nippon Med. Coll. Shinagawa, Tôkyô (Japan), 1441 Asamadai. *Compar. Histol. of Nervous System of Vertebrates and experim. Neuro-Pathol. (Rats).* ① Tôkyô Imper. Univ. 1905. ② Serial sections of Japanese fishes.

Saizawa, Kôzô, Prof. Imper. Univ. Tôkyô (Japan), Bact. Inst., Fac. of Med. *Bact.*

Sakai, Kikuo, Ass. Prof. Tôhoku Imper. Univ. Sendai (Japan), Bact. Inst., Coll. of Med. *Bact.*

Sakai, R., Ass. Prof. Hokkaidô Imper. Univ. Sapporo (Japan), Pharmacol. Inst., Fac. of Med. *Pharmacol.* o

Sakai, Takuzô, Prof. Med. Univ. Chiba near Tôkyô (Japan), Physiol. Inst. *Physiol.*

Sakamura, Tetsu (1888), Prof. Plant Physiol. Hokkaidô Imper. Univ. Sapporo (Japan), Botan. Inst. *Plant Physiol. and Cytol.* ① Sapporo 1913.

Sakař, Jaroslav, o. Prof. der Anat. der Haustiere, Dir. Anat. Inst. Zagreb (S.H.S.), Sudnička ul. 11. o

Sakurai, Motoi, Prof. Kyôto Imper. Coll. of Sericiculture. Hanazono near Kyôto (Japan), Physiol. Inst. *Physiol. and Anat. of Silk-Worms.*

Sakurai, Tsunejirô, Prof. Kyûshû Imper. Univ. Fukuoka (Japan), Anat. Inst., Coll. of Med. *Anat.*

Sala, Luigi (1863), Prof. o. Anat. Umana Normale Univ., Dir. Ist. Anat. Pavia (Italia). *Sistema nervoso centrale e periferico. Maturazione e fecondazione, uova di Ascaris megalocephala. Sviluppo del sistema linfatico.* ① Torino 1887.

Salant, William (1870), Prof. of Physiol. and Pharmacol. Med. Dept. Univ. of Georgia. Augusta, Ga. (U.S.A.). *Effect of ions, temperature, etc. on the action of drugs.* ① Columbia 1899.

Salasar, Zacaricas, Prof. escuela esp. de ing. agron. Madrid (España). *Zootechnik. Pathol.* o

Salaskin, Sergey Sergeyewitsch (1862), Dir. der Abt. der biol. Chem. des Inst. für experim. Med., Prof. des Med. Inst. Leningrad (U.d.S.S.R.), Lopuchinskaja 12. *Physiol. des Ammoniaks, des Harnstoffs und der Verdauungssäfte.*

Salazar, Abel Lima (1889), Prof. o., Dir. Inst. d'Hist., Fac. Méd. Porto (Portugal). *Ovarium.* ① Porto 1915.

Graf von Saldern-Aplimb, Leopold (1886), Ringenwalde, Kreis Templin, Prov. Brandenburg (Deutschland). *Bakt., theoretische u. angewandte Botan.*

Salée, A., Prof. Univ. Cath. Louvain (Belgique). *Paleont.* o

Salfi, Mario (1900), Ass. Ist. Anat. e Fisiol. compar. Univ. Napoli (Italia), Via Montesilvano 30. *Morfol., Biol. e sistematica delle Ascidie del Golfo di Napoli, degli Ortotteri circummediterranei.* ① Napoli.

Salimovsky, Antonina (1901), Hydrol. Inst. Leningrad (U.d.S.S.R.), 6 Linie 17, Qu. 11. *Mikrobiol. des Wassers.* ① Leningrad 1924.

Saling, Theodor Otto (1878), wissensch. Mitarbeiter Preuß. Landesanstalt für Wasser-, Boden- und Lufthygiene (Biol.-zool. Abt.) Berlin-Dahlem. Berlin-Charlottenburg 5 (Deutschland), Sophie-Charlotte-Str. 68. *Angewandte Zool. (Schädlingsbekämpfung).* ① Marburg 1906.

Salisbury, Edward James (1886), Reader in Plant Ecol., Fellow of Univ. Coll. London (England). *Plant Ecol., Palaeobotan. and Taxonomy.*

Sallač, Wilhelm, Dr., Prof. d. Dendrol., Jagdtierbiol. u. Forstentomol, Leiter d. Inst. f. Dendrol., Jagdtierbiol. u. Forstentomol. d. čech. techn. Hochschule. Praha-Vršovice (C.S.R.), Havlíčkovo sady. o

Sallee, Roy Merrudith (1889), Ass. in Botan. Univ. of Illinois. Urbana, Ill. (U.S.A.). *Plant Ecol., Soil Relations of Plants.* ② Western Illinois State Teachers Coll. 1923.

Salmon, W. D. (1895). Auburn, Ala. (U.S.A.), Experim. Station. *Vitamin B content of plant products. Effect of minerals on growth, reproduction and body composition of the white rat.* ① Univ. of Missouri 1921.

Salopek, Marijan, Prof. Dr., Dir. des Geol.-Palaeont. Inst. Univ. Ljubljana (S.H.S.), Kongresni trg. *Ammoniten.* ①

Salter, Robert Mundhenk (1892), Chief in Agronomy, The Ohio Agric. Experim. Station. Wooster, O. (U.S.A.). *Field crop improvement.* ① Ohio State Univ. 1914.

Saltykow, Sergej (1874), o. Prof. der allgemeinen Pathol. und der pathol. Anat. der Kgl. Univ., Dir. des Pathol.-anat. Inst. Zagreb (S.H.S.), Šalata. *Entzündungslehre, Tuberkulose, Transplantation, Geschwülste, pathol. Histol. des Zentralnervensystems, experim. Atherosklerose, experim. Pathol., pathol. Anat., Bact.* ① Basel 1904, Kiew 1914.

Salus, Gottlieb, Dr., Prof. d. Hygiene (Bact.). Praha II (C.S.R.), Preslova, Hygien. Inst. d. Deutsch. Univ. o

Salvi, G., Prof. Univ. Napoli (Italia). *Anat.* o

Salvioli, Gaetano (1894), Prof. agrégé de l'Univ. et chargé de l'insegnement officielle de ,,Patol. exotique". Padova (Italia). *Bact. Immunol. Physiol. humaine. Anat. pathol.* ① 1923.

Salzmann, Maximilian (1862), o. Prof., Dir. Augenklinik Univ. Graz (Österreich), Lichtenfelsgasse 15. *Anat. Histol. des Auges. Zonula ciliaris u. ihr Verhältnis zur Umgebung.* ① Wien 1887.

Šamal, Jaromír (1900). Ass. à l'inst. zool. Univ. Praha (C.S.R.). *Entomol.: Morphol., Anat., Ökol. Entomol. agric. et forestière: Micro-lepidoptères. Insectes d'eau douce: Plecoptera, Ephemerides, Neuroptera.* ① Praha 1923.

Sambbi, N., Prof. Univ. Nation. Asunción (Paraguay). *Anat. Physiol. Hygien.* o

Sambo, Maria Cengia (1888), Dr. Prato in Toscana (Italia), Rue Firenze 16. *Biol., systématique et géographie des Lichens.*

Samec, Max (1881), o. Prof. für Chemie an der Univ. in Ljubljana (S.H.S.). *Chem. und Kolloidchem. der Pflanzenkolloide, Stärke.* ① Univ. Wien 1904.

Samofai, Savva A., Chef des travaux Inst. Forestier. Leningrad (U.d.S.S.R.). *Pathol. Forestier.* o

Samoiloff, Alexander (1867), Prof. Physiol. Kasan (U.d.S.S.R.), Physiol. Labor. Univ. *Elektrophysiol.* ① Dorpat 1892.

Samoiloff, Jhon (1900), Ass. Field Husbandry Sta. of Acclim. Detskoe Selo (U.d.S.S.R.), 2 Volpinskaja. *Wheat, Buck wheat and Tobacco plants.*

Sampson, Homer C. (1885), Prof. of Botan., The Ohio State Univ., Dept. of Botan., Columbus, O. (U.S.A.). *Physiol. and Ecol.* ① Univ. of Chicago 1914.

Sampson, Kathleen (1892), Lect. in Agric. Botan., Univ. Coll. of Wales, Aberystwyth. Chesterfield, Derbyshire (England), 6 Gladstone Road. *Diseases of cereals, clovers and grasses. Cereal smuts.* ① London 1914. ② Cultures, seed.

Sampson, Myra Melissa (1887), Associate Prof. of Zool., Smith Coll. Northampton, Mass. (U.S.A.). *Physiol.* ① Univ. of Michigan 1926.

Šamšin, Vladimir (1868), Doc. II. Univ. Moskau (U.d.S.S.R.). *Pathol. Anat.* ① Moskau 1891.

Samuel, A., Prof. Hochsch. f. Tierärzte. Konstantinopel (Türkei), Bajtar m. *Pathol.* o

Samuelsson, Gunnar (1885), Prof., Dir. der Botan. Abt. des Naturhistorischen Reichsmus. Stockholm 50 (Sverige). *Pflanzengeographie Skandinaviens. Kritische Arten der Nordischen Flora. Systematik verschiedener Pflanzengruppen, speziell der eurasiatischen Flora.* ① Upsala 1913.

Samut, R., Prof. Univ. of Malta. La Valletta (Malta). *Physiol. and Pathol.* o

Samutsevitsch, Maria (1889), Labor. Mycol., Phytopath. Labor. Leningrad (U.d.S.S.R.), Nicolajevskaja 76, log. 13. *Biol. der Pilze, Reinkulturen. Bodenpilze.* ① Petersburg 1916. ② Pilzkulturen.

Sanborn, Charles Emerson (1877), Prof. of Entomol. and Experim. Station Entomol. Oklahoma Agric. and Mechanical Coll. Stillwater, Okla, (U.S.A.). *Cotton insects, Aphididae.* ② Insects. especially Aphididae and Coleoptera.

Sanchez Roig, Mario Idelfonso (1890), Prof. Zool. Agric. School. Havanna (Cuba), Cerro 827. *Zool., Paleont.* ① Havanna 1912.

Sanchez Navarro y Neumann, Emilio, Dr. en Ciencias Nat., Prof. auxiliar en el Inst. Cádiz (España), Santa Inés. *Entomol.* o

Sánchez y Sánchez, Domingo, Dr. en Ciencias Nat. Med., Conservador, por oposición, en el Mus.

de Antropol., Prof. en la Escuela de Artes e Industrias. Madrid (España), Atocha 96. *Anat. compar.* ○
Sánchez y Sánchez, Manuel (1893), Prof. à l'Univ. de Zaragoza (España), Costa 4. *Parthénogénèse expérim.; morphol. et Physiol. de la cellule nerveuse.* ⓓ Madrid 1914.
Sand, Knud (1887), Prof. de l'Univ., Dr. med. København (Danmark), Frederik V. vej 9. *Biol. sexuelle.* ⓓ 1918.
von Sande, Karl Hermann Friedrich Heinrich (1877), Techn. Abt.-Leiter am Bact. und Serum-Inst. Landsberg a. d. W. (Deutschland), Friedeberger Straße 4. *Serol. und Bact.* ⓓ Berlin 1922.
van de Sande Bakhuyzen, Hendrik Leo (1891), Research Ass. Food Research Inst., Ass. prof. biol. Stanford Univ. California. (U.S.A.). *Physiol. phenomena of growth in plants.* ⓓ Utrecht 1920.
Sanders, George Ethelbert (1884), Entomol. to Ansbacher Insecticide Company Manufacturers of liquid and dust insecticides and fungicides. New York City (U.S.A.), 527 Fifth Ave. *Insects and diseases, insecticides and fungicides.* ⓓ Toronto, Canada 1907.
Sanders, Harold George (1898), Inst. of Animal Nutrition, School of Agric. Cambridge (England). *Agric. Physiol. Statistical Research of Milk Records.* ⓓ Cambridge 1920.
Sanders, James G. (1880), Entomol. for Sun Oil Co. and Manager of Spray Oil Dept. Philadelphia, Pa. (U.S.A.). *Homoptera. Families Coccidae, Cicadellidae.* ⓓ Ohio State Univ. 1903.
Sanders, Paul De Leon (1901), Ass. Entomol. Univ. of Maryland. College Park, Md. (U.S.A.). ⓓ Univ. of Maryland 1924.
Sandeyson, E. Dwight (1878), Prof. of Rural Social Organization, Cornell Univ. Ithaca, N.Y. (U.S.A.). *Economic Entomol.* ⓓ Univ. of Chicago 1921.
Sandground, Jack Henry (1899), Instr. in Tropical Helminthol. Boston, Mass. (U.S.A.), Dept. of Tropical Med. Harvard Med. School. *Systematics and biol. of parasit. worms. Immunity and general host-parasite relationships.* ⓓ Johns Hopkins Univ. 1925.
Sandiford, Irene (1889), Associate, Section on clinical Metabolism. Instructor Biochem. Rochester, Minn. (U.S.A.), Mayo Clinic.. *Metabolism.* ⓓ Univ. Minnesota 1919.
Sandmo, Julius Klykken (1885), Prof. der Forstnutzung, Landwirtschaftl. Hochsch. Aas (Norge). *Forstnutzung.* ⓓ Landw. Hochsch. 1911.
Sando, Charles Earl (1894), Associate Biochem., Bureau of Plant Industry, U. S. Department Agric. Washington (U.S.A.). *Coloring Matters and Lipoides.* ⓓ Univ. of Maryland 1920.
Sando, William Joseph (1892), Associate Agronomist, U. S. Dept. of Agric. Washington, D.C. (U.S.A.), 3021 So. Dak. Ave. N.E. *Agronomy, genetics and cytol.* ⓓ College Park, Md., 1921.
Sands, David Richmond, Lect. Dept. of Botan. Ontario Agric. Coll. Guelph, Ontario (Canada). ○
Sands, Harold Collender (1884), Pathol. Spaulding, Maine. (U.S.A.). *Cytol. Potato diseases; pathogenic bact. both animal, plant and med., genetics as applied to egg laying habit of chickens.* ⓓ Columbia Univ. 1923.
Sandt, Walter (1891), Dr., Priv.Doc. für Botan. an der Univ. München 38 (Deutschland), Notburgastraße 4 I. *Experim. Morphol. Entwicklungsgeschichte.* ⓓ München 1920.
Sandu-Aldea, C., Prof., Scoala super. de agric. Bucarest (România). *Pflanzenzüchtung.* ○
Sanfelice, Francesco (1861), Prof. o. Igiene. Univ. Modena (Italia). *Epidemiol. Alimentazione.* ⓟ Microorganismi patogeni.
Sanford, Arthur Hawley (1882), Head of Section, Clinical Labor., Mayo Clinic, Prof. of Clinical Pathol., Mayo Foundation (Univ. of Minn.). Rochester, Minn. (U.S.A.). *Bact., Parasit., Serol.* ⓓ Northwestern Univ. 1907.
Sanford, Guthrie B. (1890), Plant Pathol. in charge, Dominion Labor. Plant Pathol. Univ. of Saskatchewan. Saskatoon (Canada). *Actinomyces scabies. Potato scab.* ⓓ Minnesota 1920.
Sani, Luigi, Prof. incar. R. I. med. vet. Milano, Lib. doc. Ist. sup. med. vet. Bologna (Italia). *Pathol.* ○
Sankas, Sugiam Huta (1899), Instr. in Anat. Bangkok (Siam), Chulalongkorn Med. school. ⓓ Chulalongkorn Med. School 1924.
Sanotowshi, Marjan (1894), Ass. de l'univ. Kraków (Polska), rue Lubier 46. *Géographie végétale.* ⓓ Cracovie 1923.
Sanozkij, Anton Stepanovič, Prof. Inst. f. Landwirtsch. Minsk, Weißrußl. (U.d.S.S.R.). *Anat. Physiol. d. Haustiere.* ○
Sansom, G. S., Hon. Research Ass., Dept. of Anat. and Histol. at Univ. Coll. London (England). *Embryol.* ○
Sansome, Frederick Whalley (1902), Ass. Univ. Glasgow (Scotland). *Cytol. Genetics, Melandrium: Sex Determination. Brassica: Cytol. Wheat. Lolium.* ⓓ Edinburgh 1926.
Santarelli, Enrico (1879), Prof. ord. Trani, Bari (Italia), Corso Cavour 67. *Flora delle Alpi Apuane, del Serchio. Azione della luce solare, rossa e violetta, sull'allungamento del fusto delle piantine, e sulla nutrizione.* ⓓ Pisa 1920.
Santesson, Carl Gustaf (1862), Prof. der Pharmacol. am Carolinischen medico-chirurgischen Inst. Stockholm (Schweden), Pharmacol. Abt.*Experim. Pharmakodynamik.* ⓓ Upsala 1892.
Santesson, Lars Johan Henrik (1901), Ass. am Anat.-Pathol. Inst. Univ. Upsala. Stockholm (Sverige), Hantverkaregatan 34 II. *Gewebezüchtung.* ⓓ Upsala 1925.
Sántha, Ladislaus, Dr., Dir. Ampelol. Inst. d. Ung. Min. f. Landw. Budapest II (Ungarn), Debröigasse 17. ○
Santiago di Suñez, Prof. Univ. Zaragoza (España). *Fisiol. Hum.* ○
Santos, Jose K. (1889), Ass. Prof. of Botan., Univ. of the Philippines. Manila (Philippine Islands). *Morphol. Cytol. of coconut palm, rice, sugar cane.* ⓓ Chicago 1923.
Santos, S., Prof. Univ. Rio de Janeiro (Brasil). *Anat.* ○
Santos y Abren, Elias, Lic. Med., Dir. del Mus. de Hist. Nat. y Etnográfico. Santa Cruz de la Palma (Canarias). *Entomol. y Botan.* ○
Santucci, Renato (1896), Aiuto dell'Ist. di Zool. della R. Univ. di Genova (Italia), Via Balbi 5. *Endobiosi delle spugne. Sviluppo larvale dei Crostacei decapodi.* ⓓ Genova 1921.
Sanz Echeverria, Josefa, Prepar. de Osteozool. Mus. Nac. de Ciencias Nat. Madrid (España), Bravo Murillo 70. *Osteol. der Fische. Otolithen.* ○
Sanzo, Luigi, Lib. doc. Univ. Messina (Italia). *Zool.* ○
Sapěhin, Andreas A. (1883), Prof. Genetik u. Pflanzenzüchtung Landwirtsch. Hochsch., Dir. Odessaer Zentralversuchsstation. Odessa, Ukraine (U.d.S.S.R.), Institutskaja 9, Versuchsstation. *Histogenetik (Genetik auf zytol. Basis) der Weizenarten.* ⓓ Odessa 1909.
Sapěhin, Leo Andrejewitsch (1906), Aspirant d. Inst. f. Vererbungslehre. Odessa (U.d.S.S.R.), Landwirtschaftl. Versuchsstation, Puschkinskaja 44. *Genetik und Cytol. d. 28 chromosomig. Triticumarten.* ⓓ Odessa 1926. ⓟ Trit. durum, dicoccum, persicum.
Saposnikov, A., Prof. Inst. d. Volksbild. Nikolaev (U.d.S.S.R.), Nikol'ška ul. 12/2. *Zool.* ○
Šapšev, Konstantin Nik., Prof. Univ. Perm (U.d.S.S.R.), Ul. Karl Marksa 26. *Hygiene.* ○
Sarasin, Friedrich (1859), Privatgelehrter, Vorsteher des Mus. für Völkerkunde. Basel (Schweiz), Spitalstr. 22. *Anthrop., Ethnol., Zool. Botan.: Ceylon, Celebes, Neu Caledonien.* ⓓ Phil. Würzburg 1883, med. hon. causa Basel 1903, sc. hon. causa Genf 1919.

17*

Sarasin, Paul (1856), Dr. phil., Präs. Schweiz. Naturschutzkommission. Basel (Schweiz), Spitalstr. 22. *Anthrop., Zool., Botan. d. Tropen.* o

Sargent, Charles Sprague, Dir. of the Arnold Arboretum. Jamaica Plain, Mass. (U.S.A.). o

Sarin, Eduard (1876), Prof. Riga (Lettland), Univ. *Fermente.* ⓓ Petersburg 1908.

Sarkisow, S. Tomsk (U.d.S.S.R.), Technol. Inst., Geol. Kabinett. *Kreideechinoiden von Turkestan.* o

Sarmiento, Laspiur Ricardo, Prof. Univ. Nac. Buenos Aires (Argentina). *Anat.* o

Sarra, Raffaele (1861). Matera, Basilicata (Italia). *Insetti dannosi alle piante. Anarsia. Olethrentes. Cimbex. Lixus.*

Sars, Georg Ossian (1837), Prof. emer. Zool. Univ. Oslo (Norge). *Crustaceen von allen Weltteilen.* ⓓ Upsala. ⓟ Crustaceen, Süßwasserplankton.

Sartoris, George Bartholomew (1896), Ass. Pathol., Bureau of Plant Industry, U.S.D.A. Washington, D.C. (U.S.A.). *Pathol., Sugar cane. Apiosporina Collinsii.* ⓓ Univ. of Michigan 1923.

Sarvonat, François (1879), Prof. agrégé à la Fac. de Méd. de Lyon. La Demi lune, Rhône (France). *Pharmacodynamic: hypnotiques, chémothérapie, antipyrétiques.*

Sasaki, Chûjiro (1857), Prof. emer. Entomol.and Sericicol. Imper. Univ. Tôkyô (Japan), Akasakaku, Awoyama Minami-Machi, 6 Chôme N. 120. *Zool., Entomol. and Sericicol.* ⓓ Tôkyô Imper. Univ. 1881.

Sasaki, Kiichirô, Lect. Tôhoku Imper. Univ. Sendai (Japan), Zool. Inst., Coll. of Sc. *Zool. (Annelida).*

Sasaki, Madoka, Prof. Hokkaidô Imper. Univ. Sapporo (Japan), Labor. of Fishery, Coll. of Agric. *Malacol. (Cephalopoda).*

Sasaki, Takaoki, Private Labor. Nakano near Tôkyô (Japan), Uyenohara 812, Higashi-Nakano. *Chem. of Proteins.*

Sasaki, Takashi, Prof. in Horticulture and Chief of Univ. Farm, Coll. of Agric. Imper. Univ. Kyôto (Japan), Labor. of Horticulture. *Crops.*

Sassi, Moriz (1880), Dr. phil, Naturhistorisches Mus. Wien IV (Deutsch-Österreich), Schwindgasse Nr. 11. *Systematische Ornithol. (Afrika).* ⓓ Wien 1903.

Satake, Yasutarô (1884), Prof. Tôhoku Imper. Univ. Sendai (Japan), Physiol. Inst., Coll. of Med. *Physiol. der Nebenniere.*

Satina, Sophia A. (1879), Investigator, Carnegie Inst. Long Island, N.Y. (U.S.A.), Cold Spring Harbor. *Mycol.: Phycomycetes and Ascomycetes (Cytol., Morphol., Physiol.). Sexuality in Fungi.* ⓓ Moscow Univ. 1904.

Satô, Akira, Prof. Tôhoku Imper. Univ. Sendai (Japan), Inst. of Pediatrics, Fac. of Med. *Bio- and Microchem.*

Satô, Kameichi, Prof. Aichi Med. Univ. Nagoya (Japan), Anat. Inst. *Anat.*

Satô, Yatarô, Prof. Imper. Univ. Kyôto (Japan), Inst. of Forestry, Coll. of Agric. *Forestry.*

Satterthwait, Alfred Fellenberg (1879), Associate Entomol., in charge of U.S. Entomol. Labor. for Cereal and Forage Insect Invest. Webster Groves, Mo. (U.S.A.), 527 Ivanhoe Place. *Calandra, life histories, distribution, economic significance.*

Satwornitzkaja, Soja (1888). Kasan (U.d.S.S.R.), Malaja Krassnaja 3. *Mikrophysiol. der Hypophysis cerebri. Drüsen mit innerer Sekretion bei Avitaminose B.*

Saulescu, Nicolas (1898), Doc., Landw. Inst. der Univ. Jassy (Rumänien). *Genetik und landw. Pflanzenzüchtung: Weizen, Mais und Hafer.* ⓓ Berlin 1925.

Sauliak-Sawizka, M. M. (1890), Ass. Kiew (U.d. S.S.R.), Lwowska Str. 49, Wohn. 2. *Nervensystem der Blase der Ratte.*

Saunders, John Tennant (1888), Lect. Univ. Christ's Coll. Cambridge (England). *Freshwater problems.*

Sautet, Jacques Jules Jean (1903), Prép. Parasit. Fac. Méd. Paris 5 (France), Hotel de la Paix, 6, Rue Blainville. *Parasit. humaine et compar.*

Sauvayeau, Camille (1861), Prof. Fac. Sc. Bordeaux (France), 31, rue Adrien Bayssellance. *Algol.*

Savarenskij, Fedor Petrov., Doc. I. Univ. Moskau (U.d.S.S.R.). *Palaeont.* o

Savas, C., Prof. Hygiène, Microbiol. Univ., Dir. de l'Inst. d'Hygiène de Bact. Athènes (Grèce), Rue Nikis 15. o

Savelli, Roberto (1895). Rovigno (Italia), R. Stazione di Bieticoltura. *Teratol. dei fiori. Mutazioni. Ibridazioni nel tabacco e nelle Zucche. Elettrogenetica.* ⓓ Perugia.

Savicz, Lydia Ivanovna (1886), Conserv. Jardin Botan. Principale. Leningrad (U.d.S.S.R.), Pessocznaja 1/2, app. 93. *Bryol.*

Savicz, Vsevolod Pavlovicz (1885), Sécrétaire sc., Conserv. Jardin Botan. Principal. Leningrad (U.d.S.S.R.), Pessocznaja 1/2, app. 93. *Lichenol., géographie des plantes.*

Savin, Witt Nik., Prof. Staats-Univ. Tomsk (U.d.S.S.R.). *Topograph. Anat.* o

Savitsch, Nina (1894). Ass. à la Stat. Genetic Centrale, Moskau (U.d.S.S.R.), Bronnaja 15. ⓓ Moskau 1918.

Savoff, Christo (1888), Inst. agric. Univ. Nancy (France). *Pathol. végétale, Sélection de plantes.* ⓓ 1923. ⓟ Maladie des plantes.

Săvulescu, N., Prof. Scoala super. di Agric. Bucuresti (România). *System Botan.* o

Savvateev, Andrej I., Doc. II. Univ. Moskau U.d.S.S.R.). *Pathol. Anat.* o

Sawada, Taketarô (1899). Miyanoshita, Hakone (Japan). *Dendrol. japonica.* ⓓ Tôkyô imper. Univ. Econom. dept. ⓟ Plantae exsiccatae.

Sawamura, Makoto, Prof. Tôkyô Imper. Univ. Komaba near Tôkyô (Japan), Chem. Inst., Fac. of Agric. *Agric. and Physiol. Chem.*

Saweljew, S. N., Doc. Zool. Labor. Univ. Baku (U.d.S.S.R.). o

Sawenkov, W. P., Prosektor, Dr., Anat. Inst. II. Univ. Moskau (U.d.S.S.R.), Dewitschje Pole, Trubuzkoi pereulok. o

Sawicki, Bronislas, Prof., Dr. med. Warszawa (Polska), r. Bracka 16. *Pathol., Anat. pathol.* o

Sawitch, Vladimir (1874), Prof. Leningrad (U.d.S.S.R.), Povarskoj 12, 1. 3. *Endocrinol.* ⓓ Leningrad.

Sawitsch, Natalie (1894), Ass. Geobotan. Section Botan. Hauptgarten. Leningrad (U.d.S.S.R.), Pessotschnaja Str. 22, Quartier 4. *Geobotan.* ⓓ Leningrad 1917.

Sawitsch, Wladimir M., Prof. an der Univ. Wladiwostok (U.d.S.S.R.). *Geobotan.* o

Sawkin, Peter St., Dir. der Landw. Moor-Versuchsstation Nowgorod (U.d.S.S.R.), Postfach 23. *Moorkultur. Wiesenbau. Botan.* o

Sawoff, M., Dir. Landw. Abt. d. Landw. Versuchs-Stat. Sofia (Bulgarien), Plostad Solunski. *Phytopath.* o

Sawostin, Peter W., Doc. an der Univ. Tomsk (U.d.S.S.R.), Timirjasew Pr. 48, W. 2. *Pflanzenphysiol.* o

Sawyer, Chauncey E. (1891), Poultry Research Vet., Western Experim. Stat. of State Coll. Puyallup, Wash. (U.S.A.). *Etiol., prevention and control of poultry diseases, chicken pox, coccidiosis.*

Sawyer, Mary L., Ass. Prof. of Botan., Wellesley Coll. Wellesley, Mass. (U.S.A.). *Morphol., Cytol., Genetics.* o

Sawyer, Wilbur Augustus (1879), Dir. Public Health Labor. Service, International Health Board, Rockefeller Foundation. New York, N.Y. (U.S.A.), 61 Broadway. *Hookworm.* ⓓ Cambridge 1906.

Sawyer, William H. jr., Ass. Prof. of Biol., Bates Coll. Lewiston, Maine (U.S.A.), 138. Nichols St. *Mycol.* o

Sax, Karl (1892), Biol. Orono, Me. (U.S.A.), Maine Agric. Experim. Station. *Genetics and cytol. of wheat hybrids. Cytol. of fertilization, bud selection, nursery stock. Sweet corn breeding. Inheritance of size differences in Phaseolus. Genetics of ecol. and adaptation.* ⓓ Harvard Univ. 1922.

Sayle, Mary (1892), Instr. Univ. of Wisconsin. Madison, Wis. (U.S.A.), 149 W. Wilson. *Ecol., metabolism of Aeschnid, Dragon-fly Nymphs.* ⓓ Univ. of Wisconsin 1927.

Sayre, Charles Bovett (1891), Associate in Research, Horticulture. Geneva, N.Y. (U.S.A.). *Canning Crops.* ⓓ Univ. of Illinois 1924.

Sayre, Jasper Dean (1893), Ass. Prof. of Botan., Ohio State Univ. Columbus, O. (U.S.A.). *Physiol., Microchem.* ⓓ Ohio State Univ. 1922.

Sazerac, Robert (1875), Chef de labor. Inst. Pasteur. Paris (France), 28, Rue Dutot. *Syphilis, Trypanosomiasis.* ⓓ Paris 1921.

Sbarsky, Boris (1885), Prof., vice dir. d'Inst. Biochim. Moscou (U.d.S.S.R.), Worontzowo-Pole 8. *Biochim.* ⓓ Genève 1911, Leningrad 1912.

Sbitkowski, Nikolaj Alexandrowitsch (1884), Ass. Botan. Inst. Univ. Minsk (U.d.S.S.R.), Schirokaja N. 28. *Phanerogamen, Flora von Weißrußland.* ⓓ Warschau 1912.

Scaffidi, Vittorio (1877), Dir. Ist. di Patol. gen. Univ. Napoli (Italia), Sant'Andrea delle Dame 21. ⓓ Roma 1901.

Scaglia, Giuseppe (1891), Ass. Ist. anat. umana di Cagliari (Italia). *Innervazione del cuore. Costituzioni.*

Scala, Augusto César (1880), Prof., Chef du Dept. Botan. du Mus. de La Plata (Republique Argentine). *Histol. végétale (bois argent.).* ⓓ Buenos Aires. ⓟ *Plantes sèches. Prép. micr. d'histol. végétale.*

Scalfi, Antonio (1896), Aiuto Sezione „Vaccini e Diagnostici" dell'Ist. Sieroterapico Milanese. Milano 14 (Italia), Via Guastalla 3. *Microbiol. e Immunol.* ⓓ Genova 1922. ⓟ *Culture, antigeni.*

Scammon, Richard E. (1883), Prof. of Anat. Minneapolis, Minn. (U.S.A.), Inst. of Anat., Univ. of Minnesota. *Quantitative study of the growth of the human body.* ⓓ Harvard Univ. 1909.

Scarcella-Perino, Giuseppe (1899), Ass. Ist. di fisiol. Univ. Messina (Italia). ⓓ 1926.

Scarpellini, A., Prof. Univ. Padova (Italia). *Batteriol.* ○

Scarth, George W. (1881), Associate Prof. of Botan., McGill Univ. Montreal (Canada). *Cell Physiol.* ⓓ Edinburgh 1904.

Ščastnyj, S. M., Prof. Chem.-Pharmaz. Inst. Odessa (U.d.S.S.R.), Ul. Krasnoj Gvardii 17. *Bact., Serol.* ○

Šeeglova, Olga A. (1889), Labor. Sektion der Pflanzenphysiol. Botan. Garten, Ass. Botan. Hochsch. Wissenschaftl. Inst. „Lesshaft". Leningrad (U.d.S.S.R.), Moika 42. *Anat. und Physiol. der Pflanzen. Photoperiodismus.* ⓓ Leningrad 1924.

Šeegolev, Grigorij G., Doc. I. Univ. Moskau (U.d. S.S.R.). *Zool., Embryol.* ○

Ščelkanovcev, J. P., Prof. Landwirtschaftl. Inst. Voronez (U.d.S.S.R.). *Forstl. Entomol.* ○

Ščerbakov, M. T., Prof. Landwirtschaftl. Inst. Krasnodar, Kaukasus (U.d.S.S.R.), Novaja 107. *Chem. d. Weines, techn. Microbiol.* ○

Schaanning, Hans Thomas Lange (1878), Konserv. Naturh. avd., Stavanger Mus. Stavanger (Norge). *Ornithol.* ⓓ Vögel.

Schachanin, Nikolaj J. (1890), Doc. des Jaroslaw. Pädagog. Inst., Dir. Naturw. Mus. Jaroslawl (U.d. S.S.R.), Sowetskaja Platz 22. *Mycol., Geobotan.* ⓟ *Pflanzen.*

Schachner, Josef (1897), Dr., Hochschulass. Weihenstephan, Freising (Deutschland), Landwirtschaftl. Hochsch. *Gärungsorganismen.* ⓓ München 1923.

Schad, Christ (1901), Chef de trav. de botan. Ecole nation. d'Agric. Grignon, Seine et Oise (France). *Génétique et pathol.*

Schade, Friedrich Alwin (1881), Dr. phil. Dresden-A. 24 (Deutschland), Nürnberger Str. 18c. *Oekol. und Geographie der niederen Kryptogamen, Hepaticae und Lichenes.* ⓓ Jena 1911. ⓟ *Hepaticae u. Lichenes.*

Schaechtelin, Jean (1884), Ass. à l'Inst. botan. de l'Univ. de Strasbourg (France), 31, rue Oberlin. *Mycol.*

Schaede, Reinhold (1887), Priv.Doc., Ass. am Pflanzenphysiol. Inst. der Univ. Breslau 9 (Deutschland), Goeppertstr. 6/8. *Cytol.* ⓓ Breslau 1912.

Schädelin, Walter (1873), o. Prof. a. d. Eidgen. Techn. Hochsch. Zürich (Schweiz), Eleonorenstr. 26. *Waldboden, Durchforstung.* ○

Schäferna, Karel (1884), Priv.Doc. für Hydrobiol. und Fischerei, Technische Hochsch. Priv.-Doc. für Zool., Karls-Univ. Praha XVI (C.S.R.), Nábřeží legií 28. *Biol. und Systematik der Crustaceen, besonders Amphipoden und Phyllopoden. Wasserinsekten und Fische. Pathol. der Fische. Limnol. und fischereiliche Biol.* ⓓ Praha 1908.

Schaeffer, Asa Arthur (1883), Prof. of Zool., Univ. of Kansas, On Staff, Biol. Labor., Cold Spring Harbor, Long Island, New York (Summer). Lawrence, Kan. (U.S.A.), Zool. Labor. *Taxonomy, biol., physiol. of Amebas. Feeding and biol. of Protozoa. Movement of Amebas. Spiral movement of all organisms.* ⓓ Johns Hopkins Univ. 1909. ⓟ *Cultures of Chaos diffluens (Amoeba proteus).*

Schaeffer, Cornelia (1896), Ass. Entomol. Inst. der E. T. H. Zürich VII (Schweiz), Fehrenstr. 8. ⓓ Zürich 1926.

Schaeffer, M. G., Prof. Fac. Méd. Strasbourg (France). *Physiol.* ○

Schäperclaus, Ludwig August Wilhelm (1899), Ass. an der Landesanstalt für Fischerei Berlin-Friedrichshagen (Deutschland). *Hydrobiol., Fischereibiol.: Stoffwechselphysiol. der Wassertiere und Pflanzen. Fisch- und Krebspathol. (Infektionskrankheiten).* ⓓ Münster 1923.

Schaeppi, Theodor (1867). Zürich 7 (Schweiz), Sprensenbühlstr. 7. *Darmepithel. Nervensystem. Zusammenhang zw. Muskel u. Nerv.* ⓓ Phil. Jena 1893, med. Breslau 1896.

Schaetz, Georg (1887), Priv.Doc. der Univ. I. Ass. am Pathol. Inst. Halle a. d. Saale (Deutschland), Ulestr. 17/0. *Pathogenese von Mißbildungen (Meckels Divertikel).* ⓓ München 1922.

Schafer, Edwin George (1884), Prof. State Coll. Washington, D.C. (U.S.A.), Pullman. *Agronomy, farm crops.* ○

Schaffer, Josef J. F. (1861), o. ö. Prof. der Histol. an der Univ. Wien XVIII (Deutsch-Österreich), Hofstattgasse 4. *Histol. der Stützsubstanzen (chordoides, chondroides, Knorpel- und Knochengewebe). Thymus. Hautdrüsen. Epithelgewebe. Muskelgewebe. Mechanische Leistungen tierischer Gewebe. Histol. Technik.* ⓓ Graz 1886.

Schaffer, Karl, Prof. Dr., Dir. d. Inst. für Histol. des Gehirns Univ. Budapest (Ungarn). ○

Schaffner, John Henry (1866), Prof. of Botan., Ohio State Univ. Columbus, O. (U.S.A.), Dept. of Botan. *Taxonomy of Plants, nature of sex, rejuvenation in plants, flora of Ohio, genus Equisetum.* ⓓ Baker Univ. 1896. ⓟ *Species of Equisetaceae.*

Schaffnit, Ernst (1878), o. Prof. an der Landwirtschaftl. Hochsch. Bonn-Poppelsdorf und Dir. des Inst. für Pflanzenkrankheiten. Bonn a. Rhein (Deutschland), Hindenburgstraße 119. *Phytopath. (pathol. Physiol., Mycol., Entomol., Immunitätsforschung).* ⓓ Erlangen 1905.

Schalyt, Michail S., Wiss. Mitarbeiter der Ukrain. Akademie der Wissenschaften, Ass. Askanija Nowa (U.d.S.S.R.), Kreis Melitopol. *Geobotan.* ○

Schander, Richard (1873), Prof., Dr., Dir. des Inst. für Pflanzenkrankheiten und Hauptstelle für Pflanzenschutz der Landwirtschaftl. Versuchs- und

Forschungsanstalten. Landsberg a. W. (Deutschland), Theaterstr. 8. *Phytopath.* ⊕ Jena.
Schandl, Josef (1885), o. ö. Prof. an der Volkswirtschaftl. Univ. Budapest VIII (Ungarn), Eszterházygasse 3. *Tierzuchtlehre.* ⊕ Budapest 1914. ⊕ Wollproben.
Schaposchnikow, Wladimir N., Doc. I. Univ. Moskau (U.d.S.S.R.), Strastnoj bulv. 10, W. 26. *Pflanzenphysiol. u. Anat.* o
Scharr, Ernst (1875), Dr. med. vet., Dir. des Bact. Inst. der Landwirtschaftskammer für die Provinz Brandenburg und für Berlin. Berlin NW 40 (Deutschland), Kronprinzenufer 4. *Tierhygien. und Tierseuchenbekämpfung, spez. Tuberkulosebekämpfung.* ⊕ Berlin 1920.
Schaternikoff, Michael (1870), Prof., Dir. Inst. f. Ernährungsphysiol. und Stoffwechsellehre d. Volkskommissarlats. Moskau (U.d.S.S.R.), Granowsky's Str. 4, 1. *Respiration. Stoffwechsel.* ⊕ Moskau 1899.
Schatsky, P. S., Kustos Mus. Bergakademie Geol. Kabinet. Moskau (U.d.S.S.R.). *Fauna des Palaeozoicums.* o
Schauder, Wilhelm (1884), planmäß. a.o. Prof., Leiter der Abteilung für Histol. und Embryol. am Vet.-anat. Inst. der Univ. Leipzig (Deutschland), Tiroler Str. 4. *Histol., Embryol. u. Anat. d. Haustiere; Bewegungsapparat, Plazentaranat.* ⊕ Gießen 1912.
Schauinsland, Hugo Hermann (1857), Prof., Dir. Städt. Mus. für Natur-, Völker- und Handelskunde. Bremen (Deutschland), Städtisches Mus. *Vermes (Plathelminthen u. Priapuliden). Embryol. u. Anat. der Elasmobranchier u. Sauropsiden. Urgeschichte (Palaeolithicum). Anthrop. und Kultur Ostasiens.* ⊕ Königsberg 1883.
Schaxel, Julius (1887), OberReg.R., Prof. für Zool. und Vorstand der Anstalt für experim. Biol. der Thüringischen Landesuniv. Jena (Deutschland), Reichardstieg 4. *Cytol. und Histol. der Entwicklung bei den gewebebildenden Tieren; Regeneration, Transplantation und Parabiose; Tierpsychol.; theoretische Biol. und Geschichte der Biol.* ⊕ Jena 1909.
Schazillo, Boleslav (1888), Forschungskatheder f. Morphol. u. Physiol. Odessa (U.d.S.S.R.), Olgiewskaja Str. 4. *Innere Sekretion.* ⊕ Charkow 1921.
Schdanow, Sergius Wassilii (1900), Ass. Zool. Kabinett Vet. Inst. Kasan (U.d.S.S.R.), Zootom. Kabinett Univ. *Vergl. Morphol. der Wirbellosen; Biene und Bienenzucht.*
Schechtel, Edward (1886), ao. Prof. Fischerei u. Jagdkunde an d. agron.-forstlichen Fac. d. Univ. Poznań (Polska), ul. Mazowiecka 24. *Fischerei: Lachs. Jagdkunde. Hydracarinen.* ⊕ Lemberg 1912. ⊕ Hydracarinen.
Scheffelt, Ernst (1885), Dr. phil. Badenweiler in Baden (Deutschland). *Fische und Fischerei, Systematik der Coregonen, Plankton, Seenkunde, Entomostraca. Naturschutz.* ⊕ Freiburg i. Br. 1908. ⊕ Praep. von Coregonenlarven und Planktonpraep. von süddeutschen Seen.
Scheffer-Boichorst, Henning (1877), Forstmeister, Velen in Westfalen (Deutschland). *Angewandte Entomol., Bekämpfung von forstlichen Schädlingen.* ⊕ Entomol. Objekte der Schädlingsbekämpfung.
Scheidt, Walter (1895), Dr., Priv.Doc. für Anthrop.; Vorsteher der rassenkundlichen Abteilung am Mus. für Völkerkunde. Hamburg (Deutschland), Eppendorfer Landstr. 18. *Rassenkunde, insbes. Rassenkunde von Europa.* ⊕ München 1920.
Schellenberg, Adolf (1882), Kustos u. Prof. am Zool. Mus. der Univ. Berlin N 4 (Deutschland), Invalidenstr. 43. *Crustaceen: Amphipoda und Notodelphyidae.* ⊕ München 1910.
Schellenberg, Alfred (1895), Doc. Techn. Hochsch., Rebbaukommissar d. Kantons Zürich. Wädenswil, Kt. Zürich (Schweiz). *Wein- u. Obstbau.* o
Schellenberg, Gustav (1882), a.o. Prof. der Botan. Göttingen (Deutschland), Botan. Inst. *Systematik der Phanerogamen: Connaraceae, Olacaceae, Icacinaceae.* ⊕ Zürich 1910.

Scheller, Robert, Univ.Prof. Breslau (Deutschland), Vogelweide 185. *Hygiene.* o
Schembel, Stefan Ju. (1886), Prof., Staatl. Med. Inst. Astrachan (U.d.S.S.R.), Nabereshn. I. Maja 144. *Phytopath. Cucurbitaceae, Solanaceae.* ⊕ Petersburg 1912. ⊕ Parasit. Pilze.
Scheminzky, Ferdinand Ottokar (1899), Dr., Ass. am Physiol. Inst. der Wiener Univ. Wien IX (Deutsch-Österreich), Währingerstr. 25. *Technik der Verstärkerröhren. Elektronarkose. Elektrotaxis und Elektrotropismus. Elektrokultur. Phonetik. Elektrophysiol. der Muskel.* ⊕ Wien 1925.
Schemtschuschnukov, Georg A., Doc. Bergakademie. Leningrad (U.d.S.S.R.), W. O., 21 Linie 2, Wohn. 27. *Palaeont., Palaeofaunistik.* o
Schenck, Heinrich (1860), Dr. phil., Geh. Hofrat, Prof. der Botan. an der Technischen Hochsch. und Dir. des Botan. Gartens zu Darmstadt (Deutschland), Nikolaiweg 6. *Anat. und Biol. der Wassergewächse, der Lianen.* ⊕ Bonn 1884.
Schenk, Jakob (1876), Sekretär des Königl. Ungarischen Ornithol. Inst. Budapest II (Ungarn), Debröi ut 15. *Vogelzugforschung, Oekol., Vogel- u. Naturschutz.*
Schennikow, Alexander Petrovitsch (1888), Oberkonserv. Botan. Hauptgarten. Leningrad (U.d. S.S.R.), Forstinst. *Botan. Geographie, Phytosoziol.*
Schenoni, A., Prof. Univ. Nation. Asunción (Paraguay). *Anat.* o
Schepilewskaja, Nina (1893), Ass. Moskau (U.d. S.S.R.), Kusnetsky Most 15, W. 3. *Stoffwechsel. Avitaminose.*
Schepotieff, Alexander (1879), Prof. Zool. Vergl. Anat. Univ. Minsk (U.d.S.S.R.). *Würmer, spez. „Vermoidea" und Pterobranchia (Rhabdopleura, Cephalodiscus). Protozool., spez. pathogene Protozoen. Biochem. (serol. Untersuchungen, Anwendung d. Wassermannschen und Präzipitin-Reaktionen zur Phylogenie). Palaeont.: Graptolithen. Sericiculture. Bienenzucht.* ⊕ Heidelberg 1903.
Scherbakoff, Anatolie (1903), I. Univ. Inst. für Experim. Biol. Moskau (U.d.S.S.R.), Lobkowskij per. 2, W. 23. *Physikalisch-chem. Untersuchungen der Binnengewässer; Ernährung der Copepoden.* ⊕ Staatsuniv. Moskau 1926.
Scherefeddin, E., Prof. Hochsch. f. Tierärzte. Konstantinopel (Türkei), Bajtar m. *Botan.* o
Scherffel, Aladár (1865), Dr. phil. h. c. Gödöllő (Ungarn). *Flagellaten, Chytridineen und verwandte Organismen.* ⊕ Szeged (Ungarn) 1922.
Schermann, Szilárd (1895), Ass. Samenkontrollstation. Budapest II (Ungarn), Kis-Rókus-Gasse 15. *Floristik, Systematik der Blütenpflanzen. Angewandte Botan.* ⊕ Budapest 1921.
Schermer, Siegmund (1886), Dr. med. vet., o. ö. Prof., Dir. des Tierärztl. Inst. der Univ. Göttingen (Deutschland), Groner Landstr. 2. *Vet.-Haematol.* ⊕ Leipzig 1908.
Schern, Kurt (1880), Prof., Dr., Dir. des Bact. Inst. der Tierärztl. Hochsch. zu Montevideo (Uruguay), Inst. de Bact. de la Escuela de Vet. *Veterinärbact. und Serol. Tierseuchenforschung, Impfstoffe.* ⊕ Leipzig 1905. ⊕ Vaccina antirabica. Tuberculin. Erreger von Tierseuchen.
Scherpe, Richard (1868), Biol. Reichsanst f. Land- u. Forstwirtschaft, Lab. f. prakt. landw. Chem. Berlin-Dahlem (Deutschland). *Durch unglinstige Bodeneigenschaften verursachte Pflanzenkrankheiten.* ⊕ Berlin 1891.
Schertz, Frank Milton (1889), Associate Biochem. Washington, D.C. (U.S.A.), 1305 Farraqut St. N.W. *Effect of potash, nitrogen and phosphorus on the chloroplast pigments.* ⊕ Univ. of Chicago 1918.
Scheuner, Arthur (1879), Dr. phil. et med. vet., o. ö. Prof. der Tierphysiol. und Dir. des Vet.-physiol. Inst. der Univ. Leipzig (Deutschland), Tiroler Str. 6. *Physiol. der tierischen und menschlichen Ernährung und Verdauung (Vitamine, Darmflora, Bact. der Silage, Fütterungsfragen, Magenmechanik, Milz).* ⊕ Göttingen 1902.

Scheuring, Ludwig (1888), ao. Prof. für Zool. u. vergl. Anat. Univ., Kustos Bayr. Biol. Versuchsanstalt f. Fischerei. München (Deutschland), Veterinärstr. 6. *Hydrobiol., Biol. u. Physiol. der Fische. Parasit., bes. Fischkrankheiten. Physiol. v. Spermatozoen.* ⓓ Gießen 1912. ⓟ Fische von Deutschland, Fischparasit. u. Fischkrankheiten.

Scheurlen, Hertha (1900), Dr. rer. nat., Ass. Greifswald (Deutschland), Geol. Inst. der Univ. *Cephalopoden.* ⓓ Tübingen 1925.

Schewiakow, W. T., Prof., Dir. Zool. Labor. Univ. Irkutsk (U.d.S.S.R.). o

Schewki, A., Prof. Hochsch. f. Tierärzte. Konstantinopel (Türkei), Bajtar m. *Pathol. Anat.* o

Schiboni, L., Prof. Univ. Roma (Italia). *Anat. patol.* o

Schieblich, Hermann Martin (1893) Dr. med. vet., Wissenschaftl. Ass. am Vet.-Physiol. Inst. der Univ. Leipzig O 27 (Deutschland), Gletschersteinstr. 30 I. *Vitamine, insbes. Bildung von Vitamin B durch Bakterien. Darmbakterien. Bact. der Silage.* ⓓ Leipzig 1920.

Schiedt, Richard Conrad Francis (1859), Prof. emer. of biol., Lancaster, Pa. (U.S.A.), 1043 Wheatland Ave. *Animal pigmentation.* ⓓ Ph.D. Univ. of Pennsylvania 1899, Sc.D. (h. c.) Franklin and Marshall Coll. 1910.

Schlefferdecker, Ernst Friedrich Paul (1849), o. Hon.Prof. für Anat. und Anthrop. Bonn a. Rh. (Deutschland), Kaiserstr. 37. *Mikroskop.-anat. Untersuchung des Muskelsystems, über Rassenunterschiede und -ähnlichkeiten.* ⓓ Königsberg i. Pr. 1872.

Schlemann, Elisabeth (1861), Univ.Doz. Berlin-Dahlem (Deutschland), Alfr.-Thaer-Weg 6. *Botan. Genetik.* o

Schlemann, Oscar (1875), Prof. u. Abteilungsleiter am Inst. R. Koch. Berlin (Deutschland), Föhrer Str. 2. *Pneumococcen, äußere u. innere Desinfektion, Immunität.* ⓓ Königsberg 1910. ⓟ Pneumococcen u. Meningococcen u. ihre Immunsera.

Schlemenz, Friedrich (1899), Ass. Berlin-Friedrichshagen (Deutschland), Landesanstalt für Fischerei. *Fischereibiol.* ⓓ Göttingen 1923.

Schlerbeek, Abraham (1887). Haag (Holland), Verhulststr. 14. *Synthetische Faunistik und Floristik des Naturschutzgebietes. Geschichte der Biol.* ⓓ Groningen 1917.

Schiff, Fritz (1889), Dir. der Bact. Abteilung des Städtischen Krankenhauses im Friedrichshain. Berlin W 15 (Deutschland), Meierottostr. 5. *Bact. und Serol., insbesondere Paratyphus, Blutgruppen, Agglutination.* ⓓ 1914.

Schifferli, Alfred (1879), Besorger der Schweiz. Vogelwarte der Schweiz. Gesellschaft für Vogelkunde und Vogelschutz in Sempach (Schweiz). *Vogelkunde der Schweiz.*

Schiffers-Rafalowitsch, E. V., Ass. Section de Géobotan., Jardin Botan. principal. Leningrad (U.d.S.S.R.), Aptekarski Ostrov, Pessotchnaja $^1/_2$. o

Schiffner, Victor (1862), Prof. der system. Botan. an der Univ. Wien III (Deutsch-Österreich), Rennweg 14. *Systematische Botan., Kryptogamenkunde (Bryol., Meeresalgen) und exakte Biol. (allgemeine Biol. auf metaphysischer Grundlage).* ⓓ Prag, Deutsche Univ. 1885. ⓟ Hepaticae eur. exsicc., Algae marinae, Kollektionen von Lichenes aus Brasilien und Indien (Java, Sumatra) Moos-Herbar.: Hepaticae.

Schiberszky, Károly (1863), o. Prof. der Univ. Budapest VIII (Ungarn), Eszterházy-utca 3. *Teratol. und Pathol. der Pflanzen, Pilze.* ⓓ Budapest 1923. ⓟ Mykol., Phytopath. Herbarium.

Schilder, Franz Alfred (1896), Dr. phil., Ass. Biol. Reichsanstalt. Naumburg a. d. S. (Deutschland), Weißenfelser Str. 57. *Systematik der rezenten und fossilen Cypraeidae (Mollusca). Variationsstatistik an Cepaea (Mollusca) und Coleopteren (bes. Coccinellidae). Morphol., Biol. und Bekämpfung von Phylloxera vastatrix.* ⓓ Wien 1921. ⓟ Rezente und fossile Cypraeidae (Mollusca). Populationen von Cepaea hortensis und nemoralis (Mollusca).

Schilf, Erich (1892), Priv.Doc. für Physiol., Ass. am Physiol. Inst. der Univ. Berlin N 4 (Deutschland), Hessische Str. 3—4. *Autonomes Nervensystem (Physiol.).* ⓓ Berlin 1919.

Schiller, Josef (1877) Doc. an der Univ. Wien XII (Deutsch-Österreich), Tivoligasse 55. *Botan. Hydrobiol., Cytol., Entwicklungsgeschichte der Pflanzen.* ⓓ Wien 1905. ⓟ Meeres- und Süßwasseralgen, Moose und Farne in Form von Diapositiven mit ihren natürlichen unveränderlichen Farben und Formen.

Schilling, Claus (1871), a.o. Prof. an der Univ. Berlin; Dir. der Tropenabteilung am Inst. Robert Koch, Berlin N 39 (Deutschland), Föhrer Str. 2. *Tropenkrankheiten; Protozoenkrankheiten; Malaria und Trypanosen; Tuberkulose.* ⓓ 1894. ⓟ Stämme von Spirochaeta Obermeieri, Trypanosoma Brucei, Stamm Provazek.

Schilling, Ernst (1889), Vorsteher der Botan. und Züchtungsabteilung am Forschungsinst. für Bastfasern, Sorau, N.-L., Prov. Brandenburg (Deutschland). *Züchtung der Flachspflanze; Pathol. des Leins; Anat. der Bastfasern; pathol. Anat.* ⓓ Münster in Westf. 1914. ⓟ Samen von Linum.

Schimon, Otto (1879), Dr.-Ing., Ass. Weihenstephan bei München (Deutschland). *Chem. der Gärung.* ⓓ München, Techn. Hochsch., 1911.

Schindewolf, Otto H. (1896), Dr. phil., Priv.Doc. der Geol. und Palaeont. Marburg a. d. Lahn (Deutschland), Geol. Inst. der Univ. *Fossile Cephalopoden und fossile Korallen.* ⓓ Marburg 1919.

Schindler, Johann (1881), Reg.R., Dr. phil., Oberinspektor der Bundesanstalt für Pflanzenbau und Samenprüfung in Wien II, Lagerhausstr. 174. Wien II (Deutsch-Österreich), Engerthstr. 204. *Anat., Histol., Physiol. und Biol. der landwirtschaftl. Kulturpflanzen, landwirtschaftl. Samenkunde, landwirtschaftl. Pflanzenbaulehre, Futterbaulehre und Alpwirtschaft. Mikroskopische Bestimmung der Wiesengräser.* ⓓ Wien 1908.

Schingarewa, Prof., Dir. Protist. Labor. Med. Inst. Leningrad (U.d.S.S.R.). o

Schinz, Hans (1858), o. Univ.-Prof., Dir. des Botan. Gartens der Univ. Zürich 8 (Schweiz). Seefeldstr. 12. *Flora der Schweiz und afrikanische Flora.* ⓓ Zürich 1884. ⓟ Exotische Herbarpflanzen.

Schinz, Hans R. (1891), Dr. med., Priv.Doc. für Röntgenol. und Radiumforschung an der Univ. Zürich 1 (Schweiz), Krautgartengasse 2. *Röntgenbiol., Strahlenbiol. Pathol. Anat. und Endokrinol. der Geschlechtsorgane. Osteol. der Säugetiere. Vererbungslehre des Menschen.* ⓓ Zürich 1916.

Schlöler, E. Lehn, Dr. København (Danmark), 14-16 Uranievej. *Ornithol.* o

Schipczinsky, Nikolai (1886), Ass. Botan. Hauptgarten. Leningrad (U.d.S.S.R.), Pesotschnaja ulitza 8. *Systematik der Ranunculaceen, Pflanzengeographie, Biol. der Unkräuter.* ⓓ Leningrad 1915.

Schischkin, Boris K. (1886), Prof. an der Univ. Tomsk (U.d.S.S.R.), Botan. Garten. *Botan., Pflanzenmorphol. und -systematic. Caryophyllaceae.*

Schitikova, Alexandrine (1899), Prép. Inst. Jaczewski de Mycol. et Pathol. végétale. Leningrad (U.d.S.S.R.), Englischer Prosp. 29, 12. *Les charbons et les rouilles du blé.*

Schitt, Peter G., Prof., Landwirtschaftl. Akademie Timirjasew. Moskau (U.d.S.S.R.), Petrowsko-Rasumowskoje Akademie 9, W. 19. *Gartenbau. Obstbau.* o

Schittenhelm, Alfred (1874), o. Prof. innere Med. Univ., Dir. der med. Klinik. Kiel (Deutschland), Feldstr. 55a. *Bluthistol., spez. Entstehung der Monocyten und deren Beziehung zum Bindegewebe und Reticulo-Endothel. Anaphylaxie, Allergie. Physiol. und Pathol. des Purinstoffwechsels. Ernährungsphysiol., in Bezug auf Verwendung der Malzkeime, Hefe u. ähnl. vom Menschen.* ⓓ Tübingen 1898.

Schkorbatow, Leonid A., Prof., Dir. Botan. Garten. Charkow (U.d.S.S.R.), Klotschkowskaja 52. *Botan. Pflanzenmorphol. Systematik.* o

Schlachtin, Eugen (1865), Prof. d. Histol. und Embryol. med. Fac. Univ. Taschkent (U.d.S.S.R.). *Schaltstücke des Myocards. Tonsillen. Reticuloendotheliales System.* ⊕ Jaroslau 1919.

Schlaginhaufen, Otto (1879), o. Prof. der Anthrop. Univ. Zürich, Dir. Anthrop. Inst. Kilchberg b. Zürich (Schweiz), Schlimbergstr. 40. *Physische Anthrop.: Anthrop. Melanesiens und der Schweiz; Hautleistensystem der Palma und Planta der Primaten.* ⊕ Univ. Zürich 1905.

Schlegel, Paul Richard (1865). Leipzig C. 1 (Deutschland), Oststr. 56. *Ornithol.* ⊕ Palaearkt. Vogelbälge.

Schleip, Waldemar (1879), Dr. med. et phil., o. Prof. der Zool. und vergl. Anat. Würzburg (Deutschland), Zool. Inst. *Embryol. (Würmer, Wirbeltiere). Entwicklungsmechanik (Würmer, Mollusken, Wirbeltiere). Chromosomenlehre (Würmer, Arthropoden). Physiol. des Farbenwechsels (Arthropoden).* ⊕ Med. Freiburg i. Br. 1904, phil. 1906.

Schlesinger, Günther (1886), Prof., R.R., Dir. d. niederösterr. Landessammlungen, Leiter der Fachstelle für Naturschutz im österr. Bundesdenkmalamt. Wien I (Österreich), Herreng. 9. *Biol. der rezenten und palaeotypen Wirbeltiere; Naturschutz.* ⊕ Wien.

Schlesinger, Monroe J. (1892), Instr. in Pathol., Harvard Med. School, Resident in Pathol. Children's Hospital. Boston, Mass. (U.S.A.). *Pathol. Anat., Bact., Immunol.* ⊕ Ph. 1920, M. 1926.

Schleussing, Hans (1897), Ass.-Arzt. Düsseldorf a. Rhein (Deutschland), Pathol. Inst., Moorenstr. 5. *Normale und krankhafte Entwicklung, Tuberkulose.* ⊕ Leipzig 1923.

Schlienz, Walter (1896), Dir. Kühlfisch A.G. Wesermünde (Deutschland), Hindenburgstr. *Kältetechnik in der Fischwirtschaft.* ⊕ Hamburg 1922.

Schlingman, A. Stanley (1889), Compar. Bact. and Pathol. Detroit, Mich. (U.S.A.), Med. Research Labor. Parke, Davis and Co. *Blackleg Immunizing Agents; The Effect of Codliver Oil on Dogs Convalescing from Distemper; The Ej ect of Carbon Tetrachlorid on Puppies; Bact. Chauvoei.* ⊕ Ohio State Univ. 1923.

Schlittler, Emil (1879), Univ.-Poliklinik. Basel (Schweiz), Tiergartenrain 5. *Pathol. Anat., normale u. pathol. Physiol. des Ohres.* ⊕ Zürich 1904.

Schljapina, Helene V. (1897), Ass. Biol. Wolga-Station. Saratow (U.d.S.S.R.). *Planktonalgen. Diatomeae.* ⊕ Moskau 1925. ℗ Diatomeen.

Schlossberger, Hans (1887), Dr. med., Prof., wissenschaftliches Mitglied des staatlichen Inst. für experim. Therapie, Frankfurt a. M. (Deutschland), Böcklinstr. 2. *Bact., Immunitätsforschung und experim. Therapie (Serumtherapie, Chemotherapie).* ⊕ Tübingen 1913. ℗ Bakterienkulturen.

Schlossmann, Hans (1894), 1. Ass. am Pharmacol. Inst. der Med. Akademie in Düsseldorf (Deutschland), Goethestr. 16. *Experim. Pharmacol.* ⊕ Köln 1920.

Schlossmann, Karl (1885), Prof. Bact. Fac. Méd. Tartu (Eesti). *Bact. générale et expérim.* ⊕ Dorpat 1911.

Schlottke, Egon (1901), Dr. phil., wissenschaftlicher Ass. an der Hauptstelle für Pflanzenschutz der Landwirtschaftskammer für die Provinz Hannover. Göttingen (Deutschland), Gosslerstr. 5. *Angewandte Zool. und Pflanzenschutz.* ⊕ Göttingen 1925.

Schlüter, Johannes Ernst Curt (1881), Dr. phil., Zool., wissenschaftlicher Leiter der Naturwiss. Lehrmittelanstalt von Dr. Schlüter und Dr. Mass, Halle a. d. S. (Deutschland), Viktoriastr. 9. *Palaearktische Tenthrediniden.* ⊕ Leipzig 1911. ℗ Biol. Praep. und Modelle.

Schlumberger, Otto (1885), Dr. phil., Reg.R. Leiter des Labor. für Kartoffelbau an der Biol. Reichsanstalt in Berlin-Dahlem. Berlin-Wilmersdorf (Deutschland), Laubacher Str. 41. *Krankheiten der Kartoffel und ihre Bekämpfung.* ⊕ München 1910.

Schmalfuß, Hans (1894), Priv.Doz. Dr. Hamburg 36 (Deutschland), Chem. Staatsinst. *Fermentative Pigmentbildung, Pflanzensäuren.* ⊕ Hamburg 1921.

Schmalhausen, Ivan (1884), Prof. Zool. Univ. Kiew (U.d.S.S.R.), Uliza Korolenko 37. *Morphol. der Vertebraten, Vogelextremität. Entwicklungsmechanik: Quantitative Untersuchung des embryonalen Wachstums (Hühnchen); Potenzen isolierter Keimfragmente in vitro (Amphibien).*

Schmell, Otto (1860), Prof. Dr. Heidelberg (Deutschland), Schloß-Wolfsbrunnen-Weg 29. *Zool. und Botan. Süßwasser-Copepoden.* ⊕ Leipzig 1891.

Schmid, Ambr. (1880), Zentralverwalter der Eidg. landw. Versuchsanstalten, Oberleitung der Schweiz. Agrikulturchem. Anstalt Liebefeld-Bern (Schweiz). *Tierernährung, Tierzucht.* ⊕ 1916.

Schmid, Emil (1891), Gams, Kanton St. Gallen (Schweiz). *Phytocoenol., Laubmischwälder Europas.* ⊕ Zürich 1923.

Schmid, Günther (1888), Dr. phil., Priv.Doc. an der Univ. Halle a. d. S. (Deutschland), Botan. Inst., Kirchtor 1. *Physiol. des Protoplasmas und physiol. Pflanzenanat.* ⊕ Jena 1912.

Schmidt, Adam (1897), Chefarzt des Staatl. Hygien. Inst. in Sombor (S.H.S.). *Bact. und Serol.* ⊕ Graz 1922.

Schmidt, Antal, Dr., Konserv. Zool. Abt., National Mus. Budapest (Ungarn). *Lepidoptera.* o

Schmidt, Axel (1876), Dr. phil., Landesgeol. Stuttgart (Deutschland), Falkertstr. 63. *Systematik jungpaläozoischer und tertiärer Süßwasser-Zweischaler (Lamellibranchiaten).* ⊕ Breslau 1904.

Schmidt, Bruno Werner (1896), Hon.Prof. forstl. Hochsch., Leiter der Waldsamenprüfungsanstalt bei der Direkt. des förstl. Versuchswesens Preußens. Eberswalde (Deutschland). *Keimphysiol. und serodiagnostische Bestimmung der klimatischen Herkunft des Forstsaatguts, forstsamenkundl. Fermentstudien, biol. Grundlagen des Zapfenkleng- und Samenentflügelungsprozesses, Genotypen und Phänotypen von Pinus silvestris.* ⊕ Königsberg 1924. ℗ Forstliche Sämereien, Photos.

Schmidt, Carl L. A. (1885), Prof. of Biochem., Univ. of California. Berkeley, Cal. (U.S.A.), Budd Hall. *Metabolism, proteins.* ⊕ Univ. of California 1916.

Schmidt, Erich (1890). Bonn a. Rh. (Deutschland), Meckenheimer Allee 2. *Insektensystematik und Biol., Pflanzenschutz, bes. Hymenoptera excl. Formicidae et Micros, Odonata, Neuroptera.* ⊕ Bonn 1915. ℗ Libellen.

Schmidt, Fritz (1889), Bact. Inst. der Landwirtschaftskammer. Halle a. d. S. (Deutschland), Freiimfelder Str. 68. *Bact. und parasit. Erkrankungen des Geflügels u. der Kälber. Pathol. der Milch.* ⊕ Berlin 1921.

Schmidt, Georg (1896), Priv.Doc. Univ. Moskau (U.d.S.S.R.), Arbat, Kaloschiugasse 21, W. 18. *Embryol. Embryonalentwicklung von Piscicola geometra.* ⊕ Moskau 1926. ℗ Blutegel u. ihre Embryonen.

Schmidt, Hermann (1892), Dr. phil., Priv.Doc. u. Kustos. Göttingen (Deutschland), Planckstr. 6. *Palaeozoische Cephalopoden.* ⊕ Göttingen 1920. ℗ Fossilien.

Schmidt, Hermin Alexander (1888), Dr. phil. Zwickau i. Sa. (Deutschland), Lothringer Str. 1. *Wachstum des phanerogamen Vegetationspunktes.* ⊕ Leipzig 1921.

Schmidt, Johs., Prof. Dr., Vorstand der Physiol. Abteil. Carlsberg-Labor. København (Danmark), Carlsbergvej, Valby. *Pflanzenphysiol.* o

Schmidt, Karl Patterson (1890), Ass. Curator of Reptiles and Amphibians, Field Mus. of Natural History Chicago. Homewood, Ill. (U.S.A.), 205 E Cedar St. *Systematic Herpetol.* ⊕ 1916.

Schmidt, Karl Paul (1872), Dr. med., Prof. der Hygiene und Dir. des Hygien. Inst. der Univ. Halle a. d. S. (Deutschland), Magdeburger Str. 21 II. *Hygiene u. Bact.: Gewerbehygiene (Bleivergiftungen). Bact.: Influenza, Typhus und Paratyphus, insbes. praktische Typhusbekämpfung. Serol.: Anaphylaxie, Wassermannsche Reaktion auf Syphilis. Allgem. Hygiene: Hitzschlag und Sonnenstich; Bekämpfung des Alkoholismus.* ⓟ München 1897.
Schmidt, Martin (1897), Dr. phil., wissensch. Ass. bei der Hauptstelle für Pflanzenschutz in Berlin-Dahlem (Deutschland), Königin-Luise-Str. 19. *Angewandte Zool.* ⓟ Berlin 1921.
Schmidt, Martin Benno (1863), Geh.Hofrat, Dr. med., o. Prof. an der Univ., Vorstand des Pathol. Inst. Würzburg (Deutschland), Luitpoldkrankenhaus. *Knochenkrankheiten, Eisenstoffwechsel, Blutbildung.* ⓟ Leipzig 1887.
Schmidt, Otto Christian, Dr. phil., Ass. am Botan. Mus. zu Berlin-Dahlem (Deutschland). *Meeresalgen.*
Schmidt, Paul Hans Karl Konstantin (1882), Priv.Doc., Univ. Hamburg, Dr. med., Abteilungsleiter am Behring-Inst. für experim. Therapie, Marburg a. d. L. (Deutschland). *Serol. und Kolloidchem. (Biochem.).* ⓟ Freiburg 1911.
Schmidt, Peter Julievitsch (1872), Prof., Dir. Zool. Labor. Inst. für d. Landwirtschaft, zugleich Conserv. Zool. Mus. Akademie d. Wissenschaften. Leningrad (U.d.S.S.R.), Prosp. Mayoroff 6/7. *Experim. Zool., Genetic. System. Ichthyol.* ⓟ St. Petersburg 1904.
Schmidt, Wilhelm J. (1884), o. Prof. der Zool. und vergl. Anat., Dir. des Zool. Inst. der Univ. Gießen (Deutschland), Bahnhofstr. 84. *Doppelbrechung der Gewebe. (Untersuchung des Tierkörpers in polarisiertem Lichte.)* ⓟ Bonn 1908.
Schmidtgen, Otto (1879), Dir. des Naturhistorischen Mus. d. Stadt Mainz (Deutschland). *Wirbeltierpaläont.* ⓟ Gießen.
Schmidthoffer, Gyula, Dr., Dir. d. Staatsinst. f. Bact. Budapest (Ungarn). o
Schmidtmann, Martha (1894), Priv.Doc. Univ. Leipzig (Deutschland), Liebigstr. 26. *Experim. Blutdrucksteigerung, experim. Cholesterinkrankheiten, Pigmente, Gewebs- und Zellreaktion, Mikrochem. Reaktionen am frischen Gewebeschnitt.* ⓟ Marburg 1916.
Schmidt-Nielsen, Sigval (1877), Prof. an der Technischen Hochsch. Norwegens. Trondhjem (Norwegen). *Fettchem.- Enzymchem., Nahrungsmittelchem., Vitamine.*
Schmiedeknecht, Otto (1847), Prof. Dr. Bad Blankenburg, Thüringen (Deutschland). *Hymenoptera: Ichneumoniden.* ⓟ Jena. ⓟ Typensammlungen von Hymenopteren, speziell Schlupfwespen. Bestimme paläarktische Ichneumoniden und Braconiden.
Schmieder, Oscar, Prof. Univ. Nac. Córdoba (Argentina). *Geol. Palaeont.* o
Schmieder, Rudolf Gustav (1898), Instr. Zool., Univ. of Pennsylvania. Philadelphia, Pa. (U.S.A.), 4351 Pech Str. *Entomol., Histol. of Insects.* ⓟ Univ. of Penns. 1922.
Schmincke, Alexander (1877), Prof. Univ. Tübingen (Deutschland), Münzgasse 14. *Allgem. Pathol.* o
Schmit, Victor (1865), Prof., Dir. Inst. für biol. Forschung Univ. Perm (U.d.S.S.R.), Histol. Labor. *Histogenetische Studien an Wirbeltieren.* ⓟ Dorpat 1891.
Schmit-Jensen, Hans Oluf (1892), Head of Division, Serumlabor. of the Royal Vet. and Agric. Coll., Copenhagen; Foot-and-Mouth Disease Experim. Station. Stege (Danmark), Isle of Lindholm, Post Box 42. *Introduction of new and rare carbohydrates in med. bact. (micro-fermentation tests). Foot-and-Mouth Disease: Cattle experim. on the ways of spreading of the virus.*
Schmitz, Ernst (1882), Univ.Prof. Breslau (Deutschland), Ahornallee 27. *Physiol. Chem.* o

Schmitz, Henry (1892), Prof. and Chief, Division of Forestry, Univ. of Minnesota. St. Paul, Minn. (U.S.A.), Univ. Farm. *Forest Pathol.* ⓟ Washington Univ. (St. Louis) 1919.
Schmol, Pierre (1889), Méd.-adjoint au Labor. de l'Etat. Luxembourg (Luxembourg), Place du théâtre 9. *Bact., Anat. pathol.* ⓟ Luxembourg 1915.
Schmoll, Hazel M., Curator of Natural History. Denver, Col. (U.S.A.), State Mus. *Ecol. taxonomy.* o
Schmorl, Christian Georg (1861) Dr. des Pathol. Inst. am Krankenhaus Friedrichstadt. Dresden-N. 6 (Deutschland), Bettinastr. 15. *Pathol. Anat.* ⓟ 1887.
Schmucker, Theodor (1894), Dr., Ass. am Inst. für allgemeine Botan. und Pflanzenphysiol. der Univ. Göttingen (Deutschland), Nikolausbergerweg 18. *Physiol. und Entwicklungsgeschichte der Pflanzen.* ⓟ München 1922.
Schnakenbeck, Werner (1887), Kustos an der Fischereibiol. Abteilung des Zool. Staatsinst. Hamburg 5 (Deutschland), Kirchenallee 47 I. *Fischereibiol.: Pleuronectes flesus, Rassenuntersuchungen am Hering, Geschichte und Technik der deutschen Nordseefischerei, Biol. der Fische. Biol. von Eriocheir sinensis. Pigmentierung der Fische. Rassenmerkmale des Axolotl.* ⓟ Halle a. d. S. 1919.
Schnarf, Karl (1879), Priv.Doc. an der Univ. Wien VI (Deutsch-Österreich), Joanelligasse 5. *Embryol.* der *Angiospermen.* ⓟ Wien 1904.
Schneck, Henry Wm. (1889), Ass. Prof. of Vegetable Gardening. Ithaca, N.Y. (U.S.A.), Coll. of Agric., Cornell Univ. ⓟ Cornell Univ. 1914. ⓟ Plants.
Schnegg, Hans (1875), Dr., o. Hochschulprof. Hochschule Weihenstephan, Post Freising, Bayern (Deutschland). *Angewandte Gärungsbiol.* ⓟ München 1901. ⓟ Mikroorganismen, insbes. Hefen und Hyphomyceten.
Schneid, Theodor (1879), Hauptkonserv. Bamberg, Bayern (Deutschland), Naturalienkabinett. *Fossile Fauna und Flora Nordostbayerns (Weißer Jura). Insektenfauna des Bamberger Landes: Carabiden u. Dyticeiden.* ⓟ München 1913.
Schneider, Edward Christian (1874), Prof. of biol., Wesleyan Univ. Middletown, Conn. (U.S.A.), 97 Broad St. *Physiol.* ⓟ Phil. Yale Univ. 1901, sc. Denver Univ. 1914.
Schneider, Erich Gerhard (1903), Dr. phil., Ass. am Botan. Inst. der Landwirtschaftlichen Hochsch. Bonn-Poppelsdorf a. Rhein (Deutschland), Meckenheimer Allee 106. *Pflanzenanat. (Holz). Zellenphysiol.* ⓟ Gießen 1924.
Schneider, Karl Max (1887), Ass. am Zool. Garten Leipzig (Deutschland). *Verhalten von Tieren in der Gefangenschaft.* ⓟ Leipzig 1914.
Schneider, Paul (1879), Dr., Stadtprosektor, Leiter des Pathol. Inst., Stadtkrankenhaus. Darmstadt (Deutschland). *Kongenitale Syphilis. Mißbildungen des Atmungsapparates und männlichen Sexualorgan.* ⓟ Heidelberg 1903.
Schneider-Orelli, Otto (1880), Prof., Entomol. Eidgenössische Techn. Hochsch. Zürich (Schweiz), Entomol. Inst., Universitätsstr. 2. *Angewandte Entomol. und Insektenbiol., spez. land- u. forstwirtschaftliche Schädlingskunde u. Bienenzucht. Aphididae, Ipidae.* ⓟ Bern 1905.
Schneidewind, Wilhelm, Prof. Univ. Halle a. d. S. (Deutschland), Albrechtstr. 29. *Agric.-Chem.* o
Schnitzer, Robert (1894), Dr. med., Ass. am Inst. „Robert Koch". Berlin N 39 (Deutschland), Föhrer Str. 2. *Experim. Chemotherapie (Trypanosomen, Malaria). Chemotherapie bakterieller Infektionen (Streptokokken).* ⓟ Berlin 1919.
Schnitzler, Josef (1885), Dir. der Landwirtschaftlichen Stelle des deutschen Kaliwereins. Berlin SW 11 (Deutschland), Anhaltstr. 7. *Agrikulturchem. und Wasserchem.* ⓟ Jena 1913.
Schnorf, J. Paul (1902). Uetikon am See, Kanton Zürich (Schweiz). *Brutbiol. der Vögel.* ⓟ Zürich 1925.

Schnürer, Josef (1873), Dr. med., Tierarzt, o. ö. Prof. für Bact. und Tierhygiene der Tierärztlichen Hochsch. Wien III (Deutsch-Österreich), Linke Bahngasse 11. *Bact., Immunitätswissenschaft, Desinfektion.* ⓓ Wien 1898. ⓟ Bakterienpraep., -kulturen.

Schnurre, Otto (1894). Frankfurt a. M.-Eschersheim (Deutschland), Klauerstr. 2. *Ornithol. Historische Ornithol.* ⓓ Frankfurt a. M. 1920.

Schober, Alfred Ferdinand (1862), Prof. Dr. Hamburg 24 (Deutschland), Lerchenfeld 7. *Reizphysiol., speziell Geotropismus.* ⓓ Breslau 1886.

Schön, Arnold (1888), Dr. phil., Preußischer Oberfischmeister. Pillau, Ostpreußen (Deutschland). *Fischereibiol., Abwasserbiol.* ⓓ Freiburg i. Br. 1910. ⓟ Plankton, Ufer- und Bodenfauna des Frischen Haffs.

Schoen, Rudolf (1892), Dr. med., Priv.Doc. für innere Med. Leipzig (Deutschland), Med. Klinik. *Pharmacol. und experim.Pathol.Atmung,Stoffwechsel, Nervensystem.* ⓓ Heidelberg 1919.

Schöndorff, Bernhard (1865), ao. Prof., Abteilungsvorsteher physiol. Inst. Univ. Bonn (Deutschland), Beringstr. 2. *Biochem. Harn. Schilddrüse. Hungerstoffwechsel. Harnstoff. Diabetes und Zuckerausscheidung. Glykogen.* ⓟ Bonn 1890.

Schoene, William Jay (1879), Entomol. to the Experim. Station and State Entomol. Blacksburg, Va. (U.S.A.). ⓓ Chicago 1911.

Schönfeld, Anton, Dr., Priv.Doc. f. Bienenzucht. Praha XII (C.S.R.), Fochova 65. o

Schönheimer, Rudolf (1898), Ass. am Pathol. Inst. der Univ. Freiburg i. Br. (Deutschland), Albertstr. 19. *Pathol. Physiol. und pathol. Chem. der Lipoide.* ⓓ Berlin 1923.

Schoenichen, Walther (1876), Prof., Dir. der Staatlichen Stelle für Naturdenkmalpflege in Preußen. Berlin-Schöneberg (Deutschland), Grunewaldstr. 6/7. *Naturschutz.* ⓓ Halle a. d. S. 1898.

Schönland, S., Prof. Rhodes Univ. Coll. Grahamstown, Cape Col. (S. Africa). *Botan.* o

Schoevers, Timon Alexander Cornelis (1878), Phytopath. in-, substitute-chief of the Dutch Phytopath. Service. Wageningen (Holland). *Phytopath. and Economic Entomol.* ⓟ Microscopical slides and objects on plant diseases and noxious insects.

Schöyen, Thor Hiorth (1885), Statsentomol. Oslo (Norwegen). *Angewandte Entomol.*

Scholz, Eduard (1860), Reg.R. und Prof. Wien VII (Deutsch-Österreich), Lerchenfelder Str. 139. *Anat. und Physiol. der Pflanzen.*

Scholz, Wilhelm (1864), Dr. med., Univ.-Prof., Hofrat, Zentraldir. des Landeskrankenhauses. Graz, Steiermark (Österreich), Riesstr. 1. *Biochem.* ⓓ Prag 1894.

Schoonhoven, John J. (1864), President Dept. of Zool., Brooklyn Inst. of Arts and Sc. Brooklyn, N.Y. (U.S.A.), 773 Eastern Parkway. *Parasit. Insects as Carriers of Disease.* ⓓ St. Francis Coll. 1893.

Schopfer, William Henri (1900), Ass. Univ. Genève, Labor. de parasit. Eaux-Vives (Suisse), Rue Muzy 15. *Physiol. et physico-chim. des parasit.: Cestodes, Nematodes, Trematodes.*

Schorn, Edwin John (1896), Demonstrator Pharmacy and Pharmacognosy, Royal Technical Coll. Glasgow, Uddingston (Scotland), Laurel Bank, Main Street. *Quantitative Tissue Variation in Med. Plants.*

Schornagel, H., Prof. Dr., Dir. d. Pathol. Labor. der Tierärztl. Fak. der Univ. Utrecht (Holland). o

Schotte, Herbert (1897), Dr. phil. Berlin-Wilmersdorf (Deutschland), Pfalzburger Str. 33. *Chem. der biogenen Amine.* ⓓ Berlin 1920.

Schottmüller, Adolf Alfred Louis George Hugo (1867), o. ö. Prof. für med. Poliklinik an der Univ. Hamburg (Eppendorfer Krankenhaus). Hamburg 36 (Deutschland), Alsterufer 11. *Innere Med., Bact.* ⓓ Greifswald 1893.

Schourschine, Pierre I., Prép. Inst. Jaczewski. Leningrad (U.d.S.S.R.), Perspective Anglaise 29. *Maladies des pommes de terre, Peronosporées.* o

Schoute, Johannes Cornelis (1877), Dr., Prof. an der Univ. Groningen (Holland), Zuiderpark 2. *Morphol. der höheren Pflanzen (Blattstellungen: Wirtelstellungen, Blütenmorphol. von Polygonum, Stammesbildung bei Monokotylen, Verästelung bei Monokotylen u. Baumfarne). Anat. (Dickenwachstum b. Monokotylen, Stelarmorphol.).* ⓓ Groningen 1902.

Schouten, Albert Reinardus (1877), Doc. School tot Opleiding van Inlandsche Artsen. Weltevreden, Java (Ned.-O.-Indië). Salemba 11. *Pflanzenanat.* ⓓ Amsterdam 1908.

Schouten, Samuel Leonardus (1874), Priv.Doc. der Mikrobiol. an der Univ. Utrecht (Holland), Maliebaan 3. *Allgemeine Mikrobiol.: Individuelle Behandlung von Mikroorganismen (Isolierung, Mikrodissektion unter dem Mikroskop von Bacterien, Hefe, Pilze, Protozoen). Kultur der Mikroorganismen. Vererbung und Variabilität der Mikroorganismen.* ⓓ Utrecht 1901. ⓟ Kulturen von Mikroorganismen.

Schouteden, Henri (1881), Chef Section Sc. nat. Mus. du Congo. Bruxelles (Belgique). *Zool. de l'Afrique Centrale.* ⓓ Bruxelles.

Schrader, Albert Lee (1896), Dr. of Phil., Associate Pomol. College Park, Md. (U.S.A.), Horticulture Department, Univ. of Maryland. *Chem., physiol., and cultural studies of grapes, apples, peaches.* ⓓ Univ. of Maryland 1925.

Schrader, Franz (1891), Assoc. Prof. Biol. Bryn Mawr Coll., Bryn Mawr, Pa. (U.S.A.). *Cytol., parthenogenesis, sexdetermination, symbiosis.* ⓓ Columbia Univ. 1919.

Schramm, J. R. (1885), Editor, Biol. Abstracts, care of Department of Zool., Univ. of Pennsylvania. Philadelphia, Pa. (U.S.A.). *Plant Physiol.* ⓓ Washington 1913. o

Schratz, Eduard (1901), Dr. phil., Ass. Berlin-Dahlem (Deutschland), Kaiser-Wilhelm-Inst. für Biol. *Physiol. Untersuchungen an Laubmoosen.* ⓓ Münster 1924.

Schreiber, Ernst (1896), Vertreter des Kustos für Botan. an der Preußischen Biol. Anstalt auf Helgoland (Deutschland). *Lebensbedingungen des marinen Phytoplanktons. Sexualitätsstudien an Braunalgen.* ⓓ Berlin 1925.

Schreiber, Maximilian (1894), Dr.-Ingenieur, o. Ass. Hochsch. für Bodenkultur (Forstliche Studienrichtung). Wien XVIII (Deutsch-Österreich), Hochschulstr. 17. *Pflanzenphysiol.* ⓓ Wien 1921.

Schreiber, Oswald (1870), Dr., Dir. des Bact. und Serum-Inst. Landsberg a. d. Warthe (Deutschland). *Immunisierungsmethoden gegen Tierseuchen.* ⓓ Basel 1896. ⓟ Veterinärsera und Bakterienpraep.

Schreiner, Alette, Dr., Ass. Anat. Inst. Univ. Oslo (Norge). *Anat., Anthrop.* o

Schreiner, K. E., Prof. Dr., Dir. Anat. Inst. Univ. Oslo (Norge). *Anat., Anthrop.* o

Schreiner, Oswald (1875), Senior Biochem., U.S. Dept. of Agric. Washington, D.C. (U.S.A.). *Soil fertility, organic compounds in soils, oxidizing powers of roots, toxic effects of organic compounds on plants, action of fertilizers, biochem. relationships of soils and plants.* ⓓ Univ. of Wisconsin 1902.

von Schrenk, Hermann (1873), Pathol. Missouri Botan. Garden. St. Louis, Mo. (U.S.A.), Tower Grove and Flad Aves. *Diseases of trees and structural timbers.* ⓓ St. Louis, Mo., 1898.

Schridde, Herm. (18.'5) Dr. Prof., Dir. des Pathol. Inst. und Forschungsinst. für Gewerbe- und Unfallkrankheiten. Dortmund (Deutschland), Bismarckstr. 56. *Allgemeine Pathol. und pathol. Anat.* ⓓ Erlangen 1901.

Schrodt, Julius (1853), Prof., Dr. phil. Gardelegen (Deutschland). *Öffnungsmechanismus der Farnsporangien und Antheren.* ⓓ Kiel 188 5.

Schroeder, Laura Johanna (1897), Instr. of Botan., at Beloit Coll. Beloit, Wis. (U.S.A.), 501 Euclid Ave.

Ecol. of the oaks, compar. ecol. anat. Systematic botan. ⒹUniv. of Illinois 1926. ⒫ Wisconsin plants.
Schröder, Ludwig Julius Bruno (1867), Dr. phil. nat. Breslau 13 (Deutschland), Sadowastr. 88 II. *Phytoplankton des Süßwassers und des Meeres, allgemeine Hydrobiol. Characeenkunde.* Ⓓ 1901.
Schroeder, Henry (1873), o. Prof. der Botan., Landwirtsch. Hochsch. Hohenheim bei Stuttgart (Deutschland). *Chem. Pflanzenphysiol.* Ⓓ Bonn 1904.
Schröder, Olaw (1880), Dr. phil., Leiter des Zool. Mus. der Univ. Kiel (Deutschland), Niemannsweg 46. *Protozoen, Wirbellose (histol. Arbeiten). Protozoen (Radiolarien, Myxosporidien, Infusorien). Wirbellose: Würmer (Nematoden, Planarien, Augen von Eunice).* Ⓓ Heidelberg 1904.
Schröder, Vera (1897), Ass. Inst. experim. Biol. Moskau (U.d.S.S.R.), Woronzowo Polje 6. *Zellphysiol., experim. Hydrobiol. Physikalisch-chem. Blutstudien. Rolle der Eiweißkörper bei der Isohämoagglutination. Einfluß der physikalisch-chem. Milieufaktoren auf die morphol. Eigenschaften der Euglena gracilis.* Ⓓ Moskau 1920.
Schroeder, William Charles (1895), Ass. Aquatic Biol. Cambridge, Mass. (U.S.A.), Mus. of Compar. Ichthyol. *Ichthyol.: Fishes of the Atlantic Coast of North America, Gadus callarias, Gadus virens, Melanogrammus aeglefinus.* Ⓟ Scales of cod, pollock (G. virens), and Haddock for scales of European fish of same species.
Schröter, Carl (1855), Prof. emer. der Botan. an der Eidg. Technischen Hochsch. in Zürich (Schweiz), Merkurstraße 70. *Pflanzengeographie, Planktol. und Oekol., Naturschutz.* Ⓓ Zürich und Berlin.
Schrottenbach, Heinrich (1885), ao. Univ.-Prof. Graz (Österreich), Zinzendorfgasse 25. *Neurol. u. Psychiatrie u. deren Grenzgebiete (Normale u. pathol. Physiol. u. Histol. des Nervensyst.).* Ⓓ Graz 1908.
Schryver, S. B., Prof. Coll. of Sc. and Technol. London S.W. 7 (England), S. Kens. *Biochem.* ○
Schtscherback, Johannes (1880), Priv.Doc. Botan. Odessa (U.d.S.S.R.), Landwirtschaftl. Inst. *Pflanzenphysiol. Saprolegniaschwärmer. Geotropische Reaktion in gespaltenen Stengeln. Salzausscheidung durch die Blätter. Geotropismus. Ortho- u. Plagiotropismus.*
Schube, Theodor (1860), Prof. Dr. Breslau VIII (Deutschland), Clausewitzstr. 5. *Flora von Schlesien; Naturschutz in Schlesien.* Ⓓ Breslau 1885.
Schuchert, Charles (1858), Prof. of Paleont., emer., Yale Univ. New Haven, Conn. (U.S.A.), Peabody Mus. *Paleont. and Stratigraphy.*
von Schuckmann, Waldemar (1883), Dr. phil., Reg.R. im Reichsgesundheitsamt. Berlin-Lichterfelde (Deutschland), Gerichtstr. 12. *Parasitenkunde, speciell Protozool., Helminthol. und angewandte Entomol.* Ⓓ Freiburg i. Br. 1908.
Schübel, Konrad (1885), Univ.-Prof. Erlangen, Bayern (Deutschland), Pharmacol. Inst. *Protozoen- und Bakteriengifte.* Ⓓ Phil. 1909, med. 1919.
Schüller, Josef (1888), Priv.-Doc., Dr. med. et phil., Dir. des Pharmacol. Inst. Köln (Deutschland), Zülpicher Straße 47. *Pharmacol.* Ⓓ Phil. 1912, med. 1914.
Schüepp, Otto (1888), Dr., Priv.Doc. an der Univ. Basel. Reinach bei Basel (Schweiz). *Botan., Entwicklungsgeschichte und Morphol., Wachstum und Formwechsel des Vegetationspunktes, Wachstumsmessungen an Knospen, Wachstumskurven, Blattstellung.* Ⓓ Zürich 1911.
Schürhoff, Paul Norbert (1878), Dr. phil., Priv.-Doc. der Botan. Berlin SW 61 (Deutschland), Wilmsstr. 1. *Botan. Cytol., Pharmakognosie, Mikrobiol.* Ⓓ Bonn 1905.
Schütz, Franz (1887), ao. Prof., Dr. med., Ass. am Hygien. Inst. Berlin (Deutschland), Dorotheenstraße 28a. *Hygiene und Bact.* Ⓓ Berlin 1912.
Schütze, Harry L. H. (1882), Lister Inst. Bact. London S.W. 1 (England), Chelsea Gardens. *Bact.*, Ⓓ Würzburg 1907.

Schütz, Ernst (1901), Ass. an der naturkundlichen Abteilung des Provinzialmus. Hannover (Deutschland). *Zool., Ornithol.: Vogelzug, Biol., Anat., Entwicklung der Federn.* Ⓓ Berlin 1927.
Schugurensky, L., Prof. Univ. Nac. d. Litoral. Corrientes (Argentina). *Horticult.* ○
Schulemann, Werner (1888), Abteilungsvorstand Pharmazeut. wissenschaftl. Labor. I. G. Farbenindustrie A.-G., Werk Elberfeld. Vohwinkel, Rheinland (Deutschland), Bismarckstr. 99. *Arzneimittelsynthese (Hexeton, Tutokach, Phanodorm, Plasmochin). Vitalfärbung. Organische Quecksilberverbindungen. Malariatherapie.* Ⓓ Phil. Breslau 1913, Med. 1914.
Schulga Nesterenko, Marie (1891), Inst. géol. I. Univ., Ass. Paléont. Moscou (U.d.S.S.R.), Geol. Kabinett. *Ammonites: de l'étage d'Artinsk.*
Schulgin, Olga (1886), Dir. Bact. Mus. d. Abteil. d. landwirtschaftl. Mikrobiol., Inst. für Experim. Agronomie. Leningrad, (U.d.S.S.R.), Ulitza Herzena 42. *Bodenbact.*
Schulgin, Wassilij W., Ass. Landw. Prov.-Versuchsstat. Krasnojarsk (U.d.S.S.R.). *Selektion.* ○
Schulman, Hjalmar (1857). Helsinki (Finnland), Fjälldalsgatan 13. *Vergl. Anat. der Säugetiere (Monotremen, Insectivoren, Edentaten, Carnivoren, Rodentia), Verbreitung einheimischer Vögel u. Säugetiere.* Ⓓ Helsinki 1908.
Schulow, Iwan S., Prof., Dir. d. Versuchsstation für Flachsbau d. Landwirtschaftl. Akad. Timirjasew. Moskau (U.d.S.S.R.), Petrowsko-Rasumowskoje. *Agronomie. Angew. Pflanzenphysiol.* ○
Schulte, Hermann von Wechlinger (1876), Dean, Prof. Anat. Creighton Univ. Med. School. **Omaha, Neb. (U.S.A.), 306 North, 14 St.** *Development of vascular system, of salivary glands. Anat. of Cetacea.* Ⓓ New York 1902.
von Schulthess, Anton (1855), Dr. med. Zürich 6 (Schweiz), Wasserwerkstr. 53. *Entomol.; Hymenoptera.* Ⓓ Zürich 1878.
Schultz, Arthur (1890), Priv.Doc., Ass. am Pathol. Inst. der Univ. Kiel (Deutschland), Adolfplatz 12. *Pathol. Anat. (Gefäßpathol., Cholesterinesterverfettung)* Ⓓ 1917.
Schultz, C. H., Prof. Univ. of the Philip. Manila (Philippine Islands). *Pathol. u. Bact.* ○
Schultz, Edwin William (1887), Prof. of Bact. and Executive Head, Department of Bact. and Experim. Pathol. Stanford Univ., California. (U.S.A.). *Filterable viruses a. bacteriophagy.* Ⓓ Johns Hopkins Univ. 1917.
Schultz, Eugene S. (1884), Plant Pathol. Washington, D.C. (U.S.A.), Bureau of Plant Industry, U.S. Dept. of Agric. *Diseases of plants; virus diseases of potato.* Ⓓ Columbia Univ., New York 1925.
Schultz, Orville Carl, Associate Prof. of Botan. Oklahoma A. and M. Coll. Stillwater, Okla. (U.S.A.). *Morphogenesis and mechanics of development.* ○
Schultz-Allenstein, Walther (1877), Dr., Kinderarzt. Allenstein, Ostpreußen (Deutschland), Zeppelinstr. 2—3. *Entwicklungsmechanik: Ovarienverpflanzung auf Männchen und fremde Erbkomplexe. Parallele von Transplantation und Bastardierung. Schwärzung weißer und gelber Haare durch Regeneration in Kälte. Entwicklungsmechanische Manifestierung normalerweise dauernd latenter Erbanlagen.* Ⓓ Königsberg 1902. Ⓟ Kaninchenrassen.
Schultze, Walter Hans (1880), ao. Prof., Technische Hochsch., Prosektor Landeskrankenhaus, ao. Mitglied des Landes-Med.-Coll. Braunschweig (Deutschland), Petritorwall 30. *Pathol. Gewerbehygiene. Bact.* Ⓓ Freiburg i. B. 1903.
Schultzer, Poul (1895), Priv.Doc. Dr. med. København (Danmark), Amager Boulevard 2. *Lichtbiol., Radiobiol., Physiol.* Ⓓ Univ. Kopenhagen 1922.
Schulz, Friedrich Nikolaus (1871), Dr. med., o.ö. Prof. der Physiol. Chem. an der Univ. Jena (Deutschland), Sedanstr. 7. *Lebensfschem., Physiol. der Nierentätigkeit. Vergl. Physiol. der Verdauung und des Stoffwechsels.* Ⓓ Bonn 1894.

Schulze, Paul (1887), o. Prof. der Zool. und vergl. Anat., Dir. des Zool. Inst. der Univ. Rostock (Deutschland). *Anat. und Histol. der Wirbellosen. Oekol., Mikrochem. organischer Skel tteubstanzen, Biol. der Tiere Deutschlands.* ⓓ Berlin 1911.

Schulze, Werner (1890), Leiter Pflanzenzuchtstation. Ahrensburg, Bez. Hamburg (Deutschland), Waldstr. 10. *Weizen u. Haferzüchtung.* ⓓ Jena 1920. ⓟ *Getreidesorten.*

Schulze, Werner R. H. H. (1894), Dr. med., Priv.-Doc. Anat. Univ. Würzburg, z. Z. Ass. Chirurg. Univ.-Klinik Freiburg. Freiburg i. Br. (Deutschland), Chirurg. Klinik. *Einfluß der inkretorischen Drüsen auf Wachstum und Entwicklung. Histol. der lymphatischen Organe. Mikrochem. der Osteogenese. Gefäßstudien bei experim .Ileus.* ⓓ Halle 1918.

Schumacher, Josef (1885), Dr. med. Berlin N 24 (Deutschland) Friedrichstr. 122/23 I. *Histochem. u. Chemotherapie.* ⓓ Berlin 1913. ⓟ *Chem. und physiol.-chem. Praep. der Histochem. und Chemotherapie.*

Schumacher, Oskar (1894), Dr., Ass. des I. Anat. Inst. Wien IX/3 (Deutsch-Österreich), Währinger Str. 13. *Anat.* ⓓ Wien 1920.

Schumacher, Siegmund (1872), o. Univ.-Prof., Vorstand des Histol.-embryol. Inst. der Univ. Innsbruck (Deutsch-Österreich), Schmerlingstr. 2. *Biol. Histol. (besonders der Vögel).* ⓓ Wien 1898.

Schumann, Friedrich Paul (1884), Dir. des Bact. Inst. der Landwirtschaftskammer Schlesien. Breslau (Deutschland), Kaiserstr. 55. *Bact. Untersuchungen über Tierkrankheiten, Todesursachen, Tuberkulose, Behandlungen steriler Rinder und Pferde.* ⓓ 1907. ⓟ *Kulturen (Tierkrankheiten).*

Schumm, Otto Heinrich (1874), Prof., Wissensch. Rat Eppendorfer Krankenhaus, Leiter des Chem. Inst. des Eppendorfer Krankenhauses und des Physiol.-chem. Inst. der Hamburgischen Univ. Hamburg 20 (Deutschland), Tarpenbeckstr. 102 I. *Spektrochem. Analyse tierischer Farbstoffe, Blutfarbstoff, blutfarbstoffähnliche Pflanzenfarbstoffe, Muskelfarbstoffe, Porphyrine, Haematin im Blutserum, Koprotin und Koprotoporphyrin bei okkulten Blutungen, pathol. Farbstoffe bei congenitaler Haematoporphyrie, Farbstoffanalyse, Spektrographische Methoden.* ⓓ Hamburg 1920.

Schussnig, Bruno (1892), Dr. phil., o. Ass. Lehrkanzel für systematische Botan. (Botan. Inst.) und Priv.Doc. an der Univ. Wien (Deutsch-Österreich). *Entwicklungsgeschichte, Cytol. und Systematik der Proto- und Thallophyten. Mikrobiol., Hydrobiol.* ⓓ Graz 1917. ⓟ *Exsikkate und Praep. von Meeresalgen.*

Schuster, Carl E. (1889), Associate Prof. of Pomol. Horticultural Department, Oregon Agric. Coll. Corvallis, Ore. (U.S.A.). *Pollination questions concerning fruits. Fertilization of old prune trees, testing varieties.* ⓓ Oregon Agric. Coll. 1916.

Schuster, Helene (1884), Prosektor pathol. Inst. Univ. Lwów (Polen), ul. Piekarska 52. *Pathol. Histol. Neubildungen.* ⓓ Lemberg 1912. ⓟ *Weiße Mäuse.*

Schuster, Ludwig (1883) RR.. Berlin-Südende (Deutschland), Hermannstr. 15. *Ornithol. Fortpflanzungsbiol.*

Schutko, Alexander St., Wiss. Mitarbeiter des Landw. Inst. Woronesh (U.d.S.S.R.). *Agronomie. Pflanzenkunde.* o

Schutow, D. A. Botan. Kab. Univ. Saratow (U.d.S.S.R.). o

Schuurmans Stekhoven Jr., Jacobus Hermanus (1892), Dr., 1. Ass., Zool. Labor., Abt. Nierstrasz. Utrecht (Holland), Mengelberglaan 67 bis. *Die tierischen Parasiten des Menschen und der Tiere, speziell blutsaugende Arthropoden und parasitische Würmer.* ⓓ 1918. ⓟ *Tabanidae.*

Schwangart, Friedrich (1874), o. Prof. emer. der Sächs. Forstl. Hochsch. Kötzschenbroda bei Dresden (Deutschland), Carolastr. 8 I. *Embryol., Insektenbiol., Systematik der Spinnen, angewandte Entomol., Forstschutz, Aetiol. der Wild-Krankheiten,* Tierpsychol., Züchtung *(Spezialität: Die Katze).* ⓓ München. ⓟ *Rassekatzen.*

Schwann-Gurijskij, PaulW., Leiter d. Botan.Abt. d. Aserbaidschansker Staatl. Mus. Baku (U.d.S.S.R.), Freiheitstr. 7. *Geobotan. Arzneipflanzen.* o

Schwanwitsch, B. N. (1889), Ass. Zool. Univ. Leningrad (U.d.S.S.R.), Zootomical Labor. *Parasit. Gastropods. Wing-pattern in Lepidoptera.*

Schwartz, Benjamin (1889), Associate Zool., Bureau of Animal Industry, U. S. Department of Agric. Washington, D.C. (U.S.A.). *Life histories and physiol. of parasitic worms. Taxonomy and morphol. of worms, especially nematodes.* ⓓ George Washington Univ. 1920. ⓟ *Parasitic worms.*

Schwartz, Ernest J. (1869), Lect. in Botan. King's Coll., Univ. London W. C. (England). *Mycetozoa.* ⓓ London 1914.

Schwartz, Martin Karl Otto (1880), Dr. phil., OberReg.R., ständiger Stellvertreter des Dir. der Biol. Reichsanstalt für Land- und Forstwirtschaft. Berlin-Dahlem (Deutschland), Königin-Luise-Str. 19. *Angewandte Biol., Pflanzenschutz und Schädlingsbekämpfung.* ⓓ Breslau 1905.

Schwartz, Philipp (1894), Dr., Priv.Doc. Frankfurt a. M. (Deutschland), Pathol. Inst. der Univ. *Pathol. Anat. des Zentralnervensystems, insbes. Neugeborener und junger Kinder.* ⓓ Budapest 1919.

Schwartz, Wilhelm (1896), Dr. phil., Ass. am Gärungsphysiol.Inst. der Hochsch. Weihenstephan. Freising-Weihenstephan, Oberbayern (Deutschland). *Entwicklungsphysiol., Mykol., Symbiose. Intrazelluläre Symbiose von Insekten mit Pilzen. Bedingungen der Sexualität bei Pilzen.* ⓓ München 1922. ⓟ *Pilzkulturen.*

Schwartze, Erich Wilhelm (1891), Pharmacol. in Charge, Pharmacol. Labor., Bureau of Chem. Washington, D.C. (U.S.A.). *Isometric contraction of cardiac muscle. Cottonseed Poisoning (Gossypol). Tin, Cadmium and Zinc (Acute and chronic poisoning). SulphurDioxide poisoning and pharmacol. Strychnine (effect of age and species of animal). Phylogeny of Nervous system. Habituation to Arsenic. Pharm. Standards for Bio-assay.* ⓓ Univ. of Chicago 1914. ⓟ *Pharm. Standards for the bio-assay of Digitalis, Strophanthus Ergot, Pituitary preparations, Epinephrin and Cannabis.*

Schwarz, Carl (1876), o. Prof. der Physiol., Tierärztl. Hochsch., ao. Prof. Univ. Wien (Österreich). *Verdauungsphysiol.* ⓓ Innsbruck 1902.

Schwarz, Ernst (1889). Berlin N 4 (Deutschland), Zool. Mus., Invalidenstr. 43. *Mammalia, Tiergeographie. Mollusken. Wachstum. Geschwülste.* ⓓ München 1911.

Schwarz, Eugen (1895), Chem. in der General Dyestuff Corporation of America. New York City (U.S.A.), 230 fifth Ave. *Lepidoptera (Macro).* ⓓ Erlangen 1922. ⓟ *Lepidoptera.*

Schwarz, Leopold Karl (1877), Prof. Univ., Wissenschaftl. Rat am Hygien. Staatsinst. Hamburg (Deutschland). *Hygien. und Bact.* ⓓ Berlin 1904.

Schwarz, Marie Beatrice (1898), Botan. im Inst. für Pflanzenkrankheiten. Buitenzorg, Java (Nederl.-O.-Indië). *Pflanzenkrankheiten.* ⓓ Utrecht 1922.

Schwarzacher, Walter (1892), Priv.Doc., I. Ass. am Inst. für gerichtliche Med. der Univ. Graz (Österreich), Universitätsplatz 4. *Physikal. Chem. des Blutes, Leichenveränderungen.* ⓓ Graz 1920.

Schwarze, Carl Alois (1886), Jamaica High School, Jamaica; Long Island, N.Y. Woodhaven, Long Island, N.Y. (U.S.A.), 8610/85. Road. *Fungous diseases of plants; microscopic fungi.* ⓓ Univ. of Missouri, Columbia, Mo., 1909.

Schwarze, Curt (1877), Kustos, Inst. für allgemeine Botan. Hamburg 36 (Deutschland). *Pflanzenanat. und Physiol.* ⓓ Tübingen 1910.

Schwede, Rudolf (1860), ao. Prof. Techn. Hochschule. Dresden-A. 24 (Deutschland), Gutzkowstraße 28. *Angewandte Anat., Nahrungs- u. Genußmittel, Faserstoffe, holzzerstörende Pilze.* ⓓ Basel 1901.

Schweizer, Jean (1896), Dr. phil., Botan. Djember, Java (Nederl.-O.-Indië). *Selectionen und Biol.: Tabak, Kautschuk und Kaffee.* ⓓ Bern 1919.

Schwemmle, Julius (1894), Dr. rer. nat., Priv.Doc., 1. Ass. an dem Botan. Inst. der Univ. Tübingen (Deutschland), Hermann-Kurz-Str. 15. *Erblichkeitsuntersuchungen an reziprok verschiedenen Bastarden (Epilobium), Cytol., Vererbungstheorie. Vergl. Cytol. der Onagraceen.* ⓓ Tübingen 1922.

von Schwerin, Graf Friedrich (1856), Präsident Deutsch. Dendrol. Ges. Wendisch-Wilmersdorf bei Thyrow, Kr. Teltow (Deutschland). *Pflanzen-Acclimatisation. Acer und Sambucus.* ⓓ Univ. Königsberg.

Schwetzoff, Michael (1885), Doc. Univ., Bergakademie. Moskau (U.d.S.S.R.), Schabolowka 32. *Palaeozool.: Cephalopoden, Brachiopoden.* ⓓ Moskau 1910.

Schweyer, Alexander W. (1873), Doc. Univ. Leningrad (U.d.S.S.R.), Zootomisch. Inst. *Protozoen-Lebenszyklen; Formwechsel der Protozoenkerne; Fortpflanzung, Geschlechtsprozesse und Entwicklung der Protozoen. Fortpflanzung, Ontogenese u. Generationswechselbegriff bei Proto- und Metazoen. Infusoria ciliata: Konjugation und Ontogenese.* ⓓ Petersburg 1924.

Schwickerath, Mathias Friederich (1892). Aachen (Deutschland), Goethestr. 25. *Flora der Umgebung Aachens. Hydrobiol. der Bäche u. Teiche in der Umgebung Aachens.* ⓓ 1922.

Schwitalla, Alphonse Mary (1882), Prof. of Biol. and Dir. of the Department. St. Louis, Mo. (U.S.A.), 1402 South Grand Boulevard. *Protozoa; Ecol. Philosophical aspects of Biol.* ⓓ Johns Hopkins Univ. 1921.

Sciacchitano, Iginio (1897), Ass. Zool. R. Univ. di Modena (Italia). *Rigenerazione, meccanica dello sviluppo.* ⓓ Cagliari 1921.

Ščigelskaya, Maria Konstantinowna (1896), Labor. für physiol. Chem. Univ. Leningrad (U.d.S.S.R.), Wassil. Ostrow, 16, Linie 29. *Chem. des lebendigen Substrats, Biogeochem. u. Zoochem., Crustacea.* ⓓ Leningrad 1925.

Sclater, William Lutley (1863). London S.W. 3 (England), 10 Sloane Court. *Systematic Ornithol.* o

Sclavo, Achille, Prof. Regia Univ. degli Studi. Siena (Italia). *Igiene.* o

Sciawunos, G. (1868), Prof. Anat. Univ., Dir. Inst. anat. Athènes (Grèce), Rue Pindare 25. *Ovaire, Masseter.* ⓓ Würzburg 1890.

Scontos, Th., Prof. Paleont. Univ., Dir. du Mus. Paleont. Athènes (Grèce), Rue Asklepios 5.

Scott, Andrew, Naturalist for Fishery investigations Univ. of Durham Piel. Barrow-in-Furness (England), Fisheries Labor. *Copepoda.* ⓓ Edinburgh.

Scott, C. Emlen (1894), Ass. Plant Pathol., California State Dept. of Agric. Sacramento, Cal. (U.S.A.). *Plant disease.* ⓓ Stanford Univ. 1917.

Scott, Ernest L. (1877), Associate Prof. of Physiol., Columbia Univ. New York City, N.Y. (U.S.A.), 437 West 59 th Street. *Carbohydrates.* ⓓ Columbia Univ. 1914.

Scott, Flora Murray (1891), Instr. Los Angeles, Cal. (U.S.A.), Univ. of California at Los Angeles. *Limnol.; Histol.; Cytol.* ⓓ Stanford 1925.

Scott, George Gilmore (1873), Prof. of Biol. at the Coll. of the City of New York. New York City, N.Y. (U.S.A.), 138. St. and Convent Ave. ⓓ Columbia Univ., N.Y., 1913.

Scott, Gordon A., Prof. of the Division of Botan., Univ. of Saskatchewan. Saskatoon, Sask. (Canada). o

Scott, Gordon H. (1901), Ass. Prof. of Microscopic Anat., Loyola Univ., School of Med. Chicago, Ill. (U.S.A.), 706 South Lincoln St. *Studies on Growth of Man. Histol. changes taking place during growth.* ⓓ Univ. of Minnesota, Minneapolis, 1926.

Scott, J. Allen (1898), Research Ass. in Helminthol., Johns Hopkins Univ. School of Hygien. and Public Health. Baltimore, Md. (U.S.A.), 615 N. Wolfe St. *Hookworm Disease, host-parasite relationships, the dog hookworm in normal and abnormal hosts.* ⓓ Wesleyan Univ., Middletown, Conn., 1924.

Scott, J. M. Duncan, Prof. St. Bartholom. Hospit. and Coll. London E.C. 1 (England), West Smithfield. *Physiol.* o

Scott, John Terrell (1893), Instr. in Anat., Univ. of Colorado School of Med. Denver, Col. (U.S.A.), 4200 East 9 th Ave. *MicroscopicAnat.: Mitochondria, Autolysis.* ⓓ Univ. of Colorado 1922.

Scott, John William (1871), Prof. of Zool., Research Parasit., Dir. of Pre-Med. Course, Univ. of Wyoming. Laramie, Wyo. (U.S.A.). *Animal parasites and parasitic diseases of domestic animals. Life history of Sarcocystis tenella, of Thysanosoma actinioides and of Moniezia expansa.* ⓓ Univ. of Chicago 1904. ⓟ *Sarcocystis tenella. Melophagus ovinus. Polycelis coronata. Animal parasites of various kinds; fleas, lice, etc. Thysanosoma actinioides.*

Scott, Joseph Prestwich (1890), Associate Prof., Department of Pathol., Veterinary Division. Manhattan, Kan. (U.S.A.), Kansas State Agric. Coll. *Anaerobic bact., Blackleg.* ⓓ Kansas State Agric. Coll. 1924. ⓟ *Cultures of Clostridium chauvei.*

Scott, Kirk (Miss) (1900), Demonstrator, Department of Botan., Univ. of Manitoba. Winnipeg, Manitoba (Canada). *Botan.: Lichens and Mosses.* ⓓ Univ. of Manitoba 1923.

Scott, William B. (1858), Blair Prof. of Geol. and Paleont. Univ. Princeton, N.J. (U.S.A.). *Evolution. Physiographie. Mammals in the Western Hemisphere.* ⓓ Phil. Heidelberg 1880, L.L. Univ. of Pennsylvania 1906, sc. Harvard 1909, Oxford 1912. o

Scott, William Clifford Munroe (1903). Toronto (Canada), 2 Surrey Place, Toronto Psychiatric Hospital. *Anatomy, embryol. and histol. of the pancreas and Isles of Langerhans in mammalian forms. Pathol. anat. of the thyroid gland in Elasmobranchs.* ⓓ Univ. of Toronto, Canada, 1924.

Scott-Bishop, Katharine (1889), Histopathol., Hooper Foundation for Medical Research, San Francisco, Cal. Berkeley, Cal. (U.S.A.), 1508 La Loma Ave. *Histol. of otosclerosis (experim.), nutrition and reproduction. Connective tissue, endothelium, renal tissue.* ⓓ Johns Hopkins 1915.

Scotti-Foglieni, Luigi (1898), Ass. eff. Pavia (Italia), Ist. di Patol. Generale R. Univ. *Vitamine. Rigenerazione del miocardio.* ⓓ Pavia 1924.

Scott Robertson, G., Prof., D.Sc., Dir. of Chem. and animal nutrition research division. Belfast (N. Ireland), Stormont. o

Scremin, Luigi (1898), Libero doc. di Farmacol. Fellow, Rockefeller Inst. Padova (Italia). *Farmacol. generale.* ⓓ Roma 1926.

Scriban, Ion A., o. Prof. Univ. Cluj (România). *Vergl. Zool. u. Anat.* o

Scullen, Herman Austin (1887), Ass. Prof. of Entomol. and Specialist in Bee Culture. Corvallis, Ore. (U.S.A.). *Insect Ecol. and Bee Culture.* ⓟ Apidae.

Seurat, L. G., Prof. Univ. Alger (Afrique). *Zool. appliquée.* o

Scurti, Francesco (1878), Dir. Staz. chimico-agraria. Torino (Italia), Via Ormea 47. *Biochem. vegetale.* ⓓ Roma 1913.

de Seabra, Anthero Frederico (1876), Naturaliste Mus. Zool. Univ. Coïmbra (Portugal). *Entomol. et biol. agric., Hémiptères Hétéroptères du Portugal.* ⓟ Hémiptères-Hétéroptères.

Seagar, Edward Aitken (1887), Prof. Tropical Hygiene and Dir. Tropical Diseases Research Labor., Imper. Coll. Port of Spain (Trinidad),' *Parasit. of Tropical Diseases.* ⓓ 1913.

Seale, Alvin (1873), Supt. Steinhart Aquarium. San Francisco, Cal. (U.S.A.), Golden Gate Park.

Ichthyol. ③ Stanford Univ. 1896. ⑦ Tropical Fishes Live.

Sears, Harry Johnson (1885), Prof. of Bact., Univ. of Oregon Med. School. Portland, Ore. (U.S.A.). *Bact. metabolism, symbiosis.* ④ Stanford Univ. 1916.

Sears, Ogle H. (1892), Ass. Prof. of Soil Biol. Urbana, Ill. (U.S.A.), Univ. of Illinois. ④ Purdue 1917.

Sears, Paul Bigelow (1891), Associate Prof. of Botan., Univ. of Nebraska. Lincoln, Neb. (U.S.A.), 1435 A Street. *Physiol. analysis of variation.* ④ Chicago Univ. 1922. ⑦ Pollen specimens of native Nebraska plants for pollens from other parts of the world.

Seaton, Ian William (1898), Head of Plant Breeding Research Division (Ministry of Agric.: N. Irèland); Lect. in Plant Genetics, The Queen's Univ. of Belfast (Ireland), Stormont, Strandtown. *Oats, Flax, Potatoes and Grasses.* ④ Univ. of Edinburgh 1922. ⑦ Pure Line Varieties of Oats.

Seaver, Fred Jay (1877), Curator. New York (U.S.A.), Botan. Garden. *Discomycetes; tropical ascomycetes.* ④ 1910.

Sebentzoff, Boris (1891). Wissenschaftl. Mitarb. f. Fischzucht der Timirjaseff-Academie. Moskau 10 (U.d.S.S.R.), 2 Graschdanskaja 43, W. 3. *Angew. Hydrobiol.: Biol. d. Süßwassers m. Bezug auf Fische und Anopheles-Larven.* ④ Moskau 1926.

Sečerov, Slavko (1888). Beograd (S.H.S.), 53 Novopazarska. *Embryol.: Entstehung der Diplospondylie. Experimental Zool.: Farbenwechsel, Gonaden, Licht, Farbe, Pigmente. Biol. Philosophie: Neovitalismus, Regulation.* ② Phil. Vienna 1910, M. Sc. London.

Sedgroich, Paul J., Ass. Prof. of Botan., Syracuse Univ. Syracuse, N.Y. (U.S.A.). *Morphol.* ○

Sedlaczek, Walther (1872), Forstentomol., Dir. forstlich. Bundes-Versuchsanstalt. Mariabrunn (Österreich). *Forstl. Zool. und Schädlingsbekämpfung. Borkenkäfer, Nonne.* ④ Wien 1902.

Seefelder, Richard (1875), o.ö. Prof. der Augenheilkunde. Innsbruck (Deutsch-Österreich), Anichstr. 5. *Normale und pathol. Entwicklung des Auges. Normale Anat. des Auges.* ④ München 1898.

Seekles, Laurens (1897), Conserv. Labor. Vet. med. Chemie, Univ. Utrecht (Holland), Bellamystraat 12. *Physiol. Chem. und Pathol. Chem.* ④ Leiden 1922.

Seelemann, Martin (1899), Ass., Tierarzt. Kiel (Deutschland), Prüne 48. *Milchbact. und Milchhygiene, Infektionskrankheiten der Tiere.* ④ Berlin 1921.

Seeley, Ralph M. (1898), Entomol., Georgia State Board of Entomol. Atlanta, Ga. (U.S.A.), Capitol Building.

Seeliger, Rudolf Heinrich (1889), Dr. phil., Regierungsrat bei der Biol. Reichsanstalt für Land- und Forstwirtschaft, Zweigstelle Naumburg a. d. S. (Deutschland), Sedanstr. 37. *Genetik der Weinrebe. Züchtung von krankheitswiderstandsfähigen Apfel- und Rebsorten. Ampelographie.* ④ Göttingen 1911.

Šefer, Elena Jakov., Doc. Inst. f. Landwirtschaft u. Melioration. Saratov (U.d.S.S.R.), Teatral'naja Ploščad.. *Pflanzenphysiol.* ○

Segares, Arturo Cabal., Prof. Univ. Madrid (España), Alv. de Castro 16. *Botan.* ○

Segers, Clément (1899), Ingénieur agronome. Bruxelles II (Belgique), Rue Stevens Delannoy 42. *Entomol. économique.* ④ Louvain 1921. ⑦ Coléoptères.

Seibert, Florence Barbara (1897), Instr. Dept. of Pathol., Ass. the Otho S. A. Sprague Memorial Inst. Chicago, Ill.(U.S.A.), Univ. ④ Yale Univ. 1923.

Seidel, Friedrich (1897), Dr., Priv.Doc. Zool., vergl. Anat. und Entwicklungsgeschichte. Ass. Zool. Inst. Univ. Königsberg i. Pr. (Deutschland). *Experim. Embryol. (Insekten, Amphibien).* ④ 1923.

Seiffert, Walter (1893), Priv.Doc. Freiburg i. Br. (Deutschland), Dreisamstr. 37. *Hygiene, Bact.* ○

Seifriz, William (1888), Prof. of Botan. Philadelphia, Pa. (U.S.A.), Dept. of Botan., Univ. of Pennsylvania. *Physical properties of protoplasm and colloidal jellies. Physiol. of the cell. Periodicity of flowering in plants.* ④ Johns Hopkins Univ. 1920.

Seiler, Jakob (1886), ao. Prof. an der Univ. München. Schlederlohe, Post Wolfratshausen, Ob.-Bayern (Deutschland), Biol. Inst. *Experim. Untersuchungen über Parthenogenese bei Schmetterlingen. Experim. zytol. Untersuchungen an Schmetterlingbastarden.* ④ Zürich 1914.

Seitz, Arthur (1880), ao. Prof. a. d. Univ. Leipzig (Deutschland), Scheffelstr. 32. *Gewerbehygien., Bact. des Mundes, Endokrine Drüsen u. Immunität.* ④ München 1905.

Seitz, Otto (1888), Dr. phil. nat., Ass. an der Preuß. Geol. Landesanstalt. Berlin N 4 (Deutschland), Invalidenstr. 44. *Die Fauna der Kreideformation.* ④ 1914.

Seiwell, Harry Richard (1904), Investigator U.S. Bureau of Fisheries. Chapel Hill, N.C. (U.S.A.), Univ. of North Carolina. *Physiol. and taxinomy of Entomostraca, Oyster.*

Sekera, Emil (1864), Ph.D., o. Prof. für allgemeine Zool u.Parasit. a. d.Tierärztl. Hochsch. Brno (C.S.R.), Pražská 69. *Anat. und Biol. der Turbellarien und Süßwassernemertinen, Hydrobiol.* ④ Prag 1888. ⑦ Mikroskopische Praep. der Süßwasserturbellarien u. Nemertinen.

Seki, K., Ass. Prof. Hokkaidō Imper. Univ. Sapporo (Japan), Pathol. Inst., Fac. of Med. *Pathol.* ○

Sekreteva, Vera J., Doc. I. Univ. Moskau (U.d.S.S.R.). *Bact.* ○

Selenij, Jurij, Prof. Tierärztl. Hochsch. Leningrad (U.d.S.S.R.), Černigovskaja 5. *Physiol.* ○

Selenskij, Waldemar D. (1880), Prof. Univ. u. Staatl. Pädagog. „Herzen" Inst., Abt. Naturwiss. Leningrad (U.d.S.S.R.), Wassily 16 L, N 29 log 7. *Vergl. Morphol. Helminthes. Hirudinea.* ⑦ Hirudinea.

Selezniov, P. I., Doc. landwirtschaftl. Inst. Krasnodar, Kaukasus (U.d.S.S.R.), Novaja 107. *Milchwirtschaft.* ○

de Selgas y Marin, Ezequiel (1894), Prof. adjutant de l'Univ., Centrale de Madrid (Anat. Comparée). Madrid (España), Paseo de la Castellana 57. *Microphotographie et microcinématographie pour l'étude des mouvements rapides.* ④ Madrid 1920.

Seliber, G. (1877). Collaborateur sc. à la sect. de Microbiol. de l'Inst. „Lesshaft". Leningrad (U.d.S.S.R.), Prospect Anglais 32. *Microbiol. pure et appliquée. Physiol. des Plantes cultivées: influence du milieu sur différentes propriétés des levures, influence du milieu sur la formation des ferments par des microorganismes.* ④ Halle a. S. 1904.

Seligmann, Erich (1880), Prof., Dr., Dir. der wissenschaftl. Inst. im Hauptgesundheitsamt der Stadt. Berlin W 15 (Deutschland), Xantener Str. 5. *Hygien. u. Bact. Seuchenbekämpfung, Diphtherie, Desinfektion.* ④ Heidelberg 1904.

Sell, Wilhelm (1892), Wissenschaftl. Hilfsarbeiter an der B. Landesanstalt für Pflanzenbau und Pflanzenschutz München (Deutschland), Briennerstraße 8 II. *Züchtung von menschen- und tierpathogenen Bakterien.* ④ München 1919.

Sella, Massimo (1886), Prof. Capo Biol. Comitato Talassografico Ital., Dir. Ist. di Biol. Marina. Rovigno, Istria (Italia). Acquario. *Pesci: Anat. e fisiol. dello scheletro, muscolatura, nervi spinali dei pesci; biol. e sistematica degli Scomberoidi (sviluppo larvale, accrescimento, migrazioni). Gambusia. Spugne. Accrescimento delle Tartarughe. Malaria: biol. degli Anofeli.*

Selle, Raymond Martin (1898), Ass. Prof. of Zool. Los Angeles, Cal. (U.S.A.), Occidental Coll. *Oestrous Cycle in the Guinea Pig, Histol.; Parasit. Protozoa of Reptiles.*

Sellers, Thomas Fort (1891), Dir. of Georgia State Board of Health Labor. Atlanta, Ga. (U.S.A.), Capitol Building. *Rabies and Intestinal Parasit. Hymenolepis nana.* ④ Univ. of Michigan 1917.

Sellier, M., Prof. Fac. Méd. Bordeaux (France). *Physiol.* o
Selter, Hugo (1878), o. Prof. der Hygien. und Bact. und Dir. des Hygien. Inst. der Univ. Bonn a. Rhein (Deutschland), Theaterstr. *Tuberkuloseforschung, Trinkwasserversorgung.* Ⓓ 1902.
de Selys Longchamps, Marc A. G. (1875), Prof. à l'Univ. Libre de Bruxelles, Fac. des Sc. Bruxelles (Belgique), 61, Avenue Jean Linden. *Morphol. animales spec. Tuniciers, Phoronis, Petromyzon.* Ⓓ Univ. de Liège 1901.
Selz, Otto (1881), o. Prof. der Philos., Psychol. und Pädagogik, Dir. Inst. für Psychol. u. Pädag. Handels-Hochsch. Mannheim (Deutschland). *Psychol. der Denkvorgänge und der Persönlichkeit. Psychotechnik.* Ⓓ 1909.
Sembel, S. J., Prof. Staatl. Med. „Lunačarsky"-Inst. Astrachan (U.d.S.S.R.), Ul. Sverdlova 37. *Botan.* o
Sembrat, Kasimierz (1902), Dr. phil., Ass. à l'Inst. de Zool. de l'Univ. Jan Kazimierz, Lwów (Polska), 4, rue Mikołaja. *Recherches experim. sur la métamorphose des Batraciens. Recherches cytol. sur la gamétogénèse et l'appareil de Golgi.* Ⓓ Lwów 1925.
Semenović, Varvara A., Doc. II. Univ. Moskau (U.d.S.S.R.). *Biol. Chem.* o
Semenow, Viktor F., Prof. an der Landwirtsch. Akademie und des Med. Inst. Omsk (U.d.S.S.R.), Poststr. 20. *Botan., Geobotan.* o
Semenow Tian-Shandkij, Andreas Petrowitsch (1866), Zool. en chef Mus. zool. de l'Acad. d. Sc. Leningrad (U.d.S.S.R.), Vasil Ostrow 8, ligne 39, log. 2. *Biographie, Taxonomie, Théorie d'évolution. Entomol.: Coleopt., Hymenopt., Dermatopt., Orthopt., Dipt.*
Semichon, Louis (1879), Préparateur au Mus. Paris 5 (France), 55, rue de Buffon. *Morphol., Histol. comparée.* Ⓟ Paris.
Semper, Max (1870), o. Prof. der Palaeont. u. Geol. Techn. Hochsch. Aachen (Deutschland), Bachstr. 34. *Wissenschaftsgeschichte. Methodol.* Ⓓ München 1896.
Semtschenkow, Iwan N., Ass. Omsk (U.d.S.S.R.), West-Sibirische Landw. Versuchsstation, Sagorodnaja Roscha, Postfach 34. *Agronomie. Pflanzenkunde. Selektion landw. Pflanzen.* o
Sen, Basiswar, Dir., Vivekananda Labor. Bagh Bazar, Calcutta (India), 8, Bosepara Lane. *Electro-Physiol.* o
Sen, S., Prof. Dr. Belgachia, Bengalen (India). *Physiol.* o
Senevet, Georges (1891), Agrégé Fac. de Méd. d'Alger, Chef de Labor. à l'Inst. Pasteur d'Algérie. Alger (Algérie), 8, rue Borély la Sapie. *Parasit. humaine et animale, Ixodidés, Insectes piqueurs, Helminthes.* Ⓓ Paris 1920. Ⓟ Parasites du Nord de l'Afrique.
Senior, Harold Dickinson (1870), Prof. Anat., N.Y. Univ. New York (U.S.A.), 338 East 26th Street. *Development of the limb arteries, particularly in man.* Ⓓ Durham 1895.
Senjutkin, I. I., Doc. Univ. Nizny-Novgorod (U.d.S.S.R.), Sovetskaja Ploščad 8. *Mikrobiol.* o
Senkowski, Michel (1867), Doc. priv. Univ. Kraków (Pologne), Rue Radziwiłłowska 4. *Chim. biol. Fonct. de foi. Electrophys. d'éléments nerv. et musculaires.* Ⓓ Phil. Lwów 1892, med. Cracovie 1895.
Senn, Gustav, Prof. Univ. Basel (Schweiz), Schönbeinstr. 6. *Botan.* o
Senna, Angelo (1866), Prof. o. Zool. Univ. Firenze (Italia), Via Romana 19. *Larve planctoniche (Attiniari, Zoantari, Ceriantari; Sipunculidi, Echiuridi). Sistematica e anat. d. Crostacei decapodi macruri, Anfipodi iperini, Brentidi (Coleopt.), Pesci e Chirotteri. Oogenesi e spermiogenesi.* Ⓓ Pavia 1890.
Senning, William Charles (1899), Instr. in Zool. Ithaca, N.Y. (U.S.A.). *Vertebrate Morphol.* o
Senô, Hidezane, Prof. Imper. Coll. of Fishery. Ecchûjima, Tôkyô (Japan), Zool. Labor. o

Sera, Gioacchino Leone (1878), Prof. o. Fac. sc. Univ. Napoli (Italia), Via Univ. 39. *Anthrop. Osteol. Miol. Caretteri fisionomici. Denti. Razze.* Ⓓ Roma 1903.
Serdítych, Wladimir W., Vorsitzender der Phys.-Geograph. Sektion der Burjat-Mongoler Wissensch. Gesellschaft. Werchneudinsk (U.d.S.S.R.), Baikalskaja 4. *Botan., Phytosoziol.* o
Serebrovsky, Alexander Sergewich (1892), Prof. of Zootechnical Inst., Ass. Inst. Experim. Biol., Anikovo Genetical Station, near Moscow. Moscow 64 (U.d.S.S.R.), Voronzowo Pole 6, Inst. Experim. Biol. *Genetics of Poultry, of Drosophila. Biometry. Poultry Husbandry.* Ⓓ Moscow 1914.
Sereni, Enrico (1900), Libero Doc. di Fisiol. generale, Capo del Reparto di Fisiol. e Chimica zool. nella Stazione Zool. di Napoli (Italia), Stazione Zool. Acquario. *Anafilassi. Fisiol. muscolare, produzione di calore. Ossidazione e respirazione cellulare. Influenze nervose sul cuore.* Ⓓ Roma 1922.
Serés, Manuel, Prof., Catedrático de Anat. en la Fac. de Med. Barcelona (España). o
Sergeeva, Pelageja V., Doc. II. Univ. Moskau (U.d.S.S.R.). *Morphol. u. System. d. Pflanzen.* o
Sergent, Edmond (1876), Dir. Inst. Pasteur d'Algérie. Alger (Algérie). *Pathol. et hygiène des pays chauds. Paludisme, fièvre récurrente, trypanosomiases, piroplasmoses. Microbiol., Protozool.* Ⓓ Paris 1910.
Sergent, Etienne (1878), Chef du Service antipaludique à l'Inst. Pasteur d'Algérie. Alger (Algérie), 4 Rue michelet. *Antipaludisme, protozool.* Ⓓ Montpellier 1901.
Sergi, Sergio (1878), Prof. anthrop. Univ., Dir. Inst. d'Anthrop. Roma (Italia), Via Collegio Romano 26. *Anthrop. morphol.: Cerebra Hererica. Crania Habessinica. Moelle épinière du chimpanse.* Ⓓ 1925.
Sergiewskaja, Lydia P., Konserv. Univ. Tomsk (U.d.S.S.R.), Herbarium. *Geobotan., Systematik.* o
Sernander, Johan Rutger (1866), Dr. phil., Prof. der Pflanzenbiol. Univ. Upsala (Schweden). *Entwicklungsgeschichtliche Pflanzengeographie; Pflanzensoziol.; Geschichte der nordeuropäischen Vegetation; Torfmoore; Kalktuffe; Bodenkunde; Verbreitungsbiol. besonders myrmecochore Pflanzen; Oekol. der Flechten; Vegetationszonen im Meere und in den Binnengewässern;Naturdenkmalspflege.*ⒹUpsala1895.
Sernov, Sergej A .(1871), Prof. Timirjasev Akad. Moskau (U.d.S.S.R.), Petrovsko'e Razumovsko'e. *Hydrobiol. Nahrung. Fauna.*
Serow, Dmitrij K., Konserv. d. Botan. Kabinetts u. Herbariums d. Ukrain. Akademie d. Wissensch. Kiew (U.d.S.S.R.), Ul. Tolstogo 19, W. 1. *Geobotan. Pflanzensystematik. Moos.* o
Servit, Miroslav (1886), Dir. staatl. landwirtschaftl. Versuchsstation. Hořice v Podkrkonoší (C.S.R.). *Lichenol. Landw. Pflanzenzüchtung.* Ⓓ Prag 1913. Ⓟ Lichenen.
Sessous, George (1876), Dr., Saatzuchtdir. der Firma Fr. Strube, Schlanstedt, Bez. Magdeburg (Deutschland). *Pflanzenzucht.* Ⓓ Jena 1903.
Šestakova, Galina S., Doc. I. Univ. Moskau (U.d. S.S.R.). *Zool., Embryol.* o
Setchell, William A., Prof. of Botan. Univ. of California. Berkeley, Cal. (U.S.A.). *Aquatic Plants.* o
van Seters, Wouter Hendrik (1891), Dr. Leiden (Holland), Hooigracht 65. *L'Embryol. compar. du crâne.* Ⓓ Leiden 1921.
Seth, Mohaw Lal, Prof. Univ. Delhi (Br. India). *Biol.* o
Seth-Smith, David. London N.W. 8 (England), Curators House, Zool. Gardens. *Ornithol.* o
Setkov, Boris Andrewitsch (1883), Prosektor, histol. Labor. Krasnodar (U.d.S.S.R.), Borslkavskaja 33. *Histol., Röntgenstrahlenwirkung auf lebendes Protoplasma.* Ⓓ Krasnodar 1921.
Seuffert, Rudolf Wilhelm (1884), Dr. phil., ao. Prof., Priv.Doc. und Oberass. am Physiol. Inst. der

Tierärztl. Hochsch. Berlin NW 40 (Deutschland), Roonstr. 13. *Physiol. Chem., Ernährung und Stoffwechsel.* Ⓓ München 1910.
Seurat, Léon Gaston (1872), Prof. de Zool. à la Fac. des Sc. Univ. d'Alger (Algerie). *Nématodes parasites. Faune intercotidale de la Méditerranée. Hyménoptères entomophages. Nématodes de la Berbérie.* Ⓓ Paris 1921. Ⓟ Nématodes. Mollusques marins de la Tunisie.
Sevčík, Franz, Dr., Prof. der Bact., Serol. und Hygiene. Praha (C.S.R.), Pražska 67, Tierärztl. Hochschule. o
Severcev, Sergej A., Doc. I. Univ. Moskau (U.d. S.S.R.). *Zool., Embryol.* o
Severin, G., Conserv. Mus. R. d'Hist. nat. Bruxelles (Belgique), 31, rue Vautier. *Entomol.* o
Severin, Harry Charles (1885), Prof. Entomol., Zool. Dept., State Coll., Entomol., Exp. Station, State Entomol. Brookings, S.D. (U.S.A.). *Economic Entomol. Black Field cricket. Orthoptera.*
Sevringhaus, Elmer Louis (1894), Associate Prof. of Physiol. Chem., Chem. of the Wisconsin General Hospital Univ. of Wisconsin. Madison, Wis. (U.S.A.). *Carbohydrate, fat metabolism; Obesity. Muscular work.* Ⓓ Harvard 1921.
Seward, Albert Charles (1863), Prof. of Botan. Cambridge (England), The Master's Lodge, Downing Coll. *Palaeobotan. J. Wealden Flora. Jurassic Flora. Fossil Plants as test of climates.* Ⓓ Cambridge Univ. 1886.
Sewell, Malcolm Cameron (1887), Assoc. Prof. Soils. Manhattan, Kan. (U.S.A.), Dept. of Agronomy, Kansas State Agr. Coll. *Plant nutrition. Crop Ecol.* Ⓓ Univ. Chicago 1922.
Sewell, Robert Beresford Seymour (1880), Indian Med. Service, Dir. Zool. Survey of India. Calcutta (India), Indian Mus. *Oceanography and Hydrography. Marine free living Copepoda.* Ⓓ Cantab 1923.
Sewertzow, A. N., Prof., Dir. Inst. d. vergl. Anat. I. Univ. Moskau (U.d.S.S.R.), Herzenstr. 6. o
Seybold, August (1901), Dr. phil., Ass. Würzburg (Deutschland), Botan. Inst. *Morphol.: Symmetrieverhältnisse, Formgestaltung der Blätter. Physiol.: Transpiration, Phototropismus, Chemotropismus und oligodynamische Erscheinungen.* Ⓓ 1924.
Seyfarth, Paul Carly (1890), Dr. med. et phil., ao. Prof. an der Univ. (allgemeine Pathol.). Leipzig (Deutschland), Beethovenstr. 33. *Tropenkrankheiten (besonders Pathol. Anatomie der Malaria). Blut und Blutkrankheiten.* Ⓓ Phil. Leipzig 1913, med. 1915.
Seyot, P., Prof. Univ. Nancy (France). *Botan.* o
Seyrig, André (1897), El Soldado, Cordoba (España). *Hymenoptera: fam. Ichneumonidae (Systematique et biol.).* Ⓟ Ichneumonidae.
Šgenti, Voldemar, Doc. Univ. Tiflis (U.d.S.S.R.). *Pathol. Anat.* o
Shadin, Wladimir (1896), Leiter biol. Oka-Station. Murom (U.d.S.S.R.). *Bodenfauna der Binnengewässer; Süßwassermollusken.* Ⓟ Süßwassermollusken.
Shadowsky, A. (1889), Ass. Univ. Moskau (U.d. S.S.R.), Botan. Garten d. Univ. I, Mestschanskaja N. 28. *Embryol. und Zytol. d. Pflanzen (Angiospermen).*
Shadowskij, Anatolij E., Prof. Akademie f. Soziale Erziehung. Moskau (U.d.S.S.R.), Trubnikowskij per. 13, W. 8. *Botan., Geographie.* o
Shafer, George D., Ass. Prof. of Physiol., Leland Stanford Junior Univ. Palo Alto, Cal. (U.S.A.), 321 Melville Ave. o
Shamel, Archibald Dinon (1878), Dept. of Agric. Riverside, Cal. (U.S.A.), 3 Arroyo Drive. *Plant physiol. Soil fertilization. Plant selection.* o
Shaner, Ralph Faust (1893), Associate Prof. of Anat., Univ. of Alberta. Edmonton, Alberta (Canada). *Embryol. of Reptilia, Compar. anat. and development of the heart of vertebrata.* Ⓓ Harvard Univ. 1920.

Shanks, W. T., Prof. Univ. Leeds (England). *Physiol.* o
Shantz, Homer Le Roy (1876), Prof. of Botan. and Head of the Department, Univ. of Illinois. Urbana, Ill. (U.S.A.). *Plant Physiol., Plant Geography.* Ⓓ Phil. Univ. of Nebraska 1905, sc. Colorado Coll. 1926.
Shapiro, Inez (1903). Los Angeles, Cal. (U.S.A.), 377 S. Virgil Ave. *Physiol. chem. and nutrition.*
Shapovalov, Michael (1880), Pathol., United States Department of Agric. Riverside, Cal. (U.S.A.), Citrus Experim. Station. *Tomato Diseases, Cotton Diseases.* Ⓓ Univ. of Maine 1913. Ⓟ Tomato diseases and cotton diseases (dry specimens).
Sharif, Mohammed (1899), Ass. Superintendent, Zool. Survey of India. Calcutta (India), Indian Mus. *Ixodoidea, Spiders.* Ⓓ Punjab Univ., India, 1922. Ⓟ Punjab Arachanids.
Sharp, Lester Whyland (1887), Prof. of Botan. in Cornell Univ. Ithaca, N.Y. (U.S.A.). *Structure of chromosomes; Spermatogenesis in Pteridophytes; Angiosperm female gametophyte.* Ⓓ Univ. of Chicago 1912.
Shaver, Jesse M. (1888), Associate Prof. of Biol. at George Peabody Coll. for Teachers, Nashville, Tenn. (U.S.A.). *Avian ecol.* Ⓓ Vanderbilt Univ. 1921.
Shaw, Frederick W. (1882), Associate Prof. of Bact. Richmond, Va. (U.S.A.), Med. Coll. of Virginia. *Bact.* Ⓓ Univ. of Kansas 1906.
Shaw, Harry B., Phytopath., Federal Horticultural Board, U.S. Dept. of Agric. Washington, D.C. (U.S.A.). *Curly-top of beets. Beet flower. Beet nematode. Fertilization of beets.*
Shaw, Jacob Kingsley (1877), Research Prof. of Pomol., Massachusetts Agric. Coll. Amherst, Mass. (U.S.A.). *Genetics of Peach Varieties; Interrelation of Stock and Scion; Taxonomy of Fruit Varieties; Nutrition of plants.* Ⓓ Massachusetts Agric. Coll. 1911.
Shaw, Louis Agassiz (1886), Instr. in Physiol., Harvard Univ., School of Public Health. Boston, Mass. (U.S.A.), 240 Longwood Avenue. *Respiration, acid-base equilibrium of the body.* Ⓓ Harvard 1909.
Shaw, Margaret Fenton, Instr. Botan. Dept., Ohio State Univ. Columbus, O. (U.S.A.). *Microchem.*
Shea, Stephen, Prof. Univ. Coll. Galway (Ireland). *Anat.* o
Shear, Cornelius Lott (1865), Senior Pathol. in Charge of the Office of Mycol. and Plant Disease Survey, Bureau of Plant Industry U.S. Dept. of Agric. Washington, D.C. (U.S.A.). *Life histories and taxonomy of Ascomycetes diseases of small fruits.* Ⓓ George Washington Univ. 1906. Ⓟ Mycol. and Pathol. specimens.
Shear, Elmer Vanderzee, Junior (1896), Associate in Research, New York State Agr. Experim. Station. Poughkeepsie, N.Y. (U.S.A.), Vassar Coll. *Fruit diseases.*
Sheard, Charles (1883), Prof. of Biophysics and Physiological Optics, The Mayo Foundation, Univ. of Minnesota. Rochester, Minn. (U.S.A.), Mayo Clinic, Section of Biophysics. *Biophysics and physiol. optics. Dynamic, Skiametry, Dynamic Ocular Tests, Ophthalmic Glasses, photoelectric effects, bile pigments, color of skin, capillaries, ultraviolet radiation.* Ⓓ Princeton 1912.
Shearer, Amon Robert (1871). Mont Belvieu, Tex. (U.S.A.). *Ornithol., Palaeont.* Ⓓ Galveston, Texas, 1898. Ⓟ Skins, nests and eggs of Texas birds.
Shearer, Cresswell (1874), Lect. Embryol., Univ. Cambridge (England), Anat. School. The Mus. *Experim. Embryol.* Ⓓ Cambridge.
Sheehy, Edmond John (1890), Lect. Agric. Zool. Dublin (Ireland), Coll. of Sc., Univ. Coll. *Entomol., Helminthol., Protozoa, Mammalia. Animal Physiol., heredity, evolution.* Ⓓ London 1920.
Shegalow, Sergej I., Prof. Dr., Landwirtschaftl. Akademie Timirjasew. Moskau (U.d.S.S.R.), Pe-

trowsko-Rasumowskoje. *Genetic, Selektion landwirtschaftl. Pflanzen, Samen.* o

Shelford, Victor E. (1877), Biol. in Charge of Research Labor., Illinois Nat. History Survey and Associate Prof. of Zool., Univ. of Illinois. Champaign, Ill. (U.S.A.), Vivarium Building, Wright and Healey St. *Animal ecol.; climatol.; biotic interaction in animal communities; relations of the codling moth to weather and climate; light penetration into water.* ⓓ Chicago 1907.

Shellshear, J. L., Prof. Univ. Hong-kong (China). *Anat.* o

Shemtschushnikow, Eugen A., Prof., Inst. f. Landwirtschaft u. Melioration. Nowotscherkassk (U.d. S.S.R.), Masterowaja ul. 82a. *Botan., Pflanzen-Physiol.* o

Shepard, Harold H. (1898), Ass. Entomol. in Experim. Station Research. College Park, Md. (U.S.A.), Univ. of Maryland. *Hesperiidae (Lepidoptera); Biol., Morphol. of the Thorax of Lepidoptera.* ⓟ Lepidoptera; other groups of insects for Hesperiidae dried, or for Lepidoptera named and preserved in alcohol.

Sheppard, Edith Mabel (1894), Lect. in Zool. and Compar. Anat. Univ. Coll. of South Wales and Monmouthshire, Univ. of Wales. Cardif (England), Newport Rd. ⓓ Univ. of Wales 1916.

Sheppe, Williams Marco (1895), Dir. of Labor. Wheeling, W.Va. (U.S.A.). *Pathol. Anat.* ⓓ Univ. of Virginia 1921.

Sherbakoff, C. D., Plant Pathol., Agric. Experim. Stat. Knoxville, Tenn. (U.S.A.). o

Sherff, Earl Edward (1886), Instr. Botan., Normal Coll. Chicago, Ill. (U.S.A.), 1419 Stewart Avenue. *Taxonomy of Compositae (Bidens, Cosmos, Coreopsis, Isostigma, Xanthium, Taraxacum etc.), Ecol. of marsh plants.* ⓓ Univ. of Chicago 1916.

Sherman, Althea Rosina (1853). National, Ia. (U.S.A.), via McGregor, *Ornithol.: feeding, flocking, nesting habits of hole-nesting birds (Sparrow Hawk, Screech Owl, Northern Flicker, Chimney Swift, and Western House Wren). Feeding habits of the Ruby-throated Hummingbird. Mammal.: The habits of bats.* ⓓ Oberlin Coll. 1882.

Sherman, H. C. (1875), Mitchill Prof. of Chem., Columbia Univ. New York City (U.S.A.). *Organic Analysis. Food and nutrition. Vitamins.* ⓓ Columbia 1897. o

Sherman, Harley Bakwell (1894), Ass. Prof. of Biol. Gainesville, Flo. (U.S.A.), Univ. of Florida. *Compar. Vertebrate Anat. and Mammalogy of Southeastern U.S.A.* ⓓ Univ. of Mich. 1922. ⓟ Skins and skeletons of mammals. Embalmed mammals from Southeastern U.S.A.

Sherman, Hope, Ph.D., M.D. Cleveland Heights, O. (U.S.A.), 2915 Coleridge Road. *Respiration of Dormant Seeds. Fecal Flora.* ⓓ Phil. The Univ. of Chicago 1920, med. Johns Hopkins Med. School 1925.

Sherman, James Morgan (1890), Prof. of Bact. and Dairy Industry, New York State Coll. of Agric., Cornell Univ. Ithaca, N.Y. (U.S.A.). *Dairy Bact.* ⓓ Univ. of Wisconsin 1916.

Sherrington, Charles Sir, Prof. Univ. Oxford (England). *Physiol. of nervous system.* o

Sherwin, Carl P. (1885), Prof. of Biochem., Dir. of Research Fordham Univ. N.Y. City, Scarsdale, N.Y. (U.S.A.), 6 Carstensen Road. *Amino acids, intermediary metabolism.* ⓓ Sc.D. Tübingen, 1915, M.D. Fordham Univ. 1920, L.L.D. 1924. ⓟ Alpha-amino acids of protein.

Sherwood, Ross M. (1886), Chief, Division of Poultry Husbandry, Texas Agric. Experim. Station. College Station, Tex. (U.S.A.). *Nutrition and genetics of chickens.*

Shestakov, Andréas (1890), Prof. Zool. des pädagogischen Inst. Jaroslavl (U.d.S.S.R.), Sovietskaja pl. 22, Naturhistorisches Mus. *Hymenoptera: Sphegidae, Chalcididae (Leucospis), Ichneumonidae (Pimplinae, Anomalonini, Ophionini et Paniscini), Braco-*

nidae (Braconini). ⓟ Sphegidae aus Turkestan und Kaukasus.

Shevde, S. V., Prof. State Coll. Baroda (Baroda State, India). *Biol.* o

Shibata, Bunpei. Komaba near Tôkyô (Japan), Zool. Inst., Fac. of Agric., Imper. Univ. *Entomol. Aphidae.* o

Shibata, Keita (1877), Prof. Imper. Univ. Tôkyô (Japan), Koishikawa-Ku, Botan. Inst., Fac. of Sc. *Biochem. and Plant-Physiol.* ⓓ Tôkyô 1899.

Shibata, Tamotsu (1888). Nagoya (Japan), Buheicho 2. *Serol.* ⓓ Tôkyô 1914.

Shikata, Masuzô, Ass. Prof. Imper. Univ. Kyôto (Japan), Chem. Inst., Coll. of Agric. *Forest Chem.* o

Shikinami, Jûjirô, Prof. Med. Univ. Okayama (Japan), Anat. Inst. *Anat.* o

Shikiejef, Stepan (1898), Ass. Zool. Kabinett Univ. Rostov a. Don (U.d.S.S.R.), Fr.-Engels-Str. N. 37. *Protist.; Experim. Zool.*

Shillinger, Jacob Edward (1890), Associate Vet., Zool. Division, Bureau of Animal Industry, Dept. of Agric., Washington, D.C. (U.S.A.). *Parasites of domestic animals.* ⓓ Vet. med. 1922, Master of Sc. 1923.

Shimada, Yaichi, Biol. Govern. of Formosa. Shinchiku-Shû, Formosa (Japan), Bureau, of Industry and Agric. *Systematic botan. of Formosa.*

Shimamura, Torai, Prof. Tôkyô Imper. Univ. Komaba near Tôkyô (Japan), Inst. of Vet. Sc., Fac. of Agric. *Vet. Med. and Physiol.*

Shimanowich, Samson (1895), Junior ass. of selection of flax, Inst. of Applied Botan. and New Cultures, Dept. of Genetics and Plant-Breeding. Detskoje Selo near Leningrad (U.d.S.S.R.). *Connection between the morphol. characters and quality and quantity of the fibre.* ⓟ Linseed.

Shimek, Bohumil, Prof. of Botan. and Curator of the Herbarium, Iowa State Univ. Iowa City, Ia. (U.S.A.). *Ecol. of the prairie region, taxonomy, plants of the Pleistocene.* o

Shimer, Hervey Woodburn (1872), Prof. of Paleont. Cambridge, Mass. (U.S.A.), Mass: Inst. Technol. *Paleont. of Invertebrata.* ⓓ Columbia Univ. 1904.

Shimizu, Tayei (1889), Prof. Med. Univ. Okayama (Japan), Chem. Inst. *Med. Chem. Galle, Gallensäuren.* ⓓ Kyôto 1915.

Shimohara, Takeshi, Prof. Jikeikai Med. Univ. Shiba, Tôkyô (Japan), Zool. Labor. *Hydrozoa.*

Shindo, Tokuichi, Prof. Kyûshû Imper. Univ. Fukuoka (Japan), Anat. Inst., Coll. of Med.

Shinji, Orihei, Prof. Imper. Coll. of Agric. and Forestry. Morioka (Japan), Zool. Labor. *Entomol* ⓓ Tôkyô.

Shinoda, Osamu (1899), Govern. Research Scholar sent Abroad of the Department of Education, Rockefeller's International Education Fellow, Zool. Labor. Abt. Physiol. Utrecht (Holland). Kyôto (Japan), Zool. Labor., Coll. of Sc., Imper. Univ. *Biochem. and Nutrition Physiol. of Insects; Histo-Cytol. of Secretion.*

Shiperovitsh, Woldemar (1893), Ass. Zool. u. Entomol. Leningrad (U.d.S.S.R.), Leskoi Inst. (Forstinst.). *Biol. d. Insekten. Hymenopt. Tentredinidae. Forstinsekten.* ⓓ Petersburg 1919.

Shirai, Mitsutarô (1863), Prof. emer. Imper. Univ. Tôkyô (Japan), Shinogawamachi Ushigome. *Plant pathol. Monstrosities in Plants in China and Japan. Useful Plants of Japan.* ⓓ Tôkyô Imper. Univ. 1886.

Shiraki, Tokuichi, Biol. Govern., Central Experim. Stat. (Formosa). Taihoku, Formosa (Japan), Labor. of Applied Zool., Dept. of Agric. *Entomol.: Isoptera, Diptera Tabanidae, Lepidoptera, Pyralidae, Orthoptera.*

Shirk, Claude J. (1875), Prof. of Biol., Nebraska Wesleyan Univ. Lincoln, Neb. (U.S.A.). *Plant Physiol. and Ecol.* ⓓ Univ. of Nebraska 1924.

Shishido, Ichirô, Dr. Kyôto (Japan), Kawaramachi-Hirokôji-kita. *Zool.*

Shitkow, Boris Michailowitsch (1872), Prof. Zool. (Vertebr.) I. u. II. Univ. Moskau (U.d.S.S.R.), Zentr. I. Univ., Zool. Mus. *Allgem. Biol., Ornithol., Mammalogie, Angewandte Zool. (Pelztiere, Forstzool., Vertebrata). Histol. u. Physiol. d. Haut u. d. Haare d. Säugetiere. Insectivora, Rodentia u. Carnivora d. paläarkt. Fauna (Anat. u. System). Chitin d. Endoskeletts d. Krebse.*

Shive, John W. (1877), Prof. of Plant Physiol., Rutgers Univ., Plant Physiol. N.J. Agr. Exp. Sta. New Brunswick, N.J. (U.S.A.). *Plant nutrition, solution cultures.* Ⓓ Carnegie Inst. of Washington, Tucson, Ariz. 1913.

Shiwago, Peter J. (1883), Priv.Doc. I. Univ., Ass. Inst. f. experim. Biol. Moskau (U.d.S.S.R.), Worontzowo Pole 6, Inst. f. experim. Biol. *Ruhekern. Chromosomen.* Ⓓ Moskau 1911.

Shoemaker, James Sheldon (1898), Ass. in Horticulture. Wooster, Ohio (U.S.A.), Experim. Station. *Stone and small fruits.* Ⓓ Univ. Minnesota 1925. Ⓟ Varieties of stone and small fruits.

Shôji, Rinnosuke (1886), Prof. Imper. Univ. Kyôto (Japan), Physiol. Labor., Coll. of Med. *Human Physiol., Physico-Chem. Blood. Biocolloid.* Ⓓ Tôkyô 1911.

Shor, Bernice Catherine (1902), Instr. of Biol., Rollins Coll., Winter Park, Orlando, Fla. (U.S.A.), 20 Pershing Place. *Mycol., Pathol., Plant Physiol.* Ⓓ Rollins Coll., Winter Park, Florida, 1924.

Shore, Clarence Albert (1878), Dir. State Labor. of Hygiene. Raleigh, N.C. (U.S.A.). Ⓓ Johns Hopkins 1907.

Shore, L. E., Lect. St. John's Coll. Cambridge (England). *Physiol.* ○

Shore, L. R., Prof. St. Bartholom. Hospit. and Coll. London E.C. 1 (England), West Smithsfield. *Anat.* ○

Shoup, George R. (1873), Poultryman and Contest Superintendent Puyallup, Wash. (U.S.A.), Western Wash. Experim. Sta., State Coll. of Wash.

Showalter, Amos M., Nation. Research Fellow, Stanford Univ., Cal. (U.S.A.). *Cytol., morphol. Bryophytes.* ○

Shreve, Edith B. Tucson, Ariz. (U.S.A.). *Plant Physiol. Regulation of transpiration by internal conditions.* Ⓓ Chicago.

Shriner, Ralph Lloyd (1899), Associate in Research. Geneva, N.Y. (U.S.A.), New York Agric. Experim. Station. *Biochem. of plants. Carbohydrates.* Ⓓ Univ. of Illinois, Urbana, 1925.

Shukowskij, Alexander W., Prof. Pädagog. Inst. Jaroslawl (U.d.S.S.R.), Kotorostnaja 26, W. 3. *Botan., Systematik.* ○

Shull, A. Franklin (1881), Prof. of Zool., Chairman of the Department of Zool., and Dir. of the Zool. Labor., Univ. of Michigan. Ann Arbor, Mich. (U.S.A.), 431 Highland Road. *Heredity and sex determination in aphids and rotifers, parthenogenetic life cycles in rotifers and insects, heredity in man.* Ⓓ Columbia Univ. 1911.

Shull, Charles Albert (1879), Prof. of Plant Physiol., Department of Botan., Univ. of Chicago, Ill. Homewood, Ill. (U.S.A.) 122 East Cedar St. *Plant Physiol. Membrane permeability, Respiration, Membrane potentials.* Ⓓ Univ. of Chicago 1915.

Shull, George Harrison (1874), Prof. of Botan. and Genetics, Princeton Univ. Princeton, N.J. (U.S.A.), 60 Jefferson Road. *Plant genetics, hybrid vigor or heterosis; genetical relations of the sexes in Lychnis (Melandrium); the first sex-linked character in plants; duplication of factors in Bursa (Capsella); self-sterility in Bursa grandiflora; species hybridizations; linkage and crossing over in Oenothera; new flower-colors, double-flowers and other new mutations, lethal factors, Mendelian inheritance in Oenothera.* Ⓓ Univ. of Chicago 1904.

Shumway, Waldo (1891), Associate Prof. of Zool., Ass. Dean of the Coll. of Liberal Arts and Sc., Univ. of Illinois. Urbana, Ill. (U.S.A.). *Vertebrate Embryol., Protozool.* Ⓓ Columbia 1916.

Shutt, Donald Bethune (1893), Lect., Bact. Department. Guelph (Canada), Ontario Agric. Coll. *Dairy Bact.*

Shwartzman, Gregory (1896), Bact. to Mount Sinai Hospital, Prof. of Bact. and Public Health of the N. Y. H. Med. Coll. and Flower Hospital. New York, N.Y. (U.S.A.), 664 West 163 Str. *Bacteriophage.* Ⓓ Univ. of Brussels, Belgium, 1920.

Šiaulis, Benedict (1889), Ass. Kaunas (Litauen), Lietuvos Univ., Med. Fac., Farm. Kat. *Properties of plain muscle.* Ⓓ Moscow 1912. Ⓟ Algae (Chlorophyceae).

Sibi, Marie (1896), Chef de Travaux à l'Inst. de Chimie biol. de Cluj (Roumanie), Strada Pasteur. *Gels colloïdaux.*

Sibilia, Cesare (1895), Ass. Stazione di Patol. vegetale. Roma 30 (Italia), Via S. Susanna 13. *Funghi parassiti.* Ⓓ Roma 1926.

Sickenberg, Otto (1901), Ass. Paläobiol. Inst. Univ. Wien I (Deutsch-Österreich). *Allgem. Paläobiol., Säugetiere, Phylogenie, Biol.* Ⓓ Wien 1925.

Sideris, Christos Plutarch (1891), Physiol.-Pathol. Honolulu (Hawaii), Experim. Station, Association of Hawaiian Pineapple Canners, Univ. of Hawaii. *Physiol. of Parasitism of the subterranean parts of Plants. Inter-reactions between soil colloids, protoplasmic colloids of the host and colloidal substances (enzymes and toxins) of the pathogene.* Ⓓ Univ. of California 1924. Ⓟ Fungi: Fusarium.

Sidgwick, N. V., Lect. Univ. Oxford (England). *Nat. Sc.* ○

Sidorin, Michail I., Doc., Moskauer Inst. f. Geodäsie. Moskau (U.d.S.S.R.), Elochowskaja, Bogojawlenskij prosp. 1, W. 10. *Pflanzen-Physiol.* ○

Sidorowa, Sophie Grigorjewna (1902), Vet.-Arzt u. Ass. der Pharm., Vet.-Inst. Pharm. Labor. Kasan (U.d.S.S.R.).

Siebeck, Richard (1883), o. Prof. der inneren Med. und Dir. der Med. Poliklinik. Bonn (Deutschland), Schedestr. 11. *Physiol. und klinische Pathol. der Atmung, der Nieren, des Wasserhaushaltes.* Ⓓ Heidelberg 1910.

Siebers, H. C., Konserv., Zool. Mus. Buitenzorg Java (Nederl.-O.-Indië). *Ornithol.* ○

Sieburg, Ernst (1885), ao. Prof. für Pharmacol. an der Univ. Hamburg 13 (Deutschland), Mittelweg 30. *Pharm., Toxikol, physiol. Chem.* Ⓓ Phil. 1913, med. 1913.

Sieden, Fritz (1875), Dr. phil., Vorsteher der Agrikulturchem. Versuchsstation der Landwirtschaftskammer für die Prov. Schleswig-Holstein. Kiel (Deutschland), Kronshagener Weg 3. *Agrikulturchem.* Ⓓ Kiel.

Siedlecki, Michel (1873), Dr. en Phil. Prof. ord. de Zool. à l'Univ. de Kraków (Polska), 6 R. St. Anny. *Protozoa, Cytol., Anat. et embryol. des vertébrés inférieurs, Structure et biol. des animaux à vol plané, biol. des animaux marins.* Ⓓ Kraków 1895.

Sieglbauer, Felix (1877), o. Univ.Prof., Dir. anat. Inst. Innsbruck (Deutsch-Österreich), Müllerstr. 59. *Anat. der Extremitäten.* Ⓓ Wien 1902.

Siegler, Edouard Horace (1888), Associate Entomol., United States Department of Agric. Washington, D.C. (U.S.A.), United Staates Bureau of Entomol. *Deciduous Fruit Insects; Insecticide.*

Siegler, Eugene A. (1891), Assoc. Pathol. Washington, D.C. (U.S.A.), Bureau Plant Industry. *Fruit disease.* Ⓓ Pennsylvania State Coll. 1918.

Sieglinger, John B. (1893), Associate Agronomist, Office of Cereal Investigations, Bureau of Plant Industry, U.S. Department of Agric. Woodward, Okla. (U.S.A.), Woodward Field Station. *Grain Sorghums and Broomcorn.* Ⓓ Kansas State Agric. Coll. 1915. Ⓟ Grain sorghums and broomcorn.

Siegmund, Herbert (1892), Dr. med., Prof., Oberarzt am Pathol. Inst. der Univ. Köln-Lindenthal (Deutschland), Franzstr. 60. *Pathol. Anat. der Mundhöhle, der Infektionskrankheiten. Biol. des Reticulo-Endothels.* Ⓓ München 1916.

Siegrist, Rudolf (1886), Dr. Aarau (Schweiz). *Pflanzengeographische Untersuchungen in Auenwäldern.* Ⓓ 1913.

Siemaszko, Vincent (Wincenty) (1887), Dr. Ph., Prof. of Plant Pathol., Coll. of Agric., Warsaw, Dir. of the Phytopath. Inst. Skierniewice (Polska). *Plant Pathol. Mycol.* Ⓓ Cracow (Poland). Ⓟ Fungi.

Siemens, Hermann Werner (1891), Dr. med., Priv.Doc., Ass. der Dermatol. Univ.-Poliklinik. München (Deutschland), Bavariaring 47. *Vererbungspathol., Hautkrankheiten, Konstitutionspathol.* Ⓓ München 1917.

Siemiradzki, Joseph (1858), Prof. Palaeont. Univ. Lwów (Pologne), Długosza 8. *Faune jurassique, Ammonites.* Ⓓ Dorpat 1885. Ⓟ Fossiles du silurien et devonien.

Siengalewicz, S., Dr.¿ Prof. Wilno (Polska), Univ., Inst. d. med. leg. *Histol. Embryol.* o

Siepi, Jule ,Dr., Conserv. Jardin zool. Marseille (France). o

Sierp, Hermann (1885), o. ö. Prof. an der Univ. München-Nymphenburg (Deutschland), Menzingerstr. 9. *Experim. Pflanzenphysiol.* Ⓓ Münster 1913.

Sievers, Fred J. (1880), Prof. of Soils and Head of Department of Soils, State Coll. of Washington. Pullman, Wash. (U.S.A.). *Soil physics, soil Chem., and Soil Biol.* Ⓓ Univ. of Wisconsin 1910.

Sifton, Harold Boyd (1889), Ass. Prof. of Botan. Toronto (Canada), Department of Botan., Univ. of Toronto. *Anat. of vascular plants, Seeds and seed germination.* Ⓓ Toronto 1923.

Sigalas, Clément Marie (1866), Doyen de la Fac. de Méd. de Bordeaux (France), Prof. de Physique méd. *Physique biol. et physiol.: Respiration, chaleur animale. Gaz du sang. Calorimétrie.*

Sigalas, Raymond Michel Marie (1892), Prof.agr., Fac. Méd., Dir. Station Biol. d'Arcachon. Bordeaux (France), 99, rue de St. Genès. Arcachon, Gironde (France), 1, rue du Prof. Jolyet. *Zool. marine et Parasit.* Ⓓ Bordeaux. Ⓟ Poissons et animaux marins du Golfe de Gascogne.

Sigerfoos, Charles P. (1865), Prof. of Zool., Univ. of Minnesota. Minneapolis,Minn. (U.S.A.). Ⓓ Johns Hopkins 1897. o

Siggers, Paul Victor, Plant Pathol., U. Fruit Comp. Port Lincoln (Australia). o

Sigmond, Josef, Dr., Prof. Ecole Polytechnique Tchèque, Chef d'Inst. d'Economie forestière. Praha-Vinohrady (C.S.R.), Havlíčkovy sady. *Dendrol. forestière.* o

Sigmund, Wilhelm, Dr., Prof. d. Biochem. Praha XII (C.S.R.), Šafaříkova 3. o

Sigot, André (1900), Ass. Biol. gén. Fac. Sc. Strasbourg (France), Inst. de Zool. et Biol. générale, Rue de l'univ. *Protist.: Flagellés.*

Sigrianski, Alexandre M., Chef des trav. Section Central de Défense des Plantes (O.Z.R.A.). Moscou (U.d.S.S.R.), Boiarski Dvor, chambre 59. o

Sihvonen, Kerttu (1893), Ass. an d. physiol. Inst. Helsinki (Finland). *Biochem.*

Šikl, Hermann (1888), M.U.Dr., Priv.Doc. der pathol. Anat. Tschechische Univ., I. Ass. am Pathol.-anat. Inst. Praha (C.S.R.). *Pathol. Histol. der Nieren und des Centralnervensystems.* Ⓓ Prag 1913.

Silber, Lew (1894), Priv.Doc. I. Univ., Ass. Mikrobiol. Inst. Moskau (U.d.S.S.R.), Pagodikskaja 10. *Mikrobiol.* Ⓓ Moskau 1919.

Silberschmidt, W., Prof. Dr. Zürich 7 (Schweiz), Zürichbergstr. 54. *Hygiene.* o

da Silva, Piraja, Prof. Fac. d. med. Bahia (Brasil). *Parasit.* o

Da Silva Baltasar Brites, Geraldino, o. Prof. Univ. Coïmbra (Portugal). *Histol.* o

Silva e Sousa, Alberto, Ass. Inst. d'Anat. Fac. méd. Porto (Portugal), R. Santa Catarina 54. Ⓓ Porto 1921.

Silva Tavares, Joaquín de, Colegio del Pasaje. La Guardia, Pontevedra (España). *Zoocecidias.* o

Silveira, Henri (1901). Lisbôa (Portugal), Inst. Rocha Cabral, 37, Rue de l'Ecole Politecnique. *Microbiol. des sols*₁

Silvestri, Filippo (1873), Prof. de zool. générale et agric. Portici, Napoli (Italia), R. Istituto superiore agrario. *Morphol. des Arthropodes; Myriapodes; Termitidae; Embiidae; Zoraptera; Thysomera; Protura; Insectes nuisibles.* Ⓓ Palermo 1896.

Siman, Karel, Dr., Dir. général des forêts d'Etat. Praha-Bubeneč (C.S.R.), Bendlova 433. *Economie forestière, dendrol.* o

Simanovič, Vladimir Fl., Prof. Univ. Perm (U.d. S.S.R.), Ul. Karl Marksa 26. *Pathol.* o

Simanovsky, Nicolas (1879), Dir. Station éxperim. de cultivateurs de tabac. Driazgin, Gouv. Woronesch (U.d.S.S.R.). *Microorganismen der Krimschen Halbinsel. Biol. von ,,Heliophrys variabilis".*

Šimec, Malka, Dr., Vorsteherin der Bact. Stat. Ljubljana (S.H.S.). o

Simić, Čedomir, Dr. Skoplje (S.H.S.), Zavod za Tropske Bolesti. *Parasit.* o

Simionescu, Jon (1873), Prof. à l'Univ. de Jassy (Roumanie), Str. Sărărie 104. *Palaeont.* Ⓓ Univ. de Vienne. Ⓟ Fossiles triasiques, jurassiques, tertiaires, devoniques de Roumanie.

Simkins, Cleveland Sylvester (1889), Associate Prof. Anat., Histol., Embryol., Univ. of Tennessee, Med. School. Memphis Tenn. (U.S.A.) 875 Monroe Ave. *Development of sex organs, of the human female. Germ cells in vertebrates. Origin of the sex-cells in the male Homo.* Ⓓ Harvard 1921.

Simm, Kazimierz Jan (1884), Dr. phil., Prof. de Ecole Agricult. Cieszyn (Polska). *Eponges d'eau doux, Econom. entomol.* Ⓓ Poznań 1920.

Simmons, Herman Georg (1866), Prof. am Landwirtsch. Inst. Ultuna, Upsala (Sverige). *Pflanzengeographie.* Ⓓ Lund 1906.

Simmons, James Stevens (1890), Instr. in Bact., Army Med. School, Washington, D.C. (U.S.A.). Ⓓ Univ. of Pennsylvania Med. School 1915.

Simmons, Joseph E. (1891). Corvallis, Ore. (U.S.A.). *Med. Bact., Food Bact.*

Simmons, Margaret P. (1894). Bact. for Earp-Thomas Cultures Corporation. Long Island City, N.Y. (U.S.A.), Nelson Ave. and Hill Street. *Bact. radicicola.* Ⓟ Humus and Agar Cultures of Bact. radicicola.

Simmons, Perez (1892), Associate Entomol., U. S. Dept. of Agric. Washington, D.C. (U.S.A.). *Household insects. Sitotroga cerealella Oliv.*

Simms, Henry Swain (1896), Associate of the Rockefeller Inst. for Med. Research, Head of Division of Chem., Department of Animal Pathol. Princeton, N.J. (U.S.A.), Rockefeller Inst. *Relation between physic. properties, such as ionization and polarity, and the chem. structure and behavior of biol. substances.* Ⓓ Coll. of Physicians and Surgeons, Columbia, 1922.

Simnitzky, Wladimir (1901), Histol. Labor. Univ. Kasan (U.d.S.S.R.), Wosnessenskaja Str. 21. *Avitaminose B. Mikrophysiol. der Drüsen mit innerer Sekretion.*

Simoes-Raposo, Luis (1898), Chargé de cours Pathol. gén. Fac. Méd. Inst. Rocha Cabral. Lisbôa (Portugal). *Immunité anti-cancéreuse. Greffe du cancer et cancer du goudron. Régénération des tissus (nerveux et osseux). Embryogénèse du système nerveux.* Ⓓ Lisbonne 1921.

Simon, Dr. phil. Zool. Mus. Berlin N 4 (Deutschland), Invalidenstr. 43. *Crustacea.* o

Simon, Ch., Prof. Ecole prép. de méd. et de pharm. Reims (France). *Pathol.* o

Simon, Charles E. (1866), Resident Lect. Baltimore, Md. (U.S.A.), 615 N. Wolfe Str. School of Hygien. *Filterable viruses.* Ⓓ Univ. of Maryland 1890. Ⓟ Filterable viruses; pathol. specimens from virus diseases.

Simon, Heinrich Joseph (1871), Prof., Dir. Pflanzenphysiol. Versuchsstat. Dresden-A. (Deutschland), Wintergartenstr. 19. *Wurzelbact. der Leguminosen. Stickstoffsammelnde Kleinwesen der Ackererde. Kohlensäuredüngung der Kulturpflanzen. Spezielle Ernährungsbedingungen der Lupinenarten. Stimulationsversuche an den Samen verschiedener Kulturpflanzen.* ⒹErlangen 1897. Ⓟ Azotogen.

Simon, Italo (1878), Prof. o., Dir. Ist. di Farm. Univ. Pavia (Italia), Palazzo Botta. *Assorbimento dei farmaci, azione coagulante degli alcooli sulle proteine, azione dei metalli pesanti sulle proteine, azione della glicerina, composti di guaiacolo con iodio, peristalsi intestinale, pressione osmotica degli organi per azione di svariati farmaci, atropina, funzione renale.*

Šimon, Jaroslav (1895). Adj. Sect. für Samenprüfung d. landw. Landesversuchsanst. Brno (C.S.R.), Květná 19.

Simon, L. J., Prof. Univ. Paris (France). *Phanérog.* o

Simon, Siegfried Veit (1877), Dr. phil., o. Prof. der Botan. an der Univ. Bonn (Deutschland), Botan. Inst. *Allgemeine Botan., Vererbung (Pflanzenphysiol.).* Ⓓ Leipzig 1903.

Simonena y Zabalegui, Antonio, Prof. Univ. Madrid (España). *Pathol.* o

Simonnet, Henri (1891). Paris 7 (France), 54, Av. Bosquet. *Nutrition animale dans leurs rapports avec les vitamines. Endocrinol. en général.* Ⓓ Paris 1925.

Simons, Hellmuth Carl Rudolf (1893), Dr. phil. nat., Laboratoriumsvorstand. Düsseldorf (Deutschland), Königsallee 62 III. *Nagana-Trypanosomen (morphol., Pathol., Biochem.). Malariaplasmodien des Menschen. Alterserscheinungen an freilebenden Nematoden. Saprophytische Oscillarien des Menschen und der Tiere. Parasitische Eugleneideen.* Ⓓ Freiburg i. B. 1917.

Simpson, George Eric (1889), Ass. Prof. of Physiol. Chem., Univ. of Pennsylvania. Philadelphia, Pa. (U.S.A.). *Water exchange of the body.* Ⓓ Yale 1920.

Simunovi, Stevan (1887), Chef Bact. Station. Mostar (S.H.S.). *Malaria.* Ⓓ Prag 1919. Ⓟ Pflanzen oder Tiere aus den Sumpfgebieten.

Sinden, James W. (1902), Instr., Cornell Univ. Ithaca, N.Y. (U.S.A.). *Plant Pathol.* Ⓓ Univ. of Kansas 1924.

Sinelnikow, Nicolaus (1885), Conserv. Anthrop. Mus. I. Univ. Moskau (U.d.S.S.R.), Mochowaja 11. *Kraniol. Anthropometrie.*

Singer, Ernst (1899), Dr. med., Ass. Hygien. Inst. Deutsche Univ. Praha VI (C.S.R.), Presslová 7. *Bact., Experim. Pathol. der Infektionskrankheiten. Reticuloendothel.* Ⓓ Deutsche Univ. Prag 1923.

Singh Pruthi, Hem (1897), Ass. Superint. Zool. Survey of India. Calcutta (India), Indian Mus. *Influence of the physical and chem. conditions of water on the bionomics of insects. Taxonomy of the Indian Jassidae (Rhynchota-insects).* Ⓓ Cambridge, England, 1924.

Sinizin, Dimitrij Theodorovič, Prof. Weißrussisches Inst. f. Landwirtschaft. Minsk, Weißrußl. (U.d.S.S.R.). *Zool.* o

Sinnott, Edmund W. (1888), Prof. of Botan. and Genetics, Connecticut Agric. Coll. Storrs, Conn. (U.S.A.). *Plant genetics, inheritance of shape; morphogenesis.* Ⓓ Harvard Univ. 1913.

Sinon, S., Dr., Prof. Pflanzenphysiol. Versuchsstat. Dresden (Deutschland), Stübelallee 2. o

Sinotô, Yoshito, Dr., Ass. Imper. Univ. Tôkyô (Japan), Botan. Inst., Fac. of Sc., Koishikawa-Ku. *Genetics and Cytol.*

Sinova, Elena Stephurofna (1874), Ass. Hauptbotan. Garten. Leningrad (U.d.S.S.R.). *Algae Marinae.* Ⓓ 1908.

Siouzev, Paul V., Chef des trav. Botan. Univ. Perm (U.d.S.S.R.), Oural Septentrional. *Champignons de l'Oural.* o

Sipe, Frank Perry (1888), Instr. in Botan. Corvallis, Ore. (U.S.A.). *Lichens.* Ⓟ Lichens of the Pacific Northwest.

Siracusa, Vittorio (1896), Ass. alla Cattedra di Med. Legale, Univ. di Messina (Italia). *Traumatol. Sierol. (Individualità del sangue). Istol.* Ⓓ Messina 1921.

Sirks, Marius Jacob (1889), Research geneticist at the State Inst. for Plantbreeding at the Agric. Coll. Wageningen (Holland), Otto van Gelreweg 2. *Genetics (Selfincompatibility, quantitative inheritance, agric. crops).* Ⓓ Utrecht 1915.

Sirotin, E. E. (1888), Prof. Univ. Minsk (U.d.S.S.R.), Sovetskaja 33. *Ionic theory of excitation: theory of the adaptation of eye, excitation of cells in an electrical field, the surface tention of electrolytes and nonelectrolytes.* Ⓓ Moscow 1922.

Sirotinina, Olga N. (1892), Leiter Chem. Abt. Bact. Stadtlabor. Saratow (U.d.S.S.R.), Biol. Stat. Chem. *Nahrungsanalyse. Rhynchota.* Ⓓ Moskau 1916. Ⓟ Rhynchota, spez. Corixidae.

Sirotinine, Nicolaus (1896), Ass. Inst. physiol. pathol. II. Univ., collabor. Inst. clinique de la Diagnostic fonctionelle et de l'Inst. de l'encéphale. Moskau (U.d.S.S.R.). *L'immunité.*

Sirotkin, Aleksander Kor., Doc. II. Univ. Moskau (U.d.S.S.R.). *Anat.* o

Sirvent, Louis (1876), Ass. au Mus. océanographique de Monaco (Principauté).

Sitton, Benjamin Gaillard (1897), Ass. in Horticulture, Michigan State Coll. of Agric. and Applied Sc. East Lansing, Mich. (U.S.A.), Department of Horticulture. *Black Walnuts. Pomol.*

Sitowski, Ludwik (1880), Dr. phil., Prof. der Zool. und Entomol. an der Univ. Poznań (Polska), Sołacz Kujawska 15. *Experim., systematische und angewandte Entomol. Lepidopteren und parasitische Hymenopteren.* Ⓓ Kraków 1907.

Sittig, Walter (1896), Dr. phil., Wissenschaftlicher Mitarbeiter der Deutschen Gesellschaft für Schädlingsbekämpfung, Leiter der Landwirtschaftlichen Beratungsstelle. Frankfurt a. M.-West 13 (Deutschland), Große Seestr. 33 III. *Pflanzenschutz und Schädlingsbekämpfung. Pflanzliche und tierische Schädlinge. Rostkrankheiten.* Ⓓ Gießen 1921.

Sivickis, P. B., Prof. Zool. Head Dept. of Zool. Univ. Philippines. Manila (Philippine Islands). *Physiol. Zool. Marine Zool.* Ⓓ Univ. of Chicago 1922. Ⓟ Lingula, Corals, Mollusks, Decapod crustaceans.

Sjemjonow, Gregor (1890), Oberass. histol. Inst. med. Fac. Univ. Taschkent (U.d.S.S.R.). *Histol. und Parasit.*

Sjöqvist, J. A., Dr., Prof. Karolinska Inst. Stockholm (Sverige). *Pharmacol.* o

Sjöstedt, Yngve (1866), Prof. am Naturhistorischen Reichsmus. u. Kustos der entomol. Abt. Stockholm 50 (Schweden). *Termit. afr., Orthoptera, Odonata, Aves afr.* Ⓓ Uppsala.

Sjövall, Per Gustaf Einar (1879), Prof. der pathol. Anat. und gerichtlichen Med. an der Univ. zu Lund (Schweden), Svanegatan 7. *Pathol. Anat.* Ⓓ Lund 1906.

Sjollema, Bouwe (1868), Prof., Dir. Labor. für Vet. med. Chem. Utrecht (Holland), Maliebaan 4. *Physiol. Chem. und pathol. Chem.* Ⓓ Jena 1893.

Sjusew, Paul W., Doc. a. d. Univ. Perm (U.d.S.S.R.), Trudowaja 12, Provinz. Mus. *Botan., Biol., Systematik.* o

Skaar, Torvald (1898), Ass. Oslo (Norwegen), Erl. Skjalgssønsgate 26. *Stoffwechsel von Ca und P bei experim. Rachitis bei Hunden.*

Skadowsky, Sergius (1886), Ass. Inst. experim. Biol., Vorstand Hydrophysiol. Station des Inst. (U.d.S.S.R.), Zwenigorod. Moskau (U.d.S.S.R.), Woronkowo polje 6. *Permeabilität, Hydrophysiol. Elektrolytwirkungen auf die Copepodennauplien.* Ⓓ Moskau 1912.

Skalińska, Marja (1890), Dr. phil., Doc. de l'Univ. de Cracovie, Prof. de Botan. et Dir. du Labor. Botan.

à l'Univ. Libre de Pologne à Varsovie, Chargée de cours à l'Ecole Supérieure d'Agric. à Varsovie (Pologne), 13, rue Bagatela. *Génétique: Races polymorphes, Mendélisme. Cytol. des plantes supérieures.* ⓓ Berne 1912.

Skarić, Josip (1887), Chef der Bact. Anstalt. Split (S.H.S.). *Bact. und Serol.* ⓓ 1913.

Skarzynska, Marie (1890), Ass. Univ. Inst. de Physiol. Warszawa (Pologne), Krak. Przed. 26/28. *Physiol. normale, Digestion, Serol., Vitamine.* ⓓ Kiew 1912, Varsovie 1925.

Skaskin, Fedor (1900), Ass. Botan. Labor. der Donschen Hochsch. Landwirtsch. u. Melioration. Novotscherkassk (U.d.S.S.R.), Potschtovaja 65. *Assimilation der Kohlensäure. Physiol. der Keimung der Chlamydosporen der Brandpilze des Getreides.*

Skeats, E. W., Prof. Univ. Melbourne (Australia). *Palaeont.* o

Skeen, John Robsin (1899), Instr. in Botan., Univ. of Pennsylvania. Philadelphia, Pa. (U.S.A.). *Phytophysiol. of Soil.* ⓓ Univ. of Pennsylvania 1926.

Skene, Macgregor, Senior Lect. Botan. Univ. Bristol (England), 36 Lawrence Grove Henleaze. *Plant Physiol. Biol. of Flowering Plants.* ⓟ Aberdeen 1906.

Skijabin, Konstantin, Prof. Tierärztl. Hochsch. Leningrad (U.d.S.S.R.), Černigovskaja 5. *Parasit.* o

Skinner, Joshua John (1882), Biochem., U.S. Dept. of Agric. Washington, D.C. (U.S.A.). *Plant, soil, and fertilizer Chem.* ⓓ Ph.D. American Univ. 1917.

Sklawunos, Constantin (1895), o. Prof. für Holzmeßkunde und Forststatik. Forsthochsch. Athen (Griechenland), Schisti Petra. ⓓ München 1923.

Sklower, Alfred (1901), Zool. Inst. u. Mus. Univ. Königsberg i. Pr. (Deutschland). *Entwicklungsgeschichte, Morphol. u. Physiol. der inkretorischen Organe der Wirbeltiere, Morphol. u. Physiol. der Metamorphose der Wirbeltiere.* ⓓ Phil. Königsberg 1924, med. 1927.

Skoblionok, Sulamite (1894), Histol. Labor. Staatl. Inst. für Med. Wissenschaften. Leningrad (U.d. S.S.R.), Wilensky per. 3 u. 4. *Histogenese der Zwischenwirbelscheiben bei Säugetieren. Respiratorisches Lungenepithel.* ⓓ Leningrad 1924.

Skoda, Karl (1872), o. Prof. für Anat. an der Tierärztlichen Hochsch. in Wien III (Deutsch-Österreich), Linke Bahngasse 11. *Anat. der Haustiere.* ⓓ Wien 1900.

Skoqsberg, Tage (1887), Instr. at Stanford Univ. Pacific Grove, Cal. (U.S.A.), Hopkins Marine Stat. *Taxonomy and compar. Morphol., Ostracodes, Dinoflagellata.* ⓓ Uppsala 1920.

Skorodumowa, Alexandra (1895), Lehrass. Wologda (U.d.S.S.R.), Milchwirtschaftl. Inst. *Microbiol. der Käse. Milchsäurebakterien.*

Skottsberg, Carl Johan Fredrik (1880), Prof., Dir. Botan. Garten. Göteborg (Sverige). *Algol., Systematik u. Geographie der Phanerogamen (spez. Pazifik, Antarktis, Südamerika).* ⓓ Upsala 1907. ⓓ Herbarpflanzen, Samen und lebende Pflanzen.

Skovsted, Aa. Thorsen (1903), Ass. Labor. für Erblichkeitslehre Kgl. Landw. Hochsch. Regensen København (Danmark). *Thelephoraceae. Erblichkeitslehre.*

Skowron, Stanislaw (1900), Ass. Univ. Biol. Univ. Cracow (Poland), Biol. Inst., Jana 20. *Cytol., Genetics. Bioluminescence.* ⓓ Cracow 1923. ⓟ Axolotl.

Skramlík, Emil, Ritter vòn (1886), ao.Univ.-Prof., 1. Ass. am Physiol. Inst. der Univ. Freiburg i. Br. (Deutschland), Hebelstr. 33/1. *Physiol. der inneren Organe (Herz, Leber und Milz), der Sinneswerkzeuge, Orientierung im Raume.* ⓓ Prag 1911.

Skriptshinsky, Georg (1895), Laborant. Leningrad (U.d.S.S R.), Chalturinstr. 23. *Angewandte Entomol. Biol. parasit. Schlupfwespen. Systematik Fam. Proctotrypidae.* ⓓ Leningrad.

Skrjabin, K. I., Prof., Dir. Inst. d. experim. Vet. Moskau (U.d.S.S.R.), Pimenovski per. 5. *Zool. Embryol.* o

Skubiszewski, Louis (1886), Dr. med., Prof. Univ. anat. pathol. Poznań (Pologne), 35, rue Dąbrowskiego. *Myositis. Dermatomyositis interstitialis chronica. Microphysiol. de l'hypophyse.* ⓓ Kiev 1914.

Skuja, Andrejs, Dr., Priv.Doc. Univ. Riga (Latvija), Brīvības ielā 23. *Pathol.* o

Skuja, Heinrichs (1892), Hilfsass. Botan. Labor. Univ. Riga (Latvija), Valdemāra ielā 69a, dz. 21. *Algenflora Lettlands. Rot- und Braunalgen des Süßwassers.*

Skupieński, François-Xavier (1888), Dr. ès Sc., Ass. de la chaire de Botan. générale de l'Univ., Doc. à l'Ecole Polytechnique de Varsovie (Pologne), Krak.-Przedm. 26/28. *Biol., cytol. et systématique des Myxomycètes.* ⓓ Paris 1920. ⓟ Myxomycètes de Pologne.

Skutch, Alexander F. (1904), Ass. in Botan., Johns Hopkins Univ. Baltimore, Md. (U.S.A.). *Ecol. (Botan.), succession following forest-fires, littoral plants. Experim. Morphol. of Plants, stimuli and plant hormones. Anat. of Monocotyledons, Musa.*

Skvortzow, B. W., Manchuria Research Society. Harbin (China). *Botan.* o

Skwarza, Elisabeth (1886), Dr. phil., wissenschaftliche Hilfsarbeiterin in der Bernsteinabteilung (Bernsteininclusen) des Geol. Inst. der Univ. Königsberg, Pr. (Deutschland), Mozartstr. 10. *Freilebende Nematoden und Ameisen sowie Hochmoorbiocönotik.* ⓓ Königsberg, Pr. 1921. ⓟ Ameisen. Freilebende Nematoden.

Skwirsky, P. (1883), Priv.Doc. Leningrad (U.d. S.S.R.), Prospekt Wolodarsky 49. *Serol.* ⓓ Berlin 1910.

Skwortzoff, Anatolij (1883), Doc. Univ. Nijny-Nowgorod (U.d.S.S.R.), Tichonowskaja Str. 58. *Helminthol. Trematoden. Diclibothrium.* ⓓ Moskau. ⓟ *Parasit.* Würmer der Fische und Amphibien.

Skwortzoff, Ekaterina (1881), Doc. Univ. Nijny-Nowgorod (U.d.S.S.R.), Tichonowskajy Str. 58. *Zool. Hydrobiol.*

Skworzoff, Wladislaw (1879), Prof. II. Univ. Moskau (U.d.S.S.R.), Pogodinskaja 6. *Pharmacol. und Biochem.* ⓓ Moskau 1909.

Skworzow, Michael (1876), Arzt. Ass. d. pathol.-anat. Inst. der II. Staatsuniv. Moskau (U.d.S.S.R.), Nikitsy Buleward 5. *Pathol. Anat. der Infektionskrankheiten des Kindesalters.* ⓓ Moskau 1909.

Sladen, Alexander George Lambart. London, S.W. 1 (England), Kingswood House, The Lee-Bucks. *Ornithol.* o

Slagg, Charles Mervyn (1890), Chief, Tobacco Division, Dominion Dept. of Agric. Ottawa (Canada) Central Experim. Farm. *Pathol. of the tobacco platn,* ⓓ Univ. of Wisconsin 1919. ⓟ Tobacco seeds. tobacco leaves (cured), tobacco disease specimens.

Slagsvold, Abt.Dir. Vet.-Inst. Oslo (Norge). *Pathol. Anat.* o

Slate, George Lewis (1899), Ass. Geneva, N.Y. (U.S.A.), N.Y. Agric. Experim. Station. *Small fruits.* ⓓ Bussey Inst. 1926.

Slate, William L. (1884), Dir. Conn. Agric. Experim. Stat. New Haven, Conn. (U.S.A.), 123 Huntington St. o

Slater, William Kershaw (1893), Beit Memorial Med. Research Fellow. London W. C. 1 (England), Univ. Coll., Gower St. *Biochem.* ⓓ Manchester 1914.

Slatineano, Ali, Prof. Univ. Jassy (Rumänien). *Bact.* o

Slăvoacă, Titu, Chef de Travaux de l'Inst. Pasteur. Cluj (Romania), Str. Pasteur 6. *Bact., Sérol., Parasit.* o

Slemmons, Wilbert S. (1896), District Superintendent, Carnation Milk Products Co. Oconomowoc, Wis. (U.S.A.). *Dairy Bact., Thermal death time studies with bact. spores. Bact. flora of bovine endo-metritis.*

van Slogteren, E., Dr., Prof. Lisse (Holland). *Phytopath. der Zwiebelgewächse.* o

Slonaker, James Rollin (1866), Associate Prof. Physiol., Stanford Univ., Cal. Palo Alto, Cal. (U.S.A.), 334 Kingsley Avenue. *Eye. Diet, Activity and Sex Problems of the Albino Rat.* ℹ Clark, Worcester, Mass. 1896. ℗ *Vertebrate Eye Sections.*

Slonevskij, Sigizmund I., Doc. I. Univ. Moskau (U.d.S.S.R.). *Hygiene.* o

Stonimski, Piotr (1893), Dr. phil., Oberass. an der Univ. Warszawa (Polska), Chałubińskiego 5. *Vitale Färbungen, Peroxydasenreaktionen, Einfluß der Thyroidea auf Wirbellose, Ekol. und Systematik der Rotatoria.* ℹ Warszawa 1922.

van Slooten, Dirk Fok (1891), Dr. Sc., Ass. at the Herbarium of the Botan. Gardens at Buitenzorg, Java (Nederl.-O.-Indië). *Taxonomy of the Angiospermae. Combretaceae, Flacourtiaceae, Stylidiaceae. Dipterocarpaceae.* ℹ Utrecht, Holland, 1919.

Sluiter, Carel Philip (1854), Prof. emeritus der Zool. zu Amsterdam. Eerbeek, Geld. (Holland). *Tunicaten, Gephyreen, Echinodermen. Tierparasiten der Säugetiere.* ℹ Amsterdam 1878.

Sluzkij, Nikolaj F., Prof. Botan. I. Univ. Moskau (U.d.S.S.R.), Ostoshenka 4, W. 4. *Botan. Phytopath. der Holzpflanzen.* o

van Slyke, Donald D. (1883), Member of Rockefeller Inst. for Med. Research; Chem. of the Hospital of the Rockefeller Inst. New York (U.S.A.), 66 St. and Ave A. *Chem. of proteins and of the blood.* ℹ Univ. of Michigan, Ann Arbor, Mich., 1905.

van Slyke, Lucius S. (1859), Chief Chem. in Research, New York Agric. Experim. Station. Geneva, N.Y. (U.S.A.). *Chem. of Milk and Milk Products.* ℹ 1882.

Small, James (1889), Prof. of Botan. Queen's Univ. Belfast (Northern Ireland). *Compositae,* *Pharm. (Botan.), Tropisms. Hydrogen ion concentration in plants and its effects.* ℹ Sc. London 1919.

Small, James Craig (1888), Bact., Philadelphia General Hospital. Ass. Prof. of Bact., Graduate Sch. Med., Univ. of Pennsylvania. Philadelphia, Pa. (U.S.A.), 2021 Pine Street. ℹ Univ. of Illinois, Chicago, 1917.

Smallwood, William Martin (1873), Prof. compar. Anat. and Head of Zool. Dept. Liberal Arts Coll., Syracuse Univ. Syracuse, N.Y. (U.S.A.), 525 Euclid Ave. *Neurol.* ℹ Syracuse Univ. 1896.

Smarods, Julius (1884), Mykol. Pflanzenschutzinst. Riga (Latvija), Kirchenstr. 4a, W. 10. *Kulturpflanzenkrankheiten (Getreide u. Beerensträucher).* ℗ Parasit. Pilze.

Smart, W. A. M. (1879), Demonstrator of Physiol. and Pharmacol. The London Hospital. London E. 1 (England). *Mathematical Aspects of Physiol. and Biol. Sc.* ℹ London Univ 1910.

Smedley, Norman (1900), Ass. Curator, Raffles Mus. Singapore (Fed. Mal. States). *Zool. of Malaysia.* ℹ Cambridge 1926.

Smellie, I. M., Lect. Univ. Birmingham (England). *Pharmacol.* o

Šmelov, K. A., Prof. N. G. Černyševskij Univ. Saratov (U.d.S.S.R.), Ploščad imeni Lenina. *Pharmacol.* o

Smetana, H., Dr., Lect. Dept. of Pathol., Union Med. Coll. Peking (China). o

Smetánka, Franz, Dr., o. Prof. der Physiol., Dir. Physiol. Inst. Šalata. Zagreb (S.H.S.). *Muskel- u. Gefäßphysiol.* o

Smetanin, N. I., Doc. Univ. Nizny-Novgorod (U.d.S.S.R.), Sovetskaja Ploščad 8. *Pathol., Histol.* o

Šmidt, Eduard Vik., Doc. I. Univ. Moskau (U.d. S.S.R.). *Histol.* o

Smillie, Wilson George (1886), Ass. Dir. for the U.S., International Health Board, Rockefeller Foundation. New York City (U.S.A.), 61 Broadway. *Parasit., Hygiene.* ℹ Harvard Med. School, Boston, Mass., U.S.A. 1912. ℗ *Intestinal parasit. of man.*

Smirnov, Eugen Serg. (1898), Ass., Lect. f. Ökol. Zool. Mus. Univ., Mitarbeiter Timirjaseff-Inst. Moskau (U.d.S.S.R.). *Abstammungslehre. Systematik u. Morphol. der Syrphidae (Diptera); experim. ökol. Studien an Calliphora und Drosophila (Dipt).; mathematische Studien am Dipterenflügel.* ℹ Moskau 1920. ℗ *Insekten (Diptera, spez. Syrphidae).*

Smirnova, Olga (1892), Ass. Botan. Garten (Herbarium). Leningrad (U.d.S.S.R.), Pessotschnaia $^1/_2$. *Wichtige Arzneikräuter des Orenburgschen Gouvernements.*

Smirnova, Valentine (1886), Labor. zool. Acad. des Sc. Leningrad (U.d.S.S.R.), 17, rue Torgovaya. *Zool. experim.: régénération et culture des tissus in vitro.* ℹ Petersburg 1914.

Smirnova, Zoë Nicolajewna (1898), Ass. Landwirtsch. Inst. Botan. Labor. Leningrad (U.d. S.S.R.), Bolschaja Zelenina 31, qu. 5. *Systematik und Oecol. der Moose und der Vascularpflanzen. Flora des Gouv. Leningrad und des Urals. Geobotan.*

Smirnow, Alexander I., Prof. d. Kuban. Landwirtsch. Inst. Krasnodar (U.d.S.S.R.), Inst. f. Tabakzucht. *Pflanzenphysiol.* o

Smirnow, Peter (1878), Prof. Gorsky Pädagogisches Inst. Wladikawkas (U.d.S.S.R.). *Pflanzenphysiol., Microbiol.*

Smirnow, Valentin I., Prof. a. d. Irkutsker Staatl. Univ. Irkutsk (U.d.S.S.R.). *Botan. Geobotan.* o

Smit, Jan (1885), Ass. am Hygien. Inst. der Univ. Amsterdam; Priv.Doc. der Allgemeinen Mikrobiol. Amsterdam (Holland), Reynier Vinkeleskade 41. *Sarcinen des Bodens und der Getreidearten. Neuere Methoden d. Abwasserreinigung (Aktivierter Schlamm).* ℹ Amsterdam 1913.

Smith, Alfred (1888), Ass. Prof. of Soil Technol. Davis, Cal. (U.S.A.), Branch of the Coll. of Agric. *Soil physics, survey, crop adaptation.* ℹ Univ. of Wisconsin.

Smith, A. Nimmo, Lect. School of Med. R. Coll. Edinburgh (Scotland). *Bact.* o

Smith, Arthur Henry (1893), Ass. Prof. Physiol. Chem. Yale Univ. New Haven, Conn. (U.S.A.), 333 Cedar Str. *Physiol. Chem., Nutrition.* ℹ Yale Univ. 1920.

Smith, Bertram Garner (1876), Associate Prof. of Anat., New York Univ. New York (U.S.A.), 338 East 26 Street. *Amphibians and fishes; life histories of urodeles; embryol. of Cryptobranchus alleghaniensis; history of the nucleus during cleavage of the egg of Cryptobranchus.* ℹ Columbia Univ 1912. ℗ *Eggs, embryos and larvae of Cryptobranchus alleghaniensis.*

Smith, Charles E. (1887), Ass. Entomol., U.S. Bureau, Truck Crop Insects. Baton Rouge, La. (U.S.A.), Louisiana Experim. Station. *Insects affecting truck crops.*

Smith, Christianna (1893), Associate Prof. of Zool. South Hadley, Mass. (U.S.A.), Mount Holyoke Coll. *Histol., Med. Zool.* ℹ Cornell Univ. 1923.

Smith, David Tillerson (1898), Bact. in Charge of the Labor. of the New York State Hospital for Incipient Tuberculosis. Ray Brook, N.Y. (U.S.A.). *Spirochetal and fungus infections of the lung.* ℹ Johns Hopkins 1922. ℗ *Cultures of fungi isolated from sputum of Cases of pulmonary disease.*

Smith, Dorothea Egleston (1890), Research Ass. Bryn Mawr, Pa. (U.S.A.), Dalton Hall, Bryn Mawr Coll. *Biochem., Bact., Immunol.* ℹ Radcliffe Coll. 1926.

Smith, Erwin F. (1854), Pathol., in charge Labor. of Plant Pathol., U.S. Dept. of Agric. Washington, D.Col. (U.S.A.), 1474 Belmont St. *Bacterial diseases of plants.* ℹ Sc. Michigan 1889, Wisconsin 1914, L.L. Michigan 1924. o

Smith, Folmer Dyrlund (1885), Vorsteher des Labor. für Elektrokultur, Norges Landbrukshøiskole. Aas (Norge). *Einfluß der Elektrizität und des Lichtes auf die Pflanzen. Korrelationen zwischen Irritationsphänomenen und Stoffproduktion bei Pflanzen.*

Smith, Frances G., Associate Prof. of Botan., Smith Coll. Northampton, Mass. (U.S.A.). *Morphol. of cycads, ecol.* o

Smith, Frank (1857), Prof. emer. of Zool., Urbana, Ill. (U.S.A.), Univ. of Illinois. *Systematic and morphol. of Oligochaeta and of fresh-water sponges. Migration of birds, gastrulation of Aurelia flavidula; variation in the sacral ribs of Necturus.* ⓓ Sc. hon. Hillsdale Coll. 1923. ⓟ Oligochaeta, fresh-water sponges.
Smith, Frank Campbell (1898), Demonstrator in Physiol. London E. 1 (England), London Hospital. *Ultraviolet absorption, spectra of certain biol. fluids.* ⓓ London Univ. 1922.
Smith, George Hathorn (1885), Prof. of Immunol., Yale Univ., School of Med. New Haven, Conn. (U.S.A.). *Bact., Serol., Immunol.* ⓓ Brown Univ., Providence, R.I., 1914.
Smith, George Hume (1896), Instr. in Botan., Univ. of Illinois. Urbana, Ill. (U.S.A.), 308 Natural History Building. *Anat. and histol., vascular anat. of flowers.* ⓓ Cornell Univ., Ithaca, N.Y., 1923.
Smith, Gilbert Morgan (1885), Prof. of Botan. Stanford Univ. California (U.S.A.). *Freshwater Algae, Morphol. of Angiosperms.* ⓓ Wisconsin 1913.
Smith, Grafton Elliot (1871) Prof. of Anat., Univ. Coll. London W.C. I (England), 24 Gordon Mansions. *Compar. Anat. of the brain, fossil remains of man.* ⓓ Sydney 1892, Cambridge 1898.
Smith, Harry Scott (1883), Associate Prof. of Entomol., Univ. of California. Riverside, Cal. (U.S.A.), Citrus Experim. Station. *Biol. control of Insect Pests. Parasit. insects, especially Hymenoptera.* ⓓ Univ. of Nebraska 1908.
Smith, Herbert Dwight (1894), Junior Entomol., Bureau of Entomol., U. S. Dept. of Agric. Carlisle, Pa. (U.S.A.), 337 Franklin Street. *Economic Entomol., insect pests of wheat with their parasites.* ⓓ Amherst, Mass. 1917.
Smith, Hugh McCormick (1865), Adviser in Fisheries to H. Siamese Maj. Govern., Dir. Siamese Dept. of Fisheries. Bangkok (Siam); *Ichthyol., economic fisheries and pisciculture.* ⓓ M. Georgetown Univ. 1888, L.L. Dickenson 1908.
Smith, Isabel S., Prof. of Biol., Illinois Coll. Jacksonville, Ill. (U.S.A.), 1120 W. College Ave. *Vascular anat.* ○
Smith, James Lindsay Salmond, Lect., Forest Botan., Indian and Colonial Forest Tress. Edinburgh (Scotland), Botan. Department University.
Smith, J. Henderson, Rothamsted Experim. Stat. Harpenden, Herts. (England). *Mycol.* ○
Smith, Johannes Jacobus (1867). Utrecht (Holland), Willem Barentzstr. 83. *Systematik: Orchideae, Triuridaceae, Burmanniaceae, Ericaceae, Epacridaceae, Euphorbiaceae, Teratol.* ⓓ Utrecht 1910.
Smith, Joseph Russell (1874), Prof. economic geography Columbia Univ. Swarthmore, Pa. (U.S.A.), 319 Cedar Lane. ○
Smith, K. A. H., Doc., Dr. Uppsala (Sverige). *Pflanzenbiol.* ○
Smith, Kenneth Manley (1892), Senior Lect. in Agric. Entomol., The Univ. Manchester (England), Dept. of Agric. Entomol. *Insects and their connection with diseases of agric. crops, virus diseases of potatoes.* ⓓ Manchester 1926.
Smith, Lawrence Weld (1895), Ass. Prof. pathol. Harvard Univ. Med. School. Boston, Mass. (U.S.A.). *Whooping-cough, bact. and pathol. Thyroid gland pathol., histo-pathol.* ⓓ Harvard-Univ. 1920.
Smith, Loren Bartlett (1890), Entomol. in charge, Japanese Bulb Project; Dir. of Research U.S. Dept. of Agric. Riverton, N.J. (U.S.A.).
Smith, Louie Henrie (1872), Soil Survey, Dept. of Agronomy Univ. of Illinois. Urbana, Ill. (U.S.A.). ⓓ Halle 1907.
Smith, Marion R. (1894), Ass. Entomol. and Ant Specialist, Mississippi State Plant Board. A. and M. College, Miss. (U.S.A.). *Ants, Biol., Taxonomy and control.* ⓓ Univ. of Illinois, Urbana 1927. ⓟ Ants.

Smith, May (1879), Investigator to the industrial fatigue research board. London (England), 15 York Buildings, Adelphi. *Fatigue. Experim. Psychol.* ⓓ Manchester 1903.
Smith, Nathan Raymond (1888), Associate Bact. Washington (U.S.A.), U.S. Dept. of Agric. *Soil bact., relations between crop growth and the soil flora; decomposition and efficiency of organic manures; soil bact., cellulose destroying and nitrogen fixing bacteria; formation of mineral deposits by bacteria.* ⓓ Univ. of Vermont, Burlington, Vt., 1911.
Smith, Ora (1900), Instr. Ames, Ia. (U.S.A.), Dept. of Horticulture, Iowa State Coll. *Plant histol., embryol. and anat.* ⓓ Iowa State Coll. 1924.
Smith, Philip E. (1884), Associate Prof. of Anat. Stanford Univ. California (U.S.A.). *Endocrine glands.* ⓓ Cornell Univ. 1912.
Smith, Ralph Eliot (1874). Berkeley, Cal. (U.S.A.), 2721 Hillegass Ave. *Plant pathol.* ○
Smith, Ralph Henry (1888), Ass. Entomol. Citrus Experim. Stat. Univ. of California. Riverside, Cal. (U.S.A.). *Carpocapsa pomonella. Spraying of citrus trees to control citrus scale insects.* ⓓ Univ. of California 1925.
Smith, Ralph Waldo (1877), Associate Agronomist. Dickinson, N.D. (U.S.A.), Substation. *Experim. with cereal crops. Cereal varieties by crossing and selection. Seeding, methods of winter protection.*
Smith, Robert O. (1903), Jr. Aquatic Biol., U.S. Bureau of Fisheries. Beaufort, N.C. (U.S.A.). *Fisheries, ecol. factors which affect fisheries; Jumping or gray mullet, Mugilcephalus.*
Smith, Roger Cletus (1888), Prof. of Entomol. and Ass. Entomol. in the Experim. Station, Kansas State Agric. Coll. Manhattan, Kan. (U.S.A.). *Relation of climate upon Insect life. Alfalfa Insects.* ⓓ Cornell Univ., Ithaca, N.Y., 1917. ⓟ Neuroptera and Mecoptera.
Smith, Septima Cecilia (1891), Fellow, The Johns Hopkins School of Hygiene and Public Health, Baltimore, Md. (U.S.A.), 615 N. Wolfe St. *Parasit. amoebae.*
Smith, Theobald (1859), Dir., Dept. of Animal Pathol., Rockefeller Inst. for Med. Research, Princeton, N.J. (U.S.A.). *Pathol. and Bact.* ⓓ 1883.
Smith, Wendal D. Ass. Prof. of Histol. and Embryol. Norman, Okla. (U.S.A.), Fac. Exchange. *Mitochondria of adrenal cortex.* ⓓ Iowa State Coll. 1926.
Smith, William, Curator of the Marine Biol. Stat. Port Erin, Isle of Man (England). *Sea Fisheries, Biol. of Fishes.* ○
Smith, William Wright (1875), Regius Prof. of Botany, Univ. of Edinburgh and Regius Keeper, Royal Botan. Garden, Edinburgh (England). *Systematic Botan. of Central and Eastern Asia.* ⓓ Edinburgh.
Smith Medina, Guillermo (1897). Cadiz (España), Manuel Rances No. 24. *Entomol. (Coleoptera).* ⓟ Coleoptera, Lepidoptera (Chrysalidae), Reptilia.
Smits van Burgst, Constant August Louis (1858), Entomol. de l'Inst. phytopath. à Wageningen. Beek chèz Breda (Holland). *Ichneumonol.*
Smitt, Anton (1883), Abteilungsvorsteher an der Forstlichen Versuchsanstalt für West-Norwegen. Bergen (Norge), Bergens Mus.
Smolák, Jaroslav, Dr., Priv.Doc. f. Phytopath. Landwirtsch. Hochsch. Brno (C.S.R.), Černa pole N. 102. *Phytopath.* ○
Smorodincev, Ivan Andr., Prof. II. Univ. Moskau (U.d.S.S.R.). *Biochem.* ○
Smotlacha, František, Prof. Dr. Praha-Vinohrady (C.S.R.), Kolínská 18. *Mycol.* ○
Smreczyński, Stanisław (1899), Dr. phil., Ass. à l'Univ. de Cracovie. Kraków (Polska), Smoleńska 23. *Embryogénie des insectes, Systématique des paléarct. Curculionidae (Coleoptera).* ⓓ Cracovie 1924. ⓟ Curculionides Paléarctiques.

Smreczyński, Stanisław (1872). Kraków (Polska), Smoleńska-Gasse 23. *Orthopteren und Hemipteren der paläarktischen Fauna.* Ⓟ Hemiptera der paläarktischen Fauna.

Smyth, Henry Field (1875), Ass. Prof. of Industrial Hygiene, Univ. of Pennsylvania Philadelphia, Pa. (U.S.A.). Ⓓ Philadelphia 1897.

Snapp, Oliver I. (1895), In charge U.S. Peach Insect Labor. Fort Valley, Ga. (U.S.A.), Box 445. *Insects attacking peaches; curculio, peach borer, San Jose scale, Oriental peach moth.* Ⓓ Mississippi A. and M. Coll., Miss., 1925. Ⓟ Insects of the Southern U.S.

Snell, Karl (1881), Dr. phil., Labor.-Vorsteher an der Biol. Reichsanstalt für Land- und Forstwirtschaft. Berlin-Dahlem (Deutschland). *Morphol. und Physiol. der Kulturpflanzen, der Kartoffel.* Ⓓ München 1907.

Snell, Walter Henry (1889), Ass. Prof. of Botan., Brown Univ., Ass. Forest Pathol., N.Y. State Conservation Commission. Providence, R.I. (U.S.A.). *Timber Decay and Wood Preservation.* Ⓓ Univ. of Wisconsin, Madison 1920. Ⓟ Wood destroying fungi.

Snider, Howard I. (1884), Ass. Prof. in Field Research in the Univ. of Illinois. Urbana, Ill. (U.S.A.), Coll. of Agric. Ⓓ Univ. of Illinois 1921.

Snikers, Pēters, Dr., Prof. Univ. Rīga (Latvija), Brīvības bulv. 3, dz. 8. *Pathol.* o

Snodgrass, Robert Evans (1875), Entomol. U. S. Bureau of Entomol. Washington, D.C. (U.S.A.). *Morphol. of insects.* Ⓓ Stanford Univ. 1901.

Baron Snouckaert van Schauburg, René Charles Edouard Georges Jean (1857). Territet (Suisse), Hotel les Terrasses. *Ornithol. de la région paléarctique et de l'archipel Malais.* Ⓓ Leiden 1880.

Snow, Julia Warner (1863), Associate Prof. of Botan. Smith Coll. Northampton, Mass. (U.S.A.), 36 Bedford Terrace. *Morphol. of the Chlorophyceae.* Ⓓ Zürich 1893.

Snow, Laetitia M. (1874), Associate Prof. of Botan. Wellesley, Mass. (U.S.A.), Wellesley Coll. *Plant Ecol. and Soil Bact.* Ⓓ Univ. of Chicago 1904.

Snow, Robert (1897), Research Fellow of Magdalen Coll., Oxford (England). *The conduction of excitation in plants; conduction of nutritive substances in plants; correlations of growing parts in plants; role of hormones in plants, inheritance and viability of pollen in Matthiola.* Ⓓ Oxford 1921.

Snyder, Charles David (1871), Prof. of Physiol., Johns Hopkins Univ. Baltimore, Md. (U.S.A.), 309 Forest Road. *Animal physiol., electro-physiol., bio-physical fields.* Ⓓ Univ. of California 1905.

Snyder, Franklin Faust (1897), Ass. Prof. of Anat., Univ. of Rochester, School of Med. and Dentistry. Rochester, N.Y. (U.S.A.), Crittenden Boul. *Histol. and Physiol. of Reproductive Organs. Embryol* Ⓓ 1923.

Snyder, Harry (1867), Prof. of agric. chem. and soils Univ. of Minnesota. Minneapolis, Minn. (U.S.A.), 1806 Summit Av. *Chem. of Plant and Animal Life. Dairy chem.* o

Snyder, John Otterbein (1867), Prof. Zool., Dir. Natural History Mus. Stanford University, Cal. (U.S.A.). *Ichthyol.; Geographic distribution of species. Habits. Growth.* Ⓓ Stanford Univ. 1899. Ⓟ Specimens west coast fishes, reptiles, amphibians, birds and mammals.

Snyder, Laurence H. (1901), Associate Prof. of Zool., N.C. State Coll. Raleigh, N.C. (U.S.A.). *Genetics. The inheritance and racial significance of the blood groups. The modification of the germ plasm. Selective coupling on the basis of somatic characters. The inheritance of mental ability.* Ⓓ Harvard 1926.

Snyder, Thomas Elliott (1885), Entomol., Branch of Forest Insect Investigations, Bureau of Entomol. U. S. Dept. of Agric., Washington, D.C. (U.S.A.). *Termites, biol. and taxonomic. Powder post beetles.* Ⓓ Washington Univ. 1920. Ⓟ Termites.

Sō, Masao, Prof. Imper. Univ. Komaba near Tōkyō (Japan), Zool. Inst., Fac. of Agric. *Applied Zool., Entomol., Sericiculture.*

Sobernheim, Georg (1865), o. Prof. der Hygien. und Bact. an der Univ. Bern, Dir. des Inst. zur Erforschung der Infektionskrankheiten Bern (Schweiz), Friedbühlstr. 22. *Immunität bei Menschen- und Tierpocken. Herpes-Encephalitisätiol. Milzbrand- und Rauschbrandätiol. Desinfektion. Diphtherieimmunität und Diphtheriebekämpfung. Theoretische Probleme der Immunität (Antikörperentstehung und -wirkung). Infektionswege, speziell Infektion (und Immunität) per os.* Ⓓ Berlin 1890. Ⓟ Mikroskopische Bakterienpraep., Diapositive, Kulturen.

Sobolew, A. Krasnojarsk (U.d.S.S.R.), Naturhistorisches Mus. *Palaeont.; Vertebrata.* o

Sobolew, Demetrius N., Prof. Geol. Inst. f. Volksbildung. Charkow (U.d.S.S.R.). *Palaeozoische Palaeont. Theorie und Evolution.* o

Sobotta, Johannes (1869), o. Prof., Dir. anat. Inst. Univ. Bonn (Deutschland). *Allgemeine und vergl. Entwicklungsgeschichte. Histol. Anat. Anthrop.* Ⓓ Berlin 1891.

Sobrado Maestro, César, Prof. Catedrático en la Fac. de Farmacie. Santiago (España). *Botán.* o

Sochański, H., Dr. med., Doc. Lwów (Polognc), Univ., Fac. méd., r. Piekarska 52. *Biochim.* o

Söderlund, Hans (1876), Vet.-Anat. der Univ. Helsinki (Finnland). *Histol. der serösen Häute. Embryonale Entwicklung des elast. Gewebes.* Ⓓ Med. vet. Leipzig 1912, agronom., forstliche Univ. Helsingfors 1926.

Söderström, Adolf Fredrik (1888), Doc. a. d. Univ. Uppsala (Schweden), N. Rudbecksgatan 8. *Evolutionistische Morphol. Spionidae (Polychäta). Polydora ciliata. Dorsale Sinnesorgane bei Nerine fuliginosa. Metamorphose der Polygordius-Endolarve. Homol., Homogenie und Homoplasie. Verwandtschaftsbeziehungen der Mollusken. Gastrula und Protostoma, Planula und Blastopor, Oroproctula und Oroproctus. Bau der Zellen.* Ⓓ Uppsala 1920.

Söding, Hans (1898), Dr. rer. nat., Ass. am Botan. Garten der Univ. Münster i. W. (Deutschland). *Wachstumsphysiol.* Ⓓ Hamburg 1923.

Söhngen, N. L., Prof., Dr., Landbouw Hoogeschool Wageningen (Holland). o

Sörrensen, Georg (1897), Ass. am Bact. Inst. der Landwirtschaftskammer Oldenburg i. O. (Deutschland). *Tierheilkunde, Unfruchtbarkeit der Haustiere.* Ⓓ Berlin 1922.

Sokhey, Sahib Singh (1887), Ass. Dir., in charge Biochem. Unit. Parel, Bombay (India), Haffkine Inst. *Biochem.* Ⓓ Edinburgh Univ. 1911.

Sokolnikov, O. I., Doc. II. Univ. Moskau (U.d.S.S.R.). *Biochem.* o

Sokoloff, Arkadij (1887), Doc. Univ. Nijny-Nowgorod (U.d.S.S.R.). *Forst und Forstwirtschaft, geographische Verbreitung der Baumarten, Wachstum.* Ⓓ Petersburg 1913. Ⓟ Lignin.

Sokolova, Lidia (1894), Specialist of Zool. Taschkent, Turkestan (U.d.S.S.R.), Putinsevskai 7. *Pests vertebrate: rodents.* Ⓓ Taschkent 1926.

Sokolov, Vladimir Michajlovič, Prof. Univ. Kazan, Tatar. Rep. (U.d.S.S.R.). *Pharmacol.* o

Sokolow, Alexey Jakob (1888), Ass. Kasan (U.d.S.S.R.), Zool. Kabinett Univ. *Morphol. der Wanderheuschrecken.*

Sokolow, Boris P. Sinelnikowo (U.d.S.S.R.) Jekaterinoslawer Landwirtsch. Provinz.-Versuchsstation. *Selektion landw. Pflanzen.* o

Sokolow, Démétrius W. (1884), Géol.-en-chef Comité Géol. Moscou (U.d.S.S.R.), Jaouzskaja 1, la section du Comité Géol. *Céphalopodes des terrains jurassique et crétacé en Crimée et les pélécypodes du crétacé de l'île de Sachaline.* Ⓓ Moscou 1909.

Sokolow, Iwan (1885), Doc. Univ. Leningrad (U.d.S.S.R.), Labor. d. spezielle u. experim. Zool. *Vererbungscytol. Spermatogenese. Hydracarina: Systematik u. Biol.* Ⓓ Leningrad 1917.

Sokolow, Peter A. (1900), Ass. der Anat. Rostow, Don (U.d.S.S.R.), Anat. Inst. *Anastomosen d. N. medianus u. ulnaris. Hermaphroditismus externus. Topographie d. Duodenum. Pancreas.* ⓓ Rostow, Don, 1923.

Sokołowski, Jan Bogumił (1899), Dr. phil. Poznań (Polska), ul. Wjazdowa 3; Zakład Biol. i Anat. Porówn. *Systematik und Biol.; Aves, Orthoptera von Polen.* ⓓ Poznań 1926. ⓟ Vögel, Geradflügler.

Sokołowski, Stanisław (1865), Prof. Univ. Kraków (Polen), Aleja Mickiewicza 17. *Holzpflanzen. Wald.* ⓟ Mitteleurop. Hölzer.

Sokołowskij, Alexander I., Doc. a. Kiewschen Inst. f. Volksbildung., Kiew (U.d.S.S.R.), Gogolstraße 19, W. 2. *Botan.* o

Sokólska, Julja (1893), Ass. Lwów (Polska), Inst. Zool. Politechnik. *Cytol.: Golgischer Apparat u. Chromosomalverhältnisse bei den Spinnen.* ⓓ 1925.

Sokownina, Nina I., Ass. I. Univ. Moskau (U.d.S.S.R.), Ostoshenka I. Uschakowskij per. 11, W. 2. *Botan. Pflanzen-Physiol.* o

Soldatov, Vladimir (1875), Prof. of Ichthyol. Agric. Acad., Sc. Ichthyol. Marine Floating Sc. Inst. Moscow (U.d.S.S.R.), Academy of Agric., Petrovsko-Rasoumovskoje. *Biol. of Fishes, System. Fishery.* ⓓ Peterburg 1906. ⓟ Fishes of european and asiatic Russia.

Soler, Franch L., Prof. Univ. Nac. Buenos Aires (Argentina). *Anat. e Physiol.* o

Soler y Batlle, Enrique, Prof. Catedrático de la Fac. de Farmacie. Sarriá, Barcelona (España), Mayor 51. *Botan.* o

Solger, Bernhard (1849), Dr. med., a.o. Prof. emer. der Anat. Neiße, Oberschlesien (Deutschland), Bismarckstr. 15. *Menschliche und tierische Morphol.* ⓓ München 1872.

Solheim, W. G., Ass. in Botan., Univ. of Illinois. Urbana, Ill. (U.S.A.), 209 Natural History Bldg. o

Soli, Ugo, Prof. Univ. Bari (Italia). *Anat. Pathol.* o

Soliman, Labeeb Boutros (1895), Ass. Entomol. Cairo (Egypten), Bureau of Entomol., Ministry of Agric. *Entomol.: Aphididae; Insect Resistant Plants.* ⓓ Univ. of California 1926.

Soljanucha, Ignatij F., Dir. Belaja Zerkow (U.d.S.S.R.), Weißruss. Selektions-Stat. *Selektion d. Zuckerrübe.* o

Sollmann, Torald Hermann (1874), Prof. of Pharmacol., Western Reserve Univ., Cleveland, O. (U.S.A.), *Pharmacol.* ⓓWestern Reserve Univ. 1896.

Sololov, B. F., Prof. assoc. acad. Russe. Paris (France), Boulevard Raspail 96. *Zool.* o

Solotnitsky, Wssewolod (1891), Abteilung Pflanzenzüchtung, Amur Landwirtschaftl. Versuchsstation. Blagoweschtschensk (U.d.S.S.R.). *Getreidezüchtung: Weizen.* ⓟ Weizen-Sorten. Wilde Pflanzen d. Amur-Distrikts.

Soloviev, Nicolaus (1889), Pathol.-Anat., Eugenist. Jaroslavl (U.d.S.S.R.), Krasnaja 12, Log. 3. *Konstitution, Cancer.*

Solowiow, Paul (1876), Prof. Gorki Bielarus (U.d.S.S.R.), Landwirtsch. Akademie. *Systematik und Biol. der Anophelini, Biol. der Trematoden, Phänol. Beobachtungen.* ⓓ Warschau 1901, Moskau 1904.

Solowjewa, Anna N., Doc. II. Univ. Moskau (U.d.S.S.R.), Samotjek, I. Wolkonskij per. 10, W. 3. *Botan.* o

van Someren, Victor Gurnet Logan, Dr., Nairobi, Kenya Colony (Tanjanika Terr.), Med. Dept., Box 140. *Ornithol.* o

Sommer, A., Prof. Univ. Tartu (Estland), Toomemäel. *Anat.* o

Sommer, Hermann (1899); The George Williams Hooper Foundation for med. Research, Univ. of California. San Francisco (U.S.A.), Second and Parnassus Avenues. *Synthese von biogenen Aminen, Hydrierung. Chem. Untersuchung von Bakterientoxinen, physiol. Untersuchung von Farbstoffen und Derivaten.* ⓓ Univ. Zürich 1923.

Sommer, Karl Robert (1864), o. Prof. der Psychiatrie und Dir. der Psychiatrischen Klinik, Univ. Gießen (Deutschland), Frankfurter Str. 97. *Psychopath. Untersuchungsmethoden, Familienforschung und Vererbungslehre, experim. Psychol., Kriminalpsychol., Tierpsychol.* ⓓ Phil. Berlin 1887, med. Würzburg 1891.

Sommerfeld, Paul (1869), Prof. Dr., Dir. des Labor. am städtischen Kaiser- und Kaiserin-Friedrich-Kinderkrankenhaus. Berlin W 15 (Deutschland), Kaiserallee 201. *Bact. Serol.* ⓓ Berlin 1891.

Sommermeyer, Viola Louise (1896), Dir. of Labor. San Diego, Cal. (U.S.A.), 1200 First National Bank Bldg. *Blood Chem. and Protozool. of Humans.*

Sonan, Jinhaku, Entomol. Governm. Central Research Inst. Taihoku, Formosa (Japan), Entomol. Labor., Dept. of Agric. *Entomol. Cicadidae.*

Sonne, Carl Olaf (1882), Dir. Labor. Finsens med. Lichtinst. København (Danmark). *Physiol., Lichtbiol.: physiol. und therapeutische Wirkung der leuchtenden und ultravioletten Strahlen.* ⓓ Kopenhagen 1913.

Sonobe, Ichirô, Ass. Prof. Tôkyô Imper. Univ. Komaba near Tôkyô (Japan), Inst. of Forestry, Fac. of Agric. *Forestry.* o

Soós, Lajos, Dr. Zool. Abt. Ung. National Mus. Budapest (Ungarn). *Malakol.*

Soot-Ryen, Iron (1896), Curator. Tromsö (Norge). Mus. *Fauna arctica, sp. Mollusca marina and Diptera.* ⓟ Mollusca marina and Diptera.

Soó von Bere, Károly Rezső (1903), Coll. Hungaricum. Berlin NW 6 (Deutschland), Marienstr. 5. (Budapest I [Ungarn], Ménesi ut 11—13.) *Pflanzensystematik (spez. Orchideen, Melampyrum). Pflanzensoziol. u. Geographie (bes. Siebenbürgen).* ⓓ Budapest 1925. ⓟ Exsiccata aus Ungarn u. Siebenbürgen.

Sopp, O. J., Dr., Dir. Labor. Kapp, Toten (Norge). *Botan.* o

Sor, G. V., Röntgenol. u. Radiol. Inst. Leningrad (U.d.S.S.R.), Ul. Röntgen'a 6. *Pathol. Anat.* o

Sordelli, Alfredo (1891), Dir. Inst. Bact. du Depart. Nac. de Higiene, Prof. Microbiol. Fac. Sc. Buenos Aires (Argentina), Calle Velez Sarsfield 563. *Bact.: anaërobes. Immunité. Anticorps et Antigènes hétérogénétiques. Coagulation du sang. Chim. Physiol. Sécrétions internes.* ⓓ Buenos Aires 1912.

Sorenson, Charles J. (1884), Ass. Entomol., Agric. Experim. Station. Logan, Utah (U.S.A.). *Chalcis-Fly in Alfalfa seed.* ⓓ Utah Agric. Coll. 1924. ⓟ Chalcis-Flies and Tree hoppers.

Sorin, Eugen (1886), Ass. Experim. Inst. der Ernährungsphysiol. Moskau 34 (U.d.S.S.R.), Sintzev-Wrashek 41. *Physiol. der Ernährung, Endokrinol., Tuberkulose.*

Sorokin, Helen P. (1894), Research Worker, Univ. of Minnesota. Department of Botany. Minneapolis, Minn. (U.S.A.). *Plant cytol. Individuality of chromosomes and variability of the nucleus in certain Ranunculaceae. Role of nucleolus. Pathol. cytol. Mosaic disease and other pathol. problems.* ⓓ Univ. of Minnesota 1925

Sorokin, Iwan R., Leiter d. Dendrol. Abt. d. Nikitskij Botan. Gartens. Jalta, Krim (U.d.S.S.R.). *Gartenbau. Dendrol.* o

Soschkina, Elizaveta D. (1889), Doc., Ass. geol. I. Univ. Moskau (U.d.S.S.R.), Cabinet geol. *Coraux paléozoïques (étage d'Ardinsk).*

Soška, Theodor (1876), Inspektor des Botan. Gartens, Univ. Beograd (S.H.S.), Takovska 41. *Floristik Mazedoniens u. Altserbiens.*

Sosnowskij, Dmitrij I., Botan., Prof. b. Botan. Garten. Tiflis (U.d.S.S.R.). *Systematik u. Pflanzengeographie.* o

Sosnowski, Jan, Dr. Prof. Warszawa (Pologne), Ecole Sup. Agric., R. Miodowa 23. *Physiol.* o

Souèges, Etienne Charles René (1876), Chef des Travaux micrographiques Fac. Pharmacie. Paris VI (France), 4, Avenue de l'Observatoire. *Embryol. végétale.*

Soula, Louis-Camille (1888), Prof. de Pharmacol. Toulouse (France), 50, Rue Montaudran. *Physiol., Pharmacodynamie, Chimisme des centres nerveux, Anaphylaxie, Physiol. de la Rate, Poisons du cœur, Inscription électromagnétique.* ⑨ Toulouse 1911.

Soule, Malcolm Herman, Ass. Prof. of Bact., Univ. of Michigan, Hygienic Labor. Ann Arbor, Mich. (U.S.A.). *Bact. metabolism, dissociation and chem. of bacteria.* ⑨ Univ. of Michigan 1924.

de Sousa Junior, Ant. J., Prof. Univ. Porto (Portugal). *Anat. Pathol.* o

Southern, Rowland (1882), Ass. Inspector Fisheries. Dublin (Ireland), 3 Kildare Place, Department of Fisheries. *Limnol., Salmon and Trout.* ⑨ London Univ. 1908.

de Souza, O., Prof. Univ. Rio de Janeiro (Brasil). *Physiol.* o

Sowerby, Arthur de Carle (1885), Zool. Field Collector in China for the U. S. National Mus., Curator of the Mus. of the Royal Asiatic Society, North China Branch. Shanghai (China), 8 Museum Road. *Mammals, Birds, Reptiles, Amphibians, Fishes, Crustaceans, Mollusks, Insects, Aracnids, Myriapods of China and neighbouring regions.*

Sowetkina, Margarete M., Doc., Mittelasiat. Staatl. Univ. Taschkent (U.d.S.S.R.), Obuchow Per. 14. *Geobotan.* o

Sowinsky, Wadim (1881), Ass. Zool. und Anat., Mitarbeiter d. Biol. Dnjepr-Station. Kiew, Ukraine (U.d.S.S.R.), Turgenewskaja Str. 19, W. 6. *Zool.-Entomol.* (*Zoogeographie, Systematik der Lepidopteren*). ⑨ Macro- u. Microlepidoptera d. palaearkt. Region.

Spadolini, Igino (1887), Dr.Prof., Dir.Labor.Fisiol., Univ. di Cagliari. Firenze (Italia), Via Cavour 28. *Fisiol. delle ghiandole a secrezione interna (paratiroide). Fisiol. del sistema nervoso simpatico.*

Spärck, Hakon Ragnar Gisiko (1896), Dr. phil., Priv.Doc., Ass. d. Invertebratenabt. d. Zool. Mus. Kopenhagen, K. (Dänemark), Krystalgade. *Biol. der marinen Lamellibranchiata, Heteropoda, Parasit. Schnecken, Metamorphose und Oekol. der Diptera Nematocera, Allg. Tiergeographie und Oekol. der nordischen Meere.* ⑨ Kopenhagen 1919.

Spät, Wilhelm, Dr., Priv.Doc. d. Hygiene (Bact.). Praha II (C.S.R.), Preslova, Hygien. Inst. d. Deutsch. Univ. o

Späth, Ernst (1886), o. Univ.Prof., Dir. II. chem. Inst. Univ. Wien IX (Österreich), Währinger Str. 38. *Konstitution und Synthese von Naturstoffen, namentlich Alkaloiden.* ⑨ Wien 1910.

Späth, Hellmut L. (1885). Berlin-Baumschulenweg (Deutschland), Späthstr. 1. *Johannistrieb. Periodizität und Jahresringbildung sommergrüner Holzgewächse.* ⑨ Berlin 1912. ⑰ Gehölze, Gehölzsämereien.

Spaeth, John Nelson (1896), Ass. Research Prof. of Forestry, Cornell Univ. Ithaca, N. Y. (U.S.A.). *Ecol.* (*Forest*). ⑨ Harvard 1920.

Spagnol, Girolamo (1897), Fellow Rockefeller Found. Padova (Italia), Ist. di materia med. e farmacol., Univ. *Farm.* ⑨ Padova 1922.

Spagnolio, Giuseppe (1875), Prof. Lib. doc. Patol. Speciale Med. Dim. Univ., Dir. Ist. di Terapia Fisica. Messina (Italia), Piazza Roma (Piazza Stazione). *Radiol. Leishmannia. Malaria. Istol. patol., fisiopatol., ematol.*

Spalteholz, Werner (1861), Dr. med. et LLD., ao. Prof. der Anat., 1. Prosektor am Anat. Inst. Leipzig C 1 (Deutschland), Mozartstr. 21. *Blutgefäßversorgung der Organe; Bindesubstanzen.* ⑨ Leipzig 1886.

Spangenberg, Georges E., Chef des trav. Stat. Expér. Agron. Charkow (U.d.S.S.R.), rue Devitschia 5. *Ustilaginées.* o

Spangler, Robert Clifton (1884), Prof. of Botan. at West Virginia Univ. Morgantown, W.Va. (U.S.A.). *Morphol. of Plants. Cladoporium. Fem. Gametophyte of Trillium. Bact.* ⑨ Univ. of Chicago 1925.

Spann, Joseph Karl (1879), o. Hochsch.-Prof. Freising bei München (Deutschland). *Tierzuchtlehre, Alpwirtschaft, Veterinärwesen.* ⑨ Med. vet. Univ. Gießen 1905; Techn. Hochsch. München 1919.

Spanner, Rudolf (1895), II. Prosektor u. Priv.Doc. Kiel (Deutschland), Hegewischstr. 1, Anat. Inst. *Pfortaderkreislauf in der Vogel- und Reptilienniere. Nierenkreislauf an lebenden Vögeln und Reptilien. Injektionen. Entwicklung der Gefäße in der Niere.* ⑨ Köln 1920.

Sparks, M. Irving (1902), Dept. of Pharmacol. Univ. Toronto (Canada), 472 W. Marion St. *Pharmacol.* ⑨ Univ. of Toronto 1926.

Spasskij, Nikolaj, Prof. Univ. Irkutsk (U.d. S.S.R.). *Physiol.* o

Spassky, L. G., Cons. le Mus., Jardin Botan. princ. Leningrad (U.d.S.S.R.), Aptekarski Ostrov, Pessotchnaia $^1/_2$. *Chim. des produits végétaux.* o

Spassky, Sergius, Prof. Zool. und Entomol. Donsche Hochsch. für Landwirtschaft und Melioration. Novotscherkassk (U.d.S.S.R.), Comitetskaja 47-a. *Biol. und Systematik der Spinnen (Araneida).* ⑰ Araneida Rußlands: Don-Gebiet, Krim, Kaukasus, Turkestan, Sibirien.

Spatz, Hugo (1888), Priv.Doc., Dr., Leiter des Anat. Labor. der Psychiatrischen und Nervenklinik. München (Deutschland), Nußbaumstr. 7. *Allgemeine und spezielle Histopath. des Centralnervensystems. Ontogenese des Gehirns. Untersuchung von Stoffwechselprozessen im Gehirn mit morphol. Methoden.* ⑨ Heidelberg 1914.

Spaul, Eric Arthur (1895), Lect. Zool. Birkbeck Coll. Univ. London, E.C. 4 (England), Breams Buildings, Fetter Lane. *Experim. Zool. Endocrinol.* ⑨ London 1924.

Spaulding, Perley (1878), Pathol., Bureau of Plant Industry, U. S. Department of Agric. Amherst, Mass.(U.S.A.), Northeastern ForestExperim. Station. *Forest tree diseases.* ⑨ Washington Univ. 1906.

Speck, Frank G. (1881), Prof. of Anthrop., Univ. of Pennsylvania. Philadelphia, Pa. (U.S.A.), College Hall. *American ethnol.* ⑨ Univ. of Pennsylvania 1908. ⑰ American ethnol. and archaeol. specimens.

Spee, Ferd. Graf von (1855), o. Univ.Prof. emer. Kiel (Deutschland), Propensdorf. *Anat.* o

Spehl, Paul Emile (1887), Agrégé, Adjoint à la Fondation Reine Elisabeth. Bruxelles (Belgique), Hôpital Brugmann. *Bact., hématol.* ⑨ Bruxelles 1909. ⑰ Préparations bact. ou hématol.

Speidel, Carl Caskey (1893), Assoc. Prof. of Anat., Univ. of Virginia Med. School. Charlottesville, Va. (U.S.A.). *Hematol. and endocrinol.* ⑨ Princeton 1918.

Speight, R., Prof. M.Sc. Canterbury Coll. Christchurch (N. Zealand). *Palaeont.* o

Speiser, Felix (1880), ao. Prof. Univ. Basel (Schweiz), St. Alban Vorstadt 108. *Ethnographie, Anthrop.* ⑨ Basel.

Spek, Josef (1895), ao. Prof., Ass. am Zool. Inst. Heidelberg (Deutschland). *Mechanismus der Zellteilung, Gastrulation, Faltenbildung, Protoplasmabewegung. Salzphysiol. (Abdichtungstheorie). Entwicklungsmechanik des Beroëeies. Rhopalien. Radula. Eizellen d. Nematoden.* ⑨ Heidelberg 1916.

Spemann, Hans (1869), Dr. phil., o.ö. Prof. der Zool., Geh.R., Dir. des Zool. Inst. Freiburg i. B. (Deutschland). *Entwicklungsphysiol., Amphibien.* ⑨ Würzburg 1894.

Spencer, Baldwin, Hon. Dir. Nat. Mus. of Natur. Hist. Melbourne (Australia). o

Spencer, George J. (1888), Ass. Prof. of Zool., Univ. of British Columbia. Vancouver (British Columbia). *Insect pests of home, garden and field; Ecol. and anat. of Thermobia domestica. Regeneration of insect appendages. Marine zool.: Mollusca and Arthropoda. Animal histol.* ⑨ Illinois 1924.

Spencer, Herbert (1894), Ass. Entomol., Louisiana Experim. Station. Baton Rouge, La. (U.S.A.). *In-*

sects affecting Crops, especially Vegetables, and Field Crops, Cotton, Corn, and Sugar Cane. ⓓ Ohio State Univ., Columbus, O., 1922.

Spencer-Lister, Sir T., Univ. Johannesburg, Transvaal (S. Africa). *Bact.* o

Speranskaja-Stepanowa, Katherine, Ass. Section of Pharm., Inst. of Experim. Med. Leningrad (U.d.S.S.R.). *Sympathetic Nervous Systeme.* ⓓ Leningrad.

Speranskij, Wassilij G., Prof. a. Iwanowo-Wosnessensker Inst. Moskau (U.d.S.S.R.), Nowoslobodskaja 65. *Gartenbau, Gemüsebau. Biol. der Obstbäume.* o

Sperlich, Adolf (1879), Dr. phil., ao. Prof. der Botan. der Univ. Innsbruck, Tirol (Deutsch-Österreich), Salurner Str. 16. *Tropismen und Nastien; die phyletische Potenz.* ⓓ Innsbruck 1902.

Sperov, Michael (1892), Dr., Ass. Prof. Anat. in Pedagogical Inst. Leningrad (U.d.S.S.R.). *Human Anat.* ⓓ Moscow 1917.

Sperrin-Johnson, J. C., Prof. Univ. Coll. Auckland (New Zealand). *Zool.* o

Speschilow, Paul (1877), Prosektor Anat. Inst. Univ. Saratow (U.d.S.S.R.). *Histol. Bau künstlicher Geschwülste.*

Spessard, Earle Augustus (1887), Prof. of Botan. and Head of Biol. Department, Quachita Coll. Arkadelphia, Ark. (U.S.A.). *Plant morphol. Fresh water Chlorophyceae of Arkansas and middle west. Variation.* ⓓ Univ. of Chicago 1924. ⓟ *Prothallia of lycopodium. Chlorophyceae.*

Spessivtseff, Paul (1866), Oberförster, Dr., Ass. an der entomol. Abt. der K. Schwed. forstl. Versuchsanstalt. Statens Skogsförsöksanstalt Experimentalfältet (Sverige). *Forstinsekten, paläarktische Borkenkäfer (Systematik, Biol. und geographische Verbreitung).* ⓓ Kais. Forstinst., St. Petersburg 1897. ⓟ Paläarktische Borkenkäfer.

Spett, Georg (1905), Ass. Zool. und Anat. Inst. Univ. Kiew, Ukraine (U.d.S.S.R.), Dmitrijevskaja Straße 57/1. *Entomol.* o

Speyer, Walter (1889), Dr. phil., Reg.R. bei der Biol. Reichsanstalt für Land- und Forstwirtschaft. Stade, Hann. (Deutschland). *Tierische Schädlinge der Obstgewächse: Carpocapsa, Anthonomus, Cheimatobia, Psylla (Morphol., Biol., Bekämpfung); Parasiten der Insekten.* ⓓ Königsberg i. Pr. 1919.

Spiegel, Ernst Adolf (1895), Doc. normale, pathol. Anat., Physiol. des Nervensystems, Ass. am Neurol. Inst. Wien I (Österreich), Falkestr. 3. *Experim. Physiol., Tonus, vegetatives Nervensystem, Labyrinth.* ⓓ Wien.

Spiegl, Anton (1886), Obertierarzt am Bact. Inst. der Landwirtschaftskammer für die Prov. Sachsen. Halle a. d. S. (Deutschland). *Bact. Fleischuntersuchung. Schlafkrankheiten (Aetiol., pathol. Anat. und Histol.).* ⓓ München 1920.

Spiekerkötter, Heinz (1899). Bielefeld (Deutschland), Melanchthonstr. 26 I. *Entgiftung von Zyanwasserstoff durch Traubenzucker bei Warmblütlern.* ⓓ Göttingen 1923.

Spierenburg, Dina (1880), Phytopath. beim Pflanzenkrankheiten-Dienst in Wageningen (Holland), „De Zuider-Eng". *Pflanzenpathol.*

Spinner, Henri (1875), Prof. de Botanique à l'Univ. de Neuchâtel (Suisse), Champ-Bougin 40. *Anat. et histol. des Carex et des végétaux de hautes altitudes. Etudes phytogéographiques sur le Jura neuchâtelois; tourbières du Jura neuchâtelois.* ⓓ Zürich 1902.

Spiridonow, Maxim D., Leiter d. Botan. Gartens. Omsk (U.d.S.S.R.), Sibirisches Inst. f. Land- u. Forstwirtschaft. *Geobotan.* o

Spiro, Karl (1867), o. Prof. f. physiol. Chem. Univ. Basel (Schweiz), Vesalianum. ⓓ Phil. Würzburg 1889, med. Leipzig 1893.

Spisar, Karl (1879). Ivanovice na Hané (C.S.R.), Drysice No. 36 u Vyskova. *Anat., Physiol., Pathol., Ökol., Biol., Teratol. der Pflanzen. Kulturpflanzen.* ⓓ Prag 1906. ⓟ Teratol. Photographien.

Spitta, Oscar (1870), Dr. med., Geh.Reg.R., ao. Prof. der Hygien., Vorsteher des Hygien. Labor. des Reichsgesundheitsamts in Berlin (Deutschland). *Hygien. und Bact.; im besonderen Abwasserbact. und -biol. Flußverunreinigung und Abwasserfragen. Nahrungsmittelbiol.* ⓓ Marburg 1895.

Spoehr, Herman A. (1885), Ass. Dir. in Charge, Coastul Labor., Carnegie Inst. of Washington. Carmel, Cal. (U.S.A.). *Photosynthesis and plant respiration.* ⓓ Chicago 1909.

Spöttel, Adolf Wilhelm Walter (1892), Dr. phil., Priv.Doc. für angew. Zool. und Beurteilungslehre. Halle a. d. S. (Deutschland), Landwehrstr. 7. *Spezielle Vererbung bei Haussäugetieren. Konstitution. Domestikation. Histol. und sonstige Eigenschaften von Haaren.* ⓓ Halle 1913.

Spohr, Edmund (1887), Doc. allgemeine u. system. Botan., Dir. des Pflanzenmorphol. u. -systemat. Labor. u. d. Botan. Gartens Univ. Tartu (Estland). *Entwicklungsgeschichte u. Ökol. des Blattes. Floristik u. Phytosoziol., insbesondere der Pflanzendecke Estlands. Geographie u. Ökol. der Wasser- u. Sumpfpflanzen Estlands u. Rußlands.* ⓓ Tartu 1920. ⓟ Herbarpflanzen.

Sponholz, Gerhard (1896), Ass. am Pathol. Inst. der Univ. Tartu (Estland), Gartenstr. 43, W. 1.

Sprehn, Curt (1892), Dr., Helminthol. Darmstadt (Deutschland), Moosbergstr. 65. *Parasitische Nematoden, Trematoden-Entwicklung.* ⓓ Berlin 1923.

Sprengel, Luise (1894), Dr. phil., Ass. an der Staatl. Versuchsanstalt für Wein- u. Obstbau, Zool. Abt. Neustadt a. d. Haardt (Deutschland), Maximilianstraße 20. *Erforschung u. Bekämpfung der tierischen Schädlinge im Obst- u. Weinbau.* ⓓ München 1923.

Sprenger, Albrecht Marinus (1881), Prof., Univ. of Agric. Wageningen (Holland), Hoogstraat 4. *Selection and crossing of horticultural plants. Pruning. Vegetative propagation of horticultural plants. Fertility.* ⓓ Wageningen 1905. ⓟ Plants.

Sprenger, Emil (1892). Liboch a. d. Elbe (C.S.R.). *Diatomeen (Bacillariales).* ⓟ Diatomeen-Material und Praep.

Spring, Samuel Newton (1875), Prof. Silviculture N.Y. State Coll. Agr. Cornell Univ. Ithaca, N.Y. (U.S.A.), 508 Highland Road. o

Sprytschin, Iwan I., Dir. d. Staatl. Sapowednik. Pensa (U.d.S.S.R.). *Botan. Pflanzen-Geographie.* o

Spuler, Anthony (1889) Associate Entomol., Wash. Agric. Experim. Station. Pullman, Wash. (U.S.A.), 611 Michigan Ave. *Economic insects. Borboridae (Diptera).* ⓓ Washington State Coll. 1919.

Spuler, Arnold (1869), o. Univ.Prof. d. Histiol. und Embryol. Erlangen (Deutschland), Hindenburgstr. Nr. 57. *Vertebrata; Mesenchym-Histiogenese, Entwicklung des Uro-Genitalapparates; Vergl. Myol.; Anthropogenese, Haarregeneration. Lepidoptera, insbes. Systematik, Phylogenie, Ontogenie und Morphol.* ⓓ Phil. Freiburg i. Br. 1891 med. Erlangen 1895. ⓟ Lepidoptera paläarctica.

Srdinko, Otakar (1875), Prof. histol. et embryol., Dir. de l'Inst. Praha II (C.S.R.), Albertov 4. *Histol. et embryol. de la glande surrénale. L'architecture fonctionelle du cartilage hyalin, des flexures des tendons, de tissu conjonctif, d'éndoneurium.* ⓓ Prague 1899.

Sreenivasaya, M. (1895), Research Ass. Indian Inst. of Sc. Bangalore (India), Hebbal. *Sc. Study of Lac Industry. Physiol. of the lac insect and its secretion products. Symbiosis.* ⓓ Madras 1918. ⓟ Symbiotic organisms. Coccids. Lac insects.

Ssacharow, Michail M., Doc. Landwirtschaftl. Akademie Timirjasew. Moskau (U.d.S.S.R.), Tichwinskij per. 9, W. 46. *Botan., Systematik, Pflanzen-Anat. u. Morphol.* o

Ssadikow, Wladimir Sserggeiewitsch (1874), Prof. physiol. Chem., Dir. d. Labor. Univ. Leningrad (U.d.S.S.R.), Wassily Ostrow, Linie 2, H. 3, W. 4. *Aminosäuren und Eiweißstoffe (speziell Kollagen). Proteol. Fermente. Chem. Elemente als Bestandteile*

des lebendigen Substrats verschiedener Organismen (Biogeochem.). Katalytische Hydrierung unter Druck. ⑲ Petersburg. ⓟ Sehnen und Haut-Kollagene. Kollagenase. Homogenisation und Fraktionierung von Seetieren ev. einzelne Fraktionen derselben.

Ssawitsch, Woldemar (1890), Ass. Inst. f. experim. Biol., Ass. Hydrophysiol. Station des Inst. zu Zwenigorod. Moskau (U.d.S.S.R.), Worontzowo polje 6. *Biol. des Süßwasserplanktons im Zusammenhang mit den physikal.-chem. Faktoren des Außenmediums. Ionenwirkung auf die Süßwasser-Ciliaten (speziell auf die Encystierung).* ⑲ Moskau 1913.

Ssawron, Elena (1899), Ass. Biochem. Inst. Charkow (U.d.S.S.R.), K. Liebknecht-Str. 104. *Biochem. des Stoffwechsels.* ⑲ Charkow 1923.

Sschelkanowzew, J. P., Prof. Zool. Labor. Univ. Rostow a. Don (U.d.S.S.R.). o

Ssumakov, Grigori (1862), Staatsrat, Wissenschaftlicher Hilfsarbeiter. Tartu-Jurjew (Estland), Alleestraße 64. *Entomol. System. Coleoptera.* ⓟ Käfer.

Stabe, Henry A. (1899), Ass. Prof. in Entomol. Baton Rouge, La. (U.S.A.), Box 376, Univ. Sta. *Rate of Growth of Worker, Queen and Drone Larvae of Apis. Nectar and Pollen Plants.* ⑲ Iowa State Coll. 1925.

Stach, Jan Wacław (1877), Dir. d. Physiographischen Mus. d. Polnischen Akademie der Wissenschaften. Kraków (Polska), Sławkowska 17, Akademja. Privatwohn.: Dietla 90. *Systematik u. Zoogeographie der Apterygogenea. Odontol.: Entwicklung, Histol. und Phylogenie der Zähne.*

Stach, Stefan Jan (1900). Kraków (Polska), Dietla 90. *Systematik u. Zoogeographie d. Lepidopteren.*

Stachrowsky, Woldemar Heinrichowitsch (1897), Specialist d. Biol. Anstalt d. Centralen Forststation. Moskau 13 (U.d.S.S.R.), Pogonno-Lossing Ostrow 3. *Angewandte Forst-Ornithol.*

Staderini, Rutilio, Prof. Univ. Siena (Italia). *Anat.* o

Stadler, Hans Friedrich (1875), Dr. med., wissenschaftl. Mitarbeiter am Zool. Inst. Würzburg. Lohr a. Main (Deutschland), Gartenstr. 514. *Vogelstimmen. Tierwelt von Unter- und Oberfranken. Wespenbauten. Züchtung von Köcherfliegen, Stein- und Eintagsfliegen, Trichoptera, Plecoptera, Ephemeroptera.* ⑲ Erlangen 1898. ⓟ Larven und Nymphen unterfränkischer Wasserkerfe.

Stadler, Lewis John (1896), Associate Prof. of Field Crops, Univ. of Missouri. Columbia, Mo. (U.S.A.). *Genetics of Maize.* ⑲ Univ. of Missouri 1921.

Stadnickij, Nikolaj Grig., Prof. N. G. Černyševskij Univ. Saratow (U.d.S.S.R.), Ploščad imeni Lenina. *Anat.* o

Stádnik, Jaroslav (1889) Chef de l'Inst. de Biol., Inst. d'Etat de Recherches. Praha-Vinohrady (C.S.R.), Mánesova 88. *Amélioration des plantes, Vererbung, Züchtung.* ⑲ Praha 1925.

Stadtmüller, Franz (1889), Prof. Dr. med., Prosektor am Anat. Inst. der Univ. Göttingen (Deutschland), Kirchweg 18. *Vergl. Anat. u. vergl. Entwicklungsgeschichte (vergl. Schädelmorphol., Filterfortsätze der Ichthyoiden), Anthrop.* ⑲ Freiburg i. Br. 1914.

Stäger, Robert (1867), Dr. med. Bern (Schweiz), Sonnenberg 14. *Biol. der Pflanzen und Tiere. Pflanzengeographie.* ⑲ Basel.

Staemmler, Martin (1890), ao. Prof., Dir. Pathol.-hygien. Inst. Chemnitz (Deutschland). *Pathol. Anat. des sympathischen Nervensystems. Fermentforschung. Geschwülste. Neubildungen d. Darmes.* ⑲ Berlin 1916.

Staff, Fr., Dr. Prof. Warszawa (Pologne), Ecole Sup. d'Agric., Miodowa 23. *Zool.* o

Staffe, Adolf (1888), Dr., Priv.Doc. a. d. Hochsch. f. Bodenkultur, Wien. Trautmannsdorf a. L. bei Wien (Deutsch-Österreich). *Tierzucht und Grenzgebiete (Fütterung, Milchwirtschaft).* ⑲ Wien 1913.

Stafford, Ethelbert Withrow (1887), Associate Prof., Entomol. and Zool. A. and M. College, P.O. Mississippi (U.S.A.). *Acalyptrate Muscoidea. Trematoda in Insects.* ⑲ Univ. of Toronto 1909.

Stafseth, Henrik Joakim (1890), Associate prof. and research associate in bact. East Lansing, Mich. (U.S.A.), Michigan State Coll. *Diseases of poultry.*

Stahel, Gerold (1887), Dir. der Regierungs-Proefstation. Paramaribo, Surinam (Südamerika). *Tropische Gewächse.* ⑲ Basel 1911.

Stahl, Horatio Seymour, Prof. of Botan. Blacksburg, Va. (U.S.A.). *Peach Yellows and diseases of beans, effects on chlorophyll formation in the former and in physical effects on sap of latter by rusts.*

Stahr, Hermann (1868), Prof. Dr. med., Dir. des Pathol. Inst., Städt.Krankenhaus. Danzig (Freistaat Danzig), Gralathstraße 5. *Bau und Entstehung der Geschwülste, Krebsforschung.* ⑲ Berlin 1893.

Staig, R. A., Lect. Univ., Department of Zool. Glasgow (Scotland). o

Staiger, Gottfried (1881), Nahrungsmittelchemiker, Doc. Berlin-Niederschönhausen (Deutschland), Blücherstr. 20. *Hefe.* ⑲ Frankfurt a. M. 1920. ⓟ Hefe.

Stainbrook, Merrill A. (1897), Instr., Zool. Department of Zool., Univ. of Tennessee. Knoxville, Tenn. (U.S.A.). *Invertebrate Paleont. Devonian Brachiopods and Corals.* ⓟ Devonian Brachiopoda and Corals, other Paleozoic Fossils.

Stakman, Elvin C. (1885), Prof., Head of the Section of plant pathol., Univ. of Minnesota, Pathol., Office of Cereal Investigations, Bureau of Plant Industry, U. S. Dept. of Agric. St. Paul, Minn. (U.S.A.), Univ. Farm. *Plant Pathol., mycol.* ⑲ Univ. of Minnesota 1913. ⓟ Cultures of cereal smut fungi.

Stålberg, Georges (1899). Lund (Sverige). *Hydrobiol., quant. u. qual. Unters. d. Bodenfauna Vätterns. Physiol.: Nahrungsphysiol., Phototaxis etc. bei Mysis relicta, Pallasea und Pontoporeia.* ⑲ Lund.

Stålfelt, Martin Gottfrid (1891), Priv.Doc. Botan. Inst. der Univ. Stockholm (Sverige). *Pflanzenphysiol.* ⑲ Stockholm 1921.

Štambuk, Dinko, Dr., Leiter Bact. Stanica. Bitolj (S.H.S.). o

Stamm, Johannes (1881), o. Prof. Pharmakognosie, gerichtl. Chem. Univ. Tartu (Eesti), Ritterstr. 2. *Gehalt der kultivierten Arzneipflanzen an ätherischem Oel. Ranzidität der fetten Oele und festen Fette. Variabilität der Mikroorganismen.* ⑲ Univ. Tartu 1912.

Stamm, Robert Hutzen (1877), Doc. a. d. Univ. Kopenhagen, B. (Dänemark), Stormgade 5. *Histol.* ⑲ Kopenhagen 1901.

Stammer, Ernst (1890), Inhaber der Fa. H. Bosse & Co., Konservenfabrik, Hecklingen in Anhalt (Deutschland). *Pflanzenkrankheiten.* ⑲ Breslau 1913.

Stammer, Hans Jürgen (1899), Dr., Ass. Greifswald i. Pommern (Deutschland), Zool. Inst. *Dipterenlarven, Brackwasserbiol.* ⑲ Greifswald 1923.

Stamp, Laurence Dudley (1898), Reader Economic Geography Univ. London, W.C. 2 (England), London School of Economics. *Plant Ecol. and Biography. Ecol. of Fossils.* ⑲ London 1917.

Standley, Paul C. (1884), Associate Curator, Division of Plants, U. S. National Mus. Washington, D.C. (U.S.A.). *Systematic Botan. Flora of New Mexico, of Columbia. Trees and Shrubs of Mexico.*

Staněk, Stanislaus (1903). Korytná p. Nivnice (C.S.R.), *Floristik d. mährischen Karpathen.*

Stănescu, Paul P. (1889), Chef de travaux au Labor. d'Anat. et Physiol. végétales de l'Univ. de Bucarest (România), str. Cotroceni 38. *Variations des produits de la photosynthèse dans les feuilles.* ⑲ Bucarest 1925.

Stanford, E.E., Prof. of pharmacognosy, Western Reserve Univ. Clevel. Cleveland Heights. Ohio (U.S.A.), 3217 Sycomore Road. *Systematic and pharm. botan.* o

Stang, Valentin (1876), Prof. der Tierzucht an der Tierärztl. Hochsch. Berlin NW 6 (Deutschland), Luisenstr. 56. ⑲ Bern 1900.

Stanković, Siniša, Dr., ao. Prof. d. Zool. u. vergl. Anat. d. Zool. Inst. d. Novi Univ. Beograd (S.H.S.). *Hydrobiol. u. Zoogeogr.* o

Stankow, Sergej S. (1892), Prof. a. d. Univ. Nishnij-Nowgorod, Ul. Swerdlowa 47, W. 7. *Botan., Systematik u. Pflanzen-Geographie.* Ⓓ Moskau. Ⓟ Herbarpflanzen.

Stanton, Thomas Ray (1885), Agronomist in Charge of Oat Investigations. Office of Cereal Crops and Diseases, Bureau of Plant Industry, U. S. Department of Agric. Washington, D.C. (U.S.A.). *Experim. with oats.* Ⓟ Smut- and rust-resistant strains of oats.

Stantschinsky, Wladimir (1882), Prof. d. Zool. u. Vergl. Anat. Univ., Dir. d. Zool. Inst. Univ. u. d. Biol. Station. Smolensk (U.d.S.S.R.). *Ornithol. der paläarkt. Fauna, Fam. Paridae. Oekol. Tiergeographie. Malakol. (Landmollusken).* Ⓓ Heidelberg 1906. Ⓟ Vogelbälge.

Stapledon, Reginald George (1882), Prof., Dir. Welsh Plant Breeding Sta., Prof. of Agric. Botan. Univ. Coll. of Wales. Aberystwyth (England). *Genetics. Plant Breeding.* Ⓓ Cambridge 1904.

Stapp, Carl (1888), Dr. phil., Vorsteher des Bact. Labor. der Biol. Reichsanstalt für Land- und Forstwirtschaft Berlin-Dahlem. Berlin-Steglitz (Deutschland), Mariendorfer Str. 17. *Bodenbact., Stickstoffbindende Bacterien. Bacterielle Pflanzenkrankheiten.* Ⓓ Marburg. Ⓟ Bakterienreinkulturen.

Stark, Peter (1888), Univ.-Prof., Dir. d. Botan. Gartens und des Pflanzenphysiol. Inst. Breslau (Deutschland), Botan. Inst. *Reizphysiol. (Traumatotropismus, Haptotropismus, Phototropismus). Moorkunde (Torfuntersuchung, Pollenanalyse).* Freiburg i. Br. 1911.

Stark, Mary B., Ph.D., Prof. of Embryol. and Histol., Hospital Med. Coll. and Flower Hospital. New York (U.S.A.), 64 St. and Ave. A. o

Starkenstein, Emil (1884), Dr. med., ao. Prof. der Pharmakol. und Pharmakognosie an der Deutschen Univ. Praha II (C.S.R.), Ječná 43. *Pharmakol. anorganischer Stoffe (Eisen, Kalk). Wasserhaushalt. Pharm. der Entzündung. Kombinierte Arzneiwirkung.* Ⓓ Prag 1909.

Starkey, Robert Lyman (1899), Associate Soil Microbiol. New Brunswick, N.J. (U.S.A.), Agric. Experim. Station. *Cycle of sulfur in nature.* Ⓓ Rutgers Univ. 1924.

Starmach, Karol (1900), Ass. Labor. Botan. Janczewskianum U.J. Kraków (Polen), Aleja Mickiewicza 21. *Algol.*

Starobinski, Aron (1893), Ass. Labor. de Bact. du Service d'Hygiène de Genève (Suisse). *Bact., Sérol.*

Starr, Anna M. (1867), Associate Prof. of Botan. South Hadley, Mass. (U.S.A.), Mt, Holyoke Coll. *Ecol. and Anat. of Plants.* Ⓓ Univ. of Chicago 1911.

Startschewskaja, Klara P., Ass., Inst. f. Volksbildg. Dnepropetrowsk (U.d.S.S.R.), Nagornaja 22. *Botan. Pflanzen-Physiol. u. Microbiol.* o

Stary, Zdenko (1899), Dr. med et rer. nat., Ass. am Deutschen Med.-chem. Univ.-Inst. Praha II (C.S.R), Salmovská 3. *Eiweißchemie, Stoffwechsel.* Ⓓ Med. Prag 1923, rer. nat. 1925.

Starygin, Zidia P. (1901), Ass· Bact.-Agronomical Station, Moscow (U.d.S.S.R.), Koniuschkovska 31. *Milk bact.* Ⓓ Petrovsky 1925.

Starzs, Kārlis (1897), Inst. für Planzenschutz. Riga (Lettland). *Unkräuter, Floristik Lettlands (Bäume und Sträucher), Betulaceae.* Ⓟ Blütenpflanzen Lettlands, Betulaceae, besonders Gattung Betula.

Staudacher, Josef (1876). Ljubljana (S.H.S.), Nunska ulica 8. *Coleopterol.* Ⓓ Graz 1904. Ⓟ Coleopteren, Lepidopteren.

Stauffacher, Heinrich (1865), Dr. phil., Prof. Frauenfeld (Schweiz). *Zellenforschung, Parasit. Biochem. Erreger der Maul- und Klauenseuche.* Ⓓ Zürich 1893.

Stawinski, Vitold (1888), Dr. phil., Adjoint du Labor. de Sociol. et Systématique des plantes de l'Univ. à Poznań (Polska). *Sociol. et géographie des plantes.* Ⓓ Poznań 1925.

Stchégolev, Grégoire (1882), Ass. Cabinet Embryol. et Histol. I. Univ. Moscou (U.d.S.S.R.), Mochovaja 11. *Spermatogenèse, Ovogenèse, Fécondation. Hirudinées (Morphol., Systématique).* Ⓓ Moscou 1913. Ⓟ Hirudinées.

Stearn, Allen Edwin (1894), Associate Prof. of Chem., Univ. of Missouri. Columbia, Mo. (U.S.A.), 1509 Rosemary Lane. *Bacterio-physics, Bacteriostasis, Bacterial Physiol.* Ⓓ Stanford Univ. 1916.

Stearn, Esther Wagner (1897), In charge of chem. at Christian Coll. Columbia, Mo. (U.S.A.), 1509 Rosemary Lane. *Variation and Mutation of Bacteria. Chem. Aspect of Antiseptics and Bacteriostatic Agents and Staining Reactions.* Ⓓ Univ. of Illinois 1920.

Stearns, Louis Agassiz (1892), Ass. Entomol. New Brunswick, N.J. (U.S.A.), Agric. Experim. Station. *Laspeyresia molesta.*

Stebut, Alex., Prof. Dr., Dir. d. Inst. für allg. Pflanzenzucht u. Genetik. Beograd (S.H.S.) Dobračina ul. 16. *Pflanzenzüchtung, Genetik.* o

Steche, Otto (1879), ao. Prof. Univ. Frankfurt a. Main. Hochwaldhausen (Deutschland), Post Herbstein. *Vergl. Physiol. d. Tiere.* o

Stechow, Eberhard (1883), Prof., Dr. phil., Zool. Staatssammlung München, Mitarbeiter der Deutsch. Tiefsee-Exp. (Valdivia) und versch. anderer Exp. München (Deutschland), Neuhauser Str. 51. *Marine Zool. und Biol., besonders Hydrozoen. Tierverbreitung, Temperatureinflüsse, Symbiosen. Iguanodon-Fährten.* Ⓓ München 1908.

Steck, F. W., Prof. Transvaal Univ. Coll. Pretoria (South Africa). *Physiol.* o

Steckbeck, D. Walter (1880), Ass. Prof. of Botan. Philadelphia, Pa. (U.S.A.), Univ. of Pennsylvania. *Morphol. and Taxonomy of Angiosperms.* Ⓓ Univ. of Pennsylvania 1915.

Steckl, Konstanty (1885), ao. Prof. Univ. Poznań (Polen), Sołacz. Dwór. *Forstbotan. Flora der Tatra.* Ⓓ Kraków 1920.

Stedje, Per (1884), Dir. at Norway State Fruit Experim. Station. Hermansverk, Sogn (Norway).

Steenberg, Carl Marinus (1882), Dr. phil., Lect. der Zool. an der Univ. zu Kopenhagen (Dänemark), Søborg, Silene Allé 9. *Molluskenanat. und -systematik, Anat. der Pulmonaten. Vergl. Anat. der Wirbeltiere.* Ⓓ Kopenhagen 1925. Ⓟ Molluskenschalen. In Alkohol konservierte Pupilliden.

Stefanescu, Dumitru, Prof. Academ. d. Agric. Cluj (România). *Dendrol* o

Stefanescu, Sabba, Prof. Univ. Bucarest (România). *Palaeont.* o

De Stefani, T., Prof. Univ. Palermo (Italia). *Entomol.* o

De Stefano, Giuseppe, Lib. doc. Univ. Bologna (Italia). *Palaeont.* o

Stefanoff, Boris (1894), Ass. Sofia (Bulgarien), Inst. für landwirtschaftl. Botan., Landwirtschaftl. Fak. der Univ. *Systematik, Oekol. und Pflanzengeographie, Flora der Balkanländer.* Ⓓ Sofia 1919.

Stefanowa, M., Leiter der Agric. Abt. d. Landw. Versuchs-Stat. Sofia (Bulgarien), Plostad Solunski. *Biochem.* o

Stefanski, Witold (1891), Dr. ès Sc. nat., Prof. extraord., Dir. du Labor. de Zool. et Parasit. Varsovie (Pologne), Univ. *Nématodes libres et parasit.* Ⓓ Genève 1914.

Steffen, Edwin (1889), Head, Dept. Forestry and Range Management State Coll. of Washington. Pullman, Wash. (U.S.A.). *Ecol. in relation to silviculture.* Ⓓ Iowa State Coll., Ames, Ia. 1922.

Stefko, Woldemar (1890), ao. Prof. I. Univ. Moskau (U.d.S.S.R.), Mockowaja 11, Anthrop. Inst. *Anat. und phys. Anthrop. Konstitution, Variation.* Ⓓ Moskau 1915.

Steggerda, Morris (1900), Field Investigator, Eugenics Record Office, Cold Spring Harbor, Long Island, N.Y. (U.S.A.). *Negro-white hybrids.* ⓟ Illinois 1922.

Stehle, Raymond Louis (1888), Prof. of Pharmacol. Montreal (Canada), McGill Univ. *Uric acid metabolism; effect of HCl on the mineral excretion of dogs and man; catalase; acidosis of anesthesia; action of pure ether; action of ethylene oxide; mechanism of pituitary action on urine secretion; gasometric determination of urea, non-protein nitrogen and total nitrogen.* ⓟ Yale Univ. 1913.

Stehli, Georg Jakob (1883), Dr. phil. nat., Stuttgart (Deutschland), Pfizerstraße 5. *Angewandte Entomol., landwirtschaftl. Schädlingsbekämpfung.* ⓟ Jena 1910.

Steidle, Hans (1893), Priv.Doc. Würzburg (Deutschland), Pharmacol. Inst. *Toxikol. der Pilze. Vitamine.* ⓟ Würzburg 1921.

Steil, Wm. N., Prof. of Boṭan., Marquette Univ. Milwaukee, Wis. (U.S.A.). *Morphol., cytol.* o

Stein, Alexander Konstantinowitsch (1874), Prof. Staatl. Inst. Med. Wissensch. Leningrad (U.d.S.S.R.) Ulitza Plechanowa 52, Qu. 6. *Giftige Wirkung der Insekten auf die Menschenhaut (Milben, Paederus).* ⓟ Leningrad 1911.

Stein, Wladimir W., Botan., Landw. Versuchsstation. Sotschi (U.d.S.S.R.), Cludowskaja 43. *Floristik, Angew. Botan.* o

Steinach, Eugen (1862), Univ.Prof. Wien (Deutsch-Österreich), Böcklinstr. 52. *Physiol. Experim. Biol. Keimdrüsentransplantation.* o

Steinberg, Bernhard (1897), Fellow in Pathol. Western Reserve Univ., School of Med. Cleveland, Ohio (U.S.A.). *Pathol., physiol. pathol. Soluble toxic products of colon bacilli. Supra-vital and intra-vital reactions of cells.* ⓟ Boston Univ. 1922.

Steinberg, Elisabeth (1884). Leningrad (U.d. S.S.R.), Wass. Ostrov 16, Lin. 75, Qu. 5. *Experim. Morphol. u. Biol. v. Barbarea arenata. Flora der Republik Jacutia, des Gouv. Leningrad und des Jenissei-Gebiets. Geobotan. von Leningrad u. Jenissei.*

Steinberg, Robert A. (1890), Ass. Physiol. U. S. Dept. of Agric., Bureau of Plant Industry, Office of Tobacco and Plant Nutrition. Washington, D.C. (U.S.A.). *Effect of daylength on plants under artificial light.* ⓟ Columbia Univ. 1919.

Steinbiß, Walter (1872), Dir. pathol. Inst. Auguste-Viktoria-Krankenhaus. Berlin-Schöneberg (Deutschland). *Spezielle u. experim. Pathol.* ⓟ Genf 1901.

Steinböck, Otto (1893). Graz (Deutsch-Österreich), Goethestr. 6 I. *Vergl. Anat., Histol., Ökol., geographische Verbreitung der Turbellarien.* ⓟ Graz 1923.

Steinecke, Fritz (1892), Dr. phil., Priv.Doc. an der Univ. Königsberg, Preußen (Deutschland), Botan. Garten. *Biol. und Systematik der Algen. Oekol. der Moore.* ⓟ Königsberg, Pr. 1914.

Steiner, Gotthold (1886), Nematol., Bureau of Plant Industry U. S. Dept. of Agric. Washington, D.C. (U.S.A.). *Nematodes, Gordiides, Tardigrades, Kinorhyncha, Rotifera, Gastrotricha, Soil Biol., freshwater Biol.* ⓟ Bern 1910.

Steinhausen, Wilhelm (1887), Dr. phil. et med., Ass., Priv.Doc. für Physiol. Frankfurt a. Main (Deutschland), Wolfsgangstr. 152. *Elastizität des Muskels. Reflexzeit. Elektrophysiol. Physikalische Theorie des Labyrinths.* ⓟ Phil. Gießen 1914, med. Frankfurt a. M. 1920.

Steinmann, Alfred (1892), Dr. phil., Phytopath. bei der Versuchsstation für Tee. Buitenzorg, Java (Nederl.-O.-Indië), Panaragan 18. *Mykol., pilzliche Teeschädlinge.* ⓟ Zürich 1917.

Steinmann, Gustav (1856), Prof. emer. für Geol. und Palaeont. an der Univ. Bonn (Deutschland), Colmanstr. 20. *Entwicklungslehre auf palaeont. Grundlage. Ammoniten.* ⓟ München 1877.

Steinmann, Paul (1885), Dr., Prof., Dir. des Mus. für Natur- und Heimatkunde. Aarau (Schweiz) *Tricladen, Regeneration der Tricladen, Fischereibiol.* ⓟ 1907.

Steinmetzer, Karl (1894), Ass. Tierärztliche Hochschule, Physiol. Inst. Wien III (Deutsch-Österreich), Linke Bahngasse 11. *Pharmacol. der Gallensekretion.* ⓟ Tierärztl. Hochsch. Wien 1921.

Steinsleger, M., Prof. Univ. Rosario (Argentina). *Anat.* o

Stejneger, Leonhard (1851), Head Curator of Biol. U. S. National Mus. Washington, D.C. (U.S.A.). *Classification and geographical distribution of vertebrates; bionomics of northern fur-seal; descriptive herpetology and ornithol.; zool. nomenclature.* ⓟ Oslo 1875. ⓟ Amphibians and Reptiles.

Stejskal, V., Conseiller. Hostomice pod Brdy (C.S.R.). *Mycol.* o

Stella, Giulio (1899), Ass., R. Ist. di Fisiol., Padova Rockefeller Foundation, Ist. di Farmacol. di Padova. Venezia (Italia), S. Provolo No. 4711. *Chim. fisiol.* ⓟ Padova 1923.

Stellwaag, Fritz (1886), Vorstand der Zool. Station der Weinbauversuchsanstalt Neustadt a. d. H., Pfalz (Deutschland). *Weinbauschädlinge, Obstbauschädlinge, Parasit.* ⓟ 1909.

Stempell, Walter (1869), Dr. phil., o. Prof. der Zool., vergl. Anat. und vergl. Physiol., Dir. des Zool. Inst. der Westf. Wilhelms-Univ., Mitglied des Beirats der preuß. Hauptstelle für den naturwiss. Unterricht. Münster i. W. (Deutschland), Gertrudenstraße 31. *Vergl. Anat. der Lamellibranchien, Schalenbildung der Mollusken, Morphol. und Physiol. der Protozoen (Ciliaten, Flagellaten, Cnidosporidien und Haplosporidien), Tierbilder der Mayahandschriften. Zool. Unterrichtstechnik und Arbeitsmethoden, Lehrbücher der Zool. und vergl. Physiol., wissenschaftl. Photographie.* ⓟ Berlin 1897.

Stenhouse, John Hutton, Caledonian United Service Club. Edinburgh (Scotland). *Ornithol.* o

Stenhouse, Williams R., Prof. Univ. Reading (England). *Bact.* o

Stensiö, Erik Helge Osvald (1891), Prof., Dir. palaeozool. dept. of the Royal State Mus. of nat. hist. Stockholm 50 (Sverige). *Fossil fishes, mesozoic and palaeozoic, Ostracoderms.* ⓟ Upsala 1921. ⓟ Fossil animals, from the Kambrian, Ordovicium and Silurian strata of Sweden.

Stenta, Mario, Doc. R. Univ. degli Studi. Padova (Italia). *Zool.* o

Štěpán, Václav Josef (1873), Prof., Dir. der Fischerei-Schule und Leiter d. Fischerei- und Hydrobiol. Station Vodňany (C.S.R.). *Wasserkäfer u. ihre Larven. Fischkrankheiten. Hydrobiol.* ⓟ Wasserkäfer u. Wasserfauna. Fischkrankheiten (Serienschnitte).

Stépanoff, Jean (1883), Chef du labor. antipesteux. Ouralsk (U.d.S.S.R.), Sowietskaja 134. *Bact.: peste.* ⓟ Paris 1919.

Stepanov, Paul J., Obergeol., Dir. Mus. Comité Géol. de Russie, Prof. für Geol. am Berginst. Leningrad (U.d.S.S.R.), W. O., Sredny Pr. 72-B. *Geol. (Nichterze); speziell: Steinkohlenablagerungen Südrußlands. Palaeont. (Bryozoen, Korallen).* o

Stepanow, Nikolaj N., Prof. Landwirtschaftl. Akad. Timirjasew. Moskau III (U.d.S.S.R.), Grashdanskaja 47, W. 36. *Waldbau. Wald-Melioration. Waldbau in der Steppe.* o

Stephen, Alexander Charles (1893), Ass. Natural History Dept. Royal Scottish Mus. Edinburgh (Scotland). *Quantitative Marine Zool., Mollusca.* ⓟ Aberdeen 1919.

Stephensen, Knud Hensch (1882), Museumsamanuensis. København (Danmark), Zool. Mus. *Crustacea Malacostraca, Amphipoda.* ⓟ Crustacea (and Pycnogonida), especially Malacostraca.

Stephenson, L.W. (1876), Geol. U. S. Geol. Survey. Washington, D.Col. (U.S.A.). *Palaeont. of the Atlantic and Gulf Costal Plein of U.S.A., Mexico and Venezuela.* ⓟ John Hopkins 1907. o

Stephenson, Thomas Alan (1898), Senior Ass. Zool. Department Univ. Coll. London W. C. 1

(England). *General morphol., bionomics of Actiniaria and Haliotis. British Orchids.* Ⓓ Aberystwyth 1923.
Stepnickij, S. O., Prof. Med. Inst. Moskau (U.d.S.S.R.), ul. Spartaka 2. *Anat.* ○
Steppun, Oskar Avg., Prof. II. Univ. Moskau (U.d.S.S.R.), *Biol. Chem.* ○
Stern, Curt (1902), Ass. Berlin-Dahlem (Deutschland), Kaiser-Wilhelm-Inst. für Biol. *Vererbung von Drosophila.* Ⓓ Berlin 1923.
Stern, Kurt (1892), Dr. phil. Frankfurt a. M.-Niederrad (Deutschland), Deutschordenstr. 78. *Physikalisch-chem. Pflanzenphysiol. Elektrophysiol. der Pflanzen.* Ⓓ München 1914.
Stern, Lina Solomon, Prof. II. Univ. Moskau (U.d.S.S.R.), *Physiol. d. Tiere.* ○
Sternberg, Hermann (1900), Dr. med., Ass. am Embryol. Inst. der Univ. Wien IX/3 (Deutsch-Österreich), Schwarzspanierstr. 17. *Entwicklungsmechanik. Entwicklungsgeschichte und Mißbildungslehre des Menschen, frühe Embryonalstadien.* Ⓓ Wien 1925.
Sternon, Fernand Lucien Auguste (1895), Dr., Ingénieur agricole. Pharmacien. Virton (Belgique), Grand'place 7. *Mycol. Systématique des Champignons inférieurs (Imperfecti).* Ⓓ Nancy 1925.
Sterzinger, Othmar (1879), Priv.Doc. Univ. Graz (Deutsch-Österreich), Krenngasse 28 III. *Experim. Psychol. Psychische Energie.* Ⓓ Gießen 1912.
Steudel, Hermann (1871), a.o. Prof. und Abteilungsvorsteher am Physiol. Inst. der Univ. Berlin N 4 (Deutschland), Hessische Str. 3/4. *Physiol. Chem. (Zellkern, Nucleinsäuren, Harnsäureproblem).* Ⓓ Kiel 1895.
Steuer, Adolf (1871), o. Univ.-Prof., Dir. des Zool. Inst. Innsbruck (Deutsch-Österreich), Universitätsstraße 4. *Marine Biol., Planktonkunde, Copepoden.* Ⓓ Wien 1896.
Steuer, Alexander (1867), Oberbergrat, o. Prof. Techn. Hochsch., Dir. der Geol. palaeont.-mineral. Abt. des Landesmus. Darmstadt (Deutschland), Geol. Inst. *Palaeont., Stratigraphie.* Ⓓ 1894.
Stevens, Asa Neiley (1901), Biochem., Eli Lilly & Co., Biol. Labor. Greenfield, Ind. (U.S.A.), 410 E. Main Street. *Concentration of Antitoxin of Blood, Plasma (Horse).*
Stevens, Clark Leavitt (1895), Ass. Prof. Forestry. Univ. New Hampshire. Durham, N.H. (U.S.A.). *Rate of growth of tree roots, Pinus strobus and P. resinosa.*
Stevens, Frank Lincoln (1871), Prof. of Plant Pathol. Urbana, Ill. (U.S.A.), Univ. of Illinois. *Mycol. and Pathol.* Ⓓ Chicago 1900.
Stevens, Harold Edwin (1880), Plant Pathol. Ft. Myers, Flo. (U.S.A.). *Diseases of the Palms, Oreodoxa regia, Citrus and Avocado diseases.* Ⓓ Univ. of Illinois 1910.
Stevens, Neil E. (1887), Pathol. U.S. Bureau of Plant Industry. Washington, D.C. (U.S.A.). *Diseases of Strawberries and Cranberries. Fungi: Physalospora and Botryosphaeria.* Ⓓ Yale 1911.
Stevens, Orin Alva (1885), Deputy Seed Commissioner and Ass. Prof. of Botan., North Dakota Agric. Coll. Fargo, N.D. (U.S.A.), State Coll. Station. *Seed testing; weeds. Ecol., floral and aculeate Hymenoptera.* Ⓓ Kansas State Agric. Coll. 1907. Ⓟ North Dakota Hymenoptera Aculeata.
Stevens, William Ch., Prof. of Botan., Univ. of Kansas. Lawrence, Kan. (U.S.A.). ○
Stevenson, Archibald Campbell, Pathol. to the Wellcome Bureau of Sc. Research. London W.C. 1 (England), 25/28 Endsleigh Gardens. *Pathol. Anat., Parasit.* Ⓓ London 1894.
Stevenson, Fred J. (1886), Ass. Prof. of Plant Breeding, Univ. of Minnesota. St. Paul, Minn. (U.S.A.), Univ. Farm. *Plant Breeding, Cytol. in relation to plant genetics. Genetics and breeding of small grains.* (1926).
Stevenson, Paul Huston (1890), Ass. Prof. of Anat. Department of Anat., Peking Union Med. Coll. Peking (China). *Anat., Physical Anthrop., Racial and Physical Anthrop. of Asia.*

Steward, A. N., Lect. Univ. Nanking (China). *Botan.* ○
Stewart, Alban, Prof. of Botan. and Bact., Florida State Coll. for Women. Tallahassee, Fla. (U.S.A.). *Pathol. anat., classification of flowering plants, Bact.* ○
Stewart, Colin Campbell (1873), Brown Prof. of Physiol., Acting Dean, Dartmouth Coll. and Med. School. Hanover, N.H. (U.S.A.). *Muscle Physiol.* Ⓓ Clark 1897.
Stewart, Dewey (1898), Instr.: New York State Coll. of Agric. at Cornell Univ., also Junior Pathol. U.S. Dept. of Agric. Station Rocky Ford., Col. (U.S.A.). Ithaca, N.Y. (U.S.A.). *Sugar Beet Diseases.* Ⓓ Mich. State 1924.
Stewart, Fred Carlton (1868), Chief in Research Botan., New York Agric. Experim. Station. Geneva, N.Y. (U.S.A.). *Fleshy fungi and plant pathol.* Ⓓ Iowa State Coll. 1894.
Stewart, John A. Mueller (1886), Prof. of Biol. and Geol., Thiel Coll. Greenville, Pa. (U.S.A.). *Botan., Ecol.* Ⓓ Univ. of Pittsburgh 1925.
Stewart, M. J., Prof. Univ. Leeds (England). *Pathol.* ○
Stewart, Morris A. (1902), Instr., Dept. of Biol., Univ. of Rochester, N.Y. (U.S.A.). *Siphonaptera and internal parasites of mammals.*
Stewart, Ralph Randles (1890), Vice-Principal. Rawalpindi (British India), Gordon Coll. *Taxonomy of the ferns and flowering plants of the Punjab, Kashmir and the Northwest Himalayas.* Ⓓ Columbia Univ. 1916. Ⓟ *Mosses, ferns, flowering plants and fungi.*
Steyer, Karl (1877), Prof., Dr. phil., Leiter der Landwirtschaftl. Versuchsstation (Hauptstelle für Pflanzenschutz), Lübeck (Deutschland), Mengstr. 4 I. *Pflanzenpathol., Fischereibiol.* Ⓓ Leipzig 1901.
Stiasny, Gustav, Dr., Konserv. Reichsmus. Leiden (Holland), Rijnsburgerweg 56. *Salpen, Scyphomedusen, spez. Rhizostomeen, Entwicklung der Enteropneusten, Plankton des Meeres.*
Stich, Rudolf (1875), o. Prof. der Chir., Dir. chir. Univ.-Klinik. Göttingen (Deutschland). *Transplantation.* Ⓓ Erlangen 1898.
Stichel, Hans Wolfgang (1898), Ass. Forschungsstelle für Pelztierkunde. Tharandt, Sa. (Deutschland). *Heteropteren.* Ⓓ Berlin 1923.
Stickdorn, Walther (1883), Dr. med. vet., Technischer Leiter des Bact. und Seruminst. in Landsberg a. d. W. (Deutschland), Heinersdorfer Str. 14a. *Bact. und Serol. der Tierseuchen.* Ⓓ Gießen 1909.
Stickel, Paul W. (1899), Junior Forester, U.S. Dept. of Agric., Forest Service, Northeastern Forest Experim. Station, Amherst, Mass. (U.S.A.). *Relationship between forest fire hazard and meteorol. conditions. Forest soil investigations.* Ⓓ Yale Univ. 1924.
Sticker, Anton (1861), Prof., Dir. Radium-Inst. Münster i. W. (Deutschland). *Geschwulstkrankheiten.* Ⓓ Berlin 1887. Ⓟ *praep. von experim. erzeugten Geschwülsten bei Mäusen, Ratten, Hunden und Vögeln. Mikrosk. Praep.*
Stickney, Fenner (1892), Associate Entomol., U.S. Dept. of Agric. Washington, D.C. (U.S.A.) *Coccidae (in general). Date palm pests, especially the date scales.* Ⓓ Univ. of Illinois 1921.
Stickney, Malcolm E., Prof. of Botan., Denison Univ. Granville, Licking County, Ohio (U.S.A.). *Morphol., algae.* ○
Stiehr, K. F. Gustav (1867), Reg.R. d. B. Landesanstalt für Pflanzenbau u. Pflanzenschutz. München 23 (Deutschland), Ungererstr. 30 I. *Gärungschemie u. Chem. des Düngers u. der Jauche, Bact.* Ⓓ Kiel 1903.
Stieve, Hermann (1886), Dr. med. et phil., o. ö. Prof. der Anat., Vorstand der Anat. Anstalt der vereinigten Friedrichs-Univ. Halle a. d. S.-Wittenberg. Halle a. d. S. (Deutschland), Große Stein-

straße 52. *Anat., Entwicklungsgeschichte, Histol., Entwicklungsmechanik.* ⒹMünchen 1911.

Stiffler, Ethel G. (1899), Instr. in Biol., Univ. of Arizona. Tucson, Ariz. (U.S.A.), Dept. of Biol. *Cytol.* ⓟ *Desert flora.*

Stigler, Robert (1878), o. ö. Prof. der Anat. und Physiol. der Haustiere an der Hochsch. für Bodenkultur in Wien XVIII (Deutsch-Österreich). *Physiol. Optik, Mechanik des Blutkreislaufes, Rassenphysiol., Sexualphysiol., Physiol. des Wiederkäuermagens. Liquormechanik.* Ⓓ Wien 1903. ⓟ *Ethnol. Geräte zentralafrikanischer Neger. Vogelbälge, Käfer, Schmetterlinge, blutsaugende Insekten aus Uganda.*

Stiles, Charles F. (1891), Ass. Extension. Entomol. Stillwater, Okla. (U.S.A.), A. and M. Coll. *Insect and Rodent Control.*

Stiles, Charles Wardell (1867), Prof. of Zool., U. S. Public Health Service. Washington, D.C. (U.S.A.), Hygien. Labor., 25. and E. Streets, N.W. *Med. Zool., Zool. Nomenclature.* Ⓓ Leipzig 1890.

Stiles, Walter (1886), Prof. of Botan. in the Univ. of Reading (England). *Plant Physiol.: Permeability, Photosynthesis.* Ⓓ Cambridge, England, 1922.

Stille, Hans (1876), Univ.Prof. Göttingen (Deutschland), Herzberger` Landstr. 55. *Geol. Palaeont.* o

Stiller, Győző (1860). Szeged (Ungarn), Erdö-ut 9. *Höhlenkäfer.*

Stillinger, Charles Roy (1889), Associate Pathol., U. S. Department of Agric., Office of White Pine Blister Rust Control. Spokane, Wash. (U.S.A.), 618 Realty Building. Ⓓ Univ. of Idaho 1915. ⓟ *Specimens of forest tree diseases and rusts (native).*

Stillmunkés, André (1890), Préparateur de Pathol. générale et expérim., Toulouse (France), 19, rue du Taur. *La Syncope adrénalino-chloroformique.* Ⓓ Toulouse 1912.

Stirrett, George Milton (1899), Entomol. in Charge, European Corn Borer Investigations, Entomol. Branch, Dominion Department of Agric. Chatham, Ontario (Canada), Entomol. Labor. *Ecol. of Insects, General Entomol., Taxonomy and classification of Insects.* Ⓓ Purdue Univ. 1924. ⓟ Coleoptera Halticini (chrysomelidae).

Stitz, Hermann (1868),Wissensch. Hilfsarbeiter am Zool. Mus. in Berlin (Deutschland), Invalidenstr. 43. *Entomol.: Anat. der Insekten, Ameisen, Neuropteren.*

Stjernman, Robert Oskar Gustav (1899), Ass. an der Landwirtsch. Hochsch. København (Danmark), Finsensvej 73 I. *Vergl. Myol.*

Stockard, Charles Rupert (1879), Prof. of Anat., Cornell Univ. Med. Coll. New York City (U.S.A.), 477 First Avenue. *Experim. study of development and growth, Causes of abnormal development. Influence of external conditions on inheritance and development. Relation of internal secretions to ovulation and oestrous, to development and type of form.* Ⓓ Ph. Columbia Univ. 1906, M. Würzburg, Sc. (hon.) Univ. of Cincinnati 1920.

Stocker, Otto (1888), Dr. phil. nat. Bremerhaven (Deutschland), Bogenstr. 9. *Physiol. Pflanzenökol.* Ⓓ Freiburg i. Br. 1923.

Stodel, Georges (1879), Maitre de conf. Paris (France), 19, B. Delessut. *Colloides, Immunité.* Ⓓ Paris 1908.

Stoeckhert, Emil (1888). Nittenau, Bayern (Deutschland). *Hymenopteren, Bienen (Apidae).* ⓟ *Bienen der paläarktischen Region, insbesondere der Gattungen Andrena und Nomada.*

Stoeckhert, Ferdinand Karl (1889), Dr. phil. Erlangen (Deutschland), Bismarckstr. 6. *Systematik und Biol. der Apiden (Hym.).* Ⓓ Erlangen 1918.

Stöhr jr., Philipp (1891), Dr. med., a.o. Prof. d.Anat. Gießen (Deutschland), Anat. *Mikroskopische Anat. des sympathischen Nervensystems. Entwicklungsmechanik des Amphibienherzens.* Ⓓ Würzburg 1917.

Stolanoff, N. A., Dr., ao. Prof. für angew. Botan., Inst. f. landw. Botan. d. landw. Fac. Sofia (Bulgarien). *Systematik, Phytogeographie.* o

Stokes, A., Prof. Univ. London S.W. 7 (England), S. Kensington. *Pathol.* o

Stokey, Alma G. (1877), Prof. of Botan., Mount Holyoke Coll. South Hadley, Mass. (U.S.A.). *Fern prothallia.* Ⓓ Univ. of Chicago 1908. ⓟ Massachusetts ferns.

Stoklasa, Julius, Dipl.-Ing. agr., Prof. der tschechischen techn. Hochsch., Dir. staatl. Versuchsstationen, Vize-Präsident der tschechoslowakischen Akademie für Bodenkultur. Praha XII (C.S.R.), Villa Groebe. *Biol. und Biochem. der Aufbau- und Abbauprozesse in der chlorophyllosen und chlorophyllhaltigen Zelle. Einfluß der Radioaktivität auf den gesamten Stoffwechsel des pflanzlichen und tierischen Organismus. Biochem. Kreislauf des Phosphat-Ions im Boden. Physiol. Bedeutung des Kalium-Ions im Organismus der Zuckerrübe. Biophysikalische und biochem. Durchforschung des Bodens.* Ⓓ Leipzig 1892.

Stoland, O. O. (1881), Prof. of Physiol. and Pharmacol. Lawrence, Kan. (U.S.A.), Univ. of Kansas. *Internal Secretion. Thyroid and parathyroid. Pituitary, Toxicity, Anaphylaxis.* Ⓓ Chicago 1913.

Stolbowa, Marie Michael (1897), Aspirant. Kasan (U.d.S.S.R.), Zootomisches Kabinett, Univ. *Vergl. Morphol. der Wirbellosen. Polychaeta. Biocoenotik d. Meere.*

Stoll, F. E., Konserv. Rīga (Latvija), Domas muzejs. *Mykol. u. Ornithol.* o

Stoll, Norman R. (1892), Associate in Helminthol., School of Hygiene and Public Health, Johns Hopkins Univ. Baltimore, Md. (U.S.A.), 615 N. Wolfe Street. *Quantitative methods in evaluation of hookworm infestation. Free living stages of hookworm. Epidemiol. of hookworm disease (China and Tropical America). Viability of hookworm ova in stored nightsoil.* Ⓓ Johns Hopkins Univ. 1923. ⓟ *Human hookworms; eggs, larvae, adults.*

Stoller, James H., Prof. of Geol. and Biol. Shenectady, N.Y. (U.S.A.), Union Coll. o

Stolley, Ernst (1869), o. Prof. d. Geol. u. Mineral. Techn. Hochsch. Braunschweig (Deutschland), Geol. Inst. *Palaeozool. u. Palaeophytol.* Ⓓ München 1891.

Stolses, Wm. Royal (1870), Dir., Bureau of Bact., Health Dept., Prof. of Bact., Med. Dept., Univ. of Maryland. Baltimore, Md. (U.S.A.), 1639 N. Calveil St. Ⓓ Univ. of Maryland 1891.

Stolte, Hans-Adam (1888), Dr. phil., Priv.Doc., I. Ass. am Zool. Inst. Tübingen (Deutschland). *Fortpflanzungsbiol., Altern der Anneliden, Formwechsel der Protozoen, Konstitutionsbiol. der Wirbeltiere.* Ⓓ Würzburg 1920.

Stolygvo, Nikolaj (1900), Ass. Riga (Latvija), I. Rig. Pilsętas slimonia. *Bact. Flora der Mundhöhle und der Atmungsorgane. Streptokokken und Pneumokokken.* Ⓓ Riga 1926.

Stolze, Karl Viktor (1901), Dr. agr., Diplomlandwirt, wissenschaftl. Ass. an der Hauptstelle für Pflanzenschutz der Landwirtschaftskammer Schlesien in Breslau (Deutschland), Werderstr. 17 III. *Kartoffelstaudenkrankheiten.* ⒹBonn-Poppelsdorf1924.

Stomps, Theodoor Jan (1885), Dr., o. Prof. für Pflanzenanat. und -systematik, Dir. des Botan. Gartens und Inst. der Univ. Amsterdam (Holland), Plantage Middenlaan 7 *Zytol. und Vererbungslehre.* Ⓓ Amsterdam 1910. ⓟ *Pflanzen.*

Stone, Roland Elisha (1881), Dr., Associate Prof. Botan. Ont. Agr. Coll. Guelph, Ont. (Canada). *Plant physiol. and Mycol. Life histories of fungi.* Ⓓ Cornell Univ. 1912.

Stone, Witmer (1866), Dir. of Mus., Academy of Nat. Sc. of Philadelphia, Pa. (U.S.A.), Logan.Square. *Ornithol.: Geogr. Distribution, Migration and Moult.* Ⓓ Univ. of Pennsylvania 1913.

Stoner, Dayton (1883), Ass. Prof. Zool., Univ. of Iowa, Iowa City, Ia. (U.S.A.), 603 Bummit Street. *Insects, Birds, Mammals, Bionomics, Taxonomic, Distribution, Life History.* Ⓓ Univ. of Iowa 1919. ⓟ Pentatomoidea of North America.

Stookey, Stephen W., Prof. of Geol. and Botan., and Dean, Coe Coll. Cedar Rapids, Ia. (U.S.A.). *Genetics, plant breeding.* ○

Stopford, J. S. B., Prof. Vict. Univ. Manchester (England).· *Anat.* ○

Stopnizkij, Sewerin (1865), Prof., Prosektor. Moskau (U.d.S.S.R.), Anat. Inst. *Anat.* ⓟ Moskau.

Stoppel, Rose (1874), Dr. phil., Priv.Doc., Wissenschaftl. Hilfsarbeiter. Hamburg (Deutschland), Inst. für allgemeine Botan. *Reizphysiol. der Pflanzen (Pflanzenschlaf).* ⓟ Freiburg i. Br. 1909.

Stoppoloni, Giuseppe (1875), Prof. Str. Stabile di Anat. descrittiva Istol. gen. con incarico dell' Ezoognosia. Camerino (Italia), Ist. Sup. di med. vet. *Coltura dei microorganismi. Tumore melanotico. Tessuto cartilagineo delle valvole sigmoidee.* ⓟ 1899.

Storch, Otto (1886), tit. ao. Prof., o. Ass. am II. Zool. Inst. der Univ. Wien I (Österreich), Ring des 12. November. *Polychaeten, Rädertiere, Phyllopoden, Cladoceren, Copepoden, Ostracoden, Nahrungserwerb, Fangapparate der Krebse, Cytol. der Parthenogenese, Odonaten, Eiablage und Flug der Odonaten. Dipteren.* ⓟ Wien 1910.

Storer, Tracy Irwin (1889), Associate Prof. of Zool. Davis, Cal. (U.S.A.), Branch, Coll. of Agric., Univ. of California. *Vertebrate zool., habits, ecol. and distribution of land vertebrates in western North America.* ⓟ Univ. California, Berkeley, 1912.

Storey, Harold Haydon (1894), Mycol., Department of Agric. Natal Herbarium, Durban (Union of South Africa). *Virus diseases of plants, plant pathol.* ⓟ Cambridge 1920.

Storey, Thomas Andrew (1875), Coll. Prof. New York, N.Y. (U.S.A.), 99 Clermont Ave. *Physiol. Hygiene.* ○

Stork, Harvey E. (1890), Prof. of Biol. Carleton Coll. Northfield, Minn. (U.S.A.). *Cytol. of Basidiomycetes. Wood structure Am. trees. Central American plante life.* ⓟ Cornell 1920.

Storkàn, Jaroslav (1890). Dr. des sc. nat. Ass. à l'Inst. Zool. Univ. Charles IV. Praha II (C.S.R.), U Karlova 3. *Pseudoscorpiones, Isopoda, Opiliones; System. Anat., Biol.* ⓟ Praha 1923.

Stoss, Anton (1888), Univ.Prof. München (Deutschland), Halzstr. 12. *Anat. u. Histosl. d. Haustiere.* ○

Stout, Arlow B., Dir. of the Labor., New York Botan. Garden. New York City (U.S.A.). *Cytol., Genetics, Phytopath., sterility in plants.* ○

Stout, Gilbert Leonidas (1898), Field Botan. for the State Nat. History Survey State of Illinois, Collaborator Plant Disease Survey Bureau of Plant Industry, U. S. Dept. of Agric. Urbana, Ill. (U.S.A.), Nat. History Building, Univ. of Illinois. *Ascomycetes, Fungi imperfecti, Hemisphaeriaceae, parasit. Fungi, Plant Pathol.* ⓟ Ohio 1923.

Stover, Wilmer Garfield (1881), Prof. of Botan. Columbus, O. (U.S.A.), Ohio State Univ. *Plant Pathol. Mycol. Virus diseases of tomato. Scab of apple, Basidiomycetes, Agaricaceae of Ohio.* ⓟ Univ. of Wisconsin 1921.

Strachan, Archibald Sutherland (1891), Pathol., South African Inst. for Med. Research, Head of Department of Pathol., Univ. of the Witwatersrand. Johannesburg (South Africa), Box 1038. *Pathol.* ⓟ Glasgow 1913.

Strachow, Trinofej Danilowitsch (1890), Prof. am Charkower Inst. für Land- u. Forstwirtschaft. Charkow (U.1.S.S.R.), Dewitschja 5, W 2. *Botan., Phytopath. Entwicklung d. Krankheitserreger. Immunität. Ustilagineae Uredineae.* ⓟ Par. *Pilze.*

Stracke, Gerhard Joachim (1877), Dr. bot. et zool. Amsterdam (Holland), Ceintuurbaan 249. *Mikroskopie der Nahrungs- und Genußmittel.* ⓟ Amsterdam 1904.

Strada, Ferdinando, Prof. Univ. Nac. Córdoba (Argentina). *Histol., Embryol., Anat. pathol.* ○

van Straelen, Victor Emile (1889), Dir. du Mus. royal d'Histoire nat. de Belgique, Prof. de paléontol. et de géol. à l'Univ. de Gand, Bruxelles (Belgique), 31, rue Vautier. *Paléont., Crustacés.* ⓟ Sc. nat. Bruxelles 1914, sc. géol. et paléont. 1925.

Strampelli, Nazareno (1866). Dir. Staz. Sperim. di Granicoltura Rieti; Dir. Inst. Naz. di Genetica per Cerealicoltura, Roma (Italia), Piagra Porta Pia 121. ⓟ Pisa 1891. ⓟ Sementi di movi frumenti per ibridagione e selegione genealogica.

Straňák, Franz, Dr., Priv.Doc. f. Phytopath. Praha (C.S.R.), Strašnice 198. ○

Strand, August Leroy (1894), Graduate Ass., Division of Entomol. and Economic Zool., Univ. of Minnesota. St. Paul, Minn. (U.S.A.), Univ. Farm. *Insecticides.*

Strand, Embrik (1876), o. Prof. Zool. Univ., Dir. d. Systematisch-Zool. Inst. und der Hydrobiol. Station Univ. Riga (Latvija), Pasta kastite 802. *Araneol. Lepidopterol., Hymenopterol.* ⓟ Oslo 1897.

Strandskov, Herluf Haldan, Ass., Prof. of Botan., Univ. of Louisville, Ky. (U.S.A.). *Genetics, physiol.* ○

Strangeways, T. S. P., Lect. St. John's Coll. Cambridge (England). *Spec. Pathol.* ○

Stranski, Iwan T. (1886), o. Doz., Leiter des Inst. für Allgemein. Ackerbau. Sofia (Bulgarien), Landwirtschaftl. Fac. der Univ. *Unkräuter.* ⓟ Kulturpflanzen- und Unkräuter-Samen.

zur Strassen, Otto (1869), Geh.Reg.R., Dr. phil., o. Prof., Dir. des Senckenbergischen naturhistor. Mus. Frankfurt a. M. (Deutschland), Zool. Inst., Robert-Mayer-Str. 6. *Allgemeine Biol., Entwicklungsmechanik, Tierpsychol.* ⓟ Leipzig 1892.

Strasser, Hans (1852), Prof. Dir. d. Anat. Inst. d. Univ. Bern (Schweiz), Finkenhubelweg 20. *Extremitätenknorpel, Nasenhöhlen, Knochenpneumatisation, Nervensystem.* † 16. 4. 1927.

Strasser, P. Pius (1843). Benedictiner-Stift Seitenstetten (Nieder-Österreich). *Kryptogamenflora (Flechten und Pilze) von Sonntagberg und Umgebung.*

Stratton, Robert (1883), Ass. Prof. of Botan. and Plant Pathol. Oklahoma Agric. and Mechanical Coll. Stillwater, Okla. (U.S.A.), 320 Hester Street. *Theoretical and Practical Botan., Systematic Botan., Plant Pathol., Plant Physiol., Genetics.* ⓟ Plant herbarium specimens.

Straub, Hermann (1882), Univ.Prof., Dir. der med. Klinik. Greifswald (Deutschland). *Blutkreislauf; Atmung; Mineralstoffwechsel.* ⓟ.Tübingen 1909.

Straub, Johann (1893), Ass., Dr. Debrecen (Ungarn), Simonyi-ut 16. *Quantitative Mikrochem.* ⓟ Budapest 1923.

Straub, Walther (1874), Prof., Dr. med., Dir. des Pharmacol. Inst. der Univ. München (Deutschland), Nußbaumstr. 28. *Experim. Pharm.* ⓟ München 1898.

Straus, Aubrey H. (1887), Dir. of Labor., Virginia State Board of Health. Richmond, Va. (U.S.A.). *Typhoid fever, dysentery.* ⓟ William and Mary Coll. 1923.

Strausbaugh, Perry Daniel (1886), Prof. and Head of the Botan. Dept., West Virginia Univ. Morgantown, W.Va. (U.S.A.). *Dormancy in buds and microchem. of developing cell walls.* ⓟ Univ. of Chicago 1920.

Strausz, Ladislaus (1902), Pathol.-anat. Inst.Univ. Pécs (Ungarn). ⓟ Pécs 1927.

Strawiński, Konstanty (1892), Dr. phil., Ass. am Entomol. und forstschutzlichen Inst. der Hochsch. für Bodenkultur in Warschau (Polen). Skierniewice (Polen), Palac. *Hemiptera-Heteroptera der Umgegend von Skierniewice, Monographie von Mesocerus marginatus (Hem.-Heteroptera, Coreidae).* ⓟ Charkoff 1917. ⓟ Hemiptera-Heteroptera Polens.

Strecker, Friedrich (1879), ao. Prof. Univ. Breslau (Deutschland), Alsenstr. 49. *Anat. und Biol.* ⓟ Breslau 1905.

Streeter, Daniel Denison (1885). Brooklyn, N.Y. (U.S.A.), 214 Fenimore St. *Herpetol. of Borneo and Sumatra.* ⓟ Columbia 1909. ⓟ Reptiles and amphibians of the United States.

Streeter, George L. (1873), Dir., Dept. of Embryol., Carnegie Inst. of Washington. Baltimore, Md.

(U.S.A.). *Human Embryol.: Nervous System and Special Sense Organs.* ⓓ Columbia Univ. 1899.
Streets, Rubert Burley (1895), Ass. Prof. of Plant Pathol., Univ. of Arizona; Ass. Plant Pathol. Ariz. Agric. Experim. Station. Tucson, Ariz. (U.S.A.), Room 220, Agric. Bldg. *Plant Pathol. and Mycol. of Semi-arid regions. Root rot caused by Ozonium omnivorum of cotton, alfalfa and many other plants. Graphiola leaf spot of date palms. Heart rot of date palm.* ⓓ Univ. of Wisconsin 1924. ⓟ Mycol. specimens from Arizona, Wisconsin and Montana (parasit. fungi, exsiccati in packets).
Strelin, Sergius (1875), Prof. of Phytopath. Yalta, Crimea (U.d.S.S.R.), Gov. Botan. Garden, Nikita. *Biol. of Fungi.* ⓓ Univ. of Simferopol 1920. ⓟ Herbarium specimens.
Strelnikow, I. D. (1887), Ass. Leshafts Wissenschaftl. Inst., Doc. Univ. Leningrad (U.d.S.S.R.). *Adsorpt. bei Protozoa. Schwarzmeerfauna. Mollusken.*
Stremme, Hermann (1879), Dr. phil., o. Prof. für Mineral. und Geol. an der Techn. Hochsch. Danzig-Langfuhr (Deutschland). *Palaeont. der Wirbeltiere.* ⓓ Berlin 1903.
Stremouchow, D. P., Sekretär Com. Géol. de Russie. Moskau (U.d.S.S.R.), Jamskaja 1. *Juraammoniten der Krim.* o
Streng, W. Osw., Dr. med., Prof. Univ., Dir. Bact.-Serol. Labor. Helsinki (Finnland), Georgsgatan 13. *Bact. und Serol.* o
Stresemann, Erwin (1889), Dr. phil., Kustos; Generalsekretär der Deutschen Ornithol. Gesellschaft. Berlin N 4 (Deutschland), Zool. Mus. der Univ., Invalidenstr. 43. *Ornithol. (Systematik, Biol., Vererbung, Geschichte der Ornithol., geographische Verbreitung).* ⓓ München 1920.
Stříbrný, Venceslav (1853), Landwirtschaftl. Versuchsstation in Sophia (Bulgarien), ul. Oborište 29. *Angewandte Botan.* ⓓ Pomol. Inst. Troja bei Prag 1879. ⓟ Bulgarische Herbariumpflanzen.
Strickland, Edgar Harold (1889) Prof. of Entomol. Edmonton, Alberta (Canada), Univ. *Economic Entomol.* ⓓ Harvard Univ. 1912.
Strilcinc, D., Dr., Ass. Labor. de Zootechnique Fac. méd. vét. Bucarest (Romanie). o
Strőszner, Ede (1869), Doc. Fac. Med. Univ. Budapest (Ungarn). *Bact. Serol.* ⓓ Wien 1896.
Stroganow, Aleksej N., Prof. I. u. II. Univ. Moskau (U.d.S.S.R.), Kriwonikolskij per. 5, W. 19. *Botan., Pflanzen-Anat.* o
Stroganov, Sergej Nik., Doc. II. Univ. Moskau (U.d.S.S.R.). *Mikrobiol.* o
Strohbinder, Marie (1894), Ass. Abt. der allgemeinen Mikrobiol., Inst. für experim. Med. Leningrad(U.d.S.S.R.),Lopuchinskaja 9. *Biochem. u. Bact. milchwirtschaftlicher Produkte.*
Strohl, André (1887), Prof. de Physique méd. à la Fac. de Méd. de Paris 5 (France), 13, rue Pierre-Nicole. *Electro-physiol.: Résistance et polarisation électrique des tissus vivants. Physiol. des réflexes chez l'Homme. Conductibilité électrique du corps humain.* ⓓ Méd. Paris 1913, ès sc. phys. 1924.
Strohl, Johannes (1886), Prof. Dr. Zollikon bei Zürich (Schweiz), Wytellikerstr. 12. *Mollusken. Theor. Zool.* o
Strohmeyer, Heinrich (1871), Dr. phil., Ministerialrat im Reichslandwirtschafts-Ministerium. Berlin W 30 (Deutschland), Hohenstaufenstr. 43. *Forst-Insekten: Morphol., Biol. und Systematik der Platypodidae, Scolytidae (Ipidae) und Lymexylonidae der ganzen Erde.* ⓓ Marburg a. d. Lahn. ⓟ Platypodidae, Scolytidae (Ipidae) und Lymexylonidae.
Strøm, Kaare Münster (1902), Oslo (Norge), Bygdø allé 11. *Freshwater algae, rotiferi, crustacea, protozoa. The physical factors in fresh water, and their influence on the organisms. Freshwater plankton.*
Stromer von Reichenbach, Freiherr Ernst (1871), Hon.Prof., Dr. phil., Hauptkonserv. an der bayer. Staats-Sammlung für Palaeont. und histor. Geol. München (Deutschland), Alte Akademie,

palaeont. Sammlung. *Palaeozool., besonders fossile Wirbeltiere; tertiäre und kretazische Wirbeltiere Aegyptens und der deutschen Schutzgebiete in Afrika.* ⓓ München 1895.
Stromsten, Frank Albert (1872), Associate Prof. of Zool., The State Univ. of Iowa. Iowa City, Ia. (U.S.A.), Dept. of Zool. *Compar. Anat., Embryol. and Ecol. Cladocera. Copepods. Venous system of Chelonia. Lymph hearts. Habits of the turtle. Musk glands.* ⓓ Princeton 1905. ⓟ Cladocera. Copepods.
Strong, Leonell Clarence (1894), Associate Prof. Biol., Research Fellow, School of Public Health, Harvard Univ. Boston 30, Mass. (U.S.A.), 316 Hyde Park Ave. *Genetics of cancer in experim. animals.* ⓓ Columbia Univ. New York 1922. ⓟ Transplantable Tumors in Mice.
Strong, Reuben Myron (1872), Prof. and Head of Department of Anat., Loyola Univ. School of Med. Chicago, Ill. (U.S.A.), 706 South Lincoln St. *Neurol., ossification of the skeleton, pigmentation, color of integumental structures and compar. anat. of vertebrates.* ⓓ Harvard 1901.
Strong, Richard P. (1872), Prof. of Tropical Med., Harvard Univ. Med. School. Boston, Mass. (U.S.A.). *Amoebic Dysentery, Bacillary Dysentery, Tropical Intestinal Parasit., Cholera, Beriberi, Verruga Peruviana, Trypanosomiasis, Flagellate Infections.* ⓓ Med. Johns Hopkins Univ. 1897, sc. Yale Univ. (hon.) 1914, sc. Harvard Univ. (hon.) 1916.
Stross, Wilhelm (1894), Ass. des Pharmacol.-pharmakognostischen Inst. der Deutschen Univ. Prag. Praha II (C.S.R.), Albertov 7. *Pharmacol. der Analeptika, Herzpharm., Narkose, Insulin.* ⓓ Prag, Deutsche Univ. 1920.
Strubberg, Aage Carl (1881). København (Danmark), Havnegade 9. *Fisheries. Marking experim. with various food-fishes in danish waters and in the coastal waters of the Faroes and Iceland. Immigration of elvers (glass-eels) in the fresh waters of West-Europe.*
Strubell, Adolf (1861), Dr. phil., Prof. extraord. d. Zool. u. vergl. Anat. Bonn (Deutschland), Lessingstr. 13. *Embryol. d. Würmer, Arachnoiden. Vergl. Anat. u. Embryol. der Wirbeltiere. Rübenematoden.* ⓓ Leipzig 1888.
Struwe, Wladimir P., Dir. Morschansk(U.d.S.S.R.), Krasnvarmejskaja 32. *Agronomie, Selektion.* o
Strzemleński, Jan Kazimierz (1898), Ingénieuragronome; Ass. à l'Inst. de Production des Plantes, Univ. de Kraków (Polska), Aleja Mickiewiera 17. *Chlorophylle, réaction tyrasinique dans la pomme-deterre.* ⓓ Univ. de Cracovie 1923.
Strzyzowski, C., Prof. Dr. Lausanne (Suisse), Villa Lavoisier, Chemin du Signal. *Toxicol., Chim. physiol.* o
Stubblefield, Cyril James (1901), Demonstrator Geol. *Trilobita, Development and systematic Brachyura, Graptolithina.* ⓓ London 1925.
Stucchi, Alberto, Prof. Univ. Nac. Córdoba (Argentina). *Biophysik.* o
Stuckey, Henry Perkins (1880), Dir., Horticulturist Georgia Experim. Sta. Experiment Georgia (U.S.A.). *Horticulture (Genetics).* ⓓ Clemson Coll. 1906.
Studhalter, Richard A., Prof. of Biol., Texas Technol. Coll. Lubboch, Tex. (U.S.A.). *Physiol., Pathol.* o
Studnička, František Karel (1870), o. Prof. Histol., Embryol. Univ. Brno (C.S.R.), Talgasse 73. *Plasmatol., Cytol., Histol., Embryol. und vergl. Anat. der Vertebraten. Geschichte der Biol., Mikroskopie.* ⓓ Prag 1895.
Stübel, Hans (1885), Prof., Leiter physiol. Inst. Tungchi Univ. Woosung near Shanghai (China). *Experimentalphysiol. u. Histophysiol.* ⓓ Jena 1908.
Stürtz, Bernhard (1845), Dr. phil. h. c. Bonn (Deutschland), Riesstr. 2 *Palaeont., Rheindiluvium, Palaeozoische Asteroiden, Siebengebirge.* ⓓ Bonn 1920. ⓟ Mineralien, Meteoriten• Gesteine, Petrefakten, Modelle.

Stukenberg, Elisabeth K., Dir. d. Pensaer Staatl. Provinzial-Mus. Pensa (U.d.S.S.R.), Krasnaja 45. *Botan.* ○

Stunkard, Horace Wesley (1889), Prof. of Biol., and Chairman of the Dept. of Biol., New York Univ. New York (U.S.A.), Univ. Heigths. *Parasit., Flat worms, morphol., life history, classification.* ⓓ Univ. of Illinois 1916.

Sturdirant, B. Frank (1883), Dir. Research Labor. Pasadena Hospital. Pasadena, Cal. (U.S.A.). *Pathol. and Bact.* ⓓ North Western Univ. 1908.

Sturges, William Shelton (1892), Bact. for Cudahy Packing Co. Labor. S. Omaha, Neb. (U.S.A.). *Bacteria in the curing of Meats and in the spoilage of meats and meat products.* ⓓ Yale Univ. 1920. ⓟ Cultures.

Sturkie, Dana G. (1897), Ass. Prof. in Agronomy. Auburn, Ala. (U.S.A.).

Sturm, Albert (1875), Mitarbeiter biol. Abt. E. Merck, Darmstadt. Rüdesheim a. Rh. (Deutschland). *Schädlingsbekämpfung im Obst- u. Weinbau, besonders: Heu- und Sauerwurm, Peronospora, Oidium usw.* ⓓ Rostock 1904.

Sturtevant, Alfred H. (1891), Member of Staff, Carnegie Inst. of Washington. New York City (U.S.A.), Columbia Univ. *Zool. Genetics. Drosophila.* ⓓ Columbia Univ. 1914.

Stutzer, Michael (1879), Prof. Univ., Dir. Mikrobiol. Inst. Woronesch (U.d.S.S.R.). *Darmbact. des Menschen und der Tiere. Krankheiten der Insekten und Pflanzen.* ⓓ Moskau 1904.

Stypal, Zdzisław (1899), Ass. Inst. vergl. Anat. der Haustiere. Lwów (Polen), ul. Kochanowskiego N. 44. *Vergl. Anat. der Haustiere.*

Suarez, P. A., Prof. Univ. Quito (Ecuador). *Histol.* ○

Subkowa, Lia (1895), Ass. Inst. Biochim. Worontzowo-Pole 8. Moscou (U.d.S.S.R.), *Biochim.* ⓓ Charkow 1920.

Subramanian, Krishna Aiyar (1904), Ass. Lect. Biol. Univ. Coll. Rangoon (Burma), Biol. Dept. *Acanthocephala. Cestodes.* ⓓ Madras 1926.

Sudejkin, G. S., Prof. Sibir. Inst. f. Land- u. Forstwirtschaft, Usadba za Staroj Zagorodnoj Roščej. Omsk (U.d.S.S.R.). *Entomol.* ○

Sudrabs, Jānis, Lect. Univ. Rīga (Latvija), Kr. Barona ielā 114, dz. 27 u. Kronvalda bulv. 1. *Pomol.* ○

Sueeyoshi, J., Ass. Prof. Imper. Univ. Sapporo (Japan), Chem. Inst., Fac. Med. *Biochem.* ○

Süffert, Fritz Adolf (1891), Dr. phil., Ass. am Zool. Inst. der Univ. Freiburg i. Br. (Deutschland). *Tierzeichnung und -färbung (insbesondere Schmetterlinge), Morphol., Physiol. und Ökol. derselben, vergl. und experim. Biol. der Schmetterlinge. Schillerfarben bei Tieren.* ⓓ Halle a. d. S. 1921.

Süpfle, Karl (1880), Dr.med., o. Prof. für Hygiene, Tierärztl. Fac. d. Univ. München (Deutschland), Lachnerstr. 3. *Hygiene, Bact., Serol. Das sog. Arndt-Schulzsche biol. Grundgesetz; Morphol. und Biol. der Bakterien; Bact. der akuten Mittelohrentzündung; Oligodynamische Metallwirkung auf Bakterien; Prüfung der Bakterienresistenz mittels optimaler Nährböden; Serol. und Immunität; Luft; Wasser; Nahrungsmittel.* ⓓ Heidelberg 1905.

Suerroro, Leon, Prof. Univ. St. Tomas. Manila (Philippine Islands). *Botan.* ○

Süßmann, Philipp Oskar (1890), Priv. Doc. (Würzburg), Dr. med., Oberarzt der Bact. Abteilung des Städt. Krankenhauses Nürnberg (Deutschland), Marxstr. 47 I. *Bact.: Anaerobenforschung.* ⓓ Würzburg 1914.

Sueyoshi, Yûji, Ass. Prof. Keiô-Gijuku-Univ. Yotsuya, Tôkyô (Japan), Biochem. Inst., Coll. of Med. *Biochem.*

Sugimoto, Tadazô, Prof. Ôsaka Med. Univ. Ishibashi near Ôsaka (Japan), Biol. Labor., Preliminary Course of Ôsaka. *Plant Cytol.*

Sugiyama, Shigeteru (1894), Prof. of Pathol. Med. Univ. Kanazawa (Japan), 8 Kawaramachi. *Vital and Supravital Staining, Haematol. and Biometry.* ⓓ Kyôto 1920.

Suk, Vojtech (1879), Dr., Prof. der Anthrop. und Ethnol. Brno (C.S.R.), Anthrop. Inst. der Masaryk-Univ. *Anthrop. primitiver Rassen.* ⓓ Med. et phil. Prag. ⓟ Menschl. Skelettreste. Eskimozähne.

Sukatschev, Wladimír N. (1880), Prof. Botan. Forst-inst. Leningrad (U.d.S.S.R.), Botan. Kabinett. *Postpliocäne Flora. Dendrol. Moorkunde. Morphol.*

Šuklje, Fran, Dr., Kustos Kroat. Nationalmus. Zagreb (S.H.S.), Demetrova ulica. *Palaeont.* ○

Šule, Karel (1872), MUDr., Prof. ord. histol. et embryol. Academiae vet. Brno (C.S.R.). *Systematica Coccidarum, Psyllidarum. Symbiosa intracellularis; histol. et anat. glandularum insectorum, piti mammalium.* ⓓ Univ. Carolina pragensis (bohemica) 1897. ⓟ Coccidae, Psyllidae, Cicadae.

Šulga-Nestorenko, Maria Iva, Doc. I. Univ. Moskau (U.d.S.S.R.). *Palaeont.* ○

Sullivan, Joseph T. (1900), Ass. chem., State chem. Dept., Purdue Agric. Experim. Stat. Lafayette, Ind. (U.S.A.), Purdue Univ. *Plant analysis.*

Sulze, Walter (1879), Dr. med., ao. Prof., Ass. am Physiol. Inst. der Univ. Leipzig (Deutschland), Liebigstr. 16. *Physiol. des Herzens, der Resorption und Sekretion (Permeabilität).* ⓓ Leipzig 1907.

Suminokura, Kunihiko (1890), Prof. kaiserl. landwirtsch. Hochsch. Tottori (Japan), Kôtô Nôgyôgakkô. *Biochem., Enzyme, besonders Oxydasen.* ⓓ Hokkaido kais. Univ. 1915.

von Sumiński, Stanisław (1891). Oberlehrer. Warszawa (Polska), 8 Sniadeckichstr. *Morphol., Biol., Geographie u. System. d. Odonata.* ⓓ Kraków 1919. ⓟ Heimische Libellen.

Summerby, Robert (1886), Prof. of Agronomy. Quebec (Canada), Macdonald Coll. *Field Crops.* ⓓ Cornell 1921.

Sumner, Francis Bertody (1874), Prof. of Biol., Scripps Inst., Univ. of California. La Jolla, Cal. (U.S.A.). *Geographic variation and heredity in mice of the genus Peromyscus.* ⓓ Columbia 1901.

Sumner, James Batcheller (1887), Ass. Prof. of Biol. Chem. Ithaca, N.Y. (U.S.A.), Stimson Hall. *Isolation of Enzymes, Purification of Proteins, Analysis of Blood and Urine.* ⓓ Harvard Univ. 1914.

Sumulong, Manuel, Ass. Prof. Univ. of the Philipp. Manila (Philippine Islands). *Vet. Anat.* ○

Sun, T. P., Doc. Southeastern Univ. Nanking (China). *Zool.* ○

Sund, Oscar (1884), Fiskerikonsulent, Zweigvorstand im Abt. für Meeresforschung bei Norwegischen Fischereidirektorat. Bergen (Norge), Fosswinckels g. 8. *Biostatistik des Dorschbestandes, einschließlich der übrigen Gadiden. Embryol. der Teleostier. Systematik der Dekapodenkrebse (Peneoidea, Caridea, Eryonidea).*

Sundberg, Carl Gustaf (1892), Doc. der Physiol. an der Univ. Upsala (Schweden), Physiol. Inst. *Physiol.: Endokrinol., Blutzuckerregulation bei epinephrektomierten Tieren.* ⓓ Upsala 1925.

Sundberg, Carl Jonas Gustaf (1859), Prof. emer. Stockholm (Sverige), Kungsholmstorg 8. *Pathol. Anat.* ⓓ Uppsala 1892. ⓟ Praep. von Embryonen von 15, 27, 40 mm Länge in Serien. Wachsmodelle über die embryonale Entwicklung der Nieren.

Sundstroem, Edward Sigfrid (1880), Ass. Prof. Berkeley, Cal. (U.S.A.), Dept. of Biochem., Univ. *Biochem. and physiol. of acclimatization, tropical physiol.* ⓓ Helsingfors 1910.

Sundwall, John (1880), Prof. Univ. Michigan, Ann Arbor (U.S.A.), Dir. Div. of Hygiena and Physiol., Education. ⓓ Phil. Chicago 1906, med. Johns Hopkins 1912.

Sunier, Armand Louis Jean (1886), Dr. ès. sc.; Cons. au Mus. d'histoire nat. Leiden. den Haag (Holland), Frankenslag 29. *Mollusques (Vertigo, Amphidromus, Arca): caractères de la coquille; systématique; zoogéographie écol. et génétique.* ⓓ Groningen 1911.

Sunzeri, Giuseppe (1900), Ass. Fisiol. Palermo (Italia), Ist. di Fisiol., Corso Tukory 119. *Genesi dell' urea. Metabolismo intermedio dei Grassi.*

Suomalainen, Elias Wilhelm (1883), Mag. phil., Lect. Pori (Finnland). *Ornithol., spez. Lokalfaunistik, Migration, Nahrung.* ① Helsinki 1909. ⑫ Palaearktische Vogelbälge.

Supino, Felice, Prof. incar. Univ. e Scuola sup. agric. Milano (Italia). *Zool.* ○

Supplee, George Cornell (1889), Dir. of Research Dept., of The Dry Milk Comp. of New York. Bainbridge, N.Y. (U.S.A.). *Chem. and Biol. of Milk and Milk products.* ① Cornell Univ. 1919. ⑫ Milk products.

Suprunenko, Alexander J., Prof. des Landw. Inst. Charkow (U.d.S.S.R.), Kaplunowskaja 1,W. 7. *Selektion landwirtsch. Pflanzen.* ○

Suraschewskaja, Maria Alexejewna (1880), Stellv. des Chefs Biol. Abt. Sanitäts-Inst. Moskau (U.d. S.S.R.), Große Mogilz-Gasse 6, Wohn. 2. *Sanitäre Bact.* ① Petersburg 1908.

Surányi, Lajos (1889), I. Ass. bact. Inst. Univ. Budapest IX (Ungarn), Rákos-Gasse 9. *Microbiol. Biol. d. Tuberkulosebacillus. Biol. Wirkung der Lipoide.* ① Budapest 1913.

Surbeck, Georg (1875), Dr. phil.; Fischerei-Inspektor im eidg. Departement des Innern. Bern (Schweiz), Wabernstr. 14. *Angewandte Limnol. (Fischereibiol., Abwasserbiol.). Fischereiwirtschaft und Fischzucht.* ① Basel 1899.

Sures, Bednett (1891), Associate Prof. of Agric. Chem. Fayetteville, Ark. (U.S.A.), Univ. of Arkansas. ① Wisconsin 1920.

Sureya, Mi., Prof. Univ. Stambul (Türkei). *Allgem. Pathol.* ○

Surface, Frank M. (1882), Biol., Economist. Washington, D.C. *(U.S.A.), 412, 5 St. Variation and Inheritance of Egg Production in the Domestic Fowl. Diseases of Poultry. Infectious-Abortion in battle. Growth and Variation in Maize.* ① Pennsylvania Univ. 1907. ○

Surmont, Hippolyte Octave (1862), Prof. de Pathol. interne, expérim., clinique. Lille (France), 51, Rue du Molinel. *Digestio.* ① Lille 1886.

Suschko, A. A. (1899), Praep., Aspirant Med. Inst. Kiew (U.d.S.S.R.), Slatoustowska 6, 26. *Blasennerven.*

Sushkin, Peter P. (1868), Prof., Head ornithol. Dept. Zool. Mus., and Dept. of pre-jurassic reptiles, at the geol. Mus. Leningrad (U.d.S.S.R.), Academy of Sc., Zool. Mus. *Ornithol. (systematics, morphol., distribution). Zoogeography, Palaearctic. Compar. morphol. of Vertebrates (skeleton). Morphol. of pre-jurassic Tetrapoda. Evolution.* ① Moscow 1903. ⑫ Birdskins.

Susman, William (1895), Lect. in Morbid Anat. and Histol. Manchester (England), Dept. of Pathol., Victoria Univ. *Amyloidosis. Pellagra. Endocrines.* ① Queen's Univ., Kingston, Ontario, Canada, 1923.

von Sussdorf, Max (1855), Dr. med. u. Dr. med. vet. h. c., pens. Prof. der Anat., Histol. u. Embryol. Stuttgart (Deutschland), Stafflenbergstr. 22. *Vergl. Anat. der Haustiere; Physiol. u. Histol. des Zirkulations- u. Respirationsapparates.* ① Tübingen 1879.

Susta, Václav (1892), Staatl. geol. Inst. Praha (C.S.R.). *Phytopalaeont.* ① Přibram 1924. ⑫ Karbonische Pflanzenversteinerungen.

Sutton, Alan Callender (1892), Head of Dept. of Human Genetics, Inst. of Biol. Research, Johns Hopkins Univ. and Instr. in Med., Johns Hopkins Med. School. Baltimore, Md. (U.S.A.), 9 East Chase Street. *Genetic; Constitution.* ① Johns Hopkins Univ. 1916.

Sutton, C. (1903), Instr. in Biol. Dept. Battle Creek Coll., Mich. (U.S.A.). *Ornithol., Bact., compar. Vertebrate Zool.* ① Milton Coll. 1925.

Suworow, Eugen (1880), Prof. Leningrad (U.d. S.S.R.). *Ichthyol., Hydrobiol. und Hydrol. Fauna der Tschesskajabucht.* ① Petersburg 1903.

Suza, Jindřich (1890), Ass. des Inst. für allgemeine u. systematische Botan. der Masaryk-Univ. in Brno (C.S.R.), Kounicova 63. *Pflanzengeographie. Lichenes.* ① Brno 1924.

Suzuki, Genzo (1885). Prof. of Botan. Univ. Hokkaido (Japan). *Plant. biol.* ① Tokyo 1917.

Suzuki, Jun-ichi, Biol. Gunzé Silk Factory. Ayabe, Kyôto-hu (Japan), Biol. Labor. *Pathol. and Anat. of Silk-Worms.*

Suzuki, Kenkô, Prof. Kyôto Imper. Sericultural Coll., Lect. Kyôto Imp. Univ. Hanazono near Kyôto (Japan). *Sericiculture.*

Suzuki, Kiyoshi, Prof. III. High School. Yoshidachô, Kyôto (Japan). *Anat. and Histol. of Plant.*

Suzuki, Minoru, Prof. Med. Univ. Okayama (Japan), Bact. Inst. *Bact.*

Suzuki, Susumu (1894). Doc. Fac. Med. Kaiserl. Univ. Tokio (Japan). *Pathol. Anat. der Avitaminose.* ① Tokio 1919.

Suzuki, Tachio, Ass. Prof. Tôhoku Imper. Univ. Sendai (Japan), Anat. Inst., Coll. of Med. *Anat.*

Suzuki, Tatsuo, Ass. Prof. Tôhoku Imper. Univ. Sendai (Japan), Anat. Inst., Fac. of Med.

Suzuki, Umetarô, Prof. Tôkyô Imper. Univ., Sc. of Inst. of Physical and Chem. Research. Komaba near Tôkyô (Japan), Chem. Labor., Coll. of Agric. *Agri-. Chem., Vitamine.*

Svanberg, A. C. O., Doc. Univ. Stockholm (Sverige). *Biochem.* ○

Svanov, I. M., Prof. Gosudarstv. Donskoj Univ. Rostov am Don (U.d.S.S.R.). *Anat.* ○

Svarčevskij, B. A., Prof. Pädag. Inst. d. Bergrepublik. Vladikavkaz (U.d.S.S.R.), Ul. Marksa 20. *Zool. d. Wirbellosen.* ○

Svedelius, Nils Eberhard (1873), Prof. der Botan. an der Univ. Upsala (Schweden). *Zytol. und Entwicklungsgeschichte der Algen. Generationswechsel.* ① Upsala 1901.

Svenson, Henry K. (1897), Ass. Prof. of Biol. Schenectady, N.Y. (U.S.A.), Union Coll. *Geographic distribution of plants, with special reference to pleistocene period.* ① Harvard 1920.

Svenson, Sten (1880). Falkenberg (Sverige). *Plantgeography.*

Svenssen, Harry Gustaf (1894), Priv.Doc. an der Univ. Upsala (Schweden). *Pflanzenembryol. und Zytol. (Tubiflora, Santalacea). Mycol. (Basidiomycetes). Embryol. der Hydrophyllazeen, Borraginazeen und Heliotropiazeen.* ① Upsala 1925. ⑫ Fixiertes pflanzenzytol. Material.

Sverdrup, Aslang, Amanuensis Zool. Labor. d. Univ. Oslo (Norge). *Genetic, Rassenbiol.* ○

Svetlov, Pavel Grig., Doc. Univ. Perm (U.d. S.S.R.), Ul. Karla Marksa 26. *Biol., Zool.* ○

Svirtschevska, Hélène V., Conserv. Inst. Jaczewski. Leningrad (U.d.S.S.R.), Perspetive Anglaise 29. *Systématique mycol.* ○

Swain, Albert F. (1894), Entomol., California Cyanide Company. Los Angeles, Cal. (U.S.A.), Box 250, Arcade Station. *Fumigation of agric. crops, particulary citrus fruits, with Calcium Cyanide. Aphididae.* ① Univ. of California 1915.

Swann, M. B. R., Lect. Gouville and Caius Coll. Cambridge (England). *Pathol.* ○

Swanson, Arthur F. (1891), Ass. Agronomist, Office of Cereal Crops and Diseases, Dept. of Agric. Hays, Kan. (U.S.A.), Ft. Hays Experim. Station. *Grain sorghums, wheat, oats, barley and corn.* ⑫ Seeds of grain sorghums.

Swarth, Harry S. (1878), Curator of Birds, Mus. of Vertebrate Zool., Univ. of California. Berkeley Cal. (U.S.A.). *Birds and mammals of western North America.*

Swarzewsky, Prof. Zool. Labor. Univ. Irkutsk (U.d.S.S.R.). ○

Swederski, W., Dr., Vorst. Staatl. Botan. Landwirtsch. Versuchsstat. Lwów (Polen). ○

Sweet, Georgina (1875), Lect. Univ. Melbourne (Australia). *Animal parasit.* ① Melbourne 1904.

Sweet, Winfield Carey (1891), Member of Staff of International Health Board, Rockefeller Foundation. New York City (U.S.A.), 61 Broadway. *Parasit. Hookworm. Ascaris.* ① Rush Med. 1917.
Sweetman, Harvey L. (1896), Instr., Univ. of Minn. St. Paul, Minn. (U.S.A.), Univ. Farm. *Ecol. of Phyllophaga (Coleoptera, Scarabaeidae).*
Sweetser, Albert Raddin (1861), Head of the Dept. of Botany. Eugene, Ore. (U.S.A.). ① Wesleyan Univ. 1884.
Swellengrebel, Nicolaas Hendrik (1885), Prof., Dr. am Kolonialinst. und an der Univ. Amsterdam (Holland), Inst. für tropische Hygiene, Mauritskade 57a. *Zool. in Beziehung zur Hygiene. Malaria. Anopheles.* ① Zürich 1908.
Swenander, C. G., Dr., Fiskeriintendent. Lund (Sverige). *Zool.* o
Swenk, Myron Harmon (1883), Prof. of Economic Entomol., Chairman of the Department of Entomol.; Entomol. of the Nebraska Agric. Experim. Station. Lincoln, Neb. (U.S.A.), Coll. of Agric., Univ. of Nebraska. *Taxonomy of Apoidea and Dipterous parasit. of Vertebrates; North American Mammal. and Ornithol.* ① Univ. of Nebraska, Lincoln, 1908.
Swensson, Victor, Dr., Prosektor sykehus. Trondhjem (Norge). *Pathol. Anat.* o
Swett, F. H. (1893), Associate Prof. of Anat. Nashville, Tenn. (U.S.A.), Vanderbilt Univ. School of Med. *Experim. Embryol., Regeneration, Transplantation.* ① Yale Univ. 1922.
Swett, Walter W. (1892), Bureau of Dairy Industry. Washington, D.C. (U.S.A.), Dept. of Agric *Relationship between conformation and anat. of the dairy cow and her milk and butterfat producing capacity.*
Swezey, Otto Herman (1869), Entomol., Hawaiian Sugar Planter's Experim. Station. Honolulu (Hawaii), 2048 Lanihuli Drive. *Sugar cane insect pests in Hawaii, forest insects in Hawaii and insects parasitism.* ① Northwestern Univ. 1897.
Swindle, Percy Ford (1898), Prof. and Dir. of Dept. of Physiol. Milwaukee, Wis. (U.S.A.), Marquette Univ., School of Med., 638 Fourth Street. *Respiration and Circulation. Cheyne-Stokes Breathing. Quantum Reactions and Associations.* ① Univ. of Berlin 1915.
Swingle, Charles Fletcher (1899), Ass. Pomol., U.S. Dept. Agr., Bureau Plant Industry (Nursery Stock Investigations). Washington, D.C. (U.S.A.). *Vegetative propagation; graft unions; graft hybrids.*
Swingle, D. B., Prof. of Botan. and Bact., Univ. of Montana. Bozeman, Mtn. (U.S.A.). *Pathol.* o
Swingle, Homer S. (1902), Ass. Entomol. Washington, D.C. (U.S.A.), Bureau of Entomol., U.S. Dept. Agric. *Insecticide chemistry. Insect Physiol. Digestive Enzymes of Blatta orientalis.*
Swingle, Walter Tennuyson (1871), Physiol. in charge, Crop Physiol. and Breeding Investigations, Bureau of Plant Industry. Washington, D.C. (U.S.A.), U.S.Dept. of Agric. *Plant Pathol., Cytol., taxonomic Botan., date culture.* ① Kans. State Agric. Coll. 1922. o
Swinton, William Elgin (1900), Ass., Dept. of Geol. London S.W. 7 (England), British Mus. (Nat. History), Cromwell Road. *Fossil reptilia and amphibia.* ① Glasgow Univ. 1922.
Swirenko, Dmitrij O. (1888), Dir. des Botan. Gartens, Prof. am Odessaer Inst. für Volksbildung. Odessa (U.d.S.S.R.), Bulev. d. Proletarier 87. *Systematik u. Biol. d. Algen, Phytoplankton.* ① Odessa 1924.
Swiridenko, Paul Alexeewitsch (1893), Dir. Pflanzenschutzstation, Zool. Abt. Rostow-Don (U.d.S.S.R.), Skobelewskaja 90, W 30. *System d. Mammalia. Angew. Zool.*
Switzer, H. B. (1888), Bact., Central District, Bureau of Chem., U.S. Dept. of Agr. Chicago, Ill. (U.S.A.), 1625 Transportation Bldg. *Bact. of dairy products.* ① Cornell Univ. 1912. ② *Cultures of thermophiles from sour canned foods.*

Sylvén, Nils Olof Waldemar (1880), Vorstand d. Abt. der Wiesengewächse Saatzuchtanst. Svalöf (Sverige). *Floristik und Pflanzengeographie; Pflanzenmorphol.; Forstbotan.; Vererbungslehre.* ① Uppsala 1906.
Symmers, W. St. Clair, Prof. Univ. Belfast (Ireland). *Pathol.* o
Sypniewski, Joseph (1883), Chef dépt. de l'amélioration des plantes à l'Inst. National Polonais d'Economie Rurale. Puławy (Polska). *Génétique et amélioration des plantes.* ① Puławy 1913.
Syreischikow, Dmitrij P., Konserv. Herbarium d. I. Univ. Moskau (U.d.S.S.R.), Sretenskij bulv.1/6, W. 130. *Flora d. Moskauer Gouv. Botan.* o
Sysin, Aleksej Nik., Prof. I. Univ. Moskau (U.d. S.S.R.). *Hygiene.* o
Sysoev, F. F., Prof. St. Inst. f. Med. Wiss. Leningrad (U.d.S.S.R.), 2 Sovetskijstr. 4. *Pathol. Anat. u. Histol.* o
Szabó, Zoltán (1882), o. ö. Prof. an der Volkswirtschaftl. Fac. Univ. Budapest VIII (Ungarn), Eszterházy u. 3. *Landwirtschaftliche Botan. Dipsacaceae, Organographie der Pflanzen.* ① Breslau 1905. ② *Landwirtschaftl. Kulturpflanzen. Dipsacaceae.*
Szabó-Patay, Josef (1887), Dr., Kustos in d. Zool. Abteilung d. Ung. National Mus. Budapest 80 (Ungarn). *Hymenopteren (Formiciden), Syst. Insektenbiol., Histol., Anat.* ① Budapest 1911. ② *Hymenopteren.*
Szafer, Władysław (1886), Prof. Botan., Dir. Botan. Garden. Kraków (Polska), Lubicz 46. *Ecol., Taxonomy. Morphol., Systematik of Angiosperms, Algol., Plantgeography, Paleobotan.* (*Diluvium*). ① Lwów 1909. ② *Plantae polonicae exsiccatae.*
Szaferowa, Janina (1896), Botan. Inst. Univ. Kraków (Polska), Kopernika 27. *Systematik d. Gattung Betula. Anat. des Pollens.*
Szajnocha, W., Dr., Prof. Kraków (Polska), Inst. Géol. Univ. o
Szakáll, Sándor (1901), Ass. physiol. Inst. Budapest VIII (Ungarn), Eszterházy-ut 9. *Arbeitsphysiol., Stoffwechsel.* ① Budapest 1925.
Szakien, Bronisław (1890), Ass. en chef. Wilno (Polska), Zakretowa 15, Zakład Botaniczny. *Mycol., Cytol., Les rouilles de Wilna. Chromosomes hétérotypiques dans l'Osmunda regalis.* ① Wilno 1924. ② *Rouilles.*
de Szaniawski, Władysław (1891), Dr. phil., Ass. en chef d'Anat. comparée. Varsovie (Polska), Univ. *Embryol. des Oiseaux.* ① Varsovie 1922.
Szártorisz, Béla (1877), Landw. Versuchsstationsleiter. Budapest I (Ungarn), Otthon-ut 17. *Angewandte Botan.* ① Budapest 1915.
Szász, Alfréd (1877), Dr., Königl. Oberbakteriol., Priv.Doc. für Schutz- und Heilimpfungen an der kön. ung. Vet.-Hochsch. in Budapest VII (Ungarn), Erzsébet királyné-ut 15. *Seuchenkrankheiten. Bact. Impfstoff-Produkt. Bakterien und Bakterienprodukte als Schutz- und Heilmittel.* ① Budapest 1903.
Szatala, Edmund (1889), Dr. Budapest I (Ungarn), Schwartzer Ferencz-ut 3. III. 6. *Lichenol.: ungarische, balkanische und orientalische Flechten. Coniocarpineae.* ① Budapest 1916.
v. Szeliga-Mierzeyewski, Władisław Eugen Johann (1882), Prof., Dir. Inst. vergl. Anat. Univ. Wilno (Polska), Zakretowastr. 15. *Lymphgefäßsystem der Vögel. Systematik der europ. Vögel. Orthoptera europ. (System.). Odonata europ. (System.).* ① Krakau 1911.
v. Szent-Györgyi, Albert (1893), Priv.Doc. Groningen (Holland), Physiol. Labor. *Biochem. Mechanismus der biol. Oxydation.* ① Budapest 1917.
Szepesfalvy, Johann Vitéz (1882), Kustos an der botan. Abteilung des Ungarischen National-Mus. Budapest V (Ungarn), Akademia-ut 2. *Bryol.* ① Budapest 1908. ② *Moose.*
Szidat, Lothar (1892), Ass. am Zool. Inst. und Mus. zu Königsberg i. Pr., Leiter der Zool. Station

für Schädlingsforschung, Rossitten, Ostpr. Königsberg i. Pr. (Deutschland), Sternwartstr. 1/2. *Angewandte Zool.; Parasitol., Trematoden, Cestoden, Nematoden. Faunistik. Museumskunde.* ⓓ Königsberg 1920. ⓟ Parasit. bes. Trematoden.

Szilády, Zoltán, Dr. phil., Extraordinarius der Zool. an der Univ. in Debreczen (Ungarn). *Oekol., Zoogeographie, Systematik der Dipteren (Tabanidae).* o

Sztankovics, Rudolf, Dr., Priv.Doc. Tierärztl. Hochsch. Budapest (Ungarn), Rottenbillergasse. *Histol. d. Pflanzen.* o

Sztolcman, Jan (1854), Vice-dir. Mus. Polonais d'Histoire Nat. Warszawa (Polska), Krakowskie-Przedmie¸cie 26/28. *Ornithol. et Histoire Nat. du Pérou et de l'Ecuador.*

Szűcs, Endre, Dr., Priv.Doc. Tierärztl. Hochsch. Budapest (Ungarn), Rottenbillergasse. *Histol.* o

Szwarz, Adam, Prof. Warszawa (Polska), Ecole Sup. Agr., r. Miodowa 23. *Botan.* o

Szymkiewicz, Dezydery (1885), Prof. de botan. à l'Ecole Polytechnique de Léopol, Doc. à l'Univ. Jagellonienne de Cracovie. Léopol (Polska), Nabielaka 22. *Climatol. en relation avec la vie des végétaux. Bilan d'eau chez les végétaux.* ⓓ Cracovie 1923.

Szymonowicz, Władysław (1869), o. Prof., Dir. histol.-embryol. Inst. Lwów (Polska), Frańska 18. *Nebenniere, Morphol. u. Entwicklung der sensiblen Nervenendigungen.* ⓓ Kraków 1893.

Szyniewski, W., Dr. phil., Prof. Lwów (Polska), Politechnique. *Biochim.* o

Tabenitzkij, Alexander A., Doc. des Kiewschen Polytechn. Inst. Kiew (U.d.S.S.R.), Brest-Litowsker Chaussee 39. *Pflanzenanat.* o

Tabor, Richard John (1875), Ass. Prof. of Botan. London S.W. 7 (England), Imp. Coll. of Sc. and Technol. South Kensington. *Botan., Mycol.* ⓓ London.

Tabountschikowa, Alexandrine (1883), Ass. Zool. Inst. Agronomique. Leningrad (U.d.S.S.R.), Kamennoostrowsky 54/31, log. 7. *Embryol. et Morphol. des Invertébrés.* ⓟ Petrograd 1909.

Tabusso, Edmundo, Doc. Univ. Lima (Peru). *Microbiol. Pathol. des anim.* o

Tacke, Bruno (1861), Prof., Dir. preuß. Moor-Vers. Stat. Bremen (Deutschland), Neustadtswall *Kultur v. Moor-, Marsch- u. Heideboden.* o

Täckholm, Gunnar (1891), Doc. an der Univ., Prof. der Botan. an der Univ. Kairo. Stockholm (Sverige), Botan. Inst. Univ. *Zytol. und Embryol. der Pflanzen.* ⓓ Stockholm 1922.

Tadokoro, Tetsutarô, Prof. Hokkaidô Imper. Univ. Sapporo (Japan), Chem. Labor., Coll. of Agric. *Nutritional Chem.*

Tafuri, Giovan Bernardino (1902), Ass. Ist. di Chimica Biol. Univ. di Napoli (Italia), S. Andrea delle Dame 21. *Biochimica.*

Tago, Katsuya (1877), Prof. Dr. Koishikawaku, Tôkyô (Japan), Nishiharamachi 2 Chiôme 40. *General Zool. Marine Mammalia. Amphibia. Urodela. Fishes. Molluscs.* ⓓ 1900. ⓟ Japanese Molluscs. Amphibia: Urodela: adult, larve, eggs. Cetacean embryos.

Taguchi, Sekishin, Prof. Chiba Med. Univ. Chiba near Tôkyô (Japan), Anat. Inst. *Anat.*

Tai, Fang Lan (1893), Prof. Plant Pathol. Nanking (China), Dept. of Plant Pathol. and Entomol., Nat. Southeastern Univ. *Plant-Pathol., Mycol.* ⓓ Cornell Univ. 1918. ⓟ Specimens of Chinese Fungi.

Taine, C. H., Lect. National Univ. Peking (China). *Zool.*

Tainter, Maurice Lane (1899), Ass. Prof. of Pharmacol., Stanford Univ. Med. School. San Francisco, Cal. (U.S.A.). *Pharmacol. of Edema, Tyramine, Digitalis, Cocain and Insulin.* ⓓ Stanford 1925.

Talrow, Wassilij E., Dir. d. Zentralwissensch. Versuchsstation f. Weinbereitung. Odessa (U.d.S.S.R.),

Ul. Swerdlowa 19. *Agronomie. Weinbau. Weinbereitung.* o

Tait, William Chaster (1844). Oporto (Portugal), Entre Quintas 155. *Ornithol. of Portugal.*

Tait, William Dunlop (1879), Chairman, Dept. of Psychol. and Dir., Psychol. Labor. Montreal (Canada), McGill Univ. *Memory, Emotions, Learning, Tests.* ⓓ Harvard 1909.

Takagi, Kôzô, Ass. Prof. Prov. Med. Univ. Ôsaka (Japan), Anat. Inst. *Anat.*

Takahashi, Katashi, Prof. Zool. I. High School, Tôkyô. Hongô, Tôkyô (Japan), Biol. Labor.

Takahashi, Nisuke, Prof. Zool. V. High School. Kumamoto (Japan), Biol. Labor. *Ichthyol.*

Takahashi, Ryô-ichi (1898), Ass. Entomol. of the Dept. of Agric., Research Inst., Government of Formosa. Taihoku, Formosa (Japan). *Aphididae, Coccidae, biol. of insects, etc.*

Takahashi, Shô, Biol. Dept. Agric. and Forestry. Tôkyô (Japan), Bureau of Agric. *Entomol. Lepidopt.*

Takahashi, Yoshio (1891). Okayama (Japan), Pharm. Labor., Med. Univ. *Experim. Pharmacol.*

Takamatsu, M., Prof. Imper. Univ. Hokkaidô (Japan), Inst. of Vet. Sc., Fac. of Agric. *Zootechn. Hippol.*

Takamine, Jôkichi (1889), President and Dir. Takamine Labor. New York City, 120 Broadway. Clifton, N.J. (U.S.A.). *Biol. Chem.* ⓓ Yale 1913.

Takano, R., Lect. Keiô-Gijuku Univ. Igakuhakushi. Tôkyô (Japan), Bact. Inst., Coll. of Med. *Bact. Hygiene.*

Takata, Maki (1892), Prof. Koishikawa, Tôkyô (Japan), Hospital Koishikawaboyôin, Ôzuka-Nakamachi 17/19. *Biochem., Kolloidchem., innere Med., insbesondere Verdauungsphysiol., Fermentlehre, Chem. der Zerebrospinalflüssigkeit.* ⓓ Kaiserl. Univ. Sendai 1915.

Prince Takatsukaza, Nobusuke (1888), President Ornithol. Society of Japan. Tôkyô (Japan), 2 Fukuyoshicho, Akasaka. *Ornithol.* ⓓ Tôkyô Imper. Univ. 1914. ⓟ Birds' skins and eggs.

Takayanagi, Giichi, Ass. Sendai (Japan), Bact. Inst., Univ. *Agglutinatorische Analyse der Hühnertyphusgruppe.*

Takeda, Kyûkichi. Kôjimachi-ku, Tôkyô (Japan), Fujimichô 4-Chôme 6. *Alpine Plants.*

Takeda, Yosito (1896), Mycol., Section of Fermentation Industrial Dept., Government Research Inst. Formosa. Taipeh, Formosa (Japan). *Microorganisms of Fermentation.* ⓓ Tôkyô Imp. Univ. 1922. ⓟ Microorganisms of Fermentation.

Takei, Sankichi, Ass. Prof. Imper. Univ. Kyôto (Japan), Inst. of Agric. Chem., Fac. of Agric.

Takenouchi, Makoto (1894), Ass. Univ. Fukuoka (Japan), Botan. Inst., Kaiserl. Kyûshû-Univ. *Physiol. Pflanzenoekol.* ⓓ Sapporo 1918.

Takenouchi, Matsujirô, Prof. Imper. Univ. Tôkyô (Japan), Bact. Inst., Fac. of Med. *Bact.*

Takenouchi, Yoshio (1891), Prof. Taihoku, Formosa (Japan), Coll. of Agric. and Forestry. *Plant Morphol. Seed abscission of Rice-plant in Formosa.* ⓓ Sapporo 1915.

Takeshita, Takematsu, Ass. Prof. of Tôkyô Imper. Univ. Komaba near Tôkyô (Japan), Zool. Inst., Fac. of Agric. *Zool.*

Takeuchi, Kichizô (1891), Dir. of Takeuchi Entomol. Labor. Kyôto (Japan), Shinomiya Yamashina Ujigun. ⓟ Japanese Insects.

Takeya, Ozuna (1891). Kumamoto (Japan), Anat. Inst. der med. Akademie. *Bau des Zytoplasmas und Kernes. Kolloidchem. Spermiogenese. Bidderschesorgan. Chromosomen.* ⓓ Kumamoto 1913.

Takezaki, Yoshiho, Ass. Prof. in Genetics, Coll. of Sc., Imper. Univ. Kyôto (Japan), Labor. of Horticulture, Coll. of Agric. *Genetics, especially of rice-plants.*

Takiuchi, Akiji, Lect. Provincial Med. Univ. Ôsaka (Japan), Anat. Inst. *Anat.*

Talaat, Husséin, Prof. agr. Univ. Stambul (Türkei). *Pathol.*
Talalajeff, Voldemar (1886), Ass. pathol.-anat. Inst. I. Univ. Moscau (U.d.S.S.R.), M. Moltschanowka 8, Log. 8. *Sepsis lenta. Histogenese u. Pathol. der paraplastischen Substanz.* ⑲ Moskau 1912.
Talbert, George A. (1865), Prof. Physiol. and Pharm. Med. School. Grand Forks, N.D. (U.S.A.), Univ. Station. *Physiol. and Pharm. Urine and blood constituents. Blood mobilization effected by Hypnotism. Hydrogen ions of sweat.* ⑲ Ohio Wesleyan 1888.
Talbot, M. W. (1889), Botan., in charge of Weed Investigations, Bureau of Plant Ind., U.S. Dept. of Agric. Washington, D.C. (U.S.A.). *Troublesome weeds.* ⑲ Univ. of Missouri 1913.
Talha, Prof. Agr. Univ. Stambul (Türkei). *Physiol.* ○
Taliaferro, William Hay (1895), Associate Prof. of Parasitol., Univ. of Chicago, Ill. (U.S.A.), Dept. of Hygiene and Bact. *Immunology of parasites, trypanosomes and the malarial parasites. Genetics and physiol. of the parasitic protozoa.* ⑲ Johns Hopkins Univ. 1918. ⓟ Parasitic protozoa.
Talijew, Valerij I., Prof., Landwirtschaftl. Akademie Timirjasew. Moskau (U.d.S.S.R.), Petrowsko-Rasumowskoje. *Botan., Allgem. Biol. Phytopath.* ○
Taljancev, A., Prof. Med. Inst. Ekaterinoslav, Ukraine (U.d.S.S.R.). *Allg. Pathol.* ○
Tamarit Olmos, Pedro, Prof. Univ. Valencia (España). *Pathol.* ○
Tamison, Frank Stover, Instr. in Horticulture. Agric. Building, College Station, Texas (U.S.A.). *Vegetable Crops.*
Tamm, Ernst Adam (1897), Dr. der Landwirtschaft, Diplom-Landwirt, wissenschaftlicher Ass. am Inst. f. Acker- u. Pflanzenbau d. Landwirtschaftlichen Hochsch. Berlin. Berlin-Friedenau (Deutschland), Lauterstr. 16. *Acker- und Pflanzenbau. Versuchswesen. Bodenbearbeitung. Elektrokultur.* ⑲ Berlin 1922.
Tammes, Tine (1871), ao. Professorin der Botan. Groningen (Holland), Oranjesingel 4. *Variabilität, Bastardierung.* ⑲ Groningen 1911.
Tamura, Kenzô, Prof. Imper. Univ. Tôkyô (Japan), Pharm. Inst., Fac. of Med. *Pharmacol.*
Tamura, Oto, Prof. of Med. Univ. Okayama (Japan), Pathol. Inst. *Pathol.*
Tanaka, Chôzaburô (1885), Prof. Botan., Horticulture and Genetics. Miyazaki-shi (Japan), Miyazaki Coll. of Agric. *Taxonomy, genetics and pathol. of Citrus plants, Japan region. Botan. bibliography. Plant collection.* ⑲ Tôkyô 1910. ⓟ Herbarium specimens of wild and cultivated plants of Japan, Luchu and Formosa.
Tanaka, Hisashi (1902). Okayama (Japan), Pharm. Labor., Med. Univ. *Experim. Pharmacol.*
Tanaka, Shigeho, Ass. Prof. Imper. Univ. Tôkyô (Japan), Zool. Labor., Coll. of Sc. *Ichthyol.* ○
Tanaka, Shoichi (1901), Ass. Prof. Phytopath. Tottori (Japan), Agric. Coll. *Phytopath.: Fusarium diseases and Sclerotinia diseases of Crops.* ⑲ Tottori Agric. Coll. 1924.
Tanaka, Shôzô, Lect. in Biol. High School. Ôsaka (Japan), Biol. Labor., Tennoji-ku. *Myriopoda.* ○
Tanaka, Takayashi. Wakayama-Ken (Japan), Yoshiwara, Yamada-mura, Ito-gun. *Entomol.: Hemiptera.* ○
Tanaka, Yoshimaro (1884), Prof. Sericiculture and Genetics Dept. of Agric. Kyûshû Imper. Univ. Fukuoka (Japan). *Genetics, Breeding, Physiol., Heredity of the Silkworm: Mutation; Linkage; Lethal factors; Maternal inheritance; Mosaics and Gynandromorphs.* ⑲ Sapporo, Japan, 1909. ⓟ Silkworm eggs, homozygous or heterozygous for different Mendelian characters. Alcoholic or dry specimens of silkworm larvae and cocoons.
Tanassijtchuk, N. P., Doc., Ass. Lesshaft Wissenschaftl. Inst. Leningrad (U.d.S.S.R.). *Zool.* ○

Tanberg, Andreas (1873), Oslo (Norge). *Interne Secretion: Parathyreoidea.* ⑲ Oslo 1913.
Tandler, Julius (1869), Prof. Univ. Wien IX (Österreich), Währinger Str. 13. *Anat.* ○
Tandy, Geoffrey (1900), Ass., Dept. of Botan., British Mus., (N. H.). London S.W. 7 (England), S. Kensington. *Taxonomy and Cytol.* ⑲ Oxford 1921.
Tanfiljew, Gawriil Iwanowitsch (1857), Prof., Odessaer Landwirtschaftl. Inst. Odessa (U.d.S.S.R.) Tschernomorskaja 8. W. 2. *Bodenkunde, Geobotan. Tundra, Wälder, Steppen.* ⑲ Petersburg 1912.
Tang, Seng Yu, Prof. North China Baptist Coll. Weihsien, Shantung (China). *Zool.*
Tangl, Harald (1900), Ass. physiol. Inst. Univ. Budapest VIII (Ungarn), Eszterházy-ut 9. *Stoffwechsel und innere Secretion.* ⑲ Budapest 1924.
Tåning, Åge Vedel (1890), Ass. at the danish Committee for the study of the sea. København, Valby (Danmark), Carlsberg Labor. *Marine biol.: deap-sea-fishes and fishery. Mediterranean Scopilida. Sternoptychida. Life-History of the Icelandic Plaice.*
Tanna, Kahandas J. (1884), Pathol. Department. Bombay (India), Haffkine Inst., Parel. ⑲ Ahmedabad 1907.
Tannenberg, Joseph (1895), Dr., Priv.Doc. an der Univ. zu Frankfurt a. M. (Deutschland), Pathol. Inst., Eschenbachstr. 14. *Pathol. des Kreislaufes.* ⑲ Marburg 1921.
Tanner, Fred Wilbur (1888), Prof. of Bact. and Head of the Dept. of Bact., Univ. of Illinois. Urbana, Ill. (U.S.A.), 361 Chemistry Building. *General bact. food microbiol., fermentation organisms, yeasts. Thermophilic bacteria.* ⑲ Univ. of Illinois 1916. ⓟ Pure cultures of yeasts, thermophilic bacteria.
Tanner, Vasco Myron (1892), Prof. and Head of the Dept. of Zool. and Entomol., Brigham Young Univ. Provo, Utah (U.S.A.). *Insect Morphol.; Fossil Fish of the Tertiary; Animal Ecol.; Mammals.* ⑲ Stanford Univ., California, 1925. ⓟ Animals; Coleoptera; Fish; Mammals.
Tansley, Arthur George (1871). Grantchester near Cambridge (England). *The study of Vegetation and plant ecol.*
Tantz, Arthur (1884), Obertierarzt, Univ. Halle a. d. Saale (Deutschland), Freiimfelderstr. 68. *Rindertuberkulose.* ⑲ Gießen 1912.
Tantz, Jacob (1895), Associate Prof. of Biol. Coll. of William and Mary. Williamsburg, Va. (U.S.A.). *Hygiene.* ⑲ Univ. of Michigan 1926.
Tapie, J., Prof. Univ. Toulouse (France). *Anat. Pathol.* ○
Tapke, Victor F. (1890), Associate Pathol., Office of Cereal Crops and Diseases, U. S. Dept. of Agric. Washington, D.C. (U.S.A.). *Diseases of cereals caused by smut fungi.* ⑲ Ohio State Univ. 1914.
Tarabukin, Alexander Ja., Wiss. Mitarbeiter des Jakutsker Provinz.-Mus. Jakutsk (U.d.S.S.R.). *Botan., Meteorol.* ○
Tarakanow, Eugen (1899), Ass. der Anat. Rostow, Don (U.d.S.S.R.), Anat. Inst. *Synovialfalten d. Hüftgelenkes. Arterien d. Gehirnes.* ⑲ Rostow a. Don 1923.
Tardo, Gian Vito (1884), Doc. di Patol. Chirurgica, R. Univ. Palermo (Italia), Via Paolo Paternostro 1. *Apparato urinario: Anat. e fisiol. normale e patol., Batteriol.*
Tarnani, Iwan Konstantinowitsch (1864), Prof. d. Zool., vergl. Anat. Charkow, Ukraine (U.d.S.S.R.), Agronomisch. Inst., Puschkinstr. 80. *Angewandte Zool., Entomol., Teratol.* ⑲ Insekten.
Tarnogradsky, David (1891), Dir. Stat. Biol. du Caucase du Nord, Doc. Gorsky Inst. Agronomique. Wladicawcass (U.d.S.S.R.), Borodinskaja 28. *Biol. des Culicides Syst. Rotifera et Rhizopoda. Faune des hautes montagnes.* ⑲ Wladicaucase 1920. ⓟ Animaux et plantes aquatiques des hautes montagnes du Caucase.

Tarozzi, Giulio (1868), Prof. o. Anat. pathol. Modena (Italia), Rue Muratori 7. *Anatomie pathol. Bact.*

Tarussof, Boris (1900), Aspirant Forschungs-Inst. für Biol. Odessa (U.d.S.S.R.). *Osmotische Wirkungen. Permeabilität.*

Tashiro, Shiro (1883), Prof. of Biochem., Med. Coll., Univ. of Cincinnati (U.S.A.). *Nature of Nerve Impulse; Bile salts; Microchemical analysis of gases; Application of physical chemistry to Biol.* ⓓ Ph. Univ. of Chigaco 1912, nied. Kyoto, Japan 1922.

Tashiro, Zentarô, Vice-Ass. Imper. Univ. Kyôto (Japan), Botan. Inst., Coll. of Sc. *Systematic.*

Tasker, Roy C. (1896), Prof. of Vertebrate Zool. Univ. Soochow (China). ⓓ Michigan 1922.

Tatarinov, E. A., Prof. N. G. Černyševskij Univ. Saratov (U.d.S.S.R.), Ploščad imeni Lenina. *Pathol.* o

Tatewaki, Misao (1899), Lect. on Forest Botan., Coll. of Agric., Hokkaidô Imper. Univ. Sapporo (Japan), Botan. Inst. *Oecol. Flora of northern Japan.* ⓓ Sapporo 1924.

Tattenfield, Frederick, Head of Dept. of Insectides and Fungicides, Rothamsted Exper. Stat. Harpenden (England). *Insecticides. Chem. Constitution and insect. action.* ⓓ London.

Tattersall, Walter Medley (1882), Prof. of Zool. and Compar. Anat., Univ. Coll. Cardiff, Wales (England), Newport Road. *Systematics of Crustacea. Chiefly Mysidacea, Isopoda and Amphipoda.* ⓓ Manchester 1902. ⓟ *Mysidacea, Isopoda, Amphipoda.*

Tatum, Arthur Lawrie (1884), Associate Prof. of Pharmacol., Univ. of Chicago, Ill. (U.S.A.). *Narcotic drugs.* ⓓ Chicago 1913.

Taube, Erwin (1878), Doc. Zool. am Herder-Inst. Riga (Latvija). *Embryol., Hydro-Biol., Entwicklungsmechanik.* ⓓ München 1908.

Taubenhaus, Jacob Joseph (1884), Chief, Division of Plant Pathol. and Physiol., Texas Agric. Experim. Station. College Station, Tex. (U.S.A.), A. and M. Coll. *Culture and Diseases of the Sweet Pea, of Truck Crops, Greenhouse Crops, Sweet Potato, Onion, Fresh Fruits and Vegetables.* ⓓ Univ. of Pennsylvania 1913. ⓟ *Phymatotrichum omnivorum and other specimens of disease affecting economic crops.*

Taufer, Josef, Dr., Prof. der Tierzucht. Brno (C.S.R.), Cerna pole, Landwirtschaftl. Hochsch. o

Tauson, Anastasie (1890), Doc. der Zool. Univ., Inst. für biol. Forschung Univ. Perm (U.d.S.S.R.). Zaimka. *Cytol. und experim. Zool. Oogenese und Spermatogenese. Wirkung des Mediums auf das Geschlecht der Tiere.* ⓓ Perm 1926.

Tauvel, Pierre (1866), Prof. tit. de Zool. Angers (France), 12, rue du Pin. *Annelida Polychaeta. Systématique, Distribution géographique, Anat.* ⓓ Sorbonne 1897. ⓟ *Annelida Polychaeta.*

Tavares, Amândio Joaquim (1900), Ass. Anat. Fac. Méd. Porto (Portugal), R. do Rosário, 266. *Variat. musc. du Thorax.* ⓓ Porto 1924.

Tavares, Joachim da Silva (1866), Caminha (Portugal). *Cécidol.*

Tavčar, Alois (1895), ao. Prof. für Vererbungslehre und Pflanzenzüchtung an der Landwirtsch. Fak. der Univ. Zagreb, Vorstand des Inst. für Pflanzenzucht. Zagreb (S.H.S.). *Vererbung der quantitativen Merkmale an Zea Mays, Phaseolus vulgaris und Pisum sativum.* ⓓ Technische Hochschule Praha 1921.

Taverner, Percy Algernon (1875), Ornithol. Canadian National Mus. Ottawa (Canada), Victoria Memorial Mus. *Birds of Canada. Systematic and distribution.*

Tavistock, Hastings William Sackville, Marquis of Warblington House, Havant (England). *Ornithol.* o

Tawara, Masato, Prof. Tôhoku Imper. Univ. Sendai (Japan), Botan. Inst., Coll. of Sc. *Botan., Phycol.*

Tawara, Sunao, Prof. Kyûshû Imper. Univ. Fukuoka (Japan), Pathol. Inst., Coll. of Med. *Pathol. and Pathol. Anat.*

Taylor, Alonzo Englebert (1871), Univ.Prof. of physiol. Chem. Stanford Univ., Cal. Philadelphia, Pa. (U.S.A.), Univ. of Penns. *Fermentation. Physiol. chem. Food Problem.* ⓓ M. Pennsylv. 1894, L.L. Wisconsin 1920.

Taylor, Arthur L. (1897), Lect. in Pathol. Univ. of Leeds. Yorkshire (England), Burnside Bardsen. *Heterotopias of alimentary tract. Anaerobic Streptococcus.* ⓓ Leeds 1923.

Taylor, Ernest Leonard (1894), Parasit. labor. of the Vet. Research Inst. of the British Ministry of Agric. and Fisheries. New Haw Weybridge, Surrey (England), Vet. Lab. *Parasit. Vet. Helminthol.* ⓓ Liverpool 1923.

Taylor, Leland H. (1893), Ass. Prof. of Zool., West Virginia Univ. Morgantown, W.Va. (U.S.A.). *Chrysididae (Hymenoptera). Habits of Wasps and Bees.* ⓓ Harvard 1922.

Taylor, Margaret, Doc., Hackett Med. Coll. Canton (China). *Physiol.* o

Taylor, Monica (1877), Head of Biol. dept., Convent of Notre Dame. Glasgow (Scotland). *Protozool. Cytol. Embryol. of Symbranchus.* ⓓ Glasgow.

Taylor, Norman (1883), Curator, Brooklyn Botan. Garden. Brooklyn, N.Y. (U.S.A.), 1000 Washington Avenue. *Plant ecol. and plant geography.*

Taylor, Walter Penn (1888), Biol., Bureau of Biol. Survey, U. S. Dept. of Agric. Tucson, Ariz. (U.S.A.), 1746 East Fifth Street. *Mammals, rodents. Relation of mammals to forestry and forage. Ecol. of the animal and plant communities.* ⓓ Univ. of California 1914.

Taylor, William Randolph (1895), Ass. Prof. of Botan., Univ. of Pennsylvania, and member of staff for course in Algae, Marine Biol. Labor., Woods Hole, Mass. (U.S.A.). Philadelphia, Pa. (U.S.A.), Dept. of Botan., Univ. of Pennsylvania. *Chromosome structure and form, Gasteria. Alpine algae in British Columbia; marine algae of Florida and of New England.* ⓓ Univ. of Pennsylvania 1920. ⓟ *Marine algae from New England and Florida; bryophytes from British Columbia.*

Tazawa, Yoshisaburô, Ass. Tôhoku Imper. Univ. Sendai (Japan), Bact. Inst. *Agglutinatorische Analyse von paratyphoiden Bazillen. Filtrierbares Virus.* ⓓ Sendai.

Teacher, John H. (1869), St. Mungo (Notman) Prof. of Pathol., Univ. of Glasgow, Pathol. to Glasgow Royal Infirmary. Glasgow, W. 2 (England), 32, Kingsborough Gardens. *Pathol. and Bact.; Pathol. Histol. and the implantation of the human ovum.* ⓓ Glasgow 1903.

Téchoueyres, Emile (1878), Prof. Ecole Méd. de Reims, Dir. des Services d'Hygiène. Reims (France), 72, rue des Moissons. *Purification des eaux potables.*

Tedeschi, E., Prof. Univ. Padua (Italia). *Antrop.* o

Tedeschi, Virgilio, Prof. Univ. Nac. La Plata (Argentina). *Physic. biol.* o

Tedin, Hans (1860), Svalöf (Sverige). *Plant breeding Inst. Genetics of peas and barley.* ⓓ Lund 1891.

Tedin, Karl Olof (1898), Dr. of Phil. Svalöf (Sweden), Inst. of Genetics. *The genetics of Pisum, of cultivated species of Hordeum. Inheritance of quantitative characters. Relations of genetics to taxonomy.* ⓓ Lund 1925. ⓟ *Genetic material of Pisum and cultivated Hordeum.*

Tedjuschin, A. W. (1891), o. Prof., Dir. Zool. Kabinett. Minsk (U.d.S.S.R.), Univ. *Zoogeographie, Systematik der höheren Wirbeltiere (Ornithofauna).* ⓟ *Vogelbälge u. -Skelette, Mammalia (Felle mit Schädel).*

Tehon, Leo Roy (1895), Botan., Illinois State Natural History Survey, State of Illinois. Urbana, Ill (U.S.A.). *Crop diseases, mycol., flora of the inland waters, weeds and phanerogams.* ⓟ *Mycol. specimens.*

Teichert, Kurt (1868), Dr. phil., Dir. der Staatlichen Milchwirtschaftlichen Forschungsanstalt,

Landes-Ökonomierat. Wangen im Algäu, Württemberg (Deutschland). *Milchwirtschaftliche Bact.* ⓓ Lausanne 1903.

Teisseyre, W., Prof. Dr. Univ. Lwów (Polska), Politechnique. *Paléont.* o

Teissier, Georges (1900), Agrégé-Préparateur du Labor. de Zool. de l'Ecole Normale Supérieure. Paris V (France), 45, rue d'Ulm. *Embryol. des Coelentérés. Physiol. des Métamorphoses chez les Insectes. Pigments animaux.* ⓓ Paris 1924.

Teles Palhinha, Ruy, Prof. o. Univ. Lisbôa (Portugal). *Botan.* o

Telford, Clarence John (1887), Forester for Natural History Survey. Urbana, Ill. (U.S.A.), 219 Natural History Building. ⓓ Yale School of Forestry 1915.

Tello, Julio C., Prof. Univ. Lima (Peru). *Anthrop.* o

Tello y Muñoz, Jorge Francisco (1880), Prof. d'Histol. et Anat. pathol. Fac. Méd., Dir. Inst. National d'Hygiène d'Alphonse XIII. Madrid (España), Aguirre 1. *Dévelopement des terminaisons nerveuses et du système nerveux central.* ⓓ Madrid 1902.

Tellyesniczky, Koloman, Prof. Dr., Dir. des Anat. Inst. N. 2 Univ. Budapest (Ungarn). *Anat.* o

Temminck Groll, Johannes (1887), Apotheker, Lect. an der Univ. Amsterdam-Oost (Holland), Meerlaan 57. *Physiol. Chem., Fermentchem.* ⓓ Pharm. Amsterdam 1918.

Tempel, Max Willy (1901), Diplomlandwirt, Dr. phil., Wissenschaftlicher Hilfsarbeiter an der Abt. Pflanzenschutz der Staatl. Landw. Versuchsanstalt. Dresden-A. 24 (Deutschland), Nürnberger Str. 40. *Prüfung von Pflanzenschutzmitteln auf ihre Wirksamkeit gegenüber Pflanzenschädlingen und Krankheiten sowie auf die Kulturpflanzen. Pflanzenkrankheiten (spez. Kartoffelkrebs, Synchitrium), Unkräuter, tierische Schädlinge (spez. Insekten) und ihre Bekämpfung.* ⓓ Leipzig 1924.

Temple, Charles Edward (1877), Prof. of Plant Pathol. and State Plant Pathol. College Park, Md. (U.S.A.), Univ. of Maryland. *Diseases of corn.* ⓓ Univ. of Nebraska 1909.

Tendeloo, Nicolaus Philip (1864), Prof. der allgemeinen Pathol. und pathol. Anat. und Dir. des Pathol. Labor., Reichsuniv. Leiden. Oegstgeest bei Leiden (Holland), Wilhelminapark 8. *Lungenkrankheiten, Pharynxdivertikel, Wirkung örtlicher Dehnung und örtlichen Druckes und ihre Bedeutung für die Pathol.; elastische Nachwirkung und ihre Bedeutung für die Pathol.; pathol. Anat., Konstellationspathol. und Erblichkeit.* ⓓ Leiden 1894.

Tenenbaum, Szymon (1892), Physiograph. Kommission Polnische Akad. d. Wissensch. Warszawa (Polen), Poznańska 38. *Coleoptera.* ⓟ Coleoptera Palaearctica.

Tengwall, Tor Åke (1892), Ass. Dir. Proefstation voor Rubber. Buitenzorg, Java (Nederl.-O.-Indië). *Geography and ecol. of plant. Soil. Vegetation des Sarekgebietes.* ⓓ Upsala 1920.

Tennent, David Hilt (1873), Prof. of Biol., Bryn Mawr Coll. Bryn Mawr, Pa. (U.S.A.). *Experim. embryol.; organization of the egg; specificity of fertilization. Cytol.; chromosomes of cross activated and cross fertilized echinoderm eggs; cytoplasm and its inclusions.* ⓓ Johns Hopkins Univ. 1904.

Tenney, Florence Gould (1901), Research Ass. Dept. of Soil Microbiol., New Jersey Agric. Experim. Station. New Brunswick, N.J. (U.S.A.). *Decomposition of Organic Matter in Soil.* ⓓ Connecticut Agric. Coll. 1925.

Teodoresco, Emmanuel Const. (1866), Prof. Physiol. végétale Fac. Sc., Labor. de Physiol. végétale. Bucarest (România), Str. Cotroceni 38. *Influence de la lumière sur les plantes. Plantes volubiles.* ⓓ Paris 1899.

von Teppner, Wilfried (1891). Graz (Österreich), Leechgasse 30. *Säugetierpalaeont., Lamellibranchiata tertiaria.* ⓓ Graz 1923.

Tepps, M. W., Lect. Univ. Dept. Zool. Glasgow (Scotland). o

Teräsvuori, Kaarle Kustaa (1884), Prof., Doc. für Acker- und Pflanzenbau an der Univ. Helsinki (Finnland), Museokatu 25 B. *Wiesenuntersuchung, Pflanzenzüchtung, Versuchswesen des Pflanzenbaus.* ⓓ Helsinki 1919.

Teranishi, Chô (1896), Priv. entomol. Osaka (Japan), 221 Ikue-cho, Higashinari-ku. *Taxonomy of Formicidae.* ⓓ Tokio 1920. ⓟ Specimens of Ants.

Terao, Arata, Prof. Imper. Coll. of Fishery. Tôkyô (Japan), Zool. Labor., Ecchûjima. *Crustacea.* o

Terashima, S., Ass. Prof. Tôhoku Imper. Univ. Sendai (Japan), Pathol. Inst., Coll. of Med. *Pathol. and Pathol. Anat.*

Terechov, G. I., Doc. Univ. Taschkent (U.d.S.S.R.). *Pathol. Anat.* o

Tereg, Elinor Maria (1885). Hannover-Linden (Deutschland), Falkenstr. 21 A. *Grünalgen.*

Terentjew, Fedor A., Kuban-Schwarzmeer Wissenschaftl. Forschungsinst. Krasnodar (U.d.S.S.R.), Dmitrijewskaja-ul. 64, W. 1. *Arzneipflanzen.* o

Terényi, Alexander (1897), Chem. Pflanzenbiochem. Inst. Budapest II (Ungarn), Debrői-ut 15. *Biochem.* ⓓ Dipl.-Ing. Chem. Debrecen 1923, phil. 1925.

Terho, K., Dr. phil., Doc. Univ. Helsinki (Finnland), Dicksursby. *Haustierrassen.* o

Terjaev, Valentin Aleks., Doc. I. Univ. Moskau (U.d.S.S.R.). *Palaeont.* o

Terlezkij, Boris Klementievič, Prof. Weißruss. Inst. f. Landwirtsch. Minsk, Weißr. (U.d.S.S.R.). *Palaeont.* o

Terni, Tullio (1888), Prof., Dir. Ist. Istol.-Embriol. Univ. Padova (Italia), Via Loredan 6. *Spermatogénèse des Amphibiens; Histol. du système nerveux des reptiles et des oiseaux; structure des ganglions sympathiques de l'homme; Histochim. du glycogène; Structure des tissus vesiculeux; Embryol. experim. des oursins et des amphibiens (thyroide; vitesse du development en fonction de la temperature); Grandeur et hypertrophie fonctionelle de la cellule nerveuse; nerf abducent et membrane nictitante dans les vertébrés; régénération de la queue des reptiles; Cytol. des chondriosomes; Duplicitas anterior dans l'embryon de poulet.*

Ternowskij, Michail F., Doc., Sibirisches Inst. f. Land- u. Forstwirtschaft. Omsk (U.d.S.S.R.), Postfach 34. *Agronomie. Selektion. Genetic.* o

Ternowsky, Wassiliy (1888), Prof., Dir. Inst. Anat. normalen Anat. Univ. Kasan (U.d.S.S.R.), Anat. Inst. *Vegetatives Nervensystem.* ⓓ Moskau 1922. ⓟ Osteol. Praep., Skelette, Praep. der menschlichen Organe (Eingeweide u. Gehirn).

Terras, James, Lect. School of Med. R. coll. Edinburgh (Scotland). *Botan.* o

Terroine, Emile Florent (1882), Prof. Univ., Dir. Inst. Physiol. gén. Fac. Sc. Strasbourg (France), Rue de l'Université. *Métabolisme énergétique. Métabolisme agaté. Physiol. des Substances grasses et lipoïdiques. Respiration des Tissus. Digestion. Actions diastasiques. Croissance. Sécrétion pancréatique. Métabolisme de base.* ⓓ 1919.

Terry, Robert J. (1871), Prof. of Anat., School of Med., Washington Univ. St. Louis, Mo. (U.S.A.). *Human and compar. Anat. Anthrop.* ⓓ Missouri Med. Coll. 1895.

Teruuchi, Yutaka, Prof. Keiô-Gijuku Univ., Member of Kitasuto Inst. for Infect. Diseases. Yotsuya, Tôkyô (Japan), Biochem. Inst., Coll. of Med. *Med. Chem.*

Tesch, Johan Jacob (1877), Leader Sea fisheries research in: Rijksinst. voor Biol. Visscherij-onderzoek. den Haag (Holland), Binnenhof 4. *Biol. of sea fishes (development, food, growth, economical importance).* ⓓ 1906.

Teserico, Enoch, Prof. incar. Univ. Milano (Italia). *Physiol.* o

Tesson, R., Prof. Ecole prép. de Méd. et de Pharm. Angers (France). *Pathol.* o

Testoni, Piero (1897), Aiuto di Farm. nella R. Univ. di Sassari (Italia), Ist. di Farm. R. Univ. *Biochem., Chimico-fisica, Istol. patol. Fisiol.* ⓟ 1922.

Teterevnikowa, Dara Nikolaevna (1904), Prép. Stat. pathol. végétale Inst. Agronomique. Leningrad (U.d.S.S.R.), Detskoe Selo. *Mycol. et pathol. végétale: biol. des Uredinées, Fusarium nivale.* ⓟ Champignons.

Tetjaeva, M. B., Wiss. Lesshaft-Inst. Leningrad (U.d.S.S.R.), Pr. Maklina 32. *Physiol.* o

Tetjunkina, Anna (1887), Ass. Charkow, Ukraine (U.d.S.S.R.), Pflanzenphysiol. Labor. des Inst. für Volksbildung. *Bodenbact.*

Teuscher, Heinrich (1891), Botan. in Charge, Morton Arboretum. Lisle, Ill. (U.S.A.). *Woody plants in Northern Illinois.* ⓟ Seeds, cuttings or bulbs of plants.

Teutschlaender, Otto Richard (1874), Dr. med., ao. Prof., Leiter einer Abteilung des Inst. für experim.-Krebsforschung. Heidelberg (Deutschland), Werderstr. 32. *Allgemeine Pathol., Krebsforschung, Metaplasie.* ⓓ Zürich 1906.

Teyrovský, Vladimir (1898), Dr. rer. nat., Lect. (Priv.Doc.) of systematical Zool., Ass. of the zool. inst. of the Masaryk Univ., Brno (C.S.R.), Konnicova 43. *Hemiptera-Heteroptera palaearct.; Odonata palaearct.; Zoogeography; Animal behavior.* ⓓ Praha 1920. ⓟ Hemiptera-Heteroptera; Odonata (Nester, Fraßstücke).

Thakar, C. S. (1887), Prof. Bombay (India), Physiol. Labor. Grant med. Coll. *Biochem. and Physiol.* ⓓ Bombay 1910. ⓟ Histol. Sections.

Thalhammer S. I., János (1847). Pécs (Ungarn), Plus-Coll. *Dipterol. Ungarns.* ⓟ Diptera.

Tharp, Benjamin Carroll (1885), Associate Prof. of Botan., Univ. of Texas. Austin, Tex. (U.S.A.) *Plant Taxonomy and Ecol. Texas Parasitic Fungi. Commelinaceae. Texas Vegetation.* ⓓ Univ. of Texas 1925. ⓟ Native phanerogamous plants of Texas.

Thatcher, R. W., Prof. of Plant Chem., Dir. of Experim. Stations, New York Agric. Experim. Stat. Geneva, N.Y. (U.S.A.). *Wheat and flour, soil investig.; chem. studies of insecticides.* ⓓ Agr. Nebraska 1920, I.L. Hobart 1925. o

Thatcher, Roscoe Wilfred (1872), Dir. of Agric. Experim. Stations for Cornell Univ. Geneva, N.Y. (U.S.A.). *Phytochemistry.* ⓓ Univ. of Nebraska 1901.

Thaxter, Roland (1858), Prof. emer. of Cryptogamic Botan., Honorary Curator Farlow Herbarium. Cambridge, Mass. (U.S.A.). *Mycol., fungus diseases of insects.* ⓓ Harvard 1882.

Thayer, John Eliot (1862), Ornithol. Boston, Mass. (U.S.A.), 30 State St. o

Thayer, Paul (1875), Prof. of Pomol. Extension, Pennsylvania State College, Pa. (U.S.A.). *Growing and handling of fruit; Dependable Fruits. Varieties of Apples. The Strawberry, its Culture and Varieties. The Red and White Currants.* ⓓ Michigan State Coll. 1913.

Thellung, Albert (1881), Dr. phil., Titularprof. Priv.Doc. an der Univ. und Ass. am Botan. Garten und Museum. Zürich 7 (Schweiz), Forchstr. 33. *Systematik der Phanerogamen (speziell: Avena, Amarantus, Lepidium, mitteleuropäische Epilobia); Floristik von Mitteleuropa; Adventivfloristik; Abstammung der Kulturpflanzen.* ⓓ Zürich 1906.

Thery, André (1864), Adjoint au Dir. de l'Inst. Sc. chérifien, Conserv. du Mus. de l'Inst. Rabat (Maroc), Avenue Moulay Youssef. *Insectes coléoptères: Buprestidae (Systématique, Biol., Anat.).* ⓟ Buprestides.

Thibout, F., Prof. Univ. Angers (France). *Botan.* o

Thiel, Albert Frederic (1887), Associate Pathol. Bureau of Plant Industry, U. S. Dept. of Agric. State Leader of Barberry Eradication in Nebraska, Lincoln, Neb. (U.S.A.), Coll. of Agric. *Epidemia. of stem rust, determination of biol. forms. Ecol. Studies in the Tension Zone between Prairie and Woodland. Germination of the Teliospores of Pucciniol Regional occurrence of Puccinia.* ⓓ Nebraska 1917.

Thiel, Max Egon (1898), Wissenschaftl. Hilfsarbeiter am Zool. Mus. Hamburg 5 (Deutschland), Danziger Str. 18III. *Systematik und Morphol. der Coelenteraten und Tunikaten. Biol. Untersuchungen an Süßwassermuscheln.* ⓓ Hamburg 1923. ⓟ Spongien, Coelenteraten, Würmer, Bryozoen u. Tunikaten.

Thiele, Johannes (1860), Prof. Dr., Kustos am Zool. Mus. a. D. Berlin N 4 (Deutschland), Invalidenstr. 43. *Vergl. Morphol. und Systematik von Mollusken. Leptostraken, Solenogastres und Mollusken der Deutschen Südpolar- und Tiefsee-Expedition.* ⓓ Berlin 1886.

Thielmann, Marie (1889), Priv.Doc. Riga (Latvija), Gertrudstr. 3, W. 8. *Züchtung pflanzlicher Gewebe; Experim. Zellphysiol.* ⓓ Petersburg 1914.

Thiem, Hugo (1887), Reg.R. Dr., Zool. an der Zweigstelle Naumburg-S. der Biol. Reichsanstalt für Land- und Forstwirtschaft, Berlin-Dahlem (Deutschland). *Anat. niederer Mollusken. Biol., Epidemiol. und Bekämpfung der Reblaus (Phylloxera vastatrix). Biol. der Coeciden.* ⓓ 1914.

Thienemann, August (1882), Dir. der Hydrobiol. Anstalt der Kaiser-Wilhelm-Gesellschaft in Plön und o. Prof. der Hydrobiol. Univ. Kiel. Plön (Deutschland), Haus Waldkoppel. *Limnol. Chironomiden. Planarien. Coregonen.* ⓓ Greifswald 1905.

Thienemann, Johannes (1863), Prof., Leiter Vogelwarte. Rossitten (Deutschland). *Vogelzugsforschung.* ⓓ Königsberg.

Thienes, Clinton H. (1896), Ass. Prof. of Pharmacol., Univ. of Oregon Med. School. Portland, Ore. (U.S.A.). *Pharm. of smooth muscle; general pharm.* ⓓ Stanford Univ. 1926.

Thierfelder, Hans (1858), o.Prof. der physiol. Chem. Tübingen (Deutschland), Gmelinstr. 8. *Gehirnchem.* ⓓ Rostock 1883.

Thoday, David (1883), Prof. of Botan. in the Univ. Coll. of N. Wales, Bangor, Carnarvonshire (Great Britain), Department of Botan., Memorial Buildings. *Plant physiol. Assimilation; Respiration; Physiol. anat. of some S. African plants; Contractile roots of Oxalis sp.; Taxonomy and distribution in S. Africa of genus Passerina.* ⓓ Cambridge 1908.

Thörner, Walter (1886), ao. Prof. Ass. am physiol. Inst. der Univ. Bonn a. Rh. (Deutschland), Hohenzollernstr. 20. *Allgemeine Physiol. des Nervensystems. Physiol. der Leibesübungen.* ⓓ Med. Göttingen 1912.

Thom, Charles (1872), Senior Mycol., Bureau of Chem., U. S. Dept. of Agric. Washington, D.C. (U.S.A.). *Microbiol. of food, fermentation, taxonomy of saprophytic hyphomycetes, Aspergillus and Penicillium.* ⓓ Missouri 1899. ⓟ Cultures of Aspergilli and Penicillia.

Thomas, Charles A. (1895), Ass, Prof. of Economic Entomol. Bustleton, Pa. (U.S.A.), Pa. State Coll. Labor. *Insects affecting Cultivated Mushrooms and Vegetable Crops. Mushroom Insects and Mites; Wireworms; Millipedes; Insect Behavior and Physiol.* ⓟ Scarabaeidae and Elateridae, live Elaterid larvae.

Thomas, Ethel Nancy Miles, Head of Dept. of Botan., Univ. Coll. Leicester (England). *Cytol. and Seedling Anat.* ⓓ Univ. of London.

Thomas, Frank L. (1887), Chief, Division of Entomol., Texas Agric. Experim. Station; State Entomol. of Texas. College Station, Texas (U.S.A.). *Investigation of insect pests of cotton, corn, citrus, truck. Regulatory work in connection with eradication of bee diseases.* ⓓ Amherst, Massachusetts, 1910.

Thomas, H. H., Univ. Lect. Downing's Coll. Cambridge (England). *Botan.* o

Thomas, Harvey E. (1890), Ass. Prof. of Plant Pathol. Ithaca, N.Y. (U.S.A.), Dept. Plant Pathol. Cornell Univ. *Plant pathol., diseases of fruits.* ⓓ Columbia Univ. 1920.

Thomas, Karl (1883), Prof. Dr. med., Dir. des physiol.-chem. Inst. der Univ. Leipzig (Deutschland),

Liebigstr. 16. *Stoffwechsel und Ernährung, Eiweißchem., intermediärer Stoffwechsel von Eiweiß und Fett.* ⓓ Freiburg i. Br. 1906.

Thomas, Lawrence C. (1896), Ass. in Animal Breeding, Coll. of Agric., Univ. of Illinois. Urbana, Ill. (U.S.A.). *Genetics.*

Thomas, Lyell Jay (1892). Ass. Prof. of Zool., Univ. of Illinois. Urbana, Ill. (U.S.A.). *Invertebrate Zool.* ⓓ Urbana 1923.

Thomas, Maurice, M. E. C. (1882). Bruxelles (Belgique), 10, rue Rembrandt. *Psychol. animale (Instinct-intelligence) dans tous les domaines de la Zool. Transformisme. Poissons Migrateurs. Migrations des Oiseaux. Instinct et Aptitudes des Araignées.*

Thomas, Pierre-Joseph (1876), Prof. Fac. méd., Dir. Inst. Chim. biol. Cluj (România), Strada Pasteur. *Chim. et physiol. des sucres. Micro-méthodes. Substances protéiques des végétaux inférieurs.*

Thomas, Roy Curtis (1887), Associate Pathol., Ohio Agric. Experim. Station. Wooster, O. (U.S.A.). *Bact., mycol., biochem.; plant diseases.* ⓓ Coll. of Wooster 1913.

Thomas, Royle P. (1897), Ass. in Soils, Univ. of Wisconsin, Madison, Wis. (U.S.A.), Soils Building. ⓓ Iowa State Coll. 1925.

Thomas, William Andrew (1883), Ass. Entomol. U. S. Bureau of Entomol. Chadbourn, N.C. (U.S.A.). *Truck Insects.* ⓓ 1908.

Thompson, Benjamin G. (1891), Ass. Entomol. Oregon Experim. Station. Corvallis, Ore. (U.S.A.). *Injurious Insects.* ⓟ Clear wing Moths, Aegriidae (Sesiidae).

Thompson, Ian Maclaren (1896), Ass. Prof. of Anat. Montreal (Canada), Department of Anat., McGill Univ. *Autonomic Nervous System.* ⓓ Edinburgh 1920.

Thompson, Luther (1890), First Ass. Clinical Pathol., Mayo Clinic, Rochester, Minn. (U.S.A.). *Bact. Hydrogen sulphide producing bacteria. Toxin production in canned foods. Spore-forming anaerobes.* ⓓ Univ. of Illinois 1925.

Thompson, May Bel, Knowles School. Kiuhiang (China). *Zool.* o

Thompson, Noel Finley, Ass. Pathol., Bureau of Plant Industry, U. S. Dept. Agric., State Capitol Annex. Madison, Wis. (U.S.A.). o

Thompson, Walter Palmer (1889), Prof. of Biol., Univ. of Saskatchewan. Saskatoon, Sask. (Canada), Biol. Dept. *Genetics and Cytol. of wheat.* ⓓ Harvard 1914.

Thompson, William Francis (1888), Dir. International Fisheries Commission. Univ. of Washington. Seattle, Wash. (U.S.A.). *Life history age and rate of growth of Hippoglossus. Life history of albacore and sardine.* ⓓ Stanford Univ., Cal. 1911.

Thompson, William Robin (1887), Entomol., Bureau of Entomol., U. S. Dept. of Agric. Hyères, Var. (France), European Parasite Labor., Le Mont Jenouillet. *Entomol. Parasit. Philos. Biol.* ⓓ Toronto, Canada, 1909.

Thoms, Hermann (1859), Univ.Prof., Geh.Reg.R., Dir. des Pharm. Inst. der Univ. Berlin-Steglitz (Deutschland), Hohenzollernstr. 6. *Pflanzenstoffe, insbesondere ätherische Öle. Pharm. und med. Chem.* ⓓ Erlangen 1886.

Thomson, A., Prof. Univ. Oxford (England), *Anat.* o

Thomson, Arved (1862), Prof. emer. Univ. Tartu (Estland), Vene t. 76. *Phytopath., Wiesenbau.* ⓓ Dorpat 1890.

Thomsen, Axel (1887), Abteilungsleiter am Serumlabor. der kgl. Vet.- und Landwirtschaftlichen Hochschule. København V (Danmark), Bülowsvej 30 A. *Abortus beim Rinde.*

Thomson, J. G., Prof. School of Hygien. and Tropic. Med. London N.W. (England), Endsleigh Gardens. *Protozool.* o

Thomson, James Allan (1881), Dir., Dominion Mus. Wellington (New Zealand). *Brachiopoda (Recent and Tertiary).* ⓓ New Zealand.

Thomsen, Mathias (1896), Lect. of Zool., Royal Vet. and Agric. Coll., Ass. at ,,Statens plantepatol. Forsøg", Lyngby. København (Danmark). *Cytol. of Parthenogenesis, Applied Entomol.* ⓓ Copenhagen 1920.

Thomson, James Stuart (1868), Senior Lect. in Zool. Victoria Univ. of Manchester (England). *Alcyonaria; Exoskeleton of Fishes; Pteropods; Chelonia.* ⓓ Ph. Berne 1909, Sc. Manchester 1915.

Thomson, Paul William (1892), Botan. und Quartärgeol. an der Versuchsstation des Estländischen Moorvereins Tooma bei Vägera. Tartu (Estland), Alleestr. 17, W. 3. *Pollenanalytisch-stratigraphische Erforschung der Moore u. lacustrinen Ablagerungen in Estland. Regionale Moortypen Estlands. Vegetation der estländischen Moore, Verbreitung einzelner Moorpflanzen.* ⓓ Tartu 1926.

Thomson, Robert Boyd (1870), Prof. of Phanerogamic Botan. Dept. of Botan., Univ. of Toronto (Canada), 11 Queen's Park. *Anat. of Gymnosperms and Vascular Cryptogams, fossil and living.* ⓓ Univ. of Toronto 1899. ⓟ Herbarium specimens of Canadian plants.

Thomson, Wilbur F. (1873), Dir., Pathol. Department, Hotel Dieu. Beaumont, Tex. (U.S.A.), 927 San Jacinto Life Bldg. ⓓ Tulane Univ., New Orleans, La., 1906.

Thone, Frank (1891), Staff member, Sc. Service, Washington, D.C. (U.S.A.), 21. and B streets. *Physiol. plant ecol., water relations; atmometry. Plant successions.* ⓓ Univ. of Chicago 1922.

Thornber, John J., Prof. of Botan., Univ. of Arizona, Dir., Arizona Agric. Experim. Station, and Dean, Coll. of Agric. Tucson, Ariz. (U.S.A.). *Flora of Arizona, grasses.* o

Thorndike, Edward Lee (1874), Dir., Division of Educational Psychol., Inst. of Educational Research, Teacher Coll. Columbia University, N.Y. (U.S.A.). *Behavior of animals. Inheritance of mental qualities.* ⓓ Harvard Univ., Columbia Univ. 1898.

Thornton, H. G. (1892), Head Bact. Dept. Rothamsted Experim. Sta. Northampton (England), Kingsthorpe Hall. *Soil Bact. Bacillus radicicola.* ⓓ Oxford Univ.

Thorpe, Frances Jackson, Ass. Herbarium, Univ. of Michigan. Ann Arbor, Mich. (U.S.A.). *Bryophytes, taxonomy of mosses.* o

Thorpe, William Homan (1902), Demonstrator, Zool. Labor. Cambridge (England), The Mus. ⓓ Cambridge 1924.

Thro, William C. (1875), Prof. of Clinical Pathol., Cornell Univ. Med. Coll. New York City (U.S.A.), 477 First Ave. *Calcium in the blood. Bact. Immuniol.* ⓓ Cornell Univ., Ithaca, N.Y. 1907.

Thrupp, Adrian C. (1897), Forester, Dominion Forestry Branch, B. C. Reserves, Research work in Interior of British Columbia. Kamloops (British Columbia), Box 340. *Growth of pseudotsuga contorta murrayana. Association of species in different forest types. Introduction of exotic tree species.* ⓓ Toronto 1922. ⓟ Tree seeds, particularly Conifers.

Thulin, Gustav (1889), Fil. Lic., Amanuens. am Zool. Inst. Lund (Sverige). *Tardigraden, Marine Bodenfauna Nordeuropas.*

Thunberg, Torsten (1873), Prof. a. d. Univ. Lund (Sverige). Finngatan 12. *Vitale Oxydationserscheinungen, Künstliche Atmung.* ⓓ Uppsala 1900.

Thuringer, Joseph Marion (1886) Prof. of Histol. and Embryol., School of Med., Univ. of Oklahoma, Norman Okla. (U.S.A.). *Stratified squamous epithelium (human): normal proliferation of cells of the stratum spinosum and Stratum cylindricum; rate of cell division in the Stratum spinosum and St. cylindricum.* ⓓ Creighton Univ. 1911.

Thurnwald, Richard, ao. Prof. Dr. Univ. Berlin (Deutschland). *Ethnol., Völkerpsychol., Soziol.* o

Thurston, Henry W., Jr., Ass. Prof. of Plant Pathol., Pennsylvania State Coll. State College, Pa. (U.S.A.). o

Thurston, Lloyd M. (1900), Instr. and Research Ass. St. Paul, Minn. (U.S.A.), Univ. Farm. *Nutritional Studies with Dairy Cattle. Bact. of butter.* ⓟ Michigan State Coll. 1922.

ten Thye, Jan Hendrik (1895), Conserv. at the pathol. inst. of the vet. fac. of the Univ. of Utrecht (Holland), Biltstraat 166. *Animal pathol., diseases in piggs; pathol. anat. of infectious diseases.*

ten Thye, Pieter Antonie (1899), Hauptass. der Augenheilkunde, Groningen (Holland), Inrichting voor Ooglyders. *Pathol. Anat. des Auges.*

Tiberti, Nazareno, Prof. Univ. Siena (Italia). *Patol.* o

Ticehurst, Claud. B. (1881). Lowestoft (England), Grove House. *Ornithol. Palaearctic and Indian.* ⓟ Palaearctic Bird Skins.

Tichomirow, Boris Mich. (1887), Ass. Univ. Naturwiss. Inst. Peterhof. Leningrad (U.d.S.S.R.), Was. Ostr. 16 L., N. 29. *Ornithol., Vergl. Anat. Speicheldrüsen. Trachealsystem der Insekten. Phytopath.*

Tichomirnowa, Olga Alexandr. (1897), Ass. Kasan (U.d.S.S.R.), Zootomisches Kabinett, Univ. *Vergl. Morphol. der Wirbellosen. Nemertini u. Oligochaeta, Biocoenotik.*

Tichonow, Paul M., Doc. d. Kasanschen Inst. f. Land- u. Forstwirtschaft. Kasan (U.d.S.S.R.). *Pflanzenkunde. Selektion. Genetik u. Physiol.* o

Tichonow, Stepan L., Doc. Landwirtschaftl. Akademie Timirjasew. Moskau (U.d.S.S.R.), Petrowsko-Rasumowskoje, Iwanowskaja 3, W. 4. *Botan., Systematik.* o

Tichovskaja, Soja Petr., Doc. Inst. f. Landwirtschaft u. Melioration. Saratow (U.d.S.S.R.). *Physiol.* o

Tiedjens, Victor A. (1895), Ass. Research Prof. of Vegetable Gardening, Massachusetts Agric. Coll. Waltham, Mass. (U.S.A.), 240 Bearer St. *Plant breeding, varieties of vegetables.*

Tiegs, Oscar Werner (1897), Lect. Zool. Melbourne (Australia), Univ. *Insect Metamorphosis; Muscle Physiol.* ⓟ Adelaide 1922.

Tiesenhausen, Manfred, Chef de travaux de l'Inst. de Botan. générale. Cluj (Romania), Str. Gojon 5. *Mycol.* o

Tiesenhausen, Michel (1877), Dir. Abt. Pathol. Anat. Forschungskatheder f. Anat. u. Physiol. Univ. Odessa (U.d.S.S.R.), Pasteurstr. 28. *Pathol. Anat.* ⓟ Odessa 1910.

Tietz, Harrison Morton (1895), Instr. Zool. State College, Pa. (U.S.A.). *Entomol. Lepidoptera: Noctuidae.* ⓟ Mass. Agric. Coll. 1921. ⓟ Noctuidae.

Tiffany, Lewis Hanford (1894), Ass. Prof. of Botan. Department of Botan., Ohio State Univ. Columbus, O. (U.S.A.). *Plant physiol., ecol., taxonomy, ecol. and physiol. of the freshwater algae.* ⓟ Ohio State Univ. 1923. ⓟ Freshwater algae.

Tiffeneau, M., Prof. Fac. Méd. Paris (France), 12, rue Rosa Bonheur. *Biochem.* o

Tigerstedt, Carl (1882), Prof. der Physiol. an der Univ. Helsingfors (Finnland), Physiol. Inst. *Kreislauf. Stoffwechsel.* ⓟ Helsingfors 1910.

Tiitso, Max (1900), Ass. Tartu (Eesti), Physiol. Inst.

Tikhomirow, Nicolaï (1881), Vorsteher d. Zool. Gartens, Inst. der Experim. Med. Leningrad (U.d.S.S.R.), Podrezowstr. 76, Qu. 8. ⓟ Leningrad. *Pathol. Anat.* o

Tikhonovich, Nikolaus N., Obergeol. Com. Géol. de Russie. Leningrad (U.d.S.S.R.), W.O., Sredny Pr. 72-B. *Paleont.; tertiäre Pectiniden des Fernen Ostens (Sachalin).* o

Tilden, Josephine E., Prof. of Botan., Univ. of Minnesota. Minneapolis, Minn. (U.S.A.). *Algae.* o

Tilford, Paul E., Ass. Pathol., Agric. Experim. Stat. Wooster, O. (U.S.A.). o

Tillyard, Rob. John (1881), Dr., Chief Biol. Dept. Cawthron Inst. Nelson (New Zealand). *Zool., Entomol. Insect pests.* ⓟ Sydney 1917, Cambridge 1920.

Timkó, György (1881), Kustos an der botan. Abteilung des Magyar Nemzeti Muzeum Budapest V

(Ungarn), Akadémia-ut 2. *Lichenol. Ungarische und balkanische Flechten.* ⓟ Flechten.

Timm, Hermann Rudolf (1859), Prof. Dr. Hamburg 39 (Deutschland), Bussestr. 45. *Einheimische Moose (floristisch).* ⓟ Würzburg 1883. ⓟ Pflanzen.

Timofeev, Sergei, Prof. Irkutsk (U.d.S.S.R.), Univ. *Morphol. der Polychaeten. Blutgefäßsystem von Euphrosyne, Staurocephalus. Hesione. Verhalten des Golgischen Netzes bei der Histogenese.* ⓟ Petersburg 1914.

Timofejew, Alexander St., wiss. Mitarbeiter d. Polytechn. Inst. in Tiflis (U.d.S.S.R.), Alexanderstraße 41a. *Pflanzentechnol.* o

Timofejewsky, Alexander D. (1887), Prof. Pathol. Physiol. Univ. Tomsk, Siberia (U.d.S.S.R.). *Tissues cultures, tuberculosis in vitro, prospective potencies of cells, cultivation of human tumors in vitro.* ⓟ Tomsk 1912.

Timon-David, Jean (1902), Ass. de Zool. à la Fac. des Sc. de Marseille (France), Place Victor Hugo. *Diptères.* ⓟ Marseille 1926. ⓟ Diptères.

Tims, Eugene Chapel (1894), Ass. Plant Pathol. Louisiana Agric. Experim. Station. Baton Rouge, La. (U.S.A.). *Sugar Cane Diseases.* ⓟ Univ. of Wisconsin 1924.

Tiprez, Jean-Léon (1901), Préparateur de Path. interne et expérim. et clinique des maladies appareil digestif. Lille (France), 241, R. Solfernio. *Digestion.* ⓟ Lille 1926.

Tischkus, Waclowas (1888). Kaunas (Litauen), Med. Fac. *Verdauung.* ⓟ Litauische Univ. 1925.

Tischler, Friedrich (1881), Heilsberg, Ostpreußen (Deutschland). *Ornithol. Ostpreußens.*

Tischler, Georg (1878), Dr. phil., o. Prof. an der Univ., Dir. des Botan. Inst. Kiel (Deutschland), Düsternbrooker Weg 17. *Pflanzliche Cytol.: Hybridenuntersuchungen Chromosomenforschung.* ⓟ Bonn a. Rh.

Tisdale, Wendell Holmes (1892), Pathol. in charge of Cereal Smut Investigations, Office of Cereal Crops and Diseases, Bureau of Plant Industry, U. S. Department of Agric. Washington, D.C. (U.S.A.). *Smuts of cereal crops, pathol., physiol. and genetic.* ⓟ Univ. of Wisconsin 1917. ⓟ Cereal smuts.

Tisdale, William Burleigh (1890), Plant Pathol., in Charge Tobacco Investigations, Univ. of Florida. Tobacco Experim. Station. Quincy, Flo. (U.S.A.). *Diseases of tobacco.* ⓟ Univ. of Wisconsin 1920. ⓟ Specimens of tobacco diseases.

Tison, M. A., Maître de Conf. Fac. Sc. Univ. Rennes (France). *Zool.* o

Tissot, A. N. (1897), Ass. Entomol., Florida Agric. Experim. Station. Gainesville, Fla. (U.S.A.). *Aphididae: biol. and economic. Citrus and truck crops.* ⓟ Ohio State Univ. 1925.

Tissot, M. J., Prof. au Mus. Paris (France), 57, rue Cuvier. *Physiol.* o

Titajev, Alexei Assinpropowitsch (1894), Labor. f. experim. Biol. Swerdlov Univ. Moskau (U.d. S.S.R.), Petrowsko-Rasumowskoje, Ivanovsk. 3. *Biochem. der Schilddrüse, Physiol. der Blutgefäße.* ⓟ Moskau 1918.

Titov, J. T., Prof. Staatsuniv. med. Fac. Minsk, Weißrußl. (U.d.S.S.R.). *Pathol. Anat.* o

Titow, Iwan A., Doc. a. Bogorodsker Wissensch. Pädagog. Inst. f. Landeskunde. Moskau (U.d.S.S.R.), Petrowsko-Rasumowskoje, Landwirtsch. Akademie 15. *Botan. Moorforschung.* o

Titschack, Erich (1892), Dr. phil., Leiter der Entomol. Sammlungen des Zool. Staatsinst. und Zool. Mus. Hamburg 1 (Deutschland), Steintorwall 1. *Wachstum, Nahrungsverbrauch, Eiproduktion der Insekten. Häutung der Insekten. Ökol. der Kleidermotte.* ⓟ 1919.

Titus, Edward Gaige (1873), Dir. Agric. Research Dept., Utah Idaho Sugar Co. Salt Lake City, Utah (U.S.A.). Residence: 1228 Bryan Ave. Business: Utah-Idaho Sugar Co. *Eugenics. Seed-breeding: crop rotation trials, seed variety trials, soil and water*

analyses; insect and disease control. Development of sugar beet and crops. Ⓓ Harvard Univ. 1911. Ⓟ Sugar beet, Hymenoptera. Apoidea. Megachilidae Osmiidae.

Tiwary, Nand Kumar (1888), Ass. Prof. of Botan. Benares Hindu Univ., U.P. (India), Dept. of Botan. *Plant morphol. and ecol. Polyembryony in the Genus Eugenia. Nectary in Lesbania grandiflora.* Ⓓ Muir Central Coll., Allahabad 1912. Ⓟ Bryophytes. Angiosperms.

Tizzoni, Guido (1853), Prof. Patol. Gen. Batteriol. Univ., Dir. del relativo Ist. Bologna (Italia), Via Belle Arti 33. *Istol. normale e patol. Fisiopatol. Batteriol. Patol. sperim. e Sierol. Vaccinazione e sieroterapia contro il tetano. Pellagra.* Ⓓ 1878.

Tjebbes, Klaas (1886), Ph.D., Leader of the Sc. Sugar Beet Inst. of Svenska Sockerfabriks Aktiebolaget at Hilleshög, post Landskrona (Sweden). *Genetical Studies on Beta, Phaseolus, Portulacca, other plants, dogs, cats, poultry. Species crosses in Linaria. Cytol. work on plants.* Ⓓ Amsterdam 1912.

Tjurin, Alexander (1882), Prof. Landwirtschaftl. Inst. (Forst. Fak.). Woronesch (U.d.S.S.R.). *Waldbau.* Ⓓ St. Petersburg 1909.

Tkatchenko, Michael (1878), Prof. of Silvies Forestinst., Dir. of Forest Experim. Station. Leningrad (U.d.S.S.R.). *Biol. and Ecol. of Primeforests, natural replacement of pine and spruce.* Ⓓ Petersburg 1904.

Tobler, Friedrich (1879), o. Prof. d. Botan. Sächs. Techn. Hochsch. Dir. Botan. Inst. und des Staatl. Botan. Gartens. Dresden-A. 16 (Deutschland), Stübelallee 2. *Physiol. der Lichenen, Farbstoffe im Pflanzenreich, Faserstoffe. Angewandte Botan. auf experim. Grundlage. Kolonialbotan., Hedera. Symbiose.* Ⓓ Berlin 1901.

Tobler-Wolff, Gertrud (1877), Dr. phil. Dresden-A. 16 (Deutschland), Stübelallee 2. *Botan. Cytol., Lichenen, Phycomyceten.* Ⓓ Berlin 1905.

di Tocco, Roberto (1895), Ass. R. Stazione Bacologica Sperimentale. Padova (Italia). *Bombyx mori.* Ⓓ Padova 1921. Ⓟ *Preparati anat. e anat.-patol. del Bombyx mori.*

Tocco-Tocco, Luigi (1882), Prof. Farmacognosia Scuola di Farm. Univ. Messina (Italia), Via Risorgimento 199. *Insetticidi. Funzione della miofibrilla. Diffusione.* Ⓓ 1922.

Tochibara, Isamu, Lect. Keiô-Gijuku Univ. Tôkyô (Japan), Parasit. Inst., Coll. of med. *Parasit.*

Tochinai, Yoshihiko (1893), Ass. Prof. of Phytopath., Coll. of Agric. Hokkaidô Imper. Univ. Sapporo (Japan), Botan. Inst. *Phytopath.* Ⓓ Hokkaidô Imper. Univ. 1918.

Toda, Shôzô, Prof. Imper. Univ. Kyôto (Japan), Hygien. Inst., Coll. of Med. *Hygiene.*

Todaro, Francesco (1864), Dir. Ist. di Cerealicoltura, Prof. Agric. Ist. Sup. Agr. Bologna (Italia), Via Toscana 121. *Cereali, razze.*

Todd, Arthur Theodore (1888), Demonstrator of Pathol., Bristol Univ., Clinical Lect. in Med. Bristol (England). *Pathol.* Ⓓ Edinburgh 1912.

Todd, James Trior (1894), Lect. in Pharm. and Pharmacognosy, Royal Technical Coll. Glasgow C.1 (Scotland). *Habilitation of Drugs.* Ⓓ Edinburgh 1916.

Todd, Thomas Wingate (1885), Henry Willson Payne Prof. of Anat., Dir. of the Hamann Mus. Western Reserve Univ. Cleveland, O. (U.S.A.). *Anat., Human, Mammalian and Clinical, Physical Anthrop. Anat. of the Gastro-Intestinal Tract. Mammalian Dentition.* Ⓓ London 1911.

Todd, Walter Edmond Clyde (1874), Curator of Ornithol., Carnegie Mus. Pittsburgh, Pa. (U.S.A.). *Neotropical birds; birds of Labrador Peninsula; geographic distribution.*

Toedtmann, Willy (1884), Dr. phil. Hamburg 22 (Deutschland), Finkenau 19. *Spermatozoen.* Ⓓ München 1912.

Töllner, Karl Fr. (1860). Bremen (Deutschland), Schönhausenstr. 21, und Stebben, Postbez. Bremen (Deutschland). *Pharm. Theoretische und angewandte Botan.* Ⓟ Seltene Pflanzen von pharm. oder industriellem Interesse.

Toenniessen, Erich (1883), ao. Prof. für innere Med. und Oberarzt der med. Univ.-Klinik in Erlangen (Deutschland). *Vererbungs- und Variabilitätserscheinungen bei Bakterien. Tuberkulose, Stoffwechsel, insbes. Kohlehydratstoffwechsel beim Säugetier.* Ⓓ Erlangen 1910.

Töppich, Gerhard (1892), Priv.Doc. für allgemeine Pathol. und pathol. Anat., I. Ass. am Pathol. Inst. Königsberg i. Pr. (Deutschland), Kopernikusstr. 3/4. *Tuberkulose.* Ⓓ Breslau 1918.

Toëx, Edmond Etienne (1876), Dir. Station centrale de Phytopath. Paris (France), 11 bis, rue d'Alesia. *Pathol. végétale: rouilles des céréales. Frisolée. Odiopsis taurica. Conidiophores. Cancer végétale. Leptonécrose et enroulement. Ascospores de Leptosphaeria. Maladies à Virus chez les végétaux. Tumeurs végétales. Erysiphées.*

Togashi, Kôgo (1895), Prof. Plant Pathol., Imper. Coll. of Agric. and Forestry. Morioka (Japan). *Cancer diseases of fruit and forest trees caused by the species of Valsa and Cytospora.* Ⓓ Hokkaidô Imper. Univ., Sapporo, 1922. Ⓟ *Parasitic fungi of Japan.*

Toison, J., Prof. Fac. Univ. Lille (France). *Histol.* o

Tokarewa, Anna Ivan, Doc., Ass. Dr. Histol.-Embryol. Kab. II. Univ. Moskau (U.d.S.S.R.), Pogodinskaja ul. 6. *Histol. Embryol.* o

Tokugawa, Yoshichika Marquis (1886), Inhaber und Mitglied des Tokugawa Biol. Inst. Hiratsukamachi, Ebaragun, Tôkyô (Japan). *Pflanzenphysiol.* Ⓓ Tôkyô 1914.

Toldt, Karl (1875), Hofrat, zool. Dir. Mus. Ferdinandeum in Innsbruck, stellvertr. Dir. a. D. naturh. Mus. Wien, Innsbruck (Deutsch-Österreich), Müllerstraße 30. *Integument und Osteol. der Säugetiere. Milben.* Ⓓ Wien 1899. Ⓟ Fauna von Tirol.

Tolles, Goodwin S. (1895), Graduate Ass. Entomol. Michigan State Coll. East Lansing, Mich. (U.S.A.), Ent. Dept. *Insect pests.* Ⓓ Mich. State Coll. 1926.

Tolmachoff, Innokenty Pavlovich (1872), Curator in Carnegie Mus., Prof. at the Univ. Pittsburgh, Pa. (U.S.A.), Carnegie Mus. *Paleozoic Faunas (Devonian and Carboniferous).* Ⓓ St. Petersburg (Russia) 1897.

Tolmatchew, Alexander (1903), Sekretär u. ord. Mitglied der Ständigen Polarkommission d. Akad. d. Wissensch., Kurator-Ass. am Botan. Mus. d. Akademie d. Wissensch. Leningrad (U.d.S.S.R.), Universitätsquai 5. *Pflanzengeographie der Arktis; Biogeographie; Systematik der höheren Gewächse (Papaver, Draba, Polemonium, besonders arktische Arten; Monimiaceae u. Labiatae Südamerikas).*

Tolmatschew, Iwan A. Doc. d. Kiewschen Landw. Inst. Kiew (U.d.S.S.R.), Brest-Litowskar Chaussee Nr. 39. *Pflanzenphysiol.* o

Tolskij, Andrej P., Prof., Kasansches Inst. f. Landu. Forstwirtschaft. Kasan (U.d.S.S.R.), Ul. Komlewa 20, W. 6. *Waldbau.* o

Tomin, Michail P. (1883), Ass. d. Landwirtschaftl. Inst. Woronesch (U.d.S.S.R.), Chimkorpus 10. *Pflanzenanat. u. Systematik, Flechten.* Ⓓ Moskau 1913. Ⓟ Flechten aus Rußland u. Sibirien.

Tomita, Hosuke, Ass. Prof. Prov. Med. Univ. Ôsaka (Japan), Anat. Inst. *Anat.*

De Toni, Giovanni (1895). Alessandria (Italia), piazza della Lega 4. *Hématol. chim.* Ⓓ Padoue 1919.

Tonhunas, Doc., Leiter d. Inst. f. Pflanzenbau. Dotnava (Litauen). o

Toŉkov, W. N., Prof., Dir. Anat. Inst. Militärmed. Akad. Leningrad (U.d.S.S.R.), Lopuchinskaja ulica 12. o

Tonnoir, André Léon (1885), Field entomol., Cawthorn Inst. Nelson (New Zealand). *Economic Entomol. Noxious weed. Systematic and biol. of diptera.* Ⓓ Liège. Ⓟ Diptera.

Tonon, Amelia (1899), Ass. Stazione Bacol. Sperimentale. Padova (Italia). *Embriol. e morfol. nor-*

mali e patol. degli insetti (Bombyx mori). Ⓟ Padova 1923. Ⓔ Preparati anat. del Bombyx mori.

Topi, Mario (1883), Delegato tecnico antifillosserico Ass. volontario Ist. di Zool. Univ. di Siena (Italia), Delegazione tecnica antifillosserica. *Entomol. agraria. Fillosserine. Fillossera della vite. Phloetribus oleae. Ilesini dell'olivo. Tignuole della vite.* Ⓟ Pisa 1905.

Topley, W. W. C., Prof. Vict. Univ. Manchester (England). *Bact.* o

Topsent, M. E., Prof. Zool. Fac. Sc. Strasbourg (France). *Zool.* o

Torelli, Beatrice (1896), Prof. nel R. liceo sc. di Caserta (Italia). *Isopodes, Cytol.*

Tornau, Otto (1886), Dr. phil., o. Prof., Dir. des Inst. für Pflanzenbau der Univ. Göttingen (Deutschland), Nikolausberger Weg 7. *Gräser und Kartoffeln.* Ⓟ Göttingen 1910.

Tornier, Gustav (1858), II. Dir. am zool. Mus. der Univ. Berlin i. R. Berlin-Charlottenburg (Deutschland), Eosanderstr. 14 II. *Wirbeltierphylogenie und Experimentalzool. (Formumwandlungen, Regeneration).* Ⓟ Heidelberg.

Toro, Rafael A. (1897), Ass. Plant Pathol., Insular Experim. Station. Rio Piedras (Porto Rico). *Mycol., Taxonomy of Pyrenomycetes, Perisporiales. Plant Pathol.: Banana Antheracnose.* Ⓟ Cornell Univ. 1921. Ⓔ Mycol. herbarium.

de la Torre-Bueno, José Rollin (1871). White Plains, N.Y. (U.S.A.), 11 North Broadway. *Biol. and Taxonomy of Heteroptera. Aquatic species.* Ⓔ Heteroptera.

Torremocha, Lorenzo, Prof. Dr., Labor. de Fisiol. Valladolid (España). *Centres corticaux.* Ⓟ 1926.

Torres Minquez, Alejandio (1861). Barcelona (Espana), rue St. Ramon 2, St. Pablo. 67. *Malacol.: Limaces.* Ⓟ Barcelone 1882. Ⓔ Pulmonata, Limaces.

Torrey, George S., Ass. Prof. of Plant Pathol., Connecticut Agric. Coll. Storrs, Conn. (U.S.A.). *Cryptogams, fungi.* o

Torrey, Harry Beal (1873), Cornell Univ. Med. Coll. New York City, N.Y. (U.S.A.). *Experim. Biol. Physiol.* Ⓟ Berkeley, Cal. 1895.

Torrey, Ray Ethan (1887), Ass. Prof. of Botan., Massachusetts Agric. Coll. Amherst, Mass. (U.S.A.). *Evolutionary morphol. Anat. and physiol. of the seed plants.* Ⓟ Harvard 1917.

Torup, Sophus (1861), Prof. der Physiol., Dir. Univ. Fysiol. Inst. Oslo (Norge). *Stoffwechsel, Ernährung. Blutgase.* Ⓟ København 1887.

To Shue Tung (1902), Lignan Univ. Canton (China). *Raw Silk improvement.*

Tóth, Zsigmond, Prof. Dr., Dir. Anat. Inst. Univ. Pécs (Ungarn). o

Tothill, John Douglas (1888). Superintendent of agric. and Dir. of entomol. Suva (Fiji). *Control of Leuvana iridescens. Biol. control of insects.* Ⓟ Harvard 1915.

Totten, Henry R., Associate Prof. of Botan., Univ. of North Carolina. Chapel Hill, N.C. (U.S.A.). *Pharm. botan.* o

Tottingham, William Edward (1881), Associate Prof. of Agric. Chem., Univ of Wisconsin. Madison, Wis. (U.S.A.), Agric. Chem. Bldg. *Environmental effects in plant metabolism, especially of temperature. Plant nutrition, especially the mechanism of assimilation.* Ⓟ Baltimore, Md. 1917.

La Touche, John David (1861). Kiltymon, Newtownmountkennedy Co. Wicklow (Ireland). *Ornithol. Birds of Eastern China.*

Toumey, James William (1865), Prof. of Silviculture. New Haven, Conn. (U.S.A.), 459 Prospect Street. *Forest Ecol.* Ⓟ Syracuse Univ. 1915.

Tourmont, Raymond (1902), Préparateur Station de Phytogénétique et de Phytopath., Ecole Agric. Grignon. l'Hay-les-Roses, Seine (France), 82, Avenue Larroumès. *Génétique, Pathol.*

Tournade, André (1881), Prof. de Physiol. Fac. Méd. Alger (Afrique), 7, Rue Marceau. *Capsules surrénales. Système nerveux sympathique et vague.*

Circulation. Asphyxie. Innervation gastro-intestinale.

Toverud, Guttorm (1896), Doc. Oslo (Norwegen), Holtegaten 8. *Physiol. und pathol. Biochem. der Zähne. Einfluß von Salzen, Vitaminen und Hormonen.* Ⓟ Oslo 1919.

Towers, Irving L. (1892), Ass., Junior Aquatic Biol. Beauford N.C. (U.S.A.), Biol. Labor. *Growth and development of fishes.* Ⓟ Augusta, Ga. 1925. Ⓔ Marine plants and animals.

Towler, Emmet D. (1891), Principal High School. La Grande, Ore. (U.S.A.). *Marine ecol. (Puget Sound).* Ⓟ Washington 1926.

Townsend, Charles Haskins (1859), Dir. Aquarium. New York, N.Y. (U.S.A.). *Fisheries. Fur seal. Deep-sea exploration.* o

Townsend, Charles Wendell (1859). Ipswich, Mass. (U.S.A.). *Ornith.* Ⓟ Harvard 1885.

Townsend, Myron T. (1897), Associate Prof. of Biol., Saint John's Coll. Annapolis, Md. (U.S.A.). *Reactions of the Jelly-fish acquorea. Hibernation in the Codlin moth larva. Burrowing habits of the American toad.* Ⓟ Illinois 1925.

Toxopeus, Hendrik Jannes (1902), Ass. Botan. Labor. Groningen (Holland), Hooge der Aa 37a. *Erbanalyse (Faktoranalyse) von Nigella damascena.* Ⓔ Im Freien wachsende Nigella-Arten, N. damascena aus dem mediterranen Gebiet und andere Varietäten.

Toxopeus, Lambertus Johannes (1894). Amsterdam (Holland), Koloniaal Inst. *Indo-australische Lycaenidae (Lepidoptera), Systematik, Anat. und Zoogeographie.*

Toyoshima, Jôsei (1885), Curator, Ogasawara Branch of the Imper. Experim. Sta. of Forestry. Ômura, Chichijima, Ogasawarajima (Japan). *Tropical plants.* Ⓟ Morioka Agric. Coll. 1909.

Tozawa, Tomiju. Mida, Tôkyô (Japan), Biol. Inst., Keiô-Gijuku Univ. *Compar. Histol.*

Traaen, Alf Egeberg (1885), Doc. Mikrobiol. Landw. Hochsch. Aas (Norge). *Bact. und Mykol. des Bodens. Brandkrankheiten des Getreides.*

Tracy, Henry Carroll (1879), Prof. of Anat., Univ. of Kansas, Lawrence, Kan. (U.S.A.). Dir., Sc. Staff, Biol. Survey, Mt. Desert Island Marine Biol. Labor. Mt. Desert Island, Me. (U.S.A.). *Nervous system of fishes. Life histories and development of behavior in Fishes. Anat. of Ascidians.* Ⓟ Brown Univ. 1910.

Traczewski, C., Dr. Prof., Dir. Inst. Pharmacol. Univ. Wilno (Polska). o

Trägårdh, I. O. H., Prof. Dr., Statens Skogsförsöksanstalt. Experimentalfältet (Sverige). *Entomol.* o

Trambusti, Arnaldo (1863), Prof. o. Patol. Gen. Univ. Genova 18 (Italia), Ist. di Patol. Gen., Viale Benedetto 15. *Citol. Batt.* Ⓟ Univ. di Pisa 1888.

Tramontano, Vinzenzo (1891), Prof. Anat. ed Istol. Patol. Univ., Ass. Ist. di Anat. Patol. Napoli (Italia).

Tramontano-Guerritore, Giovanni (1898), Libero doc. in anat e umana normale. Aiuto nell' ist. anat., Univ., Prof. incaricato di Istol. e Fisiol. generale. Siena (Italia), Via S. Pietro 3. *Osteol.: osso occipitale e colonna vertebrale. Neurol.: organi commissurali del cervello. Istol.: lipoidie grassi.* Ⓟ Siena 1922.

Transeau, Edgar Nelson (1875), Prof. of Botan. and Chairman of the Department. The Ohio State Univ. Columbus, O. (U.S.A.). *Physiol. and Ecol.* Ⓟ Michigan 1904. Ⓔ Zygnemales.

von Transehe, Nikolai (1886), Ass. Inst. für systematische Zool. an der Univ. Riga (Latvija), Antonienstr. 7. *Ornithol.: Faunistik und Vogelzug.* Ⓟ Leipzig 1913.

Tranzschel, Woldemar (1868), Prof. des Inst. für angewandte Zool. und Phytopath. Leningrad (U.d.S.S.R.), Universitätsquai 5. *Uredineen, Systematik und Biol., speziell Heteroecie. Flora der U.d.S.S.R.*

Trappmann, Walther (1889), Reg.R. an der Biol. Reichsanstalt für Land- und Forstwirtschaft in Berlin-Dahlem (Deutschland). *Pflanzenschutzmittelprüfung.* Ⓟ Marburg a. d. L. 1915.

Trask, James Dowling (1890), Ass. Prof. of Med., Yale Univ. New Haven, Conn. (U.S.A.), New Haven Hospital. *Med. Bact.* ⓓ Cornell Univ. 1917.

Tratz, Eduard Paul (1888), Dr. phil. h. c., Dir. des Mus. für darstellende und angewandte Naturkunde in Salzburg (Österreich), Naturkunde-Mus. *Ornithol. und naturwiss. Musealwesen. Allgemeine Zool.* ⓓ Innsbruck 1923. ⓟ Paläarktische Vogelbälge.

Traub, Hamilton P. (1890), Ass. Horticulturist, Minnesota Agric. Experim. Station. St. Paul, Minn. (U.S.A.), Univ. Farm. *Translocation of food in the apple tree. Culture of vegetable crops on peat and muck soils. Dehydration and storage of vegetable crops.*

Trauner, Franz (1867), Dr., tit. o. Univ.Prof. für Zahnheilkunde, Vorstand des zahnärztlichen Inst. Graz (Deutsch-Österreich), Landeskrankenhaus. *Normale und pathol. Anat. der Zähne.* ⓓ Graz 1892.

Trausmiler, Othmar (1893), Dir. Staatl bact. Labor. Sušak (S.H.S.). *Malaria.* ⓓ Prag 1920. ⓟ Malaria (Mücken, Praep.).

Trauth, Friedrich (1883), Dr. phil., Kustos am Naturhistor. Mus., ferner Priv.Doc. für Geol. und Palaeont. an der Technischen Hochsch. Wien I (Deutsch-Österreich), Burgring 7. *Fossile Invertebrata (Foraminifera, Anthozoa, Brachiopoda, Lamellibranchiata, Cephalopoda). Historische (stratigraphische) Geol., bes. der mesozoischen Formationen.* ⓓ Wien 1908.

Trautmann, Franz Alfred (1884), Dr., Hochschul-Prof. u. Dir. des physiol. Inst. der Tierärztl. Hochschule zu Hannover (Deutschland), Bischofsholerdamm 84. *Physiol.* ⓓ Zürich 1907.

Trautmann, Róbert (1873). Budapest II (Ungarn), Eszter-utca 22. *Systematik der mitteleuropäischen Formen der Gattung Mentha.* ⓟ Mentha.

Trautwein, Kurt (1881), Dr. phil., ao. Prof. für techn. Mikrobiol. an der bayer. Hochsch. für Landwirtschaft u. Brauerei Weihenstephan b. München (Deutschland). *Techn. Mikrobiol., insbesondere Gärungsmikrobiol.* ⓓ Würzburg 1921. ⓟ Bakterien, Pilze, Hefen.

Traverso, Giov. Battista (1878), Prof. de Pathol. végétale dans le R. Ist. superiore agrario de Milano (Italia). *Mycol.; Pathol. végétale.* ⓓ Pavia 1900. ⓟ Champignons (Micromycètes).

Trawiński, Alfred (1888), Prof. der Fleischhygien. der Tierärztl. Hochsch. Lwów (Polska), Kochanowskigasse 61. *Bact. und Epidemiol. der Fleischvergifter wie auch anderer Darmbakterien der Coli-Typhusgruppe.* ⓓ Lwów 1912.

Treadwell, Aaron Louis (1866), Prof. of Zool., Vassar Coll. Poughkeepsie, N.Y. (U.S.A.). *Taxonomy of polychaetous annelids.* ⓓ Chicago 1899.

Treboux, Octave, Dr., Doc., Dir. Abt. f. Pflanzenphysiol. Botan. Labor. d. Univ. Riga (Latvija), Kronvalda bulvari 4. *Vererbung.* o

Treillard, Marc (1897), Attaché à l'Inst. Pasteur. Paris (France), 96, rue Falguière. *Biol. des Crustacés en culture pure (Cladocères). Culture des flagellés.* ⓓ Paris 1922.

Trelease, Sam F. (1892), Associate Prof. of Botan. Columbia Univ., New York, N.Y. (U.S.A.). *Plant physiol., salt requirements, toxic action, climatic relations, water relations.* ⓓ Johns Hopkins Univ. 1917.

Trelease, William (1857), Prof. of Botan. (retired), Univ. of Illinois. Urbana, Ill. (U.S.A.). *Taxonomy: Yucceae, Agaveae, Phoradendron, Quercus, Piperales. Pollination. Mycol.* ⓓ Harvard 1884.

Trella, Taddeus (1885), Commission physiographique Acad. des Sc. Cracovie. Przemyśl (Pologne), le collège I. *Entomol.: Staphylinoidea, Heteromera.*

Trendelenburg, Paul Georg (1884), o. Prof. der Pharmacol. Univ. Freiburg i. Br. (Deutschland), Katharinenstr. 29. *Innere Sekretion.* ⓓ Freiburg i. Br. 1908.

Trendelenburg, Wilhelm (1877), Univ.Prof., Vorst. physiol. Inst. Tübingen (Deutschland), Silcherstr. 8. *Nervenphysiol.: Großhirn, Rückenmark. Physiol. d. musik. Empfindens.* o

Trenkle, Rudolf (1881), Landwirtschaftsrat I. Kl., Bayer. Landesinspektor für Obst- und Gartenbau und Vorsitzender des Arbeitsausschusses der Bayer. Landesausschusses für das Versuchs- und Forschungswesen im Obstbau. München (Deutschland), Wörthstr. 13 I. *Blütenbiol. bei Gemüsepflanzen u. Obstbäumen. Vermehrung geeigneter Unterlagen für Obstgehölze.*

Trentini, Silvio (1895), Ass. Ist. di Patol. Generale. Pisa (Italia), Via S. Maria 13. *Endocrinol. Anat. Patol.* ⓓ Pisa 1921.

Treschow, Cecil (1902), Ass. at the Plantphysiol. Labor. of the Royal Vet. and Agric. Coll. København (Danmark). *Forestry.*

Tretjakoff, Dmitry (1878), Prof. Univ. Odessa, Ukraine (U.d.S.S.R.), Kominternstr. 2. *Vergl. Anat. des Nervensystems und der Hartgebilde der Haut. Endoskelett der niederen Wirbeltiere. Bindegewebe, Knorpel, nerv. Gewebe. Spätneolithische Bevölkerung der Ukraine.* ⓓ Leningrad 1910.

Tribe, Margaret (1890), Lect. Zool. Univ., King's Coll. London (England), Strand. *Vertebrate Embryol. Mammals.* ⓓ London 1913.

Triepel, Hermann (1871), Prof., Abteilungsvorsteher am anat. Inst. i. R. Breslau (Deutschland), Hansastr. 16. *Mechan. Eigenschaften der Gewebe. „Abstrahierende" und „Fiktive" Biol. Materialistische Naturphilosophie.* ⓓ 1894.

van Trigt, Herman (1884), Dr. Aerdenhout (Holland). *Physiol. of the Fresh-water Sponges (Spongillidae).* ⓓ Leyden 1919.

Trinchieri, Giulio (1881), Libero doc. di Botan. generale nella R. Univ., Capo del Servizio di Protezione delle Piante nell' Ist. Internazionale di Agric. Roma (Italia). *Micol. e Fitopat.* ⓓ Torino 1904.

Trinci, Giulio, Lib. doc. Univ. Bologna (Italia). *Anat. compar.* o

Trofimovitsch, Alexis J., Chef des trav. Phytopath. Stat. Défense des Plantes. Poltava, Ukraine (U.d.S.S.R.), *Hyménomycètes.*

Troitska-Javoronkova, Iraide P., Conserv. Leningrad (U.d.S.S.R.), Station Génétique, Tsarkoié Sielo. *Pathol. végétale, Myxomycètes.* o

Troitskaja, Olga Vasilievna (1894), Ass. Inst. of Cryptogamous Plants in the Principal Botan. Garden. Leningrad (U.d.S.S.R.). *Algae.* ⓓ Leningrad.

Troitzky, Nikolai Nikolaewitsch (1887), Bureau f. angewandte Entomol. Inst. f. experim. Agronomie, Leiter d. Versuchsstation für angewandte Entomol. Leningrad (U.d.S.S.R.), Herzenstr. 44. *Experim. Ecol.: Immunität d. Pflanzen, Schädliche Insekten; Anat. der Insekten.* ⓟ Schädliche Insekten.

Trojan, Emanuel (1878), ao. Univ.Prof. Deutsche Univ., Hon.Doc. Deutsche techn. Hochsch. Praha II (C.S.R.), Vinicná 3. *Biolumineszenz, Sinnesorgane der Tiere, Parasit. Cytol.* ⓓ Prag 1905.

Trojanović, S., Dr., Prof. d. Ethnographie u. Ethnol. Skoplje (S.H.S.). *Anthrop.* o

Troll, Wilhelm (1897), Dr. phil., Priv.Doc. für Botan. München 38 (Deutschland), Menzinger Str. 13. *Morphol. u. Biol. der Pflanzen.* ⓓ 1921.

Trommsdorff, Richard (1874), Dr. med. München (Deutschland), Bavariaring 20. *Immunität, Bact. Milchhygien.* ⓓ Göttingen 1897.

Tronchet, Antonin (1902), Ass., Labor. Botan., Fac.de Sc. Lyon (France). *Anat.d. plantes vasculaires*

Tropea, C., Doc. Univ. Palermo (Italia). *Botan.* o

Tropp, C., Dr., Chem. Abt. Inst. f. Schiffs- und Tropenkrankheiten. Hamburg 4 (Deutschland). o

Trostaneckij, M., Prof. Med. Inst. Ekaterinoslav, Ukraine (U.d.S.S.R.). *Anat.* o

Trotter, Alessandro (1874), Prof. di Patol. vegetale nel R. Ist. Superiore Agrario di Portici, Napoli (Italia). *Cecidol., Micol., Parassit. e Fitopatol., Floristica e Geografia botan., Piante economiche.* ⓓ Padova. ⓟ Galle, fanerogame.

Trotter, Spencer (1860), Coll. Prof. Swarthmore Coll. West Chester, Pa. (U.S.A.), Darlington and Miner Sta. *Ornithol. Zoogeograph.* o

Trotzkij, Nikolaj A., Doc. a. Staatl. Polytechn. Inst. Tiflis (U.d.S.S.R.), Botan. Garten. *Angew. Botan. Wiesenbau.* Ⓟ

Trouessard, M., Prof. au Mus. Paris 5 (France), 57, rue Cuvier. *Zool.* Ⓟ

Truchin, P. A., Vorsteher d. Abt. f. Wiesenbau d. Kasanschen Landw. Provinz. Versuchsstation. Kasan (U.d.S.S.R.), St. Gorschetschnaja 29, W. 4. *Agronomie. Wiesenbau.* Ⓟ

True, Rodney Howard (1866), Prof. of Botan., Dir. of Botan. Garden, Univ. of Pennsylvania. Philadelphia, Pa. (U.S.A.), Botan. Dept. *Mineral Nutrition of Plants, Taxonomy of Dicranum.* Ⓟ Leipzig 1895. Ⓟ *Dicranum*-species.

Trümpy, Eugen. Root, Kt. Luzern (Schweiz). *Biol. Physiol. u. pathol. Wirkung der Strahlen vom äußersten Ultraviolett bis zum innersten Infrarot auf das menschliche Auge.* Ⓟ Zürich 1924.

Truffi, Giovanni (1898), Ass. Ist. di Istol. ed Embriol., R. Univ. di Padova (Italia), Via Giovanni Prati 5. *Rigenerazione dell' ovario, della tube e dell' utero. Genesi della ghiandola interstiziale ovarica. Istofisiol. dell' ovario. Linfosarcomi intestinali. Decidua ectopica extragravidica.*

Truitt, Reginald van Trump (1891), Prof. of Agric., in charge of the Chesapeake Biol. Labor. College Park, Md. (U.S.A.), Univ. of Maryland. *Marine Biol.; Biol. of Ostrea virginica and Callinectes sapidus.* Univ. 1920. Ⓟ Biol. Forms of the Chesapeake Bay, Fishes and Birds of Maryland.

Truka, Rudolf (1881). Prof. de chim. agric. à l'Ecole Sup. d'Agric. Brno (C.S.R.), Augustinská 17. *Chim. végétale, chim. des engrais. Biochem. L'action biol. de sol et analyse de sol, des engrais et des végétaux. Composition chimique des végétaux.* Ⓟ Praha 1905.

Trusowa, Nina Petrowna (1884), Phytopath. Moskau (U.d.S.S.R.), Sadowaja-Triumfalnaja 10, Zimmer 42. *Kleekrankheiten.*

Truszkowski, Richard (1897), Senior Ass., Dept. of Physiol., and Physiol. Chem., Vet. Inst., Univ. Warszawa (Polska), Hoża 9, m. 7. *Biochem.; Purine metabolism and the nuclear-plasmic ratio. Clinical observations; Vaginal Acidity and blood changes in various pathol. conditions.* Ⓟ London 1921.

Trzebiński, Józe (1867), o. Prof. der Botan., Leiter der Anstalt für Systematik d. Pflanzen, Dir. d. botan. Gartens der Univ. Wilno (Polen), Zakretowastr. 1. *Phytopath., Mycol., Pflanzengeographie und Sociol.* Ⓟ Krakau 1903. Ⓟ *Parasit. Pilze, Zoocecidien.*

Trzecieski, R., Dr., Prof., Dir. Inst. Pharmacol. Univ. Poznań (Polska).

Tsai, Burg (1898), Prof. Compar. Anat., Vertebrati Zool. Shanghai (China), Fuh Tau Univ. *Osteol. and Myol.* Ⓟ Peking 1923.

Tsakalotos, A. E. (1885), Leiter Labor. f. Chem. u. Bact. Athen (Griechenland), Sokratesstr. 45. *Kapillarchem. Rote u. blaue Farbstoffe der Blätter.* Ⓟ Athen 1911.

Tschassownikow, Nikolay (1896), Doc. Histol. phys. Fac., Prosektor Histol. und Embryol. Med. Fac. Tomsk, Sibirien (U.d.S.S.R.), Histol. Inst. Univ. *Thymusforschung; Röntgenisation, Explantation in vitro bei Säugern und Vögeln.* Ⓟ Tomsk 1922.

Tschechowitsch, Konstantin, Prof. d. Samarschen Landwirtsch. Inst. Besentschuk (U.d.S.S.R.), Samar.-Slat. sh. d. Landwirtschaftl. Provinz.-Versuchsstation, Selektions-Abt. *Selektion landw. Pflanzen.* Ⓟ

Tschellustkina, M. N. (1871), Ass. anat. Inst. Odessa (U.d.S.S.R.), Pasterstr. 18. *Mißbildungen des Nervensystems.* Ⓟ Odessa 1915.

Tschenzoff, Boris (1889), Doz. d. Anat. u. Physiol., Dir. Anat. Inst. Anat. u. Physiol. der Tiere, Polytechn. Inst. Tiflis, Georgien (U.d.S.S.R.). *Protozool., Einfluß d. Hormone auf Wirbellose.* Ⓟ Freiburg i. B. 1914.

Tscherbakow, Michail F., Prof. d. Kuban. Landw. Inst. Krasnodar (U.d.S.S.R.), Krasnaja 45/47. *Weinbau. Weinbereitung.* Ⓟ

Tscherdintzew, Viktor A., Doc. Geol. Univ. Kazan (U.d.S.S.R.). *Palaeont.; speziell Foraminifera.* Ⓟ

Tscherewkoff, A. M., o. Prof. Physiol. Inst. d. Univ. Sofia (Bulgarien). Ⓟ

Tscherfass, Boris (1895). Ass. für Fischzucht, Timirjaseff-Akademie. Moskau (U.d.S.S.R.). *Fischzucht. Teich- u. Seewirtschaft.* Ⓟ Timirjaseff-Akad. 1925.

Tscherkes, Leon (1890), Prof. General Pathol. of State Inst. of Chem. and Pharm. and Priv.Doc. Med. Fac. Odessa (U.d.S.S.R.), Str. Pasteur 19. *Pathol. of metaboliem. Avitaminose.* Ⓟ Odessa 1914.

Tscherkessow, Demetrius A. Leningrad (U.d. S.S.R.), W.O., Sredny Pr. 72-B, Com. Géol. de Russie. *Palaeont. der Wirbeltiere; speziell Reptilia.* Ⓟ

Tscherkessow, Wsewolod G., Kustos Geol. Mus. des Berginst., Ass. Palaeont., Mitarbeiter am Comité Géol. de Russie. Leningrad (U.d.S.S.R.), W.O., Sredny Pr. 72-B. *Palaeozoische Korallen.* Ⓟ

Tschermak-Seysenigg, Armin (1870), o. Prof. Physiol. Deutsche Univ. Praha II (C.S.R.), Albertov 5. *Nerven- und Sinnesphysiol. Physiol. des Gehirns. Verdauung. Zellphysiol. Physik. Chem.* Ⓟ Wien 1895.

Tschernezkaja, Sinaida S., Doc. d. Landw. Inst. d. Bergrepublik. Wladikawkas (U.d.S.S.R.), Revolutionsstr. 32. *Botan. Phytopath.* Ⓟ

Tschernjachiwsky, Alexander Grigorowitsch (1869), Prof. Histol. und Embryol. Univ., Hersch. Histol. Labor. Kiew, Ukraina (U.d.S.S.R.), Herschuniestr. 31/18. *Nervensystem. Regeneration bei Verletzungen. Plasticität des Neurons. Chondriosomen, Apparato reticola Golgi.* Ⓟ Kiew 1893.

Tschernoff, N. D. (1882), Ass. Zool. Inst. Univ. Leningrad (U.d.S.S.R.), Basseinaia Str. 2, Wohn. 15. *Bau und Entwicklung des Gehörorgans der niederen Wirbeltiere; Seitenlinie der Knochenfische.* Ⓟ Petersburg 1905.

Tschernov, Sergej (1903). Aspirant d. Katheders d. Zool. Charkow (U.d.S.S.R.), Klassitschny pereulok 4. System. d. Amphibia u. Reptilia d. U.d.S.S.R.

Tschernow, Alexander A., Prof. II. Univ. Moskau (U.d.S.S.R.). *Geol. und Palaeont. der Permischen Ablagerungen Nordrußlands.* Ⓟ

Tschernyschew, Boris J., Prof. Berginst. Ekatherinoslaw, Comité Géol. de Russie. Leningrad (U.d. S.S.R.), W.O., Sredny Pr. 72-B. *Palaeont.; speziell Ostracoda, Trilobita etc.* Ⓟ

Tscherwakow, Basil (1894), Prosektor Pathol.-Anat. Inst. Weißrussische Univ. Minsk (U.d. S.S.R.), Leninskaja ul. 29/35. *Pathol. Anat. und experim. Helminthol.* Ⓟ Moskau 1917. Ⓟ *Nematoden.*

Tschirch, Alexander (1856), Univ.Prof., Dir. des pharm. Inst. der Univ. Bern (Schweiz), Kollerweg. *Grenzgebiete zwischen Chem. und Botan. Vererbungsfragen, Beziehungen zur Pharmakognosie, z. B. beim Rhabarber die durch Aufspaltung tibetanischer Samen erzielte Wiederherstellung des Rheum palmatum von Linné. Harzbildung bei den Läusen spez. den Tachardia-Larshadia-Arten Indiens. Primärer und secundärer Harzfluß bei den Pflanzen. Anpassungen der Pflanze an Klima und Standort. Das Feigenproblem.* Ⓟ Freiburg i. B. 1881.

Tschopp, Ernst (1896), 1. Ass. d. Physiol. Chem. Basel (Schweiz), Vesalianum. *Histochem. der Gewebe.* Ⓟ Basel.

Tschorbadjieff, Peter (1882), Vorstand der Entomol. Abt. Sofia (Bulgarien), Solunplatze 3. *Lepidoptera u. Ipidae.*

Tschoumakova-Baum, Catherine E., Prép. Station Phytopath. du Jardin Botan. Leningrad (U.d. S.S.R.). *Pathol. végétale.* Ⓟ

Tschubkow, Efim A., Doc. d. Kubanschen Landw. Inst. Krasnodar (U.d.S.S.R.), Zentral-Inst. d. Versuchsstation f. Tabakkultur. *Tabakkultur.* Ⓟ

Tschuefskij, Ivan Af., Prof. N. G. Černyševskij Univ. Saratov (U.d.S.S.R.). *Physiol.* Ⓟ

Tschugunow, Nikolai (1889), Doc. d. Ichthyol., Stellv. Leiter d. Asowschen Wissensch. Fischerei-Exped. Moskau (U.d.S.S.R.), Landwirtschaftliche Akademie, Petrowsko-Rasumowskoje. *Biol. d. Fische. Nahrungsuntersuchungen, Altersbestimmung. Hydrobiol. Zooplankton d. südlichen Russischen Meeres; quantitative Untersuch. d. Benthos (Produktivität).* ⓓ Moskau 1912. ⓟ Entomostraken d. Kaspischen und Asowschen Meeres.

Tschugunowa, Nina (1900), Ass. Asowsche Wissenschaftl. Fischerei-Exped. Moskau (U.d.S.S.R.), Wissenschaftl. Fischerei-Inst., Pjatnizkaja 33. *Biol. und Systematik d. Fische.* ⓓ Moskau 1925.

Tschulok, Sinai (1875), Dr. phil., Tit.Prof., Priv.Doc. für allgem. Biol. Zürich 7 (Schweiz), Plattenstr. 33. *Methodol. und Geschichte der Biol., Deszendenzlehre.* ⓓ Zürich 1908.

Tsherbakoff, Sergej (1899), Aspirant Zool. Kabinett Univ. Nijny-Nowgorod (U.d.S.S.R.). *Protozool., tropische Infektionskrankheiten, med. Entomol. Malaria.* ⓟ Charkow 1921. ⓟ Malariamücken.

Tshetverikov, Sergius S. (1880), Doc. I. Moskauer Univ., Ass. Inst. für Experim. Biol., Kustos Zool. Abteilung des Polytechn. Mus., Genetische Abteilung der K.E.P.S. der Akademie der Wissenschaft. Moskau (U.d.S.S.R.), Maschkova Ul. 17/15. *Genetic: Drosophila; Evolution; Systematik der Lepidoptera; Ökol. der Hochmoore; Zoogeographie des hohen Nordens.* ⓓ Moskau 1906. ⓟ Lepidoptera palaearctica.

Tskimanauri, Georg (1887), Ass. pharm. u. physiol. Labor. Univ. Tiflis, (U.d.S.S.R.), Tschernischewsky-Str. 10. *Einfluß der Nerven auf das Herz. Mechanismus der Herztätigkeit.* ⓓ Kiew 1916.

Tso, F. L., Lect. Southeastern Univ. Nanking (China). *Zool.* o

Tsoong, K. K., Prof. National Univ. Peking (China). *Botan.* o

Tsou, Y. Hsuwen, Dir. Bureau of Entomol. Kiangsu Province. Nanking (China), Coll. of agric. Southeastern Univ. *Rice borer. Economic entomol.*

Tsuchiya, Itaru (1896), Demonstrator in Plantphysiol. Kyôto (Japan), Botan. Inst., Coll. of Sc., Imper. Univ. *Bio-Physical Chem. of Cells and Tissues, photochem. action of light of special frequencies. Photosynthesis of carbonic acid, physicochem.* ⓓ Kyôto Imper. Univ. 1924.

Tsukaguchi, Risaburô, Prof. Prov. Med. Univ. Ôsaka (Japan), Anat. Inst. *Anat.*

Tsuno, Yeitarô, Prof. Tôkyô Imper. Univ. Komaba near Tôkyô (Japan), Inst. of Vet. Sc., Fac. of Agric. *Vet. Hygiene and Pharmacol.*

Tubangui, Marcos A. (1893), Ass. Prof. of Parasit. Coll. of Vet. Sc., Univ. of the Philippines. Los Baños (Philippine Islands). *Helminthol., Protozool.* ⓓ Cornell Univ. 1921. ⓟ *Helminths and Protozoa from Philippine domestic animals.*

Tubeuf, Freiherr Karl von (1862), Dr., o.Univ.-Prof. München (Deutschland), Habsburgerstr. 1. *Pflanzenpathol. Forstl. Kulturpflanzen.* ⓓ München 1886.

Tuchner, A. S. (1887), Dir. d. beh. gen. Lehranst. f. Nahrungs- u. Genußmittelkunde. Wien 1 (Österreich), Graben 17. *Lebensmittelkunde, Vet. Pharmacol.* ⓓ Wien 1919.

Tucker, Bernard William (1901), Demonstrator, Department of Zool. and Compar. Anat., Univ. of Oxford (England), 9 Marston Perry Road. *Vertebrate morphol. and phylogeny, sexual and age variations in the skulls of ungulate mammals. Ornithol.* ⓓ Oxford 1923.

Tuckerman, Frederick (1857). Amherst, Mass. (U.S.A.). *Anat. of the taste-organs of Vertebrates.* ⓓ Heidelberg.

Tudoranu, Georges, (1892) Chef des trav. Fac. Méd. Jassy (Roumanie). *Hématol. et Sérol.* ⓓ Jassy 1919.

Tüxen, Reinhold (1899), Dr. phil. nat., Ass. der Provinzialstelle für Naturdenkmalpflege Hannover (Deutschland), Provinzial-Mus. *Botan.: Pflanzengeographie u. Floristik.* ⓓ Heidelberg 1925.

Tuff, Per (1878), Prof. der Rassen- und Zuchtlehre an der Norwegischen Landwirtschaftl. Hochschule. Aas (Norge). *Vererbungsversuche mit Enten und Hühnern. Milchsekretion.*

Tukáts, Sándor (1893), Adj. pharmakogn. Inst. Szeged (Ungarn), Kálvaria t. 5 b. *Rhabarber.* ⓓ Szeged 1922.

Tukey, Harold B. (1896), Associate in Research (Horticulture) New York State Agric. Experim. Station, Geneva, N.Y., and Horticulturist, Hudson Valley Fruit Investigations, Hudson, N.Y. (U.S.A.). *General fruit problems; fertilizers, cover crops, pruning, variety tests, pollination problems.* ⓓ Univ. of Illinois 1918. ⓟ Fruit Plants.

Tullgren, Hugo Albert (1874), Vorsteher Entomol. Abteilung Zentralanstalt für landw. Versuchswesen, Prof. Experimentalfältet, Stockholm (Sverige). *Angewandte Entomol. Aphiden, Arachniden.*

Di Tullio, A., Doc. Univ. Napoli (Italia). *Botan.* o

Tullio, Pietro (1881), Ass. à l'Inst. de Physiol. de l'Univ., Libero doc., Chargé de Chim. Physiol. Bologna (Italia). *Physiol. de l'oreille, des muscles et des nerfs.* ⓓ Bologne 1905.

Tullis, Edgar C. (1901), Ass. in Botan. East Lansing, Mich. (U.S.A.), Michigan State Coll. *Epidemiol. and Infection of Fire Blight of Apple and Pear (Bacillus amylovorus). Pear; advisibility of summer treatment.* ⓓ Univ. of Nebraska 1924.

Tulubjew, Alexander A., Vorsteher d. Ukrain. Moor-Versuchsstation. Iwanowo-Wosnessensk (U.d.S.S.R.), Pawlowskaja 21. *Agronomie. Moorkultur und Wiesenbau.*

Tůma, Vladimír (1897), Doc. priv. de l'histol. et l'embryol., Ass. de l'Inst. pour h. et e. Praha II (C.S.R.), Albertov 4. *Trombocytes chez les oiseaux. Hypoplasie. Globulin dans le sang. Histol. du Scorpaena.* ⓓ Praha 1921.

Tumanov, Ivan I. (1894), Ass. Plant Physiol. at the Inst. of Applied Botan. and New Cultures. Leningrad (U.d.S.S.R.), Herzen str. 44. *Drought and frost resistance.* ⓓ Kiew 1923.

Tumanskaja, Olga J., Ass. Geol. Moskau (U.d.S.S.R.), I. Univ. *Ammoniten der Krim.* o

Tunnicliff, Everett A. (1897), Associate in Animal Pathol. in the Coll. of Agric. and the Agric. Experim. Station. Urbana, Ill. (U.S.A.), Univ. of Illinois. *Special phases of avian tuberculosis with relation to calves and swine; bovine abortion with relation to the agglutination test and sanitation methods; bacillary white diarrhea in chickens with reference to the agglutination test and the pullorin (intradermic) test as means of diagnosing the disease in mature fowls.* ⓓ Kansas State Agric. Coll. 1921. ⓟ Pullorin (an intradermic product for the detection of fowls infected with bacillary white diarrhea).

Tunnicliff, Ruth (1876), John McCormick Inst. for Infectious Diseases. Chicago, Ill. (U.S.A.), 637 South Wood Street. *Bact. and immunity. Phagocytosis, Fusiform Bacilli and Spirilla; Scarlet Fever and Measles.* ⓓ Rush Med. Coll., Chicago, 1903.

Tur, F. Je., Prof. Univ. Inst. f. Med. Wiss. Leningrad (U.d.S.S.R.), 2 Sovetskij Str. 4. *Physiol.* o

de Tur, Jan (1875), Prof. ord. d'Anat. comparée à la Fac. de Phil. et Dir. de l'Inst. d'Anat. Varsovie (Pologne), 9, rue Wilcza. *Embryol. comparée des Vertébrés, spécialement des Sauropsidés. Tératol. et tératogénèse.* ⓓ Lwów 1907.

Turchini, Jean (1894), Prof. agrégé d'Histol. à la Fac. de Méd. de Montpellier (France), 4, Rue de la Barralerie. *Histophysiol. Cytochim. Histophysiol. rénale. Histol. comparée de la cellule rénale.* ⓓ Paris.

Turconi, Malusio (1879), Ass. al Labor. Crittogamico de Pavia (Italia), Via G. Belli 4. *Micol. e Fitopat.* ⓓ Pavia 1899.

Turesson, Göte Wilhelm (1892), Doc. at the Univ. of Lund (Sverige). *Rassenökol. und Systematik, Geographie der Pflanzen.* ⓓ Lund 1923.

Turlais, C., Prof. Ecole prép. de Med. et de Pharm. Angers (France). *Physiol.* ○

Turnbow, Grover D. (1892), Research in Dairying. Davis, Cal. (U.S.A.), Univ. of California. *Colloidal Chem.* ⓓ Univ. of Iowa 1917.

Turnbull, H. M., Prof. Hospit. Med. Coll. and Dental School. London E. (England), Turner str., Mile End. *Pathol.* ○

Turner, Charles Wesley (1897), Ass. Prof. of Dairy Husbandry, Coll. of Agric., Univ. of Missouri. Columbia, Miss. (U.S.A.). *Dairy Cattle Breeding. Milk Secretion.* ⓓ Univ. of Missouri 1921.

Turner, Clarence Lester (1890), Prof. of Zool., Beloit Coll. Beloit, Wis. (U.S.A.). *Genetics of insects. Secondary sex characters and sex. Crayfishes.* ⓓ Univ. of Wisconsin 1918.

Turner, Edward L., Prof. American. Univ. Beirut (Syrie). *Physiol.* ○

Turner, Thomas Wyatt (1877), Head Department of Biol. Hampton Institute, Va. (U.S.A.). *Plant Physiol. and Plant Pathol. Effect of nitrogen supply and mineral salts.* ⓓ Cornell Univ. 1921.

Turner, Wendell Roscoe (1902), Junior Bact. Bureau of Chem., U. S. Department of Agric. Washington, D.C. (U.S.A.). *Bact. of food preservation, food spoilage poisoning.* ⓓ Iowa State Coll. 1925.

Turov, S. S., Prof. Pädagog. Inst. d. Bergrepublik. Vladikavkaz (U.d.S.S.R.), Ul. Marksa 20. *Zool. d. Wirbeltiere.* ○

Turutanova-Ketova, Antonina (1896), Geol. Mus. der Akad. d. Wiss. Leningrad (U.d.S.S.R.). *Palaeobotan. Filicines.* ⓓ Tomsk 1918.

Tusenko, Leonid Mikolaewitsch (1886), Ass. Histol. u. Embryol. [Odessa (U.d.S.S.R.), Podbelskiego 1. ⓓ Odessa 1913.

Tusnov, Michajl, Prof. Vet. Inst. Kazan, Tatar. Rep. (U.d.S.S.R.), Arskoe Pole 96. *Bact., allg. Pathol.* ○

Tuzson, Johann (1870), o. Prof. systematische Botan., Pflanzengeographie Univ. Budapest VIII (Ungarn), Muzeum Körut 4. *Vegetation der Sand-, Salz- und lehmigen Steppen. Arabis. Entwicklungsgeschichte. Fossile Pflanzenreste des Tertiärs des Balatonsees.* ⓓ Kolozsvár 1898. ⓟ Arabis und Blütenpflanzen der ungarischen Flora.

Twort, Charles Claud (1881), Pathol. to the Manchester Committee on Cancer. Pathol. Department, Victoria Univ. Manchester (England). *Infection and Immunity.* ⓓ Aberdeen Univ. 1909.

Tyler, John M. (1857), Prof. of Biol. Amherst Mass. (U.S.A.), College. *Evolution of Man.* ⓓ Colgate 1888. ○

Tysdal, Hewitt M. (1902), Graduate Ass. and Instr. in Agronomy. St. Paul, Minn. (U.S.A.), Univ. Farm.

Tyzzer, Ernest Edward (1875), George Fabian Prof. compar. pathol. Howard Med. School. Wakefield. Mass. (U.S.A.). *Pathol., protozoa, leucocytes, vaccina, parasit.* ○

Ubaldo, Sammartino (1887), Aiuto Ist. di Chim. Fisiol. Univ. Roma (Italia), Via Depretis 92. *Fisiopatol. del Ricambio. Insulina.*

v. Ubisch, Gerta (1882), Dr. rer. nat., Priv-Doc., Ass. am botan. Inst. Heidelberg (Deutschland). *Vererbung und Reizphysiol.* ⓓ Straßburg 1911.

v. Ubisch, Leopold (1886), ao. Prof. f. Zool. u. vergl. Anat., Dr. iur. et phil. Würzburg (Deutschland), Zool. Inst. *Entwicklungsphysiol.: Linsenbildung, Regeneration, Symmetrie, Keimblatt-Determination. Entwicklungsgeschichte: Echinodermen. Tiergeographie: Kontinental-Verschiebungs-Theorie u. Tierverbreitung.* ⓓ Würzburg 1911.

Uchida, Seinosuké (1883), Ornithol. Bureau of Animal Ind., Dept. Agric. and Forestry. Tôkyô (Japan), N. 1 7-chome Aoyama Kitamachi. *Ornithol., Entomol.: Mallophaga.* ⓓ Tôkyô Imper. Univ.

Uchida, Tôichi (1898), Ass. of Entomol. Hokkaido Imper. Univ. Sapporo (Japan). *Entomol., Hymenoptera.* ⓓ Hokkaido Imper. Univ. 1913. ⓟ Hymenoptera. Ichneumonidae.

Uchida, Tôru (1897). Tôkyô (Japan), Zool. Inst. Sc. Fac., Imper. Univ. *Embryol., Morphol. and Systematic of Coelenterata (especially Medusae).* ⓓ Imper. Univ. Tôkyô 1923.

Uchiyama, Yasushi, Ass. Prof. Tôhoku Imper. Univ. Sendai (Japan), Pathol. Inst., Coll. of Med. *Pathol.*

Uchtomskij, A. A., Prof. Staatl. Univ. Leningrad (U.d.S.S.R.). *Physiol.* ○

Ucke, Alexander (1864), o. Prof. pathol. Anat. und allgemeine Pathol., Dir. des pathol. Inst. der Univ. Tartu (Estland). *Pathol. Anat.* ⓓ Dorpat 1891.

Ude, Hermann (1860), Prof. Dr., Doc. für Zool. an der Tierärztl. Hochsch. Hannover (Deutschland), Lavesstr. 28. *Oligochaeten.* ⓓ Göttingen 1887.

Ueda, Kazuharu, Prof. Med. Univ. Kanazawa (Japan), Physiol. Inst. *Physiol.* ○

Uehara, Jakehiko (1897), Prof. physiol. Nippon Med. Coll. Tôkyô (Japan), Physiol. Labor., Fac. Med., Imper. Univ. *Physiol. problems, physical, chem. and physico-chem.* ⓓ Tôkyô Imper. Univ. 1922.

Uehi, Homiki (1882), Prof. Dendrol. and Sylviculture, Expert of the Forestry Experim. Stat. Governm. General of Chosen. Suigen, Corea (Japan), Agric. and Forestry Coll. *Dendrol. and Silviculture.* ⓓ Tôkyô 1904, M.For. Boston, Mass. 1922.

Uehlinger, Arthur Bernhard (1896), Forstm. II. Kreis Kanton Schaffhausen. Schaffhausen (Schweiz), Grubenstr. 1. *Geobotan. des Buchenwaldes.*

Uehlinger, Erwin (1899), Ass. Pathol. Inst. Univ. Zürich. Schaffhausen (Schweiz), Grubenstr. 1. *Pathol. des Herzens.* ⓓ Zürich 1924.

Ueno, Masuzô, Vice Ass. Ôtsu Lake-side Biol. Stat. Imper. Univ. Kyôto (Japan), Zool. Labor., Coll. of Sc. *Ephemerida, Crustacea, Cladocera.*

Baron von Uexküll, Jakob Johann (1864), Dr. med. h. c., Honor.Prof. der Univ. Hamburg. Hamburg-Fuhlsbüttel (Deutschland), Heinrich-Trann-Str. 3. *Vergl. Physiol.; Umweltforschung.*

Ufireduzzi Bordoni, G., Prof. Univ. Roma (Italia). *Igiene.* ○

Ugarte, Trifów, Prof. Univ. La Plata (Argentina). *Toxicol.* ○

Uglitzkih, Alexander N. (1876), Doc. des Kubanschen Landw. Inst. Krasnodar (U.d.S.S.R.), Proletarskaja 50. *Forstwirtschaft. Dendrol. Gehölzflora, Pflanzenbiol.* ⓓ Krasnodar 1925. ⓟ Gehölzflora u. Samen.

Uhlenhuth, Paul (1870), Univ.Prof., Dir. hygien. Inst. Freiburg i. Br. (Deutschland), Hebelstr. 42. *Hygiene u. Bact.*

Uhlmann, Eduard (1888), a.o. Prof. für Zool., Konserv. des Phyletischen Mus. der Univ. Jena (Deutschland), Maurerstr. 1a. *Abstammungslehre, angewandte Zool., Instinkt: Bauinstinkt der Trichopteren.* ⓓ Jena 1919. ⓟ Trichopteren und Schädlinge.

Uichanco, Leopoldo B. (1894), Associate Prof. of Entomol. and Acting Head, Dept. of Entomol., Coll. of Agric. Los Baños-College (Philippine Islands) *Entomol.: Philippine economic insects; biol., embryol., anat. and taxonomy of aphids; taxonomy of Psyllidae and Thysanoptera; biol. of philippine termites; plant golls.* ⓓ Harvard Univ. 1922. ⓟ Philippine economic insects.

Uji, Hideo, Prof. Zool. High School. Matsue (Japan), Zool. Labor.

Uklopskaja, Maria J., Doc. an der Mittelaslat. Staatl. Univ. Taschkent (U.d.S.S.R.), Dshisakskaja 28. *Reisbau. Phytotechnik.* ○

Ulbrich, Eberhard (1879), Dr. phil., Prof. u. Custos am Botan. Mus. der Univ. Berlin. Berlin-Dahlem (Deutschland), Königin-Luise-Str. 6—8. *Basi-*

diomycetes, Phycomycetes; Ranunculaceae, Astragaleae, Malvaceae, Bombacaceae. Chenopodiaceae. Floristik. ⓓ Berlin 1905.

Ulehla, Vladimir (1888), Prof. Univ. Brno (C.S.R.), Kounicova 63. *Movement of Plants. Physical Chem. of the Plant: Swelling, Negative Osmosis, Water Relations, Suction Tension, Surface Tension, Anomalous Osmosis, Plasmoptysis. Experim. Ecol.: Xerophyte Problem, Lime-Stone Problem, pH Regulation of the environment.* ⓓ Sc. Strasbourg 1911, ph. Prague 1917. ⓟ *Films about Plant Movements, Microscopic Slides.* Lantern Slides about Plant Physiol.

Ullrich, Hermann Theodor (1900), Dr. phil., Ass. beim Botan. Inst. der Univ. Leipzig C 1 (Deutschland), Johannis-Allee 1 II. *Physiol. des Eiweiß- und Säurestoffwechsels der grünen Pflanzen. Bewegung niederer Organismen.* ⓓ Leipzig 1924.

Ulmansky, Sava (1886), Dr., o.ö. Univ.Prof. Zagreb (S.H.S.), Tvornička 10. *Abstammung und Rassenkunde der Haustiere, insbesondere der Schweine. Festlegung des Sus mediterraneus-Typus. Vererbung der Wolleigenschaften beim Hausschaf.* ⓓ 1910.

Ulrich, Werner (1900), Dr. phil., Ass. am Zool. Inst. der Univ. Rostock (Deutschland). *Allgemeine Zool., Coleopteren, Hymenopteren, Strepsipteren, Brackwassertiere: Bryozoen.* ⓓ Berlin 1922.

Unamuno Yrigoyen, Louis M. (O. S. A.) (1873), Univ. de Madrid et Prof. d'Histoire Natur. dans le Collège des PP. Agustinos de Llanes, Oviedo (España). *Mycol. Systemat., Champignons microscopiques, parasit. et saprophytes des plantes supérieures.* ⓓ Madrid 1906.

Underhill, Frank Pell (1877), Prof. of Pharm. and Toxicol. New Haven, Conn. (U.S.A.), 333 Cedar Street, Sterling Hall of Med. *Physiol. of Amino-Acids.* ⓓ Yale 1903.

Underhill, Grover William (1888), Ass. Entomol. of the Virginia Agric. Experim. Station Blacksburg, Va. (U.S.A.). *Economic entomol.* ⓓ State Coll. Raleigh, N.C., 1918.

Underwood, William L. (1864), Lect. on Biol. Cambridge, Mass. (U.S.A.), Massachusetts Inst. of Technol. *Bact. Mosquitoes-, Moth and their extermination.* o

Unger, Emil (1883), Adjunct Royal Hungarian Biol. Stat. for Fisheries and for Sewage Purification. Budapest II (Hungary), Debröi-ut 15. *Pisciculture, biol.* ⓓ Budapest 1909.

Unger, W. Byers (1898), Instr. in Zool., Dartmouth Coll. Hanover, N.H., (U.S.A.). *Protozool. Rhythms in nutrition and excretion of Paramecium.* ⓓ Yale Univ. 1925.

Unna, Paul Gerson (1850), Prof., Dr., Leiter seiner Priv.-Klinik für Hautkrankheiten und seines wissenschaftl. Labor. Hamburg (Deutschland), Osterstraße 129. *Histochem. der einzelnen Formelemente der Haut. Sauerstoff- und Reduktionsorte.* ⓓ Straßburg 1875.

Untersteiner, Raimund (1894), Priv.Doc. Merano (Italia). *Neurol. und Psychiatrie: Liquor- und Serumdiagnostik. Histol. (Meningitis, Encephalitis).* ⓓ 1920.

Uphof, Johannes Cornelis Theodorus (1886), Head Dept. of Ecol., Prof. of Botan. Rollins Coll., Winter Park, Fla. Orlando, Fla. (U.S.A.), West Central Avenue, Route 3. *Ecol. and Dendrol., Purper bacteria as symbionts of Lichens. Subtropical and tropical agric. of Florida.* ⓓ Highland Univ. 1915. ⓟ Flora of Florida.

Upshall, W. Harold (1901), Pomol.; Horticultural Experim. Station, Vineland Station, Ontario (Canada). ⓓ Michigan State Coll. 1926.

Urasow, Johann (1895), Staatl. Inst. d. Med. Wissensch. Leningrad (U.d.S.S.R.), Baskow per. N. 13/15, Qu. 31. *Cytol. Hypophysis des Menschen und der Säugetiere.* ⓓ Leningrad 1925.

Urban, Ignatz (1848), Prof. Dr., Geh.Reg.R. Berlin-Dahlem (Deutschland), Botan. Mus. *Morphol., Biol., Geschichte der Botan., Pflanzengeographie, Systematik, Flora von Brasilien und Westindien.* ⓓ Berlin 1873.

Urban, Jaroslav (1896), adj. à l'inst. agric., Brno (C.S.R.). *Contrôle des semences.* ⓓ Praha 1918.

Urbanowicz, Casimire (1897), Ass. Inst. Biol. gén. Univ. Wilno (Pologne), rue Zakretowa 15. *Tardigrades: Macrobiotus Oberhaeuseri. Physiol. de croissance cellulaire.* ⓓ Vilno 1925.

Uribe, César (1896), Dir. National Hygiene Labor. Prof. of Parasit. Bogota (Colombia). *Trematodes. Parasit. amoebae, trypanosomes and blood and tissue flagellates.* ⓓ Medellin 1919. ⓟ Animal Parasit.

Urita, Tomoe, Teacher in Biol. Ôtomari Girls' High School. Ôtomari Kôtô Jôgakkô, Karafuto (Japan). *Decapoda.*

Urizar, R., Prof. Univ. National. Asunción (Paraguay). *Histol.* o

Urquhart, Alexander Lewis (1887), Lect. in Pathol. and Bact. St. Thomas' Hospital, Med. School, and Curator Pathol. Mus. St. Thomas' Hospital London, Sc. Ass. in Pathol. London Univ. (England). *General Pathol. and Bact.* ⓓ Edinburgh 1911.

D'Urso, Angelo, Lib.Doc. Univ. Catania, Sicilia (Italia). *Anat. umana.* o

Ursprung, Alfred (1876), Prof. d. Botan. a. d. Univ. Freiburg (Schweiz), Botan. Inst. *Pflanzenphysiol., physiol. Anat. Osmotische Zustandsgrößen der Pflanzenzelle.* ⓓ Basel 1900.

Urusov, Dziobrej Sarkisovič, Doc. I. Univ., Ass. Anat. Inst. Moskau (U.S.S.R.). *Anat.* o

Usakova, Aleksandra, Doc. Biol. Geograf. Naučno-Inst. Irkutsk (U.d.S.S.R.). *Histol.* o

Uschakov, Paul (1903), Ass. Inst. Hydrol. Leningrad (U.d.S.S.R.), Moika 8, log. 6. *Zoographie des mers russes du Nord. Nemertini and Hydroidea.*

Uspenskaja, W. J., Ass. Inst. d. experim. Biol. Moskau (U.d.S.S.R.), Woronzowo Pole. *Entwicklungsmechanik.* o

Uspenskij, Eugen Eug., Doc. I. Univ. Moskau (U.d.S.S.R.), Nishnjaja Presnja 1/30, W. 1. *Pflanzenphysiol. u. Anat.* o

Uspensky, Konstantin P., Doc. I. Univ., Ass. Anat. Inst. Moskau (U.d.S.S.R.). *Anat.* o

Ussow, Sergei (1867), Prof. Zool. Inst. Univ. Moskau (U.S.S.R.), Nowo-dewitschy Monastyr corp. 2. log. 6—7. *Embryol. Histol. Skelett, Muskeln.*

Uti, Keizô, Ass. Prof. Imper. Univ. Kyôto (Japan), Pharmacol Inst., Coll. of Med. *Pharm.*

Utsunomiya, Hiroaki (1891), Med. Univ. Okayama (Japan), Pharm. Inst. *Experim. Pharm.* ⓓ Med. Akad. Okayama 1915.

Uyttenboogaart, Daniel Louis (1872), Dr., Rotterdam (Holland), Parklaan 8. *Coleoptera, particulièrement Gyrinidae, Curculionidae (Biol., Zoogéografie, Systématique).* ⓓ Amsterdam 1896.

Uzel, Heinrich (1868), Prof. tschechische techn. Hochsch. Praha (C.S.R.), Kgl. Weinberge, Slovenská ul. 19. *Schädliche Insekten, Thysanoptera, Apterygota.* ⓓ Prag 1896.

Vabre, Henri Victorin Joseph (1895), Préparateur de chim. biol. et méd. à la Fac. de Méd. de Montpellier (France), 7, Rue d'Alger. *Chim. analytique et microanalytique appliquée à la Biol.* ⓓ Montpellier 1921.

Vacek, Tomáš (1899), Doc. Physiol., Dir. Physiol. Dept. Vet. Univ. Coll. Brno (C.S.R.). *Physiol. of fonctional Adaptation, Regulation of the body-temperature, Physiol. of Hibernation, smooth muscle.* ⓓ Brno 1923.

Vadász, Elemér (1885), Prof. emer. d. Palaeont. Budapest II (Ungarn), Lövöház-ut 22/a. *Foraminiferen, Ammoniten, Echinodermen.* ⓓ Budapest 1907.

Välikangas, Ilmari (1884), Dr. phil., Kustos Zool. Mus. Univ. Helsinki (Finnland), Liisank. 15. *Hydrobiol., spez. Brackwasserplankton, Euglena.*

20*

Vági, Stefan (1888), o. Prof. Sopron (Ungarn). *Bodenchem., Biochem., Holzchem.* ⑨ Budapest 1919.

Valllonis, L., Doc., Leiter Labor. f. Pflanzenanat. u. -physiol. Kaunas (Litauen), Wilnastr. 2. o

Vajnberg, Solomon Berk., Doc. I. Univ. Moskau (U.d.S.S.R.). *Pathol.-anat.* o

Vakilenko, Ivan Longin, Prof. Staats-Univ. Tomsk (U.d.S.S.R.), Prospekt Timirjazeva 3. *Biol. Chemie.* o

del Val, Davis, Prof. Univ. St. Tomas. Manila (Philippine Islands). *Anat.* o

Valckenier Suringar, Jan (1864). Wageningen (Holland), Arnh. Straatweg 83. *Dendrol.; Nomenclatur; Melocactus.* ⑨ Leiden 1898.

Valdez, F. B., Prof. Univ. Rosario (Argentina). *Anat.* o

Valdez de la Torre, Carlos, Prof. Univ. Lima (Peru). *Zool.* o

Valdivieso, José D., Doc. Univ. Lima (Peru). *Viticult.* o

Valenti, Adriano (1874), Dir. Ist. Farm., Tossicol. e Terapia sperim. Univ. Milano (Italia), Città degli studii, Via G. Strambio. *Ricambio materiale, spec. dell'acido urico e corpi xantinici. Fisiol. del vomito, della fame e della sete. Farm.* ⑨ Roma 1899.

Valenti, Giulio (1880), Prof. o. Anat. umana normale Univ. Bologna (Italia), Via Irnerio 48. *Anat. Embriol.*

Valentini, Luigi, Lib. doc. Univ. Catania, Sicilia (Italia). *Fisiol. sperim.* o

Valentino, L., Doc. Univ. Palermo (Italia). *Fisiol.* o

Valeton, Theodoric (1855). Leiden (Holland), Oegstgeesterlaan 21. *Systematische Bearbeitung der Familien der altweltlichen tropischen Zingiberaceae und Rubiaceae.* ⑨ Groningen 1886.

Val'ker, F. I., Prof. Univ. Taschkent (U.d.S.S.R.). *Topographic. Anat.* o

Della Valle, Antonio (1850), Prof. o. Univ., Dir. Ist. di Anat. Compar. Napoli (Italia), Via Salvator Rosa 259. *Gammarini. Ascidie Composte. Vertebrati. Embriol. Compar. Embriol.* ⑨ Napoli Med. 1873, Nat. 1874.

Valle, Kaarlo Johannes (1887), Ass. Zool. Labor., adjung. Doc. Zool. Univ. Turku (Finnland). *Ökol. der Boden- und Tiefenfauna der Seen. Systematik, Ökol. und Tiergeographie der Odonaten.* ⑨ Odonaten (Imagines).

Valleau, William Dorney (1891), Plant Pathol., Kentucky Agric. Experim. Station and Associate Prof. of Plant Pathol., Univ. of Kentucky. Lexington, Ky. (U.S.A.). *Corn and Tobacco root diseases. Tobacco mosaic and leaf spot diseases. Nitrogen deficiency diseases.* ⑨ Univ. of Minnesota 1917.

Vallée, Henri Pierre Michel (1874), Prof., Dir. honoraire des Ecoles vét., Dir. du Labor. de Recherches du Ministère de l'Agric., Membre de l'Académie de Méd. Alfort, Seine (France), Chemin du Fort 1. *Bact. générale et comparée. Tuberculoses, anémie infectieuse, Charbon symptomatique. Fièvre aphteuse.* ⑨ Alfort 1897.

Vallilo, Giovanni, Lib. doc. R. Ist. Med. vet. Milano (Italia). *Anat. pathol.* o

Vallisnieri, Ercolo, Lib. doc. Univ. Modena (Italia). *Pathol.* o

Vallois, Henri Victor (1889), D.M., D.S., Prof. d'Anat. à la Fac. de Méd. Toulouse (France). *Ostéol., Arthrol. et Myol. comparées des vertébrés et spécialement des mammifères. Anthrop. physique: homme et primates.* ⑨ Toulouse 1921. ⑨ Pièces squelettiques d'homme.

Vámossy, Zoltán, Prof. Dr., Dir. des Pharm. Inst. Univ. Budapest (Ungarn). o

Vandecaveye, Silvere Cyril (1888), Bact., Experim. Station. Pullman, Wash. (U.S.A.). *Soil and Dairy Bact.* ⑨ Iowa State Coll. 1923.

Vandel, Albert (1894), Maître de Conférences à la Fac. des Sc. de Toulouse (France). *Zool.: Planaires,*
Isopodes, Formicidae. Biol.: Parthénogenèse. ⑨ Paris 1922.

Vandendorp, Francis (1900), Moniteur d'Histol. Roubaix (France), 22, rue Victor Delaunoy. *Histol. du Poumon humain.* ⑨ Lille 1921.

Vandendries, René Fr. P. (1874). Anvers (Belgique), rue de l'église 44. *Cytol. Sexualité des Basidiomycètes.* ⑨ Louvain 1898.

Vándorfy, Josef (1895), Ass. III. Med. Klinik. Budapest VIII (Ungarn), Ludoviceum-ut 2. *Physiol. des Magens.* ⑨ Budapest 1919.

Vaney, M., Prof. Fac. Sc. Lyon (France). *Zool.* o

Vanin, Stephan Ivanovitch (1891), Chef des travaux Inst. Jaczewski, Prof. Inst. Zool. appl. et Pathol. vég., Chef des trav. Phytopath. Inst. Forestier. Leningrad (U.d.S.S.R.). *Maladies des arbres forestiers, Polyporacées, Zoocécidies.* ⑨ Leningrad.

Vannini, Giuseppe. Lib. Doc. Univ. Bologna (Italia). *Pathol.* o

Vansell, George H. (1892), Junior agric. in the Experim. Station and Associate in Entomol. Davis, Cal. (U.S.A.). *Entomol. and Agric.* ⑨ Univ. of Kansas 1917.

Varga, Franz (1889), Prof. Aszód, Kom. Pest (Ungarn). *Pflanzenanat., Pilzparasitäre Krankheiten der gärtn. Kulturgewächse, gärtn. Pflanzenzüchtung.* ⑨ Budapest 1920.

Varga, Lájos (1890) Sopron (Ungarn), Rákóczi-Realschule. *Rotatorien.* ⑨ Kolozsvar 1915. ⑨ Rotatorien-Praep.

Varga, Oszkár, Dr., Priv.Doc. der Ung. Tierärztl. Hochsch. Budapest (Ungarn), Rottenbillergasse. *Botan.* o

de Varigny, M. H. Paris 16 (France), 18, rue Lalo. *Zool.* o

Varsanofieva, Vera (1889), Prof. géol. cabinet géol. II. Univ., Collab. Comité géol., Membre de l'Inst. géol. Moscou (U.d.S.S.R.). *Faune silurienne.*

Vas, Bernhard (1864). Budapest (Ungarn), Hauptstädt. hygien. und bact. Inst. *Bact., physiol. Chem.* ⑨ Budapest 1889.

Vasile, Zăhărescu (1889), Chef des trav. Morphol. Fac. Sc. Univ. Jassy (Roumanie), Laboratoire de Morphologie animale. *Atypus piceus. Syngnathus Agassizi. Gastéropodes: Pulmonés terrestres et fluviatiles.* ⑨ Jassy 1914.

Vasil'ev, Ivan Petrovič, Prof. Univ. Kazan, Tatar. Rep. (U.d.S.S.R.). *Pathol. Anat.* o

Vasil'evskij, Aleksej Dmit., Prof. Omsky Med. Inst. Omsk (U.d.S.S.R.), Ul. Lenina 9. *Pharmacol.* o

Vasiliu, Titu (1885), Prof. Anat. Pathol. Fac. Med. Univ. Cluj (Romania), Inst. Anat.-Pathol. *Histol. pathol.; Morphol. des tumeurs. Haematol. Etiol. microbienne. Maladies infectieuses.* ⑨ Bucarest 1912. ⑨ Prep. de Pathol.

Vasil'jevskaja, E. V., Doz. Univ. Taschkent (U.d.S.S.R.). *Allg. Pathol. u. pathol. Histol.* o

Vasil'jevskij, V. V., Prof. Univ. Taschkent (U.d.S.S.R.). *Allg. Pathol u. pathol. Anat.* o

Vasnecov, Vladimir Vikt. Doc. I. Univ. Moskau (U.d.S.S.R.), *Zool., Embryol.* o

Vassil'ev, A. T., P.D. d. Petrogr. Univ. Paris (France), *Zool. Histol. Embryol.* o

Vassil'ev, L. L., Prof. Herzen-Inst. Leningrad (U.d.S.S.R.), Moïka 48. *Physiol.* o

Vassiljev, Ivan (1901), Ass. Labor. Allgem. Pflanzenzucht Donsche Hochsch. f. Landwirtsch. u. Melioration. Novotscherkassk (U.d.S.S.R.), Potschtovaja 65. *Physiol. Grundlagen der Widerstandsfähigkeit gegen die Dürre. Transpiration von Weizen.*

Vassiljevskij, N. J. (1884), Ass. d. Abt. f. Pflanzenpathol. Botan. Garten. Leningrad (U.d.S.S.R.), Pessotschnaja 2. *Fungi imperfecti. Pflanzenpathol. u. Mykol.* ⑨ Leningrad 1912.

Vatova, Aristocle (1897), Ass. R. Ist. Biol. Marina. Acquario, Rovigno d'Istria (Italia). *Sistematica de Fanerogame; sistematica e distribuzione geografia della fauna bentonica adriatica.* ⑨ Torino 1924.

Vaughan, Richard E. (1884), Prof. of Plant Pathol. Univ. of Wisconsin. Madison, Wis. (U.S.A.). Ⓓ Univ. of Wisconsin 1912.

Vaughan, Thomas Wayland (1870), Prof. of Oceanography and Dir. Scripps Inst. of Oceanography. La Jolla, Cal. (U.S.A.). *Geol., paleont., ocrals, foraminifera.* Ⓓ Harvard 1903.

Vavilov, N. V., Doc. Gosudarstv. Donskoj Univ. Rostov a. Don (U.d.S.S.R.). *Pharmacol.* o

Vavra, Wenzel, Dr., Dir. der Zool. Abt. Praha II (C.S.R.), Vaclavské, Nationalmus. o

Vayssière, A., Prof. Labor. de Zool. marine d'Endoume. Marseille (France). o

Vaza, David Lvovič, Doc. II. Univ. Moskau (U.d.S.S.R.). *Anat.* o

Vázquez Sans, Juan, Prof. ayudante de la Fac. Med., Barcelona (España), Paseo de Gracia, 117. *Anat. compar. y Embryol.* o

Vecchi, Anita (1893). Aiuto incaricato in Zoocoltura nel R. Ist. Sup. Agrar. Bologna (Italia), Ist. Zool. *Genetica: Conigli. Ecologia: Api, Molluschi, Bachi da seta. Zoogeogr.: Api. Sistem.: Crostacei Insetti.* Ⓓ Bologna 1918.

de Vecchi, Bindo (1877), Prof. St. Anat. patol. Firenze (Italia), Via Niccolini 5. *Amebiasi intestinale. Cuore da gozzo e cuor e tireotossico.* Ⓓ Perugia 1920.

Vedeneeva, Zinaide S., Conserv. Stat. Expérim. Agron. d'Ouzbekistan, Taschkent (U.d.S.S.R.), rue Pouschkinskaia 37. *Pathol. végétale.* o

Véfik, Ahmed, Prof. agr. Univ. Stambul (Türkei). *Pharmacol.* o

Végh, Franz (1896), Ass. pharm. Inst. Szeged (Ungarn), Kálvária ter 5. *Wirkung des Insulins auf die Atmung der Schildkröte. Diuresestudien mit verschiedenen Quecksilberpraep.* Ⓓ Szeged 1926.

Véghelyi, Ludwig (1895), Ass. a. d. Kön. Ung. Entomol. Inst. Budapest II (Ungarn), Kitaibel Pàl Gasse 1. *Angewandte Entomol. und Mammal.* Ⓓ Ungarische Hemipteren, Micromammalien.

Velhmeyer, Frank I. (1886), Ass. Prof. Irrigation Investigations, Coll. Agric. Univ. of California. Davis, Cal. (U.S.A.). *Soil physics, plant physiol., water relations of plants.* Ⓓ Johns Hopkins 1927.

Vejdovský, František (1849), Prof., Univ. Charles Prague. Praha II (C.S.R.), Riegrovo náb řeží 28. *Cytol.* Ⓓ Praha 1876.

van de Velde, A. J. J., Dir. Labor. chim. et bact. Dir. de l'Inst. sup. des fermentations. Gent (Belgique), 20, rue du Chantier. o

von den Velden, Reinhard (1880), a. o. Prof. d. Med. Fac. Univ., dirigierender Arzt Städt. Krankenhaus Berlin-Wilmersdorf. Berlin W 30 (Deutschland), Bamberger Str. 49. *Angew. Pharm., experim. Pathol., Blutgerinnung, Konstitutionslehre, innere Sekretion, Kreislauffragen, Brustkorbmechanik.* Ⓓ Heidelberg 1904.

Velenovský, Josef (1858), o. Prof. Univ. Praha II (C.S.R.), Na Slupi 433. *Floristik: Phytopalaeont. (Kreideformation), Mykol. Bryol. Balkanflora; Morphol.* Ⓓ Praha.

Velich, Alois, Dr., Prof. der Anat., Physiol. der Haustiere, Landwirtsch. Bact., Dir. Inst. für Anat. u. Physiol. d. Haustiere der tschech. techn. Hochsch. Praha XII (C.S.R.), Kopernikova 7. *Microorganismes thermophiles.* o

Velich, Vratisav, Dr.-Ing., Prliv.Doc. für landwirtschaftl. Mykol. Praha II (C.S.R.), Na Bojisti, Untersuchungsanstalt für Lebensmittel. *Mycol. agricole.* o

Velluda, Constantin (1893), Chef de trav. anat. Cluj (România), Rue Mico 3. *Anat. et embryol. du système vasculaire et nerveux.* Ⓓ Cluj 1921.

Veluet, M., Prof. Univ. Poitiers (France). *Histol.* o

Veneziani, Arnoldo (1878), Prof., Lib. doc. Univ. Ferrara (Italia). Ⓓ Genova 1918.

Ventschikov, Anatol (1897). Taschkent (U.d. S.S.R.), Kuropatkinskaja 50. *Nahrungsphysiol.* Ⓓ Taschkent 1925.

Ventura, Maria (1897), Ass. Ist. Botan. Univ. Roma (Italia). *Embriol. vegetale.* o

Venulet, Franz (1878), Dr. med., o. Prof. für allgemeine und experim. Pathol. an der Univ. Warschau (Polen). *Experim. Untersuchungen über Abwehrfermente. Serol. Luesuntersuchungen: Wassermann mit aktivem Serum. Der Einfluß von Galle auf Infektion. Beziehungen der Milz zum Kohlehydratstoffwechsel.* Ⓓ Warschau 1904.

Vera, J. P., Prof. Univ. Nation. Asunción (Paraguay). *Biol.* o

Veratti, Emilio (1872), Prof. incaricato di Batteriol. nella R. Univ. di Pavia (Italia), Via U. Foscolo 21. *Istol. patol., Batteriol.; Culture dei Tessuti.* Ⓓ Pavia 1921. Ⓟ Culture di germi patogeni.

Vercellana, Joseph (1897), Prof. Univ., Dir. Labor. Bact. Parma (Italia), rue Cavour 107. Ⓓ Parma 1921.

Verdun, M. P., Prof., Fac. Méd. Pharm. Lille (France). *Zool.* o

Vereščagin, Nikolaj Konst., Doc. I. Univ. Moskau (U.d.S.S.R.) *Physiol.* o

Veress, Elemér (1876), o. ö. Prof. der Physiol., Dir. des Physiol. Inst., Szeged (Ungarn), Kálvária-Platz 5. *Physiol. der Sinne, Muskelphysiol.* Ⓓ Kolozsvár (Ungarn) 1899.

Verga, Pietro (1894), Aiuto Anat. Patol. Univ. Parma (Italia), Viale Umberto I. 34. *Cisti della dura mater spinale e dei gangli spinali.*

Verguin, Jacques (1901), Ingénieur agronome, préparateur à l'Inst. des Recherches Agronomiques (Station Entomol.) Rouey (France), 16, rue Dufay. *Biol. des insectes et animaux utiles et nuisibles à l'agric.*

Verhoeff, Frederick Herman (1874), Prof. ophthalmic Research, Harvard Univ., Ophthalmic Chief of Research, Pathol. and Ophthalmic Surgeon, Massachusetts Eye and Ear Infirmary. Boston, Mass. (U.S.A.), 82 Commonwealth Ave. *Pathol. of the eye.* Ⓓ Johns Hopkins 1899.

Verhoeff, Karl Wilhelm (1867), Dr. phil. Pasing bei München (Deutschland), Otilostr. 11a. *Myriapoda, Isopoda, Thysanura, Coleopteren, Metamorphose und vergl. Morphol.* Ⓓ Bonn 1893. Ⓟ Myriapoden, Isopoden, Insectenlarven.

Verhoeven, Willem Boudewijn Leeuwenburg (1890), Agric. Engineer Phytopath. of the Phytopath. Service. Wageningen (Holland). *Phytopath. and economical entomol.*

Verlaine, Louis Joseph Auguste (1889), Prof. de Biol. à l'Univ. Coloniale d'Anvers, Chargé du cours de Physiol. animale au Doctorat en Zool. à l'Univ. Bruxelles. Schaerbeek, Bruxelles (Belgique), 31, rue Joseph Coosemans. *Psychol. comparée. Hyménoptères. Ethol.* Ⓓ Bruxelles 1913.

Vermeulen, Hendrik Adriaan (1870), Dr.med.vet., Conserv.-Lect. an der Reichsuniv. Vet. Fak. Utrecht (Holland), Nieuwe Gracht 165. *Vergl. Anat., speziell Haustiere. Neurol., speziell Vagus und Hypoglossusareal der Säugetiere. Organe der inneren Sekretion.*

Verne, Claude Marie Jean (1890), Prof. agrégé d'Histol. à la Fac. de Méd. de Paris 7 (France), 38, Rue de Varenne, *Histol., histochim.* Ⓓ Paris 1923.

Verneau, René (1852), Prof. d'Anthrop. au Mus. national d'Histoire nat. et à l'Inst. de Paléont. humaine, Conserv. du Mus. d'Ethnographie. Paris XIV (France), 72, Avenue d'Orléans. *Anthrop. physique, Préhistoire et Ethnographie.* Ⓓ Paris 1875.

Verney, Ernest Basil (1894), Prof. of Pharm., Univ. of London W. C. 1 (England), Univ. Coll., Gower Street. Ⓓ Cambridge 1916.

Verney, Ruth Eden Couway (1894), Ass., Physiol. Dept. Univ. Coll. London W.C. 1; London N. 3 28 Clifton Avenue. *The determination of the H-ion concentration of the blood.* Ⓓ Manchester 1923.

Vernhout, Johannes Hendrik (1866), Dr. zool. et botan. Middelburg (Holland). *Embryol., Spongiol., Malacol., Bact.* Ⓓ Utrecht 1894.

Vernon, Horace Middleton (1870), Investigator Industr. Fatigue Research Board, London. Oxford

(England), 5 Park Town. *Industrial Fatigue and Efficiency.* ⊕ Oxford 1899.

Vernoni, Guido (1881), Prof. di Patol. generale R. Univ. Sassari, Sardegna (Italia). *Fisio-patol. dei vasi tumori maligni.* ⊕ Bologna 1906.

Verona, Onorato (1904), Ass. Patol. vegetale e Batt. Agraria, Ist. Superiore Agrario. Pisa (Italia). *Batt. agraria.* ⊕ Pisa 1926.

Verrier, Marie Louise (1904), Boursière de doctorat au Mus. national d'Histoire nat. Paris 14 (France), 21, Boulevard Jourdan. *Histol. de la rétine chez les Poissons et mécanisme de la vision chez les Poissons. Morphol. du tube digestif des Poissons.* ⊕ Paris 1924.

Verrill, Addison E. (1839), Prof. emer. of Zool., Yale Univ. Leihu, Kauai Island (Hawaiian Islands). *Deep-sea fauna, Marine fauna of Bermuda, W. Indies, Brazil and Panama, Crustacea, Geol. and Palaeont. of Bermuda. Alcyonaria. Starfishes.* ⊕ Yale 1867. o

Versari, R., Prof. Univ. Roma (Italia). *Anat. umana norm.* o

Verschuer, Otmar, Freiherr von (1896), Dr. med., 1. Ass.-Arzt der Med. Univ.-Poliklinik Tübingen (Deutschland), Melanchthonstr. 25. *Menschliche Erblichkeitslehre: vererbungsbiol. Zwillingsforschung. Konstitution und Rasse.* ⊕ München 1923.

Versé, Max (1877), Prof Univ. Marburg a. d. L. (Deutschland), Weißenburger Str. 15. *Pathol. Cholesterinstoffwechsel.* o

Versinin, N. V., Prof. Univ. Tomsk (U.d.S.S.R.), Prospekt Timirjazeva 3. *Pharmacol.* o

Versluys, Jan (1873), o. ö. Prof. an der Univ. Wien XIX (Deutsch-Österreich), Grinzinger Allee 18. *Vergl. Anat. der Wirbeltiere. Phylogenie der Tiere.* ⊕ Gießen 1899.

Verwey, Jan (1899), Ass. Zool. Labor. Univ. Leiden (Holland). *Aves: Biol., Systematik, Physiol. Lebensgewohnheiten (Biol. im Sinne von Uexkülls) von Evertebraten. Coccidia.* ⊕ Leiden 1926. ⊕ Aves.

Verzár, Fritz (1886), Dr., Prof. für Physiol. und allg. Pathol. an der Univ. in Debrecen (Ungarn). *Muskel- und Nervenphysiol. Actionsströme der Nerven. Blutgaswechsel.* ⊕ Budapest 1909.

Vesclov, Vasilij Sokr., Prof. Med. Inst. Omsk (U.d.S.S.R.), Ul. Lenina 9. *Bact.* o

Vestal, Arthur G. (1888), Ass. Prof. of Biol., Stanford Univ. California (U.S.A.), Box 147. *Ecol. of plants and animals. Geographic distribution. Grassland vegetation of Western North America, associations (communities) of plants and animals.* ⊕ Univ. of Chicago 1915.

Vialas, Prof. Ecole Colon. d'Agric. Tunis (Afrique). *Botan.* o

Viale, Gaetano (1889), Dir. de l'Inst. de Physiol. de la Fac. de Méd. de Rosario de S. Fé (Rep. Argentina). *Physiol. des capsules surrénales: electrophysiol., les substances photodynamiques, la physiol. du travail, insuline.* ⊕ Turin 1926.

Vialleton, Louis (1859), Prof. d'histol. à la Fac. de Méd. de Montpellier (France), 1 Boulevard du Peyron. *Histol., Embryol. et Anat. comparées.* ⊕ Lyon 1885.

Vialli, Maffo (1897), Lib. Doc., Aiuto Ist. Anat. e Fisiol. compar. Univ. Pavia (Italia), Palazzo Botta. *Anat. compar. dei Vertebrati, Fisiol. degli Invertebrati.* ⊕ Pavia 1922.

Vianna, Aug., Prof. Univ. Bahia (Brasil). *Bact.* o

Vianna, G , Prof. Univ. Porto Alegre (Brasil). *Anat.* o

Vickery, Hubert Bradford (1893), Research Chem. of the Carnegie Inst. of Washington, D.C. New Haven, Conn. (U.S.A.), Connecticut Agric. Experim. Station. *Chem. of Proteins. Nitrogenous Constituents of Plants.* ⊕ Yale 1922.

Vickery, Roy Albian (1883), Ass. Entomol., Bureau of Entomol., U. S. Dept. of Agric. Stratford, Conn. (U.S.A.), 205 Beardsley Ave. *Insects infesting cereal and forage crops, life histories and habits of Aphididae. European corn borer, Pyrausta nubilalis.* ⊕ Univ. of Minnesota 1906.

Vidal, Joaquim (1891). Ass. Labor. d'Anat. pathol. Fac. med. Univ. Rio de Janeiro (Brasil). *Organes des sens. Dosage et élimination des médicaments dans le sang.* ⊕ Rio de Janeiro 1913.

Vidovic, Josef C. (1862), Kustos Zool.-botan. Abteilung, Steiermärkisches Landes-Mus. Waldhaus, Post Wetzelsdorf bei Graz (Österreich). *Vögel der Steiermark.*

Viehvever, Arno (1885), Curator, Dir. microsc. labor. and botan. gardens., Phil. Coll. of Pharm. and Sc. Glenolden, Pa. (U.S.A.), 112 N. Scotl. Av. *Pharmacognosy. Biol.* o

Vielwerth, Vladimir (1891), Chef Inst. phytopath. des Inst. pour les recherches agron. Bratislava (C.S.R.), Matuškova 934. *Phytopath.* ⊕ Prague 1916. o

Vierhapper, Friedrich (1876), Dr. phil., a.o. Univ.-Prof. und Hon.Doc. an der Botan. Lehrkanzel der Tierärztl. Hochsch. in Wien III (Deutsch-Österreich), Fasangasse 38. *Systematische Botan. und Pflanzengeographie.* ⊕ Wien 1899.

Vietinghoff-Riesch, Arnold, Freiherr v. (1895), Dr. oec. publ., Forstassessor. Schloß Neschwitz i. Sachsen (Deutschland). *Lokalavifaunistische Forschungen, bes. Nahrungsbiol. Das Verhalten der Vögel zu den forstschädlichen Insekten.* ⊕ 1923. ⊕ Vogelbälge der palaearktischen Fauna.

Vieweger, Feodor (1889), Prof. de physiol. générale à l'Univ. Libre de Pologne; Dir. du labor. de physiol. génér. Varsovie (Pologne), rue Polna 30. *Protistophysiol., métabolisme du système nerveux.* ⊕ Bruxelles 1912.

Vigano, Luigi (1877), Capo della Sezione Vaccini umani e Diagnostica Ist. Sieroterapico. Milano (Italia), Via Borgonuovo, 19. *Sierol., Immunol., Batt., Igiene.* ⊕ Pavia 1901.

Vignes, Henri (1884) Accoucheur des Hôpitaux de Paris; Dir. de l'Inst. International d'anthrop. Paris (France), 52, Rue de Vaugirard. *Physiol. de la femme enceinte, de l'ovaire et de la menstruation. Mariages consanguins, Médicaments agissant sur l'appareil génital de la femme, Anat. pathol. et physiol. de la placenta.* ⊕ Paris 1914.

Viguier, René (1886), Dr., Prof. à la Fac. des Sc. de l'Univ. de Caen, Calvados (France), Inst. Botan. Jardin des Plantes. *Flore de Madagascar. Araliacées. Anat. Systématique. Paléont. végétale.* ⊕ Plantes de Madagascar.

Viktorow, Konstantin (1878), Prof., Dir. Physiol. Labor. des Vet.-Inst. Kazan (U.d.S.S.R.), Arskoje Pole 96. *Physiol. d. Herzens, Muskelsystem, Nervensystem.* ⊕ Kazan 1903.

Vila Gómez, Michel (1884), Prof. ajudant du l'Inst. Nacional de 2 Enseñance del Valencia (España), Rue Boix 6, 2. *Biol. agricole.*

Vilde, Jānis, Dr., Ass. Univ. Riga (Latvija), Brunineku ielā 5/7. *Anthropometrie u. Anat. d. Menschen.* o

Vilhelm, Jan (1876), Prof. à l'Univ. Charles Praha VI (C.S.R.), Benátská 433. *Botan. théoretique (Systématique). Cryptogamol. générale et systématique (Charophyta, Bryophyta), Teratol., Morphol. et Biol. de la fleur. Thermalvegetation.* ⊕ Prague 1901.

de Vilhena, Henrique (1879), Prof. Anat. Fac. Méd., Prof. d'Anat. artistique Ecole des Beaux-Arts. Lisbôa (Portugal), Avenida Republica, 102, 3. *Myol. Variations anat.*

de la Villa y Sanz, D. Julian, Prof. Univ. Madrid (España). *Anat. descr. Embryol.* o

Villalba, Aquiles S., Prof. Univ. Nac. Córdoba (Argentina). *Hygien.* o

Villani, A., Prof. Univ. Parma (Italia). *Botan.* o

Villaret, Prof. agrég. Univ. Paris (France). *Pathol. méd.* o

Villarreal, Aureliano González, Prof. Esc. de Vet. León (España). *Allg. Anat.* o

Villavicencio, R. P., Prof. Univ. Quito (Ecuador). *Patol.* o

Vilicaitis, Vincas (1892), Doc. der Botan. an der landwirtsch. Hochsch. Dotnuva (Litauen). *Desmidiaceenflora Litauens.* ⒟ 1924. ⓟ *Pilze, bes. pathol.*

Villegas, Valente (1893), Ass. Prof. Animal Husbandry. Los Baños College (Philippine Islands). *Horses, cattle, water buffaloes, sheep and goats: feeding and management.* ⓟ Ames, Iowa, 1921.

de Villiero, Cornelius Gerhardus Stephanus (1894), Prof. Zool., Univ. Stellenbosch (South Africa), van-Riebeekstraat. *Sternum and Skeletogenous strata of Amphibia.* ⒟ Zürich 1921. ⓟ *Anura.*

Vimmer, Antonin (1864), Dir. und Inspektor, President der Societas Entomol. Cechosloveniae. Prag XII (C.S.R.), Palackého tř., č. 37. *Die Systematik der Dipteren, Metamorphose der Dipteren, Anat. und Morphol. der Dipterenlarven.*

Vimtrup, Bjovulf Jensen (1890), Dr. med., Prosector pathol. Kopenhagen (Dänemark), Frederik den Femtes Vej 11. *Histol., Histopathol. und Histophysiol. der Capillaren, insbesondere die der Nieren.* ⒟ Kopenhagen 1922.

Vinall, Harry Nelson (1880), Agronomist. Washington, D.C. (U.S.A.), Bureau Plant Industry, U.S. Dept. Agric. *Culture, improvement and utilization of sorghums, millets, and field peas; forage production in the semiarid regions; improvement of pastures and hay lands by the use of introduced grasses and legumes; history, botan., and genetics of the sorghums (Sorghum vulgare Pers.).* ⒟ Cornell Univ. 1912. ⓟ *Sorghums (Sorghum vulgare Pers.).*

Vinassa de Regny, Paolo (1871), Prof. o. Geol., Paleont. Pavia (Italia), Univ. *Briozoi, Idroidi, Coralli, Spugne. Paleobiol.* ⒟ Pisa 1893. ⓟ *Fossili mesozoici e terziari italiani.*

Vincent, Emile, Dr. sc., Conserv. au Mus. R. d'Hist. nat. Bruxelles (Belgique), 35, rue de Pascale. *Conchylia fossilia.*

Vincent, Swale (1868), Prof. Physiol., Univ. Middlesex Hospital Med. School. London (England). *Secretion. Internal Secretion. Physiol., Comp. Anat. Histol.* ⒟ London 1896.

Vincenti, R., Ass. Prof. Univ. of Malta. La Valletta (Malta). *Pathol.* ○

Vinci, Gaetano (1869), Prof. stabile di farm. nella R. Univ. di Messina (Italia), Via Risorgimento 199. *Diuretici, derivati della purina, adrenalina, ipnotici, funghi. Anestetici locali.*

Vinegradov, Paul (1901), Ass. Entomol. Plant Protection Station. Stawropol, Caucasus (U.d.S.S.R.), Okrzu. *Entomol.: Odonata and Coleoptera (Tenebrionidae). Economic Entomol.*

Vinson, Albert Earl (1873). Chem. Service Techn. d'Agric.Port au Prince (Haiti). Tuscon, Ariz.(U.S.A.). *Chem. of the date and soils.* ⒟ Göttingen 1904.

Vinson, Carl George (1891), Biochem., Boyce Thompson Inst. for Plant Research, Inc. Yonkers, N. Y. (U.S.A.). *Biochem., horticulture.* ⒟ Univ. of Minnesota 1926.

Vintemberger, M. P., Chef de trav., Fac. Méd. Strasbourg (France). *Embryol.* ○

Vinter, Sergej, Prof. Landw. Inst. Kiev (U.d.S.S.R.), Brest-Litovskoe-Chaussee 39. *Forstkvnde.* ○

Virchow, Hans (1852), Dr., Prof. o. honor. emer. Univ. Berlin. Berlin-Charlottenburg 2 (Deutschland), Knesébeckstr. 78/79. *Anat.* ⒟ 1875.

Virnich, Doc. Univ. Woosung (China). *Physiol.* ○

Virsaladze, Spiridon, Prof. Univ. Tiflis (U.d.S.S.R.). *Pathol.* ○

Virtanen, A. J., Dr. phil., Doc. Univ. Helsinki (Finnland), Repslagareg. 7 A. *Biochem.* ○

Vischer, Wilhelm (1890), Priv.Doc. für Botan., Univ. Basel (Schweiz), Rittergasse 31. *Experim. Morphol., bes. der Chlorophyceen in Reinkultur. Tropische Agric. (Hevea brasiliensis, Kautschuk).* *Vegetation von Paraguay.* ⒟ München 1914.

Visher, Stephen (1887), Prof. of Geography, Indiana Univ. Bloomington, Ind. (U.S.A.). *Biogeography of the Great Plains of NorthAmerica. Influence of climatic eruditions (changes of climate, storms,* *especially cyclones, and rainfall) upon plants and animals.* ⒟ Univ. of Chicago 1914.

Visscher, John Paul, Associate Prof. of Biol., Western Reserve Univ. Cleveland, Ohio. (U.S.A.). *Fouling of Ships' Bottoms, Protozoa.* ⒟ Johns Hopkins Univ. 1924.

Vitali, Giovanni, o. Prof. Univ. Cagliari, Sardinia (Italia). *Anat., Istol.* ○

Vitek, Eugen, Ing., Chef de la Station de Contrôle des Graines. Praha II (C.S.R.), Václavské náměstí 47. *Botan. appliquée.* ○

Vitzthum, Hermann, Graf (1876), Dr., Wissenschaftl. Hilfsarbeiter an der Tierärztl. Hochsch. Berlin W 15 (Deutschland), Ludwigskirchstr. 9 a. *Acarol.* ⒟ Jena.

Vivian, Alfred (1867), Dean Coll. of Agric. Columbus, O. (U.S.A.), 1778 N. High St. *Agric. chem. Pharm.* ○

Vjekoslav, Magdić (1898), Ass. der agrochem. Abteilung der landw. Versuchsstation Osijek IV, 40 (S.H.S.). *Nährstoffaufnahme durch die Pflanzen.* ⒟ Prag 1925.

Vládescu, M., Prof. Univ. Bucarest (România). *Botan.* ○

Vladescu, Radu (1886), Prof. chim. biol. Fac. Méd. vét. Bucarest (Romania), Splaiul Davila 8. ⒟ Méd. vét. Bucarest 1911, sc. phys. Paris 1921.

Vladimirov, A. A., Prof. Inst. f. exper. Med. Leningrad (U.d.S.S.R.), Lopuchinskaja ulica 12. *Vergl. Pathol.* ○

Vlès, Fred (1885), Prof. à la Fac. des Sc., Chargé de Cours à la Fac. de Méd. Strasbourg (France), Inst. de Physique biol. Univ. *Physico-chim. biol.; Radiations. Propriétés optiques des muscles. Physico-Chim. des Pétroles. Physique biol. et Chim.-physique.* ⒟ Paris 1911.

van der Vliet, Adrianus Cornelis (1879). Rotterdam (Holland), Graaf Florisstr. 81a. *Physiol. als Theorie der Gärtnerei.*

Vlissidès, G., Prof. à l'Ecole Forestière. Athènes (Grèce) Rue Acharnon. *Botan.* ○

van Vloten, Heinrich (1895), Boschbouwkundig-Ingenieur bei der holl. forstl. Versuchsanstalt. Wageningen (Holland), Belmontelaan 5. *Durchforstungsversuche in Kiefernbeständen.*

Vluka, Josef (1869), Privatgelehrter, Besitzer einer Zuchtstation für Arzneipflanzen und landwirtschaftliche Gewächse. Slavkov u Opavy (C.S.R.), *Pflanzenveredlung, Phytopath., Pharm. und Pharmakoergasie, Krebsforschung.* ⓟ *Samen von Arzneipflanzen Mitteleuropas; Käfer.*

Vodrážka, Otakar (1883), ao. Prof. Botan. Hochsch. für Bodenkultur. Brno (C.S.R.), Černá pole. *Anat. der Pflanzen, des Holzes. Statolithenorgane der Saugorgane, der Hemiparasit., der Grasblätter.* ⒟ Prag 1909. ⓟ *Botan. mikrosk.* Praep.: *Holz, Stärke, Textilpflanzen.*

Voegtlin, Carl (1879), Chief, Division of Pharm., Hygien. Labor., U.S. Public Health Service. Washington, D.C. (U.S.A.). *Nutrition. Beri-beri. Pellagra. Function of the parathyroid gland. Anaphylaxis. Liver function. Industrial poisoning.* ⒟ Freiburg i. Br. 1903. ○

Voelkel, Hermann (1888), Wissenschaftl. Hilfsarbeiter an der Biol. Reichsanstalt zu Berlin-Dahlem. Berlin-Steglitz (Deutschland), Külzerstraße 2. *Physiol. der Insectenatmung, vergl. Physiol. der niederen Sinne. Angewandte Entomol. Ungeziefer des Menschen.* ⒟ Marburg a. d. Lahn 1919.

Völker, Ottomar, Dr., Prof. der Anat. Brno (C.S.R.). Anat. Inst. der Masaryk-Univ. ○

Völki, Wilhelm (1872), Univ.Prof. Königsberg i. Pr. (Deutschland), Hammerweg 16. *Tierzucht.* ○

Vogd, Irwin, H. (1887), Ass. manager of J. M. Lupton and Son. Mattituck, N.Y. (U.S.A.). *Breeding and reproduction, seed growing. Pathol. phases of seed production.* ⒟ Iowa State Coll. 1916.

Vogel, Leonhard (1863), Dr. phil. et med. vet h. c. Geh. Hofrat, o. Prof. an der Tierärztl. Fak. der

Univ. München (Deutschland), Veterinärstr. 6. *Landwirtschaftl. Tierzucht, Organisationsfragen (staatliche und öffentliche Maßnahmen zur Förderung der landwirtschaftl. Tierzucht), Wachstum bei landwirtschaftl. Haustieren.* Ⓓ Erlangen 1887.

Vogel, Martin (1887), Dr. med., Dir. Hygien. Mus. Dresden-A. (Deutschland), Zirkusstr. 38. *Hygiene.* ○

Vogel, Richard (1881), Prof. ,Dr., phil., Konserv. an der Württ. Naturaliensammlung und Prof. an der Landwirtschaftl. Hochsch. Hohenheim. Stuttgart (Deutschland), Harenbergstr. 56. *Vergl. Anat. und Entwicklungsgeschichte. Insektenanat. Angewandte Entomol.* Ⓓ Jena 1909.

Vogelsang, Heinrich W. (1897), Oberass. und Prof. im Pathol. und Parasit. Inst. Montevideo (Uruguay), Larranaga 572. *Parasit.* Ⓓ Montevideo 1920.

Voglino, P., Doc. Univ. Torino (Italia). *Botan.* I

Vogt, Ernst (1890), Dr., Wissenschaftl. Mitarbeiter am Badischen Weinbauinst. Freiburg i. Br.(Deutschland), Kaiserstr. 18. *Pflanzenphysiol., Mykol., Pflanzenschutz; Wirkung chem. Mittel auf Parasit. und Wirtspflanze.* Ⓓ Straßburg i. Elsaß 1914.

Vogt, Oskar (1870), Dir. Neurobiol. Inst. Univ. u. Kaiser-Wilhelm-Inst. für Hirnforschung. Berlin W 35 (Deutschland), Magdeburger Str. 16. *Hirnforschung u. Allg. Biol.*

Vogt, Walther (1888), ao. Prof., Leiter Abt. für Histol. und Embryol. Anat. Anstalt. München (Deutschland), Pettenkoferstr. *Anat., Histol., Embryol., Entwicklungsmechanik Anat. und Entwicklungsgeschichte des Bauch-Situs; Morphol. und Physiol. der Primitiventwicklung der Amphibien und ihre experimentelle Bearbeitung.* Ⓓ Marburg 1912.

Voigt, Alban (1857), Kustos Botan. Inst. Techn. Hochsch. Dresden (Deutschland), Hübnerstr. 14. *Systematik und Pflanzengeographie, besonders mediterrane Floren.*

Voigt, Alfred, Univ.Prof. Hamburg (Deutschland), Wandsbeker Stieg 13. *Botan.* ○

Voigt, Walter (1856), Univ.Prof. im Ruhestand. Bonn (Deutschland), Maarflach 4. *Turbellaria, Oligochaeta.* Ⓓ 1884.

Volnar, Alexis (1904), Forschungs-Katheder f. Morphol. u. Physiol. Odessa (U.d.S.S.R.), Metschnikoffstr. 4. *Wärmeregulation bei Kaninchen.* Ⓓ 1925.

Voinow, D. N., Prof. Univ. Bucuresti (Romănia). *Allg. Zool. u. Morphol.* ○

Voit, Erwin (1852), a.o. Univ.-Prof. München (Deutschland), Bauerstr. 28 III. *Physiol., Stoffwechsel und Ernährung.* Ⓓ München 1877.

Voit, Max (1876), o. Prof. an der Univ. und Abteilungsvorsteher am anat. Inst. Göttingen (Deutschland), Planckstr. 7. *Vergl. Aant. und Entwicklungsgeschichte des Wirbeltier-, speziell des Säugetierschädels.* Ⓓ München 1903.

Vojnovski-Krieger, Tamara (1902), Ass. Leningrad (U.d.S.S.R.), Kowenski Str. 4. *Angewandte Entomol. Biol. paraist. Schlupfwespen. Systematik: Braconidae.* Ⓓ Leningrad.

Volák, Jan (1902), Doubrava (C.S.R.). *Mikrohydrobiol.: Rhizopoda, Sporozoa, Flagellata, Cytoidea. Crustacea: Entomostraca. Vergl. Morphol.: Phanerogamae.*

Volcanezkij, Elias (1895), Doc. Saratow (U.d. S.S.R.), Prijutskajastr. 31. *Vogelfauna des Nieder-Wolga-Gebietes. Zeichnung und Färbung der Vogelbefiederung.* Ⓓ Vögel d. Nieder-Wolga-Gebletes.

Volhard, Franz (1872), o. Univ.Prof. und Dir. der Med. Klinik. Frankfurt-Main (Deutschland). *Erforschung der Nieren und Herzkrankheiten.* Ⓓ Halle a. d. S. 1897. Ⓟ Paratfinherzen.

Volhard, Justus (1881), Wissenschaftl. Ass. an der Versuchsanstalt Möckern. Leipzig-Gohlis (Deutschland), Kirchweg 10. *Agriculturchem., Pflanzen- un Tierernährung.* Ⓓ Halle a. d. S. 1895.

Volkart, Albert (1873), Vorsteher der Schweiz, landwirtschaftl. Versuchsanstalt Oerlikon und Prof. des Pflanzenbaues und der Pflanzenanat. an der eidgen. techn. Hochsch. Zürich 6 (Schweiz), Frohburgstr. 67. *Angewandte Botan.* Ⓓ Zürich 1899.

v. **Volkmann,** Rüdiger (1894), Ass. am Anat. Inst. der Univ. Tübingen (Deutschland). *Sekretorische Prozesse der Epiphyse; Cytoarchitektonik der Hirnrinde, spez. motorische und Sehregion und Beziehung zur Funktion.* Ⓓ Jena 1923.

Volodina, Sofia Nik., Doc. I. Univ. Moskau (U.d.S.S.R.). *Zool., Embryol.* ○

Voloschinova, Bina A., Conserv. Stat. Défense d. Plantes. Charkov (U.d.S.S.R.). *Pathol. végétale.* ○

Volskaja, Olga Nik., Doc. I. Univ. Moskau (U.d.S.S.R.). *Agric., Chem.* ○

Volterra D'Ancona, Luisa (1902), Ist. di Anat. e Fisiol. Compar. Univ. Roma (Italia), Via Depretis 91. *Plancton d'acqua dolce, Biol. delle Dafnie, Variabilità e ciclo dei Cladoceri, Rotiferi, Peridinei.* Ⓓ Roma 1924.

Volz, Emil C. (1891), Associate Prof. of Horticulture. Ames, Ia. (U.S.A.), Iowa State Coll.

Vonk, Herbertus Johannes (1897), Hauptass. an der Abt. für Vergl. Physiol. des Zool. Labor. Utrecht (Holland), 47 Witte Vrouwensingel. *Phagozytäre Verdauung bei Lamellibranchiaten. Enzyme der Fische. Ph-Bestimmungen im Darmkanal von Fischen und Evertebraten. Verdauung bei den Fischen.* Ⓓ Utrecht 1927.

Vonwiller, Paul (1885), Dr. med. et phil., Priv.Doc. für Anat. an der Univ. Zürich (Schweiz), Frohburgstraße 69. *Vitalfärbung, Lebendbeobachtung von Zellen und Geweben in situ an höheren Tieren und Pflanzen mittelst der Mikroskopie im auffallenden Licht. Kurvengemäße Mikrotomie regelmäßig gebogener Körper. Osteol., Anat. des Blut- und Lymphgefäßsystems, spec. Injectionen.* Ⓓ Zürich 1910.

Voorhies, Edwin Coblentz (1892), Ass. Prof. of Agric. Economics and Associate Agric. Economist in the California Agric. Experim. Station. Berkeley, Cal. (U.S.A.), 2615 Shasta Road. *Economic studies of the crops of the animal industries.*

Vorbrodt, Carl (1865). Bern (Schweiz), Zeughaus Lyss. *Schweizerische Macrolepidoptera: Biol. u. Verbreitung.*

Vorbrodt, Ladislas, Dr. Prof., Dir. Univ. Kraków (Pologne), R. Mickiewicza 17. *Botan.* ○

Vorhies, Charles Taylor (1879), Prof. of Entomol.; Agric. Experim. Station, Univ. of Arizona. Tucson, Ariz. (U.S.A.), Bin C, Univ. Station. *Economic Entomol.* Ⓓ Wisconsin 1908. Ⓟ Gila Monsters, Heloderma suspectum.

Voris, Ralph (1902), Tutor Dept. of Zool., Indiana Univ. Bloomington, Ind. (U.S.A.), Biol. Building rm 36. *Taxonomy and Biol. of Beetles (Staphylinidae).* Ⓟ Staphylinidae.

Vorms, Vladimir Aleks., Prof. N. G. Cernysevskij Univ. Saratov (U.d.S.S.R.), Ploscad imeni Lenina. *Biol. Chem.* ○

Voronin, Voldemar, Prof. Univ. Tiflis (U.d.S.S.R.). *Allg. Pathol.* ○

Voronoff, Serge (1866), Dir. du labor. de Chirurgie Expérim. du Coll. de France, Dir. adj. du labor. de Biol. de l'Ecole des Hautes Etudes. 1. Mai à 15. Oct.: Paris (France); 15. Oct. à 1. Mai: Menton (France), Château de Grimaldi. *Glnades endocrines: testicule, ovarie, thyroide, parathyroide, pancreas.* Ⓟ Paris1893.

Voronov, Georges N., Conserv. Jardin Botan. Leningrad (U.d.S.S.R.). *Flore macol. du Caucase.* ○

de **Vos,** Anna Petronella Cornelia (1893), Biol. am Reichsinst. für Fischereiuntersuchungen. Helder (Holland). *Praktische Fischereiuntersuchungen. Biol. und Systematik der aquatischen Insektenlarven.*

Voss, Friedrich (1877), Dr. phil., a.o Prof. an der Univ. Göttingen mit staatl. Lehrauftrag für ,,Landwirtschaftl. Zool. und Landesfauna''. Göttingen (Deutschland). *Morphol. und Kinematik der Insekten. Kinematische Biol., insbesondere Tierflug. Vögel, Vogelbastarde, Systematik der europäischen Gallinsekten, Ameisen, Libellen, Orthopteren.* Fau-

nistik Deutschlands, Hydrobiol., Schädlingskunde. ⓓ Neapel 1904. ⓟ *Biol. Herbar, Pflanzenbeschädigungen, Geradflügler (Orthoptera), Libellen, Ameisen, Gallinsekten, Turbellarien.*

Voss, Hermann (1894), Dr. med., Priv.Doc. Rostock (Deutschland), Anat. Inst. *Traumatische Parthenogenese des Froscheies. Entwicklungsphysiol. des Froscheies. Oxydasereaktion. Nuclealreaktion. Makroskopisch-anat. Technik.* ⓓ Rostock 1919.

Voss, Hermann Emil (1888), Vorstand d. Zool.-histol. Abteilung des Pharm. Inst., Leiter histol. Labor. der Univ.-Nervenklinik. Tartu (Eesti), Univ. *Cytol. der Keimzellen. Experim. Hermaphroditismus, Histol. der Geschlechtsdrüsen, Sexualhormone.* ⓓ Freiburg i. B. 1910.

Vosseler, Julius (1861), Dir. des Zool. Gartens. Hamburg (Deutschland), Tiergartenstr. 1. *Histol., Entomol., Amphipoden, Copepoden, angewandte Zool., Tiergärtnerei.* ⓓ Tübingen 1886.

Votsch, Wilhelm (1879), Dr. phil. Delitzsch (Deutschland), Eilenburger Str. 4II. *Systematische Anat. der Pflanzen. Algen. Süßwasserplankton.* ⓓ Erlangen 1903.

Vouk, Valentin (1886), o. ö. Prof. der Botan. an der philosophischen Fak. der Univ. in Zagreb, Vorstand des Botan. Inst. und Gartens, Sekretär der Südslavischen Akademie der Wissenschaften in Zagreb (S.H.S.), Botan. Inst. der Univ., Mažuranićev trg 29. *Anat. und Physiol. der Pflanzen Morphol. und Biol. der Thallophyten (Myxomyzeten, Cyanophyceen).* ⓓ Wien1 908.

Voukassovitch, Parlé (1893), Biol. de l'inst. central d'Hygiène Beograd. (S.H.S.), Vardarska ul. 17. *Insectes entomophages parasit., Hymenoptera.* ⓓ Toulouse, France, 1924. ⓟ *Insectes entomophages parasites.*

Vrabely, Vera (1898), Ass. Pécs (Ungarn), Chem. Inst. Univ. *Analyse organischer Naturstoffe.* ⓓ Univ. Pécs.

Vraný, Václav, Custode et sécrétaire du Mus. Tisovec (C.S.R.), Sladkovičova ulice 131. *Phanérogames.* o

de Vregille, P., Prof. St. Josephs-Univ. Beirut (Syrie). *Bact.* o

Vrevskij, M. S., Prof. Wiss. Lesshaft-Inst. Leningrad (U.d.S.S.R.), Pr. Maklina 32. *Biochem.* o

Vrgoč, Anton (1881), o. Prof. d. Pharm., Dir. d. pharmakogn. Inst. Univ. Zagreb (S.H.S.). *Pflanzenanat., Trennungsgewebe der Pflanzen, Pflanzenchem. und Pharmakognosie. Pharmakognosie.* ⓓ 1910. ⓟ *Offizinelle Pflanzen des illyrisch-kroatischen Pflanzengebietes.*

De Vries, Ernst (1883), Associate in Neurol. Peking (China), Peking Union Med. Coll. *Pathol. anat. und histol. of the nervous system. Neuroglia bei sekundärer Degeneration grauer Substanz.* ⓓ Zürich 1909. ⓟ *Specimen of pathol. human nervous tissue.*

de Vries, Hugo (1848), Emeritus-Prof. der Botan. Lunteren (Holland). *Erblichkeitslehre (Oenothera Lamarckiana). Die Lehre von den Mutationen.* ⓓ Leiden 1870.

de Vries, Otto (1881), Dir. Proefstation voor Rubber. Buitenzorg, Java (Niederl.-O.-Indien), Tjikeumeuhweg 16. *Hevea bras. Gummi.* ⓓ Leiden 1908.

de Vries, W. M., Prof. Dr., Patholog; Labor. Binnengasthuis. Amsterdam (Holland). o

Vuillemin, M. P., Prof. Fac. Méd. Nancy (France). *Mycol.* o

Vuyck, Laurens (1862), Dir. Mittl. Koloniale Landwirtschaftl. Schule in Deventer (Holland), Schalkhaar (O.). *Niederländische Flora und Hymenopterol.* ⓓ Leiden 1898. ⓟ *Hummeln.*

Waaler, Georg Henrik Magnus (1895), Prosektor. Oslo (Norge), Rikshospitalet. *Erblichkeitsverhältnisse bei den verschiedenen Arten von angeborener Farbenblindheit.*

Wachholder, Kurt (1893), Dr. med., Priv.Doc. der Physiol. Breslau 16 (Deutschland), Elsterweg 23. *Physiol. der willkürlichen Bewegung, Reflexe.* ⓓ Bonn 1920.

Wachs, Horst (1888), Dr., Priv.Doc. und Prof. der Zool. an der Univ. Rostock i. M. (Deutschland), Orleansstr. 9. *Experim. Biol. (Entwicklungsmechanik), Ornithol.* ⓓ Rostock 1914. ⓟ *Mikroskop. Praep., Vögel, Eier.*

Wachter, Willem Hendrik (1882). Rotterdam (Holland), Vierambachtsstraat 81a. *Floristik und Systematik der Phanerogamen und Musci.* ⓓ Utrecht. ⓟ *Salicaceae, Polygonoceae, Amorantaceae, Musci.*

Wacker, Johann (1868), Prof. für landwirtschaftlichen Pflanzenbau. Hohenheim bei Stuttgart (Deutschland), Landw. Hochsch. *Züchtung landw. Kulturpflanzen.* ⓓ Leipzig 1897.

Wada, Kametoshi (1894), Ass. Prof. Pathol. Kanazawa (Japan), 48 Imamachi. *Pathol. Anat., experim. Pathol.*

Waddell, James (1898), Instr. Agric. Chem. Madison, Wis. (U.S.A.), 227 Clifford Court. *Nutrition: Iron metabolism and Vitamin.* ⓓ Univ. of Wisconsin 1926.

Waddell, Susan S., Prof. Shantung Christian Univ. Tsinan (China). *Physiol., Pharm.* o

Wade, Bryan L. (1902). Hillsboro, W.Va. (U.S.A.) *Genetics of corn (maize).*

Wade, Joseph Sanford (1880), Associate Entomol., U. S. Dept. of Agric., Bureau of Entomol. Washington, D.C. (U.S.A.). *Cereal and Forage Insect. Coleoptera Tenebrionidae.*

Wadi, Woldemar (1891), Doc. innere Med., Dir. d. inneren Univ.-Hospitalklinik. Tartu (Estland), Schloßstr. 13. *Stoffwechsel: Wasser- u. Mineralstoffwechsel, Pathol. Physiol.* ⓓ Dorpat 1924.

Wadsworth, Augustus Baldwin (1872), Dir. div. of labor. and research N.Y. State Dept. of Health. Albany, N.Y. (U.S.A.), 327 State St. *Bact. pathol. hygiene.* o

Wächter, Wilhelm (1870), Dr. phil. München (Deutschland), Augustenstr. 64. *Pflanzenphysiol., Arzneipflanzen, Nutzpflanzen.* ⓓ Rostock 1897.

Wächtler, Walter (1901). Leipzig (Deutschland), Eilenburger Str. 31 I. *Anat., Biol. und Systematik der Pulmonaten.*

Wähner, Franz, Dr., Prof. der Geol. und Palaeont. Praha II (C.S.R.), Vincena 3, Geol. Inst. der Deutschen Univ. o

De Waele, Henri P. L. (1876), Prof. de Physiol. à l'Univ. de Gand (Belgique), rue du pain-perdu 13. *Choc anaphylactique et phénomènes concommittants, vasomotrical, coagulation.* ⓓ Gand 1900.

Wätzold, Paul (1875), Dr., Prof.an derUniv. Berlin. Berlin-Lichterfelde (Deutschland), Hortensienstr. Nr. 12A. *Pathol. Anat. des Auges.* ⓓ Leipzig 1902.

von Wagenen, Gertrude (1891), Instr. Anat., Med. School. Berkeley, Cal. (U.S.A.), Anat. Department, Univ. of California. *Endocrinol. Neurol.* ⓓ Univ. Iowa 1920.

Wagener, Willis W. (1892), Ass. Pathol., Bureau of Plant Industry, U. S. Dept. of Agric. San Francisco, Cal. (U.S.A.), Forest Service, Ferry Bldg. *Forest Pathol.* ⓓ Leland Stanford J. Univ. 1918.

Wagenseil, Doc. Univ. Woosung (China). *Anat.* o

Wager, Harold, D.Sc. Leeds (England), 4 Bank View, Chapel Allerton. *Botan. generally. Mycol.*

Wagler, Alexander Erich (1884), Dr. phil., Prof., Leiter des Inst. für Seenforschung und Seenbewirtschaftung in Langenargen am Bodensee (Deutschland). *Süßwasserbiol., Crustaceen (Cladoceren). Meeresbiol.* ⓓ Leipzig 1910. ⓟ *Material aus dem Bodensee.*

Wagner, A. C. Wilhelm (1866), Hymenopterol. am Zool. Staatsinst. u. Zool. Mus. Hamburg. Hamburg-Fuhlsbüttel (Deutschland), Farnstr. 36. *Fauna des Niederelbgebiets. Crabronidengattung Diphlebus.*

Wagner, Adolf, Univ.Prof. Innsbruck (Österreich). *Botan.* o

Wagner, Antoni (1860), Dir. Mus. Polonais d'Histoire Nat. Warszawa (Polska). *Anat., systématique et distribution géographique des Mollusques, terrestres et fluviatiles (Helicinidae, Clausiliidae), Zoogéographie.* ① Vienne 1886.

Wagner, Gerhard, Dr., Priv.Doc., Techn. Hochschule. Danzig. *Hyg. Bact.* o

Wagner, George (1873), Assoc. Prof. of Zool., Univ. of Wisc. Madison, Wis. (U.S.A.), 68, Biol. Bdg. *Vertebrate Zool. Compar. Anat.* ① Michigan 1903. ② Vertebrates.

Wagner, Johann (1870), Dir. der Lehrerbildungsanstalten. Budapest VI (Ungarn), Nagy Jánosgasse 37 III. *Gattung Tilia.* ② Ungarische Phanerogamen. Tilia-, Centaurea-, Pulsatilla- und Quercusarten.

Wagner, Julius (1865), o. Prof. der Belgrader Univ. Beograd (S.H.S.), Entomol. Inst. *Aphaniptera (Suctoria) und Diptera. Entwicklungsgeschichte der Arthropoden. Angewandte Entomol.* ① St. Petersburg 1896. ② Aphanipteren.

Wagner, Karl (1884), Emerit. o. Prof. der Histol. u. vergl. Anat. der litauischen Univ. in Kowno (Litauen), Mickevičiaus g. 3. *Cytol. und Histol. der Geschlechtszellen und -drüsen. Bestimmung des Geschlechtes bei Amphibien. Histol. und Cytol. des Pigmentes und der Pigmentzellen.* ① Leipzig 1909.

Wagner, Nikolaj (1893). Praha (C.S.R.), Pflanzenphysiol. Inst. der Böhmischen Univ. *Pflanzenzytol.*

Wagner, Richard (1893), Priv.Doc. Physiol. Tübingen (Deutschland), Physiol. Inst. *Bewegungsphysiol.: Innervierung antagonistischer Muskeln. Muskelphysiol.: Temperaturkoeffizient der Muskelkraft.* ① München 1919.

Wagner, Wladimir Alexandrowitsch (1849), Prof. Biol., Vergl. Psychol. Leningrad (U.d.S.S.R.), Tschernyschew pereulok, 9, Q. 6. *Zoopsychol. Evolution d. psych. Begabung. Psych. Typen.* ① Moskau 1901.

Wagstaffe, W. W., Ass. Univ. Oxford (England). *Anat.* o

Wahby, Aly (1877), Dr. Prof. Zool. Fac. Sc. et Dir. Inst. de Zool. Kadi Keny-Constantinople (Turquie), 38, Rue Cherki bey. *Toxine. Embryol. Paléont.*

Wahl, Bruno (1876), Dir. der Bundesanstalt für Pflanzenschutz, Priv.-u. Hon.Doc. Univ. Wien II (Deutsch-Österreich), Trunnerstr. 1. *Pflanzenschutz, Zool. Anat. u. Entwicklungsgeschichte der Dipteren. Turbellaria.* ① Wien 1900.

von Wahl, Carl (1869), Vorstand der botan. Abteilung der Badischen Staatlichen Landwirtschaftlichen Versuchsanstalt Augustenberg bei Karlsruhe i. B. (Deutschland). *Samenkontrolle, mikroskop. Futtermittelkontrolle, Hefereinzucht, Landw. Bact., Wiesenbestandsaufnahmen.* ① Berlin 1896.

Wail, Salomon (1898), Med.-Adj. I. Univ. Moskau (U.d.S.S.R.), Pokrowka 29, Log. 26. *Intracellulärer Fettstoffwechsel. Experim. Zytol. Mitochondrien. Strahlenwirkung auf d. Gewebe.* ① Moskau 1920.

Wailes, George Herbert (1862). Vancouver (British Columbia), 4337 Willow St. *Protozoa: Marine and Freshwater Micro-Plankton. Algae.* ② Marine microplankton.

Wainwright, Colbran Joseph (1867). Birmingham (England), Daylesford Handsworth Wood. *Diptera: Tachinidae.* ② Tachinidae.

Waite, Frederick Clayton (1870), Prof. of Histol. and Embryol., Western Reserve Univ., Cleveland, O. (U.S.A.), 2109 Adelbert Road. *Histol. and Embryol.* ① Harvard Univ. 1896 und 1898.

Waitkus, Wladas (1888), Doc. der Landwirtsch. Akademie. Dotnuwa (Litauen). *Forstbotan.*

Wakabayashi, Steich, Secretary of the Japanese Association of North America. Seattle, Wash. (U.S.A.), 216 Fifth Ave., S. *Physiol., genetics.* o

Wakiya, Yōjirō, Dir. Govern. Fishery Experim. Stat. Fusan, Korea (Japan). *Ichthyol.*

Waksman, Selman A. (1888), Prof., Microbiol. New Brunswick, N.J. (U.S.A.), N.J. Agric. Exper. Stat. *Soil micro-flora and micro-fauna, soil fungi and actinomyces, soil bact., sulfur oxydizing forms. Enzymes.* ① Univ. of California 1917. ② Actinomyces, various groups of soil Bact. and Fungi.

Walbum, Ludvig Emil (1879), Adj.-Vorsteher Statens Seruminst. København (Danmark), Frederiksberg Allee 23. *Bact. Toxine und Antitoxine (spez. Diphtherie). Stimulationseinfluß der Metallsalze auf die antitoxinbildenden u. ähnlichen Prozesse.* ② Univ. in Kopenhagen.

Waldo, George, Junior Pomol., U.S. Dept. of Agric. Washington, D.C. (U.S.A.). *Small fruits. Plant anat. Physiol.*

Waldron, Lawrence Root (1875), North Dakota Agric. Experim. Stat. Fargo, N.D. (U.S.A.), State Coll. *Genetics of plants. Production of improved varieties of farm crops.* ② New varieties of farm-crops adapted to climatic conditions of N.D.

Walkden, Herbert H. (1893), Junior Entomol., Bureau of Entomol., U. S. Dept. of Agric. Wichita, Kan. (U.S.A.), 126 S. Minneapolis Ave. *Cereal and Forage Crop Insect.* ① Mass. Agric. Coll. 1916.

Walker, Albert Trincano (1898). Oakland, Cal. (U.S.A.), 413 Siebert St. *Endocrinol.: anterior lobe of the pituitary body.* ① Univ. of California 1926.

Walker, Charles Ralph (1897), Assoc. Prof. Biol. Western State Coll. of Colorado. Gunnison, Col. (U.S.A.). *Parasit.: Cercaria opaca.*

Walker, E. W. A., Lect. Univ. Oxford (England). *Pathol.* o

Walker, Elda R. (1877), Associate Prof. of Botan. Lincoln, Neb. (U.S.A.), The Univ. of Nebraska. *Gametophytes of Equisetum. General Morphol. Algae of Sandhill lakes. Structure of grass pistils.* ① Univ. of Nebraska 1907. ② Gametophytes of Equisetum.

Walker, Hilda (1878), Senior Lect. in Physiol., Univ. of Birmingham (Engl.), 345 Gillott Road, Edgbaston. *Physiol., Physiol. Chem.* ① Birmingham 1918.

Walker, John Charles (1893), Assoc. Prof. Plant Pathol., Univ. of Wisconsin, and Pathol., Office of Vegetable and Forage Diseases, Bureau Plant Industry, U.S. Dept. of Agric. Madison, Wis. (U.S.A.), Coll. of Agric. *Diseases of vegetable crops, of cabbage and onion; breeding of Fusarium resistant varieties of cabbage; genetic studies and nature of disease resistance in cabbage.* ① Univ. of Wisconsin 1918.

Walker, Leva B. (1878), Assoc. Prof. of Botan., Univ. Lincoln, Neb. (U.S.A.). *Morphol. and development of Basidiomycetes: Sphaeropsis, Sphaerobolus, Endogone, Mucoraceas, Pluteus, Tubaria, Cyathus, Crucibulum.* ① Pacific Univ. 1901.

Walker, Marion Newman (1900), Ass. Physiol. Gainesville, Fla. (U.S.A.), Fla. St. Experim. Station. *Physiol., pathol. of cotton. Mosaic diseases of plants: cucumber, Physalis, tomato, Phytolacca, and sugar cane.* ① Auburn, Alabama, 1921.

Walker, Norman, Lect. Univ. Leeds (England). *Plant Morphol.* o

Walker, Ruth, I., Instructor Botan. Madison, Wis. (U.S.A.), Dept. of Botan., Univ. of Wis. *Cytol., Morphol. of Angiosperms, Histol.* ① Univ. Wisconsin 1926.

Wall, Arnold, Prof. M.S. Canterbury Univ. Coll. Christchurch (New Zealand). *Botan.* o

Wall, Sven Daniel (1877), Prof., Dir. Vet.-Bact. Staatsinst. Experimentalfältet Stockholm (Sverige). *Haustierseuchen. Seuchenhafter Abortus beim Rind.* ① Leipzig 1927. ② Sera und Vaccine, Tuberkulin u. Mallein, Bakterienkulturen.

Wallace, George B. (1876), Prof. of Pharm., Univ., and Bellevue Hospital Med. Coll. New York (U.S.A.), 3386 E., 26 St. ① Michigan 1897.

Wallace, Louise Baird, Lect. in Zool. South Hadley, Mass. (U.S.A.), Mount Holyoke Coll. *Cytol., Spermatogenesis and oogenesis of the spider. Axillary gland of Batrachus. Germ ring in the egg.* ① Mount Holyoke Coll. 1919.

Wallengren, Hans F. S. (1864), Univ.Prof., Dir. Zool. Inst. Lund (Sverige). *Infusoria, Allgem. Physiol.* ⑨ Lund 1897.

Waller, Adolph Edward (1892), Ass. Prof. of Botan., The Ohio State Univ., Columbus, O.(U.S.A.). *General and economic botan., ecol., plant geography.* ⑨ Ohio State Univ. 1918.

Wallgren, Axel (1869), o. Prof. Patol. Anat. Univ. Helsingfors (Finnland), Bergmansgatan 11. *Struktur, Morphogenese, Physiol., Pathol. der Zelle.* ⑨ Helsingfors 1899.

Wallin, Ivan E. (1883), Prof., Head Dept. of Anat., Univ. of Colorado, School of Med. Denver, Col. (U.S.A.). *Cytol.: The nature of Mitochondria. Bact. Evolution. Symbionticism. Thymus-like structures in Ammocoetes.* ⑨ New York Univ. 1916.

Wallis, Grace Hite, Head, Dept. of Botan., East St. Louis High School. East St. Louis, Ill. (U.S.A.). *Algae.* ○

Walmsley, T., Prof. Univ. Belfast (Ireland). *Anat.*

De Walsche, Louis (1904). Bruxelles Midi (Belgique), Rue Sallaert 13. *Embryol. et Anat. comparée des Vertébrés et Embryol. expérim.* ⑦ Lépidoptères et Coléoptères.

van Walsem, Gerard Christiaan (1863), Prof., Dir. emer. Pathol. Labor. Leiden. Haarlem (Holland), Julianastraat 47. *Mikroskopische Technik der tierischen (menschlichen pathol.) Morphol. Pathol. Anat. des Centralnervensystems. Morphol. Blutuntersuchung.* ⑨ Leiden 1889.

Walsh, T., Prof. Univ. Coll. Galway (Ireland). *Pathol.* ○

Walter, Emil (1868), Dr. phil., Betriebsleiter der teichwirtschaftlichen Versuchsstation (Hofer-Inst.) in Wielenbach, Oberbayern (Deutschland). *Versuche auf dem Gebiet der Teichdüngung, Fütterung und Rassenkunde der Teichfische.* ⑨ Halle.

Walter, Heinrich (1898), Dr. phil., Priv.Doc. der Botan. an der Univ. Heidelberg (Deutschland), Botan. Inst. *Wasserhaushalt der Pflanzen, Xerophytenproblem, Pflanzengeographie von Deutschland.* ⑨ Jena 1919.

Walter, Herbert Eugene (1867), Prof. of Biol., Brown Univ. Providence, R.I. (U.S.A.). *Genetics. The Human Skeleton. Birds.* ⑨ Harvard 1906.

Walther, Adolf Richard (1885), Dr. phil. et med. vet., Prof. der Tierzuchtlehre. Hohenheim bei Stuttgart (Deutschland). *Vererbungsfragen und Rassebildung bei Haustieren.* ⑨ Med. vet. Leipzig 1910, phil. Gießen 1912. ⑦ Skelette von Haustierrassen.

Walther, Johannes (1860), Geh. R.R., Prof. der Geol. und Paleont. Univ. Halle a. d. S. (Deutschland), Fasanenstr. 4. *Fossile Meerestiere. Bionomie des Meeres.* ⑨ phil. Jena 1882, Sc. Melbourne 1914, med. h. c. Halle a. d. S. 1925.

Walther, Oskar A. (1884), Prof. Landwirtschaftl. Inst., Dir. Pflanzenphysiol. Versuchsstat., Abtl.-Vorstand Station für Akklimatisation der Pflanzen Detskoje Sseló. Leningrad (U.d.S.S.R.), Bolschaja Possadskaja 9, W. 9. *Einfluß der Wasserstoffionenkonzentration auf die Entwicklung u. Mineralstoffaufnahme der Pflanzen. Permeabilität pflanzlichen Plasmas für Wasser und für Elektrolyte. Temperaturabhängigkeit der Enzymwirkung. Physiol. Differenzierung des Gramineenwurzelsystems. Chlorbedürfnis des Buchweizens.*

Walton, John (1895), Lect. in Botan. Manchester (England), The Victoria Univ. *Palaeobotan.: Carboniferous period.* ⑨ Cambridge 1922.

Walton, Lee Barker (1871), Prof. of Biol. Gambier, O. (U.S.A.), Kenyon Coll. *Evolution, Phil. Biol. Euglenoids. Anat. of Insects. Taxonomy of the Endomychidae (Coleoptera). Land Planarians (Nearctic Region).* ⑨ Cornell Univ. 1902.

Walton, William Randolph (1873), Assoc. Chief Bur. Entomol. Washington, D.C. (U.S.A.), U.S. Bureau of Entomol. *Dipterol. Syrphidae. Asilidae. Tachinidae. Muscoideae.* ○

Waly, Mohammed (1887), Lect. in Zool. of the Fac. of Sc. Cairo (Egypt.), Egyptian Univ. ⑨ Med. Cairo 1907, Lic. sc. nat. Lyon 1911.

Walz, Karl (1866), Obermed.R., Vorstand der Prosektur des Katharinenhospitals in Stuttgart (Deutschland). *Geschwulstlehre.* ⑨ Tübingen 1891.

Wang, C. Y., Prof. Univ. Hong-Kong (China). *Pathol.* ○

Wang, Chi Che (1891), Head of Department of Chem., Nelson Morris Memorial Inst. for Med. Research, Michael Reese Hospital. Chicago, Ill. (U.S.A.), 2900 Ellis Ave. *Food Chem., Blood Chem., Energy Metabolism. and Mineral and Nitrogen Metabolism.* ⑨ Univ. of Chicago 1918.

Wang, T. K., Prof. Coll. Shanghai (China). *Botan.* ○

Wangerin, Walther (1884), ao. Prof. der Botan. Mus. für Naturkunde und Vorgeschichte. Danzig-Langfuhr (Freistaat), Kastanienweg 7. *Systematik der Phanerogamen, Pflanzengeographie (speziell Pflanzensoziol. unter besonderer Berücksichtigung von Mitteleuropa, Biol. der Blütenpflanzen.* ⑨ Halle 1906. ⑦ Phanerogamen-Exsikkaten.

Wang Hwei-Wen, Prof. Shantung Christian Univ. Tsinan (China). *Anat.* ○

Wann, Frank Burkett (1892), Associate Prof. of Botan. Logan, Utah (U.S.A.), Utah Agric. Coll. *Plant Physiol. Nutrition of algae. Chlorosis.* ⑨ Cornell Univ. 1920. ⑦ Myxomycetes. Pure cultures of green algae.

Wanner, Johann (1878), Dr. phil., Univ.Prof. Bonn a. Rh. (Deutschland), Geol. Inst., Nußallee 2. *Fossile Echinodermen und Foraminiferen.* ⑨ München 1901.

Warburg, Elsa, Doc. Univ. Upsala (Sverige). *Palaeont.* ○

Warburg, Erik Johan (1892), Reservelaege. København (Danmark), Kommunehospitalet. *Blood, Heart, Kidney and Metabolism. Physico-chem. Biol. of the Blood.* ⑨ Copenhagen 1922.

Warburg, Otto (1859), Dr. phil., ao. Prof. an der Univ. Berlin, Dir. of the Inst. of Agric. Tel-Aviv Palestine, and of the Inst. of Natural History of Palestine, Hebrew Univ. (Jerusalem). Berlin W (Deutschland), Uhlandstr. 175. Tel-Aviv (Palestine). *Botan. (Systematik, Pflanzengeographie), Kolonialbotan. (angewandte) und koloniale Landwirtschaft.* ⑨ Straßburg i. E. 1883.

Warburton, Clyde William (1879), Pres. Am. Soc. Agronomy. Chevy Chase, Md. (U.S.A.), 20. W. Lenox St. *Cereal agronomy.* ○

Ward, Archibald Robinson (1875), Ass. Chief, Dairy Research Division of the Mathews Industries Inc. Detroit, Mich. (U.S.A.), 685 Mullett Street, Box 834. *Bact. of milk.* ⑨ Cornell Univ. 1901.

Ward, Henry Baldwin (1865), Prof. of Zool. and Head of Dept., Univ. of Illinois. Urbana, Ill. (U.S.A.). *Helminthol.; Med. Zool.; Limnobiol.; Parasit. of the Eye.* ⑨ Harvard Univ., Cambridge, Mass. 1892.

Warden, Carl John (1890), Instr. compar. psychol., in charge of the animal labor., dept. of psychol., Columbia Univ. New York City (U.S.A.), 1009 Physics Bldg. *Animal behavior; theoretical zool.; ecol. Mental life of animals; tropisms, instinctive activities, sensory capacities; imitation, learning and „rational" behavior.* ⑨ Univ. of Chicago 1922.

Wardlaw, Claude Wilson (1901), Lect. Botan. Univ. Glasgow (Scotland). *Mycol., and Morphol. of Cryptogams.* ⑨ Glasgow 1925.

Wardle, Robert Arnold (1890), Lect. in Economic Zool., Univ. of Manchester (England). *Entomol. of Cotton Plant.* ⑨ Univ. of Manchester 1912.

Warén, Harry Ilmari (1893), Stellvertreter Prof. der Botan. an der Finnischen Univ. Turku (Finnland), Yliopistonkatu 9. *Moorforschung (Entwicklung und Pflanzengesellschaften der Moore, botan. und chem. Zusammensetzung des Torfes); Nahrungs-*

physiol. der Algen (Flechtengonidien, Desmidiaceen); Sphagnumarten. ⓓ Helsinki 1923. ⓟ Sphagnumarten und Phanerogamen.

Warga, Kálmán, Dr., Dir. der Landwirtsch. Akad. Debreczen (Ungarn), Jallag. *Pflanzenkultur.* ○

Waring, J. H. (1889), Prof. of Horticulture, Univ. of Maine. Orono, Me. (U.S.A.). *Pomol.*

Warington, Katherine (1897), Ass. botan. Harpenden, Herts. (England), Rothamsted Experim. Station. *Plant nutrition, effect of particular elements on growth.* ⓓ London 1920.

Warnecke, Georg (1883), Landgerichtsrat. Altona, Elbe (Deutschland), Moltkestr. 72. *Zoogeographie der palaearktischen Makrolepidopteren, Systematik der palaearktischen Geometriden (Lepidoptera).* ⓓ Palaearktische Geometriden (Lepidoptera).

Warren, Don C. (1890), Poultry Geneticist, Res. Prof. Manhattan, Kan. (U.S.A.), Kansas State Agric. Coll. *Genetics of poultry. Drosophila. Mutatio. Genetic.* ⓓ Columbia Univ. 1923

Warren, E., Prof. Natal Univ. Coll. Pietermaritzburg (South Africa). *Zool.* ○

Warren, Edward Royal (1860) Colorado Springs, Col. (U.S.A.), 1511 Wood Avenue. *Mammal. and Ornithol.* ⓓ Boston, Mass. 1881.

Warren, Herbert Stetson (1887), Ass. Prof. of Zool., Univ. of Idaho. Moscow, Id. (U.S.A.). *Embryol. of the Crustacea.* ⓓ Stanford Univ., California, 1926.

Warren, Howard Crosby (1867), Stuart Prof. Psychol., Univ. Princeton, N.J. (U.S.A.), Eno Hall. ⓓ 1917.

Warren, Paul Alanson, Associate Prof. of biol., Coll. of William and Mary, Prof. of botan. and Head of Dept. of Botan. and Pharm., School of Pharm., Med. Coll. of Virginia, Richmond, Va. Williamsburg, Va. (U.S.A.), 104 Chandler Court. *Genetics.* ○

Warthin, Alfred Scott (1866), Prof. and Dir. of the Pathol. Labor., Med. Dept. Ann Arbor, Mich. (U.S.A.), 1020 Ferdon Road, Univ. of Michigan. *Anat. and pathol. of haemolymph glands, of the blood and blood forming organs, toxic action of mustard gas.* ⓓ Univ. of Michigan 1893.

Warwick, Bruce L. (1893), Ass., Dept. of Animal Industry, Ohio Agric. Experim. Station. Wooster, O. (U.S.A.). *Inheritance of hernia, color and other characters in swine. Inheritance of various characters in sheep. Prevention of Haemonchus contortus and Oesophagostomum of sheep.* ⓓ Univ. of Wisconsin 1925.

Waser, Ernst B. H. (1887), Prof., Abtl.-Vorsteher chem. Inst. Univ. Zürich 7 (Schweiz), 5 Freiestr. *Organische Chem. Arzneimittelsynthese. Biol. wichtige Aminosäuren. Methodik des Wärmestiches. Temperaturmessung mit Thermoelementen.* ⓓ Eidg. Techn. Hochsch. Zürich 1911.

Washburn, Frederic L. (1860), Prof. of Economic Vertebrate Zool., Univ. of Minnesota. Minneapolis, Minn. (U.S.A.). *Biol., Entomol.* ⓓ Harvard 1895. ○

Washburn, Margaret F. (1871), Prof. of Psychol., Vassar Coll. Poughkeepsie, N.Y. (U.S.A.). *Animal Mind.* ⓓ Cornell Univ. 1894. ○

Wasicky, Richard (1884), o.ö. Prof. für Pharmakognosie, Vorstand des pharmakognost. Inst. der Univ. Wien IX (Deutsch-Österreich), Währingerstraße 13a. *Pharm. und Biochem. der Saponine und Saponinpflanzen, Pharm. und Biochem. der Digitalis, Probleme der Arzneipflanzenkultur.* ⓓ Wien 1911.

von Wasielewski, Theodor (1868), o. Prof. für Hygien., Dir. d. Hygien. Inst. u. d. Landes-Lebensmittel-Untersuchungsanstalt Rostock (Deutschland), Buchbinderstr. 9. *Zellschmarotzende Mikroven, Blutzellschmarotzer, Variola-Vakzine-Erreger und berwandte Lebewesen.* ⓓ Berlin 1892. ⓟ Amoebenkulturen, Praep. von Amoebenkernteilungen, Mikrophotogramme.

Wasilewska, Stanislawa (1893), Ass. Lwów (Polen), Inst. Zool. Politechnik. *Cytol.: Golgischer Apparat der Myriapoden.*

Wasmann, Erich (1859), S. I. Aachen (Deutschland), Kurbrunnenstr. 42, oder: Valkenburg, L (Holland), Ignatiuskolleg. *Myrmecophilen und Termitophilen (Gäste der Ameisen und der Termiten): biol. (Anpassungserscheinungen) und systematisch für bestimmte Coleopterenfamilien (Paussiden, Clavigeriden, Staphyliniden), vergl. psychol. und stammesgeschichtlich.* ⓓ Phil. h. c. Freiburg i. d. Schw. 1921.

Wasmund, Erich (1902), Dr. phil., Leiter an der Biol. Station Mooslachen, Wasserburg a. Bodensee (Deutschland). *Limnol., Biogeographie und -soziol. Geol. und Palaeont. der Quartärzeit.* ⓓ 1925.

Waśniewski, Stanislas (1886), Dir. ferme experim. agric., Univ. de Cracovie (Pologne). *Botan. appl., alimentation et élevage de plantes.* ⓓ Cracovie 1914.

Wassermann, Friedrich (1884), Dr. med., a.o. Univ.Prof., Prosektor und Konserv. des anat. Inst. der Univ. München (Deutschland), Pettenkoferstraße 11. *Cytol., Chromosomenfragen und Zellteilung; Histogenese und Histophysiol. des Fettgewebes.* ⓓ München 1909.

Wassilieff, Vladimir (1900), Ass. in Botan. Yalta, Crimea (U.d.S.S.R.), Botan. Garden, Nikita. *Geobotan.: Flora of the Crimea.*

Wassiliewsky, Michael M., Obergeol. am Comité Géol. de Russie. Leningrad W.O. (U.d.S.S.R.), Sredny Pr., 72-B. *Palaeozoicum; devonische Trilobiten, Brachiopoden.* ○

Wassiljewa, Vera (1897), Ass. Kasan (U.d. S.S.R.), Physiol. Inst. d. Univ. *Elektrophysiol.*

Wassilkov, Joseph (1879), Prof. d. Botan. Akad. für Landwirtschaft, Vorsteher des Botan. Gartens d. Akademie. Gorki, Weißrußland (U.d.S.S.R.). *Floristik. Unkräuter. System. Pflanzenanat.* ⓟ Exsiccaten d. weißrussisch. Flora und Samen.

Wassjutotschkin, Artemy (1892), Ass. cytol. Labor. Univ. Leningrad (U.d.S.S.R.), Wass. Ostr. 5, Linie 4. *Histol. d. Muskelfasern.* ⓓ Leningrad 1921.

Wastl, Josef (1892), Dr. phil. Wien VIII (Deutsch-Österreich), Lerchengasse 27. *Rassenanthrop. Untersuchungen an turkotatarischen Völkern (Baschkiren, Krimtataren, Kasantataren, Nogaier, Tschuwaschen, Mischeren, Tipteren). Körperbaustudien an Sportschwimmern.* ⓓ Wien 1925.

Watanabe, Hotori, Ass. Prof. Imper. Univ. Kyôto (Japan), Inst. of Microbiol., Fac. of Med. ○

Watanabe, Kiyohiko (1900), Ass. botan. Inst. d. Kaiserl. Univ. Tôkyô (Japan), Koishikawa. *Pflanzenkrankheiten, verursacht durch Protisten* ⓓ Tôkyô 1924. ⓟ Cyanophyceen, Flagellaten und andere Protist. von Japan.

Watanabe, M. (1890), Ass. Prof. School of Fishery, Hokkaido Imper. Univ. Sapporo (Japan). *Hydrobiol., water animals, fishes.* ⓓ Hokkaido 1912.

Watari, Masami (1897). Tôkyô (Japan), Nakanegishi 100, Shitayaku. *Rhopalocera.* ⓓ Tôkyô 1921. ⓟ Rhopalocera and Grypocera from Japan and Formosa.

Watase, Shôzaburô, Prof. emer. Tôkyô Imper. Univ. Setagaya near Tôkyô (Japan), Sanshuku 111.

Waterhouse, Walter Lawry (1887), Lect. and Demonstrator in Plant Pathol., Plant Breeding and Genetics, and Agric. Botan. Sydney (Australia), Fac. of Agric., the Univ. of Sydney. *Cereal Diseases, especially Rusts, and General Mycol. Cereal Breeding.*

Waterman, Hein Israël (1889), Prof. in der Chem. Technol. an der Technischen Hochsch. Delft (Holland), Nieuwe Plantage 120. *Biochem. Methoden zur Bestimmung von Salicylsäure. Stoffwechsel und Mutationen von Penicillium, Aspergillus, Permeabilität. Physiol. der Essigbakterien. Stickstoffnahrung der Preßhefe. Kartoffeln. Amygdalin als Nahrung niederer Pflanzen.* ⓓ Delft 1913.

Waterman, Warren G., Associate Prof. of Botan., Northwestern Univ. Evanston, Ill. (U.S.A.). *Ecol.* ○

Waters, Charles W., Ass. Prof. of Botan., Miami Univ. Oxford, O. (U.S.A.). ○

Waterston, David (1871), Bute Prof. of Anat. Univ. StAndrews, Fife (Scotland), 2 Howard Place.

Human Anat., and Embryol.; Heart. Physical Anthrop. ⒟ Edinburgh 1895.

Watkins, John H. (1900), Ass. Bact., Engineering Experim. Station, Iowa State Coll. Bact. Dept. Ames, Ia. (U.S.A.), *Bact.*

Watkins-Pitchford, W., Prof. Univ. Johannesburg, Transvaal (S. Africa). *Pathol. and Bact.* o

Watrin, Prof. agr. Histol. Fac. Méd. Nancy (France). *Histol.* o

Watson, Alexander McLeod (1880), Lect. Glasgow (Scotland), Inst. of Physiol. Univ. *Histol.* ⒟ Glasgow 1924.

Watson, Clarence Wilford (1895), Ass. Prof. Moscow, Id. (U.S.A.), School of Forestry, Univ. of Idaho. *Silviculture. Forest ecol.* ⒟ Yale Univ. 1920.

Watson, David Meredith Seares (1886), Jodrell Prof. of Zool. and Compar. Anat., Univ. Coll., Univ. of London W. C. 1 (England). *Vertebrate Morphol., Vertebrate Palaeont.: Reptiles, Amphibia and Fishes.* ⒟ Manchester 1904.

Watson, Joseph Ralph (1874), Entomol. Florida Agric. Experim. Station, Univ. of Florida. Gainesville, Fla. (U.S.A.). *Insects and other animals destructive to plants and animals in Florida, truck crops and citrus. Thysanoptera, classification, control of aphids, aleurodids, and scale insects on citrus; root knot nematodes (Heterodera radicicola) on vegetables.* ⓟ Slides of Thysanoptera mounted in balsam.

Watson, Katharine Margarite (1891), Honorary Research Ass., Dept. of Zool., Univ. Coll. London N.W. 6 (England), 115 Greencroft Gdns. *Vertebrate Embryol.* ⒟ Univ. Coll., London, 1912.

Watt, James Crawford (1886), Associate Prof. of Anat. Toronto (Canada), Dept. of Anat., Univ. *Embryol., monstrosities and various anomalies. Development of bone, ossification. Calcification* ⒟ Univ. of Toronto 1918.

Watt, John Mitchell (1892), Prof., Head of the Dept. of Pharm. Johannesburg (South Africa), Pharm. Labor., Univ. of the Witwatersrand. *Action and utility of South African indigenous med. and poisonous plants.* ⒟ Edinburgh 1916.

Watt, Morris N., Lect. Med. School. Dunedin (N. Zealand). *Entomol.* o

Watts, Ralph L. (1869), Dir. School of Agric. and Expt. Station. State Coll., Pa. (U.S.A.), 215 Foster Av. *Horticulture.* o

Waugh, Frank Albert (1869), Prof. Mass. Agric. Coll. Amherst, Mass. (U.S.A.). *Horticulture.* o

Waugh, Theodore Rogers (1890), Ass. Prof. of Pathol., McGill Univ.; Ass. Pathol., Royal Victoria Hospital, Montreal (Canada), Pathol. Inst., McGill Univ. *Haematol.* ⒟ McGill Univ. 1920.

Waung, Chia Lin (1902), Academy, Head Dept. of Biol. Soochow (China). *Genetica.* ⒟ Soochow Univ. 1925. ⓟ Elementary species of silk-worm.

Wawilow, Nikolaj I., Solonzowo-Meliorat. Inst. Moskau (U.d.S.S.R.), Srednjaja Presnja 15 *Botan.* o

Wawilow, Nikolai Iwanowitsch (1887), Prof., Dir. Inst. für angewandte Botan. Leningrad (U.d.S.S.R.), Herzenstr. 44. *Genetik, angewandte Botan., Pflanzenzüchtung und Agronomie.* ⒟ Moskau 1911.

Wazenko, Aleksej A., Jekaterinoslawer Landwirtschaftl. Versuchsstation. Sinelnikowo, Jekaterinoslawsk. gub. *Selektion d. Weizens.* o

Wearn, Joseph Treloar (1893), Ass. Physican, Thorndike Memorial Labor., Boston City Hospital; Ass. Prof. of Med., Harvard Med. School. Boston, Mass. (U.S.A.). *Physiol. of Coronary Circulation. Concentration of Chlorides in the Glomerular Urine.* ⒟ Harvard Med. School 1917.

Weatherby, Charles A. (1875), Ass., Gray Herbarium of Harvard Univ. Cambridge, Mass. (U.S.A.). *Taxonomy of pteridophytes and phanerogams.* ⒟ Harvard Univ. 1898. ⓟ Herbarium specimens of flowering plants and ferns.

Weatherwax, Paul (1888), Associate Prof. of Botan., Indiana Univ. Bloomington, Ind. (U.S.A.). *Systematic morphol. of Grasses. Zea Mays L.* ⒟ Indiana Univ. 1918. ⓟ Herbarium specimens, seeds, and histol. material of grasses (Gramineae).

Weaver, John E. (1884), Prof. of Plant Ecol., Univ. of Nebraska. Lincoln, Neb. (U.S.A.), Station A. *Ecol., physiol.* ⒟ Univ. of Minnesota 1916.

Weaver, Ralph H. (1903), Instr., Univ. Kentucky. Lexington, Ky. (U.S.A.), 347 Ironsylvania Park. *Bact. of Meat and Meat Products.* ⒟ Michigan State Coll. 1926.

Webb, Robert Alexander (1891), Demonstrator Pathol., Univ. Cambridge (England), 12, Storey's Way. *Cytol. of the Blood, Serum Precipitation Reaction.* ⒟ M. Johns Hopkins 1917, Ph. Cambridge (Engl.) 1925.

Webber, Herbert John (1865), Prof. of Subtropical Horticulture and Dir. of Citrus Experim. Station and Graduate School of Tropical Agric., Univ. of California. Riverside, Cal. (U.S.A.). *Varieties and root-stocks of Citrus. Spermatogenesis and Fecundation of Zamia. Sooty mold of the Orange.* ⒟ Agr. Univ. of Nebraska 1913, phil. Washington Univ. 1900.

Webber, J. Milton (1897), Ass., Botan. Dept. Botan., Univ. of California. Berkeley, Cal. (U.S.A.). *Cytol. and Genetical Studies of Interspecific. Hybrids of the Genus Nicotiana.* ⒟ Coll. of Agric., Univ. of California 1925.

Weber, Amédée (1877), Prof. d'anat. humaine normale à l'Univ. de Genève (Suisse), Ecole de Méd. *Embryol. expérim. Implantations de zygotes sur Batraciens adultes. Développement des membres des Batraciens.* ⒟ Nancy 1917.

Weber, Anna (1852). Eerbeck (Hollande). *Algol. marine.* ⓟ h. c. Utrecht. ⓟ Dasycladiaceae: Bornetella, Neomeris, Acicularia, pour des algues marines des côtes de l'Afrique.

Weber, Anna (1893), Ass. bei ,,Statens plantepatologiske Forsøg". Lyngby (Danmark). *Krankheiten bei Tomaten, Gurken, Obst (Malus, Pirus, Prunus, Ribes, Rubus).*

Weber, Basil V., Comité Géol. de Russie. Leningrad (U.d.S.S.R.), Wassily Ostrow, Sredny Prospekt 72-B. *Palaeont., spez. Palaeozoische Korallen.* o

Weber, Carl Albert (1856), Prof. Dr. phil. Bremen (Deutschland), Friedrich-Wilhelm-Str. 24. *Biol. der Pflanzenformationen, Grasfluren und Moore. Entwicklungsgeschichte der Moore und ihrer Vegetation. Palaeophytol. der Quartaerformation. Landwirtschaftl. Botan.* ⒟ Würzburg 1879.

Weber, Elisabeth (1901), Diplomlandwirt. Berlin-Nikolassee (Deutschland), Lückhoffstr. 19. *Pflanzenschutz.*

Weber, Friedl (1886), Dr., a.o. Prof., Ass. Graz (Österreich), Schubertstr. 53. *Pflanzenphysiol.: Physikalische Chem. des Protoplasmas. Frühtreiben. Physiol. der Stomata.* ⒟ Wien 1910.

Weber, Gertrud-Agnes (1889), Comité Géol. de Russie. Leningrad (U.d.S.S.R.), Sredny Pr. 72-B. *Jura- und Kreide-Echiniden, Kreide-Gastropoden und Pelecipoden. Psychiatrische Familienforschung.* ⓟ Kreide- und Paleocän-Fauna der Krim.

Weber, H., Assistent am Inst. f. Pflanzenkrankheiten. Bonn-Rh.-Poppelsdorf (Deutschland), Nußallee 7. *Sinnesorgane, Mollusken. Entomol.*

Weber, L., Doc. Tierärztl. Akad. Lwów (Polska). *Bienenzucht.*

Weber, Rudolf (1894), Dr. med. et dent., Priv.Doc. Köln a. Rh. (Deutschland), Moltkestr. 137. *Pathol. und vergl. Histol. der Mundhöhle.* ⒟ Med. Würzburg 1919, med. dent. Würzburg 1920.

Weber, Ulrich (1898), Dr. phil., Priv.Doc. für Botan. in Würzburg (Deutschland), Botan. Inst. der Univ. *Reizphysiol., Geotropismus.* ⒟ München 1920.

Weber, Valerian N., Obergeol. Com. Géol. de Russie. Leningrad, W.O. (U.d.S.S.R.), Sredny Pr., 72-B. *Palaeont.; Silurische und Carbonische Trilobiten.* o

Weberman, E. Tallinn (Estland), Kreutzwaldi 15/17. *Fischzool.* o

Webster, Charles, Prof. American Univ. Beirut (Syrie). *Anat. Eye and ear diseases.* o

Webster, Leslie T. (1894), Associate, Dept. of Pathol., Rockefeller Inst. Med.Res. New York (U.S.A.), 66 St. and Avenue A. *Experim. epidemiol.: mode of spread of infectious disease under the controlled conditions of the labor.* ⓓ Johns Hopkins Univ. 1919.

Webster, Dr. Robert L. (1885), Prof., Head Dept. Zool., Washington State Coll. Pullman, Wash. (U.S.A.). *Entomol. Insect Control.* ⓓ Cornell Univ. 1921.

Wechoff, Nikolai (1887), Hauptass. Inst. f. angewandte Botan., Dir. Tula-Akklimatisations-Stations. Hijustim, Tula-Gouvernement (U.d.S.S.R.). *Naturalisation ausländischer Holzpflanzen.* ⓟ Gehölzpflanzen.

Wedekind, Rudolf (1883), Univ.Prof., Dir. paläont. geol. Inst. Marburg (Deutschland), Friedrichspl. 3. *Paläont. d. Invertebraten.* o

Wedenskij, Nikolaj P., Dir. d. Abtlg. f. Naturwissenschaft d. Kubaner Wissenschaftl. Mus. Krasnodar (U.d.S.S.R.), Raschpilewskaja 3. *Botan. Geobotan.* o

Wedgworth, Herman H. (1901), Associate Plant Pathol. of the Mississippi State Plant Board and Experim. Station. A. and M. College, Miss. (U.S.A.). *Irish Potato Scab.*

Weed, Alfred (1902). Dept. Economic Entomol., Univ. of Wisconsin. Madison, Wis. (U.S.A.). *Insect Physiol. and Ecol.* ⓓ Univ. of Wisconsin 1923.

Weed, Lewis Hill (1886), Prof. of Anat. and Dean of Med. Fac., Johns Hopkins Univ. Baltimore, Md. (U.S.A.), Johns Hopkins Med. School. *Anat. of nuclear masses in medulla; decerebrate rigidity; anat. of mammalian meninges; physiol. and anat. of circulation of cerebro-spinal fluid; experim. alteration of brain volume; development of excitatory areas in mammalian cerebral cortex.* ⓓ Johns Hopkins 1912.

Weeks, Andrew Gray (1861). Boston, Mass. (U.S.A.), 15 State St. *Diurnal Lepidoptera.* ⓓ Harvard Univ. Cambridge 1883.

Weese, Asa Orrin (1885), Prof. Zool., Univ. of Oklahoma. Norman, Okla. (U.S.A.). *Animal Ecol., Bio-ecol.* ⓓ Univ. of Illinois 1922.

Weese, Hellmut (1897), Ass. am pharmacol. Inst. der Univ. München (Deutschland), Biedersteiner Str. Nr. 6 II. *Standardisierung von Histamin. Phosphorvergiftung.* ⓓ München 1924.

Weese, Josef (1888), o. Prof. d. Techn. Hochsch., Dir. Botan. Inst. und Inst. für techn. Mikroskopie und organ. Rohstofflehre der Techn. Hochsch. Wien VII/2 (Österreich), Neustiftgasse 36a/13. *Mikroskopie der technisch verwerteten organischen Rohstoffe (speziell Fasern, Samen, Früchte, Kryptogamen) u. der vegetab. Nahungs- u. Genußmittel, mykol. Pflanzenpathol., Pilzsystematik (speziell Ascomyceten), Pflanzenanat., techn. Botan.* ⓓ Graz. ⓟ Pilze.

Weevers, Theodorus (1875), Dr., Prof. der Pflanzenphysiol. Univ. Amsterdam Amersfooit (Holland), *Physiol. des Stoffwechsels der Pflanze.* ⓓ Amsterdam 1902.

Wegelin, Carl (1879), o. Prof. für allgemeine Pathol. und pathol. Anat. an der Univ. Bern (Schweiz), Pathol. Inst. *Pathol. der Schilddrüse, Kropf.* ⓓ Bern 1904.

Weger, Václav (1875), Sektions-Chef Ministerium für Landwirtschaft. Praha XII (C.S.R.), Korunní tř. č. 47. *Tierernährung.*

Wegerko, Jacob (1890), Ass. II. Clinique des mal, internes Univ. Warszawa (Polska). *Etat colloidal du sang; propriétés chim. et physiques du plasma et sérum; échange de hydro-carbones, corps acétoniques.* ⓓ Bern 1913.

Węgłowski, R., Dr. Prof. Lwów (Polska), Hôpital Mil., R. Łyczakowska. *Histol. Embryol.* o

Wegner, Richard Nikolaus (1884), Prof. Dr. med. et phil., Abteilungsvorsteher am Anat. Inst. der Univ., Vorsitzender der Frankfurter Gesellschaft für Anthrop., Ethnol. und Urgeschichte. Frankfurt a. M. (Deutschland), Gartenstr. 95. *Vergl. Anat., Palaeont. und Embryol. der Wirbeltiere.* ⓓ Phil. Breslau 1910, med. München 1913. ⓟ Anat. Praep., Wirbeltierschädel, embryol. Material, mikroskopische Praep.

Wegner, Theodor (1880), o. Prof. Geol. u. Paläont. Münster, Westf. (Deutschland), Pferdegasse 3. *Reptilien der Kreide.* ⓓ Berlin 1905. ⓟ Fossilien der Kreide.

Wehmer, Carl (1858), o. Honorarprof. für Techn. Bact. und Botan., Vorstand des Bact.-Chem. Labor. an der Techn. Hochsch. zu Hannover (Deutschland), Alleestr. 35. *Holzzersetzende Pilze; Aspergillusarten; Pflanzenchem.; Gärungschem.* ⓓ Göttingen 1886. ⓟ Lebende Reinkulturen von Pilzen.

Wehmeyer, Lewis E. (1898), National Research Fellow in Biol. Sc., Harvard Univ. Cambridge, Mass. (U.S.A.), 68 Oxford St. *Taxonomy, Morphol. and Cytol. of the Pyrenomycetes.* ⓓ Univ. of Michigan 1925.

Wehr, Everett E. (1895), Ass. Zool., Univ. of Illinois. Urbana, Ill. (U.S.A.), 301 Natural History. *Parasit.: Tabanidae, Syrphidae of Nebraska.* ⓓ Univ. of California 1921.

Wehrhahn, Heinr. Rudolf (1887), Gartendir. Schorbus, Post Leuthen bei Cottbus (Deutschland). *Geschichte der Gartenpflanzen, Züchtung von Zierpflanzen, Systematische Bearbeitung aller in die Kultur eingeführten Stauden.* ⓟ Lebende Stauden und Samen, Neuzüchtungen und Neueinführungen.

Wehrle, Lawrence Paul (1887), Instr. Entomol. Cornell Univ. Ithaca, N.Y. (U.S.A.). ⓓ Kansas State Agric. Coll. 1914.

Wehtra, Nikolai (1897), Ass. an d. Katheder d. Med. Mikrobiol., Univ. Lettland. Riga (Lettland), Terbatas iel. 57/61, dz. 14. *Kultivierung der Tuberkelbazillen. Bact. Untersuchung des Blutes. Epidemiol. des Abdominaltyphus und Paratyphus.* ⓓ Riga 1923.

Weichardt, Wolfgang (1875), Univ.Prof. Erlangen (Deutschland), Löwenichstr. 24. *Hygiene, Bact., experim. Therapie.* ⓓ Breslau 1900. ⓟ Mikroorganismen und biologisch-chem. Praep.

Weidemann, Gunnar (1898), Amanuensis. Oslo (Norge), Inst. of Pharmacol. *Vitamin A and D.*

Weidenreich, Franz (1873), Dr., Prof. an der Univ. Heidelberg. Leiter des Biochem. Inst. der von Portheim-Stiftung in Heidelberg. Mannheim (Deutschland), P 7, 21. *Allgemeine Biol. und Abstammungslehre, Morphol. Anthrop., Konstitutionslehre, Lymphgefäßsystem, Histol. des Blutes, des Bindegewebes, des Knochens und der Zähne, einschließlich der Histogenese.* ⓓ Straßburg 1899. ⓟ Mikrophotographien (Diapositive) histol. Praep. von Knochen- und Zahngewebe.

Weidinger, Philipp (1870), Reg.R., Leiter der techn. U.-Abt. der Landesanstalt für Pflanzenbau u.-schutz. München (Deutschland), Liebigstr. 25. *Bekämpfung landwirtschaftlicher tierischer und bakterieller Schädlinge.*

Weidman, Frederick Deforest (1881), Prof. Dermatol. Research, Univ. of Pennsylvania. Philadelphia, Pa. (U.S.A.), Medical Hall. *Mycol., Animal Parasit.* ⓓ Univ. of Pennsylvania 1908. ⓟ Cultures of Pathogenic Fungi, Histol. sections of dermatoses.

Weidman, Robert H. (1886), Dir. N. Rocky Mountain Forest Experim. Sta. Missoula, Mtn. (U.S.A.). *Silricult.* ⓓ Ann Arbor 1914.

Weigel, Charles Adolph (1887), Entomol., Greenhouse Insect Investigations. Bureau of Entomol., U. S. Department of Agric. Washington, D.C. (U.S.A.). *Insect pests in greenhouses.* ⓓ Ohio State Univ. 1925.

Weigel, Theodor Oswald (1884), Dr. h. c. Leipzig (Deutschland), Königstraße 1. *Systematische Botan.* ⓓ h. c. Gießen 1921. ⓟ Pflanzen in Herbarform.

Weigelt, Johannes (1890), a.o. Prof. für Geol. und Palaeont. an der Univ. Halle a. d. S. (Deutschland), Bernburger Str. *Strand- und Küstenbiol., Palaeont.* ⓓ Halle a. d. S. 1913.

Weigert, Josef (1888), Leiter der Abteilung für Pflanzenbau, R.R. München 22 (Deutschland), Landesanstalt für Pflanzenbau u. Pflanzenschutz. *Bodenkunde u. Pflanzenernährung.*

Weigl, Rudolf Stefan (1883), o. Prof. med. Fac. Univ. Lwów (Polska), Nikołaja 4. *Cytol , Experim. Biol. Mikrobiol.* ⓓ Lemberg 1906. ⓟ Zucht normaler Läuse, Rikketsia Provazekii (Fleckfiebervirus).

Weigner, Karl (1874), o. Prof. der Anat. an der Karls-Univ. zu Praha-II-Budovy (C.S.R.), lék. fakulty 1660. *Topographische Anat., Anat. des periphären Nervensystems.* ⓓ Praha 1899.

Weigold, Max Hugo (1886), Dr. rer. nat., Dir. der Naturwissenschaftl. Abteilung des Provinzialmus. zu Hannover (Deutschland), R.-v.-Bennigsenstr. 1. *Ornithol., besonders Vogelzugsforschung. Faunistische und zoogeographische Erforschung Westchinas und Südosttibets. Fauna Hannovers. Museumskunde.* ⓓ Leipzig 1909. ⓟ Tier- und Pflanzenpraep., Käfer.

Weil, Arthur (1887). New York, N.Y. (U.S.A.), 130 Gun Hill Road. *Endocrinol. Neuropathol. Proteine der Tierwelt. Darstellung der Aminosäuren.* ⓓ Berlin 1914

Weil, Konrad (1893), Dr., Stellvertretendes Vorstandsmitglied der Firma C. A. F. Kahlbaum, Chem. Fabrik G. m. b. H., Berlin-Adlershof (Deutschland), Glienicker Weg 15. *Innere Sekretion.* ⓓ Göttingen 1921.

Weiler, Wilhelm (1890). Worms-Pfifflighelm (Deutschland), Westendstr. 81. *Fossile Fische.* ⓓ Gießen 1918. ⓟ Fossilien aus dem Mainzer Becken.

Weill, Robert (1902), Ass. Stat. Zool. Strasbourg (France), 11, Quai Koch. *Invertébrés marins.*

Weimer, James Le Roy (1887), Assoc. Pathol., U.S. Dept. of Agric. Manhattan, Kan. (U.S.A.), Dept. of Botan., Kansas State Agric. Coll. *Forage Crop Diseases* ⓓ Cornell Univ. 1916.

Weinard, Frederick Francis (1893), Assoc. Floric. Physiol., Univ. Illinois Coll. of Agric., Exp. Stat. Urbana, Ill. (U.S.A.), 203 Floriculture Bldg. *Physiol., Pathol.* ⓓ Univ. Illinois, 1922.

Weinberg, Ernst (1896), Dr. med., Ass. of the Neurol. Clinic, Univ. at Tartu (Eesti). *Neurol. Pathol. Anat. of the Nervous System.* ⓓ Microscopical preparations.

Weinberg, Michel (1868), Prof. à l'Inst. Pasteur de Paris (France), 25, rue Dutot. *Anat. pathol., Helminthiol., Hématol. Bact.: étude des Microbes anaérobies et de leur rôle en Pathol. générale.* ⓓ Paris 1921.

Weinberg, Wilhelm (1862), San.R., Labor. für Statistik und Kataster der württembergischen Geisteskranken. Stuttgart (Deutschland), Rotebühlstraße 51. *Biol. und med. Statistik.* ⓓ München 1886.

Weiner, Peter Arkadiewitsch (1904), Alt-Peterhofsches Naturw. Inst., Ass. am Pädagog. Inst. Leningrad (U.d.S.S.R.), Histol Labor. der Univ. *Cytol.: Eizellen (Plasmabestandteile), Fettbildung und Fettresorption. Regeneration des Epithels.* ⓓ Leningrad 1924.

Weinland, Ernst Friedrich (1869), Dr. phil. (zool.) et med., o. Prof. der Physiol. und Dir. des Physiol. Inst. der Univ. Erlangen (Deutschland), Bohlenplatz 21. *Chem. Physiol., Vergl. Physiol., Sinnesphysiol.* ⓓ Phil. Berlin, med. Leipzig.

Weir, James Robert (1882), Pathol. in Charge of Mycol. Collections. Bureau of Plant Industry, U. S. Dept. of Agric. Washington, D.C. (U.S.A.). *Fiber and oil plants, cacao, rubber, timber trees; ecol., distribution and taxonomy of tropical fungi.* ⓓ Univ. München 1911. ⓟ Loranthaceae, fungi.

Weis, Frederick Anton (1871), Prof. of physiol. of plants at the Royal Vet. and Agric. Coll. København (Danmark), Rolighedsvej 23. *Physiol. of plants, microbiol., biol. and biochem. of soils.* ⓓ Kopenhagen 1902.

Weis, Johann Alfred (1897), Ass. Leipzig C 1 (Deutschland), Nostitzstr. 43 II. *Botan. Physikalische Chem. d. Zelle. Plasmahaut, Permeabilität.* ⓓ Leipzig 1924.

Weisbach, Walter (1889), Dr. med., a.o. Prof. der Hygiene an der Univ. Halle a. D. S., wissenschaftl. Dir. des Deutschen Hygien. Mus. Dresden-N. 6 (Deutschland), Bautzner Str. 14. *Serol. der Lues; soziale Hygiene und Gewerbehygiene; pathogene Protozoen.* ⓓ Freiburg i. Br. 1913.

Weiser, Stefan, Dr., Priv.Doc., Ung. Tierärztl. Hochsch. Budapest (Ungarn), Rottenbillergasse. *Chem. d. Lebensmittel.*

Weiss, Charles (1894), Associate in Bact., Columbia Univ., School of Tropical Med. (in Porto Rico) and Dir. of Labor. of the Presbyterian Hospital. San Juan (Porto Rico). *Chemotherapy of filariasis and on enzymes of pleural exsudates.* ⓓ Phil. Univ. of Pennsylvania 1918, med. 1924.

Weiss, Emil (1893), Ass. Prof. Bact. and Pathol., Loyola Univ., School of Med. Chicago, Ill. (U.S.A.), 706 S. Lincoln Street. *Immunol., Bact. and Pathol.* ⓓ Praha (C.S.R.) 1919.

Weiss, Frederick Ernest (1865), Prof. der Botan. an der Univ. Manchester. Disley, Cheshire (England), Easedale. *Palaeobotan. des Carbons. Genetica, Pfropfbastarde.* ⓓ London 1888.

Weiss, Freeman (1892), Associate Pathol., U. S. Dept. Agric. Washington, D.C. (U.S.A.). *Diseases of potatoes, of greenhouse and garden plants. Physiol. of Genus Fusarium.* ⓓ Cornell 1923.

Weiss, Jules Adolphe Georges (1859), Prof. de Physique biol. Strasbourg (France), 10, Rue St. Elisabeth.

Weiss, Harry B. (1883), Chief, Bureau of Statistics and Inspection, New Jersey State, Department of Agric. New Brunswick, N.J. (U.S.A.), 19 North 7 Ave. High and Park. *Greenhouse insects, food habits of insects, Insect distribution, Fungous insects.*

Weiß, Otto (1871), Pro. Prof. Physiol. Univ. Königsberg i. Pr. (Deutschland), Copernicusstr. 1/2. *Sinnesphysiol. Nerven- und Muskelphysiol. Stimme und Sprache.* ⓓ Göttingen 1896.

Weiss, Paul Alfred (1898), Adjunkt d. Biol. Versuchsanstalt d. Akademie der Wissenschaften. Wien XIX (Deutsch-Österreich), Strassergasse 13. *Regeneration und Transplantation bei Amphibien und Reptilien. „Feldtheorie" der Entwicklung. „Resonanztheorie" der motorischen Nerventätigkeit.* ⓓ Wien 1922.

Weisse, Arthur (1861), Prof. Dr. phil., Wissenschaftl. Mitarbeiter am Botan. Mus. in Berlin-Dahlem. Berlin-Steglitz (Deutschland), Sachsenwaldstr. 30 II. *Botan. Morphol., insbesondere Phyllotaxis und Teratol.* ⓓ Berlin 1889.

Weissenberg, Richard (1882), Dr. med., nichtbeamteter a.o. Prof., Ass. am anat. biol. Inst. der Univ. Berlin NW 6 (Deutschland), Luisenstr. 56. *Parasit. Protozoen, insbesondere Microsporidien und Zellparasit. Petromyzonten. Lokale Vitalfärbung von Embryonen. Parasit. Hymenopteren.* ⓓ 1907.

Weitz, Wilhelm (1881), Dr. med., Univ.Prof. Tübingen (Deutschland), Wildermuthstr. *Herz; Vererbung; Verdauungsorgane.* ⓓ Kiel 1905.

Welzsäcker, Viktor, Freiherr v. (1886), a.o. Prof. an der Univ. Heidelberg, Leiter der Nervenabteilung. Heidelberg (Deutschland), Plöck 68. *Physiol. und Pathol. der Muskeln, des Herzens, der Sinne, insbes. der Hautsinne. Koordination des menschlichen Bewegungsapparates. Allgemeine physiol. und Klinik der Nervenkrankheiten und Psychotherapie.* ⓓ Heidelberg 1910.

Welch, Donald Stuart (1894), Ass. Prof. Plant Pathol. Ithaca, N.Y. (U.S.A.), N.Y. State Coll. of Agric. *Plant and Forest Pathol.* ⓓ Cornell 1925.

Welch, Paul S. (1882), Associate Prof. of Zool., Department of Zool., Univ. of Michigan. Ann Arbor,

Mich. (U.S.A.) *Biol. of Aquatic Insects; Respiration in Aquatic Insects; Parasit. of Insects. Limnol. Survey of Lakes; Physical, Chemical and Biol. Factors in luencingbiol. Productivity in Lakes; Thermal and Chem. Stratification in Lakes.* ⑨ Univ. of Illinois 1913.

Welch, William Henry (1871), Dir. School of Hygiene and Public Health John Hopkins. Baltimore, Md. (U.S.A.), 807 St. Paul St. *Pathol. anat. Bact.* ⑨ Chic. Vet. Coll. 1892. o

Weleminsky, Friedrich (1868), Dr., Priv.Doc., Abteilungsvorstand am Hygien. Inst. der deutschen Univ. Prag-Smichow (C.S.R.), Zborovskà 58. *Experim. Tuberkulose.* ⑨ Deutsche Univ. Prag 1893.

Welker, William Henry (1879), Prof. of Physiol. Chem., Coll. of Med., Univ. of Illinois. Chicago, Ill. (U.S.A.), 1817 W. Polk St. ⑨ Columbia Univ. 1908.

Wellensiek, Susan Jacobus (1899), Plant Pathol. at the Agric. Coll. Wageningen (Holland). *Diseaseresistance in Plants. Genetics of immunity from wart in potatoes. Corn-rust. Genetics of Pisum.* ⑨ Wageningen (Holland) 1924. ⑨ Pure culture of Gloeosporium Caulivorum Kirchner. Lines of peas.

Weller, D. M., Histol. Honolulu (Hawaii), Exper. Station, H.S.P.A. *"Root-rot" of Sugar Cane: fungi and bact. Histol. of the Eye spot disease.* ⑨ Microscope slides of sugar cane tissue.

Wellhouse, Walter Housley (1890), Assoc. Prof. Entomol. Ames, Ia. (U.S.A.), Iowa State Coll. *Insects infesting wild pomaceous fruits.* ⑨ Cornell Univ. 1920.

Wellington, Richard (1884), Associate in Research (Horticulture). Geneva, N.Y. (U.S.A.). *Plant Breeding: Grapes, Plums, Cherries.* ⑨ Harvard Univ., Cambridge, Mass., 1911.

Wellman, Frederick L., Ass. Plant Pathol. Univ. of Wisc. Madison, Wis. (U.S.A.). o

Wellmann, Oskar (1876), o. ö. Prof. an der Tierärztl. Hochsch., Dir. des Zootechn. Inst. Budapest VII (Ungarn), Rottenbiller-ut 23. *Energie- und Stoffwechselumsatz, biol. Blutuntersuchungen, Kreuzungsstudium an kleinen Tieren* (*Hunde, Kaninchen, Geflügel*). ⑨ 1907.

Wells, Bertram Whittier (1884), Prof. of Botan., Head of Department, North Carolina State Coll. Raleigh, N.C. (U.S.A.). *Cecidol.* ⑨ Chicago 1917.

Wells, Harry Gideon (1875), Prof. and Chairman of the Department of Pathol., Dir. of the Otho S.A. Sprague Memorial Inst. Univ. of Chicago, Ill. (U.S.A.), *General pathol., its chem. aspects. Tuberculosis.* ⑨ Med. Rush Medical College 1898, phil. Univ. of Chicago 1903.

Wells, Howard Mitchell (1896), Superintendent Graham Horticultural Experim. Station. Grand Rapids, Mich. (U.S.A.). *Pruning, cultivation, fertilization.*

Welo, Lars A. (1888), Physicist, Rockefeller Inst. for Med. Research. New Rork City (U.S.A.), 66 St. and Avenue A. *Biol. action of Iron, particulary of the oxides and the complex salts. Magnetic Properties of Complex Salts of Iron. Catalytic action of Iron Oxides.* ⑨ Univ. of California 1918.

Welsh, D. A., Prof. Univ. Sydney (Australia). *Pathol.* o

Welsh, Mark Frederick (1895), Ass. Prof. of Bact. and Pathol. College Park, Md. (U.S.A.). *Pathogenic Bact.*

Wendnagel, Ad., Dir. d. zool. Gartens. Basel (Schweiz). o

von Wendt, Georg Willehad (1876), o. Prof. Univ., Dir. Inst. für Haustierlehre. Helsinki (Finland), Regerignsgatan 13. *Stoffwechsel und Ernährung: Mineralstoffwechsel, Immunität und Stoffwechsel. Vitamine.* ⑨ Med. u. chir. Helsingfors 1905, phil. Leipzig 1908.

Wendtlandt, Willi (1884), Dr., Marine-Oberstabsarzt, Vorstand der Hygien. Untersuchungsstation des Sanitätsamts. Wilhelmshaven (Deutschland). *Bact., Serol.* ⑨ Berlin 1909.

Wenig, Jaromir (1877), Prof. de Zool. à l'Univ. Praha II (C.S.R.), u. Karlova 3. *Embryol. et organogénèse des Vertébrés.* ⑨ Prague 1901.

Wenrich, David Henry (1885), Ass. Prof. of Zool. Zool. Labor. ,Univ. of Pennsylvania. Philadelphia, Pa. (U.S.A.). *Protozool.: Protozoa of Tadpoles; Flagellates of Rats and Mice: Flagellates of Amphibia. Spermatogenesis of Phrynotettix. Reaction of Bivalve Mollusks to Changes in light intensity. Synapsis and Chromosomes. Organis. in Chorthippus.* ⑨ Harvard Univ. 1915.

Wenstrup, Edward (1894), Prof. Biol. Beatty, Pa. (U.S.A.), St. Vincent Coll. *Lepidoptera.* ⑨ St. Vincent Coll. Beatty 1893. ⑨ Lepidoptera.

Went, F. A. F. C., Prof. Dr., Dir. Botan. Labor., Praes. d. Niederl. Akad. d. Wissensch. Utrecht (Holland), Lange Nieuwstr. *Pflanzenphysiol., Enzyme, Phototropismus. Podostemonaceae. Phytopath.* ⑨ Marburg a. d. L. 1909. ⑨ Fossile Land- und Süßwassermollusken.

Wentz, John B. (1891), Assoc. Prof. Farm Crops. Ames, Ia. (U.S.A.), Iowa State Coll. *Crop breeding.* ⑨ Cornell Univ. 1916.

Wenyon, Charles Morley, Dir.-in-Chief Wellcome Bureau of Sc., Research. London, W.C. 1 (England), 25 28, Endsleigh Gardens. *Protozool. Human Intestinal Protozoa.* ⑨ Univ. of London 1905.

Wenz, Wilhelm (1886), Dr. phil. Frankfurt a. M. (Deutschland), Gwinnerstr. 19. *Mollusken, besonders Land- und Süßwassermollusken, rezent und fossil.* ⑨ Marburg a. d. L. 1909. ⑨ Fossile Land- und Süßwassermollusken.

Wepfer, Emil (1883), Prof.Dr.phil., Württemberg. Landesgeologe. Stuttgart (Deutschland), Hackländer Str. 43I. *Palaeont., Trias-Stegocephalen.* ⑨ Königsberg i. Pr. 1908.

Wercholanzewa, Maria P., Doc. II. Univ. Moskau (U.d.S.S.R.), Chlebnij per. 9, W. 12. *Botan., Physiol., Anat. d. Pflanzen, Microchem.* o

Wercholanzewa, Nadeshda P., Doc. I. Univ. Moskau (U.d.S.S.R.), Chlebnij per. 9, W. 12. *Botan. Microbiol.*

Wereschagin,Viktor I. (1874),Wissensch. Mitarbeiter d. Altajer Staatl. Mus. Barnaul (U.d.S.S.R.), Puschinskaja 51. *Botan. Floristik des Altai.*

Wereschtschagina, Warwara (1895), Ass. Wologda (U.d.S.S.R.), Milchwirtschaftl. Inst. *Mikrobiol. des Käses und der Butter. Milchsäure-Bakterien.*

Werestschagin, Iléb (1889), Kustos Zool. Mus. d. Akademie d. Wissenschaften. Leningrad (U.d. S.S.R.), Wasil. Ostrow. Sredny Prospekt 48; 25. *Limnol. Hydrobiol. Cladoceren.* ⑨ Warschau 1913.

Werkman, Chester H. (1893), Ass. Prof., Bact. Iowa State Coll. Ames, Ia. (U.S.A.). ⑨ Iowa State. Coll. 1923.

Wermel, Eugen (1904), Forschungs-Inst. für Zool. I. Univ. Physiko-Mathemat. Fac. Moskau 19 (U.d. S.S.R.), Wolchonka, B. Znamensky 4, 9. *Zytol. des Zellkernes. Histol. der Schilddrüse. Hydra. Flageaten. Actinosphaerium.* ⑨ Moskau 1925.

Werner, Bruno Clemens Fritz (1896), Ass. an der Univ.-Ohrenklinik im Eppendorfer Krankenhaus. Hamburg 20 (Deutschland), Hellwigstr. 108. *Labyrinth der Amphibia und Pisces. Cladoceren.* ⑨ Leipzig 1924.

Werner, F. Felix (1885), Dr. med., Leiter des Forschungsinst. Bad Mergentheim (Deutschland), Edelfinger Str. 14. *Physiol. und physiol. Chem. Pharm. Physiol. und pharm. Studien an der Atmung der Kaltblüter. Druckpuls der Arter. car. und Elektrokardiogramm. Moschus. Calciumwirkung am isolierten Krötenherzen.* ⑨ Gießen 1919.

Werner, Franz (1867), Prof. Zool. Univ. Wien V (Österreich), Margaretenhof 12. *Systematik und Ethol. der Reptilien uns Amphibien. Orthopteren* (*spez. Mantiden*). *Skorpione und Solifugen. Tiergeographie spez. Fauna Oesterreichs, der Balkanhalbinsel, Kleinasiens, Nord- u. Nordostafrikas.* ⑨ Wien 1890.

Werner, Harvey O. (1893), Associate Horticulturist, Univ. of Nebraska. Lincoln, Neb. (U.S.A.).

Cultural and physiol. lines with the Potato. *Olericulture.* Ⓓ Univ. of Nebraska 1923.
Werth, Emil (1869), Prof. Dr., Reg.R., Mitglied der Biol. Reichsanstalt in Berlin-Dahlem, Labor.-Vorsteher des Labor. für Meteorol. und Phänol. der Biol. Reichsanstalt. Berlin-Wilmersdorf (Deutschland), Bingerstr. 17. *Phytopath., Pflanzengeographie und -oekol., Phänol., Palaeont. (des Menschen).* Ⓓ Bern 1900.
Wertheimer, Ernst (1893), Dr., Priv.Doc., Ass. am Physiol. Inst. Halle a. d. Saale (Deutschland), Magdeburger Str. 36. *Physiol. Ernährungs- und Stoffwechselphysiol., Zellpermeabilität.* Ⓓ Heidelberg 1920.
Werthemann, Andreas (1897), Dr. med., Prosektor und I. Ass. am Pathol. Inst. Basel (Schweiz), Hebelstraße 24. *Locus minoris resistentiae. Histol. d. kleinsten Arterien.* Ⓓ 1921.
Wesenberg, Georg (1871), Dr., Leiter des Bact. Labor. der I. G. Farbenindustrie A.-G., Werk Elberfeld (Deutschland), Müllerstr. 137. *Med. und technische Bact.* Ⓓ Marburg a. d. L. 1924.
Wesenberg-Lund, Carl Jörgen, Prof. Univ. Copenhagen, Dir. of the Freshwater biol. labor. of the Univ. København. Villa alba Hilleröd (Danmark). *Limnol. Culicidae, Zoothamnium, Rotifera and Crustacea. Insects. Planctoninvestigations.* Ⓓ Copenhagen 1899. Ⓟ Freshwater fauna.
Wesselkina, Walentine (1879), Abtlg. d. experim. Pathol. des wissensch. Inst. Lesshaft, Ass. Med. Hochsch. Leningrad (U.d.S.S.R.), Torgowaja 25-A, W. 40. *Stoffwechsel nach Leberexstirpation.* Ⓓ Med. Hochsch. f. Frauen S. Petersburg 1910.
West, Cecil McLaren (1893), Univ. Anat. School of Anat. Trinity Coll., Dublin (Ireland). *Human Anat. and Embryol.* Ⓓ Dublin 1915.
West, Cyril (1887), Research in plant physiol. under the Food Investigation Board of the Department of Sc. and Industrial Research. Cambridge (England), Low Temperature Research Station, Downing Street. *Plant growth and senescence; Seed germination and dormancy; Fruit and vegetable storage.* Ⓓ London 1917.
West, Edward Staunton (1896), Ass. Prof. Washington Univ., School of Med. St. Louis, Miss. (U.S.A.). *Condensation products of acetoacetic ester and acid, preparation and metabolism in the animal body. Oxidation-Reduction potentials of organic compounds.* Ⓓ Univ. of Chicago 1923. Ⓓ Condensation products of glucose and acetoacetic ester.
West, Luther Shirley (1899), Prof. of Biol. and Eugenics. Battle Creek Coll. (Dept. of Biol.). Battle Creek, Mich. (U.S.A.). *Invertebrate Zool., Parasit., Med. Entomol. Systematic entomol., Vertebrate Histol.* Ⓓ Cornell Univ. 1925.
Westberg, Paul (1862). Riga (Lettland), Friedenstraße 47, W. 5. *Netz der Webespinnen.*
Westcott, Cynthia (1898), Heckscher Research Ass. Plant Pathol. Ithaca, N.Y. (U.S.A.), Cornell Univ. *Sclerotinia and Botrytis.*
Westenhöfer, Max (1871), Dr., ao. Prof. und Kustos des Pathol. Mus. der Univ. Berlin. Zepernick b. Berlin (Deutschland), Post Röntgental. *Pathol. Anat. und Anthrop.* Ⓓ Berlin 1894.
Wester, Jurjen (1869), Dr. méd., Prof. an der Vet.-med. Fac. Utrecht (Holland), Frederik Hendrikstraat 86. *Tierheilkunde. Eierstock und Ei; Befruchtung und Unfruchtbarkeit bei den Haustieren. Physiol. und Pathol. der Vormägen beim Rinde.* Ⓓ Utrecht 1923.
Wester, Peter Johnson (1877). Ballston, Va. (U.S.A.), Box 204. *Economic botan., the culture, propagation and improvement of tropical crops.* Ⓓ Gefleborgs Lans Folkhogskola 1895.
Westerdijk, Johanna (1883), Prof. Dr., Dir. Phytopath. Labor. Willie Commelin Scholten, Baarn, ao. Prof. Univ. Utrecht (Holland). *Pflanzenpathol., Mycol.* Ⓓ Zürich 1906. Ⓟ Pilzkulturen.

Westerlund, C. A. S., Doc. Univ. Lund (Sverige). *Anat.* o
Weston, Hubert Claude (1894), Investigator, Industrial Fatigue Research Board, Med. Research Council. Orpington, Kent. (England), Inverclyde, The Avenue. *Fatigue and the efficiency of the worker.*
Weston, William H. jr., Ass. Prof. of Cryptogamic Botan., Harvard Univ. Cambridge 38, Mass. (U.S.A.), 20 Divinity Ave. *Mycol., plant pathol.*
Westveld, Marinus (1889), Associate Silviculturist. Northeastern Forest Experim. Station. Amherst, Mass. (U.S.A.).
Wetmore, Alexander (1886), Ass. Secretary, Smithsonian Inst., with direction of the U. S. National Mus., and oversight over the National Zool. Park and the National Gallery of Art. Washington, D.C. (U.S.A.). *Ornithol.: systematic, anat., osteol. and paleont. lines.* Ⓓ George Washington Univ. 1920.
Wetmore, Ralph Hartley (1892), Ass. Prof. of Botan., Harvard Univ. Cambridge, Mass. (U.S.A.), Botan. Labor., Univ. Mus., Oxford St. *Plant Morphol., Cytol. of Hybrids, cytol. of Aster and Solidago.* Ⓓ Harvard Univ. 1924.
Wetochin, Ivan (1884), Priv.Doc., erster Ass. Kasan (U.d.S.S.R.), Physiol. Labor. Univ. *Sympathisches Nerven-System. Vergl. Physiol.* Ⓓ Kasan.
von Wettstein, Fritz (1895), o. Univ. Prof., Dir. d. Botan. Inst. d. Univ. Göttingen (Deutschland), Nikolausberger Weg 18. *Heteroploidie, Geschlechtsbestimmung, Entwicklungsphysiol. auf genetischer Grundlage.* Ⓓ Wien 1919.
Wettstein-Westersheim, Otto (1892), Kustos am Naturhistorischen Mus., Zool. Abt., Leiter der Herpetol. Sammlung. Wien I (Deutsch-Österreich), Burgring 7. *Systematik der Lacertilia. Wirbeltierfauna Österreichs. Faunistik der europäischen rezenten und fossilen Vertebraten, besonders Micromammalia. Eiszeitforschung, Spelaeol. Artenwandel auf Inseln, bes. adriatische Insel-Eidechsen.* Ⓓ Wien 1915. Ⓟ Europäische Micromammalia.
Wettstein-Westersheim, Richard (1863), Prof. der Botan. an der Univ., Dir. des botan. Gartens und Inst. Wien III (Deutsch-Österreich), Rennweg 14. *Phylogenie des Pflanzenreiches. Systematik der Anthophyten. Spezielle Untersuchungen über Sempervivum, Euphrasia, Gentiana.* Ⓓ Wien 1884. Ⓟ Lebende sukkulente Pflanzen (Kakteen, Mesembryanthemum, Sempervivum usw.).
Wetzel, Arno Willy (1890), Dr. phil., Ass. Leipzig (Deutschland), Zool. Inst., Talstr. 33. *Ciliata, Ökol. und Morphol.* Ⓓ Leipzig 1924.
Wetzel, Georg (1871), o. Prof. Greifswald (Deutschland), Steinstr. 11. *Entwicklungsmechanik u. Anat. des Kindesalters.* Ⓓ Berlin 1896,
Wetzel, Karl (1893), Dr. rer. nat., Ass. Leipzig-Plagwitz (Deutschland), Elisabeth-Allee 21a I. *Stickstoff- und Säurestoffwechsel der Pflanzen. Wasserökonomie der Pflanzen.* Ⓓ Tübingen 1921.
Wetzel, Robert (1898), Dr. med., Priv.Doc. und Prosektor der Anat. Würzburg (Deutschland). *Frühe Entwicklungsgeschichte des Hühnchens.* Ⓓ München 1923.
Weysse, Arthur W. (1867), Prof. of Biol. Univ. Boston, Mass. (U.S.A.), 688 Boylston St. *Anat., Embryol., Physiol.* Ⓓ Ph.D. Harvard 1894. o Basel (Schweiz) 1907. o
Wheatley, Bertha (1885), Demonstrator in Bact., Univ. of Leeds (England), The Med. School. *Coccal organisms.* Ⓓ Leeds 1922.
Wheeler, Ruth (1877), Prof. and Head Dept. of nutrition, Coll. of Med. Iowa City, Ia. (U.S.A.). *Physiol. Chem.* o
Wheeler, William Morton (1865), Prof. of Entomol., Harvard Univ., Dean of the Bussey Inst. for Applied Biol. Boston, Mass. (U.S.A.), Bussey Institution Forest Hills. *Entomol. Social Insects (especially Formicidae). Compar. Psychol.* Ⓓ Worcester, Mass., 1892.

Whelan, Don Bion (1887), Ass. Prof., Department of Entomol., Univ. of Nebraska. Lincoln, Neb. (U.S.A.). *Entomol.* ① Manhattan, Kan., 1914.

Wherry, Edgar T. (1885), Senior Chem., Bureau of Chem., U. S. Department of Agric. Washington, D.C. (U.S.A.). *Ecol.; distribution, growth and chem. composition of native and cultivated plants in relation to their soil conditions, to soil reaction, acidity or alkalinity.* ① Univ. of Pennsylvania 1909.

Whetzel, Herbert H., Prof. of Plant Pathol., New York State Coll. of Agric., Cornell Univ. Ithaca, N.Y. (U.S.A.). *Pathol., mycol.* o

Whipple, George Hoyt (1878), Dean and Prof. of pathol. School of med. Univ. Rochester, N.Y. (U.S.A.), 320 Westminster Rd. o

Whistler, Hugh (1889). Battle, Sussex (England), Caldbec House. *Punjab and Indian Ornithol. Palaearctic area generally. Migration and distribution of birds.* ⑫ Birdskins or eggs.

Whitaker, Joseph I. S. Palermo (Italia), Malfitano. *Ornithol.* o

Whitaker, J. Ryland, Lect. School of Med. N. Coll. Edinburgh (Scotland). *Anat.* o

Whitcomb, Warren D. (1895), Ass. Research Prof. in Entomol., Massachusetts Agric. Coll. Experim. Station. Waltham, Mass. (U.S.A.), Market Garden Field Station, 240 Beaver St. *Fruit and Garden Insects. Greenhouse Red Spider. Plum Curculio in Apples. Garden Cutworms.*

White, Adam Cairns (1901), Lect. Edinburgh (Scotland), Pharm. Dept. Univ. *Adrenaline action on splanchnic circulation.* ① Edinburgh 1925.

White, Benjamin (1879), Dir., Division of Biol. Labor., Department of Public Health, Commonwealth of Massachusetts. Ass. Prof. of Bact. and Immunol. and Preventive Med. and Hygiene, Harvard Med. School and Harvard School of Public Health. Jamaica Plain, Mass. (U.S.A.), Antitoxin and Vaccine Labor., 375 South Street. *Antitoxin and Vaccine. Fixation tests for syphilis, gonorrhea, glanders. Immunology.* ① Yale Univ. 1903. ⑫ Various serums and vaccines.

White, Charles David (1862), Senior Geol., U. S. Geol. Survey; associate in paleont., Smithsonian Inst. Washington, D.C. (U.S.A.). *Stratigraphic paleobotan. of the Paleozoic formations. Origin and constitution of coal and other carbonaceous rocks and petroleum.* ① Cornell Univ. 1886, sc. (Hon. Degree) Univ. of Cincinnati and Univ. of Rochester 1924, Williams Coll. 1925. ⑫ Paleozoic fossil plants.

White, Edward Albert (1872), Prof., Head of Department of Floriculture. Ithaca, N.Y. (U.S.A.), Coll. of Agric. ① Amherst, Mass., 1895.

White, Errol Ivor (1901), Ass., Dept. of Geol. London S.W. 7 (England), Brit. Mus. (Nat. Hist.). *Eocene Fishes of Nigeria.* ① London 1921.

White, Harvey Lester (1896), Associate Prof. of Physiol. and Instr. in Pediatrics. Washington Univ. St. Louis, Mo. (U.S.A.), Scott and Euclid. *Kidney function, Circulation.* ① Washington Univ., St. Louis, Mo. 1920.

White, Orland Emile (1885), Curator of Plant Breeding and Economic Plants. Brooklyn, N.Y. (U.S.A.), Brooklyn Botan. Garden, 1000 Washington Avenue. *Pisum, Nicotiana, Ricinus, Zea, Althaea; Occurrence and cause of fasciation; Heat- and cold-resistance in plants; Tropical economic plants and tropical botan. exploration.* ① Harvard 1913.

White, Philip R. (1901), Special Investigator, United Fruit Company. Baltimore, Md. (U.S.A.), 1010 Roland Avenue. *Mycorrhiza and Anat. of Fragaria. Sterility in Musa. Gametophytes of Pteridophyta, particularly of the tropical epiphytes. Plant tissue cultures and virus diseases.*

White, Philipp, Prof.Univ.Bangor(England).*Zool.* o

White, Richard P. (1896), Ass. Plant Pathol., Kansas Agric. Experim. Station. Manhattan, Kan. (U.S.A.). *Plant Pathol, Fruit and vegetable diseases.* ① Cornell Univ. 1926.

White, Samuel Albert. Wetunga, Fulham (South Australia). *Ornithol.* o

Whitehead, F. E. (1891), Associate Prof. of Entomol. and Extension Entomol. Univ. of Idaho. Moscow, Id. (U.S.A.). *Entomol.*

White Hineline, Gertrude M. New York City, N.Y. (U.S.A.), 1316 Riverside Drive. *Animal Behavior.* ① Univ. of Wisconsin 1918.

Whiteside, Beatrice (1891), Associate Prof. of Histol., Detroit Coll. of Med. and Surgery. Detroit, Mich. (U.S.A.). *Gustatory Apparatus.* ① Univ. Zürich 1920.

Whiting, Albert Lemuel (1885), Associate Prof. of Agric. Bact. Univ. of Wisconsin. Madison, Wis. (U.S.A.). *Nitrification in soils, soluble nitrogen, total nitrogen. Root-nodule bacteria of Leguminosae, effect on yield and quality of special crops (peas, beans, soybeans, alfalfa and clovers). Strains by agglutination reactors. Cross inoculation studies among nodule bacteria. Fermentation of straw and corn stalks.* ① Univ. Illinois 1912.

Whiting, Phineas Wescott (1887), Prof. of Biol., Univ. of Maine. Orono, Me. (U.S.A.). *Genetics of parasit. Hymenoptera. Sex-determination and parthenogenesis.* ① Univ. of Pennsylvania 1916.

Whiting, William A., Prof. of Biol., Southern Coll. Birmingham, Ala. (U.S.A.). o

Whitley, E., Demonstr. Univ. Oxford (England). *Biochem.* o

Whitmer, Harriet H., Prof. Ginling Coll. Nanking (China) *Zool.* o

Whitmore, Eugene Rudolph (1874), Prof. Georgetown Univ. Washington, D.C, (U.S.A.), 2139 Wyoming Av. N. W. *Bact. Parasit. Pathol.* o

Whitney, David Day (1878), Prof. of Zool., Univ. of Nebraska. Lincoln, Neb. (U.S.A.). *Sex in Rotifers.* ① Columbia Univ. 1909.

Whitney, Will Alvah (1902), Junior Pathol., U. S. Department of Agric., Bureau of Plant Industry. Washington, D.C. (U.S.A.). *Diseases of beans (Phaseolus vulgaris) and sweet potatoes (Ipomoea batatas).*

Whitson, Andrew Robeson (1876), Prof. Univ. of Wisc. Madison, Wis. (U.S.A.), Morningside Nook R. 7. *Soils and soil fertility.* o

Wiant, James Stewart, Fellow in Pathol. Cornell Univ., Ithaca, N.Y. (U.S.A.), 214 ThurstonAvenue. o

Wiaschlinsky, Dimitry Michailowitsch (1902), Labor. Biol. Anstalt Zentr. Forststation. Moskau (U.d.S.S.R.), Troubnikovsky 11, log. 5. *Mammalia. Pelz und Mauser d. Säugetiere. Veränderlichkeit des Haares d. Pelztiere.* ⑫ Insectivora und Rodentia aus Mittelrußland.

Wibaut-Isebree Moens, Neeltje Louwrina (1884), Biol. Gemeentelijken Geneeskundige en Gesondheidsdienst. Amsterdam (Holland), Harmoniehof 68. *Biol. u. Hygiene des Wassers, Plankton, Insektenbekämpfung. Rotatoria, Protozoa.* ① Amsterdam 1911.

Wick, Hans Hermann (1898), Dr. phil., Diplom-Landwirt, Ass. am Inst. für Pflanzenbau der Univ. Göttingen (Deutschland), Nikolausberger Weg 7. *Kartoffeln.* ① Göttingen 1925.

Wickham, Henry Frederick (1866), Prof. State Univ. of Ia. Iowa City, Ia. (U.S.A.). *Entomol.* o

Widakowich, Victor, Prof. Univ. Nac. La Plata (Argentina). *Embryol. Histol.* o

Widal, Prof. Paris 8 (France), 153, Boulevard Haussmann. *Pathol.* o

Widder, Felix Josef (1892), Priv.Doc. für systemat. Botan.; o. Univ.-Ass. Graz, Steiermark (Österreich), Holteigasse 6. *Morphol. und Systematik der Blütenpflanzen (Doronicum, Leontodon, Xanthium); Flora der Ostalpen.* ① Graz 1919. ⑫ Pflanzen.

Widdicombe, J. H., Lect. Downing Coll. Cambridge (England). *Physiol.* o

Widmark, E. M. P., Prof., Dr. Lund (Sverige). *Physiol. Chem.* o

Wiebe, Abraham H. (1892), Sc. Investigator, U. S. Bureau of Fisheries. Fairport, Ia., or Homer, Minnesota (U.S.A.). *Limnol. and Entomol. Artificial Fertilization in Aquatic Insects. Freshwater Snail Gonibasis.* ⑨ Ohio State 1924.

Wiebe, Gustav A. (1899), Ass. Agronomist, U. S. Department of Agric. Aberdeen, Id. (U.S.A.), Substation. *Breeding barley, oats and wheat for disease resistance, genetic studies on cereals.* ⑫ Cereals, Wheat Oats and Barley.

Wiechmann, Ernst (1894), Dr. med., Priv.-Doc. für innere Med. an der Univ. Köln, Oberarzt der med. Univ.-Klinik Köln-Lindenburg. Köln-Lindenthal, (Deutschland), Kinkelstr. 5. *Permeabilität der Zellen und Gewebe. Blut, Kreislauf, Diabetes mellitus.* ⑨ Kiel 1918.

Wiechowski, Wilhelm, Dr., Prof. d. Pharmacol. Praha II (C.S.R.), Albertov 5, Pharmacol. Inst. d. Deutschen Univ. o

Wiegand, Karl McKay, Prof. of Botan., New York State Coll. of Agric., Cornell Univ. Ithaca, N.Y. (U.S.A.). *Morphol. and taxonomy of vascular plants.* o

Wiegner, Georg (1883), o. Prof. für Agrikulturchem. an der Eidg. Technischen Hochsch. Zürich (Schweiz), Universitätsstr. 2. *Ernährung der landwirtschaftl. Nutztiere. Konservierung der Futtermittel.* ⑨ 1906.

Wiehr, Robert (1902), wissenschaftl. Hilfsarbeiter. Kiel (Deutschland), Preuß. Versuchs- u. Forschungsanstalt für Milchwirtschaft. *Milcherzeugung.*

van der Wiel, Pieter (1893). Amsterdam (Holland), Cornelis van der Lindenstraat 20. *Entomol.: Systematik und Biol. von Coleopteren (Käfer) und Formiciden (Ameisen); speziell die Mittel- und Nordeuropäischen Arten.* ⑫ Mittel- und Nordeuropäische Coleopteren und Formiciden.

Wieland, George Reber (1865), Associate, Carnegie Inst. of Washington; Associate Prof., Paleobotan., Yale Univ. New Haven, Conn. (U.S.A.). *Petrified Plants. Fossil Vertebrates, Testudinata, Dinosauria. Evolution. Cycadeoids. Fossil insects. American Fossil Cycads.* ⑨ Yale 1900.

Wieland, Hermann (1885), o. Prof. der Pharmacol. an d. Univ., Dir. d. pharmacol. Inst. Heidelberg (Deutschland), Bergstr. 64. *Experim. Pharm. der Narkose und Lokalanästhesie, der Atmung, des Kohlendioxyds, der Säurewirkung überhaupt; Anwendung physikalischer Chemie auf die Wirkungsmechanismus von Giften.* ⑨ Straßburg i. E. 1909.

Wieler, Arwed Ludwig (1858), Dr. nat., a.o. Prof. für Botan. an der Technischen Hochsch. Aachen (Deutschland), Nizza-Allee 71. *Anat. und Physiol. der Pflanzen; Rohstofflehre; Rauchschadenforschung.* ⑨ Tübingen 1883.

Wieman, Harry Lewis (1883), Prof. of Zool., Univ. of Cincinnati, O. (U.S.A.). *Experim. embryol.* ⑨ Univ. of Chicago 1909.

Wiener, Hugo, Dr., Prof. experim. Pathol. Praha II (C.S.R.), Salmova 3, Experim. pathol. Inst. d. Deutsch. Univ. o

Wierdak, Sz., Prof. Techn.Hochsch. Lwów (Polska). *Forstbotan. Baumkrankheiten.* o

Wieringa, Gerrit (1883), Nederlandsch Landbouwkundige, Botan. an der Reichsversuchsstation für Samenkontrolle in Wageningen (Holland), Rijksstraatweg 43. *Keimungsuntersuchung und Keimfähigkeitsbestimmung von Samen.* ⑨ Wageningen.

Wierzchowski, Zenon (1890), Adjunkt landwirtschaftl. Inst. Puławy (Polen). *B-Vitamine der Weizenkörner, Verdaulichkeit der Nahrungsstoffe bei Vögeln (Hühner, Tauben).* ⑨ Lwów 1912. ⑫ Vitaminpräparate aus Weizenkleie.

Wierzuchowski, Mieczislas (1895), Fellow in Med. of the Rockefeller Foundation. Krakóv 14 (Polska), Lwowska 29. *Intermediary carbohydrate metabolism, Hypoglycemia with convulsions in phlorhizinized dogs. Metabolism of levulose and glucose. Respiratory metabolism.* ⑨ Kraków 1920.

von Wiese, Werner (1889), Dr. agr., Saatzuchtleiter. Knehden, Post Templin (Deutschland). *Züchtung von Xero- und Hygrophytentypen bei Sommerweizen, Futterrüben und Mais durch künstliche Kreuzung unter Benutzung einer Beregnungsanlage zur Regelung des Wasserhaushalts der Pflanzen.* ⑨ Berlin 1925. ⑫ Rassen von Mais, Sommerweizen und Futterrüben.

Wiesmann, Robert (1899), Dr. phil., Entomol. der chem. Fabr. Dr. Maag, Dielsdorf (Schweiz). *Angewandte Entomol. (Obst und Weinbauschädlinge).* ⑨ Zürich 1925.

Wiesner, Richard (1875), a.ö. Univ.Prof., Prosektor. Wien IX (Deutsch-Österreich), Lackierergasse 1a. *Pathol. Anat. u. Histol., Bact.* ⑨ Wien 1902.

Wigdor, Meyer (1896), Acting Ass. Surgeon, U. S. Public Health Service, also General Practice of Med. New York City (U.S.A.), 2080 Grand Avenue, Bronx. *Parasit.* ⑨ Cornell Univ. 1923.

Wiggans, Cleo Claude (1889), Prof. of Horticulture, Univ. of Nebraska, Horticulturist, Nebraska Experim. Station. Lincoln, Neb. (U.S.A.). *Pomol., fruitfulness in the apple.* ⑨ Univ. of Missouri, Columbia 1918.

Wiggans, Clifford Barnes (1900), Ass. Horticulturist, Univ. of Arkansas. Fayetteville, Ark. (U.S.A.). *Peach and grape nutrition.* ⑨ Univ. of California 1923.

Wiggans, Roy Glen (1891), Ass. Prof. of Plant Breeding. Department of Plant Breeding, N. Y. State Coll. of Agric. Ithaca, N.Y. (U.S.A.). *Genetic studies in maize. Varietal studies of maize, soybeans, sunflowers for silage purposes. Development of desirable types of sunflowers. Seed source tests of red clover and alfalfa.* ⑨ Cornell Univ. 1919.

Wiggers, Carl John (1883), Prof. Head dept. physiol. Western Reserve Univ., School of Med. Cleveland, O. (U.S.A.), 3357 Euclid, Hts. Boul. o

Wigglesworth, Grace (1877), Ass. Keeper Botan. Manchester Mus., The Univ. Wiltrington (England), 18 Oak Road. ⑨ Manchester 1906.

Wigham, Joseph F., Prof. Univ. Trin. Coll. Dublin (Ireland). *Pathol.* o

Wight, Toynbee (1871), Pathol., U. S. V. B. Hospital. Palo Alto, Cal. (U.S.A.). *Pathol. and serol., human parasit., Protozoa, tuberculosis.* ⑨ Harvard Univ. 1901. ⑫ Protozool. preparations.

van Wijhe, S. W., Prof. em. der Anat. Univ. Groningen (Holland). *Kopfproblem der Vertebraten, Amphioxus.* o

Wijnhoff (Stiasny-), Gerarda (1885). Leiden (Holland), Rijnsburgerweg 38. *Anat. und Systematik der Nemertinen. Geographische Verbreitung der marinen Invertebraten.* ⑨ Utrecht 1910.

Wilckens, Otto (1876), Dr. phil., o. Prof. der Univ. Straßburg, beauftragt mit Vorlesungen an der Univ. Bonn (Deutschland), Scharnhorststr. 4. *Mollusken der oberen Kreideformation, fossile Faunen der südpacifischen Region.* ⑨ Freiburg i. B. 1903.

Wilcox, Alice W., Prof. of Biol., Brenan Coll. Gainesville, Ga. (U.S.A.). *Physiol., systematic botan., biol., genetics.* o

Wilcox, Edwin Mead (1876). East Lansing, Mich. (U.S.A.). *Plant Pathol.* ⑨ Harvard Univ.

Wilcox, Raymond Boorman, Ass. Pathol., Bureau of Plant Industry. U. S. Dept. of Agric., Ohio Agric. Experim. Sta. Wooster, O. (U.S.A.). o

Wilcoxon, F., Biochem. Res. in Plants, Boyce Thompson Inst. for Plant Res. Yonkers, N.Y. (U.S.A.). o

Wilczek, Ernest (1867), Prof. Botan. Pharmacognosie Univ. Lausanne (Suisse). *Botan. systématique et pharmaceutique.* ⑨ Zürich 1892.

Wilczyński, Jan Zygmunt (1891), Prof. extrao., Dir. Inst. de Biol. génér., Dir. Station hydroboil. sur les ,,Lacs Verts" (Zielone Jeziora). Wilno (Polska), 15, Rue Zakretowa. *Physiol. des Sipunculidae; Bonellia; Turbellaries et sa position morphol. et phylogenetique; Transformations des*

21*

Vorticelles (*Epistylis*). *Energétique de la vie. Dégradation morphol.* ① Univ. Cracovie 1914. ⑫ Invertebrés d'eaux douces des environs de Wilno.

Wilde, Earle Irving (1888), Prof. of Floriculture, Pennsylvania State College, Pa. (U.S.A.). *Genus Delphinum.* ① Massachusetts Agric. Coll. 1912.

Wilde, Johannes (1887), Ass. des hygien. Inst. der Univ. Tartu (Eesti), Aia tän. 46. *Nahrungsmittelchem.*

De Wildeman, Emile (1866), Dir. du Jardin Botan. de l'Etat, Prof. aux Univ. de Gand et d'Anvers (Un. coloniale). Bruxelles (Belgique), Rue des Confidérés 122. *Botan. systématique (Flore d'Afrique), Botan. coloniale.* ① Bruxelles.

Wilder, George Durand (1869), Head Dept. of Birds Peking Labor. of Natural History. Tunghsien (China). *Birds of North China.* ① Yale 1894. ⑫ Bird skins.

Wilder, Harris Hawthorne (1864), Prof. of Zool. Northampton, Mass. (U.S.A.), Smith Coll. *Vertebrate Anat. Physical Anthropol. Amphibians, Epidermic markings of human and simian palms.* ① Freiburg (Breisgau) 1891. o

Wilder, Russell Morse (1885), Prof. of Med., Mayo Foundation, Univ. of Minnesota. Rochester, Minn. (U.S.A.), Mayo Clinic. *Nutrition, carbohydrate metabolism.* ① Phil. Univ. of Chicago 1912, med. 1912.

Wildervanck, L. S. (1902), Ass. Botan. Labor. Groningen (Holland), Martinikerkhof 29. *Pflanzenphysiol.: Elastizität, Saugkraft, osmot. Druck der Zelle.*

Wiley, Harvey Washington (1844), Dir., Bureau of Foods, Sanitation and Health of Good Housekeeping Magazine. Washington, D.C. (U.S.A.), Cosmos Club. *Foods, Growth, Nutrition, Hygiene.* ① 1873.

Wilhelm, Karl (1848), Hofrat¹ Hochsch.Prof. im Ruhestande. Wien XIX (Deutsch-Österreich), Dionysius-Andrássy-Str. 5. *Anat. und Morphol. der Pflanzen; Naturgeschichte der Forstgewächse; Bau der Hölzer.* ① Straßburg i. Elsaß 1877.

Wilhelmi, Hedwig (1889), Dr. phil. München (Deutschland), Anat. Inst. *Entwicklungsmechanik: Symmetrieproblem.* ① Rostock 1917.

Wilhelmi, Julius (1880), Prof., Dr., Preuß. Landesanstalt für Wasser-, Boden- und Lufthygien., Biol.-zool. Abt. u. Mus., Berlin-Dahlem. Berlin-Lichterfelde (Deutschland), Stubenrauchstr. 4. *Med. Zool. (tierische Gesundheitsschädlinge und hygien. Hydrobiol.). Biol. des Trink-, Brauch- und Abwassers; Meerwasserverunreinigung. Wohnungs- und Körperungeziefer als Krankheitsübertrager und seine Bekämpfung.* ① Marburg 1904.

Wilke, Siegfried (1898), Ass. an der Biol. Reichsanstalt für Land- und Forstwirtschaft, Berlin-Dahlem. Berlin-Lichterfelde-W. (Deutschland), Moltkestr. 51. *Angewandte Entomol., entomol. Systematik (Coleoptera).* ① Berlin 1921.

Wilkie, A. A. W., Dir. Zool. Gardens (R. Park). Melbourne (Australia). o

Wilkie, David (1903), Ass. Edinburgh (Scotland), Pharm. Dept. Univ. *Relation of dosage to effect, illustrated by adrenaline.* ① Edinburgh 1926.

Wilkins, F. Scott (1889), Ass. Chief Forage Crop Investigations. Ames, Ia. (U.S.A.), Iowa Agric. Experim. Station. *Plants used as forage.* ① Iowa State Coll. 1915.

Wilkinson, Albert Edmund (1879), Extension Prof. of Vegetable Growing, Vegetable Specialist, Connecticut Agric. Coll. Storrs, Conn. (U.S.A.). *Vegetable growing.* ① Rhode Island State Coll. 1916.

Wilcox, Sir Wm. H., Prof. St. Mary's Hospit. Med. School. London W 2 (England), Paddington. *Pathol. Chem.* o

Will, Ludwig (1861), Dr. phil., o. Hon.Prof. der Zool., Univ. Rostock (Deutschland), Haedgestraße 35. *Vergl. Embryol. (Reptilien, Insekten); Histol. (Eibildung, Bau und Funktion der Nessel-*

kapseln); Experim. Zool. (Coelenteraten). ① Würzburg 1883.

Willaman, John James (1889), Associate Prof., Agric. Biochem., Univ. of Minnesota. St. Paul, Minn. (U.S.A.), Univ. Farm. *Biochem. of normal and abnormal plants; resistance to disease in plants; winter hardiness.* ① Univ. of Chicago 1919.

Willard, Charles J. (1889), Prof. of Farm Crops, The Ohio State Univ. Columbus, O. (U.S.A.). *Sweet clover (Melilotus sp.). Life history. Root development.* ① Ohio State Univ. 1926.

Willard, Harold Francis (1884), Entomol., Bureau of Entomol., U. S. Dept. of Agric. Honolulu (Hawaii), Box 340. *Biol. of the Mediterranean fruit fly Ceratitis capitata and the melon fly Bactrocera cucurbitae, and their parasit.; the biol. of Bruchidae attacking the algaroba bean Prosopis juliflora and their parasit..* ① Massachusetts Agric. Coll. 1911. ⑫ Fruit flies, bruchids, and their parasit.

Willard, William Albert (1873), Prof. of Anat., Univ. of Nebraska, Coll. of Med. Omaha, Neb. (U.S.A.), 42 and Dewey Ave. *Compar. anat. and physiol. of the nervous system of vertebrates.* ① Harvard Univ. 1910.

Wille, Fritz (1888), Dr. phil., ing. agr., Biol. der Aluminium-Industrie AG. Neuhausen. Siders, Wallis (Schweiz). *Pflanzenpathol., Rauchschadenfrage, Pflanzenanat.* ① Bern 1915.

Willem, Victor (1866), Prof. d. Zool. Univ. Gand (Belgique), 57 Rue du Jardin. *Respiration et circulation sur les Vertébrés inférieurs.*

Willemse, Cornelis Josef Maria (1888). Eygelshovers Z. L., (Holland). *Entomol.: Orthoptera speciel Indo-austral. Fauna.* ① Amsterdam 1913. ⑫ Orthoptera.

Willer, Alfred (1889), Dr. med. et phil., a.o. Prof. und Dir. des Fischerei-Inst. der Univ. Königsberg i. Pr., Oberfischmeister für die Prov. Ostpreußen. Königsberg i. Pr. (Deutschland), Tragheimer Kirchenstr. 74. *Hydrobiol. (Aufwuchsstudien), Fischereibiol. (Biol. der planktonfressenden Fische und wirtschaftliche Anwendung der entsprechenden Ergebnisse), Vererbungsstudien an Teichfischen.* ① Phil. Jena 1912, med. Berlin 1914.

Williams, Carrington Bonsor (1889), Dir. Plant Protection Section, Ministry of agric. Cairo (Egypt). *Applied Entomol. Insect Ecol. in relation to cereals. Migration of Insects.* Cambridge 1916.

Williams, Charles Burgess (1871), Dean N.C. Agric. and Mich. Coll., Chief div. of agronomy. Raleigh, N.C. (U.S.A.), 1405 Hillsboro St. o

Williams, Charles Laval (1887), Labor. at the New York Quarantine Station. New York City (U.S.A.), Rosebank, S. I. *The bionomics and the geographical distribution of rats and rat fleas as related to the spread of bubonic plague.* ① Univ. of Virginia 1911.

Williams, Oscar B., Instr. in Botan., Univ. of Texas. Austin, Tex. (U.S.A.). *Bact.* o

Williams, Paul Springer (1895), Ass. Prof. of Dairy Husbandry. Pennsylvania State College, Pa. (U.S.A.). ① Pennsylvania State Coll. 1920.

Williams, Rhys David (1889). Aberystwyth (England), Welsh Plant Breeding Station, Univ. Coll. of Wales. *Leguminous forage crops.*

Williams, Robert Stenhouse (1871), Research Prof. Dairy Bact. Univ. of Reading (England). ① Univ. of Edinburgh.

Williams, Samuel (1897), Lect. in Botan. in the Univ. Glasgow (Scotland). *Morphol. of Cryptogams.* ① Glasgow 1926.

Williams, Vester V. (1898), Junior Entomol. of U. S. Bureau of Entomol. in Cotton Insects Investigations. Jalluloh, La. (U.S.A.). *Cotton insects.* ① Alabama Polytecnic Inst. 1921.

Williamson, Edward Bruce (1877), Honorary Curator Dragonflies Univ. of Michigan. Bluffton, Ind. (U.S.A.). *Odonata.* ① Ohio State Univ. ⑫ Dragonflies.

Williamson, Henry Charles (1871), Zool. Nanaimo (Brit.Columbia), Pacific Biol.Station. *Fishes general; Gadidae; Salmonidae. Decapod Crustacea, Mollusc.* ⊕ St. Andrews 1893.

Williamson-Chambers, Helen Stuart (1884), Research Ass. London, W.C. 1 (England), 3 Verulam Buildings Gray's Inn. *Mycol.* ⊕ London 1906.

Willans, Maud (1883), Lect. in Bio-Physics, Chelsea Polytechnic. London S.W.5. (England), 42 Nevern Sq. *Influence of immersion in Electrolytes on plant cells. Effects of X Rays and Radiations from Radium upon plant cells.* ⊕ London 1905.

Willier, Benjamin H. (1890), Ass. Prof. Zool. Univ. Chicago, Ill. (U.S.A.), Hull Zool. Labor. *Experim. embryol.: Chick embryos. Cytol. of intracellular digestion in Planaria.* ⊕ Univ. of Chicago 1920.

Willis, Henry Stuart Kendall (1891), Associate in Med., Johns Hopkins Hospital. Baltimore, Md. (U.S.A.). *Tuberculosis. Immunity.* ⊕ Johns Hopkins Univ. 1919.

Willmer, Edward Nevill (1902), Demonstrator in Experim. Physiol. The Victoria Univ. Manchester (England). *Tissue Culture, in relation to growth.*

Wilmott, Alfred James (1888), Ass. Keeper, Dept. of Botan., British Mus., Nat. Hist. London S.W. 7 (England), Cromwell Road. *European Phanerogams (Systematic).* ⊕ Cambridge 1909.

Wilsdorf, Max Georg (1871), Dr. phil., Tierzuchtdirektor. Berlin-Halensee (Deutschland), Paulsborner Str. 25. *Allgemeine Vererbungswissenschaft. Schafzucht. Wollerzeugung.* ⊕ Leipzig 1895.

Wilsie, Carroll P. (1902), Research Ass. Urbana, Ill. (U.S.A.), Dept. of Horticulture, Univ. of Illinois. *Sweet corn, tomatoes.* ⊕ Madison, Wisconsin, 1926.

Wilson, Carl Louis (1897), Instr. in Botan., Dartmouth Coll., Hanover, N.H. (U.S.A.). *Anat. and Morphol. of Vascular Plants. Phylogeny of the Angiosperms.* ⊕ Cornell Univ. 1923.

Wilson, Cecil Cline (1896), Junior Entomol. Sacramento, Cal. (U.S.A.), 600, 26 Street. *Biol. of destructive grasshoppers in the Pacific Coast states.* ⊕ Orthoptera of California.

Wilson, David Wright (1889), Benjamin Rush Prof. of Physiol. Chem. Philadelphia, Pa. (U.S.A.), School of Med., Univ. of Pa. *Blood and Tissue Chem.* ⊕ Yale Univ. 1914.

Wilson, Edmund B. (1856), Da Costa Prof. of Zool. New York (U.S.A.), Columbia Univ. *Embryol. Cellular Biol. Karyokinesis, Heredity.* ⊕ John Hopkins Univ.: Ph.D. 1881, L.L.D. 1902; L.L.D. Yale 1901, Chicago 1901; D.Sc. Cambridge 1909; Med. h. c. Leipzig 1909. ○

Wilson, Edward Elmer (1900), Research Ass. in Plant Pathol. Madison, Wis. (U.S.A.), Dept. of Plant Pathol., Coll. of Agric. *Diseases of orchard fruits. The apple scab fungus. Venturia inaequalis.*

Wilson, Ernest Henry (1876), Ass. Dir., Arnold Arboretum, Harvard Univ., Boston, Mass. (U.S.A.), 380 South Street, Jamaica Plain. *Cherries, Conifers and Taxaceae of Japan.* ⊕ Hon. Harvard Coll. 1916.

Wilson, Everett Victor Boyle (1907), Ass. Plant Breeding Research Division, Ministry of Agric. N. Ireland. Belfast (Ireland), Stormont, Strandtown. *Agric. Plants.* ⊕ Belfast 1926. ℗ Oats, flax.

Wilson, Francis H. (1902), Acting Prof. of Botan., Univ. of Richmond, Va. (U.S.A.). *Mallophaga.* ⊕ Cornell Univ. 1925.

Wilson, Frank Norman (1890), Prof. of Med., Univ. of Michigan Med. School, Ann Arbor, Mich. (U.S.A.), Univ. Hospital. *Electrocardiography.* ⊕ Univ. of Mich. 1913.

Wilson, Gregg, Prof. Univ. Belfast (Ireland). *Zool.* ○

Wilson, Guy West, Prof. of Biol., Upper Iowa Univ. Fayette, Ia. (U.S.A.). *Parasit. fungi, Phycomycetes.* ○

Wilson, Harley F. (1883), Prof. of Economic Entomol., Univ. of Wisconsin (U.S.A.). *Apiculture.* *Behavior of the bee colony in winter.* ⊕ Oregon Agric. Coll. 1913.

Wilson, Harold Kirby (1900), Instr. Iowa State Teachers Coll., Cedar Falls, Iowa. Urbana, Ill. (U.S.A.), Dept. of Agronomy, Univ. of Illinois. *Crop production, Plant Physiol., Germination as effected by temperature, (Giberella saubenetii) and other parasit. fungi.* ⊕ Illinois 1927. ℗ Cereals.

Wilson, Henry Van Peters (1863), Kenan Prof. of Zool., Univ. of North Carolina. Chapel Hill, N.C. (U.S.A.). *Embryol. and regeneration; systematic zool. of sponges.* ⊕ Johns Hopkins 1888.

Wilson, Ira T. (1895), Prof. of Biol., Heidelberg Coll. Tiffin, O. (U.S.A.), 89 Clinton Ave. *Ecol. of fresh water lakes and streams: Mollusca. Genetics of tumor-like growths in the fruit fly Drosophila melanogaster.* ⊕ Indiana Univ. 1923.

Wilson, Irl Donaker (1888), Head, Dept. of Zool and Animal Pathol. Blacksburg, Va. (U.S.A.). ⊕ Iowa State Coll. 1914.

Wilson, J. F., Prof. St. John's Coll. Cambridge (England). *Anat.* ○

Wilson, J. Walter (1896), Ass. Prof. of Biol., Brown Univ. Providence, R. I. (U.S.A.). *Experim. Zool., Regeneration.* ⊕ Brown Univ. 1921.

Wilson, Minnie A. (1877), Bact.; Supervisor of Serol. Labor., Dept. of Health. New York City, N.Y. (U.S.A.), Foot of East 16 Street. *Bact., serol.*

Wilson, Orville Turner (1886), Associate Prof. of Botan., Univ. of Cincinnati, O. (U.S.A.). *Pathol. and teratol. Diatoms. Anat. of Alphalpha. Osmosis. Regeneration of plant structures. Ecol. Flora of Cincinnati.* ⊕ Wisconsin 1915. ℗ Plant material.

Wilson, Sidney Ramson (1882), Lect. applied Physiol. and Anaesthetics. Manchester Univ. Rusholme, Manchester (England), Birch Croft, Dickenson Road. ⊕ London.

Wilson, T. Dean (1895), Ass. Pathol., Ohio Agric. Experim. Station. Wooster, O. (U.S.A.). *Vegetable Diseases.* ⊕ 1926.

Wilson, William B., Prof. of Biol. and Dean of Ottawa Univ., Summer Session. Ottawa (Canada). *Morphol., pathol.* ○

Wilson, William Powell (1844), Dir. Mus. Philadelphia, Pa. (U.S.A.), 34th. Str. below Spruce. ⊕ Tübingen 1880. ○

Wiltshire, Samuel Paul (1891), Ass. Dir. Imper. Bureau of Mycol. Kew, Surrey (England), 17, The Green. *Plant pathol., mycol.* ⊕ Bristol 1913.

Wiman, C. J. J. E., Prof. Dr. Upsala (Sverige). *Palaeont.*

Wimmer, Emil G. (1877), o. Prof. der Forstwirtschafts-Wissenschaft an der Univ. Gießen. Freiburg i. Br. (Deutschland), Sternwaldstr. 31. *Forstliche Produktionslehre: Waldbau und Forstschutz; angewandte (Forst-) Botan., Zool. und Bodenkunde.* ⊕ München 1907.

Windle, William Frederick (1898), Ass. Prof. of Anat., Northwestern Univ., Med. School. Chicago, Ill. (U.S.A.), 303 East Chicago Av. *Histol. of the Nervous System.* ⊕ Northwestern Univ. 1926.

Winfield, George, Lect. Univ. Leeds (England). *Physiol.* ○

Wing, Henry Hiram (1850), Prof. Cornell Univ. Ithaca, N.Y. (U.S.A.). *Dairy husbandry.* ○

Wingard, Samuel Andrew (1895), Associate Plant Pathol. of the Virginia Agric. Experim. Station. Blacksburg, Va. (U.S.A.). ⊕ Columbia Univ. 1925.

Winge, Öjvind (1886), Prof. of Genetics at the Royal Vet. and Agric. Coll. København V (Danmark), Rolighedsvej 23. *Genetic-experim. and cytol. investigations on: Species hybrids in plants, coupling in higher plants and fish, sex-linked and one-sided inheritance in higher plants and in fish, sex-chromosomes investigations, chromosome numbers, cytol. of malignant tumours in plants and animals.* ⊕ Copenhagen 1917.

Winkel, A. J. (1879), Bact., Ryks-Seruminst. Rotterdam. Voorburg (Holland), Kon. Wilhelminalaan 800. *Veterinärbact.* ① Bern, Schweiz, 1912.

Winkelmann, August (1899), Dr. phil., Wissenschaftl. Hilfsarbeiter bei der Biol. Reichsanstalt für Land- und Forstwirtschaft, Berlin-Dahlem (Deutschland), Königin-Luise-Str. 19. *Phytopath.* ① Münster i. Westf. 1924.

Winkler, Albert Julius (1894), Ass. Viticulturist in the Agric. Experim. Station at the Univ. of California. Davis, Cal. (U.S.A.). *Physiol. of Viticulture.* ① Univ. of California 1921. ② Vitis vinifera varieties.

Winkler, Artur (1890), Dr., Priv.Doc. an der Univ., Sektionsgeol. an der Geol. Bundesanstalt. Wien III (Deutsch-Österreich), Rasumofskyg. 23. *Palaeont. (Tertiärconchylien). Sarmatische obermiozäne Conchylien der steirischen Tertiärablagerungen.* ① Wien 1914. ② Sarmatische (obermiozäne) Conchylien aus dem steirischen Tertiärbecken.

Winkler, Cornelis (1855), Retired Prof. in Psychiatry and Neurol. at the Univ. Utrecht and Amsterdam. Utrecht (Holland), Heerenstraat 35. *Anat. of the brain. Pathol. and experim. anat. of the nervous system.* ① Utrecht 1879.

Winkler, Hans (1877), Dr phil , o. Prof. der Botan. an der Univ., Dir. des Inst. für allgemeine Botan. und des Botan. Gartens in Hamburg 36 (Deutschland), Jungiusstr. 6. *Botan. und Vererbungslehre. Apogamie und Parthenogenesis im Pflanzenreiche. Pfropfbastarde.* ① Leipzig 1898.

Winkler, Hubert (1875), Dr. phil., a.o. Prof. der Botan. an der Univ. Breslau 9 (Deutschland), Goepperstr. 4. *Botan. Systematik und Pflanzengeographie, Morphol., Ökol., trop. Nutzpflanzen; Ficaria.* ① Breslau 1901. ② Herbarmaterial v.Ficaria.

Winkler, Wolfgang Friedrich (1890), Dr. med., Priv.Doc. für Hygiene. Rostock i. M. (Deutschland), Hygien. Inst. *Pockenvirus und -immunität. d'Hérellesches Phaenomen. Serodiagnostik der Lues.* ① Leipzig 1920.

Winogradoff, Michael (1891), Ass.Prof. Zool. et Genetic Inst. Agronomique. Leningrad (U.d.S.S.R.), Nijegorodskaja 12, 12. *Anabiosis.*

Winogradow, Alexander P. (1871), Ass. Pharmacol. Inst. Odessa (U.d.S.S.R.), Olgiewskaja 4. *Physiol. u. Pharm. d. Hepar.* ① St. Petersburg 1900.

Winogradow, B. S. (1891), Curator, Zool. Mus., Akad.d.Wissensch. Leningrad(U.d.S.S.R.). *Mammal.*

Winogradow, Sergej I., Doc. Landw. Inst. d. Bergrepublik. Wladikawkas (U.d.S.S.R.), Moskowskaja 24/14. *Pflanzen-Physiol.* ○

Winogradow-Nikitin, Paul S., Prof. Staatl. Polytechn. Inst. Tiflis (U.d.S.S.R.), Ul. Perowskoj 20. *Dendrol. Forst-Biol.* ○

Winogradsky, Serge, (1856), Prof. à l'Inst. Pasteur, Dir. du Labor. de Microbiol. agric. à Brie-Comte-Robert (France), SAM. *Microbiol. Agric., Microbiol. du Sol. Sulfobact. Ferrobact. Nitrification et l'assimilation gazeux dans le Sol.* ① 1890.

Winokuroff, Sergius (1899), Biochem. Inst. Charkow (U.d.S.S.R.), Pouschkine Str. 86. *Biochem. d. Ernährungsstörungen.* ① Odessa 1922.

Winokurowa, Maria F., Doc. II. Univ. Moskau (U.d.S.S.R.), I. Spasso-Naliwkinskij per. 14, W. 2. *Pflanzen-Physiol.*

Winslow, Charles-Edward A., Prof. of Public Health. New Haven, Conn. (U.S.A.), Yale Univ. *Water bact. Systematic relationship of the Coccaceae.* ① M.S. Mass. Inst. Techn. 1899; P.H. New York 1918. ○

Winston, J. R., Assoc. Pathol., Bureau of Plant Industry, U.S. Dept. of Agric. Washington, D.C. (U.S.A.). ○

Winteler, Emil (1878), Prof. allgem. Pathol. u. pathol. Anat. Kaunas (Litauen), Univ., Pathol.-anat. Inst. *Rhinosklerom. Zysten des Wurmfortsatzes.* ① Charkow 1909.

Winter, Floyd L. (1898), Instr. and First Ass. in Plant Breeding. Urbana, Ill. (U.S.A.), Agric. Experim. Station, Univ. of Illinois. *Plant breeding of maize plant, inheritance.* ① Univ. of Illinois 1924. ② Maize seed from different varieties.

Winters, Lawrence Merriam (1891), Prof. of Animal Husbandry. Saskatoon, Sask. (Canada), Univ. of Saskatchewan. *Animal Breeding. Wool studies of purebred and crossbred sheep.* ① Iowa State Coll. 1920. ② Samples of wool from purebred and grade sheep.

Winterstein, Ernst Heinrich (1865), Prof. für Chem. an der Eidg.Technischen Hochsch. in Zürich 7 (Schweiz), Physikstr. 4. *Physiol. und Pflanzenchem.* ① Zürich 1893.

Winterstein, Hans (1879), Dr. med., Univ.Prof. Rostock (Deutschland), Am Reifergraben 3. *Regulierung der Atmung, Stoffwechsel des Nervensystems, vergl. Physiol.* ① Prag 1903.

Winton, Frank Robert (1894), Ass. in the Dept. of Pharmacol.,Univ.Coll.,London(England). *Physiol. and Pharm. of unstriated Muscle. Pharm. action of Squills on Rats. The Influence of the Pituitary on Frog Melanophores.* ① Cambridge 1926.

Winton, Will McClean (1885), Prof. Palaeont. Fort Worth, Tex. (U.S.A.), Texas Christian Univ. ○

Wintrebert, Paul Marie Joseph (1867), Prof. d'Anat. et Histol. comparées à la Fac. des Sc. de Paris, en Sorbonne. Paris V (France), 41, rue de Jussieu. *Fonctions transitoires et spéciales de l'embryon (chez les Vertébrés anamniotes).* ① Paris 1922.

Wirgin, Germund (1868), Prof. der Hygiene und Bact. der Königl. Univ. zu Upsala (Sverige). ① Upsala 1902.

Wirth, David (1885), o. ö. Prof. an der Tierärztl. Hochsch. in Wien III (Deutsch-Österreich), linke Bahngasse 11. *Spezielle Pathol. und Therapie der inneren und Hautkrankheiten der Equiden, Carnivoren und der Kleintiere (Geflügel, Kaninchen). Hämatol. der Haustiere.* ① Wien 1908. ② Diapositive kranker Equiden, Carnivoren und kleiner Haustiere.

Wirtner, Modestus (1861). St. Bonifacius, Pa. (U.S.A.). *Hemiptera.* ① St. Vincent Coll., Beatty, Pa. 1880. ② Hemiptera.

Wishart, George Macfeat (1895), Lect. Univ. of Glasgow (England), Inst. of Physiol. *Chemical Physiol.* ① Glasgow 1918.

Wislocki, George (1892), Associate Prof. of Anat., Johns Hopkins Univ. Baltimore, Md. (U.S.A.), Johns Hopkins Med. School. *Placentation.* ① Johns Hopkins Univ. 1916.

Wislouch, Stanislas (1875), Doc. Univ. Warszawa (Polska), Brzozowa-Str. 12. *Algol. (Phytoplankton, Diatomeen), Hydrobiol. (benthonische Mikroorganismen, Schwefelbakterien), Biol. Wasseranalyse.* ① Forst-Inst. St. Petersburg 1898.

Wiśniowski, Piotr (1881), Prof. der allgemeinen Botan. Univ.Wilno (Polska), Botan. Inst., 15 Zakretowastr. *Experim. Morphol. der Pflanzen. Die Ruheperiode.* ① Lwów 1910. '

Wiśniewski, Tadeusz (1905), Bibliothècaire adj. de la Société Botan. de Pologne. Warszawa (Polska), Wspólna 47a. *Phytosociol., Ecol. et Géographie des Muscinées, Flore diluviale.* ② Bryophytes et préparations des plantes de la flore diluviale.

Wiśniowski, T., Dr. Prof. Warszawa (Polska), Kolonja Lubeckiego. *Palaeont.* ○

Wissler, Clark (1870), Curator of Anthropol American Mus. of Nat. Hist. New York, N.Y. (U.S.A.). *American Indians.* ① Columbia 1901. ○

Witanowski, Witold (1899). Warszawa (Polen), Złota 3/6. *Biochem. der Nerven: Vagus, Sympathicus, Humorale Übertragbarkeit der Herznervwirkung. Natriumsalze: Wirkung auf Herzautomatie, auf Nervenerregbarkeit.* ① Warschau 1924.

Witenberg, George (1895), Ass. Helminthol. Microbiol. Inst. Hebrew Univ. Jerusalem (Pa-

Witherby, Harry Forbes. London, N.W. 3 (England), 12 Chesterford Gardens, Hampstead. *Ornithol.* o

Withers, Thomas Henry (1883), Ass. London, S.W. 7. (England), British Mus., Nat. History. *Fossil Arthropods: Cirripedia.*

Witkowski, Władysław (1899), Ass. Inst. Vergl. Anat. der Haustiere. Lwów (Polska), ul. Tarnowskiego 69.

Witschi, Emil (1890), Priv.Doc. für Zool., Lekt. für Genetik. Basel (Schweiz), Zool. Inst. der Univ. *Geschlechtsbestimmung, Experim. Embryol., Parabiotische Zwillinge, Zytol. der Amphibienkeimzellen, Physiol. der Überreife der Eier. Farbvererbung bei Kaninchen.* ⑨ München 1913.

Witte, Hernfrid (1877), Dir. Swedish State Seed Testing Stat. Stocksund (Sverige). *Breeding of Meadow Plants. Sc. Seed Control.* ⑨ 1906.

Witte, Jürgen Johannes (1896), Ass. am Vet.-hygien. und Tierseucheninst., Abteilungsvorsteher. Gießen (Deutschland). *Bornasche Krankheit des Pferdes. Tuberkulose. Paratyphus. Rauschbrand.* ⑨ Gießen 1923.

von Wittenburg, Paul W., Oberkustos Geol. Mus. der Akademie der Wissenschaften. Leningrad (U.d.S.S.R.). *Triadische Fauna des Fernen Ostens und des Kaukasus.* o

Wittmack, Ludwig (1839), Dr. phil., Geh.Reg.R., Dr. agric. h. c., Dr. med. vet. h. c., o. Prof. an der Landwirtschaftl. Hochsch. und o. Hon.Prof. an der Univ. Berlin-Lichterfelde-O. (Deutschland), Hobrechtstr. 10. *Systematische Botan., angewandte Botan., besonders landwirtschaftl.* ⑨ Göttingen 1867.

Wittnich Carrisso, Luis, o. Prof. Univ. Coimbra (Portugal). *Biol.* o

Witzemann, Edgar J. (1884), Research Chem., Rochester, Minn. (U.S.A.), The Mayo Foundation. *Biol. Oxidation, Carbohydrates, Colloids, Insulin.* ⑨ Ohio State Univ., Columbus, Ohio, 1912.

Wize, Kazimierz (1873). Sędzino poste Buk (Polska). *Parasit., Bact., Mycol., Entomol.* ⑨ Leipzig: Med. 1899, Phil. 1907.

Wladimiroff, Georges (1901), Labor. Central Psycho-Physiol. Labor. Moskau 20 (U.d.S.S.R.), Hospital-Platz. *Biochem., Embryochem.* ⑨ Moskau 1923.

Wladimirsky, Alexander (1886), Doc. Zootomisch. Inst. Univ. Leningrad (U.d.S.S.R.). *Experim. Zool. und Ökol. Wirkung der Umgebung auf die Färbung der Schmetterlingspuppen.* ⑨ Leningrad.

Wlassow, Cyrill (1896), Ass. Nordwestl. Milchwirtschaftl. Untersuchungs-Labor. Leningrad (U.d.S.S.R.), Prosp. Wolodarskij 39. *Chem. und Technol. der Milchprodukte.*

Wlastowa-Minenkowa, Natalija W., Doc. Moskauer Vet.-Inst. Moskau (U.d.S.S.R.), Ul. Worowskogo 10, W. 21. *Geobotan. Pflanzen-Oekol.* o

Włodek, Jan (1885), a.o. Prof. Pflanzen- u. Ackerbaulehre Univ. Kraków (Polska), ul. Pędzichów-boczna 5. *Verhältnis von Pflanzen zu Boden. Einfluß der Ernährung der Pflanze auf Chlorophyll der Blätter, Acidität u. Kalkbedarf der Böden, Einfluß von Tag u. Nacht auf Zusammensetzung der Blätter.* ⑨ Berlin 1910.

Wodehouse, Roger Philip (1889), Dir. of Protein Labor., The Arlington Chem. Co. Yonkers, N.Y. (U.S.A.), 10 Stone St. *Morphol. of Pollen grains in relation to plant classification. Chem. of Animal and Vegetable proteins. Botan. of Hayfever plants.* ⑨ Harvard Univ. 1915. ⑨ Herbarium of Eastern U. S.

Wodinskaja, Kleopatra (1879), Ass. Bureau applied Entomol. of State Inst. of experim. Agron. Leningrad (U.d.S.S.R.), Herzenstr. 44. *Angew. Entomol., Histol. der Dipteren.* ⑨ Dipteren, Anthomyiden.

Wodsedalek, Jerry Edward (1884), Prof. of Zool. Univ. of Idaho. Moscow, Id. (U.S.A.), 215 IV Van Buren St. *Animal behavior, cytol., starvation experim., economic, entomol., genetics.* ⑨ Wisconsin 1913. o

Wodzicki, Kazimierz Antoni, Graf (1900), Dr.phil., Landw. Ingenieur, Ass. für vergl. Anat. der Univ. Kraków (Polska), 6 sw. Anny. *Anat. und Histol. der Vögel, intersexuelle und Vererbungsprobleme der Vögel, Kleintierzucht.* ⑨ Kraków 1925.

Wodziczko, Adam (1887), Dr. phil., Prof. à l'Univ., Dir. de l'Inst. d'Anat. et de Physiol. Végétales. Poznań (Polska), Inst. de Botan., rue Słowackiego 4—6. *Botan. théorique, Mycol. (Anat. végétale; localisation des ferments oxydants; champignons.)* ⑨ Cracovie 1916.

Woehrel, Th., Inst. Pasteur. Lille (France), 20, Boulevard Louis XIV. o

Woerdeman, Martinus Willem (1892), Prof. Dr. med., Vorstand des Anat.-embryol. Inst. der Univ. Groningen (Holland), Oostersingel. *Experim. Embryol.: Entstehung der Zellpolarität. Bewegungslehre.* ⑨ Amsterdam 1921.

Woglom, William Henry (1879), Prof. Inst. of Cancer Research. New York (U.S.A.), Columbia Univ. *Pathol. Bact. Cancer.* ⑨ Coll. Phys. and Surg. Columbia 1901.

Woglum, Russell S. (1882), Entomol., California Fruit Growers Exchange. Los Angeles, Cal.(U.S.A.), Box 530, Station C. *Insect pests of citrus fruits in California.*

Wohlgemut, Julius (1874), Prof., Dir. Chem. Abteilung Rudolf-Virchow-Krankenhaus. Berlin N (Deutschland), Augustenburger Platz. *Fermentmethoden, Stoffwechsel, Untersuchung des menschlichen Pankreassekrets, Pankreassekretion beim Menschen, Pankreasfunktion. Endokriner Stoffwechsel. Fermentbestand und Stoffwechsel der Haut.* ⑨ Freiburg i. B. 1898.

Wohlgemuth, Richard (1887), Dr., Landesfischereirat. Dresden-A. (Deutschland), Sidonienstraße 14, Landwirtschaftskammer, Abt. Fischerei. *Fischereibiol.; Teichdüngung; Abwasser.* ⑨ 1913.

Wohlwill, Friedrich (1881), Prof. Dr., Prosektor am Pathol. Inst. des allgemeinen Krankenhauses St. Georg. Hamburg 37 (Deutschland), Werderstraße 70. *Pathol. Anat., speciell des Nervensystems.* ⑨ Straßburg i. E. 1906. ⑨ Pathol.-histol. Praep.

Woinow, Georg W., Nikitskij Botan. Garten. Jalta (U.d.S.S.R.). *Botan., Dendrol.* o

Woit, Konstantin V., Prof., Landw.-Inst. Woronesch (U.d.S.S.R.), Professorskij korpus, W. 20. *Waldbau. Oekol. Phytosoziol.* o

Wojcicki, Stanislas (1896), Academie Méd. Vét. Lwów (Polska), Rue Kochanowskiego 63. *Elevage de bétail, culture des plantes fourragères et des prairies.*

Wojciechowski, Adolf (1886), Dr. med., Priv.Doc. der Chirurgie an der med. Fak. der Univ. Warschau (Polska), Nowogrodzka 59. *Anat., Embryol. und Physiol. des Omentum maius. Gewebekultur in vitro. Anat. der Lymphgefäße und des sympatischen Nervensystems.* ⑨ Würzburg 1910. ⑨ Omenten.

Wojtkiewicz, Anton (1876), Dir. Bact.-agron. Station. Moskau (U.d.S.S.R.), Koniuschkowskaga 31. *Boden- und Milchbakt.* ⑨ Milchsäurebakterien-Kulturen, Bact. Danysz, Knöllchenbakterien.

Wojtkewicz, Olga W. (1881), Ass. Bact.-Agron. Stat. Moskau (U.d.S.S.R.), Koniuschkovska 31. *Milk bact.* ⑨ Timirjaseveky Agric. Acad. Moskau 1921.

Woker, Gertrud (1878), Dr. phil. (Sc.), Priv.Doc. und Leiterin des Labor. für physik.-chem. Biol. an der Univ. Bern, Freie Straße 3 (Chemiegebäude). Oberstaufbach b. Uerligen, Thunersee (Schweiz). *Grenzgebiet zwischen Biol., Chem. und physikalischer Chem. Biol. Fragen, die einer chem. Behandlung zugänglich sind, z. B.: Biol. Katalysen, Fermentwirkungen, osmotische und Permeabilitätsprobleme. Carbohydrase- und Paroxydasemodelle als Teilgebiet Biol. Katalysen.* ⑨ Bern 1904.

Wolcott, George N. (1889), Chief Entomol., Service Technique, Port-au-Prince (Haïti), Barneveld, N.Y. (U.S.A.). *Economic Entomol., Insectae Portoricensis.* ① Cornell 1925.

Wolcott, Robert Henry (1868), Prof. of Zool., Head of Dept., Univ. of Nebraska. Lincoln, Neb. (U.S.A.). *Zoogeography, Ecol., Mites, Insects, Birds.* ① Univ. of Mich. 1893.

Wolda, Gerrit (1869), Ornithol. am Phytopath. Dienst. Wageningen (Holland), Bowlespark 10. *Physiol. der Singvögel: Akklimatisierung und Deklimatisierung. Morphol.: Bau, Reduction der Feder, ihre Symmetrie. Geburtsstatistik: Periodizität der Geburten in Holland und Schweden. Tuberkulose im Zusammenhang mit der Periodizität der Geburten.*

Wolf, Egbert (1860), Doc. Forst-Inst., Haupt-Botan. Garten. Leningrad (U.d.S.S.R.). *Dendrol.*

Wolf, Ernst (1902), Dr. phil. nat. Heidelberg (Deutschland), Sternwarte. *Orientierungsvermögen von Bienen.* ① Heidelberg 1924.

Wolf, Jan, Priv.Doc., Histol. et embryol. Praha II (C.S.R.), Albertov 4. *Histol. du cartilage (Origine de la substance élastique, Canaux vasculaires). Histogenèse du Pancreas (Ilets de Langerhans). Genèse des fibres collagènes. Bact.* ① Praha 1919.

Wolfe, Herbert Snow (1898), Prof., Ass. Prof. Botan., West Virginia Univ. Morgantown, W.Va. (U.S.A.). *Plant Physiol., Metabolic Changes in Stimulation from Dormancy.* ① Chicago 1925.

Wolfe, Thomas Kennerly (1892), Agronomist, Virginia Agric. Experim. Stat., Prof. of Agronomy, Virginia Polytechnic Inst. Blacksburg,Va. (U.S.A.). *Factors governing the growth and maturity of corn; variations in orchard grass.* ① Cornell Univ. 1921.

Wolff, Bruno (1893), Dr. med. et phil. Neuzelle, Kreis Guben (Deutschland). *Insektenmetamorphose, speziell Dipterenmetamorphose.* ③ Greifswald phil. 1920, med. 1925.

Wolff, E., Doc. Leningrad (U.d.S.S.R.), Forstinst. *Botan.* ○

Wolff, Erich K. (1893), Priv.Doc., Ass. am Pathol. Inst. der Univ. Berlin W50 (Deutschland), Prager Str. 35. *Allgemeine Pathol. und pathol. Anat., Bact., Serol. Infectionspathol.: Fettstoffwechsel.* ① 1919.

Wolff, Gustav (1865), o. Prof. allgem. Biol. u. biol. Psychol. Univ. Riehen-Basel (Schweiz). *Entwicklungsphysiol., theoretische Biol., Psychol.* ① Phil. München 1890, Med. Würzburg 1896.

Wolff, Max (1879), o. Prof. der Zool. und Leiter des II. Zool. Inst. der Forstlichen Hochsch. in Eberswalde (Deutschland), Moltkestr. 19 II. *Angewandte Zool.; vorzüglich Forstzool., besonders Forstentomol; Wissenschaftl. Photographie.* ① Jena 1903. ② Forstschädlinge.

Wolff, Paul (1894), Dr. med. et phil., Berlin NW87 (Deutschland), Altonaer Str. 7. *Pharm., Pharmakotherapie.* ① Med. 1920, phil. 1921.

Wolgounoff, Dimitri (1890), Dir. Sect. des plantes nuisibles, Stat. de la défense des plantes. Essentonky, Gouv. Terek (U.d.S.S.R.), 8 r. Batalinskaja. *Flore.*

Wolkow, Luka I., Doc. Nord-Kaukasische Univ. Rostow a. D. (U.d.S.S.R.), Ul. Engelsa 37, Univ. Botan. Kabinett. *Botan. Hydrobiol.* ○

Wollaston, Alexander Frederick Richmond. Dursley, Gloucestershire (England), Bencombe House, Uley. *Ornithol.* ○

Wollebek, A., Konserv. Zool. Mus. Oslo (Norge), Töien. *Zool.* ○

Wollenweber, Hans Wilhelm (1879), Dr. phil., Vorst. Mykol. Labor. der Biol. Reichsanstalt, Berlin-Dahlem (Deutschland). *Pilzforschung (Pyrenomyzeten und Nachweis ihrer Konidienpilze, Fusarien etc., fungi imperfecti). Algenforschung (Haematococcus, Chlamydomonas).* ① Berlin 1908.

Wollman, Eugène (1883) Chef de Labor. à l'Inst. Pasteur Paris (France). *Microbiol.: Vie sans microbes (élevages aseptiques), Bactériophage, Biol. des bactéries.* ① Liége 1909.

Wologdin, Alexander G., Bergingenieur, Com.Géol. de Russie. Leningrad W.O. (U.d.S.S.R.), Sredny Pr. 72-B. *Palaeont.; Archaeocyaten des Sibirischen Cambriums.* ○

Wolstenholme, Harry. Sydney (Australia), „Maybanke", Junction Road, Wahroonga. *Ornithol.* ○

Woltereck, Richard (1877), Etatsmäßiger ao. Prof. der Zool. an der Univ. Leipzig (Deutschland), Zool. Inst., Talstr. 33, und Leiter des Biol. Labor. in Seeon, Chiemgau (Deutschland). *Kausale Embryol. und kausale Hydrobiol. im Hinblick auf die Entstehung und Veränderung von Rassen und „Arten". Funktion der „Schwebefortsätze" bei pelagischen Crustaceen. Sexualität der Cladoceren. Variation und Artbildung.* ① Freiburg 1898. ② Seenplankton.

Wolters, Karl Ludwig (1892), Dr., Dir. des Bact. Inst. der Anhaltischen Kreise, Dessau (Deutschland), Herzogin-Marie-Platz. *Bact. und Serol.* ① 1920.

Wolterstorff, Willy (1864), Dr., Kustos am Mus. für Natur- und Heimatkunde der Stadt Magdeburg (Deutschland), Domplatz 5. *Urodelen, insbesondere der Alten Welt. Systematik, Lebensweise, Pflege, Zucht, Vererbung. Reptilien und Froschlurche, Fische, allgemeine Zool., Palaeont.* ① Erlangen 1898. ② Urodelen (Molche und Salamander).

Wolvekamp, Hendrik, Pieter (1904), Ass. bei der Abteilung für Vergl. Physiol. des Zool. Inst. der Reichs-Univ. zu Utrecht (Holland), Nicolaas-Beetsstraat 10bis.

Wong, Chaak Po (1900), Instr. Sericiculture. Canton (China), Ling Nan Univ. *Cocoon improvement.* ① Canton Christian Coll. 1922.

Wong Man, Hackett Med. Coll. Lect. Canton (China). *Anat., Dermatol.* ○

Wong, San Yin (1894), Lect. Physiol., Biochem. and Pharm. Hongkong (China), Univ. *Methods of Biochem. Analysis; Nitrogenous Metabolism of Chinese; Chinese Drugs.* ① Columbia Univ., 1924.

Wonsblein, Michail N., Prof. Dr., Zootechn. Inst. Moskau (U.d.S.S.R.), Tschernyschewskij 10. *Landwirtschaft, Forstwirtschaft.* ○

Wood, Horatio Charles (1874), Prof. Pharm., Therapeutics Univ. of Pennsylvania; Coll. of Pharmacy and Sc. Philadelphia, Pa. (U.S.A.), 319 South 41st St. ① Univ. of Penns. 1896.

Wood, Jessie I., Junior Pathol., Bureau of Plant Industry, U.S. Dept. of Agric. Washington, D.C. (U.S.A.). ○

Wood, Milo Nelson (1883), Pomol. U.S. Dept. of Agric. Sacramento, Cal. (U.S.A.), 409 Native Sons Bldg *Nut breeding, fruit tree sterility and fertility.* ① Univ. of California, Berkeley 1917.

Wood, Norman Asa (1857), Curator of Birds, Mus. of Zool. Univ. of Michigan. Ann Arbor, Mich. (U.S.A.). *Birds and Mammals of U.S.A.* ② Birdskins.

Wood, W. H., Prof. Univ. Liverpool (England). *Anat.* ○

Woodard, John (1882). Columbia, Mo. (U.S.A.), 117 Anderson Ave. *Plant Physiol., Ecol., Horticulture. Origin of Prairies.* ① Univ. of Chicago 1923.

Woodcock, Edward Fred (1885), Assoc. Prof. Botan. East Lansing, Mich. (U.S.A.), Michigan State Coll. ① Yale Univ. 1917.

Woodcock, Harold Mellor (1879), Protozool. Dept., Lister Inst. of Preventive Med., Grantee of Med. Research Council. London, S.W. 1 (England), Chelsea Gardens. *Protozool., Cell-biol., Haematol.* ① London 1905.

Woodger, J. H., Prof. Middlesex Hosp. Med. School. London (England), Mortimer Str., Oxford Str. W. *Biol.* ○

Woodhead, Arthur E. (1888), Instr. Zool. Ann Arbor, Mich. (U.S.A.), Dept. of Zool., Univ. of Michigan. *Parasit.* ① Univ. of Michigan 1926.

Woodhead, Norman (1903), Demonstrator Botan. Bangor (Great Britain), Botan. Dept., Memorial

Buildings, Univ. Coll. of North Wales. *Growth and differentiation of Kleinia articulata*. *Ecol. of the highland lakes of North Wales*. ⓓ Victoria Univ. of Manchester 1924.

Woodland, William Norton Ferrier, Dr. sc. Wellcome Bureau of Sc. Research. London, W.C. 1 (England), 25/28, Endsleigh Gardens. *Helminthol*. ⓓ London 1903.

Woodroof, Jasper Guy (1900), Assoc. Horticulturist. Georgia Agricultural Experiment Station, Ga. (U.S.A.). *Pest control, varietal adaptation, morphol. activities of the peach, pear, pecan fruits*. ⓓ Univ. of Georgia 1926.

Woodroof, Naomi Chapman (1900), Ass. Botanist Georgia Agric. Expt. Station, Ga. (U.S.A.). *Plant Pathol. with some additional work in the morphol. of the pistillate flower of the pecan. Cotton root rots and methods of control*. ⓓ Univ. of Idaho 1924.

Woodruff, Lorande Loss (1879), Prof. Protozool. New Haven, Conn. (U.S.A.), Yale Univ. *Protozoa. History of biol.* ⓓ A.M. honorary Yale Univ.; Ph.D. Columbia 1905.

Woods, Albert Fred, President, Univ. of Maryland. College Park, Md. (U.S.A.). *Phytopathol*. o

Woods, Farris Hardin (1898), Instr., Univ. Missouri .Columbia, Mo. (U.S.A.). *Embryol., Germcell studies on Sphaerium*. ⓓ Univ. of Missouri 1926.

Woods, Frederic A., M.D. Lect. on Biol. Mass. Inst. of Technology. Brookline, Mass. (U.S.A.), 1006 Beacon St. o

Woods, William Colcord (1893), Master Biol., Kent School. Kent, Conn. (U.S.A.), St. Andrew's Rectory. *Biol. and Histol. of Leaf Beetles (Coleoptera: Chrysomelidae)*. ⓓ Cornell Univ. 1917.

Woodward, Karl Wilson (1881), Prof. of Forestry, Univ. of New Hampshire. Durham, N.H. (U.S.A.). ⓓ Yale Univ. 1905.

Woodward, Robert Cecil (1898), Mycol. Advisory Officer, Ministry of Agric. and Fisheries. Oxford (England), School of Rural Economy, Univ. *Erysiphaceae*. ⓓ British Columbia 1921.

Woodworth, Charles William (1865), Prof. of Entomol. Berkeley, Cal. (U.S.A.), Univ. of California. *Economic Entomol. Geometrical Optics*. ⓓ Univ. of Illinois 1886.

Woodworth, Clyde Melvin (1888), Assoc. Prof. Plant Breeding, Univ. of Illinois. Urbana, Ill. (U.S.A.), 110 Agr. Bildg. *Field crops, corn, wheat, oats, borley, soybeans. Genetics of the soybean*. ⓓ Univ. of Wis. 1920.

Woollard, H. H., M.D., Ass. Prof. Dept. of Anat. and Histol. at Univ. Coll. London (England). o

Work, Paul (1886), Prof. Vegetable Gardening. Ithaca, N.Y. (U.S.A.), N.Y. St. Coll. of Agric. *Tomato nutrition, Systematic Vegetable Crops*. ⓓ Univ. of Minnesota 1921.

Worobiev, W. P., Prof. Anat. Labor. d. Med. Inst. Charkow (U.d.S.S.R.). o

Worobjew, Fedor I., Doc. Landw. Inst. Stawropol (U.d.S.S.R.). *Botan*. o

Worobjew, Semen O., Prof., Landw. Inst. Odessa (U.d.S.S.R.), Ul. Swerdlowa 99. *Phytobiol*. o

Woronichin, Nikolaj Nikolajevitsch (1882), Leiter Hydrobiol. Abtl. Botan. Garten, Mitarbeiter Botan. Mus. d. Akad. d. Wissensch. Leningrad (U.d.S.S.R.), Karpowka 19, log. 48. *Oekol. und Geographie d. Süßwasseralgen. Süßwasseralgen d. Kaukasus. Phytoplankton d. Flusses Neva, Algenvegetation d. Mineralquellen und salzhaltigen Binnengewässer: Algen d. Schwarzen Meeres. Pilze d. Kaukasus; Capnodiales*. ⓟ *Pilze d. Kaukasus, Algen d. Schwarzen Meeres*.

Worontzoff, Alexander (1860), Forstentomol. Forst-Abtlg. des Gouvernement, Doc. der prakt. Entomol. Univ. Nijny-Nowgorod (U.d.S.S.R.), Belinsky-Str. 33, I. *Cephus pygmaeus. Myclophilus piniperda. Tomicus curvidens. Polygraphus pubescens. Fichtenborkenkäfer. Grapholita nigricana. Tannenborkenkäfer*.

Woronzow, Daniel (1886), Prof. d. Physiol. Univ. Smolensk (U.d.S.S.R.), Physiol. Labor. *Physiol. der peripheren Nerven, physiko-chem. Analyse der Nervenerregung; Nerventonus; Physiol. des Nervenquerschnittes; Einwirkung d. Elektrolyten und anderen Substanzen und d. elektrischen Stromes auf den Nerven. Elektrokardiogramm*. ⓓ Odessa 1917.

Woronzow, Wassilij Nikolaewitsch (1877), Prof. Pharmacol., Dir. pharm. Inst. Univ. Woronesch (U.d.S.S.R.), Prospekt Revoluzii 3. *Überlebende Organe*. ⓓ Jurjew (Dorpat) 1910.

Woronzowa, Mary A. (1902), Labor. of Experim. Biol. of the Zoopark. Moscow (U.d.S.S.R.). *Mechanics of development. Influence of internal factors on the morphogenesis*.

Woskobolnikow, M. M., Prof., Dir. Zool. Labor. Pädagogische Hochsch. (V.I.N.O.). Kiew (U.d.S.S.R.). o

Woskressensky, Nicolaus M. (1889), Dir. Labor. Experim. Biol. des Röntgeninst., Ass. Zool. Kabinett, Univ. Kiew (U.d.S.S.R.), Leo-Tolstoi-Straße 7. *Wirkung der Röntgenstrahlen auf die Keimzellen der Drosophila, die wachsenden Embryonen d. Axolotes u. auf Flagellaten*.

Wottschal, Eugen F., Ukrainische Akademie d. Wissenschaften. Kiew (U.d.S.S.R.). *Landwirtsch. Biol*. o

Wóycicki, Zygmunt (1872), Prof. de Botan. Générale de l'Univ. de Varsovie (Polska), Wspólna Nr. 16. *Cytol., Histol.; Embryol. chez Larix et Caryocinèse chez Jucca*. ⓓ Varsovie 1899.

Wóycik, K., Dr. Prof. Poznań (Polska), Inst. Géol. Univ. *Palaeont*. o

von Wrangell, Margarete (1877), Prof. Agric.-Chem., Dir. Pflanzenernährungs-Inst. landwirtsch. Hochsch. Hohenheim, Württemberg (Deutschland). *Aufnahme der Phosphorsäure durch die Pflanzen und das Verhalten der Phosphorsäure im Boden*. ⓓ Tübingen 1909. ⓟ *Bodenproben*.

Wrede, Fritz (1891), Priv.Doc. Greifswald (Deutschland). *Schwefel- und selenhaltige Zucker. Thioglucose, Pyocyanin, Spermin, Atmung der Insekten*. ⓓ Phil. 1914, Med. 1915.

Wriedt, Christian (1883), Staatskonsulent in Genetic. Ski (Norge). *Vererbungsuntersuchungen in Tierzucht: Pferde, Rindvieh, Schafe, Ziegen, Schweine. Experim. Untersuchungen von Rindvieh, Schafen, Ziegen, Schweinen, Hunden, Tauben und Hühnern*.

Wright, Albert Hazen (1879), Prof. Zool. Cornell Univ. Ithaca, N.Y. (U.S.A.), 113 E. Upland Rd. o

Wright, Sir Almroth E., Prof. Univ. London, S.W. 7 (England), South Kensington. *Experim. Pathol*. o

Wright, H. D., Prof. Univ. Coll. Hospit. Med. School. London, W.C. (England). Univ. Street, Gower St. *Bact*. o

Wright, Samson (1899), Senior Demonstrator Physiol., Middlesex Hospital Med. School. London, W. 1 (England). *Experim. Physiol*. ⓓ London 1920.

Wright, Sewall (1889), Assoc. Prof. of Zool., Univ. Chicago, Ill. (U.S.A.). *Genetics of mammals, (guinea pig), coat color and other coat characters, polydactyly, abnormalities, growth and fecundity. Effects of inbreeding, crossbreeding and selection. Biometry*. ⓓ Harvard Univ. 1915.

Wright, Stillman (1898), Fellow Zool. Univ. of Wisconsin. Madison, Wis. (U.S.A.), Biol. Building. *Limnol., Taxonomy of Copepods (Fresh-water)*. ⓓ Beloit Coll. 1921. ⓟ Fresh-water Copepods.

Wright, William, Prof. Univ. London, S.W. (England), South Kensington. *Anat*. o

Wright, William Harmon (1885), Assoc. Prof. Agric. Bact. Univ. of Wisconsin. Madison, Wis. (U.S.A.) Coll. of Agric. *Soil and dairy bact., microbic physiol., legume bact. and bact. technique*. ⓓ Univ. of Wisconsin 1924. ⓟ *Bact. cultures*.

Wrzosek, Adam, Prof. Dr. med., Dir. Inst. de pathol. générale et expérim. Poznań (Polska), Univ. o

Wseswjatskij, Boris W., Vorsteher d. Biol. Station Timirjasew. Moskau (U.d.S.S.R.), Sokolniki, Rostokinskij pr. 7. *Biol., Zool.* o

Wu, Chenfu Francis (1896), Assoc. Prof. Zool., Yenching Univ. Peking (China), Department of Biol. *Invertebrate Zool. and Entomol. Morphol., Anat. and Ethol. of Nemura.* ③ Cornell Univ. 1922. ⑫ Entomostraca (Crustacea). Insects: Mallophaga, Orthoptera, Coleoptera.

Wu, Hsien Wu (1893), Associate Prof. of Biochem. Peking (China), Dept. of Biochem., Peking Union Med. Coll. *Biochem.* ① Mass. Inst. of Tech. 1916.

Wülker, Gerhard (1885), Dr. phil., Priv.Doc. an der Univ. Frankfurt a M. (Deutschland), Klettenbergstr. 7. *Parasit., angewandte Zool., Protozool., zool. Systematik. Blutprotozoen, besonders Haemosporidia. Nematoden, besonders parasit. Blutsaugende Insekten. Insekten als Schädlinge, besonders Curculionidae, Tenthredinidae. Cephalopoden.* ① Leipzig 1910. ⑫ Nematoden und andere parasit. Würmer.

Wünn, Hermann (1866), Kirn a. d. Nahe, Rheinprovinz (Deutschland). *Coccidae.* ⑫ Cocciden an Pflanzenteilen.

Wünschová, Amalie, M.U.Dr., Ass. Labor. de Fysiol. Univ. Masaryk. Brno (C.S.R.), Údolní 73. o

Wüst, Ewald (1875), o. Prof. der Geol. und Palaeont., Dir. des Geol.-Palaeont. Inst. der Univ. Kiel (Deutschland), Schwanenweg 20a. *Quartärfauna, Säugetiere, Mollusken (Mollusca extramarina palaearctica).* ① Halle a. d. S. 1901.

Wüstenfeld, Hermann (1879), Dr., Abteilungsleiter Inst. für Gärungsgewerbe, Berlin, Seestr. 13. Neufinkenkrug, Osthavelland (Deutschland). *Bact. der Essiggärung.* ① Univ Berlin 1908.

Wukolow, Sergej M., Prof. Kubaner Landw. Inst. Krasnodar (U.d.S.S.R.). *Agronomie. Gemüsebau.* o

Wulf, Fanni D., Doc. II. Univ. Moskau (U.d. S.S.R.). *Pathol. Anat.* o

Wulff, Alfred (1888), Dr. phil., Kustos für Seefischerei an der Biol. Anstalt Helgoland (Deutschland). *Fischereibiol., Hummer, Nannoplankton.* ① Kiel 1915.

Wulff, Eugen (1885), Prof. Univ. in d. Krim, Inst. für angewandte Botan. Leningrad (U.d. S.S.R.), Herzen-Str. 44. *Genetische Pflanzengeographie. Flora der Krim u. angrenzender Länder. Scrophulariaceae.* ① Wien 1910.

Wulzen, Rosalind (1886), Assoc. in Physiol. Berkeley (U.S.A.), Department of Physiol., Univ. of California. *Problems of nutrition of the Invertebrates. Pituitary gland, its effect on the growth of Planaria maculata.* ① Univ. of California 1914.

Wunder, Wilhelm (1898), Dr. phil., Priv.Doc. und Ass. am Zool. Inst. der Univ. Breslau 9 (Deutschland), Sternstr. 21. *Fischauge (Netzhautanat. und -physiol.). Trematoden (Cercarien).* ① München 1921.

Wunderlich, Ludwig (1859), Dr. phil., Dir. des Zool. Gartens in Köln-Riehl (Deutschland). *Tiergärtnerei.* ① Leipzig 1884.

Wurmbach, Hermann (1903), Ass. Zool. Inst. Univ. Bonn. Bonn (Deutschland), Troschelstr. 5, oder Marburg (Deutschland), Biegenstr. 20^1/$_2$. *Regeneration bei Wirbeltieren, Entzündung.*

Wurmser, René (1890), Ass. au Coll. de France. Paris (France), 9 place Marcelin Berthelot. *Photosynthèse. Chim. physique des oxydations.* ① Paris 1921.

Wwedensky, Boris (1891), Ass. Katheder der Zool. Univ. Nijny-Nowgorod (U.d.S.S.R.), Oscharskaja 28, 3. *Hydrobiol. (Limnol.).* ① Warschau 1914.

Wyatt, Stanley (1890), Investigator Industrial Fatigue Research Board London, Lect. in Psychol. Univ. of Manchester. Chapel-en-le-Frith, Derbyshire (England), ,,Wyenoft", Eccles Road. ① Manchester 1912.

Wylie, Robert B., Head Dept. and Prof. of Botan., State Univ. of Iowa. Iowa City, Ia. (U.S.A.). *Aquatic seed plants, studies on foliage leaves, morphol.* o

Wyman, Lenthall (1888), Assoc. Silvicult., U.S. Forest Service. Starke, Fla. (U.S.A.). *Silviculture. Factors affecting yield of oleoresin and effect of chipping on vitality and growth of trees.* ① Harvard 1914.

Wymore, Floyd Howard (1892), Research Ass., Univ. of California, Coll. of Agr., Agr. Experim. Stat. Davis, Cal. (U.S.A.), Univ. Farm. *Truck crop insect pests. Scutigerella immaculata.* ① Univ. of California 1923.

Wyneken, Karl (1884), Dr. phil. Leer, Ostfriesland (Deutschland), Heisfelder Str. 143. *Wundheilung an Pflanzen, speziell Blättern; Geschichte der Zool. und Botan.* ① Göttingen 1908.

Wyschinsky, Léon (1876), Chef de la Station Séricicole. Fergana (Turkestan). *Sericicult.* ① 1921.

Wysogórski, Johann (1875), Prof. Dr., Hauptkustos des Mineral.-geol. Staatsinst. in Hamburg 5 (Deutschland), Lübecker Tor 22. *Palaeont., Palaeozoische Brachiopoden.* ① Breslau 1900.

Wyssotskij, Michail M., Doc. Landw. Inst. Gorki (U.d.S.S.R.), Landwirtschaftl. Inst. *Pflanzen-Biol.* o

Yabuta, Teijirô, Prof. Tôkyô Imper. Univ. Komaba near Tôkyô (Japan), Chem. Labor., Coll. of Agric. *Fermentation Chem.*

Yagi, Nobumasa, Ass. Prof. Imper. Univ. Kyôto (Japan), Entomol. Labor., Coll. of Agric. *Insect Physiol.* ① Kyôto 1926.

Yagi, Seiichi, Prof. Tôhoku Imper. Univ. Sendai (Japan), Pharmacol. Inst., Coll. of Med. *Pharm.*

Yagita, Kun-ichirô, Prof. Med. Univ. Okayama (Japan), Anat. Inst. *Anat.*

Yago, Masatoshi, Biol. in Provincial Agric. Experim. Stat. Shizuoka (Japan), Kajima-mura, Fuji-gun. *Entomol.: Lepidoptera.*

Yakimoff, Wassily (1870), Prof., Chef du Service antipiroplasmique de l'Inst. vét. expérim. Léningrade (U.d.S.S.R.), Ordinarnaia, 5, app-t. 18. *Protozool., Hématol., Trypanos. Leishmania. Piroplasma.* ① 1915. ⑫ Piroplasma, Trypanosomes, Tiques russes.

Yakowlew, Nikolaus N., confer Jakowlew.

Yamada, Gentarô (1873), Dir. Prof. of the Fottari Agric. Coll. (Japan). *Parasit. Fungi, Gymnosporangium, Sclerospora.* ① Hokkaido Univ. Sapporo 1898.

Yamada, Shin-ichirô, Biol. Imper. Inst. for Infectious Diseases. Tôkyô (Japan), Shiba. *Diptera.*

Yamada, Yukio (1900). 2. Ass., Botan. Inst. Imper. Univ. Tôkyô (Japan). *System. Botan.: Phytol.* ① Tôkyô 1924.

Yamaguchi, Sachô, Dr. med., Lect. Zool. Labor. of Imper. Univ. Kyôto (Japan), Coll. of Sc. *Parasit. Nematol.*

Yamaguti, Yasuke (1888), Mitglied Ôhara Inst. landwirtschaftl. Forschungen. Kurasiki (Japan), Okayama-Ken. *Genetik der Kulturpflanzen, Reis.* ① Tôkyô 1914.

Yamaha, Gihei (1895), Lect. Botan., Sc. Fac., Imper. Univ. Tôkyô (Japan), Botan. Inst., Koishikawa. *Plant Cytol., experim. Plant Caryol. Fixation of cellular structures. Artificial Culture of cellular elements, Plastides, Nuclei, Chromosomes.* ① Imper. Univ. Tôkyô 1919.

Yamakawa, Makoto, Prof. Tôkyô Imper. Univ. Komaba near Tôkyô (Japan), Inst. of Hydrobiol., Fac. of Agric. *Hydro-biol. Chem.*

Yamamoto, Seriji (1889), Research work in a private labor. Uji near Kyôto (Japan), Hanayasiki. *Amphibian Spermatogenesis. Eugenics. Human sexual physiol. and psychol.* ① Tôkyô 1921.

Yamane, Jinshin (1889), Prof. Kaiserl. Hokkaido Univ. Sapporo (Japan).*Züchtungsbiol.*①Sapporo1916.

Yamanouchi, Gendô, Biol. Gunze Silk Factory. Ayabe near Kyôto (Japan), Biol. Labor. *Cytol. of Insects.*

Yamanouchi, Shigeo, Prof. High Seminary. Tôkyô (Japan), Biol. Labor. (Kôtô Shîhan Gakkô). *Cytol. and Genetics of Plant.*

Yamanouchi, Suehiko, Ass. Zool. Labor., Imper. Univ. Kyôto (Japan), Coll. of Sc. *Reizphysiol.*
Yamauchi, Masashi (1900). Okayama (Japan), Pharm. Inst. *Experim. Pharm.* ⓓ Okayama 1923.
Yanischewsky, Michael E., Prof. Palaeont. Univ., Obergeol. am Com. Géol. de Russie. Leningrad W.O. (U.d.S.S.R.), Sredny Pr. 72-B. *Palaeont. des Palaeozoicums. Carbon-Brachiopoden, Pelecypoden, Ostracoden, Cambrium-Trilobiten und Vermes.*
Yapp, Richard Henry (1871), Prof. of Botan. Birmingham (England), The Univ. *Plant Ecol.: salt marsh vegetation, water relations of plants.* ⓓ Univ. of Cambridge 1898.
Yasuda, Sadao (1895), Prof. Imper. Coll. of Agric. and Forestry. Morioka (Japan). *Physiol. studies on the fertilisation in flowering plants.* ⓓ Kyushu Imper. Univ. 1924.
Yatsu, Naohide, Prof. Imper. Univ. Tôkyô (Japan), Zool. Labor., Coll. of Sc. *Embryol., Experim. Biol.* o
Yazaki, Yoshio, Prof. in Bact. of Fikeikwai med. Univ. Tôkyô (Japan), Bact. Inst., Shiba. o
Yeager, Albert I. (1892), Horticult., North Dakota Experim. Stat. Fargo, N.D. (U.S.A.), State Coll. Stat. *Breeding fruits and vegetables.*
Yee, Martin (1892), Ass. Biochem. Peking (China), Union Med. Coll. *Biochem., Food analysis.* ⓓ Toledo, Ohio 1922.
Yenikomshian, H. A., Prof. Americ. Univ. Beirut (Syrie). *Pathol.* o
Yeolekar, F. G., Prof. Univ. of Bombay. Poona (Br. India). *Botan., Zool.* o
Yerkes, Robert Mearns (1876), Prof. Psychol., Inst. of Psychol., Yale Univ. New Haven, Conn. (U.S.A.), Kent Hall. *Psycho-biol. research, infra-human primates. Dancing Mouse. Vision in Animals, Chimpanzee intelligence and its vocal expressions.* ⓓ Harvard Univ. 1902.
Yocom, Harry Barclay (1888), Prof. Zool. Eugene, Ore. (U.S.A.), Univ. of Oregon. *Protozool.; Cytol. of animals. Spermatogenesis in the mouse. Neuromotor apparatus of Euplotes patella. Luteal Cells in the gonad of the Phalarope, sexual dimorphism of the feathering of wild birds.* ⓓ Univ. of California 1918.
Yocum, L. Edwin, Prof. of Botan., State Coll. for Women. Greensboro, N.C. (U.S.A.). *Plant physiol.* o
Yoder, Peter A. (1867), Assoc. Technol., Bureau of Plant Industry, U.S. Dept. of Agric. Washington, D.C. (U.S.A.). *Agric. Chem. Technol. Crop sugarcane.* ⓓ Univ. of Göttingen 1901.
Yokota, Takezô, Prof. Med. Univ. Niigata (Japan), Physiol. Inst. *Physiol.*
Yokoyama, Kirio, Entomol., Imper. Seric. Exper. Stat. Nakano near Tôkyô (Japan). *Entomol. Lepidoptera. Coleoptera.*
Yokoyama, Tadao, Entomol. Imper. Seric. Experim. Stat. Nakano near Tôkyô (Japan). *Anat. of Lepidoptera.*
Yonge, Charles Maurice (1899), Ass. Naturalist, Marine Biol. Assoc. United Kingdom. Plymouth (England), The Labor., Citadel Hill. *Compar. physiol. of digestion in Invertebrates. Feeding mechanisms. Marine wood-boring animals (Teredo, Limnoria etc.).* ⓓ Edinburgh 1924.
Yorke, Warrington, Prof. Univ., The Scool of Tropic. Med. Liverpool (England). *Parasit.* o
Yoshida, Sadao (1878), Prof. Zool., Parasit. Osaka (Japan), Pathol. Department, Med. Coll. *Helminthol.: Development, Morphol., pathogenity, Distribution. Prevention etc.* ⓓ Tôkyô Imper. Univ. 1906. ⓔ Specimens of Parasites.
Yoshii, Yoshitsugu, Ass. Prof. Tôhoku Imper. Univ. Sendai (Japan), Botan. Inst., Coll. of Sc. *Botan.*
Yoshioka, Toshisuke, Ass. Anat. Labor. of Imper. Univ. Tôkyô (Japan), Coll. of Med. *Cytol.*
Yothers, Merrill Arthur (1883), Assoc. Entomol., Bureau of Entomol., U.S. Dept. Agric. Yakima.

Wash. (U.S.A.), Box 243. *Life history and control of Apple insects: codling moth, red spiders, tree hoppers.* ⓓ State Coll. of Washington 1914.
Yothers, William Walter (1879), Assoc. Entomol., Bureau of Entomol., U.S. Dept. of Agric. Orlando, Fla. (U.S.A.), Post office Box 491. *Insects attacking citrus trees in Florida; Oil emulsions; sulphur sprays.*
Youden, William John (1900), Physical-chem., Boyce Thompson Inst. for Plant Research. Yonkers, N.Y. (U.S.A.), 1086 North Broadway. ⓓ Columbia Univ. 1924.
Young, Albert Gayland (1898), Ass. Prof. Pharm. Ann Arbor, Mich. (U.S.A.), Univ. of Michigan, Dept. Pharm. *Experim. Therapeutics and Chemotherapy.* ⓓ Univ. Wisconsin 1924.
Young, Donnell Brooks (1888), Prof. Biol., Univ. Arizona. Tucson, Ariz. (U.S.A.). *Protozoa: free living ciliates.* ⓓ Columbia Univ. 1923. ⓔ Gila Monsters, Heloderma suspecta. Horned Toads, Pyrosoma solare, Tarantulas. Desert animals.
Young, E. Gordon (1897, Prof. Dalhousie Univ. Halifax, Nova Scotia (Canada). *Biochem. Proteins. Carbohydr. metabolism of Bact.* ⓓ Cambridge. 1921.
Young, Harry C. (1888), Chief, Dept. of Botan. and Plant Pathol., Ohio Agric. Experim. Stat. Wooster, O. (U.S.A.). *Fruit diseases, parasitism of the Fusarium group.* ⓓ Washington Univ., St. Louis, Mo.
Young, Martin Tranthan (1894), Ass. Entomol. Tallulah, La. (U.S.A.). *Boll Weevil Control, Cotton Insects.* ⓓ Miss. A. and M. Coll. 1917.
Young, Maxwell Williamson (1898), Govern. Marine Biol. Portobello (New Zealand), Marine Fisheries Stat. *Fishery.*
Young, Paul A. (1898), Research Plant Pathol. Bozeman, Mtn. (U.S.A.), 109 Lewis Hall, Agric. Experim. Station. *Mycol., Ecol.* ⓓ Univ. of Illinois 1925.
Young, Vive Hall (1887), Prof. Plant Pathol., Coll. of Agric., Univ. of Arkansas; and Plant Pathol., Arkansas Agric. Experim. Stat. Fayetteville, Ark. (U.S.A.). *Physiol. of the fungi* ⓓ Univ. of Wisconsin 1916. ⓔ Fungal cultures, diseases of cotton and rice.
Youngken, Heber. Wilkinson (1885), Prof. of Botan., Pharmacognosy, Massachusetts Coll. of Pharmacy. Boston, Mass. (U.S.A.), 179 Longwood Av. *Compar. anat. of species· of the genus Aconitum and Viburnum. Pharmacoporial drugs.* ⓓ Ph.D. Univ. of Pennsylvania 1915; Ph.M. hon. Philadelphia, Coll. of Pharmacy 1919.
Yuasa, Hachirô (1890), Prof. Entomol. Kyôto (Japan), Coll. of Agric., Imper. Univ. *Entomol; morphol. and ecol. of insects. Larvae of the Tenthredinoidea.* ⓓ K. S. A. C. 1915; U. of I. 1920.
Yuasa, Keion, Entomol. Imper. Agric. Experim. Stat. Tôkyô (Japan), Entomol. Labor., Nishigahara.
Yuncker, Truman G. (1891), Prof. Botan., De Pauw Univ. Greencastle, Ind. (U.S.A.). *Taxonomic botan., Convolvulaceae (Cuscuta).* ⓓ Univ. of Illinois 1919.

Zabłocka, Wanda (1900), Ass. Labor. Bod. Jancrewskianum. Kraków (Polska), Al. Mickiewicza 17. *Phytopath.* ⓓ Cracovie, Univ. Jag. 1925.
Zablocki, Jan (1894), Ass. an der Univ. Kraków (Polska), Al. Mickiewicza 17. *Palaeobotan. d.Tertiärs und der Kreide. Systematik der Phanerogamen, spez. der Gattungen: Carex, Viola und Gentiana. Gallen der Pflanzen.* ⓓ Kraków 1926. ⓔ Herbarmaterial von Carex, Viola und Gentiana und andere Gattungen der polnischen Flora.
Zabolotnov, Petr. Pavl, Prof. Univ. Ploščad Imeni Lenina. Saratov (U.d.S.S.R.). *Pathol. Anat.* o
Zabolotny, D. M., Prof. Med. Inst. Leningrad (U.d.S.S.R.), Petr. st., ul. L'va Tolstogo 6—8. *Mikrobiol.* o

Zach, Franz (1878), Prof., Wien VIII/2 (Deutsch-Österreich), Lerchenfelder Str. 88/90. *Mykol., Phytopath.* ⓟ Wien 1918. ⓟ Pilzkulturen und Praep.

Zacharov, Vladimir (1893), Labor. f. experim. Biol. Swerdlov Univ. Moskau (U.d.S.S.R.), Arbat, Glasovsky per. 9/5, W. 6. *Bedingte Reflexe, Einfluß des Schilddrüsenhormons. Verdauungssekretion.* ⓟ Moskau 1926.

Žacharova, Taisya Mich., Doc. I. Univ. Moskau (U.d.S.S.R.). *Botan.* ○

Zacher, Friedrich (1884), Vorsteher des Labor. für Vorrats- und Speicherschädlinge der Biol. Reichsanstalt für Land- und Forstwirtschaft. Berlin-Steglitz (Deutschland), Schildhornstr. 9. *Angewandte Entomol., besonders Vorratsschädlinge, Spinnmilben (Tetranychidae). Orthopteren.* ⓟ Breslau 1910. ⓟ Orthopteren.

Zaćwilichowski, Jan (1890), Physiographische Kommission Akad. der Wissensch. Kraków (Polska), w. Anny 6, Zool. Inst. Univ. *Allgemeine und Experim. Entomol.; Faunistik: Odonata, Neuroptera, Panorpatae, Tenthredinoidea.* ⓟ Kraków 1926. ⓟ Polnische Odonata, Neuroptera, Panorpatae, Tenthredinoidea.

Zade, Adolf (1880), Dr., o. Prof. an der Univ., Dir. des Inst. für Pflanzenbau und Pflanzenzüchtung. Leipzig (Deutschland), Philipp-Rosenthal-Straße 25. *Pflanzenpathol., Versuchstechnik, Unkrautbiol. und Unkrautbekämpfung. Hafer und Gräser.* ⓟ Jena 1909.

Zadovskij, Anatolij E., Doc. I. Univ. Moskau (U.d.S.S.R.). *Botan.* ○

Zaewloschin, Michail (1886), Prof. Histol. u. Embryol. Odessa (U.S.S.R.), Karetnij Pereulok 18, Qu. 14. *Innere Sekretion.* ⓟ Odessa 1919.

Zagami, Vittorio (1902), Ass. Ist. di Fisiol. Univ. Messina (Italia). ⓟ Messina 1926.

Zagni, Luigi (1894), Aiuto Patol. Speciale Chirurgica Univ. Bologna 21 (Italia), Via Malaguti 29. *Storia della Medicina. Istol. Patol., Batt.*

Zagorowsky, Nicolas (1893), Prof. Inst. für Volksaufklärung. Odessa (U.d.S.S.R.), Straße Tchitcherin 24. *Hydrobiol.: Seewasserfauna und Fauna der Salzgewässer; Zool. und vergl. Anat.* ⓟ Odessa 1916.

Zaitzev, Philippe (1877), Prof. et Recteur de l'Inst. Polytechn. Tiflis, Géorgie (U.d.S.S.R.). *Entomol.: coléoptères aquatiques, les moustiques de Russie, entomogéographie du Caucase, insectes nuisibles (spéc. Scolytides).* ⓟ Leningrad 1900.

Zakolska, Zenobja (1892), Ass. de l'Univ. à Varsovie (Polska), Inst. Histol., r. Chałubińskiego 5. *Etudes cytol. sur le développement des Batraciens, des Insectes. Orthoptères, Dixippus morosus, Coleoptères (Tenebrio molitor).* ⓟ Varsovie 1923.

Zakrzewski, Aleksander (1894), Dr. med. und Tierarzt, Leiter des Pathol.-anat. Inst. der Tierärztl. Hochsch. in Lwów (Polen), ul. Kochanowskiego 59. *Lyssa der Fleischfresser. Pathol. Stoffwechsel der Vogelniere. Nierensystem.* ⓟ Lwów 1920.

Załęski, Edouard, Prof. Kraków (Polska), R. Czapskich 5. *Botan.* ○

Zaleski, Jean (1868), Prof. chim. pharmac. à l'Univ. de Varsovie (Pologne). *La matière colorante du sang.*

Zaleski-Mart, Elisabeth (1891), Ass. Charkow, Ukraine (U.d.S.S.R.), Pflanzenphysiol. Labor. des Inst. für Volksbildung. *Stoffwechselprozesse bei den Pflanzen.*

Zaleski, Karol Wilhelm (1890), Ass. Inst. de Botan. et de Phytopath. Univ. Poznań-Sołacz (Polska), ul. Śląska 1, 5. *Phytopath., Mycoflora des terres incultivées, Penicillium.* ⓟ 1926.

Zaleski, Wjatscheslow (1871), Prof. Charkow, Ukraine (U.d.S.S.R.), Pflanzenphysiol. Labor. des Inst. für Volksbildung. *Biochem., Stoffwechselprozesse der Pflanzen. Eiweißbildung in Phosphorverbindungen in den Pflanzen. Bodenbact.* ⓟ Nowo-Alexandria 1898.

Zalessky, Michael D. (1877), Prof. Palaeobotan.; Obergeologe am Com. Géol. de Russie. Orel (U.d.S.S.R.), Borissoglebskaja 12, oder Leningrad (U.d.S.S.R.), Wassily Ostrow, Sredny Pr. 72-B. *Palaeozoische Flora; speziell Permische und Carbonische Flora Asiens. Kaustobiolithe. Insects fossiles.*

Zambrino, G., Doc. Univ. Napoli (Italia). *Botan.* ○

Zāmels, Aleksanders (1897), Priv.Doc. Ass. Univ. Riga (Latvija), Botan. Labor., Kronwald-Boulev. 4. *Pulsatilla.* ⓟ Pulsatilla, Alchemilla.

Zamfirescu, Georges (1884), Prosector de l'Inst. d'Anat. Fac. de Méd. Univ. Jassy (România). *Anat. Ostéol. Embryol. hum.* ⓟ 1919.

Zander, Enoch, Univ.Prof. Erlangen (Deutschland), Rathsbergerstr. 7. *Zool.* ○

Zander, Robert (1892), Verwalter der Kustosstelle am Botan. Inst. Halle a. d. S. (Deutschland), Cröllwitzer Str. 25. ⓟ Halle 1923.

Zandt, Ferdinand (1892), Prof., Anstalt für Bodenseeforschung. Konstanz a. Bodensee (Deutschland). *Parasit., Fischkrankheiten, Bact.* ⓟ Freiburg i. Br. 1924.

Zanella, Baccio (1899), Ass. Ist. Farmacol., Tossicol. e Terapia sperim. Univ. Milano (Italia), Città degli studii, Via G. Strambio. *Fisico-chim.: Fototropia, Campi di forza negli elettroliti. Farm.: Olio di Chaulmoogra, Sedatiri.* ⓟ Bologna 1922.

Zanetti, Giovanni (1890), Ass. Padova (Italia), Ist. di Anat. Patol. Univ. *Neuropatol., micol.* ⓟ 1922.

Zange, Johannes Carl August Edmund (1880), Dr. med., o. ö. Prof. für Hals-, Nasen-, Ohrenheilkunde, Vorstand der Univ.-Klinik in Graz (Österreich), Beethovenstr. 25. *Pathol. Anat. und Physiol. des Ohres, der Nase und des Kehlkopfes.* ⓟ 1908.

Zanolli, V., Doc. Univ. Padua (Italia). *Anthropol.* ○

Zao, C. C., National Normal Univ. Peking (China). *Zool.* ○

Žapanić, Niko, Dr., Dir. d. Ethnograph. Mus. Ljubljana (S.H.S.), Bleiweisova cesta 24. *Anthropol.* ○

Zaprometov, Nicolaj (1893), Doc. Univ., Vorsteher der Phytopath. Section Uzbekische Exper. Stat. für Pflanzenschutz. Taschkent (U.d.S.S.R.), Uzbekistan. *Mykol. und Phytopath.* ⓟ Taschkent. ⓟ Fungi, Herbarium.

Zarevsky, S. F., Leiter der Wiss. Abt. Zool. Mus. Leningrad (U.d.S.S.R.), Univ. nab. 1. *Herpetol.* ○

Zarnik, Boris (1883), o. Prof. allg. Biol., Histol. u. Embryol. Univ., Med. Fac.; Vorstand des Morphol.-Biol. Inst. Zagreb (S.H.S.), Salata. *Anat. u. Entwicklungsgesch. v. Amphioxus, Vergl. Anat. der Exkretionsorgane d. Wirbeltiere, Cytol. der Mollusken (Chromosomen), Geschlechtsbestimmung.* ⓟ Würzburg 1904.

Zaslavsky, Abram (1897), Aspirant wissenschaftl. Forschungsinst. Odessa, Ukraine (U.d.S.S.R.), I. N. O. Kominternstr. 2. *Bact. Microbiol. der Rieselfelder, des schwarzen Schlammes der Odessaer Limane.*

Zattler, Fritz (1900), Dr. phil. z. Zt. München (Deutschland), Ainmillerstr. 29 III. *Vererbungserscheinungen und Sexualität bei Hutpilzen (Basidiomyzeten).* ⓟ Würzburg 1924.

Zavadskij, Alexander M., Prof. Univ. Kazan, Tatar Rep. (U.d.S.S.R.). *Zool.* ○

Zavattari, Edoardo (1883), Prof., Dir. dell' Ist. di Anat. e Fisiol. Compar. Univ. Pavia (Italia), Palazzo Botta. *Anat. dei Vertebrati.* ⓟ Pavia 1925.

Zavialoff, Vassily V. (1873), Prof. Physiol., physiol. Chem., med. Fac. Univ. Sofia (Bulgarien), Physiol. Inst. *Physiol.-chem. Studien über das Plastein, Unitar-Theorie der Magenfermente, Chondroitinschwefelsäure, Biochem. der Wasserstoffgärung.* ⓟ Dorpat 1899.

Zavialoff, Vsevolod V. (1895), Ass. physiol. Inst. med. Fac. Univ. Sofia (Bulgarien). *Allgemeine Physiol. des Herzens, Elektrokardiographie, Funktionelle Störungen der Herztätigkeit.* ⓟ Odessa 1919.

Zavřel, Jan (1879), Dr., prof. de zool. Brno (C.S.R.), Kounicova 63. *Métamorphose des insectes, Chironomidae.* ⓓ Praha 1903.

Zavrnik, Fran I. (1888), o. Prof. Univ. Zagreb (S.H.S.), Srebrnjak 87. *Histol. und Embryol. der Haustiere. Angiol. und Hämatol.* ⓓ Wien 1914. ⓟ Histol. Präp. aller Organe der Haustiere.

Zawadovsky, Boris Michailovitsch (1895), Prof., Dir. d. Labor. f. Exp. Biol. a. d. Swerdlov Univ., d. K. A. Timirjasev Biomus. I. Staatsuniv. Moskau 55 (U.d.S.S.R.), Miusskaja Platz 3. *Exp. Physiol. endocr. Drüsen, Morphogen. Bedeutung d. Schilddrüse d. Vögel, Schilddrüsenhormon in d. Geweben, Wechselbeziehungen d. endocr. Drüsen u. d. von ihnen bedingten Reflexe b. Säugetieren u. Vögeln.* ⓓ Odessa 1920. ⓟ Kollektionen russischer Fauna und Flora.

Zawadowsky, Michael M. (1891), Prof., Dir. Zool. Park, Dir. Inst. of General Biol. II. Univ. Moscow (U.d.S.S.R.), Zoopark. *Experim. Zool., Morphogenetics, Significance of the sexual glands, External factors of development of the eggs of Ascaris megalocephala.* ⓓ Moscau 1914.

Zawarzin, Alexej Alexejewicz (1886), Prof. Histol. und Embryol. an der Med. Kriegsakademie. Leningrad (U.d.S.S.R.), Grosse Sampsonlewskij pr. 5, W. 2. *Vergl.Histol.des Nervensystems Wirbelloser. Blut und Bindegewebe. Biol. Wirkung der Strahlenenergie. Sensibles Nervensystem. OptischeGanglien derInsekten.*

Zawisch-Ossenitz, Caroline (1888). Wien I (Österreich), Singerstr. 22. *Exp. Histol., Osteol.* ⓓ Wien.

Zawistowsky, I. A. (1876), Prosector Histol. und Embryol., Histol. Labor. Kiew, Ukraine (U.d.S.S.R.), Leningrad. 37. *Bindegewebe.* ⓓ Kiew 1904.

Zawrentjew, Boris (1892), 1. Prosektor d. histol. Labor. d. Staatsuniv. Kasan (U.d.S.S.R.). *Morphol. d. peripheren Nervensystems. Sensible und motorische Nervenendigungen in d. Urethra-Schleimhaut, Froschmusculatur, Grandry-Körperchen. Morphol. d. autonomen Nervensystems. Bau der sympathischen Ganglien. Regeneration d. sympath. Nervenfasern und ihrer Endigungen. Morphol. der Vagusendigungen. Herznerven.*

Zay, Carlo Edvardo (1860), V.-Dir. Staz. Chimicoagr. Torino (Italia), Via Ormea 47. *Chimica agraria.* ⓓ Torins 1915.

Zbóray, Adalbert (1899), Ass. Pharmakognost. Inst. Budapest VIII (Ungarn), Üllői-ut 26. ⓓ Budapest 1926.

Zdravko, Arnold (1898), Ass. des botan. Inst. der Univ. Zagreb (S.H.S.), Mažuranićev trg 29 II. *Cytol. der Samen von Getreidearten, Keimungsphysiol., Hortikultur.* ⓓ Zagreb 1921.

Zdravosmyslov, Vladimir Mich., Prof. Univ. Perm (U.d.S.S.R.), Ul. Karl Marksa 26. *Mikrobiol.* o

Zdrodowski, Paul-Pantaleon (1890), Prof. de microbiol. et d'épidémiol. Fac. méd., Dir. Inst. de microbiol. et d'hygiène. Bakou (U.d.S.S.R.), 39, Chemakhinskaia. *Bact. expér.: cholera, melitococcie, scarlatine. Immunité: réaction d'Abderhalden; anatoxine diphthérique, vaccination contre la diphthérie et la scarlatine; immunité, para-anaphylaxie. Biol.: Anopheles et Aëdes caspius en Aserbaïdjan.* ⓓ Rostoff 1920. ⓟ Cultures de micrococcus melitensis; prép. microscopiques et macroscopiques: melitococcie expérim. chez les cobayes; Anopheles. Moustiques.

Zebedeff, Wladimir (1882), Subdir. Inst. Experim. Biol., Prof. der Zool. H. Univ. Moscau (U.d.S.S.R.), Woronzowo Pole 6. *Biol., Genetic.* ⓓ München 1909; Moscau 1910.

Zebrikow, Woldemar M., Prof. Geol. II. Univ. Moskau (U.d.S.S.R.). *Palaeont.* o

Zebrowski, George (1895), Prof. of Biol., Coll. Villanova, Pa. (U.S.A.), Permanent Address is Buck Creek, Ind. (U.S.A.). *Invertebrate morphol. and Parasit. Hog Lung worm, American Dog Tick.* ⓓ Purdue Univ. 1923. ⓟ Microscopic slides of parasit. Invertebrates, Cestoda, Nematoda, Trematoda, Acarina and Insecta.

Zechmeister, László (1889), Dr. ing., o. ö. Univ.-Prof., Dir. des Chem. Inst. Pécs (Ungarn). *Pflanzenchem., Pflanzenfarbstoffe, Synthetisch-organ. Chem.* ⓓ Zürich 1913.

Zederbauer, Emmer., Univ.Prof. Wien (Österreich). *Botan.* o

Zedlitz-Trützschler, Graf von, Otto (1874).Tovhult b. Kalvsjöholm (Sverige). *Palaearktische u. afrikanisch-tropische Ornithol., Europäische u. afrikanische Säugetiere, Arktische Vögel u. Sänger.*

Zee, Harvey N., Prof. Kwang Hua Univ. Shanghai (China). *Zool.* o

de Zeeuw, Richard (1880), Assoc. Prof., Michigan State Coll. East Lansing, Mich. (U.S.A.), 533 Evergreen Ave. *Cytol. Botan.* ⓓ Univ. of Michigan 1909.

Zegola, S. I., Prof. Timirjasev-Akad. Moskau (U.d.S.S.R.), Petrovskoe-Razumovskoe. *Genetik, Pflanzenselektion.* o

Zeiger, Karl (1895), Dr. med., Priv.Doc. für Anat. und Entwicklungsgeschichte, 1. Ass. an der Dr. Senckenbergischen Anat. der Univ. Frankfurt a. M. (Deutschland), Theodor-Stern-Kai 36/37. *Vergl. und makroskopische Anat. des Menschen und der Wirbeltiere, der Hautmuskulatur der Säugetiere. Histol. und mikroskopische Anat. Theorie der Färbung und Fixierung. Histol. der quergestreiften Muskulatur.* ⓓ Frankfurt a. M. 1921.

Zeissler, Johannes Karl (1883), Vorstand des Bact. Untersuchungsamtes der Stadt Altona und Oberarzt am Städt. Krankenhaus. Altona a. d. Elbe (Deutschland), Bebelallee 29. *Die anaeroben Bakterien. Gasödeme, Tetanus, Botulinus, Rauschbrand.* ⓓ Leipzig 1909. ⓟ Anaerobe Bakterien.

Zelada, M., Doc. Univ. Tucumán (Argentina). *Botan.* o

Zeldowitsch, J. B., Prof. Dr., Dir. Anat. Labor. Inst. für Med. Wissensch.. Leningrad (U.d. S.S.R.). o

Zeleny, Charles (1878), Prof. Zool., Univ. of Illinois. Urbana, Ill. (U.S.A.). *Genetics, Experim. Embryol., Regeneration.* ⓓ Chicago 1904.

Zelift, Clarke Courson (1897), Instr. in Biol. Toledo,O.(U.S.A.), Univ. *Botan.* ⓓ Philadelphia 1923.

Zelinka, Karl, Univ.Prof. Wien III (Österreich), Siegelgasse 3. *Zool. Biol. Evertebraten.*

Zeliony, Georges (1879), Prof. Physiol. Ecole vet. sup., Dir. Labor. de physiol., Conserv. Inst. de Physiol. à l'Académie des Sc. Leningrad (U.d. S.S.R.). *Physiol. du cerveau, reflexes conditionnés, Associations cerebrales chez l'homme et les animaux. Reflexes rythmiques conditionnés.* ⓓ Leningrad 1907. ⓟ Filme cinematographique d'un chien sans hemisphères cérébraux.

Željko, Kovačević (1893), Chef d. entomol.-phytopath. Abteilung d. landwirtschaftl. Station. Osijek (S.H.S.). *Angewandte u. systematische Entomol., Coleoptera, Ichneumonidae.* ⓓ Zagreb 1922.

Zeller, Hermann (1882), ORR. Reichsgesundheitsamt, Berlin-Dahlem (Deutschland), Unter den Eichen 82/84. *Veterinärbact.* ⓓ Leipzig 1908.

Zeller, Sanford Myron (1885), Plant Pathol., Oregon Agric. Coll. and Experim. Stat. Corvallis, Ore. (U.S.A.). *Fruit trees, small fruits: Rubus and Fragaria. Mycol., taxonomy of the Hymenogastrales and other gastromycetes.* ⓓ Univ. of Washington1917.

Zeman, Francis (1881), Inspector of the Dendrol. Society. Průhonice near Praha (C.S.R.). *Dendrol.* ⓟPlants hardy in the open and seeds of gardenplants.

Zeman, Victor, Prof. Univ. Nac. d. Litoral. Corrientes (Argentina). *Histol., Physiol.* o

Žemčužnikov, Eugen (1891), Prof. d. Donschen Hochsch. für Landwirtsch. und Melioration. Nowotscherkassk (U.d.S.S.R.), Potschtowaja 65. *Transpiration des Pflanzen und Kohlensäureassimilation.*

Zemplén, Géza (1883), Prof. organ. Chem. Techn. Hochsch., Priv.Doc. Univ. Budapest I (Ungarn), Gellérttér 4. *Kohlenhydrate, Aminosäuren.* ⓓ Budapest 1904. ⓟ Seltene Zucker- und höhere Kohlenhydrate, ev. Aminosäuren.

Zenari, Silvia (1896), Ass. Ist. Botan. R. Univ. Padova (Italia). *Sistematica, Floristica, Fitogeografia, Genetica.* ℗ Padova 1918.

Zenkewitsch, Lew (1889), Priv.Doc. Univ., Leiterd. Hydrobiol. Abteilung des Schwimmenden Wissenschaftl. Meeresinst. Moskau (U.d.S.S.R.), Univ., Zool. Mus. ul. Herzena 6. *Morphol. u. Histol. des Nephridialsystems der Polychaeten. Süßwasserpolychaeten (Baikal-See). Quantitative Bodenuntersuchungen der Meere.* ℗ Moskau 1925. ℗ Süßwasserpolychaeten und Brackwasserpolychaeten.

Zenoni, Costanco, Prof. incar. Univ. Milano (Italia). *Histol. pathol.* o

Zepeda, R. B., Prof. Univ. Tegucigalpa, Honduras (Zentralamerika), Plaza de la Merced. *Parasit.*o

Zernow, S. A., Prof. Dir. Ichtyol. Labor., Landwirtschaftl. Akad., I. Univ. Moskau (U.d.S.S.R.), Petrowsko-Rasumowskoje. *Embryol.* o

Zerny, Hans (1887), Dr. phil., Kustos am Naturhistorischen Mus. in Wien I (Deutsch-Österreich), Burgring 7. *Systematik und geogr. Verbreitung der Lepidoptera, Diptera, Neuroptera. Floristik der ganzen Erde (nur Pteridophyta und Phanerogamae).* ℗ Wien 1911. ℗ Herbarpflanzen.

Zerow, Demetrius (1895), Kustos Botan. Mus. Ukrainische Akademie der Wissensch., Ass. am Inst. der Volksbildung. Kiew, Ukraine (U.d.S.S.R.), Tarassowskaja 1—3. *Bryophyta (Sphagnaceae) der Ukraine und des Kaukasus.* ℗ Bryophyta.

Zetek, James (1886), Entomol., Curator of Barro Colorado Island Labor. of Tropical Biol. of the Inst. for Research in Tropical America. Ancon (Canal Zone), Box 245. *Tropical Insects; Med. Entomol.; Central American Mollusks.* ℗ Illinois 1911. ℗ Panama Mollusks.

Zettler, Hofr. Prof. Dr., Kustos Nat. Kabinett. Mannheim (Deutschland). o

Zeuner, Heinrich (1885), Dr. Würzburg (Deutschland), Riemenschneiderstr. 9. *Systematik, Biol., Verbreitung der höheren Pilze.* ℗ Würzburg 1922.

v. Zeynek, Richard (1869), o.ö. Prof. f. med. Chem. a.d. deutschen Univ. u. Vorstand des med.-chem. Inst. Prag (C.S.R.), Salmg. 3. *Physiol. Chem.* ℗ 1893.

Ziegenspeck, Hermann (1891), Priv.Doc. und Ass. an den Botan. Inst. der Univ. Königsberg i. Pr. (Deutschland) *Physiol. Chem. der pflanzlichen Membranine. Botan. Serodiagnostik. Endotrophe Mycorrhiza. Bodenbact., Physiol. Anat. Mitteleuropäische Orchideen. Öle. Analytische Chem. im Dienste der botan. Physiol. Milchsäfte und Schleime. Schleudermechanismen von Ascomyceten und Bewegungsmechanismen. Casparyscher Streifen, Salz-Basenäquivalent. Sparstärke.* ℗ Jena 1916.

Ziegler, August (1885), Ökonomie- und Landwirtschaftsrat, Leiter der Bayer. Hauptstelle für Rebenzüchtung Würzburg (Deutschland), Schönleinstr. 3. *Züchterische Verbesserung der Weinrebe durch vegetative Züchtung und Sämlingszüchtung. Keimfähigkeit und Stimulation der Traubenkerne. Blütenbiol. Untersuchungen der Rebblüte. Züchterische Verbesserung der Obstsorten und Blütenbiol. Untersuchungen über Sterilität und Fertilität.* ℗ München 1911. ℗ Literatur über Rebenzüchtung und Obstzüchtung.

Zielstorff, Willy, Univ.Prof. Königsberg i. Pr. (Deutschland), Kirchenstr. 83. *Agric.*

Zietzschmann, Reinhold Otto (1879), o. Prof. der Anat. an der Tierärztl. Hochsch. zu Hannover (Deutschland), Bischofsholerdamm 86. *Anat., Histol. und Entwicklungsgeschichte der Haut und ihrer Anhänge. Anat. und Histol. des Auges. Entwicklungsgeschichte des Skeletts. Bau und Entwicklung der Milchdrüse.* ℗ Zürich 1902.

Ziganow, Sergej Wasilijewitsch (1889), Ass. Pharm. Inst. Odessa (U.d.S.S.R.), Olgijewskaja 4. *Experim. Pharmacol. u. Toxikol.* ℗ Moscau 1913.

Zikes, Heinrich (1860), Dr. phil., o. ö. Prof. an der Technischen Hochsch. und an der Univ. Wien IX (Deutsch-Österreich), Währingerstr. 41. *Technische Mykol. Biochem. der Gärung.* ℗ Wien 1885. ℗ Hefen, Schimmelpilze, Bakterien.

Žila, Vladimír Loro (1889), Chef de service Inst. des Fermentations, Ecole polytechnique tchèque. Brno (C.S.R.). *Microbiol. technique.*

Zilber, Lev. Alek., Prof. I. Univ. Moskau (U.d. S.S.R.). *Bact.* o

Zile, Mārtiņš, Dr. Prof. Univ. Rīga (Latvija), Elizabetes ielā 63, dz. 7. *Pathol.* o

Žilinskas, Jurgis (1885), a.o. Prof. Kaunas (Litauen), Anat. Inst. *Anat. und Anthropol.* ℗ Kaunas 1922. ℗ Schädel der Litauer.

Zillig, Hermann (1893), Dr. phil., Leiter der Zweigstelle der Biol. Reichsanstalt für Land- und Forstwirtschaft Berncastel-Cues a.d. Mosel (Deutschland). *Rebenkrankheiten und -schädlinge, Ustilagineen.* ℗ Würzburg 1920. ℗ Ustilagineen.

Zimmer, Carl (1873), o. Prof. der system. Zool. an der Univ., Dir. des Zool. Mus. Berlin N 4 (Deutschland), Invalidenstr. 43. *Crustacea, Deutsche Fauna.* ℗ Breslau 1897.

Zimmer, John Todd (1889). Ass. Curator of Birds at Field Mus. of Nat. Hist. Chicago, Ill. (U.S.A.). *Ornithol.* ℗ Nebraska 1910.

Zimmermann, Albrecht Wilhelm Philipp (1860), Geh.Reg.R. Prof. Dr., pensioniert als OberReg.R. an der Biol. Reichsanstalt für Land- und Forstwirtschaft in Berlin-Dahlem. Berlin-Zehlendorf-W. (Deutschland), am Heidehof 24. *Pflanzl. Zytol. Mikrotechnik. Tropische Pflanzenkrankheiten. Tropische Agric. Physiol. der Pilze.* ℗ Leipzig 1882.

Zimmermann, August (1875), Dr., o. Prof., Dir. des Anat. Inst. der königl. ung. Tierärztl. Hochsch., Doc. der Univ. Budapest VII (Ungarn), Rottenbiller-ut 23. *Vergl. Anat. und Embryol. der Wirbeltiere; Myol., Angiol., Hautgebilde.* ℗ Budapest 1900.

Zimmermann, Ernst H. (1860), Preußisch. Landesgeolog a. D., Prof., G.BergR. Berlin N 4 (Deutschland), Invalidenstr. 44. *Ceratites, Nautilus, Prospondylus, Dictyodora, Chondrites. Problematische Fossilien.* ℗ Jena 1884. ℗ Phycodes circinnatum u. Dictyodora Liebeana aus Thüringen.

Zimmermann, Hans (1871), Dr. phil., Landesökonomierat, Leiter der Hauptstelle für Pflanzenschutz für Mecklenburg-Schwerin und Mecklenburg-Strelitz.Rostock,Meckl.(Deutschl.),Graf Lippe-Str. 1. *Angew. Botan., Phytopath.* ℗ Erlangen 1898.

Zimmermann, Karl Wilhelm (1861), a.o. Prof., Prosektor am Anat. Inst. der Hochsch. Bern (Schweiz), Friedeckweg 20 I. *Histol.: Zelle im allgemeinen, Epithelien, Speicheldrüsen, Magen-Darmkanal, Niere, Gefäßsystem, Gehörapparat. Embryol.: Kopfentwicklung.* ℗ Berlin 1886.

Zimmermann, Percy White (1884), Prof. Botan. Univ. of Maryland, College Park,Md.(U.S.A.). *Plant Physiol. Veg. plant Propagation. Seed Germination. Root Growth in Cuttings.* ℗ Univ. of Chicago 1915.

Zimmermann, Sergius Emil (1882), Prof. Anat. des Menschen, Dir. anat. Inst. Univ. Taschkent (U.d.S.S.R.), Med. Fac. *Biomechanik; Anthropol. (Gehirn u. Schädel).* ℗ Moskau 1909.

Zimmermann, Walter (1892), Priv.Doc. und Ass. am Botan. Inst. der Univ. Tübingen (Deutschland), Naukierstr. 31. *Algol. (spec. Cytol.), Reizphysiol., Palaeobotan. und Pflanzengeographie, Euphorbiaceae.* ℗ Freiburg i. Br. 1920.

Zinserling, Iwrij Dmitrievitsch (1894), Konserv. Botan. Garten. Leningrad (U.d.S.S.R.). *Pflanzengeographie, Palaeobotan. (posttertiär), Systematik: Sorbus, Cotoneaster, Heleocharis, Spiraeoideae.*

Zinsser, Hans (1878), Prof. Bact. and Immunol. Harvard Univ. Boston, Mass. (U.S.A.), Department of Bact., Harvard Med. School. *Bact. of Infectious Diseases, Immunity. Hygiene.* ℗ Columbia 1903.

Ziobrowski,Stefan (1890).Kraków14(Polska),Józefińska 23. *AngewandteBotan. Dendrol.* ℗ Kraków 1926.

Zipf, Karl (1895), Dr. med., Priv.Doc. für Pharmacol. Münster i. W. (Deutschland), Westring 12. *Kohlehydratstoffwechsel der Leber. Aufnahme und*

Abgabe von Stoffen durch die Zelle. Vitalfärbung. Narcose. ⓓ Heidelberg 1921.

Zirkle, Conway (1895), National Research Fellow Biol. Sc.; Applied Biol., Harvard Univ. Boston,Mass. (U.S.A.), Bussey Inst. *Structure Development and Inheritance of Plastids.* ⓓ Johns Hopkins Univ. 1925.

Zirn'ts, János (1893), Entomol. Pflanzenschutzinst., Leiter d. Abteil. Priekuli. Priekulu selekcijas Stac. C. Césim (Latvija). *Angewandte Entomol., Aphididae.* ⓟ Aphididae.

Zironi, Amilcare, Prof. Univ. Milano (Italia). *Microbiol.* o

Zirpolo, Giuseppe (1887), Lib. Doc. Zool. e Lib. Doc. Anat. e Fisiol. comp. Univ. Napoli (Italia), Via Duomo 193. *Briozoi: morfol., biol., sistematica. Echinodermi: rigenerazione, anomalie. Batteri luminosi: bioluminescenza batt.*

Zitowitsch, B., Dr. med., Labor. Reval, Tartin (Eesti). *Bact. Fermente.*

Zitowitsch, Ivan (1876), Prof. Pharmacol. Univ. Rostov a. Don (U.d.S.S.R.). *Wirkung d. Alkohols auf die Magenverdauung. Bedingte Reflexe. Physiol. und Pharm. d. Nieren.* ⓓ 1911.

Zlataroff, Assen, Dr. a.o. Prof. Chem. Inst. d. Univ. Sofia (Bulgarien). *Phytobiochem.* o

Zlatnik, Alois (1902). Dvůr Králové n. Lab. (C.S.R.). *Géobotan.: Ecol. et sociol. de Sesleria calcarea. Ecol. forestière: Fagus. Botan. systématique: Hieracium sg. Archieracium.* ⓓ Prague 1925. ⓟ Hieracium sg. Archieracium.

Zlobin, Alek. R., Doc. II. Univ. Moskau (U.d. S.S.R.). *Pathol. Anat.* o

Zodda, G., Doc. Univ. Napoli (Italia). *Botan.* o

Żołka, Josef (1890), Prof. Kladno (C.S.R.). *Pilze in Schächten und Bergwerken, Kellern.* ⓓ Prag 1924. ⓟ Photographien von Höhlenpilzen; auch getrocknete und präparierte Exemplare.

Żółciński, Dr. phil., Prof. Lwów (Polska), Dublany. *Biochem.* o

Zoljesoft, Jurij (1888), Ass. Inst. für Mikrobiol. landw. Fac. Zagreb (S.H.S.), Wilsonplatz 4. *Bodenbakterien. Kefyr u. Kumys.* ⓓ Charkow 1918, Zagreb 1926.

Zollikofer, Clara (1881), Dr. phil., Priv.Doc. für Botan. an der Univ. Zürich 7 (Schweiz), Bergstr. 118. *Pflanzenphysiol., Reizbewegungen.* ⓓ Berlin 1918.

Zon, Raphael (1874), Dir., Lake States Forest Experim. Station. Univ. Farm. St. Paul, Minn. (U.S.A.). *Forest research.* ⓓ Cornell Univ. 1901. o

Zondek, S. G. (1894), Dr., Priv.Doc. für Pharmacol. Univ. Berlin NW (Deutschland), Siegmundshof 7. *Die Bedeutung der Ionenlehre für Physiol. und Pathol.* ⓓ Berlin 1919.

Zorn, Wilhelm (1884), Prof. Dr. Tschechnitz (Deutschland). *Tierzucht und Züchtungsbiol., Grünland.* ⓓ 1911.

Zorzi, P., Prof. Univ. Padua (Italia). *Fisiol.* o

Żossimow, Wsewolod Wladimir (1899), Ass. Kasan (U.d.S.S.R.), Zootomisches Kabinett, Univ. *Morphol. d. Wirbellosen (Oligochaeten), Hydrobiol. (Acarina), Biocoenotik.*

Zoth, Oskar (1864), o. Prof. Physiol. Univ. Graz 3 (Österreich), Harrach-Gasse 21. *Haemoglobin. Muskelphysiol. Energieumwandlungen der Netzhaut. Herzbewegung. Blutgasanalyse. Objektiv darstellbare Spektralfarben. Drehmomente der Augenmuskeln. Gesichtswahrnehmungen. Hören.* ⓓ Graz 1888.

Zotta, Georges (1886), Prof. de Parasit. Fac. Méd., Chef de la Section de Parasit. Inst. de Sérothérapie ,,Prof. Dr. J. Cantacuzène". Bucarest (Romania). *Protistol.: Flagellés et Rhizopodes pathogène; Sporozoaires: Haemosporidies. Insectes: Diptera Culicidae, systématique, oecol.* ⓓ Jassy 1912. ⓟ Culicidae.

Zriakovski, Basile (1888). Dir. Sect. entomol., Stat. de la défense des plantes à Essentuky, Gouv.Terek. (U.d.S.S.R.) rue Batalinskaja 8. *La faune du gouv. Terek, nuisible aux plantes cultivées. Orthoptera, Lepidoptera, Coleoptera du gouv.Terek.* ⓓ Warszawa 1912. ⓟ Lepidoptera, Orthoptera, Coleoptera du gouv.Terek.

Zschokke, Achilles (1867), Oberstudiendir., Prof., Dr., Dir. der Staatlichen Lehr- und Versuchsanstalt für Wein- und Obstbau in Neustadt a. d. Haardt (Deutschland), Maximilianstr. 45. *Pflanzenbau und Pflanzenschutz: Reben, Obstbäume u. Beerenobst. Gärungsbiol. v. Obst- u. Traubenweinen.* ⓓ Zürich 1898.

Zschokke, Fritz (1860), o. Prof. Zool. u. vergl. Anat. Univ. Basel (Schweiz), Zool. Anstalt. *Hydrobiol. Parasiten.* ⓓ Genf 1884.

Zubina, Esther M. (1899), Labor. of Experim. Biol. of the Zoopark. Moscow (U.d.S.S.R.), Maroseika, Starosadsky per. 4, Ap. 4. *Sexual gland in fowl.* ⓓ Kazan 1922.

Zühdi-Bey, Mehmed (1894), Arzt u. Tierarzt, Prof. agrégé de Bact. Konstantinopel (Türkei), Tierärztl. Hochsch., Skutarie-Sélimiée. *Anaërobier u. Serol. Geflügelcholera.* ⓓ Berlin 1919.

Zuitin, Adolf (1891), Ass. Labor. f. Genetik u. Experim. Zool. d. Naturwissenschaftl. Inst. Univ. Leningrad (U.d.S.S.R.). *Abänderung der Variabilität nach Alter und Geschlecht. Vererbungszytol., Chromosomen.*

Zulueta, Antonio de (1885), Prof. Mus. Nacional de Ciencias Naturales. Madrid 6 (España), Palacio del Hipódromo. *Hérédité chez les Insectes.* ⓓ Madrid 1910.

Zundel, George Lorenzo Ingram (1885), Investigator. New Haven, Conn. (U.S.A.), Osborn Botan. Labor. Yale Univ. *Ustilaginales. Cereal and Potato diseases.* ⓓ Cornell Univ. 1915. ⓟ Ustilaginales.

Zunz, Edgard-Victor (1874), Prof.pharmacol.expér. Fac. méd. Univ., Dir. Inst. thérapeutique Univ. Bruxelles (Belgique), 67 Rue des Deux-Eglises. *Coagulation du sang, anaphylaxie, tession superficielle du plasma, action des alcaloïdes sur le tube digestif. Métabolisme de base.* ⓓ Bruxelles 1897.

Zuwerkalow, Demetrius (1892), Ass. Biochem. Inst. Charkow (U.d.S.S.R.), Joumowskaja 5. *Vitamine, Aminosäurenstoffwechsel.* ⓓ Charkow 1922.

Zwaardemaker, Hendrik (1857), o. ö. Prof. der Physiol. Univ. Utrecht (Holland), Van Wijckkade Nr. 28. *Olfactol., Akustik, experim. Phonetik, Reflexlehre, Bioradioaktivität.* ⓓ Amsterdam 1883.

van Zwaluwenburg, Reyer Herman (1891), Ass. Entomol., H. S. P. A. Experim. Stat. Honolulu (Hawaï), P. O. Box 411. *Taxonomy of Elateridae, Soil fauna.* ⓓ Mass. Agr. Coll. 1913.

Zweibaum, Jules (1887), Dr., Prof. agrégé, Adjoint à la Fac. de Med. Univ. d'Histol. et d'Embryol. Univ. de Varsovie (Polska), Chałubińskiego 5. *Protozool., Coloration vitale, Régénération, Culture des tissus.* ⓓ Bologna 1913.

Zweigelt, Fritz (1888), Leiter der Bundes-Rebenzüchtungsstation. Klosterneuburg (Österreich), Feldgasse 7 A. *Rebenzüchtung. Maikäferbiol., Gallenforschung, Blattlausgallen.* ⓓ Graz 1911.

Zwetkowa, Eugenia Sergeewna (1892), Ass. Physiol. und Anat. d. Pflanzen Forstinst. Leningrad (U.d.S.S.R.). *Transpiration und Wasserbewegung in d. Pflanze.*

Zwick, Wilhelm (1871), o. Prof. Univ., Dir. Vet.-hygien. und Tierseuchen-Inst. Gießen, Hessen (Deutschland), Leihgesterner Weg 20. *Infektiöse Krankheiten der Haustiere und Vet.-Hygiene.* ⓓ Sc. nat. Tübingen 1896; med. vet. h. c. Gießen 1919.

Zwitkiss, Isaak (1868), Vorsteher Pasteurschen Abteilung Bact. Inst. Kiew, Ukraine (U.d.S.S.R.), Paschkinska 2. ⓓ Kiew 1893.

Zwölfer, Wilhelm (1897), Biol. Reichsanstalt für Land- u. Forstwirtschaft, fliegende Station. Rastatt i. Baden (Deutschland), Saatzuchtanstalt d. Bad. Landwirtschaftskammer. *Angewandte Entomol.; Infektionskrankheiten der Insekten* ⓓ Tübingen 1924.

Zworykin, Sergej (1895), Biol. Ona-Station. Murom, Gouv. Wladimir (U.d.S.S.R.), Sowetskaja 45. *Hydrochem.*

Zybine, Sophie (1885), Ass.-Pathol. Univ. Stat. Protection des Plantes du Gouv. N.-Nowgorod (U.d.S.S.R.), Univ. Cabinet Botan. *Maladies de lin: Melampsora lini, Colletotrichum linicolum, Polispora lini.* ⓟ Prép. d. maladies parasit. d. plantes.

II. LABORATORIA

Laboratoria biologiae generalis.
Institutiones biologicae generales.

Abisko, Lappland (Sverige).
Naturwissenschaftliche Station Abisko (Abisko Naturvetenskapliga). Dir. Dr. Bror Hedemo. — Leiter der biol. Abt.: Carl G. Alm, Upsala; Leiter der geophysischen Abt.: Dr. Bruno Rolf, Stockholm. *Sommer 8 Gastplätze. 4 Wohnzimmer, jedes für 2 Gäste, stehen unentgeltlich zur Verfügung. Biol., tier- und pflanzengeographische, faunistische und floristische Untersuchungen. Großes Laboratoriumszimmer. Vollständiges meteorologisches Observatorium. Kleiner botan. Versuchsgarten.*

Alma, Mich. (U. S. A.).
Dept. of Biol. and Geol. Alma Coll. Dir. H. M. MacCurdy.

Aluschta, Krim (U. d. S. S. R.).
Wissenschaftliche Station des Staats-Natur-Reservats in der Krim. Dir. Nikolaus Troitzkij. — Henryka Poplawska, Botan.; Ivan Pusanow, Prof. d. Zool. *5—6 Plätze. Flora und Fauna der Krim.*

Amherst, Mass. (U. S. A.).
Dept. of Biol. Amherst Coll. Prof. H. H. Plough (Biol.).

Annandale-on-Hudson, N.Y. (U. S. A.).
Dept. of Biol. Saint-Stephan's Coll. Dir V. Obreshkove.

Antigonish, N. S. (Canada).
Dept. of Biol. Univ. of St. Francis Xavier's Coll, Dir. C. J. Connolly, Ph.D. — A. F. Chaisson, M.A., B.Sc.; J. C. Chisholm, B.A.

Arkadelphia, Ark. (U. S. A.).
Dept. of Biol. Onachita Coll. Dir. E. A. Spessard.

Auckland (New Zealand).
Auckland Inst. and Mus. Dir. Gilbert Archey, O.B.E., M.A., F.Z.S. — L. T. Griffin, F.Z.S., Ass. Curator. *1 working-place for guests.*

Baldwin City, Kan. (U. S. A.).
Dept. of Biol. Baker-Univ. Dir. G. H. Bretnall.

Baltimore, Md. (U. S. A.).
Inst. for Biol. Research Johns Hopkins Univ. Dir. Prof. R. Pearl. — A. L. Allen.

Battle Creek, Mich. (U. S. A.).
Dept. of Biol. Battle Creak Coll. C. F. Sulton (Instr. in Biol.).

Beirut (Syria).
Biol. Labor. American Univ. Dir. Alfred Ely Day. — Dr. Wm. T. Van Dyck, Zool.; J. F. Crawford. ⊕ Herbarium of Syrian Plants. *A few animal skins. — The Herbarium labor.*

Berlin (Deutschland).
Staatliche Hauptstelle für den naturwissenschaftlichen Unterricht, Biol. Abt., W 35, Potsdamer Straße 120. Dir. Dr. Martin Herberg, Studienrat. — Ehrenamtliche Mitarbeiter: Prof. Dr. phil. et med. h. c. R. Kolkwitz (Botan.); Dr. phil. H. Bethge (Botan.); Dr. phil. O. Curio (Tierphysiol.); W. Fuhrmeister (Tierkunde). — Zweigstelle Geestemünde der Staatlichen Hauptstelle für den naturwiss. Unterricht (Hochseefischerei).

Berlin-Dahlem (Deutschland).
Kaiser-Wilhelm-Inst. für Biol. I. Dir. Prof. Dr. C. Correns; II. Dir. Prof. Dr. R. Goldschmidt. — I. Abt. Correns: Vererbungslehre u. allgem. Biol., Botan.; II. Abt. Goldschmidt; Vererbungslehre u. allgem. Biol., Zool.; III. Abt. Prof. Dr. Max Hartmann: Protisten; IV. Abt. Prof. Dr. Otto Warburg: Physiol.; V. Abt. Dr. Mangold: Entwickelungsmechanik; VI. Abt. Prof. Dr. Otto Meyerhoff: Physiol. *Etwa 8 Ass. Etwa 15—20 Plätze.*

Besse, Puy de Dôme (France).
Station Biol. Dir. M. Moreau, Prof. Botan. Fac. Sc. Univ. Clermont. — M. Denis, Ass. Fac. Sc. Univ. Clermont, Adjoint au Dir. *Systématique, organisation, biol. des animaux et des plantes des montagne; organismes des tourbieres et des lacs; flores anciennes; travaux myol., bryol., lichénol.; questions forestières, études sur les pâturages et les troupeaux, la maturation des fromages, l'amélioration des plantes cultivées en montagne, les plantes médicinales etc. — Sur le flanc Sud des Monts Dore, à proximité de la région des lacs du massif. central; altitude: 1000 M. 11 lits, 1 salle à manger pour les travailleurs et leur famille; une salle de recherches, une salle de pisciculture, un jardin d'expériences; un cours d'eau et un îlot pour observations en milieu naturel.*

Biol. Board of Canada (Canada).
Dir. A. P. Knight, Alice St., Kingston, Ont. — E. E. Prince, 206 O'Connor St., Ottawa, Ont.; A. G. Huntsman, Atlantic Biol. Station, St. Andrews, N.B.; A. H. Leim, Atlantic Biol. Station, St. Andrews, N.B.; W. A. Clemens, Pacific Biol. Station, Nanaimo, B.C.; J. C. Forbes, Atlantic Experim. Station, Halifax, N.S.; H. E. Tanner, Atlantic Experim. Station, Halifax, N.S.; Hess, Atlantic Experim. Station, Halifax, N.S.; R. E. Foerster, Vedder River Cressing, B.C.; D. B. Finn, Pacific Experim. Station, Prince Rupert, B.C.; F. B. Adamstone, Ph.D., Univ. of Manitoba, Winnipeg, Man.

Bolschewo, an der Nordbahn (U. d. S. S. R.).
Biol. Station (Eigentum d. ,,Ges. der Freunde von Naturforschung, Anthrop. u. Ethnogr."), Sewern. sh. d. b. Dorf Burkowo, Gouv. Moskau. Dir. Prof. N. V. Bogajavlenskij.

Brandon, Man. (Canada).
Biol. Dept. of Brandon Coll. Dir. J. W. Hill, M.A.

Bratislava — Preßburg (C. S. R.).
Inst. für allgemeine Biol., Schiffbeck. Dir. Prof. J. Babor. — Dr. B. Krajník, Ass. Ⓟ Mollusca. — Psychophysiol. Labor. (geplant).

Braunsberg, Ostpr. (Deutschland).
Naturwiss. Kabinett der Staatl. Akademie. Dir. Prof. Dr. med. et phil. Johannes Baron. — Ⓟ Anschauungsmaterial aus der Vererbungslehre.

Brisbane, Queensland (Australia).
Dept. of Biol. Unv. Dir. Prof. E. J. F. Goddard.

Brno — Brünn (C. S. R.).
Inst. f. allgem. Biol. Univ., Udolni 73. Prof. Dr. K. F. Studnička. — Doc. Jan Bělehrádek, Ass.; Ludw. Drastich, Ass.
Biol. Forschungsinst. d. Landw. Hochsch., Lednice. Dir. Em. Bayer.: cf. Lednice.
L'Inst. des Recherches Biol., Černá pole. Dir. Dr.-Ing. Jaroslav Krizenecky. — Dr.-Ing. J. Podhradský, Ass.; Dr.-Ing. J. Kostomárov, Ass.; Dr -Ing. J. Chomkovič, Ass.

Bucureşti de la Sinaia (România).
Staţinnea biol. a Univ. Dir. Prof. A. Popovici Băznoşann.

Buenos Aires (Argentina).
Gabinete de Biol. Univ. Fac. phil. Dir. Prof. Chr. Jakob.

Changsha, Hunan (China).
Biol. Dept. Coll. of Yale. Dir. H. B. Bender.

Charkow, Ukraine (U. d. S. S. R.).
Biol. Labor. der Kommunistischen Univ., 54. Artiom-Str. Dir. Prof. Vitaly Rischkow. — E. Finkelstein, Lector; E. Poliakow, Lector; S. Schapiro, Ass.; Bulanoka, Ass. Ⓟ Buntblättrige Pflanzen.

Cherson, Ukraine (U. d. S. S. R.).
Biol. Kabinett u. Labor. des Inst. für Volkserziehung namens N. K. Krupska. Dir. Lektor A. V. Myrnenko.

Christchurch (New Zealand).
Biol. Labor. Canterbury Coll. Dir. Dr. Chas. Chilton, Prof. of Biol. — Chas. E. Foweraker, Lect. in Charge Forestry Dept.; J. E. Hutchinson, Lect. in Forest Utilisation; Miss Elizabeth M. Herriott, Ass. Lect. in Biol. (Botan.); Miss Flora B. Murray, Ass. Lect. in Biol. (Zool.). *3 or more places. Researches on New Zealand Zool. and Botan. Collections of New Zealand plants and animals. Special collection of New Zealand and Subantarctic Crustacea, particularly, Amphipoda and Isopoda.* Ⓟ New Zealand plants, seaweeds Herbarium and spirit specimens. New Zealand animals, especially Crustacea; large collection of microslides of Amphipoda and Isopoda. Forestry specimens. — Mountain Biol. Station Cass, 600 m above sea-level. *Specially fitted for researches on subalpine and alpine vegetation and lacustrine fauna and flora.*

Concepción (Chile).
Labor. für allgem. Biol. d. Univ. Dir. Prof. O. Wilhelm. *Innere Sekretion.*

College Station, Tex. (U. S. A.).
Dept. of Biol. Agric. Coll. Dir. O. M. Ball, Ph.D., Head Prof. — J. E. Adams, M.S., Instr. Biol.; E. H. Gibbons, B.S., Ass. Prof. Biol.; E. H. Harper, Ph.D., Assoc. Prof. Biol.; P. A. Miller, M.S., Ass. Prof. Biol.; L. J. Pessin, Ph.D., Ass. Prof. Biol.; D. J. Pratt, A.M., Assoc. Prof. Biol.

Cork (Ireland).
Labor. of biol. of the Univ. Coll.

Decatur, Ill. (U. S. A.).
The Biol. Labor. of Millikin Univ. Dir. Prof. Helmer Parell von Wold Kjerschow Agersborg, B.S., M.S., A.M., Ph.D. — Dr. Thomas D. Howe, Instr. in Biol.; Miss Myrna Jones, B.A., M.A., Instr. in Biol. *I would be very glad to accommodate 1 or 2 persons to study with me the fauna and flora of Lake Decatur, a new lake made as a result of damming up a river (Sangamon) which has produced a lake circa 25 miles long and 3 miles wide. Necessary working facilities will be provided.*

Disko (Greenland).
Danisch Arctic Sta. Dir. Morten P. Porsild. *Biol. research in arctic science. Two working places (addressed to Danish Government, Dept. of the Interior of Prof. C. H. Ostenfeld, Pres. of committee for the Sta. København). Arctic plants, material culture of the Eskimo.*

Dmitriew, Kursk. Gouv. (U. d. S. S. R.).
Staro-Perschinski Biol. Station d. Moskauer Gesellsch. d. Naturforscher a. d. I. Moskauer Univ.

Durham, N.C. (U. S. A.).
Duke Univ. Biol. Labor. Dir. Prof. Bert Cunningham, Ph.D. — A. S. Pearse, Ph.D. (Zool., Ecol.), H. L. Blomquist, Ph.D. (Botan.); F. G. Hall, Ph.D. (Physiol.).

Fredericton, N. B. (Canada).
Dept. of Biol. Univ. of New Brunswick. Dir. Philip Cox, B.A., Ph.D.

Gainesville, Fla. (U. S. A.).
Dept. of Biol. Univ. of Florida. Ass. Prof. T. H. Hubbell (Biol.).

Geneva, N.Y. (U. S. A.).
Biol. Labor. Hobart Coll. Dir. Prof. Elon Howard Eaton. — Theodore Tellefsen Odell, Instr. *We should be glad to cooperate with anyone wishing to use our general equipment.* Ⓟ *Flora and fauna of the immediate vicinity.*

Gent — Gand (Belgie).
Labor. der allg. levensverrichtingen. Dir. Prof. H. de Waele.

Glasgow (Scotland).
Biol. Dept. Convent of Notre Dame. Dir. M. Taylor.

Grand Rapids, Mich. (U. S. A.).
Biol. Labor. of the Calvin Coll. Dir. J. P. van Haitsma.

Hamamatsu (Japan).
Biol. Labor. Hamamatsu Imper. Coll. of Technol.

Hong-Kong (China).
Labor. of biol., Univ.

Irkutsk (U. d. S. S. R.).
Biol.-Geograph. Forschungs-Inst. Univ., Naberežnaja 20. Dir. Prof. Timofeev. — Wiss. Sekr. Prof. Smirnov. Vier Sektionen: 1. Biol. (Zool., Botan., Histol.); 2. Geographische (Geographie, Ethnographie, Archäol.); 3. Geophysische; 4. Biochem. — Biol. Mitglieder: Die Prof.: Vitalij Dorogostajskij (Zool.); Valentin Smirnov (Botan.); Sergej Timofeev (Zool.,Histol.); Albert Frank-Kamenecki (Biochem., Hydrochem.); Viktor Burov, Ass. (Zool.); Ivan Molodych,Ass.(Hydrographie); Aleksandra Ušakova, Doz. (Histol.) Vladislav Jasnickij, Doz. (Botan.). — Zoofarm am Baikalsee (Füchse, Silberfüchse, Polarfüchse, Zobel), Wiss. Leiter: Prof. V. Dorogostajskij. Biol. Station am Baikalsee, Dir. V. Jasnickij.
Inst. f. allgem. Biol. u. Parasit. Dir. Prof. V. Ševjakov.

Jaroslavl (U. d. S. S. R.)
Kabinett für Biol. des Pädagogischen Inst.

Kiew, Ukraine (U. d. S. S. R.).
Inst. Ukranien sc. de biol., Puskinskaja 8, W. 8. Dir. J. Hežcnko.
Labor. für Biol. des Kiewer Inst. für Volkswirtschaft.

Kingston, Ont. (Canada).
Dept. of Biol. Queen's Univ. Dir. W. T. MacClement, M.A., D.Sc. — A. B. Klugh, B.A., Ph.D.; R. O. Earl, B.A., M.Sc.

Kraków (Polska).
Inst. de Biol. et Embryol., Św. Jana 20. Dir. Prof. E. Godlewki. — Dr. J. Bury; Dr. T. Marchlewski; Dr. M. Krahelska.

Lafayette, Ind. (U. S. A.).
Dept. of Biol. in building named „Stanley Coulter Hall of Biol." of Purdue Univ. Dir. Howard Edwin Enders. — Charles A. Behrens, Bact.; Philip A. Tetrault, Assoc. Bact.; Charley L. Porter, Botan.; Edwin J. Kohl, Ass. Botan.; Oliver P. Terry, Physiol.; Charles M. James, Ass Physiol.; Eleven Instr. and Ass. *Open to students regularly registered.*

La Laguna, Tenerife (Canarias).
Labor. de Biol., Fac. de ciencias, Seccion Univ. Establecida en la Laguna.

Lednice — Eisgrub (C. S. R.).
Biol. Versuchsstation der čech. Hochsch. in Brno. Dir. Prof. Emil Bayer. — Ass. Bajkov. *5 Plätze. Mikroskopische Zool. u. Botan., speziell Morphol. u. Systematik. Planktonuntersuchungen. Systematische Entomol. u. Ornithol. Binokulare, Mikromanipulator. Groß. binokular. Fernrohr für Ornithol. Vollständige Apparatur für Limnol. (Plankton) u. Entomol.* ℗ Entomol. Studienmaterial, Wasserfauna, Wasseru. Sumpfflora.

Leicester (England).
Biol. Labor. of Univ. Coll. Dir. Prof. Dr. Ethel N. Miles Thomas. — *2 working-places for guests. Plant Anat. Botan. and Zool.* ℗ Serial preparations of Seedling Anat.

Leningrad (U. d. S. S. R.).
Labor. für allgemeine Biol. des Staatl. Inst. für med. Wissenschaften, Sovietski 4. Dir. Prof. Dr. P. P. Ivanov. — Dr. V. A. Pavlov, I. Ass. u. Priv.-Doc.; Dr. P. E. Butning, II. Ass.; Hr. S. I. Rubaschev, II. Ass. *2—3 Plätze für morphol., experim.-morphol. u. spez. physik.-chem. Fragestellungen.*
Biol. Abt. des Röntgenol. und Radiol. Staats-Inst. ul. Röntgena (ehem. Licejskaja) 6.
Leningrader Zentral-Pädagog. Biostation, Demidow Per. 1. Dir. B. E. Rajkow.
Leningrader Wissenschaftl. Institut namens P. F. Leshaft, Pr. Maklina 32. Dir. N. A. Morosow (cf. die Abteilungen bei den Fachgruppen).
Naturwiss. Abt. des Leningrader Staatl. Pädagogischen Inst., Mojka 48. Dir. F. E. Tur.
Biol. Kabinett des Zentr. experim. Technikums „N. A. Nekrasov", Pr. Karla Marksa 84. Dir. I. Poljanskij. — G. N. Boč.

Lexington, Va. (U. S. A.).
Labor. of Biol. of the Washington and Lee Univ. Dir. Prof. W. D. Hoyt.

Lincoln, Chester County, Pa. (U. S. A.).
Dept. of Biol. of the Lincoln Univ. Dir. H. F. Grim.

Ljubljana — Laibach (S. H. S.).
Biol. Inst. d. Univ., Zološka cesta. Dir. Dr. P. Grošelj.

London (England).
Dept. of Biol., Middlesex Hosp. Med. School of the Univ., Mortimer Str., W.; Oxford Str. Dir. J. H. Woodger.
Wellcome Bureau of sc. Research, W. C. 1, 25/28, Endsleigh Gardens. Dir. C. M. Wenyon, C.M.G. C.B.E., M.B., B.S., B.Sc. — H. C. Brown, Bact.; S. H. Daukes, Dir. of Mus.; L. J. Davis, Ass. Bact.; C. A. Hoare, Protozool.; B. Jobling, General Sc. Worker; M. E. MacGregor, Entomol.; A. C. Stevenson, Pathol.; W. N. F. Woodland, Helminthol. — Wellcome Mus. of Med. Sc.; Wellcome Physiol. Research Labor.; Wellcome Chemical Research Labor.; Wellcome Entomol. Field Labor.; Wellcome Historical Medical Mus.
Biol. Labor. of the South-Eastern Agric. Coll. of the Univ., Wye, Kent.
Dept. of Biol., London School of Med. for Women of the Univ., Royal Free Hospital, 8 Hunter Str., Brunswick Square, W. C. Dir. G. P. Mudge.
Dept. of Biol. Kings Coll. for Women, Univ. of London, W. C. 2, Strand. Dir. P. C. Esdaile.

Los Angeles, Cal. (U. S. A.).
Biol. Labor. of the Univ. of Southern California. Dir. Prof. C.V. Beers.Prof.A.D.Howard;Prof.I.McColloch.

Lynchburg Va. (U.S.A.)
Biol. Dept. of the Lynchburg College. Head Prof. R. S. Freer. *Pteridophyta, Plant ecol.*

Madrid (España).
Labor. de Investigaciones Biol., Paseo de Atocha13. Dir. Prof. Dr. S. Ramon y Cajal.

Mexico D. F. (Mexico).
Dirección de Estudios Biol., 7/a calle de Balderas, núm. 94. Departamento de Exploración de la Fauna y de la Flora; Departamento de Museos; Parque Zoológico; Jardin Botánico y Acuario de Chapultepec; Sección de Administración y Archivo. Dir. Prof. Alfonso L. Herrera. — Dr. Leopoldo Flores (Exploraciones y Encargado de la Secretaria); Prof. Max. Martinez (Botánica); Prof. M. García Tunco (Quimica); Dr. J. Solis (Fisiol. Compar., Biol. Méd. y Microbiol.); Prof. V. Santiago (Ornitol.); Dr. Alejandro Ruelas (Entomol.); Prof. F. Contreras (Invertebrados); J. Garduño (Taxidermia); Fco. Moctezuma (Panoramas y Modelado); Prof. J. Cancino Gómez (Peces, Reptiles y Batracios); Prof. A. del Río (Mineral., Geol. y Paleont.); José A. Durán (Parque Zool. de Chapultepec); Ing. Octavio Solís (Jardin Botán. de Chapultepec); Ing.Daniel Ruiz Benitez (Acuario del Parque Zool. de Chapultepec).

Milano (Italia).
Labor. di biol., Univ. Dir. L. Necchi.

Montreal, Que. (Canada).
Dept. of Biol. Univ. of Montreal. Dir. L. J. Dalbis, D. ès S.

Moskau (U. d. S. S. R.).
Dir. des Facteurs biol. de la Soc. Inst. de recherches sc. de Timirjaseff. Dir. Prof. G. Bossé.
Inst. f. experim. Biol., Woronzowo Pole 6. Dir. Prof. W. Lebedeff.
Labor. für allgemeine Biol. der 2. Univ. Dir. M. M. Zavodovskij.
Moskauer Pädagogisch-Biol. Station, Sadowaja Kudrinskaja 7.
Naturwiss.-Pädagogisches Zentral-Inst., Kitajskij proesd 3, Polytechn. Mus
Staatl. wissenschaftl. biol. Forschungsinst. „Timirjaseff", Pjatnickaja ul. 48. Dir. Prof. S. G. Navašin. — 84 wiss. Mitarb.

Nagasaki (Japan).
Biol. Labor. Nagasaki Imper. Med. Univ.

New York (U. S. A.).
Dept. of Biol. New York Univ., Heightr. Dir. H. W. Stunkard. — A. B. Dawson, Prof. of Biol.; R. P. Hull, Microbiol.
Dept. of Biol. Coll. of the City. Dir. A. L. Melander.

Nikolaev (U. d. S. S. R.).
Biol. Kabinett am Inst. f. Volksbildung.

Nottingham (England).
Biol. Inst. Univ. Coll. Dir. Prof. Dr. J. W. Carr.

Odessa (U. d. S. S. R.).
Labor. für Morphol. u. Physiol. des Wissensch. Forschungsinst., Olgievskaja 4. Dir. Prof. N. K. Lyssenkov. — Leiter der Sektion f. pathol. Anat.: Prof. M. M. Tizengauzen (Tiesenhausen), Wirkl. Mitgl.: Prof. N. S. Kondrat'ev; Prof. F. M. Žmajlovič; Prof. B. A. Šacillo; Prof. A. M. Melik-Megrabov. 5 wiss. Mitarbeiter, 5 Aspiranten.

Inst. für Biol. Med. Inst. Dir. Prof. D. K. Tretjakov.

Katedra biol., Kominternstr. 2. Sektionen: Mikrobiol.; Anat. u. Physiol. der Pflanzen; Morphol. u. Systematik d. Pflanzen; Zootomie; Zool. Dir. Prof. F. M. Porodko. — Leiter der Sektionen: Prof. Ja. Ju. Bardach (Mikrobiol.); Prof. D. O. Svirenko (Morphol. u. Systematik d. Pfl.); Prof. D. K. Tretjakov (Zootomie); Prof. N. Lignau (Zool.). Wirkl. Mitgl.: D. L. Rubinstein. 5 wiss. Mitarbeiter, 10 Aspiranten.

Omsk (U. d. S. S. R.).
Inst. f. Biol. d. Sib. Inst. f. Land- u. Forstwirtschaft. Dir. Prof. S. D. Lavrov.

Orono, Me. (U. S. A.).
Dept. of Biol. Coll. of Agric. Dir. P. W. Whiting, Ph.D., Prof. in Biol. — J. W. Gowen, Ph.D., Biol.; Karl Sax, D.Sc., Biol.; C. H. Batchelder, M.S., Assoc. Prof. Biol.; I. H. Blake, A.M., Assoc. Prof. Biol.; W. H. Eyster, Ph.D., Assoc. Prof. Biol.; Marjorie E. Gooch, M.S., Ass. in Biol.; P. A. Harriman, B.A., Instr. Biol.; E. D. Hull, M.S., Instr. Biol.; Beatrice W. Johnson, B.A., Ass. in Biol.; E. Elizabeth Jones, M.A., Instr. Biol.; Iva A. Merchant, B.S., As. Aid.; H. B. Smith, M.S., Ass. Biol.; Emmeline D. Wilson, Lab. Ass. in Biol.; Helen Woodbridge, M.S., Instr. Biol.

Oslo (Norge).
Biol. Labor. Univ. Dir. Prof. Johan Hjort.

Ottawa, Ont. (Canada).
Dept. of Biol. of the Victoria Memorial Mus. Dir. R. M. Andersen, Ph.D. — E. M. Kindle, Ph.D.; P. A. Taverner.

Dept. of Biol. Univ. of Ottawa. Dir. J. A. Lajeunesse, O.M.I., M.A., Ph.L.

Oxford (England).
School of Rural Economy, Univ. of Oxford, Parks Rd. Dir. Prof. J. A. Scott Watson. — Helen Bancroft, D.Sc., Botan.; Norman Cunliffe, M.A., Entomol.; R. C. Woodward, B.Sc. Agric., Ph.D., Mycol. *3 places for Botan., Zool., Mycol. work.*

Paris (France).
Labor. de Recherches Biol. à l'Ecole des Hautes Etudes. Dir. E. Rabaud.

Labor. Biol. générale, cf.: Physiol. gen. fac. med.

Pecs (Ungarn).
Biol. Inst. der Univ. (Med. Fac.), Rákóci-Str. 80. Dir. Dr. Alexander v. Gorka. — Eugen Hanzséros, Prakt.; Ladislaus Kanizsay, Prakt.

Peking (China).
Biol. Inst. Tsinghua Univ. Dir. Prof. P. W. Claassen.

Labor. of Biol. Yenching Univ. Dir. A. M. Boring.

Perm (U. d. S. S. R.).
Inst. des recherches biol. à l'Univ. avec Station biol., Zaimka. Dir. Prof. Viktor Karlovitch Schmidt. — Wirkliche Mitglieder: D. Alexeew (Chem.);

W. Beklemischew (Zool.); A. Henckel (Botan.); W. Zdrawosmyslow (Bact.); Lubischew (Zool.); A. Perichanianz (Physiol.); D. Sabinine (Physiol. des Plantes); Sjusew (Botan.); A. Tuline (Agrochem.); W. Schmidt (Histol.); Collaborateurs sc.: E. Danini (Histol.); E. Karnauchowa (Bact.); N. Kostromine (Bact.); P. Krasowski (Botan.); P. Henckel (Botan.); E. Lazarenko (Histol.); W. Nikitine (Pédol.); W. Petropawlowski (Physiol.); M. Polukarow (Chem.); A. Tausson (Zool.); T. Temnikowa (Chem.); L. Trefilowa (Botan.); D. Charitonow (Zool.); A. Trifonow (Chém.); Aspiranten: W. Baskina (Zool.); G. Fridmann (Zool.); K. Igoshina (Botan.); O. Muchina (Histol.); L. Litwinow (Physiol. des plantes); L. Sabinina (Chém.). *Environs 12 places.* ⓟ *Parmi les animaux on étude des rotifères, araignées, turbellaries, Halticini (Coléoptères).* — *Station biol. sur les Kama.*

Perth (W.-Australia).
Biol. Labor. Univ. Dir. Prof. George E. Nichols.

Peru, Neb. (U. S. A.).
Biol. Dept. Nebr. State Teachers Coll. Dir. A. E. Holch.

Poltawa, Ukraine (U. d. S. S. R.).
Biol. Kabinett der Univ.

Porto (Portugal).
Biol. Inst. Univ. Dir. Prof. Americo P. de Lima. — A. L. M. Guimarães; A. P. Nobre; G. A. Ferreira da Silva.

Poznań — Posen (Polska).
Inst. der allgem. Biol. der Univ. Dir. Prof. Dr. Edward Lubicz Niezabitowski. — Dr. Bugochwał Kalocsay Kalusza, Ass. *3 Arbeitsplätze für Gäste.*

Praha — Prag (C. S. R.).
Inst. für allgemeine Biol. und experim. Morphol. der med. Fac., II, Kateřinská 32. Dir. Prof. Dr. Vladislav Růžička. — MUDr. Vladimir Bergauer, Ass.; MUC. Bohumil Hluchovský, Ass.; MUC. Franz Patočka, wiss. Hilfskraft. *3 Plätze für physikal.-chem. Untersuchungen der lebendigen Substanz.*

Princeton, N. J. (U. S. A.).
Dept. of Biol. of the Univ. D. Causey, Instr.

Providence, R. I. (U. S. A.).
Biol. Dept. Brown Univ. Dir. F. P. Gorham.

Put-in Bay, O. (U. S. A.).
The Franz Theodore Stone Labor. of Ohio State Univ. (Formerly The Lake Labor.). Dir. Dr. Raymond C. Osburn. — Frederick H. Krecker, Ecol.; Stephen R. Williams, Zool.; Clarence H. Kennedy, Entomol.; Lewis H. Tiffany, Algol.; Malcolm E. Stickney, Botan. *60 persons can be accommodated in the new labor. which is now being built. Work on all phases of fresh-water biol. — Boats (gasoline launches and row-boats), all kinds of collecting apparatus, aquaria etc.*

Quebec (Canada).
Dept. of Biol. and Anat. Laval Univ. Dir. C. Végina, M.D. — R. Potvin, M.D.

Rabat (Maroc).
Inst. sc. chérifien, Avenue Moulay Youssef. Dir. Dr. Jacques Liouville. — A. Théry, Entomologiste, Conserv. de Mus.; L. Emberger-Flahault, Botaniste; R. Ph. Dollfus, Parasitologiste, Office de faunistique et parasit. marocaines; J. Granjon de Lépiney, Entomologiste agric.; P. Regnier, Conserv. honoraire de l'Herbier du Maroc; J. Bourcart, Chef de la mission géol. permanente; G. Dedebant, Chef du Service de Météorol. générale. *6 places pour Zool. et Faunistique marocaines, Ornithol, Parasit. locale, Entomol agric., Paléont. nord africaine, botan., phytopath.* ⓟ *Toute la flore et faune marocaine,*

compris la faune marine du plateau continental et la paléont. locale.

Rangoon, Burma (Br.-India).
Dept. of Biol. Judson Coll. Dir. F. E. Northup.

Reno, Nev. (U. S. A.).
Dept. of Biol. Coll. of Agric. Dir. Peter Frandsen, A.M., LL.D., Prof. Biol. — C. L. Brown, A.M., Instr. Biol.; P.A. Lehenbauer, Ph.D., Assoc. Prof. Biol.; Margaret E. Mack, A.M., Assoc. Prof. Biol.

Rochester, N.Y. (U. S. A.).
Dept. of Biol. Univ. of Rochester. M. A. Stewart, Instr.

Sackville, N. B. (Canada).
Dept. of Biol. Mount Allison Univ. Roy Fraser, M.A., B.S.A.

St. Louis, Miss. (U. S. A.).
Dept. of Biol. Dir. A. M. Schwitalla.

Salisburg Cove, Me. (U.S.A.)
Biol. Labor. Mount Desert Esland. Dir. U. Dahlgren.

Saskatoon, Sask. (Canada).
Dept. of Biol. Univ. of Saskatchewan. Dir. W. P. Thompson, Ph.D. — A. E. Cameron, D.Sc.; W. P. Fraser, M.A.; L. G. Saunders, Ph.D.

Schenectady, N. Y. (U.S.A.)
Dept. of Biol. Union College. Dir. James W. Mavor, Ph. D. — Henry K. Svenson, A. M. David McC. De Forest, A. B.

Sherman, Tex. (U. S. A.).
Dept. of Biol. Austin Coll. Dir. P. E. Reid.

Sofia (Bulgarien).
Biol. Inst. d. Univ. Med. Fac. Dir. Prof. Dr. Popoff.

Soochow (China).
Biol. Supply Service Univ. Station. Dir. Chenfu F. Wu.
Dept. of Biol. Soochow Academy. Dir. Prof. Jos. William Dyson. *Flora, plant morphol.*

Stambul (Türkei).
Labor. of biol. of the Robert Coll. Dir. Prof. V. Besnard.

State College, N.M. (U. S. A.).
Dept. of Biol. Coll. of Agric., Mesilla Park. Dir. R. F. Crawford, M.S., Prof. Biol., Biol. — P. M. Gilmer, Ph.D., Assoc. Prof. Biol., Ent.

Sydney (Australia).
Biol. Branch Dept. of Agric. Dir. Prof. R. D. Watt. — H. J. Hynes, Ass.

Toronto, Ont. (Canada).
Dept. of Biol. McMaster Univ. R. W. Smith, B.A., Ph.D.
Biol. Dept. of the Ontario Coll. of Pharm. Dir. Paul L. Scott, M.B.
Biol. Dept. of the Ontario Provincial Mus., St. James Square. Dir. C. W. Nash.

Toulouse (France).
Labor. biol., Inst. catholique. Dir. Prof. L. Boule.

Tsinan, Shantung (China).
Biol. Labor. Shantung Christian Univ. Dir. Arthur Paul Jacot. *2 or 3 working-places for guests. Life history, anat., histol., in plant or animal biol., land or freshwater* ⓟ *Plants and animals of the province (and of northeast China). Exchange only for publications on fauna and flora of north or central China.*

Tucson, Ariz. (U. S. A.).
Bureau of Biol. Survey Dept. of Agric. W. P. Taylor, Biol.
Dept. of Biol. Coll. of Agric. Dir. D. B. Young, Ph.D., Head of Dept. — C. T. Vorhies, Ph.D., Prof. Entomol.; J. G. Brown, Ph.D., Prof. Plant Pathol.; G.T. Caldwell, M.S., Assoc. Prof. Biol.; D. A. Gilchrist, B.S., Rodent Control Specialist (Coop. U.S.D.A.: P. O., Phoenix); A. F. Hemenway, Ph.D., Assoc. Prof. Biol.; A. A. Nichol, B.S., Ass. Entomol.; O. L. Raber, Ph.D. Ass. Prof. Biol.; R. B. Streets, Ph.D., Ass. Prof. and Ass. Plant Pathol.

Urbana, Ill. (U. S. A.).
Illinois Labor. of Natural History.

Valparaiso, Ind. (U. S. A.).
Dept. of Biol. Univ. Dir. L. F. Heimlich.

Varna (Bulgarien).
Bulgar. Inst. zur Erforschung des Schwarzen Meeres. Das Inst. wird die Aufgabe haben, das Schwarze Meer in naturgeschichtl., phys., ethnograph., wirtschaftl., polit. u. histor. Beziehung zu erforschen.

Virginia (U. S. A.).
Dept. of Biol. Hampton Inst. Dir. T. W. Turner.

Vjatka (U. d. S. S. R.).
Wiss. Forschungsinst. f. Heimatkunde, Vjatstij; Pedagog. Inst., Ul. Lenina III. Dir. Prof. Bori, Sozontovič Lukaš. — Sekr. Arkadij Iosifovič Šernins Abt.-Leiter: 1. Abt. f. Bodenkunde: St. L. Ščeklejn; 2. Abt. f. Agronomie: P. T. Rešetnikov; 3. Abt. f. angew. Zool. u. Selektion: B. S. Lukaš; 4. Abt. f. angew. Chem.: P. A. Bobrov; 5. Abt. f. Wirtschaft des Heimatgebiets: V. A. Tanaevskij; 6. Abt. f. Gesch. des Heimatgebiets: P. N. Luppov; 7. Abt. f. Elektrotechnik: P. K. Meyer — Inst.: Labor. f. Selektions- u. zool. Arbeiten; Labor. f. Vegetationsversuche u. bodenchem. Untersuchungen.

Waco, Tex. (U. S. A.).
Biol. Inst. Baylor Univ. Dir. Prof. W. A. Buice.

Warszawa (Polska).
Biol. Inst. Marcell Nenckl's Societatis sc. Varszaviensis. Präs. Kazimierz Białaszewicz. — 1. Physiol. Labor., Dir. K. Białaszewicz; 2. Biol. Labor., Dir. Romuald Minkiewicz; 3. Labor. f. experim. Embryol., Dir. Józef Eysmond; 4. Hydrobiol. Station am See Wigry, Leiter Alfred Lityński.
Biol. Inst. der Univ. Dir. J. Eismond.

Washington, D. C. (U. S. A.).
Biol. Surveys U. S. Dept. of Agric. Dir. E. W. Nelson, Chief. — A. B. Howell, Myol. and Osteol.; F. C. Lincoln, Assoc. Biol.; Waldo L. McAtee; A. R. Kellogg, Assoc. Biol.; Sernon Bailey, Biol.; H. H. T. Jackson, Biol.

Waxahachie, Tex. (U. S. A.).
Dept. of Biol. Trinity Univ. Dir. C. D. Day.

Wilno (Polska).
Inst. de Biol. générale Univ., 15, Rue Zakretowa. Dir. Prof. Dr. Jan Wilczyński. — Dr. Jan Bowkiewicz, I. Ass.; Dr. Casimire Urbanowicz, II. Ass.; Leonard Dąbrowski, Ass. suppléant. *2—3 places. Microscopie, recherches hydrobiol.* ⓟ *Les animaux de la faune d'eaux douces, spécialement Invertebrés.* — Station hydrobiol. (aux environs de Wilno) sur „les Lacs Verts" (Zielone Jeziora).

Winnipeg, Man. (Canada).
Dept. of Biol. Univ. of Manitoba. Dir. V. W. Jackson, Sc.M.

Wolfville, N. S. (Canada).
Dept. of Biol. Arcadia Univ. H. G. Perry, M.A. — R. H. Wetmore, Ph.D.

Wooster, O. (U. S. A.).
Dept. of Biol. Coll. of Wooster. Dir. R. V. Bangham.

Woosung b. Shanghai (China).
Inst. f. Zool. u. Botan. Tungchi Univ. Dir. Prof. Dr. Klautke.

Woronesch. (U. d. S. S. R.)
Labor. d. zool. Mus. d. Univ. Dir. Prof. C. Saint-Hilaire. — W. Buchalowa, Doc. u. Ass.; L. Ignatowa, Laborant; W. Lichnitzkaja, Laborant. ⊕ Insekten und Wassertiere.
Wissenschaftl. Forschungs-Inst. a. d. Staatl. Univ.
Wissenschaft. Forschungs-Inst. namens Timirjasew, Ul. 9 Janvarja 21, W. 1.

Laboratoria evolutionis et geneticorum.

Confer: *Seminologia* (pag. 519), *Agricultura* (pag. 507), *Botania* (pag. 390), *Zoologia* (pag. 356), *Cultura animalium* (pag. 523).

Åkarp b. Alnarp (Sverige).
Institutionen för ärftlighetsforskning. Dir. N. H. Nilsson-Ehle.

Ames, Ia. (U.S.A.).
Dept. of Genetics, State Coll. of Agric. Dir. E. W. Lindstrom, Ph.D., Head of Dept., Chief Geneticist. — M. R. Irwin, M. S., Ass. in Genetics; W. V. Lambert, M. S., Instr. and Asst. in Genetics.

Amsterdam (Holland).
Labor. für angewandte Zool. u. Erblichkeitslehre. Confer: Parasit. generalis.

Antibes, Alpes Marit. (France).
Stat. de génétique et de pathol. végétale de l'Inst. d. Rech. agron. Confer: Phytopath. generalis.

Baltimore, Md. (U.S.A.).
Dept. of Human Genetics, Inst. for Biol. Research. Dir. A. C. Sutton.

Berkeley, Cal. (U.S.A.).
Division of Genetics of the Department of Agric. of the Univ. of California. Dir. E. B. Babcock, Prof. of Genetics and Geneticist in the Experim. Stat. — R. E. Clausen, Associate Prof. of Genetics; J. L. Collins, Ass. Prof. of Genetics; one Instr. in Genetics and Cytol. in the Experim. Stat.; two research Ass. (part time). *5 cytol., 6 drosophilists, 8 plant genetists (labor., greenhouse and garden). — Complete cytol. equipment. Complete arrangements for work on Drosophila. Labor., greenhouse and garden facilities for plant genetics.* ⊕ Seeds of many species of Crepis. Seeds of most species of Nicotiana.

Berlin (Deutschland).
Inst. für Vererbungsforschung der Landwirtschaftlichen Hochsch., -Dahlem, Schorlemerallee. Dir. Prof. Dr. Erwin Baur. — Dr. Paula Hertwig, Ass. Zool. Abt.; Bernhard Husfeld, Außen-Ass.; Prof. Dr. Hans Nachtsheim, Abt.-Vorst. der Zool. Abt. u. Ass.; Dr. Elisabeth Schiemann, Ass.; Dr. Emmy Stein, Ass. *4 Arbeitsplätze für Gäste. Genetik. — Hilfsmittel zu botan.- u. zool.-genetischer Experimentalarbeit.* ⊕ Getreide, Antirrhinum, Ratten, Mäuse, Kaninchen, Cyprinodentien.
Kaiser-Wilhelm-Inst. f. Biol. Confer Institutiones generales.

Borsuki, poste Lanowic, Volynie (Polska).
Station de sélection Société d'Amélioration des Plantes „Ouditsch" à Varsovie, rue Hoza 66. Dir. Edward Kostecki. — W. Bagiński, Ass. *Croisements imprévus par insectes provoqués dans les homotypes voisins de trifolium pratense.* — Station de selection Kwasów, poste Pacanów; Station de selection à Dolne, poste Przeworsk.

Boston, Mass., (U.S.A.).
Bussey Inst., Harvard Univ. Dir. W. M. Wheeler (Dean). — Prof. E. M. East (Genetics).

Breslau (Deutschland).
Abt. f. Entwicklungsmech. u. Vererbung d. Anat. Inst. Univ. XVI, Maxstr. 6. Dir. Prof. Dr. B. Dürken.

Burlington, Vt. (U.S.A.).
Dept. of Genetic, Univ. of Vermont. Dir. A. Gershoy.

Cold Spring Harbor, Long Island, N.Y. (U.S.A.).
Carnegie Institution of Washington, Station for Experimental Evolution. Dir. Chas. B. Davenport; Ass. directors: A. F. Blakeslee, H. H. Laughlin. — A. F. Blakeslee; John Belling; A. M. Banta; H. J. Banker; M. Demerec; E. C. MacDowell; C. W. Metz; H. H. Laughlin; Oscar Riddle; A. H. Estabrook. *2 working-places for guests. Eugenics research; experimental breeding. — Statistical machinery.* ⊕ Seeds.

College Station, Tex. (U.S.A.).
Dept. of Genetics, Agric. College. Dir. E. P. Humbert, Ph. D., Head Prof.; W. R. Horlacher, M. S., Assoc. Prof. Genetics.

Dolne, poste Przeworsk (Polska).
Station de sélection à Société d'Amélioration des Plantes „Ouditsch" à Varsovie, rue Hoza 66. Dir. Dr. Edward Kostecki. — Ingen. J. Dziedzic. *Modes d'ensemencement des champs d'essais comparatifs avec des variétés de céréales.* — Station de selection Kwasów, poste Pacanów; Station de selection Borsuki, poste Lanowce, Volynie.

Hamburg (Deutschland).
Labor. für Erblichkeitsforschung der Staatskrankenanstalt, Friedrichsberg.

Helsinki (Finnland).
Genetisches Inst., N. Järnvägsgatan 13. Dir. Prof. Harry Federley.

Jassy (România).
Labor. d'amélioration des plantes cultivées, Univ. Dir. N. Saulescu, Maître de Conférences. — *Génétique et amélioration des plantes cultivées.*

Kamenetz-Podolsk, Ukraine (U. d. S. S. R.).
Labor. für Selektionslehre d. Inst. für Landwirtschaft K. Marx. Dir. Prof. A. Mychaylowski.

Kiew (U. d. S. S. R.).
Labor. für Genetik u. Selektion des Landw. Inst.

København (Danmark).
Genetic Labor. of the Royal Vet. and Agric. Coll. V, Rolighedsvej 23. Dir. Prof. Dr. phil. O. Winge. — Ass.: Dr. phil. J. Clausen; Mag. scient. O. Hagerup; A. F. Skovsted. *2—3 working-places for guests. Cytol. and genetical work.* — A field labor. with experim. field in Lyngby, 12 km from Copenhagen, applied for experim. work during the summer.

Krasnodar, Kuban (U. d. S. S. R.).
Selektionsstation „Kruglik" des Landw. Inst. Dir. A. St. Pustowoj.

Kurashiki, near Okayama (Japan).
Genetical Labor., Ohara Inst. of Agric. Research. Dir. Yasuke Yamaguchi. — 3 Ass.

Kwasów, poste Pacanów (Polska).
Station de selection, Société d'Amélioration des Plantes. Dir. Dr. Edward Kostecki. — Ass.: T. Zaleski; Ing. A. Godek. *Influence de perte des racines dans des champs comparatifs avec les variétés de betteraves à sucre sur les résultats obtenus.* — Station de selection à Dolne, poste Przeworsk; Station de selection Borsuki, poste Lanowce, Volynie.

Kyôto (Japan).
Labor. of Genetics, Coll. of Agric., Kyôto Imper. Univ. Dir. Prof. Dr. Hitoshi Kihara. — Dr. I. Nishiyama.
Labor. of Genetics, Kyôto Imperial Sericicultural Coll., Homasono near Kyôto. Dir. Prof. Dr. Sadaharu Minami.

Leningrad (U.d.S.S.R.).
Labor. für Genetik der Univ. Dir. Prof. Dr. Jar. Philiptschenko. — Staatl. Doc. I. I. Sokolow; Ass.: Th. G. Dobzhansky; X. A. Fermor-Adrianowa; I. I. Lus.
Bureau für Genetik und Eugenik bei der Russischen Akademie der Wissenschaften. Dir. Prof. Dr. Jar. Philiptschenko. — Th. G. Dobzhansky; T. K. Liepin; J. J. Lus.
Abt. f. angew. Botan. u. Selektion d. Staatl. Inst. f. experim. Agronomie. Dir. N. I. Wawilow.

Lexington, Ky. (U.S.A.).
Dept. of Genetics of the Univ. of Kentucky. Dir. W. S. Anderson.

Madison, Wis. (U.S.A.).
Dept. of Genetics, Coll. of Agric. Dir. L. J. Cole, Ph.D., Chair. of Dept. — F. A. Abegg, M.S., Asst. in Genetics; R. A. Brink, D. Sc., Ass. Prof. Genetics; C. R. Burnham, B.S., Ass. in Genetics; D. G. Steele, M.S., Ass. in Genetics; E. E. Van Lone, M.S., Asst. in Genetics.

Matsudo (Japan).
Genetical Labor., Imperial Coll. of Horticulture, Chiba-Ken.

Massukylä (Finnland).
Experim. Stat. Dir. Dr. Mauri Otavi Meurmann. — *Plantcytol.; chromosome behaviour and genetical investigations by Avena u. Pisum.*

Morschansk (U.d.S.S.R.).
Selektions-Versuchsstation f. Wiesenbau „Marussino", Krasnoarmejskaja 32.

Moskau (U.d.S.S.R.).
Genetische Abt. der K. E. P. S. (Kommission zum Studium der natürlichen Produktivkräfte) der Akademie der Wiss. U.d.S.S.R., Voronzovo Pole 6, Inst. für Experim. Biol. Dir. N. K. Koltzoff. — Ass.: Prof. W. N. Lebedeff; S. S. Tschetverikoff; M. S. Avdéeva; P. A. Kosmiusky; B. L. Astauroff; E. I. Balkaschina; I. G. Kohau; M. G. Lobaczeva.

Nazarievo b. Moskau (U.d.S.S.R.).
Zentrale Station zur Untersuchung der Genetik d. landwirtschaftlichen Tiere, Gavoronki. Dir. Prof. Dr. N. K. Koltzoff. — W. A. Razieborsky, Stellvertreter des Dir. Abt. für Genetik der Hühner und für allgem. Genetik: Prof. Dr. A. S. Serebrovsky, Leiter d. Abt.; Ass.: D. D. Romaschev; Kupczenko; Petroff. Abt. für Genetik der Schafe: B. N. Wassin; K. T. Popóva-Wassina; P. A. Göptner. Abt. für Genetik des Rindes: O. A. Ivanova; Blés-Kovarsky. Chem. Abt.: N. G. Savitsch. Anat.-Histol. Abt.: S. N. Bogolioubsky. *5 Arbeitsplätze für genetische Untersuchungen; 5 Arbeitsplätze für cytol.-genetische Untersuchungen.*

New Haven, Conn. (U.S.A.)
Department of Genetics of the Connecticut Agric. Experim. Stat., 123 Huntington St. Dir. D. F. Jones. — P. C. Mangelsdorf, Ass. Geneticist; H. R. Murray.

Nikita Yalta, Krim (U.d.S.S.R.).
Labor. of Genetics Botan. Garten.

Odessa (U.d.S.S.R.).
Odessaer Inst. für Vererbungslehre u. Pflanzenzüchtung, Versuchsstation, 29/60. Dir. Prof. Dr. A. A. Sapěhin. — Wiss. Mitglied D. I. Baranskij, Ass.; Agronom L. A. Sapěhin, Aspirant; Agronom A. I. Worobjow, Aspirant; Ass.-Agronom A. M. Negrul, Aspirant. *2—3 Plätze. Zytol. und Genetik d. Kulturpflanzen. — Besondere Sääpparate. Große Räume für Aufbewahrung d. Materials. Mikromanipulator.* ℗ Die verschiedensten Sippen von Getreidearten, Sonnenblume, Tomaten, Kenaph usw.

Omsk (U.d.S.S.R.).
Labor. f. allgem. Zootechn. Vererbungs- u. Fortpflanzungslehre. Dir. Prof. S. M. Kočergin.

Oslo (Norge).
Inst. for Arvelighetsforskning (Inst. für Erblichkeitsforschung), Univ. Dir. Dr. phil. Kristine Bonnevir. — Amannensis: cand. real. Aslang Soerdrup; Ass.: cand. mag. T. Quelprud.

Paris (France).
Labor. d'Evolution des êtres organisés et Embryol. général, Fac. d. Sc. nat. Univ., 107, Bvd. Raspail. Dir. Prof. M. Caullery. — Prof. François Picard; Marcel Abeloos, Ass.; Marcel Avel, Ass. *10 places de travail. — La Station Zool. de Wimereux.*
Labor. de culture du Museum national d'Histoire naturelle, Ve, 61, rue de Buffon. Dir. Prof. D. Bois. — André Guillaumin, Dr.sc., Ass.; Robert Franquet, Lic.sc., Préparateur. *Génétique, physiol. végétale, systématique et anat. végétales, plantes utiles (classification, selection et essais).* ℗ Herbier de plantes cultivées, graines et fruits.
Station génétique Inst. d. recherches agron., VIIe, 42 bis, Rue de Bourgogne.

Perm (U.d.S.S.R.).
Kabinett für Selektion. Dir. Ass. A. P. Gorin.

Praha — Prag (C.S.R.).
Inst. f. Pflanzenveredlung u. spezielle Pflanzenproduktion d. čech. techn. Hochsch., -Vršovice, Jablonského 163. Dir. Prof. Dr. Johann Jelinek.

Shinkwa, Formosa (Japan).
Hereditological Labor., Branh Experim. St. f. Sugar Indust., Government Central Research Inst., Tainan-shû.

Skaane (Norge).
Experim.-biol. Inst. für Vererbungswissenschaften.

Stellenbosch (South Africa).
Department of Genetics, Univ. Dir. J. H. Neethling, MSc. Prof.

Stende (Latvija).
Selektionsstation landwirtschaftl. Ministeriums, Stendes muiza. Dir. Agr. Lielmans.

Svalöf (Sverige).
Institutionen för Ärftlighetsforskning. Dir. Prof. H. Nilsson-Ehle-Fil. Kand. A. Müntzing, Fil. Lic. E. Hellerström. *Small number of working places. Plant genetics.* Collabor. with the Swedish Plant breed. Assoc.

Tiflis (U.d.S.S.R.).
Selectionskabinett im Botan. Garten.

Tôkyô (Japan).
Genetical Laboratory Imperial Agric. Experim. Station, Nishigahara.
Genetical Laboratory Tôkyô Imperial Sericult. Coll., Nishigahara.

Tübingen (Deutschland).
Rassenbiol. Inst. Dir. Prof. A. Basler.

Upsala (Sverige).
Statens Inst. för Rasbiol. (Staatliches Inst. für Rassenbiol.). Dir. Prof. Dr. med. Herman Lundborg. — Doz. Dr. phil. F. J. Linders, Vizedir. u. Statistiker; Dr. med. T. Sjögren, Ass.-Arzt; Dr. med. J. Edwardson, Anthropologe; Magister A. Ljung, Genealoge; Dr. phil. W. W. Krauß, männl. Reise-Ass.; Frau G. Sandgren, weibl. Reise-Ass. u. Photograph; Kand. phil. S. Wahlund, Amanuens für Statistik; Kand. med. G. von Sydow, ao. Amanuens für Vererbungsforschung. *2 Arbeitsplätze f. Gäste. Rassenbiol. Anthropometrische und photographische Apparate. Anthropometrisches und genealogisches Material.* ⊕ *Photographien von Rassentypen (Schweden, Finnen, Lappen u. a.) aus Schweden.*

Utsunomiya (Japan).
Hereditological Labor. Utsunomiya Imper. Coll. of Agric. and Forestry.

Vinderen b. Oslo (Norge).
Winderen Labor. Dir. Dr. Jon Alfred Mjöen. — Dr. med. Fridtjof Mjöen, Vererbungsforschung; cand. math. Hans Koch, Mathematik, Statistik; Dr. A. Funke, Med.; 1 Hilfs-Ass., Biol. *2 Arbeitsplätze. Psychotechnische Messungen, besonders für die Basaleigenschaften der musikalischen Fähigkeit; sozialbiol. Aufklärungsabteil; Gesetzvorschläge etc. — Psychotechnische und anthropometrische Meßapparate.*

Vjatka (U. d. S. S. R.).
Abt. f. angew. Zool. u. Selektion. Dir. Prof. B. S. Lukaš.
Labor. f. Biol. am Pädag. Inst. Dir. Prof. B. S. Lukaš. *Genetik und Ichtyol.*
Labor. f. Selektions- u. zool. Arbeiten, Wiss. Forschungs-Inst. für Heimatkunde.

Warszawa (Polska).
Inst. f. Genetik u. Melioration d. Pflanzen, Hochsch. f. Land u. Forstwirtsch. Dir. Prof. E. Malinowski.

Wien (Österreich).
Forschungs-Inst. für Rassen- u. Konstitutions-Anthropol., Naturhistorisches Staatsmus., I, Burgring 7. Dir. J. Bayer. — Sekr.: Viktor Lebzelter.

Zürich (Schweiz).
Anthrop. Inst., Univ. Confer: Anthropol.

Laboratoria hydrobiologiae.
1. Laboratoria marina.
Confer: *Cultura piscium* (pag. 527,) *Zoologia* (pag. 356).

Aberdeen (Scotland).
Marine Labor. of the Fishery Board for Scotland, Torry, Wood Street. Dir. Dr. Bowman, Scientific Superintendent. — Dr. R. S. Clark, Plankton, Young Fish; Mr. H. Thompson, M.A.B.Sc., Haddock, Tunlcata, Biochemistry; Mr. H. Wood, M.A., Herring Investigations; Mr. G. C. Gibbons, B.Sc., Plankton general; Mr. J. B. Tait, B.Sc., Hydrography; Mr. D. G. Raitt, B. Sc., Bottom Fauna; Miss H. Ogilvie, B.Sc., Micro-Plankton. *Several places can be accommodated. Fisheries research. Investigations in connection with the sea fisheries and salmon fisheries of Scotland, and they are also participants in the programmes of the International Council for the Exploration of the Sea. — Research Vessel ,,Explorer'' (Mersey class trawler, adapted and fitted for marine research work). Motor Boat ,,Enid'' (Inshore work).* ⊕ *Marine fauna of the Scottish area collected. Type series kept. No general exchange, but exchanges could be effected with specialists working on particular groups.* — *Aquarium at Bay of Nigg, Aberdeen.*

Agar's Island (Bermuda).
Bermuda Biol. Station for Research. Dir. E. L. Mark. — *The Station being open for only a few weeks in summer. The Station is being reorganized on an international basis under the name: ,,The Bermuda Biol. Station for Research, Incorporated.'' When the reorganization is completed the Station will be open throughout the whole year. 12 to 15 working-places. Marine biol. Work on fundamental problems in physiol. (chiefly of a physico-chemical nature) are being carried on throughout the year, under the auspices of the ,,Rockefeller Inst. for Medical Research'' of New York, by Prof. J. W. V. Osterhout in a sub-laboratory of the Station. — Gasolene Launch, row boats, self closing plankton net, tow nets, etc. dredges, tangles, etc. Running salt water, and fresh water. Photographic dark room.*

Alexandrien (Egypt).
Inst. Royal d'Hydrobiol.

Alexandrovsk, Murmansk,
Nordküste Lapplands (U. d. S. S. R.).
Murman Bio-Stat. d. Leningrader Vereins d. Naturforscher. Dir. G. A. Kluge.

Amoy (China).
Amoy Marine Station. Dir. Prof. Dr. Chenfu F. Wu, Soochow Univ.

Arcachon et Guéthary (France).
Station Biol. d'Arcachon, Gironde et annexe de Guéthary (Basses Pyrénès). Dir. Prof. agrégé Dr. R. Sigalas. — *10 places. Toutes les recherches concernant la biol. marine et plus particulièrement la physiol.* — *A Guéthary: aucune installation spéciale. A Arcachon: eau douce et eau de mer courantes. Outillage complet de Physiol.-Mus. et Bibliothèque.* ⊕ *Faune et la flore marines du Golfe de Gascogne.*

Arendal (Norge).
Flødevigens Utklekningsanstalt, Biol. Station. Dir. Alf Dannevig. — *Wissenschaftliche Untersuchungen über Fischkultur und Biol. der Fische.*

Asamushi, near Aomori, Aomoriken (Japan).
Asamushi Marine Biol. Labor. (of Tôhoku Imperial Univ.). Dir. Prof. Shinkishi Hatai. — Seijirô Kokubo, Ass.-Prof. Plankton. *Physiol.*

Baltimore (Ireland).
Biol. Stat. on the South Coast of Cork.

Banyuls sur mer, Pyr. Orles (France).
Labor. Arago, Inst. nation. de Biol. mar. Dir. Prof. O. Duboscq (Sorbonne). — Ass. de zool.: Dr. Migot; R. Denis. *20 places de travail pour recherches de Zool., Histol., Physiol.*

Barsebäck am Öresund (Sverige).
Biol. Labor. am Barsebäck-Hafen d. Zool. Inst. d. Univ. Lund.

Batavia, Java (Ned.-O.-Indië).
Labor. for marine investigations and Aquarium of the Botanical Gardens. Dir. Dr. H. C. Delsman (embryol. of fishes). — *5 tables for foreign guests.* — *Research vessel: ,,Dog'', Captain H. L. Claessen, and motor boat: ,,Max Weber''.*

Beaufort, N. C. (U. S. A.).
U. S. Fisheries Biol. Station. Staff: Dir. 2 junior aquat. biol. 1 ass. aquat. biol. *12 Places. Life history studies of fishes; experim. terrapin breeding; life history study of the scallop.* — *Large tanks for retai-*

ning living specimens, fresh and salt running water. ⚲ Materials for special study, frequently are collected for investigators upon special request.

Beaulieu (France).
Labor. de Biol. Maritime (École prat. des Hautes Études, Paris).

Blakeney Point, Norfolk (England).
Field Stat. for maritime Ecology Dept. of Botany, Univ. Coll. London.

Bornö b. Holma (Sverige).
Hydrographische Station d. schwed. Abt. d. Int. Kommission f. Meeresforschung. Dir. O. Pettersson. — 1. Ass. Dr. Hans Pettersson; 2. Ass. Dr. A. Molander; Dr. Nybelin, Spez. f. Planktonarb. *Stationsarbeiten: Bearbeitg. des durch die Terminfahrten erhaltenen hydrograph. Materials; autom. Aufzeichnungen d. Wasserbewegungen in allen Tiefen des Fjordes.*

Büsum, Holstein (Deutschland).
Zool. Station. Dir. Sebastian Müllegger. — *30 Arbeitsplätze.*

Cagliari (Italia).
Stazione Biol. Univ. Dir. L. Granata.

Cau-da par Nha-trang, Annam (Indochine).
Station Maritime, Service Océanographique des Pêches de l'Indochine. Dir. Armand Krempf. — 2 ass. *Océanographie biol. Utilisation industrielle des Produits de la pêche. Farines de Poisson. Huiles de Poisson. Autolysats de poisson riches en acides amidis.* ⚲ Faune et flore marine.

Cette, Herault (France).
Station zool. marine. Dir. Prof. M. Bataillon (Montpellier). — M. Benoît, sous-dir. (Montpellier). *4 places. Biol. marine.*

Concarneau, Finistère (France).
Labor. maritime du Coll. de France. Dir.: les prof. de sc. nat. du Coll. de France. — Sous-Dir. R. Legendre. *10 places pour Zool. et physiol. marines. — Eau douce et eau de mer. Aquaria. Embarcations et bateau mixte. Equipement pour la microscopie, la microphotographie, la physiol. et la chimie.*

Dafundo (Portugal).
Aquário Vasco da Gama Station de Biol. maritime. Dir. Dr. A. Ramalho. — Ass. A. Candeias. *2—3 places. Travaux d'anat. et microscopie. — Instruments ou reactifs pour les travaux d'anat. et microscopie.* ⚲ Poissons. Copépodes. Plankton marin.

Dröbak (Norge).
Die biol. Meeresstation der Univ. Dir. Dr. Hj. Broch (Oslo). *10 Arbeitsplätze, zum Studium der Tierwelt des Oslo-Fjordes (auch f. Histol.).*

Dublin (Ireland).
Zool. Dept., Univ. Coll. Confer: Zool. generalis.

Dunedin (New Zealand).
Portobello Marine Labor. Dir. Maxwell W. Young, Zool.

Dún Laoghaire (Ireland).
Marine zool. Labor., Univ. of Dublin. Dir.: Prof. J. Bayley Butler. *4 working places for guests.* ⚲ Marine specimens from the Irish Sea.

Eugene, Ore. (U.S.A.).
Labor. of Physiol. and Zool. Confer: Physiol. comparata.

Fiskebäckskil (Sverige).
Klubbans Biol. Station d. Univ. Uppsala. Dir. Dr. Sixten Bock. *15 Arbeitsplätze (die Station ist nur während des Sommers offen). Biol. Untersuchungen des Meeres.*

Kristinebergs Zool. Station.
Dir. Prof. E. Lönnberg, Prefekt, Stockholm 50. — M. Aurivillius, Lic. Phil., Vorstand. *Im Sommer 25 Plätze, im Winter 2—3.*

Friday Harbor, Wash. (U.S.A.).
Puget Sound Biol. Station. Dir. Dr. T. C. Frye. — Dr. T. C. D. Kincaid, Morphol.; Dr. V. E. Shelford, Ecol.; Dr. A. O. Weese, Ecol.; Dr. M. T. Harman, Embryol.; Dr. N. L. Gardner, Algae; Dr. J. E. Guberlet, Parasit.; Dr. E. J. Lund, Physiol. *6 places, 2 specially for ecol. — Chiefly apparatus for the exploration of the sea bottom. — There is a special labor. for animal ecol.*

Halifax, N.S. (Canada).
Atlantic Experimental Stat. of the Biol. Board. Dir. J. C. Forbes. — H. E. Tanner; Hess.

Den Helder (Holland).
Zoöl. Station der Nederl. Dierkundigen Vereeniging. Dir. H. C. Redeke. — A. P. C. de Vos, Adj.-Dir. *6 Plätze für faunistische, morphol. und allgemein biol. Untersuchungen. — Arbeitsaquarien. Motorboot.*

Helgoland (Deutschland).
Staatliche Biol. Anstalt. Dir. Prof. Dr. W. Mielck. — Prof. Dr. A. Hagmeier, Zool.; Prof. Dr. A. Wulff, Seefischerei; Prof. Dr. R. Drost, Ornithol., Vogelwarte; Prof. Dr. E. Schreiber, Botan.; Prof. Dr. H. Hertling, Ass.; Prof. Dr. R. Kändler, Austernforschung. Im Labor. der Deutsch. Wissenschaftl. Kommission für Meeresforschung: Prof. Dr. Fr. Heincke; Dr. A. Bückmann, Ass.; Dr. Cl. Künne, Ass. *Insgesamt ca. 50 Plätze, davon 25 vornehmlich für selbständige Arbeiten, 25 vornehmlich für Lehrkurse. Gesamte marine Naturforschung. In erster Linie marine Zool. u. Botan. u. Vogelzugsforschung sowie Fischereibiol. (Ausrüstung für anat., histol., embryol., physiol., bakteriol., systematische, faunistische, quantitative Plankton- u. Bodenforschungsarbeiten), ferner auch hydrographische, meteorol. und geol. Arbeiten. Die Plätze für Lehrkurse werden auch solchen Doz. zur Verfügung gestellt, welche die Kurse selbst zu leiten wünschen. — Schauaquarium, Nordseemus. (marine Biol. u. Ornithol.). Exkursionen: 1 Motorschiff (80 Ps., 27 m lang), 2 Motorboote, mehrere Segel- u. Ruderboote. Ausgerüstet mit den gebräuchlichen Geräten für Meeresforschung. Untersuchungen: Die Bibliothek. Wissensch. Aquarien mit fließ. unfiltrierten u. filtrierten Seewasser und Durchlüftung. (Alle Leitungen ohne Metall, statt dessen Celluloidrohre u. Steingutpumpen, bes. Eignung für physiol. Arbeiten). Rührapparate u. kl. Pumpen für Aquarien. Kymographion. Sterilisationsapp. Mikrophot. App. Dunkelräume. Kulturräume im Keller und mit Oberlicht. Die Vogelwarte (ornithol. Abt.) mit Beobachtungs- und Fanggarten (Markierungen). Meteorol. Station.* — Zweiglabor. für Wattenmeerforschung in List (Insel Sylt). Beobachtungsstelle der ornithol. Abt. (Vogelwarte) auf Insel Mellum (Jade-Weser-Mündung).

Helsinki (Finnland).
Inst. für Meeresforschung, Konstantinsgatan 8. Dir. Prof. R. Witting. — Abt.-Vorst. Dr. H. Renqvist, Wasserstand; Mag. phil. G. Granqvist, Eis u. sonst. Beobachtungen. *Hydrographisch-biol. Untersuchungen.*

Herdla b. Bergen (Norge).
Die biol. Station des Mus. zu Bergen. Dir. Prof. Dr. A. Brinkmann. — Dr. Sven Runnström. *Im Sommer bis 15, im Winter 5 Arbeitsplätze. Marin-biologische Untersuchungen. — Motorschiff „Hermann Friele", ein kleines Motorboot und Ruderboote, die mit einer reichlichen Einrichtung mit Fanggeräten zu Arbeiten bis zu 1200 m Tiefe versehen sind. Große Bibliothek.* ⚲ Marine Organismen jeder Art.

Honolulu (Hawaii).
Marine Biol. Labor. Dir. C. H. Edmondson, Ph.D.

København (Danmark).
Den Danske Biol. Station (Danish Biol. Station). Strandvej 34, Hellerup. Dir. Dr. phil. A. C. Johansen. — Dr. phil. H. Blegvad. Ass. 1.4.—15.9.: Nyborg, 15.9.—1.4.: København.

The Plankton Labor. of the „Kommissionen for Danmarks Fiskeri-og Havundersøgelser", Strandvej 34, Hellerup. Dir. Prof. Dr. C. H. Ostenfeld. — P. Jespersen, mag. scient., Ass. Naturalist. *1—2 guests, investigations on marine. Plankton from Danish waters, Iceland, Faroers and the Atlantic. — Implements for collection of plankton.* ⓟ *Marine plankton organisms: Zooplankton and Phytoplankton from danish waters, Iceland, Faroers and the Atlantic.*

Kuusnõmme, Insel Oesel (Eesti).
Biol. Stat. d. Zool. Inst. Univ. Tartu.

Laguna Beach, Cal. (U.S.A.).
Laguna Marine Labor. Dir. W. A. Hilton. — Ass. *8—10 places. General Zool.; Marine Zool.; Zool. of insects and other land forms.*

La Hougue, Manche (France).
Labor. Maritime, Mus. Nat. d'Histoire naturelle. Dir. L. Mangin.

La Jolla, Cal. (U.S.A.).
Scripps Institution of Oceanography of the Univ. of Calif. Dir. Dr. T. Wayland Vaughan. — G. F. McEwen, Physical Oceanographer; F. B. Sumner, Biologist; W. E. Alleen, Biologist; E. G. Moberg, Chemist; P. S. Barnhart, Curator; C. O. Esterly, Zoologist. *4 labor. rooms, providing for about 8 guests. Biol., zool., chem. research.*

Le Croisic, Loir Infer. (France).
Labor. de Biol. marine annexe du Labor. de Zool. de l'Ecole de méd. de Nantes. Dir. Prof. Dr. Alphonse Labbé (Nantes). — Prof. Pelvies, Anat. comparée; Prof. Guéguen, Algol. *6 places. Tous travaux de biol. marine (zool., cytol., algol.). Plus spécialement études sur la faune et la flore des marais salants.*

Le Havre (France).
Inst. Océanographique du Havre au Mus., Place du Vieux-Marché. Dir. Dr. Adrien Loir. — Henri Legangneux, Chef de Labor.; Etienne Peau, Chef des Travaux biol. *Geol. Paléont. Zool.* ⓟ *Tous animaux et végétaux marins.*

Liverpool (England).
Department of Oceanography, Univ. of Liverpool. Dir. J. Johnstone. — R. T. Daniel, M.Sc.; J. R. Bruce, M.Sc.; H. C. Chadwick, M.Sc.; W. C. Smith. — Marine Biol. Station Port Erin, Isle of Man, England.

Los Angeles, Cal. (U.S.A.).
The Marine Biol. Stat. of the Univ. of Southern California. Dir. A. B. Ulrey.

Zool. Dept. of the Scripps Institution of oceanography. Dir. Prof. C. O. Esterly.

Luc-sur-Mer (France).
Labor. Maritime, Univ. de Caen.

Madrid (España).
Départ. de Biol. de l'Inst. Espagnol d'Océanographie, Calle Alcalá 31. Dir. Dr. Fernando de Buen, Chef du Dépt. de Biol. — Dr. Victoriano Rivera, Prof. de l'Inst. d'Huesca; Dr. Alfonso Gandolfi, Prof. agrégé de l'Inst. espagnol d'Océanographie; Dr. Luis Alaejos, Dir. du Labor. de Santander de l'Inst. espagnol d'Océanographie; Dr. Juan Cuesta, Ass. du Labor. de Santander de l'Inst. espagnol d'Océanographie; Licenciés en Sciences: Alvaro de Miranda, Dir. du Labor. de Málaga de l'Inst. Esp. d'Océanographie; Francisco de P. Navarro, Dir. du Labor. de Palma de Mallorque; Luis Bellón; Emma Bardán; Miguel Massutí; 1 Chef et 1 Ass. *Biol. marine appliquée à la pêche.*

Málaga (España).
Labor. Biol.-Marino (Inst. Español de Oceanografía), Paseo de la Farola 47. Dir. Dr. Alvaro de Miranda. — Dr. Luis Bellón Uriarte. *En période de réorganisation. Il faut se renseigner et demander permission à la Direccion general de Pesca, Alcalá 31, Madrid. — Appareils océanographiques; chimie de la Mer; Biol. générale.* ⓟ *Crustacés, Poissons, Algues, Plankton.*

Marseille-Eudoume, Bouches du Rhône (France).
Labor. Marion, Rue Batterie des Lions. Dir. M. Kollmann, Prof. à la Fac. d. Sc. — Dr. Van Gaver (F.); J. Timon-David. *4 places.* ⓟ *Animaux du golfe de Marseille.*

Messina (Italia).
R. Istituto centrale di biol. marina, S. Ranieri. Dir. Prof. L. Sanzo.

Millport, Isle of Cumbrae (Scotland).
Millport Marine Biol. Station, Labor. of the Scottish Marine Biol. Association. Dir. Superintendent R. Elmhirst. — A. P. Orr, Bio-chemist; Miss S. Marshall, Naturalist. *8 places and class room. The Labor. exists primarily for the investigation of the fauna and flora of the Clyde Sea Area (Firth of Clyde). — Electric plant giving 110 volts for lighting and experimental purposes.*

Misaki (Japan).
Misaki Marine Biol. Labor. (of Tōkyō Imperial Univ.), Kanagawa-Ken. Dir. Prof. Naohide Yatsu. *Systematics.*

Monaco (Principauté).
Labor. du Mus. océanographique le Monaco. Dir. Dr. J. Richard. — L. Sirvent, Ass.; M. Oxner, Ass.; Giauffret, Préparateur. *8—10 places.*

Moskau (U.d.S.S.R.).
Schwimmende Wiss. Meeresinst., ul. Herzena 6, Univ. zool. Mus. Dir. Prof. A. J. Rossolimo. — Expeditionsleiter: I. I. Mesjacev; L. A. Zenkevitsch; Prof. S. A. Zernov; Prof. V. K. Soldatov; Prof. I. V. Samojlav; Prof. V. S. Butkevič; Prof. V. V. Sulejkin; V. A. Jašnov; V. V. Alpatov; M. E. Makušok; B. K. Flerov; Frl. V. S. Malinina; Frl. I. I. Gorškova; Frl. V. A. Brockaja; Frl. M. K. Klenova; A. O. Starostin.

Nanaimo (British Columbia).
Pacific Biol. Station. Dir. Wilbert A. Clemens, M.A., Ph.D. — R. E. Foerster, M.A., Ph.D., Biologist for salmon investigations — H. C. Williamson, M.A., D.Sc., Zoologist for marine fishery investigations; L. W. S. Clemens, Gen. Marine and Fisheries investigations. *Approximately 16 places for Marine investigations. — General biol., bact., chem. and oceanographical; fresh and salt water; gas and electricity; boats.*

Napoli (Italia).
Stazione Zool. Dir. Prof. Dr. R. Dohrn. — Prof. M. Fedele, Capo del Reparto Zool.; Prof. E. Sereni, Capo del Reparto Fisiol.; Dr. J. Gross, Bibliotecario; Dott. S. Ranzi, Ass. al Reparto Zool. *Ca. 50 posti. Per tutte le ricerche biol., sia morfol. che sperimentali, sulla fauna e flora marina del Golfo di Napoli. — Diversi tipi di Galvanometri (Einthoven, Broca, ed altri); Kimografi, Segnali, leve, orologi Jacquet, slitte; Interruttore rotativo a Mercurio; Centrifughe, ultrafiltrazione, Colorimetro, Potentiometro, Determinazione PH secondo Michaelis, Refrattometro ecc.*

New Brunswick, N.J. (U.S.A.).
Oyster Investigation Labor., N.J. Agr. Expt. Station, Dept. of Biol. Houseboat; located at various points on N.J. Coast. Dir. T. C. Nelson. — In past. Dr. George W. Martin. *1 place. Physiol. of shellfish and of plankton Organisms. — Plankton collecting, titration and pH determination. Photomicrography Collecting boats and apparatus.* ℗ Oyster in all its stages, other molluscs, plankton of brackishwater estuaries. Teredo and Bankia. — Bivalve on the Maurice River, N.J., for investigation of special problems in connection with the marketing of oysters. Is in charge of permanent investigator.

Noworossijsk (U.d.S.S.R.).
Noworossijsker Biol. Station am Schwarzen Meer „Arnoldi", Slepzowskaja 3.

Ostende (Belgique).
Labor. Maritime, Inst. d'Etudes maritimes, Phare. Dir. Prof. G. Gilson.

Otaru, Hokkaidô (Japan).
Oshoro Marine Biol. Station (of Hokkaidô Imperial Univ.).

Pacific Grove, Cal. (U.S.A.).
Hopkins Marine Station of Stanford Univ. Dir.: Walter K. Fisher, Prof. of Zool. — Charles Vincent Taylor, Associate Dir., Prof. of Biol.; Harold Heath, Prof. of Zool.; Lawrence B. Becking, Prof. of Economic Biol.; Tage Skoqsberg, Instr. in Zool. *25 places. Zool., Physiol., Marine Botan., Biochem., Biophysics.*

Palma de Mallorca, Iles Baléares (España).
Labor. Biol. Marino, Aigo Dolça, Terreno. Dir.: Francisco de P. Navarro Martin. *Le labor. est (1925) en cours d'une nouvelle installation; 6 au moins places. Biol. marine.*

Plymouth (England).
Labor. of the Marine Biol. Association of the United Kingdom, Citadell Hill. Dir. Dr. E. J. Allen, F.R.S. — Head of Department of General Physiol.: W. R. G. Atkins, Esq., Sc.D., F.I.C., F.R.S.; Chief Naturalist: J. H. Orton, Esq., D.Sc.; Naturalists: E. Ford, Esq., A.R.C.Sc.; Miss M. V. Lebour, D.Sc.; Physiologist: C. F. A. Pantin, Esq., M.A.; Dir. Research Ass.: Mrs. E. W. Sexton, F.L.S.; Hydrographical and Administrative Ass.: H. W. Harvey, Esq., M.A.; Ass. Naturalists: F. S. Russell, D.S.C., D.F.C., B.A.; O. D. Hunt, Esq., B.Sc.; Temporary Ass. Naturalist: C. M. Yonge, Esq., B.Sc., Ph.D.; Student Probationers: R. Palmer, B.Sc.; H. O. Bull. *30 places for research work. 20 places for students. Fitted for Zool., botan., physiol. and bio-chem. work. — Steam trawler drifter „Salpa" with crew of six. 25 foot motor boat „Gammarus" with crew of two. Physiol. and chem. departments.* Sub-labor. at Cawsand Bay.

Port Erin, Isle of Man (England).
Marine Biol. Station. Dir. Prof. James Johnstone, D.Sc. — J. R. Bruce, M.Sc., Bio-chemist; H. C. Chadwick, M.Sc., Research Zool.; W. C. Smith, Sea-Fisheries Research; Ass.-Curator and Fish-Culturist; Junior Ass. *14 places: Marine Biol. 10, Chem.-Physiol. 4.*

Port Eynon, Gower, near Swansea (England)
Ecol. and Marine Biol. Labor.

Puerto Galera, Mindoro (Philippine Islands)
Marine Biol. Stat., Dept. of Zool., Univ. of the Philippines. Djr. P. B. Sivickis. *Sessions in April and May. Tropical Marine and shore Fauna. Guests welcome. Common biol. apparatus will be provided by the Dept. of Zool., Univ. of the Philippines.*

Riga (Latvija).
Hydrobiol. Station der Lettländischen Univ. (Latvijas Univ. Centralā Hidrobiol. stacija), Baznicas ielā No. 5, dz. 9. Dir. Prof. E. Strand. — Viktor Osolin, Adjunkt der Station. *12 Arbeitsplätze.* — Zweiglabor., hauptsächlich für Sommerarbeiten, am Usma-See in Kurland.

Roscoff, Finistère (France).
Station Biol. de Roscoff (Labor. Lacaze-Duthiers). Dir. Prof. Ch. Pérez (Paris). — Prof. Edgard Hérouard, Sous-Dir. Paris; Dr. M. Prenant, Chef des Travaux; Dr. M. Hérubel, Ass. *25 grands cabinets de travail, 10 petits. Labor. de physiol., chim., physique (claire), physique (obscur). Salles de travail et de travaux-pratiques pour les étudiants. — Bateaux pour excursions et opérations de pêche. Automobiles. Outillage général de labor. biol. Eau de mer et eau douce. Vide. Glacière. Aquariums à circulation d'eau.*

Rovigno d'Istria (Italia).
Ist. di Biol. Marina (R. Comitato Talasso grafico Italiano). Dir. Prof. Dott. Massimo Sella. — Dott. Aristocle Vatova, Ass. *N. 10 posti di lavoro. Biol. marina (ricerche sistematiche, anat., microscopiche) — Acquario (50 vasche, 34 mc), piccoli acquari nelle stanze di lavoro, motoscafo.* ℗ *Fauna bentonica, alghe.*

Saint Andrews (Scotland).
Gatty Marine Labor. and Bell-Pettigrew Mus., Univ. Dir. Prof. d'Arcy Wentworth Thompson.

St. Andrews, N.B. (Canada).
Atlantic Biol. Station of the Biol. Board. Dir. A. G. Huntsman. — A. H. Leim, Ass.-Dir. *Fisheries. Marine and fresh Water biol.*

St. James (South Africa).
The St. James Aquarium. Dir. Dr. Cecil von Bonde. — General Ass. *1 place. Marine Biol.* ℗ *Marine Fauna.*

St. Servane, Ille-et-Vilaine (France).
Labor. maritime.

Salammbô, Régence de Tunis (Protect. français).
Station océanographique du Golfe de Tunis. Dir. Dr. Henri Heldt. — Mm. Heldt, Ass. *Aquarium, labor. de photographie et microphotogr., de chimie. Musée. — L',,Andre Choleski", pinasse automobile. Le ,,Courlie" voilier mixte grée en dunder. Bâtiments à Tabarka, La Galite, Bizerte, Sousse, Sfax, Houmt-Souk. Renseignements pratiques aux pêches maritimes. Hydrobiol. et biol. des lacs de Tunisie. Etudes sur les poissons migrateurs.*

Salisbury Cove, Me. (U.S.A.).
Mount Desert Island Marine Biol. Labor. Dir. H. C. Tracy. — A. Dahlgren.

San Bartolomeo di Cagliari (Italia).
Ist. di biol. marina per il Tirreno. Dir. Prof. E. Giglio-Tos.

Santander (España).
Inst. Español de Oceanografia. Dir. L. Alaejos y Sanz.

Seto-Kanayama, Wakayama-Ken (Japan).
Seto Marine Biol. Labor. (of Kyôtô Imper. Univ.). Nishi-Muro-Gun. Dir. Prof. Kôzô Akatsuka. — Jirô Ikari, Ass. (Plankton esp. diatom.). *Several working places is wholly dedicated for foreign guests. The Labor. is fitted chiefly systematics esp. Planktonresearches.*

Sewastopol (U.d.S.S.R.).
Sewastopoler Biol. Station d. Akad. d. Wiss., Primorski bulv. Dir. Prof. N. V. Nasonov. — Prof. V. N. Nikitin, Oberzoologe; Frl. L. Jakubova; V. Popov; M. Galad'ev; N. Crigirin.

Tamaris-sur-Mer (France).
Inst. maritime de biol. de l'Univ. de Lyon. Dir. Prof. H. Cardot (Lyon). — A. Bonnet (Zool.). Julius (Physiol.). *Questions de zool., de botan. et de physiol. marines, les pêcheries, la pisciculture, l'ostréiculture, la mytiliculture, etc. — Quatre chambres. Instrument de pêche.* ⓟ Algues et animaux marins.

Taranto (Italia).
Regio Labor. di Biol. marina, Corso due mari 12. Dir. Dr. Prof. A. Cerruti. — Prof. I. Pierpaoli; Prof. G. Cossu; Prof. A. Palombi; Dr. A. Agostini. *3 posti: Biol., Oceanografia, Zool., Botan. — Motobarca; barche a remi. Apparecchi per lo studio dell' oceanografia.* ⓟ Molluschi.

Tortugas by Key West, Fla. (U.S.A.).
Tortugas Labor., Carnegie Inst. of Washington. *Meeresbiol. — Labor., Aquarium, Schiffe, Boote.*

Travemünde b. Lübeck (Deutschland).
Fischereibiol. Labor. d. Landwirtschaftl. Versuchsstation. Dir. Prof. Dr. Karl Steyer.

Tsuyasaki, near Fukuoka (Japan).
Tsuasaki Marine Biol. Station (of Kyûshû Imperial Univ.), Fukuoka-Ken.

Tvärminne (Finnland).
Zool. Station der Univ. Helsingfors, 1. Juni bis 1. Sept. Tvärminne, 1. Sept. bis 31. Mai Helsingfors, Zool. Inst. d. Univ. Dir. Alex. Luther. Jeden Sommer werden etwa 3 zool. u. botan. Unterrichtskurse für Studenten von Helsingforser Universitätslehrern abgehalten (in finnisch u. schwedischer Sprache). *Etwa 15 Arbeitsplätze. Ausrüstung in erster Linie für Hydrobiol. (Brackwasser des Finnischen Meerbusens).*

Valparaiso (Chile).
Officina Hydrographica.

Varna (Bulgarien).
Zool. Station des Zool. Inst. der Univ. Sofia. Dir. Prof. Dr. G. Chichkoff. — Dr. G. Paspaleff, Doz. *4 Arbeitsplätze.* ⓟ Die im Schwarzen Meer vorkommenden Tiere und Pflanzen.
Bulgar. Inst. zur Erforschung d. Schwarzen Meeres. Confer: Institutiones generales.

Veracruz (Mexico).
Station Biol. du Golfe. Dir. E. Beltrán.

Vladivostok, Ostsibirien (U.d.S.S.R.).
Biol. Station d. Vladivost. Sektion d. Staatl. Geograph. Ges. Dir. G. N. Gassovskij, Zool.

Wimereux, Pas-de-Calais (France).
Stat. Zool. maritime (attachée à la chaire d'Evolution, de la Fac. Sc. Paris). Dir. Prof. M. Caullery. — M. R. Weill, Ass.; Prof. E. L. Ch. Guyenot, Dir. adjoint. *10 places.* ⓟ Faune marine de la Manche.

Woods Hole, Mass. (U.S.A.).
Marine Biol. Labor. Dir. Merkel H. Jacobs, Prof. of general physiol., Univ. of Pennsylvania. — G. A. Drew, Ass. Dir. Zoöl.: I. Investigation: Gary N. Calkins, Prof. of Protozoöl., Columbia Univ.; E. G. Conklin, Prof. of Zoöl., Princeton Univ.; Caswell Grave, Prof. of Zoöl., Washington Univ.; H. S. Jennings, Prof. of Zoöl., Johns Hopkins Univ.; Frank R. Lillie, Prof. of Embryol., The Univ. of Chicago; C. E. McClung, Prof. of Zoöl., Univ. of Pennsylvania; S. O. Mast, Prof. of Zoöl., Johns Hopkins Univ.; T. H. Morgan, Prof. of Experim. Zoöl., Columbia Univ.; G. H. Parker, Prof. of Zoöl., Harvard Univ.; E. B. Wilson, Prof. of Zoöl., Columbia Univ. II. Instruction: J. A. Dawson, Instr. in Zoöl., Harvard Univ.; Horace B. Baker, Instr. in Zoöl., Univ. of Pennsylvania; E. C. Cole, Ass. Prof. of Zöol., Williams Coll.; Rudolf Bennitt, Instr. in Biol., Tufts Coll.; T. H. Bissonnette, Prof. of Biol., Trinity Coll.; Madeleine P. Grant, Ass. Prof. of Zoöl., Mount Holyoke Coll.; B. H. Willier, Ass. Prof. of Zoöl., The Univ. of Chicago; Donnel B. Young, Assoc. Prof. of Biol., Univ. of Arizona. Protozoöl.: G. N. Calkins, Prof. of Protozoöl., Columbia Univ.; Woolford B. Baker, Ass. Prof. of Biol., Embry Univ.; Mary Stuart MacDonald, Prof. of Zoöl., Agnes Scott Coll. Embryol.: Hubert B. Goodrich, Prof. of Biol., Wesleyan Univ.; Benjamin H. Grave, Prof. of Biol., Wabash Coll.; Charles Packard, Associate in the Inst. of Cancer Research, Columbia Univ.; Harold H. Plough, Prof. of Biol., Amherst Coll.; Charles G. Rogers, Prof. of Compar. Physiol., Oberlin Coll. Physiol.: I. Investigation: Harold C. Bradley, Prof. of Physiol. Chem., Univ. of Wisconsin; Walter E. Garrey, Prof. of Physiol., Vanderbilt Univ. Med. School; Ralph S. Lillie, Prof. of General Physiol., The Univ. of Chicago; Albert P. Mathews, Prof. of Biochem., The Univ. of Cincinnati; II. Instruction: Merkel H. Jacobs, Prof. of General Physiol., Univ. of Pennsylvania; William R. Amberson, Ass. Prof. of Physiol., Univ. of Pennsylvania; Wallace O. Fenn, Prof. of Physiol., Univ. of Rochester; Frank P. Knowlton, Prof. of Physiol., Syracuse Univ. Botan.: I. Investigation: B. M. Duggar, Prof. of Plant Physiol., Washington Univ.; C. E. Allen, Prof. of Botan., Univ. of Wisconsin; S. C. Brooks, Department of Public Health, Washington, D.C.; Wm. J. Robbins, Prof. of Botan., Univ. of Missouri; J. R. Schramm, Editor-in-Chief, Biol. Abstracts, Univ. of Pennsylvania; II. Instruction: Ivey F. Lewis, Prof. of Biol., Univ. of Virginia; Tracy E. Hazen, Ass. Prof. of Botan., Barnard Coll. Columbia Univ.; William Randolph Taylor, Ass. Prof. of Botan., Univ. of Pennsylvania. *Approximately 150 private research rooms with space in the general labor. for about 60 additional workers. The Labor. is equipped for work in the fields of Biol., Physiol. and Physiol. Chem. with especial facilities for the study of marine organisms. — Dark rooms, constant temperature rooms, X-ray apparatus, string galvanometer and a variety of physico-chem. instruments.* ⓟ Chiefly marine animals and plants from the vicinity of Woods Hole but other biol. material including microscopical slides can be supplied.

2. Laboratoria limnologiae.

(Confer: *Zoologia* (pag. 356), *Cultura piscium* (pag. 527).

Aneboda (Sverige).
Schwedische Station für Süßwasserbiol. Dir. Dr. H. Nordqvist. — Dr. E. Naumann (Lund). *Limnol. u. Fischereibiol.*

Balaschew, Saratow-Gouv. (U.d.S.S.R.).
Biol. Station.

Beppu (Japan).
Beppu Hot-Spring Research Labor. of Kyôtô Imperial Univ., Ôita-Kén. Dir. Prof. Dr. T. Shida. — Prof. Masamichi Suzuki, Ass. *Thermophile Tiere u. Pflanzen. Eisenbakterien.*

Berlin (Deutschland).
Hydrobiol. Gruppe der Biol. Abt. d. Preuss. Landesanstalt für Wasser-, Boden- u. Lufthygiene, -Dahlem, Ehrenbergstr. 38—40—42. Dir. Prof. Dr. R. Kolkwitz. — Mitglieder: Prof. Dr. J. Wilhelmi, Wasser- u. Abwasserfragen; Dr. E. Tiegs, Untersuchung von Wasser-, Abwasser- u. Vorflutproben; Dr. H. Helfer, Fischereibiol.: Beziehungen zwischen Fischen u. Pflanzen in hygien. Beziehung.

Mus. d. Preuß. Landesanst. für Wasser-, Boden- und Lufthygiene, -Dahlem, Ehrenbergstr. 38—42. Dir. Prof. Dr. J. Wilhelmi. — Dr. H. Beger, Mitarbeiter.

Bogorodsk, Moskauer Gouv. (U.d.S.S.R.).
Biol. Station.

Bolschewo, Gouv. Moskau (U.d.S.S.R.).
Stat. biol. de la Soc. des Amis des Sc. nat. Dir. Prof. N. V. Bogviavlensky. — Ve. Nic. Davyeloff (Biol. et histol. des Moll. et Insecta); A.S. Bugoslovsky (Biol. exper. plancton); S. J. Konlaeff (Cyclogenie des Hirudinea). *30 places Spongiae, Bryozoa, Plancton.*

Brno — Brünn (C.S.R.).
Hydrobiol. Labor. of the Department of Veter. Univ. Coll. Confer: Physiol. comparata.

Cheboygan, Mich. (U.S.A.).
Biol. Stat. of the Univ. of Michigan. Dir. Prof. G. R. La Rue.

Clermont-Ferrand, Puy-de-Dôme (France).
Station biol. de Besse-en-Chaudesse. Dir. L. Calvet.

Doksy (Hirschberg) (C.S.R.).
Forschungsanstalt f. Fischzucht u. Hydrobiol. Dir. Prof. Dr. V. Langhans. — Prof. A. A. Pascher, Botan. Abt. *Angew. Fischereibiol. Fischkrankheiten, Fischsterben.*

Drozdowice (Polska).
Biol. Station (Eigentum der Polnischen Copernicus-Naturforscher-Gesellschaft), Postamt Gródek Jagiellonski bei Lwów. Dir. Prof. Dr. Jan Hirschler. — Ludwik Monné, Ass. *3 Arbeitsplätze für hydrobiol. Untersuchungen.*

Friday Harbor, Wash. (U.S.A.).
Puget Sound Biol. Station. Confer: Labor. marina.

Genève — Genf (Suisse).
Labor. de zool. lacustre, de protist. et de parasit., Rue de l'Ecole de Méd. Dir. Prof. Dr. Emile André. — Ass. W. Schopfer, lic. ès sc. *8 places.* ⓟ Parasites. Faune des eaux douces.

Gnesdilowo, Smolensker Gouv. (U.d.S.S.R.).
Biol. Station.

Grenoble (France).
Labor. de Zool. et Hydrobiol. de l'Univ. Confer: Zool. generalis.

Hakkwaishô, Formosa (Japan).
Fresh-water Fishery Experim. St. (Bureau of Industry, Governm. of Formosa), Tô-en-gun, Shinchiku-shû.

Hillerød (Danmark).
Freshwater biol. Labor. of the Univ. Copenhagen. Dir. Prof. Dr. C. Wesenberg-Lund. — Kand. mag. Kai Berg, Ass. *1 place. Limnol.* ⓟ Freshwater organisms. — Freshwater labor. at Suserup Scriv. Bustruplake, Seeland.

Irkutsk (U.d.S.S.R.).
Biol. Station am Baikalsee, Nabereshnaja 8, W. 2. Dir. Prof. V. Jasnickij.

Kiew, Ukraine (U.d.S.S.R.).
Biol. Dnjepr-Station, Korolenkostr. 37. Dir. Prof. D. Beling, Hydrobiol., Ichthyol. — Prof. N. Cholondny, Leiter der Botan. Abt. der Station; Ass. W. Sowinsky, Zool., Entomol.; I. Markowsky, Hydrobiol., Cladocera; D. Radzimowsky, Phytoplankton; M. Miklucha-Maklay, Zool., Hirudinci; A. Miroschnitschenko, Zooplankton, Copepoda; M. Grimajlowskaja, Hydrobiol., Bentos. *3—4 Arbeitsplätze. Hydrobiol. Untersuchungen.* ⓟ Hydrobiol. Material, speziell Fische. — Sommerabt. der Station in Starosselje am Dnjepr (15 km von Kiew).

Konstanz, Baden (Deutschland).
Anstalt für Bodensee-Forschung der Stadt Konstanz. Dir. Prof. Dr. M. Auerbach, Karlsruhe (Zool., Hydrographie, Fischerei). — Prof. Dr. J. Schmalz, Konstanz, stellvertr. Dir. (Zool., Hydrographie, Chem., Fischerei); Prof. Dr. F. Landt, Konstanz (Fischerei, Fischparasiten); Prof. Dr. W. Maerker, Konstanz (Botan.); Prof. Dr. J. W. Feldmann, Schaffhausen (Zool., Fischerei, Abwasserfragen); Prof. Dr. K. Hummel, Gießen (Geol.); Prof. Dr. H. Leininger, Karlsruhe (Entomol.); Dr. H. Noll, Glarisegg (Ornithol.); F. Kiefer, Dielsberg (Zool., Copepoden); Dr. Spuls, Karlsruhe (Photometrie). *4 Plätze. Zool., Botan., Hydrographie, Chem.*

H. Kuria près Perm, Uralsk. obl. (U.d.S.S.R).
Station Biol. du fleuve Kama. Dir. W. Beklemischev. — Dr. P. Krassowski (Géobotan.); Dr. A. Triphonow (Chim.); A. Warow (Chim.); Dr. W. Baskina (Hydrobiol.); Dr. I. Tschetyskina (Géozool.) *6 places. Géobotan., Géozool., Hydrobiol.* ⓟ Flore et faune locale.

Langenargen a. Bodensee (Deutschland).
Inst. für Seenforschung und Seenbewirtschaftung. Dir. Prof. Dr. R. Demoll (München). — Dr. Erich Wagler, Leiter des Inst.; Dr. Hermann Lechler, Ass. *12 Plätze. Süßwasser- u. Fischereibiol. — Netze, Wasserschöpfer, Tiefenthermometer, Planktonpumpe, Dredgen, Interferometer, Apparate zur Bestimmung der PH.* ⓟ Bodenseematerial für wissenschaftliche Zwecke.

Lednice — Eisgrub (C.S.R.).
Biol. Versuchsstation d. tschech. Hochschulen Brno. Confer: Institutiones generales.

Leningrad — Alt-Peterhof (U.d.S.S.R.).
Hydrobiol. Labor., Naturwiss. Inst. Dir. Prof. K. Derjugin. — Dr. W. Rylow: eigene wissenschaftliche Arbeit, Leitung der speziellen Studentenarbeiten; Dr. P. Reswoy, Dr. I. Kisselew, Dr. M. Sokolowa: eigene wissensch. Arbeit u. Leitung der praktischen Arbeiten der Studenten; Dr. E. Gurjanova, Dr. M. Wirketiss, Dr. V. Mossewitsch: Aspiranten. *10 Arbeitsplätze. Plankton, Bentos und hydrologische Untersuchungen.* ⓟ Süßwassertiere. — Hydrochem. Zimmer.

Lunz am See (Nieder-Österreich).
Biol. Station (Kupelwieser'sche Stiftung). Dir. Franz Ruttner. *20 Arbeitsplätze. Mikroskopische, physiol. und hydrographische Untersuchungen. — 2 Glashäuser (Warm- und Kalthaus) mit zahlreichen Zementaquarien, Freilandbecken und -teiche, Einrichtungen f. Mikroskopie, Bakt., chem. Analyse, Apparate zur Bestimmung d. elektrolyt. Leitvermögens, der Wasserstoffionenkonzentration etc.; Bibliothek; Unterkunfträume für die Gäste.* ⓟ Lebende, biologisch interessante Wasserpflanzen. Konserviertes Material (Plankton- und Litoralproben) aus Binnenseen. — Zweiglabor. Obersee bei Lunz.

Madison, Wis. (U.S.A.).
Zool. Dept. of the Wisconsin Geol. and Natural History Survey, Biol. Building. C. Juday (Limnol.).

Madrid (España).
Seccion de Hidrobiol. Mus. Nac. de Ciencias, Palacio del Hippodromo. Dir. Arévalo Carretero.

Milano (Italia).
Stazione di Biol. e di Idrobiol. applicata, Via Gadio 2. Dir. Pr. Felice Supino. — Dr. Erminio Schleppati, Ass.; Dr. Paola Manfredi, Ass. *Appareils pour la microscopie; et appareils spéciaux pour les recherches hydrobiol.* ⓟ Annexé à l'Inst. il y a un Aquarium d'eau douce.

Militsch, Schles. (Deutschland).
Labor. für Teichwirtschaft u. Hydrobiol. Dir. Friedrich John. *Teichwirtschaft u. Hydrobiol.*

Monte del Lago, Umbria (Italia).
R. Stationes Idrobiol. del Lago Trasimeno. Dir. Prof. O. Polimanti. — 1 Ass. *3 Plätze für Limnol. und allgemeine Biol. — Apparate f. limnol. Untersuchungen.* ⊕ Sammlung von zool. u. botan. Präparaten vom Trasimenosee und Umgebung.

Mooslachen b. Wasserburg, Bodensee (Deutschland).
Biogeol. Station Mooslachen. Dir. Dr. Helmut Gams, Leiter. *Biocönotik, Systematik, Oekol. u. Geograph. der Gefäßpflanzen, Bryophyten u. Algen; Stratigraphie u. Palaeont. des Quartärs; Hydrobiol.*

Moskau (U. d. S. S. R.).
Hydrobiol. Inst. der 2. Univ. Dir. S. N. Stroganov.
Hydrobiol. Station am See „Glubokoje", 1. Univ., Zool. Mus. Dir. Dr.A.Rumjantzew. — S. N. Duplakoff, Hydrobiol.; G. S. Karsinkin, Hydrobiol.; S.G. Krischanowsky, Zool.; A. P. Scherbakoff, Chemie. *6 Plätze für hydrobiol. Untersuchungen.* ⊕ Planktonproben.
Hydrobiol. Abt. des Labor. für exper. Biol. I. Univ. Leiter Prof. S. A. Sernov. — W. A. Jaschnov; L. W. Arnoldi; A. J. Vesselovskoja. *1—2 Plätze.*
Hydrobiol. Kabinett d. Timirjasef Landw. Akademie. Dir. Prof. S. A. Sernov. — N. S. Gojevskoja-Sokolova. *3—4 Plätze.*

München (Deutschland).
Abwasserstation, staatl. Bayer. Biol. Versuchsanstalt, 34, Veterinärstr. 6. Dir. Prof. Dr. Fr. Graf. — *2 Ass. 2 Plätze für Doktoranden und Gäste. Abwasserchemie und Hydrobiol. — Hydrobiol. und chemisch-technische Laboratoriumseinrichtung.* — Zweiglaboratoria: Wielenbach und Langenargen für Teichwirtschaft und Seenforschung.

Murom (U. d. S. S. R.).
Biol. Oka-Station. Dir. W. I. Shadin. — K. Neiswestnowa-Shadina, Zoologe; N.Kabanov, Botaniker; Const. J. Meyer, Botaniker; S. Zworykin, Chemiker; D. Sassuchin, Zoologe. *5 Arbeitsplätze. Biol. der Tiere und Pflanzen des Süßwassers.* ⊕ Süßwasserfauna und -flora.

Obora b. Blatna (C. S. R.).
Hydrobiol. Station.

Onega See (U. d. S. S. R.).
Borodin Bio-Station des Leningrader Vereins der Naturforscher. Dir. B. V. Perfiljev.

Oster, Kr. Tschernigow (U. d. S. S. R.).
Biol. Labor. des Päd. Technikums. Dir. A. G. Rosanow. — G. G. Tschernogolowko, Planktonkunde; B. S. Popowitschenko, Entomol. *2 Arbeitsplätze für hydrobiol. Untersuchungen. — Planktonnetzen (typ. Zeppelin und Kolwitz).* ⊕ Süßwasserfauna.

Ōtsu (Japan).
Ôtsu Lake-side Biol. Station (of Kyôto Imperial Univ.), Mihogasaki, Shiga-Ken. Dir. Prof. Tamiji Kawamura. — Denzaburô Miyaji, Lect., Ecol., Physiol.; Yasuji Kondô, Ass., Ecol.; Masuzô Ueno, Vice-Ass., Hydrobiol. esp. Cladocera, Ephemerida.

Ottawa, Ont. (Canada).
Dept. of Marine and Fisheries. Confer: Producta lactaria.

Paris (France).
Labor. Pêcher et Productions coloniales d'origine animale du Mus. d'Histoire natur., Ve, 57, Rue Cuvier. Dir. Prof. A. Gruvel. — Préparateur au Mus.: G.Petit; Th. Mouod; R. Dollfus; P.Chabauoud. *4 places. Toutes recherches de zool. marine coloniale ou océanographiques. Crustacés. Poissons. Parasites (Trematodes). Technique de la pêche coloniale. Procédés de conservations des produits de la pêche. Sousproduits de la pêche, etc.* ⊕ Animaux marines ou d'eaux douces des Colonies françaises.

Plön, Holstein (Deutschland).
Hydrobiol. Anstalt der Kaiser-Wilhelm-Gesellschaft. Dir. Prof. A. Thienemann. — Dr. Fr. Lenz, 1. Ass.; Dr. H. Utermöhl, Volontärass. *6 Arbeitsplätze. Limnol. Untersuchungen. — Limnol. Apparatur, Motorboot, Ruderboot.*

Praha — Prag (C. S. R.).
Hydrobiol. und Fischereistation an den Luárě-Teichen, XII, Ul. 1211. Dir. Doc.Dr.Karel Schäferna. — Prof. Dr. J. Wenig; Prof. Dr. J. Komárek; Prof. Dr. B. Němee; Prof. Dr. K. Kairna; Dir. Dr. V. Vaira; Prof. J. Jauda. *2 Laboratoriumräume mit 5 Arbeitsplätzen. Ausstattung zu Freilanduntersuchungen. 2 Ruderboote. — Limnol. und Fischereiuntersuchungen. Biol. und Pathol. der Fische.*

Puckovo a. d. Weißrussischen Eisenbahn (U.d.S.S.R.).
Biol. Station „Am tiefen See".

Put-in-Bay, Ohio (U. S. A.).
The Franz Theodore Stone Labor. of Ohio State Univ. (Formerly the Lake Labor.). Confer: Institutiones generales.

Rostow a. Don (U. d. S. S. R.).
Biol. Stat., Axaje. Dir. Prof. V. N. Verškovskij.

Rouge-Cloître, Brabant, près de Bruxelles (Belgique).
Labor. de Biol. lacustre.

Saratow (U. d. S. S. R.).
Biol. Wolga-Station. Dir. Prof. Dr. A. L. Behning. — 1. Ass. M. M. Lewaschoff; 2. Ass. A. N. Popowa; 2. Ass. F. F. Djakonow; 2. Ass. H. V. Schljapina; Chemiker V. P. Radischtschev. Außeretatmäßig arbeitende Mitglieder: Zoologe N. B. Medwedewa; Hydrophysiologe N. W. Jermakow; Zoologe K. V. Schljapina; Entomologe O. N. Sirotinina. *Für wissenschaftl. Arbeiten 3 Plätze, für Kursteilnehmer 15. Potamoplankton, Acipenserembryol., Studium der Altwässer, künstl. Fischzucht. — Planktonnetze, Dredschen, Motordampfer, Fischzuchtapparate, Aquarien.* ⊕ Wasserinsekten, Kaspiseecrustaceen, Sterletentwicklung, Wolgaplankton, Fischparasiten, Polypodium hydriforme Uss.

Seeon, Chiemgau (Deutschland).
Biol. Labor. Dir Prof. Dr. Richard Woltereck. *Kausale Hydrobiol. im Hinblick auf die Entstehung und Veränderung von Rassen und „Arbeiten". Funktion der „Schwebefortsätze" bei pelagischen Crustaceen. Sexualität der Cladoceren. Variation und Artbildung.* ⊕ Plankton aller europäischen und außereuropäischen Seen.

Smolensk (U. d. S. S. R.).
Biol. Station d. Univ.Smolensk im G. „Vonlarowo", Universitätsstr. Dir. Prof. Dr. Wl. Stantschinsky. — Ass. S. J. Koschkin, Entomol.; Wiss. Mitarbeiter M. A. Emeljanoff, Hydrobiol.; W. P. Schwansky, Malacol. ⊕ Vögel d. palaearktischen Fauna. Tiere v. Westrußland (alle Klassen).

Stabler, Wash. (U. S. A.).
Wind River Experiment Station.

Starosillja, Ukraine (U. d. S. S. R.).
Biol. Station des Dnjeprflusses.

Wigry près Suwalki (Polska).
Station hydrobiol. Dir. Dr. A. Litynski. — Ass. Zygmunt Koźminski. *4 places.* — *Machines à sonder, tubes sondeurs, bathomètres, bathythermomètres, filets planctoniques, appareils pour explorer le fond des lacs, bateau à moteur, canots.* ⓟ Poissons, crustacés Planctoniques, algues, larves des Chironomides.

Swenigorod, Moskauer Gouv. (U. d. S. S. R.).
Hydrobiol. Station d. Wissenschaftl. Inst. d. „Narkomsdraw". Dir. Prof. S. N. Skadovsky. *Ca. 30 Arbeitsplätze.*

Taschkent, Usbekistan (U. d. S. S. R.).
Labor. des zool. des invertebrés et d'hydrobiol., Univ. des Etats de l'Asie Centrale. Dir. Prof. Abrahamm Brodsky. — Nicolas Keiser, Ass.; Victor Gourvitsch. *2—3 places. Pour les travaux sur la faune aquatique (hydrobiol.), edaphon (pedobiol.).* ⓟ *Les araignées, les insects aquatiques (des hautes montagnes); les crustacés des eaux douces et salines.*

Tihany (Ungarn).
Biol. Balatonsee Station (Magyar Nemzeti Muz.) (bis 1927 in Révfülöp). Dir. Prof. Dr. B. Hankó. *Großer Neubau. 6 Arbeitsplätze für Gäste. Zool. und botan., morphol., histol., ökol., experim.-morphol. Untersuchungen. — Alle Apparate zur genannten Arbeit, mit Ausnahme von Mikroskopen, die mitzubringen sind.* ⓟ *Fauna und Flora des Balatonsees.*

Toulouse (France).
Inst. de Hydrobiol. u. Pisciculture. Dir. Prof. Dr. L. Jammes.

Tsinan, Shantung (China).
Biol. Labor. Shantung Christian Univ. Confer: Institutiones generales.

Wielenbach (Deutschland).
Bayrische teichwirtschaftliche Versuchsstation (Hofer-Inst.). Dir. Prof. Dr. R. Demoll in München. — Dr. E. Walter, Betriebsleiter in Wielenbach. *Teichwirtschaftliche Versuche zur Düngung der Teiche und Aufzucht und Fütterung der Fische.*

Wien (Österreich).
Hydrobiol. Donaustation, II, Kaisermühlen. Dir. Prof. Adolf Cerny. — Priv.Doc. Dr. Otto Pesta; Priv.Doc. Prof. Dr. J. Schiller; Dr. Otto Koller, Regierungsrat Dr. Neresheimer. *4 Arbeitsplätze für Gäste für hydrobiol. Untersuchungen. — Apparate und Geräte zur Untersuchung des Planktons und der Bodenorganismen: Mikroskope, Planktonnetze, Dredschen, Bodengreifer, Wasserschöpfer, Zentrifugen usw. Einrichtungen zur hydrographischen und hydrochemischen Untersuchung.* ⓟ Tausch von Planktonproben. Die Station gibt Plankton der Donaugewässer und des Neusiedlersees ab.

Wilno (Polska).
Inst. de Biol. générale. Confer: Institutiones generales.

Wladikawkas (U. d. S. S. R.).
Nord-Kaukas. Hydrobiol. Station, 7. Wersta Woenno-Grusinskoj dorogi.

Zielone Jeziora près de Wilno (Polska).
Station hydrobiol. sur les „Lacs Verts". Dir. J. Z. Wilczynski.

Musea biologica generalia.

Aarau (Schweiz).
Mus. für Natur- u. Heimatkunde. Dir. Prof. Dr. Paul Steinmann. *Tricladen, Fischereibiol.*

Alabama (U. S. A.).
Alabama Mus. of Natural History Univ. Dir. E. A. Smith.

Albany, N.J. (U. S. A.).
New York State Mus. Dir. J. M. Clarke. — S. C. Bishop, Zoologist; H. D. House, Botanist; E. P. Felt, Entomologist.

Altona, Elbe (Deutschland).
Altonaer Mus., Museumstr. Dir. Prof. Dr. Lehmann. — Direktorial-Ass. Dr. Hubert Stierling; Konserv. Rudolf Schmitt. *2 Plätze.* ⓟ *Fauna und Flora Schleswig-Holsteins.*

Aschabad, Turkmenien (U.d.S.S.R.).
Turkmenisches Reichsmus. Dir. Stan. Bilkewicz. *Fauna und Flora d. Transkaspischen Gebietes, Buchara, China, Nord- und Ostpersien, Amu-Darja-Becken.*

Auckland (New Zealand).
Inst. and Mus. Confer: Institutiones generales.

Bamberg (Deutschland).
Naturalienkabinett der Hochsch. Dir. Dr. Theodor Schneid, Hauptkonserv. ⓟ *Sammlungen aller Naturreiche. Vorwiegend einheimische Insektenfauna und die fossile Fauna des Heimatgebietes.*

Barcelona (España).
Mus. de Biol. Dir. J. B. de Aguilar-Amat.
Mus. municip. de Ciencias Nat. Dir. Prof. F. Serdillo. — *Zool.; Malacol.; Ostéol.; Paléont.; Entomol.; Botan.; Flora iberica.*

Barnaul, West-Sibirien (U. d. S. S. R.).
Naturwissenschaftliches Mus., Respublikstr. 50. Dir. Michel Surin. — V. Werestschagin, Botaniker; Wachter, Sekretär, Präparator. *Mus. hat keine Plätze für Gäste.* ⓟ *Mineralien, getrocknete Pflanzen, Insekten, Vögel und ausgestopfte Tiere, archäol. und ethnographische Gegenstände.*

Beograd (S. H. S.).
Serbisches Landesmus., Miloša Velikog ulica. Provis. Dir. Petar S. Pavlović, Kustos der Geol.-Palaeont. Abt — Dušan Stojčević, Kustos der zool. Abt.; Dr. Danilo Kalić, Kustos d. botan. Abt.

Berlin (Deutschland).
Inst. u. Mus. f. Meereskunde, NW7, Georgenstr.34. Dir. Prof. Dr. Krumbach Vorsteher der biol. Abt.

Besançon, Doubs (France).
Mus. d'Histoire natur. Dir. E. Fournier, Paléont. — M. Marceau, Zool.

Bombay (Br.-India).
Mus. of Bombay Natural History Society, 6, Apollo Street. Dir. S. H. Prater, Curator. ⓟ *Collection of Mammals, Birds, Reptiles, Insects, etc.*

Bordeaux (France).
Mus. d'histoire naturelle. Dir. J. Chaine.

Braunschweig (Deutschland).
Naturhistorisches Mus., Schloß. Dir. Dr. von Frankenberg. *5 Arbeitsplätze.* ⓟ *Abgabe hauptsächlich von Vögeln und Säugern.*

Bremen (Deutschland).
Städtisches Mus. für Natur-, Völker- und Handelskunde, Bahnhofsplatz. Dir. Prof. Dr. H. H. Schauinsland. — Dr. Ludwig Cohn, Zool. und Anthrop.; Dr. Johannes Weißenborn, Völkerkunde; Dr. Herman Farenholtz, Botan.; Dietrich Alfken, Entomol.;

August Jordan, Palaeont.; Arnold Böhne, Conchyliol.; Hermann Brakenhoff, Botan.

Brest (France).
Mus. d'Hist. Nat.

Brno — Brünn (C. S. R.).
Zool. and palaeont. Abt. Mähr. Landesmus. K. Absolon, Kustos.

Bruxelles — Brussel (Belgique).
Mus. royal d'Histoire nat. de Belgique, 31, rue Vautier. Dir. Prof. Dr. V. Van Itraelen. — G. Severin, entomol.; E. Vincent, invertébrés tertiairs; E. Maillieux, invertébrés primairs; P. Duputs, mollusques récents; L. Giltay, Vertébrés; A. Ball, Entomol.; 26 chefs de labor. *10 places de travail. Systématique, anat., paléont. animales, paléont. végétale, pétrographie, minéral, préhistoire. — Importante collection ostéologique. Importants collections de la faune vivante et fossile de la Belgique. Collection de plants fossile et de la Belgique.* ⓟ Faune vivante et fossile de la Belgique. Flore fossile de la Belgique.
Section Sc. nat., Mus. du Congo. Dir. Dr. Henri Schouteden.

Bucuresti (România).
Mus. Zool., 1. Şoseana Kisselev. Dir. Prof. Gregori Antipa. — Prof. J. Popa-Burca, Ass.; Rich. Canisius, Ass. *Zool., Palaeont., Botan., Geol., Mineralogie, Anthrop., Ethnographie, Fauna Rumäniens.*

Bulawayo (Southern Rhodesia).
Rhodesia Mus. Dir. Dr. G. Arnold. — A. Frost, A.R.C.Sc.; B.Sc. London, Geologist. *Zool., Botan., Geol., Ethnol. objects.*

Caen, Calvados (France).
Mus. d'histoire nat. Fac. Sc. nat. Univ., rue Pasteur. Dir. A. Bigot.

Calcutta (Br.-India).
Indian Mus., 27, Chowringhee-Road und Sudder Street. — Zool. and Anthrop. Section (Zool. Survey of India). Geol. Section (Geol. Survey of India). Archaeol. Section.

Chabarowsk (U. d. S. S. R.).
Landes-Mus., Ul. Schewtschenko. Botan. u. Zool. Abt.

Chambéry, Savoie (France).
Mus. d'histoire nat., Route de Lyon.

Charkow (U. d. S. S. R.).
Biol. Mus. des Zootechnikums.

Chicago, Ill. (U. S. A.).
Field Mus. of Nat. History. Dir. W. H. Osgood, Curator of Dept. of Zool. — K. S. Schmidt, Ass. curator of Reptiles and Amphibeans; J. B. McNair, Assoc. in Economic Botany.

Christchurch (New Zealand).
Canterbury Mus. Dir. Dr. A. Tonnoir. — Pres. Cur. Prof. R. Speight. *Zool. Botany.*

Cuzco (Peru).
Mus. d'histoire nat. Univ. Dir. F. L. Herrera.

Dieppe, Seine Inf. (France).
Mus. des Sc. nat., Vieux château.

Dublin (Ireland).
Mus. of Natural Philosophy, Univ.

Ferrara (Italia).
Mus. e Labor. di Storia Naturale Univ., Via Buonporto n. 18. Dir. Prof. G. A. Barbieri, Mineral.; Prof. L. Giannelli, Zool.

Frankfurt a. M. (Deutschland).
Senckenberg-Mus. Dir. Geh.Reg.Rat Prof. Dr. O. zur Straßen. — Prof. Dr. O. zur Straßen, Leiter der Zool.; Prof. Dr. F. Drevermann, Leiter der Geol.-Palaeont.; Prof. Dr. R. Nacken, Leiter der Mineral.; Dr. R. Mertens, Kustos der Wirbeltiere; Dr. F. Haas, Kustos der wirbellosen Tiere; Prof. Dr. A. Seitz, Kustos der Entomol. *Arbeitsplätze und Ausrüstung nach Bedarf.*

Genève — Genf (Suisse).
Mus. d'Histoire nat. Dir. Dr. M. Bedof. — J. Carl, 1. Ass.

Genova (Italia).
Mus. Civico di Storia nat. „Giacomo Doria", Via Brigata Liguria 9. Dir. O. de Beaun.

Graz (Österreich).
Zool.-botan. Abt. des Steiermärkischen Landes-Mus. Joanneum, I, Raubergasse 10. Dir. Dr. Adolf Meixner, Vorstand. — Schulrat Josef Vidovic, Kustos. *5 Arbeitsplätze für systematische Studien: Zool., Botan. u. Phytopalaeont.* ⓟ Conchylien, Phanerogamen.

Greenwich, Conn. (U. S. A.).
Bruce Memorial Mus. of nat. Hist., Bruce Park. Dir. P. G. Howes, Curator.

Halifax, N.S. (Canada).
Provincial Mus. of Nova Scotia. Dir. Harry Piers, Curator.

Honolulu (Hawaii).
Bernice P. Bishop Mus. Dir. Herbert E. Gregory, Ph.D., Geologist. — William H. Dall, Ph.D., Naturalist; Elmer D. Merrill, M.S., Botanist; Otto H. Swezey, M.S., Entomologist; Clark Wissler, Ph.D., Anthropologist; Forest B. H. Brown, Ph.D., Botanist; Elizabeth W. Brown, Ph.D., Botan.; Edwin H. Bryan, Jr., M.S., Entomologist; C. Montague Cooke, Jr., Ph.D., Malacologist; Charles H. Edmondson, Ph.D., Zoologist; Kennith P. Emory, M.A., Anthrop.; E. S. Craighill Handy, Ph.D., Anthrop.; J. F. Illingworth, Ph.D., Entomol.; Marie C. Neal, M.S., Malacologist; John F. G. Stokes, Anthrop.: Gerrit P. Wilder, Botan.

Irkutsk (U. d. S. S. R.).
Irkutsker Wissenschaftl. Mus., Ul. K. Marksa.

Jakutsk (U.d.S.S.R.)
Jakutsker Gebiets-Mus.

Jaroslawl (U.d.S.S.R.).
Naturwissenschaftl. Mus. d. Jaroslawer Naturwissenschaftl. Gesellschaft u. d. Ges. f. Landeskunde, Sowetskaja Ploschad 22.

Jefferson City, Mo. (U. S. A.).
Missouri State Resources Mus. Dir. A. C. Burrill, Curator.

Klagenfurt (Deutsch-Österreich).
Naturhist. Landesmus. Dir. Dr. Fr. Lex. — Th. Posse (Botan. Garten); Mag. A. Pokorny, Kustos; Prof. E. Bendl, Kustos. *Zool.; Palaeont., Botan.*

Köln a. Rh. (Deutschland).
Städt. Mus. für Naturkunde, Stapelhaus. Dir. Univ.Prof. Dr. Otto Janson. — *Zool., mineral., geol. Sammlungen; botan. Abt. in Vorbereitung.*

Koršnu-Sevčenko, Ukraine (U.d. S. S. R.).
Naturwissenschaftliches Mus. Dir. A. Feščenko, Verw. Lektor.

Kraków (Polska).
Physiograph. Mus. der Poln. Akad. der Wiss., Slawkowska 17, Akademja. Dir. Jan Waclaw Stach. — Witold Niesiolowski, Kustos, Lepidoterol. Abt.

Kuala Lumpur (Fed. Malay States).
Selanger Mus. Dir. C. B. Kloss. *Zool., Botan.*

La Habana (Cuba).
Mus. de Historia Nat., Inst. de Segunda Enseñanza. Naturwiss. Mus., Acad. de ciencias med. fisicas y nat. de la Habana, de la Academia-Cuba 84a. *Anthrop., prähist., zool., mineral., botan. u. teratol. Sammlung.*

Mus. „Sanchez Roig", Cerro 827. Dir. Prof. M. J. Mario Sanchez Roig. *Zool., paléont.*

La Jolla, Cal. (U.S.A.).
Mus. of Scripps Inst. of Oceanography of the Univ. of Calif. *Seeflora u. Fauna. Aquarium.*

La Rochelle, Charente-Inférieure (France).
Mus. Lafaille et Mus. Fleurian, 28, rue Albert 1er. Dir. Dr. E. Lappé. — Conserv.: Darde; Dr. Bourrian. *Naturgesch., exot. Ethnographie; prähistor. Funde d. Umgegend; Ozeanographie; Fischereikunde.*

La Valletta (Malta).
Mus. of Nat. History, Univ. of Malta.

Le Hâvre (France).
Inst. Océanographique du Hâvre au Mus. Confer: Labor. marina.

Leipzig (Deutschland).
Naturkundl. Heimatmus. d. Leipziger Lehrervereins, Lortzingstr. 3. Dir. R. Buch. *Heimatl. Geol., Botan.; Zool.; photogr. Heimatmus.; Archiv d. naturwiss. Heimatkunde Sachsens.*

Le Mans, Sarthe (France).
Mus. municipal (Préfecture).

Leningrad (U. d. S. S. R.).
Zentralmus. f. Meereskunde.

Lille (France).
Mus. d'Histoire nat., rue de Bruxelles Dir. A. Malaquin, Conserv.

Lima (Peru).
Mus. de Historia Nat., Univ. Dir. Prof. C. Rospigliosi Vigil.

Linz a. D. (Österreich).
Oberösterreichisches Landesmus. Dir. Dr. Herm. Ubell. — Dr. Theod. Kerschner, Kustos, Leiter der naturhistorischen Abt. *Zool. Heimatsammlung, geol. und mineral. Sammlungen Oberösterreichs, botan. Sammlung, Bälgesammlung, entomol. Forschersammlung, im besonderen Oberösterreichs.*

Ljubljana — Laibach (S.H.S.).
Narodni muz. (National Mus.), Bleiweisova 24. Dir. Dr. Josip Mal, Leiter des Mus. — Dr. Franc Kos für Zool.; Prof. Dr. Angela Piskernik für Botan.

London (England).
British Mus. (Natural History), S.W. 7, Cromwell Road. Dir. Sir S. F. Harmer, K.B.E., Sc.D., F.R.S. — G. J. Arrow (Coleoptera); Major E. E. Austen, D.S.O. (Diptera); F. A. Bather, M.A., D.Sc., F.R.S. (Fossil Echinoderms, etc.); H. A. Baylis, M.A., D.Sc. (Parasitic Worms); K. G. Blair, B.Sc. (Coleoptera); M. Burton, M.Sc. (Sponges); C. T. Calman, D.Sc., F.R.S. (Crustacea); W. E. China, B.A. (Hemiptera); L. R. Cox, B.A. (Mollusca); J. G. Dollman, B.A. (Mammals); J. H. Durrant (Pterophorma Tineina); F. W. Edwards, B.A. (Diptera); W. N. Edwards, B.A. (Fossil Plants); A. W. Exell, B.A. (Flowering Plants); A. Gepp, M. A. (Mosses); R. D. O. Good, B.A. (Flowering Plants); M. A. C. Hinton (Mammals) A. T. Hopwood, M.Sc. (Fossil Mammals); N. B. Kinnear (Birds); F. Laing, M.A., B.Sc. (Economic Entomology); W. D. Lang, Sc.D. (Fossil Corals); P. R. Lowe, O.B.E., M.B. (Birds); C. C. A. Monro, B.A. (Worms); J. R. Norman (Fishes); H. W. Parker, B.A. (Reptiles); W. P. Pycraft (Cetacea and Osteol.); J. Ramsbottom, O.B.E., M.A. (Fungi); C. Tate Regan, M.A., F.R.S. (Fishes); A. B. Rendle, M.A., D.Sc., F.R.S. (Flowering Plants); N. D. Riley (Butterflies); G. C. Robson, M.A. (Mollusca); W. E. Swinton, B.Sc. (Fossil Reptiles); W. H. T. Tams (Moths); G. Tandy, B.A. (Flowering Plants); A. K. Totton, M.C. (Corals); J. Waterston, D.Sc. (Hymenoptera); E. I. White, B.Sc. (Fossil Fishes); A. J. Wilmott, B.A. (European and British Flowering Plants); T. H. Withers (Fossil Arthropods). *Facilities for work are given, as far as possible, to duly qualified specialists. — The apparatus and accessories are those usually employed in connexion with Mus. work.* ⑨ The animals and plants of the world. Exchanges are made with other Mus. and with private Workers.

Mus. of Nat. History King's Coll. of the Univ., Strand W.C. 2.

Lons-le-Saunier, Jura (France).
Mus. municipal (Sciences Nat.), Hôtel de Ville. Dir. L. A. Girardot, Conserv.

Los Angeles, Cal. (U.S.A.).
Los Angeles Mus., Exposition Park. Dir. W. A. Bryan.

Lübeck (Deutschland).
Naturhistorisches Mus. L. Benick, Konserv.

Lwów (Polska).
Muz. im. Dzieduszyckich, ul. Rutowskiego 18. Dir. Prof. Jarosław Łomnicki. — Dr. Jan Kinel; Prof. Dr. Józef Sicmiradski (Palaeont.); Prof. Dr. Leon Koztowski (Vorgeschichte); Dr. Tadenk Wilczyński (Pflanzenkunde); Dr. Adam Krasucki (Zool.). *Einige Arbeitsplätze für beschreibende Zool., Botan. u. Palaeont.* ⑨ *Gegenwärtige und fossile Fauna und Flora ausschließlich des polnischen Gebietes.*

Naturwiss. Mus. d. Russ. National Inst. „Volkshaus", Rutoswkiego 22. Dir. Dr. B. Lahola. *Naturwiss. u. physik. Objekte.*

Lyskov, Gouv. Nizegorod (U.d.S.S.R.).
Städt. Histor. u. Naturkundl. Mus. Dir. V. J. Malinovskij, Konserv.

Madrid (España).
Nacional de ciencias nat. Mus., Palacio del Hippodromo. Dir. T. Bolivaz. Sections: Palaeont., Malacol. et animales inférieures, Entomol., Osteozool., Hydrobiol., Microbiol.

Magdeburg (Deutschland).
Städt. Mus. f. Heimatkunde, Domplatz 4. Dir. Prof. A. Mertens. — Conserv. Dr. W. Wolterstorff (Amphibia). *Zool., Botan., Palaeont., Geol., Mineral.*

Mainz (Deutschland).
Naturhistorisches Mus. der Stadt. Dir. Prof. Dr. Otto Schmidtgen. — Dr. Fritz Ohaus, Insektenabt. ⑨ *Palaeont. u. geol. des Mainzer Beckens und des Rheindiluviums.*

Manchester (England).
Manchester Mus. Univ. Dir. G. H. Carpenter. — J. W. Jackson (Senior Ass. Keeper).

Mannheim (Deutschland).
Naturalienkabinett, Schloß. H. R. Prof. Zettler, Kustos.

Marseille (France).
Mus. d'Histoire nat., Palais de Longchamps. Dir. Prof. A. Vayssière.

Melbourne (Australia).
National Mus. of Nat. History, Geol. and Ethnol. Dir. Prof. Sir Baldwin Spencer. — Curator: J. A. Kershaw; Palaeontologist: F. Chapman; Entomologist: S. F. Hill.

Mons (Belgique).
Mus. d'Histoire nat., Rue de Houdain. Dir. Dr. Emile Hublard.

Montevideo (Uruguay).
Mus. de Historia Nat., Casilla Correo 399. Dir. Dr. G. J. Devincenzi. — Dr. Fl. Felippone; P. Cautera; L. Barattini.

Moskau (U.d.S.S.R.).
K. A. Timirjasev Biomus. an der Kommun. J. M. Swerdlov Univ. Dir. B. M. Zawadovsky. — S. J. Bessmertny, Vertreter d. Dir.; M. N. Kischkin, Leiter der Demonstrationsabt. und der Werkstätten, Kustos d. Mus.; N. P. Krenke, Leiter der botan. Abt. und der Abt. f. biol. Grundlagen der Landwirtschaft; M. J. Lomatsch, Leiter der Abt. der lebenden Natur; L. P. Liptschina (experim. Abt.); N. J. Dubrovitzkaja (Botan.); T. N. Belskaja (biol. Grundlagen d. Landwirtschaft); P. N. Chrapov (Excursionen); M. A. Voronzova (physiol. Grundlagen der Sozialhygiene und der Kultur d. menschlichen Körpers). *Grundaufgabe des Museums: die Exposition der Arbeitenresultate experim. Anstalten, Aquarien und Terrarien.* — An der botan. Abt. des Mus. hat eine Gruppe v. Mitarbeitern u. Externen unter Leitung von N. P. Krenke ein experim. Labor. organisiert, wo Arbeiten auf dem Gebiete des Mechanismus der Organenentwicklung in natürlichen sowie auch in experim. Bedingungen ausgeführt werden.
Darwinsches Mus. d. „Narkomprossa", Dewitschje Pole, B. Trubnikowskij 7.

München (Deutschland).
Alpines Mus. u. alp. Pflanzengarten, Praterinsel 5. Dir. Dr. Karl Müller.

Nairobi (British East Africa).
Mus. of the East Africa and Uganda Nat. History Society. Dir. of the Mus.: Dr. V. G. L. van Someren.

New York (U.S.A.).
The American Mus. of Nat. History, 77th Street and Central Park West. Pres.: Henry Fairfield Osborn. — Honorary Dir.: Frederic A. Lucas; Acting Dir. and Executive Secretary: George H. Sherwood. 1. Division of Mineral., Geol. and Geography, Cur.-in-Chief: W. D. Matthew; Geol. and Invertebrate Palaeont. (Acting Curator: W. D. Matthew); Mineral. (Cur.: H. P. Whitlock); Vertebrate Palaeont. (Hon. Cur.: H. F. Osborn, Cur.-in-Chief: W. D. Matthew). 2. Division of Zool. and Zoogeography. (Confer Mus. Zool.) 3. Division of Anthrop.: Cur.-in-Chief: C. Wissler; Cur. of Ethnol.: P. E. Goddard; Comparative Physiol.: Cur.: R. W. Tower. 4. Division of Asiatic Exploration and Research: Cur.-in-Chief: R. C. Andrews.

Nîmes (France).
Mus. d'Histoire nat., 17, Grand' Rue.

Nottingham (England).
Nat. History Mus.

Odessa (U.d.S.S.R.).
Naturwissenschaftl. Mus., Bulv. Feldmana.

Oldenburg i. Old. (Deutschland).
Naturhistorisches Mus. u. Vor- u. Frühgeschichtl. Sammlung, Damm 40. Dir. Prof. Dr. H. v. Buttel-Reepen. *Das Mus. umfaßt Zool., Botan., Palaeont., Mineral., Ethnographie, Prähistorie, Geol.*

Orléans (France).
Mus. d'Histoire nat., Place Louis-Roguet.

Ottawa, Ont. (Canada).
Canadian National Mus. J. H. Fleming (Hon. curator of ornithol.).

Victoria Memorial Mus. Biol.: R. M. Andersen, Ph.D.; E. M. Kindle, Ph.D.; P. A. Taverner. Botan.: M. O. Malte, Ph.D.

Paris (France).
Inst. Océanographique, 195, Rue Saint-Jacques. Dir. L. Mayer. — Prof.: A. Berget (Océanographie physique); Prof. L. Joubin (Océanographie biol.); Prof. P. Portier (Physiol. des êtres marines).
Mus. National d'Histoire nat. au Jardin des Plantes; Administration: (V°) 57, Rue Cuvier. Dir. Louis Mangin. — Secrét.: Moine.; Surveillant général: Peyselongue; Prof. R. Anthony (Anat. comparée); Prof. Jean Becquerel (Physique appl. à l'hist. nat.); Prof. A. Lacroix (Minéral.); Prof. Paul Lemoine (Géol.); Prof. L. J. Simon (Chimie organique); Prof. L. Mangin (Cryptogamie); Prof. H. Lecomte (Phanérogamie); Prof. R. Verneau (Anthrop.); Prof. L. Roule (Erpétol.); Prof. C. Gravier (Vers et Crustacès); Prof. E. Bouvier (Entomol.); Prof. V. Joubin (Malacol.); Prof. M. Boule (Paléont.); Prof. D. Bois (Culture); Prof. J. Costantin (Organographie); Prof. A. Gruvel (Pêches et produits coloniaux d'origine animale); Prof. J. Tissot (Physiol.).

Peking (China).
Peking Labor. of Nat. Historie, 11 Kaka Hutung. Dir. A. W. Graban, S.M., S.D. — Sohtsu G. King, Custodian and honorary Secretary; F. N. Kolarova, Ph.D. (Pelecypoda); N. Gist Gee, M.A. (Fresh-water and terrestrial invertebrata); C. Ping, Ph.D. (Fish, Amphibians and Reptiles); G. D. Wilder, D.D. (Birds); Bernard E. Read, Ph.C., Ph.D. (Plants). *It is a voluntary association of sc. men engaged in the study of the fauna and flora of China, and has for its purpose the systematic survey, and monographic publication of the animals and plants of China. In this work the Labor. stands ready to cooperate with, and if it seems advisable, divide the field with other organizations.* — Thus it is already affiliated with the Chinese Geol. Survey and supplements the work of the Palaeont. division of that Survey, by taking over the study of the Pleistocene Mollusca — to be published, however, in the Palaeont. Sinica, and by continuing the work in the modern fauna and flora. Branch labor. may be established, if it is deemed desirable, at other places and centers, the central labor. acting as a clearing house for correlation of the work of all the branches. The labor. aims to associate with itself active workers in the systematic zool. and botan. of China, in any part of the country, whose work is in the line of that aimed at by the labor.

Pekle b. Maribor (S.H.S.).
Bosnisches Landesmus. Dr. Otmar Reiser, Kustos.

Pensa (U.d.S.S.R.).
Staatl. Provinz.-Mus., Krasnaja 45/10. Dir. E. K. Stuckenberg.

Perak (Fed. Malay States).
Perak State Mus. *Zool., Geol., Mineral., Ethnogr., Archäol.*

Perth (Scotland).
Mus. Dir. John Ritchie, F.R.A.I. *Biol. Work.*

Philadelphia, Pa. (U.S.A.).
Mus. Acad. of Nat. Sc. Dir. W. Stone.

Pittsburgh, Pa. (U.S.A.).
Carnegie Mus. Dir. D. Stewart. — A. E. Ortmann (Curator of Invertebrate Zool.); J. P. Tolmachoff (Curator); W. E. E. Todd (Curator of Ornithol.).

Poltorack, Mittelasien (U.d.S.S.R.).
Turkmenisches Staatsmus. Dir. St. Bilkievič. *Abt. für Geol., Palaeont., Botan., Zool., Ethnographie, Geschichte, Archäol., Landwirtschaft.*

Port-Elizabeth, Transvaal (South Africa).
The Port Elizabeth Mus., Bird Street. Dir. Dr. Frederick W. Fitz Simons. *Nat. Hist., Botan., Geol., Ethnol.*

Poznań — Posen (Polska).
Naturwissensch. Abt. des Groß-Polnischen Mus. Dir. Prof. N. Niezabitowski. — 1 Ass.

Pretoria, Transvaal (South Africa).
State Mus. V. Fitz Simons (Sc. Ass. zool.)

Quebec (Canada).
Mus. of the Dept. of Public Instruction. Dir.: Curator Rev. Canon V. A. Huard, Sc.D.

Reykjavik (Island).
Mus. der Isländischen Naturforschenden Ges. Dir. Lekt. Bjarni Saemundson.

Rouen (France).
Mus. d'Histoire Nat., 98, Rue Beauvoisine. Dir. Dr. R. Regnier.

Rybinsk, Gouv. Tver (U. d. S. S. R.).
Naturwiss. Mus., Puschkinskaja 30. Dir. L. W. Wassiljewa; Praeses d. Ryb. Wiss. Gesellsch. A. A. Solotarjoff. — L. A. Albitskij; A. A. Barajeff. *2 bis 3 Plätze. Allg. biol. Sammlung. Fossilien. Lokalfauna, -flora.* ⓟ Fossilien des Jura der oberen Wolga.

Salem, Mass. (U. S. A.).
Peabody Mus. Dir. A. P. Morse (Curator of Nat. Hist.).

Samarkand (U. d. S. S. R.).
Samarkander Gebiets-Mus., Tschernjawskaja 18.

San Diego, Cal. (U. S. A.).
Nat. History Mus. Balboa Park. Dir. C. G. Abbott.

San Francisco, Cal. (U. S. A.).
Mus. of California Academy of Sc. Dir. Dr. Barton Warren Evermann. *Div. of Zool., Entomol., Ornithol., Botan., Herpetol., Palaeont., Mammal.*

Santo Cruz de la Palma (Canarias).
Mus. de Hist. Nat. y Etnográfico. Dir. E. Santos y Abren.

Sarajevo, Bosnien (S. H. S.).
Bosnisch-herzegowisches Landesmus. Hofrat Viktor Apfelbeck, Kustos der Zool. Abt.; Dr. Stjepan Bolkay, Kustos-Adjunkt der Zool. Abt.; Karlo Maly, Kustos der Botan. Abt.

Sarawak, Sarawak (British Borneo).
Nat. Hist. Mus. *Ethnographie, Zool. u. Botan. von Borneo.*

Simferopol (U. d. S. S. R.).
Naturwissensch. Abt. am taurischen Zentralmus., Karl-Liebknecht-Str. 35. Dir. Prof. A. Deutsch.

Sofia (Bulgarien).
Naturhistorisches Mus. S. M. des Königs von Bulgarien, Tzar Osswoboditel 1. Dir. Dr. Iwan Buresch. — Pentscho Dreńsky (Entomol.); Božimir Davidoff (Botan.); Nenko Radeff (Zool.). *3 Arbeitsplätze für Fauna und Flora Bulgariens.*

Split (S. H. S.).
Naturwissenschaftliches Stadtmus. u. Zool. Garten, Dir. G. Umberto.

Stanford University, Cal. (U. S. A.).
Nat. History Mus. Dir. J. O. Snyder.

Stavanger (Norge).
Naturhist. Abt. des Mus. Dir. Jan Petersen (Archäol.). — Konservator: H. Tho. L. Schaaming (Zool.).

Stawropol, Caucasus (U. d. S. S. R.).
Pravelan Mus., Lunatsharsky Place. Dir. Victor Lutshnik. — Vice-Direktor: O. Prave; Chief of the School Dept.; Z. Borisova; Curator: E. Chodarin. *4 places.*

Stockholm (Sverige).
Naturhistoriska Riksmus, 50; die ethnographische Sammlung: Vallingatan 1, C.). 8 Abt., jede einem Intendent (mit dem Titel Prof.) unterstellt. Mineral. Abt.: Indentend: Dr. Gregori Aminoff; Ass.: Lic. phil. Nils Zenzén. Paläobotan. Abt.: Indentent: Dr. Thore G. Halle; Ass.: Lic. phil. Rudolf Florin. Paläozool. Abt.: Intendent: Dr. Erik H. O. Stensiö; Ass.: Lic. phil. Richard Hägg. Botan. Abt.: Intendent: Dr. Gunnar Samuelsson; Ass.: Dr. Erik Asplund. Abt. für Evertebraten: Intendent: Dr. N. J. Teodor Odhner; Ass.: Dr. Nils Hj. Odhner. Entomol. Abt.: Intendent: Dr. B. Yngve Sjöstedt; Ass.: Dr. Abraham Roman. Abt. für Vertebraten: Intendent: Dr. Einar Lönnberg; Ass.: Dr. Nils C. G. Fersen Graf Gyldenstolpe. Ethnographische Abt.: Intendent: Carl Vilhelm Hartman; Ass.: Dr. Karl Gerhard Lindblom.

Stuttgart (Deutschland).
Württembergische Naturaliensammlung, Archivstr. 3. Dir. Prof. Dr. Max Rauther. — Hauptkonservator Dr. Erwin Lindner (Entomol); Hauptkonservator Dr. Fritz Berckhemer (Palaeont. und Geol.); Konservator Prof. Dr. Richard Vogel (Zool.); Konservator Dr. Reinhold Seemann (Geol. und Mineral.); Ass. Dr. Wilhelm Goetz (Zool.); Hofgartendir. a. D. Alwin Berger (Botan.). *3—4 Plätze für systematischzool. bzw. botan. und für palaeont. Untersuchungen.* ⓟ Besondere Pflege erfahren die Sammlungen von rezenten Tieren und Pflanzen aus dem zum Lande gewordenen Württemberg. In der zool. Abt. wird den Vögeln, Teleosteern und Insekten überhaupt besondere Sorgfalt gewidmet. Mit allen diesen Objekten evtl. Tausch möglich.

Sucre (Bolivia).
Mus. f. Anat. u. Naturgesch. Inst. Med.

Swerdlowsk (U. d. S. S. R.).
Staatl. Ural-Provinz-Mus., Ul. Lenina 28.

Sydney (Australia).
Macleay Mus. of Nat. Hist. Dir. John Shewan. Division of Nat. History of the Australian Mus.

Szeged (Ungarn).
Naturgeschichtl. Abt. d. Städt. Mus. Dir. Dr. K. Czógler. *Zool., Botan., Mineral.*

Taschkent (U. d. S. S. R.).
Zentralmus. für Naturkunde, Ethnographie und Archäol. Mittelasiens. Dir. J. Jankovskij. — Kustos: M. E. Masson; I. I. Bezdeva. *Abt.: A. Die Natur Mittelasiens. a) Das Land (physikal. Geographie, dynam. und histor. Geol. mit Palaeont., Mineral.); b) Fauna; c) Flora. B. Der Mensch in Mittelasien in Gegenwart und Vergangenheit. a) Anthrop. und Ethnographie; b) Archäol. mit Numismatik.*

Tegucigalpa (Honduras).
Mus. Nacional. *Zool. Garten; Botan. Garten; Industrie- u. Ackerbauprodukte; mineral. Ausstellung; archäol. Ausstellung.*

Temrjuk, Kuban-Gebiet (U. d. S. S. R.).
Naturwissenschaftliches Mus. Dir. S. Vojcehovskij *Erforschung der Taman-Halbinsel; Medizin. Pflanzen; Die Heuschrecke u. der Kampf mit ihr im Kuban-Gebiet.*

Toulouse (France).
Mus. d'histoire nat. et Jardin zool., au Jardin des Plantes. Prof. Lécaillon (Zool.); Prof. Mengaud (Paléont., Botan., Mineral); Lacomme (Conserv.).

Trichinopoly (Br. India).
Naturhistorisches Mus. au St. Josephs Coll. Dir. Curator C. Leigh, S.J.

Trivandrum, Madras Pres. (Br. India).
Government Mus. and Public Gardens. Dir. Dr. J. Pryde. — Superintendent in charge: A. Narayanan Nair.

Tromsö (Norge).
Tromsö Mus. Dir. Konservator T. Soot. Ryen. — Ass. Frl. O. Mathisen. 2 —3 *Plätze. Marine Flora u. Fauna.* ℗ *Arktische Tiere und Pflanzen.*

Toronto, Ont. (Canada).
Provincial Mus., St. James Square. Dir. R. B. Orr. — C. W. Nash (Biol.).

Tucumán (Argentina).
Mus. de Sc. nat. Univ. Dir. Prof. M. Lillo.

Tuscaloosa, Ala. (U.S.A.).
Mus. of Nat. History.

Uppsala (Sverige).
Biol. Mus. d. Stadt. Dir. Dr. G. Kolthoff.

Valparaiso (Chile).
Mus. de Historia Nat., Calle Hospital.

Varese, Prov. Como (Italia).
Mus. all'Isolino, Insel Virginia, Lago di Varese. *Prähistorische Pfahlbauten, Ornithol., Fischerei.*

Victoria (Br.-Columbia).
Provinz. Mus. G. A. Hardy, Ass. (Biologist). *Naturwissenschaftl. u. anthrop. Sammlungen.*

Vjatka (U.d.S.S.R.).
„Darwin" **Mus. f. Naturkunde.** Dir. Prof. B. S. Lukaš. — Mykologe: A. D. Phokine.

Vladivostok (U.d.S.S.R.).
Mus. f. Naturgeschichte u. Ethnographie. Dir. Prof. A. M. Čepurkovskij (Anthrop.). — G. N. Lindberg (Zool.); M. A. Firsov (Ornithol.). *Abt.: Anthrop. u. Ethnogr.; Geol. u. Palaeont.; Botan., Zool.; Ichthyol.*

Walla Walla, Wash. (U.S.A.).
Whitman Coll. Mus., Alderstreet 433 E. Dir. H. S. Brode (Curator).

Waren, Mecklbg.-Schwerin (Deutschland).
Von Maltzansches Naturhistorisches Mus. für Mecklenburg.

Warszawa (Polska).
Mus. für Gewerbe u. Landwirtschaft (Muz. Przemyslu i Rolnictwa), Krak.-Przedm. 66. Haupt-Dir. St. Leśniowski. — Verw.-Dir. L. Janikowski. *Sammlungen: ethnogr., archäol., mineral., entomol., zool., landwirtsch.* — An d. Mus. sind folg. Labor. u. Inst. angeschlossen: Inst. f. Gärungsgewerbe u. landwirtsch. Bakteriol., Chem. Labor., Physisches Inst. mit d. Magnet. Observatorium, Inst. d. wissenschaftl. Organisation, Inst. f. Erforschung d. wirtschaftl. Zustandes d. östl. Provinzen, Ethnol. Bureau, Station für Samenschätzung.

Mus. Polonais d'Histoire Nat. Section de Zool., Krakowskie Przedmiescie 26—28. Dir. Dr. Antoni Wagner. — Jan Sztolcman, vice-dir.; Doc. Dr. W. Poliński (Invertébrés exc. les Insectes); Dr. T. Jaczewski (Entomol.); Dr. J. Kremky (Lépidoptères); Mieczysław Węgrzecki. *Recherches sur la systématique l'anat. et la zoogéographie.* ℗ *Animaux du monde entier, particulièrement: Pologne, Sibérie Orientale, Pérou, Brésil. Collections principales: Oiseaux (coll. Taczanowski, Jelski, Sztolcman, Kalinowski, Domaniewski, Chrostowski etc.); Mollusques (coll. Lubomirski, Wagner); Araignées (coll. Kulczynski); Diptères (coll. Dziedzicki); Hémiptères (coll. Zaczewski).*

Washington (U.S.A.).
Coleman Mus. *Naturhist. Sammlungen, bes. Palaeont.*

United States National Mus. of Smithsonian Institution. Dir. Alexander Wetmore, Ph.D., Ass. Secretary. — G. S. Miller (Curator, Division of Mammals); W. R. Maxon (Assoc. Curator, Division of Plants); J. M. Aldrich (Assoc. Curator, Division of Insects); H. W. Krieger (Curator of Ethnol.); M.J. Rathbun (Assoc. in Zool.); R. Ridgway (Curator; Division of Birds); L. Stejneger (Head Curator of Biol.); B. C. Standley (Assoc. Curator, Division of Plants). *Facilities provided for outside workers in all labor. All branches of systematic biol.*

Wellington (New Zealand).
Dominion Mus. Dir. Dr. J. A. Thomson, M.A. (Palaeont.). — W. R. B. Oliver (Zool.); W. J. Phillips (Zool.); Harold Hamilton, A.O.S.M. (Zool.); Miss Amy Castle (Zool.); Miss M. B. Mestayer (Zool.); H. Farquhar (Gen. Zool.).

Wien (Österreich).
Naturhistor. Mus., I, Burgring 7. 1. Dir. Hofrat Prof. Dr. H. Rebel. — Zool. Abt.: Dir. Prof. Dr. H. Rebel; Kustos Reg.R. Dr. K. Attems; Kustos Reg.R. Dr. K. Holdhaus; Kustos Reg.R. Dr. V. Pietschmann; Kustos Dr. O. Pesta; Kustos Dr. H. Zerny; Kustos Dr. Fr. Maidl; Kustos Dr. M. Sassi; Kustos Dr. O. Wellstein; Kustos Dr. O. Koller. Geol.-palaeont. Abt.: Dir. Hofrat Prof. Dr. F. X. Schaffer; Kustos Dr. Fr. Trauth; Kustos Dr. Jul. Pia. Botan. Abt.: Dir. Hofrat Dr. K. Keißler; Kustos Dr. H. Handel-Mazetti. *Arbeitsmöglichkeit für Gäste besteht in jeder der zahlreichen Spezialsammlungen.*

Niederösterr. Landessammlungen. Dir. G. Schlesinger.

Wiesbaden (Deutschland).
Städt. Naturhist. Mus., Rheinstr. 10. Dir. Kustos Chr. Fetzer. — Entomolog: W. Roth; Präparatoren: J. Burger, P. Zimmermann; Vorst. d. Geol. Abt.: O.-Stud.-Dir. D. Heineck. *Zool., mineral., petrograph., palaeont. u. Pflanzensammlungen.*

Yellowstone, Wyo. (U.S.A.).
Mus. of the National Park Service.

Zerbst (Deutschland).
Naturwissensch. Abt. d. Mus. Dir. Dr. Gustav Hinze.

Institutiones historiae biologiae.

Berlin (Deutschland).
Inst. für Geschichte der Naturwissenschaft. Dir. Prof. Dr. Julius Ruska. *Geschichte der Naturwissenschaft im Orient.*

Jena (Deutschland).
Ernst-Haeckel-Mus. u. Archiv, Berggasse 7. Dir. Prof. Heinrich Schmidt. — *Zahlreiche Dokumente zur Gesch. u. Theorie der Entwicklungslehre, die E. H. gesammelt hat.*

Leipzig (Deutschland).
Inst. für Geschichte der Med., Univ. Med. Fac. Dir. Prof. Dr. H. E. Sigerist.

Laboratoria zoologica.

Laboratoria zoologica generalia et Laboratoria morphologiae zoologicae (Anatomiae comparatae).

Confer: *Musea biol. gen.* (pag. 350), *Musea zoologica* (pag. 370), *Institutiones generales* (pag. 336), *Laboratoria hydrobiologiae* (pag. 343) *et parasitologica* (pag. 379).

Aberystwyth, Wales (England).
Dept. of Zool. Univ. Coll. of Wales, Aberystwyth. Dir. Prof. R. Douglas Laurie. — T. Travis Jenkins, D.Sc., Ph.D., Hon. Lecturer in Oceanography; E. Emrys Watkin, B.Sc., Ass. Lecturer; E. Aneurin Lewis, M.Sc., Demonstrator, and Field Officer in Helminthology to the Inst. of Animal Parasit., Ministry of Agric. and Fisheries; Blodwen Fox, M.Sc., Research Fellow of the Univ. of Wales; E. E. Edwards, M.Sc., Research under Scholarship from Ministry; Louise Beanland, M.Sc, Research under grant from the Dept. of Scientific and Industrial Research; Kathleen E. Carpenter, Ph.D., Research on Lead pollution of rivers, under a ,,personal" grant from the Dept. of Sc. and Ind. Res. *1 or 2 places*. *Marine and Fresh Water Ecol.*, *Fisheries*, *Helminthol.*, *Entomol.*

Adelaide (South Australia).
Dept. of Zool. Univ. Dir. T. B. Robertson.

Alger (Algérie).
Labor. de Zool., Fac. Sc. Dir. Prof. Boutan.

Allahabad (Br.-India).
Dept. of Zool. Fac. of Sc. Dir. D. R. Bhattacharya.

Amherst, Mass. (U.S.A.).
Dept. of Zool. and Geol. Coll. of Agric. C. E. Gordon, Ph.D., Head of Dept.; G. C. Ring, M.A., Instr. Zool.

Amsterdam (Holland).
Zool. Labor. Univ., Pl. Doklaan 44. Dir. Prof. Dr. J. E. W. Ihle. — Ass.: Dr. C. J. v. d. Horst; Dr. J. K. de Jong; Fräulein A. van Dam; J. Tensen. ℗ Parasitische Würmer.

Angers, Maine et Loire (France).
Labor. de Zool. de l'Univ. Cathol., 2, rue Voluey. Dir. Prof. P. Teauvel. — Prof. Manquat. *1 place*. *Vermes (Polychaeta)*. ℗ *Insecta (Lepidoptera, Coleoptera)*; *Mollusca (Coquilles)*; *Annelida Polychaeta*.

Ann Arbor, Mich. (U.S.A.).
Zool. Labor., Univ. of Michigan. Dir. A. Franklin Shull. — Jacob Reighard, Prof.; Alexander G. Ruthven, Prof., Dir. of Mus.; George R. La Rue, Assoc. Prof., Dir. of Biol. Station; Paul S. Welch, Assoc. Prof.; Peter O. Okkelberg, Ass. Prof.; Lewis V. Heilbrunn, Ass. Prof.; Frank N. Blanchard, Ass. Prof.; Harry T. Folger, Instr.; T. C. Byerly, Instr.; Arthur E. Woodhead, Junior Instr.; Frank E. Eggleton, Junior Instr. 12 student assistants who teach in the labor. under supervision of an instr. or prof. *2 or 3 places*. *General Physiol.*, *Embryol.*, *Cytol.*, *Ecol.*, *Entomol.*, *Parasit.*, *Animal Behavior*. — Constant temperature chambers, aquaria, light reaction room, photographic equipment, incubator, type Kapotentiometer, large Freas thermostat. — The Biol. Station of the Univ., conducted in summer at Cheboygan, Michigan, is directed by member of the Zool. Labor. staff, Prof. George R. La Rue. The Mus. of Zool., in Ann Arbor, Michigan, is directed by a member of the Zool. Labor. staff, Prof. A. G. Ruthven.

Askania Nowa, Gouv. Dnepro-Petrowsk (U.d.S.S.R.).
Wissenschaftl. Steppen-Station a. d. I. Staatl. eppen-Sapowednik, Melitopol. okr.

Astrachan (U.d.S.S.R.).
Astrachaner Staatl. Sapowednik, Ul. Ryleewa 2/4.

Athènes (Grèce).
Labor. de Zool. Univ., Edifice de l'Univ. Dir. Prof. Skuphos.

Ayabe (Japan).
Morphol. Labor., Sericicultural Experim. Stat. of Gunse Silk Factory, Kyôto-fu. Dir. Dr. G. Yamanouchi.

Baku, Azerbaijan (U.d.S.S.R.).
Zool. Labor. of Univ. of Azerbaijan, Mal. Morskaja, 4. Dir. Prof. V. Elpatievsky. — Dr. S. I. Veißig, Ass.; Dr. S. N. Saveliev, Ass. ℗ Parasitical worms. Plankton of Caspian Sea and its southern tributaries.

Baltimore, Md. (U.S.A.).
Zool. Labor. John Hopkins Univ. Dir. H. S. Jennings. — C. D. Beers.

Bangor, N.-Wales (England).
Dept. of Zool. Dir. Prof. Ph. J. White.

Barcelona (España).
Inst. de Zool. Dir. José Fuset Fabiá.

Bargusin am Baikal-See (U.d.S.S.R.).
Bargusiner Staatl. Sapowednik d. Volkskommissariats f. Landwirtschaft (Narkomsem), Burjat-Mongoler.

Bari (Italia).
Gabinetto di zool. e anat. compar. della R. Univ. Dir. L. C. De Martiis.

Basel (Schweiz).
Zool. Anst. Univ., Rheinsprung 9. Dir. Prof. F. Zschokke.

Baton Rouge, La. (U.S.A.).
Zool. and Entomol. Labor. Dir. Wm. H Gates. — Dr. Ellinor Behre (Physiol. and Anat.); Dr. R. L. Mayhew (Physiol. and Parasit.); Prof. O. W. Rosewall (Entomol.); Prof. H. A. Stabe (Entomol.); Mr. J. R. Fowler (General Zool.). *Zool.*, *Parasit.*

Belfast (Ireland).
Dept. of Zool. Queens Univ. Dir. G. Wilson.

Beograd (S.H.S.).
Zool. Inst. Univ., Phil. Fac. Dir. Prof. Dr. Ž. Georgévitch. — Prof. Dr. B. D. Milojević (Zool. u. vergl. Anat.); Prof. Dr. S. Stanković (Zool. u. vergl. Anat.).

Bergen (Norge).
C. Sundts zool. Lehrkanzel des Mus. Dir. Prof. Dr. Aug. Brinkmann. — Amanuensis Johan Huus.

Berkeley, Cal. (U.S.A.).
Zool. Labor., Univ. of California. Dir. Prof. Charles A. Kofoid, Chairman (Protozool., Parasit.). — Prof. S. J. Holmes (Evolution and Genetics); Prof. J. F. Daniel (Experim. Zool.); Prof. J. Grinnell (Vertebrate Zool.); J. A. Long, Associate Prof. of Embryol.; S. F. Light (Invertebrate Zool.); C. L. Camp (Comparative Anat.); H. C. Hinshaw; E. L. Lazier. *2 in addition to 20 for graduate students*. *Zool.*, *embryol.*, *experim.*, *cytol.*, *protozool.*, *cultural research*. — High grade microscopical equipment,

animals in mammal house, incubators, autoclaves. Ⓟ *Marine life of Pacific Coast, Protozool. and parasit.* — *Parasit. Labor. of the California State Board of Health, C. A. Kopold, Consulting Parasit., Dir.*

Berlin (Deutschland).
Zool. Inst. Univ., N 4, Invalidenstr 43. Dir. Prof. Dr. Richard Hesse. — Prof. Dr. W. Berndt, Abt.-Vorst.; Prof. Dr. P. Krüger, Ass.; Dr. E. Marcus, Ass.
Zool. Inst. der Landwirtschaftlichen Hochsch., N 4, Invalidenstr. 42. Dir. Prof. Dr. R. Heymons. — Ass.: Prof. Dr. H. von Lengerken; *2—3 Plätze für entomol. Untersuchungen* Ⓟ *Haustiere. Ökonomisch wichtige Insekten. Pentastomida (Linguatulida).*

Bern (Schweiz).
Zool. Inst. der Univ., Bollwerk 10. Dir. Prof. Dr. F. Baltzer. — Ass.: ao. Prof. Dr. F. Baumann; Dr. G. Fankhauser.

Besançon, Doubs (France).
Labor. de Zool. Univ. Dir. Prof. Marceau.

Birmingham (England).
Dept. of Zool. Univ. Dir. Prof. Dr. F. W. Gamble.

Bloemfontein (South Africa).
Dept. of Zool. Grey Univ. Coll. Dir. Prof. T. F. Dreyer.

Bloomington, Ind. (U.S.A.).
Dept. of Zool., Indiana Univ. R. Voris (Tutor).

Bologna (Italia).
Labor. di Anat. compar. Univ. Dir. Prof. Ercole Giacomini. — Dott. Ettore Remotti, Aiuto; Dott. Edoardo Benedetti, Tecnico. *4. Anat. e Istofisiol. compar. e Biol. sperim.* — *Apparecchi di oscillazioni elettriche. Apparecchi per chim. fisiol. Apparecchi per microfotografia e Apparecchi per disegnare.* Ⓟ *Preparati istol. dei vari organi dei Vertebrati. Serie embriol. di vari gruppi di Vertebrati. Serie di cervelli di Vertebrati inferiori.*
Ist. di zool. Univ. Dir. Prof. A. Ghigi. — Dott. A. Vecchi, Aiuto. 1 Ass. *Genetica, Sistematica, Zoogeografia, Ecol., Zool. applicata, Avicoltura e Zoocultura in genere.* Ⓟ *Gallinacei: Fagiani.*

Bonn (Deutschland).
Zool. u. vergl. anat. Inst. Univ., Poppelsdorfer Schloß. Stellv. Dir. Prof. Dr. A. Borgert. — Ass.: Dr. V. Bauer; Dr. C. Heidermanns; H. Wurmbach. *20 Arbeitsplätze für zool., anat., physiol. Unters.*

Bordeaux (France).
Labor. de Zool. et Physiol. animale, Fac. Sc. Univ., Cours de la Marne, 149. Dir. Prof. Bounhiol. — Dr. J. Feytaud, Maître de Conférences. *3 places. Histol. et embryol. des animaux.* — *Microphotographie, physiol. respiratoire.*

Bou Saada (Algérie).
Labor. de Zool. Générale, Univ. d'Alger.

Bozeman, Mtn. (U.S.A.).
Dept. Zool. Coll. of Agric. M. H. Spaulding, A.M., Head of Dept.

Breslau (Deutschland).
Zool. Inst. u. Mus. Univ., Sternstr. Dir. Prof. Dr. P. E. C. Buchner. — Prof. Ferd. Pax, Leiter des Mus.; Dr. Giersberg, Priv.Doc. (Physiol.).

Brno — Brünn (C.S.R.).
Inst. de zool. Univ., Kounicova 63. Dir. Prof. Dr. Jan Zavřel. — Dr. V. Teyrovský, Agrégé; Dr. V. Vrtiš. *2 places pour morphol., histol., embryol.*
Zool. Inst. d. Landw. Hochsch. Dir. Prof. Em. Bayer.

Brookings, S.D. (U.S.A.).
Dept. of Zool. and Entomol. State Coll. of Agric. H. C. Severin, M.A., Prof. Zool. and Entomol.; R. D. Bulger, B.S., State Leader Barberry Erad. (Detailed by U.S.D.A.); M. D. Farrar, B.S., Instr. Zool. and Entomol.; George Gilbertson, M.S., Ass. Prof. and Ass. Entomol.; E. C. O'Roke, M.A., Ass. Prof. Zool.

Brooklyn, N.Y. (U.S.A.).
Dept. of Zool. Brooklyn Inst. of Arts and Sciences. Dir. J. J. Schoonhoven (President).

Bruxelles — Brussel (Belgien).
Inst. zool. Torley-Rousseau, Univ. (jadis Station biol. d'Overmeire). Dir. Aug. Lameere. *Inst. fondé pour le développement de la Zool. théorique pure.* — Un labor. de Biol. lacustre à Rouge-Cloître (Brabant, près de Bruxelles).
Labor. de zool. et d'anat. compar. Dir. Prof. A. Lameere. — Dr. Frechkop; P. Brien, Ass.

Budapest (Ungarn).
Zool. und compar. anat. Inst., Univ. Dir. Prof. Lajos Méhely. — Dr. A. Abrahám; Dr. G. Mödlinger; A. Wolsky; Dr. S. Abonji.
Inst. für Zool. und Parasitol. Tierärztl. Hochsch., Rottenbiller Gasse. Dir. extraord. Dr. Alex. Kotlán.

Buenos Aires (Argentina).
Inst. de Zool. Fac. sc. Univ. Dir. Prof. A. Gallardo.
Inst. de Zool. agric. Fac. veter. et agric. Univ. Dir. Prof. Y. M. Huergo. — Prof. F. Lahille.

Buitenzorg, Java (Ned.-O.-Indië).
Zool. Mus. and Labor. of the Botan. Gardens. Dir. Dr. K. W. Dammermann, Chief (Zoogeography and research of the new fauna of Krakatau). — Ass.: Dr. H. H. Karny (Entomol., Orthoptera, Thripsidae, Psocidae, Psyllidae); H. C. Siebers (Ornithol.). 1 Taxidermist.

Burlington, Vt. (U.S.A.).
Dept. of Zool. Coll. of Agric. H. F. Perlins, Ph.D., Prof. Zool.; Ruth J. Ball, M.S., Ass. Prof. Zool.; Rhoda A. Hartwell, M.S., Instr. Zool.

Caen, Calvados (France).
Labor. de Zool. et Physiol. animale Fac. Sc. nat. Univ. Dir. Prof. L. Mercier. — E. Audigé.

Cagliari, Sardinia (Italia).
Ist. di zool., fisiol. e anat. compar. della R. Univ. Dir. Ermanno Giglio-Tos.

Cairo (Egypt).
Dept. of Zool. Univ. Dir. Prof. V. Jollos. *Experim. Vererbungslehre u. Geschlechtsbestimmung, Protistenkunde.*

Calcutta (Br.-India).
The Zool. Survey of India, Indian Mus. Dir. Major R. B. Seymour Sewell, I.M.S. — Dr. Baini Prashad; Dr. S. L. Hora; Dr. B. N. Chopra; Dr. H. S. Rao; Dr. Hem Singh Pruthi; Mr. Mohammed Sharif (Temporary). *4 places for Zool. for taxonomic and systematic investigations. 2 places for Physical anthrop.* — *Special apparatus for Micro-photography and for physical anthrop.* Ⓟ *A representative collection of the fauna of India and Indian Seas. The Marine collections of the R.I.M.S. ,,Investigator,,.*

Cambridge (England).
Zool. Labor., The Mus. Dir. Prof. J. Stanley Gardiner, F.R.S. — Cyril Crossland; Herbert Caldwell James, Entomol. Research; Harold Hulme Brindley, Demonstrator in Elementary Biol.; John Tennant Saunders, Lect.; James Gray, Lect. in Experim. Zool.; Joseph Omer Cooper, Demonstrator; William Homan Thorpe, Demonstrator;

H. F. Gadow, Reader. *Equipment for all classes of morphol. and experim. researches.* — Subdepartments of Experim. Zool. and Entomol.

Cambridge, Mass. (U.S.A.).
Dept. of Biol. and Public Health of the Massachusetts Inst. of Technol., 222 Charles River Road. Dir. Prof. Samuel C. Prescott, Sc.D. — Prof. Robert P. Bigelow, Ph.D. (Zool. and Parasitol.); Associate Prof. Clair E. Turner, A.M., C.P.H. (Biol. and Public Health); Associate Prof. John W. M. Bunker, Ph.D. (Biochem. and Physiol.); Ass. Prof. Murray P. Horwood, Ph.D. (Biol. and Public Health); Ass. Prof. Francis H. Slack, M.D. (Public Health Labor. Methods); Ass. Prof. C. H. Blake; 9 special lect.; 2 Instr.; 1 Research Associate; 2 Ass.; 2 Research Ass. *2 or more places for compar. Anat., Bact., Biochem.* — *Bact. and Biochem. Apparatus.*

Harvard Zool. Labor., Mus. Compar. Zool. Dir. Prof. G. H. Parker. — W. M. Wheeler, Prof. of Entomol.; W. E. Castle, Prof. of Genetics; H. W. Rand, Associate Prof. of Zool.; C. T. Brues, Associate Prof. of Economic Entomol.; G. M. Allen, Lect. on Zool.; H. B. Bigelow, Lect. on Zool.; T. Barbour, Lect. on Zool.; S. R. Detwiler, Associate Prof. of Zool.; J. A. Dawson, Instr. in Zool.; J. Wyman, Jr., Instr. in Zool. *4 places for General Zool.*

Camerino (Italia).
Gabinetti di zool., anat. e fisiol. compar. della Libera Univ. Dir. R. Grandori.

Ist. di zootomia e istol. generale Libera Univ. Dir. G. Stoppolino.

Canton (China).
Dept. of Compar. Anat. Univ. Dir. Prof. Kwaan Seung Wo.

Cape Town (South Africa).
Dept. of Zool. Univ. Dir. Prof. I. D. F. Gilchrist.

Caracas (Venezuela).
Labor. de Zool. Univ. Dir. Prof. F. M. Martinez.

Cardiff, Wales (England).
Labor. of Zool. and Compar. Anat. Univ. Coll., Newport Road. Dir. Prof. W. M. Tattersall. — J. H. Lloyd, M.Sc., Lect.; Miss E. M. Sheppard, M.Sc., Lect. ⊕ Crustacea, mainly Mysidacea, Isopoda and Amphipoda.

Catania (Italia).
Ist. di Zool. e di Anat. e Fisiol. compar. Dir. A. Russo.

Cernăuţi — Czernowitz (România)
Zool. Inst. Univ. Dir. Prof. Dr. E. Botezat. — Prof. Cehovschi Constantin, Chef der praktischen Arbeiten; Dr. Marcu Orest, Ass. *Histol.*

Chapel Hill, N.C. (U.S.A.).
Zool. Labor. of the Univ. Dir. Henry Van Peters Wilson, Kenan Prof. of Zool. — Robert Ervin Coker, Ph.D., Prof. of Zool. *2 or 3 places for the usual morphol. work.*

Charkow, Ukraine (U.d.S.S.R.).
Zootomisches Kabinett, Gotpitolny per. N. 5. Dir. Akademiker Prof. A. M. Nikolsky. — Ass. T. V. Radionowa; E. Umontky; S. Schermow. *10 Arbeitsplätze für zootomische und systematische Untersuchungen ausgerüstet.* ⊕ Das Labor. hat 1. eine bedeutende Kollektion Reptilien, Amphibien, Fische (russische und asiatische Arten), Vögel (russische Fauna). 2. Osteol. Mus.

Zool. Labor., Inst. f. Land- u. Forstwiss. Dir. Prof. I. K. Tarnani. — Ass.: A. A. Ustinow; A. N. Rachmaninow; A. S. Belikowa; Aspiranten: G. Ch. Kipritsch; E. M. Klokow; R. G. Pusyzny; A. I. Wolkowa. *2 Arbeitsplätze.* ⊕ Tiere, angewandte Zool. — Zweiglabor. für Entomol.

Chicago, Ill. (U.S.A.).
Hull Zool. Labor. and Whitman Labor. of Experim. Zool. Dir. Prof. Frank Rattray Lillie. — Charles Manning Child, Prof.; Horatio Hackett Newman, Prof.; Warder Clyde Allee, Associate Prof.; Sewall Wright, Associate Prof.; Carl Richard Moore, Associate Prof.; Benjamin Harrison Willier, Ass. Prof.; L. H. Hyman. *Several places for experim. Zool., Biol. of Sex, Embryol., Genetics, Ecol.* ⊕ Genetical Records, Sex records guinea pigs, rats, free-martins, fowl. — Sub-labor.: Whitman Labor. of Experim. Zool.

Clemson College, S.C. (U.S.A.).
Dept. of Zool. and Entomol. Clemson Agric. Coll. Franklin Sherman, M.S., Prof. Zool. and Entomol., Entomol. (State Entomol.); E. J. Anderson, M.S., Ass. Prof. Zool. and Entomol.; J. A. Berly, B.S., Ass. State Entomol.; O. L. Cartwright, M.S., Ass. Entomol. (P.O., Columbia); C.O. Eddy, M.S., Assoc. Prof. and Assoc. Entomol.; C. B. Nickels, M.S., Research Ass. Entomol.; J. O. Pepper, M.S., Ext. Entomol.; E. S. Prevost, Beekeeping Specialist.

Clermont-Ferrand, Puy-de-Dôme (France).
Labor. de Zool. Fac. Sc. Univ. Dir. Prof. L. Calvet.

Cluj — Klausenburg (România).
Inst. de Zool. Inst. didactique et de recherches, Str. Miko 5. Dir. Prof. I. Scriban. — Th. Buşniţă, Chef de travaux.

Cold Spring Harbor, N.Y. (U.S.A.).
The Biol. Labor. of the Long Island Biol. Association. Dir. Reginald G. Harris. — H. E. Walter, H. M. Parshley, G. F. Sykes, Zool. Instr.; H. S. Pratt (Compar. Anat.); W. W. Swingle (Endocrinol.); J. S. Nicholas (Experim. Surgery); J. M. Andrews (Parasitol.); Asa A. Schaeffer (Protozool.); N. M. Grier (Botan.). 3 Ass. in Zool., 1 Ass. in compar. Anat., 1 Ass. in Botan., 1 Ass. in Experim. Surgery. *30 working places for guests. Fitted for marine work and general zool. Cytol., protozool., endocrinol., genetics, embryol., special equipment for work on mammals, histol., physiol.* — *Aquaria, marine and fresh water, dredge, motor boat, operating room, animal cages, microscopes, microtome, sterilizer, dark room, centrifuge, distilling apparatus, hydrogenion determining apparatus, refrigerator.*

College Park, Md. (U.S.A.).
Dept. of Zool. Univ. of Maryland. Dir. C. I. Pierson.

Columbia, Mo. (U.S.A.).
Dept. of Zool. Univ. of Missouri.
Dept. of Zool. Coll. of Agric. W. C. Curtis, Ph.D., Prof. and Chair. of Dept.; Mary J. Guthrie, Ph. D., Ass. Prof. Zool.; W. R. B. Robertson, Ph.D., Assoc. Prof. Zool.; G. W. Tannreuther, Ph.D., Assoc. Prof. Zool.

Cork (Ireland).
Zool. Labor., Univ. Coll. Dir. Prof. Louis P. W. Renouf, B.A., Dip. Agric. — Miss Marjorie M. Murphy, B.Sc., Dip. in Educ. *The main work of the Baltimore Station (see Below) is concerned with the local fauna and flora. Especially the marine fauna — and from this a great many species can be obtained.* Baltimore biol. Station on the South Coast of Cork.

Corrientes (Argentina).
Labor. de Zool. Fac. agric. Univ. Dir. Prof. M. Iglesia.

Corvallis, Ore. (U.S.A.).
Dept. of Zool. and Physiol. Stat. Agric. Coll. Nathan Fasten, Ph.D., Prof. Zool. and Physiol.; W. D. Courtney, B.S., Instr. Zool. and Physiol.; Laura Garnjobst, B.S., Tech. in Zool. and Physiol.;

Florence S. Hague, Ph.D., Ass. Prof. Zool. and Physiol.; J. L. Osborn, A.M., Instr. Zool. and Physiol.; H. M. Wight, M.S., Ass. Prof. Zool. and Physiol., Ass. Zool.

Debreczen (Ungarn).
Zool. Inst. Univ. Dir. Dr. Z. von Szilády.

Děčín-Libverd — Tetschen-Liebwerd (C.S.R.).
Lehrkanzel f. Zool., Fischzucht u. Teichwirtschaft d. Landw. Abt. d. Deutschen Techn. Hochsch. Prag. Dir. ao. Prof. Dr. V. Langhans.

Delhi (Br.-India).
Dept. of Zool. Univ. B. Das; M. L. Seth; H. Lal.

Detroit, Mich. (U.S.A.).
Dept. of Biol. Univ. of Detroit. Dir. R. A. Muttkowski. *Ecol. of aquatic animals, Arthropoda. Insect Physiol. and morphol. Taxonomy of Odonata.*

Dijon (France).
Labor. de Zool. et Physiol., Fac. Sc. Dir. Prof. Hesse.

Dublin (Ireland).
Zool. Dept., Univ. Coll., Upr. Merrion Street. Dir. Prof. J. Bayley Butler. — Thomas Dinan; Lect. Edmond John Sheehy. *4 places in General Zool., 4 in Marine Zool., 4 in Marine Zool. — An agricultural Coll. farm, with Orchards and Greenhouse, and a small Marine Zool. Labor.* ℗ *Marine specimens from the Irish Sea.* — A small Marine Zool. Labor. at Dûn Laoghaire, 7 miles from the Coll. and an Agric. Coll. and Farm at Glasnevin, 6 miles from the Coll.
Mus. and Dept. of Zool. Univ. Trinity Coll. Dir. I. B. Gatenby.

Dunedin (New Zealand).
Dept. of Zool. Univ. Dir. Prof. W. B. Benham.

Durham, N.H. (U.S.A.).
Zool. labor. Univ. of New Hampshire. Dir. C. F. Jackson. — Donald G. Barton (Anat.); Edythe May Tingley (Genetics and Geol.); A. D. Jackson (Compar. Embryol. and Histol.). *3 places. Marine or fresh-water Ecol. or invertebrate Embryol. Compar. Anat. of invertebrates.*

Eberswalde (Deutschland).
I. Labor. für Zool. Forstl. Hochsch. Dir. Prof. K. Eckstein.
II. Labor. für Zool. d. Forstl. Hochsch. Dir. Prof. M. Wolff. *Angewandte Zool., vorzüglich Forstzool., besonders Forstentomol.*

Edinburgh (Scotland).
Dept. of Zool. Univ. Dir. Prof. J. C. Ewart. — Prof. J. H. Ashworth.

Edmonton (Canada).
Dept. of Zool. Univ. of Alberta. Wm. Rowan, M.Sc.; W. Hughes, M.A.

Erlangen (Deutschland).
Zool. Inst., Universitätsstr. 18. Dir. Prof. A. Fleischmann. — Ass. Dr. F. Schwarz. *Morphogenet. Untersuchungen.* ℗ *Embryonen der Amnioten.*

Eugene, Ore. (U.S.A.).
Labor. of Physiol. and Zool. Confer: Physiol. comparata.

Fargo, N.D. (U.S.A.).
Dept. of Botan., Bact. and Zool., State Coll. of Agric. Confer Bact. agric.

Firenze (Italia).
Ist. di Zool. Univ., Via Romana 19. Dir. Prof. Dott. A. Senna. — Dott. E. Calabresi, Aiuto; Dott. L. di Caporiacco, Ass.
Labor. di Anat. e Fisiol. compar, Via Romana 19. Dir. Prof. N. Beccari. — Dott. L. Lurini, Ass. *4 places. Ricerche microscopiche.* ℗ *Preparati macroe microscopici di Anat. compar. dei vertebrati.*

Frankfurt (Deutschland).
Zool. Inst. Dir. Prof. O. zur Strassen. — Dr. Kuhn, Ass.

Freiburg i. Br. (Deutschland).
Zool. Inst. Dir. Prof. Dr. Hans Spemann. — Dr. Bruno Geinitz, Ass.; Dr. Fritz Sieffert, Ass.; Frau Dr. Else Bautzmann, Hilfsass. *6 Plätze für experim. Embryol.*
Forstzool. Inst. d. Univ., Schänzleweg 9. Dir. Prof. Dr. R. Lauterborn. — Dr. W. Bischoff, Ass. ℗ *Forstinsekten, heimische Tierwelt, Süßwasserplankton, Dipterenmetamorphosen.*

Freiburg (Schweiz).
Zool. Inst. d. Univ. Dir. Prof. Dr. A. Reichensperger. — Priv. Doc. Dr. Gandolfi-Horngold; Priv. Doc. P. Dr. G. Rahm.

Fukuoka (Japan).
Zool. Labor., Dept. of Agric., Kyūshū Imperial Univ. Dir. Prof. Hiroshi Ohshima, D.Sc. — Ass. Prof. Teiso Esaki (Entomol.); Ass. Prof. Junji Oyama (Zool.). *Invertebrate and Vertebrate Embryol. Fauna and biol. of Kyūshū, especially of marine life.* ℗ *Anything occurring in this part of Japan, especially of the groups interested by the members of the labor., and if duplicates of specimens are available, we are willing to offer them in exchange.* — Marine Biol. Labor. of the Kyūshū Imperial Univ., Tsuyasaki, Fukuoka Prefecture.

Gand — Gent (Belgique).
Inst. d'Anat. comparée Univ., 14, Rue longue du marais. Dir. Prof. Hect. Lebrun. — Prof. V. Willem. *Cytol.*

Genève — Genf (Suisse).
Inst. de Zool. et Anat. comparée Station de Zool. expérim. Dir. Prof. Dr. Emile Guyénot. — Dr. A. Naville, Chef des travaux; Dr. Mlle. K. Ponse, Ass. à la Station de Zool. expérim.; Dr. O. Schotté, Ass. *50 places pour travaux pratiques de Zool. générale. 20 places pour recherches de Zool. expérim. (régénération, sexualité, hérédité, variation, greffes et transplantation) d'Histol. et de Protistol. (Sporozoaires).* — *Installations de rayons X et ultra violets, microphotographie, étuves, glacières. Salle d'aquariums, 16 bassins, 1 grand bassin, 5 terrariums, 10 terrariums avec bassins, 2 bâtiments d'élevages, 1 grand jardin.* Labor. de Protistol. et Histol. Station de Zool. expérim. (16, chemin Sautter).

Genova (Italia).
Ist. di Anat. compar. Univ., Via Balbi 5. Dir. Prof. Luigi Cognetti de Martiis. — Libero Doc. Raffaello Anselmi, Aiuto (Anat. compar.). *Anat., Istol.*
Labor. di Zool. Univ., Via Balbi 5. Dir. Prof. Raffaele Issel. — Aiuto: Dott. Renato Santucci; Aiuto volontario: Dott. Alessandro Brian. *Il lavoro è ristrettissimo, sta per essere iniziata la costruzione di un labor. nuovo, specialmente adatto per lo studio (anche sperimentle) delle variazioni locali degli animali e con numerosi posti di studio.* ℗ *Collezione di preparati di parasitol. umana per la Scuola. Larve e adulti di cefalopodi ed eteropodi.*

Gießen (Deutschland).
Zool. Inst., Bahnhofstr. 84 I. Dir. Prof. Dr. W. I. Schmidt. — Priv. Doc. Dr. E. Merker; Dr. W. E. Ankel. *Histol. u. Cytol. im polarisierten Lichte.*

Gihu (Japan).
Zool. Labor. Gihu Imper. Coll. of the Agric. and Forestry.

Giza (Egypt).

Egyptian Government Zool. Service, Zoological Gardens. Dir. Major Frank William Borman, F.Z.S., M.B.O.U. — Dr. Ibrahim Kadry (Sub-Dir.); Helmi Elsamma Effendi (Inspector); Youssef Sabit Effendi (Inspector). ℗ A valuable collection of birds' skins (Egyptian) is maintained in the Giza Zool. Mus. (situated in the Giza Zool. Gardens) together with the Anderson collection of Reptilia. These collections are available to students. — Giza Zool. Mus. Gezira Aquarium.

Göttingen (Deutschland).

Zool. Inst. der Univ., Bahnhofstr. 28. Dir. Prof. Dr. Alfred Kühn. — Prof. Dr. R. W. Hoffmann; Dr. Kröning; Dr. Otto Kuhn; Dr. Henke, Ass. *2 Arbeitsplätze für Untersuchungen über Lichtsinn der Tiere, für Tierzucht, bes. Modifikationsversuche bei Insekten. — Spektralapparate, auch für ultraviolettes Licht; photometrische Einrichtung; Einrichtungen zur Zucht bei konstanten Temperaturen.*

Grahamstown (South Africa).

Dept. of Zool. Rhodes Univ. Coll. Dir. Prof. J. E. Duerden.

Graz (Österreich).

Zool.-zootomisches Inst. Phil. Fac. Univ. Dir. Prof. Dr. Ludwig Böhmig.

Greifswald (Deutschland).

Zool. Inst. der Univ., Bismarckstr. 11/12. Dir. Prof. Dr. E. Matthes. — Dr. G. Just, Priv.Doc. u. Ass. (Genetik); Dr. H. J. Stammer, Ass.; Dr. A. Koch, Hilfs-Ass. *Mehrere Plätze für mikroskopische u. hydrobiol. Untersuchungen (Biol. des Brackwassers); Vergl. Anat., Genetik. — Boot, Aquarien, Dunkelzimmer, chemisches Zimmer.* ℗ *Tierwelt der benachbarten Ostseegebiete und Brackwassergebiete.*

Grenoble (France).

Lab. de Zool. et Hydrobiol. Univ., Rue Hébert. Dir. Prof. L. Léger. — Dr. Perrin, Chef des travaux; Melle M. Gauthier, Préparateur d'enseignement; M. A. Dorier, Aide préparateur d'Hydrobiol. *6 places pour Parasitol., Protist., Hydrobiol. générale; Pisciculture des Salmonides. — Tous engins pour les recherches hydrobiol. et la Pisciculture des Salmonides.* ℗ Collections relatives à la Faune des eaux douces.

Groningen (Holland).

Zool. Labor. der Rijks-Univ., Reitemakersrijge 14. Dir. Prof. Dr. J. F. van Bemmelen. — Dr. A. E. van Giffen, Konserv.; J. P. Otto, Ass. *2 Plätze für anat. u. entomol. Untersuchungen. — Kollektion Schmetterlinge.*

Gunnison, Col. (U.S.A.).

Dept. of Zool. Rocky Mountain Biol. Station Western State Coll. of Col. Dir. Prof. J. C. Johnson.

Halifax, N.S. (Canada).

Dept. of Zool. Dalhousie Univ. Dir. J. N. Gowanloch, B.Sc.

Halle a. d. S. (Deutschland).

Zool. Inst. der Univ., Domplatz 4. Dir. Prof. Dr. V. Haecker. — Kustos: Prof. Dr. L. Brüel; Ass.: Prof. Dr. F. Alverdes; Dr. V. Ziehen. *Plätze nach Bedarf.* ℗ Insekten.

Hamburg (Deutschland).

Das Zool. Staats-Inst. und Zool. Mus., Steinthorwall. Dir. Prof. Dr. H. Lohmann. — Wissenschaftliches Mitglied: Prof. Dr. Wilh. Michaelsen. Ständ. Mitarbeiter: Prof. Dr. Ernst Ehrenbaum (Leiter der fischereibiol. Abt.); Prof. Dr. Ludw. Reh; Dr. Georg Duncker; Prof. Dr. Ernst Hentschel (Leiter der hydrobiol. Abt.); Prof. Dr. Berthold Klatt; Dr. Erich Titschak; Dr. Degner. 4 wissenschaftl. Hilfsarbeiter. *Die Anstalt dient dem Universitätsbetrieb, dem Museumsbetrieb und den wissenschaftlichen Arbeiten der wissenschaftlichen Kräfte der Anstalt.*

Hannover (Deutschland).

Zool. Inst. Tierärztl. Hochsch. Dir. Prof. H. Ude.

Heidelberg (Deutschland).

Zool. Inst. Univ., Plöck. Dir. Prof. Dr. Curt Herbst. — Dr. Clara Hamburger, Kustos; Prof. Josef Spek, Ass.; Dr. Karl Baldus, Hilfs-Ass. *Einige Plätze mit Ausrüstung nach Bedarf.*

Helsinki (Finnland).

Zool. Labor. u. Mus. d. Univ., N. Järnvagsgatan 13. Dir. Prof. Dr. E. Reuter. — Mag. phil. V. Korvenkontio, Amanuensis; H. Välikangas (Kustos).

Hirosaki (Japan).

Zool. Labor., Hirosaki High School. Dir. Prof. K. Izumi. — Prof. H. Negishi.

Hiroshima (Japan).

Zool. Labor., Hiroshima High Normal School. Dir. Prof. Y. Abe.

Innsbruck (Österreich).

Zool. Inst. Univ. Dir. Prof. Dr. Steuer. *Mit Sammlung adriatischer u. alpin. Tiere.*

Iowa City Ia. (U.S.A.).

Dept. of Zool. Univ. of Iowa. O. M. Helff; Prof. F. A. Stromsten.

Ithaca, N.Y. (U.S.A.).

Zool. Labor., Cornell Univ. B. P. Yonny, Ass. Prof. (Zool.); A. H. Wright, Prof. (Zool.); A. A. Allen, Prof. (Zool.); Instructors in Zool.: W. C. Jenning; R. P. Hunter; A. S. Hazzard; M. D. Pernie; Lillian A. Phelps; John Greeley; A. Grace Mekeel; Elanore McMullen. *6 or more places. Any Zool. research.* ℗ Preserved Specimens, Embryos, Larval.

Iwanowo-Wossnesensk (U.d.S.S.R.).

Zool. Labor. des Polytechnischen Inst. Dir. Prof. D. A. Lastočkin. — N. W. Korde, Ass.; L. J. Jarowitzina, Privat-Ass. *1 Platz für hydrol. und hydrobiol. Untersuchungen. — Apparate für hydrol. und hydrobiol. Untersuchungen (speziell bio-soziol).* ℗ Die hiesige Fauna (speziell Hydrofauna).

Jaroslavl (U.d.S.S.R.).

Kabinett für Zool. des Pädagogischen Inst. Dir. Prof. B. S. Cjreze.

Jassy (România).

Labor.et Mus. de Zool. descriptive, Str. Buzdugan 3. Dir. Prof. J. Borcea. — Const. Motăs, Chef des travaux. *Aphides, Zoocicidées, Bostrychides, Hydracariens.*

Labor. de Morphol. animale Univ. Dir. Prof. Paul Bujor. — Vasile Zaharesco, Chef des travaux pratiques; Georges Dornesco, Ass.; Vasile Radu, Préparateurs. *3 ou 4 quatre places pour Anat. comparée, Zool., Histol.*

Jena (Deutschland).

Zool. Anstalt Univ., Neugasse 25. Dir. Prof. Dr. Ludwig Plate. — Prof. Dr. V. Franz, Kustos; Priv-Doc. Dr. H. Hoffmann, 1 Ass.

Anstalt für experim. Biol. Univ., Dornburger Str. 25, II. Dir. Prof. Dr. Julius Schaxel. — Ein Ass. z. Z. unbesetzt. *3 Arbeitsplätze für Gäste.*

Johannesburg (South Africa).

Inst. f. Zool. u. vergl. Anat. Dir. Prof. H. B. Fantham.

Kagoshima (Japan).

Zool. Labor. Kagoshima Imperial Coll. of Agric. and Forestry.

Kajana, Suvenniemi (Finnland).

Zool. Stat. Dir. F. Rammler.

Kamenetz-Podolsk, Ukraine (U.d.S.S.R.).
Zool. Inst. d. Inst. f. Volksbildung. Dir. Prof. Kožuchiv.
Zootomisches Kabinett des Inst. f. Volksbildung. Dir. Dir. Prof. Chytkiv.
Inst. für Zool. u. Entomol. des Inst. für Landwirtschaft „Karl Marx". Dir. Prof. V. Chranevič.

Kansas City, Kan. (U.S.A.).
Zool. Division Bureau of Animal Industry U.S. Dept. of Agric. M. Imes (Veterinarian).

Karlsruhe, Baden (Deutschland).
Zool. Inst. Techn. Hochsch. Dir. Prof. W. May. — Prof. M. Auerbach.

Kaunas (Litauen).
Vergl. Anat. Inst., Wilnastr. 2. Dir. Dr. Ellisonas.
Zool. Inst., Wilnastr. 2. Dir. Prof. Ivananskas.
Zootomicum. Dir. Doc. Z. Mockus. — W. Starostinas, Laborant. ⊕ Skelette der Haustiere.

Kazan (U.d.S.S.R.).
Zootomisches Kabinett Univ., Waskressenskaja. Dir. Prof. N. Livanow. — B. G. Fedorow, Ass.; W. W. Isossimow, Ass.; Z. H. Sabussowa, Ass.; O. A. Tichomirnewa, Ass.; D. W. Bjelichow, Aspirant; M. M. Stolbowa, Aspirant; S. W. Schdanow. *20 Arbeitsplätze für morphologische und hydrobiol. (Biocoenotik) Untersuchungen über die Wirbellosen.* ⊕ Polypodium, Amphilina, Bajkal-Gammariden. — Biol. Wolgastation der naturforschenden Gesellschaft an der Univ. Kasan.
Zool. Kabinett der Univ. A. J. Sokolow, Ass.; M. W. Ostroumowa, Ass.; H. K. Holzmayer, Ass. *20 Arbeitsplätze für Morphol., Systematik u. Verbreitung d. Wirbeltiere. — System. Mus. der Wirbeltiere.* ⊕ Embryol. Material des Sterlets u. einheimischer Vögel.

Keijô, Korea (Japan).
Zool. Labor. Coll. of Med. Keijô Imperial Univ.

Kiel (Deutschland).
Zool. Inst., Hegewischstr. Dir. Prof. Frhr. v. Buddenbrock. —Priv.Doc. Dr. Eggers; Priv.Doc. Dr. Remane. *Fauna der Ostsee, Sinnesphysiol.*

Kiew, Ukraine (U.d.S.S.R.).
Zool. Labor. des Inst. für Volksbildung (I.N.O.), Ul. Korolenko 58. Dir. Prof. J. Schmalhausen. — J. Semenkewitsch; N. Woskressensky. *Zool.-Morphol. Untersuchungen.*
Zool. und Anat. Labor. des Inst. für Volksaufklärung (I.N.O), Korolenkostr. 58. Dir. Prof. Dr. Beling. — Ass.: W. Sowinsky; G. Spett. Aspiranten: S. Paremonoff; A. Miroschnitschenko; M. Grimailowska. *8 Arbeitsplätze für Gäste. Systematische Zool., Vergleichende Anat., Hydrobiol.* ⊕ *Tiere der Ukraine, besonders Fische und Evertebrata.*
Biol. Inst. der Ukrainischen Akademie der Wissenschaften, Uliza Korolenko 37. Dir. Prof. I. Schmalhausen. — J. Stepanowa, Laborant; N. Bordzilowskaja; A. Skoworoda-Zatschinjajeff. *2 Arbeitsplätze für beschreibend-morphol. Untersuchungen (vergleichende Anat., Embryol., Histol.). Teilweise auch für experim.-morphol. Untersuchungen (Entwicklungsmechanik).* ⊕ Embryonen der Vertebraten.
Zool. Labor. d. Landwirtsch. Inst. Dir. Prof. A. G. Lebedev. — M. Levitt. ⊕ Landwirtschaftliche und forstliche Insekten und Fraßstücke.

Kingston, R. I. (U.S.A.).
Dept. of Zool. State Coll. of Agric. John Barlow, M.A., Prof. (Zool.); Frederick Bauer, M.S., Instr. (Zool.).

Knoxville, Tenn. (U.S.A.).
Dept. of Zool. Univ. of Tennessee. Dir. Prof. E. B. Powers. — M. A. Stainbrook (Instr. Zool.); J. C. Jones (Ass. Prof.).

Dept. of Zool. and Entomol. Coll. of Agric. E. B. Powers, Ph.D., Prof. Zool.; G. M. Bentley, M.A., Assoc. Prof. Entomol.; Simon Marcovitsch, M.S., Entomol.; B. C. V. Ressler, M.S., Ass. Prof. Zool.

København (Danmark).
Univ. Zool. Studiesamling og Labor., Norregade 10. Dir. Prof. Dr. Ad. S. Jensen. — Lect. Dr. C. M. Steenberg; Mag. sc. Hj. Ditlevsen; Mag. sc. Ingvald Lieberkind. ⊕ Objects for education in zool.
Den K. Veterinaer- og Landbohøjskoles Zool. Labor. og Samling, Bülowsvej 13. Dir. Prof. Dr. M. Thomsen. — Fil. Kand. Stjernman. ⊕ Tiere aller Art, spez. Objekte, die auf Forstzool., landwirtschaftl. und gärtnerische Zool. Bezug haben.

Köln a. Rh. (Deutschland).
Zool. Inst. Univ., Phil. Fak. Dir. Prof. E. Bresslau. — Dr. O. Harnisch.

Königsberg i. Pr. (Deutschland).
Zool. Inst. und Mus., Sternwartenstr. 1. Dir. Prof. Dr. Otto Koehler. —Priv.Doc. Dr. Fr. Seidel, 1. Ass.; Dr. Lothar Szidat, 2. Ass.; Dr. Erich Murr, 3. Ass.; Dr. Riech und Dr. Sulower, Sammlungshilfs-Ass. *3 Arbeitsplätze für Gäste für Sinnesphysiol., Entwicklungsmechanik, Histol., Museumstechnik. — Neben histolcytol. Ausrüstung Spektralvorrichtungen, entwicklungsmechanisches Gerät, Kymographion und sonstige einfachste physiol. Apparatur.* ⊕ Das zool. Mus. sammelt Typen des ganzen Tierreichs mit gewisser Bevorzugung der Provinzialfauna Ostpreußens. — Schädlingsforschungsanstalt Rossitten (Haus Löns), Kurische Nehrung. Dr. Lothar Szidat. 2 Gastplätze. Gelegenheit zum Fang besonders im Haff, jedoch auch in der See.

Kraków (Polska).
Inst. Zool. de l'Uuniv., R. St. Anny 6. Dir. Prof. Dr. M. Siedlecki. — Ass.: Dr. M. Ramult; Dr. S. Kolodziejski. *6 places de travail pour les travaux de zool. expérim. de morphol. et de cytol.*

Krasnodar, Kuban (U.d.S.S.R.).
Zool. Labor. des Kubanisch-Landwirtsch. Inst. Dir. Prof. C. A. Meyer.

Kumamoto (Japan).
Zool. Labor., 5th High School. Dir. Prof. N. Takahashi. *Ichthyol.*

Kyôto (Japan).
Zool. Labor. Kyôto Provincial Medical Univ., Hanazono near Kyôto. Dir. Prof. Dr. Tadanaru Minoura. — Lect. O. Minouchi.
Zool. Inst. Dept. of Sc., Kyôto Imper. Univ. Dir. Prof. T. Kawamura and Prof. T. Komai in alternate years. — Prof. Dr. Taku Komat (Morphol., Embryol., Genetics); Prof. Tamiji Kawamura (Ecol., Physiol.); Ass. Prof. Kôzô Akatsuka (Hydrobiol.); Lect. Osamu Minouchi (Histol., Cytol.); Lect. Seigo Hareyama (Systematik); Lect. Dr. Sachû Yamaguchi (Parasitol.); Ass. Suehiko Yamanouchi (Physiol.).
Zool. Labor., 3rd High School. Dir. Prof. E. Ishibashi.

Lansing, Mich. (U.S.A.).
Dept. of Zool., Michigan State Coll. Dir. Harrison R. Hunt, Ph.D. — Joseph W. Stack, Assoc. Prof. of Zool. *2 places. Genetics, Eugenics, Ornithol.*

La Plata (Argentina).
Inst. f. Vergleich. Anat. Dir. Prof. M. Fernandez.
Zool. Inst. Univ. Dir. Prof. H. Arditi.

Lausanne, Canton de Vaud (Suisse).
Labor. de Zool. et d'anat.-comparée de l'Univ., Palais de Rumine. Dir. Prof. Dr. H. Blanc. — Chef des travaux: Dr. Paul Murisier; 1 Ass.

Lautaret, Hautes-Alpes (France).
Lab. de zool. de l'Inst. Alpin. Dir. M. Mirande.

La Valletta (Malta).
Labor. of Zool., Univ. of Malta. Dir. Prof. J. Borg. — Prof. O. J. Fogarty.

Lawrence, Kan. (U.S.A.).
Zool. Labor., Univ. of Kansas. Dir. Prof. H. H. Lane. — Prof. A. A. Schaeffer (Zool.).; Assoc. Prof. W. J. Baumgartner.

Leeds (England).
Dept. of Zool. of the Fac. of sc., Univ. Dir. Prof. W. Garstang.

Leiden (Holland).
Zoöl. Labor. der Rijksuniv., Sterrewachtslaan 10. Dir. Prof. Dr. P. N. van Kampen. — Dr. A. B. Droogleever Fortuyn, Lect. (Histol.); Dr. C. J. van der Klaauw, Conserv., Priv.Doc. (vergl. Anat.); Dr. H. Boschma, Haupt-Ass. Priv.Doc. (Invertebr. u. Embryol.); Dr. W. H. van Seters, Priv.Doc..(Histol.); Frl. K. Schijfsma, Ass.; Dr. J. Verweij, Ass. (Ornithol.). ⊕ Evertebrata aus den Niederlanden, Rhizocephala, Embryonen u. Neonaten von Vertebraten.

Leipzig (Deutschland).
Zool. Inst. der Univ., Talstr. 33. Dir. Prof. Dr. Johannes Meisenheimer. — Prof. Dr. Richard Woltereck; Prof. Dr. Friedrich Hempelmann; Priv.Doc. Dr. Georg Grimpe; Dr. Arno Wetzel; Dr. Wilhelm Ludwig.

Leningrad (U.d.S.S.R.).
Zool. Labor. der Univ., W.O. Dir. Prof. Dr. K. Derjugin. — 1. Ass.: P. Rezvoy, A. Gawrilenko, N. Tschernoff, B. Tichomiroff. Ass.: P. Kalaschnikoff. Doc.: W. Rylow. *24 Arbeitsplätze für mikroskopische Untersuchungen. — Mikromanipulator. —* Zool. Mus. und Labor. für Systematik d. Wirbeltiere und Hydrobiol.
Zootomisches Labor. Univ. Dir. Prof. V. Dogiel. — Doc.: A. W. Schweyer, A. P. Wladimirsky. Ass.: B. N. Schwanwitch, W. A. Pawlow, Frl. N. N. Pulikowskaja, A. P. Rimsky-Korsakow. Aspiranten: A. W. Farsenko, A. S. Montchadsky, E. S. Rammelmeyer. *2—3 Arbeitsplätze für Gäste besonders für zytol. und morphol. Arbeiten an Evertebraten.* ⊕ Praeparate der Protozoa.
Zool. Labor. der Akademie der Wissenschaft, Toutschkova Naberejnaja N 2a. Dir. Akad. N. W. Nassonov. — D. M. Fedotov (Sektion der Morphol.); P. G. Svetlov (Sektion der Zool., experim.); W. S. Kasanzeff; N. W. Okunev; W. M. Smirnova; W. J. de Lodijensky. *4 Plätze für Morphol. der Tiere und experim. Zool. — Operationssäle, Apparate und Hilfsmittel für physico-chemische Studien.* Sektion Morphol. und Sektion experim. Zool.
Zool. Inst. an der Militär-Medizinisch. Akademie, Nijegorodskaja N 4. Dir. Prof. Dr. E. N. Pawlowsky. — Ober-Ass. Dr. N. N. Kostylew; Ass. P. P. Perfiliew; Präparator Dr. G. G. Smirnow. *10 Arbeitsplätze für Gäste. Vergl. Anat. der Arthropoden. Ektoparasiten: Anat., Wirkung auf den Menschen und die Tiere, parasitische Würmer, Acanthocephalus. Wirkung der Endoparasiten auf ihren Wirt. Parasitol. experim., Anat. der Gifttiere.* ⊕ Parasiten, Gifttiere; Microsk. Praepar., Museumspecim.; Biol. Objekte. Photographien.
Zool. Labor. des Inst. für d. Landwirtschaft, Kamenni Ostrov Beresovaya Allee. Dir. Prof. J. Schmidt. — Ass.: A. V. Tabunschtschikova, M. P. Vinogradov, T. V. Vinogradova.
Zool. Kabinett d. Forstinstitutes. Dir. Prof. Dr. M. Rimsky-Korsakow. — Dr. J. N. Filipjev (Lepidoptera, Zoogeographie u. Oecol., angewandte Entomol., frei lebende Nematoden); A. W. Jacentkowsky Entomol., Forstinsekten, Borkenkäfer); W. J. Schiperowitsch (Forstinsekten, Forstliche Blattwespen (spez. Lophyrus), Panorpiden). *1 Platz für Biol. u. Systematik d. Forstinsekten. — Sammlungen der russischen Forstschädlinge.* ⊕ Forstliche Insekten, spez. Borkenkäfer.
Sect. Zool. de l'Inst. Lesshaft, Anglyisky pr. 32. Dir. P. P. Sushkin, Membre de l'Academie des Sc. — I. D. Strelnikov, Chef du Mus. d'anat. comparée et de Zool.; M. A. Boino-Rodzewitsch, Conserv.; E. M. Nepenine, Conserv.; L. K. Losinal-Losinsky, Ass.; N. P. Panassitschuk, Ass. *12 places pour recherches d'anat. comparée et zool. expérim.* ⊕ Les collections des préparations d'anat. comparée, et de Zool. ethol. — Mus. d'anat. comparée et Zool.

Leningrad-Alt-Peterhof, Sergiewka (U.d.S.S.R.).
Zool. d. Wirbeltiere d. Naturwissenschaftl. Inst. Peterhof. Dir. Prof. Dr. K. Derjugin. — B. Tichomiroff; N. Chranlioff. *12 Plätze.*
Labor. der Zool. der Wirbellosen des Naturw. Inst. Peterhof. Dir. Prof. V. A. Dogiel. — Ass.: A. P. Wladimirsky, J. J. Sokolow, A. W. Fursenko. Aspiranten: J. I. Poljansky, Frl. I. R. Malkina. *5 Plätze jährlich für morphol. (besonders für protistol.) Untersuchungen. Arbeitszeit für Gäste Mai bis September.*

Lexington, Ky. (U.S.A.).
Dept. of Zool. of the Univ. of Kentucky. Dir. Prof. W. R. Allen. — Prof. W. D. Funkhouser.

Liège (Belgique).
Inst. Ed. v. Beneden d'Anat. comp. et Embryol., Univ., Quai Ed. v. Beneden. Dir. Prof. Dr. Damas.
Collect. et Labor. d'anat. comp. et d'embryol. Dir. Prof. Dr. Ch. Julin.

Lille (France).
Labor. de Zool., Fac. Sc. cathol. de l'Univ., 13, Rue de Toul. Dir. Prof. St. Van Oye.
Labor. de Zool. générale et appliquée, Fac. Sc. Univ. Dir. Prof. A. Malaquin.

Lima (Peru).
Labor. de Anat. y Fisiol., Fac. de ciencias, Univ., Parque Universitario. Dir. Dr. W. Molina. — Dr. L. Huapaya; Dr. C. Valdez de la Torre.

Lincoln, Neb. (U.S.A.).
Zool. Labor., Univ. of Nebraska. Dir. Prof. Dr. Robert H. Wolcott. — Prof. Dr. David D. Whitney; Prof. Dr. H. W. Manter; Prof. Dr. Irving H. Blake; Mr. Eugene F. Powell; Mr. T. W. Andersen; Mr. Otis Wade; Mr. J. F. Schuett. *Places can be arranged for guests.*

Lisbôa (Portugal).
Mus. e Labor. zool. Fac. sc. Univ. Dir.: Dr. B. da Cunha Osorio. *2 Naturalistas adjuntos.*

Lisle, Ill. (U.S.A.).
Dept. of Biol., St. Procopius Coll. Dir. H. S. Jurica. — E. J. Jurica (Prof. of Zool.).

Liverpool (England).
Zool. Labor. Univ. Dir. Prof. W. J. Dakin. — S. T. Burfield, M.A. (Lect. Vertebrate anat.); R. Bisbee, M.Sc. (Lect. Genetics and Cytol.); M. Fordham, M.Sc. (Lect. Invertebrata, Parasit.). *Marine and fresh water zool., also comparative physiol. Special arrangements may be made for research workers in addition to those working for degrees.*

Ljubljana — Laibach (S.H.S.).
Zool. Inst. Univ., Kongresni teg. Dir. Prof. Dr. Jovan Hadži. — Dr. Roman Kenk, Doc. ⊕ Höhlentiere (Proteus) sowie andere heimische Tiere des Festlandes, des Süßwassers sowie der Adria.

Logan, Utah (U.S.A.).
Dept. of Zool. and Entomol. Agric. Coll. I. M. Hawley, Ph.D., Prof. Zool., Entomol.; H. J. Pack,

Ph.D., Assoc. Prof. Zool., Entomol., Assoc. Entomol.; C. J. Sorenson, B.S., Ass. Entomol.

London, Ontario (Canada).
Dept. of Zool. of the Univ. of Western Ontario. Dir. A. D. Robertson. — J. D. Detweiler; Miss H. I. Battle; C. M. McCallum; Miss Nelda Wright.

London (England).
Zool. Labor. Univ. Coll., W.C. 1. Dir. Prof. D. M. S. Watson, F.R.S. — Dr. E. A. Traser, Senior Lect.; Dr. T. A. Stephenson, Senior Ass.; H. S. Pearson, M.Sc. Ass.; S. L. Garstang, B.A., Ass.; N. S. Berrill, B.Sc., Ass.; Dr. K. M. Watson, Research Ass.; Dr. H. P. Hacker, Research Ass. *5 working places for general microscopical work, Vertebrate palaeont., experim. work on fresh water and marine animals.*
Zool. Labor. King's Coll., Univ., Strand, W.C. 2. Dir. Prof. Julian S. Huxley. — Dr. Doris L. Mackinnon (Reader); Dr. Margaret Tribe (Lect.); Miss M. E. Shaw (Demonstrator); Mr. Billinghurst (Demonstrator). *4 or 5 places for Protozoan, Vertebrate Embryol, General experim. Zool.*
Zool. Dept. Birkbeck Coll. of the Univ., Breams Buildings, E.C. 4. Dir. H. G. Jackson.
Zool. Labor., East London Coll. of the Univ., Mile End Road, E. 1. Dir. Prof. G. P. Mudge.
Zool. Dep. Bedford Coll. of the Univ., Inner Circle, Regents Park, N.W. 1. Dir. Prof. C. L. Boulenger. — Miss M. Hett, Lect.; Miss G. Faulkner, Demonstrator.
Dept. of Zool., Imperial Coll. of Sc. and Technol. School of the Univ., Imperial Inst. Road, S. Kensington, S.W. 7. Dir. Prof. E. W. McBride.
Zool. Labor. Chelsea Polytechnic and St. Mary's Hospital Medical School, H.W. 2, Paddington. Dir. Prof. W. H. Leigh-Sharpe. *2 or 3 places for General Zool.*
Prosecutorium of the Zool. Society, Regents Park, N.W. 8. Dir. The Prosectorial Committee. — Prof. R. T. Leiper, M.D., D.Sc., F.R.S (Helminthol.); Dr. C. M. Wruyon, C.B.E., C.M.G., M.D. (Protozool); H. Harold Scok, M.D., F.R.C.P., F.R.S.L., D.T.M.H. (Pathologist); John Beattie, M.B., M.Sc. (Anatomist).

Los Angeles, Cal. (U.S.A.).
Zool. Labor. of the Univ. of Southern California. Dir. Prof. M. L. Fossler.
The Comparative Anat. Labor. of the Univ. of Southern California.

Louisville, Ky. (U.S.A.).
Biol. Labor., Univ. Dir. A. R. Middleton (Zool.). — Prof. A. L. Eddy (Applied Biol.); Ass. Prof. H. H. Shanskov (Biol.).
Dept. of Anat. Histol. and Embryol. School of Med. Univ. Dir. Prof. S. J. Kornhauser.

Louvain — Leuven (Belgique).
Labor. d'Anat. comparée et de Zool. Marine, 95, rue de Namur. Dir. Prof. G. Gilson. *Histol. comparée, Anat. des animaux, Zool. marine, Crustacés-Annelida.* ℗ Faune marine.

Lucknow (Br.-India).
Zool. Labor. of the Univ. Dir. Prof. K. N. Bahl, DSc., DPhil. — Dr. G. S. Thapar, PhD.; Mr. M. L. Bhatia, MSc.; Mr. J. Dayal, MSc. *2 places for Morphol. and Embryol.* ℗ Earthworms, Leeches and Insects.

Lund (Sverige).
Zool. Inst. d. Univ. Dir. Prof. Hans Wallengren. — Vorstand der Abt. für Vertebraten: Prof. Hans Wallengren. Vorstand der Abt. für Evertebraten: Prof. Oscar Carlgren. Vorstand der Abt. für Insekten: Dr. S. Bengtsson. Ass.: G. Thulin, H. Bergman, E. Dahr. Konserv.: Dr. H. Berlin. *2 oder 3 Plätze für morphol. Untersuchungen.* — *Kleines Labor. am Barsebäck-Hafen am Öresund.*

Lwów (Polska).
Inst. f. vergl. Anat. d. Univ. Dir. K. Kwietniewski.
Zool. Inst. an der Jan-Kazimierz-Univ., St. Mikolaj-Gasse 4. Dir. Prof. Dr. Jan Hirschler. — Dr. Gustaw Poluszyński, Adjunkt; Dr. Kazimierz Sembrat, Älterer Ass.; Stefan Drzewicki, Jüngerer Ass.; Ludwik Monné, Demonstrator. *2—3 Arbeitsplätze für Gäste, für morphol.-zool. und experim.-zool. Untersuchungen.*— Zool. Mus. von Prof. Dr. Benedykt Dybowski gegr.; sibirische Fauna.
Inst. de zool. et biol. générale, Akadem. Medyc. Weter, r. Kochanowski 67. Dir. Prof. Dr. L. J. Bykowski. — Dr. L. Fedak, Ass. *4 places pour zool. et parasit.* ℗ Parasites.
Zool. Inst. am Polytechnicum. Dir. Prof. Dr. Benedykt Fuliński. — Ass.: Dr. Julja Sokólska, Stanisława Wasilewska. *6 Plätze für anat., embryol., histol. und faunistische Untersuchungen.*

Lyon (France).
Labor. de Zool., Fac. Sc. Univ. Dir. Prof. Koehler.

Madison, Wis. (U.S.A.).
Dept. of Zool. Univ. of Wisconsin. Ass. Prof. C. A. Herrick (Helminthol.).

Manchester (England).
Beyer Zool. Labor. Dir. Prof. J. S. Dunkerly. — Dr. J. S. Thomson, Senior Lect.; Dr. K. M. Smith, Agric. Entomol.; Dr. G. Lapage, Lect.; Mr. R. A. Wardle, Economic Zool. *8 places for general zool. research.* — *Special apparatus and hot-houses in connection with Cotton research.* ℗ Coelenterata (especially Alcyonaria). Protozoa (especially parasit. forms). Platyhelminthia and Nemathelminthia. Economic Insects. — Oaks Estate Experim. Station, with hot-house and labor.
Dept. of Zool. Univ. Dir. Prof. S. J. Hickson.

Manhattan, Kan. (U.S.A.).
Dept. of Zool. State Coll. of Agric. R. K. Nabours, Ph.D., Head of Dept.; J. E. Ackert, Ph.D., Prof. Zool., Parasit.; Bertha L. Danheim, M.S., Ass. in Parasit.; C. A. Gunns, Instr. Zool.; Mary T. Harman, Ph.D., Prof. Zool.; G. E. Johnson, Ph.D., Ass. Prof. Zool., Mammalogist; Carrie I. Potter, M.S., Ass. in Genetics; T. B. Williams, Ph.D., Ass. Prof. Geol.

Manila (Philippine Islands).
Dept. of Zool., Univ. of the Philippines. Dir. P. B. Sivicks. — Dr. L. Clemente; Dr. H. A. Roxas; Mr. F. V. Santos; Mr. J. S. Domantay; Dr. A. Feliciano and Mrs. R. S. Filoteo. *1 or 2 places by special arrangement.* ℗ Decapod crustaceans, corals, brachiopods, mollusks, etc. — Marine Biol. Station at Puerto Galera, Mindoro.

Marburg a. d. L. (Deutschland).
Zool. Inst. der Univ. Dir. Prof. Dr. Eugen Korschelt. — Prof. Dr. Karl Tonniger, 1. Ass.; Dr. Konstantin von Haffner, 2. Ass.; Dr. Otto Mattes, 3. Ass.; Georg Wiese, Konserv.

Marseille (France).
Labor. de Zool. générale Fac. Sc. Univ. Dir. M. Kollmann. — M. Van Gaver, Chef des Travaux pratiques; M. Timon-Dand, Ass. *4 places.*

Matsue (Japan).
Zool. Labor., Matsue High School. Dir. Prof. Hideo Uji.

Melbourne (Australia).
Dept. of Zool. Univ. Dir. Prof. W. E. Agar.

Messina (Italia).
Gabinetto di zool. ed anat. compar. Univ. Dir. G. Mazzarelli. — Lib. doc. Dt. A. Misuri

Milano (Italia).

Labor. di Anat. ed fisiol. compar. Univ., via Gadio, No 2. Dir. Prof. Dr. Rina Monti. — Dr. Carlo Maglio, Ass. piazzi. *Istol. sperim.*

Milwaukee, Wis. (U.S.A.).

Dept. of Zool. Marquette Univ. Dir. E. J. von Komorowski Menge.

Minneapolis, Minn. (U.S.A.).

Labor. of Animal Biol. of the Univ. of Minnesota. Dir. Dr. William A. Riley. (Med. Zool.). — Royal N. Chapman, Ph.D. (Ecol.); C. P. Sigerfoos, Ph.D. (General Zool.); Hal Downey, Ph.D. (Haematol.); Dwight E. Minnich, Ph.D. (Experim. Zool.); C. E. Mickel, Ph.D. (Entomol.); Adolph R. Ringoen, Ph.D. (Embryol.); Maynard S. Johnson, Ph.D. (Economic Zool.). 16 teaching ass. *6 places for haematol, histol. and physiol. work.*

Minsk (U.d.S.S.R.).

Labor. d. Biol. und d. vergl. Anat., Univ. Dir. Prof. A. Schepotieff. — Sophie Gussewa, 1. Ass. (Amphioxus und Tardigraden); Andreas Pigulewskij, 2. Ass. (Spermotogenese bei Lumbricus). *Mikroskopische Fauna der Moose und Plankton (spez. Süßwasserplankton).* ⱷ Süßwasserplankton; mikroskopische Fauna der Moore.

Kabinett für Zool. d. Univ. Dir. A. V. Fedinschin. — A. N. Rojdestwenskii; U. S. Swirskaja. ⱷ Wirbeltieren: Bälge (Felle) mit Schädel. Vögelbälge (Exotische Fauna).

Zool. Labor. Inst. f. Landw. Dir. Prof. D. Th. Sinizin.

Modena (Italia).

Ist. di Zool. ed Anat. compar. Univ. Dir. Prof. Daniele Rosa. — Ass.: Dr. Iginio Sciacchitano. *5 Arbeitsplätze. Morfol.*

Montpellier (France).

Labor. de Zool., Fac. Sc. Dir. Prof. Bataillon. — Prof. A. Soulier.

Montreal, Quebec (Canada).

Dept. of Zool. Mc.Gill. Univ. Dir. A. Willey, M.A., D.Sc. — J. K. Breitenbecher (Lect. in Zool.); L. T. Hogben, M.A., D.Sc.; Mrs. K. F. Pinhey; Miss J. T. Henderson, B.A.; M. Notkin, M.D.

Morioka (Japan).

Zool. Labor. Morioka Imper. Coll. of Agric. and Forestry. Dir. Prof. Dr. O. Shinji.

Moscow, Id. (U.S.A.).

Dept. of Zool. Coll. of Agric. J. E. Wodsedalek, Ph.D., Prof. Zool.

Moskau (U.d.S.S.R.).

Labor. de Zool. expérim. de 1. Univ. de Moscou, Mochovaja. Dir. Prof. Dr. N. K. Koltzoff. — Prof. agrégé: Dr. S. S. Tschetverikoff, Dr. S. N. Skadovsky; Ass.: Dr. M. P. Sadovnikova-Koltzova, Dr. G. I. Roskin, Dr. P. I. Givago, Dr. S. L. Frolova, Dr. V. N. Schroeder; Libre doc.: Dr. A. S. Serebrovsky, Dr. P. A. Kosminsky. *7 places pour Cytol. expérim., Génétique, Physico-chem.*

Labor. de Zool. de la 2. Univ. (Fac. pedagogique et chemio - pharmaceutique). Pirogovskaja, Dir. Prof. Dr. N. K. Koltzoff. — Prof. Dr. W. N. Lebedeff (Fac. chem. pharmaceutique); Ass.: E. T. Schaposchnikova, A. T. Jazenko, S. L. Frolova.

Labor. of Experim. Biol. of the Zoopark of Moscow, Zoopark. Dir. Prof. Dr. M. M. Zawadowsky. — Dr. L. J. Blacher, Ass.; Dr. N. A. Iljin, Ass.; E. M. Zubina, Laborant; Dr. L. M. Grigorieff, Prep.; Dr. M. P. Lubimoff, Parasitologist. *17 places. Investigations of the influence of external and internal factors upon the morphogenesis and questions of formal genetics. Mechanics of the individual development of an organism in connection with the glands of internal secretion and the questions about the hereditary transmission of character from one generation to another. — Apparatus for the definition of the concentration of hydrogene iones, apparatus for the definition of gazeous exchange, surgical instruments.* ⱷ Guinea pigs, rabbits, rats, white mice, poultry, ducks, pheasants, deer, roes, hybrides (wolves and dogs, zebu and yak), arctic foxes, axolotles, tritons, lebistes, gold fishes, drosophila, etc.

Inst. für Experim. Biol. (Volkskommissariat f. öff. Gesundheitsschutz), Voronzovo Pole 6. Dir. Prof. Dr. N. K. Koltzoff. — Stellvertr. des Dir.: Prof. Dr. W. N. Lebedeff; Ass.: Dr. S. N. Skadovsky, Dr. Tshetverikoff, S. N. Prof. Dr. V. V. Bunak, Prof. Dr. A. S. Serebrovsky, Dr. D. P. Filatow, Dr. M. P. Sadovnikova-Koltzova, Dr. A. W. Rumianzew, Dr. G. I. Roskin, Dr. P. I. Givago, Dr. L. S. Pleshkovskaia, Dr. N. W. Popoff, Dr. W. G. Savitsch, Dr. W. N. Schröder, Dr. A. T. Jazenko, Dr. G. A. Jegoroff, Dr. I. G. Kohan; Laboranten: Dr. G. W. Soboleva, Dr. M. G. Lobaceva. *22 Arbeitsplätze. Physiol.-chem. Biol., experim. Chirurgie, Blutuntersuchung, Cytol., Gewebekulturen, Zellkulturen (Drosophila, Bombyx mori, Laboratoriumsäugetiere), Anthropometrie und Rassenbiol., Zoopsychol. (Animal Behavior). — 8 selbständige Abt.:* Chem. Labor. Labor. für physikalische Chem. *2 Operationssäle.* Röntgenapparat, Ultraviolettstrahlen. Photographie, Mikrophotographie, Kinematographie (auch Mikrokinoaufnahmen. Mikroskope, Mikrotome und Ausrüstung für cytol. Arbeiten. *Spezielle Ausrüstung für Gewebekulturen. Anthropometrische Apparate. Apparate für zoopsychol. Untersuchungen. Vivarium für Laboratoriumsäugetiere, Hühner, Schafe, Affen. Aquaria und Terraria mit Luftpumpe. Termostaten für Drosophila; Apparatur für Seidenraupenzucht. Gartenanlagen.* — Hydrobiol. Station bei Svenigorod.

München (Deutschland).

Zool. Inst. Univ., Neuhauser Str. 51, Alte Akad. Dir. Prof. Dr. K. v. Frisch. — Priv.Doc. Prof. W. Goetsch; Priv.Doc. Prof. J. Seiler; Priv.Doc. E. Basler; Ass. Dr. W. Jacobs. *Physiol. der Sinnesorgane (Bienen). Zellforschung. Genetik.*

Zool. Inst. d. Tierärztl. Fac. Dir. Prof. Dr. Reinhard Demoll.

Münster i. W. (Deutschland).

Zool. Inst. der Westfälischen Wilhelms-Univ., Johannisstr. 12—17. Dir. Dr. Prof. W. Stempell, z. Z. beurlaubt; mit seiner Vertretung beauftragt: Priv.Doc. Dr. H. I. Feuerborn. — Dr. phil. H. Kemper; Th. Schräder; Fr. Peus. *20 Plätze für Morphol., Biol., Physiol. d. Tiere. — Optische u. physiol. Apparate und Instrumente.* ⱷ Insekten, spez. westfälische Fauna.

Mukden, Manchuria (China).

Zool. Labor. South Manchurian Med. Univ. Lect. Genkichi Nishimura.

Nagoya (Japan).

Zool. Labor. Aichi Med. Univ. Dir. Prof. Dr. Bun-ichirô Aoki.

Zool. Labor., 8th High School. Dir. Prof. U. Kawano.

Nancy (France).

Labor. de Zool. Pharmacol. Fac. de Pharmacol. de l'Univ. Dir. Prof. L. Bruntz.

Sect. Zool. de l'Ecole nation. des Eaux et Forêts. Inst. Zool. Fac. Sc. Univ. Prof. Maurice Bouin (Zool. appliquée); Prof. Lucien Cuénot.

Nanking (China).

Biol. Labor. Sc. Society of China. Dir. Chi Ping. — Ass. T. H. Chang (Compar. Anat.); C. L. Chang (Neurol. and. embryol.); Collector L. T. Cheng (Taxidermist.). *4 places for worms morphol. and taxon.*

Napoli (Italia).
Ist. di Zool. Univ. Dir. F. S. Monticelli.
Ist. di Anat. e Fisiol. compar. della R. Univ., Palazzo Medievale a Via Mezzocannone. Dir. Prof. Umberto Pierantoni. — Prof. Ermete Marcucci, Aiuto; Dott. Mario Salfi, Ass. — Labor. per ricerche di microbiol. applicata alle simbiosi fisiol.

Nara (Japan).
Zool. Labor., Nara Girls' High Seminary. Dir. Prof. M. Eri.

Neuchâtel (Suisse).
Labor. Zool. Univ. Dir. Dr. Otto Fuhrmann. — Ass. Th. Delachaux. *2—3 places pour Hydrobiol., Parasitol.* (*Platodes*). ⓟ Plankton, Cestoden.

New Brunswick, N.J. (U.S.A.).
Dept. of Zool. State Coll. of Agric. A. A. Boyden, Ph.D. (Instr. Zool.); L. A. Hausmann, Ph.D. (Ass. Prof. Zool.); A. R. Moore, Ph.D. (Prof. Physiol. and Biochem.); T. C. Nelson, Ph.D. (Assoc. Prof. Zool.).

New Haven, Conn. (U.S.A.).
Osborn Zool. Labor. Yale Univ. Dir. Prof. Ross G. Harrison. — Prof. Coe; Prof. Woodruff; Prof. Petrunkevitch; Assoc. Prof. Baitsell; Ass. Prof. Buchanan; Ass. Prof. Ball; Ass. Prof. Nicholas; Instr. Kirby; Instr. Steele; 13 Ass. *No room has been assigned definitely for guests, but accommodations have always be found for all qualified persons, who have desired to work in the labor. — Incubators, refrigerators, constant temperature rooms, animal rooms, operating room, photographic room and apparatus.*

New York City (U.S.A.).
Dept. of Zool. Columbia Univ. Dir. Prof. H. F. Osborn; I. H. McGregor. — Prof. B. Dean (Vertebrate Zool.); Ass. Prof. R. H. Bowen (Zool.); Ass. Prof. A. F. Huettner; E. B. Wilson (Da Costa Prof. of Zool.); Ass. Prof. D. E. Lancefield.
Dept. of Zool. Barnard Coll. Columbia Univ., 119th Street 8 Broadway. Dir. Prof. Henry E. Crampton, Head of Dept. — Louise H. Gregory, Assoc. Prof.; Florence de L. Lowther, Ass. Prof.; M. Grace Springer Forber, Instr.; Mary L. Austin, Lect.; Loir Te Winkel, Ass.; S. J. Hook, Ass.

Nijny-Nowgorod (U.d.S.S.R.).
Zool. Kabinett. Dir. Prof. W. Miliutin. — Doc. B. Wwedensky, Ass. *10 Plätze für Embryol. u. vergl. Anat.* ⓟ Örtliche Fauna. — Hydro-Biol. Abt. (wird organisiert).

Nikolaev, Ukraine (U.d.S.S.R.).
Zool. Inst. am Inst. f. Volksbildung. Dir. A. Prof. Sapošnikov.

Norman, Okla. (U.S.A.).
Zool. Labor. Univ. of Oklahoma. Dir. Dr. A. Richards, Prof. of Zool. and Dir. of Mus. of Zool. — Dr. A. O. Weese, Prof. of Zool.; Dr. A. I. Ortenburger, Ass. Prof. of Zool.; Miss Dixie Young, Instr. in Zool.

Northampton, Mass. (U.S.A.).
Zool. Labor. of Smith Coll. Dir. H. H. Wilder (Ethnol. and Physical Anthrop.). — Inez W. Wilder (Mammalian Zool.); H. M. Parshley (Genetics and Entomol.); Myra M. Sampson (Physiol.); E. R. Dunn (Compar. Anat. and Herpetol.). *4 places for subjects given above.* ⓟ Ethnol. material, Hemiptera, Amphibia.

Novočerkassk (U.d.S.S.R.).
Zool. Inst. Donskoj Polytechn. Inst. Dir. Prof. D. D. Pedašenko.

Odessa (U.d.S.S.R.).
Zool. Labor. d. Inst. f. Volksunterricht (I.N.O.), Comminternstr. 2. Dir. Prof. Nic. Lignau. — Sergius, Morin Ass.; Demetrius Snoiko, Aspirant; Sergius Nikitin, Aspirant; Barbara Zwjetkowa, Konserv. u. Fauna (*Myriopoda, Arachnoidea, Insecta, Blutparasiten*), *Hydrobiol.*
Zool. Sect. d. biol. Katheders d. Wissenschaftl. Forschungsinst. Dir. Prof. N. Lignau.
Zootom. Labor. u. Mus. Pädag. Hochsch. Dir. Prof. Tretjakow.

Omsk (U.d.S.S.R.).
Zool. Abt. d. med. Inst. Dir. Prof. S. D. Lavron.

Ôsaka (Japan).
Zool. Labor. Ôsaka Provincial Med. Univ., Ishibashi near Osaka.
Zool. Labor., Ôsaka High School, Tennôji. Dir. Prof. J. Morita. — S. Tanaka, Lect. *Cytol.*

Osijek, Kroatien (S.H.S.).
Zool. Abt. Landw. Versuchsstat. Dir. Dr. Želislav Kovačević.

Oslo (Norge).
Univ. Zool. Labor. Dir. Prof. Dr. Kristine Bonnevie. — Doc. Dr. H. Broch; Amanuensis: Gudrun Rured; Ass.: Björn Föyn.

Oxford (England).
Zool. Labor. Univ. Prof. E. S. Goodrich; Prof. E. B. Poulton.

Padova (Italia).
Inst. Zool. et Anat. compar., Via Loredan 6. Dir. Prof. Paolo Enriquez. — Dott. F. Bertolini, Aiuto; Dott. G. Cipria, Ass. *10 posti.*

Palermo (Italia).
Ist. di zool. e anat. compar.

Paris (France).
Labor. de Zool. Fac. d. Sc. de l'Univ., Ve, Sorbonne, 1 rue Victor Cousin. Dir. Charles Pérez, Prof. de Zool. — Chef d. travaux: Dr. L. Dehorne; Ass.: Dr. Hérubel, Lic. Azéma, Dr. M. Goldsmith. *10 places.*
Labor. d'Anat. et d'Histol. compar. Fac. d. Sc. Univ., 1, rue Victor Cousin. Dir. Prof. P. Wintrebert. — Ass. Dr. M. Parat. *10 places pour Cytol.; Physiol. de l'embryon; Histo-Physiol.; Colorations vitales; le Vacuome animal; Cytophysique et Cyto-chim. — Microdissection. Salle d'aquariums et d'élevage. Potentiomètre.* ⓟ Préparations microscopiques.
Labor. de Zool. Fac. de pharmacol. Univ. Dir. Prof. H. Coutière.
Labor. d' Enseignement de Zool. (P.C.N.), Fac. Sc. 12, rue Cuvier. Dir. M. Rémy Perroir. — M. M. Georges Bohn, Chef des Travaux pratiques; Delphy, préparator; Ass.: Anglus, Loisel, de Ribaucourt, Dauphin, Matisse, Bertin.
Labor. de Malacol. du Mus. d'histoire nat., 55, rue de Buffon. Dir. Dr. Joubin. — Ass.: Germain, Lamy; Préparateurs: Ranson, Boudarel. *10 places pour Anat., Microscopie, Specification, Photographie.* ⓟ Mollusques, Echinodermes, Coelentérés, Tuniciers, Brachiopodes.
Labor. de Zool. (mammifères et oiseaux) Mus. d'Histoire Nat., Ve, 57, rue Cuvier. Dir. Prof. Trouessart.
Labor. d'Ichthyol et Erpétol. du Mus. National d'Histoire Nat., Ve, 57, rue Cuvier. Dir. Prof. Louis Roule — Sous-Dir. Pellegrin; Ass.: Angel, Chevey, Vaillant. *4 places pour les études sur les Poissons et les Reptiles.* ⓟ Poissons, Reptiles.
Labor. d'Anat. compar. du Mus. national d'Histoire nat., 55, rue de Buffon. Dir. R. Anthony. — H. Neuville, Ass.; L. Semichon, Préparateur; F. Coupin, Préparateur. *5 ou 6 places pour Anat. compar. des Vertébrés.*

Labor. Morphol. expérim., Ecole prat. des Hautes Etudes. Dir. Prof. L. Magnan.
Labor. de Zool. de l'Ecole Normale Supérieure, 45, Rue d'Ulm. Dir. Prof. R. Lévy. — Georges Teissier, Agrégé-préparateur. *1 place pour Histol. et d'histophysiol.* — *Physiol. des invertébrés.*

Parma (Italia).
Gabinetto di Zool. ed Anat. compar., Palazzo dell'Ateneo. Dir. Prof. Dott. A. Andres. — Dott. V. Depardo, Ass. *5 posti.. Ricerche anat.-istol.; lavori sistematici.*

Pavia (Italia).
Inst. de Zool. Univ. Dir. Prof. Cesare Artom. — Dr. Emilio Corti, Adjoint; Dr. Franca Cavallini, Ass. *Recherches de Cytol., d'Histol., d'Embryol.*

Perm (U.d.S.S.R.).
Zool. Inst. Univ., Zaimka. Dir. Prof. W. Beklemischev. — Ass.: A. O. Tauson, D. E. Charitonov, N. J. Oparina; Aspiranten: W. P. Baskina, G. M. Friedmann. *Morphol. der Tiere, Systematik.* ⓟ Tiere und Präp. für Vorlesungsdemonstrationen.
Zootom. Kabinett. Dir. Prof. B. V. Beklemischev.

Perugia (Italia).
Ist. ed Mus. di zool., anat. ed fisiol. compar. Univ. Dir. G. Trinci.

Philadelphia, Pa. (U.S.A.).
Zool. Labor., Univ. of Pennsylvania, 34th and Woodland Ave. Dir. Prof. Clarence E. McClung. — Dr. Edwin Linton (Physiol.); Dr. David E. Fink (Entomol.); Dr. Martha Bunting. *10 places for Physiol., Cytol., Ecol.* — *Ultra-microscope. Ultra-violet apparatus, photographic apparatus, fresh and salt water aquaria.*

Pietermaritzburg (South Africa).
Zool. Labor. Dir. Prof. E. Warren.

Pisa (Italia).
Ist. de Zool. Univ. Dir. Prof. V. Diamare.

Pittsburgh, Pa. (U.S.A.).
Dept. of Zool. Univ. Assoc. Prof. A. E. Emerson (Zool.).

Poitiers (France).
Labor. de Zool. Fac. d. Sc. Dir. Prof. A. Billard. — Dr. Coulongeat, Ass. *1 place pour Zool. générale.* ⓟ Hydroïdes.

Poznań — Posen (Polska).
Inst. Zool. Univ. Dir. Prof. Jan Grochmalicki. — Dr. Juljan Rzóska; Dr. Ambroży Moszyński. ⓟ Palaearktische Wirbeltierfauna, Mollusken, Crustaceen.
Inst. Zool. u. Entomol., Sotacz, Sotacka 3. Dir. o. Prof. Dr. Ludwik Sitowski. — Dr. Jan Ruszkowski, Univ.-Adjunkt; Tadeusz Zoli, Ass.; Tomasz Serafiński, Ass. *6 Arbeitsplätze über wald-und landwirtschaftliche Schädlinge.* ⓟ Schädlinge und parasitische Insekten.
Inst. f. Biol. u. vergl. Anat., ul. Wjazdowa 3. Dir. Prof. Dr. Antoni Jakubski. — Ass.: Dr. Marie Dyrdowska, Dr. Jean Sokołowski. *2 places pour morphol.* ⓟ Mollusques terrestres de Pologne, Margarodes polonicus en divers stades.

Praha — Prag (C.S.R.).
Zool. Inst. der Deutschen Univ., II, Vinična 3. Dir. Prof. Dr. Carl J. Cori. — Ass.: Dr. Friedrich Eckert, Karl Pfleger; wissenschaftl. Hilfskräfte: Dr. Friedrich Egerer, Dr. Anton Meyer. *12 Plätze für morphol., entwicklungsgeschichtliche, histol. und biol. Untersuchungen.* — *Ein Limnetikon für Materiallieferung und Experim. im Institutsgarten. Eine Aquariumseinrichtung. Künstliche Sonne für Kulturzwecke. Eine umfangtechnische Einrichtung für biol. Wasseruntersuchung.* ⓟ Mikrofauna des Süßwassers Böhmens.

Labor. zool. ústavu Karlovy Univ., II, ul. Karlova 3. Dir.: Prof. Dr. Jaromír Wenig; Prof. Dr. Jul. Komárek. — Ass.: Dr. Jar. Hahn, Dr. Jar. Šámal, Dr. Jar. Štorkan, R. N. C. O. Jírovec. *2 places.*

Pretoria (South Africa).
Zool. Inst. Univ. Prof. Dir. D. E. Malan.

Princeton, N. J. (U.S.A.).
Princeton Univ., Biol. Labor. Dir. Prof. E. G. Conklin. — Chas. F. W. McClure, Prof. of Compar. Anat. and Embryol.; Ulric Dahlgren, Prof. of Histol.; George H. Shull, Prof. of Genetics and Botan.; E. Newton Harvey, Prof. of General Physiol. and Bio-Chem.; Edwin G. Conklin, Prof. of General Biol. and Cytol.; Lewis R. Cary, Ass. Prof. of Biol. of Parisit.; Kenneth P. Stevens, Instr. in Biol. *4 or 5 private rooms and firstclass facilities. Experim. Morphol. and Physiol.* — *Large Vivarium with animal rooms and aquaria of freshwater and sea water. Experim. Garden and Greenhouses.* ⓟ Anat. dissections of all classes of Vertebrates, Microscopical prep., Vertebrate Embryos, etc. — Confer Mt Desert Marine Biol. Labor., Woods Hole Marine Biol. Labor.

Providence, R. I. (U.S.A.).
Zool. Dept. Brown Univ. Prof. L. Hoodley.

Provo, Utah (U.S.A.).
Dept. of Zool. and Entomol. Brigham Young Univ. Dir. V. M. Tanner.

Puławy (Polska).
Abt. f. experim. Morphol. des Staatl. Wissensch. Inst. f. Landwirtschaft. Dir. Prof. S. Kopec.

Pullman, Wash. (U.S.A.).
Dept. of Zool. State Coll. R. L. Webster, Ph.D. Head of Dept., Prof. Entomol.; B. A. Slocum B.S., Specialist in Apiculture; Anthony Spuler, M.S. Ass. Entomol.

Quebec (Canada).
Dept. of Zool. MacDonald Coll. Dir. Wm. Lochhead, B.A., M.Sc. — E. M. Du Porte, Ph.D.; A. D. Baker, B.S.A.

Quito (Ecuador).
Zool. Inst. Univ. Dir. Prof. H. Borja.

Rangoon, Burma (Br.-India).
Biol. Dept., Univ. Coll. Dir. Prof. F. S. Meggitt. — Dr. Ghose, M.Sc., Ph.D., F.L.S., Lect. in Botan.; Mr. Handa, M.Sc., Ass. in Botan.; Mr. Subramanian, B.A., Ass. Lect. in Zool.; Prof. C. L. Boulenger, Bedford Coll., London, England (Nematoda); Dr. Chapin, Bureau of Animal Industry, Washington, U.S.A. (Coleoptera). *The department specialises in helminths and algae.* ⓟ Cestoda, Trematoda, Nematoda, Acanthocephala and Algae.

Reading (England).
Zool. Labor. of the Univ. Dir. Prof. F. J. Cole.

Reims (France).
Labor. de Zool. et Parasit. Dir. Prof. Perrin.

Rennes (France).
Labor. de Zool., Fac. Sc. Univ. Dir. Prof. L. Bordas.

Riga (Latvija).
Vergl.-anat. und experim.-zool. Inst. der Latvijas-Univ., Alberta iela 10. Dir. Prof. Dr. phil. N. G. Lebedinsky. — A. Dauwart; V. Melders. *2 Plätze für Embryol. u. Histol.* — Vergl.-physiol. Abt. Vorst.: Priv.Doc. L. Abolins.
Systematisch-Zool. Inst. der Univ. (Sistem.-Zool. Inst.), Alberta ielā 10. Dir. Prof. E. Strand. — Dr. N. von Transehe, 1. Ass.; O. Trauberg, Ass. *Arbeitsplätze für systematische, faunistische,*

zoogeographische Untersuchungen. ⓟ Lettländische Fauna. — Hydrobiol. Station der Univ. Riga.
Zool. Abt. d. Naturwissensch.-Math. Fac. d. Herder-Inst. Dir. Prof. E. Taube.

Roma (Italia).
Ist. zool. Univ., Via Ulisse Aldrovandi 18. Dir. Prof. Federico Raffaele. — Ass. Dr. P. Pasquini. *10 a 12 posti. Ricerche microscopiche in genere. Embriol. Istol.*
Ist. di Anat. e Fisiol. Compar. Univ., Via Depretis No 91. Dir.: vacante. — Prof. Umberto D'Ancona, Aiuto; Prof. Lidia La Face, Ass. *10 posti. Anat. e Fisiol. compar., Morfol. sperimentale, Istol., Entomol.* ⓟ Prep. di Anat. compar., Entomol. — Labor. di Entomol. Agraria.

Rosario (Argentina).
Inst. für spez. Zool. Dir. Prof. M. Faulin.

Rostock i. M. (Deutschland).
Zool. Inst. der Univ., Blücherplatz. Dir. Prof. Dr. Paul Schulze. — Dr. Werner Ulrich, Ass.

Rostow am Don (U.d.S.S.R.).
Zool. Kabinett der Nordkaukasischen Univ., Fr. Engels-Str. 37. Dir. Prof. A. N. Bartenef. — Vasiljef Leonid Ivanovitsch, Ass.; Alexis Erschof (Ichtyol.), Ass. — Biol. (ichthyol.) Station wird in Stanitza (Dorf) Elizabetinskaja eingerichtet.

St. Louis, Mo. (U.S.A.).
Dept. of Zool., Washington Univ. Prof. F. Blair Hanson (Zool.); Ass. F. M. Heys (Zool.).

Salt Lake City, Utah (U.S.A.).
Dept. of Zool. Univ. of Utah. Dir. H. R. Hagan.

Sapporo (Japan).
Zool. Inst., Coll. of Agric., Hokkaido Imperial Univ. Dir. Saburo Hatta. — Ass. Prof. Kan Oguma; Ass. Prof. Tetsuo Inukai; Ass. Teijirô Hayashi.

Saratow (U.d.S.S.R.).
Zool. Inst. und Zool. Mus. d. Univ., Leninsplatz. Dir.: vacat; Vertreter: E. Volcanezkij, Doc. ⓟ Vögel, auch Säuger, Reptilien u. Insekten d. Nieder-Wolgagebiets.
Labor. für Zool. des Inst. für Landwirtsch. u. Melioration. Dir. Prof. V. S. Elpatjewskij.

Sassari (Italia).
Ist. di Zool., Anat. e Fisiol. compar. Univ.

Seattle, Wash. (U.S.A.).
Dept. of Zool. Univ. of Washington. Dir. F. Kincaid (executive Chairman).

Sendai (Japan).
Zool. Inst. Fac. of Sc., Tôhoku Imper. Univ. Dir. Prof. Dr. S. Hatai. — Prof. Shinkishi Hatai (Physi.); Prof. Sanji Hôzawa (Morphol.); Prof. Ekitaro Nomura (Physiol.); Ass. Prof. Shichiroku Nomura (experim. Biol.); Lect. Kiichirô Sasaki (Morphol.).

Sheffield (England).
Zool. Dept. Univ.

Shizuoka (Japan).
Zool. Labor., Shizuoka High School. Dir. Prof. T. Hukui. *Helminthol. Hemiptera.*

Siena (Italia).
Ist. di Zool. e Anat. compar., Piazza Giordano Bruno. Dir. Prof. Giuseppe Colosi. — Dr. Angelo Caroli, Ass.; Dr. Mario Zopi, Ass. volontario.

Simferopol, Krim (U.d.S.S.R.).
Zool. Labor. des Pädagogischen Inst., Leninstr. 17. Dir. Prof. J. Pusanow. — Doc.: W. K. Popoff (Parasitol.), M. A. Galadschieff (Protistol.), A. N. Kasanski (Entomol.), K. W. Dahl (Zootechnik); Ass. Jakob Zeeb. *1—2 Plätze für vergl. Anat. der Wirbeltiere. Hydrobiol.* ⓟ Wirbeltiere der taurschen Fauna, besonders Vögel. Landmollusken. Süßwasserplankton.

Sloboda Borissowka, Gouv. Kursk (U.d.S.S.R.).
Verwaltung d. Sapowednik „Less na Worskle".

Smolensk (U.d.S.S.R.).
Zool. Labor. der Univ., Universitätskaja 6. Dir. Prof. Dr. Wladimir Stantschinsky. — G. L. Grawe, Ober-Ass.; W. A. Melander, Ass.; W. D. Semjonoff, Ass. *1—2 Arbeitsplätze.* ⓟ Säugetiere, Vögel, Land- und Süßwassermollusken, Insekten. — Zweiglabor.: Biol. Station der Univ. Smolensk.

Sofia (Bulgarien).
Zool. Inst. Univ., Ul. Oboriste 13. Dir.: Prof. Georg Chichkoff; Prof. Dr. Theodor Moroff. — ao. Prof. Dr. St. Konsuloff; Dr. Georg Paspaleff; A. Jeliaskowa; Dr. A. Dimitrowa, Priv.Doc.; Dr. Joakimoff; Dr. A. Chranowa, Assistenten. *3 bis 4 Plätze.* ⓟ Verschiedene Insekten und Fische der Balkanhalbinsel. Copepoda und Turbelaria. — Zool. Station in Varuce am Schwarzen Meer.

South Hadley, Mass. (U.S.A.).
Zool. Labor., Clapp Labor., Mount Holyoke Coll. Dir. Ann Morgan. — A. Elizabeth Adams, Assoc. Prof.; Christianna Smith, Assoc. Prof.; 1 Ass. Prof.; 1 Lect.; 2 Instr.; 2 Ass. (full time); 3 Ass. (half time, graduate students). ⓟ Ephemerida (Insecta).

Stambul (Türkei).
Labor. de Zool. Fac. de Sc. Univ. Dir. A. Prof. Wahby. — Prof. Sabri (Anat.).
Labor. f. vergl. u. topograph. Anat. u. med. Physik, Tierärztl. Hochsch., Skutarie-Sélimcie. Dir. Prof. Ahmed Hamdi.
Labor. de Zool. Ecole de Pharmacol. Dir. Prof. Houloussi.

State College, Pa. (U.S.A.).
Dept. of Zool. and Entomol. School of Agric. E. H. Dusham, Ph.D., Head of Dept.; R. D. Casselberry, M.S., Ass. Prof. Zool.; P. A. Frost, M.S., Ass. Prof. Zool.; S. W. Frost, M.S., Assoc. Prof. Econ. Entomol. (P. O., Arendtsville); V. R. Haber, Ph.D., Ass. Prof. Zool.; H. E. Hodgkiss, B.S., Prof. Entomol. Ext.; G. F. MacLeod, B.S., Ass. Prof. Entomol. Ext.; G. B. Newman, M.S., Assoc. Prof. Zool.; G. H. Rea, Ass. Prof. Apiculture Ext.; C. A. Thomas, B.S., Ass. Prof. Econ. Entomol. (P.O., Bustleton); H.N. Worthley, M.S., Ass. Prof. Entomol. Ext.

Stellenbosch (South Africa).
Soölogiese afdeling van die Uniw., Biol. Geb. Dir. Prof. Cornelius G. S. de Villiers. — Dr. C. S. Grobbelaar, Lect. *20 places, specially fitted for research in embryol. and micro-technic. — Complete outfit for the study of embryol. and micro-technic. Also a local mus., with good collection of South African animals.* ⓟ Embryol. and compar. anat. material of all kinds, particularly of Anura.

Stillwater, Okla. (U.S.A.).
Zool. Dept. A. and M. Coll. G. A. Moore (Instr. Zool.).

Stockholm (Sverige).
Stockholms Högskolas Zootomiska Inst., Drottninggatan 116. Dir. Prof. Dr. Nils Holmgren. — Prosektor: Doc. Dr. John Runnström; Doc.: Dr. Gösta Grönberg, Dr. Gert Bonnier, Dr. Hjalmar Rendahl; Ass.: Dr. Kåre Bäckström. *2 Plätze für anat.-histol.-embryol. Untersuchungen, Tieroperationen. — Hilfsmittel für anat. und histol. Untersuchungen. Operationszimmer, chirurgische Instrumente, Zimmer für Versuchstiere etc. Mikromanipulator.* ⓟ Anat. u. embryol. Untersuchungsmaterial.

Inst. f. Zool., Jagd u. Fischerei, Forsthochsch. Dir. Doc. G. Gröneberg.

Stoke on Trent (England).
Zool. Sekt. of the North-Staffordshire Field Club.

Storrs, Conn. (U.S.A.).
Dept. of Zool. Coll. of Agric. G. H. Lamson, jr., M.S., Prof. Zool. and Entomol., Zool.; L. B. Crandall, B.S., Prof. and Apiculture Specialist; J. A. Manter, B.S., Instr. Entomol.; A. F. Schulze, M.S., Instr. and Ass. Zool.

Strasbourg — Straßburg (France).
Labor. de Biol. générale. Dir. Prof. Edouard Chatton. — André Sigot, Ass. *16 places de travail pour les travaux morphologiques, les places pour les travaux physiol. et biochim. — Tout l'outillage pour l'anat., l'histol. et la cytol., la bact. et la protist. Pour les stagiaires une salle de morphol., une salle de cultures, une salle de chim. Locaux pour l'élevage des animaux aériens. Jardin, pièce d'eau et bassins pour la faune d'eau douce.*
Inst. de Zool. et Mus. zool. Univ. et de la Ville, Boulevard de la Victorie. Dir. Prof. E. Topsent.
Inst. d'Histol. et d'Anat. générale. Dir. Prof. P. Bouin. — Max Aron, Chargé de Cours; M. Jacques Benoit, 1er Prép.; Marc Klein, Chargé des fonctions de 2e Prép.; Melle. Lacour, Aide Prép. *3 places. Dans un nouvel inst. en construction: 6 places. Histol. expérim. Histophysiol. de la reproduction.* ⑨ *Prép. macroscopiques d'organes sexuels (Mammifères et autres Vertébrés) et prép. microscopiques des glandes sexuelles et organes de la reproduction dans la série des Vertébrés.*

Sverdlowsk (U.d.S.S.R.).
Zool. Inst. Uralisches Polytechn. Inst. Dir. Prof. Kler.

Swarthmore, Pa. (U.S.A.).
Zool. Labor. Univ. of Pennsylvania. Dir. C. E. McClung.

Sydney (Australia).
Dept. of Zool. Dir. Prof. L. Harrison.

Szeged (Ungarn).
Zool. u. Kompar.-Anat. Inst. Univ., Tisza Lajos-Körut 6. Dir. Prof. Dr. Josef v. Gelei. — Jenö Màtyàs, dr. Adjunkt; Mihàly Rotarides, dr. Ass.; Gàbor Kolosvàry, dr. Hilfs-Ass.; Lajos Boros, Praktikant; Irén Magoss, Hilfsarbeiterin f. Mikrotechnik. *2 Arbeitsplätze für cytol.-histol. Untersuchungen.* ⑨ *Spinnen, Schnecken, Turbellarien, Süßwasserpolypen, Protisten. Mikroskopische Anat. der Vertebratenknochen.*
Zool. Labor. II, Szukováthy tér 1. Dir. Prof. Dr. B. Farkas. — M. Vasvári, Ass.; *2 Plätze für histol. u. hydrobiol. Arbeiten.* ⑨ *Tiere der ungarischen Tiefebene und histol. Praep.*

Taihoku, Formosa (Japan).
Zool. Labor. Taihoku Governmental Coll. of Agric. and Forestry.

Tartu — Dorpat (Eesti).
Zool. Inst. Univ. Dir. Prof. J. Piiper. — Prof. H. Riikoja; Frau L. Poska-Teiss; Magister E. Reinwaldt. *4 Plätze für Embryol. u. Planktol.* ⑨ *Embryol. (Vögel) u. planktol. Material. — Die Biol. Station zu Kuusnõmme auf Insel Oesel.*
Zootomisches Inst. der Vet.-med. Fac. Univ., Vene tän 32. Dir. Prof. Dr. Hans Richter. — Prosektor A. Mahlmann. *1 Platz für makroskopische u. mikroskopische u. embryol. anat. Untersuchungen.*

Tiflis (U.d.S.S.R.).
Zool. Kabinett Univ. Dir. Doc. G. Džavachišvili.
Inst. f. Zool. u. Forstentomol., Techn. Hochsch. Dir. Prof. F. A. Zajcev.

Tôkyô (Japan).
Zool. Inst., Sc. Fac., Tôkyô Imper. Univ. Dir. Prof. Seitaro Goto. — Prof. Naohide Yatsu; Ass. Prof. Shigeho Tanaka; Lect. Tokuzo Goda; Ass. Yoshimasa Ozaki; Ass. Takeo Kamada; Ass. Osamu Hattori; Ass. Junichi Asada. ⑨ *Trematoda. Cestoda. Aves. Pisces.*
Zool. Inst., Fac. of Agric., Tôkyô Imper. Univ., Komaba near Tôkyô. Prof. Masao Sô; Ass. Prof. Tokio Kaburaki; Ass. Prof. Takematsu Takeshita.
Zool. Labor., Imper. Agric. Experim. Station, Nishigahara. Dir. Dr. S. Uchida. — Dr. Seinosuke Uchida (Ornithol.); Hisakichi Kishida (Mammal., Arachnida).
Anat. Labor., Tôkyô Imper. Sericicult. Coll., Nishigahara. Dir. Prof. H. Itô. — Prof. Hiroo Itô.
Zool. Labor. Jikeikwai Med. Univ., Shiba. Dir. Prof. Takeshi Shinohara. — Lect. J. Miyashita.
Zool. Labor. Imper. Fishery Inst., Hukagawa. Dir. Prof. Dr. K. Kishinoue. — Prof. Arata Terao.
Zool. Labor., 1st High School, Hongô. Dir. K. Takajaschi. *Histol.*
Zool. Labor., Tôkyô High Normal School, Koischikawa. Dir. Prof. Dr. A. Oka. *Annelida, Tunikata.*
Zool. Labor., Gakushûin. Dir. Prof. Dr. A. Izuka. *Annelida.*
Zool. Labor., Seijô High School. Y. Miyashita, Lect.

Tomsk, Westsibirien (U.d.S.S.R.).
Kabinett der vergl. Anat. Univ. Dir. Prof. Hermann Johansen. — Ass. Dr. H. Johansen. *2 Arbeitsplätze.* ⑨ *Vertebrata Sibiriens.*
Zool. Kabinett Univ. Dir. Prof. Michael Ruzky.
Labor. of Experim. Zool. Dir. Prof. Vitaly A. Khakhlof. — *1 collaborator. 2 places for Genetics.* ⑨ *Guineapigs, mice, fowls, bull-finches.*

Torino (Italia).
Ist. di Anat. e di fisiol. compar., Palazzo Cariguano. Dir. Prof. Alfredo Corti. — Dott. Mario Benazzi, aiuto. *Una diecina posti per ricerche di istol.*

Toronto (Canada).
Dept. of Biol. of the Univ. of Toronto. Dir. Prof. B. A. Bensley, Head of Dept., Dir. of the Royal Ontario Mus. of Zool. — R. R. Wright, M.A., B.Sc., LL.D., Prof. Emeritus; B. A. Bensley, B.A., Ph.D., Prof. of Zool.; W. H. Piersol, B.A., M.B., Prof. of Histol. and Embryol.; E. M. Walker, B.A., M.B., Assoc. Prof.; A. G. Huntsman, B.A., M.B., Assoc. Prof. of Marine Biol.; A. F. Coventry, B.A., Ass. Prof. of Vertebrate Embryol.; J. R. Dymond, M.A., Ass. Prof. of Systematic Zool.; J. W. MacArthur, M.A., Ph.D., Ass. Prof. of Genetics; W. H. T. Baillie, M.A., M.B., Ass. Prof. of Mammalian Anat.; E. H. Craigie, B.A., Ph.D., Ass. Prof. of Compar. Anat. and Neurol.; W. J. K. Harkness, M.A., Lect. in Limnobiol.; N. H. C. Ford, Ph.D. *Ample labor. facilities for General zool., Canadian taxonomy, compar. neurol., general histol., vertebrate and experim. embryol., fresh-water fisheries and marine fisheries investigation.* ⑨ *All taxonomic material prepared and exchanged by the Royal Ontario Mus. of Zool., or by private workers in particular branches especially entomol. and aquatic biol.* — *The Ontario Fisheries Research Labor. is organized within the Dept. The labor. of the Biol. Board of Canada, maintained by the Government of Canada for Research on marine and fresh-water fishes, is housed in the Dept. The Royal Ontario Mus. of Zool. is under the direction of the Dept.*

Tottori (Japan).
Zool. Labor., Tottori Imper. Coll. of Agric. Dir. Prof. S. Inomata.

Toulouse (France).

Labor. de Zool. (P. C. N.) Fac. Sc., Allées Saint-Michel. Dir. A. Vandel. — R. Delmas, Ass.
Labor. de Zool. Fac. des Sc. Dir. Albert Lecaillon. — Mr. Bonnet (Pierre), Ass. *10 places pour Entomol., Histol., Embryogenie.*

Touro, N. S. (Canada).

Dept. of Zool. and Entomol Coll. of Agric. Dir. W. H. Brittain, Ph.D. — W. E. Whitehead.

Toyama (Japan).

Zool. Labor. Toyama Imperial Coll. of Pharmacy.
Zool. Labor., Toyama High School. Dir. Prof. A. Hujita.

Trinidad (Br.-W.-India).

Inst. f. Zool. u. Entomol. Inst. f. tropical. Agric. Port of Spain. Dir. Prof. H. A. Ballon.

Tsu (Japan).

Zool. Labor. Mie Imperial Coll. of Agric. and Forestry.

Tübingen (Deutschland).

Zool. Inst. der Univ., Hölderlinstr. 12. Dir. Prof. Dr. Harms. — Priv.Doc. Dr. H. A. Stolte, Ass.; Dr. Fr. Bock, Ass.; Dr. Br. Eggert, Ass. *4 Plätze für vergl. Anat., Embryol., experim. Morphol. — Micromanipulator. Hirnmikrotom.*

Tunis (Protectorat Français).

Labor. de Zool. générale et Entomol. Ecole Coloniale d'Agric. Dir. Th. Ch. L. Pagliano. *Parasit. agric., Apiculture, Sériciculture.*

Turku (Finnland).

Zool. Inst. Dir. Prof. Dr. Walter M. Linnaniemi. — K. J. Valle, Ass. *Entomol. Systematik. (Sahlbergsche Sammlungen, sowie die Sammlg. v. Ehnberg u. Starck.)* ⓟ Coleoptera, Lepidoptera, Collembola.

Ueda (Japan).

Anat. Labor. Ueda Imperical Serie Coll., Nagano-Ken. Dir. Prof. T. Gamô.

Upsala (Sverige).

Zool. Inst. Univ. (mit Zool. Mus., Zootom. Labor. und Physiol. Labor.). Dir. Prof. Dr. N. von Hofsten. — Priv.Doc. Dr. I. Arwidsson, Konserv. am Zool. Mus.; Prof. ad int. Dr. S. Bock, Leiter des Physiol. Labor.; Fil. mag. G. Gustafsen, Amanuensis Zool. Mus.; Fil. mag. E. Rehman, Amanuensis am Zootom. Labor. *15 Arbeitsplätze. Morphol. und physiol. Untersuchungen. — Physiol. Apparate.* ⓟ Schwedische Fauna; europäische und außereuropäische Fauna, besonders makroskopische marine Evertebraten; osteol. Sammlungen; anat. Praep. und mikroskopische Praep. (Anat., Histol., Embryol.). — Die marine Biol. Station der Univ. Upsala, Klubban in Bohuslän (Vorstand Dr. S. Bock).

Urbana, Ill. (U. S. A.).

Zool. Labor., Univ. of Illinois. Dir. Prof. Dr. H. B. Ward. — Full Prof.: Zeleny, Shelford, Van Cleave, Shumway; Ass. Prof.: Adams, Kudo, Thomas; Instr.: Cahn, Hartmann, Hann, Adamstone, Essex. 2 Research Ass. *Indefinite workingplaces for guests. Most types of biol. research.* ⓟ Animal Parasites.

Utrecht (Holland).

Zoöl. Labor. der Rijks Univ., Janskerkhof 3. Dir. Prof. Dr. H. F. Nierstrasz. — Prof. Dr. H. J. Jordan (vergl. Physiol.); Dr. G. C. Hirsch, Lect. (experim. Histol.); Dr. G. Entz (Protist.); Dr. G. J. v. Oordt (experim. Morphol.); Dr. Schurmans Stekhoven (Parasit.). *Mehrere Plätze für vergl. Anat. und Physiol. (confer Ph. comp.), Protist., experim. Histol. (confer Histol.).* ⓟ Solenogastres, Chitonen, Nemertinen, parasit. Gastropoden, Isopoden. — 2 Unterabt. für vergl. Physiol. und experim. Histol.

Valencia (España).

Inst. zool. Univ. Dir. Prof. B. Bigorra.

Vancouver (Br.-Columbia).

Dept. of Zool. Univ. of Br.-Col. Dir. C. McLean Fraser, B.A., Ph.D. — G. J. Spencer, B.S.A., M.Sc.; G. Van Wilby, M.A.

Vermillion, S. D. (U. S. A.).

Zool. Labor., Univ. South Dakota. Dir. Dr. E. P. Churchill. — Prof. W. H. Over, Curator of Mus. 4 Ass. ⓟ Mussels, birds, eggs, fish.

Villafranca del Bierzo-Leon (España).

Labor. Biol. de S. Vicente de Paul. Dir. Salustiano Diéguez. — El P. Pujiula del Labor. Biol. de Sarriá (Barcelona). *Morfol., Histol., Citol., Fisiol., Biogénia y Bionomía.*

Vladikavkaz (U. d. S. S. R.).

Kabinett f. Zool. d. Wirbeltiere, Pädag. Inst.
Kabinett f. Zool. d. Wirbellosen, Pädag. Inst.

Warszawa (Polska).

Labor. de Zool. et Parasit. Dir. Prof. Dr. Witold Stefanski.
Inst. de Zool. Univ., Krakowskie-Przedmieście 26. Dir. Prof. Dr. Constantin Janicki. — Ass.: Dr. Jerzy Ruszkowski; Dr. Kazimierz Gajl; Dr. Jerzy Jarocki. Volontaire: Zygmunt Koźmiński.
Zool. Labor. der Freien Univ. Polens, Polnastr. 30. Dir. Prof. Dr. Richard Błędowski. — Ältere Ass.: Fr. K. Krainska. 1 jüngerer Ass. *1 Platz für Embryol. der Insekten.* ⓟ Parasit. Insekten, ganz besonders Ichneumoniden, Praeparate des Insektenparasitismus.
Inst. der vergl. Anat., r. Chalubińskiego 5. Dir. Prof. Dr. Jan de Tur. — Dr. D. Krzyżanowski; St. Trojanowski. *Embryol. comparée des Vertébrés, spécialement des Sauropsidés. Tératol. et tératogénèse.*

Washington, D.C. (U. S. A.).

Zool. Division, Bureau of Animal Industry. Dir. M. C. Hall. — J. E. Shillinger (Assoc. Veterinarian); E. W. Price (Assoc. Parasit); A. Hassall (Zoologist); E. B. Cram (Assoc. Zool).

Weihenstephan a. Freising, Oberbayern (Deutschland).

Zool. Inst. der Hochschule f. Landwirtschaft und Brauerei. Dir. Prof. Dr. K. Th. Andersen.

Wellesley, Mass. (U. S. A.).

Labor. of Zool. and Physiol., Wellesley Coll. Dir. Miss Marian E. Hubbard (Present Chairman). — Dr. Julia E. Moody, Prof. of Zool.; Dr. Margaret A. Hayden, Ass. Prof. of Zool.; Miss Janet Williamson, Instr. in Zool.; Miss Helen Avery, Instr. in Zool.; Miss Marion Lewis, Instr. in Zoll.; Miss Jean Walker, Instr. in Zool.; Miss Verz Goddard, Instr. in Physiol.; Mr. A. P. Morse, Curator of Mus.; Dr. Elizabeth MacNaughton, Instr. in Zool. 3 Labor. Ass.

Wien (Österreich).

I. Zool. Inst. Univ. Dir.: vacant. — Konserv.: Prof. Th. Pintner (Ichthyol.). *Systematik der Tiere.*
II. Zool. Inst. Univ. Dir. Prof. J. Versluys. — Dr. H. Joseph; Prof. O. Storch. *Vergl. Anat., Reptilien, Embryol., Ernährungsbiol.*
Inst. f. Zool. und Parasitenkunde der Tierärztl. Hochsch. Wien III. Dir. Prof. L. K. Böhm.

Wilno (Polska).

Inst. d'Anat. comp. Univ., Dir. Prof. Wl. Mierzejewski.
Zool. Inst. of the Univ. of Wilno, 15, Zakretowa 15. Dir. Dr. Jan Prüffer. — Dr. Hieronim Jenotowski; Marja Raeiseka; Marja Ostreyko. *2 places for systematical and ecol. research with Insects and research concerning their sense organs.* ⓟ Insecta and Myriapoda.

Winnipeg, Man. (Canada).
Dept. of Zool. Univ. of Manitoba. Dir. C. H. O'Donoghue. D.Sc. — C. F. C. Riley, M.A.; F. Neave, B.Sc.; R. D. Bird, B.Sc.; Ruby Bere, B.Sc.; Harry Morgan.

Wolfsville, N. S. (Canada).
Dept. of Osteol. and Physiol. Arcadia Univ. C. E. A. De Witt, B.A., M.D.

Woronesch (U.d.S.S.R.).
Inst. vergl. Anat. u. Embryol. Univ., Fac. paed. Dir. Ds. N. Nikoljukin.
Kabinett f. Zool. u. Entomol., Landwirtsch. Hochschule.

Würzburg (Deutschland).
Zool.-zootomisches Inst. Univ. Vorstand: Prof. Dr. W. Schleip. — Prof. Dr. L. von Ubisch, 1. Ass.; Priv.Doc. Dr. A. Penners, 2. Ass. *1—2 Plätze. — Strahlenstichapparat nach Tschacholin. Mikromanipulator nach Peterfi.*

Yamaguchi (Japan).
Zool. Labor. Yamaguchi High School.

Zagreb — Agram (S.H.S.).
Morfol.-biol. Inst. kr. Sveučilišta u Zagrebu (Morphol.-Biol. Inst. Univ.), Šalata. Dir. Prof. Dr. Boris Zarnik (allg. Biol., Histol. u. Embryol.). — Dr. Vladimir Plješakov, Prosektor für Histol.; Dr. Wilhelm Mršić, Adjunkt für Biol.; Dr. Dragutin Damaška, Ass. f. Biol.; Daniel Smekal, Ass. f. Histol. 6 Hilfs-Ass. *9 Arbeitsplätze. Morphol. der Tiere, experim. Morphol., Gewebekultur, vergl. Physiol. — Große Aquarien, Terrarien, Tierställe, Mikrophotographie, Apparate für Gewebekultur, chem. Labor.* ⊕ Cytol. Praep. d. Tiere u. Pflanzen, histol. Praep., Embryonen insbes. v. Wirbeltieren. — Labor. f. Gewebekultur, chem. Labor. (Physiol. Labor. in Vorbereitung.)
Inst. für vergl. Anat. Philosoph. Fac. (Komparativno-anat. zavod), Wilsonov trg 10. Dir. Prof. Car.
Zool.-zootomisches Inst. der Univ., Demetrova 1. Dir. Prof. Dr. Langhoffer. — Dr. Nikola Fink, Adjunkt.

Zaragoza (España).
Inst. Zool. générale. Dir. Prof. F. Aranda y Millán.

Zürich (Schweiz).
Zool.-vergl. anat. Inst. d. Univ. u. E.T.H. Dir. Prof. Dr. Karl Hescheler. — Prof. Dr. Jean Strohl, Leiter der physiol. Abt.; Frl. Prof. Dr. Marie Daiber, Prosektor des Inst.; Dr. Walter Knopfli, Ass. — Zool. Mus. der Univ. Zürich.

Musea zoologica.

Confer: *Musea biologica generalia* (pag. 350), *Laboratoria Zoologica* (pag. 356).

Amoy (China).
Zool. Mus. of the Univ. Dir. Prof. S. F. Light.

Amsterdam (Holland).
Zool. Mus., Plantage Middenlaan. Dir. Dr. L. F. de Beaufort. — Miss Tera van Benthem Jutting (Molluscs, Crustacea); Dr. H. Engel (Echinoderms, Nudibranchiate Molluscs); J. Corporaal (Entomol.). *A few Places for Sc. Mus.-Workers.*
Abt. Handelsmus. des Konigl. Kolonialinst. Mauritskade 57a. Dir. Prof. Dr. L. Cosquino de Bussy, — Dr. E. C. Jul. Mohr; Dr. W. L. Utermark; P. Hondius; Dr. Hindrik Hajo Ganssonius. *Tierische u. pflanzliche Produkte der Kolonien.*

Ann Arbor, Mich. (U.S.A.).
Mus. of Zool. Univ. of Michigan. Dir. Alexander G. Ruthven. — Bryant Walker, Honorary Curator Molluscs; E. B. Williamson, Assoc. Curator of Diptera; Dr. Howard A. Kelly, Honorary Curator of Reptiles; Bradshaw H. Swales, Assoc. Curator Birds; William W. Newcomb, Honorary Curator Odonata; James Speed Rogers, Assoc. Curator Diptera; Arthur W. Andrews, Assoc. Curator Coleoptera; Arthur S. Pearse, Honorary Curator Crustaceans; Dora S. Lemon, Custodian of Protozoa; George R. LaRue, Honorary Curator Parasitic. Worms; Walter E. Hastings, Custodian Birds'Eggs; Jan Metzelaar, Custodian Michigan Fishes; L. R. Dice, Curator of Mammals; W. A. Wood, Curator of Birds; C. L. Hubbs, Curator of Fishes. 1 Curator of Reptiles and Amphibians, 1 Curator of 1 Curator of Insects. *4 working-places for guests. Insects, Reptiles and Amphibians, Birds, Molluscs.* ⊕ *Fishes, Birds, Reptiles and Amphibians, Molluscs, Mammals, Insects.*

Aschabad, Turkmenische Republik (U.d.S.S.R.)
Turkmenisches Reichs-Mus. Dir. Stanislaw Bilkewicz. — Akademik. Andreas Karelin Kunst, Archäol. u. Etnographie; Oscar Wiesel, Etnograph; Woldemar Godecky; Andreas Smirnoff, Archäol. Technisches Personal: Alesandroff Sergius, Praeparator; Sofie Michajloff, Laborant; Karlieff Durdy, Praktikant. *2 Arbeitsplätze für biol. Untersuchungen. Zoogeographie der Vögel. Vertebraten der Insel Nowaja Semlja. — Die Sammlungen der wissenschaftlichen Fac.* ⊕ Mammalia, Ornis, Reptilia usw.

Beirut (Syrie).
Zool. Mus. Dir. William van Dyck.

Bergen (Norge).
Die zool. Abt. des Mus. zu Bergen (Bergens Mus. zool. avdeling). Dir. Prof. Dr. August Brinkmann. — James A. Grieg, Kustos der Evertebratensammlung; Sigurd Johnsen, Kustos der Vertebratensammlung. *Es können 2 Gäste untergebracht werden. Ausrüstung für systematische und anat. Untersuchungen. Instrumente für osteometrische Untersuchungen.* ⊕ *Allgemein zool. Sammlung, besonders die nordische und arktische Fauna. Sammlungen mariner Organismen. Besonders große Sammlung von Walen und Haustiersammlung. — Bergens Mus. biol. station. (Die biol. Station des Mus. zu Bergen confer Labor. marina.)*

Berkeley, Cal. (U.S.A.).
Mus. of Vertebrate Zool. Univ. of California. Dir. J. Grinnell. — H. C. Bryant (Economic Ornithol.); E. R. Hall (Research Ass.); H. S. Swarth (Curator of Birds).

Berlin (Deutschland).
Zool. Mus. der Univ., N 4, Invalidenstr. 43. 1. Dir. Prof. Dr. C. Zimmer; 2. Dir. Prof. Dr. P. Pappenheim. — Kustoden: Prof. Dr. G. Enderlein, Prof. Dr. E. Hesse, Prof. Dr. A. Schellenberg, Dr. H. Kuntzen, Dr. W. Ramme, Dr. H. Bischoff, Dr. J. Moser, Dr. E. Stresemann, Dr. W. Arndt, Dr. H. Pohle, Dr. M. Hering. Ass.: Dr. E. Ahl und Dr. G. Rensch. Wissenschaftl. Hilfsarbeiter: Dr. E. Bannwarth, Dr. M. Eisentraut, Dr. E. Mayr, Herr H. Stitz. *Arbeitsplätze nach Bedarf für Untersuchungen auf dem Gebiete der systematisch-faunistischen Zool. — Apparate und Instrumente für systematisch-faunistische Zool. Präsenzbibliothek mit zahlreichen Einzelwerken und 600 Nummern laufender Periodica.* ⊕ *Alle Gegenstände aus dem Gesamtgebiet der Zool.*

Naturwissensch. Abt. d. Märk. Mus., S 14. Dir. Dr. Max Hilzheimer. *Rezente und fossile Säugetiere, besonders Haustiere, Faunistik und Tiergeographie.*

Bern (Schweiz).
Städtisches naturhistor. Mus., Waisenhausstr. Dir. d. zool. Sammlg.: Prof. Baumann. Konserv. d. entomol. Sammlg.: Dr. Ch. Ferrière.

Budapest (Ungarn).
Zool. Abt. des Ungarischen Nationalmus. Dir. Ernest Csiki (Coleopterol.). — Mitarbeiter: Dr. Zoltán von Szilády (Dipterol., bes. Tabanidae, Stratiomyidae); Dr. Ludwig Soós (Malakol.); Dr. Anton Schmidt (Lepidopterol.); Dr. Alexander Pongrácz (Orthopterol., Neuropterol., Fossile Ins.); Dr. Baron Géza von Fejérváry (Herpetol., rec. und fossil); Dr. Josef von Szabó-Patay (Hymenopterol., bes. Formicidae); Dr. Ludwig Biró (Hymenopterol., bes. mikros.); Dr. Julius Éhik (Mammalol.); Dr. Béla Hankó (Ichthyol.); Dr. Andreas Dudich (Crustaceol.), niedere Tiere); Dr. Ladislaus von Szalay (Arachnol., bes. Acariden); Dr. Eugen Greschik (Anat. der Vögel); Dr. Stefan Gaál (fossile Malakol.); Dr. Ludwig Véghelyi (Hemipterol.) Dir. im Ruhestand als Volontär: Dr. Géza von Horváth (Hemipterol.). Volontärer Mitarbeiter: Dr. Teodor Kormos (Vertebraten-Palaeont.).

Cairo (Egypt).
Giza Zool. Mus., Zool. Survey of Egypt.

Cambridge, Mass. (U.S.A.).
Mus. of Compar. Ichthyol. W. C. Schroeder (Ass. Aquatic Biol.).
Mus. of compar. Zool. Dir. S. Henshaw. — H. L. Clark (Curator of Echinodermes); R. V. Chamberlin (Curator of Myriapoda); H. B. Bigelow (Research Curator); W. J. Clench (Curator); T. Barbour (Curator).

Cambridge (England).
Univ. Mus. of Zool. Dir. C. Forster Cooper, M.A. — Hugh Scott, Curator in Entomol.; H. F. Gadow, Curator Ornithol. Curator of Invertebrates. In connection with Dept. of Zool.

Coimbra (Portugal).
Labor. du Mus. Zool. de l'Univ., Largo do Marquez de Pombal. Dir. Dr. Bernardo Ayres. — A. F. de Seabra, ancien Ass. de l'Univ. de Lisbonne, nat. du Mus. Zool. de Coimbra; Barros e Cunha, Ass. de l'Univ. de Coimbra. *L'étude de la faune du Pays.* ⓟ Coléoptères et Hémiptères du Portugal.

Cordoba (Argentina).
Mus. de Botan. y Zool. Univ. Confer Musea botan. et Herbaria.

Cork (Irland).
Mus. of Zool. Dir. Prof. Louis P. W. Renouf, B.A., Dipl. Agric. — ⓟ Local material, which is very rich, is obtained mainly from the Baltimore (Cork) Biol. Station.

Darmstadt (Deutschland).
Zool. Abt. des Hessischen Landesmus. Dir. Prof. Th. List. — Kustos: Dr. A. Schwarz.

Dubrovnik, Dalmatien (S. H. S.).
Städt. Mus. mit zool. Sammlung.

Edinburgh (Scotland).
Nat. History Dept., Royal Scottish Mus. Dir. Dr. James Ritchie. — P. H. Grimshaw, Ass.-Keeper and Entomologist; A. C. Stephen, Ass. in charge of Fishes and certain groups of Invertebrates. *3 tables in Study Room. Systematic and morphological investigation of the zool. material in the Mus. exhibited and cabinet collections. — Measuring apparatus etc.* ⓟ Collections of Zool. material, notably bird skins and marine invertebrates from Antarctic and Arctic regions, and palaeozool. collections, notably of fishes.

Firenze (Italia).
Mus. Zool. della R. Univ., Via Romana 19. Dir. Commissione Direttiva. — Baldasseroni Prof. Vincenzo, Zool. aggiunto. ⓟ Collezione zool. generali. Collezione di prep. ceroplastiche di anat. umana.

Hamburg (Deutschland).
Hydrobiol. Abt. des Zool. Staatsinst. und Zool. Mus., Steintorwall 1. Leiter: Prof. Dr. E. Hentschel.

Hannover (Deutschland).
Naturwiss. Abt. am Provinzialmus. Dir. Dr. Max Hugo Weigold. *Ornithol., besonders Vogelzugforschung. Faunistische und zoogeographische Erforschung Westchinas und Südosttibets. Fauna Hannovers. Museumskunde.*

Hobart (Tasmania).
The Tasmanian Mus. Dir. C. E. Lord. *Vertebrate Fauna of Tasmania.*

Innsbruck (Österreich).
Zool. Sammlung des Mus. Ferdinandeum, Museumstraße. Dir. Hofrat Dr. Karl Toldt. ⓟ Fauna von Tirol.

Jena (Deutschland).
Phyletisches Mus., Neugasse 22. Dir. Prof. Dr. Ludwig Plate. — Prof. Dr. V. Franz, Kustos; Prof. Dr. E. Uhlmann, Konserv. ⓟ Objekte für Phylogenie.

Karlsruhe, Baden (Deutschland).
Zool. Abt. Landessammlungen für Naturkunde. Dir. Prof. Dr. M. Auerbach. — Prof. Dr. H. Leininger, Kustos für Entomol.; A. Kneucker, Kustos für Botan. ⓟ Deutsche u. speziell badische Fauna u. Flora. — Anstalt für Bodenseeforschung, Staad b. Konstanz (confer Limologia).

Kiel (Deutschland).
Zool. Mus. d. Univ., Hegewischstr. Dir. Prof. J. Reibisch. — Dir. Prof. Dr. Olaw Schröder.

Kiew (U. d. S. S. R.).
Zool. Mus. der Ukrainischen Akademie der Wissenschaften, Tschudnowskystr. 2. Dir. W. Karawajew. — Zoologen: N. Charlemagne, S. Paramonow, V. Dirsch. *3 Arbeitsplätze für systematische Zool.*
Zool. Mus. des Landw. Inst.

Kingston (Jamaica).
Zool. Dept. of the Mus. of Inst. of Jamaica.

København (Danmark).
Det zool. Mus., Krystalgade 23. 1. Abt. (Wirbeltiere): Prof. Dr. phil. Ad. S. Jensen, Mag. sc. R. Hørring, Cand. mag. M. Degerbøl; 2. Abt.: Dr. phil. O. Th. J. Mortensen, Mag. sc. Th. Hj. Ditlevsen, Mag. sc. P. L. Kramp, Dr. phil. H. R. G. Spärck; 3. Abt.: Mag. sc. W. Lundbeck, Cand. mag. K. Stephensen, Mag. sc. K. L. Henriksen.

La Habana (Cuba).
Mus. zool. Antonio Modesto del Vallec Iznoga e Mus. de Gundlach, Inst. de Segunda Enseñanza. *Allg. Zool., Vögel.*

La Plata (Argentina).
Zool. Abt. d. Mus. de La Plata.

Lausanne (Suisse).
Mus. zool. de l'Univ., Palais de Rumine. Conserv.: Prof. Dr. H. Blanc.

Leiden (Holland).
's Rijks Mus. van Nat. Historie, v. d. Werffpark. Dir. Prof. E. D. van Oost. — Conserv.: Dr. C. M. L. Popta, R. van Eecke, Dr. G. Stiasny, Priv.Doc.; Ass.: M. A. Koekkoek.

Leipzig (Deutschland).
Zool. Mus. d. Univ. Dir. Prof. J. Meisenheimer.

Leningrad (U.d.S.S.R.).

Mus. der Wirbeltiere u. Systematik, Univ., Was. Ost. 16 L., 29. Dir. Prof. Dr. K. Derjugin. — B. M. Tichonuroff, Ass.; I. A. Alexejew, Praeparator. *3 bis 5 Arbeitsplätze für Ichthyol., Herpetol. u. Ornithol.* Ⓟ Pisces, Amphibia, Reptilia, Aves.

Mus. d. wirbellosen Tiere u. Parasit. Labor., W.O. 16, Linie 29.

Zool. Mus. der Akademie der Wissenschaften, Wasili-Ostroo Universitäts-Quai, 1. Stv. Dir. A. Bialynicki-Birula. — Abt. I: Mammal.: Vorst. A. B. Birula; Ass. B. S. Winogradow. II: Aves: Vorst. Ak. P. P. Suschkin; Ass. P. W. Serebrovski. III: Rept.-Amph.: Vorst. S. Th. Izarevski. IV: Pisces: Vorst. P. I. Schmidt. V: Entomol.: G. G. Jacobson (Coleopt.); N. I. Kusnezov (Lepidpt.); A. N. Kiritschenko (Hemipt.); W. W. Barovski (cf Asiat. Coleopt.); Th. D. Pleske (Dipt.); A. W. Martynov (Orthopt.-Triehopt.); A. S. Skorikov (Hymenopt.). Ass.: A. P. Semenov-Tianschanski (Coleopt.), N. N. Filipjev (Lepidt.), E. Th. Miram (Orthopt.), A.A. Stackelberg (Dipt.). VI—VII: Invertebr.: A. K. Mordvilko (Vermes intest.Bryozoa);W.W.Redikorzev (Arach. Myr. Tunic.); A. M. Djakonov (Echinoder.); W. A. Lindholm (Mollus.); G. I. Werestschagin (Crust.); W. M. Rylov (Coelent. Porif.); N. P. Annenkowa (Vermes Chetop.); Ass.P. D. Resvoj (Porifera). *25 Arbeitsplätze für Gäste. Morphol., Systematik, Zoogeographie.* Ⓟ Abt.: Vertreter des Tierreiches, besonders die Tiere des Palaearctischen Faunengebietes. — Kommission und Labor. zur Erforschung d. Malaria und Ektoparas. Kommission und Labor. zur Erforschung Vermes intestinales Rußlands.

Mus. d'anat. compar. et Zool. de l'Inst. Lesshaft, Angliyski pz. 32. Dir. I. D. Strelnikov, Chef du Mus. — M. A. Boïno-Rodzewitsch; E. M. Nepenine. Confer Section Zool. de l'Inst. Lesshaft (Zool. generalis). Ⓟ Les collections, des prép. d'anat. comparée, spécialement.

Liège (Belgique).

Collections de Zool. d. l'Univ. F. M. J. Carpentier (Conserv.).

Logan, Utah (U.S.A.).

Mus. of Zool. of the Agric. Coll.

London (England).

Dept. of Zool., British Mus. of Nat. History, Keeper C. Tate Regan, F.R.S. — W. T. Calman, F. R. S. (Crustacea); R. Kirkpatrick (Tunicata, Protozoa); W. P. Pycraft (Osteol.); S. Hirst (Arachnida); J. G. Dellman (Mammalia); P. R. Lowe (Aves); G. C. Robson (Mollusca); N. B. Kinnear (Aves); H. A. Baylis (Parasit. Worms); A. K. Potton (Coelenterata); M. A. C. Hinton (Mammalia); J. R. Norman (Pisces); C. C. A. Monro (Annelida, Echinotermata); H. W. Parker (Reptilia, Baliachia); M. Burton (Porifera). *Systematic work.*

Mus. of Zool. of the Univ. Coll., Gower Street, W.C. Dir. Prof. D. M. S. Watson.

The Horniman Mus. (Zool. Anthrop.), Forest Hill, S.E. 23. Curator H. S. Harrison, D.Sc., A.R.C.Sc. F.R.A.I. — H. N. Milligan (Zool.

Lyon (France).

Collections du labor. de zool. de l'univ. Dir. Prof. René Koehler.

Maiko (Japan).

Chonchyological Mus., Maiko near Kobe.

Melbourne (Australia).

Mus. of the Royal Australasian Ornithol. Union, 346, Flinders St. Dir. J. A. Ross. — Secretary D. J. Dickison.

Mexico, D. F. (Mexico).

Invertebrat. Sect. d. Dir. de Estudios Biol., 7/a calle de Balderas 94. Dir. Prof. F. Contreras.

Sect. f. Fische, Reptilien u. Amphibien d. Dir. der Estudios Biol., 7/a calle de Balderas 94. Dir. Prof. J. Canzino Gómez.

Minneapolis, Minn. (U.S.A.).

Zool. Mus. Dir. T. S. Roberts.

Des Moines, Iowa (U.S.A.).

Brown Marine Mus. Dir. R. A. Brown (Collector of Marine specimens).

Monaco (Principauté).

Mus. océanographique de Monaco. Confer Labor. marina.

Moskau (U.d.S.S.R.).

Kabinett f. allgemeine Biol. der Kommun. J. M. Swerdlov Univ., 55,, Miusskaja Platz 3. Dir. Boris Zawadovsky (allgemeiner Kursusleiter). — G. J. Asimoff; I. I. Agol; N. V. Kirillowa (Leiter d. Kabinetts); S. A. Miletzkaja-Asimova; V. N. Slepkoff; O. V. Kozulina; Z. I. Senilova; Z. M. Perelmutter; M. N. Lapiner (Laborant d. Kabinetts).

München (Deutschland).

Zool. Museum, Neuhauser Str. 51. Dir. Prof. Dr. Ludwig Döderlein. — Hauptkonservator u. Abteilungsleiter: Prof. Dr. Wilh. Leisewitz; Konservatoren: Prof. Lorenz Müller, Prof. Dr. Heinrich Balß, Dr. Kurt von Rosen.

New Haven, Conn. (U.S.A.).

Zool. Coll. Yale Univ. Dir. W. R. Coe (Curator).

New York City (U.S.A.).

Division of Zool. of the Amer. Mus. of Nat. Hist. 77th st. and Central park west. Dir.: R. C. Murphy (Ass. Dir.); H. M. Chapmann (Curator in Birds). — J. Dwight (Research Ass.); W. G. van Name (Assoc. Curator); W. G. Matthew (Curator of vertebrate paleont. chief division I, earth Sc.); R. W. Tower (Curator compar. Physiol.); C. Wissler (Curator of anthrop.); C. Fisher (Curator in Visual Instructions); E.W. Gudger (Ichthyol.); G. K. Noble (Dept. of Herpetol.); J. F. Nichols (Assoc. Curator of recent fishes); F. E. Lutz (Curator of Entomol.).

Dept. of Tropical Research New York Zool. Society, Zool. Park. Dir. W. Beebe. — *Evolution of Fish and Birds. Oceanography.*

Norman, Okla. (U.S.A.).

Zool. Mus. Univ. of Oklahama. Dir. Ante Richards.

Odessa (U.d.S.S.R.).

Mus. f. Flora u. Fauna, Shukowskaja 35.

Oslo (Norge).

Zool. Mus. Univ. Dir. A. Wollebaek.

Paris (France).

Biol. Marine, Inst. Oceanographique, 195, rue Saint Jacques. Dir. Prof. L. Joubin. — Germain Louis, Ass. *4 places pour Biol. marine. — Microscopie, Anat., Photographie etc.*

Princeton, N.J. (U.S.A.).

Mus. of Zool. C. H. Rogers (Curator).

Rybinsk (U.d.S.S.R.).

Zool. Abt. Naturwissensch. Mus.

Saint Andrews (Scotland).

Gatty Marine Labor. and Bell-Pettigrew Mus. Univ. Confer: Labor. marina.

Salzburg (Österreich).

Mus. für darstellende u. angewandte Naturkunde. Dir. Dr. Paul Eduard Tratz. *Ornithol. Allgem. Zool.*

Samarkand (U.d.S.S.R.).

Zool. Abt. d. Samarkander Gebiets-Mus.

San Francisco, Cal. (U.S.A.).

Mus. of Mammal., California Acad. of Sc. Dir. Dr. B. W. Evermann.

Mus. of Zool., California Acad. of Sc. Dir. Dr. B. W. Evermann.

Santiago (Chile).
Museo y Labor. de Zool. aplicada. Dir. Prof. C. Porter.

Singapore (Straits Settlements).
Zool. Dept. Raffles Mus. Dir. C. B. Kloss.

Skoplje (S. H. S.).
Zool. Mus., Prosvetni Dom. Dir. Dr. Stanko Karaman.

Sofia (Bulgarien).
Zool. Abt. d. Naturhist. Mus. S. M. des Königs, Tsar Oswoboditel 1. Dir. Dr. N. Radeff.

Stockholm (Sverige).
Abt. f. Evertebraten, Naturhist. Reichsmus. Dir. Prof. N. J. T. Odhner. — Ass. Dr. N. H. Odhner. **Abt. für Vertebraten, Naturhist. Reichsmus.** Dir. Prof. E. Lönnberg. — Ass. Dr. N. C. G. Fersen Graf Gyldenstolpe.

Stuttgart (Deutschland).
Zool. Abt. d. Württembergischen Naturaliensammlung. Dir. Prof. Dr. R. Vogel. — Ass. Dr. W. Götz.

Sydney (Australia).
Zool. Abt. d. Sydney Technol. Mus.

Taschkent (U. d. S. S. R.).
Faunistische Abt. d. Zentralmus. f. Naturkunde, Ethnographie u. Archäol. Mittelasiens.

Tomsk (U. d. S. S. R.).
Zool. Mus. Univ. Dir. H. E. Johansen.

Torino (Italia).
Mus di Zool. Univ. Dir. U. Pierantoni.

Tours (France).
Mus d'Anat. compar.

Tring (England).
Zool. Mus. Dir. Dr. E. J. O. Hartert. — Custos d. Entomol. Abt.: Dr. K. Jordan.

Trondhjem (Norge).
Zool. Abt. Det Kgl. Norske Videnskabers Selskab. Mus. Dir. Kustos O. Nordgård (Vögel, Fische, Evertebraten). — Kustos Carl Dons (Säugetiere, Vertebraten).

Tschimkent, Turkestan (U. d. S. S. R.).
Syrdarja Mus., Abt. Zool. und Botan. *Zool. Botan. Ethnographie.*

Vjatka (U. d. S. S. R.).
Zool. Labor. d. Bezirksmus.

Washington (U. S. A.).
U. S. Nation. Mus. of Smithsonian Inst. Confer Mus. biol. generalia.

Wellesley, Mass. (U. S. A.).
Zool. Mus., Wellesley Coll. A. P. Morse (Curator of Zool.).

Wien (Österreich).
Zool. Sammlung, Hochsch. f. Bodenkultur. Dir. Doc. Dr. L. Lorenz.
Zool.-vergl. Anat. Sammlung. Dir. Prof. J. Versluys.
Zootomisches Mus. der med. Fac., IX. Schwarzspanierstr. 7.

Zagreb — Agram (S. H. S.).
Zool. Abt. d. National-Mus., Demetrova ul. 1. Dir. Prof. August Langhoffer. — Dr. Krunoslav Babić, Kustos.

Zürich (Schweiz).
Zool. Mus. Univ. Dir. Prof. Dr. Karl Hescheler.

Hortus zoologici (Vivaria).

Adelaide (S. Australia).
Zool. Gardens, Irome Road. Dir. Alfr. C. Minchin.

Alipore near Calcutta (Br.-India).
Zool. garden.

Amsterdam (Holland).
Koninklijk Zoölogisch genootschap „Natura Artis Magistra", Plantage Middenlaan. Dir. Dr. A. L. Sunier. — Diergaarde en Aquarium, Inspecteur der lebenden Tiere: A. F. J. Portielje. Insectarium, Volontair-Entomol.: R. A. Polak.

Anvers — Antwerpen (Belgique).
Jardin Zool., 20, Place de la Gare. Dir. Dr. M. L'hoëst. ℗ Le Jardin Zool. dispose pour la vente ou pour l'échange de tous les duplicata de mammifères, oiseaux et reptiles.

Basel (Schweiz).
Zool. Garten. Dir. Ad. Wendnagel. *Ca. 1000 Tiere.*

Berlin (Deutschland).
Zool. Garten, W 62, Kurfürstendamm. Dir. Prof. Dr. L. Heck. — Ass. Dr. L. Heck.
Zool. Garten, Abt. Aquarium, W 62, Kurfürstendamm 9. Dir. Dr. Oskar Heinroth. *Auf Wunsch kann im Aquarium wissenschaftlich gearbeitet werden.*

Bruxelles — Brussel (Belgique).
Aquarium et Mus. de pisciculture, Avenue Louise 525. Dir. J. A. Lestage.

Budapest (Ungarn).
Zool. botan. Garten der Hauptstadt. F. A. Cerva, Oberinspektor; Dr. Rajcsics.

Buenos Aires (Argentina).
Jardin zool. Dir. Dr. A. D. Holmberg.

Chapultepec (Mexico).
Zool. Garten u. Aquarium d. Dir. de Estudios Biol. Dir. José A. Duraz. — D. R. Benitez.

Dresden (Deutschland).
Zool. Garten, Aquarium u. Insektarium. Dir. Dr. Gust. Brandes.

Dublin (Ireland).
Zool. gardens, Phoenix-Park.

Düsseldorf (Deutschland).
Zool. Garten, Brehmplatz 1. Dir. Dr. G. Aulmann.

Edinburgh (Scotland).
Zool. Park. Dir. T. H. Gillespie.

Frankfurt a. M. (Deutschland).
Zool. Garten der Stadt Frankfurt a. M. mit Aquarium, Insektenhaus u. Abt. f. Schädlingskunde, Am Schützenbrunnen 16. Dir. Dr. phil. Kurt Priemel. — Entomol. u. Aquariumsvorsteher: Gustav Lederer; Volontär: Richard Wieschke.

Halle a. d. S. (Deutschland).
Zool. Garten, Tiergartenstr. Dir. Dr. Friedrich Hauchecorne.

Hamburg (Deutschland).
Zool. Garten u. Aquarium, Tiergartenstr. 1. Dir. Prof. J. Vosseler. — Leiter des Aquariums: Prof. J. Baron v. Uexküll.

Hannover (Deutschland).
Zool. Garten. Dir. Prof. Fritze.

Högholmen (Finnland).
Tiergarten. Dir. Mag. phil. Rolf Palmgren.

Irkutsk (U.d.S.S.R.).
Zoofarm am Baikalsee. Dir. Prof. V. Dorogostajskij.

Köln a. Rh. (Deutschland).
Zool. Garten, -Riehl. Dir. Dr. L. Wunderlich.

Königsberg i. Pr. (Deutschland).
Zool. Garten. Dir. Dr. Max Meissner.

Kyôto (Japan).
Kyôto Municipal Zool. Garden, Okasaki-chô. Dir. Minamiôji. — Minamiôji; Suzuka.

Leipzig (Deutschland).
Zool. Garten. Dir. Dr. Johannes Gebbing.

Leningrad (U.d.S.S.R.).
Zool. Garten, Lenin Park 1. Dir. I. N. Danilow.

Lisbôa (Portugal).
Jardin zool. e de acclimatação, Parque das Laranjeiras. Dir. A. D. Ramada Curto. *300 Säugetiere, 800 Vögel, 15 Reptilien.*

London (England).
Zool. Gardens, Zool. Society of London, Regent's Park, N.W. 8. Superintendent: Dr. G. M. Vevers. — Curator of Mammals and Birds: D. Seth-Smith; Curator of Reptils: Miss Joan B. Procter; Dir. of the Aquarium: E. G. Boulenger.

Jardin zool. de la ville. Parc de la Tête d'Or. Dir. Pierre Augustin Didier.

Madrid (España).
Jardin zool.

Maksimir b. Zagreb — Agram (S.H.S.).
Zool. Garten.

Marseille (France).
Jardin zool. municipal. Dir. Michel Fernand. — Conserv.: Jule Siepi.

Melbourne (Australia).
Zool. Gardens, Royal Park. Dir. A. A. W. Wilkie.

Münster (Deutschland).
Westfälischer Zool. Garten.

New York (U.S.A.).
New York Zool. Park, Bronx Park 185th Street and Southern Boulevard. Dir. W. T. Hornhaday. — Curator of Birds: L. S. Crandall.

New York Aquarium, Battery Park. C. M. Breder (Ichthyol.).

Ôsaka (Japan).
Municipal Zool. Garden, Tennôji.

Paris (France).
Jardin zool. d'acclimation, Bois de Boulogne.
Vivarium, Mus. nat. d'Hist. Nat., Jardin des Plantes. Dir. R. G. Jeannel.

Pretoria (South Africa).
National Zool. Gardens. Dir. A. K. Haagner.

Rotterdam (Holland).
Zool. Gardens, Kruisstraat 21. Dir. Dr. K. Kuiper. — 1 Curator of the botan. dept. ⊕ All kinds of animals and plants in the collection may be sold or given in exchange.

San Francisco, Cal. (U.S.A.).
Steinhart Aquarium. Dir. Dr. B. W. Evermann. — A. Seale (Supt.).

Sofia (Bulgarien).
Königl. Zool. Garten, Bulevard Christo Botew 47. Dir. B. Kurzius. — Ad. Schumann, Inspektor.
Aquarium d. Univ.

Split (S.H.S.).
Naturwissenschaftl. Stadtmuseum u. Zool. Garten. Confer Mus. biol. generalia.

Stellingen-Hamburg (Deutschland).
Carl Hagenbecks Tierpark. Besitzer: Heinrich u. Lorenz Hagenbeck. — Zool.: Heinz Heck u. Ludwig Zukowsky. ⊕ Wirbeltiere aller Erdteile. — Tierfangstationen in allen Erdteilen.

Stockholm (Sverige).
Zool. Garten „Skansen". Dir. M. Behm.

Sydney (Australia).
Zool. gardens, Moore Park.

Tegucigalpa (Honduras).
Zool. Garten d. Mus. nation.

Toulouse (France).
Mus. d'histoire nat. et Jardin Zool. Confer Mus. biol. generalia.

Varna (Bulgarien).
Aquarium d. Univ. Sofia.

Washington (U.S.A.).
National Zool. Park, Smithsonian Institution. Superintendent: W. M. Mann.

Laboratoria specialia embryologiae.

Confer: *Anatomia generalis et experim.* (pag. 423), *Histol.* (pag. 375), *Zoologia generalis* (pag. 356), *Anatomia animalium dom.* (pag. 433).

Baltimore, Md. (U.S.A.).
Dept. of Embryol. Carnegie Inst. of Washington Wolfe and Madison Streets. Dir. G. Streeter. — C. G. Hartmann (Research assoc.); C. H. Heuser; M. R. Lewis; W. H. Lewis. *Menschl. Embryol. Große embryol. Sammlung v. Dr. Franklin P. Mall.*

Bordeaux (France).
Labor. d'Embryogénie, Fac. Sc. Dir. Prof. Kunstler.

Bratislava — Preßburg (C.S.R.).
Histol. embryol. Inst. Univ., Prayova 5. Dir. Prof. Zd. Frankenberger.

Bruxelles — Brussel (Belgique).
Labor. d'Embryol. Dir. de Selys Longchamps.

Graz (Österreich).
Inst. für Histol. u. Embryol. Med. Fac. Univ. Dir. Prof. Hans Rabl.

Kraków (Polska).
Inst. de Biol. et Embryol. Confer Inst. gen.

Lausanne (Suisse).
Labor. d'Embryol. (École de Méd. de l'Univ.). Dir. Prof. N. Ropoff.

Leningrad (U.d.S.S.R.).
Anat.-Histol. Labor. der Univ. Dir. Prof. Dr. D. I. Deineka. — Dr. D. N. Nassonov; Dr. N. N. Rojdestwensky.

Embryol. Kabinett, Staatl. Univ.
Lisbôa (Portugal).
Inst. Bento da Rocha Cabral pour la recherche Sc. Confer: Physiol. gen. Fac. med.
London (England).
Dept. of Embryol. of the Univ. Coll., Gower Street, W.C. Dir. Prof. J. P. Hill.
London, Ont. (Canada).
Dept. of Histol. and Embryol. of the Univ. of Western Ontario. Dir. C. C. Macklin, M.D., Ph.D. — M. T. Macklin, M.D.
Lwów (Polska).
Histol. u. embryol. Inst. Univ., Piekarska 52. Dir. Prof. Dr W. Szymonowicz. — Ass.: Dr. med. Bernard Kalwaryjski, Dr. med. Helene Frank-Pitta, Abs. fil. Anna Herzig. *6 Plätze für histol. und embryol.-morphol. Untersuchungen.* ⓟ Praep. der sensiblen Nervenendigungen, der Hautgebilde u. d. drüsigen Organe werden gesammelt und auf Wunsch getauscht.
Lyon (France).
Inst. d'Histol. Fac. Méd. Univ. Dir. Prof. A. Policard. — Prof. agrégé Robert Noël, Chef d. trav. prat. Ass.: Dr. Doubrow, Dr. Chaix, Dr. Paupert-Ravault, Dr. Jeannin, Madame D. Pillet, Melles. Pallot et Boucharlat. *9 à 10 places pour stagiaires pour recherches d'histol. générale, de cytol. et d'histochimie. — Installation pour l'histochimie (microincineration, etc.), pour l'étude des tissus aux rayons ultra-violets, pour la microdissection (appareil de Peterfi), pour explantations in vitro de tissus. —* Un département d'Histol. expérim. spécialement destiné aux recherches chirurgicales et méd.
Paris (France).
Labor. d'Embryogénie compar. et Cytol. (Hautes Etudes), V, Collège de France. Dir. Prof. Henneguy.

— Fauré-Fremiet, Melle Garrault. *8 places. —Appar. à microdissection, Install. pour les rech. de chimie biol. et pour la culture des tissus.*
Porto Alegre (Brasil).
Histol. u. embryol. Inst. Dir. Prof. M. Pereira.
Quebec (Canada).
Dept. of Embryol. Laval Univ. Dir. C. Dagneau, M.D.
Stockholm (Sverige).
Anat.-Histol. Inst. Veterin. Hochsch. Dir. Prof. Agduhr.
Strasbourg — Straßburg (France).
Labor. d'Embryol. Fac. Méd. Dir. Prof. Ancel.
Toronto (Canada).
Dept. of Histol. and Embryol. Univ. W. H. Piersol (Prof. of Histol. and Embryol.).
Tsinan, Shantung (China).
Biol.-Labor. Shantung Christian Univ. Confer Institutiones generales.
Utrecht (Holland).
Embryol. Inst. van het Hubrechtfonds, Janskerkhof 2. Dir. Dr. Dan. de Lange jr. — Ass. J. D. F. Hardenberg, biol. doct. *12 Arbeitsplätze. — Sammlung von ca. 50000 Objektträgern (3000 Schnittserien darstellend) mit Schnitten durch embryol. Objekte (nur Vertebraten, der Hauptsache nach Mammalia, Selachier und Petromyzon, wiewohl auch Vögel, Reptilien- und Amphibienserien anwesend sind).* ⓟ Schönes Tauschmaterial seltener Mammalia (Tarsius, Galeopithecus, Manis usw.).
Wien (Österreich).
Embryol. Inst. Univ., 9/3. Dir. Prof. Alfred Fischel. *Experim. Embryol.*
Inst. f. Embryol. u. Histol. Veterin. Hochsch. Dir. Prof. Dr. J. Fiebiger.

Laboratoria histologiae.

Confer: *Laboratoria anatomiae gen. et experim. med.* (pag. 423), *Zoologia generalis* (pag. 356), *Embryologia* (pag. 374), *Physiologia generalis* (pag. 444).

Alfort, Seine (France).
Labor. d'Anat. de l'Ecole vétérin. Confer: Anat. animalium Fac. veterin.
Alger (Algérie).
Labor. d'Histol. et Embryol. Univ. Fac. méd. Dir. J. Weber.
Amsterdam (Holland).
Histol. labor., J. D. Meijerplein 3. Dir. Prof. Dr. Heringa. *Physik.-chem. Unters. des Bindegewebes.*
Athènes (Grèce).
Labor. d'Histol. et d'Embryol. Univ, Rue de l'Academie. Dir. Prof. G. F. Cosmetatos. *Histol. générale et spéciale. Recherches embryol.*
Bahia (Brasil).
Histol. Labor. Dir. Ad. Gordilho.
Barcelona (España).
Inst. d'Histol. Dir. E. Fernandez Galiano.
Beograd (S. H. S.).
Inst. d'Histol. Fac. méd., Hôpital civil, Resavsko ulica. Dir. Prof. A. Kostitch. — Dr. M. Danitch, Chef de Trav. prat. Ass.: Dr. G. Pastely, Dr. S. Ivanovitch, A. Telebakovitch, S. Vasoyévitch, S. Popovitch. *Histol. et la biol. expérim. culture de tissus. — Microphotographie, microcinématographie, chromophotographie.* ⓟ Mus. d'Histol. et d'Embryol.,

où sont logés les planches murales microphotographiques (plus de 400), les diapositifs et les préparation.
Bordeaux (France).
Labor. d' Anat. générale et histol. Univ., Place de la Victoire. Dir. G. Dubreuil, Prof. — Dr. A. Lacoste, Prof.; Dr. M. Beylot, Chef des trav.; Dr. Baudrimont, Prép.; Mlle Escudier, Prép. adj.; M. Merlet. *6 places pour Histol. générale et spéciale.* ⓟ Pièces d'histol. humaine.
Brno — Brünn (C. S. R.).
Histol.-embryol. Inst. Univ., Talgasse 73. Dir. Prof. F. K. Studnička.—MUDr. Jan Florian, 1. Ass.; MUDr. Otmar Olšovský, 2. Ass. *4 Arbeitsplätze. Cytol., Histol., Embryol., vergl. Anat. — Mikromanipulator, Raum für Gewebskulturen.*
Inst. histol.-embryol. Academiae, veterin. Pražská 69. Dir. Prof. MUDr. Karel Šulc. — MVC. J. Svoboda, Ass.; Arabadži, prép. *2 démonstrateurs 2 places pour l'Anat. microscopique des insects et des mammifères (vertébrés).*
Inst. für Pathol. Anat. u. Histol. d. Tierärztl. Hochsch. Confer Anat. pathol.
Bruxelles — Brussel (Belgique).
Labor. d'Histol. Fac. méd., 3, rue du Maelbeek. Dir. Prof. P. Gerard. — Dr. R. Cordier, Ass. *1 place.*
București (România).
Lab. d'Histol. et Embryol. Fac. méd. Univ. Dir. Prof. St. Besnea.

Lab. de Physiol. et Histol. Fac. méd. vétérin. Univ. Dir. Prof. I. Athanasin.

Burlington, Vt. (U.S.A.).
Dept. of Cytol., Univ. of Vermont. Dir. R. Bamford.

Charkow, Ukraine (U.d.S.S.R.).
Histol. Labor. d. Med. Inst., Str. K. Liebknecht 41. Dir. Prof. W. I. Rubaschkin. — Ass. Dr. W. W. Schmelzer; Ass. Dr. W. K. Bezuglaia; Ass. Dr. G. L. Derman; Dr. S. D. Schachov; Ass. Dr. Z. Z. Zelikowskaja.

Cleveland, O. (U.S.A.).
Labor. of Histol. and Embryol. School of Med. Western Reserve Univ., 2109 Adelbert Road. Dir. Frederick C. Waite. — Dr. Bradley M. Patten, Assoc. Prof.; Dr. Samuel W. Chase, Ass. Prof.; Dr. Theodore S. Eliot, Instr. *Working-places for 3 guests. — Full photographic outfit. Constant temperature rooms. Apparatus for grinding sections of bone and teeth.*

Cluj — Klausenburg (România).
Inst. d'Histol. et d'Embryol. Fac. Méd., rue Pasteur 6. Dir. Prof. J. Dragoiu. - Dr. C. Crişan, Chef des travaux pratiques d'Histol. Ass.: Dr. E. Pop, Dr. Th. Buşnitza. Préparateurs: Mme F. Popp, Mr C. Sava, Mr V. Dragoş, Mlle I. Zagony. *2—3 places.*

Coïmbra (Portugal).
Inst. d'Histol. et d'Embryol. Dir. Prof. Dr. G. Brites. — Dr. J. Oliveira Reis, Ass. *25 places pour Histol.* ⊕ Préparations de tissus humains.

Ekatherinoslav, Ukraine (U.d.S.S.R.).
Histol. Labor. Med. Inst. Dir. Prof. V. Karpov.

Ellensburg, Wash. (U.S.A.).
Dept. of Biol. State Teachers Coll. Dir. J. P. Munson. *Minute structure of the Chelonian brain; Origin of germ cells; cell division and cell differentiation; Cytol. of Colleterial glands; cancer; centrosome.*

Genève — Genf (Suisse).
Labor. d'Histol. et d'Embryol. Univ. Dir. Prof. E. Bujard.

Gent — Gand (Belgie).
Labor. v. Weefsel- en ontwikkelingsleer. Dir. Prof. H. Lamo.

Halifax, N.S. (Canada).
Dept. of Histol. and Embryol. Dalhousie Univ. Dir. J. N. Lyons.

Hartford, Conn. (U.S.A.).
Biol. Labor. Trinity Coll., Boardman Hall. Dir. T. H. Bissonnette. *1 working-place for Embryol. and Histol.* ⊕ Cattle Embryos, treemortin Material.

Heidelberg (Deutschland).
Biomech. Inst. der von-Portheim-Stiftung. Dir. Prof. Dr. Franz Weidenreich. *Lymphgefäßsystem, Histol. des Blutes, des Bindegewebes, des Knochen und der Zähne, einschließlich der Histogenese.* ⊕ Mikrophotographien (Diapositive) histol. Praep. von Knochen- u. Zahngewebe.

Innsbruck (Österreich).
Histol. u. Embryol. Inst. Univ. Dir. Prof. Sigmund Schumacher.

Irkutsk (U.d.S.S.R.).
Histol. Labor. Univ., Nabereschnaja 20. Dir. Prof. S. Timofeev. — S. Frank-Kamenetzki, Ass.; V. Burow, Ass.

Jassy (România).
Labor. d'Histol. Fac. de Méd. Dir. Prof. E. Puscariu. — Dr. Jacques Goldner, Chef des Travaux. *10 places de travail.*

Kaunas (Litauen).
Histol.-Embryol. Labor., Gediminogue N 29. Dir. Prof. Dr. E. Landau. — Dr. J. Rubenaite, 1. Ass. 1 Sub-Ass. *2 Arbeitsplätze für Anat. des Zentralnervensystems.*

Kazan (U.d.S.S.R.).
Histol. Labor. Univ. Dir. Prof. A. N. Mislawsky. — Dr. J. M. Lazowsky, Ass.
Histol. u. Embryol. Anst. d. Veterinär-Inst., Arskoe Pole 96. Dir. Prof. Victor Johanovicz Loginoff. — Klennyzkii, Ass.; Spasschii, Ass. *1 Platz für Embryol.: Entwicklung des Rindsembryonen.* ⊕ Rindsembryonen, Embryonen der Fleischfresser und der Einhufer.

Kiev, Ukraine (U.d.S.S.R.).
Histol. Labor. des Med. Inst., Str. Lenin N 37. Dir. Prof. A. G. Tschernlachowsky. — Dr. N. A. Romankewitsch; Dr J. A. Zawistowsky, Ass.; Dr. B. I. Deikun, Ass.; Dr. M. I. Kutscherenko-Rawizkaja, Ass.

København (Danmark).
Histol.-embryol. Labor. Univ., Stormgade 5. Dir. Priv.Doc. R. H. Stamm. — Dr. phil. C. M. Steenberg, Ass. *Gäste nur ausnahmsweise.*

Krasnodar, Kuban-Gebiet (U.d.S.S.R.).
Histol. und Embryol. Labor., Kubanischen Med. Inst., Str. Kotlarewskaja 4. Dir. Prof. P. Sa. Lachowski. — Dr. B. A. Setkow, Prosektor; Dr. A. A. Ovtscharow, Ass. *1—2 Plätze für Ultramikroskopie und gewöhnl. histol. Arbeiten.*

La Plata (Argentina).
Embryol. et Histol. Escola d. sc. med. Dir. Prof. V. Widakowich.

Lausanne (Suisse).
Labor. d'Histol. de l'Univ., Solitude 19. Dir. Prof. Loewenthal.

Leipzig (Deutschland).
Inst. f. pathol. Anat. Confer: Anat. pathol. generalis humana.

Leningrad (U.d.S.S.R.).
Labor. für Cytol. und Physiol. Histol. der Reichsuniv., W. O. Uniwersitetsk. Naberesschnaja 7/9. Dir. Prof. Dr. A. W. Nemiloff — Dr. A. M. Wasjutockin, 1. Ass.; Dr. L. I. Iwanow, 2. Ass. ⊕ Cytol. Praep.; Praep. von Geschlechtsdrüsen und Inkretionsorganen.
Inst. für Histol. und Embryol., med. Kriegs-Academie, Nischegorodskaja Str. 4. Dir. Prof. Alexius Zawarzin. — Dr. Sophie P. Alfejew, Prosektor; Dr. Nikolaus G. Chlopin, Priv.Doc.; Dr. Jurij Orlow, Ass. *4 Plätze für Gäste. Allgemeine vergl. Histol. und experim. Gewebeforschung. — Abt. für Explantation (Gewebezüchtung).*

Leningrad — Alt-Peterhof (U.d.S.S.R.).
Vergl.-Histol. Labor. des Peterhofschen Naturwissenschaftlichen Inst. Dir. Prof. D. I. Deineka. — Dr. D. N. Nassonov, Ass. Mitarbeiter: Dr. P. A. Weiner, P. W. Makarov. *2 Arbeitsplätze für vergl. histol. Untersuchungen.*

Liège (Belgique).
Inst. Auguste Swaen, Labor. d'Anat. normale, Univ., 16, rue de Pitteurs. Dir. Prof. Ch. Julin J. Duesberg et H. de Winiwarter. — 3 Ass. *2 places pour anat. macroscopique et microscopique.*

Lille (France).
Labor. d'Histol. et d'Embryol., Fac. de Méd. de l'Univ., rue Jean Bart 1. Dir. Prof. Dr. E. Laguesse. — Dr. Debeyre, Agrégé, Chef des travaux pratiques; Dr. Morel, Aide-préparateur; Monsieur Vandendorpe, Moniteur (interne des hôpitaux). *3 places pour*

technique histol. courante. ℗ Organes de l'homme; Animaux de labor.
Labor. d'Histol., Fac. cathol. de Méd. de l'Univ., 56, Rue du Port et 11, Rue de Toul. Dir. Prof. J. Toison. — Prof. Cappe de Bailleu.
Labor. d'Histol. comparée et biol. maritime, Fac. des Sc., de l'Univ. Dir. Prof. Dehorne.

Lisbôa (Portugal).
Inst. de Histol e Embryol., Fac. de Med., Univ. Dir. Prof. Dr. A. Celestino da Costa. — Prof. libre Dr. P. Roberto Chaves, Premier Ass.; Dr. Luiz Dias-Amado, deuxième Ass. *4 places pour Histol. et Embryol.* ℗ Collections de coupes d'embryol. des Mammifères.

London (England).
Histol. Labor., Royal Dental Hosp. of London and School of Dental Surgery of the Univ., 32 Leicester Square, W.C.

Louvain — Leuven (Belgique).
Labor. d'Histol. et Embryol. Dir. Prof. J. Havet.

Lund (Sverige).
Histol. Labor.

Lwów (Polska).
Histol. Embryol. Labor. d. Tierärztl. Akad. Dir. Prof. S. Czerski.

Madrid (España).
Labor. de Anat. microscópica (Residencia de Estudiantes).
Labor. de Histol. normal y patol. Inst. Nacional de ciencias, Residencia de Estudiantes. Dir. D. Santiago Ramón y Cajal.
Inst. d'Histol. veg. et anim. Fac. sc. Univ. Dir. Prof. J. Madrid y Moreno.

Marseille (France).
Labor. d'Histol. École méd. Univ.

Minsk (U. d. S. S. R.).
Histol. Labor. Univ. Dir. Prof. P. A. Maurodiadi.

Modena (Italia).
Ist. di anat. e istol. Univ. Dir. Prof. G. Tarozzi.

Montpellier (France).
Labor. d'Histol. Fac. Méd., Rue de l'école de Médecine. Dir. Prof. L. Vialleton. — Dr. Jean Turchini, Agrégé, Chef des travaux; Dr. François Granel, Chef de labor. *Places du travail pour les recherches histol.*

Montreal, Que. (Canada).
Dept. of Histol. and Embryol. Univ. Dir. L. J. Jutras, M.D.

Moskau (U.d.S.S.R.).
Histol. Inst. der 1. Univ., Mochovaja 11. Dir. Prof. Dr. A. Gurwitsch. — Prosektor Dr. W. Fomin; Priv.Doc. Dr. L. Gurwitsch; Ass. Dr. E. Schmidt; Ass. Dr. G. Petchasky; Ass. Dr. G. Chruschtschoff; Ass. Dr. B. Kedrowsky; Ass. Frau Dr. M. Kisliak; Aspiranten: Frank, Salkind, Sorin. *2—3 Plätze für experim. Untersuchungen über Zellteilung (Mitogenetische Strahlungen). — Induktorien zur Untersuchung mitogenetischer Strahlen. Quarzspektrograph.*
Kabinett für Histol. und Embryol. der Physik-Mathem. Fac., 1. Univ., Herzenstr. 6. Dir. Prof. Dr. N. W. Bogojawlesky.—Doc. Dr. A. Rumjantzew; Ass. Dr. G. G. Schegoleff; Ass. Dr. H. W. Epstein; Ass. Dr. G. A. Schmidt; Ass. Dr. K. S. Bogojawlesky. — Labor. für Protistol. Leiter Priv.Doc. Ass. H. Epstein.
Histol.-Embryol. Kabinett der 2. Univ., Pogodinskaja ul., 6. Dir. Prof. W. P. Karpov. — M. N. Rosanowa-Liachowetzkaja; A. M. Liachowetzky; G. M. Petschersky; N. G. Logwinowitsch; A. I. Tokarewa; M. A. Kotliarewskaja; N. N. Kuznezov; N. L. Gerbilsky; N. A. Maslov.

Histol. Abt. des staatl. Timirjaseff-Inst. für wissenschaftl. Forschung, Piatnizkaja 48. Dir. Prof. Dr. Olga Borisowna Lepeschinskaja. — Ass.: Valentina Petrowka Smirnowa; Mina Lwowna Eidinowa (Aspirant); Rewekka Korakowskaja.

München (Deutschland).
Abt. für Histol. and Embryol. Anat. Anstalt. Dir. W. Vogt.

Nancy (France).
Histol. Inst. Univ. Fac. méd. Dir. Prof. Dr. Remy Collin.

Napoli (Italia).
Inst. Hist. et Physiol. gen. Fac. Sc. nat. Univ. Dir. Prof. Vincenzo Diamare.

Nijny-Nowgorod (U. d. S. S. R.).
Kabinett der Histol. u. Embryol. d. Anat. Inst. Dir. Prof. W. Miliutin. — Doc. Kisseleff; Doc. E. Miliutina. *5 Plätze für Histol. u. Embryol.* ℗ Praep.

Nova Goa (Portugies.-Indien).
Kab. f. norm. u. pathol. Histol. Escola Medico-Chir.

Odessa (U. d. S. S. R.).
Histol. u. embryol. Labor. Med. Inst., Olgierskaja 4. Dir. Prof. M. N. Zaewloschin. — Ass.: E. E. Maloritschko; B. I. Kardasewitsch, L. N. Tuzenko. *10 Plätze für morphol. Arbeiten.*

Omsk (U.d.S.S.R.).
Histol. Abt. d. med. Inst. Dir. Prof. Al. E. Efimov.

Padova (Italia).
Inst. d'Histol. d'Embryol. generale. Dir. Prof. Tullio Terni. — Dott. Bernardino Pamzza; Dott. Achille Francescon; Dott. Giovanni Truffi. *6 Places de travail, pour Recherches d'histol., d'histochimie, embryol. déscriptive et expérim. — Ultramicroscope Leitz, Aquarium pour élever les œufs des Amphibiens. Microdissection.* ℗ Preparations microscopiques des ganglions sympathiques de l'homme et des animaux; centres préganglionnaires spinaux des oiseaux; caryocinèses spermatocytaires des Amphibiens; condriosomes dans les Spermatocytes et dans les mitoses Embryons de reptiles, des oiseaux et des Mammifères préparé avec la méth. Cajal pour le système nerveux. — Microdissection.

Paris (France).
Labor. d'Histol. Méd., VIe, 15 rue de l'École de Médecin. Dir. Prof. Auguste Prenant. — A. Branca, Prof. sans chair, Chef des travaux; P. Mulon, Agrégé, Chef des travaux adjoint; C. Champy, Agrégé; J. Verne, Agrégé; J. Millot, Prép. du cours. Prép. des travaux pratiques: De Kervily, Lelièvre, Bulliard, Mme Laroche, Giroud. Aide-prép. des travaux pratiques: Mlle Duboc, Mlle Demay, MmeParat, Mer, Laroche, Isidor. *6 places pour Histol.*
Labor. d'Histophysiol. Coll. de France, Ve, place Marcellin Berthelot. Dir. T. Tolly. — H. Lieure, Préparateur; M. Férester. *5 places de travailleurs pour des recherches d'Histophysiol.*

Perm (U.d.S.S.R.).
Histol. Labor. der Univ. Dir. Prof. Dr. med. V. Schmidt. — E. S. Danini, Ass. Priv.Doc.; Th. M. Lazarenko, Ass.; M. S. Rogosina.
Inst. des Recherches biol. Confer Institutiones generales.

Poitiers (France).
Inst. d'Histol. Fac. méd. Dir. Prof. M. Veluet.

Porto (Portugal).
Inst. d'Histol. Fac. Méd. Dir. Prof. Dr. A. L. Salasar. — J. Pinto Nunes; Paulo Gonçalves, Adelaide Estrada. *Travaux spéciaux sur l'ovaire.* ℗ Collections de préparations histol.

Praha — Prag II (C. S. R.).

Histol. Inst. d. Deutsch. Univ., Salmovská 5. Dir. Prof. Dr. Alfred Kohn. — Fr. Köhler; Max Watzka. *3 Plätze. Histol. d. Epithelkörper, chromaffinen Gewebes, d. sympath. Nervensystems.*
Inst. pour histol. et embryol., Albertov 4. Dir. Dr. Otakar Srdinko. — Priv.Doc. Dr. J. Wolf, 1. Ass.; Priv.Doc. Dr. Ve Tâma, 2. Ass.; Dr. I. Janatka, 3. Ass.; Melle C. R. Prchalova, 4. Ass. *6 places de travail pour tous les travaux microscopiques et expérim. — Aquarium, l'établissement pour les grenouilles, terraria, la chambre microchimique, le labor. neurol., deux chambres particulières pour les cultures des tissus, l'appareil frigozifique, app. microcinématografique, labor. pour la reconstruction des moulages, la chambre aseptique pour les opérations sur les animaux.*

Quito (Ecuador).
Histol. u. Hygien. Inst. Dir. Prof. P. A. Suarez.

Rio de Janeiro (Brasil).
Inst. f. Histol. Dir. Prof. D. de Barros.

Roma (Italia).
Ist. Histol. e Physiol. gener. Dir. Prof. G. Fano.

Rosario (Argentina).
Inst. de Histol., Corrientes 728. Dir. Prof. G. Kaminsky. — Dr. A. Malamud; Dr. Prof. G. Schlieper; E. Apfelbaum; Prof. T. Cerruti.

Rostow am Don (U. d. S. S. R.).
Histol. Labor. der Nordkaukasischen Univ., Krasnoarmeiskaja 67. Dir. Prof. Al. Kolossow. — Dr. N. Sasybin, Al. Winogradow. ℗ Histol. Praep.

Rouen (France).
Inst. d'Histol. Dir. Prof. A. Halipré.

Santiago (España).
Inst. f. Histol., Histochem. u. pathol. Anat. Dir. Prof. V. Goyanes Cedrón.

Saratow (U. d. S. S. R.).
Labor. f. Histol. Univ. Dir. Prof. V. A. Pavlov.

Sevilla (España).
Inst. f. Histol. u. pathol. Anat. Dir. Prof. M. Domínguez.

Smolensk (U. d. S. S. R.).
Histol. Kabinett. Dir. Prof. Michalovskij.

Sofia (Bulgarien).
Inst. d'Histol. et d'Embryol., Fac. de Méd. Univ. Dir. Prof. Dr. A. Mankowski. — Prosecteur: Dr. M. Angelova; Ass.: Dr. M. Mankowska. *Dans le nouveau Inst., qui est en construction, le nombre de places sera 8—10 pour les travaux mikroscopiques. — L'appareil ultramicroscopique. — Un labor. pour faire les modèles embryologiques.*

Stambul (Türkei).
Histol. u. Embryol. Inst. Univ. Dir. Prof. Tewfik Redjib.
Histol. Inst. Tierärztl. Hochsch. Dir.Prof. M.Hadi.

Stockholm (Sverige).
Histol. Inst. Karolinska Inst. Dir. Prof. G. P. E. Häggqvist.
Anat. u. Histol. Inst. Tierärztl. Hochsch. Confer Anat. animalium Fac. veterin.

Tartu — Dorpat (Eesti).
Histol. Labor. Univ.-Nervenklinik. Dir. H. E. Voss.
Inst. d'Histol. de l'Univ. Dir. Prof. Dr. med. Harry A. Kull. — Dr. med. Eduard Aunap. *2—3 places du travail. Recherches cytol. et histol.*

Zool. histol. Abt. des Pharmacol. Inst. Dir. H. E. Voss.

Taschkent (U. d. S. S. R.).
Histol. Inst. der med. Fac. Univ., Allgemeines klinisches Krankenhaus. Dir. Prof. E. Schlachtin. — G. Sjemjonow, Prosektor; S. Pigulewskij. *2 Arbeitsplätze für allgemeine und spezielle Histol.* ℗ Praep. der frühzeitigen embryonalen Entwicklung des Menschen. — Embryol. Labor.

Tegucigalpa (Honduras).
Inst. f. Histol. u. Anat. Dir. Prof. Dr. J. M. Hernandez.

Tomsk (U. d. S. S. R.).
Histol.-Embryol. Labor. Univ. Dir. Prof. S. V. Mjassojedoff. — Dr. A. S. Tschassownikov, Prosektor.

Toulouse (France).
Labor. d'Histol. Fac. de Méd. Dir. Dr. Argaud.

Tsinan (China).
Dept. of Histol. and Embryol. Dir. Prof. R. T. Shields.

Utrecht (Holland).
Labor. für Embryol. u. Histol., Nic. Beetsstraat 22. Dir. Prof. Dr. J. Boeke. — Konserv.: R. Berkelbach v. d. Sprenkel. *Nerven, Muskeln, Sinnesorgane.*
Abt. für prop. Zool. u. experim. Histol. des Zool. Labor. Univ., Janskerkhof 3. Leiter: Lect. Dr. G. C. Hirsch. — Dr. G. J. van Oordt, Konserv. (experim. Morphol.); N. J. ten Cate Hoedemaker, Ass.; L. Bretschneider, Ass. *3 Plätze für Histophysiol. der Sekretion, Gewebeatmung, Enzymforschung. — Mikromanipulator.* ℗ Histol. Praep.

Valencia (España).
Inst. f. Histol. u. Histochem. Dir. Prof. J. Bartual y Moret.

Warszawa (Polska).
Labor. d'Histol. et d'Embryol. Fac. de Méd., rue Chałubińskiego 5. Dir. Prof. Dr. M. Konopacki. — Dr. J. Zweibaum, Adjoint, Prof. agrégé; Dr. P. Słonimski; Dr. Z. Zakolska; Dr. A. Elkner. *12 places pour Histol. et Embryol. morphol. et expérim. et la culture de tissus. — Chambre à temperature constante, Micromanipulateur, Opakiluminateur, Ultramicroscopie.* ℗ Collection macroscopique et microscopique des embryons humains.
Inst. de Cytol. de l'Univ. Dir. V. B. Baron de Baehr.

Wien (Österreich).
1. Histol. Inst. der Univ., Währinger Str. Dir. Hofrat Prof. Dr. Josef Schaffer. — Priv.Doc. Dr. med. Josef Lehner; Priv.Doc. Dr. med. Victor Patzelt; Priv.Doc. Dr. phil. et med. Hanns Plenk. *20 Arbeitsplätze für selbständige Arbeiter auf histol. Gebiete.* ℗ Alles vergl.-anat. Material von Wirbeltieren.
2. Histol. Labor. der Univ., Schwarzspanierstr. Dir. Prof. W. A. Kolmer.

Wilno (Polska).
Inst. d'Histol. à l'Univ. Dir. J. S. Alexandrowitz. — S. E. Baginski, Ass.

Woronesch (U. d. S. S. R.).
Histol. Inst. Univ. Dir. Prof. S. E. Pučkofskij.

Zagreb — Agram (S. H. S.).
Histol.-embryol. Inst. Vet. Fac. Univ. Dir. Prof. Zavrnik. — Priv.Doc. Vinko Marochino (Parasit.). ℗ Histol.-mikroskopische u. embryol. Praep. aller Haustiere.

Laboratoria parasitologica.
Laboratoria parasitologiae generalis (Zoologiae applicatae).
Confer: *Bacteriologia* (pag. 479), *Zoologia generalis* (pag. 356), *Entomologia* (pag. 381), *Anatomia pathologica* (pag. 459), *Phytopathologia* (pag. 415).

Alfort, Seine (France).
Labor. parasit. de l'Ecole vét. Dir. Prof. A. C. Henry.

Alger (Algérie).
Labor. de parasit. et pathol. exotique de l'Inst. Pasteur. Dir. L. M. Parrot.
Labor. Zool. appl., Fac. Sc. Dir. Prof. L. G. Seurat.

Amsterdam (Holland).
Inst. voor Tropische Hygiene (Abt. d. Königl. Kolonial-Inst.), Mauritskade 57. Dir. Prof. Dr. W. Schüffner (Tropenhygiene u. Epidemiol.). — Prof. Dr. N. H. Swellengrebel (med. Zool.); Prof. Dr. E. P. Snijders (Bact.). *Etwa 6 Arbeitsplätze für Gäste für Malaria, andere Protozoenkrankheiten (Dysenterie, Spirochaetosen usw.), Helminthen, Blutkrankheiten, med. Bact.* ⊕ Pathogene Protozoen, Helminthen, Insekten als Überträger von Krankheiten, Bakterienstämme, spez. für tropische Krankheiten.
Labor. für angewandte Zool. und Vererbungslehre, Doklaan 44. Dir. Prof. Dr. J. C. U. de Meyere. — Ass.: B. de Vos de Wilde. ⊕ Schädliche Tiere, namentlich Insekten, holländische Dipteren im allgemeinen.

Beirut (Syria).
Dept. of Pathol., Bact., Hygiene, Parasit. American Univ. Confer Pathol. generalis humana.

Beograd (S. H. S.).
Labor. für Tropenkrankheiten. Vorstand: Dr. E. Džunkovski.

Berlin (Deutschland).
Biol. Reichsanstalt für Land- u. Forstwirtschaft, Labor. für Speicher- u. Vorratsschädlinge, -Dahlem, Königin-Luise-Str. 19. Dir. R. R. Dr. Fr. Zacher. — Ass. Dr. E. Janisch.
Zool. Gruppe der biol. Abt. der Preuß. Landesanst. für Wasser-, Boden- u. Lufthygiene, -Dahlem, Ehrenbergstr. 38—42. Dir. Prof. Dr. J. Wilhelmi (hygien. Zool. u. Parasit., Meeresfauna). — Mitglied: Dr. H. Helfer (Fischereibiol. u. Beziehungen zwischen Fischen und anderen Tieren in hygien. Hinsicht); Mitarbeiter: Dr. F. Roch (Holz- u. Steinschädlinge der Meeresküsten, festsitzende Fauna des Süßwassers), Tierarzt Dr. Th. Saling (Würmer u. Eiskriebelmücken), Dr. G. Kunicke (Fliegen, Flöhe, Wanzen, Läuse, Schaben und andere Gesundheitsschädlinge).
Tropenabt. des Inst. „Robert Koch", N 39, Föhrer-Str. 2. Abt.-Leiter Prof. Dr. Claus Schilling.

Bonn (Deutschland).
Parasit. Labor. Univ., Endenicher Allee 19. Dir. Prof. Dr. Gräfin M. von Linden. — L. Zenneck. *1 Platz für Parasit., Bact.* ⊕ Praep. über Lungenstrongylose.

Bordeaux (France).
Labor. de Zool. et Parasit. Fac. méd., rue Leytein. Dir. Prof. Dr. Mandoul. — Sigalas; Girau. *6 places pour Parasit. humaine.*

Brazzaville (Afrique équatoriale française).
Inst. Pasteur de Brazzaville. Dri. Dr. G. Ledentu. — Dr. M. Vaucel. *1 ou 2 chercheurs pourraient, le cas échéant, trouver place dans les labor. de la trypanosomiase et de microbiol. tropicale.* ⊕ Ornithodorus moubata. Glossines. Préparations de Trypanosoma Gambiense et de Spirocheta Duttoni.

Brno — Brünn (C. S. R.).
Inst. für Zool. u. Parasit. an der Tierärztl. Hochsch., Pražská 69. Dir. Prof. Dr. E. Sekera. — MVDr. K. Rašín (Parasiten bei den Haustieren, evtl. bei den Fischen und Amphibien). ⊕ Mikroskopische Praep. sowie Formol- u. Alkoholpraep. der Parasiten verschiedener Gruppen.

București (România).
Labor. de Parasit. Fac. de Med. Univ. Dr. Nikulesco; Dr. Aubert; Dr. Radacovici; Dr. Vasiliu; Dr. C. Manole.
Section de Parasit. l'Inst. de Sérothérapie. Dir. G. Zotta.

Buitenzorg, Java (Ned.-O.-Indië).
Veeartsenijkundig Inst., Boeboelak. Dir. C. Bubberman, D. V. Sc. — F. L. Huber, DVSc.; L. W. M. Lobel, Vet. Surg.; F. C. Kraneveld, Vet. Surg.; W. K. Picard, D.V.Sc.; O. Ch. Nieschulz, Dr. Nat. Sc. *Studying animal diseases.* ⊕ Pathol.-anat. praep., bloodsucking insects.

Catania, Sicilia (Italia).
Ist. di Parasit. med. Dir. M. Condorelli.

Chicago, Ill. (U.S.A.).
Ricketts Labor. Univ. of Chicago. M. A. Jacobson.

Entebbe, Uganda (Br.-East-Africa).
Commission Internationale pour l'étude de la Maladie du sommeil au Afrique central. Labor.-Dir. Dr. Duke.

Fort Collins, Col. (U.S.A.).
Dept. of Zool. and Entomol. of the Agric. Coll. Dir. Prof. Dr. C. P. Gillette. — S. A. Johnson, Assoc. Prof.; C. R. Jones, Assoc. Prof.; George M. List, Ass. Prof.; M. A. Palmer, Ass. Prof.; J. H. Newton, Ass. Prof.; John L. Hoerner, Instr.; W. L. Burnett, Curator, Mus.; C. P. Gillette, State Entomologist; George M. List, Chief Deputy; W. L. Burnett, Deputy, Rodent Control; R. G. Richmond, Deputy, Apiary Inspection; J. H. Newton, Deputy, Alfalfa Weevil Control; WM. P. Yetter, Jr., Deputy; George S. Langford, Deputy; Chas. F. Rogers, Deputy, Weed Control; S. C. McCampbell, Deputy; E. Roberts, Clerk.

Genève — Genf (Suisse).
Labor. de Zool. lacustre de protistol. et de parasit. Confer Labor. Limnol.

Halle a. d. S. (Deutschland).
Labor. f. Schädlingsbekämpfung des Bact. Inst. d. Landwirtschaftskammer. Dir. Prof. Dr. Raebiger. — Frl. G. Haas, Ass.

Irkutsk (U. d. S. S. R.).
Inst. f. allg. Biol. u. Parasit. Confer Institutiones generales.

Jassy (România).
Labor. d'Histoire nat. méd. Fac. méd. Dir. Prof. N. Leon.

Johannesburg (South Africa).
Protozool. and Parasit. Dept. S. Afric. Inst. for med. Research. Dir. Prof. H. B. Fantham. — Prof. A. Porter.

Kartabo (British-Guiana).
New York Zool. Society, Dept. of Tropical Research.

Kazan (U.d.S.S.R.).
Parasit. Labor. Vet.-Inst., Arskoe Pole 96. Dir. Prof. M. Arnoldov.

Khartum (Egypt).
Wellcome Tropical Research Labor. of the Gordon Memorial Coll.

Königsberg, Pr. (Deutschland).
Forschungsstation für Schädlingsbekämpfung, Univ. Confer Rossitten.

La Plata (Argentina).
Labor. de Parasit. Fac. de méd. vet. et d'Ecole méd. Univ. Dir. Prof. J. Mendy. — Prof. D. Greenway.

Laramie, Wyo. (U.S.A.).
Bureau of Biol. Survey, U.S. Dept. of Agric. A. M. Day (Rodent control).
Dept. of Zool., Coll. of Agric., Univ. of Wyoming. J. W. Scott, Ph.D., Prof. Zool., Research Parasit.; W. H. Carrington, Ass. in Rodent Control Work (Coop. U.S.D.A.); C. L. Corkins, M.S., Prof. Entomol., Entomol. (State Entomol.); A. M. Day, B.S., Rodent Control Specialist (Detailed by U.S.D.A.); H. M. Smith, Ph.D., Instr. Zool. and Physiol.; E. C. Harrah, Assoc. Prof.

Leiden (Holland).
Inst. für Tropenkrankheiten, Rapenburg 33. Dir. Prof. Dr. P. C. Flu. — Dr. P. H. van Thiel, Konserv. *Parasit. der Kolonien, Bact.*

Leningrad (U.d.S.S.R.).
Microbiol. Labor. der Staatl. Univ., W. O. Dir. Prof. Dr. B. Issatschenko. — A. N Moschkowa, Haupt-Ass.; R. K. Mutafowa, Ass.; N. B. Netschaewa, Ass. *8 Arbeitsplätze für Gäste.*
Inst. f. angew. Zool. u. Phytopath., Staatl. Univ., ul. Tschaikowskaja 7. Dir. N. N. Bogdanow-Katjkow.
Parasit. Labor. d. Univ., Wassilystr. 16, L. 28. Dir. Prof. W. D. Selensky. — S. M. Rosanow; T. W. Fedorowa-Winogradowa. *Helminthol.*
Service de Protozool. de l'Inst. bact. vét., Zabalkansky persp., 83A. Chef Prof. W. L. Yakimoff. — E. N. Markoff-Petraschewsky; E. F. Rastegaieff; J. W. Klimas. *5 places pour les recherches de la Protozool., Chimiothérapie, Hématol. et Tiques.* ℗ *Les tiques. Les frottis des protozocières.*
Epizootool. Abt. des Staatl. Inst. f. experim. Med., Lopuchinskaja ulica 12. Dir. V. N. Matreev.
Inst. für Mikrobiol. des Staatl. Inst für med. Wissenschaften, 2. Sovetskij Str. 4. Dir. Prof. B. P. Ebert.
Mus. d. wirbellosen Tiere u. Parasit. Labor. Confer Mus. zool.

León (España).
Labor. de Parasit., Bact., Sueros y Vaccinas, Escuela de Vet. Dir. Prof. J. V. Fernandez.

Leverkusen, Rheinland (Deutschland).
Zool. Labor. der I. G. Farbenindustrie A. G. Dir. Dr. Adolf Herfs.

Lille (France).
Labor. de Zool. méd. et pharm., Fac. Méd. Univ. Dir. Prof. P. Desoil.

Lima (Peru).
Labor. de Parasit. Fac. Méd. Dir. Dr. Ramon Ribeyro. 1 Chef de travaux et 1 prep. *4—6 places pour parasit. tropicale.* ℗ Prep. et objets de Parasit.

Lincoln, Neb. (U.S.A.).
Dept. of Med. Zool. and Parasit. of the Univ. of Nebraska. Dir. Prof. F. D. Barker.

Lisbôa (Portugal).
Inst. de Parasit. e Entomol. méd., Escola de Méd. Trop. Dir. Prof. A. Kopke.

Liverpool (England).
Dept. of Parasit., Fac. Med. Univ. Dir. Prof. W. Yorke.

London (England).
Dept. of Protozool., and Dept. of Trop. Pathol. London School of Hygiene and trop. Med. Univ., Endsleigh Gardens, N.W. Dir. J. G. Thomson. — A. Robertson.
Dept. of Protozool., Lister Inst. of preventive Med. of the Univ., Chelsea, Bridge Rd. S.W. 1. Dir. H. M. Woodcock.
Dept. of Parasit. of the Royal Vet. Coll., Univ. of London, Gt. Coll. St., Camden Town, N.W. 1. Principal: Sir John McFadyean.

Los Baños (Philippine Islands).
Coll. of Vet. Sc. Univ. of the Philippines, Coll. Laguna. M. A. Tubangui (Ass. Prof. Parasit.); A. K. Gomez (Ass. Prof. Pathol.); R. Q. Javiez (Instr. in Bact.); Z. de Jesus (Instr. Vet. Hygiene); F. M. Fronda (Ass. Prof. of Poultry Husbandry).

Lyon (France).
Labor. de Parasit. Fac. Méd. Univ., quai Claude-Bernard. Dir. Prof. J. Guiart. — Dr. Ch. Garin, Prof. agrégé; Dr. Massia, Chef des travaux pratiques; Dr. Morenas, Préparateur. *6 places pour Parasit.* ℗ Parasites animaux et végétaux.
Mus. de Parasit. de l'Univ. Dir. Prof. P. Guiart.

Madrid (España).
Labor. de Parasit. Ecole vet. Dir. Prof. V. Colomo y Amarillas.

Manhattan, Kan. (U.S.A.).
Biol.Survey, Care State Coll. A.E. Oman, Ass. Biol.

Manila (Philippine Islands).
Dept. of Parasit. Univ. Coll. of Med. Sc. Dir. Prof. Luis Guerroro. — L. Leiva; O. Africa.
Division of Zool. Bur. of Sc. Dept. of Agric. Dir. R. C. McGregor.
Labor. of Vet. Parasit. Coll. of Vet. Confer Los Baños.

Marseille (France).
Labor. de Parasit. et d'Hist. nat. de l'Inst. de Méd. et de Pharmacie coloniale. Dir. Prof. J. de Cordomoy.

Milano (Italia).
Ist. di Microbiol. Univ. Dir. Prof. A. Zironi.

Montpellier (France).
Station de Zool. Ecole d'Agric.

Moskau (U.d.S.S.R.).
Vet. parasit. Abt. des Tropen-Inst.
Protozool. section of the Tropical Inst, Pogodinskaja, 10. Dir. Prof. Dr. E. J. Marzinowski. — Ass.: Dr. Moskovski, Dr. Pikoul, Dr. Popoff, Dr. Shukoff, Dr. Shourenkoff. *3 places for human protozool., esp. Malarial parasites, Haemoflagellata, Intestinal parasites.* — *Köhler's apparatus for microscoping work with ultraviolet light. Micromanipulator nach Péterfi.* ℗ Pathogenic protozoa and histol. prep. of lesions caused by them.
Protozool. Abt. d. Metschnikow-Inst. für Infectionskrankheiten, Pokrowka 44. Dir. Prof. S. Koschun. — Abt.-Leiter: Priv.Doc. H. Epstein; A. P. Muratowa; K. A. Plewako. *2 Plätze. Micromanipulator.* ℗ Blut- u. Protozoenpraep.

Ôsaka (Japan).
Parasit. Inst. Ôsaka Provincial Med. Univ. Dir. Prof. Dr. Sadao Yoshida.

Paris (France).
Labor. coloniale. Dir. Dr. Achalme.
Labor. de Parasit. Fac. Méd. Univ., 19, rue école de Méd. Dir. Prof. E. Brumpt. — Prof. agrégé M. Neveu-Lemaire; Prof. agrégé Ch. Joyeux; Dr. M.

Langerou, Chef de Labor., Chef des travaux pratiques pour l'Inst. de méd. Coloniale; Dr. Larrousse, Dr. Galliard, Mr. Sautet, Préparateurs. *Toutes recherches de Parasit. humaine et comparée, Collections. Protozool. Helminthol. Entomol. Mycol. Malariol.* ⑨ Tous objets concernant la Parasit. humaine et comparée.
Labor. zool. appl. d'Inst. National Agronomique, (Ve). Ecole Sup. de l'Agric., 16, Rue Claude-Bernard. Dir. Prof. Marchal.
Labor. de Protozool., Inst. Pasteur, 96, rue Falguière. Dir. Prof. F. Mesnil. — Ass.: Mlle Ch. Pérard, A. Lwoff. Attachés: Mlle M. Treillard, J. Colas-Belcourt, Mlle Marguerite Lwoff, M. R. Deschiens. *Protozool. et Microbiol. coloniale.* ⑨ Trypanosomes pathogènes divers. Spirochètes de fièvres récurrentes. Préparations de Protozoaires parasites.
Service de microbiol. de l'Inst. Pasteur. Dir. A. Calmette. — A. Borrel; M. Nicolle; A. Marie; A. Besredka.
Service de microbiol. coloniale de l'Inst. Pasteur. Dir. F. Mesnil. — E. Marchoux.
Travaux de parasit. de l'Inst. méd. colon. Dir. M. C. P. Langeron.

Port of Spain, Trinidad (Br.-W.-India).
Tropical Diseases Research Labor. of the Imperial Coll. Dir. E. A. Seagar.

Quebec (Canada).
Labor. de Biol. Univ. Laval, Ministère de l'agriculture. Dir. Georges Maheux (Entomologiste). — Omer Caron (Botaniste); Moïse Gagnon (Phytopathologiste); Lionel Daviault (Entomologiste); Richard Bordeleau (Phytopathologiste); Joseph Belleau (biologiste). *2 places pour l'étude des parasit. agric. en général et des abeilles en particulier.* ⑨ Insectes; invertébrates; plantes; cryptogames; organismes bactériens des maladies des plantes et des abeilles.

Rabat (Maroc).
Div. parasit. de l'Inst. sc. chérifien. Dir. R. Ph. Dollfuss.

Rio de Janeiro (Brasil).
Inst. Oswaldo Cruz. Dir. Prof. Dr. Carlos Chagas. — C. M. Torres (Ass. anat. Pathol.); A. du Costa Lima (Ass. Entomol.); Cezar Pinto (Ass. med. Entomol.).

Rossitten, Kurische Nehrung (Deutschland).
Zool. Station für Schädlingsforschung. Dir. Prof. Dr. Otto Koehler. — Dr. Lothar Szidat, Stationsleiter. *3 Plätze für Hydrobiol. u. Schädlingskunde.* ⑨ Helminthen und sonstige Schädlinge, Fische.

Saigon (Indo-Chine).
Section microbiol. animale, Inst. Pasteur.

Santiago (España).
Inst. f. Pharmacol. Zool. Dir. Prof. C. Sobrado Maestro.

São Paulo (Brasil).
Dept. of Parasitol., Caixa Postal 1985. Dir. Dr. Samuel B. Pessôa (Chief of Labor.). — Dr. Alcides Prado (Med. entomol. and malaria); Dr. Clovis Corrêa (Mycol. and protozool.). *2 or 3 places for research of Malaria, Ancylostomosis and other intestinal parasit. General biol. of worms.* ⑨ Worms: Arthropoda and Protozoa.
Dept. of Microbiol. *Fully equiped for investigations.*

Saratow (U. d. S. S. R.).
Protozool. Abt. d. Staatsinst. f. Mikrobiol. u. Epedemiol. Dir. Dr. J. G. Hoff.

Serajewo, Bosnien (S. H. S.).
Parasit. Inst. Dir. Viktor Apfelbeck.

Skoplje, Mazedonien (S. H. S.).
Zool. Abt. Inst. f. Tropenhygiene. Dir. Dr. Stanko Karaman.

Stade, Prov. Hannover (Deutschland).
Zweigstelle der Biol. Reichsanstalt für Land- und Forstwirtschaft, Harsefelder Str. 57a. Leiter: Prof. Dr. K. Braun, Reg.R. ⑨ Vergleichsmaterial: Schädlinge, Fraßbeschädigungen usw. Pilze und Beschädigungen derselben. Schäden durch Witterungseinflüsse.

Stambul (Türkei).
Parasit. Inst. d. Univ. u. d. Tierärztl. Hochsch. Dir. Prof. J. Hakki.

Tegucigalpa (Honduras).
Parasit. Inst. Univ. Dir. Prof. B. R. Zepeda.

Tikkurila (Finnland).
Staatl. Landw. Versuchsanstalt, Abt. f. Schädlinge (Maatalouskoelaitos, Tuhoeläinosasto). Yrjö Hukkinen, 1. Ass.; Jaakko Listo, 2. Ass.; Niilo A. Vappula, 2. Ass. ⑨ Schädliche Tiere, beschädigte Pflanzen, Gallen, Minen.

Tôkyô (Japan).
Labor. of Med. Zool. Kitasato Inst. of Infect. Diseases, Shirokane-Sankô-Cliô. Dir. Dr. Mikinosake Miyazima.
Parasit. Inst. Coll. of Med., Keiô-Gijutu Univ. Dir. Prof. Dr. T. Koizumi. — Lect. I. Tochihara; Lect. Dr. M. Miyazaki.

Toulouse (France).
Inst. f. Parasit., Ecole nation. Vétérin. Dir. Prof. A. Martin. *Botan. méd. et fourragère, Zool. méd., Maladies parasit.*

Utrecht (Holland).
Inst. für parasit. u. Infektionskrankheiten der Tierärztl. Fac., Biltstraat 168. Dir. Prof. Dr. L. de Blieck. — Conserv.: Dr. T. van Heelsbergen, Dr. E. A. R. T. Baudet; B. J. Krijgsman, Ass. *2 Plätze für Bakt., Serol., Parasit.*

Laboratoria entomologica specialia.

Confer: *Phytopathologia* (pag. 415), *Zoologia generalis* (pag. 356), *Parasitologia generalis* (pag. 379), *Laboratoria culturae plantarum et animalium* (pag. 497).

A. and M. College, Miss. (U. S. A.).
Dept. of Zool. and Entomol. Agric. Coll. Dir. R. W. Harned, B.S., Prof. Zool., Entomol., Entomol. (State Entomol.).—H. W. Allen, M.S., Assoc. Entomol.; R. N. Lobdell, M.S., Assoc. Prof. Zool., Zool.; E. W. Stafford, M.S., Assoc. Prof. Zool., Entomol.; A. L. Hamner, M.S., Ass. Entomol.; J. M. Langston, M.S., Assoc. Entomol. (Coop. U.S.D.A.).
Entomol. Dept. of the Mississippi State Plant Board. M. R. Smith (Ass. Entomologist).

Alhambra, Cal. (U. S. A.).
Bureau of Entomol., P. O. Box 297. Roy E. Campbell (Entomologist); C. K. Fisher (Entomologist); J. E. Clifford (Entomologist).

Ames, Ia. (U. S. A.).
Dept. of Zool. and Entomol. State Coll. of Agric. Dir.: C.J.Drake, Ph.D.,Head of Dept.,Chief Entomol. (State Entomol.). — F. M. Baldwin, Ph.D., Prof. Physiol.; F. D. Butcher, A.B., Ext. Entomol.; B. B.

Fulton, M.S., Instr. Zool. and Entomol., Ass. Entomol.; J. E. Guthrie, M.S., Prof. Zool.; B. M. Harrison, Ph.D., Ass. Prof. Zool.; Sarah Hoke, M.S., Instr. Zool. and Entomol.; H. S. Hopkins, Ph.D., Ass. Prof. Physiol.; H. H. Knight, Ph.D., Ass. Prof. Entomol.; Henry Ness, B.S., Ass. State Entomol.; F. B. Paddock, M.S., Assoc. Prof. Apiculture, Apiculturist (State Apiarist); R. L. Parker, M.S., Ass. Chief in Apiculture; W. H. Wellhouse, Ph.D., Assoc. Prof. Entomol.; A. D. Worthington, Ext. Ass. in Apiculture.

Amherst, Mass. (U.S.A.).

Dept. of Entomol.State Coll. of Agric. Dir.: H.T.Fernald, Ph.D., Head of Dept. — C.P.Alexander, Ph.D. Ass. Prof. Entomol.; A. I. Bourne, B.A., Ass. Research Prof. Entomol.; M. H. Cassidy, B.S., Ass. Prof. Beekeeping; G. C. Crampton, Ph.D., Prof. Insect Morphol.; W. D. Whitcomb, B.S., Ass. Research Prof. Entomol. (P. O., Waltham).

Amsterdam (Holland).

Insektarium der Univ. Dir. Prof. Dr. J. C. H. de Meijere.

Arlington, Mass. (U.S.A.).

European Corn Borer Labor. B. E. Hodgson (Entomologist).

Aschersleben (Deutschland).

Biol. Reichsanstalt für Land- u. Forstwirtschaft, Zweigstelle. Dir. Reg.R. Dr. L. Peters. — Ass. Dr. F. Dykerhoff. *Tierische Schädlinge der Zier- und Gemüsepflanzen.*

Astrachan (U.d.S.S.R.).

Astrachaner Pflanzenschutzstation, Straße 1. Mai 144. Dir. Prof. S. J. Schembel. — W. D. Wodolagin (Entomol.); H. M. Stiepanow (Phytopath.). *3 bis 4 Arbeitsplätze für die Untersuchungen schädlicher Insekten, Zieselmäuse und Pflanzenkrankheiten.* ⊕ Schädliche Insekten und Pilze, Zieselmäuse und andere Nagetiere, Vögel, Pflanzen, Insekten, Astrachaner Flora und Fauna. — Zweiglabor. ist eine experim. Parzelle in Astrachan.

Atlanta, Ga. (U.S.A.).

State Board of Entomol., Capitol Building. R. M. Seeley (Entomologist).

Auburn, Ala. (U.S.A.).

Dept. of Zool. and Entomol.Coll.ofAgric. Dir.: J.M. Robinson, M.A., Actg.Head ofDept. — F.E.Guyton, M.S., Ass. Prof. Zool. and Entomol., Ass. Entomol.; H. G. Good, M.S., Ass. Prof. Zool. and Entomol., Ass. Entomol.; W. A. Ruffin, M.S., Ext. Entomol.

Baton Rouge, La. (U.S.A.).

Dept. of Zool. and Entomol. Coll. of Agric. W. H. Gates, D.Sc., Prof. Zool. and Entomol.; W. E. Hinds, Ph.D., Entomol.; Ellinore H. Behre, Ph.D., Assoc. Prof. Zool.; E. C. Davis, Specialist in Bee Culture; Mrs. Morris Faures, B.A., Instr. Zool.; J. R. Fowler, B.S., Instr. Zool.; Corinne Keaty, B.A., Instr. Zool.; H. W. Manter, Ph.D., Instr. Zool., Parasit.; O. W. Rosewall, M.S., Prof. Entomol.; Herbert Spencer, Ph.D., Ass. Entomol.; H.A. Stabe, M.S., Instr. Entomol.

Beograd (S.H.S.).

Entomol. Inst. der Univ., Krunska, 77e. Dir. Prof. Dr. J. Wagner. — Ass. Wladimir Martino. ⊕ Balkanische Insekten.
Entomol. Abt. Landwirtschaftl. Versuchsstation, Topčíder. — Nikola Beranov, Ass.

Berkeley, Cal. (U.S.A.).

Dept. of Entomol. Coll. of Agric. Dir. W.B. Herms, M.A., Head of Div. Prof. Parasit., Entomol. — A.J. Basinger, M.S., Ass. Entomol. (P. O., Riverside); J. C. Chamberlin, M.A., Ass. Entomol. (P. O., Riverside); Harold Compere, Ass. Entomol. (P. O., Riverside); E. R. de Ong, M.S., Ass. Entomol.; E. O. Essig, M.S., Assoc. Prof. and Assoc. Entomol.; S. B. Freeborn, Ph.D., Assoc. Prof. and Assoc. Entomol. (P. O., Davis); H. J. Quayle, M.S., Prof. Entomol., Entomol. (P. O., Riverside); E. W. Rust, A.B., Ass. Entomol. (P. O., Riverside); H. H. P. Severin, Ph.D., Instr. and Ass. Entomol.; Filippo Silvestri, D.Sc., Entomol. Explorer (P. O., Hongkong, China); H. S. Smith, M.A., Assoc. Prof. Entomol., Entomol. (P. O., Riverside); T. I. Storer, Ph.D., Ass. Prof. and Ass. Zool. (P. O., Davis); P. H. Timberlake, B.A., Ass. Prof. and Assoc. Entomol. (P. O., Riverside); E. C. Van Dyke, B.S., M.D., Assoc. Prof. Entomol.; G. H. Vansell, A.M., Assoc. in Entomol., Junior Apiculturist (P. O., Davis); C. W. Woodworth, M.S., Prof. Entomol.

Berlin (Deutschland).

Labor. f. physiol. Zool. d. Biol. Reichsanst. f. Land- u. Forstwirtschaft., -Dahlem, Königin-Luise-Str. 19. Dir. R. R. Prof. Dr. A. Hase.
Deutsches Entomol. Inst. der Kaiser-Wilhelm-Gesellschaft, -Dahlem, Goßlerstr. 20. Dir. Dr. Walther Horn. *Ca. 6 Plätze mit notwendiger Ausrüstung.* ⊕ Alle Insekten u. ihre biol. Objekte.

Billings, Mtn. (U.S.A.).

Billings Labor. U.S. Dept. of Agric, Box 1094. A. T. Stewart Lockwood (Assoc. Entomologist).

Blacksburg, Va. (U.S.A.).

Dept. of Entomol. of the Virginia, Experim. Station Dir. W. J. Schoene, State Entomologist .— W. S. Hough, Assoc. Entomologist; G. W. Underhill, Ass. Entomologist; L. R. Cagle, Ass. Entomologist. — 2 sublabor.

Bologna (Italia).

Labor. di Entomol., Ist. Superiore Agrario. Dir. Prof. Dr. G. Grandi. — Dr. Anna Firoi, Ass.; Dr. Lavinio Baldassarri, Ass. borsista; Athos Goidanich, Allievo interno; Dr. Dina Lombardi, Ospite. *Biol., morfol. e sistematica degli Insetti dei Fichi di tutto il mondo. Biol. e morfol. degli Imenotteri melliferi e predatori. Biol. degli Afidi, specialmente di quelli dannosi. — Il Labor., di recente formazione, si trova nel periodo di assestamento.* ⊕ Insetti di ogni ordine.

Bozeman, Mtn. (U.S.A.).

Dept. Entomol. Coll. of Agric. Dir. R.A.Cooley,B.S., Head of Dept. (State Entomol.).—W. C. Cook, Ph.D., Ass. Entomol.; J. R. Parker, M.S., Assoc. Entomol.

Brookings, S.D. (U.S.A.).

Entomol.-zool. lab. S. Dak. State Coll. Dir. H. C. Severne. — G. Gilbertson (Entomol.); M.H. Farrer (Keeping instr.); E. C. O. Roke (Zool.); A. L. Ford (Extension Entomologist). *5 to 10 places for Systematic Work in Entomol. Life history Work in Entomol. and Helminthol. Economic Entomol.* ⊕ Insects of all orders. Amphibia. Reptiles. Birds nests and eggs.

Brownwood, Tex. (U.S.A.).

Bureau of Entomol. of the U.S. Dept. of Agric. H. S. Adair (Junior Entomologist); B. M. Broadbent (Junior Entomologist).

Budapest (Ungarn).

Entomol. Station, Kitaibel-Gasse 1. Dir. Josef Jablonowski. — Mitarbeiter: Eugen Györffy, Gabriel Bakó, Julius Kadocsa.

Buitenzorg, Java (Ned.-O.-Indië).

Zool. Labor. Inst. for plant diseases. Dir. S. Leefmans. — Dr. P. V. d. Goot (Entomologist); Jr. L. Kalshoven (Entomologist).

Canton, N.Y. (U.S.A.).

Biol. Dept. St. Lawrence Univ. Dir. J. L. Buys. *Cicadellidae of the order Homoptera. Systematic.*

Carlisle, Pa. (U.S.A.).
Entomol. Labor., 337, Franklin Street. Ass. C. C. Hill (Entomologist); H. W. Smith (Junior Entomologist).

Cesim (Latvija).
Entomol. Abt. des Pflanzenschutzinst. Dir. J. Zirnits.

Chateauneuf Sion, Wallis (Suisse).
Station cantonale d'entomol. appliquée. Dir. Dr. Hans Leuzinger. Entomol. adhibita. Blutlaus, Phylloxora. Traubenwickler, Simaethis pariana. Entwicklungsgeschichte v. Carausius morosus.

Chatham, Ontario (Canada).
Entomol. Labor. Dept. of Agric., 261, Victoria Ave. A. B. Baird (Entomologist); G. M. Stirrett (Entomologist).

Clemson College, S. C. (U.S.A.).
Division of Entomol. and Zool. Clemson, Agric. Coll. Dir. Franklin Sherman, Chief. — Clifford O. Eddy, Assoc. in Research; Oscar L. Cartwright, Ass. in Research; David Dunavan, Ass. in Teaching; Joseph A. Berly, Ass. in Inspection Work; John O. Pepper, Assoc. in Extension Work; Edward S. Prevost, Beekeeping Extension; Harold J. Henderson, Ass. in Teaching; L. C. McAlister, Ass. in Research. ⓟ Insects: Birds, Reptiles, Amphibians.

College Park, Md. (U.S.A.).
Dept. of Entomol., Univ. of Maryland. Dir. Ernest N. Cory. — H. S. McConnell (Experim. Station); P. D. Sanders; H. H. Shepard; P. X. Peltier; Paul Knight. 1 or 2 workers may be accommodated for microscopical or biol. research on insects. ⓟ Economic Insects.

College Station, Tex. (U.S.A.).
Dept. of Entomol. Agric. Coll. F. L. Thomas, Ph.D., Chief of Div. (State Entomol.); S. W. Bilsing, Ph.D., Head Prof. Entomol.; A. H. Alex, B.S., Queen Breeder, State Apicultural Research Labor., P.O., San Antonio; R. K. Fletcher, M.A., Assoc. Prof. Entomol.; C. E. Heard, B.S., Foulbrood Insp. (P.O., Hebbronville); V. A. Little, M.S., Ass. Prof. Entomol.; S. E. McGregor, jr., State Apiary Insp.; H. B. Parks, B.S., Apiculturist in charge State Apicultural. Research Labor. P.O., San Antonio; H. J. Reinhard, B.S., Entomol.; R. R. Reppert, Entomol.

Columbia, Mo. (U.S.A.).
Dept. of Entomol. Coll. of Agric. Leonard Haseman, Ph.D., Prof. and Chair. of Dept., Chief Nursery Insp.; K. C. Sullivan, A.M., Ass Prof Entomol., Deputy Nursery Insp.

Columbus, Ohio (U.S.A.).
Dept. of Zool. and Entomol. State Coll. of Agric. R. C. Osburn, Ph.D., Head of Dept.; W. M. Barrows. D.Sc., Prof. Zool.; D. M. DeLong, Ph.D., Prof. Entomol.; J. S. Hine, B.S., Assoc. Prof. Zool. and Entomol.; C. H. Kennedy, Ph.D., Ass. Prof. Entomol.; W. J. Kostir, M.S., Ass. Prof. Zool ; F. H. Krecker, Ph.D., Prof. Zool.; R. N. McCormick, M.S., Instr. Zool.; D. F. Miller, M.A., Instr. Zool.; Herbert Osborn, D.Sc., Research Prof. Zool. and Entomol.; T. H. Parks, M.S., Entomol. Specialist; Hugh Setterfield, M.S., Instr. Zool ; Mary Waters, A.B., Ass. in Zool.

Como, Miss. (U.S.A.).
State Plant Board of Mississippi. G. Hoke (Ass. Entomologist).

Corvallis, Ore. (U.S.A.).
Entomol. Dept. and Oregon Experim. Stat., Oregon Agric. Coll. Dir. Don C. Mote. — H. A. Scullen, Ass. Prof. Apic. and Hymenoptera; W. J. Chamberlin, Ass. Prof. (Forest Insects, Coleoptera); B. G. Thompson; J. Wilcox (Investigation of crop insect pests, Aegeriidae). ⓟ Insects of Oregon and Western U.S. especially Coleoptera, Hymenoptera and Aegeriidae.

Dallas, Tex. (U.S.A.).
Research Bureau of Entomol. U.S. Dept. of Agric., Box 205. W. E. Dove (Parasit.).

Delaware (U.S.A.).
Dept. of Entomol. of Agric. Experim. Stat. Bionomics and Control of the Codling Moth, Bionomics and Control of the Grape Leaf Hopper, A Systematic, Biol., and Ecol. Study of the Hemiptera of Delaware.

Durham, N.H. (U.S.A.).
Agric. Experim. Station of the Univ. of New Hampshire. Dir. of Experim. Station: J. C. Kendall; Head of Dept. of Entomol.: W. C. O'Kane. — P. R. Lowry, Ass. Prof. ⓟ Economic insects.

Edmonton (Canada).
Dept. of Entomol. Univ. of Alberta. E. H. Shickland, M.Sc.

Estancia, N.M. (U.S.A.)
Bureau of Entomol. U.S. Dept. of Agric., P.O. Box 353. J. R. Douglas (Entomologist).

Fargo, N.D. (U.S.A.).
Dept. of Entomol. State Coll. of Agric. J. A. Munro M.S. (Entomologist).

Fayetteville, Ark. (U.S.A.).
Dept. of Entomol. Coll. of Agric. Dir.: W. J. Baerg, Ph.D., Head of Dept. — Dwight Isely, M.A., Assoc. Prof. and Assoc. Entomol.; Dr. G. Holl.

Firenze (Italia).
Stazione Entomol. Agraria, Museo di Storia nat. Via Romana 19.

Florence, S.C. (U.S.A.).
Division of Boll Weevil Control Clemson Coll. Pee Dee Experim. Station. Dir. George Miller Armstrong. — F. A. Fenton (Entomologist).

Forest Grove, Ore. (U.S.A.).
U.S. Entomol. Labor. S. E. Keen (Junior Entomologist); L. P. Rockwood (Assoc. Entomologist); M. M. Recher (Ass. Entomologist).

Fort Collins, Col. (U.S.A.).
Dept. of Entomol. und Zool. State Coll. of Agric. Dir. Gillette in charge. — S. A. Johnson, M.S., Assoc. Prof. Entomol. and Zool.; W. L. Burnett, Deputy State Entomol.; J. L. Hoerner, B.S., Instr. and Ass. Entomol.; C. R. Jones, M.S., Assoc. Prof. Entomol. and Zool., Assoc. Entomol.; G. S. Langford, M.S., Ass. Deputy State Entomol.; G. M. List, M.S., Assoc. Entomol., Chief Deputy State Entomol.; J. H. Newton, B.S., Ass. Entomol. (Deputy State Entomol., P.O., Paonia); Miriam A. Palmer, M.A., Ass. Prof. and Ass. Entomol.; R. G. Richmond, B.S., Deputy State Entomol., Inspr. of Apiaries; W. P. Yetter, jr., B.S., Ass. Entomol. (P.O., Grand Junction).

Fort Valley, Ga. (U.S.A.).
Peach Insect Labor. O. I. Snapp (Entomologist).

Fredonia, N.Y. (U.S.A.).
Vineyard Labor. of the N.Y. Agric. Dir. Dr. R. W. Thatcher, Geneva, N.Y. — Fred E. Gladwin (Horticulture); Deril M. Daniel (Entomol.).

Fresno, Cal. (U.S.A.).
Dried-Fruit Labor. Bureau of Entomol., U.S. Dept. of Agric. J. C. Hamlin, Assoc. Entomol.

Gainesville, Fla. (U.S.A.).
Dept. of Entomol. Coll. of Agric. J. R. Watson, M.A., Entomol.; H. E. Bratley, M.S., Labor. Ass. (Pecan Invest.); E. F. Grossmann, M.A., Ass. Ento-

mol. (Cotton Invest.); A. N. Tissot, M.S., Ass. Entomol.

Gatun Lake (Panama Canal Zone).
Barro Colorado Island Labor. of Entomol., Canal Zone, Box 245 Ancon Canal Zone. Dir. James Zetek, Curator. — Dr. Thomas Barbour; Dr. Alex. G. Ruthren; Dr. David Farichild; Dr. John R. Johnston; Dr. Vernon Kellogg; Dr. Wm. M. Wheeler. *40 places. — Excellent screened buildings, dormitories (fully equiped) and dining room service. Animal cages and usual labor. equipment. Scientists are expected to bring with them special equipment. 25 miles of coast line; almost 6 sq. miles of virgin jungle. 15 miles of trails through jungle.*

Gembloux (Belgique).
Station Entomol. de l'Etat. Dir. Prof. R. Mayné. — V. Lathouwers, Chef de tr.

Geneva, N.Y. (U.S.A.).
Dept. of Entomol. of the State Agricultural Experim. Stat. Percival J. Parrott, M.A., Chief in Research (Entomol.); Hugh Glasgow, Ph.D.; Fred Z. Hartzell, M.A. (Fredonia); Hugh C. Huckett, Ph.D. (Riverhead); Frederick G. Mundinger, M.S. (Poughkeepsie), Assoc. in Research (Entomol.); S. Willard Harman, B.S.; Foster L. Gambrell, M.S.; Derrill M. Daniel, B.S., Ass. in Research (Entomol.). *Studies on apple insects and pear insects. Control of sucking insects with dust mixtures. Insects in relation to peach culture. Insects injourius to canning crops. Studies with grape insects. Study and control of cucumber beetles. Control of aphids in cauliflower seedbeds. Spraying vs. dusting potatoes.*

Gihu (Japan).
Nawa Entomol. Research Labor.

Glendora, Cal. (U.S.A.).
California Spray-Chemical Co., Box 111. H. Knight (Research Entomologist). *Fumigation with HCN gas. Oil sprays. Insect pests infesting citrus trees.*

Guelph, Ont. (Canada).
Dept. of Entomol. Agric. Coll. Dir. A. W. Baker, B.S.A. — R. D. Colquette, B.S.A.; L. Caesar, B.A.; B.S.A.; J. A. Flock, B.S.A.

Halle a. d. S. (Deutschland).
Sammlung pflanzenpathol. Insekten d.Landwirtsch. Inst., Univ.

Hann.-Münden (Deutschland).
Zool. Inst. der forstlichen Hochsch., Schloß. Dir. Prof. Dr. L. Rhumbler. — Prof. Dr. H. Frhr. Geyr von Schweppenburg (Ornithol. u. Forstschutz). *Insektenschädlinge aller Art. — Moderne Apparatur für forstliche Schädlingsuntersuchungen aller Art.* ℗ Forstl. Schädlingspraep.

Hayling Island, Hants. (England).
British Mosquito Control Inst. Dir. John F. Marshall. *Photomicrography of insects.*

Helsinki (Finnland).
Landwirtschafts- und Forstzool. Inst., P. Rautatiek 13 F. Dir. Uunio Saalas, Fil. Dr., Prof. — 1 Ass.

Honolulu (Hawaii).
Bureau of Entomol. U.S.D.A. A. C. Mason (Ass. Entomologist); H. F. Willard (Entomologist).
Dept. of Entomol. of the Div. of Agric. and Home Economics. Dir. Prof. D. L. Crawford, M. A. — E. H. Bryan jr., M.S., Instr.
Dept. of Entomol. of the Experim. Stat. Dir. Dr. J. F. Illingworth, Entomologist.

Hood River, Ore. (U.S.A.).
Branch Experim. Station of Phytopathol. Leroy Childs, Ass. *Entomol. and Plant pathol., control. of insect pests and plant diseases of apples and pears.*

Ithaca, N.Y. (U.S.A.).
Dept. of Entomol. of the Cornell Univ. New York State Coll. of Agric. J. H. Comstock, Prof. emeritus; J. G. Needham, Prof.; Glenn W. Herrick, Prof.; C. R. Crosby, Prof.; O. A. Johannsen, Prof.; J. C. Bradley, Prof. and Curator; G. C. Embody, Prof.; Robert Matheson, Prof.; E. F. Phillips, Prof.; P. W. Claassen, Ass. Prof.; L. P. Wehrle, Instr.; R. B. Willson, Specialist in Apiculture; W. T. M. Forbes, Instr.; P. P. Babiy, Ass. Curator; Crace H. Griswold, Instr.; P. J. Chapman, Instr.; Paul R. Needham, Instr.; N. L. Cutler, Instr.; F. C. Fletcher, Teaching Ass. in systematic entomol.

Iwanovo-Voznessensk (U.d.S.S.R.).
Lab. f. Entomol. Polytechn. Inst. Dir. Prof. N. Kasanskij.

Jalta, Crimée (U.d.S.S.R.).
Cabinets d'entomol. et de phytopath. de la Station de la défense des Plantes sur les côtes meridionales de la Crimée. Dir. S. M. Fedorov. — Prof. N. N. Dekenbach (Phytopath.). *8 places pour la biol. des insectes et des maladies de cultures spéciales, préférablement de la vigne et de tabac et des moyens de la lutte.* ℗ *Les insects nuisibles de la vigne et de tabac et un herbier de maladies d'eux.*

Johannesburg (South Africa).
Entomol. Dept. S. Afric. Inst. for med. Research. Dir. Dr. A. Ingram.

Kagoshima (Japan).
Entomol. Labor. Kagoshima Imperial Coll. of Agric. and Forestry.

Kazan (U.d.S.S.R.).
La chaire de la Parasit. et la Pathol. et la Térapie des maladies invasionnées, Arskoe Pole, 96. Dir. Prof. B. Massino.— I. A. Solonitzin; W. I. Karokhin; M. Petroff. *4 places pour les études de l'helminthofauna de l'homme et des animaux vertébrés. — Les microscopes, les thermostates, les centrofougues et tous les appareils spéciaux pour les analyses du sang.* ℗ *Des collections, des préparations des vers parasit. (de la region Volga-Kama) de toutes les classes des vertébrés.* — 1 laboratoire helminthologique.

Kiev (U.d.S.S.R.).
Entomol. Labor. des Landw. Inst.

Kraków (Polska).
Lepidopterol. Abt. im Physiograph. Mus. der Poln. Akad. d. Wiss. Kustos: Witold Niesiolowski.

Krasnodar, Kubangebiet (U.d.S.S.R.).
Entomol. Labor. des Kuban. Landw. Inst. Dir Doc. Dr. V. O. Pickel.

Kurashiki near Okayama (Japan).
Entomol. Labor. Ôkaka Inst. of Agric. Research. Dir. Dr. Chûkichi Harukawa. — 1 Ass.

Kyôto (Japan).
Entomol. Labor. Kyôto Imper. Sericicultural Coll., Hanazono near Kyôto. Dir. Prof. Seiji Itô.
Entomol. Labor., Coll. of Agric., Kyôto Imper. Univ. Dir. Prof. Dr. Hachirô Yuasa. — Ass. Prof. Dr. Nobumasa Yagi (Physiol.); Ass. Yasuji Yamada (Systematic).

Lafayette, Ind. (U.S.A.).
Dept. of Entomol. of the School of Agric., Purdue Univ. J. J. Davis, B.S., in charge; C. R. Cleveland, B.A., Ass. Entomol.; C. O. Dirks, M.S., Instr. Entomol.; G. C. Oderkirk, B.S., Rodent Control Work (Detailed by U.S.D.A.); W. A. Price, M.S., Assoc. Prof. Entomol.; James Troop, M.S., Prof. Entomol.

East Lansing, Mich. (U.S.A.).
Entomol. Labor. of Michigan State Coll. Dir. R. H. Pettit, Head of Department. — E. I. McDaniel (Assoc. Prof. Entomol.); C. G. Gentner; Instructors: Don Reis, C. B. Dibble, K. Arbuthnot; R. H. Kelty (Ass. Prof. Agric.). ⓟ Insects.

Lawrence, Kan. (U.S.A.).
Dept. of Entomol., Univ. of Kansas. Dir. Prof. H. B. Hungerford. — Prof. P. B. Lawson; Ass. Prof. P. A. Readio; Ass. Prof. R. J. Beamer.

Leningrad (U.d.S.S.R.).
Bureau of Applied Entomol. of State Inst. of Experim. Agronomy Ges. Inst., Ul. Hertzena 44. Dir. Prof. Vladimir Petrovitsh Pospelov. — Learned specialists: J. I. Baeckmann (Mus. and Bureau of Determinations, Coleoptera), A. J. Dobrodeev (Forest pests), J. N. Filipjev (Statistics), A. N. Kiritshenko (Hemiptera), N. F. Meyer (Parasitic Insects), A. K. Mordvilko (Aphidodea), N. N. Troitzkij (Experim. Station); Ass.: S. A. Predtetshenskij (Orthoptera), A.N.Reichardt (Coleoptera), M. A. Rjabov (Lepidoptera), A. A. Stackelberg (Diptera); Laborants: M. T. Aristov (Garden Rests), A. E. Skriptshinskij (Chalcididae); Ass.: E. A. Krejler, K. I. Vodinskaja, M. J. Konstantinova, T. G. Kriger-Vojnovskaja (Parasites). *5—6 places for research of all kind in applied Entomol., systematic and experim.* ⓟ Insects, especially economic ones, their transformation stages, and damaged objects. — Experim. Station at Detskoje Selo Leader: N. N. Troitzky.
Entomol. Kabinett d. Staatl. Univ., Wassily Ostrow. Dir. Prof. Dr. M. Rimsky-Korsakow. — S. J. Malyshew (Doc. f. nützliche Insekten); J. N. Filipjev (Doc. f. schädliche Insekten). *1 Platz für Biol. d. Insekten.* ⓟ Beschädigungen d. Pflanzen d. Insekten aus Rußland; Insekten aus d. Umgegend von Leningrad.

Lexington, Ky. (U.S.A.).
Dept. of Entomol. and Botan., Coll. of Agric., Univ.of Kentucky. Dir.: HarrisonGarman,D.Sc.,Head of Dept., Prof. Entomol. — Mary LeG. Didlake, M.S., Ass. Entomol. and Botan.; Carrie L. Hathaway, Seed Analyst; Marie Jackson, Seed Analyst; H. H. Jewett, M.A., Research Ass. in Entomol. and Botan.

Lincoln, Neb. (U.S.A.).
Dept. of Entomol., Coll. of Agric. of the Univ. of Nebraska. M. H. Swenk, A.M., Chair. of Dept. (State Entomol.); Raymond Roberts, B.S., Instr. Entomol.; D. B. Whelan, M.S., Ass. Prof. and Ass. Entomol.; Prof. L. Bruner.

Liverpool (England).
Labor. of Entomol., Fac. of Sc. of the Univ.

London (England).
Dept. of Entomol., British Mus. of Nat. History, Cromwell Road, S.W. Keeper: C. J. Gohan. — Deputy Keeper: E. E. Austen; 4 Ass. Keepers; 5 Ass.
Dept. of Entomol., London School of Hygien. and Trop. Med. of the Univ., Endsleigh Gardens, N.W. Dir. E. A. Buxton.
Imperial Bureau of Entomol., S.W. 7, Cromwell Road. Dir. Dr. G. A. K. Marshall. — Ass. Dir.: S. A. Neave; Chairman: The Earl Buxton.

Los Angeles, Cal. (U.S.A.).
Entomol. Station of the California Cyanide Co. U. I. Safio (Chief); A. F. Swain.

Los Baños (Philippine Islands).
Dept. of Entomol., Coll. of Agric. Dir. L. B Nichanco.

Lourenço Marques (Port. East Africa).
Div. of Entomol., Dept. of Agric. Dir. Christian B. Hardenberg.

Madison, Wis. (U.S.A.).
Dept. of Economic Entomol., Coll. of Agric., Univ. of Wisconsin, 1532 University Ave. Dir. Prof. H. F. Wilson. — C. L. Fluke, M.S., Ass. Prof. and Ass. Econ. Entomol.; A. A. Granovsky, M.S., Instr. and Ass. Econ. Entomol.; G. E. Marvin, B.S., Instr. Econ. Entomol. *Arrangements made upon application. Apicultural Investigations and Insects as carriers of plant disease organisms.* — Collection of beekeeping literature.

Madrid (España).
Labor. d'Entomol. du Mus. nat. d'Hist. nat., Palacio del Hippodromo. M. Martinez de la Escalera (Prof. agrégé).

Malang (Ned.-O.-Indië).
Labor. for the Study of the Beetleborer of Coffee Berries, Experim. Station building. Dr. H. Begemann (Entomol.).

Manhattan, Kan. (U.S.A.).
Dept. of Entomol. and Zool. Kansas State Agric. Coll. President F. D. Fanell; Prof. Geo A. Dean, Head Dept. of Entomol.; Dr. R. K. Nabours, Head Dept. of Zool. — Dr. Roger C. Smith, Prof. of Entomol.; Prof. J. W. McColloch, Prof. of Entomol.; Dr. R. L. Parker, Assoc. Prof. of Entomol.; Dr. R. N. Pamter, Ass. Prof. of Entomol.; Mr. H. R. Bryson, Ass.; Dr. J. A. Ackert, Prof. of Zool.; Dr. Mary T. Hannon, Prof. of Zool.; Dr. Geo J. Johnson, Assoc. Prof. of Zool.; Dr. Minnie Jewel, Ass. Prof. of Zool.; Miss Naomi Zimmerman, Instr. *Perhaps a dozen places in the two departments.* ⓟ Insects, especially Nemoptera, Coleoptera, Diptera and Orthoptera, Lizards, snakes, fish, birds and parasites.

Melrose Highlands, Mass. (U.S.A.).
Bureau of Entomol. U.S. Dept. of Agric. P. B. Dowden; C. F. W. Muesebeck; C. W. Collins.

Menton (France).
Insectarium de Menton (Station de Zool. agric.), Avenue Cernuschi. Dir. Raymond Poutiers. — Paul Genleys; Lucien Molinari. *2 places pour Biol. entomol. ou Recherches entomol. appliquée.* ⓟ Insectes, plus spécialement insects nuisibles et utiles à l'agric.

Mexico, D.F. (Mexico).
Entomol. Sekt. d. Dir. de Estudios Biol., 7/a calle de Balderas 94. Dir. Dr. Alejandro Ruelas.

Minsk (U.d.S.S.R.).
Entomol. Labor. Inst. f. Landw. Prof. E. V. Jazentkowski.

Monroe, Mich. (U.S.A.).
Corn Borer Station, Drawer 359. P. Luginbill (Assoc. Entomol.).

Montpellier (France).
Station Entomol. Dir. Prof. J. Lichtenstein.

Morgantown, W.Va. (U.S.A.).
Dept. of Entomol. Coll. of Agric. L. M. Peairs, Ph.D., Head of Dept.; W. E. Rumsey, B.S., Entomol. (State Entomol.).

Morioka (Japan).
Entomol. Labor. Morioka Imperial Coll. of Agric. and Forestry. Dir. Prof. N. Ishimori.

Moscow, Id. (U.S.A.).
Dept. of Entomol. Coll. of Agric. W. E. Crouch, B.S., Rodent Control Specialist (P. O., Boise); R. W. Haegele, A.B., Ass. Entomol. (P. O., Twin

Falls); Claude Wakeland, M.S., Entomol. (P. O.,
Parma); F. E. Whitehead, M.S., Ass. Prof. Entomol.

Moskau (U.d.S.S.R.).
Labor. der angewandten Zool. Landwirtschaftl. Akademie. Dir. V. F. Boldyrev.

Mound, La. (U.S.A.).
U.S. Bureau of Entomol. G. H. Bradley; W. V. King.

München (Deutschland).
Inst. für angewandte Zool., Amalienstr. 52. Dir. Prof. Dr. K. Escherich. — Dr. H. Eidmann, Ass. *3 Plätze für Biol. der Insekten.* ℗ Schädliche Insekten und ihre Fraßstücke.

Nanking (China).
Bureau of Entomol. (Kiangsu Prov.), Coll. of Agric. South Eastern Univ. Dir. Y. H. Tsou. — H. S. Chang.

Napoli (Italia).
Labor. relativi e delle annesse Stazione di microbiol. Industriale e Staz. Agric. Antimalarico. Dir. G. M. M. Rossi.

Naumburg a. d. S. (Deutschland).
Zweigstelle der Biol. Reichsanstalt für Land- und Forstwirtschaft, Weißenfelser Str. 57. Leiter: Oberreg.R. Dr. Carl Börner. — Reg.R. Dr. R. Seeliger (Rebenzüchtung); Reg.R. Dr. H. Thiem (Reblausbekämpfung); Dr. Fr. Schilder, Ass.; Dr. Jancke (Obstbauschädlinge); Dr. G. Lorbeer, Ass. *2—3 Plätze für angewandt-entomol. Arbeiten, insbesondere der Wein- und Obstbauschädlinge (spezieller Reblaus, Blutlaus, Aphiden, Cocciden, biol. Mittelprüfung), für reben- und obstzüchterische Arbeiten.* ℗ Phytophthires, Apterygoten, Otiorrhynchen, Coccinelliden, parasitische Mikrohymenopteren; lebende Samen und Vermehrungsholz von Vitis und Malus. — Je 1 Prüfstelle bei Ingelfingen (Württemberg) und bei Iphofen (Franken) zur Untersuchung von Rebsorten auf Reblausbefall.

Nelson (New Zealand).
Cawthron Inst. Dir. Prof. F. H. Easterfield, M.A., Ph.D. (Chemist). — 2 Departments, (1.) Agric. Chem. and (2.) Biol.; under Biol. the staff is: A. Division of Entomol.: Chief Biol., Field Entomol.: Dr. R. J. Tillyard; first and second Ass. Entomol.: A. Philpott, B. Division of Mycol.: Mycol.: Dr. Kathleen, Ass. Mycol.: M. Curtis. *2 places for Entomol.* ℗ New Zealand Insects and Fungi.

Neustadt a. d. H. (Deutschland).
Zool. Abt. der Staatl. Lehr- u. Versuchsanst. für Wein- u. Obstbau. Dir. Prof. Dr. Fritz Stellwaag. *Weinbauschädlinge, Obstbauschädlinge.*

Newark, Del. (U.S.A.).
Dept. of Entomol. School of Agric. H. L. Dozier, Ph.D. (Entomol.).

New Brunswick, N.J. (U.S.A.).
Dept. of Entomol. State Agric. Experim. Station. T. J. Headlee, Ph.D., Entomol. (State Entomol.); C. S. Beckwith, M.S., Assoc. Entomol.; J. M. Ginsburg, Ph.D., Biochem. in Entomol. Invest.; C. C. Hamilton, M.S., Assoc. Entomol.; Ray Hutson, B.S. Ass. Entomol. in Apicult.; Carl Ilg, Labor. Ass.; F. W. Miller, Ass. Entomol.; L. A. Stearns, M.S., Ass. Entomol.; A. E. Merke, Ass. Entomol.

New Haven, Conn. (U.S.A.).
Dept. of Entomol. State Agric. Experim. Station. W. E. Britton. Ph.D., Entomol. in charge (State Entomol.); J. T..Ashworth, Deputy in charge Gypsy and Brown-tail Moth Work; R. C. Botsford, Deputy in charge Mosquito Elimination; R. B. Friend, B.S., Ass. Entomol.; Philip Garman, Ph.D., Ass. Entomol.; B. H..Walden, B.Agric., Ass. Entomol.; M. P. Zappe, B.S., Ass. Entomol.

Okitsu near Shizuoka (Japan).
Entomol. Labor. Okitsu Branch of Imper. Agric. Experim. Sta.

Omsk (U.d.S.S.R.).
Labor. f. Allgem. u. Angew. Forst-Entomol. d. Sibirischen Inst. f. Land- u. Forstwirtschaft. Dir. Prof. G. S. Sudejkin.

Orono, Me. (U.S.A.).
Dept. of Entomol. Coll. of Agric. Edith M. Patch, Ph.D., Entomol.; Alice W. Averill, Labor. Ass. Entomol.

Osijek I (S.H.S.).
Entomol. Fitopath. Abt. Dir. Kovačević F. Željko.

Ottawa, Ont. (Canada).
Labor. of Entomol., Dept. of Agric. Dir. A. Gibson. — J. M. Swaine; L. S. MacLaine; J. D. McDunnough; H. G. Crawford; C. H. Curran.

Oxford (England).
Dept. Forest Zool. Dir. R. N. Chrystal.

Paris (France).
Service des Epiphyties, VII, 42bis Rue de Bourgogne. 1. Stations Entomol.: Paris, Meudon, Saint-Genis-Laval, Montpellier, Bordeaux, Rouen, Chalette-Montargis. 2. Stations de Pathol. végétale: Paris, Antibes, Brive, Grignon, Villen Ave d'Ornon.
Stat. Entomol. de l'Inst. des Recherches Agron., 16, rue Claude Bernard. Dir. Prof. Dr. Paul Marchal. — Paul Vayssière; Bernard Trouvelos; Fernand Willaume.

Perugia (Italia).
Entomol. Inst. d. R. Inst. sup. Agrario. Dir. Prof. Dr. Carlo Fuschini.

Philadelphia, Pa. (U.S.A.).
The Philadelphia Mus., Commercial Mus., 34th Street below Spruce. Dir. W. P. Wilson. — Wilfred H. Schoff, Secretary. *Entomol. Samml. d. American Entomol. Society.*

Port-au-Prince (Haiti)
Dept. of Entomol. Technical Service, Dept. of Agric. Dir. George N. Wolcott. *Economic. Entomol. Ecol. Insects of Sugar-Cane.*

Pretoria (South Africa).
Dept. of Entomol., Univ. Dir. Prof. J. C. Faure.

Pulawy (Polska).
Abt. f. Entomol. des Staatl. wissensch. Inst. f. Landwirtsch. Dir. Prof. S. Minkiewicz.

Rabat (Maroc).
Division d'agric. entomol. Inst. sc. chér. Dir. Prof. I. Granjon de Lépiney.

Raleigh, N.C. (U.S.A.).
Dept. of Zool. and Entomol. State Coll. of Agric. Dir. Z.P. Metcalf, D.Sc., Head of Dept. — C.H.Brannon, B.S., Ass. Entomol. Specialist; C. S. Brimley, Ass. Entomol.; J. C. Crawford, M.S., Ass. Entomol.; Elizabeth Haban, Labor. Ass.; J. A. Harris, B.S., Ass. Entomol.; R. W. Leiby, Ph.D., Entomol.; W. B. Mabee, B.S., Entomol. Specialist; F. B. Meacham, M.S., Instr. Apicult.; T. B. Mitchell, M.S., Ass. Prof. Zool. and Entomol.; C. L. Sams, Beekeeping Specialist; L. H. Snyder, M.S., Assoc. Prof. Zool.

Rennes (France).
Station entomol. de Bretagne. Dir. Prof. Dr. L. Bordas.

Riga (Latvija).
Entomol. Kabinett Univ., Kronvalda bulvarī No 1. Leiter: Laimons Gailīts. — Edgārs Ozols, Sub-Ass. ℗ Insekten.

Rio de Janeiro (Brasil).
Inst. Biol. de Defesa Agric. Dir. Carlos Moreira. — Eugenio dos Santos Rangel (eng. Agric.); Angelo Moreira da Costa Lima (Med.); Luiz Augusto de Azevedo Marques; Agesislau Antonio Bittencourt (eng. Agric.); 1 Chef de Entomol. Agric.; 1 Ass. de Entomol. Agric.; 1 Chef de Phytopath.; 1 Ass. de Phytopath.; 1 Chef de Vigilance Sanitaire Végétale; 1 Ass. de Vigilance Sanitaire Végétale. *Entomol., Agric., Phytopath.* Ⓟ Insectes et Fungi.

Riverton, N. J. (U.S.A.).
Japanese Beetle Labor. Dir. L. B. Smith. — Dr. P. A. Van-der-Meulen (Chem.); Prof. O. G. Anderson (Horticulture); H. W. Allen (Entomol.). *Parasit.; Soil Insecticides; Toxicol.; Ecol.; Chaemotoxy; Biol.; Insecticides; Physiol.* Ⓟ Insects: Diptera, Coleoptera and Hymenoptera. Especially Tachinidae, Scuotaeidae and Scoliids.
Insect. Stat. of the U.S. Dept. of Agric. J. L. King (Entomol.). *Insect Parasit.*

Rosenthal b. Breslau (Deutschland).
Fliegende Station der Biol. Reichsanstalt für Land- und Forstwirtschaft, Zuckerfabrik. Leiter: Dr. O. Kaufmann. *Lebensgeschichte und Bekämpfung der Rübenfliege.*

Rostow am Don (U.d.S.S.R.)
Zool. Abt. d. Nordkaukasischen Pflanzenschutzstation. Dir. P. M. Swiridenko.

Rouen (France).
Station entomol. du Nord Ouest, 16, rue Dufay. Dir. Robert Regnier. — Jacques Vergouis; Roger Pussard; Alexandre Fillotre. *2 Places pour Entomol. agric. Apiculture.* Ⓟ Insectes. Cécidies.

Sacramento, Cal. (U.S.A.).
Entomol. Labor. of the California Spray Chem. Co., Box 1184. A. C. Browne (Economic Entomol., Plant Pathol.).
U.S. Bureau of Entomol., 600, 26th Street. W. B. Cartwright (Entomol.).

Saint-Genis-Laval, Rhône (France).
Station entomol. du Sud-Est. Dir. A. Paillot. *1 place pour recherches de biol. (Insectes) ou de parasit. agric.* Ⓟ Collections biol. d'Insectes nuisibles à l'Agric.

Saint Paul, Minn. (U.S.A.).
Division of Entomol. and Economic Zool., Univ. Farm. Dir. Dr.R.N.Chapman. — S. A. Graham, Ph. D., Forest Entomol.; M.S.Johnson, Ph.D., Economic Zool.; A. H. MacAndrews, M.S., Ass. Forest Entomol.; G. A. Mail, B.S., Ass. Entomol.; C. E. Mickel, Ph.D., Systematic and Extension Entomol.; T. A. Olson, A.B., Instr. in Entomol.; W. A. Riley, Ph.D., Parasit.; Wm. Robinson, Ph.D., Ass. Entomol.; A. G. Ruggles, M.A., State Entomol.; A. L Strand, M.S., Research Ass. in Insecricides; H. L. Sweetman, M.S., Ass. Entomol. *Research in ecol. and especially on the relationship of temperature and humidity to the biol. of insects; insecticides; insect taxonomy; forest entomol.; parasit. and general economic entomol. — Equipment for maintaining constant temperature and humidity, large insect collection for taxonomic studies.* Ⓟ All orders of insects and parasit. specimens.

Ste. Anne de la Pocatiere, Que. (Canada).
Dept. of Entomol. of the Agric. School. Dir. M. G. Bouchard.

Sapporo (Japan).
Entomol. Labor. Governm. Agric. Experim. Stat., Kotoni-mura near Sapporo. Satoru Kuwayama (Lepidopt.).
Entomol. Inst. Coll. of Agric., Hokkaidô Imper. Univ. Dir. Prof. Dr. Shônen Matsumura.

Sandusky, Ohio (U.S.A.).
Corn Borer Stat. of the U.S. Bureau of Entomol. L. H. Patch (Ass. Entomol.). *European Corn Borer.*

Sanford, Fla. (U.S.A.).
Entomol. Station of the State Plant Board of Florida. E. D. Ball; J. A. Reeves.

San Francisco, Cal. (U.S.A.).
Dept. of Entomol. Cal. Acad. of Sc. Dir. E. P. Van Duzee (Curator).
Mus. of Entomol., California Acad. of Sc. Dir. Dr. B. W. Evermann.

Saratow (U.d.S.S.R.).
Entomol. Abt. Inst. f. Landwirtschaft u. Meliorat. Dir. Doc. N. L. Sacharov.

Saskatoon, Sask. (Canada).
Dominion Entomol. Labor. K. M. King (Entomol.).

Shinkwa, Formosa (Japan).
Entomol. Labor., Branch Experim. Stat. for Sugar-Industry, Government Central Research Inst., Tainan-shû. Dir. Hidezô Takano.

Sofia (Bulgarien).
Entomol. Station S. M. des Königs v. Bulgarien, Buleward Ewlogi Georgieff No 60. Dir. Pentscho Drensky. *2 Arbeitsplätze für Erforschung der entomol. Fauna Bulgariens. — Sammlungen über entomol. Fauna Bulgariens.* Ⓟ Insekten aus Bulgarien.
Abt. f. angew. Entomol. Zentralinst. f. landwirtsch. Versuchswesen, Solunplatz 3. Dir. Dr. P. Čorbadev.

Sopron — Ödenburg (Ungarn).
Labor. für forstliche Entomol. Dir. Arthur Kelle. — Ass. Josef Tóth.

Stellenbosch (South Africa).
Dept. of Entomol. Univ. Dir. Prof. C. K. Brain, M.A., D.Sc.

Stillwater, Okla. (U.S.A.).
Oklahoma Agric. Experim. Stat. Dir. Prof. C. E. Sanborn. — Prof. Williamson J. Brown, Ass. Entomol.; Prof. Gustav A. Bieberdorf, Ass. Entomol. *Sufficient for 2 places. Systematic and biol. work. — Microscopes, collecting apparatus, photograph apparatus, insect breeding cages, imbedding and sectioning apparatus.* Ⓟ Insects, especially Aphididae and Coleoptera.

Stockholm (Sverige).
Centralanstalten för försöksväsendet på jordbruksområdetê: Lantbruksentomol. avdelingen, Experimentalfältet. Dir. Prof. Albert Tullgren. — Laborator: Dr. N. A. Kemner; Ass.: Dr. O. Lundblad, Cand. phil. O. Ahlberg, Agronom A. Lindblom. Ⓟ Tierische Schädlinge, schwedische Insekten.
Entomol. Abt. Naturhist. Reichsmus. Dir. Prof. Dr. B. Y. Sjöstedt. — Ass. Dr. A. Roman.
Inst. f. Forstentomol. Forsthochsch. Dir. Doc. I. Trägårdh.

Stoke on Trent (England).
Entomol. Sekt. of the North-Staffordshire Field Club.

Stuttgart (Deutschland).
Entomol. Abt. der Württembergischen Naturaliensammlung. Dir. Dr. Erwin Lindner.

Sverdlowsk (U.d.S.S.R.).
Entomol. Inst. Ural. Polytechn. Inst. Dir. Doc. Kolosov.

Taihoku, Formosa (Japan).
Entomol. Labor. Dept. of Agric., Government Central Research Inst. Dir. Dr. T. Shiraki. — Ryôichi Takahashi.
Entomol. Labor., Taihoku Governmental Coll. of Agric. and Forestry. Dir. Prof. Shûchi Isshiki.

Tallulah, La. (U.S.A.).
Delta Labor., U.S. Dept. of Agric., Bureau of Entomol. K. P. Ewing (Entomol.). *New cotton insect pest, the cotton flea hopper (Psallus seriatus).*

Tartu — Dorpat (Eesti).
Kab. für angew. Zool., Aia tän 46. Dir. Prof. J. Mägi. — Ass. K. Zolk; A. Määr.

Taschkent (U.d.S.S.R.).
Uzbekistan Stat. of Plant Protection, Orchard's Division. Dir. W. P. Nevskij. *System. of Aphididae and Elateridae. Applied Entomol.*

Tharandt (Deutschland).
Zool. Inst. der Forstl. Hochsch. Dir. Prof. Dr. H. Prell. — William Baer, etatmäßiger Ass.; Dr. phil. A. Prell, Ass. für Bienenkunde; Dr. phil. W. Stichel, Ass. für Pelztierkunde; Dr. phil. H. Hrabowski, Ass. für die Hauptstelle des Forstl. Pflanzenschutzes. *Entomol. Untersuchungen, bes. Determinationen.* ℗ Sämtliche Fraßbeschädigungen durch Tiere an einheimischen Holzgewächsen. — Forschungs-Inst. für Bienenkunde. Forschungs-Inst. für Pelztierkunde. Hauptstelle des Forstl. Pflanzenschutzes.

Tiflis (U.d.S.S.R.).
Entomol. Kabinett. Univ. Dir. Lect. Ph. Zajcev.

Tôkyô (Japan).
Anat. (of Insect) Labor., Imperial Sericicultural Experim. Station, Suginamichô near Tôkyô.
Entomol. Labor. Imper. Sericicult. Experim. St., Suginamichô near Tôkyô. Dir. T. Yokoyama.
Entomol. Inst. Nôgyô Daigaku (Agric. Univ.), Komazawa near Tôkyô. Lect. Dr. S. Ishiwata.
Entomol. Labor. Imper. Agric. Experim. Sta., Nishigahara. Dir. Shûta Kinoshita. (Thysanura); — Ass. S. Shinkai.
Entomol. Inst. Fac. of Agric. Tôkyô Imp. Univ., Komaba near Tôkyô. Ass. Prof. Nobukatsu Marumo; Ass. Prof. Sôkan Yano; Ass. Prof. Jirô Machida.
Entomol. Labor. Imperial Forest Experim. St., Meguro near Tôkyô. Dir. Dr. Sôkan Yano.
Parasit. Labor. Imper. Inst. f. Insect. Dis.

Tottori (Japan).
Entomol. Labor. Tottori Imper. Coll. of Agric.

Urbana, Ill. (U.S.A.).
Crop Protection Inst., State Entomology Bldg. L. L. English.
Dept. of Entomol. Univ. of Illinois.

Utsunomiya (Japan).
Entomol. Labor. Utsunomiya Imper. Coll. of Agric. and Forestry.

Uvalde, Tex. (U.S.A.).
Bureau of Entomol., U.S. Dept. of Agric. D. C. Parman (Ass. Entomologist).

Vernon (British-Columbia).
Dominion Entomol. Branch, Court House. E. R. Buckell.
Entomol. Labor., Dir. Max Hermann Ruhmann. *Entomol. Research.* ℗ *Entomol. specimens.*

Victoria (British-Columbia).
Dominion Entomol. Labor. W. Downes.

Vienna, Va. (U.S.A.).
I. and F. Board U.S. Dept. of Agric. W.M.Davidson (Entomologist).

Vincennes, Ind. (U.S.A.).
Bureau of Entomol. U.S.D.A. A. Porter.

Wageningen (Holland).
Labor. f. Entomol. des Inst. f. Phytopath., Landbouwhoogeschool. Dir. Prof. D. W. Roepke. — C.A. L. Smits van Burgst, Betrem.

Warszawa (Polska).
Entomol. Labor. a. d. Hochsch. f. Land- u. Forstwirtsch. Dir. Prof. Z. Mokrzecki.

Washington, D. (U.S.A.).
Bureau of Entomol. U.S. Dept. of Agric, Dir. L. O. Howard. — R. T. Cotton (Entomologist); M. Colcord (Librarian); J. A. Hyslop (Insect pest survey); N. E. McIndoo (Entomologist); E. W. Laake (Assoc. Entomologist); J. W. Mason (Assoc. Entomologist); S. A. Rohwer (Entomologist); V. H. Chittenden (Entomologist); F. H. Lathrop (Entomologist); M. C. Lane (Ass. Entomologist); J. S. Wade (Assoc. Entomologist); H. S. Swingle (Ass. Entomologist); C. A. Weigel (Ass. Entomologist); J. W. Bulger (Insect Physiol.).
Cereal and Forage Insect Investigations U.S. Dept. of Agric. W. H. Larrimer (Senior Entomologist).
Division of Forest Insect Investigations Bureau of Entomology, U.S. Dept. of Agric. Ass. R. A. St. George (Entomologist).
Insecticide and Fungicide Board, U.S. Dept. of Agric. Dir. J. K. Haywood, Chairman.

Webster Groves, Mo. (U.S.A.).
U.S. Entomol. Labor. A. F. Satterthwait (Ass. Entomologist).

West Lafayette, Ind. (U.S.A.).
Cereal and Forage Insect Division, Bureau of Entomol., U.S. Dept. of Agric. W. B. Noble (Ass. Entomologist).

Wichita, Kan. (U.S.A.).
Cereal and Forage Insect Investigations Wichita Labor. J. R. Horten (Assoc. Entomologist); H. H. Walkden (Junior Entomologist).

Winnipeg, Man. (Canada).
Dept. of Entomol. Univ. of Manitoba. Dir. A. V. Mitchener, B.A., B.S.A.

Wisley, Ripley, Surrey (England).
Wellcome Entomol. Field Labor. (Wellcome Bureau of Sc. Research). Dir. Malcolm E. MacGregor. *3 or 4 working-places. Entomol. Research. — Apparatus for the field study of entomol. problems.* ℗ *Various insects, but principally mosquitoes at present.* — The Wellcome Field Entomol. Labor. is a special branch of the Wellcome Bureau of Sc. Research.

Wolfen, Kr. Bitterfeld (Deutschland).
Biol. Labor. der I. G. Farbenindustrie A. G. Wolfen, Kreis Bitterfeld. Dir. Dr. Werner Ext. *Schädlingsbekämpfung, Saatgutbeizung, Insektenphysiol. u. -toxiol.*

Wooster, Ohio (U.S.A.).
Dept. of Entomol. Ohio Agric. Experim. Station. Dir. J.S.Houser,M.S.,Chief.—C.R.Cutright,Ph.D., Ass. Entomol.; L. L. Huber, Ph.D., Ass. Entomol.; C. R. Neiswander, M.S., Ass. Entomol. (P. O., Oak Harbor); Herbert Osborn, D.Sc., Assoc. Entomol. (P. O., Columbus); Joseph Polivka, M.S., Ass. Entomol. (P. O., Oak Harbor).

Zagreb — Agram (S.H.S.).
Inst. für angewandte Zool., Kačićgasse 9. Dr. Prof. Dr. Erwin Rössler. — Prof. Josef Plančić, Ass. ℗ Ingluvialien von Vögeln, Plankton u. Bodenfauna.
— Teichwirtschaftliche Versuchstation Crua Uclaka.

Zürich (Schweiz).
Entomol. Inst. der Eidgenössischen Technischen Hochsch., Universitätsstr. 2. Leitung: Prof. Dr. O. Schneider-Orelli. — Dr. C. Schaeffer. *1—2 Plätze für Zuchtversuche mit Obstbaumschädlingen u. Forstinsekten. Systematische Lepidopteren-, Coleopteren- u. Orthopterenuntersuchungen. — Versuchsbienenstand. Topfbäume für Zuchtversuche. Treibhaus. Palaearktische Schmetterlings- u. Käfersammlungen.*

Laboratoria malariae.

Confer: *Bacteriologia hominis* (pag. 479), *Hygiena hominis* (pag. 487), *Parasitologia generalis* (pag. 379).

Burgas (Bulgarien).
Inst. f. d. Bekämpfung der Malaria. Dir. Dr. Konstantin Drensky. *8 Plätze für in- u. ausländische Forscher.*

Struga na Ohridskomjezern, Macedonija (S.H.S.).
Malariastation (Stanica za Malariju). Dir. Dr. Nikola Nezlobinski.

Sušak (S.H.S.).
Bakteriol. stanica. Dir. Dr. O. Trausmiler. — Dr. H. Emili. *2 Plätze. Malariastudien. Malariabekämpfung im Terrain* ℗ Malariapraep.

Woronesch (U.d.S.S.R.).
Malaria-Stat. Dir. Dr. Paul P. Muffel.

Laboratoria helminthologiae.

Confer: *Zoologia generalis* (pag. 356), *Bacteriologia* (pag. 479), *Parasitologia generalis* (pag. 379).

Baku, Aserbeijan (U.d.S.S.R.).
Helminthol. Labor. at the Inst. of Microbiol., Shemachinskaja 39. Dir. Prof. V. Elpatievsky.

Calcutta (Br.-India).
Hookworm Research Labor., School of Tropical Med. Rice Inst. A. C. Chandler (Biol.).

Honolulu (Hawaii).
Dept. of Nematol. of the Experim. Stat. of the Association of Hawaiian Pineapple Canners. Dir. Dr. G. H. Godfrey. — Miss H. T. Morita, Ass.

Houston, Tex. (U.S.A.).
Hookworm Research Labor., School of Tropical Med. Prof. A. C. Chandler (Biol.).

Jerusalem (Palestine).
Microbiol. Inst. Hebrew. Univ. G. Witenberg (Ass. in Helminthol.).

London (England).
Dept. of Helminthol., London School of Hygiene and Tropical Med. of the Univ., Endsleigh Gardens, N.W. Dir. Prof. R. T. Leiper.

Moskau (U.d.S.S.R.).
Helminthol. Abt., Inst. f. Tropenkrankh., Pogodinskaja 10. Dir. Prof. K. I. Skrjabin.

Stambul (Türkei).
Inst. f. pathol. Anat. u. Fleischbeschau Tierärztl. Hochsch. Confer: Pathol. anat. animal.

Institutiones ornithologiae.

Budapest (Ungarn).
Königlich Ungarisches Ornithol. Inst., II, Debroigasse 15. Dir. Titus Csörgey (Ökonomische Ornithol.) — Mitarbeiter: Jakob Schenk (Vogelzug). Volontäre Mitarbeiter: Koloman Warga; Ladislaus Szemere; Nikolaus Vasvári.

Burg Seebach, Kr. Langensalza (Deutschland).
Versuchs- und Musterstation für Vogelschutz. Dir. Dr. phil. h. c. Hans Freiherr von Berlepsch.

Cleveland, Ohio (U.S.A.).
Baldwin Bird Research Labor., S. 17, Williamson Building. Dir. S. P. Baldwin.

Cold Spring Harbor, Long Island, N. Y. (U.S.A.).
American Mus. Nat. History. W. R. Boulton (Ornithologist).

Gates Mills, Ohio (U.S.A.).
Baldwin Bird Research Labor.

Helgoland (Deutschland).
Vogelwarte der Biol. Anstalt. Confer: Labor. marina.

Langres, Haute-Marne (France).
Mus. des Sc. Nat. *Vogelsammlg.*

Liběchov — Liboch a. d. Elbe (C.S.R.).
Ornithol. Station des „Lotos". Leiter: Ing. Kurt Loos, Forstmeister. *Beringung d. Zugvögel.*

Mexico, D. F. (Mexico).
Sect. Ornithol. d. Dir. de Estudios Biol., 7/a calle de Balderas 94. Dir. Prof. V. Santiago.

Oberlin, Ohio (U.S.A.).
Baldwin Bird Research Labor. S. C. Kendeigh (Research Ass.). *Life Histories of Birds, Physiol. Processes in the living Bird.*

Ottawa, Ont. (Canada).
Ornithol. Dominion Observ. R. E. De Lury (Bird banding).

Pullman, Wash. (U.S.A.).
Charles R. Conner Mus. State Coll. of Washington. Dir. D. J. Leffingwell (Curator).

Rossitten (Deutschland).
Vogelwarte Rossitten, Kurische Nehrung Ostpreußen. Dir. Prof. Dr. J. Thienemann. — 1 Ass., 1 Hilfsarbeiterin, 1 Diener. *Die Vogelzugsbeobachtungen finden hauptsächlich in der zur Vogelwarte gehörigen Beobachtungshütte Ulmenhorst statt.* ℗ Lokalsammlung der Nehrungsornis.

San Francisco, Cal. (U.S.A.).
Mus. of Ornithol., California Acad. of Sc. Dir. Dr. B. W. Evermann.

Sempach (Schweiz).
Schweiz. Vogelwarte der Schweiz. Gesellschaft für Vogelkunde u. Vogelschutz. Besorger: Alfred Schifferli.

Tunghsien (China).
Dept. of Birds Lab. of Nat. Hist. Dir. G. D. Wilder.

Warmbrunn, Schlesien (Deutschland).
Ornithol. Sammlung mit Insekten-, Schmetterlings- u. ethnograph. Kabinett. Dir. Kustos G. Martini. — Konserv.: K. Martini.

Woronesch (U.d.S.S.R.).
Versuchsstat. f. Vogelzucht, Landwirtsch. Hochsch.

Laboratoria spelaeologiae.

Cluj — Klausenburg (România).
Inst. de Speol. Case postale No. 158. Dir. E. G. Racovitza. — R. Jeannel, Sous-Dir.; P. A. Chappuis, Adjoint à la Direction; V. Puscarin, Ass. *3 places de travail. Recherches d'histoire nat.*, *spécialement zool.* ⓟ Faune et Flore souterraines. Faune de Transylvanie.

Wien (Österreich).
Speleol. Inst., VIII, Auerspergstr. 1. Vorst. Prof. Dr. Georg Kyrle.

Laboratoria botanica.

Laboratoria botaniae generalis.

Confer: *Institutiones generalis* (pag. 336), *Hortus botanici* (pag. 408).

Aachen (Deutschland).
Botan. Inst. d. Techn. Hochsch. Dir. Prof. A. Wieler.

A. and M. College, Miss. (U.S.A.).
Dept. of Botan. Agric. Coll. J. M. Beal, M.S., Head of Dept.; J. C. McKee, M.S., Assoc. Prof. Botan.; D. C. Neal, M.S., Plant Pathol.

Aas b. Oslo (Norge).
Bot. Inst. Landw. Hochsch. (Norges landbrukshøiskoles Botan. Inst.), Dir. Prof. Dr. Henrik Printz. — Elias Mork, Forstkandidat, Ass. *5—10 Plätze für Pflanzenphysiol., Bodenbiol. und Algol. ausgerüstet.* —Zweiglabor. für Pflanzenkrankheiten unter Leitung von Doc. A. E. Traaen.

Aberystwyth, Wales (England).
Botan. Dept., Univ. Coll. of Wales. Dir. Prof. Wilfrid Robinson, D.Sc. — 1 Lect., 1 Ass. Lect. *2 or 3 places for Marine Algol., Mycol.* ⓟ Marine Algae. Fungi.

Alger (Algérie).
Labor. de Botan. générale et appliquée, Fac. des Sc., Univ. Dir. Prof. Dr. René Maire. — Charles Killian, Maître de conférences; Henri Humbert, Chef de travaux; Lucienne Gauthier-Lièvre, Ass. *3 places pour Mycol., Anat. et Cytol., Flore de l'Afrique du Nord.* ⓟ Plantes et Champignons de l'Afrique du Nord.

Allahabad (British-India).
Botan. Labor., Univ. Dr. J. H. Mitter, Reader; Mr. S. Ranjan, Reader; Mr. R. K. Saksena, Lect.; Mr. G. D. Srivastava, Mr. S. P. Naithani, Demonstrators; Mr. K. Prasad, Labor. Ass. ⓟ Fungi.

Amherst, Mass. (U.S.A.).
Dept. of Botan. State Agric. Coll. A. V. Osmun, M.S., Head of Dept.; A. S. Ball, Labor. Ass. in Botan.; O. L. Clark, B.S., Ass. Prof. Botan.; W. H. Davis, Ph.D., Ass. Prof. Botan.; W. L. Doran, M.S. Ass. Research Prof. Botan.; E. F. Guba, Ph.D., Ass. Research Prof. Botan. (P.O.), Waltham); F. A. McLaughlin, B.S., Ass. Prof. Botan.; Gladys I. Miner, Curator, Dept. of Botan.; R. E. Torrey, Ph.D., Ass. Prof. Botan.

Amsterdam (Holland).
Botan. Inst. (Abt. für Pflanzenanat. u. Systematik), Plantage Middenlaan 2. Dir. Prof. Dr. Theo J. Stomps. — Conserv. herbarii: Frl. B. Polak. Ass.: J. Heimans, Dr. H. Dulfer, W. van Dieren. *Zytol. und Erblichkeitslehre. — Quarzlampen.* — Pflanzenphysiol. Abt. unter Prof. Dr. Th. Weevers.

Angers (France).
Labor. de Botan. de l'Univ. Cathol. Dir. G. Bioret. *2 places pour Anat. et Systematique des Lichens. Systematique et écol. des Algues d'eau douce.* — ⓟ Lichens et Algues d'eau douce (plancton).

Ann Arbor, Mich. (U.S.A.).
Dept. of Botan. Univ. of Michigan. P.E. Dale (Botan.); S. H. Emerson (Botan.); Ass. Prof. C. D. La Rue (Botan.).

Athen (Griechenland).
Botan. Labor. d. landwirtschaftl. Hochsch. Dir. Prof. Dr. N. Montesantos. — Prof. Dr. P. Koutsomitopulos. Dr. S. Malakates. 2 Ass. *20 Plätze. Physiol. Apparate. Sammlungen v. Pflanzen.*
Labor. de Botan. Univ., Edifice de Chemie, rue Solon. Dir. Prof. Politis.

Auburn, Ala. (U.S.A.).
Dept. of Botan. Coll. of Agric. W. A. Gardner, Ph.D., Head Prof. Botan., Plant Physiol., Botan.; W. L. Blain, M.A., Assoc. Plant Pathol.; L. E. Miles. Ph.D., Plant Pathol.; Martin Palmer, B.S., Instr. and Ass. in Botan.

Baltimore, Md. (U.S.A.).
Dept. of Botan. School of Pharmacy, Univ. of Maryland. Prof. C. C. Plitt.

Bangor, Wales (England).
Dept. of Botan. Memorial Buildings Univ. Coll. of North Wales, Carnarvonshire. Dir. Prof. D. Thoday. — Miss A. J. Davey, M.Sc., Lect.; Mr. N. Woodhead, B.Sc., Demonstrator.

Barcelona (España).
Inst. d. botan. Univ. Dir. A. Vila Nadal.

Bari (Italia).
Gab. di Botan. Dir. V. Rivera.

Basel (Schweiz).
Botan. Anstalt Univ., Schönbeinstr. 6. Dir. Prof. G. Senn.

Baton Rouge, La. (U.S.A.).
Dept. of Botan. and Bact. Coll. of Agric. C. W. Edgerton, Ph D., Prof. Botan., Plant Pathol.; W. L. Owen, B.S., Research Bact.; E. V. Abbott, Ph.D., Ass. Plant Pathol.; W. N. Christopher, B.S., Instr. Bact.; C. F. Moreland, M.S., Assoc. Prof. Botan. and Bact.; E. C. Tims, Ph.D., Ass. Plant Pathol.

Belfast (Ireland).
Dept. of Botan. Queens Univ. Dir. Prof. J. Small.

Beograd (S.H.S.).
Botan. Labor., Studenicka 54. Dir. Dr. Pierre Georgevitch. — Ass.: Dr. Frank Loschnigg, Dr. Roko Vuković. *2 Plätze für cytol. und bakt. Unters.* ⓟ Einheimische Pflanzen.
Botan. Inst. Philosoph. Fac. Univ., Dalmatinska ul. 11. Dir. Prof. Dr. Nedeljko Košanin. — Dr. Ljubiša Glišić, Doc. für Botan.; Ass. Stevan Jakooljević (Anat. d. Pfl.). *Geobotan., Embryol. d. Pflanzen*.

Bergen (Norge).
Botan. Labor. des Bergens Mus. Dir. Prof. Dr. Oscar Hagem. *3—4 Arbeitsplätze für bodenbiol. Untersuchungen, Bact. und Mykol.*

Berkeley, Cal. (U.S.A.).
Dept. of Botan., Univ. of California. Chairman: T. H. Goodspeed. — W. A. Setchell, Prof. of Botan. (Cryptogamic Botan. and Geographic Distribution); W. L. Jepson, Prof. of Botan. (Phaenogamic Botan.); N. L. Gardner, Assoc. Prof. of Botan. and Curator of the Herbarium (Cryptogamic Botan.); T. H. Goodspeed, Assoc. Prof. of Botan. and Curator of the Botan. Garden (Cytol., especially of hybrid plants); R. M. Holman, Assoc. Prof. of Botan. (Plant Physiol.); Lee Bonar, Ass. Prof. of Botan. (Physiol. of the Fungi). *10 places for Taxonomic botan., cytol., physiol., genetics.* ⓟ *Flowering plants and marine algae of the Pacific area.* — Many divisions of the Coll. of Agric. of the Univ. of California Berkeley, are engaged in research in the plant sciences.
Dept. of Botan., Coll. of Agric. W. A. Setchell, Ph.D., Prof. Botan.; H. S. Reed, Ph.D., Prof. Plant Physiol. (P.O., Riverside); H. A. Borthwick, M.A., Ass. in Botan. (P.O., Davis); A. R. C. Haas, Ph.D., Ass. Prof. Plant Physiol. (P.O., Riverside); F. F. Halma, B.S., Research Assoc. in Plant Physiol. (P.O., Riverside); R. M. Holman, Ph.D., Ass. Prof. Botan.; W. W. Robbins, Ph.D., Assoc. Prof. Botan. (P.O., Davis).

Berlin (Deutschland).
Inst. für Botan. d. Landwirtschaftlichen Hochsch., N 4, Invalidenstr. 42. Dir. Prof. Dr. H. Miehe. — Ass.: Dr. G. Mäckel; Dipl.-Landwirt E. Egglhuber. *2 Plätze für mikroskopische, bakt. und pflanzenphysiol. Arbeit.* — *Mikroskop mit Quarzoptik für Beleuchtung mit ultraviolettem Licht. Großes Mikrotom für Holzschnitte.*
Biol. Reichsanst. für Land- u. Forstwirtschaft, — Dahlem, Königin-Luise-Str. 19. Labor. für Pflanzenanat.: Dir. Dr. W. v. Brehmer. Labor. für Phänol.: Dir. Reg.R. Prof. Dr. E. Werth; Ass. Reg.R. Dr. H. Pape; Ass. Dr. S. Wilke. Labor. für Botan.: Dir. OberReg.R. Prof. Dr. O. Appel.
Botan. Gruppe der biol. Abt. d. Preuß. Landesanst. für Wasser-, Boden- u. Lufthygiene, Dahlem, Ehrenbergstr. 38—40—42. Dir. Prof. Dr. R. Kolkwitz. — Mitglied: Dr. E. Tiegs (Rauchschäden, Lufthygienefragen). Mitarbeiter: Prof. Dr. R. Malguth (Abwasserpilze); R. W. Kolbe (Meeresflora, Diatomeen); Dr. H. Bethge (Plankton, Schlamm); Dr. W. Dörries (Wasserpilze); Dr. K. Gemeinhardt (Algen und Pilze); Dr. E. Huth (Algen und Bodenpilze).

Bern (Schweiz).
Botan. Inst. u. Garten der Univ. Dir. Prof. Dr. Eduard Fischer. — Konserv. der Herbarien: Prof. Dr. Walther Rytz. Ass.: Dr. Günther von Büren, Priv.Doc. *4—6 Plätze für Biol. der parasit. Pilze. Mikrotomarbeiten.* ⓟ *Parasit. Pilze.*

Besançon (France).
Botan. générale Univ. Dir. Prof. Parmentier. — Jacques Pottier, Maître de conf. adj. *4 places pour Anat. végétale.* — *Riches herbiers.*

Birmingham (England).
Botan. Labor. Univ. Dir. Prof. R. H. Yapp. — Dr. J. S. Elliott, Lect; W. Leach, Lect.; C. G. C. Chesters, Ass. Lect.; W. B. Grove, Honorary Curator of the Fungus Herbarium. Also 1 ,,Teaching Scholar", who holds office for 2 years. *Usually 2 or 3 places for physiol. ecol., mycol. and general work.*

Blacksburg, Va. (U.S.A.).
Dept. of Botan. and Plant Pathol. Dir. F. D. Fromme. — James Nodkin, Ass.; R. H. Hurt, Ass.; C. N. Priode, Ass.; A. B. Massey, Assoc.; H. S. Stahl, Assoc.; S. C. Wingard, Assoc.; F. J. Schneiderhau, Ass. *Botan., Plant Pathol., Mycol., Bact.* — Field Labor. for research in diseases of apple and peach: 1. at Winchester, Va.; 2. at Staunton, Va.

Bloemfontein (South Africa).
Dept. of Botan. Grey Univ. Coll. Dir. Prof. G. Potts.

Bologna (Italia).
Ist. di botan. Univ. Dir. F. Morini.

Bonn a. Rh. (Deutschland).
Botan. Inst. u. Botan. Garten, Univ., Poppelsdorfer Schloß. Dir. Prof. Dr. H. Fitting. — Dr. R. Bode (Wasserhaushalt d. Pflanze); Dr. P. Dahm (Ernährungsphysiol. d. Pflanze); Dr. H. Zycha (Permeabilitätsbeeinflussung durch Licht). *Einige Plätze für Pflanzenphysiol.*
Botan. Inst. der Landwirtsch. Hochsch., Poppelsdorf, Meckenheimer Allee 106. Dir. Prof. Dr. Max Koernicke. — Ass.: Priv.Doc. Dr. Wilhelm Riede, Dr. Erich Schneider, Dr. Hubert Iven. *3 Plätze für cytol. und pflanzenphysiol. Untersuchungen.* — *Ein pflanzenphysiol. Versuchshaus mit Abt. zum Untersuchen des Wurzelwachstums. Dunkelraum. Bakterienkulturraum. Zimmer mit konstanter Temperatur. Ein Doppelkulturhaus für Kohlensäure- u. Elektrokulturforschung. Ein Gewächshaus mit Kalt- und Warmabt.*

Bordeaux (France).
Labor. de botan. et m. méd. Fac. méd. Univ. Dir. Prof. Beille.
Inst. Botan. Fac. Sc. Univ., 20, Cours Pasteur. Dir. Prof. Sauvageau. — M. Gard, Chef des travaux pratiques; A. de Puymaly, Ass.

Bourg St. Pierre, Valais (Suisse).
Labor. et Jardin alpin de la Linnaea (Annexe de l'Inst. botan. de l'Univ. de Genève). Dir. Prof. Dr. Robert Chodat. — Dr. Fernand Chodat. *12 places for paying guests (120 fs. p. mois). Soil acidity (pH); atmometrie; écol., biol. florale. Réactions du sol.* — *Loupes à dissection; atmomètres; réactifs et appareils pour mesurer les réactions du sol; appareils pour l'écol.* ⓟ *Alpine plants.*

Bozeman, Mtn. (U.S.A.).
Dept. of Botan. and Bact., Coll. of Agric. Confer: Bact. agris.

Braunschweig (Deutschland).
Botan. Inst. u. Garten, Humboldtstr. 1. Dir. Prof. Dr. G. Gassner. — Dr. J. Esdorn, Ass.; Dr. O. Appel, Ass.; Apotheker H. Rabien, Ass. *2 Plätze für physiol. Untersuchungen.* — *Botan. Garten, Gewächshäuser, Versuchsfeld für Pflanzenbau und Pflanzenschutz.*

Bristol (England).
Botan. Dept. Univ. (Botan. Labor. and Univ. Gardens). Dir. Prof. O. V. Darbishire. — 1 Prof., 2 Lect., 1 Ass. Lect. *Not limited working-places.* ⓟ *A seed exchange list is published every year and can be had on application, in February.*

Brno — Brünn (C.S.R.).
Botan. Inst. Univ., Kounicova 63. Dir. Prof. Dr. Jos. Podpěra. — Ass.: R. N. Dr. Jindrich Suza und R. N. C. Otto Mrkos. Demonstrator: R. N. C. Ludmila Pásková. Doc.: Gymn. Prof. Dr. František Nábělek. Wissensch. Mitarbeiter: Mag. d. Botan. Ivan Širjaev Grig (aus Charkow in Rußland). *2 Plätze für systematische, morphol. und geobotan. Arbeiten.* ⓟ *Botan. Objekte, Herbarien. Exsikkatenwerk: Plantae rei publicae bohemo-slovenicae exsiccatae.*
Inst. de botan. Ecole polytechnique Tcheque, Veveři ul. Čes. technika. Dir. Prof. Dr. Jan. Macků.
Labor. de botan., l'Inst. des Recherches agric. et Forestière, Květná ul. Dir. Dr. Ing. Ladislav Smolik. — Ing. Jan Appl, Ass.

Botan. Inst. Hochsch. für Bodenkultur (Landw. Hochsch.). Dir. Prof. Dr. O. Vodrážka. — Ing. R. Ille Ass.; Ing. C. V. Zajtschek, Demonstrator. *1 Arbeitsplatz für Anat. der Pflanzen, Mikrophotographie, kulturelle Versuche mit Pflanzen, Untersuchungen im polar. Licht. — Mikroskopie im polarisierten Lichte und Kinematographie.* ⓟ Praep. von Holz, Samen, Gräser, Praep. von pflanzlichen Parasiten des Holzes, Praep. der Textilpflanzen, Praep. der pflanzlichen Objekte im polar. Licht. — Biol. Station in Eisgrub (Lednice, C.S.R.) für Biol. der Wasserpflanzen; Dir. Prof. E. Bayer (Zool.). Labor. im bot. Garten der Hochsch. für Bodenkultur in Brno (C.S.R.) Dir. Prof. Dr. Otto Vodrážka (für Versuchszwecke).

Inst. botan. de l'Ecole Vétérinaire, Pražská 69. Dir. Prof. Dr. Rudolf Dostál. — Méd. Vét. Dr. Vl. Kubeš; Méd. Vét. Cand.V.Slezák; Hortus Inspector J. Hlavatý. ⓟ *Les plantes méd. et vénéneuses.*

Lehrkanzel für Botan., Warenkunde, techn. Mikroskopie u. Mykol. der Deutsch. techn. Hochsch., Komenckého nám. Dir. Prof. Dr. Ph. Osw. Richter.

Brookings, S.D. (U.S.A.).

Dept. of Botan. and Plant Pathol. State Coll. of Agric. A. T. Evans, Ph.D., Prof. Botan., Plant Pathol.; W. G. Houk, B.S., Ass. in Botan.; L. M. Pultz, B.S., Ass. in Botan.

Bruxelles — Brussel (Belgique).

Labor. de Morphol. et de Botan. systématique Univ. (Inst. Léo Essera), rue Botanique 40. Dir. Prof. Dr. Ch. Bommer. — Ass. P. V. D. Ledoux.

București (România).

Botan. Labor. Polytechn. Schule. Dir. Prof. Jacobescu.

Budapest (Ungarn).

Inst. für systematische Botan. und Pflanzengeographie der Univ., VIII, Muzeumkörut. 4. Dir. Prof. Dr. Johann von Tuzson. — Dr. P. Palik, Adjunkt; Dr. Baron G. Andreánszky, Ass.; Dr. J. Bernatsky; Dr. E. Gombocz. *4 Plätze für systematiseh botan. und pflanzengeographische Arbeiten. — Generalherbarium bestehend aus ca. 200000 Blättern von Blütenpflanzen und 50000 Blättern Kryptogamen; Herbarium Borbásianum, bestehend aus ca. 120000 Blättern von Blütenpflanzen; Warnstorfs Moosherbarium; phytopalaeont. Sammlung, enthaltend ca. 4000 Stück fossile Pflanzenreste, hauptsächlich aus dem Tertiär Ungarns.* ⓟ Herbarexemplare von Cryptogamen; Blütenpflanzen; sowie fossile Pflanzenreste, besonders aus dem Tertiär.

Allgemeines Botan. Inst. (Inst. für Pflanzenmorphol. und Physiol.), VIII, Museumring 4. Dir. Dr. Sándor Màgocsy-Dietz, Univ.-Prof. — Priv.Doc.: Dr. Jenö Bernátsky (Anat.), Dr. Endre Gombocz (Hybridisation), Dr. Arpád Paál (Reizphysiol.); Dr. Franz Hollendonner (vergl. Pflanzenanat.). *Für Anat. 3 Plätze, für Physiol. 2. — Neben Mikroskopen Apparate für physiol. Untersuchungen.* ⓟ Ökol., terratol. Gegenstände.

Botan. Inst. Technische Hochsch., Müegyetem. Dir. Prof. Dr. G. Istrauffi. — Dr., F. Hollendonner.

Buenos Aires (Argentina).

Inst. de Botan. méd. Fac. Univ. Dir. Prof. L. Durñaoua.

Botan. Labor. d. Fac. de ciencias, Univ. Dir. C. M. Hicken.

Inst. de Botan. Pathol. végét. et Microbiol. Fac. vét. Univ. Dir. Prof. L. Hauman.

Burlington, Vt. (U.S.A.).

Dept. of Botan., Univ. of Vermont. Dir. George P. Burns. — B. F. Lutman (Plant Pathol.); A. H. Gilbert (Pathol.); E. J. Dole (Systematic and Keeper of Herbarium); A. Gershoy (Genetics); R. Bamford (Cytol.); W. R. Adams (Forestry). *5 places for Genetics, Silvic., Pathol. — One Herbarium of Mexican Plants. Ample forest plantation and forest nurseries.*

Caen, Calvados (France).

Inst. Botan. de l'Univ., Jardin des plantes. Dir. Prof. Dr. René Viguier. — Dr. H. Bouygues, Maître de Conférences; Dr. P. Bugnon, Chef des Travaux pratiques; M. Lortet, Conserv. des Collections. *6 places pour travailleurs.* ⓟ Collections considérables comprenant de nombreux Herbiers spéciaux avec des plantes du monde entier et de tous les groupes. L'Herbier Lenormand compte plus de 2000 paquets. L'Herbier Vieillard est spécial à la Nouvelle Calédonie. HerbierLamouroux comprenant de nombreux types d'Algues. Echanges de plantes.

Cagliari, Sardinia (Italia).

R. Ist. Botan. Univ., R. Orto Botanico. Dir. Prof. Eva Mameli Calvino. *Sistematique, Microchimie, Anat. végétales.* ⓟ Un herbier générale. Un herbier de l'île de Sardaigne. Un herbier de l'île d'Elba.

Cairo (Egypt).

Botan. Dept., Univ., Zaafaran Palace. Dir. Prof. Gunnar Täckholm, D.Sc. — M. T. Hefnawy, B.S., Lect.; M. A. Mustafa, B.A., Lect.; Y. S. Sabet, B.Sc., Lect. 6 Demonstrators. *Working-places for guests not limited for Cytol. and Physiol. and Egyptian Flora.* ⓟ Egyptian Herbarium Specimens.

Calcutta (Br.-India).

Dept. of Botan. Univ. Dir. Prof. S. P. Agharkar. — Prof. P. Brühl.

Cambridge (England).

Botan. School, Dowing St. Dir. Prof. A. C. Seward. —F. F. Blackman, Reader in Botan. (Plant Physiol.); F. T. Brooks, Univ.Lect. (Mycol., Cytol.); H. H. Thomas, Univ.Lect. (Morphol., Palaeobotan.); H. Godwin, Demonstrator; G. E. Briggs (Plant Physiol.); H. Gilbert-Carlei, Curator of Herbarium and Dir. of Botan. Garden (Systematic Botan.). 4 Ass. *15—20 places for Physiol., Mycol., Morphol. etc.*

Cambridge, Mass. (U.S.A.).

Botan. Labor. of the Univ. Mus., Havard Univ. Ass. Prof. R. H. Wetmore (Botany).

Camerino (Italia).

Ist. di botan. ed orto botan. della Liberia Univ. Dir. R. Grandori.

Canton (China).

Dept. of Botan. and Herbarium Christian Coll. Dir. Prof. F. A. McClure.

Cape Town (South Africa).

Dept. of Botany Univ. Dir. Prof. R. S. Adamson. — Prof. R. Compton.

Cardiff, Wales (England).

Botan. Labor., Univ. Coll. of South Wales, Newport Road. Dir. Prof. R. C. McLean, M.A., D.Sc. — Dr. W. B. Crow, Ph.D., M.Sc., Lect.; Miss C. E Quinlan, M.Sc., Ass. Lect. *Room for 1 or 2 outside workers. — Fine microscopical work.* — A labor. for ecol. and marine biol. work at Port Eynon, Gower, near Swansea.

Cernăuți — Czernowitz (România).

Botan. Inst. und botan. Garten, strada Universității n. 11. Dir. o. Prof. Dr. M. Guşuleac. — Dr. A. Mühldorf, Conférencier, 2 Ass. — ⓟ Flora Rumänlens.

Inst. de Fiziol. și Anat. vegetala al Univ. Dir. Prof. Dr. med. F Netolitzky. — Ass. Lic. Radu Popovici. *2 Plätze für Mikroskopie.* ⓟ Heilpflanzen, Folklore.

Chapel Hill, N.C. (U.S.A.).

Botan. Labor. of the Univ. of North Carolina. Dir. Dr. William Chambers Coker. *About 6 places. Most of our equipment is especially suited for research on fungi.* ⓟ Fungi.

Charkow, Ukraine (U.d.S.S.R.).
Botan. Kabinett des Technol. Inst. Dir. Prof. Al. Korschikoff.
Botan. Labor. d. Inst. d. Volksunterrichts, Klotchkowskaja 50. Dir. Al. Korschikoff. — N. Dedussenko, Ass.; An. Lawrenko, Ass. Ⓟ Süßwasseralgen- und Lichenensammlungen aus vielen verschiedenen Regionen von U.d.S.S.R.
Botan. labor. of the Inst. for Agric. and of Technicum of Pharmacy, Poushkinsjaja, 80. Dir. J. Rolle. — T. Michajlowkaja (Inst. of Agric.); N. Shoslenko (Techn. of Pharmacol.) *The Inst. of Agric. may give 2 places and the Technicum of Pharmacol. 1 place. The place may be employed for the study of spore and flower plants. — In the Inst. of Agric. there is herbarium of the highest plants of the district of Kharkow; herbarium of the flora of Wologda district and other regions. Algol. collections from the district of Kharkow (marshes, ponds, lakes and rivers); plancton of r. Toretz, r. Dniepr, lake Siliger, different basins of Archangelskaja and Olonezkaja government. In the Technicum of Pharm. a herbarium of local flora and drug plants. At the labor. of Inst. of Agric. there is a botan. garden and a small hot-house. At the disposal of the Technicum of Pharmacol. is a nursery of drug plants.*

Charlottetown, P.E.I. (Canada).
Dept. of Botan. of the Prince of Wales Coll. Dir. J. F. McMillan, B.A.

Chicago, Ill. (U.S.A.).
Dept. of Botan. Univ. of Chicago. Ass. Prof. C. V. Eaton (Plant physiol.); Assoc. Prof. G. D. Fuller (Plant Ecol.); Prof. C. A. Shull (Plant Physiol.); C. J. Chamberlin (Morphol., Cytol.).

Clemson College, S.C. (U.S.A.).
Dept. of Botan. and Bact. Clemson Agric. Coll. Dir. Barre in charge (State Pathol.). — W. B. Aull, B.S., Assoc. Prof. and Assoc. Bact.; L. M. Fenner, M.S., Ass. State Pathol.; J.H. Hunter, M.S., Instr. Botan.; C. A. Ludwig, Ph.D., Assoc. Botan. and Plant Pathol.; W. D. Moore, Ph.D., Plant Pathol. (P.O., Beaufort); D. B. Rosenkrans, M.A., Assoc. Prof. Botan.; H. H. Tryon, A.B., M.F., For. Specialist (P.O., Aiken).

Clermont-Ferrand, Puy-de-Dôme (France).
Labor. de Botan. Fac. Univ. Dir. Prof. Moreau. — M. Denis, Ass. *8 places pour Systématique, Cytol., Cultures pures.* Une annexe du Labor. existe en montagne, sous le nom de Stations biol. de Besse (Puy-de-Dôme). Elle est particulièrement outillée pour l'étude des questions relatives à la biol. des organismes de montagne à celle des organismes des lacs et tourbières. Elle peut recevoir (coucher et nourrir) à la fois une douzaine de travailleurs.

Cluj — Klausenburg (România).
Labor. Inst. de Botan. Univ. générale. Dir. Prof. I. Grinescu — Manfred Tiesenhausen, Chef des travaux.

Coïmbra (Portugal).
Inst. Botan. Univ., Instituto Botânico. Dir. Prof. Dr. L. W. Carrisso. — Dr. Júlio A. Henriques, Naturaliste, Ancien Dir. de l'Inst., Dir. de l'Herbier; Dr. A. Quintanilha, 1. Ass.; F. d'Ascensão Mendonça, Inspecteur du Jardin; A. L. Franco, 2. Ass.; A. Gonsalves da Cunha. *3 places au labor. (étude de cytol. végétale). 3 à l'herbier (flore portugaise, phytogeographie). — Cytol. A l'herbier, l'herbier de Willkomm (auteur du ,,Prodromus Florae hispanica"), l'herbier portugais, l'herbier général, dans un total de plus de 100000 échantillons.* Ⓟ Des graines et des échantillons d'herbier et des prép. microscopiques.

College Park, Md. (U.S.A.).
Dept. of Botan. Coll. of Agric. J. B. S. Norton, D.Sc., Botan. and Plant Pathol.; P. W. Zimmerman,
Ph.D., Prof. Botan. and Ecol., Assoc. Dean Coll. of Agric.; F. S. Holmes, M.S., Seed Insp.; Ellen Emack, Ass. in Seed Insp.; Anna M. Hook, Ass. in Seed Insp.; Olive M. Kelk, Ass. in Seed Insp.; Ruth M. Mostyn, Ass. in Seed Insp.; Katherine Smith, Ass. in Seed Insp.

College Station, Tex. (U.S.A.).
Dept. of Botan. Agric. Coll. Helge Ness, M.S.; Chief of Div. (Berry Breeder).

Columbia, Mo. (U.S.A.).
Dept. of Botan., Univ. of Missouri. Dir. William J. Robbins. — Harold W. Rickett, Ass. Prof.; Willis E. Maneval, Ass. Prof.; Irl T. Scott, Plant Pathologist; Ernest E. Naylor, Instr. *2 places fitted for Investigation in Plant Physiol., Plant Pathol., Systematic Mycol. or Cytol. — Large herbarium parasit. and saprophytic fungi. Muffle furnace, transfer rooms.*

Columbus, Ohio (U.S.A.).
Botan. Labor. of the Ohio State Univ. Dir. Prof. Edgar N. Iranseau. — Prof.: Dr. J. H. Schaffner, Dr. H. C. Sampson, Dr. W. G. Stover. Ass. Prof.: Dr. A. E. Waller, Dr. L. H. Tiffany, Dr. J. D. Sayre. Instr.: Dr. B. S. Meyer, Dr. L. Lampe. 10 Ass. *About 15 places for taxonomy, plant physiol., morphol., ecol. and pathol. — Chemical and physical measurements, microchemistry.* Ⓟ Flower Plants and freshwater Algae.

Cordoba (Argentina).
Labor. de Botan. Fac. Sc. nat. Univ. Dir. Prof. C. C. Hosseus.

Corvallis, Ore. (U.S.A.).
Dept. Botan. and Plant Pathol. State Agric. Coll. H. P. Barss, M.S., Prof. and Chief in Botan. and Plant Pathol.; W. M. Atwood, Ph.D., Prof. Plant. Physiol.; Helen M. Gilkey, Ph.D., Ass. Prof. Botan., Curator of Herbarium; Bertha C. Hite, B.A., Seed Analyst (Detailed by U.S.D.A.); W. E. Lawrence, B.S., Assoc. Prof. Plant Pathol.; M. B. McKay, M.S., Plant Pathol.; C. E. Owens, M.A., Assoc. Prof. Plant Pathol.; F. P. Sipe, M.S., Instr. Botan.; Margaret Stason, M.S., Instr. Botan.; S. M. Zeller, Ph.D., Plant Pathol.

Danzig (Danzig).
Botan. Inst. Techn. Hochsch. Dir. Dr. Wangerin.

Darmstadt (Deutschland).
Botan. Garten u. Inst., Roßdörferstr. 40. Dir. Prof. Dr. Schenck. *Anat. u. Biol. der Wassergewächse, der Lianen.*

Debreczen (Ungarn).
Botan. Inst. Landw. Akademie. Dir. Prof. A. Gulyás.

Děčin-Libverd—Tetschen-Liebwerd (Č.S.R.).
Lehrkanzel f. Botan. u. Pflanzenschutz d. Landwirtsch. Abt. d. Deutsch. Techn. Hochsch. Prag. Dir. Prof. Dr. Karl Boresch, Prof. d. Pflanzenphysiol., Agrikulturchemie, chem. Technol.

Delft (Holland).
Labor. für Technische Botan., Poortlandlaan 35. Dir. Prof. Dr. G. van Iterson. — Frl. A. Kleinhonte, Konserv.; Frl. Dr. A. Hartsema, 1. Ass.; Frl. Ir. A. C. Sloep, Ass.; Ir. H. Eilers, Ass. *Maschinen für halbtechnische Gewinnung von Pflanzenprodukten. Extraktionsapparate, Zerkleinerungsmaschinen, Trokkenapparate, Autoklaven, Zentrifugen, Ölpresse, Versuchspapiermaschine.* Ⓟ Pflanzliche Rohstoffe, Spezialität Faserstoffe. — Das Labor. arbeitet öfters zusammen mit dem Prüfungsamt für Kautschukindustrie und -handel im selben Gebäude, sowie mit dem Prüfungsamt für Faserindustrie und -handel.

Dijon (France).
Inst. et Jardin botan. d'univ. (Hortus Botan. Divionensis). Dir. Prof. C. Oueva.

Dnepro-Petrowsk (U.d.S.S.R.).
Berginst., Poltawskaja 1.

Dresden (Deutschland).
Botan. Inst. Techn. Hochsch., A. 24, Bismarckplatz. Dir. Prof. Dr. Fr. Tobler. — Prof. Dr. R. Schwede, Ass.; Alban Voigt, Kustos des Herbariums; Dr. F. Mattick. *2 Plätze für Herbarium, Mykol. Textiltechnol. — Mikroskopische und physiol., auch mykol. Apparatur, photographische Ausrüstung, pflanzliche Rohstoffe.*

Dublin (Ireland).
School of Botan., Trinity Coll. Dir. Henry H. Dixon. — Thomas A. Bennet-Clark, Ass. *10 places for Physiol., Anat., Systematic Botan., Cytol. — Physiol. apparatus, Botan. Garden, Herbarium, Apparatus for microscopic research.* ⊕ Pteridophytes, Seed-Plants
Botan. Dept. Univ. Coll. Dir. Prof. Joseph Doyle. — P. O'Connor, B.Sc., General Ass.; Miss Phyllis Clinch, M.Sc. (Plant Physiol.). *2 places for Cytol. and compar. Morphol.; Biochemistry; experim. Physiol.* — Experim. Farm connected with the School of Agric.

Durham, N.H. (U.S.A.).
Dept. of Botan. Coll. of Agric. O. R. Butler, Ph.D., Prof. Botan., Botan.; Mabel Brown, Ph.D., Ass. Prof. Botan.; F. R. Clark, M.S., Ass. Prof. and Ass. Botan.

Durlach, Baden (Deutschland).
Botan. Abt. Staatl. Badische Landwirtschaftl. Versuchsanst. Augustenburg. Dir. Prof. C. v. Wahl.

East St. Louis, Ill. (U.S.A.).
Dept. of Botan., East St. Louis High School. Dir. H. G. Wallis.

Eberswalde (Deutschland).
Botan. Inst. Forstliche Hochsch. Dir. Prof. Schwarz.

Edinburgh (Scotland).
Botan. Labor. of R. Veterin. Coll. Dir. Prof. R. St. Macdoregall.

Edmonton, Alberta (Canada).
Dept. of Botan. Univ. of Alberta. F. R. Lewis, D.Sc.

Erlangen (Deutschland).
Botan. Inst. Dir. Prof. Dr. Kurt Noack. — 2 Ass. *Plätze für Pflanzenphysiol. und Biochemie. — Botan. Garten mit Kalt- und Warmhäusern. Gute Sukkulenten- u. Orchideenbestände.* ⊕ Lebende Sukkulenten

Eugene, Ore. (U.S.A.).
Dept. of Botan. Dir. A. R. Sweetser.

Evanstown, Ill. (U.S.A.).
Dept. of Botan., Coll. of Liberal Arts, Northwestern Univ. Dir. Prof. Charles Beach Atwell.

Exeter (England).
Biol. Labor. Univ. Coll. of the South West. Dir. J. L. Sager, M.A., F.L.S. — 1 Ass. *3 places for Cytol., Plant Anat., Plant Physiol. — Greenhouses, Hot-house and open-air plots for Plant Physiol.*

Fayetteville, Ark. (U.S.A.).
Botan. Dept. Univ. of Arkansas. Dir. D. M. Moore.

Feilding (New Zealand).
Dept. of Botan. Agric. High. School. H. H. Allan (Botan.).

Firenze (Italia).
Ist, Botan. Univ., 14, Via Lamarmora 4. Dir. Prof. Dott. Giovanni Negri. — Prof. R. Pampanini, Aiuto e Conserv. del Mus.; Dr. A. Chiarugi, 1. Ass.; Dr. G. Micatovoch, 2. Ass. *Embriol. e Morfol. delle Cormofite. Geografia Botan. e Sistematica.*

Fort Collins, Col. (U.S.A).
Dept. of Botan. State Coll. of Agric. L. W. Durrell, Ph.D., Prof. Botan., Botan. and Plant Pathol.; C. D. Learn, M.A., Ass. Prof. and Ass. Botan.; E. L. LeClerg, M.S., Ass. Plant Pathol.; Anna M. Lute, B.S., Instr. Botan.; Seed Analyst.; E. C. McCarty, B.S., Ass. Prof. and Ass. Botan.; Caroline M. Preston.

Frankfurt a. M. (Deutschland).
Botan. Inst. der Univ., Victoria-Allee 9. Dir. Geh.R. Prof. Dr. M. Möbius (1926 emeritiert). — Priv.Doc. Dr. F. Overbeck, Ass.

Freiburg i. Br. (Deutschland).
Botan. Inst., Schänzleweg 9. Dir. Geh.R. Prof. Dr. Fr. Oltmanns (Algen, Pflanzengeographie). — Dr. F. Rawitscher. ao. Prof. für Forstbotan. (Reizphysiol.); Dr. B. Huber, 1. Ass., Priv.Doc. (Physiol. und Ökol.). *Arbeitsplätze für Gäste beliebig.*

Freiburg (Schweiz).
Botan. Inst. Dir. A. Ursprung. — Priv.Doc. Dr. G. Blum. *Mehrere Plätze. — Saugkraftmessung.*

Fukuoka (Japan).
Labor. of Plant Morphol. Coll. of Agric., Kyûshû Imper. Univ.

Gand — Gent (Belgique).
Inst. botan. de l'univ. Dir. Prof. C. de Bruyne.

Geneva, N. Y. (U.S.A.).
Div. of Botan. of the New York Agric. Experim. Stat. Fred C. Stewart, M.S., Chief in Research (Botan.); Mancel T. Munn, M.S., Assoc. in Research (Botan.); Elizabeth F. Hopkins, A.B., Ass. in Research (Botan.); Walter O. Gloyer, M.A.; W. Howard Rankin, Ph.D.; Edward E. Clayton, Ph.D. (Riverhead); Elmer V. Shear, Jr., M.S. (Poughkeepsie); Leon K. Jones, Ph.D., Assoc. in Research (Plant Pathol.). *Raspberry disease investigations, Diseases of canning crops, Crown-gall on apple trees, Apple fruit diseases, Root diseases of fruit trees, Diseases of cruciferous crops, Dust treatment for seed potatoes, Mushroom studies, Seed testing and seed studies.*

Genève — Genf (Suisse).
Inst. de Botan. de l'Univ. Dir. Prof. Dr. Robert Chodat. — Dr. Marcel Minod; Dr. Fernand Chodat; Dr. Gustave Beauverd. *5 places pour Taxonomie (travail d'Herbier), Géobotan., Anat., Cytol., Physiol., Microbiol. générale, Physiol. de la nutrition, électrophysiol., Fermentations (ferments oxydants), Cultures pures d'algues, Bact., champignons.—Mesures électriques, Microdissecteurs, Labor. de chimie et Installations bact.* ⊕ Cultures pures d'algues, Mucorinées, Levures de vin. — Labor. et Jardin botan. de la Linnaea (Cours et Labor. de vacance du 8. juillet au 5. septembre). *Ecol. expérim., Biol. florale, Algol.*

Genova (Italia).
Ist. Botan. R. Orto Botan., dell'Univ., Corso Dogali 1B. Prof. Dott. Ubaldo Ricca, Aiuto. *6 posti: Anat. vegetale, Fisiol. vegetale.*

Gießen (Deutschland).
Botan. Inst. der Univ., Brandpl. 4. Dir. Prof. Dr. Ernst Küster. — H. Timmel, 1. Ass. *1 Platz für Zellenphysiol.*

Gihu (Japan).
Botan. Labor. Gihu Imperial Coll. of Agric. and Forestry.

Glasgow (Scotland).
Botan. Dept. the Univ. Dir. J. M. F. Drummond, M.A. (Regius Prof. of Botan.).—S. Williams, M.Sc.,

Ph.D., Lect. (Morphol.); C. W. Wardlaw, B.Sc., Ph.D., Lect. (Mycol.); L. J. F. Brimble, B.Sc., Lect (Physiol.); F. W. Sansome, B.Sc., Ph.D., Ass. (Cytol. and Genetics); I. M. Case, B.Sc., Ass. (Mus., Herbarium and Library). *The Dept. has charge of the Kidston Collection of Fossil Slides and of the Kidston Library. Room could be found for one guest for Palaeobotan. work, or for Morphol. or Mycol.* ⓟ Material for Plant Morphol. generally (especially Pteridophyta and Angiosperms). Cultures of Fungi.

Botan. Labor., Bact. Labor. Royal Technical Coll. Dir. Prof. David Ellis, D.Sc., Ph.D., F.R.S.E. — Blodwen Loyld, M.Sc. (Ass. Lect. in Botan.); I. B Wilson (Demonstrator in Botan.); Jas. Mull Lectch, B.Sc. (Demonstrator in Bact.).

Göttingen (Deutschland).

Botan. Inst. und Botan. Garten der Univ., Untere Karspüle 2. Dir. Prof. Dr. Gg. Ritter (Solanaceae, Rosaceae, Crassulaceae, Lichenes). — Prof. Dr. G. Schellenberg, Ass. (Connaraceae); Dr. O. Schwartz, Hilfs-Ass. (Cytol. u. Systematik besonders Pontederiaceae). *Größere Laboratoriumsräume für anat. und systematische Untersuchungen.* ⓟ Herbarien (besonders Mittel- und Südamerika), Samen- und Fruchtsammlung, Trocken- und Spirituspraep. (es besteht ein umfangreiches botan. und pharmakognostisches Mus.). — Versuchsgarten auf dem Brocken im Harz.

Inst. für allgemeine Botan. und Pflanzenphysiol., Nikolausberger Weg 18. Dir. Prof. Dr. Fr. v. Wettstein. — Dr. Th. Schmucker, Ass. *Allgemeine Botan., besonders Entwicklungsphysiol. und Genetik.*

Grahamstown (South Africa).

Labor. of botan., Rhodes Univ. Coll. Dir. Prof. S. Schönland.

Granada (España).

Labor. de Botan. Fac. pharmacol. Univ. Dir. Prof. J. L. Diez de Tortosa.

Graz (Österreich).

Labor. für Botan. Warenkunde, Chemie der Nahrungs u. Genußmittel, techn. Mikroskopie u. Mykol. d. Techn. Hochsch., Dir. Prof. Dr. F. Reinitzer.

Inst. für systematische Botan. mit dem botan. Universitätsgarten, Philos. Fak. Univ. Dir. Prof. Dr. Karl Fritsch.

Greifswald (Deutschland).

Botan. Inst. der Univ., Grimmerstr. 86/88. Dir. Prof. Dr. Joh. Buder. — Prof. Dr. E. Leick, ao. Prof.; Dr. S. Lange, Ass.; Priv.Doc. Dr. B. Huber, Ass. *2—3 Plätze für reizphysiol. Untersuchungen. — Mehrere Dunkelzimmer mit konstanter Temperatur. Lumineszensmikroskopie.*

Grenoble, Isère (France).

Labor. de Botan. Fac. Sc. Univ. et Inst. Botan. Alpin du Lautaret (Hautes-Alpes) à 2110 m d'altitude. Dir. Prof. Dr. M. Mirande. — J. Offner, Chef des travaux; M. Guéraud, Ass. *Travaux de physiol., Anat. Systématique et Chimie végétales. — Polarimètres, Labor. bact., Herbiers (riches collections).* ⓟ Collections botan.: Herbiers, plantes fossiles, graines et objets végétaux divers. — Botan. alpine avec l'Institut botan. Alpin du Lautaret.

Grignon, Seine et Oise (France).

Labor. de Botan. et Pathol. végétale, Ecole nationale d'agriculture. Dir. V. Ducomet. — Ch. Schad, Chef de travaux.

Groningen (Holland).

Botan. Labor. der Reichs-Univ., Groote Rozenstraat 31. Dir. Prof. Dr. J. C. Schoute. — Prof. Dr. Tine Tammes; Prof. Dr. W. H. Arisz; Dr. J. J. Beyer; Dr. H. J. Toxopeus; Dr. L. S. Wildervanck; Cand. H. de Haan. *1 Platz.* ⓟ Samen, cf. Samenliste.

Botan. Abt. der Rijkslandbouwproeptation, Eemskanaal Z.Z. 1. Dir. Dr. K. Zijlstra. — M. A. J. Goedewaagen.

Guelph, Ont. (Canada).

Dept. of Botan. Agric. Coll. Dir. A. L. Gibson, B.S.A. — D. A. Kimball, D.S.A.; D. R. Sands, B.S.A., M.S.

Halifax, N. S. (Canada).

Botan. Labor. Dalhousie Univ. H. P. Bell (Assoc. Prof. Botan.).

Halle a. d. S. (Deutschland).

Botan. Inst. der Univ., Am Kirchtor 1. Dir. Prof. Dr. G. Karsten. — Prof. Dr. C. Mantfort, 2. Prof.; Dr. Günther Schmid, 1. Ass.; Dr. Kurt Mathes, 2. Ass.

Hamburg (Deutschland).

Inst. für allgemeine Botan. u. Botan. Garten, 36, Jungiusstr. 6. Dir. Prof. Dr. Hans Winkler. — Prof. Dr. Edgar Irmscher, Kustos des Herbar; Dr. Curt Schwarze; Dr. Ernst Heinsen; Frl. Dr. Rose Stoppel.

Hannover (Deutschland).

Botan. Inst. Techn. Hochsch. Dir. O. Gerke.

Hann.-Münden (Deutschland).

Botan. Inst. u. Botan. Garten, Forstliche Hochsch. Dir. Prof. Dr. Jahn.

Harpenden, Herts (England.)

Dept. of Botan. of the Inst. of Plant Nutrition and Soil Problems. Dir. W. E. Brenchley. — K. Warington.

Heidelberg (Deutschland).

Botan. Inst. u. Botan. Garten. Dir. Prof. Dr. L. Jost.

Helsinki (Finnland).

Botan. Labor. der Univ. Dir. Prof. Dr. K. Linkola. — Dr. Ernst Hayrén, Adjunkt; Dr. P. R. Collander, Ass. *2 Arbeitsplätze für Gäste. — Dunkelzimmer mit regulierbarer Temperatur; elektrometrischer Apparat für Azidititätsbestimmungen.*

Hohenheim b. Stuttgart (Deutschland).

Botan. Inst. der landwirtschaftl. Hochsch. verbunden mit **W. Landesanstalt für Pflanzenschutz** (Leiter: Dr. Lang) **und Samenprüfungsanstalt** (Leiter Prof. Dr. Lakon). Dir. Prof. Dr. H. Schroeder. — Mehrere Ass.

Innsbruck (Österreich).

Botan. Inst. der Univ. u. botan. Garten, Hötting, Sternwartstr. 13. Dir. Hofrat Prof. Dr. Emil Heinricher. — o. Prof. Dr. Ad. Wagner; ao. Prof. Dr. Ad. Sperlich; Dr. Arth. Pisek, Ass. *1—2 Plätze.* ⓟ Kultur parasit. Samenpflanzen.

Ithaca, N. Y. (U. S. A.).

Dept. of Botan. Coll. of Agric. K. M. Wiegand, Ph.D., Prof. and Head of Dept. — S. H. Burnham, B.S., Ass. Curator; O. F. Curtis, Ph.D., Prof. Botan.; A. J. Eames, Ph.D., Prof. Botan.; E. F. Hopkins, Ph. D., Ass. Prof. Botan.; Lewis Knudson, Ph.D., Prof. Botan., Plant Physiol.; W. E. Manning, A.B., Instr. Botan.; W. C .Muenscher, Ph.D., Ass. Prof. Econ. Botan.; L. C. Petry, Ph.D., Prof. Botan.; Donald Reddick, Ph.D., Prof. Plant Pathol.; L. W. Sharp, Ph.D., Prof. Botan.

Iwanowo-Wosnessensk (U. d. S. S. R.).

Botan. Kabinett des Polytechnischen Inst. Dir. V. Miller. — Xenie Aljawdina, Ass. ⓟ Pilze.

Jaroslavl (U. d. S. S. R.).

Kabinett für Botan. des Pädagogischen Inst. Dir. Dr. A. V. Zukovskij.

Jena (Deutschland).
Botan. Anstalt der Univ. u. Botan. Garten. Dir. Otto Renner. — Prof. Dr. Theodor Herzog, Abteilungsvorsteher; Dr. Leo Brauner, 1. Ass.; Dr. Peter Michaelis, 2. Ass.

Jerusalem (Palestine).
Section of Botan. of the Inst. of Nat. History of the Hebrew Univ. Dir. Prof. Dr. Otto Warburg. Botan. (*system. Pflanzengeogr.*); *angew. Kolonialbotan. u. koloniale Landwirtschaft.*

Jassy (România).
Labor. de Botan. de l'Univ. Dir. Prof. Dr. Al. Popovici. — C. Petrescu, Ass.Chef des traveaux; Hélène Popocivi, Ass.; Dr. C. Popp, Ass. *30 places pour Anat. et systématique.* ⓟ Champignons: Ascomycetes et Basidiomycetes, Phanerogames en association avec des champignons, mousses et hépatiques.

Johannesburg (South Africa).
Botan. Dept. Univ. Dir. Prof. C. E. Moss.

Iowa City, Ia. (U.S.A.).
Dept. of Botan., Univ. of Iowa. Dir. Robert B. Wylie. — Dr. Robert B. Wylie, Morphol.; Dr. Bohumil Shimek, Taxonomy and Ecol. of vascular plants; Dr. G. W. Martin, Mycol and Pathol.; Dr. Walter F. Loehwing, Physiol. Research Ass. *5—20 places depending upon the nature of the problem. Taxonomy and ecol. of the prairies and border assoc. Mycol., pathol. Physiol. researches. Morphol. and Cytol of vascular plants.*

Kagoshima (Japan).
Botan. Labor. Kagoshima Imperial Coll. of Agric. and Forestry.

Kamenetz-Podolsk, Ukraine (U.d.S.S.R.).
Vereinigtes botan. Labor. des Landwirtschaftlichen Inst., des Inst. für Volksbildung sowie des Forschungskatheders für Podolien (Sektion der angewandten Botan.), Zatonskistr. 31. Dir. Prof. Dr. N. Hamorak. —Fedir S. Panasiuk, Ass. u.Lect.; Mykola A.Ljubynśkyj, Aspirant; Wossyl I. Paltschewśkyj, Aspirant. *Wasserhaushalt der Pflanzen. — Selbstregistrierende Apparate zur Messung der Transpiration und der Absorption.* ⓟ Blütenpflanzen u. Pilze und speziell Uredineae Podoliens.

Kanazawa (Japan).
Bot. Labor. 4. High School. Dir. Prof. T. Ichikawa.

Kaunas (Litauen).
Inst. f. Pflanzenanat. u. Physiol., Wilnastr. 2. Dir. Dr. L. Vailionis.

Kazan (U.d.S.S.R.).
Botan. Labor. der Univ., des Inst. f. Land- u. Forstwirtschaft u. des Vet. Inst., Ug. Tschernyschewskogo. Dir. Prof. A. G. Ponomaren.

Kew, Surrey (England).
The Jodrell Labor., Royal Botanic Gardens. Dir. Royal Botan. Gardens, Kew; T. F. Chipp (Ass. Dir.). — L. A. Boodle (Ass. Keeper). *3 or 4 places for Botan. work.* ⓟ Plant-material is supplied from the Botan. Gardens as far as possible.

Keijô, Korea (Japan).
Botan. Labor. Coll. of Med., Keijô Imper. Univ.

Kiel (Deutschland).
Botan. Inst. u. Botan. Garten, Univ. Dir. Prof. Dr. Georg Tischler. — Prof. Dr. Wilhelm Nienburg (Abteilungsvorsteher); Dr. Curt Hoffmann (Ass.); Dr. Robert Jaretzky (außerplanmäßiger Ass.); *Algenforschung.*

Kiew (U.d.S.S.R.).
Labor. der Morphol. und der Systematik am Inst. der Landwirtschaft, Brest-Litowskoje Chaussee 39.

Dir. Prof. Wladimir W. Finn. — Ass. P. F. Oksijuk. *Morphol., Systematik, Embryol. und Cytol.*

Kingston, R. I. (U.S.A.)
Dept. of Botan. State Coll. of Agric. H. W. Browning, Ph.D., Prof. Botan.; F. T. McLean, Ph.D., Plant Physiol.; Marian E. Deats, A.M., Instr. Botan.

Knoxville, Tenn. (U.S.A.).
Dept. of Botan. Coll. of Agric. L. R. Hesler, Ph.D., Prof. Botan.; S. H. Essary, M.S., Botan.; C. D. Sherbakoff, Ph.D., Plant Pathol.; J. O. Andes, M.A. Ass. Plant Pathol.; E. S. Brown, Ass. Plant Pathol.; Mrs. Helen M. Hutchens, A.B., Ass. Plant Pathol.; H. M. Jennison, Ph.D., Prof. Botan.

København (Danmark).
Univ. Botan. Labor., Gotherkade 130. Dir. Prof. Dr. C. H. Ostenfeld. — Dr. Henning E. Petersen, Sc. Ass. *2 places for general botan. work.*
Botan. Labor. Bülowsvej. Dir. Prof. Dr. phil. Aug. Mentz. — Einar Larsen, Ass. mag. sc.

Köln a. Rh. (Deutschland).
Botan. Inst. Univ., Zollstock, Vorgebirgstr. Dir. Prof. Dr. Esser. — Ass. Dr. André, Priv.Doc.

Königsberg i. Pr. (Deutschland).
Botan. Inst., Univ., Besselplatz 3. Dir. Prof. Dr. Carl Mez. — Dr. Hermann Ziegenspeck, Ass., Priv.-Doc.; Dr. Fritz Steinecke, Stud.R., Priv.Doc.

Krasnodar, Kubangebiet (U.d.S.S.R.).
Botan. Labor. u. Mus. d. Kuban. Landw. Inst. Dir. Prof. P. S. Mistschenko.

Kraków (Polska).
Botan. Inst., Lubicz 46. Dir. Prof. Dr. W. Srafer. — Dr. J. Wołosryńska, Ass.; Dr. M. Sokołowski, Ass. *1 place for Paleobotan., Diluvium.* — Botan. Mus.; *Herbarium*. ⓟ Paleobotan. objects (Diluvium).
Labor. Botan. Janczewskianum Univ. (avec "Hortus agronomico-botan."), Aleja Mickiewicza 21. Dir. Prof. Dr. Kazimier Stefan Rouppert. — Adjoint: Dr. phil. K. Piech. Ass.: Dr. phil. J. Zabłocki, Dr. phil. W. Zabłocka, K. Starmach. *4 places pour Anat. physiol., Cytol., Phytopath., Cryptogames.*

Kyôto (Japan).
Botan. Inst. Coll. of Sc., Kyôto Imperial Univ. Dir. Prof. Dr. Yoshinari Kuwada. — Prof. Dr. Y. Kuwada (Histo-Cytol.); Prof. Dr. Kavan Kôriba (Physiol.); Ass. Prof. Dr. Gen-ichi Koizumi (Systematik); Lect. Kôki Masni (Physiol.); Lect. Itaru Tsuchiya (Physiol.); Ass. Takeshige Maeda (Cytol.); Ass. Shigeru Miki (Ecol.); Vice-Ass. Zentarô Tashiro (Systematik). 2 Ass.
Botan. Labor. 3. High School. Dir. Prof. K. Suzuki. *Pflanzenhistol.*

Lafayette, Ind. (U.S.A.).
Dept. of Botan. of the School of Agric., Purdue Univ. H. S. Jackson, A.B., in charge; E. G. Campbell, Ph.D., Ass. Prof. Botan.; L. E. Compton, B.S., Field Ass. in Plant Pathol. (Detailed by U.S.D.A.); F. W. Gardner, Ph.D., Assoc. in Botan.; C. T. Gregory, Ph.D., Assoc. in Botan.; A. A. Hansen, M.S., Assoc. in Botan.; G. N. Hoffer, D.Sc., Assoc. in Botan. (Coop. U.S.D.A.); J. B. Kendrick, Ph.D., Ass. in Botan.; E. J. Kohl, M.S., Instr. Botan.; W. E. Leer, M.S., State Leader of Barberry Erad; E. B. Mains, Ph.D.,Assoc. in Botan.(Coop.U.S.D.A.); C. L. Porter, Ph.D., Ass. Prof. Plant Pathol. and Physiol.; G. M. Smith, M.S., Ass. Pathol. (Detailed by U.S.D.A.); R. R. St. John, B.S., Ass. in Corn Disease Invest.; Dorothy M. Thompson, B.S., Ass. in Botan; R. A. Weaver, B.S.Ch.E., Ass. in Botan. (Coop. U.S.D.A.); J. F. Trost, M.S., Ass. Pathol. in Botan.

East Lansing, Mich. (U.S.A.).
Beal Botan. Labor. of Michigan State Coll., East Lansing, Michigan U.S.A. Dir. Ernst A. Bessey, Prof. of Botan. — George H. Coons (Plant Pathol.); Carlyle W. Bennett (Plant Pathol.); Richard de Zeeuw (Botan. Microtechnique); Henry T. Darlington (Curator of herbarium and Dir. of Beal Botan. Garden); Edward F. Woodcock (Plant anat.); Rufus P. Hibbard (Plant physiol.); Harry F. Clements (Plant physiol.); Ray Nelson (Plant Pathol.); John E. Kotila (Plant Pathol.); Miriam Carpenter (Plant Pathol.); F. G. Larmer; H. A. Elcock; R. A. Diettert. *2 or 3 places. Preferably arrangements should be made several weeks beforehand Research may be undertaken in Plant Pathol., Plant Physoil., Plant Anat., etc. — The labor. has ample chemical equipment, apparatus for physiol. researches, herbarium small botan. garden, limited greenhouse space, culture chambers.* ⓟ *Herbarium material, especially plants from Michigan. — It is under consideration to establish a substation in the northern part of Michigan.*

Lausanne (Suisse).
Labor. de botan. et de pharmacognosie de l'Univ., Palais de Rumine. Dir. Prof. Ernest Wilczek. — Daniel Dutoit, Ass. *Environ 10 places pour Botan. systématique, Anat. pharmaceutique.* ⓟ *Plantes alpines cultivées dans les deux jardins de Lausanne et de Pont de Nant.*

Lautaret, Hautes-Alpes (France).
Labor. de Botan. de l'Inst. Alpin. Dir. M. Mirande.

La Plata (Argentina).
Inst. f. agric. Botan. Dir. Prof. L. Parodi.
Labor. de Botan. du Mus. Dir. Augusto Cesar Scala. *10 places pour Phytohistol.* ⓟ *Plantes indigènes; matérial histol.*

Laramie, Wyo. (U.S.A.).
Dept. of Botan., Coll. of Agric. Univ. of Wyoming. Aven Nelson, Ph.D., Prof. Botan., Botan. and Hort.; E. B. Payson, Ph.D., Assoc. Prof. Botan., Assoc. Botan. and Hort.; R. U. Cotter, B.S., Specialist in Barberry Erad. Work (Detailed by U.S.D.A.).

Lawrence, Kan. (U.S.A.).
Dept. of Botan. of the Univ. of Kansas. Dir. Prof. W. C. Stevens. — Assoc. Prof. A. J. Mix; Ass. Prof. S. M. Charles; Ass. Prof. C. N. Sterling; Ass. Prof. W. H. Horr.

Leeds (England).
Dept. of Botan. Fac. of sc. Univ. of Leeds. Dir. Prof. J. H. Priestley. — Reader W. H. Pearsall.

Leiden (Holland).
Botan. Labor., Nonnensteeg 3. Dir. Prof. Dr. J. M. Janse.

Leipzig (Deutschland).
Botan. Inst. der Univ., Linnéstr. 1. Dir. Prof. Dr. Wilhelm Ruhland. — Dr. Rudolf Gießler, Kustos und Ober-Ass.; Dr. Fritz Bachmann, Ass.; Dr. Karl Wetzel, Ass.; Dr. Hermann Ullrich, Ass. *20 Plätze für das Gesamtgebiet der allgemeinen Botanik. — Zimmer für konstante Temperatur, physikalisch-chemische Apparatur, chemisches Labor., Dunkelzimmer, Experimentalgewächshaus, Apparatur für gasometrisches Arbeiten, Einrichtungen für physiol. Arbeiten.*

Leningrad (U.d.S.S.R.).
Botan. Abt. Roentgenol. u. Radiol. Staats-Inst. ul. Röntgen'a (ehem. Licejskaja) 6. Dir. G. A. Nadson.
Botan. Labor. der Militär-Med. Akademie, Viborger Seite, Nischegorodski Str. 6. Dir. V. N. Lubimenko, Prof. der Botan.
Labor. der Systematik und der Geographie der Pflanzen, Univ., Universitätsquai 11. Dir. Prof. Dr. N. A. Busch. — A. P. Schénnikow, Doc. der Geobotan.; Ober-Ass. M. A. Rósanowa; Ober-Ass. A. P. Iljinskij. *2 Plätze für systematische, phytogeographische und geobotan. Untersuchungen.* ⓟ *Pflanzen.*
Kabinett der Pflanzensystematik und Deudrol. des Leningrad. Forst-Inst. mit dem Dendrol. und Botan. Garten, Lesnoje bei Leningrad. Dir. Prof. W. Sukatschew. — E. Wolf, Ass., Verwalter des Dendrol. Garten; G. Anufriew, Doc.; A. Schénnikow, Ass.; S. Sokolow, Ass.; W. Powarnitzin, Ass.; P. Bogdanow, Wissenschaftlicher Mitarbeiter. ⓟ *Pflanzensamen.*
Labor. für Anat. und Physiol. d. Pflanzen d. Forst-Inst. Dir. Prof. Dr. L. A. Iwanoff. — W. P. Maltschewsky, Ass.; E. G. Zwetkowa, Assistentin; N. Z. Kossowitsch, wiss. Mitarbeiterin. *4 Plätze für pflanzenökol. Untersuchungen, besonders des Lichtgenusses und Wasserregims der Pflanzen.*
Botan. und Microbiol. Labor., Technol. Inst. Dir. Prof. Nicolai N. Iwanoff. — V. Czastuchin, Ass.
Labor. d. Morphol. und der Systematik d. Pflanzen des Landwirtschaftlichen Inst., Kamenkyj Ostrolodie 2. Birnen-Aliee Nr. 2. Dir. Prof. Dr. Nicolai Adolfowitsch Busch. — Olga Wassiljewna Troïtzkaja, Ober-Ass.; Zoë Nicolajewna Smirnowa, Ass.; Olga Fedorowna Hase, Ass. *2 Arbeitsplätze für die systematischen, experimentell-morphol., geobotan. und algol. Untersuchungen.* ⓟ *Pflanzen.* — *Eine Sonderabt. des Labor. in Detskoje (vormals Tzarkoje) Sselo, 25 km von Leningrad.*
Labor. für Anatomie und Physiol. der Pflanzen (am Landwirtsch. Inst.), Nab. B. Newki 18. Dir. Prof. Oskar A. Walther. — M. K. Ostrowskaja, 1. Ass.; E. I. Lowtschinowskaja, Ass.; L. M. Pinewitsch, Ass.; O. P. Kamenogradskaja, Ass.-Gehilfe u. Präparator; Z. A. Tschijewskaja, Aspirant. ⓟ *Pflanzliches Material über anat.-ökol. Anpassungen.*
Labor. für Anat. und Physiol. der Pflanzen an der Russischen Akademie der Wissenschaft, Vassili Ostrov, Tutchkov Quai, No 2a. Dir. Akademiker S. P. Kostyitchev.
Sect. botan. de l'Inst. sc. de Lesshaft, 25 rue Petshatnikov (ancienne Torgovaja). Dir. Dr. Vladimir Lubimenko. — Mlle E. R. Hübbenet, Chef du Labor. botan.; Mlle O. A. Sžeglova, Prép. du Labor. *1 place pour les recherches microscopiques et spectroscopiques. Dosages des pigments des plantes par la méthode spectrocolorimétrique. — Appareils pour la spectroscopie et la spectrocolorimétrie.*
Labor. de Botan. de l'Inst. de Méd., Tolstoj-Str. Nr. 6—8. Dir. Prof. G. A. Nadson.
Labor. of Botan. of the State Pedagogical Inst., Moika 50. Dir. Prof. Nicolas A. Maximow. — Ass.: Eugenie I. Lovčinovskaja, Nadeshda W. Stark. *4 places for physiol. and anat. research.*

Leningrad — Alt-Peterhof (U.d.S.S.R.).
Labor. der Morphol. und der Systematik der Pflanzen des Naturwissenschaftlichen Inst., Sergijewka. Dir. Prof. Dr. Nicolai Adolfowitsch Busch. — Wissenschaftliche Mitarbeiter: Zoë Nicolajewna Smirnowa, Elisabeth Iwanowna Steinberg, Eustolie Iwanowna Lapschina. *2 Plätze für systematische, experim.-morphol., pflanzengeographische und geobotan. Untersuchungen.*

Lexington, Ky. (U.SA.).
Dept. of Botan. of the Univ. of Kentucky. Dir. Prof. F. T. McFarland.
Dept. of Entomol. and Botan. Confer: Entomol.

Liège (Belgique).
Inst. botan. univ. Dir. Prof. Dr. A. Gravis.

Lille (France).
Labor. de Botan., Fac. Sc. de l'Univ. Dir. Prof. L. Maige.
Labor. de Botan., Fac. Sc. cathol. de l'Univ., 12, Rue de Toul. Dir. Prof. A. Carpentier.

Lima (Peru).
Labor. Botán., Patol. Vegetal y Física, Escuela Nacional de Agric. y Veterinaria, Santa Beatriz. Dir. Dr. J. Gaudron.
Labor. de Botán., Fac. de ciencias, Univ., Parque Universitario. Dir. Dr. E. G. Hernández. — Dr. A. R. Dulanto; Dr. F. Herrera.

Lincoln, Neb. (U.S.A.).
Dept. of Botan. Coll. of Agric. of the Univ. of Nebraska. R. J. Pool, Ph.D., Chair. of Dept.; G. L. Peltier, Ph.D., Prof. Plant Pathol., Plant Pathol.; Emma N. Andersen, A.M., Ass. Prof. Botan.; R. W. Goss, Ph.D., Assoc. Prof. and Assoc. Plant Pathol.; Leva B. Walker, A.M., Assoc. Prof. Botan.; J. E. Weaver, Ph.D., Prof. Plant Ecology; P. B. Sears, Assoc. Prof. Botan.; C. R. Walker, Assoc. Prof. Botan.; E. N. Anderson, Ass. Prof.

Lisbôa (Portugal).
Inst. de Botan., Fac. de Farmacia, Univ. Dir. Prof. R. Teles Palhinha.

Liverpool (England).
Labor. of Botan., Fac. of Sc. of the Univ. Dir. Prof. J. McLean Thompson.

Ljubljana — Laibach (S.H.S.).
Botan. Inst. d. Univ., Kongresni trg. Dir. Prof. F. Jesenko.

Logan, Utah (U.S.A.).
Dept. of Botan. and Plant Pathol. Agric. Coll. of Utah. B. L. Richards, Ph.D., Prof. Botan. and Plant Pathol., Botan. and Plant Pathol.; H. L. Blood, Ass. in Botan. and Plant Pathol.; L. F. Nuffer, M.A., Ass. Prof. and Ass. Botan.

London (England).
Dept. of Botan., Univ. Coll., Gower Street, W.C. Dir. Prof. J. W. Oliver. — Dr. T. G. Hill, Plant Physiol.; Dr. P. Haas, Plant Chemistry; Dr. E. J. Salisbury, Ecol.; E. M. Cuttine, Mycol.; Miss B. Russell-Wells, Plant Chem.; Miss V. L. Anderson, Ecol. *8—10 places. — Soil investigation, Bio-Chemistry of Plants. Ecol. of maritime vegetation and woodlands.* — Field station for maritime Ecol. at Blakeney Point, Norfolk.
Botan. Dept., King's Coll., Strand, W.C. Dir. Prof. R. Ruggles Gates. — Dr. E. J. Schwartz; Mr. R. E. Chapman; Dr. J. Latter; Mr. W. R. I. Cook. *Cytol.* ⊕ Cytol. preparations. — Genetical research is carried on in connection with the Royal Botan. Gardens, Regents Park, N.W. 1.
Botan. Dept., Bedford Coll. of the Univ., Inner Circle, Regent's Park, N.W. 1. Dir. Prof. W. N. Jones.
Botan. Labor., East London Coll. (Univ. of London), E. 1, Mile End Rd., Dir. Prof. Dr. F. E. Fritsch. — Dr. F. M. Haines, Plant-Physiologist; Dr. Nellie Carter, Lect.; Miss F. Rich, Research-Ass. *6 places, especially equipped for algol. research. — Considerable collection of figures and papers relating to the taxonomy of fresh-water Algae.*
The Botan. Labor., Westfield Coll. of the Univ., Finchley Road, Hampstead, N.W. 3. Dir. Dr. E. M. Delf, Head of Dept. — Miss E. J. Fry, M.Sc., Ass. Lect. *2 places for Anat. or morphol. of plant structures.* ⊕ Marine Algae: Material for anat. investigations into plant structure, Lichens.
Dept. of Botan., Birkbeck Coll., Univ. of London, Breams Buildings, E.C. 4. Dir. Prof. Helen Gwynne-Vaughan, D.B.E., LL.D., D.Sc., F.L.S. — Lect.: B. Barmes, B.Sc., F.L.S., H. Duerden, B.Sc. Part time Lect.: J. Ramsbottom, O.B.E., M.A., F.L.S. A. J. Wilmott, B.A.; B. D. Bolas, M.Sc. Prof. research Ass.: H. S. Williamson, B.Sc. *About 6 places especially fitted for cytol. and mycol. work.*
Botan. Labor. Imperial Coll. of Sc. and Technol. of the Univ., Prince Consort Rd., S.W. 7. Dir. Sir John B. Farmer. — Prof. V. H. Blackman, Prof. of Plant Physiol.; Prof. P. Groom, Prof. of the Technol. of Woods and Fibres; Prof. S. B. Schryner, Prof. of Biochemistry. Ass. Prof.: S. G. Paine, Bact.; W. Brown, Plant Pathol.; R. T. Tabor, Mycol. *About 40 places for general Botan., Cytol., Plant Physiol., Biochemistry, Mycol., Bact.*
Labor. of botan. of the Royal Holloway Coll. for women, School of the Univ., Englefield Green, Surrey.

London, Ont. (Canada).
Dept. of Botan. of the Univ. of Western Ontario. Dir. N. C. Hart, M.A. — H. V. Berdan, B.A.

Los Angeles, Cal. (U.S.A.).
The General Botan. Labor. of the Univ. of Southern California. Dir. Prof. H. de Forest. — Prof. G. R. Johnstone; Prof. A. C. Life; Prof. E. S. Spalding.

Louvain — Leuven (Belgique).
Labor. botan. Dir. Prof. P. Biourge. — Prof. V. Grégoire; Prof. P. Martens.

Lucknow (British-India).
Botan. Dept., Univ. Dir. Prof. Dr. B. Sahni. — S. K. Mukerji (Lect. in Plant Ecol. and Systematic Botan.); H. P. Chowdhary (Lect. in Plant Physiol.); S. K. Pandé (Demonstrator). *2 guests can be accommodated for Plant Morphol.* ⊕ Fossil plants.

Lugansk, Dongebiet, Ukraine (U.d.S.S.R.).
Kab. f. landwirtsch. Botan. d. Timirjasev-Technikums, Poćtoogi jaščiks 3.

Lund (Sverige).
Botan. Mus. und Garten der Univ. Dir. vacant.: Priv.Doc. C. Naumann, Doc. an der Univ. Lund.— Konserv. Otto R. Holmberg. 1 Amanuensis; mehrere ao. Aman. *Einige Plätze für Systematik, Morphol., Pflanzengeographie.* ⊕ Flora von Skandinavien.

Lwów (Polska).
Botan. Inst. für Systematik u. Pflanzenmorphol. d. Univ., Mikołaja 4. Dir. Stanisław Kulczyński. — Antoni Maczak, Ass.; Dr. Marjan Koczwara, Ass. — Botan. Garten (Długossa 4).
Biol.-Botan. Inst. d. Univ. Dir. S. Krzemieniewski.
L'Inst. de Botan. et d'Agric. de l'Académie de la Méd. Vétérinaire, rue Kochanowskiego 63. Dir. Prof. Bronislas Janowski. — Ing. St. Wojcicki. ⊕ Herbiers des plantes fourragères.
Inst. botan. de la polytechnique. Dir. Prof. D. Szymkiewicz. — Dr. Marie Matlakówna.

Lyon (France).
Labor. de Botan., Fac. Méd. Univ. Dir. Prof. Bretin.
Labor. de Botan. Fac. des Sc. Univ. Dir. Prof. J. Beauverie. — R. Douin, Maître de Conférences; H. des Gayets, Chef des travaux; L. Faucheron, Ass.; A. Tronchet, Ass.; Rosset Martin, Chef de Labor. *4 places pour Cytol., Phytopathol., etc. Les grands herbiers généraux: Roland Bonaparte et Gandoger, ainsi que l'herbier de France de Rouy, permettent d'entreprendre des recherches de Biogéographie, etc.* ⊕ Herbiers généraux Roland Bonaparte, Rouy, Gandoger, etc.

Machatsch-Kala (U.d.S.S.R.).
Botan. Abt. des Dagestaner Wissenschaftl. Forschungs-Inst., Ingenernaja 45.

Madison, Wis. (U.S.A.).
Botan. Labor., Univ. of Wisconsin, Biology Bldg. Dir. James Bertram Overton (Chairman). — Prof., Assoc. Prof., Ass. Prof.: James Bertram Overton (Chairman), Charles Elmer Allen, Benjamin Minge Duggar, Edward Martinus Gilbert, George Smith Bryan, John Jefferson Davis, Rollin Henry Denniston, Emma Luella Fisk, Charles Herbert Otis. *Can accommodate 8 or 10 guests. — Complete outfits*

for morphol. and cytol. botany. Chemical and physical apparatus and labor. for general and research work in plant physiol. lines. Greenhouses, herbarium, etc. Ⓟ Herbarium material for exchange.

Madrid (España).
Inst. de Botan. Fac. Sc. Univ. Dir. A. Garcia Varela (Organogr.). — Prof. A. Cab. y Segares (Phytogr. et geogr. botan.); Prof. J. Gogorza y Gonzalez (Physiol.).

Magyaróvár (Ungarn).
Botan. Labor., Landw. Akademie Magyaróvár. Dir. Prof. Franz Üzouyi. — Desiderius Révy, Ass. Ⓟ Phanerogame Pflanzen; parasit. Pilze der höheren Pflanzen.

Manchester (England).
Botan. Labor. of the Victoria Univ. Dir. Prof. F. E. Weiss. — Prof. W. H. Lang, Barker Prof. of Cryptogamie Botan.; W. O. Howarth, Lect. in Botan.; J. Walon, Lect. in Botan.; L. J. F. Brimble, Lect. in Botan.; Miss Kathleen Drew, Ass. Lect. in Botan. *6 working places for guests.*

Manhattan, Kan. (U. S. A.).
Dept. of Botan. and Plant Pathol. Head of Dept.: Prof. L. E. Melchers. — Prof. E. C. Miller, Physiologist; Assoc. Prof. F. C. Cates, Taxonomist and Ecologist; Ass. Prof. R. P. White, Ass. Plant Pathologist; Dr. H. Fellows, Assoc. Pathologist; Dr. J. L. Weimer, Assoc. Pathologist; C. O. Johnston, Ass. Pathologist; Prof. W. E. Davis, Plant Physiol. (seed germination); Assoc. Prof. H. H. Haymaker (Botan., Plant Pathol.); Ass. Prof. N. E. Dalbey (Botan.); D. J. Cashen (Instr. Botan.); D. R. Porter (Plant Pathol. Spec.). *1 Place for plant Pathol., Plant Physiol., Taxonomy, Ecol. — Fully equipped labor. Greenhouse with control apparatus for plant pathol.* Ⓟ Fungi, cultures and other plant material.

Manila (Philippine Islands).
Dept. of Botan. Univ. of the Philippines. Dir. José K. Santos. — W. H. Brown; J. Marañon.
Labor. of Botan. Univ. Santo Tomas. Dir. M. Fernandez Gonzalez. — Leon Guerrero.
Division of Botan., Bureau of Sc. Dept. of Agric. Dir. Dr. W. H. Brown.

Manitou, Col. (U. S. A.).
Botan. Dept. of the Alpine Labor. F. E. Clements (Ass. in Ecol., Mycol.).

Marburg a. d. L. (Deutschland).
Botan. u. pharmacognostisches Inst., Pilgrimstein 3. Dir. Prof. Dr. Peter Claußen. — Ass.: Karl Dening, Hilde Wenderoth. Im Inst. ist tätig: Prof. Dr. Max Nordhausen. *Bact., Mycol., Anat. der Pflanzen. — Zimmer f. konstante Temperatur, Zentrifugenraum, bact. Raum, Versuchsgewächshaus.*

Marseille (France).
Labor. de Botan. Fac. Sc. Univ. Dir. Prof. J. de Cordemoy. — Prof. H. Jumelle.

Melbourne (Australia).
Labor. de Botan. and Plant Physiol. Univ. Dir. Prof. A. J. Ewart.

Mexico, D. F. (Mexico).
Sect. botan. der Dir. de Estudios Biol., 7/a. calle de Balderas 94. Dir. Prof. Max Martinez.

Milano (Italia).
Ist. botan. Univ. Dir. Prof. G. B. Traverso. — Prof. U. Brizi.

Milwaukee, Wis. (U. S. A.).
Dept. of Botan. Marquette Univ. J. A. Lounsbury (Ass. Prof. of Botan.).

Minneapolis, Minn. (U. S. A.).
Dept. of Botan. Univ. of Minnesota. Dir. J. A. Harris. — H. P. Sorokin (Research worker).

Minsk (U. d. S. S. R.).
Botan. Inst. d. Weißrussischen Univ. Dir. Prof. N. M. Gaidakov. — Ass.: N. A. Isitkovski, Frl. O. D. Akimowa. Ⓟ Samen.
Labor. de Botan. de la Stat. Expérim. Paludéenne (Lect. Botan.-Lysimetrique), 4, Lagerny per. (Corps Physique de l'Univ.). Dir. V. V. Adamoff. — V. V. Gorbounoff (Botan.). Specialistes en cultures: S. P. Garkovy, S. S. Letkovsky, N. M. Polotchanine. Spec. forestier: I. I. Soboleff; I. St. Tomachevitsch, A. Ph. Khmyza. *4 places Pavillons de végétation, terrain cultivé et vierge pour expérim. Collections des plantes vivantes, Collections dendrol. Acclimatations, introduction, hybridisation. Etudes des mauvaises herbes, Carex, Hieracium, Salix, Cryptogames.* Ⓟ Plantes vivantes.

Missoula, Mtn. (U. S. A.).
Dept. of Botan. State Univ. Prof. J. E. Kirkwood (Ecol. of the Rocky Mountain flora).

Modena (Italia).
Inst. Botan. Univ., Viale Margherita. Dir. Prof. Augusto Béguinot. — Dott. Francesco Panini, *8 places pour Anat. et Biol. des plantes. Classification des plantes surtout med. et industrielles. Cryptogamie.*

Montpellier (France).
Labor. de Botan. Fac. Sc. Dir. Prof. Flahault. — Prof. J. Pavillard.

Montreal, Quebec (Canada).
Dept. of Botan. Mc. Gill. Univ. Dir. F. E. Lloyd, M.A., F.R.S.C. — Miss C. M. Derick, M.A.; G. W. Scarth, M.A.
Dept. of Botan. Univ. of Montreal, 1265, Rue Saint Denis. Dir. Frère Marie-Victorin, M.R.S.C.

Morgantown, W.Va. (U. S. A.).
Dept. of Botan. Dir. P. D. Strausbaugh.

Morioka (Japan).
Botan. Labor. Morioka Imper. Coll. of Agric. and Forestry.

Moskau (U. d. S. S. R.).
Labor. d. Botan. Gartens d. Univ., 10, Mestscharskaja 28. Dir. Prof. Dr. M. I. Galenkin. — K. I. Meyer; N. J. Katz; P. A. Smirnaff; A. E. Schadowsky.
Wissenschaftl. Forschungsinstitut f. Botan. a. d. Physik.-Mathemat. Fac. d. 1. Moskauer Staatl. Univ., Mochowaja 11. Dir. Prof. L. J. Kursakow. — Prof. Al. Kiesel; Doc. Dr. M. P. Vercholanzeva.
Botan. Labor. der 2. Univ. Dir. S. J. Ivanov.
Labor. für Anat. der Pflanzen der 2. Univ. Dir. A. N. Stroganov.
Kabinet der Morphol. und Systematik der Pflanzen, Gr. Nikitskaya 6. Ass.: Dr. N. Komarnitzi, Dr. B. Flerov, Dr. A. Rasumov. Custos des Herbariums: Dr. D. Sireystschikov. *2—3 Gäste können arbeiten über Morphol. und Cytol. der Algen und Pilze. Systematik der Phanerogamen und Floristik der U.S.S.R.* (Herbarium).
Botan. Labor. d. Mendeleev'schen Chem.-Technol. Inst., Ploscad Iljica.
Labor. für allgemeine Botan., Landw.,,,Timirjasev"-Akad., Petrowsko-Rasumowskoje. Prof. Al. Kiesel.

Moscow, Id. (U. S. A.).
Dept. of Botan. Univ. Coll. of Agric. F. W. Gail, Ph.D., Prof. Botan., Botan.; Lois Clark, Ph.D. Ass. Prof. Botan.

München (Deutschland).
Botan. Inst. u. Garten, Menzingerstr. 13. Dir. Prof. Dr. C. von Goebel. — Hauptkonserv.: Prof. Dr. Walter Kupper. Konserv.: Dr. Karl von Schoenau.
Botan. Inst. der Techn. Hochsch., 2 NW., Arcisstr. 21. Dir. Geh.Reg.R. Prof. K. Giesenhagen. — Prof. Dr. G. Dunzinger, Konserv. *2—4 Plätze für*

mikroskopisch-morphol. Untersuchungen. ⊕ Technisch wichtige Rohstoffe aus dem Pflanzenreich.

Münster (Deutschland).
Botan. Inst. und Garten. Dir. Prof. Dr. W. Benecke.

Mukden, Manchuria (China).
Botan. Labor. South Manchurian Med. Univ. Dir. Prof. Yasona Hukuda.

Nancy (France).
Sect. Botan. forestière d'Ecole Nation. des Eaux et Forêts.
Labor. du Jardin alpin Vosgien de l'Univ. de Nancy (au Hohneck). Dir. Prof. Edmond Gain. — Avant 1914 il y avait: 1 Dir. en résidence à Nancy; 1 Conserv., 1 Jardinier chef des Cultures, en résidence à Gérardmer (Vosges). ⊕ Graines de plantes des Hautes-Vosges.—Il y avait une exposition de plantes alpines dans des cadres vitrés et une exposition de plants des Hts. Vosges dans des cadres vitrés.

Nanking (China).
Dept. of Botan. Biol. Labor. Sc. Society, Wonder Lane. Dir. Hu Hsen-Hsu. — Ching-Yueh Chang (Prof. of Plant Morphol.).

Napoli (Italia).
Royal Inst. et Jardin Botan., Via Foria 223. Dir. Prof. Dr. Fridiano Cavara. — Dr. Gaetano Rodio, Aide; Dr. Rosa Parisi, Ass.; Dr. Maria Fiore, Ass. adjoint. *50 places pour Anat., Physiol., Mycol. Génétique.* ⊕ Collection de graines et fruits et d'Exsiccata.

Neuchâtel (Schweiz).
Labor. f. Botan. u. Pflanzenphysiol. Dir. Prof. Dr. Henri Spinner.

New Brunswick, N.J. (U.S.A.).
Botan. Dept. Coll. for Women. M. E. Brumfield (Ass. in Botany).
Dept. of Botany and Plant Pathol. Univ. State Coll. of Agric. J. W. Shive, Ph.D., Chief of Div., Prof. Plant Physiol.; L. G. Campbell, Field Ass. in Plant Pathol.; M. A. Chrysler, Ph.D., Assoc. Prof. Botan.; Elizabeth Clark, M.S., Research Ass. in Plant Pathol.; Jessie G. Fiske, M.S., Ass. Prof. Botan. and Physiol.; B. R. Fudge, M.S., Research Ass. in Plant Physiol.; C. M. Haenseler, Ph.D., Instr. Botan.; A. P. Kelley, Ph.D., Instr. Botan.; W. H. Martin, Ph.D., Ass. Prof. and Plant Pathol.; W. R. Robbins, M.S., Research Ass. in Plant Physiol.; J. A. Sutfin, Research Ass. in Plant Pathol.

New Haven, Conn. (U.S.A.).
Dept. Botan. State Agric. Experim. Station. G. P. Clinton, Sc.D., Botan. in charge; W. R. Hunt, Ph.D., Ass. Botan.; Florence A. McCormick, Ph.D., Pathol.; E. M. Stoddard, B.S., Pomol.
Osborn Botan. Labor. Yale Univ. C. G. Deuber (Plant Physiologist); R. P. Marshall (Ass. Pathol.); G. L. I. Zundel (Investigator).

New York City, N.Y. (U.S.A.).
Dept. of Botan. Columbia Univ.
Labor. of the New York Botan. Garden, Bronx Park, New York City, U.S.A. N. L. Britton, Ph.D., Sc.D., LL.D., Dir.-in-Chief—Dir. A. B. Stout, Ph.D. Marshall A. Howe, Ph.D., Sc.D., Ass. Dir.; John K. Small, Ph.D., Sc.D., Head Curator of the Mus.; A. B. Stout, Ph.D., Dir. of the Labor.; P. A. Rydberg, Ph.D., Curator; H. A. Gleason, Ph.D., Curator; Fred J. Seaver, Ph.D., Curator; Arthur Hollick, Ph.D., Paleobotanist; Percy Wilson, Ass. Curator; Palmyre de C. Mitchell, Assoc. Curator; John Hendley Barnhart, A.M., M.D., Bibliographer; Sarah H. Harlow, A.M., Librarian; H. H. Rusby, M.D., Honorary Curator of the Economic Collections; Elizabeth G. Britton, Honrary Curator of Mosses; Mary E. Eaton, Artist; Kenneth R. Boynton, B.S., Head Gardener; Robert S. Williams, Administrative Ass.; Torasaburo Susa, M.S.Ag., Technical Ass.; H. M. Denslow, A.M., D.D., Honorary Custodian of Local Herbarium; E. B. Southwick, Ph.D., Custodian of Herbaceous Grounds. *10 places for plant taxonomy, morphol., and genetics.* ⊕ About 15000 species and varieties of plants are growing on the grounds and in the greenhouses.

Nijny-Nowgorod (U.d.S.S.R.).
Kabinett der Morphol. und der Systematik der Pflanzen, Univ. Dir. Prof. S. S. Stankoff. — Ass.: D. S. Awerkieff, S. P. Sybina, A. A. Sokoloff, W. P. Nogteff. *14 Plätze für Histol. und Anat. der Pflanzen Phytopathol., Geographie der Pflanzen.* ⊕ Herbaria, Pflanzen aus Mittel-Rußland. — Forstlabor. und Labor. der Phytopathol.

Nikolaev, Ukraine (U.d.S.S.R.).
Botan. Inst. am Inst. f. Volksbildung. Dir. Prof. F. Geist.

Nikolsko-Ussurijsk (U.d.S.S.R.).
Ständiges Botan. Kabinett z. Erforschung d. Flora d. Ussurijschen Gebiets, Krasnoarmejskaja, Haus d. Geograph.-Gesellsch.

Norman, Okla. (U.S.A.).
Dept. of Botan. Univ. of Oklahoma.

Northampton, Mass. (U.S.A.).
Botan. Labor. of Smith Coll. Dir. Prof. W. F. Ganong. — A staff of 9 persons engaged in: 1 Prof. in plant physiol., 1 Assoc. Prof. in morphol., 1 Assoc. Prof. in ecol. and classification, 1 Ass. Prof. in horticulture and landscape gardening, 1 Ass. Prof. in bact. and pathol., 1 Instr. in botan. history and education, 1 Ass. in the physiol. labor. *Gardens, greenhouses, special labor. These facilities are open for competent investigators.*

Novočerkassk (U.d.S.S.R.).
Botan. Kabinett am Don. Polytechnischen Inst. Dir. Prof. Dr. A. Th. Fleroff. — N. Hyp. Andrejev; W. N. Balardin. *5 Plätze für mykol., mikrobiol. und botan. Untersuchungen, speziell technische Mykol. und Gärungsorganismen, Hydrobiol.* ⊕ Pflanzen vom Nordkaukasus. — Mykotechnische Abt.

Noworossijsk (U.d.S.S.R.).
Staatl. Inst. z. Erforschung der Inundationsgebiete, Archangelskaja 15.

Odessa, Ukraine (U.d.S.S.R.).
Botan. Labor. u. Mus. INO, Kominternstr. 2. Dir. Prof. Dr. Th. M. Porodko. — I. D. Stscherback, wissensch. Mitarbeiter; B. N. Axentiew, Ass.; V. V. Abramowitsch, Aspirant; A. A. Iwanowskaja, Aspirant. *3 Plätze für Untersuchungen auf dem Gebiet des Wachstums und der Tropismen der Pflanzen.*
Inst. f. Pflanzenanat. u. system. Botan. am Chem.-Pharm. Inst.
Botan. Labor. d. Landwirtschaftlichen Inst. Dir. Prof. G. A. Borowikow. — Ass.: L. S. Ǧakulin. Stipendiaten: A. N. Schumakow, K. K. Wolf, Gojko.
Sect. f. Morphol. u. System. d. Pflanzen des Katheders f. Biol. d. Wissenschaftl. Forschungsinst. Dir. Prof. Dr. O. Svirenko.

Oberlin, Ohio (U.S.A.).
Botan. Labor. of Oberlin College. Dir. Fr. Orv. Grover. — Doctor Susan P. Nichols, Prof.; George T. Jones, Instructor. *2 working-places for guests. Morphol.* ⊕ Nord Amer. Phanerogams and Ferns, esp. from the Rocky Mountains, Ohio, New England. Fungi from Ohio.

Omsk (U.d.S.S.R.).
Botan. Abt. d. med. Inst. Dir. Prof. V. F. Semenov.

Orel (U. d. S. S. R.).
Muratowsche Botan. Station, Abt. d. Schatilowschen Landwirtsch. Bezirks-Versuchsstation, Gubsemuprawlenie.

Ôsaka (Japan).
Botan. Labor. Ôsaka Provincial Med. Univ., Ishibashi near Osaka. Dir. T. Sugimoto.

Oslo (Norge).
Botan. Labor. Univ. Dir. Prof. H. H. Gran.

Ottawa (Canada).
Division of Botan., Central Experim. Farm. John Adams; T. G. Major (Diseases of tobacco); C. E. Saunders, Ph. D. (Biol.).
Labor. of Botan., Dept. of Agric. Dir. H. T. Gussow.

Otusy, Krim (U. d. S. S. R.).
Wissenschaftl. Station v. Karadatsch, Feodossijsk. ujesd.

Padova (Italia).
Ist. botan. et Jardin botan. Univ. Dir. Prof. Giuseppe Gola. — Aiuto: Dr. C. Cappelletti (Anat., Cytol., Mycol., Hepatiques). Ass.: Dr. S. Zenari (Floristica). Allievo interno: S. Tonzig (Floristique, Systématique, Phytogéographie).

Palermo (Italia).
R. Ist. botan. dell'Univ. Dir. Prof. L. Buscalioni. — Prof. L. Domenico (Conserv. dell Erbario); Dott. B. Francesis (Vicedir. del giardino coloniale); Prof. G. Catalano (Ass. al R. Orto Botan.); Dott. G. Cultrera (Aiuto al R. Orto Botan.); Dott. S. Fel ce (Chimia). *Botan. generale e sistematica.*

Paris (France).
Labor. de Botan. 1. Fac. Sc. nat. Univ., 1, rue Victor Cousin. Dir. Prof. P. A. Dangeard. — M. Coupin, Chef des Travaux; M. Buchet; M. Pierre Dangeard. *1 douzaine place pour Anat., Histol., Développement des Algues et des Champignons.*
Labor. de Botan. général Fac. de Pharmacol. Dir. Prof. J. Guignard. — Prof. P. Guérin.
Labor. de Botan., Ecole pr. des Hautes Etudes. Dir. Prof. L. Mangin. — G. Brückner.
Labor. de Botan. de l'Ecole Normale Supérieure, 45, rue d'Ulm, Ve. Dir. Prof. Blaringhem. — Mr. Maresquelle, Agrégé-Préparateur. *Mycol.*
Labor. d'Anat. végétale, Mus. d'Histoire Nat., Ve, 57, rue Cuvier. Dir. Prof. Costantin.

Pavia (Italia).
Ist. Botan. Labor. Crittogamio. Dir. Prof. Luigi Montemartini. — Dr. L. Maffei; Dr. F. Givelli; Dr. M. Curgi; Dr. M. Turconi; Dr. M. Barbarini. *Crittogamia, Anat., Fisiol.* ⑨ Piante e funghi. — Crittogamia e parasitologia.

Peking (China).
Botan. Inst. Tsinghua Univ. Dir. Prof. S. S. Chien.

Perm (U. d. S. S. R.).
Botan. Labor. mit Versuchsstat., Univ. Dir. Prof. Dr. D. A. Sabinin.
Botan. Inst. und Hortus botan., Univ. (Morphol. Teil u. Garten). Dir. Prof. Dr. A. Henckel .— P. N. Krassowski, Ass.; K. N. Wischnewetzkaja, Ass.; R. N. Igoschin, wiss. Aspirant; Prof. A. A. Chrebtow; A. A. Drotschnewa. ⑨ Mikroskopische Planktonpraep. (Diatomeen des Schwarzen Meeres), Pilz- u. Bakterienkulturen.

Perugia (Italia).
Gabinetto e Giardino botan., R. Univ. Dir. Prof. O. Kruch.

Philadelphia, Pa. (U. S. A.).
Dept. of Botan. Univ. of Pennsylvania. Ass. Prof. W. R. Taylor; Ass. Prof. W. Seifriz.

Pietermaritzburg (South Africa).
Botan. Labor. Dir. Prof. J. W. Bews.

Poitiers (France).
Stat. de Botan. Dir. Prof. H. Ricôme.

Port au Prince (Haiti).
Dept. of Botan. and Plant Pathol. Service techn. Dir. H. D. Barker.

Poughkeepsie, N. Y. (U. S. A.).
Dept. of Botan. Vassar Coll. Dir. E. A. Roberts (Chairman).

Poznań — Posen (Polska).
Botan. Inst. Univ. Dir. A. Wodźiczko. *Mycol. (Anat. végétal); localisation des ferments oxydants; champignons.*
Inst. f. Botan. u. Phytopath., Univ. Dir. Prof. Dr. B. Namyslowski. — Dr. B. Liebetanz, Ass.

Praha — Prag (C. S. R.).
Botan. Inst. d. Čech. techn. Hochsch., XII, Haóličkovy sady 58. Dir. Prof. Dr. Jaroslav Peklo.
Lehrkanzel f. Botan., Warenkunde u. Mikroskopie d. Deutsch. Techn. Hochsch., II, Ostrovni 24.
Botan. Inst. und Botan. Garten der Deutschen Univ., II, Viničná 3a. Dir. Prof. Dr. Fritz Knoll.— Dr. Franz Pohl, Ass.; Dr. Franz Peterschilka, Ass.; Dr. Franz Schatanek, Demonstr. Am Inst. tätig: Prof. Dr. Adolf Pascher, Prof. Dr. Karl Rudolph; Dr. Franz Firbas, Prof. Dr. Günther Beck von Mannagetta. *2 Plätze für systematisch-botan., mikroskopische, floristische Arbeiten (Herbarium). Derzeit hauptsächlich Blütenforschung verschiedener Richtung Unter der Leitung von Prof. Pascher vorwiegend Kryptogamenforschung, unter der Leitung von Prof. Rudolph hauptsächlich Moorforschung im Sinne einer Floristik der Postglazialzeit (Pollenanalyse), ferner Pflanzengeographie.* ⑨ Herbarpflanzen, botan. Praep. Tausch lebender Pflanzen (Samen, Knollen, Zwiebeln u. a.) zum Anbau im Botanischen Garten.
Botan. Inst. der Karls-Univ. (Botan. ústav'a botan. zahrada), II, Na Slupi 433. Dir. Prof. Dr. J. Velenovský. — Dr. Ladislaw Viniklář; Dr. Karel Cejp; Dr. Albert Pilát. ⑨ Pilze, palaeont. Material.

Pretoria, Transvaal (South Africa).
Capetown Branch Labor. *South African Flowering Plants.*
Labor. of the Dept. of Botan., Transvaal Univ. Coll. Dir Prof. Dr. C. E. B. Bremekamp. — Mr. B. Elbrecht, Senior Lect. 2 Demonstrators. *Temporary arrangements might be made for the reception of 1 or 2 users interested in plant-physiol. research. — Apparatus for the study of growth and movement in plants.* — ⑨ Transvaal Vascular Plants. Transvaal Algae.

Providence, R. I. (U. S. A.).
Dept. of Botan. Brown Univ. Dir. Walter H. Snell, Ph.D. — Dr. N. O. Howard; Mr. W. G. Hutchinson. *Places can be provided as necessary for plant physiol., plant pathol., mycol. — Good photographic equipment.* ⑨ Wood destroying fungi, forest tree fungi.

Pullman, Wash. (U. S. A.).
Dept. of Botan. State Coll. F. L. Pickett, Ph.D., Head of Dept.; Hannah C. Aase, Ph.D., Ass. Prof. Botan., Ass. Seed Analyst; Mildred E. Manuel, B.S., Instr. Bot.; Harold St. John, Ph.D., Assoc. Prof. Botan.

Quebec (Canada).
Dept. of Botan. Mac Donald Coll. Dir. B. T. Dickson, B.A., Ph.D. — J. G. Coulson, M.A.; T. C. Vanterpool, B.S.A.
Dept. of Botan. Laval Univ. Dir. A. Robitaille, L.Ph.

Quito (Ecuador).
Botan. Inst. Univ. Dir. Prof. G. N. Paredas.

Rabat (Maroc).
Division botan. d'Inst. sc. chérifien. Dir. Prof. L. Emberger.

Raleigh, N. C. (U. S. A.).
Botan. Labor. of the North Carolina State Coll. Dir. B. W. Wells. — Dr. S. G. Lehman, Dr. R. F. Poole; Prof. I. V. Shunk; Dr. D. B. Anderson; Mr. L. A. Whitford; G. W. Fant. *3 places for Bact., Plant Pathol., Plant Physiol.* ⊕ Plant pathol. specimens.

Reading (England).
Botan. Labor. Univ. Dir. Prof. W. Stiles. — T. L. Prankerd (Lect.); W. T. Saxton (Lect.). *5 places for Plant Physiol. and Morphol.*

Reims (France).
Inst. botan. Univ. Dir. Prof. Ch. Mire.

Rennes (France).
Labor. botan. Univ. Dir. Prof. L. Daniel; Prof. P. Lesage.

Riga (Latvija).
Die morphol.-sytematische Abt. des Botan. Labor. der Univ., Kronvalda bulv. 4. Dir. Doc. Dr. N. Malta. — Aleksanders Zāmels; Paulis Galenieks; Heinrichs Skuja.
Botan. Abt. d. naturwissensch.-math. Fac. d. Herderinst. Dir. Prof. V. R. Kupfer.

Roma (Italia).
Labor. de Bot. et Hortus botan., Via Milano, 75. Dir. Prof. Dr. Pietro Romualdo Pirotta.—Dr. Enrico Carano, Prof. de Botan.; Dr. Fabrizi Cortesi; Dr. Valeria Bambacioni; Dr. Maria Ventura; Dr. Giuseppina Dragina. *Recherches de microscopie surtout, mais aussi pour la détermination etc. des plantes.*

Rosario (Argentina).
Inst. f. spez. Botan. Dir. Prof. C. di Pascal.

Rostock i. M. (Deutschland).
Botan. Inst. der Univ., Botan. Garten der Univ. Dir. Prof. Hermann von Guttenberg. — Dr. R. Bauch, Ass. *Mikroskopie, reizphysiol. Arbeiten.*

Rostow a. Don (U. d. S. S. R.).
Botan. Labor. Univ. Dir.Prof.E.K., Gemčujnikov.
Kabinett f. Botan. Dir. Prof. V. N. Verškovskij.

Rouen (France).
Ecole de Botan., 114, rue d'Elbeuf. Dir. Le Graverend. — Bourel, Jardinier-Chef. *La collection de 9000 plantes en culture.*

Ste. Anne de la Pocatière, Que. (Canada).
Dept. of Botan. of the Agric. School. Dir. Elzéar Campagna, B.A., B.S.A.

Salt Lake City, Utah (U. S. A.).
Dept. of Biol. East High School. Dir. A. O. Garrett.

San Francisco, Cal. (U. S. A.).
Dept. of Botan., Coll. of Pharmacy, Univ. of California. Dir. Prof. H. B. Casey.

Santiago (España).
Inst. f. beschreibende Botan. Dir. Prof. C. Sobrado.

Sapporo (Japan).
Botan. Inst. Col. of Agric., Hokkaidô Imper. Univ. Dir. Prof. K. Miyabe. — Prof. M. Hanzawa (Mykol.); Prof. T. Sakamorra (Physiol.); Ass. Prof. Y Toehinai (Pathol.); Ass. Prof. S. Kudô (System.); Ass. Prof. S. Kamei (Dendropathol.); Ass. S. Imai (Pathol.); Ass. S. Enomoto (Pathol.).

Saratov (U. d. S. S. R.).
Botan. Inst. der Univ. Dir. Prof. D. E. Janiševskij.

Seattle, Wash. (U. S. A.).
Botan. Labor. Univ. of Wash. Dir. Dr. F. C. Frye. — G. B. Rigg (Ass. Prof. of Botan.). — The dir. of this labor. is also dir. of the Puget Sound Biol. Station at Friday Harbor, Wash. (confer: B ologia Marina), and the 2 labor. are correlated.

Sendai (Japan).
Botan. Inst. Fac. of Sc., Tôhoku Imper. Univ. Prof. Masato Tahara; Ass. Prof. Yônosuke Okada; Ass. Prof. Shin-ichi Hibino; Lect. Tokutarô Itô.

Siena (Italia).
Ist. ed orto botan. Univ. Dir. G. Pollaci.

Simferopol (U. d. S. S. R.).
Botan. Kabinett Univ., Ul. Lenina 17.

Singapore (Br.-India).
Botan. garden and Labor. Dir. vacat. Ass. Dir. R. E. Holttum.

Smolensk (U. d. S. S. R.).
Botan. Kabinett u. Garten der Univ. Dir. Prof. J. J. Alekseev.

Sofia (Bulgarien).
Inst. f. Landwirtschaftliche Botan. N. A. A. Stoianoff, Prof. f. angew. Botan. (Systematik und Phytogeographie); Boris Stefanoff, Ass. (Systematik und Phytogeographie); Theodor Georgieff, Ass. (Systematik).
Botan. Labor. d. Physico-matemat. Fac. Dir. Prof. N. Arnaudow. — D. Jordanow, Ass.; B. Barsakow, Ass.; K. Paliewa, Ass. A. Popnikolow, Kustos.

Sopron — Ödenburg (Ungarn).
Botan. Inst. der k. ung. Hochschl. für Berg- und Forstingenieure. Dir. Prof. Dr. Dániel Fehér. — Rudolf Bokor, 1. Ass.; Géza Sommer, 2. Ass. *2 Arbeitsplätze für Pflanzenpathol. — Versuchsfelder, Apparate zur Bestimmung des CO_2-Gehalts der Luft.* — Bakt. Labor.
Inst. f. allgem. Botan. der forstlichen Versuchsstation. Dir. Prof. Dr. F. Kövessy.

Sorau, N.-L. (Deutschland).
Botan. u. Züchtungs-Abt. d. Forschungs-Inst. für Bastfasern. Dir. Prof. Dr. Ernst Schilling. — Ass. Dipl. agr. M. v. Schelika. *3 Plätze.* ⊕ Flachspflanzen, Fasern, Saatgut. — Labor. für Anat., Mycol., Vererbung von Flachs u. a. Bastfaserpflanzen. Versuchsgärten.

Southampton (England).
Botan. Labor. Univ. Coll. Dir. Prof. S. Mangham.

Stambul (Türkei).
Botan. Inst. der Univ. und der Tierärztl. Hochschule. Dir. Prof. E. Scherefeddin.

State College, Pa. (U. S. A.).
Botan. and Plant Pathol. Labor. the Pennsylvania State Coll. Dir. Frank D. Kern. — C. R. Orton, Prof. of Plant Pathol.; J. B. Hill, Prof. of Plant Morphol.; J. P. Kelly, Prof. of Botan. (Genetics); L. O. Overholts, Prof. of Botan. (Mycol.); W. S. Beach, Prof. of Plant Pathol.; H. W. Thurston, Prof. of Plant Pathol.; H. W. Popp, Prof. of Botan. (Plant Physiol.); E. L. Nixon, Prof. of Plant Pathol. Ext.; W. S. Krout, Assoc. Prof. of Plant Pathol.Ext.; G. F. Miles, Ass. Prof. of Plant Pathol. Ext.; R. S. Kirby, Ass. Prof. of Plant Pathol. Ext.; R. C. Walton, Assoc. Prof. Plant Pathol. Res.; P. Acquarone, Instr. Botan. *2 or 3 places for Plant Pathol., Uredinol., Genetics.* ⊕ Herbarium specimens of various groups. — 1 Labor. for diseases of vegetables in Bustleton, Pa.; 1 for diseases of vegetables in Arendtsville, Pa.

Stellenbosch (South Africa).
Dept. of Botan. of Univ. Dir. G. C. Nel, B.A., Ph.D., Prof. — Miss A. V. Duthie, M.A., Lect.

Stillwater, Okla. (U.S.A.).
Dept. of Botan. and Plant Pathol. State Agric. Coll. F. M. Rolfs, Ph.D., Prof. Botan. and Plant Pathol., Plant Pathol.; O. C. Schultz, B.S., Assoc. Prof. Botan. and Plant Pathol.; Robert Stratton, M.A., Ass. Prof. and Ass. Botan.; H. J. Featherly, M. S., Ass. Prof. Botan. and Plant Pathol.

Stockholm (Sverige).
Botan. Inst. der Univ. Dir. Prof. Otto Rosenberg, Ph.D. — Priv.Doc.: Dr. M. G. Stalfelt, Dr. J. A. Holmgren, Dr. G. Täckholm, Dr. L. G. Romell, Dr. K. Afzelius, Dr. O. Heilborn. 2 Ass. *3—4 Plätze für zytol. und pflanzenphysiol. Unters. — Warburgs Assimilationsapparat.* Ⓟ Anat. und zytol. Praep.
Botan. Inst. der Forsthochsch. Dir. Prof. T. Lagerberg.
Mykol. Labor. der Forstlichen Hochsch. sowie allg. forstbotan. Labor. Experimentalfältet. Dir. Prof. T. Lagerberg. — Elias Melin, Phil. Dr. Doc. *Einige Plätze für Mykol., Organographie und Anat. der Holzgewächse. — Apparate für pH-Bestimmungen.* Ⓟ Praep. der Biol. u. Morphol. der Holzgewächse; Samen und Früchte; allerlei forstschädliche Pilze, die von ihnen hervorgerufenen Krankheiten u. Fäulen.

Storrs, Conn. (U.S.A.).
Dept. of Botan. Coll. of Agric. E. W. Sinnott, Ph.D., Prof. Botan. and Genetics, Dean Div. of Agric. Sc.; G. S. Torrey, A.M., Ass. Prof. Plant Pathol.

Strasbourg — Straßburg (France).
Inst. Botan. et Jardin Botan. Fac. d. Sc., 7, rue de l'Université. Dir. Prof. Clodomir Houard. — M. Lagarde, Maître de Conférences; M. Chermezon, Chef de travaux pratiques; Mme Houard, Conserv. des Collections; M. Heé, Ass.; M. Schaechtelin, Ass.; M. Muller, Jardinier en Chef. *4 places de travail pour Recherches de Botan. et de Biol. végétale.*

Stuttgart (Deutschland).
Botan. Inst. u. Garten der techn. Hochsch. Dir. Prof. Dr. N. R. Harder.

Sverdlowsk (U.d.S.S.R.).
Botan. Labor. des Uralischen Polytechn. Inst. Dir. Prof. Kazanskij.

Swansea (England).
Botan. Dept. Univ. Coll. Dir. Prof. F. A. Mockeridge.

Sydney (Australia).
Botan. Labor. of the Univ. Dir. Prof. A. A. Lawson.

Syracuse, N.Y. (U.S.A.).
Dept. of Botan. Univ. Dir. W. L. Bray.

Szeged (Ungarn).
Botan. Inst. u. Botan. Garten, Univ., Tisza Ring 6I. Dir. Prof. Dr. I. Györffy. — Frl. Dr. E. Pákh, Ass.; Frl. Dr. E. Kol, Ass.; Zoltán Eber, Suppl. Ass.; Ant. Scheitz, Demonstrator. *2 Plätze für kryptogamische Studien.* Ⓟ Kryptogamen.

Taihoku, Formosa (Japan).
Botan. Labor. Taihoku Governm. Coll. of Agric. and Forestry.

Tartu — Dorpat (Eesti).
Pflanzenmorphol. u. Systematisches Labor., Univ. Dir. Doc. Dr. E. Spohr. — Mag. T. Lippmaa, älterer Ass.; M. Mändmetz, Hilfskraft. *3 Plätze für floristische u. systematische Untersuchg.* Ⓟ Herbarpflanzen insbesondere aus: Estland, Fennoskandien, W.- u. S.-Europa, Rußland, Asien.

Botan., Inst. Univ. Dir. Prof. H. Kaho (Pflanzenphysiol.).

Taschkent (U.d.S.S.R.).
Botan. Inst. der Mittelas. Staatsuniv. Dir. Prof. Paul Baranov. — P. Baronav, Dir. des Inst. u. Leiter der Abt. für Zytol. u. Cryptogamenpflanzen; M. Kultiassov, Leiter des Botan. Gartens; W. Drobov, Leiter der Abt. für Phanerogamenpfl.; M. Popov, Leiter der Abt. für Geographie der Pfl.; A. Wedensky, Custos des Herbariums; E. Korowin; H. Raikowa; O. Radkewitsch; E. Mokejewa. *Plätze zum Studium der mittelasiatischen Flora. — Herbarium der mittelasiatischen Flora (40000 Bl.);* Ⓟ *Pflanzen (lebende u. Exiccata), Samen und mikroskopische Praep.* — Die Botan. Gebirgsstation in Tschimgan (West-Tyanschan).

Tharandt i. Sa. (Deutschland).
Botan. Abt. der Forstlichen Hochsch. Dir. Prof. Dr. Ernst Münch. — Dr. phil. W. Bavendamm, 1. Ass.; Forstreferendar H. Lehmann, 2. Ass. *2 Plätze für forstbotan. und dendrol. Untersuchungen. — Forstbotan. Garten (1200 Nummern) und Holzsammlungen.* Ⓟ Dendrol. und Baumkrankheiten.

Tiflis (U.d.S.S.R.).
Botan. Kabinett der Univ. Dir. Lect. Z. Kančaveli.

Tôkyô (Japan).
Botan. Inst., Fac. of Sc., Tôkyô Imperial Univ. Koishikawa. Dir. Prof. K. Fujii, D. Sc. — Prof. K. Shibata (Biochemistry and Plant-Physiol.); Ass. Prof. H. Nakano (Ecol.); Prof. B. Hayata (Systematic Botan.); Ass. Prof. T. Nakai (Cryptogamic Botan.); Prof. K. Fujii (Plant-Morphol., especially Cytol., and Genetic); Lect. Dr. Ogura (Anat.); Lect. Dr. G. Yamaha (Cytol.); Ass. Dr. Sinoto (Genetics and Cytol.); Ass. Dr. M. Honda (Systematic Botan.); Ass. Dr. J. Kitasato (Biochemistry); Dr. S. Hattori (Biochemistry); Lect. T. Makino (Systematic Botan.); Lect. H. Hattori (Bact.); Ass. K. Watanabe (Plant Physiol.); Demonstrator Miss K. Yasui (Genetics); Demonstrator H. Kinoshita (Plant-Physiol.); Demonstrators: Dr. Y. Yamamoto, K. Kiyohara, Dr. A. Kimura, Dr. Y. Yamada. Ⓟ *Drie Specimens of plants. — Our Botan. Inst. includes following Labor.:* Labor. for Plant-Morphol. (Cytol. and Anat. including fossil plants labor.); Labor. for Plant-Physiol. and Ecol.; Labor. for Systematic Botan.; Labor. for Genetics; Labor. for Biochemistry (Plant-Chemistry).
Botan. Inst. Fac. of Agric., Tôkyô Imper. Univ., Komaba near Tôkyô. Dir. Prof. Dr. S. Ikeno. — Prof. Dr. S. Ikeno (Genetic); Em. Prof. Dr. M. Shirai (Pathol.); Ass. Prof. Dr. S. Kusano; Ass. Prof. K. Miyake.
Botan. Labor. Imperial Fishery Inst., Hukagawa. Dir. Prof. Dr. K. Okamura. — Prof. Dr. Kintarō Okamura (Phycol., Plancton).
Botan. Labor. 1. High School. Hongô. Dir. Dr. Ishikawa. *Zytol.*
Botan. Labor. High Seminary. (Kôtô Shihan Gakkô), Koishikawa. Dir. Prof. S Yamanouchi. *Genitik.*

Tomsk (U.d.S.S.R.).
Botan. Kabinett, Technol. Inst., Timirjazevskij Prosp. 9. Dir. Prof. N. N. Lavrov.

Torino (Italia).
R. Ist. botan. Univ. Dir. Prof. O. Mattivolo. — Ajuto D.sse Ritei Raineri; Conserv. Pietro Fontana, Ass. *Systematique, Mycol., Morphol., Physiol.* Ⓟ Podakinées, Sclerodermatacées, Hymenogastrées, Tuberacées.

Toronto (Canada).
Dept. of Botan. Univ. of Toronto. Dir. J. H. Faull, Ph.D. — R. B. Thomson, B.A.; H. B. Sifton, Ph.D.; G. H. Duff, Ph.D.; L. C. Cohman (Plant Pathol.).

Toulouse (France).
Inst. Botan. Univ. Fac. Sc. Dir. Prof. P. Dop. **Labor. de Botan. Fac. méd. Univ.** Dir. Prof. C. Gerber.

Touro, N. S. (Canada).
Dept. of Botan. Coll. of Agric. Dir. H. S. Cunningham, M.S.A.

Toyama (Japan).
Botan. Labor. Toyama Imperial Coll. of Pharmacy.

Trinidad (Br.-W.-India).
Dept. of Botan. and Genetics Coll. of tropical Agric.

Tschimgan, West-Tianschan (U.d.S.S.R.).
Botan. Gebirgsstat. d. Botan. Inst. d. Mittelas. Staatsuniv. Taschkent. Dir. Prof. P. Baranov.

Tsu (Japan).
Botan. Inst. Mie Imperial Coll. of Agric. and Forestry.

Tucumán (Argentina).
Inst f. Botan. Dir. Prof. M. Zelada.

Tübingen (Deutschland).
Botan. Inst. u. Botan. Garten. Dir. Prof. Dr. E. Lehmann.

Tunis (Tunis).
Service Botan. de la Dir. générale de l'agric., Ariana près Tunis. Dir. Prof. Vialas (Botan., Viticulture, Silviculture).

Turku (Finnland).
Botan. Labor. Dir. Harry Warén. — Mag. phil. Lauri E. Karl, Ass.; Dr. phil. E. Wainio. ,,*Herbarium Wainio*", *ein umfangreiches Flechtenherbarium, enthält Material von allen Weltteilen.* ⊕ Höhere und niedere Pflanzen (Phanerogamen, Moose, Flechten).

Upsala (Sverige).
Botan. Labor. Univ. (Botan. Institutionen). Dir. Prof. N. E. Svedelius (allg. Botan.). — Prof. Dr. H. O. Juel (System. Botan.).
Pflanzenbiol. Inst. d. Univ. (Upsala Univ. Växtbiol. Institution). Dir. Prof. R. Sernander. — Priv.Doc.: Dr. G. Einar du Rietz, Dr. Harry Smith, Dr. Hugo Osvald; Amanuensis: Carl G. Alm. *1—2 Plätze für ausländische Gäste. Bodenkunde aus biol. Gesichtspunkte. Allgemeine Kryptogamenbiol. Entwicklungsgeschichte und Verbreitung der nordeuropäischen Flora. — Spezialsammlungen von Bodenprofilen und Torfpraep., vollständiger pflanzenbiol. Mus. mit großer tiol. Flechtensammlung, Sammlung von Zooecidien, Bildungsabweichungen, Verbreitungsbiol. Haptomorphosen,*

Urbana, Ill. (U.S.A.).
Dept. of Botan. Univ. of Illinois. R. Kienholz (Instr.); J. F. Mueller (Research Ass.); Prof. H. Le Roy Shantz (Botan.).

Utrecht (Holland).
Botan. Labor. der Univ., Lange Nieuwstraat 106. Dir. Prof. Dr. F. A. F. C. Went. — Hauptass.: H. E. Dolk; Ass.: F. W. Went, S. de Boer. *2 Plätze für Untersuchungen über Reizleitung und Bewegungen bei Pflanzen. — Dunkelzimmer mit konstanter Temperatur. Apparat für Wachstumsmessung nach Koningsberger. Klinostat nach De Bouter.* Mus. u. Herbarium (Prof. Pulle), Phytopath. Labor. in Baarn bei Utrecht (Prof. Joh. Westerdijk).

Utsunomiya (Japan).
Botan. Labor. Utsunomiya Imper. Coll. of Agric. and Forestry.

Valencia (España).
Inst. f. Botan. Univ. Dir. Prof. F. Belbrán Bigorra.

Vancouver (Br.-Columbia).
Dept. of Botan. Univ. of Britisch Columbia. Dir. A. H. Hutchinson. — John Davidson, F.L.S.; Franck Dickson, B.A.

Waco, Tex. (U.S.A.).
Dept. of Botan. Baylor Univ. Dir. Prof. N. E. Jones.

Warszawa (Polska).
Inst. de Botan. Générale de l'Univ., Krakowskie-Przedmieście 26|28. Dir. Prof. Z. Wóycicki. — Dr. F. X. Skupieński; Dr. A. Luxenburg; Mme A. Mieszczańska. *22 places pour Recherches cytol. et anat.*
Botan. Labor. der freien Univ. Dir. Prof. Dr. Marja Skalińska. *Cytol. Embryol. Genetic.*
Inst. f. allgem. Botan., Hochsch. f. Land- u. Forstwirtsch. Dir. Prof. S. Dziubaltowski.

Washington, D.C. (U.S.A.).
Dept. of Botan. Howard Univ. Dir. C. S. Parker.

Weihenstephan, Bayern (Deutschland).
Botan. Inst. der Hochsch. für Landwirtschaft. Dir. Prof. Dr. Fr. Boas. — Dr. G. Claus. *4 Plätze für Pflanzenphysiologie, besonders physikalisch-chemischer Richtung.* — Labor. für Grünlandsbiol. und Pflanzenpathol.

Welikij Ustjug (U.d.S.S.R.).
Nord-Dwinsker Abt. d. Inst. f. angew. Botan., Nord-Dwinsker Gouv.

Wellesley, Mass. (U.S.A.).
Botan. Dept., Wellesley Coll. Dir. Prof. Margaret C. Ferguson, Chairman. — Margaret Clay Ferguson, Ph.D., Prof.; Howard E. Pulling, Ph.D., Prof.; Laetitia Morris Snow, Ph.D., Assoc. Prof.; Mary Campbell Bliss, Ph.D., Assoc. Prof.; Alice Maria Ottley, Ph.D., Assoc. Prof., Curator of Herbarium; Helen Isabel Davis, B.A., Ass. Prof.; Mary Louise Sawyer, Ph.D., Ass. Prof.; Helen Stillwell Thomas, M.A., Instr., Curator of Mus.; Grace Elizabeth Howard, Ph.D., Instr.; Beulah Pearl Ennis, Ph.D., Instr.; Silence Rowlee, M.A., Instr. *2 places for each subject: Anat., Bact., Cytol. and Morphol., Genetics, Physiol., Taxonomy.* ⊕ Herbarium sheets.

Wien (Österreich).
Botan. Inst. u. Garten, III/3, Rennweg 14. Dir. Prof. Dr. R. v. Wettstein. — Vice-Dir. Prof. Dr. E. Janchen.
Botan. Inst. and Inst. für Mikros. und org. Rohstofflehre, Technische Hochsch.. Dir. J. Weese. — H. Cammerloher, Ass.
Lehrkanzel f. Botan. der Hochsch. für Bodenkultur Wien, XVIII/1, Feistmantelg. 4. Dir. Prof. Dr. O. Porsch. — Dr. Viktor Folgner; Dr. Othmar Werner. *2 Plätze.* ⊕ Diapositive und Photographien.

Wilno (Polska).
Inst. d. allgemeinen Botan. d. Univ. (Inst. f. Botan. I), Zakretowa-Str. 15. Dir. Prof. Dr. P. Wisniewski. — Br. Szakien; H. Korwin-Kurkowska; I. Reniger.

Winnipeg, Man. (Canada).
Dept. of Botan. Univ. of Manitoba. Dir. A. H. R. Buller, Ph.D., D.Sc. — H. F. Roberts, B.A., LL.B., M.Sc.; C. W. Lowe; Ida K. Scott, B.A., B.Sc.

Woronesch (U.d.S.S.R.).
Botan. Inst. der Univ. (Pedag. u. Med. Fak.), Gruzowajas. Dir. Boris Mich. Kozo-Poljanski, o. Prof. der Botan. — Leon. Grig. Ramenski, Doc.; Wladislawa Lachewska; Mich. Petr. Tomin; Boris Al. Keller, e.o. Prof. für Pflanzenphysiol. ⊕ Zentralrussische Glacial-Relikte. Russ. Umbelliferae. Russ. Lichenes. — Geobotan. u. pedolog. Labor. (Doc. Ramenszl.)
Botan. Kabinett und Botan. Versuchsstation d. Landwirtsch. Hochsch. Dir. Prof. Dr. B. A. Keller.

— Dr. E. I. Proskorjakov, Ass.; Dr. M. P. Tomin, Ass.; A. F. Karelskaja; E. F. Keller; P. A. Nikitin; T. I. Popov. *4 Plätze für ökol. Anat. und Physiol. Vergl. Physiol. Geobotan. Bodenmikrobiol. — Spezielle Apparatur für ökol. Untersuchungen.* ⑨ *Pflanzen d. russischen Steppen, Halbwüsten und Wüsten.*

Würzburg (Deutschland).
Botan. Inst. Dir. H. Burgeff.

Yonkers, N. Y (U. S. A.).
Boyce Thompson Inst. for Plant Research, 1086 N. Broadway. Dir. W. Crocker. — J. M. Coulter (Plant morphol.); E. F. Davis (Biochemistry); J. M. Arthur (Biochemist); C. R. Orton, H. C. Bucha, Miles, Horsfall (seed sterilizers and insecticides); W. S. Bourn, B. Jenkins (factors, involved in the growth and maintenance of duck feeds in brackish waters); N. E. Pfeiffer (Morphologist); E. G. Vinson (Biochemist); W. J. Youden (Physical chemist); F. O. Holmes (Protozoologist); J. D. Dobrocky (Pathologist); C. M. Carlson (Microchemist); J. D. Guthrie (Ass. Biochemist); A. Hartzel (Entomologist). *5—6 places. Plant Physiol Dept., Plant Pathol. Dept., Biochemistry Dept., Microchemistry Dept., Physical chemistry Dept. — Special apparatus for controlling all conditions of plant growth on a large scale.* ⑨ *We maintain an arboretum and materials from this are subjet to exchange.*

Zagreb — Agram (S. H. S.).
Botan. Inst. der Königl. Univ. (Botan. Zavod Kr. Sveučilišta). Dir. Prof. Dr. Vale Vouk. — Dr. Ivo Pevalek, ao. Prof. der Botan. an der land- und forstwissensch. Fak., Leiter der Herbarienabt.; Dr. Ivo Horvat; Dr. Stjepan Horovatić; Ing. agr. Zdraoko Arnold. *3 Arbeitsplätze im Herbarium, 2 im Labor. für anat. bzw. zytol. und auch für physiol. Untersuchungen. Studium der kroatischen Flora, großes Herbarium (70000 Nummern).* ⑨ *Kroatische bzw. jugoslavische Flora.* — *Physiol. Labor. im Bau.*

Zaragoza (España).
Labor. botan. Fac. Sc. Univ. Dir. Prof. P. Ferrando y Mas.

Zürich (Schweiz).
Inst. f. allgemeine Botan., Univ., Künstlergasse 16. Dir. Prof. Alfred Ernst.

Syst.-botan. Inst. d. Univ. mit Botan. Garten u. Mus. Dir. Prof. Dr. Hans Schinz. — Prof. Dr. A. Thellung, Ass.; Dr. J. Bär, Kustos. ⑨ *Exotische Pflanzen.*

Musea botanica et herbaria.

Confer: *Botania generalis* (pag. 390), *Hortus botanici* (pag. 408), *Musea biologica* (pag. 350).

Aberdeen (Scotland).
Mus. of botan. Univ. Dir. Wm. G. Craib.

Amoy (China).
Botan. Mus. Univ. Dir. Prof. C. C. Chung.

Anvers — Antwerpen (Belgique).
Mus. sc. Dir. G. R. L. Maneau. *Botan. systématique et Cryptogamie. — Collection historique d'appareils de physique, spécialement d'optique microscopique.* ⑨ *Herbier de 300000 plantes,* 30000 préparations de diatomées. 10000 prép. d' anat. végétale. 10000 prép. d'anat. animale.

Ann Arbor, Mich. (U. S. A.).
Herbarium Univ. of Michigan. Dir. C. H. Kauffman. — B. B. Kanouse (Curator of Cryptogamic Collections).

Baku (U. d. S. S. R.).
Botan. Abt. Aserbaidshanser Staatl. Mus., Ul. Malygi. Dir. P. W. Schwann-Gurijsky.

Beirut (Syria).
Botan. Mus. American Univ. Dir. Prof. Alfred E. Day.

Bergen (Norge).
Abt. für System. Botan., Bergens Mus. Dir. Prof. Dr. Rolf Nordhagen.—Amanuensis: Astrid Karlsen.

Berkeley, Cal. (U. S. A.).
Herbarium Univ. S.B. Parish (Honorary Curator).

Berlin (Deutschland).
Botan.Mus.,Univ., Dahlem, Königin-Luise-Str.6/8. 1. Dir.: Prof. Dr. L. Diels; 2. Dir.: Prof. Dr. R. Pilger. — Kustoden: Dr. Burret, Prof. Dr. Gilg, Prof. Dr. Gräbner, Prof. Dr. Krause, Dr. Mattfeld, Prof. Dr. Mildbraed, Dr. Ulbrich, Dr. Vaupel; Ass.: Dr. Mansfeld, Dr. Markgraf, Dr. Melchior, Dr. Reimers, Dr. O. C. Schmidt, Dr. Werdermann. *Zahlreiche Arbeitsplätze für allgemein morphol., systematische und pflanzengeographische Untersuchungen.* — Zentralstelle für Nutzpflanzen. Leiter: Prof. Dr. Gilg und Prof. Dr. Gräbner.

Buenos Aires (Argentina).
„Darwinion", Casilla de Correo 1606. Dir. Dr. Cristobal M. Hicken. *2 Plätze für systematische*

Botan. — *Herbarien mit 45000 Arten, größte botan. Bibliothek Südamerikas.* ⑨ *Herbarmaterial.*

Buitenzorg, Java (Ned.-Oost-Indië).
Herbarium and Mus. for systematic botan. of the Botan. Gardens. Dir. Dr. J. G. B. Beumée (Plant-taxonomy in general). — Ass.: R. C. Bakhuizen van den Brink (Verbenaceae, Bombacaceae), Dr. D. F. van Slooten (Flacourtiaceae, Combretaceae, Dipterocarpaceae, Gramineae), Dr. H. J. Lam (Verbenaceae, Sapotanceae, Burseraceae), Dr. C. van Overeem (Fungi and Lichenes), Dr. B. H. Danser (Polygonaceae, plantgeography); Conserv.: Mrs. M. C. Lang and Mrs. F. F. N. Terlaak.

Burlington, Vt. (U. S. A.).
Herbarium, Univ. of Vermont. Dir. E. J. Dole

Caen, Calvados (France).
Mus. botan. (Jardin de plantes). Dir. Prof. René Vignier.—M. Lorset, Conserv. des Collections. *Herbier algol. de Lamouroux. Herbier Viellard de Nouvelle Calédonie.*

Cambridge, Mass. (U. S. A.).
Gray Herbarium of Harvard Univ. Dir. Prof. B. L. Robinson, Curator. — Prof. Merritt Lyndon Fernald; Dr. Ivan Murray Johnston, Ass.; Charles Alfred Weatherby, Ass.; Miss Ruth Dexter Sanderson, Librarian; Miss Lesley Chillingsworth Brown, Bibliographer; Miss Lily M. Perry, Ass. *About 6 tables for Monographic and floristic investigations on the taxonomy of the Spermatophyta and Pteridophyta. Also problems in plant distribution, in the relation and history of floras, on plant variation, etc. — Dissecting microscopes of powers and design appropriate to the plant group under investigation. Dissecting apparatus. Compound microscope at need.* ⑨ *Herbarium specimens of the Spermatophyta and Pteridophyta.*

Farlow Herbarium. Dir. C. W. Dodge (Curator). — R. Thaxter (Honorary Curator).

Cape Town (South Africa).
Bolus Herbar. Univ. cf.Hortus botan.Kirstenbosch.

Cluj — Klausenburg (România).
Inst. für systemat. Botan., Botan. Mus. und Botan. Garten der Univ., Grădina Botanică Str. Regală 28.

Dir. Prof. Dr. Alexander Borza. — Mus.-Kustos: E. I. Nyárády; Ass.: Emil Pop; Wissensch. Praeparator: Georg Bujorean; Chef des traveaux: vacat. *2 Plätze für system. Untersuchungen im Herbar.* ⓟ Phanerogamen der rumänischen Flora kultiviert, im Garten und getauscht auch vom Mus. — Zweiglaboratoria: Viele wissenschaftliche Reservationen und Naturparks (teilweise mit permanenten Quadraten zur Erforschung der Successionen).

Cordoba (Argentina).
Mus. de Botan. y Zool. Univ.

Corvallis, Ore. (U.S.A.).
Herbarium. Prof. H. M. Gilkey (Hypogaeous, Ascomycetes).

Detroit, Mich. (U.S.A.).
Herb arium Parke Davis and Co. Dir. O. A. Farwell (Curator).

Durban (South Africa).
Natal Herbarium and Plant Pathol. Labor., Botanic Gardens. Dir. H. H. Storey, B. A. (Mycologist in Charge). — A. P. D. McClean, M.Sc., Mycologist. *For Taxonomic Research 4 places. For Pathol. Research by arrangement. Virus diseases of plants, in particular of the Gramineae. — Herbarium of South African and exotic Phanerogams. Apparatus and Greenhouses for plant pathol. research especially virus disease research.* ⓟ Phanerogamic Herbarium material.

Genève — Genf (Suisse).
Herbier Boissier Univ., Institut Botanique Herbier Boissier. Dir. Prof. Dr. R. Chodat. — Dr. Gustave Beaurerd, Conserv. de l' herbier; Dr. Femond Chodat, 1er Ass.; Dr. Minod, Chef des Travaux. *Etudes de systématique d'après les Matériaux d'herbiers. Plusieurs places pour Lichenologie, Hépatiques, Champignons (Mucorinées etc.). Plantes supérieures du Monde entier.* ⓟ Phanérogames et Champignons. — Herbier „Müller" argoviensis, Lichens. Herbier Stephani, Hépatiques. Herbier Fackel, Champignons. Herbiers Boissier, Orient et Mediterranée et orbis terrarum, Reuler et Barbey. Herbiers généraux. Herbiers de Palégieux, Ayasse, Bouvier, Rapin, d'Europe.

Göteborg (Sverige).
Mus. des Botan. Gartens. Dir. Prof. Dr. C. Skottsberg. — Carl Blom, Ass. *3 Plätze für systematische u. anatomische, evtl. auch zytol. Arbeiten.* ⓟ Herbarpflanzen; auch Samen und lebende Gartenpflanzen.

Helsinki (Finnland).
Botan. Mus. der Univ. Dir. Prof. Dr. K. Linkola. Dr. Harald Lindberg, Kustos; Dr. Hans Buch, Amanuens. *Mehrere Plätze. — Wichtigere Herbarien: Herbarium Mus. fennici (Finnisches Nationalherbarium); Steven's Phanerogamenherbarium (mehrere Typen aus Rußland und Sibirien); Palearktisches Phanerogamenherbarium von Harald Lindberg; Moosherbarium S. O. Lindberg; Moosherbarium V. F. Brotherus; Flechtenherbarium E. Acharius; Flechtenherbarium W. Nylander; Die meisten Kryptogamenexsiccate u. a. m.* ⓟ Phanerogamen und Kryptogamen.

Indianapolis, Ind. (U.S.A.).
Herbarium Butler Univ. S. A. Cain (Curator).

Kaunas (Litauen).
Inst. für Planzensystematik an der Litauischen Univ., Botan-Garten. Dir. Prof. Dr. C. Regel. — Jurgis Kuprevičius. *2 Plätze für Flora von Litauen, Pflanzensoziol. Untersuchungen.* ⓟ Herbarmaterial aus Litauen.

Kiew, Ukraine (U.d.S.S.R.).
Botan. Mus. an der Akademie der Wissenschaften, Olgastraße. Dir. Dr. A. W. Fomin, Akademiker. — D. Zerow, Konserv., 1. Kustos; P. Oksink, Konserv., 2. Ass.; N. Pidopieczka; A. Lazarenko. *3 Plätze für Floristik.* ⓟ Herbarmaterial der Ukraine und des Kaukasus.
Inst. zur Erforschung d. Flora, Botan. Zentralsektion d. Landwirtsch. Wissensch. Kom. d. Ukraine, Korolenkostr. 21.

København (Danmark).
Botan. Garden and Botan. Mus. of the Univ. (Univ. Botan. Have og Botan. Mus.) Dir. Prof. Dr. C. H. Ostenfeld. — C. Christensen, Keeper of the Herbarium; A. Lange, Curator of the Garden; J. Boye Petersen, Sc. Ass. at the Garden; J. Gröntved, Sc. Ass. at the Mus.; F. Börgesen, Librarian. *Taxonomical studies and experim. genetic work.*
Dansk Botan. Forening, Botan. Mus. Dir. Prof. Dr. L. Kolderup Rosenvinge. — Carl Christensen, Museumsinspektor.

Krasnojarsk (U.d.S.S.R.).
Botan. Abt. Mus. des Jenissey Gebiets. Dir. A. L. Jaworskij.

La Habana (Cuba).
Botan. Mus., Inst. de Segunda Enseñanza.

Laramie, Wyo. (U.S.A.).
Rocky Mountain Herbarium, Univ. Dr. A. Nelson, Curator.

Lausanne (Suisse).
Mus. botan. et Herbier Univ., Palais de Rumine. Dir. Prof. Wilczek, Conserv.

Leiden (Holland).
's Rijks Herbarium, Nonnensteeg 1. Dir. Prof. Dr. J.W. C. Goethart. — Conserv.: W. A. Goddijn, J. Th. Henrard; Ass.: Cath. Cool; Amanuensis: C. M. Lambrechts, H. J. v. d. Hee. *2 Plätze.* ⓟ Herbar-Specimina.

Leningrad (U.d.S.S.R.).
Botan. Mus. der Akademie der Wissenschaften, Universitäts-Quai 5. Dir. Ivan Parfenjevič Borodin. Akademiker. — D. Litvinor, Conserv. des Turkestanischen Herbariums; N. Busch, Conserv. des Ostasiatischen Herbariums; W. Tränzschel, Conserv. der Kryptogamen-Abt.; S. Ganeschin, Conserv. d. Herb. des Europ. Teiles d. U.d.S.S.R.; B. Gorodkov Conserv. d. Sibirischen Herbariums; N. Kusnezov, Conserv. d. Extraeuropäischen Herbariums; Conserv.-Gehilfen: Elisab. Steinberg, Elisab. Busch, Alex. Tolmatchew, Nik. Woronichin, Lydia Savicz; Aspiranten: A. Leskov, V. Sotschava. *7 Plätze für pflanzensystematische Untersuchungen.* ⓟ Pflanzen (Phanerogamen und Cryptogamen) besonders der U.d.S.S.R.
Leshaft Inst. der physischen Erziehung, Dekabristenstr. 35. Dir. Prof. V. N. Lubimenko, Chef des Botan. Labor.

Lisbôa (Portugal).
Jardin et Mus. botan. Fac. sc. Univ. Dir. Dr. Ruy Telles Palhinha.

London (England).
Dept. of Botan., British Mus. of Nat.History, S.W.1. Cromwell Rd. Dir. Dr. A. B. Rendle. — Antony Gepp, M.A., Ass. Keeper (Cryptogams); J. Ramsbottan, M.A., Ass. Keeper (Fungi); A. J. Wilmott, B.A., Ass. Keeper (European and British Flowering Plants); R.D.O. Good, M.A., Ass. (Flowering Plants); A.W.Exell, B.A., Ass. (Flowering Plants); G. Tandy,

B.A., Ass. (Flowering Plants). *Taxonomic research.*
— *Requirements occidental to Taxonomic work on Flowering Plants and Cryptogams.* ⓟ Plants generally.
— A well equipped labor. for mycol. investigations.
Mus. of Botan. of the Univ. Coll., W. C., Gower Street. Dir. Prof. F. W. Oliver.

Lund (Sverige).
System. Abt. des Botan. Inst. der Staatsuniv. Dir.: vacat. — Konserv.: O. R. Holmberg. *Herbarium (ca. 250 000 Spannbogen, darunter das Algenherbar Ayardh's mit 50 000 Nummern).*

Madison, Wis. (U.S.A.).
Herbarium, Univ. of Wisconsin, Biology Bldg. J. J. Davis (Curator).

Marseille (France).
Mus. et Jardin botan., 105, Rue d'Edmond Rostand. Dir. Prof. Henri Jumelle.

Melbourne (Australia).
National Herbarium, South Yarra. Dir F. J. Rae. — W. Laidlaw, B.Sc., Government Botanist; J. W. Andas, Ass.; P. F. Morris, Ass. *Phanerogamen and Cryptogamen.* ⓟ Phanerogamen and Cryptogamen aus allen Teilen der Welt.

Moskau (U.d.S.S.R.).
Botan. Kabinett des Med. Inst.
Labor. für Systematik niederer Pflanzen der 2. Univ. Dir. K. M. Mejer.

München (Deutschland).
Botan. Mus., -Nymphenburg, Menzingerstr. 13. Dir. Geh.R. Prof. Dr. Ludwig Radlkofer. — Prof. Dr. Hermann Roß, Hauptkonserv. ⓟ Herbarpflanzen. — Kryptogamen-Herbarium.

Oslo (Norge).
Botan. Mus. Dir. Jens Holmboe. — Johannes Lid (Konserv.-Kustos); Bernt Lynge (Doc. u. Konserv.-Kustos). *10—15 Plätze für systematische Botan.*

Ottawa, Ont. (Canada).
Dept. of Botan. of the Victoria Memorial Mus. (Nation. Herbar.). Dir. M. O. Malte, Ph.D.

Pittsburgh, Pa. (U.S.A.).
Herbarium Carnegie Mus. E. H. Graham (Ass.).

Praha — Prag II (C.S.R.).
Dept. of Botan., National Mus., Národní Museu, Václavskénám 1700. Dir. Dr. Edvin Bayer. — Dr. K. M. Malkovský, Conserv. Adjoint; Dr. I. Kláštersky, Conserv. Ass. *5 places for plant Taxonomy and Anat.* ⓟ Herbarial plants and microscopical praep.

Pretoria (South Africa).
National Herbarium, Box 994. Dir. Dr. J. B. Pole Evans. *Systematic botan.* ⓟ General plant collections.

Rybinsk (U.d.S.S.R.).
Botan. Abt. naturwissensch. Mus.

Samarkand (U.d.S.S.R.).
Botan. Abt. d. Samarkander Gebiets-Mus.

San Francisco, Cal. (U.S.A.).
Mus. of Botan., and Herpetol. California Acad. of Sc. Dir. Dr. B. W. Evermann.

Sevastopol (U.d.S.S.R.).
Botan. Abt. Sevastopoler Mus. f. Landeskunde, Proletarskaja 5.

Sofia (Bulgarien).
Botan. Abt. d. Naturhist. Mus. S. M. d. Königs, Tsar Osvoboditel 1. Dir. Dr. Božimir. Davidoff.

Stanford University, Cal. (U.S.A.).
Dudley Herbarium. Dir. LeRoy Abrams. — Roxana S. Ferris, Ass. Curator; Gilbert T. Benson, Ass. Curator and Librarian. *Several places for taxonomic and phytogeographic work.* — *Herbarium of approximately 250 000 specimens of ferns and seed plants, also several thousand cryptogamic specimens.* ⓟ Herbarium specimens, especially of plants from western North America.

Stockholm (Sverige).
Botan. Abt. des Naturhistorischen Reichsmus. Dir. Prof. Dr. G. Samuelsson. — Dr. E. Asplund, Ass.

Strasbourg — Straßburg (France).
Collection spec. de l'Inst. botan. Univ., Rue de l'Université.

Stuttgart (Deutschland).
Botan. Abt. d. Württembergischen Naturaliensammlung. Hofgartendir. a. D. A. Berger.

Sydney (Australia).
Botan. Abt. Sydney Technol. Mus.

Taschkent (U.d.S.S.R.).
Floristische Abt. d. Zentralmus. f. Naturkunde, Ethnograph. u. Archaeol. Mittelasiens.

Tomsk (U.d.S.S.R.).
Herbarium Univ. West Siberia. Dir. Prof. Dr. P. N. Krylov. — L. P. Sergievskaja, Conserv. *2 places for working on the systematic of floral plants.* ⓟ The dry Herbarium plants.

Trondhjem (Norge).
Botan. Abt. d. Kgl. Norwegischen Gesellschaft der Wissenschaften (Trondhjems Mus.), Botanisk Avd., Museet. Dir. Ove Höeg, Konserv.

Upsala (Sverige).
Botan. Mus. Dir. Doc. Du Riek, Konserv.

Utrecht (Holland).
Botan. Mus. en Herbarium van de Rijks Univ., Lange Nieuwstraat 106. Dir. Prof. Dr. A. Pulle. — Jhr. Dr. L. H. Quarles van Ufford, Conserv.; J. Swart, Conserv.; J. Lanjouw, Ass.; C. van Steenis, Ass. *2 Plätze zu Untersuchungen über die Flora der Niederländischen Kolonien, insbesondere Niederl-Neu-Guinea und Surinam.* ⓟ Pflanzen der Niederl. Kolonien, insbesondere von Niederl.-Neu-Guinea und von Surinam.

Vjatka (U.d.S.S.R.).
Botan. Labor. d. Bezirksmus.

Wageningen (Holland).
Labor. für Pflanzengeographie u. Pflanzensystematik, Rijksstraatweg 37. Dir. Prof. J. Jeswiet. — Dr. J. T. P. Bijhouwer, Ass. *4 Plätze für mikroskopische und makroskopische Untersuchungen an Herbar- und lebenden Pflanzen für systematische Zwecke.* ⓟ Pflanzen, Samen und Diapositive.

Washington, D.C. (U.S.A.).
U.S. National Herbarium, Bureau of Plant Industry. Dir. F. V. Coville (Curator).

Wien (Österreich).
Herbar. d. Zool. Botan. Gesellschaft, III, Mechelgasse 2.

Wilno (Polska).
2. Botan. Anstalt für Pflanzensystematik u. Botan. Garten, Irakretowastr. 1. Dir. Prof. Dr. J. Trzchinski. — Ass.: Sophie Fiedarowich; Irene Sokatawska; Irene Marawska. *10—15 Plätze für mykol. und floristische Untersuchungen.* ⓟ Pilze, Zoocecidien, Phanerogamen der Wilna und Umgegend. — Botan. Garten mit einem kleinen Glashaus.

Zürich (Schweiz).

Botan. Mus. der Eldg. Techn. Hochsch., Universitätstr. 2 im Gebäude der Land- u. Forstwirtschaftlichen Hochschule. Dir. Prof. Dr. M. Rikli. — Konserv.: z.Z. vakant; Ass.: Charles Gut; Freier Ass.: Dr. Eugen Baumann. *Untersuchungen auf dem Gebiet der Systematik, der Pflanzengeographie, Planktonkunde, Pflanzengeschichte. — Große Herbarien, besonders reichhaltig das Herbarium Helveticum mit ca. 100 Faszikeln (viele Originalbelege). Große Sammlung von Polsterpflanzen. Reich vertreten: Mittelmeerflora, Flora d. Polarländer, N.-Amerika, Japan, Java. Bildergalerie hervorragender Botaniker usw. Karpologische Sammlung, das Flechtenherbarium v. Stizenberger usw.* ⓟ Mittelmeerflora, Vegetation der Arktis, Algenpflanzen. — Versuchsgarten.

Hortus botanici.

Confer: *Botania generalis* (pag. 390), *Musea et herbaria* (pag. 405).

Aberdeen (Scotland).

Cruickshank botan. garden, Chanonry, Old Aberdeen. Dir. W. G. Craib.

Alfort, Seine (France).

Jardin botan.

Amsterdam (Holland).

Hortus Botan., Plantage Middenlaan 2. Dir. Prof. Dr. Th. J. Stromps. — A. J. van Laren, Hortulanus.

Ann Arbor, Mich. (U.S.A.).

Botan. Gardens and Arboretum, Univ. Dir. H. H. Bartlett. — Frieda Cobb Blanchard, Ass. Director; E. G. Anderson, Research Fellow; Carl O. Erlanson, Research Assistant; C. G. Kulkarni, Research Fellow; Eileen W. Erlanson, *(Rosa)*; E. E. Dale *(Capsicum, Malvaceae)*; B. M. Davis, *(Oenothera)*; F. G. Gustafson *(physiol. of growth);* Fred Hermann *(Systematic Botany)*; W. Steere *(Rumex)*. ⓟ Oenothera, dried specimens and seeds. North American Wild Roses, dried specimens and seeds. Seeds of various North American perennials.

Anvers — Antwerpen (Belgique).

Jardin botan. de la ville d'Anvers, 24, Rue Léopold. Dir. F. Urom.

Asuncion (Paraguay).

Jardin Botan. Dir. Prof. Fiebrig Gertz.

Baarn (Holland).

Botan. Garten „Cantonspark", Faas Eliaslaan 49. Dir. Prof. Dr. A. Pulle. *2 Plätze. Der Garten gehört zur Univ. Utrecht. Er enthält 6 Gewöchshäuser, 1 Arboretum, eine Abt. für Planzensystematik und eine Abt. für Phytopathol.*

Bakurjani, Grusii (U.d.S.S.R.).

Bakurjansker Berg-Abt.d. Tiflisser Botan. Gartens, Gorijski ul.

Baltimore, Md. (U.S.A.).

Botan. Garden John Hopkins Univ. Dir. D. S. Johnson.

Batum, Georgien (U.d.S.S.R.).

Botan. Garten. *Subtropische Flora.*

Berlin (Deutschland).

Botan. Garten und Botan. Mus. Univ., -Dahlem, Königin-Luisen-Str. 6—8. Dir. Prof. Dr. L. Diels. — Prof. Dr. R. Pilger, 2 Dir.; Kustoden: Prof. Dr. E. Gilg, Dr. P. Graebner, Dr. J. Mildbraed, Dr. K. Krause, Dr. E. Ulbrich, Dr. Vaupel, Dr. Burret Dr. Mattfeld; Ass.: Dr. Werdermann, Dr. Melchior, Dr. Markgraf, Dr. Reimers, Dr. O. Schmidt, Dr. Mansfeld. *Ca. 20 Plätze für systematische Botan. und Pflanzengeographie, Pharmakognosie. — Herbar und botan. Garten, große beiden. Bibliothek.* ⓟ *Pflanzen, lebend und als Herbarstücke konserviert. Sowohl lebende wie getrocknete Pflanzen der ganzen Erde werden gesammelt und evtl. getauscht. — Zentralstelle für Nutzpflanzen.*

Bol. Letzy, Witebsk. Kreis (U.d.S.S.R.).

Botan. Garten.

Boston, Mass. (U.S.A.).

Arnold Arboretum Harvard Univ. Dir.: C. S. Sargent; Ass.-Dir.: E. H. Wilson. — A. Rehder (Curator of the Herbarium).

Birmingham (England).

Botan. Garden of Botan. and Hortic. Soc., Edgbaston.

Braunsberg (Deutschland).

Botan. Garten d. Staatl. Akad. Dir. Geh.Reg.R. Prof. Dr. Niedenzu.

Breslau (Deutschland).

Botan. Garten und Mus. Univ. Dir. Geh.R. Prof. Dr. Ferdinand Pax. — Prof. Dr. H. Winkler, Ass.; Dr. Al. v. Lingelsheim, Ass.

Brooklyn, N.Y. (U.S.A.).

Brooklyn Botan. Garden. Dir. C. S. Gager. — A. H. Graves (Curator of Public Instruction); M. E. Peck (Research Ass.); O. E. White (Curator of Plant Breeding). N. Taylor (Curator).

Bruxelles — Brussel (Belgique).

Jardin Botan. de l'Etat. Dir. Prof. Dr. E. de Wildeman. — Prof. Ch. Bommer, Conserv.; G. O. Boulenger. ⓟ Plantes vivantes et d'herbar.

Bucureşti (România).

Botan. Garten und Inst. Dir. Prof. M. Vladescu.

Budapest (Ungarn).

Botan. Abt. des Ungarischen National-Mus., Akademia u. 2. II. Dir. Dr. Ferd. Filarszky. — Dr. B. J. Kümmerle (Pteridol.); Dr. G. Moesz (Mycol., Cedidiol.); Dr. S. Jávorka (Floristika, ungar. und balkan. Flora); Dr. J. Szepesfalvy (Szurák) (Bryol.); Gy. Timkó (Lichenol.). *Besonders für floristische, auch mikroskopische Untersuchungen.* ⓟ *Gesammelt werden Kryptogamen (Pilze, Flechten, Algen, Moose, Gefäßkryptogamen) und Phanerogamen; dann phytopathol. und phytopalaeont. Objekte. Getauscht werden Kryptogamen und Phanerogamen, außerdem dient zum Tausche die Ausgabe des Inst.: „Flora Hungarica exsiccata", von welcher bisher 7 Centurien erschienen sind und welche Kryptogamen und Phanerogamen umfaßt.*

Zool. Botan. Garten der Hauptstadt. Confer: Hortus zoologici.

Buitenzorg, Java (Ned.-Oost-Indië).

Botan. Gardens. Dir. Dr. W. M. Docters van Leeuwen. — H. J. Wigman, Chief. *Studies on galls and plant-biol. in general. New flora of Krakatau.*

Cagliari, Sardinia (Italia).

Orto Botan. Dir. G. Negri.

Calcutta (Br.-India).

Royal Botan. Garden, Sibpore. Dir. C. C. Calder.

Cambridge (England).

Botan. Garden. Dir. H. G. Carter.

Cassel (Deutschland).

Botan. Garten, Park Schönfeld. Dir. Hermann Schulz. *Mit Herbarium.*

Catania (Italia).
Orto botan. Univ. Dir. G. Grassi-Cristaldi.

Chapultepec (Mexico).
Botan. Garten d. Dir. de Estudios Biol. Dir. Ing. Octavio Solis.

Charkow (U.d.S.S.R.).
Botan. Garten, Klotschkowskaja 52. Dir. L. A. chkorbatow.

Cork (Ireland).
Botan. garden.

Delft (Holland).
Kulturgarten für Technische Pflanzen, Poortlandlaan 35. Dir. Prof. Dr. G. van Iterson Jr. — C. H. J. Cunaeus. ⓟ Technische Pflanzen.

Dresden (Deutschland).
Staatl. Botan. Garten, A. 16, Stübelallee 2. Dir. Prof. Dr. Friedrich Tobler, o. Prof. d. Botan. a. d. Techn. Hochsch. *Garten etwa 4 ha groß mit Schauhäusern und Kulturhäusern, geographischen, systematischen und biol. Quartieren, Labor.*

Dublin (Ireland).
Botan. Gardens of Dept. of Agric. Dir. T. W. Besant.

Dunedin (New Zealand).
Botan. Gardens. Dir. D. Tannock, Botan.

Edinburgh (Scotland).
Royal Botan. Garden. Dir. Prof. W. Wright Smith, M.A. *Botan. Research.*

El Salvador (Salvador).
Botan. Garten.

Eriwan (U.d.S.S.R.).
Botan. Garten.

Evreux (France).
Jardin Botan.

Ferrara (Italia).
Orto botan. Univ., Palazzo Schifanoia. Dir. Prof. E. Baroni.

Genève — Genf (Suisse).
Conservatoire et jardin botan. de la Ville de Genève, La Console, Route de Lausanne 192. Dir. J. Briquet.

Gorki (U.d.S.S.R.).
Botan. Garten der Akad. f. Landwirtschaft. Dir. J. Wassilkov. *Delectus Seminum. Horti Botan. Akad. agron. R. P. Alboruthenicae in Gorki.*

Göteborg (Sverige).
Göteborgs Botan. Trädgård. Dir. Prof. Dr. C. Skottsberg. *Institutsgebäude mit Bibliothek, Herbarium und Labor. im Bau. Naturpark 36,8 ha. Kulturen 8,7 ha.*

's Gravenhage (Holland).
Botan. Garten. Dir. P. J. den Hertog.

Graz (Österreich).
Botan. Garten der Univ. Dir. Prof. Dr. Karl Fritsch. *Pesneriaceen; Blütenbiol. Flora von Österreich.*

Groningen (Holland).
Hortus Botan., Nieuwe Kijk-in't-Jatstr. 84. Dir. Prof. Dr. J. C. Schoute. — E. Laarman (Hortulanus); H. D. Hielkema (Amanuensis).

Helsinki (Finnland).
Botan. Garten. Dir. Kaarlo Linkola.

Jalta, Krim (U.d.S.S.R.).
Botan. Garten. Confer Nikita Yalta.

Kaunas (Litauen).
Botan. Garten an der Litauischen Univ., Freda Dir. Prof. Dr. C. Regel. — Carl Meissner, wissensch. Gärtner (Gartenin*s*pektor). *Treibhaus- und Freilandpflanzen.* ⓟ Samen und lebende Pflanzen eigener Kultur. — Phytopathol. Labor.

Kazan (U.d.S.S.R.).
Botan. Garten, Dalne-Archangelskaja ul.

Kew b. London (England).
Royal Botan. Gardens. Dir. A. W. Hill. — T. F. Chipp, Ass. Dir.; A. D. Cotton, Keeper of Herbarium; W. J. Bean, Curator; W. Dallimore (Keeper of Mus. of economic Botan.).

Kiew (U.d.S.S.R.).
Botan. Garten der Akad. d. Wissensch., Ul. Kominterna 1. Dir. V. I. Lipski. **Labor. des Botan. Gartens (I.N.O.),** Kominternstraße 1. Dir. Prof. Dr. A. W. Fomin (Pteridophyta, Akklimatisation der Pflanzen). — Prof. W. Finn (Embryol. der Pflanzen); Prof. J. Modilewski (Embryol. der Pflanzen); M. Tschernojarow (Cytol.); A. N. Oxner (Lichenes); Z. Girzitska (Mykol.); W. Kleopow (Floristik); N. Dubowik (Ass.). *4 Plätze für Embryol. der Pflanzen und Cytol., außerdem Systematik der Pflanzen. Mycol., Lichenol. und Floristik.* ⓟ Herbarium generale, Herbarium caucasicum, Herbarium Ucrainicum, Herbarium generale Lichenum, Herbarium mycol., Herbarium Filicum.

Kirstenbosch near Cape Town (South Africa).
National Botan. Gardens. Dir. Prof. Robert Harold Compton, M.A. (Cantab.). — The Bolus Herbarium (Univ. of Cape Town) is situated at the Gardens. Mrs. F. Bolus, Curator. *1 labor. seat occasionally available; for accommodation in the Bolus Herbarium reference should be made to the Curator. Special facilities for morphol. and anat. work, especially on the South African flora. Also facilities for the cultivation and distillation of essential oils .* ⓟ Seeds of South African and economic plants. Preserved material of South African plants, on request. — Sub-station, the Karoo Garden, Whitehill, near Matjesfontein, where special cultures of the succulents and other plants of the arid regions of South Africa are made.

Klagenfurt (Österreich).
Botan. Garten des Naturh. Landesmus. Dir. T. Prosser.

Köln (Deutschland).
Botan. Garten d. Stadt u. Univ. Köln, Riehl. Dir. Univ. Prof. Dr. G. Esser. — 1 Ass.

Kôshun, Formosa (Japan).
Tropical Botan. Garden, Takao-shû.

Kraków (Polska).
Botan. Garden, Kopernika 25. Dir. Prof. Dr. Wladynaw Szafer. — Dr. Bogumił Pawłowski, Adjunkt; Dr. Aniela Kozłowska, Ass. ⓟ Flora of Poland.

Kyôto (Japan).
Kyôto Provincial Botan. Garden, Shimogamo. Dir. Prof. Dr. K. Kôriba.

La Habana (Cuba).
Botan. Garten, Inst. de Segunda Ensenañza. Botan. Garten der Univ.

La Valletta (Malta).
Botan. garden, Univ. of Malta.

Leiden (Holland).
Botan. Garten der Univ. Dir. J. M. Janse.

Leningrad (U.d.S.S.R.).
Dendrol. Garten mit einer Abt. für Gräser des Leningrader Forstinst. Botan. Hauptgarten der U.d.S.S.R., Aptekarski Ostrov, Pessotchnaia ¹/₂. Dir. B. L. Issatchenko,

Prof.; Sous-Dir. V. L. Komarov, académicien. — Secrétaire-sc. V. P. Savicz et son remplaçant, Secrétaire par intérim, A. P. Iljinskij. 1. La section des Plantes vivantes: Chef de la section: V. L. Komarov, académicien (Systématisation et géographie des plantes, Flora de l'Asie Orientale, Morphol.); Sous-chef: A. P. Iljinskij (Géobotan., Morphol., Flore de la Russie Europ.); Conserv.: N. A. Maximov (Oecol. des plantes, persistance au froid, régime aquatique), A. A. Boulavkina (Systématisation: Patrinia, Flore de la Région lacustre), K. K. Kuhn (Culture des plantes); Ass.: A. I. Pojarkowa (Oecol. des plantes, Ribes de la Russie asiatique), E. V. Lebedintzeva (Oecol. des plantes). 2. L'Herbier: Chef de la Section: B. A. Fedtschenko (Systématisation et géographie des plantes: Flore du Turkestan, Papilionaceae, Liliaceae, Acanthaceae); Sous-chef: R. J. Roshevitz (Systématisation et géographie des Gramineae); Conserv.: N. V. Schipczinsky (Systématisation et géographie des plantes Ranunculaceae), I. M. Krascheninnikov (Géobotan., Systématisation: Compositeae, surtout Artemisia, la végétation des steppes et demi-deserts), M. M. Iljin (Systématisation des plantes: Compositeae et Malvaceae, Flore de Sibérie), S. V. Juzepczuk (Systématisation Rosaceae, Alchimilla, Rubus), O. I. Kuseneva-Prochorova Géobotan.: la végétation de l'Extrême Orient et de la Région arctique), A. N. Kryshtofovich (Systématisation, phytopaléont. et géographie des plantes), G. N. Woronow (Géographie des plantes: Flore du Caucase); Ass.: E. G. Czernjakowska (Systématisation et géographie des plantes de l'Asie centrale et de la Perse), O. E. Knorring (Systématisation et géographie des plantes du Turkestan), V. L. Nekrassova (Systématisation et géographie des plantes, Saxifragaceae), S. G. Gorschkova (Systématisation, Tamaricaceae), N. A. Basilevskaja (Systématisation, Astragalus). 3. Section des plantes cryptogames: Chef de Section: A. A. Elenkin (Sporol., lichenol. et algol.); Sous chef: A. N. Danilov (Biol. des algues et des champignons); Conserv.: V. P. Savicz (Lichenol.), L. I. Savicz-Ljubitzkaja (Bryol.); Ass.: L. A. Lebedeva (Mycol.), O. V. Troitzkaja (Algol. d'eau douce), E. S. Sinova (Algol. marine). 4. Le Mus.: Chef de la Section: N. A. Monteverde (Physiol. des plantes, plantes médicinales); Sous-chef: J. V. Palibin (Phytopaléont., Systématisation et géographie des plantes, plantes utiles); Conserv.: N. N. Monteverde (Plantes médicinales), L. G. Spassky (Chimie des produits végétaux); Ass.: A. F. Hammermann (Chimie des produits végétaux). 5. Section de Physiologie: Chef de la Section: V. N. Lubimenko (Physiol. des plantes); Sous-chef: S. D. Lvov (Physiologie); Ass.: V. A. Brilliant-Lerman (Physiol.), S. S. Fichtenholz (Physiol.). 6. Section de Géobotan. Chef de la Section: N. I. Kusnezov (Systématisation et géographie des plantes); Sous chef: A. P. Schennikov (Phytosociol.); Conserv.: Ju. D. Zynzerling (Systématisation et géobotan.); Ass.: E. V. Schiffers (Géobotan.), O. S. Poljanska (Géobotan.). 7. Section de semences et Station d'essais de Semences: Chef de la Section: B. L. Issatchenko (Physiol. des organismes inférieurs et étude des semences); Sous-chef: K. V. Kamenskij (Etude des semences); Conserv.: D. N. Neloubov (Physiol. des plantes); Ass.: M. I. Zavodcztkova (Etude de semences), A. A. Egorova (Physiol. des plantes). 8. Section de Phytopath.: Chef de la Section: A. S. Bondarzew (Phytopath.); Sous-chef: B. P. Karakoulin (Phytopath.); Conserv.: N. J. Vasiljevsky (Phytopath.); Ass.: G. A. Bourgvitz (Bact.), V. N. Bondarzewa (Phytopath.). 9 Section d'Acclimatation: Chef de Section: V. N. Sukaczev (Phytosociol., Dendrol., Géographie botan.); Sous-chef: E. L. Wolf (Dendrol., Acclimatation); Conserv.: M. R. Muller, Leningrad (Dendrol.), W. M. Savitsch, Vladivostok (Géobotan., acclimatation, Dendrol.), M. D. Spiridonow, Omsk (Géographie botan., acclimatation). Labor. Hydrobiol.: Chef de la Section: N. N. Woronichin (Hydrobiol., Algol., Mycol.); Ass.: V. S. Poretzky (Algol. et Diatomie). Bibliothèque: Chef de la Section: G. A. Nadson (Microbiol.); Sous-chef: J. A. Ohl (Phytopath. et Bibliographie). *Es können in allen Abt. ungefähr 40 Arbeitsplätze zur Verfügung der Gäste für wissenschaftliche Arbeit gestellt werden. — Bibliothek des Gartens (über 50000 Bände).* ⓟ Der Botan. Garten sammelt Phanerogamen- und Cryptogamenpflanzen, sowie deren Samen und Museumsexemplare der Pflanzenprodukte. Tauscht eventuell Dubletten von Exsiccaten und Herbarexemplaren. — 3 Acclimatisationsabt., nämlich in Leningrad (vormals Pomologischer Garten von Dr. E. Regel und L. P. Kesselring), Vorstand M. R. Müller; in Omsk (Vorstand M. D. Spiridonow) und in Vladivostok (Vorstand W. M. Savitch).

Botan. Garten d. Leningrader Staatl. Univ., W.O., Universitäts-Quai 7/9. Dir. W. L. Komarow.

Lima (Peru).

Jardin botan., Univ. Dir. Prof. Dr. A. Weberbauer. *Botan. pharmac.*

Lisle, Ill. (U. S. A.).

Morton Arboretum. H. Teuscher (Botanist).

Ljubljana — Laibach (S. H. S.).

Botan. Garten d. Univ., Ižanskacesta 3. Dir. Prof. Alfons Paulin. *Pflanzensystematische Untersuchungen.* ⓟ Früchte, Samen u. lebende Pflanzen.

London (England).

Botan. Gardens, Royal Botan. Soc. of London, Regent's Park, N.W. Dir. Viscount Lascelles. *Mus. Library*.

Lwów (Polska).

Botan. Garten d. Univ. Dir. St. Kulczyński.

Lyon (France).

Jardin botan. de la fac. de méd. de l'université.
Jardin et collections botan. et horticoles de la ville de Lyon, Au Parc de la Tête d'Or. Dir. Dr. Louis Marie Faucheron.

Madrid (España).

Jardin Botan., Plaza de Murillo 2. Dir. Prof. D. J. Bolivar y Mantia. — Herbaria: D. A. Cab. y Segares, E. Balguerias y Quesada; Cultivaciones: D. A. Garcia Valera.

Marburg a. d. L. (Deutschland).

Botan. Garten. Dir. Prof. Dr. Peter Claußen. *Versuchshaus für pflanzenphysiol. Arbeiten.* ⓟ Insektenfressende Pflanzen.

Messina (Italia).

Ist. ed Orto Botan. Univ. Dir. Prof. Giovanni Ettore Mattei. — 2 Aide-Botanistes. *4 places pour Anat., Physiol. et Biol. végétal.* ⓟ Plantes, fruits, graines, et tous les produits végétaux.

Migeja, Bez. Pervomajski, Ukraine (U.d.S.S.R.).

Botan. Garten.

Milano (Italia).

Orto botan. di Brera.

Mologa, Gouv. Jaroslavl (U. d. S. S. R.).

Botan. Garten.

Moskau (U.d.S.S.R.).

Botan. Garten d. 1. Moskauer Staatl. Univ., 1, Meschanskaja 28. Dir. Prof. M. I. Golenkin.
Labor. der Geobotan. der 2. Univ. Dir. M. P. Grigoriev.

München (Deutschland).

Alpines Mus. u. alp. Pflanzengarten. Confer : Mus. biol. generalia.

Muratowo, Orlovsches Gouv. (U.d.S.S.R.).
Botan. Garten. Dir. W. H. Chitrow.

Nancy (France).
Jardin botan. de la ville, Rue St. Catherine. Dir. Prof. Edmond Gain.

Nikita Yalta, Crimea (U.d.S.S.R.).
Government Botan. Garden. Dir. F. K. Kalaida. — G. Gunjko (Spec. in Culture of Technical Plants); I. Riakoff (Ass. in Pomol.); V. Maleeff (Ass. in Geobotan., Dendrol.); W. Niloff (Ass. of Chemistry, essential Oils); W. Williams (Ass. in Chemistry, essential Oils); W. Wassilieff (Ass. in Geobotan.); L. Sergeeff (Ass. in Microbiol.); V. Illuvieff (Dr. of Chemistry, Soil Technol.); M. Gerassimoff (Dr. of Chemistry, Enochemistry); N. Busin (Specialist in Grape-culture); T. Tsirina (Specialist in Botan.). *Fo. Work in Soil Improvement 3 places, Microbiol. 1 place, Pomol. 1 place, Enochemistry 2 places, essential Oils 1 place.* ⓟ Seeds, Herbarium specimens, plants etc. — Labor. of Enochemistry, of Soil Improvement, of Pomol., of Genetics.

Nikkô (Japan).
Nikkô Branch of Botan. Garden Tôkyô Imperial Univ., Tochigi-ken.

Odessa (U.d.S.S.R.).
Labor. des Botan. Gartens (I.N.O.). Dir. Prof. Dr. D. O. Swirenko. — K. Pazowsky; M. Gordienko; P. Schirschow; E. Buschenko; A. Andrejew. *3 Plätze für Systematik d. Pflanzen.*

Omsk (U.d.S.S.R.).
Abt. des Botan. Gartens Leningrad.
Botan. Garten u. Sib. Inst. f. Land. u. Forstwirtschaft. Dir. Prof. V. F. Semeonov.

Ootacamund (India).
Government Botan. Gardens and Parks, Nilgiris, Madras Presidency. Dir. F. H. Butcher, Curator.

Oslo (Norge).
Botan. Garten der Univ. (Univ. botan. have). Dir. Prof. Jens Holmboe. — Dr. phil. Erling Christophersen, Ass. *Ca. 10 Plätze.* ⓟ Sämereien, lebendige Pflanzen.

Oxford (England).
Botan. garden, Herbarium u. Mus. Dir. of the botan. Garden: Sir. F. Keeble. — Curator of Herbarium: G. C. Druce.

Palermo (Italia).
R. Orto Botan. Dir. Prof. L. Buscalioni. — Prof. D. Lanza, Conserv.; Prof. G. Catalano, Ass.; Dr. G. Cultrera- Aiuto. *Appar. di Fisiol. vegetale. Erbari per la botan. sistematica.* Giardino Coloniale.
R. Giardino Coloniale. Dir. F. Bruno.

Paris (France).
Jardin botan.

Parma (Italia).
Orto Botan. della R. Univ. Dir. C. Avetta.

Pavia (Italia).
Hortus botan. Univ. Dir. Prof. L. Montemartini.

Penang (India).
Botan. Gardens, Straits Settlements, Malaga. F. Flippavre, Curator.

Peradeniya (Ceylon).
Royal Botan. Gardens. *Zentralstelle f. d. floristisch-system. Durchforschung v. Ceylon u. f. botan. u. landw. Untersuchungen in den Tropen.*

Philadelphia, Pa. (U.S.A.).
Botan. Gardens and Greenhouses Temple Univ., Broad str. and Montgomery Ave. Dir. R. H. True. — H. F. Bernhardt.

Pisa (Italia).
Orto Botan. Univ., Via Solferino e Via S. Maria. Dir. Prof. Dott. Biagio Longo. — Dott. A. Bottini, Aiuto; Dott. L. Amadori, Ass.; Dott. U. Martelli, Ass. volontario. *2 posti per le ricerche in qualsiasi campo della Botan. — Tutti gli apparecchi necessarii alle ricerche incogni campo della Botan.* ⓟ Piante e semi in cambio.

Poznań — Posen (Polska).
Inst. f. Heilpflanzenbaulehre. Dir. Prof. Jan Dobrowolski.

Riga (Latvija).
Botan. Garten der Lettländischen Univ. (Latvijas Univ. Botan. Dārzs), Kandavas iela 2. Dir. Doc. Nikolajs Malta, Dr. rer. nat. Priv.Doc. Pauls Galenieks, Ass.

Rio de Janeiro (Brasil).
Jardim Botan.

Rouen (France).
Jardin Botan. de la Ville. Dir. E. Le Graverend. *Cataloque annuel des graines recoltées dans le Jardin Botan.*

Saharanpur (British India).
Government botan. gardens. Experim. and Educ. School of Hortic.

St. Louis, Mo. (U.S.A.).
Missouri Botan. Garden. Dir. G. F. Moore; J. M. Greenman (Curator Herbarium). — C. E. Kobuski (Rufus I. Lackland Research Fellow); H. von Schrenk (Pathologist); Ass. Prof. C. Anderson (Botan.); B. M. Duggar (Physiologist).

Santiago (Chile).
Botan. Garten.

Sassari, Sardinia (Italia).
Orto botan. Univ. Dir. V. Martelli.

Sibolangit, Sumatra (Ned.-Oost-Indië).
Botan. Gardens (altitude 500 M. about 20 H.A.), near Medan, East Sumatra. Ass. Gardener: J. A. Lörzing. — Annexed to this garden there is a nature reservation of about 300 H.A.

Sofia (Bulgarien).
Botan. Garten S. M. d. Königs. Dir. J. Kellerer, Insp.

Stockholm (Sverige).
Bergianischer Garten (Hortus Bergianus). Dir. Rob. E. Fries. — Erik Söderberg, Ass. ⓟ Lebende Pflanzen.

Sverdlovsk (U.d.S.S.R.).
Botan. Garten.

Sydney (Australia).
Botan. gardens.

Tarbes, Haut Pyrénées (France).
Jardin des plantes.

Tartu — Dorpat (Eesti).
Botan. Garten der Univ. Dir. Dr. Edm. Spohr. — Fr. Boerner. ⓟ Pflanzensamen u. Früchte.

Tegucigalpa (Honduras).
Botan. Garten d. Mus. national.

Tiflis (U.d.S.S.R.).
Botan. Garten. Herbarium; Sammlg. lebender kaukasischer Pflanzen; Selektionskabinett.

Tjibodas, Java (Ned.-Oost-Indië).
Botan. Gardens at Tjibodas. Ass. Gardener: M. L. A. Bruggeman. *Altitude 1500 m, about 20 H.A.*

Tomsk (U.d.S.S.R.).
Botan. Garten. Dir. Prof. P. N. Krylov.

Toulouse (France).
Jardin des Plantes.

Tours (France).
Jardin botan.

Urbino (Italia).
Orto botan. Univ. Dir. Inc.: G. Speranzini.

Utrecht (Holland).
Botan. Garten der Univ., Lange Nieuwstraat 106. Dir. Prof. Dr. F. A. F. C. Went. — Der Garten enthält 5 Gewächshäuser.

Vancouver (British-Columbia).
Botan. Garden.

Vjatka (U. d. S. S. R.).
Botan. Garten.

Vladivostok (U. d. S. S. R.).
Botan. Garten. Dir. Prof. V. M. Savič.

Wageningen (Holland).
Arboretum. Dir. Prof. Dr. Jakob Juswiet.

Warszawa (Polska).
Jardin botan. de l'Univ., r. Al. Ujazdonski 6/8. Dir. Prof. Dr. Boleslas Hryniewiecki.
Jardin de Saxe. Dir. Vladislas Danielewicz.

Washington (U. S. A.).
M. S. Botan. Garden. Dir. G. W. Hess. — W. J. Paget, Ass. Dir.

Wien (Österreich).
Demonstr.- u. Versuchsgarten beim Hochschulgebäude d. Hochsch. f. Bodenkultur. 4 Abt. Dir.: Professoren O. Porsch, Tschermak, Hecke.

Witebsk (U. d. S. S. R.).
Botan. Garten d. Witebsker Veterinär-Inst., Woropaewskaja 29 a.

Woronesch (U. d. S. S. R.).
Woronescher Abt. d. Russischen Botan. Gartens, Landwirt. Institut.

Laboratoria et hortus botaniae pharmacognosiae.

Amsterdam (Holland).
Labor. f. Pflanzenphysiol. u. Pharmakognosie. Confer: Physiol. plantarum.

Budapest (Ungarn).
Kgl. Ung. Drogenversuchsstation (Versuchsstat. f. Heilpflanzen), II, Debröi ut 15. Dir. Dr. Béla Augustin. — Chemiker: Dr. János Kuntz, Dr. Ilona Kelp, János Kabay; Botaniker: Dr. Adám Boros, Dr. Boriska Kemenes, Dr. Jolán Murányi; Gärtner: Géza Szathmáry. 1—2 Arbeitsplätze. Pharmakochemische, ätherische Öle, alkaloide Forschungen. Arzneipflanzenanbau und Züchtung, Samenuntersuchungen. — Destillations- und Extraktionsapparate, optische Instrumente. ℗ Arzneipflanzen. — Versuchsgärten in verschiedenen Gegenden.

Cluj — Klausenburg (România).
Arzneipflanzenversuchsstat. Dir. Bela Pater.

Johannesburg (South Africa).
Pharmacol. Labor. Confer: Pharmacol.

Lausanne (Suisse).
Labor. de botan. et de pharmacognosie, Univ. Confer: Botan. generalis.

Mogilew (U. d. S. S. R.).
Versuchsstation f. Arzneipflanzen.

Odessa (U. d. S. S. R.).
Versuchsfeld f. Arzneipflanzenkultur.

Oster, Gouv. Tschernigov (U. d. S. S. R.).
Versuchsfeld f. Arzneipflanzen.

Praha — Prag (C. S. R.).
Abt. pharmacol. Botan. und Kryptogamenkunde, Deutsche Univ. Dir. Prof. A. A. Pascher.
Pharmazeutisch-Botan. Inst. d. Karls-Univ., Benatska 433. Dir. Prof. Dr. Varl Domin.

Stockholm (Sverige).
Botan. pharmakognost. Sammlung d. K. Pharmacol. Inst., Kungstengatan 49.

Warszawa (Polska).
Pharmacogn. u. Med.-Botan. Inst. Univ., r. Brzozowa 12. Dir. Prof. Dr. Ladislas Mazurkiewicz.

Wien (Österreich).
Komitee zur staatl. Förderung d. Kultur von Arzneipflanzen in Österreich, Botan. Labor.: II, Trunnerstr. 1—3; Chem. Labor.: IX, Währinger Str. 13. Vorsitzender: Univ.Prof. Dr. Richard Wasicky. — Priv.Doc. Dr. Wolfgang Himmelbaur, Geschäftsführer, Leiter des botan. Labor. u. der Kulturen in Korneuburg bei Wien; Priv.Doc. Dr. phil. et. jur. Otto Dafert, Leiter des chemischen Labor. Botan. Labor. 6 Plätze. Chem. Labor. 12 Plätze. Botan. (anat., pharmakognostische), chemische (pharmaz.-chem., mikrochem., agrikulturchem.) Untersuchungen. ℗ Drogenprodukte.

Wilno (Polska).
Inst. f. Pharmakognosie u. med. Pflanzenkultur, Univ. Dir. Prof. J. Muszynski.

Laboratoria physiologiae plantarum.

Confer: *Laboratoria botaniae generalis* (pag. 390), *Phytopathologia* (pag. 415), *Plantae agris* (pag. 507), *Fermenta* (pag. 443).

Amsterdam (Holland).
Labor. f. Pflanzenphysiologie und Pharmakognosie, Hortus Botanicus. Dir. Prof. Dr. H. Weevers. — Dr. M. Pinkhof, 1. Ass.; Frl. C. A. Gomventak, 2. Ass.

Baltimore, Md. (U. S. A.).
Labor. of Plant Physiol. John Hopkins Univ. Dir. B. E. Livingston.

Berlin (Deutschland).
Pflanzenphysiol. Inst. Univ., -Dahlem, Königin-Luisen-Str. 1—3. Dir. Prof. Dr. H. Kniep. — Ass.:
Dr. Fr. Herrig, Priv.Doc. Dr. P. Metzner, Priv.Doc. Dr. A. Th. Czaja. Pflanzenanat., -physiol., mykol. bakt. Untersuchungen.
Pflanzenphysiol. Versuchsstat. der Lehr-u. Forsch.-Anst. für Gartenbau, -Dahlem. Dir. Doc. Dr. Gustav Höstermann. Anat., Organographie, physiol. Pflanzenphysiol. (Obstbaumphysiol.). Krankheiten der gärtnerischen Kulturgewächse, gärtnerischer Pflanzenschutz und Pflanzenzüchtung.

Breslau (Deutschland).
Pflanzenphysiol. Inst. Univ., 9, Goeppertstr. 6/8. Stelle des Dir. zur Zeit nicht besetzt; Vertreter: Dr. Schaede, Priv.Doc.

Brno — Brünn (C. S. R.).
Inst. for Plant Physiol. Univ. (Ústav pro fysiol. rostlin), 63. Kounicova. Dir. Prof. Dr. Vladimír Úlehla. — Dr. Vladimír Morávek; Ing.Vojtěch Juha; Jan Calábek; Ferdinand Herčík. *5 complete places.* — *12 Auxographs, Potentiometer, Hydrogene-Electrode, Electric Clinostats, Moving Picture equipments, Microanalysis, Gas-analysis, Growth-apparatus Complete physiol. equipment in together 35 rooms for special use or guests. 1 Experim. Greenhouse, 1 Veranda, 1 constant Temperature Room, 1 Moving Picture Room, 2 Dark Rooms, chemical and physical Labor. a. s. o.* ⊕ *Green Algae, lower Fungi, Collections for physiol. work.* — The Inst. is connected with the Biol. Station in Lednice, Moravia, Czechoslovakia. Opportunity is also given of joint research with theAgric. and with the Foresty Schools in Brno.

Bruxelles — Brussel (Belgique).
Labor. Intercommunal de Chim. et de Bact. de l'Agglomeration Bruxelloise. Dir. Prof. Dr. Hubert Kufferath. *Bact. du lait et des denrées alimentaires. Levures. Microscopie des denrées alimentaires. Algol. systématique, géographique. Cultures pures des algues.*

Budapest (Ungarn).
Versuchstation für Pflanzenphysiol. und Pathol., 11, Debröi-ut 15.—17. Dir. OberReg.R. Prof. Hermann Kern. — Prof. Ladislaus Beke; Alexander Csete, Adjunkt; Dr. Árpád v. Paál, Adjunkt, Priv.-Doc.; Dr. Béla Husz, Adjunkt; Dr. Jósef A.Krenner, Ass.; Koloman Tabajdi; Ladislaus Szemere (höhere Pilze); Julius Korponay (Obstbau-Insp.). ⊕ *Pilze.*

Buitenzorg, Java (Ned.-Oost-Indië).
Treub-labor. of the Botan. Gardens. Dir. Dr. F. C. von Faber (Plant Physiol.), especially mangrove- and crater-plants). — Ass. Ch. Coster (Plant-physiol., especially of trees). *In this labor. 6 tables for foreign botan. are available.*
Botan. labor. of the General Agric. Experim. Sta. Dir. Dr. P. van der Elst. *Physiol. of rice plant. „Mentek" disease root rot of rice. Influence of external factors, planting time, lack of necessary mineral plant foods, weather conditions on failure of crops.*

Carmel, Cal. (U.S.A.).
Coastal Labor., Carnegie Inst. of Washington. Ass. Dir. H. A. Spoehr, *Photosynthesis and plant respiration.* Confer etiam Tuscon, Ariz.

Charkow (U.d.S.S.R.).
Pflanzenphysiol. Labor. Inst. f. Volksbildung, Hospitalgasse 5. Dir. Prof. Zaleski. — Zaleski (geb. Marx); Tetjunkina; Kucharkowa; Eisarchewskaja. *3 Plätze.* — *Für Biochemie.*

College Park, Md. (U.S.A.).
Labor. of Plant Physiol. of the Univ. of Md. Dir. Prof. C. O. Appleman, Ph.D., Plant Physiol. — C. H. Conrad, Ph.D., Ass. Plant Physiol.; C. S. Johnston, Ph.D., Assoc. Prof. and Assoc. Plant Physiol.; C. L. Smith, B.S., Ass. in Plant Physiol. *Plant Biochemistry, Plant Biophysics Respiration, Salt Nutrition.*

Detskoje Seló (U.d.S.S.R.).
Pflanzenphysiol.Versuchsstation (am Landwirtsch. Inst.), Ulza Truda 1. Dir. Prof. Oskar A. Walther. — Wiss. Mitarbeiter: M. K. Ostrowskaja, I. W. Krassowskaja, M. S. Miller, L. S. Katschioni-Walther. *2 Plätze . Wasserkulturversuche, chemische u. physikalisch-chemische Untersuchungsmethoden.* ⊕ *Praep. (Alkohol u. Herbar) des Wurzelsystems der Gramineen, primäre und sekundäre Wurzeln.* — Im Winter in Leningrad, Nab. Bolsch. Newki 18.

Dresden (Deutschland).
Pflanzenphysiol. Versuchsstation. Dir. Prof. H. J. Simon.

Eberswalde (Deutschland).
Versuchsanstalt für Holz- u. Zellstoffchemie. Dir. Prof. Schwalbe.

Fontainebleau, Seine et Marne (France).
Biol. végétale. Dir. M. Molliard, Membr de l'Inst., Prof. de Physiol. végétale à la Sorbonne. — Michel-Surand, Dir.-adjoint; L. Dufour, Dir.-adjt. honoraire.

Gainesville, Fla. (U.S.A.).
Dept. of Plant Physiol. Coll. of Agric. A. F. Camp, Ph.D., Plant Physiol.(Cotton Invest.);W.A. Carver, Ph.D., Ass. Cotton Specialist; R. M. Crown, B.S., Field Ass. in Cotton Invest.

Geisenheim a. Rh. (Deutschland).
Pflanzenphysiol. Versuchsstat. der Lehr- u. Forschungsanst. für Wein-, Obst- u. Gartenbau. Dir. Prof. Dr. Karl Kroemer.

Graz (Österreich).
Pflanzenphysiol. Inst. Univ., Phil. Fak. Univ. Dir. Prof. R. Linsbauer. — E. Bersa, Ass.

Grignon, Seine et Oise (France).
Stat. de Physiol., Ecole nation. d'agric.

Harrisonburg, Va. (U.S.A.).
Dept. of Biol. State Teachers Coll. Dir. G. W. Chappelear. *Physiol. Botan.*

Hohenheim, Württbg. (Deutschland).
Pflanzenernährungs-Inst. Landw. Hochsch. Dir. Prof. Margarete von Wrangell.

Homewood, Baltimore, Md. (U.S.A.).
Labor. of Plant Physiol. of the Johns Hopkins. Univ. Dir. Burton E. Livingston. — Grace Lubin, Research Ass.; Ferdinand W. Haasis, Research and Teaching Ass. *1 or more places for research in plant physiol.* — *Chemical, physical, and physiol. equipment; controlled, unlighted temperature chambers, for seven different temperatures; greenhouses.* ⊕ *Atmometers and other physiol. and ecol. apparatus for research.*

Honolulu (Hawaii).
Dept. of Physiol. of the Experim. Stat. of the Assoc. of Hawaiian Pineapple Canners. Dr. C. P. Sideris, Physiologist; Miss B. Krauss, Ass. Physiologist.

Fukuoka (Japan).
Botan. (Pflanzenphysiol.) Labor. Kaiserl. Kyushu Univ. Dir. Rīichiro Kōketsu. — Prof. Rīichiro Kōketsu, (Elektrophysiol. u. Wasserhaushalt d. Pflanzen); Hitoshi Kojima, Ass. Prof. (Physiol. Zytol.); Makoto Takenouchi, Ass. (Physiol. Ökol.); Sadayoshi Fukaki, Ass. (Ernährung der Pflanzen). *Bedeutung von Antocyan, Beziehung zwischen den Pflanzen und dem Boden, physiol. Pflanzenanat. usw.*

Ithaca, N. Y. (U.S.A.).
Labor. of Plant Physiol. Cornell Univ. Ass. Prof E. F. Hopkins (Botan.).

Jassy (România).
Labor. de Physiol. végétale Univ., Institut de Botan. Dir. Prof. C. J. Constantineanu.

Iwanowo-Wosnessensk (U.d.S.S.R.).
Labor. für Pflanzenphysiol. der Technischen Hochsch., Landwirtsch. Fak. Dir. Prof. T. Godneff. — Frau Arzt E. Prochoroff, Ass. *Chemie der Pflanzenpigmente.*

Kiew, Ukraine (U.d.S.S.R.).
Pflanzenphysiol. Labor. (I.N.O.). Dir. Prof. Dr. N. Cholodny. — Frl. M. Moissejewa, Ass. *2 Arbeitsplätze f. Gäste. Studien über die Energetik der Photosynthesis.* — *Heliostat, Spektrophotometer, Kalorimeter, Recorder of Callendar etc.* ⊕ *Mikroskopische*

Praep. der Bakterien (spez. der Eisen- und Schwefelbakterien).

København (Danmark.)

Univ. plantefysiol. Labor., Gotersgade 140. Dir. Prof. Dr. W. Johannsen. — Lect. Dr. Boysen Jensen. *Phototropismus. Zymasegärung. Variationsstatistik.— Zur Zeit Installation f. Genetik u. Raumhygiene.*

Kgl. Veter- og Landbøhojskoles plantefysiol.Labor. Dir. Prof. Dr. Fr. Weis. — Ass.: Jakob H. Blom, Ph.D., Niels Nielsen, M.Sc., Cecil Treschow, M.F, *Above 5 working-places for guests. Anat. and physiol. of plants, microbiol. and especially biol. and biochem. researches of soil. — Appar. for plantphysiol., microbiol. and biochem., potentiometer and other equipment for measuring of pH, for volumetric analysis of gases, for mechanical and chem. analysis of soils.*

Krasnodar (U.d.S.S.R.).

Pflanzenphysiol. Labor. d. Kuban. Landw. Inst.

Landsberg a. d. Warthe (Deutschland).

Inst. f. Rodenkunde u. Pflanzenernährg. Confer: Chemia agris.

Lausanne (Suisse).

Labor. de physiol. végétale et de génétique Univ., Palais de Rumine. Dir. Prof. Arthur Maillefer. *Environ 10 places: Tropismes, Anat. végétale, Genétique. — Les appareils sont construits en vue de chaque recherche spéciale.*

Leningrad (U.d.S.S.R.).

Labor. de Physiol. Végétale à l'Univ., Vassili Ostrov, Tutchkov Quai 2a. Dir. Prof. S. P. Kostytschev,

Labor. of plant physiol. of the Inst. of Applied Botan. and New Cultures, 44 Herzen Str. and Division in Detskoje Selo near Leningrad. Dir. Prof. Nicolas A. Maximow. — Ass. Prof. Assia W. Doroshenko; Ass. Prof. Masimow Tatiane A. Krasnosselsky; Ass.: Irene W. Krassovsky, Sergius P. Kusmin, Ivan I. Tumanov; Laborant: Marie A. Krotkina, Vera L. Popova, Victor I. Rasumov. *5 places for questions applied physiol., as water-relations of plants, rootsystem in vestigations, photoperiodism, culture in artificial light, frostresistance and so on. — Experim. greenhouses (one in Detskoje Selo and the second in Voronesh Gov.). Light Chambre for artificial light. Refrigriation chambre and greenhouse (in construction).* Sub-labor.: in Kamennaja Step, Gov.Woronesch; in Mardakjani near Baku,Caucasus.

Section de Physiol. Végétale ou Jardin botan. principal. Dir. Dr. Vladimir Nikolajewitsh Lubimenko. — Dr.Serguei Dmitrievitsh Lvoff,Sous-Dir.de la Section de Physiol. végétale; Dr. Varnara Alexandrovna Lermann-Brilliante, Dr. Sophia Semenovna Fichtenholz, Ass. au Labor. de Physiol. végétale; Lydia Grigorievna Gavriloff, Olga Alexejevna Szeglova, Préparateurs du premier rang; Abraham Jakovlevitsh Kokine, Préparateur du sécond rang. *2 places pour la spectroscopie, la spectrocolorimétrie, les expériments sur la photosynthèse, le verdissement et sur l'absorption de l'eau par les racines. Il y a une serre spéciale pour les cultures des plantes. — Deux spectrocolorimètres spéciaux pour les dosages des pigments, une chambre spéciale pour les expériences sur la photosynthèse, un photomètre spécial pour les expériments sur l'absorption de l'eau par les racines.*

Leningrad — Alt-Peterhof (U.d.S.S.R.).

Lab. f. Physiol. d. Pflanzen. Dir. Prof. S. P. Kostyćev. — 1 Ass.

London (England).

Research Inst. of Plant Physiol. Imperial Coll. of Sc. of the Univ., Prince Consort Road, South Kensington. Dir. Prof. V. H. Blackman. — Prof. W. Brown (Plant Pathol.); Dr. F. C. Gregory, Ass. in Plant Physiol.; Dr. R. C. Knight, Ass. in Plant Physiol.; Dr. F. Y. Henderson, Ass. in Plant Physiol.; W. B. B. Bolas, Ass. in Plant Physiol.; Prof. S. G. Paine (Plant Bact.). *Plant Physiol., Plant Pathol. and Biochemistry.*

Los Angeles, Cal. (U.S.A.).

The Plant Physiol. Labor. of the Univ. of Southern California.

Lund (Sverige).

Physiol. Abt. d. Botan. Inst. d. Univ. Dir. Prof. J. H. Kylin.

Manila (Philippine Islands).

Labor. of Plant Physiol. and Pathol. Dir. Prof. R. B. Espino. — Ass. Prof. G. O. Octemia.

Minneapolis, Minn. (U.S.A.).

Labor. of Plant Physiol., Dept. of Botan., Univ. Dir. R. B. Harvey. — V. A. Young, M.S., Instr. in Plant Physiol.; G. P. Steinbauer, B.S., Ass. Plant Physiol. *Several desks and special equipment for work on Hight Effects on plant growth. — Spectrometer. Illuminometer recording potentiometer for light intensity records by photo electric cells, controlled temperature greenhouse, incandescent and are lighting for plant growth. Special work on light in relation to forest problems provided for at superior national Forest at Cloquet, Minn.*

Moskau (U.d.S.S.R.).

Lehrstuhl u. zugleich Versuchsstation f. Pflanzenernährung und Düngung d. Landwirtsch. Akad. Dir. Prof. Dr. Prianischnikow. — Ass.: D. W. Druginin, P. A. Golubew, T. T. Demidenko, A. N. Trojizky, P. R. Kuprejenok, M. K. Domontowisch; Z. I. Żurbicky, S. N. Rosanow, M. D. Bachulin, O. W. Sarubina; stagiäres Personal (einjährig bis dreijährig). *30 Plätze für chemische Arbeit. Vegetationsversuche. — Vegetationshaus (Glashaus) für 3000 Gefäße.*

München (Deutschland).

Pflanzenphysiol. Inst. Dir. Prof. Dr. Karl Ritter v. Goebel.

New Brunswick, N. J. (U.S.A.).

Rutgers Univ. and New Jersey Agric. Experim. Stat. Dir. Jacob Goodale Lipman. *Special provision for research in soil sc., plant physiol., biochemistry, and pomol. Facilities are available for about 25 students from foreign countries.* ⓟ *Cultures of soil microorganisms.*

Nowotscherkask (U.d.S.S.R.).

Labor. für Pflanzenphysiol. und Microbiol. der Donschen Hochsch. für Landwirtschaft und Melioration, Potschtowaja 65. Dir. Prof. E. Żmčužniko. F. D. Skaskim, Ass. *Transpiration und Kohlensäureassimilation der landwirtschaftlichen Pflanzen, die Saatgutstimulation und Bodenmikrobiol.*

Paris (France).

Labor. de Physiol. végétale Fac. Sc., Sorbonne. Dir. Prof. Molliard.

Station de biol. végétale Inst. d. recherches agron., VIIe, 42 bis, Rue de Bourgogne.

Labor. de la Biol. des végétaux, Inst. Nation. agron., Ve, 16 Rue Claud Bernard.

Organographie et Physiol. végétale du Mus. d'histoire nat., Ve, 61 Rue de Buffon. Dir. Julien Costantin. — Fritel, Ass.; Loubière, Préparateur. *1 place.* — Anat. et Paléont. végétale.

St. Paul, Minn. (U.S.A.).

Labor. of Plant Physiol., Minn. Agr. Experim. Sta. Univ. Farm. Dir. R. B. Harvey. — Mr. L. O. Regeimbal, M.S., Instr. Plant Physiol.; Mr. E. T. Erickson, B.S., Ass. Plant Physiol.; Mr. A. M. Verral, Studentass. Plant Physiol. *3 special research rooms, for work on low temperature, on light and related effects on plants, and on ripening fruits. — Auto-*

matically controlled low temperature chambers, apparatus for producing and measuring low temperatures, potentiometers, conductivity apparatus, constant temperature baths, enzyme apparatus, extensive apparatus for chemical analysis. — Special work on light in relation to forest problems provided for at Superior National Forest, Cloquet, Minn.

Poznán — Posen (Polska).
Inst. f. Pflanzenphysiol. u. landwirtschaftl. Chemie. Dir. Prof. Br. Niklewski.

Praha — Prag (C. S. R.).
Pflanzenphysiol. Inst. der deutschen Univ., II Vinična 3a. Dir. Prof. Dr. Ernst G. Pringsheim. — Ass.: Dr. Victor Czurda (Conjugaten), Dr. Felix Mainx (Algenkultur, insbesondere Euglenen und Chlamyd.), Dr. Anneliese Niethammer (Stimulation), Dr. Franz Jedlitschka. *Reizphysiol. und Mikrobiol., insbesondere Algenkultur. Vielseitige optische Ausrüstung, Vorrichtungen zur künstlichen Beleuchtung von Kulturen, bakt. Arbeitsgerät, Mikromanipulator, Dunkelkammer für Reizphysiologie, Klinostaten (nach Wiesner, Pfeffer und de Bouter).* ℗ Algenreinkulturen.
Pflanzenphysiol. Inst. der Karls-Univ., II, 433. Dir. Prof. Dr. Bohumil Němec. — Priv.Doc. Dr. S. Prát; Priv.Doc. Dr. I. Kořínek; Prof. Dr. V. S. Iljin; Dr. L. Müllerova. *5 Plätze für Cytol., Physiol. d. Ernährung, Physik-Chemie des Protoplasmas, Mikrobiol.* ℗ Kulturen von Mikroorganismen. — Abt. für Mikrobiol. (Priv.Doc. Dr. I. Kořínek); Abt. für Biochemie (Prof. Dr. V. S. Iljin).

Riga (Latvija).
Physiol. Abt. des Botan. Labor. Univ., Kronvalda bulv. 4. Leiter: Dr. phil. O. Treboux. — 2 Ass.: Priv.Doc. Marie Thielmann; Priv.Doc. Dr. phil. K. Abele.

Salt Lake City, Utah (U. S. A.).
Dept. of Agric. Research American Smelting and Refining Comp., 700, McCornick Building. Dir. G. R. Hill. *Plant Physiol. Effects smelter gases and fumes on plants, crops and animals.*

Saratow (U. d. S. S. R.).
Labor. f. angewandte Botan. d. landw. Versuchsstation. Dir. Prof. Dr. A. Richter. — Leonid Kasakewitsch; Anatol Nitschiporowitsch; Frl. Helene Dworezkaja. ℗ Samen und Pflanzen Süd-Ost-Rußlands.

Taschkent (U. d. S. S. R.).
Labor. of Plant Physiol., State Univ. Dir. Prof. A. V. Blagoveschenski. — N. D. Leonov; M. I. Kurbatov; A. G. Toschevikova; A. N. Bieloserski; W. A. Bogolinbova; N. I. Sossiedov. *2 places for bio-chemical and physiol. works.* — Mountain botan. Station at Chimgan (West Tian-Shan).

Tiflis (U. d. S. S. R.).
Inst. f. Pflanzenanat. u. -physiol. Dir. Lekt. V. Aleksandrov.

Physiol. Lab. des Botan. Gartens. Dir. Dr. W. G. Alexandrov.

Tôkyô (Japan).
Tokugawa Inst. for Biol. Research, Hiratsukamachi Ebara-gun. Dir. Marquis Y. Tokugawa. — Teijirô Kishitani (Plant Physiol.); Marquis Yoshiehika Tokugawa (Physiol. of Plant); Tadao Jimbo (Mykol.); Yoshikazu Emoto (Mykol., Bakt.).

Tucson, Ariz. (U.S.A.).
Labor. for Plant Physiol. Desert Labor., Tucson, Arizona, Coastal Labor., Carmel, California. Dir. Dr. D. T. MacDougal; Ass.-Dir. Dr. Hermann Spoehr (Coastal Labor.); Ass.-Dir. Dr. Forrest Shreve (Desert Labor.). — Forrest Shreve, Ass.-Dir. (Phytogeographer); Hermann Spoehr, Ass.-Dir. (Biochemist); James H. C. Smith (Physiologist); William G. Young (Physiologist); Godfrey Sykes (Geographer). *All collaborators by special arrangement. — The labor. are equiped for researches in biochemistry, physiol. and phytogeography.* — Several temporary special field stations have been established in California, Arizona and Sonora.

Vladikavkaz (U.d.S.S.R.).
Labor. f. Phytophysiol. Landw. Inst. Dir. Doc. S. Vinogradoff.
Kabinett f. Pflanzenanat. u. Physiol. Pädag. Inst. Labor. d. Pflanzenphysiol. und Microbiol., Gorsky Pädagogisches Institut. Dir. Prof. P. Smirnow. *Permeabilität.*

Warszawa (Polska).
Inst. für Pflanzenphysiol. der Univ. Dir. K. Bassalik.
Inst. f. Pflanzenphysiol. Hochsch. f. Land- u. Forstwirtsch. Dir. Prof. M. Korczewski.

Washington, D.C. (U.S.A.).
Experim. Stat. of Tobacco and Plant Nutrition Bureau of Plant Industry, U.S. Dept. of Agric. W. W. Garner (Physiologist); J. E. McMurtrey (Ass. Physiologist); R. A. Steinberg (Ass. Physiologist).

Wien (Österreich).
Biol. Versuchsanst. der Akad. der Wissenschaften, II, Prater, Vivarium. Dir. Hans Przibram (Zool.), Leopold Portheim (Botan.). — Wilhelm Figdor, Vorst. der pflanzenphysiol. Abt.; Eugen Steinach, Vorst. der physiol. Abt.; Paul Weiß, Adj. *20 Arbeitsplätze. — Temperaturkammern und sonstige Einrichtungen für Einwirkung äußerer Faktoren. Käfige, Aquarien, Glashäuser.* ℗ Belegexemplare entwicklungsmechanischer Versuche. — 4 Abt. (zool., botan., pflanzenphysiol., physiol.).
Pflanzenphysiol. Inst., I, Dir. Prof. Dr. Hans Molisch.

Zürich (Schweiz).
Pflanzenphysiol. Inst. der Eidg. Techn. Hochsch., Universitätstr. 2. Dir. Prof. Dr. Paul Jaccard. — Dr. Albert Frey, Ass.

Laboratoria phytopathologiae et mycologiae.

Confer: *Entomologia* (pag. 381), *Botania generalis* (pag. 390), *Microbiologia agris* (pag. 533), *Agricultura* (pag. 507), *Parasitologia generalis* (pag. 379), *Physiologia plantarum* (pag. 412).

Ames, Ia. (U.S.A.).
Dept. of Botan. and Plant Pathol. Stat. Coll. of Agric. L. H. Pammel, Ph.D., D.S., Head of Dept.; Chief Botan.; I. E. Melhus, Ph.D., Prof. and Chief in Plant Pathol.; A. L. Bakke, Ph.D., Prof. Plant Physiol., in charge Plant Physiol.; E. F. Castetter, Ph.D., Ass. Prof. Botan.; R. I. Cratty, Curator; Stuart Dunn, M.S., Instr. Plant Pathol.; O. H. Elmer, Ph.D., Ass. Plant Pathol.; V. C. Fisk, B.S., Instr. Botan.; J. C. Gilman, Ph.D., Assoc. Prof. Plant Pathol.; Ada Hayden, Ph.D., Ass.Prof.Botan.; Charlotte M. King, Ass. Chief Botan.; J. N. Martin, Ph.D., Prof. Plant Morphol. and Cytol., Ass. Botan.; R. G. Reeves, M.S., Instr. Botan.

Antibes, Alpes-Marit. (France).
Stat. de génétique et de pathol. végétale de l'Inst. d. rech. agron. Dir. G. Poirault.

Baarn (Holland).

Phytopath. Labor. Willie Commelin Scholten, Javalaan 4. Dir. Prof. J. Westerdijk. — A. van Luyk, Ass.; Christine J. Buisman; Margaretha Mes. *Pflanzenpathol., Mycol. — Sammlung von 2000 Pilzkulturen des Centralbureaus Schimmelkulturen Baarn.*

Beograd (S.H.S.).

Labor. de Pathol. végétale Fac. Agric. et Silvicult., Studenička 54. Dir. Dr. Mladen Josifovič, Doc. ⓟ Collection des maladies cryptogamiques des plantes cultivées et des arbres, trouvées dans la Royaume des S.H.S.

Berkeley, Cal. (U.S.A.).

Dept. of Plant Pathol. Coll. of Agric. R. E. Smith, B.S., Head of Div.; J. T. Barrett, Ph.D., Prof. Plant Pathol., Assoc. Dir. Citrus. Expt. Sta. (P.O., Riverside); H.S. Fawcett, Ph.D., Prof. Plant Pathol. (P.O., Riverside); E. T. Bartholomew, Ph.D., Ass. Prof. Plant Pathol. (P.O., Riverside); W. T. Horne, B.S., Assoc. Prof. Plant Pathol.; B. A. Rudolph, M.S., Research Assoc. in Plant Pathol. (P.O., Mountain View); C. O. Smith, M.S., Research Assoc. in Plant Pathol. (P.O., Riverside); Elizabeth H. Smith, M.S., Ass. Prof. Plant Pathol.

Berlin (Deutschland).

Biol. Reichsanst. für Land- u. Forstwirtschaft, -Dahlem, Königin-Luise-Str. 19. Prüfstelle für Pflanzenschutzmittel: Reg.R. Dr. E. Riehm; Ass.: Reg.R. Dr. G. Hilgendorff, Reg.R. Dr. W. Trappmann, Dr. A. Winkelmann; Labor. f. allgem. Pflanzenschutz: OberReg.R. Dr. M. Schwartz; Labor. f. Mykol.: Dr. W. Wollenweber.
Hauptstelle für Pflanzenschutz der Landwirtschaftskammer für die Provinz Brandenburg und für Berlin westlich der Oder, -Dahlem. Dir. Prof. Dr. K. Ludwigs. — Dr. M. Schmidt, Ass.
Versuchsstelle für Pflanzenschutz d. Chem. Fabrik auf Aktien (vorm. E. Schering), Zehlendorf-Schönau, Görzallee 3. Dir. Dr. Karl Görnitz.

Bonn a. Rh. (Deutschland).

Hauptstelle für Pflanzenschutz der Landwirtschaftskammer für die Rheinprovinz, Endenicher Allee 60. Leiter: Dr. B. Kessler.

Bonn-Poppelsdorf (Deutschland).

Inst. für Pflanzenkrankheiten, Landw. Hochsch., Nußallee 7. Dir. Prof. Dr. E. Schaffnit. — Dr. Volk, Dr. Wieben, Mycol. und pathol. Physiol.; Dr. Weber, Entomol. *Phytopath. (Physiol., Cytol., Biochemie, Mycol., Entomol.). Labor., Gewächshäuser, Versuchsfelder, für 12 Gäste.*

Bordeaux (France).

Station de Pathol. Végétale, 20, Cours Pasteur. Dir. Dr. M. Gard. — M. Raymond, Ass.

Bratislava — Preßburg (C.S.R.).

L'Inst. phytopathol. des Inst. d'Etat pour les recherches agronomiques. Dir. Ing. V. Vielwerth. ⓟ Collections de préparations phytopath. et entomol.

Breslau (Deutschland).

Hauptstelle für Pflanzenschutz b. d. Landwirtschaftskammer Schlesien, Matthiasplatz 5. Leiter der Hauptstelle für Pflanzenschutz: Dr. Karl Laske. — Wiss. Ass.: Dr. Lohmann, Dr. Köstlin, Dr. Stolze. *1 Gastplatz..* ⓟ Tierische Schädlinge und Krankheitsbilder von Kulturpflanzen. — Eine Versuchsabt. mit 10 Morgen Versuchsgelände.

Brives, Corrèze (France).

Station de pathol. végétale. Dir. J. Dufrénoy.

Brno — Brünn (C.S.R.).

Inst. Dendrol. et phytopath. de l'Ecole Supérieure d'Agric., Černá Pole. Dir. Dr. August Bayer. — 1 Ass. *4 places du travail. Anat. du bois, Dendrol., Phytopath. forest., Bact., Botan. spéciale. — La grande forêt domaniale de l'Ecole (ca. 8000 Hectares) est à la disposition aux études des stagiaires.* ⓟ Fruits, semences, herbiers, surtout plantes ligneuses. Semences pour échange. Bois de toutes sortes, év. pour échange. Maladies de plantes lign., champignons parasit., Défauts des bois. — 1 labor. rural (transportable) sera établi pendant 1927 dans le domaine forestier de l'Ecole Sup. d'Agric. à Adamov près de Brno.
Inst. für landw. Phytopath. und Inst. f. Pflanzenhygiene f. Mähren (Inst. des Recherches agron. et forest.), Ústav pro zdravovědu rostlin, Černá Pole. Dir. Doc. Dr.-Ing. Eduard Baudyš, Landwirtsch. Landesrat. — Ing. Aug. Kalandra (auf staat. Inst.); Dr.-Ing. K. Dvořák; Ing. Oct. Farský; Dr.-Ing. Jos. Novák, Dr. Rich. Picbauer; Ing. Jar. Rašek; Ing. Jan Rozsypal (in Landes-Inst.). *1 Platz. Landw. Entomol., Mykol., phytopath. Bakt., phytopathol. Chemie, Ornithol., Forstschutz, Gartenschutz, Kartoffelhygiene.* ⓟ Zoocecidien und alle Pilze.

Budapest (Ungarn).

Inst. für Phytopath., Volkswirtschaftl. Univ. Dir. Prof. K. Schilbersky.

Buitenzorg, Java (Ned.-O.-Indië).

Botan. Labor. Inst. for plant diseases. Dir. P. J. S. Cramer. — Dr. B. M. Schwarz.
Plant Quarantine Service. Dir. P. J. S. Cramer. — W. C. van Heurn (Zoologist).

Burlington, Vt. (U.S.A.).

Dept. of Botan. and Plant Pathol. Coll. of Agric. Univ. G. P. Burns, Ph.D., Prof. Botan., Botan.; B. F. Lutman, Ph.D., Prof. Plant Pathol., Plant Pathol.; A. H. Gilbert, M.S., Ass. Prof. Botan., Assoc. Plant Pathol.; E. J. Dole, Ph.D., Ass. Prof. and Ass. Botan.; Alexander Gershoy, B.S., Instr. and Ass. Botan.; Anna S. Lutman, B.S., Seed Analyst.

Bustleton, Pa. (U.S.A.).

Field Labor. Pennsylvania State Coll. W. S. Beach (Plant Pathol.).

Bydgoszcz — Bromberg (Polska).

Section des Maladies des Plantes à l'Inst. Agronomique de l'Etat, rue Zacisze 8. Dir. L. Gyrbowski. — S. Kéler (Entomologiste); W.Mazaraki(Chimiste); P. Leszczenko (Agronome). ⓟ Les champignons, parasit. des plantes cultivées de la region, les insectes nuisibles et leurs parasites.

Cairo (Egypt).

Plant Protection Service Ministry of Agric. Dir. C. B. Williams. — J. E. M. Mellor, Entomologist; F. Shaw, Entomologist; Hassan Efflatour, Sub-Dir.; A. Alfien, Ass. Entomol.; A. Anders, Ass. Entomol.; T. Fahmy, Mycologist.

Chabarowsk (U.d.S.S.R.).

Station f. Pflanzenschutz d. Russisch. Fern-Osten, K.-Marx-Str. 29. Dir. V. M. Engelhardt, Leiter d. Entomol. Abt. — I. N. Abramoff, Leiter d. phytopath. Abt.; L. L. Pronitschewa, Phytopathologe; N.N. Maslowsky, Entomol.; K.A. Platter-Plochozky, Mammalog.-Instr.; M. W. Krylowa, Dipl.-Agronom; N. P. Mewsos, Ass.-Laborant; J. A. G. Konowaloff, Spezialist f. angewandte Entomol., Leiter d. Filialstation in Transbaikalien (Tschita). *Im Sommer können 2—3 Gäste arbeiten. Die Station ist speziell für angewandte phyto- und entomol. Arbeiten ausgerüstet.* ⓟ Mycoflora d.Fern. Ostens. Schädliche Insekten, Rodentia. — Die Filialstation in Tschita (Transbaikalien).

Charkow, Ukraine (U.d.S.S.R.).

Phytopath. Abt. d. landwirtsch. Versuchsstation, Dewitschja ulica N. 5. Dir. Prof. T. D. Strachow. — Spezialisten: P. A. Proida, A. N. Iwachnenko, A. D. Maslowsky; Ass.: O. N. Bessonowa, K. P.

Kulschinskaja, O. W. Demianevno, G. A. Trunow, E. E. Fomin, E. A. Fialkowskaja. *3—4 Plätze für mikroskop. Unters. und Arbeit mit reinen Kulturen.* ⓟ *Parasit., Pilze.* **Forschungsinst. für angewandte Botan.** Dir. Prof. O. Janata.

Charlottetown, P. E. I. (Canada).
Experim. Farm. R. R. Hurst (Plant Pathologist). *General Plant pathol. Potato, cereal and fruit diseases, General systematic.*

Cheriban, Java (Ned.-O.-Indië).
Experim. Station. Dir. Dr. V. J. Koningsberger. — Dr. S. G. Wilbrück, P. C. Bolle (Pathologist); A. E. Berkhout (Analyste).

Chiavari, Prov. di Genova (Italia).
R. Osservatorio di Fitopat. per la Liguria, Corso Italia 11. Dir. Prof. Dr. Guido Paoli. — Carlo Menozzi. ⓟ *Insetti agrari, parti di piante alterate da insetti.*

Cluj — Klausenburg (România).
Labor. f. Pflanzenkrankheiten. Dir. Prof. I. Prodan.

College Park, Md. (U. S. A.).
Dept. of Plant Pathol. Coll. of Agric. C. E. Temple, M.A., Prof. Plant Pathol. (State Plant Pathol.); H. A. Hunter, B.S., Ass. Plant Pathol.; R. A. Jehle, Ph.D., Assoc. and Specialist in Plant Pathol.

College Station, Tex. (U. S. A.).
Division of Plant Pathology and Physiology of the Texas Agric. Exper. Sta. Chief: J. J. Taubenhaus.

Davis, Cal. (U. S. A.).
Office of Cereal Crops and Diseases Univ. Farm U.S. Dept. of Agric. V. H. Florell (Agronomist).

Detskoje Selo (U.d.S.S.R.).
Station de Pathol. végétale de l'Inst. Agronomique. Dir. Prof. N. Naoumoff. — T. L. Dobrozrakowa, Chef de travaux; M. F. Markowa, Aspirant; D. N. Teterewnikowa, Préparateur. *3 places pour l'étude de la biol. des cryptogames parasit. et des moyens de les combattre.* ⓟ *Champignons parasit. et saprophytes en herbier et cultures.*

Dresden (Deutschland).
Abt. Pflanzenschutz der Staatl. Landw. Versuchsanstalt Dresden, Hauptstelle für Pflanzenschutz für den Freistaat Sachsen, -A. 16, Stübeallee 2. Dir. Abteilungsvorstand Dr. Baunacke, Angew. Entomol. u. Zool., Botan., allg. Pflanzenschutz. — Dr. F. Esmarch, Angew. Botan., 1. Ass.; Dr. W. Tempel, Diplomlandwirt, 2. Ass.; Vorstreferendar H. Ulbrich, 3. Ass. *Parasit. Pilze, Getreidebeizung. Angew. Entomol. u. diesbezügl. wiss. Untersuchungen, Prüfung von Pflanzenschutzmitteln. Bisamrattenbekämpfung, angew. Entomol. — 1 Serienthermostat für niedere Temperaturen von 2—20⁰ C, 1 Keimschrank, Versuchsfeld m. gr. Treibhausanlage mit Zucht- und Räucherkabinen, Obstanlage m. schädlings- und krankheitsanfälligen Sorten, 1 großer Erdsterilisator.*

East Falls Church, Va. (U. S. A.).
Eastern Field Station U.S. Dept. of Agric. Bureau of Plant Industry. J. R. Christie (Assoc. Nematologist).

Eberswalde (Deutschland).
Waldbau- u. mykol. Inst. Forstl. Hochsch. Dir. Prof. Dengler.

Ekatherinoslav, Ukraine (U.d.S.S.R.).
Station de la défense des plantes. Stat. expérim. Agron. Dir. A. J. Borggardt. — Phytopathol.: Z. A. Demidova.

Essentouky, Caucase du Nord (U.d.S.S.R.).
Station de la Défense des Plantes. Dir. Alexis Lobik. — Mr. B. Zriakovski (Section entomol.); A. Lobik (Phytopathol.); Mr. D. Wolgounoff (Plantes nuisibles); B. Lobik; N. Solorieff. *4 places préparées pour l'étude de la pathol. des plantes, la systématisation des champignons, des insectes et des plantes supérieures. 1 section phytopath. avec 1 labor. bact., 1 section entomol. et 1 section pour l'étude des plantes nuisibles.* ⓟ *Échantillons de la flore mycol., graines des herbes nuisibles et papillons.*

Eutin (Deutschland).
Hauptstelle für Pflanzenschutz. Dir. J. H. Becker.

Fano (Italia).
R. Osservatorio Regionale di Fitopat. (Malattie delle piante), Via Montevecchio 8. Dir. Prof. G. Cecconi (Ispettore per le malattie delle piante). *Malattie delle piante forestali e agrarie. — Insetti dannosi e utili alle piante forestali.*

Fargo, N. D. (U. S. A.).
Dept. of Plant Pathol. State Coll. Station. W. W. Brentzel (Cereal Diseases).

Fort Valley, Ga. (U. S. A.).
Office of Fruit Diseases Investigations. J. C. Dunegan (Ass. Pathologist).

Fukuoka (Japan).
Phytopath. Inst., Coll. of Agric., Kyûshû Imperial Univ.

Gainesville, Fla. (U. S. A.).
Dept. Plant Pathol. Coll. of Agric. O. F. Burger, Sc.D., Plant Pathol.; A. N. Brooks, Ph.D., Ass. Plant Pathol. (Strawberry Invest.: P.O., Plant City); E. F. DeBusk, B.S., Ext. Citrus Pathol.; M. R. Ensign, M.S., Ext. Entomol. and Plant Pathol.; L. O. Gratz, Ph.D., Ass. Plant Pathol. (P.O., Hastings); John Gray, M.S., Ass. Prof. Plant Pathol. and Econ. Entomol.; D. G. A. Kelbert, Field Ass. in Plant Pathol.; W. A. Kuntz, Ph.D., Ass. Plant Pathol. (Coop. U.S.D.A.,; P.O., Lake Alfred); K. W. Loucks, B.S., Ass. Plant Pathol.; A. S. Rhoads, Ph.D., Ass. Plant Pathol. (P.O., Cocoa); J. L. Seal, M.S., Ass. Plant Pathol.; G. F. Weber, Ph.D., Assoc. Plant Pathol.; Erdman West, B.S., Ass. Plant Pathol.

Geisenheim a. Rh. (Deutschland).
Pflanzenpathol. Versuchsstat. der Lehr- und Forschungsanstalt. Dir. Prof. Dr. Gustav Lüstner. — Dr. Th. Gante, Ass. *6 Plätze.* ⓟ *Alle tierischen und pflanzlichen Krankheitserreger der Pflanzen.*

Gembloux (Belgique).
Station de Phytopath. de l'Etat. Dir. E. J. J. Marchal.

Gihu (Japan).
Phytopath. Labor. Gihu Imper. Coll. of Agric. and Forestry.

Göttingen (Deutschland).
Hauptstelle für Pflanzenschutz der Landwirtschaftskammer f. d. Prov. Hannover, Nikolausberger Weg 7. Dir. Dr. W. Fischer. — Dr. W. Fischer (Pflanzenschutz, angew. Botan., Landwirtschaft); Dr. E. Schlottek, Ass. (Pflanzenschutz, angew. Zool.).

Grignon, Seine et Oise (France).
Station de Phytogénétique at Phytopathol., Ecole nationale d'agriculture. Dir. V. Ducomet. — R. Fourmont, préparateur.

Halle a. d. S. (Deutschland).
Versuchsstat. f. Pflanzenkrankheiten d. Inst. f. Pflanzenbau u. Pflanzenzüchtung. Dir. Prof. Hollrung.

Hamburg (Deutschland).

Inst. f. angewandte Botan., 36, Bei d. Kirchhöfen Nr. 14. Dir.: vacat. — Ständige Mitarbeiter: Dr. Carl Brunner, Dr. Leonh. Lindinger, Dr. Johs. Meyer, Dr. Kurt Hahmann, Dr. Otto Nieser. 2 wiss. Hilfsarbeiter. *Schausammlungen und wissenschaftliche Vergleichssammlungen für alle Zweige der angewandten Botan., Warenkunde, Pharmakognosie und Landwirtschaft, Versuchsfelder, Drogen, Saatwaren, Textilfasern, Hölzer, Pflanzenschutz (Hauptstelle f. d. hamburg. Staatsgebiet).*

Hann.-Münden (Deutschland).

Mykol. Inst. der Forstlichen Hochsch. Dir. Prof. Dr. Richard Falck. — Dr. Werner Coordt, Chemiker, Institutsass.; Dr. Leopold Landauer, Chemiker, wissensch. Mitarbeiter. *3 Plätze für Untersuchungen der Holz- und Baumkrankheiten und -schutz. Mykol. Chem.-physiol. u. kulturelle Untersuchungen. — Holzbearbeitungsvorrichtungen, Zentrifugen, Luftpumpen, chem. Einrichtung, Autoklaven, phot. Einrichtung. App. für Holzschutzmittelprüfung.* ⊕ Holzzersetzende Pilze, die Formen der Holzzersetzung. — Mykol. Versuchswald mit Rellew.

Harpenden, Herts (England).

Mycol. Labor. Rothamsted Experim. Station. Dir. Dr. Wm. B. Brierley, D.Sc., F.L.S. — S. H. Smith; M. D. Glynne; B. M. Roach. *10 places for research in connection with plant-disease. — All technical apparatus for bact. and mycol. researches; Ice chambers and cold storage plant; constant temperature chambers; electrical power room; underground physiol. labor. for light control; photographic room and apparatus; special glasshouses, differential chambers for light, humidity and temperature control, insect proof chambers etc. Special physiol. and microscopical equipment. Experim. plots and fields under controlled treatment.* ⊕ Fungi, viruses and disease. — Sub-department of Algol. Sub-department for diseases of tropical crops.

Inst. of Plant. Pathol. Dir. Sir J. Russel, D.Sc. — Ass. Dir. B. A. Keen. Entomol.: A. D. Imms, J. Davidson, H. M. Morris, D. M. T. Moorland; Mycol.: W. B. Brierley, J. H. Smith, M. D. Glynne.

Hays, Kan. (U.S.A.).

Office of Cereal Crops and Diseases Ft. Hays Experim. Station. A.F. Swanson (Ass. Agronomist).

Helsinki (Finnland).

Pflanzenpathol. Inst. der Univ., Riddaregatan 6. Dir. Prof. Dr. I. J. Liro.

Herradura (Cuba).

Tropical Plant Research Foundation. F. S. Earle (Sugar cane).

Hohenheim b. Stuttgart (Deutschland).

Landesanstalt für Pflanzenschutz. Dir. Dr. Wilhelm Lang. — Dr. Jos. Krauß, Chem., Prüfung von Pflanzenschutzmitteln, Beizfragen, Bodenreaktionen; Dr. Paul Faßbender, Bodenschädlinge; Dipl.-Landwirt R. Supper, Feldversuche; Dipl.-Landwirt H. Arker, Hopfenperonospora. *Platz für 2—3 Praktikanten. Mikroskop. Untersuchungen; mykol. Arbeiten; chem. Untersuchungen. — Vegetationshaus, Sterilisiereinrichtungen, Trockenraum; Versuchsgarten u. Versuchsfeld. Vollständig ausgerüstetes chemisches Labor. für alle analytischen Arbeiten.*

Honolulu (Hawaii).

Physiol. and Pathol. Labor. of the Experim. Station of the Assoc. of Hawaiian Pineapple Canners Univ. of Hawaii. Dir. Dr. A. L. Dean. — a) Dept. of Physiol. (Plant): Dr. C. P. Sideris, Physiologist; Miss Beatrice Krauss, Ass. Physiologist; b) Dept. of Pathol. (Plant): C. P. Sideris, Pathologist; Mrs. G. C. Waldron, Ass. Pathologist; c) Dept. of Nematol.: Dr. G. H. Godfrey, Nematologist; Miss H. T. Morita, Ass. Nematologist; d) Dept. of Entomol.: Dr. J. F. Illingworth, Entomologist. *Pineapple Physiologists, Pineapple Pathol., Pineapple Nematol., Pineapple Entomol. 1 or 2 places for Physiol. and Pathol. — Potentiometers, Conductivity Cells, Calorimeters, Thermostat, Centrifuge, Surface Tension, Colorimeter and Nephelometer.* ⊕ Fungi, Bact., Nematodes and Insects. — The Chemistry Dept. and Agric. Dept. of the same institution.

Ithaca, N. Y. (U.S.A.).

Dept. of Plant Pathol., New York State Coll. of Agric. at Cornell Univ. Dir. Prof. L. M. Massey, A.B., Ph.D. — Prof. H. H. Whetzel, in charge of elementary teaching; Prof. M. F. Barrus, in charge of extension; Prof. H. M. Fitzpatrik, in charge of mycol.; Ass. Prof. F. M. Blodgett, Research; Ass. Prof. W. H. Burkholder, Research; Ass. Prof. H. E. Thomas, Research; Ass. Prof. C. Chupp, Extension; Ass. Prof. K. H. Fernow, Extension; Ass. Prof. D. S. Welch, Teaching; Instr. J. F. Flynn, Teaching; Instr. J. W. Sinden, Teaching; Instr. Dewey Stewart, Teaching; Instr. W.D. Mills, Extension; Ass.; J. G. Horsfall, S. E. A. McCallan, C. E. F. Guterman, Grace E. Peterson. ⊕ Diseased plants are collected for teaching, demonstration and research. — Four field stations.

Kagoshima (Japan).

Phytopath. Labor. Kagoshima Imperial Coll. of Agric. and Forestry.

Kamenetz-Podolsk (U.d.S.S.R.).

Phytopath. Labor. des Landwirtschaftl. Inst. K. Marx, Schewtschenkowskaja 25. Dir. Prof. Dr. D. Plevako.

Kaunas (Litauen).

Phytopath. Labor. am Botan. Garten der Univ. Dir. Prof. Dr. C. Regel. — Antanas Minkevičius. *Herbarium von Litauen.* ⊕ Moose, Pilze, Flechten.

Kazan (U.d.S.S.R.).

Labor. de Pathol. Végétale à l'Inst. Forestier. Dir. Prof. A. A. Jounitski

Ketzin a. d. Havel (Deutschland).

Pflanzenphysiol. u. Inst. für Baumschulkrankheiten. Dir. Dr. Walter Gleisberg.

Kentville, N. S. (Canada).

Labor. of Plant Pathol. J. F. Hockey.

Kew, Surrey (England.)

Imperial Bureau of Mycol., 17, The Green. Dir. E. J. Butler. — S. P. Wiltshire, Ass. Dir.; S. F. Ashby, Mycologist; E. W. Mason, Ass. Mycologist. *Several places for research in plant pathol. and mycol., especially as concerns the tropics.* ⊕ Fungi of pathol. importance, especially tropical species.

Kiel (Deutschland).

Biol. Reichsanstalt, Zweigstelle, Niemannsweg 11 bis 13. Leiter: Reg.R. Dr. Hans Blunck, Priv.Doc. — Dr. phil. Hans Bremer, Phytopathologe; Dr. phil. Karl Ludewig, Phytopathologe; Dr. phil. Fritz Merkenschlager, Agrikulturbotaniker, Priv.Doc. in Kiel; Dr. phil. Hans Hähne, Diplom-Landwirt. *3 Arbeitsplätze für Gäste. Phytopath.* Fliegende Station zur Erforschung der Rübenfliege (Pegomyia hyoxyami) in Rosenthal bei Breslau (Leiter: Dr. phil. Otto Kaufmann).

Kielung, Formosa (Japan).

Plant Quarantine Service Custom.

Kiew, Ukraine (U.d.S.S.R.).

Station expérim. agron. Section Phytopath. de l'Inst. agron., Ul. Lenina 46. Dir. G. E. Spangenberg. — H. S. Nevodskyj.

Kôbe (Japan).

Plant Quarantine Service Kôbe Custom.

København (Danmark).
Pflanzenpath. Abt. der Kgl. Veterinär- und Landwirtschaftlichen Hochsch., N., Rolighedsvej 23. Dir. Prof. Dr. C. Ferdinandsen. — N. F. Buchwald, Ass.; Ove Rostrup, Konserv. *3—4 Plätze für Mikroskopie u. Reinzüchtung.* Ⓟ Pflanzenpath. Praep.

Kokand, Fergana (U.d.S.S.R.).
Station de la défense des plantes. Phytopath.: A. G. Pospelov.

Krasnodar (U.d.S.S.R.).
Phytopath. Labor. des Kuban. Landw. Inst. Dir. Prof. J. Ed. Kuschke.

Krasnojarsk, Sibérie (U.d.S.S.R.).
Station de la défense des plantes. Phytopath.: Javorski.

Kurashiki (Japan).
Phytopath. Labor. Ôhara Inst. of Agric. Research, Kurashiki near Okayama. Dir. Yoshikazu Nishikado.

Kyôto (Japan).
Labor. of Phytopath. and Mycol., Biol. Inst., Dept. of Agric., Kyôto Imperial Univ. Dir. Prof. Dr. Takewo Hemmi — Prof. (Ass. Prof., Lect.) and Ass. Ⓟ Parasit. fungi.

Lafayette, Ind. (U.S.A.).
Dept. of Plant Pathol. and Physiol, Purdue Univ. Assoc. Prof. C. L. Porter.

Landsberg a. d. W. (Deutschland).
Inst. für Pflanzenkrankheiten, der Landwirtschaftl. Versuchs- u. Forschungs-Anst. Dir. Prof. Dr. Schander. — Götze (Zool.); Dr. Bielert (Botan., Anat.); Dipl.-Landw. Mestel (Leiter des Versuchsfeldes). *10 Plätze. — Bakt., mikroskopische, physiol., anat. Apparate, Kälteräume, Gewächshäuser und Versuchsfelder.* Ⓟ Phytopath. Objekte.

Leningrad (U.d.S.S.R.).
Inst. Jaczewski de Mycol. et de Phytopath., Perspective Anglaise. 29. Dir. Prof. A. de Jaczewski. — G. N. Doroguine, Vice-Dir.; N. Naoumov; S. Vanine; B. Bachtine; L. Roussakov; M. Antokolska; N. Rojdestvenski; L. Lebedieva; Ch. Benois; Karpova-Benois; H. Svirtchevska; E. Boudrina. *15 places pour les spécialistes travailler pour quelque temps.* Ⓟ Champignons en échantillons d'herbiers et en culture pure. Echantillons de Phytopath. — Succursale à Korenevo, Gouvernement de Moscou, à la Station centrale de sélection de pommes de terre. **Bureau of Introduction of the U.d.S.S.R. Inst. of Applied Botan. and New Cultures**, Herzenstr. 44. Dir. Prof. N. I. Wawilow. — A. N. Smyrnowa Serch (Introduction); Vorsitzender: N. P. Gorbunow; Abt. f. Feldkulturen: N. I. Wawilow; Abt. f. Obst- u. Gemüsebau: W. Paschkewitsch; Abt. f. Genetik u. Selektion: W. E. Pisarew. Ⓟ Seeds of Russian agric. crops.
Abt. f. Mykol. u. Phytopath. d. Staatl. Inst. f. experim. Agronomie. Dir. A. de Jaczewski. — A. A. Bezdolna (Phytopathol.).
Stat. phytopathol. du Jardin Botan. Dir. A. S. Bondartsev.
Inst. Cryptogamiques du Jardin Botan. Dir. Prof. A. A. Elenkine.
Mus. für Pflanzenschutz, pr. Volodarskogo 39.

Lexington, Ky. (U.S.A.).
Dept. of Plant Pathol. of the Univ. of Kentucky. Dir. Prof. W. D. Valleau.

Lima (Peru).
Sección de Botán. Aplicada la Estacion Central Agronómica. Jefe: Dr. J. Gaudron.

Lincoln, Neb. (U.S.A.).
Dept. of Plant Pathol. of the Univ. of Nebraska. Dir. Prof. G. L. Peltier. — Assoc. Prof. R. W. Goss.

Lisbôa (Portugal).
Labor. de Patol. vegetal., Inst. Superiôr de Agronomia, Tapada da Ajuda. Dir. Prof. M. da Sousa Camara. — 4 Naturalistes et Ass.

Louvain — Leuven (Belgique).
Microbiol. technique et Phytopathol, Institut Carnoy. Dir. Prof. Dr. Ph. Biourge. — Georges L. Dropsy, R.N.C.; Adrien Musquin, Pharmacien.

Lwów (Polska).
Inst. de pathol. générale et expérim. Univ. Dir. Prof. Dr. Marjan Franke. — Priv.Doc. Dr. W. Czernecki.
Labor. f. Forstbotan. u. Baumkrankheiten der Techn. Hochsch. Prof. Lz. Wierdak.
Labor. f. techn. Mykol. u. chem. Technol. d. Techn. Hochsch. Dir. Prof. W. Syniewski.

Lübeck (Deutschland).
Landwirtschaftliche Versuchsstation (Hauptstelle für Pflanzenschutz), Mengstr. 4. Dir. Prof. Dr. Karl Steyer. — G. Staude, Ass.; Dr. G. Eberle, Ass.; Dipl.-Landw. Dr. Eichler, Versuchsleiter. *1 Gastplatz in Travemünde.* — Fischereibiol. Labor. in Travemünde.

Lyngby b. København (Danmark).
Statens plantepat. Forsog. Dir. Ernst Gram.— C. M. Jørgensen, Head Botanist; S. Mrs. Rostrup, Zoologist; M. Thomsen, Zoologist; A. Miss Weber, Ass. *Guests welcomed for study of special problems.*

Madison, Wis. (U.S.A.).
Dept. of Plant Pathol. Coll. of Agric. L. R. Jones, Ph.D., D.Sc., Chair. of Dept.; J. W. Brann, M.S., Ass. Prof. Plant Pathol; J. G. Dickson, Ph.D., Assoc. Prof. and Plant Pathol. (Coop. U.S.D.A.); E. M. Gilbert, Ph.D., Prof. Plant. Pathol.; G. W. Keitt, Ph.D., Prof. Plant Pathol., Plant Pathol.; Florence L. Markin, B.S., Ass. in Plant Pathol.; A. J. Riker, Ph.D., Ass. Prof. Plant Pathol.; Mrs. Regina S. Riker, Ph.D., Ass. in Plant Pathol.; R.E. Vaughan, M.S., Assoc. Prof. Plant Pathol.; J. C. Walker, Ph.D., Assoc. Prof. Plant Pathol. (Coop. U.S.D.A.); F. L. Wellman, M.S., Ass. in Plant Pathol. (Coop. U.S.D.A.).

Madrid (España).
Estación Central de Patol. Vegetal (Fitopat. Agric.), 8, Moncloa. Dir. Prof. Miguel Benlloch. — D. Pedro Herce, Ing. Agr.; D. José del Cañizo, Ing. Agr. Lic. Sc.; D. Julian Alonso, Lic. Sc.; D. Juan Rodriguez Sardiña, Ing. Agr.; D. Santiago Blanco, Lic. Sc.; F. Lopez Garvia, Ing. Agr. (Phytopathologist); M. Arenillas, Ing. Agr. (Phytopathologist). *1 place dans chacun de labor. Entomol. agricole, Cryptogamie agricole, Maladies bact. des plantes cultivées, Thérapeutique Végétale. — Matériel entomol., micrographique, bact., photographique et microphotographique.* Ⓟ Insects ravageurs des plantes cultivées. Herbier phytopath.

Matsudo (Japan).
Phytopath. Labor. Imperial Coll. of Horticulture, Chiba-ken.

Mayaguez (Porto-Rico).
Coll. of Agric. Univ. Confer: Inst. generales culturae plantarum et animalium.

Milano (Italia).
Labor. di Patol. vegetale Ist. superiore agrario, Via G. Celoria 2. Dir. Prof. G. B. Traverso. — Prof. Luigi Jenaroli, Ass. *1 posto Patol. vegetale; Micol.* Ⓟ Micromiceti, Patol. vegetale.

Minsk (U.d.S.S.R.).
Pflanzenschutz-Stat. Dir. Prof. E. Jazentkowsky. — E. Patajuk-Jarentkowskaj, Bakteriolog; M. Pilko, Entomolog. *Plätze für bakt., entomol. und phytopath.*

Zwecke. ⓟ Fauna Weißrußlands, phytopath. Erkrankungen.

Montpellier (France).
Station de Physiol. et Pathol. végétales et chaire de Botan. agric. Ecole d'agric. Dir. G. Kühnholtz-Lordat. — Mr. Rivier, Chef des travaux. *1 ou 2 places pour Géographie botan. et Pathol. méditerranéenne.* ⓟ Plants nuisibles aux cultures. Végétation littorale. Pathol. générale et surtout méditerranéenne.

Morgantown, W.Va. (U.S.A.).
Dept. of Plant Pathol., Coll. of Agric. N. J. Giddings, Ph.D., Head of Dept.; Anthony Berg, M.S., Ass. Plant Pathol.; L. H. Leonian, Ph.D., Ass. Prof. and Ass.; Plant Pathol.; E. C. Sherwood, M.S., Ass. Plant Pathol., Specialist in Plant Diseases.

Morioka (Japan).
Phytopath. Labor. Morioka Imper. Coll. of Agric. and Forestry.

Moscow, Id. (U.S.A.).
Dept. of Plant Pathol., Coll. of Agric. C. W. Hungerford, Ph.D., Prof. Plant Pathol., Plant Pathol. (Coop. U.S.D.A.); J. M. Raeder, M.S., Ass. Plant Pathol. (Coop. U.S.D.A.).

Moskau (U.d.S.S.R.).
Stat. phytopath. de l'Inst. Agron. Dir. M. S. Outkine.
Station de la défense des plantes, Rue Sadovaia-Trioumphalnaia, 10. Dir. S. S. Bourov.
Allruss. Mikrobiol. Sammlg. Kustos: Prof. H. Zeiss.
Labor. f. Phytopath. d. L.-V.-Plechanov-Inst. f. Volkswirtschaft. Dir. N. I. Kozin.
Branche de la Section Mycol. Univ. Dir. Prof. L. I. Koursanov.

München (Deutschland).
Pflanzenschutzabt., Bayer. Landesanst. für Pflanzenbau u. Pflanzenschutz. Dir. Prof. Dr. Gustav Korff.
Inst. f. Pflanzenpathol. u. forstl. Botan. a. d. forstl. Versuchsanst., Amalienstr. 52. Dir. Prof. Freiherr v. Tubeuf.

Münster (Deutschland).
Anst. f. Pflanzenschutz.

Nagasaki (Japan).
Plant Quarantine Service Nagasaki Custom. Tei Ishii.

Nancy (France).
Labor. de Mycol., Fac. de Méd. Dir. Prof.Vuillemin.

Nataschino, près Yalta, Crimée (U.d.S.S.R.).
Stat. de la défense des plantes. Section phytopath. Dir. C. N. Dekenbach.

Natchez, Miss. (U.S.A.).
State Plant Board. W. L. Gray (Inspector).

Newark, Del. (U.S.A.).
Dept. of Plant Pathol. of the Agric. Experim. Sta. T. F. Manns, Ph.D., Prof. Plant Pathol. and Soil Bact., Plant Pathol. and Soil Bact., Vice Dir. of Sta.; F. Adams, Ph.D., Assoc. Plant Pathol. and Specialist in Plant Pathol. *Diseases of Sweet Potatoes in Field and Storage. Wilt Resistance Tests. Bact. Leaf Spot. The Value of Bees and Transparent Pollen in Increasing the Yield of Early Ripe. The Sterility and Cross-Pollination of the Hale Peach. Diseases of Cucurbits and their Control in Delaware. Diseases of Soy Beans. Diseases of Impatiens. Dahlia Diseases. Longevity of Corn Root Diseases.*

New Brunswick, N.J. (U.S.A.).
Dept. of Plant Pathol. State Agric. Experim. Sta. W. H. Martin, Ph.D., Plant Pathol.; C. M. Haenseler, Ph. D., Assoc. Plant Pathol.

Nikita — Yalta, Crimée (U.d.S.S.R.).
Stat. Phytopath. du Jardin Botan. Dir. S. L. Streline.

Odessa (U.d.S.S.R.).
Stat. Expérim. du nom de Tairov de Viticult. Section phytopath. Dir. N. A. Naoumova (Phytopath.).

Okitsu near Shizuoka (Japan).
Phytopath. Labor. Okitsu Branch of Imperial Agric. Experim. St.

Omsk (U.d.S.S.R.).
Pflanzenphysiol. u. Landwirtsch. Bakt. Inst. d. Sibirischen Inst. f. Land- u. Forstwirtsch., Vzoz 22. Dir.: Prof. G. G. Petrov, Prof. C. E. Muraschkinski. — P. U. Davydov.

Orono, Me. (U.S.A.).
Dept. of Plant Pathol. Coll. of Agric. Donald Folsom, Ph.D., Plant Pathol.; Reiner Bonde, B.S., Ass. Plant Pathol.; Louise M. Baker, Labor. Ass. in Plant Pathol.

Paramaribo (Ned.-West-Indië).
Regierungs „Proefstation". Dir. Dr. Gerold Stahel. — Dr. D. S. Fernandes.

Paris (France).
Labor. de Pathol. végétale Inst. National agronomique, Ve, 16, Rue Claude Bernard. Dir. Prof. M. Fron. — M. Maublanc, Chef des travaux. *2 places pour recherches de Cryptogamie et Pathol. végétale.*
Station Centrale de Phytopath. et de Parasit. Végétale, 11bis, rue d'Alésia Paris 14. Dir. Etienne Toëx. — Gabriel Arnaud, Dir. Adjoint; Jean Dufrenoy, Chef des Travaux; Préparateurs: Mlle Anne Marie Barbier, Mlle Marguerite Gaudineau, Robert Lemesle, Jean Barthelet, Jean Millasseau. *Phytopath.* ⓟ Objects Phytopathol.
Labor. de Cryptogamie Fac. de Pharmacie, 4, Avenue de l'Observatoire. Dir. Prof. Radais. — Prof. H. Lutz.

Peradeniya (Ceylon).
Dept. of Agric. Dir. F. A. Stockdall, C.E.B.M.A. — W. Small, M.B.E., M.A., B.Sc., Ph.D (Mycol.); J. C. Hutson, B.A., Ph.D. (Entomol.); F. P. Jepson, M.A., M.S.E.A.C. (Ass. Entomol.). M. Park, A.R. C.S., (Ass. Mycol.); A. W. R. Joachim, B.Sc., A.I.C. (Agric. Chem.); A.H.G. Alston, B.A. (System. Botan.); L. Lord, M.A. (Economic Botan.); J. C. Haigh, B.Sc., A.R.C.S.; (Ass, Mycol.). *Provision for two or three guests in each section. A special visitors laboratory is also available. Mycological, Entomological, Chemical and Botanical investigations relating to agric. problems in the tropics in connection with tropical crops and plants.* Small Labor. is maintained at the Botanic. Gardens Hakgala; elevation 5581 feet.

Plant City, Fla. (U.S.A.).
Shawberry Disease Investigations. A. N. Brooks.

Poltoratsk, Ashabad, Turkmenistan (U.d.S.S.R.).
Stat. de la défense des plantes. Phytopath.: P. G. Evstifeev.

Praha — Prag (C.S.R.).
Phytopath. Inst. d. Čech. techn. Hochsch. Vršovice, Havličkovy sady. Dir. Prof. Dr. Franz Bubak.
Phytopath. Inst. Staatl. Versuchsanst. f. Pflanzenproduktion. Dir. C. E. M. K. Blattny.

Pretoria (South Africa).
Phytopath. Labor. Division of Botan., Agric. Dept., P.O. Box 994. Dir. Dr. J. B. Pole Evans, C.M.Gr. — Plant Pathologists, Mycologists, Lay

Ass. ⓟ Fungi (South African) for exchange. — Nata Herbarium, Durban.

Puławy (Polska).
Abt. f. Pflanzenschutz des staatl. Inst. f. Landwirtsch. Dir.: vacat.

Pullman, Wash. (U.S.A.).
Dept. of Plant Pathol., State Coll. F. D. Heald, Ph.D., Head of Dept.; B. F. Dana, M.S., Ass. Prof. and Ass. Plant Pathol.; E. E. Honey, Ph.D., Instr. and Ass. Plant Pathol.; G. L. Zundel, M.S., Ext. Ass. Prof. Plant Pathol.

Riga (Latvija).
Inst. f. Pflanzenpathol. u. Parasit. Dir. Prof. J. Bitkis.
Inst. für Pflanzenschutz, Baznicas iela 4a. Dir. Maksis Eglits. — Edgars Ozols, Entomologe; Julijs Smarods, Phytopatologe; Kārlis Starzs, Spez. der Unkräuter u. Blütenpflanzen; Peters Pētersons, Mycologe; Laimons Gailits, Konsultant d. Entomol.; Jānis Zirnīts, Entomologe. ⓟ Pflanzenpathol. Material, Phanerogamen. — Zweiglaboratoria: Priekuļi à. Cēsis; Cirava; Malnava; Stende; Kaucminde.

Rio Piedras (Porto Rico).
Insular Experim. Sta. *Mycol. Plant Pathol.* ⓟ Mycol. herbarium.

Roma (Italia).
R. Stazione di Patol. vegetale, 30, Via S. Susanna Nr. 13. Dir. Prof. Lionello Petri. — Vicedir. Prof. Peyronel Beniamino; Prof. Rivera Campanile Giulia; Prof. Sibilia Cesare. *Quattro posti per il perfezionamento nella Patol. vegetale (fitopat.).*

Rostock i. M. (Deutschland).
Hauptstelle für Pflanzenschutz für Mecklenburg-Schwerin und Mecklenburg-Strelitz (mit Ausnahme von Ratzeburg) Landwirtschaftliche Versuchsstation, Abt. für Pflanzenschutz, Graf-Zippe-Str. 1. Leiter der Hauptstelle: Landesökonomierat Dr. H. Zimmermann. — Dipl.-Landw. E. Reinmuth. ⓟ Sämtliche phytopath. Gegenstände parasit. u. nichtparasit. Art.

Rostow am Don (U.d.S.S.R.).
Nord-Kaukasische Pflanzenschutzstation, Budjenow-Prosp. 105. Dir. P. A. Swiridenko (gleichzeitig Leitender der Zool. Abt.). — N. N. Archangelsky, Leitender der entomol. Abt.; N. I. Andreelo, Leitender der phytopath. Abt.; Spezialisten: D. P. Downas-Zapolsky, G. I. Lappin, L. Z. Zacharow, P. N. Nowizky, O. P. Kasanskaja. ⓟ Objekte der Entomol., Mykol. und Mammalia-Codentia. — Die Zweigabt. für Entomol., Phytopathol., Zool.

Saint Catharines, Ont. (Canada).
Dominion Research Labor. G. H. Berkeley. *Fruit Diseases, vegetable Diseases.*

Sainte Anne de la Pocatière, Quebec (Canada).
Dominion Field Labor. of Plant Pathol. Dir. H. N. Racicot. — B. Baribeau. ⓟ Specimens of Plant Diseases.

St. Paul, Minn. (U.S.A.).
Dept. of Pathol. and Botan. Dept. of Agric. Univ. Faun. Dean Freeman in charge; E.C. Stakman, Ph.D., Prof. and Assoc. Plant Pathol. (Coop. U.S.D.A.); J. J. Christensen, Ph.D., Ass. Prof. and Ass. Plant Pathol.; Louise Dosdall, Ph.D., Instr. Plant Pathol., Mycol.; Helen Hart, M.A., Ass. Plant Pathol.; R. B. Harvey, Ph.D., Assoc. Prof. and Assoc. Plant Physiol.; A. W. Henry, Ph.D., Ass. Plant Pathol.; A. H. Larson, B.S., Ass. Prof. Plant Pathol., Seed Analyst; J. G. Leach, Ph.D., Ass. Prof. and Ass. Plant Pathol. and Botan.; P. D. Peterson, B.S., Instr. Plant Pathol.; R. M. Nelson, M.S., Instr. and Ass. in For. Pathol.; L. O. Regeimbal, M.S., Experim. Labor. Ass.; H.A. Rodenhiser,

M.S., Instr. and Ass. Plant Pathol.; R. C. Rose, M.S., Plant Pathol. Specialist; Mrs. Ruby U. Crouley, Ass. Seed Analyst.; C. G. Anderson.
Office of Cereal Crops and Diseases Univ. Farm. O. S. Aamodt (Assoc. Pathologist); R. Cotter (Ass. Pathologist).

San José, Cal. (U.S.A.).
Deciduous Fruit Station. B. A. Rudolph.

Samarkand, Ouzbekistan (U.d.S.S.R.).
Stat. de la défense des plantes. Phytopath.: V. P. Schagaev.

Santiago (España).
Inst. f. veget. Pharmakol. Dir. Prof. A. Eleizegui y López.

Santiago de los Vegos (Cuba).
Dept. of Phytopathol. and Entomol. Agric. Experim. Sta. Dir. S. C. Bruner.

Sapporo (Japan).
Phytopath. Labor. Governm. Agric. Experim. Sta., Kotoni-mura near Sapporo.

Sarajevo (S.H.S.).
Labor. der phytopath. Anstalt, Zemaljski Muzej. Dir. Ing. Jovo Popović. — Dr. G. Protić (Pflanzl. Anat. und Physiol., Mykol.). *2 Arbeitsplätze für Entomol. und Mykol. — Reiche entomol. und botan. Sammlungen und große Bibliothek über Balkanhalbinsel.* ⓟ Tierische und pflanzliche Schädlinge der Land- und Forstwirtschaft.

Saskatoon (Canada).
Dominion Labor. Plant Pathol. Univ. G. B. Sanford (Plant Pathologist).

Sewero — Dwiner Gouv. (U.d.S.S.R.).
Sewero-Dwiner Pflanzenschutzstation, Velyki-Oustiug. Dir. S. I. Asow. — Asow, Entomolog; Rothers, Mykolog.

Shimonoseki (Japan).
Plant Quarantine Service Shimonoseki Custom.

Shinkwa, Formosa (Japan).
Phytopath. Labor. Branch Experim. Sta. f. Sugar-Indust., Government Central Research Inst., Tainanshû.

Skierniewice (Polska).
Phytopathological Institute. Dir. Prof. Dr. V. Siemaszko. — Miss Wanda Konopacka. Miss Zofja Zweigbaumowna. *Biological and Systematical Studies of Jungi. Herbarium of Polish Mycological Flora.* ⓟ Fungi, exsiccata; culture of fungi.

Sofia (Bulgaria).
Inst. of Phytopath. Fac. of Agric., The Univ. Dir. Dr. D. Atanasoff. *4 places for plant pathol., cytol. and bact. researches. — Air pump for compressed air and vacuum.* ⓟ Parasit. fungi and bact.
Pflanzenschutz Labor., Solunplatz 3. Dir. Dr. Bor. Iwanoff. — P. Pateff (bakt. Pflanzenkrankheiten); P. Tschorbadjeff (Entomologe). ⓟ Pilzliche und bakt. Pflanzenkrankheiten.

Hermansverk, Sogn. (Norge).
Statens forsöksstation for fruktdyrkning. Dir. Per Stedje.

Spokane, Wash. (U.S.A.).
Office of White Pine Blister Rust Control U.S. Bureaus of Plant Industry. Thomas Large (Animal Ecol. of the „Inland Empire"); C. R. Stillinger (Assoc. Pathologist).

Stauropol, Caucase (U.d.S.S.R.).
Stauropolian Plant Protection Station, Okrzu. Dir. V. Lutshnik. — B. Morozov, Learned Specialist, Chief of the Division of Phytopathol.; I. Gavalov

(Entomol.); Ass. Entomologists: V. Belizin, V. Belousov, P. Vinogradov. ⊕ Entomol., Mycol.

Stellenbosch (South Africa).
Dept. of Plant Pathol. and Mycol. incl. Bact. Univ. Dir. Prof. P. A. van der Byl, M.A., D.Sc.

Samara (U.d.S.S.R.).
Stat. de défense des plantes. Phytopath.: P. I. Balahonov.

Summerland (British-Columbia).
Dominion Labor. of Plant Pathol. H. R. McLarty (Pathol.).

Taihoku, Formosa (Japan).
Phytopathol. Labor. Taihoku Governm. Coll. of Agric. and Forestry.
Phytopath. Labor. Dept. of Agric., Governm. Central Research Inst. Dir. Tsutome Miyake.

Taranto (Italia).
R. Osservatorio di Fitopat. per le Puglie, Piazza Ebalia 1. Dir. Dottor G. Martelli. — Ass.: V. Dottor Di Cairano, Dottor A. Alemanno, Dottor C. B. Cassano, Dottor G. Quarta. *Biol. insetti e rimedi contro i medesimi.* ⊕ Insetti dannosi senza cambio. — Cocciniglie.

Tartu — Dorpat (Eesti).
Phytopath. Versuchsstation der Univ., Raadi mõis. Dir. Cand. Nikolai Rootsi. — Vicedir. Ants Käsebier; Elmar Lepik; Toomas Takjas. ⊕ Schädliche Pilze und andere phytopath. Praep.

Taschkent (U.d.S.S.R.).
Phytopath. Section Usbekische Experim. Pflanzenschutz, Puskinskaia 37. Dir. N. G. Zaprometov. — Z. G. Vedeneeva; P. P. Matzkevitz. *Krankheiten der Baumwollenstaude und der Weintraube.* ⊕ Herbarium Fungorum.
Stat. für Pflanzenschutz. Dir. Dr. Wassily Plotnikov.

Tiflis (U.d.S.S.R.).
Station Phytopath. du Jardin Botan. Dir. P. I. Nagorny.

Tôkyô (Japan).
Phytopath. Labor. Imperial Agric. Experim. Sta., Nishigahara.
Phytopath. Inst. Nôgyô Daigaku (Agric. Univ.), Komasawa near Tôkyô.

Tottori (Japan).
Phytopath. Labor. Tottori Imperial Coll. of Agric.

Tsu (Japan).
Phytopath. Labor. Mie Imperial Coll. of Agric. and Forestry.

Tsuruga (Japan).
Plant Quarantine Service Tsuruga Custom.

Tula (U.d.S.S.R.).
Stat. de la défense des plantes. Phytopath.: L. S. Guitman.

Utsunomiya (Japan).
Phytopath. Labor. Imper. Coll. of Agric. and Forestry.

Veliki-Oustioug, Sewero-Dwiner Gouv. (U.d.S.S.R.).
Pflanzenschutzstation. Confer: Sewero-Dwiner Gouv.

Verona (Italia).
R. Osservatorio regionale di Fitopat., Via G. Mameli, 3. Dir. Prof. Dott. Ettore Malenotti. *2 posti. Entomol. agraria, Distribuzione di rametti di melo con Eriosoma lanigerum contenente il suo speciale endofago, Aphelinus mali.* ⊕ Parti di piante rovinate da insetti, montate a secco su cartoncini.

Villenave d'Ornon (France).
Stat. de Pathol. végétale d'Inst. des Recherches agronomiques Paris.

Vinnitsa (U.d.S.S.R.).
Station de la défense des plantes. Phytopath.: G. F. Borisevitsch.

Vladikavkaz (U.d.S.S.R.).
Stat. f. Schutz d. Pflanzen gegen schädl. Tiere.

Wageningen (Holland).
Phytopath. Service. Dir. Ir. N. van Poeteren. — T. A. C. Schoevers; Miss B. G. Spierenburg; Ir. H. Maarschalk; Ir. W. B. L. Verhoeven; Ir. P. Hus; G. Wolda (Ornithologist); Dr. J. de Hoogh. *1 Platz for Phytopath. or economic entomol.* ⊕ Insects; Fungi; Plant diseases.

Warszawa (Polska).
Phytopath. Inst. Hochsch. f. Land- u. Forstwirtsch. Confer: Sciernewiče.

Washington, D.C. (U.S.A.).
Labor. of Plant Pathol. Bureau of Plant Industry, U.S. Dept. of Agric. L. McCulloch (Phytopath.).
Foreign Plant Quarantines, Federal Horticultural Board, U.S. Dept. of Agric., 2518, 17th str. N.W. R. K. Beattie (Pathologist).
Office of Vegatable and Forage Diseases Bureau of Plant Industry, U.S. Dept. of Agric. H. A. Edson (Pathologist).
Office of Forest Pathol., U.S. Dept. of Agric. G. F. Gravatt (Pathologist); A. R. Gravatt (Pathologist).
Mycol. Collections U.S. Dept. of Agric. J. R. Weiz (Pathologist).

Wien (Österreich).
Bundesanstalt für Pflanzenschutz, II, Trunnerstr. 1. Dir. Dr. Bruno Wahl. — Dr. Gustav Köck; Dr. Leopold Fulmek; Dr. Karl Miestinger; Dr. Friedrich Pichler; Dr. Robert Fischer; Dr. Otto Watzl; Dr. Franz Hengl. *Wiss. u. prakt. Pflanzenschutz, landwirtsch. Bakt., erzeugt Mäusebazillen, Rattenbazillen sowie Knöllchenbakt. für Leguminosen.* ⊕ Pflanzenkrankheiten, Pflanzenschädlinge, kranke und beschädigte Pflanzen. — Entomol., botan., bakt. u. chemische Labor.
Labor. f. Phytopathol. Hochsch. f. Bodenkultur, XVIII. Dir. H. Bleier.

William Head (British Columbia).
Quarantine Station. J. E. Cornwall, F.C.S.J.

Wilno (Polska).
Inst. de pathol. générale et expérim. Univ. Dir. Prof. St. Trzebinski. — Dr. E. Czarnecki.

Winnipeg, Man. (Canada).
Dom. Rust Research Labor. M nitoba Agric. Coll. Dir. G. R. Bisby, Ph.D. — C. H. Goulden (Cereal Specialist); N. M. Newton (Plant Pathologist).

Wooster, O. (U.S.A.).
Dept. of Botan. and Plant Pathol. Ohio Agric. Experim. Station. H. C. Young, Ph.D., Chief; Freda Detmers, Ph.D., Ass. Botan.; Grace Gilmor, M.S., Ass. Technician; Curtis May, M.A., Ass. Plant Pathol.; A. G. Newhall, M.S., Ass. Plant Pathol.; R. C. Thomas, M.A., Assoc. Plant Pathol.; P. E. Tilford, B.S., Ass. Plant Pathol.; R. B. Wilcox, M.S., Assoc. Plant Pathol.

Woronesch (U.d.S.S.R.).
Station de la défense des plantes, Stat. expérim. agron., Sect. Phytopath. Dir. J. G. Beiline. — Phytopath.: V. A. Kouprianov.

Yokohama (Japan).
Plant Quarantee Service Yokohama Custom. Dir. Dr. I. Kuwana.

Laboratoria oecologiae (phytosociologiae).

Confer: *Physiologia plantarum* (pag. 412).

Hallands Väderö im Kattegat (Sverige).
Ekol. Stationen, Torekov. Dir. Prof. Dr. Henrik Lundegårdh. — Ass. *6 Arbeitsplätze nur für selbständige Forscher, die die Arbeitsmethoden der experim. Ökol. kennenlernen wollen. Für ökol. Feldarbeit sehr günstige Bedingungen infolge der größtenteils unberührten Natur der Insel (Waldtypen, Strandtypen, Sumpftypen, Meeresvegetation). — Spezialapparate für Untersuchungen der Kohlensäureassimilation, Analyse des Kohlensäuregehalts der Luft, Registrierung von Lichtfaktor und anderen ökol. Faktoren, Bestimmung der Kohlensäureproduktion und Durchlüftung des Bodens. Kleines Versuchsgewächshaus. Starkstrom fehlt.*—Die Station arbeitet zusammen mit dem Botan. Inst. der ,,Centralanstalten för Jordbruksförsök'' in Stockholm-Experimentalfältet.

Odessa (U.d.S.S.R.).
Sect. f. Ökol. d. Landwirtschaftspflanzen d. Forschungsinst. f. Pflanzenbau, Ul. Swerdlowa 99. Dir. Prof. G. A. Borovikov.

Omsk (U.d.S.S.R.).
Inst. f. Ökol. u. Geograph. d. Pflanzen, Sibir. Inst. f. Land- u. Forstwirtsch. Dir. Prof. V. I. Baranov.

Poznań — Posen (Polska).
Labor. de Sociol. et Systématique des Plantes, rue Słowackiego 4—6. Dir. Prof. Dr. Joseph Paczoski. — Dr. Vitold Sławinski, Adjoint du labor. ⓟ *Les herbiers.*

Repetek at the Central Asia railway, Turkmenistan (U.d.S.S.R.).
Sand-Desert Labor. of the Inst. of Sta. Applied Botan. Dir. Dr. Vladimir Doubiansky. — Boris Orlov(Climate, microclimate and moisture of sands); Nina Basilevskaya (sandflora and Ecol. of psammophytes). *2—3 working-places. Arrangement of the Labor., not sufficient for the present, but its position in the centre of sand-desert Kara-Kum, between moving sands, very suitable of studying the endemical sandfauna and flora (psammophytes).* ⓟ *Herbarium of Psammophytes and a collection of their seeds, in exchange for seeds of Psammophytes of other countries.*

Santa Barbara, Cal. (U.S.A.).
Labor. of Ecol., Carnegie Inst. of Washington. F. E. Clements (Mycol.).

Stockholm (Sverige).
Ökologische Station der Hallands Väderö. Confer: Hallands Väderö.

Tomsk (U.d.S.S.R.).
Inst. f. Geobotan. Dir. Prof. V. V. Reverdatto.

Tucson, Ariz. (U.S.A.).
Desert Labor. Dir. D. T. MacDougal. — Forest Shreve; Godfrey Sykes; Frances Long; R. M. Fraps; Edith B. Shreve. *2 places for Physiol., Ecol.*

Valdres (Norge).
Statens forsöksstasjon for fjellbygdene (State Experim. Sta. for the Mountain Districts), Volbu. Dir. Haakon Foss. — Yngvar Vigerust, Ass. *In summer season one student-worker. 1 place. At the time only occasionally. Observations and experim. on the relation of plants to climatic conditions. — All necessaries for field experim. at the station and in the mountains. The chief apparatus for meteorol. observations.* ⓟ *Plants. Crop plants, especially hay and pasture plants, including as possible all native forms of grasses and legumes.*

Warszawa (Polska).
Inst. f. Systematik u. Geographie d. Pflanzen. Dir. Prof. B. Hryniewiecki.

Woronesch (U.d.S.S.R.).
Kabinett f. Ökolog. u. Geographie. d. Pflanzen. Dir. Doc. L. G. Ramenskij.

Zürich (Schweiz).
Geobotan. Inst. Rübel, Zürichbergstr. 30. Dir. Prof. Dr. Eduard Rübel. — Dr. Josias Braun-Blanquet, Konserv. *1—2 Plätze. Herbar und Ökol. — Herbar, besonders der Schweiz, Mittelmeer, Europa. Sammlung pflanzenökol. Apparate.*

Laboratoria anatomica.

Laboratoria anatomiae generalis et experimentalis, fac. medicinae.

Confer: *Zoologia generalis* (pag. 356), *Histologia* (pag. 375), *Institutiones generales* (pag. 386).

Aberdeen (Scotland).
Anat. Dept., The Univ. Dir. Prof. Al. Low, M.A., M.D. — Robert D. Lockhart, Ch.M., Lect. in Anat.; G. Leslie Purser, M.A., Lect. in Embryol.; Al. Galloway, M.A., M.B., Ch.B., Ass. in Anat. *6 places for Embryol., Physical Anthropol. — Apparatus for Embryol. Technique and PhysicalAnthropol.*

Adelaide (S. Australia).
Dept. of Anat. Univ. Dir. F. Wood Jones.

Alger (Algérie).
Labor. d'Anat., Fac. Méd. Dir. Prof. E. Leblanc.

Amsterdam (Holland).
Anat. Labor., Mauritskade 61. Dir. Prof. L. Bolte. — Dr. A. Dabelow, Prosector. *Zähne, Schädel.*

Ann Arbor, Mich. (U.S.A.).
Anat. Labor. Univ. of Michigan. Dir. S. C. Huber.

Asunción (Paraguay).
Anat. Inst. Univ. Dir. A. Scheuoni. — J. T. Decoud; R. Urizar.

Athènes (Grèce).
Labor. d'Anat., Rue de l'Académie. Dir. Prof. Dr. G. Sclawounos.—Dr. G. Aportolakis, Prosecteur. *6 places pour l'Anat. macroscopique et l'Histol.* ⓟ *Préparations d'Anat. macroscopique et microscopique.*

Bahia (Brasil).
Anat. Inst. Dir. Ed. Diniz.

Bangkok (Siam).
Dept. of Anat. Chulalongkara med. School. J. H. Sankas.

Barcelona (España).
Inst. d'Anat. Dir. Al. Planellas y Llanos. — A. A. Perrer Cajigal (Histol.).

Bari (Italia).
Gab. di anat. umana normale Univ. Dir. G. Favaro.

Basel (Schweiz).
Anat. Anstalt Univ., Pestalozzistr. 20. Dir. Prof. H. K. Corning.

Beirut (Syria).
Labor. anat. Univ. Dir. Prof. M. Negre. — Prof. P. de Brun (Histol.).
Anat. Labor. Americ. Univ. Y. Hitti (Anat.); N. Nucho (Histol.).

Belfast (Ireland).
Dept. of Anat. Queens Univ. Dir. Prof. T. Walmsley.

Beograd (S. H. S.).
Anat. Inst., Miloša, Poccua ulica Kasarna. Dir. Prof. Dr. N. Miljanil.

Berkeley, Cal. (U. S. A.).
Dept. of Anat. Univ. of California. Dir. Prof. H. McLean Evan. — Assoc. Prof. R. O. Moody.

Berlin (Deutschland).
Anat. Anstalt Univ., NW 6, Luisenstr. 25. Dir. Prof. Dr. Rudolf Fick (Gelenk-Muskelmechanik, allg. Vererbungslehre).—Hans Virchow, 1. Prosektor emeritus; Prof.Fr.Kopsch, 1.Prosektor; Prof.Graf Haller, Ass.; Dr. Friedel, Ass.; Dr. Mau,Ass.; Dr. Keyl, Ass.
Anat. biol. Inst. Univ., NW 6, Luisenstr. 56. Dir. Prof. Dr. Franz Keibel.— Prof. Dr. Rudolf Krause, Prosektor; Prof. Dr. Richard Weißenberg, 1. Ass.; Dr. Horst Boenig, 2. Ass. *Histol. u. entwicklungsgeschichtliche Arbeit.*

Bern (Schweiz).
Anat. Inst. Univ.

Besançon, Doubs (France).
Labor. d'Anat. Univ. Dir. Prof. L. Mandereau. — Prof. F. Prieur (Histol.).

Birmingham (England).
Dept. of Anat., Univ. of Birmingham. Dir. Prof. James C. Brash, M.C., M.A., M.D., B.Sc. — C. G. Payton, Lect. and Senior Demonstrator; W. C. O. Hill, Ass. Lect. and Demonstrator.

Bologna (Italia).
Ist. di Anat. Umana Normale Univ., Via Irnerio 48. Dir. Prof. G. Valentini. — Dr. Bertelli Buggero; Dr. Suzzi Dino; Dr. Marelli Angelo. *2 posti. Istol. normale. Dissezione sol cadavere.*

Bonn a. Rh. (Deutschland).
Anat. Inst. Univ. Dir. Prof. F. Sobotta. — Prof. Fr. Heiderich, Abt.-Vorst.

Bordeaux (France).
Labor. d'Anat., Fac. Méd. Univ. Dir. Prof. Picqué.

Bratislava — Preßburg (C. S. R.).
L'Inst. d'anat. humaine, Sasinkova 2. Dir. Dr. A. Frank. — MUDr. Leonard Bučko, Ass. *2 places.* — *L'appareil de congélation des cadavres entières pour de sections du tronc adulte.*

Breslau (Deutschland).
Anat. Inst. Univ., Maxstr. 6. Dir. Prof. Dr. H. von Eggeling. — Prof. Dr. B. Dürken, Vorsteher der entwicklungsmechanischen Abt.; Dr. E. Heidsieck, Prosektor; Dr. A. Andresen, Ass.; Dr. B. Lange, Ass. *Ca. 15 Pätze für jede Art morphol. sowie entwicklungsmechanische Arbeit.*

Bristol (England).
Anat. Dept. Univ. Dir. Prof. E. Fawcett.

Brno — Brünn (C.S.R.).
Inst. f. normale Anat. Univ., Udolni 73. Dir. Prof. Dr. Otomar Völker. — Karel Hora, Ass.; Jan Leviček, Ass.
Anat. Inst. der Tierärztl. Hochsch., Udolni 73.

Bruxelles — Brussel (Belgique).
Labor. d'anat. humaine systématique et topographique et d'embryol. Dir. Prof. Dr. A. Brachet. — A. Daleg, Chef de traveaux.

Budapest (Ungarn).
Anat. Inst. I, Univ. Dir. Prof. Mihály Lenhossék. — Dr. F. Kiss; Dr. Z. Szabó; Dr. G. Renyi; Dr. L. Walter; Dr. G. Fabinyi; T. Csepely; P. Szende; A. Egergenyi.
Anat. Inst. II, Univ. Dir. Prof. K. Tellyesniczky. — Dr. E. Barta; Dr. L. Petrovits; E. Salamon; St. Dékány; E. Schiffbeck; N. Laping; E. Missura; A. Parcsetich; A. Varga.

Buenos Aires (Argentina).
Inst. d'Anat. Fac. méd. Univ. Dir. Prof. P. Belon. — Prof. A. Gutierrez; Prof. J. Lopez Figueroa; Prof. L. R. Sarmiento.

Buffalo, N.Y. (U.S.A.).
Dept. of Anat. Univ. Dir. W. J. Atwell.

Bülowsvej-København (Danmark).
Anat. Labor. Dir. Prof. Dr.S.Paulli.—H. Moltzen Nielsen, Ass., Tierarzt; V. J. Larsen, Prosektor.

București (România).
Labor. d'Anat. normale Fac. med. Univ. Dir. Prof. Fr. Rainer.

Cadiz (España).
Labor. d'Anat. Fac. méd. Univ. Dir. Prof. F. Nuñez Romero. — Prof. J. Ceballos Bonet; Prof. E. Alcina y Quesada.

Caen, Calvados (France).
Inst. d'Anat. Fac. méd. Univ. Dir. Prof. A. Charbounier. — Prof. F. Gidon (Histol.).

Cagliari, Sardegna (Italia).
Ist. anat. umana, Via Genovesi 45. Dir. Prof. Luigi Castaldi. — 1 Aiuto, 1 Ass. ⊕ Prep. in cera de Susini di Firenze (fine sec. XVIII, principio XIX).

Cambridge (England).
Anat. Labor. Univ. Dir. Prof. J. T. Wilson. — Prof. W. L. H. Duckworth; C. Shearer, Lect. of Embryol.

Camerino (Italia).
Ist. anat. Dir. P. Dorello.

Capetown (South Africa).
Anat. Labor. of the Univ. of Capetown. Dir. Prof. M. R. Drennan. — Dr. Lennox Gordon (Surgical Anat.); Dr. R. Wolff (Regional Anat.). *Facilities are offered for the study of the anat. of the native races and for embryol. and anthropol. research. — Anthropol. apparatus.*

Caracas (Venezuela).
Labor. d'Anat. normal Univ. Dir.Prof.J. Izquierdo. — Prof. M. J. Rivero (Histol.).

Cardiff, Wales (England).
Anat. Labor. Univ. Coll. Dir. Prof. D. Hepburn.

Catania (Italia).
Ist. di Anat. umana normale, via Biblioteca▼4. Dir. Prof. Gaetano Cutore. — 1 Settore Aiuto, 1 Settore Ass.

Chiba (Japan).
Anat. Inst. Chiba Med. Univ.

Chicago (U.S.A.).
Dept. of Anat. Univ. of Chicago. Dir. Prof. R. R. Bensley. — C. J. Herrick (Prof. of Neurol.); J. B. Obenchain (Res. Assoc. in Anat.); Prof. A. Maximow (Prof. of Anat.); Prof. B. C. H. Harvey; W. Bloom (Douglas Smith Foundation Fellow).

Division of Anatomy, Northwestern Univ. Med. School, 303 E. Chicago Ave. Chairman: L. B. Avey, Rob. Langhlin Rea, Prof. of Anatomy. *Histol. Embryol. Experim. Anat. 10 working places for guests. Modern equipment.*

Clermont-Ferrand, Puy de Dôme (France).
Labor. d'Anat. Fac. méd. Univ. Dir. Prof. J. Buy. — Prof. Merle (Histol.).

Cleveland, Ohio (U. S. A.).
Anat. Labor., Western Reserve Univ., 2109, Adelbert Rd. Dir. Prof. T. Wingate Todd, M.B., Ch.B. — Dr.N. W. Ingalls. *6 places for Physical Anthrop.: Mammalian Skeleton anat. — Shop for making instruments. Photographic studio. X-ray labor..*

Cluj — Klausenburg (România).
Labor. d'Anat. déscriptive et topographique, SW. Mico 3—5. Dir. Dr. V. Papilian. — Dr. C. Velluda, Doc. et Chef de traveaux; Dr. Ecaterina Papilian et Dr. Constanta Mitrea, Ass.; St. Jianu, Liviu Funariu, Vartolomen Margineanu, Daghie, Maria Jianu, Préparateurs. *10 places.* ⓟ *Des pièces anat. et anthropol.*

Coïmbra (Portugal).
Labor. de Anat. descritiva e topografica. Univ. Dir. B. A. da Costa Freire.

Cordoba (Argentina).
Labor. d'Anat. Fac. méd. Univ. Dir. Prof. J. M. Aliaga. — Prof. F. Strada (Histol.); Prof. H. Fracassi (Anat. topogr.).

Cork (Ireland).
Anat. and Histol. Labor. Univ. Coll. Dir. Prof. D. P. Fitzgerald.

Corrientes (Argentina).
Labor. d'Anat. et Histol. Fac. agric. Univ. Dir. Prof. Cl. Benitez.

Carityba, Paraná (Brasil).
Labor. d'Anat. Fac. méd. Univ. Dir. Prof. J. Pereira de Macedo. — Prof. A. P. Carneiro (Histol.).

Debreczen (Ungarn).
Anat.-biol. Inst. Univ. Dir. Prof. Dr. Th. Huzella, — Mitarbeiter: Dr. Karl Kovács, Emerich Törö, Adalbert Éltetö, Jolán Petreczky, Elisabet Garzó. *5 Arbeitsplätze. Anat., Histol., Mikrooperation, Entwicklungsmechanik.*

Denver, Col. (U. S. A.).
Dept. of Anat. Univ. of Colorado. Dir. J. E. Wallin.

Dublin (Ireland).
Mus. and Dept. of Anat. Univ. Trinity Coll. Dir. Prof. A. F. Dixon.

Dunedin (New Zealand).
Anat. Dept., Univ. of Otago. Dir. Prof. W. P. Gowland.

Edinburgh (Scotland).
Dept. of Anat. Univ. Fac. Med. Dir. Prof. A. Robinson.

Edmonton (Canada).
Dept. of Anat. Univ. of Alberta. D. C. Revell, B.A., M.B.; R. F. Shaner, Ph.D.; E. Greene, M.D., C.M.; N. J. Minish, M.D.

Ekaterinoslav, Ukraine (U. d. S. S. R.).
Anat. Labor. Med. Inst. Dir. Prof. M. Trostanecki.

Erlangen (Deutschland).
Anat. Inst. Dir. Prof. A. Hasselwander. — Prof. A. Spuler, Vorst. d. histol. Abt.

Ferrara (Italia).
Ist. di Anat. Normale, Univ., Palazzo Schifanoia. Dir. Prof. L. Giannelli.

Firenze (Italia).
Ist. Anat. della R. Univ., Via Alfani, 33. Dir. Prof. G. Chiarugi. — Dr. G. Bozza, Ajuto; Dr. M. Calabresi, Ass.; Dr. A. Langer, Ass.; Dr. P. Francejchini, Aiuto onorario. *Anat., Istol., Embriol.* ⓟ *Preparazioni anat., istol. ed embriol.*

Frankfurt a. M. (Deutschland).
Dr. Senckenbergische Anat. der Univ., Theodor-Stern-Kai 36—37. Dir. Prof. Dr. Hans Bluntschli. — Prof. Dr. R. N. Wegner, Abt.-Vorsteher u. Prosektor; Priv.Doc. Dr. Karl Zeiger, 1. Ass.; Dr. Hans Schreiber, 2. Ass. *8 Laboratoriumsplätze: 4 für makroskopische Arbeiten auf dem Gebiet der systemat. u. topogr. Anat., vergl. Anat. und 4 für histol. u. embryol. Arbeiten.* ⓟ *Sehr gute Materialsammlung von Säugetier-, namentlich Primatenkadavern, inkl. embryol. Material von Westaffen.*

Freiburg i. Br. (Deutschland).
Anat. Inst. der Univ. Dir. Prof. Dr. Eugen Fischer. — Prof. Dr. Hans Böker, 1. Prosektor; Priv.Doc. Dr. Otto Henckel, 2. Prosektor 3 Ass. *20 Plätze für vergl. Anat., Anthrop. — Mikroskopisches, makroskopisch-anat. Labor., anthrop. Labor.*

Fukuoka (Japan).
Anat. Inst. Coll. of Med., Kyûshû Imper. Univ.

Galway (Ireland).
Dept. of Anat. Univ. Coll. Dir. Prof. St. Shea.

Genève — Genf (Suisse).
Labor. d'Anat. Univ. Dir. Prof. J. A. Weber.

Genova (Italia).
Ist. di Anat. Umana Norm. Univ. Dir. P. Lachi.

Gent — Gand (Belgie).
Labor. v. Menschelijke Ontleedkunde. Dir. Prof. G. Lebourq.

Gießen (Deutschland).
Anat. Inst. Dir. Prof. Br. Henneberg.

Glasgow (Scotland).
Anat. Dept., Univ. of Glasgow. Dir. Thomas H. Bryce, M.A., M.D. — Norman Maclaren, Ph.D.; Duncan Maccallum Blair, M.B., Ch.B. *3 or 4 places for Embryol. and Histol. — General Apparatus for Embryol. Research.*

Göttingen (Deutschland).
Anat. Inst. Dir. Prof. H. Fuchs.

Granada (España).
Labor. d'Anat. et Embryol. Fac. Med. Univ. Dir. Prof. M. Guiaro Gea. — Prof. E. Gomez Entralla; Prof. G. Sanchez Aguilera; Prof. V. Escritano y Garcia.

Graz (Österreich).
Anat. Inst. Med. Fak. Univ. Dir. Prof. F. W. Müller.

Greifswald (Deutschland).
Anat. Inst. Dir. Prof. Peter. — Prof. Dragendorff, Abt.-Vorsteher; Prof. Dr. Pfuhl, Prosektor.

Groningen (Holland).
Anat.-Embryol. Labor. der Rijksuniv., Oostersingel 69. Dir. Prof. Dr. M. W. Woerdeman. — G. Groeneveld, Prosektor; Frl. J. H. Bijtel, Ass. *6 Plätze für experim. Morphol., Embryol. compar. (Vertebraten). — Experimentierteich, Graben für Fischzucht, Mikromanipulator (Zeiss), mikrokinematograph. Apparat, Sammlung zahlreicher Vertebratenembryonen mit durchgefärbter Skelettanlage.* ⓟ Vertebratenembryonen.

Halifax, N.S. (Canada).
Dept. of Anat. Dalhousie Univ. Dir. John Cameron, M.D., D.S.C. — A. L. Curry, M.D.; V. O. Mader; G. W. Grant, M.B.

Halle a. d. S. (Deutschland).
Anat. Anstalt. der Univ, Große Steinstr. 52. Dir. Prof. Dr. med. et phil. H. Stieve. — Priv.Doc. Dr. J. Hett, Ober-Ass.; Dr. E. Hintzsche, Ass.; Dr. E. Pfeiffer, Ass. *10 Plätze für anatomische, entwicklungsgeschichtliche, histol., entwicklungsmechanische Untersuchungen. — Apparate zur Anfertigung von Plattenmodellen.*
Wilhelm-Roux-Sammlung für Entwicklungsmechanik. Dir. Prof. Stieve.

Hamburg (Deutschland).
Anat. Inst., Ericastr. 4. Dir. Prof. Dr. H. Poll. — J. H. Brodersen, Prof. extraord.; W. Blotevogel, 2. Prosektor.

Heidelberg (Deutschland).
Anat. Inst. Dir. Geh.Med.R. Prof. E. Kallius. — H. L. Hoepke, Abt.-Vorsteher; Dr. H. Münter, Leiter d. anthrop. Inst.

Helsinki (Finnland).
Anat. Inst. der Univ., Fabiansgatan 35. Dir. Prof. Dr. Hjalmar Grönroos.

Innsbruck (Österreich).
Anat. Inst. d. Univ. Dir. F. Sieglbauer.

Irkutsk (U. d. S. S. R.).
Anat. Inst. Univ. Dir. Prof. N. D. Buschmakin.

Jassy (România).
Inst. Anat. (Institutul Anatomic). Dir. Prof. Dr. V. Răşcanu. — 2 Prosecteurs. *200 places pour l'anat. descriptive, anat. topograph, méd. operatoire et embryol. humaine.*
Labor. d'Anat. topograph, Str. Vasile Conta 22. Dir. Prof. I. Tănăsescu. *App. lymphatique.*
Labor. d'anat. descriptive Fac. de méd. Dir. Prof. Dr. G. Zamfirescu. — 4 Préparateurs, 6 Aide-préparateurs (Aspirants). *200 places pour l'anat. humaine et embryol.*

Jena (Deutschland).
Anat. Anstalt der Univ. m. Anthrop. u. zootom. Samml. Dir. Geh.Hofrat. Prof. F. Maurer. — Prof. Graeper, Prosektor. *Schilddrüse, Thymus u. Kiemenspaltenderivate der Wirbeltiere. Rumpfmuskelsystem der Wirbeltiere. Integument der Wirbeltiere (Säugetierhaare). Histol. der Wirbellosen u. der Wirbeltiere.*

Johannesburg (South Africa).
Labor. of Anat. Dir. Prof. R. A. Darb.

Kanasawa (Japan).
Anat. Inst. Kanasawa Med. Univ.

Kaunas (Litauen).
Anat. Inst. Dir. Prof. Zilinskas. — 2 Ass. *1 Platz für Anthrop. — Apparatur für korsionische, trockene Praep. und Anthrop. Kabinett.* ℗ Schädelaustausch der Schädel der Litauer gegen solche der anderen Völker.

Kazan (U. d. S. S. R.).
Inst. d. normalen Anat., Univ., Universitätskaja Str. 12. Dir. Prof. Ternowsky. — A. N. Gennadiew, Prosektor; W. I. Bik, 1. Ass.; A. N. Kiparissowa; E. R. Lutzkendorff; D. M. Lapkow; A. N. Posipkin; W. M. Inkowa. *10 Plätze für Untersuchungen d. Nerven- u. Gefäßsystems.* ℗ Osteol. Praep., Skelette, flache in Agar eingebettete Praep. der menschlichen Organe (Eingeweide u. Gehirn).

Keijô, Korea (Japan).
Anat. Inst. Coll. of Med., Keijô Med. Univ.

Kiel (Deutschland).
Anat. Inst. der Univ., Hegewischstr. 1. Dir. Prof. Dr. Wilh. von Möllendorff. — Prof. Dr. Alfred Benninghoff, 1. Prosektor; Priv.Doc.Dr.Rudolf Spanner, 2. Prosektor. *4 und evtl. mehr Plätze für Histol., vitale Färbung, Gewebezüchtung,Theorie der Färbung, anat. Untersuchungen besonders mit der Technik der Gefäßinjektion.*

Kiew, Ukraine (U. d. S. S. R.).
Anat. Theater u. Inst., Katheder Normaler Anat., Str. Lewaschowskaja 39. Dir. Prof. Al. Iwakin. — P. I. Konaschko; W. P. Kibaltschitsch; M. M. Sauliak-Sawizka; I. G. Marderstein. *5 Plätze für mikroanat. Untersuchungen, besonders Lymphinjektionsuntersuchungen. — 1 Apparat des Prof. Stephanis für Injektion der Lymphgefäße.*

Kingston, Ont. (Canada).
Dept. of Anat. Queen's Univ. Dir. D. C. Matheson, M.B.

København (Danmark).
Normal-anat. Mus,, Bredgade 62. Confer: Bülowsvej. — Prof. Dr. med. Fr. C. C. Hansen; Dr. med. August Chr. Jurisch; Cand. med. Marius Hou Jensen.
Inst. Méd.-Légal de l'Univ. Copenhague (Univ. Retsmedicinske Inst.), Frederik V, vej 9. Dir. Prof. Dr. med. Knud Sand. — Prosecteurs: Dr. J. Fog, Dr. W. Munck. *3 places pour Biol. générale, Physiol., Culture des tissus, Mus. de méd. légale, Préparations de la Biol. sexuelle.*

Köln a. Rh. (Deutschland).
Anat. Inst. Univ. Med. Fak. Dir.: Prof. Dr. Veit. Prof. W. Brandt.

Königsberg i. Pr. (Deutschland).
Anat. Inst. Dir. Prof. R. Heiß.

Kraków (Polska).
Inst. d'Anat. Univ., rue Kopernika 13. Dir. Prof. Dr. K. Kostanecki. — Doc. Dr. T. Rogalski; Dr. E. Rosenhauch.

Krasnodar, Kuban (U. d. S. S. R.).
Kubansches Med. Inst., Anat. Labor., Str. Sedir Nr. 4. Dir. Dr. W. W. Bobin. — Dr. W. W. Kolesnikov; Ass. Dr. P. W. Nikolaev; Ass. Dr. N. W. Kolesnikov; Ass. Dr. W. N. Anufriev.

Kumamoto (Japan).
Anat. Inst. Kumamoto Med. Univ. Ass. Prof. M. Nakadai (Cytol.).

Kyôto (Japan).
Anat. Inst.Kyôto Provincial Med. Univ., Hiro-Kôji.
Anat. Inst. Coll. of Med., Kyôto Imperial Univ. Prof. Dr. Chikanosuke Ogawa; Prof. Dr. Seigo Funaoka; Ass. Prof. Takusaburô Kihara (z. Z. in Europa); Ass. Prof. Kakeo Kanesaki; Lect. Shumpei Kobayashi.

La Plata (Argentina).
Inst. f. allgem. Anat. Pathol. u. Hygiene. Dir. Prof. E. Blomberg.
Labor. d'Anat. topogr. et déscript. Ecole des Sc. méd. Dir. Prof. E. Galli. — Prof. F. Rophille.

Lausanne (Suisse).
Amphithéatre d'Anat. (Ecole de Méd. de l'Univ.). Dir. Prof. A. Roud.

La Valletta (Malta).
Labor. of Anat. and Histol., Univ. of Malta. Dir. Prof. R. Busuttill. — Ass. Prof. J. Briffa.

Lawrence, Kan. (U. S. A.).
Dept. of Anat. of the Univ. of Kansas. Dir. Prof. G. E. Coghill. — Prof. H. C. Tracy.

Leeds (England).
Dept. of Anat. of the Fac. of med., Univ. of Leeds. Dir. Prof. J. Kay Jamieson. — Demonstrator in Anat.: Dr. A. J. E. Cave, Dr. R. G. Inkster; Honorary Dem.: Dr. W. C. Morton, Mr. R. B. Tasker. *Lymphatic System.* ⓟ Costo-vertebral anomalies (human), anat. specimens exhibition, variations.

Leiden (Holland).
Anat. Labor. Dir. Prof. Dr. J. A. J. Barge. — Dr. A. B. Droogleever Fortuyn, Lect. der Histol.

Leipzig (Deutschland).
Anat. Inst., Univ. Dir. Prof. H. Held. — Prof. W. Spalteholz, Prosektor.

Leningrad (U. d. S. S. R.).
Anat. Abt. im Wissenschaftl. Inst. von Lesshaft. Pr. Maklina N. 32 (früher Angliisky pr.). Dir. Prof. A. Krassuskaja. — A. K. Koveschnikowa; E. A. Kotikowa.
Anat. Inst. Militär Med. Akad. Dir. Prof. W. N. Tońkov. — Dr. A. P. Lubomudrov, Ass.

Lexington, Ky. (U. S. A.).
Dept. of Anat. and Physiol. of the Univ. of Kentucky. Dir. Prof. J. W. Pryor.

Lille (France).
Labor. d'Anat., Fac. Méd. de l'Univ. Dir. Prof. C. Debierre. — Olivier.
Labor. de Anat., Fac. de Méd. cathol. de l'Univ., 56, Rue du Port et 11, Rue de Toul. Dir. Prof. H. Billet.

Lima (Peru).
Labor. de Anat., Fac. de Med., Univ., Alameda Gran. Dir. Dr. R. Palma. — Dr. A. Dammert; Dr. D. E. Lavoreria (Histol.); Dr. F. Onesada (Anat. topogr.).

Lincoln, Neb. (U. S. A.).
Dept. of Anat. of the Univ. of Nebraska. Dir. Prof. H. B. Latimer. — Prof. C. W. M. Poynter; Prof. W. A. Willard; Assoc. Prof. J. S. Latta.

Lisbôa (Portugal).
Inst. d'Anat. Fac. Méd. Univ. Dir. Prof. H. de Vilhena. — Dr. M. B. Barbosa Sueiro, 1. Ass.; Dr. Luis Guerreiro, Chef de Services; Dr. Victor Fontes, Ass. libre; Dr. Brito Fontes, 2. Ass. *3 places pour les recherches d'anat. macroscopique et Anthropometrie.* ⓟ Des variations anat. humaines. Grande collection de variations musculaires.
Inst. de Anat. topogr., Fac. de Med., Univ. Dir. Prof. A. de Almeida Vasconcelos Correia.

Liverpool (England).
Dept. of Anat., Fac. of Med. of the Univ. Dir. Prof. W. H. Wood.

Ljubljana — Laibach (S. H. S.).
Anat. Inst. d. Univ., Zaloška cesta. Dir. Prof. J. Plečnik. — Dr. A. Košir, Doc. für Histol.

London (England).
Dept. of Anat. and Histol. at Univ. Coll. Dir. G. Elliot Smith, Prof. of Anat. — H. H. Woollard, M.D., B.S. (Sub-Dean) Ass.-Prof.; H. A. Harris, M.B., B.S., B.Sc., Senior Demonstrator and Curator of the Mus.; R. W. A. Salmond, O.B.E., M.D., Ch.M., D.P.H., Hon. Lect. in Radiol.; Major C. E. S. Phillips, O.B.E., Hon. Lect. in Radiol.; E. Wolff, M.B., B.S., Demonstrator; F. Davies, M.B., B.S., Demonstrator; Una Fielding, M.B,. Ch.M., Demonstrator; J. Beattie, M.Sc., M.B., B.Ch., Demonstrator; L. Reuvid, M.B., Ch.B., Junior Demonstrator; E. R. P. Williams, M.B., Ch.B., Junior Demonstrator; R. J. Ludford, D.Sc., Ph.D., Hon. Lect. in Cytol.; G. S. Sansom, D.Sc., Hon. Research Ass. (Embryol.); J. T. Carter, Hon. Research Ass. (Histol.); F. T. Reagan, Ph.D., Hon. Research Ass. (Embryol.); Catherine J. Hill, B.Sc., Ass. (Embryol. and Histol.).
Dept. of Anat., Guy's Hosp. Med. School of the Univ., S.E. 1, St. Thomas St., Borough. Dir. T. B. Johnston.
Dept. and Mus. of Anat., London Hosp. Med. Coll. of the Univ., E., Turner Str., Mile End. Dir. Prof. W. Wright.
Dept. of Anat. of the King's Coll., Univ.. W.C. 2, Strand. Dir. Prof. E. Barclay-Smith.
Dept. of Anat., St. Bartholomew's Hospital Med. Coll. of the Univ., E.C. 1, West Smithfield. Dir. J. B. Hume. — H. E. Griffiths; L. R. Shore.
Anat. Dept. St. Thomas's Hosp., S.E. 1, Albert Embankment, Lambeth. Dir. F. G. Parsons. — J. H. Mulligan, M.B., Ch.B., Full Time Demonstrator.
Dept. of Anat., Middlesex Hosp. Med. School of the Univ., W., Mortimer Str., Oxford Str. Dir. Prof. T. Yeates.
Dept. of Anat., St. Mary's Hosp. Med. School of the Univ., W. 2, Paddington. Dir. Prof. J. E. S. Frazer.
Dept. of Anat., King's Hosp. Med. School of the Univ., S.E. 5, Denmark Hill. Dir. Prof. T. B. Johnston. — 1 Senior Ass., 2 Junior Ass. *3 places for Histol. and Embryol. — Electric Band Saw for frozen section work.*
Anat. Dept. of the London (Royal free Hospital) School of Med. for Women of the Univ., W.C. 1, 8 Hunter St. Dir. Mary F. Lucas Keene. ⓟ Human Embryoes and human Foetuses.
Dept. of Dental Anat., Univ. Coll. Hosp. Med. School, W.C., Univ. Street, Gower St. Dir. A. B. G. Underwood.
The Shatloch Mus. of Human Compar. Anat. and Pathol., St. Thomas's Hospital Med. School of the Univ., S.E., Albert Embankment, Lambeth. Dir. Prof. Leonard Stanley Dudgeon. — Curator A. L. Urquhart.
Mus. of Anat. of the Univ. Coll., W.C., Gower Street. Dir. Prof. G. E. Smith.

London, Ont. (Canada).
Dept. of Anat. of the Univ. of Western Ontario. Dir. P. S. McKibben, B.S., Ph.D. — R. P. I. Mougall, M.D.; R. A. Johnston, M.D.; H. E. Schaef, M.D.; H. M. Simpson, M.D., M.Sc.; L. W. Pritchett, M.D.; C. A. Lockwood, M.D.

Louisville, Ky. (U. S. A.).
Dept. of Anat. School of Dentistry Univ. Dir. Prof. S. E. Johnson. — Ass. Prof. S. J. Hathaway.

Louvain — Leuven (Belgique).
Labor. d'Anatomie. Dir. Prof. C. Nelis.

Lund (Sverige).
Anat. Inst. d. Univ. (Anatomiska Institutionen). Dir. Prof. Dr. med. Ivar Broman. — Doc. Dr. med. David Hohndahl, 1. Ass.; Dr. med. Sture A. Siwe, 2. Ass. *2 Plätze für embryol. Untersuchungen (Rekonstruktionen).* ⓟ Wirbeltierembryonen.
Labor. der mikroskopischen Anat. Dir. Prof. Dr. Torsten Hellman. — Sture Siwe, Ass.; Lars Essen-Möller, Amanuens; Gösta Glimstadt, Amanuens; Erik Moberg, Amanuens. *Im Herbstsemester 3, im Frühlingssemester mehrere Plätze für mikroskopische Untersuchungen.* ⓟ Lymphatisches Gewebe.

Lwów (Polska).
Inst. d'Anat. de l'Univ., rue Piekarska 52. Dir. Dr. J. Markowski.

Lyon (France).
Inst. d'Anat. Fac. méd. de l'Univ. Dir. Prof. André Latarjèt. — Gabrielle, Prof. agrégé; Gallois, Chef des travaux; Amomoux, Préparateur; Comte, Mallet-Guy, Prosecteurs; Cibert, Desacques, Dechaume, Pegalon, Aides d'Anat. *8 places pour anat. humaine descriptive et medico-chirurgicale.* —

Musée d'Anat. Salle de radiographie anat. ⑦ Ostéol. humaine.
Mus. d'anat. de l'Univ. Dir. Prof. A. Latarjet.

Madrid (España).
Labor. d'Anat. et Embryol. Fac. méd. Univ. Dir. Prof. J. de la Villa y Sanz. — Prof. Fl. Porpetay Llorente.

Manchester (England).
Anat. Labor. Univ. Dir. Prof. J. S. B. Stopford.

Manila (Philippine Islands).
Dept. of Anat. Univ. Coll. of Med. Dir. Prof. A. Garcia. — M. Cañizarez; M. Limson.
Labor. of Anat. Univ. Santo Tomas. Dir. Prof. A. Anguita. — Prof. R. Molina.

Marburg (Deutschland).
Anat. Inst. Dir. Prof. E. Göppert.

Melbourne (Australia).
Dept. of Anat. Univ. Dir. Prof. R. J. A. Berry.

Messina (Italia).
Ist. di anat. umana Univ. Dir. Prof. L. De Gaetani.

Milano (Italia).
Ist. di Anat. Umana Normale Univ., Via Strambio 3, Città degli Studi. Dir. Prof. F. Livini. — Dott. A. Defrise, Aiuto; Dott. P. Pattarin, Ass. della Sezione di Anat., macroscopica; Dott. G. Anelli, Ass. della Sezione Istol. *6 posti per Istol. ed Embriol.* ⑦ Collezione di embrioni di animali e di uomo. Gran numero di organi di feti umani a vario periodo di sviluppo, alcuni già sezionati, colorati e collezionati.

Milwaukee, Wis. (U.S.A.).
Dept. of Anat. Marquette Univ. Dir. E. J. Carey.

Minneapolis, Minn. (U.S.A.).
Inst. of Anat. of the Univ. of Minnesota. Dir. Prof. Clarence M. Jackson. — Richard R. Scammon, Prof. of Anat.; Thomas G. Lee, Prof. of Compar. Anat.; Andrew T. Rasmussen, Prof. of Neurol.; S. P. Miller, Instr. of Anat. *3 places for histol., embryol., biometry, experim. morphol. — Albino rats for experim. work; collection of sectional embryos, human and compar.; special equipment for biometric computations.*

Minsk (U.d.S.S.R.).
Inst. der Normalen Anat., Univ. Dir. Prof. Dr. S. Lebedkin. — Prosektoren: M. Boruchin, J. Irger; Hilfsarbeiter: O. Plissan und D. Golub.

Modena (Italia).
Ist. Anat. Univ., Via Foro Boario, n. 5. Dir. o. Prof. G. Favaro. — Dott. C. Bozzolo, Aiuto; Dott. G. Pancrazi, Ass.

Montpellier (France).
Labor. d'Anat. normale. Dir. Prof. P. Gilis.

Montreal, Que. (Canada).
Dept. of Anat., Univ. of Montreal. Dir. L. D. Mighault, M.D. — E. Virolle, M.D.; L. D. Delmorme, M.D.
Dept. of Anat. McGill Univ. Dir. S. E. Whitnall, M.D.; J. C. Simpson, B.Sc.; I. M. Thompson, B.Sc., M.B., Ch.B.; W. M. Fisk, M.D.; H. E. MacDermot, M.D.

Moskau (U.d.S.S.R.).
Anat. Inst. der 1. Moskauer Staats-Univ., Mochowaja, 11. Dir. Prof. Dr. P. I. Karusin. — Prof. Dr. S. O. Stopnizky, Prosektor; Prof. Dr. B. F. Adler, Priv.Doc.; Priv.Doc. Dr. M. F. Ivanizky, Ass.; G. O. Greilich, Ass.; K. P. Uspensky, Ass.; B. K. Hindze, Ass.; D. B. Urusow, Ass.; Wissenschaftl. Mitarbeiter: B. N. Uskoff, S. W. Mikolskaja, F. A. Iachimowitsch; I. S. Rijew, S. L. Dratsch, M. M. Kuzepina (Biol.). *5 Plätze. Untersuchungen der makro- und makromikroskopischen Gebiete. Neubau 1927.* ⑧ Schädel, Gehirne, Nieren.
Anat. Inst. der 2. Moskauer Univ., Dewitschje Pole, Trubezkoi pereulok. Dir. Prof. A. A. Deschin. — Dr. W. P. Sawenkov, Prosektor; Dr. N. S. Melik-Paschaev, Ass.; Dr. A. G. Korotaewa, Ass.; Dr. A. K. Diakonowa, Ass.; Dr. O. W. Jachontowa, Ass.; Dr. A. E. Dinzer, Ass.; Dr. O. D. Denissowa; Dr. W. A. Nassedkin.

Mukden, Manchuria (China).
Anat. Inst. South Manchurian Med. Univ.

München (Deutschland).
Anat. Anstalt u. Sammlung. Dir. Prof. S. Mollier. — Abt.-Leiter f. Histol. u. Embryol.: Prof. W. Vogt; Abt.-Leiter f. experim. Biol.: Prof. B. Romeis.

Münster i. W. (Deutschland).
Anat. Inst. Univ. Dir. Prof. Dr. Friedrich Heidrich. *Histol., Anat. des Kindesalters.*

Nagasaki (Japan).
Anat. Inst. Nagasaki Imperial Med. Univ. Dir. Prof. Kanai Kunitomo.

Nagoya (Japan).
Anat. Inst. Aichi Med. Univ.

Napoli (Italia).
Ist. anat. umana norm. Fac. med. Univ. Dir. Giunto Salvi.

Nancy (France).
Labor. d'Anat. Univ. Fac. méd. Prof. Dr. Adolphe Hoche, Anat. pathol.; Prof. Dr. Maurice Lucien, Anat. descript.

Nashville, Tenn. (U.S.A.).
Dept. of Anat. van der Bilt Medical School. K. E. Mason (Instr.).

New Orleans, La. (U.S.A.).
Dept. of Anat. Tulane Univ. Assoc. Prof. H. Cummins (Anat.).

New York City (U.S.A.).
Anat. Labor. of the Cornell Univ. Med. Coll. Dir. Prof. Charles R. Stockard. — Dr. Robert Chambers, Prof. of Microscopic Anat.; Dr. Charles V. Morrill, Assoc. Prof. of Anat.; Dr. George Papanicolaou, Ass. Prof. of Anat.; Dr. Jose F. Nonidez, Assoc.; Dr. Paul Reznikoff; Dr. Halsey I. Bagg, Assoc. in Anat.; Dr. Philip B. Armstrong; Dr. Louis Hausman; Dr. William L. Sneed, Instr. in Anat. *10 places for guests in gross and applied Anat. 8 places for guests research in experim. morphol. and cytol. — Microscopes and apparates for microscopical technical work. Microdissection and injection apparatus. All facilities for gross and compar. anat. — A sub-labor. of Experim. Cytol. and an experim. farm for Study of inheritance and growth in constitution and types.*
Dept. of compar. Anat. Mus. of Nat. Hist. Dir. W. K. Gregory.

Niigata (Japan).
Anat. Inst. Niigata Med. Univ.

Odessa (U.d.S.S.R.).
Abt. d. Normal. Anat. d. morphol. physiol. Katheders d. Wissensch. Forschungsinst., Olgiewskaja 4. Dir. N. Lyssenkow. — Aktive Mitglieder: N. Kondratjew, M. Gajewloschin; wissenschaftl. Mitarbeiter: W. Buschkowitsch, A. Lawrentjew; Aspiranten: N. Dowgiallo, A. Gabinsky. ⑦ Anat. Mus.; kraniol. Sammlungen.
Anat. Inst. Med. Inst., Olgiewskaja 4. Dir. Prof. N. Kondratjew. — Prosektoren: Dr. A. P. Lawrentjew, Dr. W. I. Buschkowitsch, Dr. M. I. Tschelustkina, Dr. L. I. Kostinowitsch (Ass.); Aspir.: Dr. Dowgiallo, Dr. Gabinsky. *2 Plätze für makro-*

skopische Untersuchungen des peripherischen Nervensystems.
Anat. Kabinett u. Mus. der Pädag. Hochsch. Dir. Prof. Lignau.

Okayama (Japan).
Anat. Inst. Okayama Imperial Med. Univ. Prof. Yûshô Kôsaka; Prof. Kun-ichirô Yagita; Prof. Jûjirô Shikinami.

Omaha, Neb. (U.S.A.).
Anat. Dept. Univ. of Nebraska. Dir. C. W. M. Poynter.

Omsk (U.d.S.S.R.).
Abt. für normale Anat. d. med. Inst. Dir. Doc. K. V. Romodanowskij.

Ôsaka (Japan).
Anat. Inst. Ôsaka Provincial Med. Univ. Dir. Prof. Risaburô Tsukaguchi. — Prof. Risaburô Tsukaguchi; Ass. Prof.: Kôzô Takagi, Hosuke Tomita; Lect.: Katsuhisa Saitô, Kikutaro Ogushi, Tadashi Yao, Akiji Takiuchi.

Oslo (Norge).
Anat. Inst. Univ. Dir. Prof. K. C. Schreiner.

Oxford (England).
Anat. Labor. Univ. Dir. Prof. A. Thomson.

Paris (France).
Labor. d'Anat. normale Fac. méd. Dir. Prof. A. Nicolas.

Parma (Italia).
Ist. di Anat. umana normale R. Univ. Dir. M. C. Lignière.

Palermo (Italia).
Ist. di Anat. umana normal, Bastione di Porta Carini. Dir. Prof. Lorenzo Lune. — Prof. C. La Rocca; Dott. E. Fazzari; Dott. A. Rindoni; Dott. A. Portio. *2 posti. Culture dei tessuti in vitro.*

Pavia (Italia).
Ist. di Anat. umana normale della R. Univ. Dir. Prof. Luigi Sala. — Dott. Giacomo Quarti, Aiuto; Dott. Gino Nicoli, Ass. *2 posti per ricerche in generale di Anat., istol., embriol.*
Ist. di Anat. e Fisiol. compar. della R. Univ., Palazzo Botta. Dir. Prof. Edoardo Zavattari. Dott. Prof. Maffo Vialli, Aiuto; Dott. Vittorio Citterio, Ass. *14 tavoli da lavoro per ricerche di Istol. e Anat. dei Vertebrati. — Grande Acquario con incubatorio per qualsiasi ricerca di biol. di animali d'acqua dolce. Mus. con ricche collezioni di Anat. dei Vertebrati sia a scopo didattico che di ricerche scientifiche.*

Pecs (Ungarn).
Anat. Inst. Univ. Dir. Prof. Dr. Sigismund Tóth. — Mitarbeiter: Dr. Stefan Héjj, Dr. Karl Röhlich, Koloman Gelei, Katharine Jankovich, Dr. Medardus Kováts, Luiza Haspel, Margarete Mittag, Pongrac Feniczy, Stefan Molnár, Béla Resch, Scharlotte Simon, Koloman Kisfaludy, Ludwig Baló, Emerich Cseh, Desiderius Király, Alexander Kovács, Eugen Kovács, Ludwig Kun, Maria Pánics, Elisabeth Soóky.

Peking (China).
Dept. of Anat. Union Med. Coll. Head: Prof. Davidson Black. — Prof. P. H. Stevenson; Dr. W. C. Ma; Dr. M. T. Pan.

Perm (U.d.S.S.R.).
Anat. Inst Univ., Ul. Karl Marksa 26. Dir. Prof. N. I. Anserov. — Dr. I. W. Matjuschev, Ass.; Dr. I. M. Iwanov, Ass.

Perugia (Italia).
Ist. Anat. norm. Univ. Dir. Prof. L. Caskaldi.

Philadelphia, Pa. (U.S.A.).
The Wistar Inst. of Anat. and Biol., 36th Street and Woodland Avenue. Dir. Milton J. Greenman. — Prof. Henry H. Donaldson (Neurol.); Prof. S. Hatai (Neurol.); Prof. John A. Detlefsen (Genetics); Prof. G. E. Coghill (Nervous system); Ass. Prof. Helen D. King (Embryol.); Ass. Prof. Frederick S. Hammett (Biochemistry). *Neurol. of Vertebr., Embryol.*
Daniel Baugh Inst. of Anat. and Biol. H. E. Radasch (Prof. of Histol. and Embryol.).

Pisa (Italia).
Ist. Anat. umana norm. Univ. Dir. Prof. G. Romiti.

Poitiers (France).
Labor. d'Anat. normale. Dir. Prof. B. Barnsby.

Portland, Ore. (U.S.A.).
Dept. of Anat. Oregon Med. School. Dir. W. F. Allen.

Porto (Portugal).
Inst. d'Anat. Fac. méd. Dir. Prof. Dr. J. A. Pires de Lima. — Prof. Dr. Hernani Monteiro, Sous-Dir.; Ass.: Dr. Amandio Tavares, Dr. Alberto de Sousa, Dr. Constancio Mascarenhas, Dr. Silva Leal, Dr. Esbregneira Mendes. *Mus. d'Anat. humaine et comparée, et d'Anthrop. (environ trois milles objets).*

Porto Alegre (Brasil).
Anat. Inst. Univ. 1. Dir. Prof. M. Menezes. 2. Dir. Prof. L. Leite.

Poznań — Posen (Polska).
Inst. d'Anat. Univ., Dolna Wilda. Dir. Prof. Dr. St. Różycki.

Praha — Prag (C.S.R.).
Anat. Inst. d. Karls-Univ., II, Kateřinska 32. Dir. Prof. Dr. Johann Janošik. *Abt. für topogr. u. chirurg. Anat.;* Vorstand Prof. Dr. K. Weigner.
Anat. Inst. der deutschen Univ., II, Salmovska 5. Dir. Prof. Dr. O. Grosser. — Ass.: Dr. Felix Fritschek, Otto Adam.

Quebec (Canada).
Dept. of Anat. Laval Univ. Dir. Albert Pâquet, M.D.

Quito (Ecuador).
Inst. für allgem. u. beschreibende Anat. Dir. Prof. J. G. Torres.

Reims (France).
Labor. d'Anat. normale Fac. méd. Dir. Prof. Bouvied.

Reykjavik (Island).
Anat. u. Hygien. Inst. Dir. Prof. G. Hannesson.

Riga (Latvija).
Anat. Mus. Med. Fak. Univ., Kronvalda bulv. 9. Dir. Prof. A. Starkow.

Rosario (Argentina).
Inst. f. Anat. Med. Fak. Univ. Dir. Prof. B. dell'Oro. — Prof. F. B. Valdez.

Rostock i. M. (Deutschland).
Anat. Inst. d. Univ., Gertrudenstr. Dir. Prof. Elze. — Prof. Günther Hertwig.

Rostow a. Don (U.d.S.S.R.).
Inst. d. Normalen Anat., Fr.-Engels-Str. 141. Dir. Prof. K. Jazuta. — Prosektor K. Boschoestwensky; Ass.: W. Popow, S. Danilow, P. Ssokolow, R. Chelmer, E. Melichowa, I. Iwanow, N. Odnoralow; Aspiranten: S. Booschanow und A. Chanamirow.

Rouen (France).
Labor. d'Anat. et Physiol. Dir. Prof. Guillonet; L. Longuet.

St. Andrews, Fife (Scotland).
Anat. Labor. Bute Med. School. H. J. R. Kirkpatrick, M.B., Lect. on Anat. *2 places.* — *Human Embryol. Collection.* ⓟ Reconstructions of Human Embryok.

Anat. Labor. United Coll. of St. Leonhard u. St. Salvator. Dir. Prof. D. Waterston.

San Francisco, Cal. (U.S.A.).
Dept. of Anat., Med. School Univ. of California. Dir. Prof. H. M. Evans.

Santiago (España).
Anat. u. Embryol. Labor. Dir. Prof. A. Rodriguez Cadarso. — Prof. L. Blanco y Rivero (topograph. Anat.).

Saratow (U.d.S.S.R.).
Labor. f. normale Anat. Univ. Dir. Prof. G. N. Stadnikij.

Sassari, Sardinia (Italia).
Anat. Inst. Dir. Prof. C. Ganfini.

Sendai (Japan).
Anat. Inst. Fac. of Med., Tôhoku Imper. Univ. Dir. Gennosuke Huse. — Prof. Dr. Gennosuke Huse; Prof. Dr. Kotondo Hasebe; Ass. Prof. Dr. Tatsuo Suzuki; Ass. Prof. Sakuzaemon Kodama.

Sevilla (España).
Anat. Inst. Univ. Dir. Prof. J. Gonzales y Jiménez de Meneses; Prof. D. Mesquito Moreno.

Sheffield (England).
Anat. Dept. Univ. Dir. Prof. C. J. Patten.

Siena (Italia).
Ist. di Anat. umana normale R. Univ., Laterino. Dir. Prof. Rutilio Staderini. — Prof. G. Tramontano-Guerritore, Aiuto incaricato di istol.; Dott. Q. Vischia, Ass. *2 posti. Istol.* ⓟ Collezioni osteol., specilamente crani. Collezioni nevrol. (specilamente cervelli). Collezioni istol. ed embril. — Craniologia.

Singapore (Straits Settlements).
Anat. Labor. of the Coll. of Med. Dir. Prof. G. Harrower. *5 places for guests.*

Smolensk (U.d.S.S.R.).
Anat. Inst. Univ. Dir. Prof. Juden.

Sofia (Bulgarien).
Anat. Inst. Dir. Prof. Elias Chapchal. — Dr. M. Mineff, Doc. der topographischen Anat.; Dr. M. Balan, Prosektor; Dr. D. Halatscheff, Ass.; Dr. I. Dreksel, Ass.

Stamboul (Turquie).
Labor. d'Anat., Fac. méd. Univ. Dir. Prof. Noureddine.

Stockholm (Sverige).
Anat. Abt. des Karolinska Inst. Dir. Prof. Hesser.

Strasbourg — Straßburg (France).
Labor. d'Anat., Fac. Méd. Dir. Prof. Forster. *Mus. d'anat. et d'anthrop.*

Szeged (Ungarn).
Anat. und histol. Inst. Univ., Kossuth S. sugárut 40. Dir. Dr. Eugen Davida. — Dr. Adalbert Gellért (Systemat. Anat.); Dr. Fredericus Münz (Topogr. Anat.); Dr. Andreas Veress (Histol.) ⓟ Mehrfarbige anat. Alkoholpraep.

Taihoku, Formosa (Japan).
Anat. Inst. Taihoku Governmental Coll. of Med.

Tartu — Dorpat (Eesti).
Anat. Inst. Univ., Toomemäel. Dir. Prof. A. Sommer. — R. Villems, Prosektor; W. Pärtelpoeg, Ass.

Taschkent (U.d.S.S.R.).
Anat. Inst. med. Fak. Univ. Dir. Prof. S. Zimmermann. — Dr. A. Lepechin, Prosektor u. 1. Ass.; Dr. W. Matwejew; Dr. D. Kirlka; Dr. I. Rotenberg. ⓟ Anthrop. Material (Gehirne, Schädel).

Tiflis (U.d.S.S.R.).
Anat. Inst. Univ. Dir. Prof. Nathišvili.

Tôkyô (Japan).
Anat. Inst. Fac. of Med., Tôkyô Imper. Univ. Dir. Prof. Dr. M. Inoue. — Prof. Dr. Michio Inoue; Prof. Dr. Seiho Nishi; Ass. Prof. Dr. Otto Mori.
Anat. Inst. Nippon Med. Univ.
Anat. Inst. Nippon Girls Med. Univ.
Anat. Inst. Coll. of Med., Keiô-Gijuku Univ. Dir. Prof. Dr. K. Okajima. — Prof. Dr. K. Okajima; Prof. Dr. S. Mochizuki; Ass. Prof. S. Esaki; Lect. Dr. K. Ogushi; Lect. Dr. H. Kuhlenbeck.
Anat. Inst. Jikeikwai Med. Univ., Shiba.
Anat. Inst. Military Med. Coll. (Gun-i Gakkô).

Tomsk (U.d.S.S.R.).
Inst. f. normale Anat. Dir. Prof. A. P. Asbukin.

Torino (Italia).
Inst. d'Anat. de l'Univ. Dir. G. Levi.

Toronto (Canada).
Dept. of Anat. Univ. Dir. J. P. McMurrich, M.A., Ph.D. — J. C. Watt, M.B.; H. A. Cates, M.B.; E. A. Linell, M.D.; H. G. Willson, B.A., M.B., M.D.; Miss M. I. Tom, B.A., M.B.

Toulouse (France).
Labor. d'Anat. de la Fac. de Méd., Allées saint Michel. Dir. Prof. Henri V. Vallois. — D. Dieulafé, Prof. agrégé; D. Clermont, Prof. agrégé; L. Thomas, Préparateur; A. Viéla, Prosecteur. *8 places pour anat. normale, embryol. et anthrop.* ⓟ Primates divers; pièces squélettiques appartenant aux diverses races de France.
Labor. d'anat., Ecole nation. vétérin. Dir. Prof. C. Bressou. *Anat. descriptive, systématique et topographique des Mammifères domestiques, Histol. et Embryol. des Mammifères domestiques, Tératol.*

Tsinan (China).
Dept. of Anat. Shantung Chr. Univ. Dir. Prof. L. M. Ingle.

Tübingen (Deutschland).
Anat. Inst. Dir. Prof. M. Heidenhain. — 1. Prosektor Dr. Otto Oertel, Prof. ord. ad personam. Ass.: Dr. Walter Jacobj, Dr. Kurt Neubert, Dr. Rüdiger von Volkmann. *Mikroskop. Anat. Histol. Organentstehung.* ⓟ Große Sammlungen mikroskop. Praep. vom Menschen.

Upsala (Sverige).
Anat. Institutionen. Dir. Prof. J. V. Hultkrantz. — Prof. Dr. O. M. Ramström. *Makroskopische Anat. (spez. Osteol. u. Artheol.). Rassenhygiene.*

Utrecht (Holland).
Anat. Inst., Janskerkhof 3. Dir. Prof. Dr. A. J. P. van den Broek. — Dr. H. M. de Burlet, Prosector. *Genitalorgane. Labyrinth.*

Valencia (España).
Labor. d'Anat. Fac. méd. Univ. Dir. Prof. P. Ava Sarria (Anat. descript. et Embryol.). — Prof. J. Bartrina Capella (Anat. descript. Embryol.); Prof. V. Navarro y Gil (Anat. topogr.).
Mus. de Anat. de la Fac. de Med. Dir. Espinosa Ventura.

Valladolid (España).
Inst. Anat. Sierra. Dir. Dr. S. Sierra. — Dr. R. Lopez Pietro.

Vladikavkaz (U.d.S.S.R.).
Kabinett f. Anat. u. Physiol. d. Menschen, Pädag. Inst.

Warszawa (Polska).
Inst. d. Anat. descriptive fac. Med., Chalubinskiego 5. Dir. Prof. Dr. med. et Dr. sc. E. Loth. — Dr. L. Dzwonkowski, Ass. *2 places pour travaux morphol. (surtout les primates). — Appareil radioscopique pour les travaux anat.* ℗ Primates.
Inst. d'Anat. topogr., r. Chałubinskiego 5. Dir. Prof. Dr. L. Krynski.

Wien (Österreich).
I. Anat. Lehrkanzel Univ., Währinger Str. Dir. Prof. Tandler.
II. Anat. Lehrkanzel Univ., Schwarzspanierstr. Dir. Prof. Hochstetter.
Anat. Mus. Univ. Dir. Dr. W. Ruppricht.

Wilno (Polska).
Inst. f. topograph. Anat. Fac. med.

Woosung b. Shanghai (China).
Anat. Inst. Tungchi Univ. Dir. Prof. Dr. Wagenseil.

Woronesch (U.d.S.S.R.).
Anat. Inst. Univ., Prospect Revolutii N. 2. Dir. Prof. Dr. med. G. M. Iosifov. — Oschkaderow, Prosector; Kurdiumov, Ass.

Würzburg (Deutschland).
Anat. Anstalt der Univ. Dir Prof. Dr. Hans Petersen. — Abt. f. topograph. Anat.: Dr. W. Lubosch; Abt. f. Histol. u. Embryol.: Prosektor Vogt. *Histol. des Knochens u. der Skelettorgane, Mechanik, physikal. Anat.*

Zagreb — Agram (S.H.S.).
Anat. Inst. fac. med., Voćarska cesta 97. Dir. Prof. Dr. Drago Perović.

Zaragoza (España).
Dept. Anát. Fac. Med. Dir. Prof. Baldomero Berbiela; Prof. Joaquin Gascón y Marin. — J. Conde Andreu (Prof. auxil.); A. Lorente Sanz (Prof. auxil.); 3 Ass.: N. Cifrián, R. Ipiens, A. Poch.

Laboratoria anatomiae cerebri et nervorum.

Confer: *Anatomia generalis* (pag. 423).

Amsterdam (Holland).
Centraal-Inst. voor Herzenonderzock (Gehirnforschung), Mauritskade 61. Dir. Prof. Dr. C. U. Ariens Kappers. — Dr. C. J. van der Horst.

Berlin (Deutschland).
Neurobiol. Labor. Univ., W 35, Magdeburger Str. Nr. 16. Dir. Prof. O. Vogt.

Budapest (Ungarn).
Inst. für Histol. des Gehirns, Univ. Dir. Prof. K. Schaffer. — Dr. D. Miskolozi; Dr. T. Lehóczky; Dr. L. Meduna; Dr. St. Környei.

Charkow (U.d.S.S.R.).
Ukrainische Staats-Psychoneurol. Inst. Dir. Prof. Dr. A. J. Heymanowitsch. — Morphol. Gruppe. Abt. d. topographischen Anat. d. Nervensystems, Leiter: Dr. Z. J. Heymanowitsch; Abt. d. Neuro-. histol., Leiter: Dr. N. A. Zolotowa; Abt. d. pathol. Anat. d. Nervensystems, Leiter: Prof. N. F. Melnikow-Razwedenkow, Dr. E. M. Chajet; Abt. d. Anthrop., Leiter: Prof. L. P. Nikolaew. Experim.-biol. Gruppe. Abt. d. Neurophysiol., Leiter: Prof. N. G. Ponizowsky, Ass.: f. Physiol. Dr. W. S. Ischunina, f. klin. Analyse Dr. M. M. Lomikowskaja, f. physiol. Chemie F. Berenstein; Abt. d. experim. Pathol. d. Nervensystems u. endokrin. Systems, Leiter: Priv.Doc. D. E. Alpern, Ass.: Dr. Tutkewitsch; Abt. d. Elektrophysiol. u. Biophysik, Leiter: Akademiker Prof. W. J. Danilewski, Ass.: Dr. A. M. Worosjew; Abt. f. Arbeitsphysiol., Leiter: Dr. N. N. Kadrjawzew, Ass.: Dr. I. N. Schürawlew, Dr. Denissenko; Abt. d. bedingten Reflexe, Leiter: Prof. G. W. Volborth, Ass.: Dr. Lindberg.

Hamburg-Friedrichsberg (Deutschland).
Hirnanat. Abt. der Staatskrankenanstalt u. psychiatr. Univ. Klinik. Dir. Prof. Dr. Alfons Jakob. — Priv.Doc. Dr. H. Josephy. *15 Plätze für Ärzte des In- und Auslandes für histol. u. histopathol. Arbeiten des Zentralnervensystems.* ℗ Praep. des Nervensystems.

København (Danmark).
Univ. psykiatriske Labor., Kommunehospitalet. Dir. Prof. Dr. A. Wimmer. — A. V. Neel. *Spinalflüssigkeitsuntersuchungen. Mikroskopische Untersuchungen d. Zentralnervensystems.*

Leningrad (U.d.S.S.R.).
Inst. z. Studium des Gehirnes u. der psych. Tätigkeit.

London (England).
Central Labor. County of London Mental Hospitals of the Univ., S.E.5, Mandsley Hospital, Denmark Hill. Dir. F. Golla, F.R.C.P. — S. Mann, Biochemist; C. Geary, Histologist; F. Partner, Bacteriologist. *8 places for Pathol. and Physiol. of Nervous System, physiol. Psychol. — Post mortem material. Electrical recording apparatus. Biochemical apparatus.* ℗ Pathol. specimens of Central nervous System. — 8 sub-labor. in various mental hospitals.

München (Deutschland).
Anat. Labor. d. psychiatr. u. Nervenklinik, Nußbaumstr. 7. Dir. Priv.Doc. Dr. Hugo Spatz. *Allgemeine und spezielle Histopathol. des Zentralnervensystems. Ontogenese des Gehirns. Untersuchung von Stoffwechselprozessen im Gehirn mit morphol. Methoden.*

Santpoort (Holland).
Pathol.-anat. Labor. Provinciaal Ziekenhuis. Dir. Priv.Doc. Abraham Gans. *Lokalisatorische u. histol. Befunde des Gehirns.*

Tôkyô (Japan).
Tôkyô Neurol.-Biol. Station, 1441 Asamadai Shinagawa. Dir. Dr. Tamao Saito. *Neuroglia.*

Zürich (Schweiz).
Hirnanat. Inst. der Univ., Schönberggasse 2. Dir. Prof. C. von Monakow. — Prof. M. Minkowski, Oberass. am Inst., Priv.Doc. a. d. Univ. *7—8 Plätze für normal-anat., vergl.-anat., histol., entwicklungsgeschichtl., experim.-anat. und pathol.-anat. Untersuchungen am zentralen Nervensystem. — Zahlreiche Schnittserien von Gehirn und Rückenmark für mikroskopische Untersuchungen. Makroskopische Objekte vom Nervensystem.* ℗ Makro- und mikroskopische Praep. vom zentralen Nervensystem.

Laboratoria et musea anthropologiae.

Confer: *Musea biologica* (pag. 350) *et zoologica* (pag. 370), *Anatomia* (pag. 423), *Zoologia generalis* (pag. 356), *Evolutio et genetica* (pag. 341).

Amsterdam (Holland).
Abt. Völkerkunde des Königl. Kolonial-Inst., Mauritskade 57a. Dir. Prof. Dr. van Eerde. — Prof. Dr. Kleiweg de Zwaan (Anthrop.); G. Gonggryp; B. M. Goslings. *Ethnol. u. anthrop. Studien d. Malayischen Archipels.*

Athènes (Grèce).
Mus. d'Anthrop. Univ., Académie. Dir. Prof. Jean Koumaris. ⓟ Objets préhistoriques. Crânes et squélettes. Moulages.

Berlin (Deutschland).
Inst. f. Anthrop., Erblichkeitsforschung u. Eugenik d. Kaiser-Wilhelm-Ges. zur Förderung d. Wissenschaften. Dir. Prof. Dr. E. Fischer (bisher Freiburg i. Br.). — Leiter d. Abt. f. Eugenik Dr. H. Muckermann. Leiter der Abt. f. menschliche Vererbungslehre Prof. v. Virschner.

Berlin-Charlottenburg (Deutschland).
Anthropometrisches Labor. der Deutschen Hochsch. für Leibesübungen, Deutsches Stadion. Leiter: Dr. med. Wolfgang Kohlrausch. *1 Platz.* — *Anthropometrisches Meßgerät.*

Bologna (Italia).
Ist. di antrop. Univ. Dir. F. Frassetto.

Breslau (Deutschland).
Anthrop. Inst. u. ethnol.-anthrop. Sammlung Univ.

Brno — Brünn (C.S.R.).
Anthrop. Inst. Univ., Kounieova 38. Dir. Prof. V. Suk, M.D., Ph.D. *2 places for Anthrop. (Anthropometry, Osteometry), Physical Development of Schoo Children.* ⓟ Sceletal remains human, animal. Anat. specimens. For exchange: plaster carts of Eskimofaces, and Eskimo-teeth.

Budapest (Ungarn).
Anthrop. Inst. Univ. Dir. Prof. Lajos Méhelij. — M. Malán.

Coïmbra (Portugal).
Mus. et labor. antrop. Univ.

Cold Spring Harbor, Long Island, N.Y. (U.S.A.).
Eugenics Record office Carnegie Institution of Washington. Dir. H. H. Laughlin (Ass. Dir.).— A. H. Estabrook; H. J. Banker.

Dublin (Ireland).
Mus. of anthrop. of the Univ.

Dunedin (New Zealand).
Otago Univ. Mus. Curator: Prof. W. B. Benham. — Keeper in Ethnography: Lect. H. D. Skinner.

Firenze (Italia).
Labor. d'Anthrop. de l'Univ., 3. Via del Proconsolo, 12. Dir. Prof. Aldobrandino Mochi. — Doct. Lidio Cipriani. *Recherches Anthrop. et ethnol.* ⓟ Matériaux Anthrop. et ethnographiques.

Genève — Genf (Suisse).
Labor. d'Anthrop. Univ. Dir. Prof. Eugène Pittard.

Georgetown (Br.-Guaiana).
Mus. of R. Agric. and Comm. Soc. of Brit.-Guiana, Waterstr. *Anthrop. Zool.*

Hamburg (Deutschland).
Anthropometrisches Labor. der Staatskrankenanstalt., Univ, Friedrichsberg.
Mus. für Völkerkunde, Rothenbaumchaussee 64. Dir. o. Prof. Dr. Georg Thilenius (Allgem. Abt.). — Abt.-Vorsteher: Prof. Dr. Karl Hagen (Asiat. Abt.), Dr. Arthur Byhan (Eurasiat. Abt.), Priv.Doc. Dr. Theodor Wilhelm Danzel (Afrik. Abt.), Priv.Doc. Prof. Dr. Paul Hambruch (Indoozean. Abt.), Dr. Gustav Antze (Amerik. Abt.), Dr. Gustav Schwantes (Vorgeschichtl. Abt.), Priv.Doc. Dr. Walter Scheidt (Anthrop. Abt.). *Sammlungen: Vorgeschichtl. Altertümer des hamburg. Staatsgebiets, anthrop. Sammlung, ethnograph. Sammlung. Aufstellung in Schausammlungen, Studiensammlung.*

Kiel (Deutschland).
Anthrop. Inst. der Univ., Hegewischstr. Dir. Prof. Dr. phil. et med. Otto Aichel. — Dr. phil. et med. Karl Saller, Ass. *12 Plätze für physische Anthrop., vergl. Anat., Vererbungslehre.*

Kiew (U.d.S.S.R.).
Inst. of Physical Culture in the Academie of the Science, str. Tereschenkovskaj N. 2. Dir. Prof. Woldemar Pidgaetzky. — Prof. S. Jaroslaw (Physiologist); Prof. M. Gazenuk (Biochemist); N. Kudrizky (Physiol. of children's age); St. Rodkevich (Applied-Physiol. and Sport). *3 places for anthropometrical mesures, guaze-exchange of the blood at work, ferments.— The Kata-thermometer by Hill. Anthropometrical apparatures. Establishment for the study of the study of the guaze-exchange. Chemical establishments.* ⓟ Collections of skeletons and preparations of the organs of the children of the various ages.

Kingston (Jamaica).
Anthrop. Dept. of the Mus. of Inst. of Jamaica.

Kraków (Polska).
l'Inst. d'Anthrop. de l'Univ. Dir. J. Talko-H.;ncewicz.

La Plata (Argentina).
Anthrop. Abt. des Mus. La Plata. Dir. Prof. R. Lehmann Nitsche.

Leningrad (U.d.S.S.R.).
Kabinett für ethnische Anthrop., Univ.
Kabinett für somatische Anthrop., Univ.
Mus. f. Anthrop. u. Ethnographie d. Akademie d. Wissenschaften, W.O., Tamoshennyj Per. 1. Dir. E. F. Karskij.

Liège (Belgique).
Ecole libre d'anthrop., 37, Mont St. Martin.

Lima (Peru).
Labor. de Antrop., Fac. de ciencias, Univ., Parque Universitario. Dir. Dr. J. C. Tello.

Ljubljana — Laibach (S.H.S.).
Ethnographisches Mus., Bleiweisova cesta 24. Dir. Dr. N. Županić. *Anthrop. Kraniol. Sammlg.*

London (England).
The Biometric Labor. the Francis Galton Labor. for Natural Eugenics, Univ. Coll., W.C. 1, Gower Street. Dir. Karl Pearson, F.R.S. — Ethel M. Elderton, Ass. Prof.; Percy Stocks, M.D., Med. officer; E. S. Pearson, D.Sc., Senior Ass.; G. M. Morant, D.Sc., Anthrop. and Craniometry; Ass.: M. Moul, M. N. Karn, Ida McLearn; Hon. Research Ass.: Julia Bell, M.R., C.P., M.R.C.S.; J. Wishart, M.A., B.Sc. *Postgraduate Courses in Mathematical Theory of Statistics and its applications to the biol. sc., Eugenics, Epidemiol., Courses on Vital Statistics etc. Tables for 18 postgraduate research workers. Statistical Investigations into biol. Problems. Anthropometric Labor. — Computing machines, Data in possession of the 2 Labor.; special arrangements for experim. work. There is an Anthropometric*

Labor., a small *Animal Houses, Mus. of Heredity and large osteol. and craniol. collections* (man). ⦾ Human Crania, some 7000. Skeletons, several hundreds; Prehistorie Artefacts, and many data and records of measurements on man and other animals. No exchanges.

Lwów (Polska).
Anthrop. ethnol. Inst. der Univ., Długosza 8. Dir. Prof. Dr. Jan Czekanowski. — Dr. Bolesław Rosinski, Priv.Doz.; Stanisław Klimek; Marta Gryglaszewska. *12 places pour méthode quantitative, descendance, systématique. — Arithmomètre.* ⦾ Ossements humains.

Madrid (España).
Mus. de Anthrop. Fac. d. cienc. Univ., Paseo de Atocha 13. Dir. D. M. Anton y Ferrándiz.

Moskau (U. d. S. S. R.).
Anthrop. Labor. der Univ. (ein Teil des Anthrop. Inst.), Mochovaja, 11. Dir. Prof. Dr. Victor Bunak. — Extr. Ordin. Prof.: W. Stefko; Doc.: B. Jukoff, B. Kuftin; Ass.: M. Gremiatzky, N. Sinelnikow; Laborant: P. Senkewitsch; Aspiranten u. Volontär-Ass.: A. Jarcho, J. Roginski, M. Lewin, G. Sobolew (Frau). *5—7 Plätze für vergl. Morphol. des Menschen (Kraniol., Osteol., Embryonen, Haut- und Haaruntersuchungen), Anthropometrie, Biometrie. — Osteometrisches, kraniometrisches, anthropometrisches Instrumentarium, rechnerische Apparate. 8000 Schädel, Skelette, Gehirne, Gipsmasken usw.; Kollektion von Primatenkrania.* ⦾ Menschen- und Affenschädel, Skelette, Gehirne, Embryonen.

München (Deutschland).
Anthrop. Inst. Dir. Prof. Mollison.
Anthrop.-prähistorische Sammlung, Neuhauser Str. 51. Dir.: vacat. Hauptkonserv. u. Abt.-Leiter: ao. Prof. Dr. F. Birkner. Konserv.: Dr. Friedr. Wagner.

Oxford (England).
Physical Anthrop. Dept. Univ. Lect. L. H. D. Buxton.

Padova (Italia).
Ist. di antrop. Univ. Dir. E. Tedeschi.

Paris (France).
Labor. d'Anthrop., Ecole pr. des Hautes Etudes. Dir. Prof. L. Manouvrier (Anthrop. physiol.); Dir. Adj. G. Papillault (Sociol.). — Prof. L. Capitan (Anthrop. préhist.); Prof. P. G. Mahoudean (Anthrop. zool.).
Inst. internat. d'anthrop., VIe, 15, Rue de l'Ecole de Medicine. Dir. L. Marin. — D. L. Capitan; Dr. G. Papillault; Comte Bégouen; Dr. Henri Vignes. *Physiol. de la femme enceinte, de l'ovaire et de la menstruation, Mariages consanguins, Anat. pathol. et physiol. de la placenta.*

Pavia (Italia).
Ist. di Antrop. della R. Univ., Palazzo Botta. Dir. Prof. Nello Puccioni. *Antropometria, craniologia, etnol. e paletnol.* ⦾ Crani umani, oggetti etnografici e di archeol. preistorica.

Poznań — Posen (Polska).
I. und II. Anthrop. Inst. Univ. Dir. Prof. A. Wrzosek u. Prof. J. Grochmalicki.

Praha — Prag (C. S. R.).
Anthrop. demograph. Inst. d. Karls-Univ., II, u. Karlova 3. Dir. Prof. Dr. Heinrich Maliegka.

Rio de Janeiro (Brasil).
Anthrop. Abt. d. National Mus. Dir. Prof. E. Roquette-Pinto.

Roma (Italia).
Inst. d'Anthrop., Via Collegio Romano 26. Dir. Prof Sergio Sergi. — Prof. Giuseppe Genna; Dott. Maria Genna, Ass. *L'Anthrop. physique. Anthrop. pathol. — Appareils pour la craniométrie et anthropométrie.* ⦾ Collections des cranes et des squelettes humains.

Rybinsk (U. d. S. S. R.).
Anthrop. Abt. Naturwissensch. Mus.

San Francisco, Cal. (U. S. A.).
Mus. of Anthrop., Univ. of California.

Taschkent (U. d. S. S. R.).
Anthrop. Abt. d. Zentralmus. f. Naturkunde, Ethnographie u. Archaeol. Mittelasiens.

Warszawa (Polska).
Anthrop. Labor. Societatis Sc. Varsoviensis. Dir. K. Stołyhwo. *Morphol. Abt. u. Anthrop. Mus. Abt. f. Militäranthrop.*, Leiter: Jan Mydlarski.

Wien (Österreich).
Anthrop.-ethnographisches Inst. der Univ., 9. Bezirk, Van-Swieten Gasse 1. Dir. Prof. Dr. Otto Reche. — Dr. Michael Hesch; Dr. Eberhard Geyer. *Bis zu 20 Plätze für rassen- und konstitutionsmorphol. Untersuchungen (Isohaemagglutinationsgruppen, Konst. u. Rasse an Tuberkulosen, Formmerkmale des Ohres). — Sammlungen von Lichtbildern, Skelettsammlungen aus Australien, Neuguinea, von den Buschmännern, Uganda, vorgeschichtliche Funde in Abgüssen, früh- und vorgeschichtliche Skelettfunde aus Österreich, vergl. anat. und pathol.-anat. Mat., Originalgipsabgüsse europäischer Typen, ethnol. Material.* ⦾ Gipsabgüsse von europäischen und afrikanischen Typen, eigene Originalaufnahmen.

Zagreb — Agram (S. H. S.).
Ethnographische Abt. des Kroatischen Nationalmus., Mažuranićev trg 27.

Zürich (Schweiz).
Anthrop. Inst. der Univ., VII, Plattenstr. 9. Dir. Prof. Dr. Otto Schlaginhaufen. — Frl. Gertrud Grützner, Ass. *Anthropometrische, craniometrische und osteometrische Untersuchungen.* ⦾ Schädel und Skelette von Menschen aller Rassen und von Affen und Halbaffen; Haarproben des Menschen (Rassen, Familien, besondere Varianten). — Institut steht in Beziehung zur Julius-Klaus-Stiftung für Vererbungsforschung, Sozialanthrop. und Rassenhygiene Zürich.

Laboratoria anatomiae animalium domesticorum (fac. veterin.).

Confer: *Zoologia generalis* (pag. 356).

Alfort, Seine (France).
Labor. d'Anat. de l'Ecole vét. Dir. Prof. C. Bressou. — Chef des travaux pratiques: vacat. *Recherches d'Anat., Histol. et Embryol. compar. — Un appareil à rayons X. Divers appareils spéciaux pour étudier la mécanique animale.* ⦾ Tous les objets concernant l'Anat. des animaux domestiques (Mammifères et Oiseaux).

Ames, Ia. (U. S. A.).
Veterinary division State Coll. Assoc. Prof. W. A. Aitken (Anat.).

Beograd (S. H. S.).
Inst. f. Anat. u. Physiol. d. Haustiere, Dobrašina ulica 16. Dir. Prof. Dr. J. P. Markov. — Alex. Gjaja, Ass.

Berlin (Deutschland).
Anat. Inst. Tierärztl. Hochsch., NW, Luisenstr. 56. Dir. Prof. Schmaltz.

Bern (Schweiz).
Veterinär-anat. Inst. Univ., Engehaldenstr. 6. Dir. Prof. Dr. Theodor Rubeli.

Bologna (Italia).
Inst. Anat. veterinaria, Univ. Dir. F. Negrini.

Bonn-Poppelsdorf (Deutschland).
Inst. f. Anat. u. Physiol. d. Haussäugetiere, Landw. Hochsch. Dir. vacat.

Brno — Brünn (C. S. R.).
Inst. f. Anat. u. Physiol. der Haustiere d. Landwirtsch. Hochsch., Cerna pole. Dir. Dr. Theodor Dohnal.
Inst. histol.-embryol. Acad. vet. Confer: Histol.

Budapest (Ungarn).
Anat. Inst. der Tierärztl. Hochsch., VII., Rottenbiller u. 23. Dir. Prof. Dr. A. Zimmermann. — Ass. Dr. K. Karpfer; Ass. T. Szepesi; Praktikant K. Kovács; Dr. E. Szuts; Dr. Gg. Leidenfrost. *5 Plätze für vergl. Anat. und Embryol. der Haustiere.*

Buenos Aires (Argentina).
Inst. d'Anat. Fac. méd. vet. Univ. Dir. Prof. G. Cassai.

București (România).
Labor. d'Anat. Fac. méd. vet. Univ. Dir. Prof. C. Gavrilescu.

Děčín-Libverd—Tetschen-Liebwerd(C. S. R.).
Lehrkanzel f. Anat. u. Biol. der Haustiere d. Landwirtsch. Abt. d. Deutsch. Techn. Hochsch. Praha. Dir. vacat.

Edinburgh (Scotland).
Anat. Labor. of R. Vet. Coll. Dir. Prof. O. Charnock Bradley.

Gießen (Deutschland).
Veterinär-Anat. Inst. der Univ., Frankfurter Str. Nr. 94. Dir. Geh.Med.R. Prof. Dr. Paul Martin. — Dr. R. Süppel, Prosektor; Dr. J. Kapp, Ass.

Halle a. d. S. (Deutschland).
Inst. für Anat. u. Physiol. der Haustiere der Univ., Wilhelmstr. 27|28. Dir. Prof. Ulrich Gerhardt. — Dr. Wulfsberg, Ass. *1—2 Plätze für anat. (mikroskopische) Untersuchungen, Beobachtung lebender Arthropoden.* Ⓟ Spinnen.
Skelett-, Woll- u. Fellsammlung der Haustierrassen; Futtermittel. Dir. Prof. Fröhlich.

Hannover (Deutschland).
Anat. Inst. Tierärztl. Hochsch. Dir. Prof. P. Zietschmann.

Kamenetz-Podolsk, Ukraine (U. d. S. S. R.).
Labor. für Anat. u. Physiol. der Haussäugetiere des Inst. f. Landw. „K. Marx". Dir. Prof. M. Heraschenko.

Kaunas (Litauen).
Zootomisches Inst. der Veterinärabt., Gedimino Gatve 29. Dir. Doc. Mockus.

Kazan (U. d. S. S. R.).
Anat. Inst. u. Mus. des Veterinär. Inst. Dir. Prof. K. Viktoroff.

Lafayette, Ind. (U. S. A.).
Veterinary Mus. of the Purdue Univ.

Leipzig (Deutschland).
Vet.-Anat. Inst. d. Univ. Dir. Prof. Baum.

Leningrad (U. d. S. S. R.).
Labor. für Anat. und Histol. der Haussäugetiere des Landwirtschaftl. Inst., Kamennii Ostrow, 2. Beresowaja Alleja 2. Dir. Prof. A. Nemiloff. — I. Richter; N. Nemiloff; A. Newenglowskaja. Ⓟ Histol. Praep. der Milchdrüsen und Inkretionsorgane von verschiedenen Säugern.

León (España).
Labor. de Anat. general, Escuela de Veterinaria. Dir. Prof. A. G. Villareal.

Lima (Peru).
Labor. Anat., Fisiol. Animal y Zool., Escuela Nacional de Agric y Veterinaria, Santa Beatriz. Dir. Dr. E. G. Aguinaga.

Liverpool (England).
Dept. of Vet. Anat., Fac. of Med. of the Univ. Dir. Prof. J. S. Jones.

Lexington, Ky. (U. S. A.).
Dept. of Veterinary Sc. of the Univ. of Kentucky. Prof. Dir. W. W. Dimock.

London (England).
Veterinary Labor., South Eastern Agric. Coll. of the Univ., Wye, Kent.

Lwów (Polska).
Inst. u. Mus. f. vergl. Anat. der Haustiere an der Tierärztl. Akademie, ul. Kochanowskiego 67. Dir. Prof. Dr. W. Kulcrycki. — Ass.: Stypal Zdristaw, Wladyslaw Witkowski. *Injektionspraep.* (Teichmannsche Methode) und Korrosionspraep.

Lyon (France).
Labor. d'Anat., Ecole Nationale Vét. Dir. Prof. Lesbre.

Madrid (España).
Labor. d'Anat. Ecole vét. Dir. Prof. J. Gonzalez Garcia. — Prof. G. Tejero Muñoz.

Manhattan, Kan. (U. S. A.).
Labor. of animal anat., Kansas State Agric. Coll.

Manila (Philippine Islands).
Labor. of Vet. Anat., Coll. of Vet. Sc. Dir. Prof. M. Sumulong.

Milano (Italia).
Ist. Anat. del R. Ist. sup. di Med. vet., Città degli Studi. Dir. Prof. A. C. Bruni. Ass. Dt. V. Chiodi. *6 posti Anat., Istol., Embriol.*

Minsk (U. d. S. S. R.).
Anat. Labor. Inst. f. Landw. Dir. Prof. A. St. Sanozkij.

Montevideo (Uruguay).
Inst. de Anat. Patol. y Parasit. de la Escuela de Veterinaria, Larrañaga 572. ao. Prof. Dr. Enrique G. Vogelsang. *Mehrere Plätze für pathol. Anat. u. Parasit.*

Moskau (U. d. S. S. R.).
Vet. Inst., Tverskaja, Pimenovskij per 5. Dir. Prof. A. R. Evgrafov.

München (Deutschland).
Anat. Inst. d. Tierärztl. Fac. Univ. Dir. Prof. Otto Stoß.

Novočerkassk (U. d. S. S. R.).
Donskoj Vet. Inst.

Paris (France).
Labor. d'Anat. et physiol. Inst. Nation agric., Ve, 16, Rue Claude Bernard.

Perugia (Italia).
Gabin. di Anat. normale vet. Dir. Prof. G. B. Caradonna.

Praha — Prag (C.S.R.).
Inst. f. Anat. u. Physiol. der Haustiere u. Bact. d. čech. techn. Hochsch., XII, Kopernikova 7. Dir. Prof. Dr. Alois Velich.

Pretoria (South Africa).
Dept. of Veterin. Anat. Univ. Dir. Prof. R. W. Mettam.

Riga (Latvija).
Anat. Mus. Vet.-med. Fac. Univ., Kronvalda bulv. 9. Dir. Prof. Dr. L. Kundziņš.

Rio de Janeiro (Brasil).
Inst. Experim. de Vet., R. de Rosario 134.

Sainte Anne de la Pocatière,
Quebec (Canada).
Dept. of Anat. of the Agric. School. Dir. M. Louis de Gongagne Fortin, B.S.A.

Sapporo (Japan).
Vet. Inst. Coll. of Agric., Hokkaidô Imper. Univ.

Saratow (U.d.S.S.R.).
Abt. f. Anat. u. Histol. d. Tiere, Inst. f. Landwirtsch. u. Meliorat. Dir. Prof. I. I. Kadykov.

Sofia (Bulgarien).
Inst. f. Anat., Histol. u. Embryol. d. Haustiere, Univ. Dir. Doc. Heinrich Bittner.

Stambul (Türkei).
Labor. f. Anat. u. Embryol. Tierärztl. Hochsch. Skutarie-Sélimicé. Dir. Prof. Mehmed Hilmi.

Stockholm (Sverige).
Anat. und Histol. Inst. Tierärztl. Hochsch. Dir. E. Agduhr.

Tiflis (U.d.S.S.R.).
Inst. f. Anat. u. Physiol. d. Tiere am Polytechn. Inst. Dir. Dr. Boris Tschenzoff.

Tôkyô (Japan).
Veterinary Inst. Fac. of Agric., Tôkyô Imper. Univ., Komaba near Tôkyô. Prof. Keitarô Tsuno; Prof. Naoshi Nitta; Prof. Torai Shimamura; Ass. Prof.: Kujoshi Masui, Shirô Itagaki, Osamu Emoto, Takemaro Suzuki.
Veterinary Inst. Nôgyô Daigaku (agric. Univ.), Komasawa near Tôkyô.

Utrecht (Holland).
Vet.-anat. Inst. der Univ. Dir. Prof. Dr. G. Krediet. — Dr. H. A. Vermeulen, Lect.-Konserv.; W. H. Schultze, Konserv. (Histol.). 1 Ass. *Einige Plätze für histol. u. makrosk. Unters. u. experim. Morphol. der Genitalorgane.* ⊕ Hermaphroditische Tiere und Praep.

Warszawa (Polska).
Inst. f. Anat. d. Haustiere. Dir. Prof. W. Roszkowski.
Inst. f. Tieranat., Hochsch. f. Land- u. Forstwirtsch. Dir. Prof. L. Dobrzański.

Weihenstephan b. Freising (Deutschland).
Inst. für Anat., Physiol. u. Pathol. der Haustiere der Hochsch. f. Landw. Dir. Prof. Dr. Höflich.

Wien (Österreich).
Inst. f. system. u. topograph. Anat. d. Haustiere Vet. Hochsch. Dir. Prof. Dr. K. Skoda.

Woronesch (U.d.S.S.R.).
Kab. f. Anat. u. Physiol. d. Tiere mit Labor. Landwirtsch. Hochsch.

Zürich (Schweiz).
Anat. Inst. Univ. Fac. med. vet., 7, Plattenstr. 9. Dir. Prof. Dr. Ackerknecht.

Laboratoria chemo-physicalia biologica.
Laboratoria biochemiae generalis.

Confer: *Physiologia generalis* (pag. 444) *et plantarum* (pag. 412), *Hygiena* (pag. 487), *Bacteriologia* (pag. 479), *Zoologia* (pag. 356) *et Botania generalis* (pag. 390).

Adelaide (South Australia).
Darling Labor. of Biochemistry and general Physiol. Dir. T. B. Robertson. — H. R. Marston, Demonstrator; M. C. Dawbarn, Demonstrator; Mss. Fl. McCoy Hill, Cancer Research Ass.; W. Fuller, Lect. on Histol. *4 places for Research on Animal Growth and Nutrition. Applications of Physical Chemistry to Biol. Problems. — Potentiometer, Conductivity Apparatus, Colorimeters, Spectrophotometer, Metabolimeter, Calorimeter for foodstuffs and General apparatus* Commonwealth Labor. for Nutrition Research.

Alger (Algérie).
Labor. Chimie Biol. et Toxicol., Fac. Méd. Dir. Prof. Maillard.

Athènes (Grèce).
Labor. de Chimie biol. Univ.

Ayabe (Japan).
Sericicult. Biochem. Labor., Experim. St. of Gunse Silk Factory, Kyôto-fu. Dir. Suehiko Iwaoka.

Bahia (Brasil).
Labor. d. Med. Chemie Univ. Dir. E. Dinit.

Baltimore, Md. (U.S.A.).
Dept. of physiol. chemistry, John Hopkins Med. School. M. van Reusselaer Buell.

Bangalore, (Br.-India).
Dept. of Bio-Chemistry, Indian Inst. of Science, Hebbal P.O. Dir. Dr. M. O. Forster, Ph.D. — Prof. T. K. Calterson-Smith (Electrical Technol.); Prof. Roland V. Norris, D.Sc. (Biochemistry); Prof. T. L. Simonsen, D.Sc. (Organic Chemistry); Prof. H. E. Watson (General Chemistry). 3 Ass. in the Biochemical Dept. A Senior Lecturer. *About 10 research workers are admitted annually. A deposit of Rs 150 (about $ 10) is payable by each student. Apparatus and Chemicals supplied free. The labor. is specially equipped for work on Bact. Chemistry, the Chemistry of enzyme action, fermentation problems, lac and certain plant diseases such as spike disease of sandal wood.* — Sub-labor. for micro-analysis.

Basel (Schweiz).
Physiol.-chem. Anstalt Univ. Dir. Prof. R. Spiro.

Bergen (Norge).
Bergens Mus. Biokemiske labor. Dir. Dr. phil. Torbjørn Gaarder.

Berkeley, Cal. (U.S.A.).
Dept. of Biochemistry and Pharmacol., Univ. of California. Ass. Prof.: G. W. Clark (Prof. Biochem., Pharmacol.), E. S. Sundstroem (Biochemist).

Berlin (Deutschland).

Forschungsinst. für Stärkefabrikation und Kartoffeltrocknung Landw. Hochsch., N 65, Seestr. 13. Dir. Prof. Dr. Edmund Parow. — Dr. A. Stirum, Ober-Ass.; Dr. W. Ekhard, wissenschaftl. Ass. *6 Plätze für chemische Untersuchungen. — Interferometer, Polarisationsapparat, Heleoidlampe, Colorimeter nach Ostwald, Versuchsapparat zur Herstellung von Stärke, Stärkesirup, Stärkezucker, Dextrin, Trockenkartoffeln.*

Kaiser-Wilhelm-Inst.f.Faserstoffchemie, -Dahlem, Faradayweg 16. Dir. Prof. Dr. R. O. Herzog. — 8 wissenschaftl. Mitarbeiter. *Bis zu 6 Arbeitsplätzen für physikalische u. chemische Untersuchungen an biochemischem Material, insbesondere organischen Gerüststoffen, Zellulose, Skleroproteine usw. — Apparate für physikalische und chemische Untersuchungen.* ⊕ *Faserstoffe, Gerüststoffe.*

Kaiser-Wilhelm-Inst. für Biochemie, -Dahlem, Thielallee 69/73. Dir. Prof. Dr. phil. et med. h. c. Carl Neuberg. — Ass.: Frl. Dr. M. Kobel, Dr. E. Simon; Dr. J. Wagner. *4 Plätze für allgemeine Biochemie.*

Chemisch-hygien. Abt. d. Reichsgesundheitsamtes, NW 23, Klopstockstr. 18. Dir. Dr. W. Kerp. — *Chem., hygien., physiol.-pharmacol., pharmaz. Labor.*

Pharmazeutisches Inst. Univ., -Dahlem, Königin-Luise-Str. Dir. Prof. Dr. Hermann Thoms, Geh.-Reg.R. — Priv.Doc. Dr. Dieterle, Ober-Ass. des Inst.; Ass.: Dr. Kuhnhenn, Dr. Böhm, Dr. Thimann, Dr. Heynen; Hilfs-Ass.: Dr. Unger, B. Schneider, Dr. Bull, Loth, Dr. med. Wolfgang Thoms. *10 Arbeitsplätze für Gäste für chemische, physiol.-chemische, elektrochemische und pharmacol. Untersuchungen. — Alle Apparate, welche für rein chemische sowie physikalisch-chemische und elektrochemische Arbeiten in Betracht kommen.*

Bio-chem. Abt. d. Krankenhauses Moabit, W 35. Dir. Prof. Dr. Martin Jakoby. *Fermentforschung. Pharmakol. u. Toxikol. spez. der Antigene.*

Chem. Abt. d. Preuss. Inst. f. Infectionskrankheiten „Robert Koch", N 33, Föhrerstr. 2. Dir. Geh.Reg.R. Prof. Dr. Georg Lockemann. *Arsennachweis. Adsorptionsvorgänge. Desinfektionsprüfung. Wachstumsverhältnisse von Tuberkelbacillen auf eiweißfreien Nährlösungen.*

Labor. für Biochemie, Biol. Reichsanst. für Land- u. Forstwirtschaft, -Dahlem, Königin-Luise-Str. 19. Dir. Ober-Reg.R. Prof. Dr. J. Houben; Dr. E. Pfankuch, Ass. Labor. für prakt. Chemie, Dir. Reg.R. Dr. R. Scherpe.

Biol.-chem.Abt. d. städt.Krankenhauses am Urban. Dir. Dr. Ludwig Pincussen. *Biochem. des Stoffwechsels, Fermentmethoden, Biol. der Lichtwirkung.*

Chem. Abt. Rudolf-Virchow-Krankenhaus. Dir. Prof. J.Wohlgemuth. *Fermente spez. Carbohydrasen u. Proteasen.*

Bern (Schweiz).

Med.-chemisches Inst. Univ. Dir. Prof. Dr. Bürgi.

Birmingham (England).

Dept. of Brewing and of the Biol. and Chemistry of Fermentation. Dir. Prof. A. R. Ling.

Bordeaux (France).

Labor. de Chemie Biol., Fac. Méd. Univ. Dir. Prof. Denigès.

Bratislava — Preßburg (C.S.R.).

Med.-chemisches Inst. Univ., Staatl. Krankenhaus, Mickiewiczova 13. Dir. Prof. Dr. Johann Buchtala.

Breslau (Deutschland).

Inst. für Biochemie und landwirtschaftliche Technol., Univ., 16, Hansastr. 25. Dir. Prof. Dr. F. Ehrlich. — Dr. K. Rehorst; Dr. F. Schubert; I. Bender. *30 Plätze für chemische, biochemische und technische Untersuchungen. — Lebende Mikroorganismen (Hefen, Schimmelpilze, Bakt.), etwa 200 verschiedene Rassen.*

Brno — Brünn (C.S.R.).

Med. chemisches Inst. Univ., Údolní 73. Dir. Prof. Dr. Antonín Hamsík. — MUDr. A. F. Richter; MUDr. E. Lecian.

Biochem. Inst. Tierärztl. Hochsch., Pražská 69. Dir. Ph. et MUDr. Jan Bečka. — MVDr. Emil Přibyl. 3 Demonstr. *2 Plätze für Tropfenmethode in Eiweißkörperkoagulation, Interferometrische und refraktometrische Maßanalyse, toxikol. Studien, Mikrochemie, Einrichtung für die Herstellung der wirksamen Substanzen aus Organen mit innerer Sekretion.*

Labor. de Biochemie, l'Inst. des Recherches agric. et Forestière, Květná ul. Dir. Doc. Ing. Otto Kyas. — Ing. Valíček Jan; Ing. Jan Racek; Ing. Eliška Pallatová; Ing. Rudolf Sánka; Dr. Ing. Bedřich Frodl; Dr. Ing. Hugo Volejníček; Ing. Jos. Haas; Dr. Ing. Josef Hampel; Ing. Emil Marek.

București (România).

Labor. de chemie biol. Fac. med. vét. Univ. Dir. R. Vladescu.

Budapest (Ungarn).

Kgl. Ung. Pflanzenbiochemisches Inst., II., Debröi-út 15. Dir. Dr. Johann Bodnár, o. ö. Univ.-Prof. — Ass.: Dr. Alexander Terényi, Dipl.-Ing. Chem.; Emil Gubányi, Dipl.-Ing. Chem.; wissenschaftl. Hilfsarbeiterinnen: Dr. phil. Lili Eveline Roth, Jolantha Solty, Wilhelmina Gervay. *Pflanzenchemie und -biochemie und analytische Chemie besonders für Pflanzenschutzmittel. — 1—2 Arbeitsplätze, unentgeltlich.* ⊕ *Kranke Pflanzen, besonders brandige Getreidearten.*

Chemisches Inst. der Kgl. Ungarischen Tierärztl. Hochsch., VII, Rottenbiller utca 23. Dir. Prof. Dr. Julius Gróh. — Ass.: Julius Szelestey, Chemiker-Ing. *6 Plätze für physikalische Chemie, Kolloidchemie, Spektrochemie, Reaktionskinetik, Arbeiten mit radioaktiven Elementen, usw. — Spektroskope, Spektrographe (auch im Ultraviolett), Spektralphotometer, elektrische Meßinstrumente, Messungen der Radioaktivität, Interferometrie.*

Physiol.-chemisches Inst. der königl. ungar. Pázmány Péter Univ., VIII, Eszterházygasse 9. Dir. Prof. Dr. Paul Hári. — Dr. Zoltán Aszódi, Adjunkt; Béla Róhny, Praktikant. *13 Plätze für tierische Kalorimetrie, Stoffwechsel, Spektrophotometrie, Spektrometrie, Versuche am überlebenden Herzen. — Respirationskalorimeter für kleine Tiere; Stoffwechselversuchseinrichtung für kleine und größere Tiere; Königsches Spektrophotometer; Zeissches Gitterspektrometer; Refraktometer; Kolorimeter; Apparate zur Gasanalyse; Van Slykescher Apparat zur Bestimmung des Aminostickstoffes.* ⊕ *Physiol.-chemische Praep.*

Buitenzorg, Java (Ned.-O.-Indië).

Chemical labor. the General Agric. Experim. Stat. Chief: Dr. C. van Rossem. — F. W. Weber, Ass. *Analyses of chemical constitution of crop plants at different stages of development. Experiment with chemical fertilizers for native crops. Chemical analysis of food plants, green manures, fertilizers, irrigation water, soil samples.*

Cambridge (England).

Dept. of Biochemistry Univ. Dir. Sir F. G. Hopkins. — T. S. Hele, Lect.; F. J. W. Roughton, Lect.

Camerino (Italia).

Labor. Chimica-Farmaceutica Univ. Dir. P. Saccardi. — V. Gazzi; Avanzolini.

Cape Town (South Africa).

Dept. of Biochemistry Univ. Dir. Prof. E. S. Edie.

Charkow (U.d.S.S.R.).

Ukrainisches Biochemisches Inst., Veterinärplatz. Dir. Prof. Dr. Al. Palladin. — Ass.: Dr. E. Ssawron; Dr. D. Zuwerkalow; Mitglieder: Dr H. Gowdissky,

Dr. S. Winokurow, Dr. L. Palladin, Dr. A. Kudrjawzewa, Dr. P. Normark. *15 Arbeitsplätze für Untersuchungen über Biochemie d. intermediären Stoffwechsels, Biochemie der Avitaminosen, Biochemie des Zentralnervensystems.*

Chiba (Japan).
Biochemical Inst. Chiba Med. Univ. Dir. Prof. Dr. S. Akamatou.

Chicago (U.S.A.).
Dept. of Physiol. Chemistry Univ. of Chicago. Dir. F. C. Koch (Chairman).

Cluj — Klausenburg (România).
Inst. de Chimie biol., Strada Pasteur. Dir. Prof. Dr. Pierre Thomas (de l'Inst. Pasteur de Paris). — M. Sibi, Chef de trav. prat.; Dr. R. Imas, Ass.; Dr. E. Maftei, Ass.; G. Benetato, Préparateur; Purge, Préparateur. *6 à 8 places préparées pour recherches de chimie courante, 2 places peuvent être disposées pour mesures physiques (potentiomètre) 2 pour microanalyses cliniques.— Potentiomètre de Mislowitzer. Colorimètre, néphélomètre, spectrophotomètre de Baudoies-Bénard. Appareil de van Slyke. Appareils divers pour microdosages(azote,acétone, sucre,etc.).Prochainement, un labor. de microanalyse selon Pregl sera terminé.*

Debrecen (Ungarn).
Med.-chemisches Inst. der Tisza István Univ., Simonyi ut 16. Dir. Prof. Dr. J. Bodnár. — Ass.: Dr. J. Straub, N. Dreguss, A. Karell, P. Dömsödy; Praktikanten: L. Nagy, B. Kubinyi, E. Faltin, E. Szép. — *Enzymol., med. u. analytische Chemie, Mikrochemie, gerichtliche Chemie. — 1—2 unentgeltliche Arbeitsplätze.*

Dresden, Weißer Hirsch (Deutschland).
Labor. für physiol. Chemie und Ernährungsforschung an Dr. Lahmanns Sanatorium, Lahmannring 15. Dir. Priv.Doc. Dr. Ernst Komm. — Dr.-Ing. W. Dietsch, wissenschaftl. Mitarbeiter (Chemiker); Dipl.-Ing. Käthe Sonntag, Hilfs-Ass. *3 Plätze für physiol.-chemische Untersuchungen. — Tierlabor. für Vitamin- und Stoffwechselforschung, Gaswechselapparaturen, Apparaturen für klinisch-chem. und physiol.-chem. Untersuchungen (Spektrophotometer, Polarimeter, verschiedene Kolorimeter, Mikrowage, Kymographion, Apparate für physikal.-chem. Untersuchungen, bakt. Apparaturen usw.).*

Edmonton (Canada).
Dept. of Biochemistry Univ. of Alberta. J. B. Collip, M.A., Ph.D.; J. W. Scott, M.D., C.M.

Erlangen (Deutschland).
Inst. für angewandte Chemie Univ. Dir. Prof. Dr. M. Busch. — Prof. Dr. Günter Scheibe; Prof. Dr. H. Apitzsch. *130 Arbeitsplätze für Chemiker, Pharmazeuten und Nahrungsmittelchemiker.*

Frankfurt a. M. (Deutschland).
Univ.-Inst. für Nahrungsmittelchemie, Paul-Ehrlich-Str. 40. Dir. Prof. Dr. Tillmans. — Dr. Riffart, Dr. Strohecker, Dr. Hirsch. *20 Plätze für Untersuchung von Lebensmitteln, Wasser und Abwasser.*

Freiburg i. Br. (Deutschland).
Physiol.-chem. Inst. der Univ., Sautierstr. 2. Dir. Prof. Fr. Knoop. — Dr. M. Oesterlin; Dr. H. Oesterlin; Dr. Rebel. *12 Plätze. Intermediärer Stoffwechsel, tierische Oxydationen, Reduktionen, Synthesen.*

Fukuoka (Japan).
Biochem. Inst. Coll. of Med., Kyûshû Imper. Univ.

Geneva, N.Y. (U.S.A.).
Dept. of Biochemistry State Agric. Experim. Sta. R. J. Anderson, Ph.D., Chief in Research (Biochem.); R. L. Shriner, Ph.D., Ass. in Research (Biochem.).

Div. of Chemistry of the New York Agric. Experim. Stat.
Lucius L. Van Slyke, Ph.D., Chief in Research (Chemistry); Dwight C. Carpenter, Ph.D.; Arthur W. Clark, B.S., Assoc. in Research (Chemistry); Morgan P. Sweeney, A.M.; William F. Walsh, B.S.; Millard G. Moore, B.S.; Leon R. Streeter, M.S.; R. Bruce Dayton, B.S.,Ass. in Research(Chemistry). *Chemistry of proteins.*

Göttingen (Deutschland).
Inst. f. med. Chemie u. Hygiene. Dir. Prof. Hans Reichenbach.

Halifax, N.S. (Canada).
Dept. of Biochemistry Dalhousie Univ. Dir. E. G. Young.

Hamburg (Deutschland).
Physiol.-chem. Inst. Univ., Eppendorfer Krankenhaus. Dir. O. H. Schumm.

Harpenden, Herts (England).
Dept. of Chem. of the Inst. of Plant Nutrition and Soil Problems. Dir. H. J. Page. — G. C. Sawyer; T. Eden; E. J. Maskell; R. G. Warren.

Heidelberg (Deutschland).
Inst. für Eiweißforschung, Bergheimer Str. 58. Dir. Geh.R. Prof. Dr. Albrecht Kossel.

Helsinki (Finnland).
Med.-chem. Inst. d. Univ.

Hutchinson, Kan. (U.S.A.).
Dupray Labor. of exper. Med. Confer: Therapia gener.

Iowa City, Ia. (U.S.A.).
Dept. of Biochemistry State Univ. Dir. V. C. Myers.

Irkutsk (U.d.S.S.R.).
Biochem. Sekt. des Biol. Geogr. Forschungsinst. Dir. Prof. A. Frank-Kameneckij.
Inst. f. Physiol. Chem. u. Kolloidchem. Dir. Doc. V. Ivanov.

Ithaca, N.Y. (U.S.A.).
Baker Labor. of Chemistry Cornell. Univ. Prof. E. M. Chamot (Chem. microscopy).

Jena (Deutschland).
Anstalt für pharmazeutische und Lebensmittelchemie der Univ. Dir. Prof. Oskar Keller. — Prof. Dr. phil. H. P. Kaufmann, Abt.-Vorst. (Sondergebiete der pharmaz. Chemie). *Chemie der Pflanzenalkaloide (Alkaloide der Uvagoga Ipecacuanna, Alkaloide der Delphiniumarten, Alkaloide der Helleborusarten).*

Jerusalem (Palästina).
Inst. f. Biochem. u. Kolloidchem. Hebräische Univ. Dir. Prof. Andor Fodor.

Johannesburg (South Africa).
Bio-chem. Dept. S. Afric. Inst. for med. Research. Dir. Dr. W. Fox.

Kagoshima (Japan).
Biochemical Labor. Kagoshima Imperial Coll. of Agric. and Forestry. Dir. Prof. Dr. Kiyohisa Yoshimura.

Kanazawa (Japan).
Biochemical Inst. Kanasawa Med. Univ. Dir. Prof. Dr. Kenzô Sudô.

Kazan (U.d.S.S.R.).
Labor. d. biol. Chemie, Vet.-institut, Arskoje Pole 96. Dir. Doc. S. Afonskiy. — P. Nušdin, Ass. *Physikalische Chemie, Fermente.*

Keijô, Korea (Japan).
Biochemical Inst. Coll. of Med., Keijô Imper. Univ.

Kingston, Ont. (Canada).
Dept. Biol. Chemistry Queen's Univ. Dir. J. F. Logan, Ph.D. — S. A. Beatty, B.A.

København (Danmark).
Bioteknisk-kemisk Labor., Den polytekniskeLaereanstalt, Sølvtorvet. Dir. Prof. Dr. Orla-Jensen. — Anna D. Orla-Jensen; Bernhard Spur, Carl le Dous. *1 Platz September—Februar.* ⊕ Kulturen von Milchsäurebakterien.

Köln a. Rh. (Deutschland).
Chemische Abt. des Physiol.Inst. der Univ., Lindental, Lindenburg. Abt.-Vorst. Prof. Dr. Bruno Kisch. *2 Plätze für physiol.-chemische, physikalisch-chemische Untersuchungen.*

Kraków (Polska).
Travaux pratiques à l'Inst. Chimie méd. Dir. J. Z. Robel.

Kumamoto (Japan).
Biochemical Inst. Kumamoto (Imperial) Med. Univ. Dir. Prof. Shichozô Katô.

Kyôto (Japan).
Biochemical Labor. Kyôto Imperial Sericicultural Coll., Hanazono near Kyôto. Prof. S. Hata; Prof. S. Nakane.
Biochemical Labor., Coll. of Sc., Kyôto Imper. Univ. Dir. Prof. Dr. Shigeru Komatsu. — Prof. Dr. S. Komatsu; Prof. Ass. Bunkichi Masumoto; Lect. Risaburô Nakai; Lect. Chôji Tanaka; Ass. Masao Kurata; Vice-Ass. Chûichic Okinaka; Vice-Ass. Takehiko Yukitomo; Vice-Ass. Shinsaku Ozawa.
Biochemical Inst. Coll. of Med., Kyôto Imperial Univ. Dir. Prof. Dr. Kanae Maeda. — Prof. Dr. Kanae Maeda; Ass. Prof. Senji Uchino; Lect. Sakisaburô Wada.

Lausanne (Suisse).
Labor. de Chimie physiol. et pharmaceutique de l'Univ., Place du Château. Dir. Prof. Strzyzowski.

Lawrence, Kan. (U.S.A.).
Dept. of Bio-Chemistry of the Univ. of Kansas. Dir. Prof. C. F. Nelson.

Leiden (Holland).
Labor. f. Med. Chemie. Dir. Prof. Dr. H. G. Bungenberg de Jong.

Leipzig (Deutschland).
Physiol.-chemisches Inst. der Univ. Dir. Prof. Dr. med. Karl Thomas. — Priv.Doc. Dr. Kopfhammer; Priv.Doc. Dr. Flaschenträger; Dr. med. Suger; Dr. phil. Bettzieche. *Etwa 35 Plätze. Mikroanalytische Apparatur.*
Physiol.-chem. Abt. des Vet.-physiol. Inst. der Univ., Tiroler Str. 4. Abt.-Vorst.: Prof. Dr. Martin Schenck.

Leningrad (U.d.S.S.R.).
Labor. für physiol. Chemie, Biol. Inst. d. Univ., Wassily Ostrow Linie 16 Haus 29. Dir. Prof. W. S. Ssadikow.—A. T. Risskaltschuk; M. K. Ščigelskaya; R. A. Gutner. *Neues Labor., 2—3 Plätze für Gäste.* ⊕ Kollagene, Kollagenase.
Labor. für Biochemie der Pflanzen des Inst. der angewandten Botan., Rue Herzen, 44. Dir. Prof. Nicolai N. Iwanoff. — A. M. Isaikin; O. J. Grünberg; M. I. Lischkewitsch; M. N. Lawrowa; B. S. Alexandrowa; W. F. Grigoriewa; K. N. Woroschilowa.
Labor. f. Biochemie u. Physiol. d. Pflanzen d. Akademie d. Wissensch., Tutschkowa Naberejnaia, 2a. Dir. Prof. S. Kostytschew. — Paul Eliasberg, Marie Korsakowa, Alexander Scheloumow, Georg Medwedew, Vera Berg. *10 Plätze f. Gäste f. Untersuchungen auf d. Gebiete d. Pflanzenchemie, Biochemie des Bodens und Fermentforschung. — Apparate f.*
elektrometrische Messungen, Mikroanalyse, Ultrafilter, elektrische Verbrennungsöfen.
Abt. für biol. Chemie des Inst. für experim. Med., Lopuchinskaja 12. Dir. Prof. Sergey S. Salaskin. — Ass.: P. Aschmarin, P. Astanini; M. L. Petrunkin, Anna Petrunkina, H. L. Glinka-Tschernorutzki, W. M. Wesselkina. *16 Plätze. — Alle Apparate und Hilfsmittel für chemische, biochemische und physicochemische Untersuchungen.*
Biogeoschemisches Labor. des Radium Inst. der Akademie der Wissenschaften, Röntgenstr. 1. Dir. des Radium-Inst. Akademiker W. J. Vernadsky, Leiter des biogeochemischen Labor.: Prof. W. S. Ssadikow. — M. A. Naryschkina, L. E. Kaufmann, G. G. Bergmann, A. P. Winogradow, M. K. Ičigelskaya, R. A. Gutner. ⊕ Homogenisierte Tiere und einzelne Fraktionen (Seetiere). (Nach Verfahren von Prof. W. S. Ssadikow.) — Enge Beziehungen mit biol. Stationen am Murman, Sebastopol, Astrachan und mit dem Inst. für Meeresuntersuchung.
The section of the general Microbiol. of the State Inst. of Experim. Med., Aptekarsky Island, Lopouchinskaja Street 12. Dir. W. L. Omeliansky. — Ass.: I. A. Macrinoff, W. P. Neioloff; Sc. collaborators: M. Strohbinder, I. Oulrich. *10 workingplaces for biochemistry of microorganisms.* ⊕The cultures of microbes and their preparations.
Chemisch-Bakt. Gouvern. Zentral-Labor., Pl.Kommunarow 1. Dir. P. I. Lewin.

Lille (France).
Labor. de Chimie biol., Fac. de Méd. cathol. de l'Univ., 56, Rue du Port et 11, Rue de Toul. Dir. Prof. C. Carrez.

Lincoln, Neb. (U.S.A.).
Dept. of Biochemistry Coll. of Med. Univ. of Nebraska. Confer: Omaha.

Lisbôa (Portugal).
Inst. de Chimica farmac. e chimica biol., Fac. de Farmacia, Univ. Dir. Prof. R. Lupi Nogueira.

Liverpool (England).
Johnston Labor., Univ. of Liverpool. Dir. Prof. W. Ramsden (Biochem.). — Lect. in Clinical Chemistry. *1 place for Biochemical research.*

London (England).
Dept. of Biochemistry of the Univ. Coll., Gower Street, W.C. Dir. Prof. J. C. Drummond.
Dept. of Biochem., Middlesex Hosp. Med. School, Univ., Mortimer Str., Oxford Str., W. Dir. Prof. E. C. Dodds.
Dept. of Chem. Pathol., Guy's Hosp. Med. School of the Univ., St. Thomas St., Borough, S. E. A. Dir. Prof. J. H. Ryffel.
Dept. of Chem. Pathol., St. Bartholomew's Hospital Med. Coll. of the Univ., West Smithfield, E.C. 1. Dir. Prof. R. L. M. Wallis.
Dept. of Physiol. Chem., St. George's Hosp. Med. School. of the Univ., Hyde Park Corner, S.W. Dir. Prof. J. A. Gardner.
Biochemical Dept., Imperial Coll. of Sc. and Technol. of the Univ., Imp. Inst. Road, S. Kensington, S.W. 7. Dir. Prof. S. B. Schryrer. — H. W. Buska, Demonstrator; A. B. P. Page; B. W. Town, Research Ass. *About 6 places for biochemistry generally, but especially for the biochemistry of plants. — The apparatus of the labor. consists plant machinery for working relatively large amounts of material, and also apparatus for finer work in organic and physical chemistry. Available for all workers in the labor.*
Dept. of Biochem. and Pharmacol. Nation. Inst. for Med. Research. Hampstead, W.C. 2, Dir. H. H. Dale. — H. W. Dudley; H. King,; J. H. Burn.
Dept. of Biochemistry, Lister Inst. of preventive Med. of the Univ., Chelsea Bridge Rd. S.W. 1. Dir. Prof. A. Harden.

London, Ont. (Canada).
Dept. of Biochemistry of the Univ. of Western Ontario. Dir. A. B. Macallum, M.B., M.D. — A. A. James, M.D.

Louisville, Ky. (U.S.A.).
Dept. of Physiol. Chemistry of the Univ. of Louisville. Dir. Prof. A. W. Homberger. — Prof. H. O. Calvery, Ass.; Prof. R. W. Jackson, Ass.

Lyon (France).
Labor. de Chimie Biol. de l'Univ. Dir. Prof. L. Hugounenq. — Dr. Gabriel Florence, Agrégé; Dr. Emile Couture, Préparateur-Ass. *8 places pour chimie biol. générale et chimie med.* ⓟ Produits chimiques d'origine animale.
Labor. de Chimie Biol., Ecole Nation. Vétérin. Dir. Prof. C. Porcher.
Chimie organique et Toxicol. Fac. Méd. et de Pharmacie de l'Univ., 1 rue Raulin. Dir. Prof. Dr. A. Morel. — Dr. G. Florence, Prof. agrégé, Chef des travaux, Chargé du cours de Toxicol.; Dr. M. Chambon; A. Siméon. *10 places en vue de recherches de Chimie appliquée à la Pharmacie et à la Biol.: Pharmacol., Toxicol. experim., Etude de la constitution et des propriétés des protides. Fermentations. La constitution des protides, la préparation et les propriétés des medicaments antiparasit., arsénicaux et organo-métalliques. — Etudes électriques. Pompe à vide rotative. Appareillage pour la microanalyse. etc.*

Lwów (Polska).
Med. Chem. Inst. der Univ., Piekarska 52. Dir. Prof. I. K. Parnas. — Dr. med. J. Dadlez; Dr. Phil. W. Jankowska; Dr. med. W. Mozołowski; Dr. med. J. Heller; St. Chrząszczewski. *Etwa 6 Arbeitsplätze für analytische Untersuchungen über Zusammenhang der Chemie des Blutes mit dem Gewebe.*
Inst. de Chimie de l'Acad. de la Méd. Vétérin., Kochanowskiego 61. Dir. Dr. W. Moraczewski. — M. Łaszczewski; Ed. Hamerski. *3 places.*

Mannheim (Deutschland).
Labor. der städtischen Krankenanstalten. Dir. Dr. Ernst Joseph Lesser. *Kohlehydratstoffwechsel. Anoxybiose. Fermente.*

Manila (Philippine Islands).
Labor. of Biochemistry, Univ. Santo Tomas. Dir. Prof. C. Potenciano.

Melbourne (Australia).
Dept. of Biochem. Univ. Dir. Prof. W. J. Young.

Mexico, D. F. (Mexico).
Sekt. chem. d. Dir. de Estudias Biol., 7|a calle de Balderas 94. Dir. Prof. M. García Tunco, M. Beristain.

Milano (Italia).
Labor. Chimica biol. Univ. Dir. Prof. Foà.

Minsk (U.d.S.S.R.).
Inst. für biol. Chemie der Univ. Dir. Prof. A. P. Bestužev.

Montpellier (France).
Labor. de Chimie Biol. et Méd. Fac. Méd. Univ. Dir. Prof. Eugène Derrien. — Dr. P. Cristol, Chef des Travaux, chargé de cours; H. Vabre, Ingénieurchimiste, préparateur; Dr. Charles Benoît, Chef du Labor. de Chimie de l'Hopital Général; Lang, Chef du Labor. de Chimie de l'Hopital Suburbain; Dr. Hertzel Béraha; Jacques Trivas. *4 places dans les labor. des Hopitaux pour les recherches de Chimie appliquée à la clinique. 4 places dans les labor. de Chimie biol. de l'Inst. de Biol. pour les recherches de Biochemie générale et comparée.*

Montreal, Que. (Canada).
Dept. of Biochemistry, McGill Univ. Dir. A. B. Macallum, Ph.D., D.Sc. — S. W. Bliss, M.A., Ph.D.

Dept. Biol. and organic chemistry McGill Univ. Dir. R. F. Ruttan, M.D., D.Sc.

Moskau (U.d.S.S.R.).
Labor. de Chimie biol. fac. méd. II. Univ., Devitchje Pole, Malaja Pirogovskaja 1. Dir. Prof. Dr. J. A. Smorodintzev. — Chef de labor.: Dr. A. N. Adova; Adjoint: Dr. E. A. Jljina; Dr. M. I. Ravitch-Stcherbo, Dr. E. A. Swechnikova, S. E. Menchoutin. *1 place pour récherches férmentol. — Appareil pour la détermination du pH.* ⓟ Collection speciale de préparations concernant la Chimie biol. (matiéres extractives des tissus, ferments).
Labor. für Fettchemie und Fettechnol. d. Chemisch-Technol. Inst. Mendeleew, Platz Iliyča, 5/2. Dir. Prof. Sergius Ivanow. — Alexandra Ivanowa-Schicharewa (Botan.); Fräulein Zinaida Alissowa; Wladimir Schadin (Technol.). *2 Arbeitsplätze für Gäste: physiol.-chemische Untersuchungen der Ölbildungsprozesse bei Samen, Fettsynthese, Untersuchungen der neuen Fettarten mit der Anwendung der Evolutionstheorie zur Biochemie. Firnis- und Seifenbereitung und Theorie der Prozesse.* ⓟ *Samen der Pflanzen in ihrer Abhängigkeit von den Klimaten der Erde.*
Labor. der Organischen und Biol. Chemie der Pädagogischen Fak. der II. Staatsuniv., 21, M. Pirogowskaia 57. Dir. Prof. Dr. Sergey Iakowlewitsh Demianowsky. — Al. Nik. Smolin; Mich. Wlad. Kortschagin; M. Fed. Liosin. *3 Plätze. Gärung und intermediärer Kohlenstoffwechsel im tierischen Organismus.*
Biochemisches Inst. des Kommissariats für Volksgesundheit, 64, Woronzowo Pole, 8. Dir. Prof. A. N. Bach; Vice-Dir. B. Sbarsky. — Ass.: W. A. Engelhardt, S. W. Ermoljewa, K. W. Nikolajeff. Wissenschaftl. Mitarbeiter: D. M. Michlin, S. R. Zubkowa, S. D. Balachowsky, S. S. Wassiljew W. A. Wilensky *4—6 Plätze.*
Chem.-Pharmaz. Wiss. Inst., Nikolskaja 15. Dir. P. Kaminskij.

Mukden, Manchuria (China).
Biochemical Inst. South Manchurian Med. Univ.

Nagasaki (Japan).
Biochemical Inst. Nagasaki Imperial Med. Univ.

Nagoya (Japan).
Biochemical Inst. Aichi Med. Univ. Dir. Prof. Dr. L. Michaelis.

Nancy (France).
Bio-chem. Inst., Univ. Fac. pharm. Dir. Prof. R. Douris.

Napoli (Italia).
Ist. di Chimica Biol., Sant'Andrea delle Dame 2. Dir. Prof. G. Quagliariello. — Tafuri Gian Bernardino, Ass. *Chimica e chimica-fisica biol.*
Ist. di Chimica Fisiol., Fac. med. Univ. Dir. Prof. F. Bottazzi.

New Haven, Conn. (U.S.A.).
Dept. of Biochemistry, State Agric. Exper. Experim. Sta. T. B. Osborne, Ph.D., Sc.D., Chief Research Chem. in charge; Helen Cannon, B.S., Ass. Chem.; C. S. Leavenworth, Ph.B., Ass. Chem.; H. B. Vickery, Ph.D., Ass. Chem.; A. J. Wakeman, Ph.D., Ass. Chem.

New York City (U.S.A.).
Dept. of Biol. Chemistry Columbia Univ. Dir. Prof. W. J. Gies. — Assoc. Prof. E. G. Miller jr.; Ass. Prof. M. Karshan.

New York (U.S.A.).
Biochemical Dept. H. A. Metz Labor. Dir. H. E. Dubin. — H. B. Corbilt; L. Feedman.

Niigata (Japan).
Biochemical Inst. Niigata Med. Univ.

Norman, Okla. (U.S.A.).
Dept. of Biochemistry and Pharmacol. Univ. Dir. M. Everett.

Odessa (U.d.S.S.R.).
Inst. f. Physiol. Chemie, Med. Inst. Dir. Prof. F. M. Porodko.

Okayama (Japan).
Physiol.-chem. Inst. der Univ. Dir. Prof. Dr. Tayei Shimidzu. — Shigeru Toda, Ass. Prof.; Sadatomo Yonemura, Priv.Doc.; Ass.: Keizo Misaki, Richo Karasawa. Kozoo Kaziro. *18 Arbeitsplätze. Physiol. und chemische Unters.* ⊕ *Die Gallensäuren und ihre Derivate vieler Tiere.*

Omaha, Nebr. (U.S.A.).
Dept. of Biochemistry, Univ. of Nebraska. Dir. S. Morgulis.

Ôsaka (Japan).
Biochemical Inst. Ôsaka Provincial Med. Univ. Dir. Prof. Dr. Yashirô Kotake. — Prof. Dr. Y. Kotake; Ass. Prof. Zenji Matsuoka.

Oxford (England).
Rockefeller Dept. of Biochemistry. Dir. Rudolph Albert Peters. *15 places for Subjects, related to Biochemistry.*

Paris (France).
Labor. de Chimie Biol. Fac. Sc. et de l'Inst. Pasteur, XVe, 28 Rue Dutot. Dir. Prof. Gabriel Bertrand. — M. Jarillier, Prof. adjoint; M. Rosenblatt, Ass.; L. de Sauët Rat, Ass.; M. Macheboeuf, Ass.; R. Sazerac, Chef de labor. *20 places pour Chimie Biol. pure et appliquée: Les Infinements petits chimiques (éléments oligosynergiques, diastases, vitamines, hormones, venins, etc.) Sucres et tissus végétaux, glucosides, etc. Microbes et fermentations. Chimie agric., etc. — Machines diverses (broyeurs, pilons mécaniques, presses hydrauliques, appareils à distiller et à dessécher dans le vide, etc.). Chauche Thermostat, labor. souterrain, appareil frigorifique, spectrographe, labor. la microanalyse, etc.* ⊕ *Drogues, principes immédiats naturels et autres matériaux propres aux recherches de Chimie Biol.*

Labor. de la Chimie biol. Fac. Pharm. 4 Avenue de l'Observatoire. Dir. Prof. L. Grimbert. — Paul Fleury, Chef de labor. *5 places pour l'études chimiques des produits physiologiques et pathol.*

Labor. de la Biol. comparée, Ecole Pratique des Hautes Etudes, 12 rue Cuvier. Dir. Prof. Georges Bohr. — Mme. Bohr-Drzewina, Dir. Adjoint (Ecole des Hautes Etudes); Mrs. Loisel, Anglas, de Ribaucourt, Dauphin, Matisse, Bertin, Teissier, Fischer (Paul). *Application de la physique et de la chimie à la biol.*

Services de Chimie de l'Inst. Pasteur. Dir. G. Bertrand. — A. Fernbach.

Pécs (Ungarn).
Chemisches Inst. der Univ. Dir. Prof. Dr. László Zechmeister. — Ass.: Dr. László v. Cholnoky, Dr. Johann Csabay, Dr. Vera Vrabély, Dr. Paul Rom. *4 Plätze für Pflanzenchemie. Spektroskopie von Farbstoffen. Synthetisch-organisch-chemische Arbeiten.* — *Gitterspektroskop Zeiß.*

Peking (China).
Labor. of Biochemistry P.N.M.C. Dir. Hsien Wu. — Kuo-hao Lin, Instr.; Martin Yee, Ass.; Chao-Chi Chen, Ass. — *Jood labor. Nutrition labor., rat room.*

Perm (U.d.S.S.R.).
Biochem. Labor. Univ.

Philadelphia, Pa. (U.S.A.).
Biochemistry Labor. General Hospital. J. G. Reinhold (Ass.).

Praha — Prag (C.S.R.).
Inst. f. med. Chemie d. Karls-Univ., II, Katerinska N. 32. Dir. Prof. Dr. Emanuel Formanek.

Med.-chemisches Labor. der deutschen Univ., Salmg. 3. Dir. Prof. Dr. R. Zeynek. — Ass.: Doc. Dr. Felix Haurowitz, Dr. Zdenko Stary; Dr. Victor Sellner; 1 Demonstrator. *5 Plätze für physiol.-chem. und klinisch-chemische Arbeiten.*

Lehrkanzel f. Biochemie d. Deutsch. Techn. Hochschule. I, Karlova 30.

Inst. Biochimique des Inst. pour la Production des Plantes, XII, Havličkovy sady 58. Dir. Dr. Ing. A. Němec. *Biochimie des Plantes, Enzymol. des graines. Biochimie du sol agric. et forestier, détermination de l'exigences pour engrais des sols agric. par voie microchim. Acidité du sol agric. et forestier. Nutrition des arbres forestiers.*

Pretoria (South Africa).
Biochem. Labor. Dir. Prof. H. H. Green.

Quito (Ecuador).
Bio-chem. Inst. Dir. Prof. J. M. F. Corral.

Rio de Janeiro (Brasil).
Labor. f. Bio-chem. d. Hospital Nation de alienados.

Rochester, N.Y. (U.S.A.).
Dept. of Vital Economics Univ. Prof. H. A. Mattill (Biochemistry).

Roma (Italia).
Ist. Chimica Fisiol. Univ. Dir. Prof. Domenico Lo Monaco.

Rosario, Santa Fé (Argentina).
Inst. de Quimica biol. Dir. Prof. Ventura Morera.

Rostow a. Don (U.d.S.S.R.).
Inst. f. med. Chem. u. Mikrobiol. Dir. Prof. St. M. Maksimowitsch.

Saigon (Indo-Chine).
Section de Chimie biol. et fraudes alimentaires. Inst. Pasteur.

San Francisco (U.S.A.).
Dept. of Biochemistry, Med. School, Univ. of California. Dir. Prof. C. L. A. Schmidt.

São Paulo (Brasil).
Dept. of Biol. Chemistry of the Fac. of med. chir. *Apparatus for metabolic researches, for examination of water, milk, etc.*

Saratow (U.d.S.S.R.).
Labor. für biol. Chemie der Univ. Dir. Prof. V. V. Vorms.

Chem. Abt. d. Staatsinst. f. Microbiol. u. Epidemiol. Dir. Dr. N. J. Grjasnov.

Seattle, Wash. (U.S.A.).
Dept. of Chemistry, Univ. of Washington. Ass. Prof. L. C. Boynton (Biochemist).

Sendai (Japan).
Biochemical Inst. Fac. f. Med., Tôhoku Imper. Univ. Dir. Prof. K. Inoue. — Prof. Dr. Katsuji Inoue; Lect. Takashi Hayashi.

Singapore (Straits Settlements).
Dept. of Biochemistry, Coll. of Med. Dir.: Prof. J. L. Rosedale. — G. McOwan, Reader in Chem. E. Madgwich, Reader in Physics. *4 working places for guests.*

Smolensk (U.d.S.S.R.).
Labor. f. physiol. Chemie. Dir. Prof. Demjanovskij.

Sofia (Bulgarien).
Bio-chemisches Inst. d. Univ. Dir. Prof. Dr. Assen A. O. Zlataroff. — Ass.: M. Andreitschewa, D. Kaltschewa. *Enzymochemie, Bromatol.*

Stambul (Türkei).
Biochemisches Labor., Univ. Jéré botan. Dir. Prof. Dr. Djevad Mashar. — Ass. Mehmed Ali. *Gastaufnahme durch das Prof.-Kollegium. Physiol. und pathol. Chemie. Secretion.*

State College, P. (U. S. A.).
Dept. of Chemistry (Agric. and Biol.) School of Agric. R. A. Dutcher, M.S., Head of Dept.; D. E. Haley, Ph.D., Prof. Soil and Phytochem.; A. K. Anderson, Ph.D., Assoc. Prof. Physiol. Chem.; E. S. Erb, M.S., Assoc. Prof. in charge Analyses, Met.; Hannah E. Honeywell, Ph.D., Ass. Prof. Biol. Chem.; M. W. Lisse, M.S., Ass. Prof. Biophys. Chem.; R. J. Miller, Ph.D., Assoc. Prof. Biol. Chem., Research in Poultry Nutr.; H. B. Pierce, M.S., Ass. Prof. Food and Dairy Chem.; E. S. Reider, Phar. G., Instr. Agr. Chem.; G. A. Shuey, B.S., Instr. Agric. and Biol. Chem.; Walter Thomas, B. S., Assoc. Research Prof. Phytochem.

Stellenbosch (South Africa).
Dept. of Chemistry Univ. B. de St. J. van der Riet, M.A., Ph.D. Prof. of organic chemistry, and at present gives the lect. in Biochemistry.

Stockholm (Sverige).
Physiol.-chem. Inst. Veterin. Hochsch.

Stormont, Belfast (Ireland).
Devision of chem. and animal nutrition. Dir. Prof. Dr. G. Scott Robertson. — R. G. Baskett, B.Sc.; J. Dickinson, B.Sc.; Houston, B.Sc.

Strasbourg — Straßburg (France).
Labor. de Chim. Biol. Fac. Méd. Dir. Prof. Nicloux.

Szeged (Ungarn).
Med.-chemisches Inst., Kalvária tér 3. Dir. Prof. B. v. Reinbold. — Dr. Tibor Maros; Ass.: Dr. Ida Sófalvy, Joh. Völgyessy.

Taihoku, Formosa (Japan).
Biochemical Inst. Taihoku Govern. Coll. of Med.
Biochemical Labor. Dept. of Agric., Government Central Research Inst. Otosaburô Okumura (Phytochemistry).

Tartu — Dorpat (Eesti).
Inst. für pharmac. Chemie, Rüütli t. 2. Dir. Prof. H. Parts. — Dr. E. Labi, Ass.
Chemisches u. biochemisches Labor. Univ., Innere Universitätshospitalklinik. Dir. Doc. Dr. med. W. Wadi. — A. Arrak, P. Teas, M. Illison und W. Pert. *2 Arbeitsplätze für klinische Chemie (Mikrochemie).*

Tegucigalpa (Honduras).
Bio-Chem. Inst. Univ. Dir. Prof. Dr. A. Erazo.

Tharandt b. Dresden (Deutschland).
Pflanzenchemisches Inst. der Forstlichen Hochsch. Dir. Prof. Dr. Hans Wislicenius. — Priv.Doc. Dr. Walther Gierisch, Ass.; Priv.Doc. Dr. Rudolf Lorenz, Ass. *Chemische, teilweise technisch-chemische Untersuchungen. — Spezialapparate für kolloidchemische Arbeitsverfahren (f. Pflanze und Boden), für Untersuchungen von technischen Abgasen und für experim. Rauchschäden, Rauchluft.* ⓟ *Rauchbeschädigte Pflanzenteile.* Pflanzenchemische Praep. (Drogen, Extraktstoffe, Zellulosefaser und deren Spinnerei, Weberei, Verfilzung (Papiere), Ester und Lösungen, Lignin, Harze, ferner Trockendestillate, Holzimprägnierungspraep., Gerbstoffe usw.

Tôkyô (Japan).
Sasaki's Research Labor. (Biochemistry), Uenohara 812, Higashi-Nakano, by Tôkyô. Dir. Dr. Takaoki Sasaki, Prof. emer. — Dir. Dr. T. Sasaki; Ass. S. Ueda.
Biochemical Inst. Military Med. Coll. (Gun-i Gakkô).
Biochemical Inst. Jikeikwai Med. Univ.
Biochemical Labor. Imperial Fishery Inst., Hukagawa. Dir. Prof. Dr. M. Yamakawa. — Prof. Dr. Makoto Yamakawa.
Biochemical Labor. Imperial Agric. Experim. Str., Nishigahara.
Biochemical Labor. Kitasato Inst. for Infectious Diseases, Shirokane-sankô-chô. Dir. Dr. Yutaka Teruuchi.
Biochemical Inst. Coll. of Med., Keiô-Gijuku Univ. Dir. Prof. Dr. Y. Teruuchi. — Prof. Dr. Y. Teruuchi; Ass. Prof. S. Sueyoshi; Ass. Prof. Dr. H. Kumakawa.
Biochemical Inst. Nippon Girls' Med. Univ.
Biochemical Inst. Nippon Med. Univ.
Biochemical Inst. Coll. of Med., Tôkyô Imper. Univ. Dir. Prof. Dr. S.Kakiuchi. — Prof. Saburo Kakiuchi. Ass. Prof. Teisuke Komoto.
Biochemical Labor. Tôkyô Imper. Sericicult. Coll., Nishigahara. Dir. Prof. Y. Mimuroto. — Prof. Yoshimitsu Mimuroto.
Biochemical Labor. Imperial Sericicultural Experim. Station, Suginami-chô near Tôkyô. Dir. Dr. Eikichi Hiratsuka.
Majima's Labor. Confer: Biophysica.

Tomsk (U. d. S. S. R.).
Inst. f. Bio-Chem. Dir. Prof. I. L. Vakienko.

Torino (Italia).
Inst. f. pharm. Chem. et Toxicol. Dir. Prof. L. Mascarelli.

Toronto, Ont. (Canada).
Dept. of Biochemistry Univ. of Toronto. Dir. Andrew Hunter, M.A., B.Sc., M.B. — H. Wasseneys, Ph.D.; H. B. Speakman, D.Sc.; Miss J. MacFarlane, M.A.; J. M. Luck, Ph.D.; J. A. Dauphinee, M.A.; C. C. Benson, B.A., Ph.D.; A. H. Gee, Ph.D.; A. M. Wynne, M.A., Ph.D.
Dept. of Pathol. Chemistry Univ. of Toronto. Dir. V. J. Harding, D.Sc.; G. Hunter, M.A., B.Sc.

Toulouse (France).
Labor. de Chimie biol. Fac. méd. Univ. Dir. Prof. J. Aloy.

Tsinan (China).
Dept. of Bio-Chem., Shantung Christian Coll. Dir. Prof. P. Ch'ing Kiang.

Tübingen (Deutschland).
Physiol.-chemisches Inst. Dir. Prof. H. Thierfelder. — Dr. E. Klark; Dr. E. Walz; Dr. O. Merz. *2—3 Arbeitsplätze für Gehirnchemie.* ⓟ *Physiol.-chemische Praep.*

Tunis (Tunis).
Labor. et Service de Chem. méd. et biol. Inst. Pasteur.

Ueda (Japan).
Biochemical Labor. Ueda Imperial Sericicult. Coll., Nagano-ken. Dir. Prof. Dr. R. Inoue.

Urbino (Italia).
Kabinett f. Pharmacol. Chem. Dir. Prof. Angelo Agrertini.

Utrecht (Holland).
Labor. für Physiol. Chemie, Maliebaan 50. Dir. Prof. Dr. W. E. Ringer. — Conserv. P. G. F. H. M. A. Vermast. *Enzyme.*
Labor. für Med.-Veterin. Chemie, Biltstraat 172. Dir. Prof. Dr.B. Sjollema. — Dr. L. Seekles, Conserv.

Vohwinkel, Rheinl. (Deutschland).
Pharmazeut. Wissenschaftl. Labor. J. G. Farbenindustrie A. G. Werk Elberfeld. Dir. W. Schulemann.

Vologda (U. d. S. S. R.).
Biochem. Versuchsstat. des Milchwirtschaftl. Inst. Dir. Prof. S. S. Inichoff.

Warszawa (Polska).
Physiol. Chem. Inst. Univ. Dir. Prof. St. Badzyński.
Physiol. chem. Inst. Univ. Fac. Veterin. Dir. Prof. S. Przytecki.

Washington, D.Col. (U.S.A.).
Protein Investigation Labor. of the Bureau of Chemistry, M.S. Dept. of Agric. D. B. Jones (Chemist).

Weihenstephan b. Freising (Deutschland).
Chemisches Inst. der Bayr. Hochsch. für Landwirtschaft und Brauerei. Dir. Prof. Dr. B. Bleyer. — 1 staatlich angestellter Ass.; 2—4 wissenschaftl. Mitarbeiter. *2—3 Gastplätze für biochemische Arbeiten aller Art.* ℗ Biochemische Praep.

Wien (Österreich).
Inst. f. physiol. Chemie Univ. Schwarzspanierstr. Dir. Prof. O. v. Fürth.
Inst. für angew. med. Chem. d. med. Fak. Univ. Dir. Prof. Dr. Emil Fromm.
Labor. f. pharmaz. Chemie. Dir. Prof. Faltis.
Inst. f. med. Chem. Veterin. Hochsch. Dir. Prof. Dr. H. Jansch.

Wilno (Polska).
Dept. of Physiol. Chemistry Stefan Batory Univ. Dir. Prof. J. M. Retinger.

Winnipeg, Man. (Canada).
Dept. of Biochemistry Univ. of Manitoba. Dir. A. T. Cameron, D.Sc.; — F. D. White, Ph.D.; H. D. Kitchen, M.D., C.M.
Dept. of Biochemistry Med. Coll. Prof. A. F. Cameron.

Woronesch (U. d. S. S. R.).
Inst. f. Biol. Chemie d. Univ. Dir. Prof. P. H. Nikiforovskij.

Würzburg (Deutschland).
Physiol. chemisches Inst. Univ. Dir. Prof. Dankwart Ackermann.

Yonkers, N.Y. (U.S.A.).
Protein Labor. Arlington Chemical Co. Dir. R. P. Wodehouse.

Zagreb — Agram (S.H.S.).
Inst. f. angew. med. Chem. Dir. Prof. Bubanowič.

Zürich (Schweiz).
Biochem. Inst. Univ. Dir. Prof. E. B. H. Waser.

Laboratoria biophysica.

Confer: *Biochemia* (pag. 435) et *Physiologia generalis* (pag. 444).

Bern (Schweiz).
Inst. für physikalisch-chem. Biol., Chemiegebäude der Universität, Freie Str. 3. Dir. Dr. Gertrud Wokers. *8 Plätze: 6 für chemische, 2 für mikroskopische Untersuchungen. — Chemische und mikroskopische, mikrochemische und kapillaranalytische Apparate.*

Brno — Brünn (C.S.R.).
Biophysisches Labor. Univ., Veveři 95. Dir. Prof. Dr. Vladimír Novák. — MUC. Alois Lednický, Ass.

Frankfurt a. M. (Deutschland).
Inst. für Kolloidforschung, Theodor-Stern-Kai. Dir. Prof. H. Bechhold. — 1 Ass., 1 wissenschaftl. Hilfsarbeiterin. *Kolloidforschung, chemische, biol.-med. Abt.*
Univ.-Inst. für physikalische Grundlagen der Med. (Oswalt-Stiftung). Vors.: Prof. Dr. Friedr. Dessauer. — Mitgl. d. Inst.: Dr. R. E. Liesegang. — Ass. Drs. Brenzinger, A. Janitzky, E. Lorenz, B. Rajewsky, P. Happel, R. E. Liesegang. *10 Plätze. Elektrotechn. Abt., Physik. Abt., Med.-biol. Abt. Apparate zur Erzeugung und Messung der Röntgenstrahlen u. sämtlicher Wellenlängen. Hochvakuumanlage. Ultramikroskope. Lehr- u. Forschungsinst. für die med. Anwendungsgebiete d. Physik.*

Hamburg (Deutschland).
Kolloidbiol. Station am Eppendorfer Krankenhaus, 20, Martinistr. Dir. Dr. Friedrich-Vincenz v. Hahn. — Dr. H. Junker (Zool. Ass.); Dr. Magda Wieben (Botan. Ass.); die med. Assistentenstelle ist z. Z. nicht besetzt. *2 Arbeitsplätze. Dispersoidanalyse. Zellstimulation. Vitaminforschung, allgem. Kolloidbiol.: spez. Lehre der Oberflächenaktivität.* ℗ Kolloidbiol. Praep.

Iwanovo-Voznessensk (U.d.S.S.R.).
Lab. f. Kolloidchem. Dir. Prof. Peskov.

Lauham, Md. (U.S.A.).
Biophysical Labor., Bureau of Plant Industry. G. H. Collins.

La Plata (Argentina).
Inst. f. biol. Physik, Escola de sc. med. Dir. Prof. V. Tedeschi.

Leningrad (U. d. S. S. R.).
Labor. f. allg. Biol. d. Staatl. Inst. f. med. Wiss. Confer: Institutiones generales.

Lille (France).
Labor. de Physique biol., Fac. cathol. de Méd. de l'Univ., 56, Rue du Post et 11, Rue de Toul. Dir. Prof. M. d'Halluin.

Ljubljana — Laibach (S.H.S.).
Chemisches Inst. der Univ. Dir. Dr. Max Samec. — Ass.: Richard Klemen, Janko Kavčič. *6 Plätze für kolloidchemische und physikochemische Untersuchung der Pflanzenkolloide.* ℗ *Verschiedene Pflanzenkolloide, insbesondere Stärkearten und ihre Derivate.*

Lyon (France).
Labor. Auguste Lumière, 49. 51 rue Villan. Dir. Auguste Lumière. — Chevrotier, Dr. ès-sciences; Conturier, Licencié ès-science; Mad. Montoloy, Licencié ès-sciences; F. Perrin, Chimiste; Lesbros; Grange. *10 places. Phénomènes biol. colloidanes. — Appareils habituellement utilisés pour les recherches.*
Labor. de Physique Biol., Fac. Méd. de l'Univ. Dir. Prof. Cluzet.

Montpellier (France).
Labor. de physique biol. Fac. Méd. Univ., rue Montels. Dir. Prof. Dr. J. L. Tech. — Prof. agrégé: Dr. Lamarque; Préparateur: Imbert. *Etude des fluorescences des êtres vivants et tissus provenant de ces êtres. Etude de l'action des variations du champ-électrique ambiant sur les végétaux. Etude des potentiels de contact (indice de nutrition) entre les tissus de l'homme et des animaux ou végétaux vivants et les milieux extérieurs liquides, solides ou gazeux.*

Moskau (U.d.S.S.R.).
Inst. für Physik und Biophysik d. Volkskommiss. f. Volksgesundheit, (VI) Miusskaja 3. Dir. Prof. Dr. Peter Lasareff. — N. Stschodro; S. Wawilow; P. Pawlow; W. Shuleikin; A. Predvoditelev; A. Trapeznikov. *5 Plätze für Photochemie, Molekularphysik, Biophysik.*

Paris (France).
Labor. de Physique méd. Fac. méd., 12, rue de l'Ecole de Médecine. Dir. Prof. André Strohl. — Turchini,Chef des travaux; Préparateur des travaux pratiques: Haas, Desgrez; Ph. Fabre, Préparateur du labor. de recherches; Couvreux, Préparateur du Cours. *5 places pour Electrophysiol., Radiobiol., Optique physiol. — Appareils à rayons X. Electrocardiographe. Appareils de mesures électriques, d'optique, d'enregistrement graphiques, etc.*
Labor. Physique biol. Ecole pr. des Hautes Etudes. Dir. Prof. A. d'Arsonval. — A. Mayer; B. Roussy.
Labor. de Chimie Physique biol. de la Sorbonne, 87, Boulevard Saint Michel. Dir. Dr. Pierre Girard. *Facteurs électrique de l'Osmose. La perméabilité selective des parois vivantes et des parois inertes polarisées aux différents ions. Modifications colloidales du plasma sanguin. Choc anaphylactique.*

Praha — Prag (C. S. R.).
Inst. f. allg. Biol. med. Fak. Confer: Institutiones gen.

Rosario (Argentina).
Labor. de Fisica biol. Escuela de Med. Univ. Dir. Prof. M. Neuschloß.

Strasbourg — Straßburg (France).
Inst. de Physique biol. Fac. méd. et Fac. Sc. Dir. F. Vlès. — Préparateurs: A. Dognon, M. Gex; Moniteur: G. Achard. *10 places, principalement pour physicochimie biol.; optique (radiations). Physico-chimie des matières colorantes (spectres). Physico-chimie du sol. Physico-chimie cellulaire. — Spectrophotomètres ultra-violets. Réfractomètres ultra-violets. Spectromètres infra-Rouges. Rayons X. Mesures électrométriques de p_H. Polarimétrie. Viscosimétrie. Colloïdes. Cataphorèse.*

Tomsk (U. d. S. S. R.).
Kolloidchem. Labor. Technikum. Dir. Prof. A. M. Černuchin.

Vologda (U. d. S. S. R.).
Stat. f. Kolloidchem. des Milchwirtschaftl. Inst. Dir. Prof. S. S. Prevov.

Laboratoria nutrimentorum specialia.

Confer: *Biochemia generalis* (pag. 435), *cultura plantarum* (pag. 499).

Berlin (Deutschland).
Inst. f. Nahrungsmittelkunde der Tierärztl. Hochschule. Prof. J. Bongert.

Braunschweig (Deutschland).
Labor. f. Nahrungsmittelchemie, Pharmazeutisches Inst. Dir. Prof. Paul Horrmann.

Jena (Deutschland).
Staatl. Lebensmitteluntersuchungsamt. Dir. Prof. Dr. phil. Oskar Keller. — Dr. Omar Schmidt, Abt.-Vorst.; Dr. Lothar Streicher; Dr. Schmitz. 2 Unterrichts-Ass.

København (Danmark).
Labor. f. nutrition research. Dir. M. Hindhede.

Landsberg a. d. W. (Deutschland).
Chem., Nahrungsmittelchem. Abt. d. Staatl. Hygiene-Inst., Zechower Str. 48. Dir. Prof. Dr. R. Hilgermann.

Lausanne (Suisse).
Labor. d'analyse chimique et bact. des denrées et boissons de l'Univ., Solitude 19. Dir. C. Arragon.

Lexington, Ky. (U. S. A.).
Dept. of Feed Control of the Agric. Experim. Sta. W. A. Anderson, Jr., Microscopist.

München (Deutschland).
Untersuchungsanstalt f. Nahrungs- u. Genußmittel, Karlstr. 29. I. Dir.: Dr. Theodor Paul; II. Dir.: Prof. Dr. Wilhelm Arnold. — Abt.-Leiter: Prof. Dr. Theodor Merl, ORChemiker; Prof. Dr. Konrad Amberger, ORChemiker.
Labor. f. angew. Chemie (Lebensmittelchemie u. landwirtschaftlich-techn. Gewerbe). Vorst.: Prof. Dr. Lüers.

Praha — Prag (C. S. R.).
Lebensmittel-Untersuchungs-Anstalt, II. Preslova, Dir. Prof. Dr. Anton Nestler, Prof. d. allg. Botan.

Riga (Latvija).
Inst. f. Chem. d. Nahrungs- u. Genußmittel. Dir. Prof. E. Sarinsch.

Stockholm (Sverige).
Inst. f. Lebensmittelhygiene, Veterin. Hochsch. Dir. Prof. Hülphers.

Tomsk (U. d. S. S. R.).
Inst. f. Chem. Technol. d. Nahrungsstoffe. Dir. Prof. D. S. Saratovkin.

Warszawa (Polska).
Inst. f. Untersuchung d. menschl. Nahrungs- u. Genußmittel. Dir. Prof. S. Przybytek.

Washington (U. S. A.).
Nutrition Labor. Confer: Digestio et nutritio.

Laboratoria fermentorum.

Confer: *Physiol. plantarum* (pag. 412) *et Labor. digestionis* (pag. 454), *Microbiologia agris* (pag. 533), *Biochemia generalis* (pag. 435).

Berlin (Deutschland).
Inst. für Gärungsgewerbe und Stärkefabrikation Landw. Hochsch., N 65, Seestr. 13|15. Dir. Prof. Dr. F. Hayduck. — Prof. Dr. P. Lindner; Prof. Dr. F. Schönfeld; Dr. W. Rommel; Dr. F. Stockhausen, M. Glaubitz; Dr. H. Haehn. Abt. für Brauerei, Brennerei, Hefe u. Essig in Versuchsfabriken des gärungsgewerblichen Gebietes. *Gärungsbiol. Untersuchungen.* ⊕ Kulturen von Hefen und Schimmelpilzen.

Brno — Brünn (C. S. R.).
Inst. des Fermentations. Dir. Prof. Dr. Francois Ducháček. — Ing. L. V. Žíla, Chef de service; Ing. F. Měštan, Ass.; Ing. F. Javová, Ass. 3 *places. Fermentations.*

Dairen, South Manchuria (China).
Fermentol. Labor. Central Research Inst. of South Manchurian Railway Co. Dir. Dr. Kendô Saitô.

Geisenheim a. Rh. (Deutschland).
Weinchemische Versuchsstat. Dir. Prof. Dr. von der Heide.

Graz (Österreich).
Med.-Chemisches Inst. Med. Fak. Univ., Ul. K. Marksa 26. Dir. Prof. F. Pregl.

Harpenden, Herts (England).
Dept. of Fermentation of the Inst. of Plant Nutrition and Soil Problems. Dir. E. H. Richards. — R. L. Amoore.

Keijô, Korea (Japan).
Fermentol. Labor. Keijô Governm. Technol. Coll.

København (Danmark).
Carlsberg Labor., Carlsbergvej, Valby. Dir. Prof. Vald. Henriques. — Prof. J. Hjelmslev; Prof. C. Hansen Ostenfeld. Pflanzenphysiol. Abt.: Dir. Dr. Johs. Schmidt; Ass. C. Olsen; Mag. sc. A. V. Tånnig. Chem. Abt.: Dir. Prof. Dr. S. P. Sörensen mit 5 Ass. *Gärungsgewerbe, Pflanzenphysiol.*

Krasnodar, Kuban (U. d. S. S. R.).
Gärungsphysiol. Inst. d. Kuban. Landwirtsch. Inst.

München (Deutschland).
Wissenschaftliche Station für Brauerei in München, Ohlmüllerstr. 42a. Dir. Prof. Dr. H. Lüers. — 6 wissenschaftl. Mitarbeiter. *Ausbau der wiss. Grundlage d. Brauwesens durch systemat. durchgeführte Forschungen. Chem., physiol. u. chem.-techn. Labor.*

Neustadt a. d. H. (Deutschland).
Chem. Abt. d. staatl. Lehr- u. Versuchsanst. für Wein- u. Obstbau. Dir. Prof. Dr. Schätzlein.

Odessa (U. d. S. S. R.).
Wissenschaftl. Zentral-Versuchsstation f. Weinbereitung, Ul. Swerdlowa 19.

Ôsaka (Japan).
Fermentol. Inst. Ôsaka Imperial Technol. Coll. Dir. Prof. Yasukichi Nishiwaki.

Riga (Latvija).
Gärungslabor. Landwirtsch. Fak. Univ., Kronvalda bulv. 1. Dir. Doc. P. Delle.

Taihoku, Formosa (Japan).
Governmental Central Research Inst. Dir. Ryôji Nakazawa (Fermentol. and Microbiol.).

Tunis (Tunis).
Labor. de vinification, Inst. Pasteur.

Warszawa (Polska).
Inst. f. Gärungsgewerbe u. landwirtsch. Bakt., Ul. Krakowskie Przedmieście 66. Leiter: B. Moroz. Abt. f. Reinkulturen, f. Kontrolle d. Branntweinbrennereien, analyt. Labor.

Weihenstephan b. Freising (Deutschland).
Gärungsphysiol. Inst. Vorst. der Abt. für angewandte Gärungsbiol.: Prof. Dr. Schnegg; Vorst. für theoretische Gärungsbiol.: Prof. Dr. Trautwein.

Laboratoria physiologica.
Laboratoria physiologiae generalis (fac. medicinae).
Confer: *Anatomia generalis* (pag. 423), *Hygiena hominis* (pag. 487).

Aberdeen (Scotland).
Physiol. Dept. Univ. Dir. Prof. J. A. MacWilliam.

Adelaide (South Australia).
Darling Labor. of Human Physiol. and Pharmacol., Univ. of Adelaide. Dir. C. S. Hicks.

Alger (Algérie).
Labor. de Physiol. Fac. Méd. Univ. Dir. Prof. A. Tournade. — Prof. agrégé H. Hermann; Dr. M. Chalrot, Chef des Travaux; Dr. J. Malmejac, Préparateur de recherches. *3 à 4 places. Etude de toutes les fonctions, mais particulièrement celles des glandes endocrines et de la circulation. La technique des anastomoses vasculaires entre deux animaux pour l'étude particulière des actions humorales.*

Amsterdam (Holland).
Physiol. labor., Rapenburgerstraat 136. Dir. Prof. Dr. G. v. Rijnberk. *Centralnervensystem.*

Asunción (Paraguay).
Physiol. Inst. Univ. P. B. Guggiari; J. P. Vera; J. O. de Finis.

Athènes (Grèce).
Inst. de Physiol., Univ., Rue Sina 1. Dir. Prof. S. Dontas — A. Kotsaftes. *3–5 Plätze f. Nerv- u. Muskelphysiol.*

Bahia (Brazil).
Labor. de Physiol., Fac. méd. Dir. Prof. A. Novis.

Barcelona (España).
Inst. de Physiol. Fac. Méd., Calle de Casanova. Dir. Prof. Auguste Pi-Suñer. — Prof. Jesus M. Bellido, Sous-Dir.; Dr. José PucheAlvarez; Dr. Rosendo Carrasco Formiguera; Dr. Jaime Pi-Suñer Bayo; Jaime Raventós; Ismal Bofarull, Juan Bofill. *3 places. Sensibilité interne. Physicochimique des humain metabolimetrie et électrocardiographie.*

Bari (Italia).
Gab. di fisiol. Univ. Dir.: M. Camis.

Basel (Schweiz).
Physiol. Anstalt Univ., Vesalianum, Vesalgasse 1. Dir. Prof. Dr. Ph. Broemser. — Dr. M. Duràn, 1. Ass.; Dr. H. Fasold, 2. Ass. *Auf persönliche Anfrage werden Gäste zur Bearbeitung von Fragen, die in das Arbeitsgebiet des Inst. fallen, aufgenommen.*

Belfast (Ireland).
Dept. of Physiol. Queens Univ. Fac. med. Dir. Prof. Th. H. Milroy. — Prof. John Milroy (Biochem.).

Beograd (S. H. S.).
Physiol. Inst. d. med. Fak., 92, Zrinskoga ulica. Dir. Prof. Dr. R. Burian. — Dr. Milutin Nešković, Doc.; Dr. Ilija Djuricić, Ass. *20—25 Gastplätze. — Gesamte Apparatur für physiol. Arbeiten mit graphischer und chemischer Methodik (inkl. der physiol.-mikrochemischen Methoden), für elektrophysiol. Arbeiten (auch kl. u. großes Edelmannsches Saitengalvanometer), für Arbeiten mit photographischer u. mikro- wie makrokinematographischer Methodik (in Gemeinschaft mit d. histol. Inst.) usw.*

Berlin (Deutschland).
Physiol. Univ., N 4, Hessische Str. 3–4. Dir. Geh.Med.R. Prof. Dr. Hofmann †. — Abt.-Vorst.: Prof.Dr. R. du Bois-Reymond, Prof. Steudel.
Kaiser-Wilhelm-Inst. für Arbeitsphysiol., N 4, Invalidenstr. 103a. Dir. Prof. Dr. E. Atzler. — Dr. Gunther Lehmann; Dr. Robert Herbst; Dr. Erich Müller; Dr. Wilhelm Bickert; Dr. Burkhard Kommerell. *12 Arbeitsplätze für Physiol. und Arbeitsphysiol., insbesondere für Respirationsversuche, Blutgasuntersuchungen, Haemodynamik, physikal. Chemie, Ermüdungsforschung, physiol. Eignungsprüfungen, physiol. Arbeitsrationalisierung. — Außer der gesamten Einrichtung des Inst. stehen zur Verfügung Respirationsapparate nach Pettenkofer, Benedict, Douglas-Haldane, Zuntz-Geppert, Krogh. Blutgasanalyseapparate nach Barcroft, van Slyke. Superpositionschronozyklograph. Zweiglabor.: Physiol.*

Labor. der Deutschen Hochsch. für Leibesübungen. Deutsches Stadion, Berlin-Charlottenburg 9. **Kaiser-Wilhelm-Inst. f. Biol.** Confer: Institutiones generales.

Bern (Schweiz).
Physiol. Labor. (Hallerianum) Univ. Dir. Prof. Leon Asher.

Besançon (France).
Labor. de Physiol. Univ. Dir. Prof. A. Bolot.

Birmingham (England).
Dept. of Physiol. Univ., Edmund Street. Dir. Prof. E. Wace Carlier, M.Sc., M.D. — Dr. Hilda Walker, Senior Lect.; Dr. H. I. C. Pfister, Junior Lect.

Bologna (Italia).
Ist. Fisiol., Univ. Dir. Prof. L. M. Patrizi. — G. Borgatti, Ass.
Ist. Istol. e Fisiol. generale, Univ. Dir. A. Ruffini.

Bombay (Br.-India).
Physiol. Labor., Grant Medical College. Dir. Dr. C. S. Thakar. — Dr. T. G. Paymaster, Ass. Prof.; Dr. Telang, Demonstrator; Dr. Kamat, Demonstrator. 3 Tutors, 7 Demonstrators. *2 places for Biochemical and Physiol. research. — Electro-Cardiograph.Colorimeters. Optical Apparatus. Autoclaves. Galvanometers.* Ⓟ Histol. Sections. Frogs (Indian).

Bonn a. Rh. (Deutschland).
Physiol. Inst. Univ., Nußallee 11. Dir. Prof. Dr. U. Ebbecke. — Ass.: Prof. Dr. phil. et med. P. Junkersdorf, Prof. Dr. W. Thörner, Priv.Doc. Dr. R. Matthaei; Abt.-Vorst.: B. Schöndorff (Biochemie). *2 Plätze für Elektrophysiol., physiol. Optik, Zuckerstoffwechsel. — Elektrische Meßapparate, Registrierapparate, Mikromethodik.*

Bordeaux (France).
Labor. de Physiol., Fac. Méd. Univ. Dir. Prof. Pachon.

Boston, Mass. (U. S. A.).
Labor. of Physiol. in the Harvard Medical, School and the Harvard School of Public Health, 240 Longwood Ave. The Labor. being governed by an informal committee composed of the senior members of the associated sublabor. — Walter B. Cannon, M.D., S.D.; George Higginson, Prof. of Physiol.; Alexander Forbes, M.D., Assoc. Prof. of Physiol.; Percy G. Stiles, Ph.D., Ass. Prof. of Physiol.; Alfred C. Redfield, Ph.D., Ass. Prof. of Physiol.; William T. Porter, M.D., LL.D., D.Sc., Prof. of Compar. Physiol.; Lawrence J. Henderson, M.D., Prof. of Biol. Chem.; Edwin J. Cohn, Ph.D., Ass. Prof. of Physiol. Chem.; Ronald M. Ferry, M.D., Instr. in Physiol. Chem.; Cecil K. Drinker, M.D., Prof. of Physiol.; Lawrence T. Fairhall, S.M., Ph.D., Instr. in Physiol. *The Labor. are equipped for investigation in general, mammalian and human physiol. and for physical chem. in its application to biol. problems.* The following departments of the Univ. are associated in the building accupied by the Labor. of Physiol.: The Department of Physiol. in the Harvard Med. School, the Department of Compar. Physiol. in the Harvard Med. School, the Department of Physical Chem. in the Harvard Med. School and the Department of Physiol. in the Harvard School of Public Health.

Boston City, Mass. (U. S. A.).
Thorndike Memorial Labor., Boston City Hospital. J. T. Wearn. *Physiol. of Coronary Circulation. Concentration of Chlorides in the Glomerular Mine.*

Bratislava — Preßburg (C. S. R.).
Physiol. Labor. der Comenius-Univ., Dobytci, 13 Dir. Prof. Dr. A. Blairále. — Melka, Ass.

Breslau (Deutschland).
Physiol. Inst. Univ., 16, Maxstr. 10. Dir. (Neu zu besetzen, bisher Prof. Karl Hürthle).— Prof. Dr.

Schmitz, Vorst. der physiol.-chem. Abt.; Prof. Dr. Fuchs; Priv.Doc. Dr. Wachholder. *Apparate zur Untersuchung des Blutkreislaufs, 2 komplette Saitengalvanometer, Apparate für Mikrobestimmungen.*

Brno — Brünn (C. S. R.).
Physiol. Inst. Univ., Udolní 73. Interim-Dir.: Prof. Dr. Josef Petřík. — Dr. Anna Goldfederová, Ass. *2 Arbeitsplätze. Mechanik u. Regulierung der Atmung, Respirometrie (inkl. Mikro-), biol. X-Strahlenforschung. — Möglichkeit von Neukonstruktionen in eigener mechanischer Werkstätte. Respirometer nach Zuntz-Geppert, Krogh, Mikrorespirometer verschied. Konstruktionen, Gasanalyt. Apparat nach Bohr-Tobiesen, Mikroapparate nach Krogh. Röntgeneinrichtung, Quarzlampe.* Labor. f. Neurobiol.

Bruxelles — Brussel (Belgique).
Inst. de physiol. Dir. Prof. Dr. J. Demoor et Aug. Slosse.

București (România).
Labor. de Physiol. Fac. méd. Univ. Dir. Prof. N. Paulescu.
Inst. de Physiol. Prof. Jon Athanasiu.

Budapest (Ungarn).
Physiol. und histol. Inst. der königl. ungarischen tierärztl. Hochsch., VII, Rottenbiller Str. 23. Dir. in Vertretung: Dr. med. Dezsö von Deseö. — Johann Uhrin, Ass.; Andreas Hofhauser, Ass. *2 Plätze*.
Physiol. Inst. der Péter Pázmány Univ., Eszterhazy u. 9. Dir. Prof. Dr. Géra Farkas. — Mitarbeiter: Dr. Cornelius Körössy, Dr. Lorand Jendrassik, Josef Minich, Dr. Johann Mosonyi, Desiderius Hattyasy, Dionys Klobusitezky, Josef Faubl, Alexander Lángh, Franz Leövey, Nikolaus Julesz, Dr. Harald Tangl, Dr. Irene Incze, Johann Geldrich. *2 Gastplätze, Stoffwechsel u. Arbeitsphysiol. — Apparate f. Arbeitsphysiol., Stoffwechsel, Elektrokardiographie.*

Buenos Aires (Argentina).
Inst. de Fisiol. Fac. Med. Dir. Prof. B. A. Honssay. — Raul Wernicke, Prof. Physique biol.; Narcesco Laclau, Prof. Chim. biol.; Octavio M. Rico, Prof. substitut; M. A. Magenta, Chef de travaux; P. Mazzocco, Chimiste; Chefs de travaux: L. Solari, H. Rubio, H. Mascheroni, A. Marenzi, I. Rossignoli, L. Aquino, C. T. Rietti, E. Savino. *Places pour Physiol., Physique, Chimie, Physico-Chimie biol. — Des accessoires spéciaux sont à la disposition des travailleurs. Chimie biol., Nutrition, Physico-Chimie biol., Physiol. etc. 4 étages de labor.*

Buffalo, N.Y. (U. S. A.).
Dept. of Physiol., Univ. of Buffalo. Dir. F. A. Hartman.

Cádiz (España).
Labor. de Fisiol. Fac. Méd. Dir. Prof. Dr. Leonardo Rodrigo Lavín.—Dr. Pedro Rodrigo Sabalette, Directeur des travaux appliqués à la Physiol. humaine et à la Clinique. Dr. Adolfe Vila Rodriguez, Prof. aux. de la Faculté, Directeur des travaux de Physiologie des mammifères. *Physiol. des mammifères. Physiol. humaine, et Physiol. appliquée à la pratique clinique. 45 places.*

Caen, Calvados (France).
Inst. de Physiol. Fac. méd. Univ. Dir. Prof. G. Desbouis. — Prof. S. Marais.

Cagliari, Sardinia (Italia).
Labor. di Fisiol. Univ. Dir. Prof. I. Spadolini.
Ist. di Fisiol. sperim. Dir. Tullio Gayda.

Calcutta (Br.-India).
Vivekananda Labor., 8, Bosepara Lane. Sen Boshi. *Electric changes in living tissues.*

Cambridge (England).
Physiol. Labor. Univ. Dir. Joseph Barcroft. — Drs. H. Hartridge, L. E. Shore, Sir W. B. Hardy,

E. D. Adrian, G. V. Anrep, H. E. Tunnicliffe, H. Hartree (Honorary). *25 places. Blood and respiration work. Study of Circulatory Physiol. Study of Heal-formation in muscle. Study of Electrical phenomena in muscle and nerve. All oxidation-reduction systems. Sulphure metabolism. Study of ovarian hormone.*

Cambridge, Mass. (U.S.A.).
Labor. of General Physiol., Botan. Mus. Dir. Prof. W. J. Crozier. — O. W. Richards.

Camerino (Italia).
Labor. di Fisiol. sperim. Dir. G. Gallerani.

Canton (China).
Dept. of Physiol. Univ. Dir. M. Taylor.

Cape Town (South Africa).
Labor. of Physiol. and Physiol. Chem. of the Univ. of Cape Town. Dir.: Prof. W. A. Jolly and Prof. E. S. Edie. — Dr. H. Zwarenstein; Dr. L. P. Bosman.

Cardiff, Wales (England).
Labor. of Physiol. Univ. Coll. Dir. Prof. T. Graham Brown.

Catania, Sicilia (Italia).
Ist. di Fisiol. sperim. Dir. Gaetano Onagliarello.

Chiba (Japan).
Physiol. Inst. Chiba Med. Univ.

Chicago (U.S.A.).
Hull Physiol. Labor. Dir. Prof. A. J. Cearlson. — Prof. A. B. Luckhardt, Mammalian Physiol.; Prof. R. S. Lillie, General Physiol.; Instr.: N. Kleitman, Nervous physiol.; Th. Koppanyi, Regeneration; M. Hinrichs, Physiol. of Light Action. *10—15 places. — X-ray, string Galvanometers, Instruments for metabolism investigations, aquaria, quarters for all types of animals, constant temperature rooms etc.*

Cluj — Klausenburg (România).
Inst. de Physiol. Fac. méd. Univ., Str. Miko 1. Dir. Prof. J. J. Niţescu.
Inst. de Physiol. générale, Univ., Str. Moţilor 40. Dir. Prof. D. Călugăreanu. — N. Cavrilescu, Chef des travaux.

Coïmbra (Portugal).
Labor. de Fisiol. Univ. Dir. G. B. Brites.

Columbus, Ohio (U.S.A.).
Dept. of Physiol., Ohio State Univ. Dir. Prof. A. M. Bleile. — Prof. C. S. Smith; O. L. Milton.

Concepcion (Chili).
Inst. de Fisiol. Univ., Caupolican 17. Dir. Prof. Alexander Lipschütz. — *2 chefs des travaux pratiques; 2 ass.*

Cordoba (Argentina).
Labor. de Physiol. Fac. méd. Univ. Dir. Prof. A. Stucchi (Biophys.); Prof. G. Stuckert (Biochem.).

Cork (Ireland).
Labor. of Physiol. Univ. Coll. Dir. Prof. D. T. Barry.

Corrientes (Argentina).
Labor. de Physiol. et Histol. Fac. agric. Univ. Dir. Prof. V. Zeman. — Prof. S. Castillo Odena.

Curityba, Parana (Brasil).
Labor. de Physiol. Fac. méd. Univ. Dir. Prof. C. Silveira da Mota (Biophys.). Prof. J. de Azevedo Macedo (Biochem.); Prof. M. L. Carrão (Physiol.).

Dallas, Texas (U.S.A.).
Dept. of Biol. Southern Methodist Univ. Dir. S.W. Geiser. *Physiol. problems of sex. Animal orientation to stimuli of a directive sort.*

Debrecen (Ungarn).
Physiol. und allg. pathol. Inst. der Stefan-Tisza-Univ. Dir. Prof. Dr. Fritz Verzar. — Dr. A. Zih; Dr. A. Beznák; F. Péter; A. Arvay; E. Kokas; B. Fridrik. *10 Plätze. Blutgasanalyse. Respirationsmethodik. Elektrographie. Vivisektionen. Bact. — Haldane u. Atwater Benedict Respirationsapparat. Einthoven Saitengalvanometer. Blutgasapparate verschiedenster Konstruktion. Physikalisch-chem. Apparate verschiedener Art.*

Dnepro-Petrowsk, Ukraine (U.d.S.S.R.).
The labor. of Physiol. and Biochem. Dir. Prof. W. M. Archangelsky. — Dr. H. F. Tumiantzev; Dr. N. W. Martinova; N. W. Kudriavtzev; M. P. Bochval; A. M. Kashpoor; A. M. Soobenco; M. I. Archangelskaja. *4 places. Conditioned reflexes.*

Dublin (Ireland).
Dept. of Physiol. Univ. Trinity Coll. Dir. Prof. H. Pringle.
Physiol. Dept. of Univ. Coll. Dir. Prof. I. M. O'Conor.

Dunedin (New Zealand).
Dept. of Physiol. of Univ. Dir. Prof. I. Malcolm.

Edinburgh (Scotland).
Physiol. Labor. of R. Vet. Coll. Dir. Prof. H. Dryerre.

Edmonton, Alta. (Canada).
Dept. of Physiol. Univ. of Alberta. N. B. Eddy, Ass. Prof.; A. W. Downes.

Erlangen (Deutschland).
Physiol. Inst. Dir. Prof. E. Weinland. — T. v. Brand, Ass. *Chem. Physiol. Vergl. Physiol. Sinnesphysiol. Stoffwechsel.*

Ferrara (Italia).
Ist. di Fisiol. Univ. Dir. Prof. L. Beccari.

Firenze, Toscana (Italia).
Ist. di Fisiol. Univ. Dir. Prof. G. Rossi.

Frankfurt a. M. (Deutschland).
Inst. für animalische Physiol., Süd, Weigertstr. 3. Dir. Prof. Dr. Albrecht Bethe. — Priv.Doc. Dr. Wilhelm Steinhausen, 1. Ass.; Dr. Ernst Fischer, 2. Ass. *10 Plätze. Muskel-, Nerven- und Sinnesphysiol., auch für physikal.-chem. und für histol. Arbeiten. — Spezialapparate für Muskelstudien und Kraftmeßapparate. Gut eingerichtete, aseptische Operationsräume und gute Tierställe.*
Inst. f. vegetative Physiol. Dir. Prof. G. Embden.

Freiburg i. Br. (Deutschland).
Physiol. Inst. der Univ., Hebelstr. 33. Dir. Prof. Dr. Karl Hoffmann. — Prof. Dr. E. v. Skramlik; Priv.Doc. Dr. H. Stein, Ass. *Animalische Physiol.*

Fribourg — Freiburg (Suisse).
Labor. de Physiol., Univ. Dir. Prof. Charles Dhéré.

Fukuoka (Japan).
Physiol. Inst. Coll. of Med., Kyûshû Imper. Univ

Galway (Ireland).
Dept. of Physiol. Univ. Coll. Dir. Prof. J. F. Donegan.

Genève — Genf (Suisse).
Inst. de Physiol. Univ. Dir. Prof. Frédéric Battelli.

Genova (Italia).
Ist. di Fisiol. Univ. Dir. V. Grandis.

Gießen (Deutschland).
Physiol. Inst. der Univ., Friedrichstr. 24. Dir. Prof. Dr. med. et sc. nat. K. Bürker. — Prof. Dr. R. Feulgen, Abt.-Vorst. f. Physiol. Chem.; 3 Ass. *10 Plätze. 5 für biophysikal. und 5 für biochem. Arbeiten. — Apparate zur Blutuntersuchung, zur*

Muskel- und Nervenphysiol., speziell myothermische Apparate, Apparatur und Methoden zur Biochem. u. Mikrochem. der Nukleinstoffe und spezieller Lipoide.

Glasgow (Scotland).
Inst. of Physiol. Univ. Dir. Dr. Diarmid Noël-Paton, Regius Prof. — Prof. E. P. Cathcarts-Gardiner (Physiol. Chem.); Dr. A. M. Watson (Lect. on Histol.); Dr. George M. Wishart Grieve, Lect. (Physiol. Chem.); Dr. R. C. Garry, Lect. (Experim. Physiol.); Dr. H. E. C. Wilson, Ass. (Physiol. Chem.); Dr. W. A. Burnett, Ass. (Experim. Physiol.). *About 10 places for Experim., 12 for Chem., 10 for Histol. work.*

Göttingen (Deutschland).
Physiol. Inst. der Univ., Burgstr. 51. Dir. Prof. Dr. Paul Jensen. — Prof. Dr. R. Ehrenberg, Ass.; Dr. P. Schulze, Ass. *6—10 Plätze für allgemeine Physiol., Muskel- und Nervenphysiol. Kleines chem. Labor.* ⑨ Schädel, besonders Vogelschädel, in denen die Bogengänge präpariert sind. Die Praep. stammen noch von Prof. Meissner, besonders aus den Jahren 1875—1878.

Granada (España).
Labor. de Physiol. humaine, Fac. méd. Univ.

Graz (Österreich).
Physiol. Inst. Med. Fac. Univ. Dir. Prof. O. Zoth.

Greifswald (Deutschland).
Physiol. Inst. Dir. Prof. F. Wrede, Dr. med. et phil. — Dr. E. Strack, Ass. *Konstitution von Zuckerarten, schwefel- und selenhaltige Zucker, Thioglukose, Glykoside, Bakterienfarbstoffe, Spermin, Insektenatmung.*

Groningen (Holland).
Physiol. Inst. Univ., Bloemsingel. Dir. Prof. Dr. F. J. J. Buytendijk. — Dr. J. de Haan, Lector; Dr. R. Brinkman, Conserv.; M. M. J. Dirken, Ass. *Physico-chem. Unters., Respiration und Gaswechsel. Tierpsychol.* — Abt. f. Histol.

Guelph, Ont. (Canada).
Dept. of Physiol. Agric. Coll. Dir. A. Ross, M.D., C.M.

Halifax, N.S. (Canada).
Dept. of Physiol., Dalhousie Univ. Dir. Boris Babkin, M.D., D.Sc. — N. B. Dreyer, M.A.

Halle a. d. S. (Deutschland).
Physiol. Inst. der Univ., Magdeburger Str. 21. Dir. Geh.M.R. Prof. Dr. Emil Abderhalden. — Prof. Dr. Gellhorn; Prof. Dr. Wertheimer; 8 andere wiss. Mitarbeiter. *Physiol. Chem. Interne Sekretion.*

Hamburg (Deutschland).
Physiol. Inst., Allg. Krankenhaus Eppendorf, Martinistr. 52. Dir. Prof. Dr. Otto Kestner. — Prof. Dr. Franz Groebbels; Dr. Rahel Liebeschütz-Plaut. *Stoffwechselphysiol. Flug der Vögel. Histol. Physiol.*

Hannover (Deutschland).
Physiol. Inst. Tierärztl. Hochsch., Misburger Damm 16. Dir. Prof. Dr. Alfred Trautmann. — Ass.: Dr. Thur, Dr. Luy. *Innere Sekretion und Verdauung.*

Heidelberg (Deutschland).
Physiol. Inst. Univ. Dir. Prof. August Pütter. *Physiol. der Ernährung. Physiol. der Reizbeantwortungen.*

Helsinki (Finnland).
Physiol. Inst der Univ. Dir. Prof. Carl Tigerstedt. — Dodo Rancken; Yrjö Renqvist; Ragnar Granit; Kerttu Sihvonen. *Kreislauf.*

Hong Kong (China).
Labor. of Physiol. Univ. Dir. Prof. H. G. Earle.

Innsbruck (Österreich).
Physiol. Inst. Univ. Dir. Prof. E. Brücke.

Iowa City, Ia. (U.S.A.).
Labor. of Animal Biol., State Univ. of Iowa. Dir. G. L. Houser.

Irkutsk (U.d.S.S.R.).
Physiol. Inst. Univ. Dir. Prof. N. Spasskij.

Iwanovo-Voznessensk (U.d.S.S.R.).
Kabinett f. Physiol. Dir. Prof. Lavrov.

Jassy (România).
Labor. de Physiol. animale Univ., Str. Sărăriei 76. Dir. Prof. Dr. L. N. Cosmovici. — Hélène Lupu, Chef des travaux. *Physiol. comparée chez les invertébrés: coeur, toxines. Chimie physiol., sang, lait. Histophysiol. des poissons: Cobetidiens.*
Labor. de Physiol. Fac. de Méd. Dir. Prof. V. Râşcanu. — Leon Baliff, Chef des travaux du labor. *Electrophysiol. Physiol. des muscles. Neurol. Endocrinol.*

Jena (Deutschland).
Physiol. Inst. Univ. Dir. Prof. W. Biedermann. *Fermenta, Elektrophysiol. Ernährung Wirbelloser. Muskelbau u. -physiol.*

Johannesburg (South Africa).
Physiol. Inst. Univ. Dir. Prof. E. H. Cluver.

Kanazawa (Japan).
Physiol. Inst. Kanazawa Med. Univ.

Kaunas (Litauen).
Labor. für Physiol. und physiol. Chem., Med. fac. (Fiziol. ir fiziol.chem. labor.). Dir. Prof. V. Lašas. — Dr. V. Tiškus; Dr. I. Zubkus; I. Krukonis. *7 Plätze für Mikromethodik, Tieroperationen, Blutuntersuchungen.* — Labor. für Kolloidchem.

Kazan (U.d.S.S.R.).
Physiol. Labor. d. Veterinärinst., Arskoje Pole 96. Dir. Prof. K. Viktorow. — Dr. E. Pawlowskij, Prosektor; Dr. T. Agelskij, Ass. *4 Plätze. Herz, Nervensystem, Muskelsystem, Verdauung.*

Keijô, Korea (Japan).
Physiol. Inst. Coll. of Med., Keijô Imper. Univ.

Kiel (Deutschland).
Physiol. Inst. Univ. Dir. Prof. Dr. med. Rudolf Höber. — Priv.Doc. Dr. Rudolf Mond; Dr. Hans Netter. *20 Plätze für physikal.-chem. Untersuchungen, auch an Meerestieren.*

Kingston, Ont. (Canada).
Dept. of Physiol. Queen's Univ. Dir. Prof. Spencer Melvin, M.D. — G. H. Ettinger, B.A., M.D., C.M.

København (Danmark).
Fysiol. Labor., Bredgade Nr. 62. Dir. Prof. Dr. med. V. Henriques. — Ass.: Dr. phil. R. R. Ege, H. Dam, E. Lundsgaard.

Köln a. Rh. (Deutschland).
Normal- u. pathol.-physiol. Inst. Univ. Dir. Prof. Dr. Hering. — Prof. Dr. B. Kisch.

Königsberg i. Pr. (Deutschland).
Physiol. Inst., Copernicusstr. 1|2. Dir. Prof. Dr. O. Weiß. — Priv.Doc. Dr. H. Lullies; Priv.Doc. Dr. H. Müller. *4 Plätze für Untersuchungen in Biophysik und Biochem.*

Kraków (Polska).
Physiol. Inst. an der Jagiell. Univ., Grzegorzecka Str. 16. Dir. Prof. Dr. Ernest Maydell. — Dr. med. et phil. J. Kaulbersch; Dr. med. B. Schabuniewicz; Dr. med. M. Obtułowicz; B. Skarzynski. *2 Arbeitsplätze für Elektrophysiol. und Physiol. d. Verdauungsorgane.* — *Saitengalvanometer n. Einthoven mit d.*

photographischen Registrierapparat nach Edelmann. Aseptisches Operationszimmer und Hundeklinik.

Kumamoto (Japan).
Physiol. Inst. Kumamoto Imper. Med. Univ.

Kyôto (Japan).
Physiol. Inst. Coll. of Med., Kyôto Imper. Univ. Dir. Prof. Dr. Hidezuramaru Ishikawa. — Prof. Dr. H. Ishikawa (Nervenphysiol.); Prof. Dr. Rinnosuke Shôji (physico-chem. Physiol.); Ass. Prof. Dr. Naomi Kitamura (Ernährungsphysiol.); Lect. Yoshitaka Katsu (physico-chem. Physiol.).
Physiol. Inst. Kyôto Prov. Med. Univ., Hirokôji.

La Habana (Cuba).
Physiol. Labor. d. Univ.

La Plata (Argentina).
Physiol. Inst. Escola de sc. med. Dir. Prof. Frank Soler.
Inst. f. Physiol. Dir. Prof. G. Pacella.

Lausanne (Suisse).
Inst. de Physiol., Champ de l'Air. Dir. Prof. Dr. M. Arthus. — André Arthus; Jean Favre.

La Valletta (Malta).
Labor. of Physiol. and Pathol., Univ. of Malta. Dir. Prof. R. Samut. — Ass. Prof. R. Vincenti.

Lawrence, Kan. (U.S.A.).
Dept. of Physiol. of the Univ. of Kansas. Dir. Prof. I. H. Hyde. — Prof. O. O. Stoland; Assoc.Prof. C. I. Reed; Ass. Prof. L. V. Walling.

Leeds (England).
Dept. of Physiol. of the Fac. of med., Univ. of Leeds. Dir. Prof. W. F. Skanks. — Lect. G. Winfield.

Leiden (Holland).
Physiol. Labor., Zonneveldstraat 18 a. Dir. Prof. Dr.W.Einthoven.† — Dr. F. L.Bergansius, Conserv.; Dr. S. Hoogerwerf, Hauptass.; Dr. J. Roos, W. van der Horst, J. J. Krijnen, Ass. *Elektrophysiol. Untersuchungen.*

Leipzig (Deutschland).
Physiol. Inst. d. Univ., C 1, Liebigstr. 16. Dir. Prof. Dr. med. M. Gildemeister. — Prof. Dr. W. Sulze, Oberass.; Priv.Doc. Dr. F. Kleinknecht, Dr. J.-D. Achelis, Dr. R. Krüger, Dr. H. Ballin, Ass. *Animalisch-physiol. Unters. — Vivisektorische, elektrophysiol., optische, akustische Instrumente; verschiedene Registrierapparate.*

Leningrad (U.d.S.S.R.).
Physiol. Inst. a. d. Med. Kriegsakademie, Nishegorodskaja 6.
Labor. für Physikal. Physiol. des Staatl. Inst. f. experim. Med., Lopuchinskaja ulica 12. Dir. E. A. Hanicke.
Physiol. Abt. Leningrader Wiss. „P. F. Leshaft" Inst., Pr. Maklina (ehem. Anglijskij) 32. Dir. L. A. Orbeli. — C. I. Kunstman; M. B. Tetjaeva.

Leningrad — Alt-Peterhof (U.d.S.S.R.).
Labor. f. Physiol. d. Tiere. Dir. Prof. A. A. Uchtomskij. — 1 Ass.

Liège (Belgique).
Inst. Léon Fredericq, Physiol., Univ. de Liège, Place Delcour 17. Dir. Prof. Henri Fredericq. — Dr. Jean De Jace, Ass.; 2 Elèves-Ass. (Etudiants). *Electrocardiographie, mesure de la chronaxie. Vivisection en général.*

Lille (France).
Labor. de Physiol., Fac. Méd. de l'Univ. Dir. Prof. Dubois. — P. Combemale.
Labor. de Physiol., Fac. de Méd. catholiques de l'Univ., 56, Rue du Port 11, Rue de Toul. Dir. Prof. A. Legrand.

Lima (Peru).
Labor. de Fisiol., Fac. de Med., Univ., Alameda Gran. Dir. Dr. M. E. Tabusso.

Lincoln, Neb. (U.S.A.).
Dept. of Physiol. and Pharmacol. of the Univ. Dir. Prof. A. E. Guenther. — Assoc. Prof. O. M. Cope.

Lisbôa (Portugal).
Inst. de Fisiol. e chim. fisiol., Fac. de Med., Univ. Dir. Prof. M. Athias.
Inst. Bento da Rocha Cabral pour la recherche sc., Calçada Bento da Rocha Cabral 28. Dir. M. Ferreira de Mira. — Dr. Simoes Raposo; Prof. Lopo de Carracho; Ferreira de Mira, fils; Prof. Celestino da Costa; Dr. Silveira; Prof. Joaquim Fontes; Prof. Egas Moniz; Dr. Cordato de Noronha. *Par son organisation l'Inst. reçoit des chercheurs, subsidiés ou non, sans d'autre limitation, que celle, érigée par ses revenus. Physiol. expérim., chim. physiol., histol. normale et pathol., bactl. et microbiol. agraire.*

Liverpool (England).
Dept. of Physiol. and Histol., Fac. of Med. of the Univ. Dir. Prof. J. S. Macdonald.

Ljubljana — Laibach (S.H.S.).
Physiol. Inst. d. Univ., Zološka cesta. Dir. Prof. E. Kansky. — Dr. I. Seliškar, Ass.

London (England).
Inst. of Physiol. Univ. Coll., Gower St., E.W.C. 1. Dir. Prof. C. A. Lovatt Evans, D.Sċ., Jodrell Prof. — J. C. Drummond, D.Sc., Prof. (Biochem.); D. T. Harris, M.B., B.S., Ch.B., B.Sc., Ass. Prof.; R. K. Cannan, M.Sc., A. Hemingway, M.Sc., M.B., Ch.B., Senior Ass.; G. P. Crowden, B.Sc., Ruth Conway Verney, M.D., Phyllis M. Kerridge, M.Sc.; L. E. Bayliss, B.A., G. F. Marrian, B.Sc., Ass.; H. J. Channon, B.A., M.Sc., A. C. Downing, R. J. Lythgoe, M.A., B.Ch., W. K. Slater, M.Sc., A. S. Parkes, B.A., Ph.D., A. C. Chibnall, Ph.D., Hon. Research Ass.; E. H. Starling, C.M.G., M.D., B.S., Sc.D., Fou erton Prof.; A. V. Hill, M.A., Sc.D., Fuulerton Prof. and Univ. Prof. *About 30 places for experim. Physiol. and Biochem.*
Dept. of Applied Physiol., National Inst. of med. Research, Mount Vernon Hampstead, N.W. 3. Dir. Prof. L. E. Hill. — J. A. Campbell.
Dept. of Physiol., The London Hospital Med. Coll. of the Univ., The London Hospital, Turner Str., Mile End, E. Dir. Prof. H. E. Roaf. — Dr. W. A. M. Smart; F. Campbell Smith Esq.
Dept. of Physiol., King's Coll., Univ. of London, Strand W.C. 2. Dir. Prof. R. J. McDowall.
Dept. of Physiol., St. Mary's Hosp., Med. School of the Univ., Paddington, W. 2. Dir. Prof. B. J. Collingwood.
Dept. of Physiol., Guy's Hosp. Med. School of the Univ., St. Thomas St., Borough, S.E. 1. Dir. M. S. Pembrey.
Dept. of Physiol., London School of Med. for Women of the Univ., Royal Free Hospital, 8, Hunter Str., Brunswick Square, W.C. Dir. Prof. W. Cullis.
The Harvey Physiol. Labor. St. Bartholomew's Med. Coll. of the Univ., 6, Giltspur Street, E.C. Dir.: vacant (formerly Prof. C. Lovatt Evans). — H. P. Gilding; H. Gordon Reeves. *4 places for experim. Physiol.*
Physiol. Labor., Middlesex Hospital Med. School of the Univ., Mortimer Str., Oxford Str. Dir. Prof. Vincent Swale. — T. Izod Bennett, Ass. to the Prof.; Samson Wright, Demonstrator; F. R. Curtis, Research Ass.; 3 Junior Demonstrators. *2 or 3 places for experim. Physiol., Biochem., Histol., experim. Surgery, Pharmacol.*
The Wellcome Physiol. Research Labor., Langley Court, Beckenham, Kent. Dir. Dr. R. A. O'Brien.

London, Ont. (Canada).
Dept. of Physiol. of the Univ. of Western Ontario. Dir. F. R. Miller, M.A., M.D. — N. B. Laughton, B.A., M.Sc., Ph.D.; R. A. Waud, M.D.

Los Angeles, Cal. (U.S.A.).
Labor. for Bact. and Physiol. Confer: Bact. hominis.

Louisville, Ky. (U.S.A.).
Dept. of Physiol. and Pharmacol., Univ., 101, W. Chestnutstr. Prof. H. G. Barbour (Physiol., Pharmacol.); Ass. Prof. W. F. Hamilton (Physiol.); Ass. Prof. R. Beutner (Pharmacol.).

Louvain — Leuven (Belgique).
Inst. de Physiol. Dir. Prof. A. Noyons.

Lund (Sverige).
Physiol. Institution of the Univ. Dir. Prof. Torsten Thunberg. — Ass. Prof. J. Lehmann.

Lwów (Polska).
Inst. of Physiol. of the Univ., 52 Piekara. Dir. Prof. Dr. A. Beck. — Dr. Jan Klisiecki, Dr. Viktor Tychowski, Dr. Jozef Wysocki, Ass. *6—8 working places for guests, especially fitted for research work on the central nervous system, electrophysiol., physiol. of circulation. — Galvanometer of Einthoven, microdissection apparatus of Peterfi.*

Lyon (France).
Labor. de Physiol., Fac. Méd. de l'Univ. Dir. Prof. Doyon.
Labor. de Physiol., Ecole Nation. Vét. Dir. Prof. Jung.

Madrid (España).
Labor. de Physiol. Fac. Méd. Univ. Dir. Prof. Dr. Juan Negrin. — Dr. Domingo H. Guerra.
Labor. de Physiol. Ecole vet. Dir. Prof. J. M. Diaz Villar Martinez Matamoros.
Labor. de Fisiol. de la Residencia de Estudiantes, Pinar 17. Dir. Dr. José Blanco. — Prof. Dr. Juan Negrin; Dr. Juan Sopeña.

Manchester (England).
Physiol. Labor. Univ. Dir. Prof. H. S. Raper. — Dr. F. W. Lamb, Reader in Human Physiol.; Dr. A D. Macdonald, Lect. in Experim. Physiol.; Mr. A. D. Ritchie, Lect. in Physiol. Chem.; P. W. Clutterbuck, Demonstrator in Physiol. Chem.; Dr. T. F. Brunlin, Demonstrator in Human Physiol.; E. N. Willmer, Demonstrator in Experim. Physiol. *10 places for Chem., Human and Experim. Physiol.*

Manila (Philippine Islands).
Labor. of Physiol. and Pharmacol. Univ. Santo Tomas. Dir. Prof. J. Paredes.
Dept. of Physiol. Univ. Coll. of Med. Dir. Prof. E. Bulatao. — S. Concepcion; N. Cordero.

Marburg a. d. L. (Deutschland).
Physiol. Inst. Dir. Prof. R. Dittler.

Marseille (France).
Chaire de Physiol. de l'Ecole de Méd. Dir. Prof. J. Cotte. — Chef de travaux: M. Aubert; Prép.: M. Commeret, Etudiant en Méd. *Des places, en très petit nombre, sont accordées pour des recherches.*

Messina (Italia).
Ist. di Fisiol. Sperimentale, Via risorgimento. Dir. Prof. Giuseppe Amantea. — Dott. Gaetano Martino, Aiuto; Dott. Vittorio Zagami; Dott. Gius. Scarcella Perino. *1 posto specialmente prep. per ricerche sulla fisiol. del sistema nervoso o dell'apparato genitale. — Gli apparecchi e i mezzi adatti per indagini sulla funzione dei centri nervosi col metodo chimico o farmacol., e sulla funzione dell'apparato genitale (maschile e femminile), e sulla biol. degli spermatozoi. L'Ist. funziona momentaneamente in locali pro-*visorii, e potrá assumere il suo completo sviluppo non appena pronti i locali definitivi, che sono in costruzione.

Milano (Italia).
Labor. Farmacol., Tossicol. e Terapia sperimentale Univ., Via Gaetano Strambio-Citta degli Studii. Dir. Prof. Adr. Valenti. — Dr. Celso Provinciali, Med.; Dr. Baccio Zanella, Chim. *Ricerche prevalentemente ad indirizzo fisiol. e chim. — Miografi, grandi centrifughe, spettrofotometro, potenziometro, presse, apparecchi per analisi.* ⓟ *Droghe, erbario, preparati microscopici diversi.*
Ist. di Fisiol. Univ., Via G. Strambio. Dir. Prof. Carlo Foà. — E. Peserico, Aiuto; Ass.: E. Ghirardi, C. Cantoni. *Fisiol. sperim. e chimica biol., electrofisiol.*

Milwaukee, Wis. (U.S.A.).
Dept. of Physiol. Marquette Univ. School of Med. Dir. Prof. P. F. Swindle. *Respiration and Circulation. Cheyne-Stokes Breathing.*

Minsk (U.d.S.S.R.).
Physiol. Inst. der Weißrussischen Staats-Univ., Schirokaja, 28. Dir. Prof. Leo P. Rosenow. — Ass.: Dr. Helene Nikolaewa; Chemiker-Praep.: Boris A. Ganscha. *3 Plätze. Verdauung (äußere Sekretion). Bedingungsreflexe.*

Modena (Italia).
Labor. di Fisiol. sperim. Univ., Piazza S. Eufemia No 4. Dir. Prof. Alb. Aggazzotti. — Dott. G. Bucciardi. *2 posti di studio per ricerche di fisiol. e di chim. fisiol.*

Montpellier (France).
Labor. de Physiol. Fac. Méd. Univ., Inst. de Biol., Rue Montels. Dir. Prof. E. Hédon. — L. Hédon, Prof. agrégé. *2 places équipées pour des recherches de métabolisme respiratoire et des recherches histophysiol. — Appareil à échanges gazeux respiratoires.*

Montreal, Que. (Canada).
Dept. of Physiol. Univ. of Montreal. Dir. Prof. E. Asselin, M.D.

Mukden, Manchuria (China).
Physiol. Inst. South Manchurian Med. Univ.

Münster i. W. (Deutschland).
Physiol. Inst. der Univ., Krummer Timpen 24|25. Dir. Prof. Dr. R. Rosemann. — Prof. Dr. Krummacher, Abt.-Vorst.; Dr. Weber, Ass.

Nagasaki (Japan).
Physiol. Inst. Nagasaki Imper. Med. Univ.

Nagoya (Japan).
Physiol. Inst. Aichi Med. Univ.

Nancy (France).
Inst. de Physiol. Fac. Méd., Rue Lionnois 30. Dir. Prof. M. Lambert. — Louis Merklen. Louise Hennequin, Prép. *4 places. Coeur et Circulation (méthode graphique). Physiol. de la nutrition.*

Napoli (Italia).
Ist. di Fisiol. della R. Univ. di Fac. Méd., S. Andrea delle Dame, 21. Dir. Prof. Filippo Bottazzi. — Dr. Luigi De Caro; Dr. Antonio Jappelli; Dr. Gino Bergami. *Fisiol. normale, Biochim., chim. fisica fisiol.*

New Brunswick, N.J. (U.S.A.).
Research Labor. of the New Jersey Coll. of Pharmacy. Dir. of Research: L. K. Riggs. *Pharmacol. of unsaturated Hydrocarbons. Physiol. Action of Anesthetics and Antiseptics.*

New Orleans, La. (U.S.A.).
Dept. of Physiol. Tulane Univ. Prof. H. Laurens. *Reactions of animals to light. Physiol. and anat. of the heart. Melanophores. Spectral sensitivity and*

visibility. Radiation on metabolism and growth, blood etc.

Niigata (Japan).
Physiol. Inst. Niigata Med. Univ.

Odessa (U. d. S. S. R.).
Abt. Physiol. des morphol.-physiol. Katheders d. Wissensch. Forschungsinst., Olgiewskaja 4. Dir. A. Melik-Megrabow. — E. Sinelnikoff; E. Goldenberg; W. Majewsky; A. Woinar. *Physiol. Apparate und Instrumente.*

Okayama (Japan).
Physiol. Inst. Okayama Imperial Med. Univ. Dir. Prof. Sôroku Oinuma.

Omsk (U. d. S. S. R.).
Physiol. Abt. d. med. Inst. Dir. Prof. M. P. Kalmykov.

Ôsaka (Japan).
Physiol. Inst. Ôsaka Provincial Med. Univ. Dir. Prof. Tomoichi Nakagawa. — Prof. Tomoichi Nakagawa; Lect. Tôsaku Kinoshita.

Oslo (Norge).
Univ. Fysiol. Inst. Dir.: Prof. Dr. Sophus Torup (Physiol.), Prof. Dr. Emar Langfeldt (Physiol. Chem.). — Dr. K. Chr. Geelmuyden; Dr. Ley Poulsson; Dr. Th. Skaar. *Physiol. u. physiol. Chem.*

Oxford (England).
Physiol. Inst. Univ. Dir. Prof. Sir Ch. Sherrington. *Physiol. of the nervous system.*

Padova (Italia).
Ist. di Fisiol., Viale Loredan, 6. Dir. Prof. Virgilio Ducceschi. — Prof. Achille Roncato, Libero Doc. ed Aiuto; Dott. Mario Rigoni; Dott. Bernardino Panizza. *4 posti per praticanti di Fisiol. e Chim. fisiol.* — L'Ist. ha una sezione di Chim. biol.

Palermo (Italia).
Ist. di Fisiol. R. Univ., Corso Tukory. Dir. Ugo Lombroso. — Dr. Prof. Camillo Artom, Aiuto; Dr. Giuseppe Sunzeri, Ass. *Per ricerche di chim. fisiol.*

Paris (France).
Labor. de Biol. expérim. de l'Ecole des Hautes-Etudes. Dir J. Gautrelet. — R. Bouhey. *Physiol.*
Labor. de la Biol. générale Coll. de France, Ve, place Marcellin-Berthelot. Dir. Prof. E. Gley. — Prép.: A. Quinquand, J. Cheymol, P. Gley. *2 places pour études physiol. sur le sécrétions internes.*
Labor. de physiol. Fac. de Méd. Dir.: Prof. H. Roger, Prof. H. Cardot. — Prof. F. Richet; Prof. Binet. *Physiol. gén. et comparée. Mollusques terrestres et fluviatiles de la France et de pays voisins. Lois polaire de l'excitation, Réflexe linguo-maxillaire, variations et hérédité, Coeur.*
Labor. de Physiol. à l'Ecole pratique des Hautes-Etudes, 1, rue Victor Cousin. Dir. A. E. Chauchard. — B. E. M. Chauchard, Dir.-Adjoint; G. Pouchet.
Service de Physiol. de l'Inst. Pasteur, 21—25, Rue Dutot. Dir. Prof. C. Delezenne. — Prof. W. Mestrezat.
Labor. Physiol. Inst. Cath. Ecole Sup. des Sc. Dir. Prof. A. Briot.
L'Inst. Marcy. Dir. Prof. C. Richet. *Physiol. et Psycho-Physiol.*

Parma (Italia).
Labor. di Fisiol. della R. Univ. Dir. Prof. M. Camis. — Dott. G. Pupilli, Aiuto; Dott. V. Bolcato, Ass.

Pavia (Italia).
Ist. di Fisiol. della R. Univ. Dir. F. Gayda.

Pécs (Ungarn).
Physiol. Inst. Univ., Rákóczystr. 80. Dir. Prof. Michael Pekar. — Dr. med. R. Bodó, Priv.Doc., Ass. Prof.; Dr. med. D. v. Kloburitzky, Ass. 1 Ass. u. 2—3 Volontäre.

Peking (China).
Dept. of Physiol., Union Med. Coll. Dir. Robert K. S. Lim. — Heinrich Necheles, M.D., Ph.D.; Tsang-gi Ni, M.D., D.Sc.; Hsiang-ch'uan Hou, M.D. *3—4 places.*

Perm (U. d. S. S. R.).
Physiol. Labor. Dir. Doc. W. P. Petropawlowsky. — Ass.: M. W. Mouchin, W. Ph. Tischankin; wiss. Praep.: P. Starkow, W. Tschernigowsky, T. Olenewa, A. Kolosowa, Ditsch, Ulitsky.

Perugia (Italia).
Ist. di Fisiol. Univ. Dir. Prof. Osvaldo Polimanti. — Dr. Vita Pietro. *4 Plätze für animalische Physiol.* — Monte del Lago (Umbria), R. Stazione Idrobiol. del Lago Trasimeno.

Pisa (Italia).
Ist. di Fisiol. Fac. Med. Univ. Dir. Prof. V. Aducco.

Poitiers (France).
Labor. de Physiol. Ecole réorganisée de méd. et de pharmacol. Univ. Dir. Prof. H. Delaunay.

Porto (Portugal).
Labor. de Physiol. Fac. méd. Univ. Dir. Prof. A. P. Pinto de Aguiar.

Porto Alegre (Brasil).
Labor. de Physiol. Fac. méd. 1. Dir.: Prof. R. Pilla; 2. Dir.: Prof. F. de Barros.

Poznań — Posen (Polska).
Inst. Physiol. Univ. Fac. Med. Dir. Dr. L. Zbyszewski. — Dr. Marceli Jakuliak, T. Skalmowski. *2 places pour la physiol. du système nerveux, 2 places pour chimie physique.*

Praha — Prag (C. S. R.).
Physiol. Inst. d. Karls-Univ., II, Katerinska 32. Dir. Prof. Dr. Franz Mareš.
Physiol. Inst. der Deutschen Univ., Albertov 5. Dir. Prof. Dr. Armin Tschermak-Seysenegg. — ao. Prof. Dr. R. H. Kahn; Priv.Doc. Dr. M. H. Fischer; Dr. G. Schubert; dazu 3. Ass. u. 2 Demonstratoren. *Etwa 10 Plätze für Tieroperationen, Gastranometrie, physiol. Optik. — Aseptischer Operationssaal, Vivisektorium, Saitengalvanometer, Röntgeneinrichtung, optische Apparate.*

Pretoria (South Africa).
Dept. of Physiol. Fac. Sc. Univ. Dir. Prof. F. W. Steck.

Quebec (Canada).
Dept. of Physiol. Laval Univ. Dir. J. B. Lecroix, M.D.

Quito (Ecuador).
Labor. de Physiol. Fac. méd. Univ. Dir. L. A. Rivadenaira Garcia.

Reims (France).
Labor. de Physiol. Ecole prép. de méd. et de pharmacol. Dir. Prof. Quinquaud.

Reykjavik (Island).
Bact., Physiol. Inst. Confer: Bact. hominis.

Riga (Latvija).
Physiol. Inst. Med. Fac. Univ., Kronvalda bulvari 9. Dir. Prof.- Dr. R. Krimbergs.

Rio de Janeiro (Brasil).
Inst. Physiol. Fac. med. Univ. Dir. Prof. O. de Souza.

Roma (Italia).
Ist. di fisiol. sperimentale Fac. med. Univ. Dir. Prof. S. Baglioni.

Rosario, Santa Fé (Argentina).
Labor. di Fisiol. Dir. Viale Gaetano. — Simon Neuschloß (Physique biol.); Teodoro Combes et Arturo Bruno (Chim. physiol.).

Rostock i. M. (Deutschland).
Physiol. Inst. der Univ. Dir. Prof. Dr. Hans Winterstein. — Frl. Else Hirschberg, Laborantin; Prof. Dr. F. v. Krüger, Ass., Vorst. der physiol.-chem. Abt. *6—8 Plätze für physikal.-chem. u. mikrochem. Untersuchungen. Mikrorespirometrie, Gewebsatmung.*

Rostow a. Don (U. d. S. S. R.).
Physiol. Labor. der Nord-Kaukasischen Univ., Suworowskajastr. 41. Dir. Prof. Dr. N. Rožansky. — Ass.: Dr. L. Gorschkowa, Dr. N. Danilow. *Einrichtung zum Operieren warmblütiger Tiere und Nachbehandlung. Flimmerbewegung in der Trachea; Darmperistaltik, Bluteiweißregeneration; Speichelabsonderung; Blutbewegung. — Originelle Einrichtung zur graphischen Registrierung der Flimmerbewegung; gasometrische Bestimmung der Catalasewirkung; Bestimmung der Blasendruck-Oberflächenspannung.*

Saint Andrews (Scotland).
Physiol. Dept. of the Univ. Coll. Dundee and Med. School. Dir. Prof. E. W. Reid.

San Francisco, Cal. (U. S. A.).
Dept. of Physiol., Med. School, Univ. of California. Dir. Prof. S. S. Maxwell.

São Paulo (Brasil).
Dept. of Physiol. and Psychol., Brigadeiro Tobias No 45. Sub-Dir. Benjamin Alves Ribeiro. — Dr. Octavio Martins de Camargo (Engineer). *Probably 1 or 2 places in the new building. Physiol. as applied to hygien., Professional selection.*

Saratow (U. d. S. S. R.).
Physiol. Labor. Univ. Dir. Prof. J. A. Čuevskij.

Sassari (Italia).
Ist. di Fisiol. Fac. med. Univ. Dir. Prof. G. Viale.

Sendai (Japan).
Physiol. Inst. (1. Abt.) der Reichsuniv., Tôhoku, Reichsuniversität. Dir. Prof. Y. Sataké. — Ass. Prof. Sakuji Kodama (Biol.-physikal. Chem); Ass. u. Mitarbeiter: Tadashi Sugawara, Masanosuké Watanabé, Hiroshi Tachi, Takeo Kojima, Shidzuka Saito, Mamoru Nemoto, Hiroshi Sato, Taisuke Suzuki, Hyozo Tada, Bunkichi Kamei u. Ryko Kaiwa.
Physiol. Inst. (2. Abt.) der Reichsuniv. Dir. Prof. T. Fujita. — Ass. Prof. Y. Hosoya; Ass. u. Mitarbeiter: Kisuo Sugai, Sikô Yosida, Seiiti Mori, Keizo Hashimoto.

Sevilla (España).
Lab. de Fisiol Fac. Med. Dir. Prof. Dr. Estanislao del Campo y Lopez.

Sheffield (England).
Physiol. Labor. Univ. Dir. J. B. Leather. — Elizabeth C. Eaves, M.D., Lect. (Histol. and the Physiol. of the Nervous System); Cyril Gray Imrie, M.D., Lect. (Chemical Physiol.); Georg A. Clark, M.D., Lect. (Experim. Physiol.); Robert Plate, M.D., (in charge of Physiol.Labor.Royal Infirmary). *Three places at the Univ.: others at the Physiol. Labor. at the Royal Hospital. Chemical and General Physiol. — Special facilities for study in the junctions, chemical, respiratory and circulatory in disordered conditions on patients, at the Labor. of the Dept. at the Hospitals associated with the Univ. — Labor. under the same direction and the same staff at the Royal Hospital, Royal Infirmary and Jessop Hospital.*

Siena (Italia).
Labor. di Fisiol. della R. Univ di Siena. Dir. Prof. Balduino Bocci. — Dott. Prof. L. Bellucci. *Lavori di Microscopia, Biofisica, Biochimica, Otica fisiol., Acustica, Psicol. sperimentale.*

Singapore (Straits Settlements).
Physiol. Dept. Coll. of Med. Dir. Prof. T. R. Kay-Mouat. A. J. Copeland, Lect. in Pharmacol.; C. T. Oliveiro, Ass. *5 working places for guests.*

Smolensk (U. d. S. S. R.).
Labor. f. Physiol. Med. Fac. Univ. Dir. Prof. Voroncov.

Sofia (Bulgarien).
Physiol. Inst. Univ. Dir. Prof. Dr. W. Sawjaloff. — Ass. Dr. D. Orachowatz; Ass. Dr. Ws. Sawjaloff; Ass. Dr. E. Nikoloff.

Stambul (Türkei).
Labor. de Physiol., à Haidar Pacha Constantinople. Dir. Prof. Dr. Kémal-Djénab. — Prof. Talha Saadi-Nazim, Ass.

Stellenbosch (South Africa).
Dept. of Physiol. of the Univ. Dir. Prof. P. Battaerd, M.D.

Stockholm (Sverige).
Physiol. Labor., Karolinska Inst. Dir. Prof. I. E. Johanson.

Strasbourg — Straßburg (France).
Inst. de Physiol. Fac. Méd., 1, Place de l'Hôpital. Dir. Prof. Dr. Georges Schaeffer. — Mlle. Eliane Le Bretery, Chef des travaux; Dr. Théophile Cahn, Ass.; Dr. Charles Kayser, Ass. *3 places pour physiol, 3 pour biochim. Pour la physiol.: Métabolisme: échanges respiratoires; méthodes graphiques. Pour la Biochimie: Installation complète. Physiol. de l'alimentation et de la croissance.— Tous appareils concernant la méthode graphique; salle d'opérations aseptiques pour animaux; tous les appareils nécessaires aux êtres des biochimiques: viscosimètres; appareils pour mesure de la conductivité électrique; potentiomètre pour mesure du sole; bombe calorimétrique de Mahler; ,,Portable apparatus" de Benedict; Spiromètre de Tissot; appareil de Haldane pour mesure des échanges des petits animaux; dispositifs pour étude de la respiration des tissus; appareils de Van Slyke; capacité respiratoire du sang etc.*

Sydney (Australia).
Physiol. Inst. Univ. Dir. H. G. Chapman.

Szeged (Ungarn).
Physiol. Inst. Univ. Dir. Prof. Dr. med. Elemér Veress. — Dr. med. Eugen Csinády, Ass.

Taihoku, Formosa (Japan).
Physiol. Inst. Taihoku Governm. Coll. of Med.

Tartu — Dorpat (Eesti).
Physiol. Inst. der Univ. Dir. Prof. Dr. med. Alfred Fleisch. — Dr. L. Adamberg, Ass.; Dr. M. Tiitso, Ass. *2 Plätze für Physiol. und physiol. Chemie.*

Taschkent (U. d. S. S. R.).
Labor. de Physiol. animale de l'Univ. Dir. Prof. E. Poyarkoff. — Israel, Doc.

Tegucigalpa (Honduras).
Physiol. Inst. Univ. Dir. Prof. Dr. R. R. Ramires.

Tôkyô (Japan).
Physiol. Inst. Military Med. Coll. (Gun-i Gakkô).
Physiol. Inst. Jikeikwai Med. Univ., Shiba.
Physiol. Inst. Nippon Girls' Med. Univ.
Physiol. Inst. Coll. of Med., Geiô-Gijuku Univ. Dir. Prof. G. Katô. — Ass. Prof.: Dr. M. Kubo; Lect. T. Nakazawa.

Physiol. Inst. Nippon Med. Univ. Prof. Tokehiko Uehara.
Physiol. Inst. Fac. of Med., Tôkyô Imper. Univ. Prof. Dr. Hisomu Nagai; Prof. Dr. Kunihiko Hashida Ass. Prof. Kunigô Fukuda.

Tomsk (U. d. S. S. R.).
Inst. f. Physiol. Univ. Dir. Prof. N. A. Popoff.

Torino (Italia).
Ist. di Fisiol. della R. Univ., Corso Raffaello 30. Dir. Prof. A. Herlitzka. — Dr. Rodolfo Margaria; Dr. Silvia Colla; Dr. Antonio Chiatellino; Dr. Marcello Comel. *Per qualunque ricerca di Fisiol. sperimentale. Impianti speciali per elettrofisiol. e per fisiol. colla pressione diminuita.* Sublabor.: 1. All' Ist. Mosso al Col D'Olen. 2 Al Mare a Trieste. Nedi le due schede relative.

Toronto, Ont. (Canada).
Dept. of Physiol. Univ. of Toronto. Dir. J. J. R. Macleod, M.B., Ch.B., D.Sc., LL.D. — Olmated, M.A., Ph.D., N. B. Taylor, M.B.; J. Markowitz, M.B.

Toulouse (France).
Labor. de la Physiol. Fac. méd. Univ. Dir. Prof. E. Abelous.

Tsinan (China).
Dept. of Physiol., Shantung Christian Univ. Dir. Prof. Jr. P. S. Evans.

Tübingen (Deutschland).
Physiol. Inst. Dir. Prof. W. Trendelenburg. — Dr. Wagner. *Nervenphysiol.*

Ueda (Japan).
Physiol. Labor. Ueda Imperial Sericicultural Coll., Nagano-ken.

Uppsala (Sverige).
Univ. Labor. of Physiol., Slottsgränd 3. Dir. Prof. Gustaf Fr. Göthlin. — 1 Ass. Prof. of experim. physiol. (vacat); 1 Lect. on Physiol.: Carl Gustaf Sundberg, M.D. *At most 2 places for researches in colour sense, normal or defective. — Liminespectroscope of Göthlin. Gullstrands apparatus for mixing monochromatic lights (1905). Polarization Anomaloscope of Göthlin.*

Utrecht (Holland).
Physiol. Inst. der Univ. Med. Fak., v. Wyckskade N. 28. Dir. Prof. Dr. Noyons. — Dr. T. P. Feenstra, Conserv.; Dr. J. G. Dusser de Barenne, Priv. Doc. *3 Plätze. Bioradioaktivität, experim. Phonetik, Akustik, Olfaktol. — Strahlungsapparate (Radioaktivität, Kathodenstrahlen, Licht). Apparate für experim. Phonetik. Phonimeter, komplette Reservatorenreihe, sowohl für Töne als Geräusche, Camera silenta. Olfactometer, Camera odore carens.*

Valencia (España).
Labor. Physiol. humana. Dir. Prof. A. Gily Morte.

Valladolid (España).
Labor. de Fisiol. Dir. Lorenzo Torremocha. — Prof. Auxiliar: Dr. Emilio Romo. *Reactiones psicomotrices. Centros corticales.*

Warszawa (Polska).
Labor. de Physiol. de l'Inst. Nencki, 8, rue Śniadecki. Dir. Prof. K. Bialaszewicz. — Bogucki M.; R. Szretter; S. Kuczkowski; A. Wojtczak. *12 places spécialement préparées pour les recherches: sur le métabolisme nutritif et énergétique des animaux inférieurs; sur la physiol. de la fécondation et du développement embryonnaire; sur l'échange des principes minéraux dans l'organisme. — Les appareils pour les recherches sur la réspiration et la calorimetrie des animaux de petite taille, pour les méthodes physicochimiques et microchimiques, pour la registration graphique; Cabinet chirurgique.*
Inst. de Physiol. de l'Univ., Krakowskie Przedm. N. 26|28. Dir. Prof. Fr. Czubalski. — Dr. B. Gutowski; Dr. Marie Skarzynska; Dr. J. Walawski; S. Gartkiewicz. *4—5 places. — Un électro-, cardiographe (Simens); salle d'opération aséptique (montée). Appareils pour l'étude de la circulation du sang. Accessoires (d'après Pawlow) pour l'étude de l'acte digestif.*

Wien (Österreich).
Physiol. Inst. der Tierärztl. Hochsch., III, Linke Bahngasse 11. Dir. Prof. C. Schwarz. — Dr. K. Steinmetzer. *Ca. 10 Plätze. Physiol. des Blutes u. der Verdauung.*
Physiol. Inst. Univ., Schwarzspanierstr. Dir. Prof. Dr. Durig.

Wilno (Polska).
Physiol. Inst., Zakretowastr. 15. Dir. M. Eiger. — 4 Ass.: Ing. Felix Großmann, Dr. Wladyslaw Zemojtel, Dr. Władysław Łobza und Dr. Michel Jagodowski. *15 Plätze für Elektrokardiographie, Ultramikroskopie, Interferometrie, Vivisektionen. Chemische.*

Winnipeg, Man. (Canada).
Dept. of Physiol. and Pharmacol. Univ. of Manitoba Dir. V. H. K. Moorhouse, B.A., M.B. — K. J. Austmann, M.A., M.D.; M. S. Lougheed, B.A., M.D., C.M., B.Sc.; W. G. Mackenzie, M.B., B.Sc.

Woosung b. Shanghai, Kiangsu (China).
Physiol. Inst. der Tungchi-Univ. Dir. Prof. Dr. Stübel. — Liang dsche yän. *3 Plätze für physiol. und physiol.-chemische, insbesondere auch histophysiol. Untersuchungen.*

Woronesh (U. d. S. S. R.).
Physiol. Inst. Univ. Dir. Prof. P. M. Nikiforovskij.

Würzburg (Deutschland).
Physiol. Inst. Univ. Dir. Prof. M V. Frey. *Herz.*

Zagreb — Agram (S. H. S.).
Physiol. Inst. Vet. Med. Fak., Savska c. 1. Dir. Prof. Dr. F. Smetánka. — Ass., 2 Demonstratoren.

Zaragoza (España).
Labor. de Fisiol. Fac. Med. Dir. Prof. Dr. Santiago Pi Suñer. — Dr. Máximo Muniesa.

Zürich 1 (Schweiz).
Physiol. Inst. Univ., Rämistr. 69. Dir. Prof. Dr. W. K. Hess.

Laboratoria physiologiae comparatae (fac. scientiarum natur., fac. veterin. et fac. agricult.).

Confer: *Zoologia generalis* (pag. 356), *Anatomia animalium* (pag. 467), *Histologia* (pag. 375), *Physiologia generalis* (pag. 444), *Labor. marina* (pag. 343).

Belfast (Ireland).
Dept. of Physiol., Queens Univ. Fac. Sc. Dir. Prof. Th. H. Milroy.

Beograd (S. H. S.).
Inst. de Physiol. générale Fac. philos. Univ., Kraljev trg. Dir. Prof. I. Gjaja. — B. Maleš, Ass.

2—3 places. Métabolisme énergetique, Spécialement métabolisme de sommet. Section pour l'étude de la nutrition des animaux domestiques. — Appareils pour les études des échanges gazeux de l'homme et des animaux.

Berlin (Deutschland).

Labor. f. physiol. Zool., Biol. Reichsanst. f. Land- u. Forstwirtschaft, -Dahlem, Königin-Luise-Str. 19, Dir. Reg.R. Prof. Dr. A. Hase. — Dr. H. Völkel, Ass.
Physiol. Inst. Tierärztl. Hochsch., NW 6, Luisenstr. 56. Dir. Prof. Max Cremer.
Tierphysiol. Inst. d. Landwirtschaftlichen Hochsch., N 4, Invalidenstr. 42. Dir. Prof. Dr. med. et phil. E. Mangold. — Ass.: Dr. C. Brahm, Dr. C. Schmitt-Krahmer, Dr. H. Meltzer, M. Steuber. *Physiol. und physiol.-chemische Untersuchungen: Herz-, Nervmuskel, Verdauung, Stoffwechsel, vgl. Physiol. — Respirationsapparate.*

Brno — Brünn (C. S. R.).

Physiol.Dept. ofVeterinary Univ.Coll., Fysiol. ústav vys. šk. zverolékařské. Dir. Dr. Tomáš Vacek. — Doc. Dr. O. V. Hykeš (Biol. and Pathol. of Fish and Bees). *About 4 places for animal Ecol., Hydrobiol. and other sorts of experim. Physiol. — Register apparatus, for Mikrophotography, for histol. examinations, for muscle physiol. Mercury-vapour lamp etc.* Hydrobiol. labor.

Bruxelles — Brussel (Belgique).

Labor. de physiol. animale (Inst. Solvay). Dir. Prof. Dr. M. Philippson.

Detskoje Selo b. Leningrad (U.d. S. S. R.).

Tierphysiol. Labor. des Agronomischen Inst., Dir. Prof. Dr. med. K. N. Krzyszkowsky. — Dr. med. Z. M. Rabinkowa, I. Ass.; G. N. Pawzow, Agronom II. Ass. *10 Plätze für Vogelphysiol. (Verdauung), Geschlechtsphysiol. (Fistelmethode), Physiol. der Verdauung (bei Wiederkäuern). Bedingte Reflexe, bedingte Reaktionen (Vögeln, Hunde, Katzen).*

Eugene, Ore. (U.S.A.).

Labor. of Physiol. and Zool. of the Univ. of Oregon Dir. Dr. A. R. Moore, Prof. of Gen. Physiol. — Dr. H. B. Yocom, Prof. of Zool.; Dr. R. R. Huestis, Ass. Prof. of Zool. *4 working places for guests. Gen. Physiol. and Zool. of Marine Forms.* ⑨ Marine invertebrates. — Marine station on the shore open during spring and summer.

Gent — Gand (Belgie).

Labor. d. alg. levensverrichtingen. Confer: Institutiones generales.

Kazan (U.d. S. S. R.).

Physiol. Labor. d. physiko-math. Fak. Staatsuniv. Dir. Prof. A. Samojloff. — I. A. Wetochin, Priv.Doc. und I. Ass.; W. A. Wassiljewa; M. A. Kisselew; W. W. Parin. *4 Plätze für elektrophysiol. Untersuchungen — Saitengalvanometer.*

Kiew (U.d. S. S. R.).

Tierphysiol. Labor. d. Inst. für Landwirtsch.

København (Danmark).

Labor. of Zoophysiol., B, 11 Ny Vestergade. Dir. Prof. A. Krogh. — Dr. med. Marie Krogh (Pharmacol., Human Metabolism); Dr. phil. P. Brandt Rehberg (Physiol., Courses in compar. Physiol.); O. Kasmussen (Chemistry). *2—3 places for study of capillary circulation and for biol. microchemistry. — Apparatus for microscopic and photographic study of circulation. Microtitrition and microanalysis of gases.* ⑨ The labor. workshop supplies recording metabolism apparatus microburettes and other apparatus.
Physiol. labor. of the agric. experim. Station. (Dyrefysiol. Labor.), V., 25 Rolighedsvig. Dir. Prof. Holger Møllegaard. — C. A. Lund, Chief of dept. of metabolism; Ing. Grasterlen, Chief of chemical dept· *Gasanalysis. Kalorimetric. Analytic Chemistry· Aseptic Operations. Bakt. Physiol. of metabolism and secretion. Experim. surgery. Chemotherapy. — Respiration apparatus for large animals.*

Leningrad (U. d. S. S. R.).

Labor. de Physiol. de l'Ecole supérieure vétérinaire, Tschernigowskaja, 5. Dir. Prof. Dr. med. G. Zeliony. — Valentine Adlerberg, Ass.; Georges Prokofiew Préparateur. *5 places pour recherches sur le système nerveux central; reflexes conditionnés (associatifs). — Les appareils pour les recherches mentionnés plus haut.*
Abt. f. Physiol. Len. Landw. Inst., Kamennyj Ostrov 2, Birken-Allee 28. Dir. Prof. K. N. Kržiškovsky.

León (España).

Labor. de Fisiol. e Higiene, Escuela de Veterinaria. Dir. Prof. C. S. de la Calzada.

Lwów (Polska).

Physiol. Labor. d. Tierärztl. Akad.

Lyon (France).

Labor. Auguste Lumière. Confer: Biophysica.
Labor. de Physiol., Fac. sc. de l'Univ., 16 quai Claude Bernard. Dir. Prof. H. Cardot. — M. Clement, Chargé de cours (Centrifugation). Mr. Jullien, Ass. (Histophysiol.). *4 places pour Physiol. générale et comparée et notamment physiol. du système nerveux. App. pour la mesure de la chronaxie, pour la perfusion d'organes etc.*

Leipzig (Deutschland).

Vet.-physiol. Inst. d. Univ. Dir. Prof. Dr. Arthur Scheunert. — Wissenschaftl. Ass.: Dr. Hermann Martin Schleblich; Abt.-Leiter der physiol.-chem. Abt.: Prof. Dr. Martin Schenk. *Physiol. der tierischen u. menschl. Verdauung (Vitamine, Darmflora, Bakt. der Silage, Fütterungsfragen, Magenmechanik. Milz). Bildung von Vitamin B durch Bakt.*

Marseille (France).

Labor. de Physiol. générale Fac. Sc. Univ., Place Victor Hugo. Dir. Prof. H. Bierry. — 2 Ass. *2 places physiol.chimique(Digestion et métabolisme des matières sucrees). — Appareils de chemie physiol.*

Mexico, D. F. (Mexico).

Sect. physiol. compar., d. Dir. de Estudios Biol., 71a calle de Balderas 94. Dir. Dr. J. Solis.

Milano (Italia).

Labor. di Fisiol. Sperimentale Ist. Sup. de Med. veterin. Dir. Prof. Angelo Pugliese. — Dott. F. Usuelli, Ass.; Prof. Fasoli; Prof. Cuneo; Dott. Foster; Dott. Brusotto. *3 posti per fisiol. e chimica fisiol.*

Moskau (U.d. S. S. R.).

Physiol. Cabinett des Moskauer Landwirtsch. Inst. (jetzt Timirjasew'sche Landwirt. Akademie), 8, Petrowsko-Razumovskoje. Dir. Prof. Dr. A. W. Leontowitsch. — V. W. Kudrjaschow, Ass. *3 Plätze. Vitale Methylenblaufärbung der Nerven nach Leontowitsch.*
Labor. der Tierphysiol. des Zootechnische Inst., 2, Smolenskij Boulv. 57. Dir. Prof. Boris A. Lawrow. — Frl. Natalie S. Jarussowa; Dr. Konstantin M. Michajlow.

München (Deutschland).

Tierphysiol. Inst. Tierärztl. Fac. Univ. Dir. Prof. Joh. Paechtner.

Odessa (U.d. S. S. R.).

Zoo-physiol. Labor. der Pädagogischen Hochsch. „Ino". Dir. Prof. Sinel'nikov.

Omsk (U.d.S.S.R.).
Labor. f. Physiol. u. Anat. der Haustiere am Sibirischen Inst. f. Land- u. Forstwirtsch. Dir. Prof. M. P. Kalmykov.

Paris (France).
Labor. de Physiol. générale Fac. Sc. nat. Univ., 1, rue Victor Cousin. Dir. Prof. Louis Lapicque. — P. Portier, Prof.; G. Stodel, Maître de conférences, à l'Ecole des Hautes Etudes; Laugier, Chef de travaux; Chaussin; Dessoille. *15 places pour Physiol. générale, et spécialement du système nerveux.*
Labor. de Physiol. des êtres marins, Inst. Océanographique. 195, rue St.-Jacques, (5ème). Dir. Prof. Portier.
Labor. de Physiol. compar., l'Ecole des Hautes Etudes. Dir. R. A. Legendre.

Pavia (Italia).
Ist. di Anat. e Fisiol. compar. Confer: Anat. humana experim.

Riga (Latvija).
Inst. f. Tierphysiol. u. Tierzucht. Dir. Prof. A. Buschmann.

Saratow (U.d.S.S.R.).
Tierphysiol. Labor. am Inst. f. Landw. u. Melioration. Dir. Prof. G. A. Čuevskij.

Stambul (Türkei).
Labor. de Physiol. Ecole vétérin. Dir. Prof. Dr. Hussein Sabri. — Prof. Dr. Kémal Djenab. *4 places.* — *Outillage courant de Physiol.*

Strasbourg — Straßburg (France).
Labor. de Physiol. Générale, Fac. Sc. Dir. Prof. Terroine.

Taschkent (U.d.S.S.R.).
Inst. de la Physiol. de l'Univ. de l'Asie centrale, Rue Kaufmanskaja les édifices de Fac. de Méd. Dir. Prof. Dr. I. P. Michajlovskij. — A. A. Danilov; Prosekteur A. I. Wentschikow; G. A. Iwanowskij. *2 places. La vie durant des animaux et le relever à norme.* — *"Supra-centrifuge"*. Ⓟ Artificiellement chimiques "mumiae" (les momies) des variables animaux (avec la conservation des leurs toutes intestins).
Inst. f. Physiol. d. Haustiere. Dir. Prof. G. L. Radzivilovskij.

Tiflis, Georgien (U.d.S.S.R.).
Physiol. Labor. der pädagogischen Fak. der Staatsuniv., Wake. Dir. Prof. Ivane Beritoff (Beritaschwili). — Dr. Georg Watzadse, Ass.; Alexandr Bregadse; Wladimir Sulaquelidse; Alexandr Goziridse; Schalwa Thopuria. *1 Platz am Saitengalvanometer, 2 für Nerven- und Muskelphysiol. an Kaltblütern, 1 für dasselbe an Warmblütern und für individuell erworbene (bedingte) Reflexe. — Saitengalvanometer (großes Modell) von Einthoven; alle Apparate für myographische Untersuchung an Kaltblütern und an Warmblütern; spezielle Vorrichtung für d. Arbeit mit individuell erworbenen (bedingten) Reflexen.*

Tôkyô (Japan).
Labor. of Physiol. (of silk worms) Tôkyô Imperial Sericicult. Coll., Nishigahara.
Physiol. Labor. (of Insect) Imperial Sericicult. Experim. Sta., Suginami-chô near Tôkyô.

Toulouse (France).
Labor. de la Physiol., Ecole nation. Vet. Dir. J. Lafon.

Trieste (Italia).
Sezione marina del Labor. di Fisiol. di Torino, Lazzaretto San Bartolomeo, Muggia. Dir. Prof. Amedeo Herlitzka. *Ca. 10 posti per ricerche di fisiol. umana in rapporto al mare (bagni, clima, nuoto, sports nautici, irradiazione, palombari, sottomarini).*
— *Gli apparecchi vengono messi a disposizione secondo le necessità sperimentali, secondo le ricerche degli studioti, venendo in gran parte portati dall'Ist. di Torino.*

Utrecht (Holland).
Abt. für vergl. Physiol., Zool. Labor. Utrecht, Janskerkhof 3. Dir. Prof. Dr. Hermann J. Jordan. — Dr. H. J. Vonk jr., Haupt-Ass.; N. Postma, Ass.; H. P. Wolvekamp, Ass. *4—6 Arbeitsplätze für Gäste. Vergl. Physiol. der Verdauung, der Atmung, des Blutes und hauptsächlich des Nervenmuskelsystems. — Verschiedene Respirometer, mehrere Blutgaspumpen und Gasanalyseapparate, 16 Kymographien, App. zum Studium des Tonus glatter Muskeln. App. zur quantitativen Enzymforschung u. pH-Bestimmung. Saitengalvanometer. Spezielle Arbeitsräume für physik. u. chem. Physiol.*
Labor. of veterin. physiol. and physiol. chemistry, Kliniek voor kleine Huisdieren, Einde Alex Numankade. Dir.: vacat. — Ass.: Dr. R. Toman.

Vladikavkaz (U.d.S.S.R.).
Labor. f. Tierphysiol. Landwirtsch. Inst. Dir. Prof. N. Rjazancev.

Warszawa (Polska).
Labor. d. allgem. Physiol. a. d. freien Univ. Polens. Dir. Prof. Dr. Feodor Vieweger. *Protistophysiol., métabolisme du système nerveux.*
Inst. für Tierphysiol. der Hochsch. f. Land- u. Forstwirtsch. Dir. Prof. J. Sosnowski.

Wien (Österreich).
Inst. f. allgem. u. vergl. Physiol. Dir. Prof. A. Kreidl. — Prof. Dr. Nirenstein.
Labor. f. Tierphysiol. Hochsch. f. Bodenkultur. Dir. Prof. Dr. R. Stigler.

Laboratoria physiologica specialia.
1. Digestio et nutritio.
Confer: *Nutrimenta et Fermenta* (pag. 443).

Battle Creek, Mich. (U.S.A.).
Dept. of Nutrition Research at Battle Creek Coll. Dir. Prof. H. S. Mitchel.

Berkeley, Cal. (U.S.A.).
Dept. of Nutrition, Coll. of Agric. Dir. M. E. Jaffa, Head of Div. — Hilda Faust, M.A., Specialist; Harold Goss, B.S., Instr. Nutr.

Cambridge (England).
Inst. of Animal Nutrition School of Agric. Cambridge. Dir. of Physiol. Section: F. H. K. Marshall. —S. Hammond; S. A. Asdell; H. G. Sanders. *3 places. Researchs on farm and labor. animals.* — The Field Labor. and the Univ. Farm.

Dresden-Weißer Hirsch (Deutschland).
Labor. f. physiol. Chem. u. Ernährungsforschung an Dr. Lahmanns Sanatorium. Confer: Biochemia generalis.

Göttingen (Deutschland).
Inst. f. Tierernährungslehre Univ. Dir. Prof. F. Lehmann.

Moskau (U.d.S.S.R.).
Inst. für Ernährungsphysiol. des Volksgesundheitskommissariats, Ssiwtzew-Wrajex 41. Dir. Prof. Dr.

M. N. Schaternikoff. — Dr. D. I. Romascheff, Vize-Dir.; Dr. W. M. Rodionoff, Abt.-Vorst.; Dr. O. P. Moltschanova, Abt.-Vorst.; Dr. B. I. Ilyn-Kaknef; Dr. S.W. Sorin; Dr. N. P. Rjabuschinsky; Dr. L. W. Rodina; Dr. N. S. Schepilenskaja; Dr. N. S. Jarussowa; Dr. S. W. Sschowa; S. W. Matyko. *4 Plätze für Respirations- und Stoffwechseluntersuchungen. — Respirationsapparate.*

New York (U.S.A.).
Food Research Labor., 39 W. 38th St. Harold Levine (Vitamins).

Palo Alto, Cal. (U.S.A.).
Food Research Inst. of Agric.

Paris (France).
Station centrale de Recherches sur l'Alimentation Labor. à l'Inst. des Recherches agronomiques. Dir. Prof. Dr. Jean Maurice Javillier. *Les catalysateurs biochim. (Diastases, Vitamines); les éléments catalytiques (Zinc, manganèse); métabolisme du phosphor.*

Puławy (Polska).
Abt. für die Fütterung der Tiere des landwirtschaftl. Inst. Dir. Dr. Henryk Malarski. — Dr. Zenon Wierzchowski, Adjunkt (physiol.-chem. Arbeiten speziell über Vitamine); Tadeusz Wyszyński (analytisch-chem. Arbeiten). *2—3 Plätze besonders für die chem. Arbeiten ausgerüstet. — Mikroanalyse.* ℗ Die polnische Rasse der Hühner sog. „Zielononózki".

Quincy, Ill. (U.S.A.).
Moorman Experim. Station, Moorman MF.G. Co. Dir. A. R. Lamb. *Animal nutrition and fermentation.*

Stanford University, Cal. (U.S.A.).
Food Research Inst. Dir.: Carl L. Alsberg, A.M., M.D.; Joseph S. Davis, Ph.D.; Alonzo E. Taylor, M.D., LL.D. — Economists: Holbrook Working, Ph.D., Louis B. Zapoleon; Research Assoc.: Harold Hotelling, Ph.D., Katherine Snodgrass, A.M.; Junior Research Assoc.: Merrill Bennett, A.M., Robert D. Calkins, B.S., Elizabeth P.Griffing, A.M.,

Adelaide Hobe, B.S., Margaret Milliken, B.S.
Economic and technol. research in production, distribution and consumption of food.

Stormont, Belfast (Ireland).
Div. of chem. and anim. nutrition. Confer: Biochem. gener.

Tôkyô (Japan).
Diet Research Inst. Coll. of Med.Keiô-Gijuku-Univ. Suzuki's Labor. Inst. of Physical and Chem. Res. (Kikwayaku Kenkyûsho). Dir. Dr. U. Suzuki. *Vitamine.*
Imper. Nutrition Research Inst. Dir. Dr. Saeki.

Wageningen (Holland).
Rijkslandbouwproefstation voor Veevoederonderzoek, Diuvendaal 10. Dir. Dr. B. Rae Bruyn. — C. J. Kole; Mejuffrouw W. M. van der Myll Dekker.

Washington, D.C. (U.S.A.).
Bureau of Foods Sanitation and Health of Good Housekeeping Magazine, Cosmos Club. Dir. H. W. Wiley. *Foods, Growth, Nutrition, Hygiena.*
Nutrition Labor. of the Carnegie Inst. of Washington, 29 Vila Street. Dir. Dr. Francis G. Benedict. — Dr. Thorne M. Carpenter, Physiol. Chem. *Labor. workers in respiration experim. on animals and humans; gas analysts; chemists; physicist.*

Wien (Österreich).
Ernährungsphysiol. Labor. (Josephinum). Dir. László Berczeller.

Zürich (Schweiz).
Inst. für Haustierernährung an der Eidg. Techn. Hochsch., Universitätsstr. 2. Dir. an. Prof. Dr. Georg Wiegner. — Dr. Edgar Crasemann, Ass. für chem. Arbeiten; Dr. Max Kleiber, Ass. für Respirationsversuche und Gasanalyse. *3—4 Plätze für Verdauungsversuche an kleinen und großen Tieren, für Respirationsversuche und Kolloidchem. — Verdauungskästen für kleine und große Tiere. Geflügelhof. Respirationsapparate.* Zweiglabor. ist das Agrikulturchem. Labor. der Eidg. Techn. Hochsch., das unter dem gleichen Dir. steht.

2. Physiologia exercitationis corporis.

Berlin (Deutschland).
Sportphysiol. Labor. Deutsche Hochsch. f. Leibesübungen, Deutsches Stadion. Dir. Dr. Kohlrausch.
Kaiser-Wilhelm-Inst. f. Arbeitsphysiol. Confer: Physiol. gen. fac. med.

Boulogne-sur-Seine (France).
Stat. physiol. du Coll. de France, Parc du Prince, Bois de Boulogne. Dir. Prof. E. Gley. — Dir.-adjoint: A. Pizard; Préparateur: F. Caridwil. *2 places pour recherches expérim. sur la morphogenie (influence des sécretions internes).*

Charkow, Ukraine (U.d.S.S.R.).
Arbeitsphysiol. Labor. d. Psycho-Neurol. Inst., Karl-Liebkbecht-Str. 4. Dir. N. Kudrjawzew. — Dr. med. Worobjeff; Dr. med. Gurjeew; Dr. med. Wjalkowa; Dr. med. Feldman; Dr. med. Djenisenko; Cand. med. Semernina. *Für phys. Untersuchung über Nervenermüdung und das Zentralnervensystem.*

København (Danmark).
Turntheoretisches Labor. d. Univ. (Univ. gymnastikteoretiske Labor.), K, Studiestraede 6. Dir. Prof. Dr. med. J. Lindhard.

London (England).
Inst. the Industrial Fatigue Research Board, W. C. 2, 15 York Buildings. The work of the Board is directed by a series of Committees. Secretary: D. R. Wilson, M.A. — The Senior Investigators are as follows: H. M. Vernon, M.D.; E. Farmer, M.A.; S. Wyatt, M.Sc.; May Smith, M.A.; H. C. Weston, M.J. Inst.E. There are in addition about 10 Ass. Investigators, as well as several research workers. *The Board is a Government body instituted to carry out investigations dealing with the human factor in industry. It has no laboratory of its own, but gives grants for researches in the labor. of different univ. in the country.*
Labour Research Dept., 162, Buckingham Palace Road, Westminster S.W.1.

Ôsaka (Japan).
Inst. of Physiol. of Labour, Ôsaka Prov. Med. Univ. Dir. Prof. Yasunaga Masai. — Prof. Y. Masai.

3. Physiologia humana, altitudine montium impressa.

Alagna Sesia, Provincia di Vercelli (Italia).
Ist. Angelo Mosso sul Coll. d'Olen (2900 m sul livello del mare). Non ha personale proprio, ma vi prestano servizio ass. del Labor. di Fisiol. di Torino. *25 posti. Ricerche di fisiol. normale e patol., di zool., botan., igiene e fisica terrestre in alta montagna. —* 2 stanze alla Capanna Regina Margherita sul Monte Rosa (4565 m sul mare).

Arosa (Schweiz).
Labor. der Bündner Heilstätte. Dr. W. Knoll. *Haematol. des Höhenklimas, Botan., Zool.*

Davos (Schweiz).
Schweizerisches Inst. für Hochgebirgsphysiol. und Tuberkuloseforschung. Dir. Prof. Dr. A. Loewy. *10 Plätze für physiol., chem., histol.; bact. Unters.*

4. Psychologia experimentalis.

Confer: *Neurologia* (pag. 458), *Anatomia cerebri* (pag. 431), *Physiologia generalis* (pag. 444).

Berlin (Deutschland).
Psychotechn. Inst. Leitung: Doc. Dr. Rob. Werner Schulte, Spandau, Schönwalder Allee 62.
Psychotechn. Hauptprüfstelle f. Sport u. Berufskunde, Oberwallstr.
Psychol. Inst. der Preuß. Hochsch. für Leibesübungen, Spandau, Radelandstr. Forschungsstelle für Turn- und Sportwissenschaft.
Sportpsychol. Labor. der Deutschen Hochsch. für Leibesübungen, B.-Charlottenburg, Deutsches Stadion, Hochschulgebäude. Forschungsstelle für die psychol. Sportwissenschaft.
Psychol. Abt. der Arbeitsstätte für Menschheitskunde. Dir. Prof. Dr. med. Friedenthal, Univ. Berlin.
Inst. f. angewandte Psychol., SW 68, Schützenstraße 26. Dir. Dr. Otto Lißmann.
Psychol. Inst. Univ. Dir. Prof. W. Köhler. — Prof. v. Hornbostel (Phonogramm-Archiv).

Borissovka, Kursk gov. (U.d.S.S.R.).
Zoopsychol. experim. Station (the branch of P. F. Lesshaft Inst. of Sc., Leningrad), Sloboda Borissevka. Dir. Sergius I. Malyshev. — Leon E. Arens, the substitute of director. 1 prosector. *1 or 2 places for research of the behavior of bees and wasps.* ℗ The bees (Apoidea) and wasps and their nests. — By the Station there is an forest prohibited as the movement of nature ,,The Wood on Worsklie''.

Braunschweig (Deutschland).
Inst. f. Philosophie, Pädagogik u. Psychol. Dir. Prof. Moog. — Dr. Hernig, Abt.-Vorst. d. psychol.-psychotechn. Abt.

Brisbane, Queensland (Australia).
Dept. of Psychol. Univ. Dir. Prof. J. P. Lowson.

Bruxelles — Brussel (Belgique).
Labor. de Psycho-Physiol. appliquée. Dir. Prof. Dr. A. Ley.

Cambridge (England).
Dept. of experim. Psychol. Univ. Dir. Prof. F. C. Bartlett.

Cluj — Klausenburg (România).
Labor. f. experim. u. vergl. Psychol., Univ. Dir. Fl. Stefanescu-Goangǎ.

Coïmbra (Portugal).
Inst. de Psicol. experim., Univ. Dir. J. de Carvalho.

Danzig (Danzig).
Psychol. Inst. u. Sammlung, Techn. Hochsch. Dir. Prof. Dr. Henning.

Darmstadt (Deutschland).
Psychotechn. Inst. d. Techn. Hochsch. Dir. Dr. Bramesfeld.

Dijon, Côte-d'or (France).
Labor. de psychol. Dir. Jean Lépine.

Frankfurt a. M. (Deutschland).
Psychol. Inst. Univ. Dir. Prof. Schumann.

Firenze, Toscana (Italia).
Ist. di Psicol. sperim., Univ. Dir. Prof. E. Bonaventura.

Gent — Gand (Belgie).
Labor. voor experim. Zielkunde. Dir. Prof. Dr. v. Biervliet.

Graz (Österreich).
Psychol. Labor., Phil. Fac. Univ. Interim. Leiter: Prof. Dr. Ernst Mally.

Greencastle, Ind. (U.S.A.).
Biol. Dept. of De Pauw Univ. Dir. Prof. W. N. Hess. *Physiol., Animal Behavior.*

Groningen (Holland).
Psychol. Inst. Dir. Prof. Dr. G. Heijmans.

Hamburg (Deutschland).
Psychol. Labor. der Staatskrankenanstalt, Univ., Friedrichsberg.
Psychol. Seminar u. Labor., Domstr. 9. Dir. Prof. Dr. W. Stern. — 3 wiss. Hilfsarbeiter. Dem Labor. angegliedert ist eine Abt. f. prakt. Psychol., die von der ,,Gesellschaft zur Förderung der praktischen Psychol., E.V.'' unterhalten wird. Wiss. Hilfsarbeiter: Dr. Wunderlich.

Jena (Deutschland).
Psychol. Anst. der Univ. Dir. Prof. Dr. Wilhelm Peters. *Grenzfragen der Psychol. u. Biol.: Vererbung psychol. Eigenschaften, psychol. Entwicklung, psychol. Konstitution. Probleme der experim. Psychol.: Gedächtnis, Intelligenz, Gefühlsleben, Entwickelung der Leistungsfähigkeit des Kindes.*

Kaunas (Litauen).
Labor. für Experimentalpsychol. Univ. Dir. Doc. A. Gylys.

Kiel (Deutschland).
Psychol. Inst. Univ. Phil. Fac. Dir. Prof. Wittmann.

Kiew, Ukraine (U.d.S.S.R.).
Psychol. Labor. des Inst. f. Volksbildung (I. N. O). Dir. Prof. A. Ra'evski.

København (Danmark).
Psykol. Labor., Studiestr. 6. Dir. Prof. Dr. E. Rubin. — Lect. Dr. R. H. Petersen.

Köln a. Rh. (Deutschland).
Psychol. Inst. Univ. Philos. Fac. Dir. Prof. Lindworsky.

Lawrence, Kan. (U.S.A.).
Dept. of Psychol. of the Univ. of Kansas. Dir. Prof. W. S. Hunter. — Assoc. Prof. C. Rosenow.

Leipzig (Deutschland).
Inst. für experim. Pädagogik und Psychol. des Leipziger Lehrervereins, Kramerstr. 4. Dir. Doc. Max Döring. — Felix Schlotte, wiss. Ass. *Experim. Untersuchung pädagog. u. psychol. Fragen. — Psychol. und anthropometrische Apparate zur Untersuchung von Kindern.*
Staatl. Forschungsinst. für Psychol., Universitätsstraße 7—9. Dir. Prof. F. Krueger.
Inst. f. experim. Psychol. d. Univ. Dir. Prof. Krueger.

Lexington, Ky. (U.S.A.).
Dept. of Psychol. of the Univ. Dir. Prof. J. B. Miner. — Prof. P. L. Boynton.

Lincoln, Neb. (U.S.A.).
Dept. of Psychol. of the Univ. Dir. Prof. W. F. Hyde.

London (England).
National Inst. of Industrial Psychol., 329, High Holborn, W.C. 1. Dir. Chas S. Myers, F.R.S. — Pres.: The Earl of Balfour.

Los Angeles, Cal. (U.S.A.).
Dept. of Psychol. of the Univ. of Southern California. Dir. Prof. J. W. Todd. — Prof. K. T. Waugh.

Lublin (Polska).
Labor. für Psychol. u. Biol. u. Seminarium für Naturpsychol., Human. Fac. d. Univ. Lubelski. Dir. Prof. B. Rutkiewicz.

Lund (Sverige).
Psychol. Inst. d. Univ. Dir. Prof. Herrlin.

Lwów (Polska).
Inst. für experim. Psychol., Ukrain. Ševčenko-Gesellsch. der Wiss. Vorst.: Dr. Stephan Bałej.
Psychol. Inst. d. Univ. Dir. K. Twardowski.

Mainz (Deutschland).
Psychol. Labor. Staatl. Hess. Pädag. Inst., Petersstraße 2.

Mannheim (Deutschland).
Inst. für Psychol. und Pädagogik der Handels-Hochsch. Dir. o. Prof. Dr. Otto Selz. — Dr. Eduard Meyer, Ass.; Hauptlehrer Lämmermann, Psychol. Schulberater der Stadt Mannheim; Diplom-Kaufm. Hall, Psychol. Berater der Handelssch. Mannheim. *10 Plätze für Untersuchungen auf dem Gebiet der allgemeinen, der pädagog. Psychol. u. d. Psychotechnik.*

Milano (Italia).
Ist. di Psicol. sperim. Fac. med. R. Univ. Dir. Prof. Doniselli.
Labor. di psicol. sperim. Univ. catt. del S. Cuore. Dir. Prof. A. Gemelli.

Montreal (Canada).
McGill Psychol. Labor. Dir. Prof. William D. Tait. — 1 Assoc. Prof. Educational and Compar. Psychol. 2 Ass. for general duties. *10 Research Rooms. Any sphere of Experim. Psychol. — Workshop to make special apparatus. Sound-proof Room; Dark Room; interconnection between rooms.*

Moskau (U.d.S.S.R.).
Mosk. psycho-neurol. Inst., Sadovaja-Kudrinskaja 1. Dir. A. P. Nečaer. — Vizedir.: V. A. Giliarovskij, V. V. Kramer.

München (Deutschland).
Psychol. Inst. Univ. Dir.: Profs. Becher u. Geyser.

Münster i. W. (Deutschland).
Inst. für prakt. Psychol., Landeshaus. Dir. Dr. Jos. Weber. — 1 Ass. *Begabungsforschung, Intelligenzprüfungen, persönliche u. berufliche Eignungsfeststellung, psychotechn. Untersuchungen. — Psychol. u. psychotechnisches Prüf- u. Untersuchungs-Labor., etwa 40 Apparate.*

Nikolaev, Ukraine (U.d.S.S.R.).
Inst. f. Psychol. Inst. f. Volksbildung. Dir. Prof. V. Fidrovskij.

New Haven, Conn. (U.S.A.).
Inst. of Psychol., Yale Univ. Prof. R. M. Yerkes.

New York City (U.S.A.).
Division of Psychol., Inst. of Educational Research, Teachers Coll., Columbia Univ. Dir. Prof. E. L. Thorndike.
Dept. of Psychol. Columbia Univ. C. J. Warden, Instr. in Compar. psychol.

Odessa (U.d.S.S.R.).
Psychol. Labor. Pädag. Hochsch. Dir. Prof. Sevalev.

Oslo (Norge).
Psychol. Inst. Univ. Dir.: Prof. Aall u. Prof. Schjelderup.

Padova (Italia).
Labor. di Psicol. R. Univ. Dir. Prof. V. Benussi.

Paris (France).
Labor. de la Psychol. et Physiol. des Sensations, Ecole pr. des Hautes Etudes., Sorbonne, 6, Rue St. Jacques. Dir. Prof. Henri Piéron. — I. Meyerson, Dir.-adjoint; Alfred Fessard, Chef des travaux; Marcel François, Ass.-préparateur. *5 places.*
Labor. Psychol. expérim. Ecole pr. des Hautes Etudes. Dir. Prof. P. Toulouse.
Inst. de Psychol. Fac. des Lettres Univ., 46, Rue St. Jacques. Sections: 1. Psychol. (Psychol. générale, P. pathol., P. expérim. et comparée, P. physiol., P. zool.). 2. Pédagogie. 3. Psychol. appliquée (Application au travail et à l'industrie; Sélection et orientation profess.). Prés. L. Lapique. Secr.: J. Meyerson, Dir.-adj. du Labor. de Psychol. à la Sorbonne.
Ecole de Psychol., 49, Rue Saint-André d. A.

Pécs (Ungarn).
Inst. f. Paedagogik u. psychotechn. Labor. Dir. Prof. E. Weszely.

Peking (China).
Psychol. Inst. Tsing Hua Univ. Dir. Prof. C. H. Chuang.

Perm (U.d.S.S.R.).
Kabinett für Psychotechnik Univ. Dir. Prof. I. A. Syrkov.

Poznań — Posen (Polska).
Psychol. Inst. Dir. Prof. Stefan Borowiecki.

Princeton, N.J. (U.S.A.).
Princeton Psychol. Labor., Eno Hall. Dir. Herber Sidney Langfeld, Ph.D. — Howard C. Warren, Ph.D.[t] Chairman and Prof.; Henry C. McComas, Ph.D.., Assoc. Prof.; Carl C. Brigham, Ph.D., Assoc. Prof.; Leonard Carmichael, Ph.D., Ass. Prof.; Wilbur Hulin, Ph.D., Instr.; Henry L. Eno, A.B., H. B., Research assoc.; Henry A. Cotton, A.M., M. D., Lect. *15 places. Sound-proof room, Dark rooms, well lighted rooms. — Almost all of the modern instruments for research in experimental psychol. and a fund for purchase of new instruments, when required.*

Rio de Janeiro (Brasil).
Assistencia a Alienados do Rio de Janeiro (Psychiatrisches Inst.). General-Dir.: Prof. Dr. Juliano Moreira.
Hospital Nacional de aliendos. Labor. f. pathol. Anat., biol. Chem., Bact., experim. Psychol.

Roma (Italia).
Inst. f. experim. Psychol. Dir. Prof. Sante de Sanctis.

Rosario (Argentina).
Inst. f. experim. Psychol. Dir. Prof. A. Mo.

Rostock i. M. (Deutschland).
Psychol. Inst. Dir. Prof. Katz.

Rostow a. Don (U.d.S.S.R.).
Kabinett f. experim. Psychol. Dir. Prof. I. I. Jagodinsky.

São Paulo (Brasil).
Dept. of Psycho-Technique.

Saratow (U.d.S.S.R.).
Psychol. Labor. der Staatsuniv. Dir. Prof. August Krogius. — Dr. Georg Iwanow; Frau Wera Worms-

Iwanow; Frau Marie Pustowoitow. *5 Plätze für psychol. und psychotechnische Unters. — 2 Chronoskope, 1 Tachistoskop, Müllers Gedächtnisapparat, 2 Ergographe, Dynamometer, Farbenmischapparat, Kymographion, Kardiograph, Pneumograph, Sphygmograph, Terbeux' Wage, Spirometer.* ℗ Psychol. und psychotechnische Apparate und Tests.

Sigmaringen (Deutschland).
Labor. f. experim. Psychol. d. philosoph.-theol. Lehranst., Ordinis Fratrum Minorum. Dir. Prof. D. P. Scheller.

Stanford Univ., Cal. (U.S.A.).
Psychol. Labor. Prof. W. R. Miles.

Stellenbosch (South Africa).
Dept. of Psychol. Univ. Dir. Prof. R. W. Wilcocks, B.A., Ph.D.

Stuttgart (Deutschland).
Psychol. Inst. Techn. Hochsch. Dir. Doc. F. Giese.
Inst. für Persönlichkeitsforschung, Kanonenweg 26. Dir. Dr. Römer.

Tiflis (U.d.S.S.R.).
Kabinett f. experim. Psychol. Dir. Prof. Usnadze.

Torino (Italia).
Inst. f. experim. Psychol. Dir. Prof. F. Kiesow.

Utrecht (Holland).
Psychol. Labor., Wittevrouwenstraat 9. Dir. Prof. Dr. F. M. J. A. Roels.
Bureau für Berufskunde, Wittevrouwensingel 89. Dir. Jhr. D. I. van Lennep. *Eigenungsprüfungen.*

Waco, Tex. (U.S.A.).
Psychol. Inst. Baylor Univ. Dir. Prof. Dr. A. J. Hall.

Warszawa (Polska).
Inst. f. experim. Psychol. Dir. Prof. W. Witwicki.
Psychol. Labor. d. freien Univ. Dir. Prof. Dr. Segał.
Labor. de Biol. générale à l'Inst. Neuchi, 8, Rue Sniadechich. Dir. Romuald Minkiewicz. — J. Dembowski; Mme W. S. Dembowska; Z. Czernieuwski. *5 places pour Zoopsychol. aquatique, micromorphol., influence de la lumière colorée. — Etuves à régulation différente. Etuves en verre transparent à chauffage et réglage électrique. Microchirurgie. Monochromateur Jobin-Yvon à graduation en spirale. Installation pour éclairage d'en bas: rayon en verre poli et grand miroir à inclinaison de 45°. Filtres chromatiques et catalogue des couleurs.*

Wien (Österreich).
Psychol. Inst. Univ. Dir. Prof. Dr. Bühler.
Parapsych. Inst., XVIII, Gentzg. 132. Dir. R. R. M. Tartaruga. — Präs.: Univ. Prof. K. C. Schneider (Bio- u. Zool.); Vizepräs.: Prof. R. Schmid (Phys., Chem.).

Woronesch (U.d.S.S.R.).
Psychol. Kabinett, Univ. Dir. Prof. P. L. Zagorskij.

Zagreb — Agram (S.H.S.).
Inst. f. Psychol. (Kgl. pädag. Schule).

Zürich (Schweiz).
Psychol. Inst. fac. phil. d. Univ. Dir. Prof. Dr. G. Lips.

5. Neurologia physiologica.

Confer: *Anatomia cerebri et nervorum* (pag. 431), *Psychologia experim.* (pag. 456), *Physiologia generalis* (pag. 444) *et comparata* (pag. 452).

Amsterdam (Holland).
Neurol. Labor. Dir. Prof. Dr. B. Brouwer.

Battle Creek, Mich. (U.S.A.).
Pavlov Physiol. Inst. Dir. V. N. Boldyreff.

Berlin (Deutschland).
Kais.-Wilh.-Inst. f. Hirnforschung. Dir. Prof. O. Vogt. — Prof. M. Bielschowsky; Dr. C. Vogt.

Frankfurt a. M. (Deutschland).
Neurol. Inst. Univ. Dir. Prof. Goldstein.

Leningrad (U.d.S.S.R.).
Physiol. Inst. of the Russian Akademy of Sc., Wassiljewsky Ostrow, Tuchkowa nab., 2A. Dir. Ivan Petrovich Pavlov. — N. A. Podcopaev, Senior Physiol.; I. R. Prorocov, Physiol.; A. P. Selesnev, Ingeneer-physic; A. M. Pawlowna; W. J. Pawlowna; M. K. Petrowa; W. W. Rikman. *7—8 places for the researches in the physiol. of the central nervous system by the method of conditioned reflexes.*

Lisbôa (Portugal).
Inst. de Neurol., Fac. de Med. Univ. Dir. Prof. A. C. de Abren Freire Egas Moniz.

Ljubljana — Laibach (S.H.S.).
Neurol. Labor. d. Med. Fac. Univ. Dir. Prof. A. Šerko.

London (England).
Dept. of Neurol., St. Thomas's Hospital, Med. School of the Univ., Albert Embankment, Lambeth, S.E.
Dept. of Neurol., Middlesex Hosp. Med. School of the Univ., Mortimer Str., Oxford Str., 10. Dir. D. McAlpine.

San Francisco, Cal. (U.S.A.).
Dept. of Neurol., Med. School of the Univ. of California. Dir. Prof. M. B. Lennon.

Utrecht (Holland).
Pharmacol. Labor. Confer: Pharmacol.

6. Endocrinologia (secretio interna).

Confer: *Institutiones biologicae generales* (pag. 336).

Frankfurt a. M. (Deutschland).
Biolog. Forschungsinst. Dir. Prof. Dr. F. Blum. — 1 med., 1 chem. Ass. *Erforschung endokriner Organe.*

Hellerup (Danmark).
Insulin-Labor., Onsgaardsvej 12. Dir. Dr. med. H. C. Hagedorn; M. sc. A. Hemmingen. K. Erik Jensen. *Apparatus for production and investigation of hormones. Research Labor.*

Moskau (U.d.S.S.R.).
Labor. für experim. Biol. an der Kommun. J. M. Swerdlov-Univ., Miusskaja Platz 3 „55". Dir. B. M. Zawadovsky. — G. J. Asimoff (Morphogenetische Bedeut. d. Schilddrüse u. bedingte Reflexe); A. G. Kratinoff (Hungerproblem und motorische Tätigkeit d. Verdauungstraktus i. Zusammenhange m. d. Funktionen der endokrinen Organe); A. A. Titaeff (Biochem. d. Schilddrüse); A. L. Sack (Bedingte

Reflexe, Schilddr. u. bedingte Reflexe, Bedingte Reflexe auf pharmakol. Basis); V. R. Zacharov (Schilddr. u. äußere Sekretion d. Verdauungsdrüsen, Schilddr. u. bedingte Refl. beim Hunde); M. N. Lapiner (Biochem. des Schilddrüsenhormons); Aspiranten: P. J. Dobrovitzky (Histol. innersekretorischer Drüsen); M. A. Novikova (Experim. Physiol. der Schilddrüse); T. P. Rolitsch (Histol. u. vergl. Anat. der Schilddrüse); M. L. Rochlina (Die morphogenetische Bedeutung der Schilddrüse u. die bedingten Reflexe bei Hühnern); M. S. Slotov (Schilddrüse u. bedingte Reflexe, Magensekretion); Volontäre Mitarbeiter: 1. I. Agoll (Problem der äußerl. u. innerl. Faktoren der Morphogenese und der Vererbung erworbener Merkmale); N. A. Vvedensky (Schild- u. Verdauungsdrüsen, Physikal.-chem. Charakteristik der physiol. Säfte); N. V. Kirillowa (Morphogenetische Bedeutung der Schild- u. Keimdrüsen); A. S. Lieberfarb (Physiol. der Schilddrüse u. des Ovariums); V. N. Slepkoff (Problem der äußerl. u. innerl. Faktoren der Morphogenese u. der Vererbung erworbener Merkmale). — Im Gange der Arbeit geformte Abt.: 1. Morphogenetische, 2. Experim.-physiol., 3. Biochem., 4. Reflexol. Abt. An der Akademie der Kommunist. Erziehung ist noch ein selbständiges physiol. Labor. als Filiale.

Inst. f. experim. Endokrinol. d. Volkskommissariats f. Gesundheitswesen, Voroncovo Pole, Bols. Nikolo-Vowbinskij per 14. Dir. Prof. V. D. Servinskij. — 38 wiss. Mitarb.

Laboratoria anatomiae pathologicae.

Anatomia pathologica humana experim. (fac. med.).

Confer: *Bacteriologia hominis* (pag. 479), *Histologia* (pag. 375), *Physiologia pathologica* (pag. 469).

Adelaide (South Australia).
Dept. of Pathol. Univ., North Tenace. Dir. Prof. J. B. Cleland, M.D. — Dr. Helen Mayo (Vaccine Therapy); Dr. T. Grant, Hon. Ass. (Pathol.); 9 Ass. for Labor. and Mus. *2 places for Pathol. investigations.*

Akron, Ohio (U. S. A.).
Pathol. Labor. Peoples Hospital. Dir. F. Potter.

Alger (Algèrie).
Labor. d'Anat. Pathol., Fac. mixte de Méd. et de Pharmacie de l'Univ., Dir. Prof. Dr. G. Poujol. — Dr. J. Montpellier, chef des trav. pratiques; Mlle. Lemaire, Dr. prép. *Le labor. peut recevoir un ou deux travailleurs désireux de se perfectionner dans l'étude générale de l'anat. pathol.* ⓟ Un certain nombre de blocs de paraffine, provenant des pièces étudiées sont conservés. Une importante collection le diapositifs destinés à être projetés pour l'illustration des cours a été constituée.

Amsterdam (Holland).
Ziektekundig-Ontleedkundig labor., Binnengasthuis, Grimburgwal 10. Dir. Prof. W. M. de Vries. **Labor. f. Pathol. u. Pharmacodynamie.** Dir. Prof. L. Snapper.

Ann Arbor, Mich. (U. S. A.).
Pathol. Labor. Med. Dept. Dir. A. S. Warthin.

Athènes (Grèce).
Inst. d'Anat. pathol. de l'Univ., rue de Marseille. Dir. Prof. C. Melissinos. — B. Photacis.

Bahia (Brasil).
Labor. d'Anat. pathol. Fac. méd. Dir. Prof. M. Andréa.

Bari (Italia).
Gab. di Anat. patol. Univ. Dir. Ugo Soli.

Basel (Schweiz).
Pathol. Anstalt Univ., Hebelstr. 24. Dir. Prof. Dr. Robert Rößle. — 1 Prosektor, 3 Ass. *6 Arbeitsplätze.*

Beirut (Syria).
Dept. of Pathol., Bact., Hyg., Parasit., American Univ. Dir. L. W. Parr.

Belfast (Ireland).
Dept. of Pathol. anat. Queens Univ. Dir. Prof. W. St. Clair Symmers.

Beograd (S. H. S.).
Inst. für allgemeine Pathol. u. pathol. Anat., Med. Fac. d. Univ., Višegradska ulica. Dir. Prof. Dr. S. Joannović. — Prof. Dr. S. Šahovič.

Bergen (Norge).
Dr. med. F. G. Gade's pathol. Inst. Dir. Dr. med. K. M. Haaland. — 2 Ärzte als Ass., z. Z.: cand. med. Frau Margit Haaland, Epidemiearzt des westlichen Norwegens; cand. med. Olaf Römcke, pathol. Ass. (II. Ass.). *Pathol. Anat., Bact., experim. Patho*

Berlin (Deutschland).
Pathol. Inst. u. Mus., NW. 6, Schumannstr. 20|21. Dir. Geh.R. Prof. Dr. O. Lubarsch. — Prof. Dr. Watjen, Prosektor; Prof. Dr. Bickel, Vorst. d. experim. biol. Abt.; Prof. Dr. Kuczinski, Vorst. d. parasit. u. vergl. pathol. Abt.; Prof. Dr. Rona, Vorst. d. chem. Abt.; Prof. Dr. Westenhöfer, Kustos des Mus. — Ass.: Dr. Plenge, Dr. Guillery, Dr. Borchard, Dr. de Biari, Dr. Bork, Dr. Wille, Dr. Wolff, Priv.Doc., Dr. Schwarz, Dr. van Eweyk, Dr. Mislowitzer, Priv.Doc., Dr. Kleinmann; 10 freiwillige Hilfsärzte. *50 Plätze für pathol. Anat. u. Histol., experim. Pathol., Parasit. u. Zellforschung, vergl. Pathol., pathol. u. Kolloidchem.* ⓟ Pathol. Anat. u. mikroskop. Praep., Spaltpilzkulturen, vergl. Pathol. Praep.
Pathol.-anat. u. bact. Abt. d. städt. Krankenhauses B.-Neukölln. Dir. Dr. Heinrich Wilhelm Ewald Ehlers. *Geschwulstlehre.*
Pathol.-anat. Abt., Rudolf-Virchow-Krankenhaus, N 65. Dir. Dr. Erwin Christeller. — 1 Ass. *Mehrere Arbeitsplätze für Sektionstechnik, histol. Technik, Spezialuntersuchungen.* ⓟ Histotopographische Organgefrierschnitte, Diapositive zur pathol. Anat. d. Histol., pathol.-histol. Praep.
Pathol. Inst. Augusta-Victoria-Krankenhaus. Dir. Prof. W. Steinbiß.
Pathol. Abt. Städt. Krankenhaus am Friedrichshain. Dir. Prof. L. Fick.
Pathol. Inst. des Krankenhauses Berlin-Westend. Dir. Prof. W. Koch.
Pathol.-anat. Abt. Krankenhaus Moabit, NW 40. Dir. Prof. Dr. Rudolf Jaffé. *Pathol. Anat., speziell Endokrines System, Lipoidstoffwechsel.*
Pathol.-anat. Labor. d. Reichsgesundheitsamtes, NW 40, Scharnhorststr. 65.

Bern (Schweiz).
Pathol. Inst. der Univ. Dir. Prof. Dr. C. Wegelin. — 3 Ass., 2 Volontärass. *3 Plätze für pathol. Histol.* ⓟ Anat. u. histol. Praep.

Birmingham (England).
Dept. of Pathol. Anat. Univ. Dir. Prof. G. Haswell Wilson. — C. W. Maguire; F. W. M. Lamb; E. Baylis Ash.
Pathol. Mus. Dir. C. W. Maguire. — H. H. Sampson; J. S. M. Connell.

Bologna (Italia).

Ist. Anat. patol., Univ. Dir. G. Martinotti.
Ist. di Patol. generale e Batteriol. Univ., Piazza di Porta S. Donato 3. Dir. Prof. Guido Tizzoni. — Prof. Giovanni de Angelis, Aiuto; Dott. Fráncesco Rezzesi, Ass.; Prof. Plinio Carlo Bardelli, Maggiore Vet., Ass. Onorario. *4 posti per ricerche microscopiche e 1 posto per ricerche batteriol.* ℗ Le collezioni di colture e di pezzi e prep. microscopici.

Bonn a. Rh. (Deutschland).

Pathol. Inst. Univ., Theaterstr. Dir. vakat. Stellvertr.: Prof. Dr. P. Prym. — Dr. A. Lauche, Priv.-Doc. u. Ass.; Dr. O. Schultz, Ass. *4—6 Plätze für pathol. Histol.*
Inst. f. gerichtl. u. soz. Med. der Univ., Theaterstraße 52. Dir. Prof. Dr. med. Müller-Heß. — Priv.-Doc. Dr. med. Hey; Dr. med. et phil. André; Dr. med. Wiethold; Frl. Dr. med. Nau; Frl. Dr. med. Ludwig. *6 Plätze für pathol.-anat., chem., optische Unters.* ℗ Giftpflanzen, pathol.-anat. Praep., gerichtsärztl. Praep.

Bordeaux (France).

Labor. d'Anat. pathol. et Microscopie chim., Fac. de Méd. place de la Victoire, Labor. annexe Hôpital St. André, Hôpital des Contagieux de Pellegrin-Bordeaux. Dir. Prof. Jean Sabrazès. — Dr. Léon Muratet, Prof. agrégé d'anat. pathol., chef des travaux; Henri Bonnin, Prof. agrégé, chef de labor.; Dr. Pauzat, chef de labor.; Blougtet, Interne des hôpitaux; Desmanchach, prép.; 1 prép. adjoint, 3 Ass. *4 places pour l'anat. pathol., Microscopie chim.* (*Hématol. surtout*).
Labor. de Pathol. et thérapeutique générale Fac. méd. Univ. Dir. Prof. Cruchef.

Bratislava — Preßburg (C. S. R.).

Inst. f. allg. u. experim. Pathol. d. Komensky-Univ. Staatl. Krankenhaus, Mickiewiczova 13. Dir. Prof. Dr. Milos Netonšek.
Inst. f. pathol. Anat., Histol. u.Bact. d. Komensky-Univ., Sasiukova 4. Dir. Prof. Dr. Franz Lukeš.

Bremen (Deutschland).

Pathol. Inst. der Krankenanstalt. Dir. Prof. K. A. R. Borrmann.

Breslau (Deutschland).

Pathol. Inst. Univ., Maxstr. 11. Dir. Prof. Dr. med. Friedrich Henke. — Prof. Dr. med. Ernst Mathias, 1. Ass.; Priv.Doc. Dr. med. Ernst Roesner, Dr. med. Martin Silberberg, planmäßige Ass.; Dr. med. Hans Bettinger, Dr. med. Hans Loewenstaedt, Dr. med. Gerhard Schrader, außerplanmäßige Ass. *3 Plätze für sämtl. Untersuchungen auf dem Gebiet der spez. pathol. Anat. und der allgemeinen Pathol. — Apparatur zur Zellzüchtung* (*Explantation*). ℗ Praep. aus dem Gebiet der pathol. Anat., makroskopisch und mikróskopisch. — Im Breslauer städtischen Krankenhaus „Allerheiligen Hospital" eine Prosektur mit dem zugehörigen histol. Labor.

Bristol (England).

Pathol. Dept. Univ. Dir. Prof. I. Walker Hall. — Dr. A. T. Todd; Dr. A. D. Fraser; Dr. G. Hadfield. *2 places for Pathol. and Bact.*

Brno — Brünn (C. S. R.).

Pathol.-anat. Inst. Univ., Pathologicko anat. ústav. Dir. Dr. Vádav Neumann. — Dr. Vádav Tomásek; Dr. Fr. Pavlica; Dr. I. Bouček. *5 Plätze f. Anat. pathol., bact., histol.*
Inst. f. pathol. Anat. u. Histol. d. Tierärztl. Hochschule, Prazska 67.

Bruxelles — Brussel (Belgique).

Labor. d'Anat. pathol. de l'Univ., 3, rue du Maelbeek. Dir. Prof A. P. Dustin. — Dr. R. Loy, Prof. agrégé, Chef. *6 places pour Canceriol. expérim.* — *Radiobiol. Toutes recherches de pathol. expérim., de neurol., d'Anat. pathol. — Installations très complètes de radiol. à grande puissance. Disposition de quantités importantes de matières radio-actives.* ℗ Pièces anat.-pathol. humaines et animales. Nombreuses tumeurs. Abondant matériel de thymus humains et animaux. — A l'hôpital Brugman (Bruxelles) le Centre radiobiol. et anticancereux de l'Univ.

București (România).

Labor. d'Anat. pathol. Fac. méd. Univ. Dir. Prof. V. Babes.

Budapest (Ungarn).

Allg. Pathol. Inst., Peter-Pázmany-Univ. Dir. Prof. H. Preisz. — Dr. F. Skrop; Dr. J. Tomcsik; J. Putnoky.
Pathol.-anat. Inst. No. 1, Univ. Dir. Prof. Dr. Koloman Buday. — Mitarbeiter: Dr. Béla Johann; Dr. Ernest Balogh; Dr. Josef Baló; Dr. Edmund Nachtnebel; Dr. Stefan Bézi; Dr. Edmund Zalka.
Pathol.-anat. Inst. No. 2, Univ. Mitarbeiter: Dr. Ludwig Puhr; Dr. Ladislaus Karoliny; Dr. Dionys Szüle; Dr. Andreas Korényi; Paul Hering.

Buenos Aires (Argentina).

Inst. de Anat. Pathol. Dir. Prof. J. Llambias.

Cagliari, Sardinia (Italia).

Ist. di Anat. patol. Dir. Bernardino Lunghetti.

Cambridge (England).

Dept. of Pathol. Dir. Prof. H. R. Dean. — Dr. G. S. Graham-Smith; Dr. L. Cobbett; Dr. E. G. Murray; Dr. R. A. Webb. *12 places for General Pathol., Serol., Morbid Histol., Bact.*
Bonnet Labor. of Pathol. Addenbrook's Hospital. Dir. Dr. John Joster Gaskell. *The Pathol. of Pneumonia. Pathol. of diseases of the Kidmay. Previous researches on the origin of the vascular sympathetic nervous systems. Malaria, blackwater fever, cystinaria, the action of X-rays on growing tissues. Cerebrospinal Fever.*

Camerino (Italia).

Labor. di Anat. patol. e Patol. generale. Dir. G. Pacinotti.

Cardiff, Wales (England).

Dept. of Pathol. and Bact. Nat. School of Med. Dir. Prof. E. H. Kettle.

Catania, Sicilia (Italia).

Ist. di Anat. Patol. Univ. Dir. Prof. Umberto Parodi. — Dr. Riccardo Reitani, aiuto; N. N., Ass. *8 posti. Cancer.*

Charkow (U. d. S. S. R.).

Ukrainisches Pathol.-anat. Inst., Liebknechtstr. 41. Dir. Prof. Dr. med. N. F. Melnikow-Raswedenkow. — Prof. Dr. med. K. F. Elenevsky (Leiter d. onkol. Abt. d. Inst.); Dr. W. R. Meyer (Leiter d. tanatol. Abt.); Dr. A. M. Azariew-Melnikowa; Dr. med. G. L. Derman.

Chemnitz (Deutschland).

Pathol. hyg. Inst. Dir. Prof. M. Staemmler.

Chicago, Ill. (U. S. A.).

Dept. of Pathol. Univ. of Chicago. Dir. H. G. Wells hairman). — Assoc. Prof. E. R. Long.

Chiba (Japan).

Pathol. Inst. Chiba Med. Univ.

Cluj — Klausenburg (România).

Inst. d'Anat. pathol., Str. Minerva 9. Dir. Prof. Titu Vasiliu. — Dr. R. Popa; Dr. M. Kernbach. *Histol. pathol., Bact. et parasit.*
Inst. de Pathol. générale, Inst. Pasteur, Str. Pasteur 6. Dir. Prof. M. Botez.

Coïmbra (Portugal).

Inst. de Anat. Patol., Univ. Dir. L. dos Santos Viegas.

Cork (Ireland).
Pathol. Labor. Univ. Dir. Prof. A. E. Moore.

Curityba, Parana (Brasil).
Labor. d'Anat. pathol. Fac. méd. Univ. Dir. Prof. Cairo da Silva. — Prof. I. T. Mutel.

Danzig (Danzig).
Pathol. Inst. am städt. Krankenhaus. Dir. Prof. Dr. Hermann Stahr. *Geschwülste. Krebsforschung.*

Debreczen (Ungarn).
Pathol.-anat. Inst. Univ. Dir. Dr. Fr. Orsós. — 2 Ass., 2 Prakt., 2 Demonstratoren für pathol. Anat.; 2 Ass., 2 Prakt. für gerichtl. Med.; I. Ass. für pathol. Anat.: Dr. Julius Jáki. *2 Plätze für pathol. Anat. u. Histol., Bact. — Mikrophotographische Einrichtung für gewöhnliches und ultraviolettes Licht.*

Dortmund (Deutschland).
Pathol. Inst. u. Forschungsinst. für Gewerbe- u. Unfallkrankheiten. Dir. Prof. Dr. Hermann Schridde.

Dresden (Deutschland).
Pathol. Inst. am Krankenhaus Friedrichstadt. Dir. Dr. Christian Georg Schmorl.

Dublin (Ireland).
Dept. of Pathol. Univ. Trinity Coll. Dir. Prof. J. T. Wigham.

Dunedin (New Zealand).
Dept. of Pathol. Univ. Dir. Prof. A. M Drennan.

Düsseldorf (Deutschland).
Pathol. Inst., Med. Acad. Dir. Prof. Dr. P. K. K. Huebschmann.

Edinburgh (Scotland).
Dept. of Pathol. Univ. Dir. Prof. J. L. Smith.

Edmonton (Canada).
Dept. of Pathol. Univ. of Alberta. J. J. Ower. M.D.

Ekatherinoslav, Ukraine (U.d.S.S.R.).
Pathol.-anat. Inst. Med. Inst. Dir. Prof. I. Korovin.

Erlangen (Deutschland).
Pathol.-anat. Inst. Dir. Prof. G. Hauser.

Firenze (Italia).
R. Ist. di Anat. Patol. R. Univ., 33 via Alfani. Dir. Prof. Bindo de Vecchi. — Lib. Doc. Luigi Picchi, Aiuto (Istol. Patol.); Dott. Claudio Natali; Dott. Antonio Costa; Dott. Montagnani; Dott. Pescatori. *Dai 7 ai 10 posti. Ricerche di Istol. patol., di Batt., di Patol. sperim.* ⓟ Sono collezionate e fotografate la maggior quantità possibile dei pezzi tratti dal ricco materiale necroscopico sezionato nell'Ist.

Frankfurt a. M. (Deutschland).
Senckenbergisches Pathol. Inst., Gartenstr. 229. Dir. Prof. Dr. Bernh. Fischer. — Prof. Dr. E. Goldschmid, Prosektor; Ass.: Priv.Doc. Dr. Schwartz, Priv.Doc. Dr. Tannenberg, Dr. Büngeler, Dr. Dassel, Dr. Hermann, Dr. Hirsch. *10 Plätze für Pathol., Histol.* ⓟ Pathol.-anat. Praep.

Freiburg i. Br. (Deutschland).
Pathol. Inst. der Univ., Albertstr. 19. Dir. Prof. Dr. med. Ludwig Aschoff. — Priv.Doc. Dr. H. E. Anders; Dr. med. Franz Büchner; Dr. med. Otto Ranke; Dr. med. R. Schönheimer. *10 wissenschaftl. Arbeitsplätze für experim. Pathol., pathol. Anat. und Histol., pathol. Chem.* ⓟ Praep. der menschlichen Pathol. (für den Austausch), Kropfpraep. Es besteht eine besondere Sammlung der Kriegspathol.

Fukuoka (Japan).
Pathol. Inst. Coll. of Med., Kyûshû Imper. Univ.

Galway (Ireland).
Dept. of Pathol. Univ. Coll. Dir. Prof. T. Walsh.

Gand — Gent (Belgique).
Labor. d'Anat. pathol. Fac. Med. Univ., 5, Quai de la Biloque. Dir. N. Goormaghtigh. — Amerlin A. André, Ass.; Elaut, Ass. *travaux le recherche;* Berte, Prép. *1 places pour histopathol. des glandes endocrines.* ⓟ Collection prép. de Surrénales pathol. humaines et animales.

Genève — Genf (Suisse).
Inst. Pathol. de l'Univ., Boulevard de la Cluse 40. Dir. Prof. Dr. M. Askanazy. — 4 Ass. *10 à 15 places pour travaux anat.-pathol., histol. et expérim. — 2 salles d'autopsie, 1 salle d'opération (pour animaux), 1 chambre bact.* ⓟ Prép. histol., pièces anat.-pathol. Parasites de l'homme.

Genova (Italia).
Ist. di Anat. Patol. Univ. Dir. A. Fabris.

Gießen (Deutschland).
Pathol. Inst. Med. Fac. Univ. Dir. Prof. G. Herzog.

Glasgow (Scotland).
Pathol. Inst. Royal Infirmary of Univ. Dir. Prof. John Hammond Teacher, Dept. of Pathol. — Ass.: David P. Cuthbertson (Lect. on Bio-Chem.), J. A. C. Burton, Alice J. Marshall, Helen F. Wingate, J. Steven Faulds, Elinor D. Jackson. *3 places for Histol.*
Pathol. Inst. Western Infirmary. Dir. Prof. Muir.
Dept. of Pathol. Royal Victoria Infirmary.
Dept. of Pathol. Royal Hosp. of Children.

Göttingen (Deutschland).
Pathol. Inst. Dir. Prof. E. Kaufmann.

Graz (Österreich).
Pathol.-anat. Inst. u. Mus. Med. Fac. Univ. Dir. Prof. H. Beitzke.
Inst. für allgem. u. experim. Pathol. Med. Fac. Univ. Dir. Prof. H. Pfeiffer.

Greifswald (Deutschland).
Pathol. Inst. d. Univ. Dir. Prof. Dr. Ernst Leupold. — Dr. H. Guillery, Ass.

Groningen (Holland).
Pathol. Anat. Labor. Univ., Oostersingel 63. Dir. Dr. H. T. Deelman. — Dr. O. H. Dykstra; 3 Ass.
Pathol. Labor. Dir. Prof. Dr. L. Polak Daniels

Halifax, N. S. (Canada).
Dept. of Pathol. and Bact. Dalhousie Univ. Dir. A. G. Nichols, M.D., C.M. — D. J. MacKenzie.

Halle a. d. S. (Deutschland).
Pathol. Inst. Univ. Dir. Prof. Rud. Beneke.

Hamburg (Deutschland).
Pathol. Inst. Univ., Allg. Krankenhaus Eppendorf. Prof. Dr. Theodor Fahr; Prof. Dr. Friedrich Wohlwill.

Heidelberg (Deutschland).
Pathol. Inst. Univ. Dir. Prof. Dr. Paul Ernst. — Prof. Dr. Otto Teutschlaender (Krebsforschung, Gerichtl. Med.); Prof. Dr. Siegfried Gräff (Tuberkuloseforschung, Oxydase); Dr. Curt Froboese (Lipoidstoffwechsel, Säuglingsdarm); Dr. Ernst Herzog (Nervensystem, Sympathicus). *1 großes Tauchmikrotom von R. Jung für Gehirnschnitte.*

Helsinki (Finnland).
Pathol.-anat. Inst., Nikolaigatan 10. Dir. Prof. A. L. Wallgreen; H. Castrén (stellvertr. Prof.).

Hong-Kong (China).
Labor. of Pathol. Univ. Dir. Prof. C. Y. Wang.

Innsbruck (Österreich).
Pathol.-anat. Inst., Müllerstr. 44. Dir. Prof. Dr. Georg B. Gruber. — Prof. Dr. Franz Joseph Lang; Dr. Neviny. *3 Plätze für pathol. Histol.* ℗ Menschliche Mißbildungen.
Inst. f. allgem. u. experim. Pathol. Univ. Dir. Prof. Gustav Bayer.

Irkutsk (U. d. S. S. R.).
Inst. f. allgem. u. experim. Pathol. Dir. Prof. A. Melknich.
Inst. f. Pathol. Anat. Dir. Prof. V. Donskov.

Ithaca, N. Y. (U. S. A.).
Dept. of Pathol. Cornell Univ. Ass. Prof. C. M. Carpenter.

Jassy (România).
Labor. d'Anat. pathol. Fac. Méd. Univ., Str. Muzelor 18. Dir. Prof. P. Gălăşescu. — P. N. Balan, Chef de travaux. *Histol. physio l. et pathol.* (*Hématol.*). *Anat. pathol. expérim.*
Labor. de la Clinique des maladies des yeux, Hopital St. Spiridou. Dir. Prof. Dr. Helène Puşcariu. — Dr. J. Nittulescu, Chef de labor.; Dr. Eug. Lazarescu, Ass.; Dr. E. Triaudaf, Dr. G. Vartic, Dr. Gr. Samandi, Prep. ℗ Des prép. microscopiques d'anat. pathol. et ce qui concerne les maladies des yeux.

Jena (Deutschland).
Pathol.-Anat. Inst. u. Sammlung, Univ. Dir. Prof. W. Berblinger.

Johannesburg (South Africa).
Dept. of Pathol. and Bact. of the Univ. Dir. Prof. W. Walkins-Pitchford.

Kanazawa (Japan).
Pathol. Inst. der Univ., Tsurumamachi. Dir. Prof. Dr. Hachitaro Nakamura. — Prof. Dr. Shigeteru Sugiyama; Ass. Prof. Kametoshi Wada; Ass.: Dr. Ryo Chadani, Dr. Ryosuke Okada, Dr. Kikuo Mori, Dr. Shichiichro Fuse. *23 Plätze für pathol.-anat. u. experim.-pathol. Arbeiten.*

Kaunas (Litauen).
Pathol.-anat. und pathol.-histol. Labor. der Univ. Dir. Prof. E. Winteler. — J. Mackevičolte-Lašiene, Ass.; Frl. P. Kalvaityte, Prosektor. *2 Plätze für pathol.-anat. und histol. Tumordiagnose u. sonstige krankhaft veränderte Organe oder Gewebe.* ℗ Praep. (pathol.-anat. u. histol.).

Keijô, Korea (Japan).
Pathol. Inst. Coll. of Med., Keijô Imper. Univ.

Kiel (Deutschland).
Pathol. Inst. der Univ., Hospitalstr. 20. Dir. Prof. Dr. Jores. — Priv.Doc. Dr. Arthur Schultz, Dr. Wagner, Dr. Petersen, Ass. *2 Plätze für pathol.-histol. Untersuchungen.* ℗ Menschliche Praep.

Kingston, Ont. (Canada).
Dept. of Pathol. Queen's Univ. Dir. Jas. Miller, B.Sc., M.D.

København (Danmark).
Patol.-anat. Inst., Frederik den 5tes Vej 11. Dir. Prof. Dr. med. J. Fibiger. — Prosektorer: Dr. med. Poul Møller; Dr. med. B. Vimtrup.
Patol.-Anat. Labor. Dir. Prof. A. F. Folger. — A. Møller Sørensen, Prosektor; Tierarzt Th. Primgaard, Ass.
Inst. für allgemeine Pathol., Juliane Maries Vej 22. Dir. Prof. Dr. Oluf Thomsen. — Lect. Dr. Vilh. P. H. Jensen; Cand. med. Tage Kemp; Cand. med. V. Friedenreich.

Köln a. Rh. (Deutschland).
Pathol. Inst. Univ. Dir. Prof. Alb. Dietrich.
Pathol. Inst. am Augusta-Hospital. Dir. Priv. Doc. Dr. Anton Frank.

Königsberg i. Pr. (Deutschland).
Pathol. Inst. der Univ., Kopernikusstr. 3|4. Dir. Prof. Dr. Kaiserling. — Dr. Gerhard Töppich, Priv.Doc.; Dr. Alfred Gromelski; Dr. Artur Weiß. *3 Plätze.* ℗ Farbig konservierte pathol.-anat. Praep.

Kraków (Polska).
Inst. f. pathol. Anat. der Univ., Grzegorzeckagasse 16. Dir. Prof. Dr. St. Ciechanowski. — Ass.: Dr. Skibniewski, Dr. Sciesiński, Dr. Malewski, Dr. Roman, Dr. Mróz, Kiełczewski; Prosektorsstellvertr.: Frau Dr. Łukowa. *2—3 Plätze für pathol. Histol.*

Kumamoto (Japan).
Pathol. Inst. Kumamoto (Imperial) Med. Univ.

Kyôto (Japan).
Pathol. Inst. Kyôto Provincial Med. Univ., Hirokôji.
Pathol. Inst. Coll. of Med., Kyôto Imperial Univ. Dir. Prof. Dr. Akira Fujinami. — Prof. Dr. A. Fujinami (Pathol., Parasit.); Prof. Dr. Kenji Kiyono (pathol. Anat., Mikrobiol.); Lect. Takuji Yamada, Harno Kokita, Hajime Kobayashi, Boku Nakarai.

La Plata (Argentina).
Inst. Anat. patol. espec. Embriol. e Histol., Fac. med. vet. Univ. Dir. Prof. F. Malenchini.
Inst. Anat. y Fisiol. patol. Escuela de sc. med. Dir. Prof. C. Jakob.

Lausanne (Suisse).
Inst. pathol. de l'Univ., Hôpital cantonal. Dir. Prof. Dr. J. L. Nicod. — A. de Coulon (Cancer); 3 Ass. méd. *Cancer expérim. Anat. et histol. pathol.*
Labor. de Pathol. experim. de l'Univ., Solitude 19. Dir. Prof. B. Galli-Valerio.

Lawrence, Kan. (U. S. A.).
Dept. of Pathol. and Bact. of the Univ. Dir. Prof. H. R. Wahl. — Prof. N. P. Sherwood; Ass. Prof. E. L. Treece; Ass. Prof. C. N. Downs.

Leeds (England).
Dept. of Pathol. of the Fac. of med., Univ. Dir. Prof. M. J. Stewart. — Lect.: P. L. Sutherland (Pathol. of Industrial Diseases).

Leiden (Holland).
Labor. f. allgem. Pathol. u. pathol. Anat. Dir. Prof. Dr. N. Ph. Tendeloo.

Leipzig (Deutschland).
Inst. für pathol. Anat. d. Univ., Liebigstr. 26. Dir. Prof. Dr. med. Werner Hueck. — Dr. Fritz. Klinge, Prosektor; Dr. Karl Krauspe, Ass. für Bact.; Dr. Martha Schmidtmann, Ass. für experim. Pathol.; außerdem 3 Ass. und 2 Volontär-Ass., die nach 1—2 Jahren wechseln. *6 Plätze für histol. Pathol., speziell Histol. des Mesenchyms.* ℗ Praep. für pathol. Anat. d. Menschen.
Pathol. Inst. am Krankenhaus St. Georg. Dir. Dr. Adolf Reinhardt. *Geschwülste. Secretio interna.*

Leningrad (U. d. S. S. R.).
Abt. für pathol. Anat. des Staatl. Inst. f. experim. Med.

Liège (Belgique).
Inst. d'Anat. pathol., Univ. Dir. Prof. Ch. Firket.

Lille (France).
Labor. d'Anat. pathol., Fac. de Méd. de l'Univ. Dir. Prof. F. Curtis. — Pellissier (Agrégés et exercice).
Labor. de Anat. pathol., Fac. de Méd. catholique de l'Univ., 56, Rue du Port et 11, Rue de Toul. Dir. Prof. A. Delathe.

Lima (Peru).
Labor. d'Anat. pathol. et Pathol. general, Alameda Gran. Dir. Dr. O. Hercelles; Dr. H. F. Delgado.

Lincoln, Neb. (U.S.A.).
Dept. of Anat. Pathol. and Hygien., Univ. of Nebraska. Ass. Prof. H. M. Martin (Anat. Pathol. and Hygiene); Prof. S. R. Towne.

Lisbôa (Portugal).
Inst. d'Anat. pathol. et Pathol. général, Fac. de Méd., Univ. Dir. Prof. Domingues H. F. Pereira.

Liverpool (England).
Pathol. Labor., Fac. of Med. of the Univ. Dir. Prof. E. E. Glynn.

London (England).
Inst. of Pathol. and Mus., Charing Cross Hospital, Med. School of the Univ., 62—65 Chandos Str., Strand W.C. 2. Dir. A. Piney. — A. B. Rosher (Bact.); A. R. Berrie (Bact.); J. Patterson (Biochem.); I. Muende (Pathol.). *1 place for Haematol. — Haematol. apparatus.* ⓟ Haematol. specimens.
Dept. of Pathol. and Bact., The London School. of Med. for Women of the Univ., Royal Free Hospital, W.C. 1, 8, Hunter Str., Brunswick Square. Dir. Prof. J. Henry Dible. — Joan Ross; Lucy Wills; Ceicly Wetherall; K. M. Lankester; K. Armstrong.
Pathol. Dept. St. Bartholomew's Hosp., Univ. Med. School, West Smithfield, E.C. 1. Dir. Sir Frederick Andrewes. — Dr. R. G. Canti (Bact.); Sir Bernard H. Spilsbury (Morbid Anat.); Dr. G. A. Harrison (Pathol. Chem.).
Dept. of Pathol. and Bact., St. Thomas's Hosp. Med. School of the Univ., Albert Embankment, Lambeth, S.E. Dir. Prof. L. S. Dudgeon.
Dept. and Mus. of Pathol. and Morbid Anat., and Pathol. Inst. for advance research, London Hosp. Med. Coll. of the Univ., Turner Str., Mile End, E. Dir. H. M. Turnbull.
Dept. of Pathol., Westminster Hosp. Med. School of the Univ., 12 Caxton Str., S.W. 1. Dir. J. A. B. Hicks.
Dept. of Pathol., Middlesex Hosp. Med. School of the Univ., Mortimer Str., Oxford Str., W. Dir. C. E. Lakin. — Prof. J. McIntosh.
Dept. of Pathol., St. Mary's Hosp. Med. School of the Univ., Paddington, W. 2. Dir. W. D. Newcomb. — Prof. A. E. Wright (experim. Pathol.).
Charles Graham Labor., Univ. Coll. Hospital Med. School, Univ. Str., Gower St. W.C.
Mus. of Pathol. Anat. Univ. Coll. Hosp. Med. School, Univ. Str., Gower St. W.C. Dir. Cur. W. G. Barnard.
Dept. of Pathol., Univ. Coll. Hosp. Med. School, Univ. Str., Gower St., W.C. Dir. F. H. Teale. — G. W. Goodhart.
Pathol. Labor. and Mus., Royal Army Med. Coll. of the Univ., Grosvenor Rd., S.W.
Mus. of Anat. and Pathol., King's Coll. of the Univ., Strand, W.C. 2.
Dept. of Pathol. Lister Inst. of preventive Med. of the Univ., Chelsea Bridge Rd., S.W. 1. Dir. Prof. Dr C. J. Martin.
Dept. of Pathol., Guy's Hosp. Med. School of the Univ., St. Thomas St., Borough, S.C. 1. Dir. Prof. A. Stokes. — Prof. G. W. Nicholson (pathol. Anat.).
Bland-Sutton Inst. of Pathol. Dir. Prof. James McIntosh, M.D. — S. L. Baker; L. W. Whitby; R. F. Scarff; L. Prodger; J. Morris. *3 places for Pathol. and Bact.*

London, Ont. (Canada).
Dept. of Pathol. and Bact. of the Univ. of Western Ontario. Dir. H. H. Bullard, Ph.D., M.D. — E. N. Ballantyne, M.D.; J. A. Ferguson, M.D.; C. G. Fletcher, M.D.; J. A. Lamont, M.D.; J. H. Fisher, M.D.

Louisville, Ky. (U.S.A.).
Dept. of Pathol. and Bact. of the Univ. Dir. Prof. St. Graves. — Prof. H. M. Weeter.

Louvain — Leuven (Belgique).
Labor. de l'Anat. pathol. Dir. Prof. J. Maisin. — J. Bouchaert.

Lund (Sverige).
Pathol. Inst. der Univ. Dir. Prof. Dr. med. John Forssman. — Prof. o. Einar Sjövall, Dr. med.; Priv.Doc. Dr. med. Arvis Lindau; Ass. Dr. Gulli Svarsson; 2 Ass. *2 oder 3 Plätze für pathol. und bact. Untersuchungen.* ⓟ Praep. für pathol.-anat. Sammlungen. — 2 Abt.: eine pathol.-anat. und eine bact.

Lwów (Polska).
Inst. d'Anat. pathol. de l'Univ., rue Piekarska 52. Dir.: Prof. Dr. Witold Nowicki; Prof. Dr. Th. Ostrowski. — Dr. Witold Grabowski; Dr. M. Seidler; Dr. Elène Szuster; Dr. Thaddée Wilczyński.

Lyon (France).
Labor. d'anat. pathol. Fac. méd. Univ. 18, Quai C. Bernard. Dir. Prof. J. Paviot. — Dr. J. F. Martin, Agrégé chef de travaux; Dr. J. Dechaume, Prép. de recherches; Dr. Croizat, Prép. adjoint. *6 places pour Diagnostics histopathol., Recherches expérim., Cancerol.* ⓟ Prép. histol. Autopsies. Biopsies.

Manchester (England).
Dept. o f Pathol., Victoria Univ. Dir. Prof. John Shaw Dunn. — William Susman, M.D., Lect. in Morbid Histol.; Allan Watt Downie, Ass. Lect. in Bact.; Cyril J. Polson, Ass. Lect. in Chem. Pathol. *4 places for Histol. and experim. work.*

Manila (Philippine Islands).
Dept. of Pathol. and Bact., Coll. of Med. Dir. Prof. Liborio Gomez. — Prof. Dr. Maria Paz Mendoza-Guazon; Assoc. Prof. Dr. Walfrido de Leon; Assoc. Prof. Dr. Carlos Montserrat; Ass. Prof. Dr. Regino Navarro; Ass. Prof. Dr. Juan Z. Sta Cruz; Dr. Jose Nolasco. *2 places. Later with extension of building will have place for more. Specially fitted for pathol. anat., on account of good amount of material, about 800 autopsies a year, that is 2 or 3 a day.*

Mannheim (Deutschland).
Pathol., Bact. u. Serol. Inst. der städt. Krankenanstalten. Dir. Dr. Hermann Loeschke. *Lungenpathol. Secretio interna. Knochenpathol. Mißbildungen.*

Marburg a. d. L. (Deutschland).
Pathol.-Anat. Inst. Dir. Prof. Max Versé. — H. J. Arndt, 1. Ass.

Marseille (France).
Labor. d'Anat. pathol. Ecole méd. Univ.
Labor. central des cliniques à l'Ecole méd. Dir. Dr. Albert Rouslacroix. *Anat. Pathol.*

Melbourne (Australia).
Dept. of Pathol. Univ. Dir. Prof. P. MacCallum.

Messina (Italia).
Ist. di anat. patol. Univ. Dir. Prof. P. Ferraro.

Milano (Italia).
R. Ist. di anat. patol. (R. Univ.), Ospedale Magg. Dir. Prof. Alberto Pepere. — Dr. Filippo Battaglia (Aiuto); Dr. Agatino Longhitano (1. Ass.); Dr. Rosario Marziani (2. Ass.); Dr. Alessandro Bilello (Ass. volontario); Dr. Giulio Radaell (Ass. volontario); Alfredo Fontana, Giorgio Boattini, Giuseppe Di Mattei, Vincenzo Cavallaro (Fellows Rockefeller Foundation). *16 posti. Ricerche in anat. e istol. patol., patol. sperim. Sezione Batteriol. Si sta provvedendo all'impianto di una Sezione per cultura dei tessuti.* ⓟ *Materiale anat.-patol.; Collezione istopatol.; Culture di germi patogeni.*
Ist. di Patol. Comparata Univ., Via L. Spallanzani 26. Dir. Prof. Guido Guerrini. — Vice-Dir.: Prof.

Luigi Leinati. 2 Ass. *L'Ist. è diviso in quattro Sezione: microscopia, chimica, fisiol., batteriol. Ogni sezione dispone di 4 posti per praticanti.* ℗ Parasit. animali.

Minsk (U. d. S. S. R.).
Pathol.-anat. Inst., Lenin-Straße 27. Dir. Prof. I. T. Titow. — Prosector: W. T. Tshervakow; Ass.: W. I. Glod-Werstonk, G. A. Stoljarow. *3 Plätze für pathol.-anat. Arbeiten und besonders Helminthiasis.* ℗ Pathol.-anat. Praep. (makro- und mikroskopische). Tiere: Kaninchen, Mäuse, Meerschweinchen, Ratten.

Modena (Italia).
Inst. d'Anat. pathol. Univ. Dir. Prof. Giulio Tarazzo. — Dott. Raul Barbanto, Aide. *1 posto. Anat. e histol. pathol. e bact.*

Montpellier (France).
D'Anat. pathol. Fac. Méd. Univ. Dir. Prof. E. Grynfeltt. 1 Chef des travaux pratique: Dr. E. Bosc; 1 Chef de labor. des cliniques: Dr. H. J. Guibert; 2 Moniteurs de travaux pratiques: Me le Dr. Simon-Rambaud, Mr. Buisson. *5 à 6 places pour des travaux d'histopathol., tissu conjonctif et la névroglie.* ℗ Pièces anat.-pathol. diverses.

Montreal, Que. (Canada).
Dept. Pathol. Anat. Univ. of Montreal. Dir. Latreille, M.D. — A. Bellerose, M.D.
Dept. of Pathol. and Bact. McGill Univ. Dir. H. Oertel, M.D.; L. J. Rhea, B.Sc., M.D.; A. A. Bruère, M.D.; T. R. Waugh, M.A., M.D.

Moskau (U. d. S. S. R.).
Pathol.-anat. Inst. I. Univ., Bolschaja Pirogoffskaja uliza, Dewitschje Pole. Dir. Prof. Dr. A. Abrikossoff. — Prov.Doc. I. Dawydowskie, Prosektor; Priv.Doc. W. Talalajeff. 1. Ass.; Ass.: Dr. Helene Herzenberg, Dr. M. Alexejeff, Dr. S. Weinberg; Adjunkten: Dr. S. Waïl, Dr. P. Dwijkoff, Dr. A. Kestner, Dr. D. Wyropajeff. *5 Plätze, besonders für Pathol., anat. Morphol.* ℗ Makro- und mikroskopische Sammlungspraep., typische und seltene Objekte.
Pathol.-anat. Inst. II. Univ., M. Pirogoffstr. 57. Dir. Prof. W. I. Kedrowsky. — Ass.: M. A. Skworzóff, A. I. Sawatéeff, I. L. Rapopórt, A. I. Waranoff, W. I. Schámschin; Aspiranten: M. L. Wirjukoff, W. N. Kartaschewa, L. M. Duchównikowa.

München (Deutschland).
Pathol. Inst. u. pathol. anat. Sammlung, Nußbaumstr. 26. Dir. Prof. Dr. Max Borst. — Prof. Dr. Hermann Groll (Prosektor u. Konserv.); Prof. Dr. Leonhard Wacker (Chem. Abt.); Anat. Ass.: Dr. Karl Fahrig, Dr. Ernst Dormanns, Dr. Matthias Beck, Dr. Gustav Borger. *25 Plätze für histol. u. chemische Arbeiten, speziell Geschwülste, Regeneration, Transplantation, Entzündung.* ℗ Pathol.-anat. u. histol. Praep.
Pathol. Inst. des städt. Krankenhauses München rechts der Isar. Dir. Geh.Med.R. Herrmann Dürck.

Münster i. W. (Deutschland).
Pathol. Inst. Med. Fak. Univ. Dir. Prof. W. Gross. *Vitale Färbung, bes. pathol. Zellveränderungen. Pathol. d. Niere. Pathol. Histol. d. Nervensystems.*

Mukden, Manchuria (China).
Pathol. Inst. South Manchurian Med. Univ.

Nagasaki (Japan).
Pathol. Inst. Nagasaki Imperial Med. Univ. Dir. Prof. Ikuhiko Hayashi.

Nagoya (Japan).
Pathol. Inst. Aichi Med. Univ.

Niigata (Japan).
Pathol. Inst. Niigata Med. Univ.

Nova Goa (Portuguese-India).
Dept. and Mus. of pathol. Anat. School of med.

Odessa, Ukraine (U. d. S. S. R.).
Pathol. anat. Instit. der Med. Fakultät. Olgiewskaja, 4. Dir. Prof. M. Tiesenhausen. — N. Busni Prosector; D. Chajutin, N. Glöckler, J. Medwedeff, Ass. — Mitglieder: L. Buchstab, I. Imailswitsch, D. Chaütin, R. Rublewa. *6 Plätze.* — Bakt. Abt. Morphol. Abt., experim. Abt.

Okayama (Japan).
Pathol. Inst. Okayama Imperial Med. Univ. Dir. Prof. Oto Tamura.

Omsk (U. d. S. S. R.).
Pathol. Hist. Abt. d. med. Inst. Dir. Prof. A. E. Efimov.
Pathol.-Anat. Abt. d. med. Inst. Dir. Doc. K. V. Romodanowski.

Ôsaka (Japan).
Pathol. Inst. Ôsaka Provincial Med. Univ. Emer. Prof. Naruhiko Sata (General Pathol. and Tubercul.); Prof. Miyakichi Murata (Pathol. and pathol. Anat.); Prof. Awashi Katase (Pathol. and Experim. Pathol.).

Oslo (Norge).
Pathol.-Anat. Inst. Univ. Dir. Prof. Harbitz.

Oxford (England).
Dept. of Pathol. Fac. Med. Univ. Dir. Prof. G. Dreyer. — Lect. E. W. A. Walker.

Padova (Italia).
Ist. di Anat. Patol., Via S. Massimo 18 (provisorio). Dir. Prof. Giovanni Cagnetto. — Dr. A. Fabris, Aiuto; Dr. G. Zanetti, Ass. *Richerche di anat. patol. col corredo di ricerche di istol. batt. e biochimica. Posti di lavoro: 8 per studenti ed 6 per med.* ℗ Collezione di osteo-patol. — Il Labor. è attualmente in sede provisoria. Nel prossimo anno si trasferirà nei nuovi locali dove potrà ricevere una dotazione di mezzi e di apparecchiche gli consentano di creare la possibilità di studi specializzati.

Palermo (Italia).
Ist. di Anat. patol. Fac. med. Univ. Dir. Prof. M. Soli.

Paris (France).
Labor. d'Anat. pathol. Fac. méd. Univ. Dir. Prof. A. Roussy.
Labor. de Pathol. Ext. Ecole Prat. des Hautes, Etudes, place de l'Ecole de Méd. Dir. Prof. Paul Lecène. — Dr. Pierre Moulonguet, Chirurgien des hôpitaux, Préparateur; Dr.Wolfram; Dr. Chevrillon; Dr. Sophia Dobkievitch; Dr. Pavie, Prépar. *4 places. Histol. pathol. chirurgicale, Pathol. chirurgicale expérim.* ℗ Préparations d'histol. pathol. Photographies macroscopiques et microscopiques de pathol. chirurgicale.
Mus. Dupuytren de l'Anat. pathol., 15, Rue de l'Ecole de Med. Dr. A. Herrenschmidt (Conserv.). *Histol. pathol. du Cancer.*

Parma (Italia).
Ist. di Anat. Patol. Dir. Prof. Pietro Guizzetti. — Prof. Dr. Pietro Verga, Aiuto; Dott. Emlide Forlini, Ass. *Nuovo Ist. 1927.* ℗ Mus. di Anat. Patol.

Pavia (Italia).
Ist. di Anat. Patol., Labor. di Anat. Patol. Dir. Prof. Achille Monti. — Dott. Piero Redaelli, Aiuto e libero Doc.; Dott. Luigi Nicoli, Ass. *Istol. Blatt. Microl. Micropatol.*
Ist. Camillo Goigi, Labor. di Patol. Generale ed Istol. della R. Univ., Palazzo Botta. Dir. Prof. Aldo Perroncito — Dott. Piera Locatelli, Aiuto; Dott.

Luigi Scotti Foglieni, Ass.; Dott. Umberto Collevati; Dott. Giovanni Truffi; Dott. Mario Lapidari. *48 posti per allievi interni. L'attrezzatura del Labor. riguar da la Patol. sperim., l'Istol. normale e Patol., la Parassit. e la Batt.*

Pécs (Ungarn).
Pathol.-anat. Inst. Univ. Dir. Prof. Dr. Béla v. Entz. — Ass.: Dr. Gedeon Erös, Dr. Dionys Göröng, Dr. Konrád Beöthy, Dr. Magdalena Frank, Dr. Andreas Kovács; Praktikanten: Dr. Koloman Krigl, Carl Ruszkó, Koloman Péchy, Dr. Martha Dömmel, Josef Soós, Ladislaus Strausz. *10 Plätze für pathol. u. bakt. Arbeiten.* ⓟ *Pathol.-anat. und histol. Praep. vom Menschen und von Tieren.*
Inst. f. allgem. Pathol. Dir. Prof. Dr. Géza Mannsfeld. — Mitarbeiter: Dr. Ernest Geiger, Julia Szirmay, Ladislaus Müller, Ladislaus Orosz.

Peking (China).
Dept. of Pathol. of the Peking Union Med. Coll. Head: Carl Ten Broeck.—Dr. James R. Cash, Assoc., Prof. of Pathol.; Dr. Ernest C. Faust, Assoc. Prof. of Parasit.; Dr. C. E. Lim, Ass. Prof. of Bact.; Dr. C. H. Hu, Assoc. in Pathol. *3 places for Pathol. anat., bact., and parasit.* ⓟ *Animal parasit., cultures of bact.*

Perm (U.d.S.S.R.).
Pathol. Anat. Inst. Univ.

Perugia (Italia).
Ist. di Patol. gen. Fac. med. Univ. Dir. Prof. A. Businco.

Philadelphia, Pa. (U.S.A.).
Henry Phipps Institute of the Univ. of Pennsylvania, 7th and Lombard Sts. Dir. Dr. Eugene L. Opie. — Dr. Joseph D. Aronson, Assoc.; Dr. J. Stuart Mudd, Assoc.; Dr. J. Freund (Pathol); Dr. Max Lurie; Dr. Caroline E. Whitney. *6 places for Pathol. and Bact.*

Pisa (Italia).
Ist. di Anat. Patol. Univ. Dir. A. Cesaris Demel.

Porto Alegre (Brasil).
Inst. f. pathol. Anat. u. Physiol. Dir. Prof. G. Vianna.
Inst. f. allgem. Pathol. Dir. Prof. M. Totta.

Poznań — Posen (Polska).
Inst. d'anat. pathol., 9 Kozia. Dir. Prof. Dr. med. Louis Skubiszewski. — Dr. Zeyland, Chef des trav.; Dr. Bederski, Ass.; Dr. Winter, Ass. *Pour les recherches de neoplasme.* ⓟ *Prép. macroscopiques collectionnées en mus. (environ 1400 prép.).*

Praha — Prag (C.S.R.).
Pathol.-anat. Inst. d. Karls-Univ., II, Preslova 2039, Hlávůo ústav. Dir. Prof. Dr. Rudolf Kimla. — Prof. Dr. Inan Honl; Prof. d. Bakt. u. Serol.
Inst. f. allg. u. experim. Pathol. d. Karls-Univ., II, Kateřinska 32.
Pathol.-anat. Inst. d. Deutsch. Univ., II, u nemocuice 497. Dir. Prof. Dr. Anton Ghon.
Inst. f. experim. Pathol. d. Deutsch. Univ., II, Salmova 3. Dir. Prof. Dr. Hugo Wiener. — Prof. Dr. Julius Riehl.

Quebec (Canada).
Dept. of Pathol., Anat. and Bact., Laval-Univ. Dir. A. Vallée, M.D.

Quito (Ecuador).
Labor. di Anat. Patol. y Parasit. Fac. med. Univ. Dir. Prof. L. G. Davila.
Labor. di Patol. General y Pediatria Fac. med. Univ. Dir. Prof. R. Sánchez.

Riga (Latvija).
Pathol. Inst. der Univ. Dir. Prof. Dr. med. Roman Adelheim. — Dr. med. M. Brandt, Prosektor; Dr. med. E. Kaktin, Ass. *1 Platz f. Pathol. Histol.*

Rio de Janeiro (Brasil).
Labor. Anat.-patol., Faculdade de Med. do Rio de Janeiro. Dir. Prof. Raul Leitão da Cunha. — Prof. Gustavo Hasselmann, Ass. et Chef du Labor.; Dr. Joaquim Vidal, Ass.; Dr. Augusto Duarte Pinto, Ass.; Dr. Alberto Canejo; Dr. José Pereira Roças. *1 place. Anat. pathol. Protozool. Organes des sens. Microbiol. Biometrie. Immunité. Paralysis.*

Roma (Italia).
Ist. di Anat. patol. Fac. med. e chirurgia Univ. Dir. Prof. A. Dianisi.

Rosario (Argentina).
Inst. f. pathol. Anat. u. Physiol. Dir. Prof. F. Ruiz.

Rostock i. M. (Deutschland).
Pathol. Inst. Univ., Gertrudenstr. Dir. Prof. Dr. Walther Fischer. — Prosektor: Prof. Dr. Pol; Priv.-Doc. Dr. Heine; Ass.: Dr. von Gusnar, Dr. Mayser. *Etwa 2 Plätze für pathol.-anat., pathol.-histol. Untersuchungen.*

Rostow a. Don (U.d.S.S.R.).
Inst. f. allgem. Pathol. Dir. K. R. Miram.
Inst. f. pathol. Anat. Dir. Prof. Ch. I. Krinizkij.

Rouen (France).
Labor. de Pathol., Ecole prép. de Méd. et de Pharmacie. Dir. Prof. Née. — Prof. F. Hue (Pathol. externe, Anat. topogr.).
Mus. d'Anat. pathol. Ecole prép. de méd.

Saint Andrews (Scotland).
Pathol. Dept. Univ. Coll. Dundee and Med. School. Dir. Prof. L. R. Sutherland.

St. Louis, Mo. (U.S.A.).
Dept. of Pathol., Washington Univ. Med. School. Dir. Prof. Leo Loeb. — Assoc. Prof. Frank A. McJunkin; Instr. Dr. S. H. Gray; Ass.: Dr. Walter Siebert, Dr. Walter Peterson, Dr. William Kountz; Research Ass.: Frances Lelia Haven. *Approximately 5 places. Research in general and special pathol. (experim. and microscopic).*

Samara (U.d.S.S.R.).
Pathol. anat. Inst. Univ. Dir. Prof. E. L. Kaveckij.

San Francisco, Cal. (U.S.A.).
Dept. of Pathol., Med. School, Univ. of California. Dir. Prof. G. Y. Rusk.

Saratow (U.d.S.S.R.).
Labor. f. pathol. Anat. Univ. Dir. Prof. P. P. Zabolotnov.

Sassari, Sardinia (Italia).
Ist. di Anat. patol. Fac. med. Univ. Dir. Prof. P. Mariconda.

Sendai (Japan).
Pathol. Inst. der Tôhoku Kaiserlichen Univ., Kita 4-Bancho. Dir. Prof. Dr. med. O. Kimura. — Prof. Dr. med. S. Nasu; Ass. Prof. Dr. med. T. Uchiyama (z. Z. in Marburg a.d.L.); Ass. Prof. Dr. med. T. Katsurashima; Ass. Prof. Dr. med. S. Terashima. *6 Ass. 3 Plätze; außer gewöhnlich bakteriologischen am Menschenleichen verschiedene Explantationsversuche (unter Prof. Dr. Nasu) und auch biol.-röntgenol. Untersuchungen. — Mikromanipulator, großer mikrophotographischer Apparat, Röntgenapparat.* ⓟ *Makro- sowie mikroskopische Praep. der pathol. Menschenorgane und speziell mikroskopische Praep. von Tiernerven, Parasiten des Menschen.*

Sheffield (England).
Labor. of Pathol. Fac. med. Univ. Dir. Prof. J. S. C. Douglas.

Siena (Italia).

Labor. di Anat. patol., Ist. biol., Juori Porta laterina. Dir. Prof. Ottone Barbacci. 1 Aiuto. *6 posti per allievi interni destinati essenzialmente alle ricerche di istol. patol.* ⊕ Prep. di Anat. patol. per uso del mus.

Smolensk (U. d. S. S. R.).

Pathol.-anat. Kabinett. Dir. Prof. Aljakritskij.

Sofia (Bulgarien).

Inst. für Pathol. Anat. Dir. Prof. Dr. D. O. Kryloff. — Dr. A. Pentscheff, Prosektor; Dr. M. L. Lebel, Ass.; Dr. A. Prodanoff, Ass.

Stambul (Türkei).

Pathol.-anat. Labor. Univ., Haidar-Pascha. Dir. Prof. Dr. med. H. Hamdi. — Prof. Dr. Satin Ali u. 2 Ass. *8 Arbeitsplätze für Histopathol. Cancer.* ⊕ *Mikroskopische Praep.* — *Für Krebsforschung ab 1927.*
Labor. de Pathol. gén. Fac. méd. Univ. Dir. Prof. Sureya Ali.

Stockholm (Sverige).

Pathol. Inst. Karolinska-Inst. Dir. Prof. Henschen.

Strasbourg — Straßburg (France).

Inst. d'Anat. pathol. Fac. Méd. Dir. Prof. Pierre Masson. — Dr. Louis Gery, Chargé de Cours; Dr. Charles Oberling, Chef de travaux; Dr. Charles Honette, Prép.; Dr. Josef Stolz, Prép. *6 places pour Histopathol. Matériel très riche (Mus., Collections histopathol., 1000 autopies et 2000 pièces chirurgicales par an). Il peut recevoir 8 à 10 travailleurs bénévoles.* ⊕ Prép. microscopiques de lésions humaines.

Sydney (Australia).

Dept. of Pathol. Univ. Dir. Prof. D. A. Welsh.

Szeged (Ungarn).

Pathol.-anat. u. pathohistol. Inst. d. Königl. ung. Franz-Joseph-Univ., Kossuth Street 40. Dir. Prof. Dr. Ernst von Balogh. — Dr. A. von Kálló, 1. Ass.; Dr. J. Putnoky, 2. Ass.; Dr. E. Kubányi, Vol.-Ass.; Dr. D. J. Kup. *1 Platz für pathol. Histol., Bakt. u. Gewebezüchtung.* ⊕ Gesammelt werden pathol.-anat. Musealpraep. (ca. 1000 Praep.) und pathohistol. Schnittpraep. (ca. 7000).
Allgem. Pathol. Inst. der Univ. Dir. Prof. Dr. Josef Lőte. — Mitarbeiter: Dr. Andreas Jeney, Michael Veress, Martin Nagy.

Taihoku, Formosa (Japan).

Pathol. Inst. Taihoku Governmental Coll. of Med.

Tartu — Dorpat (Eesti).

Pathol. Inst. der Univ., Lehmstr. 5. Dir. Prof. Dr. med. Alexander Ucke. — Dr. med. Albert Valdes, Prosektor; Dr. med. Rudolf Sääsk, 1. Ass.; Arzt Gerhard Sponholz, 2. Ass. *1 Platz für histol. Unters.*

Taschkent (U. d. S. S. R.).

Inst. f. Allgem. Pathol. u. pathol. Anat. Dir. Prof. V. V. Vasil'jevskij.

Tegucigalpa (Honduras).

Inst. f. Tropenpathol. Dir. Prof. B. R. Zepeda.

Tôkyô (Japan).

Pathol. Inst. Med. Fak. der Kaiserlichen Univ., Hongo. Dir. Matarô Nagayu und Tomosaburô Ogata. — Doc.: S. Fukushi und S. Suzuki; Ass.: T. Oonuma, T. Goh, Y. Kaneko, Y. Hiki, T. Azuma, I. Wake und S. Tsukahara. *20 Plätze für pathol. Anat. spez. Cancer.* ⊕ Praep.
Pathol. Inst. Nippon Girls' Med. Univ.
Pathol. Inst. Nippon Med. Univ.
Pathol. Inst. Military Med. Coll. (Guno-i Gakkô).
Pathol. Inst. Fikeikwai Med. Univ., Shiba.
Pathol. Inst. Fac. of med., Tôkyô Imper. Univ. Prof. Matoro Nagayo (Pathol., pathol. Anat.); Prof. Tomosarubô Ogata (Pathol., pathol. Anat.); Ass. Prof. Tokushirô Mitamura (Experim. Pathol.).
Pathol. Labor. Imper. Inst. of Infect. Diseases. Dir. Dr. M. Nagayo.
Pathol. Inst. College of Med., Kelô-Gijuku Univ. Dir. Prof. Dr. Z. Kawakami. — Prof. Dr. Z. Kawakami; Prof. Dr. S. Kusama; Lect. K. Kurokawa; Lect. Dr. R. Hukamachi.

Tomsk (U. d. S. S. R.).

Inst. f. pathol. Anat. Univ. Dir. Prof. V. P. Miroljuboff.

Torino (Italia).

Labor. di Patol. generale, Corso Raffaello 30. Dir. Prof. Benedetto Morpurgo. — Dr. Claudio Pulcher, Ajuto. *20 posti per ricerche microscopiche e batt. 5 posti per ricerche chimiche.* ⊕ Varie razze di Topi albini di diversa recettivita per tumori (stipite di sarcoma e di condrosarcoma).
Ist. di Anat. patol. Fac. méd. Univ. Dir. Prof. F. Vanzetti.

Toulouse (France).

Labor. d'Anat. pathol. Fac méd. Univ. Dir. Prof. J. Tapie.

Tsinan (China).

Inst. f. Pathol. Shantung Christian-Univ. Dir. Prof. L. H. Braafladt.

Tübingen (Deutschland).

Pathol. Inst., Silcherstr. 12. Dir. Prof. Dr. Alexander Schmincke. — Dr. Hans Wurm; Dr. Walter Pagel. *2 Arbeitsplätze für pathol.-histol. Untersuchungen.* — *Mikromanipulator v. Zeiss-Jena (Peterfi).* ⊕ Makro- und mikroskopische pathol.-anat. Praep. — Veterinär pathol.-histol. Labor.

Upsala (Sverige).

Pathol. Inst. der Univ. (Patol. Institutionen). Dir. Prof. Dr. med. U. Quensel. — Doc. W. Bosaus; Doc. J. Naslund. 2 Ass. *2—3 Plätze.*

Utrecht (Holland).

Pathol. Inst. Univ. Dir. Prof. Dr. R. de Josselin de Jong. — P. Nieuwenhuijse, Prosektor; mehrere Ass. *2 Plätze für pathol.-histol. Untersuchungen, auch experim.* ⊕ Pathol.-anat. Praep., teilweise in natürlichen Farben aufbewahrt. Sammlung von in Gelatine aufbewahrten Praep. Mikroskopische Praep. aller Art.

Warszawa (Polska).

Inst. d'Anat. pathol. Univ., rue Chalubiński 5. Dir. Prof. L. Paszkiewicz. — Dr. Dabrowska Janina; Dr. Henri Bychawski; Dr. Sophie Dobijowa; Dr. Clement Gerner; Dr. Stanislas Hrom; Dr. W. H. Melanowski; Dr. Siedlecki Eustachy.
Inst. für allgem. u. experim. Pathol. der Univ. Dir. Prof. Dr. med. Franciszek Venulet. — Dr. Henryk Gnoiński, Dr. Piotr Demant, Dr. Goldman, Ass. *2 Plätze für experim. Organopathol. u. serol. Untersuchungen.*
Labor. anat.-pathol. de la Clinique Neurol. de l'Univ., Nowogrodzka 59. Dir. Prof. Dr. Casimir Orzechowski. — Dr. Sigismond Messing, Chef des travaux. *6 places pour recherches histol.*

Wien (Österreich).

Pathol.-anat. Inst. der Univ., IX, Währinger Str. 6|8. Dir. Prof. Dr. Rudolf Maresch.
Pathol.-anat. Mus. der Univ., IX, Währinger Str. 11.
Anat.-pathol. Mus. d. allgem. Krankenhauses, IX, Alserstr. 4.
Inst. für allgem. und experim. Pathol., IX, Kinderspitalgasse 15. Dir. Prof. C. J. Rothberger. *Blutkreislauf (Elektrokardiogramm). Pharmacol. d. Gefäße*

Wilno (Polska).
Inst. d'Anat. pathol., Univ. Dir. Prof. Dr. Casimir Opoczynski.

Woronesch (U. d. S. S. R.).
Pathol.-anat. Inst. d. Staatsuniv., Grusovaja, 6. Dir. Prof. Dr. W. A. Afanassieff. — Ass.: Dr. A. Strukoff, Dr. S. Butschneff, Dr. Frau Ssinelstschikowa; Prosektor: Dr. Frl. Lentowskaja. *4 Plätze.*

Würzburg (Deutschland).
Pathol. Inst, Joseph-Schneider-Str. 2. Dir. Geh. Hofrat Prof. Dr. M. B. Schmidt. — Prof. E. Leupold, 1. Ass., Prosektor; Prof. E. Kirch, 2. Ass., Konserv.; Priv.Doc. E. Letterer, 3. Ass. *Ca. 12 Plätze für mikroskopische, pathol.-chemische u. experim. Unters.*

Zagreb — Agram (S. H. S.).
Pathol.-anat. Inst. Univ., Šalata. Dir. Prof. Dr. Sergius Saltykow. — Prosektor Dr. Goworow; Ober-Ass. Dr. Kornfeld; Ass.: Frau Dr. Pelčić, Frau Dr. Grubić, 3. Ass. unbesetzt. *10 Plätze für histol. (7) und bakt. (3) Untersuchungen.* ⑲ Pathol.-histol. Praep. Diapositive.

Zürich (Schweiz).
Pathol. Inst. Univ., Gloriastr. 3. Dir. Prof. Dr. H. v. Meyenburg. — Dr. A. v. Albertini, Prosektor; Dr. E. Uehlinger, 1. Ass. u. mehrere Ass. *6 Plätze, besonders für histol. u. tierexperim. Arbeiten.* ⑲ Pathol.-anat. Praep. — Neuro-Histopathol. Labor.

Zwickau (Deutschland).
Pathol. Inst. am staatl. Krankenstift. Dir. Dr. Paul Heilmann.

Anatomia pathologica animalium (fac. veterin.).
Confer: *Anatomia* (pag. 423) et *Bacteriologia animalium* (pag. 486);

Alfort, Seine (France).
Labor. d'Anat. pathol. de l'Ecole vét. Dir. G. Petit.

Ames, Ia. (U. S. A.).
Dept. of Vet. Pathol. Iowa State Coll. of Agric. Dir. E. A. Benbrook.

Amherst, Mass. (U. S. A.).
Dept. of Animal Pathol. Dir. G. E. Gage.

Berlin (Deutschland).
Biol. Reichsanst. f. Land- u. Forstwirtschaft Labor. f. Bekämpfung d. Bienenkrankheiten. Confer: Apides.
Pathol. Inst. d. Tierärztl. Hochsch., NW, Luisenstr. 56. Dir. Prof. Dr. W. Nöller. *Pathol. Anat. Protozoen.*
Veterin. Abt. d. Reichsgesundheitsamtes, -Dahlem, Unter den Eichen 82|89. Dir. Dr. E. Wehrle. *Tierseuchen.*

Blacksbury, Va. (U. S. A.).
Dept. of Zool. and Animal Pathol. Agric. Coll. I. D. Wilson, D.V.M., M.S., Prof. Vet. Sc.; C. J. Coon, Ass. Anat. Pathol.; R. A. Runnells, D.V.M., Assoc. Anat. Pathol.

Breslau (Deutschland).
Veterinär Inst. Univ. Dir. Prof. Casper.

București (România).
Labor. d'Anat. pathol. et Microbiol. Fac. méd. vet. Univ. Dir. Prof. P Riegler.
Labor. de Pathol. Fac. méd. veterin. Univ. Dir. Prof. Al. Ciucǎ.

Budapest (Ungarn).
Inst. für spez. Pathol. u. Therapie, Veterinär-Hochsch., Rottenbiller utera 23. Dir. Prof. Dr. József Marek.
Pathol.-Anat. Inst., Tierärztl. Hochsch. Dir. Prof. Dr. Karl Jórmai. — Julius Sal, 1. Ass.; Tiberius Stensky, 2. Ass.; Ladislaus Persa, Praktikant. *2 Plätze für pathol. Histol.*

Cairo (Egypt).
Veterinary Pathol. Labor., Veterinary Service.

Cambridge (England).
Inst. of Animal Pathol., Univ. Dir. Prof. J. B. Buxton.

College Park, Md. (U. S. A.).
Dept. of Bact. Coll. of Agric. E. M. Pickens, D.V.M., A.M., Prof. Bact., Anat. Pathol.; W. R. Crawford, D.V.M., Ass. Anat. Pathol.; M. B. Melroy, M.S., Ass. in Bact.; L. J. Poelma, D.V.M., Ass. Prof. Dairy Bact., Ass. Anat. Pathol.; R. C. Reed, Ph.B.,

Corrientes (Argentina).
Labor. de Anat. pathol. Fac. agric. Univ. Dir. Prof. J. Osamendi.

Děčin-Libverd—Tetschen-Liebwerd(C. S. R.).
Inst. f. Tierheilkunde. Dir. Prof. Dr. H. Oppitz.

Dresden (Deutschland).
Vet.-Abt. der sächs. Landwirtschaftskammer, -A., Sidonienstr. Dir. Landw.-Vet.-R. Dr. Kern.

Edinburgh (Scotland).
Pathol. Labor. of R. Vet. Coll. Dir. Prof. D. C. Matheson.
Animal Diseases Research Assoc. of Scotland, Moredun Inst. Gilmerton. Dir. W. A. Pool, M.R.C. V.S. — 3 Research Ass. (Animal Pathol.). *Animal Pathol.*

Elisabethville (Congo Belge).
Labor. de recherches Vétérinaire du Congo. Dir. R. van Saceghem.

Fayetteville, Ark. (U. S. A.).
Dept. Plant Pathol. Coll. of Agric. V. H. Young, Ph.D., Head of Dept.; H. R. Rosen, Ph.D., Assoc. Prof. and Assoc. Plant Pathol.

Gießen (Deutschland).
Veterinärpathol. Inst. Dir. Prof. A. Olt.

Hannover (Deutschland).
Inst. für pathol. Anat. Tierärztl. Hochsch. Dir. Prof. H. Rievel.

Kaunas (Litauen).
Pathol. anat. Inst. der Veterinärabt., Gedimino Gatve 29. Dir. Prof. El. Nonewitsch. — J. Butkewitsch, Oberass. ⑲ Katzen, Hunde, Kaninchen, weiße Mäuse.

Kazan (U. d. S. S. R.).
Pathol. Inst. des Veterinär-Inst., Arskoe Pole 96. Dir. Prof. C. Bohl. — Boris Bohl, Prosektor; Boris Iwanoff, Boris Michailoff; C. Suikowa; Michael Nechotjaeff. *2 Plätze.*
Pathol. Mus., Veterin. Inst. Dir. Prof. M. Tusnoff.

Königsberg i. Pr. (Deutschland).
Tierärztl. Inst. der Univ. Dir. o. ö. Prof. Dr. Erich Hieronymi. — Dr. med. vet. Alfred Kunze. *1 Platz für pathol.-histol. Untersuchungen und Obduktionen.* ⑲ Pathol.-anat. Praep. von Haustieren.

Kyôto (Japan).
Pathol. Labor. Kyôto Imperial Seric. Coll. Confer: Bombyx.

Leipzig (Deutschland).
Vet.-Pathol. Inst. d. Univ. Dir. Prof. Joest.

Leningrad (U.d.S.S.R.).
Abt. f. Pathol. Leningr. Landwirtschaftl. Inst., Kamennyj Ostrov 2, Birken-Allee 28.
Pathol.-Anat. Inst. der Tierärztl. Hochsch., Černigovskaij Str. 5. Dir. Prof. Dr. N. Ball. — Tierarzt G. Belkin, Prosektor; Tierarzt P. Moraev, Ass. *5 Plätze für histol. Arbeiten.* ⓟ Pathol.-anat. Praep. auf dem Gebiete der Pathol. Anat. der Tiere.

León (España).
Labor. de Terapéutica y Patol. general, Escuela de Veterinaria. Dir. Prof. J. M. Garcia.
Labor. de Histol. y Anat. patol., Escuela de Veterinaria. Dir. Prof. T. Rodriguez.

Lincoln, Neb. (U.S.A.).
Dept. of Animal Pathol. of the Univ. of Nebraska. Dir. Prof. L. Van Es. — Ass. Prof. H. M. Martin; Ass. Prof. L. V. Skidmore.

Liverpool (England).
Dept. of Veterinary Pathol. the Univ. of Liverpool. Dir. Prof. S. H. Gaiger. — K. D. Downham, B.V.Sc., M.R.C.V.S., D.V.H., Adviser in Veterinary Sc.(Other Appointments pending). *New buildings not yet completed.* ⓟ Pathol., Bact. and Protozool. Objects.

Los Baños, Laguna (Philippine Islands).
Dept. of Pathol. and Bact., Coll. of Veterinary Sc. Univ. Dir. Dr. A. K. Gomez, Chief of the Dept. — Dr. Ramon Q. Javier (Pathol.); Dr. Zacarias de Jesus (Bact.). ⓟ Pathol. slides and mus. specimens. Bact. cultures.

Lwów (Polska).
Pathol. Anat. Inst. der Tierärztl. Hochsch. Dir. Dr. Aleksander Zakrzewski. *Lyssa der Fleischfresser. Pathol. Stoffwechsel der Vogelniere. Nierensystem.*

Lyon (France).
Labor. d'Anat. Pathol., Ecole Nation. Vétérin. Dir. Prof. Ball.

Madrid (España).
Labor. d'Histol. et Pathol. anat. Ecole véterin. Dir. Prof. A. Gallego y Canee.

Manila (Philippine Islands).
Labor. of Pathol. and Bact. Coll. of Veterin. Dir. Prof. A. K. Gomez. Confer: Los Baños.

München (Deutschland).
Inst. f. Tierpathol. Univ. Dir. Prof. Th. Kitt.

Oslo (Norge).
Pathol. Abt. Veterinaer Inst. Dir. Prof. H. Holth.

Ottawa, Ont. (Canada).
Labor. of Pathol., Dept. of Agric. Dir. E. A. Watson.

Paris (France).
Station des recherches sur les maladies des animaux de l'Inst. des recherch. agron., VIIe, 42bis, Rue de Bourgogne.

Perugia (Italia).
Ist. Sup. di Med. Veterinaria. Dir. Gianbattista Caradonna.

Port au Prince (Haiti).
Veterinary Sc. Service technique. Dir. J. B. Bouglhton.

Poznań — Posen (Polska).
Inst. f. Tierheilkunde. Dir. Prof. St. Runge.

Praha — Prag (C.S.R.).
Tierärztl. Inst. Deutsche Univ. Dir. H. Descler.

Pretoria, Transvaal (South Africa).
Division of Veterinary Education and Research, Onderstepoort. Dir.: Sir Arnold Theiler; Deputy Dir.: Dr. P. J. du Toit; Sub-Dir.: D. T. Mitchell, Dr. H. H. Green, Dr. P. R. Viljoen. — Research Officers: Dr. G. de Kock, Dr. C. P. Neser, Dr. J. B. Quinlan, Dr. E. M. Robinson, H. H. Curson, A. O. D. Mogg, P. J. J. Fourie, M. W. Sheppard. Dr. F. Veglia, G. A. H. Bedford, Dr. J. P. van Zyl, Dr. J. Scheuber, G. Martinaglia, Dr. H. O. Monnig, Dr. M. Henrici, P. le Roux, Prof. Mettam, W. J. B. Green, J. I. Quin, J. H. R. Bisschop. *Dept. of Agric., Union of South-Africa, Fac. of Veterinary, Sc. of the Transvaal, Univ. Coll.* Sub-Labor.: Pietermaritzburg (Natal), Grahamstown (Cape Province), Vryburg (Cape Province), Ermelo (Transvaal), Besters Put (Freestate).

Stambul (Türkei).
Inst. f. allgem. u. spez. Pathol. Dir. Prof. A. Samuel.
Inst. f. pathol. Anat. u. Fleischbeschau. Dir. Prof. A. Schewki.

Stockholm (Sverige).
Pathol.-anat. Inst. Veterin. Hochsch. Dir. Prof. A. Hjarre.

Tôkyô (Japan).
Pathol. Labor. (of silk worms) Tôkyô Imper. Sericicult. Coll., Nishigahara.
Pathol. Labor. (of silk-worms) Imperial Sericicultural Experim. Station, Suginami-chô near Tôkyô.

Toulouse (France).
Labor. de Pathol. générale et Microbiol. Ecole nation. Vétérin. Dir. A. Dalile. *Pathol. gén. et Microbiol., Maladies microbiennes.*
Labor. d'Anat. pathol. Ecole nation. vétérin., Dir. Prof. D. Bimes. *Anat. pathol., Médecine légale, Inspection des Denrées alimentaires et des Etablissements classés soumis au controle vétérin.*
Labor. de Pathol. bovine Ecole nation, Vétérin., Dir. Prof. Ch. Besnoit. *Pathol. bovine, Médicine opératoire, Obstétrique, clinique.*

Ueda (Japan).
Pathol. Labor. (of silk-worms) Ueda Imperial Sericicultural Coll.

Urbana, Ill. (U.S.A.).
Dept. of Animal Pathol. and Hygiene Univ. of Illinois. E. C. McCulloch (Pathol. and Parasitol.)

Utrecht (Holland).
Pathol. Inst. der Tierärztl. Fac., Biltstraat 166. Dir. Prof. Dr. H. Schornagel. — J. H. ten Thye (Conserv.); Dr. H. J. M. Hoogland (Conserv.). *2 Plätze für pathol. Anat. und allgemeine Pathol. der Tiere.* ⓟ Pathol.-anat. Praep. der Haustiere.

Viçosa, Minas Geraes (Brazil).
Labor. de Veterinaria Escola Superior de Agric. Dir. P. H. Rolfs.

Warszawa (Polska).
Inst. Pathol. Anat. d. Haustiere. Dir. Prof. W. Sindemann.

Wien (Österreich).
Inst. f. Allgem. Pathol. u. pathol. Anat. Veterin. Hochsch. Dir. Prof. Dr. R. Hartl.

Anat.-pathol. Mus. der Tierärztl. Hochsch., III, Linke Bahngasse 11.

Winnipeg, Man. (Canada).
Dept. of animal Pathol. Univ. of Manitoba. Dir. A. Savage, B.S.A., D.V.M.

Zagreb — Agram (S. H. S.).
Inst. f. allg. Pathol. u. pathol. Anat. d. Haustiere. Gundulićeva ulica 22. Dir. Prof. Dr. Ludwig Jurak.

Zürich (Schweiz).
Vet.-Pathol. Inst. d. Univ., Selnaustr. 36. Dir. Prof. Dr. Walter Frei. — Dr. W. Pfenninger, Priv.-Doc., Ober-Ass., Lehrauftr. für Bakt. u. Milchuntersuchung; Dr. L. Riedmüller, Ass., pathol. u. bakt. Diagnostik von Tierkrankheiten. *6 Plätze für Bakt., pathol. Anat., Milchkunde, pathol. Physiol.* ⊕ Pathol.-anat. Praep. von Haustieren.

Laboratoria physiologiae pathologicae.

Laboratoria physiologiae pathologicae experimentalis generalis.

Athènes (Grèce).
Labor. de Physiol. pathol., Univ., Rue de l'Académie.
Labor. de Pathol. expérim. de l'Univ. Dir. Prof. J. Catsaras.

Baltimore, Md. (U. S. A.).
Labor. for Research Med., in the Johns Hopkins Hospital, Dept. of Med., N., Broadway. Dir. Dr. Leonor Michaelis, ao. Prof. Univ. Berlin, beurlaubt. — Dr. W. A. Perlzweig, Assoc. Prof. *4—5 Plätze. Physikalisch-chemische Apparate.*
Labor. of Physiol. Hygiene, 615 N. Wolfe St. Dir. W. H. Howell. — A. L. Meyer, Climatol. and Ventilation; J. H. Clark, Light and Illumination; A. M. Baetjer, Physiol. of fatigue; N. B. Herman, Industrial Diseases. *6 research rooms fitted for chemical and physical experim. — Gas analysis apparatus. Galvanometric, spectrum analysis, air conditioning X-ray, recording apparatus for physiol.*

Berlin (Deutschland).
Pathol. Inst. u. Mus. Univ. Confer: Pathol. anat. gen. humana.
Reichsgesundheitsamt. Präsident Dir. Geh.Reg. R. Dr. med. Hamel.—4 Dir., 17 Ober-Reg.-R. (darunter 12 Ärzte, 2 Chemiker, 1 Apotheker, 1 Tierarzt, 1 Zool.), 29 Reg.R. (darunter 5 Ärzte, 14 Chemiker, 2 Apotheker, 5 Tierärzte, 1 Botan., 2 Zool.) und 13 Hilfsarbeiter. *A. Hauptstätte, NW 23, Klopstockstr. 18. 1. Chemisch-hygienische Abt. (Dir.: Dr. W. Kerp): Chemisches-hygienisches, physiol.-pharmacol. u. pharmazeutisches Labor.; 2. Med. Abt. (Dir.: Dr. G. Frey): Medizinalstatistik, gewerbehygienisches Labor.; 3. Veterinär-Abt. (Dir.: Dr. E. Wehrle): Labor. für Tierseuchenforschung in der Zweigstätte Dahlem. B. Zweigstätte Dahlem, Berlin-Dahlem, Unter den Eichen 82—84. Bakt. Abt. (Dir.: Prof. Dr. L. Haendel): Labor. für Bakt., Zool., Chemie, für menschl. u. tier. Infektionskrankh. C. Zweigstätte Scharnhorststraße, NW 40, Scharnhorstr. 35 (frühere Kaiser-Wilhelms-Akademie). Hygienisch-bakt., chemisch., physikalisch-strahlenkundliches und pathol.-anatom. Labor. Sammlung anat.-pathol. Praep. (7000 Einzelpraep.).*

Bordeaux (France).
Labor. de Méd. Expér. Fac. Méd. Univ. Dir. Prof. Pierre Mauriac. — 1 Chef de Labor., 1 Préparateur, 1 Préparateur Adjoint. *5 places pour Bact., Méd. Expérim.*
Labor. de Biol. de l'Hopital St. André. Dir. Prof. Pierre Mauriac. — Dr. Louis Servantic, Chef de Labor.

Brno — Brünn (C. S. R.).
Inst. für allgemeine und experim. Pathol. Univ., Úvoz 33. Dir. Prof. Dr. Vilém Laufberger. — Ass.: Ing. chem. F. Drahovzal, Dr. med. K. Hora. 2 Demonstr. *Mehrere Plätze. — Pankreasdiabetische überlebende Hundeleber, respiratorischer Stoffwechsel des Kaninchens. Verschiedene Methoden zur pathol. Physiol. des Stoffwechsels.*

Bruxelles — Brussel (Belgique).
Lab. de pathol. générale. Dir. Prof. D. E. Spehl.
Inst. thérapeutique de l'Univ. Dir. Prof. E. V. Zunz.
Labor. de Recherche méd.

Chicago (U. S. A.).
Otho S. A. Sprague Memorial Inst. Dir. H. Gideon Wells. — Rollin J. Woodyatt, Experim. Med.; Karl K. Koessler, Experim. Med.; Maud Slye, Cancer Genetics; W. B. McClure, Experim. Med., Pediatrics; E. R. Long, Tuberculosis; M. T. Hanke, Chemistry; Julian H. Lewis, Experim. *About 6 places for Chemistry, Immunol., Pathol. Cancer Genetics, Labor.*

Cluj — Klausenburg (Romănia).
Allg. u. experim. Pathol. Abt., Pasteur Inst. Univ. Dir. M. Botez.

Darmstadt (Deutschland).
Biol. Abt. der Firma E. Merck. Dir. A. Sturm.

Detroit, Mich. (U.S.A.).
Med. Research Labor., Parke, Davis & Company. Dir. E. Mark Houghton, Ph.C., M.D.—L. T. Clark, B.S., Junior Dir.; W. E. King, M.A., M.D., Ass. Dir.; T. B. Aldrich, Ph.D., Research Chemist; E. P. Bugbee, B.S., M.D., Endocrinologist; A. D. Emmett, B.S., M.A., Ph.D., Nutritional Chemist; N. S. Ferry, Ph.B., M.D., Research Bacteriologist; O. M. Gruhzit, B.S., M.S., M.D., Pathologist; H. C. Hamilton, M.S., Pharmacolgist; Adelia McCrea, A.B. Mycologist; A. Noble, A.B., Research Bacteriologist; T. Ohno, Bacteriologist; L. W. Rowe, B.S., M.S., Pharmacologist; A. S. Schlingman, D.V.M., M.S., Veterinary Pathologist. *A private research labor. with no definite arrangement for guests. From time to time invited guests are allowed an opportunity to pursue research work.*

Elberfeld (Deutschland).
Chemo-therapeutische Abt. des Labor. der I. G. Farbenindustrie A.-G. Dir. Dr. Wilhelm Roehl. *Chemotherapie der Infectionskrankheiten, besonders der Tropenkrankheiten (Trypanosomiasis, Malaria).*

Firenze (Italia).
R. Ist. di Patol. generale Sperimentale e Batt., R. 8. Viale Morgagnie 18. Dir. Prof. Senatore Alessandro Lustig. — Prof. Marcello Lusena, Aiuto universitario; Ass. universitario: Dott. Giulio Rovida, Dott. Giovanni Favilli, Dott. Elisa Morelli, Dott. Alberto Manieri, Dott. Pio Gori. *10 posti per fisiopatol., batt. e immunol.*

Frankfurt a. M. (Deutschland).
Chemotherapeutisches Forschungsinst. „Georg Speyer-Haus", Paul-Ehrlich-Str. 42. Dir.Geh.Med.R. Prof. Dr. W. Kolle; experim.-biol. Abt.: Prof. Dr.

Kolle; chem. Abt.: Dr. Bauer. — Mitglieder: Dr. Bauer, Dr. Maschmann. 1 Laboratoriumsleiterin, 3 Ass., *Forschungen auf experim.-chemotherapeut. u. biol. Gebiete.* Eine biol. u. eine chemische Abt.

Staats-Inst. für experim. Therapie, Paul-Ehrlich-Str. 44. Dir. Prof. Dr. W. Kolle, o. Hon.-Prof. d. Univ. — Etatsmäßige wissenschaftl. Mitglieder und Abt.-Leiter: Prof. Dr. Hetsch, General-Oberarzt a. D., Prof. Dr. Laubenheimer, Prof. Dr. Caspari, Prof. Dr. Schloßberger, Dr. med. vet. Albrecht. 1 Verw.-Dir., 2 Ass., 1 Präparator. *Zweck des Inst.: 1. staatliche Kontrolle der verschiedenen Heilsera und Impfstoffe der humanen und der Veterinärpraxis, sowie der Reagentien für die Serumdiagnostik der Syphilis; 2. wissenschaftliche Forschungen auf dem Gebiete der experim. Therapie, vornehmlich der Immunität; 3. Untersuchungen experim. Art aus der Veterinärheilkunde; 4. Forschungen auf dem Gebiete der Krebsentstehung und -heilung.*

Freetown (Sierra Leone).

Sir Alfred Lewis Jones Research Labor. Dir. Prof. D. B. Blacklock. *Tropical Diseases of Africa.*

Genova (Italia).

Labor. di Therapia sperim. A. Dir. Bruschettini.

Hamadera near Ôsaka (Japan).

Ishigami Inst. for Infectious Disease, Dir. Dr. Takashi Matsuda.

Hamburg (Deutschland).

Inst. für Schiffs- und Tropenkrankheiten, Bernhardstraße 74. Inst.-Leiter u. Vorst. der klinisch-med. Abt.: O.Med.R. Prof. Dr. Bernhard Nocht; Vorst. der allg. tropenmed. Abt.: Prof. Dr. Friedrich Fülleborn; Vorst. der Abt. für prakt. Seuchenbekämpfung: Prof. Dr. Peter Mühlens; Vorst. der chem. Abt.: Prof. Dr. med. h. c. Gustav Giemsa; Vorst. der Protozoen-Abt.: Prof. Dr. Eduard Reichenow; Vorst. der entomol. Abt.: Prof. Dr. Erich Martini; Vorst. der pathol.-anat. Abt.: Prof. Dr. Henrique da Rocha-Lima; Vorst. der bact. Abt.: Prof. Dr. Martin Mayer. Außerdem 4 wissenschaftl. Hilfsarbeiter. Hafenarzt: Physikus Prof. Dr. Karl Sannemann. *Erforschung der Schiffs- und Tropenkrankheiten.*

Indianapolis, Ind. (U.S.A.).

Biol. Labor. of the Eli Lilly and Comp. Research Labor. Dir. W. A. Jamieson. *Immunol. Antitoxins, immune sera, bact. vaccines, viruses.*

Johannesburg (South Africa).

The South African Inst. for Med. Research, P. O. Box 1038. Dir. Sir Spencer Lister, M.R.C.S., L.R.C.P. — Officers Routine Division: Superintendent: G. Buchanan, M.D., Ch.B., D.P.H.; Senior Pathol.: J. G. Becker, M.B., Ch.B., D.P.H., D.Trop. Med.; A. Sutherland Strachan, B.Sc., M.B., Ch.B.; F. W. Simson, M.B., Ch.B.; Ass. Pathol.: J. J. de Waal, M.R.C.P., D.P.H., D.T.M.; L. W. Barlow, B.A., D.P.H.; Gordon D. Laing, B.Sc., M.B., Ch.B. Officers Research Division: Bact. and Pathol.: J. H. Harvey Pirie, B.Sc., M.D. (Deputy-Dir.); Fellow in Industrial Hygiene: A. Mavrogordato, M.A.; Parasit.: Annie Porter, D.Sc.; Honorary Protozool.: Prof. H. B. Fantham, M.A., D.Sc.; Entomol.: A. Ingram, M.D.; Biochem.: F. William Fox, D.Sc.; Ass. Bact.: D. Ordman, B.A., M.B., Ch.B. *Facilities for guests are granted, as the need arises, for approved workers in branches of med. sc., other than members of the permanent staff.*

Kiew, Ukraine (U.d.S.S.R.).

Abt. für Biol. und experim. Med. des Röntgeninst., L.-Tolstoi-Str. 7. Dir. des Inst.: Prof. G. P. Teslenko. — Abt.-Vorst. Prof. A. A. Krontowski; N. Woskressenski, Vorst. des biol. Labor. (experim. Erblichkeitsforschung, Zytol.); I. Bronstein, Dr. phil.Chem.(Mikrochem.); M.Kolomicz(Röntgenol.), N. Lazarew (Gewebskulturen, Tierversuche); M. Magath (Gewebskulturen, Tierversuche); N. Kiritschinska (Röntgenass.). *6 Plätze. Gewebskulturen (Explantation), insbesondere die Physiol. der Explantate, Biol. der Röntgenstrahlwirkung. — Apparatur für physikal.-chem. und mikro-chem. Arbeiten an Explantaten; Karzinom und Sarkomstämme; Röntgenapparatur; Mikromanipulatro. Biol. (zool.) Labor., mikrochem., physikal.-chem. und onkol. Labor., Vivarium.*

Abt. für experim. Med. des Bact. Inst. Dir. des Inst.: Prof. M. P. Neschtschadimenko. — Abt.-Vorst.: Prof. A. A. Krontowski; Dr. phil. I. Bronstein, I. Ass., Chem. (mikrochem. Untersuchungen im allgemeinen und in Anwendung auf Gewebskulturen); M. Jazimirska-Krontowska II.Ass. (Bact., Anfertigung der Gewebskulturen, Tierversuche). *6 Plätze. Gewebskulturen und deren Anwendung zur Infektions-, Immunitäts-, Krebsforschung u. dgl. — Großes Vivarium für Hunde (operiert nach Pawlow), Operationsraum für große Tiere u. dgl.*

Köln a. Rh. (Deutschland).

Normal- und pathol.-physiol. Inst. Univ. Confer: Physiol. gen. Fac. med.

Kuala Lumpur (Federated Malay States).

Inst. for Med. Research. Dir. William Hetcher. — R. W. Blair, (Chief Chem.); A. W. Young (Pathol.); R. Green (Pathol.); R. Lewthwaite; J. E. Lesilar (Pathol.); K. Kanagarayer; H. Marsden (Chem.); J. Shelton (Chem.); F. E. Byron (Chem.). ⊕ Cultures of pathogenic organisms.

Lausanne (Suisse).

Labor. de thérapeutique de l'Univ., Solitude 19. Dir. Prof. Arthus et Prof. Strzyzowski.

Leningrad (U.d.S.S.R.).

Labor. der Pathol. Physiol. am Leningrader med. Inst., Str. Leo Tolstoy 6|8. Dir. Prof. S. S. Chalatow. — I. M. Goldberg, I. Ass.; P. P. Meglitzky, II. Ass.; G. L. Frenckell, II. Ass., Leiter der klin. Abt.; R. I. Gawriloff, Aspirant. *4 Plätze. Cholesterinstoffwechsel. Experim. Chirurgie. — Cholesterinapparatur nach Dr. Krostelewskaja. Operationssaal.* Klin. Abt. für spezielle Studien auf dem Gebiete des Stoffwechsels.

Staats-Inst. für experim. Med., Lopuchinski-Str.12. Chef der Biol.-chem. Abt.: M. L. Petrunkin.

Labor. f. vergl. Pathol., Staatl. Inst. f. experim. Med., Lopuchinskaja ulica 12. Dir. Prof. A. A. Vladimirov.

Abt. für allgemeine Pathol. des Inst. für experim. Med., Lopuchinskaja 12. Dir. Prof. Dr. E. S. London. — Dr. A. Alexandry, Dr. N. Kotschneff, Ass. *5 Plätze. — Apparate und Hilfsmittel zum Studium des intermediären Stoffwechsels.*

Liège (Belgique).

Labor. de pathol. et de thérapeutique générale, Univ. Dir. Prof. P. Nolf.

Inst. de thérapeutique expérim., Univ. Dir. Prof. F. Henrijean.

Lille (France).

Labor. de Pathol. interne et expérim., Fac. de Méd., Univ. Dir. Prof. H. Surmont. — Dr. Jean Tiprez, Prép.; Dr. Henri Vauheeruversyn, Aide prép.; Mlle. R. Provino, Aide prép. *6 places. — Pathol. interne et expérim. Section de Chim., physiol., anat.-pathol. et histol., Histographie.*

Liverpool (England).

Liverpool School of Tropical Med. Dir. Prof. J. W. W. Stephens, F.R.S. — Prof. W. Yorke, Parasit.; Prof. D. B. Blacklock, Dir. of Sir Alfred Lewis Jones Research Labor., Freetown, Sierra Leone, Miß A. M. Evans, M.Sc., Entomol.; Lect. in Protozool., Helmintol., Clinical Pathol., Ass.-Dir., Research Ass. *6 places for Tropical Med.,* Parasit., Entomol. Sir Alfred Lewis Jones Research Labor., Freetown, Sierra Leone.

London (England).

Dept. of Bact. and experim. Pathol., Nat. Inst. for Med. Res., confer: Bact. hominis.
Dept. of Pathol. Chem., St. Mary's Hosp. Med. School of the Univ., Paddington, W. 2. Dir. Sir Wm. H. Willcox.
Naval Med. School, Royal Naval Coll. of the Univ., Greenwich. Dir. of Med. Studies: Surgeon Commander Th. Brown-Shaw. *General and Naval Hyg., Clinical Pathol., Tropical Med. and Micro-Biol.*
Biol. Labor. St. Bartholomew's Hospital Med. Coll. of the Univ., E.C. 1, West Smithfield. Dir. Prof. W. A. Cunnington. — C. C. Hentschel, Demonstr.
Dept. of Biol., Guy's Hosp. Med. School of the Univ., S.E. 1, St. Thomas St., Borough. Dir. Prof. T. J. Evans.

Lyon (France).

Labor. de Pathol. générale, Fac. de Méd. de l'Univ. Dir. Prof. Cade.

Manila (Philippine Islands).

Division of Biol. and Serum Labor. Bureau of Sc. Dir. O. Schöbl.

Melbourne, Victoria (Australia).

Walter and Eliza Hall Inst. of the Melbourne Hospital. Dir. C. H. Kellaway. — G. R. Cameron. *Compensatory Hypertrophy. Streptococcal Infection.*

Milano (Italia).

Ist. Sieroterapico Milanese, Via Darwin 20. Dir. Prof. Serafino Belfanti. — Sezione Diagnostici e Vaccini: Caporeparto Prof. Luigi Vigano; Aiuto, Dott. Antonio Scalfi. Sezione Sieri umani: Caporeparto Dott. Francesco Pepeu; Aiuto, Dott. Paolo Pauli. Sezione Vet.: Caporeparto, Dott. Giulio Ramazzotti; Aiuto, Dott. Mario Mazzucchi. Sezione Opoterapia-Fermentol.: Caporeparto Prof. Bice Neppi. Sezione Batteriol. Agraria e Industriale: Caporeparto Prof. Domenico Carbone; Aiuto, Dott. Carlo Arnaudi. Sezione Chemoterapia: Caporeparto Dott. Ugo Cazzani. Sezione Sc.: Caporeparto e Vicedirettore, Prof. Amilcare Zironi; Aiuto, Dott. Ettore Cuboni; Dott. Pia Latzer. *10 a 20 posti nelle varie Sezioni per ricerche di sierol., batteriol., biochim. e fermentol. — I Labor. sono tutti moderni e provvisti di apparechi perfezionati; i mezzi di lavoro, animali di esperienze ecc. sono abbondanti in ogni ramo.* ⓟ *Colture, sieri, vaccini ecc.*

Ist. di Patol. generale Univ., Via Gaetano Strambio 31. Dir. Prof. Pietro Rondoni. — Dott. Bruno Borghi, Aiuto; Dott. Carminati, Ass. volontario. *6 posti. — Ricerche istol., batteriol. e chim. L'Istituto dispone du buone e svariate bilancie di precisione, dispositivi per microanalisi, grande polarimetro di precisione, pressa di Buchner, ecc.*

Monaos (Brasil).

Liverpool School of Tropical Med. *Tropenhygiene, Erforschung der Tropenkrankheiten.*

Montreal, Que. (Canada).

Dept. of Pathol. Univ. of Montreal. Dir. A. Ferron, M.D. — A. Gagnon, M.D.; A. Paré, M.D.; B. Bourgeois, M.D.; A. Lesage, M.D.; A. Léger, M.D.

Moskau (U. d. S. S. R.).

Inst. Microbiol. d'Etat, r. Pogodinskaja 10. Dir. Prof. W. Barikine. — Dr. W. Friesé, sousdir., chef de la Section immunol.; Dr. A. Kompanezz, chef de la Section microbiol.; Dr. A. Zakharoff, ass. de la Section microbiol.; Dr. O. Barikine, chef de la Section histopathol.; Dr. W. Wygodtzikoff, ass. de la Section histopathol.; Dr. W. Koullikoff, chef de la Section de la chim. colloïdale; Dr. P. Smirnoff, ass. de la Section de la chim. colloïdale; Dr. S. Klukhine, chef de la Section des vaccins et des sérums; Dr. S. Minervine, ass. de la Section des vaccins et de sérums; Dr. L. Silber, ass. de la Section immunol. *10 places microbiol. et immunol.*

Inst. de Physiol. Pathol. de l'Univ. II, Bolchaja Kalengeskaja 22. Dir. Prof. Dr. A. A. Bogomolez. — Prof. agrégés: P. P. Averianoff, N. B. Medvedeva, N. N. Lirotinine. Les ass.: I. I. Bouratchevsky, L. N. Karlic, A. K. Pikkate, R. E. Kavetsky, I. M. Neumann, N. V. Valassik. Ass. aux travaux pratiques avec les étudiants, travaux sc. *10 places. Le problème de la specifité des centres nerveux végétatifs. le problème et la prophylaxie du fatigue neuromusculaire. La correlation iono-endocrinienne dans l'organisme. Les tumeurs malignes experim. — Les* dept.: chim., physiol., microbiol., physiko-chim., microscopique et immunol.

Moscow Tropical Inst., Entomol. and Toxicol. sections B. Tulskaja 77a, other sections Pogodinskaja No. 10. Dir. Prof. Dr. E. J. Marzinowski. — Dir. of sections: Prof. K. J. Skrjabin (Helminthol.); Prof. I. A. Smorodintzew (Chemotherapeut.); Dr. V.V. Nikolski (Entomol.); Dr. A.J. Metjolkin (Vet. parasit.); Dr. V. A. Nabokoff (Toxicol.); Prof. V. I. Kedrovski (Leprosy section). *2—3 places in each section.*

Inst. der allgemeinen Pathol. (pathol. Physiol.) der I. Moskauer Univ., Dewitschje Pole. Dir. Prof. G. P. Ssacharoff. — Dr. S. Tschetschulin, Ass. an der physiol. Abt.; Dr. N. Posanoff, Ass. an der bact. Abt.; Dr. B. Mogilnitzkie, Ass. an der pathol.-anat. Abt.; S. Kaplanskie, Ass. an der chem. Abt. *8 Plätze für Untersuchungen auf dem Gebiete der pathol. Anat. sowie auch der Pathol. der inneren Sekretion und der Verdauungsdrüsen. Experim. an den isolierten Organen nach Kravkow.* ⓟ *Die nach Pavlow operierten und zu verschiedenen Untersuchungen aus dem Gebiete der Pathol. der Verdauungsdrüsen vorbereiteten Hunde können evtl. verkauft oder getauscht werden.*

Staatsinst. f. experim. Therapie u. Serum-Impfstoff-Kontroll-Inst., Sivcev-Vražek 41. Dir. Prof. L. A. Tarassevič. — Stellv. Dir. Prof. V. A. Lubarskij; Abt.-Leiter: Dr. A. I. Togunova, Dr. I. N. Makarova, Dr. E. V. Glotova, Dr. L. A. Aleksina.

Dépt. de chimiothérapie de l'Inst. de maladies tropiques, Pogodinskaja 10. Dir. Prof. Dr. J. A. Smorodintzew. — Dr. A. N. Adowa; Dr. N. A. Demina. *2 places. L'action de Sb, As, „205'' et la Chim. sur les animaux de labor.*

Napoli (Italia).

Ist. di Patol. generale della R. Univ., S. Andrea delle Dame 21. Dir. Prof. Dr. V. Scaffidi. — Prof. G. di Macco, Aiuto; Dr. L. Califano, Dr. A. Gualdi, Dr. C. Amatucci Mallardo, Ass. *25 posti per Patol. sperim.*

New York City (U. S. A.).

Rockefeller Inst. for Med. Research, 66 Street and Avenue A. Board of Sc.Directors: William H. Welch, President; Simon Flexner, Dir. des Inst.; Theobald Smith; W. J. V. Osterhout; Francis G. Blake; John Howland. 1. The Dept. of the Labor. Dir.: Dr. Simon Flexner. Divisions: Pathol. and Bact., Chem., Experim. Surgery, General Physiol., Biophysics. 2. The Dept. of the Hospital. Dir.: Dr. Rufus Cole. 3. The Dept. of Animal Pathol. Dir.: Dr. Theobald Smith. Members of the Inst.: Simon Flexner (Pathol. and Bact.), Theobald Smith (Animal Pathol.), Rufus Cole (Med.), P. A. Levene (Chem.), Alexis Carrel (Experim. Surgery), Hideyo Noguchi (Pathol. and Bact.), W. J. V. Osterhout (General Physiol.), Donald D. Van Slyke (Chem.), Alfred E. Cohn (Med.), Karl Landsteiner (Pathol. and Bact.), Wade H. Brown (Pathol. and Bact.), Peyton Rous (Pathol.), Homer F. Swift (Med.), James B. Murphy (Biophysics), John H. Northrop (General Physiol.), Oswald T. Avery (Bact.), Walter A. Jacobs (Chemotherapy), Florence R. Sabin (Pathol.

and Bact.). Assoc. Members of the Inst.: J. J. Bronfenbrenner (Pathol. and Bact.), Harry Clark (Biophysic), E. V. Cowdry (Pathol. and Bact.), Pierre L. du Nouy (Experim. Surgery), Frederick L. Gates (Pathol. and Bact.), Michael Heidelberger (Chem.), Frederic S. Jones (Animal Pathol.), Paul A. Lewis (Animal Pathol.), Peter K. Olitsky (Pathol. and Bact.), Louise Pearce (Pathol. and Bact.), Thomas M. Rivers (Med.). 18 Assoc., 43 Ass. *Investigations in the sc. and arts of hygiene, med. and surgery, and allied subjects, in the nature and causes of disease and the methods of its prevention and treatment, and to make knowledge relating to these various subjects available for the protection of the health of the public and the improved treatment of disease and injury.*

Odessa, Ukraine (U.d.S.S.R.).
Labor. of General Pathol. of State Inst. of Chem. and Pharmacy, Str. Krasnoj Gvazdii 17. Dir. Leon A. Tscherkes. — Dr. J. Litvak, Ass., leading the practical works with students and sc. research; Dr. med. T. Kuperman, sc. collaborat. *Investigations on Avitaminose.*
Inst. f. Pathol. Physiol. Med. Inst. Dir. Prof. B. A. Sazillo.

Omsk (U.d.S.S.R.).
Allgem. Pathol. Abt. u. Labor. d. med. Inst. Dir. Prof. J. S. Pentmann.

Padova (Italia).
Ist. di Patol. generale Fac. med. Univ. Dir. Prof. I. Salvioli.

Palermo (Italia).
Ist. di Patol. generale Fac. med. Univ. Dir. Prof. A. Amato.

Paris (France).
Labor. de Physiol.-Pathol. Coll. de France, Ve, Place Marcellin-Berthelot. Dir. Prof. Nattan-Larrier.
Inst. Colonial Français, 4, Rue Volney.
Labor. général de Recherches de la Comité d'Encouragement aux recherches sc. Coloniales.
L'Inst. Pasteur, XVe, 21—25, Rue Dutot. Dir. E. Roux; Sous-Dir. A. Calmette. — Prof. Dr. Serge Metalnikov; Dr. Stéfan Mutermilch; Dr. Dujaorik de la Rivière; Dr. Bridré; Dr. A. Berthelot; Dr. Delagonne; Dr. Levadsh; Dr. Catoni; Dr. Mazé; Dr. Salimbain; Dr. Roubaud; Dr. Wollemann; Dr.Weinberg; Dr. Boquet; Dr. W. Mestrezat; Dr. Fourneau; Dr. Bezzaaka; Dr Nègre; Dr. Pozersky. Collaborateur: Michel Golodanoff; Prof. Dr. Constantin Levadisi. — Sous-labor.: Nha Trang (Indo-Chine).

Pisa (Italia).
Inst. f. allgem. Pathol. Dir. Prof. C. Sacerdotti.

Poznań — Posen (Polska).
Inst. de pathol. générale of expérim. Univ. Dir. Prof. Dr. I. Hoffman. — Prof. Dr. Adam Wrzosek.

Riga (Latvija).
Inst. d. allgemeinen Pathol. Med. Fac. Univ., Kronvalda bulv. 9. Dir. Prof. Mag. E. Paukuls.

Rio de Janeiro (Brasil).
Inst. f. pathol. Physiol. u. Anat. Dir. Prof. L. Da Cunha.

Roma (Italia).
L'Ist. Biol. 34, Dir. Prof. Guido Cremonese.—Prof. Dott. Giacomo Peroni, come collaboratore; Dott. Emilio Checconi, come ajuto. *Affermazione della scoperta dell'immun-malaria; Ricerca di altre applicazioni immunologiche col medesimo metodo e fra le altre, quella alla tubercolosi che è già un fatto realizzato.*

San Juan (Porto Rico).
Tropical Med. School of the Univ. Dir. Prof. Dr. R. A. Lambert.

Saratow (U.d.S.S.R.).
Labor. f. allgem. Pathol. Univ. Dir. Prof. E. A. Tatarinov.

Sassari (Italia).
Ist. di Patol. Generale Univ. Dir. Prof. Bruno Polettini.

Skoplje (S.H.S.).
Inst. für Tropenkrankheiten. Dir. Dr. Milivoj Rankov. — Zool. Abt.: Dr. St. Karaman.

Smolensk (U.d.S.S.R.).
Kabinet f. allgem. Pathol. Dir. Prof. Aljakritskij

Stanford University, Cal. (U.S.A.).
Dept. of Bact. and Experim. Pathol. Dir. E. W. Schultz, Prof. of Bact. *Filterable viruses and bact.*

Tôkyô (Japan).
Labor. of Experim. Pathol. and Kitasato Inst. for Infect. Diseases, Shirokane-Sankôchô. Dir. Dr. Shigeru Kusama.

Tomsk, Siberia (U.d.S.S.R.).
Labor. of the Pathol. Physiol. (General Pathol.) of the State Univ. Dir. Prof. A. D. Timofejewsky. — S. V. Benerolenskaja, Senior Ass. (Prosector); L. F. Larionov, Junior Ass.

Toulouse (France).
Labor. de Pathol. générale et Expérim. Fac. Med. Dir. Prof. E. Bardier. — Dr. André Stillmunkès; Dr. Paul Piquemal. *Physiol. pathol.*
Inst. d'Hydrol. Dir. Prof. Serr. — Prof. Serr (Therapeutique hydrol. et climatol.); Prof. Aloy (Chim.-Analyses); Prof. agrégés Moog et Valdiguié, Ass. au Service d'Analyse Chim.; Prof. Rispal (Analyses bact.); Dr. de Verbizier, Ass. au Service d'Analyse bact.; Prof. Marie (Physique); Prof. agrégé Escarde et Dr. Rabaud, Ass. au Service de Physique; Prof. Baylac (Pathol. interne appliquée à l'Hydrol.; Prof. Mengaud (Geol.); Dr. Bayer (Technique Hydrotherapique). *L'Inst. d'Hydrol. a pour objet de donner un enseignement complémentaire à celui qui est donné dans les Fac. pour permettre une étude approfondie des questions concernant l'Hydrol.*

Woosung b. Shanghai (China).
Allgem. Pathol. Inst. Tungchi Univ. Dir. Prof. Dr. Oppenheim.

Woronesch (U.d.S.S.R.).
Inst. f. allgem. Pathol. Univ. Dir. Prof. Afanassieff.

Zagreb — Agram (S.H.S.).
Inst. für allgemeine experim. Pathol. (Pathol. Physiol.) und Pharmacol. der Univ., Vočarska cesta No. 97. Dir. Prof. phil. et med. M. Mikuličić. — Dr. Petar Jurišić, Adjunkt des Inst., Leiter der Abt. für, physikal. Chem. der Zelle u. Gewebe; Dr. med. Vl. Sertić, Ass. bact. u. immunobiol. Abt.; Ing. chem. Vl. Anžlovar, Ass. chem. Abt.; Dr. med. J. Ivančević (d.Z. zu Studienzwecken im Auslande), Ass. klin. Abt.; Ass.-Stelle (d. Z. unbesetzt) Abt. für physiol. Graphik; Mag. pharm. N. Madirazza, Ass. pharmazeut. Abt. *4 klinische Plätze, 4 biochem., 4 für physiol. Chem., 4 für Immunobiol.—Biochem., Immunbiol., Bact., phys. Chem. in Anwendung auf biol. speziell pathol. physiol. Probleme, Pharmacol. inkl. Toxicol. u. Grenzgebiete, Biol., angewandte Klinik, experim. u. allg. Pathol.*
Inst. für allg. Pathol., pathol. Anat. u. gerichtl. Tiermed., Tierärztl. Fac., Savska cesta 14a. Dir. Prof. Sakař.

Laboratoria pharmacologiae experimentalis.
Confer: *Laboratoria physiologica* (pag. 444).

Amsterdam (Holland).
Pharmaco-therapeutisches Labor. der Univ., Polderweg 20 b. d. Linnaeusstr. Dir. Prof. Dr. Ernst Laqueur. — S. E. de Jongh, 1. Ass.; Mej. Dr. E. Dingemanse, 1. Chem.-Ass.; A. Grevenstuk, 2. Ass.; Dr. T. C. Hart, 3. Ass.

Ann Arbor, Mich. (U.S.A.).
Dept. of Pharmacol. Univ. of Michigan. Ass. Prof. A. G. Young.

Athènes (Grèce).
Labor. de Pharmacol. de l'Univ., rue de l'Acad. Dir. Prof. S. Dontas. — A. Phocas, Chef du Labor.

Bari (Italia).
Gab. d. farmacol. Dir. A. Baldoni.

Basel (Schweiz).
Pharmacol. Anstalt Vesalianum Univ. Dir. Prof. A. Jaquet.

Beograd (S.H.S.).
Farmakol. Inst. Med. Fac., Šumadiska 18. Dir. Prof. Dr. A. Holste. — Ass.: Dr. Radivoje A. Pavlović; Dr. Ilija N. Dimitrijević; Milan Arsenijević. *2 Plätze für experim. Pharmacol., besonders für die Arbeiten an isolierten Organen.*

Berkeley, Cal. (U.S.A.).
Dept. of Biochem. and Pharmacol. Univ. of California. Confer: Biochem. generalis.

Berlin (Deutschland).
Pharmacol. Inst. Tierärztl. Hochsch., NW, Luisenstraße 56. Dir. Prof. Frosch.
Pharmacol. Inst. Univ., Dorotheenstr. 28. Dir. i. V. Prof. Joachimoglu. — Dr. E. Keeser; Dr. Hintzelmann; Dr. Ehrismann. *12 Plätze für pharmacol., physiol., bact., chem. Unters.* ⓓ Drogen und Arzneimittel.

Bern (Schweiz).
Pharmacol. Inst. Univ. Dir. Prof. Dr. Alexander Tschirsch. *Grenzgebiete zwischen Chem. u. Botan., Harze.*

Bonn a. Rh. (Deutschland).
Pharmacol. Inst. Univ. Dir. Prof. Dr. phil. et med. Hermann Fühner. — Ass.: Dr. med. et phil. R. Labes, Dr. med. W. Blume, Prof. Dr. C. Bachem. *4 Plätze.*

Bordeaux (France).
Labor. de thérapeutique et pharmacol., Fac. méd. Univ. Dir. Prof. J. Carles.
Labor. de toxicol. et hygiène. Confer: Hygiena hominis.

Bratislava — Preßburg (C.S.R.).
Inst. für Pharmacol. und Pharmakognosie der Komensky-Univ., Staatl. Krankenhaus, Mickiewiczowa 13. Dir. Prof. Dr. Bohuslav Polák.

Breslau (Deutschland).
Pharmacol. Inst. Univ. Dir. Prof. J. Pohl.

Brno — Brünn (C.S.R.).
Inst. f. Pharmacol. u. Pharmakognosie d. Tierärztl. Hochsch., Pražska 67. Dir. Prof. Dr. Ottokar Rybak.
Labor. de Pharmacol. Univ., 33, Úvoz. Dir. Prof. Dr. B. Bouček. — Stanislav Petlach, MUDr., MrPh., Ass.

Bruxelles — Brussel (Belgique).
Labor. de pharmacognosie et de microscopie. Dir. Prof. Dr. N. Wattiez.
Labor. de pharmacol. expérim., fac. Méd. Univ. Dir. E. V. Zunz.

București (România).
Labor. de Pharmacol. Fac. de Méd. Univ. Dir. S. D. Lalou.

Budapest (Ungarn).
Pharmacol. Inst. Peter Pázmany Univ. Dir. Prof. Zoltán Vámossy. — Dr. G. Fritz; St. Biró; B. Paul.
Inst. für Pharmacol. Tierärztl. Hochsch., Rottenbiller'Gasse. Dir. Prof. Dr. Julius Magyary-Kossa. — Mitarb.: Dr. Franz Lakos.

Cadiz (España).
Labor. de Farmacol. Fac. Méd. Dir. Prof. Dr. José Benlloch.

Cambridge (England).
Dept. of Pharmacol. Univ. Dir. Prof. W. E. Dixon.

Canton (China).
Dept. of Pharmacol. Univ. Dir. Prof. E. C. Machle.

Catania, Sicilia (Italia).
Ist. di Mat. med. e farmacol. sperim. Dir. F. Foderà.

Chiba (Japan).
Pharmacol. Inst. Chiba Med. Univ.

Chicago (U.S.A.).
Dept. of Pharmacol. Univ. Prof. H. B. van Dyke.

Coïmbra (Portugal).
Labor. de Farmacol. Univ. Dir. F. A. da Cunha Guimarães.

Curityba, Paraná (Brasil).
Labor. de Pharmacol. Fac. méd. Univ. Dir. Prof. H. O. Riedel.

Debreczen (Ungarn).
Pharmacol. Inst. Univ. Dir. Prof. Dr. Alexander Belák. — Mitarb.: Dr. Julius von Mikó; Dr. Johann Siegler; Dr. Franz Sághy; Dr. Julius Mitroviczs; Dr. Stefan Hajdu; Ladislaus Cseresznyés; Eugen Szép; Stefan Gärtner; Zoltán Alföldy; Theresia Pala. *Allg. chem. u. pharmakodynamische Methoden, Refraktometrie, Interferometrie, Gasanalyse, Blutgase, Bact., Serol.*

Düsseldorf (Deutschland).
Pharmacol. Inst. Med. Akad. Dir. Prof. F. Hildebrandt.

Edinburgh (Scotland).
Pharmacol. Labor. Univ. A. C. White, M.B., Ph.D., Lect.; Andrew McFarlane, M.D.; G. H. Percival, M.B.; David Wilkie, M. B. *Room for 5 research workers. Suitable for all forms of research in general pharmacol. — General apparatus for investigation of functions of intact animals and isolated organs.*

Ekatherinoslav, Ukraine (U.d.S.S.R.).
Pharmacol. Labor. Med. Inst. Dir. Prof. N. Stru'ev.

Erlangen (Deutschland).
Pharmacol. Inst., Östl. Stadtmauerstr. 29. Dir. Prof. Dr. phil. et med. Konrad Schübel. — Dr. Walter Gehlen. *Chem. u. Biol.*

Ferrara (Italia).
Ist. di Farmacol. Univ., Palazzo Sch Dir. Prof. L. Beccari.

Frankfurt a. M. (Deutschland).
Pharmakol. Inst., Theodor-Stern-Haus, Weigertstr. 3. Dir. Prof. Werner Lipschitz. — Dr. Georg Barkan; Dr. Otto Girndt. *Ca. 6 Plätze für chem., physikal.-chem., spez. pharmakol. Tierexperimente. — Gitterspektroskop, Barcroft-App., Blutgaspumpe, Kleintier-Kalorimeter, Gaskettenapp., chem. Mikromethoden f. Körpersäfte, spez. pharmakol. App.*

Freiburg i. Br. (Deutschland).
Pharmakol. Inst., Katharinenstr. 29. Dir. Prof. Paul Trendelenburg. — S. Janssen; O. Krayer. *10 Plätze. — Saitengalvanometer.*

Fukuoka (Japan).
Pharmacol. Inst. Coll. of Med., Kyûshû Imper. Univ.

Gießen (Deutschland).
Pharmacol. Inst. Dir. Prof. Jul. Geppert.

Göttingen (Deutschland).
Pharmacol. Inst. der Univ., Geiststr. 4. Dir. Prof. W. Heubner. — Dr. med. Rolf Meier. *10 Plätze für die gebräuchlichsten Arten pharmacol. Untersuchungen, doch auch für physiol.-chem., analytische und physikalisch-chem. Arbeiten.*

Graz (Österreich).
Pharmacol.-pharmakognostisches Inst. Med. Fac. Univ. Dir. Prof. Dr. O. Loewi.

Greifswald (Deutschland).
Pharmacol. Inst. der Univ., Langefuhrstr. 23d. Dir. Prof. Dr. Otto Riesser. — Priv.Doc. Dr. med. Paul Wels, Ass.; Dr. med. Ernst Simonson, Volontär-Ass. *4 Plätze. Muskelphysiol. und Pharmacol. Physiol. und pharmacol. Chem. Mikroanalyse. Physikalische u. Kolloidchem. Röntgenwirkungen. Gaswechseluntersuchungen. Mikroanalyse. — Gaswechselapparate nach Luntz-Geppert, Ultramikroskop, Barcroft-Apparat, Leitfähigkeitsbestimmung, Viskosimetrie, Gaskette, Röntgenapparat, Ultraviolettlampe. Einrichtung für organisch-chem. Arbeiten, Mikrowage. Respirationskammer für Kaltblüter usw.*

Höchst a. M. (Deutschland).
Pharmacol.-wissenschaftl. Abt. des Labor. der I. G. Farbenindustrie A.G. Dir. Prof. Dr. Carl Ludwig Lautenschläger.

Halifax, N.S. (Canada).
Dept. of Pharmacol. Dalhousie Univ. Dir. O. S. Gibbs, M.B., Ch.B. — G. A. Burbidge.

Halle a. d. S. (Deutschland).
Pharmacol. Inst. der Univ., Magdeburger Str. 22a. Dir. Prof. Dr. M. Kochmann. — Dr. Wagner, Dr. Seel; Dr. Hessel. *3—4 Plätze für chem. u. biol. Arbeiten.*

Hamburg (Deutschland).
Pharmacol. Inst. Univ., Allg. Krankenhaus St. Georg, Lohmühlenstr. 3. Prof. Dr. Arthur Bornstein; Dr. Walter Griesbach.

Hannover (Deutschland).
Pharmacol. Inst. Tierärztl. Hochsch. Dir. Prof. O. Künnemann.

Heidelberg (Deutschland).
Pharmacol. Inst., Hauptstr. 47/51. Dir. Prof. Dr. H. Wieland. — Prof. Dr. Ph. Ellinger; Priv.Doc. Dr. B. Behrens. *4 Plätze.*

Helsinki (Finnland).
Pharmacol. Inst. der Univ., Riddaregatan 3. Dir. Prof. Dr. Y. Airila.

Innsbruck (Österreich).
Pharmacol. Inst. Univ. Dir. Prof. A. Jarisch.

Jassy (România).
Labor. de Pharmacol., Fac. de Med. Dir. Prof. J. Enescu.

Jena (Deutschland).
Pharmacol. Inst. Univ. Dir. Prof. Dr. Kionka.

Johannesburg (South Africa).
Pharmacol. Labor., Univ. of the Witwatersrand, P. O. Box 1176. Dir. Prof. J. M. Watt. — Marie G. Brandwijk, Junior Lect. (Chem. and Botan.). *2 places in experim. Pharmacol. and 1 place in Chem. Pharmacol.* ⊕ Collection of Native Med. Plants etc., most of which are being housed in the South African National Herbarium, Pretoria.

Kanazawa (Japan).
Pharmacol. Inst. Kanazawa Med. Univ.

Kaunas (Litauen).
Farmacol. Katedros Labor. Lietuvos Univ. Dir. Prof. P. Raudonikis. — Ben. Šiaulis, Ass. *The labor. is being equipped for detailed study of the properties of smooth muscle.*

Kazan (Ü.d.S.S.R.).
Pharmacol. Labor. des Vet.-Inst. Dir. P. Iv. Popoff. — Sophie Sidorowa. *2 Plätze.*

Keijô, Korea (Japan).
Pharmacol. Inst. Coll. of Med., Keijô Imper. Univ.

Kiel (Deutschland).
Pharmacol. Inst. Dir. Prof. Dr. Fritz Külz.

Kingston, Ont. (Canada).
Dept. of Pharmacol. Queen's Univ. Dir. Dr. T. W. Gibson.

København (Danmark).
Farmacol. Inst., Juliane Maries Vej 20. Dir. Prof. Dr. med. J. C. Bock. — Ass.: Cand. pharm. K. K. B. B. Larsen, Cand. med. Harald Okkels, Cand. med. L. Vøhtz.
Farmacol. Labor., Bülowsvej. Prof. C. H. Hansen.

Köln (Deutschland).
Pharmacol. Inst., Zülpicher Str. 47. Dir. Prof. Dr. med. et phil. Schüller. — 2 wissenschaftl. Ass.

Königsberg i. Pr. (Deutschland).
Pharmacol. Inst. der Univ., Copernicusstr. 3—4. Dir. Prof. Dr. F. Haffner. — Ass.: Dr. P. Pulewka; Dr. K. Kötzing. *6 Plätze.*

Kumamoto (Japan).
Pharmacol. Inst. Kumamoto (Imperial) Med. Univ.

Kyôto (Japan).
Pharmacol. Inst. Coll. of Med., Kyôto Imper. Univ. Dir. Prof. Dr. Kwata Marishima. — Prof. Dr. Yoshizumi Ozaki; Ass. Prof. Kikuo Ogyu; Ass. Kagemasa Kuwabara; Lect. Eitarô Marui.
Pharmacol. Inst. Kyôto Provincial Med. Univ., Hirokôji.

La Habana (Cuba).
Pharmacol. Labor., Univ.

Lausanne (Suisse).
Labor. de pharmacol. de l'Univ., Place du Château. Dir. Prof. Strzyzowski.

Leiden (Holland).
Pharmacotherapeutical Inst., Rapenburg 22. Dir. Prof. Dr. W. Storm van Leeuwen. — Dr. Nijk, Conserv.; Dr. Varekamp, Ass.; Dr. Niekert, Ass. *5 places for Researches on allergic diseases.* ⊕ Collection of various pollens and cultures of microorganisms producing climate-allergess.

Leipzig (Deutschland).
Pharmacol. Inst. d. Univ. Dir. Prof. Oskar Gros.
Inst. für Pharmacol. u. Toxicol. d. Univ.-Tierpoliklinik. Dir. Prof. Reinhardt.

Leningrad (U.d.S.S.R.).
Labor. für Experim. Pharmacol., Staatl. Inst. für Experim. Med.

Leverkusen a. Rh. (Deutschland).
Labor. für Pharmacol. der I. G. Farbenindustrie A. G.

Lille (France).
Labor. de Pharmacol., Fac. de Méd. de l'Univ. Dir. Prof. Bédart; Prof. E. Gérard.

Lisbôa (Portugal).
Inst. de Farmacol. e terapeutica, Fac. de Med. Univ. Dir. Prof. S. Rebelo Alves.

London (England).
Pharmacol. Labor. Univ. Coll., Gowen Str., W.C. 1. Dir. Prof. E. B. Verney. — Dr. F. R. Winton. *3 places for experim. investigation of the action of pharmacol. agents on physiol. processes. — String galvanometer.*
Dept. of Pharmacol., London School of Med. for Women of the Univ., Royal Free Hospital, 8, Hunter Str., Brunswick Square, W.C. Dir. E. Scarborough.
Pharmacol. Labor., London Hosp. Med. Coll. of the Univ., E., Turner Str., Mile End.
Pharmacol. Labor., Pharmaceutical Society of Great Britain. Dir. J. H. Burn.
Dept. of Biochem. and Pharmacol. Nation. Inst. f. Med. Research. Confer: Biochemia generalis.

London, Ont. (Canada).
Dept. of Pharmacol.of the Univ.of Western Ontario. Dir. J. W. Crane, M.B. — J. R. LeTouzel, M.D., C.M.

Louvain — Leuven (Belgique).
Labor. de Pharmacol. Dir. Prof. M. Ide. — Prof. A. Castille.

Ludwigshafen a. Rh. (Deutschland).
Pharmacol. Labor. der Knoll-A.-G. Dir. W. Friehler.

Lund (Sverige).
Farmacol. Inst. Univ. Dir. Prof. Overton.

Lyon (France).
Labor. de Pharmacol. Fac. Méd. de l'Univ. Dir. Prof. Leulier.

Lwów (Polska).
Inst. f. experim. Pharmacol. d. Univ. Dir.: vacat, in Vertretung M. Franke. — Dr. F. Kmietowicz.
Pharmacol. Labor. d. Tierärztl. Akad. Dir. Prof. A. Gizelt.

Madrid (España).
Labor. de Pharmacol. Fac. Méd. Dir. Dr. Tomás Aldan. — Prof. Dr. Teófilo Hernando; Dr. Juan Planelles.

Manila (Philippine Islands).
Dept. of Pharmacol. Dir. Prof. D. De la Paz. — F. Garcia; R. Guevara.

Marburg a. d. L. (Deutschland).
Pharmacol. Inst. Dir. Prof. Aug. Gürber. *Blut, Verdauung, Stoffwechsel, Pharmacol.*

Messina (Italia).
Ist. di Farmacol., Via Risorgimento 199. Dir. Prof. Gaetano Vinci. — Efisio Luigi Eocco, Libero Doc.; Giuseppe Carbonaro, Ass. *4 posti. Farmacodinamica. Farmacognosia. Chimica fisica.*

Minsk (U.d.S.S.R.).
Pharmacol. Inst. Univ. Dir. Prof. A. P. Bestužev.

Modena (Italia).
Inst. di Farmacol. Univ. Dir. G. M. Piccinini. — Dott. Angelina Levi. *12 posti. — Apparecchi di Fisiol. e di Chim.*

Montreal, Que. (Canada).
Dept. of Pharmacol. McGill Univ. Dir. R. L. Stehle, M.A., Ph.D.; — D. S. Lewis, M.Sc., M.D.

Moskau (U.d.S.S.R.).
Labor. der Pharmacol. 2. Staatl. Univ., Med. u. Chem.-Pharmacol. Fac., Pogodinskaja, 6. Dir. Prof. Dr. med. Wladislaw Skworzoff. — Ass.: Al. Lübrischin, Iv. Schischoff, Zoe Iljina, Ant. Langaglo; Priv.Doc. Nik. Grigorowitsch. *4 Plätze. — Für pharmacol. u. biochem. Analyse.* Ⓟ Chem., pharmakognostische und pharmaceutische Praep.

München (Deutschland).
Pharmacol. Inst. Univ. Dir. Prof. W. Straub. — Priv.Doc. Dr. Fromherz.
Pharmacol. Inst. d. tierärztl. Fac. Univ. Dir. U. Jodlbauer. *Lichtwirkungen.*

Münster (Deutschland).
Pharmacol. Inst. Univ., Westring 12. Dir. Prof. Dr. Hermann Freund.

Mukden, Manchuria (China).
Pharmacol. Inst. South Manchurian Med. Univ. Prof. Tei-ichi Masuda.

Nagasaki (Japan).
Pharmacol. Inst. Nagasaki Imperial Med. Univ.

Nagoya (Japan).
Pharmacol. Inst. Aichi Med. Univ.

Napoli (Italia).
Ist. di Farmacol. Dir. Pio Marfori.

Niigata (Japan).
Pharmacol. Inst. Niigata Med. Univ.

Norwich, N.Y. (U.S.A.).
Research Labor. of the Norwich Pharmacal Co.

Odessa, Ukraine (U.d.S.S.R.).
Pharmacol. Sektion d. Katheders für experim. u. klin. Med. d. wissenschaftl. Forschungsinst., Olgiewskaja 4. Dir. Prof. Dr. D. M. Zavrov.
Pharmacol. Labor. d. med. Inst., Olgiewskaja 4. Dir. Prof. Dr. Lawrow David Melitonowitsch. — Dr. Sergei Zyganoff; Dr. Alexei Leibenson; Dr. Al. Winogradoff. *6 Plätze.*

Okayama (Japan).
Pharmacol. Inst. der Kaiserlichen Med. Univ. Okayama. Dir. Prof. Dr. Kwanichiro Okushima. — Ass.: Dr. Masao Fujita, Dr. Masomi Nishishita. *2 Plätze für physiol., 2 für chem. Untersuchungen. — Physiol. und chem. Abteilung.*

Omsk (U.d.S.S.R.).
Pharmacol. Abt. d. Med. Inst. Dir. Nikolai V. Veršinin.

Ôsaka (Japan).
Pharmacol. Inst. Ôsaka Provincial Med. Univ. Dir. Prof. Sentarô Nagasaki. — Ass. Prof. Masayuki Okazawa.

Oslo (Norge).
Univ. farmacol. Inst. Dir. Prof. Dr. Poul Edvard Poulsson. — Dr. med. Klaus Hansen; Ing. chem. Gunnar Weidemann. *12 Plätze für chem. Arbeiten, experim. Pharmacol., Vitaminarbeiten.*

Oxford (England).
Labor. of Pharmacol. Fac. med. Univ. Dir. Prof. J.A. Gunn.

Padova (Italia).
Ist. di Farmacol. della R. Univ., 14, Viale Loredan 2. Dir. Prof. Luigi Sabbatani. — Dott. Egidio Meneghetti, Libero Doc.; Dott. Luigi Scremin, Libero Doc.; Dott. Mariano Messini; Dott. Giulio Stella; Dott. Giacomo Spagnol. *Ricerche di chim.-fisica e chim. colloidale applicata alla Farmacol.*

Palermo (Italia).
Ist. di Farmacol. sperim. Univ. Dir. Prof. C. Lazzaro.

Parel, Bombay (Br.-India).
Haffkine Inst. Dir. Lt. Col. F. P. Mackie, O.B.E., M.D., M.Sc. — Major S. S. Sokhey, M.D., I.M.S.;

Dr. S.N. Gore, L.M. and S.; Dr. K.S. Mhaskar, M.A., B.Sc., M.D.; Rev. Father Caius, S.J., M.S., C.I.; Dr. B. P. B. Naidu, M.D., M.Hy., D.P.H., D.T.M.; Dr. S. A. Kamat, Ph.D.; Dr. Margaret Balfour, M.D., C.B.E. *Plague vaccine, Anti-rabic vaccine. Biochem., pharmacol., bact. Section. Snake venom.* ⊕ Plague, snake venom, tropical disease specimens.

Parma (Italia).
Ist. di Farmacol. sperim. Univ. Dir. Prof. A. Chiztoni.

Pavia (Italia).
Ist. di Farmacol. sperim. della R. Univ., Palazzo Botta. Dir. Prof. Italo Simon. — Dr. Angelo Rabbeno, Aiuto; Dr. Ambrogio Mantegazza, Ass. *10 posti. — Sezioni dell'Ist.: Microchim., Chim., Fisiol. (grafiche), Chim.-fisica, Microscopia, Farmacognosia. Metodi fisiol., chim. e chim.-fisici. Crioscopia degli organi.*

Pécs (Ungarn).
Pharmacol. Inst. Univ. Dir. Prof. Dr. Géza Mannsfeld. — Dr. R. Bodó; Dr. El. Csillag; Dr. A. Lánczos; Dr. K. Hecht; L. Scheffer; L. Szirtes; Z. Horn; El. Szabó; E. Danielisz.

Peking (China).
Dept. of Pharmacol. Peking Union Med. Coll. Dir. Dr. Bernard E. Read. — Dr. T. Q. Chou, Chem.; Dr. C. Pak, Physiol.; Mr. J. C. Liu, Botan.; Mr. C. T. Feng, Pharmaceutic. *4 working places in any of the above subdivisions. — Division of Chem. Products, making ethyl esters and ephedrine. Botan. and Chinese materia med. collections; Good general pharmacol. equipments.* ⊕ Botanical and drug collections.

Perm (U.d.S.S.R.).
Pharmacol. Labor. Univ., Zaimka. Dir. Prof. Jakob Perichanjanz.

Perugia (Italia).
Ist. di Materia med. e Farmacol. Fac. med. Univ. Dir. Prof. E. Filippi.

Pisa (Italia).
Ist. di Materia med. e Farmacol. sperim. Fac. med. Univ. Dir. Prof. D. Baldi.

Pittsburg, Pa. (U.S.A.).
Dept. of Pharmacy Univ. of Pittsburg. Dir. Prof. L. K. Darbaker.

Poitiers (France).
Labor. de Pharmacie et Matière méd. École réorganisée de méd. Univ. Dir. Prof. M Roblin.

Porto Alegre (Brasil).
Pharmacol. Inst. Univ. Dir. Prof. A. Galvão.

Poznań — Posen (Polska).
Inst. de Pharmacol. Univ. Dir. Prof. Dr. R. Trzecieski.

Praha — Prag (C.S.R.).
Pharmacol.-pharmacognostisches Inst. d. Karls-Univ., II, Na Bojišti 3.
Pharmacol.-pharmacognostisches Inst. d. Deutschen Univ., II, Albertov 7. Dir. Prof. Dr. Wilhelm Wiechowski.

Quito (Ecuador).
Inst. f. Pharmacol. u. Toxicol. Dir. Prof. F. J. Barba.

Reykjavik (Island).
Pharmacol. Inst. Univ. Dir. Prof. G. Bjarnhjedinsson.

Riga (Latvija).
Pharmacol. Inst. der Univ., Kronwald-Boulevard 9. Dir. Prof. Dr. Cäsar Amsler. — Eduard Rentz, Ass.

Rio de Janeiro (Brasil).
Inst. f. Pharmacol. Dir. Prof. P. A. Pinto.

Roma (Italia).
Ist. di materia med. e farmacol. Fac. med. Univ. Dir. Prof. A. Bonanni.

Rosario (Argentina).
Inst. f. Pharmacol. Dir. Prof. L. Negrette.

Rostock i. M. (Deutschland).
Pharmacol. Inst. der Univ., Gertrudenstr. Dir. Prof. Dr. Ernst Frey. — Dr. Ernst Ruickoldt, Ass. *Physiol., chem., histol. Arbeiten.*

Rostow a. Don (U.d.S.S.R.).
Pharmacol. Labor. Univ. Dir. Prof. Zitovič.

San Francisco, Cal. (U.S.A.).
Dept. of Pharmacol., Med. School, Univ. of California. Dir. Prof. G. W. Clark.

Saratow (U.d.S.S.R.).
Pharmacol. Labor. Univ. Dir. Prof. K. A. Smelov.

Sassari (Italia).
Labor. di Farmacol. sperim. Dir. Prof. Mario Chiò. — Dott. P. Testoni, Ass. *Pharmacol. du cœur de hétérothermes.*

Sendai (Japan).
Pharmacol. Inst. Coll. of Med., Tôhoku Imper. Univ. Dir. Prof. Dr. Seiichi Yagi. — Ass. Prof. Taiichi Murashima.

Sevilla (España).
Labor. de Farmacol. Fac. Méd. Dir. Prof. Dr. Emilio Munoz Riverodel Olmo.

Shanghai (China).
Pharmacol. Inst. der Tung-chi Univ., 22a Burkill Road. Dir. Dr. med. A. Kessler.

Sheffield (England).
Labor. of Pharmacol. Univ. Dir. Prof. E. Mellanby.

Siena (Italia).
Ist. di Materia med. e Farmacol. Fac. med. Univ. Dir. Prof. C. Raimondi.

Smolensk (U.d.S.S.R.).
Pharmacol. Kabinett. Dir. Prof. Nikolaev.

Sofia (Bulgarien).
Inst. f. Pharmacol. u. Therapie Med. Fak. Univ. Dir. Prof. W. O. Alekseleff. — Ass.: Dr. K. Wassilewa; Dr. M. Dimitrakoff; Dr. Aleksander Nikolaeff.
Inst. für Pharmacol. Veterinärmed. Fac. Dir. Prof. Dr. Johan Närr. — P. A. Poppoff, Ass.

Stambul (Türkei).
Labor. de Pharmacol. Fac. méd. Univ. Dir. Prof. M. Neozad.
Pharmakol. Inst. der Tierärztl. Hochsch. Dir. Prof. I. Hakki.

Stockholm (Sverige).
Pharmakol. Inst. des Karolinska Inst. Dir. Prof. C. G. Santesson.

Strasbourg — Straßburg (France).
Labor. Pharmacodynamie Fac. Méd. Dir. Prof. Ambard.

Szeged (Ungarn).
Pharmacol. und Pharmakognostisches Inst. Univ., Kálvária tér 5. Dir. Prof. Béla v. Issekutz. — Dr. Al. Tukats, Adjunctus; Dr. Jul. Méhes; Dr. M. Leinzinger; Dr. F. Végh; Dr. J. Bath. *2 Plätze. Arbeiten mit überlebenden isolierten Organen. Biol. Wertbestimmungen der Arzneien. Insulinforschung. — Durchströmungsapparate. Gaswechselbestimmungsapparat. Barcroft-Warburgsche Apparate. Interferometrie.*

Taihoku, Formosa (Japan).
Pharmacol. Inst. Taihoku Governmental Coll. of Med.

Tartu — Dorpat (Eesti).
Pharmacol. Inst. der Univ., Domberg. Dir. Prof. Dr. S. Loewe. — Dr. M. Ilisson, Ass.; Dr. F. Lange, Ass. der Abt. Arzneiprüfungsamt; Dr. E. Käer; Dr. rer. nat. E. H. V. Voss, s. t. Priv.Doc. d. Zool., Vorst. der zool.-histol. Abt.; Mag. chem. A. Wähner, physik.-chem. Privat-Ass.; E. Paas. *8 Plätze für Biochem. u. tierexperim. Arbeiten. — Mechan. Energie künstl. Atmung, Zeitschreibung, graph. Registrierung, el. Reizung u. Heizung auf jedem Arbeitsplatz.* Ⓟ Pharmacol. Sammlungspraep. — Dem Inst. zugeordnet ist die pharmacol. Abt. des Staatlichen Arzneiprüfungsamtes.

Tiflis (U. d. S. S. R.).
Pharmacol. Inst. Univ. Dir. Prof. W. Mossešvili.

Tôkyô (Japan).
Pharmacol. Inst. Military Med. Coll.
Pharmacol. Inst. Jikeikwai Med. Univ., Shiba.
Pharmacol. Inst. Nippon Girls' Med. Univ.
Pharmacol. Inst. Nippon Med. Univ.
Pharmacol. Inst. Fac. of Med., Tôkyô Imper. Univ. Prof. Dr. Harno Hayashi; Prof. Dr. Kenzô Hamura.
Pharmacol. Inst. Coll. of Med., Keiô-Gijuku Univ. Dir. Prof. Dr. K. Abe. — Ass. Prof. Dr. S. Miyazaki; Lect. F. Arima.

Tomsk (U. d. S. S. R.).
Inst. f. Pharmacol. Univ. Dir. Prof. N. V. Veršinin.

Torino (Italia).
Ist. di materia med. e farmacol. sperim. Fac. med. Univ. Dir. Prof. P. Giacosa.

Toronto, Ont. (Canada).
Dept. of Pharmacol. Univ. of Toronto. Dir. V. E. Henderson, M.B. — M. J. Sparks; G. H. W. Lucas, Ph.D.

Tsinan (China).
Labor. of Materia med. and Pharmacy Shantung Christian Univ. Dir. Prof. W. P. Pailing.

Tucumán (Argentina).
Labor. de Farmacia práctica y Farmacol. Univ. Dir. Prof. A. Rovelli.

Tübingen (Deutschland).
Pharmacol. Inst. Univ. Dir. Prof. Carl Jacoby. *Pharmakol. u. Pysiol.*

Uppsala (Sverige).
Farmakol. Institutionen. Dir. Prof. Dr. E. L. Backman.

Urbino (Italia).
Gabinetto di Materia med. Univ. Dir. inc. Prof. C. Ricci.

Utrecht (Holland).
Pharmacol. Labor. der Reichsuniv., Servaasbolwerk 1a. Dir. Prof. Dr. R. Magnus † 25. 7. 27. — Dr. J. W. le Heux (Conserv.); Dr. A. P. H. A. de Kleyn, Dr. C. de Lind van Wijngaarden (Ober-Ass.); Dr. G. G. J. Rademaker (Ass.). *2—4 Plätze für experim. und operative Physiol. u. Pharmacol. — Brodi-Atmungsapparat.*

Valladolid (España).
Labor. de Farmacol. Fac. Med. Dir. Prof. Dr. Marino Monserrate Abad.

Warszawa (Polska).
Inst. de Pharmacol. experi m. de l'Univ., Krakowskie-Przedmieście 26|28. Dir. Prof. Dr. méd. Georges Modrakowski. — Ass.: Dr. Emile Leyko, Dr. Henri Sikorski, Dr. Jean Supniewski; Aides Ass.: Dr. Stanislaw Kroszczyński, Mag. Farm. Louis Glodowski. *4 places pour les recherches experim. dans le domaine de la circulation, de la digestion et des sécrétions; facilité spéciale pour les travaux sur les organes isolés. — Les appareils modernes pour l'analyse des gaz du sang, etc., et pour la détermination de p_H (Potentiomètre).*

Washington, D.C. (U.S.A.).
Pharmacol. Labor. of the Bureau of Chemistry. E. W. Schwartze.

Wien (Österreich).
Pharmacol. Inst. der Univ. Dir. Prof. Dr. Ernst P. Pick. *Diurese, Leberfunktion für den Kreislauf u. die Diurese. Angriffspunkte der Schlafmittel.*
Lehrkanzel für Pharmacol. an der Tierärztl. Hochschule. Dir. Prof. Dr. Gustav Günther. — Dr. Wilhelm Heeke, Ass. *10 Plätze.*

Wilno (Polska).
Inst. f. pharmacol. Chem. Univ. Dir: Prof. W. Karaffa-Korbutt.
Inst. Pharmacol. Univ. Dir. Prof. Dr. C. Traczewski.

Woosung b. Shanghai (China).
Inst. f. Pharmacol. u. Hygiene Tungchi Univ. Dir. Prof. Dr. Kessler.

Woronesch (U. d. S. S. R.).
Pharmacol. Inst. der staatl. Univ., Prospekt Revoluzii, N. 3. Dir. Prof. Dr. W. N. Woronzow. — Arzt I. W. Troitzky, zeitl. Prosektor; Arzt W. J. Leszinski, Ass.

Würzburg (Deutschland).
Pharmacol. Inst., Kölliker Str. 2. Dir. Prof. Ferdinand Flury. — Priv.Doc. Dr. Hans Steidle. *2 Ass. 5 Plätze für Chemie, Physiol., Pharmacol. (Gifte).*

Zürich (Schweiz).
Pharmacol. Inst. Univ., Gloriastr. 32. Dir. M. Cloetta. — *2 Ass. 2—3 Plätze.*

Laboratoria serologiae et immunologiae.

Confer: *Labor. Bacteriologiae* (pag. 479) et *Labor. Hygienae* (pag. 487), *Physiologia pathol. gen. experim.* (pag. 469).

Berlin (Deutschland).
Serumabt. des Hygien. Inst. Tierärztl. Hochsch., NW 6, Luisenstr. 56. Dir. Prof. K. Bierbaum.

Bologna (Italia).
Labor. Mil. per la Produzione del siero Antitetanico, Caserna Davia. Dir. Prof. Dott. Pl. C. Bardelli, Maggiore Veterinario. — Dott. V. Cilli, Ass. *Sierol. specialmente applicata ai sieri antitossici. Micol.* Ⓟ *Colture di anerobi e specialmente ceppi svariati di Bac. del tetano. Colture del "Cryptococcus farciminosus Rivoltae".*

Boston, Mass. (U.S.A.).
Wassermann Labor. Mass. Dept. of Public Health. Dir. W. A. Hinton.
Antitoxin and Vaccine Labor. Dir.: B. White; Ass. Dir.: E. S. Robinson. *Antitoxin and Vaccine gonorrhea, glanders fixation tests for syphilis, Immunol.*

Butanton, São Paulo (Brasil).
Serotherapeutisches Inst. Dir. Dr. V. Brazil.

Cairo (Egypt).
Vaccine Inst. Publ. Health Service.
Serum Inst. Veterinary Service.

Cambridge (England).
Dept. of Pathol. Confer: Anat. pathol. generalis humana.

Danzig (Danzig).
Serumlabor. d. westpreußischen Landwirtschaft. Dir.: Dr. Stube.

Glenolden, Pa. (U.S.A.).
Antivenin Inst. of America. Dir. Afranio do Amaral.

Hamburg (Deutschland).
Inst. f. Immunität Univ. (pathol. Biol.), Eppendorfer Krankenhaus. Dir. Prof. Hans Much. — Dr. H. Schmidt.
Serol. Labor. der Staatskrankenanstalt Univ., Friedrichsberg.

Helsinki (Finnland).
Staatl. Serumlabor. u. Pasteur-Inst. Dir. Dr. J. A. Murto. *Serol. u. Bact. Antibacterielle Vaccine. Bakterienconglutination, Theorie der Seroreaktionen bei Lues u. Lepra. Oligodynamische Untersuchungen an Bakterien.*

Höchst a. M. (Deutschland).
Sero-bact. Abt. des Labor. der I. G. Farbenindustrie A.-G. Dir. R. Bieling.

Jassy (România).
Labor. de Pathol. générale. Dir. Prof. Mihăeşti G. Ionescu. *Immunol.*

Jena (Deutschland).
Virusforschungsanstalt. Dir. Prof. Dr. Willy Pfeiler. *Maul- und Klauenseuche; septischer Abortus; Krebs; ansteckende Euterentzündung in Schleswig-Holstein; Hühnertyphus.*

Kaunas (Litauen).
Pasteur-Inst., Posko Gatve.

Kiew (U.d.S.S.R.).
Variola-Vakzine-Abt. des Bact. Inst. Dir. Prof. Dr. Iw. W. Hach.

København (Danmark).
Seruminst. d. Landwirtsch. Hochsch., V, Bülowsvej. Dir. Prof. Dr. C. O. Jensen. — Tierärzte: C. W. Andersen, A. Thomsen, V. Andersen, H. O. Schmit-Jensen.
Statens Seruminst., Amager Boulevard. Dir. Dr. Th. Madsen. — Abt.-Vorst.: Dr. med. Martin Kristensen und Dr. phil. L. Walbum.

La Plata (Argentina).
Labor. de Toxicol. Fac. de química y farmacol. Univ. Dir. Prof. T. Ugarte.
Labor. de Toxicol. Fac. med. vet. Univ. Dir. Prof. A. Candioti. *Semiol., Materia méd. y Toxicol.*

Leverkusen a. Rh. (Deutschland).
Labor. für Serol. der I. G. Farbenindustrie A.G.

London (England).
Calf Vaccine Labor. Lister Inst. of Preventive Med. of the Univ., Chelsea-Bridge-Rd., S.W. 1. Bact.: A. B. Green.

Mannheim (Deutschland).
Pathol., bact. u. serol. Inst. d. städt. Krankenanstalten. Confer: Anat. pathol. humana.

Marburg a. d. L. (Deutschland).
Inst. für experim. Therapie „Emil von Behring". Dir. Prof. Dr. Hermann Dold. — Priv. Doc. Dr. Hans Schmidt; Dr. W. Scholtz; Dr. H. Groß. *Für bact. u. serol. Untersuchungen.* ⓟ Bact. u. serol. Praep.

New York (U.S.A.).
Serol. Labor. Dept. of Health. M. A. Wilson (Bacteriologist).
Bureau of Labor. of the Dept. of Health of the City of New York, Foot of East, 16th Street. Dir. Dr. W. H. Park. — Ass. Dir.: Dr. A. W. Williams, Dr. Ch. Krumwiede. *Bact., Serol. u. Immunitätswiss. Herstellg. versch. Heilsera, Vaccine, Pockenlymphe usw. Bact. Kontrolle d. Milchwirtschaft.*

Olivet, Mich. (U.S.A.).
Dept. of Biol. Olivet Coll. Prof. G. F. Forster (Biol.). *Immunol. Precipitins. Hypersensitiveness to animal products.*

Palermo (Italia).
Ist. Neoimmunitario Italiano, Corso Calatafimi, 412. Dir. Dr. Nello Mori.

Paris (France).
Service de Sérothérapie de l'Inst. Pasteur. Dir. L. Martin.
Labor. du Séro-Diagnostic Inst. Pasteur., 21—25, Rue Dutot. Dir. Dr. Mutermilch. ⓟ *Produits servant au séro-diagnostic de la Syphilis.*
Labor. de Bact., de la Fac. de Méd. Dir. Prof. Besançon.
Labor. serol. de l'Ecole pratique des Hautes-Etudes. Dir. Prof. R. Lévy.
Inst. supérieur de Vaccine. (Inst. d. Acad. de Med.). Dir.: L. Camus. — Préparateur: L. Tanon; Préparateur du Labor.: M. Maitre.

Poitiers (France).
Labor. de Chimie et Toxicol. École réorganisée de méd. Univ. Dir. Prof. R. Sauvage.

Praha — Prag (C.S.R.).
Serol. Abt. hyg. Inst. der Deutschen Univ., II. Dir. F. Breinl.

Puławy (Polska).
Abt. f. Serol. des Staatl. Wissenschaftl. Inst. f. Landwirtschaft. Dir. Prof. T. Jaroszynski.

Reims (France).
Labor. de Chimie et Toxicol. École préparatoire de méd. Dir. Prof. Bottu.

Insel Riems,
PostMesekenhagen b. Greifswald (Deutschl.).
Staatliche Forschungsanstalten Insel Riems. Dir. Prof. Dr. Otto Waldmann. *Die Anstalt dient zur wissenschaftl. Erforschung d. Maul- u. Klauenseuche u. d. Schweinepest sowie d. Herstellung eines Hochimmunserums gegen diese Seuchen.*

Rio de Janeiro (Brasil).
Inst. Vaccinico.

Rotterdam (Holland).
Rijksseruminrichting. Dir. Dr. L. F. D. E. Lourens. — Abt.-Dir.: Dr. H. E. Reeser; Bakteriologen: Dr. H. v. Straaten, Dr. B. J. C. te Hennepe, Dr. A. J. S. v. Alphen, Dr. K. Büchli, F. Meyer Cluwen, Dr. A. J. Winkel; Landbouwkundige: J. H. Boersma. Conserv.: J. Aalders.

Saigon (Indo Chine).
Serol. Sect. Inst. Pasteur.

São Paulo (Brasil).
Dept. of Microbiol. Confer: Parasit. generalis.

Saratow (U.d.S.S.R.).
Staatsinst. f. Mikrobiol. u. Epidemiol. im Südosten d. U.d.S.S.R., Kozarmennaja 18. Serum-Abt.: Dir. Dr. I. V. Kolpakov — Vaccine-Abt.: Dir. Dr. E. I. Korobkov.

Sassari (Italia).
Ist. di Chimica farmaceutica e tossicol., Scuola di farmacia Univ. Dir. Prof. C. Gastaldi.

Sofia (Bulgarien).
Serol. Abt. des Staatl. Bakt. Inst., Buleward Makedonia. Dir. Dr. S. Breier.

Stambul (Türkei).
Section des Vaccins a l'Inst. bact., Rue Matbaa. Dir. Prof. S. Isan.
Inst. f.Toxikol. Tierärztl.Hochsch. Dir. Prof. Faik.

Tartu — Dorpat (Eesti).
Serum Labor. Univ., Vene t. 28. Dir. Dr. G. Heinrich.

Tôkyô (Japan).
Serol. Labor. Kitasato Inst. for Infect. Diseases, Shirokane-Sankô-Chô. Dir. Dr. Taiichi Kitashima.

Utrecht (Holland).
Rijks Serol. Inst., Sterrenbosch 1. Dir. Prof. Dr. H. Aldershoff; Onder-Dir.: Dr. A. B. F. A. Pondman. — W. A. Timmermann.

Valencia (España).
Labor. de Toxicol. Fac. med. Univ. Dir Prof. J. B. Peset Aleixandre.

Warszawa (Polska).
Inst. f. pharmazeut. u. toxicol. chem. Univ. Dir. Prof. J. Zaleski.
Serol. mikrobiol. Inst. Dir. Prof. R. Nitsch.

Wien (Österreich).
Staatl. Serotherapeutisches Inst., IX, Zimmermanngasse 3. Dir. Prof. Dr. Rudolf Kraus. — Prof. Michael Eisler; Prof. E. Loewenstein; Dr. St. Baecher; Dr. Teichmann; N. Kovaes. *Mikrobiol., Immunol. Erzeugung therapeut. u. diagnost. Sera u. bact. Praep. (Tuberkulin, bact. Vaccinen usw.).* Bact. Untersuchungsstelle u. Station f. Blutuntersuchung nach Wassermann.

Laboratoria bacteriologiae hominis.

Confer: *Labor. Hygienae hominis* (pag. 487), *Labor. Serologiae* (pag. 477), *Anatomia pathol. humana* (pag. 459), *Physiologia pathol. generalis* (pag. 469), *Parasitologia generalis* (pag. 379), *Biochemia generalis* (pag. 435), *Microbiologia agris* (pag. 533).

Adelaide (South Australia).
S. A. Government Labor. of Pathol. and Bact., North Terrace. Dir. L. B. Bull, D.V.Sc.

Akron, O. (U.S.A.).
Bact. research Labor. Municipal Univ. Dir. R. D. Fox. *Pathogenic Bact., Cultivation and Immunol., Pleomorphism.*

Alger (Algérie).
Inst. Pasteur. Dir. Dr. E. L. M. E. Sergent. — Dr. L. Parrot; Dr. M. Beguet; M. H. Rougebief; Dr. E. Murat; Dr. Et. Sergent; Dr. A. Catanel; Dr. H. Foley; A. Donatien; F. Lestoquard; E. Plantureux; L. Musso.

Ann Arbor, Mich. (U.S.A.).
Pasteur Inst., Univ. of Michigan. Dir. H. W. Emerson.

Athènes (Grèce).
Labor. de Microbiol. et d'Hygiène, d'Univ., rue de l'Académie. Dir. C. Savas.
Inst. Pasteur Hellenique, -Ambélokipi, 103, Avenue de Kiphissia. Dir. Dr. Georges Blanc. — Service méd.: Chefs de Labor.: Dr. J. Caminopetros (Pathol. expérim.), Dr. G. Johannides (Microbiol. générale, Analyses). Service vét.: Chef de Labor.: Dr. C. Melanidis; Chef de Labor. adjoint: Dr. M. Stylianopoulo. *4 places pour Bact., Histol., Chim. biol., Parasit.* ⊕ Mollusques d'eau douce. Prép. de Leishmania tropica et L. Dokovani. Virus filtrauts (Herpet-, Vaccine-, Variole aviaire etc.).

Bahia (Brasil).
Bact. Labor. Dir. Prof. A. Vianna.

Baku (U.d.S.S.R.).
Inst. de microbiol. et d'hygiene, 39, Chemaklimskaja. Dir. P. P. Zdrodowski.

Banjaluka (S.H.S.).
Bact. Station.

Beirut (Syria).
Inst. d. recherches bact. S. Josephs Univ.

Beograd (S.H.S.).
Bact. Station, Durmitorska ulica. Dir. Dr. Stevan Ivanić.

Berlin (Deutschland).
Hygienisch-bact. Abt. d. Preuß. Landesanst. für Wasser-, Boden- u. Lufthygiene, -Dahlem, Ehrenbergstr. 38—40—42. Abt.-Leiter: Prof. Dr. med. B. Bürger (Bact. Reinigung von Wasser mit Filterung und Sterilisation, insbesondere mit Chlor, Abwasserdesinfektion). — Mitglieder, Hilfs- und Mitarbeiter: Dr. med. H. Beger (Leitung des Bact. Labor. einschl. Nährbodenherstellung, allgemeine Bact. und Serol., bact. Wasseruntersuchungen, insbesondere von Vorflutern); Dr. med. B. Nehring (bact. Abwasser-Untersuchung, Gerbereiabwässer- und Milzbrandfragen); G.M.R. Prof. Dr. med. E. Wernicke (Lufthygien.); Dr. med. H. Dornedden (Langsame Sandfiltration und nicht überstaute Filter).
Preuß. Inst. für Infektionskrankheiten „Robert Koch", N 39, Föhrer Str. 2. Ehrenmitglieder: Prof. Dr. Martin Kirchner, Prof. Dr. Emil Zettnow. Präs.: Prof. Dr. Fred Neufeld. — Abt.-Dir.: Prof. Dr. F. K. Kleine; Prof. Dr. Georg Lockemann; Prof. Dr. Richard Otto; Prof. Dr. Claus Schilling. Abt.-Leiter: Prof. Dr. Eduard Böcker; Prof. Dr. Heinrich Gins; Prof. Joseph Koch; Prof. Dr. Bruno Lange; Prof. Dr. Oskar Schiemann. 1 wissenschaftl. Hilfsarbeiter, 11 Ass. Leiter der Krankenabt. (Infektionskab. des Rudolf-Virchow-Krankenhauses): Prof. Dr. Ulrich Friedemann.
Bact. Abt. d. Reichsgesundheitsamtes, -Dahlem, Unter d. Eichen 82—84. Dir. Prof. Dr. L. Haendel. *Bact. Zool. Menschl. u. tier. Infektionen.*
Wissenschaftl. Inst. im Hauptgesundheitsamt der Stadt. Dir. Prof. Dr. Erich Seligmann. *Hygien. u. Bact. Seuchenbekämpfung. Diphtherie; Desinfektion.*
Bact. Abt., Rudolf-Virchow-Krankenhaus. Dir. Prof. K. Meyer. *Mehrere Plätze für Gäste.*

Bitolj, Mazedonien (S.H.S.).
Bact. Station. Dir. Dr. M. Rankov: — Dr. Dinko Stambuk.

Bordeaux (France).
Pasteur-Inst. Dir. Prof. Dr. G. Dubreuil.

Brno — Brünn (C.S.R.).
Inst. f. Bact., Serol. u. Hygiene der Haustiere der Tierärztl. Hochsch., Pražska 67. Dir. Prof. Dr. Franz Ševčík.
Mikrobiol. Labor. Univ., Pekařská 53—57. Dir. Prof. Dr. Jan Kabelík. — MUC. Martin Skýba, Ass.

Bruxelles — Brussel (Belgique).
Larbo. de bact. Univ. Dir. Prof. Dr. J. Bordet.

Inst. Pasteur, 30, Rue de Remorquem. Dir. J. Bordet; Sous-dir. E. Renaux. — Le Fevre de Assic, Ass.
Labor. de Microscopie. Dir. Prof. A. Lameere.

Bucureşti (România).
Inst. de Seruri şi Vaccinuri „Dr. I. Cantacuzino". Dir. Prof. Dr. Jean Cantacuzène. — Dir. adjoint: Prof. Dr. M. Cinca; Prof. Al. Cinca; Dr. I. Nicolau; Dr. D. Combiesco; Dr. P. Condrea; Dr. M. Nasta; Prof. Dr. Ch. Zotte; Mlle Dr. A. Damboisceanu; Dr. Max Marlee. *3 labor. individuels et une vingtaine de places dans les différents sections: Bact., pathol. expérim., Sérol. et Immunol., Chim. biol. et Chim. physique, pathol. comparée, Parasit., Protist., Entomol.* ⓟ Collection de microbes, parasites etc.

Budapest (Ungarn).
Bact. Inst. der Univ. Dir. Prof. H. Preisz. — Dr. L. Smányi; Dr. L. Gózany; Dr. St. Went; Dr. G. Markos; Dr. G. Oláh; Dr. A. Forró; A. Szilágyi.
Inst. für Seuchenlehre, Tierärztl. Hochsch., Rottenbiller-u. 23. Prof. Dr. F. Hutijra; Prof. Dr. Rudolf Manninger. Mitarb.: Dr. Arpád Márcis; Priv.Doc. der Serol. Dr. Eugen Schütz.
Staatsinst. für Bact. Dir. Gyula Schmidthoffer.
Pasteur-Inst., Türoltó-Utero. Dir. Prof. Dr. Agost Székely.

Buenos Aires (Argentina).
Inst. bact. del Ministerio de Agric., La Paternal F.C.P. Dir. Prof. Dr. Friz Ruppert. — Dr. W. A. Collier, Parasitologe; Dr. A. Scasso, pathol. Anatom; Dr. J. Quiroya, Serologe; Dr. A. Rottgardt, Bakteriologe; Dr. C. Zanini, Bakteriologe. Dr. A. Riglos, Bacteriol.
Inst. Bact. del Dept. Nac. de Higiene, Velez Sarsfield 563. Dir. Dr. Alfredo Sordelli. — Dres. Juan. M. Miravente et Carlos Zanolli (Bact.); Dr. Romirio Biglieri (Vacunas); Dr. Conrado Villegas (Rabies); Dr. Arturo Poiré (Tuberculosis); Dr. Manuel V. Carbonell, et J. M. de la Barrera (Higiene); Dr. Leopoldo Uriarte et Dr. Nestor Morales Villazón (Peste); Dr. Pedro Beltrami (Carbunclo); Dr. Juan Lewis (Farmacol.); Dr. Guido Pacella (Patol.); Dr. César E. Pico et J. Negrete (Sueroterapia); Dr. Raél Wernicke (Físico- Química); Dr. Roberto Dios (Protozool.). *Supérieur à 10 places pour Bact., Immunité, Pharmacol., Physique et Chim. Biol.* ⓟ Serpents, spécialement venimeux.

Buffalo, N.Y. (U.S.A.).
State Inst. for the study of Malignant Disease. M. C. Marsh (Biologist); G. T. Cori (Ass.); C. F. Cori (Biol. chemist).

Cape Town (South Africa).
Dept. of Bact., Univ., New Med. School Buildings, Falmouth Road, Observatory. Dir. Prof. William Campbell, M.B., Ch.B. — Dr. Noran McCullough Lect.; Dr. Greenfield, Ass.; Dr. von Düring, Ass. *10 places for Bact., Serol., Immunol., Protozool. The Dept. is well equiped, especially for researches in Experim. med. Its equipment being thoroughly modern and always well cared for.* ⓟ Cultures and Prep. of South African bact. — Clinical Labor. at the new Somerset Hospital.

Caracas (Venezuela).
Labor. de Bact. Univ. Dir. Prof. J. R. Risquez.

Celje, Slovenien (S.H.S.).
Bact. Station.

Cetinje, Montenegro (S.H.S.).
Bact. Station.

Coïmbra (Portugal).
Labor. de Bact. Univ. Dir. L. Pereira da Costa.

Curityba, Paraná (Brasil).
Labor. de Bact. méd. Univ. Dir. Prof. A. de Assis Gonçalves.

Dijon, Côte-d'Or (France).
Inst. de bact. et d'Hygiène. Dir. Prof. Charpentier.

Donja Tuzla, Bosnien (S.H.S.).
Bact. Station.

Dublin (Ireland).
Dept. of Bact. Univ. Trinity Coll. Dir. Prof. J. W. Bigger.

Dubrovnik (S.H.S.).
State Bact. Labor. Dir. Dr. Igor N. Asheshov. — Inna A. Asheshova, Ass. *Working places for guests in Bact., Ultrafiltration, Ultraviruses.* ⓟ Living cultures of the Bacteriophages. Macroscopial permanent prep., demonstrating the action of the Bacteriophages. Photographs of the same.

Dunedin (New Zealand).
Dept. of Bact. Univ. Dir. Prof. C. S. Hercus.

Edinburgh (Scotland).
Bact. Dept. Univ. Dir. Prof. I. P. Mackie. —, Dr. D. G. S. McLachlan, Lect.; Dr. J. M. Alston. Dr. A. M. M. Griesson, Dr. A. Messer, Ass; M. A. Cheyne, Senior Labor. Ass. and 5 other labor. Ass. *6 places for Bact. and Immunol.* ⓟ Pathol. material etc. containing, microorganisms and cultures of microorganism.

Edmonton, Alta. (Canada).
Dept. of Bact. and Hygiene Univ. of Alberta. A. C. Rankin; W. C. Laidlan; R. M. Shaw; L. C. Harris.

Ekatherinoslav, Ukraine (U.d.S.S.R.).
Bact. Inst. Med. Inst. Dir. Prof. S. Predtetschefski. *Immunisierung gegen Lyssa. Heilsera.*

Elberfeld (Deutschland).
Bact. Labor. der I. G. Farbenindustrie A.-G. Dir. Dr. Georg Wesenberg. *Med. u. techn. Bact.*

Entebbe, Uganda (British East Africa).
Inst. of bact. Research. Dir. Dr. Duke.

Ferrara (Italia).
Ist. di Patol. generale e di Batt. Univ., Palazzo Schifanoia. Dir. A. Marassini.

Gent — Gand (Belgie).
Labor. voor Bact. Dir. Prof. A. Heuseval.

Glasgow (Scotland).
Bact. Inst., Royal Infirmary. Dir. Dr. J. A. Campbell.

Gospić, Kroatien (S.H.S.).
Bact. Station.

Halifax, N.S. (Canada).
Dept. of Bact. Dalhousie Univ. Dir. R. J. Bean, M.S. — M. E. MacKay.

Helsinki (Finnland).
Bact.-Serol. Labor. Univ., Nikolaigatan 14. Dir. Prof. Dr. W. Osw. Streng.

Hongkong (China).
Labor. of Bact. Univ.

Ishigami (Japan).
Inst. der Infektionskrankheiten. Ishigami's Krankenhaus, Hamadera Park, near Osaka. Dir. Dr. T. Matsuda. *Tuberkulose.*

Jassy (România).
Labor. de Bact. Fac. Méd. Univ., Spitalul Paşcanu 4. Dir. Prof. Al. Slătineanu. — I. Gheorghin, Chef de travaux. *Biol. de la cellule cancereuse.*

Johannesburg (South Africa).
Dept. of Pathol. and Bact. Univ. Confer: Anat. pathol. humana experim.
Dept. of Bact. and Pathol. of the S. Afric. Inst. for med. Research. Dir. Prof. J. H. Pirie.

Kanazawa (Japan).
Microbiol. Inst. Kanazawa Med. Univ.

Karlsruhe i. B. (Deutschland).
Labor. für Bact. Techn. Hochsch. Dir. Prof. E. von Gierke.

Kazan (U. d. S. S. R.).
Bact. Inst. der Univ., B. Krasnaja 59. Dir. Prof. W. M. Aristowsky. — Blagoweschtensky u. Stoeltzer, Ass. *5 Plätze*. ⓟ Heilserum, Vaccine, Kulturen.

Keijô, Korea (Japan).
Microbiol. Inst. Coll. of Med., Keijô Imper. Univ.

Kiew (U. d. S. S. R.).
Bact. Inst., Stenki Rasin Str. 4 (Bajkova gora). Dir. Prof. Dr. M. P. Nestšadimenko. — Abt.-Vorst. PD. I. V. Hach (Vaccineabt.); Dr. B. I. Klein (Serotherapeut. Abt.); Prof. Dr. A. A. Krontovskij (Abt. f. experim. Med.); Prof. Dr. M. P. Nestšadimenko (epidemiol. u. Malaria-Abt.); Dr. I. M. Zvitkij (Abt. f. Tollwutbehandlung nach Pasteur).

Kingston, Ont. (Canada).
Dept. of Bact., Queen's Univ. Dir. G. B. Reed, B.Sc., Ph.D. — J. H. Orr, M.D., C.M.

Knoxville, Tenn. (U. S. A.).
Bact. of Univ. of Tennessee, Rooms 100—110, Morrill Hall. Dir. Paul W. Allen. — Geo. Cameron; W. E. Cole. *4 places*. ⓟ General labor. animals.

Kraljevica, Kroatien (S. H. S.).
Bact. Station. Dir. Dr. Josip Trausmiler.

Krasnodar, Kubangebiet (U. d. S. S. R.).
Bact. Inst., Med. Inst., Raschpilewskaia 104. Dir. I. G. Sawtschenko. — Dr. J. Rosnatowsky.

Kumamoto (Japan).
Microbiol. Inst. Kumamoto (Imper.) Med. Univ.

Kurashiki near Okayama (Japan).
Bact. Labor. Ôhara Inst. of Agric. Research. Dir. Dr. Arao Itano. — 3 Ass.

Kyôto (Japan).
Inst. of Microbiol. Coll. of Med., Kyôto Imper. Univ. Prof. K. Kiyono (of Pathol. Inst.); Ass. Prof. Ken Kimura (z. Z. in Europa); Ass. Prof. Hotori Watanabe; Lect. Tsunesuke Miura.
Microbiol. Inst. Kyôto Provincial Med. Univ., Hirokôji.

La Habana (Cuba).
Bact. Labor. Univ.

La Plata (Argentina).
Microbiol. Inst. Escola de sc. med. Dir. Prof. H. Dasso.
Inst. f. spezielle Microbiol. Prof. Dir. Fred. Ruppert.

Landsberg a. d. W. (Deutschland).
Hygien.-Bact. Abt. d. Staatl. Hygiene-Inst. Zechoner Str. 48. Dir. Prof. Dr. R. Hilgermann.

Lansing, Mich. (U. S. A.).
Bureau of Labor., Michigan Dept. of Health. Dir.: C. C. Young, D.P.H.; R. L. Kahn, Ass. Dir. (Immunol.). — Ass. Dir. of Personnel, Research Bact., Immunol., Biol. Products Division, Research-Biol. *Limited to 6 places for Bact., Serol., Diagnostic Chem.* Branch Labor., at Houghton, Mich. and Western Michigan Division Labor., Grand Rapids, Mich.
Dept. of Bact. State Coll. Ass. Prof. W. Le Roy Mallmann (Bact.).

Lausanne (Suisse).
Labor. de Bact., Solitude 19. Dir. Prof. B. Galli-Valerio.

Leeds (England).
Bact. Labor. Med. School Univ., Thonesby Place. Dir. J. W. McLeod. — J. Gordon, M.D., Lect.; I. Happold, B.Sc., Demonstrator; B. Whealtey, M.B., Demonstrator. *2 places*. ⓟ Bact. cultures.

Leiden (Holland).
Klinisch-Pathol. Labor., Hospital-Univ. Dir. Prof. Dr. W. A. Kuenen. — Dr. F. J. H. van Deinse, Chef de labor.; W. J. Bruins Slot, Ass. *3 places pour bact. et sérol. chim. pathol.*
Labor. f. Hygiene u. Bact. (Lab. voor gezondheidsleer en bacteriologie), Boerhave-Labor. Boerhavestr. 34. Dir. Prof. Dr. R. P. van Calcar. — Dr. I. R. F. Rassers, Conserv.; C. L. G. M. Houtzager, Ass.

Leningrad (U. d. S. S. R.).
Bact. Pasteur Inst., Ul. Mira 12a. Dir. Ja. Ju. Liebermann. 4 Abt.: Diagnose, Wassermann, Bact., Chem.
Mikrobiol. Abt. des Röntgenol. u. Radiol. Staats-Inst., ul. Röntgen'a (ehem. Licejskaja) 6.
Mikrobiol. Labor. d. Staatliches Hydrol. Inst., W. O. 2 l. 23. Dir. Prof. Dr. B. Issatschenko. — R. K. Mutafowa, Hauptass.; N. B. Netschaewa; A. G. Salimowskaia; Z. G. Muliartschik. *2 Arbeitsplätze für Gäste*.
Bact. Abt. d. Staatl. Inst. für ärztliche Fortbildung, Kirochnaja 41. Dir. Prof. Dr. med. Georg Belonowski. — Dr. med. A. Miller, I. Ass.; Dr. med. A. Kalinin, II. Ass.; Dr. med. F. Schultz. *75 Plätze für Ärzte, welche zur Fortbildung (Kurse) kommen, 25 für wissenschaftl. Arbeiten. — Potentiometer für elektrometrische pH-Bestimmung.*
Labor. de Microbiol. de l'Inst. pour méd. expérim., Perspect Maklin 327. Dir. V. L. Omeliansky. — G. Seliber; G. A. Bovschik. *4—5 places pour microbiol. pure et appliquée.*
Mikrobiol. Labor. des Staatsinst. für Med. Wiss., Sowetsky Prospekt 4. Dir. Prof. B. P. Ebert. — Dr. L. H. Peretz, I. Ass.; Dr. S. N. Saschino, Prosektor; Dr. M. E. Warsiliewa, II. Ass.; Dr. F. E. Lippert, Präparator.

Léopoldville (Congo belge).
Labor. bact. Dir. J. Rodhain. — van Hoof; v. d. Branden.

Lexington, Ky. (U. S. A.).
Dept. of Bact. of the Univ. of Kentucky. Dir. Prof. M. Scherago.

Liège (Belgique).
Inst. de bact., Univ. Dir. Prof. E. Malvoz.

Lille (France).
Inst. Pasteur, 20 Bd Louis IV. Dir. adjoint Dr. L. Marmier. — Chefs de service: E. Rolants, C. Guérin, Grysez; Lemoigne, Chef de labor.; Dopter, Ass.; Paulhiac, Préparateur. *15 places*. *Labor.: Bact. méd.; Bact. vet.; Bact. agric.; Physique biol.; Hygiène appliquée; Chim. biol.*

Lima (Peru).
Labor. de Bact., Fac. de Med., Univ., Alameda Gran. Dir. Dr. R. Rebogliati.
Labor. de Bact. e Histol., Inst. de Odontol., Univ., Plaza de la Inquisición 507. Dir. Dr. M. Noriega del Aguila.

Lincoln, Neb. (U. S. A.).
Dept. of Pathol. and Bact. of the Univ. Dir. Prof. H. E. Eggers. — Prof. H. H. Waite; Assoc. Prof. J. T. Myers.

Lisbôa (Portugal).

Inst. Bact. Camara Pestana. Dir. Prof. Annibal Bettencourt, Prof. de Bact. et Parasit. Fac. Méd. (Le dir. de l'Inst. est toujours le Prof. de Bact. F. sc. Lisb.). — Nicolau A. de Bettencourt; Miguel A. Reis Martins; Ildefonso Borges; Estevão de Pereira da Silva; Annibal de Magalhaes; Luís Figueira; Ayres Kopke. *3 à 6 places. Recherches de Bact., Protozool., Helminthol., Immunol., Diphthérie, Rage.*

Inst. de Bact. e Higiène, Escola de Med. tropical. Dir. Prof. J. F. Sant'Ana.

Ljubljana — Laibach (S. H. S.).

Epidemiol. Inst. Bact. Station. Dir. Dr. M. Šimec.

Logan, Utah (U. S. A.).

Dept. of Bact. and Physiol. Chem., Agric. Coll. of Utah. J. E. Greaves, Ph.D., Prof. Bact. and Physiol. Chem., Bact. and Chem.; E. G. Carter, M.S., Dr. P.H., Ass. Prof. Bact. and Physiol., Assoc. Bact.

London (England).

Dept. of Bact., Univ. Coll., Hosp. Med. School, W.C., Univ. Street, Goner St. Dir. H. D. Wright.

Dept. of Bact., London Hosp. Med. Coll. of the Univ., E., Turner Str., Mile End Dir. Prof. W. Bulloch.

Dept. of Bact., Guy's Hosp. Med. School of the Univ., S.E. 1, St. Thomas St., Borough. Dir. J. W. H. Eyre.

Dept. of Bact., St. George's Hosp. Med. School of the Univ., S.W., Hyde Park Corner. Dir. E. L. Hunt.

Dept. of Bact., St. Mary's Hosp. Med. School of the Univ., W. 2, Paddington. Dir. A. Fleming. — J. Freeman.

Dept. of Bact., London School of Hygien. and tropical Med. of the Univ., N.W., Endsleigh Gardens. Dir. Prof. R. Hewlett.

Bact. Labor., Royal Inst. of Public Health, W.C. 1, Russell Square 37.

Dept. of Bact., Lister Inst. of preventive Med. of the Univ., S.W. 1, Chelsea-Bridge Rd. Dir. Prof. J. C. Ledingham.

Dept. of Bact. and experim. Pathol. Nat. Inst. for Med. Res., Hampstead. Dir. S. R. Douglas. — W. E. Gye; C. Dobell (Protistol.); P. P. Laidlaw; P. Hartley; W. J. Purdy; L. Colebrook; G. W. Dunkin.

Wellcome Bureau of sc. Research. Confer: Institutiones generales.

London, Ont. (Canada).

Dept. of Pathol. of the Univ. of Western Ontario. Dir. F. W. Luney, M.D., D.P.H. — Prof. E. L. Armstrong. *Diagnostic Bact. and Pathol., Parasit.*

Los Angeles, Cal. (U. S. A.).

Labor. for Bact. and Physiol. of the Univ. of Southern California.

Louvain — Leuven (Belgique).

Labor. de Bact. Dir. Prof. R. Bruynoghe.

Lwów (Polska).

Inst. de biol. générale de l'Univ., Zakład Biol. Ogolnej Univ. Ul. Mikosaja 4. Dir. R. S. Weigl. — Dr. Karoline Reis, Ass.; 2 Ass. *2 Plätze. Infektionskrankheiten, insbesondere Fleckfieberuntersuchungen, wie auch andere durch Insekten übertragbare Krankheiten. — Lauszucht. Apparate zur Untersuchung der ultravisiblen Vira.* ⊕ *Normale Läuse gezüchtet,* Rickkelsia Provazekii.

Labor. f. Bact u. Serol. d. Tierärztl. Akad. Dir. S. Legeżyński.

Bact. Inst. Ukrain. Ševčenko-Gesellschaft der Wissensch. Vorst.: Dr. Maxim Muzyka.

Lyon (France).

Labor. de Bact. Fac. Méd. de l'Univ. Dir. Prof. Arloing.

Labor. de Bact. de l'Ecole de Service de Santé Militaire.

Inst. Bact. de Lyon et Inst. Pasteur du Sud-Est, 61, rue Pasteur. Dir. Paul Courmont.

Madrid (España).

Inst. de Higien. mil., Alb. Agnilesa 56. Dir. Exc. Eduardo Semprun y Semprun. *Bact. Histol. Analysis hygien. Biol. Unters. Vaccina.*

Labor. de Bact. y Serol. Residencia de Estudiantes.

Manchester (England).

Bact. Labor. Univ. Dir. Prof. W. W. C. Topley.

Manhattan, Kan. (U. S. A.).

Dept. of Bact. Dir. L. D. Bushnell. — Prof. of Bact. P. L. Gainey, Prof. of Soil Bact.; A. C. Fay, Ass. Prof. of Daisy Bact.; W. R. Hinshaw, Ass. Prof. of Poultry Bact.; B. W. Lafene, Instr. in General Bact.; C. B. Hudson, Fellow; 4 Labor. Ass. on part time. *3 places.*

Manila (Philippine Islands).

Labor. of Bact. Med. Zool. Parasit. Univ. Santo Tomas. Dir. Prof. Luis Guerrero. — V. R. Lanuza.

Marburg a. d. L. (Deutschland).

Inst. f. experim. Therapie „Emil von Behring". Confer: Serol.

Marseille (France).

Labor. de Bact. de l'Ecol. de Méd. et de Pharmacie, Palais du Pharo. Dir. Prof. S. Costa. — Mr. Louis Boyer, Chef des Travaux; Mr. Arnoux, Prép. *3 places.*

Mexico, D. F. (Mexico).

Sect. physiol. compar., biol. med. u. Microbiol. d. Dir. de Estudios Biol. Confer: Physiol. comparata.

Mons (Belgique).

Inst. bact. prov., Boul. Sainctellette. Dir. Prof. Dr. Martin Herman.

Montpellier (France).

Labor. de Microbiol. Fac. Méd. Univ. Dir. Prof. M. Lisbonne. — Dr. Carrère, Chef de Trav. pratiques; Labraque Bordenave, Lic. sc. Prép. *2 places pour Microbiol. clinique.*

Montreal, Que. (Canada).

Dept. of Bact. Univ. of Montreal. Dir. A. Bernier, M.D. — H. Aubry, M.D.; P.-P. Gauthier, M.D.; A. Bertrand, M.D.

Moskau (U. d. S. S. R.).

Staatl. mikrobiol. Inst., Device Pole, Pogodinskaja 10. Dir. V. A. Barykin. — 12 wiss. Mitarb.

Labor. für Mikrobiol. der 2. Univ. Dir. I. L. Kricevskij.

Staatl. Bact. Inst. des Volkskommiss. f. Gesundheitswesen „Immunität", Tverskaja Blagoveščenskij 3. Dir. N. Vlassjevskij. — 3 wiss. Mitarbeiter.

Mikrobiol. Inst. des Volksunterrichtskommiss., M. Pirogowskaia N 57. Dir. Prof. Dr. med. et phil. I. L. Kritschewski. — Prof. Dr. med. et phil. I. Kritschewski, Leiter d. Bact. u. immunol. Abt.; Doc. Dr. phil. G. O. Rosnin, Leiter der Protozool. Abt.; Prof. Dr. chem. W. A. Ismailski, Leiter der chemotherap. Abt.; wiss. Mitarbeiter: Priv.Doc. Dr. med. K. A. Friede, Dr. med. A. M. Brussin, Dr. med. L. W. Kritschewski, Dr. med. R. S. Tscherikower. *26 Plätze für Bact., Protozool., Immunbiol., Chemotherapeutik. — Sehr reiche Apparatur für alle Zweige des Inst.* ⊕ *Sammlung der Bakterien, der pathogenen Spirochaeten und der pathogenen Protozoen.*

„Mečnikov"-Inst. f. Infektionskrankh. (Inst. des maladies infectieuses ElieMetschnikoff),Pokrovka 44. Dir. Prof. S. V. Koršun. — Abt. für Epidemiol., Serol., pathol. Anat., Protozool., Biochem., Antirabies nach Pasteur.

Mostar (S. H. S.).
Ständige bact. Station (Stalna bact. Stanica), Carina 32. Dir. Dr. med. Stevan Simunović. *Arbeitsplätze für alle bact.-serol. Arbeitsmethoden und auch für Malariafragen.* Eine Malaria-Hilfsstation im Orte Vitina (Bezirk Ljubuški) und eine Biol.- Entomol. Hilfsstation in Metković an der unteren Narenta, welche 1926 ausgestaltet wurde.

Mukden, Manchuria (China).
Microbiol. Inst. South Manchurian Med. Univ.

München (Deutschland).
Staatl. Bact. Untersuchungsanstalt. Dir. Prof. Dr. W. Rimpau. *Seuchenbekämpfung; Hygien.; Bact.*

Nagasaki (Japan).
Bact. Inst. Nagasaki Imperial Med. Univ.

Nagoya (Japan).
Bact. Inst. Aichi Med. Univ.

Nancy (France).
Labor. de Microbiol. Fac. de Pharmacie. Dir. Prof. Lasseur.
Labor. de Bact. Fac. Méd. Dir. Prof. Macé.

Nantes (France).
Inst. Pasteur de la Loire-Inférieure, 26, Boulevard Victor Hugo. Labor. Dept. de Bact. Station Agronomique.

Napoli (Italia).
Ist. Batt. e Parasit. Fac. med. Univ., S. Andrea delle Dame 2. Dir. Prof. Nicola Pane. — Dr. G. Marotta; Dr. S. Suggese. *8 posti. Batt. e Serol.*

New Brunswick, N. J. (U. S. A.).
Dept. of Bact. Rutgers Univ. J. A. Anderson (Instr.).

New Haven, Conn. (U. S. A.).
Dept. of Bact. Yale Univ., Sheffield Hall. Dir. Prof. L. F. Rettger. — Dr. A. H. Gee (Putrefaction of Haddock Muscle).

Nha Trang (Indo-Chine).
Inst. Pasteur. Dir. N. Bernard. *Zum Studium der epidem. Krankheiten von Indo-China sowie zur Herstellung des Pest-Serums 1895 errichtet. Bereitung von Serum gegen die Rinderpest.* Zweiganstalt des Inst. Pasteur in Paris.

Niigata (Japan).
Microbiol. Inst. Niigata Med. Univ.

Nižnij Novgorod (U. d. S. S. R.).
Biol. Abt. f. Reinkulturen, Chem. Zentral-Labor. Dir. Dr. Franz Razowski.

Novi Sad (S. H. S.).
Epidemiol. Inst., Vojvodina. Dir. Dr. Olga Aristonkova.
Staatl. Pasteur-Inst. Dir. Dr. Adolphe Hempt.

Nova Goa (Portug. India).
Bact. Labor. Escola Med.-Chirùrgica.

Novi Pazar, Serbien (S. H. S.).
Bact. Station. Dir. Dr. Niktopolin Černozubov.

Odessa (U. d. S. S. R.).
Sekt. f. Mikrobiol. des Katheders f. Biol. d. Wissenschaftl. Forschungs-Inst., Kominternstr. 2. Dir. Prof. Ja. Ju. Bardach.
Sekt. f. Mikrobiol. der Infektionskrankheiten des Wissenschaftl. Forschungs-Inst. Dir. Prof. V. Stefanskij.

Abt. f. Allgem. Mikrobiol am Med. Inst. Dir. Prof. V. L. Jelin.
Mikrobiol. Kabinett a. d. pädag. Hochsch. Dir. Prof. Bardach.

Okayama (Japan).
Bact. Inst. Okyama Imperial Med. Univ. Dir. Prof. Minoru Suzuki.

Olomouc (C. S. R.).
L'Inst. microbiol. de l'Univ. Brno. Prosecture de l'Hôpital provincial. Prof. Dr. Jan Kabelík, Dir. de l'Inst. microbiol. à Brno, de la Prosecture provinciale à Olomouc et de la Station diagnostique de l'Etat ibidem. — Dr. G. Gellner, Méd. Colonel; Dr. I. Vignati, Méd. sécondaires; 2—3 Méd. et Vét. *2—3 places pour toute la bact. et sérol. méd. et pour la plupart de la chimie colloïdale méd., principalement pour la néphélométrie et tyndallométrie des colloïdes. — Tyndallomètre de Mecklenburg, le néphélomètre de Kleinmann· et le viscosimètre sérol. de Kabelík.*

Omsk (U. d. S. S. R.).
Bact. Abt. d. med. Inst. Dir. Prof. V. S. Veselov.

Ôsaka (Japan).
Bact. Inst. Osaka Provincial Med. Univ. Prof. Yoshimoto Fukuhara (Bact. and Hygiene.).

Oslo (Norge).
Bact. Inst. d. Armee.

Osijek (S. H. S.).
Epidemiol. Inst. (Epidemiol. Zavod Osijek). Dir. Dr. Slavko Hirsch. — Dr. Marija Belavić; Dr. Dušan Kostič; Dr. Simeon Griner. *1 Arbeitsplatz. Bact., Serol., gerichtl. und med. Chem., Histol. Bekämpfung der Epidemien, Erforschung von Infektionskrankheiten, Durchführung von hygien. Maßnahmen.* Ⓑ Bact. Mus.

Parel, Bombay (Br.-India).
Haffkine Inst. Confer: Pharmacol.

Paris (France).
Inst. Prophylatique, Rue d'Assas. *Syphilis.*

Pavia (Italia).
Labor. di Batt., Ospedale di S. Matteo. Dir. Dott. E. Veratti, Prof. incarnicato. *Vi è posto per 3 o 4 studenti. — Culture dei tessuti.* Ⓑ Collezione di Batteri patogeni.

Perm (U. d. S. S. R.).
L'Inst. Bact. Dir. Prof. Wladimir Michailovicz Sdravosmislov. — Ass.: Dr. Elena Karnaouchova, Dr. L. E. Martinov, Dr. G. I. Badsiev. *Recherches bact., immunol. et serol.* Labor. de service malarique.

Porto (Portugal).
Bact. Inst. Univ. Dir. Prof. C. F. M. Ramãlhao.

Porto Alegre (Brasil).
Inst. Pasteur. Dir. Prof. Leite da Fonseka.
Mikrobiol. Inst. Dir. Prof. P. Filho.

Poznań — Posen (Polska).
Inst. f. med. Biol. Dir. Prof. Leon Padlewski.
Histol.-bact. Labor. der Chir. Klinik der Univ., Ulica Długa 1. Dir. Prof. Dr. Anton Jurasz. — Dr. Kowalski; Dr. Skubiszewski. *1 Platz.*

Praha — Prag (C. S. R.).
Bact.-serol.Inst. der Karls-Univ., II, Preslova 2039, Hlávův ústav.

Prizren, Serbien (S. H. S.).
Bact. Station.

Quebec (Canada).
Dept. of Bact. MacDonald Coll. Dir. F. C. Harrison, D.Sc. — J. R. Sanborn, M.Sc.; A. Loond, M. Sc.

Quito (Ecuador).
Bact. Inst. Dir. Prof. E. Cousin.

Rangoon (Br.-India).
Pasteur Inst.

Reims (France).
Labor. de Bact., Ecole de Méd. 51, rue Simon., Dir. Dr. Téchoueyres. *Hygiène et Bact.*

Reykjavik (Island).
Bact., Physiol. u. Pathol. Inst. Dir. Prof. G. Thoroddsen.

Riga (Latvija).
Labor. der med. Mikrobiol. Med. Fac. Univ., Ritterstr., I. Krankenhaus, Baracke 12. Dir. Prof. Dr. med. W. Klimenko. — Dr. N. Wetza, Ass.; Nik. Stoligvo.

Rio de Janeiro (Brasil).
Inst. f. Mikrobiol. Dir. Prof. B. Lobo.
Inst. de Bact. d'Hôpital Nation.
Inst. Pasteur.
Labor. bact. fédéral.

Roma (Italia).
Labor. Batt. della Sanità Pubblica, Piazza Vittorio Emanuele 13. Dir. Prof. B. Gosio.

Rosario (Argentina).
Labor. de Microbiol. y Parasit. Escuela de med. Univ. Dir. Prof. A. Gatti.

Saarbrücken (Deutschland).
Inst. für Mikrobiol., Winterbergstr. 16—20. Dir. Dr. W. Fornet. — Labor.-Vorst.: Dr. E. Christensen. *Organotherapie: eßbares Insulin. Bact.: Kultur entfetteter Tuberkelbazillen.*

Šabac, Serbien (S. H. S.).
Bact. Station.

Saigon (Indo-Chine).
Inst. Pasteur. Dir. Dr. N. Bernard. — 5 Sektionen: Bact. humaine, Microbiol. animale, Serum et Vaccine, Chim. biol. *Gewinnung d. Impfstoffes gegen Tollwut, Stud. epidem. Krankh.*

St. Andrews (Scotland).
Dept. of Bact. Univ. Coll. Dundee and Med. School. Dir. Prof. G. W. Tulloch.

Salisbury (Rhodesia).
Labor. of Bact. Research. Dir. Dr. A. Fleming.

San Francisco, Cal. (U. S. A.).
Dept. of Bact., Med. School, Univ. of California Dir. Prof. K. F. Meyer.

São Paulo (Brasil).
Dept. of Epidemiol. and Chair of Hygien. S. Paulo Med. School, Caixa Postal 1985. Chief of Labor.: Francisco Borges Vieira. — Dr. Samuel B. Pessôa, Acting Ass. 1 Ass. 2 labor. ass. *2 places in Bact. and Epidemiol.*

Sarajevo (S. H. S.).
Epidemiol. Inst. (Epidemioloski Zavod). Dir. Dr. Paul Kaunitz. — Dr. Simon Grüner; Dr. Milena Duraškovic-Jakovljevič; Dr. Desanka Milovanovič-Mileusnic; Dr. Vladimir Ljutov. *Für Gäste 1 Arbeitsplatz. Sero-bact. u. histol. Untersuchungen — Analysen-Quarzlampe.*

Saratow (U. d. S. S. R.).
Bact. Labor. Univ. Dir. S. M. Nikanorov.
Bact. Abt. Staatsinst. f. Mikrobiol. u. Epidemiol. Dir. Dr. S. J. Boril.

Saskatoon, Sask. (Canada).
Dept. of Bact. Univ. of Saskatchewan. Dir. W. S. Lindsay, M.B. — G. Rea, D.P.H.

Sendai (Japan).
Bact. Inst. Coll. of Med. Tôhoku Imper. Univ., Kita Yobancho. Dir. Prof. Dr. K. Aoki. — Dr. T. Hayashi; Dr. K. Hashimoto; Dr. J. Tazawa; Dr. G. Takayanagi. *10 Arbeitsplätze für Gäste; allgemeine Bact.* ⓟ *Sammlung für Bakterienstämme, besonders für Typhus, Hühnertyphus, Paratyphus, Dysenterie, Proteus, Pyocianeus, Prodigiosus.*

Sfax (Tunis).
Annexe de l'Inst. Pasteur de Tunis. Dir. Dr. A. Espié.

Siena (Italia).
Hygien. Inst. Prof. A. Sclavo.

Singapore (Straits Settlements).
Dept. of Bact., Coll. of Med. Dir. Prof. W. A. Young. — G. E. Bloche, Lect. in Public Health, G. C. B. Gilmore, Lect. in Infect. Diseases. *5 working places for guests.*

Smolensk (U. d. S. S. R.).
Bact. Inst., Leninstr. 6. Dir. Prof. M. Isabolinsky. — S. Brilling, Vorst. d. Chem. Abt.; W. Gitowitsch, Vorst. d. Vaccin-Abt.; A. Zeitlin, Vorst. d. Pasteur-Abt.; W. Judenitsch, Vorst. d. Serum-Abt.; R. Katzmann, Vorst. d. Untersuchungs-Abt.; 34 Ass. *10 Plätze.* ⓟ *Bact. Praep.*

Sofia (Bulgarien).
Bakt. Inst. des Staates, Buleward Makedonia. Dr. Johan Robeff; Dr. Mara Zachariewa; Dr. Nadejda Popkirowa.
Inst. für Bakt. u. Serol. Univ., Moskowska 43. Dir. Prof. Dr. Wl. N. Markow. — Dr. med. Sdrawka Jatschewa; Dr. med. Eugen Nikolow; Cand. med. Bogoja Jurukow. *Allgemeine Mikrobiol. u. Serol., Gärungsphysiol.* ⓟ *Kulturensammlung von Bakterien u. Hefen.*

Sousse (Tunis).
Annexe de l'Inst. Pasteur de Tunis. Dir. Dr. H. Diacono.

Split, Dalmatien (S. H. S.).
Bakt. Stat. Dir. Dr. Josip Škarić.

Stambul (Türkei).
Bakt. Inst. Univ. Dir. Prof. Réfik.
Inst. de Bakt. et Hygièn., Ecole de Pharm. Dir. Prof. Sever Kiamil.

Stockholm (Sverige).
Bakt. Staats Labor. Dir. Prof. Carl Kling. — Insp. Prof. Carl Sundberg. 2 Labor., 2 Ass., 1 Vet.-Ass.

Strasbourg — Straßburg (France).
Mus. Pasteur, 2, Rue Köberlé.

Subotica, Vojvodina (S. H. S.).
Bakt. Station.

Sucre (Bolivia).
Bakt. Labor. Med. Inst.

Sydney (Australia).
Microbiol. Labor. of the Univ., 93, Macquarie Street.

Taihoku, Formosa (Japan).
Microbiol. Inst. Taihoku Governm. Coll. of Med.

Tananarive (Madagascar).
Inst. Pasteur. Dir. Dr. Girard, Prof. d'hygiène à l'école de méd. (Peste, rage). — Dr. Jean Robic (Analys. bact.); René Rahoerson, Med. de l'Ass. indigène (Peste). *Service de Rage, Service de Vaccination Jennerienne. Labor. de la Peste. Labor. de Bact. (Clinique et de Recherches). Sous peu, l'établissement, qui va être rattaché à l'Inst. Pasteur de Paris, verra étendre son champ d'action.*

Tanger (Maroc).
Inst. Pasteur, Place Pasteur. Dir. P. Remlinger.

Tartu — Dorpat (Eesti).
Labor. de bact. Fac. Méd. Dir. Prof. Karl Schloßmann. — Dr. L. Leismann, Chef du labor. des recherches. 3 Ass. *3 places de travail pour les recherches bact. et sérol. — Réfractomètre, Potentiomètre.* — Inst. Pasteur (vaccinations antirabiques). Labor. de bact. vét.

Taschkent (U. d. S. S. R.).
Bakt. Inst. Univ. Dir. Prof. A. D. Grekov.

Tegucigalpa (Honduras).
Bakt. Inst. Fac. Med. Univ. Dir. Prof. Dr. J. M. Fiallos.

Teheran (Persien).
Inst. Pasteur.

Tiflis (U. d. S. S. R.).
Labor. f. Bakt. u. Hygien. Univ. Dir. Prof. S. Ameredzibi.

Tôkyô (Japan).
Bact. Labor. Coll. of Med., Keiô-Gijuku Univ. Dir. Prof. Dr. S. Hata. — Prof. Dr. R. Kobayashi; Lect.: Dr. R. Takano, T. Oguchi, T. Isawa, Dr. Y. Kusama.
Bact. Labor. Kitasato Inst. for Infect. Diseases, Shirokane-Sankô-chô. Dir. Dr. Sahachirô Hata.
Bact. Inst. Coll. of Med., Tôkyô Imper. Univ. Prof. Kôzô Saizawa; Prof. Matsujirô Takenouchi.
Bact. Inst. Nippon Med. Univ.
Bact. Inst. Nippon Girls' Med. Univ.
Bact. Inst. Yikeikevai Med. Univ., Shiba. Dir. Prof. Dr. Y. Yazaki.
Bact. Inst. Military Med. Coll.
Bact. Labor. Imperial Inst. f. Infectious Diseases.

Tomsk (U. d. S. S. R.).
Inst. f. Bakt. Dir. Prof. P. V. Butjagin.

Toronto, Ont. (Canada).
Dept. of Pathol. and Bact. Univ. Dir. O. Klotz, M.B., M.D. — W. L. Holman, M.D.; W. L. Robinson, M.B.

Toulouse (France).
Labor. de Bact. Fac. méd. Univ. Dir. Prof. L. Rispal.

Tsinan (China).
Dept. of Bact. Shantung Christian Univ. Dir. Prof. S. Cochran.

Tucumán (Argentina).
Inst. f. Bact. Dir. Prof. R. Cossio.

Tunis (Tunis).
Inst. Pasteur. Dir.: Ch. Nicolle; Sous-Dir.: Et. Burnet. — Chefs de labor.: Ch. Anderson, P. Durand, Jean Bance, G. Catouillard. *Service antirabique; Centre vaccinogène; Service de préparation du sérum antidiphtérique et des vaccins microbiens; Labor. de Chimie méd. et biol.; Labor. de vinification; Service des analyses bact., méd. et biol.; Recherches scientifiques; Service antipaludique.* — Labor. annexes: à Sousse (H. Diacono); à Sfax (A. Espié).

Tuzla (S. H. S.).
Bakt. Station (Stalua bakt. stanica). Dir. Dr. Robert Fried. ℗ Kulturen von pathogenen Bakterien. — 4 Antimalaria-Stationen: in den Bezirksstädten Bijeljina, Brcko, Bosn. Brod und Bosn. Gradiska.

Urbana, Ill. (U. S. A.).
Dept. of Bact., Univ. of Illinois, 361 Chem. Bldg. Dir. Fred Wilbur Tanner, Prof. and Head of the Dept., Dir. of the Labor. — Ass. Prof., Ass. *Several places for bact., microbiol., fermentol., etc.— Special apparatus may be purchased for special investigations. Large incubator for thermo-philic bact.* ℗ *Large collection of thermophilic bact. and yeasts.* — There are seven other labor. in which microbiol. work is done at this univ.

Varaždin, Kroatien (S. H. S.).
Bakt. Station.

Vancouver (Br.-Columbia).
Dept. of Bact. Univ. of Br.-Col. Dir. H. W. Hill, M.B., M.D., D.P.H. — Freda L. Wilson, M.A.; H. M. Matthews, B. A.

Warszawa (Polska).
Mikrobiol. Labor. der freien Univ. Dir. Prof. Ławrynowicz.

Wien (Österreich).
Bakt. Serol. Unters. Anst., III, Hauptstr. 146. Dir. Prof. Dr. V. K. v. Russ.
Prof. Dr. E. Pribram's mikrobiol. Sammlung, IX|2, Michelbeuerngasse 1a. Von Prof. Dr. Král 1888 in Prag gegr. wiss. Sammlung von Bakterien, Hefen u. Pilzen. Leiter: Prof. Dr. Ernst Přibram.

Wilno (Polska).
Hygien. Inst. Univ. Dir. Prof. W. Karaffa-Korbutt.
Bakt. Inst. fac. med. Univ. Dir. Prof. T. Gryglewicz.

Woronesch (U. d. S. S. R.).
Mikrobiol. Inst der Univ. Dir. Prof. M. I. Stutzer. — Dr. De-Gorge, Vorst. der Wutschutzimpfstation; Dr. Mufel, Vorst. der Malariastation; Dr. Adelson, Vorst. der Pockenimpfanstalt. ℗ Helmintol. Material. Kulturen der Krankheitserreger der Insekten und Pflanzen.
Bakt. Inst. Univ. Ass. Paul P. Muffel.

Zagreb — Agram (S. H. S.).
Staatl. Epidemiol. Inst., Mlinarska cesta 14. Dir. Dr. med. et med. vet. Berislav Borčić. — Dr. Berlot Josip, Ass. serol. Abt.; Dr. Stanko Miholić, Ass. chemische Abt.; Dr. Dora Filipovic, Ass. bakt. Abt.; Dr. Slavko Palmović, Ass. antirabische Abt.; Dr. Julije Rogina, Chef der Jenner-Abt.; Dr. Marianne Hermann, Ass. chem. Abt.; Ing. Božidar Rogina, Ass. chem. Abt.; Dr. Dimitrije Kalić, Ass. bakt.Abt. *Neues großes Inst. im Bau.* Bakt. Station in Sušak (spez. Malaria). Bakt. Station in Gospić. Bakt. Station in Varaždin. Bakt. Station in Nova Gradiška (spez. Malaria). Antimalarische Station in Suhopolje.

Zaječar, Serbien (S. H. S.).
Bakt. Station.

Zürich (Schweiz).
Hygien.-Bakt. Labor. der Eidgen. Techn. Hochsch., Clausiusstr. 25. Dir. Prof. W. v. Gonzenbach.

Laboratoria bacteriologiae animalium.

Confer: *Bacteriologia hominis* (pag. 479), *Microbiologia agris* (pag. 533), *Hygiena hominis* (pag. 487), *animalium* (pag. 492), *Anatomia* (pag. 423), *Anat. pathol. animalium* (pag. 467), *Parasitologia generalis* (pag. 379).

Alger (Algérie).
Labor. Microbie vét. de l'Inst. Pasteur. Dir. F. Lestoquard. — E. Planteneux.

Beppu (Japan).
Beppu-Hot Spring Research Labor. Confer: Labor. Limnol.

Berlin (Deutschland).
Bakt. Inst. d. Landwirtschaftskammer, NW 40, Kronprinzenufer 4—6. Dir. Dr. Scharr.

Bern (Schweiz).
Veterinär-pathol. und veterinär-bakt. Inst., 6. Engehaldenstr. Dir. Prof. Dr. Huguenin. — Ass. Bourgeois. Tierarzt. *6 Plätze* Ⓟ Pathol.-anat., bakt., parasitol. Praep.

Braunschweig (Deutschland).
Bakt. Anstalt der Landwirtschaftskammer, Hochstr. 17|18. Dir. Dr. A. Machens. — Dr. F. Cordes, Abt.-Vorst.; Dr. R. Frisch, Ass. *2 Arbeitsplätze für bakt. und serol. Untersuchungen und für biol. Untersuchungen.* Ⓟ Praep. über Tierkrankheiten.

Breslau (Deutschland).
Bakt. Inst. der Landwirtschaftskammer Schlesien, 16, Kaiserstr. 55. Dir. Dr. vet. Paul Schumann. Stellvertr. Dir.: Dr. med. vet. Lerche. — Obertierarzt: Dr. med. vet. Hustig; Tierarzt: Dr. med. vet. Gerstenberger, Tierarzt Dr. med. vet. Grasnick; Ass.-Tierärzte: Dr. med. vet. Müller, Dr. med. vet. Dohme, Dr. med. vet. Berenz, Dr. med. vet. Grimm, Dr. med. vet. Lübke. *1 Platz für bakt. Untersuchungen.* Ⓟ Pathol.-anat. Praep. von Krankheiten der Tiere. Kulturen von Tierkrankheiten.

Budapest (Ungarn).
Bakt. Inst. der kön. ung. Veterinär-Hochsch., VII, Hungária-Korút, 249. Dir. Prof. Dr. A. Aujeszky. — Dr. Julius Darányi, kön. Oberbakteriolog. Doc. an der Univ.; Dr. Josef Csontos, Ass., Doc. der Veterinär-Hochsch.; Dr. Ferdinand Kerbler, Ass. *2 Plätze.*

Dessau (Deutschland).
Bakt. Inst. der Anhaltischen Kreise, Herzogin-Marie-Platz. Dir. Dr. Wolters. — Dr. med. Zschucke, Abt.-Leiter der med. Abt.; Dr. med. vet. Hoffmann, Abt.-Leiter der vet.-med. Abt.; Dr. med. vet. Stange, wissenschaftl. Hilfsarbeiter; Dr. phil. Zschelinski, Leiter der chemischen Abt.

Emden, Ostfriesland (Deutschland).
Zweigstelle des Tierseucheninst., Landwirtschaftskammer d. Prov. Hannover.

Groningen (Holland).
Microbiol. labor. Agric. Experim. Station, Eemskanaal Z. z. 1. Dir. Dr. F. C. Gerretsem. — Dr. J. Sack (Bact.). *2 places for all sorts of research work along microbiol. lines. — All apparatus for microbiol. research work. Calorimetric and electrometric p_H determination.* Ⓟ Cultures of some Bact.

Halle a. d. S. (Deutschland).
Bakt. Inst. der Landwirtschaftskammer für die Provinz Sachsen. Dir. Prof. Dr. phil. et med. vet. h. c. Raebiger. — Abt.-Vorst. Dr. phil. H. Rautmann *(Tuberkuloseabt.)*; Obertierarzt Dr. med. vet. Spiegl, *Leiter der diagnostischen Labor. (Parasit. und Spezialist für Schafkrankheiten)*; Obertierarzt Dr. med. vet. Tantz, *Leiter des Außendienstes bei der Tuberkulosebekämpfung*; Dr. med. vet. Schmidt *(Spezialist für Geflügelkrankheiten)*; Ass.-Ärzte Dr. med. vet. Scheidemann und Dr. med. vet. Uhlhorn (Außendienst). *1—2 Arbeitsplätze für bakt. u. parasit. Untersuchungen. — Tuberkuloselabor., Labor. für Schädlingsbekämpfung, Fischlabor., Yoghurt- und Kefirlabor.* Ⓟ Es werden gesammelt: pathol.-anat., pathol.-histol. Praep., Stammkulturen tierpathogener Bakt., Schädlingsbekämpfungsmittel.

Hannover (Deutschland).
Bakt. Chem. Inst. Techn. Hochsch. Dir. Prof. Dr. Carl Wehmer. *Holzzersetzende Pilze; Aspergillusarten; Pflanzenchem.; Gärungschem.*

Tierseucheninst. der Landwirtschafts-Kammer. Dir. Dr. Karsten. — Abt.-Vorst.: Dr. Ehrlich; wissenschaftl. Hilfsarbeiter: Dr. Lüttschwager; Ass.: Dr. Holsing. *1—2 Arbeitsplätze für bakt.-serol. Untersuchungen.* — Tierseucheninst. der Landwirtschaftskammer Hannover, Zweigstelle Emden (Ostfriesland).

Helsinki (Finnland).
Staatl. Vet.-Bakt. Labor. Dir. Dr. Hindersson.

Kansas City, Mo. (U. S. A.).
Kinsley Vet. Biol. Labor. Dir. A. T. Kinsley.

Kaunas (Litauen).
Vet.-bakt. Inst Univ., Keistnéā gatvè 18. Dir. Prof. A. Gogelis.

Kazan (U. d. S. S. R.).
Bakt. Labor. Vet. Inst., Arskoe Pole 96. Dir. Prof. Dr. M. Tusnov.

Königsberg i. Pr. (Deutschland).
Bact. u. Serum-Inst. der Landwirtschaftskammer, Beethovenstr. 24|26. Dir. Dr. Paul Knauer. *Bakt. Tuberkuloseforschung.*

Leipzig (Deutschland).
Tierseucheninst. u. Inst. für animalische Nahrungsmittelkunde d. Univ., Linnéstr. 11. Dir. Prof. August Eber.

Leningrad (U. d. S. S. R.).
The Section of the gener. Microbiol. of the State Inst. of Experim. Med. Confer: Biochemia generalis. **Veterinär-Bakt. Inst.,** Meshdunarodnij pr. 83. Dir. N. A. Pokschischewskij.

Bakt. Kabinett beim Zool. Garten, Zoologischer Garten Lenin-Park. Dir. A. Pyrkow. — Labor. Tierarzte: Frl. P. Hoffmann; Tierarzt des Zool. Gartens: Dr. Benjamin Klimachevsky. *Physiol. Kabinett bei der Affenabt., bacteriosk opisches Kabinett. Helminthol., bact. und protozoische Untersuchungen.* Ⓟ Pathol.-anat. und helminthol. (ascaris) Praep.

Veterinär-Mikrobiol. Inst., Mojka 98. Dir. B. L. Pazewitsch.

Lima (Peru).
Labor. de Microbiol. y Patol. Animal, Escuela Nacion. de Agric. y Veterin., Santa Beatriz. Dir. Dr. E. Tabusso.

Montevideo (Uruguay).
Bakt. Inst. der Tierärztl. Hochsch. (Inst. de Bact. de la Escuela de Veterinaria). Dir. Prof. Dr. Kurt Schern. — Dr. Freire Muñoz; Dr. Carramaquaghui (Delegierter der Veterinärpolizei). Dr. Macció. *Arbeitsplätze für bakt., serol. Untersuchungen, auch für Milchuntersuchungen.* Ⓟ Tierpathogene Bact.

Oldenburg i. O. (Deutschland).
Bakt. Inst., Oldenb. Landwirtschaftskammer. Dir. Erich Lührs.

Oslo (Norge).
Bakt. Abt. des Veterinär-Inst. Dir. Prof. H. Holth.

Paris (France).
Services vét. de l'Inst. Pasteur. Dir. A. Prevot.

St. Paul, Minn. (U.S.A.).
Division of Veterinary Med. Univ. Farm. Dir. C. P. Fitch.

San Francisco, Cal. (U.S.A.).
National Canners Assoc. Western Branch, 322, Battery street. J. R. Esty. *Food poisoning investigations.*

Stambul (Türkei).
Bakt.-Serol. Labor. der Tierärztl. Hochsch., Skutarie-Sélimiće. Dir. Prof. Ismaïl Risa-Bey. — Dr. Mehmed Zühdi-Bey, Prof. agrégé. *2 Arbeitsplätze.*

Stockholm (Sverige).
Veter.-bakt. Staatsinst., Experimentalfältet. Dir. Prof. Dr. med. vet. Sven Wall. — Tierarzt Eugen Klarin, Vorst. der diagnostischen Abt.; Tierarzt Dr. med. vet. Edwin Lehnert, Vorst. der serol. Abt.; Tierarzt Nils Ohlsson, Ass. an der diagn. Abt.; Tierarzt Erik Thorstrand, Ass. an der serol. Abt. Außerdem 2 Extra-Ass. *Wenigstens 4 Plätze für Veterinärbakt. und pathol. Anat. sowie Serol. — Vollständige moderne Einrichtung. Interferometer.* ℗ Bakterienkulturen, pathol.-anat. sowie parasit. Praep. — 1 Zweiglabor. im nördlichen Schweden.

Tôkyô (Japan).
Veterinary Labor. Kitasato Inst. for Infectious Disease, Shirokane-Sankô-chô. Dir. Dr. Shinkichi Umeno.

Warszawa (Polska).
Bakt. hygien. Inst. Univ. Fac. Veterin. Dir. Prof. Z. Szymanowki.

Wien (Österreich).
Inst. f. Bakt. u. Hygien. Tierärztl. Hochsch. Dir. Prof. Dr. J. Schnürer.

Zagreb — Agram (S.H.S.).
Inst. für Seuchenlehre d. Haustiere, VI, Ciglana, gradska kuia. Dir. Prof. Dr. Stjepan Plasaj.
Inst. f. Fleisch- u. Milchhygiene.

Zaragoza (España).
Labor. de Physiol. et Hygiène, Escuela de Veterinaria. Dir. P. Moyano y Moyano.

Züllchow-Stettin (Deutschland).
Gesundheitsamt der Landwirtschaftskammer für die Provinz Pommern. Dir. Dr. Pröscholdt. — Dr. Niklas (Sterilitätsbekämpfung); Dr. Zeug (diagnostisches Labor.); Dr. Schultz (Tuberkuloselabor.); Dr. Penschuk (serol. Labor.); Dr. Nida (Milch- u. parasit. Labor.); Dr. Riedel (klinische Untersuchungen auf Tuberkulose, Trächtigkeit u. Sterilität). *1—2 Plätze. — Alle Hilfsmittel u. Apparate.* ℗ Diapositive von pathol.-anat. Veränderungen v. Tierkrankheiten u. v. klinischen Krankheitsbildern.

Laboratoria hygienae hominis.

Confer: *Laboratoria Bacteriologiae hominis* (pag. 479) *et animalium* (pag. 486), *Anatomia* (pag. 423) *et Anatomia pathologica generalis* (pag. 469), *Parasitologia generalis* (pag. 379).

Amsterdam (Holland).
Labor. f. Hygiene u. Bact. Dir. Prof. J. J. van Loghem.

Ann Arbor, Mich. (U.S.A.).
Hygien. Labor. Univ. of Michigan. Dir. Prof. J. Sundwall. — Ass. Prof. M. H. Soule (Bact.).

Athènes (Grèce).
Labor. central de l'Hygiène publique. Dir. C. Moutoussis. — K. Kyriasides.

Bari (Italia).
Labor. d'igiene Univ. Dir. G. Falco.

Basel (Schweiz).
Hygien. Anstalt Univ., Petersplatz 10. Dir. Prof. Robert Doerr.

Beograd (S.H.S.).
L'Inst. Central d'Hygiène. Dir. Dr. Stevan Z. Ivanić. — Section Bact. et epidemiol.: Chef Dr. St. Z. Ivanić (bact. et epidemiol.); Dr. Milutin Ranković, Dr. Desanka-Šamić, Dr. Bosiljka Djorić, Dr. Jezdimir Kušić (colaborateurs sc.), Dr. Momčilo Ivanić (protozool.); Section de méd. sociale: Chef Dr. Bogoljub Konstantinović, Branimir Maleš (physiol. et psychol.); Section antirabique: Dr. Obrenija Popović; Section parasit.: Chef Dr. Eugene Džunkowsky (Entomol.), Anna P. Bragina, Dr. Paul Vukasovič (Section chimique pour l'analyse de l'eau, des aliments, des produits pharmacol.; cette section n'est pas encore tout à fait organisée); Chimiste pour chimie med.: Vladimir A. Sirotkine; Section hémotérapique en projet; Section technique en projet; Section serol.: Dr. Dragojlo Popović, Dr. Dobrila Magovćević, Dr. Ramzin, Isava Šaulić. *Places pour les stagiaires dans chaque section.* ℗ L'Inst. collecte tous les objets de bact. (cultures), parasit., entomol., phytopath. etc. pour son mus.

Beuthen, O.S. (Deutschland).
Preuß. Hygien. Inst. Dir. Prof. W. v. Lingelsheim. — Prof. E. Jacobitz, Abt.-Vorst.

Berlin (Deutschland).
Hygien. Labor. des Reichsgesundheitsamtes. Dir. Geh.Reg.R. Dr. Oscar Spitta. *Hygien. und Bact.; im besonderen Wasserbact. u. -biol. Flußverunreinigung und Abwasserfragen. Nahrungsmittelbiol.*
Preuß. Landesanst. für Wasser-, Boden- u. Lufthygiene, -Dahlem, Ehrenbergstr. 38—40—42; Bahnstation: B.-Lichterfelde-West (Wannseebahn). Dir.: Geh.Med.R. Prof. Dr. M. Beninde; stellv. Dir.: Prof. Dr. K. Thumm. — Abt.: 1. Chem. Abt., Dir.: Prof. Dr. K. Thumm; 2. Wassertechn. Abt., Dir.: Prof. Dr.-ing. C. Reichle; 3. Biol. Abt., Dir.: Prof. Dr. R. Kolkwitz; a) Hydrobiol. Gruppe, Leiter: Prof. Dr. J. Wilhelmi; b) Botan. Gruppe, Leiter: Dr. E. Tiegs; c) Zool. Gruppe, Leiter: Prof. Dr. J. Wilhelmi; 4. Hygien. Bakt. Abt., Dir.: Prof. Dr. B. Bürger; 5. Mus., Dir.: Prof. Dr. J. Wilhelmi. *Aufgaben; Wasserversorgung und Beseitigung der Abwässer und Abfallstoffe. A. Wasserversorgung: 1. Die planmäßige wissenschaftliche und technische Prüfung und Durchbildung bestehender und neuer Verfahren der Wassergewinnung und der Wasserreinigung, sowie der Grundsätze für die quantitative Bestimmung und deren Sicherstellung; 2. Auskunftserteilung und sanitätstechnische Beratung auf Antrag von staatlichen und kommunalen Behörden, sowie von Privaten über bestehende oder geplante Wasserversorgungsanlagen; 3. die wissenschaftlich-technische Prüfung des Betriebes von Wasserwerken; 5. die Untersuchung von Wasserproben; Kenntnis der geol.-hydrol. Verhältnisse, sowie der Beschaffenheit des Oberflächenwassers im Reichsgebiet. B. Beseitigung von Abwässern und Abfallstoffen: 1. Wissenschaftlich-technische Prüfung der Verfahren zur Reinigung von Abwässern; 2. Entwässerungsanlagen; 3. Untersuchungen von Ab-*

wässerproben, Müll und sonstigen Abfallstoffen, Luft- und Bodenproben, Filterstoffen, Klärmitteln; 4. systematische Feststellung der Einwirkung der verschiedenartigen Wässer auf die Wasserläufe in chemischer und biol. Hinsicht (Fauna, Flora. Fischzucht); 5. Feststellung der Einwirkung der Schmutzwässer auf den Boden, Ausnutzung der Dungstoffe. — Der Landesanstalt ist angegliedert: das Mainwasser-Untersuchungsamt in Wiesbaden, Luisenstr. 26. Vorst.: G.M.R. Prof. Dr. G. Frank; Chemiker: G. Eckerlin.
Hygien. Inst. Univ., NW 7, Dorotheenstr. 28a. Dir. G.H.R. Prof. Dr. med. Martin Hahn. — Prof. Dr. Bruno Heymann, Abt.-Vorst.; Prof. Dr. Schütz, 1. Ass.; Priv.Doc. Dr. Hirsch, Priv.Doc. Dr. Strauß; Dr. Eisenberg, Ass.; Dr. Kappus, Dr. Wamoscher, Volontair-Ass. *5 Plätze für hygien.-chemische, hygien-physikalische, bact. und serol. Untersuchungen.* ⓟ *Bacterienkulturen, Organschnitte, Protozoenpraep.* — Zweiglabor.: bact. u. serol. Untersuchungsamt.
Forschungs-Inst. für Hygiene u. Immunitätslehre, -Dahlem, Thielallee 63. Dir. Prof. Dr. Ernst Friedberger. — Dr. Grünstein, Dr. Seidenberg, Ass. *6 bis 8 Plätze für Bakt., Hygien., Immunitätslehre.*

Birmingham (England).
Dept. of Hygien. Univ. Dir. Sir John Robertson; Prof. C. J. Lewis. — G. A. Aaden.

Bogota (Columbia).
National Hygiene Labor. C. Wribe (Prof. of parasit.).

Bologna (Italia).
Ist. d'Igiene Univ., Viale Filopanti N. 10. Dir. Prof. Dr. Donato Ottolenghi. — Dr. Giuseppe Brotzu, Libero Doc., Aiuto; Dr. Antonio Ceredi, Ass. *10—12 posti. Ricerche di batt., di immunol., di igiene sperim. e sociale, di educazione fisica. Vi sono attualmente 4 posti di studio sovvenzionati dalla Rockefeller Foundation per coloro che si dedicano a Studi di igiene. Sono riservati a giovani laureati.* — Il Dir. di questo Ist. ha fondato una Stazione per lo studio della malaria nelle bonifiche a Ferrara. Questa Stazione sorta alla fine del 1925 è divenuta ora una Sezione Speciale della Stazione sperim. per la lotta antimalarica, ist. a Roma dalla Direzione Generale della Sanita pubblica in accordo con l'Int. Health Board della Rockefeller Foundation: è diretta dal Dir. di questo Ist. Non ha avuto ancora personale fisso; ma vi si sono succeduti vari studiosi.

Bonn a. Rh. (Deutschland).
Hygien. Inst. Univ., Theaterstr. 32. Dir. Prof. Dr. Hugo Selter. — Prof. Dr. Hilgers; Prof. Dr. Bach; Prof. Dr. Gewecke; Priv.Doc. Dr. Blumenberg. *10 Plätze.*

Bordeaux (France).
Labor. de toxicol. et hygiène appliquée, Fac. med. Univ. Dir. Prof. Barthe.
Labor. d'hygiène, Fac. med. Univ. Dir. Prof. Anché.

Bratislava — Preßburg (C.S.R.).
Inst. d'hygiène. Un Univ. Dir. Prof. Dr. Stanislav Růžička. — 3 Ass., 1 wissenschaftl. Hilfskraft.

Braunschweig (Deutschland).
Hygien. Labor. u. Sammlung. Dir. Prof. Dr. W. H. Schultze.

Breslau (Deutschland).
Hygien. Inst. Univ., Maxstr. 4. Dir. Prof. Dr. Carl Prausnitz. — Priv.Doc. Dr. Herbert Lubinski; Dr. Gerhard Quast; Priv.Doc. Dr. Kollath; Dr. Gertrud Meißner. *10 Plätze für Bact., Serol., Hygien.* —*Interferometer.*Wutschutzabt.f. Schlesien (Pasteur-Inst.).

Brno — Brünn (C.S.R.).
Hygien. Inst. Univ., Uvoz 33. Dir. Prof. Dr. Jos. Roček. — Dr. Magdalena Kühnová, Ass.; Dr. Jaroslav Vitha, Ass.

Bruxelles — Brussel (Belgique).
Labor. Centr. de l'Administration de l'Hygièn. de l'Etat. Dir. J. J. M. van Boeckel.
Labor. d'Hygiene Dir. Prof. Dr. O. Gengou.

Budapest (Ungarn).
Hauptstädtisches hygien. u. bact. Inst. Dir. Prof. Dr. B. Vas, Königl. Ob.San.R. — Dir.-Stellvertr.: Dr. Edmund Strößner, Königl. San.R.; Ass.: Dr. Ilona v. Jóos, Dr. chem. Olga Waldbauer, Dr. Johann Kirchner. *3 — 4 Plätze für alle bact., hygien. und ärztlich-chem. Untersuchungen.*
K. Ung. Staatl. Hygien. Inst., IX, Gyáli-ut 4. Dir. Dr. B. Johan. — 6 Ärzte, 2 Chem. *Bact., Serol., Pathohistol., Parasitol., Chem.*

Cadiz (España).
Labor. d'Hygiène Fac. méd. Univ. Dir. Prof. A. Urtebey y Pastorino.

Caracas (Venezuela).
Labor. d'Hygiène Univ. Dir. Prof. L. G. Itrage.

Catania, Sicilia (Italia).
Ist. d'Igiene sperim. Dir. Eugenio di Mattei.

Charleston, W.Va. (U.S.A.).
West Virginia State Hygien. Labor. Dir. C. E. Gabel.

Chicago (U.S.A.).
Univ. of Chicago, Dept. of Hygien. and Bact., 5724 Ellis Av. Dir. Dr. Edwin O. Jordan. — Dr. John F. Norton, Bact. and Public Health; Dr. William H. Taliaferro, Parasit.; Dr. I. S. Falk, Immunol. and Vital Statistics; S. E. Branham, Bact.; F. A. Coventy, Instr.; A. B. Fisher, Ass. *Varies places, according to number of other workers in the labor. They are fitted for the special needs of the visitor.*

Cluj — Klausenburg (România).
Inst. d'Hygiène et d'Hygiène sociale. Str. Miko 1. Dir. Juliu Moldovan. — Titu W. Slăvoacă, Chef de travaux; Mihai Molog, Ass.; Lazăr W. Isaicu, Ass.

Coïmbra (Portugal).
Inst. de Hïgiene Univ. Dir. J. Serras e Silva.

Cordoba (Argentina).
Labor. de Hygiène Fac. méd. Univ. Dir. Prof. A. D. Villalba.

Danzig (Danzig).
Hygien. Inst., Techn. Hochsch. Dir. Prof. Johannes Petruschky.

Debrecen (Ungarn).
Hygien. Inst. der St.-Tisza-Univ. Dir. Prof. Dr. Alexander Belák. — Dr. Stefan Hajdu, Ass.; Dr. Ladislaus Csresznyés, Ass.; Dr. Theresia Pala. *Immunitätsforschung, Bact., Physik. Chem., demnächst auch für Luftionisierung und atmosphärische Elektrizität.* ⓟ Frösche, Kaninchen.

Dakar (Sénégal).
Inst. Pasteur de l'Afrique occidentale française. Dir. Dr. Constant Mathis. — Sections: Microbiol. humaine, Microbiol. vét., Chim. biol., Chim. agric. Dr. René Guillet; M. Boulay, Chim.; M. Baury, Microbiol.; M. Didier, Vét. de l'Armée. *Ouvert aux travailleurs, qui s'intéressent aux maladies de l'Afrique occidentale, maladies de l'homme, des animaux ou des plantes.* ⓟ Objets pathol. de l'Afrique occidentale.

Dortmund (Deutschland).
Hygien.-Bact. Inst. der Stadt. Dir. Dr. Max Löns. *Typhus u. Paratyphus.*

Dresden (Deutschland).
Deutsches Hygien. Mus. Dir. M. Vogel. Wiss. Dir.: Prof. Dr. W. Weisbach. *Serol. der Lues; soziale Hygien. u. Gewerbehygien.; pathog. Protozoen.*

Düsseldorf (Deutschland).
Hygien. Inst. Med. Akad. Dir. Prof. Th.J. Bürgers.

Ekaterinoslaw, Ukraine (U.d.S.S.R.).
Labor. d'Hygiène de l'Inst. Bact. Ass. Méd.: Garkavi; Gologorski.
Labor. d'Hygiène de l'Inst. Méd. Ass.: Goldenberg (Méd.), Gauchmann (Chim.).
Labor. de l'Inst. d'Hygiène Professionelle. Ass.: Jüdine (Méd.), Kastner (Chim.). *Hygiène expérim.*

Erlangen (Deutschland).
Hygien.-Bakt. Inst. der Univ., Wasserturmstr. $2^1/_2$ Dir. Geh. Med. R. Prof. Dr. Ludwig Heim. — 1 Ass.

Frankfurt a. M. (Deutschland).
Städtisches Hygien. Univ.-Inst., Paul-Ehrlich-Str. 40. Dir. Geh.Med.R. Prof. Dr. Max Neissner.— Abt.-Vorst. Prof. H. Braun; Bact. Frl. Dr. E. Klieneberger; 3 Ass.-Ärzte. *Etwa 3 Plätze für Bact. und Serol.*

Freiburg i. Br. (Deutschland).
Hygien. Inst. Dir. Prof. Paul Uhlenhuth.

Fribourg — Freiburg (Suisse).
Inst. d'Hygiène et de Bact. de l'Univ. Dir. Prof. Dr. Glücksmann. — Dr. Demont, Ass. *Einige Plätze.*

Gelsenkirchen (Deutschland).
Inst. f. Hygiene und Bact. Dir. Prof. Hugo Bruns.

Genève — Genf (Suisse).
Inst. d'Hygiène et de Bact. Univ. Dir. Prof. Hector Cristiani.

Gießen (Deutschland).
Hygien. Inst. d. Med. Fac. Univ.

Granada (España).
Labor. de Hygiène Fac. méd. Univ. Dir. Prof. A. Alvarez de Cienfugos Cobos.

Graz (Österreich).
Hygien. Inst. Med. Fac. Univ. Dir. Prof. Dr. Wilhelm Prausnitz.

Greifswald (Deutschland).
Hygien. Inst. der Univ., Martin-Luther-Str. 6. Dir. Prof. Dr. med. et phil. E. G. Dresel. — Dr. F. Schilf; Dr. O. Stickl; Dr. F. Pels Leusden. *2 Plätze für alle bact. und hygien. Untersuchungsmethoden.* Ⓟ Bakterienkulturen und Bakterienpraep.

Groningen (Holland).
Hygien. Labor. Univ. Dir. Prof. A. Klein.

Halle a. d. S. (Deutschland).
Hygien. Inst. Dir. Prof. Paul Schmidt. *Hygiene u. Bact.: Gewerbehygiene (Bleivergiftung). Bact.: Influenza, Typhus, Paratyphus, insbes. praktische Typhusbekämpfung. Serol.: Anaphylaxie, Wassermannsche Reaktion auf Syphilis. Allgem. Hygiene: Hitzschlag u. Sonnenstich. Bekämpfung des Alkoholismus.*

Hamburg (Deutschland).
Hygien. Staatsinst., 36, Jungiusstr. 1. Dir. GMR. Prof. Dr. med. et phil. R. O. Neumann. — Abt.: Direktorialabt. Abt. I: hygien.-bact. Untersuchungen. Abt. II: hygien.-chem. Untersuchungen. Abt. III: Nahrungsmitteluntersuchung, einschl. der Stationen für ausländ. Fleisch und Fett und für ausländ. Weine. Abt. IV: serobiol. Untersuchungen. Abt. V: Städtereinigung. Arbeitsgebiet für Bau- u. Gewerbehygiene. Arbeitsgebiet für Leibesübungen,
Ventilation u. Heizung. Ständiger Vertreter des Dir.: ao. Prof. Dr. J. Kister; Abt.:Vorst.: Prof. Dr. Kister, Prof. Dr. Noll, Prof. Dr. Lendrich, Prof. Dr. Gaethgens, Dr. Kammann; wissenschaftl. Mitgl. u. ständige Mitarbeiter: Prof. Dr. Buttenberg, Prof. Kickton, PrivDoc. Prof. Dr. Schwarz, Dr. Finsterwalder, Dr. Lorentz; Chem.: Dr. Angerhausen, Dr. Berg; Dr. Gahrtz, Dr. Hanne. Dr. Keim, Dr. Keiser, Dr. Korn, Dr. Nachtigall, von Noel, Dr. Nottbohm, Otte, Dr. Penndorf, Dr. Sudendorf, Dr. Weiß; ferner 14 nicht festangestellte Chem.; im ganzen 88 Beamte. *Plätze nach Bedarf und Raumverhältnissen. Alle Gebiete der Hygiene.* — 3 Außenstationen.

Helsinki (Finnland).
Hygien. Inst. der Univ., Fabiansgatan 35. Dir. Prof. Dr. O. J. von Hellens.

Innsbruck (Österreich).
Hygien. Inst. u. Bundesstaatl. Bact.-diagnost. Untersuchungsstelle. Dir. Prof. A. Lode.

Iowa City, Ia. (U.S.A.).
Water Labor. Division, Iowa State Hygien. Labor. Dir. J. J. Hinman.

Jaroslavl (U.d.S.S.R.).
Kabinett f. Hygiene. Dir. Doc. G. I. Kuročkin.

Jassy (Romania).
Labor. d'Hygiène Fac. Méd. Dir. Prof. Dr. M. Ciuca. — Dr. I. Balteano ,Conferencier d'Epidemiol. et Maladies contagieuses; Dr. G. Tudoranu, Chef de travaux; Ass.: Mlle E. Manolui, Mlle Moruzi, Munteanu; Toma, Préparateur. *10 places.* Ⓟ Différents produits pathol. de l'hôpital des Maladies contagieuses.

Jena (Deutschland).
Hygien. Anstalt. Dir. Prof. Dr. Abel.

Johannesburg (South Africa).
Industr. Hygien. Dept., S. Afric. Inst. for med. Research. Dir. Prof. A. Mavrogordato.

Kaunas (Litauen).
Hygien.-Bact. Inst., Gedimino Gatve 29. Dir. Prof. Jurgeliunas.

Kiel (Deutschland).
Hygien. Inst. Univ., Hospitalstr. 34. Dir. Prof. Dr. Korff-Petersen. — Prof. Dr. L. Bitter; Dr. med. Weigmann; Dr. phil. Liese; Dr. med. phil. Gundel.

København (Danmark).
Univ. hygien. Inst., B, Ny Vestergade 11. Dir. Prof. Dr. L. S. Fridericia, Dr. Skuli Gudjonsson, Ass.; Priv.Doc. Dr. Erik Begtrup (leitet Kursus in praktischer Ernährungslehre); Agnes Elgstrøm, Haushaltungslehrerin. *3 Plätze für Vitamin- und Stoffwechselversuche. — Versuchsräume mit Tierkäfigen, chem. und bact. Apparatur, Luftanalyse, Kalorimeter.*

Köln a. Rh. (Deutschland).
Hygien. Inst. d. Univ. (zugleich: **Bact. Untersuchungsamt der Stadt**), -Lindental, Gleulerstr. 77. Dir. Dr. Reiner Müller, Prof. für Hygiene u. Bact.— Priv.Doc. Dr. med. Karl L. Pesch, Oberarzt (Gruppe der Diphtheriebakterien, Serol.); Priv.Doc. Dr. med. Curt Sonnenschein, Ass.-Arzt (Bakteriophagen, Proteusbakterien, Ozänabakterien); Dr. phil. Elisabeth Sauerborn, Chemikerin; Dr. med. Hippolyt Guillery, Augenarzt (sympathische Ophthalmie und tuberkulotoxische Gewebsveränderungen). *2 Plätze für Bact. und Mikrobiol., Serol. Bakterienmutationen, Paratyphusbakterien, Geschichte der Hygiene und Bakt.* Ⓟ Große Sammlung von Bakterienkulturen und von Bakteriophagen vorhanden.
Mus. für Volkshygiene der Stadt, Im Dau Nr. 3. *Körperbau u. Körperverkrümmungen, Körperpflege, Zahnhygiene, Ernährung, Säuglingspflege, Bekämp-*

Laboratoria

jung des Alkoholismus, Bekämpfung d. ansteck. Krankheiten (insbes. Tuberkulose, Geschlechtskrankh., Kinderkrankh. usw.), Desinfektion, Schulhygiene (Wasserversorgung, Beseitigung der Abfallstoffe).

Königsberg i. Pr. (Deutschland).
Hygien. Inst. Univ. Leiter des Untersuchungsamtes f. ansteckende Krankheiten: Dr. W. Blumenberg.

Kyôto (Japan).
Inst. of Hygien. Coll. of Med., Kyôto Imperial Univ. Dir. Prof. Dr. Shozô Toda. — Ass. Prof. S. Otani; Lect. Fuji Teikichi; Lect. Jehara Takeo.

La Habana (Cuba).
Labor. für Hygiene, Univ.

La Valletta (Malta).
Labor. of Hygiene, Univ. Dir. Prof. A. V. Bernard.

Lausanne (Suisse).
Inst. d'Hygiène expérim. et de Parasit. de l'Univ., Solitude 10. Dir. Prof. B. Galli-Valerio. *3 places pour Hygiène, Parasit. animale.*

Leipzig (Deutschland).
Hygien. Inst. d. Univ. Dir. Prof. W. Kruse.

Lexington, Va. (U.S.A.).
Labor. of Hygien. of the Washington and Lee Univ. Dir. Prof. F. Fletcher.

Lexington, Ky. (U.S.A.).
Dept. of Hygiene of the Univ. of Kentucky. Dir. Prof. W. N. Lipscomb. — Prof. J. E. Rush; Prof. W. W. Zwick.

Lille (France).
Labor. d'Hygiène et Microbiol., Fac. de Méd. cathol. de l'Univ., 56, Rue du Post et 11, Rue de Toul. Dir. Prof. G. Lemière.
Labor. d'Hygiène et Bact., Fac. de Méd. de l'Univ. Dir. Prof. Pierret.

Liverpool (England).
Dept. of Hygiene, Fac. of Med. Univ. Dir. Prof. E. W. Hope.
Univ. School of Hygiene and City Labor., Mount Pleasant (Corner of Oxford Street). Dir. Prof. J. M. Beattie, Bact. Labor. — Dr. L. S. Ashergt, Chief Ass., Lect.; Dr. G. R. James, Lect. on Public Health Bact.; Dr. Thirza Redman, Lect. on Bact. Chem. *6—8 research workers.*

London (England).
Dept. of Hygiene, Westminster Hosp., Med. School of the Univ., S.W. 1, 12 Caxton Str. Dir. F. J. Allan.
Dept. of Hygien., St. George's Hosp., Med. School of the Univ., S.W., Hyde Park Corner. Dir. H. R. D. Spitta.
Hygiene Labor. and Mus., Royal Army Med. Coll. of the Univ., S.W., Grosvenor Road.
Inst. and Mus. of Hygiene, W. 1, 34. Devonshire St., Portland Place.

Lyon (France).
Mus. d'Hygiène de l'Univ. Dir. Prof. Paul Courmont.

Lwów (Polska).
Inst. f. Hygiene d. Univ. Dir. Z. Steusing.

Madison, Wis. (U.S.A.).
State Labor. of Hygiene. Ass. Prof. M. S. Nichols (Sanitary Chem.).

Madrid (España).
Inst. Nacional de Higiene de Alfonso XIII., Parque de la Moncloa. Dir. O. J. Tello y Muñoz.

Marburg a. d. L. (Deutschland).
Hygien. Inst. Dir. Prof. Heinrich Bonhoff.

Messina (Italia).
Ist. di Igiene Univ. Dir. Prof. G. Volpino.

Milano (Italia).
Ist. di Igiene Univ. Dir. Prof. E. Ronzani.

Modena (Italia).
Ist. di Igiene e Batt. Univ., Piazza S. Eufemia 4. Dir. Prof. Francesco Sanfelice. — Dott. Pier Luigi Carbonieri, Aiuto; Dott. Edgardo Barbanti. *5 posti. Ricerche biol., batt., Chim. bromatol.* ℗ Vicca collezione di microorganismi patogeni e non patogeni.

München (Deutschland).
Hygien. Inst. Univ., Pettenkoferstr. 34. Dir. Prof. Dr. Kisskalt. — Prof. Dr. v. Angerer; Prof. Dr. Ilzhöfer; Priv.Doc. Dr. Knorr. *10 Plätze für Bact. u. Chem.*

Münster i. W. (Deutschland).
Hygien. Inst. Univ. Dir. Prof. K. W. Jötten.

Napoli (Italia).
Ist. d'Igiene Fac. med. Univ. Dir. Prof. D. De Blasi.

Niš (S.H.S.).
Inst. Epidémiol. Dir. Dr. med. Dragolioub Popovitch. — Dr. Alexandre Papadopolos; Dr. Miron Maloitchitch; Dr. Radoslav Miletitch; Dr. Milan Yovanovitch; *1 place vacante d'Ing. chem. Examens microscopiques et bact. Serol. Examens chim. Service anti-rabique.*

Norton, Va. (U.S.A.).
Norton Branch Labor., State Board of Health. Dir. Mr. A. H. Straus, Richmond (Va.); H. L. Freese in charge of Norton Labor. *2 places could fit up for work in enteric diseases.* ℗ Stock cultures. — This labor. is a branch labor. of the State Board of Health, Labor., Richmond (Va.).

Odessa (U.d.S.S.R.).
Katheder für praktische Med., Olgievskaja 4. Dir. Prof. L. V. Gromaševskij. — Leiter d. Sektionen: Prof. V. Stefanskij (Mikrobiol. d. Infektionskrankh.), Prof. N. Kostjamin (Hygien.); wirkl. Mitgl.: Prof. V. Elin, Prof. S. M. Ščastnyj; 3 wiss. Mitarbeiter, 8 Aspiranten.
Abt. f. Hygiene des Med. Inst. Dir. Prof. N. N. Kostjamin.

Okmulgee, Okla. (U.S.A.).
Sanitary Dept. Dir. M. E. Bjerregaard.

Omsk (U.d.S.S.R.).
Hygien. Abt. d. med. Inst. Dir. Prof. K. M. Gretschischtscheff.

Oslo (Norge).
Hygien. Inst. Univ. Dir. Prof. A. Holst.

Padova (Italia).
Ist. d'Igiene e Polizia med., Fac. med. Univ. Dir. Prof. O. Casagrandi.

Palermo (Italia).
Ist. d'Igiene Fac. med. Univ. Dir. Prof. L. Manfredi.

Paris (France).
Labor. d'Hygiène Fac. méd. Univ., 391, Rue de Vaugirard. Dir. Prof. A. B. Marfan.
Labor. d'Hygiène générale et expérim. Ecole pr. des Hautes Études. Dir. Dr. Prof. F. Bordas.
Labor. d'Hygiène de la Ville de Paris, IVe, 1bis, Rue des Hospitalières St. Gervais.

Parma (Italia).
Ist. d'Igiene Fac. med. Univ. Dir. Prof. G. Sangiorgi. *Igiene. e polizia med.*

Pavia (Italia).
Ist. d'Igiene Fac. med. Univ. Dir. Prof. E. Bertarelli.

Pécs (Ungarn).
Hygien. Inst. Univ. Dir. Prof. Bela Fenyvessy.

Peking (China).
Inst. f. Hygiene u. Physiol. Tsinghua Univ. Dir. Prof. Kang Li.
Dept. of Hygiene Union Med. Coll. Dir. Prof. J. B. Grant.
Central Epidemic Prevention Bureau, Temple of Heaven. Technical Dept. u. Standardization Labor. Dir.: Surgeon-General Shisan C. Fang.

Perm (U. d. S. S. R.).
Hygien. Inst. mit Kabinett f. soziale Hygiene. Dir. Prof. K. N. Sapšev.

Philadelphia, Pa. (U. S. A.).
Labor. of Hygiene, Univ. Dir. Alexander Crever Abbott. — D. H. Bergey, Prof. Hygiene and Bact.; Seneca Egbert, Prof. Hygiene; Henny Field Smyth, Ass. Prof. Industrial Hygiene; Myriam S. Iszard, Instr. in Bact.; W. A. Kreidler, Ass. Instr. in Bact. *4 places.*

Pisa (Italia).
Ist. d'Igiene Fac. med. Univ. Dir. Prof. A. di Vestea.

Porto Alegre (Brasil).
Hygien. Inst. Univ. Dir. Prof. Freitas e Castro.

Poznań — Posen (Polska).
Hygien. Inst. Univ. Dir. P. Gantkowski.

Praha — Prag (C. S. R.).
Hygien. Inst. d. Karls-Univ., II, Na Bojišti 3. Dir. Prof. Dr. Gustav Kabrkel.
Hygien. Inst. d. Deutsch. Univ., II, Preslova 2098. Dir. Prof. Dr. Gottlieb Salus. — Prof. Dr. Oskar Bail; Dr. Wilhelm Spät, Priv.Doc. d. Hygiene; Dr. F. Breinl, Serol. Abt.

Raleigh, N.C. (U. S. A.).
State Labor. of Hygiene. Dir. C. A. Shore.

Richmond, Va. (U. S. A.).
Labor. of the State Board of Health. Dir. Dr. J. H. Strauß. — There is a branch labor. at Norton, Va.

Riga (Latvija).
Hygien. Inst. Med. Fak. Univ., Kronvalda bulvarī 9. Dir. Doc. E. Fehrmann.

Rio de Janeiro (Brasil).
Labor. de Hygiene Fac. med. Univ. Dir. Prof. A. Peixoto.

Rolla, Mo. (U. S. A.).
Hygiene Labor. of the Public Health Service. Dir. Prof. M. P. Ravenel. — J. A. Bengtson. *Med. Bact. and preventive Med. Trachoma.*

Roma (Italia).
Ist. d'Igiene sperim. Fac. med. Univ. Dir. Prof. G. Sanarelli.

Rosario (Argentina).
Labor. de Higien. y Bact. Escuela de Odontol. Univ. Dir. Prof. J. Jons.

Rostock i. M. (Deutschland).
Hygien. Inst., Buchbinderstr. 8|9. Dir. Prof. Dr. v. Wasielewski. — Priv.Doc. Dr. Winkler; Dr. Demme; Dr. Richtsteiger. *2 Plätze für Untersuchungen über Protozoenzüchtung und Pockenforschungen, Variola-Vakzine.* ℗ Protistenkulturen. Protozoen-, insbesondere Amoebenpraep., Mikrophotogramm e.

Mecklenburgische Landes-Lebensmittel-Unters.-Anst.
Dir. Prof. Dr. Theodor v. Wasielewski.

Rouen (France).
Labor. de Bact., Ecole prép. de méd. Dir. Prof. M. Guerbet.

San Francisco, Cal. (U. S. A.).
Dept. of preventive Med. and Hygiene, Med. School of the Univ. of California. Dir. Prof. W. H. Kellogg.

São Paulo (Brasil).
Inst. de Hygiene, Caixa postal 1985. Dir. Dr. P. H. Geraldo H. de Paula Souza, M.D. — Ass.: Dr. F. Borges Vieira, Dr. B. Alves Ribeiro, Dr. Clovis Correâ, Dr. S. B. Pessôa; Instr.: Dr. Gastão Fleury da Silveira, Dr. Alberto Santiago, Dr. Octavio Monteiro de Camargo. *In our new building about 6 places. Research work on leprosy, malaria, epidemiology, intestinal parasit. infections.* ℗ Preparations on leprosy, mycol., malaria, mosquitoes, coskroaches, intestinal parasites, etc. — Annexed to the labor. there is a dispensary, frequented daily by about 100 patients.

Saratow (U. d. S. S. R.).
Labor. f. experim. Hygien. Univ. Dir. Prof. V. A. Arnol'dov.

Sassari (Italia).
Ist. de Igiene, Univ. Dir. C. Fermi.

Shanghai (China).
Public Health Dept. Dir. Dr. N. Davis.

Sheffield, Ia. (U. S. A.).
Branch Labor. for the Iowa State Board of Health. Dir. J. B. Hudson. *Bact. of the car., staining of tubercle bacilli typhoid carriers.*

Sighişoara — Schäßburg (România).
Hygien. Mus.

Smolensk (U. d. S. S. R.).
Hygien. Inst. mit socialhygien. Mus. Dir. Prof. Dychno.

Sofia (Bulgarien).
Inst. für Hygiene Fac. med. Univ. Dir. Prof. T. Petrov.

Sombor (S. H. S.).
Staatl. Hygien. Inst. (Državni Higijenski Zavod). Dir. Dr. Adam Schmidt. *2—3 Arbeitsplätze besonders für klinisch-mikroskopische und bakt. Arbeiten ausgerüstet.*

Stockholm (Sverige).
Hygien. Inst. Karolinska Inst. Dr. Prof. Petterson.

Stuttgart (Deutschland).
Inst. f. Bakt. u. Hygien. Techn. Hochsch. Dir. Prof. A. Gastpar.

Szeged (Ungarn).
Hygien. Inst. Univ. Dir. Prof. G. Rigler.

Taihoku, Formosa (Japan).
Dept. of Hygiene Gouvernmental Central Research Inst. K. Morishita (Parasit.).

Tartu — Dorpat (Eesti).
Hygien. Inst., Aia t. 46. Dir. Prof. A. Rammul. — Ass. S. Lind; Ass. J. Wilde.

Taschkent (U. d. S. S. R.).
Hygien. Inst. Univ. Dir. Prof. G. M. Pinegin.

Tegucigalpa (Honduras).
Hygien. Inst. Univ. Dir. Prof. Dr. M. Carias.

Tiflis (U. d. S. S. R.).
Inf. f. Sanitätswesen mit Malariaabt. Univ. Dir. Prof. V. Voronin.

Torino (Italia).
Ist. del. Igiene Univ. Dir. Prof. F. Abba.

Toronto, Ont. (Canada).
Dept. of Hygiene, Univ. Dir. J. G. Fitzgerald, M.D.

Tübingen (Deutschland).
Hygien. Inst. Univ. u. Nahrungsmitteluntersuchungsamt. Dir. Prof. Kurt Wolf.

Tunis (Tunis).
Labor. d'Hygiène. Dir. Prof. Braquehaye. *Hygiène et Méd. usuelle.*

Utrecht (Holland).
Hygien. Labor., Catharijnesingel 59 boven. Dir. Prof. Dr. C. Eijkman. — Conserv. J. P. G. Hulshoff Pol; Th. H. Strengers, Haupt-Ass.

Valencia (España).
Labor. de Higiene con prácticas de Bact. sanitar. Fac. med. Univ. Dir. Prof. J. Campos Fillol.

Vermillion, S.D. (U.S.A.).
State Health Labor. Ch. A. Hunter, Ass. Dir.

Warszawa (Polska).
Hygien. Inst. Univ. Dir. Prof. S. Dzierzgowski. **Staatl. Hygiene-Inst.**, Kujarska 2. Dir. Dr. Rajchman; stellvertr. Dir.: Dr. Hirszfeld. — Dr. F. Pnezmycki; Dr. Sparrow; Dr. Sierakowski; Dr. Fejgin; Dr. Celarek; Frl. Halber; Frl. Seydel; Dr. Anigstein; Dr. Weil; Dr. Chodźko; Dr. Funk; Dr. Kołodziejska; Dr. Kacprzak; Fr. Adamoviczova; Dr. Nowakowski; Fr. Szrojnicka. *10 Plätze für Bakt., Serol., Biochemie. Bereitung d. Heilsera u. Impflymphen. Abt.: 1. allg., 2. bakt. (diagnost.), 3. Abt. f. Heilsera- u. Impflymphenbereitung, 4. Abt. f. Bereitung d. Pockenlymphe, 5. Pasteur'sche Abt. (Schutzimpfungen gegen Tollwut), 6. Schule f. Hygiene.* — Diagnostische Filialen: (nur bakt. Abt.) Kraków, Toruń, Łódź, Lublin, Wilno. Filiale in Lwów besitzt außerdem eine Pasteursche Abt.

Washington, D.C. (U.S.A.).
Hygienic. Labor. (U.S. Public Health Service), 5th and E-Streets, N.W. Surgeon General: Hugh S. Cumming; Dir. of the Labor. Corps: Surg. George W. McCoy. — Division of Pathol. and Bact.: In charge of division: Surg. George W. McCoy; E. Francis (Bacteriologist); A. C. Evans (Assoc. Bacteriologist). Division of Zool.: Chief of division: Ch. Wardell Stiles, Ph.D.; M. M. Brooks (Biologist). Division of Pharmacol.: Chief of division: Carl Voegtlin, Ph.D. Division of Chemistry: Chief of division: William Mansfield Clark, Ph.D.; B. Cohen (Chemist). *The purpose of the Labor is to study and investigate the diseases of men and conditions influencing the propagation and spread thereof, including sanitation and sewage and the pollution either directly or indirectly of the navigable streams and lakes of the United States.*

Wien (Österreich).
Lehrkanzel f. Hygiene Univ. Dir. Prof. R. Graßberger.

Wiesbaden (Deutschland).
Mainwasser-Untersuchungsamt der Preuß. Landesanst. für Wasser-, Boden- u. Lufthygiene, Luisenstr. 26. Geh. Med. R. Prof. Dr. G. Frank. — Chemiker: G. Eckerlin.

Wilhelmshaven (Deutschland).
Hygien. Untersuchungsstation des Sanitätsamtes. Dir. Marine-Oberstabsarzt Dr. Willi Wendtlandt. *Bact. Serol.*

Woronesch (U.d.S.S.R.).
Hygien. Inst. Univ. Dir. Prof. M. J. Stutzer.

Würzburg (Deutschland).
Hygien. Inst. Univ. Dir. Prof. K. B. Lehmann.

Zagreb — Agram (S.H.S.).
Hygiene-Schule der Rockefeller-Stiftung (Higijen. škola Rockefellerove zaklade), Zeleni Brijeg.
Hygien. Inst. fac. med., Gajeva ulica 40. Dir. Prof. Dr. Emil Prašek.
Hygien.-Bact. Inst. Fac. phil. Univ.

Zürich (Schweiz).
Hygien.-Inst. Univ., 7, Gloriastr. 32. Dir. Prof. Dr. Silberschmidt.

Laboratoria hygienae animalium.

Confer: *Laboratoria Bacteriologiae animalium* (pag. 486), *Parasitologia generalis* (pag. 379).

Freiburg i. Br. (Deutschland).
Tierärztl.-Hygien. Inst. Dir. Prof. Mat. Schlegel.

Gießen (Deutschland).
Veterinärhygien. und Tierseucheninst. Dir. W. Zwick.

Hannover (Deutschland).
Hygien.-Univ., Tierärztl. Hochsch. Dir. Prof. Dr. Hermann Mießner. *Aufzuchtkrankheiten; Paratyphosen; Rotz; Paratuberkulose; Gasödeme; Tollwut; Parasiten.*

Kamenetz-Podolsk, Ukraine (U.d.S.S.R.).
Labor. für Zoohygiene d. Inst. für Landwirtsch. „K. Marx."

Kiel (Deutschland).
Tierseuchen-Inst. der Landwirtschaftskammer für die Provinz Schleswig-Holstein, Gutenbergstr. 77. Dir. Dr. W. Kiessig. — Dr. P. Heinke. *4 Plätze für bact. Untersuchungen, Herstellung von Serum und Impfstoffen.* Ⓟ Praep. tierischer Erkrankungen.

Kindia (Afrique occidentale française).
Succursale de l'Inst. Pasteur. Dir. Maj. vét. Wilbert.

Landsberg a. d. W. (Deutschland).
Inst. f. Tierhygiene d. Preuß. Landw. Versuchs- u. Forschungsanst., Theaterstr. 26. Dir. Prof. Dr. phil. Paul Knuth. — Dr. David, diagnostische Untersuchungen; Dr. Zerbe, Bekämpfung der Tuberkulose bei Haustieren; Dr. Geldsetzer, Bekämpfung der Unfruchtbarkeit bei Haustieren; Dr. Brantin, Bekämpfung der Unfruchtbarkeit bei Haustieren. 2 techn. Ass. für bact., serol. u. parasit. Untersuchungen. *1 Platz.*

Leipzig (Deutschland).
Vet.-Hygien. Inst. der Univ. Dir. OberMed.R. Prof. Dr. Martin Klimmer. *Tuberkulose, Jungtierkrankheiten, Abortus, Euterkrankheiten.*

León (España).
Labor. de Fisiol. e Higiene, Escuela de Veterinaria. Confer: Physiol. compar.

Lwów (Polska).
Labor. f. Fleischhygiene d. Tierärztl. Akad. Dir. A. Trawiński.

München (Deutschland).
Tierhygien. Inst. Univ. Dir. Prof. Dr. K. Süpfle.

Muktesar, P.O. Ritani (Brit.-India).
Imperial Inst. of Veterinary Research. Dir. J. T. Edwards, D.Sc., M.R.C.V.S. — J. R. Haddow, B.Sc., Veterinary Research Officer (in charge of serum Production); Hugh Cooper, Pathologist. Officer in charge of preparation of Vaccines and Media; Officer in charge of Dairy and Animal Husbandry; Ass. in Protozool.; Ass. in Helminthol.; Ass. in Bio-chemistry; Ass. in general Labor. and Field Work; Engineer with a staff of 2 Ass.; Artist. *6 places. Dept.: I. The preparation of large quantities of sera and vaccines against rinderpest, anthrax, Haemorrhagic Septicaemia, strangles and Black quarter; II. The preparation of mallein, tuberculin, abortion vaccine and special vaccine; III. Imparting instruction in the control of epizootic diseases prevalent in India to officers of the Vety. Dept.; IV. Examination of pathol. specimens received from the Centres of infection; V. Research work on various diseases of domesticated animals.* — There is a branch labor. at Izatnagar, in the plains, about 70 miles from the main labor.

Stambul (Türkei).
Inst. f. Hygiene u. Tierzucht, Tierärztl. Hochsch. Dir. Prof. S. Abidin.

Stuttgart (Deutschland).
Württemb. Tierärztl. Landes-Untersuchungsamt. Dir.Ministerialrat Dr. Robert v. Ostertag. *Veterinärhygien.; Bact. der Tierseuchen,Fleischhygiene, Milchhygiene.*

Wien (Österreich).
Staatl. Tierimpfstoffgewinnungsanstalt und Station für Tierseuchendiagnostik, Mödling b. Wien, Friedrichstr. Dir. Dr. Franz Gerlach. *Bact., Serol. u. pathol. Anat.*

Cancer.
Confer: *Anatomia pathol. generalis* (pag. 459).

Berlin (Deutschland).
Inst. f. Krebsforschung, Univ., NW 6, Luisenstr. 9. Dir. Prof. Ferd. Blumenthal.

Bruxelles — Brussel (Belgique).
Centre radiobiol. et anticancereux de l'Univ., l'hôpital Brugman. Dir. Prof. A. P. Dustin.

Hamburg (Deutschland).
Inst. für Krebsforschung Univ. Eppend. Krankenhaus. Dir. Dr. R. Bierich.

Heidelberg (Deutschland).
Wissenschaftliche Abt. des Inst. für experim. Krebsforschung, Voßstr. 3. Dir. Prof. Dr. H. Sachs. — Dr. A. Klopstock, Ass.; Dr. E. Witebsky, Hilfs-Ass. *6 Plätze für Serol. und Immunitätsforschung.*

Jassy (România).
Labor. de Bact. Fac. méd. Confer: Bact. hominis.

London (England).
Imperial Cancer Research Fund. Dir. Dr. J. A. Murray. — Secretary: F. G. Hallett.

Marseille (France).
Labor. au centre Anti-Cancéreux. Dir. Dr. Albert Rouslacroix. *Cancerol. Accessoirement Bact. clinique.*

New York (U.S.A.).
Inst. of cancer research Columbia Univ., 1145 Amsterdam Ave. Dir. Prof. F. C. Wood.

Paris (France).
Inst. du Cancer Fac. méd. Dir. Prof. Dr. Gustave Roussy.
Labor. de Radiophysiol., Inst. Radium, Pavillon Pasteur, 1, rue Pierre Curie. Dir. Prof. Dr. Claude Regaud. — Antoine Lacassagne, Sous-Dir. du Labor.; René Ferroux, Chef des travaux de physique appliquée à la biol. *5 places. Physiol. des radiations de courte longueux d'onde, Cancer expérim.* — *Machines à rayons X. Radium et autres corps radioactifs.*

Strasbourg — Straßburg (France).
Labor. au Centre Régional de Lutte contre le Cancer. Dir. P. Reiß.

Toulouse (France).
Centre régional pour la lutte contre le cancer. Dir. Prof. T. Marie.

Laboratoria palaeontologica.

Amsterdam (Holland).
Mineralogisch-Geol. Labor., Nieuwe Prinsengracht 126. Dir. Prof. Dr. E. Dubois.

Athènes (Grèce).
Labor. de Paléontol. et de Géol. Univ., rue de l'Académie. Dir. Prof Skuphos.

Basel (Schweiz).
Geol.-paläontol. Anstalt Univ. Bernoullianum. Dir. Prof. A. Buxtorf. — 1 Ass. *4 Plätze.*

Beograd (S.H.S.).
Geol.-paläontol. Inst. (geol.-paleontol. zavod), Univ. Philosoph. Fac. Dir. Prof. Dr. Svetolik Radovanović.

Berkeley, Cal. (U.S.A.).
Dept. of Geol. Univ. of California. C. L. Camp (Curator of Reptiles and Amphibians).

Berlin (Deutschland).
Geol.-Paläontol. Inst. u. Mus. Univ., N 4, Invalidenstr. 43. Dir. Geh. Bergrat Prof. Dr. J. Pompeckj. — Dr. W. Janensch, Kustos u. Prof.; Dr. W. Dietrich, 1. Ass.; Prof. Dr. H. Reck, 2. Ass.; Dr. W. Quenstadt, 3. Ass. ⓟ *Fossile Tiere und Pflanzen.*
Geol. Paläontol. Inst. Techn. Hochsch., -Charlottenburg, Berliner Str. 170—2. Dir. Prof. Dr. A. Born.

Bologna (Italia).
Ist. di geol. e paleontol. Univ. Dir. Prof. M. Gortani.

Bonn a. Rh. (Deutschland).
Geol.-paläontol. Inst. Univ. Dir. Prof. Dr. Hans Cloos.

Breslau (Deutschland).
Geol.-palaontol. Inst. u. Mus. Univ., Krieten Siebenmorgenweg 67. Dir. Prof. Dr. Wolfg. Soergel.

Bruxelles — Brussel (Belgique).
Labor. de Géol. de l'Univ., 133, avenue Adolphe Buyl. Dir. Prof. M. Leriche. Mme Paul Ledoux, Ass. *1 place.* ⓟ *Fossiles.*
Service géol., Palais du Cinquantiaire.

Budapest (Ungarn).
Palaeontol. Inst. Univ. Dir. Prof. Károly Papp. — Dr. St. Mayer; Dr. A. Erdödy.

Caen, Calvados (France).
Inst. de Paléontol. Fac. Sc. nat. Univ. Dir. Prof. A. Bigot.

Puy de Dôme, Clermont-Ferrand (France).
Labor. de Paléontol., Fac. Sc. Univ. Dir. Prof. Glangeaud.

Cluj — Klausenburg (România).
Inst. Géol. Univ. Dir. Prof. Dr. Ion Popescu-Voiteşti. — Dr. Stefan Mateescu, Chef de travaux pratiques.

Coïmbra (Portugal).
Mus. e Labor. Geol. Dir. A. Ferraz de Carvalho.

Delft (Holland).
Labor. f. historische Geol. u. Palaeontol. Dir. Prof. Dr. H. A. Brouwer.

Ekaterinoslaw, Ukraine (U. d. S. S. R.).
Geol. Cabinet des Berg-Inst. Dir. Prof. N. I. Lebedew. — W. I. Pogodina, Ass. *3—4 Plätze besonders für das Carbon (Donetzbecken). — Sammlungen von Rußland, Westeuropa und Nordamerika.*

Erlangen (Deutschland).
Mineralogisch-geol. Inst. Dir. Geh.Med.R. Prof. Dr. Hans Lenk. — Prof. Dr. Lothar Krumbeck, 1. Ass.; Dr. Paul Dorn, 2. Ass. ⓟ *Objekte organogener Gesteinsbildung.*

Firenze, Toscana (Italia).
Ist. di Geol. e Paleontol. Univ. Dir. Prof. G. Dainelli. — Prof. D. Del Campana.

Frankfurt a. M. (Deutschland).
Geol. palaeontol. Inst. der Univ., Robert-Mayer-Str. 6. Dir. Prof. Dr. Fr. Drevermann. — Prof. Dr. Leuchs (allgemeine und angewandte Geol.); Prof. Dr. Richter (Stratigraphie und Palaeontol.); Priv.-Doc. Dr. Kräusel (Palaeontol.). *Palaeontol., mikrophotographische, chemisch-geol. Untersuchungen. — Moderne Mikrophotographie.* ⓟ *Fossilien des Palaeozoicums und Neozoicums; Radiolarienpraep.; Abgüsse fossiler Wirbeltiere.*

Freiburg i. Br. (Deutschland).
Geol. Inst. der Univ. Dir. Prof. W. Deecke. — 2 Ass.

Freiberg, Sachsen (Deutschland).
Geol. Inst. u. Palaeont. Sammlung d. Bergakad. Dir. Prof. Schuhmacher.

Genova (Italia).
Ist. di Geol. R. Univ. Dir. Prof. Gaetano Rovereto. — Prof. Paolo Principi, Aiuto; Dott. Marco Airoldi, Ass.

Gießen (Deutschland).
Geol. u. Palaeont. Inst. der Univ., Ludwigstr. 23. Dir. Prof. Dr. Hermann Harrassowitz. — Prof. Dr. Hummel, 1. Ass.; Dr. Heinrich Richter, 2. Ass.

Göttingen (Deutschland).
Geol.-Palaeont. Inst. der Univ., Bahnhofstr. 28. Dir. Prof. H. Stille. — Prof. Dr. H. Salfeld (geol. Vorlesungen); Dr. H. Schmidt, Priv.Doc., Kustos, Verwaltung der Sammlungen (palaeont. Vorlesungen); Dr. R. Brinckmann, Priv.Doc., Ass. (geol. Vorlesungen); Dr. F. Heidorn, Ass.; Dr. F. Lotze, Ass. ⓟ *Fossile Tiere und Pflanzen.*

Graz (Österreich).
Labor. für Phytopalaeont. Phil. Fac. Univ., Vorst.: Prof. Dr. Bruno Kubart.

Greifswald (Deutschland).
Geol.-paläontol. Inst. der Univ., Langefuhrstr. 23. Dir. Geh. Reg.R. Prof. Dr. Otto Jaekel. — Dr. Hans Frebold, Priv.Doc.; Dr. E. A. Zischke, Ass.

Grenoble (France).
Labor. de Paléont., Fac. Sc. Univ. Dir. Prof. Jacob.

Groningen (Holland).
Biol. Archaeol. Inst., Poststraat 6. Dir. A. E. van Giffen. *Haustiere u. Kulturpflanzen der Praehistorie u. frühen Historie für statistische Untersuchungen.*
Mineralogisch-geol. Labor., Melkweg 1. Dir. Prof. Dr. J. H. Bonnema. — J. E. van Veen (Ass.).

Halle a. d. S. (Deutschland).
Geol. palaeont. Inst. Dir. Geg.R. Prof. Dr. Johannes Walther. — Prof. Dr. Johann Weigelt; Dr. Bruno von Freyberg; Dr. Walter Roepke.

Hamburg (Deutschland).
Mineralogisch-geol. Staatsinst., 5, Lübecker Tor 22. Dir. Prof. Dr. Gurich. — Prof. Dr. Wysogórski, Hauptkustos; Dr. Koch, Dr. Ernst, Dr. Gripp, Kustoden; Dr. Wohlstadt, Dr. Müller, wissensch. Hilfsarbeiter. *1 Platz* ⓟ *Tertiärfossilien, Jura- und Kreidefossilien von Norddeutschland, Palaeozoicum von Skandinavien.*

Heidelberg (Deutschland).
Geol.-palaeont. Inst. Dir. Prof. Wilhelm Salomon-Calvi.

Innsbruck (Österreich).
Geol.-Palaeont. Inst. Univ., Universitätsstr. 4. Dir. Prof. Raimund v. Klebelsberg.

Jassy (România).
Labor. de géol. et paléont. Dir. Prof. I. Simionescu. — Th. Văscăutzeano, Chef de trav.; N. Macarovici, Ass.; Amélie C. Motas, Ass.; E. Saoulea, Dessenateur ⓟ *Fossiles de Roumanie (devon. trias. juras. crét. néogen).*

Jena (Deutschland).
Mineralogisch-Geol. Inst. der Univ., Palaeont. Abt.; Schillerstr. 12. Dir. Prof. Dr. Linck. — Abt.-Leiter: Prof. Dr. v. Seidlitz. *1—2 Plätze. — Vergleichende Sammlung der Thüringischen Schichten und Fossilien.* ⓟ *Fossile Vertebrata u. Evertebrata.*

Karlsruhe, Baden (Deutschland).
Geologisch-mineralogisches Inst. Techn. Hochsch. Dir. Prof. Wilh. Paulcke.

Kiel (Deutschland).
Geol.-Palaeont. Labor. Univ. Dir. Prof. Dr. Ewald Wüst. *Quartärfauna, Säugetiere, Mollusken (Mollusca extramarina palaearctica).*

Köln a. Rh. (Deutschland).
Geol.-Mineralogisches Inst. der Univ., Severinswall 38. Dir. Prof. Dr. H. Philipp. — Ass.: Dr. Ralb, Dr. Warneck, Dr. Müller. *2 Plätze. — Alle Apparate zu geol. u. mineralog. Untersuchungen, außerdem zu chemischen Analysen (Wasseruntersuchungen), Erzmikroskopie, Gesteinsanalysen, elektrischen und magnetischen-geophysikalischen Untersuchungen.* ⓟ *Palaeont.-geol. Material der Rheinlande.*

Königsberg i. Pr. (Deutschland).
Geol. u. palaeont. Inst. u. Bernsteinsammlung, Lange Reihe 4. Dir. Prof. Dr. Karl Andrée. 3 Ass.

Kraków (Polska).
Inst. paléont. Univ. (Zakład paleont. Univ. Jagiellońskiego), 53, rue Grodkza. Dir. Prof. Dr. Jan Nowak. — Ass.: Dr. Franciszek Bieda, Stanisław Sokołowski. *1 place.* ⓟ *Brachiopodes du Carbonifère inf., Céphalepodes du malm.*

Kyôto (Japan).
Geol. Inst. Coll. of Sc. Kyôto Imperial Univ. Dir. Prof. Dr. Takuji Ogawa (Geol.). — Prof. Dr. Naka-

mura (Paleont. and Historical Geol.); Ass. Prof. Jirô Makiyama (Palaeont.); Ass. Tokubei Kuroda (Chonchynol.).

East Lansing, Mich. (U.S.A.).
Dept. of Zool. and Geol., State Coll. of Agric. Confer: Zool. generalis.

Lausanne (Suisse).
Labor. de géol. et de paléont. de l'Univ., Palais de Rumine. Dir. Prof. M. Lugeon. — Chef des travaux: Dr. E. Gagnebin; Ass.: Dr. E. Peterhans. *10 places.* ⓟ Collection de Nummulites de Ph. de la Harpe et des Vertébrés de Samos de Forsyth-Major.

Leipzig (Deutschland).
Geol.-palaeont. Inst. u. Mus. d. Univ. Dir. Dr. Johannes Felix.

Leningrad (U.d.S.S.R.).
Labor. f. Palaeont. Univ.

Liège (Belgique).
Collections et labor. de paléont. animale et de paléont. végétale, Univ. Dir. Prof. Dr. Ch. Fraipont.

Lille (France).
Labor. de Paléobotan., Fac. des Sc., de l'Univ., 159, rue Brule-Maison. Dir. Prof. Paul Bertrand. — M. P. Corsin, Ass. *3 places. Flores houillères. (Morphol. externe et anat. Applications stratigraphiques.) Algues des Charbons. — Outillage photographique pour l'étude des empreintes houillères et pour l'étude anat. des végétaux à structure conservée.* ⓟ Flores houillères: toutes plantes de la houille.
Labor. de Géol. de l'Univ. et Mus. houiller de l'Univ., 159, rue Brule-Maison. Dir. Prof. Pierre Pruvost. — M. G. Dubois, Chargé de Cours; M. R. Dehée, Ass.; M. A. Duparque, Ass. *Paléont. des terrains primaires, structure microscopique des charbons. — Outillage photographique pour l'étude des empreintes houillères microscope de Lechatellier pour l'étude de la houille par reflexion.* ⓟ Faune et flore fossiles du terrain houiller.

Lima (Peru).
Labor. de Paleont. y Geol., Fac. de sc. Univ., Parque Universitario. Dir. Dr. C. I. Lissón.

Lincoln, Neb. (U.S.A.).
Dep. of Paleontol. and Geol. of the Univ. of Nebraska. Dir. Prof. E. H. Barbour (Geol.). Prof. E. F. Schramm (Geol.); Ass. Prof. C. H. Barbour (Paleont.); Ass. Prof. E. D. McEwan (Geol.).

Lisbôa (Portugal).
Inst. de Geol. e Paleont., Inst. Superiôr Técnico. Dir. Prof. E. Fleury.

Ljubljana — Laibach (S.H.S.).
Geol.-Palaeont. Inst. d. Univ., Kongresni trg. Dir. Prof. M. Salopek. *(Ammoniten.)*

London (England).
Dept. of Geol., Imp. coll. of Sc. and Technol. of the Univ., S.W. 7, Imp. Inst. Road, S. Kensington. Dir. Prof. W. W. Watts. — Prof. C. G. Cullis; A. M. Davies, Ass. Prof.; V. C. Illing, Ass. Prof.

Louvain — Leuven (Belgique).
Labor. géol. Dir. Prof. H. de Dorlodot. — Prof. A. Salée.

Lwów (Polska).
Inst. paléont. de l'univ. Joanneo-Casimirienne. Dir. Prof.Dr. Joseph Siemiradzki. — Dr. Władysław Zych, Ass. *Machine électrique pour préparations microscopiques des fossiles.* ⓟ Fossiles des terrains silurien, crétacé et miocenique de Pologne.

Lyon (France).
Labor. de Paléont., Fac. Sc. de l'Univ. Dir. Prof. Déperet.

Madrid (España).
Comision de Investigaciones paleont. y prehistóricas, Paseo de la Castellana. Dir. Prof. Eduardo Hernandez-Pacheco. — Sr. Conde de la Vega del Sella (Agregado); José Royo Gomez (Ayudante); Francisco Hernandez-Pacheco (Ayudante); Francisco Benitez Mellado (Dibujante). *Paléont des mammifères. Préhistorie paléolithique.* ⓟ Fossiles paléolithiques.

Marburg a. d. L. (Deutschland).
Geol. palaeont. Inst. Dir. Prof. Rudolf Wedekind.

Mexico, D.F. (Mexico).
Sekt. f. Palaeont. u. Geol. d. Dir. de Estudios Biol., 7|a calle de Balderas 94. Dir. Prof. A. del Rio.

Moskau (U.d.S.S.R.).
Geol. Inst. d. Bergakademie. Prof. D.J. Ilowaisky. *Palaeont. spec. Ammoniten des Jura u. d. Kreide.*
Geol. Kabinett der Bergakademie. Dir. P. S. Schatsky (Kustos am Mus.). *Fauna des Palaeozoicums.*

Münster i. W. (Deutschland).
Geol. Inst. der Univ., Pferdegasse 3. Dir. Prof. Dr. Wegner. — Priv.Doc. Dr. Andree, Ass. *1 bis 2 Plätze für palaeont. Untersuchungen. Chem. Labor.* ⓟ Fossilien. Vorzügl. Sammlung diluvialer Wirbeltiere (Skelette von Mammut, Bos primigenius). Pliozäne Wirbeltiere der Insel Samos. Fische aus der oberen Kreide Westfalens.

Neuchâtel (Schweiz).
Inst. f. Geol. u. Palaeont. Dir. Prof. Dr. Emile Argand.

Oslo (Norge).
Geol.-Palaeont. Inst. Univ. Tøien. Dir. Prof. O. Holtedahl.

Paraná (Argentina).
Palaeont. Inst. Prof. J. Frenguelli; Prof. F. de Aparicio.

Paris (France).
Labor. Paléont. Ecole pr. des Hautes Etudes. Dir. Prof. M. Boule.

Porto (Portugal).
Geol. Inst. Univ. Prof. A. A. M. Correia; Prof. J. A. dos Reis Castro Portugal.

Poznań — Posen (Polska).
Palaeont. Inst. der Univ., Słowackigasse 4|6. Dir. Prof. Dr. Wilhelm Friedberg. — Dr. Marja Rózkowska, Ass. *1 Platz.* ⓟ Erratische Fossilien von Westpolen.

Praha — Prag (C.S.R.).
Geol.-Palaeont. Inst. d. Deutschen Univ., II, Vinična 3. Dir. Prof. Dr. Franz Wähner.

Riga (Latvija).
Geol. u. paleont. Inst. Univ., Baznicas ielā 5. Dir. Prof. Dr. E. Kraus.

Rostow a. Don (U.d.S.S.R.).
Kabinett f. Geol. u. Palaeont. Dir. Prof. Grigorovič-Berezowskij.

Sendai (Japan).
Geol. Inst. Fac. of Science, Tôhoku Imper. Univ. Dir. Prof. Dr. H. Matsumoto (Paleont.).

Sofia (Bulgarien).
Geol. Inst. Dir. Prof. Peter A. O. Bakaloff. — Dr.Wassilie Radeff, Ass.; Peter Gotschev (Paleont.).

Stellenbosch (South Africa).
Dept. of Geol. Univ. Dir. Prof. S. J. Shand, Ph.D., D.Sc.; A. V. Krige, M.A., D.Sc., Lect.

Strasbourg — Straßburg (France).
Labor. de Paléont., Fac. d. Sc. Dir. Prof. Chaput.

Tôkyô (Japan).
Geol. Inst. Coll. of Sc., Tôkyô Imper. Univ. Prof. Dr. Takeo Katô; Prof. Dr. Seitarô Tsuboi.

Tomsk (U.d.S.S.R.).
Palaeont. Praktikum Technikum. Dir. Doc. P. M. Ryžkov.

Toulouse (France).
Labor. de Géol. Fac. Sc. Univ., Allée Saint-Michel. Dir. Prof. M. Charles Jacob. — N. Gaston Astre, Chargé de Cours et Ass. *2 places pour Géol. et Paléont. générales, surtout Foraminifères, Mollusques continentaux et Mammifères fossiles. — Tous les appareils pour la préparation des fossiles. Tour électrique à dégager les fossiles. Tour électrique horizontal, à 2 disques, pour lames minces de Roches. Machine électrique à scier les roches (utile pour l'étude des caractères internes des Rudistes). Série des Microscopes, simples, binoculairs et polarisants.* Ⓟ Fossiles des terrains pyrénéens et des bassins lacustres adjacents. Mollusques Continentaux. Mammifères.

Tübingen (Deutschland).
Geol.-palaeont. Univ.-Inst., Waldhäuserstr. 10. Dir. Prof. Dr. E. Hennig. — Prof. Dr. von Huene, Konservator; Dr. Staesche, Ass. *Mehrere Plätze. In den Sammlungen sind die fossilen Saurier und die Juraformation speziell reich vertreten. Praeparationslabor. mit geschulten Kräften. Chemisches Labor.* Ⓟ Fossile Säuger, Ammoniten, Jurafaunen, doch je nach Gelegenheit auch anderes.

Upsala (Sverige).
Inst. f. Palaeont. Dir. Prof. C. Wiman.

Utrecht (Holland).
Mineralogisch-Geol. Inst. der Univ., Ganzenmarkt N. 32. Dir. Prof. Dr. L. M. R. Rutten. — J. Druif, Conserv.; W. Nieuwenkamp, Ass.

Vancouver (British Columbia).
Geol. Labor. of the Univ. of British Columbia. Dir. Dean R. W. Brock. — S. G. Schofield, M-A., B.Sc., Ph.D.; M. Y. Williams, B.Sc., Ph.D.; T. E. Phemister, M.Sc., Ph.D.; E. M. Burvash, M.A.Sc. Ph.D. *Mesozoic and Tertiary fossils of western Canada.*

Warszawa (Polska).
Labor. f. Geol. u. Palaeont. d. freien Univ. Dir. Prof. Makowski.

Wellington (New Zealand).
New Zealand Geol. Survey, 156 Terrace. Dir. P. G. Morgan. — J. Marwick, Palaeontologist. Ⓟ Mollusca, fossil and recent.
Lands and Survey Dept. Dir. Dr. P. Marshall, M. A. (Palaeont.).

Wien (Österreich).
Palaeobiol. Inst. der Univ., I,Univ. Hauptgebäude. Dir. Prof. Dr. Othenio Abel. — Priv.Doc. Dr. Kurt Ehrenberg, 1. Ass.; Priv.Doc. Dr. Otto Sickenberg, 2. Ass. *3 Plätze. Palaeobiol.* Ⓟ Anpassungen im allgemeinen; alle Reste fossiler Organismen (bes. Tiere), vor allem solche, die in biol. Hinsicht (Anpassungen an Aufenthaltsort, Bewegungsart, Nahrung; Krankheiten; Standortsformen; Kämpfe) für die Stammesgeschichte von Interesse sind, sowie alle Arten von Lebensspuren, Pseudofossilien (Fälschungen), Stücke die zu Volksglauben u. Sage in Beziehung stehen und ähnliches; Phylogenie der Primaten u. Equiden (Spezialsammlungen).
Palaeont. Inst. Philos. Fac. Univ. Dir. Prof. Diener.

Wilno (Polska).
Inst. f. Geol. u. Palaeont. Univ. Dir. Prof. B. Rydzewski.

Zagreb — Agram (S.H.S.).
Geol.-palaeont. Inst. Philosoph. Fac., Demetrova ulica 1.

Musea palaeontologica.

Confer: *Laboratoria palaeontologica* (pag. 493), *Musea biologica generalia* (pag. 350).

Ann Arbor, Mich. (U.S.A.).
Geol. Mus., Univ. of Michigan. Dir. E. C. Case. — G. M. Ehlers, Curator invertebrate fossils; R. C. Hussey, Ass. Ⓟ Fossils of all sorts.

Athènes (Grèce).
Mus. paelont. de l'Univ. Dir. Prof. Th. Scoufos.

Bayreuth (Deutschland).
Kreisnaturaliensammlung, Neues Schloß. Dir. Dr.Theodor Schneid, Hauptkonserv. (in Bamberg). *Sammlungen aus dem Gebiete der Palaeont., Geol. und Mineralogie, und zwar vorwiegend aus Nordostbayern (Oberfranken).* Ⓟ Heimische fossile Fauna und Flora.

Bologna (Italia).
R. Mus. Geol. „G. Capellini", Via Zamboni 63. Dir. Prof. Dr. Michele Gortani. — Prof. Domenico Sangiorgi, Ass.-Conserv. Ⓟ Modelli in gesso di Vertebrati fossili (pronti per cambio). Graphol. (pronte per cambio). Fossili terziari dell'Emilia. Collezioni paleont. italiane. Collezione dei Vertebrati fossili. Collezione delle Cicadee fossili.

Bruxelles — Brussel (Belgique).
Mus. de paléont. Dir. Prof. Dr. L. Dolle.

Cairo (Egypt).
Geol. Mus., Dawawin, post office.

Chica o, Ill. (U.S.A.).
Walker Mus. Univ. of Chicago. A. S. Romer (Curator of Vertebrate Paleont.).

Cork (Ireland).
Mus. of Geol. and Mineralogy, Univ.

Darmstadt (Deutschland).
Geol.-palaeont. Abt. des Landesmus., Geol. Inst. Techn. Hochsch. Dir. Prof. A. Steuer. — Kustos: Dr. O. Haupt.

Dublin (Ireland).
Mus. of Geol., Univ.

Kiew (U.d.S.S.R.).
Palaeont. Sammlung der Akad. d. Wissensch. Dir. P. A. Tutkowski.

Klagenfurt (Österreich).
Naturhistor. Landesmus. Confer: Zool. generalis.

København (Danmark).
Det mineralogisk-geognostiske Mus., Ø. Voldgade 7. Dir. Prof. O. B. Bøggild. — Cand. phil. V. E. Hintze; Doc. Cand. mag. J. P. J. Ravn; Mag. sc. Frk. Karen Herdis Callisen.

La Habana (Cuba).
Palaeont. Mus. Inst. de Segunda Enseñanza.

La Plata (Argentina).
Dept. of Paleont. Mus. de La Plata, Calle N. 3, N. W 34. Dir. A. Cabrera (Curator in Chief).

Laramie, Wyo. (U.S.A.).
Geol. Mus., Univ. of Wyoming. Curator: Dr. S. H. Knight.

Lausanne (Suisse).
Mus. geol. et minéral de l'Univ., Palais de Rumine. Conserv.: Prof. Lugeon.

Leiden (Holland).
Rijks Geol.-Mineralogisch Mus., Garenmarkt. Dir. Prof. Dr. B. G. Escher. — Conserv.: Prof. Dr. H. Gerth. ⱷ Fossilien aus den Niederlanden und den niederländischen Kolonien. Fossilien aus dem Tertiär von Java.

Leningrad (U. d. S. S. R.).
Geol. Mus. (Geol. Muz. imeni Petra Velikogo-G.M.), Tučkova nab., 2. Dir. Akad. F. Ju. Loevinson Lessing. — Palaeont. Sektionen: 1. Sektion d. Wirbellosen: M. V. Bajarunas (Kainose); P. V. Wittenburg (Mesose); N. A. Kulik (Palaeose, Kollektionsyst.). 2. Osteol. Sektion: E. I. Beljaeva. 3. Palaeophytol. Sektion: M. F. Neuburg. 4. Nord-Düna-Galerie mit Kommission f. Nord-Düna-Ausgrabungen: Akad. P. P. Suškin.

Liège (Belgique).
Collection et labor. de geol., Univ. Dir. Prof. Dr. M. Lohest.

Lima (Peru).
Mus. Geol. y Paleont., Escuela de Ingenieros.

Logan, Utah (U. S. A.).
Mus. of Paleont. of the Agric. Coll.

London (England).
Dept. of Geol. and Palaeont., British Mus. of Nat. History, S.W. 7. The Keeper of Geol.: F. A. Bather. Ass. Keeper: W. D. Lang, Sc.D. (Foraminifera, Porifera, Coelentera, Polyzoa); Ass.: L. R. Cox, B.A. (Mollusca excl. Cephalopoda); E. I. White, B.Sc. (Pisces); A. T. Hopwood, M.Sc. (Mammalia); W. E. Swinton, B.Sc. (Reptilia); T. H. Withers (Arthropoda, Echinoderma). Attached Worker: L. F. Spath, D.Sc. (Cephalopoda); D. M. A. Bate (Mammalia pleistocaenica, Aves); H. M. Muir-Wood (Brachiopoda). *About a dozen places. The research is examination of the fossils, preserved in the Mus. — The Dept. is supplied with all apparatus, required for its own studies; but private workers cannot rely on being able to borrow this. The dept. library is available for qualified students.* ⱷ Fossils.

Lwów (Polska).
Palaeont. Mus. d. Univ., ul. Długosza 8. Dir. J. Siemiradzki.

Lyon (France).
Collections du labor. de géol., Fac. des Sc. d. Univ. Dir. Prof. Ch. Depéret.

München (Deutschland).
Inst. für Palaeont. u. Histor. Geol., Neuhauser Str. 51. Dir. Prof. Dr. F. Broili, o. Prof. für Palaeont. u. histor. Geol. a. d. Univ. München. — Prof. Dr. v. Stromer, Hauptkonserv.; Prof. Dr. Dacqué, Konserv.; Dr. J. Schröder, Ass.

New Haven, Conn. (U. S. A.).
Peabody Mus. C. Schuckert (Paleont.).

Oslo (Norge).
Palaeont. Mus. Inst. Tøien. Dir. Prof. Kiær.

Paris (France).
Mus. de Paléont. de l'Ecole Nationale Supérieure des Mines, 5 e, 60 Boulevard St.-Michel. Dir. G. J. Painvin, Conserv. de la Collection. — Pebelier, Chef des Travaux Pratiques de Paléont. de l'Ecole Nationale Supérieure des Mines; Laville, Ancien Chef des Travaux Pratiques, en retraite.
Inst. de Paléont. humaine (Fondation Albert Ier, Prince de Monaco), 13e. 1, rue René Panhard. Dir. Prof. Marcellin Boule. — Dr. René Verneau, Prof. au Mus. national d'Histoire nat., Conserv. de Mus. d'Ethnographie du Trocadéro Paris), Prof. d'Anthrol. préhistorique à l'Inst. de Paléont. humaine; Henri Breuil, Prof. d'Ethnographie préhistorique à l'Inst. de Paléont. humaine; Henri Neuville, Sous-Dir. du Service d'Anat. compar. au Mus. national d'Histoire nat., Secrétaire de l'Inst. de Paléont. humaine.

Rybinsk (U. d. S. S. R.).
Geol. Abt. Naturwissensch. Mus.

San Francisco, Cal. (U. S. A.).
Mus. of Paleont., California Acad. of Sc. Dr. B. W. Evermann.

Stockholm (Sverige).
Palaeozool. Dept. Royal State Mus. of Nat. Hist. Dir. E. H. O. Stensiö. — Lic. phil. R. Hägg, Ass.
Palaeobotan. Dept., Naturhistoriska Riksmus. Dir. Prof. Dr. T. G. Halle. — Dr. R. Florin, Ass. Curator. *3 places. General facilities for palaeobotan. work.* ⱷ Fossil plants of all formations collected. In exchange are offered fossil plants especially from the Rhaet-Lias of Sweden, from the Arctic regions and from China.

Stuttgart (Deutschland).
Geol. Abt. Württ. Naturaliensammlung. Dir. F. Berckhemer.

Sydney (Australia).
Palaeont. Sammlung d. Australian Mus.

Tananarivo (Madagascar).
Mus. d. l'académie Malgache. Dr. Ch. Lamberton.

Taschkent (U. d. S. S. R.).
Palaeont. Abt. d. Zentralmus. f. Naturkunde, Ethnographie u. Archaeol. Mittelasiens.

Torino (Italia).
Mus. di geol. e paleont. Fac. Sc. Univ., Palazzo Carignano. Dir. Prof. C. F. Parona.

Wien (Österreich).
Geol.-Palaeont. Abt. des Naturhistorischen Mus., I, Burgring 7. Dir. Hofrat Prof. Dr. Franz X. Schaffer. — Kustoden: Priv. Doc. Dr. Friedrich Trauth, Priv. Doc. Dr. Julius Pia; Direktionsadjunkt: Lotte Adametz. *10 Plätze. Palaeont. u. geol. Arbeiten im weitesten Sinne. Sammlung: 70000 Nrn., darunter ca. 63000 der Tertiärsamml., 15000 fossile Pflanzen, 11000 Foraminiferen, 6000 fossile Säugetiere u. 1300 fossile Fische).* ⱷ Fossile Pflanzen, Tiere, Sedimentgesteine, Praep., Abgüsse, Abbildungen.

Zagreb — Agram (S. H. S.).
Geol.-palaeont. Abt. (Geol.-paleont. odio) des Kroatischen Nationalmus. (Hrvatski narodni muz.), Demetrova ulica 1. Dir. Dr. Franz Suklje, Kustos.

Laboratoria culturae plantarum et animalium.
Institutiones generales culturae plantarum et animalium.

Alcorn, Miss. (U. S. A.).
Agric. and Mechanical Coll. L. J. Rowan, Ph.D., President; P. S. Bowles, B.S., Prof. Agric. Ed.; A. T. Busby, Ass. Prof. Agric. Ed. and Dairying; E. R. Correll, B.S., Instr. Farm Mech.; Sarah P. Holmes, Prof. Home Econ. Ed.; Selina C. Jackson. Instr. Nurse Training; Mrs. B. M. Johnson, Instr. Housekeeping; L. J. McKay, B.S., Farm Supt.; W. S. Nelson, Instr. Laundering; Mrs. Eunice M.

Powell, B.S., Prof. Home Econ. Ed.; O. W. Sanders, B.S., Ass. Prof. Voc. Ed.; Mrs. Alice L. Tanner, Instr. Dom. Art.; H. T. Tanner, Prof. Hort.; J. B. Wright, B.S., Ass. Prof. Agric. Ed.

Barro, Colorado Islands (Cuba).
Inst. for Research in Tropical America.

Berkeley, Cal. (U.S.A.).
Agric. Experim. Station Univ. of California, 124 Hilgard Hall. Dir. E. D. Merrill. — F. N. Briggs (Pathologist); Prof. E. B. Babcock (Genetics); Prof. F. T. Bioletti (Viticulturist); F. W. Allen (Ass. Pomologist).

Cairo (Egypt).
Ägyptische landwirtschaftliche Versuchsstation, in Bachtim, Post Matarieh; Bahnstation: Schubra; Hauptbureau Cairo. Dir. V. Mosséri. *2 Sektionen: I. Technische Sektion mit 4 Abt.; II. Tierzuchtsektion (Cattle Section) mit 2 Abt. Die Versuchsfelder (ca. 125 Feddans à 0,4200 ha) dienen hauptsächlich zu Versuchen mit Baumwolle, Getreidearten, Klee, Mais usw.*

Davis, Cal. (U.S.A.).
Univ. Farm. Prof. E. H. Hughes (Animal Husbandry); W. P. Duruz (Pomologist); Assoc. Prof. S. B. Freeborn (Entomol.); Assoc. Prof. F. M. Hayes (Veterinary Sc.); Ass. Prof. E. L. Proebstong (Pomol.); A. H. Hendrickson (Assoc. Pomologist); G. L. Philp (Ass. Pomologist); M. J. Heppner (Ass. Pomologist); F. H. Wymore (Research Ass.).

Experiment, Ga. (U.S.A.).
Georgia Agric. Experim. Station. Dir. H. P. Stuckey, B.S., Hort. — R. P. Bledsoe, M.S., Agron.; F. R. Edwards, M.S., An. Husb.; W. G. Friedemann Ph.D., Chem.; B. B. Higgins, Ph.D., Botan., Plant Pathol.; W. L. Brown, B.S., Ass. Chem.; Naomi Chapman, M.S., Ass. Botan., Plant Pathol.; Sarah L. Kilpatrick, Libr., Secy.; Susan J. Mathews, B.S., Nutr. Specialist; Catherine Newton, M.S., Nutr. Specialist; Frank Van Haltern, M.S., Assoc. Botan. (P.O. Tifton); M. A. Willis, M.S., Ass. Agron.; J. G. Woodroof, B.S., Ass. Hort.

Geneva, N.Y. (U.S.A.).
New York Agric. Experim. Stat. Dir. Roscoe W. Thatcher, D.Agric., LL.D. — George W. Churchill Agric.; Dir. of Agronomy, Animal Industry, Bact., Biochemistry, Botan., Chemistry, Dairying, Entomol., Horticulture.

Grignon, Seine-et-Oise (France).
Ecole nationale d'Agric. *Stations de Recherches: Station de recherches de grande culture; Station agronomique; Station de Physique et Chimie biol.; Station de Physiol. et Pathol. des Plantes cultivées; Labor. de Zootechnie; Station de l'Hydraulique.*

Honolulu (Hawaii).
Divisions of Agric. and Home Economics. A. L. Dean, Ph.D., President of Univ.; A. R. Keller, C.E., M.S., Dean Coll. of Applied Sc. — H. F. Bergman, Ph.D., Prof. Botan.; E. M. Bilger, Ph.D., Ass. Prof. Chem.; E. H. Bryan, jr., M.S., Instr. Entomol.; Minnie E. Chipman, Prof. Household Art.; D. L. Crawford, M.A., Prof. Entomol., Dir. of Ext.; Anna von B. Dahl, Ass. Prof. Textiles; J. O. Dale, B.S., Instr. Poultry Husb.; F. T. Dillingham, M.A., Prof. Chem.; C. H. Edmondson, Ph.D., Prof. Zool. and Dir. Marine Biol. Labor.; Giichi Fujimoto, M.S., Instr. Chem.; Lawrence Gay, Marketing Specialist; Clara F. Hemenway, Libr.; L. A. Henke, M.S., Prof. Agr.; Paul Kirkpatrick, Ph.D., Prof. Physiol.; F. G. Krauss, D.Sc., Prof. Agron. and Genetics; W. R. McAllep, Lect. on Sugar Tech.; M. C. Magarian, Instr. Physiol.; Carey D. Miller, M.S., Ass. Prof. Household Sc.; Leonora Neuffer, Ph.D., Prof. Chem.; J. M. Ostergaard, Ass. in Zool.; Robert Pahau, B.S., Supt. Waiakea Farm, P.O., Hilo; H.

S. Palmer, Ph.D., Prof. Geol.; Richard Wrenshall, Ph.D., Prof. Chem.

Kagoshima (Japan).
Kagoshima Imperial Coll. of Agric. and Forestry.

Leningrad (U.d.S.S.R.).
Leningrader Landwirtschaftl. Inst., Insel d. Arbeitenden, 2. Beresowaja Alleja 28. Rektor: K. D. Glinka. — Dekan N. N. Bogdanow-Katjkow, Abt. f. Landwirtschaft; Dekan M. I. Djakow, Abt. für Tierzucht; N. N. Kashanow, Abt. f. landw. Ökonomie u. Politik.
Staatl. Inst. f. experim. Agronomie, ul. Herzena N. 42|44. Dir. N. I. Wawilow. — W. P. Pospelow, Abt. f. angew. Entomol.; A. A. Jatschewskij, Abt. f. Mykol. u. Phytopathol.; N. W. Kopjew, Abt. f. Landw. Maschinenkunde; N. I. Wawilow, Abt. f. angew. Botan. u. Selektion; S. P. Kostytschew, Abt. f. landw. Microbiol.; M. I. Djakow, Abt. f. Zootechnik; L. S.Berg, Abt. f. angew. Ichthyol.; K. D. Glinka, Abt. f. Bodenkunde; M. E. Tkatschenko, Abt. f. Forstwissenschaft.

Lérida (España).
Servicio Agron. Nacional. Dir. Prof. Ramon Blanco.

Lexington, Ky. (U.S.A.).
Coll. of Agric. of the Univ. of Kentucky. Dir. Prof. T. P. Cooper.

Lyallpur, Panjab (Br.-India).
Agric. Coll. Affiliated to the Univ. of the Panjab, Lahore.

Lyon (France).
Inst. Agronomique. Sublabor: Stat. Maritime de Biol. de Tamaris-sur-Mer.

Mayaguez (Porto Rico).
Coll. of Agric. and Mechanic Arts of the Univ. of Porto Rico. T. E. Benner, Ph.D., Chancellor of Univ. (P.O., San Juan); C. E. Horne, Ph.D., Dean of Coll. (P.O., Mayaguez); A. A. Alvarez, V.M.D., Prof. Vet. Sc.; B. A. Bourne, M.S., Prof. Veg. Pathol.; H. T. Cowles, B.S., Prof. Hort., Chair of Agr. Dept.; R. E. Danforth, M.S., Prof. Biol.; M. R. Diaz, B.S., Ass. Sugar Chem · T. E Dunlap, M.S., Prof. Chem.; W. D. Durland, M.F. Prof. For.; J. W. Elliott, M.S., Prof. Agron.; A. A. Fix, B.S., Prof. An. Husb.; L. C. Monzón B.S., Ass. Prof. Chem.; M. A. del Valle, B.S. in Ch.E., Prof. Sugar Chem.; R. G. Wagner, M.A., Prof. Agr. Econ.

Newark, Del. (U.S.A.).
Agric. Experim. Station of the Univ. Charles A. McCue, Dir. and Horticulturist; Louis R. Detjen, Assoc. Horticulturist; G. F. Gray Ass. Horticulturist; F.S. Lagassé, Research Horticulturist; Thomas F. Manns, Plant Pathologist; James F. Adams, Assoc. Plant Pathologist; George L. Schuster, Agronomist; C. R. Runk, Ass. Agronomist; J. W. Graham, Ass. Agronomist; Philip B. Myers, Chemist; G. M. Gilligan, Ass. Chemist; G. L. Baker, Ass. Chemist; H. L. Dozler, Entomologist; C. C. Palmer. Consulting Veterin.; A. E. Tomhave, Animal Husbandman; H. R. Baker, Poultry Disease Investigator.

Omsk (U.d.S.S.R.).
Sibirisches Inst. f. Land- u. Forstwirtschaft, Sa staroj sagorodnoj roschej. Dir. Prof. Nikita Ivanovitsch Gribanov. — S. D. Lavrov (Biol.); G. G. Petrov (Physiol. der Pflanzen u. landw. Bakt.); V. F. Semeonov (allg. Botan.); G. S. Sudejkin (allg. u. angew. landw. u. Forstentomol.); V. I. Baranov (Ökol. u. Geographie der Pflanzen); K. P. Goršenin (Bodenkunde); M. P. Kalmykov (Physiol. u. Anat. der Haustiere); S. M. Kočergin (Milchwirtschaft, allg. Zootechnik, Vererbungs- u. Fortpflanzungslehre).

Osijek (S.H.S.).
Landwirtschaftliche Versuchs- und Kontrollstation. Dir. Prof. Dr. Nikola M. Savič. — Dr. Zeljko Kovalevic, Chef der entomol.-phytopathol. Sektion; Dr. Ing. Michail Graianin, Chef der agriculturchem. Sektion. *3 Arbeitsplätze für agriculturchemische und mikroskopische Untersuchungen. — Agriculturchemische Apparate, sowie verschiedene Schädlinge.* ⓟ Pflanzenschädlinge.

Paris (France).
Inst. des Recherches Agronomiques, VIIe 42 bis Rue de Bourgogne. Dir. M. Roux. *30 stations agronomiques 7 stations oenol., 8 stations d'entomol. agric., 7 stations de pathol. végétale, 3 stations d'essais de semences, 3 stations de génétique, 2 stations de biol. végétale, 3 stations de mécanique agric., 1 station de recherches sur l'alimentation des hommes et des animaux, 2 stations de zootechnie, 1 station de recherches sur les maladies des animaux, 2 stations de recherches viticoles, 2 stations séricicoles, 1 station de recherches horticoles.*

Pretoria (South Africa).
School of Agric. and Experim. Station. Dir. E. Parish. — 9 Lect.; 1 Ass.

Puyallup, Wash. (U.S.A.).
Western Washington Experim. Sta. of the State Coll. of Washington. Dir. Julius Wilbur Kalkus. — M. E. Mc Collam, Agronomist; H. D. Locklin, Horticulturist; C. E. Sauger, Poultry Pathologist; Geo. R. Shoup, Poultry Husbandman; Mrs. Geo. R. Shoup, Poultry Specialist; S. S. Worley, Junior Veterinarian, Plant Pathologist. *Entomol. Economic biol. Horticult. Dairy. Poultry. Agron. — Incubators. Babcock Machine. General Labor. equipment for Pathol. and Bact. Work. Dairy Herd. Poultry, Labor. Animals etc.* ⓟ Pathol. specimens of plants and animals. Seeds.

Riga (Latvija).
Landwirtschaftl. u. Tierzuchtl. Abt. d. mat.-naturwissensch. Fac. d. Herderinst. Dir. Prof. W. v. Knieriem. — **Naturwiss.-mathem. Abt. Herder-Inst.,** Antonijas ielā. Rektor: P. Sokolowski. — Biol.: Friedr. Ferle (Pflanzenbau); W. v. Knieriem (Landw. u. Tierzucht); Karl Reinh. Kupffer (Botan.); Erwin Traube (Zool.).

Siverskaja Belogorka, Gouv. Leningrad (U.d.S.S.R.).
Landwirtschaftl. Versuchsstation der Nord-Westl. Provinzen. Dir. Prof. Victor Dechewoy. — Administrateurs des sections: Petroff (d'agric.), Rganovsky (praticulture), Benoi (zootéchnie), Moliboga (météorologie), Socoloff (chimie agronomique), Kousminn (l'Economie rurale); Ass.: Smirnoff, Mortensenn, Vorobieff, Balandinn, Kroupsky, Jaroslavsky, Vitalsky, Gélninn; Manipulateurs: Horochavina, Resdolnaya, Ivanova, Koconéff, Veriga, Rasoumova, Tolstova. *12 places pour tous les sections.* ⓟ Une collection de plantes du pays de Mourmann, une collection des plantes caractéristique pour les marais, les prairies et les prairies des forêts. — Le labor. agronomique-chimique en a des sections: du sol, de la zootéchnie, d'agric. et de la praticulture.

Speyer a. Rh. (Deutschland).
Landwirtschaftliche Kreisversuchsstation und öffentliche Untersuchungsanstalt für Nahrungs- und Genußmittel, Obere Langstr. 40. Dir. Prof. Dr. Otto Krug. — a) Landwirtschaftliche Versuchsstation: Prof. Dr. Max Kling, Abt.-Vorst.; Prof. Dr. Otto Engels, Oberregierungschemiker; Dr. Wilh. Jürgens und Dr. Heinr. Holl, Regierungschemiker. b) Öffentliche Untersuchungsanstalt: Dr. Georg Fiesselmann, Oberregierungschemiker; Dr. August Gompf, Dr. Max Gareis, Wilhelm Poller, Dr. Otto Reichard, Regierungschemiker. ⓟ Gesammelt: Düngemittel, Futtermittel, Pflanzenschutzmittel, Bodenproben. Nicht getauscht.

Trinidad (Br.-W.-India).
The imperial coll. of tropical agric., Principal: Hugh Martin Leake. — Henry Arthur Ballou (Zool. and Entomol.); Sydney Francis Ashby (Mycol. and Bact.) Sydney Cross Harland (Botan. and Genetics); Frederick Hardy (Chemistry and Soil Sc.); John Sydney Dash (Agric.); vacat (Sugar Technol.); Cecil Yaxley Shephard (Econom.). Außerdem Lect.

Urbana, Ill. (U.S.A.).
Natural History Survey. Dir. S. A. Forbes. — A. E. Miller (Research Entomol.); L. R. Tehon (Botanist); J. H. Bigger (Entomol.); T. H. Frison (syst. Entomologist and Curator); C. J. Telford (Forester); S. C. Chandler (Field Entomologist); W. V. Balduf (Entomologist).

Wellington (New Zealand).
Agric. Dept. Dir. Alfred H. Cockayne, B.Sc., Botan. — E. F. Northcroft, M.Sc., Botan; H. H. Allen, Feilding, Botan.; G. H. Cunningham, Mycol.; J. G. Myers, B.Sc., F.E.S., Entomologist; D. Miller, M.Sc., Entomologist.

Laboratoria culturae plantarum generalia.

Confer: *Institutiones generales culturae plantarum et animalium* (pag. 497), *Phytopathologia* (pag. 415).

Aas (Norge).
„**Elektrokulturforsøkene**". (Dieser Name wird aber wahrscheinlich in kurzem verändert.) **Inst. für Elektrokulturversuche,** Landbrukshoiskolen. Dir. Folmer Smith.

Aberystwyth, Wales (England).
Welsh Plant Breeding Station, Alexandre Road. Dir. N. G. Stapledon. — T. J. Jenkin, M.Sc., Plant Breeder; R. D. Williams, M.Sc., Plant Breeder; E. T. Jones, M.Sc., Plant Breeder; M. G. Jones, M.Sc., Agronomist; W. Davies, M.Sc., Agronomist; Miss K. Sampson, M.Sc., Mycologist; M. A. H. Tinsher, M.Sc., Physiologist; D. W. Davies, Mycologist. *2 places for Plant-Breeding. — The Trial Grounds, Gardens and Labor. of the Welsh Plant Breeding Station.*

Adelaide (South Australia).
Waite Research Inst. Univ. of Adelaide. J. F. Phipps (Genetics). *Plant Breeding.*

Aluschta, Krim (U.d.S.S.R.).
Krim. Staatl. Sapowednik.

Anapa, Nord-Kaukas. Gebiet (U.d.S.S.R.)
Landwirtschaftl. Versuchsstation.

Anjô (Japan).
Provincial Agric. Experim. Station, Aichi-ken.

Augustenberg, Post Grötzingen, Baden (Deutschland).
Staatl. Landwirtschaftliche Versuchsanstalt. Dir. Prof. Dr. F. Mach. — Dr. A. Stang, Oberregierungschemiker; Dr. C. v. Wahl, Oberregierungsbotan.;

Dr. J. Schaller, Oberregierungschemiker; M. Fischler, Oberregierungschemiker. Reglerungschemiker: P. Lederle, Dr. W. Lepper, Dr. R. Herrmann, Dr. F. Sindlinger. ℗ Pflanzen und Samen.

Beauvais, Oise (France).
Inst. Agric. Section d'Enseignement Supérieur d'Agric. de l'Inst. Cathol. de Paris.

Beograd (S.H.S.).
Inst. für allg. Pflanzenzucht u. Genetik d. Univ., Dobracina ul. 16. Dir. Prof. A. Stebut. *Pflanzenzüchtung u. Genetik.*

Berlin (Deutschland).
Biol. Reichsanstalt für Land- u. Forstwirtschaft, -Dahlem, Königin-Luise-Str. 19. Dir. Geh.Reg.R. Prof. Dr. O. Appel. — 25 planmäßige, 6 außerplanmäßige Beamte, 13 Angestellte. Labor. f. allg. Pflanzenschutz: Vorst. OberReg.R. Dr. M. Schwartz, Ass. Dr. H. Goffart; Labor. f. allg. Sortenkunde: Vorst. Dr. K. Snell; Labor. f. Kartoffelbau: Vorst. Reg.R. Dr. O. Schlumberger, Ass. Dr. E. Köhler; Sammlung Gesetze u. Verordnungen: Dr. M. Noack; Labor. f. Forstzool.: Dr, H. Sachtleben; Labor. z. Erforschung u. Bekämpfung d. Nonnenplage: Dr. E. Knoche; Labor. f. Speicher- u. Vorratsschädlinge: Reg.R. Dr. Fr. Zacher, Ass. Dr. E. Janisch; Labor. f. prakt. Bodenbakt.: Vorst. Reg.R. Dr. H. Behn, Ass. Dr. E. Pfeil; Labor. f. prakt. landw. Chemie: Reg.R. Dr. R. Scherpe; Labor. f. Phänol. u. Meteorol. Auskunftsstelle: Vorst. Reg.R. Prof. Dr. E. Werth, Ass.: Reg.R. Dr. H. Pape; Dr. S. Wilke; Labor. f. d. Bekämpf. d. Bienenkrankheiten: Vorst. Reg.R. ao. Prof. Dr. A. Borchert; Prüfstelle f. Pflanzenschutzmittel: Reg.R. Dr. A. Riehm, Ass.: Reg.R. Dr. R. Laubert; Labor. f. Botan.: Vorst. OberReg.R. Prof. Dr. O. Appel; Labor. f. physiol. Zool.: Vorst. Reg.R. Prof. Dr. A. Hase, Ass. Dr. H. Völkel; Labor. f. Bakt.: Vorst. Dr. K. Stapp; Labor. f. Chemie: Vorst. OberReg.R. Prof. Dr. J. Houben, Ass.: Dr. E. Pfankuch, Dr. W. Fischer; Labor. f. Pflanzenzüchtung: Vorst. Dr. K. O. Müller, Ass. Dr. H. Braun; Labor. f. Mykol.: Vorst. Dr. W. Wollenweber; Pflanzenanat. Labor.: Vorst. Dr. W. v. Brehmer, Ass. Dr. E. Konstanty; Zweigstelle Naumburg: Leiter OberReg.R. Dr. C. Börner; Labor. f. Rebenzüchtung: Vorst. Reg.R. Dr. R. Seeliger; Labor. f. Reblausbekämpfung: Reg.R. Dr. H. Thiem, Ass.: Dr. Fr. Schilder, Dr. O. Jancke; Zweigstelle Aschersleben: Leiter Reg.R. Dr. L. Peters, Ass. Dr. F. Dyckerhoff, Zweigstelle Stade: Leiter Reg.R. Prof. Dr. K. Braun, Ass. Reg.R. Dr. W. Speyer; Zweigstelle Berncastel-Cues: Leiter Dr. H. Zillig, Ass: Dr. L. Niemeyer, Dr. K. Pfeilsticker; Zweigstelle Kiel: Leiter Reg.R. Dr. H. Blunck, Ass.: Dr. H. Bremer, Dr. O. Kaufmann, Dr. F. Merkenschlager, Dr. K. Ludewig, Dr. H. Hähne, Dr. W. Zwölfer, Rastatt, Bekämpfung d. Maiszünslers; Labor. f. ang. Landwirtschaft: Vorst. Reg.R. Dr. G. Schneider; Bücherei: Vorst.. Reg.R. Prof. Dr. H. Morstatt. *In jedem Labor. können Gäste nach Vereinbarung aufgenommen werden.* ℗ Pflanzenkrankheiten, parasit. Pilze, schädliche Insekten und Wirbeltiere, Bienenkrankheiten. — Zweigstelle in Naumburg-Saale für Wein- u. Obstbau. Zweigstelle in Aschersleben für Gemüse- u. Samenbau. Zweigstelle in Berncastel-Cues f. Rebenkrankheiten. Zweigstelle in Kiel für Getreide u. Futterpflanzen. Zweigstelle in Stade für Obstbau.

Inst. f. Acker- u. Pflanzenbau Landw. Hochsch., N 4, Invalidenstr. 42. Dir. Prof. Dr. C. Opitz. *Sorten- und Saatgutfrage; Abbau der Kartoffel; physikalische Bodenuntersuchungen im Anschluß an verschiedene Bodenbearbeitungsmethoden; Bestimmung des Düngebedürfnisses der Böden; Flachszüchtung.*

Landwirtschaftliche Versuchsstation der Landwirtschaftskammer für die Provinz Brandenburg und für Berlin, -Dahlem, Lentze-Allee 55/7. Dir. Prof. Dr. O. Lemmermann. — Dr. L. Fresenius, Abt.-Vorst.; Dr. P. Hasse. *Vegetationshaus, Versuchsfeld, Labor.*

Besentschuk (U.d.S.S.R.).
Landwirtschaftl. Provinz-Versuchsstation, Samaro Slat. sh. d.

Béthune, Pas de Calais (France).
Stations agric.

Bitley, Mich. (U.S.A.).
Siffey-Farms. Dir. O. Lloyd-Jones.

Blagoweschtschensk (U.d.S.S.R.).
Abt. Pflanzenzüchtung, Amur Landw. Versuchs-Stat. Dir. W. Solotnitsky.

Bonn-Poppelsdorf (Deutschland).
Inst. für Boden- u. Pflanzenbaulehre Landw. Hochsch., Katzenburgweg 5. Dir. Geh.Reg.R. Prof. Dr. Th. Remy. — DDr.: Weiske, Vasters, Christopeit, Steinberg, Liesegang; Diplomlandwirte: Witte, Hinckers, Ohly. *Bodenkundliche und pflanzenbauliche Untersuchungen.* ℗ Gesammelte Objekte, die in das Lehr- und Untersuchungsgebiet einschlagen.

Breslau (Deutschland).
Inst. für Pflanzenbau und Pflanzenzüchtung Univ. XVI, Hansastr. 25. Dir. Prof. Dr. Berkner. — Ass.: Priv.Doc. Dr. Christiansen-Weniger (Pflanzenzüchtung); Dr. Schleusener (Versuchswesen); Diplomlandwirt Schröder (Pflanzenzüchtung); Frl. Dr. Freter (chemische Ass.); Dr. Mane. *4 Plätze für chemische, agriculturchemische und pflanzenzüchterische Untersuchungen.* Versuchsgut in Schwoitsch bei Breslau.

Brno — Brünn (C.S.R.).
Inst. f. allg. u. spez. Pflanzenproduktion u. -veredlung d. Landwirtsch. Hochsch., Cerna pole. Dir. Prof. Dr. J. Munzar.

Budapest (Ungarn).
Inst. für landwirtschaftliche Botan., VIII, Eszterházy u. 3. II. Dir. Prof. Dr. Zoltán Szabó. — Priv.-Doc. Dr. Béla Augusztin (Medizinalpflanzen); Zoltán Biró (Forstkunde); Mathias Mahács (Gartenbau); Alex. Pettenhoffer (Weinbau). ℗ Objekte für landw. Botan., Kulturpflanzen, Samen, Ähren, Objekte der Pflanzenzüchtung und Vererbungslehre.

Buitenzorg, Java (Ned.-O.-Indië).
General Agric. Experim. Station. Dir. Dr. P. J. S. Cramer. — Dept.: 1. Labor. for research work on soils: Dir. Jr. J. Th. White; 2. Labor. for M crobiol.: Dir. J. Groenewege; 3. Labor. for Chem.: Dir. Dr. C. van Rossem; 4. Botan. Labor.: Dir. Dr. P. van der Elst; 5. Sect. for Agronom.: Dir. Jr. A. Wulff; 6. Sect. for Plantbreeding of permanent and annual crops: Dir. P. J. S. Cramer and L. Koch.

Cairo (Egypt).
Plant Breeding Sta. of Agric. Coll., El Gize near Cairo.

Cambridge (England).
Inst. of Agric. Botan. Univ. Dir. Sir. R. H. Biffen.

Cassel-Harleshausen (Deutschland).
Landw. Versuchsanst. d. Landw.-Kammer f. d. Reg.-Bez. Cassel. Confer Harleshausen.

Cesis (Latvija).
Versuchs- u. Pflanzenzucht Stat. des Lettl. Landw. Zentralvereins, Priekulu selekcijas. Dir. Dr. Edgar Eglit.

Charkow (U.d.S.S.R.).
Landwirtschaftl. Gebiets-Versuchsstation, Tschügüjewskoje Chaussee, Postfach 266.

Cherson (U.d.S.S.R.).
Landwirtsch. Versuchsstation.

Chibini, Murman-Gouv. (U.d.S.S.R.).
Murman-Abt. d. angew. Botan. u. Selektion d. Union-Inst. f. angew. Botan. u. neue Kulturen. Dir. J. G. Eichfeld.

Cincinnati, Ohio (U.S.A.).
Lloyd Brothers Agric. Research Labor.

Cluj — Klausenburg (România).
Labor. für Pflanzenzüchtung. Dir. Prof. Dr. Anastase Munteanu. — Andrei Piescu, Ass. *Pflanzenzüchtung.* ⓟ Landw. Samen und Pflanzen.
Versuchsfeld f. Pflanzenbau. ⌐ Dir. Prof. Joh. Dragan. — Prof. A. Munteanu.

Coimbatore (Br.-India).
Agric. Research Inst. Dir. R. C. Broadfoot.

College Park, Md. (U.S.A.).
Maryland Agric. Experim. Sta. Prof. C. O. Appleman (Plant physiologist).

College Station, Tex. (U.S.A.).
Texas Agric. Experim. Station, Dir. Dr. B. Youngblood. — A. B. Conner, Vice-Dir.; A. H. Leidigh, Ass.Dir.; M. Francis, Chief, Vet. Sc.; H. Schmidt, Veterinarian; V. J. Brauner, Veterinarian; G. S. Fraps, State Chem., Chief; S. E. Asbury, Ass. Chemist; W. H. Walker, Ass. Chemist; J. E. Teague, Ass. Chemist; J. K. Blum, Ass. Chemist; A. T. Potts, Chief, Horticulture; J. M. Jones, Chief, Animal Ind.; J. L. Lush, Animal Husbandman; G. R. Warren, Swine Husbandman; R. M. Sherwood, Poul. Husbandman; J. J. Hunt, Wool Grader; M. C. Tanquary, Chief, State Entomol.; H. J. Reinhard, Entomologist; C. S. Rude, Entomologist; H. S. Cavitt, Apiary Inspector; W. R. Jordan, Apiary Inspector; E. B. Reynolds, Chief, Agronomy; G. N. Stroman, Cotton Breeder; C. H. Mahoney, Ass. Cotton Breeder; J. J. Taubenhaus, Chief, Plant Pathol. and Physiol.; E. M. Peters, Superintendent; D. H. Bennett, Veterinarian; O. L. Carpenter, Shepherd; W. H. Friend, Superintendent; S. W. Bilsing, Prof. Entomol.; W. L. Stangel, Prof. Animal Husbandry; F. A.Buechel, Prof.Agric.Econ.; G.W. Adriance, Assoc. Prof. Hort.; W. E. Garnett, Prof. Rural Sociol.; G. P. Grout, Prof. Dairy Husb; R. C. White Assoc. Prof. Rural Sociol.; H. V. Geib, Ass. Prof. Agron.; E. O. Pollock, Ass. Prof. Agron.; L. P. Gabbard, Chief, Div. Farm and Ranch Economics; V. L. Cory, Grazing Res'h Botan.; H. E. Rea, Ass.; W. T. Carter, Chief, Soil Survey; H. W. Hawker, Soil Surveyor; Edward Templin, Soil Surveyor; H. Ness, Chief, Botan.; H. B. Parks, Apiculturist; A. H. Alex, Queen Breeder; D. T. Killough, Superintendent; F. D. Fuller, Chief, Feed Control Ser.; Inspectors: J. H. Rogers, W. H. Wood, J. D. Prewit, T. C. Davis, J. F. Schultz. Superintendents: R. A. Hall, W. S. Hotchkiss, V. E. Hafner, R. H. Wyche, A. B. Cron, P. B. Dunkle, R. E. Dickson, R. E. Karper, J. J. Bayless, L. J. McCall, G. T. McNess, D. L. Jones. *No special provision is made for working guests, except that there are a few part-time ass.-ships or fellowships, which pay a small stipend and permit the holder to do work toward the Master's degree in the Agric. and Mechanical Coll. of Texas.* ⓟ Only entomol. and herbarium specimens, sufficient to carry on properly the research work of the institution. — Dept.: Administration; Veterinary Sc.; Chemistry; Horticulture; Animal Industrie; Entomol.; Agronomy; Plant Pathol. and Physiol.; Farm and Ranch Economics; Soil Survey; Botan.; Publications; Feed Control Service; Main Station Farm. Substations: Beeville, Bee County; Troup, Smith County; Angleton, Brazoria County; Beaumont, Jefferson County; Temple, Bell County; Denton, Denton County; Spur, Dickens County; Lubbock, Lubbock County; Balmorhea, Reeves County; Coll. Station, Brazos County; Nacogdoces, Nacogdoches County.

Darmstadt (Deutschland).
Hessische Landwirtschaftl. Versuchsstat., Rheinstr. 91. Dir. Prof. Dr. Hubert Rössler. *Pflanzenernährung u. Düngung.*

Debreczen (Ungarn).
Pflanzenbau-Inst., Landw. Akademie. Dir. Prof. K. v. Varga.

Dickursby (Finnland).
Landwirtschaftliche Versuchsanstalt.

Dnepro-Petrowsk (U.d.S.S.R.).
Landwirtschaftl. Versuchsstation, Plosch-Demjana Bednogo.

Dotnuva (Litauen).
Dotnuvische Versuchsstation f. Pflanzenzüchtung. Dir. Prof. Dr. D. Rudzinski. — Z. Mackievič. ⓟ Verschiedene züchterische Sortensamenmuster.

Dublany (Polska).
Inst. d'Agric. Dr. Adam Krasinki.

Durham, N.H. (U.S.A.).
Agric. Experim. Station. Confer: Entomol.

Edinburgh (Scotland).
Dept. of Agric. Univ. Fac. Sc. Dir. Prof. J. A. S. Watson.

Ejsk, Nord-Kaukas. Geb. (U.d.S.S.R.).
Landwirtschaftl. Versuchsstation.

Eriwan (U.d.S.S.R.).
Landwirtschaftl. Mus., ,,Narkomsem".

Forus, Jederen (Norge).
Statens forsöksstation Vestenfjelds. Dir. Hönningstad.

Frankfurt a. M. (Deutschland).
Landw. Abt. der I. G. Farbenindustrie A.-G., Werk Leverkusen. Dir. W. Bonrath.

Gembloux (Belgique).
Station agronomique de l'Etat.

Gießen (Deutschland).
Landwirtschaftliches Inst. Univ. Dir. Gesevius. Abt.-Vorst. f. Pflanzenproduktionslehre: Dr. Adolf Kraft.

Glenolden, Pa. (U.S.A.).
H. K. Mulford Agric. Research Labor.

Guadalajara, Dalisco (Mexiko).
Inst. Biol., Apartado 370. 1. Dir. Prof. Victor A. Reko. — 2 Hilfskräfte, 1 Botaniker, 1 Chemiker. *Das Inst. steht Farmern u. anderen an botan. Fragen Interessierten kostenlos zur Verfügung, untersucht eingesandte Pflanzen, identifiziert sie, stellt ihre Indianer- u. wiss. Namen fest, ihre wirtschaftl. Verwendbarkeit u. sonstige Wirksamkeit. Fragen der Pflanzenerkrankung u. Schädlingsbekämpfung, des Anbaus in ungewohnten Gegenden (Transplantationen) werden durchgeprüft u. darüber Auskunft gegeben. Gegen geringe Gebühren sind Samen mexikan. Pflanzen zu beziehen, ebenso Kakteen oder Orchideen (auch im Austausche mit Sammlern). Über Bezugsquellen pflanzl. Rohstoffe aus Mexiko wird Auskunft gegeben. — Bibliothek, Labor., Versuchsgärten.* Versuchsgärten (eigene in Guadalajara, zur Verfügung stehende in Oaxaca und Tehuacan, Puebla) u. Labor.

Göttingen (Deutschland).
Inst. für Pflanzenbau, Nikolausberger Weg 7. Dir. Prof. Dr. Otto Tornau. — Dr. Hans Hermann Wick, landwirtschaftl. Ass.; Dr. Theodor Steche, chemischer Ass.

Gomel (U.d.S.S.R.).
Versuchsstation a. Staatl. Inst. f. experim. Agronomie.

Gorki (U.d.S.S.R.).
Agric. Versuchsstation der Staatl. Landwirtschaftl. Akademie, Orschanskij okr.

Grignon, Seine-et-Oise (France).
Station agronomique Ecole nationale d'agric.

Halle a. d. S. (Deutschland).
Inst. für Pflanzenbau u. Pflanzenzüchtung, Ludwig-Wucherer-Str. 2. Dir. Prof. Dr. Th. E Roemer. — Dr. Dirks (chem. Labor.); Dr. Blohm (Pflanzenbau); Dr. Rudorf (Pflanzenzüchtung). 6 Ass. *3 Plätze für Bodenuntersuchungen nach Neubauer u. Mitscherlich, für Immunitätszüchtungen.*

Hanadate (Japan).
Riku-u Branch of Imperial Agric.Experim. Station. Sempoku-gun, Akita-ken. T. Nibe (Ornithol).

Hanoi (Indo-Chine).
Sect. d'agric. d'Ecole sup. d'Agric. et sylvicult.

Harleshausen b. Kassel (Deutschland).
Landwirtschaftliche Versuchsanstalt. Dir. Prof. Dr. Emil Haselhoff. — 3 Chemiker, 1 Landwirt. *Bodenkunde, Pflanzenernährung.*

Harpenden (England).
Rothamsted Experim. Station. Dir.: Sir John Russell, D.Sc.; Ass. Dir.: B. A. Reen, D.Sc. — Head of the Mycol. Dept.: Wm. B. Brierley, D.Sc.

Heitô, Formosa (Japan).
Provincial Agric. Experim. St. Takao-shû.

Helsinki (Finnland).
Agronomisches Inst. der Univ., Riddaregatan 6.

Hildesheim (Deutschland).
Landwirtschaftl. Versuchsstat. Dir. Dr. Carl F. W. Dempwolff.

Hjellum, pr. Hamar (Norge).
Statens forsöksstation pa Möystad. Dir. Glaerum.

Holt, pr. Troms (Norge).
Statens forsöksstation for Troms og Finmark. Dir. Tjærvoll.

Honolulu (Hawaii).
Pan-Pacific Agric. Research Inst.

Hořice v. Podkrkonoši (C.S.R.).
Staatl. landwirtschaftl. Versuchsstation. Dir. M. Servit.

Insterburg (Deutschland).
Landwirtsch. Versuchsstat. u. Nahrungsmitteluntersuchungsamt. Dir. Dr. W. Heintz.
Versuchsstat. d. Mitscherlich-Gesellschaft. Dir. Dr. Born. *Feststellung des Düngebedürfnisses d. Bodens mittels Gefäßversuchen.*

Irkutsk (U.d.S.S.R.).
Ost-Sibirische Landwirtschaftl. Provinz-Versuchsstation, Timirjasewstr. 68.

Ithaca, N.Y. (U.S.A.).
Dept. of plant Breeding, Cornell Univ. Dir. Dr. A. Tavcar.

Iwanovo-Woznessensk (U.d.S.S.R.).
Inst. f. allgem. Ackerbau. Drl. Prof. A. N. Prochorov.
Wosnessensker Landwirtsch. Versuchsstation, Nikolajewskij okr.

Jaroslavl (U.d.S.S.R.).
Kabinett für Pflanzenzucht des Pädagogischen Inst. Dir. Prof. L. J. Moljakov.

Jena (Deutschland).
Landwirtschaftl. Inst. mit landw. Labor. u. landwirtschaftl. Garten. Dir. Prof. W. Edler.

Kagi, Formosa (Japan).
Kagi Agric. Experim. St.

Kalmar (Sverige).
Stat. of the Swedish plant breeding Assoc. Dir. O. Holmgren, Agron.

Kamenetz-Podolsk, Ukraine (U.d.S.S.R.).
Labor. für allg. Pflanzenzuchtlehre des Inst. f. Landw. „K. Marx". Dir. Prof. D. Plevako.
Landw. Labor. des Inst. für Volksbildung. Dir. Prof. Padalka.

Kazan (U.d.S.S.R.).
Landwirtschaftl. Provinz-Versuchsstation d. Ober-Wolga-Gebiets.

Kiel (Deutschland).
Inst. f. Pflanzenbau Univ. Dir. Prof. Walter Dix.

Kiew (U.d.S.S.R.).
Labor. für landw. Biol., Akad. d. Wissensch.
Katheder für landwirtschaftliche Pflanzenkunde.

Königsberg i. Pr. (Deutschland).
Pflanzenbau-Inst. der Univ., Tragheimer Kirchenstr. 74. Dir. Prof. Dr. Eilh. Alfred Mitscherlich. — Ass.: Chemiker Dr. Wagner, Landwirt Dr. Dühring; Leiter der Stationen der Mitscherlich-Gesellschaft: Dr. Born, Insterburg; Dr. Bimschas, Pr.-Holland; Dipl.-Landw. Baron v. Engelhardt, Osterode; Priv.-Doc. Dr. Krull, Königsberg; Dr. Kobbert, Königsberg. *Nur soweit Platz vorhanden ist; vorherige Anmeldung ist erforderlich. Botan.-mikroskopische, chemische, bodenkundlich-physikalische, pflanzenphysiol. Gefäß- und Feldversuche.* ⓟ Praep. von Pflanzenzüchtungen. Bodenproben. — Die Stationen der Mitscherlich-Gesellschaft in Ostpreußen, u. zwar in Pr.-Holland, in Insterburg und in Osterode. Speziell zum Zweck der Feststellung des Düngebedürfnisses des Bodens mittels Gefäßversuche.

Košice — Kaschau (C.S.R.).
L'Inst. des Recherches agric. Dir. Ing. A. Šeda.

Krasnodar (U.d.S.S.R.).
Kuban. Melioration-Versuchsstation d. „Narkomsem", Nowaja 107.
Agron. Inst. des Kuban. Landwirtschaftl. Inst., Nowaja ul. 107.
Kuban. Landwirtschaftl. Provinz-Versuchs-Station, Postfach 145.

Krasnojarsk (U.d.S.S.R.).
Jenisseisker landw. Prov. Versuchsstat. Dir. W. W. Sabaschnikow.
Mittel-Sibirische Landwirtschaftl. Versuchs-Station Landwirt. Abt. Staatl. Mus. des Jeniss. Gebiets. Dir. H. P. Miklaschewskaja.

Kuala Lumpur (Fed. Malay States).
Dept. of Agric. Secretary: A. S. Haynes.

Kun-Chu-ling, Manchuria (China).
Agric. Experim. St. South Manchurian Railway Co.

Kurashiki near Okayama (Japan).
Agronomical Labor. Ôhara Inst. for Agric. Research. Dir. Dr. Mantarô Kontô. — 3 Ass.

Kyôto (Japan).
Inst. of Agronomy, Coll. of Agric., Kyôto Imper. Univ. Dir. Prof. Dr. Takashi Sasaki. — Prof. Dr. Isao Namikawa (Horticulture); Prof. Dr. A. Kikuchi (Horticulture); Ass. Prof. Yoshinori Takesaki (Crops, Plant breeding).

Kyôto Provincial Agric. Experim. St., Shimogamo. Katsujirô Hisata (Phytopath., Agric. Chemistry).

Landsberg a. d. W. (Deutschland).
Inst. für Pflanzenzüchtung der Preuß. Landw. Versuchs- und Forschungsanst., Theaterstr. 25. Dir. Prof. Dr. G. Bredemann. — Wissenschaftl. Hilfsarbeiter: Dipl.-Landw. Max Klein, Dipl.-Landw. Joseph Mallach. *Plätze reichlich vorhanden. Für alle Untersuchungen betr. Züchtung der landw. Kulturpflanzen, Vererbungsstudien für praktische Züchtung, Vervollkommnung der bei den Züchtungsarbeiten anzuwendenden Methodik. — Alle zu genannten Arbeiten erforderlichen, dazu die rund 16 ha großen Versuchsfelder mit ihren Glas-, Vegetations- und Isolierhäusern.* ⊕ Zu tauschen sind Sämereien fast aller landw. Kulturpflanzen in ihren Züchtungssorten und Variationen.

La Plata (Argentina).
Inst. f. agric. Botan. Confer: Botan. generalis.

Laramie, Wyo. (U.S.A.).
Dept. of Agricult., Univ. of Wyoming. Dir. Prof. A. E. Bowman. — S. H. Dadisman.

Lausanne (Suisse).
Mus. agric. de l'école cantonale d'Agric. de l'Univ.

Leipzig (Deutschland).
Staatl. Sächsische landwirtsch. Versuchs-Anst., Möckern, Gustav-Kühn-Str. 8. Dir. Prof. Dr. phil. et agr. h. c. Gustav Fingerling. — Prof. Dr. Gabriel, Vorst. der Dünge- und Futtermittel-Abt.; Ass. an der Dünge- und Futtermittel-Abt.: Dr. Volhard, Dr. Strigel, Eisenkolbe; Ass. an der wissenschaftl. Abt.: Just, Dr. Behrens, Heise, Muth.
Landwirtsch. Inst. d. Univ. Dir. Prof. Falke.

Leningrad (U.d.S.S.R.).
Sapropelische Abt. der Russischen Akademie der Wissenschaft. Russische Akademie der Wissenschaft. Chef der Abt.: A. I. Gorbov.
Allrussisches Landwirtschaftl. Mus., Fontanka 10. Dir. N. N. Gassowsky.
Labor. f. angew. Botan. u. Selektion, Staatl. Inst. f. experim. Agronomie, Ul. Hercena 42 i 44. Dir. Prof. A. V. Vavilov.

Lexington Ky. (U.S.A.).
Dept. of Agronomy of the Univ. of Kentucky. Dir. Prof. E. J. Kinney. — Prof. G. Roberts.

Lima (Peru).
Sección de Agric. General, la Estación Central Agronómica. Jefe: Dr. J. Miranda y Rivera.

Mutterstadt, Rheinpfalz (Deutschland).
Landwirtsch. Versuchsst. der I. G. Farbenindustrie A.-G. Limburger Hof. Biol. Dr. H. Lösch. *(Pflanzenkrankheiten.)*

Lincoln, Neb. (U.S.A.).
Dept. of Agronomy of the Univ. of Nebraska. Dir. Prof. T. A. Kieselbach; Prof. W. W. Burr. — Prof. F. D. Keim; Assoc. Prof. J. C. Russel; Ass. Prof. A. Anderson; Ass. Prof. T. H. Gooding; N. F. Petersen (Botan. Ass.).
Dept. of Agric. of the Univ. of Nebraska. Dir. Prof. W. H. Brokaw. — Prof. C. K. Morse.
Agric. Exper. Sta. of Nebraska.

Linköping (Sverige).
Stat. of the Swedish plant breeding Assoc. Dir. S. Wålstedt, Agron.

Lisbôa (Portugal).
Mus. Nacional Agric., Tapada da Ajuda.

Ljubljana — Laibach (S.H.S.).
Landwirtschaftl. Versuchs-Stat. (Kmetijska pokusna in kontrolna postaja). Dr. Milena Perušek, Ass.

Lodejnoje Pole,
Gouv. Leningrad (U.d.S.S.R.).
Experim.-Station „Naturargij".

Logan, Utah (U.S.A.).
Agric. Experim. Sta.

Luleå (Sverige).
Stat. of the Swedish plant breeding assoc. Dir. F. Naesman, Agron.

Lwów (Polska).
Station für Pflanzenbau, Pflanzenzüchtung und Pflanzenwuchs der Technischen Hochsch. Dir. H. Gurski. — R. Borkowski, Adjunkt.

Manila (Philippine Islands).
Agric. Experim. Sta. Coll. of Agric. Dir. Prof. C. F. Baker. — Prof. B. M. Gonzales; Prof. M. L. Roxas.

Matsuyama (Japan).
Provincial Agric. Experim. Sta.

Mayaguez (Porto Rico).
Agric. Experim. Station. Dir. D. W. May, M.Agr. — R. L. Davis, A.M., Plant Breeder; Gerard Dikmans, D.V.M., Parasit.; H. C. Henricksen, B.Agr., Agric.; T. B. McClelland, A.B., Hort.; C. M. Tucker, B.S., Plant Pathol.; J. O. Carrero, B.S.Ch.E., Ass. Chem.; J. A. Saldana, B.S., Ass. in Plant Brdg.; A. Arroyo, Entomol. Peparator.

Minsk (U.d.S.S.R.).
Versuchsstation f. Agronomie, Kommunalnaja 36.

Minussinsk (U.d.S.S.R.).
Landwirtschaftl. Versuchsstation.

Mont-Calme, Lausanne (Schweiz).
Schweizer Samenuntersuchungs- u. Versuchsanst. *Samenkontrolle. Pflanzenschutzmittel f. Feldgewächse.*

Moskau (U.d.S.S.R.).
Labor. für Pflanzenzucht der II. Univ. Dir. A. A. Bauer.
Landwirtschaftl. Akademie namens Timirjasew. Confer Institutiones generales.
Versuchs-Abt. d. Volkskommissariats f. Landwirtschaft (Narkomsema), Staraja Ploschad,,,Bojarskij Dwor".
Staatl. Wissenschaftl. Forschungsinst. f. Kolonisation (Goskolonit), Ul. K. Marksa, Gorochowskij per. 4.

München (Deutschland).
Bayer. Landesanstalt für Pflanzenschutz, Liebigstr. 25. Dir. Ministerialrat Christmann. — I. Abt. f. Pflanzenbau: Reg.R. 1. Kl; J.Weigert, Reg. Räte: Dr. Stiche, Riedl, Kronberger, Dr. Boshart; Assesoren: Fürst, Dr. Hiltner, Estal (Gutsinspektor); wissenschaftl. Hilfsarbeiter: Apothekerin Frl. Bergold. II. Abt. f. Pflanzenschutz: Reg.R. 1. Kl. Dr. Korff, Reg.R. Dr. Flachs; Assessor Dr. Pustet, Reg.R.1. Kl.Weidinger; wissenschaftl. Hilfsarbeiter; Dr. Hecker, Dr. Sell. III. Abt. f. Samenkontrolle: Reg.R.1. Kl. Dr. Gentner; Reg.R. Dr. Merl. IV. Abt. f. Futtermittelkontrolle: Reg.R. 1. Kl. Dr. Kinzel, Assessor Küchler. *Versuchsgut Nederling b. München; Versuchsgarten Holzapfelkreuth; dazu Versuchslabor. usw. je nach Bedarf. Abt.: Pflanzenbau, Pflanzenschutz, Samenkontrolle, Futtermittelkontrolle, mit Labor., 2 landwirtschaftl. Versuchsgütern von rund 190 ha, Versuchsgut von rund 80 ha, einem Versuchs- u. Mustergarten für Obst u. Gemüse im Ausmaß von 3 ha.* ⊕ *Samensammlungen (Kulturpflanzen, Unkräuter usw.). Mittel zur Schädlingsbekämpfung (Mäuse- u. Rattengifte, Mäusebazillen; Beizmittel usw.), Verkauf im großen. Impfstoff zur Impfung der Leguminosen.*

Inst. f. Acker- u. Pflanzenbau. Techn. Hochsch.
Dir. Prof. L. Kießling. Konserv. Dr. Kreutz.
Bayerische Hauptversuchsanstalt für Landwirtschaft mit agric.-chem. Unterrichtslabor. an der Technischen Hochsch., Luisenstr. 36. Dir. Geh.-Reg.R. Prof. Dr. Theodor Henkel. — 3 Konserv.: Weiker, Dr. Metzger, Wittmann als Chemiker; 1. Konserv.: Dr. Baumgaertel als Bakteriologe; 3 Ass.: Dr. Hurtig, Dr. Stann, Dr. Fischer als Chemiker; 1 landw. Ass.: Stoll. *Im Agriculturchemischen Labor. 38 Arbeitsplätze. Außer chemischen Arbeiten auch bact. — Agricultur-chemisches und bakt. Labor., Versuchsfeld mit einer Anzahl von Silos zur Grünfutterkonservierung und Gespanntieren.* ⓟ Dünge- und Futtermittel. — Zweiglabor. auf dem Versuchsfeld Obermenzing.

Münster i. W. (Deutschland).
Landwirtschaftliche Versuchsstation, Südstr. 72.
Dir. Prof. Dr. A. Bömer. — Abt.-Vorst.: Dr. W, Gutthoff, Dr. E. Dinslage, Dr. J. Hasenbäumer. Dr. F. Bartschat, Dr. K. Lehmann.

Muratowo, Orlowsches Gouv. (U. d. S. S. R.).
Schapilowsche Landwirt. Prov.-Versuchsstat. Dir. W. N. Chitrow.

Mydlniki pod Krakowem (Polska).
Ferme expérim. agric. Univ. Dir. Dr. St. Wasniewski.

Nanking (China).
Agric. Experim. Sta.

Nara (Japan).
Provincial Agric. Experim. Sta.

Nogent-sur-Marne (France).
Inst. National d'Agronomie coloniale. Dir. Prof. E. Prudhomme.

Nossowskaja, Gouv. Tschernigow (U.d.S.S.R.).
Landwirtschaftl. Versuchsstation.

Nowosybkow, Gouv. Gomel (U. d. S. S. R.).
Landwirtschaftl. Versuchs-Station.

Odessa (U. d. S. S. R.).
Forschungsinst. f. Pflanzenbau, Ul. Sverdlova 99. Dir. Prof. A. A. Sapĕhin. — Leiter der Sektion f. Ökol. d. Landwirtschaftspfl.: Prof. G. A. Borovikov. 1 wiss. Mitarbeiter, 4 Aspiranten. *Sektionen: a) Vererbungslehre u. Pflanzenzüchtung; b) Ökol. der Landwirtschaftspflanzen.*
Odessaer Zentralversuchsstat., Landw. Hochsch., Institutskaja 9. Dir. Prof. Andreas A. Sapĕhin.

Oerlikon b. Zürich (Schweiz).
Schweizerische landwirtschaftliche Versuchsanstalt. Dir. Prof. Dr. A. Volkart. — B. Schmitz, Chemiker (Agrikulturchemie); Dr. A. Grisch, Botan. (Samenkontrolle); Dr. E. Neuweiler, Botan. (Pflanzenschutz); Dr. E. Gäumann, Botan. (Futtermittelkontrolle). *Samenkontrolle, Dünge- und Futtermittelkontrolle, Bodenuntersuchungen, Versuchstätigkeit.* ⓟ Futterpflanzen, Unkräuter, Pflanzenkrankheiten.

Omsk (U.d. S. S. R.).
West-Sibirische Provinz-Versuchsstat., Postfach 34.

Opava — Troppau (C. S. R.).
Station des Recherches agric.

Orel (U. d. S. S. R.).
Landwirtsch. Versuchsfeld, Postfach 33.

Ôsaka (Japan).
Provincial Agric. Experim. Station, Momoyama.

Oster, Gouv. Tschernigov (U. d. S. S. R.).
Landwirtsch. Versuchsstation Paed. Technik.

Osterode, Ostpreußen (Deutschland).
Versuchsstat. der Mitscherlich-Gesellschaft. Dir. Dipl. Landwirt Baron v. Engelhardt. *Düngungsbedürfnis des Bodens; Gefäßversuche.*

Oxford (England).
Agric. Inst. Univ. Dir. Lect. W. R. Peel.

Peking (China).
Agric. Experim. Stat. Dir. Chunjen Constant Chen.

Pensa (U. d. S. S. R.).
Verwaltung d. Pensaer Sapowednik, Krasnaja 45/10.

Perm (U. d. S. S. R.).
Kabinett f. Meliorationswesen. Dir. Doc. A. I. Širajejev.

Perth (W. Australia).
Dept. of Agric. Dir. Prof. J. W. Paterson.

Piracicaba, São Paulo (Brasil).
L'Ecole Agric. „Luiz de Quoiroz". Prof. A. A. Bitancourt (Botan.).

Poltawa (U. d. S. S. R.).
Landwirtschaftl. Versuchsstation.

Port-au-Prince (Haiti).
Service technique d'Agric. Dir. G. F. Freeman. — G. N. Wolcott (Entomologist).

Poznań — Posen (Polska).
Inst. f. allgem. Acker- u. Pflanzenbau. Dir. Prof. Z. Pietruszczynski.

Praha — Prag (C.S.R.).
Biol. Inst. des Staatlichen Versuchsinst. für Pflanzenproduktion. Dir. Ing. Dr. Jaroslav Stádník. — Ing. Sebald Krkoška, Inspektor; Ing. Dr. Karel Kamenický, Inspektor; Ing. Jindřich Nebovický, Adjunkt; Ing. Emil Kunz, Ass.; Ing. Dr. Bohumil Kafka, Ass.; Ing. Miloslav Sobotka, Ass.; Ing. František Landovský. *4 Plätze. — Biol. Versuchsinst. u. Züchtungsstation. Material von Getreide und Erbse seit vielen Jahren.* ⓟ Züchtungsmaterial, Samen von Kulturpflanzen, Futterpflanzen, Tabak, Gemüsepflanzen, Arzneipflanzen, dendrol. Material, pomol. Material. — Züchtungsstation in Uhříněvěs bei Praha. Pomol. Arboretum in Průhonice bei Praha. Dendrol. Park, Průhonice bei Praha.
L'Inst. des Recherches agric., Václavské nám, Rokoko. Dir. M. Doc. Dr. Straňák.

Pretoria (South Africa).
Agric. Experim. Stations, Groenkloof.

Preußisch-Holland, Ostpreußen (Deutschland).
Versuchsstat. der Mitscherlich-Gesellschaft. Dir. Dr. Bimschas. *Feststellung des Düngungsbedürfnisses des Bodens.*

Pulawy (Polska).
Inst. d'Agric. de l'Etat. Dir. T. Mieczyński. — Abt.: Landwirtschaftl. Abt., Leiter: E. Godlewski; Abt. f. Pflanzenbau, Leiter: J. Sypniewski; Abt. f. Gartenbau, Leiter: J. Dybowski; Abt. f. Pflanzenschutz, Leiter: vacat; Abt. f. Tierzucht, Leiter: L. Adametz; Abt.f.Bodenkultur, Leiter: T. Mieczyński; Abt. f. Tierernährung, Leiter: H. Malarski; Abt. f. Serol., Leiter: T. Jaroszyński; Abt. f. experim. Morphol., Leiter: S.Kopeć; Abt. f. Entomol., Leiter: S. Minkiewicz; Abt. f. Bact., Leiter: K. Markiewicz; Abt. f. Obstbäumebau, Leiter: J. Białobok. — Zweiglabor.: Staatl. Inst. für Landwirtsch. in Bydgoszoz; Stat. f. Fischerei, Halbinsel Hel.

Pusa, Bihar and Orissa (Br.-India).
Imperial Agric. Research Inst. and Coll.

Riga-Rāmava (Latvija).
Versuchsfarm Landwirtsch. Fac. Univ. Lettl. Dir. Doc. P. Lejiņš.

Rio Piedras (Porto Rico).
Insular Experim. Station. Dir. R. Menéndez Ramos, M.S. — M. T. Cook, Ph.D., Chief Div. Plant Pathol. and Botan.; H.L. Dozier, Ph.D., Chief Div. Entomol.; Montgomery Ellison, B.S., Chief Div. An. Husb.; F. A. Lopez, B.S., Chief Div. Chem.; Pedro Richardson, B.S., Chief Div. Agron.; R. F. García, B.S., Assoc. Chem.; J. P. Griffith, M.S., Plant Breeder; Pedro Osuna, B.S., Hort; F. M. Penneck, B.S., Flor.; J. H. Ramírez, B.S., Ass. Chem.; Alfonso Rivera, D.V.S., Vet.; A. H. Rosenfeld, M.S., Special Cane Tech.; Francisco Sein, jr., B.S., Ass. Entomol.; L. A. Serrano, B.S., Ass. Agron.; R. A. Toro, B.S., Ass. Pathol.

Roodeschool, Prov. Groningen (Nederland).
Experiment Farm d. Inst. f. Plantenveredeling, Wageningen.

Rostow a. D. (U.d.S.S.R.).
Rostow-Nachitschiwan Landwirtschaftl. Landes-Versuchsstation, Postfach 573. Dir. Prof. A. F. Lebedeff.

Rouen (France).
Station Agronomique de la Seine-Inférieure, Avenue de Caen et 36, Rue Blaise-Pascal.

Saguramo (U.d.S.S.R.).
Versuchsfelder d. Techn. Hochsch. Tiflis.

St. Paul, Minn. (U.S.A.).
Univ. Farm. Paul Donald Peterson (Ass. Plant Pathologist); C. Kennedy (Ass. Prof. agric. Biochemistry); H. L. Sweetman (Instr.); W. G. Brierley (Assoc. Prof. Horticulture); L. S. Palmer (Agric. Biochemistry); G. H. Nesom (Ass. Prof.); F. J. Stevenson (Prof. Plant Breeding); H. M. Tysdal (Instr. in Agronomy); A. C. Hildreth (Ass.Horticulture); C. E. Cary (Horticulture); C. H. Eckles (Dairy Husbandry); H. K. Hayes (Prof. ofplant Breeding); A. C. Arny (Ass.); C. H. Bailey (Biochemistry); J. H. Beaumont (Horticulture); H. E. Brewbaker (Ass. Plant Breeder); J. J. Willaman (Assoc. Prof. agric. Bioch.); W. C. Cook (Assoc. Prof. Entomol.); W. De Long (Ass. Horticulture).

Salatiga (Nd.-O.-Indië).
Experim. Station for Central Java. Dir. Dr. Ch. J. Bernard (Acting). — J. Th. de Haan (Agronomist). *Conditions for the various crops produced in Central Java, in collaboration with the technical staff of the Tea and Rubber station in Buitenzorg and the coffee station in Malang. Visits to estates and advice work on cacoa, kapok, coffee, rubber, manuring and green manuring.*

Salt Lake City, Utah (U.S.A.).
Agric. Research Dept. Utah — Idaho Sugar Co. 324—327 Vermont Building. Dir. Dr. E. G. Titus. — *Research workers welcomed. — Seed breeding. Development of Sugar Beets and Crops.* — The company works together with Farms in Utah, Idaho, Washington, Montana und South Dakota.

Salutchje, Gouv. Tver (U.d.S.S.R.).
Sapropelische Versuchsstation. Leiter V. A. Petrov.

Samara (U.d.S.S.R.).
Landwirtschaftliches Inst., Kooperatiwnaja 175.

Schatilowo, Gouv. Orlow (U.d.S.S.R.).
Landwirtschaftl. Provinz-Versuchsstation, Nowossilowskij ujesd.

Sidi-Bel-Abbès (Algérie).
Labor. Agric.

Simferopol (U.d.S.S.R.).
Landwirtsch. Krim-Inst. für spez. Kulturen, Gubernskaja 22.

Sinowjewsk (U.d.S.S.R.).
Adshamsker Landwirtschaftl. Versuchsstation, Postfach 202.

Sinelnikowo (U.d.S.S.R.).
Landwirtschaftl. Dnepro-Petrowsker Provinz-Versuchsstation.

Sofia (Bulgarien).
Inst. f. allgem. Ackerbau d. Landw. Fak. Johan Stranski, Doc. f. Landwirtschaft (Floristik); Johan Iwanoff, Doc. f. Landwirtschaft; M. Christoff, Ass. (Genetik); B. Ilieff, Ass.; B. Tiklaroff, Ass.; D. Kostoff, Ass.
Zentralinst. für landwirtschaftl. Versuchswesen, Solunplatz 3. Dir. Christo Savov. Abt.: Bodenkunde: Chr.Savov; Agriculturchemie: M. Stefanova; Agriculturbotan.: W. Stribrny; Samenkontrolle u. Samenzucht: Ch.Kazaski; Tierzucht: vacat; Phytopath.: Dr. B. Ivanov; angew. Entomol.: P. Čorbadev; landwirtschaftl. Maschinen u. Geräte: V. Kislinski.

Stambul (Türkei).
Agric. Inst. Tierärztl. Hochsch. Dir. Prof. N. Schewket.

Stockholm (Sverige).
Kungl. Landtbruks-Akademien. Praeses: Freiherr Carl Joach. Beck-Friis, Rittergutsbesitzer. — Sekretär: Prof. Dr. Paul Hellström. 6 Abt.: Landtbruksavdelningen; Vetenskapsavdelningen; Skogs- och trädgårdsavdelningen; Hushållnings-och slöjdavdelningen; Mekaniska avdelningen; Ekonomiska avdelningen.
Mus. f. Ackerbau u. Fischerei. Dir. Prof. Dr. Paul Hellström.
Abt. für Gartenkultur mit großen Baumschulen u. einer Gärtnerschule (30 Schüler). Vorsteher: Dir. Gustaf Lind.
Centralanstalten för försöksväsendet på jordbruksområdet (Zentralanstalt für das Versuchswesen auf d. Gebiete der Landwirtschaft). Direktion: Präsident: der Preses der Akademie; 6 Mitglieder. — Beamte: Abt. f. Landwirtschaft, Vorst.: Prof. Dr. Carl Axel Hjalmar von Feilitzen; Ober-Ass.: P. Bolin. Abt. f. Agriculturchemie, Vorst.: Prof. Dr. Sven Ludvig Alexander Odén; 1 Ass. Abt. f. Tierzucht u. Molkereiwesen, Vorst.: Prof. Dr. Nils Hansson; Ober-Ass. für die Molkereiversuche: E. J. Haglund; 1 Laborator, 3 Ass. Abt. f. Agriculturbotan., Vorst.: vacat; 1 Laborator, 1 Ass. Abt. f. Entomol. Vorst.: Prof. A. Tullgren. 1 Laborator, 1 Ass. Abt. f. Bact., Vorst.: Prof. Christian Barthel; 2 Ass.

Suchum (U.d.S.S.R.).
Suchumer Versuchsstation f. Landwirtschaft u. Gartenbau.

Suewka (U.d.S.S.R.).
Landwirtschaftl. Provinz-Versuchsstat. v. Vjatka Permskoj sh. d.

Sumy (U.d.S.S.R.).
Landwirtschaftl. Versuchsstation.

Svalöv (Sverige).
Swedish plant breeding Association. Dir. Prof. H. Nilsson-Ehle. — E. W. Ljung, Agron. Head of rye div.; Dr. A. Akerman, Head of wheat and oats div.; Dr. N. Sylvén, Head of grass, vlover and alfalfa div.; Dr. J. Rasmusson, Head of root crop div.; Doc. Dr. C. Hammarlund, Head of potato and pea div.; G. Sundelin Ass. of root crop div.; Dr. G. Nilsson-Leissner, Ass. of grass div.; J. Walldén, Dir. of seed control dept.; J. Lindberg, Chemist.

Working places may be put at disposal for a small number of guests. They are fitted for research in plant genetics. — Refrigerator compound. Baking labor. ℗ Agric. plants. — Their are branch labor. at Ultuna, Linköping, Kalmar, Skara, Varpnäs, Undrom, Torstea and Luleå.

Talowaja, Gouv. Woronesch (U.d.S.S.R.).
Steppen-Versuchsstation.

Tarasowka, Nord-Kaukas. (U.d.S.S.R.).
Donez-Versuchsfeld, Ju.-W. sh. d.

Tartu — Dorpat (Eesti).
Pflanzenbiol. Versuchsstation d. Univ. Dir. Doc. N. Rootsi. — Mag. agr. August Miljan, älterer Ass.; Frau Erika Miljan, Ass. *Feld- u. Vegetationsversuche.* Zweiglabor.: Das Labor. für Samenkontrolle bei dem Kabinett für Pflanzenbaulehre.

Taschkent (U.d.S.S.R.).
Labor. of General Agric. of Middle-Asiat. State Univ. Dir. Prof. N. I. Kourbatov. — Assoc. Prof. N. N. Mokin. *4 places for specialists, study of soils and plants.* ℗ Soil samples. — Labor. of Experim. Station for studying of fertilizers. Post-office Kaounchi. Taschkent Distrikt, Ousbekistan, U.d. S.S.R.
Turkestaner Abt. d. Staatl. Inst. of experim. Agronomic Univ.
Inst. f. spez. Akerbau, Univ. Dir. Prof. M. J. Uklonskaja.

Tegucigalpa (Honduras).
Abt. f. Ackerbauprodukte des Nation. Mus.

Tel-Aviv (Palestine).
Inst. of Agric. Dir. Prof. Dr. Otto Warburg. *Botan. (Systematik, Pflanzengeographie), Kolonialbotan. (angewandte) und koloniale Landwirtschaft.*

Tetschen-Liebwerd—Dečin-Libverd(C.S.R.).
Landwirtschaftl. Abt. d. Deutschen techn. Hochsch. Prag. Dir. Prof. Dr. Anton Jakanatz. *Botan. u. Pflanzenschutz.*
Lehrkanzel f. Pflanzenzüchtung, Wiesen- u. Hopfenbau d. Landwirtsch. Abt. d. Deutsch. Techn. Hochsch. Prag. Dir. Prof. Ing. Eligius Freudl.

Tôkyô (Japan).
Bureau of Agric. Dept. of Agric. and Forestry. Dr. Seinosuke Uchisa (Ornithol.); Hisakichi Kichida (Mammalogy); Susunun Takahashi (Entomol.).

Toulouse (France).
Inst. Agric. de l'Univ. Dir. Prof. Dr. Léon Marie Joseph Gustave Nicolas. *Physiol. végétal. Terratol. végétale. Amélioration des plantes cultivées.*

Tschakino, Gouv. Tambow (U.d.S.S.R.).
Tambower Landwirtschaftl. Versuchsstation. Dir. L. A. Pelzig.

Tunis (Tunis).
Labor. d'Agric. Dir. Prof. Coupin. *Cultures coloniales.*

Tucumán (Argentina).
Sta. experim. agric. Univ. Dir. Prof. W. Cross.

Uhříněves b. Praha (C.S.R.).
Züchtungsstat. d. Staatl. Versuchsinst. f. Pflanzenproduction.

Ultuna, Upsala (Sverige).
Stat. of the Swedish plant breeding Assoc. Dir. Agronom R. Torssell.

Undrom (Sverige).
Stat. of the Swedish plant breeding Assoc. Dir. G. Ericsson, Agron.

Urbana, Ill. (U.S.A.).
Dept. of Agronomy, Univ. of Illinois, 212 Old Agric. Building. Dir. W. L. Burlison. — Ass. Chief: G. H. Dungan (Crop Production).

Uspenskaja (U.d.S.S.R.).
Priasowskoje Versuchsfeld, Taganrogskij okr.

Varpnäs, Nor (Sverige).
Stat. of the Swedish plant breeding Assoc. Dir. G. Nilsson, Agron.

Vjatka (U.d.S.S.R.).
Abt. f. Agronomie d. Wissensch. Forschungsanst. f. Heimatkunde. Dir. Prof. P. T. Rešetnikov.

Voldbu (Norge).
Statens forsöksstation for fjeldbygderne. Dir. H. Foss.

Wageningen (Holland).
Inst. voor plantenveredeling (Pflanzenveredelung). Dir. Prof. Ir. C. Broekema. — Dr. M. J. Sirks (Geneticist); G. Azings Venema (Botanist); A. E. H. R. Boonstra (Ass.); Ir. J. D. Koeslag (Secretary of the Central Committee for Fieldinspection); Ir. W. A. Bosma (Botanist of the Committee for Research work on sugarbeets). *Rassenzüchtung v. Kulturgewächsen, Pflanzenveredelung; Kulturuntersuchung.; Auskunftsstelle für den Landbau auf d. Gebiete d. Veredelung von Kulturgewächsen.* — Experim. Farm at Roodeschool (Groningen).

Washington, D.C. (U.S.A.).
Bureau of Plant Industry U.S. Dept. of Agric. C. Drechsler (Morphol.); W. W. Eggleston (Botanist); W. S. Pate (Pathologist); L. M. Rutchius (Horticulturist); W. M. Diehl (Mycologist); B. Dodge (Pathologist); R. D. Rands (Pathologist); E. R. Ränker (Assoc. Physiol.); C. R. Ball (Sen. Agronomist); D. F. Black (Biochemist); P. Brierley (Pathologist); N. A. Brown (Pathologist); C. F. Clark (Horticulturist).
Systematic Agrostol., U.S. Dept. of Agric. A. S. Hitchock (Botanist).
National Canners' Assoc. Research Labor.

Weihenstephan b. Freising (Deutschland).
Inst. f. Pflanzenbau u. Bodenkunde, Hochsch. f. Landwirtsch. Dir. Prof. Dr. H. Puchner.

Wien (Österreich).
Bundesanst. f. Pflanzenbau u. Samenprüfung d. Landw.-Botan. Versuchsanst. Dir. Hofrat Ing. E. Haunalter. — Reg.R. Dr. E. Bogenhofer; Reg.R. Dr. H. Schindler; Ing. J. Schwarz; Dr. A. Hoffmann; Dr. F. Drahorad. *Selbständige Forschung auf dem Gebiete des Pflanzenbaues u. dessen Grenzgebieten, Übertragung der Forschungsresultate in die landwirtschaftl. Praxis, Pflanzenzüchtung, Samenkontrolle.* Samenzucht- u. Versuchsanlage in Melk, Gaming, Wieselburg a. d. Erlauf, Weideversuchswirtschaft Kragligut in Mitterndorf, Steiermark.
Landwirtsch. Labor. der Hochsch. f. Bodenkultur. Dir. Prof. Dr. H. Kaserer.

Williamstown, Mass. (U.S.A.).
Mount Hope Farm. H. D. Goodale (Biologist). *Inheritance and physiol. of reproduction.*

Witebsk (U.d.S.S.R.).
Landwirtschaftl. Versuchsstation. Janowskaja 29.

Wladikawkas (U.d.S.S.R.).
Landwirtschaftl. Inst. d. Bergrepublik, Alexandrowskij per. 3.

Woodward, Okla. (U.S.A.).
Woodward Field Station. J. B. Sieglinger (Assoc. Agronomist).

Wooster, Ohio (U.S.A.).
Ohio Experim. Station. R. M. Bethke (Animal Industry); P. E. Jilford (Plant Pathologist).

Woronesch (U.d.S.S.R.).
Kab. f. spez. Ackerbau mit Labor., Landwirtsch. Hochsch.
Woronescher Abt. d. Staatl. Inst. f. d. Studium d. Inundations-Gebiete, Landwirtsch. Inst.
Landwirtschaftl. Provinz-Versuchsstation Univ.
Kab. f. allg. Melioration, Landwirtsch. Hochschule.

Kabinett f. allgem. Ackerbau mit Labor. Landwirtsch. Hochsch.
Agronom. Kabinett, Univ. Dir. Prof. P. A. Savvin.

Würzburg (Deutschland).
Landwirtsch. Kreisversuchsstation, Luxburgstr. 4. Dir. Dipl.-Ing. S. Schumoefer, Oberregierungschemiker. — 3 Labor.

Zürich (Schweiz).
Landwirtschaftl. Versuchsfelder der Eidgen. Techn. Hochsch.

Zwätzen b. Jena (Deutschland).
Agric. Feld-Versuchsstation.

Laboratoria et institutiones agriculturae specialis.
Confer: *Phytopathologia* (pag. 415).

Aberdeen, Id. (U.S.A.).
Substation U.S. Dept. of Agric. G. A. Wiebe (Ass. Agronomist).

Ames, Ia. (U.S.A.).
Dept. of Farm Crops State Coll. M. F. Jenkins (Assoc. Agronomist).

Battle Creek, Mich. (U.S.A.).
Research Dept. Postum Cereal Co. Dir. M. S. Fine.

Berlin (Deutschland).
Preußische Versuchs- u. Forschungsanstalt f. Getreideverarbeitung u. Futterveredelung, N 65, Seestr. 11. Die allgemeine Verwaltung untersteht einem Verwaltungsdir. Die einzelnen Inst. werden von Dir. geleitet. Zur Beratung gemeinsamer Angelegenheiten bilden die Institutsdir. und der Verwaltungsdirektor ein Kollegium. *Wissenschaftliche und technische Grundlage der Getreideverarbeitung und -lagerung sowie der Gewinnung, Verwendung und Verbesserung der Futterstoffe. Anwendung wissenschaftlicher Forschungsergebnisse in der Praxis. Labor., Kursussäle, Arbeitsräume, 1 Mühle, 1 Bäckerei, 1 Kornlagerhaus, 1 Versuchsfeld und Vegetationshaus.*
Inst. I für Getreidelagerung und Futterveredelung. Dir. Prof. Dr. Gerlach. — Wissenschaftl. Hilfsarb.: Dr. Seidel, Dr. Günther May, Leonhardt. *5 bis 10 Arbeitsplätze. Chemische, botan. u. bact. Arbeitsplätze. Wissenschaftliche und praktische Arbeiten über die Lagerung des Getreides in Speichern und ähnlichen Vorratsräumen, über die Gewinnung, Erhaltung, Verbesserung und Verwendung der im Inlande erzeugten Futtermittel, über die Anlage von Silos und Trockeneinrichtungen für Futterstoffe, Fütterungsversuche.* — *Labor. Versuchsanstalt (Vegetationsstation), Versuchsfelder.*
Inst. II für Müllerei. Inst.-Dir. u. Prof. Dr. Joh. Buchwald. — Nahrungsmittelchemiker: Dr. Hugo Kühl; Botaniker: Dr. Gerhard Brückner. *Unterrichtslehrgänge für Müller im April und Oktober 14tägig, Übungslehrgänge in den Labor. 4wöchentlich. Das Arbeitsfeld des Instituts für Müllerei sind die Getreidemühlen aller Art, die Müllerei und der Mühlenbau, ferner der Handel mit Getreide, Mehl und Kleie und sonstigen Müllereierzeugnissen.* — *Versuchs-Roggen- und Weizenmühle, Tagesleistung 20 000 kg.*
Labor. f. Kartoffelbau, Biol. Reichsanst. für Land- u. Forstwirtschaft, -Dahlem, Königin-Luise-Str. 19. Dir. Reg.R. Dr. O. Schlumberger. — Dr. E. Koehler, Ass.
Labor. für angewandte Landwirtsch. Biol. Reichsanstalt für Land- u. Forstwirtschaft, -Dahlem, Königin-Luise-Str. 19. Dir. Reg.R. Dr. G. Schneider.

Bernburg (Deutschland).
Anhaltische Versuchsstation, Junkergasse 3. Dir. Prof. Dr. Wilhelm Krüger. — Versuchsabt.: Prof. Dr. Gustav Wimmer, Abt.-Vorst. u. stellvertr. Dir. der Versuchsstation; Dr. Hans Lüdecke, Ass. Untersuchungs-Abt.: Dr. Oscar Ringleben, Abt.Vorst.; Dr. Otto Voigt, Ass.; Dr. Otto Unverdorben, Ass.; Dr. Joseph Grimm, Ass.. Botan. Abt. u. Pflanzenschutz: Dir. Carl Ernst Becker, Abt.-Vorst. *Labor., Gewächshaus und Versuchsfeld.* ⓟ *Pflanzenkrankheiten und Schädlinge, besonders der Zuckerrübe.*

Biggs, Cal. (U.S.A.).
Biggs Rice Field Station. J. W. Jones (Assoc. Agronomist).

Bloomington, Ill. (U.S.A.).
Cereal Investigations Bureau of Plant Industry. J. R. Holbert (Agronomist).

Bologna (Italia).
Ist. di Cerealicoltura (di Allevamento vegetale per la cerealicoltura), 18, Via Toscana, 121. Dir. Prof. Francesco Todaro. — Dott. Cesare Orlandi, Aiuto. Dir.; Dott. Mario Bonvicini, Primo Ass.; Dott. Guglielmo Carboncini, Secondo Ass. *20 a 30 posti. Analisi botan. di sementi. Esercitazioni di Genetica applicata alla Cerealicoltura.* ⓟ *Semi di cereali.*

Buitenzorg, Java (Ned.-O.-Indië).
Section for Agronomy of the General Agric. Experim. Sta. Chief: Ir. A. Wulff. — Ir. I. G. Ossewaarde, Ass. *Field experim., testing different varieties, different treatments. Technic of native crop production. Agricultural implements. Plows and tools, adapted to native and plantation crops and different kinds of soils.*
Section for Plantbreeding of permanent and annual crops of the General Agric. Experim. Station. Dir.: Dr. P. J. S. Cramer and L. Koch. — Ass.: H. de Veer; Dr. I. Boldingh; Manager: F. A. Parkinson. *Introduction of wild forms and allied species of cultivated plants. New species of Hevea, coffee, kupok, etc. New crops. Experim. on grafting, influence of stock on scion with tropical crops (coffee, rubber, etc.). Selection of coconuts. Varieties of oil palms. The experim. are carried on in 2 gardens: Economic Garden, Buitenzorg (W. M. van Helten, manager). 20 H.A. Green manures from a practical point of view. Production of seed and planting material of economic plants and new crops. Experim. coffee plantation, Bangelan near Malang. 200 H.A. of different species, hybrids, and clones of coffee. Experim. garden, Buitenzor g (20 H.A.). Seed production garden, Buitenzor (28 H.A.). Seed selection of rice, corn and other native crops. Introduction of new varieties. Experim. with new forage and fibre plants. Seedling selection with cassave (Manihot). Production and distribution of better planting material. Varietal character of rice. Ear to row experim. with corn.*

Cambridge (England).
National Inst. of Agric. Botan., Huntingdon Road. Dir. Wilfred H. Parker. — F. C. Hawkes, Se cretary

and Ass. Dir.; A. Eastham, Chief Officer. *Crop Improvement Branch, Official Seed Testing Station, Cambridge. Iliam Farm, St. Ives. Potato Testing Station, Ormskirk.*

Columbus, Ohio (U.S.A.).
Dept. of Farm Crops' Ohio State Univ. Prof. J. B. Park (Farm Crops).

DetskojeSelo, Gouv.Leningrad(U.d.S.S.R.).
Labor. f. spez. Pflanzenbau der landw. und polytechnischen Hochsch., 1. Detskoje Selo. 2. Leningrad, Polytechnikum. Dir. Prof. N. Nedokučaev. — S. Nikitin, A. Schulz, L. Krasnokutskaja. *2—3 Plätze für speziellen Pflanzenbau und Vegetationsversuche. — Vegetationshaus und Versuchsfeld, chemisches Labor.* Ⓟ *Samen d. landw. Pflanzen, Kartoffeln, Körner usw.*
Akklimatisations-Station, Kolpinskaja, N. 2. Dir. Prof. S. A. Eghis. — Prof. Oscar Walther, Dir. der physiol. Abt.; Stanislaus Dawidowitsch, Ass. für Genetik des Tabaks; Wladimir Rybin, Cytol.-Ass.; Iwan Samoiloff, Ass. für Feldversuche; Lydia Pinewitsch, Ass. für Physiol.; Vera Krapiwina, Ass. für Hybridisation in Genus Nicotiana; Labor.-Agronom Burzeff, Genetik d. Nicotiana rustica; Labor.-Agronom Trunava, Genetik d. Buchweizens; Labor.-Physiologist Müller, Physiol. d. Tabaks; Labor.-Physiologist Tschischewskaja, Physiol. d. Tabaks. *Herbariums der verschiedenen Rassen d. Nicot. Tabacum und N. rustica und Hybriden zwischen diesen zwei Arten.* Ⓟ *Samen und Herbaria der Nicotiana-Spezies.* — *Physiol. Labor.* (Prof. O. A. Walther). *Cytol. Labor.* (W. A. Rybia).

Dickinson, N.D. (U.S.A.).
Dickinson Susbtation. R. W. Smith (Assoc. Agronomist). *Experim. with cereal crops.*

Geneva, N.Y. (U.S.A.).
Div. of Agronomy of the New York Agric. Experim. Stat. Reginald C. Collison, M.S., Chief in Research (Agronomy); James E. Mensching, M.S., Assoc. in Research (Agronomy); James D. Harlan, B.S.; Park V. Traphagen, Ass. in Research (Agronomy). *Lysimeter investigations. Plant experim. work. Studies of certain physiol. and chem. aspects of cereal straw. Investigations of high-nicotine tobacco. Physiol. and nutritional studies of the apple tree. Studies of certain problems on muck soils.*

Groningen (Holland).
Rijkslandbouwproefstation, Eemskanaal Z.Z. 1. Dir. Dr. K. Zijlstra (Botan.-landwirtsch. Untersuchg.). — Abt.-Dir.: J. G. Maschhaupt u. J. Hudig (Düngungsversuche), Dr. F. C. Gerretsen (Mikrobiol. Bodenuntersuchg.); Chemiker: C. W. G. Hetterschij, Dr. J. Sack; Agromon: C. Meijer; Plantkundige: Drs. M. A. J. Goedewaagen. *Acker- u. Wiesenbau.*

Groß-Enzersdorf (Österreich).
Versuchswirtschaft d. Hochsch. f. Bodenkultur Wien. Dir. d. Versuchsfeldes: Prof. Dr. Kaserer; Dir. d. Versuchsstalles: Prof. Dr. L. Adametz.

Hamburg (Deutschland).
Landwirtschaftliche Versuchsstation Hamburg-Horn, 26, Hammerlandstr. 231. Dir. Dr. C. Krügel. — A. Retter, Dipl.-Chemiker. *Großer Versuchsgarten mit Parzellen zur Vornahme v. Pflanzendüngungsversuchen. 1 Labor., in dem vorwiegend Analysen z. Begutachtung v. Rohstoffen u. Fabrikaten d. Superphosphat-Industrie ausgeführt werden. Daneben werden auch Bodenproben u. Ernteprodukte untersucht.*

Harrisburg, Pa., (U.S.A.).
Bureau of Plant Industry, Dept. of Agric. T. L. Guyton (Chief Entomologist); A. B. Champlain (Entomologist); J. N. Knull (Entomologist).

Hljustim, Tula Gouv. (U.d.S.S.R.).
Inst. f. angewandte Botan. Tula Akklimatisations-Stat. Nikolai Wechoff, Haupt-Ass.

Honolulu (Hawaii).
Experim. Station Assoc. of Hawaiian Pineapple Canners. Dir. A. L. Dean, Ph.D. — F. A. E. Abel, M.S., Ass. Chem.; F. A. Bowers, B.S., Ass. Agric.; Gwendolyn Cochrane, B.S., Ass. Pathol.; John Horner, A.B., Ass. Chem.; Helene Morita, B.S., Ass. Path.; C. P. Sideris, Ph.D., Pathol.; W. A. Wendt, B.S., Ass. Agric.; N. L. Dennison, B.S., Agr.

Ithaca, N.Y. (U.S.A.).
Dept. of Plant Breeding, Cornell Univ. Dir. Dr. A. Tavcar. — I. F. Phips (Geneticist).

Jackson, Miss. (U.S.A.).
State Plant Board of Mississippi, P.O. Box 424. O. M. Change (Inspector).

Kiew (U.d.S.S.R.).
Akklimatisations-Garten a. d. Akademie d. Wissenschaften, Dorogoshitzkaja 45. Dir. M. T. Katschenko.
Landwirtschaftl. Provinz-Versuchsstation, Nesterowskaja 31.

Krasnodar, Kuban (U.d.S.S.R.).
Inst. z. Wissenschaftl. Erforschung des Kuban- u. Schwarzmeer-Gebietes, Oktoberstr. 119. Dir. A. P. Protopopov. — F. A. Terentjew (wiss. Mitarbeiter). 10 Agronomen, 1 Med., 2 Bodenforscher. *Industriepflanzen. Zootechnik.*

Lednice-Morava (C.S.R.).
„Mendele", l'Inst. des recherches agric. de l'Inst. des Recherches agric. et Forestière. Dir. Dr. Ing. Karel Kočnar. — Ing. Jaromír Závada, Ass.; Ing. Vladimír Beneš, Ass.

Leipzig (Deutschland).
Inst. f. Pflanzenbau u. Pflanzenzüchtung d. Univ. Dir. Prof. Dr. Adolf Zade. *Pflanzenpathol. Versuchstechn. Unkrautbiol. u. Unkrautbekämpfung. Hafer u. Gräser.*

Leningrad (U.d.S.S.R.).
Pflanzenzucht-Station, Kamennyj Ostrov 2, Birkenallee 28. Dir. Prof. N. A. Busch. — Prof. O. A. Valter.

Lincoln (New Zealand).
Canterbury Agric. Coll. Dir. R. E. Alexander. — F. W. Hilgendorf, Plant Breeder; J. W. Calder, Ass. Breeder. *5 places for plant breeeding, cereals and grasses and plot technique.*

London (England).
Farm Labor. Med. Research Council, Mill Hill. Dir. G. W. Dunkin.

Magyar-Óvár (Ungarn).
Versuchsanstalt für Pflanzenkultur der Landwirtschaftsakademie. Dir. Emil Grabner.

Maksimir b. Zagreb — Agram (S.H.S.).
Facultätsgut mit Versuchswirtschaft Univ. Zagreb. Curatoren: Profs. Petračić, Jurić, Rittjg.

Malang (Ned.-O.-Indië).
Experim. Station. Dir. Dr. A. J. Ultee. — Dr. W. Bally, Botanist; Th. G. E. Hoedt (Agronomist).

Mohaltan, pr. Trondhjem (Norge).
Statens forsöksstation Nordentjelds. Dir. P. Lövö.

Moskau (U.d.S.S.R.).
Bureau f. d. Einführung u. Vermehrung neuer Feldpflanzensorten a. Staatl. Inst. f. experim. Agronomie, Granatnyj Per. 7.
Lehrstuhl u. zugleich Versuchsstation f. Pflanzenernährung u. -düngung. Confer: Physiol. plantarum.

Station Expérim. Linicole de l'Inst. Agronomique. Dir. A. N. Kletschetov.

Padova (Italia).
Inst. Econom. rurale et estimo. Dir. Prof. L. Di Muro.

Perm (U.d.S.S.R.).
Kabinett f. Vegetationsversuche. Dir. Doc. N. G. Kudrjavcev.

Perugia (Italia).
Versuchsfelder d. Inst. sup. Agrario. Dir. Dr. Oswaldo Kruch.

Praha — Präg (C.S.R.).
Inst. f. Pflanzenveredelung u. spez. Pflanzenproduktion, tschech. techn. Hochsch. Confer Evolutio et genetica.
Staatliche Versuchsstationen, XII, Villa Groebe. Dir. Prof. Dr. Julius Stoklasa. — Doc. Dr. Šebor, Dr. Pěnkava; Dr. Bareš; Ing. Strupl; Ing.Vrbenský; Ing. Strádal; Ing. Šilhavý.

Pretoria (South Africa).
Agric. Botan. Inst. Dir. Prof. J. M. Hector.

Priekuļi (Latvija).
Priekulsche Versuchs- und Pflanzenzuchtstation des Lettländischen Landwirtschaftlichen Zentralvereins, Priekulu selekcijas slacijas über Cesis. Dir. Agronom Edgar Eglit. — Agronom N. Brosch, Gehilfe; Stud. agr. Frl. E. Rosental, Ass.; Frl. I. Balod, Laborant.

Puerto Orotara (Canarias).
Jardín de aclimatación. Dir. Francisco Menéndez Martin.

Puławy (Polska).
Ab. f. Pflanzenzüchtung, Staatl. Inst. f. Landwirtschaft. Dir. Dr. L. Kaznowski.
Dept. de l'amélioration des plantes à l'Inst. nat. Polonais d'Economie Rurale. Dir. J. Sypniewski.

Riga (Latvija).
Versuchswirtschaft „Vecance". Dir. Prof. Janis Bergs.
Abt. f. Pflanzenbau d. naturwiss.-math. Fac. d. Herder-Inst. Dir. Prof. F. Ferle.

Sofia (Bulgarien).
Inst. f. spez. Pflanzenbau. Dir. Doc. I. Ivanov.

Stettin (Deutschland).
Anstalt für Pflanzenbau.

Stockholm (Sverige).
Abt. f. Agrikulturbotan. Zentralanst. f. Landw. Versuchswesen. Dir. vakat. 1 Ass.; 1 Laborator.

Ökol. Station der Hallands Väderö.
Confer: œcologia (pag. 423).

Stormont, Belfast (Ireland).
Plant Breeding Research Station (Ministry of Agric. Northern Ireland) and Plant Genetics Dept. The Queen's Univ. of Belfast, Strandtown. Dir. Ian W. Seaton. — 1 Ass. (General). ⊕ Varieties of Agric. Plants (principally oats and flex).

Tela (Honduras).
Research Dept. Tela Railroad Comp. H. S. Dean (Panama Disease of Bananes).

Tiflis (U.d.S.S.R.).
Inst. f. spez. Ackerbau, Techn. Hochsch. Dir. Prof. P. M. Žukoskij.
Kabinett f. subtropische Kulturen, Techn. Hochsch.

Tjinjiroean, Java (Ned.-O.-Indië).
Government Cinchona Experim. Station. Dir. Dr. M. Kerbosch. — Dr. C. Spruit, P. Pzu. (Botanist); W. van Dorssen (Chief analysist); E.v.d. Fos (Analysist); P. Schrijnen (Agric. Ass.).

Vjatka (U.d.S.S.R.).
Labor. f. Vegetationsversuche u. Bodenuntersuchungen d. wissensch. Forschungsinst. f. Heimatkunde.

Warszawa (Polska).
Inst. f. Pflanzenprodukte, Hochsch. f. Land- u. Forstwirtsch. Dir. Prof. W. Staniszkis.

Washington, D.C. (U.S.A.).
Office of Cereal Crops and Diseases, Bureau of Plant Industry, U.S. Dept. of Agric. Dir. W. A. Taylor, Chief. — H. H. McKinney (Pathologist); V. F. Tapke (Assoc. Pathologist); M. A. McCall (Cereal Agronomy); A. G. Johnson (Plant Pathologist); R. W. Lenkel (Assoc. Pathologist); I. R. Stanton (Agronomist).
Tropical Plant Research Foundation. Dir. W. A. Orton.

Weihenstephan b. Freising (Deutschland).
Inst. für Pflanzenzüchtung u. Pflanzenbau der Hochsch. für Landw. Dir. Prof. Dr. Raum.

Woronesch (U.d.S.S.R.).
Woronescher Versuchsfeld d. Leningrader Staatl. Inst. f. experim. Agronomie, Landwirtsch. Inst.
Versuchsstation für Weizen-, Roggen-, Hirse- u. Kartoffelzüchtung, Landwirtsch. Hochsch.

Zagreb — Agram (S.H.S.).
Labor. f. Pflanzenproduktenlehre Univ. Dir. Prof. Jurić.

Laboratoria silviculturae.

Confer: *Horticultura* (pag. 513), *Phytopathologia* (pag. 415).

Aas b. Oslo (Norge).
Labor. für Forstwissenschaft. Norges Landwirtschaftl. Hochsch. Prof. Barth; Prof. J. Böhmer. *Forstwissenschaft, Forstpflege, Taxation.*
Statens Skogforsöksstation, Landwirtschaftl.Hochschule. Dir. Eide.

Amersfoort (Holland).
Holl. forstliche Versuchsanstalt. Dir. Engbertus Hesselink. *Waldbauliche Probleme, wie Wurzelentwicklung, Einfluß der Herkunft des Kiefernsamens.*

Amherst, Mass. (U.S.A.).
Northeastern Forest Experim. Sta. U. S. Forest Service. Dir. Samuel Trask Dana. — C. E. Behre (U. S. Forest Service), Forest Mensuration; S. T. Dana (U. S. Forest Service), Forest Protection; P. R. Gast (U. S. Forest Service and Harvard Forest), Forest Ecol.: H. J. MacAlone (U. S. Bureau of Entomol.), Forest Entomol.; H. B. Peirson (Maine Forest Service), Forest Entomol.; Perley Spaulding (U. S. Bureau of Plant Industry), Forest Pathol.; P. W. Stickel (U. S. Forest Service), Forest Five Investigations; Marinus Westveld (U. S. Forest Service), Silviculture. *Guests always welcome. Forest Ecol., Silviculture, Forest Mensuration, Forest Entomol., Forest., F Patholorest Live Investigations.*

Asheville, N.C. (U.S.A.).
Appalachian Forest Experim. Station. C. R. Hursh (Assoc. Forest Ecologist).

Bergen (Norge).
Forstliche Versuchsanstalt für West-Norwegen (**Vestlandets forstlige forsøksstation**), Bergens Mus. Dir. Prof. Dr. Oscar Hagem. — Abt.-Vorst.: Anton Smitt.

Berkeley, Cal. (U.S.A.).
Dept. of Forestry Coll. of Agric. Dir. Walter Mulford, F.E., Head of Div. — Emanuel Fritz, M.E., M.F., Assoc. Prof. For.; Woodbridge Metcalf, M.S., Assoc. Prof. For.; H.E. Malmsten, B.S.F., Ass. Prof. For., Ass. Plant Ecologist; A. W. Sampson, Ph.D., Assoc. Prof. For., Plant Ecologist; F. X. Schumacher, B.S., Instr. For.

Berlin (Deutschland).
Labor. f. Forstzool. Biol. Reichsanstalt für Land- u. Forstwirtschaft, Dahlem, Königin-Luise-Str. 19. Dir. Dr. H. Sachtleben.

Berlin, N.H. (U.S.A.).
Research Dept. Brown Comp. H. I. Baldwin (Research Forester).

Brno — Brünn (C.S.R.).
Inst. f. Forstschutz d. Landwirtsch. Hochsch., Černa pole. Dir. Ing. Anton Dyk.
Inst. f. Forstproduktion d. Landwirtsch. Hochsch., Černa pole. Dir. Ing. Josef Konšel.
Inst. f. Forsteinrichtung, -taxation u. Dendrometrie d. Landwirtsch. Hochsch., Černa pole. Dir. Dr. Rudolf Haša.

Buitenzorg, Java (Ned.-O.-Indië).
Forest Research Inst. Dir. Dr. R. Wind, clerical staff. 1. Section of wood technol. Chief: Dr. L. G. den Berger. Ass.: Forest Officer A. Th. J. Bianchi. *Identification of tropical woods with a pocket lens; investigations on the mechanical, physical, and chem. properties, the microscopic structure, the durability, seasoning and preservation of wood; rules for the inspection of timber; grouping of the different woods in durability — and value classes (in collaboration with the section for botanical and wood technical exploration); distribution of authentic wood samples, investigation on minor forest products; extension service.* 2. Section for botan. and woodtechnical exploration of forests. Chief: Forest Officer F. H. Endert. Ass.: Forest Officer A. Thorenaar. *Collecting of botanical material and authentic wood samples of the principal trees of the Archipelago; collecting of data about range, frequency and vernacular namens of these trees and about the uses and the durability of the wood.* 3. Section for the investigation on the regeneration of the different kinds of trees, except teak. Chief: Forest Officer A. Noltee. *Investigation on the regeneration of the most important kinds of trees, except teak, for industrial purposes, for the production of tanning material, etc.; investigation on the selection cutting systems in mountain-forests; distribution of seeds.* 4. Section for the investigation on the regeneration of teak and forest protection. Chief: Forest Officer H. M. J. Hart. Ass.: Forest Officer J. P. Schuitemaker. *Sylvicultural investigation on the regeneration of teak and its principal associates. Forest protection against fires, cattle and different diseases.* 5. Section of thinnings- and yield-capacity. Chief: Forest Officer H. E. Wolff von Wülfing. Ass.: Forest Officers J. A. J. H. Stoutjesdijk and F. W. K. de Klerk. *Investigation on the yield of timber and financial produce of teak and other trees. Methods of measuring crops and increment, modes of treatment, calculation of yield-tables, etc.*

Burlington, Vt. (U.S.A.).
Dept. of Forestry, Univ. of Vermont. Dir. W. R. Adams.
Dept. of Forestry, Coll. of Agric. F. M. Callward, B.S., State Ext. For.

College Park, Md. (U.S.A.).
Dept. of Forestry, Coll. of Agric. F. W. Besley, M.F., D.Sc., Lect. on For. (State For.); F. B. Frenk, M.S., For. specialist.

Colorado Springs, Col. (U.S.A.).
Rocky Mountains Forest Experim. Station. Dir. C. G. Bates. — J. J. Roeser (Ass. Silviculturist).

Dehra-Dun (British-India).
Imp. Forest Research Inst. Dir. A. Rodger.

Durham, N.H. (U.S.A.).
Dept. of Forestry Coll. of Agric. Dir. K. W. Woodward, M.F., Prof. For. — C. L. Stevens, B.S., Ass. Prof. For.; E. D. Fletcher, Ext. For.; R. D. Stevens, B.S., Instr. and Ass. For.

Eberswalde (Deutschland).
Forstl. Versuchsanstalt und Samenprüfungsanstalt, Forstl. Hochsch. Dir. Prof. Schilling.

Edinburgh (Scotland).
Dept. of Forestry Univ. Fac. Sc. Dir. Prof. E. P. Stebbing.

Freiburg i. Br. (Deutschland).
Badische forstliche Versuchsanstalt Univ. Dir. Prof. Dr. Hans Hausrath. — Forstass. Crocoll. *Waldbau, historische Pflanzengeographie, Forstgeschichte.*

Gießen (Deutschland).
Forstinst. mit Forstgarten Univ. Geschäftsführ. Dir. Borgmann. — 1. Abt. f. Betriebslehre. Dir. Prof. Borgmann. 2. Abt. f. Produktionslehre. Dir. Prof. Vanselow. 3. Abt. f. Forstpolitik. Dir. Prof. Weber.

Gorki (U.d.S.S.R.).
Forstl. Zentral-Versuchsstation der Staatl. Landw. Akademie.

Grafrath b. München (Deutschland).
Forstl. Versuchsrevier. Confer München.

Hanoi (Indo-Chine).
Sect. forestière d'Ecole sup. d'Agric. et Sylvicult.

Helsinki (Finnland).
Forstwissenschaftl. Versuchsanstalt. Rauhastr. 4. Dir. Prof. Dr. O. Heikinheimo. — Abt. f. Waldbau: Prof. Dr. O. Heikinheimo, Ass. Dr. V. Kujala; Abt. f. Forsttaxation: Prof. Dr. Y. Ilvessalo, Ass. Mg.ph. M. Lappi-Seppälä; Abt. f. Bodenkunde: Prof. Dr. V. T. Aaltonen; Abt. f. Mooruntersuchungen: Doc. Dr. V. Auer. — 10 Versuchsstationen.
Landwirtschafts- u. Forstzool. Inst. Confer: Agric. gen.

Hütteldorf (Österreich).
Forstl. Demonstrations- u. Lehrgarten d. Hochsch. f. Bodenkultur in Wien. Dir. Prof. A. Cieslar.

Ithaca, N.Y. (U.S.A.).
Dept. of Forestry Coll. of Agric. Dir. R. S. Hosmer, B.A.S., M.F., Prof. and Head of Dept. — John Bentley, jr., B.S., M.F., Prof. For. Engin., For.; J. A. Cope, B.S., M.F., Ext. Ass. Prof. For.; C. H. Guise, B.S., M.F., Ass. Prof. For. Mgt.; A. B. Recknagel, B.A., M.F., Prof. For. Mgt. and Utilization; J. N. Spaeth, B.S., M.F., Ass. Prof. For.; S. N. Spring, B.A., M.F., Prof. Silviculture.

Kamenetz-Podolsk, Ukraine (U.d.S.S.R.).
Labor. für Forstwirtschaft d. Inst. für Landwirtsch. „K. Marx".

Kazan (U.d.S.S.R.).
Dendrologischer Forstgarten des Inst. der Land- und Forstwissenschaft. Dir. Prof. Jaschnoff. — Ass. Maria Stelmachowitch. Ⓟ Forstpflanzen (Samen, Keimlinge).

Kiew (U.d.S.S.R.).
Forstkundl. Versuchsstat. d. Landw. Inst. Dir. Prof. E. Alexseef. — Prof. S. Vinter; Prof. D. Seftschek; Prof. A. Fomin.

Kôryô, Korea (Japan).
Government Forest Experim. Station, Keikidô.

Krasnodar, Kuban (U.d.S.S.R.).
Forstwissensch. Inst. d. Kuban. Landw. Inst. Dir. Doc. A. N. Uglicki.

Kyôto (Japan).
Inst. of Forestry Coll of Agric., Kyôto Imper. Univ. Dir. Prof. Dr. Yatarô Satô. — Prof. Dr. Sanroku Ichikawa; Prof. Dr. Yatarô Satô; Ass. Prof. Eitarô Sekiguchi.

Lafayette, Ind. (U.S.A.).
Dept. of Forestry of the School of Agric. Purdue Univ. B. N. Prentice, B.A., M.F., Assoc. Prof. For.

Lansing, Mich. (U.S.A.).
Dept. of Forestry State Coll. of Agric. Dir. A. K. Chittenden, M.F., Head of Dept. — J. C. DeCamp, M.F., Ass. Prof. For.; P. A. Herbert, M.F., Ass. Prof. and Ass. For.; Fay Hyland, B.S., Research Ass. in For. (P.O., Sault Ste. Marie); R. F. Kroodsma, M.F., Forest Specialist.

Leipzig (Deutschland).
Abt. f. Forstwirtschaft d. Un v. Dir. Forstmeister Mehner.

Leningrad (U.d. S.S.R.).
Leningrader Forstinst., Lesnoj, Institutskij per. 5. Rektor: F. N. Dingelstedt; Prorektor: N. P. Kobranow. — G. I. Anufriew; M. M. Orlow; W. Sukatschev. N. Ja. Kusmin; S. A. Nepostaew; P. A. Kostylew.
Kabinett f. d. Forstkulturwesen, Forstinst. Dir. Prof. N. Kobranow. — A. Fomitschow, Ass., W. Kapper, Ass. (Leitung der praktischen Arbeiten und Kontrolle der wissensch. Arbeiten). *2 Plätze für die Untersuchung der Waldsamen und Forstpflanzen.* ⓟ Samen verschiedener Provenienz.
Abt. f. Forstwiss. d. Staatl. Inst. f. experim. Agronomie. Dir. M. E. Tkatschenko.

Lima (Peru).
Sección de Selvicultura, la Estación Central Agronómica. Dir. Dr. J. O. Solano.

London (England).
Dept. of Forestry Demonstration Plantation, South Eastern Agric. Coll. of the Univ., Wye, Kent.

Lwów (Polska).
Labor. f. Forstentomol. u. Forstschutz d. Techn. Hochsch. Dir. Prof. A. Kozikowski.

Madison, Wis. (U.S.A.).
Forest Products Labor. E. Gerry (Microscopist in Forests Products). *Physiol. Anat. and Microchem. of Forest Products.*

Maksimir b. Zagreb — Agram (S.H.S.).
Forstl. Demonstr. u. Versuchsgarten Univ. Zagreb. Dir. Prof. Petračic.

Manila (Philippine Islands).
Forest Experim. Sta., Forest School. Dir. Prof. A. F. Fischer. — Prof. H. Cuzner; Ass. Prof. F. R. Amos; Ass. Prof. A. P. Racelis.

Mariabrunn (Österreich).
Forstliche Landes-Versuchsanstalt. Dir. W. Sedlaczek.

Melbourne (Australia).
Forest Examination Board. Dir. A. J. Ewart.

Minsk (U.d.S.S.R.).
Forst-Versuchsstation Inst. f. Landw., Wassiljewskij Per. Dir. Prof. W. J. Perechod.

Missoula, Mtn. (U.S.A.).
Northern Rocky Mountain Forest Experim. Sta. of the U.S. Forest Service. Dir. R. H. Weidman. — *Windfall Problems, Forest Succession.*

Mont Alto, Pa. (U.S.A.).
Pennsylvania State Forest School. Dir. E. A. Ziegler. — Illo. Hein, Prof. of Biol. *1 place for Forest Ecol., Dendrol., Forest Entomol., Pathol.*

Montreal (Canada).
Forest Products Labor. of Canada, 100, Univ. St. Dir. W. Kynoch, B.Sc.F., F.E. (Toronto). — Dr. Claser Fritz; Mr. I. D. Hale; R. Levick. Timber tests division. Wood preservation division. Timber physics division. Pulp and Paper division. *Research into the chem., physical, anat. and other properties of woods and into the pathol. of woods.* Branch labor. at Vancouver, British-Columbia, in connection with the Univ. of British-Columbia.

Morgantown, W.Va. (U.S.A.).
Dept. of Forestry Coll. of Agric. Dir. T. W. Skuce, B. S., Forest Specialist.

Moscow, Id. (U.S.A.).
Dept. of Forestry, Coll. of Agric. Dir. F. G. Miller, M.F., Dean School of For. — E. E. Hubert, Ph.D., Prof. For. Products; H. I. Nettleton, B.S.F., Instr. For.; C. W. Watson, M.F., Ass. Prof. Silviculture
School of Forestry, Univ. of Idaho. Prof. E. E. Hubert (Forest pathol.).

Moskau (U.d.S.S.R.).
Moskauer Wissenschaftl.-Techn. Forstgesellschaft a. Moskauer Forstwissenschaftl. Inst., Wolchonka 14.

München (Deutschland).
Forstliche Versuchsanstalt, Amalienstr. 52. Geschäftsf.: Dr. Eugen Lukinger, Forstamtmann. — Forstliches Lehr- u. Versuchsrevier (Grafrath b. München). Leiter: Proff. Schüpfer und Fabricius. Forstlicher Versuchsgarten (Grafrath). Leiter: Prof. Fabricius.
Inst. für forstl. Betriebslehre, Amalienstr. 52. Dir. Prof. Schüpfer.
Inst. f. Waldbau u. Forstbenutzung, Amalienstr. 52. Vorst.: Prof. Fabricius.

Nancy (France).
Ecole nationale des eaux et fôrets de Nancy. Station de Recherches et Expériences. Labor. de Botan., Zool. et d'essais de bois, établissement de pisciculture etc. — 4 Sections: Sc. forestières; Botan. forestière et étude du bois; Zool. forestière et Pisciculture.
Sect. des Sc. forestières de l'Ecole nation des Eaux et Fôrets.

Nanking (China).
Dept. of Forestry, Coll. of Agric. and Forestry, Univ. Research-Prof.: W. C. Lowdermilk. *Methods in forest conservation.*

New Haven, Conn. (U.S.A.).
Yale School of Forestry, Yale Univ. Dir. Prof. Henry Solon Graers. — Prof. James Th. Tourney, Research in Silviculture; Prof. R. C. Hawley, Research in Silviculture; Prof. S. J. Ricord, Research in Wood Technol.; Ass. Prof. Geo. S. Garralt, Research in Forest Utilization. Usually 3 or 4 Ass. *There is room for 3 or 4 visiting researchers at our technol. labor. and at our field stations in silviculture. — One large collection of wood specimens, particulary of tropical woods, suited to labor. use Full privileges in school forests.* ⓟ Chiefly wood specimens.

Dept. of Forestry State Agric. Experim. Sta. W. O. Filley, For. in charge; H. W. Hicock, M.F., Ass. For.

New Orleans, La. (U.S.A.).
Southern Forest Experim. Station U.S. Forest Service. Dir. R. D. Forbes.

Ogasawara Island (Japan).
Ogasawara Branch of Imperial Forest Exper. Sta. *Trop. plant.*

Orono, Me. (U.S.A.).
Dept. of Forestry Coll. of Agric. J. M. Briscoe, M.F., Prof. For.; C. W. L. Chapman, M.S., Ass. Prof. For.; D. B. Demeritt, M.F., Ass. Prof. For.; M. E. Watson, B.S., For. Specialist.

Oxford (England).
Dept. of forest Botan. at School of Forestry. Dir. Prof. R. Day.

State College, Pa. (U.S.A.).
Dept. of Forestry, School of Agric. J. A. Ferguson, M.A., M.F., Head of Dept.; C. R. Anderson, M.F., Assoc. Prof. For. Ext.; W. G. Edwards, M.S., Prof. Lumbering; W. B. MacMillan, B.S., M.F., Ass. For.; F. T. Murphey, B.S., Ass. Prof. For. Ext.; H. S. Newins, M.F., Prof. Wood Utilization; G. F. Rupp, B.S., Instr. For.

Perm (U.d.S.S.R.).
Kabinett f. Forstwirtschaft. Dir. Doc. S. I. Kuprijanoff.

Petersham, Mass. (U.S.A.).
Harvard Forest Experim. Sta. Dir. R. T. Fisher.

Portland, Ore. (U.S.A.).
Pacific Northwest Forest Experim. Station, 514, Lewis Bldg. Dir. Thornton T. Munger. — Walter H. Meyer, Forest Mensurationist; Ruthford H. Westveld, Silviculturist; Leo A. Isaac, Silviculturist; Richard E. McArdle, Forest Protectionist; A. Gael Simson, Forest Protectionist. *Silviculture, botan. and dendrol.* — *Arboretum, Nursery, and a great area of virgin forest, immature timber, and cutover land on which experiments are under way.* ⊕ Tree seeds and dendrol. specimens. — The principal field center is the Wind River Branch Station located in the Columbia National Forest, Washington, U.S.A.

Poznań — Posen (Polska).
Inst. d. Forstbotan.-Univ. mit Dendrol. Garten, Sołacz, Dwór. Dir. Dr. Konstanty Stecki, ao. Prof. Univ. — Dr. Witold Kulesza, Priv.Doc. Adjunkt; Bohdan Podkhajski, Ass. ⊕ Im dendrol. Garten werden frei wachsende Laub- u. Nadelhölzer Mitteleuropas gesammelt. Zum Tausch können wir eventl. einige seltenere osteurop. Arten, die noch in Polen vorkommen, anbieten.
Inst. f. Forstbauwesen. Dir. Prof. J. Rafalski.
Inst. f. Dendrometrie u. Forststatik. Dir. Prof. T. Wielgosz.
Inst. f. allgem. Forstwesen. Dir. Prof. J. Rivoli.
Inst. f. spez. Waldbau. Dir. Prof. R. Biehler.

Praha — Prag (C.S.R.).
Inst. f. Dendrol., Jagtierbiol. u. Forstentomol. d. čech. techn. Hochsch., Vršovice, Havlickovo sady. Dir. Prof. Dr. Wilhelm Sallac.

Pullman, Wash. (U.S.A.).
Dept. of Forestry State Coll. Dir. E. H. Steffen, M.F. Head of Dept.

Raleigh, N.C. (U.S.A.).
Dept. of Forest State Coll. of Agric. F. H. Claridge B.S. M.F. Instr. For.; R. W. Graeber B.S., For. Specialist.

Rangoon, Burma (Br.-India).
Forestry Dept. Univ. Coll. Dir. Prof. R. Unwin.

Riga (Latvija).
Inst. f. Forstbotan. Dir. Doc. K. Melders.

Rouen (France).
Ecole municipale d'Arboriculture, Jardin-des-Plantes.

St. Paul, Minn. (U.S.A.).
Lake States Forest Experim. Sta. Univ. Farm. J. Kittredge (Silviculturist).
Dept. of Forestry Dept. of Agric. Univ. Farm. Dir. Henry Schmitz, Ph.D., Chief. — J. H. Allison, Ph.B., M.F., Prof. and Assoc. For.; P. O. Anderson, B.S., For. Specialist; H. E. Bartelt, B.S., Instr. For.; S. S. Burton, B.S., Instr. and Ass. For.; E. G. Cheyney, B.A., Prof. For.; D. A. Kribs, B.S., Instr. For.; E. E. Probstfield, B.S., Instr. and Ass. For.; J. P. Wentling, M.A., Assoc. Prof. and Assoc. For.

Sapporo (Japan).
Inst. of Forestry, Coll. of Agric., Hokkaidô Imper. Univ.

Sofia (Bulgarien).
Labor. auprès de l'Inst. de Sylviculture à la Fac. agronomique de l'Univ. Dir. Th. Dimitroff. — M. D. Rouskoff, Ass. *Pour de recherches sur les semences forestières.* ⊕ Semences d'essences forestières, particulièrement de résineux des forêts bulgares.

Sopron — Ödenburg (Ungarn).
Kön. ung. forstliche Versuchsstation, Förstkola Hochsch. Dir. Julius Roth. — N. Erdödy. *Püspökladány Szik-Versuchsfeld. Gödöllo Arboretum.* ⊕ Forstliche Schädlinge.

Springforbi (Danmark).
Das forstliche Versuchswesen in Dänemark (Statens forstlige Forsøgsvœsen). Dir. Prof. Dr. h. c. A. Oppermann. — C. H. Bornebusch, Laborant. 2 Forst-Ass., 2—3 Forstassessoren.

Stockholm (Sverige).
Landwirtsch. Inst. der Forsthochsch. Dir. Doc. J. Nathorst.
Inst. f. Waldbau Forsthochsch. Dir. Prof. A. Wahlgren.
Forstliche Versuchsanstalt. Experimentalfältet. Dir. O. A. H. W. Hesselman.

Storrs, Conn. (U.S.A.).
Dept. of Forestry Coll. of Agric. A. E. Moss, M.F., Ass. Prof. For.; A. A. Doppel, M.S., Instr. and For. Specialist.

Sverdlowsk (U.d.S.S.R.).
Inst. f. Forstkunde Ural. Polytechn. Inst. Dir. Doc. Demidoff.

Syracuse, N.Y. (U.S.A.).
Roosevelt Wild Life Forest Experim. Station (N.Y. State Coll. of Forestry). Dir. C. E. Johnson.
State Coll. of Forestry, New York. H. P. Brown (Wood technol.).

Taihoku, Formosa (Japan).
Dept. of Forestry Government Central Research Inst.

Tartu — Dorpat (Eesti).
Kab. für Waldbau, Aia t. 46. Dir. Dr. O. Daniel. — V. Matiisen, Ass.
Universitätsforst, Kastre-Peravallas. Dir. K. Werberg. — Dr. O. Daniel; R. Rüsberg; L. Puksov; K. J. Kull.

Taschkent (U.d.S.S.R.).
Inst. f. Forstwirtschaft Univ. Dir. Prof. A. G. Fedorov.

Tharandt, Sa. (Deutschland).
Lehrforst d. Forstl. Hochsch. Dir. Prof. Gross.

Acad. Forstgarten Forstl. Hochsch. Dir. Prof. E. Münch.

Tiflis (U.d.S.S.R.).
Inst. f. Forstdendrol., Techn. Hochsch. Dir. Prof. P. Z. Vinogradov-Nikitin.
Forstmus. Univ. Dir. Prof. S. Kurdiani.

Tôkyô (Japan).
Inst. of Forestry Fac. of Agric., Tôkyô Imp. Univ., Komaba near Tôkyô. Prof. Seiroku Honda; Prof. Shitarô Kawai; Prof. Hanshirô Migita; Prof. Kitarô Moroto; Ass. Prof. Ichirô Sonobe; Ass. Prof. Ihachiro Miura; Ass. Prof. Toshio Maki; Ass. Prof. Masao Yoshida.
Labor. of Forestry Imperial Forestry Experim. Sta., Meguro near Tôkyô. Dir. Dr. Hanshirô Migita. — Dr. H. Migita; Noritake Takashima.

Vancouver (British-Columbia).
Dept. of Forestry of the Univ. of British Columbia. Dir. H. R. Christie, B.Sc.F. — F. M. Knapp, B.S.F., M.S.F.
Vancouver labor. (Forest products labor.) in connection with the Univ. of British Columbia. Dir. T. A. McElhanney. — 2 timber tests engineers, 2 timber testers, 1 forestry Ass.

Victoria (British-Columbia).
Forest Service. Forest Experim. Sta. P. M. Bare (Ass. Forester). *Forest Research work in the Northen Interior of Britisch Columbia.*

Wageningen (Holland).
Reichs-Forst-Inst. (Rijksboschbouwproefstation), Rijksstraatweg 3. Dir. E. Hesselink. — H. van Vloten, Boschbouwkundige.

Warszawa (Polska).
Inst. f. Forstwirtsch. Hochsch. f. Land- u. Forstwirtsch. Dir. Prof. W. Jedliński.

Washington (U.S.A.).
Forest Service, U.S. Dept. of Agric. Dir. W. B. Greeley, Chief.

Wien (Österreich).
Samml. f. Forstl. Produktionslehre Hochsch. f. Bodenkultur. Dir. Prof. Dr. A. Cieslar.
Labor. f. Waldbau mit Lehrmittelsammlung, Hochschule f. Bodenkultur. Dir. Prof. Cieslar.
Samml. f. Forstzool. u. Forstschutz, Hochsch. f. Bodenkultur. Dir. Prof. Seitner.

Wooster, Ohio (U.S.A.).
Dept. of Forestry, Ohio State Agric. Experim. Station. Dir. of Sta. (State For.): Edmund Secrest, B.S., Chief For. and Assoc. — O. A. Alderman, M.F., Ass. For.; J. J. Crumley, Ph.D., Assoc. For. (P.O., Athens); F. W. Dean, B.S., Ass. For.; B. E. Leete, M.F., Ass. For. (P.O., Portsmouth); L. J. Leffelman, M. F., Ass. For.

Woronesch (U.d.S.S.R.).
Kabinett f. Spez. Forstkultur Landwirtsch. Hochschule.
Labor. f. Forsttechnol. Landwirtsch. Hochsch. Dir. Prof. N. A. Rozanov. *Forstwirtschaftl. chem. Technol.*
Kabinett f. Dendrol. Landwirtsch. Hochsch. Dir. Doc. O. G. Kapper.

Zagreb — Agram (S.H.S.).
Inst. f. Forstnutzung. Dir. Prof. Ugrenowič.
Inst. f. Waldbau Univ. Prof. Dir. Petračić.
Forstl. Versuchsanstalt. Dir. Prof. Lewakowič.

Institutiones horticulturae.

Confer: *Cultura plantarum generalis* (pag. 499), *Agricultura generalis* (pag. 507), *Phytopathologia* (pag. 415).

Aas b. Oslo (Norge).
Labor. für Gartenbau. Norges Landwirtschaftl. Hochsch. Prof. Knut Vik; Prof. Hans Misvœr; Prof. Olav Moen. *Pflanzenkultur. Gartenbau. Fruchtgarten. Gemüse.*

Acireale (Italia).
R. Staz. sperim. di agron. e frutticoltura.

Ames, Ia. (U.S.A.).
Dept. of Horticulture and Forestry, State Coll. of Agric. B. S. Pickett, M.S., Head of Dept., Chief in Hort. and For. — G. B. MacDonald, M.F., Prof. and Chief in For.; P. H. Elwood, jr., B.S., Prof. and Chief in Landsc. Arch.; A. T. Erwin, M.S., Chief in Veg. Crops; T. J. Maney, B.S., Chief in Pomol.; I. T. Bode, M.S., Ext. Assoc. Prof. For.; Perkins Coville, M.F., Instr. For.; J. C. Cunningham, B.S., Prof. Hort. and Botan., Supvr. of Noncollegiate Agr.; C. H. Diggs, Ext. Assoc. Prof. Landsc. Arch.; C. L. Fitch, M.A., Ext. Prof. Veg. Crops; S. E. Haber, M.S., Ass. Chief in Veg. Crops; C. V. Holsinger, B.S., Ext. Prof. Hort.; D. S. Jeffers, M.F., Assoc. Prof. For.; H. F. Kenney, B.S., L.A., Instr. Landsc. Arch.; H. L. Lantz, M.S., Ass. Chief in Pomol.; J. A. Larsen. M.F., Ass. Prof. For.; H. E. Nichols, B.S., Ext. Assoc. Prof. Hort.; H. H. Plagge, M.S., Ass. Pomol.; H. W. Richey, B.S., Prof. Pomol.; R. R. Rothacker, M.S., Ass. Prof. Landsc. Arch.; J. C. Schilletter, M.S., Instr. Hort.; Ora Smith, M.S., Instr. Veg. Crops; E. C. Volz, M.S., Assoc. Prof. Flor. and Veg.; Crops; W. B. Ward, B.S., Ass. Prof. Hort.

Amherst, Mass. (U.S.A.).
Dept. of Horticulture, State Coll. of Agric. F. A. Waugh, M.S., Head Div. of Hort. and of Dept. of Landsc. Gard. — W.W. Chenoweth, M.S., Head Dept. of Hort. Mfrs.; F. C. Sears, M.S., Head Dept. of Pomol.; J. K. Shaw, Ph.D., Research Prof. Pomol.; C. L. Thayer, B.S., Head Dept. of Flor.; L. B. Arrington, B.S., Instr. Hort.; J. S. Bailey, M.S., Invest. in Pomol.; W. R. Cole, Ext. Ass. Prof. Hort. Mfrs.; W. L. Cutler, Labor. Ass. in Pomol.; L. S. Dickinson, B.S., Ass. Prof. Hort.; B. D. Drain, S.M., Ass. Prof. Pomol.; C. R. Fellers, Ph.D., Research Prof. Hort. Mfrs.; A. P. French, M.S., Instr. Pomol.; A. K. Harrison, Ass. Prof. Landsc. Gard.; R. M. Koon, M.S., Ext. Prof. Veg. Gard. (P.O., Waltham); R. T. Muller, M.S., Ass. Prof. Flor.; L. R. Quinlan, M.L.A., Ass. Prof. Landsc. Gard.; G. J. Raleigh, M.S., Instr. Pomol.; W. F. Robertson, B.S., Instr. Hort. Mfrs.; G. B. Snyder, B.S., Instr. Veg. Gard.; W. H. Thies, M.S., Ext. Ass. Prof. Pomol.; C. H. Thompson, M.S., Prof. Hort.; V. A. Tiedjens, M.S., Ass. Research Prof. Veg. Gard. (P.O., Waltham); R. A. Van Meter, B.S., Prof. Pomol.

Athens, Ga. (U.S.A.).
Dept. of Horticulture, State Coll. of Agric. T. H. McHatton, D.Sc., M.H., Prof. Hort.; G. H. Firor, B.S., Field Agt. in Hort.; H. W. Harvey, B.S., Landsc. Gard. Specialist; R. L. Keener, B.S., Assoc Prof. Hort.; H. M. McKay, M.S., Field Agt. in Hort.

Auburn, Ala. (U.S.A.).
Dept. of Horticulture, Coll. of Agric. Dir. C.L.Isbell, M.S., Actg. Head of Dept. — Otto Brown, M.S., For. Specialist; S. H. Gibbons, B.S., Hort. Specialist; D. W. Kimbrough, Ph.D., Ass. Hort.; R. W. Taylor, B.S., Ass. in Hort.; L. M. Ware, M.S., Ass. Prof. Hort.

Berkeley, Cal. (U.S.A.).
Dept. of Horticulture, Coll. of Agric. H. J.Webber, Ph.D., Head Div. of Subtrop. Hort.; F. T. Bioletti, M.S., Head Div. of Vit. and Fruit Products; J. W. Gregg, B.S., Head Div. of Landsc. Design.; W. L. Howard, Ph.D., Head Div. of Pomol. (P.O., Davis); H.:A. Jones, Ph.D., Head Div. of Truck Crops; Plant Breeder (P.O., Davis); L. D. Batchelor, Ph.D., Prof. Orchard Mgt.; Hort. (P.O., Riverside); W. H. Chandler, Ph.D., Prof. Pomol.; F. W. Allen, M.S., Ass. Pomol.; J. P. Bennett, Ph.D., Ass. Prof. and Assoc. Pomol.; L. O. Bonnet, I.A., Assoc. in Vit. (P.O., Davis); S. H. Cameron, M.S., Research Ass. in Subtrop. Hort.; A. W. Christie, M.S., Ass. Prof. Fruit Prod., Assoc. Chem.; I. J. Condit, B.S., Assoc. in Subtrop. Hort.; W. V. Cruess, B.S., Assoc. Prof. Fruit Prod., Chem.; L. H. Day, B.S., Research Ass. in Hort. (P.O., Davis); W. P. Duruz, M.S., Ass. Pomol. (P.O., Davis); H. B. Frost, Ph.D., Ass. Plant Breeder (P.O., Riverside); A. H. Hendrickson, M.S., Assoc. Pomol. (P.O., Davis); M. J. Heppner, M.S., Research Ass. in Pomol. (P.O., Davis); R. W. Hodgson, M.S., Assoc. Prof. Subtrop. Hort., Assoc. Citriculturist; W. B. Hooper, B.S., Specialist in Walnut Culture (P.O., Riverside); J. H. Irish, B.S., Junior Chem. in Hort.; H. E. Jacob, M.S., Junior Vit. (P.O., Davis); Katherine D. Jones, B.S., Instr. Landsc. Design.; E. L. Overholser, M.A., Ass. Prof. and Assoc. Pomol.; E. R. Parker, M.S., Research Ass. in Orchard Mgt. (P.O., Riverside); G. L. Philp, M.S., Ass. Pomol. (P.O., Davis); E. L. Proebsting, Ph.D., Instr. and Junior Pomol. (P.O., Davis); J. T. Rosa, jr., Ph.D., Ass. Prof. Truck Crops, Assoc. Plant Breeeder (P.O., Davis); Edna M. Russ, B.S., Ass. in Pomol. (P.O., Davis); W. R. Schoonover, M.S., Citriculture Specialist (P.O., Riverside); H. W. Shepherd, B.S., Ass. Prof. Landsc. Design.; J. L. Stahl, B.S., Assoc. in Pomol. (P.O., Davis); J. G. Surr, Field Ass. (P.O., Riverside); W. P. Tufts, M.S., Ass. Prof. and Assoc. Pomol. (P.O., Davis); A. J. Winkler, Ph.D., Ass. Viticulturist. (P.O., Davis).

Berlin (Deutschland).
Lehr- u. Forschungsanst. f. Gartenbau, -Dahlem, Königin-Luise-Str. 22. Dir. Prof. Echtermeyer.

Blacksburg, Va. (U.S.A.).
Dept. of Horticulture, Agric. Coll. Dean Price in charge; T. C. Johnson, A.M., Prof. Veg. Gard. (P.O., Norfolk); Kent Apperson, Ass. Hort.; L. C. Beamer, Ass. Specialist in Veg. Gard.; L. B. Deatrick, B.S., Ass. Prof. and Specialist in Veg. Gard.; Mrs. Mary C. McBryde, Ass. Specialist in Landsc. Design.; R. C. Moore, B.S., Ass. Hort.; F. A. Motz, B.S., Assoc. Prof. Pomol., Field Hort.; C. B. Price, Instr. Hort.; D. A. Tucker, B.S., Instr. and Ass. Hort.

Bologna (Italia).
Labor. di Arboricoltura ed Orticoltura, Ist. Sup. agron., Via Filippo Re, 4. Dir. Prof. Angelo Manaresi. *Environ 30 places pour tous les recherches d'Arboriculture et d'Horticulture. — Rasoirs, caisses de végétation, appareils pour l'analyse chimique.*

Boston, Mass. (U.S.A.).
Agric. Research Labor. of the United Fruit Compagny. Dir. J. R. Johnston. — O. A. Reinking (Plant Pathologist).

Brookings, S.D. (U.S.A.).
Dept. of Horticulture, State Coll. of Agric. N. E. Hansen, Sc.D., Prof. Hort. and For., Hort. and Vice Dir. of Sta.; A. L. Ford, M.S., Hort. and Entomol. Specialist; P. L. Keene, M.S., Ass. Prof. Hort.

Clemson College, S.C. (U.S.A.).
Dept. of Horticulture, Clemson Agric. Coll. C. C. Newman, B.S., Prof. Hort.; G. P. Hoffmann, M.S., Assoc. Prof. Hort.; A. M. Musser, B.S., Assoc. Hort.; E. H. Rawl, M.S., Ass. Hort. (P.O., Aiken); A. E. Schilletter, B. S., Ext. Hort.

College Park, Md. (U.S.A.).
Dept. of Horticulture, Coll. of Agric. E. C. Auchter, Ph.D., Prof. Hort., Hort.; W. R. Ballard, B.S., Landsc. and Veg. Gard. Specialist; J. B. Blandford, Instr. and Hort. Supt.; V. R. Boswell, M.S., Instr. and Ass. Hort.; F. W. Geise, M.S., Prof. Oler., Veg. Brdg.; L. M. Goodwin, Canning Crops Specialist; A. L. Schrader, Ph.D., Ass. Pomol.; A. S. Thurston, M.S., Prof. Flor. and Landsc. Gard.; A. F. Vierheller, M.S., Hort. Specialist; T. H. White, M.S., Veg. and Flor. Work; W. E. Whitehouse, M.S., Ass. Prof. Pomol.

College Station, Tex. (U.S.A.).
Dept. of Horticulture, Agric. Coll. W. B. Lanham, M.S., Chief of Div.; E. J. Kyle, M.S., Head Prof. Hort.; G. W. Adriance, M.S., Assoc. Prof. Hort.

Columbia, Mo. (U.S.A.).
Dept. of Horticulture, Coll. of Agric. Dir. T. J. Talbert A. M., Prof. and Chair of Dept. — E. A. Bierbaum, B.S., Hort. Specialist; H. D. Hooker, jr., Ph.D., Assoc. Prof. Hort.; H. F. Major, B.S., Assoc. Prof. Landsc. Gard. and Supt. of Grounde; A. E. Murneek, Ph.D., Ass. Prof. Hort.; E. M. Page, B.S., Ext. Instr. Prof. Hort.; J. T. Quinn, A.M., Ass. Prof. Hort.; H. G. Swartwout, M.A., Ass. Prof. Hort.

Columbus, Ohio (U.S.A.).
Dept. of Horticulture, State Coll. of Agric. Dir. Wendell Paddock, M.S., Head of Dept. — F. H. Beach, B.S., Hort. Specialist; F. G. Charles, B.S., Instr. Hort.; F. W. Dean, B.S., For. Specialist; C. S. Holland, B.S., Hort. Specialist; A. C. Hottes, M.S., Prof. Flor.; Elusina Lazenby, B.A., Instr. Landsc. Arch.; G. L. Lynch, M.L.A., Ass. Prof. Landsc. Arch.; L. M. Montgomery, M.S., Prof. Hort.; John Morrison, Ass. Hort.; N. W. Scherer, B.S., Ass. Prof. For.; W. R. Sears, B.S., M.L.A., Prof. Landsc. Arch.; A. D. Taylor, M.S., Prof. Landsc. Arch. (nonresident); E. B. Tussing, B.S., Veg. Gard. Specialist.

Corvallis, Ore. (U.S.A.).
Dept. of Horticulture, State Agric. Coll. Dir. W. S. Brown, M.S., Prof. and Hort. — J. C. Bell, M.S., Instr. Hort. Products; A. G. B. Bouquet, B.S., Prof. Veg. Gard. (Veg. Gard.); Oliver Ham, Ass. in Veg. Gard. Henry Hartman, M.S., Assoc. Prof. and Assoc. Hort. (Pomol.); E. M. Harvey, Ph.D., Prof. and Research Hort. (Physiol.); C. L. Long, M.S., Hort. Specialist; A. L. Peck, B.S., Prof. Landsc. Gard. and Flor.; C. E. Schuster, M.S., Assoc. Prof. and Assoc. Hort. (Pomol.).

Durham, N. H. (U.S.A.).
Dept. of Horticulture, Coll. of Agric. G. F. Potter, M.S., Prof. Hort., Hort.; J. R. Hepler, M.S., Ass. Prof. and Ass. Hort.; James Macfarlane, Instr. Flor., Florist; H. A. Rollins, M.S., Ext. Hort.; S. W. Wentworth, B.S., Ass. Prof. and Ass. Hort.

Fayetteville, Ark. (U.S.A.).
Dept. of Horticulture, Coll. of Agric. Dir. J. R. Cooper, M.S., Head of Dept. — W. E. Loomis, Ph.D., Assoc. Prof. and Ass. Hort.; C. B. Wiggans, M.S., Instr. and Ass. Hort.; Claude Woolsey, M.S., Ext. Hort.

Fort Collins, Col. (U.S.A.).
Dept. of Horticulture, State Coll. of Agric. E. P. Sandsten, Ph.D., Prof. Hort., Hort. (State Hort.); F. M. Green, B.S., Ass. Hort. (Chief Deputy State Hort.: P.O., Delta); R. A. McGinty, A.M., Assoc. Prof. and Assoc. Hort.; Edgar Tubby, Florist.

Gainesville, Fla. (U.S.A.).
Dept. Horticulture and Botan., Coll. of Agric. W. L. Floyd, M.S., Prof. Hort. and Botan., Ass. Dean Coll. of Agr.; C. E. Abbott, B.S., Instr. Hort.; G.

H. Blackmon, B.S., Pecan Culturist; M. D. Cody, M.A., Prof. Botan. and Bact.; A. M. Funnell, B.S., Foreman Greenhouse and Grounds; E. L. Lord, B.S., Ass. Prof. Hort. and Botan.; Harold Mowry, Ass. Hort.; J. S. Rogers, A.M., Prof. Biol. and Geol. (Agric.).

Geisenheim a. Rh. (Deutschland).
Labor. u. Forschungsanst. f. Wein-, Obst- u. Gartenbau. Dir. Prof. Fr. Muth. — Prof. G. Lüstner (Pflanzenpathol.); Prof. K. Kroemer (Pflanzenphysiol.); Prof. K. v. d. Heide (Weinchemie). — Pflanzenpathol. Vers.-Anst. Pflanzenphysiol. Vers.-Anst. Weinchemische Vers.-Anst. Hefereinzuchtstation.

Geneva, N.Y. (U.S.A.).
Div. of Horticulture of the New York Agric. Experim. Stat. Ulysses P. Hedrick, Sc.D., Vice-Dir., Chief in Research (Horticulture); Assoc. in Research: Fred E. Gladwin, B.S., George H. Howe, B.S., Richard Wellington, M.S., Frank H. Hall, B.S., Harold B. Tukey, M.S., Charles B. Sayre, M.S., Alwin Berger, Ph.D., O. M. Taylor; Ass. in Research (Horticulture): George L. Slate, M.S., Olav Einset, B. Agr., Leslie R. Hawthorn, B.S., Lewis M. Van Alstyne, B.S. *Tests of varieties. Breeding experim. Fertilizer experim. with fruits. Pruning experim. Propagation experim. Strains of the Baldwin apple Stock experim. A study of sex in fruits. Work with vegetables.*

Grand Rapids, Mich. (U.S.A.).
Graham Horticultural Experim. Sta., W. Bridge Road. Dir. H. M. Wells. — 2 Ass.

Ithaca, N.Y. (U.S.A.).
Dept. of Horticulture, Coll. of Agric, Cornell Univ. E. A. White, B.S., Prof. and Head Dept. of Flor. and Ornamental Hort.; H. C. Thompson, M.S., Prof. and Head Dept. of Veg. Gard.; A. J. Heinicke, Ph.D., Prof. and Head Dept. of Pomol.; R. M. Adams, B.S., Ass. Ext. Prof. Veg. Gard.; A. C. Beal, Ph.D., Prof. Flor.; D. B. Carrick, Ph.D., Prof. and Research Pomol.; R. W. Curtis, M.S., Prof. Ornamental Hort.; E. G. Davis, B.S., Prof. Landsc. Arch.; E. V. Hardenburg, Ph.D., Ass. Prof. Veg. Gard.; L. H. MacDaniels, Ph.D., Prof. Pomol.; H. S. Mills, M.S., Instr. Veg. Gard.; Lua A. Minns, M.S., Instr. Flor.; A. H. Nehrling, Prof. Flor.; Joseph Oskamp, B.S., Ext. Prof. Pomol.; G. W. Peck, M.S., Ext. Prof. Pomol.; J. P. Porter, M. S., M.L.D., Ass. Ext. Prof. Ornamental Hort.; H. W. Schneck, M.S., Ass. Prof. Veg. Gard.; F. O. Underwood, B.S., Ass. Ext. Prof. Veg. Gard.; H. Wessels, M.S., Research Prof. Veg. Gard. (P.O., Riverhead); Paul Work, Ph.D., Prof. Veg. Gard.; A. M. S. Pridham, B.S., Instr. Flor.; Paul Yashin, Ass. Hort.; J. E. Knott, M.S., Instr. Veg. Gard.

Kamenetz-Podolsk, Ukraine (U.d.S.S.R.).
Labor. für Gartenbaulehre d. Inst. f. Landw. „K. Marx". Dir. Prof. A. Volostschuk.

Knoxville, Tenn. (U.S.A.).
Dept. of Horticulture, Coll. of Agric. J. A. McClintock, M.S., Hort., Assoc. Plant Pathol.; N. D. Peacock, M.S., Assoc. Prof. Hort.; Otha Ogle, Foreman of Fruit Farm; H. F. Helfenbein, B.S., Instr. Hort.; W. C. Pelton, M.S., Hort. Specialist; G. B. Shivery, For. Specialist.

Lafayette, Ind. (U.S.A.).
Dept. of Horticulture of the School of Agric., Purdue Univ. Laurenz Greene, M.S., in charge; O. G. Anderson, B.S., Prof. Hort.; C. E. Baker, B.S., Ass. in Pomol.; H. D. Brown, M.S., Assoc. Prof. and Research Assoc. in Veg. Gard.; C. L. Burkholder, B.S., Assoc. in Hort.; F. P. Cullinan, B.S., Research Assoc. in Pomol.; L. D. Davis, B.S., Ass. in Hort.; F. C. Gaylord, B.S., Assoc. in Hort.; I. C. Hoffman, M.S., Ass. in Veg. Gard.; E. R. Lancashire, M.S., Ass. in Hort.; W. E. Lommel, B.S., Assoc. Prof. Hort.; J. H. MacGillivray, Ph.D., Research Ass. in Hort.; V. H.Ries, M.S., Ass. Prof. and Ass. in Flor., E. C. Stair, M.S., Instr. Hort.; A. H. Watson, B.S.; Research Ass. in Hort.

Lansing, Mich. (U.S.A.).
Dept. of Horticulture, State Coll. of Agric. Dir. V. R. Gardner, M.S., Head of Dept.; — O. P. Halligan, B.S., Head of Landsc. Gard.; F. C. Bradford, M.S., Assoc. Prof. and Research Assoc. Hort.; H. A. Cardinell, B.S., Hort. Specialist; J. W. Crist, Ph.D., Ass. Prof. and Research Ass. in Hort.; W. C. Dutton, B.S., Research Assoc. in Hort.; J. B. Edmond, M.S., Instr. Hort.; H. D. Hootman, Hort. Specialist; Alex Laurie, M.A., Instr. and Ass. in Hort.; R. E. Loree, M.S., Ass. Prof. and Ass. in Hort.; R. E. Marshall, M.S., Assoc. Prof. and Research Assoc. Hort.; Harvard Norton, M.L.A., Instr. Landsc. Arch.; N. L. Partridge, Ph.D., Ass. Prof. and Research Ass. in Pomol.; G. E. Starr, Research Assoc. Hort.

Lednice — Eisgrub (C.S.R.).
Fürst Liechtenstein Pflanzenzüchtungs-Inst. (Mendel-Inst.). Dir. Hofrat W. Lauche. — Leiter Dr. Franz Frimmel. *Arbeitsplätze für wiss. u. prakt. Arbeit auf dem Gebiete gärtnerischer Pflanzenzüchtung.* ⊕ Neuzüchtungen und Sorten von gärtnerischen Blumenrassen, Gemüserassen, Obstrassen.

Leipzig (Deutschland).
Abt. f. Gartenbau d. Univ. Lektor: H. Grabbe.

Leningrad (U.d.S.S.R.).
Inst. für Gartenwesen des Leningrader Landwirtsch. Inst.

Lexington, Ky. (U.S.A.).
Dept. of Horticulture, Coll. of Agric., Univ. of Kentucky. Dir. C. W. Mathews, B.S., Head of Dept. — J. S. Gardner, M.A., Field Agt. in Hort.; W. W. Magill, B.S., Field Agt. in Hort.; A. J. Olney, M.H., Prof. and Ass. Hort.; C. S. Waltman, B.S., Instr. and Ass. Hort.

Lima (Peru).
Labor. de Arboricultura, Selvicultura y Horticultura. Confer: Labor. Silviculturae.

Lincoln, Neb. (U.S.A.).
Dept. of Horticulture, Coll. of Agric. of the Univ. of Nebraska. Dir. C. C. Wiggans, Ph.D., Chair. of Dept. — F. M. Coe, M.S., Instr. and Ass. Hort.; E. H. Hoppert, B.S., State Ext. Agt. in Hort.; H. O. Werner, M. S., Assoc. Prof. and Assoc. Hort.

Logan, Utah (U.S.A.).
Dept. of Horticulture, Agric. Coll. of Utah. T. H. Abell, M.S., Ass. Prof. in charge. — Emil Hansen, Instr. and Landsc. Gard. Specialist.

London (England).
John Innes Horticultural Inst., S.W. 19, Mostyn Rd., Merton. Dir. Sir Daniel Hall, K.C.B., Ll.D., F.R.S. — Miss A. Andersson; Miss D. Cayley; R. J. Chittenden; Dr. E. J. Collins; M. V. Crane; C. D. Darlington; Miss D. Delvinton; Miss A. E. Gairdner; C. L. Huskins; W. C. F. Newton; Miss C. Pellew; W. Lawrence; H. D. Bennett. *Genetics and Cytol.*

Long Ashton near Bristol (England).
Agric. and Horticult. Research Sta. Confer: Microbiol. agris.

Louisiana, Miss. (U.S.A.).
Research Dept., Stark Bro's Nurseries. Dir. J. T. Bregger. *Fruit varieties; propagation; Scion-rooting. Diseases of fruits. Propagation of Citrus fruits (California) and compar. value of stocks (citrus).*

Madison, Wis. (U.S.A.).
Horticultural Dept., Univ. of Wisconsin. Dir. James G. Moore. — H. L. Russell, Dean of Coll.; J. G.

Milward, Prof.; F. A. Aust, Assoc. Prof.; J. W. **Brann**, Ass. Prof.; D. Bradbury, Ass.; R. H. **Roberts**, Pomologist; J. Johnson, Tobacco culture; V. Watte, Olericulture. *Facilties for from 6 to 10 guests for chem., botan., pomol. workers. — Gardens and orchards and greenhouses in addition to usual, table facilities, with necessary microscopical and other equipment. — This in itself is apost of the Univ. of Wisconsin and other labor. In the Univ. are available for use of students in Pomol. and Olericulture.*

Manhattan, Kan. (U. S. A.).
Dept. of Horticulture, State Coll. of Agric. Dir. Albert Dickens, M.S., Head of Dept. — R. J. Barnett, M.S., Prof. Hort., Pomol.; W. B. Balch, M.S., Ass. Prof. Hort., Greenhouse Foreman; A. H. Helder, M.S., M.L.A., Ass. Prof. in charge Landsc.Gard,; E. M. Litwiller, B.S., Ass. Prof. Hort., Home Study Service; W. R. Martin, jr., B.S., Hort. Specialist; W. F. Pickett, M.S., Ass. Prof. Hort., Orchard Invest.; L. C. Williams, B.S., Hort. Specialist.

Morgantown, W.Va. (U. S. A.).
Dept. of Horticulture, Coll. of Agric. Dir. Ernest Angelo, M.S., Instr. and Junior Hort. — Dee Crane, Potato Specialist; H. L. Crane, M.S., Assoc. Prof. and Assoc. Hort.; T. M. Currence, B.S., Instr. and Junior Hort.; T. D. Gray, B.S., Hort. Specialist; H. K. Knowlton, Ph.D., Assoc. Prof. and Assoc. Hort.; H. W. Prettyman, B.S., Supt. Packing School (P. O., Inwood); K. C. Westover, M.S., Ass. Prof. and Ass. Hort.

Moscow, Id. (U. S. A.).
Dept. of Horticulture Coll, of Agric. Dir. C. C. Vincent, M.S., Prof. Hort., Hort. — E. R. Bennett, M.H., Field Hort. (P. O., Boise); L. E. Longley, M.S., Assoc. Prof. and Assoc. Hort.; C. V. Schrack, B.S., Gardener.

Newark, Del. (U. S. A.).
Dir. Horticultural Dept. of the Agric. Experim. Sta. Dean McCue, Head of Dept. — L. R. Detjen, M.S., Assoc. Prof. and Assoc. Hort.; G. F. Gray, M.S., Instr. and Ass. Hort.; F. S. Lagassé, M.S., Ass. Hort. *Study of the Physiol. Drop of Fruits in Delaware. 1. Physiol. dropping of fruits in relation to pollination, fertilization, embryo development and embryo abortion. 2. Physiol. dropping of fruits in relation to genetic relationship of plants. 3. Physiol. dropping of fruits in relation to environment differences.*

New Brunswick, N. J. (U. S. A.).
Dept. of Horticulture, State Agric. Experim. Sta. Dr. M. A. Blake, B.S., Hort. — H. G. Bailey, Veg. Gard. Foreman; H. M. Biekart, B.S., Florist.; J. H. Clark, M.S., Ass. Pomol.; C. H. Connors, B.S., Assoc. in Plant Brdg.; A. J. Farley, B.S., Pomol.; G. T. Nightingale, Ph.D., Biochem. in Hort.; L. G. Schermerhorn, B.S., Oler.; C. H. Steelman, Orchard Foreman.
Dept. of Horticulture, Coll. of Agric. Dir. M. A. Blake, B.S., Hort. — H. M. Biekart, B.S., Florist; J. H. Clark, M.S., Ass. Pomol.; C. H. Connors, B.S., Assoc. in Plant Brdg.; A. J. Farley, B.S., Pomol.; H. F. Huber, B.S., Ass. Olericulturist; Nico Mogendorff, B.S., Research Ass. Pomol.; T. C. Rogers, B.S., Research Ass. Pomol.; L. G. Schermerhorn, B.C., Olericulturist.

Norfolk, Va. (U. S. A.).
Virginia Truck Experim. Sta. Dir. T. C. Johnson, Horticulturist. — H. H. Zimmerley, Horticulture and Plant Breeding; Frank P. McWhorter, Plant Pathol.; F. W. Poos, Insects affecting truck crops; Harold S. Peters, Insects affecting potatoes.

Okitsu (Japan).
Imper. Horticultural Experim. Station, Shizuoka Ken. Dir. Dr. Agr. Hirotarô Andô — Keizô Nagai (Physiol. and Genetics); Toshiyoshi Tanikawa (Fruit Breeding and Variety Test); Tetsuji Ishii (Fruit Propagation); Kazuyori Inoue (Fruit Growing); Jôji Kishima (Vegetable Growing). *Places for Kaki and Unshiu growing.*

Paris (France).
Station de recherches d'horticulture de l'Inst. des recherches agron., VIIe, Rue de Bourgogne.

Placerville, Cal. (U. S. A.).
Eddy Tree Breeding Station. Dir. Lloyd Austin. ⊕ Pinus ponderosa, lambertiana jeffreiji, monticola. Pseudotsuga latifolio. Abies concolor, magnifica. Sequoia sempervirens, gigantea.

Pretoria (South Africa).
Inst. of Horticulture. Dir. Prof. H. Clark Powell.

Puławy (Polska).
Abt. f. Gartenbau des Staatl. wissensch. Inst. f. Landwirtsch. Dir. J. Dybowski.

Pullman, Wash. (U. S. A.).
Dept. of Horticulture, State Coll. Dir. O. M. Morris, M.S., Head of Dept. — M. D. Armstrong, B.S., Hort. Specialist; W. A. Luce, B.S., Ass. Hort. (P. O. Wenatchee); F. L. Overley, M.S., Ass. Prof. Hort.; C. L. Vincent, M.S., Ass. Prof. Hort.

Raleigh, N. C. (U. S. A.).
Dept. of Horticulture, State Coll. of Agric. Dir. C. D. Matthews, B.S., Head of Dept. — E. B. Morrow, M.S., Hort. Specialist; H. R. Niswonger, M.S., Hort. Specialist (P. O. Asheville); J. P. Pillsbury, B.S.. Prof. Hort.; W. A. Radspinner, M.S., Ass. Research Pomol.; G. O. Randall, M.S., Ass. Prof. and Ass. Hort.; Robert Schmidt, B.S., Ass. Hort.; C. F. Williams, M.S., Ass. Hort.

Riga (Latvija).
Abt. f. Obst- u. Gemüsebau. Dir. Lekt. J. Sudrabs.

Riverhead, N. Y. (U. S. A.).
Long Island Vegetable Research Farm. H. C. Huckett, Ass.

Sacramento, Cal. (U. S. A.).
Dept. of Horticulture, Agric. Dept. State of Cal. C. E. Scott, Plant pathologist.

St. Louis, Mo. (U. S. A.).
Federal Horticultural Board, U. S. Dept. of Agric. A. C. Hill, Inspector.

St. Paul, Minn. (U. S. A.).
Section of Vegetable Gardening, Univ. Farm. B. I. Burrell, Ass. Horticulturist.
Dept. of Horticulture, Dept. of Agric., Univ. Farm. Chief: W. H. Alderman, B.S. — J. H. Beaumont, Ph.D., Ass. Prof. and Ass. Hort.; W. G. Brierley, M.S., Assoc. Prof. and Assoc. Hort.; B. I. Burrell, B.S., Instr. and Ass. Hort.; C. E. Cary, B.S., Ass. Prof. Hort., in charge Landsc. Gard.; A. C. Hildreth, B.S., Instr. and Ass. Hort.; F. A. Krantz, Ph.D., Ass. Prof. Hort., in charge Veg. Gard.; R. S. Mackintosh, M.S., Hort. Specialist; H. P. Traub, M.S., Instr. and Ass. in Hort.; A. N. Wilcox, M.S., Instr. and Ass. Hort.

Saratow (U. d. S. S. R.).
Abt. f. Gartenkunde des Inst. f. Landwirtsch. u. Meliorat. Dir. Doc. V. L. Kusmič.

Sofia (Bulgarien).
Inst. f. Gartenbau.

Sotschi, Schwarzmeer-Gebiet (U. d. S. S. R.).
Provinz-Station f. Landwirtschaft u. Gartenbau.

State College, N. M. (U. S. A.).
Dept. of Horticulture, Coll. of Agric. Dir. Garcia in charge. — A. B. Fite, M.S., Assoc. Prof. and Ass. Hort.; J. W. Rigney, B.S., Ass. Prof. Hort.

State College, Pa. (U.S.A.).
Dept. of Horticulture, School of Agric. Dir. S. W. Fletcher, Ph.D., Head of Dept. — R. D. Anthony, Ph.D., Prof. Pomol.; A. W. Cowell, B.S., Prof. Landsc. Arch.; F. N. Fagan, M.S., Prof. Pomol.; L. M. Marble, B.S., Prof. Farm Storage Research (P.O., Canton); C. Emory Myers, Ph.D., Prof. Plant Brdg.; W. T. Tapley, M.S., Prof. Veg. Gard.; E. I. Wilde, M.S., Prof. Flor.; J. R. Bracken, B.S., Ass. Prof. in charge, Landsc. Gard. Ext.; H. E. Dahl, M.S., Ass. Prof. Landsc. Arch.; R. W. Evans, B.S., Instr. Hort.; F. W. Haller, Ass. in Hort.; M. T. Lewis, M.S., Ass. in Plant Brdg.; W. B. Mack, M.S., Ass. Prof. Veg. Gard.; C. R. Mason, M.S., Ass. Prof. Veg. Gard. Ext.; R.B.Maxwell, M.S., Instr. Storage Research (P.O., Canton); W. B. Nissley, B.S., Prof. Veg. Gard. Ext.; J. U. Ruef, M.S., Ass. in Pomol. Ext.; M. E. Smith, B.S., Ass. in Storage Research (P.O., Canton); R. S. Snyder, B.S., Ass. in Pomol. Ext.; R. H. Sudds, B.S., Ass. in Pomol.; Paul Thayer, M.S., Prof. Pomol. Ext.

Stillwater, Okla. (U.S.A.).
Dept. of Horticulture and Forestry, State Coll. of Agric. Dir. G. W. Cochran, M.S., Prof. Hort., Hort. — F. B. Cross, B.S., Assoc. Prof. and Assoc. Hort.; Christian Jensen, B.F., For. and Landsc. Specialist; E. D. Markwell, B.S., Ass. Prof. Hort. and For.; D. C. Mooring, M.S., Ext. Hort.; D. V. Shuhart, B.S., Ass. Prof. and Ass. Hort.

Storrs, Conn. (U.S.A.).
Dept. of Horticulture, Coll. of Agric. Dir. S. P. Hollister, B.S., Prof. Hort. — W. H. Darrow, M.S., Fruit Specialist; R. H. Patch, M.S., Ass. Prof. Flor.; A. T. Stevens, M.S., Prof. Veg. Gard.; A. E. Wilkinson, M.S., Veg. Gard. Specialist.

Taihoku (Japan).
Governmental Horticultural Experim. Station, Shirin near Taihoku.

Tiflis (U.d.S.S.R.).
Kabinett f. Gartenbau, Techn. Hochsch. Dir. Prof. A. Ch. Rolloff.

Tôkyô (Japan).
Horticult. Inst., Fac. of Agric.,Tôkyô Imper.Univ., Komaba near Tôkyô. Prof. Hiroshi Hara; Prof. Suketeru Yoshikawa; Prof. Hirotarô Andô.

Tucson, Ariz. (U.S.A.).
Dept. of Horticulture, Coll. of Agric. Dir. A. F. Kinnison, B.S., Actg. Head of Dept., Assoc. Prof. Hort., Citriculturist. — D. W. Albert, B.S., Ass. Prof. and Ass. Hort.; M. F. Wharton, M.S., Instr. and Ass. Hort.

Urbana, Ill. (U.S.A.).
Dept. of Horticulture, Coll. of Agric., Univ. of Illinois. Dir. J. C. Blair, Sc.D., Head of Dept., Prof. and Chief in Hort. — H. W. Anderson, Ph.D., Assoc. Prof. and Assoc. Chief in Pomol. Path.; Harland Bartholomew, C.E., Ass. Prof. Civic Design; W. S. Brock, B.S., Ass. Prof. and Ass. Chief in Systematic Pomol.; A. S. Colby, Ph.D., Assoc. Prof. and Assoc. Chief in Pomol.; C. S. Crandall, M.S., Prof. Pomol., Chief in Plant Brdg.; S. W. Decker, B.S., Ass. in Flor.; H. B. Dorner, M.S., Prof. and Chief in Flor.; M. J. Dorsey, Ph. D., Prof. and Chief in Pomol.; C. A. Garner, M.S., Assoc. in Oler.; M. C. Gillis, Ph.D., Research Ass. in Oler.; S. W. Hall, B.S., Ass. Prof. and Ass. Chief in Flor.; W. A. Huelsen, B.S., Assoc. in Oler.; James Hutchinson, Assoc. in Flor.; V. W. Kelley, M.S., Assoc. in Pomol.; L. A. Koritz, B.S., Research Ass. in Oler.; E. P. Lewis, M.S., Assoc. in Oler. (P.O., Des Plaines); J. W. Lloyd, Ph.D., Prof. and Chief in Oler.; K. B. Lohmann, M.L.A., Assoc. Prof. Landsc. Gard.; May E. McAdams, B.S., Assoc. in Landsc. Gard.; R. L. McMunn, B.S., Research Ass. in Pomol.; R. S. Marsh, Ph.D., Ass. Prof. Hort. Ext.; H. M. Newell, B.S., Ass. in Fruit and Veg. Marketing; L. I. Peterson, B.S., Assoc. in Landsc. Gard.; W. A. Ruth, Ph.D., Assoc. Prof. and Assoc. Chief in Pomol. Physiol.; O. G. Schaffer, B.S., Assoc. in Landsc. Gard.; B. L. Weaver, B.S., Ass. in Veg. Gard. Ext.; F. F. Weinard, Ph.D., Assoc. in Flor. Physiol.; S. H. White, M.L.A., Ass. Prof. Landsc. Gard.

Vineland Station, Ont. (Canada).
Ontario Horticultural Experim. Sta. Dir. E: T. Palmer. *Experim. and Plant Breeding*.

Waedenswil b. Zürich (Schweiz).
Eidgenössische Versuchsanst. f. Obst-, Wein- u. Gartenbau. Dir. Dr. K. Meier. *Bekämpfung von Schädlingen und Krankheiten der Obstbäume, Reben- u. Gartengewächse. Kontrollgebiet: Deutsche Schweiz.*

Wageningen (Holland).
Labor. f. Blumenzwiebeluntersuchung d. Inst. f Phytopath., Landbouwhoogeschool. Dir. Prof. E. van Slogteren.

Waltham, Mass. (U.S.A.).
Market garden Field Station Mass. Agric. Coll., 240, Beaver St. Ass. Research Prof. E. F. Guba (Botan.); Ass. Research Prof. W. D. Whitcomb (Entomol.).

Warszawa (Polska).
Inst. f. Gemüsepflanzenkultur, Hochsch. f. Land- u. Forstwirtsch. Dir. Prof. F. Kotowski.

Washington, D.C. (U.S.A.).
Federal Horticultural Board, U.S. Dept. of Agric. Dir. C. L. Marlatt, Chairman. — N. R. Hunt, Pathologist.

Wooster, O. (U.S.A.).
Dept. of Horticulture, Ohio Agric. Experim. Sta. Dir. J. H. Gourley, M.S., Chief.—F.H.Ballou, Assoc. Hort. (P.O., Newark); J. W. Bushnell, M.S., Ass. Hort.; C. W. Ellenwood, Ass. Hort.; Donald Conin, Consult. Hort.; F. S. Howlett, B.S., Ass. Hort.; I. P. Lewis, B.S., Ass. in Hort.; Roy Magruder, B.S., Ass. Hort.; J. S. Shoemaker, B.S., Ass. Hort.

Woronesch (U.d.S.S.R.).
Kabinett f. Garten- u. Gemüsebau, Landwirtsch. Hochsch.

Institutiones praticulturae.

Beaverlodge, Alta. (Canada).
Experim. Sub-Station for the Grande, Prairie District. Dominion Experim. Farms. Dir. W. D. Albright.

Kazan (U.d.S.S.R.).
Abt. f.Wiesenbau der Kazanschen Landw.Provinz.- Versuchsstat. Dir. P. A. Truchin.

Lobnja, Mosk. Gouv. (U.d.S.S.R.).
Staatl. Inst. f. Wiesenbau, Sawel. sh. d.

Morschansk (U.d.S.S.R.).
Selektions-Versuchsstat. f.Wiesenbau „Marussino" Confer: Evolutio et Genetica.

Siverskaya Bologorska (U.d.S.S.R.).
Landwirtschaftl. Versuchsstation d. nordwestlichen Provinzen. Confer: Inst. generales culturae plantarum et animalium.

Tetschen — Děčín (C.S.R.).
Lehrkanzel f. Pflanzenzüchtung, Wiesen- und Hopfenbau. Confer: Inst. gen. cult. plant. et anim.

Tiflis (U.d.S.S.R.).
Kabinett für Wiesenbau des Botan. Gartens.
Vologda (U.d.S.S.R.).
Stat. f. Futterbeschaffung u. Wiesenbau mit Lehr-Labor., Milchwirtschaftl. Inst. Dir. Prof. L.S.Moljakov.

Woronesch (U.d.S.S.R.).
Inst. f. Selektion u. Wiesenbau Landsch. Hochsch. Versuchsstation f. Weizen usw. Confer: Labor. et Inst. agric. spec.

Institutiones culturae plantarum specialium.
1. Viticultura.

Confer: *Biochemia* (pag. 435), *Fermenta* (pag. 443), *Horticultura* (pag. 513), *Phytopathologia* (pag. 415).

Anapa, Nord-Kaukas. Gebiet (U.d.S.S.R.).
Anapskaja Bezirks-Versuchsstation f. Weinbau u. Weinbereitung, Nishneje Dshemete.

Beaune, Côte d'Or (France).
Station Oenol. de Bourgogne (Fac. Sc. Univ. Dijon). *Entomol. agric. Repression des fraudes.*

Berncastel-Cues (Deutschland).
Biol. Reichsanstalt für Land- u. Forstwirtschaft. Zweigstelle. Dir. Dr. H. Zillig. — Ass.: Dr. L. Niemeyer. *Rebenkrankheiten.*

Brno — Brünn (C.S.R.).
Wein- u. Obstbau-Sektion d. Mähr. Landw. Versuchsanstalt (Inst. des Recherches agric. et forest.), Schwarze Felder. Dir. Dr.-Ing. Karel Neoral. — Ing. J. Blaha, Ing. F. Hanzelka, Ing. V. Ondrousek, Ass.; Ing. A. Chrobak, Adj. *1 Arbeitsplatz für Analysen der Wein- u. Obstindustrie, Mykol. d. Weinbereitung u. Haltbarmachung des Obstes.* ⊕ Mikroorganismen d. Weines u. Obstes.

Budapest (Ungarn).
Ampelol. Inst., II, Debröigasse 17. Dir. Desiderius Dicenty. — Mitarb.: Dr. Ladislaus Sántha (Ampelol., Lichenol.); Dr. Josef Andrásovsky (Ampelographie).

Freiburg i. Br. (Deutschland).
Badisches Weinbauinst. mit angeschlossener Hauptstelle für Pflanzenschutz in Baden, Peterhof. Dir. Dr. Karl Müller. — Dr. Gessner, Regierungsbotaniker; Dr. Kotte, Regierungsbotaniker; Dr. Vogt, Chemiker; Dümmler, Weinbauoberinspektor; Röder, Weinbauinspektor; Meinke, Durlach (Baden), Weinbauinspektor. ⊕ Rebpraep., Praep. von erkrankten Pflanzen, z. B. Kartoffeln, Getreide, Gemüse, Obst u. dgl. — Rebenveredlungsanstalt in Durlach (Baden). Rebenzüchtungsanstalt Jesuitenschloß bei Freiburg.

Geisenheim a. Rh. (Deutschland).
Rebenveredelungsstation der Lehr- u. Forschungsanstalt für Wein-, Obst- u. Gartenbau. Wissenschaftl. Abt.: Dir. Prof. Dr. Karl Kroemer. *Physiol. der Rebe. Mykol. der Weinbereitung.*

Guasti, Cal. (U.S.A.).
Labor. of the Italian Vinegard Company. L. O. Bonnet (Viticulturist).

Klosterneuburg (Österreich).
Bundes-Rebenzüchtungsstation a. d. Bundes-Lehr- u. Versuchsanstalt für Wein-, Obst- u. Gartenbau. Dir. Oberinspektor Dr. Zweigelt. — Paul Steingruber, Ass. *Hauptsächl. blütenbiol. Untersuchungen zur Rebblüte.* ⊕ Rebsorten.

Košice — Kaschau (C.S.R.).
Abt. für Wein- u. Obstbau bei der Landw. Versuchsstation. Dir. Ing. A. Fiala. — Ing. J. Jandačová, Ass. *Wein- u. Obstanalyse.* ⊕ Die Reinhefekulturen (Weinhefe), neue Weinrebensorten.

Krasnodar, Kuban (U.d.S.S.R.).
Weinbau-Inst. des Kuban. Landwirtsch. Inst. Dir. Prof. M. F. Stscherbakoff.

Lausanne (Suisse).
Station fédérale d'essais viticoles et arboricoles, Montagibert et Pully. Dir. Dr. H. Faes (Pathol. végét.). — P. Tonduz, G. Piguet, M. Staehelin et P.Castan, Ass. *Division de physiol. et pathol. végétale. Division de chim. et bact.* ⊕ Insectes, champignons, bact. et levures (bact. et levures sont échangées).

Lima (Peru).
Sección de Viticultura y Enol., La Estación Central Agronómica. Jefe: J. D. Valdivieso.

Montpellier (France).
Station viticult., Ecole d'agric.

Neustadt a. d. H. (Deutschland).
Staatl. Lehr- u. Versuchsanstalt für Wein- u. Obstbau. Dir. Prof. Dr. A. Zschokke. — Prof. Dr. Zschokke, Leiter der Botan. Abt.; Prof. Dr. Schätzlein, Leiter der chem. Abt.; Prof. Dr. Stellwaag, Leiter der zool. Abt. Jeder Abt.-Leiter hat je 1—2 wissenschaftl. Ass. *4 Arbeitsplätze. Schädlinge des Weinbaus und des Obstbaus. Zusammenarbeiten des Botanikers, Chemikers und Zoologen über die Biol. der Reben und Obstbäume, der Schädlinge und Krankheiten der Reben und Obstbäume, Züchtung von Reben und Obstbäumen. Ferner Biol. der Gärungsorganismen und Weinkrankheiten.* ⊕ Tierische Schädlinge an Reben, Obstbäumen, Beerenobststräuchern. Pilze als Krankheitserreger an Reben, Obstbäumen, Beerenobststräuchern. — 1 fliegende Station zur Erforschung der Schädlingsbiol. an Obstbäumen.

Oppenheim a. Rh. (Deutschland).
Lehr- u. Versuchsanst. f. Wein- u. Obstbau. Confer: Pomologia.

Paris (France).
Station de recherches viticoles de l'Inst. des recherches agron., VIIe, 42bis, Rue de Bourgogne.
Station Viticulture Inst. Nation. agron., Ve, 16, Rue Claude Bernard. Dir. P. Viala.
Station oenol. de l'Inst. f. Rech. agron., VIIe, 42bis, Rue de Bourgogne.

Pretoria (South Africa).
Div. of Viticulture, Elsenburg, Muldersvlei, C.P. Governm. Viticulturist: S. W. van Nieskesk.

Sofia (Bulgarien).
Inst. f. Weinbau. Dir. Doc. Nedelčev.

Stellenbosch (South Africa).
Dept. of Viticulture. Prof. A. I. Perold.

Weinsberg i. W. (Deutschland).
Württemberg. Versuchsanstalt für Wein- u. Obstbau. Confer: Pomologia.

2. Pomologia.

Confer: *Horticultura* (pag. 513).

Ames, Ia. (U.S.A.).
Pomol. Section, Agric. Experim. Sta. Dir. T. J. Maney. — H. L. Lantz, Ass. chief.

Berkeley, Cal. (U.S.A.).
Division of Pomol., Univ. of Cal. Prof. W. H. Chandler (Pomol.).

Caen, Calvados (France).
Stat. Pomol.

Davis, Cal. (U.S.A.).
Branch of Pomol., Coll. of Agric., Univ. of Cal. Dir. W. L. Howard.

Ithaca, N.Y. (U.S.A.).
Dept. of Pomol., Cornell Univ. Prof. A. J. Heinicke (Pomol.).

Ketzin a. d. H. (Deutschland).
Pflanzenphysiol. u. Inst. f. Baumschulkrankh. Confer: Phytopathologia.

Košice — Kaschau (C.S.R.).
Abt. f. Wein- u. Obstbau b. d. Landw. Versuchsstation. Confer: Viticultura.

Mountain Grove, Mo. (U.S.A.).
Missouri State Fruit Experim. Station. Dir. F. W. Faurot, B.S. — J. W. Hitt, Mgr.; C. M. Williams, Ext. Hort.

Nikita Yalta (U.d.S.S.R.).
Labor. of pomol., Botan. Garden.

Okitsu (Japan).
Pomol. Labor., Okitsu Branch of Imper. Agric. Experim. Station, near Shizuoka.

Oppenheim a. Rh. (Deutschland).
Lehr- u. Versuchsanstalt f. Wein. u. Obstbau.

Průhonice b. Praha — Prag (C.S.R.).
Pomol. Arboretum. d. Staatl. Versuchsinst. f. Pflanzenproduction. Dir. C. Kamenický.

Puławy (Polska).
Abt. f. Obstbäumebau d. Staatl. Wissensch. Inst. f. Landwirtsch. Dir. Prof. J. Bialobok.

Simferopol (U.d.S.S.R.).
Salgirsker Wissensch. Versuchsstation f. Obstbau, Woronzowskij Sad. Dir. P. J. Jakowlew.

Stellenbosch (South Africa).
Inst. for Fruitculture. Dir. Prof. O. S. H. Reinecke.

Urbana, Ill. (U.S.A.).
Dept. of Pomol. Pathol. Univ. of Illinois. Ass.Chief: H. W. Anderson.

Warszawa (Polska).
Inst. f. Obstbau. Hochsch. f. Land- u. Forstw. Dir. Prof. W. Gorjackowski.

Weinsberg i. W. (Deutschland).
Württemberg. Versuchsanst. für Wein- u. Obstbau. Dir. Dr. Otto Kramer. — 1 wissenschaftl. Ass.

3. Cultura mori.

Ascoli Piceno (Italia).
R. Stazione sperim. di gelsicoltura e bachicoltura. Dir. Acqua Prof. Camillo. — P. Lorenza Lombardi, Vice-Dir.; Michele della Corte, Ass. 2 Ass. *Patol. del filugello. Formazione di nuove razze del baco da seta. Preparazione seme bachi per secondi allevamenti. Studi sulle principali varietà di gelsi.* ℗ Collezioni di bozzoli.

Kyôto (Japan).
Labor. of Mulberry-Culture Kyôto Imperial Sericicultural Coll., Hanazono near Kyôto. Dir. Prof. Mitsuya.

Tôkyô (Japan).
Labor. of Mulberry Culture Imperial Sericicultural Experim. Sta., Suginami-chô near Tôkyô.
Labor. of Mulberry Culture Tôkyô Imperial Sericicult. Coll., Nishigahara.

Ueda (Japan).
Labor. of Mulberry Culture Ueda Imperial Sericicultural Coll., Nagano-ken. Dir. Prof. Dr. Yasutarô Endô.

4. Institutiones seminologiae.

Confer: *Genetica* (pag. 341).

Baku (U.d.S.S.R.).
Staatl. Samenkontrollstation d. „Narkompross", Kooperatiwnaja 5, Landwirtschaftl. Mus.

Belfast (Ireland).
Dept. of Agric. Botan. Seed Testing and Plant Disease Division Queen's Univ. Dir. Prof. S. P. Mercer, B.Sc. — Mr. A. E. Muskett, B.Sc. (in charge of Plant Pathol. Labor.); Mr. David Clouston, M.A., B.Sc. (in charge of Seed Labor.). 10 Female Ass. in Seed Labor. *The Division consists of 2 main sections: Seed Control; Plant Pathol.* ℗ Weed and Crop seeds (may be exchanged). Mus. of Plant Pathol. (not at present for exchange).

Berlin (Deutschland).
Biol. Reichsanstalt für Land- u. Forstwirtschaft. Labor. für allgem. Sortenkunde, -Dahlem, Königin-Luise-Str. 19. Dir. Dr. K. Snell.

Botan. Abt. d. Landwirtsch. Kontrollstation d. Landwirtschaftskammer. Dir. Dr. Paul Filter. *Samenkontrolle. Futtermittelmikroskopie.*

Braunschweig (Deutschland).
Landwirtschaftliche Versuchsstation. Confer: Chemia agris.

Brno — Brünn (C.S.R.).
Sektion für Samenprüfung, Mährische Landwirtschaftliche Landesanstalt (Inst. des Recherches agric. et forest.), Květná 19. Dir. Doc. Dr. Ing. František Chmelař. — Ing. Jos. Nádvorník; Ing. Jar. Urban; Ing. Jar. Šimon; Ing. Fr. Mikolášek. 2 *Plätze für Samen- u. Sortenprüfung. — Modernes Vegetationsglashaus; Keimlabor.* ℗ Samen, Pflanzen u. Knollen der in C.S.R. gebauten Sorten. — Zweiglabor. für Futterpflanzen in Rožnov u. in den Beskiden.

Budapest (Ungarn).
Samenkontroll-Station, II, Kis Rokusgasse15. Dir. Dr. Árpád von Degen. — Stationsleiter: Ludwig

von Baán, Dr. Béla Szartorisz, Guido von Gerhardt, Dr. Géza Lengyel, Dr. Ottó Bocskay; Adjunkte: Dr. Julius Butujás, Josef von Horváth, Stefan Vigh, Dr. Zoltan Zsák, Dr. Leonhard Papp, Georg Schnabel, Dr. Johann von Cziáky; Ass.: Dr. Konstantin Schermann. *Samenkunde, Mikroskopie von Futtermitteln.—Optische Apparate, Keimlabor.* ⓟ Pflanzen, Pflanzensamen, Keimlingpraep., Pflanzenwurzeln.

Buitenzorg, Java (Ned.-O.-Indië).
Section for Plantbreeding of permanent and annual Crops of the Gen. Agric. Experim. Station. Confer: Agricultura specialis.

Cluj — Klausenburg (România).
Samenkontroll-Station. Dir. G. Tordai.

Corvallis, Ore. (U.S.A.).
Branch Seed Labor. B. C. Hite (Ass. Botanist).

Eberswalde (Deutschland).
Forst. Versuchsanst. und Samenprüfungsanstalt. Confer: Silvicultura.

Fort Collins, Col. (U.S.A.).
Colorado Seed Labor. Dir. Dr. L. W. Durrell. — Anna M. Lute, Mildred R. Lyon: Investigations in seeds. *Purity analysis, germination testing.* ⓟ Seeds.

Hohenheim b. Stuttgart (Deutschland).
Landessaatzuchtanstalt der württembergischen landwirtschaftlichen Hochsch. Dir. Prof. Dr. J. Wacker. — Dr. G. Baur, Abt.Vorst., Futterpflanzenzüchtung; J. Nuding, Ass., Sortenversuchswesen; Dr. E. Koch, chemischer Ass. ⓟ Körnerproben von Sorten der verschiedenen landwirtschaftlichen Kulturpflanzenarten.
Württ. Landesanstalt für Samenprüfung, Abt. des Botan. Inst. d. Landwirtsch. Hochsch. Dir. Prof. Dr. Georg Lakon. *Aufgabe: Untersuchung v. Kultursämereien auf ihre Beschaffenheit u. Ausführung wissenschaftl. Forschungen auf d. Gebiete d. Samenkunde, ferner Belehrungs- u. Gutachtertätigkeit.*

Königsberg i. Pr. (Deutschland).
Samenuntersuchungsamt und Pflanzenschutzstelle der Landwirtschaftskammer Ostpreußen, Beethovenstr. 24—26. Dir. Dr. Alfred Lemcke. — Praeparatorinnen: Fräulein Hertha Perlbach u. Fräulein Käthe Lau. *Saatuntersuchungen.* ⓟ Pflanzliche und tierische Schädlinge.

Košice — Kaschau (C.S.R.).
Samenkontroll-Anstalt, Barca-Košice. Dir. Dr. Stephan Bódis. — A. Horník, Ass.; J. Novy, Ass.; M. Maloch, Ass. (Leiter der agronom. Versuchsstation in Barca bei Košice). *Samenkontrolle (Kleesamen).*

Krasnodar, Kuban (U.d.S.S.R.).
Samenzucht-Inst. d. Kuban. Landw. Inst.

Landskrona (Sverige).
Saatzuchtanstalt Weibullsholm (Weibullsholms Växtförädlingsanstalt). Dir. Harry Weibull. — Prof. Fil. Dr. Nils Heubert-Nilsson; Priv.Doc. Fil. Dr. Carl Hallqvist; Prov.Doc. Fil.Dr. Carl Hammarlund; Fil. Dr. Karl B. Kristofferson. 4 Ass. *Die wissenschaftlichen Versuche und die praktischen Resultate d.Anstalt werden gern demonstriert. Geeignetste Zeit ist Juni—August.* ⓟ Proben von Züchtungsprodukten oder Landsorten werden getauscht Von eigenen Züchtungsprodukten von Getreide, Rüben, Wiesengräsern und Gemüsen werden kleine Proben für Museumszwecke oder für Probeaussaat gern gesandt.

Leningrad (U.d.S.S.R.).
Selektions Station Leningr. Landwirtschaftl. Inst., Kamennyj Ostrov. 2, Birken-Allee 28. Dir. Prof. N. I. Vavilov.

Lwów (Polska).
Staatl. Botan.-Landwirtsch. Versuchsstation, ul. Zyblikiewicza 40. Vorst.: Dr. W. Swederski. — 2 Adjunkt., 1 Ass. *Abt. für Samenschätzung, für Pflanzenschutz und Hochgebirgs-Vers.-Stat. auf d. Polonina Pożyzewska in den Ost-Karpathen.*

Moskau (U.d.S.S.R.).
Moskauer Samenkontrollstation Timirjasew, Samarskij per. 5.
Samen-Versuchs- u. Kontrollstation (M.O.S.Ch.A.), Smolenskij bulv. 8.

München (Deutschland).
Abt. f. Samenkontrolle a.d. Landesanst. f. Pflanzenbau u. Pflanzenschutz. Dir. Dr. Georg Gentner.

New Brunswick, N.J. (U.S.A.).
State Seed Labor. M. E. Brumfield (Ass. Seed Analysist).

Odessa (U.d.S.S.R.).
Provinz-Samenkontroll-Station, Institutskaja 9.

Oslo (Norge).
Saatzuchtwirtschaft des Falleskjøpet. Hjelium. Dir. Prof. W. H. Christie. *Pflanzenzüchtung; Vererbungslehre.*

Paris (France).
Station des Semences, Inst. Nation. agron., Ve 16, Rue Claude Bernard.

Perm (U.d.S.S.R.).
Kabinett f. Samenkunde. Dir. Ass. N. M. Bubnov.

Riga (Latvija).
Samenkontrollstation Landwirtschaftl. Fac. Univ., Kronvalda bulv. 1.

Sofia (Bulgarien).
Samenkontrollabor. beim Inst. für Allgemeinen Ackerbau an der Landwirtschaftlichen Fac. der Univ. Dir. o. Doc. Iwan T. Stranski. — Ewgeni Levensson, Ass. *2 Arbeitsplätze für Samenkontrolle. — Sammlung von Samen in Bulgarien wachsender Pflanzen.* ⓟ Samen.
Abt. f. Samenkontrolle u. Samenzucht Zentral-Inst. f. landwirtsch. Versuchswesen, Solunplatz 3. Dir. Dr. Chr. Kazaski.

Stettin (Deutschland).
Samenprüfungsstelle u. Hauptstelle für Pflanzenschutz. Dir. Richard Kleine. *Fraßbildstudien. Biol. der Brenthidae. Angewandte Entomol.*

Svalöf (Sverige).
Inst. d. Schwed. Saatzuchtvereins (Sveriges Utsädesförening). Dir. Prof. St. Nilsson-Ehle. Confer: Cultura plantarum generalis.

Tiflis (U.d.S.S.R.).
Stat. f. Samenkontrolle des Botan. Gartens.

Twer (U.d.S.S.R.).
Samenzucht u. Samenkontroll-Station, Archirejskaja Datscha.

Wageningen (Holland).
Ryksproefstation voor Zaadcontrole (Samenkontrolle). Binnenhaven 1A. Dir. Dr. W. J. Franck. — Botaniker: G. Wieringa, Dr. H. Bos, Ir. K. Leendertz, J. A. Nieuwland; Mykologe: Dr. L. C. Doyer. *Physiol., anat., und mykol. Untersuchungen von Samen. — Eine Installation für Keimung bei niedriger Temperatur (in der Nähe des Gefrierpunktes).* ⓟ Zur Verfügung stehende Samen. — Zweiglabor.: ein besonderes Labor. für die Kulturkontrolle der Gewächse.

Weihenstephan b. Freising (Deutschland).
Bayer. Landessaatzuchtanstalt. Dir. Prof. Dr. L. Kießling (München). — Reg.R.: Scharnagel, Hampp, Gaßner, Weller, Müller; Ass. bzw. Assessoren: Siebert, Ziegler, Honnecker, Biéchy. *Plätze für Gäste nach besonderer Vereinbarung. — Einrichtungen für botan., chemische und technische Untersuchung bei Pflanzenzüchtungs- u. Vererbungsstudien.* ℗ Landw. Kulturpflanzen, Züchtungs- u. Sonderformen.

Wien (Österreich).
Bundesanst. f. Pflanzenbau u. Samenprüfung. Confer: Labor. cult. plant. gen.

Wjatka (U.d.S.S.R.).
Samenkontrollstation, Gumbsemuprawlenie.

Woronesch (U.d.S.S.R.).
Versuchsstat. f. Sortenprüfung Landwirtsch. Hochschule.

5. Cultura sacchari.

Baraguá, Province of Camagüey (Cuba).
Cuba Sugar Club Experim. Station of the Tropical Plant Research Foundation, Central Baraguá. Dir. and General Manager Washington Office: W. A. Orton; Local-Dir.: D. L. Van Dine. — C. F. Stahl, H. K. Plank, T. S. Roß, Entomologists; F. S. Earle, Sugar Cane, Technologist; S. Byall, Chemist. *3 places for Pathol., entomol. and chemistry. — Experim. gardens and plant houses.* Branch station for sugar cane varieties, Herradura, Cuba. Branch station for sugar cane stalk borer investigations, Central Jaronú.

Baton Rouge, La. (U.S.A.).
Sugar Experim. Station. W. L. Owen (Research Bacteriologist).

Belaja Zerkow (U.d.S.S.R.).
Belozerkowskaja Selektions-Station d. „Sacharotresta" (Zuckertrust), Alexandrijskaja 41. Dir. I. F. Soljanucha.

Berlin (Deutschland).
Inst. für Zucker-Industrie der Landw. Hochsch., N 65, Amrumer Str. Dir. Dr. O. Spengler. — Dr. C. Brendel, Vorst. d. Versuchsabt., Dr. G. Dorfmüller, Vorst. d. wissenschaftl.-chem. Abt.; Dr. R. Weidenhagen, Vorst. der Forschungsabt.; Dr. W. Paar, Vorst. der Unterrichtsabt.; Dr. A. Traegel, Unterrichtsass. *2—10 Plätze. Zahlreiche Apparate und Apparaturen.* ℗ Alle das Zuckerrohr und die Zuckerrübe betreffenden Schädlinge und Krankheitserreger.

Bernburg a. d. S. (Deutschland).
Anhaltische Versuchsstation. Confer: Labor. et Inst. agric. spec.

Honolulu (Hawaii).
Experim. Station, Hawaiian Sugar Planters' Assoc. Dir. H. P. Agee, B.S. — C. C. Barnum, B.S., Ass. Pathol.; T. K. Beveridge, B.S., Ass. Agr.; F. W. Broadbent, Ass. Agr.; A. Brodie, M.A.S., Techn. Chem.; L. W. Bryan, B.S., For. Supvr. (Hilo Project); Raymond Conant, Ass. Agr.; H. A. Cook, A.B., Ass. Chem.; C. L. Crutchfield, Ass. Chem.; F. C. Denison, Ass. Agr.; Allister Forbes, Ass. Agr.; Donald Forbes, Supt. Vineyard Street Nursery; Fred Hansson, Ph.B., Ass. Chem.; W. C. Jennings, Ass. Agr.; R. H. King, B.S., Ass. Chem.; Y. Kutsunai, B.S., Ass. Agr.; H. A. Lee, Plant Pathol.; H. L. Lyon, Ph.D., in charge Dept. of Botan. and For.; W. R. McAllep, Sugar Tech.; W. L. McCleery, Ass. Sugar Tech.; G. A. McEldowney, For. Supvr. (Oahu Project); W. T. McGeorge, M.S., Assoc. Chem.; O. C. Markwell, M.S., Ass. Agr.; J. P. Martin, B.S., Ass. Plant Pathol.; F. Muir, Entomol.; H. T. Osborn, M.A., Ass. Entomol.; F. A. Paris, A.E., Ass. Agr.; C. E. Pemberton, Ass. Entomol.; R. C.L. Perkins, D.Sc., Consult. Entomol.; Twigg Smith, Illustrator; W. E. Smith, Ass. Sugar Tech.; H. K. Stender, B.S., Ass. Agr.; G. R. Stewart, B.S., Chem.; O. H. Swezey, M. S., Entomol.; F. R. Van Brocklin, B.Ch., Ass. Chem.; R. H. Van Zwaluwenburg, B.S., Ass. Entomol.; J. A. Verret, B.S., Agr.; Neil Webster Ass. Agr.; J. A. H. Wilder, A.B., Ass. Agr.; F. X. Williams, Sc.D., Ass. Entomol.

Krasnodar, Kuban (U.S.S.R.).
Station f. d. Studium d. Zuckerrübe, Zuckerfabrik.

Mount Edgecombe, Natal (South Africa).
Natal Sugar Experim. Stat., South African Sugar Assoc. Dir. H. H. Dodds. — H. H. Storey, Government Mycologist; A. P. D. McClean, Acting Government Mycologist; C. P. van der Merve, Government Entomologist. *2 places suitable for general chemical and bio-chemical work in sugar manufacture and agric. and in soil sc.* ℗ Varieties of sugar cane. Seeds of plants suitable for green manuring.

Pasoeroean (Ned.-O.-Indië).
Sugar Experim. Station. Dir. Dr. Viktor Jakob Koningsberger. — The staff of the Station is formed by 45 Europeans, 4 Chinese and about 130 natives. 1 Ass.-Dir. head of agric. dept. Secretary: J. van Harreveld, Editor of the „Archief voor de Java-Suikerindustrie"; head of statistical work. 1. Section for cane breeding and botan. of cane: Dr. P. Posthumus *(Morphol. of cane, taxonomy of Saccharum, identification methods of cane varieties);* Dr. G. Bremer *(Cytol. studies, botan. work, study of rootsystem, investigations of diseases; sereh, mosaic disease, fungus diseases);* C. A. Backer *(Identification of cane varieties, investigation of weed-flora in Java cane fields),* Chief of experim. garden with native staff. 2. Section for fieldtrials, inspection work of plantation, studies in cultivation and irrigation: Dr. P. J. van Breemen, Chief of the fieldtrialservice *(Theory of fieldtrials, probable error; entomol. work in Aphis mavidis and Oregma lanigera);* Dr. T. A. Tengwall and C. E. van der Eyl, Field Inspectors *(Cultures of cane, irrigation, seed nurseries, fertilizers; influence of climate on crop, advices for improvement of yield of plantations).* 15 group-advisers, officers with botan. or agric. education, are spread over the cane districts of Java, doing the daily advising work, studying agrogeol. of their district, laying out and controling the fieldtrials, in close connection with the central station. 3. Section for geol., agrogeol., meteorol.: Mrs. C. H. van Harreveld-Lako *(Chemical analysis of cane soils, properties of soil, water-capacity, nitrogen retaining power of soil, soil reaction, advices on suitability of soil for cane cultivating, record on rainfall, sunshine, wind, moisture content of air);* F. Haverkamp, Chief of analytical labor. *(Fertilizers and soils).* 4. Section for engineering work: Ir. N. Nobel; Ir. L. D. Teutelink, Ass.; G. E. Ferguson; Ir. C. J. Schott; Ing. I. Hes, Ass., Chief of drawing office with native staff, Chief of constructionshop, with native staff, Chief of instrumentshop with native staff. 5. Section for chemical work: Studies on clarification, boilingprocess, evaporation, exhaustion of molasses and sugars: V. Khainovsky *(Extraction of juice from bagasse, analysis of bagasse);* Ir. J. W. L. van Ligten, Ir. G. E. van Ness, Consulting engi-

neers; C. Sylmans, Chief of chemical control; J. G. Smits, Chief of sugar labor.; Miss M. van de Kreke, Chief of analytical labor. *for molasses, limestone, lubricating oil, calory value of fuel.*

Praha — Prag (C.S.R.).
Versuchsanstalt der Zuckerindustrie, Střešovice Vořechovka. Vorst. d. phytopath. Abt.: Dr. Franz Rambousek.

Salt Lake City, Utah (U.S.A.).
Agric. Research Dept. Utah-Idaho Sugar Co. Dir. E. G. Titus. Confer: Cultura plantarum generalis.

Washington, D.C. (U.S.A.).
Sugar-Plant Investigations, U.S. Dept. of Agric. E. W. Brandes (Pathologist).

6. Cultura heveae.

Boenoet, Sumatra O.K. (Ned.-O.-Indië).
Plantation Research Dept., U.S. Rubber Plantations, Inc., Kisaran, Sumatra, O. K. Dir. James Grantham, M.A. (also Soil Chemist). — Dr. H. S. Yates, Botanist; C. Barclay, B.A., Geneticist; E. E. Probstfield, B.S., M.F., Forester; J. O. Shank, B.S., Forester. *Investigations carried out on genetics, disease, physiol. of Hevea brasiliensis, soil work and field experiment work on tapping, thinning and other estate procedures.*

Buitenzorg, Java (Ned.-O.-Indië).
Proefstation voor Rubber (Central Rubberstation and Rubberproefsation West-Java). Dir. Dr. O. de Vries. — Dr. T. A. Tengwall, Ass. Dir., Botanist Dr. W. Bobilioff, Botanist, Physiologist; Dr. Brouwer, Botanist, Selectionist; Dr. Dillewyn, Botanist Phytopathologist; M. Vrolyk, Agriculturist; J. S. Vollema, Agriculturist; W. F. Ostendorf, Agriculturist; Dr. R. Riebl, Chemist; Dr. G. M. Kraay, Chemist; Mrs. N. Beumee-Nicuwland, Chemist. ⓟ Hevea brasiliensis, taxonomic objects (seeds, leaves etc.). Fungi and insects on Hevea brasiliensis. Herbarium and seeds of Leguminosae.

Galang, Sumatra O.K. (Ned.-O.-Indië).
Wetenschappelyke Dienst der Rubber Cultuur Maatschappy Amsterdam. Dir. Dr. P. Arens. —

Dr. F. Arens; Dr. E. Fickendey; Ir. C. J. Vergeer. *Botan., Chemie, Bact.* Pflanzung Poeloe Radja, Pflanzung Nykerk.

Medan, Sumatra (Ned.-O.-Indië).
General Experim. Sta. of A.V.R.O.S. (Algem. Vereenig. van Rubberplanters Oostkust v. Sumatra). Dir. Dr. A. W. K. de Jong. — 1. Botan. division: Chief: Dr. C. Heusser, Dr. Boudijin *(Hevea selection; Experim. in the selection garden Soengei Pantjoer; Research of the cytol. of flower and fruit of the oilpalm; Research work on plant diseases)*. 2. Agric. division: Chief: Dr. J. F. Schmöle *(Manuring experim.; Tapping and thinning out experim. with rubber; Oilpalm selection; Introduction of new plants; Soil treatment experim.; Roselle fibre experim.);* Iv. Kostleven, Sec. Agric. 3. Chemical division: Chief: Ir. H. N. Blommendaal, Ir. van Harpen *(Experim. with preserved latex; Research work on the preparation of rubber, palmoil and other crops; Analysis of chemicals used in the preparation of these products; Chemical analyses of manures, fertilisers; Testing of prep. palmoil, rubber, etc.; Testing new machinery for palmoil preparation).* Branch Labor. at Soengei Pantjoer.

Salatiga (Ned.-O.-Indië).
Experim. Sta. Confer: Agric. spec.

7. Cultura nicotianae.

Confer: *Entomologia* (pag. 381), *Horticultura* (pag. 513).

Djember, Java (Ned.-O.-Indië).
Het Besoekisch Proefstation. Confer Thea.

Driazgui, Gouv. Woronesch (U.d.S.S.R.)
Stat. expérim. de cultivateurs de tabac. Phytopath.: N. Simanovsky.

Klaten (Ned.-O.-Indië).
Experim. Station for Tobacco. Dir. Dr. A. d'Angremond. — Agriculturists: A. N. J. Beets, P. M. Bartels; Analyst: E. J. Her; Ass.: T.E. van Wijlen, J. F. Haenzgen, Jagoos, M. G. Gervais. 35 Javanese labor. helpers. Dr. A. d'Angremond *(Breeding; Fighting Pytophthora Nicotianae, Br. de Haan, Fighting Oidium of tobacco;* A. N. J. Beets *(Drying and fermentation; Manuring experim.; Combustibility of tobacco);* P. M. Bartels *(Problem during which number of years stable manure can be replaced by artificial fertilizer without causing a going down of the quantity or quality of the tobacco).*

Krasnodar, Kuban (U.d.S.S.R.).
Zentral-Versuchsinst. f. Tabakkultur, Postfach 55. Dir. Prof. A. A. Schmuck. — Dir. d. Versuchsfeldes „Kruglik" bei Kr.: V. S. Pustovoit.

Medan, Sumatra (Ned.-O.-Indië).
Deli Experim. Station for Tobacco. Dir. Dr. J. Kuyper (General direction and advice service); — Secretary: C. H. ten Cate (Office work and library). Division of Chemistry: Ir. N. Noerngali *(Experim. on drying and fermentation of tobacco; Soil analyses);* B. Ph. M. de Groot, 2nd Chemist *(Analyses of samples of fertilizers, insecticides, etc.).* Division of Botan.: Dr. S. C. J. Jochems *(Plantbreeding and phytopath. work with tobacco);* A. R. R. F. Koelen, Ass. *(Breeding experim.).* Division of Agronomy: Ir. E. Sidenius *(Experim. on manuring, training, green manuring, crop rotation);* Ir. van der Poel, 2nd Agronomist; R. Noerngali, Field Ass. Division of Zool.: Dr. L. Fulmek *(Fighting caterpillars in the field and the sheds; Fighting lice; Biol. researches in connection with the jungle on tobacco soils);* J. C. van der Meer Mohr, 2nd Zoologist. *The station is working in the interest of the Sumatra Tobacco cultivation and maintained by the Deli Planters Vereeniging formed by the tobacco companies.*

Ottawa (Canada).
Tobacco Division of the Central Experim. Farm. C. M. Slagg.

Quincy, Fla. (U.S.A.).
Tobacco Experim. Station. W. B. Tisdale (Plan Pathologist).

Scafati, Salerno (Italia).
R. Ist. Sperim. per le Coltivazioni dei Tabacchi „Leonardo Angeloni" Dir. Superiore attuale: Dr. Manlio Donadoni. — L'attuale Dir. Tecnico dell'Ist.: Dr. Giacomo Trojano. *Tutto il personale tecnico dir. dell'Amministrazione delle Privative ed eventualmente, avendone i requisiti, qualche funzionario del personale tecnico esecutivo. L'Ist. offre ospitalità a chimique, autorizzato dalla Dir. Generale delle Privative, Ministero Finanze, Roma Studie ricerche sulla tabacchicultura. Reparti: Agrario,*

Industriale, Didattico. Labor. di chimica e microscopia, Gabinetto entomol. ⊕ Sementi di tabacchi. — Recentemente (1926) e stata creata in Verona una Sezione di questo Ist.

Washington, D.C. (U.S.A.).
Experim. Sta. of tobacco and plant Nutrition U.S. dept. of Agric. Confer: Physiol. plantarum.

8. Cultura theae.

Buitenzorg, Java (Ned.-O.-Indië).
Algemeen Proefstation voor Thee. Dir. Dr. Charles J. Bernard, Botaniste, Phytopath. — Dr. J. J. B. Deuss, Chimiste; Dr. C. P. Cohen Stuart, Botaniste Sélectioniste; Dr. R. Menzel, Entomologiste; Dr. Steinmann, Mycologiste; Dr. Vageler, Agrogeologue; Agronomes: A. J. Garretsen, A. Keuchenius, P. Prillwitz. *La ,,Proefstation" s'occupe de tout ce qui concerne la culture du thé: agronomie, maladies, sélection, engrais, engrais verts, fabrication. Elle comprend un labor. d'agrogéol. qui s'intéresse à toutes les cultures tropicales. Elle s'occupe également des soidisant ,,petites cultures" dans l'ouest de Java: Cacao, Coca, Palmiers à huile, fibres etc.* ⊕ *Tout ce qui concerne la plante de thé et les engrais verts (Légumineuses) et les plants de cultures secondaires; les insectes utiles ou nuisibles des plantations de thé; échantillons de thé.*

Djember, Java (Ned.-O.-Indië).
Het Besoekisch Proefstation. Dir. Johannes Gandrup, Phytopathologist; Jean Schweizer, Botanist; L. R. van Dillen, Chemist; G. A. Reydon, Agriculturist.

Malabar (Ned.-O.-Indië).
Experim. Station for tea. Situated at the Malabar Estate, Pengalengan near Bandeng. Dir. K. A. R. Bosscha. — 1 Chemist. *Plant breeding with tea and cinchona. Field experim. with artificial fertilizers and green manures for tea and Cinchona. Research work on a process to prepare chinine and kina-salts out of the fresh met bark. Daily control of the preparation of tea for the market in the labor.*

Tôen, Formosa (Japan).
Branch Experim. St. for Tea-Plantation Governm. Central Research Inst.

9. Cultura solani tuberosi.

Berlin (Deutschland).
Forschungs-Inst. f. Stärkefabrikation u. Kartoffeltrocknung Landw. Hochsch. Confer: Biochemia gen.

Korenevo, Gouv. Moscou (U.d.S.S.R.).
Section Filiale de l'Inst. Jaczewski à la Station Expérim. de Pomme de Terre. Chef de la Section: N. A. Rojdestvenski.

Wageningen (Holland).
Labor. für Mycol. en aardoppelonderzoek (Kartoffeluntersuchung) d. Inst. f. Phytopath., Landbouwhoogeschool, Binnenhaven 4. Dir. Prof. Dr. H. M. Quanjer. — Phytopathologen: Mej. H. L. G. de Bruyn, Mej. Dr. J. H. H. van der Meer, Dr. S. J. Weelensick; Ir. T. H. Thung, ass. *Phytopath. Untersuchungen.* ⊕ *Kranke Pflanzen.* — Dieses Labor. bildet zusammen mit dem Labor. für Entomol. und mit dem Labor. für Blumenzwiebeluntersuchung das zu der Landwirtschaftlichen Hochsch. gehörige Inst. für Phytopath.

10. Cultura coffeae.

Port au Prince (Haïti).
Coffee Experim. Station Service Technique. Dir. C. H. Arndt. — H. D. Barker (Botanist, Plant Pathologist).

Salatiga (Ned.-O.-Indië).
Experim. Sta. for Centr. Java. Confer: Agric. spec.

11. Cultura citri.

Lake Alfred, Fla. (U.S.A.).
Citrus Experim. Station of the Florida Agric. Experim. Stat. R. L. Miller (Ass. Entomologist). *Citrus Aphid and other citrus insects. Morphol. of Pentatomidae.*

Louisiana, Mo. (U.S.A.).
Research Dept. Confer: Inst. horticult.

Pretoria (South Africa).
Dept. of Citriculture Dept. of Agric.

Riverside, Cal. (U.S.A.).
Graduate school of Tropical agric. and Citrus experim. Sta. Univ. of Cal. Dir.: H.J.Webber; Assoc. Dir.: J. T. Barrett. — W. B. Hooper (Walnut culture); H. Compere (Entomologist); C. F. Lackey (Plant Pathologist); S. H. Reed (Prof. of Plant physiol.); M. Shapovalov (Pathologist); P. A. Miller (Ass. in Plant Pathol.); E. T. Bartholomew (Ass.); A. T. Basinger (Entomol.); J. W. Lesley (Ass. Genetics); H. S. Smith (Prof. of Entomol.); L. J. Klotz (Research Assoc. Plant Pathol.); R. H. Smith (Ass. Entomologist). *Arrangement for guests made on application for research on Plant Pathol. Plant Physiol. Genetics. Agric. Chemistry. Entomol.* — *Farm of 700 acres for plant experim.*

Laboratoria culturae animalium generalia.

Confer: *Institutiones generales culturae plantarum et animalium* (pag. 497), *Zoologia generalis* (pag. 356).

Aas b. Oslo (Norge).
Zool. Labor., Norges Landwirtschaftliche Hochsch. Dir. Prof. Isachsen. *Haustierkunde, Fütterungsversuche.*

Ames, Ia. (U.S.A.).
Bureau of Animal Industry U.S. Dept. of Agric., P.O. Box 175. C. N. McBryde (Bacteriologist).

Amherst, Mass. (U.S.A.).
Dept. of Animal and Dairy Husbandry, Mass. Agric. Coll. Dir. J. H. Frandsen.

Askania Nova, Ukraine (U.d.S.S.R.).
Zootechnisches Labor. Dir. Prof. Michael F. Iwanoff. — Ass.: Leonid Greben, Leiter der Versuche mit Landwirtsch. Tieren; Peter Belechoff, Leiter der Arbeiten über Wolluntersuchungen. Tierärzte. *2 bis 4 Plätze für Wolluntersuchungen und Arbeiten über Vererbung der Eigenschaften bei Landwirtsch. Tieren (Rinder, Schafen und Schweine).* ℗ Rinder, Schafe, Schweine.

Athens, Ga. (U.S.A.).
Dept. of Animal Industry, State Coll. of Agric. M. P. Jarnagin, D.Sc., Prof. An. Husb.; J. H. Wood, B.S., Prof. Poultry Husb.; F. W. Bennett, B.S., Assoc. Prof. Dairy Husb.; F. W. Fitch, B.S., Field Agt. in Dairy Husb.; C. E. Kellogg, M.S., Assoc. Prof. An. Husb.; J. G. Liddell, B.S., Field Agt. in Swine Indus.; L. H. Marlatt, Field Agt. in Cheese Prod. (Coop. U.S.D.A.); F. E. Mitchell, B.S., Poultry Specialist; W. S. Rice, B.S., Assoc. Prof. An. Husb.; H. O. Woodward, B.S., Adjunct Prof. Poultry Husb.

Auburn, Ala. (U.S.A.).
Dept. of Animal Industry, Coll. of agric. Head Prof. J. C. Grimes, M.S.; W. H. Eaton, B.S., Assoc. Prof. Dairying, Dairyman; J. E. Ivey, M.S., Prof. Poultry Husb., Poultry Husb.; F. W. Burns, B.S., Ass. Prof. An. Indus.; S. R. Doughty, B.S., Poultry Specialist; H. A. Gardner, M.S., Poultry Specialist; William Hardie, B.S., Dairy Specialist; L. G. Pearson, B.S., Supt. National Egg Laying Contest.; W. D. Salmon, M.S., Ass. Prof. An. Indus., Ass. in An. Indus. Research; J. D. Sykes, B.S., Poultry Specialist; G. A. Trollope, B.S., Poultry Specialist.

Baton Rouge, La. (U.S.A.).
Dept. of Animal Industry, State Agric. Coll. E. L. Jordan, B.S., Prof. An. Indus., An. Indus.; J. B. Francioni, B.S., Ass. Prof. An. Husb.; M. M. La Croix, B.S., Swine Husb. Specialist; L. E. Morgan, M.D., Livestock and Tick Erad. Specialist; E. W. Neasham, B.S., Dairy Specialist; F. C. Old, B.S., Poultry Specialist; G. H. Reymond, B.S., Ass. Dairy Specialist; C. H. Staples, B.S., Prof. Dairying; Elsmer Wilson, B.S., Ass. Poultry Specialist.

Berkeley, Cal. (U.S.A.).
Dept. of Animal Industry, Coll. of Agric. G. H. True, B.S., Head of Div. (P. O., Davis). — H. R. Guilbert, M.S., Junior An. Husb. (P.O., Davis); C. E. Howell, M.S., Ass. Prof. and Ass. An. Husb. (P.O., Davis); E. H. Hughes, M.S., Assoc. Prof. and Assoc. An. Husb. (P. O., Davis); S. W. Mead, M.S., Ass. Prof. and Ass. An. Husb. (P.O., Davis); R. F. Miller, M.S., Assoc. Prof. and Assoc. An. Husb. (P. O., Davis); W. M. Regan, A.M., Assoc. Prof. and Assoc. An. Husb. (P. O., Davis); E. C. Voorhies, B.S., Ass. Prof. An. Husb.; J. F. Wilson, M.A., Ass. Prof. and Ass. An. Husb. (P.O., Davis).

Berlin (Deutschland).
Inst. für Tierzucht Landw. Hochsch., N 4, Invalidenstr. 42. Dir. Prof. Dr. Hansen.

Bern (Schweiz).
Zootechnisches u. vet.-hygien. Inst. Univ. Dir. Prof. Duerst.

Bern-Liebefeld (Schweiz).
Centralverwaltung der Schweiz. Landw. Versuchs- und Untersuchungsanstalten. Dir. Dr. G. Schmid, Ing. agr. — Ing. agr. I. Landis, Ass.; Ing. agr. G. Gutknecht, Hilfsass. *Tierzucht, Fütterungslehre.*

Blacksburg, Va. (U.S.A.).
Dept. of Animal Industry, Agric. Coll. W. D. Saunders, Milk Invest., Cheese Specialist; R. E. Hunt, B.S., Prof. An.Husb.; C. W. Holdaway, M.S., Prof. Dairy Husb., Dairy Husb., Supt. of Creamery, Dairy, and Advanced Registry; E. J. Albert, Instr. Poultry Husb.; G. L. Booker, B.S., Instr. Poultry Husb.; F. A. Buchanan, B.S., Assoc. Prof. and Ext. Dairy Husb.; G. L. Carey, Ass. Dairy Husb. L. I. Case, B.S., An. Husb.; C. T. Cornman, Ass. Prof. Poultry Husb.; A. L. Dean, Assoc. Prof. Poultry Husb.; G. C. Herring, B.S., Ass. An. Husb.; Bessie M. Hodsden, Ass. Poultry Specialist; H. G. Iddings, B.S., Dairy Mfrs. Specialist; E. W. Lawson, B.S., Sheep Specialist; C. R. Nobles, B.S., Ass. An. Husb.; T. E. Starnes, Instr. Poultry Husb.; Rachel F. Treakle, Ass. in Poultry Husb.

Bologna (Italia).
Ist. di Zootechnica ed Ezoognosia Univ. Dir. A. Campus.

Bonn-Poppelsdorf (Deutschland).
Inst. für Tierzucht und Molkereiwesen Landw. Hochsch. Dir. Prof. Dr. A. Richardsen. *Milchuntersuchungen und Fettbestimmungen.* Landwirtschaftl. Versuchswirtschaft.

Bozeman, Mtn. (U.S.A.).
Dept. of Animal Industry, Coll. of Agric. C. N. Arnett, B.S., Head of Dept., Vice Dean of Agr.; Harriette E. Cushman, B.S., Poultry Specialist.— George Ford, Ass. in Poultry Husb.; W. E. Joseph, Ph.D., Assoc. An. Husb.; R. C. McChord, B.S., Assoc. Prof. An. Husb.; J. A. Nelson, M.S., Ass. Prof. Dairy Mfg.; J. O. Tretsven, B.S., in charg. School of Agr., Dairy Specialist; Louis Vinke, M.S., Ass. in An. Husb.; R. L. Waddell, B.S., Livestock Specialist.

Breslau (Deutschland).
Inst. für Tierzucht und Milchwirtschaft Univ., Hansastr. 25. Dir. Prof. Dr. W. Zorn. — Priv.Doc. Dr. J. Gärtner. 1 planmäßiger Ass.; 1 außerplanmäßiger Ass. *Tierphysiol., histol., biol. und milchwirtschaftliche Untersuchungen.* ℗ Skelette; Haut-, Haar-, Wollproben; mikroskopische Praep. des tierischen Körpers und seiner Organe insgesamt.

Brno — Brünn (C.S.R.).
Zootechn. Inst. der Tierärztl. Hochsch. Dir. Prof. Dr. M. V. Ant. Hcůra. — Doc. M.V. Dr. u. Dr. Ing. agron. Cyril Kućera; Ass. M.V. Dr. Jaromír Fiala; Ass. M.V. Dr. Frant. Habartík; Ass. Ing. M.V.C. Mich. Cholevćuk.
Inst. f. Biotechnol. d. Tiere d. Landwirtschaftl. Hochsch., Černa pole. Dir. Prof. J. Taufer.
Inst. f. allg. u. spezielle Zootechnik der Landwirtschaftl. Hochsch., Černa pole. Dir. Prof. Dr. Em. Bayer. — Abt.-L. f. Züchtungsbiol.: Doc. Dr. J. Kříženecký.

București (România).
Labor. de Zootechnique Fac. de méd. vét. Univ. Dir. Prof. Dr. Georges K. Constantinesco. — Dr. D. Contescu, Ass.; Dr. D. Strilcine, Ass.

Budapest (Ungarn).
Zootechn. Inst. der königl. ungarischen Tierärztl. Hochsch., VII, Rottenbillerstr. 23. Dir. Prof. Dr. Oskar Wellmann. — Ass.: Dr. Géza Fetth, Ludwig v. Bérczy. *1—2 Plätze für biol. Blutuntersuchungen, Stoffwechseluntersuchungen.*
Inst. für Tierzucht Volkswirtschaftl. Univ. Dir. Prof. K. Schandl.

Burlington, Vt. (U.S.A.).
Dept. of Animal Industry, Coll. of Agric. H. B. Ellenberger, Ph.D., Prof. An. and Dairy Husb., An. and Dairy Husb.; H. A. D. Leggett, B.S., Ass. Prof. Poultry Husb., Poultry Specialist; E. H. Loveland, B.S., Dairy Specialist; J. A. Newlander, M.S., Instr. and Assoc. Dairy Husb.; A. H. Robertson, M.S., Research Dairy Bact.; W. B. Silcox, M.S., Instr. Dairy Husb.

Cluj — Klausenburg (România).
Labor. f. Tierzucht u. Molkerei. Dir. Prof. J. Oțolu.

College Station, Tex. (U.S.A.).
Dept. of Animal Industry, Agric. Coll. J. M. Jones, A.M., Chief Div. of Range An. Husb. (Sheep and Goat Research).— D. W. Williams, M.S., Head Prof. An. Husb.; G. P. Grout, M.S., Head Prof. Dairy Husb.; R. M. Sherwood, M.S., Chief Div. Poultry Husb.; D. H. Reid, M.S., Head Prof. Poultry Husb.; Fred Hale, M.S., Chief, Div. of Swine Husb.; G. W. Barnes, B.S., Beef Cattle Specialist; D. S. Buchanan, B.S., Ass. Prof. An. Husb.; J. A. Clutter, jr., M.S., Prof. Dairy Husb.; A. L. Darnell, M.A., Prof. Dairy Husb.; V. R. Glazener, B.S., Poultry Specialist; A. H. Groth, M.S., Ass. Prof. An. Husb. (Horses); D. F. Irving, B.S., Ass. Prof. Poultry Husb.; J. L. Lush, Ph.D., An. Breeder.; A. K. Mackey, M.S., Assoc. Prof. An. Husb.; E. M. Regenbrecht, M.S., Assoc. Prof. An. Husb.; J. L. Thomas, Dairy Specialist; A. L. Ward, B.S., Swine Husb.; R. H. Williams, Ph.D., Prof. An. Husb.

Děčin-Libverd —Tetschen-Liebwerd (C.S.R.).
Inst. f. Tierzucht u. Milchwirtsch. Dir. Prof. Dr. M. Nitsche.

Erastovka, Ukraine (U.d.S.S.R.).
Zootechnisches Labor. des Landwirtsch. Technikums, Erastover Postabt. d. Ekaterinoslav. Gouv., Kreis Krivorag.

Fort Collins, Col. (U.S.A.).
Dept. of Animal Industry, State Coll. Agric. G. E. Morton, M.L., B.S., Head Dept. of An. Husb. (State Dairy Comr.); C. I. Bray, Ph.D., Assoc. Prof. An. Husb., Assoc. in An. Invest.; B. W. Fairbanks, B.S., An. Husb. Specialist; O. C. Krum, B.S., Poultry Specialist; L. P. McCann, M.S., Assoc. Prof. An. Husb.; E. J. Maynard, M.S., Assoc. in An. Husb.; C. N. Shepardson, M.S., Assoc. Prof. and in charge Dairy Mfrs.; H. H. Smith, B.S., Ass. Prof. and Ass. An. Husb.; O. C. Ufford, B.S., Ass. Prof. and Ass. Poultry Husb.

Gainesville, Fla. (U.S.A.).
Dept. of Animal Industry, Coll. of Agric. J. M. Scott, B.S., An. Indus. and Vice Dir. of Sta.; C. H. Willoughby, B.Agr., M.A., Prof. An. Husb. and Dairying; N. W. Sanborn, M.D., Prof. Poultry Husb.; H. L. Brown, M.S., Ext. Dairyman; N. R. Mehrhof, B.S., M.Agr., Poultry Specialist; Pence Peterson, Dairyman.

Geneva, N.Y. (U.S.A.).
Div. of Animal Industry of the New York Agric. Experim. Sta. Dir. William P. Wheeler, Assoc. in Research (Animal Industry). *Feeding experim. with poultry. Breeding experim. with poultry. A study of soil requirements.*

Göttingen (Deutschland).
Inst. f. Tierzucht. Dir. Prof. Jonas Schmidt.

Gomplitz b. Tetschen a. d. E. (C.S.R.).
Lehrkanzel f. Biol. u. Züchtung der Haustiere der Landwirtsch. Abt. der Deutschen Techn. Hochsch. Prag.

Grignon, Seine et Oise (France).
Labor. de Zootechnie, Ecole nation. d'agric.

Halle a. d. S. (Deutschland).
Inst. für Tierzucht und Molkereiwesen Univ., Sophienstr. 35. Dir. Prof. Dr. G. Frölich. — Priv.-Doc. Dr. Spöttel für Unterricht (Forschung u. Verwaltung der Sammlungen); Ass. Dr. Dalchau (Verwaltung des Haustiergartens); Ass. Dr. Tänzer (Unterricht, Forschung und Verwaltung der Bücherei); Ass. Dr. Wilcke (chem.Untersuchungen); Ass. Diplomlandwirt Thiele (Versuche auf dem Versuchsgut der Domäne Lettin); Ass. Dr. Lüthge (Herdbuchwesen). *2 Plätze. Chem. u. mikroskopische Untersuchungen. — Apparatur f. chem. Untersuchungen, für Blut- u. serol. Untersuchungen.* ℗ Skelette, Felle, Haar- bzw. Wollproben von Haustieren und ihren verwandten Wildformen. — *Versuchsgut auf Domäne Lettin.*

Hannover (Deutschland).
Inst. für Tierzucht u. Vererbungsforschung, Tierärztl. Hochsch. Dir. Prof. Dr. Carl Kronacher. *Haar- u. Wolleforschung; Haustiergenetik; Konstitutionsforschung; Inzuchtforschung, Körperbaulehre, Pferde-, Schweine-, Ziegenzucht.*

Helsinki (Finnland).
Inst. der Haustierkunde, 3 A, Regeringsgatan. Dir. Dr. Rich. Hinderson.

Hohenheim b. Stuttgart (Deutschland).
Inst. für Tierzuchtlehre der Landwirtschaftl. Hochschule. Dir. Prof. Dr. phil, et med. vet. Adolf Richard Walther. — Dr. Albrecht Walther, Diplomlandwirt Otto Sommer, Dr. Otto v. Bolzano, Ass. *2 Plätze. Untersuchungen über Rassenkunde und Vererbung. — Geflügelzucht- und Kleintierzuchtanstalt.* ℗ Skelette von Haustierrassen.

Iwanowo-Voznessensk (U.d.S.S.R.).
Zootechn. Inst. a. d. Polytechn. Inst. Dir. Prof. P. G. Borissov.

Jassy (România).
Labor. de Zootechnic Univ. Dir. Prof. Agricola Cardaș. *Zootechnik. Bovides.*

Jena (Deutschland).
Anstalt f. Tierzuchtlehre. Dir. Prof. Dr. Pritzwald-Stegmann.

Kamenetz-Podolsk, Ukraine (U.d.S.S.R.).
Labor. für allg. Tierzuchtlehre des Inst. für Landwirtschaft „K. Marx". Dir. Prof. O. Melnyk.

Kingston, R.I. (U.S.A.).
Dept. of Animal Industry, State Coll. of Agric. J. E. Ladd, M.S., Prof. An. Husb.; C. E. Brett, B.S., Instr. Poultry Husb.; Howland Burdick, B.S., Ass. Prof. Dairying, Mgr. of Farm; K. H. Goodner, M.A., Instr. Bact., Ass. in An. Brdg. and Pathol.; H.G.May, Ph.D., Prof. Bact., An. Brdg. and Pathol.; N. F. Waters, B.S., Ass. in An. Brdg. and Pathol.

Königsberg i. Pr. (Deutschland).
Inst. f. Betriebslehre u. Tierzucht. Dir. Prof. Br. Skalweit u. Prof. W. Völtz.

Leipzig (Deutschland).
Inst. f. Tierzucht u. Geburtskunde d. Univ. Dir. Prof. J. Richter.
Inst. f. Tierzucht u. Molkereiwesen d. Univ. Dir. Prof. Golf.

Leningrad (U.d.S.S.R.).
Abt. f. Zootechnik d. Staatl. Inst. f. experim. Agronomie. Dir. M. J. Djakow.

León (España).
Labor. de Morfol., Zootecnica, Agric., Escuela de Vet. Dir. Prof. F. Pedro Gonzalez.

Lexington, Ky. (U.S.A.).
Dept. of Animal Industry, Coll. of Agric., Univ. of Kentucky. E. S. Good, M.S., Chair. An. Indus. Group. — J. J. Hooper, M.S., Prof. Dairy Husb., in charge Dairy Invest.; Jos. H. Martin M.S., Assoc. Prof. and in charge Poultry Husb.; W. S. Anderson, M.A., Prof. Genetics, in charge Horse Husb.; J. O. Barkman, B.S., Ass. Prof. Dairy Mfrs. and Field Agt. in Dairying; J. D. Foster, M.S., Insp. in charge Creameries; P. H. Goodwin, Field Agt. in Poultry Husb.; Amanda H. Harms, B.S., Ass. Pathogenic Bact. in An. Husb.; C. E. Harris, B.S., Field Agt. in Poultry Husb.; W. J. Harris, B.S., Ass. Prof.

and Ass. An. Husb.; L. J. Horlacher, M.S., Assoc. Prof. and in charge Sheep Invest.; B.S., Field Agt. in Poultry Husb.; R. C. Miller, M.S., Field Agt. in An. Husb.; J. W. Nutter, Supt. Dairy Farm.; E. M. Prewitt, B.S., Field Agt. in Dairying; Wayland Rhoads, M.S., Field Agt. in An. Husb.; H. G. Sellards, B.S., Field Agt. in An. Husb.; J. R. Smyth, B.S., Field Agt. in Poultry Husb.; E. J. Wilford, M.S., Assoc. Prof. and Ass. An. Husb.; W. W. Dimock, D.V.M., Prof. and Head Dept. of Vet. Sc.; J. F. Bullard, D.V.M., Ass. Vet.; P. R. Edwards, Ph.D., Ass. Bact., F. E. Hull, D.V.M., Instr. Vet. Sc., Ass. Vet.; T. P. Polk, D.V.M., Field Agt. in Vet. Sc. *Etiol. and Control of Infectious abortion of mares cows and sows.* ⱷ Guinea pigs and rabbits.

Lincoln, Neb. (U.S.A.).
Dept. of Animal Husbandry of the Univ. of Nebraska. Dir. Prof. H. J. Gramlich. — Assoc. Prof. W. J. Loeffel.

Logan, Utah (U.S.A.).
Dept. of Animal and Poultry Husbandry, Agric. Coll. of Utah. K. C. Ikeler, M.S., Head Dept. of An. Husb., Livestock Specialist; Byron Alder, B.S., Assoc. Prof. Poultry Husb., Poultryman; R. J. Becraft, M.S., Ass. Prof. in charge Range Mgt.; A. C. Esplin, B.S., Ass. Prof. and Ass. An. Husb. Wool and Sheep Specialist; W. H. Warner, Ass. in Poultry Husb.

Lugansk, Dongebiet, Ukraine (U.d.S.S.R.).
Kabinett f. Zootechnik d. Landwirtsch. Timirjazew Technikums, Počtovyi jačšikz 3.

Manhattan, Kan. (U.S.A.).
Animal Husbandry Dept., Kansas, State Agric. Coll. H. L. Ibsen (Prof. of Genetics).

Montpellier (France).
Labor. de Zootechnie de l'Ecole nationale d'agric. Dir. Mr. Henri Cottier. — Mr. Dedieu, Chef des travaux. *2 places pour alimentation.*

Moskau (U.d.S.S.R.).
Zootechn. Versuchsstation, Landw. Timirjazew, Akad. Dir. E. A. Bogdanow.

München (Deutschland).
Inst. f. Tierzucht Univ., 2, NO 6, Veterinärstr. 6. Dir. Geh.H.R. Prof. Dr.Dr.h. c. Leonhard Vogel. — Ass.: Dr. Franz Kugler, Dr. Walter Koch. *Versuchsstall für Züchtung landwirtschaftlicher Haustiere.*
Inst. u. Labor. für Tierzucht u. Züchtungsbiol. Dir. Dr. H. Henseler, o. Prof. der Techn. Hochsch. — Dr. Amschler, Priv.Doc.; L. Krüger, Ass.; außerdem weit mehrere Hilfsass. *20—30 Plätze für alle biol. Untersuchungen. — Das Inst. verfügt über chem. und physikal.-mechan. sowie techn. Labor., die mit neuesten Apparaten und Hilfsmitteln vielseitig ausgestattet sind. Engste Arbeitsgemeinschaft mit den großen chem., physikal., techn. Labor. und Inst. der Techn. Hochsch. München.* ⱷ Vor allem alle Gegenstände, die das Gebiet der Haustierzucht (insbesondere Methoden der Züchtung) betreffen und berühren.

Newark, Del. (U.S.A.).
Dept. of Animal Industry, School of Agric. C. C. Palmer, D.V.M.S., Prof. An. Industr.; T. A. Baker, B.S., Prof. of An. Husb.; A. E. Tomhave, M.S., An. Husb.; H. R. Baker, M.S., Ass. Prof. Biol. Research assoc. in Poultry Diseases; M. Clarihew, A.B., Ass. in Biol.; C. W. Mumford, Supt. of Poultry; H. S. Palmer, B.S., Poultry specialist.

Oberer Hardthof b. Gießen (Deutschland).
Tierzuchtsinst. der Univ. Gießen. Dir. Prof. Dr. Hermann Kraemer. — Dr. Heinrich Lang, Ass. (Verwaltung und wissenschaftl. Versuche). *Tierzucht,*

Beurteilung der Tiere, Vererbungsfragen, Abstammungsfragen, Rassengeschichte der Haustiere.

Orono, Me. (U.S.A.).
Dept. of Animal Industry, Coll. of Agric. L. S. Corbett, M.S., Prof. An. Indus.; L. M. Dorsey, M.S., Assoc. Prof. An. Indus.; L. P. Gardner, M.S., Ass. Prof. Poultry Husb.; H. W. Hall, B.S., Instr. An. Indus.; R. F. Talbot, B.S., Dairy Specialist; O. M. Wilbur, M.S., Poultry Specialist.

Perm (U.d.S.S.R.).
Kabinett für allgem. u. spez. Zootechnik. Dir. Doc. V. P. Sergovancev.

Perugia (Italia).
Gabinetto di zootecnica della R. Univ. Dir. A. Cugnini.

Poznań — Posen (Polska).
Inst. f. spez. Tierzuchtlehre. Dir. Prof. M. Pankowski.
Zootechnisches Inst. der Univ. Dir. Z. Moczarski.

Praha — Prag (C.S.R.).
Inst. f. animal. Biotechnol. d. čech. techn. Hochschule, Vršovice, Havličkovy sady. Dir. Prof. Dr. Jarolav Just.
Inst. f. allg. u. spezielle Zootechnik d. čech. techn. Hochsch., Vršovice, Havličkovy sady. Personal: confer institutio sequens.
Inst. f. Tierzuchtbiol. d. Ackerbauminist., Vršovice, Havličkovy Sady. Dir. Prof. MUDr. et Ph. Dr. Fr. Bilek. — a) Ass.: Ing. B. Tichota, Ing. Ungermann; b) Insp.: Ing. Dr. M. Tumlířová, Ing. F. Valento. ⱷ Skelette, Schädel der Rassentiere.

Puławy (Polska).
Abt. f. Tierzucht des Staatl. Wissensch. Inst. f. Landwirtsch. Dir. Prof. L. Adametz.

Raleigh, N.C. (U.S.A.).
Dept. of Poultry Labor. of the State Coll. Station. Dir. B. F. Kaupp.

Reno, Nev. (U.S.A.).
Dept. of Animal Industry, Coll. of Agric. C. E. Fleming, B.S., Range Mgt.; V. E. Scott, B.S., Prof. Dairying, Dairy and Poultry Specialist; F. W. Wilson, M.S., Prof. An. Husb.

Riga (Latvija).
Inst. f. Tierphysiol. u. Tierzucht. Confer: Physiol. compar.

Roma (Italia).
Ist. Sperim. Zootechnico. Dir. M. Bartolo.

Saratow (U.d.S.S.R.).
Inst. f. Allgem. Tierzucht des Inst. f. Landwirtsch. u. Melioration. Dir. Doc. N. V. Gorjainova.

St. Paul, Minn. (U.S.A.).
Univ. Farm. Confer: Labor. cult. plant gen.

Siverskaya Balogorka,
Gouv. Leningrad (U.d.S.S.R.).
Landwirtschaftl. Versuchsstation der nord-westlichen Provinzen. Confer: Inst. gen. cult. plant. et animalium.

Sofia (Bulgarien).
Inst. f. allg. Tierzucht d. Landw. Fac. Dir. Prof. Jelü A. Gantscheff. — Georg S. Chlebaroff, Doc.; Peter Petkoff, Doc. für Anat. u. Physiol. der Haustiere, Histol., Insektenbiol.; Ass.: A. Kanettarjieff, Johan Popoff, Al. Petroff.
Inst. für spezielle Tierzucht, Zentral-Versuchsstation für Geflügelzucht. Dir. G. Chlebaroff. *Präzise Schädelmeßinstr., anthrop. Meßinstr., Hypogoniometer nach Rolf Dürst. Vollständige Einrichtung für Wolluntersuchungen nach der Kronacher Methode. Einrichtung für Stoffwechselversuche mit großen und kleinen Haustieren.*

Stambul (Türkei).
Inst. f. Hygiene u. Tierzucht Tierärztl. Hochsch. Confer: Hygenia animal.

State College, Pa. (U.S.A.).
Dept. of Dairy Husbandry and Bact., School of Agric. A. A. Borland, M.S., Head of Dept. — A. L. Beam, M.S., Prof. Dairy Prod.; S. I. Bechdel,Ph.D., Prof. Dairy Prod.; S. J. Brownell, B.S., M.Agr., Ass. Prof. Dairy Husb. Ext.; C. D. Dahle, M.S., Assoc. Prof. Dairy Mfrs.; F. J. Doan, M.S., Ass. Prof. Dairy Mfrs.; E. B. Fitts, B.S., Prof. Dairy Husb. Ext.; C. R. Gearhart, B.S., Ass. Prof. Dairy Husb. Ext.; F. P. Knoll, Ass. in Dairy Mfrs.; M. H. Knutsen, M.S., Assoc. Prof. Bact.; R. H. Olmstead, M.S., Ass. Prof. Dairy Husb. Ext.; I. O. Sidelmann, Ass. in Dairy Husb. Ext.; A. C. Simpson, B.S., Ass. Prof. Bact.; W. D. Swope, M.S., Assoc. Prof. Dairy Husb.; G. A. Taylor, B.S., Ass. Prof. Dairy Mfg. Ext.; R. P. Tittsler, B.S., Instr. Bact.; P. S. Williams, M.S., Ass. Prof. Dairy Husb.

Stellenbosch (South Africa).
Dept. of Animal Husbandry Fac. agric. Univ. Dir. Prof. J. H. W. Th. Reimers.

Stockholm (Sverige).
Abt. f. Tierzucht Vet. Hochsch. Dir. Prof. Nyström.
Abt. f. Tierzucht Zentralanst. f. landwirtsch. Versuchswesen. Dir. Prof. Dr. N. Hansson. — J. E. Haglund, Oberass. f. Molkereiversuche; 1 Laborator, 3 Ass.

Tartu — Dorpat (Eesti).
Kabinett f. Tierzucht, Peterburi t. 76. Dir. Dot. N. Rootsi. — Ass. A. Munga.

Taschkent (U.d.S.S.R.).
Inst. f. Viehzucht. Dir. Prof. M. G. Cislov.

Tiflis (U.d.S.S.R.).
Kabinett f. Zootechnik Techn. Hochsch.

Tschechnitz, Schles. (Deutschland).
Preuß. Versuchs- und Forschungsanstalt für Tierzucht. Dir. Prof. Dr. Zorn. — Wissenschaftl. Hilfsarbeiter: Tierzuchtinspektoren Dr. Richter, Dr. Mosig, Dr. Kroll. *3 Plätze für Züchtungsbiol. und Tierzüchtung.* — *Labor., große Viehbestände.* ⓟ Haustiere. — Größere private Zuchtbetriebe sind angeschlossen.

Tunis (Tunis).
Labor. d'Anat. et Physiol. des animaux et Zootechnie. Dir. Prof. Duvanchelle.

Utrecht (Holland).
Inst. voor zoötechniek derVet.Fac.Univ., Biltstraat No. 172. Dir. Prof. Dr. H. M. Kroon. — Conserv.: Dr. G. M. van der Plank.

Vec-Auce (Latvija).
Versuchsfarm Landwirtsch. Fac. Univ. Dir. Doc. J. Apstīs.

Vologda (U.d.S.S.R.).
Zootechn. Inst. mit Lehrlabor. Milchwirtschaftl. Inst. Dir. Prof. N. N. Pelechov.

Wageningen (Holland).
Rijkslandbouwproefstation voor veevoedercontrôle en hulpmiddelen (Viehfutterkontrolle u. Hilfsmittel). Dir. Dr. B. R. de Bruyn. — Abt.-Dir.: Dr. J. A. Ezendam (Pflanzenkunde); Jr. G. B. v. Kampen (Chemie); R. W. Tuinzing (Chemie); C. J. Kole, W. M. v. d. Mijll Dekker (Pflanzenkunde).
Zootechn. Labor. Landwirtschaftl. Hochsch. Dir. D. L. Bakker.

Warszawa (Polska).
Inst. f.Viehzucht, Hochsch. f. Land- u. Forstwirtschaft. Dir. Prof. J. Rostafinski.

Washington, D.C. (U.S.A.).
Animal Husbandry Division, U.S. Dept. of Agric. P. E. Howe (Biol. Chem.). *Biol. Chem.: nutrition, fasting and proteins.*
Bureau of Animal Industry, U.S. Dept. of Agric. Dir. J. R. Mohler, Chief.

Weihenstephan b. Freising (Deutschland).
Inst. für Tierzuchtlehre, Hochsch. Dir. Prof. Dr. I. Spann. — Dr. Eugen Probst, Ass. ⓟ Skelette von Haustieren.

Wheaton, Ill. (U.S.A.)
Dept. of Animal Biol., Wheaton Coll. Prof. C. W. Howart *(Parasitologist, Sericiculturist).*

Wien (Österreich).
Inst. f. Tierzucht, Vet. Hochsch. Dir. Prof. Dr. K. Keller.
Inst. f. Tierzuchtlehre, Hochsch. f. Bodenkultur. Dir. Prof. Dr. L. Adametz.

Wooster, Ohio (U.S.A.).
Dept. of Animal Industry, Ohio Agric. Experim. Sta. Gustav Bohstedt, Ph.D., Chief. — D. S. Bell, M.S., Ass. (Sheep Invest.); R. M. Bethke, Ph. D., Ass. (Nutr.); B. H. Edgington, M.D.C., D.V.M., Assoc. (Path.); C. H. Hunt, A.M., Ass. (Nutr.); D. C. Kennard, Ph.C., B.S., Assoc. (Poultry Invest.); L. B. Nettleton, Ass. (Poultry); W. L. Robison, M.S., Ass. (Swine Invest.); H. L. Sassaman, M.S., Ass. (Nutr.); A. R. Winter, M.S., Ass. (Nutr.).

Woronesch (U.d.S.S.R.).
Kabinett f. allgem. Zootechnik mit Labor.
Labor. f. spez Zootechnik, Landwirtsch. Hochsch.

Zagreb — Agram (S.H.S.).
Inst. für Tierzucht und angewandte Biol. der land- und forstw. Fac. der Univ., Tvornička 10. Dir. Prof. Dr. Sava Ulimansky. — Prof. Dr. Albert Ogrizek, 1. Ass.; Ing. Milutin Petrović, 2. Ass. *Derzeit 10 Plätze für Wolluntersuchung und Vererbungsstudien.* — *Alle neuen Wolluntersuchungsapparate, Kinoaufnahme- und Wiedergabeapparat usw.* ⓟ Schädel der Haustiere, Wollproben. — Experim. Inst. mit Versuchsstall am Fac.-Versuchsgut Maksimir.

Laboratoria culturae animalium specialia.
1. Cultura piscium.

Confer: *Laboratoria marina* (pag. 343) *et Limnologia* (pag. 347).

Alger (Algérie).
Service général des Pêches maritimes. Dir. J. P. Bounhiol.

Altona, Elbe (Deutschland).
Forschungsinst. f. d. Fischindustrie e. V. Dir. Dr. Hanns Lengerich. — Dr. H. Metzner. *2 Arbeitsplätze für Gäste. Einrichtung für Bact., Biochem., physiol. Chem., Nahrungsmitteltechnik.*

Bangkok (Siam).
Siamese Dept. of Fisheries. Dir. H. McCormick Smith.

Bergen (Norge).
Abt. f. Meeresunters. des Fischereidirektorats (Fiskeridirektorens Kontor, Avdeling for Havundersøkelser). Fischerei-Dir. Sigurd Asserson. — Paul Bjerkan, Biol. der Sprotte, der Pleuronectiden, des

Hummers, der Krabben usw.; Einar Koefoed, Jungfische, Eier usw., Bibliothekar; Einar Lea, Biol. des Herings; Oscar Sund, Biol. der Gadiden, insbes. des Dorsches; (z. Z. ledig) Ozeanographie. *Die Abt. verfügt über ein speziell gebautes Motorschiff, den „Johan Hjort" das für biol. Untersuchungsfahrten innerhalb des Norwegischen Meeres verwendet wird und für ozeanographische Arbeiten daselbst.* ⓟ Schuppen von Heringen, Sprotten, Gadiden usw. Plankton von den eigenen Untersuchungsfahrten.

Berlin (Deutschland).

Preußische Landesanstalt für Fischerei, Landw. Hochsch., Friedrichshagen. Dir. Prof. Dr. H. H. Wundsch. — Dr. R. Czensny, Chemiker; Dr. F. Schiemenz, biol. Ass.; Dr. W. Schäperclaus, biol. u. bact. Volontär-Ass.; Dr. H. Potonié, Biol. Volontär-Ass.; Dr. H. Fritzsche, Biol. Volontär-Ass. *Plätze nach Bedarf in größerer Zahl für hydrobiol. und fischereibiol. Untersuchungen. — 2 Ruderkähne auf dem großen Müggelsee, kleine Teiche, Zement- u. Glasaquarien, hydrobiol. Fanggeräte, Geräte für bact. Untersuchung, chemisches Labor.* ⓟ Fischereibiol. wichtige Gegenstände Deutschlands. — Versuchs- und Lehrwirtschaft „Jägerhof" am Sakrowsee bei Cladow (Bez. Potsdam).

Deutsche Wissenschaftliche Kommission für Meeresforschung. Vors.: Dr. C. Heinrici, früher Staatssekretär im Reichsministerium für Ernährung und Landwirtschaft. — Stellvertr. Vors. der Kommission: Geh.R. Heincke, Helgoland. Mitglieder der Deutschen Wissenschaftlichen Kommission für Meeresforschung sind die Herren Geh.R. Brandt, Henking; Professoren Ehrenbaum, Schott, Mielck, v. Buddenbrock, staatliche Fischerei-Dir. Lübbert und Präsident des Deutschen Seefischerei-Vereins Freiherr v. Maltzahn; ao. Mitglied Herr Studien-Dir. Dr. Strodtmann.

Boulogne-sur-Mer (France).

Station Aquicole (Labor. de l'Office Sc. et Technique des PêchesMaritimes),17,Boulevard de Châtillon. Dir. Jean Le Gall, Prof. Agrégé de l'Univ. — Albert Bride-Etivant, Préparateur; Madame Jean Le Gall Préparatrice. *2 places pour biol. des animaux marins comestibles. Industrie de la pêche.* ⓟ Faune marine. Plancton.

Brescia (Italia).

R. Stabilimento Ittiogenice. Dir. Prof. Cav. Lo Giudice Pietro. *Ricerche i Idrobiol. applicata alla pesco. Piscicultura.*

Budapest (Ungarn).

Royal Hungarian Biol. Station for Fisheries and for sewage purification, II, Debröi ut' 15. Dir. E. Unger.

Cagliari (Italia).

Scuola di pesca R. liceo sc. Dir. M. Pasquale.

Castiglione (Algérie).

Stat. expérim. des Pêches. Dir. J. P. Bounhiol.

Chapel Hill, N.C. (U.S.A.)

U.S. Bureau of Fisheries. H. R. Leiwell (Investigator).

Děčín-Libverd—Tetschen-Liebwerd (C.S.R.).

Lehrkanzel f. Zool., Fischzucht u. Teichwirtsch. Confer: Zool. generalis.

Dublin (Ireland).

Dept. of Fisheries, 3, Kildare Place. Dir. Chief Inspector of Fisheries.

Eberswalde (Deutschland).

Fischzuchtanstalt, Forstl. Hochsch. Dir. Prof. Eckstein.

Edinburgh (Scotland).

Fishery Board for Scotland. Chairman: D. T. Jones, C.B.E., J.R.S.E.
Survey of British Fresh-Water Lakes, VillaMedusa, Bosnell Road.

Fairport, Ia. (U.S.A.).

U.S. Bureau of Fisheries. A. H. Wiebe.

Fusan, Korea (Japan).

Governm. Fishery Experim. Station, Makinoshima. Dir. Dr. Yôjirô Wakiya.

Halle a. d. S. (Deutschland).

Fischlabor. des Bact. Inst. der Landwirtschaftskammer Prov. Sachsen. Dir. Prof. Dr. Räbiger.

Hamburg (Deutschland).

Fischereibiol. Abt. des Zool. Staatsinst. u. Zool. Mus., 5, Kirchenallee 47. Dir. Prof. Dr. Ehrenbaum. — Dr. W. Schnakenbeck, Kustos; Dr. H. Lissner, wissensch. Ass. der Deutschen Wissensch.; Dr. H. Heidrich, Kommission f. Meeresforschung. *3 Plätze für Fischereibiol. Arbeiten u. literarische Studien.* ⓟ Alles die Biol. der Fische im In- u. Auslande Betreffende.

den Helder (Holland).

Rijksinst. voor biol. Visscherijonderzoek. Dir. Dr. H. C. Redeke.

Helsinki (Finnland).

Inst. f. Fischereiuntersuchungen. Dir. Prof. Dr. T. H. Järvi. — Fischereibiologe: Viljo Jääskeläinen, Mag. phil.

Hutami (Japan).

Hutami Branch Experim. Sta. of Imperial Fishery Inst., Hutami-mura, Kakogun, Hyôgo-ken.

Ithaca, N.Y. (U.S.A.).

Fish Cultural Experim. Station of Cornell Univ. Dir. Dr. George C. Embody, Prof. of Aquic. *A fish hatchery and ponds.* ⓟ Fishes and forage organisms for fishes.

Kiew (U.d.S.S.R.).

Ichthyol. Versuchsstat. d. Landw. Inst. Dir. Prof. D. Belling.

Kizaki (Japan).

Kizaki Branch Experim. Sta. of Imperial Fishery Inst., by Lake Kizaki Nagano-ken.

Königsberg i. Pr. (Deutschland).

Fischerei-Inst. der Univ., Tragheimer Kirchenstr. 74. Dir. Univ.Prof. Dr. med. et phil. Alfred Willer. — Dr. phil. Johannes Lundbeck. *4 Plätze im Inst., 4 in der Seefischereistation, 1 in der Teichwirtschaft für hydrobiol. und fischereibiol. Untersuchungen. — Moderne hydrobiol. Apparatur.* ⓟ Tierwelt des Süßwassers und der Ostsee. Fische insbesondere. — Zweiglabor.: Versuchsteichwirtschaft Perteltnicken (Samland). Seefischereistation Neukuhren (Samland).

Leningrad (U.d.S.S.R.).

Bureau für angewandte Ichthyol., Morskaja, 42. Dir. Prof. Dr. Leo Berg. — Ichthyologen: J. N. Arnold, J. F. Prawdin, N. A. Smirnov, M. J. Tichji, N. M. Knipowitsch, A. J. Nedoschiwin, A. I. Rabinerson, V. S. Michin, A. N. Probatov, M. D. Ilijn, M. J. Markun, V. V. Petrov. ⓟ Fische und im Wasser lebende Evertebraten.

Manila (Philippine Islands).

Division of Fisheries and Marinebiol., Bureau of Sc., Dept. of Agric. Dir. Dr. A. W. Herre.

Miyazu (Japan).

Miyazu Provincial Fishery Experim. Sta., Kyôto-fu.

Moskau (U.d.S.S.R.).

Inst. f. Fischereiwirtschaft, Pjatnitzkaja Str. 33. Dir. V.I. Meissner.— S. Awerinzew (Ichthyol. Abt.); N. Alexandrow (Ökonomist. Abt.); G. Drucker (Technol. Abt.); A. Lipin (Hydrobiol. Abt.). Außerdem 3 Spezialisten, 18 Ass. und 4 Laboranten.
Labor. f. Fischzucht u. Fischerei Landw. Akad. „Timirjasev" Dir. Prof. F. Baranov. — Stellv. Prof. A. N. Eleonskij.

München (Deutschland).

Bayer. Biol. Versuchsanstalt für Fischerei, Veterinärstr. 6. Dir. Prof. Dr. Reinhard Demoll. — Chemiker: Prof. Dr. Franz Graf, OberReg.Chemiker; Leiter der Abwasserstation; Prof. Dr. Martin Strell, Reg.Chemiker; Dr. Ludwig Walz, Bodenchemiker; Biologen: Prof. Dr. Marianne Plehn, Konserv.; Prof. Dr. Ludwig Scheuring, Ass.; Dr. Otto Gaschott, Ass. *3 Plätze für Fischerei-Untersuchungen. — Apparate und Hilfsmittel für hydrobiol. und Fischerei-Untersuchungen. — Durch Personal-Union der Direktion verbunden mit dem ,,Hofer-Institut" Teichwirtschaftl.Versuchsanstalt in Wielenbach und dem Inst. für Seenforschung und Seenbewirtschaftung in Langenargen am Bodensee.*

Münster (Deutschland).

Fischereibiol. Inst. d. Landwirtschaftskammer für die Provinz Westfalen. Dir. Dr. Conrad Lehmann.

Nancy (France).

Sect. de Pisciculture de l'Ecole Nation. des Eaux et des Fôrets.

Neukuhren, Samland (Deutschland).

Seefischereistation. Dir. Dr. J. Lundbeck. *4 Plätze für Gäste. Hydrobiol. — Moderne Apparatur.*

Ôchô (Japan).

Ôchô Branch Experim. Sta. of Imperial Fishery Inst., Ochô-mura, Toyota-gun, Hiroshima-ken. Dir. Eiji Kajiyama.

Oslo (Norge).

Statens Forsøksvirksomhet for Ferskvandsfiskeri, Adr. Zool. Mus. Tøien. Dir. Prof. Dr. Knut Dahl. — Dr. O. Olstad, Ass.

Otaru, Hokkaidô (Japan).

Takashima Fishery Experim. Station, Takashima near Otaru.

Ottawa, Ont. (Canada).

Biol. Labor. of the Dept. of Marine and Fisheries. Dir.: A. Halkett, Frits Johansen.

Perteltnicken, Samland (Deutschland).

Versuchsteichwirtschaft d. Univ. Königsberg. *1 Platz für Gäste.*

Portobello (New Zealand).

Marine Fisheries Investigation Station. Dir. by a Board. Chairman: Hon. G. M. Thomson. — Maxwell Young, F.C.S., Biologist. *2 places. Marine Zool. ⓟ Exchange list of fauna available on enquiry.*

Poznań — Posen (Polska).

Inst. f. Jagd u. Fischerei. Dir. Prof. E. Schechtel.

Prince Rupert (British-Columbia).

Pacific Experim. Sta. for Fisheries (Biol. Board). Dir. D. B. Finn.

Puławy (Polska).

Inst. d'Agric. de l'Etat. Confer: Agric. spec.

Riga (Latvija).

Ichtyol. Labor. beim Fischereiamt des Landwirtschaftsministeriums. Dir. Wilhelm Mannsfeld. ⓟ Plankton Lettländischer Gewässer. Hirudinea Lettlands.
Inst. f. Fischzucht. Dir. Lect. J. Rubbards.

Ruda Maleniecka (Polska).

Experim. Stat. f. Fischzüchtung. Dir. Dr. Boris Dixon. *Angewandte Ichtyol.*

Sapporo (Japan).

Fishery Inst., Coll. of Agric., Hokkaidô Imper. Univ. Prof. Dr. Madoka Sasaki; Prof. Tsunenobu Hujita.

Seattle, Wash. (U.S.A.).

Labor. for study of halibut (Hippoglossus), International Fisheries Commission of United States and Great Britian, Univ. of Washington. Dir. William F. Thompson. — H. A. Dunlop, Ass. Dir. (Age and Growth); F. H. Bell (Racial Studies of halibut); L. L. Bolton (Spawning and early life history of halibut); R. V. Butt (Studies of Growth; gear used); W. F. Thompson (Tagging and migration, growth). *Research is entirely upon halibut and is carried on partly at the labor. and partly at sea. — Guests specially invited and necessary special apparatus arranged to suit.*
Coll. of Fisheries, Univ. of Washington.

Stanford Univ., Cal. (U.S.A.).

Salmon Fisheries U.S.Board of Fisheries. W. H. Rich (Chief investigator); H. B. Holmes (Junior Aquatic Biologist). *Life-histories and ecol. of the Pacific salmons, especially of the genus oncorynchus.*

Stockholm (Sverige).

Inst. f. Zool. Jagd u. Fischeuer Forsthochsch. Confer: Zool. generalis.
Mus. für Ackerbau u. Fischerei. Inspektor: Prof. Dr. Paul Hellström.

Tanabe (Japan).

Provincial Fishery Experim. Sta., Wakayama-Ken.

Tateyama (Japan).

Takashima Branch Experim. Sta. of Imper. Fishery Inst., Chiba-Ken.

Terminal near San Pedro, Cal. (U.S.A.).

California State Fisheries Labor. In charge: W. L. Scofield. — Sc. Ass.: J. A. Craig, Dr. Frances N. Clark; Ruth Thompson (Librarian); Temporary sc. Ass. 5 to 10. *A few marine fish and invertebrates from local waters for study.*

Toba (Japan).

Provincial Fishery Experim. Stat., Mieken.

Tokohashi (Japan).

Imper. Piscicult. Experim. Sta., Muroyoshida near Toyohashi Aichi-Ken. Dir. Yashiichi Matsui. — Ass.: A. Kumada.

Tôkyô (Japan).

Zool. Labor., Imperial Fishery Inst. Confer: Zool. generalis.
Bureau of Fishery Dept. of Agric. and Forestry.
Inst. of Fishery Fac. of Agric., Tôkyô Imp. Univ., Komaba near Tôkyô. Prof. Dr. K. Kishinoue (Ichthyol.); Prof. Dr. J. Hara (Econom. Oceanogr.); Prof. Dr. S. Machida (Fishery); Prof. Dr. M. Yamakawa (Biochemistry); Ass. Prof. Ikusaku Amemiya.

Toronto (Canada).

Dept. of Biol. Univ. Confer: Zool. generalis.

Toyotagun, Hiroshima Ken (Japan).

Ohcho Fish-culturing Station. Dir. Eiji Kajiyama. *Sea-fish Culturing. Sea-fish Hatching.*

Trondhjem (Norge).

Trondhjems biol. Station. Dir. O. Nordgård. *Eine Hauptaufgabe für die Station ist Schollenzucht und für diesen Zweck hat der Dir. 4 praktische Ass. oder Mitarbeiter. 6 Arbeitsplätze für Gäste. — Arbeitsaquarien im Labor. und Gelegenheit für Exkursionen mit Motorkutter ,,Gunnerus".*

Vedder Crossing (Britisch Columbia).
Pacific Salmon Research Station. R. E. Foerster (Biologist).

Villa do Conde, b. Porto (Portugal).
Staatl. Stat. f. Fischzucht. Dir. A. Pereiro Nobre. *Fischzucht in den süßen Gewässern Portugals u. wiss. Stud. über die Süß- u. Salzwasserfauna* — *5 Labor., 1 Mus., 1 Aquarium, 1 Vivarium.*

Vladivostok (U. d. S. S. R.).
The Pacific Ocean Sc. Fishery Research Station, 1st May Street-Dalriba. Staff of workers: Dir. Dr. A. N. Dershovin; Sub-Dir. M. L. Pyatakov. — Branch of Fisheries: A. N. Dr. Dershavin, Superintendent und Chief Ichthyologist; S. J. Berezofskii, Chief Ichthyologist; G. S. Lindberg, Ichthyologist; A. S. Ambros, Ichthyologist; A. D. Baturin, Zoologist; B. P. Pentegov, Chemist; V. J. Krivoborskii, Economist; N. V. Milovidova, Ass.; Branch of Fish-Culture: S. S. Kuznetsoff, Superintendent; N. N. Belov, Ichthyologist; Branch of Hydrobiol.: M. L. Pyatakov, Superintendent and Chief Zoologist;

K. O. Gomoyunov, Chief Hydrologist; S. S. Zachs, Zoologist.

Vodňany (C. S. R.).
Fischerei und Hydrobiol. Station. Dir. V. J. Štěpán.

Warszawa (Polska).
Inst. f. Ichtyobiol. u. Fischzucht, Hochsch. f. Landu. Forstwirtsch. Dir. Prof. F. Staff.

Washington, D. C. (U. S. A.).
U.S. Bureau of Fisheries, S.W. Corner 6 and B streets. L. Radcliffe (Deputy Commissioner).

Wellington (New Zealand).
Fisheries Dept. Dir. A. C. Hefford, M.Sc.

Wielenbach (Deutschland).
„Hofer Inst.", Teichwirtschaftl. Versuchsanst. Dir. Prof. R. Demoll.

Wien (Österreich).
Inst. f. Ichtyol., Veterin. Hochsch. Dir. Prof. Dr. G. Fiebiger.

2. Sericicultura (Cultura bombycis).

Confer: *Zoologia generalis* (pag. 356) et *Biochemia generalis* (pag. 435), *Entomologia* (pag. 381), *Anat. pathol. animalium* (pag. 467), *Evolutio* (pag. 341).

Ayabe (Japan).
Jôtan Sericicultural Inst., Kyôto-fu.

Canton (China).
Government Bureau of Sericiculture. Dir. Prof. Dr. Charles W. Howard. *Sericiculture, med., entomol. and parasit.*

Dairen, South Manchuria (China).
Sericicol. Labor., Central Research Inst. of South Manchurian Railway Co.

Ferghana, Turkestan (U. d. S. S. R.).
Station Séricicole. Dir. L. Wyschinsky.

Fukuoka (Japan).
Sericicol. Inst., Coll. of Agric., Kyûshû Imperial Univ. Dir. Prof. Dr. Yoshimaro Tanaka. — Ass. Prof. Eisaku Kawaguchi.

Hanazono near Kyôto (Japan).
Sericicultural Labor., Kyôto Imperial Sericicultural Coll. Dir. Prof. H. Suzuki.

Kumamoto (Japan).
Provincial Sericicultural Experim. Station, Demizumura near Kumamoto.

Kyôto (Japan).
Labor. Physiol. and Anat. of Silk Worms, Kyôto Imperial Sericicultural Coll., Hanazono near Kyôto. Dir. Prof. Motoi Sakurai (z. Zt. in Europa).
Pathol. Labor. (of Silk-worms), Kyôto Imperial Sericicultural Coll., Hanazono near Kyôto.

Montpellier (France).
Station de Sériciculture, Ecole d'agric.

Moskau (U. d. S.S.R.)
Forschungsstation f. Seidenraupenzucht. Bakuninstr. 84. Dir. Doz. B. N. Michin. Abt. f. experim. Biol.: Prof. A. W. Anutschin *(Biolog. u. Genetik.);* Prof. S. A. Ussow *(Histol u. Embryol.).* Abt. f. Seidenraupenzucht: Leiter: A. D. Platowa. *Labor. f. experim. Seidenraupenzucht. Labor. f. Kokon- u. Seidenforschung. Versuchsgarten (Pflanzung v. Morus alba). Museum.* Ass. M. G. Sserebrennimow, B. S. Nikolskaja. Wissensch. Mitarbeiter: A. W. Maslennikowa, K. M. Hartulari, E. S. Kartaschewa, T. M. Gurskaja, H. S. Lotterstein. *3 Plätze für Gäste,*

(*Tische mit vollständiger Ausrüstung).* ℗ Systematische u. biol. Sammlungen, sowie wissenschaftl. Material über Bombyx mori.

Nagano (Japan).
Provincial Sericicultural Experim. Sta.

Nantung (China).
Labor. de Sériciculture, Textile Coll. Dir. Darwin Wen.

Padova (Italia).
R. Stazione Bacol. Sperimentale, Brusegana. Dir. Pigorini Luciano. — T. Gennaro, Vice-Dir.; Dr. Roberto di Tocco, Ass.; Dr. Amelia Tonon, Ass. *10 posti per morfol. e 1stol.; fisiol. e. chimica fisiol.; microbiol.; allevamenti (anche a scopo di ricerche di genetica).* ℗ Preparati morfol., anat., istol. del Bombyx mori sano e malato. Diapositive riguardanti il Bombyx mori.

Paris (France).
Station séricicole de l'Inst. des recherches agron., VIIe, 42bis, Rue de Bourgogne.

Taschkent (U. d. S. S. R.).
Station Centrale de Sériciculture d'Asie Centrale, rue de Kafanow, 1. Dir.-Sc. Prof. Eraste Poyarkoff. — E. Gavriil Pokrowsky, Ass.; Cathérine Gorpintschenko, Ass.; Nicolas Jwirbliss, Chef de la Station de Technol. de Cocons et de la Soie; Bschirulla Khodjaew, Ass.; Léon Wyschinski, Chef de la Station filiale de Fergana; Géorge Propatschikh, Ass. *10 à 18 places pendant la saison de la vie du ver à soie. Observations et expér. sur le ver à soie: morphol., anat. histol., physiol., bact.* ℗ Animaux, producteurs de la soie, races du ver à soies. Plantes servant comme nourriture à ces animaux. — La Station se divise en 2 section: Section de biol. du ver à soie et Section de technol. des cocons et de la soie.

Tôkyô (Japan).
Labor. of Mulberry Culture, Tôkyô Imperial Sericult. Coll. Confer: Cultura mori.
Physiol. Labor. (of Insects), Imperial Sericicult. Experim. St. Confer: Physiol. compar.

Ueda (Japan).
Sericicultural Labor., Ueda Imperial Sericicultural Coll., Nagano-Ken.

3. Apides.

Confer: *Entomologia* (pag. 381).

Berlin (Deutschland).
Inst. für Bienenkunde Landw. Hochsch, -Dahlem, Lentze-Allee 86. Dir. Prof. Dr. Ludwig Armbruster. *3 Plätze für Histol., Chemie, Kinematographie. — Temperatur-Registrierapparat nach Siemens. Honigsammlung, Wachssammlung, apistische Bildersammlung, Bienenzuchtmuseum, Bienenwirtschaft.* ☽ Apides, Hymenoptera. — 2 Belegplätze, 1 Außenbienenwirtschaft.
Biol. Reichsanst. für Land- u. Forstwirtschaft, Labor. für die Bekämpf. d. Bienenkrankh., -Dahlem, Königin-Luise-Str. 19. Dir. Reg.R. Priv.Doc. Dr. A. Borchert.

College Station, Tex. (U.S.A.).
Agric. Experim.Sta. Confer: Labor.cult. plant. gen.

Erlangen (Deutschland).
Anstalt f. Bienenzucht, Univ. Dir. Prof. A. Fleischmann; E. Zander.

Jena (Deutschland).
Univ.-Bienenstand. Dir. Pfarrer Ludwig.

Krasnodar, Kuban (U.d.S.S.R.).
Labor. für Bienenzucht des Kuban. Landw. Inst. Dir. Doc. Dr. H. St. Kalaitan.

Leipzig (Deutschland).
Abt. f. Bienenzucht d. Univ. Dir. Prof. O. Krancher.

Quebec (Canada).
Labor. de Biol. Confer: Parasit. generalis.

Stettin (Deutschland).
Versuchs- u. Lehranstalt f. Bienenzucht d. Landwirtschaftskammer. Dir. Dr. phil. Joachim Evenius. — Frau Dr. phil. Christa Evenius. *1 Labor. soll in nächster Zeit erbaut werden.* ☽ Alle die Biol. u. Zucht von Apis mellifica betreffenden Objekte, einschl. Verwandten von Apis Mellifica u. einschl. Bienennährpflanzen. Versuchsbienenstand.

Wien (Österreich).
Inst. f. Bienenkunde u. Pathol., Vet. Hochsch. Dir. Doc. O. Muck.

4. Animalia pelicea.

Charlottetown (Canada).
Fox Research Sta. Dept. of Agric. Dominion of Canada, Health of animals branch. Dir. F. Torrance, B.A., D.V.Sc. Section 1: Sanitation and Diseases of Foxes: J. A. Allen, V.S., B.V.Sc., Animal Pathologist; Section 2: Sc. Nutrition of Foxes: G. Ennis Smith, B.A.Sc., Biochemist.

Irkutsk (U.d.S.S.R.)
Zoofarm am Baikalsee. Confer: Hortus Zoologici Vivaria).

Leipzig (Deutschland).
Reichs-Zentrale für Pelztier- und Rauchwarenforschung. C 1. Nicolaistr. 28/32. Kustos: Dr. W. Stichel. *Zweck der Zentrale ist Untersuchungen über Pelztierfelle. Verbreitung und Erforschung der Hege, Haltung und Zucht von Pelztieren. Forschungen auf dem Gebiete der Pelzveredelung.* Museum für Pelztierkunde.

Tharandt (Deutschland).
Forschungsstelle f. Pelztierkunde. Dir. Prof. H. B. Prell. *Vererbungslehre, speziell bei Pelztieren.*

Laboratoria productionis lactariae.

Confer: *Labor. culturae animalium generalia* (pag. 523) *et Biochemia fermentorum* (pag. 443), *Bacteriologia animalium* (pag. 486), *Chemia agris* (pag. 536).

Ames, Ia. (U.S.A.).
Dairy Dept., Iowa State Coll. W. A. Cordes; M. P. Baker.

Bainbridge, N.Y. (U.S.A.).
Research Dept. the Dry Milk Comp. of New York. Dir. G. C. Supplee.

Beograd (S.H.S.).
Inst. f. Milchwirtschaft, Dobračina ulica 16. — Dir. Dr. T. Mihovič.

Berkeley, Cal. (U.S.A.).
Dept. of Dairy Industry, Coll. of Agric. C. L. Roadhouse, D.V.M., Head of Div. (P.O., Davis). — F. H. Abbot, B.S., Assoc. in Dairy Indus. (P.O., Davis); A. H. Folger, B.S., in charge Advanced Registry Work (P.O., Davis); G. E. Gordon, B.S., Dairy Specialist; C. S. Mudge, Ph.D., Assoc. Prof. Dairy Indus., Assoc. Bact. (P.O., Davis); D. H. Nelson, M.S., Assoc. in Dairy Indus., Junior Dairy Tech. (P.O., Davis); C. A. Phillips, B.A., Junior Dairy Tech. (P.O., Davis); G. D. Turnbow, M.S., Ass. Prof. Dairy Indus., Dairy Tech. (P.O., Davis); F. R. Wilson, B.S., Junior Dairy Tech. (P.O., Davis).

Cleve, Rheinprov. (Deutschland).
Molkerei-Lehr- und Versuchsanstalt der Landwirtschaftskammer für die Rheinprovinz. Dir. L. Müller. — Dr. Johannes Frahm, Ass. *3—4 Plätze für che-* mische und bact. Untersuchungen von Milch und Molkereiprodukten.

Columbia, Mo. (U.S.A.).
Dairy Dept., Univ. of Missouri. E. C. Elting.

Davis, Cal. (U.S.A.).
Division of Dairy Industry, Univ. of Cal. D. H. Nelson.

Detroit, Mich. (U.S.A.).
Dairy Research Division, Mathews Industries. Dir. H. A. Harding. — A. R. Ward.

Geneva, N.Y. (U.S.A.).
Div. of Dairying of the New York Agric. Experim. Sta. Arthur C. Dahlberg, M.S., Chief in Research (Dairying). — Julius C. Marquardt, M.S.; J. Courtenay Hening, M.S., Ass. in Research (Dairying). *Dairy herd management. Dairy products: Icecream investigations. Cheese investigations. Analytical investigations.*

Halle a. d. S. (Deutschland).
Molkereilabor. d. Inst. f. Tierzucht u. Molkereiwesen. Dir.: vacat.

Hoorn (Holland).
Landwirtschaftl. Versuchsstation d. Rijkslandbouwproefstation. Abt.-Dir.: F. W. J. Boekhout, Dr. W. v. Dam, Dr. E. Hekma (Physiol. Abt.). — Land-

bouwkundige: Dr. J. C. de Ruijter de Wildt; Chemiker: Dr. H. A. Sirks, Dr. B. J. Holwerda; Bacteriolog: J. v. Beynum; Physiolog: Dr. E. Brouwer; Techniker: W. J. Arends jr., J. Th. Smiers, H. v. d. Kool. *Physiol. der Milch u. der Milchprod.*

Ithaca, N.Y. (U.S.A.).
Dept. of Dairy Industry, Cornell Univ. W. V. Price.

Kempten, Allgäu (Deutschland).
Milchwirtschaftliche Untersuchungsanstalt im Allgäu. Dir. Dr. phil. H. Martin, Chemiker. — Dr. O. Freiesleben, Chemiker, Ass. *4 Plätze für chem.-physikal. u. bact. Untersuchungen von Milch und ihren Produkten.*

Kiel (Deutschland).
Preuß. Versuchs- u. Forschungsanstalt für Milchwirtschaft. 6 selbständige wissenschaftl. Inst.: Chemisches Inst., Dir.: Prof. Dr. Burr; Bact. Inst., Dir.: Prof. Dr. Henneberg; Physikalisches Inst., Dir.: Prof. Dr. Rahn; Inst. für Milcherzeugung, Dir.: Prof. Dr Bünger; Inst. für Milchverwertung, Dir.: Prof. Dr. Westphal; Inst. für Maschinenwesen, Dir.: Prof. Dr. Lichtenberger. Kuratorium: Vors.: Geh.Ober-Reg.R. Dr. Oldenburg; Verwaltungs-Dir. der Gesamtanstalt: Prof. Dr. Rahn; 9 wissensch. Hilfsarbeiter, 20 techn. Hilfsarbeiterinnen. Mit der Anstalt ist verbunden: Versuchs- und Lehrmolkerei, wiss. Oberleitung: Prof. Dr. Westphal; Molkereilehranstalt, Dir.: Prof. Dr. Westphal; Maschinenprüfungsamt, Dir.: Prof. Dr. Lichtenberger. *Wissenschaftl. Erforschung des gesamten Gebietes der Milchwirtschaft u. des Molkereiwesens sowie Förderung der milchwirtschaftl. Praxis.*
Physikalisches Inst. der Preußischen Versuchs- und Forschungsanstalt für Milchwirtschaft in Kiel, Prüne Nr. 48. Dir. bis 31. Dez. 1926 Prof. Dr. Otto Rahn. — Dr. phil. Eichstädt. *Etwa 6 Plätze. Milchwirtschaftliche Physik. — Ultramikroskop, Pontentiometer, Mikrokinoapparate usw. Apparate der Molkereipraxis (Versuchsbutterfässer, Homogenisiermaschine).* ℗ Mikrophotographien von Fettkügelchen in Milch unter verschiedenen Bedingungen. Mikrophotographien von Schlagsahne, von der Entstehung und dem Gefüge der Butter. Mikrophotographien von Kondensmilch. Mikrokinofilms von der Aufrahmung roher und gekochter Milch, von der Einwirkung des Salzes auf das Wasser der Butter.
Inst. für Milcherzeugung der PreußischenVersuchs- und Forschungsanstalt für Milchwirtschaft. Dir. Prof. Dr. Bünger. — Dipl.-Landwirt Lamprecht; Wiehr; Neuhaus. *Futterbau, Futterwerbung, Futterkonservierung, Fütterung des Milchviehs, Aufzucht des Milchviehs. Einrichtung des Milchviehstalles. Gewinnung und Behandlung der Milch. Aufzucht und Mästung des Schweines. — Versuchsgut Friedrichsort bei Kiel.* ℗ *Futtermittel. Futterpflanzen.* Mikrophotographien von Fettkügelchen, von verschiedenen Stadien des Butterungsvorgangs, von Milchzuckerkristallen in Kondensmilch, von Bildern über die Struktur der Butter. Mikrofilms von der Aufrahmung roher u. gekochter Milch; Mikrofilms von der Einwirkung des Salzes auf die Butter.
Bact. Inst. der Preußischen Versuchs- und Forschungsanstalt für Milchwirtschaft, Prüne 48. Dir. Prof. Dr. Wilhelm Henneberg. — Dr. Richter (Bakteriologe); Dr. Christiansen (Bakteriologe); Dr. Seelemann (Tierarzt). Wissenschaftl. Untersuchungen, biol. Analysen. *Etwa 40 Plätze für bact. Untersuchungen, Mikroskopieren.* ℗ Sämtliche Pilze der Milch, der Butter und des Käses.

Kiew (U.d.S.S.R.).
Labor. für Milchwirtschaft des Landw. Inst.

København (Danmark).
Bioteknisk-kemisk Labor. Confer: Biochemia gen.

Königsberg i. Pr. (Deutschland).
Milchwirtschaftliches Inst. der Univ., Tragheimer Kirchenstr. 83. Dir. Prof. Dr. Grimmer. — Dr. Georg Schwarz, Ass. *6 Plätze für Biochemie.*

Lafayette, Ind. (U.S.A.).
Dairy Dept. of the Agric. Experim. Sta. W. T. Epple; G. Spitzer; S. M. Hauge.

Leningrad (U.d.S.S.R.).
Milchwirtsch. Untersuch. Labor. des Nord-West-Distrikts, Prosp. Wolodarskij 37. Dir. Prof. Simeon Paraschtschuk. — Ass.: O. Paladina, M. Strohbinder, K. Wlassoff. *2 Plätze für chemisch-bact. Untersuchungen, besonders Milchsäurebact.* ℗ Bact. Praep., besonders Butterschimmelpilze und Milchsäurebact.

Lexington, Ky. (U.S.A.).
Dept. of Dairy of the Univ. of Kentucky. Dir. Prof. J. J. Hooper.

Liebefeld b. Bern (Schweiz).
Schweiz. Milchwirtschaftliche und Bact. Anstalt. Dir. Prof. R. Burri. — Chem. Adjunkt: Dr. G. Koestler; Bact. Adjunkt: Dr. J. Kürsteiner. 4 Chemiker u. 4 Bacteriologen. *Milchkontrolle in sämtlichen Kantonen.*

Lwów (Polska).
Inst. f. Milchwirtschaft. d. Tierärztl. Akad. Dir. S. Niemczycki.

Magyaróvár (Ungarn).
Milchwirtschaftl. Forschungsstation. Dir.Otto Graz.

Muktesar (Br.-India).
Imperial Inst. of Veterinary Research. Confer: Hygiena animalium.

Münster i. W. (Deutschland).
Molkereilehr- und Versuchsanstalt für Westfalen und Lippe. Dir. H. Pflugradt. — M. Christoffer. *8 Plätze für bact. und chemische Untersuchungen auf dem gesamten Gebiete des Molkereiwesens.*

Oldenburg i. O. (Deutschland).
Versuchs- und Kontrollstation der oldenburgischen Landwirtschaftskammer, Marslatourstr. 4. Dir. Prof. Dr. Otto Max Popp. — Dr. J. Contzen, Stellvertr. des Dir., Labor.-Vorst.; Dr. W. Riedel, Vorst. der milchwirtschaftl. Abt.; Dipl.Landw. Suhren, landwirtschaftl. Ass.; Dipl.-Landw. Nieschlag, landwirtschaftl. Ass.; H. Mentz, Chemiker, Ass.; S. Gericke.

Perm (U.d.S.S.R.).
Milchwirtschaftl. Labor. Dir. Doc. J. A. Behrsin.

Pretoria (South Africa).
Milchkundl. Inst. Univ. Dir. Prof. H. B. Davel.

Reading (England).
British Dairy Inst., Univ.

Riga (Latvija).
Inst. f. Milchwirtschaft. Dir. Prof. F. Neulands.

St. Paul, Minn. (U.S.A.).
Division of Dairy Husbandry, Univ. Farm. Ass. Prof. H. Macy (Dairy Bact.); Ass. Prof. W. E. Petersen (Dairy Husbandry).

State College, Pa. (U.S.A.).
Dairy Dept. Prof. A. A. Borland (Dairy Husbandry); Prof. S. A. Bechdel (Dairy Production); Prof. F. J. Doan (Dairy Manufacturing).

Storrs, Conn. (U.S.A.).
Bact. Labor. of the Connecticut Agric. Coll. Dir. W. M. Esten. — Miss C. J. Mason, Instr.; Vinton E. White, Instr.; Wright D. Gifford. *35 places for Dairy-, soil- and water Bact. — 1 large, 4 smaller electric incubators. 3 Autoclavs. Live steam for stenti-*

gation. *Investigations of Silage fermentation, cheese fermentation and clean milk production. Sources of sour milk organisms. Experiments in soil fertility on 4 acres of land.* ⊕ 250 specimens of bact. or strains.
Dairy Manufacturing Experim. Station. E. O. Anderson (Ass. Prof.). *Milk and its products. Some of the factors (physical and chemical) which influence the stability of casein.*

Urbana, Ill. (U.S.A.).
Dairy Dept., Univ. of Illinois. Dir. H. A. Ruehe.

Utrecht (Holland).
Labor. für Nahrungsmittel tierischen Ursprungs, Veterinäre Fac. d. Univ. (Labor. voor de kennis der menschelijke voedingsmiddelen van dierlijken oorsprong), Biltstraat 166. Dir. Prof. C. F. van Oyen.

Vancouver (Br.-Columbia).
Dept. of Dairying, Univ. Dir. Wilfrid Sadler, B.S.A., M.Sc.

Vologda (U.d.S.S.R.).
Station f. d. Techn. d. Milchbearbeitung mit Lehrlabor., Milchwirtsch. Inst. Dir. Prof. A. S. Krylow.
Bact. Versuchsstation d. Vologodski Molotschno-Chosjaistwennyi Inst. (Milchwirtschaftl. Inst. in Vologda). Dir. Prof. Sergei Alexandrowitsch Korolew. — Ass. A. M. Skorodumowa, S. B. Panfilow, W. I. Wereschtschagina. *3—4 Plätze für Mikrobiol. der Milch, Käse, Butter u. dergl.* ⊕ Die Kulturen d. Mikroorganismen der Milch (Bakterien, Schimmelpilze, Hefen).

Wangen i. Allgäu (Deutschland).
Milchwirtsch. Lehr- und Forschungsanstalt. Dir. LandesökonomieR. Dr. phil. Kurt Teichert. — Dr. rer. nat. Mundinger; Dipl. agr. Stocker; Chemiker Kotterer. *6 Plätze für chem.-bact. Arbeiten.*

Washington, D.C. (U.S.A.).
Research Labor., Bureau of Dairy Industry U.S. Dept. of Agric. Dir. C. W. Larson, Chief. — G. E. Holm; D. H. Warren; A. J. Boyer; K. J. Matheson; G. R. Greenband; P. A. Wright; W. W. Swett. *Fermentation.*

Weihenstephan b. Freising (Deutschland).
Versuchs- und Forschungsanstalt für Milchwirtschaft der Hochsch. f. Landwirtschaft. Dir.: Prof. A. Fehr; stellv. Vorst.: Landwirtschaftsrat Zeller.

Weiler im bayr. Allg. (Deutschland).
Lehr- und Versuchsanstalt für Emmentaler Käserei, Lehrsennerei. Dir. Dr. Wilhelm Doll. — Ldw. Assessor Franz Hofer. Technische und bact. Kontrolle in der Rundkäserei. *2 Plätze für chemische und bact. Arbeiten auf dem Gebiet der Emmentaler-Käserei.*

Wien (Österreich).
Inst. f. Milchhygiene, Lebens- u. Futtermittelkunde, Tierärztl. Hochsch. Dir. ao. Prof. Dr. F. Zaribnicky.
Inst. f. Molkereiwesen, Hochsch. f. Bodenkultur. Dir. Prof. W. Winkler.

Laboratoria microbiologiae agris (Pedologiae).
(Bacteria et protozoa).

Confer: *Institutiones generales culturae plantarum et animalium* (pag. 497), *Botania generalis* (pag. 390), *Bacteriologia hominis* (pag. 479) et *animalium* (pag. 486), *Parasitologia generalis* (pag. 379), *Chemia agris*, (pag. 536).

Aas (Norge).
Mikrobiol. Labor. der Landwirtschaftlichen Hochschule. Dir. A. E. Traaen. — ⊕ Praep. der Pflanzenkrankheiten für den Unterricht an der Hochsch.

Alcorn, Miss. (U.S.A.).
Dept. of Bact. Agric. Coll. C. F. Briscoe, Ph.D., Head of Dept.; — H. H. Harned, M.S., Assoc. Prof. and Assoc. Bact.

Ames, Ia. (U.S.A.).
Soils Dept. Iowa State Coll. Dir. P. E. Brown. — L. W. Erdman (Ass. Chief soil Bact.).
Dept. of Bact. State Coll. of Agric. R. E. Buchanan, Ph.D., Head of Dept. and Dean of Grad. Coll.; Chief Bact. — Max Levine, Ph.D., Prof. Bact.; J. C. Weldin, M.S., Instr. Bact.; C. H. Werkman. Ph.D., Ass. Prof. and Ass. Bact.; L. A. Burkey.

Amherst, Mass. (U.S.A.).
Microbiol. Labor. of Massachusetts Agric. Coll. Dir. Charles E. Marshall. — Ass. Prof. Leon A. Bradley; James E. Fuller, Instr.; Mary E. M. Garvey, Instr.; W. Brooks Hamilton, Graduate Ass.; Ralph L. France, Graduate Ass. *10 places for Microbiol. agric. problems. — Well equiped in microbiol. field, temperature rooms, vacuum and pressure pumps, shakers, etc.*

Beograd (S.H.S.).
Pädol. Inst. Landwirtschaftl. Fac. d. Univ., Dobračina ul. 16. Dr. D. Todorović.
Inst. für Landwirtsch. Bact. u. Tierseuchenlehre, Birčaninova ulica 36. Dir. Prof. Dr. D. Konjev.

Berlin (Deutschland).
Inst. für Agriculturchemie und Bact. der Landwirtschaftlichen Hochsch., -Dahlem, Lentzeallee 55|57. Dir. Prof. Dr. O. Lemmermann. — Dr. Jessen, Chemiker; Dr. Lesch, Chemiker; Dr. Gerdum, Versuchsfeld-Ass.
Labor. für prakt. Bodenbact. Biol. Reichsanstalt für Land- u. Forstwirtschaft, -Dahlem, Königin-Luise-Str. 19. Dir. Reg.R. Dr. H. Behn. — Dr. E. Pfeil, Ass.
Labor. für Bact. Biol. Reichsanstalt für Land- u. Forstwirtschaft, -Dahlem, Königin-Luise-Str. 19. Dir. Dr. K. Stapp.

Blagoweschensk (U.d.S.S.R.).
Bact. Abt. d. Landwirtschaftl. Amur-Versuchsstation, Remeslennaja 59.

Bonn a. Rh. (Deutschland).
Bact. Inst. d. Landwirtschaftskammer, Rheindorfer Str. 12. Dir. Dr. Eickmann.
Chem. Inst. Landw. Hochsch., Poppelsdorf. Dir. Prof. Hubert Kappen. *Bodenchemie.*

Bozeman, Mtn. (U.S.A.).
Dept. of Botan. and Bact. Coll. of Agric. D. B. Swingle, M.S., Prof. Botan. and Bact.; F. B. Cotner, A.M., Assoc. Prof. Botan. and Bact.; H. E. Morris, M.S., Assoc. Botan. and Bact.; P. A. Young, Ph.D., Ass. Botan. and Bact.

Brie-Comte-Robert, S. et M. (France).
Inst. Pasteur. Dir. S. Winogradsky, Membre de l'Institut. — Mlle X. Nikitine, Préparateur. *2 places.*

Brno — Brünn (C.S.R.).
Inst. f. Landw. Phytopath. Confer: Phytopath. generalis.

Buitenzorg, Java (Ned.-O.-Indië).
Labor. for Microbiol. of the General Agric. Experim. Sta. Chief: J. Groenewege. *Bact. of the soil.*

Fermentation process; silage. Biol. purification of water.

Chiba (Japan).
Microbiol. Inst., Chiba Medical Univ.

Columbia, Mo. (U.S.A.).
Soils Dept., Coll. of Agric., Univ. of Missouri. W. A. Albrecht.

Corvallis, Ore. (U.S.A.).
Dept. of Bact., State Agric. Coll. G. V. Copson, M.S., Prof. Bact., Bact.; J. A. Berry, M.S., Instr. Bact.; W. V. Halversen, Ph.D., Assoc. Prof. and Assoc. Bact.; J. E. Simmons, M.S., Ass. Prof. Bact.

Delft (Holland).
Labor. für Mikrobiol. der Technischen Hochsch., Nieuwe Laan 3. Dir. Prof. A. J. Kluyver. — Ing. C. B. van Niel, Konserv.; Ing. K. W. H. Leeflang, Ass. *Mehrere Plätze für Stoffwechselversuche an Mikroorganismen. — Zimmer mit konstanter Temperatur, semi-technischer Gärungsapparat, Installation für kontinuierliche Aeration, reichhaltige Mikrobensammlung usw.* ⊕ Kulturen von Hefen- und Bakterienarten.

Fargo, N.D. (U.S.A.).
Dept. of Botan., Bact. and Zool., State Coll. of Agric. H. L. Bolley, M.S., Dean of Biol., Prof. Botan. and Plant Pathol., Botan. and Plant Pathol. — W. E. Brentzel, M.A., Plant Pathol. (Coop. U.S.D.A.); W. G. Couey, B.S., Seed Insp., and State Agt. in Plant Pathok; Mayme Dworak, A.M., Ass. Prof. Botan. and Bact.; Bact. and Plant Physiol.; G. J. Ikenberry, M.S., Instr. Botan.; H. D. Long, Field Crop Insp.; G. C. Mayoue, M.S., Ext. Plant Pathol.; G. E. Miller, M.A., Prof. Biol.; C. I. Nelson, A.B., Prof. Bact., Bact. and Soil Biol.; E. S. Reynolds, Ph.D., Prof. Botan.; O. A. Stevens, M.S., Ass. Prof. Botan.; Deputy Seed Comr.; A. D. Whedon, Ph.D., Prof. Zool. and Physiol.

Fayetteville, Ark. (U.S.A.).
Dept. of Bact., Coll. of Agric. Dir. W. L. Bleecker, D.V.M., Head of Dept.

Fort Collins, Col. (U.S.A.).
Dept. of Bact., State Coll. of Agric. W. G. Sackett, Ph.D., Bact.; Ida W. Ferguson, R.N., Ass. in Bact.

Fukuoka (Japan).
Microbiol. Inst. Coll. of Med., Kyûshû Imper. Univ.

Geneva, N.Y. (U.S.A.).
Div. of Bact. of the New York Agric. Experim. Stat. Robert S. Breed, Ph.D., Chief in Research (Bact.); Harold J. Conn, Ph.D., Chief in Research (Soil Bact.); George J. Hucker, Ph.D., Assoc. in Research (Bact.); Carl S. Pederson, M.S.; Paul S. Prickett, M.S., Ass. in Research (Bact.). *Classification and identification of bact. Standardization of biol. stains. Effect of straw on plants. Sauerkraut investigations. Tomato products investigations. Market milk problems. Cheddar cheese studies.*

Göttingen (Deutschland).
Inst. für landwirtschaftliche Bact. der Univ., Gosslerstr. 16. Dir. Prof. Dr. August Rippel. — Dr. Oskar Ludwig, Ass. *Für alle bact. Arbeiten. Stoffwechselphysiol. höherer und niederer Pflanzen. — Elektrometrische Messung der Wasserstoffionenkonzentration. Gasanalyse.*

Guelph, Ont. (Canada).
Dept. of Bact. Agric. Coll. Dir. O. J. Stevenson, M.A., D. Paed. — R. E. Stone, B.Sc., Ph.D.; D. H. Jones, B.S.A.; A. Davey, B.S.; D. R. Shutt, B.S.A.

Halle a. d. S. (Deutschland).
Bact. Abt. d. Versuchsstat. f. Pflanzenkrankheiten u.Pflanzenschutz d. Landwirtschaftskammer für die Provinz Sachsen. Dir. Dr. Berthold Heinze.

Harpenden, Herts (England).
Dept. of Protozool. of the Inst. of Plant Nutrition and Soil Problems. Dir. D.W. Cutler. — L.M. Crump; H. Sandon; A. Dixon.
Dept. of Bact. of the Inst. of Plant Nutrition and Soil Problems. Dir. H. G. Thornton. — P. H. H. Gray.

Honolulu (Hawaii).
Dept. of Bact. of the Board of Agric. and Forestry. B. A. Gallagher (Bacteriologist).

Jaroslavl (U.d.S.S.R.).
Kab. f. Bodenkunde. Dir. Doc. Ja. Je. Subbotin.

Kamenetz-Podolsk, Ukraine (U.d.S.S.R.).
Microbiol. Labor. des Inst. f. Landwirtsch. „Karl Marx". Dir. Prof. M. Bernadzki.

Kiew (U.d.S.S.R.).
Labor. für Microbiol. des Landw. Inst.

Knoxville, Tenn. (U.S.A.).
Dept. of Bact. Coll. of Agric. P. W. Allen, Ph.D., Prof. Bact. B. W. Hess, A.B., Instr. Bact.

København (Danmark).
Microbiol. Labor. (Landbau-Hochsch.), Rolighedsevej. 23. Dir. Hj. Jensen.

Lafayette, Ind. (U.S.A.).
Dept. of Biol. of the School of Agric., Purdue Univ. Dir. Stanley Coulter, Ph.D., LL.D., in charge. I. L. Baldwin, M.S., Ass. Prof. Bact.; C. A. Behrens, Ph.D., Prof. Bact.; D. W. Creel, B.S., Ass. in Bact.; K. L. Dickens, B.S., Ass. in Biol.; D. H. Dunham, M.S., Instr. Biol.; H. E. Enders, Ph.D., Prof. Zool. in charge Gen. Biol.; Elizabeth O. Hassenzahl, M.S., Instr. Biol.; C. M. James, M.S., Instr. Anat. and Physiol.; G. D. Kennedy, B.S., Ass. in Bact.

Leipzig (Deutschland).
Inst. für landwirtschaftl. Bact. u. Bodenkunde d. Univ. Dir. Prof. Dr. Felix Löhnis. *Landwirtsch.Bact.*

Leningrad (U.d.S.S.R.).
Microbiol. Labor. d. Landwirtschaftlichen Inst., Kam. ostrow. Dir. Prof. Dr. B. Issatschenko. — A. A. Egorowa, Haupt-Ass.; M. M. Ontschukowa, Ass. *8 Arbeitsplätze.*
Bureau f. landwirtsch. Mikrobiol. d. Inst. f. experim. Landwirtschaft, Ulica Herzena 42. Dir. S. Kostytschew, o. Mitglied d. Akad. d. Wiss. v. U.d. S.S.R. — Vorst. d. einz. Labor.: W. Omeliansky, V. Dogjel, M. Korsakowa, S. Mereshkowsky; 1. Ass.: G. Burgwitz, N. Lazarew, A. Scheloumow, O. Schulgina, I. Ulrich; 2. Ass.: O. Beressnewa, O. Schwezowa, M. Stepanowa, L. Sturm; Laboranten: W. Bylinkina, L. Harder. S. Kovrovzewa, N. Krassilnikoff, G. Lopatina, E. Nikitina, E. Rammelmeyer, M. Schukowa, N. Strelkow. *5 Plätze für Gäste, eingerichtet für Untersuchungen auf dem Gebiete d. Mikrobiol. d. Bodens und der landwirtsch. Produkte. — 1. Abt. f. allgemeine Mikrobiol. und Gärungsmikrobiol.,* Vorst.: W. Omeliansky, Mitgl. d. Akad. d. Wiss. v. U.d.S.S.R.; *2. Abt. f. Bodenbact.,* Vorst.: M. Korsakowa; *3. Abt. f. Bekämpfung d. Schädlinge d. Landwirtsch.,* Vorst.: S. Mereshkowsky; *4. Abt. f. Protozool.,* Vorst.: V. Dogjel; *5. Mus. d. lebend. Kulturen. — Apparaturen für mikrobiol. und biochemische Untersuchungen nach den im Bureau ausgearbeiteten Methoden.* ⊕ Reinkulturen verschiedener landwirtschaftlich wichtiger Mikroorganismen.

Lexington, Ky. (U.S.A.).
Dept. of Agric. Bact. of the Univ. of Kentucky. Dir. Prof. D. J. Healy.

Lima (Peru).
Sección de Microbiol. Agric. Sueros y Vacunas. La Estación Central Agronomica. Jefe: Dr. E. Tabusso.

Lisbôa (Portugal).
Labor. de Microbiol. agric., Inst. Superiôr de Agronomia, Tapada da Ajuda. Dir. Prof. A. A. B. Ramires.

Long Ashton near Bristol (England).
Agric. and Horticult. Research Sta. of the Univ. of Bristol. Dir. Prof. B. T. P. Barker.

London (England).
Bact. Labor. of the South-Eastern Agric. Coll. of the Univ., Wye, Kent.

Lugansk, Dongebiet, Ukraine (U. d. S. S. R.).
Kab. f. Bodenkunde d. Landwirtsch. Timirjazew-Technikums, Počtovyi jaščikz 3.

Madison, Wis. (U. S. A.).
Dept. of Bact., Coll. of Agric. E. G. Hastings, M.S., Chair. of Dept. — E. B.Fred, Ph.D., Prof. Agr. Bact., Agr. Bact.; W. D. Frost, Ph.D., Dr.P.H., Prof. Agr. Bact., Agr. Bact.; Edith Haynes, M.S., Instr. and Ass. in Agr. Bact.; G. E. Helz, B.S., Ass. Bact.; Harriet L. Mansfield, B.S., Ass. Agr. Bact.; Myrtle A. Shaw, M.S., Ass. Agr. Bact.; A. L. Whiting, Ph.D., Assoc. Prof. Agr. Bact.; W. H. Wright, Ph.D., Assoc. Prof. Agr. Bact., Agr. Bact.

Madrid (España).
Labor. de Microbiol. Mus. nation. d'Hist. Nat., Palacio del Hippodromo.

Manila (Philippine Islands).
Division of Soils and Fertilizers, Bureau of Sc., Dept. of Agric. Dir. A. S. Arguelles.

Marseille (France).
Labor. de Zool. agric., Fac. des Sc. Dir. J. Cotte. *2 places pour Recherches microscopiques.*

Milano (Italia).
Labor. di Batt. agraria, Ist. Sup. agrar. Dir. Prof. C. Gorini.

Minsk (U. d. S. S. R.).
Inst. für Mikrobiol. der Univ. Dir. Dr. B. I. Elberth.

Moscow, Id. (U. S. A.).
Dept. of Bact., Coll. of Agric. W. M. Gibbs, Ph.D., Prof. Bact., Bact.; C. C. Prouty, M.S., Ass. Bact.

Moskau (U. d. S. S. R.).
Section Soil Bact., Agronom. Sta. Dir. K.Rudakov. **Station of bact. agronom.** Dir. A. Woitkiewicz.

Münster (Deutschland).
Bact. Inst., Landwirtschaftskammer Prov. Westfalen, Kronprinzenstr. 15. Dir. Dr. Paul Sachweh.

Newark, Del. (U. S. A.).
Dept. of Soil Bact. of the Agric. Experim. Sta. *Soil Flora in Relation to Crop Production.*

New Brunswick, N. J. (U. S. A.).
Dept. of Bact. State Coll. of Agric. Dir. T. J. Murray, M.S., Prof. Bact.
Dept. of Sewage Disposal, N.J. Agr. Experim. Sta. and Dept. ofWater Supplies and Sewage Disposal, State Univ. of New Jersey. Dir. Dr. Jacob G. Lipman and Dr. J. M. Thomas. — Dr. H. Heukelekian, Research Bacteriologist; Daggmar Peterson, Research Zoologist; A. J. Fischer, Research Fellow; P. J. A. Zeller, Chemist. *2 places for Biol. of Sewage Disposal. — Working facilities and apparatus in connection with these specialized fields.*
Dept. of Microbiol., Agric. Experim. Sta. Dir. Dr. S. A. Waksman. — Dr. R. L. Starkey, Assoc. Microbiol.; Ass. in Microbiol.: L. J. Dubos, Florence G. Tenney. *5—12 places for research on microorganisms of the soil and their biochemical processes in the soil. Soil fungi, soil actinomyces, soil bacteria and soil protozoa.*

Northampton (England).
Bact. Dept., Rothamsted Experim. Sta. Dir. H. G. Thornton.

Omsk (U. d.·S. S. R.).
Pflanzenphysiol. u. Landwirtsch. Bact. Inst. Confer: Phytopathologia.

Orono, Me. (U. S. A.).
Dept. of Bact., Coll. of Agric. F. L. Russell, B.S., V.S., Prof. Bact. and Vet. Sc.; E. R. Hitchner, M.S., Assoc. Prof. Bact.

Ottawa (Canada).
Division of Bact. of the Central Experim. Farm. Dir. A. Grant Lochhead. — *2 Ass. Agric. Bact.*

Paris (France).
Station de Microbiol. Inst. Nation. Agron., Ve, 16, Rue Claude Bernard.

Perugia (Italia).
Microbiol. Inst. d. R. Inst. sup. Agrario. Dir. Prof. Dr. de Rossi.

Pisa (Italia).
Labor. f. Bact. agrar. et Pathol. végét. Dir. Prof. R. Perotti.

Puławy (Polska).
Abt. f. Bact. d. staatl. wissenschaftl. Inst. f. Landwirtsch. Dir. Prof. K. Markiewicz.

Pullman, Wash. (U. S. A.).
Bact. Division of the Experim. Station. Dir. Prof. E. C. Johnson. — S. C. Vandecaveye, Ph.D., Bact.; Victor Burke, Ph.D., Prof.Bact.; L. A. Burkey, M.S., Instr. Bact.; Vera E. Lautenschlager, M.A., Instr. Bact. *1 place for Soil and Dairy Bact. — Ordinary apparatus for bact. research.*

Riga (Latvija).
Mikrobiol. Labor. Landwirtschaftl. Fac. Univ. Kronvalda bulvari 1. Dir. Prof. Dr. A. Kirchenšteins.

Saratow (U. d. S. S. R.).
Abt. f. landw. Mikrobiol., Inst. f. Landwirtsch. u. Meliorat.

Sofia (Bulgarien).
Mikrobiol. Abt. d. Staatl. Bact. Inst., Bulevard Makedonia. Dir. Dr. Al. Michaelow.
Bodenkundl. Abt. Zentralinst. f. landwirtsch. Versuchswesen, Solunplatz 3. Dir. Prof. Chr. Sawoff.

Stoke on Trent (England).
Mikroskopic. Sect. of the North Staffordshire Field-Club.

Stockholm (Sverige).
Abt. f. Bact. Zentralanst. f. landwirtschaftl. Versuchswesen. Dir. Prof. Chr. Barthel. — *2 Ass.*

Storrs, Conn. (U. S. A.).
Dept. of Bact.-Coll. of Agric. W. M. Esten, M.S., Prof. Bact.; Christie J. Mason, B.Agr., Instr. Bact., L. F. Rettger, Ph.D., Bact. in Animal Diseases (P. O. Yale Station); J. G. McAlpine, Ph.D., Ass. Bact. in Animal Diseases.

Taihoku, Formosa (Japan).
Governmental Central Research Inst. Confer: Fermenta.

Trinidad (Br.-W.-India).
Inst. f. Mykol. u. Bact. Inst. f. tropical. Agric. Dir. Prof. S. F. Ashby.

Tunis (Tunis).
Inst. f. Mikrobiol. Dir. Rousseau, *Microbiol. agric., Technol. agric.*

Vjatka (U.d.S.S.R.).
Labor. f. Vegetationsversuche. Confer: Labor. et Inst. agric. spec.

Vladikavkaz (U.d.S.S.R.).
Labor. f. Pedol. mit Mus. Landwirtsch. Inst. Dir. Prof. A. Pankov.

Warszawa (Polska).
Inst. f. Mikrobiol. a. d. Hochsch. f. Land- u. Forstwirtsch. Dir. Prof. W. Dąbrowski.

Washington, D.C. (U.S.A.).
Microbiol. Labor., Bureau of Chemistry, U.S.Dept. of Agric. Dir. Dr. Charles Thom. *1 or 2 places cultural equipment for work in bact. and mycol.* Ⓟ Pure cultures of molds and bact.

Weihenstephan b. Freising (Deutschland).
Bact. Abt. and d. Süddeutsch. Forschungsanst. f. Milchwirtschaft. Dir. Dr. Karl Joseph Demeter. *Landwirtschaftl. Microbiol. spez. Molkereibact.*

Woronesch (U.d.S.S.R.).
Kabinett f. Bodenkunde mit Labor., Landwirtsch. Hochsch.

Zagreb — Agram (S.H.S.).
Inst. f. Bodenkunde, Univ. Dir. Prof. Seiwerth.
Inst. für Mikrobiol. an der landwirtschaftl. Fac. d. Univ., Wilsonplatz 2. Dir. Prof. Dr. Ludwig Gutschy. — 1 Ass. *10 Arbeitsplätze. Landwirtschaftliche u. technische Mikrobiol. — Versuchsfeld und Garten in Maksimir bei Zagreb.*

Zürich (Schweiz).
Landwirtschaftl. Bact. Labor. der Eidgen. Techn. Hochsch., Universitätsstr. 2. Dir. Prof. M. Düggeli.

Laboratoria terrae uliginosae.

Archangelsk (U.d.S.S.R.).
Archangelsker Moor-Versuchsfeld, Gubsemprawlenije.

Bremen (Deutschland).
Preuß. Moor-Versuchsstation, Neustadtswall. Dir. Geh.Reg.R. Prof. Bruno Tacke. — 3 Abt.-Vorst., 1 Botaniker, 7 chem. Ass *Förderung d. Kultur von Moor-, Sand- u. Marschböden durch wiss. Forschung.*

Hannover (Deutschland).
Versuchsanstalt f. techn. Moorverwertung d. Techn. Hochsch. Vorst.: Prof. Gustav Keppeler.

Iwanowo-Wosnessensk (U.d.S.S.R.).
Ukrain. Moor-Versuchstat. Dir. A. A. Tulubjew.

Landsberg a. d. W. (Deutschland).
Inst. f. Meliorationswesen u. Moorkultur d. Preuß. Landw. Versuchs- u. Forschungsanst.

Minsk (U.d.S.S.R.).
Moor-Versuchsstation des Inst. f. Landw., Sowetskaja 31.

Moskau (U.d.S.S.R.).
Geo-botan. Kabinett d. wissensch.-experim. Torf-Inst., Nikolskaja Str. 10|2. Dir. Prof. W. S. Dokturowsky. — D. Serassimow; W. Matjuschenko; D. Begak (Mikrobiol.). *Geobotan. Untersuchungen, stationäre Beobachtungen in Redkino, Gouv. Twer, Torfstation.* Ⓟ Versch. Praep. zur botan. Analyse d. Torfes (Pollenarten, Radizellen, Moose u. a.). — Torfstat. Redkino, Gouv. Twer.

München (Deutschland).
Bayerische Landesanstalt für Moorwirtschaft, Königinstr. 36. Dir. OberReg. R. Prof. Theodor Mayer. — Reg.R. I. Kl.: Dr. E. Gyllu, Dr. H. Paul, F. Bader, M. Harttung, Dr. J. Ibele, F. Klinger, G. Schmetzer; Reg.R.: W. Spengler, L. Schindler, Dr. H. Gümbel, L. Hartel, Dr. H. Haxpointner, L. Angermaier, Landwirtschaftsrat R. Deichstetter. 4 Assessoren, 4 wiss. Hilfsarbeiter. *Abt.: botan., chemische u. technische Abt. Zweiginst.:* 9 Moorwirtschaftsstellen in Bernau, Schleißheim, Karolinenfeld, Benediktbeuren i. O.-B., Karlshuld u. Günzburg i. Schwaben, Dingolfing i. N.-B., Weiden i. d. Oberpfalz, Landstuhl (Pfalz) u. 28 Dienststellen, die den Moorwirtschaftsstellen nachgeordnet sind. 2 Staatsgüter: Benediktbeuren u. Pfrentsch.

Nowgorod (U.d.S.S.R.).
Landwirtschaftl.Moorversuchsstation, Postfach23.

Riga (Latvija).
Abt. f. Moorkultur. Dir. Doc. E. Ansons.
Inst. f. Moorforschung u. Verwertung d. Torfs. Dir. Doc. Peter Nomals.

Rogatschew (U.d.S.S.R.).
Moor-Versuchswirtschaft, Bobrijskij okr.

Tooma b. Vägera (Eesti).
Versuchsstation d. Estländischen Moorversuchsvereins. Dir. P.W.Thomson (Botan. u. Quartärgeol.) *Pollenanalytisch-stratigraphische Erforschung der Moore u. lacustrinen Ablagerungen in Estland. Regionale Moortypen Estlands. Vegetation der estländischen Moore. Verbreitung einzelner Moorpflanzen.*

Wien (Österreich).
Landwirtschaftl. Chem. Bundesversuchsanstalt. Confer: Chemia agris.

Laboratoria chemiae et physicae agris.

Confer: *Biochemia generalis* (pag. 435) *et Biophysica* (pag. 442), *Microbiologia agris* (pag. 533), *Cultura plantarum generalis* (pag. 499).

Bern-Liebefeld (Schweiz).
Schweiz. agrikulturchemische Anstalt. Dir. Dr. A. Schmid, Liebefeld. — Dr. E. Ritter; Dr. E. Truninger; Dr. K. Keller; Dr. E. Weidmann; Dr. M. Perestein. *Plätze für agric.-chem. Versuche verfügbar.* Ⓟ Futtermittel, Düngemittel, Bodenproben.

Bordeaux (France).
Ecole de Chimie appl. à Agric. Fac. Sc. Univ.

Braunschweig (Deutschland).
Landwirtschaftliche Versuchsstation, Hochstr. 17 u. 18. Dir. Dr. A. Gehring. — Dr. E. Pommer, Fütterungslehre; Dr. A. Peggau, Dr. O. Wehrmann, Bodenchemie und -bact. *Bodenchemie und Bact. Samenkontrolle, Feldversuche.*

Breslau (Deutschland).
Agrikulturchemisches und Bact. Inst. Univ., 16, Hansastr. 25—27. Dir. Prof. Dr. Paul Ehrenberg. — Priv.Doc. Dr. E. Ungerer, Ass.; Dr. K. Maiwald, Ass.; Ch. Pfotenhauer, technische Ass. *Einige Plätze für Arbeiten über Tier- und Pflanzenernährung, sowie bact., physikalische und kolloidchemische Bodenforschung. — Ultramikroskop,* 2 Vegetationsanlagen, *Versuchsstallungen, usw. — Es ist außer einer großen*

Vegetationsanlage mit großem Zubehör und besonderem Labor. außerhalb der Stadt im Inst. ein bact. sowie ein chemisches Labor. mit kleiner Vegetationsanlage vorhanden, ferner Räume für andere Arbeiten.
Inst. für Pflanzenbau und Pflanzenzüchtung Univ. Confer: Labor. et Inst. agric. spec.

Brno — Brünn (C.S.R.).
Inst. für Agric.-Chemie der Landw. Hochsch. Dir. Prof. Dr. R. Truka.
Bodenkundl. Sektion Landw. Versuchs-Anst. (Inst. des Recherches agron. et forest.). Dir. Prof. Václav Novák.

Budapest (Ungarn).
Agrochemisches Inst. der Kgl. Ung. Volkswirtschaftlichen Univ.-Fac., IV, Szerb-utca 23. Dir. Prof. Dr. Géza Doby. — Dr. G. Eperjessy; F. Snassel; A. Medveczky. *1 Platz.*

Buitenzorg, Java (Ned.-O.-Indië).
Labor. for research work on soils of the General Agric. Experim. Sta. Chief: Ir. J. Th. White. — Dr. E. E. Scheibener, Ass. *Physical analysis of soils. Chemical analysis of different fractions of soils. Reactions of soils of known constitution on manures. Volcanic ashes. Atterberg figures. Classification of soils of newly explored regions in relations to crop production. Petrographical survey of Java in tracing the origin of the soils. Mineralogical analysis of soils. Fossils.*

Eberswalde (Deutschland).
Chemisch-bodenkundliches Inst. Forstl. Hochsch. Dir. Prof. Albert.

Gießen (Deutschland).
Agrikulturchemisches Labor. Univ. Leiter: Prof. Kleberger.

Göttingen (Deutschland).
Agriculturchem. u. bodenkundl. Inst. der Univ., Nikolausberger Weg 7. Dir. Prof. Dr. E. Blanck. — Dipl.-Ing. Dr. F. Giesecke, planm. Ass.; Dr. A. Reiser, Ass. *Agrikulturchemie u. Bodenlehre.*

Gorizia (Italia).
Ist. chimico-agrario sperimentale di Gorizia, Via Trieste 43. Dir. Prof. ing. Arturo de Varda. — Dott. in chimica Arturo Viano *(Sostanze alimentari e scorte agrarie);* Dott. ing. chimico Rodolfo Petronio *(Analisi delle terre e prove colturali);* Dott. Edmondo Happacher *(Fitopat. e bachicoltura e batt. agrarie). 1 posti per i lavori di fitopat. e bachicoltura.* ⓟ Tutti gli oggetti (piante, insetti ecc.) che servirono a studi speciali.

Grignon, Seine-et-Oise (France).
Station de Physique et Chimie biol. Ecole nationale d'agric.

Halle a. d. S. (Deutschland).
Agric. chem. Kontrollstat. der Landwirtschaftskammer für die Prov. Sachsen. Dir. Prof. H. C. Müller. — G. Metge (Abt.-Vorst.).

Hann.-Münden (Deutschland).
Forstchemisches Labor. Forstl. Hochsch. Dir. Prof. H. Süchting.

Helsinki (Finnland).
Agrikulturchemisches Labor. Univ. Dir. J. Valmari. — M. W. Breumer.

Hohenheim b. Stuttgart (Deutschland).
Landesversuchsanstalt für landw. Chemie, Landw. Hochsch. Vorst.: Prof. Dr. Brigl. — Abt. Vorst. b. d. Kontroll-Abt.: Dr. Lacour; Abt.-Vorst. b. d. wissenschaftl. Abt.: Dr. Windheuser.

Honolulu (Hawaii).
Dept. of Chemistry of the Div. of Agric. and Home Economics. Dir. Prof. F. T. Dillingham, M.A.

Jassy (România).
Labor. de Chim. agric. Dir. Prof. H. Vasiliu.

Jena (Deutschland).
Agrikulturchemisches Labor. Dir. Prof. H. Immendorf.

Kamenetz-Podolsk, Ukraine (U.d.S.S.R.).
Bodenkundl. Labor. des Inst. f. Landw. „Karl Marx". Dir. O. Krassivki.

Kiel (Deutschland).
Landwirtsch. Versuchsanstalt der Landwirtschaftskammer für die Prov. Schlesw.-Holstein, Kronshagener Weg 3. Dir. Dr. Sieden. Stellvertr.Vorst.: Dr. Blohm; Abt. Vorst.: Dr. Trieschmann, Dr. Wolff; Chemiker: Dr. Beeth, Dr. Hauptfleisch.

Königsberg i. Pr. (Deutschland).
LandwirtschaftlicheVersuchsstation und Nahrungsmitteluntersuchungsamt der Landwirtschaftskammer für Ostpreußen, Lange Reihe 3. Dir. Prof. Dr. Goy. — Dr. Köhler; Dr. Rudolph; Dr. Bodschwinna; Chemiker Müller. *3 Plätze für Agriculturchemie und Nahrungsmittelchemie.* ⓟ Futtermittel.
Agrikulturchemisches Inst. Univ. Dir. Prof. W. Zielstorff.

Köslin (Deutschland).
Agrikulturchemische Versuchs-Station.

Krasnodar, Kuban (U.d.S.S.R.).
Labor. für Agrochemie d. Kuban. Landw. Inst.

Kyôto (Japan).
Inst. of Agric. Chemistry Coll. of Agric., Kyôto Imper. Univ. Dir. Prof. Dr. Shigeru Osugi. — Prof. Dr. Bunsuke Suzuki (Biochemistry); Prof. Dr. Kinsuke Kondô (Nutrition Chemistry); Prof. Masuzô Shikata (Forestry Chemistry); Ass. Prof. Kametarô Konishi; Ass. Prof. Hideo Katagiri; Ass. Prof. Sankichi Takei.

Landsberg a. d. W. (Deutschland).
Inst. für Bodenkunde und Pflanzenernährung d. Preuß. landwirtsch. Versuchs- und Forschungsanst., Theaterstr. 25. Dir. Prof. Dr. Densch. — Dr. Hunnius, Chem. Ass.; Dr. Groh, Landwirtsch. Ass. Dr. Pfaff, Chem. Ass. *Chemische, in beschränktem Maße bodenbact. Untersuchungen.*

Laramie, Wyo. (U.S.A.).
Dept. of Agronomy, Univ. of Wyoming. Dir. Prof. A. F. Vass. — Ass. Prof. G. Hartman.

Lausanne (Suisse).
Stat. Fédérale de Chimie Agric., Montagibert. Dir. C. Duserre. *Boden-, Dünger- und Futtermittelkontrolle.*

Leningrad (U.d.S.S.R.).
Abt. f. Bodenkunde d. Staatl. Inst. f. experim. Agronomie. Dir. K. D. Glinka.

Liebefeld b. Bern (Schweiz).
Schweiz. Agriculturchem. Anstalt. Confer Bern.

Linz (Österreich).
Landwirtsch.-Chem. Bundesversuchs-Anst., Promenade 37. Dir. Reg.R. Dr.-Ing. R. Hönigschmid. — Reg.R. Dr. F. Wochack; Insp. Dr. R. Skutezky; Adjunkt: Dr.-Ihg. H. Werneck.

Lisbôa (Portugal).
Labor. de Fisica agric., Inst. Superiôr de Agronomia, Tapada da Ajuda. Dir. Prof. F. de Almeida Figneredo.
Labor. de Chimica agric., Inst. Superiôr de Agronomia, Tapada da Ajuda. Dir. Prof. L. A. Rebelo da Silva.

Lyngby (Danmark).
Statens Planteavls-Labor. Dir. K. A. Bondorff. — Sub-Dir.: Erik F. Petersen *(CO_2-Production in*

the soil. Bact., Nodule bact.), R. K. Kristensen *(Agric. Chemistry);* Ass.: F. Find Poulsen, S. Tovborg Jensen. 7 Ass. *6 places for soil chemistry and soil microbiol. Agric. chemistry in relation to cultivation of plants. — Apparatus for electrometric determinations of hydrogen-ion concentration.* ⓟ Plants in relation to soil. Microorganisms.

Magyaróvár (Ungarn).
Agric.-chem- Versuchsanstalt d. Landw. Akad.

Minsk (U.d.S.S.R.).
Station Expérim. Paludéenne. Dir. en Chef: Prof. A. T. Kirssanoff; Dir.: M. V. Dokoukine. — Experim. agron., Dir.: M. V. Dokoukine; Specialistes: A. Belkievitch, S. Garkavy, P. Gavrilovitch, N. Lebedévitch, V. Linévitch, S. Letkovsky, P. Parakhnievitch, N. Polotchanine, I. Soboleff, I. Tropachko, A. Zénink. I. Section de chimie agronomique, Vice-Dir.: B. A. Gangea (Chimiste pédologue) T. I. Rodzevitch (Chimiste); V. Sciepourgynsky; I. Tropachkó. II. Section Botan.-Lysimétrique et cabinet de botan.: V. V. Adamoff (Dendrol. sylvicoles); V. V. Gorbonnoff (Cultures); Garkavy; Letkovsky; Soboleff (Dendrol. et forestier); I. Tomachevitch; A. Khimza. III. Section d'économie agric.: M. A. Doubrovsky; Linévitch. IV. Section domaine et métairie pour économie rurale expérim.: E. J. Schuperko; K. J. Makhnatch; N. V. Petrovsky. Stagiaires spécialistes au labor. chimique de la Station: F. M. Snejkine (Zootechnique); T. I. Zenkévitch (Chimie); M. V. Bitch (Chimie); G. G. Maslakovets (Physique); V. M. Pilko (Chimie) *12 places 30 roubles par moi, pendant trois mois de l'été. Phytochimie. Physiol. des plantes, chimie pédol. Etude de la Tourbe botan., culture des plantes tourbière, Agronomie, Etude dendrol. et sylvicoles sur marais, Introduction des plantes de provenience étrangère. — Calorimètre spécial pour étudier l'effect calorique des tourbes (Bombe calorifique). Labor. chimiques, physiques.* ⓟ Specim. de diverses tourbes pour analyses. Plantes vivantes de collections (Herbarium vivum). Graines. Plantes desséchées pour herbiers et autres objets sur commande. — Succursale de labor. dans la ferme au marais Komarovsky (dans la ville même de Minsk) où se trouvent les pépinières et les terres destinées aux études et expériences, champs, prairies, bois, marais. Herbarium vivum et les pépinières forestières dendrol. ainsi, que des cultures potogères.

Montpellier (France).
Station de Chemie agric., Ecole d'agric.

Moskau (U.d.S.S.R.).
Bact.-agronomische Station, 69, Koniuschkowskaja, 31. Dir. Prof. A. Wojtkiewitsch.—K. Rudakow, W. Islailsky, Abt.-Vorst.; E. Mischustin, E. Runow, O. Wojtkewitsch, L. Starygin. *2 Arbeitsplätze für bact. Boden- und Milchuntersuchungen und Bakteriosen der Pflanzen.* ⓟ Mannigfaltige bact. Praep. und Kulturen.
Wissenschaftl. Inst. f. Dünger, Nowo-Slobodskaja, Tichwinskij per. 10.
Staatl. wissenschaftl. Forschungsinst. f. Melioration mit Salzen, Starokonjuschennij 12.

München (Deutschland).
Inst. für Bodenkunde a. d. forstl. Versuchsanstalt, Amalienstr. 52.

Nelson (New Zealand).
Agric. Chem. Dept. of the Cawthorn Inst.

New Brunswick, N.J. (U.S.A.).
Labor. of Agric. Biochemistry of the New Jersey Agric. Experim. Station. Dir. Walter C. Russell. *1 place. — Investigations in nutrition and biochemistry. An albino rat colony is maintained for vitamin research and farm animals of known history are available for research purposes.*
Rutgers Univ. and New Jersey Agric. Experim. Sta. Confer: Physiol. plantarum.

Nikita-Yalta, Krim (U.d.S.S.R.).
Labor. of Soil Improvement, Botan. Garden.

Oerlikon b. Zürich (Schweiz).
Agriculturchem. Abt. d. Schweizerischen landwirtschaftl Versuchsanst. *Futtermittel-, Boden- und Düngemittelkontrolle in der Ost- und Zentralschweiz.*

Oldenburg i. O. (Deutschland).
Versuchs- u. Kontrolistat. d. Oldenburg. Landwirtschaftskammer. Confer: Productio lactaria.

Perm (U.d.S.S.R.).
Landwirtschaftl.-chem. Kabinett. Dir. Prof. A. I. Mošev.
Kabinett f. Bodenkunde. Dir. Prof. V. V. Nikitin

Pisa (Italia).
Labor. f. Agrarchem. d. R. Inst. sup. Agrario. Dir. Prof. C. Ravenna.

Poznań — Posen (Polska).
Inst. f. Bodenkunde. Dir. Prof. F. Terlikowski.
Inst. f. Pflanzenphysiol. u. landwirtsch. Chemie. Confer: Physiol. plantarum.

Pretoria (South Africa).
Agric.-Chem. Inst. Univ. Dir. Prof. J. C. Ross

Puławy (Polska).
Abt. f. Bodenkultur des Staatl. Wissensch. Inst. f. Landwirtsch. Dir. Prof. T. Miecsynski.

Pullman, Wash. (U.S.A.).
Dept. of Soils. Dir. F. J. Sievers.

Riga (Latvija).
Inst. f. Bodenbau. Dir. Doc. Janis Balodis.
Abt. f. Agrikulturchem. u. Bodenkunde. Dir. Doc. P. Kulitaus.

St. Paul, Minn. (U.S.A.).
Division of Agric. Biochemistry, Univ. of Minnesota. Dir. R. A. Gortner.

Salzburg (Österreich).
Bodenkundliches Labor. des Landes-Meliorations-Amtes, Müllner-Hauptstr. 54. Leiter: Ing. Dr. Bernhard Ramsauer. — 1 Laborant. *Dient der kulturtechn. sowie der land- u. forstwirtsch. Bodenuntersuchung. Seit 1923 fakultative Bodenuntersuchungen, seit 1925 obligate Bodenuntersuchung für sämtl. Meliorationsprojekte. 2 Kopecky-, 1 Krauß-, 2 Junkerapparate.*

Sofia (Bulgarien).
Inst. für Agriculturchem. Dir. Doc. M. Hadčiev.
Agric.-chem. Abt. Zentralinst. f. landwirtsch. Versuchswesen, Solunplatz 3. Dir. Dr. M. Stephanova.

Sopron — Ödenburg (Ungarn).
Forstchemisches Labor. Dir. Stefan Vági. — Karl Schuhmacher, Ass. *2 Plätze für bodenkundliche und organische Chemie. — Analytische Apparate.*

Split (S.H.S.).
Landw. Kontroll- und Versuchsstation, Frankopanska ulica 1. Dir. Ing. Anaklet Gazzari. — Peter Novak, Entomol. 2 Ass. *Agro-chemische Untersuchungen und für entomol. Forschungen.* ⓟ Schädliche Insekten.

Stalingrad (U.d.S.S.R.).
Stalingrader Abt. d. Staatl. Inst. f. Melioration m. Salzen.

Stockholm (Sverige).
Abt. f. landwirtschaftl. Chem. d. Zentralanst. f. landwirtschaftl. Versuchswesen. Dir. Prof. Dr. S. L. A. Odén. 1 Ass.

Taschkent (U.d.S.S.R.).
Inst. f. Bodenkunde u. Geobotan., Obuchow-Str. 10. Dir. Prof. N. A. Dimo.

Tiflis (U. d. S. S. R.).
Labor. f. landwirtsch. Chem. Dir. Prof. Melikoff.

Tôkyô (Japan).
Chemical Inst. Fac. of Agric., Tôkyô Imper. Univ., Komaba near Tôkyô. Dir. Prof. Dr. U. Suzuki (Biochem.). Prof. Dr. Makoto Sawamura (Nutrition Chem.); Prof. Dr.Keijirô Asabu (Agric. Chem.); Prof. Dr. Sôjirô Kawase (Phytochem.); Ass. Prof. Teijirô Yabutra (Fermentol.); Ass. Prof. Yoshihiko Matsuyama; Ass. Prof. Issni Kawamura; Ass. Prof. Shin-ichirô Kasugai.

Torino (Italia).
R. Stazione Chimico-Agraria Sperim., Via Ormea No 47. Dir. Prof. Francesco Scurti. — Edoardo Dr. Carlo Zay, Vice-Dir.; Dr. Giovanni Piano, Ass. Chimico; Dr. Giulia Drogoul, Ass. Chimico; Dr. Antonino Di-Giovanni, Ass. Agronomo. *18 posti. Labor. (compresi i sotterranei nei quali sono installati gli apparecchi a tipo industriale). Chimica agraria, Biochemica vegetale, Industrie Agrarie. Applicazioni Agrarie del freddo artificiale. — L'Ist. è divisio in cinque Sezioni che sono: Sezione Ricerche; Sezione Analisi; Sezione Frodi; Sezione Industrie Agrarie; Sezione Agronomica.*

Trinidad (Br.-W.-India).
Dept. of chem. and soil sc., Coll. of tropical Agric. Dir. Prof. F. Hardy.

Vjatka (U. d. S. S. R.).
Abt. f. Bodenkunde d. Wissensch. Forschungsinst. f. Heimatkunde. Dir. Prof. St. L. Ščeklejn.

Warszawa (Polska).
Inst. für Bodendüngung u. Bodenbearbeitung. — Skiernewice Dir. Prof. Dr. Marjan Gorski. — Adjunkt Dr. Ing. Zygmunt Golonka; Ass. Ing. Olga Dąbrowska; Ing. Janina Krotowicz. *5 Plätze für chemische und physikalische Bodenuntersuchung. — Versuchsfeld 35 ha und Vegetationshalle.*
Inst. f. Agrarchem. a. d. Hochsch. f. Land- u. Forstwirtsch. Dir. Prof. G. Mikułowski-Pomorski.

Washington (U. S. A.).
Bureau of Soils, U.S. Dept. of Agric. Chief: Milton Whitney.
Bureau of Chemistry, U.S. Dept. of Agric. Chief: C. A. Browne.

Weihenstephan b. Freising (Deutschland).
Agrikulturchemische Inst. der Hochsch. f. Landw. Dir. Prof. Dr. Niklas.

Wien (Deutschland).
Landwirtschaftlich - Chemische Bundesversuchsanstalt, II, Trunnerstr. 1. Dir. Reg.R. Dipl.-Ldw. Ing. Dr. Viktor Zailer. — Oberinspekt.: H. R. Ing. Aug. Füger; Dr. Ferd. Pilz, Ing. Leopold Wilk, Dr. Rud. Miklauz, Dr. Vinzenz Fritsch, Ing. Rud. Waschata, Dr. Helmuth Müller, Dr. Jos. Mayrhofer, Dr. Franz Wobisch, Ing. Jul. Heisig, Ing. Richard Wagner, Dr. Wolfgang Himmelbaur, Dr. phil. et jur. Otto v. Dafert-Gensel-Timmer. Dr. Alfred Uhl. Ing. Alfred Welch. Dr. Klementine Fiala-Mrazek. *Pflanzenernährung u. Düngung. Molkerei. Weinbau. Chem.-techn. Untersuchungen. Moorkultur. Zollu. Steuerangelegenheiten. Fischerei u. Abwässer. Fütterung u. Ernährung. Kultur von Heilpflanzen u. Drogen, Versuchswirtschaft in Admont. Vegetationsstat. u. Versuchsfelder in Korneuburg Nr. 6.*
Sammlung f. Geol. u. Bodenkunde Hochsch. f. Bodenkultur. Dir. Prof. Dr. A. Himmelbaur.

Wilno (Polska).
Inst. de Chimie agric. et Microbiol. à l'Univ., Objazdowa L. Dir. Prof. Dr. Stephan Bazarewski. — Ass.: Ing. agronom Mlle Janina Turska et Mr. Witold Zarnowski. ⊕ *Les cultures des microbes.*

Woronesch (U. d. S. S. R.).
Versuchsstat. f. Düngung, Landwirtsch. Hochsch.

Zagreb — Agram (S. H. S.).
Agrochem. Labor. Univ. Dir. Prof. Solaja.

Zürich (Schweiz).
Agrikulturchemisches Labor. der Eidgen. Techn. Hochsch., 6, Universitätsstr. 2. Dir. Prof. G. Wiegner. — Prof. E. Winterstein.

III. PERIODICA

Abhandlungen, Botanische. C. von Goebel.
Abhandlungen aus dem geologisch-paläontologischen Institut der Universität Greifswald. O. Jaekel. *L. Bamberg, Greifswald.*
Abhandlungen des Naturwissenschaftlichen Vereins, Bremen. H. Duncker. *Illing & Lüken, Bremen.*
Abstracts, Botanical. R. G. Hoskins. *I. H. Ehlers, Ann Arbor, Mich.*
Acta botanica Instituti Botanici Universitatis Zagrebensis. V. Vouk. *Clavoni botaniškog zavoda weučilišta Zagreb; Julius Springer, Wien.*
Acta Instituti et Horti Botanici Tartuensis. H. Kaho. *Botan. Institut, Tartu.*
Acta Musei Moraviensis. K. Absolon. *Mor. zemské Museum, Brno, C.S.R.*
Acta pathologica et microbiologica Scandinavica. O. Thomsen. *Levin & Munksgaard, Kopenhagen.*
Acta Societatis Botanicorum Poloniae. D. Szymkiewicz. *Société Botanique de Pologne, Varsovie.*
Acta Societatis Entomologicae Cechosloveniae. A. Vimmer.
Acta Zoologica. Internationell tidskrift för zoologi. N. Holmgren. *A. Bonnier, Stockholm.*
Annalen des Naturhistorischen Museums in Wien. K. Keißler.
Annales d'Anatomie pathologique et normale médico-chirurgicale. B. Cunéo, R. Gregoire, P. Legéne, P. Masson, A. Policard, G. Roussay. L. Cornil, P. Moulonguet. *Masson & Cie., Paris.*
Annales de Biologie lacustre. A. Lameere, I. A. Lestage. *I. Wodon-Rousseau, Office de Publicité, Bruxelles.*
Annales de l'Institut Pasteur. E. Duclaux. A. Calmette, L. Martin. *Masson & Cie., Paris.*
Annales du Jardin botanique de Buitenzorg. Ch. J. Bernard, W. M. Docters van Leeuwen. *Brill & Co., Leiden.*
Annales du Musée Colonial de Marseille. A. Jumelle. *Musée Colonial de Marseille.*
Annales de Paléontologie. P. M. Boule. *Masson & Cie., Paris.*
Annales de Physiologie et de Physico-chimie biologique. P. Portier, L. Lapicque, C. Delezenne, A. Mayer. *G. Doin, Paris.*
Annales de la Société de Zymologie pure et appliquée. M. H. van Laer. *Société de chimie industrielle, Paris.*
Annales de l'Université de Lyon. I. I. Beauverie.
Annali di Botanica. P. R. Pirotta, E. Carano.
Annals of applied Biology. D. W. Cutler. *Cambridge University Press, London.*
Annals of Botany. R. Thaxter. *Oxford University Press.*
Annals of the entomological Society of America. H. Osborn. *Entomol. Soc. of America, Columbus, Ohio.*
Année, L', biologique. *Presses Universitaires de France, Paris.*
Annotationes Zoologicae Japonenses. Zool. Society. *Zool. Inst., University, Tokyo.*
Annuaire agricole de la Suisse. H. Faes. *Berne.*
Anthropologie, L'. P. M. Boule, R. Verneau. *Masson & Cie., Paris.*
Anzeiger, Anatomischer. H. von Eggeling. *G. Fischer, Jena.*
Anzeiger, Anthropologischer. Th. Mollison, W. Gieseler. *E. Schweizerbart, Stuttgart.*

Anzeiger, Ethnologischer. M. Heydrich, H. Buschan. *E. Schweizerbart, Stuttgart.*
Anzeiger für Schädlingskunde. Zugleich Nachrichtenblatt der Deutschen Gesellschaft für angewandte Entomologie e. V. K. Escherich, F. Stellwaag. *Paul Parey, Berlin.*
Anzeiger, Zoologischer. E. Korschelt. *Akad. Verlagsgesellschaft, Leipzig.*
Arbeiten der Biologischen Dnjeprstation. D. Beling. *Phys. Mathemat. Abteilung d. Ukrainischen Akademie der Wissenschaften.*
Arbeiten der Biologischen Wolga-Station Saratow. A. Behning.
Arbeiten aus dem Pathologischen Institut der Universität Helsingfors. A. Wallgren. *G. Fischer, Jena.*
Arbeiten des phyto-paläontologischen Laboratoriums der Universität Graz. B. Kubart. *Leuschner & Lubensky, Graz.*
Arbeiten aus dem Zoologischen Institut der Universität Innsbruck. A. Steuer, W. Junk. *W. Junk, Berlin.*
Archiv, Virchows, für pathologische Anatomie und Physiologie und für klinische Medizin. O. Lubarsch. *Julius Springer, Berlin.*
Archiv für Anthropologie. G. Thilenius. *F. Vieweg & Sohn, Braunschweig.*
Archiv für Bienenkunde. L. Armbruster. *Karl Wachholtz, Neumünster (Holstein).*
Archiv, Botanisches. Zeitschrift für die gesamte Botanik. Carl Mez. *Verlag des Repertoriums, Berlin-Dahlem.*
Archiv, Internationales, für Ethnographie. A. W. Nieuwenhuis, J. van Goyenkade. *C. J. Brill, Leiden.*
Archiv für Hydrobiologie. August Thienemann. *E. Schweizerbart, Stuttgart.*
Archiv der Julius-Klaus-Stiftung für Vererbungsforschung, Sozialanthropologie und Rassenhygiene. O. Schlaginhaufen. *Orell Füssli, Zürich.*
Archiv für Molluskenkunde. F. Haas, W. Wenz. *Deutsche malakozoologische Gesellschaft, Frankfurt a. M.*
Archiv für Naturgeschichte. E. Strand. *Nicolaische Verlagsbuchhandlung, Berlin.*
Archiv für experimentelle Pathologie und Pharmakologie. W. Straub, L. Krehl. *F. W. Vogel, Leipzig.*
Archiv, Pflügers, für die gesamte Physiologie des Menschen und der Tiere. E. Abderhalden, A. Bethe, R. Höber. *Julius Springer, Berlin.*
Archiv, Skandinavisches, für Physiologie. C. G. Santesson. *W. de Gruyter, Berlin.*
Archiv für Protistenkunde. M. Hartmann, A. Pascher. *G. Fischer, Jena.*
Archiv für Rassen- und Gesellschaftsbiologie einschließlich Rassen- und Gesellschaftshygiene. A. Ploetz, F. Lenz. *J. F. Lehmann, München.*
Archiv für experimentelle Zellforschung, besonders Gewebezüchtung (Explantation). Rhoda Erdmann. *G. Fischer, Jena.*
Archives d'Anatomie microscopique. L. F. Henneguy. *Masson & Cie., Paris.*
Archives d'Anatomie, d'Histologie et d'Embryologie. Ancel, R. Anthony, Bouin, Brachet, Forster. *Vix, Strasbourg.*

Archives Russes d'Anatomie, d'Histologie et d'Embryologie. D. Deineka. *Staatl. Verlag Gosisdat, Leningrad.*
Archives de Biologie. E. van Beneden, Ch. van Bambeke. *Masson & Cie., Paris.*
Archives de Botanique. R. Viguier. *R. Viguier, Caen.*
Archives d'Hydrobiologie et Ichthyologie. A. Litynski. *Société des Sciences de Varsovie.*
Archives internationales de médecine expérimentale. R. Bruynoghe, A. P. Dustin. *Masson & Cie., Paris.*
Archives internationales de Pharmacodynamie et de Thérapie. E. Gley, J. F. Heymans; Inst. d. Pharmacodynamie Gand. *A. Doin, Paris.*
Archives de l'Institut Pasteur d'Algérie. G. Senevet. *Institut Pasteur, Alger.*
Archives de Morphologie générale et expérimentale. R. Anthony, Bataillon, Bouin, Latarjet, Policard, Prenant. *G. Doin, Paris.*
Archives du Muséum d'Histoire naturelle. R. Anthony. *Masson & Cie., Paris.*
Archives de Physique biologique et de Chimiephysique des corps organisés. F. Vlès. *Vigot, Paris.*
Archives internationales de Physiologie. L. Fredericq, P. Heger. *Doin, Paris.*
Archives Neerlandaises de Physiologie de l'homme et des animaux. Buytendyk, W. Einthoven, W. E. Ringer, G. van Rijnberk, H. Zwaardemaker. *Nijhoff, 's Gravenhage.*
Archives Portugaises des Sciences Biologiques. Athias, A. Celestino da Costa.
Archives de Zoologie expérimentale et générale. E. Racovitza, L. Fage, O. Duboscq. *H. Le Soudier.*
Archivi di Biologia applicata alla Patologia, alla Clinica e all'Igiene. E. Maragliano, A. Trambusti, P. Canalis, F. Figari. M. Capocaccia, Sivori, Rebaudi, Bonino, Pulgher. *Istituto Maragliano, Genova.*
Archivio per l'Antropologia e la Etnologia. L. Cipriani. *Società Italiana di Antropologia e Etnologia.*
Archivio botanico per la Sistematica, Fitogeografia e Genetica. A. Béguinot. *Tipografia Valbonesi, Forli.*
Archivio Italiano di Anatomia ed Embriologia. G. Chiarugi, G. Bruno. *Niccolai, Firenze.*
Archivio di Scienze biologiche. F. Bottazzi.
Archivum Balatonicum. E. Csiki, B. Hankó. *Ungar. Nat.-Museum.*
Ardea. Tydschrift der Nederlandsche Ornithologische Vereeniging. L. F. de Beaufort, G. J. van Oordt, J. Verwey. *E. J. Brill, Leiden.*
Beiträge zur Biologie der Pflanzen. R. Schaede, F. Rosen. *J. U. Kern, Breslau.*
Beiträge zur Naturdenkmalpflege. W. Schoenichen. *Gebr. Bornträger, Berlin.*
Beiträge zur pathologischen Anatomie und zur allgemeinen Pathologie. L. Aschoff. *G. Fischer, Jena.*
Beobachter, Der ornithologische. A. Hess. *Schweizer Gesellschaft f. Vogelkunde, Basel.*
Bericht, Anatomischer. H. von Eggeling. *G. Fischer, Jena.*
Bericht, Zoologischer. C. Apstein, E. Korschelt. *Akad. Verlagsges., Leipzig.*
Berichte über die gesamte Biologie. Abteilung A. Berichte über die wissenschaftliche Biologie. M. Hartmann, F. v. Wettstein. T. Péterfi. *Julius Springer, Berlin.*
Berichte über die gesamte Biologie. Abteilung B. Berichte über die gesamte Physiologie und experimentelle Pharmakologie. P. Rona. *Julius Springer, Berlin.*
Berichte der Deutschen Botanischen Gesellschaft. E. Tiegs. B. Leisering. *G. Fischer, Jena.*
Berichte der Schweizerischen Botanischen Gesellschaft. W. Rytz. *Rascher & Co., Zürich.*
Bibliographia genetica. J. P. Lotsy, H. N. Kooiman. *M. Nijhoff, s'Gravenhage.*
Bibliographia zoologica. *Concilium Bibliographicum, Zürich.*
Bibliotheca Genetica. E. Baur. *Gebr. Bornträger, Berlin.*
Biologia generalis (International Journal of general biology. Archives internationales de biologie générale). Internationale Zeitschrift für allgemeine Biologie. L. Löhner, R. Pearl, V. Ruzicka. O. Abel, O. Porsch, C. Schwarz, J. Versluys, R. Wasicky. *E. Haim & Co., Wien.*
Biologia Hungarica. Z. von Szilády. *Szilady, Budapest.*
Blätter für Aquarien- und Terrarienkunde. W. Wolterstorff. *J. E. G. Wegner, Stuttgart.*
Blätter, entomologische. Zeitschrift für Biologie und Systematik der Käfer. R. Kleine. *F. Pfenningstorff, Berlin.*
Blätter, Praktische, für Pflanzenbau und Pflanzenschutz. G. Christmann. *F. P. Datterer & Co., Freising.*
Boletin de la Dirección de Estudios Biologicos. A. L. Herrera. *Dir. de Estud. Biol., Mexico.*
Bolletino del laboratorio di Zoologia generale e agraria della R. Scuola superiore d'Agricoltura, Portici. F. Silvestri.
Bollettino dei Musei di Zoologia e Anatomia comparata della R. Università di Genova. R. Issel, L. Cognetti de Martiis.
Botanik, Angewandte. Zeitschrift für Erforschung der Nutzpflanzen. K. Snell. *Gebr. Bornträger, Berlin.*
Botanikai, Magyar, Lapok. A. von Degen.
Bryologist, The. O. E. Jennings. *Sullivan Moss Soc., New York.*
Bulletin of the Antivenin Institute of America. R. H. Hutchison. *Antivenin Institute of America, Glenolden, Pa., U.S.A.*
Bulletin, Biological. *Marine Biological Laboratory, Woods Hole, Mass., U.S.A.*
Bulletin biologique de la France et de la Belgique. L. F. G. Blaringhem, G. Bohn, Caullery, Ch. Julin, F. Mesnil, P. Pelseneer, Ch. Péres, Et. Rabaud. *Presses Universitaires, Paris.*
Bulletin du Jardin botanique. W. M. Docters van Leeuwen. *'s Lands Plantentuin, Buitenzorg.*
Bulletin de la Société Colombienne de Sciences naturelles. Fr. Apollinaire-Marie. *Bogotà.*
Bulletin de la Société vaudoise des Sciences naturelles. A. Maillefer.
Bulletins et mémoires de la Société d'Anthropologie de Paris. R. Anthony. *Masson, Paris.*
Bulletins of the New York Agricultural Experiment Station. W. O. Gloyer. *Geneva, N. Y.*
Bulletins, Storrs Experiment Station. W. M. Estem.
Candollea. Organe du conservatoire et du jardin botaniques de la ville de Genève. J. J. Briquet. *Conservat. botanique, Genève.*
Cellule, La. G. Gilson. *A. Wystpruyst, Louvain.*
Cereal Chemistry. C. H. Bailey. *American Association of Cereal Chemists.*
Chemie der Erde. E. Blanck, G. Linck. *G. Fischer, Jena.*
Chemie der Zelle und Gewebe. Zeitschrift für die Probleme der Gärung, Atmung und Vitaminforschung. H. Haehn. *Gebr. Bornträger, Berlin.*
Comptes rendus de l'Association des Anatomistes. R. Collin. *Société d'Impressions typographiques, Nancy.*
Comptes rendus de la Société de Biologie. J. M. Jolly. *Masson, Paris.*
Contributions from the Bermuda Biological Station for Research. E. L. Mark.
Contributions from the Zoological Laboratory, Harvard University. E. L. Mark.
Cytology, General. E. V. Cowdry. *Univ. of Chicago Press.*
Entomologica Americana. F. H. Chittenden.
Ergebnisse der allgemeinen Pathologie und pathologischen Anatomie des Menschen und der Tiere. O. Lubarsch, R. v. Ostertag, M. Frei. *J. F. Bergmann, München.*

Ergebnisse der Physiologie. L. Asher, K. Spiro. *J. F. Bergmann, München.*
Ergebnisse und Fortschritte der Zoologie. M. Hartmann. *G. Fischer, Jena.*
Fermentforschung. E. Abderhalden. *Urban & Schwarzenberg, Berlin-Wien.*
Field Naturalist, Canadian. H. F. Lewis. *Ottawa Field Naturalists Club, Ottawa, Canada.*
Fischerei-Zeitung. E. Walter. *L. Neumann, Neudamm.*
Fischerei-Zeitung, Allgemeine. H. N. Maier. *Gebr. Reichel, Augsburg.*
Fischerei-Zeitung, Badische. W. Koch. *Bad. Landes-Fischereiverein, Karlsruhe.*
Fischerei-Zeitung, Schweizer. G. Surbeck. *H. Kunz, Pfäffikon.*
Fiskerier, Finlands. T. H. Järvi.
Flora oder Allgemeine botanische Zeitung. C. von Goebel. *G. Fischer, Jena.*
Folia anatomica Japonica. K. Okajima. *Maruzen & Co., Tokyo.*
Flora Batava. L. Vuyck. *Martinus Nijhoff, Haag.*
Folia Cryptogamica. J. Györffy. *Szeged.*
Folia entomologica hungarica. J. Jablonowski.
Folia myrmecologica et termitologica. A. Krausse. *Hussiten-Druckerei (A. Höhne), Bernau b. Berlin.*
Forschungen, Milchwirtschaftliche. W. Grimmer. *Julius Springer, Berlin.*
Fortschritte der Landwirtschaft. H. Kaserer, R. Miklauz. *J. Springer, Wien.*
Gazette, Botanical. J. M. Coulter. *University of Chicago Press.*
Genetica. H. N. Koolman, J. P. Lotsy. *M. Nijhoff, Haag.*
Genetics. A periodical record of investigations in heredity and variation. W. Castle, E. Conklin, C. Davenport, B. Davis, E. East, R. Emmerson, H. Jennings, H. Morgan, R. Pearl. G. H. Shull. D. F. Jones. *Brooklyn Botanic Garden.*
Jahrbuch für Morphologie und mikroskopische Anatomie. Abteilung I: Gegenbaurs Morphologisches Jahrbuch. E. Göppert. *Akad. Verlagsges., Leipzig.*
Jahrbuch für Morphologie und mikroskopische Anatomie. Abteilung II: Zeitschrift für mikroskopisch-anatomische Forschung. H. Stieve. *Akad. Verlagsges., Leipzig.*
Jahrbuch, Neues, für Mineralogie, Geologie und Palaeontologie. R. Brauns, E. Hennig, E. Kaiser. *Schweizerbart, Stuttgart.*
Jahrbücher, Botanische, für Systematik, Pflanzengeschichte und Pflanzengeographie. A. Engler. *Max Weg, Leipzig.*
Jahrbücher für wissenschaftliche Botanik. H. Fitting. *Gebr. Bornträger, Berlin.*
Jahrbücher, Zoologische. Abteilung für Systematik, Geographie und Biologie der Tiere. Abteilung für Anatomie und Ontogenie der Tiere. Abteilung für allgemeine Zoologie und Physiologie der Tiere. M. Hartmann, R. Hesse. *G. Fischer, Jena.*
Jahresbericht für Agriculturchemie. F. Mach. *P. Parey, Berlin.*
Jahresbericht, Just's Botanischer. Systematisch geordnetes Repertorium der botanischen Literatur aller Länder. F. Fedde. *Gebr. Bornträger, Berlin.*
Jahresbericht über die gesamte Physiologie und experimentelle Pharmakologie mit vollständiger Bibliographie. Zugleich Fortsetzung des Hermann Weißschen Jahresberichts über die Fortschritte der animalischen Physiologie und des Maly-Andreasch-Spiroschen Jahresberichts über die Fortschritte der Tierchemie oder der physiologischen, pathologischen und Immuno-Chemie und der Pharmakologie. P. Rona, K. Spiro. *Julius Springer, Berlin.*
Jahresbericht über die Leistungen auf dem Gebiete der Veterinärmedizin. W. Ellenberger, K. Neumann-Kleinpaul, O. Zietzschmann. *Julius Springer, Berlin.*
Journal of Agricultural Research. W. Crocker.

Journal, American, of Anatomy. C. R. Stockard. *Wistar Institute of Anatomy, Philadelphia.*
Journal, American, of Physical Anthropology. A. Hrdlicka. *Wistar Institute, Philadelphia.*
Journal of Bacteriology. C. E. A. Winslow. *Williams & Wilkins, Baltimore.*
Journal of Biochemistry. S. Kakiuchi. *Coll. of Med., University, Tokyo.*
Journal, Australian, of experimental Biology and medical Science. J. B. Cleland, T. B. Robertson, C. S. Hicks. *Univ. of Adelaide.*
Journal, British, of experimental Biology. J. Gray. *Oliver & Boyd, Edinburgh.*
Journal of Botany. A. B. Rendle. *Taylor & Francis, London.*
Journal, American, of Botany, Ecology and Genetics. C. E. Allen. *Botanic Garden, Brooklyn.*
Journal, Japanese, of Botany. S. Ikeno. *National Research Council of Japan.*
Journal of the New York Botanical Garden. M. A. Howe. *New York Botanical Garden.*
Journal of the Botanical Society of South Africa. R. H. Compton. *Specialty Press, Cape Town.*
Journal of biological Chemistry. Stanley R. Benedict, Dakin, B. Mendel, Van Slyke. *Rockefeller Institute for medical research., New York.*
Journal of Ecology. A. G. Tansley. *Cambridge University Press.*
Journal of Economic Entomology. E. P. Felt. *Amer. Assoc. of Economic Entomologists, Geneva, New York.*
Journal of Entomology and Zoology. W. A. Hilton. *Dep. of Zoology, Pomona College, Claremont, Calif.*
Journal of Genetics. R. C. Punnett. *Cambridge University Press.*
Journal of Heredity. O. Olson. *Amer. Genetic Assoc., Washington.*
Journal of Immunology. A. F. Coca. *Williams & Wilkins, Baltimore.*
Journal of the Marine Biological Association of the United Kingdom. E. J. Allen. *Brendon & Son, Plymouth.*
Journal of Morphology and Physiology. C. E. Mc Clung. *Wistar Institute, Philadelphia.*
Journal of comparative Neurology. Donaldson, Johnston, Coghill, Meyer, Strong, Herrick. *Wistar Institute of Anatomy, Philadelphia.*
Journal für Ornithologie. E. Stresemann. *Friedländer & Sohn, Berlin.*
Journal of Parasitology. H. Ward. *Wistar Institute, Philadelphia.*
Journal of the Pacific Research Institution. F. G. Krauss. *Pan-Pacific Union, Honolulu, Hawai.*
Journal, American, of Pathology. James W. Jolling, How. T. Karsner, Paul L. Lewis, F. Parker jr., G. Wells, G. H. Whipple. T. B. Mallory. *Boston City Hospital.*
Journal, British, of experimental Pathology. E. C. Doddts; C. L. Evans, W. E. Gye, E. H. Kettles, J. McIntosh, J. W. McLeod, J. A. Murray, W. J. Tulloch. P. Fildes. *Lewis & Co., London.*
Journal of Pharmacology and Experimental Therapeutics. J. J. Abel, A. R. Cushny. *Williams & Wilkins, Baltimore, Md.*
Journal of Physiology. Adrian, A. V. Hill, J. B. Leathes, C. S. Sherrington. *Cambridge University Press.*
Journal, American, of Physiology. D. R. Hooker, *Amer. Physiol. Soc., Baltimore.*
Journal, Chinese, of Physiology. R. Kho-Seng Lim, H. G. Earle, B. E. Read, Hsien Wu. *Chinese Physiological Society, Peking.*
Journal de Physiologie et de Pathologie générale E. Gley, Ch. Richet, P. J. Teissier. Henry Cardot Robert Pierret. *Masson & Cie., Paris.*
Journal of general Physiology. W. J. Crouzier, J. H. Northrop, W. J. van Leuven Osterhout. *Williams & Wilkins, Baltimore.*

Journal of comparative Psychology. K. Dunlap, R. M. Yerkes. *Williams & Wilkins, Baltimore.*
Journal of metabolic research. R. G. Hoskins. *Physiatric Inst. Morristown, New Yersey, U.S.A.*
Journal, Jowa State College, of Science. R. E. Buchanan.
Journal, Philippine, of Science. E. Quisumbing, M. A. Tubangui. *Bureau of Science, Manila, P.J.*
Journal New York Entomological Society. W. Beutenmüller.
Journal of the Royal Horticultural Society of England. F. J. Chittenden. *Royal Horticult. Soc., London.*
Journal de la Société Botanique de Russie. J. Borodin. *Musée Botanique, Leningrad.*
Journal of the Washington Academy of Sciences. *Washington Academy of Sciences.*
Journal, Japanese, of Zoology. S. Goto. *National Research Council of Japan.*
Journal of Experimental Zoology. R. G. Harrison. *Wistar Institute, Philadelphia.*
Konowia. Zeitschrift für systematische Insektenkunde. R. Meyer. F. Wagner. *F. Wagner, Wien.*
Korrespondenzblatt der wirtschaftlichen Schädlingsbekämpfung. H. Lehmann. *H. Lehmann, Luxemburg.*
Kosmos. Organ der Polnischen Kopernikus-Naturforscher-Gesellschaft. J. Hirschler. *Verl. d. Gesellschaft, Lwów.*
Lambillionea. Revue mensuelle de l'Union des Entomologistes Belges. Fr. Derenne.
Listy Biologické. V. Laufberger, K. Šulc, F. K. Studnicka. *Spolek Ceskych Lekaru, Prag.*
Lotos. Naturwissenschaftliche Zeitschrift. L. Freund. *Deutscher naturwiss.-med. Verein f. Böhmen, Prag.*
Lyon horticole et Horticulture nouvelle réunis. L. M. Faucheron. *Jardin botanique, Lyon.*
Magazin, Nyt, for Naturvidenskaberne. B. Lynge. *A. W. Brögger, Oslo.*
Meddelanden af Societas pro Fauna et Flora Fennica. E. F. Häyrén.
Meddelelser, Videnskabelige, fra Dansk Naturhistorisk Forening. A. S. Jensen. *O. A. Reitzel, København.*
Meddelelser, Entomologiske. K. L. Henriksen. *Entomol. Forening, København.*
Mededeelingen van het Phytopathologisch Laboratorium Willie Commelin Scholten. J. Westerdijk.
Mededeelingen van het Proefstation voor Thee. Ch. J. Bernard. *Thee proefstation, Buitenzorg, Java.*
Mémoires de l'Institut national Polonais d'Economie rurale à Pulawy. St. Kopeć. *Institut d'Economie rurale, Pulawy, Pologne.*
Mémoires du Musée royal d'Histoire naturelle de Belgique. V. E. van Straelen.
Mikrokosmos. Zeitschrift für angewandte Mikroskopie, Mikrobiologie, Mikrochemie und mikroskopische Technik. G. J. Stehli. *Franckscher Verlagsbuchhandlung, Stuttgart.*
Mitteilungen der Anthropologischen Gesellschaft in Wien. L. Bouchal, O. Menghin, O. Reche. *Anthrop. Gesellschaft, Wien.*
Mitteilungen, Entomologische. W. Horn. *Deutsches Entomologisches Institut, Berlin-Dahlem.*
Mitteilungen der Fischerei-Vereine für die Provinzen Brandenburg, Ostpreußen, Pommern, Grenzmark. K. Eckstein. *Verl. des Fischerei-Vereins für die Provinz Brandenburg, Berlin.*
Mitteilungen zur Geschichte der Medizin und der Naturwissenschaften. H. E. Sigerist. *L. Voß, Leipzig.*
Mitteilungen aus dem Institut für allgemeine Botanik in Hamburg. H. Winkler. *Verlag des Institutes, Hamburg.*
Mitteilungen aus dem geologischen Institut der Universität Greifswald. O. Jaekel. *L. Bamberg, Greifswald.*
Mitteilungen der Aargauischen Naturforschenden Gesellschaft. P. Steinmann. *H. R. Sauerländer, Aarau.*
Mitteilungen aus dem Zoologischen Museum in Berlin. W. Ramme. *R. Friedländer & Sohn, Berlin.*
Mitteilungen aus dem Zoologischen Staatsinstitut und Zoologischen Museum in Hamburg. H. Lohmann.
Monatsberichte, Ornithologische. E. Stresemann. *R. Friedländer & Sohn, Berlin.*
Monatshefte, Naturwissenschaftliche, für den biologischen, chemischen, geographischen und geologischen Unterricht. R. Rein. *B. G. Teubner, Leipzig.*
Monatsschrift, Ornithologische. C. R. Hennicke. *Geraer Verlagsanstalt, Gera.*
Monatsschrift, Wiener Tierärztliche. Günther, Hartl, Keller, Reisinger, Schmidt, Schnürer, D. Wirth, L. K. Böhm. *Wilh. Braumüller, Wien.*
Monitore zoologico Italiano. A. Senna, Chiarugi. *Firenze.*
Nachrichtenblatt für Naturdenkmalpflege. W. Schoenichen. *H. Bermühler, Berlin.*
Natur, Die. Naturwissenschaftlich-pädagogische Zeitschrift. V. Pollak, O. Kühn. *Deutscher Verlag für Jugend und Volk, Wien.*
Naturalist, American Midland. N. M. Grier. *University Notre Dame, Indiana, U.S.A.*
Naturaliste, Le, canadien. C. V. Huard.
Naturalist, The Scottish. P. H. Grimshaw, J. Ritchie. *Oliver & Boyd, Edinburgh.*
Naturforscher, Der. W. Schoenichen. *H. Bermühler, Berlin.*
Naturwissenschaften, Die. A. Berliner. *Julius Springer, Berlin.*
Neuheiten auf dem Gebiete des Pflanzenschutzes. G. Köck. *Bundesanstalt für Pflanzenschutz, Wien.*
News, Entomological. P. Calvert. *Amer. Entomol. Society, Philadelphia.*
News, Eugenical. Ch. B. Davenport. *Science Press Lancaster, Pa., U.S.A.*
Notulae entomologicae. R. Forsius. *Societas Entomologica Helsingforsiensis, Helsingfors.*
Nova Guinea. Résultats des expéditions scientifiques à la Nouvelle Guinée. L. F. de Beaufort. *E. J. Brill, Leiden.*
Novitates Zoologicae. Rothschild, Jordan, E. Hartert. *Tring Museum, Tring, Engl.*
Onderzoekings verricht in het Zoologisch Laboratorium der Rijks Universiteit Groningen. J. F. van Bemmelen.
Ornis fennica. Zeitschrift für Vogelforschung und Vogelschutz. I. J. Hortling, E. F. Merikallio. *Ornithol. Vereinigung in Finnland.*
Pathologica. Rivista mensile di Patologia. A. Cesaris Demel.
Phytologist, The new. A. G. Tansley. *Wheldon & Wesley, London.*
Plant Physiology. C. A. Shull. *Science Press Printing Co., Lancaster, Pa., U.S.A.*
Pflanze, Die kranke. Volkstümliches Fachblatt für Pflanzenheilkunde. W. Baunacke. *Sächsische Pflanzenschutzgesellschaft, Dresden.*
Pflanzenareale, Die. H. Winkler, E. Hannig. *G. Fischer, Jena.*
Phytopathology. W. O. Orton, E. C. Stakman, *Phytopathological Society, Lancaster, Pa.*
Protoplasma. Internationale Zeitschrift für physikalische Chemie des Protoplasmas. J. Spek, F. Weber. *Gebr. Borntraeger, Berlin.*
Psyche, a Journal of Entomology. Ch. T. Brues. *Cambr. Entomological Club, Boston, Mass.*
Pubblicazioni della Stazione Zoologica di Napoli. M. Fedele. *Friedländer & Sohn, Berlin.*
Record, Anatomical. J. L. Bremer. *Wistar Institute, Philadelphia.*
Repertorium entomologicum. H. Hedicke. *Dtsch. entomol. Gesellschaft, Berlin.*

Repertorium specierum novarum regni vegetabilis. F. Fedde. *F. Fedde, Berlin-Dahlem.*
Reports of Oxford Ornithological Society. F. C. R. Jourdain, B. W. Tucker.
Resumptio Genetica. J. P. Lotsy, H. N. Kooiman. *M. Nijhoff, Haag.*
Review, Physiological. D. R. Hooker. *Amer. Physiol. Soc., Baltimore, Md.*
Revista Argentina de Biologia. B. A. Houssay. *Asociacion Medica Argentina, Santa Fe.*
Revista Mexicana de Biologia. J. Ochoterena. *Sociedad Mexicana de Biologia, Mexico.*
Revue d'Anthropologie. P. Topinard. *Masson, Paris.*
Revue générale agronomique de Louvain. E. Hegh.
Revue générale de la Botanique. M. Marin Molliard, *Libr. Générale d. l'Enseignement, Paris.*
Revue Française d'endocrinologie. M. Lucien, J. Parisot, G. Richard. *G. Doin, Paris.*
Revue générale d'Histologie. C. Regaud, J. Renaut. *Masson & Cie., Paris.*
Revue, Internationale, der gesamten Hydrobiologie und Hydrographie. R. Wolterek. *Akad. Verlagsgesellschaft, Leipzig.*
Revue Russe d'Entomologie. A. P. Semenov-Tiau-Shandky. *Société entomologique de Russie, Leningrad.*
Revue Suisse de Zoologie. E. Guyénot. *Bedot, Genève.*
Revue Zoologique Russe. P. Kosminsky. *Staatsverlag, Moskau.*
Rivista di Antropologia. S. Sergi. *Società Romana di Antropologia, Roma.*
Rivista di Biologia. O. Polimanti. *Istituto Editoriale Scientifico, Milano.*
Rivista Italiana d'Ornitologia. Arrigoni degli Oddi. *Ant. Trischitta, Messina.*
Rivista di Patologia sperimentale. G. di Macco, J. Jacono, L. Avellone, L. Califano, A. Gualdi, P. Formicola, C. Amatucci-Mallerdo, G. Sollazzo. V. Scaffidi. *V. Idelson, Napoli.*
Rivista di Patologia vegetale. L. Montemartini. *Tipografia cooperativa, Pavia.*
Rundschau, Koleopterologische. F. Heikertinger. *Albert Winkler, Wien.*
Senckenbergiana. R. Richter. *Senckenbergische Naturforschende Gesellschaft, Frankfurt a. M.*
Sitzungsberichte der Gesellschaft für Morphologie und Physiologie in München. *J. F. Lehmann's Verlag, München.*
Sitzungsberichte und Abhandlungen der Naturwissenschaftlichen Gesellschaft Isis in Dresden. F. A. Schade. *H. Burdach, Dresden.*
Supplementa Entomologica. W. Horn. *Verlag W. Horn, Berlin-Dahlem.*
Tidskrift, Svensk Botanisk. T. Lagerberg.
Tierzucht, Deutsche landwirtschaftliche. L. Vogel, Hoesch. Gatermann. *M. u. H. Schaper, Hannover.*
Tierzucht, Süddeutsche landwirtschaftliche. L. Vogel. Stockklausner. *M. u. H. Schaper, Hannover.*
Tijdschrift voor Hydrobiologie en Visscherij. H. C. Redeke, J. A. Heymann. *C. de Boer jr., den Helder.*
Tijdschrift, Naturwetenschappelijk. G. Naveau.
Tijdschrift der Nederlandsche Dierkundige Vereeniging. van Bemmelen, Ihle, Loman. *E. J. Brill, Leiden.*
Tijdschrift, Nederlandsch, voor Hygiene, Microbiologie en Serologie. W. C. de Graaff. *S. C. van Doesburgh, Leiden.*
Tijdschrift over Plantenziekten. N. van Poeteren, J. Ritzema Bos, H. M. Quanjer, H. M. de Koning. *H. Veenman & Zonen, Wageningen.*
Transactions American Microscopical Society. H. J. van Cleave. *Banta, Menasha, Wis.*
Travaux du laboratoire d'Anatomie normale de la Faculté de Médecine d'Alger. E. Leblanc.
Travaux de l'Institut Zoologique et de la Station de Cette. J. E. Bataillon.

Treubia. Recueil de travaux zoologiques, hydrobiologiques et océanographiques. W. M. Docters van Leeuwen. *'s Lands Plantentuin, Buitenzorg.*
Umschau, Die. Wochenschrift über die Fortschritte in Wissenschaft und Technik. H. Bechhold. *H. Bechhold, Frankfurt a. M.*
Untersuchungen, Mykologische, und Berichte. R. Falck. *A.-G. für Druck und Verlag, Kassel.*
Vakblad voor Biologen. A. B. Droogleever Fortuyn, M. J. Sirks. *C. de Boer, Helder.*
Verhandlungen der Deutschen Anatomischen Gesellschaft. H. von Eggeling. *G. Fischer, Jena.*
Verhandlungen des Botanischen Vereins der Provinz Brandenburg. Th. Loesener. *Selbstverl. des Vereins, Berlin-Dahlem.*
Verhandlungen der Internationalen Vereinigung für theoretische und angewandte Limnologie. Fr. Lenz. *E. Schweizerbart, Stuttgart.*
Verhandlungen der Deutschen Zoologischen Gesellschaft. C. Apstein.
Vierteljahrsschrift der Naturforschenden Gesellschaft in Zürich. H. Schinz. *Beer & Co., Zürich.*
Versuchs-Stationen, die landwirtschaftlichen. Organ für naturwissenschaftliche Forschungen auf dem Gebiete der Landwirtschaft. G. Fingerling. *P. Parey, Berlin.*
Wisconsin Horticulture. F. Cranefield. *Wisconsin State Horticultural Society, Madison, Wis.*
Zeitschrift für induktive Abstammungs- und Vererbungslehre. Correns, Haecker, Steinmann, v. Wettstein. E. Baur. *Gebr. Bornträger, Berlin.*
Zeitschrift für die gesamte Anatomie. Abteilung I: Zeitschrift für Anatomie und Entwicklungsgeschichte. E. Kallius. *Julius Springer, Berlin.*
Zeitschrift für die gesamte Anatomie. Abteilung II: Zeitschrift für Konstitutionslehre. J. Tandler. *Julius Springer, Berlin.*
Zeitschrift für die gesamte Anatomie. Abteilung III: Ergebnisse der Anatomie und Entwicklungsgeschichte. E. Kallius. *Julius Springer, Berlin.*
Zeitschrift, Biochemische. C. Neuberg. *Julius Springer, Berlin.*
Zeitschrift für Biologie. O. Frank, M. von Frey, E. Voit. *J. F. Lehmann, München.*
Zeitschrift für wissenschaftliche Biologie. Abteilung A. Zeitschrift für Morphologie und Ökologie der Tiere. P. Buchner, P. Schulze. *Julius Springer, Berlin.*
Zeitschrift für wissenschaftliche Biologie. Abteilung B. Zeitschrift für Zellforschung und mikroskopische Anatomie. R. Goldschmidt, W. von Möllendorff. *Julius Springer, Berlin.*
Zeitschrift für wissenschaftliche Biologie. Abteilung C. Zeitschrift für vergleichende Physiologie. K. von Frisch, A. Kühn. *Julius Springer, Berlin.*
Zeitschrift für wissenschaftliche Biologie. Abteilung D. Wilhelm Roux' Archiv für Entwicklungsmechanik der Organismen. Organ für die gesamte kausale Morphologie. H. Spemann, W. Vogt, B. Romeis. *Julius Springer, Berlin.*
Zeitschrift für wissenschaftliche Biologie. Abteilung E. Planta. Archiv für wissenschaftliche Botanik. W. Ruhland, H. Winkler. *Julius Springer, Berlin.*
Zeitschrift für Botanik. H. Kniep, F. Oltmanns. *G. Fischer, Jena.*
Zeitschrift, Allgemeine botanische, für Systematik, Floristik, Pflanzengeographie usw. A. Kneucker, H. Leininger. *G. Braun, Karlsruhe.*
Zeitschrift, Österreichische botanische. R. Wettstein, E. Janchen, G. Klein. *Julius Springer, Wien.*
Zeitschrift, Hoppe-Seylers, für physiologische Chemie. A. Kossel. *W. de Gruyter, Berlin.*
Zeitschrift für angewandte Entomologie. K. Escherich. *P. Parey, Berlin.*
Zeitschrift, Deutsche Entomologische. H. Hedicke. *R. Friedländer & Sohn, Berlin.*

Zeitschrift für Ethnologie. Organ der Berliner Gesellschaft für Anthropologie, Ethnologie und Urgeschichte. *Julius Springer, Berlin.*

Zeitschrift für Hygiene und Infektionskrankheiten. F. Neufeld, M. Hahn, R. Doerr. *Julius Springer, Berlin.*

Zeitschrift für Immunitätsforschung und experimentelle Therapie. R. Kraus, E. Friedberger, Sachs, Uhlenhuth. *G. Fischer, Jena.*

Zeitschrift für wissenschaftliche Insektenbiologie. H. u. W. Stichel. *Verl. Naturwissenschaftl. Publikationen, Berlin-Hermsdorf.*

Zeitschrift für wissenschaftliche Mikroskopie und für mikroskopische Technik. E. Küster. *S. Hirzel, Leipzig.*

Zeitschrift für Morphologie und Anthropologie. E. Fischer. *E. Schweizerbart, Stuttgart.*

Zeitschrift für Naturwissenschaften. A. Japha, G. Schmid, Scupin. *L. Hofstetter, Halle a. d. S.*

Zeitschrift, Jenaische, für Naturwissenschaften. Med. Naturwissenschaftl. Ges., F. Maurer. *G. Fischer, Jena.*

Zeitschrift, Palaeontologische. F. Drevermann. *Gebr. Bornträger, Berlin.*

Zeitschrift, Frankfurter, für Pathologie. B. Fischer-Wasels. *J. F. Bergmann, München.*

Zeitschrift für Pflanzenernährung und Düngung. P. Ehrenberg, O. Lemmermann. *Verlag Chemie, Leipzig.*

Zeitschrift für Pflanzenkrankheiten und Pflanzenschutz, mit besonderer Berücksichtigung der Krankheiten von forstlichen, landwirtschaftlichen und gärtnerischen Kulturpflanzen. Carl von Tubeuf. *E. Ulmer, Stuttgart.*

Zeitschrift für Pflanzenzüchtung. C. Fruwirth. *P. Parey, Berlin.*

Zeitschrift für Pilzkunde. H. Kniep, F. Kallenbach, A. Zeuner. *W. Klinkhardt, Leipzig.*

Zeitschrift für Psychologie und Physiologie der Sinnesorgane. Abteilung I: Zeitschrift für Psychologie. F. Schumann. Abteilung II: Zeitschrift für Sinnesphysiologie. M. Gildemeister. *J. A. Barth, Leipzig.*

Zeitschrift für Tierzüchtung und Züchtungsbiologie. C. Kronacher. *P. Parey, Berlin.*

Zeitschrift für Untersuchung der Lebensmittel. A. Börner, A. Juckenack, J. König. *Julius Springer, Berlin.*

Zeitschrift für wissenschaftliche Zoologie. J. W. Harms, W. Schleip. *Akad. Verlagsges., Leipzig.*

Zeitung, Wiener entomologische. A. Hetschko, F. Heikertinger. *R. Friedländer & Sohn, Berlin.*

Zell-Stimulations-Forschungen. M. Popoff, W. Gleisberg. *P. Parey, Berlin.*

Zentralblatt für Bakteriologie, Parasitenkunde und Infektionskrankheiten. Abteilung I. Medizinischhygienische Bakteriologie und tierische Parasitenkunde. O. Uhlworm, A. Weber, E. Gildemeister. Abteilung II. Allgemeine landwirtschaftlich-technische Bakteriologie, Gärungsphysiologie, Pflanzenpathologie und Pflanzenschutz. K. Friederichs, F. Löhnis. *G. Fischer, Jena.*

Zentralblatt, Biologisches. C. Correns, R. Goldschmidt, O. Warburg. *G. Thieme, Leipzig.*

Zentralblatt, Botanisches. S. V. Simon. *G. Fischer, Jena.*

Zentralblatt, Botanisches. Beihefte. A. Pascher. *C. Heinrich, Dresden.*

Zentralblatt für die gesamte Hygiene und ihre Grenzgebiete. M. Rubner, C. Günther. *Julius Springer, Berlin.*

Zentralblatt für Mineralogie, Geologie und Paläontologie. R. Brauns, E. Hennig. *Schweizerbart, Stuttgart.*

Zentralblatt für allgemeine Pathologie und pathologische Anatomie. W. Berblinger, M. B. Schmidt. *G. Fischer, Jena.*

Zoologica. R. Hesse. *E. Schweizerbart, Stuttgart.*

C. G. Röder G. m. b. H., Leipzig

MIX
Papier aus verantwortungsvollen Quellen
Paper from responsible sources
FSC® C105338

If you have any concerns about our products,
you can contact us on
ProductSafety@springernature.com

In case Publisher is established outside the EU,
the EU authorized representative is:
**Springer Nature Customer Service Center GmbH
Europaplatz 3, 69115 Heidelberg, Germany**

Printed by Libri Plureos GmbH
in Hamburg, Germany